the **Ratings** suggest the priority that this Measure deserves in your ov[...] conservation program, in typical situations.

for **New Facilities**:

A Do it wherever it applies. It costs little, and it has no significant disadvantages.

B Do it in most cases. Modest cost. Pays back quickly. Does not need special skill or increased staffing.

C It is very expensive. Or, the payback period is relatively long. Or, operation may require substantial effort, special skill, or continuing management attention.

D It provides only a small benefit in relation to its cost. Or, it may have high risk because it is novel, unreliable, difficult to install, or difficult to maintain.

for **Retrofit**:

A Do it wherever it applies. Simple and quick. Costs little in comparison with its benefits. The risks can be managed easily by the present staff.

B Do it in most facilities where it applies. Pays back quickly. Easy to accomplish. Requires a modest amount of money, effort, and/or training. May have pitfalls that require special attention.

C Expensive or difficult. Or, the saving is small in relation to the money, effort, skill, or management attention required. The risks are clear and manageable.

D Expensive, and provides only little benefit. Or, exceptionally risky because it is difficult to accomplish correctly, or difficult to maintain, or unproven, or unpredictable.

for **Operation & Maintenance**:

A Simple, quick, and foolproof. Or, it must be done to prevent damage or major efficiency loss.

B Will be done in a well-managed facility. Pays back quickly. Fairly easy to accomplish. Not too risky. Requires a modest amount of money, effort, and/or training. Or, it is a less critical maintenance activity.

C Requires substantial money, effort, special skill, and/or management attention. Or, the benefit is small.

D The benefit is small in relation to cost. Or, it is exceptionally difficult to accomplish. Or, it has potential for serious adverse side effects.

the **Selection Scorecard** rates the financial and human factors that are most important for deciding whether to exploit the Measure in your application. The scores are for typical commercial applications. *Shaded symbols indicate a range of scores.*

Savings Potential is expressed as a percentage of the facility's total utility cost.

$ $ $ $	over 5%
$ $ $	0.5% to 5%
$ $	0.1% to 0.5%
$	less than 0.1%

Rate of Return estimates the percent of the initial cost that is saved each year.

% % % %	over 100%
% % %	30% to 100%
% %	10% to 30%
%	less than 10%

Reliability indicates the likelihood that the Measure will remain effective throughout its promised service life.

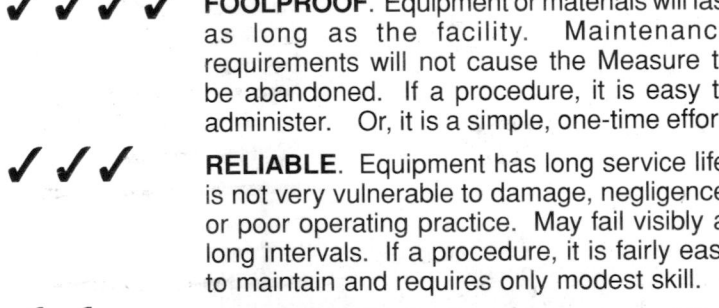

✓ ✓ ✓ ✓   **FOOLPROOF**. Equipment or materials will last as long as the facility. Maintenance requirements will not cause the Measure to be abandoned. If a procedure, it is easy to administer. Or, it is a simple, one-time effort.

✓ ✓ ✓   **RELIABLE**. Equipment has long service life, is not very vulnerable to damage, negligence, or poor operating practice. May fail visibly at long intervals. If a procedure, it is fairly easy to maintain and requires only modest skill.

✓ ✓   **FAILURE PRONE**. Equipment needs skilled maintenance, or it is vulnerable to damage or poor operating practice. Fails invisibly. If a procedure, it is easily forgotten or requires continuing supervision.

✓   **VERY RISKY**. Equipment has poor or unknown reliability. Or, it needs frequent maintenance. If a procedure, it is difficult to learn or it may easily cause damage.

Ease of Retrofit or **Ease of Initiation** indicates how easy it is for the people involved to accomplish the Measure properly.

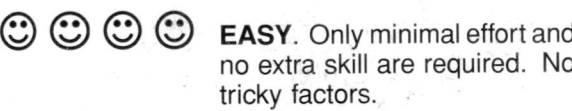   **EASY**. Only minimal effort and no extra skill are required. No tricky factors.

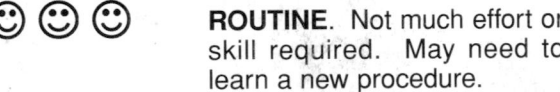   **ROUTINE**. Not much effort or skill required. May need to learn a new procedure.

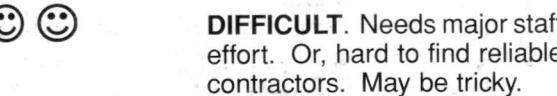   **DIFFICULT**. Needs major staff effort. Or, hard to find reliable contractors. May be tricky.

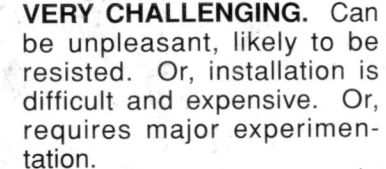   **VERY CHALLENGING**. Can be unpleasant, likely to be resisted. Or, installation is difficult and expensive. Or, requires major experimentation.

How to Use the *Energy Efficiency Manual*

The *Energy Efficiency Manual* is your primary tool for improving energy efficiency and reducing your utility costs. It is a comprehensive, step-by-step guide that is designed to help you manage your activities effectively and with confidence.

The core of the *Energy Efficiency Manual* is 400 energy efficiency "Measures." The Measures have a standard format that makes it easy to organize them into an optimum efficiency program for your facility. Refer to the inside of the front cover to learn how to exploit the Measures.

The Measures are grouped into Sections and Subsections. These correspond to ***types of energy systems*** (for example, boilers, chillers, or lighting) or to ***energy waste in specific components*** (for example, air leakage through doors, or solar heat gain through windows). This arrangement lets you quickly identify whole groups of Measures that may or may not apply to your facility. For example, if your boilers are fueled by natural gas, you can bypass the Subsection that deals with fuel oil systems. Use the Table of Contents to find the Sections and Subsections that apply to your situation.

The Reference Notes, the last Section of the book, serve you in two important ways. They support the Measures with additional explanation, which may be more basic or more advanced than the "working" information in the Measures. Also, you can read each Reference Note by itself for a concise overview of an important energy conservation topic.

Use the Index to find specific topics that interest you, or to find definitions of terms.

❑ **If you are involved in new construction** — if you are an architect, an engineer, a construction manager, a contractor, or a code official — use the *Energy Efficiency Manual* as a design review guide. As you develop your design, continually check the *Manual* for efficiency features that you can exploit. Use it to find where the design wastes energy, and to find better ways of saving energy.

❑ **If you own, manage, or operate facilities** — anything from a private house to an office complex or hospital or paper mill — use the *Energy Efficiency Manual* to find all your opportunities for savings. Then, use it to prioritize your activities. Finally, let it guide you in accomplishing and preserving your improvements.

❑ **If you are a specialist in energy efficiency**, use the *Energy Efficiency Manual* as a designer or facility manager would, depending on whether you deal with new or existing facilities. It will improve the quality of your work and reduce the time you need to provide the best service to your clients.

❑ **If you are a student or teacher**, start with the Reference Notes to learn fundamental principles. With each Reference Note, use the related Measures as examples of practical applications.

❑ **If you are an advocate for efficiency or the environment**, use the *Energy Efficiency Manual* to learn the real-world aspects of the conservation activities that interest you. The *Manual* will help you to promote resource conservation that produces credible results.

Now, please read "A Personal Note: the Right Way to Do Energy Conservation."

ENERGY EFFICIENCY MANUAL

Donald R. Wulfinghoff

for everyone who uses energy,
pays for utilities,
controls energy usage,
designs and builds,
is interested in energy and
environmental preservation

ENERGY INSTITUTE PRESS

Wheaton, Maryland   U.S.A.

Energy Efficiency Manual
by Donald R. Wulfinghoff

published by:
Energy Institute Press
3936 Lantern Drive
Wheaton, Maryland 20902
U.S.A.

301-946-1196
888-280-2665 (orders only)

Copyright © 1999 Donald R. Wulfinghoff

Custom excerpts and course packs from the *Energy Efficiency Manual* are available for purchase. Please contact the publisher for selections and prices.

Library of Congress Catalog Card Number 99-22242

ISBN 0-9657926-7-6

Library of Congress Cataloging-in-Publication Data

Wulfinghoff, Donald R.
    Energy efficiency manual : for everyone who uses energy, pays for utilities, ... / Donald R. Wulfinghoff.
        p. cm.
    ISBN 0-9657926-7-6 (alk. paper)
    1. Energy conservation Handbooks, manuals, etc.
    2. Energy consumption--Handbooks, manuals, etc.
    I. Title

    TJ163.3.W85 1999
    697--dc21   99-22242
                                          CIP

Printed in the United States of America

TABLE OF CONTENTS

A PERSONAL NOTE: THE RIGHT WAY TO DO ENERGY CONSERVATION

Improving energy efficiency may be the most profitable thing that you can do in the short term. How much you will actually benefit from this opportunity depends on how you approach it. Please take a few minutes to read the following suggestions about using the *Energy Efficiency Manual* and about your role in energy conservation. Invest a little time in learning how to use the *Manual*, and it will reward you with years of savings and achievement.

If you are involved in new construction — if you are an architect, an engineer, a construction manager, a contractor, or a code official — use the *Energy Efficiency Manual* as a design review guide. As you develop your design, continually check the *Manual* for efficiency features that you can exploit. Use it to find where the design wastes energy, and to find new ways of saving energy.

If you own, manage, or operate facilities — anything from a private house to an office complex or hospital or steel mill — use the *Energy Efficiency Manual* first to find all your opportunities for savings. Then, use it to prioritize your activities. Finally, let it guide you in accomplishing and preserving your improvements.

If you are a specialist in energy efficiency — if you are an energy consultant, a utility energy specialist, or an energy services provider — use the *Energy Efficiency Manual* in the same way, depending on whether you deal with new or existing facilities. You will find that it greatly improves the quality of your work and reduces the time you need to provide service of top quality to your clients.

If you are a student preparing to enter any of these important fields, or if you are a teacher, you will use the *Energy Efficiency Manual* in a different way. Start with the Reference Notes to learn fundamental principles. With each Reference Note, use the related Measures as examples of practical applications.

If your job or your vocation is to advocate efficiency — for example, if you are a government energy official or an environmental advocate — use the *Energy Efficiency Manual* to learn the real-world aspects of the conservation activities that interest you. Both governments and advocacy groups have played an invaluable role in promoting efficiency. At the same time, naive enthusiasm sets the stage for failures,

which undermine public confidence in energy conservation and actually waste energy. The *Energy Efficiency Manual* will help you to promote resource conservation that produces credible results.

How to Use the Energy Efficiency Manual

The *Energy Efficiency Manual* is designed to be your primary tool for improving energy efficiency and reducing your utility costs. It is a comprehensive, step-by-step technical guide, and it also helps you manage your activities efficiently. Learning to use this tool proficiently will take only a few moments.

The core of the *Energy Efficiency Manual* consists of four hundred energy efficiency "Measures." Each Measure is a specific energy efficiency improvement or cost saving activity. Each Measure gives you the information you need to plan the activity efficiently and accomplish it successfully.

All the Measures have a standard format. This includes special features, Ratings and a Selection Scorecard, that help you to quickly judge the value of each Measure for your applications. Other features, the Summary, Economics, and Traps & Tricks, give you the main features of each Measure. To become familiar with these features, refer to the key to the Measures, inside the front cover, as you browse through the Measures.

The Measures are grouped into Sections and Subsections. These correspond to *types of energy systems* (e.g., boilers, chillers, lighting) or to *energy waste in specific components* (e.g., air leakage through doors, solar heat gain through windows). This lets you quickly identify whole groups of Measures that may or may not apply to your facility. For example, if your boilers are fueled by natural gas, you can bypass the Subsection that deals with fuel oil systems. Use the Table of Contents to select the Sections and Subsections that apply to your facility.

First, find all your opportunities.

Resist the temptation to rush into energy conservation projects without considering all your opportunities first. You may be eager to get started after attending a seminar, or reading an article, or getting a sales pitch. Those are good ways to get an introduction to new concepts, but they are no substitute for knowing all your opportunities.

If you grab at opportunities randomly, you will miss many good ones and waste money. In a facility of any size, there will be many things that you can do to reduce your utility costs. Every building and plant wastes energy in hundreds or thousands of places. Find them all.

There is no way to find the best opportunities first. It is like an Easter egg hunt. You can't tell how big the prizes are until you have searched everywhere and found all the eggs. By the same token, don't expect to find a "short list" of improvements that are best for your facility. Each building and plant wastes energy in different ways.

Your search for efficiency improvements will be time-consuming. (In existing facilities, this search is often called an "energy audit.") Typically, it requires weeks or months. In a large, diverse facility, it may require more than a year. Demand the time to do it right.

A false concept that came out of the popular energy conservation movement of the 1970's is the "walk-through" or "one-day" energy audit. According to this notion, whizzing through a facility reveals energy conservation opportunities by a mystical kind of inspiration. Reject this ouija board approach, even as a starting point. Quickie surveys fool you into believing that you know your options when you really don't.

Budget your time as wisely as your money.

When you complete your list of potential efficiency improvements, your next job is to decide the most effective sequence for accomplishing them. You want to produce the greatest payoff in the shortest time. Be shrewd about managing your program's two most important resources, money and personal capabilities.

The *Energy Efficiency Manual* helps you make the best use of both these resources. The Ratings in each Measure suggest its overall priority, taking into account the economics of the Measure, the difficulty of accomplishing it, and the degree of risk. To refine your ranking, the Selection Scorecard, just below the title, rates these factors individually. At the end of each Measure, the Economics gives you general estimates of the potential savings, the cost, and the rate of return.

Recognize that your time is a more precious resource than the money needed to make the improvements. Energy efficiency is a profit maker. So, you could borrow money to fund any project that you know will pay off. The skills and effort of the people involved are the real limiting factors. Traps & Tricks, located right after Economics, alert you to aspects of the Measure that will challenge the people involved.

Give priority to the Measures, or groups of Measures, that will produce the largest savings, even though they may not pay off most quickly. Don't divert your time to minor activities while there are more important things to be done. On the other hand, if you see that you can accomplish a Measure quickly and reliably, go ahead and do it. Don't waste time analyzing small improvements in detail.

Try to accomplish groups of related Measures together. For example, make all the control improvements to your air handling systems as a single activity. This avoids duplication of effort, saves money in contracting, and produces a better overall system. The *Energy Efficiency Manual* is organized to make this easy for you.

Most important, don't get in over your head at the beginning with a large project that demands all your attention. If a Measure seems overwhelming, defer it until you have more time to study it. Don't start any Measure until you are ready to complete it *successfully*.

Don't expect instant gratification.

The desire for quick and effortless results has ruined more energy conservation projects than any other cause. Rushing into a project blindly is unprofessional. You would not want your surgeon to rush through your operation just to prove how quickly he can do it.

You have heard expressions like "no-cost energy conservation measure," "pick the low fruit," and so forth, to describe retrofit projects that are supposed to be "easy" or "simple." These notions are illusions that lure you into being too hasty. Every opportunity for saving energy requires significant effort, if it is going to work and to endure.

Your willingness to invest the needed effort and time is what guarantees the success of your projects. The *Energy Efficiency Manual* will show you how to make your improvements as quickly and easily as possible.

Rely on proven equipment and methods.

Energy conservation is not a license to use the owner as a guinea pig. In most cases, rely on conventional equipment and methods. Contrary to popular opinion, energy efficiency does not require exotic technology. That's good news. The bad news is that fads in energy conservation have strong appeal, distracting people from proven profit makers. The only good reason to do energy conservation is to produce predictable, certain savings.

Everyone is fascinated by innovation. Innovation drives progress. But, the price of innovation is a big chance of failure. Most owners can't afford that risk. Leave unproven equipment and methods to those who develop new products and have a laboratory budget.

On the other hand, if you are in a position to work at the frontiers of energy efficiency, the *Energy Efficiency Manual* will help you survive as a pioneer. You will find many Measures at the leading edge of energy efficiency (and a few that are just on the outer fringe). These too can be profitable if you give them the attention they need. Riskier Measures have a Rating of "C" or "D", and their Traps & Tricks warn you of the dangers of unexplored territory.

Why is there so much stress on reliability?

The *Energy Efficiency Manual* devotes a lot of attention to the details that make the difference between a reliable system and one that is riddled with problems. This emphasis on avoiding pitfalls and dealing with tricky factors is intended to alert you, not to frighten you. Energy conservation is still a new subject. The blunt truth is that many energy conservation projects have failed, almost always because people ignored vital issues at the outset. These issues are often simple. For example, a common cause of energy waste is failing to mark controls so that people know how to use them.

Only successful projects pay off. We want you to contribute to the successes, not to the failures. The Measures spell out the issues that you need to consider. It's like driving around potholes. Keep your eyes open and don't rush.

Why all the explanations?

A large part of the *Energy Efficiency Manual* is devoted to explaining how things work. There are several important reasons for this. If you understand the principles, you are much less likely to make mistakes. Knowing the principles also enables you to keep up with changes in technology. And, knowing what you are doing at a basic level turns the work into fun.

The "theory" is located in two places. Each Measure offers the basic information that you need, and if necessary, it suggests where to get more information. Often, a Measure will refer you to one or more Reference Notes. Each Reference Note is a self-contained explanation of a specific topic.

Don't let mere words get in your way.

Each area of design, construction, and facility operation has a separate vocabulary. Architects have one set of jargon, mechanical engineers have another, electrical contractors still another, and so forth. Don't let this deter you from making efficiency improvements in each of these areas. The principles are important, not knowing particular words.

The *Energy Efficiency Manual* keeps the language as simple as possible. For example, we say "lamp" or "light fixture" instead of "luminaire." We say "window" or "skylight" instead of "fenestration." To help you communicate with specialists who may be fussy about language, the *Manual* explains specialized terms in the places where you need to know them.

Fortunately, each area has only a few specialized terms that are important. If you find a word that is unfamiliar, the Index will steer you to a concise, practical explanation.

You don't need much math, but be comfortable with numbers.

You will probably be happy to see that the *Energy Efficiency Manual* uses little mathematics. There are only a few simple formulas, and you need only arithmetic to use them.

Even so, energy efficiency is all about numbers. In most cases, you are not doing something that is fundamentally new. Instead, you are doing something *better*. To judge whether the improvement is worth the cost, you have to be able estimate the benefit in terms of numbers. If you are not comfortable doing the math, of if you need a calculation that requires specialized knowledge, get a specialist to make the calculations for you.

Recognize that energy savings are uncertain to some extent. They are subject to conditions that you cannot predict, including future energy costs, operating schedules, weather, and human behavior. Make your estimates of savings for a reasonable range of conditions.

Keep your facility efficient for its entire life.

When energy conservation became a public issue during the 1970's, it was promoted by many well-intentioned people who lacked experience in keeping things working. Energy conservation was treated as a magic pill that would cure the disease of energy waste once and for all. In reality, energy waste is a degenerative condition that keeps trying to return.

Maintaining efficiency is like maintaining your physical fitness. You have to keep it up. Design your efficiency improvements to survive as long as the facility. Each Measure that requires maintenance tells you how to keep it profitable.

Let all your information sources work for you.

Capable professionals depend primarily on a few well-worn references. But, they also know how to get information from other sources quickly. Whether you are a professional or not, the *Energy Efficiency Manual* is your primary reference for energy efficiency. However, no single book can tell you everything you need to know. To do battle with energy waste, assemble an armory of information that is appropriate for the level of improvements that you plan to make.

You will see that the *Energy Efficiency Manual* is not cluttered with formulas and tables. When you need detailed engineering data, get it from the appropriate reference books. Fortunately, you need only a few of these. If you are involved at a professional level with heating, air conditioning, refrigeration, or designing a building's skin, you should have the four-volume *ASHRAE Handbook* on your shelf. For electric lighting, the prime reference source is the *IESNA Handbook*.

Many books are available on specialized aspects of energy conservation, such as solar energy, cogeneration, and residential insulation. Don't hesitate to get another book to expand your knowledge about a subject. There is no better bargain. A good book costs almost nothing in comparison with your utility expenses, and it protects your most valuable assets, which are your time and your professional reputation.

Once you decide to use a particular type of equipment, study the catalogs and equipment manuals of different manufacturers. These are a treasure of important details, and they are your most current source of information. But, beware. The big weakness of manufacturers' literature is a selective rendition of the truth. Knowing potential problems beforehand is critical to success, but manufacturers tend to omit or minimize this vital information.

Talk to others.

Two heads are better than one. Seek other people's opinions before you get involved with unfamiliar equipment or procedures. You can get practical advice from books, trade magazines, professional organizations, consultants, colleagues, and vendors. Talk to facility operators for their opinions about how well something really works.

As you do this, take everything with a grain of salt. People's perceptions are distorted by wishful thinking, embarrassment about disappointing outcomes, and inability to measure actual performance. I have listened to experienced plant operators brag about big efficiency improvements that they were convinced they had achieved with gadgets that were purely bogus.

Don't try to do everything yourself.

If you have a big facility, you will not live long enough to make it efficient by yourself. If you try, energy and money will bleed away while valuable efficiency improvements wait to be made.

Spread the work effectively. In a big facility, your main job is to decide which Measures to accomplish, and to make sure that they get done correctly. Use engineers, architects, contractors, specialized consultants, along with the facility staff. As your program gains momentum, you will have your hands full making sure that others do their work correctly.

Many Measures straddle the boundaries of the established design and construction disciplines. For example, successful daylighting requires close coordination between the architect, the lighting designer, the electrical engineer, and the mechanical engineer. You have to bring all these people together and require them to address all the issues that are critical for success. This is not always easy. Select your people for their willingness to listen and learn.

Seize the opportunity!

The most important point is to get started. At every moment, motors and fans are running, lights are turned on, boilers are burning fuel, and other equipment is consuming energy. Some of this energy is being wasted, and it is probably more expensive than you realize. Remember that cost savings are pure profit. You would have to sell a lot more of your product or service to make as much profit as you can from energy efficiency. Start tapping this resource.

On an industry-wide basis, the efficiency of your facilities will increasingly determine whether your organization can continue to survive and compete. On a global scale, improving efficiency is the most satisfactory way for civilization to adapt to declining energy resources and to minimize harm to the environment.

Enjoy yourself.

At this point, you may feel that you got into more than you bargained for. Don't worry. Energy conservation is a bigger challenge than most people expect, but the *Energy Efficiency Manual* breaks it down into easy steps. Set a comfortable pace, and stick with it. Your energy savings will soon show up on your utility bills, and those saving will continue to grow and accumulate.

Your energy efficiency program can be the most interesting and rewarding part of your career. It will give you an opportunity to become involved in every aspect of your industry. There is probably no other way that you can have as much fun while doing something of fundamental importance.

Donald Wulfinghoff
Wheaton, Maryland, USA

Expression of Gratitude

This book aspires to bring order and understanding to the vast field of energy efficiency. It organizes what I have learned about the subject during a career that has spanned the most exciting years of energy conservation in the United States and the world. Almost everything that I know was learned from others in one way or another. I would like to begin the book by recognizing those who contributed generously and specifically to the book, and also to recognize several persons and organizations who contributed more generally to my education in energy efficiency. This book is largely their achievement. The following brief acknowledgments cannot adequately recognize the individuals who made important contributions. However, I hope that these mentions will be accepted as a token of my deep gratitude.

Clinton W. Phillips, a figure revered in the air conditioning industry for his limitless contributions, erudition, and charm, meticulously reviewed two separate drafts of the material that deals with cooling systems. In addition to checking the text, he made important comments on both the theory of refrigeration and the lore of practical applications.

Henry Borger, a leader in construction research as well as a talented writer on diverse subjects, reviewed the entire book, suggesting improvements in structure and content.

Charles Wood reviewed the text that deals with boiler systems, providing valuable comments on this technical area and on the editorial approach.

Jim Crawford of the Trane Company contributed extensive and detailed information about the fast-changing world of refrigerants. Dave Molin of the Trane Company reviewed the Reference Note on energy analysis computer programs.

Richard Ertinger and Edward Huenniger of Carrier Corporation provided valuable information about the most recent advances in cooling technology.

Ken Fonstad, of the Graham Division of Danfoss, Inc., wrote lucid explanations of the electrical subtleties of variable-frequency motor drives, accompanied by extensive oscilloscope traces that he made. He also contributed a number of illustrations.

Sean Gallagher shared his experience with the practical aspects of lighting retrofits and with utility purchasing in this era of rapid change in the utility industry.

Don Warfield of Solarex provided information about the current state of photovoltaic technology, and made several illustrations available.

Many others contributed information during the twenty years of the book's preparation. It is impossible now to recall all the valuable discussions and presentations. I hope that the individuals will approve of the way that the book reflects their expertise.

Many organizations contributed illustrations that help to achieve the book's goal of bringing to life many unfamiliar and subtle concepts. These organizations are listed in the back of the book. The individuals who were especially helpful in providing the illustrations include Pat McDermott and Claudia Urmoneit of Osram Sylvania; Eric Johnson, Dave McDevitt, and Pat Barbagallo of Carrier Corporation; Jake Delwiche and Dick Figgie of the Trane Company; Thomas Henry and Pam Blasius of Armstrong International; Andrew Olson and Jim Baker of Rite-Hite Corporation; Tania Davero of Advance Transformer Company; Leight Murray and Diane Iaderosa of the Airolite Company; Doby Byers of American Mill Sales; Peter DeMarco of American Standard; John Figan of Bacharach; William McCloskey of Baltimore Aircoil Company; Steve Hill of Blender Products; Roy Nathan of Calmac Manufacturing Corporation; Bob Agnew of Celotex Corporation; Sharon Quint and Bill Garratt of Cleaver-Brooks; Paul Moulton of Construction Specialties, Inc.; Dewey Boggs of Coyote Electronics; Sherri Snow of Danfoss Automatic Controls; Lynn Hamrick of Donlee Technologies; Trish Steele of Dow Chemical Company; Linda Byam of Duo-Gard Industries; Herman Knapp and Keith Knapp of Fuel Efficiency, Inc.; Chris Van Name of Goodway

Tools Corporation; William Bakalich of Hi-Fold Door Corporation; Eric Huffman of Huvco, LLC; Joachim Harasko of IMR Environmental Equipment; Rob Carter of Industrial Combustion; Bruce Keller of Kalwall Corporation; Bob Hanson of Kentube; Judy Kuczynski of Leeson Electric Corporation; Jennifer Vizvary of Lennox Industries; Lee Webster of LightScience Corporation; Don Betts of Ludell Manufacturing Company; Dipti Datta of M&I Heat Transfer Products; Wayne Toenjes and Susie Toenjes of Major Industries; Bob Rank of Paragon Electric Company; Mike Leeming of Parker Boiler Company; Daniel Manna and Juli Stovall of Paul Mueller Company; Steve McNeil of Pennsylvania Separator Company; Larry Wilton of Philips Lighting; Ken Brooks of Preferred Utilities Manufacturing Corporation; Leon Siwek of Pure Water/Clean Air Group; Tom Hilty of Reliance Electric; Eileen Moran of Resources Conservation Inc.; Rick Wirth of Robicon; Henry Warner and Deb Jamour of Ruud Lighting; Mike Schweiss of Schweiss Distributing; Jean Posbic and Cindy Axline of Solarex; Jeff Sommer of Spirax Sarco; Brian Edwards of Sun Tunnel Skylights; James Satterwhite and Cindy Selig of Super Sky Products; Mike Williams of Todd Combustion; Gerry Denza of Unenco Electronics; Jim Carney of Vaughn Manufacturing; Michael Boyd of Vistawall Architectural Products; Jochen Schiwietz of Water Technology of Pensacola; Dorothea Rynearson of WaterFurnace International, and Klaus Reichardt of the Waterless Company.

The archives of Wulfinghoff Energy Services, Inc. yielded many of the figures, found among thousands of photographs that were originally taken as field notes in energy efficiency projects, and among illustrations made for courses and seminars.

Like all who are involved with energy efficiency, I owe a great debt to the American Society of Heating, Refrigerating, and Air Conditioning Engineers, known to the world as ASHRAE. This book's many referrals to the *ASHRAE Handbook* attest to its role as the primary reference for the practice of refrigeration and building design. The Society gave me the opportunity to serve with several committees that defined the course of energy conservation in response to energy crises of the 1970's. Among these responsibilities were helping to organize and to serve as a judge of the ASHRAE Energy Awards, which provided exposure to the energy conservation philosophies of the most innovative engineers of that era. ASHRAE also provided the impetus to write *Managing Your Energy*, the tutorial on the management aspects of energy conservation that became the basis of the Energy Management chapter of the *ASHRAE Handbook*. Another committee assignment gave me the opportunity to investigate how the different types of buildings use energy, then and now a topic that is rife with misconceptions.

I am particularly indebted to the National Capital Chapter of ASHRAE. During the years that I have been a member, the Chapter presented several hundred technical presentations, from each of which I learned something new and valuable. Two individuals stand out for their accomplishments in the Chapter. Jim Wolf, the president when I first began to serve in its offices, created the model of disciplined organization that keeps the Chapter effective to this day. Jim also facilitated technical reviews and illustrations from the Trane Company and American Standard. Jose Reig was the mainstay of the Energy Management Committee when I served as its early chairman, and later supported me when I became responsible for ASHRAE's energy conservation programs in the mid-Atlantic States. His intense dedication to everything he undertakes has been rewarded by the success of the engineering firm that he built.

The George Washington University provided my first platform for teaching energy efficiency to professionals, starting during the late 1970's. This gave me the occasion to consolidate the lessons that were being learned in those heady days of intense interest in energy conservation. The notes of those courses became the early structure of the *Energy Efficiency Manual*.

The U.S. Navy Engineer Officers School, San Diego, provided my first serious introduction to the machinery of energy systems. The School was a model of effective instruction that should be copied by all engineering schools.

My education in energy efficiency would have been inadequate for this task without the practical experience gained while working for the clients of my energy efficiency firm. Improving their facilities taught me the lessons of energy efficiency in the real world, including the diversity of ways that energy is wasted, the importance of details at every step, and the need for relentless maintenance and management attention. I always tried to spare our clients from the fads that were rampant during the infancy of the energy conservation movement. Still, those forward looking managers were the experimental subjects who made progress possible. And, they provided the living that financed the long years of writing.

Among our clients who became good friends, Michael Whitcomb deserves special mention as an extraordinary facility energy manager who aggressively and successfully pioneers important areas of energy efficiency. Our discussions about the practical realities of managing energy systems continue to be instructive.

The *Energy Efficiency Manual* benefited immensely from the editorial review of two extraordinary individuals. Nancy Dashiell, the original and veteran writer and technical editor of U.S. Pharmacopeia *Drug Information* (published by Consumers Union under the title *Complete Drug Reference*), edited the crucial final manuscript of the book and made valuable suggestions about earlier versions. Felicity Evans, whose experience includes service as a government energy official, insisted on essential changes to the early structure and style of the book. Among other important improvements, her suggestions led to the creation of the Reference Notes.

Dan Poynter, renowned parachutist and publishing mentor, made penetrating comments on an early draft that led to a complete rewrite to make the book easy to use by a broad audience.

In the production of the book, one individual stands out. Mark Dorbert, the proprietor of Wet Ink Printing & Graphics, shepherded the book from manuscript to press, rendering the interior design, accomplishing the composition and typesetting, and electronically processing the illustrations. An inspired artist, he also designed the end material. He did an enormous amount of work that would normally require a large team of individuals, mastering the range of prepress skills during a period in which the technology of publishing is changing from month to month. Nothing daunts him, he never slackens the pace, and his humor keeps the work enjoyable.

Cindy Fowler, of Graves Fowler Associates, rendered the cover design flawlessly and quickly, patiently dealing with many details.

Steve Dolan, of Scanners LLC, made the electronic renderings of the author's drawings, putting in many hours of work to meet a short deadline.

— Donald Wulfinghoff

Section 1. BOILER PLANT

INTRODUCTION

A large fraction of a facility's total energy usage begins in the boiler plant. The cost of boiler fuel is typically the largest energy cost of a facility, or the second largest. For this reason, a relatively small efficiency improvement in the boiler plant may produce greater overall savings than much larger efficiency improvements in individual end users of energy. Also, most boiler plants offer significant opportunities for improving efficiency. These reasons make the boiler plant a good place to start the search for savings.

How to Use Section 1

This Section is organized primarily by the subsystems of the boiler plant. Table 1, on the next page, relates the ways that boiler plants waste energy to these subsystems. Use Table 1 as the first step in identifying the areas for improving your plant. The first two Subsections apply to all boiler plants, and you should start with them. The last Subsection applies only to larger facilities. You can ignore Subsections that do not apply to your particular type of plant.

Before Getting Started ...

Read Reference Note 30, Boiler Types and Ratings, for a quick overview of boiler types and the way they are rated. Read Reference Note 20, Fossil Fuels, to understand the role of boiler fuels in efficiency, operating practices, maintenance, and cost.

Table 1. Boiler Plant Energy Losses

Energy Requirement / Loss	Percent Increase in Plant Energy	Most Improvement Possible in:	Improve with Measures in Subsections:
Unnecessary operation of equipment	0 to 200	facilities with non-continuous operations	1.1, 1.2
Improper air-fuel ratio	0 to 20	all boilers	1.2, 1.3, 1.5
Burner operation	0 to 1	large heavy oil-burners	1.4
Forced-draft & induced-draft fans	0 to 0.7	large fans	1.4
Standby losses	0 to 10	atmospheric burners	1.5
Flue losses due to fouling	0.1 to 10	all boilers	1.6, 1.8
Flue lossers due to inadequate heat transfer surface	0 to 40	old, cheap, & overdriven boilers	1.7, 1.12
Flue losses due to unrecovered latent heat of water in flue gas	2 to 10	boilers with high blowdown rates	1.8
Condensate loss	0 to 10	facilities with old or missing condensate systems	1.8
Condensate system operation	0 to 0.2	systems with large pumps	1.8
Defective vacuum condensate system	0 to 50	vacuum condensate systems	1.8
Fuel oil heating	0 to 0.5	heavy-oil burners	1.9
Fuel oil transfer	0 to 0.1	odd cases	
Steam/hot water loss	0 to 50	old, large underground distribution systems	1.10, 1.12
Steam trap leakage	0 to 20	all systems with steam traps	1.10
Combustion air leakage	0 to 0.1	boilers with seperate fans	1.11
Boiler plan radiation & conductive loss	0.3 to 4	medium-sized & older plants	1.11
Distribution system conduction loss	0.5 to 30	old distribution systems in damp soil	1.11, 1.12

Equipment Scheduling and Operating Practices

This is the place to start in reducing your boiler plant's energy consumption. The Measures in this Subsection all deal with control of your boiler plant equipment. They recommend operating practices and automatic controls that limit operation of the heating plant and its major components to conform to need. These Measures may provide large savings at modest cost. Most of them are easy to accomplish.

MEASURE 1.1.1 Minimize the duration of boiler plant operation.

A primary principle of energy conservation is turning off equipment when it is not needed. Apply this principle to your boilers and their auxiliary equipment. The subsidiary Measures give you the basic techniques. Select from these to fit your particular situation. Combine the methods, if appropriate.

Energy Saving Potential

If you minimize the operation of the boilers and their auxiliary equipment, you will save energy in these areas:

- *plant and distribution system losses.* Conductive heat losses continue as long as the boiler plant is hot, and leaks continue as long as the boiler plant is kept under pressure. Losses are greatest in plants that are large, that have long distribution systems, and that are poorly maintained.

- *unnecessary energy consumption by boiler plant auxiliary equipment.* Even if the heat demand on a boiler is minimal, some equipment may continue to operate, such as hot water distribution pumps and feedwater pumps that serve larger boilers. The amount of such energy waste depends on the design of the boiler plant.

- *unnecessary operation of user equipment that is not separately controlled.* For example, old steam radiators with manual control valves operate as long as they are provided with steam, regardless of need. The amount of energy waste depends on the amount of user equipment that lacks individual control. (This issue is discussed below.)

Where are the Best Locations to Control Heating?

The first step is to develop a consistent plan of control for heat-using equipment throughout the facility. Decide whether to control heating operation at the plant, at the end-use equipment, or in some combination of these two. Each approach has advantages and disadvantages:

- *minimizing steam consumption by turning off the end-use equipment.* You can install shutdown controls on each item of end-use equipment. From an overall efficiency standpoint, this method is good because each piece of end-use equipment can shut down in accordance with its individual heating requirements. The major disadvantage of this method is the expense and maintenance of having separate controls at each item of end-use equipment. The boiler senses the disappearance of heating load when the end-use equipment shuts off. Boiler output falls as the end-use equipment shuts down,

and no separate boiler plant controls may be required. (Individual control of heating units is covered in Section 5.) However, this method does not eliminate the energy consumed to keep the boiler system warmed up, to replace distribution system losses, and to operate boiler plant auxiliary equipment. To avoid these losses, you need additional controls to shut down the boiler plant itself.

- *minimizing steam consumption by shutting down the boiler plant.* At the other extreme, you may shut down the boiler system, which will shut down all the equipment served by the system. This method is much less expensive than installing individual shutdown controls on the end-use equipment.

This method wastes some energy unless all end-use equipment operates on the same schedule. For example, if the boiler plant serves radiators, operating a boiler to provide heat to one room keeps every radiator in the facility working.

Another disadvantage of this method is that it does not stop the energy consumption of non-heating components of end-use equipment, such as the fans in fan-coil units.

- *minimizing steam consumption when end-use equipment operates on a variety of schedules.* In situations where some end-use equipment operates on a shorter schedule than the boiler plant, you can provide separate shutdown controls just for the equipment that operates on shorter schedules. For example, you might install such controls for the administrative spaces in a hotel, while the guest rooms have heating available continuously or seasonally.

- *minimizing steam consumption using different control criteria.* You can use different methods of shutdown for the boiler plant and the end-use equipment. For example, you might shut down end-use equipment with timeclocks and shut down the boiler plant with an outside air temperature sensor.

There may be many ways to distribute shutdown control of heating within a facility. Your objective is to make the boiler plant respond efficiently to the range of conditions that may occur, including the operating schedule of heating equipment, outside temperature, etc., and to satisfy the constraints of cost, reliability, and maintenance requirements.

What is the Shortest Practical Shutdown Interval?

Shutting down a boiler causes thermal stresses in the boiler itself and in plant equipment, shortening equipment life and hastening the formation of leaks. In steam systems, allowing the plant to cool causes the steam throughout the system to condense, creating a vacuum that draws air into the system. The degree of adverse effect varies widely, depending on the type of system.

In general, you can shut down smaller boiler plants for shorter periods of time. Starting up and shutting down large boiler plants are major operations, requiring hours to accomplish. Whether to shut down a boiler plant overnight or over a weekend is a decision that you make by comparing all the costs of leaving the system in operation to all the costs of shutting it down. These costs should include labor and maintenance, as well as energy.

Don't Forget to Limit Operation of Auxiliary Equipment

Limit the operation of boiler plant auxiliary equipment along with the operation of the boilers themselves. This equipment includes heating water distribution pumps, feedwater and condensate pumps, control air compressors, and a variety of less energy-intensive equipment. See Measures 1.1.3 ff for details.

Coordinate with Control of Chiller Plant Equipment

Most of the methods given by the subsidiary Measures for minimizing the operation of a boiler plant can also be used to minimize the operation of a chiller plant. If the boiler plant is located near the chiller plant, consider combining the control installations. See Measures 2.1.1 ff for comparison.

Automatic Control is More Efficient than Manual Control

People are not well adapted for starting and stopping equipment at precise times. People are especially unreliable if action is needed at irregular times, and as conditions change. For this reason, use automatic controls wherever possible. The subsidiary Measures emphasize automation for starting and stopping boiler plant equipment. For situations where automatic operation is not advisable, Measure 1.1.1.4 provides a hybrid approach that improves reliability while retaining manual control.

Explain It!

Make it easy for plant operators to understand the desired operating schedule and the operation of automatic controls. Install effective placards at locations that are visible to the persons responsible for controlling the operation of equipment. See Reference Note 12, Placards, for details of effective placard design and installation.

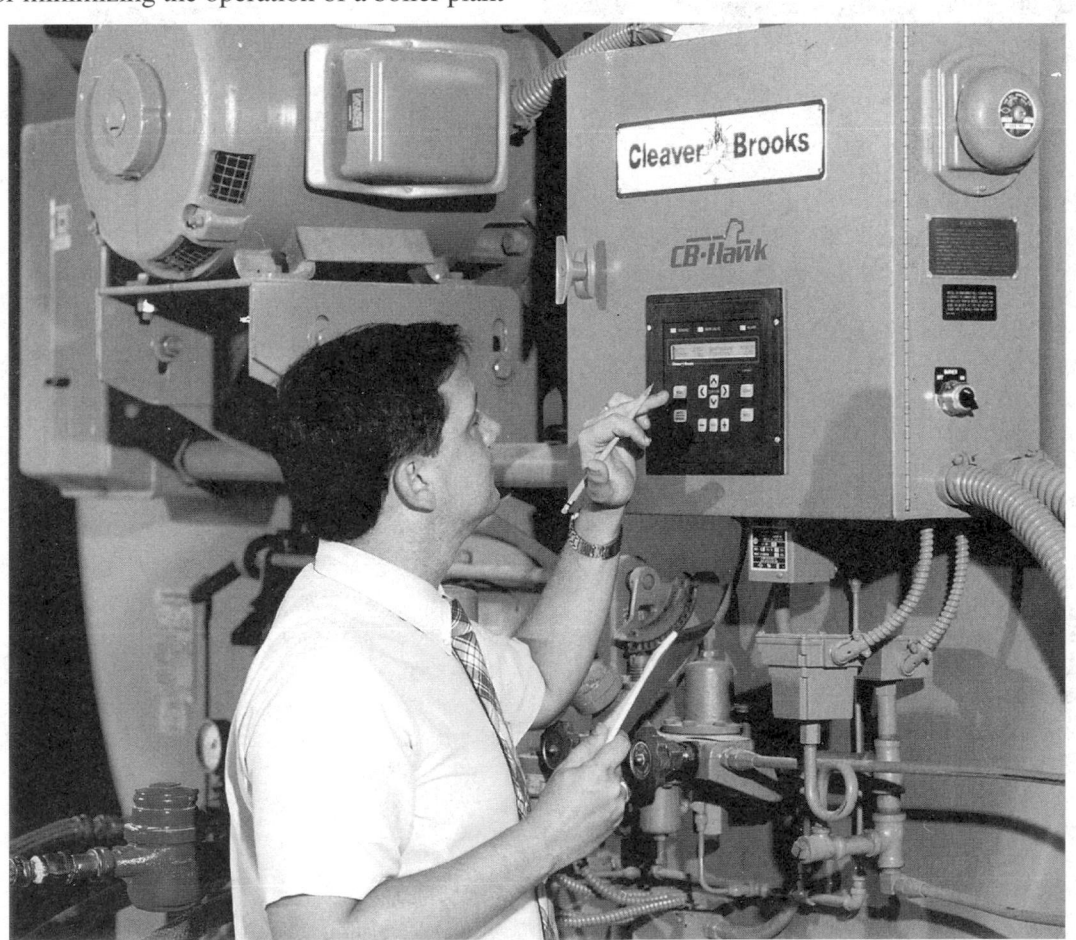

Cleaver-Brooks

MEASURE 1.1.1.1 For applications with regular schedules, install clock controls to start and stop boilers.

In many facilities, the best way to minimize unnecessary boiler plant operation is to use a timeclock to limit operation of the equipment. This method is appropriate for facilities where the need for the boiler plant is limited to specific times.

This is the first of four Measures that provide methods of limiting equipment operation. Expect to use time controls in combination with other methods of control for better tailoring of boiler operation to need.

Preliminary Decisions

As a first step, inventory the heat requirements of each space and application that is served by the boiler plant. From this, decide on the best distribution of operating controls, as discussed under Measure 1.1.1, above.

It is not always simple to know when operation of the boiler plant is actually needed. In addition to operating on a routine schedule, boilers need to be available during unusual periods of facility operation. If a time control prevents a boiler from operating when it is needed, somebody will eventually disable the time control or reset it to a less efficient schedule. This is why effective overrides are an essential part of time control. See Reference Note 10, Clock Controls, for details of overrides.

Selecting Time Control Equipment

Fortunately, you now have a rich field of choice in time controls, and prices are modest. But, don't be hasty in making your selection. Select your time controls to provide reliability, ease of use, scheduling flexibility, and easy override. Use Reference Note 10, Clock Controls and Programmable Thermostats, as a checklist for equipment features, installation practices, and operating procedures.

If your facility has old-style electro-mechanical timeclocks, consider replacing them with newer electronic models to take advantage of internal power backup, multiple schedules, and other features that are important in most timeclock applications.

If you have an energy management control system, you might use it for starting and stopping boiler plant equipment. See Reference Note 13, Energy Management Control Systems, for the advantages and disadvantages of this approach.

Optimum-Start Controls

If the boiler plant needs a variable warm-up period to adapt to changes in the outside temperature, consider

SUMMARY

An inexpensive and powerful method of reducing unnecessary boiler operation. Limited to applications where the facility operates on a rigid time schedule. Requires user-friendly overrides to reduce the likelihood of sabotage by after-hours occupants.

SELECTION SCORECARD

Savings Potential .... \$ \$ \$ \$

Rate of Return, New Facilities % % % %

Rate of Return, Retrofit.......... % % % %

Reliability .... ✓ ✓

Ease of Retrofit .... ☺ ☺ ☺

optimum-start controls, which are a special feature of time controls. See Measure 1.1.1.2, next, for the details.

ECONOMICS

SAVINGS POTENTIAL: Varies greatly, depending primarily on the amount of individual control that exists over the operation of end-use equipment. Savings within the boiler plant itself depend on the extent of plant losses and the amount of energy consumed by auxiliary equipment.

COST: Timeclocks having the full range of desirable features are available for several hundred dollars. Installation and training may cost several thousand dollars per unit.

PAYBACK PERIOD: Less than one year, to several years.

TRAPS & TRICKS

SELECTING THE EQUIPMENT: Make sure that you get all the features you need. A timeclock will soon become junk if it is difficult to use, or if it lacks a critical feature.

INSTALLATION: Install time controls where they are obvious and easily accessible to the people who are supposed to use them. Make them inaccessible to everyone else.

OPERATION: Time controls invite tampering. Make sure that operators know how to use them properly.

MONITOR PERFORMANCE: Improper operation of time controls is easy to overlook. Check them periodically to ensure that they are set properly.

MEASURE **1.1.1.2 In applications that require a warm-up period, control boiler operation using an optimum-start controller.**

This is the second of a series of Measures that limit the operation of boilers and their auxiliary equipment. The first Measure covers timers that start and stop equipment at fixed times. This Measure uses time control that adapts automatically to changes in weather. This feature is called "optimum-start."

Description

Optimum-start control is a supplemental feature of time controls that automatically adjusts the equipment start time in response to conditions. The start time usually is based on outside air temperature, and in more sophisticated models, is also based on the inside temperature.

The simplest optimum-start controllers vary the start time based on a fixed relationship to the outside air temperature (and perhaps the inside temperature). This relationship is set manually, and you have to find the optimum setting by trial-and-error. More sophisticated units have the capability to establish an optimized schedule automatically by "learning" from experience.

SUMMARY

A refinement of time control that adapts to changes in weather conditions. Savings are proportional to the variability of the climate.

SELECTION SCORECARD

Savings Potential .... $ $ $ $

Rate of Return, New Facilities % % % %

Rate of Return, Retrofit.......... % % % %

Reliability .... ✓ ✓

Ease of Retrofit .... ☺ ☺ ☺

Time controls that offer an optimum-start feature typically offer several channels of control, each of which can be operated on a separate schedule. For example, the controller could provide different turn-on times for different boilers.

Where to Use Optimum-Start Controls

At the present time, commercially available optimum-start controls are designed for applications where the start time depends on the weather. For example, it may take several hours to warm a building that has been shut down over a cold weekend. This requires starting up the boiler plant well before occupancy of the building begins. The same building may require a much shorter warm-up period the following day. If the boilers are started using simple time controls, operators tend to set the start time for the worst conceivable condition, which wastes heating energy under less severe circumstances.

You can extend the optimum-start principle to any application where there is a definable relationship between the start time and some condition that requires the boiler plant to operate. For example, a boiler plant that provides heat for a manufacturing process might be optimized to start at a time that depends on the size of the batch of product being produced. However, you will have to do this on a custom basis, typically by using a programmable logic controller (PLC).

Low-Temperature Protection

Where freezing or other damage from low temperature is a potential problem, provide supplemental controls to start the boiler plant to avoid this risk. Usually, the best approach is to include a low-temperature override that restarts the boiler plant if the

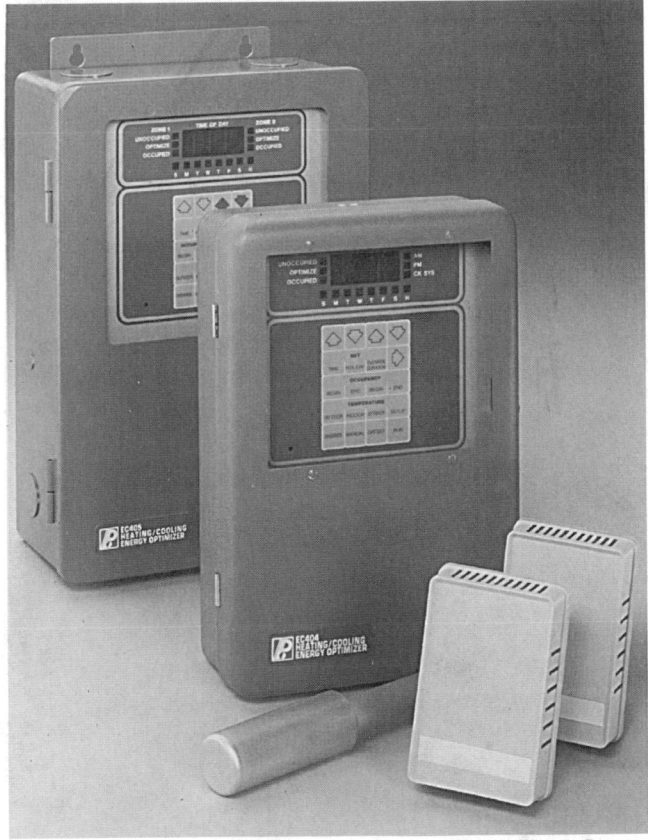

Paragon Electric Company

Fig. 1 Typical optimum-start controllers These are two fairly sophisticated time controls that include optimum-start. Note the inside and outside temperature sensors.

temperature at some critical location in the facility falls below a set level.

Methods of Accomplishing Optimum-Start

The simplest and most reliable way to install optimum-start capability is to install an optimum-start controller for the individual items of equipment that you wish to control. Figure 1 shows typical units that have a wide range of capabilities.

You can't retrofit optimum-start capability to existing time controls. You don't need to. Timeclocks with optimum start have become so inexpensive that it is often economical to replace existing timeclocks in order to acquire optimum-start capability. If you do buy a new timeclock to take advantage of optimum start, select the unit for other desirable features. The features of time controls are covered in Reference Note 10, Clock Controls and Programmable Thermostats.

Most centralized energy management control systems offer optimum-start capability. However, this approach is less reliable than using a dedicated optimum-start controller. Also, the cost of a dedicated unit may be lower than the cost of re-programming an existing centralized computer system and connecting it to the boiler plant equipment. See Reference Note 13, Energy Management Control Systems, for the reasons why.

Combine with Optimum-Start for Cooling

Many models of optimum-start controls (but not all) provide optimum start for both heating and cooling. Thus, you can use a single controller for both the chiller plant and the boiler plant. You can also use it with air handling systems, and with other heating and cooling equipment. See Measure 2.1.2.4 for optimum-start with chillers, and Measure 4.1.1.2 for optimum-start with air handling systems.

How to Install the Outside Air Temperature Sensor

Efficient operation of an optimum-start control depends on getting true outside and inside temperature signals. The placement and shielding of the outside air temperature sensor are critical. Install the sensor so that:

- it is exposed to the prevailing outdoor air temperature
- the penetration through the envelope is sealed so that positive pressure inside the building cannot force inside air over the sensor
- sunlight does not heat the sensor, either directly or by heating surrounding surfaces
- precipitation cannot reach the sensor, either directly or by splashing
- it is protected from physical impact.

ECONOMICS

SAVINGS POTENTIAL: *0.1 to 10 percent of heating costs, the percentage being larger in colder climates.*

COST: *Electronic time controls with optimum-start cost from $100 to several thousand dollars. Labor costs depend on the difficulty of making connections to the boiler and auxiliaries. The cost of programming an existing computer energy management system for optimum-start is typically several thousand dollars.*

PAYBACK PERIOD: *Less than one year, to several years.*

TRAPS & TRICKS

In addition to the cautions in Measure 1.1.1.1, make sure that the outside air temperature sensor is installed properly.

MEASURE **1.1.1.3 If the boiler plant is used only for comfort heating, limit the operation of the boiler plant based on the outside air temperature.**

RATINGS
New Facilities Retrofit O&M

This is the third of a series of Measures that prevent unnecessary operation of the boiler plant. It is limited to boilers or boiler plants that are used only to serve loads that respond to weather. Space heating is the typical application.

The technique is to control boiler plant operation simply by sensing the outside air temperature. This method is cheap, reliable, and effortless. It gives more precise response to weather than the seasonal shutdown that is commonly used for heating plants. Figures 1 and 2 show a couple of typical control installations.

Use the outside air temperature sensor to control plant auxiliary equipment as well as the boilers themselves. Refer to Measures 1.1.3 ff for this.

Where to Use This Method

The usual application for this method is a boiler plant that is used only for comfort heating. The loads of all the facilities served by the boiler plant must depend on weather in the same way.

SUMMARY

A simple, reliable method of avoiding unnecessary boiler operation. Limited to boilers that serve only space heating loads, or other loads that are related to the outside air temperature.

SELECTION SCORECARD

Savings Potential .... $ $ $ $

Rate of Return, New Facilities % % % %

Rate of Return, Retrofit.......... % % % %

Reliability .... ✓ ✓ ✓

Ease of Retrofit .... ☺ ☺ ☺

If a small fraction of the boiler plant load is not related to weather in the same way as the rest of the load, consider separate heating equipment for those applications. For example, you might install localized water heaters to allow you to shut down the boiler plant during warm weather. This approach is recommended by Measure 1.12.2.

Energy Saving Potential

Using outside air temperature to control the operation of the boiler plant saves energy in these ways:

• *by shutting off heating to end uses that tend to waste energy during mild weather.* Common

WESINC

Fig. 1 Boiler water temperature reset This control installation resets the heating water temperature based on the outside air temperature. This enhances comfort during less severe weather, and improves control valve performance.

WESINC

Fig. 2 A simple on-off control This thermostat simply turns the boiler on or off. It is connected to a temperature sensor installed outside the window. The labeling is crude but effective, at least until the next paint job.

examples are apartment complexes, military bases, and campuses where tenants do not pay directly for their utilities and may be negligent in controlling heating. Buildings with openable windows are especially vulnerable to this waste of heating energy. On the other hand, facilities with good end-user thermostatic control have little loss of this type.

- *by reducing standby and distribution losses.* A boiler system suffers heat loss even when there is no end-use heating. Large central heating systems with distribution piping in poor condition may have large losses even when there is no end-use heating load. On the other hand, typical single-building steam heating systems have little loss of this type.

- *by reducing operation of auxiliary equipment.* Heating plants continue to use pump energy, regardless of load, unless the entire plant is turned off. Steam heating systems typically have less auxiliary load during mild weather.

How to Install the Outside Air Temperature Sensor

The only critical aspect of this control method is installing the outdoor temperature sensor. Refer to Measure 1.1.1.2, previous, for details.

ECONOMICS

SAVINGS POTENTIAL: *0.2 to 5 percent of annual heating energy consumption.*

COST: *Several hundred to several thousand dollars, depending on the existing controls and the difficulty of installing the outside air thermostat.*

PAYBACK PERIOD: *One year to several years.*

TRAPS & TRICKS

INSTALLATION: *Make sure that the outside air temperature sensor is installed as described in Measure 1.1.1.2. A temperature reading error of even a few degrees can cause discomfort, leading to abandonment of the energy saving control.*

MONITOR PERFORMANCE: *Inspect the controls at scheduled intervals to ensure that they are working properly.*

MEASURE 1.1.1.4 In applications where automatic starting and stopping of boilers is not desirable, use automatic controls to signal the starting and shutdown sequence to operators.

RATINGS

New Facilities | Retrofit | O&M

A B ☐

This is the fourth of a series of Measures that prevent unnecessary operation of the boiler plant. It is useful for plants where it is not acceptable to start or stop the plant automatically. This is typical in large boiler plants where starting may require a sequence of manual operations, such as draining steam lines and preheating feedwater. From a safety standpoint, it may be necessary to have staff present to take action in the event of equipment failure at start-up. Also, it may be dangerous to turn off boilers automatically if they serve a critical function.

Unfortunately, manual starting and stopping is not reliable in terms of minimizing the operation of equipment to conserve energy. Even when operators stand watch at equipment control stations, they may not be able to judge the optimum starting times for the boilers.

A solution is to use the appropriate methods from the previous Measures to indicate when equipment should be turned on and off, rather than operating the systems directly. For example, you can use an optimum-start timeclock to flash a light on an indicator panel when the boilers should be turned on.

Similarly, you can use a series of timers to ensure that the proper auxiliaries are turned on before the boiler starts. For example, the timers can indicate the need for feedwater heating, fuel oil heating, starting of feedwater and condensate pumps, etc., at the appropriate times in advance of signalling the boiler to be lighted.

SUMMARY

A simple method of minimizing unnecessary operation in boiler plants where the equipment must be started and stopped manually.

SELECTION SCORECARD

Savings Potential .... $ $ $ $

Rate of Return, New Facilities % % % %

Rate of Return, Retrofit .... % % % %

Reliability .... ✓ ✓

Ease of Retrofit .... ☺ ☺ ☺

This type of control can be built into any large control panel, such as the one shown in Figure 1. The controls alert operators with indicator lights and audible signals, preferably associated with a diagram of the plant equipment being controlled. In effect, the panel serves as an automated checklist for start-up and shutdown. An elegantly executed panel invites people to use it.

This method is less expensive than connecting automatic controls to the equipment itself. The connections are simple and do not require specialized knowledge of the equipment controls.

Control Other Plant Equipment with the Same Panel

For greatest efficiency, use this same method to control chillers (see Measure 2.1.2.5), air handling units (see Measure 4.1.1.6), and other equipment that must be started and stopped manually.

ECONOMICS

SAVINGS POTENTIAL: *1 to 5 percent of system operating cost.*

COST: *Several hundred to several thousand dollars.*

PAYBACK PERIOD: *Less than one year, typically.*

TRAPS & TRICKS

DESIGN AND INSTALLATION: *Design and install the panel so that it grabs the attention of the operators, with bright indicator lights and loud audible signals. But, don't make it obnoxious, or the operators will disable it.*

MAINTENANCE: *Test the panel periodically to ensure that it is working properly. Replace the burned-out bulbs, etc.*

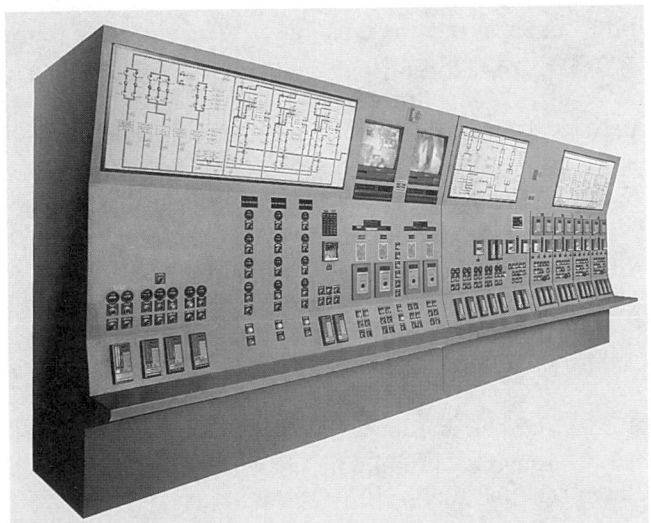

Preferred Utilities Manufacturing Corp.

Fig. 1 Control panel for a large boiler plant It is easy to include annunciators in this type of panel to alert operators when to start and stop equipment.

MEASURE 1.1.2 With multi-fuel boilers, select the most economical fuel on a moment-to-moment basis.

If you have the capability of using more than one type of fuel or energy source, use the most economical source at all times. This may provide significant savings with minimal effort. It is not uncommon for one of the fuels to be more economical at some times, and less economical at others. For example, a common dual-fuel combination is oil and natural gas. Natural gas may be less expensive than oil in summer, but more expensive in winter.

You expect plant operators to be competent in the technical aspects of plant operation, but don't expect them to know the details of energy pricing. You have to translate your energy pricing and other costs into specific instructions for fuel changeover that the plant operators

SUMMARY

An important cost saving practice in plants that can burn several fuels. Requires analysis and constant attention.

SELECTION SCORECARD

Savings Potential	$ $ $ $
Rate of Return	% % % %
Reliability	✓ ✓
Ease of Initiation	☺ ☺

can follow. See Fgure 1. Do this in the four steps explained here.

This Measure shows you how to decide when to change from one energy source to another. The subsidiary Measure recommends automatic changeover, which is possible only under certain conditions.

Step 1: Determine Your Instantaneous Energy Costs

Calculate the cost of all your fuels on a common baseline, such as dollars per million BTU. To do this, get out every one of your boiler fuel contracts and rate schedules. Study them until you understand them completely.

Fuels that you purchase in batches and store on your premises, such as oil and propane, have the simplest pricing. Base your fuel switching decision on how much your fuel is worth at the time. This is not what you paid for the fuel, but what you will have to pay when the time comes to buy the next batch. You will have to use some judgement in deciding this.

Rates for pipeline gas are usually constant over long periods of time, but check your rate schedule to make sure.

Electricity pricing can be complex. See Reference Note 21, Electricity Pricing, to help you understand your electricity rate schedule. Don't be reluctant to call your utility for clarification.

By the time you finish this step, you may learn ways to buy your energy sources more economically.

Step 2: Calculate Your Other Fuel-Related Costs

Your choice of fuel will have cost effects other than the cost of the fuel itself. Take into account:

- *maintenance costs.* For example, burning oil instead of gas increases the need for cleaning firesides, burners, fuel oil heaters, etc.

Cleaver-Brooks

Fig. 1 This boiler can burn oil or gas, but which is it burning? As long as you are paying people to staff the boiler plant, have them optimize the fuel choice at all times.

• *effect on boiler efficiency.* Different fuels have different combustion efficiencies in a particular boiler. Measure the efficiencies using the methods explained in Measure 1.2.1.

• *auxiliary equipment operation.* For example, burning heavy fuel oil instead of light oil increases the cost of soot blowing and fuel tank heating. Burning fuel oil instead of gas requires the operation of fuel oil pumps.

Do not include amortization of equipment that is not affected by your choice of fuel. For example, fuel oil requires storage tanks, but since they already exist, their cost is not a factor in your fuel selection. (Costs of this type are called "sunk costs", which means that they should have no effect on future economic decisions.)

Express these costs in the same way that you calculated your fuel costs, for example, dollars per million BTU.

Step 3: Decide on the Changeover Conditions

In most cases, you want to change to each of the available fuels as soon as it becomes the least expensive choice, considering the cost factors that you calculated in Steps 1 and 2. Now, you have to decide how and when to decide to change fuels.

If all your fuel prices remain constant for long intervals, review your fuel selection whenever the price of one of the fuels changes. For example, if you can burn light oil or natural gas, you might shift to fuel oil when its price drops to $0.50 per million BTU less than the price of natural gas. The differential accounts for differences in efficiency, maintenance costs, and auxiliary equipment. You should review the relative prices whenever you buy a new batch of fuel oil or whenever the price of gas changes.

If a fuel price depends on conditions, such as the time of day or the month of the year, you need to take action whenever these conditions change. For example,

if electricity is your least expensive energy source during off-peak hours, switch to the electric boiler during these hours. For predictable changeovers like this, consider automatic changeover, which is recommended by the subsidiary Measure 1.1.2.1.

Step 4: Give the Instructions

The fourth step is effective writing. Compose clear and specific instructions that explain when to switch from one energy source to another. If necessary, include a worksheet for any calculations that need to be done.

If you assign responsibility for deciding changeover to the boiler operators, recognize that they have a strong bias in favor of fuels that require the least equipment maintenance, regardless of cost. Operators prefer to burn gas over oil, and light oil over heavy oil. Unfortunately, the fuels preferred for low maintenance are usually the more expensive. Explain the economic consequences to the operators.

Review the procedure periodically to make sure that it is staying abreast of the latest energy price structures. Make this a permanent item in your operations calendar.

ECONOMICS

SAVINGS POTENTIAL: Up to 30 percent of fuel cost, based on typical differences in fuel prices.

COST: Minimal. A certain amount of managerial work is required to analyze fuel costs and rates, and to establish the appropriate procedures.

PAYBACK PERIOD: Immediate.

TRAPS & TRICKS

OPERATION: The main weakness of this Measure is the continuous attention that is needed to be sure that the best fuel is selected. This is a training and communication challenge. Schedule periodic checks to verify that the procedure is being followed.

MEASURE   **1.1.2.1 Install automatic fuel changeover.**

This subsidiary Measure recommends switching fuels automatically in multi-fuel installations. The previous Measure showed you how to decide when to switch.

Automatic fuel changeover is most important when it is desirable to switch fuels frequently or unexpectedly. For example, some gas suppliers offer a discount if gas service can be terminated whenever the outside temperature falls to a certain level. To exploit such opportunities, you need to be able to respond quickly and with certainty. Manual changeover may be unreliable, especially if the boiler plant is unattended.

If electricity is used as an alternate energy source, it is especially important to stop using electricity during the times specified in the electric rate schedule. Failure to do so even one time typically voids the discount that applies to off-peak electricity.

How to Install the Equipment

To switch fuels automatically, you need the right controls to shut down the burner for one fuel safely and start the burner for the other fuel. Dual-fuel boilers almost always have such controls.

To make the changeover automatic, the controls must be able to sense the conditions that determine when to change fuels. For example, if the changeover is based on the time of day, install a timeclock to provide the changeover signal to the burner controls. Similarly, the control signal might be based on outside air temperature or the facility's total electric demand. Or, the signal might be transmitted by a fuel supplier.

Boiler controls have major safety considerations, so have the work done by a boiler controls specialist.

SUMMARY

A way of ensuring that the most economical fuel is used without relying on continuous operator attention. Still requires periodic management review. Feasible only if fuel changeover relates to a condition that can be sensed automatically.

SELECTION SCORECARD

Savings Potential	$ $ $ $
Rate of Return, New Facilities	% % % %
Rate of Return, Retrofit..........	% % % %
Reliability	✓ ✓ ✓
Ease of Retrofit	☺ ☺

ECONOMICS

SAVINGS POTENTIAL: *Up to 30 percent of fuel cost, based on typical differences in fuel prices.*

COST: *Several thousand dollars, typically.*

PAYBACK PERIOD: *Varies widely, depending mainly on the rate of fuel consumption and on the difference in fuel prices.*

TRAPS & TRICKS

INSTALLATION: *Work with a vendor who is experienced with the type of changeover controls that you need.*

MONITOR PERFORMANCE: *Check the changeover operation periodically. It could fail for years without being noticed.*

MEASURE **1.1.3 Operate boiler auxiliary equipment consistent with boiler operation and load.**

RATINGS

New Facilities | Retrofit | O&M

B

In many boiler plants, you can save energy just by being careful how you operate your plant auxiliary equipment. For example, see Figures 1 and 2.

Auxiliary equipment wastes energy when it consumes more energy than is needed to handle the existing level of boiler output. There are two main causes of this waste:

- *operating unnecessary equipment.* A boiler plant typically has duplicate pumps, as in Figure 1, and it may have other duplicate equipment. These units are installed to provide increased reliability, or to accommodate different boiler loads. The plant staff may operate more of the units than are necessary. You can correct this cause of energy waste by improving operating procedures and by installing automatic controls.

- *throttling control of equipment output.* Some auxiliary units are driven at constant speed under all conditions, with their output throttled to follow the boiler load. For example, this is the mode of

SUMMARY

Another basic operating practice that saves money and reduces equipment maintenance. Mostly a matter of effective communication with the operating staff.

SELECTION SCORECARD

Savings Potential .... $ $

Rate of Return .... % % % %

Reliability .... ✓ ✓

Ease of Initiation .... ☺ ☺ ☺

operation for combustion air fans on smaller boilers, and for feedwater pumps on many larger boilers. To correct this kind of energy waste, you need to modify the auxiliary equipment or get different kinds.

This Measure and the two subsidiary Measures reduce energy waste from the first of these two causes. This Measure suggests general procedures and the subsidiary Measures recommend control improvements.

Energy waste from the second cause is covered by Measures in Subsection 1.5 (for combustion air fans),

WESINC

Fig. 1 Fuel oil pumps These do not need to operate unless the burner is firing. Control them to turn off when they are not needed, but provide for keeping them primed.

WESINC

Fig. 2 Stack draft inducer This retrofitted unit, intended to solve a draft problem, was found running continuously. It should be controlled to turn off when the boiler is not firing. Otherwise, it increases standby losses and cools the flue, which kills the draft and causes flue corrosion.

Subsection 1.8 (for water pumps and water treatment equipment), and Subsection 1.9 (for fuel oil pumps).

Energy Saving Potential

The amount of energy you can save, and the potential for wasting energy, varies widely between boiler plants. These factors depend on the amount of energy that is consumed by auxiliary equipment and on how effectively the auxiliary equipment is controlled.

For example, a large boiler plant burning heavy oil may expend five to ten percent of its total energy input to operate auxiliary equipment, and even more under certain load conditions. At the other extreme, a gas-fired boiler with atmospheric burners has virtually no auxiliary energy consumption.

As a rule, energy waste usually does not occur with auxiliary equipment that is controlled in an on-off manner in response to the boiler load. For example, there is no unnecessary operation of an on-off feedwater pump that operates in response to a boiler water level control. If the boiler is not operating, the boiler water level remains constant, and the feedwater pump remains off.

In contrast, one feedwater pump may be used to serve several boilers through solenoid valves. In most such installations, the pump runs continuously, regardless of the needs of the boilers. It is easy to overlook the fact that the pump keeps running even when all the boilers are turned off.

Decide How to Control the Equipment

Start by identifying all the auxiliary equipment in the boiler plant, including:
- heating water circulating pumps
- feedwater pumps
- condensate pumps
- fuel oil pumps
- fuel oil heaters
- equipment for handling solid fuels
- forced draft fans

- induced draft fans
- air compressors for soot blowing
- air compressors for control air
- water treatment chemical feeders
- and, other specialized equipment.

Determine the conditions under which each item is needed. See how each item is controlled to respond to boiler load.

In cases where an item of auxiliary equipment can operate unnecessarily, decide whether you want to control the auxiliary equipment manually or automatically. Manual control is satisfactory in facilities where you can maintain adequate operator training and discipline, but manual control is likely to fail in other environments. If you decide to use automatic control, refer to the subsidiary Measures.

Give the Instructions

Where you decide to retain manual control, give appropriate instructions for operating each unit of auxiliary equipment that has the potential of being operated unnecessarily. Make your instructions clear and permanent. To make the instructions permanent, install an effective placard that explains the relationship between boiler operation and auxiliary equipment operation. See Reference Note 12, Placards, for details of effective placard design and installation.

ECONOMICS

SAVINGS POTENTIAL: *Up to 40 percent of the cost of operating the boiler auxiliary equipment.*

COST: *Minimal, if the procedures are accomplished manually.*

PAYBACK PERIOD: *Immediate.*

TRAPS & TRICKS

TRAINING AND COMMUNICATION: *This activity is primarily a matter of effective training and supervision. Keep up with it. Install effective placards at the equipment that needs to be controlled.*

MEASURE 1.1.3.1 Interlock auxiliary equipment with the boilers it serves.

RATINGS

New Facilities | Retrofit | O&M

A | B |

Measure 1.1.3 recommended minimizing the operation of auxiliary equipment. This is the first of two subsidiary Measures that provide control techniques to make this foolproof.

The most effective method of minimizing the operation of auxiliary equipment is to interlock it with the operation of the boiler. To do this, provide power to the auxiliary equipment through the boiler itself.

This simple technique works even if a particular item of equipment serves several boilers. For example, consider a boiler plant where a single feedwater pump that runs continuously serves several boilers through solenoid valves. There may be long periods when none of the boilers is firing. To shut down the feedwater pump whenever possible, provide power to the feedwater pump through the solenoid valves, using relays. In this way, the feedwater pumps remain off unless any one of the boilers calls for feedwater.

How Much Cycling Can Equipment Tolerate?

This technique may cause some auxiliary equipment to cycle, rather than running continuously. Cycling equipment too often increases maintenance cost and reduces reliability. For example, the life of a motor is shortened if it is turned on and off frequently, say, every minute or two. This effect may negate the value of the energy saving. Therefore, when interlocking auxiliary equipment to boilers, avoid situations in which the equipment is cycled too frequently.

This is a judgement call. For example, the fan in a power burner cycles along with the flame, and service life is satisfactory. By the same token, you should be able to cycle a feedwater pump of modest size. Most pumps and smaller fans tolerate cycling because they do not have much inertia to resist starting. In general, it is practical to cycle smaller equipment more frequently than larger equipment.

Excessively short cycles are likely to be caused by setting the burners to fire over too small a range of temperature or pressure. (See Measure 1.5.3.4 about increasing this range.)

SUMMARY

Simple control interlocks that ensure that auxiliary equipment does not operate unnecessarily.

SELECTION SCORECARD

Savings Potential .... $ $

Rate of Return, New Facilities % % % %

Rate of Return, Retrofit.......... % % % %

Reliability .... ✓ ✓ ✓ ✓

Ease of Retrofit .... ☺ ☺ ☺

Operation Through Boiler Sequencing Controls

Measure 1.1.4.1 recommends automatic controls to sequence boilers efficiently in multi-boiler installations. Boiler sequencing controls may be able to limit the operation of auxiliary equipment as well. However, you would not install a sequencing controller primarily for this purpose. The method given here is simple, obvious, reliable, and inexpensive. Not every function has to be performed by a microprocessor.

ECONOMICS

SAVINGS POTENTIAL: Up to 40 percent of the cost of operating the boiler auxiliary equipment.

COST: Several hundred to several thousand dollars per connection.

PAYBACK PERIOD: One year to several years, typically.

TRAPS & TRICKS

DESIGN: This technique is simple and fairly foolproof. Be sure that equipment will not cycle excessively under any load conditions.

EXPLAIN IT: Record the modification in the plant's engineering records. Otherwise, it will confuse maintenance personnel when repairs are required.

MEASURE **1.1.3.2 Install power switching that prevents unnecessary operation of spare pumps.**

RATINGS

This is the second of two subsidiary Measures that eliminate unnecessary operation of auxiliary equipment. This technique avoids unnecessary operation of duplicate units.

A boiler plant typically has equipment that is installed in pairs for the sake of reliability. Common examples are pumps for heating water, condensate, feedwater, fuel oil transfer, and fuel oil pressurization. You can avoid unnecessary operation of this equipment simply by installing a transfer switch in the equipment's power circuits. See Measure 10.4.1 for the details.

MEASURE **1.1.4 Distribute the heating load among boilers in the manner that minimizes total plant operating cost.**

RATINGS
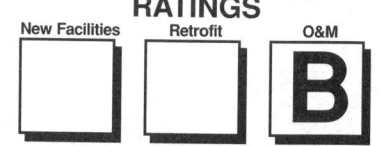

If you have more than one boiler, you may be able to reduce your operating costs by paying attention to the sequence in which you start the boilers, and by the way you distribute the load among them. Here is why:

- boilers of different types, models, and sizes may differ in efficiency by as much as 30%.
- the efficiency of a given boiler may vary significantly at different levels of load. For example, consider a pair of identical boilers whose peak efficiency occurs at 60%, and whose efficiency drops off at higher and lower loads. If the total plant load equals 120% of the full load of one boiler, the two boilers handle the plant load with best efficiency if each operates at 60% load. In contrast, if one boiler were operated at 100% load and the other boiler were operated at 20%, both would be operating well away from their points of maximum efficiency.
- boilers may differ in maintenance cost and in the cost of operating their auxiliary equipment
- different boilers may use different fuels. (This issue was covered by Measures 1.1.2 ff.)

In a nutshell, you want to operate the cheapest combination of boilers and boiler load under all conditions. Doing this involves three steps: (1) calculate the total operating cost of each boiler at all of its firing levels, (2) determine the optimum combination of boilers and boiler loading for each level of overall heating load, and (3) issue effective instructions to operate the boilers at the optimum loading.

The subsidiary Measure changes the third step by recommending boiler sequencing controls that distribute the boiler load automatically.

SUMMARY

An operating procedure that may save a significant amount of energy at minimal cost. Requires some preliminary calculations.

SELECTION SCORECARD

Savings Potential .... $ $

Rate of Return .... % % % %

Reliability .... ✓ ✓ ✓

Ease of Initiation .... ☺ ☺ ☺ ☺

Boiler Efficiency Characteristics and Energy Saving Potential

To provide background for accomplishing this Measure, let's quickly review the efficiency characteristics of boilers. This will give you an idea of the benefit that you can expect from sequencing your boilers most efficiently.

There is a distinct difference in efficiency between modern boilers and many older boilers. Some older boilers may have efficiencies as low as 70%, or even lower. When an older boiler is inefficient, the major cause usually is insufficient heat transfer. In such boilers, efficiency drops substantially as load is increased. Also, there is a drop in efficiency at the low end of the load range.

Standby losses may be the most important source of inefficiency in some boilers. The most vulnerable type, boilers with open atmospheric combustion chambers, may average standby losses of 10% over the course of a year. Standby losses are explained in Measure 1.5.3. There are large differences in the standby losses of different types of boilers. In summary, (1) boilers with open combustion chambers have much higher standby loss than boilers with closed combustion chambers, (2) single-stage burners have more standby loss than staged burners, and (3) burners that modulate over a wide range have little standby loss. In a given boiler, standby losses increase as the load is reduced.

Among the best available industrial-sized boilers, there is little difference in efficiency, for a given fuel. The best modern boilers commonly approach the theoretical maximum efficiency that is practical. Good oil-fired boilers have peak combustion efficiencies ranging from 82% to 85%, depending on the grade of oil for which the boiler is designed. Good gas-fired boilers have peak combustion efficiencies around 81%.

In boilers that are designed for high efficiency, the efficiency does not vary much across the load range. Typically, the efficiency of such boilers is almost constant in the upper half of the their load range, falling off a few percent at low loads. The reason for the drop at low load is that losses from such boilers (such as surface radiation) remain constant, so that they assume greater relative significance at lower loads.

There is one significant efficiency difference among modern boilers. Some boilers recover the heat of flue gas condensation, which gives them substantially higher efficiencies than other boilers. These boilers typically have total efficiencies exceeding 90%. They are usually fired by gas.

In general, boilers with low peak efficiency also tend to lose more efficiency if they are not maintained well. For example, if a boiler has inadequate heat transfer surface, it loses efficiency rapidly as fireside and waterside fouling accumulate.

■ How Burner Modulation Affects Efficiency

The ability of a boiler to vary its output is determined by the burner design. Most burners are either single-stage, two-stage, or modulating. A boiler's efficiency changes with output, so the way the burner changes its output determines the boiler's average efficiency.

If the boiler has a single-stage burner, the only boiler efficiency that matters is the efficiency at full output, i.e., with the burner operating. Standby losses are the only influence that can vary the boiler's average efficiency.

With two-stage burners, you are interested in the efficiency at both firing rates. The "low-fire" output is typically about 60% of the "high-fire" (maximum) output, although this may vary. In an efficient boiler, the efficiency at both stages is about the same, since modern boilers are typically designed for peak efficiency at about 80% of full load. In boilers with inadequate heat transfer, the first stage may be significantly more efficient than the second stage.

Modulating burners can adjust output over a continuous range, extending from full load down to a minimum output. Boilers with multiple burners may have a minimum output close to zero. Single modulating burners typically have a minimum output between 25% and 50% of full load.

If a boiler has an inefficient load range, a modulating burner will allow the boiler to operate in that range. The amount of energy waste depends on the length of time that the boiler is fired in the inefficient load range. This depends on the load profile, on the size distribution of the boilers, and on the way the load is shifted between boilers.

Step 1: Calculate Operating Cost vs. Load for Each Boiler

You need to work with hard numbers, not with assumptions. For example, it is not always true that a load is served most efficiently with the smallest number of boilers, or that a smaller boiler should be shut down when a larger one is started.

The starting point is to find out your actual boiler efficiencies at all their firing rates. Do this by performing combustion efficiency tests, using the methods spelled out in Subsection 1.2.

Combustion efficiency tests do not reveal standby losses. You have to estimate your standby losses from the type of boiler and its typical load profile.

After you have collected your efficiency figures, calculate the fuel cost per unit of output energy for each boiler, at each firing rate.

If your boilers have separate auxiliary equipment, calculate the costs of operating the equipment as a function of load.

If your boilers differ in maintenance cost, calculate the maintenance cost as a function of load.

Now, add all these costs together. This gives you the total operating cost of each of your boilers as a function of load.

(This procedure is similar to the one in Measure 1.1.2 for selecting the most economical fuel in multi-fuel installations. If you have multi-fuel boilers, save effort by doing both sets of calculations at the same time.)

Step 2: Decide the Start-Stop Sequence and Load Sharing

Now that you know the individual cost characteristics of your boilers, you have to figure out how to combine the boilers most economically for each level of overall load. In some cases, this is simple. In other cases, it requires a few hours of time with graph paper and a calculator. You have to do the most figuring if your boilers are not identical, and especially if their efficiencies overlap in different parts of their load ranges.

The optimum method of distributing the load depends on the type of burner modulation. To avoid confusion, we will discuss the load distribution procedures separately for each type burner.

■ Distributing the Load with Single-Stage Burners

If all the boilers have single-stage burners, the only step necessary is to sequence the boilers in the order of their operating costs.

In normal operation, this requires no special controls and no special effort. The firing of a steam boiler normally is controlled by the steam pressure in the discharge header, and the firing of a hot water boiler normally is controlled by its discharge water temperature. To get the most efficient boiler to operate first, adjust the burner control to start at the highest pressure or temperature, with the other boilers adjusted to start at successively lower pressures or temperatures, in order of their increasing operating cost.

Starting a cold plant needs special attention. Simple firing controls try to start all the boilers together. Ordinarily, you want to warm up the plant using only the most efficient boiler or boilers. Therefore, you have to lock out all the boilers except the most efficient during the warm-up phase.

However, if there is a heavy heating load, the most efficient boilers may not be able to satisfy the load alone, or they may take too long to warm up the system. In this case, the operators must decide which boilers to operate, based on the expected load.

Alternatively, you can use an automatic controller for this purpose. The controller must be able to determine the load, or to predict the load. If the boiler plant is used primarily for comfort heating, you can use an optimum-start controller that responds to changes in the outside temperature. See Measure 1.1.1.2 for details of this equipment.

Don't worry about the efficiency of starting a cold plant if the plant runs continuously, or for an entire season. On the other hand, if a plant is started every day, this can be a major factor.

■ Distributing the Load with Two-Stage Burners

Setting the most efficient firing sequence for boilers with multi-stage burners is only slightly more complex. The trick is to think of each stage of firing as a separate boiler to be sequenced efficiently.

For example, consider a boiler with a two-stage burner that is firing its first stage. As the load increases, should the boiler fire on its second stage, or should a second boiler be started on its first stage? In the form of a table, here are the two choices:

Cut-In Pressure	Boiler	Stage		Boiler	Stage
13 PSI	#1	1st		#1	1st
12 PSI	#2	1st	OR	#1	2nd
11 PSI	#1	2nd		#2	1st
10 PSI	#2	2nd		#2	2nd

If both boilers have the same efficiency characteristics, the answer depends on the efficiency of each stage of firing. For example, if both boilers are 80% efficient in the first stage and 76% efficient in the second stage, then start both boilers at the first stage before firing any boiler at second stage.

Conversely, if the first boiler is 80% efficient at both stages, and the second boiler is 76% efficient at both stages, then fire the first boiler to full output before starting the second boiler.

In normal operation, distribute the load between firing stages by setting the turn-on pressures or temperatures for each stage, as with single-stage boilers.

Starting a cold plant that has multi-stage boilers may be somewhat more complex than with single-stage boilers. Virtually all multi-stage boilers are controlled to start at high fire. You will probably need a boiler scheduling controller to let you select the firing sequence of boiler stages between different boilers.

In other respects, starting a cold plant involves the same considerations as with single-stage boilers.

■ Distributing the Load with Modulating Burners

If your boilers have similar cost-vs.-load characteristics, operate all the boilers at equal percentages of load. This general rule applies even if the boilers are different sizes. (For those who are interested in mathematical proof of this, it derives from the fact that the efficiency curves of most boilers are convex upward at all points. The same is likely to be true of the cost curves.)

If your boilers fit this description, how many should you operate? The answer: operate the number that keeps the cost per unit of output as low as possible. For example, if the boilers have the lowest operating cost at 70% load, operate the number that keeps the load on each boiler nearest to this percentage.

With modulating burners, it is easy to keep all the operating boilers near the same percentage of load. Simply set the firing controller of each boiler to the same steam pressure or water temperature.

The situation is more complex if the boilers have different efficiency profiles. For example, if the first boiler is very efficient, do not start the second boiler until the first boiler reaches full load. Conversely, if the first boiler has a serious drop in efficiency at higher loads, it might be more economical to start the second boiler before the first boiler reaches full load.

If the boilers do not have identical efficiency profiles, you will have to install a specialized controller that modulates the output of each boiler according to a schedule that is programmed into the controller. Each installation must be customized to the characteristics of the boilers. See subsidiary Measure 1.1.4.1 for details.

Warming up a cold plant efficiently involves the same considerations as for single-stage and multi-stage burners, discussed previously.

Step 3: Instruct or Automate

The final challenge is to educate the boiler operators about the operating schedule and to make sure that the procedure is followed. If the boiler load changes continuously, this activity can be tedious. People are involved, so the process may be unreliable. Except in the simplest cases, it is far more reliable to automate this process, as recommended in the subsidiary Measure.

Related Boiler Efficiency Improvements

Some of the other Measures may affect the way you shift load between boilers. For example, if you have multi-stage boilers that suffer low efficiency at high fire, you may eliminate the high-fire setting in one or more of the boilers. (Measure 1.4.3 recommends this.) That modification converts the boiler to a single-stage boiler of lower capacity, which is easier to coordinate with other boilers.

Keep in mind that the energy consumption of boiler auxiliary equipment may be an important cost factor. See Measures 1.1.3 ff, above, to coordinate auxiliary equipment operation efficiently with boiler operation.

ECONOMICS

SAVINGS POTENTIAL: *Typically, 1 to 10 percent of the cost of boiler operation, depending on the relative efficiency characteristics of the boilers that are installed.*

COST: *Small. The main cost is for the analysis required to create the optimum operating schedule, including boiler efficiency testing.*

PAYBACK PERIOD: *Less than one year.*

TRAPS & TRICKS

CALCULATIONS AND EFFICIENCY MEASURE-MENTS: *Don't make any assumptions about boiler efficiency. To distribute the load efficiently, you need to know the real efficiencies of all your boilers at all firing levels. Measure these efficiencies yourself.*

TRAINING AND COMMUNICATION: *The big challenge is keeping the plant operators on top of the procedure. This requires effective training and supervision.*

MEASURE 1.1.4.1 Install an automatic boiler scheduling controller.

The manual control procedure recommended by Measure 1.1.4 has a number of weaknesses in all but the simplest boiler installations: it requires operators to be present in the boiler plant whenever boilers are operating, it requires constant attention by the operators, and it requires resetting the boiler firing response whenever an additional boiler is started or stopped.

This subsidiary Measure avoids these shortcomings by using a boiler scheduling controller to automatically sequence boiler firing and to distribute the load among the boilers.

Where to Install a Boiler Scheduling Controller

The additional benefit provided by a scheduling controller depends on the type of burner modulation, on the relative efficiencies of the boilers, and on how often you start a cold plant.

Boilers with single-stage burners can be sequenced satisfactorily simply by sensing the system temperature or pressure, for the reasons discussed in Measure 1.1.4. If all your boilers are single-stage, a sequencing controller is not likely to be useful except when starting a cold plant. In the latter case, a sequencing controller is most valuable if the boilers differ in efficiency.

The same is usually true of boilers with multi-stage burners. Make sure that the burner staging controls can

SUMMARY

Eliminates the need for continuous operator attention to keep boiler loading optimized. Requires accurate programming.

SELECTION SCORECARD

Savings Potential .... **$ $**

Rate of Return, New Facilities **% %** % %

Rate of Return, Retrofit.......... **% %** % %

Reliability .... ✓ ✓ ✓

Ease of Retrofit .... ☺ ☺ ☺

stagger the firing of different boilers in an efficient sequence. This should not be a problem if the boilers are similar in efficiency characteristics. However, if there are more than two or three boilers with multi-stage burners, conventional staging controls that sense system temperature or pressure independently may cause unstable cycling. In this case, get a sequencing controller.

A controller is usually necessary for optimum automatic sequencing if the boilers have modulating controls, especially if the boilers start and stop automatically.

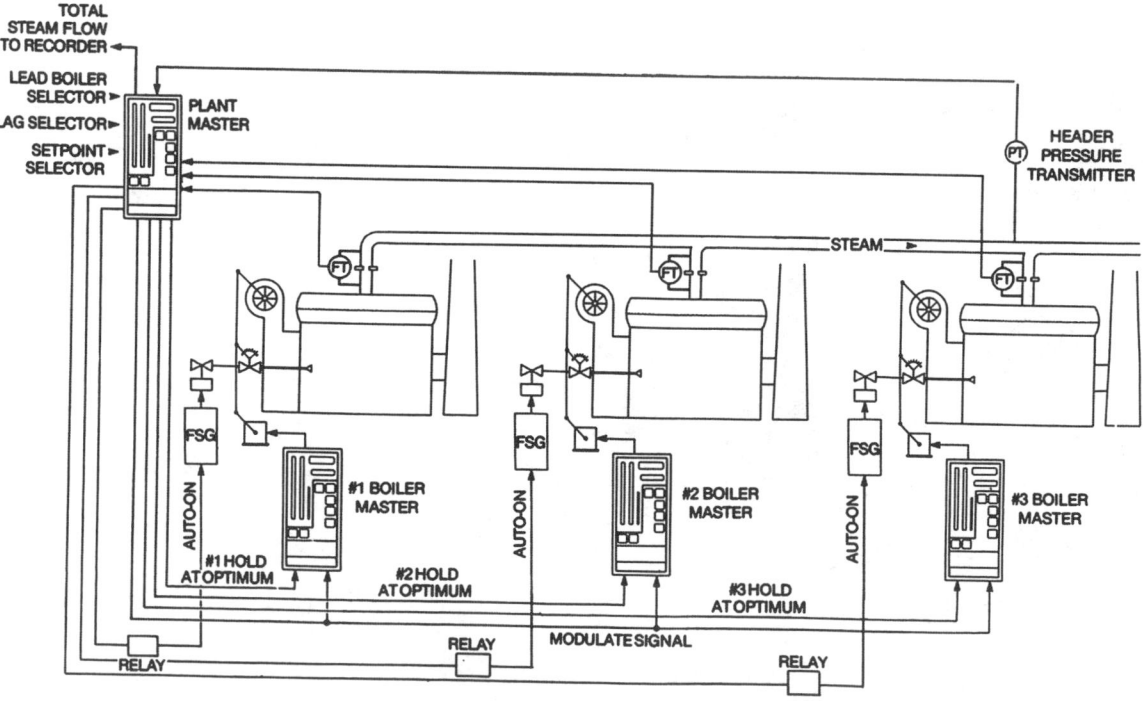

Preferred Utilities Manufacturing Corp.

Fig. 1 Typical control system for optimizing boiler load

How to Select the Controller

A boiler scheduling controller is a simple computer that senses the water temperature or steam pressure and sends the appropriate firing instructions to the burners. Various models are available that differ in sophistication and cost. Figure 1 shows a typical system installation.

Select a controller that has sufficient capability to accommodate differences between the boilers. If modulating burners are used in boilers with different efficiency characteristics, the controller should be able to shift load between the boilers based on the individual control response of each burner.

If the boiler plant is shut down frequently, be sure that the controller can warm up a cold system efficiently.

These functions could be performed with an energy management computer system, or with a general purpose computer that is properly interfaced to the boiler firing controls. However, both approaches require a substantial amount of custom work. Unless the control requirements are unusual, it is quickest and most economical to use a controller that is packaged for this purpose.

■ Combination with Other Functions

The functions of this equipment are sometimes combined with combustion controls and combustion trim systems (Measure 1.3.2). When selecting a controller, determine whether it also makes sense to use it for fuel changeover in multi-fuel systems (Measure 1.1.2.1) or for control of auxiliary equipment (Measures 1.1.3 ff).

Programming the Controller

As with any computer, the accuracy of the control signals sent to the burners depends on the accuracy of the input information. You have to input the burner firing sequences you want. With modulating burners, you have to input the burner firing rates at each pressure or temperature. To determine what these inputs should be,

make the same analysis that is outlined in Measure 1.1.4. Do this before you select the controller. It will help you to decide which controller to buy.

ECONOMICS

SAVINGS POTENTIAL: *Typically, 1 to 10 percent of the cost of boiler operation, depending on the relative efficiency characteristics of the boilers that are installed.*

COST: *A specialized boiler load controller typically costs $8,000 to $30,000, depending on the size and complexity of the boiler plant. You may be able to do the job with a programmable logic controller (PLC), which typically costs $1,000 to $5,000, but then you have to program the controller. Also, include the cost of boiler efficiency testing.*

PAYBACK PERIOD: *Several years or longer.*

TRAPS & TRICKS

SELECTING THE EQUIPMENT: *Make sure to select a controller with enough flexibility to handle your combination of boilers. If the plant has boilers of different types, a relatively complex and expensive controller may be needed.*

Resist the temptation to perform this function with a general-purpose energy management computer unless you really understand how to program the computer system. Also, such a computer is likely to become obsolete long before the boilers wear out.

PROGRAMMING: *Selecting the optimum schedule is not as easy as it may seem, especially if the boilers are of mixed types. Be sure to account for all load conditions.*

MONITOR PERFORMANCE: *Schedule periodic checks to ensure that the controller is doing what you want, and that no one has tampered with the scheduling sequence.*

MEASURE 1.1.5 In steam systems, keep steam pressure at the minimum that satisfies equipment and distribution requirements.

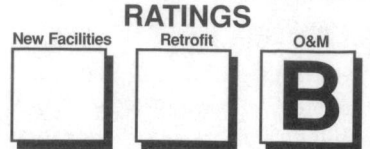

You can save energy and gain other benefits by keeping the boiler and steam distribution pressures to a minimum as the steam load varies. The greatest energy savings occur in extended distribution systems, especially if the system has developed a significant amount of leakage.

You have the greatest ability for reducing pressure at low loads in steam systems that have long pipe runs. This is because a substantial part of the boiler pressure is needed to overcome flow resistance. The resistance is proportional to the square of the flow rate, so it drops considerably at low flow rates.

In steam systems that operate at lower pressures, minimizing steam pressure may allow boilers to operate unattended that otherwise require continuous operator presence, or the boilers may be tended by operators with a lower class of license. This may save labor cost if you can reliably predict the times when steam pressure can be reduced.

Energy Saving Potential and Other Benefits

The benefits of minimizing pressure in a steam system are:

- *reduced leakage throughout the system.* In larger steam systems, leakage can steal a large fraction of the steam produced by the boiler plant. Leakage wastes the heat carried by the steam, increases water consumption, and incurs the cost of treating the replaced water. The extent of the saving is related to the size and condition of the distribution system.

  By the same token, reducing pressure also significantly reduces the growth of leaks, a relentless process responsible for major maintenance and replacement costs.

- *reduced conductive heat loss throughout the system.* If the steam is distributed in a saturated state (that is, without superheat), the temperature of the steam is proportional to its pressure. Conductive heat loss is directly proportional to the temperature, so reducing the pressure reduces the heat loss. The potential saving is related to the effectiveness of the system's insulation.

- *improved boiler efficiency.* Reducing the steam temperature increases the temperature differential between the combustion gases and the water, which increases heat transfer and efficiency. The maximum boiler efficiency improvement is in the range of 1 or 2 percent.

- *reduced scaling and oxygen corrosion.* Scale formation and oxygen corrosion rise dramatically

SUMMARY

An easy adjustment that may save a significant amount of energy, especially in steam systems that are large or leaky. Requires continuous operator attention. In some cases, may allow unattended firing of boilers or reduced operator licensing requirements. May cause problems if overdone.

SELECTION SCORECARD

Savings Potential	$ $ $
Rate of Return	% % % %
Reliability	✓ ✓
Initiation	☺ ☺ ☺

with boiler pressure. Scale reduces heat transfer and efficiency, and it is difficult to remove. Oxygen corrosion in high-pressure systems causes pitting of the boiler tubes, which ultimately leads to pinhole leaks. Reducing steam pressure is no substitute for scrupulous water treatment, but it does provide additional protection. (See Subsection 1.8 about water treatment.)

- *reduced control valve erosion.* Control valves on steam equipment suffer accelerated erosion of the disk and seat surfaces when the valves are almost closed. Fluid passing through valves that are barely open achieves high velocity and turbulent flow because of the pressure difference across the valve. Abrasion by the steam erodes the valve surfaces. Reducing the system pressure lessens this problem in two ways. First, there is less pressure to be converted to steam velocity. Second, the valve is opened wider for a given load, reducing the pressure differential across the valve itself.

- *improved control valve stability at low loads.* Steam control valves tend to chatter or cycle when they are barely open. Reducing the steam pressure causes control valves to open wider, reducing this effect.

The last two benefits occur with pressure reducing valves and with equipment inlet valves that operate directly from the main steam line. They do not occur with valves that are installed downstream of a pressure reducing valve, because the latter holds the downstream pressure constant.

How Low Can You Reduce the Pressure?

A steam system is designed to operate at a certain pressure in order to provide sufficient delivery of steam, sufficient pressure at the user equipment, and sufficient temperature at the user equipment. Do not reduce steam pressure to the point that any of these requirements are violated. For example, steam equipment such as sterilizers and cookers requires a minimum steam temperature, regardless of load.

On the other hand, in a steam system designed for space heating, space heating equipment may require only a fraction of the equipment's design pressure during mild weather. This allows the pressure at the boiler to be reduced more than if distribution pressure losses were the only factor.

Develop a Schedule of Pressure vs. Load

Your objective is to operate the boilers at the lowest acceptable pressure under all load conditions. As the steam output of the plant is reduced, reduce steam pressure correspondingly. To do this rationally, develop a schedule of boiler steam pressure as a function of steam output.

Do this by observing the performance of the user equipment as the pressure at the plant is reduced, to determine the minimum acceptable boiler pressure. Repeat this over the range of plant load conditions, as the boiler goes from low output to high output. If the plant output is related to weather, it may take a period of months to accurately determine the minimum steam pressure that is acceptable under different load conditions.

Different boiler pressures may be required for a given rate of steam output, depending on which loads are being served. For example, a campus may have dormitories that need a large amount of steam at night for heating, and a cafeteria that needs a smaller amount of steam during the daytime for cooking. Even so, the daytime pressure may need to be higher to accommodate the higher pressure requirements of the cafeteria cooking equipment. Similarly, a part of a facility that operates on a distinct schedule may require higher pressure because its distribution system has higher losses.

Modify Steam Users to Lower System Pressure Requirements

If a relatively minor steam user has a larger pressure requirement than the rest of the system, it forces the entire system to operate at higher pressure. In such cases, consider separating this steam user from the rest of the system, or modifying it to operate at lower pressure. See Measure 1.12.3 for details.

Possible Boiler Damage at Reduced Pressure

A boiler is designed to operate within a certain range of pressures. Odd as it may seem, a boiler may have a minimum safe pressure as well as a maximum safe pressure. In case of doubt, contact the boiler manufacturer to find out whether this may be a problem.

Your primary concern is excessive formation of steam bubbles within the boiler. The boiler water cools the steel parts of the boiler against damage from the higher temperature of the combustion gases. Steam does not provide this protection. When pressure is reduced, the steam bubbles in the water expand to take up more volume. This leaves metal surfaces with less cooling. This is most likely to be a problem in watertube boilers, where the enlarged steam bubbles may drive water out of the tubes.

If you reduce pressure only when boiler load is low, this problem becomes less critical because there is less steam formation and less heat transfer through the boiler tubes. However, it is not prudent to depend on this. It is likely that the boiler will sometimes be fired at a high rate even when the pressure is low. For example, this might happen when starting a cold system, or when shutting down one of two boilers.

Another possible problem is that the larger bubbles formed at lower pressure may cause increased carryover of liquid into the steam, along with the impurities and chemicals in the boiler water. This problem is also less likely to occur at low load.

Possible Steam Metering Error

Changing steam pressure may cause steam meters to read erroneously. Error occurs in both orifice meters and pitot tubes, which are calibrated for a given steam density. If metering is critical, you can compensate for pressure changes by using more complex metering equipment. Microcomputers in some steam metering systems can adjust for pressure changes.

Safety Issues if Staffing is Reduced

It is never prudent to skimp on safety, so consider whether it is appropriate to reduce the level of staffing when reducing pressure, even though this may be allowed by local laws. Consider the general condition of the boiler, the nature of the control system, the effectiveness of installed safety devices, and the vulnerability of the facility to the types of boiler failures that might occur.

ECONOMICS

SAVINGS POTENTIAL: *If the distribution system is large and in poor condition, minimizing pressure may save 5 to 20 percent of steam energy. In facilities that have short, well maintained distribution systems operating at low pressures, the benefit of reducing pressure is minimal.*

COST: *Minimal. The main cost is the manpower required to develop the steam pressure schedule.*

PAYBACK PERIOD: *Less than one year.*

TRAPS & TRICKS

PREPARATION: *Check to be sure that reducing pressure will not cause boiler problems. Have the patience to refine your steam pressure schedule until it is optimum. You may need a year of observations to determine the minimum acceptable steam pressures under all conditions of steam demand and weather.*

OPERATION: *You need to stay on top of this activity. Schedule periodic checks to verify that the procedure is being followed and that it is not causing problems. This is a training and communication challenge.*

METERING ACCURACY: *If you need accurate steam metering, get steam meters that can compensate for changes in supply pressure.*

Boiler Plant Efficiency Measurement

Measure the efficiency of your boilers on a regular basis. This is the first step in keeping them efficient. Efficiency testing is important for two reasons. First, a large fraction of the total energy consumed by a facility passes through its boilers, so even a small drop in boiler efficiency represents a large loss of energy. Second, boiler efficiency declines continuously over time, so you need to test efficiency regularly to know when to perform maintenance.

Fortunately, boiler efficiency testing is easy. The Measures in this Subsection show you how to do it, and how to keep your testing program effective.

INDEX OF MEASURES

1.2.1 Test boiler efficiency on a continuing basis.

1.2.2 Install efficiency instrumentation appropriate for the boiler plant.

1.2.3 Calibrate boiler plant instruments at appropriate intervals.

1.2.4 Keep operators proficient in using instrumentation to maximize boiler plant efficiency.

RELATED MEASURES

- Subsection 1.3, for adjusting air-fuel ratio for best efficiency
- Other Subsections in Section 1, for improving efficiency in individual boiler plant components

MEASURE 1.2.1 Test boiler efficiency on a continuing basis.

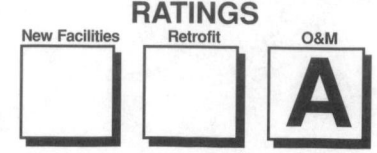

SUMMARY

The basic step in maintaining boiler efficiency. Know the theory well enough so that you understand which tests to use, and what they are telling you.

SELECTION SCORECARD

Savings Potential .... $ $ $

Rate of Return .... % % % %

Reliability .... ✓ ✓

Ease of Initiation .... ☺ ☺ ☺

Periodic efficiency testing is an important part of the management of any boiler plant, for these reasons:

- a large fraction of a facility's total energy consumption flows through its boilers, so even a small drop in efficiency represents a large amount of energy and cost in absolute terms. In facilities where boilers have operated for long periods of time without being tested, it is not uncommon to find that efficiency has fallen by five or ten percent, which may represent a very large cost.

- efficiency testing is the most accurate indicator for adjusting the boiler and its auxiliary equipment, such as adjusting the air-fuel ratio (covered in Subsection 1.3). Efficiency testing tells you the "bottom line" economic performance of the boiler, unlike indirect clues, such as flame color.

- you can localize most boiler problems by knowing how to exploit the full range of efficiency test methods. Efficiency tests are to boilers what blood tests are to human beings.

- efficiency testing is the first step in estimating the benefit of potential boiler improvements, such as adding an economizer or an air-fuel control system.

Boilers are vulnerable to conditions, such as tube fouling, that may reduce their efficiency over relatively short periods of time. Therefore, you should test boiler efficiency on a regular basis. Fortunately, boiler testing is easy. It is practical for most facilities to have at least one person who is proficient in testing. This individual need not be a boiler operator, or even a member of the physical plant staff.

You may need several types of efficiency tests to get a complete picture of boiler performance. There is one main type of test, called a "combustion efficiency" test. You may have to supplement this with one or two specialized tests, such as a test for carbon monoxide, to get a complete picture of your boilers' performance. This Measure tells you how to select the appropriate tests for your boilers, how to do them, and the weaknesses of each method. In a short time, you can be testing your boiler efficiency with confidence.

What "Efficiency Testing" Means in Boilers

Efficiency is defined as the ratio of the useful output energy produced by a system to the raw energy input into the system. In principle, it is possible to test the overall efficiency of a boiler plant. You could do this by measuring the energy content of the steam or hot water that is exported by the boiler plant and the total quantity of energy that is input to the boiler plant, including fuel for the boilers and other energy for the auxiliary equipment.

In actual boiler plants, such a comprehensive test is impractical. Measuring steam or hot water energy with reasonable accuracy, say within one percent, is possible only under laboratory conditions. Field measurements of flow are vulnerable to error, even if they are done by expensive consultants. Measuring input energy is also subject to error. You would have to analyze the energy content of your fuels, make corrections for ambient conditions, and create very stable test conditions. This is not possible under realistic conditions.

Fortunately, you don't need to measure the total efficiency of a boiler plant if your objective is to tune up the plant's efficiency, rather than running a testing laboratory. You can use an easy procedure, called a "combustion efficiency" test, to measure the aspects of boiler efficiency that cause most losses in normal operation. The "combustion efficiency" test determines how completely the fuel is burned, and how effectively the heat of the combustion products is transferred to the steam or water.

The efficiency of the other boiler plant components, such as pumps, fans, and motors, tends to remain constant, so you don't need to test their efficiency on a continuing basis.

Limitations of the Combustion Efficiency Test

The combustion efficiency test is your primary tool for monitoring boiler efficiency. You can achieve accuracy in the range of one percent of efficiency if you do the testing carefully and use equipment of good quality. However, be sure that you understand what the test is telling you. The combustion efficiency test does not account for:

• *standby losses.* You perform a combustion efficiency test when the boiler is operating under a steady load. Therefore, the combustion efficiency test does not reveal standby losses, which occur between firing intervals. You cannot measure standby losses directly. You have to estimate them from the type of boiler and the firing schedule. See Measure 1.5.3 for an explanation of standby losses.

• *heat loss from the surface of the boiler to the surrounding space.* As a practical matter, you cannot measure this loss. Typical estimates state that the loss from surface radiation is about two percent of the boiler's full load energy consumption. (Subsection 1.11 shows you how to reduce surface heat loss and recover the heat.)

• *blowdown loss.* The amount of energy wasted by blowdown varies over a wide range. (Subsection 1.8 shows you how to minimize blowdown loss and how to recover heat from blowdown.)

• *soot blower steam.* The amount of steam used by soot blowers is a variable that depends on the type of fuel and the judgement of the staff. (Subsection 1.6 shows you how to minimize soot blower steam consumption.)

• *auxiliary equipment energy consumption.* The combustion efficiency test does not account for the energy use by auxiliary equipment, such as burners, fans, and fuel pumps. (To minimize this energy consumption, see the Subsections that cover the specific types of auxiliary equipment.)

While the combustion efficiency test does not give the overall efficiency of the boiler plant, it is by far the easiest method of tracking moment-to-moment, day-to-day, and season-to-season variations of boiler efficiency. Combustion efficiency testing tells you how far boiler efficiency drifts away from the best efficiency that you can achieve when the boiler is fully tuned up.

How to Do a Combustion Efficiency Test

You test combustion efficiency by measuring either the oxygen content or the carbon dioxide content in the flue gases. The oxygen test can provide much better accuracy in the range of air-fuel ratio where the boiler is supposed to operate. In recent years, oxygen testers

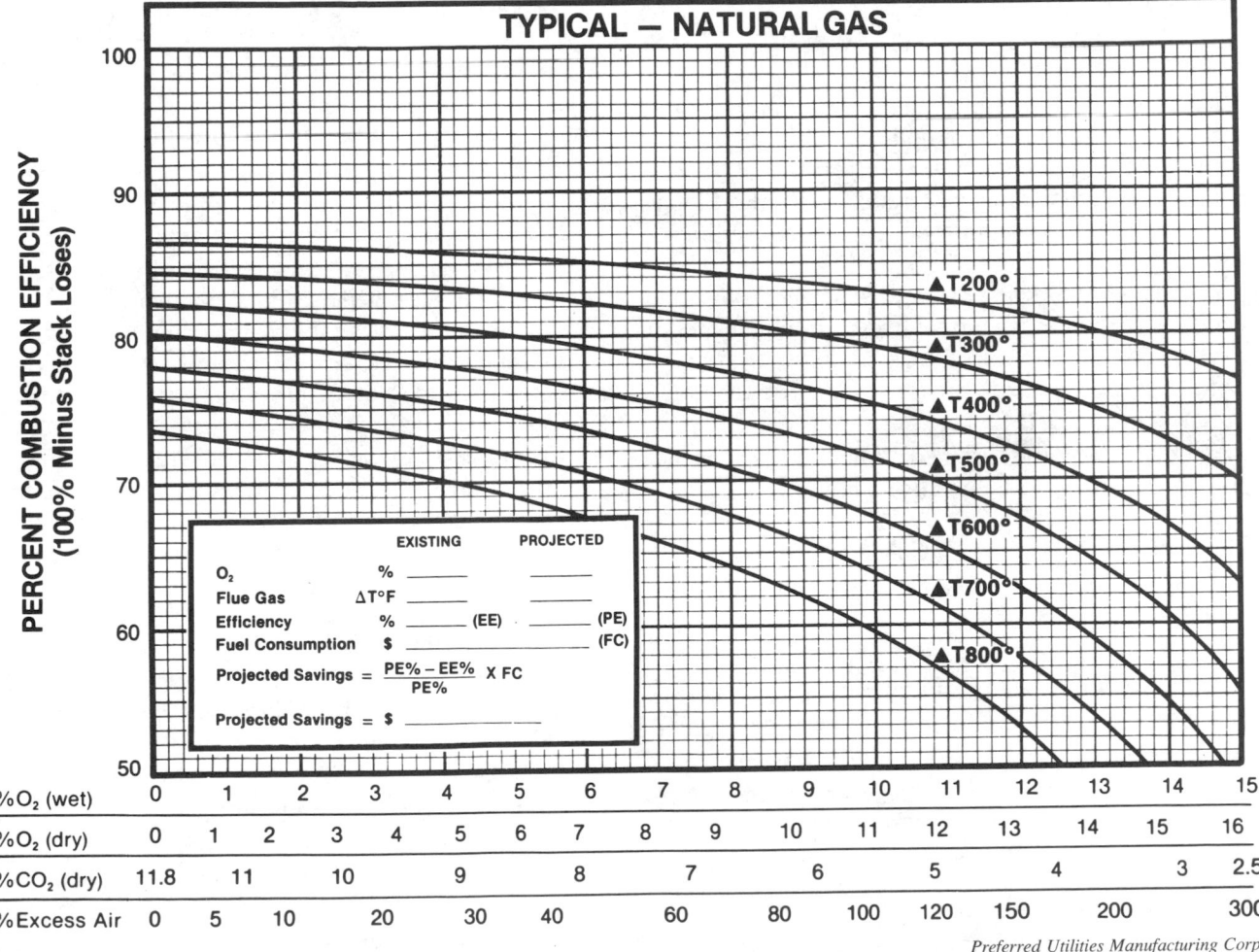

Preferred Utilities Manufacturing Corp.

Fig. 1 Form for calculating combustion efficiency This form works with either oxygen or carbon dioxide measurements. You don't need this if you have an electronic efficiency tester that calculates efficiency automatically. The graph nicely illustrates the fixed relationships between the amounts of oxygen, carbon dioxide, and excess air in the flue gas, for a particular type of fuel.

of good quality have become available at modest prices. Therefore, use the oxygen test as your fundamental test for combustion efficiency.

Testing for combustion efficiency using the oxygen method consists of three easy measurements:

- the percentage of oxygen in the flue gases
- the temperature of the flue gases
- the temperature of the air going into the boiler

Make the oxygen measurement with a specialized tester that is described below. Make the two temperature measurements with ordinary thermometers. To make the flue gas measurement, you may have to drill a small hole in the flue or breeching, as close to the outlet of the boiler as possible. (Check first. Unless the facility is brand new, somebody probably made a hole already.)

Some combustion efficiency testers have an internal computer that calculates the efficiency directly from these measurements. If you don't have such a unit, use the three measurements with an efficiency table or graph for the specific type of fuel. These are provided with the test equipment, or you can find them in reference books. A typical graph for calculating combustion efficiency is reproduced in Figure 1.

For example, using the graph in Figure 1, say that the oxygen tester gives a reading of 5% oxygen in the flue gases. (This is the "wet" value, which means that the water vapor has not been removed from the flue gas sample.) If the flue thermometer reads 480°F and the air going into the boiler has a temperature of 80°F, the flue-to-inlet temperature difference is 400°F. Entering the table with these numbers gives a combustion efficiency of 80%.

It's that easy. The carbon dioxide test is done in the same way, except that a different tester is used to measure carbon dioxide instead of oxygen, and the efficiency table or graph is based on carbon dioxide percentage.

You may do other tests, explained below, to supplement the combustion efficiency test or to measure pollutants. These tests are even simpler.

The Logic of Combustion Efficiency Tests

Your boiler burns fuel efficiently if it satisfies these conditions:

- it burns the fuel completely
- it uses as little excess air as possible to do it
- it extracts as much heat as possible from the combustion gases.

The combustion efficiency test analyzes the flue gases to tell how well the boiler meets these conditions. The test is essentially a test for excess air, combined with a flue gas temperature measurement. The only purpose of bringing air into the boiler is to provide oxygen for combustion. Bringing in too much air reduces efficiency because the excess air absorbs some of the heat of combustion, and because it reduces the

temperature of the combustion gases, which reduces heat transfer. The temperature of the flue gas indicates how much energy is being thrown away to the atmosphere.

All the calculations needed to perform the efficiency test are contained in the efficiency table, so no mathematics is needed. The efficiency table uses this logic:

- The amount of oxygen in the flue gas indicates directly the amount of excess air that is flowing through the boiler. The atmosphere contains about 21% oxygen. All this oxygen would be consumed if combustion were perfect. Therefore, the presence of oxygen in the flue gas indicates that more air is being used than is necessary for perfect combustion.
- The efficiency table is based on a certain heat content for the fuel. This is why there are different tables for different fuels. The tables also assume that the fuel is burned completely. These assumptions provide the amount of heat released in combustion.
- The volume of flue gas is known from the assumed chemical composition of the fuel and the amount of excess air. Knowing the volume of the flue gas and its temperature indicates how much energy is contained in the flue gas. This energy is wasted.
- Knowing how much energy is being produced and how much is being lost gives the combustion efficiency.

If your efficiency tester has a built-in efficiency computer, it follows the same logic, and you don't even have to look up numbers in the table.

You can also test combustion efficiency by measuring the carbon dioxide content of the flue gases instead of measuring the oxygen content. Although the test procedure is the same as for oxygen testing, the principles of the carbon dioxide test are different. Carbon dioxide content indicates both the amount of fuel burned and the amount of energy released because it is a principal residue of fuel combustion. (The other major residue is water vapor.) If there were no excess air, the amount of carbon dioxide in the flue gases would be a certain amount, determined by the chemical composition of the fuel. A lower concentration of carbon dioxide therefore indicates excess air.

In other words, the oxygen test indicates the amount of excess air directly, whereas the carbon dioxide test indicates excess air indirectly, based on an assumption about the chemical composition of the fuel.

■ Assumptions of Combustion Efficiency Testing

Be aware that the efficiency tables used in combustion efficiency testing make some assumptions. Most importantly, they assume a certain average chemical composition of the fuel. This makes it important to use the specific table for the fuel being burned. If you use a tester that calculates the efficiency, you have to set the type of fuel in the tester.

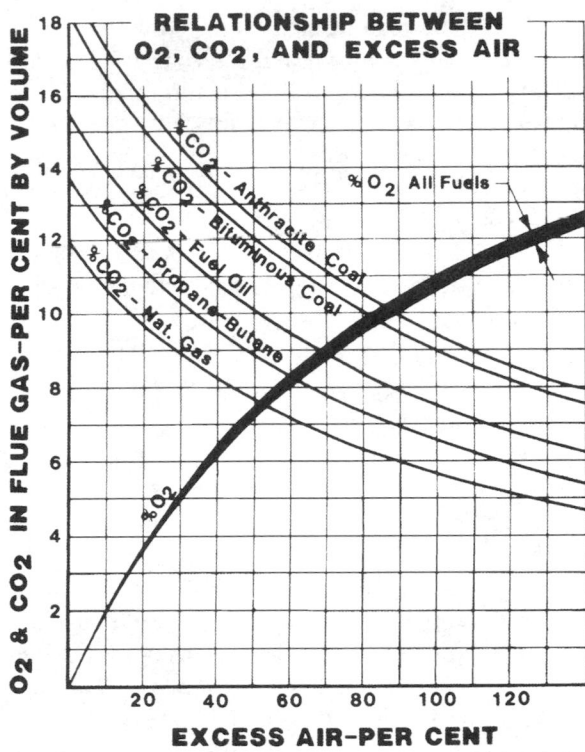

Fig. 2 Why flue gas oxygen indicates combustion efficiency more accurately than carbon dioxide First, the oxygen content changes radically with excess air. Second, the relationship between oxygen and excess air is almost independent of the fuel type.

The chemical composition of a fuel varies considerably even within a given type (such as "No. 2 oil"). However, the effect of such differences is small enough to ignore for most practical purposes.

Another assumption of the tables is that the fuel is burned completely. As long as the burners are operating properly and the air-fuel ratio is not grossly out of range, this is a reasonable assumption. However, burner malfunctions and insufficient air cause incomplete combustion. These conditions cause fuel to be thrown into the boiler without producing heat. The unburned fuel is not detected by the oxygen or carbon dioxide tests, so these tests yield artificially high efficiency figures if combustion is incomplete. This is one reason why you need an additional test to check for incomplete combustion.

To be extremely accurate, you would have to measure the temperature of the fuel, to take into account the sensible heat carried by the fuel. As a practical matter, this energy is minor in comparison with the chemical energy of the fuel. Furthermore, the temperature of the fuel that is delivered to the boiler usually varies little from one time to another. So, you can forget about fuel temperature in most cases. (The temperature of heavy fuel oil is important for viscosity control. That aspect is covered in Subsection 1.9.)

Why the Oxygen Test is Better than the Carbon Dioxide Test

For any given fuel, there is a fixed relationship between the amounts of oxygen and carbon dioxide in the flue gas. This is why you can measure either one to determine combustion efficiency. The fixed relationship between oxygen, carbon dioxide, and excess air is shown in Figure 1.

The oxygen test is more accurate than the carbon dioxide test. The reason is apparent from the graph in Figure 2. When you approach the optimum excess air setting, the relative change in oxygen is much greater than the relative change in carbon dioxide for a given change in excess air. For example, with No. 2 oil, an increase in excess air from 2% to 10% causes oxygen in the flue gas to increase by a factor of five, a change that you can measure easily. On the other hand, the same increase in excess air causes carbon dioxide to drop by only 10%, a difference that is more difficult to measure accurately.

Another advantage of the oxygen test is that the results are much less sensitive to variations in the chemical composition of the fuel. Figure 2 shows this dramatically. The amount of carbon dioxide in the flue gas depends on the amount of carbon in the fuel, and the amount of excess air is calculated from this carbon dioxide value. There are large differences in the chemical composition of some fuels, such as industrial by-product gases. All liquid and gas fuels have some variation.

In contrast, the oxygen test provides a direct indication of excess air. Variations in carbon content do not affect the results of the oxygen test at all, and variations in the total energy content of the fuel affect the oxygen content much less than they affect the carbon dioxide content.

Unlike the carbon dioxide test, the oxygen test works only in the region of excess air. There is no oxygen to measure when there is no excess air. This is not a problem in normal testing, because you should always operate boilers with a small amount of excess air.

Check the Oxygen Test with the Carbon Dioxide Test

Boiler efficiency testing is so important that it is worth a small additional effort to check the oxygen test with a carbon dioxide test. Test equipment for both tests is readily available. The cost in time and money for the additional test is minor. If continuous-reading oxygen test equipment is installed in your boiler plant, check this equipment occasionally with portable test equipment that checks for both oxygen and carbon dioxide.

If the carbon dioxide test does not give the same results as the oxygen test, something is wrong. One (or both) of the tests could be erroneous, perhaps because of stale chemicals or drifting instrument calibration.

Another possibility is that outside air is being picked up along with the flue gas. This occurs if the combustion gas area operates under negative pressure and there are leaks in the boiler casing. Watch out for this problem in boilers with induced draft fans and in boilers that have atmospheric burners.

A discrepancy between the two tests could also be caused by a defect in the burner assembly, such as a clogged nozzle. Remember that *combustion efficiency testers assume complete combustion of the fuel*. Any condition that interferes with complete combustion renders the tests invalid, and also breaks the correlation between oxygen content and carbon dioxide content.

Beware of electronic efficiency testers that claim to "measure" both oxygen and carbon dioxide. Many such units really measure only the oxygen percentage. They calculate the carbon dioxide percentage from the oxygen percentage, using assumptions about the fuel composition. *The carbon dioxide indications of such units are meaningless, and you cannot use them for cross checking purposes.*

Tests for Incomplete Combustion

No boiler is capable of burning fuel without some amount of excess air, although the percentage of excess air may be small. As the excess air is reduced toward zero, there is some fuel that is not burned completely. This partially burned fuel creates smoke, leaves deposits on firesides, and creates environmental problems.

Unburned fuel may also represent a significant waste of energy. The amount of waste depends on the energy content of the unburned fuel components. For example, the unburned components of heavy oil are mostly organic compounds that have a high energy content. On the other hand, the unburned components of coal may consist largely of foreign matter that has much lower energy content than coal itself. One source estimates that each 0.1 percent of unburned combustibles in flue gas typically represents between 0.3 and 0.6 percent of the energy content of the fuel. *This waste of energy is not measured by the combustion efficiency test.*

Therefore, optimum efficiency requires some positive amount of excess air, even though the combustion efficiency test indicates maximum efficiency with no excess air. If the boiler is working properly, the optimum amount of excess air is small.

To fine-tune the excess air, you may need an additional test that detects small amounts of incompletely burned combustion products. Two common tests for this purpose are smoke density and carbon monoxide in the flue gas.

■ Smoke Opacity Test

Before combustion efficiency test equipment became available, the amount of air was adjusted by observing the smoke emerging from the stack. For example, boilers burning heavy oil used the rule that the flue gases should be a "light brown haze." This is no longer satisfactory as a primary test, but it continues to be a useful check. If there is too much smoke when the excess air is set to a reasonably low figure, something is wrong. Therefore, measuring smoke density ("opacity," to be precise) continues to be a valuable diagnostic test.

You can use the smoke density test with heavier grades of oil and with solid fuels. Smoke density is not a reliable indicator with gaseous fuels and with light oils. The unburned residue of these fuels is not visible unless air is very deficient.

■ Carbon Monoxide Test

The carbon monoxide content of flue gas is a good indicator of incomplete combustion with all types of fuels, as long as they contain carbon. Carbon monoxide in the flue is minimal with ordinary amounts of excess air, but it rises abruptly as soon as fuel combustion starts to be incomplete. This makes it an excellent indicator when making your final adjustments of the air-fuel ratio.

An excessive level of carbon monoxide that occurs in the normal region of the air-fuel ratio indicates trouble within the boiler. Carbon monoxide rises excessively if any defect in the boiler causes incomplete combustion, even with excess air. This makes carbon monoxide testing an excellent tool for discovering combustion problems, especially if it is used in combination with oxygen testing. For example, the carbon monoxide test might reveal a fouled burner. It might also point toward a more subtle problem, such as a poor match of the burner assembly to the firebox, causing a portion of the flame to strike a surrounding surface. (Cooling the flame interrupts the combustion process, leaving carbon monoxide and other intermediate products of combustion in the flue gases.)

Carbon monoxide also forms if there is a great excess of air. This is not a matter of practical significance. Once you set the air-fuel ratio properly, the carbon monoxide content falls into the proper range if there are no other problems.

Flame Appearance

The practice of adjusting burners on the basis of the color and shape of the flame is no longer acceptable, now that more accurate tests are readily available. However, you should still check the color and shape of the flame periodically to verify that there are no combustion problems, such as a clogged burner or water in the fuel.

Tests for Specific Environmental Pollutants

Environmental regulations may require you to test flue gases to ensure that emission of certain pollutants, such as nitrogen oxides, sulfur oxides, hydrogen sulfide, and chlorine, does not exceed specified limits. These tests are usually simple, requiring nothing more than sucking a flue gas sample into a specialized chemical or electronic tester.

Controlling pollutants requires specialized procedures, such as recirculating the flue gases, increasing air flow, or adjusting the burner flame. These procedures are likely to affect efficiency. Do efficiency testing as part of any change you make to control pollutants. Keep your pollutants within limits by using methods that reduce efficiency the least.

How to Select Test Equipment

Efficiency test equipment must be accurate. The objective is to measure efficiency within a fraction of a percent, because a difference this small can correspond to a large amount of energy. The entire range of possible boiler efficiencies corresponds to a range of flue gas oxygen content from zero to twenty percent. At the high end of the efficiency range, an error of two percent in flue gas oxygen results in a half percent error in combustion efficiency.

You don't need the skill of a surgeon to do combustion efficiency testing, but you do have to be careful. Some types of test equipment require more finesse than others. Let's look at what is available.

■ Oxygen and Carbon Dioxide Testers

The pioneer of combustion efficiency testers is the Orsat analyzer, a chemical testing apparatus in which flue gases are mixed with a liquid reagent that changes volume as a result of the reaction. The apparatus displays the volume change accurately, providing an accurate indication of combustion efficiency. Oxygen, carbon dioxide, carbon monoxide, and other gases can be analyzed.

The Orsat analyzer is simple and accurate. A portable version is available at moderate price. However, the apparatus is built of delicate glass tubing and requires a fine touch. Modern testers are much more rugged and much easier to use, so the Orsat analyzer is now a museum piece.

A similar chemical approach is used by the vintage Bacharach "Fyrite" tester, which is compact and made of plastic. One version of this unit tests for oxygen, and a different version tests for carbon dioxide. They are shown in Figure 3. The "Fyrite" is less accurate than the Orsat, but it is rugged and fairly easy to operate. It is the least inexpensive of all combustion gas analyzers. You need to read a table to use this device.

These testers require replacing their chemicals after a certain number of tests, and the chemicals have a limited storage life.

In recent years, electronic testers have become available for analyzing flue gases. Some typical units are shown in Figure 4. Most electronic testers use an electro-chemical cell as the gas sensing element. A different cell is required for each gas being measured.

The most common sensor used in oxygen testers is a zirconium oxide element that develops a voltage difference across two sides if there is a difference in oxygen concentration. The zirconium oxide element has the advantage that it can be exposed to flue gases for thousands of hours.

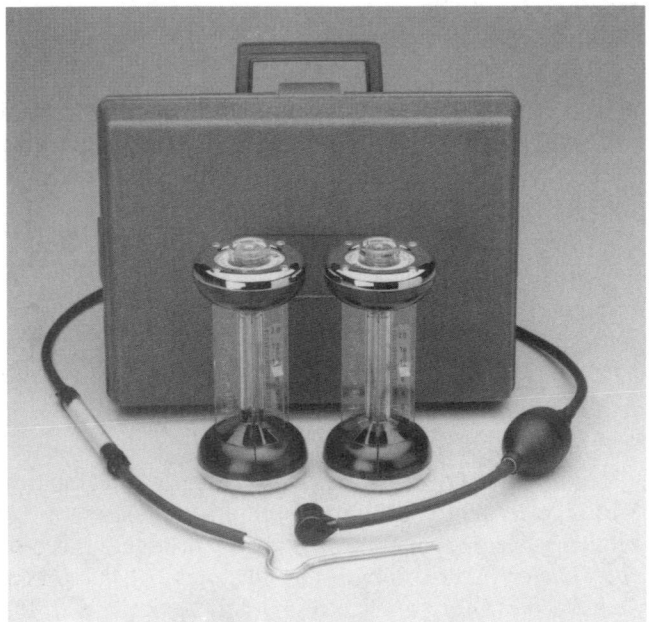

Bacharach, Inc.

Fig. 3 Chemical combustion efficiency testers One tests for flue gas oxygen, the other for carbon dioxide. The rubber hose draws a flue gas sample through a liquid in the tester. A chemical reaction changes the liquid volume, revealing the amount of gas.

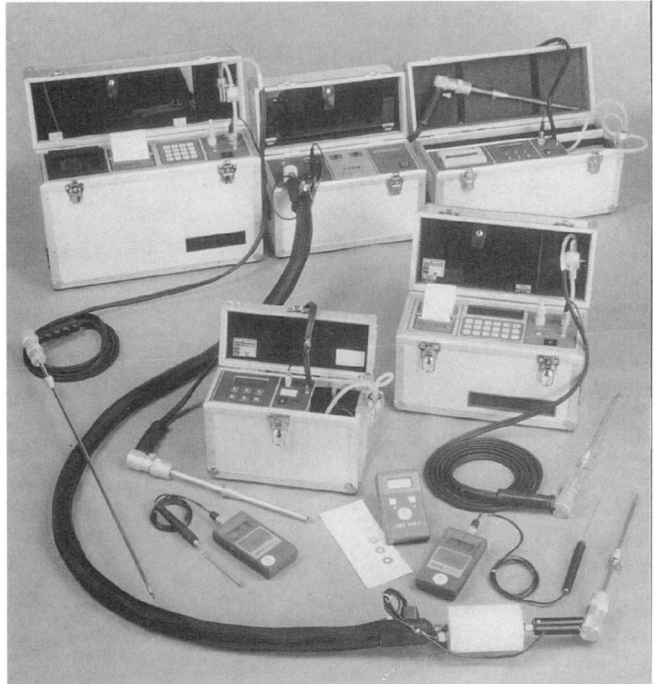

IMR Environmental Equipment, Inc.

Fig. 4 Electronic flue gas analyzers These test for a variety of gases. Oxygen and carbon dioxide testers can calculate combustion efficiency directly.

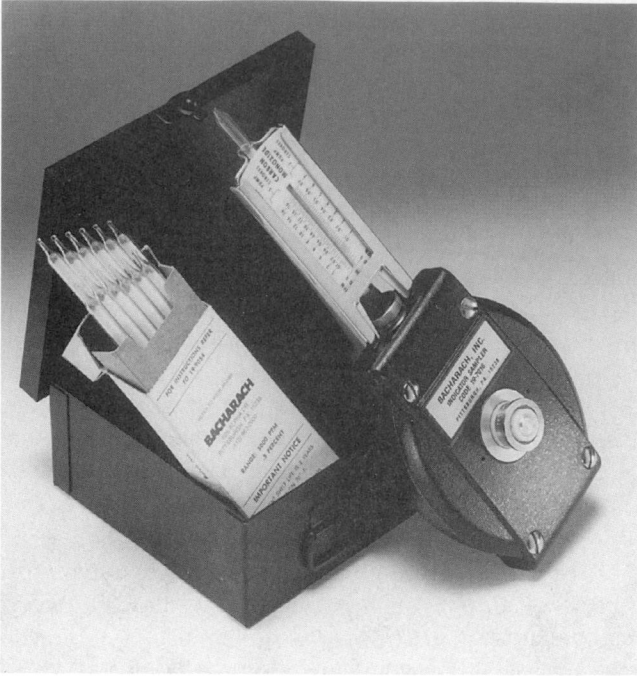

Bacharach, Inc.

Fig. 5 Chemical carbon monoxide tester The simple device on the right draws a flue gas sample through a glass tube containing a chemical that reacts with carbon monoxide to change color. Each tube is used one time. By using different chemicals, the device can test for other flue gases.

The overwhelming advantage of electronic testers is that they provide an immediate and continuous readout simply by inserting a probe into the flue gases. This allows you to see the effect your of boiler adjustments immediately.

Some combustion efficiency testers avoid the need to use a table to calculate efficiency. They perform the efficiency calculation with a microcomputer and read out combustion efficiency directly. There is a potential for error with such testers because efficiency depends on the temperature of the inlet air as well as the temperature of the flue gas. Some testers use the temperature of the air surrounding the tester as the inlet

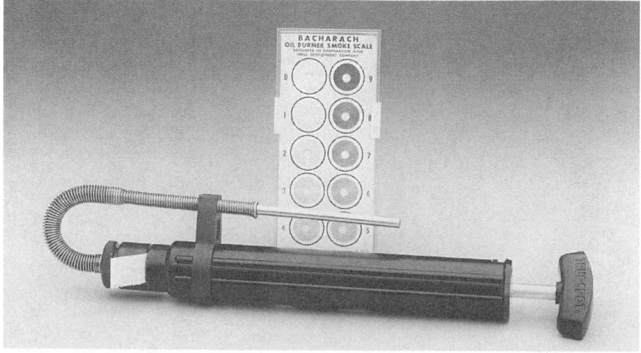

Bacharach, Inc.

Fig. 6 Smoke density tester This device, similar to a bicycle tire pump, sucks a flue gas sample through a piece of filter paper. The resulting smoke spot is compared with the chart behind the pump.

air temperature. This introduces error because the air near the flue may be much warmer than the air entering the boiler. Better testers provide a separate remote thermometer for measuring the air inlet temperature while you take the flue gas sample.

Electronic testers of good quality are substantially more expensive than chemical testers. Still, the price is small compared to the benefit of keeping boiler efficiency optimized (unless your boilers are very small).

As with most electronic equipment, the quality of electronic flue gas analyzers varies widely. The best of the lot may be capable of better accuracy than the inexpensive chemical testers. The worst of the lot are incapable of providing useful accuracy.

A weakness of all electronic gas analyzers is that they must be re-calibrated frequently because the sensor output drifts. Fortunately, oxygen testers may be self-calibrating by using the oxygen content of the atmosphere as a calibration point. Testers for other gases require calibration using pre-mixed gases. Since the inner workings of these instruments are invisible, calibration is an act of faith in the calibration gases.

Most inexpensive electronic combustion analyzers do not measure carbon dioxide directly. Some units provide a carbon dioxide reading, but this is derived by internal calculation from the oxygen measurement. If you want to use a carbon dioxide test to check the oxygen test, a value that is calculated from the oxygen measurement is useless. Measure carbon dioxide with a separate instrument designed for that purpose.

■ Testers for Carbon Monoxide and Specialized Gases

You can find electronic testers for gases other than oxygen or carbon dioxide, but these tend to be expensive and specialized. The most economical testers for carbon monoxide and specific pollutant gases use disposable chemical capsules for each type of gas. These units are reliable and easy to use. See Figure 5.

For example, one type of tester uses disposable chemical capsules in the form of slender tubes. The tester consists of a holder for the capsule that is equipped with a pump that draws the gas sample through the capsule. The gas concentration is indicated by the length of a color change inside the capsule.

■ Smoke Testers

You can measure smoke density cheaply and reliably with a tester that is constructed like a bicycle pump. It pulls a measured amount of flue gas through a piece of filter paper. You compare the blackness of the smoke spot on the filter paper to a chart that comes with the tester. See Figure 6.

Another method of testing smoke density is to install an optical densitometer in the flue. This is a simple device that shines a light beam across the flue. The amount of light that is absorbed by the smoke is

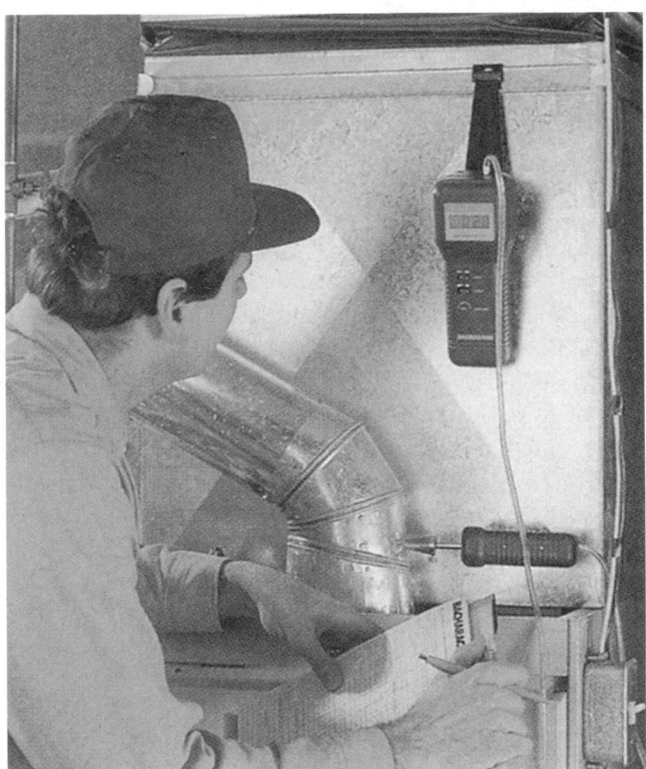

Bacharach, Inc.

Fig. 7 Small furnace, small tester Combustion efficiency testing is easy. It requires only a small hole in the flue. However, it is critical to keep air from leaking through the hole into the probe.

measured by a photocell. This device provides a continuous reading. In fact, some older oil-fired boilers use this device as the primary means of adjusting air-fuel ratio.

■ **Thermometers**

The thermometer that you use for flue temperature measurements must be accurate. An error of 40°F in the flue temperature measurement can result in a one percent error in boiler efficiency. Combustion efficiency testing is no place for a shirt pocket thermometer. Use a laboratory grade liquid-in-glass thermometer. Mercury-in-glass thermometers are available with maximum temperatures up to 1,200°F. To read the thermometer most accurately, get one that has a maximum temperature close to the maximum temperature of the flue gases. A 500°F (or 300°C) thermometer is appropriate for efficient boilers. A 750°F (or 400°C) thermometer is adequate for most others.

Good mercury thermometers are inexpensive, but they are fragile and difficult to read. Buy a couple of spares, along with a protective storage case.

Dial thermometers use a bimetallic sensing element that may drift with time. The only merit of dial thermometers is that they are easy to read. In a pinch, you can calibrate or check the low temperature end of a bimetallic thermometer by placing it in boiling water, which provides a reliable standard of 212°F (100°C) at sea level. If you use this method, correct for your

altitude, because water boils at lower temperature at higher altitudes.

Electronic thermometers have high-tech appeal, but they may be inaccurate, just a toy dressed up with a digital output. Electronic temperature sensing elements are inherently non-linear. Most of the cost of the electronic thermometer goes to compensating for the weaknesses of the sensing element. Some electronic thermometers are accurate, but you typically have no way of distinguishing these units from the toys. Even with an electronic unit of high quality, be aware of the guaranteed accuracy range. For example, a one percent error in absolute temperature corresponds to an error of about 10°F at typical flue temperatures.

Avoid Error Due to Extraneous Air

The combustion efficiency test involves measuring a small quantity of residual oxygen in the combustion gases. In a well tuned boiler, the gases contain less than one percent of free oxygen. Therefore, the test is seriously falsified by any outside air that enters the test probe.

Be careful to avoid getting outside air into the probe at the probe hole. Sloppy technique may allow outside air to travel along the shaft of the probe and enter the tip. To avoid this, aim the probe toward the boiler, preferably by bending the probe to that it faces into the flue gas stream well inside the flue. Wrap a piece of cloth around the probe to act as a gasket between the probe and the edge of the hole. It's the same general idea as safe sex. See Figures 7 and 8.

IMR Environmental Equipment, Inc.

Fig. 8 Large boiler, small tester Combustion efficiency testing is done the same way on all boilers and furnaces. Here, a test is conducted high on a large stack. The probe is located at the arrow, the tester is in the foreground. All these people are relaxing. It takes only one person to run the test.

If the flue contains a draft hood or atmospheric damper, tap the gas sample well below the point of entry of outside air. Turbulence in the flue may cause ambient air to backflow to the probe. Such backflow is especially likely to occur near bends in the ductwork.

Leakage of extraneous air into the flue is likely to occur if the boiler has an induced-draft fan, i.e., a fan that sucks combustion gases through the boiler rather than blowing them through. If you insert the probe at a point between the boiler outlet and the induced-draft fan, the induced-draft fan will draw outside air into the test hole. Air will enter at any leaks that exists in the flue ahead of the test probe. To avoid this problem, examine the boiler breeching ahead of the test hole, and plug any leaks. Drill the test hole as close to the boiler or economizer outlet as possible.

Errors from Sensing the Wrong Gases

A particular type of sensor may respond to more than one substance in the flue gas. This leads to a false reading if you are expecting to test for one substance but your tester registers another. The most common problem is that hydrogen and hydrogen sulfide register as carbon monoxide with some electrochemical carbon monoxide sensors. This is not a problem with clean fuels, which produce only carbon dioxide and water vapor as their principal end products. Beware of false readings when burning fuels, such as high-sulfur oil and coal, which produce gases other than carbon dioxide and water.

Errors Due to Internal Steam Leaks

If a boiler has an internal steam leak, the steam mixes with the flue gases. The steam is cooler than the combustion gases and lowers the temperature at the test point, giving an erroneously high efficiency calculation. If the steam leak is large, it may become apparent from an efficiency reading that seems abnormally high, especially at low loads. The rate of steam leakage remains constant at all firing rates, as long as the boiler pressure remains constant. Therefore, the error caused by the escaping steam is larger at low firing rates. If a boiler has a steam leak large enough to be detected by an efficiency test, it should be shut down immediately and repaired.

How to Test Boilers with Flue Gas Recirculation

To reduce the production of nitrogen oxides during the combustion process, many boilers now recirculate a portion of the flue gas. Exhaust gas is drawn from the flue with a fan and it is injected back into the boiler at the burner. The flue gas contains little oxygen and it is relatively cool. The purpose of this is to reduce the peak combustion temperature, since nitrogen oxides are formed mainly at high combustion temperatures.

Flue gas recirculation does not cause any problems with combustion efficiency testing. Efficiency testing works by measuring what goes into the boiler and what comes out. Recirculation is a process that occurs entirely within the boiler. Therefore, any effect that recirculation may have on efficiency is included in the flue gas analysis. Recirculation does not affect where you should tap your flue gas sample.

How to Test Boilers with Economizers and Air Preheaters

Economizers and air preheaters (covered in Subsection 1.7) are devices for capturing additional heat from the flue gases after they leave the boiler. The recovered heat typically is used to heat feedwater or combustion air, respectively. These devices lower the flue gas temperature, so they affect the efficiency test.

With both types of units, the proper place to measure the flue gas temperature is at the outlet of the heat recovery device, rather than at the outlet of the boiler.

With a combustion air preheater, the place to measure inlet air temperature is at the inlet to the air preheater, not the inlet to the burner. In other words, do not measure the temperature of the heated air produced by the preheater.

With a combustion air preheater of the heat wheel type, the flue gas test may yield an erroneously high efficiency. This is because the rotation of the wheel carries a certain amount of cool ambient air into the exhaust stream. Some types of heat wheels have purge sections that deliberately recirculate a portion of the incoming air into the exhaust from the heat wheel. Furthermore, different portions of the heat wheel are at different temperatures, so the flue gas temperature leaving the heat wheel is non-uniform. There is no practical way to make an accurate correction for these factors.

Preparation and Testing Conditions

When you first start your efficiency testing program, do a test under optimum conditions to establish a baseline for future tests. Do the baseline test with the boiler in peak condition, and across the entire load range of the boiler. Compare later tests to this baseline. When the boiler efficiency falls too far below your measured baseline, it is time to take corrective action.

For this baseline test, expect to spend considerably more effort and time inspecting the boiler plant and correcting deficiencies than in doing the testing itself. Adjust, clean, and repair all parts of the boiler that affect efficiency. Clean firesides and watersides. Clean and adjust burners. Remove all looseness from air and fuel control linkages. Check the combustion air supply. Adjust flue draft. And so forth. Refer to the other Subsections for the Measures that optimize the efficiency of your particular types of boilers.

Do the tests while the boiler has a steady load. If the system is used primarily for comfort heating, you may not have enough load to test your boilers during warm weather. It is easier to conduct tests in a plant that has more than one boiler, so the boiler being tested can shift load to and from the other boilers.

Do not attempt to create a load on the boiler by warming it from a cold state. This produces an efficiency reading that is artificially high, because the flue gases are being cooled by the mass of the boiler and water.

You have to make certain adjustments to the boiler plant while the boiler is operating. For example, determine the optimum fuel oil heater temperature by adjusting the fuel oil temperature for peak boiler efficiency. Make these adjustments as the test progresses.

How Often to Repeat Efficiency Tests

The most common reason for a decline in efficiency is accumulation of fouling on firesides, watersides, and burners. Fouling accumulates at a fairly predictable rate, if the boiler has no defects. The appropriate interval for combustion efficiency testing depends on the type of boiler and the type of fuel. A simple gas-fired boiler with atmospheric burners may hold its efficiency for years without adjustment, and with minimal maintenance. On the other hand, a large pressure-fired boiler burning a variety of fuels may have a noticeable drift in efficiency over a period of days or weeks.

Even if the boiler has recently been cleaned, test for efficiency and make adjustments whenever anything changes that can reduce efficiency. For example, test efficiency when changing from one type of fuel to another, when changing batches of fuel, and when slack in the linkages controlling air-fuel ratio becomes apparent. If controls are subject to drift, determine the length of time that it takes a significant error to accumulate. (If drift is serious, improve the control system as recommended by Measures 1.3.2 and 1.3.3.)

ECONOMICS

SAVINGS POTENTIAL: *2 to 10 percent of fuel cost, typically. Credit the savings to the specific actions that you take to improve efficiency, such as optimizing the air-fuel ratio, optimizing fuel oil viscosity, etc.*

COST: *You can purchase a good chemical combustion efficiency test kit that measures oxygen, carbon dioxide, and smoke for less than $500. A chemical carbon monoxide tester costs less than $300. Electronic testers of reasonable quality cost from $1,000 to several thousand dollars. Also, consider the cost of the actions that you take to improve efficiency.*

PAYBACK PERIOD: *Very short, to several years, depending on the amount of fuel consumed by the boiler and its tendency to lose efficiency.*

TRAPS & TRICKS

TRAINING AND ADMINISTRATION: *Efficiency testing does not require a high level of skill, but it does require practice and strict adherence to the instructions. If you delegate this responsibility to others, make sure that they are well trained.*

SELECTING THE EQUIPMENT: *Don't skimp on the quality of the test equipment. If your budget is limited, get a proven chemical test kit. The quality of electronic test equipment varies. Higher price does not necessarily mean better accuracy. High-quality electronic testers are easy to use and reduce the opportunity for mistakes.*

KEEP IT UP: *Efficiency testing is easy to forget. Schedule periodic review of testing procedures. Verify that testing is done properly, that results are recorded, and that appropriate action is taken.*

MEASURE 1.2.2 Install efficiency instrumentation appropriate for the boiler plant.

You can do a good job of testing boiler efficiency with portable instruments that are relatively inexpensive. However, it may be a good idea to install some efficiency-related instruments permanently. This can make it easier to monitor efficiency and energy consumption.

Permanently Installed or Portable?

To decide whether to install permanent instruments, and which to install, consider these factors:

- *the amount of money at stake.* Portable instruments are less expensive, it is easier to service and calibrate them, and you can use them for any number of boilers. It may not be economical to install and maintain fixed instruments in a small boiler plant. In contrast, the cost of instruments for a large boiler plant is almost trivial compared to fuel cost.
- *the types of boiler plant equipment.* For example, simple gas boilers with atmospheric burners, serving a single customer, may require no fixed instrumentation, except for a boiler gas meter and a flue thermometer. With such boilers, efficiency does not deteriorate over short periods of time, and there are few things to can go wrong in a way that affects efficiency. At the other extreme, a large, complex boiler plant serving multiple customers deserves an extensive array of instruments and data recording devices.
- *the sophistication of the boiler operators.* If there are no operators on duty in the plant, or if the boiler operators do not have the skills to exploit the instrumentation, there is no use in installing it. (See Measure 1.2.4 about training.) If you hire someone to test the boilers periodically, that person will bring his own instruments.
- *the improvement in ease of use.* Permanent instruments may be much easier to use, and this may motivate people to monitor conditions better. For example, nobody likes to clamber up a ladder to stick a sampling tube into hot boiler breeching. Portable instruments tell you nothing as long as they are locked up in the chief engineer's office.
- *instrument accuracy and longevity.* Exposing sensors to flue gases on a continuous basis causes calibration to drift and requires periodic replacement of sensor elements. Instruments that are kept inside a padded box most of the time may last much longer. However, well designed permanent instruments make it easy to replace elements, and the newer types of sensors have greatly improved longevity.

SUMMARY

Makes it easier to make important efficiency measurements. May improve motivation for efficient operation.

SELECTION SCORECARD

Savings Potential	$ $
Rate of Return, New Facilities	% % % %
Rate of Return, Retrofit	% % %
Reliability	✓ ✓
Ease of Retrofit	☺ ☺ ☺

Types of Instruments to Consider

The main types of efficiency-related instrumentation to consider for permanent installation in the boiler plant are:

- *combustion efficiency indicators.* These days, permanently installed efficiency instruments are electronic. They operate on the principles explained in Measure 1.2.1.
- *stack thermometers.* A glance at a stack thermometer can reveal that efficiency is dropping, provided that the operators have a table indicating the desired stack temperature at each level of load. The thermometer must be accurate. The typical cheap units that have bimetallic elements are not stable, so they must be checked and re-calibrated periodically. Newer electronic thermometers designed specifically for this purpose may be much more accurate.
- *smoke opacity sensors.* Boilers that burn heavy liquid fuels and solid fuels should have a smoke opacity indicator. Excess smoke density is an immediate indication of trouble that causes rapid fireside fouling. Also, smoke density is related to pollution emissions that are regulated in many jurisdictions. Figures 1 and 2 show the elements of permanently installed smoke opacity indicators.
- *pollutant sensors.* Special sensors for nitrogen oxides, sulfur oxides, chlorine, mercury, and other pollutants may be required to satisfy regulatory limits on emissions.
- *fuel flowmeters* are the most accurate way of measuring energy input. Flowmeters for individual boilers are important for making efficiency measurements, and for shifting load among boilers of different sizes. Measuring the total facility fuel

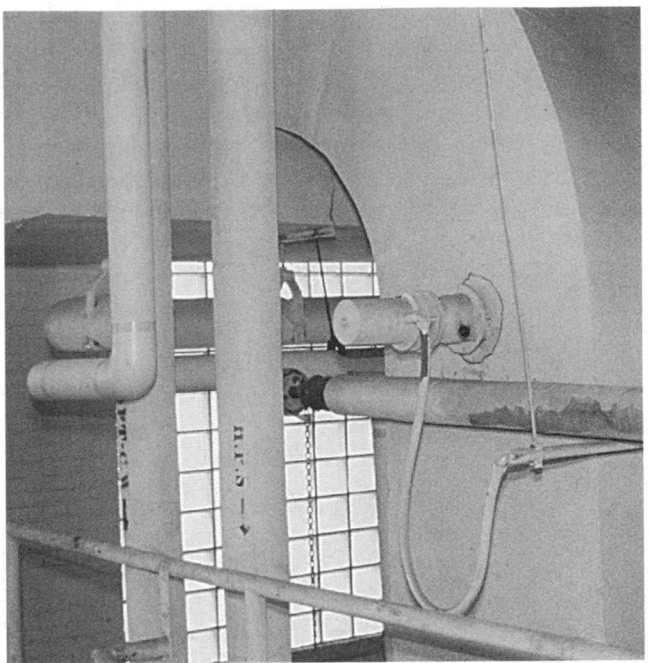

WESINC

Fig. 1 Smoke opacity monitor This is simply a bright lamp that shines across the flue toward a light sensor on the other side.

Preferred Utilities Manufacturing Corp.

Fig. 2 Remote smoke opacity indicator This gives a digital indication of smoke density at any convenient location.

flow rate gives you an instantaneous indication of facility energy consumption.

- *fuel quantity meters* are needed to indicate total fuel consumption over periods of time. Mechanical fuel meters measure consumption accurately. See Figure 3. Alternatively, you can use the same electronic instrument to measure fuel flow rate and total consumption. Metering individual boilers is not necessary unless they serve separate functions.

- *steam or BTU meters* are installed if the plant is serving multiple customers who are separately accountable for energy usage. Steam and BTU meters are limited in accuracy and reliability. Don't expect that you can use them to determine the overall efficiency of your boilers. However, you will find it informative to compare your combustion efficiency tests to the efficiency you calculate using the steam or BTU meters. True combustion efficiency is somewhat higher than true overall efficiency. This is because combustion efficiency testing does not measure all losses.

- *fuel oil thermometers.* Fuel oil temperature is a factor in burner efficiency, as discussed in Measure 1.9.1.

- *fuel oil and gas pressure gauges.* Fuel pressure is a factor in burner efficiency.

- *air casing pressure gauges.* Monitor air casing pressure when the flow of air into the casing from the forced-draft fans is adjustable. You can use this information to minimize the energy consumption of variable-flow fans, as recommended by Measure 1.4.7.

Data Displays and Recorders

Chart recorders provide a valuable display of trends in measurements. For example, a strip chart recording of oxygen in the flue gases indicates how well the air-fuel controls are working. Even if the oxygen content stays within acceptable limits, a recording might show

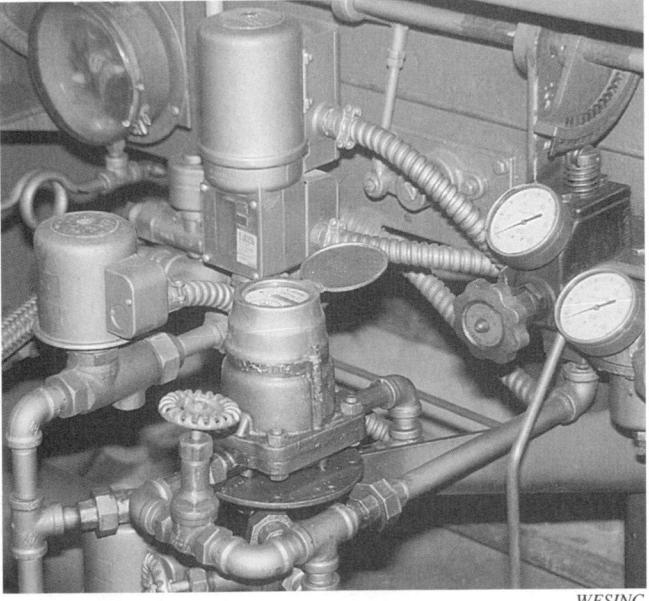

WESINC

Fig. 3 Fuel meter This accurate, inexpensive device, installed in the fuel oil line to the burner, tells you how much energy is going into your boiler. If combined with an accurate measurement of heat output, it also tells you the average efficiency of the boiler.

that the controls are starting to behave erratically, suggesting loose control linkages, a failing sensor, etc.

This does not mean that you should attach a chart recorder to every gauge in the boiler plant. Use recording instruments where it is useful to see the long-term trends or the short-term stability of measurements.

Multiple-channel recorders are valuable for seeing the relationship between different measurements. For example, there is no point in recording stack temperature alone because it fluctuates with load. But, using a two-channel recorder that records the stack temperature and the fuel flow rate simultaneously makes it possible to track relative efficiency.

The latest development in data recording is digital logging. This is much more convenient than using messy paper charts (no more ink pens!), and it gives you much more capability to analyze the data and produce reports. With a digital recording system, the sensor output is converted to a digital record. You can save virtually unlimited amounts of data on tiny digital storage media. You can perform any kind of calculation with the data, and display the output in any form that is convenient (a computer monitor, a printer, etc.). Software that allows you to take advantage of this capability is now becoming available at modest cost.

Be aware that collecting data in digital form does not improve its accuracy. Accuracy depends mainly on the sensors. Digital recording adds an extra step of conversion from the sensor output to a digital signal, which may increase the measurement error. On the other hand, digital data collection allows more accurate calculations from the data, such as totalling the steam flow that is measured by steam meters.

Coordinate with Your Other Efficiency Instrumentation

Take a broad view as you plan your permanent efficiency instruments. You will also want appropriate instruments to optimize the efficiency of your chiller plant, air handling systems, industrial processes, and other equipment. You need different types of instruments for each of those other purposes, but consider the ways that you can save effort and cost by planning them all together. As a minimum, you will probably want all your fixed instruments to provide uniform formats for collecting and analyzing data for energy management purposes.

ECONOMICS

SAVINGS POTENTIAL: *In the most favorable cases, using installed efficiency instrumentation to keep a boiler tuned may save several percent of efficiency, compared to testing and adjustment at long intervals. Credit these savings to the specific actions you take to improve efficiency.*

COST: *Less than one hundred dollars (for a good quality stack thermometer, for example) to tens of thousands of dollars for elaborate data logging equipment. The cost of appropriate instrumentation is roughly proportional to the size and complexity of the plant. The cost is minor compared to overall plant operating costs.*

PAYBACK PERIOD: *Less than one year, to several years.*

TRAPS & TRICKS

SELECTING THE EQUIPMENT: *First, figure out what instrumentation you need, and how you will use it. Don't skimp on quality, which varies widely. Unreliable instruments are worse than none.*

INSTALLATION: *Installation is usually not difficult, but follow the instructions meticulously to prevent erroneous readings.*

MAINTENANCE: *Most instrumentation related to boiler efficiency requires periodic calibration and/or maintenance to maintain accuracy. See Measure 1.2.3.*

MEASURE 1.2.3 Calibrate boiler plant instruments at appropriate intervals.

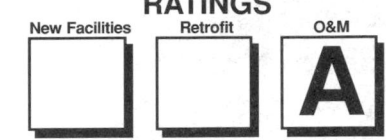

RATINGS

New Facilities	Retrofit	O&M
		A

Boiler plant instruments must be accurate in order to be useful. An instrument may have to be calibrated when it is first installed, or it may have to be re-calibrated periodically, or both. Re-calibrate your instruments often enough to keep them within the range of accuracy that you need to maintain efficiency.

Each type of instrument requires different calibration procedures, and some are more difficult to calibrate than others. For example, you need to test and/or calibrate temperature sensors against a reliable standard thermometer, or by testing them in boiling water.

Some instruments cannot be re-calibrated. Sealed units, such as liquid and capillary thermometers, must be replaced if they are inaccurate.

Some instruments cannot be tested for accuracy once they are installed. For example, there is no way of checking the accuracy of flow measuring devices without setting up a virtual test laboratory. Such equipment depends on factory calibration, which must be taken on faith. This is also true of some control elements, e.g., square root extractors for pitot tubes.

Coordinate with Your Other Efficiency Instrumentation

Develop a single calibration schedule for all the instruments in your facility, including instruments in

SUMMARY

Necessary maintenance to keep instruments useful. Tends to be neglected.

SELECTION SCORECARD

Savings Potential .... **$** $

Rate of Return .... **% % % %**

Reliability .... ✓ ✓

Ease of Initiation .... ☺ ☺ ☺

chiller plants, air handling systems, and elsewhere. This will save effort, and make the activity more habitual.

ECONOMICS

SAVINGS POTENTIAL: *Same as for Measure 1.2.2.*

COST: *Usually minor.*

PAYBACK PERIOD: *Immediate.*

TRAPS & TRICKS

SCHEDULING: *Instrument calibration is a type of activity that tends to be neglected. Have an effective system to schedule it.*

MEASURE **1.2.4 Keep operators proficient in using instrumentation to maximize boiler plant efficiency.**

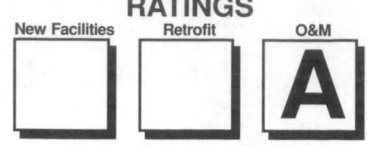

Many boiler plants are beautifully equipped with instruments to measure virtually every parameter of the plant's operation. But in many such plants, you can see that the instruments are out of calibration and no one is paying any attention to them. Or, someone takes readings every hour, and no one ever reviews the log sheets. Is this happening in Figure 1?

Instrumentation is useless unless the operating staff have the ability and the instructions to exploit it properly. Provide formal training for your boiler personnel in using efficiency instrumentation. The key elements in doing this are an effective instructor, effective training procedures, and an operating staff that is capable and motivated.

In large plants, operators may be specialized. For example, only a single individual may be trained and authorized to adjust the combustion controls. On the other hand, all operators should know how to use

---

SUMMARY

Efficiency instrumentation is useless unless the plant operators use it and act on the information.

SELECTION SCORECARD

Savings Potential	**\$ \$ \$**
Rate of Return	**% % % %**
Reliability	✓ ✓
Ease of Initiation	☺ ☺ ☺

---

instrumentation for efficient sequencing of equipment, as recommended by Measures 1.1.4 ff.

Even where boilers do not have sophisticated instrumentation, operators should know how to exploit the simple instruments that are available. Boiler efficiency instrumentation can be simple to use, as described in Measure 1.2.1. Also, the lowly stack gas thermometer provides a continuing check on efficiency, provided that the boiler load is indicated, for example, by a fuel flowmeter or a pointer on the fuel linkage.

Coordinate with Other Efficiency Training

Coordinate this Measure with staff training for efficient operation of chiller plants, process equipment, air handling systems, and other facility equipment.

ECONOMICS

SAVINGS POTENTIAL: *Same as for Measure 1.2.2.*

COST: *Typical training costs are several hundred to several thousand dollars, repeated as necessary.*

PAYBACK PERIOD: *Same as for Measure 1.2.2.*

TRAPS & TRICKS

TRAINING AND COMMUNICATION: *Good instruction is a rare art, especially when dealing with abstract concepts. Don't rest until you are sure that the plant staff understand how to monitor efficiency. Make sure that new boiler operators are trained to optimize efficiency. Schedule periodic refresher training.*

Preferred Utilities Manufacturing Corp.

Fig. 1 Instruments are useless without people Plant operators must know what the instruments are telling them, and they must act on the information.

The ratio of combustion air to fuel is the most important adjustment on a boiler. You can often improve efficiency significantly by keeping the air-fuel ratio optimized. The Measures in this Subsection optimize the air-fuel ratio, improve air-fuel controls, and correct defects in the controls.

INDEX OF MEASURES

1.3.1 Optimize the air-fuel ratio.

1.3.2 Install automatic air-fuel mixture controls.

1.3.3 Adjust and repair air-fuel ratio controls.

RELATED MEASURES

- Subsection 1.2, for methods of measuring combustion efficiency
- Measure 1.4.4, installing burners that require less excess air
- Measures 1.9.1 ff, adjustment of fuel oil viscosity to maintain proper air-fuel ratio

MEASURE 1.3.1 Optimize the air-fuel ratio.

Air-fuel ratio is by far the most important routine adjustment that is made to boilers. Of all the adjustments that plant operators can make, it has the greatest influence on efficiency. Furthermore, failure to set the air-fuel ratio properly can create serious maintenance and environmental problems.

If you have automatic combustion controls, adjusting the air-fuel ratio is easy. Using the methods explained in Measure 1.2.1, measure combustion efficiency while setting the combustion controls to the optimum air-fuel ratio. The combustion controls will then maintain this ratio under all load conditions.

Adjusting the air-fuel ratio is not much more difficult if you have burners that fire at one or more fixed firing rates. On the other hand, adjusting modulating burners can be tedious.

The basic steps are described here. See the burner maintenance manual for details of how to adjust your particular model of burner.

The Optimum Air-Fuel Ratio

A perfect boiler would use just enough air to burn all the fuel completely, with no oxygen left over in the flue gas. (The ratio of air to fuel that achieves this ideal result is called a "stoichiometric mixture" by chemists and advanced boiler people.) With real boilers, achieving reasonably complete combustion requires a certain amount of air in excess of the stoichiometric ratio. The excess air is needed to ensure that all the fuel comes in contact with sufficient oxygen for complete combustion within the flame area.

The minimum amount of excess air that is necessary for clean combustion depends on the type of fuel and on the type of burner. Table 1 in Measure 1.4.4 lists typical ranges of excess air for different fuels and burner types. More excess air is needed for fuels that are heavier and dirtier. Also, burners in smaller equipment tend to have substantially higher excess air requirements. Modern, high-efficiency burners minimize the amount of excess air required. The best modern burners do a much better job of preparing the fuel for combustion and of bringing the proper amount of air into the combustion zone.

The design of the boiler's combustion chamber may also affect the excess air requirement. The design of the combustion chamber becomes an issue in existing boilers if you plan to retrofit a new burner.

Determine the optimum air-fuel ratio for each of your boilers individually, using the tests recommended below.

SUMMARY

The most important routine adjustment for boilers. Has a major effect on efficiency, maintenance, and pollution. The trick is to set the air-fuel ratio as close to optimum as possible, while avoiding insufficient air.

SELECTION SCORECARD

Savings Potential .... $ $ $

Rate of Return .... % % % %

Reliability .... ✓ ✓

Initiation .... ☺ ☺ ☺

Efficiency Loss from Incorrect Air-Fuel Ratio

Efficiency suffers from too much air, and from too little. Efficiency declines rapidly as the amount of air is reduced below the point of best efficiency. Efficiency declines much more slowly above the point of best efficiency. This is because insufficient air and excess air waste energy in two different ways.

With insufficient air, efficiency falls primarily because combustion is incomplete. The incompletely burned portion of the fuel is being thrown away through the flue, taking along its unused energy.

With excess air, the fuel is being burned almost completely, but a portion of the combustion energy is wasted in heating the excess air. The heated excess air is carried through the boiler as useless baggage. Also, mixing the combustion gases with excess air lowers the temperature of the gases, which reduces heat transfer. See the effect in the graph of Figure 1 in Measure 1.2.1.

If the amount of excess air is extreme, the large volume of cool air can quench the combustion process, causing fuel to be burned incompletely. However, this effect does not become significant until efficiency has already been lowered drastically by the previous effect.

Other Problems Caused by Insufficient Air

To repeat, insufficient air wastes much more energy than an equal percentage of excess air. In addition, insufficient air causes these problems:

- *air pollution.* For any hydrocarbon fuel (gas, oil, coal, wood, peat, etc.), the end products of complete combustion are carbon dioxide and water, along with a small amount of additional compounds that may be formed from other substances in the fuel.

The combustion process is complex, and includes a myriad of intermediate organic compounds, especially with oil and solid fuels. Many of these partially burned combustion products are noxious compounds that comprise an air pollution problem if they escape unburned. If the air-fuel ratio is too low, the chain of combustion reactions is incomplete, greatly increasing the emission of pollutants.

- *fireside fouling.* With insufficient air, some of the products of incomplete combustion adhere to the heat transfer surfaces in the form of soot. Soot is a porous material that is mostly carbon, combined with other products of incomplete combustion that tend to be sticky. Its porous nature makes it an effective insulator, resisting the flow of heat from the combustion gases to the boiler water. Even small amounts of soot can significantly reduce efficiency. Soot accumulates rapidly when there is too little air for complete combustion.

- *explosion hazard.* When combustion is incomplete, the unburned fuel components make the flue gas potentially explosive. If the gases reach a part of the boiler or flue where they can be mixed with additional air (for example, above an atmospheric flue damper), they could explode if ignited.

- *flame instability.* Flame stability can be a problem when pushing the limits of low excess air with conventional burners. Minimizing excess air can produce noticeable pulsations in the flame. This can become dangerous if the flame is briefly extinguished and then re-ignites. This problem can occur with some burners even though the amount of air still exceeds the stoichiometric ratio. With such burners, the most direct solution is to replace them with burners that are designed to operate with low excess air. See Measure 1.4.2 for this Measure.

Insufficient air creates highly visible smoke and it deposits soot that requires hard work to remove, so boiler operators tend to avoid insufficient air. They may carry this to the extreme of carrying far too much excess air.

Incomplete Combustion Caused by Extreme Excess Air

When excess air is carried to an extreme, it cools the gases enough to prevent complete combustion. The amount of excess air that will cause incomplete combustion depends on the type of fuel, the burner design, and the firebox conditions. The main problems caused by having too much excess air are inefficiency, air pollution, and fireside fouling.

The compounds that form in excess air are different from the compounds that form when there is insufficient air. You can see this easily by observing the smoke.

For example, burning petroleum with insufficient air produces dense black smoke, while an extreme amount of excess air produces dense white smoke.

General Procedure for Adjusting Air-Fuel Ratio

Adjusting the air-fuel ratio consists of testing the combustion efficiency of the boiler and adjusting the air-fuel ratio until you find the optimum air-fuel ratio. Measure 1.2.1 gives the details of the tools you need to measure the air-fuel ratio. In summary, the test sequence is:

- *Set the air-fuel ratio by using the oxygen test.*

- *Refine the adjustment by setting carbon monoxide.* As you reduce excess air and get close to the point of optimum efficiency, the carbon monoxide in the flue gas will start to rise rapidly. This rapid increase occurs at about 200 to 400 parts per million carbon monoxide, with most fuels. Set the air-fuel ratio just short of this rapid rise.

 You will not be able to set the carbon monoxide level properly if there is a defect in the boiler that causes incomplete combustion. In such cases, the amount of carbon monoxide remains high near the optimum air-fuel ratio. In fact, this condition is a reliable indication of trouble. Typical causes of high carbon monoxide are a fouled burner or an improper flame pattern that allows the flame to contact a combustion chamber surface.

- *With dirty fuels, conduct a smoke test.* Add air as necessary to reduce soot and smoke to acceptable levels. Use a smoke test if the fuels used produce too much soot and smoke even at low carbon monoxide concentrations. Dirty fuels require additional excess air to burn up combustible material that is embedded in inorganic particles that are released when the fuel is burned.

- *Repeat the oxygen test and check it with a separate carbon dioxide test.* The oxygen test is based on some assumptions (spelled out in Measure 1.2.1) that break down under certain conditions, so it is a good idea to check it with a test of a different type. If the carbon dioxide test does not correlate to the oxygen test, there is something wrong with either the test or the boiler. If you find that you cannot reduce excess air without producing excessive smoke or carbon monoxide, the problem is probably in the boiler. Either the burners need cleaning or there is an equipment defect in the boiler, the flue, or the air delivery system.

The procedure for adjusting the burner assembly or the air-fuel controls differs from one unit to another. The only reliable guide is the burner manufacturer's instructions. Follow the instructions exactly.

WESINC

Fig. 1 Atmospheric gas burner The primary air to each venturi is adjusted on this furnace by rotating a plate with holes in it. As you can see, the settings are random, wasting energy and increasing air pollution.

WESINC

Fig. 2 Another atmospheric gas burner On this furnace, the primary air for each venturi is adjusted by a disk that moves along a threaded rod, providing fine adjustment.

Use a Little Additional Excess Air to Allow for Changing Conditions

You can set the air-fuel ratio with almost perfect precision by using the methods just described. However, the air-fuel ratio can wander after it is set, from these causes:

- changes in atmospheric conditions
- imperfect calibration of throttling burners across the load range
- wear of mechanical linkages
- friction in pneumatic actuators
- fouling of sensors in automatic combustion controls
- burner fouling
- differences in fuel composition between batches
- changes in fuel oil or gas pressure
- changes in fuel oil temperature.

With mechanical air-fuel controls, these factors make it advisable to provide an additional margin of excess air to avoid the problems caused by insufficient air. If the maintenance of the plant is competent and energetic, it is probably not necessary to add more than a few percent of excess air to account for these factors. In plants with less sophisticated controls, and in plants that are not monitored closely, you may have to widen the margin of excess air. But don't overdo it.

Even with automatic air-fuel controls, it may be desirable to add a small margin of excess air. The amount depends on the accuracy of the controls. You will get a feeling for the right amount of excess air from regular combustion efficiency testing and from monitoring the trend of fireside fouling.

How to Adjust the Air-Fuel Ratio Mechanically

If you do not have automatic combustion controls, you need to set the air-fuel ratio by making mechanical adjustments to the burners or the control linkages. You need the burner manufacturer's maintenance manual and the latest service bulletins to show you precisely what to do. There may be critical adjustments that do not seem important from their appearance. If you are not sure that you have all the information you need, call the factory. Details matter, so don't try to fake it.

Also, you should have a combustion efficiency tester for this job that provides a continuous, instantaneous readout. (See Measure 1.2.1 for tips on selecting the equipment.)

It helps to have two persons doing this work, especially if the burner adjustments are not close to the point where the flue gas sample is taken for the combustion efficiency tests. One person stays at the boiler breeching with the test equipment and calls out the readings, while the other person adjusts the burner.

Try to hold the boiler load as steady as possible during the adjustments. If the burner operates at different firing rates, you may have to set the air-fuel ratio for each firing rate at different times, as the load changes. If the load on the boiler plant is light, it is practically impossible to set the fuel-air ratio for high firing rates. Do not create a load by warming up a cold boiler, because this would produce erroneous efficiency readings and air-fuel settings.

■ Atmospheric Gas Burners

Most atmospheric gas burners have two sources of air, "primary" air and "secondary" air. Primary air enters the venturi and mixes with the fuel there. Secondary air enters around the outside of the burner, and mixes with the flame at the burner.

In packaged boilers where the burner is an integral part of the boiler, there is usually no adjustment for secondary air. In such boilers, setting the air-fuel ratio involves little more than changing the position of a shutter that controls the flow of primary air. Figures 1 and 2 show two examples.

In contrast, generic burners designed for retrofit installations have a separate adjustment for secondary air, and the proper secondary air setting is important in such installations. See Measure 1.4.2 for more about atmospheric burners.

You may find that the primary air setting has little effect on boiler efficiency or on the excess air requirement. But, the primary air setting may seriously affect the carbon monoxide concentration in the flue gas. Therefore, test for carbon monoxide when setting atmospheric burners.

■ Modulating Burners

Modulating burners with mechanical air-fuel controls typically have a single control motor that moves in response to the boiler's output pressure or temperature. The control motor has a linkage connected to an air control, which typically is a damper. The motor also has linkage to the fuel control, which is a valve of some kind. If the burner can burn more than one type of fuel, there is a separate linkage for each type of fuel.

The control linkages are non-linear, which means that a certain amount of motion in the air control linkage does not produce the same relative change as the motion in the fuel linkage. To compensate for this, the fuel linkage typically has an adjustable cam, which makes the fuel flow proportional to the air flow.

Boilers with mechanically modulating burners commonly link the air and fuel controls with a common shaft, called a "jackshaft." The load control motor drives the jackshaft. The jackshaft typically has an arm that drives a linkage to the air damper, and it has one or

more cams to depress plunger-type fuel valves. Figure 3 shows a typical jackshaft arrangement for a dual-fuel boiler, and Figure 4 shows a detail of a typical cam.

The air-fuel linkages are normally adjusted at the factory, with little provision to make it easy to adjust them in the field. For example, the typical fuel adjustment cam has a shape that is set by adjusting a large number of screws. This makes it tedious to adjust the air-fuel ratio.

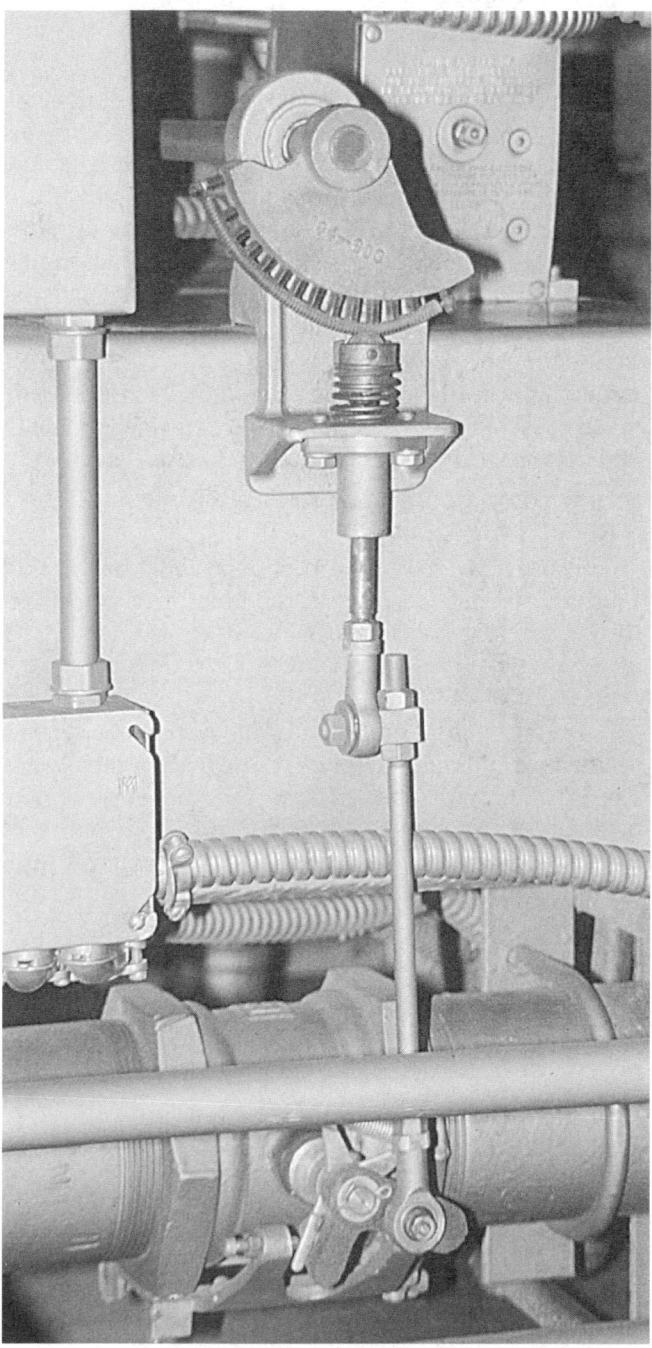

WESINC

Fig. 4 Typical cam on jackshaft air-fuel linkage The shape of the cam determines the air-fuel ratio at all boiler loads. The shape can be adjusted by turning the screws that you see between the cam body and the curved spring that rides on the plunger. This is a tedious job.

WESINC

Fig. 3 Typical jackshaft air-fuel control The jackshaft is mounted horizontally, moved by the control motor on the upper left. A cam on the left end of the jackshaft controls a fuel oil valve, and a cam on the right end of the shaft controls a gas valve. An arm to the left of the shaft's center support moves the combustion air damper. The air-fuel ratio depends on the relative positions of these mechanical linkages.

The procedure consists of holding the firing rate steady, measuring the efficiency, performing the necessary adjustment to the air-fuel ratio, and repeating the sequence until the point of best efficiency is found. Then, repeat this process for the next increment of load. The number of adjustments depends on the design of the mechanism that controls air and fuel. With a variable cam, you have to change the boiler loading in small increments that allow you to set each of the adjustment screws. Regardless of the linkage design, make the adjustments at sufficiently close intervals to keep the boiler efficient at all loads.

It may be difficult to hold the boiler output steady at each percentage of load as you make the adjustments. If you have more than one boiler, you can shift the load between boilers by changing the relative pressure or temperature settings of the boilers. You can use three people for this job. One adjusts the boiler load, another performs the combustion efficiency tests and calls out the readings, and the third adjusts the air-fuel controls.

After you go through this exercise, you will understand the advantages of combustion air "trim" systems, which adjust the air-fuel ratio automatically and continuously. See Measure 1.3.2 for details.

■ How Often Should You Adjust the Air-Fuel Ratio?

Inspect the air-fuel ratio on a periodic basis. With mechanical air-fuel controls, the ratio drifts because of the factors discussed previously. You need less excess air as a safety factor if you test and adjust the air-fuel ratio more often.

The appropriate interval between settings depends on the type of boiler and the type of fuel. A simple gas-fired boiler with atmospheric burners may hold its efficiency for years without adjustment and with minimal maintenance. On the other hand, a large

pressure-fired boiler burning a variety of fuels may have a noticeable drift in air-fuel ratio over a period of days or weeks.

Adjust Flue Draft First

Adjust the flue draft before you set the air-fuel ratio. The principles of flue draft and the methods of adjusting it are explained in Measure 1.5.3.

ECONOMICS

SAVINGS POTENTIAL: *1 to 10 percent of fuel cost, typically.*

COST: *The cost of efficiency test equipment is covered in Measure 1.2.1. The amount of labor required to set air-fuel ratio can be less than one man-hour for a boiler with a single-stage burner, to several man-days for a boiler with throttling burners and difficulty in maintaining a steady load.*

PAYBACK PERIOD: *Immediate, to one year.*

TRAPS & TRICKS

SKILLS: *Adjusting air-fuel ratio requires two skills, efficiency testing and setting the boiler's air-fuel controls. See Measure 1.2.4 about the former. Make sure that the person adjusting the boiler knows how to do it correctly.*

TEST EQUIPMENT: *The right test equipment makes the work much easier. See Measure 1.2.1 about selecting test equipment.*

BOILER CONDITION: *You can't set the air-fuel ratio properly if the boiler's controls are sloppy or defective. See Measure 1.3.3 about maintenance.*

SCHEDULING: *Repeat the procedure periodically. Make sure that you have an effective method of scheduling it.*

MEASURE 1.3.2 Install automatic air-fuel mixture controls.

RATINGS
New Facilities | Retrofit | O&M

A C ☐

Automatic air-fuel controls provide the ultimate in efficiency by continuously monitoring the flue gas composition and readjusting the air-fuel ratio. These controls are common on larger boilers. You can retrofit them to existing boilers.

Environmental regulation of boiler operation has expanded enormously in recent years, and continues to expand. If you decide to install automatic air-fuel controls, consider the effect of regulation on the capabilities that the controls should have. Try to satisfy foreseeable environmental regulations as well as maximizing efficiency.

The simplest type of automatic air-fuel control is an "oxygen trim" system. Oxygen trim systems typically sense only oxygen in the flue gases. The output of the oxygen trim system is a small adjustment in the air-fuel ratio to move the measured oxygen content in the desired direction. Figure 1 shows the control box for a typical oxygen trim system.

At the other extreme are complex air-fuel control systems for large boilers that measure a variety of flue gases and combustion conditions. Virtually all of these use flue gas oxygen as the primary measurement. They may also sense carbon monoxide, nitrogen oxides, other gases, and smoke. Such systems provide accurate control of boiler output, precise control of air-fuel ratio, and control of various functions related to plant emissions, such as stack gas recirculation and scrubbing.

SUMMARY

Keeps the air-fuel ratio optimum. Available with a wide range of features and prices. Economical for larger boilers. May not be worth the additional maintenance and failure modes in smaller boilers.

SELECTION SCORECARD

Savings Potential .... $ $

Rate of Return, New Facilities % % %

Rate of Return, Retrofit.......... % %

Reliability .... ✓ ✓ ✓

Ease of Retrofit .... ☺ ☺ ☺

Potential Savings

The improvement in efficiency that you can expect from automatic air-fuel control system varies widely from one boiler plant to another. As an example, a boiler burning heavy oil might have an optimum excess air of 15%. In this boiler, increasing the excess air to 30% would cause an efficiency loss of one percent. In some plants, excess air may drift from 15% to 30% under realistic operating conditions. Therefore, you may be able to save about one percent of fuel energy percent in such a plant. In plants where maintenance has been neglected, the saving may be as high as several percent of the fuel input.

On the other hand, a well maintained boiler that burns natural gas of stable characteristics may benefit little from automatic air-fuel adjustment. In this case, a simple mechanical linkage that regulates the air-fuel ratio may keep efficiency optimized within close tolerances. With such a boiler, the efficiency improvement would typically be only a fraction of one percent.

Take the claims of efficiency improvements made by combustion control manufacturers with a grain of salt. Before-and-after comparisons, even if they are monitored by a disinterested party, may be misleading. Much of the improvement that is measured when a trim system is installed may actually result from the cleaning and tune-up that occurs along with the installation.

Most Favorable Applications

Larger boilers are candidates for automatic air-fuel controls because small differences in efficiency can produce large differences in absolute energy consumption. For example, a boiler that produces 100

Cleaver-Brooks

Fig. 1 Oxygen trim control

million BTU's per hour of steam for 3,000 hours per year while burning heavy oil at $1.00 per gallon has an annual fuel cost of approximately $2,500,000. Improving efficiency by one half of one percent would save approximately $12,000 per year.

An automatic air-fuel control system is more valuable if the factors that affect the air-fuel ratio are subject to change. Some of the conditions that can cause the air-fuel ratio to drift are listed in Measure 1.3.1, above.

Boilers that burn solid fuels or heavy oils need to adjust the air-fuel ratio continually to account for changes in fuel characteristics, such as the content of carbon, hydrogen, oxygen, sulfur, moisture, and dirt. With coal, these factors can change as fuel is brought in from different parts of the coal pile. With oil, they can change as different strata of oil are tapped from the tanks.

Automatic air-fuel controls correct for drift, hysteresis, and non-linearity in control linkages. For example, where pneumatic actuators are used to control the fuel valves and/or fan dampers, friction in the actuators and linkages can cause errors in the air-fuel ratio. Linkage geometry in modulating burners is typically non-linear, and requires tedious adjustments of cams to compensate. Automatic controls can eliminate the need to adjust for these factors.

Staff Skill and Control Reliability

To a certain extent, automatic air-fuel controls can compensate for deficient maintenance. In particular, it can compensate for lack of tuning and for looseness in the controls. However, automatic combustion controls are not a proper substitute for skills, salary, and supervision. Installing a complex control system in an environment with poor maintenance trades continuing poor efficiency for the possibility of much worse efficiency if the control system fails.

By the same token, you need controls that are especially reliable in plants that are not continuously monitored by skilled operators. This situation is common in non-industrial facilities, where the boiler plant is often treated as an inscrutable nuisance.

Any automatic combustion control should continuously self-test, and it should sound an alarm if it senses failure or an out-of-range condition.

System Components and Installation Issues

Automatic air-fuel mixture controls vary in complexity, control accuracy, and versatility. Their performance depends on their installation as well as their components. These are the basic components, and some important issues to consider in installing them:

- *flue gas probe(s) or sensor(s).* The mechanics of installing sensors is usually easy. Be sure that the location where you install the sensor(s) provides a representative sampling of flue gases, and that it

provides a reliable average flue gas temperature. Be careful to prevent outside air from mixing with the flue gas sample. See Measure 1.2.1 for more about this.

- *analysis hardware and logic circuitry.* This usually comes in a single box. If the sensor output is electrical (for example, from a zirconium oxide oxygen sensor), the box can be installed almost anywhere. If gas samples are conveyed to the box through tubing, install the box within a specified distance from the sampling point.

- *displays.* You should install gauges or recorders that display gas content, efficiency, etc. Install them where they are mostly likely to be monitored by the operators.

- *actuators.* Actuators translate the output signal of the combustion control system to motion of the fuel valves and air dampers. Retrofitting actuators on a preexisting control system can be challenging, unless the retrofit system is custom designed for the particular boiler.

With boilers that use a common mechanical linkage to control both air and fuel, the usual practice is to insert a variable-length link in a part of the original linkage that controls either the air or the fuel. The trim system works by controlling the length of the variable link.

Larger boilers typically use pneumatic, electric analog, or digital signals to control the motion of separate actuators on the fuel valve and fan control. With such controls, you have to match the output signals of the combustion control system with the signals needed by the actuators. For example, if the original controls move the fuel and air actuators with 3-to-15 PSI air pressure, the trim system must have pneumatic outputs in this range. Alternatively, you can change the actuators.

- *failure alarms.* To repeat, the control system should test itself continuously. If a failure occurs, or if the air-fuel ratio gets out of the designated range, the controls should indicate an alarm. Make the alarm able to get attention, in a manner appropriate to the staffing. For example, if the boiler plant is not staffed continuously, have the alarm trigger a loud horn, or automatically dial the telephone of the appropriate manager, or whatever is needed to initiate immediate repairs.

How to Select Air-Fuel Controls

Air-fuel control systems differ widely in cost and in the range of features that they offer. Here are the main features to consider when shopping for a system.

■ What is Sensed

Sensing the oxygen content of the flue gas provides the most direct and accurate indication of combustion efficiency. The least sophisticated combustion air trim systems measure only oxygen, and these systems are called "oxygen trim" systems.

More sophisticated combustion air trim systems adjust the oxygen concentration by sensing unburned combustibles. These systems typically sense carbon monoxide as an indicator of unburned combustibles. Carbon monoxide plays this role because it is the second from the last combustion product in the long chain of reactions by which the carbon in petroleum or other complex fuels is converted to carbon dioxide. Hydrogen is sometimes used as an indicator of unburned combustibles, but this practice is limited to specialized applications.

Smoke density, or opacity, is commonly used as a separate indicator of unburned combustibles when burning dirty fuels, like heavy oil or coal. A separate indicator is desirable because carbon monoxide does not correlate well to the concentration of large unburned particulate matter in the flue gases. Monitoring smoke opacity is necessary in its own right because visible smoke may be an environmental or public relations issue.

Still another approach is measuring total unburned combustibles, using one of the sensor types described below. At the present time, this approach is uncommon in smaller boilers.

The combustion air trim system may also have to control other gases not directly related to efficiency. Environmental laws restrict emission of nitrogen oxides and sulfur oxides from larger boilers. Specialized boilers, such as refuse burners, may have to control other flue gas constituents, such as chlorine and mercury. Such gases are controlled by boiler design and by the addition of specialized equipment, such as flue gas scrubbers and flue gas recirculators. The operation of such equipment affects combustion air control.

■ Sensor Types

Oxygen is commonly measured with a zirconium oxide sensor. This is a solid state device that develops a voltage in response to a difference in oxygen concentration across two sides.

Carbon monoxide is measured with a variety of devices, including:

- electrochemical cells, which produce electric current when carbon monoxide is involved in specific chemical reactions
- infrared absorption detectors, which send a beam of infrared light through the flue gas. The light is tuned to an oscillation frequency of the carbon monoxide molecule, which causes carbon monoxide to absorb and scatter the light.
- polarization detectors, which function by sending a beam of polarized light through the flue gas, and measuring rotation of the direction of polarization

by the gas. This is done with polarizing filters similar to polarized sunglasses.

Unburned combustibles can be measured directly with catalytic units that work by burning up the combustibles and measuring the heat produced. In order for this type of sensor to work, there must be some oxygen in the flue gas. Unburned combustibles can also be measured with flame ionization detectors.

Smoke opacity, which may indicate both combustible and non-combustible particulate matter, is usually sensed by measuring the absorption of a beam of light in the flue.

Flue gases that are not directly related to combustion efficiency, such as nitrogen and sulfur oxides, are measured with a variety of technologies. For example, oxides of nitrogen can be sensed by chemiluminescence, which is the emission of certain wavelengths of light when nitrogen oxides undergo specific chemical reactions. Sulfur oxides and chlorine can be detected with polarized light sensing. And so forth.

With some types of sensors, one type of gas may register as another. For example, hydrogen and hydrogen sulfide register as carbon monoxide on some electrochemical carbon monoxide sensors. Guard against such cross readings if the flue gases contain large concentrations of combustion products other than carbon dioxide and water vapor.

Sensor technology is a rich and growing field. Check the latest offerings when selecting a system. The best types for air-fuel control will probably change with time as existing sensor types evolve and new types are discovered.

■ Sensor Calibration

The sensors used in combustion air controls are exposed continuously to the harsh environment of the flue. This causes the output of the sensors to drift. If not corrected, sensor drift causes serious inaccuracy in control. The only solution to drift is recalibrating the sensors at appropriate intervals. Some types of sensors can be recalibrated with the boiler in operation, and others cannot be. Sensor accuracy is important, so get sensors that can be recalibrated while the boiler is operating.

Even if the sensors can be calibrated during boiler operation, you may have to manually initiate a calibration cycle. This has the disadvantage that the procedure can be neglected, but it has the advantage of reminding operators of the need for checking the condition of sensors. Automatic recalibration cannot repair a sensor that becomes defective.

Oxygen sensors have the important advantage that they can be calibrated using ambient air. The atmosphere contains a fixed percentage of oxygen, so the sensor can be calibrated by diverting boiler room air into the sensor while it remains in place. Most zirconium oxide sensors used in combustion control systems include this feature.

Carbon monoxide sensors that use the principles of infrared absorption or polarization can be calibrated continuously by using reference beams of light. Sensors that use electrochemical cells must be recalibrated periodically by passing calibration gases through the sensor.

Instruments that need calibration gases require you to keep a stock of the appropriate gases on hand. Also, using calibration gases requires more skill.

Smoke opacity sensors can be recalibrated continuously by using a reference light beam.

Sensors for gases that are not directly related to combustion efficiency operate on a variety of principles. Some of them can be continuously recalibrated, and some cannot be.

■ Ability to Change the Air-Fuel Ratio with Load

The simplest oxygen trim systems establish a constant percentage of oxygen at all boiler loads. A constant oxygen percentage is less than optimum because it allows the percentage of unburned fuel products to rise at lower loads. (This occurs because the combustion zone is smaller and cooler at low loads, requiring more oxygen to ensure complete combustion.)

The appropriate variation in oxygen content with load depends on the fuel and the boiler design. For this reason, most oxygen trim systems include provision to manually adjust the amount of excess oxygen at different boiler loads. This is called "characterization" of the controls. Do it when the controls are installed, and when sensor elements are replaced.

More sophisticated combustion air trim systems eliminate the need for manual characterization of excess air by sensing unburned fuel components in addition to sensing oxygen. Even so, complex systems may still have a manual characterization capability to allow for control of smoke or environmental pollutants.

■ Multi-Fuel Capability

In boilers that burn more than one type of fuel, account for the fact that the optimum air-fuel ratio differs from one fuel to another. Heavier and dirtier fuels require more excess air than gaseous or light fuels. Therefore, get automatic air-fuel controls that provide separate characterization for each fuel burned.

■ Failure Modes and Overrides

Combustion controls will fail occasionally. For this reason, the system should have a reliable self-test feature, and it should respond to failures by reverting to a fail-safe setting. Typically, the fail-safe setting is a fixed percentage of excess oxygen.

A particularly reliable method of fail-safe control is to have mechanical stops in the control linkages so that high and low limits in the air-fuel ratio cannot be exceeded. Such stops are possible only if the air damper and the fuel valve are connected by a common mechanical linkage. This type of fail-safe limit is not possible with control systems in which the air and fuel are controlled separately by pneumatic, electrical, or electronic signals.

■ Alarms

The controls should have effective alarms to alert operators of any undesirable condition, especially a failure of the combustion control system or an excess of undesirable gases in the flue.

■ Displays and Data Logging

The control system should include convenient displays of all the measurements being taken, such as percentage of gas concentration and flue temperature. If you can afford it, get data logging equipment to record and analyze long-term trends in the measurements.

Some systems display derived values, such as percent efficiency and boiler load. Understand which outputs are being measured directly, and which are derived. For example, percent carbon dioxide may be calculated from the measured percent oxygen and the flue temperature. Such a derived value is dubious and can be misleading. Boiler load indications may be derived from stack temperature. Such a load indication is only approximate, because other factors can affect flue temperature.

The output of the combustion control system must be matched to the input of the recording and analysis equipment. There are a variety of signal conventions, such as 4-to-20 milliamperes or 0-to-100 millivolts. If the combustion control system includes integral data logging, you don't have to worry about this.

■ Actuator Connections

It can be tricky to make the mechanical connections to the existing air-fuel control system. Select a system that makes this job as easy as possible.

■ PID Control Characteristics

It is difficult to maintain control of the air-fuel ratio when the boiler load fluctuates rapidly, as occurs in many applications. To achieve accurate tracking, sophisticated control systems have special features to reduce error, eliminate overshoot, and prevent oscillations. These are usually called "proportional-integral-derivative" (PID) control characteristics. See Reference Note 37, Control Characteristics, for details.

ECONOMICS

SAVINGS POTENTIAL: 0.2 to 3 percent of fuel consumption.

COST: A simple oxygen trim system for a boiler that uses a mechanical linkage for control of air-fuel ratio may cost less than $10,000. Installing a complex state-of-the-art control system for a large boiler may cost several hundred thousand dollars.

TRAPS & TRICKS

SELECTING THE EQUIPMENT: *Get equipment that is well proven. Be sure to match the features of the control system to the characteristics of your boilers.*

INSTALLATION: *Follow installation instructions precisely. The details matter.*

MAINTENANCE: *Be prepared to perform repetitive maintenance, such as replacing sensors. This requires effective training and supervision. Alternatively, arrange for maintenance to be done by the manufacturer or a qualified service representative. Schedule periodic tests of the equipment.*

MEASURE 1.3.3 Adjust and repair air-fuel ratio controls.

RATINGS

New Facilities	Retrofit	O&M
		A

The air-fuel control system needs maintenance on a regular basis. The mechanical components that control air and fuel are continually moving in response to load changes. Something as simple as a loose connection can cause a major error in the air-fuel ratio. Combustion control systems have sensors that are exposed to hot flue gases, so they require periodic calibration and replacement.

Looseness in Mechanical Linkages

The air and fuel controls of most boilers have some mechanical linkages. Most small and medium sized boilers control the air-fuel ratio entirely with mechanical linkages, which may include a rotating shaft, one or more bellcranks, pushrods, and an adjustable cam. These linkages are simple, accessible, and easy to understand. Figure 1 shows a typical example.

This is fortunate, because the adjustment of the linkages is critical. Even a small amount of play may cause significant inefficiency because fuel and air controls typically have a small range of movement. Fuel valves are usually the most sensitive element. For example, movement of a millimeter in a plunger-type fuel valve may cause a significant change in fuel flow.

Air-fuel linkages have a number of wear points and opportunities for components to come loose. To discover play in a control linkage, hold one end of the linkage in a fixed position. Then, grasp the linkage at various places and attempt to move it while feeling for looseness. Inspect each bearing and joint in the linkage as you do this. If there is enough play in the linkage to allow any significant amount of motion, repair the linkage or tighten it. Replace all worn components.

Unfortunately, many burners have air-fuel linkages made of cheap stamped parts, so they are vulnerable to wear and working loose. Some common components,

SUMMARY

Basic maintenance that keeps the air-fuel controls working properly.

SELECTION SCORECARD

Savings Potential .... **$** $ $

Rate of Return .... **% % %** %

Reliability .... ✓ ✓

Ease of Initiation .... ☺ ☺ ☺

such as slotted sheetmetal control arms, are unbelievably flimsy. Linkages are often adjusted with ordinary nuts, which are virtually certain to work loose in time. Improve such features wherever you can. For example, if your linkages have slotted control arms, install friction washers on each side of the slot, and use an all-metal stop nut.

Industrial Combustion

Fig. 1 Typical dual-fuel burner assembly All the linkages that control the air-fuel ratio of this burner are easily accessible for inspection, adjustment, and repair.

■ Mechanical Components of Combustion Air Trim Controls

In air-fuel controls that have combustion air trim systems (covered in Measure 1.3.2), the trim system can compensate for wear in the mechanical components up to a point. But don't rely on this. When the movement of the linkage reverses direction, the slack in the control linkage must be taken up in the reverse direction to get the air-fuel ratio back within limits. Such control reversals may occur frequently, and the hammering that they produce in the linkage aggravates the looseness. Beyond a certain point, looseness in the control system can trigger major control oscillations so that the air-fuel ratio never stabilizes. For this reason, keep your combustion control systems just as tight as mechanical air-fuel controls.

Sensors in Combustion Control Systems

Combustion control systems continuously readjust the air-fuel ratio by sensing the combustion products in the flue gases. When they function properly, automatic combustion controls provide the ultimate in efficiency. However, their accuracy depends on proper operation and calibration of their gas sensors and associated circuitry.

The combustion gas sensors are the most vulnerable element of automatic air-fuel control systems. Most combustion control systems operate by sensing oxygen with a zirconium oxide element. Zirconium oxide elements are more tolerant of the flue environment than other types of sensors, but they eventually foul. The sensor typically must be replaced every year or two. Between replacements, the sensor and associated instrumentation require periodic recalibration. This may be done automatically by the control system, or an operator may have to initiate a recalibration cycle. Oxygen sensors typically are recalibrated with ambient air, which contains a known quantity of oxygen.

More sophisticated combustion control systems operate by sensing other flue gases in addition to oxygen, such as carbon monoxide. Sensors for other gases typically require frequent replacement. Also, calibration of the sensors may require more elaborate procedures than for calibrating oxygen sensors. Some types of sensors require calibration gases for recalibration. These are gases that have a known percentage of the type of gas that is being measured in the flue.

Test, maintain, and calibrate sensors using the manufacturer's procedures. Calibrate often enough to avoid substantial drift in control accuracy.

Upgrading

The technology of air-fuel controls has been advancing rapidly. Newer equipment may be significantly superior to older combustion control installations, justifying upgrading. See Measure 1.3.2 for the features of newer equipment.

ECONOMICS

SAVINGS POTENTIAL: *Up to 10 percent of fuel cost, in extreme cases. The potential for savings is proportional to the opportunity for error in the air-fuel controls.*

COST: *Minor. The cost of efficiency testing to re-adjust the air-fuel ratio is covered in Measure 1.3.1.*

PAYBACK PERIOD: *Immediate.*

TRAPS & TRICKS

SKILLS: *Basic repair of loose linkages is easy. However, you have to recalibrate mechanical air-fuel controls after you repair them. See Measure 1.3.1 for details.*

SCHEDULING: *This activity needs to be repeated periodically. Include this in your maintenance scheduling system.*

Burner and Fan Systems

A burner assembly prepares the fuel for burning and mixes the fuel efficiently with combustion air. The fans that move the air and flue gases are an integral part of the system.

Modern burner systems are capable of burning fuel almost completely. Even so, you can save energy in the burner system by reducing the amount of excess air, by reducing the amount of energy needed to operate the burner system, by improving the efficiency of heat transfer, by avoiding standby losses, and by burning the most economical fuels. At the same time, you want to minimize pollution, minimize maintenance, and avoid damage to the boiler.

The Measures in this Subsection exploit all these opportunities. Some are routine maintenance activities, some are simple adjustments, and some are equipment changes.

INDEX OF MEASURES

1.4.1 Clean, adjust, and repair burner assemblies at appropriate intervals.

1.4.2 Eliminate air leaks in air casings, blower housings, and connecting ducts.

1.4.3 In boilers that are fired at inefficiently high output, reduce the maximum firing rate.

1.4.4 Install burner systems that provide the best efficiency and other features.

1.4.5 Replace the motors in burners and fans with models having the highest economical efficiency.

1.4.6 Replace continuous pilot flames with electrical ignition.

1.4.7 Install variable-output fan drives on large forced-draft and induced-draft fans.

RELATED MEASURES

- Measures 1.1.2 ff, selecting the most economical fuel to burn
- Measure 1.1.4.1, installing an automatic burner sequencing controller
- Measure 1.5.3.4, minimizing burner cycling
- Subsection 1.2, for methods of measuring combustion efficiency
- Subsection 1.3, for Measures related to setting the air-fuel ratio of burners
- Measures 1.9.2 ff, stabilizing viscosity of fuel oil delivered to burners

MEASURE 1.4.1 Clean, adjust, and repair burner assemblies at appropriate intervals.

SUMMARY

Essential maintenance to keep boilers efficient.

SELECTION SCORECARD

Savings Potential .... $ $ $

Rate of Return .... % % % %

Reliability .... ✓ ✓

Ease of Initiation .... ☺ ☺ ☺

Lack of routine maintenance is the most common cause of inefficiency in burners. Fortunately, most burners are relatively simple devices, and most routine maintenance is fairly easy. Just make sure that the maintenance gets done.

Maintenance Procedures

Use the manufacturer's operating manual as your primary guide in maintaining your burners. The instructions for most burners include an exploded diagram of the burner, showing all its components. Use it as a guide.

If you cannot find adequate maintenance instructions (usually because the manufacturer has gone out of business), or if the instructions are inadequate, concentrate on these items:

• *Keep all the fuel passages scrupulously clean.* Virtually all types of burners for gas or liquid fuels prepare the fuel for combustion by dispersing it through small orifices or an array of small orifices. (The rotary cup burner is the single significant exception.) In many burners, a swirling motion of the fuel created by small vanes or flutes in the burner head helps to atomize the fuel. Recirculating oil burners contain narrow passages to control return of oil to the pump.

Burners for light oil tend to be kept clear by the solvent properties of the oil, but the parts exposed to high temperature may form coke and clog. Furthermore, particles in the oil may clog the burner.

With heavy fuel oil, distillation and coking of the fuel may leave residues that clog narrow passages and bake on to hot metal of the burner head. As a result, burners for heavy oil require periodic disassembly and cleaning. Cleaning is done with strong solvents and mechanical removal of deposits. Take care to avoid reaming or scoring the components when cleaning them mechanically. Cleaning tools should be softer than the burner materials. Copper wire and toothpicks are used for cleaning nozzles and burner head components. A dental pick is handy for getting carbon residue out of grooves, but be careful because the steel is hard.

Gas burners tend to stay clean because gas combustion does not produce products that can clog the burner. As a result, gas burners may operate for years without losing efficiency. But, don't ignore them. Check and clean gas burners periodically. Burner manifolds are made of iron, and rust may clog the nozzles over time. Atmospheric gas burners have been clogged by spider cocoons in the primary air passages.

• *Replace worn orifices.* The atomizing orifices in oil burners enlarge by erosion. Some manufacturers recommend replacing nozzles every year or two.

• *Repair leaking seals.* Inspect the burner for leaks while it is under pressure, but with the fuel valve or solenoid valve turned off. Fuel leakage may be caused by leaking fuel valves, leaking recirculation valves, or defective seals, depending on the burner design. Unfortunately, this does not reveal any leaks downstream of the fuel valve, so you have to carefully check the discharge end of the burner assembly.

Seals and sealing surfaces are vulnerable when burner tips are removed and replaced. With the high pressures that are used in oil burners, a small scratch on a sealing surface or on a copper gasket can cause a significant leak. Cross threading also causes serious leaks.

• *Keep moving parts operating freely and without play.* Lubricate linkages, damper blades, and other components that need it. Inspect thoroughly to find any points of wear. Correct for wear by adjusting or replacing the worn part.

• *Make sure that air shutters close tightly.* This is important for minimizing standby losses, which occur when the burners are not firing.

• *Keep remote forced-draft and induced-draft fans operating properly.* The maintenance required for combustion air fans is generally the same as for any centrifugal fan. The main items are keeping the impeller blades clean and the bearings lubricated.

Diagnosing Burner Problems

Burners are subject to a variety of problems that may develop quickly and reduce burner efficiency. Make sure that all boiler operators know how to spot problems.

Burner problem diagnosis charts that are provided by some manufacturers are valuable for diagnosing problems that are peculiar to a specific model of burner. However, operators must first notice that a problem exists. These are common indications of trouble:

- *smoke indicates incomplete combustion.* Check for smoke continuously. However, don't depend on smoke to indicate a problem. Not all burner defects produce smoke. A partially clogged burner results in a high air-fuel ratio that may burn off smoke before it leaves the combustion chamber. With gas fuels, incomplete combustion may not generate enough smoke to be noticeable.
- *a drop in measured combustion efficiency* could be due to many causes. You should be testing combustion efficiency on a regular basis, as recommended by Measure 1.2.1. A defect in the burner is likely if carbon monoxide and/or smoke are higher than normal in relation to the oxygen content.
- *irregular flame appearance,* including a change in the color, shape, intensity, or other appearance characteristics of the flame. Check the appearance of the flame regularly. Sparks or flashes may indicate incomplete combustion, or they may be caused by water or other contaminants in the fuel. It takes experience and a practiced eye to know what a change in flame appearance means, but any change should prompt further inspection.
- *unusual sound and vibration.* A burner defect may change the sound of the flame enough to be noticeable to an experienced operator. However, this clue is too subtle to be useful to most operators. A burner defect that causes non-uniform mixing of air and fuel may cause instability of the flame, creating a noticeable sound or even severe pulsing of the boiler. The flame may also become unstable if you attempt to minimize excess air with a burner that is not designed for low excess air. An unstable flame is dangerous. If it causes the flame to blow out, it can cause a firebox fuel explosion.
- *low flue temperature.* With non-modulating burners, you may be able to detect a drop in flue temperature that indicates a reduction in burner output. This could be caused by a clogged burner. However, a burner may be partially fouled without a noticeable drop in flue temperature. Also, a drop in flue temperature caused by a fouled burner could be masked by a rise in temperature due to soot accumulation on the tubes.

- *dirty burner appearance.* Inspect your burners regularly. Larger boilers may have multiple burner barrels that can be removed individually for inspection and cleaning while the boiler is operating. Other types of burners can be inspected only when the boiler is cold.

Burner assemblies should look clean and be tight. The critical factor in atomizing is the cleanliness of the small passages through which the fuel passes. The only way to find an obstruction in this area is to disassemble the burner.
- *soot accumulation.* Soot anywhere within the boiler, whether on the tubes, the refractories, or the burner itself, is an indication of incomplete combustion. Gas that is burned properly produces virtually no soot. Light oil should produce only a light accumulation of soot over a long period of time. Heavy oil produces a normal accumulation of soot that must be cleaned away periodically.

A change in the way soot accumulates indicates trouble. If abnormal sooting occurs within a boiler, the pattern of accumulation usually indicates whether the problem is due to the burner or to another cause. Opening the boiler to check for soot and other conditions is dirty work, but it needs to be done regularly.
- *irregular appearance of refractories.* Irregular appearance of the fire brick and other refractories usually indicates that the flame is coming too close to the refractory or that material is being thrown on the refractory by the burner. The symptoms are localized discoloration, melting, and spalling. A partially clogged burner could distort the flame shape enough to cause these symptoms. However, these symptoms are more likely to be caused by a burner that is mismatched to the boiler or is being operated at the wrong pressure. Fuel contaminants sometimes cause unusual deposits on refractories.
- *loose control linkages* can cause radically improper air-fuel mixtures. This is a common problem. For example, a jam nut may come loose on a jackshaft assembly. Such defects are usually easy to spot.

Check and Adjust the Air-Fuel Ratio

Whenever you work on a burner or burner assembly, adjust the air-fuel ratio. Measure 1.3.1 explains the procedures.

ECONOMICS

SAVINGS POTENTIAL: 0.5 to 10 percent of fuel cost.

COST: Usually minor. Facilities that have a boiler staff should be able to accomplish burner maintenance as part of the normal operating routine.

PAYBACK PERIOD: Immediate. Appropriate maintenance is always cheaper than neglecting it.

MEASURE **1.4.2 Eliminate air leaks in air casings, blower housings, and connecting ducts.**

RATINGS

New Facilities	Retrofit	O&M
		A

The boiler system's fans expend energy to move air into the boiler and to move flue gas out of the boiler. The potential energy waste is greatest if a fan is separated from the boiler by a length of ducting or flue, and if the boiler is surrounded by an air casing. The energy waste typically is small, but you can prevent it easily.

If the boiler has an induced-draft fan, air leaks into the flue may cause errors in combustion controls and combustion efficiency tests that can lead to serious problems.

Leakage typically occurs as a result of flexible duct couplings that break from vibration, sloppy joints in casings, and cover plates being left off. Repairs usually are obvious and cheap.

Finding leaks is simply a matter of careful inspection. In parts of the system that are under positive pressure, you can find leaks by running a hand around areas where it may be occurring. Finding leaks in parts of the boiler that are under negative pressure is more difficult. A smoke pencil is a good tool for finding leaks into a duct or flue.

Waste of Fan Energy

The amount of fan energy that is wasted is small in relation to other boiler plant losses. For a given amount of leak area in the duct system, the amount of air leakage depends on how the flow of air into the boiler is regulated. The loss is highest if air flow is throttled at the point where the air enters the firebox, so that the entire air passage is kept under maximum pressure by the fan. The loss is lowest if air flow is regulated at the fan, because there is less pressure in the air passage.

In boilers that have induced-draft fans, the air passages are under a slightly negative pressure. The pressure difference usually is less than with forced-draft

SUMMARY

A basic item to check with boilers that have combustion air fans separate from the burners, and with boilers that have induced-draft fans. Leaks are usually easy to fix.

SELECTION SCORECARD

Savings Potential	**$ $**
Rate of Return	**% % % %**
Reliability	**✓ ✓**
Ease of Initiation	☺ ☺ ☺

fans. Again, the amount of air leakage depends on how the flue draft is regulated.

Adverse Effects of Air Leakage into the Flue

With induced-draft fans, air may leak into the flue. This air, which is not involved in combustion, cools the flue gas. This gives a false reading of higher efficiency. It also falsely indicates high excess air, which may motivate operators to set the air-fuel ratio too rich. This causes rapid fireside fouling, a significant drop in efficiency, and other problems, which are covered in Measure 1.3.1.

Leakage in a forced-draft system usually has no ill effects other than the waste of fan energy. Air leakage may upset the air-fuel ratio in some boilers, but this problem is uncommon. Boilers large enough to have separate forced-draft fans usually have combustion controls that continuously adjust the air-fuel ratio.

ECONOMICS

SAVINGS POTENTIAL: *Perhaps a maximum of several hundred dollars per year, for typical small leaks in a large* boiler system. *A major leak in an induced-draft system may lead to incorrect air-fuel ratio settings that waste much more energy than the leak itself and that cause other problems.*

COST: *Usually minor.*

PAYBACK PERIOD: *Usually short.*

TRAPS & TRICKS

SCHEDULING: *Air leaks grow over long periods of time. Schedule inspections at appropriate intervals, such as yearly.*

MEASURE 1.4.3 In boilers that are fired at inefficiently high output, reduce the maximum firing rate.

The flue gas temperature increases with firing rate, for a given boiler. Efficiency falls as the flue gas temperature rises. In a boiler that has a high firing rate in relation to the amount of heat transfer, the flue gas temperature rises rapidly near full load, wasting energy.

If you have a boiler that fits this description, you can improve average operating efficiency by limiting the maximum firing rate of the boiler. You may have to reduce the maximum output by 20% to 40% to achieve much efficiency improvement. The question is whether you can afford to sacrifice this much capacity. You may be able to do this if you have plenty of reserve capacity, or if you take other steps to reduce your peak heating load.

Why would a boiler lack sufficient heat transfer? The reason is usually economic. The cost of a boiler is proportional to its heat transfer surface area, so the facility may have tried to save money by purchasing a smaller boiler and using a larger burner in it. The traditional rules that relate the heating surface area of a boiler to its output are somewhat arbitrary, so capacity may be stretched at the expense of efficiency.

Many older boilers would benefit from a reduction of burner output even if they are not burdened with an oversized burner. Older boilers may have less sophisticated heat transfer design, so their efficiency tapers off at higher loads, even though they may be adequate in terms of tube surface area.

There is another source of energy waste to suspect in boilers that are driven too hard. Combustion may be incomplete because the flame is being pushed too close to the surfaces of the combustion chamber. You can detect this condition by inspecting the inside of the boiler, and by finding a high level of carbon monoxide in the flue gas.

SUMMARY

May save energy in older boilers, and perhaps in newer boilers with oversized burners. To do this, you need enough reserve capacity to tolerate the reduction in output.

SELECTION SCORECARD

Savings Potential	**$ $**
Rate of Return	**% % %** %
Reliability	✓ ✓ ✓ ✓
Ease of Retrofit	☺ ☺ ☺

Energy Saving Potential

To decide whether to limit the firing rate, consider the boiler's efficiency characteristics, the turndown characteristics of the burners, and the reserve capacity of the boiler plant.

The greatest energy waste from overdriving a boiler occurs if the burner has only a single firing level. In this case, the boiler is operating at reduced efficiency whenever it is firing. If an inefficient boiler is being overdriven by a single-stage burner, the efficiency loss may be as high as 10 percent, compared to the boiler's peak efficiency.

The least energy waste occurs with a modulating burner that has a turndown ratio that is large enough to follow variations in the boiler load. In typical applications, the boiler spends only a small amount of time near full load, typically when warming up the facility after a period of shutdown.

Staged burners are an intermediate case. The improvement that a 2-stage burner offers compared to a single-stage burner depends on the load profile of the boiler, the ratio between the high-fire and low-fire outputs, and the sequence of firing stages.

Test the Potential Benefit

By this point, you should have tested the efficiency of your boilers at all their firing rates. If not, see Measure 1.2.1 for an explanation of combustion efficiency testing. In short, if the efficiency of a boiler is substantially lower at higher firing rates, you can save energy by limiting the firing rate. For example, if the boiler has a two-stage burner, reducing the firing rate is desirable if the high-fire efficiency is substantially lower than the low-fire efficiency.

Even if you do not have an efficiency tester handy, you can tell that the boiler is losing efficiency at higher firing rates if the flue temperature rises dramatically. If your boiler fires only at a single rate, calculate the difference between the flue temperature and the acid dew-point temperature for the fuel you are using. (See Measure 1.7.1.1 for an explanation of acid dew point.) If the difference is greater than necessary to protect the boiler, reducing the firing rate will probably improve efficiency.

An increasing carbon monoxide reading at higher firing rates suggests that the combustion chamber is being fired beyond its capacity. However, rising carbon monoxide could also be caused by burner defects.

Can You Afford to Lose the Boiler Capacity?

If you limit the firing rate of a boiler, and then discover that you miss the capacity, you can restore the higher firing rate. However, you may not be able to do so quickly. Therefore, calculate how much spare capacity your boiler plant presently has, including reserve capacity needed to deal with maintenance and unplanned outages. Spare capacity is common in boiler plants because design engineers tend to specify large amounts of reserve capacity to compensate for uncertainties and changing conditions. Also, the load may have been reduced by energy conservation measures, abandoning some facilities, etc.

Before reducing burner output, make note of the average boiler load under peak load conditions. Also, note how long the boiler must operate at high fire after a long period of shutdown, such as a cold Monday morning after a weekend shutdown. Then, estimate how much reduction in capacity you can tolerate, considering the amount of reserve capacity you want to keep.

Plants with Boilers of Different Efficiencies

Measure 1.1.4 recommends that you operate your most efficient boilers first. If you have boilers of different types, do not reduce the firing rate of your most efficient boilers if doing so would force less efficient boilers to operate for longer periods of time.

Methods of Reducing the Maximum Firing Rate

First of all, do not reduce the firing rate to an extent that risks acid condensation. You have to maintain a certain margin of temperature between the flue gas temperature in the coldest part of the boiler and the acid dew-point temperature of the fuel. If you are not clear about this point, review Measure 1.7.1.1.

The best method of reducing the maximum firing rate depends on the type of burner and burner controls. You may be able to use more than one of the following methods.

■ Reduce the Size of the Fuel Nozzle

You can reduce the maximum firing rate of most types of burners by replacing the existing fuel nozzle with a smaller nozzle, and adjusting the air flow accordingly. Check the manufacturer's literature or contact the manufacturer to determine the appropriate procedure. This modification is usually inexpensive.

■ Deactivate the "High Fire" Setting of Staged Burners

Most staged burners have two levels of firing, "high fire" and "low fire". The conventional method of control has the burner go immediately to high fire after starting. This provides the fastest warm-up when starting from a cold state. Also, this sequence ensures that the burner will satisfy the load.

After the set pressure or temperature is reached, the burner goes to low fire. If the load is greater than the low-fire setting is able to satisfy, the burner cycles between high fire and low fire. If the load is low, firing eventually turns off, and the cycle starts again with high fire. This is because the temperature or pressure is allowed to drop so far between firing cycles that the burner restarts at high fire. This reduces cycling, and thereby reduces standby losses.

As a result of this method of control, staged burners typically spend a large fraction of the time at high fire. One way to deal with this is to modify the burner controls to lock out the high-fire setting. The low-fire output typically is about 60% of the high-fire output. This makes the low-fire setting a good firing rate for improved efficiency in many applications.

Measure the boiler output at high and low fire before proceeding, so you won't be surprised by losing more capacity than you expected. The easiest way to estimate the boiler output is to take readings from the fuel meter.

If you want to retain the high-fire setting for warm-ups periods or during extreme loads, you might use a timeclock or an outdoor temperature sensor to allow high fire under these conditions.

The burner control sequence is built into the firing controls. Only a qualified specialist should modify these controls to bypass the high-fire setting. Contact the burner manufacturer to discuss whether this is practical.

■ Lower the High Limit of Modulating Burners

Modulating burners follow the boiler load, so they do not suffer from excessive firing during normal operation. However, modulating burners normally operate at full output during a warm-up period. If the boiler has electronic controls, you may be able to limit the maximum firing rate with a control setting. Another method is to limit the mechanical motion of the air-fuel actuators. Or, install a smaller burner nozzle, as suggested previously. Check the manufacturer's literature or contact the manufacturer to determine the appropriate procedure to limit the maximum burner output.

Limiting the maximum firing rate may also be beneficial when several boilers are operating together. This prevents one boiler from firing at an excessive level while the others are at low load. See Measures 1.1.4 ff about distributing the boiler load in the most efficient manner.

Reduce Auxiliary Energy Consumption at the Same Time

If you reduce the output of a boiler permanently, reduce the output of the boiler's auxiliaries in an efficient manner also. This is not necessary if the auxiliaries have the capability to turn down efficiently. See Measures 1.1.3 ff for efficient control of auxiliary equipment.

Leave a Record of the Change

Label or tag the burner control housing permanently to indicate the changes. See Reference Note 12, Placards, about placard materials and installation. Keep your notes about the change with the rest of the reference material for the boiler.

Alternative Methods

Increasing the heat transfer of the boiler is an alternative to reducing burner output. Methods of increasing heat transfer are covered in Subsection 1.7.

If you measure low combustion efficiencies at all firing rates, your boiler is an energy hog. In that case, consider replacing the boiler. See Subsection 1.12 about your options.

ECONOMICS

SAVINGS POTENTIAL: 0.1 to 5 percent of energy consumption.

COST: Materials costs are usually small, except for burner control circuit components, which may cost hundreds of dollars. Several hours to several days of labor are required for modification and testing.

PAYBACK PERIOD: Less than one year, to several years.

TRAPS & TRICKS

PLANNING: Make sure that your efficiency tests are accurate. If this Measure applies to your boiler, it has inadequate heat transfer surface. Consider all the Measures that might improve this situation, although this Measure and Measure 1.7.2 are the only cheap ones. If you decide to use this approach, make sure that limiting the firing rate will not cause a shortage of capacity under heavy load conditions.

MAKING THE CHANGES: Any work on burners involves safety. Make sure that you know what you are doing, or hire a specialist.

MEASURE 1.4.4 Install burner systems that provide the best efficiency and other features.

A burner costs much less than the fuel that it burns during its service life. Therefore, even a small improvement in burner efficiency is usually worth the additional cost when buying a new boiler. With existing boilers, the efficiency improvement may justify upgrading the existing burners.

Some burners are inherently less efficient than others, mainly because they require more excess air. Also, some types of burners that have high efficiency on paper may tend to be inefficient in practice because they are sensitive to lack of maintenance.

Another reason to upgrade burners is to comply with increasingly stringent pollution abatement requirements, especially those requiring reduced emission of nitrogen oxides. If you have to upgrade burners to satisfy environmental regulations, exploit the opportunity to select burners that also maximize efficiency. A low-emissions burner is not necessarily a highly efficient burner.

In selecting a replacement burner assembly, do not be satisfied just with a claimed high efficiency, or with efficiency that is measured by testing under laboratory conditions. In addition to having a good peak efficiency, the burner should remain efficient by resisting fouling and being easy to maintain.

Burner Characteristics

A burner assembly has two main functions: to prepare the fuel for burning, and to mix air with the prepared fuel. The ideal burner would completely burn all the fuel, require no excess air, and require no energy to operate. No burner system achieves this ideal, but some come close. The goal of designing the ideal burner has led to a great variety of types, each with advantages and disadvantages, and each best adapted to a particular class of boilers. The overall efficiency of a boiler depends on a number of burner characteristics. As a result, you can't select a burner on the basis of a single efficiency figure. Furthermore, there are issues other than efficiency, especially air pollution, that you have to consider. Make your choice based on an understanding of the following burner characteristics.

■ Fuel Burning Efficiency

Historically, high fuel burning efficiency was one of the first goals of boiler design, and almost complete success has been achieved. With all conventional fuels, a properly functioning burner burns more than 99% of the fuel. If incomplete burning occurs, the cause is a defect in the burner or improper matching of the burner to the boiler.

SUMMARY

In new construction, this is just a matter of smart shopping. As a retrofit, this is a major modification that requires a thorough knowledge of how burners are matched to boilers. Advisable if the boiler has an efficiency limitation related to its present burners, or if the modification provides other benefits, such as reduced maintenance or multi-fuel capability.

SELECTION SCORECARD

Savings Potential	$ $ $
Rate of Return, New Facilities	% % % %
Rate of Return, Retrofit	% % %
Reliability	✓ ✓ ✓ ✓
Ease of Retrofit	☺ ☺ ☺

■ Excess Air Requirement

Among modern burners, the main characteristic that makes one burner more efficient than another is the amount of excess air that the burner requires. Excess air absorbs heat, reduces heat transfer, and increases fan power. With current technology, all burners require some excess air, but some types require considerably less than others.

Efficient burners limit excess air by careful metering of fuel and air, and by aggressive mixing of the two. In addition, burners for liquid fuels must prepare the fuel for burning by atomizing it, which is a process of breaking the fuel into tiny droplets. Atomizing enormously increases the fuel's surface area, so that the fuel vaporizes quickly and can be mixed with air while it is still within the controlled environment of the burner.

The flame tends to become unstable when pushing the lower limits of excess air, so the design of low-excess-air burners must provide for exceptional flame stability.

■ Flame Pattern

The burner assembly should produce a flame shape that fits properly within the firebox. The burner should also distribute the hot gas efficiently through the heat transfer surfaces. An improper flame pattern may cause incomplete combustion and damage the boiler. This is explained in greater detail below. Good retrofit burners typically are designed to provide a range of flame shapes.

■ Combustion Chamber Insulation Requirements

If an oil burner projects the oil spray into the combustion chamber, the area of the spray must be kept hot enough to complete the vaporization of the oil. Some types of burners start the flame well inside the head of the burner, and these are relatively insensitive to combustion chamber temperature. These are called "retention head" or "flame retention" burners.

Other types, such as rotary cup burners, require that the front of the boiler combustion chamber be well insulated to create a hot zone around the burner. Most modern replacement burners are less sensitive to combustion chamber insulation than older types.

■ Standby Losses

Standby losses are explained in Measure 1.5.3. They occur only if the boiler cycles on and off. Modulating burners create much less standby loss than on-off burners because they spend more time firing continuously. A burner reduces standby losses by using an air register that closes tightly when the burner is not firing. Most larger power burners use air registers, but atmospheric burners do not. Even an air register cannot avoid the large standby losses that occur with purge cycles. Standby losses can be reduced by other means, such as installing a damper in the flue, but the other methods are less effective.

■ Turndown Ratio

Turndown ratio is the ratio of the maximum heat output of a burner to the minimum heat output. With boilers that operate unattended, the turndown ratio is assumed to mean the ratio that can be achieved without operator action.

Having multiple burners provides a large increase in turndown ratio. For example, if a boiler has three burners, each of which has a 4:1 turndown ratio, the overall turndown ratio is 12:1. Even greater turndown ratios are possible in boilers with multiple burners if the burner orifices can be changed while the boiler is in operation. Such boilers have essentially unlimited turndown.

A large turndown ratio minimizes standby losses by allowing the boiler to fire continuously. It also reduces the thermal stresses that occur when firing begins and ends. The importance of a large turndown ratio depends on the application. It is especially important in boilers that operate over a wide range of loads, for example, in boilers that provide comfort heating.

Turndown is achieved by using either staged burners or modulating burners. Staged burners are simpler and easier to adjust, but they have a much smaller turndown ratio than modulating burners. Staged burners eliminate standby losses within their turndown range, but they create more thermal stress because of their sudden change of firing rate.

A large turndown ratio involves design compromises. It may not be prudent to operate a modulating burner near the low end of its advertised turndown range because it may not be able to provide accurate control of the air-fuel ratio in that region. Also, it is more difficult to adjust the air-fuel ratio of a modulating burner throughout its turndown range than to adjust a staged burner at its fixed firing rates. However, these limitations can be avoided if the boiler has automatic air-fuel ratio controls.

■ Auxiliary Energy Requirements

Burner assemblies require varying amounts of energy to operate. Oil burners require pump energy. All power burners require fan energy, whether the fan is an integral part of the burner assembly or is installed remotely. Atomizing with compressed air requires compressor energy, and atomizing with steam requires additional fuel to make the steam. Only atmospheric burners do not require extra energy to operate.

■ Pollutant Emissions

Even high-efficiency burners may emit an objectionable amount of pollutants. Currently, there is emphasis on reducing the production of nitrogen oxides, which are formed when high combustion temperatures allow the oxygen in the air to react with the nitrogen in the air. Therefore, low-emission burners attempt to reduce combustion temperature, or to accomplish part of the combustion in an oxygen-poor environment, or both.

Some burners achieve these objectives by recycling combustion gases inside the burner. Other burner systems recirculate a portion of the exhaust gas from the flue back to the burner. Figure 9 shows the latter approach.

Any burner can reduce smoke and particulate emissions by using excess air, but this approach has an efficiency penalty. As with automobile engines, reconciling high efficiency with low pollutant emission involves compromises.

■ Draft Requirements

Measure 1.5.1 explains the role of draft. Burners that use forced-draft and/or induced-draft fans control draft inside the boiler by overwhelming fan power, although a stack is still required for removal of exhaust gases. At the other extreme, atmospheric burners are entirely dependent on convective draft. Small gun-type burners provide only enough draft for proportioning the air with the fuel, and they require either stack draft or a flue gas fan to move the combustion gases through the boiler.

■ Maintenance Requirements and Ease of Maintenance

Lack of maintenance may seriously reduce burner efficiency. Select burners for which maintenance is less critical, and which need maintenance less often. Burners for gas typically need little maintenance. Oil burners vary in their maintenance requirements, especially for cleaning parts exposed to the flame. For example, rotary cup burners became notorious because of their tendency to lose efficiency if they are not cleaned frequently.

Ease of maintenance is important with burner assemblies that require periodic servicing. Needed maintenance is more likely to occur if it is easy to accomplish. Select a burner assembly that allows maintenance people to reach all the components easily. Look for burners that swing out on hinges, or that slide out on rails. At the same time, consider ease of access to the interior of the boiler.

■ Initial Cost

In all but the smallest boilers, the cost of the burner is almost trivial compared to the cost of the fuel that it burns. A small loss of efficiency costs much more in additional fuel cost than the premium for a more efficient burner.

Multi-Fuel Capability

Consider adding multi-fuel capability. Having the ability to burn more than one type of fuel allows the facility to exploit price differences in fuels. Similarly, multi-fuel capability provides protection against shortages of individual types of fuels. In many facilities, these advantages may be even more important than the efficiency improvement.

Installing multi-fuel capability involves costs that may be much larger than the costs of the burners themselves. If you add the ability to use stored fuels, such as oil or compressed gas, you will need equipment for receiving the fuel, storing it, and conveying it to the boilers. If pipeline fuels are added, installing the pipelines may be a major undertaking. Selecting alternative fuels requires careful analysis of pricing policies, and shrewd guessing about how the energy market will change in the future. If you go this far, extend your analysis to cover all the fuels that may be used throughout the facility.

Especially in unattended boiler plants, a multi-fuel burner should provide the capability of switching fuels automatically, in response to a remote signal. (See Measure 1.1.2.1 about this.) This is not possible with all multi-fuel burners, because some types must be manually retracted when they are turned off to protect them from the flame in the firebox.

Table 1.   BURNER CHARACTERISTICS

	Capacity Range	Excess Air (percent)	Standby Loss	Turndown Ratio	Operating Energy	Maintenance
Gas Burners						
atmospheric	unlimited, usually small	30 - 50	very high	none	none	very low
power on-off	unlimited, typ. below 1,000 MBH	5 - 30	mostly from purge	none	low	low
power staged	unlimited, typ. 200-2,000 MBH	5 - 30	mostly from purge	1.5:1 to 3:1	low	low
power modulating	typ. over 200 MBH	5 - 30	low	typ. 7:1	low	low
Oil Burners						
low-draft gun burners	typ. below 500 MBH	typ. 50	high, if no damper	none	low	low
pressure atomizing on-off	unlimited, typ. below 1000 MBH	10 - 25	mostly from purge	none	low	low
pressure atomizing staged	unlimited, typ. 200-2,000 MBH	10 - 25	mostly from purge	1.5:1 to 3:1	low	low
pressure atomizing modulating	typ. over 200 MBH	10 - 25	low	3:1 to 8:1	moderate	low to moderate
air atomizing modulating	typ. over 3,000 MBH	5 - 25	low	typ. 7:1	high	moderate
steam atomizing modulating	typ. over 10,000 MBH	5 - 25	low	typ. 7:1	high	moderate
rotary cup	unlimited, typ. below 10,000 MBH	15 - 25	mostly from purge	typ. 4:1	moderate	high
Coal Burners						
pulverized	very large	10 - 15	low	large	very high	very high
stoker	unlimited	20 - 40	low	large	moderate	high
fluidized bed	evolving, typ. large	2 - 10	low	evolving	very high	very high

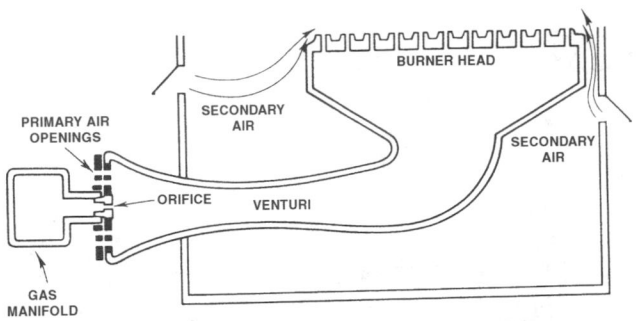

Bacharach, Inc.

Fig. 1 Atmospheric gas burner In this type of burner, air is mixed with the fuel gas in two stages. Both the primary and secondary air paths remain open, allowing serious standby losses when the burner is not firing.

Types of Burners

Attempts to combine desirable burner characteristics have given rise to a rich variety of burner designs. You have to sift through this wealth of choices. There is no formal system of classification for burners. The following groupings cover the most common types. Once you narrow the choice to a particular type, shop the market to find the best models that are currently available. Table 1 lists each of these types and their related characteristics.

■ Atmospheric Gas Burners

Atmospheric gas burners are used on the smallest gas-fired boilers, and also on some surprisingly large ones. Figure 1 shows a cross section of this type of burner. Figures 1 and 2, in Measure 1.3.1, show typical atmospheric burners in small boilers..

Atmospheric burners have two major efficiency weaknesses: (1) a relatively large amount of excess air, which reduces combustion efficiency, and (2) large standby losses, which result from the fact that the combustion chamber is open to the surrounding atmosphere. The need for excess air results from a lack of vigorous mixing of the air and fuel, and from lack of control of combustion air. The burner is surrounded by ambient air, and a large amount of non-combustion air is heated and carried through the boiler.

To make the air mixing as efficient as possible, most atmospheric burners pre-mix a certain amount of air with the fuel before it is ignited. This air is called "primary" air. Burners usually mix gas with air by venturi action. The gas, under pressure, is injected into the throat of one or more venturis, and this creates a suction that draws air into the venturis from the surrounding space. The mixture of gas and primary air is then passed through an array of small burner orifices that are open to the combustion chamber, where the gas is ignited. Enough additional, or "secondary," air is drawn from the combustion chamber to complete burning.

Primary air is controlled with a shutter at the inlet of the venturi or other mixing device. With burners that are designed for a specific boiler or furnace, the flow of secondary air to the burner is controlled by the design of openings in the boiler casing. Atmospheric burners designed for retrofit applications may have enclosures that provide a separate adjustment for secondary air.

Openings for secondary air need to be large. This makes it impossible to accurately control secondary air if there are pressure fluctuations of any magnitude in the boiler room air supply.

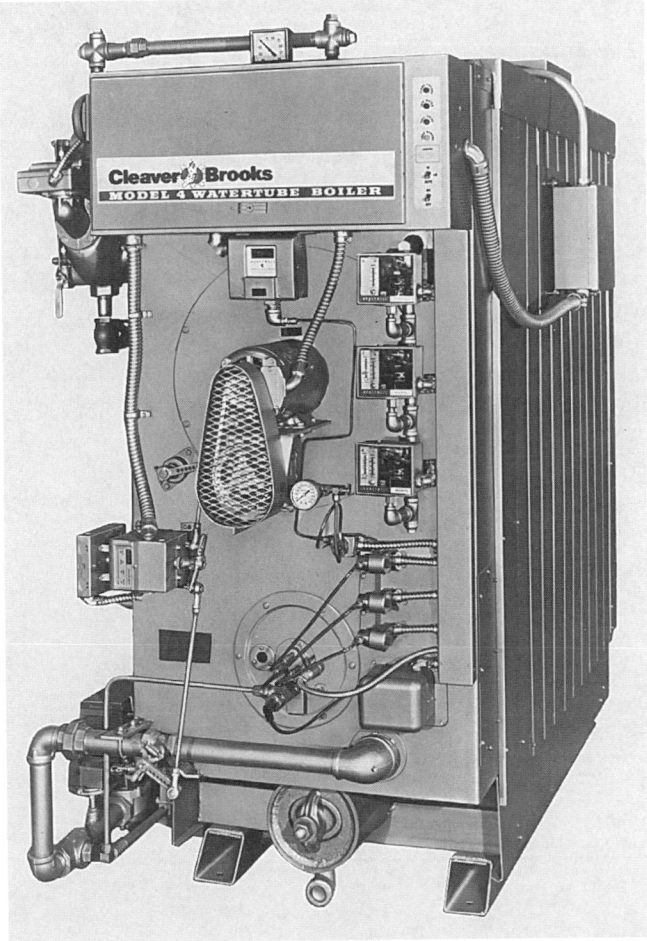

Cleaver-Brooks

Fig. 3 Boiler with staged gas power burner The three solenoid valves and thermostats on the right side control three separate burner nozzles inside the boiler.

Industrial Combustion

Fig. 2 Typical single-stage gas power burner This unit has a damper that closes the combustion air path when the burner is not firing.

Preferred Utilities Manufacturing Corp.

Fig. 4 A large modulating gas burner assembly

Parker Boiler Co.

Fig. 5 A less common type of modulating gas power burner   Here, the gas is distributed to an array of burner heads under the boiler tubes, in an arrangement similar to an atmospheric burner.

Atmospheric burners cannot adapt to changes in fuel flow. For this reason, they are controlled in an on-off mode. Multiple levels of firing can be provided by installing more than one burner assembly in the boiler.

This family of burners is called "atmospheric" because they must be kept at atmospheric pressure to function properly. The air and fuel are metered only by the low gas pressure and weak convective forces, so the air-fuel ratio is sensitive to small pressure differences in the combustion chamber and in the surrounding space. This makes it necessary to have reliable draft control,

typically provided by a draft hood or barometric damper that is installed at the flue gas outlet from the boiler. By the same token, flaws in boiler room ventilation or flue draft can alter the pressure at the burners enough to disrupt proper mixing of air and fuel.

Atmospheric burners usually do not have air inlet registers. (There is no inherent reason for this. The practical reason is that units with atmospheric burners are designed to be cheap, and adding a damper would increase cost and complication.) As a result, atmospheric burners have large standby losses. A possible remedy is to install a damper in the flue that closes when the burner is not operating. This is recommended by

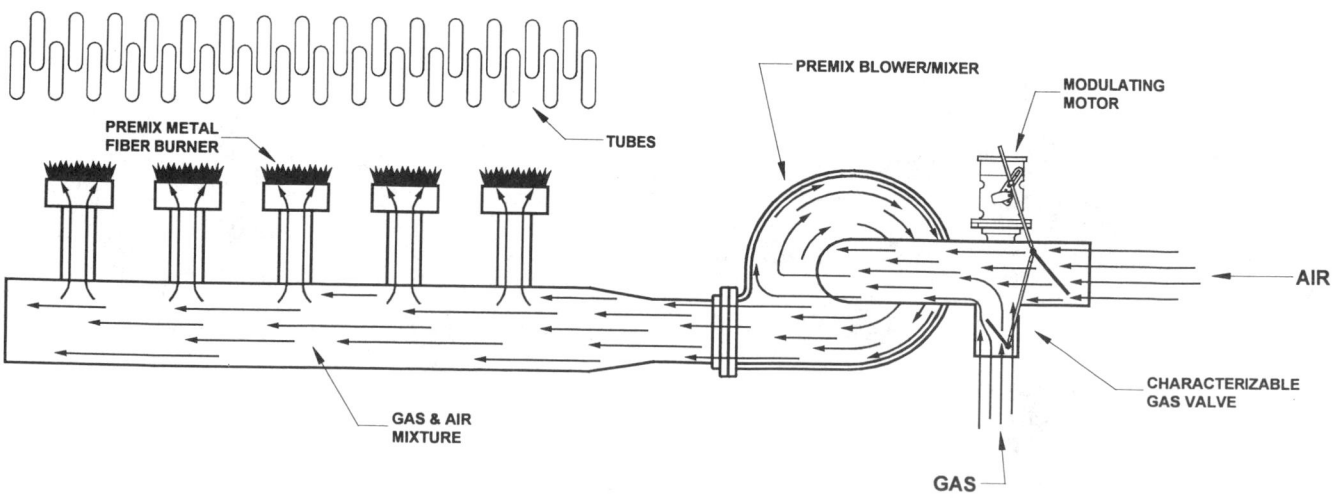

Fig. 6 Diagram of the burner shown in Figure 5

Parker Boiler Co.

Measure 1.5.3.2. However, flue dampers do not close tightly, so they do not eliminate standby losses completely.

Don't expect to retrofit a power burner in a boiler that is designed for atmospheric burners. Atmospheric burners discharge their gases upward at low velocity over a large surface area. In contrast, power burners discharge their gases at high velocity in a compact flame, and usually in a horizontal direction.

■ Single-Stage Gas Power Burners

This type of burner consists of one or more gas orifices installed in a diffuser assembly. An integral fan forces air through the assembly, which provides vigorous mixing. The fan meters the amount of combustion air, providing more positive control of air-fuel ratio than is possible with an atmospheric burner. Most units have a damper that closes when the burner is not firing. Figure 2 shows a typical unit.

■ Staged Gas Power Burners

The firing rate in a gas power burner can be varied in several ways. One type of burner has two or three separate gas nozzles, with a solenoid valve controlling each nozzle. The air flow is adjusted correspondingly by changing the damper setting of the fan, or by operating the fan at different speeds. This type has a fairly good turndown ratio because each nozzle can be optimized for its own gas flow rate. Figure 3 shows a boiler with a three-stage gas burner.

Another type has a single nozzle, typically with two settings, "high fire" and "low fire." The firing rate is changed with a two-position controller that partially closes the gas valve and the air register.

■ Modulating Gas Power Burners

This type typically has an array of gas orifices installed in a diffuser assembly. To provide modulation, a gas valve is installed to control the fuel flow and a damper is installed to control the air flow. The fan may be integral with the burner assembly, or in large boilers that have an air casing, the fan is installed remotely. Figure 4 shows a large modulating gas burner assembly.

Figure 5 shows a less common type of modulating gas burner, which uses an array of burner heads similar to those found in atmospheric burners. Figure 6 shows a cross section of the burner.

A problem with all modulating burners is maintaining the correct air-fuel ratio at all loads. Smaller boilers usually accomplish this with a mechanical linkage between the fuel valve and the air damper. Unfortunately, fuel valves and air dampers are non-linear in flow, so the two cannot be coordinated with a simple mechanical connection. A common solution is to include an adjustable mechanical cam in the linkage to control the fuel valve. The cam is set at the factory based on average conditions. Adjusting the air-fuel ratio in the field requires a tedious setting of the cam across the

entire load range, which may be impossible to accomplish accurately under realistic conditions.

The limitations of a mechanical linkage are avoided in larger boilers by controlling fuel and air separately with a control system that senses the excess air in the flue gases. "Oxygen trim" systems can be retrofitted to boilers that have mechanical air-fuel linkages, and these systems compensate for errors in the linkage. See Measure 1.3.2 for details.

■ Single-Stage Pressure-Atomizing Oil Burners

This type is used with light grades of fuel oil. All pressure-atomizing burners operate by forcing the oil through a small orifice under high pressure. The abrupt change in pressure as the oil passes through the orifice creates shearing forces in the oil that break it up into small droplets. The smallest single-stage atomizing burners are called "gun" burners because they have a barrel that "shoots" into the firebox. Actually, they look more like hair dryers than guns.

■ Multi-Stage Pressure-Atomizing Oil Burners

This type has a "high fire" and a "low fire" setting. To change from high fire to low fire, the oil pressure is reduced and the damper is partially closed. These burners are used on small and medium sized boilers, and may burn oil of various weights.

The turndown of common 2-stage oil burners is limited to a ratio of about 1.6 to 1. The reason is that the simple nozzles used in these burners cannot maintain proper atomizing if the pressure is changed over a wide range.

■ Modulating Pressure-Atomizing Oil Burners

This broad category of burner types may modulate output by varying oil pressure, or by recirculation, or both. With recirculating burners, the pressurized oil at the burner nozzle has two paths to follow, either out the burner orifice or back into the recirculation line. Burner

Preferred Utilities Manufacturing Corp.

Fig. 7 Modulating pressure-atomizing oil burner
Oil pressure for atomizing is provided by a gear pump, seen here at the lower center. Modulation is accomplished by recirculating unused fuel oil back to the tank.

output is controlled by a valve in the recirculation line that dumps unburned fuel back into the fuel tank. Figure 7 shows a burner of this type. Another type of atomizing burner modulates by changing the orifice area, for example, by inserting a tapered needle into the orifice.

The turndown ratio of a typical modulating oil burner with a fixed nozzle orifice is about 4:1. With larger boilers that have several burners, the orifice sizes of individual burners can be changed while the boiler is in operation, providing an almost unlimited modulation range.

The problems of setting the air-fuel ratio accurately in a modulating oil burner are the same as for modulating gas burners, discussed previously.

■ Air-Atomizing Oil Burners

Air-atomizing burners are used with all grades of fuel oil. Within the nozzle tip, high-pressure air is aimed at a thin stream of oil. The force of the collision breaks up the oil into droplets. Modulation is accomplished by recirculation. Turndown ratios range from about 3:1 to about 7:1.

Air-atomizing systems require a hefty air compressor and a big air storage tank. The power consumed by the compressor is typically a fraction of one percent of the energy flow through the boiler plant.

■ Steam-Atomizing Oil Burners

Steam-atomizing burners are used with heavy grades of oil. Their action is similar to that of air-atomizing burners except that steam is used instead of compressed air. The high temperature of the steam helps to break up and vaporize the oil. Modulation is accomplished by recirculation. Turndown ratios of up to 8:1 are claimed. An older type of steam-atomizing burner uses a venturi to draw oil into a steam jet, which discharges into the boiler through a nozzle.

The steam requirement of modern steam-atomizing burners is approximately one percent of boiler output. The heat of the steam is not recovered because the steam starts at the same temperature as the water in the boiler and then cools by expansion. In a steam system that has little water loss, the steam required for atomizing may add noticeably to the makeup water requirement.

■ Rotary Cup Oil Burners

This is an older type that is used with light and heavy oils. Oil is atomized by a combination of centrifugal force and collision with an air stream. The atomizing element is a rapidly rotating cylinder or cup. Oil is fed into the interior of the cup, and is thrown outward when it reaches the edge. High-velocity air flows around the outside of the cup, breaking up the oil film that is being thrown off by centrifugal force. The fan that creates the

Industrial Combustion

Fig. 8 Modulating dual-fuel burner This compact unit burns oil and gas through separate nozzles.

high-velocity air stream may be integral with the burner assembly, or it may be installed remotely, blowing air around the burner. The turndown ratio of rotary cup burners is the lowest of the modulating types, with a maximum of about 4:1.

The fuel oil is not vaporized until it has been flung beyond the cup assembly, so a high-temperature combustion zone must be maintained around the burner. This requires insulating the entry portion of the combustion chamber with refractory material. Also, matching the burner to the geometry of the combustion chamber is critical, as explained below.

Efficient atomizing depends on smooth flow of oil over the cup. Unfortunately, the cup has a large surface area within the combustion chamber, which leads to carbonizing and sludging of oil on the cup. Therefore, the burner assembly requires frequent and scrupulous cleaning. Rotary cup burners became notorious for their inefficiency when they are dirty or improperly matched to the boiler. This even led some government agencies to initiate witch hunts against rotary cup burners during the 1970's.

■ Dual-Fuel Burners

Dual-fuel burners for gas and oil have entirely separate burners for the different fuels, along with a common fan system. In oil-gas assemblies, the gas is discharged through a relatively large cylindrical array of nozzles, and the oil is discharged through a nozzle at the end of a barrel that is inserted through the center of the gas burner. Figure 8 shows a typical unit of this type.

■ Pulverized Coal Burners

These burners are used only in large boilers. The coal is pulverized into fine particles outside the boiler and is then carried through the burner barrel suspended in a high-velocity air stream. The inside end of the burner has an impeller that distributes the coal into the combustion chamber. Combustion air is provided through the space surrounding the burner by a separate fan. The particles of coal burn as they travel through the combustion chamber, so the boiler must be designed specifically for this type of burner.

Compared to other types of coal burners, pulverized coal burners have the ability to change load quickly, like oil and gas burners. Moving the air at relatively high pressure for conveying the coal is a major cost factor, because it takes about two pounds of air to carry a pound of coal. Pulverizing and other coal preparation adds considerably to the cost of boiler operation.

■ Stokers for Solid Fuels

Stokers for coal and other solid fuels, such as refuse, consist of a grate on which the solid fuel sits while air is

Industrial Combustion

Fig. 9 Flue gas recirculation A portion of the gas leaving the boiler is recirculated back to the combustion air fan, where it mixes with incoming air. The flue gas is inert, so it lowers the average temperature of the combustion gases, reducing the formation of nitrogen oxides.

fed from underneath. There are many variations of stokers. The boiler is built around the stoker, so it is not possible to retrofit a stoker to an existing boiler. On the other hand, stoker boilers have been successfully converted to burn oil and/or gas by installing conventional oil and gas burners.

■ Fluidized Bed Combustion

Fluidized-bed burning is used primarily for coal. Every aspect of the boiler is built around this method of firing, so it is not adaptable to other existing boilers. Fluidized bed boilers emerged during the 1980's in response to concern about air pollution.

Crushed coal is fed into the boiler along with granulated limestone. The "bed" of coal and limestone is suspended by air fed from underneath. This requires large amounts of fan energy. Boiler tubes are located within the bed, providing good heat transfer by physical contact between the tubes and the heated particles. The abrasion of the particles keeps the tubes clean.

Sulfur dioxide formed in the process of combustion is immediately converted to calcium sulfate, or gypsum, by reaction with the limestone. This reduces sulfur dioxide emissions, at the expense of having to dispose of the calcium sulfate, which is dry and inert. Nitrogen oxide emissions also are low because combustion occurs at lower temperature than in conventional boilers.

Match the Burner to the Combustion Chamber

The term "combustion chamber" means the region surrounding the burner where the fuel is burned. One function of the combustion chamber is to serve as a region of high temperature to sustain combustion until it is complete. With liquid and solid fuels, the combustion chamber provides for evaporation of the fuel prior to combustion. All combustion must be completed within the combustion chamber. Beyond the combustion chamber, the gas temperature falls rapidly as heat is given up to the heat transfer surfaces, stopping combustion.

Retrofit burners for gas and oil are available that provide complete combustion in a wide variety of boilers. These burners mix fuel and air inside the burner assembly to avoid external effects. The sensitivity of burner efficiency to combustion chamber temperature is reduced by using a "flame retention" head, which is basically a small combustion chamber enclosed by the burner housing to ensure that combustion is initiated properly.

A replacement burner must have a flame pattern that keeps the flame away from any of the surfaces of the combustion chamber. Flame that touches any surface is quenched by the lower temperature of the surface, with the result that combustion stops and unburned fuel,

WESINC

Fig. 10 A modern burner retrofitted to an old brick-set watertube boiler   Whether this is a prudent upgrade depends on the efficiency and condition of the boiler. Matching a modern burner to an old boiler involves a number of issues that affect efficiency and reliability.

mixed with incombustible impurities, is deposited as soot. Flame contact with refractory materials also damages the refractories. Many retrofit burners, especially larger models, allow the flame shape to be tailored to the combustion chamber by changing nozzles and diffusers.

Keeping the flame away from surfaces requires a minimum combustion chamber volume that is related to the burner output. Empirical rules for combustion chamber volume have evolved over the years. If the combustion chamber volume is not large enough for the intended burner output, efficiency suffers and damage occurs. See Measure 1.4.3 for more about this.

Another function of the combustion chamber is to feed the hot combustion gases into the heat exchange surfaces in the proper direction. Fan-forced burners give the combustion gases a significant amount of inertia. The makes the discharge pattern of the burner a dominant factor in gas distribution from the combustion chamber. For example, if an upflow atmospheric burner is being replaced with a horizontal-flow power burner, it may be difficult or impossible to keep the combustion gases from bypassing a portion of the heat exchange surfaces, or backing up in one end of the combustion chamber. A bad gas flow pattern can negate the intended benefit of installing a more efficient burner.

Especially in large boilers, burner openings may be lined with refractory to maintain high temperature adjacent to the burner. The refractory is cooled by embedded waterwall tubes to prevent it from melting. Make sure that your retrofit burner adapts properly to such features.

You may have to be inventive if your burner mounting locations are awkward. For example, if you have to install the burner through an existing door at one side of the combustion chamber, you can aim the burner toward the center of the combustion chamber by using an angled mounting plate.

How to Decide Whether to Upgrade Burners

Every boiler deserves a look to determine whether you should improve the burner efficiency. Figure 10 shows how a modern burner assembly has been retrofitted to an old boiler.

Use Table 1 to compare your existing burners to others that are available. This sequence of steps will help you decide whether to replace your burners.

(1) Consider the type and age of the boiler.

Most boilers of modern design have high efficiency. Inefficient burners tend to be found in older boilers and in newer boilers of old design. If testing shows that the boiler is not efficient, the next question is whether the problem is high excess air or high stack temperature.

High excess air is a burner problem. High stack temperature indicates a lack of heat transfer capability, which cannot be cured by a new burner. To correct high stack temperature, you need to reduce the maximum firing rate (see Measure 1.4.3), or increase the heat transfer of the existing boiler (see Subsection 1.7), or install a more efficient boiler (see Subsection 1.12).

Age is a major factor in the economic decision. It is risky to spend a lot of money converting a boiler that may be near the end of its service life.

(2) Observe whether the boiler cycles on and off frequently.

Even if the measured combustion efficiency of a boiler is high, the boiler may suffer from efficiency loss when the burner cycles on and off. One adverse effect of cycling is standby loss. Another is inefficient air-fuel ratio during the period of time that the boiler is stabilizing after it starts firing. Cycling also places stress on a boiler that causes leakage, metal fracturing, and crumbling of refractories.

Standby loss is greatly reduced if the burner has a damper that closes tightly when the burner is off. Most on-off burners have a damper, except for atmospheric burners and the smallest power burners. If the boiler does not have a combustion air damper, installing a flue damper (Measure 1.5.3.2) is an alternative. Dampers do not reduce standby loss that occurs during the purge cycles, when all dampers must be open.

You can minimize cycling by installing a modulating burner that has a high turndown ratio. A less effective solution is installing a 2-stage burner. Unfortunately, you cannot estimate standby losses accurately. Judge the seriousness of cycling losses by observing how often the boiler cycles under typical loads.

(3) Measure the boiler efficiency.

Tune up the boiler to its best air-fuel ratio, using the procedures of Measure 1.3.1. Then, measure the minimum amount of excess air required by the burners for stable operation. Compare this to the amount of excess air that is required by other burners that could be used in the boiler. Use a combustion efficiency table to calculate how much energy you could save by installing a burner that operates with less excess air.

(4) Consider the operating energy of the burners.

If the boiler uses a large amount of fan energy, there may be room to reduce fan power, especially by selecting a burner assembly that reduces its fan power requirement in proportion to the load. An alternative is to retrofit the existing fan with a variable-speed drive, as recommended by Measure 1.4.7, below.

The energy that oil burners require for atomizing may be in the form of pressure, steam, compressed air, or centrifugal force. The energy requirement for atomizing is small, the greatest being about one percent of fuel energy content, in the case of steam atomizing. Within this range, there may be an opportunity for saving energy or money.

The economic choice between air atomizing and steam atomizing depends on the relative cost of electricity and boiler fuel. One authority has pointed out that it takes about 130 BTU to produce a pound of atomizing air, while it takes about 1,300 BTU to produce a pound of steam. However, the BTU's used in air atomizing are expensive, being in the form of electricity. Steam atomizing is economically favorable only when the cost of the boiler fuel is much lower per unit of energy than the cost of electricity. This might be the case for a boiler burning refuse or wood waste.

Pressure atomizing requires less energy than steam or air atomizing. It is the most common method of atomizing light oils. Pressure atomizing has also been used extensively with heavy oils in larger boiler installations. At present, retrofit burners for heavy oils almost all use steam or air atomizing. This reflects the difficulty of atomizing heavy oil with pressure alone. However, burner technology keeps changing, so consider pressure atomizing heavy oil along with the other methods.

For solid fuels, it is usually difficult to retrofit a new type of burner because the overall design of the boiler is based on a particular method of burning. For example, a boiler with pulverized coal burners has a radically different configuration from a boiler that uses stoker firing. Modifying solid fuel burners requires case-by-case analysis.

(5) Consider how much maintenance the burners require.

Burners that require continual maintenance are likely to waste energy. Even if maintenance is good at present, it is likely that the boiler will go through periods of inadequate maintenance. It is especially important to install low-maintenance burners in boilers that are not monitored. Some types of oil burners require much more maintenance than others. The extreme example is rotary cup burners, which are notorious for becoming inefficient when they are not kept clean.

Labor cost is an economic factor if you have to hire extra people in order to accomplish adequate burner maintenance. Labor cost is not an economic factor if operators are on duty and have nothing else to do.

(6) Determine whether there are burner compatibility problems.

Start by inspecting the entire firebox environment, including the burner cones, the back wall, and all visible heat exchange surfaces. There should be no evidence of flame impingement, which causes incomplete combustion, sooting, damage to refractories, and metal stress.

Atmospheric burners and other burners that depend on convective draft are vulnerable to problems in the flue system. For example, wind descending on a stack from a tall adjacent building may upset combustion. This problem could be controlled by installing a burner assembly that gets its draft from a powerful fan. But first, fix any air flow problems. See Measure 1.5.2 about this.

Consider whether your existing burners limit opportunities to improve efficiency. For example, steam atomizing burners require high steam pressure. Using them may make it impossible to reduce system pressure to conserve energy, as recommended by Measure 1.1.5.

(7) Consider whether upgrading burners can provide other advantages.

Multi-fuel burners may provide considerable savings in fuel cost. However, consider all the costs of being able to burn multiple fuels, including new piping, tanks, pumps, heaters, etc.

With existing multi-fuel boiler installations, consider installing burners that can change fuels automatically to minimize cost. Automatic changeover is especially valuable with unsupervised boilers, although the increased reliability of fuel changeover is valuable in any plant. See Measures 1.1.2 ff for control procedures to optimize the benefit of fuel switching.

ECONOMICS

SAVINGS POTENTIAL: Varies widely. Boiler efficiency sometimes can be improved as much as 10 percent by improving burners, but the improvement is generally much less if the original burners have been operated properly. Upgrading burners may provide additional important benefits, such as reduced maintenance, multi-fuel capability, etc.

COST: Gas and oil burners costs about $5 per thousand BTUH of capacity in the smallest commercial sizes, to less than $1 per thousand BTUH in the largest sizes. Dual oil/gas capability adds a premium of about 20% in larger sizes. The cost of coal stokers exceeds $20 per thousand BTUH in the smallest sizes (about 1 million BTUH), to less than $4 per thousand BTUH in the largest sizes.

PAYBACK PERIOD: Less than one year, to many years.

TRAPS & TRICKS

SELECTING THE EQUIPMENT: Consider all eligible types of burners, and understand the advantages and disadvantages of each type. Also, be sure to understand how a new burner system must interface with your boilers.

INSTALLATION: Unless you are an expert, have the installation done by a specialist who knows how to fit the new burners properly.

EXPLAIN IT: New burners have unfamiliar operating and maintenance requirements. Inform and train the people concerned. Set up an appropriate maintenance schedule.

MEASURE 1.4.5 Replace the motors in burners and fans with models having the highest economical efficiency.

RATINGS
New Facilities | Retrofit | O&M
[ ] [ ] [B]

Be prepared to replace a failed motor with a model that has the highest practical efficiency. If the motor is exceptionally large and inefficient, you may even want to replace it before it fails. See Measure 10.1.1 for the details.

MEASURE 1.4.6 Replace continuous pilot flames with electrical ignition.

RATINGS
New Facilities | Retrofit | O&M
[ ] [D] [ ]

You can get retrofit kits to replace pilot lights with electronic igniters in many models of boilers and furnaces. Continuous flame pilot lights waste energy by consuming fuel when the boiler is not needed. They may also aggravate standby losses by enhancing convection through the boiler while the burner is not firing.

Continuous flame pilots are found only in boilers and furnaces with atmospheric natural gas burners. (A continuous pilot flame would not be safe in a boiler where fuel vapors can accumulate.) A pilot light burns fuel at a very low rate compared to the main burner, but it operates continuously. The energy waste may be significant if the pilot light continues to operate while the boiler is unused for long periods, for example, in a heating boiler that does not fire during the warm season.

Exceptions

Getting rid of the pilot light is not economical or advisable if:
- the boiler operates for a large fraction of the time
- the pilot flame is needed to maintain flue draft
- the pilot light is needed to prevent condensation on the tubes and other metal surfaces of the boiler.

Alternative: Turn off the Pilot Light Seasonally

If the heating unit is needed only on a seasonal basis, you can avoid much of the energy waste of the pilot light by turning off the pilot light at the end of the heating season.

> **SUMMARY**
>
> Saves fuel in some older gas-fired atmospheric burners. Undesirable in some applications.
>
> **SELECTION SCORECARD**
>
> Savings Potential .... $ $
>
> Rate of Return .... % %
>
> Reliability .... ✓ ✓ ✓
>
> Ease of Retrofit .... ☺ ☺ ☺

ECONOMICS

SAVINGS POTENTIAL: *0.2 to 5 percent of fuel consumption*

COST: *Several hundred dollars.*

PAYBACK PERIOD: *Several years, or longer.*

TRAPS & TRICKS

SELECTING THE EQUIPMENT: *Electronic igniters are relatively simple items that may vary in quality from one manufacturer to another. Select one that is made by a major manufacturer. It may be difficult to find a unit that you can install easily.*

INSTALLATION: *Anything related to burner ignition has serious safety considerations, so make sure that the unit is installed properly. This change probably requires a burner specialist.*

MEASURE 1.4.7 Install variable-output fan drives on large forced-draft and induced-draft fans.

If you are buying a new boiler or designing a boiler system, consider getting variable-speed fan drives to reduce fan energy consumption. You may be able to retrofit your existing boiler fans with variable-speed drives. Connecting the drive to the fan motor itself is usually straightforward. The challenge is to control the fan drive so that it maintains the proper air-fuel ratio at all loads.

Where to Install Variable-Speed Fan Drives

Variable-speed fan drives are most likely to be economical with larger boilers that have modulating burners, or perhaps staged burners. The price of variable-frequency drives per kilowatt drops rapidly with size, improving the economics of the modification for larger fans. With fans that operate continuously, the modification starts to become attractive at a fan size of about 50 KW (about 50 HP).

Combustion of hydrocarbon fuels requires about 10 to 20 pounds of air per pound of fuel, including an allowance for excess air. The low end of this range applies to coal, and the upper end to natural gas. For a fuel oil that lies in the middle of this range, burned in a boiler of average efficiency, the combustion air flow rate is roughly 250 CFM per million BTU of boiler heat output.

Assuming a typical flow resistance of 6" water gauge and fan efficiency of 60%, this translates to an average fan power requirement around 0.4 horsepower per million BTU. Thus, variable-speed fan drives start to be economical with boilers rated around 200 million

SUMMARY

Consider this for large boiler fans where the boilers operate for long periods at reduced loads. May be a technical challenge to retrofit.

SELECTION SCORECARD

Savings Potential .... $ $

Rate of Return, New Facilities % % %

Rate of Return, Retrofit .... % %

Reliability .... ✓ ✓ ✓

Ease of Retrofit .... ☺ ☺ ☺

BTU per hour. This figure is approximate. As the price of electronic drives continues to fall, they will become economical in smaller sizes.

If air flow is presently controlled with an inlet vortex damper, the potential saving offered by a variable-speed fan drive is much less. Inlet vortex dampers are a moderately efficient means of reducing fan output down to about 50 percent of full flow. Replacing an inlet vortex damper would be economical only if the boiler operates for long periods at loads well below half.

■ Verify Retrofit Burner Compatibility

Check with the burner or boiler manufacturer to make sure that this modification will work with your boiler. You can skip this step only if the fan and the damper are presently remote from the boiler.

You may not be able to retrofit a variable-speed drive if the original air control damper plays a role in mixing the air with the fuel. If the damper that throttles the flow of air is an integral part of the burner assembly, the full fan pressure may be needed at all times for proper operation.

Energy Saving Potential

Forced-draft fans and induced-draft fans are typically the most energy intensive boiler accessories. Modulating or staged burner systems may waste a large fraction of the fan energy by using dampers to throttle air flow. When a damper is used to reduce air flow, the energy consumption of the fan does not drop in proportion to the reduction in flow. Instead, the fan must expend energy to force the air through the resistance of the damper.

The energy saving also depends on the load profile of the boiler. Damper control is especially wasteful in

WESINC

Fig. 1 Vortex damper on a large forced-draft fan This method of modulating combustion air flow is less efficient than a variable-speed fan drive.

boiler applications such as space heating, where the boiler operates at low output for much of the time.

The amount of power required to move the combustion air is approximately proportional to the third power of the boiler load. For example, if the boiler load is reduced to half, the power needed to move the combustion air is reduced to about one eighth. Fan power in excess of this amount is wasted. In applications where the boiler load varies over a wide range, the saving averages from 20 to 50 percent of full fan power.

To estimate the amount of energy your boiler fans are presently wasting, measure the power consumed by the fan under different load conditions. Use a wattmeter to measure the fan power. (You can't just measure the current and multiply by the voltage. This would neglect the changing power factor of the motor, resulting in an exaggerated estimate of power consumption.)

As you measure the fan power, you can determine the boiler load from the fuel flow rate. Estimate the total length of time that the boiler operates at each percentage of load.

Calculate the saving you would get from a variable-speed drive at each percentage of load. Request certified performance data from prospective suppliers of the variable-speed drives. The usual curves found in the glossy brochures are marketing daydreams.

How to Select and Install a Retrofit Fan Drive

Study Reference Note 36, Variable-Speed Motors and Drives. In the current state of technology, the only type of drive that you need to consider is the variable-frequency electronic drive. In addition to being efficient, this type does not require any additional space at the fan itself. It does require a bulky control cabinet, but you can install the cabinet anywhere that is convenient.

Electronic variable-frequency fan drives do have some problems, especially electrical noise in the power supply system. Also, you may have to replace the present fan motor to avoid overheating. Reference Note 36 covers all these issues.

Synchronizing Air and Fuel after Retrofit

Before proceeding, get a clear idea of how you intend to synchronize the air and fuel flows. This may be simple or difficult, depending on the nature of your combustion controls. Work out the details of the control design, including detailed control schematics, before purchasing equipment.

The variable-speed motor drives offered by major manufacturers have been engineered to accept a variety of input signals, including pneumatic, electric, and electronic. For example, the fan may go from minimum to maximum output based on a pneumatic input signal ranging from 3 to 15 PSI. Typically, the fan speed

responds to the input signal in a fairly linear manner, and air flow is proportional to fan speed.

The modification is simplified if the boiler has a feedback control system that adjusts the air flow independently of the fuel flow by sensing the flue gas composition. With luck, it may not be necessary to modify the existing air-fuel controls at all, except that the control signals originally sent to the fan damper actuator are now sent to the fan drive.

However, do not take this for granted. The existing control system may not have enough control range to adapt to a large difference in response characteristics between the original damper and the new fan motor controller. Discuss this with the manufacturer of the existing air-fuel control system.

The modification is more difficult if the air-fuel ratio is presently controlled by a mechanical linkage that connects the fan damper to the fuel valve. The linkage is driven in response to the boiler output (steam pressure or water temperature). The linkage typically uses a cam to move the fuel valve. The shape of the cam is set at the factory, and compensates for the fuel valve's non-linear output behavior and for the geometry of the linkage.

When installing a variable-speed motor drive in a system that uses a mechanical air-fuel linkage, disconnect the part of the linkage controlling the air damper. See Figure 1, for example. Send the signal from the boiler output (steam pressure or hot water temperature) sensor to the new fan control as well as to the original mechanical positioner. Expect to make additional changes to get the air and fuel to track properly, since the variable-speed drive regulates air flow differently from a damper.

You may be able to get the air and fuel to track properly by adjusting the shape of the fuel cam. However, the adjustment range of the cam is limited, and resetting the shape of the cam over the entire load range of the boiler can be tedious, as explained in Measure 1.3.1.

Another possibility is changing the response characteristic of the motor speed controller. This is a specialized feature that may be available only from certain drive manufacturers. Adjusting the air-fuel ratio properly across the entire load range requires skill and patience, regardless of the method used.

How to Control Induced-Draft Fans

Variable-output control of induced-draft fans is easier to retrofit. Induced-draft fans are usually controlled to maintain a constant draft in response to a signal from the draft sensor. This is a simple control application. Just match the input signal range of the fan control to the output signal range of the draft sensor.

Test and Adjust the Air-Fuel Ratio

After the equipment changes have been made, conduct a combustion efficiency test across the boiler's full load range. Adjust the air-fuel ratio as necessary.

ECONOMICS

SAVINGS POTENTIAL: *20 to 60 percent of fan energy, depending on the average load on the boiler. Less, if the fan presently has an inlet vortex damper.*

COST: *See Reference Note 36, Variable-Speed Motors and Drives, for the costs of variable-speed drives. The additional control costs depend on the type of combustion control that is presently used.*

PAYBACK PERIOD: *Less than one year, to many years, depending on fan size, fuel cost, boiler loading, and duration of operation.*

TRAPS & TRICKS

DESIGN: *Retrofitting variable-output drives to boiler fans is still unusual. This is a major modification that cannot be done by a conventional boiler technician. Work out the details with the burner or boiler manufacturer.*

SELECTING THE EQUIPMENT: *Study Reference Note 36, Variable-Speed Motors and Drives.*

EXPLAIN IT: *The modification is unconventional. Document the design, equipment, operating procedures, and maintenance procedures. Without a clear explanation, the maintenance staff will be baffled when the system requires adjustment or maintenance.*

Boiler draft is necessary for accurate mixing of air and fuel, for moving flue gases through the boiler, and for expelling exhaust gases. Draft needs to be controlled accurately for each of these purposes. Draft should be stopped when the boiler is not firing, to prevent standby losses and to minimize thermal stresses. The Measures in this Subsection control draft for these purposes and correct problems that interfere with proper control of draft.

INDEX OF MEASURES

1.5.1 Adjust draft for maximum efficiency.

1.5.2 Correct defects in flue systems and boiler room ventilation that cause draft problems.

1.5.3 Minimize standby losses.

1.5.3.1 Control all fans in the combustion air path to stop, and all dampers to close, when the burner is not firing.

1.5.3.2 Install an automatic flue damper.

1.5.3.3 Install a burner assembly or boiler that minimizes standby losses.

1.5.3.4 With cycling burners, adjust the controls to minimize the frequency of firing cycles.

RELATED MEASURES

- Subsection 1.3, for adjusting air-fuel ratio
- Subsection 1.4, for actions related to burners and combustion air fans

MEASURE 1.5.1 Adjust draft for maximum efficiency.

SUMMARY

An important boiler efficiency adjustment that is easy to overlook.

SELECTION SCORECARD

Savings Potential	$ $
Rate of Return	% % % %
Reliability	✓ ✓
Ease of Initiation	☺ ☺

Make sure that draft is regulated properly in every boiler and furnace. In some cases, draft regulation is permanently built into the system, and it requires no special attention. However, many boilers require draft adjustments to achieve optimum efficiency. The adjustments you need to make depend on the design of the boiler, the burner system, the flue, and the draft regulating devices.

What is Draft?

In the context of boilers, draft is a difference in pressure between one part of the boiler system and another. These pressure differences move air and combustion gases through the boiler system to:

- move the proper amount of air into the boiler for combustion
- mix air with fuel at the burner in the proper ratio
- move the combustion gases through the boiler at the proper rate
- expel exhaust gases to the outside, usually through a stack.

Draft for different parts of the boiler system may be provided by convection or by fans. Both sources of draft must be regulated.

Other sources of draft have been used in specialized applications. For example, the puffing of a steam locomotive is caused by exhaust steam passing through a venturi located at the base of the stack to create draft in the firebox. Draft from sources other than convection and fans is rare today.

■ Convective Draft

Convective draft is created by the buoyancy of the combustion gases. High temperature causes them to expand, and hence to become lighter. The buoyant "lifting" force is an illusion. The actual force is provided by gravity. Gravity has a greater effect on the heavier surrounding air than on the lighter combustion gases, so the combustion gases are displaced upward.

The intensity of convective draft is proportional to the height of the stack or other structure in which the heated gases are contained. The height of the boiler itself and the breeching also creates draft. A stack is necessary if the boiler is not tall enough to create sufficient draft internally, or if the internal passages of the boiler offer too much resistance to gas flow. The stack also has the more visible function of discharging the exhaust at a high elevation, so the fumes are carried away from the site.

The intensity of convective draft increases with the temperature of the combustion gases, increasing rapidly as the boiler first warms up, and leveling off as the gas temperature increases further. The temperature of the combustion gas drops as it gives up heat to the water, but the gas temperature inside the boiler remains high enough to provide strong convection. The temperature of the gases drops slowly once they leave the heat transfer portion of the boiler, so the draft produced by a stack is essentially proportional to its height.

Convection was the original source of draft for boilers, and still plays an important role in most boiler installations. In boilers with atmospheric burners, all four of the functions of draft listed above are accomplished by convection alone. At the other extreme, boilers with forced-draft fans perform the first three functions primarily by fan power, but they rely on convection to expel exhaust out the stack. The burners used in some small and medium sized boilers have integral fans to draw combustion air into the firebox and to regulate mixing air with the fuel, but they rely on convection to move combustion gases through the boiler.

■ Fan-Induced Draft

Any or all of the four functions of draft can be accomplished with fans instead of with convective draft. Using fans allowed the development of burners with higher output and better control of air-fuel ratio than is possible with burners that operate by convection. Using fans to force combustion air through the boiler eliminates the need for a tall stack. Fans provide strong draft, so they can overpower any irregularities in gas flow caused by convection within the boiler.

Fans themselves serve as an effective draft regulator since they have stable output characteristics. However, using fans to create draft does not always eliminate the need for other forms of draft regulation. A separate stack draft regulator may be necessary if a tall stack is installed.

The fans used in a boiler may be "forced draft" or "induced draft." Forced-draft fans blow air into the boiler at the burner openings, and they create a positive pressure inside the combustion air ducting and the boiler. Induced-draft fans suck air out of the boiler from the outlet end of the boiler, and they create a negative pressure inside the boiler and the breeching.

Most small and medium sized boilers use forced-draft fans exclusively. Some larger boilers used induced-draft fans in addition to forced-draft fans. Induced-draft fans are usually installed in combination with forced-draft fans. An exception is older marine boilers operated with induced draft fans alone to prevent leakage of combustion gases into the confined spaces below decks.

One important class of boilers and furnaces does not rely on convection at all. These are units that use large heat transfer surfaces to reduce the flue gas temperature below the condensation temperature of water. The flue gases are too cool to create strong draft. This equipment relies entirely on one or more fans to move the combustion air and flue gases, and it operates satisfactorily without a stack.

Draft Regulators

Adjusting draft consists of adjusting a draft regulator. So, start by finding out what kind of draft regulator you have and understanding how it works. You want to make sure that the type of draft regulator you have is appropriate and that it is working properly.

In boiler systems that use convective draft, the stack is the primary source of draft. A problem is that stack draft changes with boiler loading, outside air temperature, wind, and other factors. A common solution is to design the stack to produce more draft than necessary, and to install a draft regulator between the boiler and the stack to limit the draft at the boiler outlet.

Boilers that use fans may have no draft regulator other than the fans themselves. Fans are effective flow regulators as long as they work against constant pressure. Forced-draft fans are strong enough to overpower small fluctuations due to convection within the boiler. However, stack draft may suffer from strong fluctuations. With a tall stack, a draft regulator may be needed to compensate for the fluctuations even though the boiler has powerful fans.

A similar situation exists in boilers that have induced-draft fans that operate at constant output. A draft regulator is placed between the fan and the outlet of the boiler to throttle the draft created by the fan.

The fact that a boiler has a fan in the burner system does not necessarily mean that the fan is providing all the draft required. The fan may be intended only for proportioning the air and fuel, and it may not have the ability to move the combustion gases through the resistance of the boiler. This is typical of gun-type oil burners, for example. In such boilers, convective draft is required to overcome the resistance of the boiler, and a draft regulator is required to control the convective draft.

WESINC

Fig. 1 Draft hoods Air from the boiler room freely enters the flues from underneath the hoods. These boilers are unusually large to have draft hoods. They are used because the boilers have atmospheric burners.

Parker Boiler Co.

Fig. 2 A boiler with a barometric flue damper This damper uses a delicately balanced blade to maintain a flue pressure that is slightly below the pressure in the boiler room. This boiler has atmospheric burners, but barometric dampers may also be used with power burners.

■ Types of Draft Regulators

A draft regulator is a device that is installed between the boiler and the stack to provide a controlled negative pressure at the boiler flue gas outlet. These are the types of draft regulators:

- *draft hoods.* A draft hood is essentially a missing section of flue inside the boiler room. The opening in the flue makes the pressure at that point equal to the boiler room pressure. A collecting hood is attached to the upper portion of the flue to keep the flue gases from spilling into the boiler room. Figure 1 shows typical draft hoods.

The effect of a draft hood is to isolate the boiler from the stack. The boiler draft is determined by the difference in height between the burner and the bottom of the draft hood. To increase the draft in the boiler, the draft hood can be installed at a higher point in the flue. The available boiler draft is usually small because of height limitations inside the boiler room.

At low firing rates, the stack draft is satisfied largely by air from inside the boiler room, which also acts to dilute the flue gases and reduce the stack draft. Draft hoods are cheap and foolproof, and they do not require adjustment. They have several disadvantages. They provide little negative pressure at the boiler outlet, they are not adjustable, and they

WESINC

Fig. 4 An air casing pressure gauge   This is on a large boiler with a forced-draft fan. Draft is just as important in big boilers as in small boilers.

cannot prevent standby losses. Also, they cannot protect against backflow from the stack. This may occur when a boiler starts firing into a cold stack, or if wind conditions cause the stack to backflow. For these reasons, draft hoods generally are limited to boilers with atmospheric burners, or to boilers with small gun-type burners.

- *barometric dampers.* A barometric flue damper is a controlled opening installed in the side of the flue. The damper allows boiler room air to enter the flue, partially killing the stack draft. The resistance of the damper maintains a small, fixed negative pressure with respect to the surrounding space. The most common type of barometric damper has a single blade with a small balance weight. You can adjust the amount of negative pressure by moving the balance weight. Figure 2 shows a boiler with a barometric damper.

Barometric dampers are used instead of draft hoods with boilers that need greater draft to overcome internal resistance to gas flow. Barometric dampers have the advantage that they can close to prevent flue gases from flowing into the boiler room if the flue pressure becomes positive.

Barometric dampers are less reliable than draft hoods. The force moving the damper blade is too weak to overcome any tendency of the blade to stick. Cheap or improperly installed barometric dampers may not maintain a constant draft as firing levels change, and they may oscillate, upsetting the air-fuel ratio. Barometric dampers cannot prevent standby losses. They are generally limited to boilers of small and medium size.

- *motorized flue dampers* function by partially blocking the flow of flue gas from the boiler. The

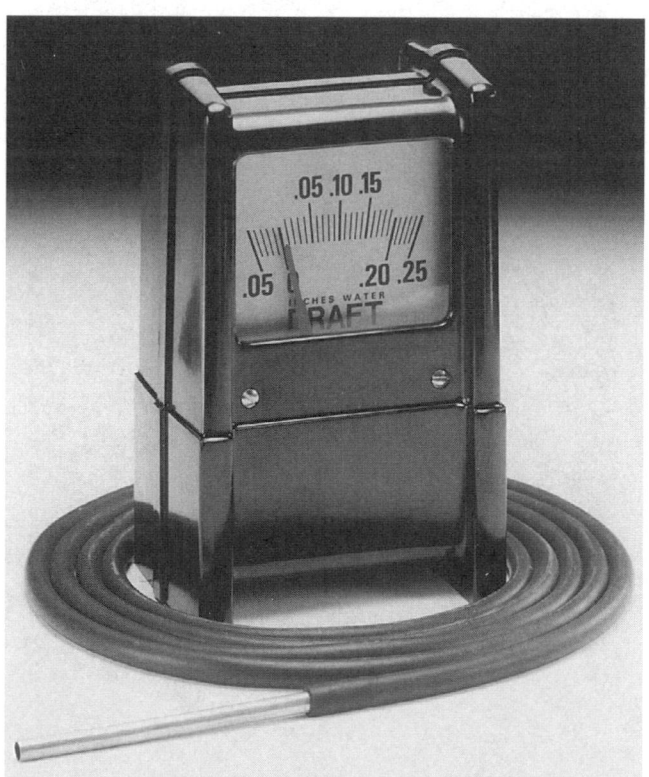

Bacharach, Inc.

Fig. 3 A portable draft gauge   This inexpensive but sensitive unit can be used to measure draft at any point in the boiler system.

damper is controlled by a draft sensor located near the burner. The damper is installed inside the boiler breeching or at some other location inside the flue close to the boiler outlet.

A significant advantage of a flue damper is that it can be controlled to close when firing is off. This reduces standby losses, as recommended by Measures 1.5.3 ff, below. Motorized draft dampers generally are limited by their cost and complexity to medium and large sized boilers.

The burner controls should open the damper fully during purge cycles. The controls should open the damper partially during burner start-up. If the damper fails in any position other than fully open, combustion gases could back up, risking explosion and asphyxiation. Therefore, the burner controls should include a low-draft interlock that stops the burner if the damper fails to open fully. The boiler should also have an easily visible draft gauge that operators can use to verify proper operation.

- *variable-speed induced-draft fan.* If a boiler uses an induced-draft fan, you can regulate the draft at the boiler outlet by modulating the speed of the fan. This method is covered by Measure 1.4.7.

Draft hoods and barometric dampers may draw substantial quantities of air into the flue from the boiler room. Normally, this is not a problem, because the boiler room has heat to spare. However, air loss through the draft regulator may interfere with arrangements that improve boiler efficiency by drawing combustion air from the warmest parts of the boiler room. See Measure 1.11.3 for details.

How to Measure and Adjust Draft

Your primary objective is to set the draft at the burner, which is called "overfire" draft. This affects the efficiency of the burner. All burners should be operated within draft limits that are specified by the boiler or burner manufacturer. You set the overfire draft by adjusting the draft regulator.

You may also measure draft at the draft regulator and at the stack, to make sure that these two items are working properly.

Measure draft with a draft gauge, which is a sensitive pressure gauge. Insert a flexible hose from the draft gauge into the area where you want to measure the pressure. The draft gauge registers the difference in pressure between the tip of the probe and the location of the gauge housing. Draft gauges are inexpensive and readily available from heating supply houses. Figure 3 shows a typical draft gauge.

Large boilers that have air casings may have permanently installed draft gauges. An example is shown in Figure 4. The casings may be under positive or negative pressure. Maintaining the correct draft is just as important in large boilers as it is in smaller boilers.

To get correct readings, make sure that all access doors, cover plates, etc., are installed in their normal positions.

■ Overfire Draft

From the standpoint of combustion efficiency, the draft at the burner is most important. It affects the air-fuel ratio and the burner's capacity. If the draft at the burner is correct, the combustion gases flow through the boiler properly.

Overfire draft with atmospheric burners should be slightly negative. Only a small gas pressure is feeding the burner and any positive pressure will prevent the flow of primary or secondary air into the combustion chamber. A boiler with atmospheric burners typically should have a negative pressure at the burner of about 0.02 inches of water.

Overfire draft should also be negative for many smaller burners that have built-in fans. The function of the fans in such burners is limited to proportioning the air and fuel properly. Stack draft moves the combustion gases through the boiler by convection.

Overfire draft in boilers with power burners is positive. The fans are the principal force driving the flow of gas through the boiler. This does not mean that power burners are independent of draft. The capacity of the burner, and possibly the air-fuel ratio, is affected by the firebox pressure.

Measuring overfire draft may require drilling a hole through the boiler front, since all doors and access panels should be closed. If possible, remove a bolt or screw from a flange or faceplate near the burner, and insert the probe through the bolt hole.

■ Draft at the Draft Regulator

The second draft measurement is taken at the draft hood or draft regulator. The purpose of this measurement is to verify that the draft control device is functioning and adjusted properly. The draft should be negative and steady, so that air flows inward, from the boiler room to the flue. You can also check this draft by dangling a lightweight sheet of paper near the draft regulator to see if it is drawn toward the flue.

If the boiler uses a draft hood, there is no adjustment to make. The appropriate overfire draft is created by the height difference between the burner and the draft hood.

If the boiler uses a barometric damper, you adjust the damper to provide the proper overfire draft. The damper is adjusted typically by changing the position of a balance weight.

Check a barometric damper to be sure that it moves freely but remains steady. The damper should settle down quickly when the firing level changes.

If the boiler uses a motorized damper, adjust the damper setting until the overfire draft is correct. There is no need to measure the draft at the damper itself unless you suspect a problem.

If the boiler uses a variable-speed induced-draft fan, adjust it to maintain the desired overfire draft.

■ Stack Draft

Stack draft is determined primarily by the stack height, and it usually is not adjustable. The purpose of measuring stack draft is to verify that stack draft is adequate, i.e., that the pressure inside the bottom of the stack is negative under all operating conditions. The stack draft itself does not have to remain at a specific value, but it should be adequate to remove flue gases under all operating conditions.

The stack may be subject to wind effects, such as a downdraft from an adjacent building, that cause the flow in the stack to reverse. Measure the stack draft during periods when you suspect trouble, to find out if such problems exist. Even though the stack draft may vary over a wide range, it should not have erratic fluctuations.

You can measure stack draft at any convenient location near the base of the stack. If a draft regulator is installed, measure the stack draft above the draft regulator at a distance equal to several flue diameters. You can insert the probe through the draft regulator, if it is long enough to extend a sufficient distance above the opening. If this is not possible, drill a small hole in the stack.

When to Check Draft

Check draft, and adjust it as necessary, every time you do efficiency testing and burner maintenance.

Measure 1.4.1 covers appropriate burner maintenance intervals. Also, measure draft when diagnosing burner problems.

Solving Flue Draft Problems

If you cannot adjust draft properly, or if there are problems that create erratic draft, you need to make changes to the draft system. In such cases, refer to Measure 1.5.2, next.

ECONOMICS

SAVINGS POTENTIAL: *Up to 2 percent of fuel cost, typically.*

COST: *The cost of a draft gauge is less than $100. The amount of labor required to adjust draft is minor.*

PAYBACK PERIOD: *Short. Adjusting draft should be a routine part of boiler operation.*

TRAPS & TRICKS

PREPARATION: *Make sure that you have the right kind of draft regulator for the boiler and the flue system, and that it is installed and functioning properly.*

SKILLS: *Adjusting draft is not difficult, but it does require specialized knowledge. Make sure that the person doing the work understands the principles.*

KEEP IT UP: *Repeat draft measurements and adjustments whenever you work on the burner. Also, check barometric dampers and motorized flue dampers at regular intervals. Put this in your maintenance calendar.*

MEASURE **1.5.2 Correct defects in flue systems and boiler room ventilation that cause draft problems.**

Improper draft reduces boiler efficiency by disturbing the air-fuel ratio. It also increases maintenance by causing smoke and sooting. These problems are explained in Measure 1.3.1. More severe draft problems can cause serious safety problems, especially backflow of flue gases into the boiler room. In the worst cases, this can smother the flame and cause an explosion.

This Measure explains the common causes of draft problems, and suggests appropriate remedies. Draft problems can arise anywhere in the path of combustion air or flue gases, including the entry into the building, the inside of the boiler, and the entire exhaust path. You may find serious draft defects that have persisted for years.

A draft problem may have more than one cause. Take corrective action that eliminates all the problems.

Relationship to Burners and Fans

The sensitivity of a boiler to draft depends on the design of the burner system. Burners and fans that operate at high pressure overpower fluctuations in draft, reducing their effect. At the other extreme, atmospheric burners are very vulnerable to fluctuations in draft. Other burner designs lie between these two extremes.

Problem: Defective, Improper, or Incorrectly Adjusted Draft Regulator

If the stack draft is adequate but burner draft is too weak, the problem probably is in the draft regulator. The draft regulator may be defective, or it may be improperly adjusted, or it may be the wrong type for the boiler installation. Refer to Measure 1.5.1 about selecting and adjusting draft regulators. Draft regulators can limit excessive draft, but they cannot compensate for inadequate draft.

Problem: Excessive Flue Draft

The usual cause of excess draft is that the stack or flue is taller than needed. This is unavoidable in some applications, for example, with basement boiler plants in tall buildings, where the flue rises to the roof. The usual remedy is to add an appropriate draft regulator. The types of draft regulators and their characteristics are covered in Measure 1.5.1.

Excessive draft can also be reduced by placing a restriction in the stack or flue. However, a fixed restriction cannot adapt well to changing boiler loads or other conditions. You may have to deal with a tall stack or flue by using both a stack restriction and a draft regulator.

Problem: Insufficient Flue Draft

Insufficient stack draft may be caused by a stack that is too short, or by obstructions in the ducting leading to the stack. The ways to deal with insufficient flue draft are:

- *remove any restrictions that exist in the flue.* If the connection between the boiler and the stack has repeated bends, it can create considerable resistance to flow. Eliminate right-angle bends and use smooth, long-radius bends.

- *install a draft-inducing fan.* Small boiler systems can use an inexpensive draft inducer that fits through the side of the flue like a paddle wheel. Larger systems may require a proper induced-draft fan that moves the entire volume of flue gas. The fan should be interlocked with burner firing to conserve energy and to avoid cooling the flue between firing cycles. Control larger fans with a variable-speed drive, as recommended by Measure 1.4.7.

- *install a taller stack*

- *replace the burner with a unit that operates efficiently with the existing draft.* This approach is especially desirable if you can substitute a more efficient burner. Upgrading burners is covered by Measure 1.4.2.

Problem: Wind Effects on the Flue

Rapid and severe fluctuation in draft is usually caused by wind effects. A stack is not affected seriously by wind that blows horizontally, but this is an ideal condition. Adjacent buildings or terrain may cause wind to have a strong downward or upward component at the top of the stack. This is a common problem when the top of the flue is below an adjacent structure.

WESINC

Fig. 1 Blocked combustion air intake This air intake to a large boiler plant was blocked to keep the cold outside air from causing discomfort inside the plant.

A downward wind component into the stack inhibits the flow of flue gases out of the boiler. If this condition is serious, the fumes tend to smother the flame, or they spill into the boiler room through the draft regulator or other openings in the flue. A backdraft may even blow out the fire in a boiler that has atmospheric burners, posing a risk of explosion when unburned fuel is ignited by contact with the heated parts of the boiler.

Usually, the best solution is to install a broad cap at the top of the stack that blocks any vertical wind component. You can design this fixture to diffuse wind and reduce its velocity.

Don't expect a draft regulator to solve a backflow problem. It cannot even do a good job of smoothing out smaller fluctuations caused by wind, which tend to be gusty in nature. Similarly, draft inducing fans do not adequately smooth out fluctuations caused by wind.

Replacing the present burner with one that operates at higher fan pressure reduces the sensitivity of the burner to draft fluctuations. However, replacing the burner is not a substitute for fixing the draft problem.

Problem: Wind Pressure on the Boiler Room

When wind strikes a surface, it is capable of creating pressures that are high in comparison with the normal pressure differences inside a boiler system. Boiler buildings that have many ventilation openings act as a shield against wind pressure. This is because the pressure inside the boiler room is an average of the pressures on the upwind and downwind sides.

However, unbalanced boiler room pressure is likely to occur if there are ventilation openings only on one side of the building. You can minimize this problem by installing carefully designed wind deflectors over the

openings. Before making any changes, refer to Measure 1.11.3, which relates to air flow through the boiler space.

Problem: Building Chimney Effect

If a boiler room is located in the basement of a tall building, the chimney effect of the building may create a negative pressure inside the boiler room. The solution to this problem is to provide a combustion air intake that leads directly to the outside of the building at grade level, and to minimize openings between the boiler room and the interior of the building. The outside air intake must be large enough to prevent drastic changes in boiler room pressure when interior doors are opened.

Problem: Inadequate Outside Air Intake Area

The space around the boiler should be at the same pressure as the air outside the building. To achieve this,

WESINC

Fig. 2 A collection of combustion air problems This air intake for a small boiler is very large. However, it has an intake fan that upsets the air-fuel ratio of the atmospheric burner. The bottom louvers of the intake have been closed, and the top is being clogged by leaves.

air must enter the boiler room through one or more openings in the boiler room enclosure. These openings must be large enough to avoid any measurable pressure drop as air enters the boiler room. Insufficient intake area causes negative pressure inside the boiler room. The pressure drop gets worse at higher firing rates.

Experience has provided rules of thumb for air intake size. A common guideline calls for one square inch of intake area for every 1,000 BTU/HR of boiler capacity. Fine mesh screens (such as insect screens, but not bird screens) over the intakes double or triple the inlet area that is advisable.

Boiler operators may reduce the inlet area to avoid discomfort from exposure to cold air during cold weather, or to protect against a perceived freezing problem. Figure 1 shows an example. To solve the combustion air problem, you must resolve the comfort problem. If necessary, relocate the outside air intake to avoid discomfort, or duct the combustion air directly to the burner area. Or, provide an enclosure for persons who work in the space. This also provides a haven from boiler room noise. However, this solution is acceptable only if plant operations permit the staff to spend most of their time inside the enclosure.

If possible, try to route outside air through the warmest parts of the boiler room to pick up heat before the air enters the boiler. This increases boiler efficiency while cooling the space. See Measure 1.11.3 for details.

■ Provide Sufficient Air for Draft Hoods and Dampers

If the boilers have barometric flue dampers or draft hoods, the boiler room needs additional inlet area for the air that is drawn into these devices. Size the inlet for the possibility that much more air may flow into the draft hood or damper than is needed for combustion. Otherwise, the boiler room may develop negative pressure, and flue gas flow may become unstable.

Problem: Fans in the Boiler Room Envelope

Fans move air by creating a pressure difference. Therefore, if a fan is used to move air into or out of a boiler room, it creates a pressure change in the boiler room. Fans may be installed in the combustion air intakes by operators seeking relief from high space temperature. Also, fans are sometimes installed because of the misconception that they "help" combustion. Figure 2 shows an example.

The original design of boiler plants sometimes includes ventilation exhaust fans, especially if the space is used for other equipment, such as chillers and air handling equipment. The designer may not have recognized the disturbance of boiler draft caused in this way.

The solution in all cases is to remove fans that create irregular combustion draft. As a guideline, if a ventilation fan creates enough of a pressure differential to cause a noticeable breeze through any air inlet opening, the fan is too large or the total inlet area is too small. In principle, you can minimize the effect of fans in the boiler room envelope by making the passive openings much larger than the combined area of all the fans in the envelope. This may be impractical to achieve. Furthermore, as we have seen, plant operators tend to obstruct large openings to avoid discomfort during cold weather.

Spot cooling of personnel by using outside ventilation air is acceptable if the total volume of air is small in relation to the capacity of properly sized outside air intakes. However, in climates that have cold weather, spot cooling is uncomfortable when the outside air is too cool. As suggested previously, if operators usually stay in one area of the boiler room, you can reduce the amount of cooling ventilation by providing an enclosure for the operators.

Ventilation fans that recirculate air entirely within the boiler space do not cause a problem, regardless of their size, unless they blow on the burners or combustion fan inlets.

Problem: Soot Obstruction

In some types of boilers, especially cast iron boilers with narrow passages, sluggish burner draft could be caused by heavy soot accumulation. In most boiler designs, the amount of soot would have to be monumental to restrict gas flow noticeably.

ECONOMICS

SAVINGS POTENTIAL: *0.1 to 5 percent of fuel consumption; plus, reduced fireside cleaning costs. Eliminating a safety problem may be the primary consideration.*

COST: *Varies widely, depending on the nature of the problem.*

PAYBACK PERIOD: *As an energy conservation measure, the payback period may vary widely. A draft problem may be a serious condition that needs to be fixed as a safety measure.*

TRAPS & TRICKS

DIAGNOSING THE PROBLEM: *Draft problems can be tricky to diagnose. Be sure you understand the problem before you choose the solution.*

MEASURE 1.5.3 Minimize standby losses.

Standby loss is a major cause of energy waste in many boilers. If your boilers suffer from this problem, you may be able to reduce the energy loss. The four subsidiary Measures provide a variety of ways to reduce standby losses, including equipment changes and simple adjustments.

What is "Standby Loss?"

When a boiler is firing, heat is transferred to the internal surfaces of the boiler because the combustion gases are hotter than the surfaces. Conversely, the boiler loses heat to the cold air inside the boiler when the burners are not firing. If air can circulate through the boiler when firing is off, a great deal of heat is lost this way. This loss of heat is called "standby loss." There are two reasons why air flows through the combustion area while the boiler is not firing: convection and purging.

■ Convection Losses

Convection circulates cold ambient air through a boiler, carrying away heat up the flue. The high temperature inside a hot boiler induces a convective force through the boiler by making the air inside the boiler less dense, and therefore lighter. If a boiler does not have a damper to prevent convection when the burner is not firing, the convection continues as long as there is heat in the boiler. Standby loss from convection increases with the length of time that a boiler is not firing.

The worst convection occurs in boilers with combustion chambers that are open to the atmosphere. Such boilers have no means to restrict convection, and the flow resistance of the gas passages is low.

The least convection occurs in boilers with closed combustion chambers and fan-forced combustion. To limit convection, most larger power burners have an integral damper that closes when the burner is not operating.

Some boiler systems use dampers in the flue that close when the boiler is not firing. In boilers that do not have dampers, the amount of convection depends largely on the boilers' internal resistance to gas flow. Boilers that use atmospheric burners typically have little internal resistance, so their convection losses tend to be high, perhaps as much as 10% of fuel consumption. At the other extreme, multi-pass firetube boilers have a high resistance to convection because of the circuitous route that the gas must follow.

Convection losses are reduced by burners that have a high turndown ratio. These allow the burner to fire for a much larger fraction of the time, reducing the period of time during which convection loss can occur.

■ Purging Losses

Purging is a safety feature of forced-draft boilers. The purpose of purging is to ensure that nothing but air is inside the combustion space when the burners are fired. If the combustion space were filled with a mixture of air and fuel vapors, an explosion could occur. To prevent explosions, the burner controls are programmed to operate the fan for a period of time prior to firing in order to clear out the entire volume of air within the boiler. This air is replaced with cold ambient air that absorbs a considerable amount of heat from the boiler.

Some burner systems also carry out a purging cycle when firing stops. (This is called "post-purge.") The purpose of post-purge is to blow all combustion products out of the system after firing stops. This may be necessary because the flue gases contract when firing stops, which may cause the draft to reverse and carry flue gases into into boiler space. Post-purge is especially wasteful because the firesides are at maximum temperature immediately following a firing cycle.

Standby loss from purging is proportional to the rate at which a burner cycles on and off. For this reason, purging losses are reduced by burners that have a high turndown ratio. These allow the burner to satisfy low loads without turning off the flame.

Reducing Standby Losses Reduces Boiler Damage

Reducing standby losses will also reduce boiler problems, such as tube sheet leakage and refractory breakage. These problems are aggravated by abrupt changes in boiler temperature, which is associated with convection and purge cycles.

When a burner turns on or off, there is a large, abrupt change in the temperature of the boiler components. The change in temperature is increased by convection of outside air through the boiler. Boilers are designed to tolerate this abuse fairly well, but there is a correlation between certain kinds of boiler damage and the frequency of firing cycles.

Purging creates the greatest cycling stress on a boiler. The sudden flow of cold air causes rapid contraction of the boiler's components. This temperature change is especially severe because it occurs just before firing commences, creating an abrupt reversal of temperature from cold to hot. Post-purge cycles change the temperature abruptly in the opposite direction.

Safety Considerations

Combustion air and flue gas must be able to move freely as long as the burner is firing. All fans in the combustion gas path must be operating when the burner is firing. All dampers must operate properly during firing, and they must open fully during pre-purge and post-purge cycles. A fan or damper failure may cause combustion gases to back up, risking explosion and asphyxiation. Therefore, any modifications should be accomplished only by a technician who is competent to deal with the safety issues.

MEASURE 1.5.3.1 Control all fans in the combustion air path to stop, and all dampers to close, when the burner is not firing.

This method of reducing standby loss applies to boiler systems that have fans and/or dampers to control draft. The procedure is simply to ensure that the burner controls turn off all the fans, and close all the dampers, whenever the burners are not firing.

Control of Fans

Boiler systems may have one or more fans in the combustion gas path. If any one continues to operate when the boiler is off, it can induce a significant amount of air flow through the boiler, even if dampers are closed. Forced-draft fans and induced-draft fans normally are controlled to turn on and off as part of the firing sequence. Verify that this is true in your plant.

Draft inducers, which are small fans installed in the flue itself, may be installed to solve a stack draft problem. Letting such fans operate continuously is a mistake. If the draft inducer can draw boiler room air into the stack through the draft regulator between firing cycles, it will cool the stack. Since the stack is cold when the boiler fires, the draft has to be restarted every time the boiler fires. Also, letting the stack cool causes corrosion by allowing acids that are contained in the flue gases to condense on the flue surfaces (see Measure 1.7.1.1 about acid condensation).

The solution is simple. Use the burner controls to control the operation of all fans in the boiler air system.

Control of Dampers

Dampers that are part of burner assemblies are almost always programmed to close when the burner stops firing. Verify this operation in your burners.

A motorized damper may be installed inside the flue to regulate flue draft. Such a damper may be controlled only in response to flue draft, not burner firing. If so, change the controls to close the damper when firing stops. This change has major safety implications. Therefore, equip the burner controls with a low-draft

interlock that stops the burner if the damper fails to open fully. The boiler should also have an easily visible draft gauge that allows the operator to verify proper damper operation.

SUMMARY

A basic control sequence for minimizing convective heat loss.

SELECTION SCORECARD

Savings Potential .... $ $

Rate of Return, New Facilities % % % %

Rate of Return, Retrofit .... % % % %

Reliability .... ✓ ✓ ✓

Ease of Retrofit .... ☺ ☺ ☺

ECONOMICS

SAVINGS POTENTIAL: *0.5 to 10 percent of fuel cost, depending on the fans and dampers that are installed.*

COST: *The additional control wiring typically costs several hundred dollars.*

PAYBACK PERIOD: *Less than one year, to several years.*

TRAPS & TRICKS

SKILLS AND APPROVALS: *The work should be done by a qualified burner controls specialist. Review the changes with the boiler and/or burner manufacturer, and if appropriate, with the applicable safety code enforcement agency.*

EXPLAIN IT: *Don't forget to record the changes in the equipment records, or else the next maintenance work on the system may restore the system to its original configuration.*

MEASURE **1.5.3.2 Install an automatic flue damper.**

A standard method of reducing standby loss is to install an automatic damper in the flue that closes when the unit is not firing. This keeps outside air from circulating through the boiler and carrying heat out the flue.

Applications and Energy Saving Potential

Standby losses range from minimal to more than 10 percent of fuel input. It is usually not practical to measure standby loss in the field. Instead, you need to judge standby losses from the configuration of the system. This will tell you whether you need a flue damper. Consider these factors:

- *burner off time.* Standby losses are directly proportional to the time that firing is off while the boiler of furnace is still hot. Continuously operating boilers do not have standby losses.

- *the amount of heat stored in the mass of the equipment.* The convective force that causes standby losses is caused by warming of air in the idle boiler or furnace. If the unit remains hot between firing cycles, as with steam boilers, standby losses are greatest. On the other hand, a hot air furnace that fires occasionally suffers standby losses only until the furnace cools down.

- *the resistance of the system to air flow.* A tight damper in the system, either at the combustion air inlet or in the flue, prevents most standby loss.

- *the strength of the convective draft.* All other things being equal, a greater convective force increases standby loss. The greatest convective force occurs if there is a tall stack, but no draft regulator.

A flue damper is the only means of limiting standby loss if the boiler or furnace has no means of closing off the air inlet. This is characteristic of all boilers with atmospheric burners. Such boilers are wide open to convection, and the flow resistance of the gas passages is low. Also, many boilers and furnaces with smaller fan-powered burners do not have inlet dampers.

A flue damper may provide some benefit even in a boiler that has an inlet damper, although the damper is unlikely to be economical. Even with the inlet damper closed, air heated inside the boiler may pass up one side of the flue while ambient air backflows down the other side. The extent of such convection is difficult to judge. Go to the top of the stack or to the draft damper after the boiler has stopped firing for a while, and see whether there is a significant flow of hot air.

Flue dampers are not useful with high-efficiency furnaces and boilers that employ forced circulation of

SUMMARY

Do this if the burner does not have an air shutter that closes when firing stops. May also be desirable in large flue systems that allow convection even with a burner damper.

SELECTION SCORECARD

Savings Potential	$ $ $
Rate of Return	% % %
Reliability	✓ ✓ ✓ ✓
Ease of Retrofit	☺ ☺ ☺

combustion gases. Such units have heat exchangers with long, narrow passages that inhibit convection. Their resistance to convective flow is so high that heat loss is negligible as long as the fan is not operating.

Installation Limitations

If the boiler has a sealed flue, installing a flue damper may be relatively easy. Just remove a short section of the flue near the boiler discharge, and insert the flue damper.

WESINC

Fig. 1 A flue damper installation challenge This boiler has atmospheric burners, several flue gas outlets, and an integral draft hood. Installing a flue damper here would require a lot of sheetmetal work.

Unfortunately, installing flue dampers is most difficult on the boilers that need them the most, which are boilers with atmospheric burners. Boilers with atmospheric burners commonly have an open discharge hood at the outlet of the boiler casing, and the flue is connected to the top of this hood. An example is shown in Figure 1.

In order to block circulation of air through the boiler, the flue damper must be installed on the boiler side of any openings in the flue. If the boiler has an integral draft hood, you must remove the hood, find a way to connect the flue damper to the boiler directly, and rebuild the draft hood and the flue connections. This job may be too difficult to be practical. In this situation, think about replacing the entire boiler with one that is more efficient.

Types of Flue Dampers

Motorized, externally powered dampers are used in boilers of all sizes. For example, one major burner manufacturer offers retrofit flue damper kits in sizes ranging from 10 inches to 42 inches in diameter. Flue dampers typically are single-blade units that have the advantages of being inexpensive and closing tightly enough to block most convection. The hardware and controls are conventional. Installation is simple, provided that a suitable length of flue is accessible for mounting the damper.

Self-powered thermostatically operated flue dampers are used in smaller units. These open in response to the high temperature of flue gases. Most thermostatic dampers have a blade made of bimetallic material that changes shape when heated. After the burner lights, hot gas reaching the damper causes the bimetallic elements to warp to an open position, allowing the gas to pass to the flue. No external power or external connections are required. The advantages of thermostatic dampers are low cost, reliability, and low maintenance.

A limitation of self-powered thermostatic dampers is that they must remain partially open so that hot gas can flow quickly to the thermostatic element as soon as the burner starts firing. Partial opening is achieved by making a hole in the damper blade, or by undersizing the damper blade. This lack of tight closure undermines the effectiveness of the damper. As a result, thermostatic dampers are limited mainly to small boilers and furnaces where their lower cost is more important.

Relationship to Draft Regulators

A flue damper must be located on the boiler side of any draft hood or barometric draft damper that is installed. Otherwise, when the flue damper closes, it would force flue gases to flow into the boiler room through the draft regulator.

If a motorized draft regulation damper is already installed inside the flue, examine it to see whether it is capable of closing tightly. If so, this damper can serve to prevent convection. See Measure 1.5.3.1, previous.

Okay with Pilot Lights

Older furnaces and boilers may use a continuous pilot light to ignite the burner. The amount of flue gas produced by the pilot light is small. Flue dampers generally have enough leakage to accommodate the pilot light.

Adjust the Air-Fuel Ratio

A flue damper restricts flow to a small extent even when it is open. This may affect the air-fuel ratio, especially with atmospheric burners. After installing a flue damper, adjust the air-fuel ratio as recommended in Measure 1.3.1.

Safety Certification

Flue dampers involve a potential safety hazard, which is the possibility of the damper remaining in the closed position while the burner is firing. If this occurs, combustion gases will back up into the space surrounding the heating unit. Approved flue dampers include safety features that deal with this possibility. Be sure to select only approved flue dampers, and to install them in a safe manner. Standards for flue dampers are set by various agencies. For example, the American Gas Association acts as an approval agency for flue dampers that are used in gas-fired heating units. Make sure that the flue damper has the seal of the appropriate organization.

ECONOMICS

SAVINGS POTENTIAL: *2 to 10 percent of fuel consumption, if the boiler does not have an inlet damper. Less than 2 percent of fuel consumption, if the boiler has an inlet damper.*

COST: *Less than $100 for small thermostatic units, to about $1,000 for large standard motorized units, installed. Customized installations for large boilers may cost thousands of dollars.*

PAYBACK PERIOD: *One year to many years.*

TRAPS & TRICKS

SELECTING THE EQUIPMENT: *Select a unit made by an experienced manufacturer. A certification label does not indicate reliability or efficient operation. Select a unit that closes as tightly as possible while satisfying safety requirements. Make sure that the unit is appropriate for your particular boiler and flue system.*

INSTALLATION: *Most installations are fairly easy, but some require ingenuity. Work out the details, or else the unit may not function properly.*

MEASURE 1.5.3.3 Install a burner assembly or boiler that minimizes standby losses.

RATINGS

New Facilities | Retrofit | O&M
A | **C** |

If the boiler has no dampers to prevent standby losses, you may be able to replace the present burner with a burner that has an integral damper. Or, you may be able to replace a cycling burner with a modulating burner. Refer to Measure 1.4.4, which covers upgrading burners in general.

Select replacement burners to provide as much improvement as possible in efficiency and ease of maintenance. Replacing the burner only to prevent standby losses would rarely be economical.

If you are installing a new boiler or furnace, select a unit that has low standby losses as well as high combustion efficiency. See Measure 1.12.1 for ideas.

MEASURE 1.5.3.4 With cycling burners, adjust the controls to minimize the frequency of firing cycles.

RATINGS

New Facilities | Retrofit | O&M
 | | **B**

If you have boilers that control their output by cycling on and off, you can reduce standby losses and boiler deterioration by reducing the burner cycling rate. However, pushing this technique too far can cause trouble.

A Simple Solution

Heat loss from purging increases with the rate of cycling because the number of purge cycles is increased. Convective heat loss increases with the rate of cycling because the combustion chamber is kept hot for a larger percentage of the "off" time.

Therefore, you can reduce standby losses by reducing the rate of burner cycling. Burner firing normally is controlled by the outlet steam pressure or outlet water temperature. You can usually set the turn-on and turn-off pressures (or temperatures) separately. Adjust the settings so that the turn-on and turn-off points are further apart.

Limitation in Multi-Boiler Installations

In a plant with several boilers, the sequence of boiler operation may be controlled by setting each boiler to start and stop at a different steam pressure or water temperature. The sequence of boiler operation affects overall plant efficiency, as explained in Measure 1.1.4. If you use this method of sequencing the boilers, you will have little opportunity for increasing the pressure or temperature range of an individual boiler.

SUMMARY

A simple adjustment that can save a small amount of energy and increase equipment life.

SELECTION SCORECARD

Savings Potential .... **$**

Rate of Return .... **% % % %**

Reliability .... **✓ ✓ ✓**

Ease of Initiation .... **☺ ☺ ☺**

Conflict with Minimizing Steam Pressure

In some steam systems, you can save energy by keeping steam pressure at the low end of the pressure range, as recommended by Measure 1.1.5. On the other hand, to minimize cycling, the boiler needs to fire to the top of its pressure range.

Analyze your system to decide which Measure provides the greater benefit. Minimizing boiler pressure provides its greatest benefit in large steam systems, which usually do not have cycling burners.

Possible Problems with User Equipment Controls

Large swings in steam pressure or water temperature may cause control problems in user equipment. This should not be a problem if the user equipment and its controls are properly designed. However, problems may arise in some marginally designed units, or problems may be triggered in units that have incipient defects.

Large swings in steam pressure may create a nuisance in manually controlled equipment, such as old steam radiators. Widely varying steam pressure makes it difficult for occupants to find a satisfactory valve setting. Changes of pressure may create cycles of noise in steam equipment, even if they have automatic controls.

Explain It!

The benefit of extending the burner firing interval is not obvious to operating and maintenance personnel. Install a permanent placard at the firing controls to indicate the proper settings and the reason for them. See Reference Note 12, Placards, for details of effective placard design and installation.

ECONOMICS

SAVINGS POTENTIAL: *0.5 to 3 percent of boiler fuel input.*

COST: *Minimal.*

PAYBACK PERIOD: *Immediate.*

TRAPS & TRICKS

CAREFUL INITIATION: *Make sure that increasing the firing interval does not interfere with boiler sequencing, cause excessive pressure or temperature fluctuations at the user equipment, or cause other problems. You may have to experiment to find the maximum desirable firing interval.*

EXPLAIN IT: *For the benefit of maintenance personnel who may work on the burner controls, install a permanent placard to indicate the firing interval that should be set.*

Firesides and Watersides

Keeping firesides and watersides clean does more to keep boilers efficient than any other routine maintenance activity. Deposits that accumulate on these surfaces impede heat transfer, forcing more of the heat of the combustion gases to be lost to the flue. The Measures in this Subsection maintain the cleanliness of heat transfer surfaces and reduce the cost of doing so.

INDEX OF MEASURES

1.6.1 Clean firesides at appropriate intervals.

1.6.2 Install soot blowers in boilers that burn sooting fuels.

1.6.3 Optimize soot blower operation.

1.6.4 Clean watersides at appropriate intervals.

1.6.5 Avoid leaving waterside deposits when deactivating boilers.

RELATED MEASURES

- Measure 1.8.1, water treatment to minimize waterside deposits and erosion
- Measure 1.9.3, fuel oil additives that reduce fireside fouling

MEASURE 1.6.1 Clean firesides at appropriate intervals.

Deposits on fireside surfaces insulate the tubes from the combustion gases, forcing more heat energy to escape into the stack. Even modest fireside fouling can waste a significant fraction of fuel energy.

Types of Fireside Deposits

The most common fireside deposit is soot. Soot consists of unburned carbon and various dirty carbon compounds that are products of partial fuel combustion. Many of these intermediate products adhere to boiler tubes. Soot has a high insulating value because it forms as a porous layer, with a large amount of dead space. Soot can be formed by any type of fuel that contains carbon. This includes all conventional fuels and most unconventional fuels. Soot forms immediately and rapidly when combustion is not complete.

Once deposits are lodged on the tube surface, the cooling provided by the tubes keeps the deposits from burning off, even if they are combustible. The heat of the furnace may bake some of the deposits on the tubes. Some noncombustible impurities contained in the fuel, such as the mineral particulates called "fly ash," may melt and stick to the boiler tubes.

How Often to Clean Firesides

Gas fuels can burn so cleanly in many boilers that little soot accumulates, provided that the air-fuel ratio is kept within proper limits. At the other extreme, heavy oils and coal produce soot in virtually any boiler, requiring periodic cleaning of firesides, perhaps several times per year. Boilers burning light oil accumulate fireside deposits more slowly, typically requiring cleaning about once a year.

How to Monitor Fireside Fouling

Visual inspection is the only reliable method of gauging fireside fouling. Most types of watertube and firetube boilers allow easy inspection of most of the fireside surface. Take the opportunity of inspecting the firesides whenever the boiler is shut down after an extended period of operation. Inspecting firesides does not require a great deal of analytical skill. Anything that is deposited on the tube surfaces should be removed.

You can use gas temperature to monitor soot accumulation between boiler shutdowns. A rise in temperature indicates fouling, although it does not distinguish between fireside and waterside fouling. Take the combustion air temperature into account when you take flue temperature readings.

SUMMARY

Important routine maintenance that tends to be neglected because it involves dirty work. Preserves boiler life as well as efficiency.

SELECTION SCORECARD

Savings Potential .... $ $

Rate of Return .... % % % %

Reliability .... ✓ ✓ ✓

Ease of Initiation .... ☺ ☺ ☺

Flue gas temperature is easy to track with on-off and two-stage burners because they always have the same heat output. Flue gas temperature is more difficult to track if the boiler has a modulating burner, because the flue temperature varies with load. To account for this, prepare a table of flue gas temperature versus load when the boiler is clean. To determine the load, use a fuel flowmeter, a steam meter, or a BTU meter. Another method is to install a position indicator on the fuel control linkage and calibrate it for load.

Cleaning Techniques

Firesides are cleaned with mechanical methods. With firetube boilers, a cylindrical brush is pushed through each tube at the end of a rotating cable. Removing heavier deposits may require a cutter head. Some machines have a vacuum cleaner attachment that disposes of the soot as it comes out of the tubes. Figure 1 shows how this type of equipment is used. Most firetube boilers allow relatively easy access to the entire fireside surface.

Cleaning watertube boilers is more tedious. Hand wire brushes and scraping tools typically are used. Access to tubes behind the first row can be a frustrating challenge. Particularly difficult cases may require air lances, sandblasting, or other messy procedures. Don't wear your Sunday clothes for this job.

Of course, you can hire contractors to clean your boilers, but you still need to monitor them. Make sure that the contractor is thorough and does not harm the boiler.

With fuels that deposit soot at a high rate, consider installing soot blowers. See Measures 1.6.2 and 1.6.3, below, about soot blowers.

Goodway Tools Corporation

Fig. 1 Cleaning the firesides of a firetube boiler This job is dirty, but not too difficult. The vacuum cleaner attachment, at right, is the largest part of the equipment.

How to Minimize Fireside Deposits

Preventing deposits is better than removing them. The way to minimize fireside deposits is to keep the air-fuel ratio adjusted properly, as recommended by Measure 1.3.1. Make sure that the boiler's air-fuel controls keep the ratio in the proper range at all times. Even a brief period of operation with insufficient air can deposit a significant amount of fouling.

The condition of burners has a major effect on fireside deposits, especially with heavier liquid fuels and with solid fuels. A partially clogged burner may cause some of the fuel to burn incompletely, creating heavy residues. For this reason, clean your burners regularly, and without fail. Follow the procedures recommended by Measure 1.4.1.

Clean Firesides Upon Shutdown

Clean firesides as soon as you shut down a boiler. During shutdown, moisture may condense on the tube surfaces. Moisture reacts with the sulfur that exists in any remaining soot to form sulfuric acid, which corrodes the tubes. For this reason, cleaning at shutdown needs to be especially thorough.

Unfortunately, the areas where soot is hardest to remove are the areas where corrosion can cause the most trouble. In firetube boilers, take special care to remove any soot that collects around the outside of the tube where it projects through the tube sheet. In watertube boilers, concentrate on the areas where tubes enter the steam and mud drums. Rear rows of watertubes are especially difficult to reach.

ECONOMICS

SAVINGS POTENTIAL: 0.5 to 10 percent of fuel cost, depending mainly on the type of fuel and how well the air-fuel ratio is kept adjusted.

COST: The cost of typical rotating brush cleaning equipment for firetube boilers ranges from $1,500 to $4,000, depending on size and features. If you hire a cleaning service, the cost is proportional to the size of the boiler and to the difficulty of reaching all the fireside surfaces. A medium-sized firetube boiler with moderate sooting can be cleaned in one day.

PAYBACK PERIOD: Several months to several years.

TRAPS & TRICKS

DILIGENCE: Cleaning firesides is dirty, tedious, and uncomfortable, which motivates people to quit the job before it is finished. Inspect the work to make sure that all fireside surfaces are cleaned. Deposits that are not removed will cause maintenance problems later. Be careful to avoid damaging tubes and brickwork.

SELECTING THE EQUIPMENT: If you buy your own cleaning equipment, select equipment that makes the job as easy and pleasant as possible. The tools should not be so aggressive that they risk damaging the metal. Get a vacuum cleaner attachment to collect the dust.

MEASURE **1.6.2 Install soot blowers in boilers that burn sooting fuels.**

If you wait until shutting down the boiler to clean firesides, the deposits will reduce the boiler's efficiency for the entire time it is operating. Soot blowers allow you to remove deposits while the boiler is operating, at any interval you choose. With dirty fuels, soot blowers cannot eliminate the need for mechanical cleaning, but they may substantially reduce the drop in efficiency between cleanings.

Whether to Install Soot Blowers

A certain amount of sooting is virtually impossible to prevent. With gas and light oil, soot may deposit on firesides so slowly that it burns off before it can accumulate. With all the heavier grades of fuel oil and with all grades of coal, soot accumulates as long as the boiler is firing. Heavier grades of oil build up soot much more quickly than do lighter oils. Soot accumulation in coal-fired boilers depends on several factors, especially the grade of coal, the burner design, and the tube design.

With heavy oils and coal, significant soot accumulates within a few hours. A boiler may operate without interruption for thousands of hours, allowing

SUMMARY

Primarily for boilers that burn heavy oil and coal. Retrofit is tricky, and is more likely to be feasible in firetube boilers than in watertube boilers. Restrictions on soot discharge into the environment may be an obstacle.

SELECTION SCORECARD

Savings Potential	$ $ $
Rate of Return, New Facilities	% % % %
Rate of Return, Retrofit..........	% % %
Reliability	✓ ✓ ✓
Ease of Retrofit	☺ ☺ ☺

enough soot to accumulate to seriously reduce efficiency. Fortunately, soot can be kept under control while the boiler is in operation by using soot blowers. These are devices that blast soot loose with jets of steam or compressed air. Soot blowers are available for both watertube and firetube boilers.

Fuel Efficiency, Inc.

Fig. 1 Soot blower for firetube boiler This model uses compressed air. The four valves at right blow different groups of tubes in sequence.

If a boiler does not already have soot blowers, estimate the potential benefit by noting how much soot accumulates between maintenance intervals. If soot accumulates rapidly enough to reduce efficiency significantly between manual cleanings, consider soot blowers. Soot blowers do not eliminate the need for periodic manual cleaning, but they may extend the cleaning interval dramatically.

Soot blowers are especially important with economizers that have gill rings or other components that trap soot. If an economizer has aluminum gill rings, soot accumulation can cause a thermite fire, which is intense and virtually impossible to extinguish.

In calculating the economics of adding soot blowers, include the savings in fuel cost and labor. Expenses include the cost of installation and the cost of the steam or compressed air that is consumed. (Minimizing the cost of soot blowing steam or air is covered by Measure 1.6.3, next.) The cost of maintaining the soot blower itself should be minimal if the unit is well designed.

Soot Blowers for Firetube Boilers

Soot blowers for firetube boilers consist of nozzles that are located just beyond the end of each tube. The nozzles are fed from manifolds that are located inside the ends of the boiler, where the combustion gases turn around to go from one pass to the next. Figure 1 shows a typical installation.

The nozzles are usually located at the burner end of the boiler. This is because the combustion gas at the end of the first pass, at the rear of the boiler, is too hot for the equipment. Therefore, the steam or compressed air will be blowing in a direction opposite the second pass (or fourth pass) of the boiler, and in the same direction as the gas in the third pass. To keep from interfering with the combustion gas flow, the nozzles are controlled in groups, so that only a fraction of the tubes in each pass are being blown at a time. This complicates the steam or air manifolds, and requires an array of valves and a sequencing controller.

The nozzles and manifolds take up a lot of space, which may interfere with gas flow. Because of the potential for problems, check on the performance of the soot blower before purchase. Compare notes with someone who has installed the same type of soot blowers in your model of boiler.

Soot blowers for firetube boilers typically are made by specialty companies.

Soot Blowers for Watertube Boilers

A typical soot blower unit for a watertube boiler consists of a pipe that extends across the width of the boiler, passing either among the tubes or in front of them. Orifices are located in the pipe to blast steam or air through the bundle so that all tubes are cleaned. The soot blower is usually rotated during operation so that

the blast reaches all the area around the blower pipe. In a typical boiler, a number of units are needed to provide complete coverage of the tubes. Figure 2 shows the typical external hardware for a steam soot blower.

Soot blowers are customized for each model of boiler. They may be made by the boiler manufacturer or by specialty companies. Each orifice or nozzle must line up properly with the adjacent tubes. Installing a soot blower assembly may be difficult or impossible in boilers that were not originally intended to have them.

Steam or Compressed Air?

High pressure is a key to soot blowing. In firetube boilers, a minimum pressure of 125 PSI is recommended, for both steam and air. Therefore, if the boiler does not produce this much steam pressure, compressed air is the only alternative. Slightly less pressure may be permissible if the soot is easily removed.

Measure 1.1.5 recommends reducing the steam pressure to reduce system leakage, and for other possible benefits. In some cases, this may conflict with steam soot blowing.

With firetube boilers, the equipment that must be installed in the boiler to blow soot with steam is more

WESINC

Fig. 2 Soot blower for watertube boiler Rotating the tube manually, using the chain and gear, opens a cam-operated steam valve under the gear, allowing steam to flow through the elbow into the blower tube.

expensive than the equipment needed to blow soot with air. However, if a separate air compressor is needed for soot blowing, the total equipment cost is about the same for steam and air.

Steam is more expensive than compressed air, unless the steam is being produced with a very cheap fuel. The loss of steam in soot blowing requires additional makeup water, but you might not notice the difference unless the steam system has low water loss.

Compressed air starts out at room temperature and it cools by expansion as it comes out of the soot blower orifices, so there is a large temperature difference between the air and the boiler steel. In firetube boilers, the air is discharged into the end of the tube, where it fits into the tube sheet. This area is sensitive to thermal stress. Whether air blowing can cause thermal damage is an issue that is not well documented.

Soot Discharge Into the Environment

Soot is dirty and corrosive, so soot blowing can create a serious problem in the vicinity of the boiler plant. For example, automobiles parked near boiler plants become covered with pinhole rust spots from soot blowing.

If soot discharge is a potential obstacle, your choice is to give up soot blowing or to install flue gas cleaning equipment.

ECONOMICS

SAVINGS POTENTIAL: *0.5 to 10 percent of fuel cost, depending on the rate of soot accumulation. Using soot blowers may substantially reduce the cost of manual fireside cleaning, and it may reduce repair costs.*

COST: *Compressed air soot blowers for firetube boilers cost about $5,000 for 100 tubes, to about $12,000 for 300 tubes. The corresponding range for steam soot blowers is about $8,000 to $18,000. Air compressors for soot blowing cost from $1,000 to $5,000, depending on the size of the boiler.*

PAYBACK PERIOD: *1 to 10 years, depending primarily on the size of the boiler, the type of fuel, and fuel costs.*

TRAPS & TRICKS

SELECTING THE EQUIPMENT: *Retrofit installation of soot blowers is not common, so performance is not well documented. The equipment must be designed to fit the particular model of boiler, to avoid interference with gas flow, and to survive without maintenance. Ask vendors for references, and talk with purchasers who have used the equipment for several years.*

INSTALLATION: *As a rule, let the vendor install the equipment, because it is specialized. Do-it-yourself installation may void the warranty or weaken later damage claims.*

MEASURE 1.6.3 Optimize soot blower operation.

Soot blowing commonly is controlled on the basis of time, for example, 30 seconds every four hours. Such a rote procedure may result in unnecessary usage of steam or compressed air. On the other hand, inadequate soot blowing wastes fuel and increases the effort needed to clean firesides. Soot blowing does not use a large amount of energy, but it uses enough to make it worthwhile to optimize the procedure.

The prudent rule is to err on the side of cleanliness, but not to use more soot blower operation than is necessary to maintain a safe margin of cleaning.

How to Schedule Soot Blowing

Start with the recommendation of the soot blower manufacturer and/or the boiler manufacturer in scheduling soot blowing. Using this as a baseline, refine the schedule over the long term by inspecting the cleanliness of firesides whenever the boiler is cold.

You may hear that soot blowing should be done when a rise in flue temperature becomes apparent. This is wrong. Too much soot can accumulate before you see an increase in flue temperature. Blow soot on a firm schedule, and adjust the schedule based on experience that you gain from inspecting the firesides.

If different parts of the boiler have different soot blowing requirements, develop an appropriate soot blowing schedule for each soot blower. For example, an economizer may have different soot blowing requirements than the tubes in the firebox.

SUMMARY

Refining soot blowing practice to minimize energy consumption while keeping tubes clean.

SELECTION SCORECARD

Savings Potential .... **$** $

Rate of Return .... **% % % %**

Reliability .... ✔ ✔

Ease of Initiation .... ☺ ☺ ☺

ECONOMICS

SAVINGS POTENTIAL: *A fraction of one percent of fuel consumption, typically.*

COST: *None. The purpose of this Measure is to reduce the cost of steam or compressed air.*

PAYBACK PERIOD: *Immediate.*

TRAPS & TRICKS

EXPLAIN IT: *Train operators to use the equipment and to recognize trouble. Establish a written schedule for soot blowing.*

MONITOR PERFORMANCE: *Include a column in the plant operating log to record the time of soot blowing, the stack temperature before blowing, and the stack temperature afterward. It may take months or years of operation to find the optimum soot blowing schedule, adjusted for firing rate and hours of operation.*

MEASURE 1.6.4 Clean watersides at appropriate intervals.

RATINGS

New Facilities	Retrofit	O&M
		A

SUMMARY

Important long-term maintenance that tends to be neglected because it is difficult, dirty, and out of sight. Preserves boiler life as well as efficiency.

SELECTION SCORECARD

Savings Potential .... **$ $** $

Rate of Return .... **% % %** %

Reliability .... **✓ ✓**

Ease of Initiation .... ☺ ☺ ☺

Deposits on the insides of boiler tubes reduce efficiency in the same way as deposits on the outside, by reducing heat transfer from the combustion gases to the water.

In extreme cases, waterside deposits may reduce boiler efficiency by as much as ten percent. A boiler with deposits this severe is on the verge of failure. In more typical cases of neglected waterside maintenance, a boiler suffers an efficiency loss of several percent.

Cleaning watersides is important to protect the boiler. The combustion temperature inside a boiler is high enough to destroy the steel of boiler materials. Damage does not occur in normal operation because the boiler water cools the steel enough to keep its temperature at a safe level. However, the insulating property of waterside deposits inhibits this cooling. If the deposits are not prevented from forming, or if they are not cleaned out, they can accumulate to the point that a tube will fail from overheating. Failure is most likely to occur while the boiler is under heavy load, which maximizes the damage and the hazard that results.

Characteristics of Waterside Fouling

Deposits on the watersides of tubes are mostly of a type called "scale", because they typically are hard and

Goodway Tools Corporation

Fig. 1 Mechanically cleaning the watersides of a large watertube boiler This is dirty, noisy, claustrophobic work, but it is essential. The steam drum must be large enough to give a man and his tools access to the tubes. Otherwise, the tubes must be cleaned chemically.

tenacious. Especially in watertube boilers, the agitation of the water washes off softer deposits, which sink to the bottom of the boiler or float to the top.

Scale and other waterside deposits conduct heat much less efficiently than boiler steel. The most common types of mineral deposits have thermal conductivities that are only two to four times higher than the conductivity of the insulating brickwork used in boilers. Silicate scale, the least conductive, has ten times the insulating value of firebrick. The causes of waterside scaling are covered in Measure 1.8.1.

When Do Watersides Need Cleaning?

Visual inspection is the only accurate way to determine the thickness of waterside deposits. Unfortunately, some boilers make it difficult to see the watersides. For example, a watertube boiler with small drums may require an inspection mirror, which can provide a look only into the ends of the tubes. Scrape some of the deposit down to bare metal to see how thick it is. The thickness will suggest the cleaning methods needed.

Flue gas temperature is not a reliable indicator of waterside deposits. The deposits accumulate too slowly to be noticeable, and the resulting increase in flue gas temperature is masked by changes in boiler load. However, it is educational to note the difference in flue gas temperature immediately before and after waterside cleaning.

Cleaning Techniques

Mechanical and chemical methods are used to clean watersides. Both methods involve a lot of dirty work and both involve a risk of damaging the boiler. Unless the boiler plant operators have the skill and the equipment required, leave this activity to a specialist.

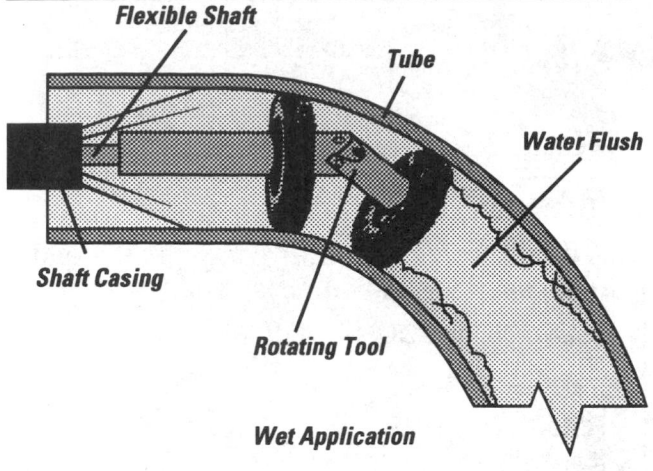

Flexible Shaft

Tube

Water Flush

Shaft Casing

Rotating Tool

Wet Application

Goodway Tools Corporation

Fig. 2 Rotating brush for watertube cleaning

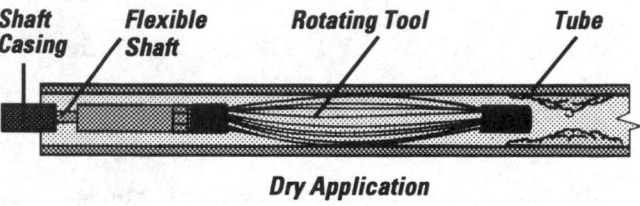

Shaft Casing

Flexible Shaft

Rotating Tool

Tube

Dry Application

Goodway Tools Corporation

Fig. 3 Rotating cutter for watertube cleaning

Large watertube boilers can be cleaned mechanically by passing a cleaning tool through each tube at the end of a rotating cable. This method is possible only if the steam drum is large enough to allow a cable to be directed into each tube. Figure 1 shows how this is done.

In order of increasing scale removing ability, the tools used are wire brushes, abrasive bead brushes, and cutter heads. Cutter heads have teeth that are hard and sharp to break up the hard scale. Even in experienced hands, it is easy for cutter heads to damage tubes. Figures 2 and 3 show typical cleaning tools.

Mechanical cleaning is difficult or impossible with many types of boilers, especially firetube boilers, because it is impossible to reach all the tube surface. Many firetube boilers have only a small handhole that exposes a small portion of the tube bundle. If deposits are soft, you may be able to remove them with a high-pressure water lance. Unfortunately, even a water lance cannot reach most of the tube surface in many boilers.

All types of boilers can be cleaned chemically. The basic technique is to use acid to eat away the mineral deposits. To keep the acid from eating away the boiler steel also, corrosion inhibitors are added to the acid. Even with inhibitors, careful control of the process is required.

Acid cleaning commonly opens up leaks at the joints between the tubes and the tube sheets, or between the tubes and the drums. This is especially likely if waterside cleaning has previously been neglected. What happens is that the acid removes mineral deposits that have plugged small leaks as they formed. The acid treatment itself may worsen the leaks once they are opened up. For this reason, be prepared to make repairs before putting the boiler back in service. You may be able to repair minor leaks by re-expanding the tube end into the tube sheet or drum. In more severe cases, the tube has to be replaced after the sealing surface is refinished.

Prevention is Better Than Cure

Preventing waterside fouling is easier, more efficient, and safer than cleaning it out. Also, it is usually cheaper, especially when considering the higher average boiler efficiency. Prevent waterside fouling with an effective water treatment program. This may radically reduce the need for waterside cleaning. See Measure 1.8.1 about water treatment.

ECONOMICS

SAVINGS POTENTIAL: *0.5 to 10 percent of fuel cost.*

COST: *The cost of typical rotating cleaning equipment for watertube boilers ranges from $1,500 to $4,000, depending on size and features. The time required for cleaning is proportional to the size of the boiler and to the difficulty of reaching all the waterside surfaces. For a typical commercial-size boiler, cleaning can be accomplished in one day. Acid cleaning requires specialists. Especially with acid cleaning, set aside additional funds to repair leaks that may appear when scale is removed from leaky tube joints.*

PAYBACK PERIOD: *Several months to several years.*

TRAPS & TRICKS

SKILL AND DILIGENCE: *By any method, waterside cleaning is dirty and tedious. Skill is required to avoid boiler damage. Make sure that the people doing the work are proficient. Inspect the work to make sure that it is thorough.*

MEASURE 1.6.5 Avoid leaving waterside deposits when deactivating boilers.

Most of the solids that are dissolved in boiler water decrease in solubility when the water temperature is lowered. Therefore, when a boiler is cooled down, dissolved material deposits on the internal surfaces of the boiler. Also, suspended material falls out of suspension. To prevent this, reduce the concentration of solids in the boiler water while the water is still hot. Do this by increasing blowdown for several days prior to shutdown.

Keep the water treatment chemicals at their normal concentration while increasing blowdown. The purpose is to eliminate the products of water treatment, not the treatment chemicals themselves.

Do not drain the boiler while it is still hot, because the heat of the boiler may bake any floating matter to the waterside surfaces, and may solidify sludge in the bottom of the boiler. As soon as the boiler is cooled and drained, wash it out vigorously.

If the boiler is being put into storage, take steps to prevent corrosion. If the boiler is stored with water in it, ask your water treatment specialist which chemicals to add during storage.

SUMMARY

A procedure that reduces scaling at the end of an operating period. Easy and free.

SELECTION SCORECARD

Savings Potential	$
Rate of Return	% % % %
Reliability	✓ ✓ ✓
Ease of Initiation	☺ ☺ ☺ ☺

ECONOMICS

SAVINGS POTENTIAL: *The primary benefit is reducing the effort required to clean the firesides.*

COST: *None.*

PAYBACK PERIOD: *Immediate.*

TRAPS & TRICKS

PLANNING: *The key to success is being ready several days before boiler shutdown. Make this a permanent item on your planning calendar.*

Combustion Gas Heat Transfer And Heat Recovery

The largest source of energy waste in a properly operating boiler is failure to transfer all the heat of combustion into the steam or hot water. The energy waste is large even in the best conventional boilers. The energy loss may be much larger in boilers that are inefficient in design or that have poorly matched burners.

The Measures of this Subsection capture additional heat from the combustion gases, either inside the boiler or after the gases have left the boiler. Accomplishing these Measures in the most efficient manner, and without doing damage to the boilers, requires a thorough understanding of the principles involved. These principles are presented in the Measures.

Some of these Measures are unusual, so approach them with caution. Their large potential savings make them worth considering by sophisticated plant operators.

INDEX OF MEASURES

1.7.1 Install a flue gas heat exchanger to recover additional heat.

  1.7.1.1 Install a conventional (non-condensing) economizer.

  1.7.1.2 Install a heat recovery air preheater.

  1.7.1.3 Install a condensing economizer.

  1.7.1.4 Install a water spray heat recovery unit.

1.7.2 In firetube boilers, install turbulators.

RELATED MEASURES

- Subsection 1.12, for replacing inefficient boilers with more efficient units
- Measure 1.4.4, reduce burner capacity in relation to boiler size

MEASURE 1.7.1 Install a flue gas heat exchanger to recover additional heat.

In an ideal boiler, all the energy of the fuel would be absorbed in the water or steam, and the flue gases exhausted from the boiler would have the same temperature as the entering air. In reality, the flue gases of efficient modern boilers have temperatures in the range of 300°F to 600°F, and even higher if they operate at high pressure. Less efficient boilers may have exhaust gas temperatures that are much higher.

These temperatures represent a large amount of energy that is being wasted. It may be practical to recover more heat from the flue gases by installing a heat exchanger in the flue gases.

The four subsidiary Measures cover the available flue gas heat recovery methods, including unconventional ones.

What Determines Flue Gas Temperature

To approach flue gas heat recovery intelligently, you need to understand the factors that determine flue gas temperature. The main factors are:

- *the boiler's heat transfer capacity.* Many older boilers have high flue gas temperatures because of insufficient heat transfer surface and unsophisticated heat transfer features. Some newer boilers may skimp on heat transfer surface as a cost cutting measure.
- *burner capacity.* Getting too much heat out of a boiler by installing an oversized burner is another cost cutting measure that raises flue temperature.
- *boiler water temperature.* The combustion gases must be hotter than the boiler water in order for heat transfer to occur. Practical limitations in heat transfer require the gas temperature to be 50°F to 100°F higher than the temperature of the boiler water. For example, a boiler operating at 600 PSI has a water temperature of about 490°F, so it cannot have a flue gas temperature leaving the boiler lower than about 550°F.
- *superheated steam temperature.* In steam boilers with superheaters, the limiting temperature usually is the temperature of the water, not the temperature of the superheated steam. Typically, boilers are designed so that combustion gases flow over the superheater tubes first, then through the tubes where water is boiled. However, if the boiler has superheaters that are fired in a separate firebox that exhausts directly to the flue, the flue gases coming

from the superheater firebox must be hotter than the superheated steam temperature.

- *preventing acid condensation.* The flue gas temperature may be kept high to avoid condensing acids in the flue. This is explained in Measure 1.7.1.1, below.
- *latent heat of water vapor.* Fuels that contain hydrogen produce water as a combustion product. When the water vapor is lost as part of the exhaust gases, it carries away about 1,000 BTU of energy per pound of vapor in the form of latent heat, i.e., the heat needed to transform liquid water into vapor. This heat can be recovered only at temperatures lower than the condensing temperature of the water vapor. This temperature depends on the concentration of water vapor in the flue gas.

A large fraction of the heat that is contained in high-temperature flue gas can be recovered by passing the flue gas through a heat exchanger after it leaves the boiler. The limitation of this technique is finding an application for the relatively low-temperature heat that is recovered. Heating feedwater and combustion air are the common applications.

Acid condensation and the lack of low-temperature recovery applications limit recovery of the lowest temperature component of flue gas heat, but methods do exist to capture this component.

Other Ways of Improving Heat Transfer

It may not be prudent to add an expensive heat recovery device to a boiler that has inferior efficiency characteristics, such as inefficient burners, insufficient combustion chamber volume, or other limitations that either cannot be improved or that would be expensive to improve. Also, it is risky to install expensive new equipment on a boiler that may be approaching the end of its service live. So, consider the other available methods of improving heat transfer.

Turbulators are an inexpensive way to improve heat transfer in firetube boilers that lack sufficient tube surface area. See Measure 1.7.2 for more about turbulators.

If a boiler is inefficient because it is being fired beyond its efficient capacity, consider reducing the burner output. See Measure 1.4.3 for this.

At the high end of the cost spectrum is purchasing a more efficient boiler. See Measure 1.12.1 if you consider this approach.

MEASURE 1.7.1.1 Install a conventional (non-condensing) economizer.

SUMMARY

A well proven but expensive method of improving the efficiency of boilers that have high flue gas temperatures. Economy of scale limits economizers to larger boilers.

SELECTION SCORECARD

Savings Potential .... $ $ $

Rate of Return .... % %

Reliability .... ✓ ✓ ✓ ✓

Ease of Retrofit .... ☺ ☺ ☺

This subsidiary Measure covers the most conventional method of recovering heat from flue gas after it leaves the boiler. This is to add an "economizer." An economizer is a heat exchanger that recovers heat from the flue gas to warm a liquid. Economizers are available from a number of manufacturers as standard items.

The conventional economizers covered by this Measure recover only the sensible heat of the flue gas.

Applications and Potential Savings

Figure 1 shows how an economizer is used in a typical feedwater heating application. In most cases, the liquid being heated is the boiler's own feedwater. Heating feedwater is a well balanced application, because the need for the recovered heat coincides with its availability. Also, only a small amount of piping is required, which makes for a compact installation and reduces cost.

In this example, the energy saving is about five percent. This is about the maximum saving that an economizer can provide. A saving much less than this usually does not justify the cost of the economizer. Therefore, economizers have a limited range of application, especially as retrofit items.

The temperature of the gas leaving the boiler is the key issue that determines whether an economizer is worthwhile. There are two general categories of applications. One consists of all boilers that operate at high pressure or high temperature. The other consists of existing boilers that are inefficient because they have inadequate heat transfer.

High-pressure boilers can benefit from economizers because the high temperature of the water or steam forces the gas to leave the boiler with a large amount of heat remaining. For example, if a boiler operates at 600 PSI, the gas temperature leaving the boiler is at least 550°F. If this boiler burns a fuel with an acid dew point (discussed below) of 250°F, and a safety margin of 100°F is needed to protect the flue, this leaves 200°F to be recovered.

For this reason, efficient high-pressure boilers usually include an economizer as an integral feature. Economizers are easy to include in the design of watertube boilers at modest cost, resulting in a compact overall package. In new construction, this is usually the best way to acquire an economizer. The main exception would be a firetube boiler operating near the maximum pressure for this type of boiler. Economizers cannot be made an integral part of firetube boilers.

With an existing boiler, if the flue gas temperature is too high, verify that the reason is inherent to the boiler, i.e., either high operating pressure or insufficient heat transfer surface. If the gas temperature is high for a reason that can be fixed, such as dirty firesides or watersides, it would be folly to install an economizer. Also, consider whether an old boiler has enough remaining life to justify the expense of an economizer. You may be able to substantially improve the efficiency of a large old boiler by adding an economizer, but the expense may not be justified if the boiler has to be replaced within a few years.

Retrofit economizers are expensive, and they have economies of scale, so they are economical only for larger boilers. The smallest standard economizers are rated at about 3 million BTU's per hour.

Efficient boilers that operate at low pressure or temperature cannot benefit from conventional economizers. The acid dew points of typical fuels range between 140°F and 250°F. If the flue gas temperature is raised about 100°F above the acid dew points to protect the flue, this results in flue gas temperatures that must be kept at 240°F to 350°F, respectively. A steam boiler operating at 15 PSI has a water temperature of 250°F, and good boiler design can reduce the flue gas temperature to within 50°F of this, or about 300°F. Thus, there is little or no opportunity to reduce flue gas temperature.

Construction of Freestanding Economizers

An economizer can be made from virtually any type of gas-to-liquid heat exchanger. A common type that is made in small sizes consists of one or more tubes coiled inside a cylindrical shell. Feedwater is passed through the tubes, and flue gases are passed through the shell. This design is simple and compact. The economizer can be inserted in-line in the flue, provided that a length of flue is accessible. Figure 2 shows an economizer of this type.

Most larger economizers consist of a bundle of straight finned tubes installed in a rectangular enclosure. These are typically installed alongside the boiler on a separate stand. This requires a large empty space adjacent to the boiler breeching. Moving things around to create this space may add significantly to the cost of the installation.

The finned tubes in this type of economizer increase heat transfer and reduce the overall bulk of the unit. The fin spacing is selected for the dirtiest fuel that the boiler burns, to allow for soot removal. A typical selection is 2 fins per inch with No. 6 oil, 3 fins per inch with No. 2 oil, and 4 to 6 fins per inch with gas.

The flue gas side of an economizer is a trap for soot, so economizers usually include integral soot blowers. Some manufacturers install soot blowers on all models, while others install soot blowers only on units that are used with oil and solid fuel. (Soot blowing is covered by Measures 1.6.2 and 1.6.3.)

Easy access to the heat transfer surfaces is a critical feature for an economizer. Even if the unit has a soot blower, you need access to the heat transfer surfaces for inspection, localized cleaning, and repair.

Acid Dew Point Limitations on Heat Recovery

Sulfur or sulfur compounds are present in virtually all petroleum and coal, and even in natural gas. The sulfur burns to become the gas sulfur dioxide. This gas can combine with water to form corrosive (sulfurous and sulfuric) acids. Water vapor is always present in the flue gas. It is created by combustion if the fuel contains hydrogen. It also enters as humidity in the air, and as a contaminant of the fuel. The acids are harmless as long as they remain in the vapor state. However, if the acid vapors contact a surface that is cold enough, they will condense and corrode steel components.

As with water, the condensation temperature of flue gas acids can be expressed as a dew point, which is called the "acid dew point." The acid dew point depends on the acid's boiling point and the amount of acid that is present in the flue gases. The boiling point of sulfuric

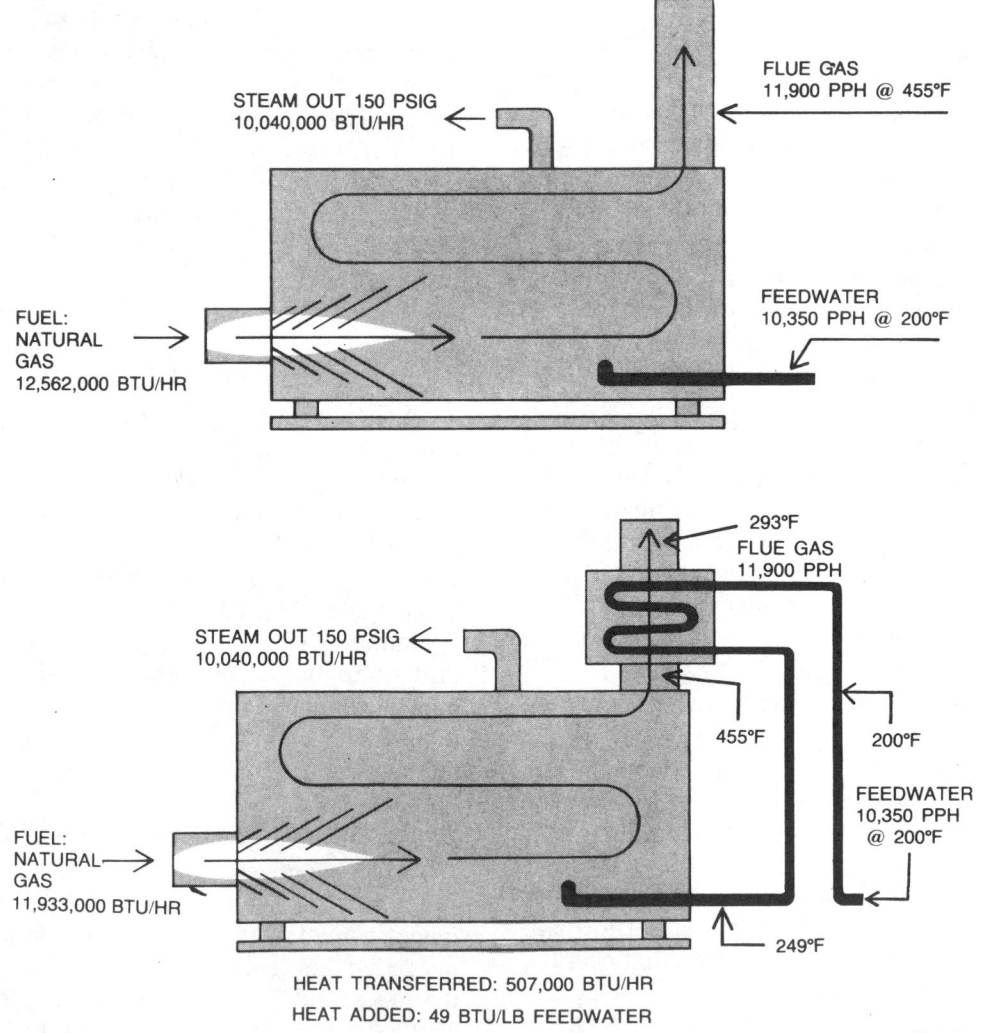

HEAT TRANSFERRED: 507,000 BTU/HR
HEAT ADDED: 49 BTU/LB FEEDWATER

Kentube

Fig. 1 Example of economizer installed to heat feedwater

acid ranges from 332°F to 626°F, depending on the degree of binding with water. Like water, acid can condense well below its boiling point. (A fogged window illustrates this.) A higher concentration allows acid to condense at a higher temperature.

The acid concentration in flue gases varies widely among different types of fuels, depending on the amount of sulfur that the fuels contain. Typical acid dew points for some common fuels are:

Fuel	Typical Acid Dew Point, °F
Natural Gas	140
No. 2 oil	180
No. 6 oil	200 - 250
High-Sulfur Coal	250

In practice, the minimum flue gas temperature coming out of the economizer should be considerably higher than these figures. Two temperatures must be controlled to prevent acid condensation. First, the temperature of the economizer surface must be kept higher than the acid dew point to prevent corrosion of the economizer itself. Second, the gases leaving the economizer should be hot enough to prevent corrosion of the flue system (the breeching and the stack).

The flue gas temperature varies widely with boiler load. The limiting condition is the flue gas temperature at the boiler's minimum load.

In the case of the economizer itself, the most vulnerable part is the end where the feedwater enters. Both the feedwater and the flue gases are coolest at this point. (The problem of corrosion at this point in an economizer is called "cold end corrosion." To maximize heat transfer, economizers are designed with counterflow circuiting, i.e., the feedwater enters at the top of the unit, where the flue gases exit.)

To estimate the inlet feedwater temperature that must be maintained to prevent corrosion, first find out the highest acid dew point that may occur with the type(s) of fuel being used. The fuel supplier should be able to provide this information.

The surface of the economizer must be kept slightly warmer than the acid dew point at all times. For example, if the economizer is used for feedwater heating, and the maximum expected acid dew point is 200°F, then the inlet feedwater temperature should not drop below about 210°F.

The temperature of the gases leaving the economizer must also be controlled. This is to protect the flue system from acid corrosion. All parts of the flue that can corrode must be kept warmer than the acid dew point. Gases cool as they travel through the flue because they lose heat to the flue surfaces. To account for this, a rule of thumb has arisen that the flue gas temperature leaving

the economizer should be at least 100°F higher than the acid dew point. Another rule of thumb calls for a 0.5°F increase in gas outlet temperature for each foot of flue.

Less temperature margin is needed if the flue and stack are completely insulated. If the stack or flue liner are made of non-corrosive materials, such as brick, the acid dew point can be ignored in these parts of the system.

How to Protect Economizers and Flues from Acid Condensation

If the economizer is sized for maximum heat recovery, there may not be enough hot gas generated at low boiler loads to keep the inlet end of the economizer warmer than the acid dew point. One solution is to preheat the feedwater before it enters the economizer. The preheater is thermostatically controlled to maintain a minimum feedwater inlet temperature, and it should operate only at low boiler loads. Boiler steam or hot water is used as the heat source. Most of the energy used by the preheater is recovered.

Kentube

Fig. 2 Economizer installed on a small boiler A larger economizer would be installed on its own stand, alongside the boiler.

Another technique for protecting the economizer is to recirculate a portion of the water leaving the economizer back to the economizer inlet. This requires a separate small pump.

The flue can also be protected at low boiler load by preheating the feedwater. The feedwater in turn heats the flue gases. The same preheater can be used for both purposes. The preheater is controlled to maintain both the appropriate stack temperature and the appropriate economizer temperature. The flue temperature sensor should be located near the highest point of the flue system that is vulnerable to corrosion.

Heat that is used to keep the flue gases warm cannot be recovered. Therefore, to provide an economical net saving, the economizer needs to recover substantially more heat, at higher loads, than is needed to keep the flue gases warm at lower loads.

There is no way to prevent acid condensation in a cold flue or stack. For this reason, make sure that any draft inducing fan turns off when the boiler is not firing. (See Measure 1.5.3.1 about this.) Corrosion of the upper few feet of a steel stack or flue liner is unavoidable because turbulence allows air to enter the top of the stack, which cools it below the acid dew point. You can observe that the tops of steel stacks are often seriously corroded.

Rain causes more stack corrosion when the flue gas temperature is reduced, because the rain is able to penetrate deeper into the stack and linger in the liquid state longer. Therefore, it may be advisable to add a rain cap to the stack when installing an economizer.

Examples

Here are some examples that illustrate the key points of controlling acid condensation:

- Oil with an acid dew point of 200°F is being burned. The inlet feedwater temperature is 160°F, and the gas temperature leaving the economizer is 300°F at low load. In this case, the feedwater must be preheated to prevent corrosion of the economizer, but the flue is safe.
- Oil with an acid dew point of 200°F is being burned. The inlet feedwater temperature is 210°F, and the gas temperature leaving the economizer at low load is 230°F. In this case, the economizer is safe. The flue is likely to corrode if it is made of corroding material or if it is not well insulated.
- Natural gas with an acid dew point of 140°F is being burned. The inlet feedwater temperature is 160°F, and the gas temperature leaving the economizer at low load is 230°F. In this case, there is no risk to either the economizer or the flue.

Consider Future Fuels

The previous discussion showed that the allowable reduction of flue gas temperature depends on the fuel,

and particularly on its sulfur content. In other words, installing an economizer commits the facility to burning only fuels that have a certain maximum sulfur content. Take this into account when considering an economizer. Depletion of low-sulfur fuels is forcing the substitution of higher-sulfur fuels. Environmental laws may require burning low-sulfur fuels, but regulations cannot override the reality of future fuel supply.

This has already been strongly evident in the depletion of low-sulfur coal, and its increased substitution with high-sulfur coal. The same trend toward higher acid dew points may occur with other fuels, especially if shortages of natural gas recur. The last of the previous examples shows that natural gas allows more aggressive heat recovery than does oil. However, if a gas-fired boiler is capable of burning oil, it is prudent to assume that it will become necessary to burn oil again at some time in the future. In summary, be realistic in considering all the types of fuels that a boiler may burn during its remaining life.

Heat Exchangers that are Not Limited by Acid Dew Point

You can avoid all the temperature limitations created by flue gas acids if you use an economizer and a flue that do not corrode. Measure 1.7.1.3, below, covers this possibility. Another possibility is condensing the flue gases using a water spray, which is covered by Measure 1.7.1.4. However, both of these techniques are limited in application by the low temperature of the recovered heat.

Increased Draft and Pump Pressure Requirements

Economizers add resistance to the flow of flue gases. Typically, the resistance of an economizer is in the range of one inch of water gauge at maximum firing rate. This may or may not require modifications to increase draft. Such modifications may include increasing the power of the forced draft fan, replacing the burner assembly with one having a stronger fan, adding an induced draft fan, or increasing the height of the stack. Sources of draft are covered in Measure 1.5.1.

Economizers have long water passages, and the flow is turbulent to maximize heat transfer. This increases resistance to feedwater flow. In most cases, the feedwater pump has ample reserve pressure to deal with this. However, verify that the pump is able to provide the maximum required feedwater flow once the economizer is installed. If not, you will have to increase the pressure output of the feedwater pumps, either by increasing the diameter of the pump impeller or by installing a new pump.

The additional fan and pump energy that is required should be factored into the cost-benefit calculation.

Using an Economizer With More Than One Boiler

Economizers are expensive, so it is tempting to consider using one economizer to serve more than one boiler. This arrangement is a candidate for serious trouble. If one boiler is operated while another is idle, some gas will flow from the active boiler to the idle boiler, causing acid corrosion in the idle boiler. Attempting to operate more than one boiler simultaneously through a single economizer would require complications to maintain the proper draft within each boiler.

In a plant with several boilers, the most economical approach is to install economizers on the boilers that can be operated most efficiently, and to use these boilers as the lead units. The other boilers without economizers are kept in reserve, or are used only when required by a heavy load.

ECONOMICS

SAVINGS POTENTIAL: *2 to 10 percent of fuel cost.*

COST: *Economizers typically cost $200 to $600 per million BTUH of boiler capacity. The unit cost declines considerably with size. Installation cost varies widely, but it will be high if it is necessary to move piping and equipment to make room for the economizer.*

PAYBACK PERIOD: *Several years or longer.*

TRAPS & TRICKS

DESIGN: *The main performance issues are heat recovery and avoiding acid condensation. Consider all fuels that may be used in the boiler. Make sure that the boiler retains adequate draft. Design the physical layout for ease of maintenance.*

SELECTING THE EQUIPMENT: *Do your shopping thoroughly. Economizers come with a range of features, and in a range of quality. Avoid units that lack adequate access for cleaning.*

OPERATION AND MAINTENANCE: *Soot blowing and cleaning must be done regularly. If the economizer has variable modes of operation, such as preheating feedwater to prevent acid condensation, monitor these features and be alert to problems. Install effective instruction placards, if appropriate.*

MEASURE 1.7.1.2 Install a heat recovery air preheater.

SUMMARY

Expensive and bulky, air preheaters are commonly installed on large boilers. They offer an additional increment of efficiency that is not available with conventional economizers. They are rarely installed on a retrofit basis.

SELECTION SCORECARD

Savings Potential .... $ $ $

Rate of Return, New Facilities % %

Rate of Return, Retrofit.......... % %

Reliability .... ✓ ✓ ✓

Ease of Retrofit .... ☺

This is the second of four subsidiary Measures devoted to recovering additional heat from flue gas after it leaves the boiler. The method covered here is installing a heat exchanger that recovers heat from the flue gas to preheat the combustion air. This is a well balanced heat recovery application, because the need for the recovered heat coincides with its availability.

Combustion air preheaters are bulky because they transfer heat from one gas to another at relatively low temperature. This makes them expensive. In the past, economies of scale have limited air preheaters to very large boilers, such as those used for power generation and large industrial processes. Newer technology, discussed below, may make air preheaters less bulky and expensive, and hence allow them to be economical in smaller sizes. Keep an eye on developments.

Applications and Potential Savings

Air preheaters compete with economizers as a method of recovering heat from flue gas. Within the range of flue gas temperatures where both apply, conventional economizers are generally less expensive and easier to install. However, air preheaters have an advantage over conventional economizers in being able to recover the bottom range of temperature from flue gas. This is because the outside air used for combustion air is typically much cooler than feedwater, so it can absorb more heat from the flue gases. As a result, a combustion air preheater may increase boiler efficiency by 1% to 4% more than a conventional economizer.

The difference is especially important in plants with high-pressure boilers, because the feedwater is especially warm as a result of passing through deaerating feedwater heaters. For example, outside air may have an annual average temperature of 50°F, whereas the feedwater entering the boiler may have a temperature of 200°F.

(You may wonder why an economizer could not be used for warming condensate before it enters the feedwater heater, while it is still relatively cool. The reason is that the condensate is heated by steam as part of the deaerating process, and more heat cannot be added because the pressure within the deaerating tank must be kept reasonably low. There are ways to use flue gases to heat condensate, but these create further complications.)

Where boilers are used to drive high-pressure steam turbines, the feedwater may be heated deliberately to much higher temperatures to increase the cycle efficiency. In such cases, the temperature of the feedwater becomes too high for recovery of flue gas heat, and air preheating then becomes the method of choice for recovering all flue gas heat.

Air preheaters are usually found in plants where boiler efficiency is being pushed to the limit. Such plants have economizers, so the role of the air preheater is limited to removing the last increment of heat from the flue gases. In principle, an air preheater could be used without an economizer, in which case it would remove much more heat from flue gases, and presumably would provide better economics. In such an application, the air from the preheater would be very hot, and a retrofit installation would probably encounter major engineering obstacles. For example, the density of the air might be too low to enter the boiler through the existing openings and it might not provide proper burner operation.

Because air preheaters can cool flue gas to low temperatures, they can condense the water out of the flue gas that is formed during combustion. This may offer a significant energy recovery advantage, because this recovers much of the latent heat of the water vapor in the fuel. For a given outside air temperature, the actual amount of latent heat recovered depends on the dew point of the water vapor in the flue gases. In turn, this depends on the concentration of water vapor, which depends on how much hydrogen is in the fuel. Gaseous fuels contain much hydrogen, so they produce a large amount of water vapor.

Petroleum produces much less water vapor than gas. Coal produces little water vapor by combustion. (However, coal may contain a large amount of free water as a contaminant.)

(Recovery of the latent heat of water vapor is also a feature of the unconventional heat recovery methods covered by Measures 1.7.1.3 and 1.7.1.4, next.)

The inherent disadvantage that goes along with condensation of the water vapor is that any sulfur in the fuel will combine with the water to form acids. The preheater itself must be able to tolerate acids, which makes it more expensive. The portion of the flue downstream of the preheater must be non-corroding, so preheaters are usually found in boiler plants that have brick or concrete stacks. The corrosion problem effectively bars combustion air preheaters from many plants that already have economizers.

Types of Air Preheaters

In principle, any type of gas-to-gas heat exchanger could be used for preheating combustion air, and many types have been proposed and tried. In the current state of technology, the following are types worth considering.

■ Tube Bundle (Tubular) Preheaters

A tube bundle, or tubular, air preheater uses metal tubes as the heat transfer surface. Usually, the flue gas flows inside the tubes, and the incoming air flows around the outside. The tubes typically are straight, and are held in parallel bundles by tube sheets. The tube sheets

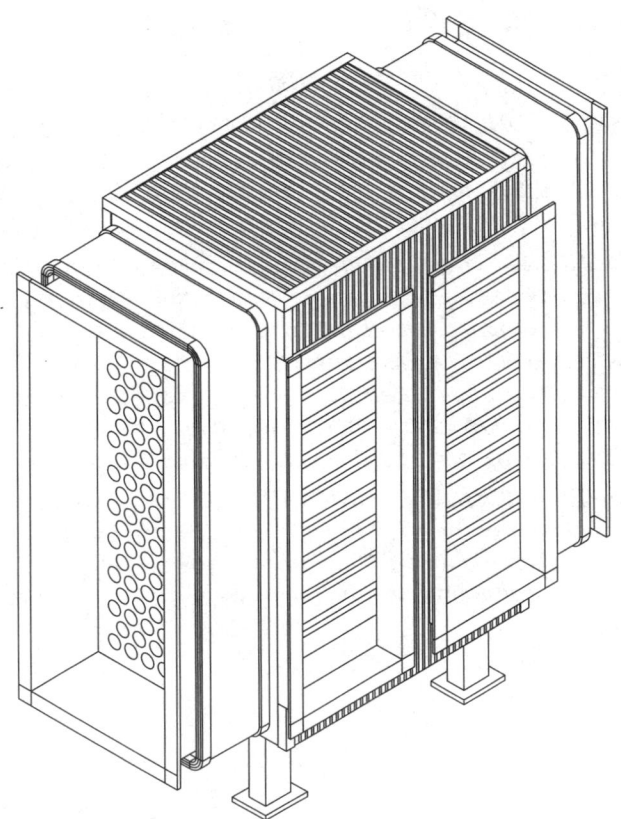

Kentube

Fig. 1 Tubular air preheater The flue gas flows through the tubes, entering at left. The incoming combustion air flows across the outside of the tubes.

also direct the flow of air around the tubes. Figure 1 shows a typical unit.

Tubular preheaters are fairly popular. They are simple, and easy to clean. Their main disadvantage is that they are rather bulky in relation to their heat transfer capacity. However, they can be made in many configurations to fit different spaces.

■ Parallel-Plate Preheaters

A parallel-plate heat exchanger is an assembly of metal plates held in parallel so that the flue gas and combustion air streams pass between alternate plates. Heat is transferred from flue gas to air by conduction through the plates.

This type of air preheater was one of the first to be used. Its advantages are simplicity, compactness, and relatively easy cleaning. It has fallen out of popularity, probably due more to effective marketing by competitors than because of any serious disadvantages. This type appears to be a leading candidate for use with smaller boilers.

■ Rotary Preheaters

A rotary air preheater is a rotating cylinder filled with air passages in the axial direction. Exhaust gas flows through one side of the rotor, and combustion air flows through the other. The material of the rotor, which is a thin metal honeycomb, absorbs the exhaust gas heat. The rotor turns slowly, at about 3 RPM, transferring heat from the exhaust stream to the incoming air stream.

Combustion air is at higher pressure than the flue gas. This causes air to leak from the air side to the flue side of the heat exchanger, wasting fan power. To minimize this loss, the unit includes seals between the frame and the cylinder to isolate the two sides. The seals require maintenance.

Rotary air heaters for boilers are specialized items made by a small number of manufacturers. They are offered in a large range of sizes. For example, one manufacturer offers units with rotor diameters ranging from 4 feet to 17 feet. (Smaller units are used in air conditioning heat recovery applications, where they are called "heat wheels." The latter are made by a different group of manufacturers.)

The main advantage of rotary preheaters is that they are compact. Their disadvantages include complexity, leakage around the seals, difficulty in cleaning, and a need for maintaining moving parts.

■ Heat Pipe Heat Exchangers

Heat pipe heat exchangers are described in Measure 4.2.8. They are commercially available as packaged units for heat recovery in air handling systems, and have been used for this purpose since the 1970's. They are not yet standard items for flue gas heat recovery. However, they may appear in a future generation of air preheaters for smaller boilers.

The potential advantage that heat pipes promise is reduced size. This is important because bulk is a limiting factor both in the cost of the heat exchanger itself, and in the feasibility of installation in an existing boiler room. Heat pipe heat exchangers are simpler than rotary preheaters, although they do have some moving parts.

Cleaning Air Preheaters

All fuels except gas produce a significant amount of dirt in the flue gas. If a preheater is used, it should have an effective means of periodically clearing the dirt from the flue gas passages. This is typically done with a device similar to a soot blower. With some fuels, it may be appropriate to have a detergent washing system.

Expect Acid Condensation

The basic principles of acid condensation from flue gas are covered in Measure 1.7.1.1. In most plants, the economic advantage of an air preheater over a conventional economizer rests on its low heat recovery temperature. Therefore, expect acid condensation in the heat recovery unit.

Some preheaters may deal with acid corrosion by periodic replacement of the corroded parts. For example, one manufacturer of rotary preheaters makes the low-temperature portion of the rotor (where the air enters and the flue gases exit) easily replaceable. Acid-resistant coatings also are used to retard corrosion. Pinhole perforations of the rotor material by corrosion are not damaging, because the rotor material is used only as a mass for heat storage, and does not separate one gas stream from another.

Other types of heat exchangers are made of non-corrosive materials, which greatly increases cost.

An air preheater reduces the flue gas temperature to the region of acid condensation, so the stack must resist corrosion. Also, the condensate drained from the stack is somewhat acidic. This may require neutralization to conform to environmental regulations dealing with acid sewage.

Increased Draft Requirement

Any type of flue gas heat recovery unit increases resistance to the flow of flue gas and combustion air. If stack draft is not adequate to overcome the added resistance, it may be necessary to increase draft by increasing the power of the combustion air fans or by increasing the height of the stack. Sources of draft are covered in Measure 1.5.1. The additional fan power reduces the overall energy saving provided by the preheater.

Plant Modifications and Installation Costs

All types of heat recovery preheaters are bulky in themselves. In addition, a great deal of space is needed for the ducting that carries the flue gas and combustion air. Expect to relocate equipment that is in the way. Such work is expensive. The cost of installation is one of the major reasons why air preheaters are usually found only in very large boiler plants.

ECONOMICS

SAVINGS POTENTIAL: 1 to 4 percent of fuel cost, assuming that an economizer is installed. Up to 10 percent of fuel cost, if no economizer is installed. Deduct the increased fan energy requirements from these savings.

COST: Tubular air preheaters typically cost from $5 to $10 per SCFM of heated air. Installation cost, including the cost of extra combustion air ducting, varies widely. In retrofit, the cost of relocating existing plant equipment and structure may be very high.

PAYBACK PERIOD: Ten years or longer, typically.

TRAPS & TRICKS

SELECTING THE EQUIPMENT: Each type of air preheater has advantages and disadvantages. Do your shopping thoroughly, study the literature, and discuss candidate equipment with owners who have used it.

DESIGN: The main performance issues are heat recovery and avoiding acid condensation. Consider all fuels that may be used in the boiler. Make sure that the boiler retains adequate draft. Design the physical layout for ease of maintenance. If the air preheater is used with high flue temperatures, the high combustion air temperature may add significant engineering problems.

OPERATION AND MAINTENANCE: Air heaters, especially rotary types, are specialized items that require understanding and skill to keep in operation. Cleaning must be punctual if the fuels are dirty. Be alert for problems.

MEASURE **1.7.1.3 Install a condensing economizer.**

SUMMARY

An improvement on conventional economizers. Requires custom design and fabrication. Most likely to be economical with gas fuels. With sulfur-bearing fuels, expect to deal with acid sewage and flue corrosion.

SELECTION SCORECARD

Savings Potential .... $ $ $

Rate of Return, New Facilities % % %

Rate of Return, Retrofit.......... % %

Reliability .... ✓ ✓ ✓ ✓

Ease of Retrofit .... ☺ ☺

This is the third of four subsidiary Measures devoted to recovering additional heat from flue gas after it leaves the boiler. This method uses a variation of the conventional type of economizer covered in Measure 1.7.1.1. Conventional economizers are limited in their heat recovery potential by the need to avoid acid condensation in the heat exchanger itself and in the flue. A condensing economizer differs primarily in using a larger heat transfer surface to condense water vapor from the flue gases, and in using non-corrosive materials for the low-temperature portion of the heat exchanger.

The first widespread use of condensing heat exchangers occurred in the 1980's with the introduction of gas-fired residential-sized condensing boilers and gas furnaces. These use stainless steel for the low-temperature portion of the heat exchanger. At present, condensing economizers for heat recovery in larger boilers are unusual items that would be built on a custom basis. You will be a pioneer if you attempt this.

Where to Consider a Condensing Economizer

The advantage of a condensing economizer is its ability to recover the latent heat of the water vapor in the gases. Significant amounts of water vapor may come from two sources: hydrogen in the fuel, and water that is included in the fuel as a contaminant. Hydrogen is an important source in gas fuels, and is less important in oil. Contaminant water is most important in coal, and there may be a substantial amount in heavy oil. See Reference Note 20, Fossil Fuels, for more on the hydrogen content of different fuels. The latent heat present in the combustion gases of common fuels is given in reference books. The energy content of unconventional fuels, such as municipal refuse and wood waste, must be analyzed on an individual basis.

The concentration of water vapor in the flue gases determines the amount of latent heat that is present, and also the fraction of the latent heat that can be recovered. Latent heat is not recovered unless the water vapor condenses. The concentration (or dew point) of the water vapor determines whether it condenses. If the flue gas has only a small amount of water vapor, little latent heat will be recovered at any useful temperature. For example, a condensing economizer may not be effective when burning oil, because oil creates too little water vapor to condense at useful heat recovery temperatures.

Here are several examples to illustrate how the hydrogen content of the fuel and the temperature at which heat is recovered determine whether a condensing economizer can be useful:

• Natural gas with an acid dew point of 140°F is being burned. The inlet feedwater temperature is 110°F. In this case, the flue gas temperature could be reduced to a range of 120°F to 160°F, depending on load. A substantial amount of water will condense, recovering latent heat, and some acid will also condense. Consider a condensing economizer for this application.

• Natural gas with an acid dew point of 140°F is being burned. The inlet feedwater temperature is 200°F. In this case, the flue gas temperature which exits the economizer cannot fall below about 210°F. No condensation will occur in the economizer, and probably none in the flue. There is no point in using a condensing economizer in this application.

• Heavy oil with an acid dew point as high as 230°F is being burned. The inlet feedwater temperature is 210°F. The stack is made of concrete with a non-corrosive liner. In this case, the flue gas temperature which exits the economizer cannot fall below about 220°F. At the minimum flue gas temperature, some condensation might form in the economizer, and condensation in the flue would be unimportant. No latent heat would be recovered. A condensing economizer would make little difference in this application. It would be more economical to use a conventional economizer and to design it to keep the flue gas temperature above 220°F during low boiler loads.

• Heavy oil with an acid dew point as high as 230°F is being burned. The inlet feedwater temperature is 110°F. The stack is made of concrete with a non-corrosive liner. In this case, the flue gas temperature could be reduced to a range of 120°F to 160°F, depending on load. A large amount of acid would condense in the economizer, as well as in the flue. Some latent heat would be recovered, limited by the amount of hydrogen in the fuel and the water dew point. The efficiency would be higher than that of a conventional economizer, but probably not enough to justify the cost of dealing with the acids.

These examples lead to the conclusion that a condensing economizer is useful only if the heat recovery temperature is low. The most favorable applications are inefficient boilers burning natural gas. For such application, the total efficiency improvement of the boiler system might be 10 to 15 percent.

The major disadvantage of condensing economizers is their high cost, which results from:

- the large surface area required in relation to the relatively small temperature reduction achieved
- the need for expensive corrosion-resistant materials
- the custom design and fabrication required.

The cost limits practical application of condensing economizers to boilers with a long remaining life. Newer boilers are likely to be more efficient than older boilers, so their savings potential is likely to be lower. Thus, condensing economizers occupy an even narrower niche than conventional economizers.

Dealing with Acid Condensation

If water is condensed out of the flue gases, the sulfur in the fuel combines with the water to form acids, mostly sulfuric acid. The acids must be drained from the economizer. The amount of acid depends on the amount of sulfur in the fuel, and the amount of water depends on the amount of hydrogen in the fuel. If the fuel has a high hydrogen content and a low sulfur content, the acid solution is weak. If the acidity is high, you may have to neutralize the condensate to protect the plant's sewage system and to conform to environmental regulations dealing with acid sewage.

A condensing economizer requires a flue that is made of non-corroding materials. After the gases exit the economizer, they cool further in the flue, where they deposit more acid and more water. If the existing flue cannot tolerate the acidity and moisture, a new flue is required. The flue used with a condensing economizer can be somewhat smaller than the original because the cooler gases occupy less volume.

In small condensing gas furnaces and boilers, plastic pipe has been used successfully as a flue material. However, this may be dangerous with larger units. An economizer fire caused by soot accumulation would burn a flue made of combustible material.

ECONOMICS

SAVINGS POTENTIAL: *5 to 15 percent of fuel cost, in feasible applications.*

COST: *$300 to $1,000 per million BTUH of boiler capacity. Installation costs vary widely, but may be very high if it is necessary to relocate piping and equipment to make room for the economizer.*

PAYBACK PERIOD: *Several years or longer.*

TRAPS & TRICKS

DESIGN: *Condensing economizers are unusual items in large boiler applications, so be sure you deal with all the issues. Provide easy access for cleaning and repair. If you burn sulfur-bearing fuels, deal with acid condensate and protect against flue corrosion.*

EQUIPMENT: *The equipment should be fabricated by a manufacturer of economizers who has successful experience in making units of corrosion-resistant materials. Check references.*

OPERATION: *Cleaning must be done regularly if dirty fuels are used.*

MEASURE **1.7.1.4 Install a water spray heat recovery unit.**

This is the last of a series of four subsidiary Measures devoted to recovering heat from flue gas after it leaves the boiler.

In this method of heat recovery, the flue gas is passed through a water spray. This lowers its temperature to between 120°F and 200°F, depending on the application. The water spray is heated by the sensible heat of the exhaust gas and also by the latent heat of the water vapor produced during combustion. The latent heat is an important energy component that is not recovered by ordinary economizers. (Latent heat recovery is explained in Measures 1.7.1.1 and 1.7.1.3.) The water spray unit takes heat recovery near its theoretical limit of efficiency.

Heat Recovery Vessel

The heat recovery unit consists of a corrosion-resistant vessel in which a water spray can pass through the flue gas. Because there is no conducting barrier that separates the flue gas from the water it is warming, this type of vessel is sometimes called a "direct contact heat exchanger."

The heat recovery unit is continuously wetted and cooled, and it does not contain any pressure. In principle, it could be cheap to manufacture, perhaps with plastic materials. However, typical commercial units have a stainless steel tank, and stainless steel surfaces for distributing the water flow, so they are expensive. A typical unit is shown in Figure 1, and the installation is shown in Figure 2.

Application Limitations

The major application limitation of this technique is that heat is recovered at a low temperature. The temperature must be kept low to maximize sensible heat recovery, to maximize the capture of latent heat, and to minimize water loss from evaporation. (All the water would be lost to evaporation if its temperature rose to 212°F, or less at higher altitudes.) By the same token, you should look for an application that uses the spray water directly. Using a heat exchanger between the spray water and the application would cost a large fraction of the available temperature of the recovered heat.

The Condition of the Water

The water from the spray unit is contaminated by anything that is in the flue gases, including particulates, fly ash, and miscellaneous chemical compounds. The degree of contamination depends on the fuel. A clean

SUMMARY

An unusual technique for heat recovery that pushes the limits of efficiency. Requires a low-temperature heat recovery application. Reduced air pollution may be a valuable bonus.

SELECTION SCORECARD

Savings Potential .... $ $ $

Rate of Return, New Facilities % % % %

Rate of Return, Retrofit .... % % %

Reliability .... ✓ ✓ ✓ ✓

Ease of Retrofit .... ☺ ☺

fuel, such as natural gas, gives clear water. The water condensed from other fuels may require filtration or other treatment.

Fuel that contains sulfur makes the water acidic. However, the acidity is mild. With natural gas, the pH may be reduced by only a tenth of a point, e.g., from a pH of 7 to a pH of 6.9. Even with high-sulfur fuels, the dilution by the large volume of water limits the pH change to no more than about one point. If necessary, this amount of acidity can be neutralized easily.

In some cases, the spray water could be used for boiler feedwater makeup. To calculate whether this is economical, consider the value of the energy recovered, the cost of treatment for the spray water, and the cost of treatment for the conventional water source. This application is practical only for boilers systems that require a large amount of makeup water.

Exploit the Environmental Benefits

The spray system acts as an air washer, removing a considerable fraction of solids and acids. The cost of the unit may substantially reduce the need for other pollution abatement equipment. Also, the acid removal benefit may allow you to burn less expensive fuels that contain larger amounts of contaminants.

Match the Heat Requirement to Heat Availability

To keep the water temperature within the narrow limits that are necessary for heat recovery, the temperature and flow rate of the water must be kept in balance with the temperature and flow rate of the exhaust gases. If the water flow rate is too low, the water overheats and evaporates, taking the recovered heat with it. If the water flow rate is too high, the water temperature is too low.

If the need for recovered heat exceeds the amount that is available, the volume of spray water is adjusted to provide the desired water temperature. Additional heat must be provided by other sources.

If the available flue gas heat exceeds the requirement, you have to limit contact between the flue gas and the water. You can do this by diverting flue gas from the heat exchanger or by changing the spray pattern within the heat exchanger. The latter method increases the temperature inside the heat exchanger, perhaps requiring more expensive materials.

ECONOMICS

SAVINGS POTENTIAL: *5 to 20 percent of the energy content of the boiler fuel, depending on the type of fuel and efficiency of the boiler. The economic value of the*

recovered heat depends on energy costs and the nature of the application.

COST: *A packaged, stainless steel unit may cost about $10,000 for a boiler of 3 million BTUH capacity, and about $30,000 for a boiler of 25 million BTUH capacity. In new construction, installation adds little to plant cost, and flue cost may be reduced. In retrofit, modifications*

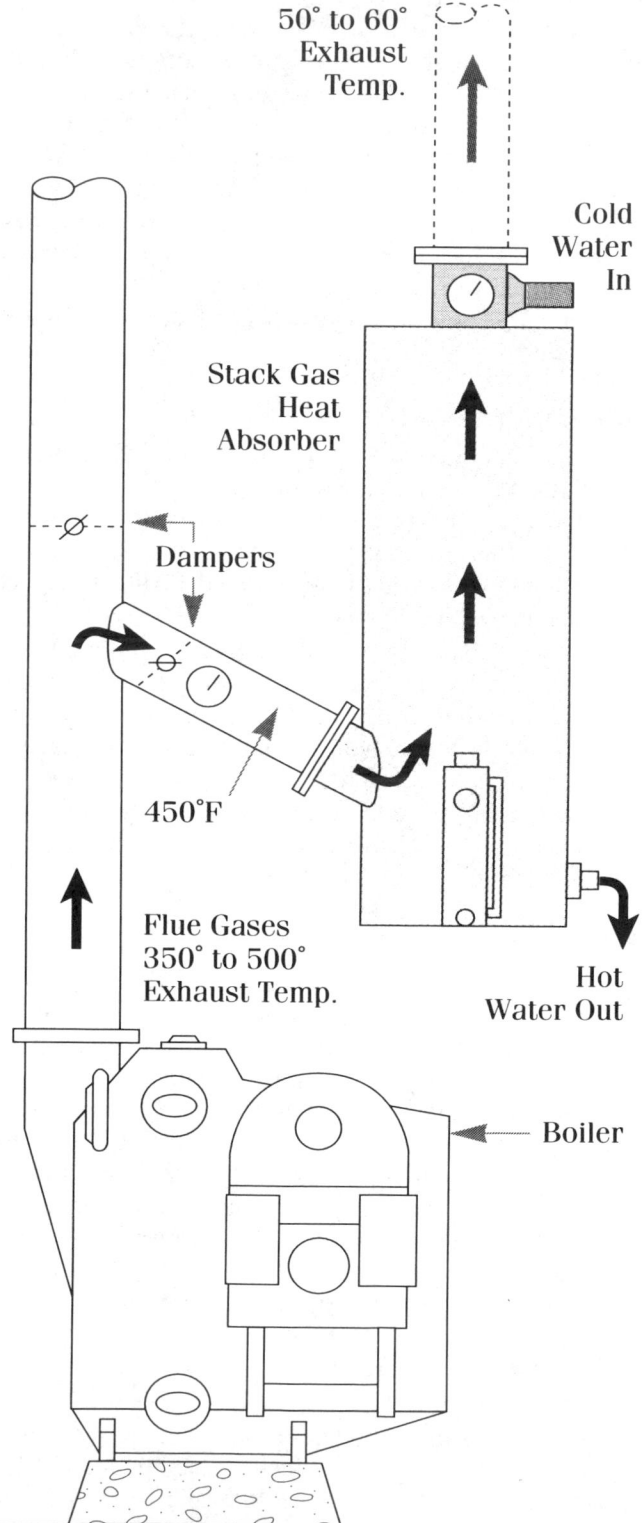

50° to 60° Exhaust Temp.

Cold Water In

Stack Gas Heat Absorber

Dampers

450°F

Flue Gases 350° to 500° Exhaust Temp.

Hot Water Out

Boiler

LOW STACK TEMPERATURE Typically within 10° of incoming process water.

INCOMING WATER distributed evenly with wide angle jet.

ALL STAINLESS STEEL CONSTRUCTION for durability, long life.

STAINLESS STEEL HEAT TRANSFER MEDIUM Maximizes flue gas heat transfer to incoming water.

FROM BOILER

DIRECT CONTACT between flue gases and water puts virtually all the heat into the water.

TO PROCESS Preheated water has absorbed virtually all the heat from 350° - 500° flue gases.

Ludell Manufacturing Company

Fig. 1 Water spray heat recovery vessel

Ludell Manufacturing Company

Fig. 2 Typical flue gas heat recovery installation

to the boiler breeching and moving existing equipment may be expensive.

PAYBACK PERIOD: *In the most favorable applications (gas fuels and the ability to use large amounts of low-temperature heat), the equipment may pay for itself in less than one year. In less favorable applications, the payback period may be several years or longer.*

TRAPS & TRICKS

DESIGN: *The method is unusual, but it has been done before. Achieving maximum efficiency requires a good grasp of the theory, combined with a talent for practical innovation.*

EQUIPMENT: *Some manufacturers make the equipment as a catalog item, typically from stainless steel. You might be able to make the heat exchanger from less expensive materials. The manufacturer may be pioneering along with you. Check his references.*

INSTALLATION: *The fact that the equipment is unconventional increases the risk of installation problems. Work closely with the installer.*

OPERATION AND MAINTENANCE: *You may have unexpected problems. Stay on top of operation until it becomes routine. Once operating procedures stabilize, install instruction placards.*

MEASURE 1.7.2 In firetube boilers, install turbulators.

RATINGS

New Facilities	Retrofit	O&M
	C	

"Turbulators" are devices inserted inside the fire tubes of boilers to increase heat transfer between the combustion gases and the tube walls. Turbulators may take many forms, such as twisted strips or coils of steel. Figures 1 and 2 show these forms.

Turbulators have been used in boilers for many years, but they are still controversial. Their actual benefit, and even the way they work, have not been well documented. Turbulator manufacturers claim typical efficiency improvements in excess of 10 percent, which is dramatic. However, more independent authorities claim that savings do not exceed 3 percent in the best of cases.

The most compelling advantage of turbulators is that they are much less expensive than any other method of improving heat transfer. This suggests that you should consider them before using more expensive methods of improving heat transfer or retiring an existing boiler to replace it with a more efficient one.

Alleged Principles of Operation

Turbulators are intended to improve heat transfer within the tubes of firetube boilers by changing the flow of the combustion gases through the tubes. The specific action, or actions, of turbulators are open to question. Manufacturers offer these explanations:

• **increased turbulence at the tube surface.** Heat transfer from a gas to a metal surface is increased by turbulence. The name "turbulator" derives from the assumption that turbulators function by

SUMMARY

A method of improving heat transfer in firetube boilers that tempts with its low cost and ease of installation. The typical candidate is an old boiler that has only one or two passes. Performance claims tend to be exaggerated. May cause boiler damage if not installed properly.

SELECTION SCORECARD

Savings Potential	**$ $** $
Rate of Return	**% %** %
Reliability	✓ ✓
Ease of Retrofit	☺ ☺ ☺

increasing the turbulence of the flue gases. However, turbulator manufacturers typically recommend installing turbulators in only a portion of the length of the tubes, and only in the last pass. This leaves a large fraction of tube area unaffected by the alleged benefit of increased turbulence.

• **improved gas distribution through the tubes.** In most firetube boilers, the tubes are arranged in several passes. Gas flow reverses direction at the ends of the boiler in going from one pass to the next. The inertia of the gas as it makes the turn may cause unequal gas flow into the following tubes. Turbulators may reduce this tendency (if it

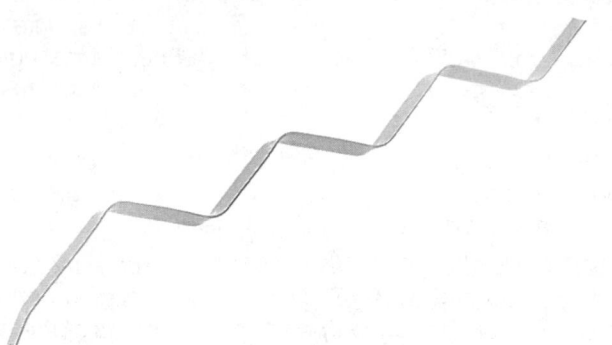

Fuel Efficiency, Inc.

Fig. 1 Typical turbulator This is simply a strip of steel, bent into a shape that holds it within the boiler tube by friction.

exists) by increasing resistance to flow, thereby causing the gas to distribute itself more evenly. Vendors typically recommend that longer strips be inserted in the upper tubes of a pass, which are the ones that might receive the most gas flow if inertia is a factor.

• *increased gas residence time.* One vendor claims that turbulators slow the flow of gases out of the firebox, thereby increasing the time that the gas is in contact with the flame and the time that the flame is in contact with the tubes. This cannot be true as a primary cause. The velocity of a given volume of flue gas is a function only of its density and the size of the tubes. Cooling reduces the density of the gas and slows it down, but this is an effect, not a cause.

In addition to these unconvincing explanations, the literature of turbulator manufacturers includes various other assertions that are overtly wrong. Nonetheless, it

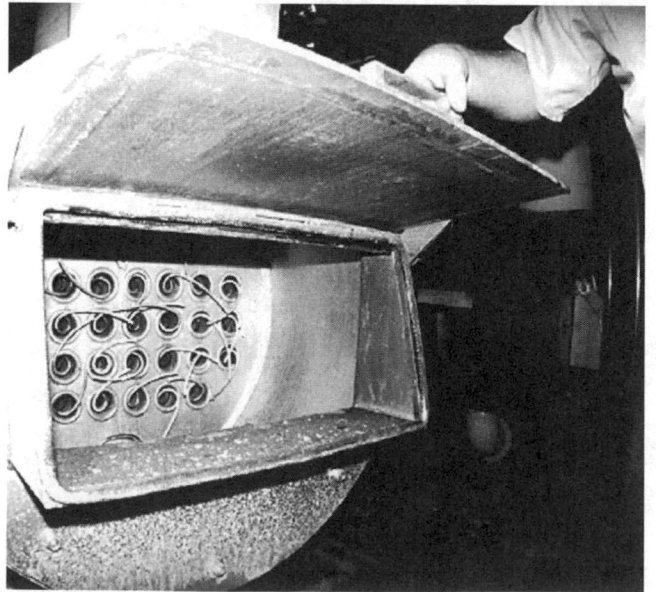

WESINC

Fig. 2 Turbulators installed in a firetube boiler These turbulators are in the form of coils. They are kept from sliding into the tubes by pigtails that extend outside the tube sheet.

is widely believed that turbulators work. One possibility is that turbulators do improve efficiency, but that nobody really knows why.

Where to Consider Turbulators

The typical candidates for turbulators are older firetube boilers that have one or two passes. High stack temperature is the symptom that indicates a need to increase heat transfer. Read the stack temperature when the boiler is clean and working perfectly. If the boiler exit temperature exceeds the acid dew-point temperature of the fuel by much more than 100°F, consider turbulators. (The relationship of stack temperature to acid dew point is explained in Measures 1.7.1 ff.)

Test combustion efficiency before and after turbulators are installed. Estimate the potential for improving efficiency from the initial efficiency test and the acid dew point of the fuel. Based on this, negotiate with the vendor for a guarantee that the turbulators will produce a stated efficiency improvement.

Turbulators have a tendency to trap soot inside the tubes. For this reason, one manufacturer of turbulators advises that they should be used only with gas and light oil fuels. However, another manufacturer states that turbulators can be used even in coal-fired boiler, but only with frequent soot blowing.

Alternative: Reduce Burner Capacity

Many boilers that are candidates for turbulators are inefficient because they have an unsophisticated design that is not capable of recovering the heat of the flue gases with great efficiency. However, some boilers are inefficient because the burner has too much capacity for the boiler. In such cases, an alternative is to reduce the capacity of the burner. Measure 1.4.3 recommends this approach.

Effect on Draft

Proper boiler draft is crucial to efficiency, for the reasons explained in Measure 1.5.1. If turbulators are installed, they will reduce the draft at the burner, upsetting the air-fuel ratio. Therefore, adjust the draft at the burner after installing turbulators.

The stack must be able to provide enough additional draft to overcome the added resistance of the turbulators. Verify this before ordering turbulators. If the stack cannot provide sufficient draft for the existing burner, you have various expensive ways to deal with the problem. In most such cases, the best choice is to install a burner that has a more powerful fan. If you take this approach, select the burner for high efficiency, as recommended in Measure 1.4.2.

Reduction of Standby Losses

The resistance to gas flow added by turbulators reduces standby loss, which results from convective flow

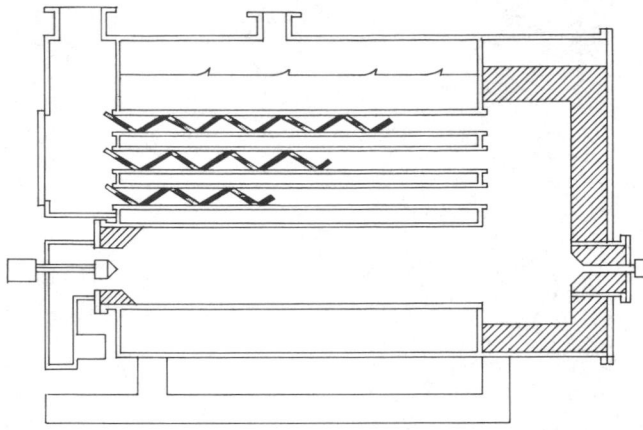

Fuel Efficiency, Inc.

Fig. 3 Turbulator installation pattern This is how several manufacturers suggest installing turbulators. The details vary with the boiler configuration.

of air through the boiler while the boiler is not firing. The reduction of standby loss depends on whether the boiler has a damper that closes when the boiler is not firing. If the boiler has such a damper, either in the burner or in the flue, turbulators provide little additional benefit. If the boiler does not have such a damper, turbulators are a poor substitute for installing one. See Measures 1.5.3 ff for methods of reducing standby loss.

Installation Procedures

Turbulators are ordered custom for each boiler. Typically, they are installed in the last pass only. It does not seem to matter which end of the tubes they are inserted into. A set of turbulator strips is supplied in various lengths. The longer strips are installed in the top row of tubes, and progressively shorter strips are installed in the lower rows, as shown in Figure 3.

The strips are held in place by spring action, and have no fasteners. The strips are slid into the tube until a tab at the outer end of the strip is stopped by the end of the tube. Installation typically takes one day. The work can be done by the facility staff, if they are sophisticated enough to perform routine maintenance on the boiler.

One manufacturer states that if the flue gas temperature is reduced excessively, risking acid condensation, the turbulators strips should be shortened to adjust the flue gas temperature.

Interference with Fireside Cleaning

Turbulators must be removed when firesides are cleaned. This should not be difficult or time consuming in most cases.

One manufacturer states that soot blowers can be operated with turbulators in place. However, if you have soot blowers, confirm compatibility with the manufacturer of the soot blower. (Soot blowers for firetube boilers are described in Measure 1.6.2.)

Manufacturers claim that turbulators reduce the need for cleaning because the increased turbulence inhibits the deposition of soot. View this claim skeptically. It is conceivable that turbulators could trap soot, at least in local areas of the tubes.

Possible Damage to Boiler

It does not appear that the performance of boiler turbulators has been studied extensively, so it is not possible to predict whether turbulators will cause boiler trouble. With respect to efficiency, a tendency of turbulators to trap fireside deposits would undercut any increase in heat transfer that they provide. With respect to maintenance, be wary of possible problems, such as tube erosion or stress cracking at the tube sheet. Discuss this with other operators of similar boilers who have used turbulators for a period of years.

ECONOMICS

SAVINGS POTENTIAL: *Who knows? Turbulator manufacturers claim efficiency improvements in excess of 10 percent, but such performance claims do not seem credible. Estimate your boiler's potential for improving efficiency. Then, require a guarantee from the vendor that installing turbulators will improve efficiency in your boiler by a stated amount.*

COST: *Typically $10 to $15 per tube. Firetube boilers have 25 to 150 tubes in the last pass, in which the turbulators are installed, for a total material cost of $250 to $2,500. Installation typically requires less than one day.*

PAYBACK PERIOD: *Maybe about one year, in favorable applications.*

TRAPS & TRICKS

CHOICE OF METHOD: *One big question is whether the equipment will produce the promised efficiency improvement. Another is whether turbulators will harm the boiler. Before buying, discuss performance with other purchasers who have operated with the same type of turbulator in your type of boiler. Ask vendors for references. Get a guarantee of efficiency improvement and a warranty against boiler damage.*

INSTALLATION: *The method of installing turbulators may be critical to maximizing efficiency, avoiding draft problems, and avoiding tube damage. Follow the manufacturer's instructions.*

MAINTENANCE: *Turbulators must be removed for boiler cleaning, and they must be reinstalled in the original pattern. This increases cleaning cost. Make sure that removal and reinstallation procedures are documented in a manner that is clear to the people who do boiler cleaning.*

Condensate, Feedwater, and Water Treatment

The condensate system is the return portion of a steam system. "Condensate" is the liquid residue of steam that is used by equipment or that condenses in steam pipes because of heat loss from the pipe.

The feedwater system is the input system to the boiler. "Feedwater" is purified and/or treated water that is pumped into the boiler.

The condensate system feeds into the feedwater system. In many cases, the two systems are merged, so that the same set of equipment performs the functions of both condensate and feedwater system.

"Makeup water" is treated water that is added to the feedwater or condensate system to compensate for losses from the system.

Functions of Condensate and Feedwater Systems

Condensate and feedwater systems perform some or all of these functions:

- filling the boiler system and replacing water that is lost
- collecting and returning condensate from steam-using equipment
- draining liquid water from steam lines and idle equipment
- injecting water into the boiler against the boiler's internal pressure
- purifying and/or treating the boiler water to prevent damage and inefficiency in the boiler and in steam-using equipment
- preheating feedwater, usually with recovered heat
- recovering heat from condensate, such as the heat of condensate flash steam
- creating a vacuum in condensate return systems to improve condensate flow and to improve the temperature control of steam heating equipment
- creating a vacuum for steam machinery condensers.

Each of these functions consumes energy or affects the efficiency of the boiler system. One of these functions, water treatment, also has a major effect on maintenance costs and equipment life.

INDEX OF MEASURES

1.8.1 Test and treat boiler water on a continuing basis.

    1.8.1.1 Hire a qualified consultant and contractor to perform water treatment.

    1.8.1.2 Install automatic water treatment equipment.

1.8.2 Control top and bottom blowdown to maintain required water quality and minimize waste of boiler water.

    1.8.2.1 Install automatic blowdown control.

1.8.3 Install blowdown heat recovery.

1.8.4 Maximize condensate return.

    1.8.4.1 Recover the heat from condensate that must be discarded.

    1.8.4.2 Recover the energy of high-temperature condensate that would be lost by flashing.

1.8.5 Keep vacuum condensate systems operating properly.

1.8.6 Replace pump motors with models having the highest economical efficiency.

RELATED MEASURES

- Measures 1.1.3 ff, minimizing the operation of installed auxiliary equipment
- Measures 1.7.1 ff, recovery of flue gas heat to preheat feedwater
- Measures 1.10.3 and 1.10.4 ff, selecting and maintaining steam traps to prevent condensate loss
- Measure 1.10.5, recovery of vent steam for return of heat and water

Condensate Systems

Condensate systems are simple. They consist of a condensate tank ("receiver") and a system of pipes that lead condensate from the steam-using equipment to the tank. In larger facilities, a number of satellite condensate receivers may be installed at distant sites, each of which pumps condensate to a primary condensate receiver located in the boiler plant. A typical condensate receiver is shown in Figure 1.

The condensate system begins at the steam traps that drain condensate from steam-using equipment and from steam lines. The steam traps keep steam from blasting through the equipment before the steam gives up all its latent heat. Steam traps operate by recognizing the difference between steam and water, and they pass only the water that has condensed inside the equipment. Steam traps are essential for efficient condensate return. See Measures 1.10.3 and 1.10.4 ff for details about them.

In many boiler systems, makeup water is introduced into the system by pouring it into the condensate receiver. Makeup water input is controlled by a float valve located in the receiver.

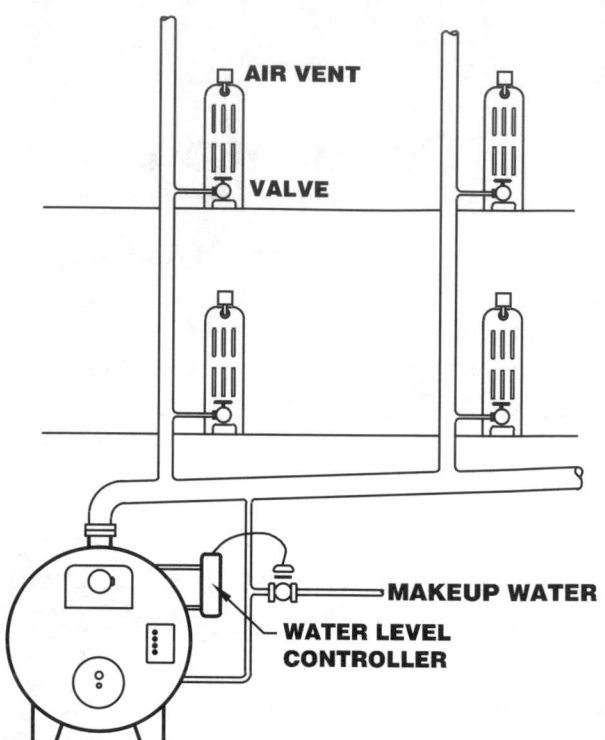

Fig. 2 One-pipe steam system This old type of steam system combined steam distribution and condensate return in single pipes. Its simplicity is offset by difficulty in controlling heating equipment at low output temperatures. It deserves more attention for high-temperature heating applications.

The details of a condensate system depend on its operating pressure. Some condensate systems operate at pressures higher than atmospheric, and others operate at pressures well below atmospheric. These types of systems are "closed," i.e., they are not open to the atmosphere. Many condensate systems operate at atmospheric pressure. These have vents to the atmosphere located at the condensate receiver or receivers. An important advantage of closed systems is that they keep air out of the water. Air is a primary cause of corrosion on all parts of the system. On the other hand, vented systems may be simpler, especially in facilities where condensate is returned from long distances.

■ High-Pressure Condensate Systems

High-pressure closed condensate systems can be used in cases where the lowest heating temperature that is required corresponds to a steam pressure that is higher than atmospheric. Typical examples are industrial process heating and drying applications. All parts of a steam system that has a high-pressure condensate system are under pressure that is higher than atmospheric. Therefore, such systems have the least amount of trouble from air entering the system.

In high-pressure condensate systems, the steam-using equipment is usually located near the boilers. Otherwise, heat loss from the condensate pipe would

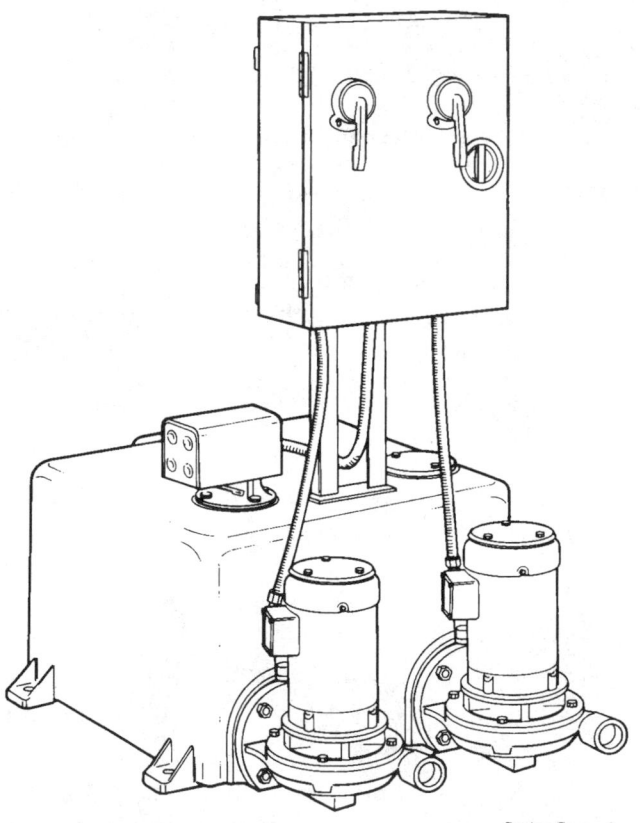

Spriax Sarco, Inc.

Fig. 1 Condensate receiver This is the basic element of a condensate system. It is just a tank with a pump that is activated by a float switch. (One pump is a spare.) The pump sends the condensate to the boiler, or to a feedwater tank, or to another receiver that is closer to the boilers. The tank is heavy because a failed steam trap may subject it to steam pressure. Also, it may be under vacuum when the condensate pipes cool down.

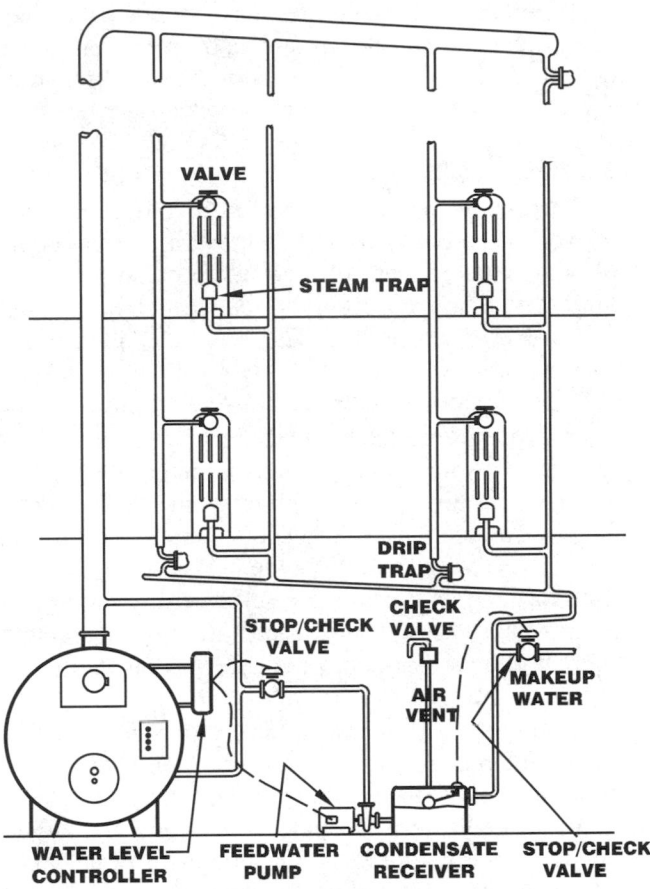

Fig. 3 Vented (atmospheric pressure) condensate system
This is the most common type of boiler water system for heating applications. It is simple and fairly reliable. Its main weakness is allowing air to contact the boiler water.

cause large variations in condensate system pressure, creating problems in pumping the condensate.

■ One-Pipe Systems

The "one-pipe" heating system is a combination of a steam distribution system and a condensate return system. Figure 2 shows a typical system. As the name implies, the system uses only a single pipe to each item of heating equipment for supplying steam and returning condensate. The condensate from the heating equipment drains back to the boiler by gravity through the steam supply pipe. The system is unique in that it does not require steam traps, and it usually requires no condensate receivers. This is an old system that was once common for space heating with cast iron radiators in multi-story buildings.

At first glance, the one-pipe system appears to be a closed system. Unfortunately, in typical space heating applications, it is not. In order for the radiators to control temperature properly, they require air vents. (Measure 5.2.7 explains why.) The air vents allow air to enter the system in large amounts.

■ Atmospheric-Pressure Condensate Systems

Most condensate systems that extend over a considerable distance are vented. The condensate tank

itself is vented, and the entire condensate system operates at atmospheric pressure. Figure 3 shows a typical system.

Heat loss from the pipe keeps most parts of the system at a temperature well below the boiling point of water. Therefore, the water vapor pressure inside the system is below atmospheric. The difference in pressure is made up by atmospheric air that enters the system through vents and leaks. Therefore, vented systems act to mix air into the condensate. The air must be removed before the water enters the boiler, if the boiler is to survive for long.

In most vented condensate systems, condensate drains from the steam-using equipment into a receiver by gravity. For this reason, the condensate receiver is usually located below the equipment it drains. If the condensate cannot flow by gravity all the way back to the boiler room, it is first collected in a local receiver. From this receiver, a float-actuated pump sends the water to a condensate receiver in the boiler plant, or to a separate feedwater system.

■ Vacuum Condensate Systems

Some closed condensate systems operate at pressures well below atmospheric. These are called "vacuum" condensate systems. A condensate system may be operated under a vacuum to aid steam flow, to improve control of low-pressure steam equipment, or to improve the efficiency of steam-powered machinery. Condensate systems for steam machinery may operate

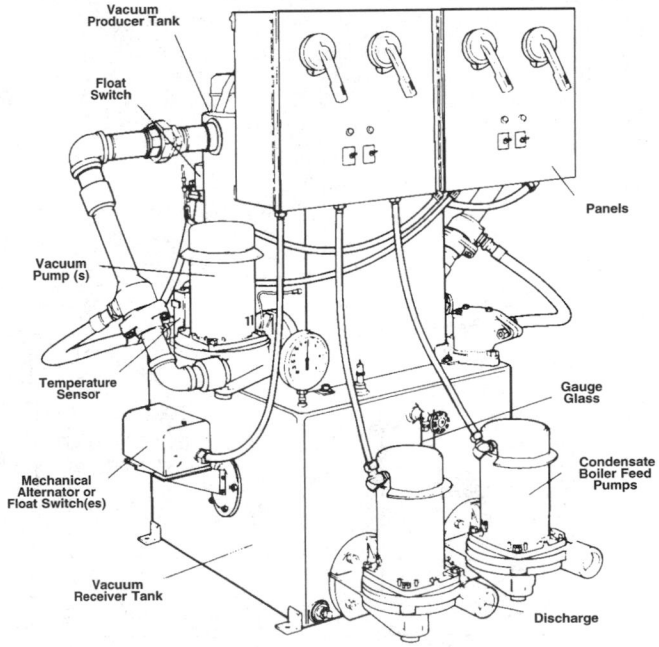

Spirax Sarco, Inc.

Fig. 4 Vacuum condensate receiver This is the heart of a vacuum condensate system. In addition to the tank and pump of a conventional condensate receiver, it has a vacuum pump that removes air and other gases. Therefore, the pressure in the system depends entirely on the condensate temperature.

WESINC

Fig. 5 Boiler water system for a large office building
The tank is the main condensate receiver. The three pumps to the right are individual feedwater pumps for three boilers. The pump in front injects water treatment chemicals into the feedwater.

at high vacuum, while systems for heating equipment typically operate at a moderate vacuum.

Vacuum condensate systems are closed. However, the fact that the system is below atmospheric pressure invites air leakage into the system. Air and other non-condensible gases are removed from these systems by various types of positive-displacement pumps or by venturi "air ejectors." As long as the system is operating properly, the amount of air that enters the boiler water is much less than with a vented system.

The main equipment difference between a vacuum condensate system and an atmospheric system is the condensate receiver, which includes the vacuum pumps. Figure 4 shows a typical vacuum condensate receiver.

Feedwater Systems

The feedwater system begins at the condensate receiver in the boiler room. In the simplest cases, the only major component of the feedwater "system" is a pump that takes water from the condensate receiver and pushes it into the boiler. The feedwater pump, or a feedwater inlet valve, is controlled by the water level in the boiler. Figure 5 shows a typical system of this type.

In more complex boiler plants, there may be a separate feedwater tank. This tank is almost always used for removing air and other gases from the feedwater. The feedwater tank is a closed unit that removes oxygen from the feedwater and heats the feedwater by blasting steam through it. Installing a separate deaerating tank requires the system to have a separate feedwater pump. Figure 6 shows a boiler water system that includes a deaerating tank.

In this system, the condensate pump or a separate pump can circulate water through the feedwater tank to deaerate it prior to starting the boiler system. The deaerating tank usually operates at a pressure higher than atmospheric, but some plants use feedwater tanks that operate at pressures below atmospheric.

The feedwater tank may also be used as a site for recovering waste heat to heat the feedwater. Waste heat may be extracted from a flue gas economizer, from

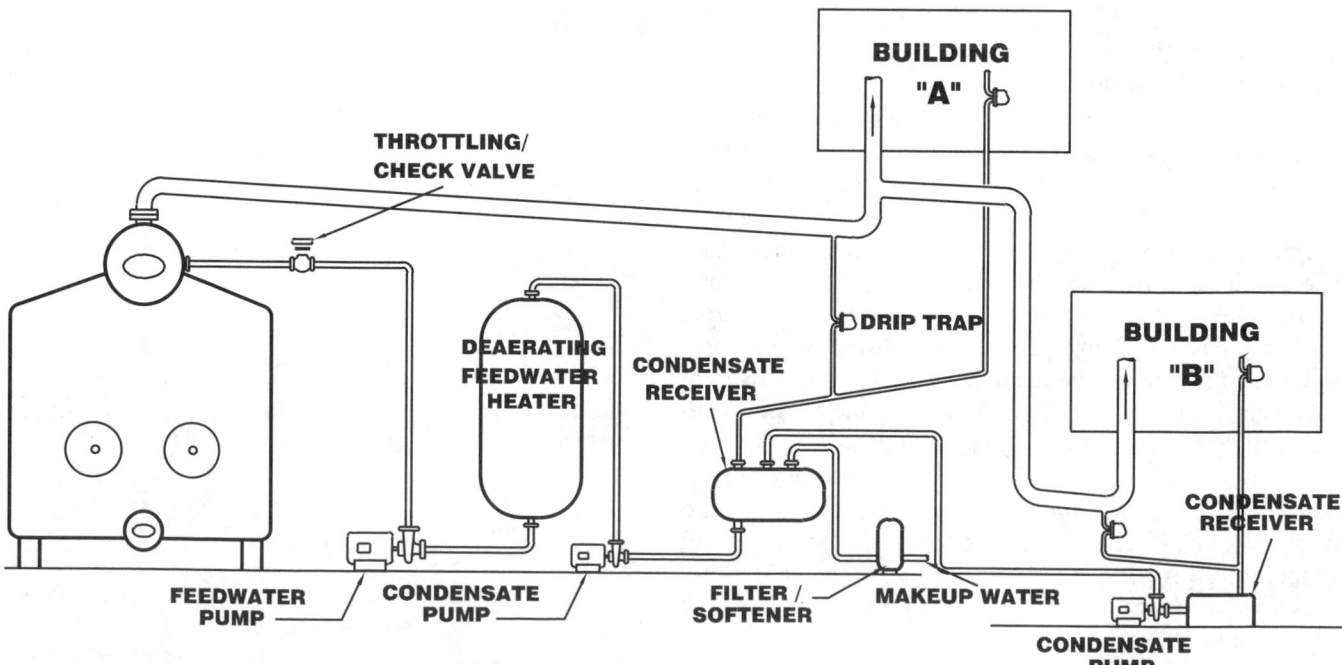

Fig. 6 Boiler water system with a deaerating feedwater heating tank This system is appropriate for higher boiler pressures and larger systems, primarily because it provides for better water treatment. Using a separate feedwater tank requires a separate feedwater pump.

blowdown, or from other sources. Makeup water may also be added at the feedwater tank, after initial treatment.

In large steam systems that drive high-pressure turbines, there may be several feedwater tanks that serve as feedwater heaters at progressively higher temperatures. This arrangement increases the efficiency of the steam turbine cycle.

Separate feedwater systems tend to be found in plants that operate at higher steam pressure. This is because the oxygen that is trapped in boiler water becomes much more damaging at higher pressures. Also, high-pressure steam systems tend to have more opportunity for heat recovery. Nonetheless, deaerating feedwater tanks are often found in steam systems that operate at relatively low pressures.

Water Treatment

Water treatment is a primary function of condensate and feedwater systems. There are many types of water treatment, just as there are many types of medical treatment, each of which is intended to solve specific conditions. Water treatment processes can be grouped into three broad categories: (1) treatment of makeup water, by "softening" and other techniques, (2) adding chemicals to the boiler water, and (3) removing oxygen and other gases by mechanical separation in a deaerating feedwater tank.

The particular combination of water treatment methods that is best for a boiler plant depends on the pressure and temperature of the system, the exposure of the system to the atmosphere, the contents of the makeup water, environmental regulations, and the preferences of the people doing the treatment.

In addition to water treatment, boiler systems use blowdown to control water contaminants. Blowdown is the controlled dumping of a certain fraction of the boiler water in order to reduce the concentration of contaminants.

Hot Water Boiler Systems

Hot water boilers do not require condensate or feedwater systems as such. In a hot water system, there is no change of state from water to steam and back. The water is moved by pumping, in a continuous loop. The boiler can be considered as a heater located on the loop.

Water treatment is important for hot water systems. However, for a given water temperature, water treatment in a hot water system is generally less elaborate and less demanding than in a steam system.

MEASURE **1.8.1** Test and treat boiler water on a continuing basis.

SUMMARY

Important maintenance needed by most boilers on a continuing basis. Maintains peak efficiency, minimizes the need for nasty waterside cleaning, and extends boiler life. Can do harm if done improperly. Requires specialized skill.

SELECTION SCORECARD

Savings Potential .... **$ $** $

Rate of Return .... **% % % %**

Reliability .... ✓ ✓

Ease of Initiation .... ☺ ☺

Water treatment is the most important day-to-day maintenance in a boiler plant. Inadequate or incorrect water treatment may reduce the efficiency of the boiler quickly. Sooner or later, it will ruin the boiler and the entire steam system. Figure 1 shows what neglect of water treatment can do to boiler tubes.

Water treatment is customized for each boiler plant. Treatment requirements may change as conditions change. Review your treatment program periodically to refine the procedures. There are many water contaminants, each of which must be considered individually. You may have a selection of water treatment methods to deal with each one. The chemical and physical actions of water contaminants and water treatment chemicals are extensive and complex. Incorrect water treatment can be harmful in two ways, by failing to prevent damage and by damaging the system itself.

This Measure gives you a good introduction to boiler water treatment. It starts by listing the reasons why water treatment is important. It describes the boiler plant characteristics that determine the need for water treatment. It tells you how neglecting water treatment can harm a boiler system, so you can keep a lookout for the signs of trouble. It summarizes the main water treatment methods. And, it outlines how to test your boiler water to adjust your treatment to conditions.

This introduction will not make you an expert. It is intended to give you enough information to deal intelligently with water treatment specialists, and to judge how well they are doing their jobs. In most cases, you will hire a water treatment consultant to lay out your water treatment program, and you will have the boiler plant staff or a contractor perform routine testing and treatment. Using water treatment consultants and contractors is tricky in itself. This aspect is covered by subsidiary Measure 1.8.1.1.

In all but the smallest plants, it helps to have automatic water treatment equipment. Subsidiary Measure 1.8.1.2 recommends this equipment.

Water Treatment Benefits and Costs

Effective water treatment provides several important benefits:

- *maintaining heat transfer efficiency.* Insufficient water treatment leads to scaling of boiler tubes, which reduces boiler efficiency significantly.
- *reducing the need for blowdown,* which wastes energy and water
- *reducing the need for boiler cleaning,* which is much more arduous than water treatment. In many boilers, water treatment can eliminate the need for periodic cleaning of watersides.

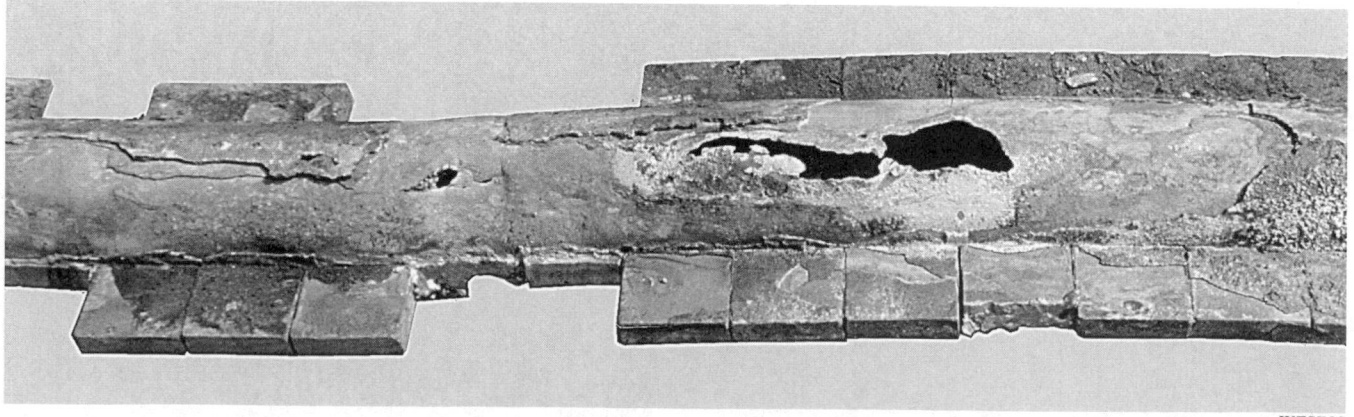

WESINC

Fig. 1 This is why boiler water treatment is important This tube from a watertube boiler has been destroyed by neglect on both the waterside and fireside. If boiler water is treated properly, the waterside will last almost indefinitely.

• *minimizing boiler damage,* especially failed tubes and leaky tube sheets.

• *minimizing distribution system and user equipment damage,* such as corrosion, valve leakage, and turbine blade fouling.

The cost of water treatment varies widely, but the cost is less than the cost of neglect. Know which water treatment costs to expect, and be prepared to justify them. These costs include consulting services, service contracts, chemicals, chemical feeders, test equipment, maintenance of water treatment equipment, staff training, and labor.

Factors that Determine the Type and Amount of Water Treatment

Tailor your water treatment program to the conditions that exist at your plant. Do all boilers need water treatment? The answer is that almost all boilers do.

About the only situation in which water treatment is not critical is operation at very low pressure or temperature, in a system with very low water loss, and with "soft" makeup water.

The main factors that determine the methods and intensity of water treatment are:

• *the rate of water replacement (makeup).* Water quality does not degrade significantly inside the boiler. Water treatment is needed almost entirely to remove the contaminants in makeup water, or to render them harmless. A totally closed steam or hot water system that has no leaks requires little water treatment, although the small amount that is required may be important.

(For this reason, eliminating water loss from the system is a principal way of reducing water treatment requirements. Measure 1.8.4 improves condensate return, and the Measures in Subsection 1.10 reduce water loss from the system.)

• *the type and amount of contaminants in the makeup water.* Water that originates underground (i.e., well water and spring water) typically is saturated with dissolved minerals. For example, in locations where water flows through limestone, it is "hard," i.e., loaded with calcium compounds that can leave hard deposits inside boilers. Also, the carbonate component of limestone forms carbonic acid, carbon dioxide, and related compounds that cause corrosion.

Surface water (i.e., river water that originates from surface runoff of rain) is saturated with oxygen because of its contact with the atmosphere. Surface water may also carry large amounts of suspended organic and inorganic matter.

• *whether the condensate system is open or closed.* An open (atmospheric) condensate system allows

air to be dissolved in the water with each pass through the system. An open system requires additional steps to remove the air and to neutralize the carbonic acid that is formed by dissolving the carbon dioxide in air. It may also require additional water treatment methods to protect the condensate system.

• *the boiler pressure and temperature.* Water treatment requirements increase radically with increasing pressure and temperature because chemical reaction rates rise exponentially with temperature and pressure. An adverse reaction that may be negligible at low temperature can become seriously destructive at high temperature. Also, the solubility of contaminants may be affected strongly by temperature. Some compounds that cause scaling are less soluble at higher temperature, so that higher pressure boilers have a greater tendency to form scale and sludge.

For example, domestic water heaters, which operate at low temperature and pressure, survive in many areas without water treatment, even though contaminants are being introduced at a high rate. (Water heaters do form considerable scale, which should be flushed out periodically.) In contrast, high-pressure shipboard boilers require exacting water treatment, even though they distill their makeup water.

• *whether the boiler produces steam or hot water.* Other factors being equal, steam boilers require considerably more water treatment than hot water boilers. There are several reasons for this:

- leakage from a steam system is primarily from the steam-carrying parts of the system, leaving contaminants behind in the water. In contrast, leakage from a hot water system carries a proportionate amount of dissolved matter along with it.

- evaporation of water in the boiler leaves the contaminants behind to become increasingly concentrated within the boiler. In a hot water system, contaminants are distributed throughout the system.

- formation of steam bubbles at the heat transfer surfaces leaves dissolved solids behind on these surfaces. This is discussed further, below.

• *the purposes for which the steam or hot water is used.* If steam or hot water is to be consumed, select water treatment methods that are compatible with the application. For example, if the steam is used for humidification of ventilation air, do not use toxic water treatment chemicals, such as amines, that can get out of the system.

• *environmental regulations.* The blowdown from boilers contains the water treatment chemicals and their by-products. Environmental regulations limit the dumping of certain types and concentrations of these products into sewage systems.

Once you define the water treatment needs of the boiler plant, decide how to satisfy these requirements. You often have options. The choices depend on relative cost and on the types of water treatment equipment already installed. For example, you can minimize oxygen in feedwater by chemical treatment or by installing a deaerating feedwater tank.

Adverse Effects of Inadequate Water Treatment

The problems caused by inadequate water treatment can be grouped into the following classes. These problems may be caused by neglect of water treatment, by inappropriate water treatment, or by failing to remove the products of water treatment.

■ Scale Formation

Scale is a general name for hard deposits that form on heat transfer surfaces. Scale typically has a stony appearance. The type of scale that may form depends on the local water conditions. Scale forms from whatever mineral and organic matter is in the water if the concentration of the material becomes too high.

The most rapidly accumulating types of scale are compounds of calcium and magnesium. Boilers that operate at high temperatures may also form a scale from compounds of silica (the ingredient of common sand), which is hard, tenacious, and has a high insulating value. Oil, grease, and dirt that are not removed from new or newly repaired boilers also contributes to scaling.

Scale has these adverse effects:

• *reduced heat transfer.* If scale accumulates to any visible extent, it inhibits the flow of heat from the combustion gases enough to cause a significant loss of efficiency. Common scale compounds have thermal conductivities that are only a few percent of the conductivity of boiler steel.

• *tube destruction by heat.* The combustion temperature inside a boiler is high enough to destroy the steel of boiler materials. Damage does not occur in normal operation because the boiler water cools the steel. Scale inhibits this cooling. In extreme cases, scale may even obstruct water flow. The reduced cooling allows the flame to heat the metal to failure. Failure is most likely to occur while the boiler is under heavy load, which maximizes the damage and the hazard that results.

• *increased need for waterside cleaning.* Scale is difficult and dirty to remove, and the removal process has the potential of damaging the boiler. (See Measure 1.6.4 for more about this.)

One way that scale forms is that the solubility of certain compounds diminishes when water temperature rises. This is the case with calcium compounds, which deposit calcium carbonate (the ingredient of limestone) and calcium sulfate (the ingredient of gypsum). If the water is saturated with a substance having this solubility characteristic, the substance deposits on the hottest surfaces, which are the tubes. This process can be prevented by keeping the concentration of the scale-forming contaminants below their saturation level.

Another process of scale formation applies mostly to steam boilers. When a steam bubble forms, the solid matter that was dissolved in the water may deposit on the tube surface. If the concentration of the material in the boiler water is below saturation, the material deposited on the boiler surface may be re-dissolved by the boiler water, but if the concentration is high, the rate of deposition exceeds the rate of dissolving.

Some materials undergo a chemical change in the boiler that results in deposition of scale. An example is the conversion of magnesium sulfate to magnesium hydroxide, a hard scaling material. This reaction also produces sulfuric acid.

■ Obstruction by Sludge

Sludge is a muddy accumulation of precipitated material and dirt that tends to collect in low points in the boiler and in areas where the water is stagnant. Sludge is a normal product of water treatment. It is removed by bottom blowdown. It becomes a problem if blowdown is not controlled properly.

Its worst effect is obstructing water flow. For example, if sludge piles up in the bottom header of a watertube boiler, it blocks convective flow of water into the bottoms of the tubes in that area, leading to failure of the tubes. Sludge may harden and plug blowdown lines if it is left in a boiler that is allowed to cool.

■ Oxygen Corrosion

Oxygen corrosion rusts away the boiler steel. Oxygen corrosion in higher pressure boilers takes the form of deep, scattered pitting that eventually leads to pinhole leaks.

There are two kinds of iron oxide, one beneficial and the other associated with corrosion. The beneficial iron oxide is black and magnetic. It forms a thin protective coating on steel, and then stops forming. One of the objectives of water treatment is to preserve this coating. The harmful iron oxide is ordinary rust, which is brown. It continues to eat away the steel as long as oxygen is present.

Water absorbs oxygen anywhere that it is exposed to air. Air dissolves in condensate that is returned through an atmospheric condensate system. Even in a closed system, oxygen can be introduced by leaks in portions of the system that are below atmospheric pressure, such as the suction of a feedwater pump. When a boiler system is shut down, the condensate and feedwater systems are flooded with air, which dissolves into the water that enters the system when it is restarted.

■ Acid and Alkali Damage to Steel

The most common acids occurring in boiler water are sulfuric, hydrochloric, and carbonic. All are formed by chemical reactions that start with minerals that enter with the makeup water, or with chemicals that are added for water treatment. In addition, carbonic acid is derived from atmospheric carbon dioxide and organic matter in makeup water. Excessive alkali concentration usually results from improper water treatment.

Some acids and alkalis attack the grain structure of steel, weakening it in a manner that is more subtle than the highly visible rusting caused by oxygen corrosion. One effect of this microscopic damage is localized cracking at points of stress, for example, where tubes are held in tube sheets.

In steam systems, lack of proper water treatment may corrode the condensate system as well as the steam system. This is mainly the result of carbon dioxide being formed in the boiler water from bicarbonate alkalinity. The carbon dioxide travels with the steam and dissolves in the condensate, where it forms carbonic acid.

■ Carryover

Nothing but pure steam is supposed to leave the boiler. However, certain conditions can cause vigorous eruption at the water surface that carries contaminants out of the boiler. This eruption is called "foaming" or "priming," depending on its nature and causes.

Carryover is caused by materials that float on or near the water surface, such as oil, suspended solids, organic matter, and some carbonates. Carryover also results from a high level of dissolved solids. The adverse effect is fouling of steam pipe, steam valves, and steam using equipment. Silicates carried in steam from high-pressure boilers may travel for long distances, depositing on components such as valves and turbine blades.

■ Destruction of Sealing Materials

Certain types of boilers, mostly cast iron sectional boilers, require gaskets or other sealing materials to prevent leakage. Some boiler treatment chemicals attack the seals, requiring major disassembly for repairs. Figure 2 shows an example of this.

General Methods of Water Treatment

In principle, you could prevent water problems in boilers by using only perfectly pure water as makeup. However, this method is too expensive for general use. (It is used on oceangoing ships, which must distill their makeup water from seawater.) The common approach is to treat makeup water by a variety of methods to remove as much harmful matter as possible, and to render the remaining contaminants harmless.

The following are the main classes of water treatment methods. Have a water treatment expert select an appropriate combination of them for your boilers. Water treatment procedures are critical. For example,

chemicals should be injected at specified points in the system, and feed rates should be precisely controlled. The boiler is literally a pressure cooker that promotes chemical reaction among all the compounds in the boiler water and with the metal of the boiler. *It is dangerous to add water treatment chemicals without understanding all the chemical consequences.*

■ Cleaning the System Before Filling

Thoroughly clean out the boilers and the rest of the system before putting the system into service for the first time, and before returning the system to service after maintenance or modifications. Preservatives used in new or mothballed boilers cause scaling and foaming. Dirt that is left in a boiler after maintenance, such as mechanical cleaning of watersides, can cause scaling and priming.

■ Removing Suspended Matter from Makeup Water

Use filtration to remove suspended organic and inorganic matter in the makeup water if the water is not provided by a clean source. Remove smaller particles by adding coagulants, such as alum.

■ Softening Makeup Water

Calcium and magnesium compounds in makeup water cause "hardness," i.e., a strong tendency to form scale. Getting rid of these compounds is called "softening."

Fig. 2 Leakage at the seals of a cast iron sectional boiler
This problem can be caused by using the wrong water treatment methods.

A common method of softening is chemical treatment that converts the original compounds into other compounds that remain dissolved instead of forming scale. For example, the "hard" makeup water can be circulated through a bed of ordinary salt (sodium chloride) to convert the calcium and magnesium compounds into sodium compounds, which are much more soluble. This general method of swapping atoms between similar molecules is called "ion exchange." A number of ion exchange processes are available commercially, most of which are based on a combination of chemicals called "zeolite." Figure 3 shows an ion exchange water softener.

Ion exchange processes may make the water corrosive. Correcting this requires additional chemical treatment.

Precipitation is another method of softening water. A combination of chemical processes is used to convert the calcium and magnesium compounds into solids that can be filtered out. The most common of these methods is the "soda-lime" process.

■ Chemical Removal of Oxygen

Oxygen can be removed from boiler water by adding a chemical to the water that combines with any oxygen to form a harmless compound. The most common chemical used for this purpose is sodium sulfite. Boilers that operate at high pressure may use hydrazine, because sodium sulfite has a number of disadvantages that become serious at high pressures.

■ Mechanical Deaeration of Feedwater

Oxygen and other gases that are dissolved in the feedwater can be removed by heating the feedwater and spraying it inside a tank. The solubility of the gases drops at high water temperature, and the agitation and large surface area of the spray causes the gases to separate out. The gases are vented to the atmosphere.

Most deaerating tanks "scrub" the entering feedwater spray with a blast of steam. Figure 4 shows a deaerating feedwater tank that operates on this principle. The steam heats the feedwater as well as scrubbing it. Little energy is lost in the process, provided that the steam is not allowed to vent to the atmosphere. (See Measure 1.10.5 about correcting this problem, if it occurs.)

Deaerating tanks are expensive, but they reduce the need for chemicals to remove oxygen. Consider installing a deaerating tank for boilers that produce more than about 10 million BTU's per hour.

■ Chemical Prevention of Scaling in Boiler Water

"Softening" makeup water may not be sufficient to prevent scaling in boilers that operate at higher pressures. You can add phosphates to the boiler water to combine with calcium and magnesium, the elements that form

Cleaver-Brooks

Fig. 3 Water softening system   This is used to treat makeup water before it enters the boiler system. It replaces minerals that cause scaling with other materials that remain dissolved in the boiler water.

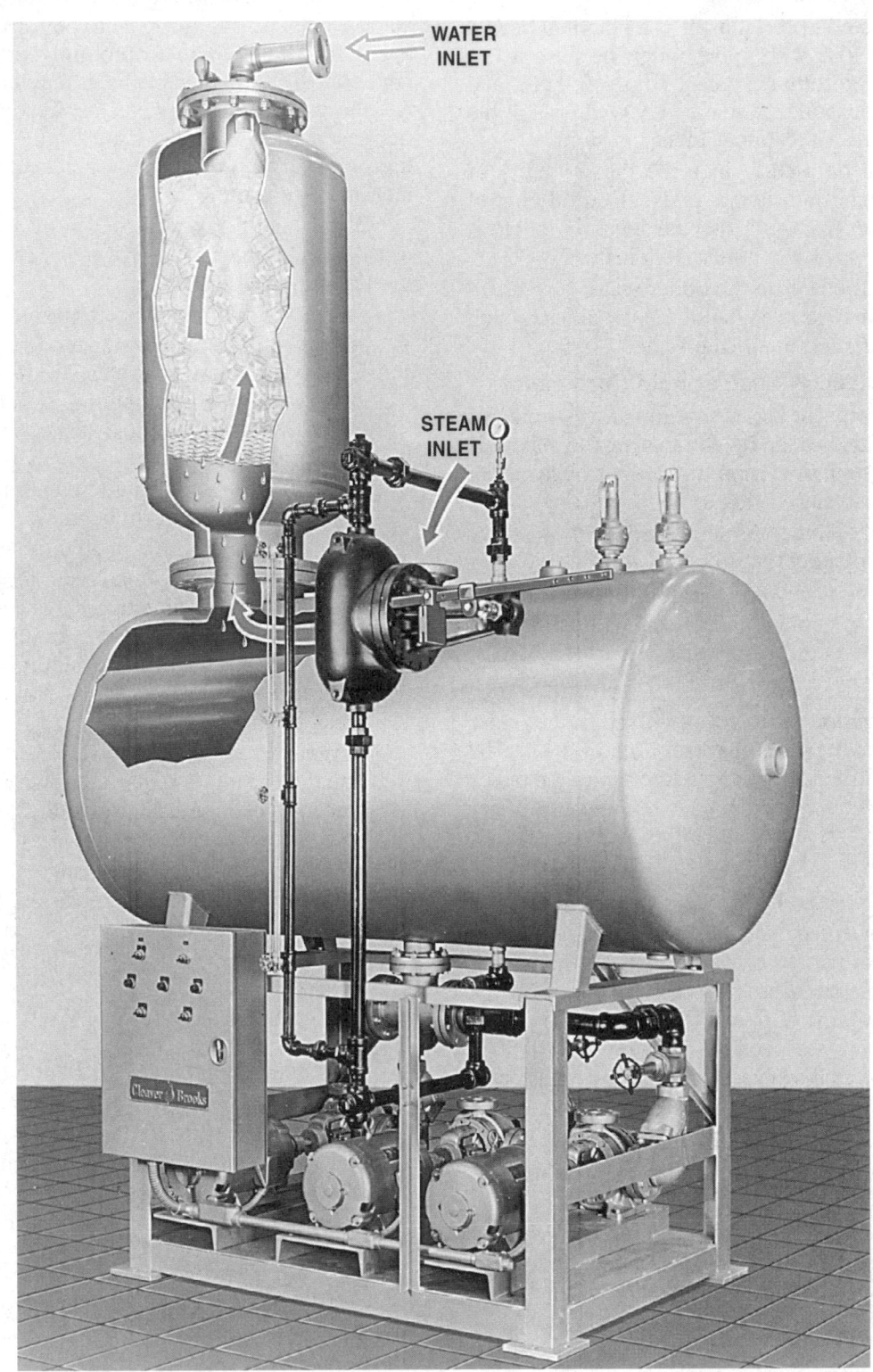

WATER
INLET

STEAM
INLET

Cleaver-Brooks

Fig. 4 Deaerating feedwater heating tank The incoming feedwater is sprayed downward into the upper tank, separating air and other gases that are dissolved in the water. The gases are further removed from the water droplets by a blast of steam from below. The gases are vented to the space. The lower tank is only for storage.

scale. The resulting compounds precipitate, and are removed by blowdown.

The precipitated material itself is a potential hazard as long as it remains within the boiler, because it can clump together and form deposits. To prevent this, add dispersants to the boiler water to keep the insoluble products of chemical treatment in suspension.

Chelants can be used to increase the solubility of calcium and magnesium compounds. Careful control of chelants is necessary. If the hardness is too low, chelants tend to attack the boiler steel instead.

Reducing carbonates in the boiler water (expressed as "alkalinity") increases the solubility of silica, which prevents the formation of silicate scale.

■ Chemical Protection Against Acid Corrosion

One way to prevent acid corrosion is to keep acids from forming. This is done by eliminating the materials that form acids, such as carbon dioxide, or by avoiding the conditions that cause acids to form.

Another method is to neutralize acids with an alkali.

Still another method is to form a surface on the steel that resists corrosion. For example, chromates and nitrites added to the boiler water react with the iron in the boiler to form a corrosion resistant surface.

■ Blowdown

The concentration of matter dissolved in the boiler water increases as steam is lost from the system. The solution is to continuously dump a fraction of the boiler water so that it can be replaced with makeup water having a lower concentration of dissolved matter. Blowdown is covered by Measures 1.8.2 ff, below.

■ Eliminating Carryover

Baffles are installed inside the steam outlets of most boilers to prevent carryover of water with the steam. Additional protection is provided by putting additives into the boiler water to suppress carryover. This saves energy by reducing the need for blowdown to reduce the concentration of dissolved solids, which is a factor in carryover. Corn starch is one material used for this purpose, among others.

■ Chemical Protection Against Corrosion of the Condensate System

Condensate systems are vulnerable to corrosion by carbonic acid. Also, air enters the condensate system when it is shut down, admitting oxygen. Dealkalizing, a process similar to water softening, keeps carbonic acid from forming. Neutralizing agents, such as amines, can be injected into the steam line to reduce the severity of both problems. When the steam condenses, it carries the agents into the condensate system.

Boiler Water Testing

As we said before, water treatment must be tailored closely to the condition of the water. To determine the condition of the water, you must perform testing. In addition, you must test the water to check the effect of your water treatment. This is a continuing process. In some plants, it is appropriate to perform chemical tests every few hours, and to monitor certain conditions continuously. In other plants, it may be satisfactory to test the water every few weeks. Testing requirements may vary. For example, if the boiler is steaming at a high rate, it requires more water treatment chemicals than if it is idling.

Water condition is indicated by a number of standard tests. Each test measures a particular characteristic. The most common of these are:

- *hardness,* a measure of the concentration of minerals that have a strong tendency to form scale. The predominant "hardness" minerals are calcium and magnesium compounds. You can measure hardness with chemical tests for these specific elements. A simple, more general test is to measure the amount of soap solution that must be added to a water sample to form bubbles.

- *pH,* the basic measure of acidity. Pure water has a pH of 7.0. pH that is too low or too high can cause damage to metal directly. pH is also a controlling factor in many chemical reactions that occur in boiler water. You can measure pH with a simple chemical test. Or, you can install a pH meter that reads continuously.

- *oxygen,* which causes corrosion in any concentration. Keep the oxygen concentration as close to zero as is economically possible. Measure oxygen content with a chemical test.

- *alkalinity,* primarily a measure of the concentration of naturally occurring bicarbonates, which are associated with several adverse boiler conditions. Alkalinity also indicates the presence of alkaline water treatment chemicals. Despite its name, the alkalinity test is not directly related to pH. Two chemical tests for different sources of alkalinity are phenolphthalein (called "P alkalinity") and methyl orange (called "M alkalinity").

- *silica,* which forms a scale in high-pressure boilers that is particularly difficult to remove. Measure silica with a chemical test.

- *iron,* which combines with phosphates and hydroxides to form scale. Measure iron with a chemical test.

- *total dissolved solids (TDS) in boiler water,* is mostly an indication of the concentration of the by-products of chemical treatment, provided that boiler water is properly treated. The key word is "dissolved." TDS remain safely dissolved in the water and do not deposit on the steel surfaces as long as they are kept within the limits that are appropriate for the boiler conditions.

Blowdown is the primary method of keeping TDS within safe limits. Blowdown wastes energy and requires makeup water treatment, so the objective is not to minimize total dissolved solids, but to let them rise to a safe maximum.

The easiest way to measure TDS is by measuring the electrical conductivity of the water with a permanently installed meter that is calibrated in terms of TDS. You can also measure TDS chemically.

• **total dissolved solids (TDS) in condensate,** which is independent of the boiler water TDS. Condensate TDS typically reflects chemical treatment to protect the condensate system, such as with amines.

Not all of these characteristics are important for all boilers. Your water treatment consultant should specify the appropriate tests, how often to perform them, and the acceptable range of results.

You should also measure the concentration of treatment chemicals in the boiler water. For example, if you use phosphates and sulfite for water treatment, measure the levels of these chemicals to determine the quantity to be added.

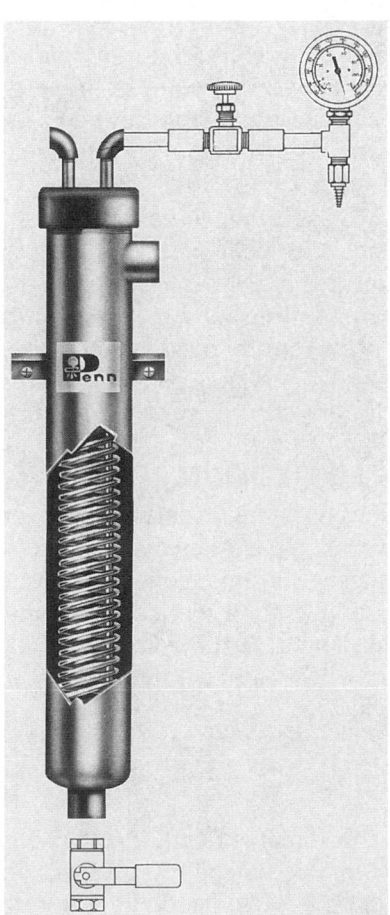

Pennsylvania Separator Co.

Fig. 5 Sample cooler This is valuable when taking water samples from a boiler that operates above boiling temperature. Without it, tests will show concentrations higher than they actually are. It also reduces the risk of scalding when taking samples.

With high-pressure boilers, install a sample cooler to take samples of boiler water for testing. Otherwise, a large part of the water sample will flash into steam, and the material dissolved in the sample will register higher concentrations than actually exist inside the boiler. Sample coolers designed to be cooled with domestic water are available commercially. Figure 5 shows a typical unit.

The visible condition of the boiler during maintenance is also a form of water testing. In particular, scaling is very visible, and it will be cursed by those who must remove it. Other defects, such as acid embrittlement of steel, may not be visible.

Coordinate with Other Water Treatment

Cooling towers need a water treatment program, as recommended by Measures 2.3.3 ff. Hydronic systems may need some water treatment, although they are much less demanding. In many locales, it is desirable to treat the domestic water. It would save effort, and perhaps increase reliability of service, if you find a water treatment specialist who can deal with all your requirements. However, do not compromise to achieve this. Each application has different water treatment requirements, and one specialist may not be competent to handle them all.

ECONOMICS

SAVINGS POTENTIAL: *Typically several percent of boiler fuel cost; plus, elimination of most waterside cleaning cost; plus, elimination of water-related boiler damage.*

COST: *Varies widely, depending on water condition, boiler pressure, and other factors.*

PAYBACK PERIOD: *Immediate. Appropriate water treatment always costs less than neglecting it.*

TRAPS & TRICKS

SETTING UP THE PROGRAM: *Water treatment is complex. Different treatment methods are required for different water conditions, different types of boilers, and different operating conditions. Have a qualified specialist set up the water treatment program. See Measure 1.8.1.1, next.*

SKILLS: *Once the water treatment program has been set up, routine water testing and treatment can be done by the facility staff. This requires training and diligence. If it is not practical to do water testing and treatment in-house, hire these services.*

MONITOR CONDITIONS: *Regular testing does no good unless it results in appropriate action. Record water test results and review them on a continuous basis. At every opportunity, inspect the watersides to judge the effectiveness of treatment and to look for adverse effects. Make corrections to the treatment program as indicated.*

MEASURE 1.8.1.1 Hire a qualified consultant and contractor to perform water treatment.

SUMMARY

A competent consultant and a competent contractor are both vital to the water treatment program. Distinguish their roles.

SELECTION SCORECARD

Savings Potential .... $ $ $

Rate of Return .... % % % %

Reliability .... ✓ ✓ ✓

Ease of Initiation .... ☺ ☺

Measure 1.8.1 makes the point that water treatment is a complex subject. To do it correctly and economically, you need to bring together the right combination of people. The participants are the in-house staff, an expert water treatment consultant, and a reliable water treatment contractor.

The ultimate check on the effectiveness of water treatment is the condition of the boiler. Examine watersides at every opportunity. This reveals how well the staff, the consultant, and the contractor are doing their jobs.

Functions of the Consultant and Contractor

The functions of the water treatment *consultant* are to:
- test the facility's water source
- specify the appropriate treatment methods, taking into account all the factors described in Measure 1.8.1
- specify the appropriate boiler water tests to be done by the facility staff (or a contractor), how often to perform them, and the acceptable ranges of results
- periodically check the performance of the in-house staff and/or the contractor in maintaining proper water conditions.

The functions of a water treatment *contractor* are to:
- provide the appropriate chemicals on a continuing basis
- provide the appropriate test equipment
- install automatic chemical feeders, if appropriate (see Measure 1.8.1.2, next)
- in a plant that does not have specialized boiler operators, accomplish testing and perhaps other routine functions.

Failing to distinguish between the roles of the consultant and the contractor can be expensive. Water treatment companies make money primarily by selling chemicals, and they have no interest in keeping your costs down. They may treat water excessively or with unnecessarily expensive methods. A particular contractor may not offer the most appropriate methods, or may not be familiar with them. For these reasons, you need a consultant who is a genuine expert in water treatment and who has no financial interest in sales of chemicals or equipment. Be prepared to pay the consultant a fee for a service that the contractor claims to offer "free."

How to Select the Consultant

Selecting a water treatment consultant is like selecting a surgeon. Since you are not an expert in the subject matter yourself, you have to judge by the results that the expert has produced elsewhere. Call other facilities that have equipment like yours, and ask for recommendations. Your consultant may be an independent expert, a professor, or a large company specializing in water consulting. If your boiler plant is at least moderately large, you can afford to bring in a water treatment consultant from anywhere in the country. In fact, don't be surprised if you have to bring in someone from out of town.

Again, as in selecting a surgeon, get an expert who has experience with your specific set of problems. An expert in municipal water systems may not be appropriate for setting up a water treatment program for a boiler plant. Furthermore, an expert in large, high-pressure utility boilers may prescribe a program that is too elaborate for your low-pressure heating boilers.

How to Select the Contractor

You want a water treatment contractor from the local area. Otherwise, your plant will not get the regular attention it needs. Unfortunately, not all water treatment companies provide good service. More than one plant has suffered damage at the hands of water treatment companies, even major ones. Water treatment companies have as much trouble training and keeping qualified staff as other organizations. This is another reason why you need a separate water treatment consultant.

If your consultant is familiar with the local area, he may be able to recommend a contractor. In any event, contact the operators of other boiler plants in the vicinity who have boilers like yours. Be aware that the opinions of other operators may not be reliable. They are not water treatment experts, either. Other boiler operators may not be competent to judge the results that they are getting, and they may be more influenced by the contractor's charm than by performance. In the end, you have to know enough about water treatment to ask the right questions.

Relationship to the Boiler Plant Staff

How much should the boiler plant staff be involved with water treatment? This depends on the complexity of treatment requirements and the sophistication of the staff. If there is at least one trained boiler operator, this person can do at least the routine chemical treatment and water testing. However, many commercial and institutional facilities do not have staff capabilities to handle water treatment reliably. In such cases, contract the water treatment functions, and ask your water treatment consultant to make periodic checks of the contractor's performance.

On the other hand, a sophisticated staff may be able to purchase its own chemicals and test equipment, eliminating the need for a water treatment contractor, provided that the staff is properly advised by the consultant.

ECONOMICS

See Measure 1.8.1.

TRAPS & TRICKS

SELECTING SERVICES: *Do not select your specialists on faith. Ask candidates for references, and check them. Then, judge results by keeping a close watch on the condition of your boilers.*

MEASURE 1.8.1.2 Install automatic water treatment equipment.

Automatic feeders for water treatment chemicals are an effective means of ensuring that chemicals are injected properly. Chemical feeders are fairly simple, consisting of a chemical storage bin, an injection pump, and controls. Enough bulk chemical is stored in the bin to avoid the need for frequent handling of chemicals. Figure 1 shows a typical chemical feeder.

An automatic feeder can inject chemicals frequently, which would be a bother for human operators. How desirable is this? It depends on the rate of water loss from the system. If the system loses only a small fraction of its water volume per day, individual small doses of treatment chemicals suffice. In such cases, the only advantage of a chemical feeder is that it automatically mixes and injects the dose.

On the other hand, a system having a high rate of water loss benefits significantly from automatic injection. If chemicals are added manually when the loss rate is high, the concentration of chemicals in the boiler water rises at the time of treatment and tapers off

Fig. 1 Automatic boiler water chemical feeder   This is simply a plastic storage bin with an injection pump installed underneath. This equipment tends to clog, so don't neglect it.

SUMMARY

Provides savings in chemicals and labor. The equipment requires regular maintenance. Does not reduce the need for monitoring water quality.

SELECTION SCORECARD

Savings Potential .... **$**

Rate of Return, New Facilities **% %** %

Rate of Return, Retrofit.......... **% %** %

Reliability .... ✓ ✓

☺ ☺ ☺

until the next treatment. This procedure requires more chemicals than if the concentration is kept steady.

Methods of Control

Chemical feeders can be controlled to inject an individual dose, to inject fixed doses at periodic intervals based on time, or to inject different quantities of treatment chemicals based on water conditions.

Timed feeding alone cannot respond to changes in boiler load. For example, in comfort heating applications, changes in load throughout the day and from one day to the next affect the rate of water loss. Feeding at a steady rate under these circumstances could lead to chemical levels in the boiler water that are too high or too low.

The best control for most purposes is based on water conditions. A sensor is installed on the boiler to control each chemical being fed. The sensor controls the operation of the pump on the feeder.

There are several hazards to automatic control. One is that a failure of the control sensor could cause too little or too much chemical to be injected, which could have serious consequences. Another hazard is that automatic operation causes human operators to neglect monitoring the boiler water conditions.

Even with automatic control, monitor the condition of the boiler water with a separate alarm having its own sensor. Continue manual water testing even with automatic chemical feeding.

Maintenance

While chemical feeders are useful in reducing routine labor, they require maintenance of their own. Many automatic chemical feeders are abandoned because of inadequate maintenance. Chemical feeders are fairly simple, but maintaining them is another bit of specialized knowledge that the staff must learn.

If the facility does not have the ability to maintain its own water treatment program, and a contractor is hired for this function, the contractor should maintain the chemical feeder. This arrangement is most likely to be successful if the contractor provides the feeder and its associated controls and alarms.

ECONOMICS

SAVINGS POTENTIAL: *Included in Measure 1.8.1. This is primarily a means of reducing the costs of chemicals and labor.*

COST: *Equipment and installation typically cost several thousand dollars. Water treatment contractors may* install the equipment "free" as part of a service contract, or the contractor may lease the equipment to the facility.

PAYBACK PERIOD: *Immediate, if the feeders are rented or provided by the water treatment contractor. Several years or longer, if the feeders are purchased. Purchasing the feeders provides a larger net saving in the long term.*

TRAPS & TRICKS

SELECTING THE EQUIPMENT: *Ask other boiler operators with similar water treatment requirements about their experience with chemical feeders. Shop thoroughly.*

SKILLS AND TRAINING: *Even though the equipment is fairly simple, it will be neglected unless operators are trained to maintain it. Automatic water treatment requires as much supervision and follow-up as manual methods.*

MEASURE **1.8.2 Control top and bottom blowdown to maintain required water quality and minimize waste of boiler water.**

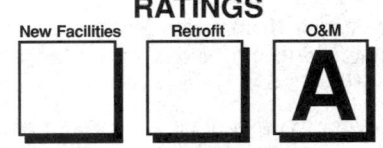

RATINGS

New Facilities | Retrofit | O&M

A

Blowdown is the deliberate dumping of boiler water to flush out accumulated contaminants. Most contaminants are brought into the system by makeup water. Chemical treatment converts contaminants that are harmful in small quantities into substances that can be tolerated in much larger quantities. Eventually, even these substances accumulate to the point that they must be removed from the system to avoid trouble.

For example, in treating "hardness" (mainly calcium and magnesium compounds), two pounds of phosphate water treatment chemicals combine with one pound of "hardness" compounds to produce three pounds of a material that sinks to the bottom of the boiler and becomes sludge. The sludge must be blown out before it hardens in place or accumulates enough to restrict water flow.

Blowdown, therefore, is an integral part of water treatment. The necessary percentage of blowdown depends on the contaminant concentration in the makeup water and on the allowable contaminant concentration in the boiler water. In boiler plants that have pure makeup water, the blowdown percentage may be very low. In applications where makeup water is heavily loaded with contaminants, the percentage of blowdown may be large.

SUMMARY

Too much blowdown wastes fuel in the short term. Too little reduces combustion efficiency in the long term and causes boiler damage.

SELECTION SCORECARD

Savings Potential .... **$** $

Rate of Return .... **% % % %**

Reliability .... ✓ ✓

Ease of Initiation .... ☺ ☺

Insufficient blowdown results in fouling of the heat transfer surfaces, which can seriously degrade efficiency. It also leads to serious damage to the boiler, ranging from various types of corrosion to obstruction of water flow that can cause tube failure.

Excessive blowdown wastes energy, increases water treatment cost, and wastes raw water.

Excessive blowdown occurs in many boiler plants because boiler operators try to stay "on the safe side." A contributing factor is the crude method of controlling blowdown on many boilers, which is to partially open a manual blowdown valve and allow an unknown quantity

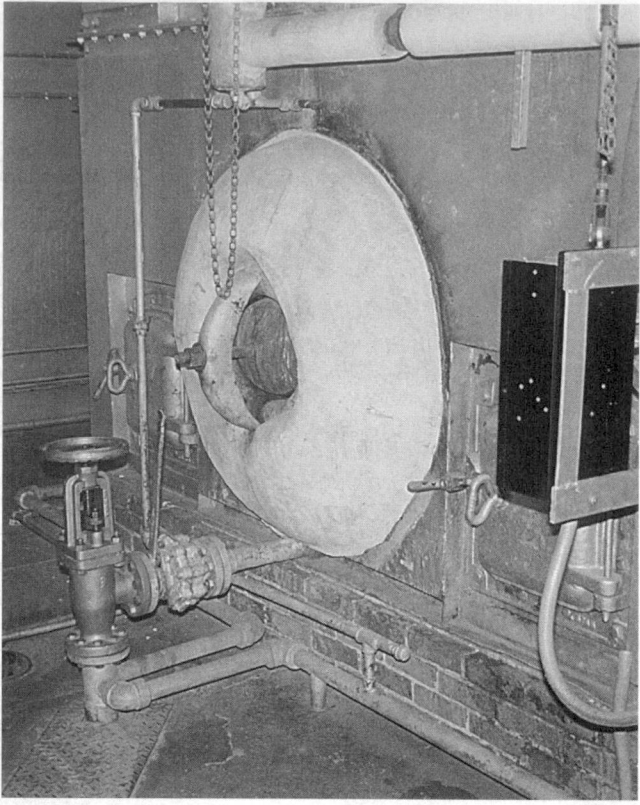

Fig. 1 Bottom blowdown connection The valve is connected to the bottom of the lower (mud) drum of a large watertube boiler. It discharges to the sewer.

Fig. 2 Top blowdown connection The insulated pipe starting just below the manhole taps surface blowdown from the steam drum of this large watertube boiler.

of water to gush out. Instead of this haphazard approach, control blowdown rationally. Monitor water conditions continuously, and adjust the blowdown rate accordingly.

This Measure gives you the principles of controlling blowdown, which you can do manually. The subsidiary Measure 1.8.2.1 recommends installing automatic blowdown control, where appropriate, to control blowdown most accurately.

Top Blow and Bottom Blow

Some solid contaminants are heavier than water, and these settle to the bottom of the boiler. These are removed by one or more blowdown drains at the bottom of the boiler. This is called "bottom" blowdown. Figure 1 shows the bottom blowdown valve of a large boiler.

The sludge that concentrates in the bottom of the boiler is in a relatively undisturbed area, and can wait

there until enough has accumulated to justify blowing it out. In boilers that have several isolated low points, such as watertube boilers with several lower headers, a blowdown drain is installed at the bottom of each. The lower headers of watertube boilers are called "mud drums" because the solids accumulate there in the form of a sludge.

Other solid contaminants are lighter than water. In steam boilers, these contaminants float to the surface. Also, dissolved contaminants come out of solution at the water's surface as the water evaporates. These contaminants are removed at a blowdown drain that is installed just below the waterline. This is called "top" or "surface" blowdown. Figure 2 shows the top blowdown connection of a large boiler.

It is better practice to control the concentration of dissolved contaminants by blowing from the top blowdown drain, rather than from the bottom drain. Contaminants and water treatment products that float to the surface carry over with the steam. Also,

---

FORMULA 1

$$\text{Blowdown rate} = \text{Makeup rate} \times \frac{\text{TDS of feedwater}}{\text{TDS in boiler} - \text{TDS of feedwater}}$$

where:
- TDS means "total dissolved solids"
- the blowdown rate and makeup rate are expressed in any identical units of mass flow, for example, pounds per hour

---

evaporation at the surface makes the contaminant concentration highest there.

Can Bottom Blow Cause Tube Burnout?

In watertube boilers, there is some concern that extended bottom blowdown may cause burnout of generating tubes. The concern is that opening the blowdown valve offers a path of least resistance so that water in the bottom drum will flow out the blowdown connection instead of up into the tubes. Reduced flow within the tubes allows the tube temperature to rise dangerously. This risk is alleged to be most severe in headers serving waterwall tubes, which are exposed to the full heat of the firebox.

It seems doubtful that blowdown connections would be installed where using them might damage the boiler, but only brief periods of blowdown may be intended. In case of doubt, check with the boiler manufacturer. It may be good practice to limit bottom blow to periods when the boiler is at a lower firing rate.

How to Calculate the Blowdown Rate

You can calculate the blowdown rate easily from the concentration of contaminants in the water by using Formula 1.

For example, if the TDS of the feedwater is 5 percent of the TDS desired in the boiler, the blowdown rate should be slightly more than 5 percent of the makeup rate. If the system is losing 10,000 pounds of steam per hour, the proper blowdown rate is 526 pounds per hour.

The Energy Penalty of Blowdown

The amount of energy carried away by blowdown is proportional to the temperature of the boiler water and to the blowdown rate. (The energy content of water inside the boiler is entirely sensible heat, which is linearly proportional to the temperature.)

For example, if the boiler pressure is 250 PSI, the boiler water temperature is 405°F. If the makeup water temperature is 65°F, then each pound of water that is lost requires another pound of makeup water to be heated 340°F, which requires about 340 BTU.

If the boiler has the same blowdown rate that we calculated in the previous example, about 180,000 BTU per hour is being lost. This is about 1.8 percent of the energy content of the steam.

In this example, the heat loss is unavoidable because blowdown is necessary to maintain proper water conditions. However, any blowdown in excess of the minimum is a waste of energy and water.

How to Control Top Blowdown

Ask your boiler manufacturer or water treatment consultant (see Measure 1.8.1.1) to specify the water conditions that you should maintain with blowdown. Usually, blowdown is controlled on the basis of maintaining the maximum allowable level of total dissolved solids (TDS).

To a large extent, the concentration of TDS in boiler water is related to "hardness," i.e., the concentration of calcium and magnesium ions. In a well treated boiler, TDS mostly consists of the residue of chemical treatment to neutralize hardness.

Unusual circumstances may make it appropriate to control blowdown on the basis of some other water characteristic, but this is rare. Other boiler contaminants or conditions, such as pH and alkalinity, are controlled by appropriate chemical treatment rather than by blowdown.

■ Should Top Blowdown be Continuous or Intermittent?

As a practical matter, blow down intermittently if you have a low blowdown rate, and blow down continuously if you have a large blowdown requirement. The limiting factor is knowing your actual blowdown rate. If you open the blowdown valve fully for a measured period of time, you can judge the amount of blowdown accurately. On the other hand, you cannot judge your blowdown rate accurately if you keep the blowdown valve open slightly. (In principle, you could meter the blowdown or calibrate your blowdown valve, but this is an unnecessary complication.)

The main issue is keeping the blowdown rate as close to optimum as possible. Blowdown requirements vary enormously from one boiler plant to another. If you need a high blowdown rate, you may need continuous blowdown to keep the TDS from drifting too far away from its ideal value. On the other hand, if the makeup rate is low and the makeup water is clean, it may be satisfactory to blow down every few hours.

■ How to Measure TDS

The easiest way to measure TDS is with a TDS meter. This is actually a meter that measures the electrical conductivity of the boiler water, but it is calibrated in terms of TDS. A TDS meter has the advantage that it provides a continuous readout. If you have a high blowdown rate, a TDS meter is practically a necessity.

The main disadvantage of the TDS meter is that it can get out of calibration. This is not likely, but a false reading can result in the wrong blowdown rate, with possibly serious consequences. If you can afford it, install two TDS meters and compare them regularly.

You can also measure TDS by using an inexpensive chemical test. This may be satisfactory in a small boiler plant that has a low blowdown requirement. The TDS should not be able to drift far from its ideal value between tests. You can also use a chemical test to check your electrical TDS meter.

■ **Manual Control Methods**

If you blow down intermittently, open the blowdown valve for a fixed period of time, such as ten seconds, and observe the effect on the TDS meter. Repeat this as necessary to keep the TDS at its proper level.

With a steam boiler, the TDS meter will register the effect of blowdown quickly because all the TDS are circulating vigorously within a confined space. In a hot water system, the TDS are circulating throughout the facility, so you have to wait for the water to circulate through the system before the effect of blowdown will register properly on the TDS meter.

If you blow down continuously, keep adjusting the blowdown valve based on the TDS meter reading. The operators have to monitor the TDS meter continuously to do this right.

If the manual procedure seems too difficult or too risky, install an automatic blowdown control, as recommended by Measure 1.8.2.1.

How to Control Bottom Blowdown

Do bottom blowdown intermittently. Determine the appropriate interval by piping the discharge into a safe container and seeing what settles out. If the bottom blow does not contain a significant amount of visible material, extend the interval.

Minimize the Need for Blowdown

The formula for blowdown rate given above states that the need for blowdown is proportional to the makeup rate and to the concentration of solids in the feedwater. Let's examine this important point more closely. First of all, if the boiler system has minimal water loss, it needs only minimal blowdown because there is little makeup to bring contaminants into the system. Second, if the makeup water is perfectly pure, the need for blowdown is minimal because the makeup water is not bringing contaminants into the system. Therefore, to reduce the need for blowdown, (1) reduce system leakage as much as possible, and (2) treat the makeup water as effectively as possible before it becomes feedwater.

To reduce system leakage, see Measures 1.10.1 and 1.10.2. Reducing "hardness" is the part of water treatment that has the greatest effect on reducing blowdown requirements. Do this with a water softening process, as described in Measure 1.8.1. Also, remove suspended solids in the makeup water by filtration and coagulation.

Recover Blowdown Heat, Maybe

If the system still requires a substantial amount of blowdown after all improvements have been made, consider recovering the heat from blowdown water. Measure 1.8.3 tells you how.

ECONOMICS

SAVINGS POTENTIAL: 0.1 to 2 percent of boiler fuel, depending mainly on the makeup water rate, the boiler pressure, and makeup water conditions.

COST: Minimal.

PAYBACK PERIOD: Immediate.

TRAPS & TRICKS

SETUP AND MONITORING: Develop the appropriate blowdown schedule based on water conditions, the type of boiler, and boiler operating conditions. Use water testing results, especially TDS, as the primary guide to blowdown.

When the boiler is cold, inspect the watersides to verify adequate blowdown. Inspect the bottom drain area for adequate bottom blowdown.

TRAINING AND SUPERVISION: Blowdown will be forgotten unless it is done on a regular basis. Establish a blowdown schedule and put it on the boiler operating log, along with appropriate instructions. Check that the procedure is being followed.

MEASURE **1.8.2.1 Install automatic blowdown control.**

RATINGS

New Facilities · Retrofit · O&M

B **C** ☐

The formula in Measure 1.8.2 shows that the blowdown rate should be proportional to the makeup rate. If the makeup rate fluctuates, the boiler operators cannot optimize the blowdown rate manually. The solution is to install an automatic valve for the top (surface) blowdown that senses the appropriate condition in the boiler water. Usually, this condition is the concentration of "total dissolved solids" (TDS). The equipment for this is a standard boiler accessory.

Equipment and Installation

To control blowdown automatically, install an automatic valve in the top blowdown line. Typically, this is a solenoid valve. The solenoid valve is controlled by a sensor that measures the electrical conductivity of the boiler water, which relates to the TDS concentration. Figure 1 shows a typical blowdown controller.

Automatic blowdown may operate continuously or intermittently, depending on the blowdown rate. If the blowdown requirement is small, intermittent blowdown is more practical. A modulating valve operating at a low flow rate would require frequent maintenance because of scoring of the valve seat.

WESINC

Fig. 1 Automatic blowdown controller The controller senses boiler water conductivity to activate surface blowdown.

SUMMARY
Provides the most accurate control of blowdown.

SELECTION SCORECARD

Savings Potential	**$**
Rate of Return, New Facilities	**% %** %
Rate of Return, Retrofit..........	**% %**
Reliability	✓ ✓
Ease of Retrofit	☺ ☺ ☺

If you use a solenoid valve, it is a good idea to control it through a cycle timer. The conductivity sensor triggers the cycle timer, which keeps the valve open for a fixed period of time, such as ten seconds. This method reduces the TDS in small increments and allows the TDS concentration to stabilize between blowdown cycles, preventing erratic operation of the blowdown valve.

Limit the capacity of the blowdown valve to keep the water level from falling to an unsafe level if the valve fails in a fully open position. Make the blowdown line, or the valve, small enough so that the feedwater pump can maintain the water level while the boiler is at full output and the blowdown valve is fully open.

Do not install the blowdown valve in a descending pipe. This would allow dirt to settle in the valve and foul it. Do not install a strainer anywhere in the blowdown line. A strainer might clog and then blow out explosively.

As a safety feature, monitor the condition of the boiler water with a separate alarm having its own sensor. The alarm should activate if the controlling parameter (usually, the TDS) is either too high or too low.

Top Blowdown Only

Usually, only top blowdown is controlled automatically. Do bottom blowdown manually, which allows you to inspect the blowdown for accumulated solids. In most plants where operators are not on duty continuously, you can defer bottom blowdown to normal working hours.

Continue Water Testing

Installing automatic blowdown does not reduce the need for regular boiler water testing. Blowdown controls only one water condition, hardness. You need to keep other water conditions within proper limits.

Furthermore, you need testing to verify that the automatic control is functioning properly.

ECONOMICS

SAVINGS POTENTIAL: *Covered in Measure 1.8.2.*

COST: *Several thousand dollars, typically.*

PAYBACK PERIOD: *Several years or longer, depending primarily on the average blowdown rate required.*

TRAPS & TRICKS

MAINTENANCE: *Conductivity sensors require periodic maintenance and/or calibration. Blowdown valves tend to foul, so clean them periodically.*

CHECKING: *Review the water test results continually, especially TDS, to make sure that the blowdown is operating properly. Failure is easy to overlook until serious damage occurs. Test the TDS alarm on a daily basis.*

MEASURE **1.8.3 Install blowdown heat recovery.**

Blowdown water contains valuable heat. You can recover this heat easily. Whether it is economical to do so depends mainly on these factors:

- the amount of heat available for recovery, which mainly depends on the blowdown rate. The blowdown rate is proportional to the makeup water requirement, which is equal to system water loss plus blowdown.
- whether an application is available that can use heat at the temperature of the blowdown
- whether the potential application is synchronized with the time of blowdown.

In brief, blowdown heat recovery is most likely to be practical with larger boilers that require a large amount of makeup. In such cases, the payback period for blowdown heat recovery may be short. On the other hand, blowdown heat recovery is unlikely to be practical with smaller boilers, or in steam systems that have only small water losses.

Recover Heat from Water, Flash Steam, or Both?

You can recover almost all the heat from blowdown water, regardless of the boiler pressure, if you cool the blowdown water enough in the process. However, raising the heat recovery temperature reduces the fraction of the heat that is recovered. This is because all the energy of the blowdown water is in the form of sensible heat (i.e., proportional to temperature).

In many applications, it would be desirable to use the recovered heat in the form of steam. For such applications, you can exploit the fact that a certain fraction of the blowdown water will flash into steam when it is exposed to the lower pressure outside the

SUMMARY

Worth considering in boilers that require a significant amount of blowdown. Heating makeup water is the most common application. High-pressure systems offer a greater variety of applications.

SELECTION SCORECARD

Savings Potential .... **$**

Rate of Return, New Facilities **% %** %

Rate of Return, Retrofit.......... **% %**

Reliability .... ✓ ✓ ✓

Ease of Retrofit .... ☺ ☺ ☺

boiler, provided that the water is not cooled first. The flashing occurs because the heat contained in the water boils part of the liquid into steam.

Flashing occurs instantly, resulting in a quantity of liquid at boiling temperature and a quantity of steam at the same temperature. Both the steam and the remaining water contain heat energy. The question is, how much of the original energy goes into the steam, and how much remains in the water?

First of all, the total amount of energy contained in blowdown water is proportional to the temperature of the boiler water, which is directly related to the boiler pressure. For different boiler pressures, this table shows the amount of water that remains as a liquid, the amount that flashes into steam, and the amount of energy in each:

Boiler Pressure, PSI gauge	Weight, lb		Energy, BTU		Percent of Energy to Steam
	Water	Steam	Water	Steam	
10	0.97	0.03	202	33	14
35	0.93	0.07	167	82	33
88	0.88	0.12	158	143	47
205	0.81	0.19	146	218	60
655	0.68	0.32	123	365	75
1310	0.58	0.42	104	485	82

(The table assumes a baseline of zero energy for water at 32°F, which is standard practice. The table also assumes that the condensate discharges to atmospheric pressure.)

This table reveals some interesting results that help to decide whether it is practical to recover the blowdown heat as flash steam, or from the liquid:

- At low boiler pressures, most of the energy remains in the water.
- Above about 100 PSI, there is more energy in the flash steam than in the remaining water. At still higher pressures, the fraction of the energy contained in flash steam rises rapidly.
- The energy content of the water actually declines as boiler pressure rises. The reason is that the increased energy content of the water causes more of it to flash into steam. The water always cools to the same temperature once it gets outside the boiler, so less water means less energy.

The conclusion is clear. Recovering blowdown heat in the form of flash steam is much more attractive at higher boiler pressures.

Equipment for Recovering Blowdown Heat

Recovering heat from the liquid is usually done with a simple heat exchanger. Figure 1 shows a typical blowdown heat recovery unit, which is available as a standard product.

A heat exchanger is usually needed when recovering heat from the liquid blowdown water, because it is contaminated with water treatment chemicals and their residue. Therefore, you usually cannot add the blowdown water directly to the liquid you are heating. The heat exchange process reduces heat recovery to some extent.

To recover blowdown heat in the form of flash steam, the water is dumped into a tank, called a "flash tank." A portion of the water flashes to steam inside the tank, and the steam is piped away to its application. The flash tank has a steam trap, or float valve. The steam trap is needed to keep the steam from flowing into the sewer along with the remaining blowdown water.

The flash tank may drain the remaining liquid to a heat exchanger to recover the heat in the liquid. The heat of the steam is recovered at steam temperature, while the heat in the liquid is recovered at lower temperature. Figure 2 shows a typical heat recovery

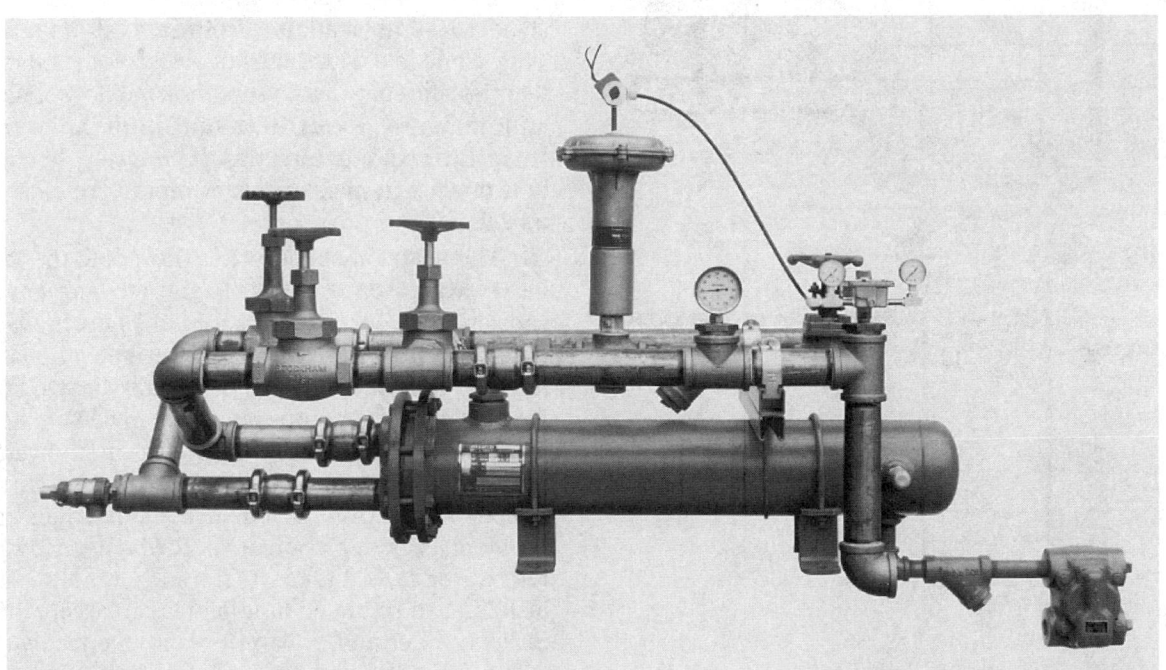

Cleaver-Brooks

Fig. 1 Liquid-to-liquid blowdown heat recovery unit In this packaged unit, the flow of blowdown water is regulated to maintain the desired temperature of the liquid being heated, as long as blowdown occurs. Heat may be recovered at any temperature up to the temperature of the boiler water. However, the amount of heat recovered is inversely proportional to the recovery temperature. The steam trap at lower right keeps any flash steam from blowing out of the unit.

vessel for recovering heat from both the steam and the liquid.

For some applications, you can recover the energy of flash steam without a heat exchanger. To do this, simply bubble the steam through the liquid or other material to be heated. This captures all the heat of the steam, and it also recovers the pure water condensed from the steam. The main limitation is that the application must be at atmospheric pressure, or at least, at a pressure much lower than boiler pressure. Of course, the heated material must tolerate the added water.

You need a heat exchanger to recover the heat from the flash steam if you want to heat a substance that cannot be exposed to the steam, or to heat a material that is under pressure.

Using Blowdown to Heat Makeup Water

A common application for blowdown heat recovery is heating makeup water. This is a good application because the outgoing flow of blowdown water is synchronized with incoming flow of makeup water. The

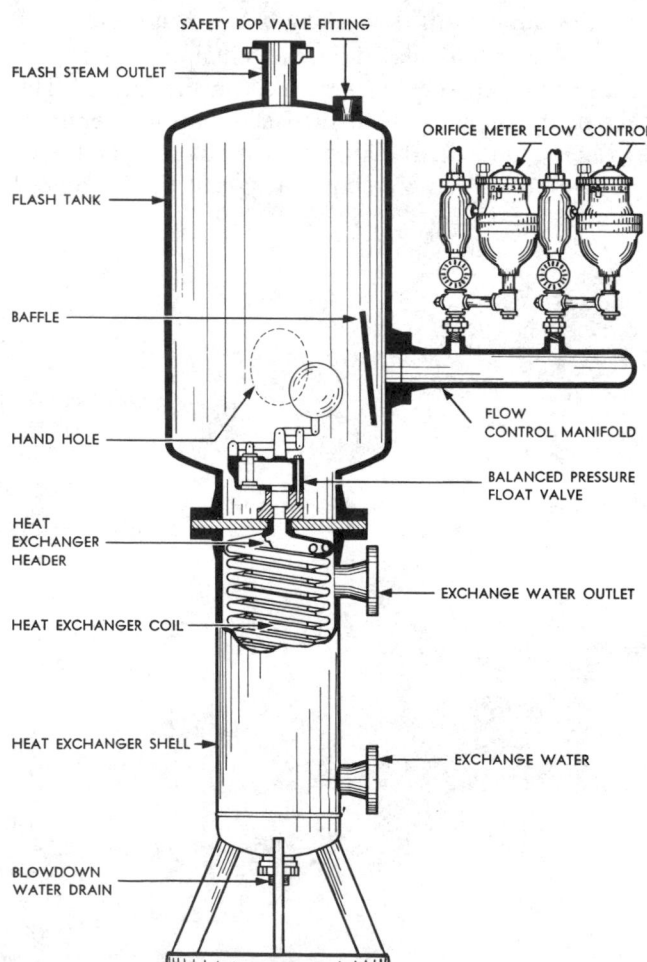

Fig. 2 Blowdown heat recovery unit providing steam and hot water The blowdown water enters through the metering valves on the right. It flashes in the upper tank. The remaining liquid drops through the float valve to give up its heat in the heat exchanger below.

quantity of makeup water to be heated is always greater than the quantity of blowdown, because the amount of makeup water equals the blowdown plus the total system water loss.

Here are some examples to illustrate makeup water heating:

- A boiler operates at 10 PSI, and the makeup water rate is five times the blowdown rate. If all the blowdown heat is recovered, it can raise the temperature of the makeup water by 43°F.
- A boiler operates at 655 PSI, and the makeup water rate is five times the blowdown rate. If all the blowdown heat is recovered, it can raise the temperature of the makeup water by 94°F.
- A boiler operates at 655 PSI, and the makeup water rate is two times the blowdown rate. If all the blowdown heat is recovered, it can in principle raise the temperature of the makeup water by 235°F. However, if the makeup water starts at 50°F, the final temperature is 285°F, which is too high for the feedwater system. Instead, some of the blowdown heat is used to heat makeup water, and the rest is sent to the deaerating tank in the form of flash steam..

These examples suggest that most boilers can use their blowdown heat entirely for heating makeup water. A possible exception is high-pressure boilers with low makeup water rates, which cannot recover all the heat.

■ Installation Considerations

Packaged blowdown-to-makeup heat exchangers are standard items available from several manufacturers. They come in a wide range of sizes. Some units include thermostatic controls to proportion the flow of blowdown and makeup water for optimum heat transfer. Installation is usually inexpensive, because the blowdown and makeup water piping are close to each other.

The heated makeup water flows into the feedwater tank. Make sure that the feedwater tank has enough capacity to accommodate rapid changes in water level during blowdown. Normally, water level changes are not a problem, because the water drawn from the feedwater tank to compensate for blowdown is replaced by heated makeup water. However, if feedwater is fed to the boiler with on-off control, the flows may temporarily get out of balance. One solution to this problem is better control of feedwater flow. If the feedwater tank lacks sufficient capacity, consider installing an insulated holding tank to accept the heated makeup water until it can flow into the feedwater tank.

A single heat recovery unit can serve any number of boilers. Simply pipe the blowdown from all the boilers to the heat recovery unit. This arrangement should be satisfactory if a common feedwater tank is used for the same boilers.

Using Blowdown to Heat Feedwater

In a boiler system that uses an open (vented to atmosphere) feedwater tank, 200°F is about the maximum practical temperature to which makeup water can be heated. If the makeup water is heated more, it will flash away through the tank vent.

If the system uses a deaerating feedwater tank, flash steam from the blowdown can contribute to the steam needed by the deaerator to "scrub" oxygen out of the feedwater. This is economical because only a flash tank is required. However, blowdown usually occurs in brief discharges, so the feedwater system must be designed to exploit the brief bursts of flash steam.

Figure 3 shows a blowdown heat recovery system that heats both feedwater and makeup water.

Other Applications

The heat contained in blowdown can be used for any application that is compatible in terms of temperature, quantity of heat, and time synchronization.

Blowdown is not a steady source of heat. Makeup water heating is the only application in which the need for blowdown is synchronized with its availability. In most other applications, consider blowdown heat as a supplemental source. Domestic water heating and fuel oil tank heating might be applications.

In high-pressure plants, most of the blowdown will flash into steam. Such plants may have a low-pressure steam system that is used primarily for heating. In such cases, simply feed the flash steam into the low-pressure system.

If the application requires higher temperatures, less heat can be extracted from the blowdown. Only higher-pressure boilers produce blowdown water that is hot enough for heating applications above the boiling point of water.

Combination with Condensate Heat Recovery

In some applications, it is necessary to dump the condensate from steam equipment, rather than return it

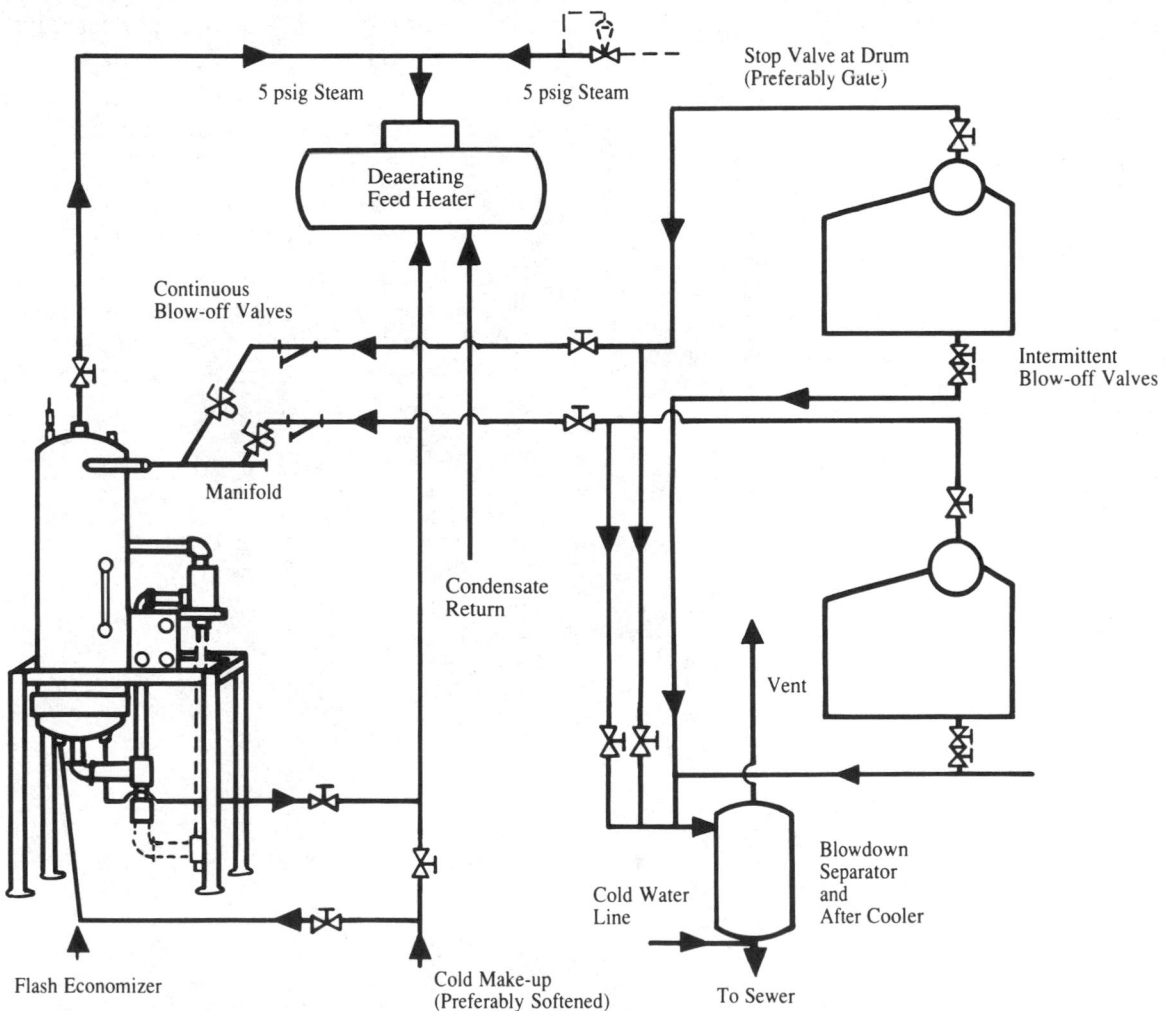

Fig. 3 Blowdown heat recovery system for makeup water and feedwater The deaerating feedwater heater needs a constant supply of steam, which is supplemented by flash steam. The makeup water can accept heat as it becomes available. This system recovers heat from both top and bottom blowdown.

to the boiler. (For example, there may be a risk that the condensate is contaminated by materials that would foul the boiler.) You can still recover the heat of the dumped condensate by using the blowdown heat recovery system. See Measure 1.8.4.1 for this application.

Heat Exchanger Fouling and Maintenance

Blowdown water is laden with dissolved materials. As the blowdown water cools, some of these materials precipitate on the surfaces of the heat exchanger. The mess that occurs in the heat exchanger consists of the substances that were removed from the boiler to avoid leaving a mess there. Plan regular cleaning of the heat exchanger to remove these deposits. Design the heat exchanger to make cleaning easy.

The blowdown heat recovery units offered by several manufacturers have stainless steel heat exchanger tubes. This is desirable because the heat exchanger is hot, wet, and open to the atmosphere. It would corrode quickly if it were made of plain steel.

Treat the makeup water before it passes through the heat exchanger. This minimizes fouling of the makeup water side of the heat exchanger.

Safety Issues

If you plan to use flash steam at a pressure higher than atmospheric, select the flash tank for the expected pressure.

Design the heat recovery system to deal with the possibility that the blowdown valve may stick in the fully open position. If the blowdown is recovered in a closed tank or heat exchanger, it should have a pressure relief valve sized for the maximum blowdown rate.

No component of the heat recovery system should be able to become so fouled that it can block the blowdown discharge. If this occurs, full boiler pressure will be applied upstream of the blockage, possibly causing a dangerous blowout. For this reason, don't install any strainers in the blowdown circuit of the heat recovery system.

ECONOMICS

SAVINGS POTENTIAL: *0.1 to 2 percent of boiler fuel, depending mainly on the blowdown percentage.*

COST: *Typically, several thousand dollars for a smaller boiler plant. Up to $20,000 for a sophisticated system that serves a larger boiler plant.*

PAYBACK PERIOD: *Less than one year, to many years, depending mainly on the blowdown rate and boiler pressure.*

TRAPS & TRICKS

CHOICE OF METHOD: *Consider all your heat recovery options. Estimate your potential savings carefully.*

SELECTING THE EQUIPMENT: *Select the equipment to tolerate heavy fouling and to be easily cleaned. Ask vendors for references. Talk with purchasers who have used the equipment for several years.*

MAINTENANCE: *Schedule cleaning at appropriate intervals. Fouling reduces heat recovery.*

EXPLAIN IT: *The purpose of a blowdown heat recovery unit is not obvious from its appearance. Install a placard on it that indicates what it does and how it works. See Reference Note 12, Placards, for details of effective placard design and installation.*

MEASURE 1.8.4 Maximize condensate return.

RATINGS

New Facilities | Retrofit | O&M

A | **C** |

Try to recover condensate wherever it is economical to do so, and keep the condensate as warm as possible. The condensate contains a considerable amount of energy, and the treated water is valuable in itself.

All steam-operated equipment generates condensate, unless the steam is used in a process that consumes the steam or releases it to the atmosphere. Even so, condensate may be dumped rather than being returned to the boiler. This may occur because of cost cutting in the original installation, or because condensate lines were abandoned instead of repairing them, or because facility additions neglected to connect to an existing condensate system. Design errors, installation errors, inadequate maintenance, or improper plant operation may prevent an installed condensate system from functioning.

Condensate is often returned over long distances, as shown in Figure 1. If you cannot afford to return condensate, try to recover its heat energy at the site. See subsidiary Measure 1.8.4.1 for this. To minimize loss of condensate by flashing from high-pressure equipment, see subsidiary Measure 1.8.4.2.

Reasons to Return Condensate

These are the reasons to return condensate:

• *to recover the heat contained in the condensate.* The energy saving is significant if the steam-using equipment is close enough to the boiler plant so that you can keep the condensate warm until it returns to the boiler. Condensate retains its heat better if there is a large condensate flow rate and the lines are well insulated. In many installations, return lines would be so long that condensate could not remain warm.

SUMMARY

Consider this any time condensate is being discarded. Usually expensive.

SELECTION SCORECARD

Savings Potential .... $ $

Rate of Return, New Facilities % % % %

Rate of Return, Retrofit.......... % %

Reliability .... ✓ ✓ ✓

Ease of Retrofit .... ☺ ☺ ☺

• *to reduce water treatment cost.* This cost is related to the quality of the raw water in the locale. Condensate return is more attractive if the makeup water is expensive to treat.

• *to conserve raw water.* Water costs are rising everywhere. Even though water is still cheap in most places, saving raw water may be a significant cost factor.

• *to satisfy environmental regulations.* Condensate may be considered to be contaminated water, although the quantity of water treatment chemicals that pass through the steam system is small. Or, there may be objections to the high temperature of the condensate if it is discharged into a body of water.

Heat Recovery Potential

The amount of energy in condensate depends on the characteristics of the steam-using equipment. In an ideal steam system, the steam-using equipment would extract all the thermal energy of the steam, and return

WESINC

Fig. 1 Steam and condensate lines on a large military base

only cold water. This is almost never possible in practice, for two reasons. One is that the application requires heat at some minimum temperature. For example, if steam is used for heating a batch of material at 300°F, the condensate is formed at a temperature of 300°F.

The second reason relates to the thermal characteristics of water and steam. Below boiling temperature, pure steam has a pressure lower than atmospheric. However, the steam inside the equipment must be above atmospheric pressure to push the condensate out of the equipment. As a result, condensate enters a vented condensate system at a temperature somewhat higher than the boiling temperature.

(Vacuum condensate systems allow steam to condense in the equipment at a much lower temperature. Hence, the condensate in a vacuum system carries much less energy. See Measure 1.8.5, below, for more about vacuum condensate systems.)

As a typical example of condensate recovery, condensate returns to the boiler at a temperature of 180°F, and the makeup water has a temperature of 60°F. In this case, each pound of condensate that is recovered saves 120 BTU. If steam is produced at 15 PSI, this represents 10.5% of the energy that would be needed to make the steam from cold makeup water.

■ Loss of Heat in Condensate Piping

If there is a long distance between the boiler and the steam-using equipment, much of the condensate's heat may be lost through the walls of the pipe on the way back to the boiler. Before spending money to install condensate piping, calculate how much of the condensate's heat will survive the trip. Your estimate should not exaggerate the effectiveness of insulation. Straight sections of pipe can be heavily insulated, but expansion rollers, pipe hangers, exposed flanges, and other features conduct heat through the insulation. On the other hand, if you can run the condensate pipe inside hot steam tunnels, heat loss is greatly reduced.

The rate of condensate flow has a major effect on the temperature of the returning condensate. The rate of heat loss from a pipe depends on its insulation and surface area. The loss is constant for a given temperature. If there is a large flow inside the pipe, the heat loss may comprise a small percentage of the total heat being transmitted, and the condensate arrives hot. However, if there is only a low flow rate, the heat loss from the pipe results in the condensate being cold by the time it arrives. This is a mathematical way of saying that a long condensate return may be worthwhile for a large steam user but not for a small one.

Flash steam forms when the condensate exits the steam-using equipment, and it keeps the liquid condensate warm. For example, condensate that is originally at 240°F cools to 212°F as soon as it leaves the trap. The energy corresponding to the 28°F temperature difference is contained in the flash steam. The steam travels through the condensate pipe, keeping the pipe warm until it all condenses. (The flash steam also displaces air from the condensate pipe, preventing corrosion and contamination of the condensate.)

■ Recovery of Condensate Heat in High-Pressure Applications

In high-pressure applications, the condensate is substantially hotter than the atmospheric boiling temperature. If the condensate is returned through a vented condensate system, a large fraction of it will flash into steam. Capturing the energy of the flash steam involves additional steps. These are covered by subsidiary Measure 1.8.4.2.

General Approach

The first step is to find out whether you are losing condensate. See Measure 1.10.1 for methods of monitoring water loss.

If you find that you are losing condensate, the second step is to find out where it is going. Find any steam-

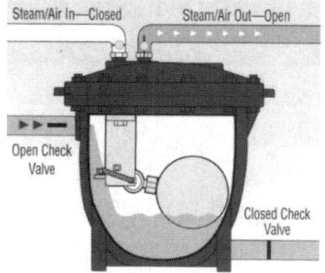 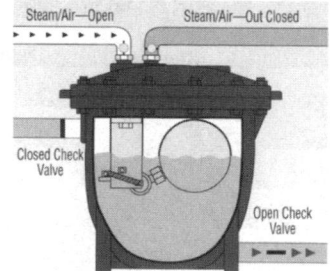 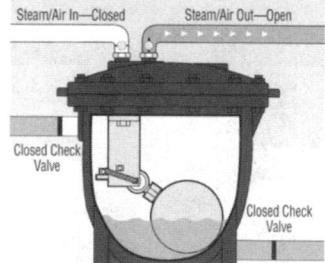 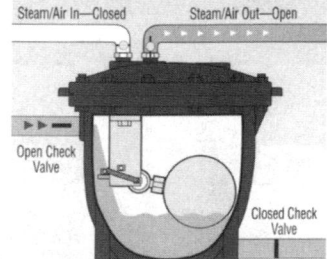

FILLING
1. During filling, the steam or air inlet and check valve on pumping trap outlet are closed. The vent and check valve on the inlet are open.

BEGIN PUMPING
2. Float rises with level of condensate until it passes trip point, and then snap action reverses the positions shown in step one.

END PUMPING
3. Float is lowered as level of condensate falls until snap action again reverses positions.

REPEAT FILLING
4. Steam or air inlet and trap outlet are again closed while vent and condensate inlet are open. Cycle begins anew.

Armstrong International, Inc.

Fig. 2 Steam powered condensate mover This is essentially a positive-displacement steam pump without a piston. It is especially useful where the condensate system is not tall enough to provide adequate suction pressure for a centrifugal pump.

using equipment that is not returning condensate. See Measure 1.10.2 for methods of detecting leakage from the condensate return system. Also, look for blowing steam vents and other places where water may be disappearing.

The third step is to take whatever action is necessary to return condensate, subject to the limitations of economics and practicality. Install condensate returns where they are missing. Repair the condensate return system where it is defective.

Condensate System Layout

The types of condensate systems are summarized in the introduction to this Subsection. Recovering condensate is a two-step process of draining by gravity into a receiving tank ("condensate receiver"), and then pumping from the receiver back to the boiler plant. To avoid complications, the receiver must be located below the equipment that it serves, and the condensate must be able to flow downhill from the receiver. All condensate piping must be pitched downward. To satisfy these requirements in a large facility, there may be many receivers in different locations. For example, each building in a facility that is served by a central steam plant might have its own receiver.

How to Pump High-Temperature Condensate

If the condensate pipe is so well insulated that condensate reaches a condensate or feedwater pump at a temperature above about 180°F, the pump may be unable to function properly. The problem is that the hot water partially flashes into steam at the reduced pressure of the pump suction. The presence of steam bubbles in the pump reduces pump capacity and efficiency, hastens deterioration of the impeller and bearings, and makes a loud rattling noise. This problem causes some facilities to deliberately cool the condensate. This wasteful practice is not necessary, because there are several ways to pump hot condensate.

One solution is to install the condensate pump below the condensate receiver. The hydrostatic pressure of the water column between the receiver and the pump prevents flashing. Calculate the required height from the suction pressure requirement of the particular pump. (This requirement is called "net positive suction head," and it is indicated on the pump curves.) A height of 8 to 15 feet usually suffices for water at the boiling point, with less height being required for cooler water.

Another solution is to use a steam-powered condensate mover instead of a centrifugal pump.

Figure 2 shows how this equipment works. It consists of a tank, a float valve, two check valves, and a steam connection. In operation, condensate is allowed to drain into the tank. Then, steam is admitted to the tank to blow the condensate out, after which the cycle repeats. The unit is essentially a reciprocating steam pump without a piston. It can be made with hardware similar to that found in steam traps, and compact units are offered by several steam trap manufacturers. In fact, these units are sometimes called "pumping traps."

Upgrade Your Water Treatment Program to Deal with Condensate

Condensate returning through an atmospheric condensate system dissolves oxygen and carbon dioxide from the air. This in turn causes the water to corrode the iron in the pipe and to dissolve iron compounds. Revise your water treatment program to deal with these contaminants. See Measure 1.8.1 about water treatment.

ECONOMICS

SAVINGS POTENTIAL: *3 to 12 percent of fuel cost for the portion of the system that does not return condensate; plus, major reduction of water treatment cost; plus, major reduction of makeup water cost.*

COST: *Condensate receivers typically cost several thousand dollars. Condensate piping costs from $10 to $20 per foot ($30 to $60 per meter) in typical sizes. Trenching, if required, costs from $0.50 to $2.00 per foot ($1.50 to $6.00 per meter). Piping in insulated conduit, if required, costs $20 to $40 per foot ($60 to $120 per meter). (U.S. installed prices.)*

PAYBACK PERIOD: *Varies widely. Typically, several years or longer.*

TRAPS & TRICKS

DESIGN: *Condensate systems are conventional, but don't overlook unusual options, such as steam-powered condensate movers. Design details have a major effect on system performance and life. Be generous with insulation. Minimize heat loss at insulation breaks, such as pipe hangers. The life of the system is determined by the rate of corrosion, both inside and outside the pipe. Keep the pipe and insulation dry. Lay out the pipe with adequate pitch to keep it drained. Use appropriate corrosion protection methods. Select the most corrosion-resistant steel that you can afford. In open condensate systems, arrange vents so they minimize entry of air into the system.*

MEASURE 1.8.4.1 Recover the heat from condensate that must be discarded.

The condensate from some equipment may be contaminated, making it undesirable to return it to the boiler. For example, a leak inside a steam heated fuel oil heater would get oil into the boiler, causing a particularly nasty form of waterside fouling. Therefore, this condensate is dumped.

Even so, you can still recover the heat from this condensate. Do this with a heat exchanger, in the usual way. Perhaps add a flash tank, if the condensate is delivered at high pressure. Since the condensate may be dirty, use a heat exchanger that is especially easy to clean.

Combine with Blowdown Heat Recovery

A blowdown heat recovery system (Measure 1.8.3) is a method of extracting heat from blowdown water that is dumped to the sewer. If the condensate is being dumped near the boiler, you may be able to use a common heat exchanger to recover heat from condensate and blowdown.

ECONOMICS

SAVINGS POTENTIAL: *Typically, a few percent of the fuel required to make steam for the heating application from which condensate is dumped.*

RATINGS

New Facilities	Retrofit	O&M
C	**C**	

SUMMARY

Even if you have to dump condensate, you may still be able to recover its heat.

SELECTION SCORECARD

Savings Potential	**$**		
Rate of Return, New Facilities	**%**	%	%
Rate of Return, Retrofit..........	**%**	%	
Reliability	✓	✓	✓
Ease of Retrofit	☺	☺	☺

COST: *Several thousand dollars, for an application of average size. May be inexpensive if the condensate can be dumped to an existing blowdown heat recovery system.*

PAYBACK PERIOD: *Several years or longer.*

TRAPS & TRICKS

DESIGN: *Consider all your heat recovery options. Be careful if you combine condensate heat recovery with blowdown recovery.*

EXPLAIN IT: *Describe the installation in the plant operating manual. The purpose of the equipment is not obvious, so install a placard on it that states what it does.*

MEASURE 1.8.4.2 Recover the energy of high-temperature condensate that would be lost by flashing.

RATINGS

New Facilities	Retrofit	O&M
C	C	

SUMMARY

Specialized methods for recovering the maximum amount of condensate heat in high-temperature steam heating applications.

SELECTION SCORECARD

Savings Potential .... $ $

Rate of Return, New Facilities % % %

Rate of Return, Retrofit.......... % % %

Reliability .... ✓ ✓ ✓

Ease of Retrofit .... ☺ ☺ ☺

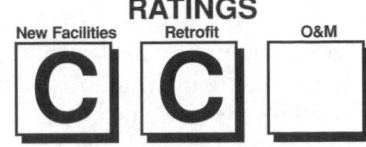

If condensate returns in a condensate system that works at atmospheric pressure, you will lose a certain percentage of the heat in the form of flash steam. Only liquid water returns to the boiler. Any flash steam that remains by the time the condensate gets back to the receiver is lost through the vent on the receiver. The amount of flash steam depends on the pressure, and hence the temperature, at the discharge of the steam-using equipment.

If steam equipment operates at high temperature, much of the energy of the condensate may be wasted. For example, if steam is used to heat a vat to 330°F, the condensate temperature is also 330°F inside the heating equipment. When the condensate drains to a condensate system operating at atmospheric pressure, approximately half of the condensate flashes into steam. Part of this flash steam condenses in the return pipe, and keeps the condensate warm. The rest is lost through the vent at the condensate receiver.

You can use a variety of techniques to recover the heat of high-temperature condensate, rather than letting it flash away to the atmosphere. These are some approaches:

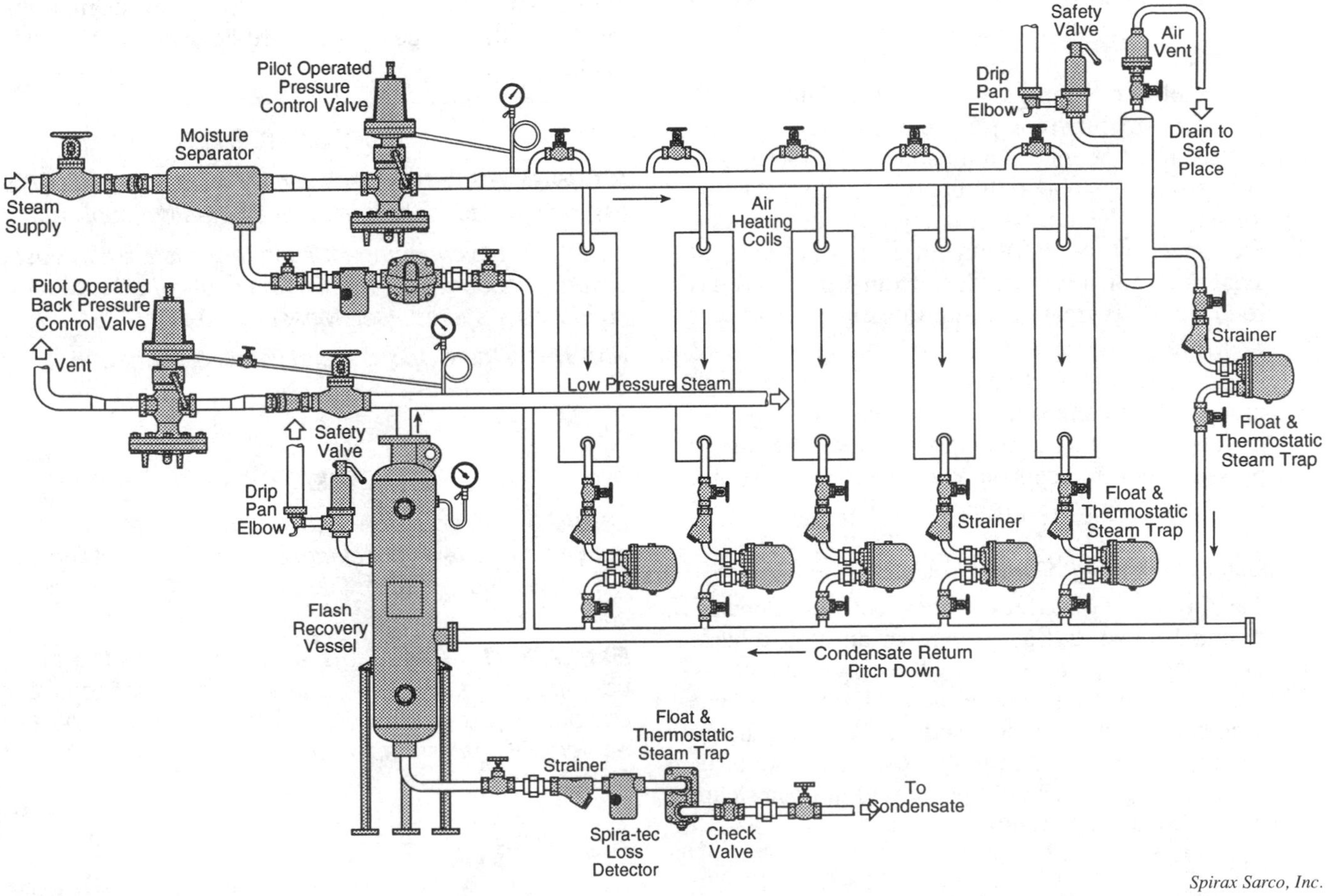

Spirax Sarco, Inc.

Fig. 1 High-temperature condensate heat recovery In this example, condensate from high-temperature air heating coils is drained to a flash tank, where it produces a large quantity of steam. This diagram illustrates a host of steam system accessories, not all of which are necessary for simpler applications.

• Recover the heat locally. If you have a possible low-temperature heating application near the point where the condensate is discharged, discharge the condensate through a heat exchanger that serves the lower-temperature process. This cools the condensate, so it will return through the atmospheric condensate system without flashing.

If you need the recovered heat in the form of low-pressure steam, use a flash tank.

The main feasibility question is whether the low-temperature application can be synchronized in time with the operation of the equipment that discharges the high-temperature condensate.

• If a low-pressure application exists elsewhere, install a low-pressure steam line and allow the condensate to flash into it. If no low-temperature heat recovery application is available, return the flash steam in a separate line. Drain the high-pressure condensate to a flash tank, which produces low-pressure steam. Figure 1 shows a typical application.

The steam is returned to the boiler plant or elsewhere for use in low-pressure applications, such as feedwater heating. This method is feasible only with large volumes of high-pressure condensate. A small quantity of low-pressure steam would condense before it is able to travel any great distance.

• Use a closed, high-pressure condensate system. There is no inherent reason why a condensate system must operate at atmospheric pressure. The condensate system could operate at the discharge pressure of the steam-using equipment, returning the condensate as high-temperature liquid. A closed system eliminates water loss from vents, and has reduced water treatment requirements.

A high-pressure condensate is restricted to applications that discharge at the same high pressure. It cannot accept condensate from any lower-temperature applications. Also, the higher pressure requirements on pipe, receivers, and other hardware may add significant cost.

Feasibility and Energy Saving Potential

The purpose of this Measure is to recover the energy that would be lost as flash steam in an atmospheric condensate system. The amount of energy that appears as flash steam depends on one factor: the temperature at which the steam is condensed in the steam-using equipment. Here is a table that shows the amount of condensate energy that would be wasted in flash steam at different condensing temperatures:

Steam Pressure (PSI)	Condensate Temperature (°F)	Percent of Condensate Energy in Flash Steam at Atmospheric Pressure	BTU's in Flash Steam, per Pound of Condensate
10	240	14	33
35	280	33	82
88	330	47	143
205	390	60	218
655	500	75	365

This table shows that the amount of condensate heat lost by flashing is modest unless the heating application requires a temperature much higher than the atmospheric boiling point. This limits the range of applications of this Measure.

Furthermore, not all the flash steam is wasted in a conventional condensate system. Measure 1.8.4 explains that flash steam maintains the temperature in the condensate system, replacing heat lost by conduction. This effect is valuable in longer condensate systems.

Providing a Steady Flow of Flash Steam

If you want to keep the flow of high-temperature condensate as steady as possible, use steam traps of the float-and-thermostatic type. Bucket traps pass condensate in intermittent slugs. Thermostatic traps, impulse traps, and disk traps release condensate in pulses, but over shorter intervals. Selecting steam traps involves other issues, which are covered in Measure 1.10.3.

ECONOMICS

SAVINGS POTENTIAL: *Typically a few percent of the fuel required to make steam for the heating application.*

COST: *A typical compact flash tank installation costs several thousand dollars. For the costs of an extended condensate system, see Measure 1.8.4.*

PAYBACK PERIOD: *Several years, or longer.*

TRAPS & TRICKS

DESIGN: *If you don't do much design of steam equipment, review your steam theory before tackling this job. Make sure that nothing can obstruct the flow of condensate.*

EXPLAIN IT: *Describe the installation in the plant operating manual. The purpose of the heat recovery equipment is not obvious, so install a placard on the equipment that states what it is and explains how it works.*

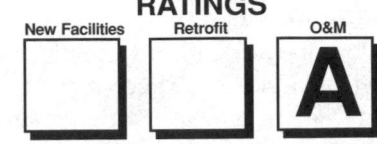

MEASURE 1.8.5 Keep vacuum condensate systems operating properly.

Vacuum condensate systems are sometimes found in facilities that have low-pressure steam heating equipment, such as steam radiators for space heating. The purpose of these systems is poorly understood, so they tend to be neglected. If a vacuum condensate system is defective, the steam-using equipment usually continues to function, but with poor control and reduced efficiency. Fortunately, vacuum condensate systems are simple and they are easy to maintain. If you have one, keep it working properly.

Vacuum condensate systems are similar to conventional atmospheric condensate systems. The piping is laid out in the same manner. The main difference is that the condensate receiver is closed and the entire system is kept below atmospheric pressure by a vacuum pump. The specialized condensate receiver is the item that needs the most attention. Figure 1 shows a typical unit.

Why Vacuum Condensate Systems Are Used

There are three main reasons to operate a condensate system under vacuum. Two of them apply to low-pressure steam heating systems, and one is important for the operation of steam machinery.

■ Improved Control of Low-Pressure Heating Equipment

If the temperature of pure steam inside a heating unit is less than 212°F, its pressure is lower than atmospheric. This has always been a vexing problem in designing heating equipment to operate at temperatures lower than 212°F.

For example, a radiator that is supplied with steam at 230°F is limited to a narrow range of stable temperature control between 230°F and about 210°F. If the steam valve is throttled to the point that the steam pressure falls below atmospheric, air backflows into the radiator from the condensate system, mixing with the steam and reducing the heating capacity of the radiator. This causes the thermostatic control or the human occupant to open the steam valve. Then, the air vent on the radiator gradually purges the air, which restores the radiator's capacity.

This sequence produces a cycle of unstable temperature control. As a result, occupants of buildings with steam radiators tend to keep the radiators turned up to maximum output and they open windows to vent the excess heat. Needless to say, this is a very wasteful way to operate a heating system.

The only practical way to achieve stable temperature control of steam equipment at low temperatures is to use a vacuum condensate system. This allows the steam

SUMMARY

Vacuum condensate systems tend to be neglected because they are poorly understood. Failure causes reduced efficiency and poor control of steam-using equipment.

SELECTION SCORECARD

Savings Potential .... $ $

Rate of Return .... % % % %

Reliability .... ✓ ✓ ✓

Ease of Initiation .... ☺ ☺ ☺

within the equipment to condense and drain at a lower temperature. For example, if the pressure in the condensate system is kept 10 PSI below atmospheric pressure, steam can condense and give up its heat down to a temperature of about 162°F.

■ Improved Steam Distribution for Low-Pressure Steam Heating

As we have seen, a vacuum condensate system allows steam heating equipment to operate at a much lower inlet pressure. Thus, there is a larger pressure differential that is available to drive steam through the system. This allows a low-pressure boiler to serve equipment in a relatively long distribution system without excessively large pipes.

Vacuum condensate systems were essential to the steam heating of early highrise buildings that used radiators. The greater pressure differential helped to overcome the gravity head of the steam in the supply pipes.

■ Improved Efficiency of Steam Machinery

Designing steam machinery to exhaust into a vacuum was one of the major advances in steam engine efficiency that occurred in the nineteenth century. In most cases, the machinery is closely coupled to a condenser, and there is no need for a vacuum condensate system to connect them.

How Vacuum Condensate Systems Work

A vacuum condensate system is simply a closed condensate system. In the absence of air, the pressure inside the condensate system is determined solely by the temperature of the water vapor and condensate. The system uses a vacuum pump to extract air and other non-condensing gases out of the condensate system.

The only unusual equipment in a typical vacuum condensate system is a specialized condensate receiver, which includes the vacuum pump and the condensate pump. There is also an air separator that recovers water vapor from the air that is pumped out of the system.

Maintenance

The two main items of maintenance for a vacuum condensate system are minimizing leakage of air into the system and keeping the vacuum pump in good operation. Air leakage reduces the vacuum inside the condensate system, and increases the energy used by the vacuum pump. Vacuum condensate receivers come in a number of designs. They are fairly easy to repair if you follow the manufacturer's instructions.

The condensate receiver has a vacuum gauge. Keep it in good condition and properly calibrated. This gauge is your primary indication of system performance. If the condensate system is spread out, it should have vacuum gauges at one or more extreme locations. These help you to identify air leakage that is distant from the condensate receiver.

Explain It!

One of the main obstacles to maintenance is that the purpose and operation of the system is not obvious to plant operators. If the system fails, they may not understand that problems with temperature control throughout the building are being caused by a failed vacuum condensate unit located in the basement. Therefore, install a placard near the equipment that states what it is and explains how it works. Be sure to state the vacuum that should be maintained. See Reference Note 12, Placards, for details of effective placard design and installation.

ECONOMICS

SAVINGS POTENTIAL: *5 to 50 percent of heating system energy. Failure of the vacuum system causes people to operate steam heating equipment in a wasteful manner.*

COST: *A few hundred dollars per year, on average.*

PAYBACK PERIOD: *Immediate. Appropriate maintenance is always cheaper than neglecting it.*

TRAPS & TRICKS

EXPLAIN IT: *Install a prominent placard at the vacuum pump that identifies the equipment, describes how it works, and states the vacuum that should be maintained in the system. Train new operators about the system.*

WESINC

Fig. 1 Vacuum condensate receiver The receiver is the main hardware difference between a vacuum condensate system and a vented system. This receiver has a vacuum pump that sucks air and other non-condensing gases out of the system, lowering the pressure to a value that is determined by the condensate temperature. Reading the vacuum gauge is the primary way of telling whether the system is operating properly.

MONITOR PERFORMANCE: *Keep checking the operation of the system. Add a column to your plant* *operating log to record condensate system vacuum. Then, review the logs.*

	RATINGS	
New Facilities	Retrofit	O&M
		B

MEASURE 1.8.6 Replace pump motors with models having the highest economical efficiency.

Be prepared to replace a failed motor with a model that has the highest practical efficiency. If a motor is exceptionally large and inefficient, you may even want to replace it before it fails. See Measure 10.1.1 for the details.

A fuel oil system may use energy to pump the oil from the storage tank to the burner, to pressurize the oil for atomization, and to heat the oil for pumping and burning. In most cases, these functions absorb only a small fraction of the boiler plant's energy consumption. Still, improving these functions may provide significant energy savings. The Measures of this Subsection exploit the potential for saving energy and cost in performing these functions.

Fuel oil that is burned in land-based boilers is commonly classified by numbered grades, ranging from Number 1 to Number 6. The properties of fuel oil can vary considerably within a grade. They depend on the location of origin, the refining and "cracking" processes used, and blending that is done by the seller. See Reference Note 20, Fossil Fuels, for an overview of the properties of fossil fuels.

MEASURE 1.9.1 Adjust fuel oil temperature to provide the optimum viscosity for burner efficiency.

Viscosity is the thickness, or resistance to flow, of a liquid. Fuel oil viscosity must stay within a particular range in order for the burners to function properly. This is a potential problem because viscosity varies widely with temperature, and from one batch of oil to another.

The viscosity of heavy oils is adjusted for combustion by heating it in a thermostatically controlled heater that is located in the fuel line ahead of the burners. Figure 1 shows a typical fuel oil heating station. (Heaters may also be installed in storage tanks to preheat the oil to make it easier to pump.)

You should control the viscosity of your heavy fuel oils, typically, Number 4 and heavier. Lighter oils have viscosities that are low enough at normal storage temperatures to atomize properly without preheating. There are exceptions. Some burners for Number 2 oil have heaters as an aid for cold starting.

You can install controls that automatically adjust fuel oil viscosity, which eliminates the need for manual testing or adjustment. See Measure 1.9.1.1, next, about installing this equipment.

Why Fuel Viscosity Matters

If the viscosity of fuel oil is too high, the burner cannot atomize it properly. This causes incomplete combustion and sooting. Trying to compensate for the incomplete combustion by increasing the amount of combustion air would significantly reduce the combustion efficiency of the boiler. Poor atomization causes a longer flame length because the fuel is thrown farther into the boiler before it is burned. In extreme cases, the flame may reach the inside surfaces of the boiler, fouling the surfaces and quenching the flame.

> **SUMMARY**
>
> Good operating practice to maximize burner efficiency when burning No.4 and heavier oils.
>
> **SELECTION SCORECARD**
>
> Savings Potential .... **$**
>
> Rate of Return .... **% % % %**
>
> Reliability .... ✓ ✓
>
> Ease of Initiation .... ☺ ☺ ☺

If fuel viscosity is too low, the fuel components that are more volatile may separate from the less volatile components within the burner and ignite prematurely. This disrupts combustion of the remaining fuel and fouls the burner. In the worst case, the flame can be blown out, creating an explosion hazard as raw fuel enters a hot firebox. Excessive oil temperature increases the risk of a boiler room fire in the event that a fuel line ruptures.

Excessive oil heating also consumes additional energy in the fuel oil heater. Most of this heat is recovered inside the boiler. If the energy source for heating is more expensive than boiler fuel (for example, if the boiler uses an electric fuel heater), excessive heating wastes money.

Expect Fuel Oil Viscosity to Vary

Heating fuel oil to a particular temperature maintains proper viscosity if the characteristics of the fuel remain constant. However, viscosity can vary dramatically within a given grade of fuel. For example, different Number 5 fuel oils may vary in viscosity from 150 to

WESINC

Fig. 1 Heater for heavy fuel oil Keep the temperature set to provide best burner efficiency.

800 SSU units at the standard test temperature. (Burners for heavy fuel are designed to operate with a fuel viscosity close to 150 SSU units.) Furthermore, if fuel is stored for a long time, viscosity may change with time within a given batch as the lighter components of the oil float toward the top of the storage tank.

Finding the Optimum Viscosity by Combustion Efficiency Testing

In principle, the most accurate way to optimize fuel oil viscosity is to adjust the fuel oil temperature until the combustion efficiency of the boiler peaks, while providing the correct amount of excess air. Combustion efficiency test methods are explained in Subsection 1.2, and the significance of excess air is explained in Subsection 1.3.

There is a tricky aspect to this adjustment. The combustion efficiency test is not able to account for unburned fuel. Improving atomization slightly increases the amount of fuel that is burned, which increases the stack temperature. The combustion efficiency test interprets the higher stack temperature as a reduction in efficiency. To correct for this, keep the carbon monoxide or smoke level constant as you perform the tests. Combustion efficiency testing also reveals any loss of efficiency that may be occurring for other reasons.

Rapid-reading test equipment makes this adjustment less tedious. See Measure 1.2.1 for guidance in selecting of boiler efficiency test equipment.

Alternative: Measure the Fuel Oil Viscosity

An alternative way to adjust for differences in fuel batches is to measure the viscosity of the new batch of fuel. A variety of fairly simple and inexpensive methods are available for testing viscosity. The simplest is just a container with a calibrated orifice. Calculate the viscosity by timing how long it takes for a sample of oil to drain out of the container. Another method measures the friction drag on a disk spinning in a sample of oil. Viscosity testing equipment is available from chemical supply houses and boiler equipment suppliers.

To achieve meaningful comparisons, conduct all viscosity tests at the same temperature. This requires a thermostatically controlled heater that consistently heats fuel oil samples within a tolerance of a few degrees. Chemical supply houses have expensive heaters for this purpose. A fondue pot with a good thermometer serves just as well.

■ Units of Viscosity

A system of viscosity units used specifically with petroleum is Saybolt Seconds. This system is based on the number of seconds required for a sample of oil to drain out of a standardized cup. To complicate matters, there are two different Saybolt systems of measurement. Saybolt Seconds Universal (SSU) is used with lighter oils, and Saybolt Seconds Furol (SSF) is used with heavier oils. SSU is measured at 100°F, and SSF normally is measured at 122°F.

The metric unit of viscosity is the "poise." Another parameter called "kinematic viscosity" is measured by a unit called the "stoke." As conversion to the metric system becomes universal, you may encounter these units in fuel oil specifications. Regardless of the units that are used, the key point is that the viscosity of the fuel must match the viscosity requirement of the burners and pumps.

How Often to Adjust Viscosity

Adjust fuel oil viscosity any time that a significant change in fuel properties may occur. Viscosity and other properties vary depending on the origin of the oil, and on the refining and "cracking" processes that it undergoes. Even if the same supplier is used, changes in blending by the supplier may significantly change viscosity. For example, fuel suppliers may change the blending of heavy oil during cold weather to provide easier pumping.

The safest practice is to adjust viscosity any time a new batch of fuel is started. This includes fuel that has sat undisturbed on the premises for a long period of time, for example, from a previous heating season. Stagnant fuel in a tank may stratify into layers of different viscosity. To compensate for stratification, you would need to adjust the fuel oil temperature as the fuel level is drawn down. To avoid this problem, consider preventing stratification with the methods recommended by Measure 1.9.3.

Batches of fuel oil may differ in a number of other important characteristics. Differences in energy content require adjusting the air-fuel ratio, which is covered in Subsection 1.3. Various adverse characteristics of fuel oil can be treated with additives and other methods, which are covered by Measure 1.9.3.

ECONOMICS

SAVINGS POTENTIAL: 0.1 to 1.0 percent of boiler fuel cost.

COST: Cup-type viscosity measurement devices are available for less than $100. Other types of viscosity test equipment can be much more expensive, but they are not necessary for this purpose. See Measure 1.2.1 for the costs of efficiency testing.

PAYBACK PERIOD: Short, in boiler plants that burn a substantial amount of heavy fuel oil.

TRAPS & TRICKS

TRAINING AND DILIGENCE: Viscosity testing is not complicated, but it is a kind of procedure that tends to be forgotten. Use the boiler operating log to tell operators when to adjust for viscosity, for example, "when changing fuel tanks."

MEASURE **1.9.1.1 Install automatic fuel oil viscosity control equipment.**

You can install controls that automatically measure and adjust fuel oil viscosity. This eliminates the need for manually readjusting the fuel oil heater temperature. Figure 1 shows a typical controller.

Automatic viscosity control is built around a viscosity sensor that is installed on the discharge side of the fuel oil heater. The viscosity measurement generates a standard pneumatic, electric, or electronic signal that is used to reset the temperature of the fuel oil heater.

Different types of viscosity sensors are available. The choice depends on the range of oil viscosities, whether the sensor is subject to high pressure, and whether the sensor is to be installed in a tank or pipe.

SUMMARY

Another automatic control that reduces the need for attention and labor, at the cost of additional plant complexity. Economical for larger boilers that burn heavy oil.

SELECTION SCORECARD

Savings Potential .... **$**

Rate of Return, New Facilities **%** %

Rate of Return, Retrofit.......... **%** %

Reliability .... ✓ ✓ ✓

Ease of Retrofit .... ☺ ☺ ☺

Don't have blind faith in an automatic control. To check on the performance of the viscosity control system, install a viscosity meter (usually included as part of the equipment) and a fuel oil thermometer where they are plainly visible to plant operators.

ECONOMICS

SAVINGS POTENTIAL: Somewhat more than with the manual methods of Measure 1.9.1.

COST: $10,000 to $20,000.

PAYBACK PERIOD: One year to many years, depending mainly on the amount of heavy oil burned by the boiler. For example, the payback period might be about one year for a boiler steaming at an average rate of 50,000 pounds per hour.

TRAPS & TRICKS

SELECTING THE EQUIPMENT: Get the equipment from a leading manufacturer of boiler equipment. Ask vendors for references, and talk with purchasers who have used the equipment for several years.

SKILLS AND MAINTENANCE: This equipment saves labor, but does not reduce skill requirements. Train appropriate operators how to adjust the equipment.

EXPLAIN IT: The purpose of the equipment is not obvious to plant operators, so install a placard that identifies it and describes how it works.

SCHEDULING: Improper operation of this kind of auxiliary equipment is easy to ignore. Schedule periodic checks of the controls in your maintenance calendar.

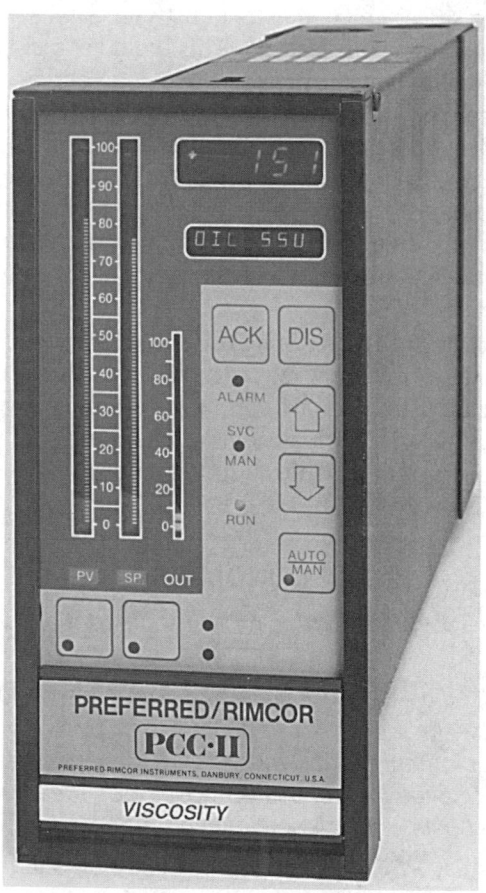

Preferred Utilities Manufacturing Corp.

Fig. 1 Automatic viscosity controller You could guess that this is simply a standard programmable logic controller (PLC) that has been programmed to maintain optimum viscosity. It needs input from an automatic viscosity measuring device.

MEASURE **1.9.2 Use the most economical heat source for fuel oil heating.**

SUMMARY

Applies to heavy oil, especially if stored in exposed tanks and pumped long distances. May be a good application for waste heat.

SELECTION SCORECARD

Savings Potential .... **$**

Rate of Return, New Facilities **% %** % %

Rate of Return, Retrofit.......... **% %** %

Reliability .... ✓ ✓ ✓ ✓

Ease of Retrofit .... ☺ ☺ ☺

Heavy oil requires heating in order to atomize and burn properly. Heating may also be needed to allow heavy oil to be pumped from the storage tank. Electricity is often used as a source of heating energy because electric fuel oil heaters are inexpensive and easy to install. Also, electric heating allows starting up a cold plant. However, if the fuel used by the boiler is significantly less expensive than electricity, it may be worthwhile to use steam or hot water for fuel oil heating.

In calculating the cost of heating with steam, consider the boiler efficiency. Also, note that condensate from fuel oil heaters usually is dumped. This is to avoid the danger of contaminating the boiler if a leak develops inside the fuel oil heater. The most economical situation is using waste steam to heat fuel oil, if enough waste steam is available.

Consider each of the three main portions of the fuel oil system:

- *storage tank heating.* This may use a substantial amount of energy if the storage tank is located above ground. The storage tank should have a hot well, which is an isolated area surrounding the discharge point that can be heated without heating all the fuel in the tank. Tank heating is a good application for steam or hot water heating, including condensate and flash steam.

- *transfer piping* is heated typically with steam tracing lines. If the piping is exposed to external conditions, the steam requirement varies widely depending on the weather. This is not a good application for waste heat unless the source is available continuously.

- *pre-combustion heating,* immediately ahead of the burner, can easily be accomplished with a steam heat exchanger.

Using Waste Heat to Heat Fuel Oil

Fuel oil heating is an attractive application for waste heat because it allows recovery of heat at low temperature. Larger boiler plants often have a considerable amount of otherwise useless low-temperature heat, such as the heat contained in blowdown water and condensate flash steam, the exhaust steam from auxiliary equipment drive turbines, etc. Another advantage of this application is that the need for fuel oil heating and the availability of waste heat both tend to be proportional to boiler load, so that availability of heat is synchronized with need.

If you have an excess of waste steam for heating the storage tank, route the steam to the hot well first. Let any remaining steam heat the bottom of the tank

through separate coils. Keeping the oil warm reduces stratification and sludge formation. See Measure 1.9.3 for details.

Back up the waste heat source with a primary source of energy, such as a low-pressure steam main. If waste heat is used to supplement another heat source, be sure to install a placard at the relevant valves or controls so that operators understand the arrangement. Otherwise, operators may valve off the waste heat source and rely solely on the back-up source. See Reference Note 12, Placards, for details of effective placard design and installation.

ECONOMICS

SAVINGS POTENTIAL: *A fraction of the cost of fuel oil heating energy. Depends on the relative costs of different energy sources that might be used for fuel oil heating.*

COST: *Varies widely, depending on the installed equipment and available options.*

PAYBACK PERIOD: *Less than one year, to many years.*

TRAPS & TRICKS

DESIGN: *Consider all potential heat sources, including waste heat. Heating fuel pipes with steam in freezing conditions is tricky. The major steam trap manufacturers can help you with this.*

EXPLAIN IT: *Describe the fuel oil heating procedures in the plant equipment records. If more than one heat source is available, use the plant operating log to remind operators to use the appropriate one. Install placards at the valves or controls used to change heat sources. Include this in your training for new plant operators.*

MONITOR PERFORMANCE: *Review the plant operating logs periodically to check that temperatures are correct and that the correct heat sources are being used.*

MEASURE **1.9.3 Use fuel oil additives to improve combustion efficiency and/or improve other fuel oil properties.**

A wide variety of fuel oil additives have been marketed over the years. Some of these provide a real benefit, some are harmful, and some have effects that are minimal or unknown. Most additives are marketed to correct problems associated with heavy oil. Light oil is generally free of troublesome conditions that require corrective action.

Figure 1 shows a fuel oil additive being injected into a boiler that burns heavy oil.

Alleged Benefits and Mechanisms of Action

Some fuel oil additives claim to correct specific conditions, while others claim a range of benefits. Here are some of the benefits claimed by various fuel oil additives.

■ Improved Atomization

Good burners are capable of burning fuel virtually completely. However, all burners require some excess air to achieve complete combustion because they cannot mix the air and fuel in perfect proportions. Some fuel oil additives claim to improve atomization, on the theory that finer atomization allows combustion to be completed closer to the burner, reducing the requirement for excess air. Whether this is true in practice is questionable. These are two ways that different additives claim to improve atomization:

• *reducing surface tension.* Some additives reduce the surface tension of the oil, allowing it to break up more easily.

• *steam formation in the oil droplets.* The oldest additive claimed to improve combustion is water. The theory is that water added to the fuel explodes into steam in the heat of the firebox, breaking apart large droplets of oil. The purpose of the additive is to hold the water in emulsion in the fuel oil.

In order for this to work in theory, the microscopic water droplets in the oil must be much smaller than the oil droplets themselves. The quantity of emulsifying agent is inversely proportional to the water droplet size, which means a lot of emulsifying agent is needed if the droplets are to be small. If less than the required amount of emulsifier is added, the water droplets are too large to be useful.

(Emulsifiers work by encapsulating water droplets in a chemical film. The volume of the droplet drops in proportion to the third power of its radius, but the surface area drops only in proportion to the second power, so encapsulating a given quantity of water in smaller droplets requires much more emulsifier.)

SUMMARY

Fuel oil additives range from useful to harmful, and some are simply useless. Understand the properties of any additive you consider. If an additive claims to solve a boiler problem, you may be able to solve the problem better in other ways.

SELECTION SCORECARD

Savings Potential .... $

Rate of Return .... % % % %

Reliability .... ✓ ✓ ✓ ✓

Ease of Initiation .... ☺ ☺

■ Enhancing Combustion by Catalytic Action

Heavy fuel oil consists of large molecules. These go through a series of chemical reactions in the process of being reduced to carbon dioxide and water. Combustion catalysts hasten the decomposition by reducing the input energy required for each of the intermediate reactions. Catalysts typically are metals or metal compounds.

Modern burners that are properly matched to the combustion chamber are capable of almost completely burning fuel, so catalysts cannot significantly improve burning efficiency. However, quicker breakdown of the fuel might reduce the need for excess air with some burners.

■ Reduced Fireside Fouling

Improving the atomization of fuel prevents the formation of large particles of combustion products that cause smoke and adhere to boiler tubes. However, a properly operating burner does not form excessively large particles of fuel. Therefore, additives cannot improve this aspect of combustion.

Additives may prevent fouling by the non-combustible fuel contaminants that are common in heavy oil. For example, it is believed that iron, sodium, and vanadium compounds that deposit on boiler tubes act as catalysts for forming corrosive compounds from the sulfur in fuel. Some additives change the melting temperature of the metal compounds so that they do not adhere to the tube surfaces. Other additives soften the fouling so that it can be removed more easily.

Abnormal fireside deposits may occur for a variety of reasons that cannot be corrected by fuel oil additives. For this reason, you should diagnose the cause of fouling before resorting to additives.

■ Preventing Fuel System Corrosion by Fuel Acids

Fuel oil contains a variety of contaminants that either are acidic or that form acids in combination with water. The acids cause corrosion throughout the fuel oil system, especially in storage tanks. Additives may protect against acids by:

- *neutralizing the acids*, with alkaline substances.
- *forming a protective film at the metal surface*. A number of different mechanisms of action are claimed.

■ Eliminating Free Water in the Fuel

Water in fuel causes a variety of problems. It corrodes the system directly, and it combines with sulfur and other matter in the fuel to form corrosive acids. It is also a factor in the formation of sludge in tanks. In large amounts, it may interfere with combustion. Additives use two opposite approaches to dealing with water:

- *keeping the water in the fuel emulsified* to minimize its contact with metal surfaces. This is done with a variety of agents that act as emulsifiers.
- *causing water to settle out.* This is done with "demulsifying" agents. If you add a demulsifier, you must promptly pump out the water that separates, or it will corrode the bottom of the tank.

■ Reducing Sludge Formation

Sludge forms in the bottom of fuel tanks. The sludge is a combination of impurities in the fuel oil, corrosion residue, heavy fractions that settle out of the oil, and chemical changes in the oil that cause semi-solid compounds to form. Water plays a part in chemical reactions that form sludge. Additives may inhibit sludge formation by these processes:

- *keeping particles from clumping together.* "Dispersants" are the class of additives that perform this function.
- *inhibiting sludge-forming chemical reactions.* Different reactions occur in different circumstances, so the additive must be tailored to the fuel conditions.
- *keeping free water out of the fuel.* Water is involved in forming sludge. The methods of getting rid of water that were mentioned previously also reduce this cause of sludge.

■ Preventing Stratification in Storage

Heavy oil separates into fractions of different weight if it sits for a long time, and the lighter fractions rise above the heavier ones. The extent of this stratification depends on the oil composition, storage temperature, and mostly, the length of time that the oil has been stagnant. The primary means of reducing stratification are rotation of the oil stock and recirculation. Some additives may reduce stratification, especially those intended to reduce sludge formation.

WESINC

Fig. 1 Fuel oil additive being injected into a boiler's fuel oil system This is a test being conducted to determine the actual improvement in performance.

Be Skeptical

The effectiveness of fuel oil additives is controversial. A particular fuel oil additive may be valuable, or useless, or harmful. Be clear about diagnosing any conditions that an additive is supposed to cure, or else the additive may correct a problem that does not exist in your plant. Judging the merits of a fuel oil additive may require outside expertise. Do not use any additive without the advice of an unbiased expert who has the ability to evaluate the claimed benefits.

Additive vendors may be secretive about the composition of their products. It is reasonable to suspect that this is because the contents are ordinary materials being sold at a high markup. You can determine the contents of an additive by sending it to a laboratory for chemical analysis, but this does not answer the question of whether the additive will actually provide a benefit in your particular boiler.

A boiler efficiency test will tell you whether there is room for improving efficiency. No additive can improve efficiency beyond the heat exchange characteristics of the boiler. (Measure 1.2.1 covers efficiency testing.) If the additive is proposed to correct a maintenance or fuel storage problem, consult an expert in boiler plant operations to find the best solution, which may or may not involve fuel additives.

Be suspicious of tests that claim to demonstrate the effectiveness of an additive. Even if an objective observer witnesses a test, any measured improvement of efficiency may be mostly the result of cleaning and adjusting the boiler prior to the test. The only reliable way to test efficiency improvement claims is to rig a test setup that allows the boiler to switch back and forth between treated and untreated fuel.

It is difficult to verify a claim that a particular additive improves maintenance. A long time must pass for an effect to be noticeable. During this time, other conditions related to the problem may change.

Alternatives to Fuel Additives

Fuel oil additives almost always address problems that can be corrected in other ways. The most direct solution to a fuel oil problem is to purchase fuel oil that has better properties, e.g., less sulfur, less water, refining by a process that has less tendency to form sludge, etc. An informed purchaser can avoid most fuel oil problems at the outset by specifying the fuel properly and checking it when it arrives.

Recirculation prevents stratification and probably reduces sludge formation and water settling.

Another approach is to keep heavy fuel warmer in storage. This reduces the solidification of heavier fractions, prevents formation of wax crystals that interfere with atomization, reduces condensation of water, and reduces sludge formation. If the bulk of the oil in a tank is kept heated, consider improving the tank

insulation to save energy. Also, see Measure 1.9.2 about using waste heat to keep the fuel oil warmed.

ECONOMICS

SAVINGS POTENTIAL: Additives can improve efficiency less than one percent in a modern, properly operating boiler. If a condition is causing a larger efficiency loss than this, fix the underlying problem. In the case of additives claimed to reduce maintenance problems, calculate the savings on an individual basis.

COST AND PAYBACK PERIOD: These vary too widely to allow a general estimate.

TRAPS & TRICKS

SELECTING THE ADDITIVE: Even legitimate additives can provide only small improvements in boiler operation, so it is difficult to measure the actual benefit provided. Discuss any additives you are considering with boiler operators who have used them. But, be skeptical. People tend to believe in bogus cures. If possible, arrange for an objective test in one of your boilers.

USAGE: If an additive is appropriate for your plant, it must be used properly to be effective and economical. Provide clear instructions to operators. Keep the injection equipment working properly. Schedule periodic checks to verify that the additive is having the intended effect, and no adverse effects.

	RATINGS		
MEASURE 1.9.4 Replace pump motors with models having the highest economical efficiency.	New Facilities	Retrofit	O&M
			B

Be prepared to replace a failed motor with a model that has the highest practical efficiency. In some cases, it may even be economical to replace a functioning motor

with a more efficient model. See Measure 10.1.1 for the details.

Steam And Water Leakage

The Measures in this Subsection prevent loss of steam and water in the heating plant and its distribution system. Leakage can be a major source of energy waste and maintenance costs, especially in plants with extensive distribution systems. Most of the Measures in this Subsection are routine maintenance functions. The exception is Measure 1.10.3, which replaces inefficient types of steam traps with more efficient types.

INDEX OF MEASURES

1.10.1 Monitor boiler system water loss.

1.10.2 Locate and repair steam and water leaks at appropriate intervals.

1.10.3 Use the most efficient type of steam trap for each application.

1.10.4 Test and repair steam traps on a continuing basis.

    1.10.4.1 Install accessory devices to assist in steam trap diagnosis.

    1.10.4.2 Hire specialists to perform periodic steam trap inspections.

1.10.5 Recover heat and water from steam vents.

RELATED MEASURES

- Measure 1.1.5, reducing system pressure to minimize system leakage
- Subsection 1.8, for recovery of steam and condensate in condensate return systems

MEASURE **1.10.1 Monitor boiler system water loss.**

Steam and water carry the energy of the boiler system. If the system leaks, the energy is lost along with the steam and water. Leakage is worst in systems that are long, in poor condition, and operating at higher pressures.

Monitoring leakage is the first step in controlling it. You can track leakage easily by installing a water meter on the makeup water line. In most systems, the amount of makeup water equals the loss, including losses both inside and outside the boiler plant. Recording makeup water is the only reliable method you have for detecting and measuring leakage that is invisible, such as from underground distribution pipe.

The Costs of Water Loss

Measuring water loss is important primarily because it indicates the amount of energy by lost from the system. For example, loss of one pound of water in the form of steam wastes about 1,000 BTU. Water loss also incurs the cost of treating makeup water, along with the cost of the replacement water itself.

Condensate loss wastes much less energy than steam loss, for a given weight of water that is lost. The makeup water reading itself cannot distinguish between the two. If the entire steam system has condensate return, it is likely that most water loss is in the form of steam, because steam is under pressure. But, this is not true of all steam systems. A defective condensate system can waste a large amount of water. See Measure 1.8.5 for improving condensate return.

Record Water Loss Patterns to Locate Leaks

Record the amount of makeup water on the boiler plant operating log. A time plot of makeup water consumption, at hourly or shorter intervals, may provide valuable clues about the causes of water loss. See Measure 1.10.2 for more about this.

Makeup water readings at the boiler plant cannot distinguish steam or water that is consumed in processes, such as steam laundry presses, aircraft catapults, etc. Also, there may be portions of the distribution system that dump condensate, rather than returning it to the boiler plant. You can distinguish these steam losses by

---

SUMMARY

An fundamental aspect of boiler plant management. A makeup water meter is the only equipment you need.

SELECTION SCORECARD

Savings Potential .... **$ $ $**

Rate of Return .... **% % % %**

Reliability .... ✓ ✓

Ease of Initiation .... ☺ ☺ ☺

---

installing steam meters at the individual items of equipment, or on the portions of the distribution system that do not return condensate. (See Measure 1.8.4 about returning condensate to the boiler plant.)

Act on the Information

Analyzing water loss is a useless effort unless you take action to reduce it. Measure 1.10.2 spells out the appropriate corrective action.

ECONOMICS

SAVINGS POTENTIAL: *Up to 20 percent of fuel cost; plus, reduced water treatment cost; plus, reduced makeup water cost. Leakage varies widely, depending on the characteristics of the system. Credit the savings to the specific actions that you take to correct the leakage.*

COST: *Minimal, if a makeup water meter is already installed. Installing a water meter typically costs several hundred dollars.*

PAYBACK PERIOD: *Cannot be stated in terms of this activity alone.*

TRAPS & TRICKS

DILIGENCE: *This kind of good plant management practice tends to be neglected. Include a column for makeup water consumption in the boiler plant operating log. Schedule periodic analysis of water consumption in your maintenance calendar. If monitoring indicates significant water loss, act on the information.*

MEASURE 1.10.2 Locate and repair steam and water leaks at appropriate intervals.

Monitor system water loss as recommended in Measure 1.10.1. If your water loss is significant, search for leaks on a regular schedule. Repair visible leaks as soon as they appear. Individual leaks grow rapidly. Once a leak starts, the flow of steam or water erodes the passage, enlarging it.

In some plants, fixing leaks may be neglected because the task looks overwhelming. A fog of steam leaking from pipe joints, valve packings, pump glands, and vents gives the false impression that leaks are everywhere. Not every situation is as bad as the one shown in Figure 1, but steam leaks always waste significant amounts of energy. The solution is to start repairs one leak at a time, just like picking up trash along a road. In less time than you expect, the leaks are fixed and losses are greatly reduced.

Potential Savings

Leakage from the distribution system continues as long as the system is under pressure. Furthermore, the rate of leakage is almost constant, regardless of the heating load, because the system pressure is what forces the steam and water out of the system. Because the leakage is continuous, even small leaks waste a considerable amount of steam or hot water. In facilities that have extensive hydronic or steam distribution, serious leakage may waste a large fraction of the total energy input. The costs incurred by leakage include:

• the energy consumed to heat the lost steam or water

SUMMARY

Important basic boiler plant maintenance. You can find leaks almost anywhere they occur, if you have the right skills and equipment.

SELECTION SCORECARD

Savings Potential .... $ $ $

Rate of Return .... % % % %

Reliability .... ✓ ✓ ✓ ✓

Ease of Initiation .... ☺ ☺ ☺

• the cost of the makeup water itself
• treating makeup water
• increased blowdown
• loss of antifreeze, if used, in hydronic systems
• increased maintenance cost for heating system components caused by contaminants in the makeup water.

Analyze Makeup Water as the First Step in Finding Leaks

Measure 1.10.1 recommends monitoring your makeup water rate. This tells you when you need to check for leaks, and it may also tell you the general areas where leaks are occurring. A time plot of makeup water consumption at hourly, daily, and monthly

WESINC

Fig. 1 Major steam leaks This government facility was wasting a lot of taxpayers' money by letting steam escape from the boiler plant and the distribution system.

intervals may provide strong evidence about the causes of water loss.

As we said before, water loss from the distribution system tends to remain constant, as long as the pressure in the system remains constant. This is an approximation. The downstream pressure in a long distribution system changes with load because of flow resistance. Also, you may want to modulate the system pressure to follow the load to reduce water and heat loss. See Measure 1.1.5 for more about this.

An increase in water loss at certain times indicates leakage or water consumption by certain equipment, or leakage from parts of the distribution system that are used only at certain times. For example, if you have laundry presses, examine your makeup water records to see how much the water loss increases during the hours when the presses are operating to estimate the amount of water they are consuming. Similarly, if water loss increases during cold weather, suspect leakage from space heating equipment.

Don't bother trying to estimate the loss from individual steam leaks that you can see. Fix all leaks, regardless of size. Little leaks become big leaks.

Leaks in Pipe Systems

Expect all sealing materials used in pipe systems to develop leaks eventually, and to require occasional replacement. Sealing materials fail because of hardening, loss of adhesion, cracking, and corrosion. Seals around valves and other moving components also wear mechanically. The most common sealing elements are gaskets and valve packings.

Leakage is hastened by thermal expansion and mechanical motion, which tend to concentrate stresses at the joints where the sealing materials are used. A system design that fails to allow for thermal expansion and ground motion invites leakage at joints. The worst design mistake is to allow bending stresses to occur at joints.

As the system ages, the pipe itself starts to leak. This is most likely to occur first at welds where flanges and fittings are attached, and may also occur along the longitudinal weld of the pipe itself.

Older steam pipe may develop pinhole leaks. These start with pits formed inside the pipe by oxygen corrosion resulting from inadequate water treatment. Pinhole leaks concentrate at localized flaws in the steel. If pinhole leaks start to proliferate, the piping system is probably ready to be scrapped.

Condensate pipe deteriorates most seriously from carbonic acid corrosion. This tends to uniformly eat away the thickness of the metal that is in contact with the liquid condensate. Therefore, the bottom of condensate pipe usually leaks first.

The outside of pipe may suffer massive corrosion if it remains wet. This is a major problem of buried pipe. Pipe in this condition cannot be repaired economically, and it must be replaced sooner or later. Repairs can be very expensive, in the multi-million dollar range for large systems. To place this in perspective, serious leakage in a large system may cost millions of dollar in energy in a few years.

If you cannot survey your pipe systems adequately yourself, consider hiring pipe surveying services. Specialists who do this work have the latest equipment for detecting leaks, eliminating the need for you to purchase the equipment and learn how to use it. As with all specialized services, check references thoroughly before signing the contract.

■ How to Find Leaks in Above-Ground Pipe Systems

Leaks are generally easy to find in parts of the system that are visible. This includes equipment in the boiler house, in steam tunnels, in pipe chases, in attics, etc. The main part of the job is tracing all the pipe.

In hot water distribution systems, leaks are usually easy to find as drips. Steam leaks may create a visible plume, but not always. Try to search for steam leaks on a humid day, as they are more visible then. In climates where the air is dry, small steam leaks may be difficult to find.

You can listen for steam leaks, but this works only in a quiet environment, or when you are very close to the leak.

You can also detect leaks by feel, passing a hand around areas where leaks may occur, such as gaskets and packings. Do not feel too closely, as this risks burns from the steam itself and from hot surfaces. Do not use this technique in high-pressure plants, where leaking steam can travel far and burn severely.

■ How to Find Leaks in Buried Pipe Systems

Underground conditions in many locations promote pipe deterioration, so buried distribution systems typically account for most steam and water loss.

Leaks may be difficult to find in buried pipe. You may be able to find them from visual clues. In cold weather, leakage may cause faster snow melting above the location of the leak. In warmer weather, the water leaking from a pipeline may stimulate the growth of vegetation above the leak.

You can use infrared thermal scanners to detect leaks in buried pipe. They work by detecting small temperature differences on surfaces. The escaping steam or water heats the material at the point of leakage, and this temperature change is detected by the scanner. Infrared sensing equipment is not reliable for detecting escaping steam or water itself. This is because gases and liquids do not radiate strongly in the infrared spectrum. Infrared scanning works best in cold weather, at night. See Reference Note 15, Infrared Thermal Scanning, for more about this method.

Abnormal surface warming over a distribution system may be caused by leakage or by defective insulation. Defective insulation usually appears on an infrared scan as an area of broad warming, whereas leakage shows up as a localized hot spot. To resolve the question, you may have to dig into the hot spot. Check insulation condition at the same time you check for leaks. See Measure 1.11.1 about improving insulation condition.

Leaks in buried low-temperature hot water systems (and in buried chilled water systems) may not be detectable from changes in surface temperature.

If you find bad pipe, determine whether the pipe deterioration is local or widespread. The most reliable inspection procedure is to remove a section of pipe in a part of the system where leakage occurs, strip the insulation off the pipe, and examine the exterior and interior surfaces in detail. If leakage appears to be a possible major problem, examine sections at different locations in the distribution system. If you suspect severe corrosion, sacrifice a length of pipe for sectioning and metallurgical examination. Specialists in this type of work can diagnose the cause of the corrosion by inspecting such samples.

You can inspect the interior of pipe with a specialized video camera that you insert into the pipe anywhere you can create an opening. However, this method does not tell you the condition of the pipe's outside surface.

■ Don't Skimp When Repairing Joints

Leaks occur most commonly at pipe joints, and these are the easiest leaks to repair. However, these repairs do require some care. Discard all the old sealing materials and components, such as gaskets. These materials are not intended to be reused, and they will fail quickly if they are. Remember that energy loss and labor, not expendable materials, are your major cost items in repairing leaks.

Refinish the metal surfaces that contact the sealing materials. You can buff out mild erosion with abrasive cloth. More severe erosion requires cutting or grinding the metal surface. Be sure to finish the metal surface flat, and make the surfaces of adjacent pipe or fittings parallel. If the sealing surface is badly worn, try to take the pipe section to a machine shop. If you have to do major refinishing in place, make sure that you have the proper tools to do the job properly.

If you cannot stop a valve packing leak by tightening the packing nut moderately, replace the entire packing. In such cases, consider removing the valve and taking it to the shop for a complete overhaul of the packing, stem, disk, seat, and gasket surfaces.

Leakage from Steam Vents

There is a tendency to overlook dramatic leaks, such as a blowing vent on a piece of equipment. People may believe that anything so obvious is intentional, so major leaks can persist for years. In fact, there is usually no valid reason to discharge visible amounts of steam. The only exceptions are industrial processes in which the steam is necessarily lost to the atmosphere.

Steam systems have vents for removing air and other gases at certain points in the system, such as at deaerating feedwater heaters and condensate receivers. A large amount of steam can be lost from such vents. If a vent is blowing heavily, there is usually a defect, such as a leaking steam trap. You may be able to save venting steam by installing a vent condenser (see Measure 1.10.5).

Internal Boiler Leakage

Major steam leaks can occur inside boilers, where they are invisible. Internal leaks commonly occur at pressure fitted joints. In firetube boilers, leaks occur where tubes enter tube sheets. In watertube boilers, leaks occur where tubes enter upper and lower drums. In cast iron sectional boilers, leaks occur between sections.

Tubes can develop leaks in a number of ways. Corrosion causes pinhole leaks. A weakness in a welded seam that is aggravated by corrosion can result in a lengthwise split that can quickly develop into a major leak. Accumulation of scale can overheat a tube, causing it to weaken or burn. The result may be a progressive leak or an abrupt catastrophic failure.

Find internal leaks by inspecting the boiler when it is cold. An exceptionally large internal leak in an operating boiler may be revealed by a small drop in stack temperature. An operator would have to be exceptionally alert to catch this.

You can often correct small leaks at tube sheets and drums by re-expanding the tube end into the tube sheet or drum. More severe leaks require removing tubes and refinishing the holes in the tube sheet or drum. Any defect in the tube itself requires the tube to be replaced.

Losses in Condensate Return

Not just steam leaks out of a steam system. Do not overlook the possibility of leaks in the condensate return system. Leakage of steam may be visible and noisy, while serious leakage of condensate may occur quietly.

The worst case of condensate loss occurs when there is no condensate return system. See Measures 1.8.4 ff about this.

Steam Trap Leakage

Leakage of steam traps is a cause of invisible steam loss that requires specialized corrective measures. Steam traps are covered in Measures 1.10.3 and 1.10.4 ff.

ECONOMICS

SAVINGS POTENTIAL: *0.1 to 50 percent of total system energy consumption. Larger losses occur in systems with extensive distribution systems, and in systems that operate at higher pressures, and in systems that have not been well maintained.*

COST: *Labor is the major cost for routine leak repair. For fixing indoor leaks, labor cost depends on whether plant operators have spare time to perform maintenance. Materials for repairing external leaks typically consist of inexpensive gaskets and packings. Repairing leaks in buried pipe involves major digging costs.*

The cost of extensive pipe replacement is high, especially if the pipe is buried.

See Reference Note 15, Infrared Thermal Scanning, for the cost of thermal scanning equipment and services.

Repairing internal boiler leaks typically costs hundreds or thousands of dollars.

PAYBACK PERIOD: *Short, if only minor repairs are needed. Several years or longer, if extensive pipe replacement is needed. Eliminating leaks generally costs less than neglecting them, until a point is reached at which major pipe replacement is required.*

TRAPS & TRICKS

DILIGENCE: *It is easy to forget this, or to let other concerns take priority. Do a leakage survey whenever the makeup water rate indicates a problem.*

WORKMANSHIP: *Workmanship makes a big difference in how long repairs last. Leak repair is tedious, which invites sloppy work, such as failure to refinish sealing surfaces adequately. Do it right the first time.*

MEASURE 1.10.3 Use the most efficient type of steam trap for each application.

Steam traps are vital components of a steam system. If they do not perform properly, they can be a source of major energy and water loss. Normal wear of traps is a major cause of steam leakage. Steam traps can fail in a way that effectively creates a hole in the steam system. Using the wrong type of trap can cause improper operation of the steam-using equipment, damage piping, and destroy the trap itself. Loss of steam through a trap is almost invisible, because the steam disappears into the condensate system.

Make sure that each steam trap is individually matched to its application in terms of type and size. About half a dozen different types of steam traps are in common use today. The variety of types results from differences in application requirements and differences in cost. Trap types differ in reliability and time-between-overhaul. Some types of steam traps have higher average leakage than others. The cost of the trap should be a lesser factor in selecting one type of trap over another, because the trap cost is dwarfed by the cost of wasted steam.

In existing systems, you may find traps that are not the best type for the application. Proper trap application requires a certain amount of specialized knowledge, so traps may have been installed without considering all the relevant factors. Also, facility operators tend to favor trap types that are cheap and compact.

This Measure gives you an introduction to steam trap selection, with an emphasis on selecting traps that have the lowest average leakage over the life of the plant. Supplement this with manufacturers' literature for the specific types of traps that apply to your facility.

SUMMARY

Inappropriate types of steam traps waste steam. Select trap sizes using manufacturers' procedures.

SELECTION SCORECARD

Savings Potential .... **$ $**

Rate of Return, New Facilities **% %** % %

Rate of Return, Retrofit.......... **% %** % %

Reliability .... ✓ ✓ ✓ ✓

Ease of Retrofit .... ☺ ☺ ☺

Once you understand trap selection, survey all the steam traps in your facility to determine whether each is the appropriate type and capacity for its application.

Finally, replace all inappropriate traps with the proper types and capacities.

This Measure deals with selecting steam traps and replacing inappropriate or less efficient types. See Measures 1.10.4 ff for trap maintenance.

The Two Applications of Steam Traps

Steam traps are devices that block the passage of steam while allowing liquid condensate to pass. Traps are used in two general applications, which are illustrated in Figure 1:

- *on the outlets of steam-using equipment*, to keep steam confined inside the equipment until it has

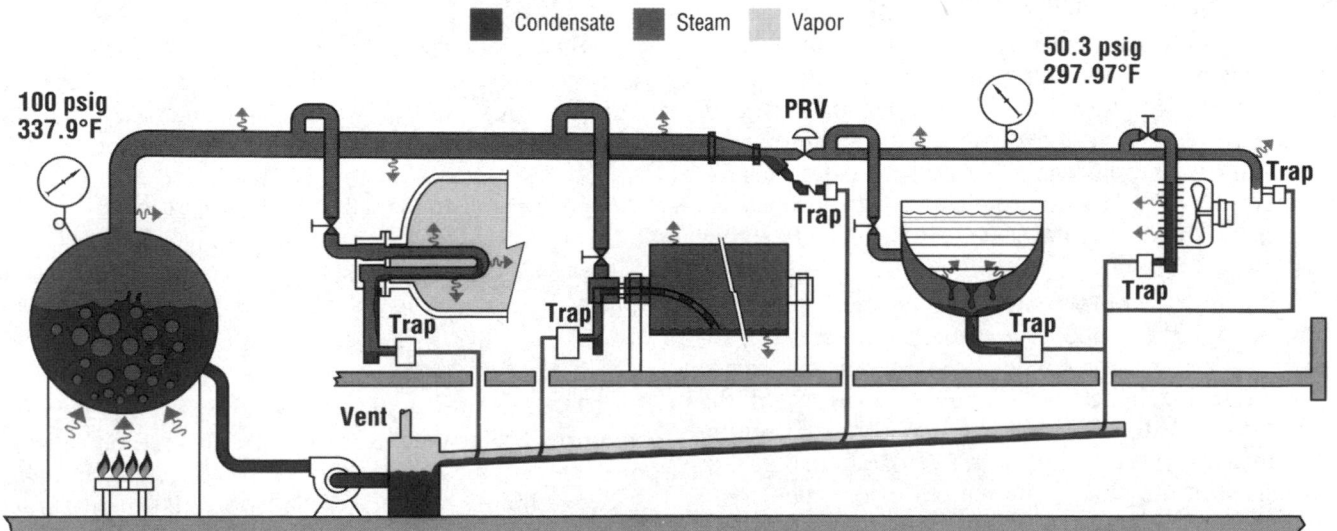

Armstrong International, Inc.

Fig. 1 The two functions of steam traps One is to keep steam inside heating equipment until it has condensed and given up its latent heat. The other is to drain condensed water from steam lines before it damages the system.

given up its heat, i.e., to keep the steam from blowing straight through the equipment

- *on steam lines*, to remove condensate that forms in the pipe. Condensate forms because of heat loss from the pipe. If allowed to accumulate, slugs of condensate propelled by steam pressure may accumulate enough kinetic energy to destroy valves, piping, and equipment. Traps used to drain steam lines are called "drip traps."

In many steam systems, the traps also serve as an important means of removing air and other non-condensible gases from the system. If these gases are not removed, they may block the flow of steam. For example, in a steam coil with many circuits, air that accumulates in the coil causes most of the steam to be routed through a few of the circuits, rendering the rest of the coil useless. Also, non-condensible gases are the source of corrosion in steam pipe and equipment. Oxygen corrodes steel directly, while carbon dioxide forms carbonic acid, which causes acid corrosion.

To minimize these problems, most applications require steam traps to have the ability to remove, or "vent," non-condensible gases (i.e., gases other than steam). Traps need a large venting capacity to clear air out of the steam system after a period of shutdown. Traps need a smaller venting capacity to vent the system continuously while it is in operation, to remove gases that are carried in the steam.

Steam traps usually vent non-condensible gases into the condensate system. The location of the traps may not allow complete venting of the system. Parts of a steam system may trap air by letting it become stagnant, so it is not carried along by steam flow. Separate air vents are installed in these parts of the system. These vents typically discharge air directly to the atmosphere, rather than to the condensate system.

How Steam Trap Leakage Wastes Energy

Most of the useful energy of steam is in the form of latent heat, which is the heat required to turn liquid water into steam inside the boiler. For a steam system to operate efficiently, all the steam must condense inside the steam-using equipment, so that the latent heat is transferred to the heating application. If steam leaks through a trap, most of the energy of the leaked steam is wasted.

Little or none of this energy is recovered on the way back through the condensate system. Steam that leaks through a trap into the condensate system is condensed by conductive heat loss through the pipes of the condensate system. In an atmospheric or vacuum condensate system, any steam leakage that is not condensed in the pipes is blown out of the condensate system vents, which are typically located at the condensate receiver.

Causes of Steam Trap Leakage

A large part of this Measure deals with the leakage characteristics of different types of steam traps. Before getting into the individual types, it is worth noting that all steam traps may leak for these reasons:

- *inherent leakage of the trap design.* In principle, all types of traps, except orifice traps, are capable of blocking steam completely. However, all traps will leak eventually. Field experience suggests that some trap types tend to operate longer before developing leakage. Also, some types of traps will leak if they are not installed in a certain way, whereas the method of installation does not cause leakage in other types.

- *sticking in an open or partially open position.* All types of traps are subject to complete failure as a result of fouling, corrosion, or mechanical failure. Failure in the open position is the equivalent of having a hole in the system the size of the trap's internal discharge passages. Failure in the closed position causes steam equipment to cease operating because it cannot discharge condensate. Failure of drip traps in the closed position is dangerous because slugs of condensate remain in steam lines. Some traps tend to fail in an open position, while others tend to fail in a closed position. This tendency may also be influenced by the characteristics of the steam system.

- *wear and fouling of sealing surfaces.* All types of steam traps (except orifice traps) block the flow of steam by metal-to-metal contact of sealing surfaces. Even if a trap continues to close properly, it will eventually develop leakage because of steam abrasion, fouling, and hammering of the sealing surfaces. Once leakage begins, it typically progresses rapidly. Some types and models of traps develop this type of leakage more quickly than others.

Steam loss also depends on the size of traps. Larger traps can waste more steam, and they cost more to replace. Survey all your traps, large and small, to make sure that each is appropriate for its application.

Steam Trap Types and Leakage Characteristics

The following is an introduction to the types of steam traps that are presently in common use, with emphasis on their steam leakage characteristics. The trap types are covered in approximate order of condensate draining capacity.

This comparison of efficiency characteristics requires broad generalizations. Information on trap leakage provided by manufacturers is sketchy and may lack credibility. One problem is that all types of traps can leak seriously under certain conditions, so any comparison depends on assumptions about the application. The major trap manufacturers tend to be

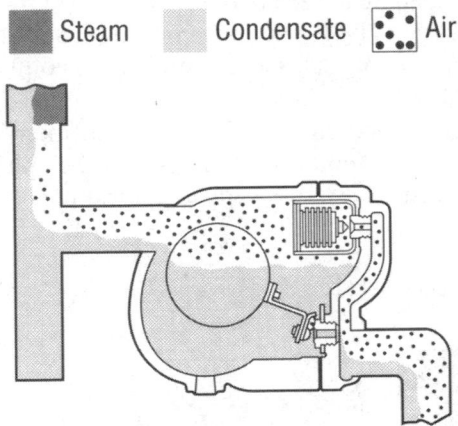

Steam ▮   Condensate ▮   Air ⁙

1. On start-up low system pressure forces air out through the thermostatic air vent. A high condensate load normally follows air venting and lifts the float which opens the main valve. The remaining air continues to discharge through the open vent.

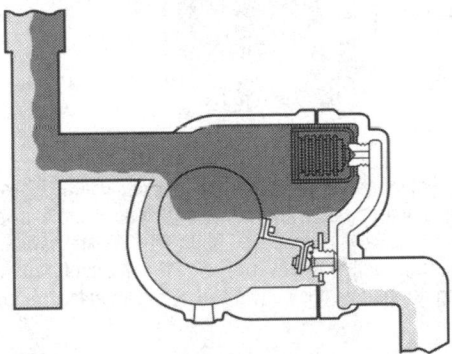

2. When steam reaches the trap, the thermostatic air vent closes in response to higher temperature. Condensate continues to flow through the main valve which is positioned by the float to discharge condensate at the same rate that it flows to the trap.

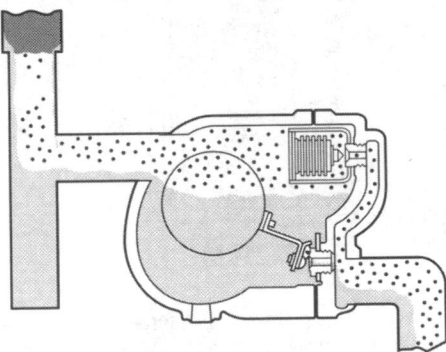

3. As air accumulates in the trap, the temperature drops below that of saturated steam. The balanced pressure thermostatic air vent opens and discharges air.

Armstrong International, Inc.

Fig. 2 How float-and-thermostatic (F&T) traps work

cautious about pointing the finger at certain trap types, probably because they make a variety of types themselves.

■ Float-and-Thermostatic (F&T) Traps

F&T traps are used for draining steam equipment and as drip traps. They are one of the most popular types, and they are used in a wide range of sizes. Figure 2 shows how they work.

As the name implies, a float-and-thermostatic trap is a combination of two separate devices, a float trap and a thermostatic vent. The float trap is the essence of simplicity. It consists of a chamber with a discharge valve at the bottom. The discharge valve is actuated by a float and lever. When the chamber is dry, the weight of the float keeps the valve closed. When the chamber fills with condensate, buoyancy lifts the float and opens the discharge valve. The float is usually spherical in shape to resist the pressure of the steam.

Float traps are efficient in separating condensate from steam because the trap directly senses the presence of condensate through its great buoyancy. Furthermore, the float ball can be made as large as needed to provide a strong operating force for the condensate valve. Steam cannot leak through the trap because the discharge is located under water.

A simple float trap cannot vent air. It would eventually fill with air and keep the valve from opening ("air bind"). To prevent this, virtually all float traps include a separate thermostatic valve near the top of the trap. This gives them the name "float-and-thermostatic." The thermostatic valve remains open until the trap heats up. F&T traps are able to vent large amounts of the air from the system, which is an important feature.

In F&T traps, the thermostatic element can become a source of steam leakage. Fortunately, the thermostatic

WESINC

Fig. 3 Typical F&T traps The chubby shape provides space for the float ball to swing up and down. The thermostatic element is on the upper left.

element in an F&T trap wears out more slowly than the element in a thermostatic trap. This is because it remains closed most of the time, opening only when a volume of air accumulates inside the trap.

Float traps are vulnerable to dirt. The discharge valve is located near the bottom of the trap, where large dirt particles may accumulate. Also, float traps can operate steadily in a partially open position, which allows debris to become lodged between the plug and seat of the discharge valve. The protection against dirt is to buy a trap that has an integral strainer, or to install a strainer ahead of the trap.

You can recognize F&T traps from their large, rounded shape, which accommodates the spherical float ball inside. Figure 3 shows typical units.

■ Steam   ▨ Condensate   ▦ Air   ▨ Flashing Condensate

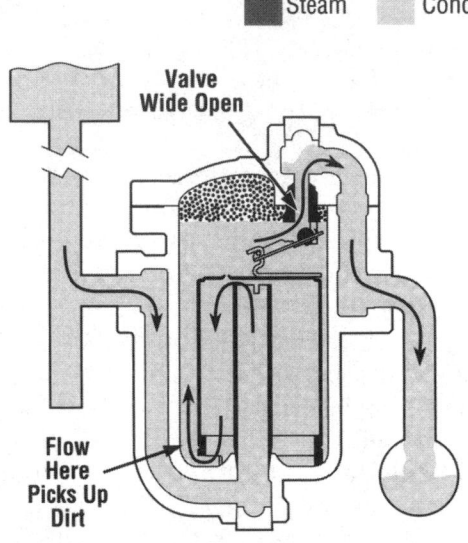

1. Steam trap is installed in drain line between steam-heated unit and condensate return header. On start-up, bucket is down and valve is wide open. As initial flood of condensate enters the trap and flows under bottom of bucket, it fills trap body and completely submerges bucket. Condensate then discharges through wide open valve to return header.

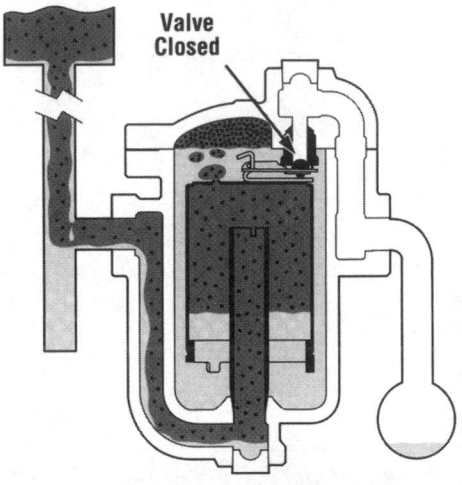

2. Steam also enters trap under bottom of bucket, where it rises and collects at top, imparting buoyancy. Bucket then rises and lifts valve toward its seat until valve is snapped tightly shut. Air and carbon dioxide continually pass through bucket vent and collect at top of trap. Any steam passing through vent is condensed by radiation from trap.

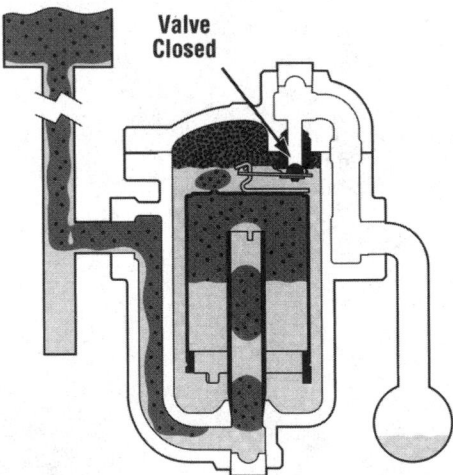

3. As the entering condensate starts to fill the bucket, the bucket begins to exert a pull on the lever. As the condensate continues to rise, more force is exerted until there is enough to open the valve against the differential pressure.

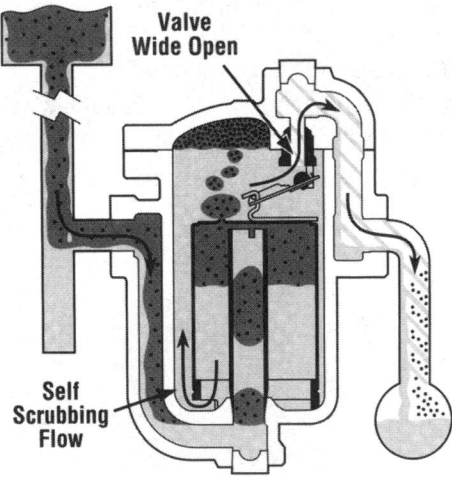

4. As the valve starts to open, the pressure force across the valve is reduced. The bucket then sinks rapidly and fully opens the valve. Accumulated air is discharged first, followed by condensate. The flow under the bottom of the bucket picks up dirt and sweeps it out of the trap. Discharge continues until more steam floats the bucket, and the cycle repeats.

Armstrong Inernational, Inc.

Fig. 4 How inverted bucket traps work

and be discharged. The vent hole also allows a small quantity of steam to escape from the bucket into the main body of the trap, where it condenses. The presence of the vent hole also allows the trap to open more quickly when condensate enters it. In normal operation, loss of steam through the vent is limited by the rate at which steam condenses in the trap, which is small.

Inverted bucket traps are resistant to dirt because the discharge valve is located at the top of the trap, away from dirt that settles in the bottom of the trap body. The discharge valve opens abruptly and fully, so dirt carried in the condensate does not become lodged in the valve seat.

Inverted bucket traps are most likely to fail in the open position. They fail in the open position if they run dry, because then the bucket cannot float. (This problem is most likely to occur if the steam is superheated.) They may also fail from misalignment of the internal mechanical linkage. When they fail in the open position, the size of the discharge orifice is the only factor that limits steam loss.

You can recognize inverted bucket traps from their cylindrical shape, which conforms to the shape of the bucket inside. Figures 5 and 6 show two different models.

Armstrong International, Inc.

Fig. 5 A small inverted bucket trap draining a steam-fired space heater

■ **Inverted Bucket Traps**

Inverted bucket traps are used for draining steam equipment and as drip traps. They are used in a wide range of sizes. Figure 4 shows how they work.

An inverted bucket trap is built around a floating bucket that has its open side facing downward. Steam entering the trap is fed into the bucket, causing it to float to the top of the surrounding pool of condensate and close the discharge valve. When only condensate enters the trap, the steam in the bucket condenses and the bucket sinks, opening the discharge valve. Thus, condensate drains in cycles.

Inverted bucket traps allow little steam leakage when they are operating properly because the buoyancy of the bucket provides strong closure of the discharge valve. As with float traps, inverted bucket traps are relatively reliable because the bucket can be made as large as needed to provide a strong operating force for the condensate valve.

The top of the bucket has a small vent hole. This allows non-condensible gases to escape from the bucket

Armstrong International, Inc.

Fig. 6 A stainless steel inverted bucket trap installed as a drip trap Note the test valve installed on the discharge (left) side of the trap.

■ Thermostatic Traps

Thermostatic traps are used for draining steam equipment and as drip traps. They are most common in smaller capacities. Figure 7 shows typical units, which are characterized by small physical size.

Thermostatic traps operate by sensing the difference between the temperature of live steam and the temperature of condensate or non-condensible gases that cool inside the trap. Their operating principle is simple: steam cannot cool below its condensation temperature, but condensate and non-condensible gases can cool. When a thermostatic element inside the trap senses a temperature lower than steam temperature, it assumes that it is surrounded by water and opens the discharge valve.

Presently, there are two main categories of thermostatic traps:

* *bimetallic traps,* which consist of a valve that is moved by a simple bimetallic element that is located on the steam side of the valve. This type has a fixed operating temperature, so it cannot respond to changes in the pressure and temperature of the steam.

Bimetallic traps are compact because they require space only for the small thermostatic element. The housing may have any shape.

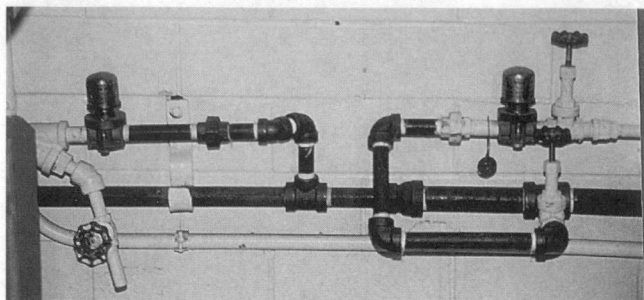

Armstrong International, Inc.

Fig. 7 A pair of thermostatic wafer steam traps The main attraction of these is their small size and low cost.

* *bellows traps* close a discharge valve using the pressure of a boiling fluid contained inside a bellows. The fluid in the bellows is selected to boil at a temperature lower than steam temperature, so the bellows wants to expand when steam is present. The pressure inside the bellows is partially balanced by the steam pressure, so the opening temperature of the bellows trap can adapt to changes in steam pressure. Figure 8 shows how a bellows trap works. This type of trap should not be exposed to superheated steam, because the pressure inside the bellows may exceed the steam pressure enough to burst the bellows.

■ Steam   ■ Condensate and Air   ■ Condensate

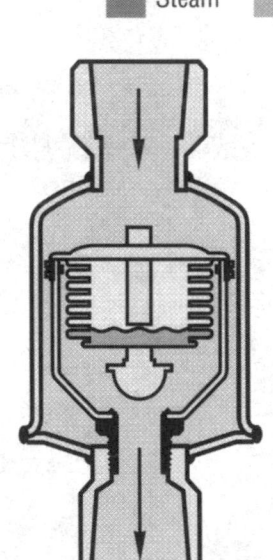

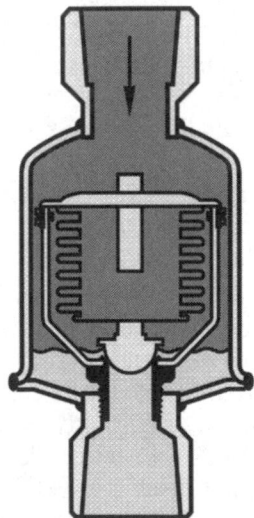

1. On start-up, condensate and air are pushed ahead of the steam directly through the trap. The thermostatic bellows element is fully contracted and the valve remains wide open until steam approaches the trap.

2. As the temperature inside the trap increases, it quickly heats the charged bellows element, increasing the vapor pressure inside. When pressure inside the element becomes balanced with system pressure in the trap body, the spring effect of the bellows causes the element to expand, closing the valve. When temperature in the trap drops a few degrees below saturated steam temperature, imbalanced pressure contracts the bellows, opening the valve.

Armstrong International, Inc.

Fig. 8 How a thermostatic bellows trap works

Bellows traps are small. Some have a cylindrical housing that conforms to a cylindrical bellows. Others have a compact housing that surrounds capsules of different shape, one of which resembles a large button.

Thermostatic traps are prone to leakage because they open and close relatively slowly. This provides time for erosion by high velocity steam and water while the sealing surfaces are barely separated. The partially open discharge valve may trap dirt, preventing tight closure.

The piping layout of thermostatic traps may be critical. The trap does not open until the condensate cools somewhat. If the condensate does not cool quickly enough, condensate may back up into the steam equipment or steam line. Therefore, install thermostatic traps so that they are surrounded by air that is substantially cooler than steam temperature. With low-pressure steam, the trap may have to be installed at a certain distance from the equipment or steam lines. Do not insulate the trap or the pipe that leads to it.

A standard thermostatic steam trap may be used as an air vent. You can recognize this application from the fact that the trap is installed at a high point on the steam equipment or pipe, rather than at a low point.

■ Disc Traps

Disc traps are used primarily as drip traps and for low steam loads, such as steam tracing lines. They are used in smaller capacities. (Disc traps are sometimes called "thermodynamic" traps. This may be a derivation of the name Thermo-Dynamic, which is a trade mark of Sarco Spirax, the original producer of disc traps.)

A disc trap consists of a flat disc resting on a circular seat that is smaller than the disc. The disc is enclosed in a chamber above the seat, where it moves freely. Figure 9 shows how simple these traps are in construction. Figure 10 shows how they work.

Condensate or steam enters the trap through the center of the seat, flows over the seat, and discharges through ports located under the perimeter of the disc. The disc is confined in a small chamber, into which it fits loosely so that steam can leak into the space above the disc. When condensate is present, it lifts the disc and exits. When steam enters the trap, Bernoulli effect reduces the pressure between the disc and seat. (Perhaps, this type of trap should be called "aerodynamic" rather than "thermodynamic.") Steam at full steam pressure leaks into the space above the disc, aiding the Bernoulli effect, and the trap snaps shut. Once the trap is shut, the disc is held down by the steam pressure above the disc.

The main advantage of disc traps is their small size in relation to their condensate capacity. You can usually identify them from the disc chamber, which is a small cylindrical housing that forms the top of the trap. Some disc traps are installed in-line, and these may be barely

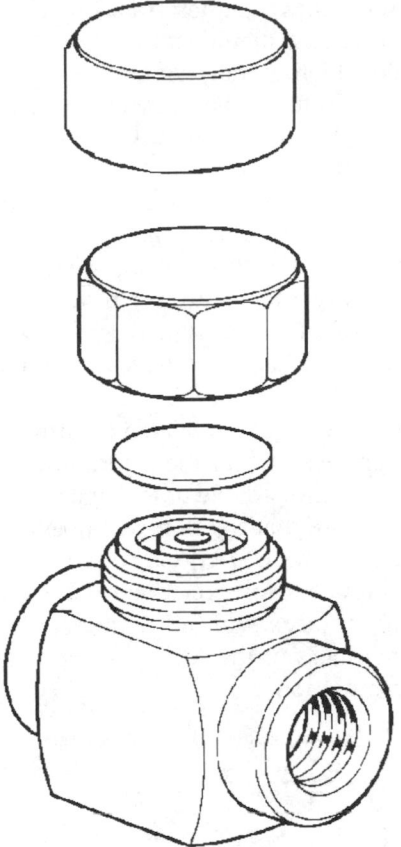

Spirax Sarco, Inc.

Fig. 9 Disc trap The thin disc sits freely on the seat. Steam and condensate rise through the center hole, turn around under the disc, and bleed off through the surrounding space. The cap over the disc forms a chamber in which the disc is enclosed. At the top is an insulating cover, which improves trap performance.

larger than the steam pipe itself. Figure 11 shows an installed unit.

The reliability of disc traps is a controversial issue. One leading trap manufacturer (who manufactures all the major types) asserts that disc traps are as reliable as other trap types. Other parties assert that leakage increases after a relatively short time because the sealing surfaces become deformed from hammering and steam abrasion. Also, the large contact area of the sealing surface makes disc traps vulnerable to leakage caused by fouling. Increasing leakage at the sealing surface allows the steam above the disc to vent quickly to the outlet, causing the trap to cycle more and more rapidly. Some manufacturers admit to this weakness, and they design their models for quick replacement of the disc and seat.

Even the way that disc traps operate is controversial. All agree that disc traps operate in a cyclic manner. Many parties say that disc traps cycle open periodically even when there is no condensate in the trap. This occurs because the steam above the disc condenses, eliminating the pressure that holds the disc closed. A puff of steam is lost with each operating cycle. However, one major manufacturer denies that their traps open in the absence of condensate. This manufacturer states that live steam never reaches the trap, but that the trap is closed by the flashing of condensate that is near steam temperature. This manufacturer insists that a water seal should be created ahead of the trap.

All seem to agree that trap cycling increases if heat loss through the trap body causes the steam above the disc to condense. For this reason, some manufacturers offer insulating caps for the disc cover. Other manufacturers offer more expensive models that surround the disc chamber with an outer chamber that is filled with inlet steam.

The design of the seat is also controversial. Some manufacturers assert that the operation of disc traps depends on a controlled rate of leakage between the disc and seat to ensure that the trap will open. They create this small leak with a tiny groove or a carefully roughened surface. However, one leading manufacturer uses sealing surfaces with no deliberate leakage. All agree that hardness of the sealing surfaces is important to resist deformation.

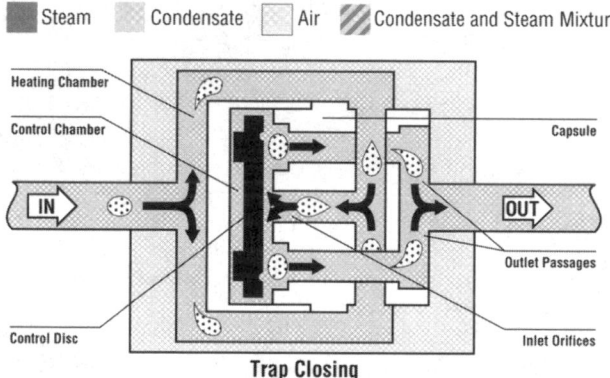

1. On start-up, condensate and air entering the trap pass through the heating chamber around the control chamber and through the inlet orifice. This flow lifts the disc off the inlet orifice, and the condensate flows through to the outlet passages.

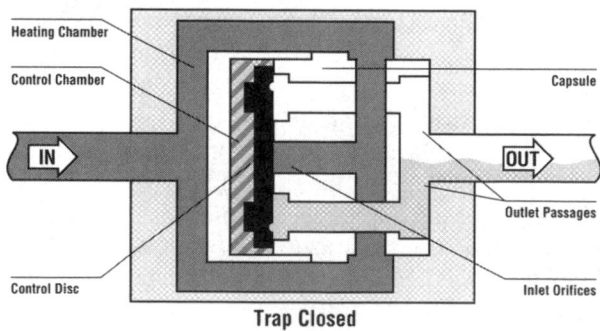

2. When steam reaches the disc, increased flow velocity across the face of the disc reduces pressure at this point, and the disc closes the orifice. Controlled bleeding of steam from the control chamber causes the trap to open. If condensate is present, it will be discharged. The trap recloses in the presence of steam and then continues to cycle at a controlled rate.

Armstrong International, Inc.

Fig. 10 How a disc trap works

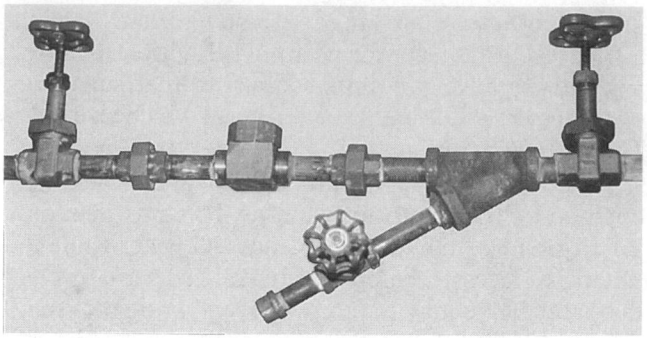

WESINC

Fig. 11 Disc trap installed   This type is very small, as you can see in comparison with the valves and strainer.

The most common failure mode of disc traps is increasingly rapid cycling that is caused by fouling or deformation of the sealing surfaces. When this occurs, steam is lost in puffs. The rate of steam leakage is limited by the interrupted flow and by the small size of the passages. If the steam plant or the application is shut down periodically, the trap might also stick shut or fully open.

Try to install disc traps so that the disc lies on its seat horizontally, equalizing the forces on the disc. Disc traps are sometimes installed with the disc vertical, but this hastens deterioration.

■ **Orifice Traps**

Orifice traps are used almost exclusively as drip traps. They are made only in smaller sizes.

An orifice trap, as its name implies, consists of nothing more than a small hole. A screen typically is installed upstream of the hole to prevent clogging. This type was introduced formally in the 1970's. The fixed orifice is essentially an adaptation of the practice of draining condensate manually by slightly opening a drain valve.

Orifice traps can be considered a controlled leak, whose principal merit is that the volume of the leak is known in advance. They are limited to use as drip traps, and they must be sized closely to match the expected condensation rate. They are not practical for steam-using equipment, which has large and variable steam consumption.

The housing of an orifice trap is very compact, typically only slightly larger than the pipe diameter. A sloppy insulation job may hide the device entirely.

The *ASHRAE Handbook* explains the operation of orifice "traps" as follows: "... an orifice of any size has much greater capacity for condensate than it does for steam because of the significant differences in their densities and because flashing condensate tends to choke the orifice ..." This explanation is not persuasive, because steam flows much more easily than water, in terms of volume. It is true that condensate chokes an orifice. However, a continuously open orifice is passing dry steam most of the time in a drip trap application, especially if the steam has any superheat. The *ASHRAE Handbook* goes on to say that "the steam loss is usually comparable to that of most cycling-type traps," but does not state which types.

Orifice traps are vulnerable to dirt because the orifices are quite small. The orifice must be protected by a screen that has a smaller mesh than the orifice size, and such a screen easily becomes clogged.

■ Other Types of Steam Traps

Various other types of traps have emerged over the years. Some, such as piston traps and open bucket traps, have become obsolete because of poor reliability, complexity, or large size. Other types of traps are limited to specialized applications, such as a variant of the inverted bucket trap designed to lift condensate. Consider replacing obsolete or inappropriate traps, rather than repairing them.

Other Selection Characteristics

From the standpoint of efficiency, ability to block steam flow is the main consideration in selecting traps. In addition, you need to consider other characteristics, especially these:

• *reliability.* Most steam is wasted by trap failure, rather than by inherent leakiness of certain types of traps. Different trap types vary in susceptibility to the three failure modes discussed previously.

All traps can be expected to fail, but the average time between failures may vary widely among different types of traps. Many people feel that F&T traps and inverted bucket traps have the longest intervals between failure. Thermostatic traps probably have shorter intervals between maintenance because of their gradual closing characteristics. The reliability of disc traps is controversial, as discussed previously. Orifice traps are very vulnerable to clogging.

All traps, except orifice traps, will eventually fail by leaking steam, but they may also stick fully open or fully closed. The latter modes are more likely in systems that shut down periodically, because this allows the mechanism to corrode into position. Inverted bucket traps tend to fail in the open position, because this is their shut-down position. Float traps tend to fail in the closed position, which is their shut-down position. A float trap can also fail by corrosion of the float ball, which closes the valve. Thermostatic traps can fail by failure of the thermostatic element. Orifice traps, of course, can fail only to a closed state by clogging.

If an F&T trap fails in the closed position, its thermostatic element will cause it to behave like a thermostatic trap. This may reduce its capacity considerably, but will make the failure difficult to diagnose.

No trap is reliable unless it is properly matched to its application. For example, float traps are vulnerable to water hammer, freezing, and dirt, whereas inverted bucket traps are resistant to these problems.

• *capacity range.* Float, bucket, and thermostatic traps block steam efficiently from zero condensate flow up to their maximum rated capacity. Disc traps adapt to different drainage rates, but they are limited to small capacities because of the way they operate. Orifice "traps" continuously leak steam when condensate is not present, so they must be sized accurately for the maximum expected condensate flow rate.

• *system pressure.* Inverted bucket traps are available for any pressure. Float traps are limited in pressure by the possibility of crushing the float. Bellows-type or encapsulated thermostatic traps are limited in pressure by the possibility of crushing the thermostatic element. Orifice and disc traps are limited in pressure by the erosion that occurs when high-pressure steam passes through narrow passages.

Some disc traps require a minimum pressure drop between the steam side and the condensate side to operate properly, typically 10 PSI or more. In addition, disc traps are vulnerable to back pressure because proper operation requires steam to be able to exit from the trap at high velocity.

• *venting a cold system at start-up.* Thermostatic traps provide rapid venting of cold systems. Thermostatic elements are included in F&T traps for cold system venting.

Inverted bucket traps do not vent air rapidly because the smallness of the vent hole in the bucket limits the flow of gases through the trap. To compensate for this, a thermostatic elements can be fitted to the bucket that increases the size of the vent hole when the bucket is cold. This added complication reduces the reliability of the trap, all other things being equal.

Disc traps vent a cold system very slowly because the trap is closed by air in the same way as by live steam.

Orifice traps are poor for venting a cold system because of the typically small size of the orifice.

You can vent a cold system by using separate air vents, which gives you greater latitude in selecting trap types.

• *venting a warmed-up system.* F&T traps, inverted bucket traps, and orifice traps all do a good job of venting non-condensible gases from a system that is operating at normal temperature.

If a thermostatic trap is kept flooded by condensate, air never has a chance to reach the thermostatic element, so the trap cannot vent a warmed-up system. This limitation is characteristic of most thermostatic traps because the thermostatic element must be set to close at some temperature lower than the steam temperature.

Disc traps vent an operating system slowly for the same reason that they vent a cold system poorly, namely, that non-condensible gases cause the trap to close rather than to open. Venting becomes impossible if the trap is installed with a water seal ahead of it to reduce cycling, as one manufacturer suggests.

- *vulnerability to freezing.* Inverted bucket traps and float traps remain partially filled with water when they are idle, which invites freezing damage. Of the two, float traps are more vulnerable because they contain a float ball and a thermostatic element (in F&T traps) that are easily crushed by ice expansion.

Bimetallic thermostatic traps, disc traps, and orifice traps are less likely to be harmed by freezing.

Installation practice is a major factor in avoiding freeze damage, both for traps and for other steam equipment. The basic principle is to completely drain the portions of the system that may be exposed to freezing temperatures. Competent steamfitters have a variety of techniques for accomplishing this. Major trap manufacturers publish guidance in avoiding freezing in steam systems.

- *operation with superheated steam.* Superheated steam can be a problem for several trap types when they are used in drip legs. If superheated steam reaches an inverted bucket trap, the steam will rush through and keep the trap dry, causing the trap to remain in the open position. Superheated steam may cause a disc trap to chatter and pass steam. Superheated steam may burst the expanding element of a bellows trap. Superheated steam may increase the loss through orifice traps.

In principle, you could use any of these types with superheated steam, provided that you install them so that liquid condensate always forms ahead of the trap. However, this is a bad gamble, especially if the trap can be damaged by contact with superheated steam. Under some circumstances, the condensate may not form as you hope.

- *vulnerability to water hammer.* A slug of condensate propelled by high steam pressure has enough energy to crush trap components, especially thermostatic elements and float balls. Float traps and bellows-type thermostatic traps are sensitive to water hammer. Other common types resist damage much better.
- *size.* Float traps and bucket traps are bulky, whereas other common types are small.
- *cost.* Float traps and bucket traps are more expensive than other common types.

Other characteristics may be significant, and putting all the selection factors into perspective requires experience. Several steam trap manufacturers publish detailed selection guides. Experienced manufacturers' representatives can offer valuable advice. Recognize that vendors are biased toward the types that they offer, especially if they are proprietary, and toward more expensive models.

Sizing and Surge Capacity

Do not select traps with excess flow capacity. The capacity of the trap determines the size of the valve orifice (or internal passages, in the case of a disc trap). If the trap fails in the open position, the orifice size determines the rate of steam loss.

Select trap sizes using the instructions in the manufacturers' catalogs. Calculate sizing from the maximum condensate load, the pressure differential across the trap, and the nature of the application (which dictates extra capacity as a safety factor). Typically, F&T traps and inverted bucket traps are offered in a range of body sizes, which are combined with a range of orifice sizes to satisfy the full range of capacity requirements.

Consider the surge capacity inside the body of the trap. Surges of condensate must be kept from backing up into steam equipment or into steam lines. Float traps and bucket traps have a significant amount of surge capacity. You can also gain surge capacity by increasing the volume in the pipe that leads to the trap.

Strainers

As we have seen, dirt is a potential problem with steam traps. It can keep the valve from closing completely (all types, except orifice traps) and it can clog the trap passages (especially orifice and disc traps). The general solution to these problems is to install a strainer ahead of the trap. Figure 11 shows a strainer that is properly installed.

The strainer itself is an item that requires periodic inspection and cleaning. Unless the steam system has a particular dirt problem, inspection is needed only at long intervals. This leads people to forget about them. Make sure that your maintenance schedule includes such multi-year inspections. Figure 12 shows a strainer that has likely been forgotten.

To simplify the installation, you can get traps with integral strainers in most types and sizes. However, these do not eliminate the need to clean out the strainers periodically.

Piping Details

Proper operation of a steam trap depends on the way it is installed. As we have seen, installation practice is important for freeze protection, and it may be a factor in providing adequate surge capacity. Piping the trap to

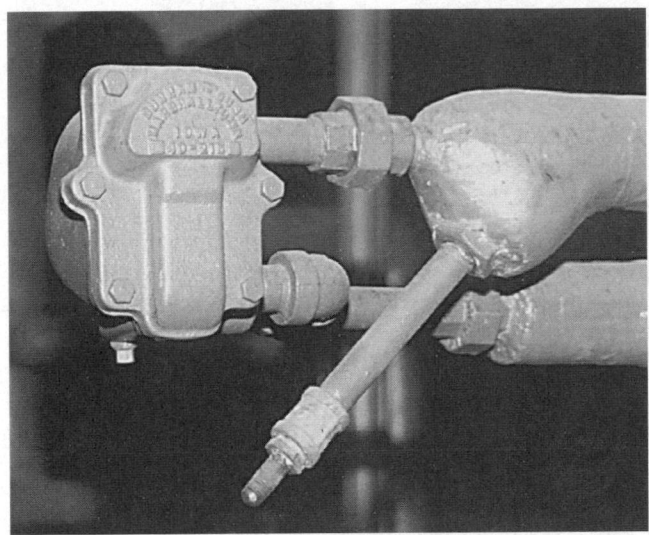

Fig. 12 A strainer that needs cleaning The strainer installed for this F&T trap is completely hidden by insulation. Still, it could be used if it had a blowdown valve, which it lacks. The crude plug in the blowdown pipe has probably never been removed.

maintain a water column ahead of the trap may be necessary for satisfactory operation and longevity of thermostatic and disc traps. Refer to the manufacturer's literature for more details.

Avoid Sharing Traps

Do not use a single trap to serve more than one item of equipment. Condensate drainage is sensitive to discharge pressure, so if two units discharge to a common line, the discharge from the higher pressure unit can block the flow of condensate to the other unit. This problem is especially severe with modulating equipment. Sharing traps makes it difficult to diagnose problems in traps or the equipment they serve.

ECONOMICS

SAVINGS POTENTIAL: *Varies widely. For traps serving equipment, the wrong types of traps may waste 1 to 30 percent of the steam flow to the equipment. For drip traps, an orifice trap or a malfunctioning trap of any type may drain much more steam than condensate.*

COST: *F&T traps and inverted bucket traps typically cost from $100 to $500 each, in sizes where less efficient types of traps might be used.*

PAYBACK PERIOD: *Several months to several years.*

TRAPS & TRICKS

SELECTING THE TRAPS: *Start by educating yourself about steam traps. Know where to use each type of trap, and how to size traps. Take one of the courses offered by major trap manufacturers. Buy your traps from reputable manufacturers.*

INSTALLATION: *Trap installation requires special practices for each type of trap and application. Do your homework.*

MEASURE 1.10.4 Test and repair steam traps on a continuing basis.

SUMMARY

Important steam system maintenance that is often neglected. Trap testing requires special skill. Trap repair requires a delicate touch.

SELECTION SCORECARD

Savings Potential .... **$ $ $ $**

Rate of Return .... **% % % %**

Reliability .... **✓ ✓**

Ease of Initiation .... ☺ ☺

All steam traps become leaky in time. The most reliable types of steam traps may operate several years before starting to leak seriously. The least reliable types may develop significant leakage after several months of service. (The characteristics of different types of steam traps are covered in Measure 1.10.3.) Therefore, all steam plants should have a continuing program of steam trap inspection and maintenance.

The techniques of steam trap testing tend to be skill-intensive and somewhat uncertain. Furthermore, testing tends to be awkward and uncomfortable, especially with traps that are installed in hot, inaccessible locations. This makes the operation of an effective inspection and maintenance program a challenge of boiler plant management.

The appropriate response to finding a defective trap may be to repair the trap, to replace the trap with a similar new unit, or to replace the trap with a more efficient type (as recommended in Measure 1.10.3).

There are two subsidiary Measures that provide alternative ways to ease trap testing. The first, Measure 1.10.4.1, recommends installing special equipment to help diagnose the condition of traps. The second, Measure 1.10.4.2, recommends hiring trap testing specialists to do this job for you.

Energy Waste from Trap Leakage

Leaking steam traps can be serious sources of energy waste. A leaking steam trap acts as a short circuit from the high-pressure portion of the system to the condensate system, and the loss continues as long as the steam system is operating. If the trap drains to a condensate line, the steam loss is invisible, so it may persist for years.

The worst energy loss occurs when the steam trap fails in a fully open position. In this case, the rate of steam loss is proportional to the size of the orifice in the trap and to the pressure difference across the trap. These same characteristics determine the rated load of the trap, so steam loss from a stuck-open trap is proportional to its rated capacity. (This is an important reason to avoid oversizing traps.)

The range of steam loss in typical applications varies enormously. For example, a leaky steam trap on a radiator in a low-pressure steam system may lose 5 pounds of steam (about 5,000 BTU) per hour. At the other extreme, a large steam trap with a valve seat diameter of 1/2", serving a process application at 300 PSI, may waste over 2,000 pounds of steam per hour (over 2 million BTU per hour) if the trap sticks fully open.

If the steam leaking through a trap is mixed with condensate, the water chokes the flow of steam through the orifice, reducing steam loss. This redeeming effect is most likely to occur in traps installed on steam-using equipment. In drip trap applications, the condensate flow may be low in relation to the trap capacity, so there is not enough condensate to choke the flow through a failed trap. In steam lines with superheat, the line may be completely dry.

Other Consequences of Steam Trap Malfunctions

Failure of a trap in the closed position may have serious consequences. If the trap is installed on steam-using equipment, a closed trap will prevent the equipment from working. If a drip trap is stuck closed, condensate can accumulate in the pipe. The condensate forms slugs that are propelled at high speed by the pressure of steam in the system, and these slugs can destroy pipe, valves, and equipment.

Causes of Steam Trap Failure

The common causes of trap failure are:

• wear of sealing surfaces from cyclic operation, and from abrasion by steam, water, and particles

• restricted motion of the valve parts because of corrosion or fouling

• inability to close fully because dirt or corrosion scale is trapped between the valve and seat

• misalignment of the sealing surfaces because of water hammer, freezing, or incorrect installation of replacement parts

• rupture or distortion of floats and thermostatic bellows by freezing, water hammer, or corrosion

• in inverted bucket traps, loss of the water seal, causing the trap to hang fully open

• in disc traps, lack of a water seal at the inlet, causing the trap to chatter.

The first two of these causes will eventually occur with any trap that stays in service long enough. The third is likely with certain types of traps, especially if inadequate water treatment has allowed the system to become corroded. The last four problems arise mainly from improper installation or from installing an inappropriate type of trap.

Expect any trap to leak eventually. If the system is shut down often, expect certain types of traps to stick fully open or fully closed because of corrosion while the system is filled with air and residual water. Float traps tend to fail fully closed in this case, and inverted bucket traps tend to fail fully open.

All types of traps, except orifice traps and disc traps, contain a condensate discharge valve that consists of a hard metal plug that seats in a hard metal orifice. The plug typically has the shape of a ball or cone, and the orifice typically has an angled shoulder against which the plug rests. This arrangement allows the plug and seat to wear into each other, maintaining a tight seal even as the components deform from usage. However, this self correcting action cannot continue indefinitely. At some point, the valve starts to leak, after which leakage accelerates. This will occur eventually in any steam trap that has a discharge valve.

All types of traps have a certain amount of snap action in opening and closing. Snap action occurs because a portion of the seat area is exposed to the lower pressure of the discharge side when the valve is closed. When the valve starts to open, this area is exposed to the higher pressure of the steam side, which assists in opening the valve. The opposite occurs as the valve closes. The snap action keeps the trap from operating in a barely open position, which would trap debris and hasten erosion. The snap action is more forceful at higher steam pressures and with larger orifices.

(Orifice traps are an exception to these behavior patterns. This type can be considered as being failed in an open position while it is working, but otherwise fails to a closed position.)

Trap Testing Methods

Over the years, many methods of testing steam traps have been tried. No method yet developed is completely reliable, and most methods require special skill. The following are the most common methods being used today.

■ Checking Condensate System Vents

With condensate systems that operate at atmospheric pressure, you can detect heavy leakage of steam traps from steam blowing at the condensate system vents. The vents are usually located at the condensate receiver. Venting steam does not indicate which trap is leaking, but it does indicate that a trap somewhere in the system is failing.

In long condensate systems, steam from a moderate trap leak is likely to condense before it reaches the condensate receiver vent. Therefore, a blowing vent indicates a leaking trap, but a quiet vent does not mean that the traps are working properly.

■ Test Valves

Steam trap test valves are described in subsidiary Measure 1.10.4.1. A test valve allows you to temporarily divert the output of the trap from the condensate system to the atmosphere, so you can see the trap output. The test valve setup should also allow you to briefly shut off the trap discharge and vent the condensate system, so you can see if any blowing steam is coming from the condensate system rather than from the trap itself.

Test valves work well with traps that alternately open and close, such as inverted bucket traps, thermostatic traps, and disc traps. With such traps, operation of the trap is apparent as the discharge valve opens and closes.

Test valves may require more skill when used with float traps, F&T traps, and thermostatic traps. These traps may discharge condensate continuously, rather than in cycles, when the condensate load is high. If the trap discharges continuously, it may be difficult to distinguish between live steam and flash steam. (Flash steam is the steam that is formed instantly from condensate as soon as the condensate passes through the trap into the lower pressure of the atmosphere.) If the system pressure is high, the condensate carries a large amount of energy that produces flash steam, which emerges from the test valve vigorously. Don't let flash steam mislead you into believing that a healthy trap is leaking.

Using test valves with disc traps may be deceiving if the condensate system operates at a pressure higher than atmospheric. This is because the performance of disc trap depends on the pressure differential across it. Opening the test valve to atmospheric pressure may give the false impression that a failing disc trap is behaving properly.

Stress safety when using test valves. If a trap fails in the open position, live steam will blast out of the test valve when it is opened.

■ Audible Sounds of Trap Operation

Each type of steam trap has characteristic sounds that indicate whether the trap is operating properly. For example, most traps snap audibly as they close. Rapid rattling of an inverted bucket trap indicates that it is stuck open. Cycling of a disc trap at too high a rate indicates that it is leaking excessively. And so forth.

You can hear the sounds of trap components opening and closing with an inexpensive stethoscope of the kind used by automobile mechanics. If you don't have one, hold the tip of a screwdriver on the trap and place the handle on your cheekbone. Some special stethoscopes amplify the sound and can highlight certain frequency ranges. Ordinary stethoscopes do not pick up the high-frequency sound of leaking steam.

Background noise is a problem in using sound to diagnose traps. Sound travels well in metal, so the sounds of the trap may be masked by other system noise. If a number of traps are located close together, their sounds may mingle. Moving the listening device from one location to another can resolve the origin of sounds.

The main weakness of this method is that it requires a detailed knowledge of how each type of trap sounds under different conditions, and a good deal of experience in listening to trap sounds.

■ Ultrasonic Listening Devices

Even if a trap is cycling properly, worn valve surfaces or misalignment may allow considerable leakage. Steam leaking through a narrow restriction emits a large amount of sound in the ultrasonic range, but little in the audible range. Therefore, listening for ultrasound is a valuable method of diagnosing leakage. This requires an ultrasonic "stethoscope," which is a device that converts ultrasound into audible sound or some other form of output. Figure 1 shows a typical unit.

Using this equipment effectively requires experience. An ultrasonic listening device does not detect ordinary trap noises very well, so use it in combination with a conventional stethoscope or other means of detecting audible noise.

■ Gauge Glasses on Traps

Some models of bucket traps and float traps are equipped with gauge glasses, as described in subsidiary Measure 1.10.4.1. These allow you to directly observe the water level inside the traps. However, if no water level is visible, you cannot tell whether the trap is full of water (stuck closed) or steam (stuck open).

■ Temperature Difference

No method of steam trap testing has raised more false hope than temperature testing, which is based on the assumption that condensate leaving a trap is cooler than steam leaking through the trap. This assumption is false. It overlooks the fact that the temperature on the discharge side of a steam trap depends on only one thing: the pressure of the condensate system.

When steam leaks into the condensate system, its pressure drops immediately to the pressure in the condensate system, and its temperature changes immediately to the corresponding saturation temperature. On the other hand, when the trap passes liquid condensate into the condensate system, the condensate flashes to steam at the same temperature. For example, if the condensate system is operating at sea level atmospheric pressure, the temperature of the trap discharge is always 212°F, whether the trap is operating properly or not.

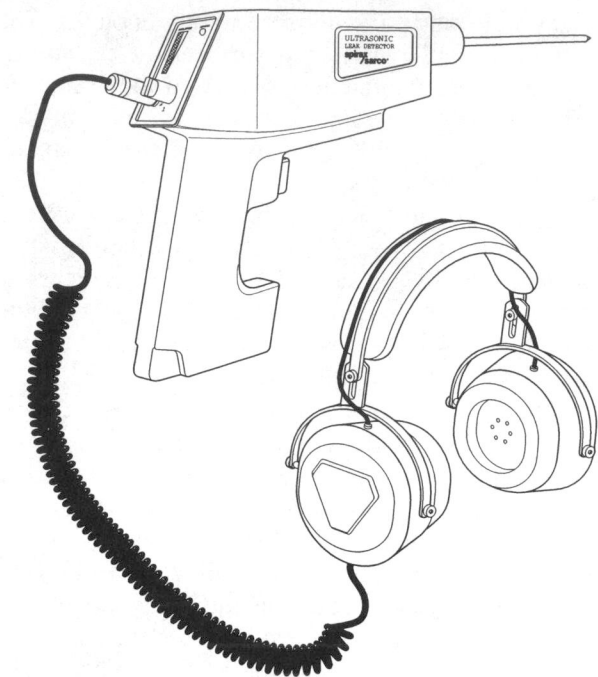

Spirax Sarco, Inc.

Fig. 1 Ultrasonic trap tester This device picks up high-frequency trap sounds with its probe, and converts them to audible sounds. It is especially useful for detecting leakage through worn valve seats.

In a system that operates at higher pressures, leaking steam may increase the temperature of the discharge for a short distance below the trap, before the steam or condensate has time to equilibrate with the condensate system pressure. However, you may not be able to tell whether a high surface temperature is caused by leaking steam or by conduction through the metal of the components. Furthermore, the condensate itself discharges from the trap at an elevated temperature.

The only exception is a trap that holds condensate long enough for it to cool below the boiling point temperature of the condensate system. This can occur only with thermostatic traps. However, to avoid backing up condensate, thermostatic traps are designed to cool condensate only a small amount below the temperature of the steam system. Some thermostatic traps are designed to deliberately "subcool" condensate, but such applications are unusual.

The trap temperature does provide a reliable indication if a trap is stuck closed or valved off. In this case, the body of the trap is substantially cooler than the system temperature. You can measure the surface temperature of the trap by holding a thermometer against the surface with a rag over it, or by using crayons that melt at calibrated temperatures, or by using a pyrometer. If the system operates at low pressure, you can tell whether the trap is near normal operating temperature by squirting water on it. Checking by feel is asking for trouble.

■ Infrared Imaging Scanners

Infrared cameras are a versatile tool that allow a user to actually see patterns of temperature variation. Although spot testing of trap and system surface temperature is a poor method of checking steam traps, for the reasons given previously, infrared imaging can be effective. If you have the right equipment and enough experience, you can observe heat patterns on the surface of the trap and adjacent pipe that indicate whether the trap is operating properly. You may also be able to detect leaks.

See Reference Note 15, Infrared Thermal Scanning, for more about this equipment. Scanning equipment is expensive, fragile, and somewhat awkward, but it may be the best method for facilities with many traps.

■ Proprietary Trap Performance Indicators

In the continuing search to find reliable methods of diagnosing steam trap performance, various devices have been invented for installation on or near steam traps to indicate whether they are operating properly. Measure 1.10.4.1 describes several of these devices that are presently on the market.

Preferred Test Methods

For most applications, consider a combination of test valves and acoustical and/or ultrasonic testing. If a trap is not discharging condensate, check its surface temperature to tell whether it is stuck closed or some other stoppage exists.

Between test periods, keep an eye on the condensate system vents.

Test Both Start-Up and Hot Operation

Steam traps perform different functions at start-up and during warmed-up operation. Therefore, test your traps under both conditions. For example, an F&T trap may appear to function properly in warmed-up operation because the float is draining condensate. However, the thermostatic element may be defective, failing to vent the system at start-up.

Drip traps are easy to check. On start-up, a drip trap should drain clearly observable amounts of condensate. After the system is warmed up, the trap should pass little condensate and no steam.

Traps installed on steam-using equipment may be more difficult to check, especially in high-pressure systems. If the equipment has a heavy load, the heavy condensate drainage may make it difficult to distinguish direct steam leakage from flash steam. If the equipment has modulating control, the steam pressure inside the equipment is reduced at lower loads. A leak that is easy to find under high pressure may not be apparent at low load.

How Often to Test

If a trap fails the day after you test it, it continues to waste energy until the next test. Therefore, test your traps as often as practical. Use these guidelines to develop an appropriate testing schedule:

- *the types of traps installed.* Inverted bucket traps and float traps are the most reliable types. In average service, they may work for years without developing problems. Disc traps are considered the least reliable, and may start to have escalated steam consumption within a few months. The less reliable types also have a broader variation in failure rate, making it less certain that an individual trap will survive for a given period of time.
- *the number of traps in the system.* The more traps there are in the system, the greater is the probability that a failed trap is leaking steam at any given time.
- *trap capacity.* The rated capacity of a steam trap reflects its orifice size and the pressure difference under which it operates. Both of these factors determine the leakage rate in the event of failure. Thus, it pays to test big traps often because they can waste a lot of energy when they fail. On the other hand, smaller traps do not have the strong opening and closing forces of larger traps, and their components are not as strong, so they may not last as long as larger traps on average.
- *availability of personnel.* Balance the cost of steam loss against labor cost. There is no reason why traps in boiler plant should not be tested daily if the plant is manned and operators have free time.
- *accessibility of traps.* Accessibility is a major factor in labor cost. If a trap is in an awkward location, consider moving it to a more accessible location.

The large cost penalty of steam trap leakage in a large, high-pressure steam plant may make it worthwhile to test steam traps daily or weekly in such plants. At the other extreme, a low-pressure steam plant with only a few traps may need trap testing only several times per year.

Group Repair and Reduced Testing Interval

After a facility has been in service for several years, the lives of the traps are no longer synchronized, so trap failures are likely at any time. This forces you to test traps more often to keep traps from leaking for long periods. Therefore, in steam systems that have a large number of similar traps, it may be more economical to overhaul all the traps of each type as a group. This provides a period of time during which there are few trap failures, so you can reduce trap inspections during this period. Later, when the failure rate starts to rise, overhaul all the traps of that type.

The optimum time-between-overhaul varies for different types of traps. Therefore, each type of trap should have its own group overhaul schedule.

Trap Repair Procedures

Good traps are designed to allow easy replacement of worn parts, and parts kits are available for common models. Replacement parts are few and simple, but training and care are needed to install them properly. Follow overhaul instructions exactly. For example, valve plugs and seats are supplied in matched sets. Replacing one without the other virtually guarantees leakage. Aligning the parts in some traps requires a delicate touch.

Where to Get Training

Trap testing and repair requires hands-on training. Trap diagnosis can be learned properly only in a setting where students can observe functioning equipment, including their sounds and other behavior when operating properly and when defective. Some trap manufacturers have excellent training facilities that provide these opportunities. Some demonstration setups have piping, steam coils, and traps made of transparent materials. These are invaluable because the operation of steam equipment is difficult to visualize through steel.

The training provided by trap manufacturers to steam users is either free or it costs only a modest amount. This training is one of the best bargains in energy conservation.

ECONOMICS

SAVINGS POTENTIAL: *1 to 20 percent of system steam production, depending largely on the size and condition of the steam distribution system.*

COST: *Most of the cost is for labor. This cost depends on the amount of free time that plant operators have available. Steam trap testing equipment typically costs a few hundred dollars. The cost of accessory devices installed on traps for diagnosis is covered in subsidiary Measure 1.10.4.1.*

See Reference Note 15, Infrared Thermal Scanning, for the cost of thermal scanning.

PAYBACK PERIOD: *Immediate. Maintaining steam traps always costs less than neglecting them.*

TRAPS & TRICKS

DILIGENCE: *Steam trap maintenance tends to be neglected because leakage is invisible and inspection may be awkward. It becomes easier with practice. Schedule trap inspections in your maintenance calendar, and give them priority.*

TRAINING AND EXPERIENCE: *Steam trap testing requires a knowledge of trap operation, experience, and a certain amount of intuition. Invest in training one or more talented individuals as trap testers. Get hands-on training with your types of traps from a trap manufacturer or other effective source of instruction. If possible, have the same individuals test and repair traps, because the two activities enhance each other. Alternatively, hire expert testing services, which are covered in subsidiary Measure 1.10.4.2.*

SELECTING TEST METHODS: *Select the most effective test methods for your types of traps. In case of doubt, discuss testing methods with operators of other facilities who have your types of traps. Install appropriate test devices, if they can speed testing or increase the certainty of diagnosis (see subsidiary Measure 1.10.4.1).*

MEASURE **1.10.4.1 Install accessory devices to assist in steam trap diagnosis.**

This is the first of two subsidiary Measures that help you test your steam traps. The method here is to install an accessory testing device at each trap that tests the condition of the trap.

The uncertainty and difficulty of steam trap diagnosis keeps people inventing new devices for testing steam traps. Some of these are installed as permanent trap accessories. The best ones may greatly increase the speed and certainty of testing when the user is familiar with them. Installed trap test devices are especially worth considering for facilities that have a large number of traps. You can use some of these accessory devices with all types of traps, but others are limited to certain trap types.

Applications and Alternatives

All permanently installed steam trap testing devices require a certain amount of space in the condensate discharge piping. This makes them impractical for traps that are installed in tight quarters, such as the traps on many steam radiators. Also, the devices are somewhat expensive, which makes them uneconomical with most small traps.

Some of these accessory devices can be a mixed blessing. Using each type requires its own specialized techniques, and requires more theory for the trap tester to learn. Furthermore, each device introduces its own failure modes and maintenance requirements. As a

SUMMARY

Some test devices may substantially speed and simplify steam trap diagnosis. They typically require additional skill. Some proprietary devices are confusing, fragile, and/ or expensive. All require some maintenance.

SELECTION SCORECARD

Savings Potential .... $ $ $

Rate of Return, New Facilities % % % %

Rate of Return, Retrofit.......... % % % %

Reliability .... ✓ ✓ ✓ ✓

Ease of Retrofit .... ☺ ☺ ☺

result, they are more applicable to plants that have skilled operators. In other facilities, they will add confusion.

Test Valves

Test valves are simple and fairly obvious to use. They can by used with all types of steam traps. They indicate some trap problems clearly, but not all.

The simplest arrangement is a single valve in the trap discharge that diverts the discharge to the atmosphere. For example, see Figure 6 in Measure 1.10.3. This allows you to observe whether steam is

Armstrong International, Inc.

Fig. 1 Steam traps with inlet and outlet test valves Note how the discharge pipe for each test valve is aimed away from the valve handle.

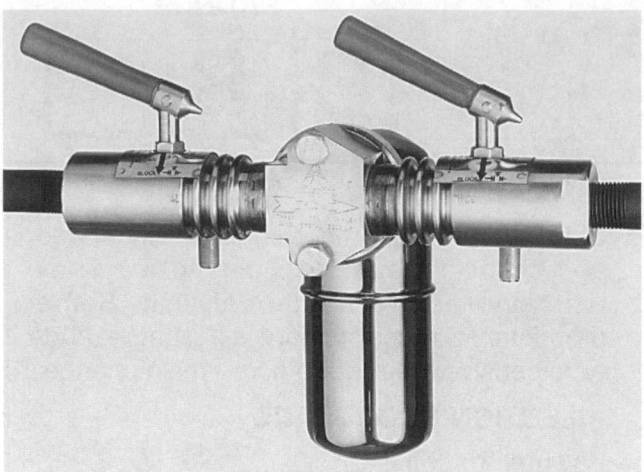

Fig. 2 A steam trap with integral test valves

leaking through the trap, or whether it is stuck in a closed position.

If a trap is used as a drip trap, the test valve cannot indicate a stuck-closed condition when the system is warmed up. This is because the steam line may be dry at that time. To test whether a drip trap is stuck closed, install a test valve in the inlet side of the trap. This will show whether water is backed up ahead of the trap. Figure 1 shows a group of traps that have both inlet and outlet test valves. This method is reliable with float traps and with inverted bucket traps. However, it is not

reliable with thermostatic, disc, and orifice traps, because water may back up ahead of these trap types in normal operation.

Figure 2 shows a steam trap that is equipped with integral inlet and outlet test valves. This simplifies installation.

It can be difficult to tell whether steam blowing from a trap discharge is the result of a steam leak in the trap, or is flash steam from the condensate, or is steam that is coming from the condensate system. The amount of flash steam from condensate increases considerably with boiler pressure.

Also, a steady flow of steam, without condensate, may come from the condensate system itself. This can occur if other traps have serious steam leaks. To determine whether blowing steam is coming from the trap or from the condensate system, install a set of three test valves, as shown in Figure 3. One valve closes the discharge line to the condensate system, one opens a sampling line from the trap discharge, and one opens a sampling line from the condensate system. Use the valves in this way:

- *normal operation.* The discharge valve from the trap to the condensate system is open. The other two are closed. This is the state shown in Figure 3.

- *trap test.* Close the discharge valve, and open the valve above it. This discharges the trap to the atmosphere, so you can observe the discharge directly. A steady flow of steam indicates a leaking trap. If nothing comes out, this may indicate that the trap is stuck in a closed position. However, this may be normal for a drip trap on a warmed-up steam line.

- *condensate system test.* Close the discharge valve, and open the valve below it. This lets you see

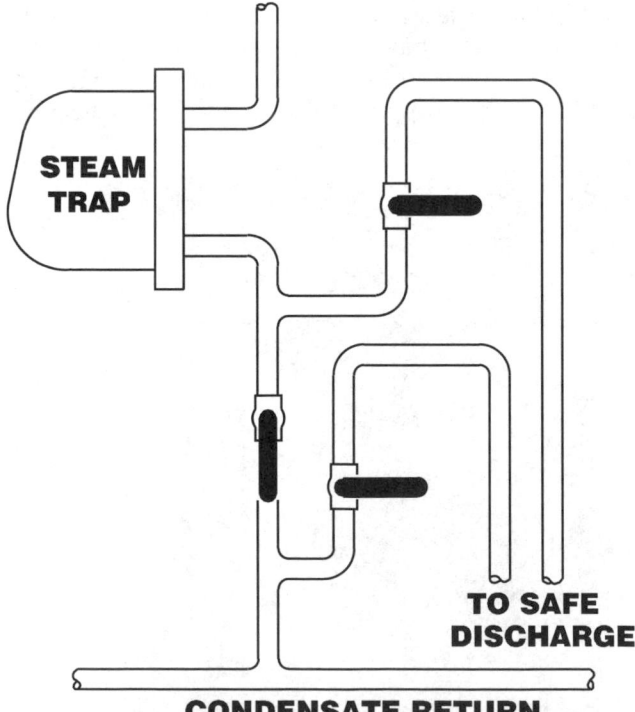

Fig. 3 Test valves that isolate the condensate system This arrangement makes it possible to tell whether steam blowing from the trap discharge is coming from the trap itself or from the condensate system. Note how the test valves are installed in vertical pipe, to keep them free of debris.

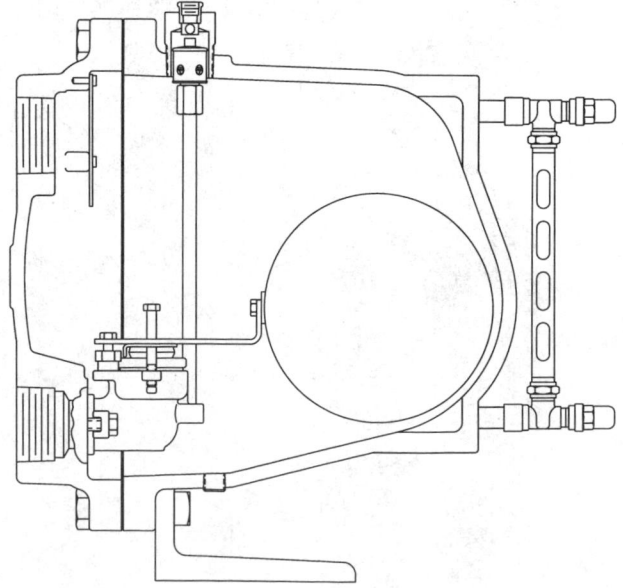

Fig. 4 Steam trap with gauge glass

whether blowing steam is coming from the condensate system, rather than from the trap itself. This indicates that there are leaks in other traps connected to the same condensate system.

At least one manufacturer makes a three-position stainless steel valve that accomplishes the same purposes as this set of three valves. It is more compact, and easier to install.

Test valves are exposed to heat, moisture, and air, so they should be made of non-corrosive materials.

Try to install test valves so that condensate and debris cannot come to rest inside the valves. The best practice is to install the test valve in a vertical pipe. Figure 3 shows how to do this.

Test valves add a potential safety hazard. Bear in mind that opening a valve ahead of the trap vents the full pressure of the steam system. Also, if the trap fails in the open position, live steam will blast out of the sampling valve with the potential to cause severe burns. For this reason, pipe the discharge from the test valve in a safe direction. The test valves and the connecting pipe should have the smallest practical diameter, to protect people and to minimize steam blast in case any of the connections breaks from corrosion or impact.

Label test valves to indicate their purpose, to indicate the valve positions for each purpose, and to warn of the safety hazard. See Reference Note 12, Placards, for placard design, materials, and installation methods.

Gauge Glasses

With float traps and inverted bucket traps, the level of water inside the trap is a reliable indication of whether the trap is working properly. You can't see the water level inside a trap, but a gauge glass makes this possible. The gauge glass used with a trap is similar to the gauge glasses used on boilers, feedwater tanks, or other pressurized vessels. Some traps are offered with gauge glasses as an option. Figure 4 shows an example.

Gauge glasses cannot be installed on steam traps unless the traps are designed to take them. The bottom of the gauge glass must be connected to the body of the trap, using a special fitting ahead of the discharge valve.

Gauge glasses add a potential safety hazard because they contain full steam pressure. For this reason, gauge glasses must be sturdy and well protected. Both top and bottom connections should include small orifices to limit steam release if the gauge glass breaks.

Gauge glasses are subject to fouling that can make it difficult to see the water level inside. Therefore, the gauge glass should have a blowdown valve at the bottom connection to allow it to be cleared out occasionally.

Proprietary Test Devices

New types of devices for testing steam traps keep emerging. For example, Armstrong International, a major trap manufacturer, offers a sensor that is installed

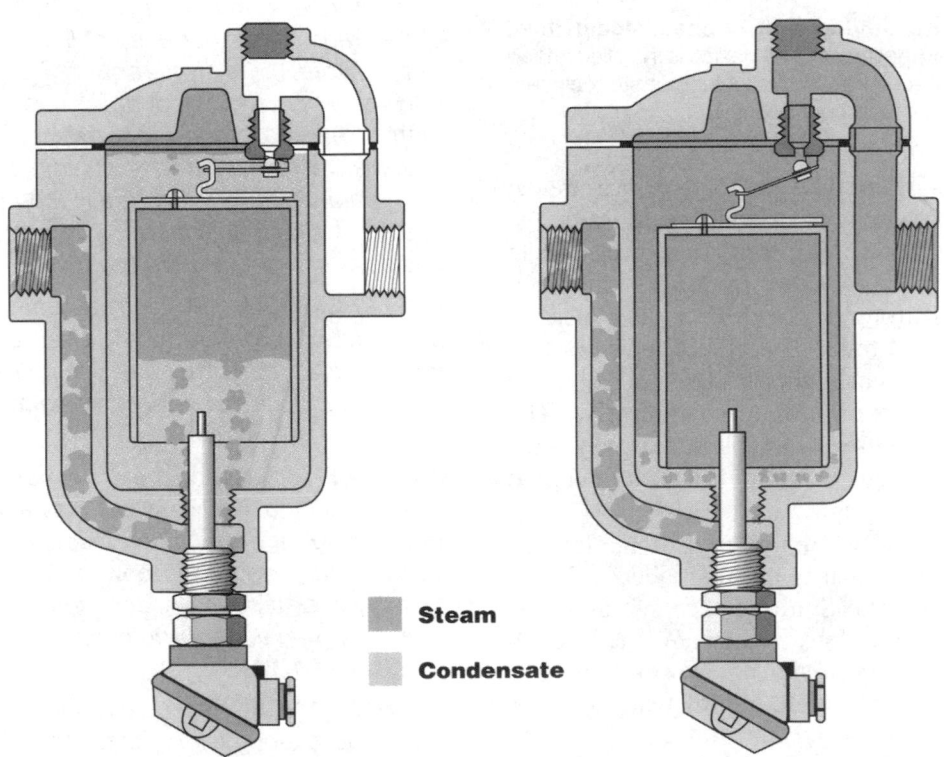

Steam

Condensate

Armstrong International, Inc.

Fig. 5 Trap testing device that senses conductivity and temperature This device can detect a trap that is stuck open or closed. It can monitor traps remotely and continuously. It requires specially designed traps.

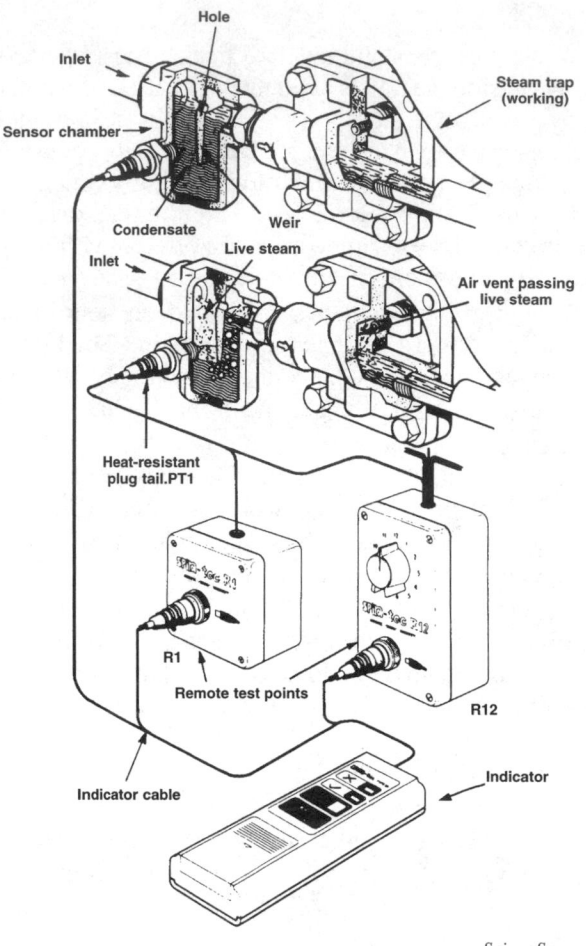

Spirax Sarco, Inc.

Fig. 6 Trap testing device that senses steam flow, conductivity, and temperature The device is installed ahead of the trap. It can detect a trap that is leaking or stuck closed. It can monitor remotely and continuously.

in the bottom of its inverted bucket traps to measure the temperature and conductivity of condensate inside the trap. See Figure 5. Using this information, an electronic logic circuit determines whether the trap is operating properly. The Armstrong device detects either a steam leak or a stuck-closed trap. The device indicates trap condition either with signal lights installed on individual traps, or the traps can be monitored remotely. The remote monitoring feature is valuable for large systems because it allows continuous monitoring with minimal labor.

Spirax Sarco, another major trap manufacturer, offers a sensing device that is installed ahead of each trap. See Figure 6. The Spirax device senses steam flow but not condensate flow, so a steam leak through the trap generates an alarm signal. Newer models of this device detect a stuck-closed trap by using a sensor that measures conductivity and temperature. The Spirax device can be used with most types and models of traps. Traps can be monitored locally or remotely.

Proprietary devices have a number of disadvantages. One is that the manufacturer's monopoly on the device raises the price. At the same time, the manufacturer is less motivated by competitive pressure to improve the product. One-of-a-kind devices that are discontinued leave users with equipment that cannot be repaired. Nonetheless, trap diagnosis is important enough to justify the effort and risk of considering new methods of doing it better.

ECONOMICS

SAVINGS POTENTIAL: The savings that result from trap testing are given in Measure 1.10.4. This Measure improves the reliability and reduces the effort of trap testing. The potential savings depend on labor costs, the skill of the staff, the size and number of traps, and the accessibility of traps.

COST: Varies widely. Test valves may cost less than $100 per trap, while a remote monitoring system for a small number of traps may cost several thousand dollars per trap.

PAYBACK PERIOD: Less than one year, to several years.

TRAPS & TRICKS

SELECTING THE METHOD: Favor test methods that provide reliable diagnosis without the need for special expertise or extensive experience. Select equipment for ruggedness and longevity in the harsh environment of the traps. Consider ease of installation and ease of use. Avoid units that require frequent or difficult maintenance. Ask vendors for references, and check them. Beware of proprietary devices that may become obsolete, especially if they contain electronics. Favor major manufacturers, especially manufacturers of traps and other steam specialties, who are likely to remain in business and support the product.

INSTALLATION: Most trap testing devices have specialized installation requirements. Follow the manufacturer's instructions. Install to avoid internal fouling and external impact damage, and for ease of use and maintenance.

EXPLAIN IT: Describe the equipment and its operation in a manner that is accessible to everyone involved with trap testing. If appropriate, install placards at the test devices. Train your staff how to use and maintain the test devices. If possible, combine this with the general training in trap testing and maintenance recommended in Measure 1.10.4.

MAINTENANCE: Schedule regular maintenance of the test devices along with maintenance of the traps.

MEASURE **1.10.4.2 Hire specialists to perform periodic steam trap inspections.**

This is the second of two subsidiary Measures that help you test your steam traps. The approach here is to hire specialists to take care of the whole job, or most of it. For example, some steam trap manufacturers offer trap inspection as a service, as in Figure 1. This may be the only practical approach for a facility that cannot maintain a skilled boiler staff.

This approach does not eliminate the uncertainties of trap testing. Inspection companies have the same problems in training and keeping personnel as other companies, and personnel assigned to a particular customer may be spotty in skills or initiative. Therefore, check the reputation of the company. Ask about the facilities and staff that the company uses for training its personnel, about the number of trained personnel, their length of service with the company, etc.

There is a conflict of interest when hiring contractors to perform trap testing. If the contractor is also given responsibility for trap repair, there is a tendency to do unnecessary work. A possible solution is to hire a contractor for testing only, and let the in-house staff do

Armstrong International, Inc.

Fig. 1 Trap inspection service   Consider a professional service to test and/or repair your steam traps if this approach would provide greater skills or consistency.

SUMMARY

Consider this approach if you cannot maintain a skilled in-house staff to do trap testing.

SELECTION SCORECARD

Savings Potential	$ $ $ $
Rate of Return	% % % %
Reliability	✓ ✓ ✓
Ease of Initiation	☺ ☺

---

the repair work. The logic in this division of labor is that testing requires more subtle skills than repair. However, this motivates the contractor to be hasty in doing the testing.

Reducing Inspection Costs with Group Overhaul

The need for frequent trap testing raises the cost of using contractors. One way to increase the inspection interval is to overhaul all traps as a group. See Measure 1.10.4 for details.

Consider In-House Trap Testing and Repair

If the facility has a skilled physical plant staff, it is usually best to do steam trap testing in-house. In-house personnel can be selected and supervised more effectively than contractor personnel. The major steam trap manufacturers offer excellent training for facility maintenance personnel, as discussed in Measure 1.10.4.

However, do not be excessively optimistic about in-house testing if conditions are not favorable for it. Trap testing will be neglected if there is high staff turnover, or if physical plant maintenance does not have the attention of upper management, or if existing staff do not have the capability to stay proficient, or if the staff are overworked. In such cases, hire a reputable contractor and hope for the best.

ECONOMICS

SAVINGS POTENTIAL: The savings that result from trap testing are given in Measure 1.10.4. This Measure improves the reliability and reduces the effort of trap testing in facilities where it is not appropriate to perform trap testing and maintenance in-house.

COST: Varies widely. For large facilities, the cost of hiring competent outside services is probably about the same as the cost of training and maintaining a skilled in-house staff for trap maintenance. With smaller non-

industrial facilities, hiring outside services may be the only practical way of keeping traps maintained.

PAYBACK PERIOD: *Immediate, since no initial investment is required. In order to be economical, the cost of using a contractor must be less than the additional savings achieved by using a contractor.*

TRAPS & TRICKS

SELECTING THE CONTRACTOR: *Look for experience. Ask contractors how many skilled test personnel they*

have, and how long they have been employed by the company. Ask what test methods they use. Ask for references, and check them.

MONITOR PERFORMANCE: *Schedule periodic checks of the contractor's work, immediately after an inspection cycle. Look for failed traps that they may have missed, and check that repaired traps are functioning properly. This requires at least one trained trap tester on your own plant staff, or available from another source.*

MEASURE **1.10.5 Recover heat and water from steam vents.**

A blowing steam vent is like a flare at an oil refinery. It is unmistakable evidence that a large amount of energy is being wasted. Even a small vent that operates continuously can waste a considerable amount of energy and money. Do not miss an opportunity to recover this energy.

Another reason to recover vented steam is to recover the pure water that comes from condensing the steam. Recovering the water saves the cost of replacing it, and also reduces water treatment costs.

Recovering the heat and the water is usually a simple matter of heat exchange. You can use any liquid, gas, or solid medium to condense the steam if its temperature is below the boiling point of water. Most of the energy of steam is latent heat, so the temperature of the condensing medium does not matter much, as long as it is cooler than the boiling temperature. For example, you can use warm condensate to condense venting steam. This warms the condensate, reducing fuel cost, and allows the condensed steam to be used as makeup water. If a large amount of steam is being vented, consider using it for heating domestic water, heating fuel tanks, process heating, etc.

Heat Recovery Methods

You can use an ordinary steam-to-liquid heat exchanger for many applications. Where the application is heating liquid in a tank, consider a steam immersion coil.

There is no need for a heat exchanger if you can mix the vented steam with the condensing medium. One method is to pipe the steam into a tank of liquid. Or, spray the liquid into a pipe that contains the vent steam.

> **SUMMARY**
>
> Consider this anywhere a significant amount of steam is venting. Usually simple.
>
> **SELECTION SCORECARD**
>
> Savings Potential .... $ $ $
>
> Rate of Return, New Facilities % % % %
>
> Rate of Return, Retrofit.......... % % %
>
> Reliability .... ✓ ✓ ✓
>
> Ease of Retrofit .... ☺ ☺

A vent condenser is a specialized heat exchanger designed specifically to recover heat and water at steam vents. A vent condenser is cooled by a liquid, and the condensed steam falls back into the vessel. Vent condensers are common on deaerating feedwater heating tanks.

If there is no application for heat recovery, but you want to recover the water, pass the steam through an air coil to condense it.

How to Find Leaking Steam Vents

There are many industrial processes that involve steam vessels with vents. To find them, look for plumes of steam. A major plume of steam may persist for a long time because everyone assumes that it is intentional or necessary.

Within the boiler plant itself, the deaerating feedwater tank (if one is installed) is the equipment most likely to be venting steam. Deaerators use steam to separate air from the feedwater, and the pressure within the tank typically ranges up to 15 PSI. At the vent where the air is ejected from the tank, a considerable amount of steam may be lost along with the air. The tank may already have a vent condenser installed. If so, check that the vent condenser is operating properly.

Sizing the Heat Exchanger

You need to estimate the amount of steam that is being wasted in order to size the heat exchanger and to estimate the potential savings. Engineering tables are available that give the rate of steam flow through a hole, based on the diameter of the hole and the pressure difference. Do not be fooled by a vent that is discharging both air and steam, where the size of the hole or the appearance of the plume exaggerates the amount of steam being wasted.

One way of measuring the amount of steam being wasted is to run a hose from the steam vent into an empty drum. Cool the drum by spraying it with cold water to make the steam condense. The rate of steam loss, by weight, equals the rate at which water condenses in the drum. Each pound of condensate represents about 1,000 BTU of recovered steam energy.

ECONOMICS

SAVINGS POTENTIAL: *Depends on the application. For deaerating feedwater heaters, the loss of vented steam can account for 0.1 to 1 percent of fuel input.*

COST: *Piping steam directly to a nearby application may cost several hundred dollars. Installing a typical vent condenser and piping to an application may cost several thousand dollars.*

PAYBACK PERIOD: *Less than one year, to several years.*

TRAPS & TRICKS

EXPLAIN IT: *The purpose of the steam recovery equipment is not obvious from its appearance. Install a placard that identifies it, states how it works, and states that there should be no steam plume coming from the equipment. See Reference Note 12, Placards, for details of effective placard design and installation.*

Conduction and Radiation Losses

The equipment in boiler plants and heat distribution systems is hot, so it loses heat by radiation and by conduction to the air in the space. Heat loss from the boiler itself and its auxiliary equipment typically wastes about two percent of the boiler's full-load energy consumption. Heat loss from the distribution system varies enormously, depending on the size, design, and condition of the distribution pipe and its accessories. In long distribution systems, a large part of boiler energy input may be lost by conduction and radiation.

The Measures in this Subsection reduce conductive heat losses from the boiler plant equipment and distribution system. The third Measure recovers part of this heat.

INDEX OF MEASURES

1.11.1 Locate and repair defective insulation on all heating plant equipment and piping.

1.11.2 Minimize cooling or ventilation of pipe tunnels and other unoccupied spaces surrounding hot distribution equipment.

1.11.3 Route combustion air to the boiler by a path that recovers heat from the boiler room.

MEASURE 1.11.1 Locate and repair defective insulation on all heating plant equipment and piping.

Insulation is what keeps the heat energy of a boiler system from leaking out of the system on its way to the end users. Inspect the condition of your insulation often enough to keep the insulation from becoming seriously deteriorated. This Measure summarizes what insulation does, what to insulate, and how to insulate. It also gives you several methods of finding defective insulation, even if it is buried.

Basics of Boiler System Insulation

■ Heat Loss of Bare Surfaces

Surfaces lose heat by radiation and convection. If the surface has little insulation value, such as the surfaces of steel pipe and fittings, the rate of heat loss depends almost entirely on the surface area and surface temperature. Convective heat loss occurs strongly even at low temperatures, but levels off to a maximum rate. In contrast, radiation heat loss continues to increase exponentially with temperature.

Radiation becomes the dominant mode of heat loss from exposed surfaces in high-temperature plants. For example, a bare pipe carrying 15 PSI steam is about 150°F hotter than typical boiler room air, and it loses about 400 BTU per hour per square foot of surface area. By comparison, a bare pipe carrying 600 PSI saturated steam loses about 1,600 BTU per hour per square foot.

The amount of bare surface in a heating plant is usually minor, so bare surface heat loss is not a major consideration. Still, it may pay to insulate localized bare surfaces, as recommended below.

■ Upgrading Intact Insulation

If a surface is insulated, the rate of heat loss is determined primarily by the surface area, the thermal resistance (the R-value, in English units) of the insulation, and the insulation thickness.

The shape or curvature of insulation is also a factor. Insulation that is curved has more outer surface area in relation to its thickness, and hence more heat loss. This is significant in the case of pipe insulation. Because of this effect, a given thickness of insulation is more effective with larger pipe diameters.

Insulation that is kept dry and free of physical damage remains effective indefinitely. In most heating plants, the insulation was installed properly at the time of construction, although certain minor areas may have been neglected. Damage over the years may lead to some loss of insulation, but incidental damage usually does not justify replacing insulation. For example, see Figure 1.

If insulation is in reasonably good condition, the remaining question is whether there is enough of it. In many heating plants, the amount of insulation may be inadequate in terms of present economics. It may be economical to supplement the original insulation, or to replace the existing insulation with new material having higher thermal resistance. Upgrading is most likely to be worthwhile in systems that have extensive piping that is easily accessible.

However, upgrading existing insulation usually has a low rate of return. Each additional increment of insulation thickness provides less benefit. Economics favors installing plentiful insulation on new equipment. A large fraction of the cost of insulation is for labor, and it does not require much additional labor to increase the insulation thickness.

Select your insulation quantity using engineering calculations. The principles and formulas are simple. Refer to ASHRAE publications or other books dealing with heating systems.

■ Heat Loss of Wet Insulation

Moisture is the worst enemy of insulation. If fibrous or open-cell insulation becomes wet, its insulation value falls drastically, and it may be permanently damaged. If buried pipe insulation of this type becomes soaked, it becomes ineffective, and the surrounding soil determines the rate of heat loss. Closed-cell insulation may lose little insulation value if wetted, provided that it is installed snugly. Even closed-cell foam may not keep the pipe itself from becoming wetted, which causes corrosion.

In indoor locations, insulation becomes wet if there is a leak, usually in the pipe or equipment that the insulation is covering. A leak may soak the insulation. Insulation should be installed so that a leak inside the insulation can drain out, but this is rarely done in practice.

The worst moisture damage to insulation occurs in buried pipe. Deterioration of buried insulation is a common, insidious source of major energy waste in large distribution systems. In the past, engineers tended to underestimate this problem, causing many facilities to suffer high heat loss from their underground pipe. Fortunately, piping system designs and equipment that preserve the integrity of the insulation have now become widely available. Unfortunately, replacing underground pipe is very expensive.

In steam tunnels and pipe trenches, as in Figure 2, the relative humidity of the air surrounding the insulation may remain at 100% if the soil is wet. This does not cause the insulation to become wet, provided that the pipe remains warmer than the air inside the tunnel. The higher temperature of the pipe keeps the relative humidity inside the insulation well below 100%. (Just the opposite situation exists with chilled water piping, which faces a serious problem of waterlogging when it

is installed inside tunnels.) If the piping system is shut down at certain times, the insulation may become wet. Whether this is a problem depends on the type of insulation.

What Should Be Insulated

Install adequate insulation on all equipment and piping that carries heat, unless the heat is being discarded. Equipment that may lack adequate insulation includes return hot water piping, condensate piping in closed systems, and fuel oil piping that carries heated oil.

Small items matter if they are very hot. In high-pressure systems, insulate system accessory equipment, such as valves and the larger types of steam traps.

■ Where Not to Insulate

Not every bare steel surface in a heating plant should be insulated. Do not insulate:

Armstrong International, Inc.

Fig. 1 The Insulation from the Black Lagoon Even insulation that looks this bad may be in reasonably good condition. The main points are whether the insulation is dry and snugly fitted, and whether there is enough of it. The main problem with the insulation of this tank is peeling paint. The lower pipe insulation is in bad condition, and should be repaired.

WESINC

Fig. 2 An ideal environment for heating system insulation
If the pipe is warm, the insulation will remain dry regardless of the humidity in the tunnel. However, this is a challenging environment for chilled water pipe insulation, especially if hot and cold pipe run in the same tunnel.

- *drain lines to sewers*
- *pipe carrying unheated fluids,* such as domestic water
- *thermostatic steam traps and the condensate pipe that leads to them,* for the reasons given in Measure 1.10.3
- *condensate pipe leading to disc traps,* in some cases. See Measure 1.10.3 for details.
- *soot blower elbows,* which are hot only briefly while soot is being blown. The heat in the elbows cannot be retained until the next operation. Metal "reflective insulation," which is explained below, may be installed primarily to act as safety guards.
- *boiler air casings,* if the casings are not hot to the touch. The air entering the boiler through the casing captures most of the heat radiated from the hot interior hardware.

You may harm some boiler room equipment by insulating it. Any hot equipment that is covered by insulation becomes much hotter. Damage can occur in ways that are not apparent, such as from thermal expansion. For example, it would not be prudent to add insulation over the sheetmetal skin of a boiler, because the skin may deform excessively from thermal expansion.

■ Whether to Insulate Condensate Systems

The question arises whether it is correct to insulate condensate return systems. The answer depends on the design of the system. Closed (pressurized) condensate systems should be insulated. In the case of open (atmospheric pressure) condensate systems, some people say that insulation is a bad idea because keeping the condensate warm causes it to evaporate away. This is wrong. Evaporation occurs only through the small vent or vents, usually located only at the condensate receiver. As long as the temperature inside the condensate system is below the boiling point, the vapor pressure of the water is below atmospheric pressure, so water vapor is not actively forced out of the system.

Condensate coolers are sometimes installed in condensate systems to prevent flashing of steam at the inlet of a condensate pump or feedwater pump. In such cases, insulating the condensate piping defeats the purpose of the condensate cooler. You may be able to avoid this obstacle by changing the pump installation or by installing a steam-powered condensate mover. See Measures 1.8.4 ff for these methods.

How to Find Defective Insulation

Visual inspection is usually good enough for insulation that is under roof and accessible, as in boiler rooms, pipe chases, and steam tunnels. If the insulation is intact, fits snugly, and remains dry, it is almost certainly doing its job. Damage that is serious enough to reduce the effectiveness of insulation is usually obvious. The only subtle case is where insulation may become wet from a leak in the equipment it surrounds.

You can use infrared cameras and other types of thermal scanners to survey large areas of insulation and inaccessible equipment. See Reference Note 15, Infrared Thermal Scanning, for details.

In a buried steam distribution system, you can estimate the rate of heat loss from the pipe insulation by measuring the rate of condensate flow from the drip traps. Each pound of condensate represents a conductive energy loss of about 1,000 BTU. If the system has an intact condensate return, pick a time when all the end user equipment can be turned off, so the condensate flow is returning only from the drip traps. Alternatively, drain the drip traps into containers and time how fast they fill.

This test method works only if the steam in the lines is saturated. Some facilities superheat the steam to keep it from condensing in the lines. To get around this, turn off the boiler's superheaters during the test, or use the desuperheater.

Steam becomes superheated in passing through a pressure reducing valve. If the line is fed from a pressure reducing valve, reduce the boiler pressure enough to allow you to bypass the pressure reducing valves for test purposes.

You can localize defects in buried insulation by using the same methods used to find leaks. See Measure 1.10.2 for details.

Insulation Methods and Materials

Extensive insulation repairs are usually done by contractors having the specific skills and equipment appropriate to the type of insulation being installed. Experience indicates that some methods of insulation are much better than others, particularly for buried pipe. Become familiar with the advantages and disadvantages of all the insulation methods before initiating the work.

Refer to the *ASHRAE Handbook* or other references for an overview of insulation materials and installation practices, and for the formulas needed to calculate heat loss. After selecting your general method of insulation, study the manufacturers' literature. The following highlights deserve particular attention.

■ Moisture Resistance

No insulation system is able to keep moisture away from the insulation in the long term. The thermal resistance of insulation can be preserved entirely only by keeping it in a well drained enclosure. The enclosure may be the boiler house, a steam tunnel, or in the case of buried pipe, a drained conduit.

If it is impossible to keep the insulation dry at all times, as with buried pipe, the insulation itself should be impermeable to moisture. But, be aware that impermeable insulation alone cannot maintain thermal integrity or protect the pipe. Joints and cracks in the insulation allow water to travel between the insulation and the pipe, causing at least localized heat loss and corrosion. No insulation material is entirely impermeable when exposed to continuous moisture.

■ Insulation Temperature Ratings

All new insulation should be rated for the highest temperature of the equipment being covered. This is true even if the insulation is being used to cover existing insulation, because a break in the original insulation can expose the new insulation to the temperature of the equipment.

With organic insulation, long-term exposure to heat reduces the strength of the material. The rate of decay rises exponentially with temperature, so organic insulation should have a temperature rating that is substantially higher than the maximum operating temperature of the equipment.

■ Workmanship

Insulation should fit snugly all around the equipment. Also, adjacent sections of insulation should fight together tightly. Otherwise, heat is lost by convection between the heated surface and the insulation.

Using tape to cover gaps or hold insulation together is not satisfactory. All tape adhesives deteriorate with time, exposing any openings in the insulation.

Make the insulation removable at locations that may require access for maintenance, such as flanges, in-line traps, etc. Anticipate that water leakage may occur at such places, and install the insulation so that leakage at these points cannot soak the insulation. Leave an opening in the bottom of a removable segment of insulation to allow any water leakage to drain out.

■ Preformed Insulation

Preformed pipe insulation is a great labor saver. If it is sized properly, is provides snug contact with the pipe, reducing convection losses.

The elegant plant operator can now purchase custom fitted insulating covers for valves, steam traps, and other items.

■ Reflective Insulation

Reflective insulation is nothing more than a sheetmetal shell surrounding a hot surface. It is a supplemental method of reducing heat loss from localized surfaces of very high temperature, such as superheater elbows, where it would be awkward to use conventional insulation. The same metal shell may also be used as a safety guard. Reflective insulation is not appropriate as general insulation. Reflective insulation has little or no application in heating plants that operate at low and medium pressures. The metal covers sometimes seen on outside pipe systems are for protection against weather, and they do not function as reflective insulation.

Reflective insulation functions simply as a heat reflector. At very high temperatures, such as the temperature of high-pressure steam, radiation from a bare surface may account for much more heat loss than convection. Heat radiation from hot surfaces is electromagnetic radiation in the infrared range, with a much longer wavelength than visible light. This radiation is reflected by a smooth metal surface. To function properly, the reflector must not touch the hot surface. The reflector should remain at low temperature so that it does not radiate heat.

The shell does not reflect perfectly, and the hot surface does not reabsorb all the reflected heat. As a result, heat radiation bounces back and forth between the pipe and the shell. The shell itself absorbs some of the radiated heat, becomes hot, and loses this heat to the environment. The efficiency of reflective insulation declines seriously as it becomes oxidized or dirty. Even at best, heat loss from reflective insulation is much higher than with conventional insulation.

Don't expect to enhance the performance of conventional insulation with reflecting insulation. The outer surface of the conventional insulation is so cool that it does not lose much heat by radiation.

ECONOMICS

SAVINGS POTENTIAL: *Up to 40 percent of boiler energy input. Savings depend on the condition of insulation, how well insulation is protected from moisture, the extent of the distribution system, soil conditions, operating temperature, and other factors.*

COST: *The cost of repairs and improvement per unit of energy saved varies enormously, depending on the* accessibility of the insulated surfaces, the size of the project, and other factors.

PAYBACK PERIOD: Less than one year, to several years, for insulating bare surfaces. Several years or longer, for replacing defective insulation.

TRAPS & TRICKS

PLANNING: Your most productive investment is the effort you spend in selecting insulation materials and installation methods. Especially with underground distribution, some insulation methods provide virtually

unlimited service, while others start to deteriorate almost as soon as they are installed. Do your homework.

INSTALLATION: Workmanship has a major effect on insulation effectiveness. Each type of insulation requires its own specialized installation techniques. Know the proper techniques before the work begins. Monitor the installation.

FOLLOWUP: Repeat insulation inspections at long intervals. Have a long-term maintenance calendar for scheduling this.

MEASURE **1.11.2 Minimize cooling or ventilation of pipe tunnels and other unoccupied spaces surrounding hot distribution equipment.**

RATINGS

New Facilities	Retrofit	O&M
A	**C**	**B**

If steam or hot water pipe runs in tunnels or chases, you can reduce heat loss from the pipe by keeping the tunnels or chases as warm as possible. This may be a simple matter of closing manhole covers and access doors. For example, see Figure 1.

If no closures presently exist, add them. Improve existing closures, if appropriate.

If a pipe chase passes through a structure that is sensitive to heat, such as the occupied portion of a building, the pipe chase may be vented to keep it from heating the adjacent structure. A more efficient alternative is to insulate the walls of the chase.

Limitations

Safety is an important consideration if people are able to enter the heated space. For example, a tunnel should not be so hot that a worker might collapse from heat prostration while walking through the tunnel. Many countries have safety regulations covering this situation, but use common sense and alertness in any case.

Do not cut off ventilation to any spaces that need air for combustion.

If the space contains equipment that is cooled, such as chilled water lines or air conditioning ducts, calculate whether there is a higher net benefit in keeping the space warm or letting it cool. This compromise may change with the seasons. For example, if a tunnel carries both steam and chilled water lines, it may be best to keep the tunnel heated in winter and to keep it vented in summer.

A certain amount of venting may be required to remove humidity. The amount of venting that is needed

SUMMARY

A simple procedure or modification that can substantially reduce heat losses from distribution piping. May have safety implications.

SELECTION SCORECARD

Savings Potential	$	$		
Rate of Return, New Facilities	%	%	%	%
Rate of Return, Retrofit	%	%	%	%
Rate of Return, O&M	%	%	%	%
Reliability of Equipment	✓	✓	✓	✓
Reliability of Procedure	✓	✓		
Ease of Retrofit	☺	☺	☺	
Ease of Initiation	☺	☺		

depends on the amount of moisture released into the space, which may come from pipe leaks, the surrounding soil, rain, etc. High humidity does not harm heated pipe and fittings, because the heat of the pipe prevents condensation of moisture on the pipe or its insulation. However, condensation occurs on unheated equipment. Warm, wet steel corrodes quickly.

Explain It!

Where doors, access covers, etc. may be left open, mark them in a bold and permanent manner to indicate that they should be kept closed when the space is

WESINC

Fig. 1 Door to steam tunnel This steam tunnel opens into the lower level of a barracks. There would be less heat loss from the steam pipe, and the barracks would remain cooler, if the door were kept closed. All it takes is a sign.

unoccupied. See Reference Note 12, Placards, for effective marking techniques.

ECONOMICS

SAVINGS POTENTIAL: *Up to 20 percent of heat loss from piping.*

COST: *Usually small.*

PAYBACK PERIOD: *Short.*

TRAPS & TRICKS

INGENUITY: *Find all the places where you can reduce heat loss by minimizing ventilation of equipment, and figure out effective ways of sealing them off.*

ACCESS AND MARKING: *Make the accesses to the heated spaces easy to close tightly. The key to maintaining this activity is effective marking of the accesses.*

MEASURE **1.11.3 Route combustion air to the boiler by a path that recovers heat from the boiler room.**

SUMMARY

An experimental concept that saves a small but worthwhile fraction of boiler fuel. May require some highly visible changes to the boiler room that are probably not as difficult as they look. Involves comfort and safety considerations.

SELECTION SCORECARD

Savings Potential .... $ $

Rate of Return, New Facilities % % %

Rate of Return, Retrofit .... % % %

Reliability .... ✓ ✓ ✓

Ease of Retrofit .... ☺ ☺

As all boiler operators know, the air in boiler plants can be quite warm, and some parts of the boiler house are much warmer than others. You can improve boiler efficiency by drawing combustion air from the warmest part of the boiler room. This may be simple to accomplish, although it may require some strange and bulky contraptions.

Energy Saving Potential

Boiler efficiency increases by about one percent for each 40°F increase in combustion air temperature. Usually, the limiting factor in the effectiveness of this Measure is the temperature rise that you find inside the boiler room, rather than the amount of waste heat available. The temperature rise is limited by heat loss from the air to the boiler room envelope, and by the need to keep most of the space cool enough for people.

In an existing boiler plant, if combustion air is drawn directly from the outside, you can increase the combustion air temperature by moving the combustion air intake to a warm area inside the boiler room. For example, if the outside air temperature is 32°F (0°C), moving the combustion air intake to the inside may easily achieve a temperature rise of 40°F (about 20°C). You may be able to achieve an even greater average temperature increase in colder climates.

If the combustion air intake is already located inside the boiler room (typically near the burner front), rearranging the air intake may be able to increase the combustion air temperature by 20°F to 40°F (about 10°C to 20°C). Even this smaller increase saves one half to one percent of boiler fuel, which is economically important in all but the smallest heating plants.

Thus, the range of possible efficiency improvement is typically from 0.5 to 2 percent, based on temperature rise. The next question is whether enough heat is present in the boiler room to sustain this temperature rise. In most cases, the answer is yes. At full load, a typical boiler loses about two percent of its energy input into the space by convection and radiation. At partial load, the percentage of energy loss is higher. This is because the exterior surface of the boiler tends to remain at the same temperature regardless of load. This is a primary reason why overall boiler efficiency drops at low loads.

(Heat loss from the boiler surface is most constant with firetube boilers, because the shell of the boiler is filled with water and steam that remains at constant temperature. With watertube boilers, radiation losses may decrease significantly as the load falls, because the combustion gases lose more of their heat to the tubes before they reach the outer surfaces of the boiler. Measure 1.1.5 recommends lowering the steam pressure or water temperature as the load drops. This may reduce the heat loss into the space significantly.)

Relative heat loss tends to decrease with larger boilers because larger boilers have less surface area in relation to their capacity. Boilers with air casings may have relatively low heat loss, because the flow of air in the casing captures the surface heat of the boiler. Thus, with larger boilers, the availability of waste heat in the space is likely to be the limiting factor in raising the combustion air temperature. Still, larger boilers provide savings that are large in absolute terms.

An important amount of heat may be lost into the boiler house by hot piping and auxiliary equipment. Furthermore, if the boiler room is connected to adjacent steam tunnels and pipe chases, it may be practical to draw heat from these spaces. However, the latter may be self-defeating, because it increases the rate of heat loss from the piping. See Measure 1.11.2 for more about this.

■ Relationship to Boiler Plant Insulation

Improving the insulation of the heated equipment in the boiler plant reduces heat loss, so it somewhat reduces the potential benefit of this Measure. Improving insulation enough to make a major difference in the total plant heat loss is expensive, unless there are serious localized defects in the insulation. In most cases, the benefit of this Measure is not reduced much by potential insulation improvements.

General Approach

The basic idea is to use the boiler room (and other connected spaces that contain hot equipment) as a preheater for the combustion air. Figure 1 sketches the concept in its basic form. Heat lost from the surfaces of the boiler and other hot equipment heats the boiler room air. The heated air rises and is confined. A duct is installed to draw the heated air into the boiler for combustion.

To keep people cool, locate the outside air intake so that outside air flows to the operator station before it flows over hot equipment.

The challenge is to maximize the amount of waste heat that is recovered. This may involve a variety of changes to the boiler room. Many different configurations are possible, and there is plenty of opportunity for ingenuity. Follow these guidelines:

• *Keep the waste heat from escaping.* Ideally, to recover the heat of boiler room air, the exterior surfaces of the boiler room should be well insulated and sealed against air leakage. In the real world, boiler rooms have little or no insulation, and they typically have copious air leakage. Give priority to minimizing air leakage, which is usually simple. It is usually not economical to insulate the boiler room, except perhaps to substitute insulating panels for useless windows or other openings. Try to minimize conduction loss by routing combustion air away from exterior surfaces.

• *Route the air through the space by a path that maximizes heat collection from hot equipment.* Don't bother trying to tailor heat collectors closely to hot sources. Simple baffles and hoods may adequately collect the warmer air by convection.

• *Avoid diluting the combustion air with non-combustion air.* To maximize the temperature of the combustion air, only combustion air should flow through the heated part of the space. If extra air is needed, for operator ventilation or to feed draft regulators, try to keep this air flow away from the heated equipment and keep it from entering the boilers.

If the job is done well, most of the air within the boiler space remains still, especially air near the exterior surfaces of the space. Only enough air is circulated around hot equipment to satisfy the combustion requirement.

This Measure creates a zone of hot air in some elevated part of the boiler plant. This zone is too hot for people to occupy. Therefore, design the air flow path so that the heated air does not traverse normally occupied space as it flows toward the collection point.

Rate of Combustion Air Flow

Combustion of hydrocarbon fuels requires about 10 to 20 pounds of air per pound of fuel, including an allowance for excess air. The low end of this range applies to coal, and the upper end to natural gas. For fuel oil that lies in the middle of this range, burned in a boiler of average efficiency, the combustion air flow is roughly 250 CFM per million BTUH of boiler heat output. Use this figure to estimate the approximate amount of air flow through the space. The air flow changes with the load on the boilers.

Fig. 1 The basic concept Use the warmest air in the boiler room as combustion air, to improve combustion efficiency.

In small boilers, the combustion air flow is barely noticeable. In larger boilers, it can create quite a draft where the outside air enters the boiler room. If the draft is noticeable, route it to provide operator comfort as well as efficient heat collection.

How Heat Moves in the Boiler Room

The way hot equipment loses heat depends on its surface temperature. Insulated surfaces lose most of their heat by convection, whereas surfaces at steam temperature lose heat primarily by radiation, with a large amount of convection loss also. In a properly insulated boiler room, most heat is lost by convection. If the waste heat reaches the cold surfaces of the boiler room, by convection or radiation, it cannot be recovered.

Convection is a two-step process. Heat is first transferred to the air by conduction. Then, the heated air rises because it is less dense than the surrounding air.

Heat radiation passes through air without stopping until it reaches a surface. For all practical purposes, air does not absorb heat radiation or radiate heat itself.

How to Channel the Waste Heat

Ideally, combustion air should flow over all the heated equipment in the space, collecting as much heat as possible, and then flow into the boiler. It should avoid cold surfaces. The typical boiler room is not laid out for this to happen, so you need to guide the heated air to the combustion chamber. This involves collecting air in a hot area of the boiler room, and then routing it to the burner through a duct. Here are the main steps.

■ Close Penetrations in the Boiler Room Envelope

Boiler rooms typically have openings in the walls and ceiling to vent heat, such as roof ventilators and high windows. Keep these closed, and provide comfort ventilation for the operating personnel as suggested below. If you cannot close the openings permanently, mark them prominently to indicate that they should be kept closed for the purpose of heat recovery. See Reference Note 12, Placards, for effective marking techniques.

■ Partition the Boiler Space

If the boilers are installed in the same space as other equipment, such as chillers or air handling units, consider installing partitions to help concentrate the heat from the boiler plant. This may also improve the efficiency and longevity of the cooling equipment by a modest amount. Partitions for this purpose can be very light, perhaps consisting of nothing more than fireproof fabric.

■ Install Heat Collecting Hoods

Heated air rises, so you can collect heated air by installing a hood over the boiler, and perhaps over other equipment that loses a significant amount of heat. The basic idea is sketched in Figure 2. Hoods keep the heated air from coming into contact with cold exterior surfaces before it is recovered. Boiler rooms typically have minimal insulation, so this can be important. It is especially important if the boiler space is much larger than the boiler itself.

Installing hoods over the boilers may allow ventilators to be installed in the ceiling or walls. This may be the best way to collect waste heat while providing adequate cooling for people.

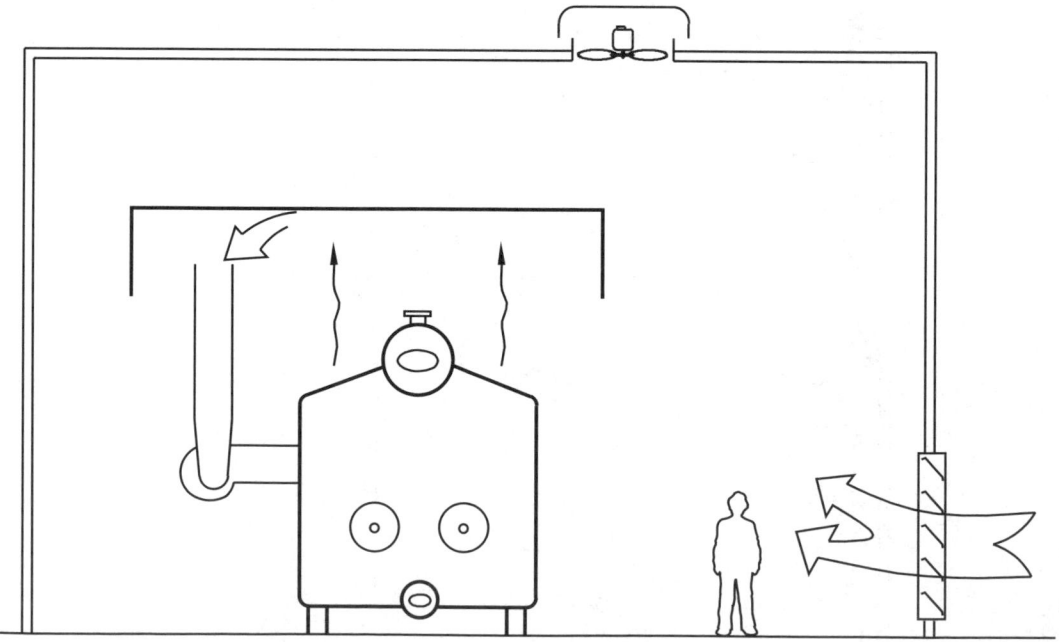

Fig. 2 Heat collecting hood   The hood captures the air heated by the boiler and other hot equipment before it is cooled by contact with exterior surfaces.

You can make the hoods of lightweight material. To improve performance slightly, add a thin layer of insulation to either the top or bottom. For example, duct board should work well. The material should be fire-resistant and have little fuel value. The major installation problem is working around the clutter of pipes and pipe hangers that exist above the equipment.

The height of the hood above the equipment does not matter much, unless there are ventilators moving air through the space. Install hoods high enough to provide space for working on the equipment.

The major disadvantage of this method is that fails to recover heat from equipment that is not located under the hood. To do this, you can install additional hoods that feed air into the main hood. The air should rise into the main hood through large slanted ducts or chutes, like a water slide turned upside down. There should be only one combustion air collection point for each boiler, because it is impractical to balance the air flow from different collecting points.

Hoods recover radiant heat only by absorbing it or by reflecting it back to the hot equipment. Radiant heat travels in all directions, so the only way to capture more radiant heat is to surround the hot equipment more completely. Any opaque material stops radiant heat, no matter how thin it is. Metal surfaces reflects a large part of heat radiation.

■ **Separate the Air Flow into Draft Regulators**

If the flue system uses draft hoods or barometric draft regulators, they may be a serious impediment. Both types of draft regulators must draw air from the interior of the boiler room in order to regulate draft properly. Unfortunately, these dampers are typically located right above the boiler, where they may draw a large portion of the heated air directly into the flue. Figure 3 illustrates the problem.

The only solution is to install a heat collecting hood below the level of the draft hood or barometric regulator. See Figures 1 and 2 in Measure 1.5.1 to see what you have to work around in order to accomplish this.

■ **Isolate Comfort Ventilation**

Operator comfort may require additional outside air, especially in smaller boiler plants where the amount of air brought in for combustion air is not sufficient to provide comfort ventilation first. If possible, keep the additional air separate from the combustion air stream. For example, consider providing an enclosure for operating personnel that has a separate fan for comfort cooling.

Where to Locate the Outside Air Intake

The air that is drawn from the space for combustion must be replaced with outside air. The location of the outside air intake determines where the flow of air through the space begins. If you want to relocate the intake in an existing plant, consider using a duct to route air from the existing outside air louver.

Bring in outside air at a low point in the space. This helps to preserve the stratification needed to keep the entering air from diluting the warmer air in the space. Also, the lower level is where people are usually located.

■ **A Layout That Won't Work**

It may seem easy to route incoming air over the boilers just by placing an opening in the ceiling over the boiler. You hope that the entering cold air blankets

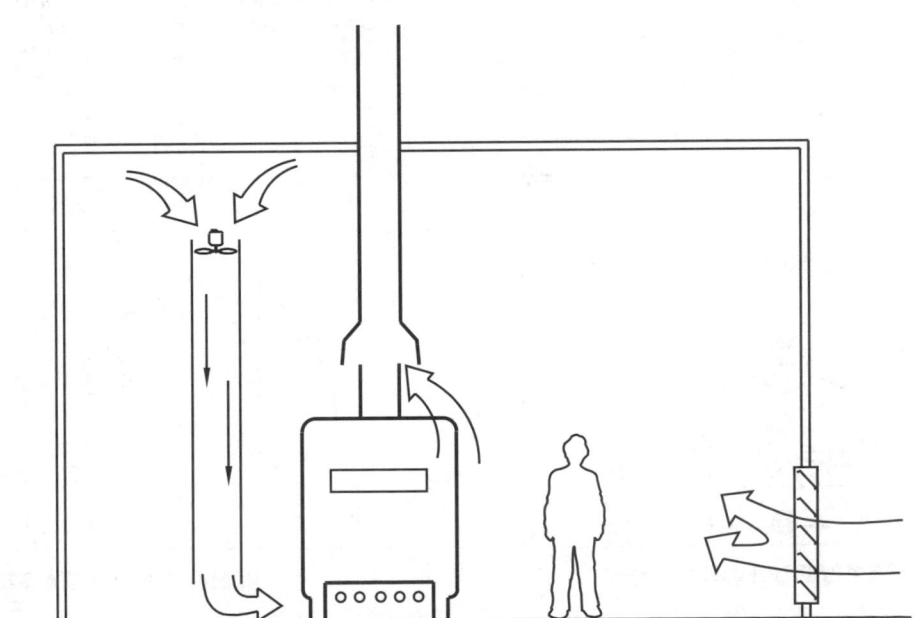

Fig. 3 Two problems If a draft hood or barometric damper is installed in the flue, it will draw the heated air out of the boiler room. If the boiler does not have an enclosed burner system, the heated air cannot be routed into the boiler reliably.

the boiler and is warmed before it gets to the burners. Unfortunately, this won't work. Figure 4 shows why. If there is a hole in the roof, the warm air in the space will rise by convection and exit the hole. Cold outside air will enter the same hole and fall to the floor in a confined stream, picking up little heat along the way.

Where to Collect Boiler Room Air

If you don't use a heat collecting hood, locate the inlet to the combustion air duct above the major heat producer, typically above the boiler itself. The height of the duct inlet is not critical, as long as it is above the top of the tallest heat producer. Do not locate the inlet immediately below the ceiling, as this would drag air along the ceiling and cool it.

It might seem clever to make a survey of the temperatures in the space, and draw air from the warmest point. This may mislead you. For example, it is very warm right above the steam drum of a watertube boiler, but the heat emitting surface is so small that the amount of heat to be recovered is also small.

Connecting the Duct to the Combustion Air Fan

Attaching the duct to the fan or burner assembly may be easy or it may be the Achilles heel of the job. If you go to a lot of trouble to collect hot air all over the boiler room, you have to be sure that the air goes into the boiler. There are three distinct possibilities, as follows.

■ Air Intake with Separate Forced-Draft Fans

If the boiler system has a separately mounted combustion air fan, you can run a duct to it from anywhere inside the boiler room. This is may be easy to accomplish.

■ Bolting the Duct to the Burner Assembly

Burner assemblies that include an integral combustion air fan may have flanges for attaching ducts or intake silencers to the combustion air intake. This makes it easy to install the combustion air duct, provided that the air intake is not in an awkward location, such as facing downward. In the latter case, see whether you can make the fan housing more accessible. Larger burner assemblies may be designed so that the fan housing can be rotated to make the intake face in a variety of directions. If this is not possible, you may be able to attach a sheetmetal transition box.

Design the duct so that it does not significantly reduce the fan capacity or increase its power requirement. Make the duct large, and use smoothly rounded bends. The construction can be very lightweight because there is only negligible pressure on the duct.

The duct should not hinder access to the burners for operation and maintenance. If the burner assembly must be removed to provide access to the inside of the boiler, design the duct to swing out of the way or to be easily removable.

An easy installation is to suspend the duct from the ceiling. This saves structural cost, and also makes it easier to swing the duct out of the way for burner maintenance.

■ If the Duct Cannot be Attached to the Burner Assembly

Making an airtight connection between the duct and the fan inlet may not be practical, especially with smaller boilers. In such cases, terminate the duct as near to the fan inlet as practical. Cover the burner assembly as much as possible to prevent air leakage. For example, you might terminate the duct with a fireproof fabric shroud over the burner.

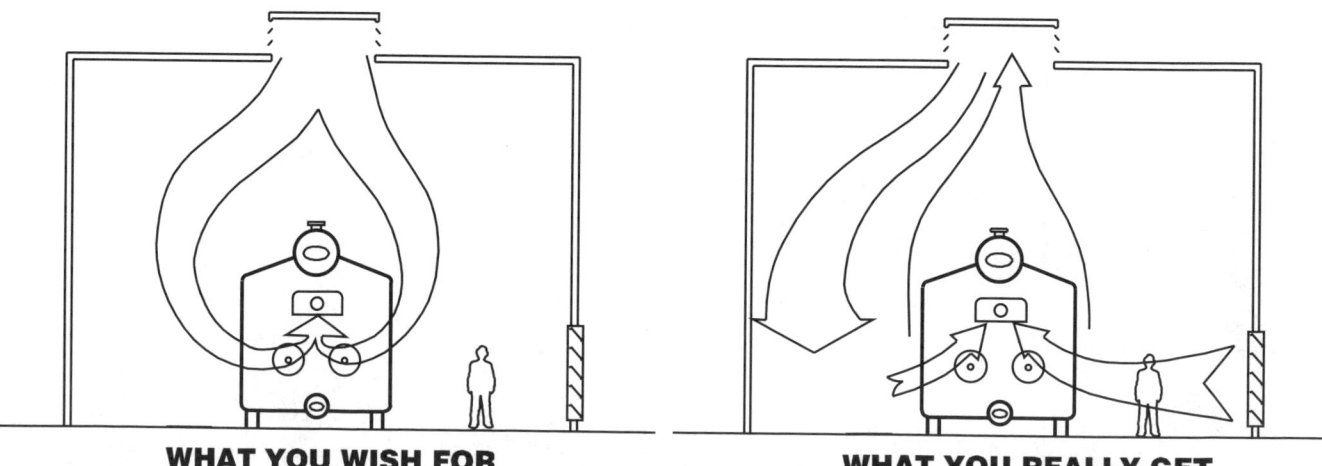

WHAT YOU WISH FOR   **WHAT YOU REALLY GET**

Fig. 4 Why the outside air intake must be located at a low point   If cold outside air enters at a high point in the space, it will drop to the floor and enter the burners without collecting much heat from the boiler. The warm air in the space will escape out of the same hole.

If the boiler does not have a combustion air fan, as with atmospheric burners that draw air from floor level, it may be impossible to achieve a good seal. Atmospheric burners are sensitive to ambient air pressure, so it is not advisable to surround the burner assembly with any sort of enclosure.

If the boiler does not have a combustion air fan, install a fan inside the duct. The location of this fan is not critical. Installing the fan at the top of the duct allows the duct to be made of lightweight material, such as fireproofed fabric. The air discharge keeps the duct inflated if it is suspended properly. Figure 3 shows how this is done. It also illustrates that dumping the warm air into the open is less effective than ducting the warm air directly into the burner.

Interlock the operation of the fan with the operation of the burners. Preferably, the fan speed should change if the burner firing rate changes. This maintains the highest combustion air temperature. The energy consumption of the fan is minor if it is selected properly.

This configuration is less efficient than a tight connection to the burner because a greater volume of air is pulled through the area at the top of the duct, lowering the temperature in that area. Also, some dilution of the hot air may be unavoidable.

Operator Comfort and Safety

■ Provide for Work in the Hot Air Collecting Area

This Measure creates one or more regions of hot air in the boiler room. Design the air flow path so that this does not create discomfort or a safety hazard for the people in the space. Any area of the boiler room may have to be entered at some time. For example, there may be light fixtures within the hot zone that require lamp replacement, or pipe joints that may develop leaks. For these occasions, make it possible to cool down the hot zone. You may be able to do this by switching boilers, or by venting the hot area, or by ventilating the hot area with a fan.

■ Don't Trap Explosive and Suffocating Gases

An important safety consideration in boiler room design is venting dangerous gases, both in normal operation and in the event of a casualty, such as a fuel line rupture. Do not let your heat recovery system concentrate or trap potentially explosive vapors.

Natural gas and some other gaseous fuels are lighter than air, so your heat collecting areas might trap them in the event of a gas leak. If your boilers can burn gas, consider fail-safe safety vents that are activated by combustible vapor detectors.

Most of the vapors of liquid fuels, the gases of incomplete combustion, and refrigerant vapors are heavier than air. If the boilers cannot burn gas fuels, your heating collecting system probably cannot cause a problem.

Explain It!

The purpose of the heat recovery system is not obvious from its appearance, so install a prominent placard in the boiler room to explain it to plant operators. Spell out any actions needed to maintain the efficiency of heat recovery, such as keeping doors and windows closed, using appropriate ventilators for comfort, etc. See Reference Note 12, Placards, for effective placard design and installation techniques.

ECONOMICS

SAVINGS POTENTIAL: *0.2 to 2 percent of fuel consumption. If fans are used, their operating energy reduces the saving by a small amount.*

COST: *Hundreds of dollars to thousands of dollars.*

PAYBACK PERIOD: *Less than one year, to several years.*

TRAPS & TRICKS

DESIGN: *You will be a pioneer. A combination of ingenuity, daring, and common sense is needed here. You will have a great conversation piece, but also try to save some energy. Be realistic about how air moves, and do not overestimate the temperature rise that you can achieve. Provide for cooling any portion of the space that people may have to enter.*

MATERIALS: *The materials can be very light and inexpensive because you are simply trying to channel air currents. Even aluminum foil or flame-retardant fabric could work if you support it properly. Furthermore, lighter materials are easier to fit around obstructions. Design to resist damage, especially from maintenance work.*

System Design for Efficient Low-Load Heating

The load on a typical boiler plant varies widely with the season and the time of day. Large boilers become inefficient at low load. Boiler auxiliary equipment, such as fans, pumps, and air compressors, may consume excessive amounts of energy at low load. The Measures of this Subsection correct these inefficiencies by matching the boiler plant's energy consumption to the facility's heat requirements. Some of the Measures also reduce distribution losses, which may be a major source of energy waste in large facilities.

These Measures may provide major benefits in addition to energy savings, especially improved reliability and reduced plant overhead cost.

These Measures are major, expensive activities. They may not add much to the cost of a new boiler plant. In some cases, they may even reduce construction cost. However, they are generally expensive to retrofit, except when replacing old equipment or expanding an existing boiler plant.

INDEX OF MEASURES

1.12.1 In facilities that operate for extended periods with low heating loads, install a small, efficient lead boiler.

1.12.2 Install localized heating units to allow shutting down the central plant during periods of low load.

1.12.3 If it is desirable to reduce the boiler operating pressure, eliminate high-pressure steam users or provide separate high-pressure steam boilers.

1.12.4 If a facility has several boiler plants, provide cross connections that allow shutting down the least efficient boilers.

RELATED MEASURES

- Subsection 1.1, for operating practices and controls to maximize plant efficiency
- Subsection 1.2, for measuring boiler efficiency
- Subsections 1.4, 1.5, 1.8, and 1.9 for efficient matching of auxiliary equipment energy consumption to boiler load

MEASURE 1.12.1 In facilities that operate for extended periods with low heating loads, install a small, efficient lead boiler.

Boiler plants commonly waste energy because the boilers and other equipment are sized for the peak load requirement, ignoring the fact that heating plants typically operate for a large fraction of the time at low output. You can correct this deficiency by installing a small, efficient boiler to satisfy low heating loads. The smaller boiler is called a "lead" unit because it is the first to operate at low loads.

How Improper Sizing Wastes Energy

A boiler plant may waste energy in these areas when the load is low:

- *the boiler itself.* Boilers of good design typically do not lose much efficiency until the load is reduced to a small fraction of full load, say 30% or less. Below this level, efficiency starts to fall rapidly. Boilers that have on-off or two-stage firing control do not suffer low-load efficiency loss in combustion efficiency because they always fire at higher load levels. However, such boilers may suffer serious standby losses at low loads, depending on their designs. Large boilers typically have modulating burners, which can turn down their firing rate into the region where overall boiler efficiency (but not necessarily combustion efficiency) drops seriously.

- *auxiliary equipment.* Large boilers typically require a large amount of energy for operation of auxiliary equipment, such as forced draft fans, feedwater pumps, fuel oil pumps, air compressors for control air, and other equipment. Auxiliary equipment is often designed in such a way that its energy consumption remains high at all loads. For example, auxiliary energy consumption may account for only 2% of plant energy consumption at full load, but may account for 5% to 10% at low loads.

  Auxiliary equipment that operates in an on-off manner suffers little efficiency loss at partial boiler loads. However, continuously operating fans and pumps typically are commonly controlled by throttling their output, which is inefficient. The type of boiler and the kind of fuel have a large influence on auxiliary energy requirements. For example, a small gas-fired boiler may have a minimal auxiliary load, whereas a fluidized bed coal boiler has an enormous energy requirement for fans and fuel handling.

- *plant overhead.* A large part of the cost of operating a boiler plant is related to the size and complexity of the systems that are kept in operation, regardless of the actual heating load. Such overhead costs

SUMMARY

Appropriate for many boiler plants. In addition to saving energy, it may provide opportunities for reducing maintenance and labor costs, and for using less expensive energy sources. In new facilities, this is basic good design that can save money as well as energy. In existing facilities, the change is expensive.

SELECTION SCORECARD

Savings Potential .... $ $ $

Rate of Return, New Facilities % % % %

Rate of Return, Retrofit.......... % % %

Reliability .... ✓ ✓ ✓

Ease of Retrofit .... ☺ ☺

include labor, maintenance, and housekeeping energy consumption, such as lighting and ventilation for the boiler house.

- *distribution losses.* The pressure and temperature within distribution systems typically remains constant as the load changes. Leakage is proportional to system pressure, and conductive heat loss is proportional to system temperature, so distribution losses are independent of boiler load. At low loads, these losses waste a relatively larger fraction of plant output.

If the existing boiler plant operates for long periods of time at low loads, you can reduce inefficiency in the first three of these areas by installing a small, efficient lead boiler system to carry low loads. This allows you to shut down the larger boilers for much of the time. (To reduce inefficiency in distribution, see the Measures in Subsections 1.10 and 1.11, as well as Measures 1.1.5, 1.12.2, and 1.12.3.)

An Example

A typical example is a college campus having an original estimated peak heating load of 200 million BTU/HR. To serve this load, the campus has two boilers of 250 million BTU/HR capacity, one of which serves as a standby unit. The boiler plant operates primarily to provide space heating, but it continues to operate in warm weather to provide heat for domestic water, cooking, and other functions. The nighttime load in summer typically is less than 5 million BTU/HR, and the daytime load in summer is always less than 30 million BTU/HR.

We could improve efficiency substantially by installing a 30 million BTU/HR boiler to handle the load during warm weather. This allows shutting down the existing 250 million BTU/HR boilers for the entire warm season.

We could go further by installing an additional 5 million BTU/HR boiler to handle the summertime night load. However, the marginal efficiency improvement provided by this boiler (compared to the 30 million BTU/HR boiler alone) would probably be much smaller than the marginal efficiency improvement provided by the 30 million BTU/HR boiler (compared to the original 250 million BTU/HR boilers).

Weather data for mid-latitude locations shows that the heating load typically is less than half the peak load for most of the year. Therefore, we could improve efficiency during the spring and autumn heating seasons by adding a boiler with a capacity of about 100 million BTU/HR. This allows the original boilers to remain shut down except for one or two months per year.

Many other combinations are possible. For example, installing a single 50 million BTU/HR boiler would allow shutting down the original oversized boilers for about half the year, and would keep them operating in a more efficient range of loads. A boiler of this capacity would handle the summer daytime load fairly efficiently, although it would still be too large for efficient handling of the summer nighttime load.

Cost vs. Efficiency

The previous example would be very expensive, especially if two new boilers are installed. This would be true of any retrofit application, unless it is necessary to replace existing boilers. In addition to buying one or more new boilers, you have to expand the boiler house.

Another apparent disadvantage in existing facilities is that this modifications will idle a large fraction of the equipment capacity most of the time. This disadvantage is psychological rather than economic. In retrofit applications, this Measure may be resisted because it leaves a large monument to the original sizing mistake in the form of the original equipment.

From a purely economic standpoint, the correct approach is to view the original investment as a sunk cost. The economic analysis should consider only the new investment and the savings that the new investment will provide, along with the salvage value of any equipment that can be sold. In retrofits, consider selling some of the equipment that is rendered unnecessary. There is a thriving market for used boilers and accessories.

If you are building a new facility, the principles illustrated by the previous example can reduce total construction cost. This is because you will need less reserve capacity.

Maximize Efficiency in the New System

The efficiency of a lead boiler depends on much more than its sizing. Optimize every aspect of the boiler plant that relates to the lead system. With respect to the boiler itself, high combustion efficiency and low standby losses are the primary criteria, so select a boiler with these characteristics.

Select the lead boiler and its auxiliary equipment to maintain high efficiency at all fractions of load. The lead boiler usually operates over a wider range of load, and for a longer time, than any of the other boilers.

For boilers that are fired by gas fuels, consider condensing boilers, which provide higher efficiency than conventional boilers. Figure 1 shows typical units. Condensing boilers are described in Reference Note 30, Boiler Types and Ratings. Flue gas condensation in general is explained in Measure 1.7.1.1.

Size, in itself, does not have a significant effect on peak efficiency. The best models of boilers of all sizes approach the theoretical limits of combustion efficiency as long as they operate within their efficient load range. At present, condensing boilers are available only in relatively small capacities, but you can successfully operate a large number of these units in parallel.

Select your new boiler to minimize auxiliary load. For example, some unusual types of boilers require large circulation pumps, and they offer no significant benefit in return. Techniques for matching auxiliary energy

WESINC

Fig. 1 A battery of condensing boilers for a large library
Each boiler operates at a fixed output that optimizes its efficiency. Flue gas condensation provides high individual efficiency. Small increments of size adapt efficiently to all load conditions. The low flue gas exit temperature allows flues to be made of plastic pipe and to exhaust from an adjacent side wall.

consumption to variable boiler loads are found in Subsection 1.4 for burner systems, Subsection 1.8 for feedwater and condensate pumps, and Subsection 1.9 for fuel oil pumps. Many of these techniques are unconventional, and a great deal of thought is needed to derive the most efficient, reliable, and economical design.

Specify your new equipment by manufacturer and model number when purchasing. Otherwise, contractors or suppliers may substitute cheaper equipment that is less desirable in efficiency, features, or reliability.

Feasibility Analysis

You need a detailed feasibility study to decide whether it is economical to add a small lead boiler or a combination of lead boilers. The analysis should emphasize the following issues.

■ Duration of Low-Load Periods

Determine the shortest intervals for which it is practical to shut down the larger boilers, and compare this to your typical operating profile. For example, a boiler plant serving a large commercial facility may have a substantial drop in load from midnight to 0500, but it may not be practical to shut down the boiler plant for such a short time. Frequent shutting down and restarting of a boiler causes leaks, deteriorates refractories, introduces oxygen into the boiler water, and promotes corrosion of flues. In addition, starting a large boiler system requires at least an hour of effort by one or more skilled operators. Balance the energy saving against these additional maintenance costs.

■ Comparative Efficiency and Cost

If your heating load fluctuates in typical fashion, calculate the cumulative energy consumption of the original plant and each potential new configuration over the course of a year of typical operation. This calculation should include the energy consumptions of the boilers, auxiliary equipment, and plant overhead functions (air compressors, lighting, etc.). To simplify this tedious calculation, use an appropriate computer program. See Reference Note 17, Energy Analysis Computer Programs, for details.

Consider different combinations of boiler sizes. If the boiler load varies widely, the analysis will not point to a combination that is clearly the best. Instead, one combination of sizes will provide the highest overall system efficiency, and a different combination will provide the best economics. In the end, you will have to make an informed judgement call.

In existing facilities, measure the actual efficiencies of the existing boilers over their entire load range, using the methods and equipment recommended in Measure 1.2.1. At the same time, use a wattmeter to measure the energy consumption of electrical auxiliary equipment having variable output, such as forced draft fans and feedwater pumps that operate against a throttling valve. Also, measure the energy consumption of equipment having a steady output, such as on-off pumps, plant lighting, and air compressors. You can measure their durations of operation by using an interval recorder.

■ Ability to Reduce Plant Overhead Costs

Try to minimize plant overhead, which consists of costs external the boiler system itself, such as labor, maintenance, and plant utilities. For example, you may be able to operate a small lead boiler automatically, even though the larger boilers must be manned.

This may provide an opportunity to reduce staffing, which could substantially increase total cost savings. However, staffing reductions usually are practical only if you can limit operation to the lead boiler for an entire shift or for an entire season. Staffing requirements in many jurisdictions are dictated by the operating pressure and/or the capacity of the boiler.

Boiler maintenance cost depends largely on the number of operating hours, so reducing the operation of larger boilers should save cost. (On the other hand, starting and stopping a boiler increases some maintenance costs.) To achieve a saving in plant overhead energy consumption, such as lighting and ventilation, you would have to shut down the space in which the main boiler equipment is located.

■ Potential Fuels

Use the feasibility analysis as an opportunity to consider using different fuels that may provide advantages in cost, plant efficiency, reliability, maintenance, or environmental factors.

Distribute the Boiler Load Efficiently

To gain the potential benefit of this modification, you need to start and shut down the boilers in the sequence that satisfies the heating load most efficiently. This requires effective operator training, well documented procedures, and continuing supervision.

Measures 1.1.4 ff gives procedures for distributing the load among boilers most efficiently. Usually, you should load the lead boiler system to full capacity before starting larger units, but this is not always true. To avoid running out of capacity, operators need to foresee the possible need to start larger boilers. This requires supervisors to keep track of weather forecasts, plant equipment operating schedules, or other factors that may require additional heating capacity.

Automatic sequencing controls for boilers are commercially available. These are an effective method of ensuring that the most efficient combination of boilers is operated as load conditions change. Measure 1.1.4.1 covers these controls.

Alternative Measures

An alternative approach is to shut down the boiler plant entirely during periods of low loads, and to install localized heating units. Measure 1.12.2, next, covers this approach. It has the advantage of eliminating distribution losses during the period of time when the boiler plant is shut down.

If a facility is served by more than one boiler plant, you may be able to handle low loads efficiently by cross connecting the plants. This allows you to increase the average load on the boilers that operate, and to select the most efficient boilers to operate with priority. Measure 1.12.4 gives this approach.

If energy waste at low load is caused mainly by inefficient operation of auxiliary equipment, rather than by inefficiency of the boiler itself, then improve the efficiency of the auxiliary equipment. See Subsection 1.4 for burner systems, Subsection 1.8 for feedwater and condensate pumps, and Subsection 1.9 for fuel oil pumps.

ECONOMICS

SAVINGS POTENTIAL: *5 to 20 percent of boiler plant operating cost, in eligible applications.*

COST: *The typical cost of efficient gas- and oil-fired boilers ranges from about $50 per thousand BTUH of capacity in residential sizes, to about $15 per thousand BTUH in large commercial sizes, to about $5 per thousand BTUH in large industrial sizes. Packaged stoker-fired coal boilers cost from $15 to $30 per thousand BTUH, depending on size. In new construction, this Measure may reduce the overall boiler plant cost.*

PAYBACK PERIOD: *May be immediate, in new construction. Several years or much longer, in retrofit applications.*

TRAPS & TRICKS

DESIGN: *Don't rush into this. Do your feasibility analysis carefully to get the sizing right. Consider all your configuration options.*

SELECTING THE EQUIPMENT: *Select a proven boiler design from a major manufacturer. Select the model with the highest efficiency. If you consider an innovative type, such as a condensing boiler, check its performance history.*

OPERATION: *It does no good to install an efficient lead boiler unless you operate it properly in relation to the other boilers. Be sure to accomplish Measure 1.1.4 or 1.1.4.1.*

MEASURE 1.12.2 Install localized heating units to allow shutting down the central plant during periods of low load.

Measure 1.12.1 explains why it is inefficient to operate a large boiler plant to serve small heating loads. If the boiler plant is serving only small items of equipment when it is operating at low load, you can shut down the plant by replacing those items with independently heated units.

For example, if a central boiler plant operates during the summer only to provide domestic hot water to various buildings, provide each of those buildings with its own hot water heater. This may allow you to shut down the central plant for a large part of the year, eliminating all its energy losses, and perhaps eliminating all its operating costs.

This Measure applies to facilities, such as college campuses and military bases, where the central plant exists primarily to provide winter heating. The boiler plant is typically operated during warm weather for domestic water heating, cooking, and other low-energy applications.

Feasibility Considerations

A major facility modification such as this deserves a detailed feasibility study before you start spending serious money. Consider the following factors.

■ The Condition of the Distribution System

The choice between this Measure and Measure 1.12.1 is likely to be determined by the amount of distribution system losses. Refer to Measures 1.10.1, 1.10.2, and 1.11.1 for methods that you can use to estimate energy loss in the distribution system. Distribution system losses may be large in systems that are extensive or old.

■ Fuel Choices for the Central Plant

If the central plant can use fuels that are significantly less expensive than the fuels that are available for localized heating units, this Measure is unlikely to be economical. Only a central plant can burn coal, heavy oil, and most waste fuels.

Don't overlook the possibility of converting the central boiler plant to use fuels that are less expensive or more desirable in other ways.

■ Fuel Availability and Storage for Localized Units

If the localized heating units burn natural gas, you have to install gas lines and metering, if these are not already present. If the localized units burn a liquid fuel, you have to install fuel tanks and associated piping. If the localized units use electricity, you probably have to increase the amperage, and maybe the voltage, of the electrical service.

SUMMARY

Appropriate for facilities where the boiler plant load is mostly seasonal. Especially desirable for reducing losses in long distribution systems. May reduce plant overhead costs. Expensive to retrofit.

SELECTION SCORECARD

Savings Potential .... **$ $ $**

Rate of Return, New Facilities **% %** %

Rate of Return, Retrofit.......... **% %** %

Reliability .... ✓ ✓ ✓

Ease of Retrofit .... ☺ ☺

■ Ability to Reduce Central Plant Operating Cost

Shutting down the central plant may eliminate most of its physical operating overhead for the season, especially boiler maintenance and housekeeping energy consumption, such as lighting and ventilation for the boiler house. The savings depend on the size and complexity of the systems that can be shut down.

Aside from energy cost, the largest plant operating cost is for labor. Shutting down the boiler plant on a seasonal basis may or may not let you reduce labor cost. It is difficult to get skilled boiler operators to work on a seasonal basis, so the boiler plant staff may have to be kept on the payroll all year. One possible way of reducing labor cost is to reassign boiler operators to seasonal duties that normally require other personnel.

■ Ability to Control Localized Units

This Measure increases the number of units that need to be controlled, and hides them from the physical plant staff. This is an invitation to excessive operation of the localized units, which can dissipate much of the potential savings. Therefore, a critical part of this Measure is providing reliable controls to limit operation of the localized units to times when they are needed. This may be feasible in some environments, but not in others. Refer to Subsection 5.1 for methods of controlling localized units.

■ Options for Localized Units

Consider alternative ways of installing localized heating units. For example, you may have a choice between installing a small number of small water heaters in a building, or installing a larger central unit to serve the whole building. Consider the relative costs of equipment, installation, connections to the existing internal distribution system, and the value of space that is taken up or that must be added.

■ Relative Heating Unit Efficiencies

In the past, good central plant boilers had higher efficiencies than small fuel-fired heating units. This is no longer a valid generalization. In particular, small furnaces and boilers that exploit the heat of condensation of flue gases have seasonal efficiencies that range from about 86% to about 96%. This is much better than the efficiencies of large conventional boilers.

Most localized units that have very high efficiencies require gas fuels. Condensing oil-fired units are now making a hesitant entry into the market. The best conventional small oil-fired heating units approach the efficiencies of the best central plant boilers. All electric heating units have an efficiency of virtually 100%.

Small high-efficiency units experienced reliability problems when they were first introduced during the 1980's. Some models have survived long enough to establish a reasonable record of reliability. Investigate the performance records of the specific models you consider.

Less efficient small heating units still remain on the market because of their lower cost. The seasonal efficiencies of such units range from less than 70% to about 80%, typically worse than the efficiency of large boilers.

With the broad range of efficiencies that are available, be specific in designating the models of equipment you want. In most countries, new heating units carry efficiency ratings. These are approximate, primarily because standby losses cannot be estimated accurately. (The most efficient units avoid standby losses by using sealed combustion chambers.)

Of course, if the boilers in an existing facility are old and inefficient, this further improves the feasibility of installing local units.

■ Boiler Plant Auxiliary Energy Consumption

Refer to Measure 1.12.1 about the energy consumption of boiler plant auxiliary equipment, which varies widely. Localized heating units usually have minimal auxiliary energy consumption.

Alternative Measures

Installing a small, efficient lead boiler in the boiler plant can improve the low-load efficiency of the central plant considerably, but it does not reduce distribution losses. See Measure 1.12.1 for this approach.

You can reduce distribution losses somewhat during low-load conditions by reducing the pressure and temperature of the distribution system. See Measure 1.1.5. You may have to make the system changes recommended by Measure 1.12.3 in order to reduce the system pressure.

You may be able to reduce low-load energy waste in the boiler plant's auxiliary equipment. See Subsection 1.4 for burners and fans, Subsection 1.8 for feedwater and condensate pumps, and Subsection 1.9 for fuel oil pumps.

If the facility is served by more than one boiler plant, you may be able to handle low loads more efficiently by cross connecting the plants. This allows you to shut down some boilers while increasing the load on others. Refer to Measure 1.12.4 for this approach.

ECONOMICS

SAVINGS POTENTIAL: *5 to 30 percent of total heating energy consumption. There may also be significant savings in maintenance and operator labor costs.*

COST: *Efficient gas-fired air heating units cost about $20 per million BTUH of capacity. Efficient gas- and oil-fired boilers cost $30 to $50 per million BTUH in residential sizes. Retrofit installation may double or triple these prices. In new construction, adding localized units may reduce the cost of the main boiler plant.*

PAYBACK PERIOD: *Several years or longer, except shorter in especially favorable situations.*

TRAPS & TRICKS

DESIGN: *You may have to invest a lot of thought and analysis to achieve the maximum potential of this activity. Investigate conditions at each of the locations where localized heating equipment would be installed, do load calculations for each location, and figure out the best method of installing the new equipment.*

SELECTING THE EQUIPMENT: *Stress efficiency when selecting small heating units. Use condensing boilers and furnaces where you can.*

OPERATION: *This activity increases the number of units that need to be controlled, and hides them from the physical plant staff. Provide reliable controls to prevent unnecessary operation. Include the appropriate control Measures from Subsection 5.1.*

MEASURE 1.12.3 If it is desirable to reduce the boiler operating pressure, eliminate high-pressure steam users or provide separate high-pressure steam boilers.

RATINGS

New Facilities   Retrofit   O&M

B **C** [ ]

Measure 1.1.5 recommends that you keep the boiler pressure as low as possible. This may provide a number of benefits, including:
- reduced steam leakage
- reduced conductive heat loss
- improved boiler efficiency
- reduced scaling and corrosion
- reduced control valve erosion
- improved control valve stability
- reduced operator licensing requirements.

(Reducing the pressure of an existing boiler has some disadvantages, including steam metering error and possible boiler circulation problems. See Measure 1.1.5.)

In some cases, you may be prevented from lowering the system pressure in the boiler plant because one or more small users of steam require a substantially higher pressure than is needed by the rest of the users. For example, sterilizers or laundry presses may require a much higher pressure than steam heating.

The limitation may also occur inside the plant. For example, a boiler may have steam atomizing burners that require a pressure of 150 PSI, while the steam pressure requirement of the facility may be lower than this for much of the year.

SUMMARY

Appropriate where a few minor users of high-pressure steam force the entire steam system to operate at high pressure. May provide a number of benefits.

SELECTION SCORECARD

Savings Potential	$ $
Rate of Return, New Facilities	% % % %
Rate of Return, Retrofit	% %
Reliability	✓ ✓ ✓ ✓
Ease of Retrofit	☺ ☺

In such cases, you can reduce the steam pressure requirement in two ways:

- ***provide a separate small boiler at the high-pressure application.*** For example, if a campus clinic needs high-pressure steam to operate sterilizers, install a small, high-pressure boiler at the clinic to serve this need. Small high-pressure boilers typically do not require operators. Figure 1 shows an example.

- ***replace the user equipment with a different type.*** An alternative solution in the previous example is to replace the steam sterilizer with an electrically heated unit. Similarly, if steam atomizing burners dictate a high boiler pressure, replace them with air atomizing burners.

ECONOMICS

SAVINGS POTENTIAL: *Applications vary too much to allow general estimates.*

COST: *Small high-pressure boilers cost $20 to $40 per MBH of capacity, and retrofit installation may double or triple these costs. In new construction, other substitutions of equipment may cost little or nothing extra.*

PAYBACK PERIOD: *Several years or longer, usually.*

TRAPS & TRICKS

PLANNING: *Consider all the potential consequences of a change. For example, if you change from steam-atomizing to air-atomizing burners, will the new burners have a satisfactory turndown ratio, will you have the space for a large air compressor, will the compressor noise be objectionable, etc.?*

WESINC

Fig. 1 High-pressure boiler in a laundry This remote boiler in a large hotel avoided the need to design the main boiler plant to provide high pressure for a specialized application.

MEASURE 1.12.4 If a facility has several boiler plants, provide cross connections that allow shutting down the least efficient boilers.

SUMMARY

Applies to facilities served by several boiler plants. Requires compatible steam pressures and/or water temperatures. Provides a number of significant advantages. Usually very expensive, unless it avoids the need to increase boiler capacity.

SELECTION SCORECARD

Savings Potential	$	$	$
Rate of Return, New Facilities	%	%	%
Rate of Return, Retrofit..........	%	%	
Reliability	✓	✓	✓
Ease of Retrofit	☺	☺	

If your facility is served by several independent boiler plants, you may be able to interconnect the heating systems. This will allow you to select the most efficient combination of boilers and auxiliary equipment for each level of heating load. The efficiency improvement is likely to be greatest at low loads. If the boilers in individual plants are too large to handle low loads efficiently, having one boiler serve more than one system allows that boiler to operate at higher load.

Potential Savings

The savings that result from improved boiler loading are explained in Measure 1.12.1. The savings that result from shutting down a boiler plant are explained in Measure 1.12.2.

On the negative side, interconnecting systems may increase distribution losses because of the added piping. This is significant only if you have to add a long run of piping to interconnect the systems, and if this additional piping serves no other purpose. If heat is distributed by means of hot water, additional pump power may be required. Be sure to include these losses in the overall savings calculation.

System Compatibility

A facility with a number of different boiler plants may have a hodgepodge of different types of distribution systems. Separate plants may have been built as a part of successive facility expansions that did not consider overall facility efficiency. The differences in system characteristics may prevent interconnections from being made, or they may limit the choices. For example, a high-pressure steam system can feed a low-pressure system, but not the reverse. In this case, connecting the systems is useful only if the higher pressure system is the one that provides more efficient operation during low-load conditions.

You can feed heat from a steam system into a hot water distribution system by using heat exchangers. You can even do the reverse, if the temperature in a hot water distribution system is hot enough. However, in both cases, the heat can move only from the hotter system into the cooler one.

An Example

Consider a facility that has two separate boiler plants. Both of the plants are intended primarily to provide space heating, but they also provide heat for domestic water, cooking, and other functions. One of the plants has two boilers rated at 500 million BTU/HR

each, and the other boiler plant has two boilers rated at 100 million BTU/HR each. Both boiler plants distribute steam at the same pressure. The heating season lasts six months, and both boiler plants operate efficiently during the heating season. However, during the warm season, the larger plant has an average daytime load of only 30 million BTU/HR, and the smaller plant has an average daytime load of only 10 million BTU/HR.

The individual load on each boiler plant during the long warm season is too low for efficient operation. But, if the two systems are connected, one of the 100 million BTU/HR boilers can serve the total average summer daytime load of about 40 million BTU/HR, which is a fairly efficient situation. Equally important, the larger boiler plant can be shut down for the entire warm season, saving a considerable amount of plant auxiliary and overhead cost.

The efficiency benefits extend into the early and late parts of the heating season. Typical heating loads in mid-latitude locations are only a fraction of the peak load for most of the heating season. In this example, one of the 100 million BTU/HR boilers can continue to supply the total facility load during the early and late parts of the heating season. During the coldest part of the year, the small boiler plant can be shut down entirely, and the entire facility can be served by the larger plant alone. A single boiler plant crew could shuttle between the two plants as the seasons change.

In this example, the great flexibility of the interconnection depends on the fact that both steam systems operate at the same pressure. If the smaller system operated at higher pressure than the larger system, an interconnection would still provide substantial savings, but it would not be possible to shut down the smaller plant in winter. In the opposite case, if the larger system operated at a higher pressure than the smaller system, there would be little benefit in connecting the two systems.

How to Minimize Piping Cost

Installing piping is the main expense of interconnecting boiler plants. Pipe diameter is the dominant factor in the materials cost of the piping and accessories. Therefore, minimizing pipe diameter is a primary way to limit cost.

In steam systems, the piping must be large enough so that steam arrives at the remote ends of the system with adequate pressure. Pressure losses increase greatly with increasing flow rate, so you may have to compromise between cost and efficiency by limiting operation of the interconnections to times when flow rates are low. This may not be a serious limitation if the primary purpose of the interconnection is to improve efficiency during low loads.

If the facility has several boiler plants, you may have a choice of several ways of connecting the systems. The most versatile, and most expensive, is a loop that allows steam or hot water to flow from any boiler plant to any user. This may be an unnecessary extravagance. If you focus on the conditions when interconnection would save the most energy and money, this may narrow down the choices.

Pumping Efficiency in Hot Water Distribution Systems

In a hot water distribution system, minimizing pump power is a major efficiency consideration. It is pointless to save energy in the boilers by connecting systems, only to lose energy in the pumps. Pumps have a hunger for energy that is insidious because they are small and operate quietly. If a pumping system operates at full flow under all load conditions, pump energy consumption can become a major part of total plant energy consumption.

The pump power required to overcome the resistance of the added piping depends on the flow velocity, and hence, on the pipe diameter. In brief, bigger pipe allows smaller pumps. The energy cost saving with smaller pumps may eventually offset the higher cost of larger pipe. Calculate the best balance between pump power and piping cost.

Consider reducing pump energy consumption by using variable-flow pumping. You may be able to avoid installing supplemental pumps for the interconnections by converting the existing pumps to variable-flow operation.

As with steam distribution systems, you may be able to limit pipe size by designing the interconnections to operate only during low-load conditions.

Use Cross Connection to Increase Effective Capacity

If you can foresee a need for additional heating capacity, interconnecting existing plants may be a less expensive alternative to purchasing new boilers. Because of diversity, an interconnected system needs less total boiler capacity than isolated boiler plants. This benefit occurs only if the different systems have their peak loads at different times.

Use Cross Connection to Improve Reliability

Interconnections provide reserve capacity in the event of a boiler failure in any of the connected plants. This reserve capacity is limited by the capacity of the interconnections. This may argue in favor of larger pipe sizes, which are also more efficient.

Condensate Return

When steam systems are interconnected, provide for returning condensate to the correct boilers. For example, you could install a level sensor in the condensate receiver of one plant that causes additional condensate to be pumped to another plant. In retrofit, the details of modifying each plant depend on how the condensate and feedwater systems of the plants are presently arranged.

Sequence the Boilers Efficiently

Interconnection requires more sophisticated control of boiler sequencing and loading. In a solitary plant, boilers can be started in a fixed sequence. When boilers at different sites are interconnected, operators have to start and stop boilers at one plant with respect to conditions that exist at other plants, and these conditions change. If the operators don't do this efficiently, the benefit of the interconnections is wasted. For this reason, make sure that plant operators remain aware of the correct sequencing and loading procedure. These are the main points:

- turning on the most efficient boilers first
- selecting the boilers that operate most efficiently at each level of load
- minimizing energy consumption by boiler auxiliaries
- controlling the interconnections, including operation of valves and pumps.

Subsection 1.1 recommends various methods for accomplishing these functions, including automatic controls for sequencing the operation of equipment. Automatic controls are better at "remembering" repetitive functions than people. However, automatic controls do not relieve managers of the responsibility of understanding the control functions and ensuring that they are accomplished. Plant operators will override sequencing controls that they do not understood.

Feasibility Analysis

A major facility modification such as this deserves a detailed feasibility study. Consider these factors:

- *pressure and temperature compatibility.* This determines whether interconnection is possible at all.

- *duration of low-load periods.* It may not be practical to shut down large boilers for short periods, for reasons explained in Measure 1.12.1. Consider whether major equipment can be shut down for periods long enough to justify interconnection.

- *fuels usable by different boiler plants.* You may gain a major cost saving with an interconnection that allows a plant burning a less expensive fuel to serve a larger load.

- *ability to reduce plant overhead costs.* If you can shut down a plant for an extended period of time each year, you may be able to gain a large saving in plant overhead costs. The main overhead factors are covered in Measure 1.12.2.

- *the value of increased reliability.* Interconnections provide increased reliability in case of equipment failures. Consider how important this is in your circumstances.

- *ability to increase capacity or to avoid adding new capacity.* An interconnection may make it unnecessary to add new boilers or to replace old ones. This can be a major economic advantage.

- *comparative efficiency and cost analyses.* If your analysis of the previous factors indicates that it may be desirable to interconnect plants, make detailed efficiency and cost calculations.

- *capacity of the interconnecting piping.* The capacity of the interconnecting piping may have a major effect on cost. It may also determine whether the interconnection can improve system reliability.

Alternative Measures

You can improve the efficiency of each boiler plant individually during low-load conditions by having a small, efficient lead boiler in each plant. See Measure 1.12.1 for this alternative.

Another approach is to shut down the boiler plants entirely during periods of low loads, and to install localized heating units. This solution eliminates distribution losses as well as avoiding inefficient operation of boiler plant equipment. See Measure 1.12.2 for this solution.

If inefficient operation of auxiliary equipment at low loads is a problem, try to improve the efficiency of auxiliary equipment. See Subsection 1.4 for burners and fans, Subsection 1.8 for feedwater and condensate pumps, and Subsection 1.9 for fuel oil pumps.

ECONOMICS

SAVINGS POTENTIAL: *2 to 20 percent of total boiler plant energy consumption.*

COST: *Varies widely, from tens of thousands of dollars for short interconnections, to millions of dollars for large facilities. This Measure may offer an immediate net saving if it provides an alternative to installing new boilers.*

PAYBACK PERIOD: *Several years or longer, typically. The payback period may be shorter in especially favorable situations.*

TRAPS & TRICKS

DESIGN: *Although the concept is simple, it is easy to be hasty and overlook opportunities for improving efficiency and reducing cost. A lot of money is at stake, so take your time.*

OPERATION: *For this expensive modification to be successful, the boilers and the system valves at different sites must be sequenced in a precisely defined manner. Operators in different locations must be well trained, coordinated, and alert. Review Measures 1.1.4 and 1.1.4.1.*

Section 2. CHILLER PLANT

INTRODUCTION

Cooling accounts for a large fraction of the energy cost of modern buildings in most locations. In warm and humid climates, cooling and dehumidifying may account for more than half of a facility's energy cost. Cooling is also a major cost of many industrial processes.

There is a rich field of opportunity for improving the cooling efficiency of existing buildings and plants. Many improvements in cooling technology have made it possible to build new facilities with improved cooling efficiency. There are also many techniques for improving the efficiency of existing cooling equipment. These range from simple adjustments to expensive plant modifications.

Many of these improvements pay back quickly. As in the case of boiler plants, a large fraction of the total energy consumption of the facility enters at the chiller plant. For this reason, an improvement in the chiller plant may save much more energy than improvements at individual items of end user equipment.

How to Use Section 2

This Section is organized by the components of the chiller plant and by particular types of modifications. The cheaper, more universal Measures are in the earlier Subsections, and the most expensive and specialized Measures are in the later Subsections. Subsection 2.1 is the place to start with any chiller plant. Go through the rest of the Subsections in their approximate order, selecting the Measures that apply to your plant.

Before Getting Started ...

To understand the real meaning of chiller efficiency ratings, see Reference Note 31, How Cooling Efficiency is Expressed.

To learn about the factors that determine the efficiency of a compression cooling unit, see Reference Note 32, Compression Cooling.

To learn about the efficiency aspects of absorption chillers, see Reference Note 33, Absorption Cooling.

To find out what is behind the current excitement regarding refrigerants, see Reference Note 34, Refrigerants.

Equipment Scheduling And Operating Practices

Turning off equipment when it is not needed is the most powerful principle of energy conservation. The Measures in this Subsection recommend operating practices and automatic controls that limit operation of the cooling plant and its major components to conform to need. In addition, the first Measure provides for selecting the most efficient combination of equipment for each set of conditions. These Measures all deal with control. They may provide large savings at modest cost. Most of them are easy to accomplish.

MEASURE 2.1.1 Distribute the cooling load among chillers in the manner that minimizes total plant operating cost.

SUMMARY

May save a significant amount of energy. Costs nothing to do, but requires continuous attention by operators. First, you have to analyze your plant operating costs precisely.

SELECTION SCORECARD

Savings Potential .... $ $ $

Rate of Return .... % % % %

Reliability .... ✓ ✓ ✓

Ease of Initiation .... ☺ ☺ ☺

If you have more than one chiller, you may be able to reduce your operating costs significantly by paying attention to the sequence in which you start the chillers, and by the way you distribute the load among the chillers. Here is why:

- if a chiller can operate at more than one level of output, its efficiency varies with load
- if you have more than one type of chiller, they may differ in their efficiency characteristics
- different chillers and combinations of chillers may have different auxiliary equipment energy requirements.

The chiller operating sequence is most important when the chillers have different efficiency characteristics. For example, if you have added newer chillers to an existing plant, the new machines are probably more efficient than the older units. Also, different types and models of chillers may have large differences in their part-load characteristics. For example, reciprocating chillers suffer less efficiency loss at low cooling loads than centrifugal chillers, even though they have lower peak efficiency.

Once you have figured out the optimum chiller loading schedule, you need to ensure that it is followed. Manual setting is satisfactory only if the chillers are tended continuously, and even this does not guarantee reliability. The subsidiary Measure recommends a controller to perform this function automatically.

How to Sequence Chillers

The most efficient way to load chillers depends on the way that they unload or modulate individually. To avoid confusion, we will discuss each type of modulation separately.

■ Optimum Sequence for Single-Stage Chillers

Single-stage chillers operate at constant output. They respond to changes in the cooling load by turning on and off, usually in response to the temperature of the returning chilled water. They usually have reciprocating or scroll compressors.

If all the chillers are single-stage and identical, the operating sequence does not matter.

If you have single-stage chillers with different efficiency characteristics, start the chillers in order of their efficiency, regardless of their capacity.

The only complication arises if different chillers have different auxiliary energy requirements. For example, a larger, more efficient chiller might require a larger condenser water pump. In this case, calculate the overall energy consumption of each chiller plus its auxiliary equipment, and operate the chillers in the sequence that uses the least energy overall.

■ Optimum Sequence for Multi-Stage Chillers

Most reciprocating chillers adjust their output in fixed stages. They usually do this by disabling, or "unloading," individual cylinders in the compressor. Older reciprocating machines had from one to four cylinders, and typically each cylinder could be unloaded separately. This provided a number of stages equal to the number of cylinders. It is becoming more common for the individual compressors of reciprocating chillers to operate in only two stages, and to achieve a larger overall number of stages by using several compressors. (For the reasons, see Reference Note 32, Compression Cooling.) For example, a modern reciprocating chiller might have three compressors, each with two stages, for a total of six steps of cooling output.

Reciprocating chillers are most efficient at full output, because friction losses are constant, and unloading the cylinders incurs some losses. Therefore, it is usually best to load the first chiller up to full output before starting the next. If a chiller has several compressors, the manufacturer arranges the compressors to start and stop in the most efficient sequence.

If you have multi-stage chillers with different efficiency characteristics, sequence the chillers in order of their efficiency, regardless of their capacity. The output of staged chillers is usually controlled by sensing the temperature of the returning chilled water.

The only complication arises if there are differences in the auxiliary energy requirements of different chillers. For example, a larger, more efficient chiller might require a larger condenser water pump. In this case, calculate the overall energy consumption of each chiller plus its auxiliary equipment, and operate the chillers in the sequence that uses the least energy overall.

■ Optimum Sequence for Modulating Chillers

Centrifugal and screw chillers adjust their cooling output by continuous modulation, down to a minimum load. At the minimum load, the compressor must turn off, or the machine must resort to false loading. (False loading is very wasteful. Measures in Subsections 2.6 and 2.8 attack this problem.) Chiller operation normally is controlled to maintain a fixed chilled water supply temperature.

Both centrifugal and screw chillers suffer serious efficiency loss at low loads. Therefore, your first guideline is to operate the minimum number of chillers.

If all the chillers are similar in efficiency characteristics but differ in size, it is generally best to start the smallest chiller first. This exploits the fact that the efficiency of modulating chillers improves with load, up to a large percentage of full load. This is especially important if the smaller chiller operates with less auxiliary power, for example, if it is served by smaller pumps.

When you have to start an additional chiller, distribute the load so that all operating chillers are at the same percentage of load. This is true even if the chillers are of different size, provided that their curves of COP versus load are similar. (This is not obvious. If you are interested in the mathematical proof, it derives from the fact that the efficiency curves of most modulating chillers are convex upward at all points.)

If the chillers are identical, it is easy to distribute the load evenly between the chillers. Simply control each chiller to maintain the same chilled water supply temperature. Be sure that the chilled water thermostat for each chiller is installed where it senses only the discharge from its own chiller.

If the efficiency characteristics of the chillers differ, operate the most efficient units first. For example, newer centrifugal chillers are more efficient than older units, so operate them first.

Control becomes complicated if there is a combination of chillers with different efficiency characteristics. For example, consider a chiller plant that has one older centrifugal chiller and a new screw chiller. The screw chiller is more efficient than the centrifugal machine at low load, but is less efficient at high load. In this case, it might be most efficient to operate the screw machine up to 60% of full capacity, then turn off the screw machine and transfer all the load to the centrifugal machine. If the load climbs to the point that both machines are needed, the load is shared between the two machines in a ratio that varies with the load.

In a complex case where dissimilar chillers operate together, develop a table that specifies which chillers to run, and the distribution of load between the chillers, at each total facility load from zero to maximum. To develop this table, you need to know the efficiencies of all the chillers throughout their load ranges. From this, calculate the energy input to the chillers at each percentage of load. You may have to communicate with the manufacturers' engineering departments to get reliable part-load efficiency data for your chillers.

Similarly, calculate the chilled water pumping energy, the condenser water pumping energy, and the cooling tower fan energy for each level of load. Fan energy can be complicated, as explained in Measure 2.2.2. You may have to make some simplifying assumptions.

With this information in hand, plot operating cost versus output for each chiller, including any significant differences in maintenance cost. Then, for each level of total plant load, tabulate the least expensive combination of chillers and auxiliary equipment.

Methods of Setting Chiller Load

If you have to distribute the load in unequal percentages to maximize efficiency, you need a reasonably accurate and convenient means of setting the load of each chiller individually. Some larger chillers have panel-mounted load limiters (commonly called "demand limiters") that you can use for this purpose. Figure 1 shows an example. Don't expect the calibration of these controls to be precise.

With water chillers, you can shift the load among chillers by setting different units to provide different chilled water temperatures. This is a relative adjustment, because resetting the chilled water temperature of one chiller will change the loads on all chillers.

This method has an efficiency penalty. It requires the most heavily loaded chillers to operate at a supply

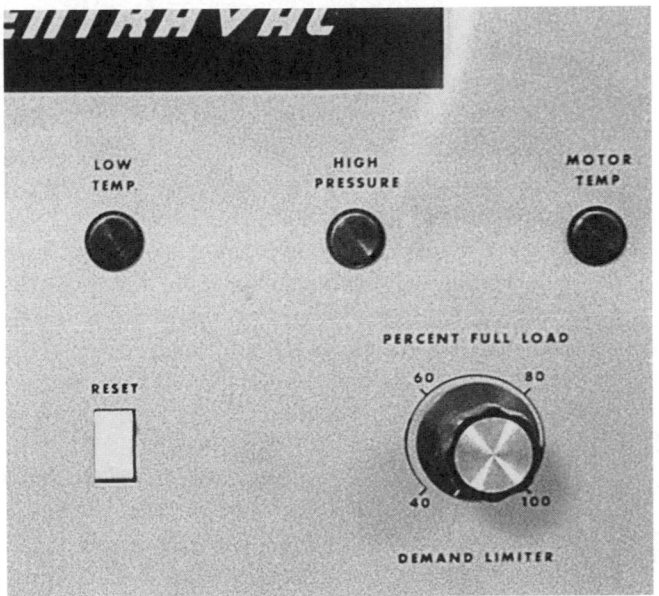

WESINC

Fig. 1 "Demand limiter" This control limits the maximum output of the chiller. It is called a demand limiter because it is intended to limit electricity demand charges. You can also use it to shift the cooling load among chillers.

temperature that is lower than average, which reduces their efficiency. The chilled water temperature setting has a strong effect on a chiller's efficiency. (See Measure 2.2.1 for details.) Therefore, if you lower the temperature setting of one chiller to make it carry more of the cooling load, raise the temperature settings of the other chillers to achieve the highest average chilled water temperature that satisfies the cooling requirement.

How to Determine Chiller Output

If your chillers do not have a direct method of setting their individual or relative outputs, you need a method of measuring the chiller output. Use one of these methods:

- *the manufacturer's chart that shows how output relates to supply and return water temperatures.* To use this method, make sure that all your supply and return water thermometers are well calibrated. This method assumes particular water flow rates through the evaporator and condenser. The actual rates may be quite different. Therefore, this method works only as a relative indication of output, and only if the water flow rates are consistent in all the chillers.

- *a BTU meter in the chilled water line of each chiller.* This is the only way to measure cooling output directly. See Reference Note 16, Measurement of Liquid, Gas, and Heat Flow, about BTU meters.

- *motor wattage.* Measuring wattage requires a wattmeter, which is not a standard item on most chiller control panels. To tell actual cooling output, find the chart in the manufacturer's catalog that shows how input wattage relates to load, as corrected for evaporator and condenser temperatures. This method tells you true output, provided that nothing has changed the efficiency characteristics of the chiller since the catalog data were measured.

- *motor amperage.* Amperage provides only a relative indication of load, so it is useful only for identical chillers. It does not provide a true indication of output, because the power factor of the motor varies with load over a wide range.

- *calibrate the position of mechanical control linkages,* such as the positioner for centrifugal chiller inlet guide vanes. This provides only a relative indication, as the output will change with evaporator and condenser temperature. Therefore, this method is useful only with identical chillers.

- *with steam-fired absorption chillers,* measure the steam flow rate or the condensate flow rate. Convert this to output by using the catalog data.

- *with direct-fired absorption chillers,* measure the fuel flow rate. Convert this to output by using the catalog data.

Sequencing Combinations of Compression and Absorption Chillers

In a chiller plant that has both absorption and compression machines operating in parallel, the procedures for distributing the cooling load are the same as above, but it may be more complicated to calculate the optimum distribution of load between the two types.

One type or the other will be more economical at a particular moment. However, don't expect the same type to be more economical all the time. Compression chillers operate with electricity (except for engine-driven units), while absorption machines operate with some form of fuel or heat input. The relative costs of the electricity and heat sources may change dramatically over the life of the facility. Therefore, you have to repeat the calculation periodically.

Another complication is that electricity prices may vary over a wide range from moment to moment. The major variable component is the "demand charge," which may be a large fraction of total electricity cost. (See Reference Note 21, Electricity Pricing, for details.) The desire to avoid high electricity demand charges is often a major factor in selecting absorption chillers. Be sure to include demand charges in the comparative cost calculations. For example, it may be most economical to operate absorption chillers during periods of peak electricity prices, and to operate compression chillers at other times.

ECONOMICS

SAVINGS POTENTIAL: *1 to 20 percent of the cost of chiller operation. Depends on the relative efficiency characteristics of the chillers that are installed.*

COST: *Small. The main cost is for the analysis required to create the optimum operating schedule.*

PAYBACK PERIOD: *Less than one year.*

TRAPS & TRICKS

CALCULATIONS AND EFFICIENCY MEASURE-MENTS: *To do this right, you need to know the real efficiencies of all your chillers, including their auxiliary equipment. Don't make assumptions. Get certified chiller data from the manufacturer, and figure the auxiliary energy inputs yourself.*

TRAINING AND COMMUNICATION: *The big challenge is keeping the plant operators on top of the procedure. This requires effective training and supervision.*

MEASURE 2.1.1.1 Install an automatic chiller scheduling controller.

Measure 2.1.1 explains how to sequence chillers to minimize energy costs. If the plant has different types and/or sizes of chillers, the manual optimization procedures described in Measure 2.1.1 may require too much attention from the plant operators to be reliable. In such cases, install an automatic controller to perform this function.

Should You Install a Scheduling Controller?

An automatic optimizing controller is most desirable for chillers that have modulating controls, because this is the most difficult case. It is especially difficult if you combine modulating chillers of different types. In the latter case, the optimum distribution of load among the chillers varies with the overall cooling load.

There may be little or no reason to use a scheduling controller with single-stage or multi-stage chillers. You can usually sequence these satisfactorily by staggering their chilled water temperature setpoints. However, a controller may still be desirable if the staged chillers differ in both size and efficiency.

Selecting and Programming the Controller

A chiller scheduling controller is a simple computer that senses the chilled water temperature and sends the appropriate starting, stopping, and loading signals to the chillers in response. A variety of chiller load schedulers are available commercially for these functions. If the controller is not designed specifically for your combination of chillers, you have to do the analysis described in Measure 2.1.1. Do this analysis before you buy the controller, as it will help you to recognize which features you need.

The less expensive units perform elementary sequencing functions. The most elementary function is to turn on successive chillers as the previous chillers load up to specified levels. Another function is shutting down smaller chillers when they become fully loaded, and transferring the load to the larger chillers. You have to input the compressor loading sequences desired. Controllers in the middle of the range can distribute the cooling load among modulating chillers at specified ratios.

You may need more sophisticated control capability than this to deal with complex combinations of chiller characteristics. If you have different types or models of chillers with modulating compressors, the controller must be able to distribute the load optimally throughout the load range. This requires a controller that can memorize and process the efficiency characteristics of

SUMMARY

Eliminates the need for continuous operator attention to keep chiller loading optimized. Usually needed only with modulating chillers.

SELECTION SCORECARD

Savings Potential .... $ $ $

Rate of Return, New Facilities % % %

Rate of Return, Retrofit.......... % % %

Reliability .... ✓ ✓ ✓

Ease of Retrofit .... ☺ ☺ ☺

each chiller in the plant. You have to input the desired load for each chiller at each chilled water temperature.

■ Don't Forget to Control the Auxiliary Equipment

If the chiller plant has separate auxiliary equipment, such as chilled water and condenser water pumps, select a chiller scheduler that can optimize the operation of these also, to provide the optimum combination of auxiliary equipment under all load conditions.

In some plants, optimum scheduling of auxiliary equipment may be complex. You have to program the controller to deal with each item of auxiliary equipment individually. To do this, you have to define the conditions under which each auxiliary should run. If the auxiliary has variable output, you have to decide the level of output for each level of cooling load.

Do not use a controller to operate auxiliary equipment if you can perform that function efficiently with a simple electrical connection between the chiller and the auxiliary equipment.

See Measure 2.1.3 about turning condenser cooling equipment on and off. See Measure 2.2.2.2 about control of cooling tower fans. See Measure 2.1.4 about turning chilled water pumps on and off. See Measure 2.5.2 about variable-flow chilled water pumping.

Alternative: Use An Energy Management Computer

If you have an energy management computer, you may be able to use it to optimize chiller sequencing and loading. This may be the most economical approach if the appropriate software is available and debugged. Otherwise, programming an energy management computer to perform this specialized function may be impractical. See Reference Note 13, Energy Management Control Systems, about the difficulties you may face.

ECONOMICS

SAVINGS POTENTIAL: *2 to 10 percent of chiller plant energy consumption. Depends on the relative efficiency characteristics of the chillers that are installed.*

COST: *$4,000 to $30,000, depending on the complexity of the plant and installation.*

PAYBACK PERIOD: *Several years, or longer.*

TRAPS & TRICKS

SELECTING THE CONTROLLER: *Be sure to select a controller that can deal with the characteristics of all your chillers. If you have chillers of different types or models, you may need a relatively complex and expensive controller. Resist the temptation to use a general-purpose energy management computer for this purpose, unless you certain that it can be programmed properly. Also, such a computer is likely to become obsolete long before the chillers wear out. If you know how to use programmable logic controllers (PLC's), you can create virtually any chiller control sequence. But, even if you use a PLC, you still have to interface the PLC to the chillers.*

PROGRAMMING: *Account for all load conditions. Selecting the optimum schedule may be tricky, especially for combinations of types or models.*

MONITOR PERFORMANCE: *Schedule periodic checks to ensure that the controller is doing what you want, and that no one has tampered with the scheduling sequence.*

MEASURE 2.1.2 Use automatic controls to shut down the entire chiller plant when there is no cooling load.

A chiller plant should not operate when there is no need for cooling. If a chiller plant operates unnecessarily, it may waste energy in a number of ways. With plants that are required on a seasonal basis, it is common practice to start the plant and shut it down manually. This results in inefficient operation at the beginning and end of the season, and inability to provide cooling when it may still be needed.

The general solution to these problems is to use automatic controls to start and stop all the chiller plant equipment in response to actual need for cooling. The five subsidiary Measures provide different methods of controlling chiller plant operation, adapted to different situations:

- Measures 2.1.2.1 and 2.1.2.2 sense the presence of an actual cooling load. The first of these senses thermostatic control signals in the cooled spaces. The second senses the outside air temperature.
- Measures 2.1.2.3 and 2.1.2.4 control chiller plant operation by time. The second of these provides for pre-cooling based on weather conditions.
- Measure 2.1.2.5 is a hybrid of automatic and manual control for situations where safety considerations make it inadvisable to control equipment automatically.

In many cases, you want to control the running of the chiller plant on the basis of both load and time. In these cases, combine the subsidiary Measures as appropriate.

Energy Saving Potential

This Measure can prevent or reduce these sources of energy waste:

- *unnecessary auxiliary equipment energy consumption.* Even if the chiller stops running, the chilled water pumps, condenser water pumps, and perhaps the cooling unit fans continue to operate. The power consumption of this equipment is typically 5 to 20 percent of the plant's full-load power consumption.
- *false loading.* In some chillers, especially older ones, the chiller does not stop when the cooling load disappears. Instead, a false load is used to keep the chiller running above its minimum load, typically by hot gas bypass. If false loading is used, it typically consumes 10 to 40 percent of the plant's full-load energy consumption.
- *distribution losses.* Chilled water distribution pipe loses some cooling energy by conduction. This loss is minor in small systems. In large central cooling systems with pipe insulation in poor condition,

conduction loss may waste as much as 10 percent of the plant's full-load power consumption.

- *waste of cooling by end users.* Certain applications tend to waste cooling if it is available during mild weather. This is a problem in apartment complexes, military bases, and campuses, where tenants do not pay directly for their own utilities. Buildings with openable windows are especially vulnerable.

Not a Substitute for Local Control of Cooling

Turning off cooling at the chiller is not an acceptable substitute for good control of cooling equipment within the spaces. For one thing, the chiller plant must operate as long as any space requires cooling, even though the other spaces may not. For example, if the chiller plant is turned on to serve one classroom in a school after hours, the entire school will be cooled. Another reason is that turning off the chiller plant does not stop the energy consumption of end-use equipment accessories, such as the fans in fan-coil units and air handling units.

The chiller plant equipment and the end-use cooling equipment each need their own appropriate controls. For example, it might be appropriate to control space cooling units with timeclocks and to control the chiller plant based on outside air temperature.

Provide Separate Cooling for Small After-Hours Loads

A common waste of energy is keeping a large chiller system in operation to serve a few small spaces that operate during hours when the rest of the facility is shut down. An example is the cooling of computer rooms in office buildings. In such cases, consider installing separate cooling units to serve the spaces with longer operating schedules. See Measure 2.8.2 for this approach.

Coordinate with Measure 2.1.3

The purpose of this Measure is to shut down all chiller plant equipment during periods when there is no cooling load. Measure 2.1.3, below, interlocks the chiller's condenser cooling equipment with the running of the compressor under all conditions. Both Measures involve common control and wiring work, so try to accomplish both as a common project.

Coordinate with Control of Boiler Plant Equipment

The subsidiary Measures recommend techniques similar to those used for limiting the operation of the boiler plant (Measures 1.1.1 ff). If the boiler plant is located near the chiller plant, consider using the same control device for both heating and cooling equipment.

MEASURE 2.1.2.1 Control chiller plant operation by sensing the end-user cooling load.

New Facilities   Retrofit   O&M

B **C**

This is the first of five subsidiary Measures that automatically shut down the cooling plant when it is not needed. In this method, the plant shutdown controls sense whether there is an actual cooling load anywhere in the system. The control system does this by monitoring the thermostatic signals that control the end-use cooling equipment, such as fan-coil units in individual spaces. If none of the space thermostats calls for cooling, all chiller plant equipment is turned off.

This method applies to any type of cooling load, including solar load, heat gain from equipment and people, etc. The key point is that the thermostatic control system must be able to recognize the presence of a cooling requirement anywhere in the building. Figure 1 shows that this can be tricky.

WESINC

Fig. 1 Why it is tricky to sense cooling load Some spaces on this face of the building are in deep shade, while the rest are exposed to strong solar gain. To turn off the chillers that cool this zone, sensors must be located where they will detect a cooling load in any part of the zone.

SUMMARY

The only accurate method of controlling the operation of the chiller plant based on actual cooling load. The cost and complexity depends on the sophistication of the present thermostatic controls.

SELECTION SCORECARD

Savings Potential	$ $
Rate of Return, New Facilities	% % %
Rate of Return, Retrofit..........	% % %
Reliability	✓ ✓ ✓
Ease of Retrofit	☺ ☺ ☺

See Reference Note 14, Control Signal Polling, for methods of collecting load information from a large number of thermostats. If a centralized system for monitoring all space temperatures is already installed, this Measure should be cheap. Otherwise, installing a system to monitor space temperatures may be considerably more expensive and complex than the subsidiary Measures that follow.

ECONOMICS

SAVINGS POTENTIAL: *Typically a few percent of total chiller plant energy consumption, but may be more in some cases.*

COST: *Several hundred to several thousand dollars, typically. Maybe much more, if the areas served by the chiller plant are dispersed.*

PAYBACK PERIOD: *Several months, to several years.*

TRAPS & TRICKS

MAINTENANCE: *This method will not work properly unless all thermostats being monitored are intact and properly calibrated.*

EXPLAIN IT: *Explain this control feature to the plant operators, so they will not be surprised when the entire plant shuts down by itself. Write a clear explanation in the plant operating manual.*

MONITOR PERFORMANCE: *Operators in a hurry to fix cooling problems will be tempted to bypass the shutdown control. Schedule periodic checks to make sure it is working properly.*

MEASURE **2.1.2.2 Limit the operation of the chiller plant based on the temperature or enthalpy of the outside air.**

This is the second in a series of five methods for shutting down the chiller plant automatically when it is not needed. In this method, the system shutdown controls sense the temperature or enthalpy of the outside air. The main virtue of this method is that it is simple and easy to retrofit. Only one sensor is needed, and it is relatively easy to install. The advantages of this method over Measure 2.1.2.1 are much lower cost and greater control reliability. However, this method is satisfactory in a much more limited range of applications.

Where to Use this Method

Limit this technique to applications where the need for cooling is related closely to the outside air temperature or to the outside air enthalpy. Unfortunately, in many applications, the need for cooling does not correlate well with the outside air temperature. The cooling load of a space is the sum of several different heat loads, including solar radiation heat gain, internal heat gains, cooling and dehumidification of ventilation air, and conduction through the building envelope. This method is desirable only in applications where sensing the outside temperature provides a fair approximation to the effect of all these heat loads.

In particular, this method is unsatisfactory for facilities that have large solar heat gains. The intensity of solar radiation can overwhelm the effect of outside air temperature on the cooling load. Also, there are large differences in solar heat gain between different parts of a building and different times of day. This variability makes it impossible to compensate for large solar gains by fudging the outside air temperature setting.

On the other hand, in facilities where the windows can be opened, the high degree of ventilation provided by windows tends to make this method feasible. For example, this method of control may be satisfactory in multi-family housing complexes.

If some of the space served by the chiller system has high internal heat gains, you can compensate by lowering the outside temperature at which the chiller plant turns on. This works even if the heat gains are fairly high, provided that the heat gain is predictable. Lighting in office buildings is a typical example of large, regular internal heat gains. (We are assuming here that the interior spaces are cooled by outside air ventilation when the outside air is cool. This is called an "economizer cycle." See Measure 4.2.5 for details.)

SUMMARY

A simple, inexpensive, and reliable method of controlling the operation of chillers for some air conditioning applications. Not appropriate for facilities with high solar heat gain.

SELECTION SCORECARD

Savings Potential .... $ $

Rate of Return, New Facilities % % % %

Rate of Return, Retrofit.......... % % % %

Reliability .... ✓ ✓ ✓

Ease of Retrofit .... ☺ ☺ ☺

Whether to Use Temperature or Enthalpy as the Controlling Factor

In climates that tend to remain dry during the transition season, ordinary (dry-bulb) temperature is satisfactory for controlling the running of the chiller system.

Where high humidity may occur during the transition seasons, consider enthalpy controls instead of simple temperature controls. Enthalpy controls start and stop the chiller in response to humidity as well as to temperature. For example, an enthalpy control might start the chiller plant at an outside temperature of 65°F if the relative humidity is 100%, and at a temperature of 76°F if the relative humidity is 30%.

Use a type of enthalpy sensor that is accurate and stable. Some cheaper types of enthalpy sensors used in HVAC work, including all the early types, require too much maintenance and re-calibration to be practical. See Measure 4.2.6 for more about enthalpy controls.

How to Install the Outside Air Sensor

The way you install the outdoor temperature sensor is critical. An improper installation can affect the temperature reading by several degrees, which may be enough to render the Measure useless. Install the sensor so that:

- it is exposed to the prevailing outdoor air temperature
- sunlight does not heat the sensor, either directly, by reflection, or by heating surrounding surfaces
- building air leaking through the wiring penetration cannot reach the sensor

- precipitation cannot reach the sensor, either directly or by splashing
- if an enthalpy sensor is used, the unit is far away from foliage, cooling towers, and other sources of moisture
- it is protected from physical impact.

A typical good location is on top of a tall mast at the upwind edge of the roof, under a shield.

Explain It!

Install a placard on the chiller controls to inform operators why the chillers do not operate on mild days. See Reference Note 12, Placards, for effective placard design and installation techniques.

ECONOMICS

SAVINGS POTENTIAL: *Typically a few percent of total chiller plant energy consumption, but may be more in some cases.*

COST: *Several thousand dollars, typically.*

PAYBACK PERIOD: *One year, to several years.*

TRAPS & TRICKS

CHOICE OF METHOD: *Do not use this method if solar load is a significant part of the cooling load. If humidity is a large part of the cooling load, consider using an enthalpy sensor rather than a thermostat.*

DETAILS: *The location and shielding of the outside air sensor are critical, whether it senses temperature or enthalpy. If you control using enthalpy, be sure to select a sensor that is stable and reliable in the long term.*

EXPLAIN IT: *Tell the plant operators what to expect. Describe the controls in the plant operating manual. Install an effective placard at the chiller controls*

MONITOR PERFORMANCE: *Operators in a hurry to fix cooling problems will be tempted to bypass the shutdown control. Schedule periodic checks to make sure it is working properly.*

MEASURE **2.1.2.3 In applications with regular schedules, use time controls.**

This is the third is a series of subsidiary Measures that automatically shut down the chiller plant when it is not needed. Consider this approach in applications where cooling is needed on a regular time schedule. It is similar to Measure 1.1.1.1 in its technical features and economics. Refer to there for those aspects. As an alternative to conventional external time controls, some

EF/EX/FA CHLR	OCC PC01S	TIME PERIOD SELECT	
PERIOD	ON	OFF	M T W T F S S H
1	0700	1800	X X X X X
2	0600	1300	X
3	0000	0300	X
4	0000	0000	X X
5	0000	0000	
6	0000	0000	
7	0000	0000	
8	0000	0000	
OVERRIDE	0 HOURS		
NEXT	PREVIOUS	SELECT	EXIT

Carrier Corporation

Fig. 1 Time scheduling option on chiller control panel
This feature of a modern chiller provides a limited degree of flexibility in scheduling chiller operation by time.

SUMMARY

An inexpensive way of limiting chiller plant operation in facilities that operate on a predictable schedule.

SELECTION SCORECARD

Savings Potential .... $ $ $

Rate of Return, New Facilities % % % %

Rate of Return, Retrofit.......... % % % %

Reliability .... ✓ ✓

Ease of Retrofit .... ☺ ☺ ☺

chillers offer time control as part of the machine's own control package. Figure 1 shows an example.

Time control alone is rarely sufficient to avoid unnecessary chiller plant operation, because it does not respond to cooling load conditions. Consider combining this Measure with either of the two preceding Measures, as appropriate.

MEASURE **2.1.2.4 In applications where pre-cooling is required, use optimum-start controllers.**

This Measure is a more efficient variation of time control, which is covered in the preceding Measure. It is similar to Measure 1.1.1.2 in its technical features and economics. Refer to there for the details.

SUMMARY

A refinement of Measure 2.1.2.3 for applications where you want to cool down the facility prior to occupancy.

SELECTION SCORECARD

Savings Potential .... $ $ $

Rate of Return, New Facilities % % % %

Rate of Return, Retrofit.......... % % % %

Reliability .... ✓ ✓

Ease of Retrofit .... ☺ ☺ ☺

MEASURE 2.1.2.5 In applications where automatic starting of chillers is undesirable, use automatic controls to alert personnel to start them manually.

RATINGS

New Facilities **B** Retrofit **B** O&M

This is the last of a series of five subsidiary Measures that eliminate unnecessary operation of the chiller plant. It is similar to Measure 1.1.1.4 in its technical features and economics. Refer to there for those aspects.

In summary, this method uses any of the methods of Measures 2.1.2.1 through 2.1.2.4, as appropriate, to signal operating personnel to start and stop the cooling plant equipment manually, rather than using automatic controls to operate the equipment directly. This method applies where automatic control may be a safety hazard.

In facilities where the boiler and chiller plants have a common control station, save cost and effort by creating a control panel to serve both sets of equipment.

SUMMARY

A way of achieving the benefit of automatic control in facilities where you need to start and stop the chiller plant manually.

SELECTION SCORECARD

Savings Potential .... $ $ $

Rate of Return, New Facilities % % % %

Rate of Return, Retrofit.......... % % % %

Reliability .... ✓ ✓

Ease of Retrofit .... ☺ ☺ ☺

MEASURE 2.1.3 Turn off and isolate heat rejection equipment when the corresponding chiller turns off.

RATINGS

New Facilities **A** Retrofit **B** O&M

Every mechanical cooling system requires heat rejection equipment to get rid of the heat that is collected in cooled spaces or processes. This equipment consumes energy for fans, pumps, or both. It does not need to operate when the chiller is not operating.

In systems that are assembled from separate components, the condenser cooling equipment may not be controlled properly to turn off when the chillers turn off. Or, if the heat rejection system is complex, the equipment may not turn off in a manner that minimizes energy consumption under all operating conditions. Figure 1 shows a typical cooling tower system where such inefficiency may arise.

Check your heat rejection systems to see if they have equipment that operates unnecessarily. If so, you can save energy by making some changes that are usually fairly simple.

Applications and Energy Saving Potential

This Measure applies to all types of chillers, unless they run continuously. The savings are proportional to the size of the equipment and to the length of time that the chiller or chillers are turned off completely.

SUMMARY

Worth considering for all chillers where the compressor does not operate continuously. Control modifications are usually simple. Saves the most energy in water-cooled systems. In multi-chiller systems, you may have to install special valves.

SELECTION SCORECARD

Savings Potential .... $ $ $

Rate of Return, New Facilities % % % %

Rate of Return, Retrofit.......... % % % %

Reliability .... ✓ ✓ ✓

Ease of Retrofit .... ☺ ☺ ☺

Chillers using reciprocating and scroll compressors are designed to change output by cycling or staging the compressors. They turn off completely during periods of low load, allowing the condenser cooling units to be turned off. Single-stage compressors turn off for the largest fraction of the time, but these are used only in smaller systems.

Centrifugal and screw compressors modulate their output. Even these types must turn off below a certain fraction of full load, unless they are kept running by false loading.

If the system has some form of "free cooling" that requires the operation of the cooling tower, then the condenser cooling equipment probably has to keep running as long as there is a cooling load. See Subsection 2.9 for details.

How to Control the Equipment

The method of control depends on the type of heat rejection equipment that the cooling system uses. The following are the usual situations.

■ Condenser Water Pumps and Condenser Isolation Valves

If a water-cooled chiller has its own cooling tower or cooling tower cell, turn off the condenser water pump when the compressor stops.

If a single cooling tower serves several water-cooled chillers, the design of the cooling tower and the configuration of the pumps determine whether you can turn off individual pumps when a chiller compressor turns off. Consider these factors:

• you have to keep the flow rate through each operating condenser high enough to maintain good heat transfer and minimize fouling. The chiller manufacturer normally specifies this flow rate. To satisfy this condition, the pump configuration must allow you to turn off individual pumps as chillers turn off, without reducing the flow rate in other condensers excessively.

• you have to valve off the condensers of idle chillers when the corresponding pumps turn off. You can operate the condenser valves manually if you start and stop the chillers manually. However, if the chillers are started and stopped automatically, you need automatic motorized valves.

• reducing the flow rate in the cooling tower may reduce its cooling ability. Gravity-flow towers typically need a certain flow rate to maintain good contact between the air and water. The manufacturer always specifies the upper limit of flow, but you may have to ask about the lower limit. The actual flow requirement may be increased by installation or maintenance defects, such as distributor trays that are not level or nozzles that are partially obstructed. If the tower has separate cells, you may be able to connect individual cells to individual chillers. This requires some piping modifications and the installation of motorized valves.

■ Cooling Tower Fans

The simplest and most common cooling tower fan controls simply operate the fans whenever the cooling water temperature is higher than the fan setting, even if the chillers served by the tower are turned off. A variation of this method operates the fans on the basis of outside air wet-bulb temperature.

It is easy to prevent unnecessary fan operation if a cooling tower serves only one chiller. Simply power the fan through an auxiliary contactor in the chiller controls.

Baltimore Aircoil Company

Fig. 1 Typical cooling tower system This system is divided into three cells, and each cell has two fans. This system could waste fan and pump energy by operating more of the cells than needed, and it could waste fan energy by inefficient control of the fans.

If a cooling tower serves more than one chiller, the modification is only slightly more complicated. In this case, energize the motor starters for the pumps and fans through relays from each of the chiller controls. The relays are wired so that operation of any chiller energizes the pumps and fans.

You don't need these methods if you install the sophisticated controls recommended by Measure 2.2.2.2. Refer to Measures 2.2.2 ff and 2.4.2 for details of efficient fan control.

■ Evaporative Condenser Fans and Circulating Pumps

An evaporative condenser uses a pump to spray water over smooth coils that contain the compressed refrigerant from the compressor. A fan circulates air over the wetted coils, cooling them by evaporation. The fan in an evaporative condenser typically runs continuously, since the unit has no cooling ability without the fan. There is no need for the pump or fan to operate when the compressor is turned off. Avoid unnecessary operation simply by powering the pump and fan through an auxiliary contactor in the chiller controls.

■ Cooling Fans in Air-Cooled Chillers

With air-cooled chillers, operation of the cooling fans is usually an integral part of the chiller control circuitry. Operation of the fan is normally controlled on the basis of compressor operation or condenser pressure, so that the fan(s) will not run unless needed. Sometimes, fan operation is also controlled by outside air temperature. In such cases, you may find that the fans run during hot weather even when the compressor

is not running. This is a control oversight. Correct it by modifying the fan controls so that the fans run only when the compressor operates.

ECONOMICS

SAVINGS POTENTIAL: *5 to 30 percent of the energy used by the condenser water pumps and cooling unit fans.*

COST: *Several hundred to several thousand dollars for simple control modifications. Motorized valves for condenser water and related piping modifications cost much more.*

PAYBACK PERIOD: *One year, to many years.*

TRAPS & TRICKS

DETAILS: *Exploit the chiller's installed features for controlling auxiliary equipment, if it has any. Be sure to provide adequate safety features to ensure chilled water flow and condenser cooling when starting the chiller. When making this change to an existing plant, be sure that you understand the existing controls for starting equipment, including the internal starting sequences of chiller-mounted controls.*

EXPLAIN IT: *Make sure that the plant operators understand how the equipment is supposed to be linked. Otherwise, the control interlocks will eventually be disconnected. Install a system diagram and an explanation of system operation at a prominent location. See Reference Note 12, Placards, for guidance.*

MONITOR PERFORMANCE: *Schedule periodic checks of the procedures and/or controls in your maintenance calendar.*

MEASURE 2.1.4 In plants with multiple water chillers, minimize the operation of chilled water pumps and isolate idle evaporators.

Figures 1 and 2 show two common chilled water pumping arrangements that are used in plants with several chillers. With either of these configurations, there are two simple things you can do to save energy as the cooling load falls: (1) turn off pumps that are not needed to carry the load, and (2) close valves to stop flow through the evaporators of idle chillers.

Both actions save pump energy. Turning off a pump saves pump energy. Shutting off flow to an idle chiller also reduces pump energy slightly, by reducing the flow rate through the pump.

Shutting off the chilled water flow to an idle chiller allows you to reduce the energy consumption of the remaining chillers. This is because the warm return water flowing through the evaporator of the idle chiller dilutes the chilled water from the chillers that are running. The mixing requires the operating chillers to produce chilled water at lower temperature.

For example, consider a low-load situation where only one of the chillers in Figure 2 is operating. The return water temperature is 48°F, and the system requires 46°F chilled water to satisfy the cooling load. The chilled water of the operating chiller is mixed with the warm return water coming through the idle chillers, so the operating chiller must produce chilled water at 42°F, instead of the system requirement of 46°F.

The lower chilled water temperature wastes energy because chiller efficiency typically declines about two percent per degree Fahrenheit that the chilled water temperature is reduced. (See Measure 2.2.1 for details.) You avoid this loss of efficiency simply by stopping the flow of water to the idle chillers.

In some systems, you will be able to accomplish both these actions at the same time. In others, you must do them at different times. Which you do depends on the sizing and connections of the pumps in relation to the chillers. It also depends on whether the end-use equipment has throttling valves or bypass valves. You can save much more energy if the system uses throttling control valves, rather than bypass valves. In many cases, you can convert a system that has bypass valves to operate with throttling valves.

Applications and Energy Saving Potential

These factors determine whether this Measure is appropriate for your plant, and how much energy it can save:

- *whether the chillers use supply water temperature control.* This Measure will not work with return water temperature control if you want to use automatic control of the pumps and isolation valves.

SUMMARY

An easy operating procedure that saves chilled water pump energy and improves chiller efficiency. Some piping and valve changes may be needed. Can damage chillers if you do it wrong. Easy to forget. In new construction, use variable-flow pumping (Measure 2.5.2) instead.

SELECTION SCORECARD

Savings Potential	$ $ $
Rate of Return, Retrofit	% % % %
Rate of Return, O&M	% % % %
Reliability of Equipment	✓ ✓ ✓ ✓
Reliability of Procedure	✓ ✓
Ease of Retrofit	☺ ☺ ☺
Ease of Initiation	☺ ☺ ☺

You can accomplish this Measure manually with return water temperature control.

- *the cooling load profile.* Savings are possible only when the load is low enough to allow one or more pumps to be turned off. Therefore, savings are greatest if the system operates for long periods at low loads.

- *the number and sizes of the chillers.* The savings are greatest if the chillers are sized so that unused chiller capacity can be turned off as the load falls. In other words, the saving potential is greatest in a plant with a larger number of chillers, or in a plant where a smaller chiller is installed to carry low loads.

- *the relationship of pump and chiller sizes.* You can achieve the greatest saving when the capacities of individual pumps correspond to the flow requirements of individual chillers. The worst case is a plant in which one pump is sized to carry the full chilled water flow of the system. In the latter case, the potential for saving pump energy is minimal, and you may not even be able to isolate the evaporators of idle chillers.

- *the types of control valves in the end-use equipment.* The savings potential is larger if the end-use equipment uses throttling valves instead of bypass valves. You can modify bypass valves on the end-use equipment, or replace them with throttling valves, to exploit this fact.

• *the length of the distribution system.* If the system uses bypass control valves, the savings potential is greater if the distribution system is long. This is because a larger fraction of pump power is devoted to overcoming piping loss, which declines radically as flow is reduced.

Dangers

The procedures in this Measure can damage equipment if they are done incorrectly. Be sure to avoid these three potential problems:

• *inadequate evaporator water flow.* Stopping the evaporator flow in a chiller that is operating will freeze the evaporator. In addition to bursting evaporator tubes, this gets water into the refrigerant circuit, which may destroy the compressor.

• *excessive evaporator water flow.* Shutting off the flow in idle chillers may accelerate the flow in the operating chillers. This may cause tube erosion. This hazard does not exist if each chiller has a dedicated chilled water pump.

• *pump cavitation.* If flow through a centrifugal pump is restricted too much, there is great turbulence in the impeller blades, which is aggravated by the formation of bubbles of water vapor. This is called "cavitation." Cavitation makes noise, and can wear out pumps.

The following discussion tells you how to avoid these problems.

How to Operate Pumps with a System Bypass Valve

In some constant-flow chilled water systems, a bypass valve is installed between the system pumps and the rest of the system. This is shown in Figure 3.

The bypass valve may have one or two functions. One function is to limit system pressure. The bypass valve dumps supply water flow back to the pumps if the system pressure becomes excessive. Another function is to maintain a minimum flow through the pumps. This is to prevent pump cavitation, which occurs when the flow through an operating pump is restricted. A bypass valve is used most commonly in systems where the end-use equipment has throttling valves.

If the system has chilled water pumps that are matched in capacity to the chillers, the procedure is simple. Operate the same number of pumps and chillers. Whenever a chiller is turned off, turn off the corresponding pump and isolate the evaporator of the chiller.

If the pump and chiller capacities are not matched, use a pressure sensing control to control the number of pumps that operate. How to determine the minimum acceptable system pressure is explained below.

This method of control should eliminate the need for the bypass valve under all or most conditions. Pump cavitation is not a problem with this Measure, because it reduces the number of pumps in operation. To make sure that the bypass valve does not waste energy by bypassing water unnecessarily, determine the maximum safe pressure for the system. Set the bypass valve to the highest allowable pressure.

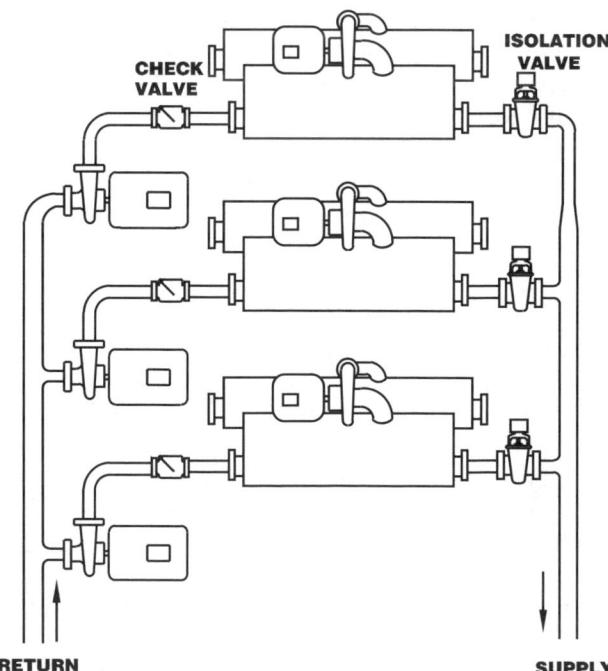

Fig. 1 Series chilled water pumps A chilled water pump is dedicated to each chiller. This is the easiest system to control efficiently. Its only disadvantage is that a pump failure will make a chiller unavailable.

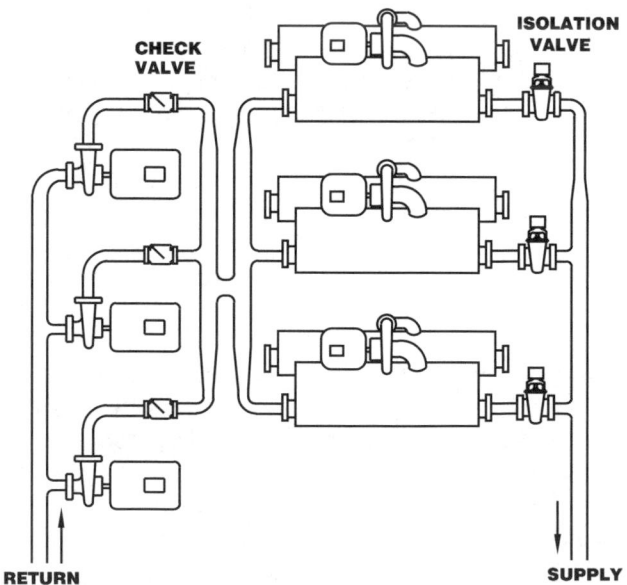

Fig. 2 Parallel chilled water pumps This arrangement is efficient only if the capacity of each pump is matched to the flow requirement of each chiller. Its only advantage is that it allows a single pump to serve as a reserve unit for any other pump.

How to Operate Pumps if the Cooling Equipment has Bypass Control Valves

In many chilled water systems, bypass control valves are installed on the end-use equipment, rather than throttling valves. Bypass valves control equipment output by splitting the flow of chilled water, passing some through the equipment and shunting the rest. As in the previous case, bypass valves are used to limit system pressure, or to maintain a minimum flow through the system pumps, or both. Also, some engineers believe, incorrectly, that bypass control valves are necessary for stable control. Figure 4 shows a chilled water distribution system that uses bypass control valves on the end-user equipment.

(We are using "bypass valve" as a broad term to contrast with "throttling valve." If a bypass valve is installed on the return side of equipment, it may be called a "mixing valve." If it is installed on the inlet side, it may be called a "diverter valve.")

Figure 5 shows a typical bypass control arrangement on a cooling coil. The sketch shows that a balancing valve is installed in addition to the bypass valve. The balancing valve is adjusted to provide the same flow resistance as the cooling coil. This arrangement provides better control by the control valve, especially at low loads. In many cases, a balancing valve is not used. As an alternative to using a balancing valve, the bypass valve is selected to provide the characteristics of the balancing valve. (With this type of balancing valve, the bypass path has a minimum flow resistance approximately equal to the resistance of the controlled equipment.)

Some equipment that uses chilled water may have no control valves at all, i.e., they are always "wide open." For example, the cooling coils of multizone air handling units sometimes have no control valves. Such equipment behaves in the same way as equipment that is controlled by bypass valves.

The most efficient approach is to get rid of bypass valves, and replace them with throttling valves, as explained below. Bypass valves limit your ability to turn off chilled water pumps because they keep the flow (almost) constant, regardless of the cooling load. If you turn off a pump, the system pressure will drop. You can turn off pumps only to the point that the pressure drop is not excessive. This seriously limits your ability to turn off chilled water pumps.

If you retain the bypass control, you will have to do some preliminary work to find out how to operate the pumps most efficiently. To do this, turn off a chilled water pump and see what happens to the system pressure. The pressure drop that occurs will depend on the characteristics of the pumps and of the system. Centrifugal pumps produce more flow as output pressure requirement is reduced. The relationship between pressure and flow depends on the characteristics of the

specific pumps used. (See Reference Note 35, Centrifugal Pumps, for more about this.)

One factor in your favor is that the pressure requirement of the end-use equipment drops significantly as load lessens. However, be careful when exploiting this fact to reduce pump power. Make sure that none of the end-use equipment requires maximum pressure while the rest of the system is at low load. For example, an air handling unit for a computer room may operate at high load continuously, even though the load on other air handling units varies with the weather.

If the chilled water system has a lot of distribution pipe resistance, you are more likely to get away with turning off a pump. Piping resistance is proportional to the third power of the flow rate, so it drops radically as flow is reduced, partially compensating for the reduced pump output. As a result, when you turn off a pump, the pressure loss at the end-use equipment is less than the pressure drop at the pump end of the system.

■ Operate Pumps on the Basis of System Pressure and Cooling Load

The objective is to minimize the number of pumps in operation under all load conditions. Or, if pumps of

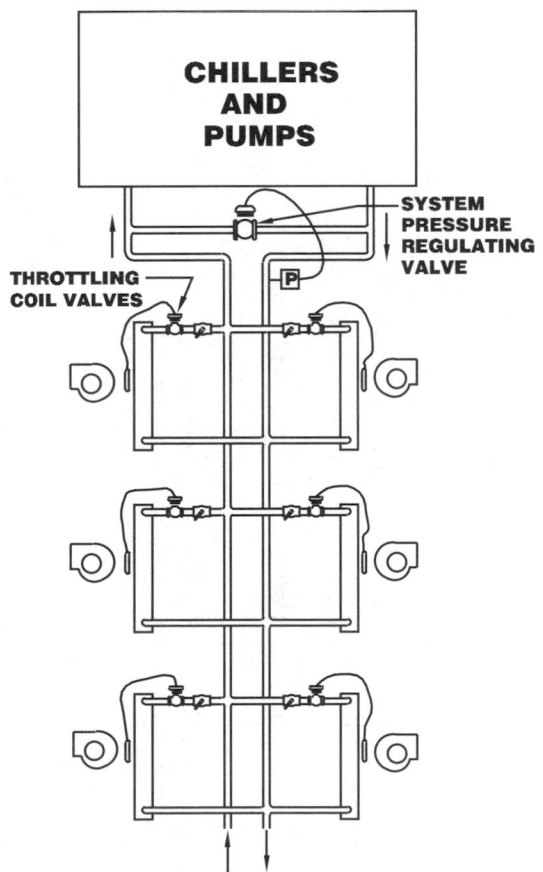

Fig. 3 System bypass valve This arrangement is wasteful if all the pumps operate continuously and the valve is used to maintain constant system pressure. However, it is a useful safety device if pumps are operated efficiently, and the bypass valve is set only to relieve excess pressure.

different sizes are installed, the objective is to minimize the total pump power. You have to judge how many pumps to operate from the system pressure and flow requirements. Stay within these limits:

- keep the system *pressure* high enough to satisfy the requirements of the end-use equipment and their control valves, but low enough to prevent control valve instability and other problems
- keep the system *flow* high enough to maintain good heat transfer in the chiller evaporators, and to avoid cavitation in the pumps.

The system must provide enough pressure to force chilled water through the end-use equipment and its control valves. So, first determine the pressure requirements of the end-use equipment. Find this information in the mechanical drawings. For example, the coils in air handling units typically have a full-load pressure requirement ranging from 2 to 15 PSI.

Additional pressure is needed for satisfactory control valve operation. A general rule is that control valves should have at least as much resistance as the controlled equipment. The actual resistance of the control valves may not be stated anywhere in the plant records, so start by assuming that the combined resistance of the

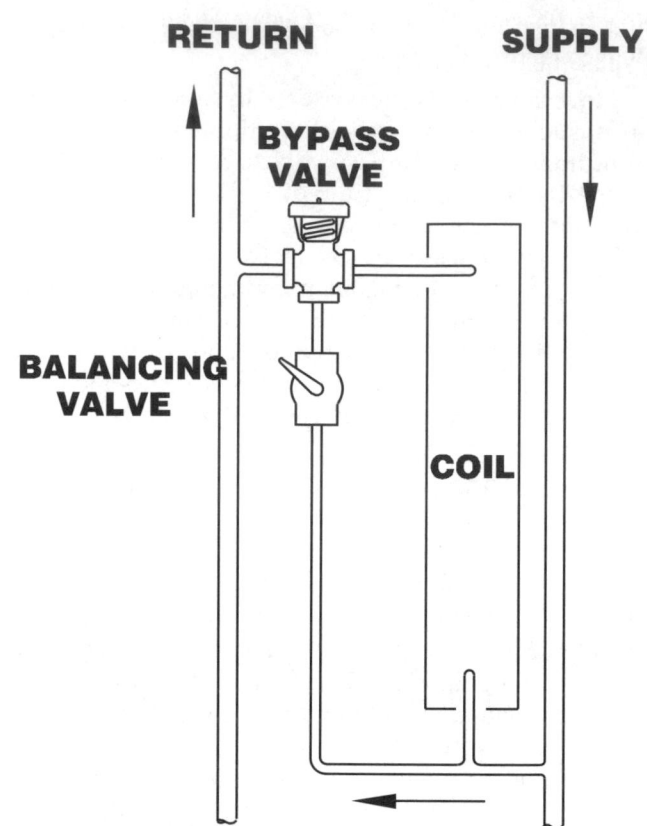

Fig. 5 Typical coil bypass control valves The bypass valve controls the flow through the coil. The balancing valve equalizes the flow resistance between the two paths. Completely closing the balancing valve converts the bypass valve to a throttling valve.

equipment and its control valves is twice the resistance of the equipment itself.

To measure the system pressure, install a pressure differential gauge between the supply and return lines near the end of the distribution system. Install the gauge as far away from the pumps as possible to account for pressure drop in the distribution piping. This gauge installation becomes the key control point for turning pumps on and off. Install the gauge where it is easy to read, or alternatively, install a pressure differential sensor that indicates remotely at the pump control location.

(Install a gauge that measures pressure differential directly, rather than using two separate pressure gauges. If you use separate gauges, errors in the gauges can combine to produce a large error in the pressure differential.)

Then, conduct a series of trials throughout the range of system loads to measure the lowest pressure differential that is satisfactory for each level of cooling load. Deficiencies in cooling capacity are likely to appear first in the end-use equipment that has the highest design pressure requirement. If there is some equipment that operates at high load while the average system load is low, pay particular attention to this equipment.

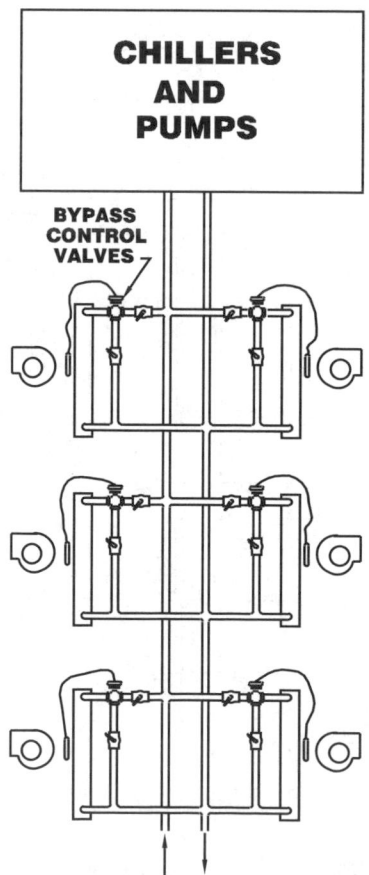

Fig. 4 Bypass control valves on cooling equipment This common arrangement is the most wasteful because it requires the maximum continuous pump output. The preferred remedy is to replace the bypass valves with throttling valves.

From these trials, prepare a schedule that tells the plant staff which pumps to operate at each level of cooling load, based on the indicated pressure differential.

Eliminate Chilled Water Bypassing to Increase the Energy Saving

To maximize your ability to turn off pumps, you have to eliminate bypass operation at the control valves. This may be easy. Also, it becomes much simpler to relate the operation of the pumps to the operation of the chillers.

If your end-user equipment has balancing valves in the bypass line, you can eliminate bypass operation simply by closing the balancing valve. In effect, the bypass valve becomes a throttling valve. The valve is likely to behave in a satisfactory manner because you are turning off pumps, which prevents excess pressure on the valve. But, check the valve performance to be sure. If you do this, tag the balancing valve in a permanent manner, stating that it should be kept in the closed position.

If your cooling equipment has bypass valves, but no balancing valves, you may be able to achieve the same effect by closing off the bypass line. Figure 6 shows an example where this has been done successfully with a very large air handling unit.

If these cheap expedients do not work well, replace the bypass control valves with throttling control valves. It may not be economical or even necessary to replace all the bypass control valves in the system. For example, in a system that serves large air handling units and small fan-coil units, it may only be economical to replace the valves on the air handling units.

If you do convert to throttling control valves, consider installing a pressure-actuated bypass valve between the supply and return legs as a safety device, as in Figure 3. This protects the equipment in case someone neglects to turn off pumps when the load is low. Analyze your system to decide whether this is necessary to prevent damage from either excessive pressure or pump cavitation. If you do use a system bypass valve, make sure that it operates only as a safety device.

How to Isolate Idle Evaporators

Ideally, you want to valve off the evaporator of each chiller that is not operating. As we discussed before, this keeps return water from flowing through the idle chillers and forces all flow through the operating chillers. In this way, there is no separate return water path to dilute the chilled water.

The simplest case is shown in Figure 1. Here, each chiller has its own evaporator circulation pump. In this case, simply turn off the pump for each chiller and close its isolation valve when the chiller turns off. Note that this assumes that you are able to make the chilled water

flow rate in the system proportional to the cooling load, as discussed previously.

The situation in Figure 2 is almost as simple. In this case, operate the same number of pumps as the number of operating chillers, and close the isolation valves of the idle chillers. Again, this assumes that you are able to make the chilled water flow rate in the system proportional to the cooling load.

The situation becomes more complicated and less efficiency if the number of chilled water pumps does not equal the number of chillers. For example, a system with three chillers may have two chilled water pumps, each of which is sized for the total capacity of the plant. In such cases, you cannot simply valve off each evaporator as its chiller turns off. This is because you may cause water flow rates through the evaporators of the operating chillers that are too high or too low, as follows.

■ Evaporator Flow Rate Limits

Keep the flow rate in operating chillers within these limits:

* *low enough to avoid tube erosion.* Valving off all the idle evaporators may cause excessive water velocity through the operating chillers, which

Fig. 6 A large bypass control valve converted to throttling operation The control valve for this enormous air handling unit was converted to throttling operation simply by eliminating the bypass path. Major energy savings result. However, it becomes more critical to operate the correct number of chilled water pumps.

Fig. 7 Automatic evaporator isolation valves These greatly
increase the reliability of efficient operation. They are not
cheap, but with chillers this size, they pay back quickly.

accelerates tube erosion. This problem is most
likely to occur in systems where the individual
pump capacities are not matched to individual
chiller flow requirements.

This problem occurs if the chilled water pumps are
sized to handle more than one chiller. For example,
if one pump is sized for the flow of two evaporators,
you may have to keep both evaporators open to flow,
whether or not both chillers are operating.

A generally accepted figure is that the water velocity
in an evaporator should not exceed 11 feet per
second (3 meters per second). Calculate whether
any combination of chillers and pumps will exceed
this flow rate. To do this, look up your chillers'
characteristics to see how their evaporator water
velocities relate to flow rate. Also, be aware that
the flow rate produced by the pumps may be
significantly higher than their rated values when
the pumps are operating against reduced system
pressure.

• *high enough to maintain proper heat transfer.*
Heat transfer depends on turbulent flow of the
chilled water. Turbulence occurs naturally at water
velocities above a threshold value, which should
be specified in the chiller catalog. For smooth
evaporator tubes, this water velocity is about 3 feet
per second (1 meter per second).

Failing to maintain proper water velocity can cause
serious trouble. If idle evaporators are not valved
off when some of the pumps are turned off, flow
through the operating chillers will be inadequate.

This reduces chiller efficiency substantially, and
may cause evaporator freezing.

In cases where you cannot simply isolate the
evaporator of a chiller when the chiller is turned off,
you have to develop a schedule that states which
evaporators to valve off as the load changes.

■ Operate Evaporator Isolation Valves Automatically, If Practical

Evaporator isolation valves are almost always
installed as part of the original equipment. However,
do not expect to operate these valves manually unless
you also start and stop the chillers manually. Starting a
chiller with the evaporator valved off can freeze the water
in the evaporator, seriously damaging the chiller.
(Chillers have safety devices to protect against this
occurrence, but don't rely on safety devices as a means
of control.)

If the chillers start and stop automatically, install
motorized valves that are interlocked with the chiller
controls. Figure 7 shows motorized chilled water valves
on an evaporator.

If you must operate the valves manually, at least
install safety interlocks that prevent chiller operation
unless the valves are open.

Regardless of whether you operate the valves
manually or automatically, install placards at the valves
explaining the operation of the system. See Reference
Note 12, Placards, for effective placard design and
installation techniques.

Keep at Least One Pump Running

You have to keep at least one chilled water pump
running as long as the chiller plant is in operation, even
if no chiller is running at the moment. This is because
chilled water temperature provides the information about
cooling load needed to start the first chiller. To avoid
the energy waste of running this last pump unnecessarily,
use one of the methods recommended in Measures
2.1.2 ff.

Potential Problems with Return Water Temperature Control

Reciprocating compressors and other compressors
that control output in steps usually control their output
on the basis of return water temperature. You cannot
use this Measure with return water temperature control
if the pumps are started and stopped automatically. In
that case, the pumps and chillers would oscillate on and
off in attempting to follow the changes in return water
temperature caused by the changes in flow rate.

If the pumps are started and stopped manually, this
Measure probably can work with return water
temperature control. Stopping a pump reduces the flow
rate, causing a higher return water temperature for a
given load. The system will stabilize with a lower
average supply water temperature.

Use Chilled Water Temperature Reset

Isolating idle evaporators allows the operating chillers to increase their chilled water supply temperature. However, you will not reduce energy cost unless you actually exploit this opportunity. Use automatic reset controls for this function, because the system load fluctuates continuously. See Measures 2.2.1 ff for details.

A More Efficient Alternative

Variable-flow chilled water pumping is the most efficient design for a chilled water system. It perfectly matches pump power to the system flow requirements under all load conditions. See Measure 2.5.2 for details.

Variable-flow pumping has the greatest advantage over this Measure in plants that use one or two large chilled water pumps. In retrofit applications, variable-flow pumping is much more expensive than this Measure. In new construction, variable-flow pumping adds relatively little cost in systems of medium and large sizes.

ECONOMICS

SAVINGS POTENTIAL: *20 to 70 percent of chilled water pump energy. 1 to 5 percent of chiller energy.*

COST: *If the right kinds of valves are already installed, the cost is limited to installing effective placards, which may cost about one hundred dollars.*

The cost of converting bypass control valves to throttling valves varies widely. If you are lucky, it will cost almost nothing.

Simple interlocks to prevent chiller operation with the evaporator valves closed may cost several thousand dollars.

Automatic chilled water reset controls may cost several thousand dollars.

If you need to install automatic evaporator isolation valves, they will cost from thousands of dollars to tens of thousands, depending mainly on the pipe size.

PAYBACK PERIOD: *Immediate, if you do everything manually. One year to many years, if you need to install automatic valves and controls.*

TRAPS & TRICKS

PLANNING AND DESIGN: *You will have to spend some time studying the characteristics of your chillers and pumps to work out the optimum schedule of pump and valve operation. Consider all the possibilities for adding or changing valves to exploit this technique as much as possible.*

OPERATION: *The plant operators will determine whether this activity succeeds or fails. So, provide clear instructions. Install highly readable placards on the valves and pump controls that explain how to operate them. Include this information in your training curriculum for new plant operators. Schedule periodic checks of proper operation.*

MEASURE **2.1.5 Install power switching that prevents unnecessary operation of spare pumps.**

RATINGS

New Facilities	Retrofit	O&M
A	**B**	

Refer to Measure 10.4.1.

WESINC

Fig. 1 Pump row of a large hotel All these pumps are in duplicate sets. Only one of each needs to be running. A transfer switch for each pair is a cheap, foolproof way to avoid unnecessary operation.

Cooling plants commonly have duplicate chilled water and condenser water pump sets for the sake of reliability. Figure 1 shows a facility with many duplicate pumps. Typically, these are installed with no controls or instructions to avoid unnecessary operation of the reserve pumps. Unnecessary operation of duplicate pumps is a common cause of energy waste, because operating personnel may not be sure how many pumps should be running.

A simple solution is to install a power switch for the pumps that keeps the spare pump from operating at the same time as the others.

Before proceeding, check the flow requirements of the system to verify that the pumps actually duplicate the required capacity. For example, if the plant has two chillers, each pump may be sized for only one chiller, in which case there is no standby pump.

You need more flexible pump control if you are able to run different numbers of pumps in response to changing load conditions. See Measure 2.1.3 for control of condenser water pumps, and Measure 2.1.4 for control of chilled water pumps.

MEASURE **2.1.6 Turn off compressor sump/crankcase heaters during extended shutdown intervals.**

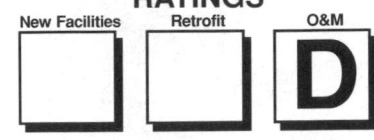

Most halocarbon refrigerants mix freely with the compressor lubricating oil. In chillers that use such refrigerants, a substantial amount of refrigerant may mix with the oil in the compressor sump (or crankcase) when the compressor is not running. When the compressor restarts, the pressure in the sump is suddenly reduced, and the refrigerant boils out of the oil. The frothing that results keeps the oil pump from delivering adequate lubrication to the machine, and the compressor may be seriously damaged.

A common accessory installed to prevent this problem is a small electric heater that warms the oil. The heater keeps the temperature of the oil high enough to drive off excessive dissolved refrigerant. The heater is usually a low-wattage electric resistance heating element. It may be wrapped around the outside of the compressor sump, or it may be installed inside the sump.

Sump heaters typically consume about 100 watts in the smallest compressors that are used in central systems, and up to several hundred watts in larger compressors. The control of sump heaters is simple. Typically, the heater turns on when the compressor is off, and vice versa.

If the chiller is shut down for an entire season, the sump heater may operate for that entire time. This is unnecessary. The sump heater has to operate for only a few hours prior to start-up to expel refrigerant from the oil. The manufacturer should recommend the minimum time required.

You can prevent unnecessary heater operation simply by disconnecting the heater power circuit during extended periods of shutdown.

The practical problem is ensuring that the heater is turned on well before restarting the chillers. Failure to do this can destroy the compressors. As a minimum,

---

SUMMARY

Save a few hundred watts in each compressor during the off season. Be careful to avoid equipment damage.

SELECTION SCORECARD

Savings Potential .... **$**

Rate of Return .... **% % % %**

Reliability .... ✓ ✓

Ease of Initiation .... ☺ ☺ ☺

---

lock the chiller power off, and install a tag that tells the operator to turn on the sump heater in advance of start-up.

ECONOMICS

SAVINGS POTENTIAL: *$20 to $200 per year per compressor, typically.*

COST: *Usually none.*

PAYBACK PERIOD: *Immediate.*

TRAPS & TRICKS

SCHEDULING: *Since this procedure occurs only once each year, nobody will remember it. Schedule it as part of your annual cooling system shutdown and start-up routines.*

SAFETY: *To keep from destroying the compressor, install a lock on the starting switch over winter. Install an effective placard where the lock is located that states how long to operate the sump heater before starting the compressor.*

Optimum Operating Temperatures

Optimizing the operating temperatures of cooling machines is one of the most important aspects of efficient operation. The COP of any machine depends strongly on the temperature difference against which it operates. (Reference Note 32, Compression Cooling, explains this issue.) Cooling plants typically keep this differential higher than it needs to be, providing you with an opportunity for major energy savings.

You can reduce the temperature differential from the bottom by increasing the evaporator temperature. You can reduce it from the top by lowering the condensing temperature. Both adjustments are physically easy to make. The appropriate temperatures depend on the cooling load, the weather, chiller design, and conflicts with the efficiency of other equipment in the cooling system. Your main challenge is figuring out the optimum temperature settings for all operating conditions. In most applications, the optimum temperatures change over a wide range as load conditions change. Therefore, try to use automatic controls to adjust the temperature settings on a continuous basis.

This Subsection has two principal Measures. The first optimizes evaporator temperature, and the second optimizes condenser temperature. The two Measures are similar in theory, and both may provide major savings. However, the two differ greatly in practice. Optimizing the evaporator temperature is usually easy, simple to understand, and safe. Optimizing the condenser temperature sometimes requires a lot of analysis, and the chiller may be damaged if the condensing temperature is set too low.

INDEX OF MEASURES

2.2.1 Keep the chilled water supply temperature as high as possible.

> **2.2.1.1 Reset chilled water temperature manually.**

> **2.2.1.2 Install an automatic chilled water temperature controller.**

2.2.2 Optimize the condensing temperature.

> **2.2.2.1 Adjust the condenser temperature manually.**

> **2.2.2.2 Install automatic condenser temperature reset controls.**

RELATED MEASURES

- Subsection 2.4, for improvements to heat rejection equipment that provide more accurate control of condenser temperature

MEASURE 2.2.1 Keep the chilled water supply temperature as high as possible.

Keeping the chilled water temperature as high as possible provides major energy savings. Manufacturers' technical literature shows a saving of approximately two percent of input energy per degree Fahrenheit (or about four percent per degree Celsius) that the chilled water temperature is raised. This number applies to all types of chillers, with minor variations. Figure 1 shows how COP improves with increasing chilled water temperature, for a typical chiller.

This simple procedure applies to virtually all chilled water systems. Raising the chilled water temperature generally does not create any risks to equipment. And, it costs little or nothing to accomplish.

The amount that you can increase the chilled water temperature is limited only by the need to satisfy the cooling load. Most of the work of accomplishing this Measure consists of determining the maximum allowable chilled water temperatures over the range of cooling loads. The two subsidiary Measures offer manual and automatic methods, respectively.

The Potential for Raising Chilled Water Temperature

Chilled water systems are commonly designed to provide full cooling load with a chilled water temperature of about 42°F. Plant operators typically leave the chilled water temperature fixed at this value or some other. This is inefficient for most applications, such as air conditioning, where the load is well below its maximum most of the time. Typically, you can raise the chilled water temperature by 5°F to 10°F for much of the time. Even at full load, the typical oversizing of airside components (air handling units, fan-coil units, etc.) usually allows some increase in chilled water temperature.

A single space or a small number of spaces may require colder chilled water than is needed by the rest of the facility. In such cases, determine the limiting factor (e.g., inadequate air flow to the space) in the offending spaces. Consider spending some money to eliminate the problem in order to reap the savings that result from higher chilled water temperature.

Compromise with Fan Power in VAV Systems

In a variable-air-volume (VAV) air handling system, space cooling is controlled by varying the supply air flow, and the supply air temperature is nominally kept constant. Raising the chilled water temperature may raise the air temperature, which will cause the fans to operate at higher power.

Typically, more energy is saved in the chiller than is lost in the fans, so the best efficiency is usually produced by raising the chilled water temperature as much as possible. However, this may not be true in all cases. In case of doubt, calculate the optimum compromise between chiller power and fan power.

Compromise with Pump Power in Variable-Flow Chilled Water Systems

A variable-flow chilled water system saves pump energy by distributing chilled water only in the quantities needed by the air handling systems and older equipment. (Older conventional design bypassed unused chilled water around the user equipment.) In a variable-flow system, increasing the chilled water temperature increases the amount of chilled water that must be pumped, for a given cooling load.

Typically, you save more energy in the chiller than you lose in the pumps, so you get the best efficiency by raising the chilled water temperature as much as possible. However, this may not be true in all cases. In case of doubt, calculate the optimum compromise between chiller power and pump power.

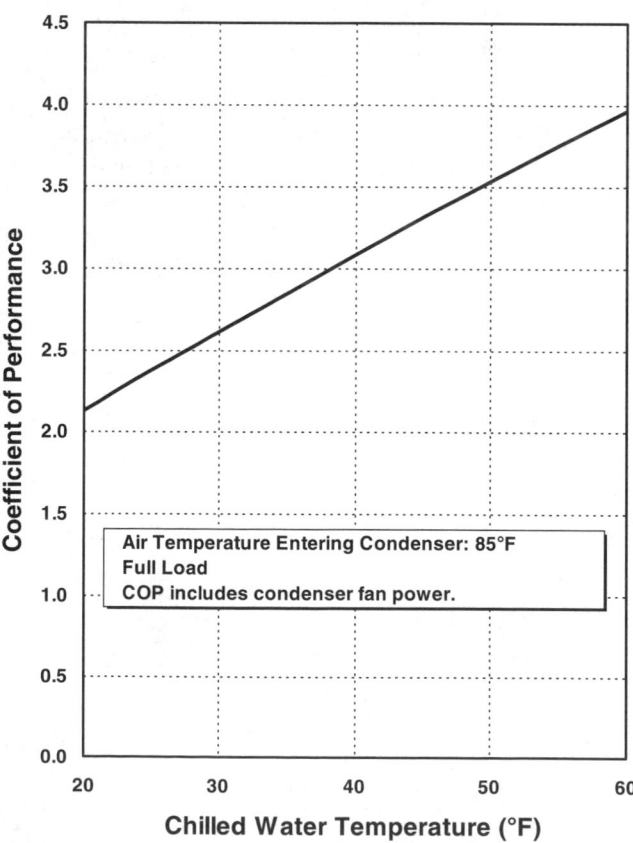

Fig. 1 Improve chiller COP dramatically by raising the chilled water temperature This curve is for a typical centrifugal chiller. Similar improvement is available with other types of chillers.

MEASURE **2.2.1.1 Reset chilled water temperature manually.**

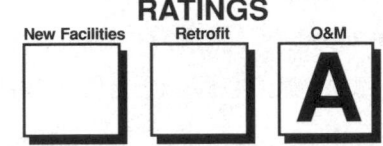

This is the first of two subsidiary Measures that optimize the chilled water temperature. The simplest and cheapest method is to set it manually at the chiller control panel. See Figure 1.

The practical limitation with manual resetting is that the cooling load changes continuously. Don't expect operators to keep the chilled water temperature fine-tuned on a continuing basis. Therefore, manual adjustment is practical primarily in response to seasonal changes in cooling load. This limits savings to perhaps less than half the potential maximum savings, which can be achieved only with automatic reset controls (recommended by subsidiary Measure 2.2.1.2, coming up next).

If your budget limits you to setting the chilled water temperature manually, give your chiller operators a table of chilled water temperature settings to follow. You have to develop this table from experience over a period of time. As conditions change, raise the chilled water temperature until cooling capacity becomes inadequate at some location in the facility, and note the conditions (the time of day, the outside air temperature, the

SUMMARY

Usually as simple as turning a knob. The main weakness of manual control is inability to track continuously changing cooling load.

SELECTION SCORECARD

Savings Potential .... $ $ $

Rate of Return .... % % % %

Reliability .... ✓ ✓

Ease of Initiation .... ☺ ☺

percentage of occupancy, etc.). Repeat this until you have covered all conditions.

For convenience, schedule chilled water temperature adjustments at the same time as condenser water temperature adjustments. See Measure 2.2.2.1 about this.

ECONOMICS

SAVINGS POTENTIAL: *4 to 10 percent of the average annual chiller energy consumption. The saving is partially reduced by increased fan power in variable-air-volume (VAV) air handling systems, and by increased pump power in variable-flow chilled water systems.*

COST: *Minimal.*

PAYBACK PERIOD: *Immediate.*

TRAPS & TRICKS

DILIGENCE: *This activity is easy to neglect. The optimum chilled water temperature takes some effort to find. Also, it is tempting to lower chilled water temperature in response to cooling complaints that have not been adequately diagnosed. Provide effective instructions, and schedule periodic checks of the procedure in your maintenance calendar.*

RUNNING TEMP CONTROL LEAVING CHILLED WATER		06-26-95 11:48 28.8 HOURS
CHW IN **55.1**	CHW OUT **44.1**	EVAP REF **40.7**
CDW IN **85.0**	CDW OUT **95.0**	COND REF **98.1**
OIL PRESS **21.8**	OIL TEMP **132.9**	MTR AMPS **93**
CCN	LOCAL RESET	MENU

Carrier Corporation

Fig. 1 Critical temperatures This display on the control panel of a modern chiller shows you the critical temperatures that determine the efficiency of the machine. However, you have to figure out the specific temperatures to set.

MEASURE **2.2.1.2 Install an automatic chilled water temperature controller.**

This is the second, and preferable, method of optimizing chilled water temperature. Wherever it is economical to do so, reset the chilled water temperature using an automatic control. This is a simple, common control function that can be accomplished with a few standard control components. You can also use your energy management computer system for this purpose, if you have one. Some manufacturers offer specialized chiller controllers that perform this function.

The challenge with automatic controls is designing them to maintain the most efficient relationship between the chilled water temperature and the cooling load. Failing to do this accurately wastes some of the savings potential of the Measure, or causes comfort problems. The cooling load relates to several conditions, including the outside air temperature, the humidity, the amount of sunshine, the number of occupants, and the heat emitted by equipment.

The most accurate way of responding to the cooling load is to use the signals from the space thermostats. When the signal from any one thermostat indicates that the airside unit is unable to satisfy the load in that space, the chilled water temperature is lowered incrementally. See Reference Note 14, Control Signal Polling, for methods of selecting the critical thermostat signal.

A cruder method of controlling chilled water is to sense the load at the chiller, for example, by sensing the difference between the supply and return chilled water temperatures. This method is less accurate than sensing space loads directly, but it is simple, cheap, and reliable.

With either method, do not let a single space or piece of equipment force the entire chiller system to operate at a much lower chilled water temperature. See Measure 2.2.1 about this.

ECONOMICS

SAVINGS POTENTIAL: *5 to 15 percent of annual chiller energy consumption. The saving is partially reduced by increased fan power in variable-air-volume (VAV) air handling systems, and by increased pump power in variable-flow chilled water systems.*

COST: *Several thousand dollars, typically, for a specialized reset controller. The cost of programming an energy management computer system to perform*

SUMMARY
The preferred way to control chilled water temperature.

SELECTION SCORECARD
Savings Potential .... $ $ $

Rate of Return, New Facilities % % % %

Rate of Return, Retrofit.......... % % % %

Reliability .... ✓ ✓

Ease of Retrofit .... ☺ ☺ ☺

this function is typically about the same, but may be much higher.

PAYBACK PERIOD: *Typically less than one year, with larger chiller systems. Up to several years, with small systems.*

TRAPS & TRICKS

DESIGN: *If the cooling system is complex, for example, if it includes variable-flow chilled water pumping or VAV air handling units, optimizing the chilled water temperature to achieve the lowest overall system energy consumption may be complex. Make sure that the control design is correct, keep the control as simple as possible, and make it easy to diagnose.*

OPERATION AND MAINTENANCE: *Select a control system that is compatible with the people who operate the chiller plant. For example, a common response to cooling complaints is to defeat any automatic chilled water controls and manually lower the chilled water temperature. Make sure that this sort of thing does not occur.*

DOCUMENTATION: *The purpose of the controls is not obvious. Put a clear description of the controls in the plant operating manual. (You do have a book of operating instructions for your plant, don't you?) Install an effective placard at the controls.*

MONITOR PERFORMANCE: *Failure of this kind of non-critical control is invisible. Schedule periodic checks in your maintenance calendar.*

MEASURE   2.2.2 Optimize the condensing temperature.

You can reduce the energy consumption of your chillers by keeping the condensing temperature as low as possible. Typically, compressor power drops by about 1.5 percent per degree Fahrenheit (about 3 percent per degree Celsius) of condensing temperature. Reference Note 32, Compression Cooling, explains the reasons.

Figure 1 shows how the coefficient of performance improves with lower condensing temperature, for a typical reciprocating chiller. However, not all cooling equipment benefits as much as the chiller represented in Figure 1. For reasons which we will discuss, the benefit provided by minimizing the condensing temperature varies widely among different types of cooling equipment.

Chiller manufacturers may be vague about condenser temperature requirements, or they may specify condenser temperatures without explaining them. As a result of confusion about chiller requirements, the condensing temperature in an operating chiller plant may be left higher than necessary. Correcting this is an important energy conservation opportunity.

There are two subsidiary Measures. Measure 2.2.2.1 recommends setting the condensing temperature manually, if the plant has no automatic controls that can perform this function. This is an operator activity that is quick and costs nothing. However, manual setting is unreliable, and it cannot maintain optimum settings at all times. Subsidiary Measure 2.2.2.2 avoids these limitations by recommending automatic controls to reset the condenser temperature.

Cautions

Unfortunately, optimizing the condenser temperature is not as simple as optimizing the chilled water temperature (which is covered by Measures 2.2.1 ff). It is complicated by these factors:

- In condensers that use fans for cooling, *lowering the condenser cooling temperature setting increases fan energy consumption*. Fans for condenser cooling typically consume 5 to 20 percent of system energy consumption. At some point, attempting to reduce the condensing temperature by increasing fan operation will cost more fan energy than is saved in compressor energy. Remember that the objective is to reduce the total energy consumption of the chiller system, not just the energy consumption of the compressor.

- Unlike ideal chillers, *real chillers do not always use less energy as the condensing temperature is reduced*. Reducing condenser temperature

improves the COP of all types of chillers at the high end of the temperature range. However, the energy consumption of some chillers may actually increase if the condensing temperature is reduced too much. This behavior occurs most commonly under low loads. Different types and models of chillers vary greatly in this regard.

- *Many chillers and chiller systems require a minimum condensing temperature* to operate properly or to avoid damage. There are a variety of reasons for this that are specific to individual chiller models.

Because of these complications, it is not appropriate to say simply, "lower the condenser temperature as much as possible." In some cases, the minimum condensing temperature is a fixed number that is easy to determine. In other cases, you have to calculate the optimum condenser temperature for each combination of load and ambient temperature.

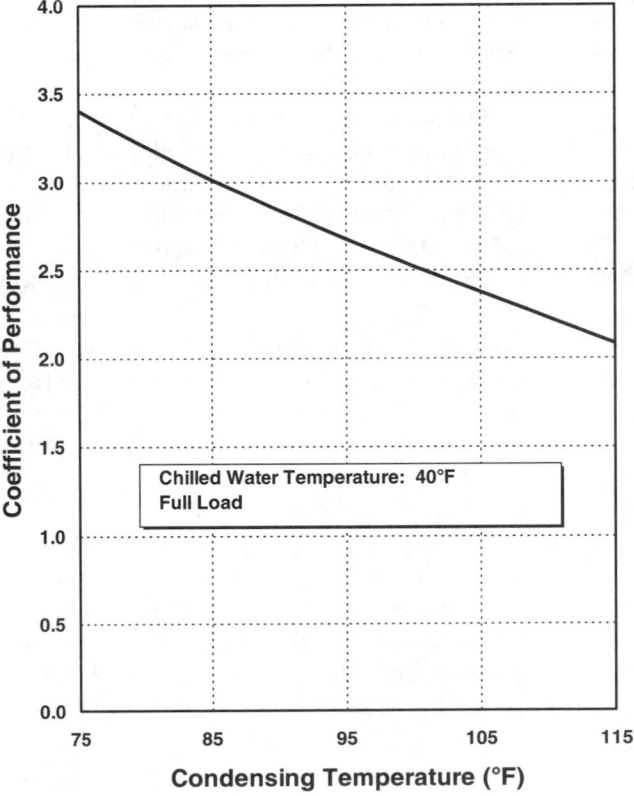

Fig. 1 Improve COP by lowering condensing temperature This curve is for a modern reciprocating chiller. Centrifugal chillers may respond less favorably, especially at low loads. Be aware that many kinds of cooling equipment have a minimum allowable condensing temperature.

How Energy Consumption Depends on Condensing Temperature

Two major energy users in a cooling system, the compressor and the heat rejection fans, are affected by the condensing temperature. The compressor power depends on the condensing temperature, which is determined by the operation of the fans. The fan power is determined by the cooling load and by the air temperature (for dry coolers) or by the wet-bulb temperature (for cooling towers and evaporative condensers).

■ Response of Compressor Power to Condenser Temperature

For most chillers under most load conditions, lowering the condensing temperature reduces the energy consumption of the compressor. However, there are exceptions in some chiller models, especially centrifugal chillers. These exceptions occur at lower condensing temperatures and at low loads.

■ Response of Condenser Temperature to Fan Power

The condensing temperature is limited by the temperature of the cooling medium. If the chiller uses a dry cooler, the fans cannot cool the condenser below the outside air dry-bulb temperature. If the chiller uses a cooling tower or evaporative condenser, the fans cannot cool the condenser below the outside air wet-bulb temperature.

Fans operate at full power if they attempt to maintain a condensing temperature lower than allowed by the ambient air temperature (dry-bulb or wet-bulb, as appropriate). Fan power drops rapidly if the fans are controlled to maintain a condensing temperature higher than ambient, but this reduces compressor COP in most cases.

Based on this behavior, a fairly good approximation in most cases is to set the fan control temperature from 5°F to 15°F higher than the ambient dry-bulb or wet-bulb temperature. The actual optimum temperature differential depends on the heat transfer efficiency of the cooling unit, and on the cooling load.

At lower cooling loads, you can set the cooling unit temperature somewhat closer to the ambient temperature, because there is less temperature loss in the condenser system. This is not a big factor. Typically, the cooling load affects the optimum condenser cooling temperature setting within a range of less than 10°F.

■ Combined Compressor and Fan Power

In view of the previous, you can usually calculate a condenser temperature setting that is close to optimum. At high loads, the curve of compressor power versus condenser temperature is steep, so keep the condenser temperature as low as possible, but slightly above the ambient (dry-bulb or wet-bulb) temperature. At high

chiller load, the optimum fan temperature setting is typically about 10°F above ambient.

This temperature differential does not change much at lower cooling loads, because two effects tend to cancel. On one hand, the curve of compressor input versus condensing temperature is less steep, making the fan energy curve more dominant, and hence tending to push the optimum condensing temperature higher to minimize fan energy. On the other hand, the fan power curves are shifted toward the left at lower loads. So, on balance, the same differential between ambient temperature and condensing temperature may be satisfactory for all load levels.

There is an obvious exception. If the compressor energy input starts to increase at lower condensing temperatures, do not reduce the condensing temperature below the point where the compressor energy requirement is lowest.

Why Cooling Machines Have Minimum Allowable Condensing Temperatures

If your cooling equipment has a minimum condenser temperature limit, be sure to find this out. Condenser temperatures that are too low may reduce capacity, reduce efficiency, or cause damage.

Most cooling machines do have such a limit. The manufacturer may express the limit in various ways. For example, the manufacturer may specify a fixed minimum condensing temperature, typically around 70°F or 85°F. Or, the manufacturer may require a minimum difference between the condensing and chilled water temperatures. In some cases, the minimum condensing temperature may depend on load, but this degree of detail is not usually specified. In case of doubt, pick up the telephone and call the manufacturer's engineering department.

Some of the reasons why a particular machine may have a minimum condensing temperature are:

• *to provide enough pressure to force refrigerant from the condenser back to the evaporator.* Both the evaporator and the condenser contain refrigerant vapor in equilibrium with liquid refrigerant. Therefore, the pressure in each vessel is proportional to its temperature. The pressure in the condenser must be higher than the pressure in the evaporator to force refrigerant through the metering device that separates the two. If the pressure difference is not great enough, the reduced refrigerant flow limits the cooling capacity of the unit.

The required temperature difference between condenser and evaporator is greatest in chillers that use refrigerant metering devices that have considerable resistance to flow, especially expansion valves. The superheat that is required in the operation of an expansion valve further increases the temperature differential.

(It is an article of faith among most engineers that the condensing temperature must always be higher than the evaporator temperature. In fact, there are ways of avoiding this requirement, but they require exotic design. For example, such designs may be used in heat pump loop systems, where the condensing temperature may be colder than the evaporator temperature. In principle, if the condenser is colder than the evaporator, the compressor could be designed to behave like a motor, providing energy recovery.)

• *with air-coil evaporators,* to keep the evaporator coil from frosting. If the evaporator is starved of refrigerant, the suction of the compressor lowers the evaporator pressure, which lowers the temperature inside the coil enough to allow frosting.

• *to keep lubricating oil that is mixed with the refrigerant from pooling in the condenser.* This would deprive the compressor of oil for lubrication. This limitation occurs only in chiller systems where the lubricant travels with the refrigerant.

• *with flooded evaporators,* to keep the refrigerant level from falling below the level of some tubes in the evaporator. This reduces heat transfer. This problem occurs in some flooded evaporators if the chiller lacks a separate refrigerant storage vessel of sufficient volume.

• *with some centrifugal chillers,* to avoid unstable gas flow through the impeller. The condensing temperature at which this occurs depends on the load.

• *in absorption chillers,* to avoid carrying salt from the generator to the condenser section of the distiller. A condenser temperature that is too low may increase the vapor velocity in the distiller to the point that the vapor carries salt over to the pure water.

• *in absorption chillers,* to avoid the risk of crystallizing the salt solution with sudden changes in load or cooling water temperatures. Low cooling water temperature increases this risk.

• *in systems that are operated during freezing weather,* to limit excessive ice formation in cooling towers and evaporative condensers. Keeping the condenser water warm melts frost.

These are all practical problems, rather than theoretical limitations. By the 1990's, some manufacturers were aggressively redesigning their chillers to avoid the need to maintain elevated condensing temperature.

How to Calculate the Optimum Condensing Temperatures

Unfortunately, there is nothing you can observe directly that tells you the optimum condenser temperature. You have to calculate the optimum temperature based on the chiller's characteristics, the cooling load, and the temperature or enthalpy of the outside air. You have to create a table or graph of the compressor power and the fan power for all operating conditions. Then, for each combination of compressor load and outside air temperature (dry-bulb or wet-bulb, as appropriate), you have to set the condenser cooling temperature that provides the lowest overall power.

You may not have all the data you need to develop these curves. If the chiller is fairly modern, you can get the full-load curve of compressor power versus condensing temperature from the chiller catalog. Getting part-load curves may require a call to the chiller manufacturer's engineering department. Get data for the cooling unit fans from the cooling unit manufacturer.

Account for All Methods of Limiting Condenser Cooling

Identify all the methods that your chiller system uses to limit condenser cooling. Make appropriate adjustments at each control. With virtually all condensers that depend on a fan for cooling, whether wet or dry, controlling fan operation is the primary method of controlling condensing temperature.

The system may use additional methods to maintain a minimum condensing temperature. These include:

• *dampers* on one or more fans of the cooling unit. This method is used most often with dry coolers.

• a *bypass valve* to divert cooling water around wet coolers

• a *throttling valve* in the cooling water line

• *flooding the refrigerant side* of the condenser to reduce surface area

• *shutting off individual sections* of condensers.

If the system uses one of the latter methods, set the fan control so that the fan does not operate in a futile attempt to reduce the cooling temperature below the temperature being set by the other method. For example, a cooling tower may serve a number of cooling units that maintain a minimum condensing temperature of 85°F by using thermostatic throttling valves in their cooling water lines. If the cooling tower serves only these units, control the cooling tower fan so that it does not attempt to reduce the cooling water temperature below 85°F.

Most of the latter methods increase fan or pump energy. Do not use any of them as a primary method of controlling condenser temperature. Generally, they should operate only when needed to maintain a minimum condenser temperature during cold weather, when turning off the fan is not adequate.

Improve Condensing Temperature Control Accuracy

If the method of controlling condenser temperature allows the temperature to swing over wide limits, the average temperature has to be higher. The method of controlling the fan air flow determines how accurately you control the condenser water temperature. For example, a single-speed fan is the worst case, because large swings in cooling temperature occur as the fan turns on and off.

You can achieve more efficient control of condensing temperature with multi-speed fan control, or multiple fans, or some combination of these. Variable-speed drives provide the greatest efficiency. See Subsection 2.4 for fan efficiency improvements.

MEASURE 2.2.2.1 Adjust the condenser temperature manually.

RATINGS		
New Facilities	Retrofit	O&M
		A

This is the first of two subsidiary Measures that offer ways of minimizing the condenser temperature. The cheap way is to adjust the condensing temperature manually. This requires virtually no effort. Just turn a screw in the thermostat that controls the condenser temperature. Or, if you have electronic controls, tap a few keys on the keyboard.

But, you have to invest in some brain work beforehand. Calculating the optimum condensing temperatures can be complicated. Unlike setting evaporator temperatures, where you can optimize by trial-and-error, you have to calculate the optimum condenser temperatures from chiller system characteristics that you measure or acquire from manufacturers. See Measure 2.2.2 for the procedures.

Once you develop the curves explained in Measure 2.2.2, you face the administrative problem of getting the plant operators to use them on a regular basis. The cooling load and the outside air conditions change from hour to hour, so it may be necessary to adjust the condenser temperature setting several times per day. This is most likely to be practical if the facility has an operator continuously on duty who can perform this task as a regular assignment.

How to Measure the Wet-Bulb Temperature

If your system uses a wet cooler, control the temperature settings of the condenser cooling equipment on the basis of the wet-bulb temperature of the outside air. The cheapest and most reliable way to measure wet-bulb temperature is with a sling psychrometer,

SUMMARY

Changing condenser water temperature is usually as simple as turning an adjustment screw. You may have to do a good deal of analysis to determine the optimum settings. Be careful to avoid equipment damage.

SELECTION SCORECARD

Savings Potential .... $ $ $

Rate of Return .... % % % %

Reliability .... ✓ ✓

Ease of Initiation .... ☺ ☺

shown in Figure 1. The only disadvantage of this method is that you have to take the psychrometer outside and whirl it whenever you want to take a reading.

An alternative is to install sensors to indicate wet-bulb temperature directly. To do this, you need a temperature sensor, a humidity sensor, and appropriate electronics to convert the readings to wet-bulb temperature. You can get this as a package. If you take this approach, be careful to install both sensors properly. See Measure 1.1.1.2 for installation tips.

How to Determine the Chiller Load

If your condenser temperature settings are supposed to change with the chiller load, the chiller operator has

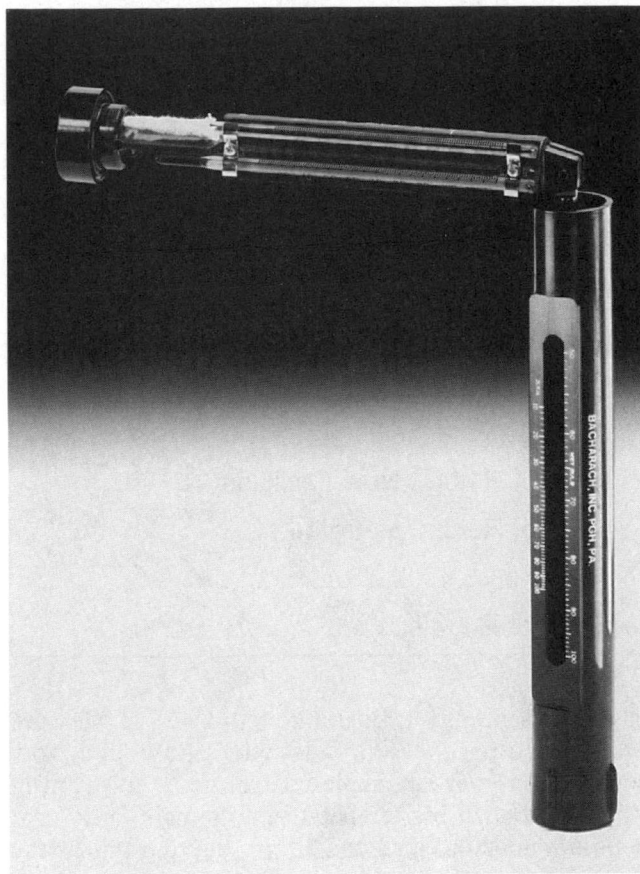

Bacharach, Inc.

Fig. 1 Sling psychrometer This consists of a thermometer with a wet wick wrapped around the bulb, mounted on a handle. You whirl it to cool the bulb by evaporation. The reading that results is the "wet-bulb" temperature.

to be able to determine the chiller load. See Measure 2.1.1 for methods you can use.

Coordinate with Adjusting the Evaporator Temperature

Both condenser temperature and evaporator temperature are affected by chiller load and weather conditions. If you set both manually, reset both on the same schedule. See Measures 2.2.1 ff about setting evaporator temperature.

ECONOMICS

SAVINGS POTENTIAL: *3 to 8 percent of the average annual chiller system energy consumption.*

COST: *Minimal.*

PAYBACK PERIOD: *Immediate.*

TRAPS & TRICKS

USING THIS METHOD: *Setting the condenser water temperature manually is probably too esoteric for most operating personnel. It is also a chore. Do it as a stopgap until you install the automatic condenser temperature control recommended by Measure 2.2.2.*

PREPARATION: *You have to be a scientist for a while to figure out the optimum condenser water temperature for each load level.*

TRAINING AND COMMUNICATION: *The big risk is neglecting the procedure. Provide clear instructions. Install an effective placard on the thermostat. Schedule periodic checks of the procedure in your maintenance calendar.*

MEASURE 2.2.2.2 Install automatic condenser temperature reset controls.

This is the more efficient of two subsidiary Measures that offer methods of optimizing condensing temperature. The manual method of control that is covered by Measure 2.2.2.1 is tedious if the optimum condensing temperature may change much with changes in the chiller load and outdoor temperature and humidity. Therefore, it is inaccurate and unreliable in these circumstances. To avoid these limitations, install automatic controls to optimize the condenser water temperature.

SUMMARY

The only way to reliably maintain accurate control of condenser water temperature. May or may not be complicated, depending on the characteristics of the chiller system.

SELECTION SCORECARD

Savings Potential	**$ $** $
Rate of Return, New Facilities	**% % % %**
Rate of Return, Retrofit	**% % %** %
Reliability	✓ ✓
Ease of Retrofit	☺ ☺

Your Control Choices

If you are buying new chillers, you may now be able to specify a standard control that performs this function. Chiller manufacturers are starting to design their controls to exploit lower condensing temperatures. (More fundamentally, be sure to select chillers that are designed to operate efficiently at low condensing temperatures.)

If a customized control is available for your specific model of chiller, this avoids the analysis chore needed to do the control design yourself. This is most likely with packaged air-cooled chillers. With water-cooled chillers, you still have to determine the characteristics of the separate condenser heat rejection equipment.

Some systems need relatively simple controls, while others may need controls that are more complex. The simplest case is maintaining a constant differential between the condensing temperature and the outside air temperature (either dry-bulb or wet-bulb). You can make a control to do this by using a signal offset relay and a few other standard control components.

For more complex control characteristics, consider using your energy management control system, if you have one. But, first review Reference Note 13, Energy Management Control Systems.

If you know how to use a programmable logic controller (PLC), you can create virtually any chiller control sequence. If you use a PLC, you have to know how to interface it to the chiller's capacity controls.

If you are going to install a chiller sequencing controller, as recommended by Measure 2.1.1.1, try to find a unit that can control condensing temperature as well. Such units may or may not be on the market by the time you read this.

How To Define Your Own Control Schedule

If a customized control is not available for your particular chiller system, you need to make the analysis of condenser water temperature explained in Measure 2.2.2. Doing the analysis will show you how complicated your automatic control needs to be. You want the condensing temperature to remain fairly near optimum under all load conditions, but don't make the control excessively complicated.

The simplest case is a control that simply keeps the condenser temperature at a fixed differential about the ambient (dry-bulb or wet-bulb) temperature. This is described in Measure 2.2.2. This method of control yields good efficiency in many chiller installations. It is likely to be satisfactory if the compressor power always declines with a reduction of condenser temperature. All chillers behave this way during warm weather, but some may not behave this way during cooler weather.

If the chiller has a minimum condenser temperature, you can accommodate this with a supplemental control that overrides the normal control when the minimum condensing temperature is reached.

Optimizing the condenser temperature becomes more complex if the compressor power increases with a drop in condenser temperature. This behavior occurs only with certain models of chillers, and only when they operate during colder weather. A custom control to deal with such characteristics is much more complex than a simple differential temperature control. If the system operates only for a limited amount of time at low loads and low temperatures, forget about trying to design a control that follows the characteristics of the chiller in this range. Instead, simply set a limit on the minimum condensing temperature at this point.

ECONOMICS

SAVINGS POTENTIAL: *5 to 15 percent of the average annual chiller energy consumption.*

COST: *Several thousand dollars, typically. The cost can be much higher in more complex chiller systems.*

PAYBACK PERIOD: *One year, to many years.*

TRAPS & TRICKS

KEEP IT SIMPLE: *If control may be complex, try to find a dedicated controller that can handle the job with sufficient accuracy. Otherwise, you may need to find exceptionally good control skills, either for designing the controls or for programming your energy management computer. Don't create a complex control monster that the operating staff cannot understand or maintain properly.*

SELECTING THE EQUIPMENT OR SOFTWARE: *Study the details of any controller that you consider buying, to make sure that the equipment or software can do what you want.*

EXPLAIN IT: *Explain the operation of the control clearly in the plant operating manual. If you use specialized hardware, identify it with an effective placard.*

MONITOR PERFORMANCE: *The control may fail without anyone ever noticing until there is equipment damage or loss of capacity. Schedule periodic checks of the control in your maintenance calendar, including the calibration of temperature and humidity sensors.*

Condenser and Evaporator Heat Transfer Efficiency

The evaporator, the condenser, and the condenser heat rejection equipment lose heat transfer efficiency as a result of fouling. Fouling increases chiller energy consumption and reduces capacity. (The reasons are explained in Reference Note 32, Compression Cooling.) The Measures in this Subsection maintain the effectiveness of heat transfer primarily by improved maintenance to minimize fouling. The last Measure improves the heat transfer efficiency of the evaporators in water chillers by installing turbulators, a technique that is old but still controversial.

INDEX OF MEASURES

2.3.1 In systems with open-loop cooling towers, clean condenser tube watersides regularly.

    2.3.1.1 Install automatic condenser tube cleaners.

2.3.2 With water chillers, clean evaporator tube watersides at appropriate intervals.

2.3.3 With wet condenser cooling systems, test and treat cooling water on a continuing basis.

    2.3.3.1 Hire a qualified consultant and contractor to perform water treatment.

    2.3.3.2 Install and maintain automatic chemical feeders.

2.3.4 In wet cooling systems, adjust the bleed rate to maintain proper water conditions with minimum water consumption.

    2.3.4.1 Install and maintain an automatic bleed control.

2.3.5 In chilled water systems, install turbulators in the evaporator tubes.

RELATED MEASURES

- Subsection 2.4, for Measures that improve the efficiency of heat rejection equipment

MEASURE 2.3.1 In systems with open-loop cooling towers, clean condenser tube watersides regularly.

Cooling towers are effective traps for dirt, and they are nutritious breeding sites for fouling organisms. Therefore, condensers that are cooled by open-loop cooling towers are susceptible to serious fouling. Under adverse conditions, fouling accumulates rapidly. Condenser fouling increases chiller energy consumption by raising the condensing temperature. Heavily fouled condensers can increase energy consumption and reduce capacity by more than 10%.

Clean condenser tubes regularly, before they can accumulate significant fouling. Use approved methods. You can remove moderate fouling with specialized tube cleaning brushes. Figure 1 shows a typical cleaning tool.

Severe fouling, or fouling of inaccessible heat transfer surfaces, may require cleaning using strong chemical solutions. Cleaning condensers is laborious, and it may hasten the development of leaks and tube cracking. Make sure that the persons who do it are well trained.

You can reduce or eliminate the need for manual tube cleaning by installing automatic tube cleaning equipment inside the condenser. This is covered by subsidiary Measure 2.3.1.1.

Goodway Tools Corporation

Fig. 1 Condenser tube cleaning The tool used here is a rotating brush at the end of a flexible drive cable. This is not brain surgery, but it does require some training and care.

SUMMARY

Routine maintenance needed to minimize the condensing temperature.

SELECTION SCORECARD

Savings Potential .... $ $ $

Rate of Return .... % % %

Reliability .... ✓ ✓

Ease of Initiation .... ☺ ☺ ☺

When to Clean Condensers

Monitor the need for condenser cleaning by watching these temperature differences in the condenser:

- the temperature difference between liquid refrigerant and the water discharged to the cooling tower should be low. This indicates effective transfer of heat from the refrigerant to the cooling water.

- the difference between the condenser water inlet and outlet temperatures should be large. This indicates that the condenser water is absorbing heat effectively.

These temperature differences are only a few degrees, so you need well calibrated thermometers to measure them with enough accuracy to be useful.

The temperature differences change with chiller load, so develop a chart that shows what they should be at each level of chiller output. If your chillers have modulating output, use one of the methods described in Measure 2.1.1 to determine the chiller output.

Take baseline readings immediately after you clean the condenser, with the chiller operating properly in all respects, including proper refrigerant charge. Clean the condenser again when the temperature differentials change by a few degrees.

Inspect the condenser visually whenever you open it up. Relate what you see to the temperature differences that you record while the chiller is operating. Over the long term, this will help you see useful patterns in your cleaning methods and water treatment. Experience should indicate the approximate interval between cleanings, although this interval may change significantly with changes in weather and other environmental factors.

Prevention First

Effective condenser water treatment should be your primary approach to keeping condensers clean. Effective water treatment can prevent biological fouling entirely, and it can greatly retard the deposit of fouling from other causes. Chemical treatment is covered in Measures 2.3.3 ff, below.

ECONOMICS

SAVINGS POTENTIAL: 2 to 10 percent of chiller energy consumption.

COST: The cost of typical rotating brush cleaning equipment for condenser and evaporator tubes ranges from $1,500 to $3,000, depending on size and features. An average chiller can be cleaned in one day.

PAYBACK PERIOD: Short. Over the long term, the savings from condenser cleaning greatly exceed the cost of doing it.

TRAPS & TRICKS

SKILLS: Wire brushing is not difficult to learn, but requires care. Chemical cleaning usually is done by specialists, and involves a significant risk of damage if done wrong.

MONITORING: In your chiller operating log, make sure to have columns for water inlet temperature, water outlet temperature, refrigerant gas temperature, and chiller load. Review these temperatures periodically and compare them to your baseline figures.

MEASURE 2.3.1.1 Install automatic condenser tube cleaners.

RATINGS

New Facilities	Retrofit	O&M
C	**C**	

An alternative to the manual condenser cleaning recommended by Measure 2.3.1 is installing automatic condenser tube cleaning equipment. In principle, continuous automatic cleaning is better than manual cleaning because the tube surfaces are always kept as clean as possible.

One such device has brushes in each tube that shuttle back and forth within the condenser tubes while the chiller is operating. The brushes have plastic bristles. To make the brushes shuttle, the flow in the condenser tubes is reversed by using a reversing valve in the condenser water line. The brushes are cycled several times per day by an automatic controller.

When the brushes are not operating, they are held in place outside the condenser tubes by a retainer basket. The idle brushes and their retainer add a flow resistance of about one PSI, or two feet of water column. This is a small fraction of the flow resistance of the tubes themselves and the system as a whole. It amounts to about 0.0015 pump horsepower per ton of cooling, in compression chillers. In retrofit applications, the additional resistance may not be noticeable. In fact, it may save a tiny amount of power by throttling excess pump capacity. (See Measure 10.2.2 about this.)

Because condenser pipe is large, the reversing valve is expensive, typically accounting for about three fourths of the cost of the installation. The brushes themselves are inexpensive, although there are many of them. The brushes typically last from one to five years, depending

SUMMARY

A way to clean condenser tubes continuously and to save some labor. Retrofit requires major work on the condenser water piping. The equipment is still uncommon, leaving doubts about effectiveness and reliability.

SELECTION SCORECARD

Savings Potential	**$ $** $
Rate of Return, New Facilities	**% %** %
Rate of Return, Retrofit.........	**% %** %
Reliability	✓ ✓ ✓
Ease of Retrofit	☺ ☺

on water conditions, and the retainer basket is expected to last for the life of the chiller.

Some condenser tubes have surface features to enhance turbulence and heat transfer, such as spiral grooves. Automatic tube brushes can accommodate some of these designs, although perhaps not all. Investigate this issue before installing an automatic tube cleaner in a chiller with condenser tubes that are not smooth cylinders.

No independent analysis of the long-term effectiveness, reliability, and possible operating problems of automatic tube cleaners has been found.

There is some concern that automatic tube cleaners may damage or wear out condenser tubes. However, plastic bristles seem unlikely to cause either corrosion or abrasion problems.

Needs Its Own Maintenance

If an automatic tube cleaner works properly, it may eliminate the need for all or most manual tube cleaning. However, it does not eliminate all labor, because it requires its own specialized maintenance. The reversing valve must be lubricated periodically, and the brushes must be replaced at intervals ranging from one year to five years. Replacing the brushes requires the condenser water box to be opened at one end. Also, an automatic tube cleaner does not eliminate the need for chemical water treatment or strainer maintenance.

ECONOMICS

SAVINGS POTENTIAL: *3 to 12 percent of chiller energy consumption. In new installations, the resistance of the cleaning equipment adds about 0.002 kilowatts of pump power per ton of cooling capacity, offsetting a small fraction of the savings. There is also a net saving in chiller maintenance cost, averaging from several hundred dollars to several thousand dollars per year.*

COST: *Typical prices of complete systems are about $30 per ton for a 1,000-ton chiller, about $50 per ton for a 300-ton chiller, and $100 per ton for a 100-ton chiller. These prices include all equipment, and installation of the brushes inside the evaporator. They do not include installation of the reversing valve and other modifications to the condenser water pipe. The brushes alone, needed for occasional replacement, cost about $2 apiece, one being needed for each condenser tube.*

PAYBACK PERIOD: *5 to 15 years, typically. Depends on energy costs, chiller size, water conditions, operating schedule, etc.*

TRAPS & TRICKS

SELECTING THE METHOD: *Automatic tube cleaners are not well proven, so you are pioneering. Talk to others who have used the particular model of equipment for several years in chillers similar to yours.*

OPERATION AND MAINTENANCE: *Although the equipment operates automatically, train the plant operators to monitor and maintain it. Insert a clear description of the equipment in the plant operating manual.*

MEASURE **2.3.2 With water chillers, clean evaporator tube watersides at appropriate intervals.**

Fouling of the waterside surfaces of the evaporator increases energy consumption by forcing the evaporator to operate at a lower temperature to provide a given cooling temperature.

All new chilled water systems need to be cleaned thoroughly to remove dirt, preservatives, and sealant materials left over from construction. Small particles are collected by the system strainers, but some of the material dissolves and deposits on the system surfaces. A new system may need to be cleaned several times, at successively longer intervals.

Chilled water systems are closed, and under normal conditions they accumulate fouling very slowly. The high velocity of the water in the tubes tends to keep the tube surfaces clean, except for a very thin film. In an older system, the appropriate interval for cleaning the chilled water system may be several years or longer. Fouling is usually not serious, so you can schedule the cleaning for convenience, for example, when the condenser watersides are being cleaned.

The evaporator is the coldest part of the chilled water system. The solubility of most solids declines as the water temperature falls, so any solids that are dissolved to the point of saturation tend to precipitate out in the evaporator. However, this is not a problem if you keep the dissolved solid content of the chilled water well below saturation.

Cleaning Methods

Evaporator watersides are usually cleaned by brushing, using the same equipment used for cleaning condenser watersides. Evaporator fouling is less serious than condenser fouling, and it is usually easier to remove. Still, be careful to avoid tube damage.

Unfortunately, if you find significant fouling inside the chiller, fouling is likely to be a worse problem throughout the chilled water system, where the same thickness of fouling can clog or greatly reduce the flow of water through the narrow passages of air coils. In that case, you may have to clean the entire chilled water system using an appropriate chemical method.

Prevention is Better Than Repeated Cleaning

After you clean the chilled water system thoroughly, it should not be necessary to clean it again for a long time. Use appropriate chemical additives to prevent

SUMMARY

Especially important for new systems. Additional cleaning may be needed occasionally, even though the system is closed.

SELECTION SCORECARD

Savings Potential .... $ $

Rate of Return .... % % %

Reliability .... ✓ ✓

Ease of Initiation .... ☺ ☺

fouling in normal operation. These stabilize the fouling at a low "fouling factor." If your system needs an antifreeze solution in the chilled water, this solution may include an appropriate fouling inhibitor. Check to be sure.

Systems with "Strainer Cycles"

If a chiller system has a "strainer cycle," the chilled water circuit may need periodic cleaning. The strainer cycle, described in Measure 2.9.3, admits filtered cooling tower water directly into the chilled water circuit.

ECONOMICS

SAVINGS POTENTIAL: 0.5 to 3 percent of chiller energy consumption.

COST: The cost of typical rotating brush cleaning equipment for condenser and evaporator tubes ranges from $1,500 to $3,000, depending on size and features. An average evaporator can be cleaned in one day.

PAYBACK PERIOD: Varies. The long-term saving substantially exceeds the cost.

TRAPS & TRICKS

SCHEDULING: Since this is something that needs to be done only at long intervals, it is likely to be forgotten. Make sure that your maintenance schedule can accommodate long-term items such as this.

SKILLS: Evaporator tubes are somewhat delicate. Make sure that cleaning is thorough, while taking care to avoid tube damage or loosening of tube joints.

MEASURE 2.3.3 With wet condenser cooling systems, test and treat cooling water on a continuing basis.

RATINGS

New Facilities   Retrofit   O&M

A

Wet condenser cooling systems tend to grow or accumulate matter that creates fouling at a rapid rate. You can eliminate or minimize fouling by using a combination of three techniques:

- *treat the water.* Add appropriate chemicals to the water to deal with the different types of contaminants that may grow or accumulate in the system. Water treatment additives are sold commercially for this purpose, as well as dispensing equipment and water treatment services. Additives should be selected by a person trained in water treatment, based on the specific water conditions that exist at the site.

- *bleed a fraction of the water from the system.* The concentration of contaminants dissolved in the water increases with time. The only way to keep the dissolved material within acceptable limits is to bleed water from the system at a controlled rate. See Measures 2.3.4 ff for details.

- *physically remove solids from the system.* Normal water treatment cannot be effective in any part of the system where solids have accumulated. The system has strainers. Clean them out periodically. Solids may also accumulate in the cooling tower sump. Scoop these out.

There are two subsidiary Measures. The first recommends hiring specialists to help with these functions. The second recommends installing automatic water treatment equipment.

Causes of Fouling and General Solutions

There are several causes of cooling water fouling, each of which requires individual corrective measures. The following are the main classes of fouling.

■ Microorganisms

Microorganisms are usually the main cause of fouling in cooling towers and evaporative condensers. The highly aerated cooling water is a rich medium for growth of microorganisms, including algae, bacteria, mold, and fungi, which it traps from the atmosphere. These grow quickly in wet cooling units if conditions permit. They become troublesome when they form matted colonies that are capable of inhibiting heat transfer and clogging parts of the system.

Microorganisms are killed with biocides. The available selection is being progressively restricted by concerns about the environmental effects of cooling tower effluent. Figure 1 shows a simple iodine injection system to kill microorganisms.

SUMMARY

Important maintenance needed by virtually all wet cooling systems. Minimizes condenser losses, minimizes the need for severe condenser cleaning methods, and preserves equipment. Requires specialized knowledge.

SELECTION SCORECARD

Savings Potential .... $ $ $

Rate of Return .... % % % %

Reliability .... ✓ ✓

Ease of Initiation .... ☺ ☺

Certain diseases, of which Legionnaires' Disease is the most notorious, have been linked to the growth of organisms in cooling towers. The danger may exist even if growth of visible colonies has not occurred. The biocides normally used in water treatment are believed to eliminate the hazard in most cases.

Algae and some other fouling organisms require sunlight for growth. For this reason, cooling units that have enclosed sides, such as evaporative condensers and upflow cooling towers, may have less trouble with biological fouling than cooling units that are open to sunlight. In principle, you could add sun shielding to the latter types, but this involves serious practical problems, such as wind forces and air recirculation.

■ Airborne Dirt and Debris

Dirt is washed out of the atmosphere by the action of wet cooling units, and the dirt is carried by the water to settle throughout the condenser system.

The treatment is to use detergent or specialized polymeric agents to keep the dirt in suspension until it is bled from the system or is trapped by the strainers.

Determine the concentration of chemical treatment needed to control dirt by checking the rate of dirt accumulation visually. Keep monitoring this, because the amount of dirt in the environment may vary considerably.

■ Waterborne Minerals

The minerals that cause trouble most commonly are compounds of calcium (dissolved from limestone) and magnesium. As water evaporates from the cooling unit, the concentration of minerals in the system rises. If the concentration is allowed to rise excessively, the minerals deposit on the surfaces of the system in the form of scale.

The content of minerals capable of depositing scale is expressed as the "hardness" of the water.

The primary method of dealing with dissolved minerals is to bleed water from the cooling unit and replace it with fresh water at a rate that keeps the mineral concentration low enough to prevent scaling. See Measures 2.3.4 ff, below, for details of bleed control.

Chemical water softening may also be advisable in locations where the water supply is very hard. In such cases, the usual treatment is ion exchange with salts, the same type of treatment used to reduce the hardness of domestic water.

Get Help from Specialists

Unless you are an expert in water treatment, hire a water treatment consultant to analyze the cause of fouling in your system. The consultant uses visual inspection and chemical tests to identify each source of fouling, for which he prescribes appropriate corrective measures.

You have to decide whether to treat the system yourself or to hire a contractor to perform this service. The answer depends on the capabilities of the facility staff. Treatment requires periodic inspection and application of chemicals, at intervals ranging from several hours to several days. Periodic functions tend to be neglected, so water treatment by the staff may not be sufficiently reliable. In many cases, it is best to hire a contractor to perform water treatment, as recommended by Measure 2.3.3.1, the first subsidiary Measure.

Automatic water treatment equipment reduces the labor required to treat condenser water, but it does not eliminate the problem of neglect. See Measure 2.3.3.2, the second subsidiary Measure.

Coordinate with Other Water Treatment

Boilers also need a carefully controlled water treatment program, as recommended by Measures 1.8.1 ff. These Measures also provide more guidance on how to set up a water treatment program.

Hydronic systems may need some water treatment, although they are less demanding. This is described in Measure 2.3.2. You may also have a need to treat your domestic water.

It saves effort to find a specialist who can handle all these aspects of water treatment. But, don't compromise to do this. You may need different specialists to deal with the different water treatment requirements.

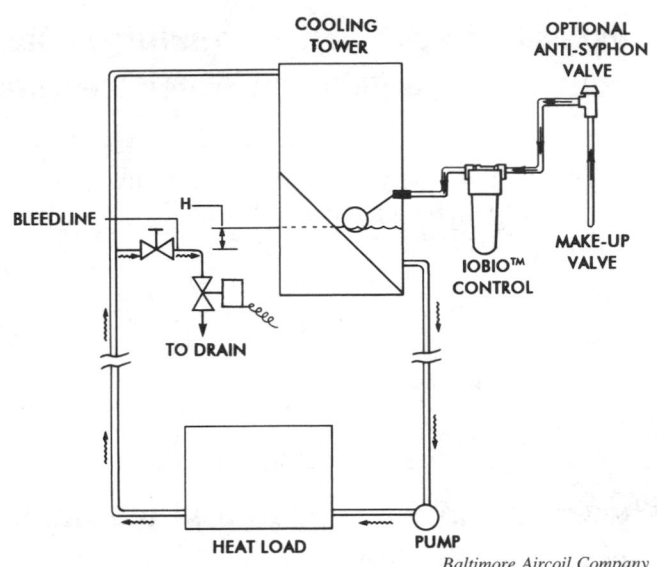

Baltimore Aircoil Company

Fig. 1 Iodine injection system to kill microorganisms in cooling tower water

ECONOMICS

SAVINGS POTENTIAL: *1 to 5 percent of chiller system operating cost, depending on the severity of water conditions. Water treatment reduces maintenance cost and lengthens equipment life.*

COST: *Several hundred to several thousand dollars per year, typically, depending on plant size and water conditions.*

PAYBACK PERIOD: *Short. Over the long term, the savings greatly exceed the cost.*

TRAPS & TRICKS

PREPARATION: *Water treatment is a specialty that requires considerable experience, especially to select the best methods for a particular system. Have your program set up by a qualified specialist. See Measure 2.3.3.1.*

TRAINING AND MANAGEMENT: *Water testing and water treatment are not difficult, but they may seem somewhat mysterious at first. Also, they are easy to neglect. If it is not practical to do water testing or treatment in-house, hire these services.*

MONITOR CONDITIONS: *At every opportunity, inspect the cooling tower and the condenser watersides to check the effectiveness of treatment and to look for adverse effects. Schedule inspections at intervals appropriate for your water conditions. Make adjustments to the water treatment program as indicated.*

MEASURE **2.3.3.1 Hire a qualified consultant and contractor to perform water treatment.**

RATINGS

New Facilities   Retrofit   O&M

                                  A

This is essentially the same as Measure 1.8.1.1, which deals with boiler water treatment. Refer to there for details.

MEASURE **2.3.3.2 Install and maintain automatic chemical feeders.**

RATINGS

New Facilities   Retrofit   O&M

B   **C**

This activity is similar to Measure 1.8.1.2, which recommends automatic chemical feeders for boiler water treatment. Refer to there for details.

Figure 1 shows a simple chemical feeder for water treatment chemicals.

In most cases, combine automatic control of chemical treatment with automatic control of cooling water bleed. Measure 2.3.4.1 covers automatic bleed control.

WESINC

Fig. 1 Automatic chemical feeder for cooling tower water   The plastic drum stores the chemicals. A pipe feeds the chemicals into the tower along with the makeup water. A solenoid valve and controller regulate the dose. The water meter reads the makeup water because water used by evaporation is deducted from the sewage bill.

MEASURE 2.3.4 With wet cooling systems, adjust the bleed rate to maintain proper water conditions with minimum water consumption.

A

SUMMARY

A careful adjustment minimizes the costs of water and water treatment.

SELECTION SCORECARD

Savings Potential .... $ $

Rate of Return .... % % % %

Reliability .... ✓ ✓

Ease of Initiation .... ☺ ☺

Water evaporates rapidly from cooling towers and evaporative condensers. This loss of water is compensated by adding makeup water from the local water supply. The makeup water contains dissolved minerals and a certain amount of particulate matter. In addition, the water spray captures significant amounts of atmospheric pollutants. If only enough water were added to the system to compensate for evaporation, the minerals and other matter would concentrate rapidly. This accumulation would soon deposit as fouling on the surfaces of the cooling unit, the chiller condenser, and the entire piping system.

To prevent this, you have to bleed water from the cooling system at a controlled rate. This carries away the contaminants that would cause fouling if they became too concentrated. Water is usually bled from a drain located in the sump. Bleed is necessary in all wet cooling systems, and it must continue as long as the system is operating.

Bleed has a significant cost. It is a significant loss of water, and it also carries away a portion of the chemicals that you add to control fouling.

Even if you hire a water treatment contractor to take care of chemical treatment, do not assume that the contractor will optimize the bleed rate. After all, he is in the business of selling chemicals, and his primary job is to prevent fouling, so his motivation leans toward setting the bleed rate too high. Therefore, deal with control of bleed as a separate cost saving issue.

You can set the bleed rate manually or automatically. Manual setting of bleed rate is always wasteful because the evaporation rate varies with the cooling load. A bleed rate that is set to maintain a safe concentration at high cooling load is wasteful under low cooling load. Figure 1 shows a cooling tower that depends on manual bleed.

The most efficient way to control bleed is to install a bleed valve that is controlled by a concentration sensor. This technique is covered by the subsidiary Measure 2.3.4.1.

How to Calculate the Bleed Rate

The appropriate bleed rate is determined by the evaporation rate and by the concentration of impurities in the makeup water. The evaporation rate is proportional to the cooling load. If the impurity level is high in the incoming water, the dissolved material soon reaches a concentration in the cooling unit that causes fouling unless the material is flushed out.

Formula 1 shows how the bleed rate relates to these factors. This formula neglects the effect of drift (loss of unevaporated water spray) and the addition of dirt washed out of the air. These factors are usually minor.

In the language of the water treatment trade, the ratio of the maximum contaminant concentration in the cooling water to the contaminant concentration in the makeup water is called "cycles of concentration." Formula 2 expresses bleed rate in these terms.

For example, your water treatment specialist may determine that three cycles of concentration are allowable. From this, the formula tells us that the bleed rate should be set to half the evaporation rate.

The Cost of Bleed

Although water is usually cheap in terms of unit cost, the bleed requirement is continuous as long as the chiller operates. Also, if you need to use water treatment chemicals, bleeding may be responsible for most of the consumption of chemicals.

Here is an example to illustrate the magnitude of the water cost. Any cooling tower evaporates about 1.5 gallons per ton-hour of heat rejected, a figure determined solely by the latent heat of water. At a water cost of 0.1 cents per gallon, this translates into a cost of 0.15 cents per ton-hour, or about 0.15 cents per kilowatt-hour input to a chiller system of typical efficiency. If the cooling water is kept at two cycles of concentration, the bleed rate equals the evaporation rate, adding another 0.15 cents per kilowatt-hour.

The bleed rate can easily get out of control, in the direction of either too little bleed or too much. The plumbing involved in controlling the bleed of many cooling towers is primitive. Bleed may be controlled with a manual bib valve, i.e., a garden hose spigot, which cannot control flow accurately. It is not extraordinary to find water being bled at a rate ten times greater than necessary. Using the previous example, this would raise chiller system water cost to 1.5 cents per kilowatt-hour. A poorly controlled cooling tower bleed may account for a large fraction of total water cost of a facility.

FORMULA 1

$$\text{Bleed rate} = \text{Evaporation rate} \times \frac{\text{Concentration (M)}}{\text{Concentration (T)} - \text{Concentration (M)}}$$

where:

* Concentration (M) is the concentration of critical contaminants in the makeup water
* Concentration (T) is the maximum allowable contaminant concentration in the cooling tower
* bleed rate and evaporation rate are expressed in identical units, for example, gallons per hour, percentage of condenser circulation rate, etc.

FORMULA 2

$$\text{Bleed rate} = \frac{\text{Evaporation Rate}}{\text{Cycles of Concentration} - 1}$$

where:

* bleed rate and evaporation rate are expressed in consistent units, for example, gallons per hour, percentage of condenser circulation rate, etc.

Test the Makeup Water

The first step in setting the bleed rate is testing your water supply. In many geographic areas, the determining factor is the "hardness" of the water, which is a measure of its calcium carbonate (dissolved limestone) content. In other locations, other substances in the water supply determine the bleed rate. Vendors of water treatment supplies offer inexpensive chemical test kits for the common types of impurities.

Your water treatment specialist will tell you which contaminants to test for in your water supply. Repeat water testing at appropriate intervals, and adjust the bleed rate as necessary.

The other thing you need to know is the maximum allowable contaminant concentration. Again, your water treatment specialist will tell you this. Or, if you don't have one, you can get this information from reference publications, or perhaps from your chiller manufacturer.

The condition of your equipment cannot tell you whether you are bleeding enough, because visible fouling may occur even at an excessive bleed rate if you do not treat the water properly.

Use Continuous Bleed Rate

In most applications, use continuous bleed as long as the tower is operating. An amount of water equal to the total volume of the cooling system may evaporate within several hours of operation. Intermittent bleed is unsatisfactory unless you find a way to do it reliably on an hourly basis. Even better, use automatic bleed control, which is recommended by the subsidiary Measure.

WESINC

Fig. 1 Cooling tower with manual bleed The middle sized pipe is the tower drain and bleed line, which has a manual valve. This is an unreliable way of controlling bleed.

ECONOMICS

SAVINGS POTENTIAL: Up to 80 percent of the cost of water and water treatment chemicals used in cooling units. Also, reduction of equipment maintenance costs.

COST: At most, several hundred dollars per year for testing supplies. Labor costs should be minimal.

PAYBACK PERIOD: Short, except in small systems. The saving is much greater than the cost over the long term.

TRAPS & TRICKS

TRAINING: Write clear instructions for controlling the bleed rate, and train the operators how to follow them.

MONITOR PERFORMANCE: Controlling bleed rate will be forgotten unless it is done on a regular basis. Add a column to the chiller plant operating log for water test results. Check the logs, and follow up to ensure that the procedure is being followed.

MEASURE 2.3.4.1 Install and maintain an automatic bleed control.

The rate of evaporation from a cooling tower changes continually with changes in cooling load. Only automatic control of bleed can follow these changes and minimize waste of cooling water and water treatment chemicals. To control bleed automatically, install an automatic valve at the bleed connection. The valve is controlled by a sensor that measures the contaminant concentration.

The equipment and installation are simple. You can use automatic control with virtually any cooling tower or evaporative condenser. However, the blowdown control becomes another item of equipment that you have to maintain. It will work reliably if you inspect it regularly and maintain it as necessary.

Equipment and Installation

■ Type of Valve

Automatic bleed may operate continuously or intermittently, depending on the bleed rate. Intermittent bleed is more practical, except for the largest systems. Use a solenoid valve for intermittent control. They are simple and reliable. A modulating valve is more complex and more sensitive to the harsh environment of a cooling tower. A modulating valve operating at a low flow rate would require frequent maintenance because of valve seat erosion.

If you use a solenoid valve, add a control feature that keeps the valve from cycling frequently. One method is to control the valve through a cycle timer. The contaminant sensor triggers the cycle timer, which keeps the valve open for a fixed period of time, such as two minutes. This method reduces the contaminant concentration in small increments and allows the concentration to stabilize between bleed cycles.

■ Valve and Drain Location

The bleed valve can be installed anywhere in the bleed line. The bleed line is usually connected to a point in the sump. In most cases, the original drain connection of the bleed line is satisfactory.

As a safety feature, locate the bleed connection at a point in the sump where it cannot drain the sump completely if it sticks open.

Water is driven out of the bleed line by gravity. Make sure that there is enough difference in height between the water level in the sump and the discharge point of the bleed line. Otherwise, there may not be enough gravity head to force the bleed water out of the system at the rate needed.

SUMMARY

Provides more accurate control of bleed rate. Another thing to maintain.

SELECTION SCORECARD

Savings Potential .... $ $

Rate of Return, New Facilities % % % %

Rate of Return, Retrofit .... % % % %

Reliability .... ✓ ✓

Ease of Retrofit .... ☺ ☺ ☺

Do not install the bleed valve in a descending pipe. This would allow dirt to settle on top of the valve when it is closed.

Do not install a strainer anywhere in the bleed line. A strainer would eventually clog, preventing bleed.

■ Type of Sensor

The type of sensor you use to control the valve depends on the nature of the contaminant that controls the blowdown rate. Your water treatment specialist should select the type of sensor. For example, if the controlling factor is hardness, you may use a sensor that measures the electrical conductivity of the cooling tower water.

■ Control Setting

To minimize the cost of makeup water and chemicals, set the sensor to control the bleed rate so that the concentration of impurities remains just below the point where they cause trouble.

■ Safety Alarm

It is good practice to monitor the contaminant concentration of the water with a separate sensor that sounds an alarm if the concentration is too high or too low. A concentration that is too high indicates that bleed is not adequate. A concentration that is too low indicates that the bleed valve is probably stuck open.

Coordinate with Automatic Chemical Feeding

This method of control is similar to automatic feeding of water treatment chemicals, and their functions are closely related. To avoid duplication of effort, install both functions at the same time. See Measure 2.3.3.2 about automatic chemical feeders.

ECONOMICS

SAVINGS POTENTIAL: *30 to 60 percent reduction in water and additive costs, compared to manual bleed control.*

COST: *One thousand dollars to several thousand dollars.*

PAYBACK PERIOD: *Less than one year, to several years, compared to manual feeding.*

TRAPS & TRICKS

EQUIPMENT SELECTION AND MAINTENANCE: *Select the equipment for reliability and ease of maintenance. A bleed valve failure is likely to occur at an awkward time, and it needs to be corrected quickly.*

MONITORING: *Review the water test results from the chiller operating log periodically to make sure that the bleed rate is correct.*

EXPLAIN IT: *Describe the automatic bleed valve in the plant operating manual. Install a placard that identifies the valve and states what it does.*

MEASURE 2.3.5 In chilled water systems, install turbulators in the evaporator tubes.

RATINGS

New Facilities	Retrofit	O&M
	D	

In evaporators, heat transfer efficiency increases with surface area. Unfortunately, evaporator surface area is also a major factor in the cost of a chiller system. For this reason, some chillers lack sufficient surface area for good efficiency.

In a chilled water system, you may be able to improve the heat transfer efficiency on the water side of the evaporator by installing "turbulators." These are inserts placed inside the water tubes of tube-and-shell heat exchangers to increase the turbulence of the water flowing through the tubes. The increased turbulence at the tube surface improves heat transfer.

(The same principle applies to condenser tubes. However, turbulators do not appear to be practical in condensers, because they would trap the dirt that is present in the cooling water.)

Turbulators have many configurations, including helical wire coils and twisted metal strips. They usually extend the full length of the tube, or almost the full length. They must be easily removable for tube cleaning.

Where to Consider Turbulators

The main candidates for turbulators are chillers that have a substantial temperature drop across the evaporator. Call the manufacturer to find out the design temperature drop of the evaporator. If the manufacturer's catalog for the chiller is available, see whether the chiller is an economy model that has an evaporator at the small end of the size range.

As a last resort, measure the difference between the temperature of the chilled water leaving the evaporator and the temperature of the liquid refrigerant, with the

SUMMARY

An old technique for improving heat transfer that is still controversial. Performance is uncertain, and tube damage is possible. Inexpensive. Proceed cautiously.

SELECTION SCORECARD

Savings Potential .... **\$ \$** \$

Rate of Return .... **% %** %

Reliability .... ✓

Ease of Retrofit .... ☺ ☺

chiller under full load. In evaporators that use an expansion valve, do not use the suction gas temperature to judge this temperature differential, because the expansion valve acts to superheat the vapor. See Reference Note 32, Compression Cooling, for details.

The evaporator tubes used in some newer chillers are made with a texture that enhances heat transfer. Turbulators are not useful in these chillers.

Energy Saving Potential

The energy saving potential of turbulators is limited. Flow is inherently turbulent above a certain velocity, and chiller systems are designed to exceed this velocity. Therefore, be skeptical of the savings claims of turbulator vendors. In the absence of a formal efficiency testing program by a competent and objective agency, it is easy to fall prey to wishful thinking.

Potential Equipment Damage

Turbulators have been used for decades, and they are controversial. The record suggests that turbulators have been involved in tube failure, perhaps by several mechanisms:

- the increased turbulence causes mechanical erosion of the tubes by the fluid itself
- vibration of the turbulators against the tube surfaces, causing mechanical damage
- electrolytic erosion that may occur as a result of dissimilarity of metals between the turbulators and the tubes.

A systematic study of the nature and prevalence of these problems has not been found, so it is impossible to predict with certainty whether turbulators will cause trouble, or which type of turbulator is least likely to cause trouble. The best approach is to check with other people who have extended experience with the types of turbulators you are considering. Also, check with the chiller manufacturer, who may have received field reports about the long-term effects of turbulators.

Secondary Effects on System Efficiency and Capacity

Turbulators increase the resistance of the evaporator to chilled water flow. The resistance arises from the reduction of the tube cross section, the friction of the turbulators, and the turbulence they cause. Adding turbulators with no other adjustments to the chilled water system will reduce chilled water flow somewhat. This throttling effect will reduce the power consumption of the chilled water pumps by a small amount. If chilled water temperature reset is used to maximize chiller efficiency (see Measures 2.2.1 ff), the reduced flow rate will require a lower chilled water temperature to serve a given cooling load, and this reduces the chiller's efficiency.

These are small effects that are not likely to be significant. Even so, you can restore the original chilled water flow rate if the system was installed with flow regulating ("balancing") valves. Open these enough to compensate for the added resistance of the turbulators. The reduced energy loss in the balancing valves compensates for the increased energy loss in the evaporator. While you are at it, optimize the chilled water system flow. See Measures 10.2.1 and 10.2.2 for details.

If the system has variable-flow chilled water pumping (Measure 2.5.2), pump power will be increased somewhat.

If the system's chilled water pumping capacity is already marginal, the reduced flow will limit system cooling capacity. Look into this before making any changes. It is not economical to modify the pumps to overcome the resistance of turbulators.

ECONOMICS

SAVINGS POTENTIAL: *Up to 3 percent of chiller energy. The saving is higher in less efficient chillers. The overall saving may be reduced by higher pump power.*

COST: *Spiral turbulators typically cost from $10 to $20 per tube. A large chiller may have several hundred tubes.*

PAYBACK PERIOD: *Several years, typically.*

TRAPS & TRICKS

CHOICE OF METHOD: *Turbulators are something of a gamble. Their efficiency improvement and potential for creating problems are not well known. Check with other plant operators who have used the kinds of turbulators that you are considering.*

INSTALLATION: *Follow the instructions. The method of installation may affect the potential benefit and the potential for problems.*

EXPLAIN IT: *Put the installation instructions in the plant operating manual. The turbulators need to be removed and reinstalled whenever the tubes are cleaned.*

Heat Rejection Equipment

Every mechanical cooling system must get rid of the heat that it collects in the process of cooling spaces or processes. The heat is dumped into the environment, either into the atmosphere, or into a body of water, or into the soil of the earth. The cooling system needs special equipment for rejecting the heat. The Measures in this Subsection maintain or improve the efficiency of heat rejection equipment that rejects heat to the atmosphere.

The heat rejection equipment is often the condenser itself, in the form of a finned coil that is aided by a fan. Such condensers are called "air-cooled condensers."

The heat rejection of an air-cooled condenser can be enhanced by keeping it wet to take advantage of evaporative cooling. This variation is called an "evaporative condenser." Figure 1, on the next page, sketches an evaporative condenser.

In many cooling systems, especially larger ones, water is cooled by evaporation in a remote "cooling tower," and the water is pumped some distance to the condenser for cooling. Figure 2 sketches a simple cooling tower, and Figure 3 shows details of a different type.

The fans in these units are major consumers of energy, accounting for typically 5 to 20 percent of chiller system energy consumption. The least fan energy is used in systems that are aided by evaporation, and the most is used by equipment that is directly air-cooled. Heat rejection equipment also has a major effect on the efficiency of the chillers, through its effect on condensing temperature. This important aspect is covered in Subsection 2.2.

Heat can also be rejected through water coils that are immersed in bodies of water, such as rivers, lakes, and underground aquifers. Or, coils can be buried in soil. Reference Note 22, Low-Temperature Heat Sources & Heat Sinks, covers these methods of heat rejection.

INDEX OF MEASURES

2.4.1 Modulate fan output in heat rejection units to follow the cooling load.

    2.4.1.1 Install variable-frequency fan drives.

    2.4.1.2 Install variable-pitch propeller fans.

    2.4.1.3 Install dual or multi-speed motors.

2.4.2 In multiple-cell cooling units, sequence the fans efficiently.

2.4.3 Clean heat rejection units at appropriate intervals.

    2.4.3.1 Install and screen heat rejection units to minimize debris accumulation.

2.4.4 In gravity-flow cooling towers, ensure proper water distribution.

2.4.5 Keep heat rejection unit housings and fittings intact.

2.4.6 Avoid recirculation of air through the same or adjacent heat rejection units.

2.4.7 Install fan and pump motors having the highest economical efficiency.

RELATED MEASURES

- Measure 2.1.3, interlocking fan operation to chiller operation

- Measures 2.2.2 ff, adjusting condensing temperature to achieve the best compromise between energy consumption in the chiller and the heat rejection equipment

- Measures 2.3.3 ff, chemical treatment of cooling water

- Measures 2.3.4 ff, adjusting cooling water bleed rate

- Subsection 2.5, for reducing cooling water pump power

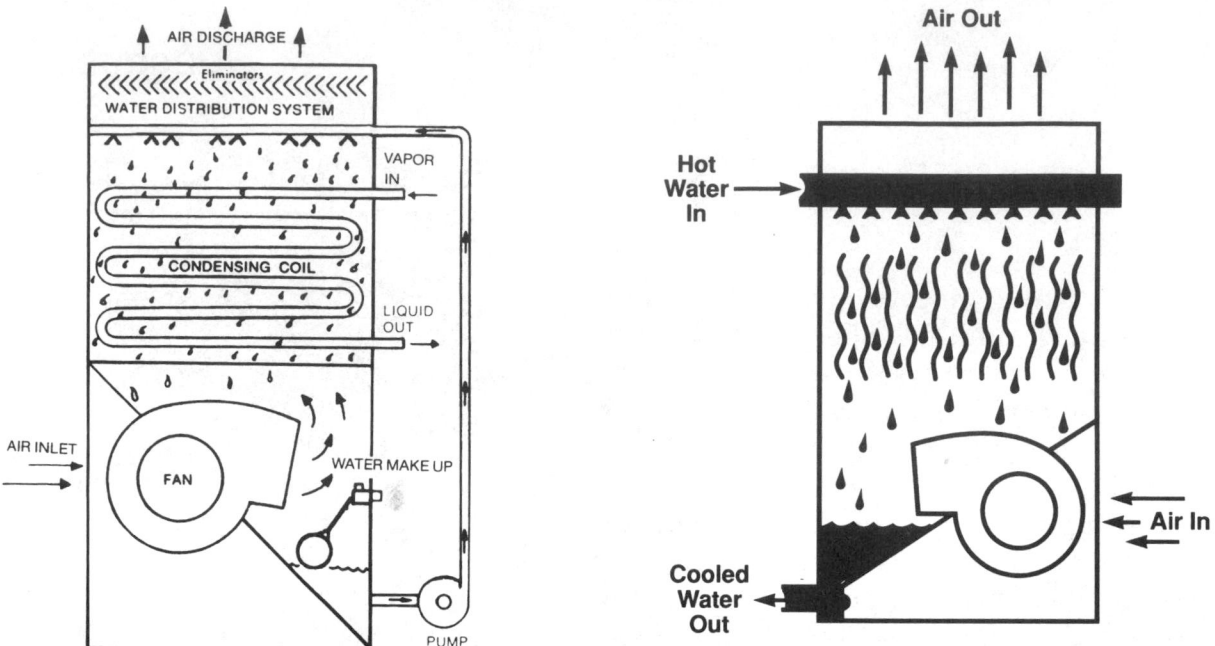

Fig. 1 Evaporative condenser Water is sprayed over the hot refrigerant condenser coil to cool it. The water is recirculated within the enclosure by a small pump. A fan blows air upward through the falling spray to stimulate evaporation and cooling.

Baltimore Aircoil Company

Fig. 2 Upflow cooling tower Water is sprayed into the top of the enclosure. A fan blows air upward to stimulate evaporation and cooling of the water. The water is pumped away to a condenser or other cooling application.

Baltimore Aircoil Company

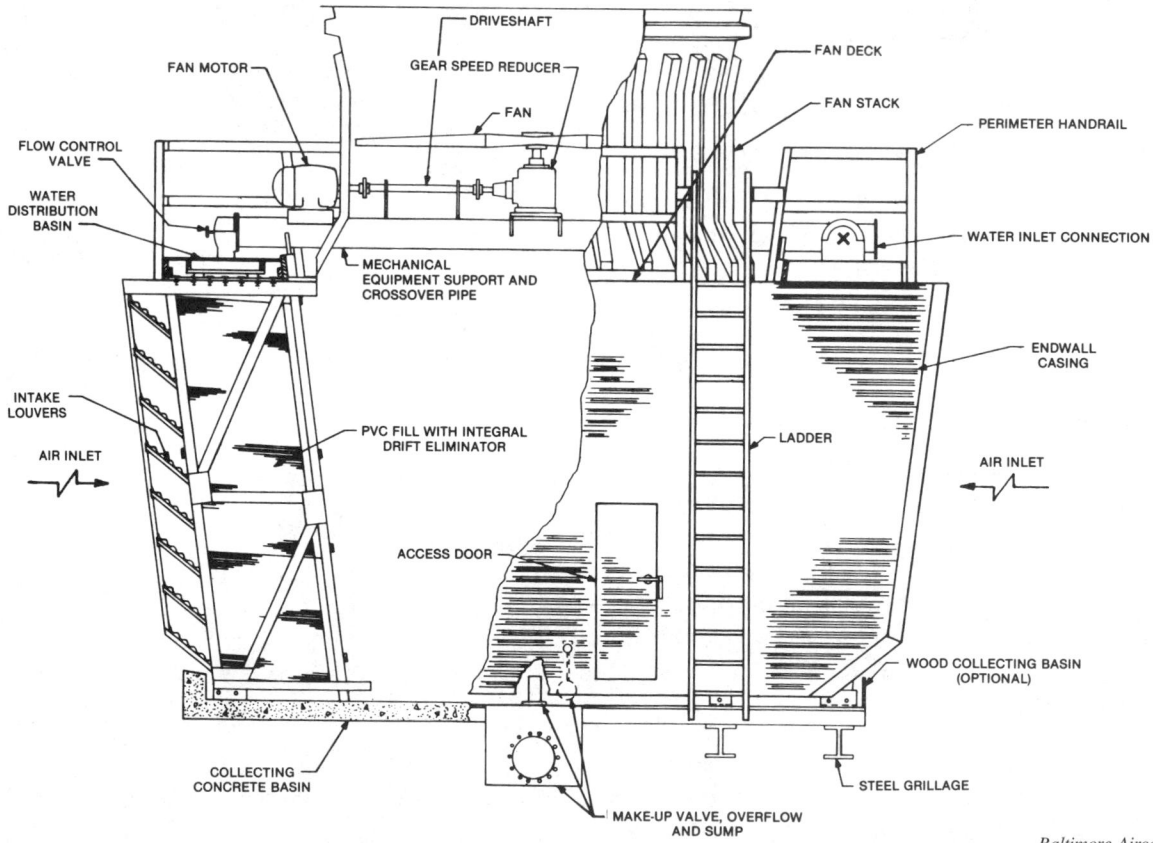

Baltimore Aircoil Company

Fig. 3 Crossflow cooling tower On each side of the tower, water falls from a distributor tray through orifices. The water drips down layers of "fill" material, which increases the surface area for evaporation. Air flows across the fill from both sides, drawn toward the center by the fan. Water lost by evaporation is replaced by a float valve that maintains the level of the water in the collecting basin. "Drift eliminators," which may be located anywhere in the air discharge path, minimize the loss of unevaporated water droplets.

MEASURE 2.4.1 Modulate fan output in heat rejection units to follow the cooling load.

This Measure eliminates or minimizes energy waste due to the cycling of fan output that occurs in cooling units with single-speed, fixed-pitch fans. Cycling fan output wastes fan power and raises the average condensing temperature. The solution is to install a fan drive that is better able to follow the actual cooling load. In the current state of technology, one of these fan drives will be best for your application:

- *electronic variable-speed fan drives,* which are covered by subsidiary Measure 2.4.1.1. These provide the best efficiency. They are easy to install, even in retrofit applications.

- *variable-pitch fans,* covered by subsidiary Measure 2.4.1.2. These provide efficiency that is comparable to the efficiency of variable-speed drives. They may be the best choice in cases where the adverse side effects of electronic drives are too expensive to overcome. They are limited to cooling units of larger size. They may be difficult to retrofit.

- *multi-speed or dual fan motors,* which are covered by subsidiary Measure 2.4.1.3. These are a compromise. They have lower average efficiency than the previous two types. They are inexpensive, easy to install, and free of serious problems. They are most likely to be the best choice in cooling units of small and medium size.

Energy Saving Potential

The best overall chiller system efficiency is achieved if the condenser cooling fans modulate their output efficiently to match the heat rejection requirements of the chiller. In contrast, if a fan must change its output in stages, fan energy is wasted and the average condenser temperature is raised.

The methods recommended by this Measure to improve fan modulation are expensive. Whether any is worthwhile depends on the potential efficiency improvement. The least efficient operation occurs when condenser cooling depends on a single fan that turns on and off. If a single-stage fan motor is larger than a few horsepower (or kilowatts), analyze whether a fan drive modification would be economical. If the fans presently operate with two or more steps of speed, upgrading to continuous modulation is likely to be economical only if the fans are very large.

■ Fan and Drive Energy

Average fan energy consumption is higher with a cycling fan because fan power is proportional to the third power of the air flow rate. For example, consider an ideal fan that requires 100 kilowatts to produce full output. If this fan operates at full output half the time, it consumes an average of 50 KW. If the same fan is run continuously and is modulated to move the same amount of air, it consumes 12.5 KW. Thus, the cycling fan uses four times as much energy as the continuously operating fan.

The energy-saving potential depends on the cooling load profile. The previous example is almost a worst case. In contrast, if the cooling load remains near maximum at all times, the single-speed fan does not waste much energy. At the other extreme, if the cooling load is usually low, the fan does not operate enough to provide a large opportunity for saving. The most energy is wasted if the fan operates between 30 to 70 percent of the time, which is typical of chiller applications.

If the cooling unit has more than one fan serving each cell, then the flow is varied in stages, which is more efficient than turning a single fan on and off. (In units with multiple fans, be sure that the fan speeds are sequenced efficiently, as recommended by Measure 2.4.2.)

Account for these factors when you estimate fan and drive energy savings:

- *drive losses.* Cycling and staged fan speeds have only minimal drive losses. Variable-pitch fans have little or no drive losses, but motor efficiency drops at low loads. Electronic variable-speed drives have average drive losses in the range of 5% to 20%, depending on the load profile, drive type, and size.
- *variations in fan efficiency with flow rate.* The fixed-pitch propeller fans used in most cooling units become more efficient as their speed is reduced. Variable-pitch fans are designed for maximum efficiency at some intermediate speed. The effect of pitch on efficiency is relatively small in most applications.
- *motor starting current.* When a cycling fan starts, it has an exceptionally high starting current. This involves energy losses, but only for a brief period during each starting cycle.
- *differences in the amount of air flow required.* For a given amount of condenser cooling that is required, cycling requires more total air flow than modulated fan operation. For example, if the cooling load is half of maximum, substantially less than half the maximum cooling air flow is required to maintain the condenser temperature.

■ Effect on Condensing Temperature

Modulating the fans provides more accurate control of condenser temperature, which can be used to improve chiller efficiency. If a fan is cycled or staged, the condenser water temperature fluctuates between the temperatures at which the fan turns on and off. This temperature difference is typically set at 3°F to 6°F to prevent excessive fan cycling. The temperature fluctuation wastes energy in itself, by raising the average condensing temperature.

If a heat rejection equipment has more than one fan, it is common to divide the cooling unit into separate cells, to avoid short-circuiting air through the fans that are not running. (This is discussed in Measure 2.4.6.) However, different portions of the heat transfer surface still receive different degrees of cooling, which reduces overall cooling efficiency. Operating all the fans together with continuous modulation eliminates this inefficiency.

To get the most benefit from fan modulation, you have to select the most efficient condenser temperatures. This aspect is covered in Measure 2.2.2.

Other Benefits

Eliminating or reducing fan cycling extends the life of the fan, the transmission, and bearings. Propeller fans have large moments of inertia, requiring motors with high starting torque. This results in a great deal of stress during starting. Modulating fan drives reduce or eliminate high-torque fan starting.

Eliminating cycling also eliminates the overt change in noise level that occurs when the fan turns on and off, which may be more objectionable than the steady noise itself.

Minimizing the air flow velocity through the cooling tower minimizes "drift," which is the escape of unevaporated water droplets. This is only a minor efficiency factor.

Comparison of Methods

In many applications, all three methods of modulating fan output may be usable. The choice is a compromise involving the following factors.

■ Efficiency

Electronic variable-speed drives and variable-pitch fans are comparable in efficiency. Both can track the cooling load perfectly, down to a small fraction of full load. Multi-speed fan motors are not able to track the cooling load as closely, so they are less efficient than fully modulating methods.

Electronic variable-speed drives have some internal electrical losses that depend on the load profile, drive type, and size. These drives also cause some electrical losses in the motor, because the power provided by the electronic drive contains harmonics that cannot be used efficiently by the motor. Losses added by the drive are around 5% at full speed. Combined drive and motor losses may exceed 30% at low speeds, but energy input has dropped to a small fraction of full load input by then. From the standpoint of the fan, speed control is efficient because aerodynamic losses are reduced as fan speed and load are reduced.

Variable-pitch fans are efficient because they add no drive losses. From the standpoint of the fan itself, variable-pitch control is aerodynamically efficient, although not as efficient as changing fan speed. The propeller remains at maximum speed, so the friction loss of the blades does not decline much at lower cooling loads. At low output, there is some waste of energy because the inner portion of the blade is moving more air than the outer portion. As the fan approaches zero output, the outer portion of the blade actually reverses. This effect sacrifices little efficiency under typical loads.

There is some loss of motor efficiency with variable-pitch fans when the load falls below about half of full load. Larger motors suffer relatively less efficiency loss at lower loads. See Measure 10.1.1 for more detail.

Multi-speed fan motors provide less accurate tracking of the cooling load than the other two methods. Therefore, this method involves some waste of fan power and excessive condensing temperature. Even so, these inefficiencies are much lower than with single-speed fans.

From the standpoint of drive efficiency alone, multi-speed motors compare well to variable-frequency drives and variable-pitch fans. Two-speed motors are somewhat less efficient than single-speed motors, especially at the lower speed, but they require no separate mechanical or electrical components. By the same token, replacing a single-speed motor with a multi-speed motor involves no additional equipment. Three-and four-speed motors are somewhat less efficient than two-speed motors. From the standpoint of the fan, varying speed is aerodynamically the most efficient means of reducing output.

Baltimore Aircoil Company

Fig. 1 Cooling tower fan drive gearbox This is a common method of matching the motor to the fan in larger cooling towers. Before retrofitting a speed reduction device for the motor, make sure that the gearbox remains properly lubricated at low speed.

If a cooling unit has more than one fan per cell, using multi-speed fan motors incurs an additional efficiency penalty because the fans do not always operate at the same speeds. See Measure 2.4.2 for details.

■ **Number of Fans**

If a cooling unit has more than one fan, the best efficiency is achieved by operating all the fans at the same speed. This is simple and economical with electronic drives, because one drive can be used to operate all the existing fans and fan motors. In contrast, a separate variable-pitch replacement would have to be installed for each fan. Similarly, a multi-speed motor would have to be substituted for each of the existing fan motors.

■ **Fan Size**

Variable-pitch fans are available only in larger sizes. Electronic drives are available for fan motors down to about one horsepower (or kilowatt), and they have no upper size limit. Efficient multi-speed fan motors are available in sizes down to about one horsepower (or kilowatt).

■ **Resonances**

Each of the subsidiary Measures gives methods of avoiding resonances. In brief, resonances are usually easy to avoid with variable-pitch fans and two-speed motors. Variable-speed drives operate fans over a wide range of speeds, so they may trigger serious mechanical vibration of the cooling tower structure. The controllers of some variable-speed drives give you the option of locking out speeds that may cause resonances.

■ **Electrical Noise**

Electronic drives may introduce an objectionable level of electrical noise throughout the electric power supply system of the facility. Correcting this problem may be expensive, especially in facilities that use equipment that is sensitive to power supply quality. Some newer models of electronic drives are designed to reduce this problem substantially.

This problem does not occur with variable-pitch fans and multi-speed motors.

■ **Gearbox Lubrication**

Many larger cooling units use a right-angle gearbox to drive the fan. Figure 1 shows a typical installation. If the gearbox uses splash lubrication, installing a variable-speed drive may slow the fan enough to cause a loss of lubrication. This problem can be avoided by maintaining a minimum fan speed, but this wastes a small amount of energy, and may not allow the fan to reduce output sufficiently at low loads. Check with the cooling tower or gearbox manufacturer before retrofitting a variable-speed motor drive.

Installing a two-speed fan usually does not cause lubrication problems, but verify this with the tower or gearbox manufacturer.

Lubrication problems do not occur with variable-pitch fans, which operate at fixed speed.

■ Maintenance

A variable-pitch fan has its critical moving components located in the hub, which is continuously exposed to a humid and corrosive atmosphere. As a result, variable-pitch fans require regular inspection and maintenance.

Electronic drives have no moving parts, and require no routine maintenance. The vulnerable electronic components can be installed remotely from the cooling tower, in an enclosure. The components are subject to occasional failures, but they typically can be replaced easily.

Multi-speed motors operate in the same environment as the motors they replace, and they have minimal maintenance requirements.

■ Starting Stress

Electronic drives are limited in starting current, which gives them an inherent "soft" starting characteristic. Additional soft starting features are available on many models.

Variable-pitch fans have high starting inertia, and hence, high starting stress. The stress can be reduced by installing a conventional "soft" motor starter. In any event, a variable-pitch fan has far fewer starting cycles than a cycling fan.

Multi-speed fans have the same starting stresses as single-speed motors, if they are controlled to start at high speed. After starting initially, multi-speed motors have lower stresses in switching from one speed to another, and these stresses occur less often than with single-speed motors.

■ Audible Noise

The noise produced by a fan increases radically with the fan speed. Lowering the fan speed to, say, 70% of full speed provides a substantial reduction in fan noise. Therefore, variable-speed drives reduce the average fan noise level considerably. In contrast, the noise produced by variable-pitch fans is always near maximum.

Multi-speed fans are much less noisy at the lower speeds. There is an overt change in sound level as the fan speed changes, but the change is less noticeable than with the original single-speed fan.

■ Ease of Retrofit

Installing an electronic drive requires no mechanical changes to the cooling tower, unless stiffening is required to avoid resonances. The electronics cabinet is relatively small and its location is not critical.

Installing a variable-pitch fan may require significant alteration of the fan mounting to accommodate the greater weight and deeper hub dimensions.

Replacing a single-speed motor with a multi-speed motor usually requires little or no modification of the cooling unit. However, the multi-speed motor may be larger than the original, which may hinder installation in tight quarters.

■ Cost

Like all electronic products, variable-speed drives have experienced a large drop in price over the years. They are now much cheaper than variable-pitch fans for most applications.

As noted previously, the relative cost of an electronic drive becomes much lower if one unit can be used to control several fans.

Multi-speed fan motors are typically much cheaper than the other two methods. The only costs are for a new motor, a modest amount of installation labor, and simple controls.

Beware of Reduced Cooling Surface Area in Some Cooling Towers

All cooling towers increase their evaporating surface area by using "fill" that forms passages over which the water flows. In some towers, the passages are so small that reducing the air flow through the tower would cause the passages to become plugged by water held in place by surface tension. This raises the water temperature, which causes the controls to increase the fan output to compensate. This involves some efficiency loss. In such cases, it may be economical to modify the passages so they remain open under all air flow rates.

MEASURE 2.4.1.1 Install variable-frequency fan drives.

This is the first of three subsidiary Measures that provide alternative methods of improving the fan modulation of cooling towers and dry coolers. Review "Comparison of Methods," in Measure 2.4.1, to help you decide which to choose.

This Measure covers electronic variable-frequency drives (VFD's). These drives operate fans at the precise speed needed to satisfy each load condition. See Reference Note 36, Variable-Speed Motors and Drives, for the characteristics of variable-frequency drives and for the selection factors that you should consider.

Operate All Fans at the Same Speed

The lowest energy consumption occurs when all the fans serving a particular cooling system are operated at the same speed (assuming that the fans are identical). This arrangement is also simple to control. Any number of fans can be controlled from a single electronic drive if it has enough power output. Figure 1 shows a typical installation.

Turndown Ratio

In principle, variable-frequency drives are able to reduce motor speed efficiently almost to zero. Some models are now available that operate very well at low speed. However, be aware that certain types of variable-frequency drives may cause a strong pulsation of the fan at very low speed, which is sometimes called "cogging." Drives with this characteristic are typically less expensive. With such units, you can avoid trouble by setting the lower speed limit of the drive above the speed at which cogging becomes troublesome. In most cooling unit applications, this compromise has only a negligible effect on efficiency.

Dealing with Mechanical Resonances

Both fans and cooling unit structures may have strong resonances at a number of frequencies. These resonances may create troublesome noise, or even damage the equipment. A unit may have several resonances within the speed range of the fan. This is especially likely in larger, loosely constructed units. The problem occurs primarily with cooling units that have propeller fans, rather than centrifugal fans. The resonant frequencies of centrifugal fans are usually too high to trigger serious vibration in heavy structures.

One solution to this problem is to lock out certain fan speeds using the fan control circuitry. This is a standard feature with some variable-speed drives. Locking out speed ranges invites the fan to "hunt" above and below these speeds when the cooling load happens

SUMMARY

Becoming the preferred method of modulating cooling unit capacity. Easy to retrofit. Has serious potential problems, usually avoidable.

SELECTION SCORECARD

Savings Potential .... $ $ $

Rate of Return, New Facilities % % %

Rate of Return, Retrofit.......... % % %

Reliability .... ✓ ✓ ✓

Ease of Retrofit .... ☺ ☺

to coincide with them. The speed control should be able to prevent the fan from hunting excessively.

Another solution is to dampen or eliminate the resonances by modifying the structure, installing braces, or pre-tensioning parts of the structure that may vibrate.

Install vibration sensors on the tower to shut down the fans if vibration becomes dangerous. Larger towers may have such sensors as part of their original equipment.

Dealing with Power System Distortion and Electrical Noise

This is the worst area of trouble for VFD's in general. It is a big subject. See Reference Note 36, Variable-Speed Motors and Drives, for the details.

Control Issues

Variable-frequency drives are now commonly equipped to accept standard pneumatic and electric control signals. This should make it easy to retrofit the drive controls to the thermostatic controls that sense the condenser cooling load. Acquire the equipment from a vendor who is able to supply a control package specifically designed for cooling tower applications.

An important factor to consider is the long time lag in control response, especially with wet coolers. When the fan speed increases in response to a rise in condensing temperature, the effect is not felt at the condenser until the water has time to circulate from the tower to the chiller. This causes oscillation of the fan controls, unless they are set to respond slowly. A more sophisticated approach is to use proportional-integral-derivative (PID) controls. These are covered in Reference Note 37, Control Characteristics.

If the drive defaults to full-speed operation in the event of a drive circuit failure, the controls should include an alarm to alert operators that the control has failed.

The most difficult part of optimizing the control settings is determining the optimum condenser temperature. Measures 2.2.2 ff describe how to do this.

Remove Existing Dampers

Some smaller heat rejection units, both wet and dry, employ dampers to throttle air flow as a means of modulating the cooling capacity of the cooling tower. Dampers should become unnecessary when this Measure is accomplished. Remove the dampers or lock them fully open.

If the dampers must be kept in operation to limit air flow at minimum fan speed, ensure that the damper controls open the dampers fully at higher fan speeds. Try to avoid this control complication, which is another potential failure mode.

Eliminate Cooling Tower Bypass, If Appropriate

Some cooling tower installations have a diverter valve that bypasses a portion of the condenser cooling water around the cooling tower. This may be done to provide more accurate control of condenser water, to maintain a minimum condenser temperature, or as part of a freeze protection scheme. Installing a fully modulating fan should eliminate the need for tower water bypass.

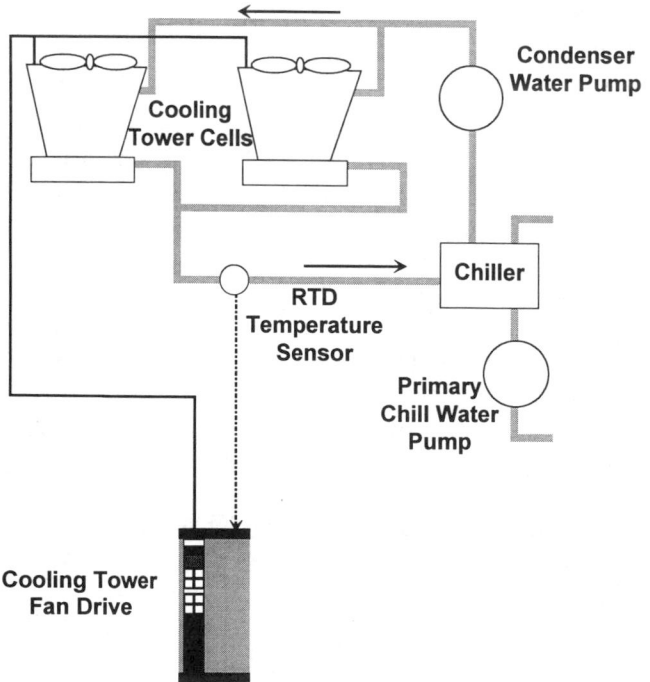

Graham Company

Fig. 1 Variable-frequency drive for cooling tower fans
This is a good application for variable-frequency drives. Retrofit is especially easy. A single drive can operate any number of fans at identical speed, which is the most efficient way of operating multiple fans.

If a cooling tower bypass is installed, find out why. If installing a variable-flow fan eliminates the need for the bypass, either remove the bypass valve entirely or lock the bypass closed. In the latter case, install a placard on the bypass valve to indicate that it is no longer used.

The effect of a bypass valve on pump energy consumption is minor, in conventional condenser cooling systems. The main reason for eliminating the bypass is to avoid unnecessary complication and possible conflict with the fan controls.

Some chillers require a minimum condenser water temperature, typically about 65°F (18°C). If the chiller is turned off during cold weather, the temperature of the water standing in the tower may fall below the minimum allowable temperature. If a bypass is installed to maintain a minimum cooling water temperature, make sure that it opens only when needed for this purpose. The warm-up interval is usually no longer than a few minutes. Also, make sure that the fan controls keep the fans turned off as long as the bypass is open.

Important Features for Cooling Unit Applications

See Reference Note 36 for the selection issues that apply to all electronic drive applications. Drive features that are important for cooling tower fans in particular are:

- the ability to lock out specific resonant speed bands
- a control package that is appropriate for condenser cooling
- if a wet cooler is used during cold weather, the ability to reverse the fans to melt ice.

Whether to Replace the Fan Motor

Electronic drives produce distorted power waveforms. The high-frequency distortion components cannot be used efficiently by motors, and they are converted to heat inside the motor. The amount of distortion varies widely among different types and models of VFD's. Not surprisingly, those with smoother waveforms cost more, other things being equal.

Newer motors that are designed for use with electronic drives suffer less efficiency loss from drive distortion. If you install a drive that does not produce especially smooth motor power, it may be advisable to replace older fan motors. See Reference Note 36 for the details.

ECONOMICS

SAVINGS POTENTIAL: *30 to 70 percent of fan energy. 0.5 to 3 percent of chiller energy, because of lower condensing temperature.*

COST: *Variable-frequency drives cost from $50 to $400 per horsepower (or kilowatt), depending on size, type, and optional features. High-efficiency motors cost $50* *to $100 per horsepower. In retrofit applications, several man-days of installation labor are typically required.*

PAYBACK PERIOD: *One year, to many years.*

TRAPS & TRICKS

CHOICE OF METHOD: *Variable-speed drives should be installed on most heat rejection units of large and*

medium size. Variable-frequency drives are probably the wave of the future for this application. Several years ago, you would have been a pioneer retrofitting a VFD to a cooling tower, but no longer. Electronic variable-speed drives are still improving and becoming cheaper. Be familiar with the selection issues explained in Reference Note 36.

MEASURE **2.4.1.2 Install variable-pitch propeller fans.**

RATINGS

New Facilities	Retrofit	O&M
B	**D**	

This is the second of three subsidiary Measures that improve the fan modulation of cooling towers and dry coolers. Review "Comparison of Methods," in Measure 2.4.1, to help you decide which to choose.

This subsidiary Measure covers variable-pitch fans. Like variable-frequency drives, which are covered in the previous Measure, variable-pitch propellers provide precisely the degree of cooling required by the load. Instead of changing the speed of the fan, they change the pitch. A mechanism in the hub of the propeller changes the pitch of the blades by any desired amount, so the air flow can be adjusted anywhere from full output to almost zero flow.

The propeller and its accessories are specially designed for the corrosive and dirty environment of cooling towers. The propeller blades are typically aluminum alloy or fiber-reinforced plastic, and the pitch mechanism usually is actuated by pneumatic pressure.

Variable-pitch fans are losing ground to variable-speed drives, primarily because of their higher cost. Even before this competition arose, variable-pitch fans were limited by economics to larger cooling towers. They are worth considering in cases where the disadvantages of electronic drives cannot be tolerated. However, these disadvantages are being lessened in newer models of electronic drives, further limiting the application of variable-pitch fans.

Retrofitting a variable-pitch fan is feasible in most cooling towers above a certain size. Installation service is offered by the major cooling tower manufacturers. The modification is straightforward, although not as simple as Measures 2.4.1.1 and 2.4.1.3. The replacement propeller fits into the same opening as the original fan, but you may have to modify the fan transmission mounting.

SUMMARY

Well proven in cooling towers. The most efficient choice if a variable-speed drive is not feasible. Expensive to retrofit.

SELECTION SCORECARD

Savings Potential	$ $ $
Rate of Return, New Facilities	% % %
Rate of Return, Retrofit..........	% % %
Reliability	✓ ✓ ✓
Ease of Retrofit	☺ ☺

Operate All Fans at the Same Speed

If a single cooling system is served by more than one variable-pitch fan, the lowest energy consumption occurs when all fans are operated at the same pitch. This arrangement is simple to control. Any number of variable-pitch fans can be controlled by providing the same control signal to all the fans.

Turndown Ratio

Variable-pitch propellers have the best turndown ability of any type of variable-output fan. They are able to reduce air circulation smoothly from full load to almost zero.

Dealing with Resonances

Resonances of the fan and the cooling tower structure are not likely to be a problem with variable-pitch fans, but consider the possibility. Resonances are relate to the fan speed and the number of propeller blades. The speed is constant, so you can avoid resonances simply by operating the fan at a speed that does not excite any resonances.

In retrofit applications, verify that there will be no resonance at the anticipated operating speed. This may occur if the variable-pitch fan has a number of blades different from the original fixed-pitch fan. If the fan has a belt drive, you can easily change the fan speed by a small amount to avoid a resonant frequency. The capacity of the heat rejection unit is not affected by a small speed adjustment because the propeller pitch changes to compensate for it.

Eliminate Cooling Tower Bypass, If Appropriate

Refer to this heading in Measure 2.4.1.1, previous. The considerations are identical when dealing with variable-pitch propellers.

Belt or Gear Drive?

Gear drives may achieve 98% efficiency, in the best of cases. Belt drives have somewhat lower efficiencies, typically in the range of 92% to 96%.

Belt drives provide great flexibility in varying the speed of the fan in relation to the motor speed. This can be valuable in avoiding resonances.

Another advantage of using a belt-driven fan is that it may avoid the need to remount the power transmission between the motor and the fan. On the other hand, a gear drive may be an integral part of a variable-pitch fan package, avoiding any problems of separate mounting.

Control Issues

The controls for a variable-pitch fan are relatively simple. They are available as a standard package from some fan vendors. The typical control senses condensing temperature to change the air pressure delivered to the variable-pitch mechanism in the hub of the fan.

The most difficult part of optimizing the control settings is calculating the optimum condenser temperature for each chiller load and outside air temperature. Measures 2.2.2 ff describe how to do this.

Standard control packages allow the designer to select a fail-safe pitch setting in the event of failure of the pitch control. For most applications, the appropriate fail-safe position is full pitch, providing maximum cooling capacity. If freezing is a potential problem, it may be more desirable to have the pitch control fail to minimum pitch.

An important factor to consider is the long time lag in control response, especially with wet coolers. When the fan speed increases in response to a rise in condensing temperature, the effect is not felt at the condenser until the water has time to circulate from the tower to the chiller. This causes oscillation of the fan controls, unless they are set to respond slowly. The control packages provided by the manufacturers of variable-pitch fans for cooling towers should deal with this problem satisfactorily.

The cooling tower will continue to operate with a failed pitch control, so include an alarm in the control package to alert operators if the pitch control fails.

ECONOMICS

SAVINGS POTENTIAL: *30 to 70 percent of fan energy. 0.5 to 3 percent of chiller energy, because of lower condensing temperature.*

COST: *Approximately $30 per ton of cooling tower capacity, for the equipment. Retrofitting a variable-pitch propeller to an existing cooling tower involves a large labor cost, which varies widely depending on circumstances.*

PAYBACK PERIOD: *One year, to many years.*

TRAPS & TRICKS

SELECTING AND INSTALLING THE EQUIPMENT: *Installing variable-pitch fans is specialized work. Buy the equipment from the people who will install it, so that there is no question about who is responsible for fixing problems. Make sure that both equipment and installer have a good reputation. Check with others who have used both.*

MAINTENANCE: *The variable-pitch mechanism operates in a harsh environment, so maintenance is critical. Make sure that you have a way of scheduling the maintenance reliably.*

MEASURE 2.4.1.3 Install dual or multi-speed motors.

SUMMARY

An alternative to Measures 2.4.1.1 and 2.4.1.2 that is easy, reliable, and less expensive, but also less efficient.

SELECTION SCORECARD

Savings Potential .... $ $ $

Rate of Return, New Facilities % % % %

Rate of Return, Retrofit .... % % %

Reliability .... ✓ ✓ ✓ ✓

Ease of Retrofit .... ☺ ☺ ☺

This is the last of three subsidiary Measures that improve the fan modulation of cooling towers and dry coolers. Review "Comparison of Methods," in Measure 2.4.1, to help you decide which to choose. This subsidiary Measure covers the use of dual motors and multi-speed motors in cooling towers and dry coolers. Both methods have virtually identical efficiency characteristics, in most applications.

Multi-speed motors offer two or more fixed speeds. Most heat rejection units equipped with multi-speed motors have only two speeds, but three or more speeds may be worth considering. Multi-speed motors may be easy to retrofit to existing single-speed heat rejection units.

In dual-motor installations, one motor at a time drives the fan, while the other motor idles. Figure 1 shows a typical installation. Dual motors offer only two speeds, but virtually any ratio of two speeds can be obtained if at least one of the motors is connected with a belt drive. Usually, multi-speed cooling tower fans operate at a 2:1 ratio, but there is no inherent reason why you should not select a different ratio. The power rating of each motor is selected for the speed at which it drives the fan. This configuration is generally limited to new equipment, as it would be difficult to retrofit.

The two methods are similar in cost. They are much less expensive than the variable-output drives covered by the previous two Measures. The main disadvantage of both methods is that they are less efficient than the two previous Measures.

Efficiency Characteristics of Multi-Speed Fan Control

If a single-speed fan would operate about 50% of the time, then installing dual motors or a multi-speed motor will provide a significant improvement.

A typical condenser cooling unit retains about 70% of its maximum cooling capacity if the fan speed is reduced to half. Therefore, a 2:1 speed ratio provides substantial fan power savings if the system operates much of the time in the vicinity of half load. The fan input power is reduced to about 20% of full-load power when it operates at half speed, which includes a motor efficiency loss at the lower speed.

Since there is a relatively small change in cooling capacity when the motor switches between the two speeds, the condensing temperature does not rise much when the fan switches to lower speed.

On the other hand, the multi-speed fan offers much less benefit if the system lingers at low load. This occurs, for example, in the air conditioning systems of many types of facilities located in temperate climates. When the load is low, the fan does not run very long with either a single-speed or multi-speed motor, so not much motor energy is saved. Also, the fluctuations in condenser temperature are large in both cases.

A multi-speed fan also offers less benefit if the cooling unit operates most of the time near full load. In this case, either a single-speed or multi-speed motor would operate at full output most of the time.

How to Select the Motor

See Reference Note 36, Variable-Speed Motors and Drives, for details of multi-speed motors and dual motor installations. In most cases, a motor with two speeds is the best choice. Motors with three or four speeds are available on a custom basis, but they are larger and more expensive, and they suffer higher efficiency losses.

Control Issues

Controlling a multi-speed motor is simple. Use a multi-stage thermostat having the same number of stages as the available speeds. With a two-speed motor, one stage of the thermostat turns the fan to low speed when the condensing temperature rises to a certain setting, and to full speed at a slightly higher temperature. In cooling towers, the thermostat is usually installed to sense the temperature of the water in the sump. With air-cooled condensers, the fan speeds are controlled by the temperature or the pressure of the refrigerant in the condenser.

An important factor to consider is the long time lag in control response, especially with wet coolers. When the fan speed increases in response to a rise in condensing temperature, the effect is not felt at the condenser until the water has time to circulate from the tower to the chiller. This causes the fan controls to hunt between fan speeds. One solution is to build a delay

Baltimore Aircoil Company

Fig. 1 Two-motor cooling tower fan drive Only one motor operates at a time. Using two fan speeds instead of one provides major advantages in some applications, but not in all. The ratio of fan speeds may strongly affect the efficiency gain. Because these motors have belt drives, the fan speed with each motor can be adjusted over a wide range.

into the starting and stopping of each stage. A more sophisticated approach is to use proportional-integral-derivative (PID) controls. These are covered in Reference Note 37, Control Characteristics.

The most difficult part of optimizing the control settings is determining the optimum condenser temperature. Measures 2.2.2 ff give the procedures for doing this.

Control of Speeds in Multi-Fan Installations

In heat rejection units where a cell has more than one fan, the manner in which the fan speeds are sequenced has a significant effect on energy consumption. See Measure 2.4.2 for the details. In this type of cooling unit, this Measure suffers a greater efficiency disadvantage when compared to the preceding two Measures.

Dealing with Resonances

Both fans and cooling unit structures may have strong resonances at a number of frequencies. These resonances may become loud enough to annoy people, and they may even damage equipment. Select all the operating speeds of the fan to avoid any resonant frequency.

In retrofit applications, you can use the original single-speed motor to check for resonances at the intended speeds of the multi-speed motor. Do this by changing pulleys to operate the original fan at the intended high and low speeds for test purposes.

If a resonance does appear at any of the intended operating speeds, several solutions are available. If a belt drive is used, you can adjust the fan pulleys to avoid the resonant speeds.

Another approach is to change the resonant frequencies of the structure or to dampen the resonances. You can do both by modifying the structure, installing braces, or pre-tensioning parts of the structure that may vibrate.

Remove Dampers in Retrofit Units

Some small heat rejection units, both wet and dry, employ dampers to throttle air flow as a means of modulating the cooling capacity of the cooling tower. For example, you may find dampers if the chiller must maintain a minimum condenser temperature. Dampers are inefficient as a means of controlling the output of a heat rejection unit, and they can usually be eliminated when multi-speed fan modulation is installed. Remove the dampers, or lock them fully open, if you provide efficient modulation.

If the dampers must retained for control at low fan speed, ensure that the damper controls open the dampers fully when the fan is operating at high speed. Avoid this control complication, if possible, because it is another potential failure mode.

Eliminate Cooling Tower Bypass, If Appropriate

Refer to this heading in Measure 2.4.1.1, above. The considerations are similar when dealing with multi-speed motors.

ECONOMICS

SAVINGS POTENTIAL: *30 to 60 percent of fan energy. 0.5 to 2 percent of chiller energy, because of lower condensing temperature. The life of all drive components is extended.*

COST: *Single-winding 2:1 motors cost about $400 per horsepower in small sizes and about $100 per horsepower in large sizes. The cost of dual motors is similar. Controls add several hundred to several thousand dollars. In retrofit applications, labor for motor replacement is a relatively small cost, if there are no installation difficulties.*

PAYBACK PERIOD: *One year, to many years.*

TRAPS & TRICKS

DO YOUR HOMEWORK: *You have more choices of multi-speed motors than you may realize. See Reference Note 36 for the full story. Figure out the best motor speed ratio for your application based on the cooling load profile.*

INSTALLATION: *This feature is so simple and reliable that there is no excuse for problems. Set the speed controls for maximum overall efficiency. In retrofit, make sure that the motor or motors are aligned properly.*

MEASURE **2.4.2 In multiple-cell cooling units, sequence the fans efficiently.**

Some heat rejection units have a number of fans, each of which operates in one or two stages. The sequence of fan operation affects the overall fan power, the condenser temperature, or both. The fan sequence that provides the lowest fan energy consumption also provides the lowest average condenser temperature. Minimizing the condenser temperature increases chiller efficiency, for reasons explained in Measure 2.2.2.

In some cases, you may find that the sequence is not the optimum one. You are most likely to find this in multi-cell cooling towers that are installed as a part of engineered (i.e., not packaged) cooling systems. Multi-fan dry coolers that serve a single chiller generally include controls that sequence the fans in an optimum manner. But, don't assume this. It takes only a brief period of observation during the cooling season to check the fan sequence. Changing the fan sequence is usually a simple control modification.

Optimum Fan Sequencing

If the fans have continuous fan speed modulation, best efficiency occurs if all the fans are kept running, and at the same speed. (If the fans do not have continuous modulation, consider installing it, as recommended by Measures 2.4.1.1 and 2.4.1.2.)

If the fans are staged, keep the loading of all the fans that serve a particular cooling circuit as close together as possible. For example, in a cooling unit with three two-stage fans, the first fan starts at low speed, then the second fan, then the third. Next, the first fan goes to high speed, then the second, then the third. The fans should unload in reverse order.

This manner of fan sequencing applies to all the fans that serve a single chiller system. A cooling tower with several cells may be connected to more than one chiller system. Optimize the fans for each chiller system as a group.

Be sure of the cooling unit circuiting before proceeding. In the case of wet cooling units, this is usually easy to trace. Be careful with multi-cell cooling towers that have piping connections that allow different cells to serve different loads. With such towers, the optimum fan sequence is different for different water flow connections.

In dry coolers and evaporative condensers, circuiting may be less obvious. For example, if a reciprocating chiller has several compressors, each compressor is likely to have its own condensing coil circuit. If the sequence of fan operation in a packaged heat rejection unit appears to differ from the sequence recommended

SUMMARY

A simple correction to the fan control sequence may save a significant amount of fan and chiller energy.

SELECTION SCORECARD

Savings Potential .... **$** $

Rate of Return, New Facilities **% % % %**

Rate of Return, Retrofit.......... **% %** %

Reliability .... ✓ ✓ ✓ ✓

Ease of Retrofit .... ☺ ☺ ☺

below, there may be a good reason. In case of doubt, contact the manufacturer before making any changes.

Why Optimum Sequencing Saves Fan Energy

All the fans that serve a particular cooling circuit should keep their loading as close together as possible. The reason is that fan energy is proportional to the third power of flow volume, rather than being directly proportional to flow. For example, two fans operating at half speed consume considerably less energy than one fan operating at full speed.

Why Optimum Sequencing Lowers Condenser Temperature

Similarly, the lowest condensing temperature is achieved when all the fans that serve a particular cooling circuit keep their loading as close together as possible. To see why, consider a cooling tower with two cells serving a single circuit, each fan having two speeds. If one fan operates at full speed and the other fan is off, the temperature reduction is limited to about half of maximum, because the water from idle cell is mixed with the water from the active cell. On the other hand, if both fans operate at half speed, the temperature drop is near maximum, because half the maximum air flow provides full cooling at low cooling load.

ECONOMICS

SAVINGS POTENTIAL: *10 to 40 percent of fan energy; 1 to 5 percent of chiller energy.*

COST: *In existing facilities, a controls specialist can typically make the change in an hour or two, at a cost of several hundred dollars.*

PAYBACK PERIOD: *Less than one year, to several years, in retrofit.*

TRAPS & TRICKS

LOOK BEFORE YOU LEAP: *Before you start tinkering with the fan controls, be sure you understand why the unit works the way it does. Especially in factory-* *assembled units, the manufacturer has a specific reason for the way the fans are sequenced. You are most likely to find an outright sequencing mistake in a field-erected cooling tower. If you find that the fan sequencing is not the most efficient, suspect that this may be related to freeze protection or other functions. If so, you have to find ways to accommodate those other functions.*

MEASURE 2.4.3 Clean heat rejection units at appropriate intervals.

RATINGS

New Facilities	Retrofit	O&M
		A

Heat rejection equipment seems to give the impression that its condition has no effect on cooling system efficiency until it physically falls apart. This is not true. If heat rejection units are not kept clean, heat transfer suffers. This reduces chiller system efficiency and capacity, and increases the fan power requirement. Also, neglecting cleaning hastens equipment deterioration.

Cleaning heat rejection equipment is straightforward. This Measure tells you what needs to be done. The subsidiary Measure recommends changes to heat rejection units that reduce their tendency to become fouled.

SUMMARY

Routine maintenance that lowers condenser temperature and maintains capacity.

SELECTION SCORECARD

Savings Potential	$ $ $
Rate of Return	% % % %
Reliability	✓ ✓
Ease of Initiation	☺ ☺ ☺

How Lack of Cleaning Wastes Energy

Lack of cleaning wastes energy in a number of ways, depending on the type of heat rejection equipment:

• *dry coolers suffer restricted air flow from dirt and debris.* Dirt that settles on the fin surfaces restricts the flow of air through the coil and reduces heat transfer efficiency. The fan uses more energy as it tries harder to maintain the condenser temperature setting.

• *evaporative condensers and closed-loop cooling towers* have heat transfer surfaces between the evaporating water and the refrigerant or condenser cooling water, respectively. Fouling of these surfaces raises the condensing temperature. Obstruction of cooling water flow by accumulation of debris in piping, strainers, and drains raises the condensing temperature. Fan energy consumption increases as the fan tries harder to maintain the condenser temperature setting.

• *open-loop cooling towers, i.e.,* units that do not use heat exchangers, do not suffer an efficiency loss themselves as a result of fouling. Water evaporates from slimy splash bars as efficiently as from clean ones. The trouble with fouling that accumulates and grows in the cooling tower is that it travels to the heat transfer surfaces of the condenser. (Condenser cleaning is covered in Measures 2.3.1 ff.) Obstruction of condenser water flow by accumulation of debris in condenser tubes, piping, strainers, and drains raises the condensing temperature.

Factors that Increase Dirt Accumulation

The height of the unit above ground, especially above surrounding foliage, is a major factor that determines how fast debris accumulates. Units located below grade, either in wells or in basement mechanical equipment rooms, may accumulate debris prodigiously. Birds may clog units with feathers, nest material, and food leftovers. The bird problem is worst when the unit is located inside a well, behind a decorative screen, or in other sheltered locations. Improving the installation to reduce debris accumulation is recommended in the subsidiary Measure.

In wet cooling units, the nature of the local water supply determines the potential for mineral scaling of heat transfer surfaces. Measures 2.3.3 ff recommend ways to control these problems.

How to Clean Dry Coolers

If a dry cooler has a coil that is horizontal or slanted, debris falls off the bottom side of the coil when the fan turns off. The air flow is usually upward, which blows debris off the top of the coil when the fan is running. If the environment is clean, rain may provide all the cleaning that is necessary. Vertical coils, including the condenser coils of packaged air handling units, do not have this advantage.

In less satisfactory installations, airborne debris is drawn into dry cooling units by wind and by the action of the fans. The debris accumulates:

• over the coil fins

• between the coil fins

• on intake screens

• on fan guard screens.

Inspect each of these areas carefully. Some deposits are not obvious. In particular, a coil may become almost totally obstructed, even though the outer fin surfaces have been cleaned by superficial washing. You have to look through a coil and see light on the other side to be sure that it is clear.

If the dry cooler is installed so that the passage for the entering air is restricted, the velocity of the entering air is increased. This increases the tendency of the air to carry debris into the cooler. Such installations may also make it difficult to inspect the coil for debris. If you are unable to inspect the coil easily, suspect a problem.

Cleaning dry coolers is usually just a matter of sweeping the debris out of the cooling unit and cleaning the area surrounding the unit. Clear the surrounding area because any loose material will be drawn into the unit by the suction of the fan.

If dirt has collected between the coil fins, it is likely to be muddy in nature, so that it tends to cement itself in place. Water washing with a garden hose nozzle may dislodge the material after dissolving it. This may require patience. If the garden hose does not work, a high-pressure water lance can usually do the job, but be careful to avoid damaging the fins.

How to Clean Wet Coolers

The water sprays used in open-loop cooling towers, closed-loop cooling towers, and evaporative condensers are effective traps for atmospheric contaminants. These include inorganic particulate matter, general debris, gases, chemical pollutants, and a large variety of microorganisms. Each of these creates its own categories of problems.

The combination of water, warmth, heavy oxygenation, and nutrient matter fosters prolific growth of microorganisms, especially algae, bacteria, mold, and fungi. These may form colonies that insulate heat transfer surfaces and clog the system. In addition, some organisms may create health problems, such as Legionnaires' Disease. These are the areas that need periodic cleaning:

Baltimore Aircoil Company

Fig. 1 Water distribution orifices in a crossflow cooling tower These are easily obstructed by debris that falls into the open top of the cooling tower. This causes uneven water flow in the tower. The easy access to the trays makes them easy to clean out.

- *intake and fan guard screens,* which are easy to inspect and clean
- *drift eliminators,* which are baffles installed in cooling towers to prevent water droplets from being blown out. A water lance is effective for cleaning these.
- *water distribution nozzles,* which typically are simple orifices in the bottoms of the distributor trays at the top of gravity-flow towers. Figure 1 shows how they are arranged. The nozzles are typically a fraction of an inch (one or two centimeters) in diameter. Although the nozzles have large diameters, it is not unusual for them to become clogged. The trays in many towers are open, making it easy for large debris, such as leaves and dead animals, to fall in. Also, the nozzles in older steel cooling towers may become obstructed by rust chips. By the same token, the trays are easy to clean out, even with the tower in operation.
- *water spray nozzles,* in induced-draft towers. In this type of tower, air flow is induced without fans by spraying the water into the tower at high pressure. The venturi effect of the spray nozzles draws air into the spray. Unfortunately, this requires small nozzles that are easily clogged. This type of tower needs continuous filtration of the cooling water and frequent inspection of the nozzles.
- *heat transfer coils,* in evaporative condensers and closed-loop cooling towers. Most fouling can be cleaned off with detergent and brushing. Tenacious fouling, especially from mineral scaling, may require more brutal treatment, including specialized chemical cleaning.
- *sumps,* from which debris should be removed using methods similar to cleaning a swimming pool
- *condenser water strainers,* which should be flushed periodically. Failure to do so makes the trap itself an obstruction to cooling water flow, and can substantially raise condensing temperature.

- *fill media,* consisting of splash bars or water film surfaces. There is generally no need to clean fill media unless debris accumulates to the point it reduces the evaporating surface area significantly. The cleaning method for heavily fouled fill media depends on the type of fill. If the surfaces are accessible, mechanical brushing or high-pressure washing may be best. Chemical cleaning may be needed for packed fill media.

How to Prevent Fouling in Wet Coolers

Eliminate biological fouling by effective water treatment, which is recommended by Measures 2.3.3 ff. Chemical treatment, combined with bleed (see Measures 2.3.4 ff), can also help to keep dirt from depositing on system surfaces. Water treatment cannot keep large debris, such as leaves and rust chips, from obstructing flow. You have to remove debris manually.

ECONOMICS

SAVINGS POTENTIAL: *0.1 to 5 percent of chiller system operating cost.*

COST: *Typically a few hours of labor every few months.*

PAYBACK PERIOD: *Short. The overall savings greatly exceed the cost.*

TRAPS & TRICKS

INSTRUCTIONS: *Provide explicit instructions and check the work. Cleaning finned units is especially tedious, so the inner coil areas may not be cleaned adequately. Care is needed to avoid progressive damage when using vigorous cleaning methods.*

SCHEDULING: *Nobody likes to clean cooling equipment, and it can always be delayed, so it tends to be forgotten. Make sure that you have an effective way of scheduling the activity.*

MEASURE **2.4.3.1 Install and screen heat rejection units to minimize debris accumulation.**

Take a look at your heat rejection units, and see whether they are picking up an excess amount of debris. If so, it may be worthwhile to modify the cooling units to reduce the rate of debris accumulation. This saves labor, and maintains higher efficiency between cleanings. There are two fairly easy solutions that cover most cases. One is to increase the height of the cooling unit above sources of debris. The other is to install screens that prevent debris from entering the cooling unit.

Install Screens

In most situations where cooling units tend to foul by larger debris, you can protect the unit by installing screens. Usually, it is best to install the screen so that it covers the entire air intake to the heat rejection unit. If you do this, install the screens so that the screens themselves do not become choked. To keep the screens clear, install them so that entering air must flow upward into them. In this way, debris is pulled away from the screens by gravity when the fan stops..

Do not let screens obstruct air flow. A coarse screen that is appropriate for stopping leaves and birds will not interfere with air flow itself. However, if the screen may become partially obstructed, compensate by making the surface area of the screen larger.

SUMMARY

Simple installation improvements that reduce the rate of condenser fouling.

SELECTION SCORECARD

Savings Potential .... **$** $ $

Rate of Return, New Facilities **%** **%** % %

Rate of Return, Retrofit .... **%** % %

Reliability .... ✓ ✓ ✓ ✓

Ease of Retrofit .... ☺ ☺ ☺

In some cases, it is more practical to install a screen surrounding the unit. Figure 1 shows an example. This arrangement does not provide complete protection, but it cannot clog the air flow.

Elevate Cooling Units

If the cooling unit is being fouled by leaves and other foliage, it may help to elevate the unit. This is because debris is typically sucked up from an adjacent surface by the high velocity of the intake air. Adding legs, or extensions to existing legs, may be fairly easy. Keep the air intake path as open as possible to minimize entering air velocity. Exploit this opportunity to improve access to the cooling unit for cleaning and maintenance.

WESINC

Fig. 1 This fence keeps debris from fouling the condensing unit   It is reasonably effective in this role, provided that there are no nearby trees and no turbulent wind currents to introduce leaves or other debris over the top of the fence.

ECONOMICS

SAVINGS POTENTIAL: 0.1 to 5 percent of chiller system operating cost.

COST: Several hundred to several thousand dollars.

PAYBACK PERIOD: Several years, typically.

TRAPS & TRICKS

OBSERVE EFFECTIVENESS: This is usually a low-risk activity. However, air moves in mysterious ways, so check your new installation periodically to make sure that it is staying clean.

MEASURE **2.4.4 In gravity-flow cooling towers, ensure proper water distribution.**

SUMMARY

A simple check that may eliminate a source of cooling tower inefficiency.

SELECTION SCORECARD

Savings Potential .... $ $

Rate of Return .... % % % %

Reliability .... ✓ ✓

Ease of Initiation .... ☺ ☺ ☺

In cooling towers where the water flows over the fill by gravity, it is important to maintain uniform distribution of water. Any gaps in the curtain of water allows air to short-circuit through the tower without cooling. This reduces cooling effect, increases fan power, and reduces tower capacity.

In gravity-flow cooling towers, water typically is pumped into distributor trays at the top of the unit. (See Figure 1.) The water falls down to the fill or splash bars through holes ("nozzles") in the bottoms of the trays. The water should be equally deep at all points in the distributor trays to provide uniform flow. The depth is typically no more than one or two inches (a few centimeters). If the tower is tilted even slightly, portions of trays may run dry or have weak flow. You can correct the tilt by adjusting the trays individually or by shimming the supports for the entire tower, depending on the design of the unit.

Cooling towers may have a number of separate trays. To achieve uniform filling, the cooling water pipe filling each tray may have a balancing valve. If so, be sure that these valves are set properly. If you find that the valves have different settings, suspect a water distribution problem that should be corrected.

If the water levels in the trays are too low to provide good distribution, even with all valves in the system fully open, check that the strainer is not clogged. If the strainer is clear, the nozzles may be too large, or there may be a problem of inadequate pump capacity (which is rare).

If the balancing valves are heavily throttled, verify that the water depths in the trays are adequate. If so, you may be able to save energy by trimming the condenser water pump capacity. See Measure 10.2.1 for this.

Keep the nozzles clear to provide uniform water flow. This is covered by Measure 2.4.3, above.

WESINC

Fig. 1 Distributor trays on crossflow cooling tower Warm condenser water is pumped into the shallow trays, from which it drains through orifices to the fill material below. If the tower is not completely level, the trays do not fill evenly, and cooling capacity and efficiency suffer. Also, note the balancing valves where the water enters the trays. If these are not adjusted properly, water flow will be uneven.

ECONOMICS

SAVINGS POTENTIAL: *0.2 to 2 percent of chiller system operating cost.*

COST: *Depends on the problems to be corrected, but usually modest.*

PAYBACK PERIOD: *Usually short.*

TRAPS & TRICKS

SCHEDULING: *Some problems, such as clogging of distributor tray nozzles, will recur. Repeat the inspection occasionally. Put this in your maintenance schedule.*

MEASURE **2.4.5 Keep heat rejection unit housings and fittings intact.**

A

The efficiency of a heat rejection unit is mainly a matter of how effectively air is brought in contact with the cooling water or the cooling coils. The air is guided by the outer enclosure and by internal structures. Defects in these components may persist unnoticed, because the unit continues to operate, although with reduced efficiency and capacity. Check your heat rejection units periodically to make sure that all the parts are intact. Combine this with the periodic cleaning recommended by Measure 2.4.3.

Defects in the enclosure waste energy in two ways: (1) inefficient air flow increases the average condensing temperature, which reduces chiller efficiency for the reasons explained in Measure 2.2.2, and (2) the fans use more energy in trying to maintain the condenser temperature setting. Defects also reduce the capacity of the cooling system.

Close Openings in the Enclosure

It is not uncommon to find large pieces missing from cooling unit enclosures. Panels may be removed for maintenance and then forgotten. Access doors and

---

SUMMARY

Routine inspection and maintenance to ensure effective air flow through the cooling unit.

SELECTION SCORECARD

Savings Potential .... **$ $**

Rate of Return .... **% % % %**

Reliability .... ✓ ✓

Ease of Initiation .... ☺ ☺ ☺

---

handholes may be left astray. Figure 1 shows that this really happens.

In old units, portions of the enclosure may have rusted or rotted away. If you find these conditions, the corrective action is usually obvious.

Insist that all the fasteners be used to secure removable panels, rather than allowing panels to hang in place with one or two bolts. This keeps panels from

WESINC

Fig. 1 Cooling tower operating with an access door removed Air is short circuited to the fan, bypassing the water to be cooled. This radically increases the condenser water temperature, reduces the cooling capacity of the tower, and keeps the fan running.

blowing loose, and motivates people to replace them properly.

Make Access Doors and Panels Easy to Secure

A common problem is neglecting to replace access panels that are not attached to the casing by a positive means. To reduce this problem, install retainers in a manner that motivates replacing the panels. For example, attach the panel to the casing with a cable in such a way that the panel dangles in front of the opening when the panel is removed. Ease of use is a key feature. Design and install access panels so that they are easy to install and easy to latch securely.

Keep Fill Intact

In older wet coolers and in wet coolers that have been rebuilt, some of the internal "fill" material that increases the evaporating surface area may be broken or missing. Serious defects are usually obvious. The corrective action consists of replacing the decayed, broken, or misplaced elements.

A more subtle problem may exist in rebuilt towers, where the geometry of the fill may be wrong or portions of the fill may be missing. When a cooling tower is rebuilt, check the geometry of the fill material against the latest version of the cooling tower drawings.

Keep Drift Eliminators Intact and Clear

Drift eliminators are baffles installed in the air discharge path to minimize the discharge of unevaporated water droplets, which are returned to the sump. Not all cooling towers have them. If drift eliminators are missing or defective, tower water consumption increases. Drift eliminators that are clogged by debris, slime, or other material reduce air flow and cooling effectiveness.

ECONOMICS

SAVINGS POTENTIAL: 0.1 to 2 percent of chiller system operating cost.

COST: Depends on the nature of the defect. Usually minor.

PAYBACK PERIOD: Usually short.

TRAPS & TRICKS

ANALYZE THE PROBLEM: When you find a big gap where none belongs, figure out what caused it. Then, fix it in a way that makes it unlikely to recur.

MEASURE **2.4.6 Avoid recirculation of air through the same or adjacent heat rejection units.**

The way a heat rejection unit is installed may cause air to circulate from the discharge back to the intake. The recirculated air has little remaining cooling capacity, so recirculation wastes fan energy, reduces the capacity of the cooling unit, and increases the increased condensing temperature.

A similar problem may occur where cooling units are installed adjacent to each other. In this case, the discharge from one cooling unit may be drawn into the intake of another.

Take a look at all your heat rejection units to see whether recirculation may be a problem. The possibility of significant recirculation varies with wind, temperature, and humidity. Consider all the possibilities.

How Much Energy is Really Being Wasted?

You may be able to spot a recirculation problem by observing the vapor plume of a heat rejection unit, but don't expect to judge the severity of the problem from the appearance alone. A few wisps of water vapor recirculating back to the intake do not necessarily indicate a significant problem, although they may. On the other hand, serious recirculation may occur invisibly during warm or dry weather, when the discharge is invisible.

Judge the potential for energy waste by examining the installation to see whether it has any of the characteristics that tend to cause recirculation, which are explained below. Then, if you suspect trouble, you can estimate the percentage of recirculation easily by taking air temperature or humidity measurements under a variety of typical wind conditions.

The air temperature and humidity at the intake of the heat rejection unit should be the same as the

temperature and humidity of free air away from the unit. If the intake air is warmer or more humid, then some recirculation is occurring.

To estimate the amount of recirculation in dry coolers, use Formula 1, making the air temperature measurements indicated.

Formula 1 is not accurate for wet coolers, because they perform a large fraction of cooling by evaporation. Formula 1 may still give you a rough estimate of recirculation under humid conditions, when evaporation is limited.

For an accurate estimate of recirculation in a wet cooler, use Formula 2. Measure the humidity with a psychrometer (see Figure 1 in Measure 2.2.2.1) or relative humidity sensor, and convert to absolute units by using a psychrometric chart.

If the cooling unit does not have a localized intake point where you can take readings, take an average of

---

SUMMARY

Corrections for common installation problems that waste fan energy, reduce chiller efficiency, and reduce capacity.

SELECTION SCORECARD

Savings Potential .... $ $

Rate of Return, New Facilities % % % %

Rate of Return, Retrofit.......... % %

Reliability .... ✓ ✓ ✓ ✓

Ease of Retrofit .... ☺ ☺

FORMULA 1

for dry coolers,

$$\text{Percent recirculation} = \frac{\text{intake temperature - ambient temperature}}{\text{discharge temperature - ambient temperature}} \times 100$$

where:
• the temperatures can be expressed in any consistent set of units

FORMULA 2

for wet coolers,

$$\text{Percent recirculation} = \frac{\text{intake humidity - ambient humidity}}{\text{discharge humidity - ambient humidity}} \times 100$$

where:
• humidity is expressed in absolute units of moisture content, for example, grains of moisture per pound of air

readings around the unit, especially upwind and downwind.

If you find that significant recirculation occurs under conditions that exist often, consider corrective measures.

What Determines the Amount of Recirculation

The design of the cooling unit and the way it is installed tell whether the unit is likely to have recirculation problems. Consider these are factors:

- *discharge velocity.* Recirculation is reduced by high discharge velocity, which throws the discharge air away from the unit. The discharge air is warmer than the ambient air, so it tends to rise clear of the unit once it is beyond the influence of air currents in the vicinity of the unit.

- *intake velocity.* Recirculation is increased by high inlet velocity, which creates a low-pressure area near the inlet that attracts recirculation. High intake velocity may be caused by a small intake size in the cooling unit. It may also be caused by a tight air passage leading to the unit.

- *fan modulation.* A variable-flow fan may increase recirculation by reducing the discharge velocity, especially at low loads. However, the intake velocity is reduced in the same proportion, which tends to reduce recirculation into the same unit.

WESINC

Fig. 1 Horizontal-flow cooling tower If wind blows opposite the direction of discharge, a certain amount of recirculation will occur. Recirculation is aggravated by the location of the tower alongside a wall. This type of tower causes serious recirculation to any other heat rejection units located in the general direction of the discharge.

Furthermore, fan speed is controlled in response to cooling load, so any loss of cooling capacity is compensated automatically by increased fan output. Fan power is wasted, but chiller efficiency may not be seriously affected. Fan modulation is not likely

WESINC

Fig. 2 Cooling tower installed in a well This installation invites serious recirculation. A stack has been installed on the fan on the right to minimize recirculation. Even so, you can see that some of the discharge is being drawn back down into the cooling tower. The discharge plume is visible because of the high relative humidity in this location.

to cause recirculation unless the air flow around the tower is confined.

- *wind speed.* Wind blows the discharge air in a downwind direction. Wind also creates a suction on the downwind side of the cooling unit because the unit is not streamlined. Therefore, discharge air tends to tumble into the downwind side of the same or adjacent units. You can reduce this effect by shielding the unit from wind, but the structure used for shielding may cause recirculation itself, as explained below.

- *orientation and wind direction.* If a heat rejection unit does not discharge vertically, its orientation is important. Figure 1 shows a unit that discharges horizontally. If the unit discharges into the wind, a significant amount of recirculation is likely. Also, a unit that discharges horizontally is likely to inject warm air into adjacent cooling units.

More generally, an air inlet located on the downwind side of any cooling tower tends to recirculate air. Wind direction is variable, so some recirculation is likely to occur for part of the time.

- *enclosures and surrounding structures.* Serious recirculation is likely if a heat rejection unit is surrounded. For example, the unit may be hidden by a decorative enclosure, or it may be installed in a well. The problem is especially severe if the enclosure rises to the top of the cooling unit or higher. Figure 2 shows an extreme case. This geometry forces the incoming air to pass near the exiting air stream, which invites mixing. This effect is aggravated by wind, which tips the discharge air column into the entering flow.

- *adjacent structures.* Tall parapets and adjacent buildings that are upwind of the cooling unit may have strong downdrafts that mix the discharge air with the entering air. In some cases, a downdraft may exist far downstream of its source.

All other factors being equal, the least recirculation occurs in induced-draft heat rejection units, i.e., units in which the fans are located at the discharge end of the unit. These units have relatively high discharge velocity and relatively low inlet velocity. Most dry coolers use an induced-draft fan arrangement, as do most crossflow cooling towers.

Updraft, or counterflow, cooling towers typically have a forced-draft fan arrangement, i.e., with the fan blowing into the tower. These units have relatively low discharge velocity and relatively high inlet velocity, so they are more susceptible to recirculation.

How to Minimize Recirculation

The following is a menu of techniques for reducing recirculation. The best solution for your equipment is a matter of layout, cost, appearance, and other factors.

Consider combining some of these methods, if appropriate.

■ Modify the Enclosure to Reduce Recirculation

As explained previously, enclosures that surround a heat rejection unit tend to cause recirculation. The severity of this effect depends on the relative height of the enclosure. If the top of the enclosure is as high as the discharge point or higher, contact between the entering and leaving air streams is almost unavoidable. You can avoid this problem or reduce it by modifying the enclosure. Here are some ways to do this:

- *install cutouts at the bottom of the enclosure* to allow air to enter, as illustrated in Figure 3. Make the cutouts large enough. If there is a prevailing wind during the cooling season, try to install the cutouts on the windward side. This technique may be inexpensive, and it preserves the decorative screening effect of the enclosure, especially where the unit is installed on a roof.

- *substitute open latticework* for unbroken panels, as illustrated in Figure 3

- *reduce the height of the enclosure* as much as possible. If the unit is located well above viewers, the enclosure does not have to be as tall as the cooling unit to hide the unit completely. The esthetics of the building may even allow a glimpse of the upper portion of the unit.

- *increase the clearance* between the enclosure and the inlet sides of the heat rejection unit.

■ Install an Exhaust Stack on the Heat Rejection Unit

Adding a stack to the cooling unit is a way of raising the discharge point without raising the heat rejection unit itself. Figure 4 sketches how this is done. Figure 2 shows an actual retrofit of a stack to an existing cooling tower.

Figure 5 shows how a stack can reduce recirculation that is caused by the cooling tower being installed close to a wall.

The cross sectional area of the stack should be the same as the cross sectional area of the original discharge of the heat rejection unit. This is important to maintain the discharge velocity. A stack that is reasonably short, i.e., no taller than several diameters, does not significantly increase fan power.

If the stack is tapered toward the outlet, the discharge velocity is raised. The higher velocity reduces recirculation. However, the fan power requirement increases because the fan must add more kinetic energy to the discharge air. The reduction of recirculation probably does not justify the increase in fan power, except in specialized circumstances.

If you install a stack, brace it well to withstand wind. You may have to brace a stack directly to the building structure, rather than to the heat rejection unit itself, if the heat rejection unit does not provide a strong attachment point.

■ Raise the Heat Rejection Unit Above Surrounding Structures

Raising a heat rejection unit may be the only available option if the unit is tightly enclosed within a well, or if you want to exploit wind to aid evaporation.

Figure 4 sketches this solution. The cost of the supporting structure and the cost of modifying the piping makes it expensive.

The change in height does not affect cooling water pumping. The increased pressure needed to pump water to the top of the tower is balanced by the increased pressure in the return leg.

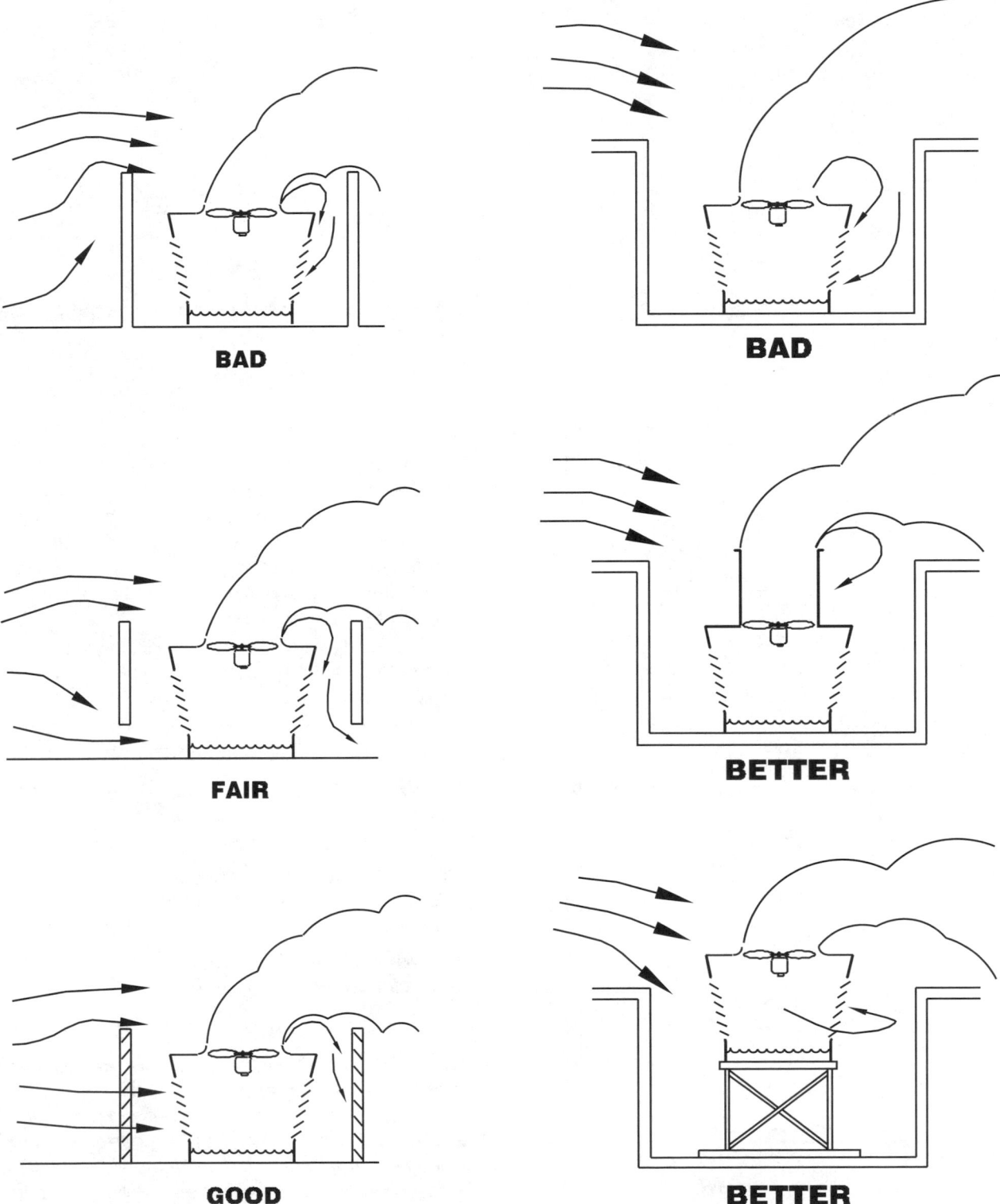

Fig. 3 How to reduce recirculation in an enclosure

Fig. 4 How to reduce recirculation in a well

■ Isolate Adjacent Heat Rejection Units

If one heat rejection unit discharges toward the intake of another, install a barrier to deflect the discharge. Figure 6 sketches an example.

If adjacent heat rejection units all discharge vertically, a recirculation problem is not likely unless there are aggravating factors, such as those discussed previously.

■ Relocate the Heat Rejection Unit

An approach of last resort is to move the heat rejection unit to some other location. This is very expensive because of the costs of rigging, piping, and fabrication of support structures.

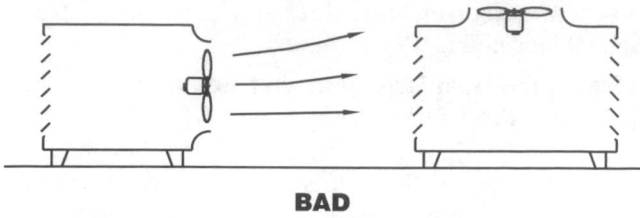

BAD

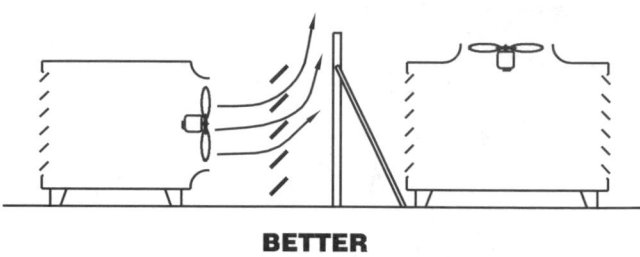

BETTER

Fig. 6 How to reduce recirculation from one heat rejection unit to another

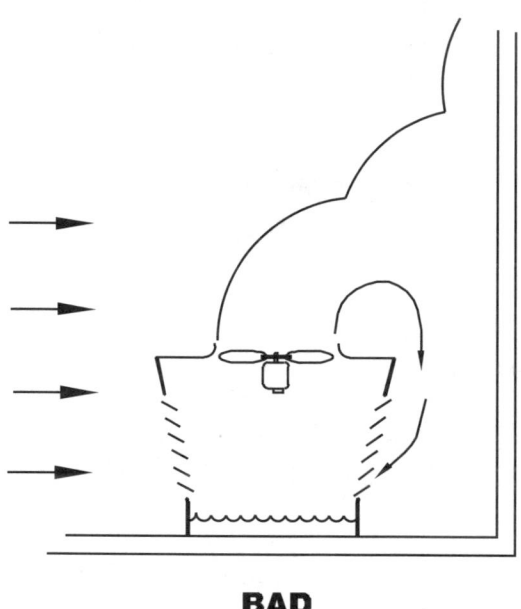

BAD

In deciding whether to move a heat rejection unit, consider whether you have an opportunity of moving the unit to a location where the ambient air is cooler, such as the shaded side of a building. This would be most valuable with dry coolers, whose condensing temperature relates directly to the dry-bulb temperature. Location is less relevant with wet coolers, whose condensing temperature depends on the wet-bulb temperature. The wet-bulb temperature of the air is not affected by shading.

Especially with wet coolers, consider any possibility of exploiting ambient wind to reduce fan power requirements.

ECONOMICS

SAVINGS POTENTIAL: *5 to 30 percent of the fan energy of the heat rejection unit. 0.1 to 2 percent of chiller energy consumption.*

COST: *In new construction, there may be little or no additional cost. In retrofit, modifications may cost less than one hundred dollars, to several thousand dollars.*

PAYBACK PERIOD: *Short, in new construction. One year to many years, in retrofit.*

TRAPS & TRICKS

ANALYZE THE PROBLEM: *Before you go to a lot of effort, make sure that you really have a problem. For example, do not worry about recirculation caused by strong north winds in winter if your chiller is loaded only during warm weather. Calculate the actual recirculation under conditions that matter. Then, if you have a problem, make sure that your intended solution will actually solve it.*

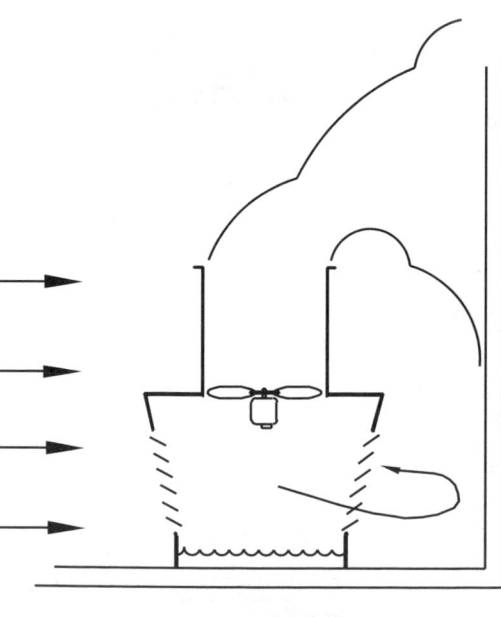

BETTER

Fig. 5 How to reduce recirculation against a wall

RATINGS

New Facilities | Retrofit | O&M

A | | **B**

MEASURE **2.4.7 Install fan and pump motors having the highest economical efficiency.**

See Measure 10.1.1 for the details. Also, consider the following points that relate specifically to motors in heat rejection units.

Electronic variable-speed drives modify the frequency, voltage, and waveform of the power provided to the motor. This makes it desirable to select motors intended specifically for variable-frequency drives. (See Measure 2.4.1.1 about installing these drives on cooling towers.)

For motors that drive variable-pitch propellers (see Measure 2.4.1.2), the part-load efficiency of the new motor is important, so check the efficiency figures for partial loads.

For single-speed motors, the only efficiency figure that matters is the efficiency at the percentage of load at which the motor operates. This percentage should be 60% to 90% of the motor's full-load rating.

Be prepared to replace a failed motor with a model that has the highest practical efficiency. If a motor is exceptionally large and inefficient, you may even want to replace it before it fails.

Pump Energy Consumption

The Measures in this Subsection are equipment modifications and operating practices that reduce the energy required to pump chilled water and condenser cooling water. These Measures apply to systems of any size, from a small building to a district cooling system that serves an entire city.

A NOTE ON UNITS

It is common practice to specify pressures in water distribution systems in terms of height of water column, such as "feet of head." One foot of water column equals 0.43 PSI. The pressure output of pumps and the flow resistance of coils and other equipment are both expressed this way. This simplifies system pressure calculations by equating pressure requirements due to gravity ("static head") with pressure requirements due to pipe friction and equipment resistance.

INDEX OF MEASURES

2.5.1 **Adjust the discharge of pumps to match system flow and/or pressure requirements.**

    2.5.1.1 **Trim pump impellers.**

    2.5.1.2 **Throttle pump discharge valves.**

2.5.2 **Install variable-flow chilled water distribution.**

2.5.3 **Install pump motors having the highest economical efficiency.**

RELATED MEASURES

- Measure 2.1.3, efficient control of condenser cooling pumps
- Measure 2.1.4, efficient control of chilled water pumps
- Measure 2.1.5, avoiding unnecessary operation of standby pumps

MEASURE 2.5.1 Adjust the discharge of pumps to match system flow and/or pressure requirements.

Pumps are used in chiller systems for distributing chilled water and for circulating condenser cooling water. If the pumps operate at constant speed, adjust their output so that they provide the minimum flow needed for efficient system operation. This reduces energy input, and in some cases, it may also reduce tube erosion and improve control valve performance.

The subsidiary Measures are two common methods of adjusting pump output. Measure 2.5.1.1 is much more efficient. Measure 2.5.1.2 is a stopgap measure that is quick and cheap.

Matching Flow Requirements

In both chilled water and condenser water systems, the flow rate has a major effect on heat transfer efficiency, and hence chiller efficiency. These flow rates are specified by the chiller manufacturer. When you make adjustments, err on the side of high flow.

The need for chilled water flow may also be determined by the cooling equipment that uses the chilled water. This occurs if the flow rate needed by the chilled water system is higher than the flow rate needed by the evaporator. Be sure to accommodate the needs of the cooling equipment.

The minimum flow rate of condenser water pumps may also be determined by the need of the cooling tower for a minimum flow rate. This would be unusual. It might occur if the cooling towers, or individual cells, are connected to chillers in odd ways.

Alternative: Adjust Impeller Speed

The output of a pump can be reduced efficiently by reducing the pump speed. The effects on pump flow, pressure, and power input are almost the same as for impeller trimming. To reduce pump speed on a one-time basis, you have to change the motor or insert a speed changing device between the motor and pump. These are unconventional modifications. In chilled water distribution (see Measure 2.5.2), using a variable-speed drive can save large amounts of energy in many applications.

MEASURE 2.5.1.1 Trim pump impellers.

Some pumps in a chiller plant may develop more pressure than needed, usually because of excess capacity that is built into the design as a safety factor. This wastes energy, so consider adjusting the pump pressure to the minimum needed for proper operation. The most efficient method of doing this is reducing the diameter of the pump impeller. See Measure 10.2.1 for the details.

RATITINGS
New Facilities · Retrofit · O&M

MEASURE **2.5.1.2 Throttle pump discharge valves.**

If the capacity of a pump is excessive, you can save a small amount of energy by partially closing a valve in the pump discharge line, or by placing some other restriction in the discharge. This method is much less efficient than impeller trimming, but you can do it easily, and it is reversible. See Measure 10.2.2 for the details.

MEASURE **2.5.2 Install variable-flow chilled water distribution.**

Variable-flow chilled water distribution is a major advance in chiller system efficiency that became popular in the late 1970's. It saves a considerable amount of the energy required for chilled water distribution, while preserving good regulation of the chilled water pressure that is supplied to the end-use equipment.

Variable-flow pumping had been possible in earlier years, but the development of efficient variable-flow pump drives was probably the main stimulus that motivated designers to attempt it. The layout is simple, but understanding it requires some study. There are traps to avoid, especially in the many variations that are possible. The basic layout may seem too simple to be true. As a result, early variable-flow systems had complications that were unnecessary and counterproductive.

In new construction, consider variable-flow pumping for all but the smallest systems.

In existing facilities, you can retrofit variable-flow chilled water pumping in existing chiller systems. The feasibility of retrofit depends on the size of the system and the amount of equipment that needs to be modified. Variable-flow pumping requires a larger number of pumps than conventional chilled water distribution, although the total pump capacity is no greater. Expect to modify piping within the chiller plant. Also, you will probably want to modify at least some of the chilled water control valves at the end-use equipment.

The Basic System

Figures 1 and 2 compare an old-style constant-flow chilled water distribution system with a basic variable-flow system. The main difference is that the variable-flow system consists of two loops, each of which has its own pumps. We will call the two loops the "chiller loop" and the "load loop," respectively. (The terminology is

> **SUMMARY**
> Saves a large fraction of chilled water distribution energy. Increases average chiller efficiency in multi-chiller installations. Reduces heating of chilled water by pumping losses. Requires additional pumps. In retrofit, requires some piping modifications, and probably, replacing cooling equipment control valves. Simple in layout. Tricky to design. Reliable, but only if operators understand it.
>
> **SELECTION SCORECARD**
> Savings Potential .... $ $ $
> Rate of Return, New Facilities % % % %
> Rate of Return, Retrofit .... % %
> Reliability .... ✓ ✓ ✓
> Ease of Retrofit .... ☺ ☺

not standardized.) The two loops share a common "bypass line."

The easiest way to understand the system is to visualize the chiller loop, including the bypass line, as an independent system. It circulates water through the chillers at a constant flow rate that is independent of the load. The load loop taps chilled water from the chiller loop as necessary to satisfy the cooling load. The number of chillers that operate in the chiller loop is controlled by the cooling load.

These are the characteristic features of a variable-flow system:

• *The evaporator circulating pumps are sized to serve individual chillers.* Each pump is sized for the flow required by the evaporator of one chiller, and the pump produces only enough pressure to overcome the resistance of the evaporator. Only one pump runs for each operating chiller.

- *The flow rate in the chiller loop is always higher than the flow rate in the load loop.* Therefore, the direction of flow in the bypass line is always as indicated. (This is true of the basic system of Figure 2. It is not true for one variant of the system, which is discussed below.) This is necessary to

maintain the chilled water temperature in the load loop. The direction and volume of water flow in the bypass line control the number of chillers that operate, as described below.

- *Each operating chiller is controlled individually,* to maintain its own chilled water supply temperature. There is no need for collective chiller control. Centrifugal and screw chillers use conventional controls. Return water temperature

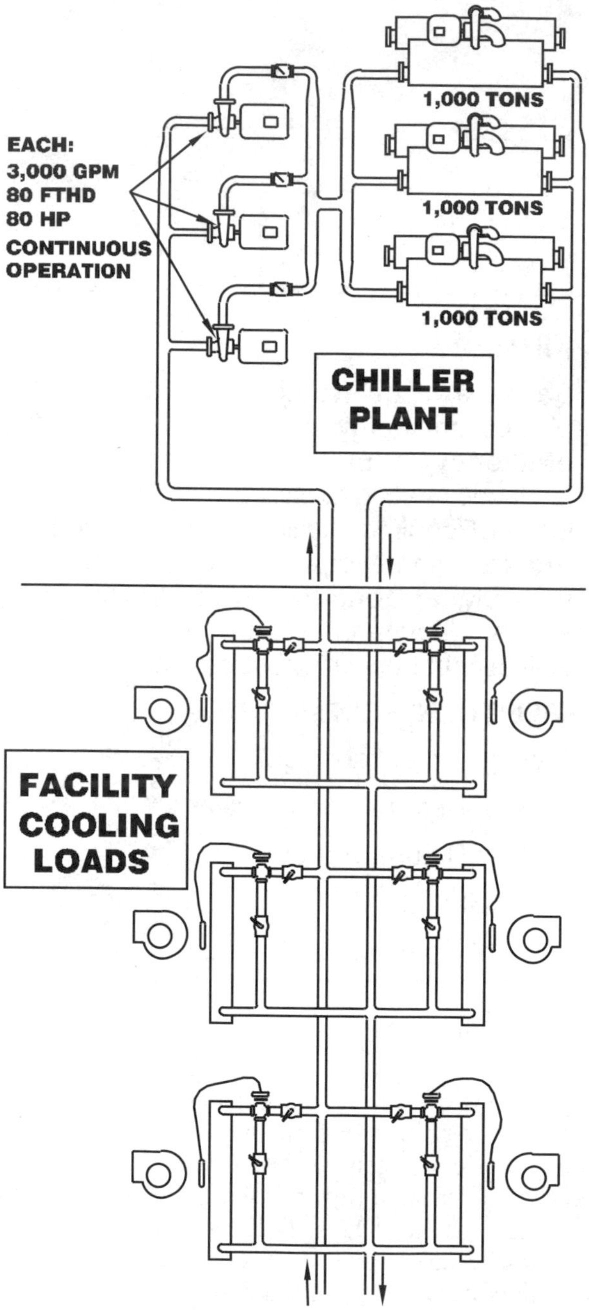

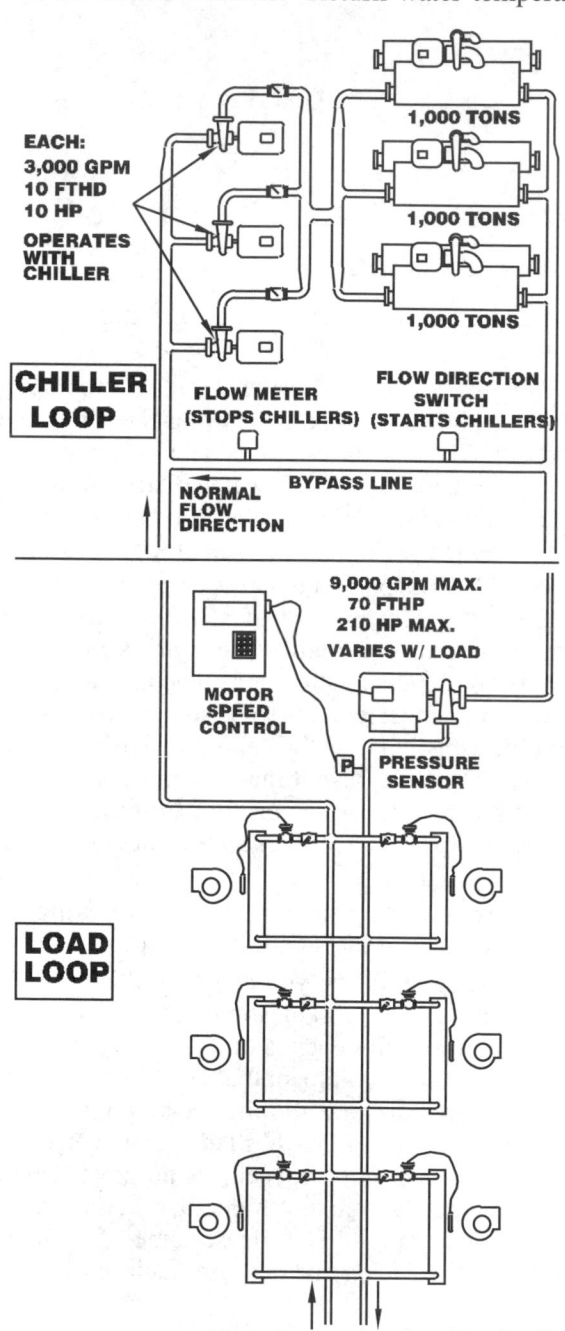

Fig. 1 Constant-flow chilled water system The main features of this system are the constant flow requirement of the end-use equipment and the series connection of the chiller plant to the chilled water distribution system. This keeps the pump load high at all times, regardless of the cooling load. It also reduces the average chiller efficiency at partial cooling loads.

Fig. 2 Variable-flow chilled water system The main feature of this system is that the flow through the chillers is uncoupled from the flow to the facility cooling equipment. This makes it possible to independently optimize pumping of water through the evaporators and pumping to the facility loads. In turn, this allows water flow through the chillers and water flow to the facility loads to be reduced in proportion to the cooling load.

control is not practical, so reciprocating and scroll compressors may need special controls to adapt them to supply water temperature control. Chilled water temperature reset (recommended by Measure 2.2.1) may be done in the usual manner.

- *All the equipment that uses chilled water is controlled by throttling (two-way) valves,* instead of by bypass valves. This allows the flow rate in the load loop to be reduced to match the cooling load, thereby saving pumping energy.
- *The load loop pump, or pumps, are selected and controlled to maintain the pressure needed by the load loop.* The load loop pumps satisfy the pressure requirements of the end-use equipment, and overcome the resistance of the distribution piping. Their characteristics are independent of the chiller characteristics.

These are the essential features. Design and installation details are explained below, along with more complicated variations of the system.

Energy Saving Potential

■ Pump Energy Savings

The power consumed by any pump is proportional to the pressure difference produced by the pump, multiplied by the flow rate through the pump. In a conventional constant-flow chilled water system, the chilled water pumps continuously deliver the flow rate and the pressure that are required at maximum load. This is wasteful because both the flow and pressure requirements decline as the load drops. Furthermore, the flow requirement varies differently from the pressure requirement.

The variable-flow system saves energy by making it possible to optimize the pump pressures and the pump flow rates separately, and to continuously adjust these values as the cooling load changes.

To illustrate the potential for saving pump energy, consider a typical constant-flow system that uses three 1000-ton chillers, as in Figure 1. The chilled water flow rate is based on 3,000 tons, because this is the capacity of the chillers. This flow rate is constant, regardless of the number of chillers in operation, because the pumps must continue to satisfy the flow requirements of the bypass valves on the end-use equipment.

If the chillers produce maximum output at a temperature differential of 8°F, the chilled water flow rate is 9,000 GPM. The pumps produce a total pressure of 80 feet head, which is needed to overcome these system pressure drops:

- 10 FTHD (12.5%) in the evaporator
- 30 FTHD (37.5%) in the end-use equipment
- 40 FTHD (50.0%) in distribution piping loss.

With pumps and motors of average efficiency, the chilled water pumping requires a total continuous power input of 300 horsepower.

Now, compare this to the variable-flow system shown in Figure 2. To calculate the savings, you need to know the load profile of the system. In this example, experience shows that the peak load is 2,400 tons and the average load is 800 tons. Also, you have recorded that one chiller operates for 100% of the cooling season, a second chiller operates for 40% of the season, and a third chiller operates for 10% of the season. From this, you calculate these savings:

- *in the evaporator.* From the length of time that each chiller operates, it is easy to calculate that the average quantity of chilled water flow through the evaporators is 50% of the maximum. The pressure drop across the evaporators is always the same. Therefore, the energy needed to pump water through the evaporators is reduced to 50% of the original amount.
- *in the end-use equipment.* The equipment end-use pressure requirement remains unchanged. The flow rate is roughly proportional to the cooling load. Since the average load is 800 tons and the original pump was sized for 3,000 tons, pumping energy is reduced to 27% of the original figure. Pump speed reduction incurs an additional 20% drive and pump loss, which increases this figure to 32%.
- *in the distribution piping.* Pumping energy to overcome pipe friction is proportional to the third power of the flow rate. You would need to know the precise load profile to estimate this precisely. You probably can't get this information, so assume that this portion of pumping energy, including drive loss, is 30% of the original.

Adding these savings together yields an overall savings of two thirds of the original pump energy, or an average saving of 200 horsepower for the duration of the cooling season.

■ Savings from Eliminating Chilled Water Temperature Mixing

Many constant-flow chilled water systems need to maintain maximum chilled water flow at all times because the end-use equipment uses bypass valves to control capacity. (Some end-use equipment, such as multizone air handling units, may have no control valves at all.) Without keeping full pump capacity running, the pressure at the end-use equipment would become too low to provide full flow to any equipment that needs it. In this situation, the pumps must maintain full evaporator flow through all chillers, even those that are turned off. Otherwise, the flow velocity through the active chillers would be excessive.

Unfortunately, this reduces the efficiency of the chillers that are running. The chilled water produced by the operating chillers is mixed with return water passing through the idle chillers. As a result, the active chillers must produce chilled water at a lower temperature than if mixing did not occur. Lowering the chilled water temperature reduces chiller efficiency, for the reasons explained in Measure 2.2.1. The loss in chiller efficiency is typically about two percent per degree Fahrenheit.

For example, consider a plant with three identical chillers in which only one chiller is operating. The return water temperature is 48°F and the system requires 46°F chilled water to satisfy the cooling load. Mixing through the idle chillers forces the operating chiller to produce chilled water at 42°F, which is 4°F lower than the water temperature needed by the end-use equipment.

Variable-flow pumping completely eliminates this inefficiency.

■ Savings from Reduced Heating of Chilled Water by Pump Losses

Most of the energy consumed by a chilled water pump eventually is converted to heat in the chilled water. With a chiller of average efficiency, this pumping heat requires an additional cooling energy input equal to about 20% of the pump power. A variable-flow pumping system reduces this energy penalty in proportion to the saving in pump energy.

Where to Use Variable-Flow Pumping

The potential benefit of variable-flow chilled water pumping depends on these factors:

- *the size of the pumps.* Variable-flow pumping is more economical in larger systems. In new construction, variable-flow pumping may be less

WESINC

Fig. 3 Bypass control valve for an air handling unit This type of control keeps the water flow requirement essentially constant, regardless of the load. Replacing bypass valves with throttling valves is a basic step in converting to variable-flow pumping.

expensive than a constant-flow system, above a certain size.

- *the cooling load profile.* Variable-flow pumping produces its benefit when the cooling load is low. Therefore, it is most economical in systems with long durations of reduced load.
- *whether the system uses bypass valves.* Constant-flow systems commonly use bypass valves on the user equipment, or they may use a single large bypass valve that is located near the pump discharge. Either way, bypass valves force all the system's pumps to operate under all conditions. If variable-flow pumping is installed, no bypass valves are needed. In retrofit, it is usually economical to replace bypass control valves on equipment. Or, you may be able to convert existing bypass control valves to act as throttling valves.

  (If a constant-flow system uses several pumps, you can save energy in a similar way by turning chilled water pumps on and off in relation to the cooling load, after modifying the valves. See Measure 2.1.4 for details. This method does not save as much energy as a variable-flow system, but it is much cheaper to retrofit.)

- *whether distribution system resistance is a major factor.* If pipe friction accounts for a large fraction of pump power, variable-flow pumping can greatly reduce the energy requirement. This part of pump power is proportional to the third power of the flow rate.
- *the potential for using chilled water temperature reset.* Chilled water supply temperature reset, which is explained below, reduces the advantage of variable-flow pumping. This is because it tends to keep the flow in the load loop more constant. The potential for chilled water temperature reset is limited if the different loads served by the system do not increase and decrease together.
- *whether it is desirable to load certain chillers preferentially.* A variable-flow chilled water system allows you to load some chillers before others. This feature can improve average chiller efficiency when recovering heat from a chiller at elevated temperatures. It may also be valuable when the chillers have differing efficiency characteristics.

Some of the variations of the basic system, which are explained below, increase the advantage of variable-flow pumping, while other variations reduce the advantage.

Design and Installation Details of the Basic System

Although variable-flow pumping is simple in layout, the details are critical. To avoid confusion, let's start by considering only the factors that relate to the basic layout of Figure 2. We will cover the variations a few pages later.

■ Install Throttling Control Valves on the End-User Equipment

The cooling equipment that is served by the chilled water system should be controlled with simple throttling valves. This keeps the total flow in the load loop proportional to the cooling load, which minimizes pump power.

Select the valve actuators to keep the valves closed against the maximum pressure that may develop in the load loop. Maximum pressure occurs when a failure of the variable-flow controls causes the pump(s) to develop maximum output under low-load conditions.

Avoid oversizing any throttling valves that you install. Oversized valves wear out rapidly at low load, and they may cause noise and control problems.

In conversion projects, you may find that the existing end-use cooling equipment is controlled by bypass valves. Constant-flow chilled water systems use bypass valves specifically to maintain a constant flow rate in the chilled water system. The control valves bypass water that is not needed for cooling to the return line. Figure 3 shows this common arrangement on an air handling unit.

(There are two main reasons for maintaining constant flow in the original distribution system. One is to keep the flow rate through the chiller evaporator high enough to maintain good heat transfer. Another is to prevent cavitation in the chilled water pumps, which wears out impellers and makes noise. These requirements do not exist in a variable-flow system, because the flow through the evaporators and pumps is not tied to the flow in the end-use equipment. Still another reason for the popularity of bypass valves is a superstition among engineers that they are needed to provide stable control.)

You may be able to convert existing bypass control valves to operate as throttling valves. If you block the bypass line, the bypass valve becomes a throttling valve. The valve is likely to behave in a satisfactory manner because the variable-flow operation of the pumps in the load loop maintains a fixed pressure on the valve. But, check the valve performance to be sure.

This is especially easy if your end-user equipment has balancing valves in the bypass line. In this case, simply close the balancing valves. Install permanent tags on the balancing valves that state that the valves should be kept in the closed position.

If this cheap expedient does not work well, replace the bypass control valves with throttling control valves. It may not be economical or even necessary to replace all the bypass control valves in the system. For example, in a system that serves large air handling units and small fan-coil units, it may only be economical to replace the valves on the air handling units.

Some types of end-use equipment do not use any chilled water control valves, so they always take maximum flow. Typically, such equipment needs to maintain a constant low temperature. A common example is multizone air handling units, which regulate air temperature by mixing chilled air from a chilled water coil with warm air from a heating coil. You can install throttling valves on such equipment, because they typically need only a fraction of maximum coil flow for much of the time. (Throttling valves also make it possible to use cold deck temperature reset with multizone air handling units. See Measures 4.6.1 ff for this.)

■ How to Install the Evaporator Circulating Pumps

In the chiller loop, flow of water through the evaporators is provided by a separate pump for each chiller. Each pump operates only when its chiller is operating, and each pump operates at constant flow.

Select the pump pressure only to overcome the resistance of the evaporator, because the pump plays no role in circulating water through the load loop. Select the pump flow rate to optimize the efficiency of the chiller, following the chiller manufacturer's specifications. Evaporator flow rate involves a compromise between pump power and evaporator temperature. Also, a flow rate that is too low increases fouling, while a flow rate that is too high hastens tube erosion.

Converting an existing system to variable-flow operation requires new evaporator pumps. The conversion reduces the pressure requirement in the chiller loop so much that the existing chilled water pumps cannot be converted efficiently for evaporator circulation.

■ Install Evaporator Check Valves

Each evaporator should have a check valve in its discharge line to prevent supply chilled water from flowing backward through the evaporator when the chiller is off. Backflow would lower the return water temperature to the running chillers, reducing their efficiency. Also, it would waste a certain amount of pump energy.

In conversions, the original system may have motorized shutoff valves that close when the chiller turns off. These can serve the same purpose as check valves. However, check valves are more reliable because they require no power source and no controls.

■ How to Start and Stop Chillers and Pumps

In a variable-flow system, chillers are started and stopped individually in response to the flow demands of the load loop. (The flow in the load loop is proportional to the cooling load.)

The flow rate in the chiller loop should always be higher than the flow rate in the load loop. When this condition is met, the direction of flow in the bypass line is as shown in Figure 2. If this condition is not met, some return chilled water is recirculated through the bypass line. This would dilute the chilled water from the chiller loop, raising the chilled water supply temperature.

Subject to the previous condition, the flow rate in the chiller loop should be kept at the minimum needed to satisfy the flow requirement of the load loop. The flow rate in the chiller loop is determined by the number of chillers in operation, since the flow through each evaporator is fixed. Operating more chiller capacity than needed increases circulation of chilled water within the chiller loop. This lowers the evaporator temperature, reducing efficiency. Also, operating excess chiller capacity is inefficient, for the reasons given in Measure 2.1.1.

The individual chillers are started and stopped by the two sensors in the bypass line, the flow direction sensor and the flow rate sensor. The flow direction sensor starts an additional chiller if it senses that the flow direction is about to reverse from its normal direction. The flow rate sensor stops a chiller when it senses that the flow in the bypass line exceeds the flow produced by the last chiller in the starting sequence.

If the chillers are not identical, start them in order of their efficiency. See Measures 2.1.1 ff for details.

■ How to Control Chiller Loading

Control of chiller output in a variable-flow system is simple for centrifugal and screw chillers. All chillers that are running are controlled to maintain a fixed chilled water supply temperature. Each operating chiller uses its individual controls in the normal manner to maintain a set chilled water supply temperature. There is no need for collective chiller control. When converting from a constant-flow system, you usually do not have to make any changes to the chiller controls.

The most common practice is to set all chillers to maintain the same supply chilled water temperature. Since all chillers receive return water at the same temperature, this approach loads all chillers to the same percentage of full load, regardless of size or type. With chillers having similar characteristics of efficiency versus load, this is the most efficient way of distributing load. With chilled water temperature set in this way, you achieve the best overall efficiency simply by starting chillers in the order of their efficiency.

To ensure uniform distribution of load, set the chilled water supply temperatures of all chillers to the same value. Small differences in the temperature settings cause significant shifts in loading among the chillers.

In some cases, you may want to load some chillers at a higher percentage of load then others. This is somewhat tricky, and usually reduces overall efficiency somewhat. Methods of preferential loading are covered below.

Reciprocating chillers and other types that control output in abrupt steps usually control their output by sensing the return water temperature. Conventional return water controls cannot maintain a fixed supply

water temperature, because they would hunt in attempting to follow the abrupt changes in supply temperature that occur as stages turn on and off. To control staged chillers on the basis of supply chilled water temperature, you need more sophisticated controls that anticipate the supply water temperature changes that occur as the chiller changes its output. Such controls are available commercially.

(Return water temperature is not feasible in variable-flow systems, because the return water temperature is not a reliable indication of the system load. A variable-flow system tends to hold the return water temperature constant, regardless of load. Also, the return water temperature is affected by the relative flow rates in the chiller and load loops, which fluctuate widely as chillers are started and stopped.)

■ How to Design the Bypass Line

The flow of chilled water through the bypass line is completely unrestricted. The line should be large enough to accommodate the full evaporator flow of the largest chiller with minimal resistance. The size of the bypass line has minimal effect on system cost because the bypass line can be very short.

The flow rate and flow direction sensors are installed in the bypass line. The configuration of the line should allow these sensors to operate accurately.

■ How to Control the Distribution (Load Loop) Pumps

Control the chilled water distribution pumps to maintain a constant pressure in the load loop. The pressure setting is determined by the needs of the cooling equipment that is served by the chilled water system.

The most efficient and precise way to control the load loop pressure is to regulate the speed of the distribution pumps. The best way to do this is to install electronic variable-speed drives for the pump motors. See Reference Note 36, Variable-Speed Motors and Drives, for details.

In a retrofit project, installing an electronic variable-speed drive allows you to use the existing pumps. A single electronic drive can operate several pumps. If extra pumps are installed as standby units, an electronic drive can easily be switched between pumps. However, you may want to upgrade the pump motors for the reasons given in Reference Note 36.

If the original chilled water distribution system uses several pumps in parallel to achieve maximum system flow, another possibility is operating these pumps in sequence to provide variable flow. The abrupt changes in pressure as an individual pump turns on or off may cause some control oscillation at the end-use equipment, but this may be tolerable.

Pressure losses in the distribution piping decline as the flow rate drops. This may be significant in a long distribution system. In such systems, you can save pump energy by reducing pressure enough to exploit the reduced piping losses. To do this, simply locate the pressure sensor that controls pump output near the end of the distribution system. In a large system, this may put the pressure sensor miles away from the pumps. The cost of the signal line is minimal in comparison with the additional energy saving.

In principle, you could save additional pump energy under reduced-load conditions by lowering the pressure setting as the cooling load drops. This exploits that fact that the end-use equipment needs less flow at low loads. However, this poses the risk of inadequate pressure for any item of end-use equipment that may need to operate at high load while the rest of the system is at low load.

Effect of Chilled Water Temperature Reset

Raising the chilled water supply temperature increases chiller efficiency. (Measure 2.2.1 explains why.) In constant-flow systems, a good rule is simply to keep the supply water temperature as high as possible. However, in a variable-flow system, this is not always the most efficient method of operation. The reason is that raising the water temperature requires more water to be pumped for a given cooling load, increasing pump power.

Calculate the optimum compromise between chilled water temperature reset and pumping power for your installation. If the distribution system does not have much resistance, you will probably find it most efficient to keep the chilled water temperature as high as possible. However, this may not be true in systems that have long distribution systems, where distribution pumps consume more power in comparison with the chillers.

If the plant has more than one chiller, the variable-flow system increases the opportunity for chilled water temperature reset, by eliminating mixing of supply water with return water. (See "Savings from Eliminating Chilled Water Temperature Mixing," above.)

Variations of the Basic Layout

Variable-flow chilled water distribution provides a great deal of flexibility for improving pumping efficiency. The following are variations that you may want to consider. Variations from the basic layout are more complex in behavior, and they may have unwanted effects. Approach them carefully.

■ Mixing Chiller Types

The basic variable-flow system that we described previously can handle any combination of chiller types and sizes without complications. This is because each chiller has its own independent evaporator pump, and because circulation of water in the load loop is independent of the chiller characteristics. This makes it easy to add chillers to increase capacity or to improve system efficiency.

A requirement when mixing chiller types is that all the chillers must produce full output at the same supply and return temperatures. To achieve maximum overall efficiency when chillers of different types are operating simultaneously, they should have similar efficiency-versus-load characteristics (but not necessarily the same efficiencies). See Measure 2.1.1 for more about this.

■ Parallel or Dedicated Evaporator Pumps

The evaporator circulation pumps can be arranged in parallel, as shown in Figure 2, or each chiller can have a dedicated evaporator pump. The latter arrangement is shown in Figure 1 of Measure 2.1.4. The choice does not affect the operation of the variable-flow system, but it has some other aspects that affect reliability and efficiency.

The parallel arrangement allows you to add a reserve pump (or more than one) in parallel with the others. This provides a high degree of pumping reliability at modest cost. However, the presence of a reserve pump increases the likelihood of running more pumps than necessary. See Measure 10.4.1 for ways of avoiding this problem.

Installing a dedicated evaporator pump for each chiller makes it easier to avoid unnecessary pump operation, but it has the disadvantage that failure of an evaporator pump will disable the chiller that it serves. This is not a serious limitation in typical installations with more than one chiller. Centrifugal pumps are reliable, and they are less likely to fail than the chiller itself. In order for a pump failure to reduce cooling output, it would have to occur at a time when the remaining chillers are incapable of carrying the load.

■ How to Accommodate Diverse Pressure Requirements Efficiently

If some chilled water users or some parts of the distribution system require higher pressure than others, pump power is wasted because the load loop pumps must operate at the highest pressure required in the system. In this case, try to refine the layout of the load loop to accommodate the different pressure requirements efficiently.

One method is to use separate distribution pumps and piping for loads having different pressure requirements. This may be economical if only minimal extra piping is needed, for example, if high chilled water pressures are needed in one piping leg and lower pressures are needed in another. Piping is expensive, so it is rarely economical to have separate piping systems running in the same direction just to carry different pressures.

Another approach is to use booster pumps feeding from the main piping to serve loads with higher pressure requirements. Variable-flow booster pumps may be economical with larger loads. For smaller loads that do not merit the cost of a variable-flow booster pump, use a small constant-flow pump. In these cases, retain bypass valves on the end-use equipment. Control the booster pumps so that they operate only when needed.

■ The Ultimate in Distribution Efficiency

All conventional distribution systems use pumps to create enough pressure to provide maximum flow through the end-use devices. The flow is regulated by valves. This is true of contemporary variable-flow systems, also. The pumping arrangements discussed previously reduce the excess energy output of the pumps, but they still waste energy in two ways. One is the dissipation of energy across all the control valves and

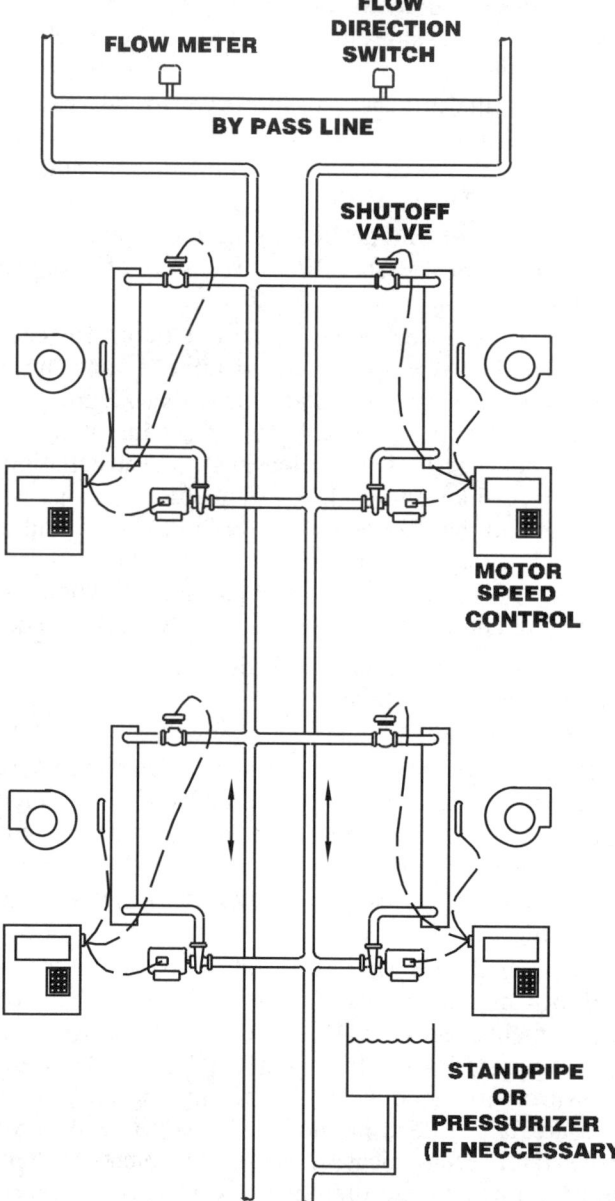

Fig. 4 The ultimate in pumping efficiency This arrangement uses a separate variable-flow pump for each item of end-use equipment. It eliminates all control valves, thereby eliminating the pump energy needed to overcome the resistance of the valves. It requires no central distribution pump. However, it may require a system pressurizing tank to ensure adequate suction pressure at each pump.

balancing valves in the system. The other is the excess energy dissipated if different end-use equipment has different pressure requirements. The previous heading dealt with the latter issue, but as we saw, it is not practical to completely eliminate this cause of energy waste if there is a great deal of diversity in pressure requirements.

However, we can completely eliminate energy waste of both types by taking an entirely different approach, which is shown in Figure 4. This is to install a separate variable-flow pump for each load, and to control each pump thermostatically. For example, the pump may be controlled to maintain a particular coil discharge temperature or a particular space temperature. This approach allows you to eliminate control valves entirely, along with their energy losses. Flow through the equipment is controlled entirely by pump speed. In effect, each item of equipment is served by its own distribution pump. For this reason, balancing valves are no longer necessary, and their energy losses are eliminated, also.

In this arrangement, there is no need for a separate pump to force water into the distribution system. Each load pump is selected to have enough pressure capacity to account for its share of the resistance in the load loop. Because each pump is sucking from the load loop distribution pipe, make sure that there is enough static pressure in the load loop to provide adequate suction to all the load pumps. Otherwise, friction in the distribution lines may restrict flow to the load pumps enough to cause pump cavitation.

Each item of end-use cooling equipment needs an external valve to stop flow completely when cooling is not needed. A centrifugal pump is an open passage when it is turned off. In most cases, a check valve should do the job. This is because the pressure on the discharge side of the equipment should always be higher than the pressure on the inlet side. (The opposite would be true if the load loop has its own distribution pump, as in the "conventional" variable-flow configuration.) Otherwise, there is no way of limiting flow in the forward direction. For this reason, the load loop cannot use a separate distribution pump to overcome the resistance of the pipe.

This idea is entirely novel, so be careful. Its time is coming because of the declining cost of variable-speed motor drives. This approach is already economical for larger items of end-use equipment, such as air handling units of medium and large size. Variable-speed motors are rapidly becoming economical in smaller sizes, such as those appropriate for fan-coil units.

■ How to Shift Chiller Loading by Adjusting the Supply Water Temperature

As discussed previously, the basic variable-flow system loads all operating chillers by the same percentage. This is usually the most efficient distribution of load, or nearly so. However, if you want to shift more of the load to a particular chiller, you can do this easily by setting the supply chilled water temperature of that chiller lower than the setting of the other chillers.

For example, consider a situation where two chillers are operating. You set the supply temperature of one at 45°F and the supply temperature of the other at 42°F. If the return temperature is 48°F, then the first chiller will carry one third of the load, and the second will carry two thirds.

Typically, you would load one chiller preferentially if it is more efficient than the rest. But, this is worthwhile only if unloading the other chillers does not make them much less efficient, which would forfeit the overall efficiency advantage. Make a careful analysis when assigning load between chillers with dissimilar efficiency characteristics. See Measures 2.1.1 ff for details.

A chiller's supply chilled water temperature setting has a strong effect on its efficiency. Therefore, when lowering the supply temperature of one chiller to take on more of the cooling load, raise the supply temperature of the other chillers. This minimizes the net energy loss due to reduced chiller efficiency.

■ **How to Load a Chiller Preferentially by Locating It Ahead of the Bypass Line**

A more radical method of shifting load to one particular chiller is to install that chiller, along with its evaporator pump, on the load side of the bypass line. We will call this chiller the "lead" chiller.

When only the lead chiller is operating, the system acts like the basic variable-flow system. The lead chiller's evaporator flow is kept constant by the chiller's evaporator pump. The load loop pump varies flow to match the cooling load, and any excess flow is recirculated through the bypass line..

The behavior of this configuration differs when any of the other chillers are operated. Because of the direction of flow in the bypass line, water returning from the load loop first enters the lead chiller. Only the return water flow that exceeds the evaporator flow of the lead chiller flows to the other chillers. As a result, the lead chiller is always loaded to 100% load, and the remaining chillers share the remainder of the load.

Typically, you might use this method of preferential loading to exploit an especially efficient chiller as the lead chiller. This is valuable if the lead chiller loses efficiency at lower loading. However, for this configuration to provide an overall benefit, the remaining chillers must operate at reasonably high efficiency at lower loads. For example, a centrifugal or screw chiller might be used as the lead chiller, while the rest are reciprocating chillers.

This method makes a permanent commitment to keeping the lead chiller loaded whenever it is running. Before installing this configuration, make sure that operating conditions of the plant will always favor it.

Alternatively, you can install some valves and short pieces of pipe to give you the option of "moving" the bypass line to the load side of the lead chiller. Before adding such a clever feature, consider that it may utterly confuse operating personnel.

This method is also used with chiller heat recovery, which is covered in Subsection 2.10. Preferential loading is valuable in cases where heat is recovered at

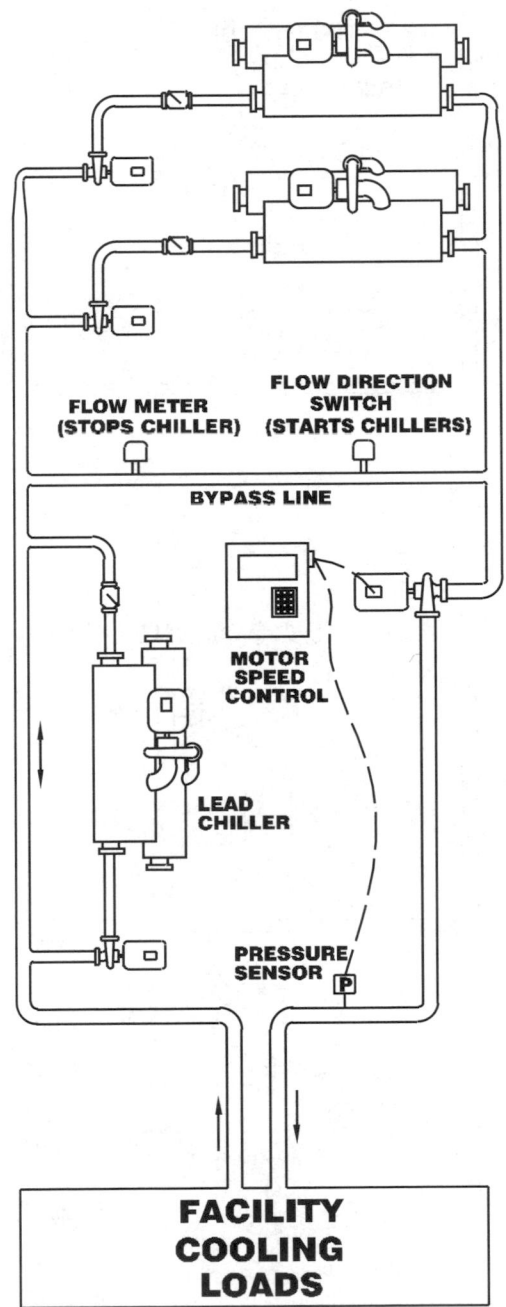

Fig. 5 Preferential loading arrangement for a heat recovery chiller The lower chiller, ahead of the bypass line, operates at a load that satisfies the heat recovery requirement. Additional cooling, when needed, is provided by the chillers above the bypass line. This provides the most efficient control of chilled water temperature.

elevated condensing temperature from a particular chiller, and you want to shift the load to this chiller.

Trouble may arise if the heat recovery chiller is controlled to satisfy the heat recovery load, rather than the cooling load. In this case, the chilled water supply temperature floats. When a second chiller is started to meet a rising cooling load, chilled water at the fixed supply temperature of the second chiller is mixed with chilled water at the floating supply temperature of the

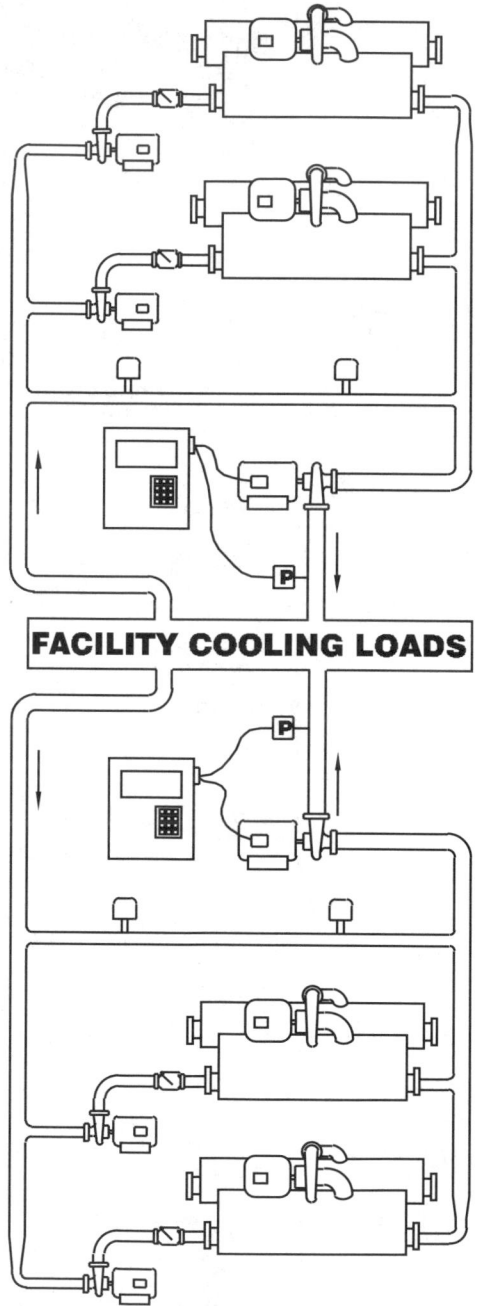

Fig. 6 Two variable-flow chiller plants serving a single facility Control is somewhat trickier than with a single system because each system may try to respond to changes created by the other. For minimum trouble, control each system to maintain a constant pressure in its own system. In this arrangement, the cooling load is distributed between the two systems by adjusting their relative output pressures.

heat recovery chiller. This reduces average chiller efficiency.

Also, if the heat recovery chiller is large compared to the other chillers, its chilled water supply will dilute the chilled water of the other chillers when the heat recovery load is low. This reduces total chiller capacity and requires an excessive number of chillers to keep operating.

■ How to Load a Chiller Preferentially by Locating It In the Return Line of the Load Loop

If you want to operate a heat recovery chiller based on the heat recovery load, rather than the cooling load, you can minimize the previous efficiency problems by installing the heat recovery chiller in the return line of the load loop, as shown in Figure 5. In this arrangement, the heat recovery chiller operates in series with the other chillers. The heat recovery chiller cools the return water, reducing the temperature differential across the other chillers.

This series arrangement affects the efficiency of all the chillers. See Reference Note 32, Compression Chillers, for more about series evaporator connections. Under some conditions, this arrangement may provide higher total plant efficiency. However, over the course of a typical range of load conditions, total plant efficiency is likely to be reduced. You have to make a detailed estimate for your particular situation to determine how much the value of the recovered heat exceeds the additional cost of cooling.

When the heat recovery chiller is able to carry the entire cooling load alone, water flows in the bypass line opposite to its normal direction. This is okay, but it complicates the control logic needed to sequence the chillers.

This arrangement could be used to preferentially load any chiller, not just a heat recovery chiller. However, it is difficult to imagine an application, other than heat recovery, for which this arrangement is best.

■ How to Interconnect Different Chiller Plants

Variable-flow pumping applies to a chilled water system that has chillers at different locations. Such a system is depicted in Figure 6. There are two distinguishing features in such a layout: (1) a separate variable-flow distribution pump draws water from each chiller loop, and (2) each distribution pump has a check valve at its discharge. Without the check valves, the flow from one chiller plant would short-circuit through the bypass lines of the other chiller plants.

Control of the system is fairly simple, but be careful to avoid unstable operation. These are the main features of the controls:

- Control of chiller output in the individual plants is not affected.
- The distribution pump for each system is set to maintain a constant output pressure.

- The load is distributed among all the operating plants by adjusting the relative pressures of the distribution pumps. Increasing the output pressure of one of the distribution pumps increases the load carried by the corresponding chiller(s).

- A distribution pump may be called upon to produce more flow than its chillers can provide. To prevent this, the reverse flow sensor in the plant's bypass line should activate control logic that limits further output by the distribution pump. This may require some complication, especially because the flow from a centrifugal pump varies with pressure, even with the speed constant.

- For given pressure settings, the distribution of load among plants depends on the pressure loss in the distribution piping between the plants. If there is minimal piping pressure loss, the plant set for higher pressure will carry all the load, up to its maximum capacity. Then, the plants set for lower pressure will carry the remainder. On the other hand, if there is large pressure loss in the distribution piping, each plant will serve the loads closest to it.

- There is a tendency for the output of the distribution pumps to oscillate because the pressure sensor controlling each pump feels the pressure output of the other pumps. This "hunting" tendency is damped by pressure drop in the distribution piping, but this damping may not be adequate. To avoid oscillation, make the distribution pump control response of the lead chiller system (the system that starts first) substantially slower than the control response of the distribution pumps of the chiller systems that follow it. Also, the pump curve of the lead chiller plant should be flatter (i.e., the pump output should be less sensitive to discharge pressure) than the pump curves of the following chiller plants.

Explain It!

By now, it should be obvious that variable-flow chilled water pumping may confuse plant operators. Make it easy for operators to understand the system layout, the purpose of the system, and the control logic. Otherwise, they may defeat the controls and operate the system inefficiently. This will take a well prepared formal training session. Install effective placards on the system components. See Reference Note 12, Placards, for tips on how to do this.

ECONOMICS

SAVINGS POTENTIAL: 40 to 70 percent of chilled water pumping energy; 0.5 to 4 percent of chiller energy, provided that the chiller has chilled water temperature reset controls.

COST: In new construction, variable-flow pumping may cost no more than constant-flow pumping, especially in larger systems. Retrofitting variable-flow pumping may cost several thousand dollars for the smallest eligible systems, and much more for large or complicated systems.

PAYBACK PERIOD: One year, to many years.

TRAPS & TRICKS

DESIGN: As with the formula $E = MC^2$, the apparent simplicity of a well-designed variable-flow system masks a lot of theory. Be sure you understand how your system will work under all operating conditions. Variations of the basic system may be tricky.

TRAINING AND DOCUMENTATION: Although the system operates automatically, operators must understand how it works. Otherwise, controls will be set or repaired incorrectly, and the system will operate wastefully. Install a large, well designed system diagram where everyone will see it, along with a clear explanation of how the system works. Put the details in the plant operating manual.

MONITOR PERFORMANCE: Keep an eye on the chiller sequencing and chilled water pump output. On the chiller plant operating log, include columns to record chiller loading (usually determined from chiller amperage) and chilled water discharge pressure. Review the logs regularly to ensure that the system is operating efficiently.

MEASURE **2.5.3 Install pump motors having the highest economical efficiency.**

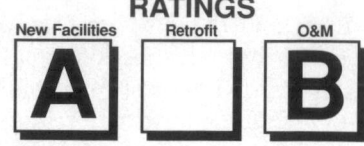

RATINGS

New Facilities	Retrofit	O&M
A		**B**

See Measure 10.1.1 for the details. Installing efficient motors throughout the system is an effortless opportunity that should not be missed.

Be prepared to replace failed motors with models that have the highest practical efficiency. If a motor is exceptionally large and inefficient, you may even want to replace it before it fails.

When replacing a pump motor, consider whether you can reduce the pump power requirement to use a smaller motor. See Measures 2.5.1 and 2.5.2 about this.

If you plan to use a motor with a variable-frequency drive, select a motor that is designed for this application. For details, see Reference Note 36, Variable-Speed Motors and Drives.

Most of the energy that is consumed for cooling goes into the compressors of the cooling units, or actually, into the motors that drive them. A single action to improve the efficiency of these components may save as much energy as many other actions to reduce the cooling load. However, compressor improvements are typically expensive, involving replacement or addition of major equipment.

There are large differences in the efficiency characteristics of different types of compressors. Also, there are large differences in the efficiency of different models of the same type. In existing facilities, be prepared to exploit the fact that compressors available today are significantly more efficient than the best models available a few decades ago.

INDEX OF MEASURES

2.6.1 If the compressor motor fails, replace it with a more efficient motor.

2.6.2 Replace inefficient compressors with efficient units.

2.6.3 In centrifugal chillers, install variable-speed compressor drives.

RELATED MEASURES

- Subsection 2.8, for Measures that minimize the use of inefficient existing chillers

MEASURE 2.6.1 If the compressor motor fails, replace it with a more efficient motor.

Chillers have large motors. In the past, when a chiller motor failed, it was usual procedure to rewind the motor, rather than replacing it. Rewinding costs about half as much as a new motor. However, the potential savings in energy cost make it worthwhile to consider replacing the failed motor with a new motor that is selected for maximum efficiency.

See Measure 10.1.1 for the factors to consider when you upgrade motors for efficiency. The following points relate specifically to chiller drive motors.

Emphasize Part-Load Efficiency

Unless the compressor has only a single stage, the motor operates at a fraction of full load for a large part of the time. If you replace the motor on a staged or modulating compressor, select the motor with an eye toward part-load efficiency.

Chillers Driven Variable-Frequency Drives

If the motor is driven by an electronic variable-frequency drive, note the special requirements for motors used with these drives. See Reference Note 36, Variable-Speed Motors and Drives. Also, see Measure 2.6.3, below, about adding a variable-frequency drive to an existing chiller.

Replacing Hermetic Motors

Motors for hermetic chillers are made specifically for those chiller models, so there are rarely any options among replacement hermetic motors. However, you may need a new motor if you are converting the chiller to a different refrigerant to satisfy environmental regulations. The insulation and other materials used in original hermetic motors may be incompatible with the new refrigerants. As a bonus, the newer motor may offer some efficiency improvement.

Refrigerant conversions typically require other changes to the chiller, such as reducing impeller

---

SUMMARY

In open-drive chillers, installing a new motor provides greater efficiency than rewinding the original motor. There may be options even in hermetic chillers, where refrigerant substitution may be an important consideration.

SELECTION SCORECARD

Savings Potential .... **$ $**

Rate of Return .... **% % %**

Reliability .... ✓ ✓ ✓ ✓

Ease of Initiation .... ☺ ☺ ☺

---

diameters, and changing sealing components. The conversion also is likely to involve some loss of capacity. See Reference Note 34, Refrigerants, for more about refrigerant conversion.

ECONOMICS

SAVINGS POTENTIAL: *1 to 5 percent of motor energy, for substituting a high-efficiency motor instead of rewinding the old motor.*

COST: *A new motor costs about twice as much as rewinding a motor. In addition, the cost of a high-efficiency motor is roughly 20% higher than the cost of an average motor, but this figure varies widely.*

PAYBACK PERIOD: *Less than one year, to several years.*

TRAPS & TRICKS

PLANNING: *First, review Measure 10.1.1. To replace an internal (hermetic) motor, contact the chiller manufacturer about options. Also, check whether there are any other manufacturers who offer upgraded motors for your model of hermetic chiller.*

MEASURE **2.6.2 Replace inefficient compressors with efficient units.**

If a chiller has a compressor that is relatively inefficient, you may be able to improve efficiency substantially by replacing the compressor with a more efficient unit. This is possible with compressors of all sizes. Furthermore, you may be able to replace a compressor of one type with a compressor of a more efficient type. Figure 1 shows a large replacement centrifugal compressor.

In the past, you would consider this option when a compressor failed or neared the end of its service life. Now, you should also consider replacing the compressor as one way of satisfying requirements to use newer types of refrigerants. See Reference Notes 32 and 34 for more about these issues.

Select the Compressor for Efficiency at All Loads

When you select a compressor, be careful to compare the efficiency characteristics of competing compressors for identical rating conditions. The conditions under which the compressor operates, especially the evaporator and condenser temperatures, can have a larger effect on performance than the characteristics of the machines themselves. See Reference Note 31, How Cooling Efficiency is Expressed, for more about this important point.

Consider the efficiency of the compressor across the entire load. A major benefit of newer compressors is their ability to operate more efficiently at low loads.

SUMMARY

Less expensive than installing a new chiller, but also less efficient. Provides an opportunity for changing refrigerant type. Difficulty varies widely, from simple to impractical.

SELECTION SCORECARD

Savings Potential .... $ $ $

Rate of Return .... % % %

Reliability .... ✓ ✓ ✓ ✓

Ease of Retrofit .... ☺ ☺ ☺

In typical HVAC applications, this may be more valuable than the improvement in full-load efficiency.

Matching the Replacement Compressor to the Chiller

The feasibility of upgrading a compressor is largely a question of how the compressor is connected mechanically to the other system components, especially to the evaporator and condenser. The easiest situation is one in which the compressor is entirely separate from the other components, connected only by ordinary piping. Reciprocating compressors, even those installed as part of a packaged chiller, are usually installed in this manner.

At the other extreme are chillers in which the compressor is mounted directly to the evaporator and/or condenser with custom fittings as part of a compact package. This is typical of centrifugal chillers. Replacing the compressor with a unit made by a different manufacturer may require considerable adaptation and additional space. Unless the replacement compressor is a custom fit, it is mounted on a separate stand and piped to the original heat exchangers.

You may be able to replace one type of compressor with a different type. For example, you might replace a small single-stage reciprocating compressor with a scroll compressor. Or, you might replace a large reciprocating compressor with a screw compressor. Such changes require specialized knowledge, because the evaporator, condenser, and refrigerant metering apparatus may have features that are intended for a particular type of compressor.

Another issue is matching the replacement compressor to the chiller system controls. A variety of schemes are used for controlling the output of chillers. If the replacement compressor does not use the same control signals as the original, special skill is needed.

Trane Company

Fig. 1 A large replacement compressor Most of the energy of the chiller plant goes into the compressor. Make it as efficient as possible. At the same time, select it for a refrigerant that will be available in the long term.

Consider Reducing the Compressor Size

In a plant that has excess chiller capacity, you may be able to reduce the cost of replacing the compressor by substituting a smaller size. Doing this makes the existing evaporator and condenser larger with respect to the compressor, which should increase the average efficiency. This is because the temperature drop across a heat exchanger increases with the load. Using a smaller compressor eliminates the least efficient operating range of the heat exchangers, which is high load.

Using this approach, you can convert one of your existing chillers to serve as a smaller lead chiller of improved efficiency. This is desirable in systems that operate for long periods at low cooling loads. See Measure 2.8.1 about lead chillers.

Reducing the compressor size may not reduce the power required for pumping water through evaporators or condensers. Do not expect to reduce system pump power along with compressor power. The evaporator and condenser pumps must continue to provide adequate flow velocity within the evaporator and condenser, even though the amount of heat rejected may be significantly reduced.

Update the Refrigerant at the Same Time

If you are presently using a refrigerant that is due to be phased out for environmental reasons, replacing the compressor makes it much easier to change to a new type of refrigerant. Most refrigerant compatibility problems occur in the compressor, so you can avoid these problems by installing a replacement compressor that is designed for the new refrigerant. In fact, the marketing of replacement compressor kits mushroomed in the late 1980's in response to concern about the atmospheric effect of refrigerants. (For details, see Reference Note 34, Refrigerants.)

Also, the new compressor is designed for maximum efficiency with the new refrigerant. Even so, the new refrigerant may not be optimum for the existing evaporator and condenser, because of differences in density, heat transfer properties, and other characteristics.

(You may be able to substitute a new refrigerant without changing the compressor, but such swaps are beset by problems. For example, HCFC-123 attacks the seals used in CFC-11 machines, and the swap results in a loss of cooling capacity and some loss of efficiency. Also, HCFC-123 attacks the insulation of hermetic motors that are not designed to operate with that refrigerant.)

Feasibility and Alternatives

This Measure is most likely to be economical when the present compressor needs to be replaced. If the original compressor is worn out, consider whether the rest of the chiller system is in sufficiently good condition to justify installing a new compressor.

Replacing a functioning compressor solely to gain improved efficiency improvement is not economical, except in unusual circumstances. A better alternative is replacing the chiller entirely with a new, highly efficient unit (see Measure 2.8.1). All aspects of chiller efficiency have been improved in recent years, including major advances in heat transfer efficiency.

In facilities that have more than one chiller plant, another alternative is connecting the chiller plants together (Measure 2.8.3). This may allow you to operate your least efficient chillers much less. Of course, this works only if the other chillers are more efficient.

ECONOMICS

SAVINGS POTENTIAL: 10 to 35 percent of chiller operating energy.

COST: Depends on the basis of comparison. This method is cheaper than purchasing an entirely new chiller system. The premium for purchasing a compressor of higher efficiency varies widely. The cost of other equipment modifications to adapt to a different compressor may be substantial.

PAYBACK PERIOD: Less than one year, to many years, depending on the circumstances.

TRAPS & TRICKS

PLANNING: Consider all your options. Compare efficiency, cost, and difficulty. This Measure is most desirable where it saves a lot of money compared to installing a new chiller, while providing comparable efficiency. If you convert to a different type of compressor, be prepared for tricky control modifications. Talk with others who have already installed the same compressor you are considering. Discuss installation problems and how well the chiller performs with the new compressor.

INSTALLING THE EQUIPMENT: It is prudent to let the compressor manufacturer be responsible for the installation. That way, you don't have to go to court to find out who is responsible if you have problems.

MEASURE 2.6.3 In centrifugal chillers, install variable-speed compressor drives.

Centrifugal chillers are the most efficient type when operating near full load, but their efficiency suffers seriously at low loads. This is due to inherent characteristics of centrifugal compressors that are explained in Reference Note 32, Compression Cooling.

Centrifugal compressors also have limited turndown ratio. This means that the minimum load at which the machine can operate is a larger fraction of full load than with other compressor types. It is not desirable to turn a centrifugal chiller on and off frequently, so there is a temptation to create a false load to keep the chiller running. This is a very wasteful practice.

Trane Company

Fig. 1 Variable-frequency chiller drive This drives the electric motor of the compressor at variable speed to follow the cooling load. It is quiet, and has no moving parts. It is a mainstream feature of new chillers, but approach retrofit with caution.

SUMMARY

Significantly improves the efficiency of centrifugal chillers at low loads, and extends the turndown range. Make the change only with the approval of the chiller manufacturer. Not well proven as a retrofit.

SELECTION SCORECARD

Savings Potential .... **$ $** $

Rate of Return .... **% %** %

Reliability .... **✓ ✓**

Ease of Retrofit .... ☺ ☺

You may be able to improve both the part-load efficiency and the turndown ratio of a centrifugal chiller by installing an electronic variable-speed compressor drive. This is a device that controls the speed of the compressor by varying the frequency of the power provided to the motor. See Reference Note 36, Variable-Speed Motors and Drives, for details of electronic drives. Figure 1 shows a unit designed to operate chillers.

A big advantage of this method of speed control is that no major changes to the chiller hardware are required. The electrical connections to the compressor drive motor are relatively easy. The electronic control package can be installed in any reasonably well protected location. Retrofit kits for centrifugal chillers are now being offered.

The modification makes speed control the primary means of controlling capacity in approximately the top half of the load range. At lower loads, the compressor operates at a fixed minimum speed, and further reduction in capacity is provided by the chiller's original inlet guide vanes. This more complex control arrangement must be interfaced with the existing chiller controls.

Energy Saving Potential and Limitations

The energy saving provided by variable-speed drives depends on the amount of time that the chiller operates in the inefficient portion of its load range. In typical air conditioning applications, the load is low for a large fraction of the time. In such cases, the distribution of chiller sizes determines how much load can be maintained on individual chillers. The opportunity for saving energy is greatest if the plant has only one or two large chillers, so they are forced to operate at low load for much of the time.

The savings are especially large if false loading is presently used to maintain a minimum load to satisfy the chiller. The false loading may occur within the chiller system itself, usually in the form of hot gas bypass. Or, the staff may create an artificial cooling load, for example, by disabling the economizer cycles of air handling units. Be sure to identify any false loading of the system, which may not be recognized if it has become habitual.

Variable-speed drives do not offer the same range of turndown with centrifugal chillers as they do with centrifugal fans and pumps. One of the compromises of centrifugal chiller design is that the impeller dimensions can be optimized only for a single speed. As a result, varying the impeller speed is efficient only in a limited range of speed reduction. This is why control at low loads requires the inlet guide vanes originally installed on the chiller.

Using an electronic drive reduces chiller efficiency somewhat at the top end of the load range, because of losses within the drive itself and because of increased heating losses in the motor.

Be skeptical of performance claims. New concepts in energy efficiency are prone to exaggeration, even if they are made by large and reputable manufacturers. A variable-speed drive cannot turn a sow's ear into a silk purse. The efficiency of a chiller depends on many factors, and this Measure improves only one aspect of chiller efficiency.

Potential for Chiller Damage

Adapting a centrifugal chiller to variable-speed drive is much more of a challenge than adding a variable-speed drive to a pump or fan. The flow of refrigerant gas in a centrifugal compressor is vulnerable to several modes of instability that occur when the speed of the impeller is not matched to the gas flow. These instabilities reduce efficiency, make noise, and may damage the compressor.

Also, changing the compressor speed may destroy an impeller or other rotating component by torsional resonance. Three-phase motors are smooth power sources, as long as the phases are balanced. However, serious torsional resonances may still occur somewhere within the speed range of the drive. Also, the irregular power waveform produced by electronic drives may introduce some torsional vibration.

Therefore, consult with the chiller manufacturer before installing a variable-speed drive. The drive should be designed as a package intended for the particular chiller model, and the work should be done by someone who can provide a credible warranty of reliability and efficient performance.

Beware of trouble when modifying a chiller with a conversion kit sold by another manufacturer. In case of trouble, each party is likely to blame the other, and any remaining chiller warranty may be voided.

Avoid Problems Outside the Chiller Plant

Variable-frequency drives have matured, and they now function reliably. However, they do have problems that you need to consider. The most serious problem is electrical noise, which can cause interference with other equipment throughout a facility. See Reference Note 36, Variable-Speed Motors and Drives, for details.

Other Ways of Improving Low-Load Efficiency

The purpose of this Measure is to improve efficiency at low cooling load. The most efficient way to do this is to install a small, efficient lead chiller to carry low loads. See Measure 2.8.1.

If a facility is served by several chiller plants, you may be able to improve efficiency at low loads by connecting the plants together. See Measure 2.8.3.

ECONOMICS

SAVINGS POTENTIAL: *10 to 40 percent of chiller input energy. Even more, if this Measure allows false loading to be reduced or eliminated.*

COST: *Retrofit variable-speed compressor drives cost about $25,000 for a 400-ton chiller, and about $35,000 for a 600-ton chiller.*

PAYBACK PERIOD: *Several years, or longer.*

TRAPS & TRICKS

CHOICE OF METHOD: *Consider all your options, even installing a more efficient chiller. Talk to the manufacturer of your chiller about the change. Talk to other operators who have made the same conversion. Be prepared to deal with control modifications that may be tricky.*

SELECTING AND INSTALLING THE EQUIPMENT: *Buy the equipment only from (or through) a manufacturer of centrifugal chillers. Anyone else lacks the expertise to avoid chiller problems. Have the installation done by the equipment manufacturer.*

Managing the refrigerant of your cooling equipment is the most important routine maintenance for keeping the equipment at maximum efficiency. You need to maintain the proper amount of refrigerant, neither too little nor too much. You also need to make sure that nothing enters the equipment but refrigerant. The Measures in this Subsection guide you in these activities. They apply to both compression cooling and absorption cooling, and they apply to cooling units of all sizes.

These Measures also reduce expensive equipment failure, and they extend equipment life. Maintaining efficiency and avoiding equipment damage are closely related aspects of refrigerant management, especially with low-pressure refrigerants.

Before getting started with these Measures, you may want to review Reference Note 34, Refrigerants. It will give you a good introduction to refrigerant properties and to the contemporary issues that affect your choice of refrigerants.

INDEX OF MEASURES

2.7.1 Repair chiller system leaks.

2.7.2 Maintain the proper refrigerant charge.

2.7.3 Operate purge units appropriately.

　2.7.3.1 Install high-efficiency purge units.

2.7.4 Install accessories that prevent air leakage into idle chillers.

2.7.5 Drain the water from the evaporators and condensers of idle chillers.

MEASURE 2.7.1 Repair chiller system leaks.

Leakage is the first cause of most problems that relate to refrigerants and the refrigerant side of the chiller system. It is not unusual for chillers to leak large amounts of refrigerant. Older chillers that are not well maintained typically lose a substantial fraction of their total charge in the course of a year. Newer chillers are designed to minimize refrigerant loss, but all chillers develop leaks that need to be repaired. Only small, truly hermetic chiller systems are immune from normal leakage, and even these may suffer failures of soldered joints that allow leakage.

You have to keep checking for leaks, and repairing them, as long as you have chillers. Schedule this as a regular maintenance item. This Measure tells you about the problems that leakage can cause, and how to find leaks. Leaks occur in defective equipment. When you find a leak, repair the equipment as recommended by the manufacturer.

Types of Leakage

There are three modes of system leakage to monitor: loss of refrigerant, leakage of air and water vapor into the system, and leakage of water into the system. Each of these can create serious problems. Which of these problems may occur depends on the pressure characteristics of the refrigerant and the design of the system.

■ Loss of Refrigerant

If the pressure of the refrigerant inside the system is higher than atmospheric pressure, the system may lose refrigerant. Loss of refrigerant causes a chiller to operate inefficiently, and it may cause equipment damage. Problems related to improper refrigerant charge are explained in Measure 2.7.2, below.

In order of increasing pressure, the common refrigerants HFC-134a, CFC-12, ammonia, HCFC-22, and CFC-502 are high-pressure refrigerants. The vapor pressures of these refrigerants are above atmospheric even on the suction side of the compressor.

Some low-pressure refrigerants, notably the common refrigerants CFC-11 and HCFC-123, operate at pressures higher than atmospheric on the discharge side of the compressor, which includes the condenser. Therefore, the high-pressure side of these systems can lose refrigerant.

In addition, low-pressure refrigerants are lost from the system by the action of purge units, which are required in all systems that use low-pressure refrigerants. This is not leakage in the literal sense, but it has the same effects. The operation of purge units is related to

SUMMARY

Basic maintenance that maintains chiller efficiency and prevents equipment damage. This is the first step in maintaining proper refrigerant charge. With low-pressure refrigerants, it is also necessary for effective purging.

SELECTION SCORECARD

Savings Potential .... $ $

Rate of Return .... % % % %

Reliability .... ✓ ✓ ✓ ✓

Ease of Initiation .... ☺ ☺

leakage of air into the system, which is discussed next. See Measures 2.7.3 ff about purge units.

The cost of replacing lost refrigerant is significant. Packaged water chillers typically use several pounds of refrigerant per ton of cooling capacity. Distributed chiller systems may use even more. If you do not monitor leakage, you may lose an expensive quantity of refrigerant before a chiller operating problem alerts you to the leakage. Refrigerant cost is escalating as more expensive refrigerants are required to satisfy environmental concerns.

In the United States and other countries, laws require the repair of significant leaks if the refrigerants are designated as environmentally harmful.

■ Leakage of Air and Water Vapor into the System

If the pressure of the refrigerant inside the system is lower than atmospheric pressure, air may leak into the system, along with the water vapor that air contains. The presence of non-refrigerant gases in the chiller reduces efficiency.

In chillers that use the common halocarbon refrigerants, air and moisture can also cause serious damage. Air and moisture commence to damage chillers as soon as they enter the system, even though dryers and purge units may be installed to remove them. The problems caused by air and moisture are explained in Measure 2.7.3, below.

Refrigerants that have low vapor pressures invite leakage of air and atmospheric water vapor into the system. The refrigerants CFC-113, CFC-11, HCFC-123, and some others have vapor pressures below atmospheric pressure at room temperature. With such refrigerants, air can leak into the chiller whether it is operating or idle.

Even if a refrigerant has a high vapor pressure at room temperature, air may leak into the evaporator of the chiller if the evaporating temperature is low enough. For example, the refrigerants ammonia, CFC-12, and HFC-134a have evaporating pressures lower than atmospheric when used in low-temperature applications, such as cold storage and making ice cream.

With any refrigerant, air and water vapor may enter the system as a result of improper refrigerant charging and oil filling procedures.

■ Leakage of Water into the System

Water chillers and chillers with water-cooled condensers face the additional problem of water leakage into the refrigerant side of the system. Water that leaks into a chiller in the liquid state causes the same kinds of damage as water that leaks into the chiller as vapor. With a liquid leak of any size, the effects may be much more severe.

The most likely route of water leakage is through the evaporator of a water chiller. This part of the system has lowest internal pressure. The critical factor is the pressure differential in the evaporator when the compressor is running. Even with some high-pressure refrigerants, the water pressure in the evaporator may be higher than the refrigerant pressure. This is because the evaporator may be subjected to large hydrostatic pressure from a tall chilled water distribution system. (If you have a tall chilled water system, be aware of whether it includes a heat exchanger to isolate the chiller from high chilled water pressure.)

Water can leak into the system through a water-cooled condenser, in some cases. This risk exists primarily in systems that use low-pressure refrigerants. Most high-pressure refrigerants have a pressure that is high enough to resist water leakage at the condenser. The maximum water pressure on the condenser is determined by the height of the cooling tower above the condenser. This height is usually too low to cause trouble with high-pressure refrigerants, but there are exceptions.

To make the story complete, be aware that water may enter any type of compression chiller as a contaminant of lubricating oil. Water vapor in the air is absorbed by refrigerant oil if the container is not kept properly sealed. This is not leakage in the literal sense, but it has the same effect. The small amount of water introduced with contaminated oil can cause harm.

Where Leaks Occur

Leaks can occur anywhere in a chiller system, but most leaks occur in these locations:

- shaft seals of compressors with external drive motors
- operating shafts for the inlet guide vanes of centrifugal chillers. Figure 1 shows this vulnerable site.
- anywhere else that mechanical motion is transmitted through the casing
- packing boxes for wiring that passes into the refrigerant side of the system for internal motors, cylinder unloaders, sensors, etc.
- charging connections
- gauge taps
- threaded fittings
- gaskets
- fins of refrigerant coils, especially at brazed connections

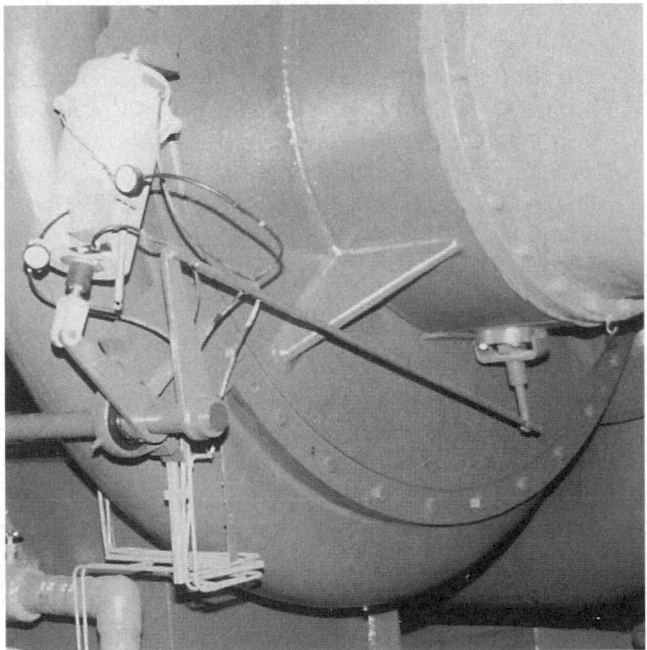

WESINC

Fig. 1 Potential leakage site in a centrifugal chiller This actuator for the machine's capacity control penetrates the casing. It is a moving component, so suspect leakage here. The refrigerant operates below atmospheric pressure, so leakage anywhere on this machine will be inward, making conventional leak detection methods ineffective. This fitting is located at the compressor suction, increasing any leakage that occurs here.

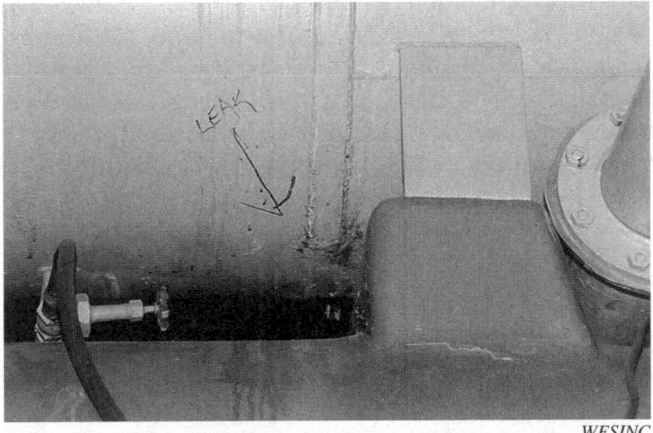

WESINC

Fig. 2 Leakage at an original weld of an old chiller

• welded joints, especially in older units where internal corrosion may have penetrated through imperfections in the weld. Figure 2 shows a leak at a welded joint.

In water chillers or in chillers that have water-cooled condensers, there may be internal leakage between the liquid side and the refrigerant side. This usually occurs at the tube sheet of shell-and-tube heat exchangers, either because of defective fit between the tubes and the tube sheet, or because of stress cracking of the tubes.

Symptoms of Leakage

The first step in managing leakage is knowing whether there is a problem. Keep a continuous record of the amount of refrigerant that you use to refill the system. In systems with positive-pressure refrigerants, loss of refrigerant indicates the existence of leaks. In systems with negative-pressure refrigerants, loss of refrigerant indicates an inefficient or malfunctioning purge unit, or a major air leak into the system.

In systems with purge units, keep track of the length of time that the purge unit operates. This indicates the rate of air leakage into the system. Some sophisticated purge units record the cumulative amount of air removed, as explained in Measure 2.7.3.1.

Air that leaks into a chiller raises the condensing pressure. This is because the partial pressure of the non-refrigerant gases adds to the partial pressure of the refrigerant gases. If you measure a condensing pressure that is higher than indicated in the table, suspect air in the system. This is a good test because the pressure rise is proportional to the efficiency loss caused by the excess air. (Measure 2.7.2 explains how to figure out what the condenser pressure should be.)

However, this symptom does not become apparent if the chiller has a purge unit that is operating properly. The purge unit does not allow enough air to accumulate to affect the condensing pressure noticeably.

If the system is undercharged, the lack of refrigerant may create a drop in pressure that masks the rise in pressure caused by the presence of air. This is a very inefficient condition.

The usual way of detecting the presence of moisture in the refrigerant circuit is to use a refrigerant dryer that contains a chemical indicator that changes color when a certain amount of moisture is absorbed. This works only in chillers that have a sight glass on the dryer, or that have a removable dryer element. Consider installing a visible dryer for this purpose if none is presently installed. If a purge unit is installed, the purge unit may indicate the presence of moisture. The performance of purge units in revealing and removing moisture varies widely, as explained in Measures 2.7.3 ff.

Eventually, the presence of air and water will be revealed during repair by their adverse effects: corrosion, acidification and sludging of oil, restriction of throttling valves, clogging of screens, etc.

Methods of Finding Leaks

Most refrigerant leaks start very slowly, which makes them difficult to detect. The methods that you can use to detect leaks depend on the type of refrigerant and on the location of the leak in the system, as follows.

■ Leakage of Refrigerant Out of the Chiller into the Surrounding Space

Check for leaks at likely leakage points everywhere that the system is under positive pressure. You can detect outward leakage of refrigerant with a variety of methods. These are the main ones:

• *visible appearance of corrosion.* Escaping refrigerant combines with water vapor in the air to create highly corrosive acids. Look for localized rust in steel components, and green corrosion of copper alloys.

• *soap bubbles.* A bottle of soap or detergent solution is cheap and easy to use. You can use it with any type of refrigerant. The procedure is to wet the suspect area and look for bubbles. Even slow leaks create bubbles if the soap solution covers the leak long enough. This method fails if the leak is located where the detergent solution drains off quickly or cannot be observed well. A clear oil may work in locations where a detergent solution is too runny.

• *the halocarbon torch.* This is an old, reliable method of detecting leaks of halocarbon refrigerants. It consists of an ordinary propane torch with an attachment that holds a small copper surface in the flame. A sniffer tube draws air into the flame from the suspect location. Leaking refrigerant breaks down in the heat of the flame and reacts with the copper to produce a distinct blue-green flame. The open flame is a fire hazard if there are fuels or flammable materials in the vicinity.

• *the electronic ionization detector,* which is a newer type of tester for halocarbon refrigerants. It functions by ionizing the halocarbon gas between charged platinum electrodes and sensing the ionization current. It is very sensitive.

• *dyes in the refrigerant oil.* This is an older method that is suitable for refrigerants that carry oil through the system. The dye travels with the oil, and appears at points of leakage. Fluorescent dyes are available that can be seen in small amounts with a portable ultraviolet lamp.

• *smell,* which works for ammonia and other gases with strong odors, if the leak is large enough.

■ **Leakage of Refrigerant into Evaporator and Condenser Water**

Your first clue that high-pressure refrigerant is leaking into the waterside of the evaporator or condenser is usually indirect. If you have to add refrigerant but you can't find any external leaks to account for the loss, suspect a leak into the waterside. To localize the leak, drain the evaporator or condenser and test for refrigerant leakage using the methods given above.

As a first check for leakage into the evaporator, use a leak detector at the vent of the chilled water system surge tank. This test cannot detect a small leak, and it cannot detect refrigerant that is entirely dissolved or chemically changed by water.

As a first check for leakage into the condenser, use a sensitive refrigerant detector at the point where cooling water is pumped into the cooling tower or at the suction point of the sump, with the cooling water pumps turned off. This test cannot reveal a small leak.

You may have to test the water itself to find leaking refrigerant in it. If your water treatment contractor cannot do this test, send a sample to a laboratory that tests for refrigerant pollution in water wells.

■ **Leakage of Air into the Refrigerant Circuit**

There is no simple, reliable method of finding the location of an air leak into a chiller system. This is a problem if the chiller uses a refrigerant that operates below atmospheric pressure. The best you can do easily is to seal suspect areas and observe whether the rate of air leakage into the system drops.

Purge units that record the amount of air removed from the chiller are handy for seeing the effect of a repair within a reasonably short time. In the case of slow leaks, this may be a few days.

Measure 2.7.4 recommends equipment that keeps chillers with low-pressure refrigerants under positive pressure while they are idle. This keeps air from leaking into idle chillers. You can use the same equipment to pressurize a chiller enough to force refrigerant out of any leaks that exist, making it possible to find the leaks with ordinary refrigerant leak testing equipment.

■ **Leakage of Water into the Refrigerant Circuit**

Leakage of water into the refrigerant is revealed in the refrigerant dryer, as mentioned above. If water accumulates in the dryer rapidly, suspect a leak from the waterside, rather than water vapor entering along with air.

If a chiller that uses a high-pressure refrigerant shows a water leak while it is running, the leak is almost certainly in the evaporator. To find the leak, drain the evaporator and search for escaping refrigerant using the test methods given previously.

If a low-pressure refrigerant has a condensing pressure higher than atmospheric, you may be able to localize a water leak to the evaporator or condenser without draining them. (CFC-11, HCFC-123, and some other low-pressure refrigerants have condensing pressures higher than atmospheric.) Use a fresh moisture indicator in the purge unit, and see if it indicates moisture while the chiller is running continuously. If there is no moisture, the leak is in the condenser.

To localize the leak, drain whichever heat exchanger has the leak, and inspect it carefully. If you have the equipment recommended by Measure 2.7.4, discussed previously, you can use it to pressurize the system above atmospheric pressure. This allows you to find the leak with conventional refrigerant leak detection equipment.

ECONOMICS

SAVINGS POTENTIAL: *Chillers that are chronically leaky may operate with an average efficiency loss of 5 to 20 percent. Preventing leakage of moisture into chillers avoids expensive maintenance problems.*

COST: *Minimal. This activity saves refrigerant cost and reduces long-term maintenance cost.*

PAYBACK PERIOD: *Immediate.*

TRAPS & TRICKS

CHOICE OF TEST METHODS: *Equipment and materials for finding refrigerant leaks are relatively cheap. Buy equipment that is sensitive, reliable, and easy to use. Maybe get several types for different leakage locations.*

TRAINING: *Provide training for anyone you assign to do leak testing. A certain amount of experience and finesse are required.*

SCHEDULING: *This kind of activity tends to be forgotten. Make leak testing a regular item on your maintenance schedule.*

MEASURE 2.7.2 Maintain the proper refrigerant charge.

The efficiency of all chillers suffers if the system has either too little or too much refrigerant charge. Also, the compressor may suffer damage if the system is overcharged. Some systems have only minimal reservoir capacity, making it important to charge the system precisely. Such systems are more vulnerable to loss of efficiency from small leaks. Other systems have a large amount of reservoir capacity. In these systems, a small leak may persist for a long time before being noticed.

Check the refrigerant charge in your cooling units often enough to keep the charge within proper limits. This Measure gives you procedures for checking and maintaining the proper refrigerant charge and explains the effects of improper refrigerant charge.

SUMMARY

A fundamental chiller maintenance procedure with a significant effect on efficiency. Finding the level of charge may be tricky. Some inexpensive accessories may help.

SELECTION SCORECARD

Savings Potential .... $ $

Rate of Return .... % % % %

Reliability .... ✓ ✓ ✓ ✓

Ease of Initiation .... ☺ ☺ ☺

Bad Effects of Incorrect Refrigerant Charge

Both the COP and the capacity of a cooling unit suffer if the refrigerant charge is too low. When that occurs, evaporator capacity is reduced because less of its surface is wetted, and the average evaporator temperature differential increases. The compressor must work harder to satisfy the same cooling load.

WESINC

Fig. 1 Refrigerant level sight glass in a large chiller The glass is located just below the center of the evaporator shell, to the right of the instrument panel. It is so small that it may not line up with the liquid level. If so, you can't tell whether the liquid level is above or below the glass. In this case, observe the glass when the machine starts.

In hermetic chillers, in which the motor is cooled by the refrigerant gas, low charge can overheat the motor, reducing its life.

If there is too much refrigerant in the system, the excess may back up in the condenser, reducing its effective surface area and increasing the average temperature differential across the condenser. In chillers that have a flooded cylindrical evaporator and no device to regulate the refrigerant level in the evaporator, high refrigerant level reduces the evaporation surface area.

In some types of systems, excess refrigerant can travel through the evaporator in the liquid state, continuing into the compressor. This can destroy a positive displacement compressor immediately, and it can destroy a centrifugal compressor gradually.

How to Measure Refrigerant Charge

The most difficult aspect of maintaining the proper refrigerant charge may be measuring the charge that is presently in the system. In some cases, this can be tricky, tedious, or both. The best method of checking the refrigerant charge depends on the type of system. Use the best method or combination of methods for your system. The following are the various methods that are available.

■ Liquid Level Indicators and Sight Glasses

Some chiller units, and some vessels in a chiller system, may have a means to indicate the refrigerant quantity directly. These work only if a predictable quantity of refrigerant remains in one part of the system. The most common liquid level indicator is a sight glass on the vessel where the refrigerant collects.

Refrigerant level sight glasses are common accessories of packaged water chillers. They are useful on these machines because all the refrigerant remains within the shell of the machine and drains freely into the evaporator. Figure 1 shows a typical sight glass. It can be used when the machine is running or turned off, although the level is more stable when the machine is not running.

Many refrigerant level sight glasses are perversely small, making it difficult to check the level if it is above or below the level of the sight glass. In such cases, it helps to look at the sight glass as the chiller is being started. If the refrigerant surface is above the sight glass, you can probably see bubbles as the chiller starts, or the refrigerant level drops to the level of the sight glass. If the refrigerant level is below the sight glass, you may be able to see splatter on the sight glass, which indicates that the charge is low.

Some older units have a liquid level test cock on the evaporator shell. However, these require venting some refrigerant to test the liquid level. This practice is now considered very bad form, for environmental reasons.

In chiller systems where the components are spread out, refrigerant quantity indicators do not work as well, or they may not work at all. The problem is that refrigerant migrates from one part of the system to another. When the chiller is running, the distribution of refrigerant in the system varies with load. When the chiller is not running, refrigerant migrates to the coldest part of the system. For example, the refrigerant might accumulate in the condenser during winter and in the evaporators during summer.

If a spread-out chiller system has a receiver (refrigerant surge tank) or a shell-and-tube evaporator, it may be practical to use a level indicator in one of these vessels. In such cases, the level indicator provides useful information only when the system is running and stabilized. Even then, the level of refrigerant varies with the cooling load. The refrigerant level indicator should be readable anywhere within the acceptable charge range. When the system is turned off, refrigerant pools in the coldest parts of the system, and the level indicator gives a false reading.

If your chiller system does not have an easy-to-read refrigerant level indicator, consider adding one, if possible. If you do, install a placard at the sight glass or gauge that indicates the normal range of readings, and the conditions under which the readings are valid. For example, the placard might say, "Refrigerant level indicator valid only if the compressor is running, or if the receiver temperature is at least 10°F colder than the outside air temperature." (See Reference Note 12, Placards, for tips on how to create an effective placard.)

Any kind of refrigerant level gauge or sight glass should be strong and well protected. A broken sight glass or gauge connection would vent the entire refrigerant charge into the surrounding space. With high-pressure refrigerants, the blowout continues at full pressure as long as there is liquid in the system, which is a dangerous situation.

■ Discharge and Suction Pressures

With all types of compression cooling equipment, you can check the state of refrigerant charge by measuring the discharge and suction pressures in the system. Do this while the compressor is operating and the system is in stable operation.

Larger machines usually have gauges installed that indicate the evaporator and condenser pressures at all times. Figure 2 shows a typical example. Use portable gauges if the machine does not have gauges installed.

The normal discharge pressure depends on the condensing temperature. To check system charge, use a table of refrigerant pressures and temperatures. This tells you what the condensing pressure should be at the current condensing temperature. If the discharge pressure is lower than it should be at that temperature, the system is low on charge.

Refrigerant pressure gauges often have the corresponding saturation temperatures printed right on the gauge dials. This saves you the trouble of finding a refrigerant pressure chart. Portable refrigerant gauges typically show the saturation pressures for several of the most common refrigerants. If you need to use a refrigerant table, you can find one in many reference books. Also, refrigeration supply houses common give away refrigerant tables that are printed on handy cards. If the type of refrigerant used in the chiller system has been changed, be sure to use a refrigerant table for the current refrigerant.

If the refrigerant charge is low, both the discharge and suction pressures will be lower than normal. The discharge pressure is low because there is not enough gas in the system for the compressor to squeeze to the normal discharge pressure. The suction pressure drops

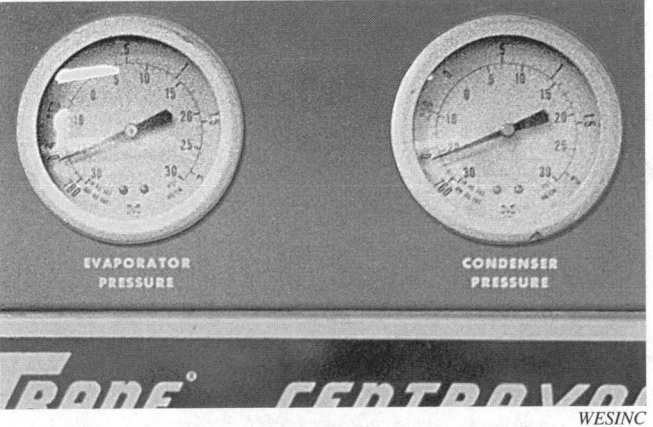

WESINC

Fig. 2 Evaporator and condenser gauges These tell you immediately whether the machine has the minimum amount of refrigerant for efficient operation. They do not tell you the actual amount. The condenser pressure provides an uncertain indication of excessive charge.

because there is not enough liquid refrigerant in the evaporator to boil off vapor at the normal vapor pressure. As a result, the vapor expands into the compressor suction, lowering its pressure. In other words, the compressor starts to act like a vacuum pump.

Low suction pressure also creates abnormally low suction temperature. This occurs because the refrigerant gas is cooled below its saturation temperature by the greater expansion. The suction temperature can eventually fall enough to freeze the evaporator coil. In a water chiller, this can cause major damage.

Suction pressure could be lower than normal for other reasons, such as obstructed air flow through an air-cooled evaporator. For example, opening the evaporator coil access panel in an air handling unit short-circuits the flow of air around the coil, causing its refrigerant pressure to drop.

Discharge pressure is much less reliable as a clue to excessive refrigerant charge. If the condenser floods from excess charge, its cooling capacity is reduced, so the discharge pressure rises. A noticeable pressure rise occurs only under high load. A condenser that is heavily flooded with excess refrigerant will also cause cooling water or cooling air temperatures that are lower than

normal, because the condenser is not rejecting as much heat. However, this symptom is subtle.

(If the discharge pressure is lower than normal and the suction pressure is higher than normal, the compressor may be worn out, or the compressor or system may have an internal leak from the discharge side to the suction side, or the system may have hot gas bypass.)

So, the suction and discharge pressures are a reliable indicator of low charge, and the discharge pressure is a less reliable indicator of excessive charge. However, system pressure cannot tell how much refrigerant is in the system within the normal range of charge. As long as there is enough liquid within the system to keep the evaporator supplied, the readings are normal. In systems without refrigerant quantity indicators, you have to check the refrigerant pressure at appropriate intervals to detect the first sign of inadequate charge. When leakage finally causes liquid starvation in the evaporator, pressures start to decline. The rate of decline depends on the leakage rate and on the volume of refrigerant in the system.

On the other hand, air in the system causes all pressures to be higher than normal. This can mask a

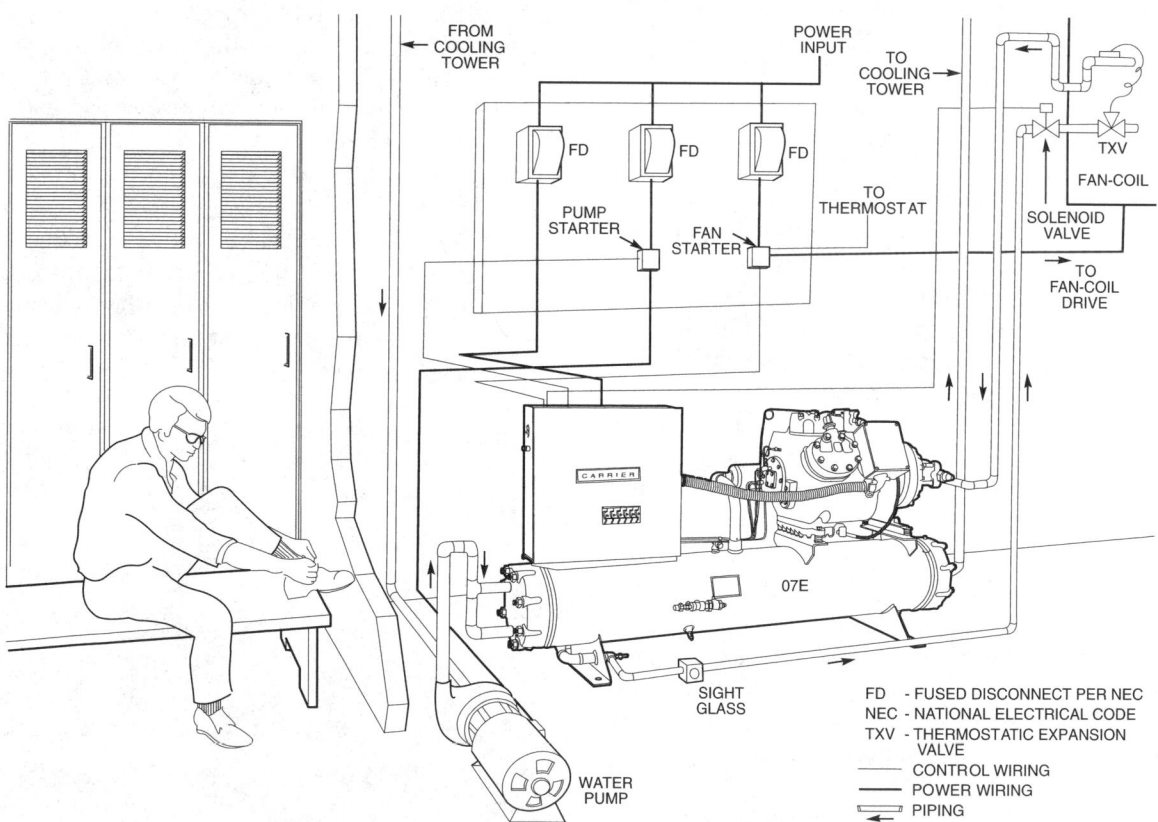

Carrier Corporation

Fig. 3 Evaporator liquid line sight glass It is generally located as shown here, close to where the refrigerant liquid enters the evaporator. Bubbles in the sight glass while the system is running probably indicate low refrigerant charge, but they may also indicate an obstruction of the refrigerant line in the direction of the condenser.

low refrigerant charge. Keep air out of the system at all times. This is covered by Measures 2.7.1 and 2.7.3.

■ Evaporator Liquid Line Sight Glass

In chillers that use a throttling type of refrigerant metering device (an "expansion valve," capillary tubes, etc.) to control the flow of refrigerant to the evaporator, a sight glass may be installed in the refrigerant line leading to the evaporator. Figure 3 shows where to look for the sight glass.

Bubbles in the sight glass indicate that there is not enough liquid in the system to keep the line filled. Bubbles first appear under high cooling load, when liquid is being drawn out of the line most rapidly. Bubbles that occur when the system first starts may be normal, and do not indicate low charge.

Bubbles in a sight glass are not a foolproof indication. If the sight glass is located upstream of a partially obstructed filter or dryer, the back pressure may keep bubbles from forming even when the charge is low. Conversely, if the sight glass is downstream of a clogged filter or dryer, the reduced pressure at the sight glass may cause bubbles to form even though the amount of refrigerant in the system is proper. Adding more refrigerant based on this false indication may overcharge the system and cause compressor damage.

A liquid line sight glass cannot reveal excessive refrigerant charge.

■ Suction Gas Superheat

In direct-expansion chiller systems (which send the refrigerant directly into the cooling coils), low charge is indicated by high superheat in the gas leaving the evaporator, especially when the compressor is operating at full load. Superheat is the excess of the gas suction temperature above the gas saturation temperature. When the evaporator becomes "starved" for refrigerant, the available refrigerant boils off quickly and the unsatisfied heat load of the evaporator superheats the refrigerant gas excessively.

In systems that use a thermostatic expansion valve, the valve is designed to maintain a fixed amount of superheat. The purpose of the superheat is to ensure that liquid refrigerant does not enter the compressor. Do not let this superheat fool you into believing that the charge is low. If the superheat setting of the valve is unknown (it is typically $10°F$ to $20°F$, or $5°C$ to $11°C$), the charge is probably not low if the superheat remains essentially the same at all loads.

■ Condensate Subcooling

In systems with air-cooled condensers, excessive charge is indicated by excessive subcooling of the refrigerant. Subcooling is cooling of the liquid refrigerant below its saturation temperature. When the system is overcharged, the condenser fills with liquid refrigerant, the condenser capacity drops, and the liquid

lingers in the condenser long enough to become excessively subcooled.

The difference in temperature between normal and subcooled refrigerant from a condenser is small. This makes the test too subtle for any but experienced technicians. Look for condenser subcooling as confirmation of excess charge if the discharge pressure is too high.

This symptom is accompanied by abnormally high condenser pressure, especially at high cooling load.

■ Bleeding Refrigerant Pressure

As a last resort, you can bleed refrigerant from the system until the operating pressures drop, and then add the recommended amount of extra refrigerant. Do not use this method with any environmentally harmful refrigerant unless you have the equipment to salvage the refrigerant.

Should You Add a Receiver?

All chiller systems have a certain amount of storage volume for liquid refrigerant, but the amount varies widely. Ample refrigerant storage capacity ensures that refrigerant is available to the evaporator. It compensates for accumulation of refrigerant in different parts of the system under different operating conditions. It prevents back-flooding of refrigerant into the condenser. And, it provides a reserve to make up for leakage.

Some chillers inherently have large storage volume. For example, packaged centrifugal water chillers store a large amount of refrigerant in their evaporator shells. On the other hand, direct-expansion chillers may have little storage capacity, because air coils have small liquid volumes. In the past, it was common practice to install a "receiver" in such systems, which is simply a storage tank, or surge tank. Different chiller system designs may have receivers in different parts of the system.

It has become commonplace to eliminate the receiver from chiller systems as a cost saving measure. In such systems, storage volume is limited to the condenser itself and to the piping downstream of the condenser. Therefore, a relatively small overcharge may cause refrigerant to back up into the condenser, and a relatively small undercharge may starve the evaporator of refrigerant. For example, a difference of a few ounces of refrigerant charge may affect chiller performance in a small split system.

In systems that lack refrigerant storage capacity, it may be desirable to add a receiver to the system. The mechanical installation is usually not complicated, but it should be done by a refrigeration specialist familiar with proper piping practices and other aspects of assembling cooling systems. Finding the proper location for the receiver in the system requires a clear understanding of chiller system design.

Installing a receiver is not a substitute for keeping the system free of leaks. If the system operates properly when it is properly charged, it probably does not need a receiver. Instead, put your emphasis on proper charging procedure and checking for leaks.

How to Add Refrigerant

Follow the refrigerant charging procedures specified by the manufacturer. If your system does let you measure the refrigerant charge directly, find the point of minimum charge as described previously. Then, add refrigerant in the amount specified by the manufacturer. If you use a large bulk container of refrigerant, put it on a portable scale as you charge the system. Calculate the amount of refrigerant added from the change in weight.

Be careful to keep air from entering the system when you recharge it. This requires great care if the refrigerant in the chiller is below atmospheric pressure. Even with high-pressure refrigerants, be careful to purge all the refrigerant gauge and filling hoses before opening the chiller service ports.

If you are filling a chiller system that has been opened to the atmosphere, you have to use a vacuum pump to remove all air and vapor from the system before recharging. Don't try this without training. Chiller servicing should be done only by technicians who fully understand what they are doing. Inadequate training of maintenance personnel is a common cause of chiller damage and inefficiency.

ECONOMICS

SAVINGS POTENTIAL: *Up to 20 percent of chiller operating cost.*

COST: *Usually minimal.*

PAYBACK PERIOD: *Short.*

TRAPS & TRICKS

SKILLS AND TRAINING: *Understand how refrigerant travels in your chiller system. Know the best methods of checking the charge in that type of system. Invest in training the right person for this responsibility. Keep unqualified people from messing with refrigerant charge. They can do a lot of harm.*

SCHEDULING: *This is another function that tends to be forgotten. If the charge in your chillers can be checked easily, put a column for refrigerant level on the chiller operating log. Otherwise, schedule checks in your maintenance calendar. (You do have a chiller operating log and a maintenance calendar, right?)*

MEASURE **2.7.3 Operate purge units appropriately.**

A

Air and moisture can leak into any chiller system if the pressure inside the system is lower than atmospheric. For this reason, chillers that operate at low internal pressure are equipped with purge units to remove the contaminants. Figure 1 shows a typical purge unit.

Proper operation of the purge unit is essential for efficient operation, and to minimize corrosion damage. This Measure tells you what purge units do, how to keep them working, and how to operate them appropriately.

Water may enter a chiller along with the air, in the form of water vapor. Also, water may leak into the system from the chilled water side of the evaporator and from the condenser side of a water-cooled condenser. Liquid water that leaks into the chiller partially vaporizes inside a low-pressure system. The vapor travels in the same manner as water vapor that enters through air leaks. Purge units can remove some of the water that remains inside the system as a liquid, but their ability to do this varies with the type of purger.

Newer models of purgers have performance that is much better than the performance of older models, and they are easier to use. Subsidiary Measure 2.7.3.1 recommends installing the best type of available purger.

Purging is not a substitute for repairing leaks. As soon as air and moisture enter a chiller, they start to combine with the refrigerants to form acids, which in turn form oil sludge. The acids and sludge are not removed by purging. The only completely effective way to prevent damage is to keep contaminants out of the chiller in the first place. See Measure 2.7.1 about this important maintenance.

Chillers That Require Purging Equipment

Leakage of air and atmospheric moisture into chillers is a problem with refrigerants that have evaporating pressures below atmospheric pressure. (The system pressure is lowest in the evaporator circuit.) Expect to find purge units on systems that use refrigerants CFC-11, CFC-113, HCFC-123, and other low-pressure refrigerants. All absorption chillers that use water and lithium bromide as the refrigerant operate under high vacuum, and these require purge units.

Even if a refrigerant has a high saturation pressure at normal temperatures, the evaporating pressure may be less than atmospheric if the evaporating temperature is low enough. For example, the refrigerants ammonia and HFC-134a have evaporating pressures lower than atmospheric when used in low-temperature applications, such as frozen food storage. Such systems may not have purge units, even though they would be desirable. See the subsidiary Measure about adding a purge unit.

SUMMARY

An important operating procedure with chillers that operate at low refrigerant pressure. Maintains efficiency and prevents chiller damage.

SELECTION SCORECARD

Savings Potential .... **$** $

Rate of Return .... **% % %** %

Reliability .... ✓ ✓

Ease of Initiation .... ☺ ☺

Energy Waste Caused by Air in the Refrigerant

In chillers that use compressors, non-refrigerant gases reduce efficiency by wasting the compressor energy that is required to compress the gases through the system. These gases have minimal refrigeration effect. This energy waste is proportional to the increase in condensing pressure.

In chiller systems that use air coil evaporators and condensers, air that accumulates in the coils can substantially reduce their heat exchange surface area, reducing efficiency and capacity. The same effect occurs in water-cooled condensers.

In absorption chillers, non-condensible gases reduce the vacuum that is necessary to achieve rapid evaporation of the refrigerant. Small amounts of air substantially reduce efficiency and capacity.

Equipment Damage Caused by Air and Moisture

Air and moisture leakage into chillers causes equipment damage and interferes with performance in a variety of ways, of which the most serious are:

• *corrosion.* In chillers that use halocarbon refrigerants (the most common types), water combines with the refrigerants to form powerful acids, primarily hydrochloric and hydrofluoric. Even small quantities of these acids can cause expensive damage and shorten chiller life. The acids preferentially attack weaknesses in steel, such as welds in compressor casings, causing microscopic leaks. Acid also preferentially attacks polished surfaces, such as valve plates.

In absorption chillers that use a lithium bromide solution (the most common type), air hastens corrosion by interaction with the salt solution.

• *electrical burnout,* in chillers that have internal motors. The same acids that cause corrosion also attack motor insulation. Eventual arcing of the insulation causes catastrophic failure, which requires expensive motor repair or replacement. The burnout itself creates contaminants that disperse throughout the chiller system, requiring tedious and expensive system cleaning.

• *damage by oil breakdown products.* The acids that are formed by reactions between the refrigerant and water react with the lubricating oil to form sludge. The sludge blocks oil filter screens and dryers, and deposits on heat transfer surfaces. The results are higher operating pressure and compressor temperature. In turn, the increased temperature hastens carbonization of the oil, which acts as a catalyst for further acid formation.

• *copper plating.* Copper is dissolved from copper alloy system components and is then deposited on hot surfaces, such as crankshaft journals. This occurs in a series of chemical reactions involving water that leaks into the system.

• *higher operating temperatures.* Air in the system causes the compressor to run hotter, because the air must be compressed along with the refrigerant,

requiring higher discharge pressure and creating more heat of compression. The higher temperature shortens compressor and motor life, and aggravates all the previous adverse effects.

• *obstruction of refrigerant flow.* Moisture that remains free freezes in the coldest components in the system, where it obstructs refrigerant flow. Freezing is especially likely to occur in the refrigerant metering device.

Operation and Types of Purge Units

Purge units expel non-condensible gases that leak into the system. To do this, they continuously draw vapor from the inside of the chiller, condense the refrigerant and return it to the chiller, and vent the gases that do not condense.

Purge units come in a variety of designs. Some are self-contained cooling units with their own compressors, condensers, and separation equipment. These units can operate whether the chiller itself is running or not. Other purge units use the liquid refrigerant or chilled water generated within the chiller as the means of condensing the refrigerant in the purge unit, so these types cannot operate unless the chiller is running. Some purge units use cold domestic water as the condensing medium to allow purge operation when the chiller is turned off. The modern high-efficiency purge units covered by Measure 2.7.3.1, below, are self-contained units that provide low condensing temperatures.

In low-temperature applications, such as ammonia food freezers, purging can be done with a simple device similar to an inverted-bucket steam trap. This device contains a refrigerant condensing coil that is operated by the chiller system itself.

■ How Purgers Remove Water Vapor

Purge units remove water by two processes. One of these is inherent in the operation of the purger, and the other is not. Water is only slightly soluble in halocarbon refrigerants. As refrigerant and water vapor condense in the purge unit, the fraction of the water that separates from the refrigerant floats to the top of the refrigerant in the purger, where it is tapped off. With some purge units, the water must be drained from the purger manually.

The solubility of water in halocarbon refrigerants decreases as the temperature of the refrigerant decreases. If the purge unit cools the condensed refrigerant below the evaporator temperature, no free water will exist inside the chiller, although some water remains dissolved in the refrigerant. On the other hand, if the purge unit does not cool the condensed refrigerant below the evaporator temperature, free water can exist in the chiller. Free water causes severe corrosion as it reacts with the refrigerant to create an acid solution. In a water chiller, especially severe corrosion occurs where water condenses on the steel surfaces of the evaporator shell.

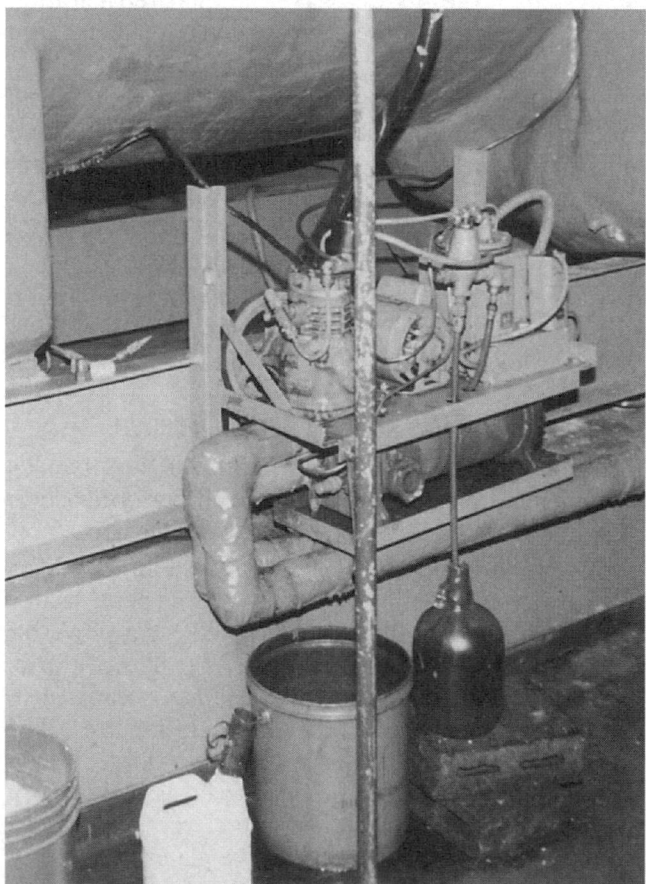

WESINC

Fig. 1 Purge unit on chiller with low-pressure refrigerant
This older purge unit discharges non-condensible gases through the jug of refrigerant oil on the floor. Bubbles in the oil show that air is leaking into the chiller. Refrigerant ejected by the purger dissolves in the oil, so it does not form bubbles.

A chemical dryer element may be installed to remove the water that is not removed by the action of the purge unit. The dryer typically consists of a disposable cartridge of hygroscopic (water-absorbing) material. The dryer element must be replaced when it becomes saturated with water. This is usually indicated by a chemical added to the dryer that changes color when the dryer is near saturation.

The dryer element is not an inherent part of the purger, but the purger is a logical location for the dryer because the refrigerant vapor is recirculated through the purger. The dryer may be built into the purge unit, or it may be installed separately in the line that returns liquid refrigerant from the purger to the chiller. Dryers may be installed in other parts of the system not related to purge operation, such as the liquid refrigerant line in chillers that use expansion valves. Some of the high-efficiency purge units covered in Measure 2.7.3.1 use a water removal process that does not require a separate chemical dryer to remove normal amounts of moisture.

Liquid water can leak into a chiller through the evaporator of a water chiller, or through the condenser of a water-cooled chiller. No purge unit can deal effectively with a water leak. This is because the purge unit draws only vapor from the inside of the chiller system. The pressures that exist inside chillers with low-pressure refrigerants are high enough to keep water in the liquid state, and the purger cannot remove water as a liquid.

For this reason, water chillers or water-cooled chillers that use low-pressure refrigerants should have a reliable method of indicating a water leak into the refrigerant circuit as soon as it occurs, along with a chemical dryer to absorb water until a repair is made.

Refrigerant Loss with Purge Operation

Purge units vent a certain amount of refrigerant along with the air they remove. In fact, an older purge unit may remove several times as much refrigerant as air. Replacing refrigerant is expensive, and the escaped refrigerant may be an environmental hazard. Therefore, older purge units are usually operated intermittently, based on experience with the rate of air leakage into the chiller.

This is not a concern with the newer high-efficiency purge units recommended by Measure 2.7.3.1, below. These units waste much less refrigerant, and their operation is controlled automatically.

How to Check Purge Operation

Different types of purgers provide different ways of checking their operation. Start by reading the purger instruction manual.

With many units, if the unit is ejecting non-refrigerant gases, you can see the gases bubble through a liquid seal, or through a bubbler that is installed specifically for this purpose. If the purge unit does not have a liquid seal that you can see, pipe the purge vent into a transparent container of compressor lubricating oil. The ejected gas forms visible bubbles.

The absence of bubbles does not necessarily mean that the purge unit is not working, because there may be only a minimal amount of air inside the chiller. On the other hand, a small quantity of bubbles emitted continually may be refrigerant, rather than non-condensible gases. This loss of refrigerant is characteristic of older, less efficient purgers.

Some purgers are constructed so that moisture eliminated from the chiller is visible in a sight glass.

You can check accumulation of moisture from the color of the dryer element, if your chiller has a chemical dryer that indicates moisture by changing color.

Check the operation of the purger indirectly by monitoring the relationship of the pressures and temperatures in the condenser and evaporator. See Measure 2.7.2 for details. If both these pressures are higher than normal, there are foreign gases in the chiller.

Purge Idle Chillers

If chillers are idle for long periods, or if the rate of leakage is substantial, operate purge units at appropriate intervals while the chillers are idle. For example, if a CFC-11 chiller is located in an unheated equipment space, the refrigerant vapor pressure remains below atmospheric pressure throughout much of the cold weather season, inducing air leakage.

An alternative for such chillers is to install equipment that prevents air leakage into idle chillers by pressurizing them above atmospheric pressure. See Measure 2.7.4, below.

Absorption chillers that use water as the refrigerant have high internal vacuum. This makes them susceptible to air leaks. Any air that enters causes rapid corrosion because of the action of the salt solution inside the chiller. For this reason, absorption chillers should be purged continually, whether they are operating or not.

If the present purge unit cannot operate unless the chiller is running, consider replacing it with a unit that can operate with an idle chiller. This is an opportunity to upgrade the purge unit, as recommended by Measure 2.7.3.1, next.

ECONOMICS

SAVINGS POTENTIAL: Up to 10 percent of chiller operating cost. In addition, proper purging reduces long-term maintenance costs.

COST: Minimal.

PAYBACK PERIOD: Immediate.

MEASURE 2.7.3.1 **Install high-efficiency purge units.**

RATINGS

New Facilities	Retrofit	O&M
A	C	

Purge units manufactured as late as the 1980's may extract several times more refrigerant than air from the system. Newer models of purge units were introduced during the 1990's specifically to reduce the emission of halocarbon refrigerants into the atmosphere. The best new purge units radically reduce loss of refrigerant, simplify use, and provide a record of leakage. Figure 1 shows a modern unit.

If your chillers with halocarbon refrigerants don't have high-efficiency purge units, consider installing them. Low-loss purgers are available as retrofit units for older chillers. Some chiller manufacturers offer them as standard equipment on new chillers.

If a chiller does not have a purge unit, but it operates with an evaporating pressure lower than atmospheric pressure, consider installing a purge unit. This will improve efficiency with all types of refrigerants. It will also prevent damage if the refrigerants have harmful reactions with air or water. Measure 2.7.3 explains purge units in detail.

Advantages of High-Efficiency Purgers

The main feature of high-efficiency purge units is that they cool the recirculated refrigerant vapor to a low temperature. This condenses most of the refrigerant, allowing it to be recovered. It also minimizes the solubility of water in the liquid refrigerant, so that the water separates from the refrigerant in the purger.

High-efficiency purge units reduce the cost of refrigerant wasted during purging, along with the presumed environmental problems caused by refrigerants. One prominent manufacturer claims to cut the loss of refrigerant to less than one percent of the quantity of air removed.

High-efficiency purgers operate only in response to an accumulation of air or water. Thus, they run whenever needed, but not more than needed. Some high-efficiency

SUMMARY

Consider replacing an old purge unit with a new one if the refrigerant is expensive or environmentally harmful. If a chiller does not presently have a purge unit, consider installing one if any part of the cooling system operates below atmospheric pressure.

SELECTION SCORECARD

Savings Potential	$ $
Rate of Return, New Facilities	% % %
Rate of Return, Retrofit..........	% % %
Reliability	✓ ✓ ✓ ✓
Ease of Retrofit	☺ ☺ ☺

purge units use a water removal process that does not require an expendable dryer cartridge.

A valuable feature of some purge units is the ability to keep track of the amount of air removed from the chiller. This is helpful for indicating whether leaks exist, and how bad they are.

Purgers for Refrigerants that do Not React with Air or Moisture

Some refrigerants do not have harmful reactions with air or water. Ammonia is the most important example. Water mixes readily with ammonia, forming a harmless solution. Water and air may still rust the steel components of the system, but not in the aggressive manner that occurs with halocarbon refrigerants.

Air that enters the system still increases compressor power and wastes energy. So, even with these refrigerants, it is still important to have a purger. In low-temperature applications, such as ammonia food freezers, purging can be done with a device similar to an inverted-bucket steam trap that contains a refrigerant condensing coil. This simpler type of purger operates well because it exploits the low-temperature operation of the system itself.

ECONOMICS

SAVINGS POTENTIAL: *Savings in refrigerant cost vary with the size and leakiness of the chiller system. Several hundred dollars per year is typical. More effective purging reduces long-term maintenance costs.*

COST: *Approximately $1,000 for the equipment. In retrofit applications, installation may cost several hundred dollars.*

PAYBACK PERIOD: *Several years, or longer. The payback period may be shorter than this if the high-*

efficiency unit is installed as a replacement for a failed unit.

TRAPS & TRICKS

SELECTING THE EQUIPMENT: *Purgers are an evolving product, so shop the market for features. Favor the equipment of major manufacturers. Check with other users of models you are considering.*

INSTALLATION: *Good installation workmanship is important to keep the purger connections from becoming an air leak. Follow the manufacturer's instructions precisely.*

OPERATION: *Train the staff in using the equipment. Provide clear instructions in the plant operating manual. Install an effective placard at the unit.*

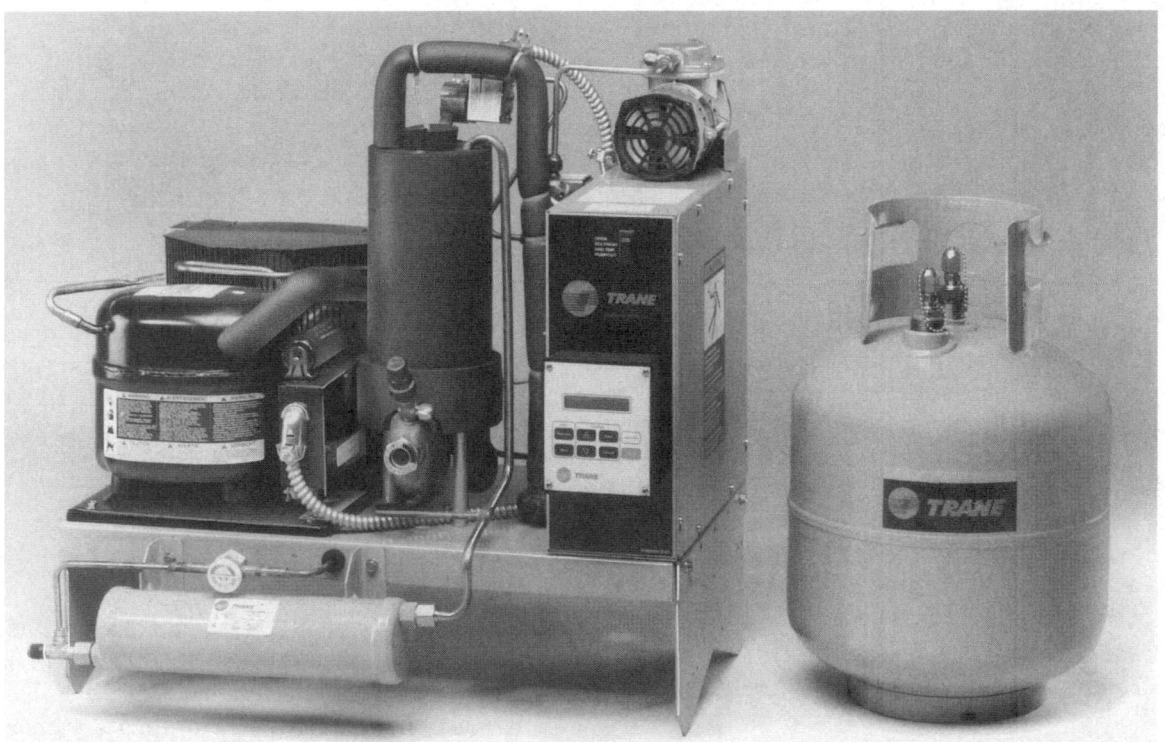

Trane Company

Fig. 1 Modern purge unit The best contemporary purge units extract much less refrigerant from the chiller than older models, and they are controlled to avoid unnecessary operation. Some also record the amount of air leakage into the chiller. The refrigerant canister on the right indicates the size of the unit.

MEASURE 2.7.4 Install accessories that prevent air leakage into idle chillers.

If you have chillers with low-pressure refrigerants that stand idle for long periods, you can prevent air leakage into the chillers by pressurizing them to keep the internal pressure higher than atmospheric. This approach has the advantage that air is never allowed to enter the chiller, while purging removes air only after it has entered the system.

There are two practical ways to pressurize an idle chiller. One is to warm the refrigerant in the chiller, and the other is to add an inert gas to the refrigerant. This Measure applies to several common refrigerants that have a saturation pressure that is only slightly below atmospheric pressure. It does not apply to absorption chillers, which have a high internal vacuum.

Don't expect these methods to prevent water leaks into the chiller from the evaporator or condenser. The amount of pressurization required would be much too high, forcing refrigerant out of the system. Instead, see Measure 2.7.5 about preventing water leakage into idle chillers.

Pressurization by Warming the Refrigerant

The newest method of pressurizing chillers is to heat the refrigerant until its vapor pressure is slightly higher than atmospheric. This method is limited to refrigerants having saturation pressures that are not much below atmospheric pressure. In practice, this limits the technique to compression chillers with the common low-pressure refrigerants. The refrigerant temperature needed is about 76°F (24°C) for CFC-11, and about 84°F (29°C) for HCFC-123.

Accessory units that warm the refrigerant are available commercially. One type uses an electric heating blanket that is wrapped around the evaporator or refrigerant sump. Another type uses a small water heater that is piped to circulate through the chilled water side of the evaporator. The heater is typically electric, but the same effect could be achieved with a heater that uses other energy sources.

Both types require a sensitive gauge to measure the difference between the inside and outside pressures, and an electronic control to regulate the heating element. Thermostatic control allows the pressure differential of the refrigerant to be controlled precisely, typically to about 0.01 PSI higher than atmospheric pressure. The smallness of the pressure differential minimizes the amount of refrigerant that is forced out of the system through any leaks.

The chilled water isolation valves of the evaporator must close tightly if the refrigerant is warmed. Otherwise, the chilled water will conduct the heat away from the chiller, requiring the heating unit to operate at a high power level and waste energy.

Pressurization by Adding Inert Gas

An older method is pressurizing chillers by using a pressurized source of dry nitrogen. Nitrogen is inert and inexpensive. A specialized pressure regulator is needed to provide only a slight pressure differential in the chiller. The nitrogen must be removed completely before the chiller is started. This is done by the chiller's purge unit.

Also Finds Refrigerant Leaks

As discussed in Measure 2.7.1, it is difficult to find air leaks into a chiller that has low-pressure refrigerant. The methods in this Measure can be used to pressurize the chiller to force refrigerant out of the chiller through any leaks that exist. Then, ordinary refrigerant leak testing methods will reveal where the air leaks are located. For finding leaks, the pressure is increased temporarily to several PSI. This is a standard feature of refrigerant warming systems.

ECONOMICS

SAVINGS POTENTIAL: Savings in refrigerant vary with the size and leakiness of the chiller system. Several hundred dollars per year is typical. The main benefit is a long-term saving in chiller maintenance costs.

COST: Approximately $2,000 to $3,000 for the equipment. Several hundred dollars worth of connecting pipe are needed. Installation typically takes about 12 hours.

PAYBACK PERIOD: Several years, or longer.

TRAPS & TRICKS

SELECTING THE EQUIPMENT: *Refrigerant warming equipment is less cumbersome to use than inert gas pressurization, and it allows precise control of the pressure differential. However, refrigerant warming is still novel, so check performance with current users before you buy. Pressurization with inert gas may be the best option if idle chillers are located in a very cold space, or if the evaporator isolation valves are not tight.*

INSTALLATION: *Let the vendor install the equipment. This avoids warranty disputes. Good workmanship is needed to avoid an air leak at the sensor connections.*

SCHEDULING: *Use your yearly maintenance schedule to remind you to follow the procedures for idling the chillers. Schedule periodic checks of operation during the season when chillers are idle. Put a place for this on the plant operating log.*

EXPLAIN IT: *Operators will have no idea what the equipment is supposed to do unless you explain it. Install a permanent, clearly written instruction placard at the equipment.*

MEASURE **2.7.5 Drain the water from the evaporators and condensers of idle chillers.**

A small leak that allows liquid water to enter a chiller from an evaporator or condenser is much more serious than a small air leak. Much more water enters the chiller this way. Purge units are much less able to remove liquid water. The hydrostatic pressure in the chilled water system or cooling tower system creates a much larger pressure differential to force the water into the system than does atmospheric pressure.

You can easily prevent water leakage into the chiller when the chiller is idle for extended periods. Simply drain the water from the evaporator and condenser. You don't have to drain the entire water system to do this. You already have isolation valves that are installed to allow maintenance on these heat exchangers. Close these valves and drain the heat exchangers.

(You cannot use the method of Measure 2.7.4, pressurizing the inside of the chiller, to prevent leakage of water into the chiller from evaporators and condensers. Even if you could pressurize the refrigerant side enough to overcome the hydrostatic pressure of the water, the high internal pressure would accelerate loss of refrigerant through any leaks in other parts of the system.)

Piping Improvements

If you do not drain the entire system when you drain the chiller, you may need some pipe and valve modifications to make draining easy and trouble-free. Expect the isolation valves to leak somewhat. The drained condenser or evaporator is filled with air, so water that leaks into it can rust the steel parts.

To prevent this, arrange the piping so that any leakage through an isolation valve is diverted to the sewer without the water entering the heat exchanger first. A simple drain tap at a low point in the water line should suffice. Visualize where water from a leaking isolation valve would pool. If this is in a portion of pipe outside the chiller, install a small drain at this point.

If water from a leaking isolation valve would drain into the evaporator or condenser, or into the water box of a larger heat exchanger, modify the pipe to keep the water outside. For example, you can weld a small dam inside the pipe. Tap a small drain in the pipe on the upper side of the dam. Or, if the pipe diameter is small, insert a small dip in the line outside the heat exchanger.

ECONOMICS

SAVINGS POTENTIAL: *The primary benefits are extended chiller life and reduced long-term maintenance.*

COST: *May cost nothing. If you have to modify the piping or add drain lines, the cost may range from several hundred dollars to several thousand dollars, depending on the pipe size.*

PAYBACK PERIOD: *Immediate, to several years.*

TRAPS & TRICKS

PREPARATION: *Spend some time looking at the water connections to the evaporator and condenser. Imagine how water would flow from leaking isolation valves. Figure out an inexpensive way of creating a low drain point outside each heat exchanger connection.*

SCHEDULING AND DOCUMENTATION: *Use your yearly maintenance schedule to remind you to follow the procedures for idling the chillers. Explain the procedure in the plant operating manual.*

System Design for Efficient Low-Load Cooling

The Measures in this Subsection are major changes to the chiller plant that increase cooling efficiency at low cooling loads. The three Measures are alternative approaches to this goal. Some of the Measures in other Sections can improve the part-load efficiency of a chiller system, but they are usually less effective.

In most applications, the cooling load is much lower than the peak load for most of the cooling season. To achieve good efficiency under all load conditions, you need chillers that are sized so that no chiller is forced to linger at an inefficiently low load. In addition, you may want a chiller of a separate type to satisfy low loads.

One class of energy waste that results from low-load operation is inefficiency of the chillers and their auxiliary equipment. Unfortunately, the types of chillers that are most efficient at full load (centrifugal and screw compressor types) suffer serious loss of efficiency at reduced loads. While modern versions of these chiller types have greatly improved part-load performance, many machines that are presently installed have poor part-load efficiency. (See Reference Note 32, Compression Cooling, for more about chiller efficiency characteristics.)

Another class of low-load energy waste is false loading. This is done to avoid cycling chillers on and off when the cooling load is less than their minimum capacities. False loading throws away the portion of the chiller output that is not required. This is an extremely wasteful way of dealing with reduced cooling loads. It is often encountered with centrifugal chillers, because of their inability to operate at low output. Especially with older centrifugal chillers, the minimum load is a large fraction of full load. False loading is also encountered with cooling machines having other types of compressors, including screw and reciprocating compressors.

INDEX OF MEASURES

2.8.1 Install chillers and auxiliary equipment of appropriate size to avoid extended operation at low load.

2.8.2 Install local cooling units to allow shutting down the central chiller plant during periods of low load.

2.8.3 If a facility has several chiller plants, provide cross connections that allow shutting down the least efficient chillers.

RELATED MEASURES

- Subsection 2.1, for controls and procedures that optimize the use of chiller plant equipment
- Measures 2.2.1 ff, chilled water temperature reset, which improves chiller efficiency at reduced loads
- Measures 2.4.1 ff, improving condenser cooling fan modulation
- Measure 2.5.2, variable-flow chilled water pumping
- Measure 2.6.2, replacing the existing compressor with a compressor that has better low-load efficiency
- Measure 2.6.3, installing variable-speed drives on existing compressors

Chillers can be false-loaded internally, which is done by using hot gas bypass. Or, chillers can be false-loaded externally, by adding unnecessary cooling load. For example, chillers can be loaded by increasing reheat in air handling systems, by increasing outside air quantities, or by cooling vacant spaces.

Defects in the chiller may restrict throttling ability, raising the load at which false loading is employed. For example, if the external linkage that actuates the inlet guide vanes of a centrifugal chiller is not adjusted properly, the vanes may not close fully, limiting turndown. Another problem is instability of the control system, so that the chiller output oscillates wastefully at low load. This can result from play in linkages, improper control installation, improper charge, and other causes.

If some parts of a facility continue to require cooling outside the normal cooling season, cooling efficiency will be especially low during this period. For example, chillers that serve multizone units must operate continuously in order to maintain a constant cold deck temperature in the air handling units. You may be able to eliminate such waste by modifying the application. For example, Subsection 4.6 recommends modifications of multizone air handling units that eliminate the need for continuous chiller operation.

Inefficient chiller operation at low load reduces the benefit of energy conservation measures that you accomplish to reduce the cooling load. For example, if you add expensive sun shades to reduce solar load, they provide no economic benefit during periods when false loading is used. When the cooling load rises above this minimum, the conservation benefit of the sun shades is limited to the difference between the actual cooling load and the false loading level.

MEASURE 2.8.1 Install chillers and auxiliary equipment of appropriate size to avoid extended operation at low load.

RATINGS
New Facilities | Retrofit | O&M
A | C |

Your primary method of avoiding inefficient operation at low loads is to select the chiller sizes carefully, and to match the sizing of the chiller auxiliary equipment to the individual chillers. Your goal is to select a combination of chiller sizes that allow the plant to operate efficiently under every level of cooling load that may occur for a significant length of time.

For example, if the chiller plant provides air conditioning, it should be able to handle spring and fall loads as efficiently as summer loads, and it should handle nighttime loads as efficiently as daytime loads. This may seem obvious, but you probably don't have to travel far to find a chiller plant that violates this principle. Commonly, you will find a plant that has two enormous chillers. One is sized for the peak cooling load, plus a large margin. The second chiller, identical to the first, sits there as a standby unit. For most of its operating hours, the one chiller that runs is operating at greatly reduced efficiency.

In both new construction and retrofits, you face a dilemma when selecting chiller sizes: the chiller types that are more efficient at full load tend to be most limited in their ability to operate efficiently at low load. This conflict between full-load and part-load efficiency is most severe with centrifugal chillers, and it is also significant with screw compressors. The best of the newer models of centrifugal and screw chillers have greatly improved their part-load efficiency, but this factor is still important.

In new construction, you can usually select an efficient combination of chiller sizes and types without adding much to the cost of the plant. You can typically maintain efficient chiller performance with only one or two chiller sizes. In a large plant with a highly variable load profile, you may wish to use three chiller sizes. You may actually save some money by using a combination of sizes. This is because you need less reserve capacity if you combine chiller sizes intelligently.

Retrofit applications are more challenging. You probably already have all the chiller capacity you need. The cost of installing another chiller to handle low loads is high. Cost usually limits you to one additional chiller. This chiller will usually be smaller than the existing chillers. It must be small enough to handle the lowest loads efficiently, but large enough to avoid operating an existing chiller at inefficiently low load. Modern chillers are substantially more efficient than older units. If the existing chillers are old, you want to make the new chiller larger so that it carry more of the cooling load.

SUMMARY

Avoids a major cause of inefficiency with centrifugal, screw, and absorption chillers. May also reduce maintenance and labor costs. May cost nothing in new construction. Expensive as a retrofit.

SELECTION SCORECARD

Savings Potential .... $ $ $ $

Rate of Return, New Facilities % % % %

Rate of Return, Retrofit.......... % % %

Reliability .... ✓ ✓ ✓

Ease of Retrofit .... ☺ ☺

In a group of chillers, the one that operates first is called the "lead" unit. This chiller is the most efficient at the low loads that it serves. It is usually the smallest unit in the plant. As the cooling load increases to the point that the lead chiller cannot carry the load alone, another chiller starts. The best distribution of the cooling load between the chillers depends on their relative efficiency characteristics. See Measure 2.1.1 for details.

Base Your Decision on the Future, Not the Past

In retrofit applications, an apparent disadvantage of this Measure is that it keeps more of the original chiller capacity idle. This is a matter of psychology rather than engineering or economics. There is a natural inhibition against correcting a chiller sizing mistake if the original equipment remains as a monument to the mistake.

However, from a purely economic standpoint, the correct approach is to view the original investment as a sunk cost, and to consider only the economics of the new investment. Your economic analysis should consider the salvage value of any chillers that you may be able to sell as a result of adding the new lead chiller.

Efficiency Features of the Main Chiller Types

See Reference Note 32, Compression Cooling, and Reference Note 33, Absorption Cooling, for the characteristics of different types of chillers. The following discussion deals primarily with the chiller characteristics that matter when selecting chiller sizes to maximize efficiency.

Although the chiller type is the main factor in part-load efficiency, there are major differences among different models of a given type. Pay attention to the differences between different models. Be specific about acceptable model numbers when acquiring a chiller by using a bid specification. Otherwise, contractors or suppliers may substitute cheaper equipment that is less desirable in efficiency, features, or reliability.

■ Centrifugal Chillers as Lead Chillers

Some centrifugal chillers now combine variable-speed drives with conventional inlet guide vane throttling. This substantially improves their COP at partial loads. This improved part-load performance may make a centrifugal chiller appropriate for use as a lead chiller, if it is small enough in relation to typical low loads. Centrifugal chillers are available in sizes as small as 100 tons, although their cost per ton is high in small sizes.

■ Screw Compressor Chillers as Lead Chillers

The peak efficiency of screw compressor chillers is less than the peak efficiency of centrifugal machines, but they are able to operate at reduced loads with less loss of efficiency. Furthermore, most modern screw chillers are tolerant of cycling on and off at minimum load, unlike centrifugal machines. Screw machines are available in sizes as small as about 25 tons, although their cost per ton is relatively high in smaller sizes.

■ Reciprocating Chillers as Lead Chillers

The full-load efficiency of reciprocating chillers is substantially less than that of centrifugal and screw compressor machines. This difference in efficiency exists for fundamental physical reasons, and it will probably widen as centrifugal and screw chillers become more efficient. Reciprocating chillers unload more efficiently than centrifugal chillers, but even their advantage at low load is disappearing. The remaining niche for reciprocating chillers is small cooling loads, less than about 100 tons. Most reciprocating chillers tolerate cycling well, and they lose relatively little efficiency down to zero load.

A difficult control problem arises if a reciprocating chiller is used as the lead chiller with centrifugal chillers or screw compressor units. Reciprocating chillers use step control of capacity, so their output is usually controlled by sensing return water temperature. In contrast, centrifugal and screw chillers typically control output by sensing supply water temperature.

One way to use a reciprocating chiller as a lead unit is to shut down the reciprocating chiller when a larger non-reciprocating chiller is started. However, this approach makes the lead chiller useless for carrying part of the load in the event that one of the larger chillers should fail.

Another approach is to use supply water temperature control with the reciprocating chiller. This requires the control system to anticipate the changes in supply chilled water temperature as the compressor stages. This is a touchy method of control for reciprocating chillers, especially if they are operating in parallel with other types of chillers.

■ Scroll Compressor Chillers as Lead Chillers

Chillers that use scroll compressors control output by cycling on and off. They are presently similar to reciprocating chillers in efficiency. They may gain an advantage in the future because they do not have all the inherent efficiency limitations of reciprocating units. Like reciprocating chillers, they generally control output based on return water temperature, and therefore they have the same problems in operating in parallel with throttling types of chillers. At present, scroll compressors occupy the bottom end of the size range, but they can be ganged together to achieve any desired capacity.

■ Absorption Chillers

Absorption chillers are unsuitable as lead units in most applications. Start-up and shutdown is slow, and requires a careful sequence to avoid crystallizing the salt absorbing solution, as well as to avoid other troubles. Absorption chillers are best used to exploit an inexpensive source of heat, and to minimize electricity demand charges.

Exploit Low Condensing Temperatures

Chiller efficiency improves radically at lower condensing temperatures, provided that the chiller is designed to exploit low condensing temperature. This feature is especially important for chillers that operate at low loads, because low loads typically occur at times when the outside air temperature is low. Make this an important factor in selecting your lead chiller. Be aware that chillers typically have limitations in their ability to exploit reduced condensing temperature. See Measure 2.2.2 for details.

How to Select Chiller Size

Sizing is most critical for the lead chiller, because it operates at the lowest percentages of load. If the lead chiller is a throttling type and if it is too large, it too will operate inefficiently at low loads. If the lead chiller is too small, the next chiller may operate at inefficiently low load when it starts. Consider these factors in sizing a lead chiller:

- *cooling load profile.* Analyze the cooling load profile in detail, and match it to the part-load efficiency profiles of the chiller types you are considering. In existing facilities, measure the cooling load profile with hourly measurements of chiller load over the course of a cooling season. The cooling load pattern varies from one season to another, so try to get data from several years. If the plant is new, or if you do not have load data, you have to calculate the load profile. In most cases, you should use an energy analysis computer program for this.

• *reserve capacity requirements.* In an existing chiller plant, adding a small lead chiller always increases reserve capacity. If you are installing a new chiller plant, cost limitations may force you to compromise between part-load efficiency and reserve capacity.

For example, with a design load of 2,000 tons, it would be common to install three 1,000 ton chillers, one of which is a spare unit. In contrast, a combination of a 400 ton lead chiller and two 1,000 ton units would be considerably more efficient, and it would cost less. The disadvantage is that failure of one of the 1,000 ton chillers would reduce the available capacity to 1,400 tons. However, with a typical load profile, this capacity would suffice most of the time. Even if a chiller failure occurred during a peak load period, capacity would probably be adequate to keep the facility operating while the failed chiller is repaired.

• *other cooling conservation activities.* Consider the effect of other energy conservation measures on the cooling load, especially measures that reduce cooling load during low-load conditions. For example, eliminating reheat in air handling systems (see Section 4) and window treatment (see Section 8) dramatically reduce the cooling load during mild weather. These improvements reduce the total plant capacity requirement and provide more freedom in selecting the lead chiller for efficient part-load performance.

• *efficiency of the other chillers.* When sizing a lead chiller for an existing facility, consider which chiller will be started second. It might be the smallest of the original chillers, or a different one that is more efficient. Select the relative sizing of the first and second chillers so that both can operate efficiently at the point where the second chiller starts, or the second can operate efficiently if the first is turned off. If one or more of the original chillers is especially inefficient, consider increasing the size of the new lead chiller to allow the inefficient machine(s) to be placed out of service either permanently or on a seasonal basis.

Ability to Reduce Auxiliary Load

In retrofit applications, adding a smaller lead chiller may or may not give you an opportunity to reduce the energy consumption of auxiliary equipment. It depends on these factors:

• *chilled water pumping.* You can reduce chilled water pumping energy only if one of the chilled water pump is sized to match the capacity of the lead chiller. Also, the end-use equipment needs to use throttling control valves instead of bypass valves. See Measure 2.5.2 for more about this.

• *condenser water pumping.* If the new lead chiller has a water-cooled condenser that is served by an existing cooling tower and condenser water system, you can save pumping energy by installing a condenser water pump sized specifically for the new chiller.

• *condenser cooling fans.* If the condensers are cooled by water, the fan energy used by the cooling towers is proportional to the cooling load. So, adding a lead chiller probably will not save much fan power. However, if you improve fan capacity modulation (see Measures 2.4.1 ff) when you install the lead chiller, this will save fan energy.

Base Your Decisions on Detailed Calculations

The cooling load fluctuates continuously in most applications. Dealing with this accurately requires you to calculate the moment-to-moment energy consumption of the plant over the course of a year. Repeat the calculation for different chiller size combinations to find the most economical combination. In a retrofit application, calculate how a system with a lead chiller would compare to the existing system. To simplify these tedious calculations, use an appropriate computer program. See Reference Note 17, Energy Analysis Computer Programs, about these programs.

Control the Chiller Loading Efficiently

Clever equipment sizing will not provide the expected benefit unless you arrange to start and stop the chillers in the sequence that satisfies the cooling load most efficiently. Doing this reliably requires operator training, effective documentation of procedures, and continuing supervision. Or, it requires reliable controls. In plants that have a variety of chiller types and/or sizes, you need to decide the proper sequence in terms of the percentage of load at which each chiller is turned on or off. See Measures 2.1.1 ff for details.

Consider Using the Lead Chiller for Heat Recovery

You may be able to use some of the heat that is rejected by the chillers. Sometimes, this is done with a small chiller that has special characteristics for heat recovery. The heat recovery chiller is usually loaded ahead of the other chillers, so consider combining the functions of a lead chiller and a heat recovery chiller. See Measure 2.10.3 about chiller heat recovery.

Comparison to Measure 2.8.2

Measure 2.8.2 deals with a particular situation in which a chiller plant is kept in operation at inefficiently low loads to serve a few small applications. To deal with this situation, it may be more efficient to shut down the chiller plant entirely during periods of low loads, and to provide local cooling units for the small applications. For a comparison of the two approaches, see Measure 2.8.2.

ECONOMICS

SAVINGS POTENTIAL: *5 to 20 percent of chiller plant operating cost, in typical applications. Savings are greatest if false loading is presently used to reduce chiller output.*

COST: *In new facilities, selecting chiller sizes for optimum efficiency may cost little or nothing extra. In retrofit applications, an additional chiller costs from $300 to $1,000 per ton of capacity, including design, auxiliary equipment, and installation.*

PAYBACK PERIOD: *In new facilities, the payback period may be short. In existing facilities, the payback period ranges from several years to many years. The payback period may be immediate if a new lead chiller replaces a chiller that needs to be replaced.*

TRAPS & TRICKS

PLANNING: *The design procedures are routine, but you need to make an exceptionally thorough analysis to select the optimum chiller sizes. If you combine different types of chillers, be sure that you can control them to operate together efficiently. Design the auxiliary equipment to minimize the auxiliary load.*

SELECTING THE EQUIPMENT: *Select the lead chiller(s) with an emphasis on high turndown ratio. Invest in efficiency options.*

TRAINING AND FOLLOWUP: *The benefit of this major aspect of plant design may be lost if the plant operators do not understand how to sequence the chillers efficiently. Make sure that they are properly trained. Put clear instructions in the plant operating manual. Check the chiller plant log periodically to verify that the chillers are being sequenced and loaded efficiently. Review Measures 2.1.1 ff.*

MEASURE 2.8.2 Install local cooling units to allow shutting down the central chiller plant during periods of low load.

SUMMARY

Appropriate where the duration of chiller operation is extended substantially to serve lesser cooling requirements. Improves overall efficiency, and may significantly reduce chiller plant overhead costs.

SELECTION SCORECARD

Savings Potential .... $ $ $

Rate of Return, New Facilities % % %

Rate of Return, Retrofit.......... % % %

Reliability .... ✓ ✓

Ease of Retrofit .... ☺ ☺

A common problem is that a few small cooling applications require the operation of a large chiller for long periods of time while the rest of the facility is shut down. For example, the computer rooms in an office complex typically need continuous cooling, even though most of the complex is shut down more than half the time.

The same problem occurs on a larger scale if a central chiller plant serves an extended facility, such as a college campus or a military base. For example, if a college campus is mostly shut down during the hot summer months, the chiller plant may still have to operate to serve one administrative building.

In such cases, consider installing separate small cooling units to serve the applications that operate on longer schedules. This allows you to shut down the central chiller plant entirely for a large part of the time.

Potential for Energy and Cost Savings

The benefit of this Measure is based on marginal differences in efficiency and energy costs. When you calculate the potential benefit, take the following factors into account.

■ Relative Energy and Demand Costs

The energy used by local cooling units may not cost the same as the energy used by the central plant. Even if the central chillers and the local cooling units both use electricity, the electricity cost may differ if the local cooling units are not served by the same electricity meter. The cost difference is greatest if the different meters are subject to different rate schedules. For example, a central chiller plant may be served under a lower rate than the other buildings in a complex.

Even if the rate is the same, differences in the amount of electricity consumed through each meter may affect the average cost. This effect does not occur if the consumption through all meters is consolidated before the rate is applied.

If the central chillers and the local cooling units use different energy sources, the energy costs are certain to be different. The most common example of this situation is that the central plant uses absorption chillers that are direct-fired or operated by steam.

Another possibility is that the local cooling units use an energy source other than electricity. For example, small cooling units operated by fuel-fired reciprocating engines have gone through several episodes of popularity in recent decades.

Comparing the cost of electricity and other energy sources is complicated by electricity demand charges, which differ by hour, day, and month. Reference Note 21, Electricity Pricing, explains demand charges. To account for the fluctuations in electricity prices, you may need to use an energy analysis computer program to calculate your total yearly costs. These programs are explained in Reference Note 17, Energy Analysis Computer Programs. Even then, you face the possibility of large future changes in the relative cost of electricity and other energy sources.

■ Relative Efficiencies of Central and Local Cooling

To calculate the comparative performance of central and localized cooling, you have to know the detailed efficiency characteristics of both sets of equipment.

Smaller, standard cooling units made in the United States and some other countries are rated for efficiency. See Reference Note 31, How Cooling Efficiency is Expressed, about the types of ratings being used. Be clear about the energy consumption that is covered by the efficiency ratings of local cooling units. For example, the Seasonal Energy Efficiency Rating of a split system may or may not include the energy consumption of the condenser cooling fans. Include the energy needed for any external equipment, for example, for a separate condenser cooling water system.

For larger, engineered cooling systems, don't depend on the efficiency figures in the catalogs. Require certified performance data from the manufacturer. Remember that the COP's given for large chillers apply only to the chiller itself. They do not include energy requirements for separate condenser cooling equipment, for chilled water distribution, or for the end-use equipment, such as air handling units and fan-coil units.

These are large energy loads, so be sure to include them in your calculation. For example, with a central chiller plant, the saving in chilled water pumping may be the major factor that justifies installing local cooling units.

In retrofit applications, you may or may not be able to get reliable chiller efficiency information from the manufacturer. Reliable data may not be available for older chillers, and older machines may have lost efficiency for a variety of reasons. Since a lot of money is at stake, make the effort to measure the efficiency of installed chillers. Install an accurate BTU meter in the chilled water output, and use a reliable method of measuring the energy input. See Reference Note 16 for methods of measuring energy output, and Measure 2.1.1 for various methods of measuring chiller energy input.

The efficiency of chillers varies with load, especially if they have modulating control. To calculate the energy consumptions of both system configurations over a yearly cycle of operation, you may need to use an energy analysis computer program.

■ Ability to Reduce Central Plant Operating Cost

Shutting down the central plant may eliminate most of its non-salary overhead costs, including chiller maintenance, water treatment, and housekeeping energy consumption, such as lighting and ventilation.

Labor is usually the largest overhead cost. It is difficult to get skilled plant operators on a seasonal basis, so the chiller plant staff may have to be kept on the payroll all year. One possible way of reducing labor cost is to reassign chiller operators to seasonal duties that normally would require other personnel.

How to Select Local Cooling Units

See Measure 5.5.3 for tips on selecting local cooling units for efficiency. See Reference Notes 32 and 33 for details of different types of cooling equipment.

For most applications, local cooling units are so small that they are limited to reciprocating or scroll compressors. In extended facilities where even the local loads are large, it may be appropriate to use a screw chiller or even a centrifugal chiller to carry local cooling loads. In this case, see Measure 2.8.1 about installing smaller chillers to serve low loads.

Installation Considerations for Local Cooling Units

Central chiller plants are popular because localized cooling units have a number of disadvantages, or at least, they are more bother. If you install local cooling units, minimize the potential problems. Consider the following points.

■ Ability to Control the Operation of Local Units

It takes more effort to monitor efficient operation of a number of distributed cooling units. Local units are less apparent to the physical plant staff, so they may operate unnecessarily for long periods of time. This can dissipate much of the potential savings. Therefore, provide reliable controls to limit operation of the local units to times when they are needed. Refer to Subsection 5.1 for methods of controlling local units.

Even cooling units that are supplied by a central chiller plant can operate when they are not needed. However, units served by the central plant cannot waste energy when the plant is not operating, whereas local cooling units can waste energy at all times.

■ Ability to Provide Condenser Cooling

Any local cooling unit requires an external condenser or heat rejection unit. Simple split systems are usually satisfactory if the cooled space is located a short distance from a suitable condenser location. For longer distances, consider water-cooled units that are served by a small cooling tower. This configuration has the advantage that a single cooling tower can serve any number of local cooling units.

In extreme cases, you may have to cool condensers using domestic water. This requires about 50 gallons of domestic water per ton-hour. If you use this method, try to save water cost by recovering the used water for further use. Even better, try to exploit the heat collected in the water. See Measure 5.5.9 for details.

■ Space and Appearance

Local cooling units take up space and affect the decor. A large variety of units are available that may minimize the space requirement, or even avoid it. For example, some units are made for installation in ceiling plenums. Others are available with attractive evaporator enclosures that are mounted on the wall above the working level. Another possibility is installing the cooling unit in an adjacent vacant space, and ducting air into the cooled space.

■ Noise

The compressor makes most of the noise, so try to select units that keep the compressor out of the occupied space. Great strides have been made in reducing the noise levels of condenser and evaporator fans, so shop for a quiet unit.

■ Wiring Amperage

Cooling units are large users of electricity compared to other equipment in the spaces they serve, so you may have to increase the amperage capacity of the wiring that serves the area you plan to cool locally.

■ Distribution of Local Units

It there is more than one candidate for localized cooling within an area, there may be several ways to provide the cooling. For example, there may be a choice between installing a number of small air conditioning units for individual spaces, or installing a larger local unit to serve several spaces.

Comparison to Measure 2.8.1

In many cases, this Measure is an alternative to installing a small lead chiller in the central plant, as recommended in Measure 2.8.1. In deciding between the two approaches, consider these issues:

- *how much the efficiency of the central plant will be improved during the main cooling season.* Careful chiller sizing (Measure 2.8.1) can virtually eliminate inefficiency caused by operation at low loading. If you shut down the chiller plant for extended periods, such as the winter season, the chillers may still operate at inefficiently low loads, although not as long. This is because periods of low load occur even during the seasons of heavy cooling. Furthermore, removing some of the load from the central chiller plant (Measure 2.8.2) worsens the mismatch between the capacity of the chillers and the cooling load they serve.

- *relative cooling efficiency during the low-load season.* Measure 2.8.2 eliminates chilled water pumping power, distribution losses, and auxiliary equipment energy consumption during the chiller plant shutdown period. On the other hand, the local cooling units may or may not be more efficient than the central plant chillers when they are operating at low load. If the local cooling units are less efficient than the central chillers, this is not important if the local cooling units are small and the low-load energy losses of the chiller plant are large. It may be important if a substantial fraction of the cooling load is converted to localized cooling.

- *relative cost.* If you need only a small number of local cooling units to shut down the central plant during low-load periods, this Measure may be much less expensive than Measure 2.8.1.

These factors balance in a complex way, so that it may not be obvious which Measure is better. Use the same computer program discussed previously to compare the two approaches.

ECONOMICS

SAVINGS POTENTIAL: 5 to 30 percent of total cooling energy consumption. There may also be significant savings in central plant maintenance and labor costs.

COST: The cost of local cooling units ranges from $5 to $15 per square foot ($50 to $150 per square meter) of conditioned space, for common applications. In new construction, local cooling reduces the capacity requirement of the central chiller plant, which partially offsets the cost of the local cooling units.

PAYBACK PERIOD: Less than one year, to many years, depending on the amount of local cooling equipment that has to be installed.

TRAPS & TRICKS

DESIGN: It may take lot of thought and analysis to get the most out of this approach. Investigate conditions at each of the locations where you would install local cooling units, do load calculations for each location, and figure out the best method of installing the new equipment. Consider whether the units should also have heating capability, and coordinate this with Measure 1.12.2.

SELECTING THE EQUIPMENT: There is a wider range of efficiencies in small cooling units than in large central chillers. Shop the market thoroughly. Stress efficiency, along with other important characteristics, such as noise level and control features.

OPERATION: This activity increases the number of units that need to be controlled, and hides them from the physical plant staff. Provide reliable controls to prevent unnecessary operation. Use the appropriate control Measures from Subsection 5.1.

MEASURE 2.8.3 If a facility has several chiller plants, provide cross connections that allow shutting down the least efficient chillers.

Larger facilities may be served by several independent chiller plants. In new facilities, this may be done to allow separate areas to operate independently, or to reduce distribution cost. In existing facilities, separate plants may have been built as the facility expanded at different times, perhaps with insufficient regard to the equipment that already existed.

In all these cases, interconnecting the separate plants may save energy by:

• allowing the most efficient chillers in any plant to serve the whole facility

• allowing large chillers to operate at higher, more efficient loading during periods of low cooling load

• reducing the total energy required for condenser cooling and plant overhead.

The savings potential of the first two points is covered in Measure 2.8.1. The savings potential of the third point is covered in Measure 2.8.2.

On the negative side, interconnecting the plants requires extending the distribution piping. This may require additional chilled water pump power, and it may increase heat loss.

An Example

Consider a facility that has two separate chiller plants. One plant has two centrifugal chillers rated at 1,000 tons each, and the other plant has two centrifugal chillers rated at 500 tons each. Both chiller plants operate efficiently during the peak cooling season. However, during the transitional season, the larger plant has an average load of only 200 tons, and the smaller plant has an average load of only 100 tons. These loads are inefficient for both plants.

If the plants are connected, the total average transition season load of 300 tons is combined to become an efficient load for one of the 500-ton chillers. The larger chiller plant can be shut down for the entire transition season, saving a considerable amount of plant auxiliary and overhead cost.

During the warmest part of the year, you may be able to shut down the small chiller plant except during peak days, and to serve the entire facility with the larger plant alone. If the plant sizes and other conditions are favorable, a single chiller plant crew might shuttle between the two plants as the seasons change.

Piping Design Issues

The major expense of interconnecting chillers is the cost of the piping, which in many installations can rival the cost of chillers themselves. Pipe diameter is the

SUMMARY

Improves overall efficiency, especially at low cooling loads. May reduce maintenance, reduce staffing costs, increase reliability, and/ or avoid the need to purchase additional chillers. Usually very expensive, unless it avoids the need to purchase additional chiller capacity.

SELECTION SCORECARD

Savings Potential .... $ $ $

Rate of Return, New Facilities % % %

Rate of Return, Retrofit.......... % % %

Reliability .... ✓ ✓

Ease of Retrofit .... ☺

dominant factor in the cost of piping. Both material and installation costs of piping are high.

Design the interconnection to minimize pump power. It is pointless to save energy in the chillers only to lose it in the pumps. Pumps have a hunger for energy that is insidious because they are small and operate quietly. If a pumping system operates at full flow under all load conditions, pump energy consumption can become a major part of total plant energy consumption. Pump energy consumption in distribution systems can be reduced by using variable-flow pumping, which is recommended by Measure 2.5.2.

The pump power needed to overcome the resistance of the added piping is sensitive to the pipe diameter. Bigger pipe allows smaller pumps. The cost saving with smaller pumps may substantially offset the higher cost of larger pipe. Do enough sketching and calculating to find the best balance between pump power and piping cost.

Use these techniques to minimize piping cost:

• *be selective about the interconnections.* In facilities where there are several chiller plants, you may be able to interconnect them in various ways. If cost were no object, the best approach would be to create a large loop header that allows chilled water from any chiller plant to flow to any user. This may be an unnecessary extravagance. Most of the benefit of this Measure occurs under low-load conditions, so the connections can be optimized to transmit water from the low-load chillers to the low-load applications.

- *limit the use of the interconnections to periods of low cooling load,* i.e., to periods when some of the chillers are turned off. Pressure losses increase rapidly as the flow rate increases, so consider limiting operation of the interconnections to times when flow rates are low. This is a practical approach because the interconnections typically are not needed during periods of high load, when all the chiller plants are operating. However, this strategy adds control complications that may be difficult for plant operators to understand.

- *minimize average flow velocity with variable-flow pumping.* The higher cost of larger pipe is a strong motivation to reduce the average flow rate. Variable-flow pumping is an effective way of doing this. See Measure 2.5.2 for details.

In retrofit applications, if the connections are fairly short, you may be able to use the existing chilled water pumps to push water through the connecting lines. Resistance to chilled water flow comes from the end-use coils, control valves, chiller evaporators, and piping. The resistance of straight pipe is relatively low. If you use variable-flow chilled water pumping, the piping resistance drops when the cooling load is low, because the resistance is proportional to the square of the flow rate.

Pipe resistance becomes a major factor if the connections are long, especially if the connections are used when the cooling load is high. In this case, you may need additional pumps. Consider adding variable-flow pumping, especially if chilled water pump modifications are required. This may substantially reduce the average pump power.

Exploit Interconnections to Improve Reliability and Capacity

In deciding how to design the interconnections, consider the value of improving the facility's overall reliability and capacity. If you foresee a need for additional cooling capacity, interconnection may be a less expensive alternative to purchasing new chillers. This possibility is enhanced if there is diversity of cooling loads, so that an interconnected system needs less total chiller capacity than isolated chiller plants.

Interconnections provide reserve capacity in the event of a chiller failure in any of the connected plants. This reserve capacity is limited by the capacity of the interconnections. This may argue in favor of larger pipe sizes.

Optimize Control of the Entire System

Interconnection requires more sophisticated control of chiller operation. Instead of simply starting chillers and letting them run, operators will have to continually start and stop chillers and control the interconnections in a manner that optimizes overall efficiency. Failure

of the operators to control the flow of energy between plants will reduce or negate the purpose of the modification. For this reason, proper training of plant operators is a vital part of this activity.

Automatic controls are better than people at "remembering" repetitive functions. Use them wherever it is practical to do so. Subsection 2.1 presents automatic methods of sequencing chiller plant equipment. However, using automatic controls does not relieve managers of the responsibility of understanding the control functions and ensuring that they are accomplished. Automatic controls that are not understood will be overridden by plant operators.

Try to Reduce Central Plant Operating Cost

Shutting down a central plant eliminates most of its non-salary overhead cost. These include chiller maintenance, water treatment, and housekeeping energy consumption, such as lighting and ventilation. Labor is usually the largest overhead cost. Shutting down a chiller plant provides a saving in labor cost only if the size of the staff can be reduced, or if the staff can be assigned to other duties during the period of time that plants are shut down.

Base Your Decisions on Accurate Calculations

This modification is very expensive, and it produces its benefit by marginal improvements in efficiency. These improvements depend on the load profiles of the facilities served by all the chiller plants, on the efficiency profiles of all the chillers, and on the other factors discussed here. Calculations that include all these factors are complex. To get the most accurate analysis, use a computer program of the type covered by Reference Note 17, Energy Analysis Computer Programs. Most of those programs are not designed to handle separate chiller plants serving a distributed load, so you may need ingenuity to get the program to model this configuration with acceptable accuracy.

ECONOMICS

SAVINGS POTENTIAL: *5 to 20 percent of total chiller plant energy consumption.*

COST: *Varies widely, from tens of thousands of dollars for short interconnections to millions of dollars for large facilities. This Measure may offer an immediate net saving if it provides an alternative to installing new chillers.*

PAYBACK PERIOD: *Several years or longer, except in especially favorable situations.*

TRAPS & TRICKS

PLANNING AND DESIGN: *Although the concept is simple, it is easy to be too hasty and overlook* opportunities for improving efficiency and reducing cost. A lot of money is at stake, so take your time.

TRAINING AND OPERATION: For this expensive modification to be successful, chillers and system valves at different sites must be sequenced in a precisely defined manner. Operators in different locations must be well trained, coordinated, and alert. Automate the controls wherever it is practical to do so. Review the methods of Measures 2.1.1 ff.

EXPLAIN IT: Make it easy for system operators to understand how the system is laid out and controlled. Install a large, well rendered diagram of the system at each operator stations. Install well designed placards on the cross connection valves and controls.

Exploiting Low Ambient Temperature for Water Chilling

When the outside air temperature is lower than the inside temperature, you can use the low outside temperature to cool the water in a chilled water system. The cooling is enhanced by the evaporative cooling effect of the system cooling towers. The Measures of this Subsection use this principle to reduce the load on the chiller, or even to allow the chiller to be turned off.

These Measures can provide only a fraction of the full-load capacity of the chiller system. However, in air conditioning applications, the cooling load is usually low when the weather allows these techniques to be used. Some of the Measures require the chiller to be turned off, while others can reduce the load on the chiller while it is running.

Where to Consider These Measures

These Measures are expensive, so they must be able to operate for many hours per year to pay for themselves. The potential running time depends primarily on the following factors, which apply to all the Measures in a similar way.

■ Favorable Cooling Load Characteristics

These Measures are useful only if the facility requires a significant amount of cooling when the weather conditions are able to provide it. In other words, look for applications that have high heat gains during cool weather.

Buildings with a large amount of glazing have high solar heat gain, and these may be good candidates. Buildings with high internal heat gains may also be good candidates. High internal heat gain may come from intense lighting, energy-consuming equipment, and concentrations of people. The interior zones of large buildings always have net internal heat gain, because they have no natural path for rejecting heat.

Industrial process cooling may also benefit from these Measures.

At the other end of the spectrum, these Measures are not justified in buildings that avoid the need for cooling in cool weather, typically by opening windows. For example, apartment complexes, typical primary and secondary schools, and older buildings of all types tend to be ineligible.

INDEX OF MEASURES

2.9.1 Install chiller "free cooling."

2.9.2 Cool chilled water with a heat exchanger in the cooling tower circuit.

2.9.3 Install a strainer system to use cooling tower water directly in the chilled water system.

2.9.4 Install a "waterside economizer" system using separate cooling coils.

2.9.5 Install a closed-loop atmospheric cooling unit in the chilled water circuit.

RELATED MEASURES

- Measure 4.2.5, outside air cooling through the air handling systems
- Measure 4.2.7, air purge cycle
- Measure 4.2.8, air-to-air heat recovery in air handling systems
- Subsections 5.6 and 5.7, for heat pump systems which can exploit ambient energy sources

■ Eligible Cooling System Characteristics

The feasibility of these Measures depends on the temperature requirement of the cooling application. In air conditioning applications, the goal is to maintain a space temperature around 75°F (about 24°C). In a typical mechanical air conditioning system, the air used to cool the space has a temperature of about 55°F (about 13°C). This air is cooled in a coil by chilled water, which typically has a temperature of about 42°F (about 6°C) at full load.

In order to use outside air to cool the chilled water, the temperature of the outside air must be lower than the temperature of the chilled water by some margin. In the case of evaporatively cooled systems, the wet bulb temperature of the air must be lower than the chilled water temperature.

From the preceding numbers, it is clear that cooling with outside air is not practical unless the temperatures in the system can be raised considerably. Fortunately, if the cooling load is low, the chilled water temperature and the supply air temperature may be much warmer than in normal operation. The feasibility of these methods depends on the extent to which higher temperatures can be used in the cooling systems.

The savings potential is greatest if the chiller plant serves only single-zone air handling systems or fan-coil units. In these types of equipment, the supply air temperature rises proportionally as the cooling load falls, so these units can use chilled water at higher temperatures. This provides the greatest opportunity to exploit the temperature differential between the outside and inside.

Variable-air-volume (VAV) systems and reheat systems provide substantially less opportunity for exploiting these methods. This is because the supply air must be held at a lower temperature to account for diversity in the cooling loads of the different spaces served by the air handling system. In these systems, it is essential to use supply air temperature reset (see Measures 4.4.1 ff) so that the systems can operate with the highest possible chilled water temperatures.

■ Whether Humidity Control is Needed

In normal operation, the humidity in spaces can be controlled by adjusting the temperature of the chilled water. Colder chilled water wrings more moisture out of the air at the cooling coils, reducing the space humidity. However, the Measures of this Subsection cannot provide much savings unless the chilled water temperature is allowed to rise as much as possible. This effectively eliminates humidity control.

In many locations, the lack of humidity control with these Measures is acceptable. Also, raising the chilled water temperature provides important energy conservation benefits, of which the most important are improving chiller efficiency and reducing reheat. High chilled water temperatures are not acceptable in locations that are humid during cool weather, such as New Orleans and London. In such climates, the dehumidification provided by low coil temperature is as important as cooling.

(Modern air conditioning seems to increase the sensitivity of people to humidity level. To a large extent, this results from poor coordination of temperature with humidity. It also relates to the nature of indoor activities, to fashions in clothing, and to other factors that are less well understood.)

■ Favorable Weather Profiles

To make these Measures pay off, there must be many hours of weather that is cold enough to cool the chilled water adequately while the facility is operating. The Measures are most likely to be economical in moderate or cooler climates. They are unlikely to be economical in consistently warm climates, such as those of Miami or Singapore. They are also uneconomical in many applications in colder climates, such as those of Minneapolis or Hamburg, because cooling may not be required for a large number of hours.

These Measures also require a climate in which the relative humidity is usually low when the outside temperature is cool. This is because of the comfort considerations discussed previously, and also because most candidate cooling systems use evaporative cooling towers. The evaporative cooling is essential to making the Measures effective with the low outside-to-inside temperature differentials under which they operate.

With evaporative cooling towers, the "wet-bulb" temperature is the relevant factor, rather than the "dry-bulb" temperature. A rule of thumb is that feasibility of these Measures requires a large number of hours during which the wet-bulb temperature is below about 45°F. In arid climates, this corresponds to outside air temperatures less than about 60°F. In general, for a given outside air temperature, these Measures are more effective if the climate is drier.

The need to avoid freezing sets a lower limit on the condenser water temperature. The fill structure of most cooling towers is not designed to carry an ice load, so the cooling tower must be controlled to keep ice from accumulating. Evaporative cooling can cause freezing at air temperatures well above 32°F. The usual minimum condenser water temperature is about 43°F with a crossflow tower, and about 40°F with a counterflow tower. (Crossflow towers have a higher temperature limit because they do not cool the water uniformly.)

These temperatures are low enough for cooling chilled water in most applications. However, achieving these low temperatures without freezing the cooling tower requires accurately modulating the cooling tower output in proportion to load. Typically, this requires a variable-output fan drive (see Measures 2.4.1 ff), and perhaps additional features, such as periodic fan reversing to de-ice the tower.

In applications that need to chill a liquid during predominantly cold weather, consider using a dry heat rejection unit and antifreeze in the cooled liquid. This system is simple, which makes it reliable. However, it is restricted to applications that need cooling during colder weather. Dry coolers suffer too much temperature loss to be effective with the higher outside air temperatures of typical applications.

■ Ability to Use Heat Sinks Other Than Outdoor Air

With all these Measures, a cooling tower is commonly used to reject heat, using the atmosphere as a heat sink. More broadly, you can adapt these Measures to use any other available heat sink, provided that the heat sink has sufficient capacity and a temperature that is low enough. See Reference Note 22, Low-Temperature Heat Sources & Sinks, about the various heat sinks that are available. Among these are ground water, rivers, and lakes.

Comparison of the Measures in this Subsection

All the Measures of this Subsection are alternatives to each other. They differ in these significant characteristics, which are summarized in Table 1:

• *whether the Measure can operate in conjunction with compressor operation.* Some of the Measures

cannot function while the chiller is operating. They are useful only when the cooling load is low enough for them to provide all the cooling that is needed. They are limited to a smaller number of hours of operation than the other Measures.

If the plant has more than one chiller, the chiller that exploits ambient cooling might operate in parallel with another chiller. However, this would require controls to limit the output of the chiller that does not have ambient cooling. Alternatively, the chiller with ambient cooling could feed chilled water to the other chiller for further cooling. However, this would require a series piping arrangement. Both of these approaches involve an additional level of complexity.

• *heat transfer steps.* The Measures of this Subsection move heat solely by the differential between the space temperature and the outside air temperature (dry-bulb or wet-bulb). This valuable temperature difference is usually small, so it is important to dissipate as little of it as possible in heat transfer. Table 1 shows the relative number of heat transfer steps needed by each Measure. The table does not include the evaporative heat transfer in the cooling tower or heat transfer from the chilled water to the end-use cooling equipment, because

Table 1. COMPARATIVE EFFICIENCY CHARACTERISTICS OF THE MEASURES IN SUBSECTION 2.9

Measure	Can augment compressor?	Can variable-flow CHW pumping improve chiller efficiency?	Heat Transfer Steps	Auxilary Equipment Load
Chiller Free Cooling (Measure 2.9.1)	No	Yes	■ Evaporator waterside ■ Evaporator refrigerant side ■ Condenser refrigerant side ■ Condenser waterside	■ CT system ■ CHW system
Cooling Tower-to-Chilled Water Heat Exchanger (Measure 2.9.2)	Yes	Yes	■ Heat exchanger CT side ■ Heat exchanger CHW side	■ CT system ■ CHW system (may require additional pump power) ■ Additional HX pump
Strainer Cycle (Measure 2.9.3)	No	Yes, but may not be feasible	■ none	■ CT system ■ CHW system
Waterside Economizer (Measure 2.9.4)	Yes	NA	■ Heat exchanger CT side ■ HX condenser side	■ CT system (may increase CT fan power) ■ Economizer system pump ■ Additional AHU fan power, at all times
Atmospheric Cooling of Chilled Water (Measure 2.9.5)	Yes	Yes	■ Heat exchanger CHW side ■ Heat exchanger waterside	■ CHW system ■ Extra cooling unit (only when providing supplemental cooling)

Abbreviations:
CT   cooling tower
CHW   chilled water
HX   heat exchanger
NA   not applicable

these are common to all the Measures. This comparison is useful for understanding the theoretical limits of each technique. However, the number of heat transfer steps is not an accurate guide to relative efficiency. For example, the refrigerant-to-liquid heat transfer steps in chiller free cooling involve less temperature loss than typical liquid-to-liquid heat transfer.

- *parasitic equipment load.* These Measures can save only a fraction of chiller energy consumption because they operate only at low cooling loads. Therefore, the system pump and fan energy consumptions are large in relation to the savings, and may offset them substantially. The table shows which auxiliary equipment must operate for each of the Measures.

- *whether variable-flow chilled water pumping increases the feasibility of the Measure.* Variable-flow chilled water pumping (see Measure 2.5.2) is growing in popularity as an energy conservation feature in chilled water plants. One of the effects of variable-flow pumping is to increase the return chilled water temperature. This increases the feasibility of the Measures because the warmer return chilled water has a greater tendency to reject heat to the outside.

  Variable-flow chilled water pumping is not an issue with one of the Measures, a "waterside economizer," because this approach does not use the chilled water system. Variable-flow pumping may be incompatible with another Measure, the "strainer cycle."

Alternatives to the Measures in This Section

You have a rich assortment of choices if you want to exploit the coolness or dryness of the outside air for cooling during mild weather. Consider all the possibilities before making a selection.

The most direct competitor to the Measures of this Subsection is the airside economizer (Measure 4.2.5), which brings outside air directly into the air handling systems. This technique offers the greatest energy saving potential because it involves no heat transfer steps. It is simple in concept, and well proven in actual experience. An airside economizer is the preferred method in most cases where it is practical to install the necessary ductwork and dampers.

The only major disadvantage of the economizer cycle is increased intake of outside air pollutants. Prevailing opinion tends to view outside air as less dangerous than inside air, so this disadvantage occurs only in a limited number of locales. Airside economizers cannot be used in multizone and dual-duct air handling systems unless you make unusual modifications to the air handling units.

A variety of other low-energy methods of cooling are available for cooling spaces directly. These include:

- *ventilating the space directly,* by installing openable windows, by using existing windows effectively, or by installing ventilation openings and fans

- *using space air circulation fans,* which cool people by "wind chill factor." Circulation fans can enhance the effect of mechanical air conditioning.

- *using room dehumidifiers.* These provide a significant cooling effect by themselves in humid climates. They may also work in combination with conventional air conditioning to reduce overall energy consumption.

- *using evaporative space coolers.* This method of cooling has much lower energy consumption than mechanical cooling. It is limited to dry climates.

Other Measures that Reduce the Effectiveness of These Measures

The benefit of the Measures in this Subsection and of the Measures mentioned previously is reduced by any actions that reduce heat gain. These include the Measures that improve envelope insulation (Section 7), that shade the spaces from solar gain (Section 8), and that improve lighting efficiency (Section 9).

The benefit of these Measures is also reduced by chiller heat recovery (Subsection 2.10), which offsets heating cost, and which may be an alternative to these Measures in some applications.

MEASURE **2.9.1 Install chiller "free cooling."**

SUMMARY

Often the easiest way to exploit cooling without compressor operation. Limited to certain types of chillers and system configurations. Cannot be used to supplement chiller operation.

SELECTION SCORECARD

Savings Potential .... $ $ $

Rate of Return, New Facilities % % %

Rate of Return, Retrofit.......... % %

Reliability .... ✓ ✓ ✓

Ease of Retrofit .... ☺ ☺

"Free cooling" is a modification of a chiller or cooling system that allows direct heat exchange between the evaporator and condenser. In effect, the chiller acts as a heat exchanger between the cooling tower and the chilled water system.

The modification consists of a pipe with a valve that is installed between the refrigerant sides of the evaporator and condenser. The pipe bypasses the compressor, allowing refrigerant vapor to migrate freely between the evaporator and condenser when the compressor is not operating. Figure 1 shows how free cooling is installed in a chiller that has a multi-stage compressor.

When the outside air wet-bulb temperature is low enough to make the condenser water colder than the chilled water, refrigerant evaporated by the warmer chilled water migrates to the colder condenser. The

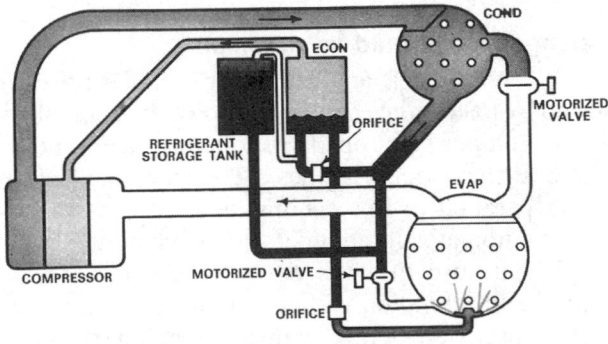

NORMAL CHILLER OPERATION

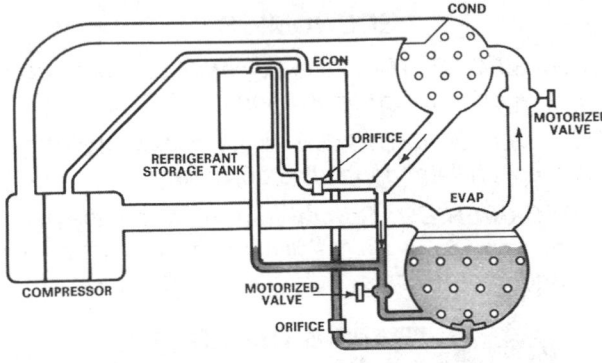

FREE COOLING

Trane Company

Fig. 1 Chiller free cooling Focus on the lower right. The weather is cold, so the condenser, on top, is colder than the evaporator, on the bottom. Therefore, refrigerant evaporates from the warmer evaporator, migrates to the condenser, and condenses. The liquid refrigerant flows back down to the evaporator, completing the cycle. The hardware on the left is for an inter-stage economizer, which is not necessary for free cooling.

refrigerant condenses in the condenser, giving up its latent heat, and the liquid refrigerant returns by gravity to the evaporator. This is a rapid and efficient means of heat transfer.

You may recognize that this is how a heat pipe works. So, you can consider free cooling as a way of using the chiller as a heat pipe between the cold outside air and the warm inside air. Heat is moved from the inside of the building to the outside solely by the differential between the space temperature and the outside air wet-bulb temperature.

To use free cooling, the compressor is turned off, the refrigerant bypass connection is opened, and the chilled water and cooling tower systems are kept running. Heat rejection from the building occurs as long as the chilled water temperature is higher than the cooling tower water temperature.

(The term "free cooling" is not unique to this technique. It may be used for any or all methods of cooling without operating the chiller compressor.)

Appropriate Applications

Refer to the Subsection introduction about the general application of free cooling and the alternative methods to consider.

This method can be used only with chillers and chiller systems that allow rapid migration of the refrigerant vapor from the evaporator to the condenser, and drainage of the condensed vapor back to the evaporator. Packaged water chillers are typically well adapted to free cooling. A chiller system with a remotely mounted evaporator and/or condenser may be adaptable to free cooling, but custom design of such an installation would require exceptional skill.

This method of free cooling with packaged centrifugal water chillers was patented by the Trane Company, which offers it as an option on some of its models. Investigate the status of patent coverage before attempting to adapt this technique to existing chillers or chiller systems.

This method does not work with air-cooled condensers or with direct-expansion evaporators. The refrigerant gas passages in the coils are too restricted to allow adequate convective circulation, and the liquid refrigerant would not flow properly without the pressure differential provided by the compressor. Exceptions are conceivable, but you would have to become an inventor to make them work.

Advantages and Disadvantages

In systems where this Measure applies, it may be easier and cheaper to accomplish than the other Measures of this Subsection. Chiller free cooling requires no modifications or additions outside the chiller itself.

The main efficiency disadvantage of free cooling is that it cannot be used in conjunction with normal chiller operation. Therefore, in order for free cooling to be useful, the cooling load must be low enough for free cooling to satisfy the entire cooling that occurs during the times when weather conditions allow free cooling to be used.

Free cooling requires the largest number of heat transfer steps of the Measures in this Subsection. On the other hand, the refrigerant heat transfer is highly efficient. This is because the high latent heat of the refrigerant provides a large amount of heat flow in relation to the mass that is evaporated.

One major chiller manufacturer claims that free cooling can provide as much as 45% of normal chiller capacity in its centrifugal machines. However, this would occur only at the lowest acceptable outside air temperature. The average capacity of free cooling under typical conditions is much less than this.

Free cooling involves a failure mode that is subtle and may waste large amounts of energy. If the refrigerant bypass valve leaks or fails to close fully when the compressor is operating, high-pressure gas is bypassed back to the compressor suction. This accidental form of hot gas bypass may waste a large amount of energy.

How to Control Free Cooling

Control of the free cooling connection itself is simple. The refrigerant bypass line must be tightly closed when the compressor is operated, and it must be open when free cooling is used.

The number of hours that free cooling can operate is determined by the chilled water temperature at which the compressor turns on. Therefore, keep the chilled water temperature setting as high as possible. Refer to Measure 2.2.1 about chilled water temperature reset.

Similarly, keep the condenser water temperature as low as possible when free cooling is used. However, many cooling units require an elevated condenser water temperature when they are running. (See Measure 2.2.2 for the reasons why.) In such cases, provide controls that raise the condenser water temperature when the compressor starts.

In order to exploit the potential of this Measure without wasting fan energy, the controls for the cooling tower fans must be fairly sophisticated. Water in cooling towers does not evaporate as well at the lower temperatures at which free cooling operates, so the cooling fans must expend energy to maintain the lowest possible cooling water temperature. Such fan usage is excessive for normal chiller operation, so the fan control system must be adaptable to the different modes of operation. Refer to Measures 2.2.2 ff for details.

If the modification is offered as a standard package by a chiller manufacturer or other vendor, make sure that the controls included in the package satisfy all these requirements, and any others required by the chiller system.

■ Changeover Should be Automatic

Free cooling is available only at irregular times, as allowed by load and weather conditions. It is unrealistic to expect chiller plant operators to remain alert enough to grab at these opportunities. The changeover from chiller operation to free cooling, and vice versa, must be done automatically in order for the technique to pay off. Make sure that the controls are designed to maximize the duration of free cooling by minimizing the condenser water temperature and keeping the chilled water temperature as high as possible.

ECONOMICS

SAVINGS POTENTIAL: 5 to 30 percent of compressor energy, in appropriate applications.

COST: $8,000 to $16,000 as an accessory of new centrifugal chillers, depending on chiller size.

PAYBACK PERIOD: Several years or longer, the period depending primarily on the climate, the facility's cooling load, and energy costs.

TRAPS & TRICKS

PLANNING: Remember, this kind of free cooling cannot operate while the chiller is running. Make sure that your load and weather conditions allow free cooling for enough hours per year to make it pay. Don't kid yourself about how much cooling capacity this modification can provide under typical conditions. Talk with plant operators who have used the equipment for several cooling seasons.

EQUIPMENT AND INSTALLATION: *The modification is a standard option of at least one chiller manufacturer. The hardware is simple, so it should be reliable. However, optimum control is somewhat complex. Let the manufacturer install the equipment, to avoid performance and warranty disputes. Then, check out performance thoroughly after installation. In particular, make sure that the chiller turns off and changes to free cooling whenever load conditions permit, and that the cooling tower water temperature is kept as low as possible during free cooling operation. Be sure to use chilled water temperature reset.*

STAFF COMMUNICATION: *Make sure that plant operators understand how free cooling works, or they will defeat it sooner or later in attempting to solve cooling problems. Explain the free cooling system in the plant operating manual. Install a placard at the bypass valve and controls. Put a column in the chiller plant log that indicates when free cooling is operating.*

MAINTENANCE: *Only minimal routine maintenance is required. A leak in the bypass valve will waste energy, and it is invisible. Schedule checks of this in your maintenance calendar.*

MEASURE 2.9.2 Cool chilled water with a heat exchanger in the cooling tower circuit.

Another method of exploiting cool weather conditions to chill water is to install a heat exchanger between the cooling tower water and the chilled water, as shown in Figure 1. This method is conceptually similar to Measure 2.9.1, but has the advantage that it can be used to pre-cool chilled water while the chiller is operating. Also, it is readily adaptable to most chilled water systems that use water-cooled chillers.

Appropriate Applications

This method applies to all water-cooled water chillers. Refer to the Subsection introduction for general application considerations and alternative methods.

Advantages and Disadvantages

The relative efficiency characteristics of the Measures in this Subsection are shown in Table 1 of the Subsection introduction. The table shows that this Measure lies in the middle of the pack. This Measure has no major advantages or disadvantages of a practical nature. Installation should be fairly easy, even in retrofit. The largest item is a compact heat exchanger. Pipe connections are short.

Heat Exchanger and Piping Installation

This method of cooling recovery can be piped in a variety of ways. Figure 1 shows a simple arrangement that is satisfactory for many applications. The major feature is a heat exchanger installed in the chilled water return line. Cooling tower water is pumped through the heat exchanger in the counterflow direction. There is

SUMMARY

Often easy to install. Has no outstanding advantages or disadvantages over the other Measures of this Subsection.

SELECTION SCORECARD

Savings Potential .... $ $ $

Rate of Return, New Facilities % % %

Rate of Return, Retrofit .... % %

Reliability .... ✓ ✓ ✓

Ease of Retrofit .... ☺ ☺ ☺

no need for a control valve in the operation of the heat exchanger.

■ Deal with the Flow Resistance of the Heat Exchanger

Installing the heat exchanger in the chilled water return line increases flow resistance. In retrofit, you may have to install a chilled water pump with greater pressure output. In some cases, the chilled water system has a throttling ("balancing") valve that can be adjusted to account for the difference in flow resistance. Figure this out beforehand, as equipment changes are expensive. After installing the equipment, verify that the flow rate is adequate.

A possible alternative is installing a bypass around the heat exchanger in the return chilled water line, so that the resistance is removed when the heat exchanger is not in use. The reduction in flow rate that occurs when the heat exchanger is used should be tolerable because the cooling load is low then.

Before changing the chilled water pump, consider installing variable-flow chilled water pumping, as recommended by Measure 2.5.2.

■ Condenser Recirculation Line

If necessary, you can add a recirculation line around the chiller condenser to maintain a minimum condensing temperature for chillers that have this requirement. See Measure 2.2.2 about chiller condensing temperature limitations. In order for the heat exchanger to be effective, the cooling water temperature must be as low as the cooling tower can provide. The condenser recirculation line resolves the conflict between the two different temperature requirements.

Some chillers tolerate low condensing temperatures, which may allow you to omit the recirculation line. However, note that the cooling tower water temperature may be lower than the chilled water temperature. Most chiller manufacturers presently do not conceive of chiller operation with the condensing temperature lower than the evaporating temperature.

How to Control the System

Control details vary among different possible configurations. The general control steps are:

- the heat exchanger pump operates whenever the temperature of the cooling tower water is lower than the temperature of the returning chilled water, and it is turned off at other times
- the condenser pump operates when the compressor runs, and is turned off at other times
- the chilled water supply temperature is reset continuously to keep it as high as possible. (Refer to Measure 2.2.1 about chilled water temperature reset.)
- if a condenser recirculating line is installed, the chiller condenser water temperature is controlled by the recirculating valve when the heat exchanger pump is operating. When the heat exchanger pump is off, the recirculating valve is shut and the condenser water temperature is controlled by the operation of the cooling tower fan. This control changeover keeps the fan from running excessively when the heat exchanger is not used.

To exploit the potential of this Measure without wasting fan energy, the cooling tower fans need controls

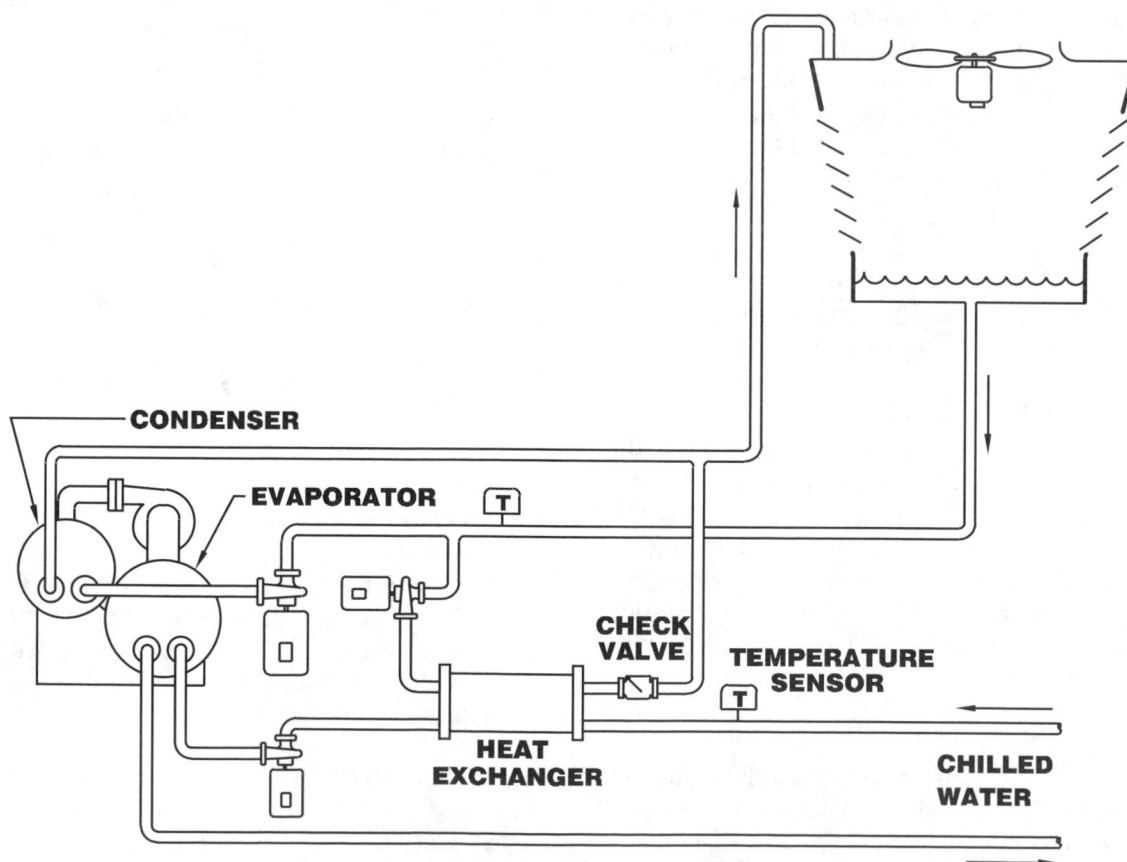

Fig. 1 Heat exchanger installed between condenser water system and chilled water system The heat exchanger circulating pump operates only when the temperature sensors indicate that the water from the cooling tower is colder than the chilled water.

that are fairly sophisticated. Water in cooling towers does not evaporate well at the lower temperatures at which the heat exchanger operates, so the cooling fans must operate aggressively to maintain minimum cooling water temperature. Such fan usage is excessive for normal chiller operation, so the fan controls should adapt to the different modes of operation. See Measures 2.2.2 ff for details.

___

ECONOMICS

SAVINGS POTENTIAL: *10 to 30 percent of compressor energy, in appropriate applications.*

COST: *Heat exchangers and a moderate amount of piping cost from $150 to $400 per ton of free cooling capacity, in typical applications.*

PAYBACK PERIOD: *3 to 15 years, depending mainly on climate and energy costs.*

TRAPS & TRICKS

YOU ARE PIONEERING: *The design is unconventional, so work out the details carefully. Achieving efficient control is challenging.*

EXPLAIN IT: *Put a clear explanation of the system in the plant operating manual. Install placards on the heat exchanger and the controls.*

MONITOR PERFORMANCE: *Include columns in the chiller plant operating log to record the outside air wet-bulb temperature and the inlet and outlet temperatures of the heat exchanger. Failure or bad calibration of the controls is invisible, so review the logs regularly to check performance.*

MEASURE 2.9.3 Install a strainer system to use cooling tower water directly in the chilled water system.

RATINGS

New Facilities	Retrofit	O&M
D	**D**	

From the standpoint of heat flow, the most direct way of cooling a building during cold weather is to use the cooling tower water directly in the chilled water system. This approach avoids most heat exchange steps. You can accomplish it with a simple arrangement of valves and crossover connections.

This type of system must resolve a serious conflict. Cooling tower water is dirty. At the same time, the water in the chilled water system must be kept clean to avoid clogging the narrow passages in the heat exchange coils throughout the facility. To deal with this problem, the system includes a high-efficiency strainer to clean the cooling tower water before it enters the chilled water system. For this reason, this approach is sometimes called a "strainer cycle."

Figure 1 shows a typical strainer cycle installation. When the strainer cycle goes into operation, the cooling tower water is connected into the chilled water system, the chilled water pump discharges through a high-efficiency strainer, rather than through the chiller evaporator, and the cooling tower water pump is turned off.

Efficiency Characteristics

The relative efficiency characteristics of strainer cycles are shown in Table 1 of the Subsection introduction. Strainer cycles eliminate the condenser

SUMMARY

Provides efficient heat exchange, but cannot supplement compressor operation. The risk of fouling the chilled water system is a major drawback.

SELECTION SCORECARD

Savings Potential .... **$ $** $

Rate of Return, New Facilities **% %** %

Rate of Return, Retrofit.......... **%** %

Reliability .... ✓

Ease of Retrofit .... ☺ ☺

and evaporator heat transfer steps, along with the operation of the cooling tower water pumps. As a result of these advantages, the strainer cycle is one of the most efficient means of exploiting low outside air temperature for cooling.

On the other hand, strainer cycles cannot operate while the chiller is running. As a result, they fail to provide savings during many hours when cool outside temperatures could be used to augment the cooling provided by chiller operation. On balance, the inability of a strainer cycle to supplement compressor operation usually makes this method less efficient overall than some of the other Measures of this Subsection.

Applications

This Measure is used with chilled water systems that employ water-cooled chillers. This Measure is not practical in all facilities, because some design requirements of strainer cycles are not adaptable to all chilled water systems. Refer to the Subsection introduction about the general application of free cooling and the specific methods to consider.

Design Issues

The equipment of a strainer cycle is easy to install, and it does not require much space. The largest additional item is the high-efficiency strainer. Piping connections and valve installations are simple. The following are the main design issues.

■ Cleaning Cooling Tower Water

Cooling tower water is dirty because cooling towers wash dirt and debris out of the atmosphere and trap it. Isolating the chilled water system from this dirt is a primary concern.

Cooling tower systems normally have strainers to prevent clogging of the system and to protect the cooling water pumps and the condensers. These functions can be accomplished with coarse strainers, which are not satisfactory for protecting the chilled water system. To protect the small passages of chilled water heat exchangers, the strainer cycle uses a high-efficiency strainer to remove small particulate material.

The strainer is an area of serious risk. A strainer failure will quickly distribute dirt throughout the chilled water system and clog the insides of coils. Fixing this requires shutting down the entire chiller system and doing tedious cleaning. Therefore, install a strainer cycle only if you are sure of a high level of maintenance discipline over the long term. The only requirement is to check and flush the strainers periodically. This is not difficult, but the procedure can never be neglected.

Install the high-efficiency strainer on the discharge side of the chilled water pump. Otherwise, the resistance of strainer may reduce pump suction pressure too much, causing pump cavitation. Select a strainer that has a large capacity in relation to flow, reducing resistance.

In retrofit applications, the additional resistance of the strainer should not be a problem. The strainer is located in the evaporator bypass line, so it does not increase flow resistance during normal chiller operation. During strainer cycle operation, the increase in resistance causes a moderate reduction of flow. This should be tolerable with the low cooling load that exists when the strainer cycle is used.

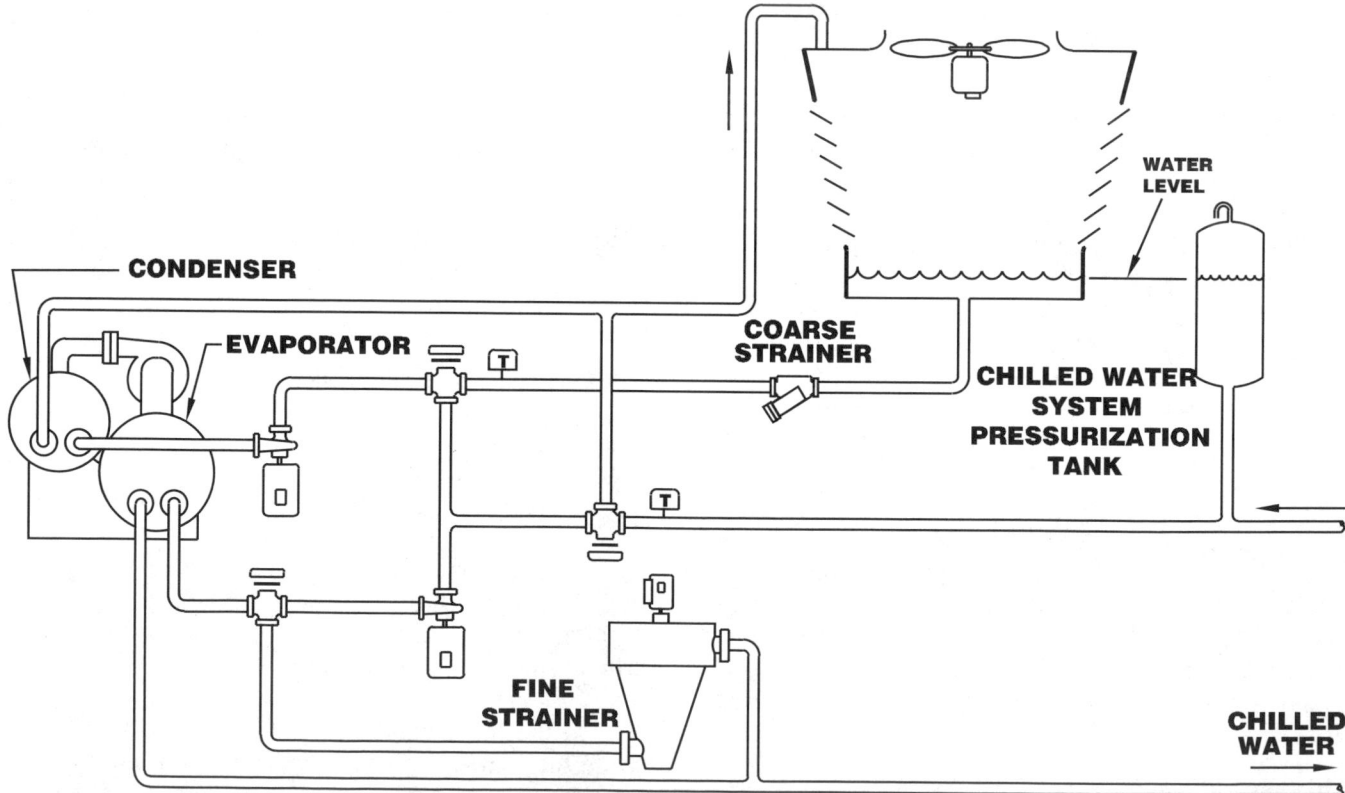

Fig. 1 Strainer cycle The main efficiency advantage of this system is that the cooling tower water is used directly in the chilled water system. This characteristic is also the system's most serious liability. A high-efficiency strainer is needed to clean the tower water sufficiently to use in the narrow coils of the cooling equipment. If the strainer fails, the entire chilled water system may be fouled.

■ **Water Treatment**

Cooling tower systems have dirty water for three broad reasons. One is that they collect dirt and debris from the surrounding atmosphere and environment. Another is that the cooling tower acts as a growth site for a wide range of biological fouling. Thirdly, since cooling towers operate by evaporation, they require continuous intake of raw water. This replacement water carries mineral fouling and a certain amount of dirt.

Dirt is removed by the strainers. These must be flushed periodically. In a strainer cycle, the operation of the strainers is so critical that you should consider installing strainers with automatic backflushing.

Biological fouling is controlled by chemical treatment of the cooling tower water. This is part of the proper operation of any cooling tower system. See Measures 2.3.3 ff about chemical treatment. Remember that cooling tower water treatment is normally designed for the condenser cooling water system, not for the chilled water system. If a strainer cycle is installed, adjust your water treatment accordingly. For example, consider using more suspension agents to keep dirt from depositing on system surfaces.

Failure of water treatment is even more likely than failure to keep strainers cleaned out, because water conditions are not directly visible and water treatment requires specialized skills. Consider it virtually certain that some lapses in water treatment will occur during the life of a facility.

Mineral fouling is prevented by keeping the concentration of minerals in the cooling tower water low enough. This requires proper control of cooling tower bleed. See Measures 2.3.4 ff about this function.

■ **Dealing with Expansion Tanks**

A chilled water system needs an expansion tank to deal with expansion and contraction of the water. In conventional systems, the expansion tank can take one of two forms. The simplest is an open, vented tank, which must be located above the highest point in the system. The other is a closed tank that is pressurized with air, which may be installed at any point in the system.

During strainer cycle operation, the cooling tower creates an opening in the chilled water system. In effect, the cooling tower sump also acts as an expansion tank. For this reason, the chilled water system's normal expansion tank must be located so that its water level is the same as the water level in the cooling tower sump. If the expansion tank were lower, water would drain out of its vent. If it were higher, water would drain out of the expansion tank into the cooling tower sump.

You cannot use a pressurized expansion tank in a system with a strainer cycle, because the pressurization tank would blow water out of the system when the strainer cycle is operating.

A strainer cycle is probably impractical if you cannot install the cooling tower sump higher than the tallest point in the chilled water distribution system. In that case, the chilled water system would tend to drain out through the cooling tower sump. Preventing this would require complex plumbing and control. The benefit would not justify the complications, especially since other Measures can achieve the same purpose without the complications.

A related issue is that the cooling tower sump must be high enough above the chilled water pump to provide good pump suction. This condition is satisfied if the chilled water pump is installed at the same elevation as the condenser water pump, or lower.

■ **Relative Cooling Tower and Chilled Water Flow Rates**

Cooling towers are designed for optimum performance at a particular flow rate. By serendipity, compression chillers require about the same chilled water and condenser water flow rates. This makes it simple to use the chilled water pump to circulate water through the cooling tower.

This lucky coincidence does not occur with absorption chillers, which require a much higher condenser water flow in relation to the chilled water flow. In an absorption chiller system, the water flow provided by the chilled water pump during strainer cycle operation is only a fraction of the normal cooling tower flow rate.

The same problem arises in a compression chiller system if the system uses variable-flow chilled water pumping (Measure 2.5.2). In a variable-flow system, the flow of chilled water is proportional to the cooling load. The strainer cycle operates when the cooling load is low, so the chilled water flow rate is greatly reduced.

There are two potential solutions for the latter cases. One is to operate the cooling tower at the reduced flow rate. Some cooling towers may provide sufficient cooling for the reduced cooling load, and some may not. Check this with the cooling tower manufacturer. Operation of cooling towers at reduced flow rates is an area of uncertainty.

The other solution is to maintain full cooling tower flow by operating the condenser water pump. A bypass valve is used to tap off the part of the flow needed by the chilled water system. The energy penalty of operating the condenser water pumps is a major factor. The additional complexity increases the possibility of operating the system in an inefficient manner.

How to Control the Strainer Cycle

Control the strainer cycle to operate as long as the temperature of the cooling tower water is low enough to satisfy the cooling load, instead of running the chiller.

As the cooling load increases, the chilled water supply temperature needs to be lower, and the chilled water return temperature increases. When the return water temperature rises to a certain point, the strainer cycle must be turned off and the chiller must be started. Refer to Measure 2.2.1.2 for controls that keep the allowable chilled water temperature as high as possible.

The cooling tower fan controls need to be sophisticated to maintain the optimum balance between saving chiller energy and not wasting fan energy. Water evaporates slowly at the lower temperatures at which free cooling operates, so the cooling tower fans must operate more aggressively to create evaporative cooling. Such fan usage is excessive for normal chiller operation, so the fan control system should be adaptable to the different modes of operation. See Measures 2.2.2 ff for details.

■ Changeover Should be Automatic

The strainer cycle can operate only at irregular times that are determined by the cooling load and the weather conditions. It is unrealistic to expect chiller plant operators to remain alert enough to grab at these opportunities. The changeover from chiller operation to strainer cycle operation, and vice versa, must be automatic.

ECONOMICS

SAVINGS POTENTIAL: *5 to 30 percent of compressor energy, in appropriate applications.*

COST: *Several thousand dollars, up to considerably more.*

PAYBACK PERIOD: *Less than one year, to many years. The payback period is shorter in larger systems, all other factors being equal.*

TRAPS & TRICKS

CHOICE OF METHOD: *Are you sure you want to do this? The simplicity of the concept and the high potential efficiency are appealing, but this technique is appropriate only for a plant that will always have reliable operation and maintenance. When calculating the potential benefit, remember that the strainer cycle cannot operate while the chiller is running. Make sure that your load and weather conditions allow strainer cycle operation for enough hours per year to make it pay.*

DETAILS: *The hardware design is fairly simple, but designing the controls to optimize the benefit of the system is challenging. Make sure that the chiller turns off and the system switches to strainer cycle operation whenever load conditions permit. Design the controls to keep the cooling tower water temperature as low as possible during strainer cycle operation.*

MAINTAIN IT: *A strainer cycle needs maintenance that never fails. A failure that allows cooling tower dirt to get into the chilled water system could require an expensive cleaning operation and an extended cooling shutdown. Failure of the strainers is easy to overlook until serious system fouling has occurred. Controls may malfunction indefinitely, or they may not be set for maximum efficiency, without the problem being noticed.*

EXPLAIN IT: *Make sure that plant operators understand how the strainer cycle works, or they will defeat it sooner or later in attempting to solve cooling problems. Explain the system in the plant operating manual. Install placards at the valves and controls. Put a column in the chiller plant log that indicates when the strainer cycle is operating.*

MEASURE **2.9.4 Install a "waterside economizer" system using separate cooling coils.**

SUMMARY

Lets the cooling tower provide direct cooling in combination with DX coils. Piping cost usually limits it to air handling units having their own compressors and condensers. Retrofit is usually not practical.

SELECTION SCORECARD

Savings Potential .... **$ $** $

Rate of Return, New Facilities **% %** %

Rate of Return, Retrofit.......... **%** %

Reliability .... ✓ ✓ ✓

Ease of Retrofit .... ☺

Another way to move low outdoor temperatures indoors is to install separate cooling coils in the air handling units or other cooling equipment that is fed by water from the cooling tower. This type of system is commonly called a "waterside economizer." (The name distinguishes this method of capturing outside air cooling from the "airside economizer," which is covered by Measure 4.2.5.)

Figure 1 shows a typical waterside economizer installation. A unique aspect of the waterside economizer is that it can use an atmospherically cooled water system to supplement direct-expansion (DX) cooling, as shown in Figure 1. In fact, this is the only common application where waterside economizers are preferable to the other methods of this Subsection.

Where to Consider a Waterside Economizer System

Refer to the Subsection introduction for a comparison of the different methods of free cooling that are available.

The waterside economizer has an exclusive niche. It is the only method that is available for exploiting direct cooling with cooling tower water that can work with direct-expansion (DX) cooling coils.

As a practical matter, waterside economizers are usually limited to DX air handling units that have their own compressors and condensers. This is because it is expensive to run water piping to numerous air handling units. If the units need water piping for condenser cooling, the additional piping cost for the economizer coils is relatively small. Typically, self-contained air handling units are installed to avoid the cost or space requirements of a central chiller plant. Figure 2 shows a typical air handling unit of this type that uses a waterside economizer.

This is primarily a method for new construction. Retrofitting economizer coils to existing air handling units would be difficult or impossible.

Comparative Efficiency Characteristics

The comparative efficiency characteristics of a waterside economizer system are shown in Table 1 of the Subsection introduction.

A unique efficiency disadvantage of the waterside economizer is that it requires a separate coil in the air stream. This coil has a significant amount of resistance, compared to other pressure losses in the air handling system, so it requires a corresponding increase in fan power. The coil is always present, so fan power is increased at all times, even when the waterside

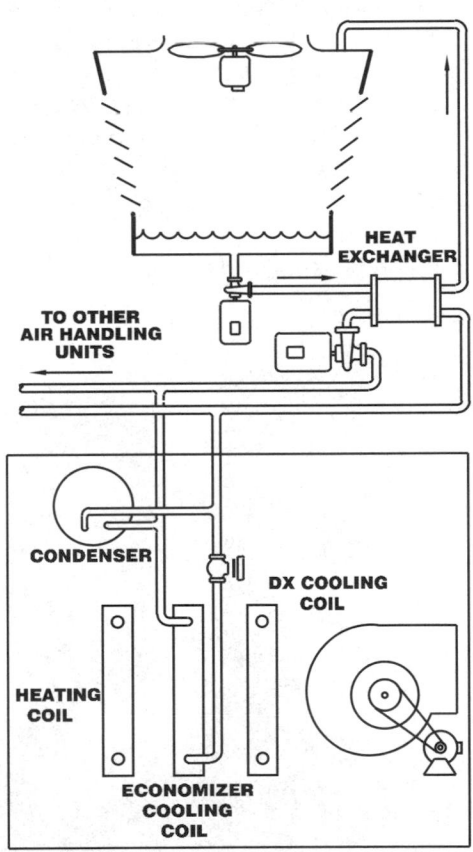

SELF-CONTAINED AIR HANDLING UNIT

Fig. 1 Waterside economizer system Water from the cooling tower is used to cool remote coils in the air handling units. These coils operate independently of the DX cooling coils operated by the compressors. A heat exchanger is usually needed to isolate the economizer coils and the condensers from the dirt picked up by the tower.

economizer is not operating. This makes it especially important to have weather conditions that favor a waterside economizer for a large fraction of the time.

A waterside economizer system may or may not provide a net saving in pump energy, depending on the cooling load profile. The system requires no chilled water pumps. The power required for pumping cooling tower water is increased. The amount of pump power depends on how the water is circuited inside the air handling units, as discussed below.

Installing an economizer system typically requires a heat exchanger for the cooling tower water. This is because the economizer coils cannot tolerate the dirt that exists in cooling tower water. The temperature differential in the heat exchanger wastes part of the cooling potential of the cooling tower. The heat exchanger also requires a separate pump, and the additional resistance in both water circuits increases total pump power.

Because the condensers in the air handling units are served by the same water circuit as the economizer coils, the heat exchanger also raises the condensing temperature, reducing the efficiency of mechanical cooling whenever the water temperature is higher than the minimum required by the compressor systems. In other words, the economizer system imposes an efficiency penalty on the compression cooling systems even when the economizer itself is not operating.

Figure 3 shows a typical plate heat exchanger for a waterside economizer. It is relatively small compared to the other components of the system.

How Waterside Economizer Coils are Connected

The economizer coil in each air handling unit is installed directly ahead of the main cooling coil. If a preheat coil is installed, the economizer coil is installed downwind of the preheat coil, since the latter is vulnerable to freezing.

In each air handling unit, the water flow to the economizer coils may be piped in parallel with the water flow to the condenser, or it may be piped in series. A parallel connection provides the lowest water temperature to both, but it increases pump power and pipe size. If the water is piped in series, it goes to the economizer coil first. The warming of the water in the economizer coil does not cost an efficiency penalty if the compressor must maintain a condensing temperature that is higher than the discharge temperature of the economizer coil. This is typically true with contemporary equipment, but it may not be true in the future, as equipment manufacturers become more aggressive about exploiting the efficiency benefit of lower condensing temperatures.

If the chiller requires an elevated condensing temperature, it also needs a recirculation circuit around the condenser. This maintains a constant flow rate through the condenser for proper heat transfer.

You can save pump energy by installing a separate pump for each air handling unit or group of air handling units that operates on a separate schedule.

An additional refinement for minimizing pump power is bypassing the water flow around the economizer coils when they are not needed. To exploit this fully, you need a variable-flow pumping system.

WESINC

Fig. 2 Self-contained air handling system that uses a waterside economizer The cooling water pipes coming from the left serve both the economizer coil and the condenser of the DX cooling unit.

This is relatively easy in an economizer system. Pump the water with a single distribution pump that is controlled to maintain a fixed pressure differential in the system. Also, install shutoff valves to stop the flow to condensers that are not in use.

How to Control a Waterside Economizer

Each economizer coil is controlled individually, as follows:

- the economizer coil and the primary cooling coil operate in sequence, with the economizer coil valve opening fully before the primary coil starts to operate
- when the cooling tower water temperature is higher than the temperature of the air entering the economizer coil, the economizer coil is shut off
- a signal to start the primary coil is generated whenever the economizer coil cannot satisfy the demand of the thermostatic control

There are differences in the control scheme that depend on the type of air handling system. In single-zone air handling units:

- control the coils in response to the space temperature or return air temperature, as appropriate
- to turn off the economizer coil when heating is desired, use the same control signal that turns off the primary cooling coil.

In air handling units that are controlled to maintain a specific supply air temperature (VAV and reheat units):

- control the coils to maintain the desired supply air temperature
- use supply air temperature reset aggressively to maximize the benefit of the economizer.

The cooling tower fans need special control, especially if the chillers are not designed to exploit minimum condensing temperatures. The economizer cycle needs as much cooling effect from the cooling tower as possible. Water in cooling towers evaporates more sluggishly at lower outside air temperatures, so the cooling fans must be controlled to produce the maximum evaporative cooling effect in this temperature range. Such fan usage is excessive for normal compressor operation. Design the fan controls to optimize overall system efficiency under both economizer and non-economizer modes of operation. See Measures 2.2.2 ff for details.

ECONOMICS

SAVINGS POTENTIAL: *10 to 30 percent of compressor energy, in appropriate applications.*

COST: *In new facilities, a waterside economizer costs $50 to $150 per ton of system capacity, typically. In retrofit, the cost of adding economizer coils to existing air handling units is probably prohibitive.*

PAYBACK PERIOD: *Several years or longer, typically.*

TRAPS & TRICKS

DESIGN: *Waterside economizers are fairly common. However, do not copycat existing designs, as they may be flawed or inefficient. Select cooling equipment that exploits the efficiency advantage of low condensing temperature.*

EXPLAIN IT: *Put a clear explanation of the system in the plant operating manual. Install placards at the economizer coils and the controls.*

MONITOR PERFORMANCE: *Record the outside air wet-bulb temperature and the supply and return temperatures of the economizer circuit in the chiller plant operating log. Review the log regularly to check performance, because failure or incorrect calibration of the controls is invisible.*

WESINC

Fig. 3 Heat exchanger for waterside economizer system
This plate heat exchanger is very compact, but expensive. It reduces the efficiency of both the waterside economizer and the DX cooling systems.

MEASURE 2.9.5 Install a closed-loop atmospheric cooling unit in the chilled water circuit.

You can use outside air to cool chilled water directly by using a conventional heat rejection unit. The heat rejection unit may be a closed-loop cooling tower, or even a dry cooler. This method is unusual. It is too expensive to compete with the other Measures in common applications, but it may be the best choice in some special cases. Figure 1 shows the arrangement.

Appropriate Applications

Refer to the Subsection introduction about the general application of free cooling and the various methods to consider.

Table 1 in the Subsection introduction shows that this method is relatively efficient compared to the other Measures. It can operate along with the normal chiller operation, providing free cooling as long as the outside air temperature (dry-bulb or wet-bulb, depending on the type of heat rejection unit) is lower than the return chilled water temperature.

The major disadvantage of this approach is high cost. In constant-flow chilled water systems, the heat rejection unit must be sized for the full flow of the chilled water circuit, but it operates with temperature differentials that are typically small. This requires a heat rejection unit that is large in relation to the quantity of the heat rejected. In chilled water systems that have cooling towers, all the other Measures of this Subsection are usually more economical.

The heat rejection unit can be much smaller if the chilled water system has variable-flow pumping, because the flow rate is much lower when the cooling load is low, as it usually is when the outside air is cold. Also, the efficiency of this Measure is higher with variable-flow pumping systems because the return chilled water temperature is higher.

This is the only Measure of this Subsection that can be used with air-cooled water chillers. However, its range of applications is still narrow. Air-cooled chillers are commonly selected for arid climates to avoid the need for cooling water, and these climates are typically too warm for this Measure. At the other temperature extreme, air-cooled chillers may be used so they can operate during cold weather without freezing the heat rejection unit. Liquid chilling during freezing weather is an uncommon application.

This Measure is relatively simple to accomplish, although expensive. The main physical limitation is finding a mounting location that can accommodate the size and weight of the heat rejection unit.

SUMMARY

An alternative that offers good efficiency, but is economical only in specialized applications. Most likely to be desirable in systems with variable-flow chilled water pumping, and with air-cooled water chillers that operate during cold weather.

SELECTION SCORECARD

Savings Potential .... $ $ $

Rate of Return, New Facilities % % %

Rate of Return, Retrofit.......... % %

Reliability .... ✓ ✓ ✓

Ease of Retrofit .... ☺ ☺

Dry Cooler or Wet Cooler?

This Measure can be accomplished with either a closed-loop evaporative cooler or a dry cooler. A quick analysis with a psychrometric chart shows that an evaporative cooler provides much greater savings. The evaporative cooling effect lowers the chilled water temperature enough to greatly increase the number of hours of non-chiller cooling, especially in climates that are dry.

Furthermore, the heat exchange loss in a dry cooler keeps the chilled water at least 15°F warmer than the outside dry-bulb temperature. Thus, for example, making 50°F chilled water with this method alone would require an outside air temperature of about 35°F.

How to Control the System

The atmospheric cooling unit is turned on, and the transfer valves connect it into the chilled water circuit, whenever the cooling unit can reduce the temperature of the returning chilled water enough to justify its operation. A dry cooler can be controlled with a switch that senses the difference in temperature between the return chilled water and the outside air. With a wet cooler, the control should sense the outside air wet-bulb temperature. For best efficiency with a wet cooler, use a sliding scale of temperature differentials for different wet-bulb temperatures.

Controls are also needed for freeze protection, unless the system uses antifreeze or is drained before the onset of freezing weather.

The potential benefit of the Measure is influenced heavily by the chilled water temperature. Keep it as high as possible. This also improves the COP of the chiller. See Measure 2.2.1 about chilled water temperature reset.

Freeze Protection

Chilled water systems are usually located entirely indoors, avoiding the need for freeze protection. However, the atmospheric cooling unit is exposed to outside conditions, where it may be damaged by freezing. A leak in this unit would shut down the entire cooling system.

The most efficient way to avoid freezing is to drain the exposed portion of the chilled water piping when freezing is a possibility. Freezing conditions come and go repeatedly, so install an automatic drain-down. Drain-down is activated when the water in vulnerable parts of the system approaches freezing temperature.

An alternative method of freeze protection is to use antifreeze solution with the chilled water system. A disadvantage of this method is that antifreeze solutions have poorer heat transfer than water. Using antifreeze reduces the efficiency of all the heat exchangers in the system.

Sump heaters are sometimes used to avoid freezing in heat rejection units. Avoid these because of their energy consumption.

ECONOMICS

SAVINGS POTENTIAL: *10 to 30 percent of compressor energy, in appropriate applications.*

COST: *$300 to $600 per ton of free cooling capacity.*

PAYBACK PERIOD: *Several years, to much longer, depending mainly on energy costs and the weather profile.*

TRAPS & TRICKS

PLANNING AND DESIGN: *Make sure that this is the best approach for your application. Do not let the simplicity blind you to the cost. The design is not particularly tricky, but it is unconventional. Efficient drain-down for freeze protection requires some cleverness. Be prepared to work out bugs in the installation.*

EXPLAIN IT: *Put a clear explanation of the system in the plant operating manual. Install placards at the heat rejection unit and the controls.*

MONITOR PERFORMANCE: *Include columns in the chiller plant operating log to record the outside air wet-bulb temperature and the supply and return temperatures of the heat rejection circuit. Review the logs regularly to check performance, because failure or incorrect calibration of the controls is invisible.*

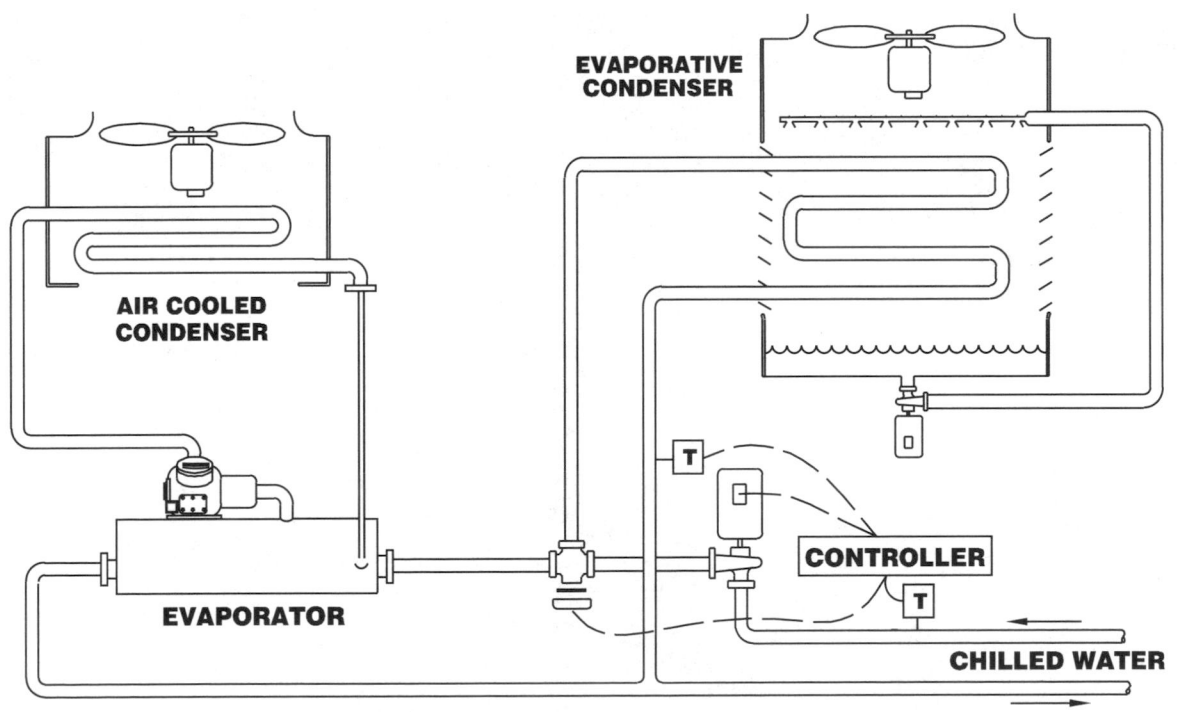

Fig. 1 Cooling the chilled water circuit with its own heat rejection unit This unusual arrangement works even with air-cooled cooling machines.

Heat Recovery From Chillers

The Measures in this Subsection recover the heat that is rejected by chillers, or increase the amount of heat that you can recover. When a chiller is working, it rejects an amount of heat that is equal to the cooling load plus the energy input to the compressors. In facilities that have a large cooling load, heat is discarded to the atmosphere in enormous amounts. This energy seems to cry out to be put to use.

Fortunately, recovering heat from chillers at their normal condensing temperatures is usually simple and free of trouble. However, the low temperature of the recovered heat is a fundamental limitation. To achieve maximum efficiency, chillers are normally designed to operate with condensing temperatures close to the temperature of the outside air. Most facilities cannot use much heat that is recovered at these low temperatures. Preheating domestic water is by far the most common application. Other applications are unusual, and they tend to be specialized to particular industries.

Increasing Heat Recovery by Increasing the Condensing Temperature

You can increase the temperature of the recovered heat as much as 20°F to 50°F (11°C to 28°C) by increasing the condensing temperature. This also increases the amount of heat you can recover. And, you may be able to find new applications for the recovered heat, including some space heating and industrial heating applications.

However, increasing the condensing temperature imposes a serious penalty on chiller efficiency. It also invites a host of complications and problems. Chillers are designed to produce a specified capacity at a specified maximum ambient temperature, which typically involves condensing temperatures in the range of 100°F to 110°F. In normal operation, the average condensing temperature is lower than this. It is usually safe to operate a chiller continuously at its design maximum temperature, or even at a somewhat higher temperature, but motor life may be reduced.

You can increase the heat recovery temperature further by using a chiller that is designed specifically for heat recovery. A heat recovery chiller is simply a conventional unit that is optimized for higher condensing temperatures.

INDEX OF MEASURES

2.10.1 Use condenser water directly for heating applications.

2.10.2 Use an auxiliary condenser or double-bundle condenser for heat recovery.

2.10.3 To recover large amounts of heat at elevated condensing temperature, install a heat recovery chiller.

2.10.4 To recover small quantities of heat at maximum temperature, install a desuperheater.

2.10.5 Improve the quantity or economics of heat recovery by adding or increasing heat storage.

■ The Effect of Higher Recovery Temperature on Chiller COP

If you increase the condensing temperature, balance the increased value of the recovered heat against the increased cost of cooling. Increasing the condensing temperature typically reduces the chiller COP by 1% to 2% per degree Fahrenheit.

Your best compromise depends on the relative value of chiller input energy and the heat that is recovered. The recovered heat is usually less valuable than the chiller input energy. The input energy for compression chillers is usually electricity or another expensive source. On the other hand, the heat pump effect of a chiller makes the amount of recovered heat several times greater than the amount of chiller input energy.

If you select a new chiller for higher condensing temperature, it will have a somewhat higher COP in heat recovery operation than a similar chiller that is selected for normal condensing temperatures. However, this "heat recovery" chiller will have a lower COP at normal condensing temperatures.

■ The Effect of Higher Heat Recovery Temperature on Cooling Capacity

Operating a chiller at increased condensing temperature reduces its capacity substantially. The effect of condensing temperature on capacity varies with the type of compressor. Centrifugal chillers suffer a more serious loss of capacity than other types because the refrigerant gas in a centrifugal compressor is not moved by positive displacement. As the gas flows through the chiller, it is slowed by the increasing discharge pressure caused by higher condensing temperature.

Whether the reduction of cooling capacity is important depends on the time relationship between the cooling and heat recovery applications. For example, preheating domestic water may have to be curtailed during periods of high cooling load. Typically, this does not reduce heat recovery much, because peak cooling conditions are occasional. You can avoid this limitation completely by using heat storage.

■ The Effect of Higher Condensing Temperature on Equipment Life

Increasing the condensing temperature for long periods of time may adversely affect compressor and motor reliability. The effect depends on the type of compressor, the type of refrigerant, and the details of the specific model. Check the chiller specifications before operating at elevated condensing temperature for long periods. Discuss the advisability of heat recovery operation with the chiller manufacturer's engineering department.

The most serious long-term threat is to the motor insulation. The life of the insulation declines rapidly as its temperature increases. Motor temperature rises because motor current increases as the motor must produce greater condensing pressure. Failure of an internal motor is especially bad because the burnout products are corrosive and travel throughout the refrigerant circuit, requiring a messy cleaning procedure. If the chiller motor is external, it may be advisable to replace it with a motor that has a higher temperature rating. See Measure 10.1.1 for details of motor temperature and reliability.

Some refrigerants have exceptionally high discharge temperatures, or superheat. Increasing the condensing temperature may raise the discharge temperature enough to break down the lubricating oil that is carried along in the refrigerant. You can deal with this by selecting a lubricating oil suitable for the higher temperatures.

■ Unloading Limitations of Centrifugal Compressors at Elevated Condensing Temperatures

Centrifugal chillers have the highest peak efficiency of all contemporary chiller types. This makes them the most popular type in large sizes. Unfortunately for heat recovery, a centrifugal compressor is less tolerant of having its condensing temperature increased than other types of compressors. The most serious problem is that centrifugal chillers lose their ability to throttle capacity as condensing temperature is increased. This is because a centrifugal chiller "surges" at some point as its output is reduced. The load at which this occurs increases with condensing temperature.

Surge is a special problem of centrifugal devices (including centrifugal fans and pumps). In brief, the higher pressure in the perimeter housing (the "volute") of the compressor tends to force gas back into the impeller, against the outward flow. The centrifugal force of the outward flow is reduced as the chiller output is reduced. When the centrifugal force can no longer overcome the discharge pressure, the outward flow stalls in the diffuser area leading to the volute, the capacity drops abruptly, and the unstable flow causes severe vibration.

Increasing the condensing temperature for heat recovery raises the discharge pressure. To avoid surge, the chiller must operate at higher load to increase the centrifugal force which drives the outward gas flow. Surge can occur even at full load if the condensing temperature is too high.

Centrifugal chillers offered as "heat recovery" models are not different from other centrifugal chillers in any basic way. They simply are designed to produce greater centrifugal force in the gas. The manufacturer does this by increasing the impeller diameter or by increasing the impeller speed.

Special Heat Recovery Modes of Certain Chiller Types

■ Oil Heat Recovery in Screw Compressors

Most screw compressors inject oil into the compression process to seal the gaps around the rotors. This oil absorbs a considerable amount of the heat of compression, and this heat is removed from the oil at a temperature that is high enough to be useful for some applications. The quantity of heat removed from the oil is limited, but don't overlook it.

■ Recovery of Heat from Discharge Cooling of Ammonia Compressors

Ammonia is limited to reciprocating and screw compressors because of its high pressure differential. Ammonia has exceptionally high superheat, so the compressor may be designed for water cooling of the discharge passages to protect the machine. The quantity of heat in this compressor cooling water is a small fraction of the total heat rejection, but it is worth circuiting the cooling water so that this heat can be collected along with the condenser coolant. To recover the superheat at a temperature higher than condensing temperature, refer to Measure 2.10.4.

■ Heat Recovery from Absorption Chillers

As a rule, do not try to increase heat recovery from absorption chillers by increasing their condensing temperature, especially if they are driven by steam or hot water. The COP of absorption chillers is low at best, and increasing the condensing temperature makes the COP even worse. Also, the operating temperatures of absorption chillers are critical. (See Measure 2.2.2 for some of the reasons.) Read the operating manual carefully before making any adjustments.

Some direct-fired absorption chillers offer an option described as "heat recovery." This is a misnomer invented to boost sales. The chiller is powered by an internal burner. The "heat recovery" consists of diverting heat from the burner to an internal boiler. This saves the expense of installing a separate boiler. This feature is a convenience that does not increase overall efficiency.

How to Match the Heat Recovery Application to the Chiller Plant

■ Heat Recovery is Usually Supplemental

Consider chiller heat recovery to be a supplemental heat source. The chiller may not be able to provide the needed amount of heat or the maximum temperature at all times. Even if it can, you need a conventional backup source for times when the heat recovery chiller is taken out of service.

■ Match the Timing of Cooling and Heat Recovery

Recovered heat must be available at the time the application requires it. Therefore, the amount of heat that can be recovered depends on the relative timing of chiller operation and the heat recovery application. For example, domestic water heating may use chiller heat recovery on a supplemental basis whenever the chiller happens to be operating.

More energy is recoverable in a warm climate, where chillers operate continuously, than in a climate

where the chiller operates for only several months per year.

If you want to use chiller heat recovery for space heating, the facility must have a substantial cooling load during cool weather to provide the heat.

You can deal with short-term differences in timing, extending as long as several days, by using heat storage. This is covered by Measure 2.10.5.

■ Match the Heat Recovery and Application Temperatures

In some applications, the amount of heat that can be recovered is proportional to the condensing temperature. If heat recovery cannot provide the maximum temperature of the application, then heat recovery is still valuable for preheating. For example, in heating domestic water from 55°F to 130°F, chiller heat recovery can preheat the water to 100°F, and a conventional water heater is used to complete the job.

In other applications, you would have to modify the application to exploit heat recovery at all. For example, to use heat recovery for space heating, the heating system must be designed for exceptionally low heating water temperature. Chiller heat recovery cannot work with conventional hydronic heating that has a typical supply water temperature of 160°F (71°C) and a typical return water temperature of 120°F (49°C). To use heat recovery, the system would have to operate with a supply temperature of 130°F (54°C) or less, and a return temperature of about 90°F (32°C). This would require special heating units, and the low supply air temperature may cause complaints about comfort.

■ Heat Exchanger Losses and Contaminant Isolation

Heat recovery usually requires a heat exchanger. Condensing temperatures are low, so the temperature drop that occurs in a heat exchanger may seriously reduce the amount of condenser heat that you can recover. Therefore, design the heat exchangers for minimum temperature loss.

If the condenser is water-cooled, you may be able to recover the heat simply by using the cooling water directly. But this is not always possible. If the system uses an open cooling tower to cool the condenser, the cooling water is dirty. Cooling towers wash pollutants out of the atmosphere and trap debris. Also, cooling tower water is usually treated with chemicals that are toxic. This may force you to use another heat exchanger to isolate the tower water from the application. For example, the dirt in the water may foul heating coils.

As an alternative, you may be able to recover heat without a heat exchanger by installing a strainer system in the cooling tower circuit. See Measure 2.9.3 for this technique, which is not foolproof.

If the condenser is air-cooled, you have to install a heat exchanger to recover the heat of the compressed refrigerant gas. In some applications, such as heating potable water, the heat exchanger may also have the purpose of blocking contamination. A leak in the heat exchanger may contaminate the heated medium with refrigerant. Halocarbon refrigerants combine with water to form hydrochloric and hydrofluoric acids, which are toxic and corrosive. Ammonia refrigerant combines with water to form window cleaner.

The most efficient way to protect against contamination is using a double-wall heat exchanger. In this type of heat exchanger, the heat transfer surface consists of two metal surfaces with a space between them that is vented to the outside. Thermal conduction between the two surfaces occurs across metal-to-metal contacts that are spaced at regular intervals.

With refrigerants having very low condensing pressure, especially R-123 and R-11 (which is being phased out for environmental reasons), you may be able to provide isolation by keeping the heated liquid at a pressure higher than the condensing pressure. For example, this may be possible in a domestic water system where pressure is maintained by a standpipe. Even with this arrangement, contamination may occur if the domestic water pressure is turned off during a period of time when a leak exists in the heat exchanger.

(Contamination in the other direction, from the water side into the refrigerant, is a serious problem that is faced by all water-cooled chillers that have single-wall heat exchangers. See Measure 2.7.1 for details.)

■ Distribute the Load Efficiently Between Heat Recovery and Non-Recovery Chillers

Raising the condensing temperature costs a penalty in chiller COP and capacity. Therefore, you want to operate the minimum amount of chiller capacity at elevated condensing temperature. At the same time, you want to avoid operating any chiller at an inefficient load.

For example, assume that you have a chiller plant with two 1000-ton chillers and a single 500-ton chiller. At a particular moment, the total cooling load is 900 tons, and 300 tons of this load are needed to satisfy the heat recovery load. If you operate only a single 1000-ton chiller, it has to operate with elevated condensing temperature, so the entire cooling load is delivered with a COP penalty. On the other hand, you could operate the 500-ton chiller at a load of 300 tons to provide heat recovery, and one of the 1000-ton chillers at 600 tons to

provide the rest of the cooling load with low condensing temperature. In this case, both chillers are loaded near their points of peak efficiency.

You can shift the cooling load between the heat recovery chiller and other chillers in several ways:

- *manually reset the chilled water temperature settings of the chillers.* Standard chiller controls adjust their loading to meet a temperature setting for the chilled water supply or return. Lowering the temperature setting of one chiller in a group increases the load on that chiller and reduces the load on the others.

- *install a chiller sequencing controller.* Chiller sequencing controllers are automatic devices that start and stop chillers in a programmed sequence in accordance with the cooling load. They also distribute the load among the chillers. See Measures 2.1.1 ff for details.

- *exploit the preferential load capability of variable-flow pumping.* In variable-flow pumping systems, a selected chiller can be loaded preferentially by the manner in which it is installed in the chilled water piping system. Using this method, the selected chiller is loaded fully before any other chillers are loaded. This method is inflexible, but it has certain advantages. Refer to Measure 2.5.2 for details of variable-flow pumping.

How This Subsection is Organized

The first three Measures of this Subsection recommend recovering heat from the condensers of chillers. They are presented in order of increasing cost and complexity. The fourth Measure captures the relatively small quantity of energy that is available at high temperature in the form of superheat. The last Measure recommends heat storage, primarily to increase the amount of heat that can be usefully recovered.

You can recover heat from cooling equipment in many ways. The Measures in this Subsection represent the basic and intermediate levels of complexity. Other techniques are more complex, such as two-stage heat recovery, combinations with heat pumps, etc. These methods require an advanced knowledge of refrigeration theory and practice, and they are more difficult to operate efficiently. These heat recovery techniques are usually specialized to particular industries, and they are likely to be integrated into the design of specialized processes.

MEASURE **2.10.1 Use condenser water directly for heating applications.**

The easiest way to recover heat from a chiller is simply to divert the condenser water through a heating application on the way to the cooling tower. The heat recovery temperature can be increased by raising the condenser water temperature setting, subject to the limitations of the chiller.

The simplest arrangement is shown in Figure 1. You can use this for an application such as tank heating, where the amount of heat available exceeds the requirement, and where the normal condensing temperature is adequate.

Another arrangement is shown in Figure 2. This one provides somewhat higher recovery temperature because it avoids the heat loss that would occur in an idle cooling tower. This advantage is most significant if heat is being recovered during cold weather.

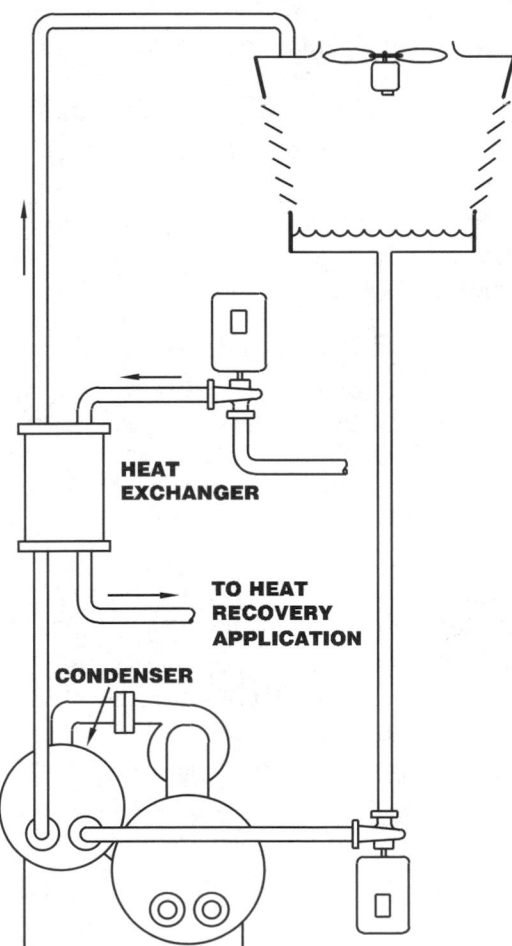

Fig. 1 Condenser heat recovery This arrangement simply taps heat from the condenser cooling water as it leaves the condenser.

SUMMARY

Inexpensive and simple. Application is limited by cooling tower water contaminants.

SELECTION SCORECARD

Savings Potential	$ $ $	
Rate of Return, New Facilities	% % % %	
Rate of Return, Retrofit	% % %	
Reliability	✓ ✓ ✓	
Ease of Retrofit	☺ ☺ ☺	

Energy and Cost Saving Potential

If you can recover heat at the chiller system's minimum condensing temperature, this Measure saves energy in three ways:

- *heat recovery.* All the heat rejected by the chiller can be recovered by this method. The application determines how much of this heat is useful.

- *reducing compressor power,* due to the cooling of the condenser water in the heat recovery process. This cooling augments the cooling tower. This saving occurs as long as the cooling tower water would be warmer than the minimum allowable temperature without heat recovery. The efficiency improvement is greatest when the chiller is heavily loaded and the cooling tower has the highest temperature differential.

- *reducing cooling tower fan power.* Heat recovery leaves less heat for the cooling tower to discard, so the tower fans do not have to work as hard. The saving depends on how the cooling water temperature is controlled.

If you raise the condensing temperature to recover more heat, the chiller will consume more energy. The cooling tower fans will use less energy because they are rejecting less heat, and because the cooling tower is more effective at higher temperature. If you have a choice in selecting the condensing temperature, make a careful analysis to find the temperature that produces the greatest overall saving.

Protecting Against Contamination

The main limitation of this method is that the cooling tower water is contaminated by dirt and water treatment chemicals. The water may also become contaminated by refrigerant. Therefore, the application must be able to tolerate the contaminants, or it must be isolated from them.

The simplest application is one in which the cooling tower water can be passed through a coil that is designed to operate with dirty water. For example, heating storage tanks for heavy oil is a good application. If the application needs protection against the possibility of a heat exchanger failure, refer to the Subsection introduction for various methods.

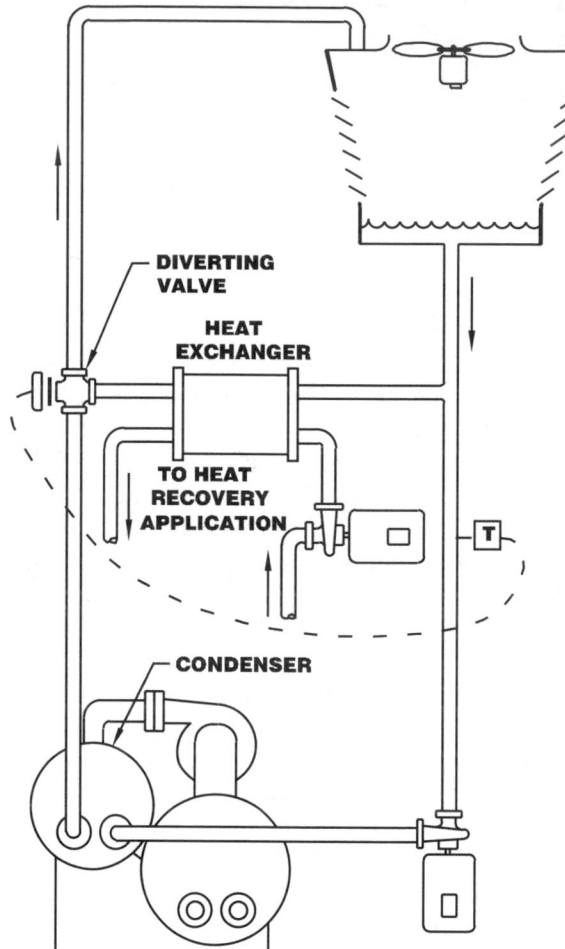

Fig. 2 Condenser heat recovery with option to bypass the cooling tower This arrangement maximizes the amount and temperature of heat recovery by avoiding heat loss from the cooling tower. This is especially useful in cold climates.

Controls

If you intend to recover heat only at normal condensing temperatures, no additional controls may be needed.

If you intend to raise the condensing temperature to increase heat recovery, install controls that limit the condensing temperature to the minimum needed to satisfy the moment-to-moment heat requirement of the application.

If the chiller must maintain a minimum condensing temperature, and the initial temperature of the substance you are heating is below this temperature, install a thermostatically controlled bypass to maintain the minimum temperature in the condenser.

ECONOMICS

SAVINGS POTENTIAL: *The amount of heat that can be recovered is equal to the cooling load plus the energy input to the chiller. The energy consumption of the cooling tower fans is reduced. If the heat is recovered at minimum condensing temperature, the chiller COP is improved by the reduction of the condenser temperature. If heat is recovered at elevated condensing temperature, compressor power is increased.*

COST: *Varies widely, typically several thousand dollars for more. Most of the cost is for piping, and perhaps, for a heat exchanger.*

PAYBACK PERIOD: *Less than one year, to many years. The economics are better in larger systems, other factors being equal.*

TRAPS & TRICKS

DESIGN: *If you plan to increase the condensing temperature to increase heat recovery, design the controls to optimize the balance between heat recovery and reduced chiller performance. This requires a detailed knowledge of the chiller's efficiency characteristics.*

MAINTENANCE: *Cooling tower water is dirty, so schedule regular cleaning of any heat exchangers.*

EXPLAIN IT: *Put a clear explanation of the heat recovery system in the plant operating manual. Install placards on any unusual items.*

MONITOR PERFORMANCE: *Schedule periodic checks of control operation and heat exchanger cleaning in your maintenance calendar.*

MEASURE **2.10.2 Use an auxiliary condenser or double-bundle condenser for heat recovery.**

If you don't want to recover heat directly from the cooling tower water (Measure 2.10.1) because of the dirt and chemical content of the water, consider using an auxiliary condenser. An auxiliary condenser is a second condenser. It is similar to the primary condenser, except that you size it for the amount of heat that you wish to recover. Clean water is circulated in a closed loop between the auxiliary condenser and the heat recovery application. Figure 1 shows a chiller with a large auxiliary condenser.

The compressor discharges freely to both the main condenser and the auxiliary condenser. The division of refrigerant flow between the main condenser and the auxiliary condenser depends on their relative temperatures and sizes.

Heat can be recovered at any temperature up to the temperature of compressor discharge. However, if the temperature of the main condenser is considerably lower than the heat recovery temperature, most of the heat will go into the main condenser. To increase heat recovery at maximum temperature, the flow of cooling water to the main condenser has to be throttled back.

A double-bundle condenser is a variation of this arrangement, in which a single condenser shell contains

SUMMARY

Requires an expensive special condenser. Avoids cooling tower water contaminants, but not the risk of refrigerant contamination. The economics are best in applications where heat recovery can occur at lower condensing temperatures. Can be used with air-cooled chillers.

SELECTION SCORECARD

Savings Potential .... $ $ $

Rate of Return, New Facilities % % %

Rate of Return, Retrofit.......... % %

Reliability .... ✓ ✓ ✓

Ease of Retrofit .... ☺ ☺

two isolated tube bundles. One of the tube bundles is cooled by cooling tower water, like a normal condenser, and the other tube bundle is used for heat recovery, like an auxiliary condenser. Figure 2 shows heat recovery from a double-bundle condenser.

Trane Company

Fig. 1 Chiller with auxiliary condenser This chiller has two condensers of approximately equal size. One is connected to a cooling tower, and the other to a heat recovery application. An auxiliary condenser of this size can recover an enormous amount of heat at low temperature for applications such as pre-heating domestic water.

Double-bundle condensers are used in new chiller installations, while auxiliary condensers are used in both new and retrofit applications. The main differences are in cost and equipment layout.

Comparison to Measure 2.10.1

The energy and cost saving factors are the same as those of Measure 2.10.1.

Heat recovery in Measure 2.10.1 occurs in series with the condenser, whereas heat recovery in this Measure occurs in parallel with the main condenser. In Measure 2.10.1, all the discharge heat enters the heat recovery process. In this Measure, a fraction of the heat bypasses the auxiliary condenser and is lost to the cooling tower.

For example, consider preheating domestic water that enters the auxiliary condenser at 55°F, while the water from the cooling tower enters the primary condenser at 90°F. If the two condensers are identical, most of the chiller's heat is rejected to the auxiliary condenser because it is colder, but some heat is lost to the cooling tower. This is not a problem if the amount of heat rejected by the chiller substantially exceeds the needs of the application.

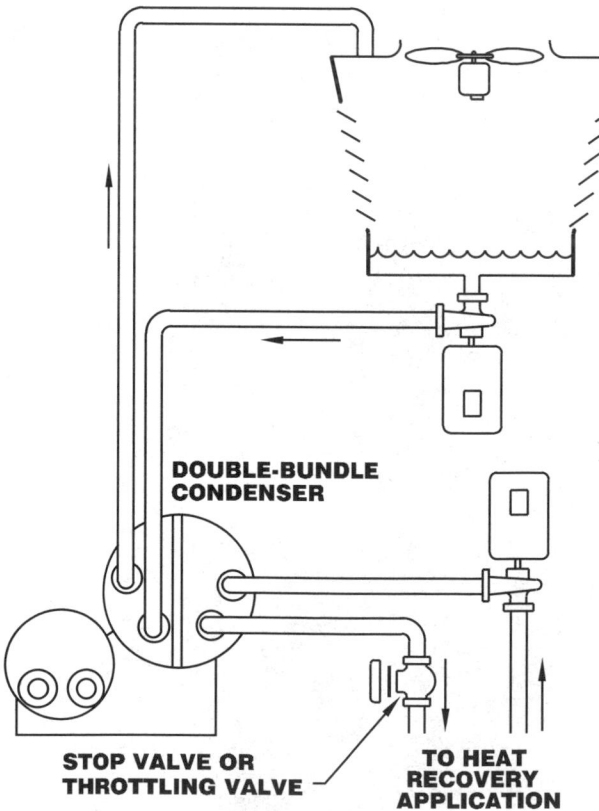

DOUBLE-BUNDLE CONDENSER

STOP VALVE OR THROTTLING VALVE

TO HEAT RECOVERY APPLICATION

Fig. 2 Heat recovery from a double-bundle condenser This arrangement has a throttling valve to regulate the heat recovery temperature. To maximize the heat recovery temperature, restrict the flow of water from the cooling tower to its tube bundle.

Contamination By Refrigerant or Leakage Into Refrigerant

The auxiliary condenser has its own closed water circuit, independent of the cooling tower water. This keeps the water clean, so it can be used in conventional heating coils.

A leak in the auxiliary condenser will contaminate the heat recovery water circuit with refrigerant, if a high-pressure refrigerant is used. Therefore, heating potable water and other critical liquids requires additional safeguards, which are explained in the introduction to this Subsection. If a low-pressure refrigerant is used, a leak in the auxiliary condenser will allow the water or other heated liquid to contaminate the refrigerant.

Control with Water-Cooled Primary Condensers

In the simplest cases, a water-cooled auxiliary condenser may require no controls. Preheating domestic water is such an application, provided that the incoming domestic water temperature is not too low.

To raise the heat recovery temperature, raise the temperature setting of the cooling tower. This causes more of the refrigerant to give up its heat in the auxiliary condenser.

The controls should optimize the system economics under all operating conditions, taking into account the value of heat recovery, chiller input energy, cooling tower fan energy, and energy consumption of heat recovery pumps. These controls may be simple or complex, depending on the heat recovery application and the amount of money at stake.

Stop the flow of liquid through the auxiliary condenser when the compressor is not running. Otherwise, the liquid may be cooled, rather than heated. As explained in Measure 2.9.1, an idle chiller acts like a heat pipe, in which heat from warmer parts flow to the coldest part by refrigerant convection. For example, if cooling tower water continues to be circulated through the main condenser, the temperature there will approach the outside wet-bulb temperature. If the weather is cool, water flowing through the auxiliary condenser may lose heat to the main condenser.

Unless the compressor operates almost continuously, install an automatic valve to shut off the auxiliary condenser. This is an added complication, and another item to maintain.

Accessories and Control with Air-Cooled Chillers

In principle, you can use an auxiliary condenser with an air-cooled chiller. This is an odd arrangement, and it may take a great deal of skill to get it operating properly. The air-cooled condenser exposes refrigerant directly to outside air temperature. As a result, the relative temperatures of the two condensers cannot be used to regulate the amount of heat recovery. Instead, it would be necessary to use backpressure valves in the compressor discharge line to distribute the flow of gas to the two condensers, along with check valves to control the flow of liquid refrigerant.

Another problem is that refrigerant would accumulate in the air-cooled condenser during cold weather. Therefore, the system would need enough extra refrigerant to make up for the refrigerant trapped in the air-cooled condenser during cold weather, and it would need a large receiver to hold the excess refrigerant during warm weather. In systems where oil travels with the refrigerant, piping technique would be critical.

ECONOMICS

SAVINGS POTENTIAL: This arrangement can recover a large fraction of the heat rejected by the chiller system. The fraction depends on the relative size and recovery temperature of the auxiliary condenser (or bundle). The energy consumption of the cooling tower fans is reduced. If the heat is recoverable at minimum condensing temperature, chiller COP is improved by reducing the condenser temperature. If the temperature of the cooling tower condenser is increased to enhance heat recovery, the compressor input energy is increased.

COST: $10 to $20 per thousand BTUH of heat recovery capacity.

PAYBACK PERIOD: Several years, or longer. Larger systems typically have shorter payback periods.

TRAPS & TRICKS

PLANNING AND DESIGN: Make sure that you can recover enough heat to justify the cost of the modification. As in all heat recovery methods, the tricky part of the design is the controls, which become more complex if you want to operate with elevated condensing temperatures. Provide safeguards against refrigerant contamination of the water, if this could be a problem. If you want to try this with an air-cooled system, treat it as a pioneering venture.

EXPLAIN IT: Put a clear explanation of the system in the plant operating manual.

MONITOR PERFORMANCE: Schedule periodic checks of control operation in your maintenance calendar.

MEASURE 2.10.3 To recover large amounts of heat at elevated condensing temperature, install a heat recovery chiller.

The main limitation of chiller heat recovery is the low temperature of the recovered heat. While conventional chillers can be operated at increased condensing temperature, they pay an increasingly severe penalty in COP, capacity, and service life as the heat recovery temperature is increased. If you have a potential heat recovery application and exploiting it would require a condensing temperature that is too high for your conventional chillers, consider installing a heat recovery chiller.

A heat recovery chiller is any type of cooling machine that is built to operate at a higher condensing temperature for the purpose of recovering its heat output at higher temperature. The difference is a matter of selection, not fundamental design. You select the condensing temperature to satisfy the temperature requirements of the heat recovery application. Typically, you select the capacity of the machine to satisfy the heating load, rather than the cooling load. By the same token, you may control the output of the machine to follow the heating load, rather than the cooling load.

A large cooling machine that is designed with a view toward recovering its heat output may be called a "heat pump," although the term is commonly reserved for small units that are designed primarily as a substitute for other types of heating equipment. On the other hand, the more theoretical members of the HVAC trade use the term "heat pump" to cover all cooling machines, regardless of whether or not the heat output is exploited. Don't be confused by the terminology. Regardless of what anyone calls the machine, the basic issues are the same.

Reasons to Install a Special Heat Recovery Chiller

A specialized heat recovery chiller may offer some or all of these advantages, when compared to chillers that are not selected specifically for heat recovery:

- it is capable of achieving higher recovery temperatures. In applications such as heating domestic water, this allows heat recovery to satisfy a larger fraction of the application's energy requirement. In some cases, it allows heat recovery for applications that would not otherwise be possible.
- it achieves higher recovery temperature without sacrificing reliability
- it has a COP that is optimized for the heat recovery temperature (but it has a lower COP than conventional chillers when operating at conventional temperatures)

SUMMARY

The most efficient way to recover large amounts of heat at an elevated condensing temperature. Requires a specialized chiller. The same chiller may serve at different times for ice thermal storage.

SELECTION SCORECARD

Savings Potential .... $ $ $

Rate of Return, New Facilities % % % %

Rate of Return, Retrofit.......... % % %

Reliability .... ✓ ✓ ✓

Ease of Retrofit .... ☺ ☺

- in the case of a centrifugal chiller, it retains better unloading capability
- it is sized so that the COP penalty resulting from the high condensing temperature is limited to the smallest possible fraction of the cooling load
- it can be operated without a cooling tower, if all condenser heat is recovered
- in existing plants, it does not cause the chiller plant to lose cooling capacity because of operation at high condensing temperature.

Energy Saving Potential

The net saving provided by a heat recovery chiller is a balance between the value of the recovered heat and the cost of the additional energy required by the chiller.

The amount of recoverable heat equals the cooling load plus the motor input energy. The application determines how much of this heat is useful. If the chiller is sized properly, most of this heat is recovered.

On the debit side, the heat recovery chiller requires substantially more energy than a cooling-only chiller that produces the same amount of cooling at lower condensing temperatures. The compressor energy input typically increases from 1.0% to 1.4% per degree Fahrenheit (1.8% to 2.5% per degree Celsius) that the design condensing temperature is increased. If the application needs an intermediate heat exchanger to isolate it from possible refrigerant contamination, the temperature differential is increased, which increases energy input.

From the standpoint of energy alone, the increase in compressor energy is recoverable as heat. From an economic standpoint, the compressor input energy usually has higher value than the recovered heat, unless the recovered heat is used to displace resistance heating. Electricity typically costs several times more, per unit of energy, than heat produced with common fuels.

The heat recovery process saves a fraction of cooling tower fan energy, because the recovered heat does not have to be rejected by the cooling tower. On the other hand, any pumps used in the intermediate heat transfer loop consume additional energy. These factors are usually minor in comparison with the first two factors.

Chiller Cost

In new construction, installing a heat recovery chiller may not add much to the chiller plant cost. The heat recovery chiller is typically a smaller unit, and it carries a part of the cooling load. The cost per unit of cooling capacity is somewhat higher because capacity drops at higher condensing temperatures.

In existing facilities, adding a specialized chiller to exploit heat recovery is expensive. To pay off, the potential heat recovery application must keep the chiller heavily loaded for long periods.

If the plant presently lacks a chiller that is small enough to handle low cooling loads efficiently, consider using a smaller heat recovery chiller to serve as a lead chiller. (See Measure 2.8.1 about installing an efficient lead chiller.) This approach makes sense if there is a large heat recovery load during periods when the cooling load is low.

The economics of adding a heat recovery chiller to an existing plant are most favorable if the chiller satisfies a need for additional cooling capacity.

Sizing, Operating Schedule, and Heat Storage

The ideal heat recovery application is one that requires a steady input of heat at times that coincide with long periods of cooling. Such ideal loads are rare. Sizing the heat recovery chiller is usually a compromise between chiller cost and the amount of heat that can be recovered.

One way to level the heat recovery load is to store recovered heat until it is needed, as recommended by Measure 2.10.5. This provides good utilization of the heat recovery chiller even with irregular heat loads, such as domestic water heating.

Electricity pricing schedules may affect the feasibility and operating schedule of a heat recovery chiller. See Reference Note 21, Electricity Pricing, for details. The heat recovery chiller has higher demand for a given amount of cooling than conventional chillers, so it may not be desirable to recover heat during the time of day when electricity demand rates are high. If there is enough cooling load to produce useful amounts of heat outside the hours of high electricity prices,

consider heat storage as a way to satisfy the heat requirement during peak-price hours.

How to Distribute the Chiller Load During Heat Recovery

The heat recovery chiller normally is throttled to produce only enough cooling to satisfy the heat recovery load, because its higher condensing temperature gives it a lower COP. The introduction to this Subsection covers the ways of distributing the cooling load among chillers.

A subtle and serious problem may arise if a particular chiller is dedicated to heat recovery at high condensing temperature. The heat recovery chiller cannot share its cooling load with the next chiller that starts, so the next chiller may operate at low load for many hours, depending on the chiller sizing and the cooling load profile. This is especially serious with centrifugal chillers, and to a lesser extent, with screw compressor chillers. Correcting this situation may require sacrificing some heat recovery to provide higher cooling load for the cooling-only chillers. See Measures 2.1.1 and 2.8.1 for more about this.

Operating the Heat Recovery Chiller at Reduced Condensing Temperature

If there are periods of time when no heat recovery is required, operate the heat recovery chiller at normal lower condensing temperatures, subject to any chiller design limitations. As with any chiller, cooling efficiency improves if you lower the condensing temperature.

A heat recovery chiller is less efficient at normal condensing temperatures than a similar chiller that is designed to operate at the lower condensing temperature. If you want the chiller to operate at both condensing temperatures, give the chiller manufacturer plenty of advance notice, as this may affect some design features.

Selecting the Type of Compressor

See Reference Notes 32 and 33 for the characteristics of different types of chillers. See the introduction to this Subsection for chiller characteristics that relate specifically to heat recovery.

Screw compressor chillers are becoming attractive for heat recovery in all sizes except the largest and smallest. Certain models of screw compressors have variable discharge ports that allow them to adapt more efficiently to different pressure differentials. The advantage of these machines increases in applications where the cooling load and/or the heat recovery load varies widely, and also where the desired heat recovery temperature may vary.

Reciprocating chillers are well proven in heat recovery applications, and they continue to be the best choice in many smaller applications. They have a great deal of flexibility in condensing pressure, throttling, and other characteristics. Their relatively low COP is less of a disadvantage in heat recovery applications, because their higher input energy is captured as heat.

For large heat recovery applications, a centrifugal heat recovery chiller offers the highest COP at its design condensing temperature. A big drawback of centrifugal chillers is limited ability to unload. Do not use a centrifugal chiller unless there is enough cooling load to keep the chiller above its minimum load during the period when heat recovery is useful. Centrifugal chillers have less ability than other types to increase their COP at condensing temperatures that are lower than their design condensing temperature. This is important if the chiller is needed for extended periods without heat recovery.

Centrifugal chillers with variable-speed drives are able to operate at reduced condensing temperatures without as much sacrifice of COP, and they are able to unload better. However, the variable-speed drive is an expensive feature that may not be justified in a dedicated heat recovery chiller.

■ **Combinations of Chiller Types**

In most cases, you can use any type of heat recovery chiller in combination with other types of chillers. For example, you might install a screw chiller for heat recovery in a plant where the other chillers are centrifugal.

You may have to deal with a control complication if you use a reciprocating chiller in combination with centrifugal or screw chillers. Reciprocating chillers use step control of capacity, so their output is usually controlled on the basis on return water temperature, while centrifugal and screw chillers typically control output on the basis of supply water temperature. Supply water temperature control is possible with a reciprocating chiller, but this requires special controls that anticipate the changes in supply chilled water temperature as the compressor unloads. This is a touchy method of control, especially where reciprocating chillers operate in parallel with other types.

Using the Same Chiller for Cooling Thermal Storage

The distinguishing feature of a heat recovery chiller is that it is designed for a larger difference between the evaporating and condensing temperatures. This same temperature differential may be appropriate for ice storage, which is covered in Subsection 2.11. For example, a heat recovery chiller may be optimized for an evaporator temperature of 40°F and a condensing temperature of 110°F. A chiller optimized for ice storage, with an evaporator temperature of 20°F and a condensing temperature of 90°F, has about the same design temperature differential.

Thus, a chiller that is used for ice storage in summer to avoid peak electricity demand rates can also be used to recover heat at elevated temperatures during the rest of the year. This combination may greatly improve the utilization and economics of the chiller. For this to work well, the ice storage load and the heat recovery cooling load should be similar in magnitude.

The chiller cannot simultaneously recover heat at elevated temperature and store cooling at reduced temperature, because the increased temperature differential of the chiller can be used for only one application at a time.

Ice storage requires the chiller to cool an antifreeze solution as a medium for freezing the ice. Avoid using antifreeze solution when the chiller is used for conventional cooling, because all antifreeze solutions have poorer heat transfer characteristics than water. Switching the evaporator from chilled water to ice freezing solution requires draining the evaporator each time the chiller is changed from one function to the other. This is not too burdensome if done only on a seasonal basis. Some clever plumbing may ease the changeover.

ECONOMICS

SAVINGS POTENTIAL: The amount of heat that can be recovered is equal to the cooling load plus the energy input to the chiller. The net saving is reduced by substantially increased compressor energy input, which usually has a higher value than the recovered heat. The net financial saving depends largely on the relative prices of electricity and heat.

A minor bonus is that the energy consumption of the cooling tower fans is reduced. However, any heat transfer pumps require additional energy.

COST: Heat recovery chillers cost between $500 and $1,000 per ton of cooling capacity. Because the cooling capacity is reduced at elevated condensing temperature, the cost averages about 20% higher than for chillers operating at minimum condensing temperatures. In terms of recovered heat, the cost of the chiller ranges from $35 to $70 per thousand BTUH.

PAYBACK PERIOD: One to three years, in new construction. Several years to much longer, in retrofit. The payback period is much shorter if the chiller is installed to satisfy a need for increased cooling capacity.

TRAPS & TRICKS

PLANNING AND DESIGN: Expect to spend time on this. A heat recovery chiller is one of the most complex ways to conserve energy. It can waste your investment or even waste energy if it is not planned properly, or if conditions change. Heat recovery design is complicated by interactions between the heat recovery chiller and the other chillers, by fluctuation in the value of the recovered heat, and by the variety of design options.

SELECTING THE EQUIPMENT: Get your chiller from a manufacturer with experience in producing chillers for heat recovery applications. Check with other plant operators about the performance of the types of chillers you are considering.

EXPLAIN IT: Even though heat recovery should be controlled automatically, make sure that plant operators understand how heat recovery is supposed to work. Put

a clear explanation of the system in the plant operating manual. Otherwise, maintenance work or attempts to respond to cooling problems may sabotage the system.

MONITOR PERFORMANCE: Improper control operation is invisible, so schedule periodic checks of heat recovery efficiency. Include columns in the chiller operating log to record condensing temperature and other parameters related to the effectiveness of heat recovery.

MEASURE 2.10.4 To recover small quantities of heat at maximum temperature, install a desuperheater.

SUMMARY

Recovers heat at higher temperatures than from condenser, without reducing chiller COP. Simple to install. Superheat is only a small fraction of total chiller heat rejection, and it depends on the type of refrigerant and compressor.

SELECTION SCORECARD

Savings Potential	$		
Rate of Return, New Facilities	%	% %	%
Rate of Return, Retrofit..........	%	% %	
Reliability	✓	✓ ✓	
Ease of Retrofit	☺	☺ ☺	

With most refrigerants (but not all), the compression process raises the refrigerant gas temperature to a level higher than the saturation temperature in the condenser. This is why the compressor discharge pipe may be much hotter than the liquid refrigerant line from the condenser. The excess heating of the refrigerant over its saturation temperature is called "superheat." The energy contained in the gas as superheat is a small fraction of the total heat rejected by the chiller.

A "desuperheater" is a heat exchanger that is installed between the compressor discharge and the condenser. The difference between the desuperheater and the condenser is that the heat removed in the desuperheater is only sensible heat, so no condensation occurs. Recovering the superheat at a temperature higher than the condensing temperature does not reduce the chiller COP, because this process does not affect the condensing temperature or the condensing pressure. (The compressor discharge pressure is determined by the condensing temperature of the refrigerant gas in the condenser, which is determined by the temperature of the cooling tower water.)

No control is needed on the refrigerant side of the desuperheater. The temperature of the liquid being heated is controlled by throttling its flow through the desuperheater. Alternatively, the liquid can pass through the desuperheater with no control, picking up as much heat as possible. Most commonly, the liquid is heated in the condenser first, and is then heated in the desuperheater to maximize its temperature.

Be aware that the term "desuperheater" may be used for any heat recovery heat exchanger that is installed in the compressor discharge line, regardless of how much it cools the gas. In this sense, a "desuperheater" is an auxiliary condenser that is connected in series with the main condenser.

Applications and Energy Saving Potential

A desuperheater is useful where the desired heat recovery temperature is higher than the condensing temperature. Desuperheaters may deliver heat at a temperature that is useful for domestic water heating, for some space heating, and for some process applications. The recovered energy is sensible heat, so increasing the heat recovery temperature also increases the quantity of heat recovered.

If all the chiller's rejected heat can be used at condensing temperature, a desuperheater cannot increase the total amount of heat that is recovered. If no desuperheater is installed, the condenser collects the superheat energy, but at the lower condensing temperature.

Superheat comprises only a small fraction of the chiller's heat rejection. Most of the energy in hot refrigerant gas is contained as latent heat, not as sensible heat. Designers who are accustomed to recovering sensible heat from liquids and solids should not be fooled by the hotness of the compressor discharge pipe into believing that a lot of high-temperature heat is available.

Superheat is sensible heat, so the gas temperature drops as more heat is extracted. This forces you to compromise between the heat recovery temperature and the amount of heat that you recover in the desuperheater.

■ Common Refrigerants with High Superheat

With most common refrigerants, the amount of superheat is too small to be worth recovering. This is true of the common old refrigerants R-11 and R-12, and of the important replacement refrigerants R-123 and 134a. Of the common halocarbon refrigerants, only R-22 produces a substantial amount of superheat. The greatest amount of superheat among conventional refrigerants is produced by ammonia, which is notorious for its high discharge temperatures.

For example, consider a theoretical compression cycle from a 40°F evaporating temperature to a 120°F condensing temperature:

- with R-134a, the superheat energy is only about 4% of total heat rejection, and the discharge temperature is only about 12°F higher than the condensing temperature
- with R-22, superheat comprises about 11% of total heat rejection, and the compressor discharge temperature is about 155°F, which is about 35°F higher than the condensing temperature
- with ammonia, superheat is about 14% of total heat rejection, and the compressor discharge temperature is about 225°F, which is about 105°F higher than the condensing temperature.

These numbers apply only to "isentropic" compression, which is a theoretical situation. The temperatures in real equipment may be much lower, for the following reasons.

■ Effect of Compressor Features on Superheat

The numbers in the previous example assume that the gas is compressed without gaining or losing any heat between the evaporator and the condenser. This assumption is not valid with most compressors where compression occurs in stages, such as the common multi-stage centrifugal machines. Most of these machines use interstage cooling or refrigerant injection to improve efficiency, and these features reduce superheat considerably.

Most screw compressors use oil to seal the gaps between the lobes, and this oil acts as a coolant, reducing superheat.

Ammonia has a discharge temperature that is so high that reciprocating ammonia compressors commonly have water-cooled cylinder heads. The water cooling reduces the superheat. Screw compressors used with ammonia may also have water-cooled discharge components.

Before buying a new compressors, get information about its potential for recovering superheat from the manufacturer's engineering department.

To estimate the potential benefit of installing a desuperheater on an existing compressor, measure the superheat under different loads and condensing temperatures. To do this, you need to measure the temperature of the discharge line. You can do this easily by wrapping the sensing element of an accurate thermometer against the discharge line with a wad of any insulating material.

■ Limitations in Recovering Superheat from Air-Cooled Chillers

In an air-cooled chiller, the refrigerant discharge gas is in direct contact with the outside air temperature at the condenser. Therefore, the condenser pressure falls radically during cold weather. (Various features can be added to air-cooled condensers to maintain a higher condensing temperature, but these cannot maintain a high pressure under all loads.) The condensing temperature depends on the discharge pressure, so the high discharge temperature disappears during cooler weather. An air-cooled chiller is usually not eligible in applications where high temperature is needed during cold weather application, as in space heating.

Superheat could be maintained by installing a backpressure valve at the gas outlet end of the desuperheater. However, this would keep the chiller operating at reduced COP, even during periods when the condensing temperature can be reduced. Also, a backpressure valves adds potential failure modes. Failure of the valve in the closed position would shut down the chiller system, and failure in the open position would disable heat recovery.

Heat Exchanger Configuration

A desuperheater is relatively small. This is because the temperature differential is high and the amount of heat recovered is small. The gas passages must be large enough to pass full gas flow, since no condensation occurs. The desuperheater should allow free drainage of any refrigerant that condenses inside it. With refrigerants that carry oil, the desuperheater should allow proper return of oil to the compressor. The easiest way to satisfy these requirements is to install the desuperheater so that its discharge line drains toward the condenser.

To achieve the highest heat recovery temperature, the desuperheater should be a true counterflow heat exchanger. That means, the heated liquid should flow in a direction opposite to the direction of the discharge gas. To achieve this, a coaxial unit is needed, rather than a shell-and-tube type.

How to Protect Against Contamination

If a leak in the desuperheater is a potential hazard, use a double-wall heat exchanger. These are readily available in desuperheaters.

ECONOMICS

SAVINGS POTENTIAL: *Recoverable superheat is a few percent of the heat rejected by the chiller, the amount depending on the refrigerant and the features of the compressor.*

COST: *Double-wall desuperheaters cost from $50 to $100 per ton of compressor capacity, depending on size. Controls are additional, and may cost from several hundred to several thousand dollars. Installation typically requires one day to several days, depending on the complexity of the system.*

PAYBACK PERIOD: *Several years, or longer.*

TRAPS & TRICKS

CHOICE OF METHOD: *Don't let the high recovery temperature fool you into thinking that there is much heat available. Temperature is one thing, heat is another. Control the flow of liquid through the desuperheater so that the high temperature is not dissipated.*

EXPLAIN IT: *Although heat recovery should be controlled automatically, make sure that the plant operators understand it. Install a permanent placard with a diagram of the heat recovery system near the system controls. Placard the control valves and the controls.*

MONITOR PERFORMANCE: *Schedule periodic checks of heat recovery operation. Include a column in the chiller plant operating log to record the liquid discharge temperature from the desuperheater.*

MEASURE 2.10.5 Improve the quantity or economics of heat recovery by adding or increasing heat storage.

In many potential applications for chiller heat recovery, the heat is available when it is not needed, and vice versa. In such applications, you can make recovered heat available when it is needed by installing heat storage as part of the heat recovery system.

You can use heat storage with any of the other Measures of this Subsection. The storage "technology" consists of an insulated tank. Liquid heated by the chiller condenser is stored in the tank. If that application already involves a tank, such as a domestic water heating tank, you can increase the storage effect by making the tank larger, or by adding another tank. Storage usually does not require any additional heat exchangers. The only energy penalty is heat loss from the storage tank, and perhaps the operation of a small additional circulating pump.

With some combinations of cooling and heating load profiles, using heat storage can reduce the size of the heat recovery chiller. This is important if you want to recover heat at an elevated condensing temperature. (Measure 2.10.3 explains why.)

Energy and Cost Saving Potential

The cost of heat storage equipment is high in relation to the value of the energy that can be stored at any one time. In order to be economical, the storage equipment must be needed to charge and discharge frequently, and to discharge fully during most cycles.

■ Applications that Benefit from Heat Storage

Heat storage for chiller heat recovery provides a benefit only if these timing relationships exist between the cooling load and the heat recovery application:

- there are periods of time when the amount of recoverable heat exceeds the immediate need for heat
- there are also periods of time when the need for heat cannot be satisfied by the available recovered heat
- these two periods alternate with each other. Recovered heat must be used fairly soon after it is stored. The stored heat continuously leaks out of the storage tank by conduction. With typical storage tanks, losses become excessive in time intervals ranging from one day to several days.

Figure 1 illustrates one typical day of operation with heat storage. On this day, heat storage is able to increase heat recovery only by a modest amount. By the same token, the storage capacity that is required to serve the heat recovery needs during this day is much less than the amount of heat that is recovered.

SUMMARY

Increases the amount of heat that can be recovered in applications where the cooling load is out of phase with the heat requirement on a daily cycle. May be able to reduce the size of the heat recovery chiller. The mechanical installation is fairly simple. Control may be complex if heat is recovered at elevated condensing temperatures.

SELECTION SCORECARD

Savings Potential	$	$	$
Rate of Return, New Facilities	%	%	
Rate of Return, Retrofit	%	%	
Reliability	✓	✓	✓
Ease of Retrofit	☺	☺	☺

■ An Example of Savings

Let's look at an application in which condenser heat is used to preheat domestic water from a supply temperature of 55°F to an average final temperature of 85°F. The storage tank has a capacity of 10,000 gallons, and it costs $10,000 to install.

Each gallon of water stores about 240 BTU, so the entire tank stores about 2.4 million BTU. If fuel-fired water heating costs $5 per million BTU, then the value of the energy stored in the tank is $12. In order to repay its cost, the storage tank must discharge completely 834 times ($10,000 divided by $12). The storage system operates on a daily cycle, and it discharges completely an average of 200 days per year. Therefore, the storage system breaks even about four and a half years after it goes into operation.

This example does not include a deduction for any additional pump energy required. Also, it assumes a very regular water usage pattern, which is unusual.

■ Effect of Electricity Demand Charges

Electricity rates do not affect your decision to use heat storage if you recover the heat at normal condensing temperatures.

If you increase the condensing temperature to recover more heat, this reduces the chiller COP and increases the demand cost of operating the chiller. Therefore, it may not be economical to recover heat during the time of day when electricity demand rates are high. (See Reference Note 21, Electricity Pricing, for an explanation of demand charges. They are becoming an increasingly large part of electricity cost.)

Heat storage lets you operate the heat recovery chiller during hours when demand rates are low. However, this works only if there is enough cooling load at those times to produce the desired amount of heat. For example, this situation may occur in a hospital, but is unlikely to occur in an office building.

Electricity demand charges typically are seasonal. If you want to use heat recovery outside the months of highest electricity prices, then the effect of demand charges is less of an impediment to heat recovery during the daytime.

How to Estimate Storage Capacity

The chiller load varies from day to day, so estimating the most economical storage equipment capacity may be tedious. You need a series of graphs, similar to Figure 1, for your typical daily load profiles. From the number of days of each type, make a graph of the total annual heat recovery potential versus storage capacity and cost. From this, decide on the heat storage capacity. This is usually a compromise. The total savings increase as you increase capacity, up to the useful maximum, but the payback period gets longer.

An appropriate computer program may reduce the drudgery of this calculation, and it can model expected loads in new installations. See Reference Note 17, Energy Analysis Computer Programs, for details.

In retrofit applications, you have the advantage that you can make your estimates with real data. Use a chart recorder or digital data logger to measure these profiles. Metering the chiller motor wattage is usually the easiest way to track the chiller load. You may have to install a flow meter to measure the usage profile of hot water or other material that is to be heated.

The Heat Storage Equipment

The heat storage unit typically is a tank that has some additional hardware. The most important aspect of the tank design is keeping the heated liquid separate from the unheated liquid. Refer to the Measure 2.11.1 for the ways that this is done in chilled water storage. The same methods apply to storing heated water.

The simplest, cheapest, and most common method of storing heated water is by exploiting the natural tendency of water to stratify in a tank. Compared to chilled water storage, hot water storage has the advantage that the change in water density with temperature is much greater, providing more reliable stratification. For example, water expands four times more in warming from 80°F to 100°F than it does in warming from 40°F to 60°F. Even so, be sure that the top and bottom connections are designed to limit water velocity and vertical mixing. This typically requires diffusers inside the top and bottom of the tank.

The major disadvantage of stratified storage is conduction of heat from the warmer water to the cooler water inside the tank. If water stands still inside the tank, the temperature of the water eventually becomes uniform. To maximize stratification and minimize the conduction problem, make the storage tank tall in relation to its width. Do not use a horizontal tank if you are depending on stratification. (A horizontal tank with a series of vertical siphon baffles is a more complicated method of storage.)

In some applications, you can exploit storage capacity that already exists. For example, if the

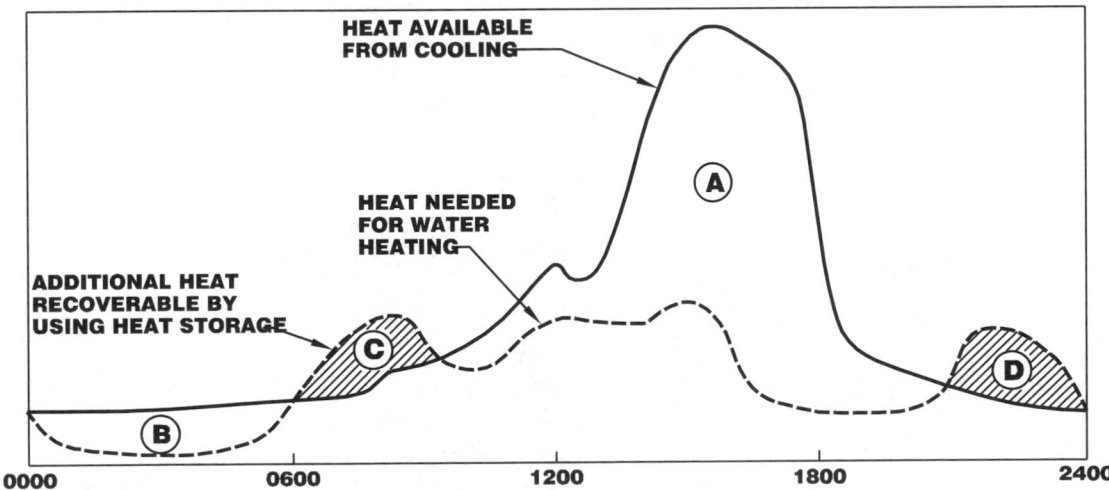

Fig. 1 Example of heat storage These are profiles of chiller heat rejection and water heating energy requirements for a large hotel on a typical day. By using heat storage, some of the excess heat produced during period "B" can be saved for use during period "C." Similarly, excess heat produced during period "A" can be saved for use during period "D." The required storage capacity is the larger of "C" or "D." To determine the storage capacity that provides the best economics, you need to make similar curves for all operating conditions.

application is fuel oil tank heating, the mass of the fuel oil itself acts as a storage medium.

Typical Domestic Water Preheating Systems

Domestic water preheating with chiller heat recovery is fairly simple. Figure 2 shows a typical installation. The heat exchanger recovers heat from the leaving condenser water, and a pump circulates water between the heat exchanger and the heat storage tank. Hot water enters or leaves the tank at the top, and cold water enters or leaves the tank at the bottom. An ordinary domestic water heater is used for final heating.

The final heating tank is typically much smaller than the storage tank, because it is sized only to provide for the maximum hot water usage rate. You can keep it small by increasing the capacity of the heater. But, if you use electricity for final heating, try to limit electric heating to periods of low electricity prices. This may require a larger final heating tank.

Control for this type of system is simple. In this example, the heat exchanger circulation pump runs whenever the condenser water entering the heat exchanger is more than a few degrees warmer than the temperature at the top of the storage tank. This single control performs these important functions:

- stops the pump when the chiller turns off. Otherwise, the flow of cooling tower water through

the heat exchanger would cool down the storage tank.
- stops the pump when the storage tank is fully charged
- prevents diluting the tank with colder condenser water if the condensing temperature is lowered to improve the COP after the tank is fully charged.

Typical Recirculating Heating Systems

If the application is a recirculating heating system, the heat storage layout is generally similar to the arrangement for domestic water heating. Although the layout is simple, some control complications are needed to match the characteristics of the storage equipment to the characteristics of the heating equipment. The design issues are similar to those described in Measure 2.11.1 for cooling with chilled water storage.

In brief, the ability of the storage unit to provide heating energy is determined by the peak temperature of the heating system, and by the differential between the heating water supply and return temperatures. A major obstacle is that most heating systems operate at temperatures higher than typical heat recovery temperatures. To bring the two temperatures together, it is necessary to lower the system heating temperature and to raise the chiller heat recovery temperature. This complicates the design of the heating system, invites comfort problems, and sacrifices chiller COP.

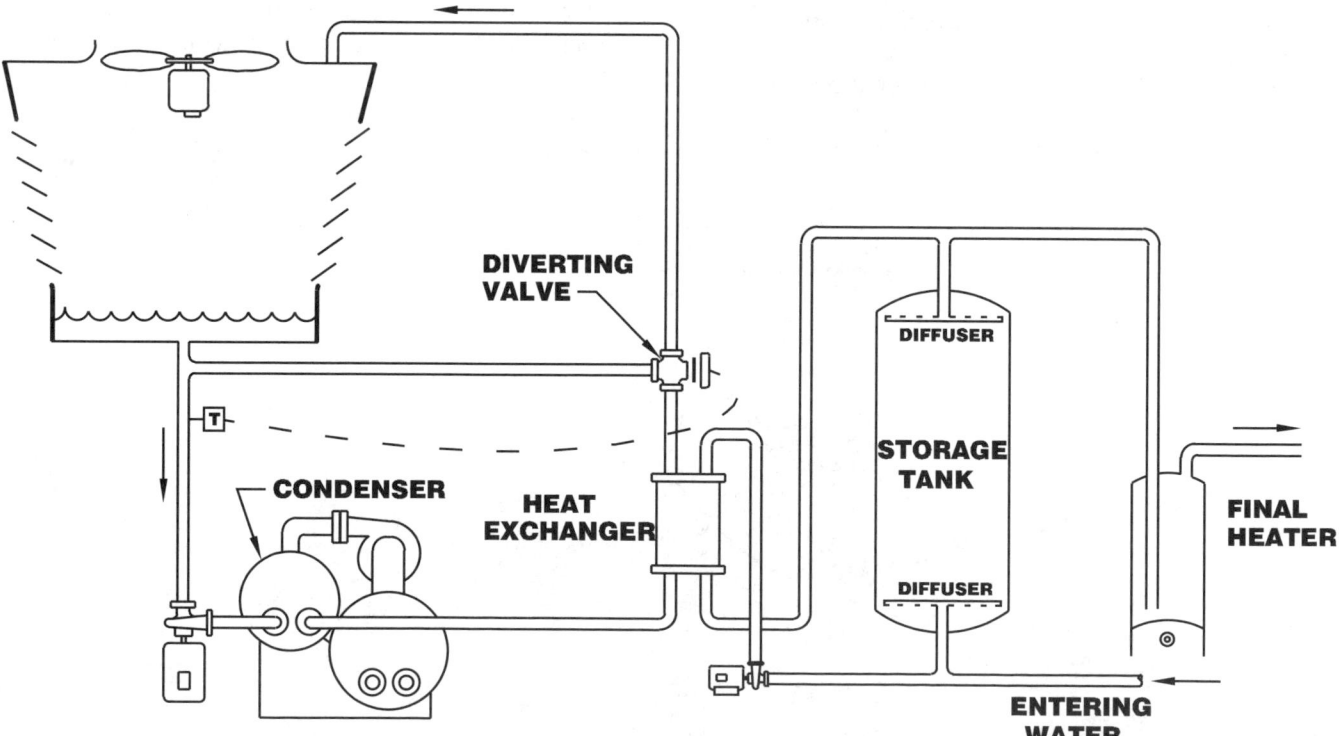

Fig. 2 Heat storage for domestic water heating   A small pump circulates water from the heat exchanger into the top of the storage tank, which depends on stratification to keep the warmest water near the top. Diffusers are installed at the top and bottom connections to preserve stratification.   The pump operates only until the storage tank is filled with enough hot water to satisfy needs during the next discharge cycle. A diverting valve is installed in the condenser water line to avoid losing heat to the cooling tower when maximum heat recovery is desired.

Another obstacle is the very large storage volume needed to serve a typical space heating application. The amount of heat that can be stored in a given volume of water is directly proportional to the temperature differential in the heating system. To minimize the storage volume, design the heating system to maintain a large temperature differential between the supply and return water. This requires special heating coils, 2-way coil control valves, and variable-flow pumping, as in chilled water storage systems.

Balancing Storage Capacity with Heat Recovery Temperature

The amount of heat that can be stored in the form of hot liquid is directly proportional to the temperature differential that can be stored. Therefore, recovering heat at higher temperature reduces tank size and cost, even if the storage temperature is higher than the application requires. On the other hand, increasing the heat recovery temperature reduces the chiller COP. You need to calculate an optimum compromise between heat recovery and cooling efficiency.

Heat storage is most economical when it is used with a chiller that is optimized for heat recovery. The benefit is mutual. The higher recovery temperature of the chiller increases heat storage, and heat storage increases the utilization of the heat recovery chiller.

Minimize Operation at Increased Condensing Temperature

If you want to increase the condensing temperature at certain times to increase heat recovery, install time controls or other automatic controls to limit the duration of operation at elevated condensing temperature. These controls simply switch between two sets of condenser thermostatic controls, one for normal operation and the other for heat recovery operation.

The controls should limit operation at elevated condensing temperature to the minimum needed to provide for the estimated heating needs of the next heating cycle. If heat usage depends on weather, the day of the week, or other variables, intelligent logic is needed to estimate the heat requirement. Computer controls are available for this function, but manual input may still be needed to account for weather changes that the computer cannot foresee.

If demand charges are a factor in scheduling heat recovery, use either time controls or demand controls to minimize condenser temperature during periods of high demand charges.

If you install a specialized heat recovery chiller to operate at higher condensing temperatures (Measure 2.10.3), you need additional controls to distribute the load between the chillers to optimize overall plant performance.

Combining Heat Storage with Solar Collectors

In locations that are suitable for solar collectors, consider using them as a source of energy to charge the heat storage system. Combining the two may be favorable for these reasons:

- *complementary time schedules.* On a seasonal basis, solar collectors can continue to produce heat during cooler weather, when the chillers are operating at reduce load, or are not operating at all. Even during the cooling season, electricity demand charges make it undesirable to operate chillers at increased condensing temperature during peak daytime hours. Solar collectors are a daytime heat source, so they can increase the temperature of the liquid after initial heat recovery.

  If the system uses cooling thermal storage (Measure 1.11.1), the chillers do not operate during periods of high demand charges, which are typically summer days. Solar collectors are most effective at these times.

- *compatible temperatures.* Solar collectors can boost the temperature of water heated by the chiller condensers. Flat-plate collectors produce higher temperatures up to about 140°F, at useful flow rates. Focussing reflectors can produce temperatures that are much higher.

- *shared equipment costs.* The storage tank and piping of a heat storage system can also serve as part of an active solar system. Active solar systems have poor economics by themselves, largely because of the cost of these components. This is a way to get an active solar system at substantially reduced cost.

See Reference Note 23, Non-Fossil Energy Sources, for more about solar energy systems.

ECONOMICS

SAVINGS POTENTIAL: *The saving consists of the additional heat that is usefully recovered by virtue of having the storage system.*

COST: *Insulated steel storage tanks cost about $1 per gallon in large sizes, up to about $4 per gallon in small sizes, including a moderate amount of piping. This translates to $4 to $20 per thousand BTUH stored, in typical applications. If storage is used to reduce the capacity of a new heat recovery chiller, the net cost of the storage equipment may be greatly reduced.*

PAYBACK PERIOD: *3 to 20 years, depending mainly on energy costs and the number of discharge cycles per year.*

TRAPS & TRICKS

CAREFUL DESIGN: *To calculate the optimum amount of storage capacity, make a range of estimates for* reasonably foreseeable conditions. Design the storage tanks for effective separation of the heated water from the unheated water. Pay special attention to the control strategies if heat is recovered at elevated condensing temperature, or if final heating is done with electricity.

EXPLAIN IT: Put a clear explanation of the storage tank in the plant operating manual as part of the explanation of the overall heat recovery system. Install a placard on the storage tank.

MONITOR PERFORMANCE: Include a column in the chiller operating log to record storage tank temperature. Review the logs regularly to check that the system is operating properly.

Cooling Thermal Storage

Cooling storage is an old technique that was originally employed to allow small cooling equipment to serve large cooling loads for short intervals. After a long period of dormancy, cooling storage experienced an explosion of interest in the 1980's that continues to the present. This time, the primary interest is in shifting the cooling load to times of day when electricity rates are lower. The difference in electricity prices between different hours of the day continues to widen. This suggests that cooling storage will continue to grow in popularity.

After a rocky start, storage equipment is becoming more reliable. Several types continue to evolve, and new types occasionally appear. The growing pains have not ended yet. Although the concept is simple, cooling thermal storage is a broad subject with many complications. A chiller system with cooling storage is a challenge to design, to install, and to operate. Many systems have failed. The initial phase of cooling storage has yielded a wealth of experience that can guide the designer who is wise enough to learn from the past.

This Subsection contains a single Measure that covers the contemporary applications and variations of this technology. It organizes the subject for you, and incorporates the important lessons of experience. It stresses planning and analysis, which were often neglected in enthusiastic haste to install systems. It will help you to decide whether to use cooling storage, and it will point you toward the best equipment for your application. Expect to do a lot of additional study and investigation before making a commitment to installing a system.

INDEX OF MEASURES

2.11.1 Install cooling thermal storage.

RELATED MEASURES

- Measure 2.10.5, storage of heat recovered from chillers

MEASURE 2.11.1 Install cooling thermal storage.

This Measure helps facility owners to decide whether to acquire cooling storage, provides an outline for the designer of a cooling storage system, and guides facility staffs in operating cooling storage systems.

Deciding whether to install cooling storage is complex. Stated most generally, cooling storage is desirable in applications where the opportunities and benefits outweigh the disadvantages. Both the benefits and the disadvantages are explained below. Following that is a sequence of steps that lead you from the initial decision to consider cooling storage to successful operation of the installed system. This is followed by a comparative review of all contemporary types of cooling storage equipment. At the end is an introduction to the major engineering issues.

Overview

A cooling thermal storage system is like a conventional chiller system, with the addition of a large container that stores cooling in ice, chilled water, or some other material. A cooling storage system is analogous to the electrical system of an automobile. The chiller is like the automobile's generator, and the cooling storage unit is like the automobile's battery. At different times, the cooling load of the facility may be served by the chiller directly, by the cooling storage unit, or by both. (Not all systems can operate in all three of these modes.)

Cooling storage has two main advantages over a system without storage. One is that cooling can be available on any desired schedule, independently of the operation of the chillers (within limits). The other is that the cooling storage unit may be able to deliver cooling at a higher rate than the chillers, or to supplement the chillers.

Cooling thermal storage is not an energy conservation measure in the strict sense. On the contrary, all forms of thermal storage involve losses, and they require additional energy for the operation of the system. Some forms of cooling storage make the chillers operate less efficiently. Thermal storage exists primarily for the benefit of the electric utility. It allows the utility to generate more of its electricity with its most efficient generators. It may also allow the utility to generate electricity with fuels that are less scarce or less critical.

From the standpoint of the facility owner, thermal storage is a means of purchasing electricity at lower rates. The electric utility rewards the facility owner who uses cooling storage by offering lower electricity rates. All parties should understand that this is a business arrangement, not a matter of engineering. The electric

SUMMARY

Currently popular as means of purchasing electricity at lower rates. Sometimes used in new plants to reduce chiller cost. Bulky, expensive, and complex. Reliability has been poor, but performance will probably improve with time. Will not operate properly without continuous monitoring. Ice storage, the most common method, seriously reduces chiller efficiency. Economic feasibility depends largely on long-term considerations that are not under the user's control.

SELECTION SCORECARD

Savings Potential	$ $ $ $
Rate of Return, New Facilities	% % %
Rate of Return, Retrofit	% %
Reliability	✓ ✓
Ease of Retrofit	☺ ☺

utility attempts to persuade the facility owner to use cooling storage for its own purposes. In turn, the facility owner must have enough savvy to negotiate the best rates with the utility.

Less frequently, the facility owner may use cooling storage as a way of reducing the chiller capacity that is needed for a new facility. By itself, this yields an economic benefit for the owner only if the cooling load is concentrated in short time periods. In many applications, cooling storage can provide both types of benefit.

At the present time, cooling storage is partly a fad among facility owners who fail to understand the underlying purposes of the technology, and the business motives behind it. Be especially cautious that cooling storage will serve your purposes in the long term. Cooling storage requires the facility owner to understand the motives and needs of the electricity utility company, to understand the theory of electricity pricing, and to be able to negotiate effectively with the electric utility.

Cooling storage has long been used successfully in a variety of specialized cooling applications. It cannot yet be considered a success in air conditioning applications, where it has experienced a large number of failures. These have been caused by deficient design, installation flaws, neglected maintenance, and failure to understand the logic of the systems' operation.

Cooling storage will undoubtedly become more successful, both technically and economically, as experience accumulates. Even so, cooling storage will always be risky. Long-term success depends largely on future conditions that cannot be predicted reliably and that the user does not control.

Benefits of Cooling Storage

The following are the benefits provided by cooling storage. The first two are the main reasons that cooling storage is usually installed. The remaining benefits are ancillary. They may improve the economics of cooling storage, but they generally do not justify it by themselves.

■ Reduced Electricity Cost

At the present time, most cooling storage is installed to reduce electricity costs. Review Reference Note 21, Electricity Pricing, for an understanding of the electricity pricing issues that are involved. In brief, the price of electricity is related to the time period when the electricity is purchased. Electrically powered chillers typically account for a large fraction of a facility's energy consumption, and cooling storage allows the operation of the chillers to be shifted to time periods when electricity is cheaper.

The key issue in reducing electricity costs is the "demand" charge, which varies with the time of day, the day of the week, and the season. Make sure that you are comfortable with this concept before studying the following examples. In brief, the demand charge is a charge based on the maximum *rate* of energy consumption (measured in kilowatts), not for the *amount* of electricity consumption (measured in kilowatt-hours). In the following examples, the objective of cooling storage is the minimize the demand charge.

Figure 1 shows how cooling storage can reduce electricity cost in a facility where the non-cooling load (lights, air handling units, computers, etc.) are essentially constant during the period of high demand charges. For example, this is typical of office buildings and retail stores. In this example, cooling is provided by storage, rather than by chiller operation, during the periods of higher demand charges. There are three demand charge periods during the day, and the objective is to operate the chillers only during the period of lowest rates.

The cooling load to be avoided during periods of higher demand charges is marked "A." The cooling storage unit can be charged at any time during the off-peak rate period, but preferably during the coolest hours of the night (to maximize chiller efficiency). This cooling load is marked as "B." During off-peak periods, the chillers can also serve the cooling load directly, if a cooling load exists then. Direct cooling is more efficient than cooling with storage, as we shall see, so we try to use direct cooling as much as possible during the off-

peak period. In our example, this direct cooling load is marked "C."

Figure 2 shows how cooling storage can reduce electricity demand charges in a facility where the non-cooling load fluctuates widely during the demand charge period. For example, this is typical of colleges and manufacturing facilities. In this example, cooling storage is controlled so that it satisfies the portion of the cooling load marked "A." The purpose of this is to limit the maximum electric load during the peak demand charge period. The cooling storage unit is charged during off-peak hours. As in the previous example, this load is marked "B." Again, direct cooling is used during off-peak hours to the maximum extent possible. The off-peak direct cooling load is marked "C."

Unlike the example in Figure 1, direct cooling is used for a portion of the cooling load during the period of higher electricity rates. This is because the demand charge during this period is determined by the highest value of the non-cooling electrical load, which occurs at 6 PM in this example. This method of using cooling storage is called "peak shaving." In this case, the cooling unit can be charged at any time, provided that the cooling load never increases the demand above that determined by the non-cooling loads.

Demand charges typically are established on a monthly or seasonal basis. In Figure 2, the non-cooling demand that is shown represents the highest demand during the measurement period, not the highest daily demand, which may be much lower.

Don't blindly follow the example of Figure 2 when you are considering cooling storage for peak shaving. The electricity rates that you pay may have an important additional complication. Some utilities penalize high demand not only by a demand charge, but also by increasing the kilowatt-hour charge during periods of higher demand rates. This makes it desirable to shift as much of the cooling load as possible to the off-peak rate period, regardless of the demand profile of the non-cooling loads.

Reducing demand benefits the electric utility, so many utilities presently offer incentives for the owner to install cooling storage equipment. These incentives typically take the form of partial rebates of the cost of installing cooling storage. Where such rebates are offered, they may be a major factor in the overall economics. Unfortunately, such rebates tend to dazzle facility owners, making them careless in assessing the overall cost of the project or the long-term operating liabilities. It is foolish to burden a facility with inappropriate equipment for the sake of a rebate.

■ Ability to Reduce Chiller Capacity in New Plants

If you are building a new chiller plant, you can use cooling storage to reduce the chiller capacity, saving overall chiller plant cost. The price of storage equipment is only about 10% to 20% of the price of conventional chiller equipment, per unit of cooling capacity. Therefore, storage may reduce equipment costs. However, the cooling load in your application must occur during predictable intervals that leave ample time for charging the storage unit. For example, cooling storage has been used to reduce the size of air conditioning equipment in churches, which need a large cooling capacity for only a few hours per week. Similarly, cooling storage has long been used in dairy operations, where rapid cooling of a large volume of milk is required for a few hours each day.

Figure 3 shows how cooling storage can be used to reduce chiller capacity. Here, the cooling load pattern is typical of an office building. In this example, the chiller is controlled on weekdays so that it operates at full output until the cooling storage unit is fully charged, at which time the chiller throttles back to handle the current cooling load. This method of control avoids the possibility of running out of cooling capacity during weekdays, although it may keep more energy in storage than necessary. During the weekend, the chiller does not store cooling until late on Sunday. This minimizes losses from the cooling unit. (Other possible control approaches are covered below.)

In applications where cooling storage can reduce chiller capacity, storage is likely to reduce electricity costs as well. Design the system to get the greatest combined benefit. In the previous example, you might wish to increase the size of the chillers and increase the storage capacity so that the chillers can be turned off during periods of higher electricity rates.

■ **Ability to Exploit Lower Condensing Temperatures**

Cooling storage operation usually shifts the operation of chillers into the night, when the outside air is considerably cooler. If the chillers are able to reduce their condensing temperatures to follow the night air temperature, their average COP improves considerably. (Many chillers cannot exploit reduced condensing temperature. See Measure 2.2.2 for the reasons.)

This is a major bonus that tends to be overlooked. With chilled water storage, the chillers may actually be able to operate with higher net COP than in conventional daytime cooling. With ice storage, the reduced condensing temperature may substantially reduce the COP penalty caused by the low evaporating temperatures needed for freezing.

■ **Reserve Capacity in the Event of Chiller Failure**

Cooling storage buys time for the repair of a failed chiller, provided that the chiller fails when the storage unit has a substantial remaining charge. However, this bonus is offset by the fact that the storage system itself increases the probability that a failure will occur in the chiller system.

■ **Smaller Water and Air Distribution Components**

In attempting to make a virtue of the lower evaporator temperatures required for ice storage, systems are being developed that exploit the lower chilled water temperatures. The main advantage of these systems is smaller size for chilled water and air distribution equipment. These systems are a mixed blessing. They are discussed below.

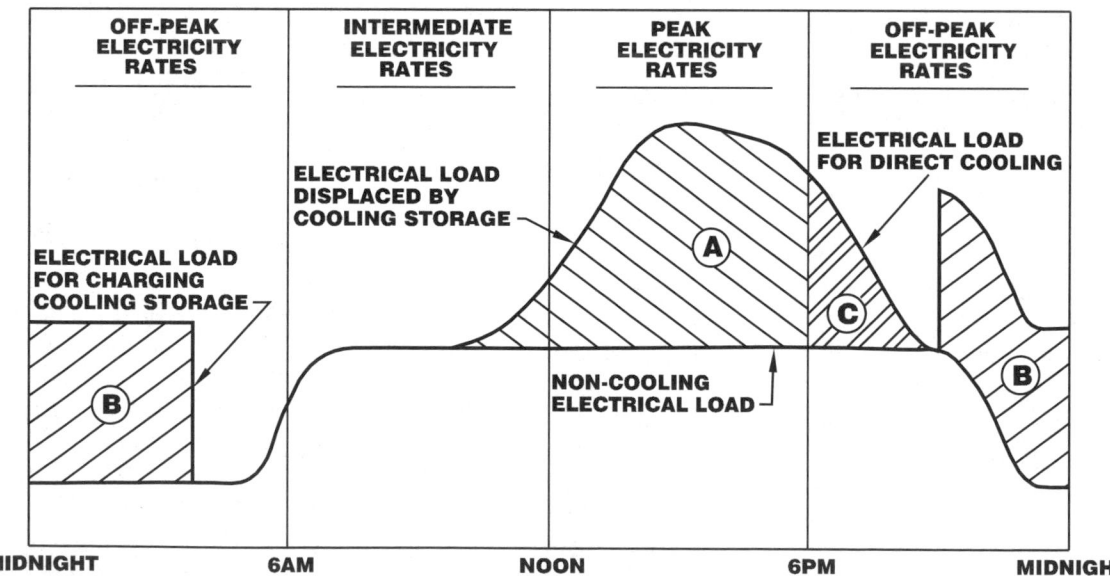

Fig. 1 Using cooling storage to minimize electricity cost (case 1): the electrical loads other than cooling are fairly constant during the period of high demand charges The load profiles are typical of an office building. This situation favors providing enough storage capacity so that all chiller operation occurs during off-peak hours. This method of control is simple and reliable.

■ Compensating for Capacity Loss with Replacement Refrigerants

Some common refrigerants used in the past must be replaced to satisfy environmental concerns. (See Reference Note 34, Refrigerants, for details.) Converting existing chillers to use new refrigerants has a capacity penalty, typically 10% or less. Most chiller plants have enough reserve capacity to make up for the lost capacity. If not, you can use cooling storage to make up for the loss. However, before you do this, look for energy conservation measures to make up the difference. (You should exploit most of your other energy conservation opportunities before getting involved with cooling storage.)

■ Increasing Heat Recovery

Cooling storage may improve the potential for recovering condenser heat from chillers. This is because cooling storage causes the chillers to operate at steady loads for longer, predictable periods of time. On the other hand, cooling storage may make heat recovery more expensive by eliminating chiller operation during periods of time when recovered heat is needed. Refer to Subsection 2.10 for details of chiller heat recovery.

Ice storage reduces chiller COP by lowering the evaporator temperature. Similarly, if you increase the condensing temperature to enhance heat recovery, this also reduces the COP. Therefore, you usually don't want to do both at the same time. On the other hand, a chiller that is designed for ice storage may also be effective for

heat recovery at elevated temperatures when it is used for direct cooling of chilled water.

Under some conditions, you can increase chiller heat recovery by installing a heat storage unit. See Measure 2.10.5 for details. In some of these cases, which are discussed below, you can use the same storage unit for both heating storage and cooling storage.

Disadvantages of Cooling Storage

Cooling storage illustrates the maxim that engineering is compromise. Cooling storage systems can be designed in many ways, each with its own advantages and disadvantages. The following are the main disadvantages of cooling storage, which vary in importance among the different types.

■ Energy Losses

Cooling storage is not an energy conservation measure. Storage systems consume more energy at the facility than conventional cooling. In the least favorable applications, the increase in electricity consumption would negate the advantage of lower electricity prices. These are the main energy losses:

- *reduced chiller efficiency,* especially with ice storage. Ice storage pays a high energy efficiency penalty because the chillers need to operate at low evaporating temperature in order to freeze the ice. This lowers the chiller COP from 30% to 50%, compared to water chilling, for a given condensing

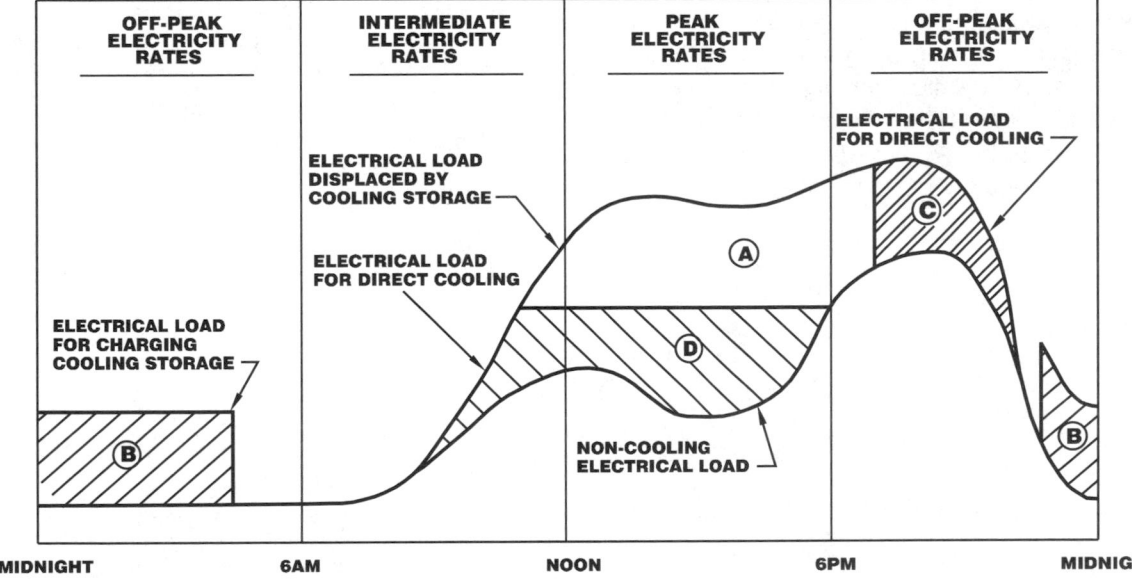

Fig. 2 Using cooling storage to minimize electricity cost (case 2): the electrical loads other than cooling vary widely during the period of high demand charges The load profiles are typical of a college campus. In this situation, there is no cost penalty if the chillers operate during the hours of high demand charges, but only up to a maximum total demand that is determined by the non-cooling electrical loads. (This part of the cooling load is marked "D.") This method of control allows the cooling storage unit to be smaller, and it exploits the fact that direct cooling is more efficient than storage cooling. However, you have to guess what the non-cooling loads will be. The chiller loading decision becomes more complicated if the kilowatt-hour charge (commodity charge) for electricity changes with the time of day.

temperature. Even storing chilled water as a liquid requires lower evaporator temperature.

A new chiller that is designed for both cooling storage and direct cooling of chilled water has reduced efficiency in both applications. Existing chillers may be especially inefficient if they are converted to ice storage, or they may not work at all.

The loss of chiller efficiency is partially compensated by the fact that the chiller spends more time operating at night, when the condensing temperature can be lower.

• *conduction loss.* All thermal storage loses energy by conduction from the storage unit. Storage systems for larger commercial applications typically lose several percent of their energy charge per day. The relative energy loss decreases as the storage vessel becomes larger. This is an important factor that weighs heavily against smaller storage systems.

• *loss of residual stored energy.* Energy that is left over in the storage unit at the end of an operating cycle suffers higher storage losses. Most types of ice storage require the storage unit to melt all ice completely at the end of the storage cycle. This energy is lost if it cannot be used during the off-peak period.

• *auxiliary equipment energy consumption.* Cooling storage also requires additional energy to put the cooling energy into storage and take it back out. This usually takes the form of energy needed to operate additional pumps. Depending on the type and design of the storage system, the additional pump load may be minor, or it may be several times greater than conventional chilled water pumping.

From a global perspective, the energy efficiency picture becomes more favorable when you consider the reasons why electric utilities promote cooling storage. Utilities promote this technology primarily to delay expensive and troublesome power plant construction, and to generate electricity with energy sources that are less expensive and less critical. It also allows utilities to minimize the operation of less-efficient generators, especially gas turbines, to handle peak loads.

Delaying power plant construction also delays the large expenditure of energy that occurs in construction. However, this advantage is lost after several generations of plant construction, because later plants must be increased in size to provide the additional energy used by the early generations of storage systems.

■ Space Requirements

A large amount of space is needed by the storage container and its accessories. The volume of the storage container is roughly proportional to the volume of the space being cooled. Chilled water storage requires about 20 to 30 cubic feet per ton-hour, ice storage requires about 3 to 4 cubic feet per ton-hour, and eutectic salts require a storage volume intermediate between the other two types.

For example, a typical office building with a floor area of 100,000 square feet, operating on a daily storage cycle, requires a minimum of about 3,000 cubic feet of ice storage, or about 100,000 gallons of chilled water storage.

■ Operation and Maintenance Burden

Cooling storage technology burdens the operating staff with additional complexity. The storage equipment is not particularly mysterious, but the control needed to optimize the operation of the system is somewhat subtle. Even if cooling storage technology were mature, many facilities would be unable to maintain and operate it reliably.

Another problem is continuity of operation. Unless you negotiate special terms with the utility, you will

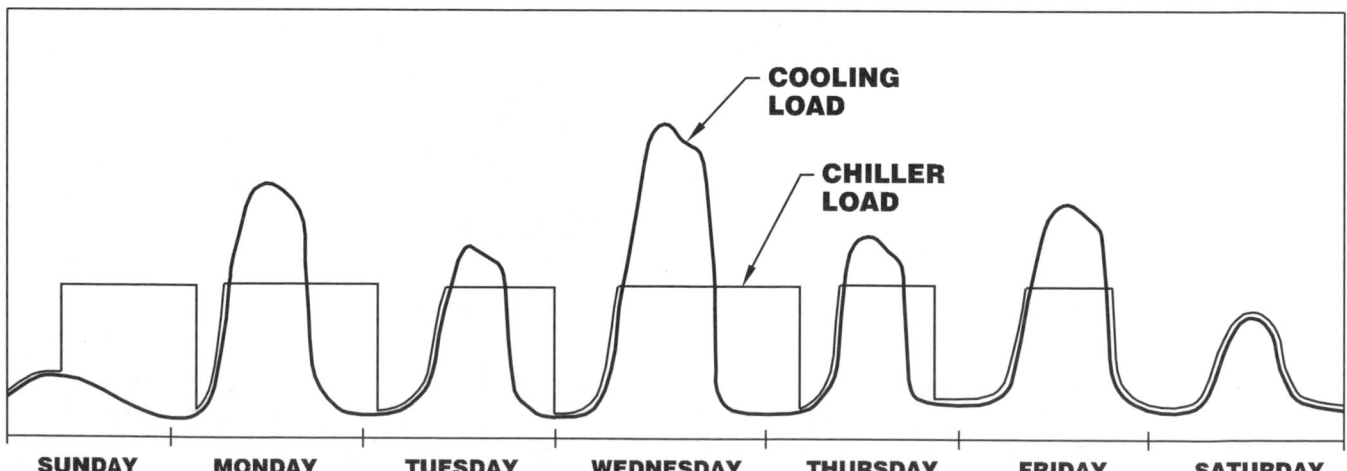

Fig. 3 Using cooling storage to reduce chiller size The load profile is typical of an office building. During weekdays, the chiller always operates at full load, long enough to keep the storage unit at full charge. On weekends, the chiller does not charge the storage unit until Sunday afternoon, to minimize storage losses. During months when the cooling load is lower, the chiller could be operated at lower output to reduce demand charges.

forfeit all or most of your savings in electricity prices if the plant fails even once during the high-demand period. Skilled staff must occupy the plant continuously during the months when peak demand occurs, and they must be able to immediately correct any failure of the storage system.

As a rule, do not install cooling storage in a facility where the system cannot be monitored continuously and repaired immediately. You may be able to relax this requirement somewhat if you can negotiate certain arrangements with the electric utility that are described below.

■ **Economic Risk**

Any saving in electricity cost that is provided by cooling storage is based entirely on the pricing policy of the electric utility. The willingness of the utility to continue providing favorable pricing will depend on the benefit that cooling storage provides to the utility. Both will change with time, depending on the factors explained in Reference Note 21, Electricity Pricing.

Several growing issues presently place the commitment of utilities to thermal storage in greater doubt. One is utility deregulation, which eliminates the marriage of individual customers to a particular utility. This may kill utility support for thermal storage, except for the incentive provided by current rate schedules. Another issue is the renewed interest in electric vehicles. If electric vehicles become widespread, they will charge their batteries mostly during off-peak periods, eliminating the need for thermal storage to level the load.

To reduce the economic risk to reasonable proportions, you need to negotiate a long-term rate contract, as outlined below. If the utility will not agree to an acceptable contract, it would be economic folly to invest much money in cooling storage.

HOW TO APPROACH STORAGE COOLING

Although the concept of cooling storage is simple, success requires the proper execution of many steps, each of which may involve a great deal of detail and special expertise. The following are the main steps in the development of a cooling storage system, from conception to birth.

Step 1: Clarify Your Motives

Many cooling storage plants have been installed as a fad, or to reap utility incentives. Plants built for such reasons alone are likely to fail. Before investing substantial effort or expense, consider all the potential benefits and disadvantages discussed previously. Decide whether the potential economic saving is worth the long-term operational commitment.

Cooling storage systems are expensive. Decide whether the initial investment and the long-term operating cost might better be invested in other cost saving measures, or in improving your product, or in the stock market. Don't get preoccupied with cooling storage until you have exploited all the possibilities for reducing your energy costs that are less expensive and more reliable.

Step 2: Consider the Facility's Ability to Operate Cooling Storage

A properly designed cooling storage plant does not require a great deal of extra labor to operate. However, it does require an operating staff with a high level of specialized skill and the ability to monitor the storage system continuously. If the plant fails for even a short time during the peak demand period, the utility rate advantage may be lost for a month, a cooling season, or an entire year.

Step 3: Develop the Facility's Cooling Load Profiles

The potential saving in electricity costs is determined by the peak cooling load that occurs during the utility's high billing periods. The capacity of the cooling storage system is determined by the duration of the peak cooling periods. High cooling loads of short duration provide the greatest saving in relation to cost, whereas high cooling loads that last all day require larger storage capacities to gain the same benefit.

In new construction, you need a sophisticated computer program to estimate future load profiles, as well as for analyzing the performance of candidate storage systems. See Reference Note 17, Energy Analysis Computer Programs, for details.

In existing facilities, get the detailed load profiles for several past years and study them. You will need to contact the electric utility for this information, unless you have an in-house energy monitoring system that provides sufficient detail. Even the electric utility may not have the historical load information you need. In that case, either set up your own metering, or use a computer simulation program.

The load profiles help you to gauge the likelihood that the utility will maintain a long-term commitment to cooling storage. If only a few days of heavy cooling load occur each year, as in Minnesota, for example, it is cheaper for the utility to operate peaking generators to cover these loads than to maintain support for cooling storage. In such locales, a utility's interest in promoting cooling storage may be transient, making future operation of storage plants uneconomical. On the other hand, in a location such as Florida, where high cooling loads occur for many days each year, the utility is more likely to continue its interest in cooling storage.

If you are installing a new chiller plant, the simulated load profiles will tell you how much you can reduce your chiller capacity by using cooling storage. The greatest savings in chiller capacity are provided by cooling loads that are sharply peaked.

Step 4: Estimate the Long-Term Saving in Electricity Cost

Estimate the potential saving in electricity cost from the cooling load profiles and from the utility's rate schedules. See Reference Note 21 for guidance in understanding rate schedules. The Big Question is whether the lower electricity rates for cooling storage will outweigh the increased consumption caused by reduced chiller efficiency. The Big Challenge is to estimate how these rates will change over the life of the cooling storage system.

It is foolhardy to assume that present electricity prices will continue, or even that they will change gradually. To convince yourself of this, review the pattern of electricity prices over the past twenty or fifty years. You can guess future rates intelligently only by assessing the utility company's future. This requires a crystal ball. The utility industry itself, with staffs of experts, has proven to be wildly wrong in predicting future conditions, even over time periods as short as a decade.

Furthermore, in the United States and some other countries, the structure of the utility industry is changing from a regulated environment to an environment in which some parts of the industry are regulated and others are not. There will be a period of chaos in which the long-term economic benefit of cooling storage is even more unpredictable.

To establish the range of possibilities, calculate for a best case and a worst case. If cooling storage looks potentially attractive, this exercise will help you to prepare for the rate contract negotiations recommended below. If cooling storage looks unrewarding or too risky at this point, forget about it and spend your money on reliable energy conservation measures. Unfortunately, the uncertainties of predicting the future are such that the best case is likely to make cooling storage look advantageous, while the worst case is likely to make cooling storage look unfeasible.

Step 5: Investigate Utility Ownership and Operation of the Cooling Plant

Most cooling storage is installed for the ultimate benefit of the electric utility, so consider a deal in which the utility operates the storage chiller plant. The cleanest arrangement is for the utility to build and operate the cooling storage plant at its own expense. It is also conceivable that the utility could own the plant while the facility operates it, or vice versa. However, splitting responsibilities invites poor performance and eventual litigation.

At present, this is not a common arrangement, but there is precedent for such arrangements. Cogeneration plants were installed and operated at customer sites by natural gas utilities during the 1960's and early 1970's.

Unfortunately, this precedent does provide a model for success, because most of those plants were abandoned before they were able to repay their costs. In that episode, the utility companies could not operate the satellite plants any better than their customers. However, the customer was not saddled with the full financial loss.

Step 6: Negotiate a Long-Term Rate Contract with the Electric Utility

By this point, it should be apparent that cooling storage is an unacceptable gamble unless you are assured of favorable electricity prices for the life of the storage plant. Electricity prices change. It is a fundamental mistake to assume that present electricity prices provide sufficient incentive for cooling storage, even if subsidies are also provided by the utility to offset the cost of construction. A storage system will cease to be profitable unless it continues to provide an electricity price advantage that compensates for the energy losses of the system, the additional operating costs, and amortization of the original investment.

Therefore, early in the development of a cooling storage project, negotiate a long-term rate contract with the utility. This is not an activity for the naive. To do it right, you need to understand rate structures. You need to tactfully investigate the utility's present motivations and to assess its future conditions. You need to assess the future of utility regulation in your jurisdiction. You need negotiating skill, along with the authority to negotiate. The negotiation should be headed by a senior facility manager who has studied the issues in depth.

Take the initiative to prepare a contract offer that protects your investment. After all, you will be the party taking the risk of owning and operating the system, or at least providing the space for it. Furthermore, even though utilities are primary sponsors of cooling storage, many are naive about the economic and technical aspects of operating these systems. Both for negotiating purposes, and to ensure the success of your system, you need to know more about cooling storage than the utility. The utility is more likely to respond favorably to a reasonable proposal than to originate an arrangement that is satisfactory to you.

Your contract proposal should cover these four critical issues:

- the specific rates
- the duration of the rates
- exemptions for demand peaks caused by occasional equipment failure
- which metered loads will be subject to the cooling storage rate.

■ The Rates

The structure of electricity rates is explained in Reference Note 21. Review this material to understand how a utility is compensated for the electricity that it produces, and how cooling storage reduces some components of the utility's costs.

The starting point for the customer is calculating the electricity price differential that is needed to make cooling storage attractive. This may require the utility to offer a more favorable price or a more favorable rate structure than it presently offers.

Remember how the utility views cooling storage. It is not a revenue producer like other electrical sales, but a means to solve a utility problem. When talking to the utility, focus on the aspects of cooling storage that help the utility. You will not get a favorable electricity rate unless your system helps the utility to satisfy its objectives. Don't be surprised if the utility itself is not clear about its objectives, especially in the long term.

■ The Duration of the Rates

As a minimum, the contract should guarantee that favorable rates will continue long enough to pay off the system and to compensate for the effort, investment, and risk involved in installing it. If the utility is unwilling to negotiate such a contract, you have no rational basis to install cooling storage. Without a long-term rate contract, all it takes is the future decision of a utility executive to turn your cooling storage plant into a white elephant.

Electricity prices depend on many factors that vary in unpredictable ways, as explained in Reference Note 21. Do not expect the utility to commit to specific prices for a long term in the face of these uncertainties. Such an agreement would be a pure gamble for both parties. Perhaps the best way to deal with this uncertainty is to negotiate long-term rates that are expressed in terms of a discount from any future non-storage rates.

For example, agree that electricity for the chiller plant, purchased during off-peak hours, will be priced at a fixed percentage of the lowest non-storage rate offered by the utility to any customer.

■ Forgiveness Clause for Occasional System Failures

There is an economic hazard for the owner of a cooling storage plant that is not yet adequately recognized. The hazard is that a brief failure of storage could wipe out the saving in electricity costs for a long period of time. This is because demand charges are based on measurements that are made throughout a specified time period.

If the cooling storage systems fails for only one day during the peak period, the demand that occurs on that day becomes the basis for the demand charge for the entire demand billing period. The demand billing period may be a month, the entire cooling season, or the entire year. While demand peaks are almost always measured on a monthly basis, "ratchet" provisions in the rate schedule may extend a demand charge that is determined during one month into the months that follow.

For example, one utility bases its yearly demand charges on the highest demand that occurs during the months of July, August, or September. If the cooling storage system fails on a hot afternoon in July, the demand registered during that period becomes the basis of demand charges for the entire year. A failure that lasts one hour can wipe out most of savings that the storage system is supposed to provide for that entire year.

To avoid this risk, negotiate a forgiveness clause in your long-term rate contract. This clause exempts demand charges resulting from occasional failures of the cooling storage system. For example, the contract might stipulate that no demand charges will be levied for the three days of highest facility demand during the cooling season, provided that none of these days is coincident with the three highest days of the utility's demand. Thus, if the storage plant suffers a failure that requires two days to repair, the facility probably will not lose its savings in electricity cost.

At first glance, it might appear that this defeats the utility's purpose in having demand charges. This is not true. From the standpoint of the electric utility, rare failures of cooling storage by individual customers are not important. This is because a utility has many customers using storage or other demand-reduction techniques. Statistically, the utility can expect that only a few storage failures, at worst, will occur during a utility load peak.

This concept is novel, so be prepared to explain and persuade. Another problem is that the utility will probably have to modify the way it calculates demand. Be prepared to argue that most large utilities now calculate demand from a central computer, which can be programmed with modest effort to accommodate different rate arrangements.

■ Electricity Metering Configuration

The distribution of metering may have a profound effect on electricity cost. The chiller plant may be served by the same meter as the entire facility, or the chiller plant may have separate meter, or the chiller plant may be combined with selected other loads. Each of these arrangements produces a different total electricity cost.

For example, if storage is used for "peak shaving," the chillers should be served by the same meter that serves the variable non-cooling loads. On the other hand, if the rates include excess-consumption charges, it may be better to have a separate meter for the chiller plant.

You will have to negotiate a change in metering with the utility. If you think that rearranging the metering would produce lower electricity costs, be careful to consider the net effect on electricity cost for all the metered areas. Be clear about which meters are covered by your negotiations, and which continue to be governed by the utility's existing rate schedules, and their successors.

Step 7: Assess the Potential for Heat Recovery from the Storage Chillers

As discussed previously, cooling storage may increase the opportunity for recovering the heat rejected by the chiller plant. On the other hand, heat recovery may reduce the efficiency of cooling operation. If the facility presents an opportunity for heat recovery, refer to Subsection 2.10 for the factors to consider.

Step 8: Select the Control Strategy for Minimizing Electricity and Equipment Cost

Controlling a cooling storage installation may be tricky. How you control storage depends on the purposes you hope to achieve. In turn, the control strategy largely determines the storage capacity that you need to install. Select the control strategy to satisfy these primary objectives:

- scheduling the chiller load to minimize electricity cost
- minimizing the amount of excess stored cooling that remains at the end of each discharge cycle
- minimizing total equipment cost. In new construction, you can reduce the cost of the chillers, and perhaps the cost of other facility cooling equipment.

You will have to compromise between these objectives.

A simple example is a typical large office building in which the non-cooling electrical loads are essentially constant throughout the working day. This example is illustrated in Figure 1. If the electric utility's demand period is from 0800 to 1800, then the chiller is controlled so that it does not run during these hours. At the end of each day, the control system estimates the amount of cooling that is needed for the next day. The chiller is operated during the coolest hours of the night to put this amount of cooling into storage, plus a reserve to cover the uncertainty in predicting the next day's cooling load.

A more complex case is one in which the electrical demand of non-cooling equipment varies widely throughout the day, as shown in Figure 2. In such cases, the chillers may operate during the demand period, but their output is limited to keep the overall demand of the facility within a set limit. Thus, chiller output declines as other electrical loads increase. Operating the chiller during the demand period allows you to reduce the storage capacity. If the demand charge changes from one month to another (which is typical), then the optimum amount of chiller operation during the demand period will vary from month to month. This particular control strategy is called "demand limiting," although all modes of chiller control should attempt to limit demand.

If the purpose of cooling storage is to minimize chiller capacity, then chillers operate as close to the usage period as possible, but still during periods of lowest electricity prices. For example, a dairy operation may have two milking cycles per day, both of which occur during the utility's demand period. The chiller operates at night to provide enough stored cooling for both of the cooling cycles. If the utility does not have a demand charge on weekends, the chiller may operate during the day on Saturday and Sunday to store cooling only for the evening cycle.

Controls are covered in greater detail at the end of this Measure.

Step 9: Estimate the Cooling Storage Capacity

The cooling storage capacity that you need is determined by the purposes of the system and by the facility's cooling load profile. For example, an office building in a mid-latitude climate typically has an afternoon cooling peak lasting only a few hours. This profile requires less storage capacity for a given demand reduction than a hotel located in a warm, humid climate, where the cooling load may remain high all day.

It may not be economical to displace all chiller operation from the period of high electricity prices. For example, if a heavy cooling load occurs only for a few weeks a year, a storage system designed for those loads will have a large amount of idle capacity for the rest of the year. If the utility has no ratchet charges, the excess capacity provides little benefit.

The actual storage capacity needs to be somewhat greater than the amount of cooling load that is displaced. As with most equipment, provide some reserve for loads greater than expected, and for equipment performance poorer than expected.

With all cooling storage methods, the chilled water temperature rises as the storage unit discharges. This reduces cooling capacity at the end of the discharge cycle. If the cooling load remains high at the end of the discharge period, you may need additional storage capacity to account for this effect. On the other hand, you may be able avoid this problem by installing multiple storage units, and discharging them in sequence.

In ice storage units, the rise in chilled water temperature toward the end of the discharge cycle may be severe, and the effect increases with discharge rate. This may not be a problem in an ice storage system that distributes chilled water at normal temperatures, because ice storage produces water that is much colder than normal chilled water.

On the other hand, if the system uses low-temperature chilled water distribution, and the cooling load remains high near the end of the storage cycle, the rise in chilled water temperature may be unacceptable. The problem is least severe with ice shedders and ice slurry systems. Internal-melt systems and ice capsules suffer a large rise in temperature, unless the discharge rate is low. (Individual types of storage units are described in greater detail below.)

If you compensate by increasing the capacity of the storage unit, this will increase that amount of cooling energy that remains in the storage unit at the end of the cooling storage period. Much of this energy is wasted. For example, if cooling storage is not needed over the weekend, some of the residual cooling left in storage at the end of Friday is lost by conduction over the weekend. Even worse, some ice storage units require all the ice to be melted at the end of each discharge cycle, whether the cooling is needed or not.

Step 10: Estimate the Heat Recovery Storage Capacity, If Used

If you want to recover heat from the chillers, installing a heat storage may substantially increase the amount of heat that you can recover. The heat storage capacity is determined largely by the chiller operating strategy. Refer to Measure 2.10.5 for details of heat recovery storage.

Step 11: Estimate the Cooling Storage Chiller Capacity

The chiller capacity is determined by the control strategy and the storage capacity. If you rely entirely on storage cooling during the peak electricity price period, the chillers must be large enough to charge the storage unit during the period when electricity prices are lowest. If the chillers serve the cooling load in parallel with the storage unit, then the chiller capacity is determined by the storage load or the direct cooling load, whichever is larger.

When you select the chiller size for charging the storage unit, recall that the chiller's charging capacity is less than its capacity for direct cooling. This is because the chiller must operate at lower evaporator temperature when charging. In ice storage systems, chiller capacity is reduced by about 40% compared to direct cooling, for the same condensing temperature. This effect is compensated somewhat by lower nighttime condensing temperatures, provided that the chiller is able to exploit lower condensing temperatures.

In new construction, if storage is being used to reduce chiller capacity, consider future changes in the cooling of the facility. Leave space for more chiller capacity if the cooling load increases. For example, an office building in a warm climate may presently be cooled only eight hours per day, but it is conceivable that future conditions will require much longer periods of cooling.

Step 12: Select the Cooling Storage Technology

The various types of cooling storage are described below. Your selection of the type of storage unit has major effects on system size, cost, complexity, and efficiency.

The first selection step is deciding whether to use ice, chilled water, or eutectic storage. Under present conditions, ice appears to be preferable for most applications. Chilled water is rarely used unless there is an existing water tank of adequate capacity. Eutectic storage probably is not the best choice unless there has been a significant improvement in the technology by the time you read this. If you select ice storage, you have another major decision in selecting the type of ice maker.

A major part of this decision is figuring out where to put the storage vessel. The vessels used with the different types of storage vary considerably in volume, height requirements, and accessibility requirements. For example, underground tanks are acceptable for some applications and storage methods, but not for others.

Step 13: Select the Type of Cooling Storage Chiller

See Reference Note 32, Compression Cooling, and Reference Note 33, Absorption Cooling, for details of chiller selection factors. As in all chiller plants, high COP should be a dominant criterion.

You face additional selection issues if the chiller makes both ice (for storage) and chilling water (for direct cooling). This requires the chiller to operate at two different pressure differentials. These issues are similar to the ones related to selecting chillers for heat recovery at elevated temperatures. See Measure 2.10.3 for details. In brief, screw compressors and reciprocating compressors are favored in this case because of their ability to adapt to different pressure differentials.

Be sure to select chillers that can exploit low ambient condensing temperatures, for the reasons explained previously.

Step 14: Select the Evaporator Configuration

The method of controlling refrigerant flow to the evaporator has a major effect on system COP. Storage systems have been built with each of the three major evaporator types: liquid overfeed, flooded evaporators, and expansion valve. These types are explained in Reference Note 32, Compression Cooling. Expansion valve evaporators are inherently less efficient than the other two types.

In ice storage systems, expansion valve evaporators require additional complications to allow them to switch efficiently between making ice and chilling water directly. One way of doing this is to install an expansion valve that maintains a fixed superheat, regardless of the evaporator temperature. Such expansion valves have been available for a number of years. Another approach is to use two expansion valves, one for making ice and one for chilling water. Solenoid valves control which expansion valve is used, based on signals from the system controls.

Step 15: Design the Connections between the Chiller Plant, the Storage Unit, and the Facility Equipment

The type of storage unit dictates major aspects of the design. These include isolating the storage unit from the pressure of the chilled water system, reconciling the supply and return water temperatures, and switching from charging to direct cooling. These topics are covered at the end of this Measure. In general, keep the overall configuration as simple and reliable as possible, while providing for efficiency in all the modes of operation.

Step 16: Design the Controls and Alarms

Control is the trickiest aspect of the design, installation, and operation of a cooling storage system. Cooling storage may require a large number of control functions because a system changes its connections, running time, and chiller loading for different operating conditions. A peculiar difficulty of cooling storage is that the control system must predict load conditions from several hours to several days in advance. Details are given below.

Step 17: Design the Physical Layout

From the standpoint of long-term performance, the most important aspect of the physical layout is easy access for maintenance. Design the layout so that any component in the system can be replaced easily. *Make it easy to maintain the system for the life of the facility.* Don't design for the short lifetimes used in economic analysis. You don't want a system to be unrepairable just because it is paid off.

Many cooling storage systems are being installed today in a manner that makes it virtually impossible to repair them economically. Burying systems underground not only makes them inaccessible, but invites deterioration by ground water. Equally bad is stacking modular units inside a building in a way that makes it impossible to remove an individual unit or to replace the components in a unit. No system on the market is sufficiently well proven to justify these practices.

Step 18: Install the System

Installing cooling storage is similar to installing other complex HVAC equipment. Effective communication between the designer and the contractor is important. Selecting packaged equipment reduces installation cost, and it lessens the opportunities for installation problems.

Step 19: Train and Preserve a Skilled Staff

Among the many failures of cooling storage systems, failure to operate the systems properly looms large as a primary cause. Operating a cooling storage plant requires a high level of skill, unconventional knowledge, and continuous monitoring. The staff must receive special training to operate the system. There can be no lapses in staff continuity as long as the system is operating.

This is a major management challenge. Staff turnover requires continual retraining. Furthermore, talented operating personnel tend to be diverted to tasks outside the chiller plant.

COOLING STORAGE EQUIPMENT

The various types of cooling storage differ markedly in size, efficiency, reliability, and other significant characteristics. Early cooling storage for HVAC applications focussed on liquid water as the storage medium. Within a few years, growing awareness of the huge volume required by water storage, along with its economic and practical problems, forced designers to switch their preference to ice storage. In turn, ice storage has posed a whole new set of challenges. The design compromises required in ice storage are stimulating development of a variety of storage vessels and system types. The following is an introduction to the types of cooling storage that are presently available, with emphasis on their comparative characteristics.

■ Common Considerations for Ice Storage Units

Using ice for thermal storage has two major advantages over using chilled water storage. One is that much more energy can be stored in a given volume. The heat required to melt ice is about 144 BTU's per pound, while liquid water can absorb less than 20 BTU's per pound in typical cooling storage applications. As a result, a typical ice storage unit requires only about one fifth the volume of a chilled water tank.

The other major advantage of ice storage is that ice always melts at a fixed temperature, regardless of the state of charge of the storage unit. This keeps the chilled water supply temperature fairly constant as long as ice remains, regardless of the return water temperature. Therefore, the cooling system served by the storage unit can be entirely conventional.

This advantage is lost in air conditioning systems that are designed to exploit the low chilled water temperature available with ice storage. This is covered in greater detail below.

The main disadvantage of ice storage is low energy efficiency. The evaporator temperature needed to freeze ice is about 12°F to 30°F (7°C to 17°C) colder than is needed to chill water in conventional applications. This imposes a severe reduction of chiller COP, typically in the range of 20% to 40%. Some types of ice storage require substantially lower evaporator temperatures than others.

A fundamental problem in ice storage design is that ice expands when it freezes. Freezing of water is one of nature's most destructive forces. Given sufficient time, it destroys mountains. In much less time, it destroys man-made constructions, such as roads and thermal storage systems. Ice also exerts force by its buoyancy. Some types of ice storage accommodate these forces with a greater margin of safety than others. With types that require coils in the tank for freezing, use well proven packaged units, rather than run the risk of repeating past mistakes with a custom design.

Another challenge in the design of ice storage units is the relatively poor thermal conductivity of ice, which is about the same as for granite, and about 0.5% the conductivity of copper pipe. Lower evaporator temperatures are needed to freeze greater thicknesses of ice, so the storage unit should minimize the thickness. Some types do this better than others.

Another problem is non-uniform flow of the chilled water or heat transfer fluid as it flows through the melting ice. Where water flows freely in a tank, it tends to melt channels through the ice, reducing heat transfer during the discharge cycle. Some types of ice storage are very vulnerable to this problem, while at least one type is immune to it.

Ice Shedders

A simple method of cooling storage is making ice, dropping it into a tank, and circulating water through the ice. The ice making machines are similar to the ones that make the ice cubes for your drinks. These storage systems are called "ice harvesters," or "ice shedders," because the ice maker is usually mounted above the ice storage tank, where it drops the ice into the tank below. Figure 4 shows the equipment layout.

Ice shedder systems are popular because they offer several significant advantages, and they are free of some of the problems of other types of ice storage. A primary advantage is the simplicity of the system connections. Ice shedders average about the same overall efficiency as other ice storage systems.

A major advantage of ice shedder systems is that they are less vulnerable to flaws in the design of the site installation. The ice making machine is usually purchased as a complete package, ready to be installed on top of the tank. If the machine is derived from well proven standard equipment, it is less likely to have crippling flaws that appear the day after the warranty expires.

The ice maker itself is relatively expensive, although its cost per unit of capacity drops radically with increasing size. Therefore, ice shedders are favored for applications that require a large storage volume in relation to chiller capacity. The storage tank is relatively inexpensive because it is simple. Commonly used materials for the tank and tank circuit include plastic, fiber-reinforced plastic (fiberglass), and concrete.

Ice shedder systems may be several stories tall, because the ice maker is mounted on top of the tank.

Figure 5 shows a typical installation. The height can be reduced by using a conveyor to move the ice to the tank, but this sacrifices some of the elegant simplicity of the system.

■ Evaporator Features

Ice is formed on the evaporator from water that is circulated from the storage tank. Many contemporary units use flat plate evaporators, from which the ice sheds in the form of sheets or large flakes. The flat sheet form is favored because it shortens the path of heat through the freezing ice, allowing higher evaporator temperatures, and hence higher COP. However, ice can be made in many shapes. Figure 6 shows a typical unit in operation.

Ice tends to stick to the evaporator as it freezes, so the method of releasing the ice from the evaporator is a major issue with ice shedders. At present, the most popular method is similar to a defrost cycle, which operates about 5% of the time. The ice is released by periodically bleeding hot gas into the evaporator while the compressor is turned off. The hot gas typically comes from a compressor discharge manifold that links several compressors. Figure 7 is a diagram of the cycle.

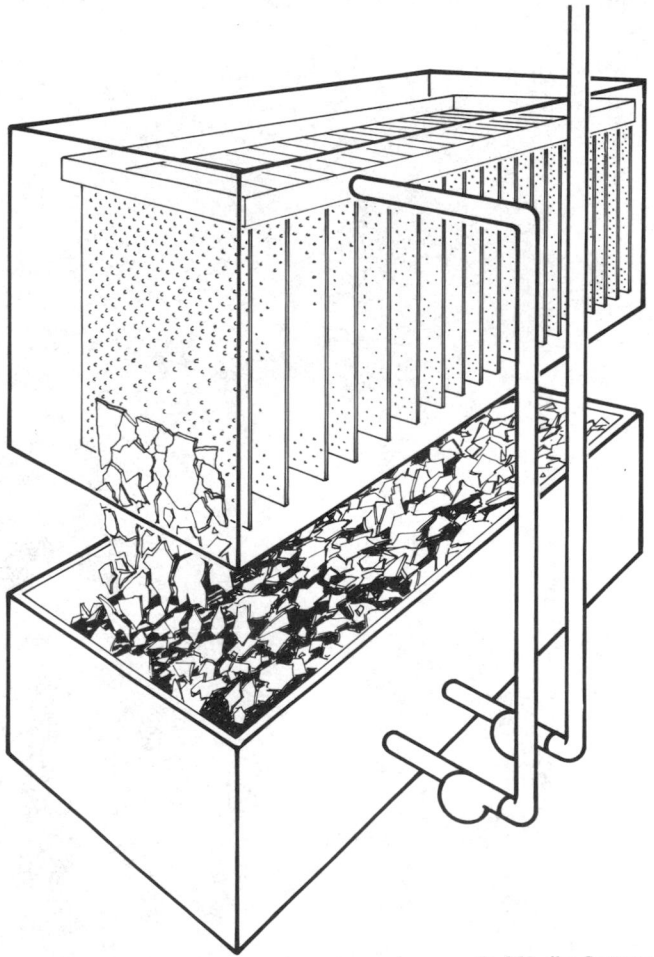

Paul Mueller Company

Fig. 4 The basic components of an ice shedder system
An ice maker dumps ice into a tank, and chilled water is circulated through the ice.

Defrosting wastes a certain amount of energy, so manufacturers search for more efficient methods of releasing the ice from the evaporator. One method is scrapers, which come in various designs. Another involves freezing the water as it is carried by a carrier liquid that acts as a release agent. The latter type are called slurry systems. Two versions of slurry systems are described below.

Some models of ice shedders are incapable of cooling chilled water directly. Such units have expansion valve evaporators that are set to operate at only one temperature for freezing.

Other models have liquid overfeed evaporators, which can vary their evaporator temperature. These units cool chilled water by pumping it over the evaporator in the same manner as for ice making.

■ Storage Tank Features

Ice shedders avoid the trickiest problems of designing ice storage units, which is dealing with the expansion of water as it freezes. The ice tank is simply a bin. The equipment inside the bin is not exposed to expansion or buoyancy forces. The return chilled water distributors may be kept above the ice, and the headers that extract water from the bottom of the tank may easily be designed to resist damage. Although the tanks are simple, tank shape and control of water level are important to keep large voids from forming in the ice. Voids can be eliminated by stirring the ice pile mechanically, but it is best to avoid the complication.

The chilled water is cooled by direct contact with ice in the form of small pieces, providing a large surface area. As a result, ice shedder system have a high peak cooling rate. By the same token, if the water flow through the tank is well designed, the water discharge temperature is kept low until most of the ice has melted.

Short-circuiting of water flow poses somewhat less of a problem than with other methods of ice storage. Minimize it by careful design of tank shape and header configuration, by control of the water level in the tank, and perhaps by using mechanical agitators to redistribute the ice. Short-circuiting is more difficult to control if the ice is stored for long periods, because the ice pieces may fuse together, channeling water flow.

The independence of the storage bin from the ice maker allows you to select the size and shape of the bin to fit available space and to meet the owner's stylistic preferences. This flexibility is not unlimited. The ice falling from the ice maker forms a conical pile, which leaves some void space. If you need to minimize the

Paul Mueller Company

Fig. 5 Ice shedder system for a military club in Florida The compressor and evaporator unit, surrounded by the railing, sits atop a rectangular concrete ice tank. The cooling tower for the unit is at upper left.

volume of the tank, make the tank round and tall. This shape also improves heat transfer and reduces heat loss.

■ Possible Need for Heat Exchanger

Ice shedders heavily aerate the water in the tank by virtue of pumping it continuously over the evaporator. The resulting high oxygen content makes the water especially corrosive to steel components. As a result, these systems commonly use a heat exchanger to isolate the tank water from the steel piping of the chilled water system.

Ice Freezing on Coil, External Melt

At first glance, the simplest way to make ice for cooling storage is to put a refrigerant coil in a tank of water and freeze the water around the coils. Cooling is recovered by circulating water around the frozen coils. This technique has been named "external melt" ice storage, to distinguish it from another method in which the ice is melted by the coil. Figure 8 shows a cutaway of an external-melt ice storage unit.

The large ice surface area gives external-melt units a high discharge rate, if needed. Another virtue of external-melt systems is that their system connections are simpler than with some other types. Figure 9 shows the basic layout. This reduces the uncertainties of engineering and makes it easier for the facility staff to understand the operation of the system.

External-melt systems tend to have lower overall efficiency than other types of cooling storage systems. An inherent problem is the thickness of ice on the coils, which typically grows to several inches with reasonable coil spacing. This forces the chiller to operate at progressively lower evaporator temperatures as the ice accumulates. The average efficiency of the system is further reduced by inability to predict the exact cooling

WESINC

Fig. 6 Ice shedder in operation Ice is being released from the vertical evaporator plates into the ice bin below, which is almost full. The thin tubes feed refrigerant to the evaporator plates from the horizontal header at the bottom. The horizontal header at the top feeds hot gas to the evaporator plates in sequence to release the ice. At lower right is the ice level sensor that tells when the ice bin is full.

load during the next cycle, the irregularity of ice formation, and other storage inefficiencies. The cost of the storage units is relatively high, because they are packed with coils, attachments, and accessories.

The simplicity of the concept made this method the first to be tried on a large scale. However, serious problems in the first generation of equipment almost led to the extinction of this method. Most of the problems have now been brought under reasonable control, and this type of ice storage has climbed back to a prominent position. External-melt storage units are now used in systems of all sizes. Figure 10 shows one module of a large system being installed.

A major failing of early external-melt storage units was that they failed to account for the enormous expansion force of freezing ice, which destroyed coils and tanks. Also, the buoyant force of the ice tore the coil assemblies loose from their attachments. These problems are now avoided by using packaged ice storage units that are carefully designed and heavily built. It would be folly to repeat the mistakes of the past by attempting to design external-melt storage units on a custom basis, unless the designer is very skilled and the customer has a large budget.

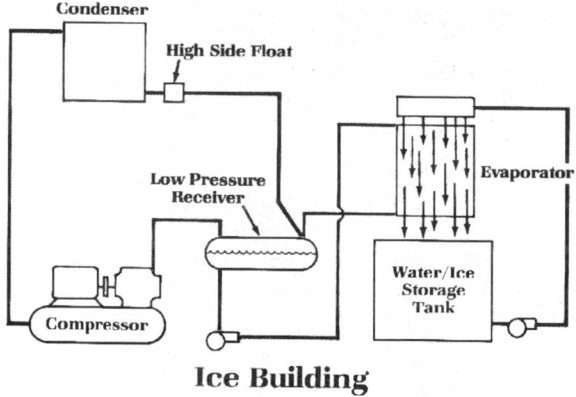

Ice Building

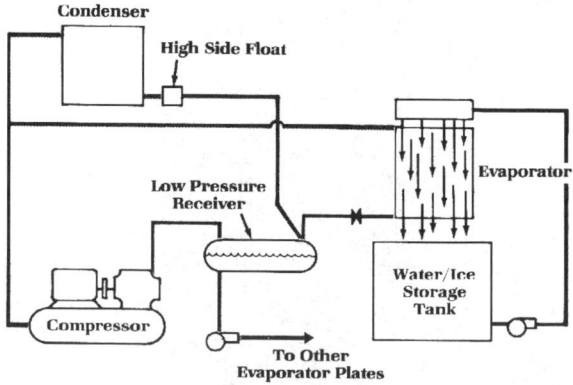

Ice Harvesting

Paul Mueller Company

Fig. 7 Ice making and release cycle of an ice shedder Ice forms on the cold evaporator plates. To release the ice, it is partially melted by feeding hot gas from the compressor into the evaporator. This process involves some energy loss.

▪ Direct-Expansion Storage Units

Ice can be formed on direct-expansion coils as well as on coils that are cooled indirectly by an antifreeze solution. Direct-expansion coils avoid heat transfer losses, but system design is more complex, mostly in the feeding of refrigerant to the coils. Coils have been made from various plastics, copper alloys, and steel. The strength of steel is probably needed for larger units. Steel coils make it possible to cool directly with ammonia, which is an excellent refrigerant for large ice storage units. (Ammonia attacks components containing copper.) Figure 11 shows a direct-expansion external-melt system that serves a small professional building.

▪ Ice Bridging

Another fundamental problem of external-melt systems is ice bridging from one coil to adjacent coils. Bridging impedes the flow of chilled water in some parts of the storage unit, forcing short-circuiting through other parts. In order to minimize storage volume and cost, the coils are packed as close as possible. As a result, bridging will occur if the nominal ice capacity of the unit is exceeded.

Bridging also occurs if the ice is not completely melted between cooling cycles. Ice grows in a somewhat irregular manner, so bridging becomes probable after several cycles of incomplete ice melting. Therefore, the controls must ensure that the coils are melted completely at the end of every operating cycle, or perhaps after several operating cycles. This takes no special action if the cooling load continues into hours of low electricity prices. This allows the storage unit to operate until all the ice has melted. However, if cooling occurs mostly during hours of peak electricity cost, some ice may remain until the next charging cycle. In this case, energy may have to be expended to melt the remaining ice prior to charging.

▪ Achieving Uniform Melting

External-melt systems have difficulty maintaining low chilled water temperature as the ice melts, especially toward the end of the discharge cycle. There is a strong tendency toward short-circuiting, and this problem has not been solved entirely. If one path through the ice enlarges faster than others, the increased water flow will further hasten the enlargement, to the point that one part of the storage unit may melt completely while the rest of the unit still has considerable ice.

Various methods are used to encourage uniform melting, including careful coil design, baffles, and headers. One major manufacturer even uses an air bubbling system to keep ice melting uniform. A disadvantage of this method is that it aerates the chilled water, making it corrosive. This requires major corrosion protection features for the storage unit, and usually, a heat exchanger to isolate the aerated water from the chilled water system.

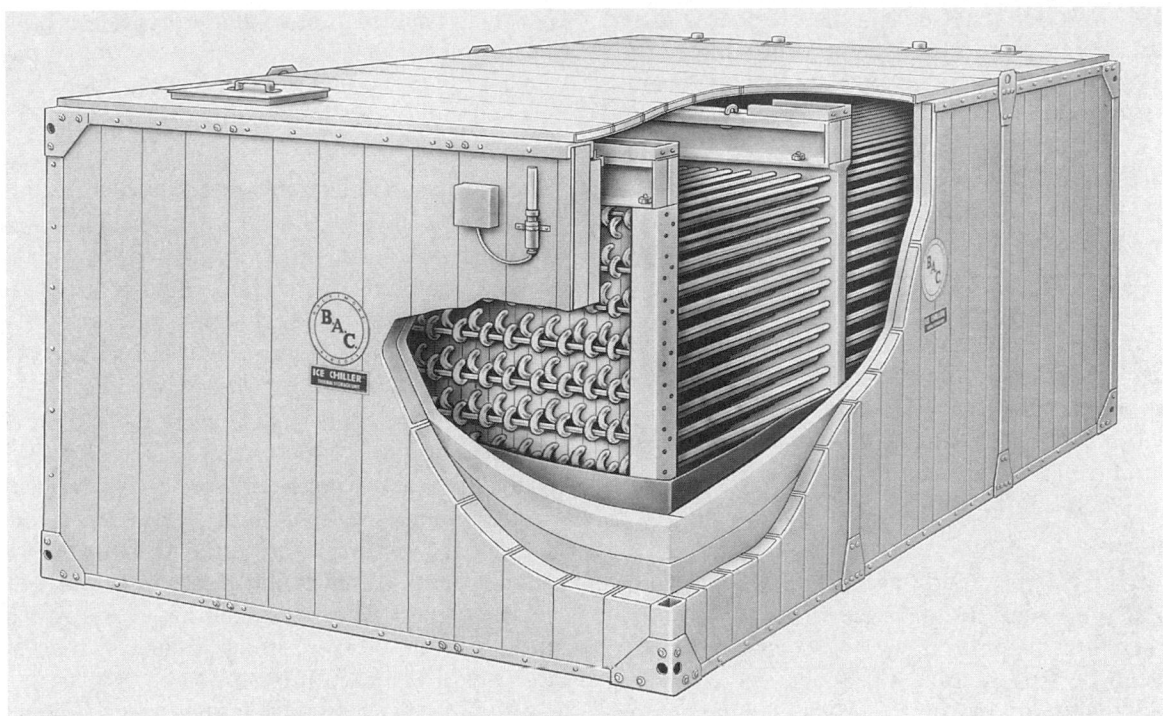

Baltimore Aircoil Company

Fig. 8 External-melt ice storage unit This is a tank full of water that is filled with freezing coils. The coils may be cooled by glycol from a chiller, or they may be cooled by direct expansion. Avoiding ice bridging between the coils and maintaining uniform water circulation are important challenges with this type of unit.

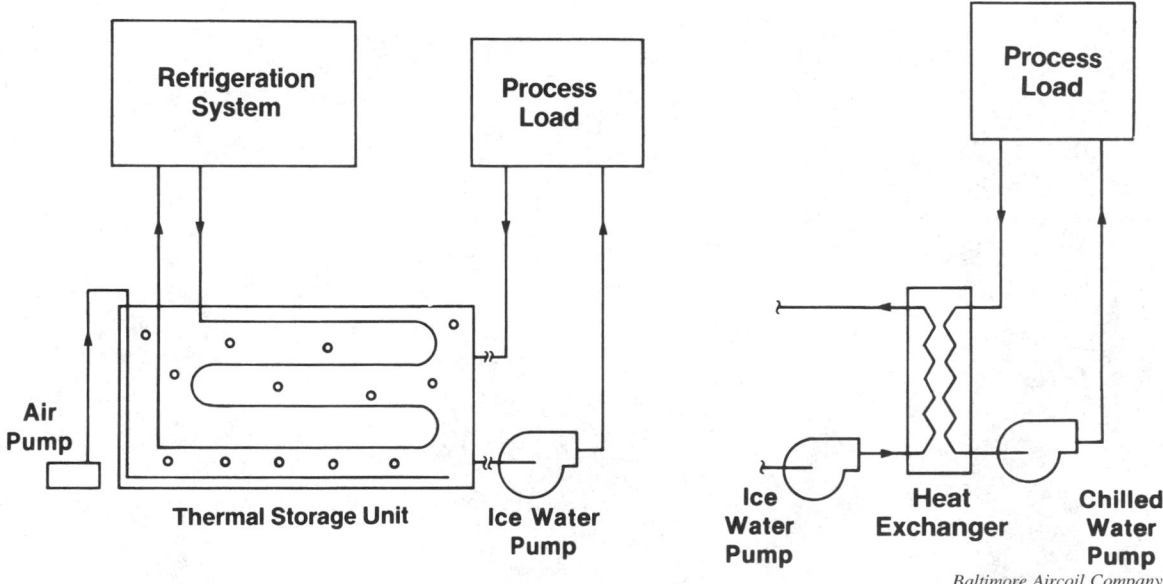

Baltimore Aircoil Company

Fig. 9 System connections with external-melt systems The chilled water can be pumped directly from the tank, or a heat exchanger can be used. A heat exchanger is needed to isolate the tank from pressure if portions of the chilled water system are located above it. This system includes an air bubbler to promote uniform water circulation. The bubbler saturates the water with oxygen, making it corrosive to steel. This is another reason to use a heat exchanger.

■ Alternating Use for Heat Storage

An interesting possibility is using the storage unit to heat and store hot water during the off-peak season. To do this, the evaporator coil connection is switched so that the coil becomes a second condenser bundle. This has been done in some small packaged systems.

Ice Freezing on Coil, Coil Melt

Another type of ice storage also freezes ice on coils, but the method of melting the ice is entirely different. Instead of melting the ice from the outside of the ice layer, the ice is melted by the same coil that freezes it. This system uses a heat transfer fluid, typically glycol, inside the coils. During the charging cycle, the glycol flows through the chiller to freeze the ice. During the discharge cycle, the ice is melted by circulating the heat transfer fluid with the chiller turned off. Figure 12 shows the system operating modes.

■ Advantages

Coil-melt systems avoid the most vexing problems of external-melt ice storage. Ice bridging is not a problem because the entire mass of water surrounding the coils can be frozen solid. This also increases the compactness of the storage unit. The water in the storage tank does not circulate, so there is no need for headers, baffles, or other complications.

It is easy to protect the storage equipment from corrosion because the water that freezes in the storage unit does not circulate, and it is not aerated. There is no corrosion problem in the chilled water system, because it is a closed system.

■ Charging and Discharging Characteristics

The performance characteristics of coil-melt systems vary during the freezing and thawing cycle to a greater extent than with other ice storage methods. The charging efficiency and charging rate are greatest at the beginning of the charge cycle because ice was cleared from the coils by the previous discharge.

The temperature rises considerably throughout the discharge cycle, for a constant charge rate. This is due to the thermal resistance of the growing water layer that forms around the tubes as the ice melts. Liquid water has about four times the thermal resistance of ice when stationary, but convection currents in the water layer probably reduce the resistance considerably. This effect is not a major worry in chilled water systems that operate at conventional temperatures. However, it may disqualify this method of ice storage if you want to exploit low-temperature chilled water distribution.

Baltimore Aircoil Company

Fig. 10 An external-melt storage module for a district cooling system   A large number of these modules are installed underneath a parking lot. The system serves a portion of downtown Baltimore.

By the same token, the maximum discharge rate also declines as the ice melts. This may be a problem if the cooling load remains high near the end of the discharge cycle.

■ Storage Unit Design

Packaged coil-melt storage units are available in a variety of sizes and shapes. The storage unit can freeze solid from one side to the other, so the tank must be designed to resist bursting or progressive leakage. (As an alternative, the system can be controlled to prevent complete freezing of the tank, but this method is less reliable.) Plastic pipe is typically used in smaller units, and steel pipe in larger units. Figure 13 shows a cutaway of a typical packaged storage unit.

Large storage systems can be assembled easily from any number of modular units. Figure 14 shows an example. Another approach is to use prefabricated coil arrays that are available for installation in large site-built tanks.

■ Connections to the Chilled Water System

The chiller and the ice storage unit can serve the cooling load individually or simultaneously, and the chiller can charge the storage unit while it serves the cooling load. It is challenging to design a system configuration that optimizes operating efficiency under all possible operating modes.

Some designs compromise by simply connecting the chiller and storage unit in series. This is fairly satisfactory, but sacrifices some efficiency or capacity, depending on whether the chiller is upstream or downstream of the storage unit.

Small systems may use the heat transfer solution directly in the chilled water system. This approach reduces heat transfer in the end-use equipment because glycol and other heat transfer fluids have worse heat transfer characteristics than water. An alternative is using a heat exchanger, which allows pure water to be used in chilled water circuit. This approach incurs a temperature rise in the heat exchanger. In larger systems, a heat exchanger becomes more desirable to reduce the quantity of glycol that is needed.

The heat exchanger, if one is used, needs freeze protection. This is because the glycol solution may be well below freezing temperature while the chilled water is unprotected. One method is to bypass the glycol around the heat exchanger when the glycol is below freezing temperature. Reliability requires multiple bypass valves and safety controls that stop glycol flow if the heat exchanger gets too cold.

Ice Capsules

In ice capsule storage systems, a tank is filled with sealed containers, or capsules, of water. The capsules are frozen by circulating a heat transfer fluid (generally glycol) through the tank. The storage unit is discharged simply by turning off the chiller and continuing to circulate the glycol through the tank, thawing the capsules.

The capsules are made of thin-walled plastic, typically polyethylene. They are shaped to survive the expansion of ice inside them. Some are dimpled spheres, typically a few inches in diameter. Others are shaped like large rectangular hot water bottles. These shapes

WESINC

Fig. 11 External-melt cooling storage system using direct-expansion coils The partially buried, round tanks are filled with spiral copper evaporator tubing. The compressor and condenser units are on the right . This system serves a small professional building.

allow the glycol to flow between the capsules. Some shapes, especially spheres, may be held in place only by dense packing. Other shapes require an external framework, or they may be held apart by spacers that are molded into the capsules.

Ice capsules are the simplest form of ice storage. The tank can be of any type that has the appropriate volume and shape. Common materials are concrete, steel, and fiberglass. Tank design is fairly simple, because the ice capsules float freely in the tank, requiring no supporting structures.

Ice capsule systems are immune to overcharging problems. They are relatively foolproof, simple to install and control, and relatively inexpensive. Their particular failure modes, ruptured capsules and leaking tanks, are non-catastrophic and fairly easy to repair. Ice capsule systems have not received as much attention as they

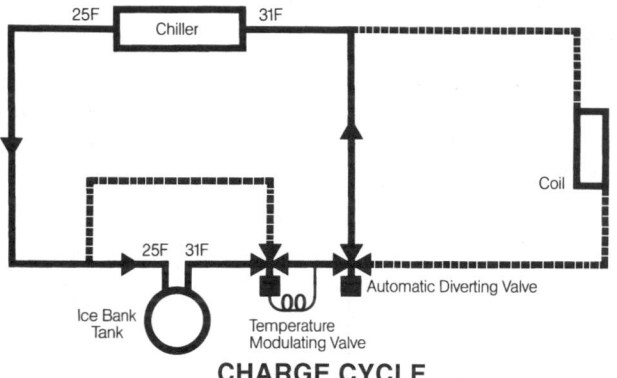

CHARGE CYCLE

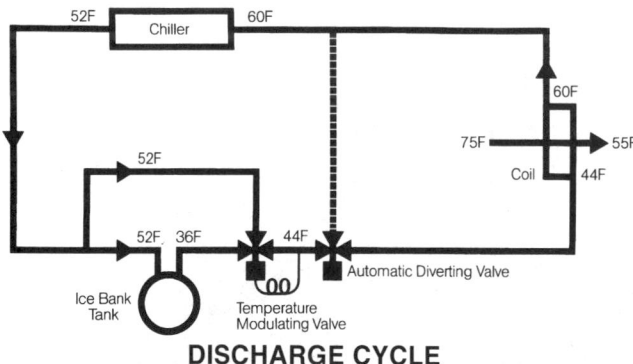

DISCHARGE CYCLE

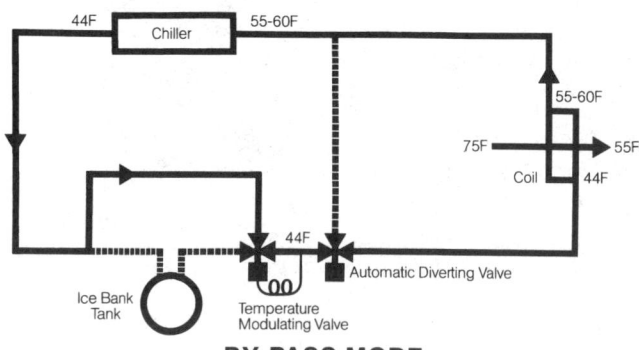

BY-PASS MODE

Calmac Manufacturing Corporation

Fig. 12 Coil-melt cooling storage system operating modes

deserve, perhaps because their price is too low to finance advertising and technical support.

The storage unit is completely independent of the chiller. This makes it simple to operate the chiller separately from the ice storage unit, at normal chilled water temperatures. The chiller can also operate in parallel with the ice storage unit. The chillers are conventional, except that the evaporator temperature is selected for efficient freezing.

■ Discharge Characteristics

The surface temperature of each ice capsule remains near freezing until the ice melts completely, because the small volume of water inside each capsule stays in equilibrium with the remaining ice. As a result, the discharge temperature remains almost constant until some of the capsules thaw completely. Beyond that point, the discharge temperature rises rapidly.

To preserve full cooling capacity at the end of the discharge cycle, design the flow of glycol in the tank so that all the ice capsules melt evenly. During the discharge cycle, the first capsules to encounter the warm entering glycol are melted most rapidly. One solution is to install a pipe network that distributes entering glycol over all the ice capsules evenly. This requires a lot of pipe, takes up space, and complicates installation. Another solution is to oversize the storage tank for the load. Such steps may be unnecessary if chilled water is used by the facility at normal temperatures, and if the cooling load tapers off at the end of the cooling period.

■ Freezing Problems Inside the Capsules

Ice capsules have an odd problem that occurs at the beginning of the freezing cycle. If water is kept still and cooled slowly, it will cool to well below 32°F (0°C) before it starts to freeze. These conditions occur inside the sealed ice capsules. Therefore, the capsules must be cooled to about 25°F (-4°C) to be sure of freezing. Once freezing starts, the first ice triggers freezing of the rest of the water at normal freezing temperature.

Any residual ice inside a capsule prevents subcooling. However, this effect is not helpful. If some capsules are completely thawed at the beginning of the next charging cycle, and some are not, this only aggravates the problem of unequal freezing.

Ice capsules that are purchased with the water permanently sealed should contain an agent that stimulates freezing. If capsules are filled on site, add an appropriate agent for this purpose.

If the glycol temperature distribution in the tank is irregular, some of the capsules may never be cooled enough to freeze. To get a full cooling charge, you may have to operate the chiller at below-normal for a short time at the end of the charging cycle.

Stored Ice Slurry Systems

The stored ice slurry system is one of the newest entries in the cooling storage market. Figure 15 shows a diagram of this type of system. It has several unusual features

One distinguishing feature of the system is a special evaporator that allows water to be frozen while combined with a carrier liquid. The carrier liquid is usually propylene glycol. The solution of water and glycol flows by gravity down a tall evaporator surface. The surface is agitated aggressively by a mechanical scraper, so that the water freezes in the form of microscopic crystals, forming a slurry with the glycol.

In one version of the system, the slurry is pumped to an insulated storage tank. The ice-bearing slurry floats to the top of the tank, leaving a solution of glycol and water at the bottom of the tank. Usually, the solution at the bottom of the tank is pumped to the application. The returning warmed liquid is sprayed over the slurry at the tope of the tank.

Another version of the system pumps the slurry itself to the application. This can be a very efficient method of transporting cooling, because the ice crystals in the slurry have a high cooling energy content in the form of latent heat of freezing.

The stored slurry system should offer a high discharge rate because the ice crystals have a large surface area.

The system usually employs a heat exchanger. This isolates the glycol mixture, which has a poor heat

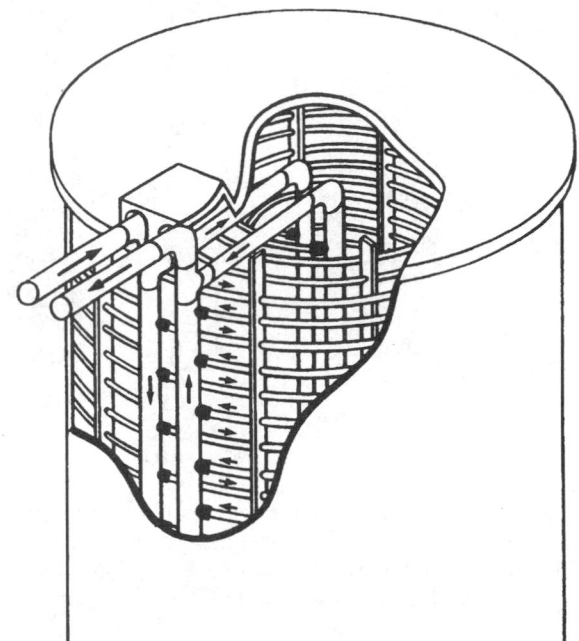

Calmac Manufacturing Corporation

Fig. 13 Coil-melt cooling storage unit The tank is full of water, which does not circulate. The water is frozen by circulating glycol from a chiller to coils within the tank. Stored cooling is extracted from the ice by the same glycol, which is now connected to the chilled water system.

Calmac Manufacturing Corporation

Fig. 14 A large modular coil-melt cooling storage system The storage units are installed on grade, outside the low mechanical equipment building. The system serves the large domed building in the rear.

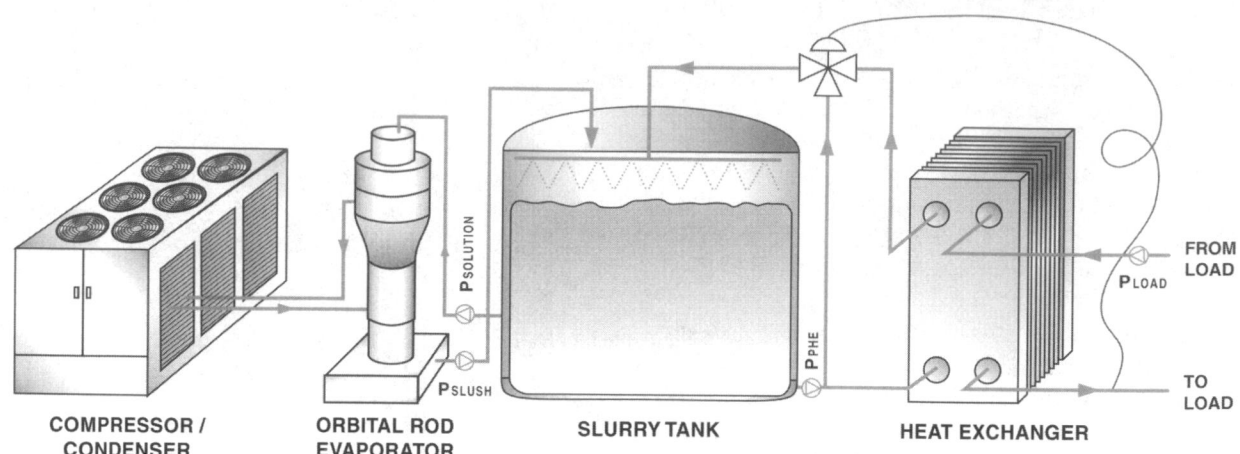

Fig. 15 Stored ice slurry system A slurry of ice in glycol is formed in the evaporator. The slurry is pumped to the storage tank. In this version of the system, a mixture of glycol and water settles out of the bottom of the slurry. This liquid is pumped to a heat exchanger to cool the chilled water system. The return liquid is sprayed over the slurry to cool it. In a different version of the system, the slurry itself is pumped to the cooling application.

transfer characteristics, from the pure water of the chilled water system. However, there are applications that do not need a heat exchanger.

The chiller, the evaporator, and the storage tank are independent of each other, except for pipe connections. This provides a great deal of flexibility in physical layout. The tank can be any shape or size, within broad limits, but it does require specialized design. The warm returning solution is sprayed uniformly over the top of the slurry mass in the tank. This is necessary to prevent channeling of flow.

Freezing the water on the surface of the evaporator avoids the heat transfer loss that occurs in ice storage systems that freeze water with glycol. However, mixing of the water with the carrier liquid reduces the freezing temperature. Also, the various circulation pumps require energy.

The big question about stored slurry systems is their actual performance. They have only recently arrived on the market. As with any equipment that is supposed to last for decades, performance can be judged fairly only after years of operation in actual applications.

Separated Ice Slurry Systems

An older type of ice slurry system freezes ice in large flakes. Water that is mixed with a carrier liquid flows by gravity down a tall plate evaporator. Ice freezes out of the stream, but does not stick to the evaporator because it is carried along in the other liquid. The ice forms a slurry with the water, which is pumped to a

Fig. 16 Separated ice slurry cooling storage system The system provides air conditioning for an elementary school. The storage equipment is surrounded by the brick enclosure in the foreground.

collecting bin. In the bin, the glycol falls free of the ice, which forms in a mass similar to a pile of snow. The separated glycol is then mixed with melt water to repeat the freezing process.

(Because water is combined with another liquid for freezing, slurry systems are sometimes called "binary" systems.)

The evaporator in a separated slurry system is much taller than the evaporator in a stored slurry system. The evaporator plate and ice collector is typically several stories high. This is necessary to allow the large flakes of ice to form. The ice bin also tends to be tall, because this is an efficient configuration for storing the ice, separating the glycol, and circulating chilled water through the ice.

Separated slurry systems are the most complex type of cooling storage, in terms of the number of large components and the complexity of piping. Figures 16, 17, and 18 show an ice slurry system used for air conditioning an elementary school in Florida.

The efficiency of a separated slurry system is probably better than the average for ice storage systems. Heat transfer on the evaporator is good, but mixing of the water with the carrier liquid reduces the freezing temperature. Distributing the chilled water efficiently with the ice mass is a challenge. Also, the additional pumps and conveyors require energy.

The system may not need a heat exchanger. If there is no heat exchanger, a certain amount of the carrier liquid will be carried along into the cooling distribution system.

The complexity of ice slurry systems outweighs their advantages in most applications. Initial cost is high. Slurry systems were never popular, and it seems that their niche in cooling storage is becoming narrower. Consider them for larger applications where the cost of custom fabrication is less of an economic penalty.

The original motivation for slurry systems was to freeze ice in such a way that it can be separated from the evaporator without a defrost cycle. In this regard, separated ice slurry systems can be considered a variation on ice shedder systems.

Liquid Chilled Water Storage

When electric utilities first started promoting commercial-scale cooling storage during the early 1980's, most designers envisioned storing the chilled water itself in large tanks. This method is conceptually simple. It also has a fundamental efficiency advantage over ice storage, namely, it does not require the chillers to operate at the low evaporator temperatures needed to freeze ice.

The storage tank is independent of the chillers, making it possible for the two to operate in parallel or separately. The connections between the chiller, the

storage units, and the chilled water system are relatively simple.

However, after a few years of analysis and experience, including some successful installations, it became clear that the enormous volume required to store water as a liquid is a serious economic disadvantage and a physical obstacle. The minimum temperature of liquid water is 32°F (0°C), and avoiding freezing in heat transfer forces the minimum temperature to be a few degrees warmer than this. In most HVAC applications, the chilled water is warmed to a temperature no higher than about 50°F (10°C). As a result, the cooling storage capacity of water is limited to about 15 BTU's per pound. This translates to about 100 gallons per ton-hour, an enormous amount of weight and volume that is difficult or impossible to accommodate in the layout of most facilities. Furthermore, the large surface area increases conduction losses.

In general, chilled water storage is not the best method of cooling storage unless storage is cheap or the storage volume is very large. Proponents of chilled water storage often pin their hopes on the availability of a large fire protection storage tank. Figure 19 shows

Fig. 17 Separated ice slurry system equipment The tall tower in the center is the evaporator. The ice storage tank is on the right. The chiller condensing unit is on the left.

a chilled water storage tank that also serves as a storage tank for fire protection.

Economy of scale benefits chilled water storage more than it benefits ice storage (except for ice capsule systems, which also have large economy of scale). The main reason is that tanks comprise most of the extra cost of chilled water storage, and the cost of the tanks per unit volume drops sharply with size. Also, relative heat loss drops substantially with larger tanks. Thus, chilled water storage may emerge as the best choice in very large applications.

■ Compatibility with the Chilled Water System

To achieve maximum storage capacity, the chilled water in the storage tanks is kept at the lowest possible temperature. This is lower than the usual chilled water supply temperature for HVAC applications, which is 42°F (6°C) or higher. Solving this problem is easy. Supply water from the storage tank is mixed with return water to produce the desired supply temperature.

A more critical problem is maintaining the highest possible chilled water return temperature. If the chilled water returns while it is still relatively cool, its remaining cooling energy will probably be unavailable for the remainder of the cooling cycle. This is because the water will be too warm to satisfy the cooling load if it is sent back through the chilled water system. If chilled water is wasted in this way, storage may be depleted before the end of the cooling cycle.

Keeping the return water temperature as high as possible requires special design of the user systems. The application equipment, typically cooling coils, should have throttling control valves, rather than bypass valves. The coils or other devices should be designed to maximize the temperature rise of the chilled water. Expect to use variable-flow chilled water distribution. This adds cost, but it also improves pumping efficiency, a valuable energy conservation measure in itself. (See Measure 2.5.2 for details.)

To keep the supply chilled water as cold as possible until the end of the storage cycle, the storage unit must keep the return water from mixing with the supply water. Any mixing dilutes the chilled water in the tank and poisons the ability of the storage unit to provide full cooling capacity. The need to separate the return chilled water somewhat increases the already huge volume requirement, and complicates the storage design. There are several methods of storing chilled water in a way that separates the return and supply water. Each of these methods, which are described next, has significant drawbacks.

■ Storage Using Multiple Tanks

One way of keep the supply water separate from the return water is to use a number of separate tanks. For example, use four tanks of equal size, three of which hold the initial charge of chilled water, while the fourth is empty. As chilled water is drawn from the first tank, return water is drained into the fourth tank. When the first tank is emptied, it becomes the return tank for the second tank, and so forth.

This method provides absolute segregation of supply and return water, which is especially valuable if chilled

WESINC

Fig. 18 Separated slurry system ice storage tank The returning chilled water is sprayed into the ice tank for cooling. The irregularity of the ice pile suggests that water distribution is less than ideal.

water is stored for longer periods, say, longer than one daily cycle. On the other hand, this method has several drawbacks. It requires extra volume. It increases tank cost per unit volume. The tanks have more surface for conductive heat loss. And, the piping and valve arrangements are somewhat complex.

■ Storage Tanks with Siphon Baffles

To avoid the disadvantages of multiple tanks, a single large tank may be divided into a labyrinth by partitions that form a series of siphon loops. Water is taken out of one end of the tank and is returned at the other end. At the point where the head of the return water stream contacts the tail of the chilled water stream, the area of contact is reduced by the baffles. Mixing is resisted by the difference in density of the supply and return water, especially in the downward flowing legs of the siphon loops.

Unfortunately, labyrinth tanks do not work as well in practice as it seems they should. The return water tends to overrun the chilled water, especially in the upward legs of the siphon loops. Also, the uninsulated steel baffles conduct heat from the return water to the chilled water. Experience teaches that a labyrinth tank should not be used unless its exact configuration has been tested and proven satisfactory.

■ Storage Tanks Using Stratification

"Stratification" is another method of keeping return water separate from the supply water. It allows water to be stored in a single large tank by exploiting the density differences of water at different temperatures. Water that is less dense floats on top of water that is denser. Cold water at supply temperature enters or leaves from the bottom of the tank, while return water enters or leaves from the top of the tank.

Water must enter and leave the tank gently, because the tendency to stratify is weak. The density difference between water at 39°F (4°C) and 50°F (10°C) is only about 0.03 percent. Therefore, the tank needs extensive diffuser trees in the top and bottom to reduce velocity and to control the flow direction.

The major disadvantage of stratified storage is its inability to separate supply and return water as the temperature approaches the point of maximum water density. The density of water is greatest at 39.2°F (4.0°C). If water is either warmer or cooler than this, it becomes lighter. If water colder than 39°F is pumped into the tank, it will float upward to mix with the water warmer than 39°F, destroying the stratification. Since a stratified tank cannot store chilled water below about 40°F, it needs more volume to compensate for the unusable temperature band.

There is much misunderstanding of this behavior, including a superstition that chilled water cannot be stored below 39°F. This is true only of tanks using stratification to separate the return and supply water. The other storage methods allow water to be stored at any temperature down to freezing.

Stratified storage has the advantage that the chillers operate at higher evaporator temperature, which increases COP. Also, using a single large tank has the advantage of minimizing the surface area for conductive heat loss.

An important efficiency drawback is that heat flows between the supply and return water by conduction within the tank. Any stratified tank has a region where the return water and supply water are of mixed temperature. This region, called the "thermocline," is typically several feet thick, regardless of tank shape. It

WESINC

Fig. 19 Chilled water storage This system provides air conditioning for a Florida prison. The huge tank was originally designed to hold water for fire protection. It functions well in both roles.

represents wasted cooling by the chiller, and it requires even more storage volume. You can minimize the thermocline problem by selecting a tank that is tall in relation to its diameter. For a system of any size, such a tank will be very tall.

You can achieve positive separation of the supply and return water within a single tank by using a moving diaphragm inside the tank. At the beginning of the discharge cycle, all the water is on one side of the diaphragm. As water returns, it enters on the other side of the diaphragm. The diaphragm itself adds some complexity, but it also eliminates the need for elaborate diffusers at each end of the tank. So far, this method has not become popular, perhaps because present diaphragm designs lack long-term reliability. Also, present diaphragm designs conduct heat from the return water to the supply water.

With stratified storage, the chillers operate at about the same temperature as they do when generating chilled water directly. Therefore, it is not necessary to make substantial efficiency compromises when selecting the chiller. For the same reason, stratified storage is usually compatible with existing chillers, except perhaps absorption units. With other types of chilled water storage, existing centrifugal chillers may seriously lose efficiency and capacity if chilled water is produced near freezing temperature. However, if charging is always done at night, the lower condensing temperature may avoid this problem.

■ How to Measure the Charge

With any of the previous types of storage, you can measure the amount of chilled water remaining in the tank fairly accurately. The usual method is installing a series of temperature sensors along the path of water flow in the storage unit. Systems with multiple tanks also need level sensors in each tank.

■ Combining Thermal Storage with Chiller Heat Recovery Storage

Chilled water storage offers the possibility of using the same tanks, at different times, for both chilled water storage and for storage of heat recovered from air conditioning condensers. This possibility becomes more attractive if the system has several storage tanks. For example, consider a system with four tanks. During hot weather, all four could be used for cooling storage. During cold weather, all four could be used for heating storage. During spring and fall, one or two of the tanks could be used for one purpose, and the rest used for the other purpose. See Measure 2.10.5 for storage of heat recovered from chillers.

Eutectic Storage Media

Eutectic cooling storage seeks to combine the efficiency advantage of chilled water storage with the compactness of ice storage. The main efficiency disadvantage of ice storage is that ice freezes at a temperature that is lower than necessary for most cooling applications, forcing ice storage chillers to operate at low evaporating temperatures. The desire to overcome this limitation has led to a search for materials that change in phase at a more desirable temperature. Such materials are called "eutectic." (This word is based on Greek words that mean, approximately, "good freezing.")

Ironically, the present eutectic alternatives to ice have phase change temperatures that are too high for optimum system performance.

■ Eutectic Materials to Date

Up to now, the best results for cooling storage have been obtained with hydrated salts, in which heat is absorbed in separating water molecules from salt molecules, a reaction that occurs at a fairly distinct temperature. (This is not a "phase change," strictly speaking, but the term is used.) The material has a phase change temperature averaging about 47°F (8°C).

Formulations of salts with other melting temperatures are being tried. Unfortunately, the phase change in hydrated salts involves substantially less heat than the melting of ice. Commercial eutectic salt storage units require somewhat more than twice the volume of ice units. This is somewhat less than half the volume of chilled water storage.

Changes in the structure of organic polymers, especially the melting of waxes, have also been investigated, but these have relatively low heat storage potential. These materials also suffer from loss of storage capacity with cycling. The melting temperature may be different from the freezing temperature.

Other exotic materials have been investigated, such as clathrates, where the heat is given off when molecules of one type enter voids in the crystal lattice of a different type of material.

Eutectics are an evolving area. Newer materials may someday improve upon the shortcomings of the present materials. A phase change temperature in the range of 38°F to 42°F (3°C to 5°C) would be ideal. Even with this property, eutectic materials would need to increase their volumetric heat capacity by at least a factor of two to make them superior to ice in most applications.

■ Tank Configuration

Storage units using eutectic salt solutions are similar to those using ice capsules. Sealed containers of the eutectic material are stacked in a tank, and water is circulated through the tank. One major advantage is that glycol is not necessary, since water can be circulated through the tank without freezing.

The design of the storage unit is simplified by the fact that eutectic materials are heavier than water, so they do not tend to move around the tank by flotation. Also, eutectic materials do not change volume appreciably with the phase change, so there is no stress on the containers.

The lack of volume change makes it impossible to measure the charge by measuring the system water volume. So far, nobody has developed a reliable method of directly measuring the cooling charge in a eutectic system.

■ Dealing with the High Phase Change Temperature

The freezing temperature of present eutectic systems is too high to provide chilled water at normal temperatures. In addition, present eutectic materials do not maintain their low temperature throughout the discharge cycle, so the chilled water temperature may exceed 50°F (10°C) toward the end of the discharge cycle. Chilled water systems designed for conventional temperatures will not be able to provide full cooling capacity at these higher temperatures. If a system is designed to operate with the warmer chilled water, it may not be able to provide adequate dehumidification.

A eutectic system can be used for pre-cooling return chilled water in conventional chilled water systems, providing about half the cooling load, while the chiller further reduces the chilled water system. This may be satisfactory in some applications where only a partial reduction of the peak cooling load is required to satisfy the purposes of the system.

Present eutectic systems cannot use heat exchangers, because the temperature rise in heat transfer would worsen the problem of high chilled water temperature. Since eutectic storage cools the chilled water directly, the system does not need a heat exchanger unless there is excessive static pressure in the chilled water system. System pressure can be handled in a variety of ways that are suggested below.

■ Unusual Opportunities

Eutectic storage at these high temperatures offers several possibilities that are not available with other methods of cooling storage. One is operation with conventional absorption chillers. Another possibility is storing "free cooling" from the operation of cooling towers alone (see Measures 2.9.2 and 2.9.3). This can be done when the outside air temperature is below the phase change temperature.

GENERAL DESIGN ISSUES

The previous discussion outlined the steps needed to put a storage cooling system into operation, and gave a comparative overview of the types of storage cooling technology that are available. We will now examine in greater detail several important engineering aspects of installing the system. These apply to most types of systems.

How to Deal with Chilled Water System Pressure

The height of a chilled water system causes hydrostatic pressure. Unlike chillers, most cooling storage units are not capable of resisting this pressure. Furthermore, most storage units are vented to the atmosphere. If they were directly connected to the chilled water system, portions of the system higher than the tank would overflow the tanks when the pumps are turned off. The following are methods of dealing with the static pressure of the distribution system.

■ Use a Heat Exchanger

The simplest method of isolating distribution system pressure is to use a heat exchanger. This method works well with ice storage systems that are designed to provide chilled water at normal temperatures. The low storage temperature easily makes up for the temperature drop in the heat exchanger. On the other hand, if an ice storage temperature is designed for low-temperature chilled water distribution, a heat exchanger would limit the minimum distribution temperature.

Heat exchangers are undesirable with chilled water storage, because their temperature loss would seriously reduce the effective storage capacity. By the same token, heat exchangers cannot be used with contemporary eutectic storage, which stores water at temperatures that are already higher than conventional chilled water temperature.

■ Elevate the Storage Unit

You can avoid the need for pressure isolation in open storage units by making the water level in the storage unit as high as the top of the distribution system. However, the enormous weight of water or ice is an obstacle to installing the storage unit high in the building structure. In existing facilities, it may be impractical to add sufficient structural support. In new construction, it may be easy to modify the structural design to provide adequate support. In fact, some huge ice storage systems have been installed in buildings above the level of the chillers, at the cost of heavily reinforced structures.

The weight of a storage system can be distributed to different parts of the structure by using a number of smaller storage modules. This practice is becoming common, especially for larger systems.

Chilled water storage is four to six times heavier than ice storage, making it generally impractical to install on the building structure. An exception is using fire protection tanks as storage units. These probably will not be large enough for a full-capacity storage system, but they may serve as valuable mini-storage units to reduce chiller size and reduce peak electricity costs.

■ Use a Backpressure Valve and Check Valves

Another approach is to install a backpressure valve at the bottom of the chilled water return leg. The valve is set to maintain a pressure somewhat higher than the hydrostatic pressure in the system. When the chilled water distribution pump turns off, the backpressure valve closes completely, trapping the water in the return leg.

Similarly, a check valve is used at the bottom of the supply leg to prevent water from draining out of the supply leg when the pump turns off.

Both backpressure valves and check valves tend to be leaky. One solution is to add a small pump to return any leaked water from the storage unit to the chilled water system. Another is to install automatic tightly closing valves to augment the backpressure and check valves when the pump turns off.

A backpressure valve may greatly increase the energy needed for pumping chilled water. In a closed system, the hydrostatic pressure in the return leg balances the hydrostatic pressure in the supply leg, so gravity plays no role in the pump power. However, using a backpressure valve effectively adds a gravity load that is equivalent to pumping all the water in the system to a height that corresponds to the pressure setting of the backpressure valve. Then, all the pumping energy is converted to heat in the return water, increasing the chiller load. This energy penalty will be imposed whether cooling is being provided from storage or by the chillers directly. In a tall building, a backpressure valve may cause pumping to absorb a large fraction of the energy consumption of the chilled water system.

■ Use a Pressurized Storage Unit

In principle, most types of cooling storage could be installed in closed tanks directly connected to the chilled water system. Expansion of ice would be accommodated by the expansion tank in the chilled water system.

The practical limitation is tank weight and cost. The wall thickness that a tank needs to withstand a given pressure is directly proportional to the tank diameter. Cooling storage tanks are large, so designing them to withstand pressure would require extreme thicknesses. In the United States, any vessel holding 15 PSI of pressure requires ASME pressure vessel certification, which further increases cost. This amount of hydrostatic pressure is produced by a height of merely three stories. Furthermore, tanks require access for inspection and maintenance. Large access panels are difficult to create and seal in pressurized tanks.

Coil-melt ice storage systems are a significant exception. In these systems, only the interior of the coils are subjected to system pressure. The coil diameters are small, so they can easily withstand large pressures. However, this is usually a moot point. The coils in coil-melt systems are filled with glycol. To avoid the need to fill the entire distribution system with glycol, it is common to use a heat exchanger with coil-melt systems. This by itself isolates the storage unit from system pressure.

Pressurized vessels cannot be used ice shedders, slurry systems, and multi-tank chilled water systems, because these types must be open to the atmosphere.

Chilled Water Supply and Return Temperature

Some cooling storage systems impose severe restrictions on chilled water supply and/or return temperatures. Using these types of storage systems requires you to modify the equipment that uses chilled water. Other types of storage systems impose no restrictions. The following are the major cases.

■ Ice Storage with Normal Chilled Water Temperatures

The simplest situation is using an ice storage system to provide chilled water to the facility at conventional chilled water temperatures. In this case, no change to the distribution system or user equipment is needed.

Using the chilled water at storage temperature would waste energy by excessive dehumidification, and might cause control instability at low cooling loads. To avoid these problems, control the chilled water supply temperature by using a mixing valve to dilute the cold water from the ice tank with return water.

■ Ice Storage with Minimum Chilled Water Temperatures

In attempting to salvage some benefit from the unnecessarily cold chilled water produced by ice storage systems, user systems are being developed that exploit the lower chilled water temperatures, especially in new construction. Colder chilled water carries more cooling energy, so pipes and pumps can be reduced in size, and the pump energy can be reduced correspondingly.

Where the low-temperature chilled water is used for air conditioning, air can be made colder. This can reduce the size of ducts, enclosures, and fans, and can reduce fan power consumption. These advantages have captured the attention of many designers. However, enthusiastic designers should be cautious of these potential problems of low-temperature air distribution:

- *cold air dumping.* Cold air "dumps" from overhead diffusers, causing serious localized discomfort. Overcoming this problem requires new types of diffusers that mix supply air with room air. It remains to be seen whether such diffusers will be satisfactory, especially in VAV applications. Terminal units with mixing fans may be required, adding cost, noise, and energy consumption.
- *duct sweating,* which may damage ducts, insulation, ceiling surfaces, and other equipment.
- *excessive dehumidification.* The low coil temperature increases dehumidification, which increases energy expenditure. Below normal air distribution temperature, the additional dehumidification usually does not improve comfort, and it may be an irritant.

Low-temperature air distribution is likely to provide more headaches than benefit, at least for the near future. This is a new area of comfort engineering, and it is likely that lessons will be learned at the cost of expensive mistakes.

Low-temperature distribution marries the facility systems to the low evaporator temperatures required for ice storage. This may be an unwise commitment, because the low chiller efficiency may cause storage cooling, at least on a continuous basis, to be abandoned at some time in the future.

If the facility is designed to use low-temperature chilled water distribution, the ice storage system must be able to sustain low water temperature throughout the discharge cycle. Ice shedders and external-melt units are best in this regard, while internal-melt units are worst. Ice capsules suffer a serious rise in chilled water temperature in the second half of the discharge cycle.

■ Maximizing Efficiency with Liquid Chilled Water Storage

Refer to the description of liquid chilled water storage in the previous summary. The main design challenge is that you have to maximize the temperature differential between supply and return chilled water in order to preserve the cooling capacity of the storage unit. As explained previously, the way to do this is to use variable-flow chilled water pumping. In retrofit, this may require major modifications of the cooling equipment, including coils, control valves, and pumps.

■ Eutectic Storage with High Chilled Water Temperatures

Present eutectic storage systems produce chilled water that is warmer than normal. If the chilled water from the storage unit is not cold enough to satisfy the cooling load, it can be cooled further by a chiller before being distributed to the facility. This method is a hybrid of direct and storage cooling.

In any event, the return water temperature must remain high enough so that melting of the eutectic material does a large fraction of the cooling. Using variable-flow chilled water pumping and 2-way throttling valves on the coils helps to maximize the return temperature.

Keep Charging Separate from Direct Cooling

Chiller COP falls along with evaporator temperature, so operate the chillers at low evaporator temperature only for charging the storage unit. Try to operate the chiller separately, at the highest possible evaporator, for direct cooling. In other words, avoid using the same chiller for both storage charging and direct cooling simultaneously.

In principle, any kind of cooling storage system can be designed to allow the chiller to switch between charging and direct cooling. This is easier with some types of systems than with others. It is easiest with a liquid chilled water storage system. It is also fairly easy if the chiller cools a glycol solution. The glycol solution can be sent either to the cooling storage unit or to a heat exchanger for direct cooling.

Using a heat exchanger forces the chiller to operate with lower evaporator temperature. In principle, the chiller could switch between chilling glycol (for ice freezing) and chilling water (for direct cooling), but the complications would not be justified by the reduced temperature loss.

There is another solution that is less conventional, but more efficient: add a second evaporator to cool the chilling water directly. The two evaporators cannot operate at the same time, because the evaporator temperatures are different. This solution may be fairly simple if the chiller system is designed for a separate evaporator, as with a liquid overfeed system. It may be difficult to accomplish with some packaged chillers. In any event, this is a job for someone who knows how to design unconventional chiller systems.

■ Separate Chillers for Storage and Direct Cooling

With any type of storage, a chiller that is optimized for one evaporator temperature is less than optimum at any other temperature. This is especially true of centrifugal chillers. Screw compressors with variable discharge ports can maximize their COP's at different evaporator temperatures. Reciprocating chillers are fairly tolerant of changes of evaporator temperature.

From the standpoint of efficiency alone, it would be desirable to install separate chillers for direct cooling and for charging the storage unit. However, this is generally not economical if it leads to duplication of equipment cost. Consider this approach if the facility has a large cooling load during the utility's off-peak periods. In this case, size one set of chillers to operate continuously during the off-peak period to charge storage, and size another set of chillers to satisfy the off-peak cooling load. For example, hospitals and hotels might benefit from this approach.

Controls and Alarms

Cooling storage systems require specialized control. You may have to do some or all of the control design separately, even with packaged storage systems. This is because the control sequences vary with the purposes of the system, with different electricity pricing structures, and with the load characteristics of the facility. You generally need a computer or a programmable logic controller (PLC) to control cooling storage for optimum efficiency. The control functions are too complex, too variable, and too repetitive for manual operation. The following is a guide to the control and alarm functions that the system may need.

■ Controlling the Amount of Storage for Each Cycle

The system controls must charge the storage unit based on the anticipated cooling requirement during the next discharge cycle. Having too little capacity will either forfeit the electricity price saving or leave the facility without cooling at the end of the day.

On the other hand, storing too much cooling wastes energy, because more energy is required to produce stored cooling than to cool the space directly. Some of the excess stored cooling is wasted between discharge cycles by conductive loss from the tank.

With some types of storage, such as ice shedders, residual ice is melted if the chiller changes to direct cooling, wasting the ice. Even if the ice storage chillers are not used for direct cooling, the storage units require periodic melting of all the ice to prevent channeling or uneven ice accumulation. Exceptions are ice capsule systems and some slurry systems.

Unfortunately, the control system cannot predict the storage load for the next cycle accurately, because of changes in weather and internal loads. If the system wastes much energy as a result of leftover cooling storage, be ready to make manual control adjustments when a weather system is approaching or other changes in the load are foreseeable. Manual adjustment is an unreliable mode of operation in the long term, so try to avoid a system design that requires it.

■ Time Control to Conform to Electricity Rate Schedules

Electricity prices are usually related to the time of purchase. For example, the utility may charge its highest demand rate between the hours of noon and 8 PM. There may be a lower demand rate that applies for several hours before and after these times, respectively. Design the controls to limit chiller operation to the times of lowest electricity cost.

Optimum scheduling depends on the electricity rates for each period, the existence of ratchet provisions in the rates, the cooling load, the storage capacity, etc. Operate the storage system only at times when electricity costs are low enough to offset its efficiency disadvantage. The exception is when storage capacity is needed to augment chiller capacity.

■ Other Control to Conform to Electricity Rate Schedules

Electricity rates may not always be based on time. For example, excess-consumption charges and "hour charges" are based on consumption. Design your controls to get the most benefit from these rates, which involves measuring the total facility load. Some rates make it difficult to achieve optimum control. Expect to use a computer that you can program to follow changes in rates and rate schedules.

■ Facility Demand Limiting While Cooling Directly

If the electric utility's demand charge is an important cost factor, balance direct cooling with cooling from the storage system to keep demand below a target level. If the non-cooling electrical load varies during the

demand period, as illustrated in Figure 2, this load usually determines the peak demand setting. A higher demand level is necessary if the storage capacity is not large enough to carry the entire cooling load. The arrangement of electric metering is a factor to consider, because different metering arrangements result in different total demand charges, as explained previously.

If the non-cooling load is unpredictable, limit chiller operation based on a conservative estimate of the non-cooling load. If the non-cooling load rises above this level, creating a higher demand charge for the period, raise the chiller load limit accordingly. Reduce the charging of the storage unit by a corresponding amount. Start all over again at the beginning of the next demand measurement period.

■ Time Control to Minimize Condensing Temperature

The controls should exploit the fact that condensing temperature has a major effect on chiller efficiency. Unless a weather system is moving through the area, the outside temperature drops continuously throughout the night. Therefore, charge the storage units as late as possible without running into a period of higher electricity rates. This control feature applies only to chillers that are able to exploit low ambient condensing temperatures.

■ Time Control to Reduce Storage Losses

The storage tank loses heat by conduction, so recharge it as late as possible. Charging at night provides an additional bonus of lower condensing temperature. If the storage tank is well insulated, small differences in the time of charging will not make much difference.

■ Control of Evaporator Temperature When Charging Storage

To save energy in charging, keep the evaporator temperature as high as possible. This requires control to correlate the evaporator temperature to the percentage of charge. An easy way to achieve this is to keep the compressor load constant throughout the charging period. With most types of storage units, this slows the rate of charging as the charge increases. Exceptions are ice shedders and some ice slurry systems, in which the charging rate is constant.

Increasing the evaporator temperature during charging lengthens the charging time. Therefore, coordinate this control function with the control function that determines the starting time for charging.

■ Maintaining Efficient Chiller Loading During Charging

The efficiency of a chiller may vary substantially with load. Set the chiller load during charging for maximum efficiency, whenever possible. The effect of loading on efficiency for different types of chillers is covered by Reference Note 32, Compression Cooling.

Depending on the type of chiller, this control consideration may conflict with minimizing the evaporator temperature differential during charging. This is because the temperature differential increases with load.

■ Distributing the Load Among Chillers

See Measures 2.1.1 ff for the most efficient ways to distribute load among chillers. The storage system controls should determine which chillers serve the cooling load directly and which chillers charge the storage unit. If more than one chiller is used for charging storage, or if more than one chiller is used for direct cooling, try to use the same controller that distributes the load between the chillers, if there is one.

■ Safety Control to Prevent Overcharging

Install an independent override control to prevent overcharging. Overcharging may burst the tank with some types of ice storage. Also, energy is wasted in making more ice than the storage unit's design capacity. This control is a safety device, so it should be entirely independent of the operating controls. It should have a separate sensor that is triggered when the tank's ice capacity is exceeded.

■ Alarms

Install an alarm for each critical control function. For example, if a tank is not accumulating ice when it should be, an alarm should indicate this as early as possible, to avoid exceeding the demand limit during the next discharge cycle.

There should also be alarms for casualty conditions that are not related to controls. For example, open tanks should have high and low level alarms. The system should have all the conventional refrigeration alarms, such as high and low head pressure.

Alarms should be independent of the control system, and they should have their own sensors. Otherwise, a failure of the control system may disable the alarm. For example, the timing of chiller operation with respect to demand periods should be monitored by an independent alarm having its own clock.

ECONOMICS

SAVINGS POTENTIAL: *Cooling storage provides cost savings, not energy savings. The savings vary widely with circumstances, depending mostly on the pricing policies of electric utilities and your ability to negotiate with the utility. The cost savings provided over the life of the system are much less predictable than with most energy conservation measures.*

COST: *Coil-type thermal storage units and large chilled water storage tanks cost from $50 to $100 per ton-hour of storage, depending on size and type. This cost does not include the chillers themselves, piping connections, or controls. Heat exchangers, if needed, cost from $100 to $200 per ton of heat exchanger capacity. The evaporator assemblies of ice harvester systems cost from $500 to $700 per ton of capacity, not including the compressors, insulated storage tank, piping, and pumps. Ice slurry evaporators cost from $300 per ton in large sizes, to $800 per ton in small sizes. At the present time, many electric utilities will defray part of the system cost.*

PAYBACK PERIOD: *Several years, to much longer.*

TRAPS & TRICKS

MOTIVES: *One of the reasons that cooling storage units have a poor performance record is that they are installed as toys. When the charm wears off, the toy is abandoned. Before you install one of these big, expensive systems, make sure that you are ready to live with its operating and maintenance requirements. Understand that thermal storage exists for the benefit of the utility, and that you will benefit only to the extent that the utility provides a rate subsidy or other incentive for operating the system. Utility representatives will give you an optimistic pitch about this, but probably will not offer to provide a rate guarantee for a period that is long enough to protect your investment. You must learn enough about utility issues to judge whether a thermal storage installation is likely to be profitable in the long term. In particular, consider the effect of utility deregulation. Also, monitor the development of electric vehicles, which may abolish utility interest in thermal storage.*

DESIGN: *Designing an optimum thermal storage system involves a broad range of subjects, some of which are outside the realm of conventional HVAC design. A single detail that you overlook can cripple your system. Don't take the building design off on tangents to accommodate the storage system. In particular, be skeptical about low-temperature chilled water and air distribution.*

SELECTING THE EQUIPMENT: *Thermal storage equipment is still evolving, and performance is spotty. Favor equipment that has been in operation long enough to demonstrate good reliability. Shop the market thoroughly. Check with other users of the equipment you are considering. Remember that the equipment has to last for 20 years or longer if it is replaceable, and for the life of the facility if it is not replaceable.*

INSTALLATION: *Storage systems are heavy and require a lot of space. Make sure that structural support is adequate. Install the equipment so that it will survive. For example, avoid burying steel tanks. Provide easy access to all parts of the equipment for repair and replacement.*

ALERTNESS: *Maintaining a high level of skill and attention is an inherent part of operating a thermal storage system. During peak load conditions, the system must operate with total reliability to avoid losing the* economic benefit. The system requires continuous surveillance during the peak load period. Install alarms to reveal all likely failure modes.

EXPLAIN IT: The logic of the system can be confusing to the operators. Put up a well designed system diagram that explains how the system works, identifies its components, and spells out operating procedures. Install placards on all parts of the system that require action or inspection.

MONITOR OPERATING EFFICIENCY: Make a continuous record of the charge status of the storage units and the loads on the chillers. Check the record often to make sure that the amount of energy stored is not too much or too little for the facility cooling load. Schedule periodic checks of controls operation.

Section 3. SERVICE WATER SYSTEMS

INTRODUCTION

Service water systems provide cold and hot water for consumption. In residential and commercial facilities, service water is used in three main areas: (1) sanitary applications, including lavatories, showers, toilets, and urinals, (2) food and beverage applications, including food preparation, dishwashing, and drinking fountains, and (3) lawn and plant watering. Industrial uses of service water are even more varied. They include washing, cooling, and use of water as an ingredient in materials being manufactured.

Virtually all service water is recycled eventually. In many industrial operations, water is recycled within the facility. In most residential and commercial facilities, service water is not recycled within the facility. Instead, it is recycled through the environment, usually after being treated by a public sewage system. Water must be purchased to replace water that is discharged to the sewer, water that evaporates, and water that becomes part of products.

How to Use Section 3

This Section is organized by the major cost components of service water systems. These are:

- *water consumption.* The Measures of Subsection 3.1 reduce water consumption by improvements to water fixtures and by helping users to conserve water.

- *water heating.* The Measures of Subsection 3.2 improve the selection and operating efficiency of water heaters.

- *service water pumping.* The Measures of Subsection 3.3 reduce pumping energy by improvements to pumps, piping, and storage.

Reducing Service Water Consumption

In many facilities, the cost of domestic water can be a surprisingly large fraction of total utility costs. Furthermore, the cost of water in many locations is rising rapidly. The Measures in this Subsection reduce consumption of service water with more efficient fixtures, more efficient use of fixtures, and better maintenance. Where the fixtures use hot water, these Measures also reduce energy consumption for water heating.

New fixtures are being developed for the purpose of using less water. However, newer fixtures are not necessarily better. Equipment that attempts to reduce water consumption may be unpleasant to use. Some items intended to conserve water, such as toilets with reduced tank capacity, may actually increase water consumption in practice. There is a place for new technology, but combine it with common sense in selecting fixtures and with motivating people to use water efficiently.

INDEX OF MEASURES

3.1.1 Repair water fixtures regularly.

3.1.2 Install efficient wash basin fixtures.

3.1.3 Install efficient shower heads.

3.1.4 Install shower valves that allow easy control of temperature and flow rate.

3.1.5 Provide instructions for efficient use of water in showers and lavatories.

3.1.6 Install efficient toilets.

3.1.7 Install efficient urinals or improve existing urinals.

MEASURE **3.1.1 Repair water fixtures regularly.**

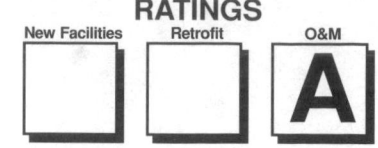

The most common control devices for domestic water are valves, faucets, faucet aerators, toilet and urinal flush valves, and shower heads. All of these will eventually leak, fail, or get lost. Leaks can waste a large amount of water over the course of time because they are continuous. Repairing water control equipment is simple, important maintenance that should be accomplished continually.

Faucets and Valves

Stop leaks early to reduce the amount of maintenance required. If a faucet leaks for a long period of time, the valve seat will erode. Replacing the soft valve disk as soon as it wears out minimizes the need to replace valve seats.

Valve seats do wear out eventually, especially in environments with acidic or gritty water. Replacing them in faucets is usually easy. Use a thread sealant to prevent leakage between the valve seat and the body of the valve. Hose connection valves ("hose bibs") may lack

Fig. 1 Typical wash basin The elements that need regular maintenance are the faucet(s), the faucet aerator, and the stopper.

WESINC

SUMMARY

Basic maintenance that reduces the water bill and slows fixture deterioration.

SELECTION SCORECARD

Savings Potential .... $ $

Rate of Return .... % % % %

Reliability .... ✓ ✓

Ease of Initiation .... ☺ ☺ ☺

---

removable seats, making it necessary to replace the valve body.

Adjust the valve stem packings to prevent leakage around the valve stem. Use a wrench that cannot mar the surface of the nut. When you need to apply too much force on the nut to stop leakage around the stem, it is time to replace the packing material. Valves commonly have a second packing that prevents leakage around the valve stem when the valve is fully open. Inspect this packing when replacing the primary packing.

Some types of valve disks last longer than others. Learn from experience which types provide the best service, and keep an ample stock of spares on hand. Also, keep on hand a stock of repair kits for your specialized faucets, such as single-lever types. Keep a few spare valve stems, as these wear out occasionally, especially in acidic water.

Faucet Aerators

Faucet aerators serve several purposes. They improve cleansing. They provide a sensation of vigorous flow that is desired by users. And, they conserve water by restricting flow. Figure 2 shows a typical aerator.

Aerators are screwed into the end of the spigot. They typically consist of a screen that breaks up the water flow. The screen is held by a fitting with air passages. This arrangement mixes air with the water stream as the stream is broken up by the screen.

Aerators must be removable to allow for occasional removal of debris that backs up behind the screen. The fact that they are removable allows them to work loose and be lost. For this reason, check faucets periodically to be sure that the aerators are present and installed tightly.

You may reduce loss of aerators by using thread locking compound when reinstalling them. However, this is not necessary if maintenance personnel are careful, and it may cause more trouble than it avoids. If you do this, select a compound that does not grab excessively or harden with age.

Basin Stoppers

In lodging and residential facilities, leaking basin stoppers not only waste water, but they annoy users who need to fill the basin with water. See Figure 1. The annoyance may motivate the occupant to abandon the stopper and to keep a continuous stream of water running into the basin.

The usual problems with mechanical basin stoppers are improper adjustment, lack of lubrication of the linkage that drops the stopper into place, and clogging by hair and other debris. Basin stoppers need frequent, regular cleaning. Make this somebody's responsibility. You may have to train maids or cleaning crews how to do this. This will radically reduce the need for maintenance calls and increase occupant satisfaction.

Gravity Tank Toilet Valves

Toilets that use gravity tanks have two important valves. The flush valve is a large stopper in the bottom of the tank that is lifted by the flush handle when someone flushes the toilet. The fill valve refills the tank as soon as the tank starts to drain. It is operated automatically by a float,

The flush valve in the bottom of the gravity tank toilets will eventually start to leak. This allows water to leak out of the tank into the bowl, where it siphons into

Fig. 2 Faucet aerator This simple faucet part is important for water conservation and user satisfaction. It must be unscrewed periodically to clear out grit. As a result, it may get lost.

the sewer. The flush valve has a large sealing area, so it can waste a lot of water when it leaks. A leak becomes progressively worse, and it will eventually start to destroy the valve seat. Replacing the valve seat is a messy job. It is easier in the long run to check for leakage and make repairs in a timely manner. At first, leakage of the flush valve is not obvious, so the trick is to identify it early.

Any easy test method is dropping a dye tablet into the tank. If the flush valve is leaking, the dye appears in the toilet bowl after several minutes. Dye tablets for this purpose are a standard hardware store item, but any dye can be used that does not stain the toilet. Food coloring works well.

Another foolproof method is to turn off the water supply to the tank when it is full. A leaking flush valve allows the tank to drain. You can usually tell that the water level has dropped within a fraction of an hour. A bonus of this method is that it exercises the shutoff valve, which typically freezes from lack of use.

A leak will start to become apparent if it is ignored too long. If the toilet is not used much, you may occasionally hear the tank valve open to refill the tank. If the water contains any staining agents, a stain may form in the bowl where water enters from the tank. But, don't wait for this. A leak may exist a long time before a stain becomes apparent. Also, the stains may be permanent.

The second important valve in a gravity tank toilet is the tank filling valve. Make sure that the fill valve shuts tightly after the tank refills. The float level needs to be adjusted properly. If the float cuts off the water supply at a level that is too low, the toilet does not flush properly. If the float is set too high, the incoming water drains continuously into the overflow pipe. If the fill valve leaks, the tank continues to fill slowly, and the excess water dumps into the overflow pipe. The operation of this valve is easy to check visually.

In time, the fill valve has to be replaced. The assemblies for most toilets are standard and inexpensive. Replacement takes about fifteen minutes with practice, if none of the coupling nuts are frozen.

Flushometer Valves

Flushometer valves on toilets and urinals eliminate the need for a gravity tank. They remain open for a fixed time interval, discharging the appropriate quantity of water into the bowl. The lack of a tank makes it difficult to test for valve leakage. In both toilets and urinals, you have to check visually for water flow into the bowl. Dry the sides of the bowl and inspect carefully.

Repairing a flushometer valve requires spare parts that are specific to the particular model.

Shower Heads

Shower heads eventually clog because they act as strainers. To clean them, simply unscrew them and dump out the grit. High-efficiency shower heads need cleaning more often because they have smaller orifices.

Shower heads that have their own shutoff valves eventually leak. When they do, replace them with shower heads that do not have shutoff valves. See Measure 3.1.3 for recommendations.

Shower valves that allow the user to adjust the spray pattern eventually wear out. Replace them with the most efficient type.

ECONOMICS

SAVINGS POTENTIAL: *Several hundred to several thousand dollars per year, in facilities of typical size and water consumption.*

COST: *Minor. Repair parts for common fixtures are inexpensive.*

PAYBACK PERIOD: *Short. The cost of the water saved is much greater than the cost of the repairs.*

TRAPS & TRICKS

SCHEDULING: *Inspect all water fixtures often enough to keep them from leaking for long periods. Make this a repetitive item on your maintenance calendar. If possible, train maids and cleaning crews to report defective fixtures.*

SPARE PARTS: *After you learn which repair items work best, keep an ample stock on hand. The bother of finding the right repair parts may deter maintenance more than the work itself.*

MEASURE **3.1.2 Install efficient wash basin fixtures.**

The use of water in wash basins is typically wasteful. The way you select faucets can reduce water consumption, although you can't expect the efficiency of a soldier shaving from his helmet. The key points are to select wash basin fixtures that minimize the flow rate and the duration of flow. This is especially important for hot water, where you are saving heating energy as well as the water itself.

In the United States, Federal law requires that faucets made after 1993 must limit water flow to these values, assuming a supply pressure of 80 PSI:

- lavatory faucets and aerators, 2.5 gallons per minute
- kitchen faucets and aerators, 2.5 gallons per minute
- self-closing faucets, 0.25 gallons per cycle.

Do not assume that all hardware will meet these criteria, or that they substitute for good judgement in selection. The flow rate of faucets may increase considerably if the aerator is lost or if the self-closing mechanism fails. Durability is an important selection factor.

Private-Use Wash Basins

Where wash basins are used for all purposes, as in private rooms and dormitories, follow these guidelines:

- *use single-spigot fixtures.* The basin should have a single spigot that mixes hot and cold water from separate taps. Using separate spigots for hot and cold water requires mixing the water in the basin to get the proper temperature. This wastes water and takes time.

- *provide effective stoppers.* The stopper should close the drain tightly. If the stopper leaks, the user must refill the basin continually. If leakage is severe, the user will leave the stopper open and use a continuous stream of water. Stoppers should resist jamming and should be easy to clean.

- *use aerators.* Aerators restrict the flow, make the stream effective for washing at a lower flow rate, and provide an increased sensation of flow.

Public-Use Wash Basins

Where wash basins are used mostly for hand washing, as in public lavatories, follow these guidelines:

- *use single-spigot fixtures,* for the same reasons as before

- *use aerators,* for the same reasons as before

- *use faucets that turn off automatically.* This feature is particularly important where wash basins are used by transients, as in airports and restaurants.

SUMMARY

Simple decisions about standard equipment can save a lot of water.

SELECTION SCORECARD

Savings Potential .... **$ $**

Rate of Return, New Facilities **% %** % %

Rate of Return, Retrofit.......... **% %**

Reliability .... ✓ ✓ ✓ ✓

Ease of Retrofit .... ☺ ☺ ☺

It is less important in facilities where the population is stable and responsible, as in office buildings.

A disadvantage of current types of automatic faucets is that they always operate with maximum faucet flow, which makes it especially important to maintain the aerators to limit flow. However, the full flow is tolerable if the faucets operate only when they are needed.

The older style of automatic faucet uses a dashpot that allows flow for a fixed time interval after a valve is depressed. This type is somewhat awkward to use, and it tends to stick.

Newer automatic faucets use infrared sensors that sense the presence of hands in the basin. This type allows water flow only when needed, and forces the user to be somewhat deliberate in hand washing. To date, this type of equipment is expensive and

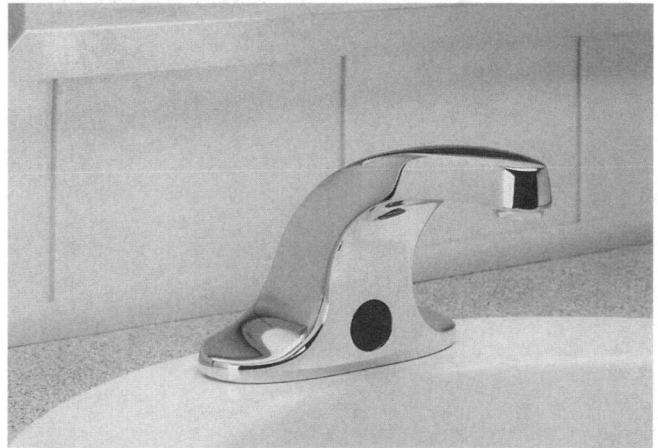

American Standard, Inc.

Fig. 1 Hand washing faucet with infrared sensor

This method of control is efficient for hand washing because it limits flow to precisely the period needed. It requires electrical power for the valve and the sensor. Do not use it in locations exposed to heat sources other than users' hands.

somewhat temperamental, but it may improve in time. Figure 1 shows a typical unit. Do not use this type in open environments where the infrared sensor may be triggered by a heat source in the background, such as sunlight through a window.

• *use single-faucet fixtures with a fixed-temperature water supply.* To avoid the waste of energy that occurs when users attempt to mix hot and cold water, install fixtures that have a single valve. This approach is more expensive because it needs a separate branch of the service water system to serve the faucets. This branch has additional piping, a mixing valve, and probably an additional recirculation pump.

In general, avoid basin fixtures that are either awkward or unpleasant to use, even if they are efficient. Figure 2 shows faucets of this nature.

Soap dispensers are a continuing maintenance nuisance, and fiddling with the soap dispenser may cause users to waste water. Figure 3 shows a compact type of basin that combines soap dispensing, washing, and drying in the same location.

■ Avoid Communal Wash Basins

Avoid communal fountain-type lavatory fixtures with multiple nozzles or spigots. It makes no sense for an individual to wash his hands with a fixture that is designed to provide water flow for a dozen persons. Also, the spray pattern of many models is particularly inefficient for hand washing.

In men's restrooms, these fixtures tend to be confused with urinals, especially by foreign visitors, sometimes leading to considerable embarrassment.

Reliability and Ease of Maintenance

Measure 3.1.1 emphasizes that maintenance is a major factor in water conservation. Select basin equipment for reliability and ease of maintenance. Some plumbing fixtures are much better in these respects than others.

ECONOMICS

SAVINGS POTENTIAL: *30 to 70 percent of the water and water heating energy used in wash basins.*

COST: *Self-closing single-temperature faucets cost about $70 each. Single-temperature faucets controlled by infrared sensors cost about $170 each, in small quantities.*

PAYBACK PERIOD: *Several years or longer.*

WESINC

Fig. 2 An efficient type of faucet that is awkward to use
These faucets require the user to push on a valve rod under the spigot. Foolproof in terms of water conservation. But, unfamiliarity and fear of contamination by other users discourage people from washing their hands.

American Mill Sales

Fig. 3 Automatic hand washing station This basin includes a soap dispenser, faucet, and hot air dryer, all operated automatically. The covers are removed to show the equipment inside.

MEASURE 3.1.3 Install efficient shower heads.

RATINGS

New Facilities	Retrofit	O&M
A	**B**	

Ordinary shower heads flood the bather with much more water than is needed for efficient bathing. Figure 1 shows an example.

A variety of "high-efficiency" shower heads are available that you can use to limit water flow. These reduce water consumption by three methods: using smaller orifices, narrowing the discharge pattern, and mixing air with the water to increase the sensation of flow.

The sensation provided by the shower head is an important subjective factor for bathers. Experience shows that a particular shower head may be pleasing to one person and annoying to another. A partial solution to this problem is to select shower heads that let the bather change the spray pattern. However, this also forfeits part of the potential saving.

Savings Potential

Present types of high-efficiency shower heads use one to two gallons per minute if the user controls the water pressure to the shower head, and about two to three gallons per minute with full-flow valves. In contrast, common types of shower heads that are not designed for efficiency may pass 4 to 8 gallons per minute. The difference is starting to narrow as manufacturers increasingly design for water conservation. Even today, selecting a shower head for efficiency may reduce water consumption by half or better.

High-efficiency shower heads may cost less than conventional shower heads. This is because they are simpler. In new construction, this makes high-efficiency shower heads pay off immediately. Figure 2 shows a typical inexpensive high-efficiency shower head.

More expensive, "deluxe" shower heads have features that tend to be wasteful. Figure 3 shows an example. Of course, it is always possible for a manufacturer to jazz up an efficient unit and raise the price.

SUMMARY

An inexpensive way to save water and water heating energy. Easy to install. Differences in personal preference are a problem.

SELECTION SCORECARD

Savings Potential	$ $ $
Rate of Return, New Facilities	% % % %
Rate of Return, Retrofit	% % %
Reliability	✓ ✓ ✓
Ease of Retrofit	☺ ☺ ☺ ☺

In retrofit, the payback period for replacing shower heads with high-efficiency units varies widely, depending on:

- ***the average number of users per shower head per day.*** High-efficiency units save more energy when installed in a busy gymnasium or in a communal dormitory shower than when installed in a hotel room or in a private residence.

- ***the habits of the bathers.*** The water consumption during each shower depends on the duration of the shower and the setting of the flow rate. This depends largely on whether the shower is used primarily for hygiene or as an esthetic experience. See Measure 3.1.5 about showering behavior.

- ***the type of valve.*** See Measure 3.1.4 about efficient shower valves.

- ***the cost of water and water heating energy***

- ***the characteristics of the shower head,*** discussed next.

Control of Spray Characteristics and Flow Rate

Many high-efficiency shower heads are adjustable to accommodate user preferences. Some people prefer a needle-like spray, while others want a fuller flow at lower velocity. In order for a shower head to work properly, the full pressure of the water system must be delivered to the orifices to break up the water into droplets.

Most adaptable shower heads adjust the spray pattern by varying the sizes of the orifices. The flow rate varies with the user's setting. With larger orifices, the stream is not broken into droplets. There is no way to avoid the fact that a coarser spray uses more water.

The least expensive models change orifice size by sliding one or more tapered plugs, or plugs with tapered grooves, to various positions in a nozzle plate. More expensive models use several sets of nozzles. Experience teaches that no single type is preferred by a majority of people.

Some models include a pulsing spray as one of the settings, as in Figure 3. This is an esthetic feature, not an energy conserving feature.

Select for Ease of Use

You will save more energy if you make it easy for bathers to use showers efficiently. If you install adjustable shower heads, select a type that is easy to use. The easiest adjustment is a large butterfly knob on the side of the shower head. Some cheap units have a knob in the center of the nozzle plate, forcing the user to reach through the spray to adjust it. This is annoying.

Shower heads that include several spray nozzles commonly require the bather to turn a large ring

Fig. 1 Water waster This shower head has large orifices, which require a high flow rate to provide a satisfactory spray.

surrounding the head, as in Figure 3. This can be confusing and awkward for unfamiliar users.

For those of European taste, high-efficiency shower heads are available in "telephone" style, installed on flexible hoses. They are awkward to use. If the shower space is confined, it is more convenient and more efficient to take "Navy" showers using a fixed shower head. See Measure 3.1.5 about this.

Install User-Friendly Shower Valves

If the valve in the shower provides only full flow, the bather must adjust the spray at the shower head to modulate the flow. This is likely to produce a shower that is not as pleasing to the user. Simple shower valves are more efficient and more pleasing. See Measure 3.1.4 about shower valves.

Avoid On-Off Valves on Shower Heads

Some high-efficiency shower heads include an on-off valve. Figure 4 shows an example.

In principle, this valve avoids the need to adjust the main hot and cold water valves. Once the main valves are set properly, the water is simply turned off at the shower head.

Don't use this type of shower head. In reality, it probably does not speed the process of getting the desired water temperature in the shower. The stagnant water must still clear out of the hot water supply line, and this takes longer if the user does not open the hot water valve to clear the line quickly.

Furthermore, this type of shower head may cause a continuous leak. If the main valves are left open with the shower head valve closed, the water line that connects the valves to the shower head is kept under pressure. This causes the valves to leak if they have leaky packings. If the shower is used with a tub, a diverter valve is installed, and this may leak. The valve in the shower head is a crude device that eventually starts leaking itself, so it has to be abandoned anyhow.

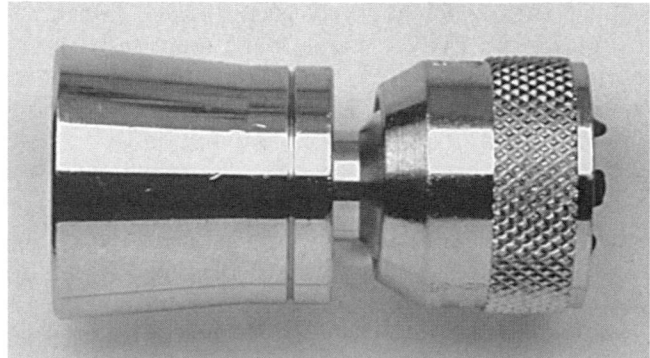

Fig. 2 Inexpensive water-saving shower head This unit saves water simply by having small orifices and a narrow spray pattern. The unit must be removed occasionally to clean out grit that backs up behind the orifices. For this reason, prefer units that have flats for a wrench instead of a knurled area.

Water Temperature Fluctuations

The bather adjusts the shower water temperature by adjusting the relative flow of hot and cold water. The hot water arrives by a separate pipe all the way from the hot water heater, which maintains the hot water pressure to the shower head. If someone turns on an adjacent cold water fixture, such as a faucet or toilet, the cold water pressure to the shower drops. This causes a sudden increase in the shower water temperature. In extreme cases, this may scald the bather.

High-efficiency shower heads may worsen this problem. If the existing shower heads are sensitive to water pressure, test this possibility before replacing them with high-efficiency units.

This problem occurs only if the cold water piping is too small, so that it cannot maintain water pressure when another fixture draws a large flow. In a well designed service water system, both the cold and hot water piping is of ample diameter to serve all fixtures. Individual fixtures are fed from lines of small diameter, to prevent them from bleeding too much pressure from the main pipe.

WESINC

Fig. 3 Versatile and amusing, but not highly efficient
This shower head provides a variety of spray patterns, including a pulsating spray. Some of the patterns have higher flow rates. The pattern is selected by rotating the large outer ring, which is somewhat awkward.

Testing and Selecting Shower Heads

In the United States, Federal law requires that shower heads made after 1993 must limit water flow to 2.5 gallons per minute with a supply pressure of 80 PSI, but do not assume that this criterion will be met. Reliable data on the water consumption of individual models is usually not available. Consumer publications, such as *Consumer Reports*, occasionally review the efficiency and esthetic factors of currently available shower heads.

If you cannot find reliable information about the performance of different models, you can test samples yourself. Measure the length of time required to fill a bucket through the shower head with the water turned on fully. In retrofit applications, start by testing your existing shower heads.

The nature of the facility environment may be a major factor in your choice. Luxury hotels and private residences typically require greater adaptability than public showers, such as those in gymnasiums and swimming centers. Simple, non-adjustable shower heads may be quite satisfactory for the latter applications, and they are more rugged than the adjustable types.

Maintenance

Non-adjustable high-efficiency shower heads have tiny orifices. These trap particles behind the nozzle plate. The particles typically are too large to clog the orifices themselves, but enough of them may accumulate to choke flow through the entire unit. Correct these problem by unscrewing the unit and dumping out the grit. If the water supply is particularly gritty, you may have to do this often.

Do Not Use Flow Restrictor Inserts

Shower flow restrictors are simple orifices that are intended to reduce shower water consumption. They usually take the form of a plastic insert that is installed inside the shower head connector. These devices have been distributed widely by utility companies, government agencies, and others to promote conservation. Unfortunately, they do not work and they are a nuisance. In order for the flow restrictor to reduce water flow, it must dissipate most of the water pressure ahead of the shower head, so the shower head does not have enough pressure to work properly.

ECONOMICS

SAVINGS POTENTIAL: 40 to 70 percent of the water consumption and water heating energy used in showering.

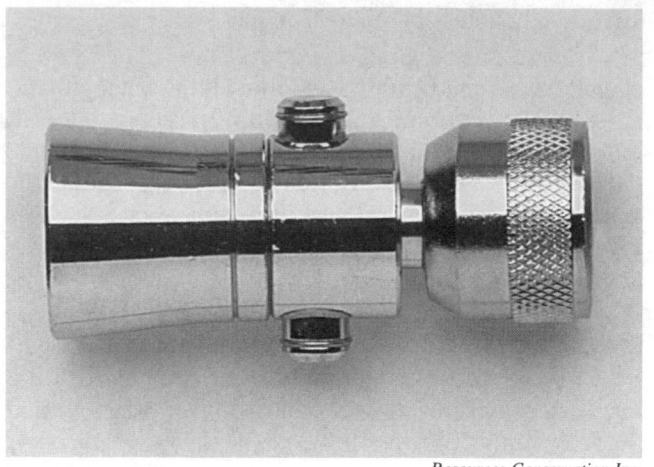

Resources Conservation Inc.

Fig. 4 Shower head with integral shutoff valve This may seem like a good idea, but it isn't, for several reasons.

COST: Low or none, in new facilities. High-efficiency shower heads are typically less expensive than standard units. Typical prices range from $5 to $20 for single-setting shower heads, and up to $50 for units with variable spray patterns.

PAYBACK PERIOD: Immediate, in new facilities. One year to several years, in retrofit.

TRAPS & TRICKS

SELECTING THE EQUIPMENT: Be aware that people differ in their preferences for shower heads. Try to find a type that most bathers like. In residential facilities, keep a few good types in stock so that you can respond to complaints by installing a different type. Install a few samples on a trial basis before buying in quantity. Test their water consumption.

MEASURE **3.1.4 Install shower valves that allow easy control of temperature and flow rate.**

RATINGS

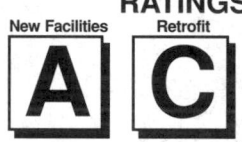

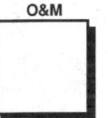

New Facilities | Retrofit | O&M

A | **C** |

The old original style of shower valves, with separate valves for cold and hot water, is the most efficient for general use. Figure 1 shows a typical installation.

Separate valves allow the bather to quickly set the desired flow rate and temperature. They are simplest and easiest to maintain. Everyone knows how to use them. In new installations, they are usually the least expensive type.

Newer types of shower valves waste energy in several ways. The worst energy waste occurs with shower valves that operate only at full flow. High-efficiency shower heads (see Measure 3.1.3) typically use about 50% more water when operating at full pressure than when the bather is able to set the flow rate. Less efficient shower heads may double their water consumption with full-flow valves.

All non-standard types of valves waste water by forcing the bather to fiddle with the valve while attempting to figure out how it works. As the bather fiddles, water is being wasted. Also, a bather who is unfamiliar with a shower valve tends to set it to a higher flow rate than he would use with a familiar valve.

Again, the worst case is full-flow shower valves, which require the water to go to full flow before they control temperature. They also annoy users because they operate in a manner that is contrary to common

SUMMARY

Conventional double faucets are the most efficient, the easiest to use, the least expensive, and the easiest to maintain.

SELECTION SCORECARD

Savings Potential .... $ $ $

Rate of Return, New Facilities % % % %

Rate of Return, Retrofit .... % %

Reliability .... ✓ ✓ ✓ ✓

Ease of Retrofit .... ☺ ☺ ☺

sense. They probably increase the risk of scalding. This type of valve is the stupidest development in the history of plumbing, an egregious case of spending extra money to irritate the user.

In facilities where inefficient shower valves are installed, try to replace them with conventional separate valves for hot and cold water.

It may not be economical to install double valves in showers where a full-flow valve is presently installed. In such cases, the best replacement is a combination valve in which total flow is controlled by moving the knob in and out, and temperature is controlled by turning the knob. This type does not regulate flow as easily as separate valves, but it is fairly easy to use after a brief period of fumbling to figure out how it works.

It is not a good idea to use an on-off valve that is part of the shower head. See Measure 3.1.3 about this.

ECONOMICS

SAVINGS POTENTIAL: *Approximately half of the cost of water and water heating that is used in showers.*

COST: *Low or none, in new facilities. Efficient shower valves typically cost less than the wasteful types. Retrofit costs range from $50 to $200 per shower.*

PAYBACK PERIOD: *Immediate, in new construction and renovation. Several years or longer, in retrofits.*

TRAPS & TRICKS

SELECTING THE EQUIPMENT: *Here again, simplest is best. The most efficient valves are the most obvious to use. Stress reliability and ease of maintenance. There is no need to install complex, expensive valves to achieve an upscale appearance.*

WESINC

Fig. 1 Anyone can understand how to use these shower valves This conventional type is simplest, cheapest, easiest to maintain, easiest to use, and probably safest. Put a high-efficiency shower head in here, and this will be an ideal shower installation.

MEASURE 3.1.5 Provide instructions for efficient use of water in showers and lavatories.

The usual practices of showering and lavatory use in prosperous countries are very wasteful. A continuous stream of water is used where only brief wetting for lathering and rinsing is required.

In environments where it is possible to influence the practices of shower users, request bathers to use "Navy showers." This procedure is used on ships where the fresh water supply is limited. It consists of turning on the shower briefly to wet the body, lathering with soap, and then turning on the shower briefly to rinse. The amount of water and water heating energy used in this way is a small fraction of the amount used in a normal shower.

In environments where it is not appropriate to request showering practices as strict as these, make a more general request to conserve water, which applies to all water usage in the lavatory.

Match the Instructions to the Environment

Strict "Navy" showering is appropriate in environments where showering is considered primarily as a hygienic measure, and to locales where water is considered a scarce commodity. Typical sites are military facilities, schools, prisons, and cities with acute water shortages.

Where showering may be considered an esthetic experience or a means of relaxation, as in hotels, it is more appropriate to post general requests to conserve water. If water is scarce, say so.

How to Convey the Message

Posting appropriate placards at the showers and lavatories is the most effective method of requesting users to conserve water. See Reference Note 12,

SUMMARY

Saves a lot of water at low cost in facilities where it is appropriate to tell people how to use water.

SELECTION SCORECARD

Savings Potential	$ $ $
Rate of Return, New Facilities	% % % %
Rate of Return, Retrofit..........	% % % %
Reliability	✓ ✓
Ease of Retrofit	☺ ☺ ☺

Placards, for details of effective placard design, materials, and installation.

In environments with stable populations, such as schools and military bases, also use existing communications channels, such as announcements, newsletters, posters, etc.

ECONOMICS

SAVINGS POTENTIAL: 50 to 90 percent of the water and water heating energy used in showers.

COST: Durable, waterproof placards typically cost several dollars or more, depending on size and quantity.

PAYBACK PERIOD: Less than one year, typically.

TRAPS & TRICKS

PERSUASION: This activity is non-technical. Hand it to the public relations or personnel department, or to the warden.

MEASURE **3.1.6 Install efficient toilets.**

The gravity tank toilet is the most common type in residential applications, while the flushometer type is the most common in commercial applications.

Typical older gravity tank toilets use five to seven gallons per flush. Figure 1 shows how this type of toilet works by siphon action. The amount of water required for effective siphoning has long been recognized as a problem. In response, a variety of toilet designs have appeared over the years that attempt to use less water. These include toilets that flush with positive pressure, toilets that grind the waste, etc. More recently, laws have been enacted that get politicians involved in designing toilets, with results that should not be surprising.

The reduction of water consumption in new toilets of conventional design has been marginal or illusory, so far. Any improvement is usually at the cost of increased annoyance. Some of the newer types of toilets do use substantially less water, but they may be complicated, messy, unreliable, or unpleasant. None of the newer types has yet proven to be clearly superior to the older types.

SUMMARY

A variety of toilets have been developed to reduce water consumption. So far, none is superior to older models in all respects.

SELECTION SCORECARD

Savings Potential .... $ $

Rate of Return, New Facilities % % % %

Rate of Return, Retrofit .... %

Reliability .... ✓ ✓ ✓

Ease of Retrofit .... ☺ ☺

A problem in using less water for flushing is that the waste is not carried as far along the horizontal portions of sewage lines. The effect of this, and possible solutions, are still being debated.

Even so, if you are planning a new facility, you will have to select a new type of toilet, as the old reliable versions have been banned. The following are the main

American Standard, Inc.

Fig. 1 Conventional siphon toilet The water level in the bowl remains even with the top of the siphon elbow until flushing. When the water in the tank is released into the bowl, a siphon action is started that draws the waste into the sewer line, aided by the swirling of water into the bowl.

American Standard, Inc.

Fig. 2 Pressure-flushing toilet Compare this to Figure 1. The water for flushing is stored in a pressurized tank. Waste enters the discharge pipe by siphon action, and is then pushed through the discharge pipe by the pressurized water, which also increases the siphon effect.

points to consider. At this time, it is still too early to replace existing toilets for the sake of water conservation.

Be Skeptical of Ratings and Performance Claims

In the United States, Federal law requires that all types of toilets made after 1996 must use no more than 1.6 gallons of water per flush. A single exception is the blowout toilet, which is a pressurized flush toilet using a large trapway. This type of toilet, used only in special applications, may use 3.5 gallons per flush.

Government-mandated ratings are not a reliable guide to water consumption, because they only designate the amount of water consumed per flush. Several flushes may be required to get rid of waste. Some "water conserving" toilets flush more effectively than conventional toilets, and some flush less effectively. Some European designs appear to be intended more to allow examination of the waste than disposing of it. The ability of a toilet to get rid of waste also depends on the amount and characteristics of the waste.

How to Select Toilets

Select toilets for flushing effectiveness, esthetics, reliability, and ease of maintenance. Any toilet requires periodic maintenance, and it will waste water without it. Beware of the newer types of toilets that require mechanisms and seals in the path of the waste. When these fail, they become unpleasant to use and to repair.

At the present time, pressure-flushing toilets appear to offer a good compromise between water consumption, flushing effectiveness, and freedom from the more unpleasant types of maintenance. A typical pressure-flushing toilet is shown in Figure 2. The distinguishing feature of this type is the pressure tank, which is shown in Figure 3. A significant disadvantage of this type is that the flushing action is very loud, although it is brief.

In facilities where toilet water consumption is substantial, be willing to consider novel types of toilets, but do your homework. In the United States, start with the Environmental Protection Agency information center on water conservation, but don't stop there. Investigate how the types that you are considering will work in a real environment. This may be challenging. The most novel units tend to be installed by enthusiasts who are blind to the practical and esthetic issues.

ECONOMICS

SAVINGS POTENTIAL: 30 to 70 percent of the water used in toilets.

COST: Toilets of conventional design, but improved allow flushing with smaller quantities of water, cost about $70 each. Pressure-flushing toilets cost about $170 each. More exotic types range widely in price.

PAYBACK PERIOD: Several years or longer.

TRAPS & TRICKS

SELECTING THE EQUIPMENT: All toilets must meet the same standards for water consumption per flush. Therefore, when selecting new toilets, the main factor affecting efficiency is flushing effectiveness. Also, select for reliability, ease of maintenance, and esthetic factors. Put yourself in the position of the persons who must use and repair the unit.

WESINC

Fig. 3 Water tank of a pressure-flushing toilet A quantity of air and water is pressurized inside the cylindrical tank by the system water pressure. The toilet is flushed by pressing the white plunger on top, allowing the air to force the water into the bowl and the discharge pipe. In this model, the pressure tank is installed inside the tank of a conventional toilet, showing how much less water is used.

MEASURE 3.1.7 Install efficient urinals or improve existing urinals.

SUMMARY

Current water consumption standards for new urinals are lax, so you have room to select more efficient types. Consider improving existing units, if possible.

SELECTION SCORECARD

Savings Potential	**$ $**
Rate of Return, New Facilities	**% %** % %
Rate of Return, Retrofit..........	**% %**
Reliability	✓ ✓ ✓ ✓
Ease of Retrofit	☺ ☺

For disposal of urine alone, urinals have the potential of being much more efficient than toilets. If you have more than a few good men, install urinals adjacent to the toilets.

In the United States, Federal law requires that all types of urinals made after 1993 must use no more than 1.0 gallon of water per flush. This is a lax standard. It is easy to see that a urinal could be designed to operate with much less water than this, perhaps with only a cupful of water per use. Some designs use no water at all.

In existing facilities, consider replacing urinals if the existing ones are particularly inefficient. A urinal is not a major water user, provided that it does not flush continuously. Also, it uses only cold water. This makes urinal efficiency a relatively minor matter. Still, try to improve bad cases.

How to Select Urinals

More efficient urinals may cost little more than other types. Some types cost less. As with all equipment, shop diligently and examine actual installations.

The only efficiency issue with urinals is the quantity of water used. Urine, being a liquid, enters the sewage system without assistance. The only purpose of the water supply is to flush out the film of urine that remains on the surface of the urinal, for sanitary and esthetic reasons. Urinals should be designed to require the smallest amount of water for this simple flushing function.

In addition, the water should flow for the shortest possible time. The common flushometer valve works well for this. To eliminate the need for people to touch the flush handle, install urinals in which the flushing is controlled by a personnel sensor. Theses sensors are similar to the ones used in automatic faucets, covered by Measure 3.1.2. Figure 1 shows a typical unit.

You may have to go to another country to find the best units, as fashion in urinals is national. In the United States, start with the Environmental Protection Agency information center on water conservation. In addition to efficient water usage, consider:

- *ease of use*
- *containment,* considering drippage and bad aim
- *effective rinsing* of the surface
- *ease of cleaning,* including removal of paper towels, cigarette butts, etc.
- *mechanical simplicity* and *ease of maintenance.*

■ Waterless Urinals

Some urinals require no water, except from the user. The simplest type is nothing more than a bowl with a

American Standard, Inc.

Fig. 1 In by the left, out by the right This urinal has an infrared sensor to detect a user. It flushes automatically when the person departs.

pipe that leads to the sewer. In most buildings, this arrangement is not acceptable, because any plumbing fixture requires a trap between the fixture and the sewer. (The purpose of the trap is to prevent gases in the sewage system from entering the space. The trap is usually a simple siphon loop. Look under a sink to see a typical example.) In a urinal with no water to flush the urine through the trap, the urine would linger in the trap, emitting odor into the space.

A refinement of the waterless urinal has a simple oil trap that controls odor. The urine passes through the oil, and then to the sewer by siphon action. The oil floats above the urine, preventing the odorous elements of the urine from evaporating into the space. Figure 2 shows a typical urinal of this type. Figure 3 is a detail of the oil trap.

The merits of this urinal, in addition to zero water consumption, are simplicity and low cost. The unit's

main disadvantage is that it does not flush the bowl surface. This allows more odor than a flushing urinal.

It appears that the best applications for waterless urinals are spaces that have a high ventilation rate or a rustic environment, such as restrooms in parks. They may also be appropriate for facilities, such as military bases, where esthetics is less of a concern and urinals are cleaned on a daily basis.

Waterless urinals require periodic replenishment of the oil in the siphon trap. They also require periodic replenishment of the filter cartridge, typically two to four times per year. The cartridge becomes clogged with small particulate matter, especially hair. Large debris, such as cigarette butts, does not pass into the cartridge, but must be cleaned out manually, as with conventional urinals.

Improve Existing Urinals that Lack Flush Valves

If existing urinals use a continuous stream of water, you may be able to eliminate most of the water consumption simply by installing a self-closing valve. A simple spring-return valve is inexpensive, reliable, and efficient. This type is satisfactory for most informal environments. For applications where a larger volume of water is required for flushing, install a flushometer valve that delivers appropriate volume.

For group urinals and for environments where users may not flush, consider using a personnel sensor to activate a flush cycle. Infrared sensors or light beams are effective for this purpose. Figure 4 shows an old communal urinal that could benefit from such a control.

Especially with urinals that were fabricated on site, you may be able to reduce water consumption by improving the distribution pattern of the flushing water.

Waterless Co.

Fig. 2 Waterless urinal This type uses no water for flushing. In the bottom, it has an oil trap that keeps the odor of the urine in the drain trap from entering the space.

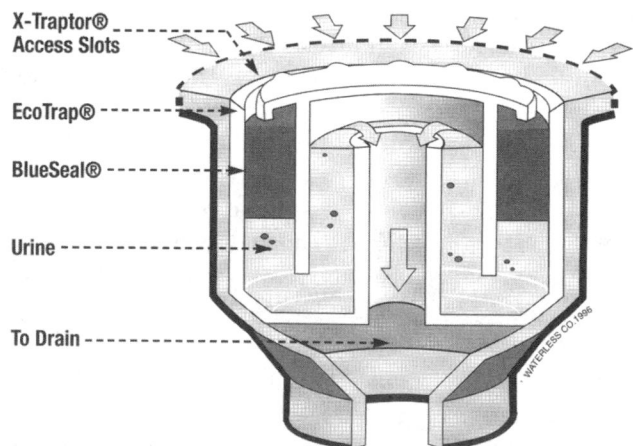

Waterless Co.

Fig. 3 Oil trap for waterless urinal This is used in the urinal shown in Figure 2. It acts as a required plumbing trap, while keeping urine from being exposed to the interior of the space.

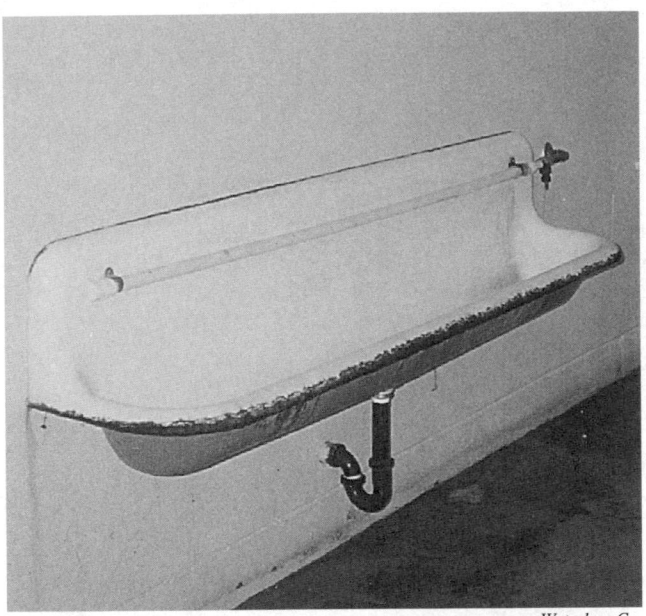

Waterless Co.

Fig. 4 Old communal urinal that flushes continuously This is a serious water waster. A valve actuated by a personnel sensor should be installed in the water line.

ECONOMICS

SAVINGS POTENTIAL: *30 to 100 percent of the water used in urinals.*

COST: *Flush valves activated by infrared personnel sensors cost about $200 each. The bowl itself costs $100 to $200, depending on size and style. Manual flushometer valves cost about $100 each. Simple self-closing valves cost about $50 each. All these prices are for small quantities. Waterless urinals with oil traps cost about $300. Replacement cartridges for waterless urinals cost about $6 apiece.*

PAYBACK PERIOD: *One to several years, when selecting more efficient units in new construction. Many years, typically, for replacing existing urinals. One to several years, for improving existing urinals.*

TRAPS & TRICKS

SELECTING THE UNITS: *When selecting new urinals, search for units that use substantially less water than prescribed by current government water consumption standards. Avoid designs that are complex or difficult to maintain.*

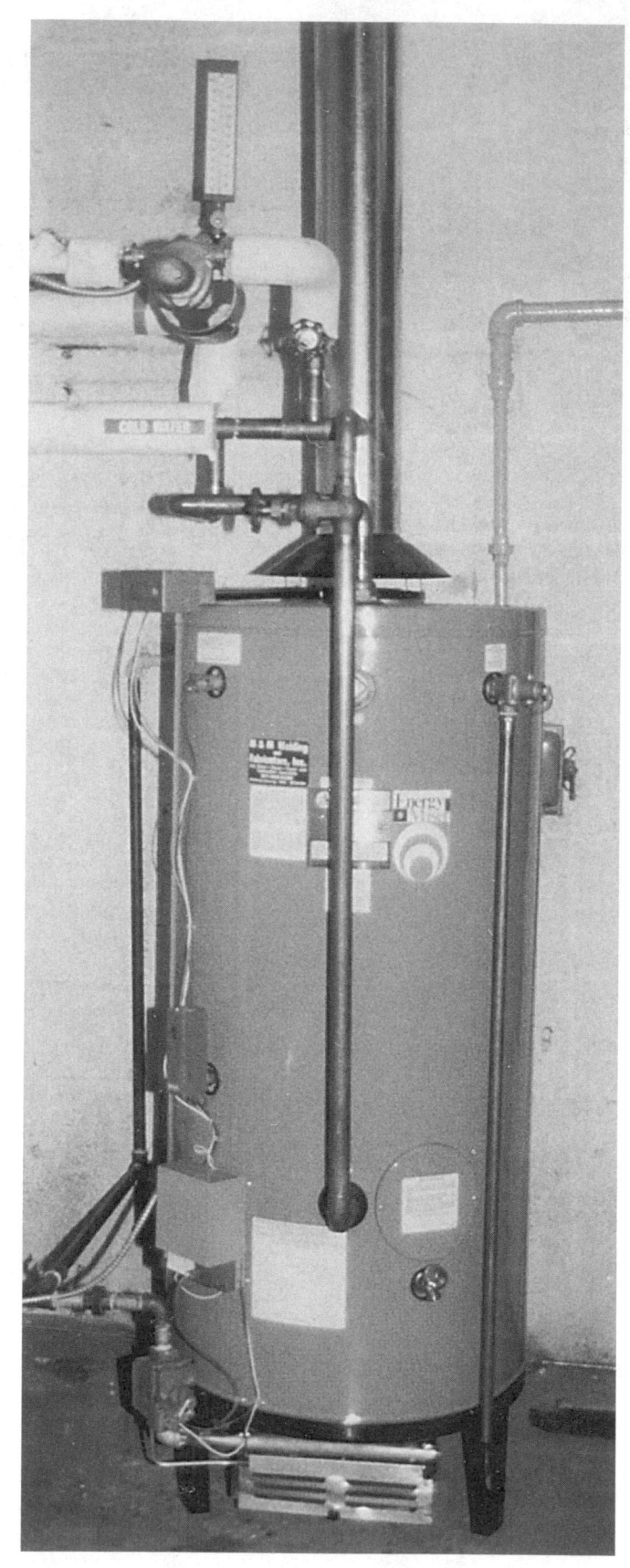

Water Heating Systems

Water heating is a significant energy user in almost all facilities. In housing, food service establishments, and some industrial operations, water heating is a major part of total energy cost. Water heating systems usually operate continuously, so their losses are continuous. Inefficiency in water heating is caused primarily by inefficiency of the heating equipment, and by heat loss from water storage tanks and distribution pipe.

The Measures in this Subsection increase the efficiency of water heating units and systems. In addition, the last two Measures reduce the cost of electric water heating.

INDEX OF MEASURES

3.2.1 Minimize the hot water temperature.

    3.2.1.1 Use low-temperature detergents.

3.2.2 Install a separate high-temperature water heater for high-temperature applications.

3.2.3 Install water heaters that have the lowest energy cost and highest efficiency.

3.2.4 Install supplemental insulation on water heaters.

3.2.5 Install automatic flue dampers on fuel-fired water heaters.

3.2.6 Clean and adjust the combustion systems of fuel-fired water heaters periodically.

3.2.7 Clean out scale from water heaters periodically.

3.2.8 Exploit interruptible or storage rates for electric water heating.

3.2.9 Control electric water heating to reduce demand charges.

RELATED MEASURES

- Measure 1.10.3, installation of the most efficient types of steam traps on steam-fired water heating equipment
- Measures 1.10.4 ff, maintenance of steam traps on steam-fired water heating equipment
- Measure 1.12.2, installing localized water heaters to allow shutting down a central heating plant during periods of low load
- Subsection 2.10, for heating of service water with heat recovered from chiller condensers

MEASURE **3.2.1 Minimize the hot water temperature.**

You can minimize energy losses by keeping the water temperature as low as practical. This a common opportunity because water heating systems are designed to produce water at a substantially higher temperature than is needed for most applications. For example, the typical domestic hot water heater is designed to produce water at temperatures as high as 180°F (82°C), but lavatory applications typically require water no hotter than 110°F (43°C).

Lowering the water temperature may reduce scalding hazard in some applications. On the other hand, be aware that lowering the water temperature may increase health hazards in certain situations. It also reduces the effective capacity of the hot water storage tank. Thus, water temperature is a compromise. You need to tailor the compromise to the particular application.

Energy Losses Related to Water Temperature

Energy consumption increases with water temperature in these ways, which are listed in approximate order of importance:

- *more heat delivered per unit of water.* In applications where the end-use water temperature is not regulated, as in residential clothes washers and dishwashers, lowering the water temperature reduces the amount of energy delivered. This is not true in applications where the end-use temperature is held constant by a thermostatic mixing valve.

- *more water consumed.* In lavatory and shower applications, excessive temperature increases the amount of water that is used at fixtures. This is because users must mix hot and cold water to achieve the desired temperature. For example, a bather who encounters high hot water temperature in a shower will spend time adjusting the mixed water temperature before stepping into the shower. Also, the bather will use a higher flow rate as a safety measure to provide good control of temperature.

- *increased surface loss.* Hot service water systems have a large amount of surface area in the tanks, distribution pipes, and recirculation pipes. Surface heat loss is directly proportional to the surface area and to the temperature difference between the water and the system environment. It is inversely proportional to the thermal resistance of the system insulation. Depending on the situation, some of the surface heat loss may be recovered as a form of

SUMMARY

Saves a modest amount of energy with minimal effort. Consider the potential health hazards.

SELECTION SCORECARD

Savings Potential	**$**
Rate of Return	**% % % %**
Reliability	✓ ✓ ✓
Ease of Initiation	☺ ☺ ☺ ☺

space heating, or it may increase the cooling energy requirement of the surrounding space.

- *increased standby loss.* In fuel-fired heaters, higher water temperature increases convective flue losses during periods when the burner is not operating.

- *increased combustion loss.* In fuel-fired heaters, higher water temperature increases stack losses by reducing the temperature differential between the combustion gases and the water.

Health Concerns Related to Low Hot Water Temperature

People are concerned that reducing domestic water temperature allows dangerous microorganisms to grow in the hot water system. This issue will continue to be the subject of research and conjecture for some time. Microorganisms are everywhere, and human beings have evolved resistance to infection. Nonetheless, be aware of the relevant safety issues for your application.

A variety of microorganisms found in hot water systems have been suspected of causing disease. These include, but are not limited to:

- *legionella pneumophila,* a widespread bacterium that grows in soil in all moderate climates. Legionella gained notoriety (and its name) in the Legionnaire's Disease outbreak. Legionella may grow rapidly in water temperatures between 85°F and 113°F. They are killed by temperatures above 131°F. Some believe that algae must be present in the water supply to provide amino acids needed for growth.

Infection occurs by respiration of an aerosol containing the bacteria. The result is pneumonia. Showers appear to be a prime infection site. Only a small fraction of persons exposed to legionella develop symptoms. However, the mortality rate is high among those who do develop symptoms.

• *Pontiac fever bacteria* have characteristics similar to those of legionella. The result is a flu, and exposure almost always results in symptoms. The mortality rate is low.

• *salmonella* are common bacteria that cause illness by food poisoning. Drinking or eating is the usual route of infection. The principal protection is proper cooking of foods in which the bacteria can proliferate.

It is not practical or necessary to eliminate all microorganisms from domestic water supplies. The key is to prevent proliferation, because infection is related to the concentration of the microorganisms. Cold water is too cool for rapid growth.

Above 140°F, most infectious microorganisms are killed. Unfortunately, scalding becomes a hazard at this temperature. Even above this temperature, some bacteria may survive by being encapsulated inside amoebae that may be present in the water supply. However, bacteria cannot spread while in this state.

Whether disease occurs depends on the susceptibility of the people who are exposed. For example, "hospital-acquired pneumonia" is a serious health problem in hospitals, and legionella in the hospital water supply is suspected. Patients are assumed to be especially susceptible because of their weakened conditions.

Relationship of Health Hazard to Type of Water Heater and System

Some research has associated infections with certain types of water heaters. All water heater tanks and hot water storage tanks accumulate debris at their bottoms. This consists of dirt that enters along the with the raw water. In water heaters, it also consists of fragments of scale that is formed by minerals in the water. The debris is presumed to serve as a breeding site for micro-organisms.

Fuel-fired water heaters that are fired from the bottom appear to be immune to growth of micro-organisms regardless of the water temperature setting. This is probably because the flame sterilizes the debris at the bottom of the tank.

Electric water heaters do not offer this protection, because the heating elements are installed well above the bottom of the tank. Incoming cold water typical enters the bottom of the tank, keeping the sediment relatively cool. The same applies to fuel-fired water heaters in which the flame does not cover the entire bottom of the tank.

In recirculating systems, this sterilization effect may protect the entire system provided that the circulating water is brought in contact with the fuel-fired surface on each pass. If the water heating system includes a storage tank without heating elements, the sterilizing

capability of the system depends on how the water flow is piped. If recirculated water is piped only through the storage tank and the storage temperature is low enough, the entire circulating water mass could grow micro-organisms.

Water Temperature is a Compromise

From the standpoint of energy conservation, set the water temperature just high enough to satisfy the highest temperature requirement of the system. For example, if the system serves only wash basins, the temperature can be low, say 100°F (38°C). If you need higher water temperature for localized purposes, such as dishwashing, consider installing an auxiliary heater for those applications. See Measure 3.2.2 for details.

High water temperature is the only reliable way of killing microorganisms in service water systems. Some microorganisms are very adaptable. For example, legionella survives throughout the range of pH found in domestic water supplies and also resists the

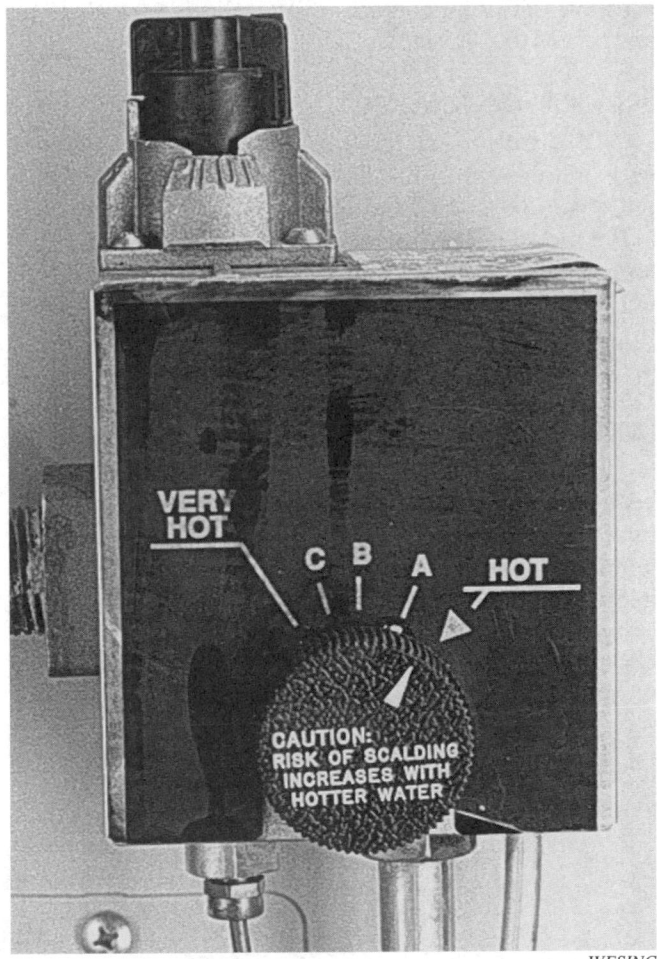

Fig. 1 Insufficient information This thermostat on a gas water heater does not state the actual water temperature, which you need to know. In any event, you should determine water temperature from a tank outlet thermometer, not from markings on the thermostat.

concentrations of fluorine and chlorine found in potable water.

This being said, keep the health aspect in perspective. Human beings do not require sterilized water, but rather water in which pathogens are kept at a safe level. All public water supplies contain a variety of pathogens. Prudent policy is to design and operate water systems so that pathogens are not allowed to multiply to dangerous levels.

The potential safety hazard from low water temperature depends on how the water is used. As discussed above, the hazard appears to lie primarily from inhalation of aerosols of heavily infested water. Water at reduced temperature may be safe for most applications if aerosols are not formed.

From a health standpoint, the water temperature setting at the heater can be deceptive. For example, in electric water heaters, the thermostat may be located immediately above the heater. The water that is below the heater may remain much cooler. For example, experimentation has shown that an electric water heater set at 140°F may contain many gallons of water at a temperature of less than 120°F.

Hot Water Thermometers

To provide an accurate guide to hot water temperature setting, install thermometers at both the tank outlet and the return end of the recirculation line. The outlet thermometer provides a direct reading of the tank temperature. You need one because the thermostat does not indicate temperature accurately. For example, see Figure 1. The return line thermometer provides a check on the effectiveness of recirculation.

Label the Thermostat

Install a placard at the thermostat of each water heater that states the temperature that should be set. State that the tank outlet thermometer should be used for temperature setting, not the indications on the thermostat itself. Include warnings about scalding hazards and growth of dangerous organisms in the water, as appropriate. See Reference Note 12, Placards, for details of effective placard design and installation.

Effect on Storage Capacity

Lowering the water temperature reduces the effective storage capacity of the hot water tank. For example, if water is needed at 100°F, hot water at 140°F is mixed with cold water at 60°F in even proportions. However, if the hot water storage temperature is reduced to 100°F, water is drawn from the tank at twice the rate, reducing its effective capacity in half.

This may not matter if the tank has a high "recovery rate," which is the heating capacity of the heating element. However, if the recovery rate is low in relation to the tank volume, lowering the water temperature may cause the system to run out of hot water occasionally. This may happen if the capacity of the storage tank was selected to exploit load diversity.

In facilities that use electric water heaters, you can reduce electricity demand charges by storing hot water. (See Measure 3.2.8 for details.) Lowering the water temperature interferes with this technique by reducing the effective storage capacity.

Relationship to Recirculation

In lavatory and shower applications, an effective recirculation system allows you to maintain lower hot water temperatures. It also makes the system more pleasant for users. See Measure 3.3.2 for more about recirculation.

ECONOMICS

SAVINGS POTENTIAL: *5 to 40 percent of water heating energy. Reducing hot water temperature may also reduce water consumption in shower and lavatory applications.*

COST: *Minimal.*

PAYBACK PERIOD: *Immediate.*

TRAPS & TRICKS

DECISION: *The big issue is deciding the temperature. Set the lowest temperature that is consistent with the needs of the application, the hazard from microorganisms, and hot water storage capacity.*

MARKING: *Install a placard at the thermostat that states the temperature that should be set.*

MEASURE **3.2.1.1 Use low-temperature detergents.**

B

SUMMARY

Lets you reduce the hot water temperature in cases where the choice of detergent is the limiting factor.

SELECTION SCORECARD

Savings Potential .... **$**

Rate of Return .... **% % % %**

Reliability .... **✓ ✓**

Ease of Initiation .... ☺ ☺ ☺ ☺

In some cases, the water temperature may be kept high to provided effective cleaning with detergents. For example, this is common in janitorial and laundry applications. You may be able to lower the water temperature if you substitute low-temperature detergents. Such detergents work well with the tepid temperatures needed for other domestic hot water applications, such as lavatories and showers.

Low-temperature detergents typically do not cost much more than conventional detergents. Even so, calculate whether any additional cost is justified by the increase in hot water system efficiency.

The effect of water temperature on the cleaning effectiveness of detergents is a matter of degree. You may have to experiment to find the lowest acceptable water temperature.

Be sure that high water temperature is not needed for its own sake, for example, for sterilization in the rinse cycle of dishwashers. If the application specifically requires high water temperature, consider installing an auxiliary heater. See Measure 3.2.2 for details.

ECONOMICS

SAVINGS POTENTIAL: *1 to 10 percent of water heating energy.*

COST: *Usually minimal.*

PAYBACK PERIOD: *Immediate.*

TRAPS & TRICKS

PLANNING: *Do not compromise with cleaning efficiency or sanitation. Balance the energy saving against any increase in detergent cost.*

KEEP IT UP: *Once you identify the appropriate detergent, make sure that the person responsible for buying cleaning supplies continues to specify this type.*

MEASURE 3.2.2 Install a separate high-temperature water heater for high-temperature applications.

Some hot water applications require exceptionally high temperatures, which may be much hotter than needed by most users. For example, a dormitory that uses most of its hot water in showers and lavatories may have a cafeteria that requires much hotter water for dishwashing. The high dishwashing temperature increases heat loss throughout the hot water system. (See Measure 3.2.1 for the ways that energy is wasted.)

The solution is to install a separate water heater to serve the high-temperature application. This approach is worth considering if the high-temperature application is localized, while usage of water at ordinary temperature is dispersed. Install the high-temperature heater as close to the user equipment as possible, in order to minimize heat loss in the connecting piping.

There are two ways to provide a separate high-temperature water supply. One is to install an entirely separate high-temperature heater that draws water from the cold water system. This approach may be best if high-temperature usage is a large fraction of total usage, or if the equipment that uses high-temperature water is clustered in a particular area.

On the other hand, if high-temperature usage is a small fraction of total hot water usage, it is usually more economical to install a "booster" heater for each application. The booster heater draws water from the lower-temperature heating system. A booster heater can have a smaller heating capacity than a heater that draws from the cold water system. Booster heaters are offered as an integral part of some equipment, such as dishwashers.

Cost of the Energy Source

The cost premium that you can tolerate for separate high-temperature water heating depends on the losses that you can avoid. This Measure rarely provides a cost saving if the energy source for the high-temperature heater is more expensive than the energy source for the primary hot water system.

Some equipment may be equipped with an integral booster heater that requires a more expensive energy source. For example, a dishwasher may have an electric booster heater, while the primary water heater is fired by less expensive gas. In such cases, it may be economical to install an external booster heater that operates with the least expensive available energy source.

SUMMARY

Consider this for applications where a fraction of hot water is needed at especially high temperature. Must use an energy source that is no more expensive than the energy source used by the primary heater.

SELECTION SCORECARD

Savings Potential .... $ $

Rate of Return, New Facilities % %

Rate of Return, Retrofit.......... % %

Reliability .... ✓ ✓ ✓ ✓

Ease of Retrofit .... ☺ ☺ ☺

Control of Auxiliary Heaters

If the high-temperature water is needed on a predictable schedule, you may be able to save additional energy by turning off the high-temperature heater when it is not required. For example, you can use a timeclock to turn on the high-temperature heater one or two hours prior to the beginning of high-temperature water usage. This refinement is worth the effort and expense only in larger systems that have significant heat losses.

If the high-temperature heater is electric, its schedule of operation may have a significant effect on electricity cost. See Measure 3.2.8 for details.

ECONOMICS

SAVINGS POTENTIAL: *1 to 30 percent of water heating energy, depending on the system layout and the types of water heating applications being served.*

COST: *Several thousand dollars, typically.*

PAYBACK PERIOD: *Several years or longer, typically.*

TRAPS & TRICKS

SELECTING THE EQUIPMENT: *Select water heaters for high efficiency. See Measure 3.2.3, next, about this. Also, be sure to use the most economical available energy source. This usually has a greater effect on operating cost than energy efficiency.*

TEMPERATURE SETTING: *The purpose of this activity is to allow you to reduce the temperature of the primary water heater. Make sure that the temperature is actually reduced. See Measures 3.1.1 ff.*

MEASURE 3.2.3 Install water heaters that have the lowest energy cost and highest efficiency.

RATINGS

New Facilities   Retrofit   O&M

A | **C** | **A**

There are substantial differences in the efficiencies of different models of water heaters. Be sure to select the most efficient models for new construction and when replacing failed units.

In existing facilities, it may even be worthwhile to replace a functioning heater with a more efficient unit.

The convenience of electric water heaters prompts people to install them even where a less expensive energy source is available. Figure 1 shows an example. Don't make this mistake. Install an electric water heater only if installing the flue and the fuel supply is too difficult, or if electricity is the least expensive energy source, taking all factors into account.

Efficiency Characteristics of Water Heaters

■ Electric Water Heaters

With electric water heaters, the efficiency of energy conversion in the water heater itself is virtually 100 percent. The only losses related to the heater itself are heat loss from the surface of the heater to the surrounding space.

As with all electrical equipment, there are large energy losses in the production and distribution of the electricity. (This is why electricity is usually expensive.) Within the boundaries of the facility, there are relatively minor losses within the transformers and power wiring.

SUMMARY

Not only for new facilities. Be ready to replace failed units with efficient models. It may even be economical to replace functioning units if they are especially inefficient. Consider converting to a less expensive energy source.

SELECTION SCORECARD

Savings Potential .... $ $ $

Rate of Return, New Facilities % % %

Rate of Return, Retrofit.......... % %

Rate of Return, O&M .... % % %

Reliability of Equipment .... ✓ ✓ ✓ ✓

Reliability of Procedure.......... ✓ ✓

Ease of Retrofit .... ☺ ☺ ☺

Ease of Initiation .... ☺ ☺ ☺

■ Steam-Powered Water Heaters

With steam-coil heaters, the efficiency of energy conversion in the water heater itself is also virtually 100 percent. Losses related to the heater are limited to heat loss from the surface of the unit.

There are substantial energy losses in the steam plant and the distribution system, including the condensate return. The latter includes leakage of the heater's steam trap. These losses are covered in Section 1.

■ Conventional Fuel-Fired Water Heaters

Fuel-fired units are subject to the same combustion efficiency limitations as boilers. These are covered in the Subsections 1.2 and 1.3. In summary, they are:

- *improper air-fuel ratio.* Setting the proper ratio is covered by Measure 3.2.6.
- *flue losses due to fouling of the heat exchange surfaces.* Fuel-fired heaters lose efficiency as fouling accumulates on the heat transfer surfaces. Cleaning heat exchange surfaces is covered by Measure 3.2.7.
- *flue losses due to unrecovered latent heat of water in the flue gas.* Water vapor is a combustion product of hydrocarbon fuels, especially of gas. If the water remains in the vapor state and escapes out the flue, it does not give up its latent heat.

Conventional water heaters may have much less effective heat transfer than boilers. This is because the heat transfer surface is limited to the bottom of the tank

WESINC

Fig. 1 Going the wrong way This old gas water heater in a government building was replaced with an electric heater, even though gas is much less expensive than electricity in this location.

and the flue pipe that rises through the tank. This limits the firing rate of a water heater, and hence its recovery rate. Some water heaters have several flue tubes, or flues of increased surface area, to increase the heat transfer area.

The portions of the tank surface that absorb heat from the flame and flue gases cannot be insulated. This creates standby losses when the unit is not firing. The heat of the water in the tank induces convection of cold air through the heater when it is not firing. If air is free to circulate, it carries away heat from the water, requiring additional firing.

You can reduce standby loss in existing water heaters by installing a flue damper, which is recommended by Measure 3.2.5.

■ High-Efficiency Fuel-Fired Water Heaters

High-efficiency fuel-fired water heaters increase efficiency in two ways, by recovering the latent heat of the water vapor in the flue gas, and by minimizing standby loss. Figure 2 shows one type of unit.

WESINC

Fig. 2 High-efficiency water heater The exhaust is so cool that plastic pipe is used for the flue. This model uses pulsed combustion, an unusual feature of one manufacturer. The small muffler in the combustion air intake pipe and the very large muffler in the exhaust pipe minimize the noise.

To recover more heat from the flue gas, including both sensible heat and some latent heat, high-efficiency water heaters have heat exchangers with more surface area. At least the cooler end of the heat exchanger is made of stainless steel, to avoid corrosion by the condensed water vapor and acid.

High-efficiency water heaters may eliminate virtually all standby loss by using one or more of these methods:

• an integral flue damper
• heating the water in an external heat exchanger, so the storage tank has no uninsulated heat transfer surfaces
• using narrow combustion air passages that require a fan to move the combustion air. These passages allow minimal convection when the fan is not running.

■ Direct-Contact Fuel-Fired Water Heaters

Direct-contact water heaters are the most efficient kind of fuel-fired water heater. They eliminate most of the energy losses that relate to the exchange of heat between the combustion gases and the water. Figure 3 shows how a typical unit is constructed.

As the name implies, the water to be heated comes in direct contact with the combustion gases. Typically, the combustion gases flow upward through a stainless steel vessel, while the water to be heated is sprayed into the vessel from the top. Various kinds of filler material, such as clusters of stainless steel shells, distribute the water flow while allowing the combustion gas to heat the water.

This arrangement acts as a counterflow heat exchanger. As a result, the flue gases are reduced to a temperature almost as low as the temperature of the incoming water. The temperature difference may be as low as 10°F (6°C).

Direct-contact water heaters also recover most of the latent heat of the water vapor in the flue gas. With gas fuels, this accounts for more than 10% of the fuel's energy.

Direct-contact water heaters require expensive non-corroding materials, usually stainless steel, because the vessel and the filler material are exposed both to hot combustion gas and to incoming water that may be saturated with air. The units become economical in large sizes for applications that require large volumes of hot water, such as concrete plants, commercial laundries, and meat processing plants. Direct-contact heaters may provide a significant cost advantage in new applications where they avoid or reduce the need for boilers and related piping.

The cleanliness of the water depends on the fuel. The water will absorb whatever is in the flue gas. With gaseous fuels, the water is clear, even potable. It will dissolve the carbon dioxide in the flue gas, forming a weak solution of carbonic acid, considerably less concentrated than in club soda. If other fuels are used, the water will not be potable, but it may be suitable for washing and for inclusion in products, such as concrete.

The exceptionally low flue gas temperature allows flues to be made of materials that cannot tolerate high temperatures, such as plastic pipe. However, the flue gas is saturated with water vapor and it picks up oxygen from the incoming water, so the flue material must be non-corroding.

Effect of Water Temperature on Efficiency

The water temperature has a major effect on the heat loss from all types of water heaters and from the rest of the hot water system. Therefore, keep the water temperature as low as practical. See Measure 3.2.1 about this.

Water Heater Insulation

The surface area of a water heater is relatively small, but the unit's insulation is important because the water

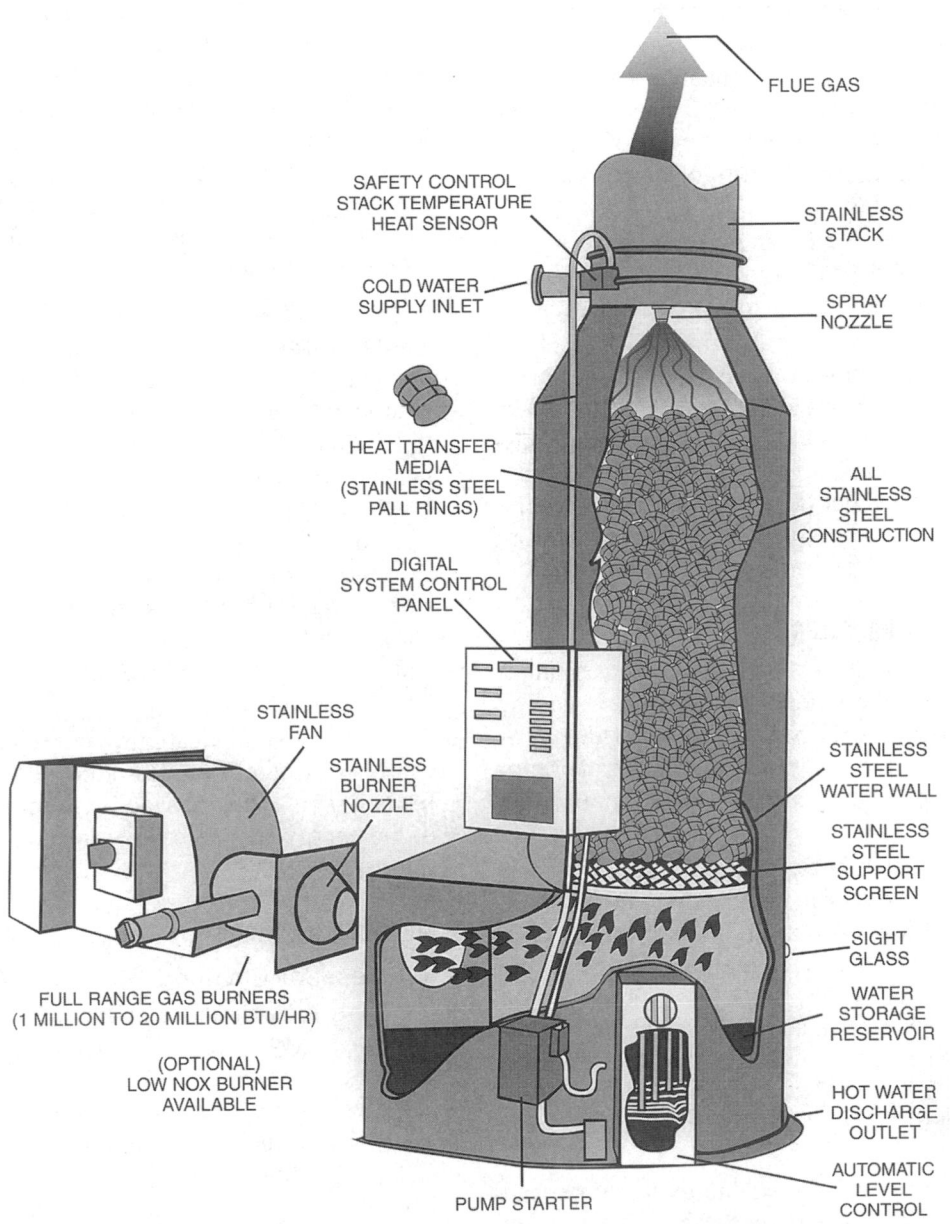

Ludell Manufacturing Company

Fig. 3 Direct-contact water heater This is the most efficient type of fuel-fired water heater because it eliminates heat exchange losses and captures a large fraction of the latent heat of the water in the flue gas. If the fuel is clean gas, the cleanliness of the water is not significantly affected, except for a slight increase in acidity. If the fuel is dirty, the dirt will go into the water, which is still suitable for applications such as industrial washing and making concrete during cold weather.

temperature is high continuously. The best new water heaters may have insulation R-values five times higher than those of some older units.

If you are installing an engineered hot water tank, calculate the best compromise between insulation effectiveness and cost. Insulation is cheap. It is hard to overdo it.

Water Heater Efficiency Ratings

In the United States, smaller residential-type water heaters are rated by using an "energy factor." This is defined as the ratio of the energy in the water to the input energy. It is measured under a standard set of usage conditions.

Contemporary electric water heaters have energy factors from about 0.80 to about 0.95. The only source of energy loss in an electric water heater is the insulation, so differences in the energy factors of electric heaters relate only to the insulation.

Contemporary gas-fired water heaters have energy factors from about 0.50 to about 0.70. These energy factors result from all the energy losses explained above.

Packaged water heaters are also rated by "R-factor", which is the average thermal resistance of the unit's insulation. Contemporary electric water heaters have R-factors ranging from 10 to 25. Contemporary gas-fired water heaters have R-factors ranging from about 6 to 18. The shape of a water heater has a minor effect on its surface heat loss. For a given R-factor, a more compact shape allows less heat loss.

Energy and Cost Saving Potential

The cost saving that you can gain by installing a water heater with better efficiency depends on the energy source, the cost of the energy source, and perhaps the amount of hot water consumption. In retrofit applications, it also depends on the efficiency of the existing equipment.

With electric water heaters, the amount of water consumption is not a factor in efficiency. The only loss is heat loss through the insulation, which depends only on the insulation and the water temperature. With an existing electric water heater, you can reduce heat loss by adding supplemental insulation, which is recommended by Measure 3.2.4. In selecting a new electric water heater, consider the "energy factor," as this takes into account both the insulation R-value and the shape of the tank.

With fuel-fired water heaters, energy losses depend on the amount of hot water that is used. If the water heater sits idle most of the time, its energy losses consist mostly of insulation heat loss and standby loss. If the unit has a high rate of water consumption, efficiency also depends heavily on the sophistication of the heat transfer surfaces. The "energy factor" covers all these

aspects. Use it as a selection guide, if it is available. If not, examine the design of the unit in light of the efficiency factors explained previously. If no hot water meter is installed, estimate consumption by recording the fuel input to the water heater.

To estimate the efficiency of an existing fuel-fired water heater, start by looking at its features, as discussed above. This is the best available indication of standby loss and insulation loss. You can do a combustion efficiency test to measure the unit's efficiency of converting fuel energy to heat in the water. See Measure 1.2.1 for details.

The value of improved insulation is reduced if heat lost from the water heater is a useful source of space heating. This happens only if the water heater is located in a heated space. Heat loss from the water heater is least important if the climate is predominantly cold, and if the energy used for space heating costs as much as the energy used for water heating. Conversely, the value of insulation increases if the water heater is located in an air conditioned space.

■ Electricity vs. Other Energy Sources

Do not assume that an electric water heater will cost much more to operate than other types. In water heating equipment, this is true less widely than with other types of heating equipment. The reason is that electric water heaters are more efficient than fuel-fired water heaters, for the reasons covered previously.

Furthermore, you may be able to improve the economics of electric water heating by installing a storage water heating system, or simply by allowing the electric utility to exercise interruptible control of your water heating. See Measures 3.2.8 and 3.2.9 for the details.

As in other applications, your selection of the energy source for water heating should consider other pertinent factors, such as maintenance requirements, the reliability of the energy supply, ease of installation, the need for a flue, etc.

Significant Options

■ Access for Cleaning

With fuel-fired water heaters, select a unit that provides easy access for cleaning the heat transfer surfaces. Separate handholes for cleaning are available in water heaters above a certain size and price.

Some of the more exotic designs may have higher efficiency than other types when they are new, but inability to clean out scale will make them progressively less efficient throughout their lives.

■ Fast-Recovery Heaters

Some water heaters have extended-surface heat exchangers to provide a higher recovery rate. The increased surface area may or may not produce higher efficiency, depending on the firing rate.

■ Power Venting

Some water heaters have power venting, which means that the combustion gases are drawn through the unit by a small fan. Power-vented units have low flue gas temperatures and low standby losses, making them more efficient than average. The low flue gas temperature permits venting through an adjacent wall, without the need for a flue. Some units can use plastic pipe for venting.

This type saves money in new construction by avoiding the construction of a flue. In an existing facility, substituting a power-vented unit may be advantageous if it allows a flue leading from a heated space to be blocked off, reducing outside air infiltration.

■ Non-Metallic Tanks

Some electric water heaters have tanks made of fiber-reinforced plastic. This avoids the need for an anode rod in areas that have corrosive water. Anode rods cause odor problems with some water sources.

ECONOMICS

SAVINGS POTENTIAL: *10 to 40 percent of water heating energy.*

COST: *Typical residential-size water heaters cost several hundred dollars. The cost premium for more efficient models of conventional water heaters ranges* from 30 to 150 percent, depending on features and pricing policy. Many utility companies have rebate programs to defray the cost of smaller high-efficiency units. At the other end of the size range, direct-contact water heaters typically cost $15,000 for a unit with a capacity of 1 million BTUH, to $75,000 for a unit with a capacity of 25 million BTUH. If you change to a water heater with a different energy source, the cost of connecting to the new energy supply may be as much as the cost of the heater.

PAYBACK PERIOD: *Several years, or longer, depending mainly on the quantity of water consumed.*

TRAPS & TRICKS

SELECTING THE EQUIPMENT: *Study the water heater market. There are many more choices than previously. Use the cheapest energy source that is readily available. Then, stress overall efficiency. Select a larger unit if this allows you to lower the water temperature. With units heated by fuel or steam, consider ease of cleaning the heat transfer surfaces. With complex high-efficiency units, check whether maintenance will be difficult. Check with others who have used your intended model for at least a few years.*

TEMPERATURE SETTING: *After installing a new water heater, minimize the water temperature. See Measures 3.2.1 ff.*

MEASURE 3.2.4 Install supplemental insulation on water heaters.

SUMMARY

An easy way to save a small amount of energy with poorly insulated water heaters. This is the only action possible to improve the efficiency of existing electric and steam-fired water heaters. Provides only a marginal improvement in fuel-fired heaters.

SELECTION SCORECARD

Savings Potential .... **$ $**

Rate of Return .... **% %** %

Reliability .... ✓ ✓ ✓ ✓

Ease of Retrofit .... ☺ ☺ ☺

Water heaters lose heat continuously through their insulation. The heat loss is independent of heater operation or water consumption. It depends only on the difference between the temperature of the water and the temperature of the surrounding space. If an existing water heater is poorly insulated, you can improve the insulation by covering the outside with a supplemental blanket of insulation. This activity is so popular that pre-cut insulation kits are available for the purpose. If a suitable kit is not available, you can do the job with batt insulation, a pair of scissors, and suitable tape.

Measure 3.2.3 points out that insulation heat loss is essentially the only source of energy waste in electric and steam-fired water heaters. Therefore, adding to the present insulation can make old heaters of these types about as efficient as new high-efficiency units.

How to Add Insulation

Insulation blankets for water heaters usually consist of glass fiber batt with an external covering. The covering may be thin plastic or fiber-reinforced foil. The insulation should have an insulation value of R-6 or more. This corresponds to a thickness of at least two inches.

Workmanship is important. The insulation should fit snugly at all points to avoid convective heat leakage between the heater and the insulation. Eliminate gaps. Tape the insulation at all edges and joints, and around pipes. Do not cover electrical connection boxes and thermostatic controls.

Fire safety is a concern. Keep the insulation and tape away from any hot areas. Be especially careful with fuel-fired heaters. Even glass fiber insulation includes a binder material that burns.

Alternative: High-Efficiency Water Heater

In fuel-fired water heaters, insulation heat loss is only one of several sources of energy waste, and it is usually not the most important one. Therefore, consider replacing a fuel-fired heater with a more efficient unit,

especially if the unit receives heavy usage. See Measure 3.2.3 about the efficiency of fuel-fired water heaters.

When Not to Insulate a Water Heater

This Measure is not economical if the water heater is installed in a space that is heated for a large part of the year. Also, the heat lost from water heaters may be useful for reducing humidity damage in basements and other humid areas.

ECONOMICS

SAVINGS POTENTIAL: *2 to 6 million BTU's per year, for water heaters of typical residential and commercial size.*

COST: *Materials cost about $20. Installation typically takes an hour or two.*

PAYBACK PERIOD: *One year, to several years.*

TRAPS & TRICKS

WORKMANSHIP: *Proper installation matters as much as the amount of insulation. Fit all the insulation snugly against the heater surface. Seal edges and joints tightly.*

MEASURE **3.2.5 Install automatic flue dampers on fuel-fired water heaters.**

RATINGS

New Facilities | Retrofit | O&M

C

You can reduce the standby loss in a fuel-fired water heater by installing an external flue damper. Flue dampers apply to water heaters in the same way as to boilers. See Measure 1.5.3.2 for the details.

The saving is typically smaller for water heaters because they burn less fuel than boilers, at least in typical applications.

Whether to Install a Flue Damper

External flue dampers are useful only with conventional water heaters that rely on convective circulation of flue gases. Modern high-efficiency water heaters have integral flue dampers or other means of reducing flue losses.

An external flue damper addresses only one of the sources of energy waste that occur in fuel-fired water heaters. If the present unit is old and/or inefficient, consider replacing it with a high-efficiency unit, as recommended by Measure 3.2.3.

Is Installation Practical?

With many water heaters, installation may be difficult or impossible. The damper must be installed between the exhaust outlet and the draft hood. However, water heaters typically have an integral draft hood mounted on the heater casing. In order to install a flue damper, you would have to relocate the draft hood away from the heater. This may be too awkward to be practical.

SUMMARY

Reduces flue losses in conventional water heaters. May be difficult to fit.

SELECTION SCORECARD

Savings Potential .... **$ $**

Rate of Return .... **% %**

Reliability .... ✓ ✓ ✓ ✓

Ease of Retrofit .... ☺ ☺ ☺

ECONOMICS

SAVINGS POTENTIAL: *Typically 5 to 10 percent of fuel consumption.*

COST: *Self-powered dampers cost from $20 to $60. Installation time varies with difficulty, averaging about one hour.*

PAYBACK PERIOD: *Several years or longer.*

TRAPS & TRICKS

SELECTING THE EQUIPMENT: *Select a unit that looks well made. Flue dampers are made by many shops, and some are flimsy. Safety certification does not indicate reliability or efficient operation. Select a unit that closes as tightly as possible while satisfying safety requirements.*

INSTALLATION: *Install the unit so that it blocks the flow of gas from the heater, ahead of the draft hood. Be sure that the flue gas flows properly into the draft hood when the damper is open and closed. Get this right or flue gas will enter the space.*

MEASURE 3.2.6 Clean and adjust the combustion systems of fuel-fired water heaters periodically.

Fuel-fired water heaters are usually reliable, and they require only a modest amount of maintenance. To keep them at peak efficiency, give these maintenance items your attention at regular intervals:

- *burner cleanliness and air-fuel ratio.* Routine burner maintenance consists of (1) cleaning, followed by (2) readjusting the air-fuel ratio. Both of these functions require specialized equipment and techniques. Burner cleaning is covered by Measure 1.4.1. Adjusting the air-fuel ratio is covered by Measure 1.3.1.

  Gas-fired heating units typically require burner maintenance only at long intervals. Inspect the flame visually on a routine basis. Make a combustion gas analysis at the beginning of each heating season. If the test indicates the proper efficiency, leave the burner alone.

  The burners of oil-fired water heaters require periodic cleaning because the fuel oil tends to form sludge and coke on the burner. Clean the burner at scheduled intervals. Refer to the heater's maintenance manual, or determine the appropriate intervals from experience.

- *fireside cleaning.* Accumulation of combustion products on the fireside surfaces of the heater increases flue losses. Clean the firesides of oil-fired heaters on the basis of soot accumulation. Typically, clean an oil-fired water heater about once a year.

  Gas-fired heaters that are adjusted properly may never require fireside cleaning. This is because soot burns off or spalls off faster than it can accumulate. If a heating unit accumulates soot or other fouling at an abnormal rate, the usual cause is a dirty or improperly adjusted burner.

- *automatic flue dampers* are important to efficiency. Failure of a flue damper can be a serious safety hazard, so they are made to be reliable. They require no routine maintenance. Check the operation of flue dampers visually at the beginning of the heating season.

SUMMARY

Basic maintenance needed to maintain the efficiency of fuel-fired hot water heaters.

SELECTION SCORECARD

Savings Potential .... $ $

Rate of Return .... % % %

Reliability .... ✓ ✓ ✓ ✓

Ease of Initiation .... ☺ ☺ ☺

- *burner fan impellers* should be cleaned to keep air flow efficient. Be prepared for the fact that cleaning a fan may make it noticeably louder.

- *fan motor lubrication.* If the fan motors are not permanently lubricated, oil them at appropriate intervals.

- *combustion air dampers* should be checked for proper closure, freedom of movement, proper adjustment of linkages, and proper control. Keep them clean and they will probably require no other maintenance.

ECONOMICS

SAVINGS POTENTIAL: *2 to 30 percent of energy input, the larger savings applying to oil-fired heaters.*

COST: *$10 to $100 per unit per year, the larger costs applying to oil-fired heaters.*

PAYBACK PERIOD: *Short. Over the long term, the savings substantially exceed the cost.*

TRAPS & TRICKS

SCHEDULING: *This activity is easy to overlook. Make sure that you have a maintenance schedule to remind you of repeated, long-interval maintenance activities such as this.*

SKILLS: *Burner maintenance and adjustment requires special equipment and skills. Make sure that the skills are available in-house, or hire the maintenance service.*

MEASURE **3.2.7 Clean out scale from water heaters periodically.**

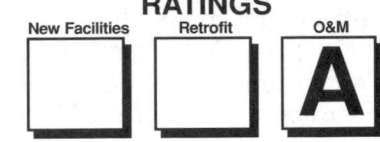
Water heaters may accumulate prodigious amounts of scale if the water supply has a high content of dissolved solids. Scale forms because the solubility of most solids drops as the water temperature rises. When the water is heated inside the heating unit to the point that it is saturated with dissolved minerals, the minerals come out of solution and deposit on the hottest surfaces of the unit.

With fuel-fired heaters, the hottest surface is the bottom of the tank, where the flame heats the water. Unfortunately, scale is an effective insulator. With fuel-fired heaters, the insulating effect of the scale causes a larger fraction of the fuel energy to be forced into the flue. In addition to dissolved minerals, a certain amount of miscellaneous dirt enters with the water supply. The dirt to the bottom of the tank, increasing the insulation effect.

In water heaters heating tanks that have electric or steam-fired heating coils, scale forms on the heating coils. In electric resistance and electrode water heaters, scale has no effect on efficiency or capacity, because the electrical energy has no place else to go. (Extreme scaling may cause an electric heating element to burn out, but it does not reduce efficiency.) Steam-fired coils may suffer a reduction of capacity. This is not important, unless the tank needs rapid recovery.

WESINC

Fig. 1 Clean-out valve on a water heater Attach a short hose to the faucet and run it to a container or sink. Open the valve wide for a few minutes. Judge how often to do this from the amount of scale that appears in the container.

SUMMARY

An easy routine procedure that reduces heat loss.

SELECTION SCORECARD

Savings Potential .... **$ $**

Rate of Return .... **% % %** %

Reliability .... ✓ ✓

Ease of Initiation .... ☺ ☺ ☺ ☺

How to Clean the Waterside

Virtually all fuel-fired water heaters have a drain connection at the bottom of the tank to remove scale and dirt. Scale breaks loose as a result of the expansion and contraction that occurs as the flame turns on and off, making it possible to blow out the scale chips.

Connect a hose to the drain connection, open the valve fully, and discharge the water to a bucket or drum. Look inside the bucket to see how much debris came out of the heater. Repeat the process until you have cleared out the heater.

Be careful to avoid scalding. Be ready to close the water supply valve that feeds the hot water tank if the blowdown valve breaks or fails to close completely.

Larger water heaters may have cleanout holes. If the heater has this feature, use it. Open it and clean the surfaces mechanically. Use a wire brush, water lance, or other appropriate tools.

Schedule this procedure often enough so that large amounts of debris do not accumulate between cleanings.

Label the Water Heater

A non-critical operation of this kind is easy to forget. Make it harder to forget by installing a placard at each fuel-fired water heater or group of water heaters. State the times when the heaters should be flushed, and when they should be mechanically cleaned. See Reference Note 12, Placards, for details of effective placard design, materials, and installation.

Reducing Scale with Water Treatment

Hard scale consists mostly of calcium compounds. The tendency of water to form scale is described in terms of "hardness," which is the quantity of dissolved minerals of calcium. You can reduce hardness by treating the incoming water supply in a water softener. This device passes the water through a bed of salt. It converts the dissolved calcium compounds into sodium compounds. The sodium compounds are far more soluble, so they do not precipitate to form scale.

Water softeners are not economical just to reduce water heater scale. They are usually installed when the facility needs large amounts of soft water for purposes such as laundry, or to reduce stains in fixtures. If you are considering a water softener for one of these purposes, remember that keeping the water heater more efficient is a bonus.

A certain amount of "soft scale" can form in water heaters even with soft water.

Possible Health Benefit

Debris that falls to the bottom of the tank is suspected of being a breeding ground for dangerous microorganisms. See Measure 3.2.1 for more about this. The problem is compounded by the fact that the water in the bottom of the tank may be much cooler than the water above the heating elements. Therefore, cleaning out tanks periodically may reduce the health risk. But, don't depend on this.

ECONOMICS

SAVINGS POTENTIAL: *A few percent of water heating energy.*

COST: *Minimal, for routine flushing. Mechanical cleaning of larger units, in tanks where this is possible, may require several hours.*

PAYBACK PERIOD: *Immediate, for periodic blowdown. Other methods pay for themselves within a few months. Over the long term, the saving is many times the cost.*

TRAPS & TRICKS

SCHEDULING: *The only trick is to remember to do it. Put it on your maintenance schedule as a repeated item.*

MEASURE **3.2.8 Exploit interruptible or storage rates for electric water heating.**

SUMMARY

May provide substantial savings in electricity cost for water heating. The savings are foolproof because interruption is under utility control. Requires some discipline in scheduling water usage to avoid shortages, and may require additional storage capacity.

SELECTION SCORECARD

Savings Potential .... $ $ $

Rate of Return .... % % % %

Reliability .... ✓ ✓ ✓ ✓

Ease of Initiation .... ☺ ☺ ☺

You may be able to get electricity at reduced rates by allowing the electric utility to install a remote control on your electric water heaters, as shown in Figure 1. The utility will turn off the heaters at certain times when they need to reduce their generating load. Reference Note 21, Electricity Pricing, explains the utility's motivation for offering this deal.

In some cases, the utility offers the lower rate only for the electricity that goes to the water heater. In small facilities, the utility may extend the discount to all the electricity. This saves the utility the cost of installing a separate meter for the water heater.

The utility installs and maintains the control equipment. Control is reliable because the utility itself controls water heater operation.

During the period of time when the power to the water heater is interrupted, the facility must get by with the amount of water that is stored in the tank.

This customer option has been offered by many utilities for a long time. There are two common variations. One is called "interruptible" water heating. The other is called "storage" water heating. Both depend on the inherent storage capacity of typical water heaters.

Under an "interruptible" arrangement, the utility has the option of turning off the water heating element for a relatively short period, typically several hours. The utility typically exercises this option only a few times per year, during periods of especially high energy consumption. The interruptible arrangement typically is used by utilities serving areas that have only occasional high peak loads. Under this arrangement, a customer with a storage tank of normal size may suffer the inconvenience of rescheduling water usage for a few hours, but only a few times per year.

Under a "storage" arrangement, the utility typically turns off the water heating element for its entire period of high demand charges, which may last from six to eighteen hours per day. The interruption may occur every day, or only during certain extended periods, such as the summer or winter months. The customer must have enough storage capacity to satisfy hot water requirements for this entire period. Typically, reheating the tank takes two or three hours.

If the utility offers both "interruptible" and "storage" arrangements, the customer needs to consider two factors in making the choice. One is the relative electricity cost, and the other is the storage capacity requirement. Utilities usually offer more favorable rates for storage water heating because it provides more benefit to the utility.

To exploit this opportunity, you need enough storage capacity to provide hot water during the periods when the power to the heating element is turned off. The amount of storage capacity depends on the length of time that the utility turns off the heating element, and it depends on the water usage pattern of the facility.

If your present water storage capacity is not adequate, you have two choices. One is to install a larger storage tank, or a supplemental tank. Many utilities provide discounts or rebates to encourage you to install more hot water storage capacity. The other approach is to reschedule the operations that require hot water. For example, a hotel could delay evening dishwashing to a time period when the water heating element is turned on.

Disadvantages

The main disadvantage is the possibility of running out of hot water during the interruption period. Deal with this by installing adequate hot water storage capacity. If you need to install a larger tank, or a supplemental tank, the additional space is typically not a problem. A modest increase in the dimensions of the tank provides much larger capacity.

How to Exploit This Opportunity

■ **Step 1: Analyze the Utility's Rate Schedules and Incentives**

Start by reviewing Reference Note 21, Electricity Pricing. Then, ask your electric utility for all the current rate schedules that apply to your facility. Compare the cost of electric water heating under all the rates, including interruptible, storage, and conventional demand rates.

Also, be aware that many electric utilities offer specialty rates. For example, some utilities offer more attractive rates if you allow the utility to control other electric loads, such as dual-fuel space heating or electric storage space heating. If your facility has a large amount of interruptible electrical load, try to negotiate a special contract. Usually, it takes a minimum of several hundred kilowatts of interruptible load to entice the utility into a special arrangement.

■ Step 2: Analyze the Facility's Hot Water Storage Requirements

The facility must have enough storage capacity to provide hot water through the period of heating interruption. To estimate the amount of storage you need, analyze the facility's water consumption pattern, especially the amount needed during the interruption period.

For example, assume that the facility uses a maximum of 1,000 gallons of hot water per day. If the tank capacity is 1,000 gallons or more, the facility is completely unaffected by the interruption schedule of the utility, provided only that the water heater has a few hours to recharge.

WESINC

Fig. 1 Interruptible water heating The arrow points to a small relay that the electric utility installed to interrupt the operation of the water heaters during periods of high electrical load. In return, the utility offers the electricity for the heaters at lower prices.

On the other hand, interruption may or may not be acceptable with the existing tank capacity. It depends on the length of the interruption and the pattern of consumption. If hot water consumption is distributed throughout the day and interruptions last only a few hours, hot water shortages will be rare. On the other hand, if an interruption extends all day, hot water shortages may occur too often to be acceptable.

Measure water consumption during typical days of high usage. You can do this by measuring the hot water flow with an external flow meter. (See Reference Note 16, Measurement of Liquid, Gas, and Heat Flow, for details.) As an alternative, you can estimate water consumption from the energy input to the water heater. The simplest way to do this is to use a portable time recorder on each heating element. Or, if the heater has a number of heating elements, use a kilowatt-hour meter to measure the total electrical input.

A simple yes-or-no approach is to simulate interruption by turning off the power to the heating element during several days of heavy water usage. This tells you whether the present storage capacity is adequate.

Be cautious about utility claims that water heating power will be interrupted "only a few times" or "only for a short while." This may be true at present, but interruptions are likely to get worse in the future, as the utility's load increases. Install enough storage to carry through the longest period of interruption stipulated in the rate schedule.

■ Step 3: Estimate Your Cost

The utility typically installs the control for your water heaters, and this part of the job costs you nothing.

If you have to add more storage capacity, the main item is the installed cost of the new tank. The utility may offer discounts or rebates for installing the additional capacity.

■ Step 4: Negotiate with the Utility

If the utility has a standard water heater control rate and your hot water consumption is not a major electrical load, simply request the standard arrangement. As stated before, if you have a large amount of demand that may be controlled, you may be able to negotiate more favorable terms.

■ Step 5: Install Additional Storage Capacity, If Needed

Install any additional hot water storage that you may need. Storage tanks are discussed below. Many utilities offer to handle the installation. Also, the utility may provide financing and charge time payments as part of the utility bill.

■ Step 6: Establish Appropriate Water Management Procedures

Do what you can to avoid water shortages during the period of interruption. In facilities that have a regular water use pattern, no special action is necessary if you have planned properly. If an unusually high demand for water may occur during an interruption period, prepare to educate water users to the problem. You may be able to this effectively with placards, as explained below.

Storage Water Heaters

Storage water heaters are the same as ordinary water heaters, except that the tanks are larger. The heating capacity is usually the same as for conventional heating. This is satisfactory because the elements operate during periods of low water consumption, when the tank can be reheated at leisure. For the same reason, you do not have to increase the capacity of the electrical circuits that serve the heaters.

Select storage water heaters that have exceptionally good insulation. Conductive heat loss from a well insulated water heater amounts to no more than a few percent per day. Heat loss does not increase much as you increase storage capacity, because surface area increases more slowly than volume.

You can also increase storage capacity by adding a second water heater in series with the original unit. To avoid increasing the total circuit amperage, recharge the downstream water heater first, and then the upstream heater. To do this, you have to power the two heaters in sequence. This not a difficult control arrangement, but it is unusual.

In larger water systems, you can increase storage capacity by installing a plain tank, without a heater, ahead of the heating tank. Install a small pump to circulate hot water from the heating tank into the supplemental tank during the heating period.

Hot Water Temperature, Storage Capacity, and Safety

The amount of heat stored in water is directly proportional to the temperature of the water. Therefore, raising the temperature of the storage tank increases its effective capacity. This fact creates the temptation to limit the cost of storage heaters by raising the storage temperature. In principle, an ordinary water heater could store water almost at boiling temperature. In practice, the safety hazard is usually unacceptable if the water can come in contact with people, as in showering.

At first glance, it may appear that you can solve both the scalding hazard problem and capacity problem by using a tempering valve. A tempering valve is a mixing valve with an internal thermostat that mixes cold water with the hot water from the heater. The mixed water is then distributed within the facility.

This is probably a bad idea. The tempering valve is a mechanical device that may fail. If it fails, the water at the point of use may be much hotter than expected. In systems where the hot water is not recirculated, the

danger is more acute. When the user turns on the tap, cold water comes out first, causing the user to open the hot water tap fully. When the cold water clears out of the line, hot water follows. Tempering valves are not a solution to this problem, because they respond too slowly to stop the initial slug of hot water.

From a safety standpoint, the best approach is to install tanks that are large enough to store an adequate quantity of water at a safe temperature. Then, you may still want to use a tempering valve to lower the water temperature further for energy conservation. See Measure 3.2.1 for details.

The appropriate temperature setting for human usage of hot water is bounded narrowly by two considerations. 140°F is commonly quoted as the maximum temperature to avoid scalding. You may want to use a tank discharge temperature almost this high to ensure that the temperature inside the tank is high enough to prevent proliferation of microorganisms. See Measure 3.2.1 for more about this issue.

Higher temperatures are mandatory in sanitation applications, such as dairy barns and restaurant dishwashing, and in some other specialized applications. In such applications, the staff must be trained to deal with the hazards of the high water temperatures.

Equipment Placards and Interruption Indicators

The operating staff will not know about the purpose of interruptible water heating unless you tell them. Therefore, install a prominent placard at the hot water tank and/or at a location where hot water usage is controlled, which says:

• power to the water heating system may be interrupted as a cost saving measure
• the times and/or durations of interruption
• a signal will be activated when interruption is in effect
• restrict hot water use during periods of interruption.

There is a bonus in making all water users in the facility aware of heating interruptions, even though hot water continues to be available from storage. It reduces the chance that a water shortage will occur, and it motivates water conservation and energy conservation in general.

The utility should provide an indicator to show when water heating is being interrupted. This could be nothing more than a red light on the utility's control box that turns on when power to the water heater is interrupted. If this is not satisfactory, you can install a remote indicator anywhere that it will get the attention of the people who need to know. The control receiver that turns off the heating elements provides a signal voltage. Use this to trigger your indicator or alarm. As an alternative, use a relay to sense when power to the heating elements is turned off.

Understand the Utility's Motivation

A utility offers discounted rates for interruptible service because this provides a benefit to the utility. The value to the utility changes with time as the utility's load and generating resources change. The highest rate incentives are offered when the utility is most desperate to reduce its demand. However, when the utility is desperate, it will seek a variety of other solutions. The utility may eventually abandon the rate incentive.

A utility may reduce or abandon a discount rate at any time, subject to approval by the relevant regulatory agencies. Before making a substantial investment to gain a reduced utility rate, make an informed judgement about how long the utility is likely to continue the favorable rate.

This Measure shifts the utility's electrical load, but it does not reduce it. The water heating load that is deleted during the hours of interruption is added to the load during the other hours of the day. Only a few hours are needed for reheating the tank. The utility wants to move this load into the least-used off-peak nighttime hours. However, the need for the water heater to recharge extends the peak demand period. Eventually, there may be no more opportunity for transferring daytime load to nighttime.

This situation is most likely to arise in areas where electric space heating creates a peak load for the utility. Ironically, space heating can be a peak load in warm climates as well as cold climates, because the low cost of electric heating units makes them popular in warm locations.

What to Do if the Utility Does Not Offer a Storage Rate

Even if the utility does not offer an interruptible or storage rate for water heating, you can still reduce demand charges by controlling water heating on your own. See Measure 3.2.9 for this alternative.

ECONOMICS

SAVINGS POTENTIAL: *The savings are almost entirely in the form of reduced electricity cost. The potential saving varies widely. Many utilities offer no special rates for interruptible service, while others may reduce the price of interruptible electricity by as much as 50%.*

COST: *The cost to the facility is minimal, if only interruption is employed. Increasing storage capacity may cost several hundred dollars to tens of thousands of dollars. Many utilities provide payments or discounts to offset the cost of installing additional storage capacity.*

PAYBACK PERIOD: *Immediate, for simple interruption. Typically several years, if you have to add storage capacity.*

TRAPS & TRICKS

PLANNING: *Estimate the storage capacity that you need to avoid hot water shortages. Base this estimate on a storage temperature that provides the optimum compromise between efficiency and equipment cost. Investigate whether it is practical to reschedule any hot water usage, as an alternative to buying more storage capacity.*

SELECTING THE EQUIPMENT: *The same criteria apply to storage water heaters as to other water heaters. See Measure 3.2.3.*

PUBLICITY: *Keep this activity from creating unpleasant surprises. Install placards that request water conservation in locations where water users will see them. If interruption does not occur on a regular basis, install an interruption alarm.*

MEASURE **3.2.9 Control electric water heating to reduce demand charges.**

SUMMARY

An alternative way of reducing demand charges if Measure 3.2.8 is not available. Requires strict operating discipline.

SELECTION SCORECARD

Savings Potential	$ $
Rate of Return, New Facilities	% % % %
Rate of Return, Retrofit..........	% % %
Reliability	✓ ✓
Ease of Retrofit	☺ ☺

Refer to Measure 3.2.8 for the basic principles. If the electric utility does not offer a special rate for interruptible or storage water heating, you can still reduce demand charges by controlling the operation of the water heaters yourself. This exploits the fact that a water heater is an energy storage device, which allows you to turn it off for a period of time. Figure 1 shows an exceptionally favorable opportunity.

This Measure is similar to Measure 3.2.8, except that responsibility for managing control of the water heater rests with the facility. This is a major difference. The activity is no longer foolproof, and the facility must pay for the necessary controls.

Give serious thought to whether your facility can manage interruption of water heating demand reliably. Although control is automatic, some discretion may be required. If the hot water storage capacity is marginal, interrupting water heating may sometimes create a hot water shortage. In that case, you must decide whether to forego the availability of the hot water or the savings in demand charges. A way out of this dilemma is to increase hot water storage capacity, but this is expensive.

How to Control Water Heating

The basic procedure is to install a demand controller to control the operation of the water heaters. A demand controller is a control device that senses the overall facility power consumption. When the facility load rises to a preset level, the controller turns off the water heater. The water heater is turned back on when the facility load drops below this level.

If the water heater has a number of heating elements, the demand controller should turn them off in sequence, so that water heating may continue to be available at a reduced level.

(Some electric water heaters have a pair of identical heating elements, one at the top of the tank and one at the bottom. These operate in sequence. The top element is intended to provide hot water quickly, after which the bottom element heats the rest of the water in the tank. This arrangement is used to limit the wiring capacity needed to serve the water heater. It also acts to limit demand.)

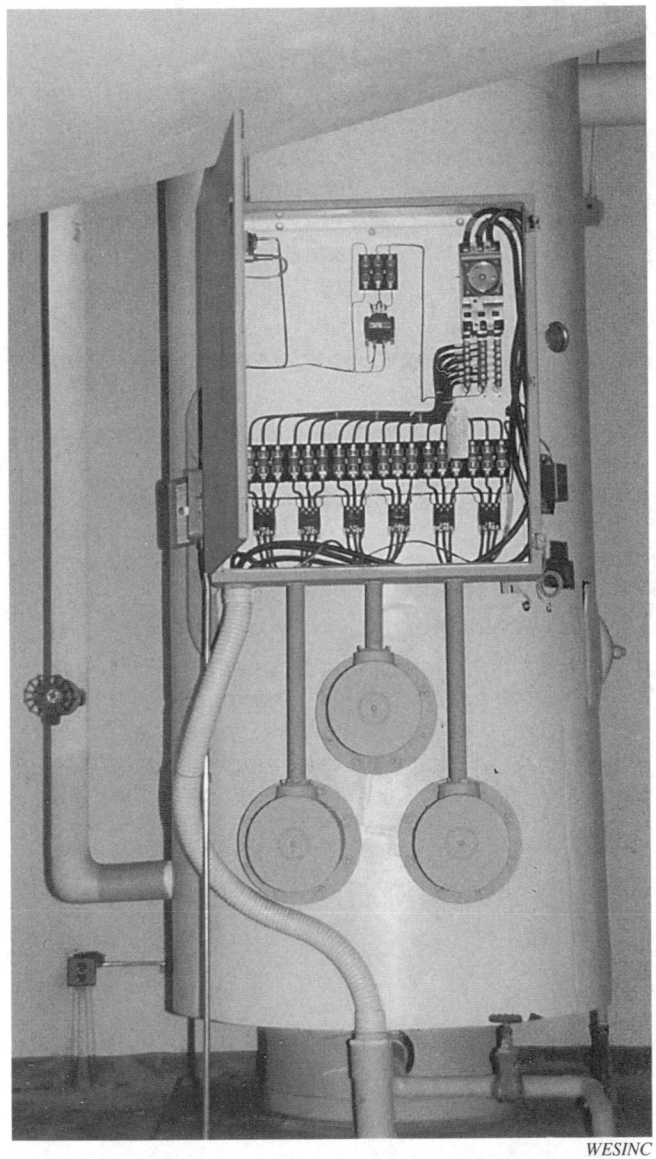

WESINC

Fig. 1 An exceptionally favorable opportunity for reducing demand charges This 100-KW water heater is greatly oversized for the facility it serves, and all six heating elements are connected in parallel. As a result, demand charges are very high. They could easily be reduced or eliminated with a reliable time control and a few relays installed in the wiring cabinet.

More sophisticated demand controllers "learn" from experience. If the facility load rises to a new high during a demand billing period, this level becomes the new threshold for demand control, reducing unnecessary interruption of water heating.

These functions can also be performed by many energy management computer systems, provided that you can program the computer to do it. See Reference Note 13, Energy Management Control Systems, for details. A dedicated controller is more reliable, and it is usually cheaper in the long run.

Control Other Eligible Equipment at the Same Time

The facility may have other equipment that can be turned off on an elective basis. Use the same demand controller to control this equipment as well as the water heater. For example, the demand controller can sequence different air conditioners to limit the overall demand. Water heating is usually the only major controllable load in commercial and residential facilities. Some industrial facilities may offer other major opportunities.

Explain It!

The controls are automatic, but they will be sabotaged sooner or later if the staff does not understand their purpose. Failing to control water heating demand even once during a demand billing period will forfeit all savings during the billing period. Therefore, install an effective placard near the water heater controls that informs the staff of the purpose of the demand control. See Reference Note 12, Placards, for details of effective placard design, materials, and installation.

ECONOMICS

SAVINGS POTENTIAL: Varies widely. Demand charges in the United States average several dollars per month per kilowatt of demand during the peak demand period.

COST: Several thousand dollars for the demand controls. The cost of adding more water storage capacity, if necessary, varies from hundreds of dollars to thousands of dollars.

PAYBACK PERIOD: Several years, or longer.

TRAPS & TRICKS

PREPARATION: Before you can do this successfully, you need to learn about electricity rates. First, request from your utility all the rate schedules that apply to your facility. Then, learn how to read them. After that, make an educated guess about the demand rates that the utility will offer in the future. Reference Note 21 will guide you. See Measure 3.2.8 about the possibility of increasing your water storage capacity.

SELECTING THE CONTROL EQUIPMENT: The most reliable approach, and probably the cheapest, is to install a single-purpose demand controller. Shop the market to see what is currently available. You can also use your energy management computer, but be prepared for programming headaches. Also, you will lose your control capability when the computer becomes obsolete in a few years.

PUBLICITY: Tell water users ahead of time when to expect water heating to be interrupted. See Measure 3.2.8 for details.

EXPLAIN IT: Plant operators usually are not the people who pay the electric bill, so this activity could fail without anyone noticing. Therefore, describe the demand limiting control clearly in the plant operating manual. Install a placard at the control that explains how to tell whether the control is working.

MONITOR PERFORMANCE: Schedule periodic checks of the controls during periods of time when demand charges are affected. Also, periodically review the current rate schedules to verify the times and conditions under which interrupting water heating saves money. Utilities continually change their rate schedules.

Service Water Pumping

The Measures in this Subsection reduce the energy consumption of service water pumps. Many facilities do not need their own pumps to provide service water pressure, because the public water utility provides adequate pressure. However, many facilities do need to provide or augment their own service water pressure, and the pumps for this function may be major consumers of energy.

Expect to find water pressurization pumps in facilities that have their own sources of water, typically from wells. Also, a facility needs to have its own pumps if the pressure provided by the water utility is not high enough, or if the public water supply is unreliable. The facility may have its own pumps for applications that need water at exceptionally high pressure. For example, tall buildings commonly have their own service water pumps to serve upper floors.

Most facilities have small pumps for recirculating hot water. These pumps typically consume only a small amount of energy. However, there are cases where the design of the service water system requires larger pumps for recirculation. You may be able to save a significant amount of energy by improving these arrangements.

MEASURE 3.3.1 In facilities that have their own service water pumps, configure the system to minimize pump energy consumption.

The purpose of this Measure is to improve the efficiency of pumping systems that provide pressure for service water. The pumps may be relatively small, perhaps a few horsepower. However, if the facility does not have a standpipe (water tower) or other type of storage tank that maintains pressure, the pumps consume a large amount of energy in relation to their size because they must operate continuously.

There are two general ways to improve pumping efficiency in such cases:

- *install pumps in increments of capacity,* so that pump power can increase in proportion to the flow rate demanded by the system. This approach is recommended by subsidiary Measure 3.3.1.1.
- *install one or more standpipes or pressurization tanks,* to limit pump operation to brief periods of running at maximum efficiency. This approach is recommended by subsidiary Measure 3.3.1.2.

Where to Consider this Measure

This Measure applies to facilities that employ their own pumps to provide service water pressure, or to augment the pressure provided by the water utility ("city water"). In-house pumps are used in tall buildings, in facilities that have their own water supply, and in locations where city water pressure is unreliable.

If an existing facility uses a standpipe for pressurization, your ability to make efficiency improvements is limited, because standpipe systems are inherently efficient. However, even if a standpipe is installed, you may be able to make improvements that are worth the effort.

The two subsidiary Measures offer a spectrum of possibilities that range from adding a single small pump to installing multiple standpipes. The more expensive changes are economical as retrofits only in facilities that have larger service water pumps.

An Example

To see how a service water system can waste pumping energy, consider the example of a large hotel in which service water pressure is produced entirely by in-house pumps. This is depicted in Figure 1. The hotel has a tall guest room tower. Meeting rooms are on the second floor. Kitchens, restaurants, laundry, ballrooms, offices, and lobby are on the first floor. The main mechanical equipment room is in the basement.

This system is inefficient because it must provide enough pressure to flush a toilet in the penthouse, even though most of the water flows to the large water-using equipment on the lower floors. The capacity of the

pressurization system must satisfy the requirements of all the showers, cooking facilities, laundry, and other water users. Even taking diversity into account, the pumping capacity must be much larger than the typical flow. If the entire system were pressurized by a single pump, the combined pressure and flow requirements result in the pump being vastly oversized.

We will return to this example in the subsidiary Measures to illustrate the various methods that we can use to reduce pump energy.

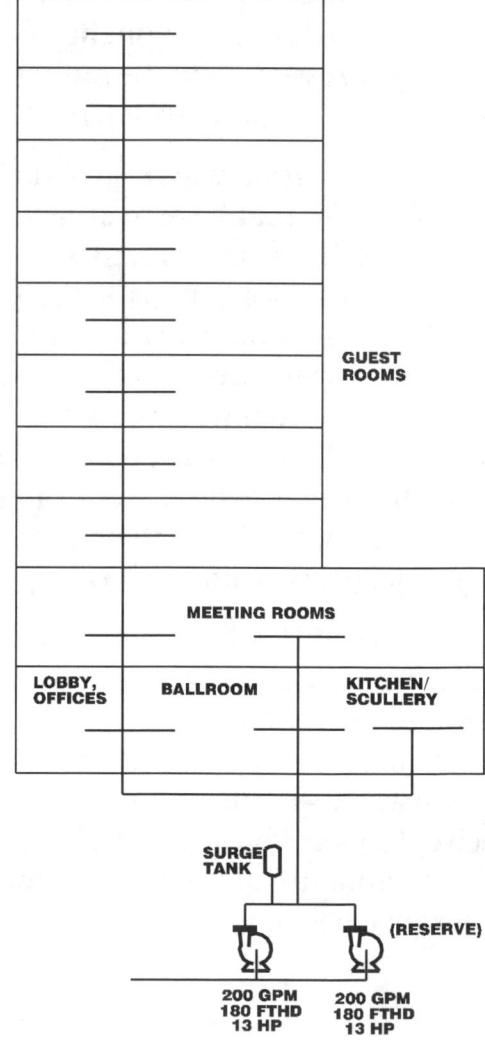

Fig. 1 How service water pressurization wastes energy
In this example of a worst case, service water pressure to a tall hotel is provided by a single pump. The pump must operate continuously. It must be sized for the maximum flow rate of the entire building, even though the average flow rate is a fraction of this amount. It must be sized to provide water pressure to the highest level in the building, even though most water consumption occurs at the lower levels.

Analysis of System Energy Consumption

(The following discussion requires an understanding of basic centrifugal pump characteristics. If you need a quick refresher, see Reference Note 35, Centrifugal Pumps.)

The pump load in service water systems is created mostly by gravity and by the pressure requirements of the water-using equipment. These factors are constant. Therefore, the pump pressure requirement is fairly constant. On the other hand, the flow requirement typically varies over a wide range.

(The amount of pump energy dissipated by pipe friction is minor, especially because the average flow rate is low. However, the resistance caused by pipe friction may become important at higher flow rates, requiring the pump pressure rating to be increased somewhat.)

These pump load characteristics cause a service water system to waste energy in several ways. Each of these has a different remedy. The following are the main causes of energy waste.

■ Extended Operation at Low Flow Rate

The key point is that a single domestic water pressurization pump spends much of the time operating at a very low fraction of its design maximum flow. Likewise, most of the energy that the pump consumes over the course of time is consumed while the pump is operating at such low flow rates. For example, in a hotel,

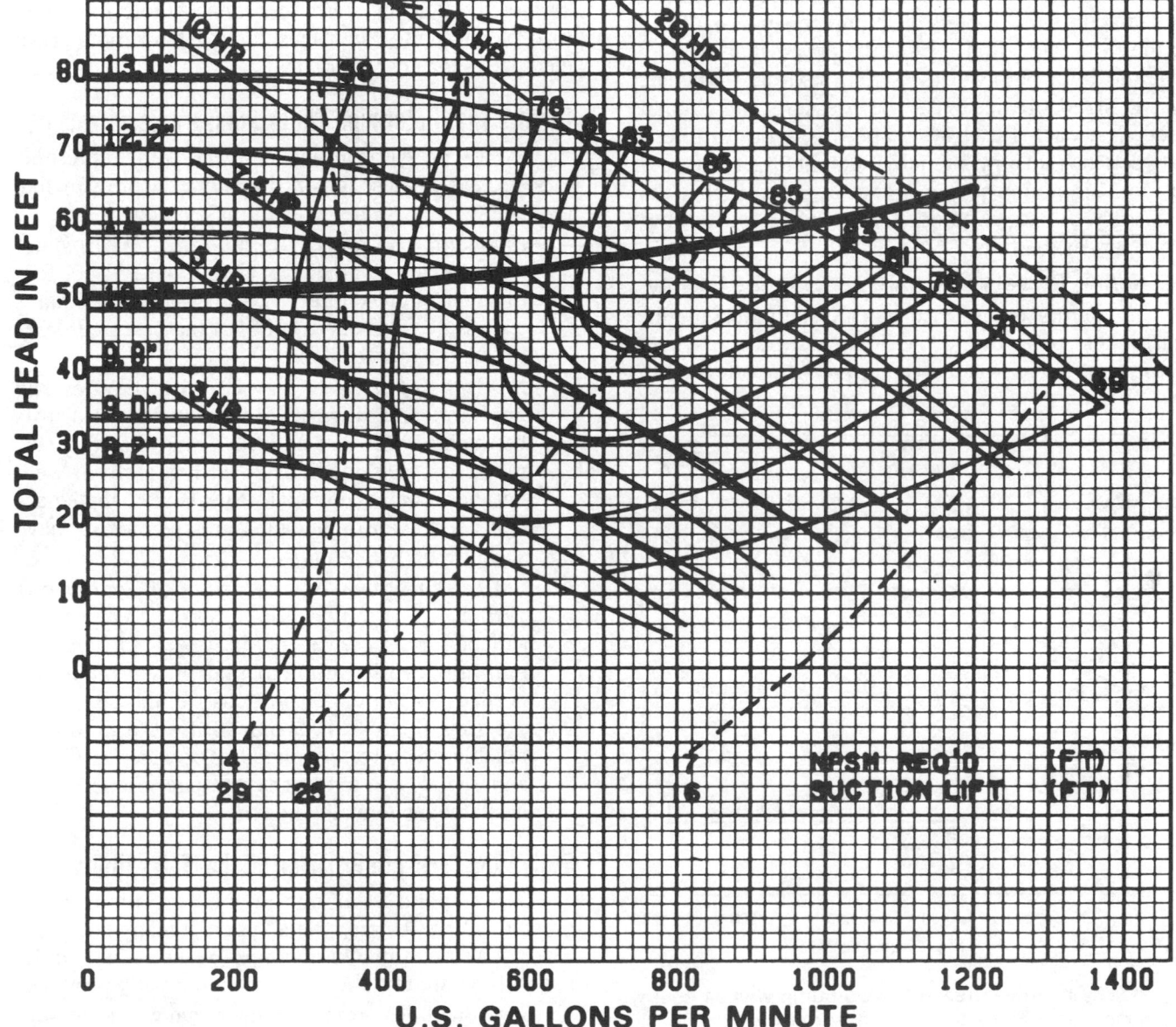

Fig. 2 Pump curve and system curve for a single pump providing service water pressure In a typical application, the pump spends most of its time operating at a very low fraction of maximum flow. The pump can operate only along its pump curve, so it produces more pressure than necessary at low flow rates. Also, the efficiency of the pump declines seriously when it operates at low flow rates.

a single pump might operate at less than 10% of maximum flow for a large fraction of the time. This wastes energy in two ways, which can be seen from the typical pump and system curves in Figure 2.

One cause of energy waste is that the pump operates at low efficiency for most of the time. Pump efficiency declines seriously when the pump operates far from its design pressure and flow rate. For example, in Figure 2, pump efficiency declines from about 85% at maximum flow to less than 50% at low flow rates.

The other cause of energy waste is that the pump can operate only along its pump curve. At low flow rates, the system curve is far from the pump curve, so the pump is producing much more pressure than needed to satisfy the flow requirement.

The graph in Figure 3 shows the overall effect of these two factors. It plots the pump power as a function of the flow rate. It also plots the power that would be required by a pump of the same maximum efficiency that is sized precisely for the pressure and capacity needed at each flow rate. There is a big difference between the two power requirements at lower flow rates. In fact, the pump power is several times higher than necessary to satisfy the system requirement during a large part of the time.

■ Differing Pressure Requirements in the Facility

The pumps in a service water system must produce the highest pressure needed anywhere in the system. Pump energy is wasted in delivering water at this maximum pressure to portions of the system that need less pressure. This is a common problem in tall buildings, as in our sample hotel.

■ Systems with Pressure-Reducing Valves

Some buildings use pressure-reducing valves to keep system pressure within acceptable limits. This arrangement is typically found in tall buildings, where pressure-reducing valves might be installed in the lower portions of a tall riser. All the pressure that is dissipated by the valves must be produced by the pumps, wasting a considerable fraction of the pump's energy input.

■ Motor Efficiency

The efficiency of electric motors starts to decline rapidly when the motor is loaded less than about 50%. The motor is always somewhat oversized for the pump, so motor efficiency starts to decline before the pump load falls to half. See Measure 10.1.1 for more about electric motor characteristics.

Factors That Determine Water Pressure Requirements

Excess pressure at any delivery point represents wasted energy. These factors determine the pressure that the facility's own water system must produce:

- ***the height of the delivery point.*** Hydrostatic pressure due to gravity is about 0.43 PSI per foot of height. For example, a 20-story building requires a pressure of about 100 PSI at the basement level to pump water to the top floor.

- ***the types of fixtures served.*** Most fixtures and appliances require minimum water pressures between 10 PSI and 20 PSI. Some shower heads may require as much as 30 PSI to perform well, but these are exceptional. Specialized appliances that need even higher pressures commonly have separate booster pumps.

- ***the reliable pressure contribution of the city system.*** The city water system may provide satisfactory pressure for the lower portion of a building, while supplemental pressurization is needed for the upper portions. If the city water pressure fluctuates, design to use the minimum reliable pressure. If reliability of water service is an issue, design to use city pressure whenever it is available, as explained next.

Exploit the Pressure Provided by the Water Utility

Don't spend your own money to create water pressure where the city water system already does this. Select your pressurization pumps to take suction at the lowest pressure that the city system normally provides.

Some facilities provide all their own water pressure from a vented tank at ground level that is filled from the city water supply. This arrangement may be used to ensure an adequate supply of water because the city water supply was considered unreliable at the time the

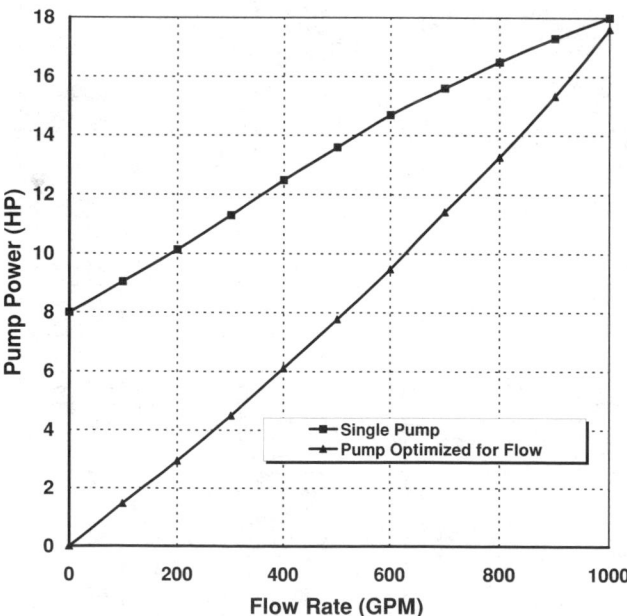

Fig. 3 Why a single pressurization pump wastes energy
The upper curve shows the actual power consumption of the single pump characterized in Figure 2. The lower curve shows the power needed to satisfy the system flow requirements by a pumping system that always operates at peak efficiency. The power differs enormously at low flow rates.

facility was built. Providing a reliable supply of water for firefighting may be one of the primary reasons for installing such a system. This arrangement duplicates the pressure of the city water system, wasting the corresponding pump energy. Figure 4 shows an example.

A solution is to connect the domestic water system directly to the city water supply, bypassing the storage tank. Fire protection pumps are separate from service water pumps, and these remain connected to the reservoir tank. Maintain a full tank of water for firefighting by using a float-actuated fill valve.

In retrofit, leave the original pumps in place for periods when the city supply fails. Install reliable controls to switch to the original pumps when needed. A simple method is to use the pressure of the city water supply to control a transfer switch that provides power to only one of the pumps. With a manual override on the transfer switch, the original pumps serve as a reliable backup for the new pumps.

With this arrangement, you have to be sure that the pump check valves do not leak. The check valves of both the new and the original pumps are under full supply pressure at all times. If the check valve of the new pump leaks, water will backflow into the city water supply. If the check valve of the old pump leaks, water will backflow into the vented tank, overflowing the tank into the sewer. Both leaks waste pump energy.

Pressurize Hot Service Water Efficiently

The hot water flow capacity is determined by the capacity of the cold water pumps. Therefore, remember the hot water requirements in each pressure zone when sizing the service water pumps for that zone.

In terms of pressure, the hot water system is connected directly to the cold water system through the water heater. In other words, hot service water is produced at the same pressure as the cold water system that feeds it. If the water system is divided into different pressure zones, the hot water usually is distributed to the same pressure zone as the cold water, even if not to

WESINC

Fig. 4 Failing to exploit city water pressure In this large hotel, city water is delivered to the vented tank in the background. All the city pressure is lost. The large pumps in the foreground must operate continuously to provide pressure within the building.

the same physical space. Exceptions might occur if hot water is needed at a different pressure than cold water.

A separate water heater is needed for each separate pressure zone in which hot water is delivered. The cost and difficulty of installing additional water heaters is a major limiting factor in subdividing the water system for efficient pumping.

In all but the smallest facilities, hot water requires recirculation to maintain adequate temperature at the water-using fixtures. To avoid complications and waste of pump energy, each pressure zone should have its own recirculation system. If recirculation is designed properly, the recirculation pumps are very small.

MEASURE 3.3.1.1 Use multiple pressurization pumps.

This subsidiary Measure explains the first of two approaches for minimizing service water pumping energy. This approach relies on using a clever combination of pumps.

There are two aspects to this approach. One is to install a group of several pumps in parallel to adapt to the changes in the water flow requirement. Another aspect is to install different groups of pumps that are matched to different pressure requirements within the facility.

The selection and grouping of pumps is largely a matter of judgement. There is a trend of diminishing returns in using larger numbers of pumps to match flow and pressure requirements more accurately. As you make the configuration more efficient, it becomes more complex and more expensive. Many pumps may be needed to efficiently match all the facility's flow and pressure conditions.

SUMMARY

Simple and reliable. Facilities with diverse pressure and flow requirements may require numerous pumps. Retrofit applications require piping changes, which may be modest. Requires a separate water heater and recirculation system for each pressure zone.

SELECTION SCORECARD

Savings Potential .... $ $

Rate of Return, New Facilities % % %

Rate of Return, Retrofit.......... % %

Reliability .... ✓ ✓ ✓ ✓

Ease of Retrofit .... ☺ ☺ ☺

How to Accommodate Changing Water Usage

To match pump power to the variations in flow requirements, install a parallel combination of pumps. Select the pump capacities so that some combination of pumps is reasonably efficient for all flow rates that may occur.

For example, consider our typical hotel, introduced in Measure 3.3.1. Assume that the hotel has these water flow requirements:

- flow is maximum between the hours of 0600 and 0800, but never more than 200 GPM
- usually below 100 GPM during lunch and dinner hours
- usually below 50 GPM during the rest of the daytime and evening
- usually below 10 GPM during late night hours.

To meet these flow requirements efficiently, we will replace the present single pump and its standby unit with a parallel group of five pumps, as shown in Figure 1.

By using three pumps rated at 100 GPM each, we satisfy the maximum flow requirement, with one pump as a spare. Also, selecting this size allows a single pump to handle the lunch and dinner hours. A single pump of 50 GPM capacity is installed to handle the lower demand during the rest of the daytime. A single pump of 10 GPM capacity handles the minimal late-night requirements.

Obviously, other combinations are possible. The example makes it clear that there is a compromise between the number of pumps and the efficiency of pumping.

In parallel groups of pumps, select all of them to operate at the same pressure, since they may operate together. In principle, pumps operating individually at the lower system flow rates could be designed for a somewhat lower pressure, since they face less piping resistance. However, this refinement is not worth the extra complication.

■ How to Control Pump Sequencing

The best way to control the operation of the pumps is to sense the flow rate in the common discharge line. Remember that the flow rate is determined by the water users, not by the pumps. A specific pump or combination of pumps is turned on in each range of flow rates.

You cannot use pressure for normal control of pump operation, because the system pressure is supposed to remain constant at all flow rates. However, you might sense the pressure in the discharge line to start an additional pump in the event that one of the pumps fails.

■ How to Select Reserve Pumps

In systems designed to operate with only one pump, it is common practice to provide a reserve pump. However, in a system that uses multiple pumps, failure of one of the pumps may not be critical. In fact, a single pump failure may have no effect except during a peak consumption period. During a peak, the loss of pressure would be most noticeable in the highest portion of the building. Even then, the characteristics of centrifugal pumps allow them to "trade" pressure for increased flow.

If several pumps of mixed size are installed in parallel, a reserve unit equal in size to the largest pump usually provides a generous guarantee of performance.

■ **How to Minimize Pressure Fluctuations**

Starting or stopping a pump causes a small but fairly abrupt change in system pressure. You can smooth these fluctuations by installing a surge absorber in the discharge line from the pumps. This device is a small tank that is filled with air at system pressure. It is connected at the bottom to the water pipe. The pressurized air acts as a cushion that absorbs sudden increases in pump pressure, and it expands to compensate for small drops in system pressure.

How to Accommodate the Differences in Pressure Requirements

The previous technique, using a group of parallel pumps, reduces the waste of energy that results from excessive flow capacity. Additional steps are needed to minimize the waste of pump energy that results from differences in the pressure requirements of different parts of the facility. In tall buildings, pressurization pumps must provide enough pressure to serve the upper levels, resulting in unnecessary pressure at lower levels. Also, some branches of the service water system may have exceptionally high pressure requirements because they serve special requirements, such as dishwashers, high-pressure hosedown, or a decorative fountain. The following are two ways to solve this problem.

■ **Method A: Install Separate Pumps and Risers for Different Pressure Zones**

If cost were no obstacle, it would be desirable to provide separate pressurization for each area of a facility that has a different pressure requirement. The cost of piping and pumps determines how close to this ideal you should try to make the system. Even in retrofit applications, you may be able to cut apart existing risers and provide separate pumps, with little new piping being required.

The modification shown in Figure 2 illustrates this approach. In this case, three separate pressure zones have been created. One serves the higher portions of the building, a second serves most of the lower portions, and a third serves the higher pressure requirements of the dishwashing area. A short riser was added to connect the new guest room pumps to the existing guest room riser. Also, the piping in the equipment room had to be rearranged to provide a separate pressure zone for the dishwashing area.

The number of pumps has been increased further, compared to Figure 1. This is unavoidable, because each pressure zone must have a variety of pumps to accommodate different flow rates. However, the total pump energy consumption is reduced, because the pumps serving the largest flow requirements operate at lower pressure.

Remember that a separate hot water heater is needed for each pressure zone. In retrofit applications, this method allows the new pumps to be installed near the old ones, which makes it easier to install the additional water heaters in a cluster.

■ **Method B: Install Booster Pumps to Serve Zones with Higher Pressure Requirements**

Another way to provide increased pressure for certain areas or equipment is to install booster pumps. Figure 3 shows this approach being used in our hotel example to serve the upper guest room floors. Also, a single booster pump has been added to serve the high-pressure applications in the dishwashing area.

This method incurs no cost for additional piping, except for connections to the booster pumps. Compared to Figure 1, the number of pumps has been increased. This is necessary because each pressure zone needs a variety of pumps to accommodate different flow rates.

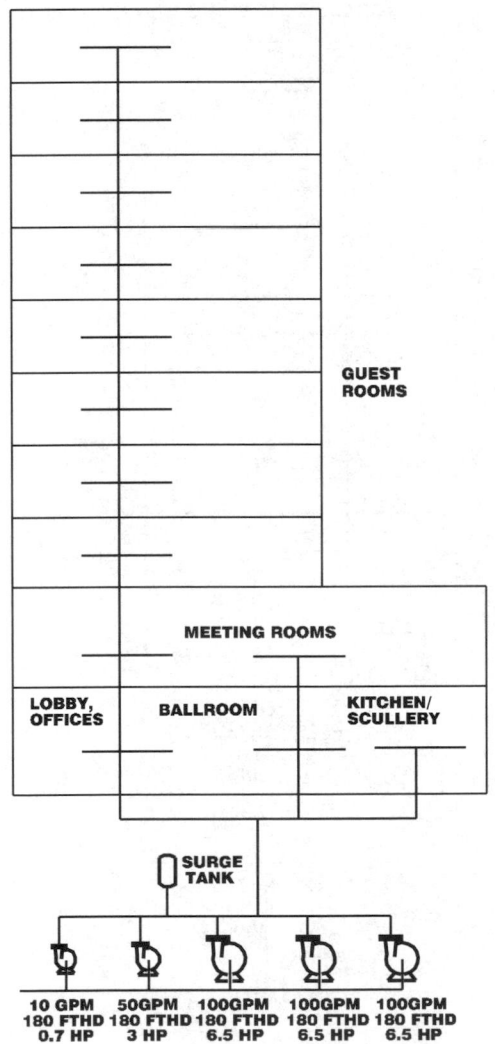

GUEST ROOMS

MEETING ROOMS

LOBBY, OFFICES | BALLROOM | KITCHEN/ SCULLERY

SURGE TANK

10 GPM	50 GPM	100 GPM	100 GPM	100 GPM
180 FTHD	180 FTHD	180 FTHD	180 FTHD	180 FTHD
0.7 HP	3 HP	6.5 HP	6.5 HP	6.5 HP

Fig. 1 Adapting efficiently to changes in water demand
The individual pumps are sized to satisfy the water demand at different times of day, minimizing excess flow capacity at any moment. Still, this arrangement does not account for the great differences in pressure needed to move water to different parts of the building.

Booster pumps are most appropriate in cases where the city water pressure is sufficient for most of the facility. In such cases, using booster pumps eliminates the need to provide in-house pumps for the portions of the facility that can be served directly by city water pressure.

Each booster pump, or set of booster pumps, is usually controlled on the basis of the flow in its discharge line, as with the other pumps. However, a booster pump might be started by the operation of high-pressure equipment in the branch that it serves. For example, in Figure 3, the booster pump that serves the dishwashing area might be started by operation of the dishwashers. The check valve installed in parallel with the booster pump allows the primary pumps to provide other water requirements when the booster pump is not operating.

Installing additional hot water heaters may be cumbersome with this method. For example, you would have to install the water heater for the upper floors in the same location as the booster pumps for the upper floors. This might require a water heater that uses a more expensive fuel than the main heater in the basement. If so, the increased fuel cost may largely offset the saving in pump energy.

Why You Can't Use a Variable-Speed Pump Instead

At first glance, installing a variable-speed pump drive seems to be a much simpler way of accommodating changes in water demand, instead of installing all those fixed-capacity pumps. Figure 4 shows that people have tried this. Unfortunately, it doesn't work. Slowing a centrifugal pump reduces the pressure output as well as the flow. Most service water applications need the water pressure to remain constant, regardless of the flow rate. Even a small reduction in pump speed would starve the system of water, starting at the highest points.

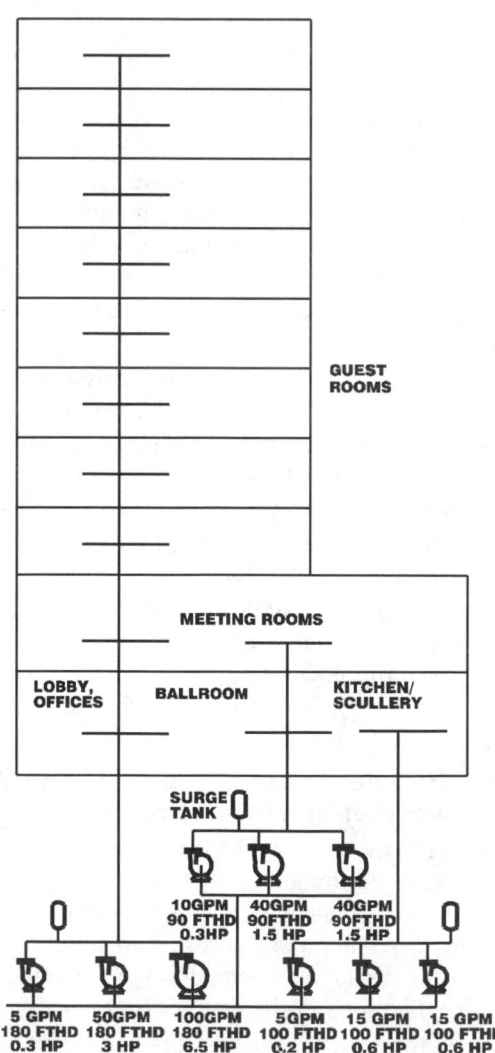

Fig. 2 Separate risers for different pressure zones This requires many pumps and extra pipe. Even so, the guest room riser still delivers water at greatly differing pressures.

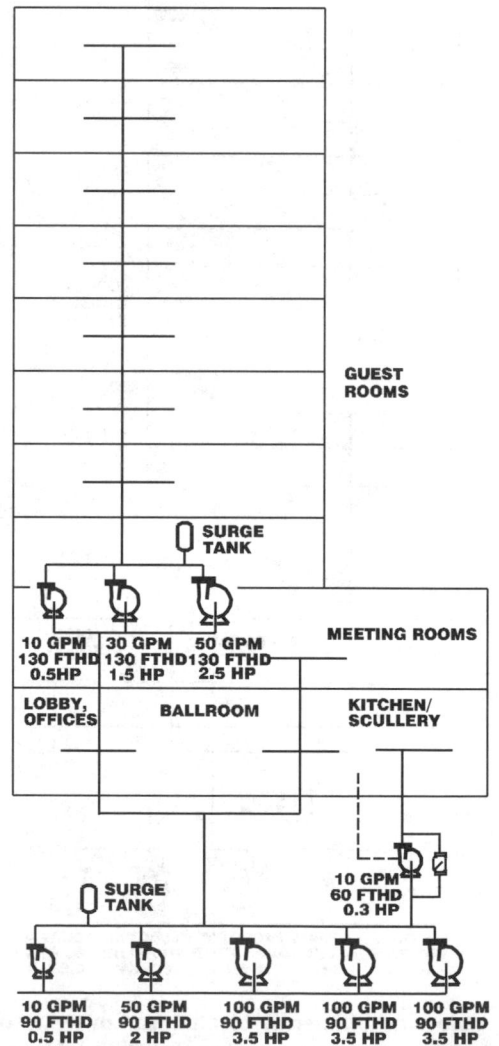

Fig. 3 Booster pumps This arrangement achieves the same purpose as the one in Figure 2. It may require less pipe, and it is easier to retrofit. The number of pumps is the same, but one group must be installed at a remote location. The overall efficiency is about the same.

ECONOMICS

SAVINGS POTENTIAL: *40 to 80 percent of service water pumping energy.*

COST: *Several thousand dollars and up.*

PAYBACK PERIOD: *Less than one year, to many years. The payback period is best with large motors and easy piping.*

TRAPS & TRICKS

DESIGN: *This is basically a matter of balancing three variables: energy cost, pump cost, and piping cost. Sketch out all the configurations that make sense, then make a judgement call. Before installing pumps in remote locations, consider that pumps make a certain amount of noise, they may take up rentable space, they may leak, and they may be less accessible for maintenance.*

EXPLAIN IT: *Proper pump sequencing is the key to efficient operation. Describe the system, the sequencing scheme, and the control settings in the plant operating manual. Install a placard at the sequencing control that states the control settings.*

WESINC

Fig. 4 Wrong application for a variable-speed pump
This variable-speed pump installation is intended to minimize the energy required to deliver service water in a tall hotel. However, the need to maintain a fixed pressure in the system dooms this approach to failure. The losses in this old hydraulic drive negate any small saving in pump power.

MEASURE 3.3.1.2 Install gravity tanks or pressurized storage tanks.

RATINGS

New Facilities   Retrofit   O&M

C   **C**   [ ]

This subsidiary Measure recommends another way to minimize the energy needed to pressurize the service water system. This approach relies on one or more tanks to maintain pressure on the system. The system's pumps are used only to refill the tanks, so they operate only intermittently.

This is usually the most efficient method. The average rate of water usage in most facilities is much less than the peak consumption, so the pumps are stopped most of the time. The pumps can be optimized for the single pressure and flow rate needed to refill the tank. By the same token, the capacity of the pumps does not affect efficiency, unlike systems where the pumps provide system pressure directly.

SUMMARY

Allows the pumps to operate at their maximum efficiency. Provides stable water pressure. Requires a separate water heater and recirculation system for each pressure zone. In retrofit, structural support for the tank(s) may be a problem, but clever design can minimize tank size.

SELECTION SCORECARD

Savings Potential .... **$ $**

Rate of Return, New Facilities **% %** %

Rate of Return, Retrofit.......... **%** %

Reliability .... ✓ ✓ ✓ ✓

Ease of Retrofit .... ☺ ☺

You have a choice of two types of tanks. You can install a simple gravity tank, which must be located above the highest water user in the system. Or, you can install a closed, pressurized tank at any elevation.

Figure 1 illustrates both types installed in the hotel that we have been using as an example in the previous Measures. The tall portion of the building is served by a standpipe located above the roof. The lower portions of the building are served by a closed pressurization tank located in the equipment room.

From the standpoint of energy efficiency, gravity tanks and pressurized storage tanks are essentially equal. The choice between the two usually does not affect the other components of the water system. The choice is guided mainly by relative cost and by the structural limitations of the building. Each tank requires only one pump, with perhaps a second reserve pump.

This method is likely to be more expensive than the methods of Measure 3.3.1.1, but this is not always true. This method adds cost for the tank(s) and supporting structure, but it may add relatively little piping cost. In new construction, the additional structural cost may be minor. However, structural support may be a large cost in retrofit, especially for standpipes on top of the building. In new construction, this method allows the pumps to be much smaller.

As in Measure 3.3.1.1, a separate hot water heater and recirculation system is needed for each pressure zone, i.e., for each pressurization tank.

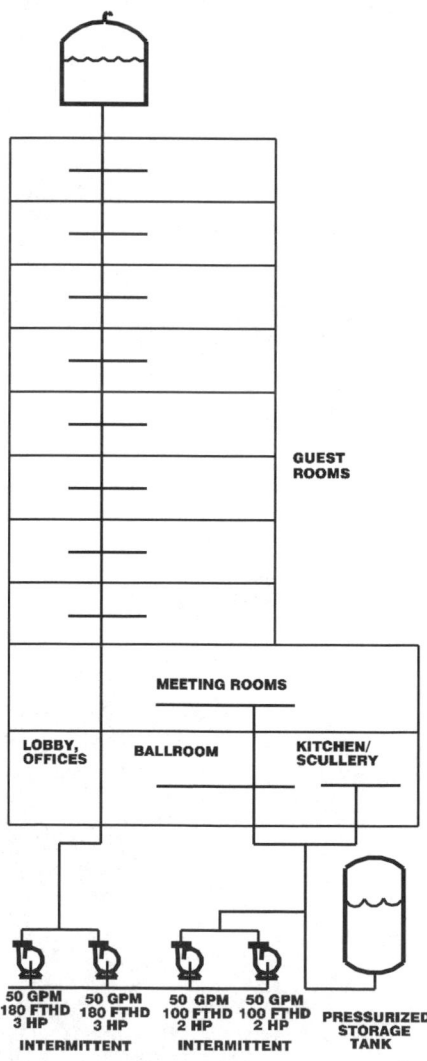

Fig. 1 Gravity tank and pressurized tank Using storage tanks eliminates excess pump operation. In addition, splitting the building into pressure zones improves pumping efficiency.

Standpipes

A "standpipe," or gravity tank, is an elevated tank that is intended to store water and to maintain a steady water pressure. In a standpipe system, water pressure is provided entirely by gravity and it is determined by the height of the tank above the user. The tank is vented and requires no accessories, except for a level switch to control the operation of the pump that fills it. Figure 2 shows a number of gravity tank installations.

Many buildings have gravity tanks for fire protection. The tank ensures both an adequate quantity of water and adequate pressure. A fire protection tank can also pressurize the service water system.

When considering a gravity tank, assess the ability of the building's structure to carry the localized weight of the tank and its contents. In new construction, this may be a minor matter. A filled service water pressurization tank may weigh less than a boiler or a chiller. You can minimize the size of the tank by selecting the pump capacities carefully, as explained below.

A gravity tank must be located well above the highest level of water usage in order to provide adequate pressure. For example, if you need to provide 15 PSI to the top floor of a building, the water level in a gravity tank must be 35 feet above the tallest fixture.

If only a small amount of water is needed at the highest level in the building, it is more practical to use a small booster pump with a separate riser that serves just the upper floors of the building, as shown in Figure 3. This makes structural support easier. It also reduces the problem of freeze protection, which is discussed below.

The cheapest way to refill the tank is through the service riser. However, this causes pressure fluctuations at the user equipment when the pump starts or stops. You can damp these fluctuations by using a surge

WESINC

Fig. 2 Gravity tanks These buildings use gravity tanks for two purposes, pressurizing the domestic water system and providing a separate source of water for fire protection. In most of these buildings, separate tanks are used for the two purposes. The elevation of the tanks above the roofs provides adequate pressure for the upper floors.

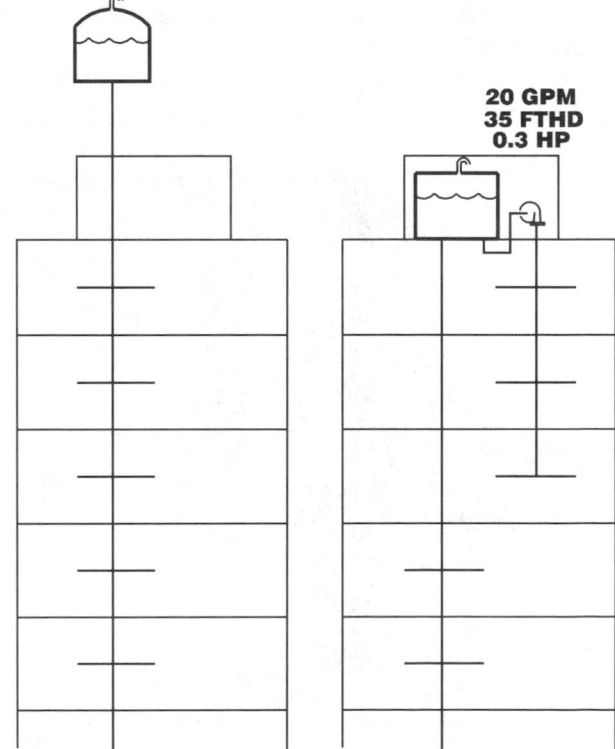

Fig. 3 How to provide pressure to the upper floors
As a third approach, you could add a small pressurization tank just to serve the upper floors.

absorber, described previously. You can eliminate the fluctuations completely by installing a separate pipe to refill the tank.

Closed Pressurization Tanks

You can also store system pressure with a closed pressurization tank. This is simply a tank that has an air space to provide pressurization. A small air compressor replenishes the air occasionally as it is absorbed slowly into the water. Figure 4 shows an example.

A pressurization tank has the advantage that it can be installed anywhere in the service water system, provided that the piping is of adequate size at that point.

The tank wall thickness is proportional to the diameter of the tank, so it is desirable to keep the tank as small as possible. You can minimize the tank capacity by selecting the pump capacities carefully, as explained below. In a tall building, you can reduce the tank's pressure rating by installing it high in the system.

A pressurization tank acts as a surge absorber. The tank is most effective in this role if it is located near the pump discharge.

How to Select Tank Capacity and Pump Size

In a standpipe system, the pump turns on and off continually. The starting current of a motor is several times the normal full-load current, so frequent starting causes the motor to run hotter than normal. Calculate the maximum cycling rate that you can expect. Then, select the motor accordingly. See Measure 10.1.1 for guidance in selecting motors.

The size of the tank is determined primarily by the need to limit pump motor cycling. Here is an example.

WESINC

Fig. 4 Service water pressurization equipment in the basement of a tall office building The closed tank on the right maintains constant water pressure. The three pumps on the left operate in sequence.

The peak flow rate is about 1,000 GPM, the pump has a capacity of 1,000 GPM, and the tank has a capacity of 2,000 gallons. The cycling behavior of the pump at different rates is:

Consumption Rate (GPM)	Pump ON Time (minutes)	Pump OFF Time (minutes)	Motor Starts per Hour
1,000	continuous	none	0
800	10	2.5	5
600	5	3.3	7
400	3.3	5	7
200	2.5	10	5
100	2.2	20	3

This example shows that the number of motor starts per hour remains fairly constant throughout a wide range of water demand rates. A large tank capacity is needed to extend the period of time that the pump remains off. The number of motor starts per hour is inversely proportional to the tank capacity.

Greater pump capacity shortens the time needed to refill the tank, somewhat increasing the number of motor starts per hour. The pump capacity should be able to satisfy the maximum continuous flow requirement, but there is no merit in adding additional pump capacity, except to add a spare pump.

This analysis suggests a way of reducing the tank capacity. Namely, divide the pump capacity among several smaller pumps. Use a series of tank level switches to operate individual pumps, so that the minimum number of pumps are operated to satisfy the current demand flow rate. This slows the rate of filling and matches pump output to the flow requirement, allowing tank capacity to be reduced without excessive pump cycling.

For example, if the maximum expected flow rate is 1,000 GPM, install three pumps, each of 350 GPM. Install four tank level switches, at 100%, 30%, 20%, and 10% of full capacity. One pump is started when the tank level falls to 30%, a second when the tank level falls to 20%, and the third when the tank level falls to 10%. All pumps are turned off when the tank is filled to 100%.

Extending this example, a fourth pump can be installed as a spare. It is controlled with a level switch located at, say, the 5% level. It starts automatically if any of the other pumps fails, or if the maximum flow rate is greater then estimated.

Using Existing Pumps

If you add a standpipe or pressurization tank to an existing facility, consider using one or more of the existing pumps for refilling the tank. You may be able to save even more energy by trimming the pumps to operate at the lowest practical pressure. See Measure 10.2.1 for details of pump impeller trimming.

Freeze Protection of Exterior Tanks

Exterior tanks have been used for years in many cities with colder climates, such as New York City. A tank located outside the building in a cold climate should be insulated. The large mass of water keeps the tank from freezing, as long as some flow through the tank occurs on a daily basis. However, heat loss from the tank increases the energy required for heating the water. Exposed connecting pipe is vulnerable to freezing. Preventing this requires heat tracing of the pipe, which uses energy.

You can avoid the freezing problem entirely by keeping the tank within a heated space, such as an equipment penthouse. If you do this, provide adequate pressure for the upper floors with a small booster pump, as suggested previously.

ECONOMICS

SAVINGS POTENTIAL: *40 to 80 percent of service water pumping energy.*

COST: *$10,000 and up.*

PAYBACK PERIOD: *Several years or longer. The payback period is best with large motors and easy piping.*

TRAPS & TRICKS

DESIGN: *The concepts are simple, but there is room to do a better job than is often found. You can substantially reduce the tank size and the corresponding structural support requirements by taking care in selecting pump capacities and designing pump controls.*

EXPLAIN IT: *Describe the system in the plant operating manual, including the pump operating sequence. Install a placard at the pump controls that states the sequence.*

MEASURE 3.3.2 Design hot water recirculation to minimize pump energy.

RATINGS

New Facilities   Retrofit   O&M

The purpose of recirculation in a hot water system is to ensure that the temperature of the water at the fixtures does not vary, and to keep hot water available without a delay. Most hot service water systems in commercial and industrial facilities have recirculation.

Recirculation systems are needed because hot water cools quickly as it sits in the pipe. Water pipe has a large surface area in relation to its volume, so the water cools in a matter of minutes, even if the pipe is insulated.

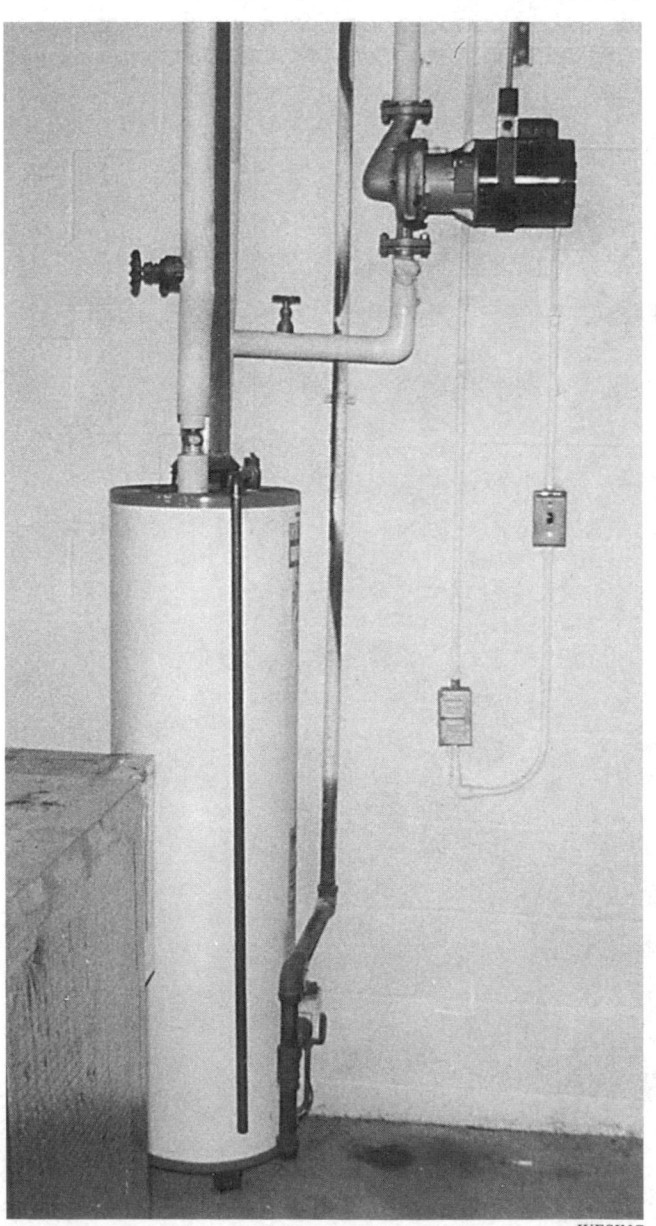

WESINC

Fig. 1 Little water heater, big recirculation pump This pump is probably much larger than needed, although it is still only a fraction of a horsepower. Note that the hot water distribution and recirculation lines are insulated.

SUMMARY

Various improvements to the recirculation system. The potential for savings is small unless a mistake was made in the original system design.

SELECTION SCORECARD

Savings Potential .... $

Rate of Return, New Facilities % % %

Rate of Return, Retrofit.......... % %

Reliability .... ✓ ✓ ✓ ✓

Ease of Retrofit .... ☺ ☺ ☺

Without recirculation, the first water drawn from the system by a fixture or appliance would be cooler than normal. Unless the hot water tank is close to the user, there would be a long delay in receiving hot water.

Recirculation is simple. Small pipes are installed in the hot water system at each area where hot water is used. These pipes draw water from the system continuously and return it to the water heater. This small flow keeps the hot water supply pipe warm. For example, each restroom typically has a recirculation line that draws from the common hot water line that serves the wash basins. A small pump operates continuously to recirculate the water.

Pump Energy Requirement

Recirculation requires little pump energy if it is designed properly. Only a low flow rate is needed to keep the pipes warm. Recirculation takes place in a closed loop, so there is no gravity load. A typical recirculation pump has a motor of one-tenth horsepower or less. However, recirculation pumps operate continuously, or at least for a large fraction of the time.

In existing facilities, recirculation systems typically have little potential for improvement. However, if a recirculation pump is larger than a fraction of a horsepower (or kilowatt), review the system layout. For example, Figure 1 looks suspicious. A design mistake is a possibility.

If you see that the recirculation pump is unusually large, consider the possibility that the pump is designed to overcome the resistance of a backpressure valve. Backpressure valves may be used in the return legs of recirculation lines at different levels in a tall building. Figure 2 shows an example. If this is the cause, consider redesigning the hot water distribution to eliminate the need for the backpressure valve, as recommended by Measures 3.3.1 ff.

The flow capacity of a recirculation pump is determined by the flow velocity that is needed to keep the hot water supply line warm. A larger hot water supply pipe requires a larger recirculation pump and larger recirculation lines.

Insulate the Pipes Well

Recirculation keeps all the supply pipes and the recirculation pipes warm. Therefore, insulate the pipes to the economical maximum.

In addition to reducing heat loss, insulating the pipes also reduces the pump power that is needed. This is because the recirculation flow and pump power is directly proportional to the rate of heat loss from the system. Recirculation occurs an in open loop, so the pump power is proportional to the third power of the flow rate. Therefore, if you reduce the heat loss by a factor of two, you reduce the pump power requirement by a factor of eight.

How to Control Recirculation Pumps

You can save a modest amount of energy by installing a control to turn off recirculation pumps when they are not needed. The amount of energy at stake is usually small, so the method of control must be inexpensive to be economical.

There are two situations when recirculation is not needed. One is when there is no hot water usage. The other is when hot water usage is high, so that normal usage maintains the water temperature in the pipes.

You can easily minimize pump operation under both conditions by controlling the pump with a temperature sensor located at the remote end of the hot water system. If the system has several branches, install a temperature sensor at each end, and connect the sensors in parallel, so the pump starts if any branch gets cool. This may require long wiring runs to the temperature sensors. To keep the pump from cycling too much, set a reasonable difference between the turn-on and turn-off temperatures, say 10°F.

Thermostatic control is sometimes done incorrectly by installing the thermostat in the return portion of the recirculation pipe, perhaps at the pump. Figure 3 shows an example. This method is undesirable because it senses the water temperature in the recirculation pipe, rather than the supply pipe. The pump simply cycles as the water in the return pipe cools off, rather than responding to the actual water temperature at the application.

If safety is not an issue and no hot water is ever required outside normal hours of operation, consider controlling the recirculation pump with a timeclock. Think hard before you do this. Hot water may be needed at odd hours. For example, an office building may be

WESINC

Fig. 2 Large hot water recirculation pumps required by pressure reducing valves This steam-fired water heater serves a tall office building. Pressure reducing valves are installed at various levels in the building to limit the pressure at the fixtures. The large recirculation pumps on the right are needed to restore the pressure dissipated in these valves.

vacated at night by the tenants, but cleaning crews may need hot water. See Reference Note 10, Clock Controls and Programmable Thermostats, for guidance in selecting timeclocks. An alternative is using your energy management computer system, which is usually more expensive and less reliable. See Reference Note 13, Energy Management Control Systems, for details.

Be especially cautious if turning off recirculation pumps may create a safety hazard. If the water in the supply pipes is allowed to cool substantially, there will be an abrupt rise in water temperature as the column of freshly heated water reaches the appliances. This can be a safety hazard in showers and in other applications where people are in contact with the water. Some appliances, including some shower valves, include thermostatic mixing valves to regulate water temperature. Do not rely on such devices if safety is an issue.

If you decide to limit operation of a recirculating pump, install a placard that explains the controls. See Reference Note 12, Placards, for details of effective placard design, materials, and installation.

WESINC

Fig. 3 Wrong way to control recirculation Here, thermostats are installed on the return ends of two recirculation lines. They control the operation of the recirculation pumps. This is incorrect. The temperature at this point bears no relationship to the temperature of the water supplied to the fixtures.

ECONOMICS

SAVINGS POTENTIAL: 40 to 70 percent of recirculation pump energy, which is usually a small amount. The saving may be larger if the recirculation design is extravagant.

COST: Several hundred dollars to several thousand dollars, typically.

PAYBACK PERIOD: Several years or longer.

TRAPS & TRICKS

RECOGNIZING PROBLEMS: The first thing to check is the size of the recirculation pumps. They should be small. If they are not, check for a pressure reducing valve in the system, and try to modify the system to get rid of it. To minimize pump operation, the only correct approach is to sense the water temperature at all the end users. Calculate whether the savings justify the cost of the wiring.

LABELING: If you install a control to limit the operation of the recirculation pump, its purpose will not be obvious to future maintenance personnel. Install a placard to explain it.

MEASURE **3.3.3 Trim pump impellers to eliminate excess system pressure.**

In an existing facility, the service water pumps may develop more pressure than needed. This wastes energy, so consider adjusting the pump pressure to the minimum needed for proper operation of the water-using equipment. The only effective method, in most cases, is reducing the diameter of the pump impeller.

See Measure 10.2.1 for the details.

This activity is limited in potential benefit. Service water systems, especially those in tall buildings, provide less potential for impeller trimming than other types of systems. This is because the pump must produce enough pressure to overcome the gravity requirement of the tallest parts of the system. Piping friction accounts for less of the pump pressure rating, so reduced friction at lower flow rates is less able to compensate for a reduction of the pump output pressure.

Unless the pump has been selected with a much higher pressure rating than needed, trimming the pump impeller risks inadequate pressure at the highest fixtures, especially when the system flow is near maximum.

■ Why Pump Throttling Won't Work Instead

Another general method of reducing pump output is throttling the pump discharge. (This is covered by Measure 10.2.2.) This method would produce little benefit in service water systems. The reason is that pump discharge throttling is effective mostly near the maximum system flow rate, whereas the typical service water system operates mostly at flow rates much lower than maximum. In other words, the fixtures in the system already throttle the pump most of the time.

MEASURE **3.3.4 Install power switching that prevents unnecessary operation of spare pumps.**

It is common practice to install spare pump sets for the sake of reliability. These are often installed with no controls or instructions to avoid unnecessary operation of the spare units. A simple, foolproof way to avoid unnecessary pump operation is to install power switching that denies power to the spare pumps as long as the regular pumps are operating.

See Measure 10.4.1 for the details.

Before proceeding, check the flow requirements of the system to verify that the pumps really are spare units.

MEASURE **3.3.5 Install pump motors having the highest economical efficiency.**

See Measure 10.1.1 for the details. When designing new pump installations, make the small additional effort to specify efficient motors for the pumps.

In existing facilities, be prepared to replace a failed motor with a model that has the highest practical efficiency. In some cases, you may even want to replace a functioning motor with a more efficient model.

Before replacing a pump motor, consider whether you can reduce the power requirement by any of the other Measures in this Subsection. If so, you may be able to reduce the size of the replacement motor.

Section 4. AIR HANDLING SYSTEMS

INTRODUCTION

Most larger buildings use centralized air handling systems to provide cooling, heating, and ventilation. Air handling systems are usually the largest user of energy in buildings that use them as their primary conditioning equipment.

Typical modern air handling systems expend much of their energy consumption for overhead functions, especially air delivery and reheat, that provide no useful heating or cooling of the spaces. Much of this energy consumption can be eliminated. As usual, the opportunities are greatest when designing new buildings, but major improvements are usually possible with existing systems that are wasteful. Make this excess energy consumption a prime target of your efforts to conserve energy. Many significant improvements to air handling systems can be made at modest cost. This Section shows how to do it.

What Air Handling Systems Do

Air handling systems may perform a surprisingly large number of functions, including these:

- *heat and/or cool space air*. Heating and cooling may be done entirely by coils contained within the air handling system, or in combination with localized heating and cooling equipment.

- *control the humidity of space air*. Most systems do this by cooling the air below its dew point. Some systems remove humidity directly by using desiccants.

- *bring in outside air to reduce the concentration of internally generated pollutants*. Most systems mix outside air with recirculated air. Some systems use outside air entirely, and exhaust all air in the first pass.

- *bring in outside air for cooling*. This function is limited to times when the outside air temperature is lower than inside temperature.

- *provide cooling by evaporation of water*. This function is limited to hot, dry climates.

- *distribute air within the spaces* in a controlled pattern. This may be done to provide effective distribution of heated and cooled air, to provide a sensation of air movement or to avoid such a sensation, to enhance cooling by perspiration, or to reduce concentrations of air pollutants by mixing air within the building.

- *remove air pollutants by filtering* incoming and/or recirculated air

- *control the interior pressure of the building with respect to the outside pressure*. Pressure differences may be minimized, to prevent drafts at windows and to avoid difficulty in opening exterior doors. Or, a positive pressure may be maintained to avoid infiltration of outside air through the envelope.

- *control pressure relationships between different areas in the building*. Pressure differences may be minimized to avoid difficulty in opening interior doors. Or, pressure differences may be created to prevent contamination of one space by the air from another.

An individual air handling system typically does not perform all these functions. Some of the functions may not be needed, or they may be overlooked in the system design. Some of these functions may be accomplished in a manner that wastes energy. As a result, many air handling systems offer opportunities both for improving functional performance and for saving energy. By the same token, when you set out to improve the efficiency of existing air handling systems, keep all these functions in mind, so you do not undermine any of the system's purposes. The Measures discuss these functions where appropriate.

Types of Air Handling Systems

It is useful to consider an air handling system as consisting of three groups of components. Figure 1 shows six different arrangements of components for the first group. We will call this group the "front end" of the air handling system. This part of the air handling system recirculates air from the spaces, controls the intake of outside air and the corresponding exhaust of space air, and largely controls the pressures inside the spaces that are served by the system.

The term "front end" is not standard, because the industry does not yet recognize the separate importance of these components, and of the need to give them distinct treatment in design and installation. This blindness to the front end is responsible for much inefficiency in air handling systems. Measure 4.2.1 explains the differences between the six front end arrangements shown in Figure 1. Some of these configurations are much more efficient and versatile than the others.

The components on the left in each panel of Figure 2 represent the second group of air handling system components. These extend from the coils and supply fan to the spaces. They heat, cool, and control the humidity of the air, they move the air into the spaces, and they regulate space temperature. These components determine the name that is given to the system, such as "multizone," "terminal reheat," etc.

Figure 2, which extends over several pages, shows eleven principal types of air handling systems. These include all the common types, along with several less common types that are important. Some types have the potential to be much more efficient than others. However, variations within each type, as well as the adjustment of a system's controls, can cause major changes in their energy and comfort characteristics. Therefore, a more efficient type of air handling system may waste more energy than a less efficient type if it is badly designed or adjusted.

In principle, any of the front end arrangements in Figure 1 can be combined with any of the configurations in Figure 2. In fact, you can find examples of most such combinations.

The third group of air handling system components consists of fans that draw air directly from the spaces. Most buildings have a variety of these fans, serving applications such as toilets, kitchen hoods, laboratory hoods, etc. These fans are symbolized by a single fan on the right side of each panel in Figure 2. Treat exhaust fans as an integral part of air handling systems, even though they may have no mechanical connection to the rest of the systems.

Individual exhaust fans may remove air from spaces that are served by several air handling units. Be aware of their effects whenever you design or make any changes to an air handling system.

Why Air Handling Systems Waste Energy

Large buildings have been heated for several thousand years, using both central and distributed heating methods. Heating design regressed after the collapse of the Roman Empire, and remained primitive until the nineteenth century. The major advance of that century was replacing fireplaces with steam radiators, the steam being provided from a central boiler plant.

Ventilation of buildings, primarily for cooling, remained equally primitive. Thousands of years ago, some cultures in Mesopotamia developed effective convective ventilation systems, but these were exceptional and depended on special climatic conditions. Throughout the later centuries, large buildings could become brutally hot during warm weather. Relief was provided primarily by ventilation through windows, and by using tall ceilings that allowed stratification of warmer air above the people in the spaces.

Central ventilation systems using fans were introduced late in the nineteenth century. They became common in the period between the two World Wars, largely because they allowed buildings to be constructed with large core areas having no direct access to the outside for ventilation. This made space cheaper, and reduced the cost of heating. At the same time, mechanical air conditioning started to appear, first in the form of small units designed for individual spaces, and later as large central systems.

After World War II, a doctrine of comfort became popular that stressed cooling, air flow, and control of humidity. This doctrine, combined with the rising demand to cool entire buildings, caused central ventilation systems to evolve into central air handling systems. These systems combined the functions of heating, cooling, and ventilation. They promised economy of scale by using a small number of large components to serve a large area. They also made it possible to keep the noisy heating and cooling machinery at a distance from the occupied spaces.

Centralized air handling systems influenced the shape of modern buildings strongly by making it possible to design buildings with large core areas. In buildings of all sizes and shapes, central air handling systems motivated using windows that cannot be opened. In turn, the non-openable windows made buildings dependent on air handling systems under all conditions.

After several decades of experience, it has become apparent that air handling systems waste energy in major ways that do not occur with smaller HVAC equipment that serves only individual spaces. The main areas of air handling system energy waste are:

- *reheat*. This is used primarily to control the temperatures of individual spaces in a system that serves spaces with a diversity of loads. Reheat also provides control of humidity.

- *increased fan power*. Central air handling systems require long duct systems. The cost of ductwork, and the need to fit it into small spaces, forces the design toward higher fan power.

- *inaccurate control of outside air intake*. Conditioning outside air is especially expensive. It is possible to control outside air intake accurately, but many systems lack this capability.

- *uneven distribution of ventilation air.* Central air handling systems that combine the function of ventilation with the functions of heating and cooling cannot distribute outside air accurately to individual spaces. If the system is to provide adequate ventilation to each space, it must provide too much ventilation to most of the other spaces.

- *inability to turn off heating, cooling, and ventilation to individual spaces*. This problem could be solved by installing a shutoff damper at each space, but hardly any systems have this feature at present.

Energy waste from reheat can be enormous, and energy waste from the other causes is substantial. The variable-air-volume (VAV) type of air handling system was developed specifically to correct the first two of these of these problems. Unfortunately, lack of sufficient analysis led to serious comfort problems in VAV systems, and crude attempts to solve the comfort problems reintroduced a large amount of energy waste.

This is where the situation presently stands. The VAV concept has become dominant in all but the smallest air handling systems, but comfort and efficiency problems remain.

In a nutshell, the efficiency and comfort problems of central air handling systems occur because the designer tries to achieve too many purposes with a single type of equipment. We will venture the prediction that the energy and comfort problems of air handling systems will not be solved until outside air ventilation is handled separately from heating and cooling. The most comfortable and efficient buildings of the future will once again separate the heating and cooling functions.

How to Use Section 4

Section 4 is organized primarily by type of air handling system. Select the appropriate Subsection from Subsections 4.3 through 4.8 that applies to each type of air handling system that you have. The Measures in each Subsection are tailored to the type of system.

The first two Subsections, 4.1 and 4.2, have Measures that apply to all types of systems. The Measures in Subsection 4.1 recommend control practices that minimize unnecessary system operation. The Measures in Subsection 4.2 recommend improvements to the "front end" of all types of air handling systems. Subsection 4.9 includes Measures specifically related to motors and motor installations, which may apply to any type of air handling system.

As a general guide, each Subsection starts with the Measures that are least expensive and most widely applicable, and progresses to Measures that are more expensive and specialized. Measures providing exceptionally large savings are moved toward the front. Technically related Measures are grouped together.

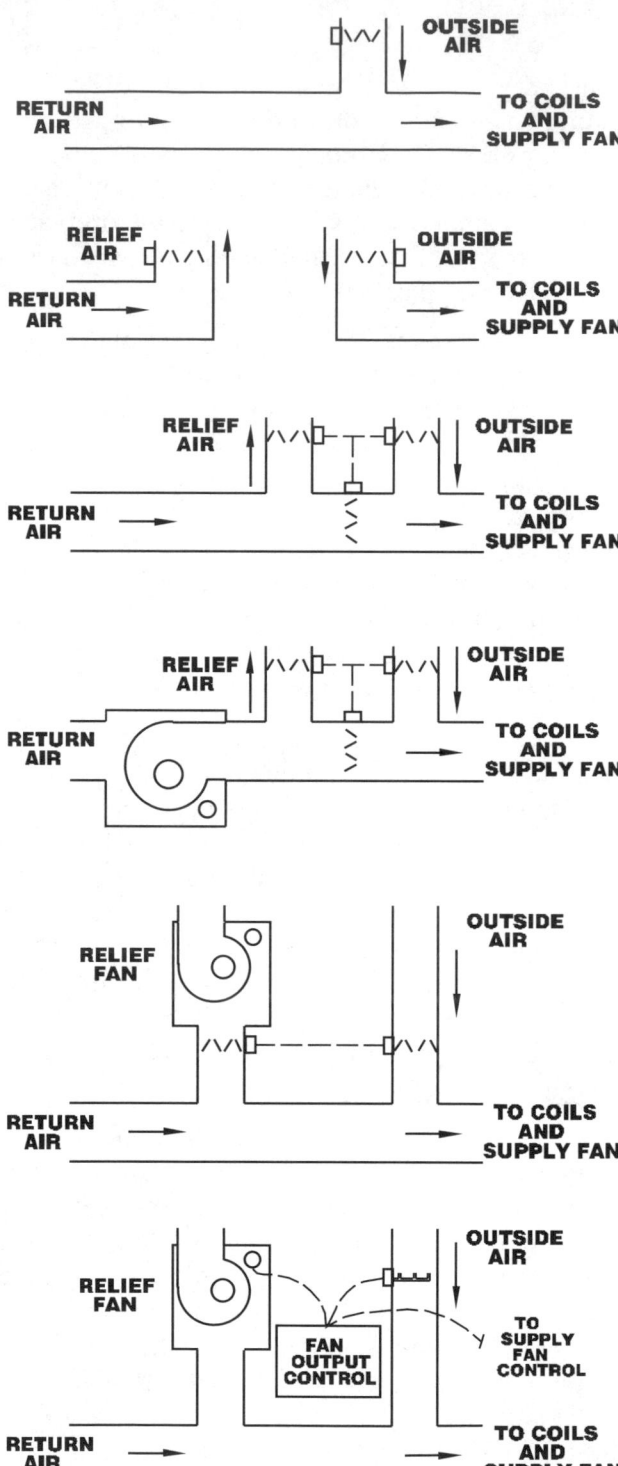

Fig. 1 Front end configurations These are different arrangements of the components that control the flow of air into the air handling system, either from the outside or from the spaces. They also have a major effect on the pressure in the spaces.

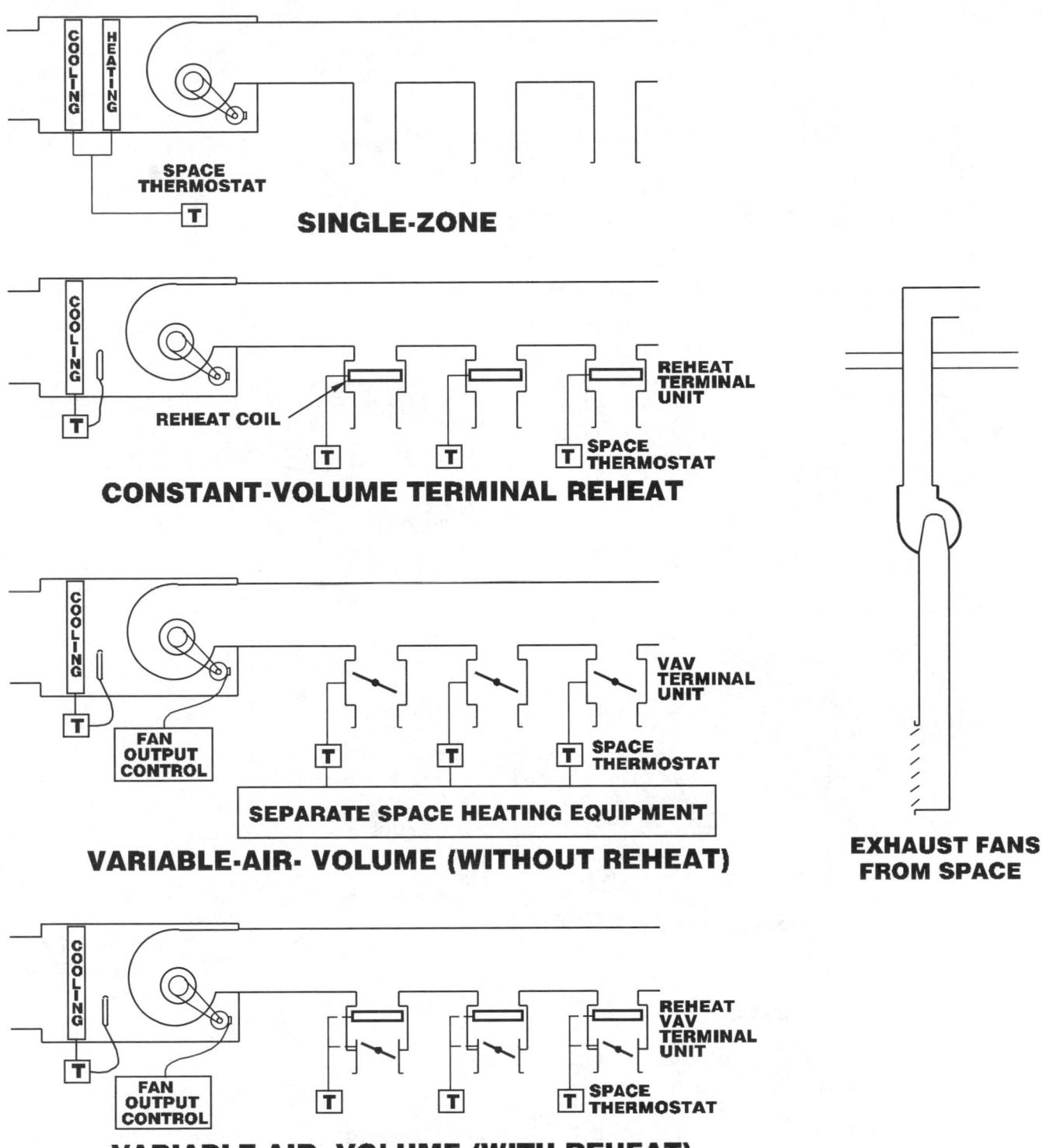

Fig. 2 Types of air handling systems The components on the left condition the air, distribute the air to the spaces, and provide space temperature control. Space exhaust fans, symbolized by the fan on the right, discard air directly to the outside.

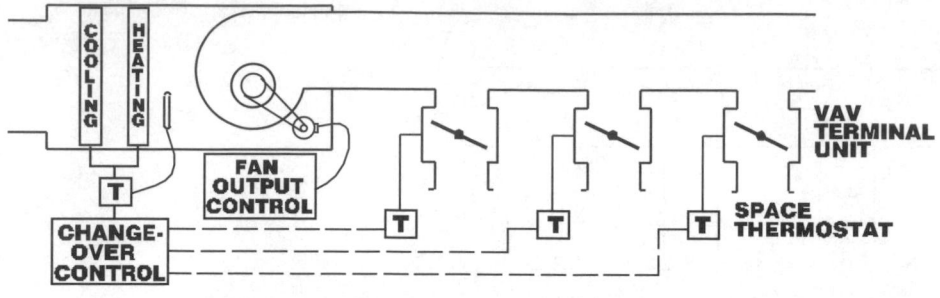

VARIABLE-AIR- VOLUME CHANGEOVER

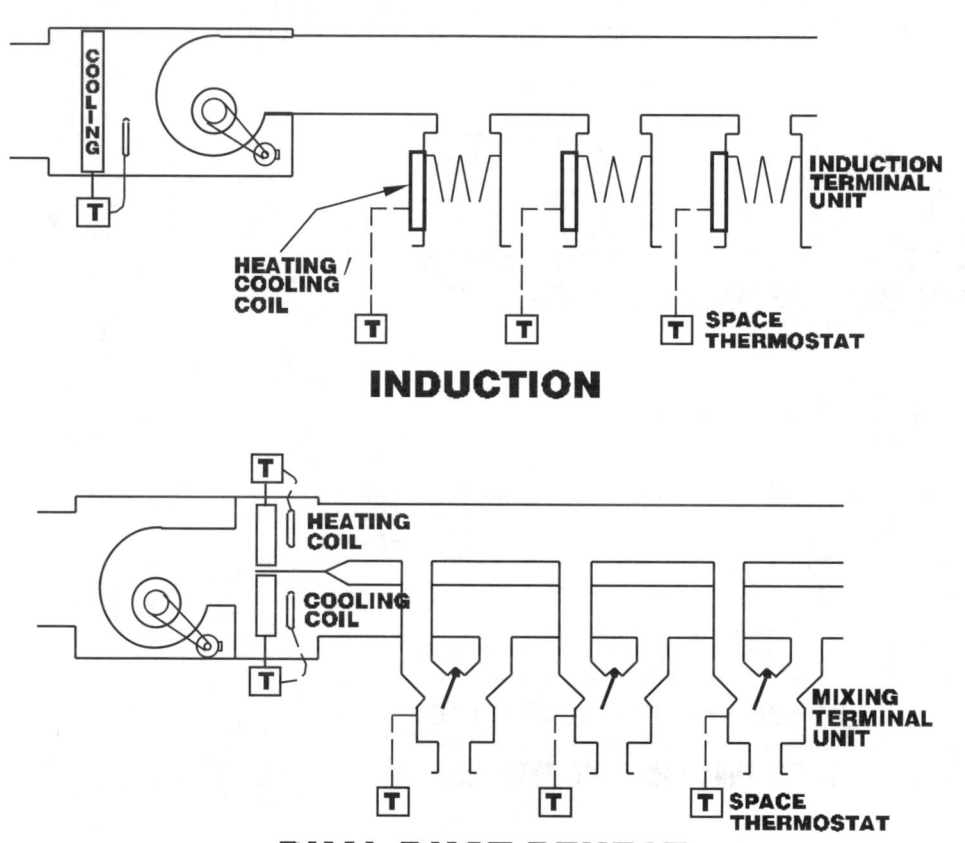

INDUCTION

DUAL-DUCT REHEAT

EXHAUST FANS FROM SPACE

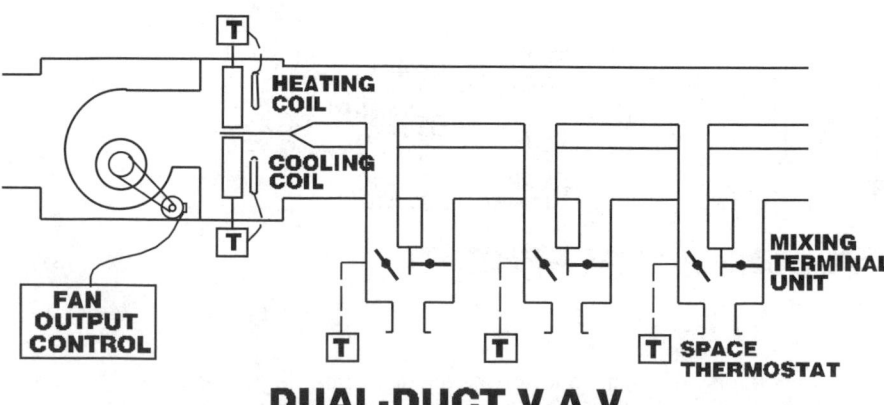

DUAL-DUCT V.A.V.

Fig. 2 (cont.) Types of air handling systems

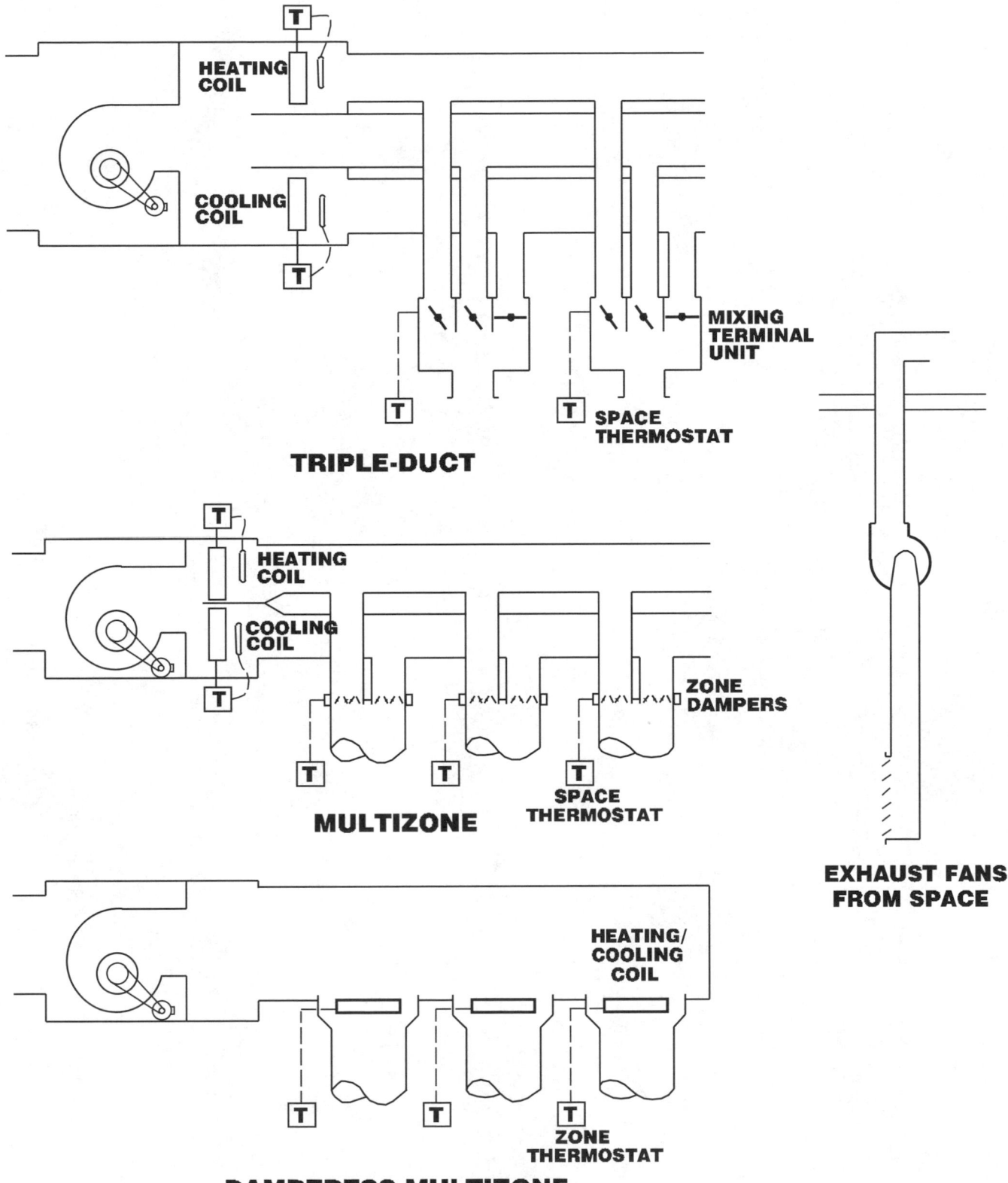

TRIPLE-DUCT

MULTIZONE

DAMPERESS MULTIZONE

EXHAUST FANS FROM SPACE

Fig. 2 (cont.) Types of air handling systems

Minimizing Duration of Operation

This is the place to start in reducing the energy consumption of air handling systems. The Measures in this Subsection tailor the running of air handling systems to actual need. These Measures often provide large savings at modest cost, and they are usually easy.

INDEX OF MEASURES

4.1.1 Turn off air handling systems when they are not needed.

4.1.1.1 Where spaces operate on regular schedules, use timeclocks to start and stop air handling equipment.

4.1.1.2 Install optimum-start controllers to adapt starting times to weather conditions.

4.1.1.3 In spaces with irregular usage, install rundown timer switches to provide user control of air handling system operation.

4.1.1.4 In spaces with irregular usage, install personnel sensors to control air handling equipment.

4.1.1.5 Assign responsibility for operating air handling systems to the personnel who administer the spaces.

4.1.1.6 In applications where automatic starting and stopping of air handling units is undesirable and operators are on duty, use automatic controls to alert operators to turn systems on and off.

RELATED MEASURES

- Subsection 4.2, for control of outside air
- Measures 1.1.1 ff, control of boiler plant equipment
- Measures 2.1.2 ff, control of chiller plant equipment
- Subsection 5.1, for control of room conditioning units and self-contained HVAC equipment

MEASURE 4.1.1 Turn off air handling systems when they are not needed.

Turning off unnecessary equipment is a fundamental principle of energy conservation. The subsidiary Measures control the running of air handling systems in response to a variety of conditions. You can combine these techniques to provide great flexibility in tailoring the operation of air handling systems to the usage of the spaces they serve.

Energy Saving Potential

Turning off air handling equipment whenever possible eliminates or reduces these causes of energy waste:

- *unnecessary conditioning.* Energy delivered by an air handling system changes space temperature and humidity. If the spaces are unoccupied, temperature and humidity usually can be allowed to drift within wide limits. Also, unoccupied spaces usually do not need outside air, which requires a lot of conditioning energy.

- *reheat losses.* Several common types of air handling systems consume energy for reheat, a method of temperature control in which heat is used to cancel cooling to regulate supply air temperature. Reheat provides no useful conditioning. It ceases when the air handling unit is turned off. (Measures that reduce reheat during system operation are in Subsections 4.4 through 4.8.)

- *unnecessary system equipment energy consumption.* Fans typically account for a significant fraction of air handling energy consumption. You can usually turn off the system fans when conditioning is not required (but not always). In some systems, you can reduce the operation of the pumps that transport hot water and chilled water from the central plant to the air handling units.

Automatic Control is More Efficient than Manual Control

Human beings are not well adapted to starting and stopping equipment at precise times. In particular, physical plant operators are likely to neglect this function because they are diverted by a variety of other responsibilities. Human control is especially unreliable if the starting and stopping times vary with changing conditions, such as weather. For these reasons, use automatic controls wherever they are practical.

Nonetheless, you may have to start and stop some air handling units manually for reasons of safety or flexibility. For situations where automatic starting and stopping of air handling equipment is not advisable, Measures 4.1.1.5 and 4.1.1.6 recommends methods that improve control reliability while retaining human control.

Make Conditioning Available Whenever It is Needed

Controlling the running of air handling systems is not as simple as it may appear. Air handling systems provide comfort, so occupants will seek to disable any type of control that threatens to turn off conditioning when they may want it. For this reason, automatic controls for starting and stopping air handling systems have experienced a poor rate of survival.

Controls that merely turn off conditioning systems on a fixed schedule are not satisfactory. The challenge is making the system available during irregular or unexpected periods of space usage. The first step is to inventory the conditioning requirements of each space and application that is served by the air handling system. From this inventory, design a combination of controls that keeps the air handling system running as long as any space that it serves needs conditioning.

For example, the controls should accommodate individuals who work late, they should provide conditioning for cleaning crews who may arrive long after the facility has closed, and they should provide enough conditioning for security personnel who patrol at all hours. The subsidiary Measures provide a variety of techniques to respond to these needs.

Other Ways to Deal with Irregular Space Usage

Consider reducing conditioning levels for certain activities. For example, it is reasonable to require cleaning crews, who are physically active, to be clothed to work at lower space temperatures than are normal for sedentary daytime work.

Also, consider installing a separate small conditioning system for people who must remain within the building during normally unoccupied hours. For example, provide separate conditioning for guard stations.

Limit Temperature Extremes During Off Hours

If appropriate, include control provisions to restart the air handling system any time the spaces become too cold or too warm. The most common concern is freezing, but problems may be caused by less drastic temperature changes. Install safety thermostats that are set for the low and high temperature limits. Install them in the locations where temperatures first become critical. For example, install a freeze protection thermostat in a poorly insulated part of a space that has water piping.

Coordinate with Control of Other Systems

You can use similar techniques to control the facility's boiler plant equipment (Measures 1.1.1 ff), chiller plant equipment (Measures 2.1.2 ff), and individual conditioning units (Subsection 5.1). Coordinate all these controls to reduce cost and increase opportunities for conservation.

Comparison to Temperature Setback

Temperature setback is an energy conservation technique that consists of lowering the space thermostat settings during off hours, while the air handling system continues to operate normally. Fans and other equipment continue to operate during the setback period, so setback is less efficient than turning off the air handling system entirely. Measure 4.3.2 gives details of temperature setback.

In any type of air handling system that uses reheat for temperature control, temperature setback may not save much energy. In fact, it may increase energy cost. The only types of air handling systems in which you can install temperature setback without careful analysis are non-reheat single-zone systems, non-reheat VAV systems, and triple-duct systems.

MEASURE **4.1.1.1 Where spaces operate on regular schedules, use timeclocks to start and stop air handling equipment.**

RATINGS

New Facilities	Retrofit	O&M
A	**A**	

Where a facility operates on a regular schedule, timeclocks are usually the best way of automatically starting and stopping equipment, at least as a starting point. Timeclocks can be combined effectively with controls that respond to other inputs, such as outdoor temperature, occupancy, etc.

Where Not to Use Time Controls

Time control does not work in facilities that operate on an irregular schedule, such as conference rooms, auditoriums, and surgical suites. Unfortunately, one often encounters vain attempts to use timeclock systems in such spaces. In these misbegotten attempts, facility managers may publish periodic activity schedules that the operating staff is supposed to follow in resetting timeclocks. The staff invariably respond by ignoring the schedules and operating the air handling systems continuously.

How to Select Time Controls

Be careful in choosing from among the large selection of time controls that are available. Each application has its own requirements for reliability, ease of use, variety in scheduling, and override control. See Reference Note 10, Clock Controls and Programmable Thermostats, for equipment features and installation practices.

Where older timeclocks that lack desirable features are presently installed, consider replacing them with newer electronic models. This can provide capabilities

SUMMARY

A basic and powerful energy conservation tool. Be sure to accommodate irregular space usage. Will fail if it is not installed in a user-friendly manner.

SELECTION SCORECARD

Savings Potential	$ $ $ $
Rate of Return, New Facilities	% % % %
Rate of Return, Retrofit	% % % %
Reliability	✓ ✓
Ease of Retrofit	☺ ☺ ☺

such as internal power backup, multiple schedules, cycling, automatic daylight saving time conversion, optimum-start, etc.

If you have an energy management computer system, consider using it to start and stop the air handling equipment. However, large-scale control systems tend to be unreliable, and the cost of connecting each air handling system is high. Simple localized controls are less expensive, they are individually much more reliable, and they are more adaptable to local conditions. The main advantage of a centralized system is its ability to re-schedule air handling systems from a central location. See Reference Note 13, Energy Management Control Systems, for details.

After-Hours Operation

In most applications, timeclocks should include a manual override feature. The primary purpose of the override is to allow conditioning in the event that the air handling system is needed during unscheduled times. This is essential to protect the time control, as well as for comfort. If an override is not easily accessible, irate occupants or operating staff will eventually disable the time control.

Overrides are not an efficient means of adapting to unscheduled operation. Overrides typically function to keep the equipment operating until the following "off" time. For example, if an air handling unit is scheduled to turn off at 6 PM and someone restarts the unit at 8 PM, the unit will continue to operate until 6 PM of the following day. Consider this type of override to be an emergency feature rather than a routine control function.

If an air handling system is needed often outside of normal hours, control it with an interval timer, a personnel sensor, or other control method that adapts to the duration of usage. You can install any of these controls as a supplement to timeclock control. They are covered in the following subsidiary Measures.

ECONOMICS

SAVINGS POTENTIAL: *10 to 70 percent of the air handling system's operating cost.*

COST: *The price of electronic timeclocks ranges from less than one hundred dollars to several hundred dollars. Installation and training may cost from several hundred dollars to several thousand dollars per unit.*

PAYBACK PERIOD: *Less than one year, to several years.*

TRAPS & TRICKS

SELECTING THE EQUIPMENT: *Get all the appropriate features. A timeclock soon becomes junk if it is difficult to use, or if it lacks any feature needed to adapt equipment operation to people's needs.*

INSTALLATION: *Install the timeclock itself so that it is obvious and accessible only to those who should set it. Install a separate override in a location where it is conspicuous to occupants of the space.*

EXPLAIN IT: *Time controls invite tampering. Make sure that operators know how to use them properly, and that they are kept set properly. Provide clear setting instructions at the unit, if these are not included with the timeclock itself. In any event, install a placard at the unit that describes what the timeclock controls, the schedule that should be set, and any additional controls for the same air handling system.*

MONITOR PERFORMANCE: *Sooner or later, all time controls fail or get out of time. Schedule periodic checks of all time controls to verify proper operation and settings.*

MEASURE **4.1.1.2 Install optimum-start controllers to adapt starting times to weather conditions.**

SUMMARY

Important for minimizing the running time of air handling systems during extreme weather. To retrofit it, upgrade the time controls.

SELECTION SCORECARD

Savings Potential ....

Rate of Return, New Facilities

Rate of Return, Retrofit ....

Reliability ....

Ease of Retrofit ....

Optimum-start control is a supplemental feature of timeclocks that automatically adjusts the equipment start time in response to weather conditions.

If air handling systems are started using simple timeclocks, facility operators typically set the start time to cover the worst weather conditions that may occur. This wastes conditioning energy, and also wastes the energy required to operate fans, pumps, and other equipment.

For example, consider a building that has been shut down over a weekend during very cold weather. This is a shutdown period of about 77 hours. The air handling systems of the building may have to operate for five hours prior to occupancy to warm the building to normal temperature. The next day, the same building may require only a one-hour warm-up period because the weather becomes warmer and the building has been shut down for only about nine hours. An optimum-start controller senses the differences and adjusts the start time accordingly.

How Optimum-Start Works

The time required for a building interior to reach normal temperature after a period of shutdown depends on the building's thermal mass and on its rate of heat loss. Thermal mass soaks up conditioning energy during the start-up interval, increasing the time required to reach normal temperature. Heat loss diverts conditioning energy from the building interior, extending the start-up interval.

Optimum-start controllers take both of these factors into account. The time required to replenish heat lost from the internal mass of the building depends on the space temperature at start-up. The controller uses an inside thermometer to sense the space temperature. The controller calculates the rate of heat loss to the outside from the difference between the outside temperature and the desired inside temperature. The controller also has an outside thermometer for this purpose.

The controller must adjust for the building's thermal storage and heat loss properties. In less sophisticated models, this is done by manual adjustments when the controller is installed. You may have to find the optimum settings by trial-and-error over a period of months. More sophisticated units can automatically "learn" an optimized schedule from experience.

Controller Options

Optimum start is a common feature of more sophisticated electronic timeclocks. Many models provide optimum start for both heating and cooling, although some models function only for heating.

Don't expect to add optimum-start capability to an existing timeclock. Instead, simply replace the timeclock with a model that has optimum-start capability. Fortunately, this equipment has become inexpensive. While you are at it, select the timeclock for the other important features that are described in Reference Note 10, Clock Controls and Programmable Thermostats.

Most computer energy management systems include optimum start as a programming option. However, these large computer systems are less reliable than dedicated optimum-start controllers. Furthermore, the cost of a dedicated unit typically is lower than the cost of re-programming an existing centralized computer system and connecting it to the air handling systems. See Reference Note 13, Energy Management Control Systems, for more about this approach.

How to Install the Outside Air Temperature Sensor

The placement and shielding of the outside air temperature sensor are critical. Install the sensor so that:

- it is exposed to the prevailing outdoor air temperature
- the penetration through the envelope is sealed so that positive pressure inside the building cannot force inside air over the sensor
- sunlight does not heat the sensor, either directly or by heating surrounding surfaces
- precipitation cannot reach the sensor, either directly or by splashing
- it is protected from physical impact.

Use One Controller for Several Applications

Time controls that offer an optimum-start feature typically offer several channels of control, each of which can be operated on a separate schedule. For example, a single controller could provide different turn-off times for different air handling systems. It could even be used to start different types of equipment, such as boilers and chillers. Using a single controller for different equipment simplifies installation, inspection, and maintenance. Inventory all your potential applications before selecting the controller.

Relationship to Temperature Setback Control

Temperature setback is an energy conservation measure that limits the temperature drop inside the building during unoccupied hours. In turn, this limits the time required to restore normal space temperature.

Optimum-start control is more efficient than temperature setback because it turns off the system completely. It also eliminates uncertainty about restoring space temperature on time. In most cases, optimum-start control is a better alternative than temperature setback.

However, one caution is to avoid possible damage from low temperatures during the off period. See Measure 4.1.1 about avoiding this danger.

ECONOMICS

SAVINGS POTENTIAL: *1 to 10 percent of the cost of operating the air handling systems, compared to using ordinary timeclocks. The greatest savings occur with long shutdown periods and severe climates.*

COST: *Optimum-start is a feature that is available with many electronic timeclocks that are priced from $100*

up. *Labor cost depends mainly on the difficulty of making connections to the controlled equipment and the difficulty of installing the outside air temperature sensor. Installation typically costs from several hundred dollars to several thousand dollars.*

The cost of connecting and programming an existing computer energy management system for optimum-start is typically several thousand dollars.

PAYBACK PERIOD: *Several months to several years, the period being shorter with larger systems and more severe climates.*

TRAPS & TRICKS

SELECTING THE EQUIPMENT: *If you use a timeclock for optimum start, make sure to get all the appropriate features. A timeclock soon becomes junk if it is difficult to use, or if it lacks a critical feature.*

INSTALLATION: *Install the control so that it is accessible only to those who should set it. Install an override that is conspicuous to occupants of the space. A certain amount of patience may be needed to optimize the settings that account for the building's thermal characteristics. Make sure that the outside air temperature sensor is installed properly.*

EXPLAIN IT: *The purpose of the optimum-start controller is not obvious from its appearance. Install a placard to explain what the controller does, and the schedule of operation. The unit itself should have clear instructions. See Reference Note 12, Placards, for tips on designing effective placards.*

MONITOR PERFORMANCE: *Improper operation is easy to overlook. Schedule periodic checks of all time controls to verify proper operation and settings.*

MEASURE 4.1.1.3 In spaces with irregular usage, install rundown timer switches to provide user control of air handling system operation.

RATINGS

Rundown timers are timing devices that turn on equipment immediately, and turn it off after a set amount to time. Rundown timers are appropriate for controlling conditioning in spaces that are used on an irregular basis, such as meeting rooms and auditoriums.

To control the air handling unit, install a rundown timer in each space that may need conditioning at odd hours. Connect the timers so that a signal from any one space will start the air handling system. Figure 1 shows a typical installation.

WESINC

Fig. 1 Rundown timer used to control conditioning in a library function room This is very effective for the purpose. The space has no exterior exposure, so warm-up time is minimal. However, there is no placard to tell unfamiliar users what the timer does, or how to use it. Also, the locked cover on the thermostat is a bad idea.

SUMMARY

Inexpensive and fairly foolproof. Somewhat less efficient than personnel sensors (Measure 4.1.1.4). Limited in application by inability to anticipate occupancy for warm-up or cool-down.

SELECTION SCORECARD

Savings Potential .... $ $ $

Rate of Return, New Facilities % % % %

Rate of Return, Retrofit .... % % % %

Reliability .... ✓ ✓ ✓

Ease of Retrofit .... ☺ ☺ ☺

Advantages

Rundown timers are inexpensive. In most applications, they are the most economical method of providing user control of conditioning.

Rundown timers are easy to use. They can be understood even by people who are unfamiliar with the space, provided that the timers are properly marked. This eliminates the need for members of the physical plant staff to get involved in scheduling the air handling systems.

Installation is usually easy. Just be sure that the timer is conspicuous and accessible to users of the space.

Limitations

The main limitation of rundown timers is that they cannot provide a warm-up or cool-down period prior to occupancy. Therefore, restrict rundown timers to applications where not much of a temperature change is required prior to occupancy. Rundown timers typically work well in interior spaces that are surrounded by other conditioned spaces.

A warm-up or cool-down period is acceptable if staff personnel are always available who can be assigned responsibility for starting the air handling system prior to occupancy. For example, rundown timers installed in hotel function rooms can be started by the staff who set up the spaces.

From an efficiency standpoint, rundown timers eliminate all system operation prior to occupancy, but they may leave the system running unnecessarily after occupancy. People have a tendency to set the timer for maximum duration. They may easily forget to turn the timer off when the space is vacated. This is relatively unimportant if the space is used only occasionally, but it becomes a more serious limitation if the space is used often.

Likely Alternative Methods

In applications where you consider rundown timers, also consider personnel sensors, which are recommended by Measure 4.1.1.4, next. Both types of controls may perform satisfactorily in a given application. Rundown timers are cheaper, simpler to comprehend, easier to install, and probably more reliable. On the other hand, personnel sensors tailor the operation of the system to occupancy more efficiently.

Types of Rundown Timer Switches

The most common type of rundown timer switch is the "dial timer," which is a spring-wound mechanical timer that is set by twisting the knob to the desired time. These are available in a wide range of maximum times, up to about 24 hours.

This type of timer switch is vulnerable to damage. The shaft is weak, and the knob is usually not attached to the shaft securely. For this reason, dial timers may not last long in a hostile environment, such as a high school. Expect to replace units occasionally. Fortunately, they are inexpensive and easy to replace.

Mechanical rundown timers make a slight noise as they operate, which can be irritating in quiet environments.

Electronic switches are available that perform the same function as mechanical rundown timers. The best of these units are much more durable than mechanical timers. Some electronic timers offer too many options, which provides the opportunity for inefficient operation. For example, most electronic units do not limit the maximum running time sufficiently. Also, they may provide an override "on" position, which will be abused. Even the simplest electronic time control will baffle some space users.

Minimize the Time Intervals

For each space, select a timer that has the shortest time interval that is appropriate for the usage of the space. For example, the timer in a public library function room should probably have a duration no longer than three hours. For the sake of efficiency, err on the side of shorter intervals. If the timer runs down, the occupants can easily reset it.

Connections to the Air Handling System

Connecting timers to the air handling system controls is usually simple. You can connect an number of timers in parallel, so that a timer in any space can start the air handling system.

Explain It!

In most applications, the rundown timers will be operated by people who are unfamiliar with the conditioning system. Therefore, success requires instructions that are conspicuous and easily understandable. Install a placard adjacent to each rundown timer that states:

- that conditioning of the space is controlled by the timer switch in order to conserve energy
- that conditioning can be provided at any time by setting the timer to the desired duration
- that the switch should be turned to "off" when the space is vacated
- whom to contact in the event of difficulty.

See Reference Note 12, Placards, for details of effective placard design and installation.

ECONOMICS

SAVINGS POTENTIAL: *80 to 95 percent of unnecessary operation of the conditioning unit. The fraction of total system energy consumption that is saved varies widely, depending on the occupancy of the space and its conditioning load.*

COST: *Mechanical rundown timers cost about $20 apiece. Electronic units cost less than $50 apiece. Wiring materials and labor typically cost several hundred dollars per space.*

PAYBACK PERIOD: *Several months to several years.*

TRAPS & TRICKS

CHOICE OF METHOD: *Do not use this method for initially warming up or cooling down a space over a large temperature range. It will take too long.*

SELECTING THE EQUIPMENT: *The timers will be subjected to abuse in almost any type of environment. Select them for ruggedness and ease of use.*

MAKE IT OBVIOUS: *The most important installation feature is locating the controls where they are accessible and obvious to transient occupants, while remaining well protected. Be sure to install an effective placard along with the timer. The activity will fail otherwise.*

FOLLOW UP: *Schedule periodic checks of the timers. Expect to replace broken units occasionally.*

MEASURE **4.1.1.4 In spaces with irregular usage, install personnel sensors to control air handling equipment.**

In appropriate applications, personnel sensors match the operation of air handling systems to space occupancy more accurately than any other method. The distinctive advantage of personnel sensors is that they respond directly to occupancy. They start the conditioning system as soon as anyone enters the space, and they stop conditioning as soon as everyone vacates the space.

To control the air handling unit, install a personnel sensor in each space that may need conditioning at odd hours. Connect the sensors so that a signal from any one space will start the air handling system. Figure 1 shows a typical installation.

Installing personnel sensors is not difficult, but it is tricky. You need to select the right location and orientation for each unit individually. See Reference Note 11, Personnel Sensors, for details of features and installation methods.

The hardware cost of the sensors themselves has become almost trivial. The main costs are for installation labor and wiring materials.

Sensor Types to Consider

With present technology, the best types of personnel sensors for HVAC control are those using either infrared or ultrasonic detection. See Reference Note 11 for details of all contemporary types of personnel sensors.

Limitations Related to Occupancy Patterns

A fundamental limitation of personnel sensors is their lack of ability to anticipate occupancy. Therefore, personnel sensors are inappropriate in applications

SUMMARY

Very efficient, fairly foolproof. Limited to spaces that do not require extended warm-up or cool-down.

SELECTION SCORECARD

Savings Potential .... $ $ $

Rate of Return, New Facilities % % % %

Rate of Return, Retrofit .... % % %

Reliability .... ✓ ✓ ✓

Ease of Retrofit .... ☺ ☺ ☺

where a significant amount of warm-up or cool-down period is needed prior to occupancy.

If the start time of the air handling system is predictable but the stop time is not, use a time control to start the system and a personnel sensor to stop it. For example, this combination is appropriate for a restaurant, because operations always begin at a certain time but the staff may remain for an unpredictable length of time after the place closes.

Personnel sensors may start an air handling system unnecessarily by sensing the transient presence of security personnel, cleaning crews, etc. This may not waste much energy, because the system will stop shortly after the personnel leave the space. However, if there is a lot of transient traffic, personnel sensors may be less efficient than other types of control, such as interval timers.

Tailor Each Sensor to the Space

You have to tailor the installation of each personnel sensor. The reasons are explained in Reference Note 11. In a given facility, you may need as many as a half dozen different types and models of personnel sensors, and many mounting configurations. You need to plan the installation of each sensor on a space-by-space basis to achieve satisfactory performance.

By the same token, personnel sensors are vulnerable to changes in space layout. Expect such changes to occur from time to time over the life of the facility. Seemingly minor changes to space configurations may force you to relocate the sensors within the space. For example, installing partitions can shield people from the view of the sensor. If a sensor is installed on a wall, placing a coat rack near the sensor may blind it to a large part of the space.

WESINC

Fig. 1 Ultrasonic sensor used to control conditioning in a conference room This, or an infrared sensor, might be installed with a turn-on delay to avoid turning on conditioning when someone passes briefly through the space.

Personnel sensors are limited in range. In large spaces, you may have to install several units in order to provide adequate coverage of the space. You can reduce the number of units by being clever about where you place them. For example, install sensors near exits or near centers of activity.

Likely Alternative Methods

In applications where you consider personnel sensors, also consider rundown timers, which are recommended by subsidiary Measure 4.1.1.3. Both types of controls may perform satisfactorily in a given application. Rundown timers are cheaper, simpler to comprehend, easier to install, and probably more reliable. On the other hand, personnel sensors tailor the operation of the system to occupancy more efficiently.

Connections to the Air Handling System

Connections to the air handling system controls are simple. You can connect any number of sensors in parallel, so that any sensor will start the air handling system.

Explain It!

Even though personnel sensors are automatic, let occupants know that conditioning is being provided. For example, if the space needs a warm-up or cool-down period, occupants need to be told that the space will reach the proper temperature in a short time.

In each space where you use personnel sensor control, install a placard at an appropriate location, e.g., adjacent to the space thermostat, stating that heating and cooling of the space is started automatically as soon as people enter the space. See Reference Note 12, Placards, for details of effective placard design and installation.

ECONOMICS

SAVINGS POTENTIAL: *A well tailored control system based on personnel sensors eliminates 70 to 95 percent of unnecessary operation of the air handling system. The cost saving varies widely among applications, depending on operating schedule, conditioning load, and other factors.*

COST: *A variety of satisfactory sensors are available for less than $100 apiece. Installation cost is the major expense. Connections to a typical air handling system may cost several hundred dollars to several thousand dollars.*

PAYBACK PERIOD: *Several months to several years.*

TRAPS & TRICKS

CHOICE OF METHOD: *Do not use this method if the space may have to be warmed up or cooled down over a large temperature range. It takes too long. For warm-up or cool-down, consider combining this method with other Measures. Do not use personnel sensors if there is frequent transient traffic through the space outside of normal hours of occupancy.*

EQUIPMENT SELECTION AND INSTALLATION: *The type of sensor needs to be tailored to the space, along with the location and orientation of the sensors. Review Reference Note 11.*

PUBLIC RELATIONS: *People need to be assured that they are getting heating or cooling. Install a prominent placard in the space telling occupants that heating or cooling started automatically when they entered the space, and that the space will soon be at the proper temperature.*

MONITOR PERFORMANCE: *Schedule periodic checks of the sensor controls, along with routine checks of all the thermostatic controls.*

MEASURE 4.1.1.5 Assign responsibility for operating air handling systems to the personnel who administer the spaces.

Although automatic control is usually more reliable than manual control, there are situations in which automatic control is not practical. For example, it may not be possible to design a satisfactory automatic controls for a hotel ballroom. The problem is that the ballroom is used on an irregular schedule, and it has transient traffic going through it during the hours when it is not used. These conditions would fool any conceivable combination of automatic controls.

In many such cases, specific people are responsible for the day-to-day "hands on" administration of the spaces. For example, the surgical suites in a hospital are administered by the chief surgical nurse, the ballrooms in a hotel are supervised by the banquet manager, etc. Such personnel can be assigned the responsibility of operating the air handling systems.

This method is reliable for turning on air handling systems, because failure to do so leaves the space uncomfortable. The big challenge is getting people to turn off the systems.

Make Controls Accessible and Explain Them

The only equipment change that may be needed is installing a large "on-off" switch in an area that is readily accessible to the manager of the space. Accessibility is a key point.

Install an effective placard at the switch that states its purpose and the desired manner of operation. See Reference Note 12, Placards, for details of effective placard design and installation.

ECONOMICS

SAVINGS POTENTIAL: *50 to 95 percent of unnecessary operation of the air handling system, depending on the effectiveness of management. The fraction of total system energy consumption that is saved varies widely among different applications, depending on operating schedule, conditioning load, and other factors.*

SUMMARY

A cheap and powerful way to save energy, but requires effective long-term management.

SELECTION SCORECARD

Savings Potential .... $ $ $ $

Rate of Return .... % % % %

Reliability .... ✓ ✓

Ease of Initiation .... ☺ ☺

COST: *Minimal, if existing switches are adequate. Installing more accessible switches may cost hundreds of dollars to thousands of dollars.*

PAYBACK PERIOD: *Immediate, if no physical changes are needed. Several months to several years, if you have to install switches that are more user-friendly.*

TRAPS & TRICKS

ADMINISTRATION AND COMMUNICATION: *This is primarily a matter of effective management and leadership. The physical plant department needs to communicate effectively with the other departments that are being asked to control the equipment. Back up the request with a well written memorandum. Show the people what needs to be done. Say please and thank you.*

MAKE IT EASY: *Make it easy and obvious for the people who are assigned to control the systems. The procedure is less likely to be forgotten if the switch controlling the system is located out in the open. Don't expect people to go looking for the switch in a locked room that is unrelated to their usual activities. Install an effective placard at the switch that explains which equipment is controlled, and when it should be started and stopped.*

FOLLOW UP: *The procedure will tend to fade away. Keep it refreshed by periodic contacts with the departments responsible. Schedule this in your maintenance calendar.*

MEASURE 4.1.1.6 In applications where automatic starting and stopping of air handling units is undesirable and operators are on duty, use automatic controls to alert operators to turn systems on and off.

RATINGS

New Facilities	Retrofit	O&M
B	**B**	

In some applications, it is not appropriate to control the running of air handling systems automatically. For example, automatic starting may be dangerous for personnel who may be working nearby. Automatic stopping could be dangerous in critical conditioning applications, such as a surgical suite. However, relying on individual initiative is an unreliable way of conserving energy.

A compromise solution is to use automatic controls to signal to system operators when equipment should be turned on and off, rather than operating the systems directly. For example, an optimum-start timeclock can be wired to flash an indicator light or sound a buzzer when the air handling units should be turned on. Use any combination of the automatic methods recommended previously, in Measures 4.1.1.1 through 4.1.1.5, as appropriate.

This method is less expensive than connecting the automatic controls directly to the equipment itself.

Staffing Requirements

For this Measure to work, equipment operators must be on duty at the times when equipment needs to be started and stopped. The most favorable situation is one in which operators are assigned to a specific location, such as a central control station or a security desk.

Use the Same Indicator Panel for Other Plant Equipment

In applications where this Measure is appropriate for controlling air handling systems, it is likely to be appropriate for controlling boiler plant and chiller plant equipment also. See Measure 4.1.1.6 for control of boiler plant equipment and Measure 2.1.2.5 for control of chiller plant equipment.

SUMMARY

The best method of minimizing system operation when manual control is required.

SELECTION SCORECARD

Savings Potential	$ $ $ $
Rate of Return, New Facilities	% % % %
Rate of Return, Retrofit	% % % %
Reliability	✓ ✓
Ease of Retrofit	☺ ☺ ☺

ECONOMICS

SAVINGS POTENTIAL: *70 to 95 percent of unnecessary operation of the air handling system, depending on the effectiveness of management. The fraction of total system energy consumption that is saved varies widely, depending on operating schedule, conditioning load, and other factors.*

COST: *Several hundred dollars to several thousand dollars.*

PAYBACK PERIOD: *Less than one year, to several years.*

TRAPS & TRICKS

CHOICE OF METHOD: *Do not rely on this method unless you are certain that operators will always be present to respond to the signals.*

MAKE IT OBVIOUS: *Design and install the panel so that it grabs the attention of the operators, with bright indicator lights and loud audible signals. But, don't make it obnoxious, or the staff will disable it.*

MONITOR PERFORMANCE: *Periodically, check that the panel is working properly and that people are responding to it.*

Outside Air Intake and Building Pressurization

The Measures in this Subsection deal with outside air intake and building pressurization. Outside air is taken into the building in large quantities, primarily to maintain air quality. Conditioning this air typically accounts for a large fraction of the energy consumption of air handling systems.

The movement of air through the building determines the pressure relationships inside the building, and between the inside of the building and the outside. In some types of buildings, these pressure relationships are important for safety, and in most buildings they have significant comfort effects.

The intake of outside air is controlled by two groups of air handling system components. One group comprises the "front end" of the air handling system. Various arrangements of these components are shown in Figure 1 in the Section 4 Introduction. The other group consists of exhaust fans and relief fans that draw air directly from the spaces in the building. These components are symbolized on the right side of Figure 2 in the Section 4 Introduction. While the space exhaust fans may be physically separate from the rest of the air handling system, consider them an integral part of the system. In many cases, a particular exhaust fan affects the operation of several air handling systems.

Air handling system design often fails to pay adequate attention to these components. The best time to initiate these Measures is when the facility is being designed. However, they can be retrofitted to existing facilities, in many cases. Many existing systems have defects in this area.

The first three primary Measures, 4.2.1 ff through 4.2.3, are concerned with avoiding unnecessary intake of outside air. At the same time, these Measures emphasize indoor air quality, a serious concern that may conflict with energy conservation. One way of resolving this conflict is to use air cleaning, which is recommended by Measure 4.2.4.

Under some weather and building load conditions, you can reduce the cost of cooling by exploiting outside air intake in a controlled manner. This principle is applied in Measures 4.2.5 through 4.2.7.

Exhaust air heat recovery is a major technique for reducing the energy required to condition outside air. It is covered by Measure 4.2.8.

The last three Measures, 4.2.9 through 4.2.11, correct common defects in outside air intake that waste energy and cause discomfort.

MEASURE **4.2.1 Adjust outside air intake to the minimum needed to satisfy comfort, health, and code requirements, and to maintain proper building pressurization.**

RATINGS

New Facilities · Retrofit · O&M

| | | A |

Conditioning outside air usually requires more energy than conditioning air that is recirculated from within the building. For this reason, reduce outside air to the minimum that is needed to maintain acceptable air quality. (An exception occurs when outside air can be used for cooling. The air handling system should be able to exploit this opportunity when it occurs. See Measure 4.2.5 for the details.)

Ventilation is presently a controversial issue. This Measure summarizes the current knowledge on the subject, tells you which ventilation standards prevail at the present time, and discusses the weaknesses of these standards.

Building pressurization is controlled by the same components of the air handling system that control outside air intake. Building pressurization is important because it affects energy consumption, comfort, and operation of doors. In some specialized facilities, creating differential pressurization between spaces is used as a primary method of preventing contamination.

To provide adequate control of the quantity of outside air and building pressure, you may have to make improvements to some parts of your air handling systems. This Measure describes the different methods that air handling systems use to control outside air intake, and it suggests improvements that you may need to make.

There are two subsidiary Measures. The first recommends special control procedures to minimize outside air ventilation during periods of reduced occupancy or activity. The second recommends controlling the outside air ventilation rate by sensing the concentration of air pollutants in the spaces.

Energy Effects of Outside Air Intake

Bringing outside air into the building increases all these classes of energy consumption:

- *heating.* Heating energy is directly proportional to the temperature rise required. For example, consider a situation in which the space temperature is 70°F, the outside temperature is 30°F, and the supply air temperature is 90°F. In this example, recirculated air is heated only 20°F, whereas outside air is heated 60°F, requiring three times more energy.

- *sensible cooling,* which is cooling that lowers the temperature of air. Sensible cooling energy is directly proportional to the temperature reduction required. For example, consider a situation in which the space temperature is 78°F, the outside

---

SUMMARY

A fundamental step that may save lots of energy. Must be balanced with health and safety implications. May be hindered by deficiencies in system design and equipment.

SELECTION SCORECARD

Savings Potential	$ $ $ $
Rate of Return	% % % %
Reliability	✓ ✓
Ease of Initiation	☺ ☺

---

temperature is 88°F, and the supply air temperature is 68°F. In this example, recirculated air is cooled only 10°F, whereas outside is cooled 20°F, requiring twice as much energy.

- *latent cooling* or *dehumidification*, which is energy that is required to remove moisture from air. Latent cooling energy is proportional to the amount of moisture removed. In dry climates, little or no energy is required for latent cooling. In humid climates, latent cooling may require as much energy as sensible cooling. In contrast, recirculated air typically requires little latent cooling.

- *humidification* of outside air while heating. This is energy intensive. For example, consider outside air at 30°F and 90% relative humidity that is heated to 70°F indoor temperature. If the air is humidified to 60% relative humidity at the indoor temperature, the energy requirement is increased by 70%.

  Recirculating air within the building saves some of this energy, but not all of it. This is because building air continuously loses humidity by diffusion through the building envelope. The rate of moisture loss depends on the surface area and permeability of the envelope. (Building air loses humidity rapidly if it is exposed to envelope surface temperatures that are below its dew point. See Reference Note 42, Vapor Barriers, for more about this.)

- *fan energy.* Outside air intake has only a minor effect on fan energy in most types of air handling systems.

Effects of Building Pressurization

Positive pressure inside the building increases loss of conditioned air. The extent of leakage depends on the tightness of the building envelope. For a given amount of envelope leakage, the savings that result from reducing building pressurization depend on the degree of recirculation. In a building that uses 100% outside air, leakage through the envelope does not reduce efficiency because the air will be discharged anyhow. On the other hand, a leaky building that recirculates a large fraction of conditioned air can lose a substantial amount of air through leakage.

Negative pressure inside the building causes outside air to leak into the building. This causes discomfort during cold weather, especially to people who are located near the leaking portions of the building envelope. It also wastes energy, because occupants increase heating to compensate for the discomfort. In leaky buildings, this can be a major problem.

Reducing the building pressure increases the flow of radon from soil into the building. The extent of the radon problem depends on the local geology, the area of soil in contact with the building interior, and the effectiveness of measures to block the flow of radon into the building.

Either positive or negative pressurization can cause difficulty in opening or closing exterior doors. Differences of pressurization between different parts of the building make it difficult to move interior doors.

If the interior temperature of the building differs from the outside temperature, the relative pressure inside the building changes with height. This is called "chimney effect." If an air handling system serves several floors, this problem is unavoidable within the zone served by the system. Vertical passages in tall buildings, such as elevators, stairwells, and pipe chases, allow chimney effect of large magnitude to develop. Air handling systems generally cannot compensate for chimney effect. (A system could be designed to do so,

Table 1. BUILDING AIR POLLUTANTS

Hazard Category	Associated with:	Solution Category
biological	■ air handling equipment type	■ equipment replacement
	■ equipment configuration defects	■ equipment modifications
	■ containment accumulation in air handling equipment	■ cleaning ■ biocidal measures ■ drying cycle ■ insulation relocation ■ downstream filtering ■ increased ventilation
	■ wet condenser coolers	■ biocides ■ relocation of equipment
chemical, emitted inside the building	■ smoking	■ isolation of smoking ■ filtering inside spaces
	■ building materials	■ increased ventilation ■ materials substitution ■ vapor barriers
	■ furnishings	■ increased ventilation ■ substitution of furnishing materials
	■ occupant equipment	■ equipment ventilation ■ increased space ventilation ■ substitution of equipment ■ isolation of equipment
	■ occupant respiration (carbon dioxide)	■ increased ventilation
pollutants in outside air	■ geographic location	■ reduced outside air intake ■ special filtering
	■ location of outside air intakes	■ relocation of outside air intakes
fibrous	■ asbestos	■ encapsulation ■ removal
	■ air distribution insulation	■ relocation of insulation, with acoustical modifications
radon	■ geographic location	■ move the facility
	■ proximity of spaces to ground	■ vapor barriers ■ increased ventilation of lower levels ■ construction above ground

but its additional energy consumption would offset any saving related to chimney effect.)

How Much Outside Air is Needed?

The pendulum of opinion has swung repeatedly regarding the question of how much outside air is necessary for occupant comfort and health. When modern conditioning systems started to evolve in the mid-1800's, there was concern that isolation from the outside environment would cause health problems. This concern receded quickly when no overt problems appeared. By the time of the first energy crises in the 1970's, outside air ventilation in excess of that provided by leakage and exhaust fans was considered largely unnecessary. The pendulum swung to the other extreme again during the 1980's, when the notion of "sick building syndrome" arose. Within a short time, many assumed that inadequate ventilation was a principal cause of this phenomenon. As a result, building codes were amended to require radical increases in ventilation rates.

To approach ventilation rationally, let's review the nature of indoor air pollution. Some pollutants are controlled by outside air ventilation, and others are not. Much of the controversy over the amount of outside air ventilation results from failure to distinguish between the various pollutant sources. Table 1 summarizes the major categories of agents that cause indoor air quality problems.

■ Where Increased Outside Air is Beneficial

The table makes it clear that outside air ventilation is desirable to clear out pollutants emitted from the building structure and furnishings (e.g., formaldehyde), user equipment (e.g., ozone from copy machines and laser printers), and people (e.g., carbon dioxide and body odor).

Increased outside air ventilation can also reduce the concentration of radon, in locations where this is a problem. The radon problem is serious but regional, and occurs because of natural discharge of radon from the soil underneath a building. In locales where radon emerges from the soil at dangerous rates, the extent of penetration into the building depends mainly on the building's subsurface and ground level architecture.

■ Where Increased Outside Air is Detrimental

Increasing outside air ventilation worsens the problem of pollutants that originate outside the building. Typical outdoor pollutants include nitrogen oxides and ozone derived from vehicle exhaust, and emissions from nearby industrial processes. In many locales, it is wrong to assume that the air outside a building is cleaner than the air inside. Especially where ventilation air is taken into the building at or below street level, the quality of outside air may be much worse than the quality of interior air. In fact, many buildings in urban areas limit outside air ventilation for this reason.

■ Where the Amount of Outside Air is Not a Primary Factor

The amount of outside air ventilation is only a secondary factor with two important categories of indoor air pollutants: microorganisms and fibers that originate inside the building. Biological air contaminants, which became notorious with the first diagnosed episode of Legionnaire's Disease, are characterized as pathogens (which cause disease) or allergens (which cause allergies).

It appears that most pathogens grow in the damp insides of cooling equipment, from which they migrate to any portions of the air distribution system that offer them food (dust is nourishing for many microorganisms) and sufficient moisture. Allergens typically consist of dust, the organisms and residue of organisms that grow in dust, and pollen (from outside the building). Allergies may also be caused by a host of non-biological materials.

Fibrous and porous materials used for insulation inside air handling systems act as a collection and breeding area for microorganisms. The same is true of air filters, if they are not kept clean.

At least some types of fibers cause lung diseases. Asbestos was the first fiber to be considered dangerous, and now other types of fibers are considered dangerous. How fibers cause disease is not well understood. Some researchers suspect that the shape of the fiber is as important as its chemical composition. Fiber hazards in most buildings are associated primarily with fibrous insulation inside air handling systems, including the exposed top surface of fibrous insulation inside ceiling plenums.

The principal defense against these latter types of hazards is to keep air distribution systems clean, a condition that is virtually impossible to achieve in many existing systems. The next best thing is to avoid knocking these contaminants loose into the air stream.

■ Relationship to "Stuffiness"

Lack of sufficient outside air ventilation is often mistakenly blamed for comfort problems, especially "stuffiness." Research in this area tends to show that such complaints are more likely to be the fault of inadequate air velocity within the space or of excessively high temperature and humidity. However, opinions will continue to differ about this subject.

Ventilation Standards

So, we see that the quantity of outside air ventilation is sometimes a matter of conflicting considerations. Code requirements serve only as a crude guide. They are usually better than no guidance at all, but they fail to provide the complete picture. Following them blindly may waste energy or fail to deal adequately with air quality. Standards for ventilation have been published by ASHRAE, building code agencies, and others.

At the present time, the dominant ventilation standard in the United States is ASHRAE Standard 62-89, which states specific ventilation rates, equipment requirements, and design practices. ASHRAE Standard 62-89 is still rudimentary, and it includes a large judgemental component that is controversial. Undoubtedly, it will continue to evolve. The basic research that supports the health issues is far from complete. Energy efficiency and health seem to pull ventilation requirements in opposite directions, and the equilibrium between the two will probably continue to shift.

ASHRAE Standard 62-89 has several different ways of defining the amount of outside air that is required, including:

- *a table of ventilation rates* that states ventilation rates for particular types of activities. For some environments, such as offices, dry cleaning shops, and theaters, the ventilation rate is stated on a per-person basis. Other environments, such as stores and garages, are covered on the basis of floor area. Bedrooms, living rooms, and baths are covered on a per-room basis.

- *control based on occupancy.* The Standard allows reduction of outside air during unoccupied periods or periods of low occupancy. (The details of reducing ventilation during unoccupied periods are covered in Measure 4.2.1.1, below.)

- *control based on measured contaminant level.* The Standard lists a small number of contaminants, for which maximum allowable concentrations are given. These concentrations are controversial, but at least they provide a measurable basis for regulating outside air ventilation. (Control of outside air based on contaminant level is recommended by Measure 4.2.1.2, below.)

- *reduction of outside air by air cleaning.* ASHRAE Standard 62-89 allows the quantity of outside air to be reduced by air cleaning, but the Standard is vague about how to determine the reduction. (Air cleaning is recommended in Measures 4.2.4 ff.)

Ideally, the ventilation rate in a building should be controlled to keep any every pollutant in the building below a concentration that is objectionable. However, this approach is far from being feasible in the most general sense. To date, maximum safe concentrations have been defined for only a limited number of specific air contaminants, such as carbon monoxide and formaldehyde. Ventilation is often done in response to contaminants, such as cigarette smoke and body odors, for which objective standards are difficult to define. Furthermore, some researchers feel that only a fraction of the pollutants that cause air quality problems have been identified.

Air for Exhaust Fans

In many cases, the outside air requirement exceeds ventilation standards because of the need to provide sufficient air flow at exhaust fans. For example, in a restaurant where the kitchen draws air from the dining area, air flow through the dining room may greatly exceed the per-person ventilation standard because a large volume of air is needed to make the kitchen hood function properly.

Check all exhaust fans that draw from the spaces served by the air handling system to ensure that their flow rates are neither too high nor too low. Exhaust fans can be elusive, so make a careful search to find them all. Review the building drawings, and also look for any additional exhaust fans that are not shown on the drawings.

Exhaust requirements tend to be inflexible. For example, it can be dangerous to throttle kitchen or laboratory hoods below their rated flow. Several techniques can reduce the energy consumption related to exhaust while maintaining adequate flow rates:

- using *unconditioned makeup air* with exhaust hoods. Hoods are commercially available that have integral outside air makeup. These units are designed to avoid discomfort by keeping the flow of makeup air away from the people working at the hoods.

- installing *separate air handling systems for areas with high exhaust requirements.* For example, in a restaurant, install separate conditioning systems for the kitchen and dining area. Air handling systems in areas with a high exhaust flow may be candidates for air heat recovery (Measure 4.2.8), which reduces the amount of energy needed to condition outside air.

- *air cleaning,* which can be used to reduce exhaust requirements in many applications, such as toilet exhaust. This technique is covered in Measure 4.2.4.

System Configurations for Controlling Outside Air

Before you start making adjustments, study your system and know its limitations. Figure 1 shows the common configurations that air handling systems use for controlling the intake of outside air.

System A in Figure 1 depicts the simplest type of air handling system, which has only a supply fan. In this system, the amount of outside air is determined almost entirely by exhaust fans that draw air from the spaces that are served by the air handling unit. The amount of outside air drawn into the system is equal to the exhaust from all the exhaust fans, plus any leakage through the envelope. Outside air usually enters through a damper located ahead of the air handling unit.

In this system, there is no way to increase the amount of outside air beyond the amount that is expelled by the exhaust fans. The amount of outside air can be reduced by partially closing the outside air damper at the air handling unit. This will also reduce the amount of air available to the exhaust fans, possibly causing problems.

In System A, the building pressure cannot be adjusted independently of the outside air quantity. The pressure in the building tends to remain negative because of the suction of the exhaust fan. Closing the outside air damper makes the building pressure more negative. The negative pressure induces infiltration throughout the building. Conversely, the intensity of the negative pressure depends on the leakiness of the building envelope.

The space pressure near the air supply may become positive under some conditions. This may occur if the outside air damper is fully open, the return air ducting has significant resistance, and the exhaust fans are turned off. In this situation, the space pressure near the air supply may become positive because the supply fan is pushing against the resistance of the return air ducting.

Systems B, C, D, and E in Figure 1 can control the outside air intake rationally. The minimum intake of outside air is determined by the needs of the space exhaust fans. In systems B, C, and E, the maximum outside air is 100% of the air handling system flow. In system D, the maximum outside air is determined by the space exhaust fans and by a relief fan.

The common feature of systems B and C is an arrangement of three dampers. One damper is in the outside air intake path, another damper is in the relief air path, and the third is in the path of the recirculated air. The recirculation damper creates resistance to the returning air, forcing some of it out the relief damper. The recirculation dampers also creates an area of reduced pressure in front of the fan, causing outside air to enter the system.

The three dampers are connected by a mechanical linkage or by a common control system. The relief and outside air dampers move in unison, while the recirculation damper moves in the opposite direction. I.e., when the relief and outside air dampers close, the recirculation damper opens, and vice versa.

In System B, both outside air intake and building pressure are controlled by the relative positions of the three dampers. For example, assume that you want to increase the pressure inside the building while taking in the same amount of outside air. First, close the recirculation and the relief dampers further to restrict the return of air from the space. This also reduces the supply fan's input into the space. To compensate for this, open the outside air damper wider.

A single-fan system, such as System B, is difficult to adjust properly because of the interaction between

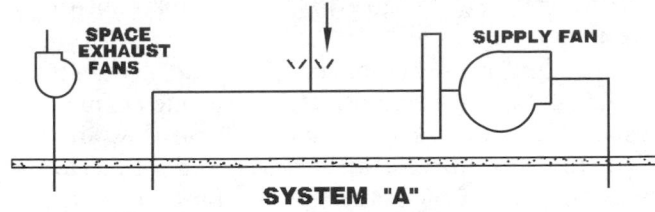

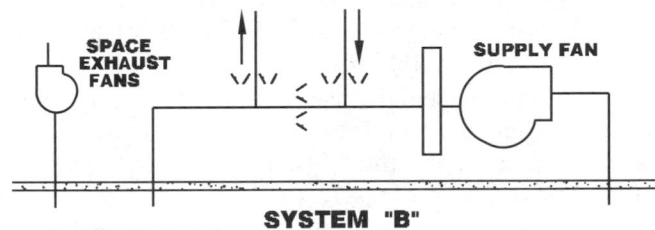

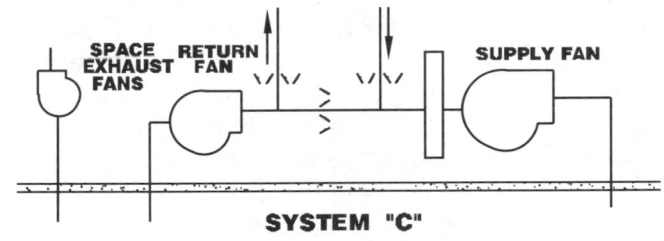

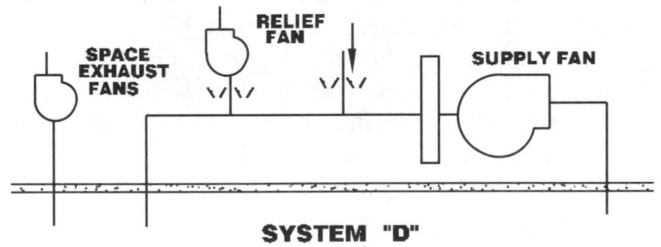

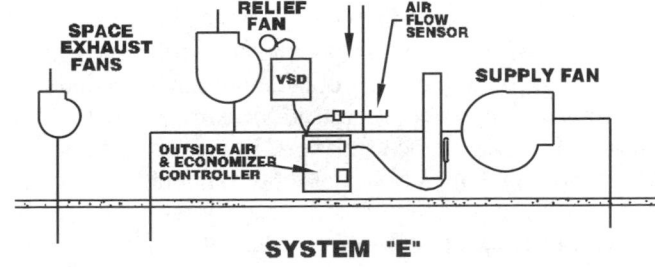

Fig. 1 Different arrangements for controlling outside air intake and space pressure All the fans and dampers are critical, including remote exhaust fans. The resistance of the return air path may also have a major influence. Some of these arrangements are much more effective than the others.

outside air quantity and building pressure. This configuration generally provides poor control of ventilation.

The distinguishing feature of System C is a return air fan. This fan is almost as large as the supply fan. Using a return fan allows the outside air flow and the building pressurization to be controlled more independently of each other. First, adjust the return fan capacity to provide the proper pressure in the spaces. Then, use the dampers to control outside air intake. If the dampers are selected and installed properly (often not the case), this system provides the most positive control of air flow and pressure.

System C is widely used in larger air handling systems, where the added cost of the return is a relatively small fraction of the total system cost. It is covered more thoroughly in Measure 4.2.2.

System D in Figure 1 is a less common approach to controlling the outside air quantity. In this arrangement, a relief fan is installed in the air return. The relief fan is sized to make up the difference between the rate of outside air intake and the rate of exhaust from exhaust fans in the spaces. The fan helps to draw return air back to the air handling unit, reducing the positive pressure on the spaces. However, if there is a great deal of resistance in the return air path, this system will still produce positive pressure in the space.

In effect, System D uses an exhaust fan as an integral part of the air handling system. The primary advantage of this arrangement is that it provides accurate control of outside air intake. This is because a centrifugal fan provides more accurate regulation of relief air flow than typical dampers. If system operation never changes, no damper is needed ahead of this fan. However, a damper is needed to prevent infiltration when the air handling system is turned off.

System D does not need a recirculated air damper. At the maximum flow rate of the relief fan, the intake and relief dampers are fully open. Under this condition, no energy is wasted in moving air across the resistance of dampers.

The main disadvantage of System D is that it is inflexible. If the relief fan operates at a fixed flow rate, it cannot provide the large outside air flows needed for economizer cycle operation (see Measure 4.2.5) and for smoke clearance in the event of a fire. To accommodate these modes of operation, the relief fan would need the same maximum capacity as the supply fan (although a lower pressure differential). Also, it would have to operate in variable-flow mode during economizer operation, and at a low fraction of its peak capacity at other times. The two dampers are poorly adapted to efficient and accurate operation over such a wide range of air flows.

System E of Figure 1 is an unusual configuration that offers advantages of efficiency combined with accurate control of outside air intake. It is now practical because of the availability of variable-speed fan drives that maintain high efficiency over a wide range of speeds. This system has no dampers, and hence it has no losses from dragging air through the resistance of dampers. It controls outside air precisely by using direct measurement of intake air flow to control the operation of a relief fan. It accommodates economizer cycle operation accurately and efficiently. It is covered more thoroughly in Measure 4.2.2.

System E needs dampers in the intake and relief air path to block outside air infiltration during periods when the air handling system is not running. However, these dampers play no role in controlling air flow during normal operation.

(All the configurations shown in Figure 1 can control outside air intake by direct measurement of the intake air flow. However, the systems differ widely in their range of outside air intake, and in the way that changing the amount of outside air affects building pressure.)

If Necessary, Improve the Systems

The previous discussion shows that the air handling system layouts represented by Systems C, D, and E are the only ones that provide good control of outside air intake and building pressurization. Consider converting other configurations into one of these. This is especially important with larger systems. See Measure 4.2.2, below, for details.

How to Measure Outside Air Quantity

You have to measure the outside air flow in order to set it. There are various methods that you can use.

You can easily measure the percentage of outside air with no equipment except a thermometer. Just use Formula 1.

To calculate the absolute quantity of outside air, multiply this percentage by the supply air flow rate. You can find the supply air flow rate in the building drawings. The actual supply air flow will differ somewhat from

---

FORMULA 1

$$\text{Percent outside air} = \frac{(\text{return air temperature}) - (\text{mixed air temperature})}{(\text{return air temperature}) - (\text{outside air temperature})} \times 100$$

---

the design figure because of variations in equipment and installation.

Stratification can cause an error in using Formula 1 by creating misleading temperature readings, especially in the mixed air stream. See Measure 4.2.11 for details.

You can measure air flow directly by using a portable flow measuring device. This is satisfactory if you are adjusting the system on a one-time basis. Pitot tubes are moderately accurate for measuring air velocity. You can insert a pitot tube through small holes in ducts and air handlers. Irregularities in flow caused by bends and connections may seriously reduce the accuracy of the measurement. Correct for this by making a careful traverse of the duct with the pitot tube.

As an alternative, you can place a portable anemometer inside the duct. Expect some difficulty in placing the unit where it can read average air flow. Get an anemometer with a remote readout, because you cannot leave the access door open to read the unit while making measurements. In large systems, you may be able to stand inside the outside air intake. Don't let your own body interfere with the measurement.

In VAV air handling systems, and in some other systems where the flow requirement varies, the system controls need to measure air flow continuously. The control system needs to adjust the supply and return fans separately, because their air flow rates differ by the amount of outside air taken into the system. To control outside air intake most accurately, install a permanent flow measuring device to measure it directly. The system also needs a flow measuring device for either supply air flow or return air flow.

Air flow measurement devices actually measure air velocity. To calculate the actual volumetric flow, multiply the measured velocity by the cross section area of the duct at the point where the velocity is measured. Fixed flow measuring devices provide good accuracy if they are installed properly and kept clean.

How to Measure Building Pressure

You can measure building pressure by using a sensitive manometer, such as the type used to measure flue draft for boiler testing. To measure the difference in pressure between the inside and the outside of the building, you need to find a small hole in an exterior wall, door, or window. Leave the manometer on one side of the hole, and pass a probe through the hole to the other side. The main practical difficulty is finding a suitable small hole. You can't work through an open door or window, because the large opening kills most of the pressure differential in the vicinity of the manometer. If necessary, try feeding the probe through a wiring penetration.

You can easily get a rough idea of the strength of pressure differences by holding a door barely open. Doors are excellent indicators of pressure differential because they are large and balanced. If there is a pressure difference, it will tend to slam to door shut or push it farther open. For better sensitivity, disconnect the door closer.

If no exterior door is available, open a window slightly. Observe the strength of the draft by holding a lightweight piece of paper near the opening, or use a smoke pencil. This method is less satisfactory than using a door.

Measure building pressure when there is no wind. The velocity pressure of a moderate wind is as strong as a typical building pressure differential caused by air handling system adjustment.

If the air handling system serves spaces that differ in height by several floors, chimney effect can have a significant effect on the pressure measurements. Chimney effect is proportional to the temperature difference between the inside of the building and the outside. It may be very strong in a tall building on a cold day. It vanishes when the outside temperature equals the inside temperature.

Monitor the Dampers

The system's dampers are the Achilles heel of effective flow control. Many buildings were built with dampers that were junk when they were new. Even the best dampers allow a substantial amount of leakage because they have extensive mating surfaces. Damper leakage gets worse with time. Blades bend, seals wear out or fall off, and linkages work loose.

Furthermore, maintenance personnel are inclined to tamper with air flow settings in response to comfort complaints, and for other reasons. It is common to find outside air dampers blocked fully closed or fully open, with the damper linkages disconnected.

Improper ventilation air settings can persist indefinitely because they have no visible effect, other than increased energy bills. For these reasons, treat dampers as equipment that needs continual maintenance.

ECONOMICS

SAVINGS POTENTIAL: *May be a large fraction of the cost of heating and cooling energy. The savings are greatest in climates that have extremes of temperature or humidity.*

COST: *Small, in cases where only adjustments are required. The cost of air handling system modifications to allow precise regulation of air intake is covered in Measure 4.2.2.*

PAYBACK PERIOD: *Less than one year.*

TRAPS & TRICKS

DECISIONS, DECISIONS: *Adjusting the percentage of outside air has become a lot more complicated than it used to be. Nowadays, you have to be familiar with legal requirements, you have to study many conflicting opinions, you have to examine evidence of air quality problems, and you have to avoid fooling yourself. In the end, the amount of outside air is a judgement call that you must make.*

ADJUSTMENT PROCEDURES: *Adjusting air flow seems to mystify a lot of people. There is no reason for this. Use common sense, understand the principles, and have good instruments. Be prepared for the possibility that your air handling systems may be so poorly designed that it cannot regulate air flow properly. In this case, you need to make hardware improvements, as recommended by Measure 4.2.2. Be wary of dealing with "air balancing" services. This trade has gained a bad reputation for preaching hokum and faking work. At least the first time, adjust the outside air quantity yourself. It will be educational.*

TRAINING AND COMMUNICATION: *Once you have figured out how to deal with outside air, train the operating staff. This will make them part of the solution, rather than part of the problem. Explain the air handling systems in the plant operating manual, and provide a list of the correct settings.*

MONITORING: *Because of air quality concerns and the high cost of conditioning outside air, outside air ventilation deserves continuing attention. Repeat air flow measurements and check the integrity of the system dampers and controls at scheduled intervals. Be alert to indications of air quality problems, but don't get hysterical. "sick building syndrome" is like many contemporary concerns. It probably exists, but it probably is not as common as generally believed. Keep yourself informed as this issue evolves.*

MEASURE **4.2.1.1 During periods of reduced occupancy, control outside air dampers and exhaust fans to reduce the quantity of ventilation air appropriately.**

RATINGS

New Facilities | Retrofit | O&M

A **A** [ ]

Measure 4.2.1 provided guidelines for adjusting the outside air ventilation rate. When the facility is not operating, it may not need any ventilation, or it may need ventilation at a reduced rate. Control the air handling system to reduce outside air intake and building exhaust to match the requirements. Make sure that each air handling system has the appropriate controls to adapt ventilation to the area it serves. This includes controlling all those exhaust fans on the roof. See Figure 1.

Consider each of the following situations. Along with checking the controls, make sure that the dampers are capable of limiting the air flow to the required amount. See Measure 4.2.1 for details.

■ **When No After-Hours Ventilation is Needed**

The controls of most air handling systems are designed to close their dampers when the fans stop. This is important because the duct system still acts as an infiltration path. The fans themselves offer little resistance to air flow when they are idle, so the only positive obstacles to infiltration are the system dampers. Make sure that the dampers are controlled to close when the fans stop.

Also, take a good look at the dampers themselves, and repair them as necessary. See Figure 2. If the dampers are not designed to close tightly, consider improving them as recommended by Measure 4.2.2. Make sure that damper linkages are connected and adjusted properly. Figure 3 shows a common problem.

SUMMARY

Basic but frequently neglected control features that save a lot of energy.

SELECTION SCORECARD

Savings Potential .... $ $ $ $

Rate of Return, New Facilities % % % %

Rate of Return, Retrofit.......... % % % %

Reliability .... ✓ ✓

Ease of Retrofit .... ☺ ☺

■ **When After-Hours Ventilation is Determined by Exhaust Fans**

After hours, the need for outside air may be determined by the need to keep some exhaust fans running, even though these fans are not explicitly part of the air handling system. For example, in a college chemistry building, the daytime ventilation rate is dictated by the occupancy of the classrooms, but night ventilation may be needed to feed the laboratory fume hoods at a reduced flow rate. Under these circumstances, the air handling unit may be turned off, unless it is needed to maintain space temperature.

Whether the air handling unit fans are running or not, the dampers should be set so that they do not force

WESINC

Fig. 1 What are all these exhaust fans and ventilation intakes, and do they need to be operating? A single exhaust fan that operates unnecessarily wastes more energy than an enormous hole in the building envelope. This is the roof of a large college laboratory building. The operation of each fan should be tailored to the operating schedule of individual spaces.

WESINC

Fig. 2 These dampers need to be replaced They were too flimsy in the first place, a bad bargain.

air into the system. The controls should close the recirculation and relief dampers. The outside air damper should be opened completely or partially.

Provide controls to limit the running of all exhaust fans, such as toilet exhaust fans and kitchen hood fans, unless they must operate continuously. Provide separate controls for exhaust fans that operate on independent schedules.

Exhaust fans should be equipped with individual dampers. Make sure that dampers are installed, and that they close reliably when the fans turn off.

If any exhaust fans must continue operating, try to satisfy their air requirements with unconditioned air. For example, hoods are available with outside air connections directly to the perimeter of the hood.

■ When After-Hours Ventilation is Determined by People

A building may never be completely unoccupied. Typically, cleaning crews and security personnel are present during unoccupied hours. If people remain in the area served by the air handling system, then ventilation is subject to requirements that are given in ASHRAE Standard 62-89 or other relevant standards. See Measure 4.2.1 for details.

The important point is that it may be satisfactory to reduce the ventilation rate during this time period. For

WESINC

Fig. 3 Common damper problem For some reason, damper manufacturers have not yet figured out how keep a simple rod connected reliably. Therefore, you have to keep checking.

example, if the ventilation standard is expressed on a per-person basis, then provide ventilation for the number of people who are actually in the building.

If the ventilation rate is determined by the presence of people, then install controls that can sense when people are actually present. For example, you might install a personnel sensor (see Measure 4.1.1.4) to control the operation of the air handling system. A personnel sensor cannot tell how many people are present, so make a reasonable assumption in setting the ventilation rate.

Ventilation may be provided for after-hours occupancy by operating one or more exhausts fans, without running the air handling system fans. For example, operating the toilet exhaust fan may provide adequate ventilation for after-hours purposes.

On the other hand, it may be necessary to operate the air handling unit to maintain space temperature. In this case, the ventilation rate may be determined either by the air handling system dampers or by the exhaust fans. Select the method that most closely tailors the ventilation rate to the after-hours occupancy.

Don't Overlook Any Dampers or Exhaust Fans

Inventory all the ways that outside air is brought into the building. There may be several dampers that connect the air handling system to the outside. Most air handling systems have an outside air intake damper. Many larger air handling systems also have a relief air damper at the air handling unit or nearby. Some air handling systems have air relief dampers located remotely in the building envelope, with no duct connections to the rest of the air handling system. Such remote relief dampers are easy to overlook.

(A note about terminology. "Relief air" is air that is discharged by an air handling system to make room for an equal quantity of outside air. Relief air carries away pollutants from the entire area served by the air handling system. "Exhaust air" is used to clear out pollutants that originate from a more localized source, such as a kitchen hood or a toilet. The distinction between the terms is not observed rigorously.)

In existing buildings, exhaust fans can be elusive. There are typically many of them, and not all of them may be shown on the original construction drawings. Make a careful search of the entire building envelope to find them all. If the building is large, it helps to use a camera in making your survey.

Provide Multiple Ventilation Rates as Needed

Ventilation requirements may vary in stages. For example, full ventilation may be required during weekdays, a lesser rate may be required during evenings and weekends, and no ventilation may be required after midnight. The system controls should accommodate these changing requirements as closely as possible. This requires multiple stages of control.

The controls must have a way of "knowing" when to switch from one ventilation rate to another. If the ventilation rate is determined by time or occupancy, use one or more of the control methods recommended by Measures 4.1.1 ff. The ultimate refinement is varying the ventilation rate continuously as pollutant levels change. This approach is recommended by Measure 4.2.1.2, next.

Anticipate the Beginning and End of Occupancy

The mass of air in a building dilutes contaminants, so it makes sense to cease outside air ventilation some time prior to the end of occupancy. By the same token, start ventilation prior to the beginning of occupancy to clear out accumulated contaminants. ASHRAE Standard 62-89 specifies permissible lead times for starting and stopping ventilation. These times are specified in terms of the air volume of spaces.

ECONOMICS

SAVINGS POTENTIAL: *Typically, all or a large fraction of the energy used to condition outside air during unoccupied periods.*

COST: *Varies widely, depending on what needs to be done. Simple adjustments are cheap. Damper repairs, or even replacement, usually do not cost much in relation to the cost of conditioning energy. Adding control capabilities may cost thousands of dollars.*

PAYBACK PERIOD: *Less than one year, to several years.*

TRAPS & TRICKS

DESIGN DETAILS: *Design the air handling system controls to accommodate each mode of ventilation that may be needed. Consider all exhaust fans to be an integral part of the air handling system. Be sure to coordinate damper positions with the operation of exhaust fans.*

EXPLAIN IT: *Describe the after-hours ventilation controls in the plant operating manual. Install explanatory placards on fans and dampers that operate on special schedules. Inform the operating staff how the system is supposed to operate, even though the controls are automatic. This way, they will not try to start fans that are supposed to be off, or vice versa.*

MONITORING: *Fan and damper operation is vulnerable to tampering and control failure. Improper operation is not obvious, so schedule periodic checks.*

MEASURE 4.2.1.2 Control outside air intake by sensing air contaminants.

Measure 4.2.1 explained why it is important to minimize the outside air ventilation rate. An emerging method of doing this is to regulate ventilation air by sensing the concentration of specific contaminants within the spaces. In principle, this is the most efficient method of regulating ventilation, because it goes directly to the purpose of ventilation.

Identifying the Critical Pollutants

This method of controlling outside air is valid only if you can identify all the pollutants that determine the needed ventilation rate. For example, in an industrial operation where the dominant hazard is known to be formaldehyde, control ventilation on the basis of the formaldehyde concentration.

ASHRAE Standard 62-89 gives maximum allowable concentrations for a short list of contaminants. Government regulations and reference books on industrial hygiene give recommended maximum exposures for a much broader range of air contaminants.

This method is already popular for controlling the operation of garage exhaust fans, such as those in Figure 1. This method of control senses the carbon monoxide concentration in the air, because carbon monoxide is the primary concern in garages.

In buildings with clean environments that are intended primarily for human occupancy, such as office buildings and theaters, you may be able to control ventilation based on the carbon dioxide concentration in the air. Carbon dioxide is the main product of human respiration (along with water vapor), so its concentration can be used to indicate the density of people in the spaces

SUMMARY

In principle, the most efficient method of controlling ventilation. Still limited by the types of pollutants that can be sensed. Unproven, except in a few specialized applications. Requires sensor maintenance.

SELECTION SCORECARD

Savings Potential .... $ $ $ $

Rate of Return, New Facilities % % % %

Rate of Return, Retrofit.......... % % % %

Reliability .... ✓

Ease of Retrofit .... ☺ ☺

served by an air handling system. Carbon dioxide sensor systems for controlling ventilation are becoming available in the HVAC trade.

Control of ventilation by contaminant monitoring is not satisfactory in applications where there are non-specific air quality problems, commonly called "sick building syndrome." This is because the causes of indoor air quality problems have not yet been adequately defined. Many substances found in air are suspected of causing discomfort and health problems, but few of these suspicions have been confirmed. Even where certain substances or classes of substances (such as "volatile organic compounds") have been identified as likely culprits, rational concentration thresholds have not been established for them.

WESINC

Fig. 1 An application for control by carbon monoxide sensors These fans ventilate a garage. Similar technology that controls on the basis of carbon dioxide can be used to control ventilation in benign occupied spaces.

Design is Unconventional

Even if you can identify all the critical pollutants, you still have to find sensors for these pollutants that are suitable for control purposes. Sensors are available for many individual chemicals, for groups of substances, and for particulates in general. You may have a variety of choices. However, control of ventilation based on contaminant concentration is still a new concept in conventional air conditioning applications. For this reason, you may have to adapt sensors that presently are designed for non-HVAC industrial or research applications. This crosses the boundaries of established disciplines, so expect this to be a challenge.

Sensor Maintenance

Sensor systems for airborne contaminants require skilled maintenance and periodic calibration. Such equipment does not have an established track record in control of indoor air pollutants. You will have to develop an effective maintenance program from experience. You can expect satisfactory operation in the long term only in a disciplined maintenance environment where control of air quality is a concern of management.

The main exceptions are carbon monoxide and carbon dioxide. Newer types of sensors have been developed for these gases that require little maintenance or calibration, or none at all. Lack of maintenance is the major failure of unconventional systems, so be sure to select sensors that need a minimum of maintenance.

Get a Code Variance, If Necessary

Most codes that specify outside air ventilation are based on tables of ventilation rates for particular types of activities, as explained in Measure 4.2.1. In jurisdictions that specify ventilation rates in this manner, request a variance from the code that allows you to control ventilation based on pollutant level.

Cost Components

The sensor systems themselves typically cost several thousand dollars. Installation costs vary widely, depending on the ease of interfacing the controls to the existing air handling system. After identifying the best type of sensor, try to find a model that provides standard output signals for interfacing with your HVAC controls. Except in systems that are packaged specifically for ventilation control, design fees will comprise a major fraction of the installation cost.

Sensors for most pollutants, other than carbon monoxide and carbon dioxide, require continuing expenditure for maintenance and calibration. You also have to keep a stock of supplies, such as test chemicals or calibration gases.

ECONOMICS

SAVINGS POTENTIAL: *30 to 70 percent of the cost of conditioning outside air.*

COST: *Simple sensor systems that control ventilation on the basis of carbon monoxide or carbon dioxide content cost several thousand dollars. Custom controls that sense other pollutants would cost many times more.*

PAYBACK PERIOD: *Several years or longer, probably.*

TRAPS & TRICKS

CHOICE OF METHOD: *You will be a pioneer if you control by sensing anything other than carbon monoxide or carbon dioxide. Experimental and unusual equipment tends to be abandoned in operational environments, so don't do this unless your facility has a maintenance staff that is technically oriented.*

PREPARATION: *Expect to study the subject. Have the time and resources available for experimentation and working out details. Before getting started, verify that your method will satisfy local code requirements.*

CHOICE OF EQUIPMENT: *As always, try to use standard equipment, and check its reputation with other users. You will be entering a whole new world if you try to adapt pollutant and chemical sensing equipment that is not designed for HVAC control.*

MAINTAIN IT: *Sensor maintenance is the Achilles heel of this control method. The effort is small, but critical. Include sensor tests and maintenance in your maintenance calendar. To minimize the risk of tampering or improper operation, include a clear description in the plant operating manual. Install a placard at the sensor that identifies and explains it.*

MEASURE **4.2.2 Provide accurate control of outside air intake and building pressurization by adding a return fan or relief fans and improving the damper configuration.**

Poor design of the "front end" of many existing air handling systems prevents accurate control of outside air intake and building pressurization. The result is either waste of energy from taking in too much air, or health hazards and discomfort from taking in too little air. Lack of accurate control of outside air intake and air exhaust can also result in building pressurization problems. These issues are covered in Measure 4.2.1.

The usual defects in "front end" design are the lack of a return air fan and ineffective damper arrangements. This Measure corrects these problems.

There are two approaches. The more satisfactory one is to improve the "front end" of the air handling system. The objective is to achieve either configuration "C" or configuration "E," as shown in Figure 1 of Measure 4.2.1. This is the approach to use in all new construction, and in retrofit wherever practical.

In retrofit, if neither of the previous approaches is practical, an alternative is to add relief fans in the spaces that are served by the existing air handling system.

The Best Fan and Damper Configurations

Figure 1 of Measure 4.2.1 shows two front end configurations for air handling systems that provide good control of outside air intake along with good control of building pressure. These are systems "C" and "E." The industry does not have standard names for these configurations, so we will call them systems C and E in this Measure.

System C has been used for a long time, and it has been well proven by experience. However, it is often installed with poor damper arrangements that void much of its benefit.

System E is an unusual configuration that offers advantages of efficiency combined with accurate control of outside air intake. It is now practical because of the availability of variable-speed fan drives that maintain high efficiency over a wide range of speeds. System E is novel. Therefore, you are pioneering if you use it. It deserves your attention because it avoids all the problems of control dampers, including their energy waste.

■ System "C" Front End Configuration

System "C," in Figure 1 of Measure 4.2.1, has become fairly standard in larger installations because it provides the best control of outside air intake and building pressurization. This configuration can induce outside air in any quantity, from a minimum that is determined by the space exhaust fans, up to 100% of system flow. It maintains fairly good regulation of

SUMMARY

Mainly a matter of selecting the most efficient front end configuration for air handling systems. Not applicable if the exhaust requirement always exceeds the outside air requirement.

SELECTION SCORECARD

Savings Potential	\$ \$ \$ \$
Rate of Return, New Facilities	% % % %
Rate of Return, Retrofit	% % %
Reliability	✓ ✓
Ease of Retrofit	☺ ☺

outside air quantity in the face of external forces, such as wind, opening exterior doors, etc. The main elements are three dampers and two fans. The three dampers are:

- an *outside air intake damper*
- a *relief damper,* which is needed to get rid of air in excess of the amount removed from the spaces by exhaust fans
- a *recirculation damper,* which creates resistance to the returning air, forcing some of it out the relief damper. The recirculation damper also creates an area of low pressure in front of the fan, which draws outside air into the system.

The three dampers are connected by a mechanical linkage or by a common control system. The relief and outside air dampers move in unison while the recirculation damper moves in the opposite direction. I.e., when the relief and outside air dampers close, the recirculation damper opens, and vice versa.

The return fan stabilizes the control provided by the dampers. The amount of air flowing through each of the three dampers is determined by the pressure difference across each damper. Using a return fan isolates the dampers from any pressure changes originating within the spaces. Such changes might be caused by opening doors and windows, or by turning exhaust fans on and off. Without a return fan, such changes could strongly affect the amount of air returning from the spaces, and hence, the amount of outside air intake.

The return fan also neutralizes the tendency of the supply fan to create excessive pressure inside the building. The return fan compensates for the resistance of the return air path, which includes the recirculation and relief dampers. If only a supply fan is used, the pressure of the supply fan backs up against the resistance of the return air path to create a positive pressure within the building. As explained in Measure 4.2.1, this wastes conditioned air through leakage, and may cause difficulty in door operation.

This combination of three dampers and two fans works well with any kind of air handling system: single-zone, multizone, dual duct, etc.

The main weakness of System C is its reliance on variable dampers ("control dampers") to regulate air flow. These regulate air flow by creating resistance. Moving air across this resistance wastes energy.

Dampers also provide poor accuracy of control if the air flow varies over a wide range, especially near minimum flow. In many damper installations, the dampers are not capable of accurately maintaining the minimum air flow. Also, control dampers are so leaky in the closed position that they allow a considerable amount of infiltration into the building. These weaknesses can be overcome by careful damper selection, and by using combinations of dampers, as discussed below. However, you have to do much better than current conventional practice.

■ System "E" Front End Configuration

System E controls outside air intake entirely by changing the speed of a relief fan. It has no control dampers. For this reason, it avoids the energy waste caused by dragging air through the resistance of dampers. Air flow is controlled by fan output, which is based on measurement of actual air flow. Therefore, the control of air flow is accurate under all conditions.

System E does need shutoff dampers in the intake and relief air path to block outside air infiltration during periods when the air handling system is not running. However, these dampers play no role in controlling air flow during normal operation. To exploit this advantage, use low-leakage dampers to block infiltration, as discussed below.

System E has one important disadvantage, compared to System C. Because System E does not have a return air fan installed in the return air duct, it cannot compensate for the resistance of the return air path. As a result, the air pressure inside the spaces served by the air handling system is always positive. If the resistance of the return air path is high, the positive pressure may also be high.

To understand this, start with the fact that the pressure inside the air handling system is zero (equal to outside air pressure) at the outside air intake. It remains approximately zero at the inlet to the relief fan. Therefore, to return air from the space, the pressure in the space must be equal to the flow resistance of the return air path. For example, if the return air path has a resistance of 0.2 inches of water, then the pressure inside

the space (near the air return point) must also be approximately 0.2 inches of water.

Where to Use These Configurations

These front end configurations involve additional costs for the return fan, the dampers, connecting ductwork, and simple controls. In variable-air-volume (VAV) systems, an air flow measuring control system is needed for proper control.

In retrofit applications, another obstacle is the extra space required to properly install the return fan and the dampers. If this space is not available, you may be able to install space relief fans, as discussed below, instead of a return fan.

This Measure is not appropriate if the amount of air exhausted from the spaces by exhaust fans always exceeds the amount of outside air needed for ventilation. This situation exists commonly in facilities with large exhaust fans. For example, a restaurant may draw air through the customer area to be exhausted through the kitchen hood.

For example, consider a system in which the supply fan provides 100,000 CFM and the exhaust fans in the spaces remove 20,000 CFM. Only 80,000 CFM returns to the supply fan from the spaces, so 20,000 CFM of outside air is drawn into the supply fan inlet. If the outside air ventilation requirement is 20,000 CFM or less, there is no need for dampers to force air into the system. In this example, it is impossible to reduce the outside air flow to less than the exhaust flow.

Even if high exhaust flow makes the dampers unnecessary, it may still be desirable to install a return fan to keep the pressure in the spaces from being too high. However, the energy saving benefit would be minor, in most cases.

Design and Installation Issues

Despite the popularity of System C, it is commonly designed and installed with mistakes that cause poor performance. System E is novel, so it involves even more opportunity for mistakes. Work out the following issues.

■ Match Return or Relief Fan Capacity to Ventilation Requirements

The capacity of the return fan or the relief fan is selected to equal the capacity of the supply fan, minus the amount of air removed from the spaces by exhaust fans. For example, if the supply fan provides 100,000 CFM and the spaces have a number of exhaust fans totalling 15,000 CFM in capacity, then the return fan should have a capacity of 85,000 CFM.

If the supply fan has variable output, the return fan must also have variable output. In VAV systems, control the return fan to follow the supply fan. Also, if exhaust fans are turned on or off while the air handling system is operating, the control system should adjust output of the return fan accordingly. Installing variable-flow fans is covered in the later Subsections that deal with the individual types of air handling systems.

■ Select Dampers for Low Leakage When Closed

Make sure the controls shut off outside air ventilation during periods when ventilation is not required. (See Measure 4.2.1.1 for details of this control function.) The outside air and relief dampers are the hardware involved. Unfortunately, the cheaper type of opposed-blade dampers leak about 5 to 10 percent of their maximum flow, for a given pressure across the damper. Figure 1 shows this weakness dramatically. But this is not the whole story. If the air handling system continues to operate while the outside air and relief dampers are closed, increased pressure differential across the dampers increases their leakage.

For example, consider an air handling system designed to deliver 100,000 CFM to the spaces, with 20,000 CFM of outside air. With these requirements, the return fan is designed for only 80,000 CFM. There is low pressure in the damper area because the supply fan is trying to draw more air than the return fan provides. With ordinary dampers, the amount of leakage could be from 5,000 CFM to 10,000 CFM. The actual amount of leakage depends on the size and design of the dampers, their condition, and the fan characteristics.

There are two basic steps in minimizing damper leakage: (1) install dampers that are designed for low leakage, and (2) make the dampers as small as possible.

There is a large range of leakiness in conventional dampers. The damper industry in the United States has established a rating system for damper leakage. The best dampers for common applications have a "Class 1" leakage rating. This means that the damper has a leakage of less than 4 CFM per square foot at a test pressure of one inch of water gauge. Use dampers having a rating of Class 1 or better for blocking outside air infiltration.

Damper leakage tends to get much worse with age, especially with cheaper dampers. To minimize long-term deterioration of the damper's sealing ability, spend the money to get these features (which are not inherently part of the Class 1 rating):

- a frame that is well reinforced to resist deformation
- extruded airfoil blades, which are very rigid and are shaped to improve air flow
- non-round blade shafts, and blades that are formed tightly to the shafts, to keep the blades in alignment
- good shaft bearings that do not bind or become sloppy with age
- strong linkages between blades, and between blades and actuators
- blade edge seals that are mechanically locked in place (not glued or clipped on)
- a method of easily replacing blade edge seals.

Good low-leakage dampers are significantly more expensive than the junk that is commonly installed, but the price premium is minimal in comparison with the energy savings.

Air leakage through a closed damper is proportional to its surface area, or more precisely, to the total length of blade edges. Conventional opposed-blade control dampers tend to be leaky because of the large ratio of blade edge length to surface area. Therefore, make intake and relief dampers as small as possible, subject to admitting the rated amount of ventilation air when the dampers are open. With typical outside air intake percentages, this area is small. The intake and relief dampers typically should be smaller than the cross section area of the return duct.

An important exception to the last statement is systems using an economizer cycle. (See Measure 4.2.5 for an explanation of economizer cycles.) In these systems, both the intake and exhaust dampers must be large enough to pass the entire system air flow. This aspect of economizer operation is discussed below.

Special low-leakage dampers have been designed that block almost all air leakage. Figure 2 shows an installation. Unfortunately, most low-leakage dampers cannot modulate air flow effectively over a wide range. One solution is to use a combination of damper sizes, as in Figure 2. Another is to install a combination of damper types. One type is selected strictly to provide positive shutoff of outside air, and the other type, which is installed in series, modulates the air flow.

■ Select Dampers for Accurate Control of Outside Air

Conventional opposed-blade dampers are crude in their control of air flow. They fail to provide reliable control accuracy at both ends of their range of motion, and they are non-linear in the middle range of travel. For a given damper position, a small change in the pressure differential across the damper has a large effect on flow.

If you decide to install System C, the basic solution is to keep the outside air intake damper as small as possible, so that the damper will operate in the portion of its operating range where it can provide accurate control. This is the same solution suggested above for minimizing leakage when the damper is closed.

Trouble arises when the amount of outside air varies over a large range. This occurs with economizer cycle operation, and perhaps in other applications. In such cases, the outside air and relief dampers must be so large that they cannot provide accurate control at low flow rates. Solutions are offered below.

■ Control by Flow Measurement and Feedback

For the reasons given previously, damper position is not an accurate indication of the amount of outside air being drawn into the building, especially in systems that have changing outside air requirements, variable exhaust fan requirements, and economizer cycles. To control outside air with any degree of accuracy, both System C and System E need controls that measure the actual air flow.

In System C, if the fans operate at fixed speeds, the dampers can be controlled with a single air flow measurement device located in the outside air intake. This simple arrangement can compensate for changes

Blender Products, Inc.

Fig. 1 Large damper array An installation like this cannot control low air flow rates reliably, and it cannot prevent major infiltration when the facility does not need outside air. Note the large unit heaters, probably installed for freeze protection. Such a high-energy alternative to good dampers is a very bad bargain.

in exhaust fan operation and changes in building pressure. Flow measurement and control become more complex if the supply and return fans have variable output, or if accurate control of building pressurization is an issue. In these cases, the system requires several flow measuring stations.

In System E, the speed of the relief fan is controlled with a single air flow measurement device located in the outside air intake. The flow measuring station responds to controls that determine the need for outside air. These might be a timeclock, an economizer cycle thermostatic control, etc. The output of the supply fan may be constant, or it may be controlled independently of the relief fan.

One of the advantages of System E, compared to System C, is that control of the relief fan's output is entirely independent of control of the supply fan's output.

■ Avoid Stratification

Eliminating stratification is a major aspect of good front end design. It is a challenge to mix the air streams in the damper section of an air handling system as damper positions and fan speeds change over a wide range. See Measure 4.2.11 for details.

■ Access to Equipment

Provide easy access to all equipment that may need maintenance or adjustment. This includes access to fans, motors, fan drives, dampers, actuators, thermostats, flow measurement devices, and other control components.

Economizer Cycle Operation

Economizer cycles are covered in detail by Measure 4.2.5. In brief, an economizer cycle is an energy conservation measure in which a large volume of outside air is taken into the building for cooling during cool weather. For the economizer cycle to function properly,

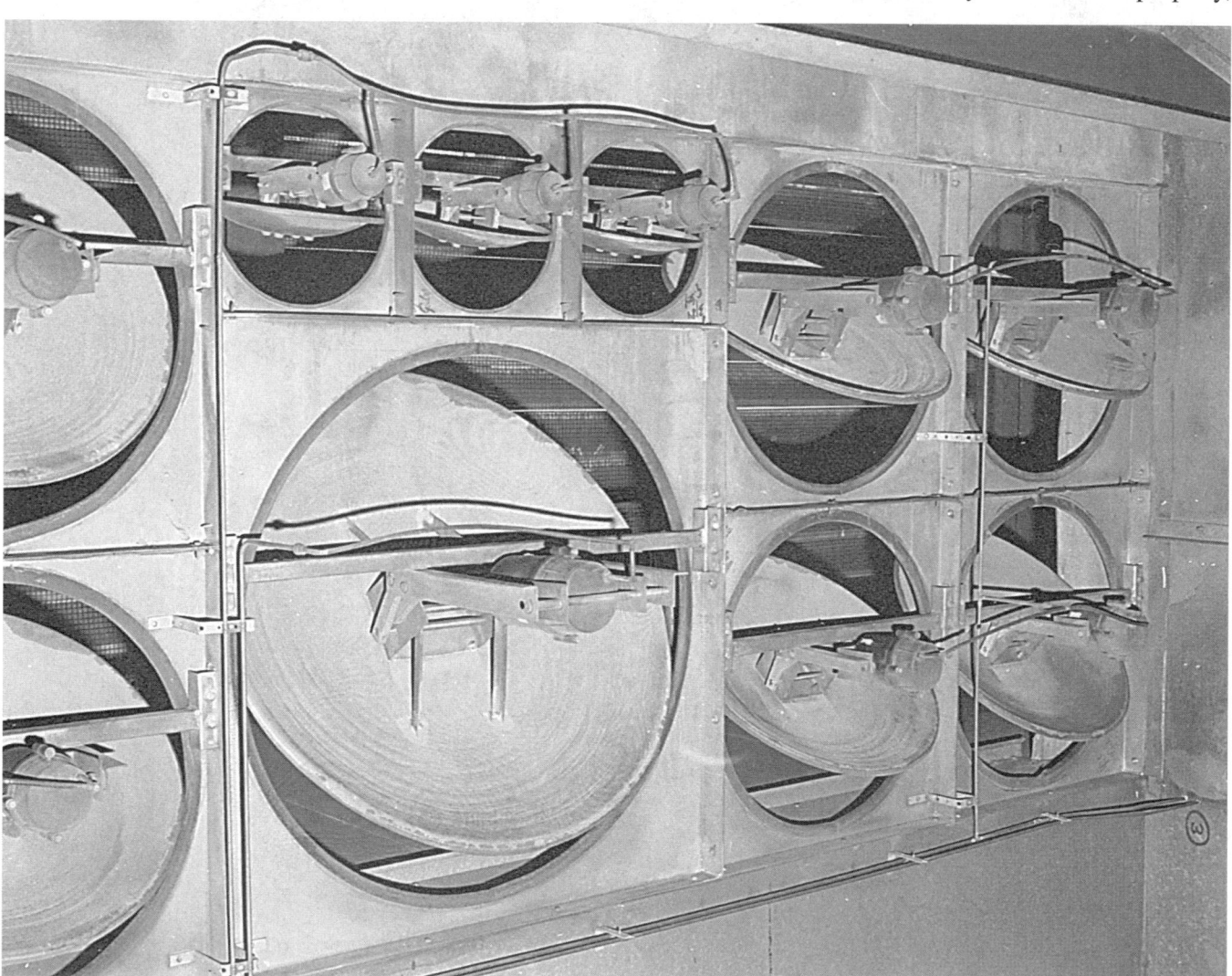

M&I Heat Transfer Products

Fig. 2 Low-leakage dampers These can seal almost hermetically, and they retain their tightness over the long term. They have the minimum possible edge length in relation to their area. However, they cannot modulate air flow over a wide range. To deal with this limitation, the array combines a variety of damper sizes. For example, this array can operate efficiently with both minimum outside air requirements and with an economizer cycle.

the air handling system must be able to take in outside air in quantities ranging from zero to 100% of system air flow.

Air infiltration during system shutdown is a common problem of economizer cycles, because a conventional damper large enough to handle 100% outside air may leak badly when it is fully closed. Also, the large outside air and relief dampers have poor control at low outside air intake rates, when the economizer cycle is not in use.

An unconventional solution is to use two sets of dampers. A large set of modulating dampers is used for economizer operation. When the economizer cycle is not in use, these large dampers are isolated by a special airtight damper. Ventilation air is then controlled by an appropriately sized smaller damper.

Flow measurement may be a problem with economizer cycles because most flow measurement devices are not accurate over a wide range of flows. One solution is to size the flow measurement devices to control outside air intake during non-economizer operation. During economizer operation, control the quantity of outside air with a thermostat that senses space temperature (in the case of single-zone systems) or supply air temperature (in the case of other types of systems).

Alternative: Space Relief Fans

An alternative to installing a return fan is installing one or more relief air fans that draw directly from the conditioned spaces. The fans draw an amount of air from the spaces equal to the amount of outside ventilation air desired for the spaces, minus any amount required by exhaust fans (for toilets, hoods, etc.). Relief fans should be controlled by the air handling system's controls in the same way as a return fan.

Relief fans may be the only choice in buildings where you cannot install a return fan and a full complement of dampers. They can also be used in buildings that lack a return air duct system. Relief fans have these disadvantages:

- *more expensive to modulate air flow.* Each relief fan serving a space with variable ventilation requirements requires dampers or a variable-flow drive.
- *more expensive control and wiring.* In a system that uses flow sensors, each relief fan requires a flow sensor. The distance of the relief fans from the supply and return fans increases wiring cost.
- *usually impractical to accommodate economizer cycle operation.* For an economizer cycle to work in a system that uses relief fans, the fans would have to be much larger than otherwise, and they would need the ability to modulate their output.

- *usually impractical to accommodate exhaust air heat recovery.* An exhaust air heat recovery system would need a connection to each relief fan.

The design and installation precautions in a relief fan system are generally the same as those discussed previously for a return fan system.

Coordinate with Other "Front End" Measures

This Measure involves the same part of the air handling system as the other Measures of this Subsection. To avoid wasting money and opportunity, decide which of the Measures is appropriate for each air handling system, and accomplish all of them together.

ECONOMICS

SAVINGS POTENTIAL: Conditioning outside air typically accounts for a major fraction of heating and cooling cost. This Measure may reduce the cost of ventilation by 20% to 70%, compared to installations that control ventilation poorly.

COST: It is much cheaper to provide effective control of outside air in new construction than in retrofit. In either case, the cost varies widely, depending on the building and system configuration. For example, an array of tight-closing dampers costs about $500 to $1000 per square meter, for the equipment.

PAYBACK PERIOD: Usually short, in new construction. One year to many years, in retrofit.

TRAPS & TRICKS

DESIGN: The reason that so many air handling systems fail to control ventilation properly is lack of attention to detail in the design. The layout may be simple on paper, but the actual physical arrangement of the fans and dampers is critical. Use some imagination here. The method of damper and fan speed control is also important.

SELECTING THE EQUIPMENT: Good equipment is available, but you have to find it among the junk, and pay the higher price. It doesn't cost much extra.

INSTALLATION: Work with the contractor. Don't expect him to work out the design details. His job is sheetmetal, not engineering. Don't let the contractor substitute cheaper equipment.

MONITOR PERFORMANCE: The front end of an air handling system contains moving machinery that must be maintained. Damper linkages come loose and controls get out of calibration. Blades warp, and blade seals wear out. Tampering occurs in response to comfort complaints, to save energy, to avoid coil freezing, or for other reasons. Schedule periodic checks of the damper section, including air flow measurements.

Once-through air handling systems are required in applications where recirculation of air is hazardous or undesirable, as in kitchens, surgical suites, and medical laboratories. However, once-through systems are sometimes found in applications where they are not needed. Conditioning outside air requires a lot of energy, so failing to recirculate air when it is possible to do so is a major waste. Examine each non-recirculating air handling system to decide whether it can exploit air recirculation for at least part of the time.

Air handling systems that could use recirculation may lack it for a variety of reasons. Sometimes, an air return is omitted to reduce installation cost. Recirculation is commonly omitted in spaces that require a high outside air ventilation rate at certain times, but not at other times. For example, 100%-outside-air systems are common in gymnasiums and locker rooms, even though they are unoccupied, or have minimal occupancy, most of the time. Sometimes, a space that originally required 100% outside air was changed to a function that does not require it.

This modification involves two main additions. One is a return air path from the conditioned spaces back to the air handling unit. The other is a set of dampers, and perhaps an additional fan, to regulate the intake of outside air.

Providing the Air Return Path

You can add an air return path in a variety of ways. The choice is governed by a variety of considerations, including fire safety, noise isolation, isolation from lighting heat gain, avoiding air quality problems, limitations in the space available to install ducts, isolation from perimeter infiltration, isolation from other air handling systems, and cost. Consider these possibilities:

- *ducted returns.* These are the most foolproof because they isolate the air handling system from external factors. They are also the most expensive. If you cannot find a way to install a return air duct inside the building, consider installing it outside the building, e.g., on the roof. Pay special attention to sealing and insulating external ducts.
- *plenum return.* This is satisfactory in many applications. Problems may arise if other air handling systems use the same plenum space as a return, or if the plenum space is in contact with a larger exterior surface. Making holes in partitions to create a plenum return increases the transfer of noise across the partitions. Also, the holes are a path for fire and smoke, although this problem can be limited with fire dampers. Plenum return picks

up the heat of light fixtures laid in suspended ceilings. Whether this makes a difference in the conditioning load, compared to ducted returns, depends on the layout of the building.

- *corridor air return.* This method is inexpensive, and it is used in many older buildings. It is least desirable from a fire safety standpoint because the air handling system circulates smoke through the corridors in the event of a fire. For this reason, corridor returns are now prohibited in many jurisdictions. If you want to use a corridor return, you have to install grilles of adequate size in the doors or walls of all the spaces that are served by the system, unless these spaces are permanently open to the corridor.
- *return through a void space* is sometimes possible. Be careful to avoid air quality problems. Expect problems if the space contains moisture, bare soil, animal residue, or dust that is light enough to be picked up by the air stream.

Design and Installation of the Outside Air Intake

The challenging part of the modification is designing the system to provide good control of outside air intake. See Measure 4.2.2 about these aspects.

Satisfying Ventilation Code Requirements

Refer to Measure 4.2.1 for ventilation requirements. If the original once-through air handling system results from a ventilation code requirement, providing air return may require a variance from the code. The designer is likely to get a variance if he can demonstrate that the system is capable of tailoring ventilation to requirements. Note that ASHRAE Standard 62-89, the basic ventilation standard for many jurisdictions, provides for reducing ventilation during periods of reduced occupancy.

Coordinate with Other "Front End" Measures

This Measure involves a major addition to the air handling system. When planning it, review the other Measures of this Subsection to see whether you can save money or effort by accomplishing any of them at the same time.

ECONOMICS

SAVINGS POTENTIAL: *Possibly a large fraction of the total heating and cooling cost of the air handling system.*

COST: *Modifying a system to provide rational recirculation is much cheaper, and more satisfactory, in new construction than in retrofit. In either case, the cost varies widely, depending on the building and system configuration.*

PAYBACK PERIOD: *A few years, or less, in new construction. Longer, in retrofit.*

TRAPS & TRICKS

CHOICE OF METHOD: *Moving air through a building is subject to a variety of factors, including structural, fire safety, noise, conflict with other systems, etc. Take the broad view and consider all your options.*

OPERATION AND MAINTENANCE: *The result of this Measure is a conventional recirculating air handling system. See Measure 4.2.1 about efficient adjustment and operation of the system.*

MEASURE **4.2.4 Use air cleaning to reduce the need for outside air ventilation.**

You may be able to save a lot of energy by cleaning and recirculating the indoor air, rather than replacing it with outside air. A variety of air cleaning methods are available. Individually or in combination, they have the capability of removing most types of indoor air pollutants.

Filtration to improve air quality is widely used in applications that are critical to health, such as surgical suites. Other air cleaning methods are less common. The idea here is to expand the application of air cleaning to take advantage of its energy saving potential. Most of the methods require specialized equipment and maintenance. Therefore, they tend to be limited to facilities with more sophisticated maintenance staffs.

(The ordinary filters used in air handling units do little to improve air quality for occupants. These filters, which are called "low efficiency" filters, serve mainly to protect the coils from becoming clogged. They trap mainly fibers and large particles. They cannot trap more than a small fraction of the dust, pollen, microorganisms, and gases that exist in building air.)

Energy Saving Potential

Measure 4.2.1 estimates the amount of energy that you can save by reducing outside air quantity. The ventilation rates specified by guidelines such as ASHRAE Standard 62-89 depend on the usage and

SUMMARY

Air cleaning may deserve more attention outside the specialized applications where it is presently used. All methods require continuing maintenance. Some are limited to narrow classes of contaminants.

SELECTION SCORECARD

Savings Potential	$	$	$
Rate of Return, New Facilities	%	%	%
Rate of Return, Retrofit..........	%	%	%
Reliability	✓	✓	
Ease of Retrofit	☺	☺	

conditions of the facility. By improving air quality conditions in the occupied spaces, you may be able to reduce the ventilation rates that are required.

■ Fan Power Penalty

Most air cleaning equipment adds considerable resistance to air flow. Additional fan power is needed to overcome this resistance. The additional fan energy substantially reduces the potential saving in conditioning energy, perhaps severely enough to make air cleaning uneconomical.

■ **An Example**

Consider an air handling system that has a flow rate of 100,000 CFM. Without air cleaning, 30,000 CFM of outside air is needed. With air cleaning, outside air can be reduced to 10,000 CFM.

Outside air is heated by an average of 30 degrees Fahrenheit for a duration of 3,000 hours per year. Air cleaning results in a saving in air heating of 1,950 million BTU per year. Air heating energy costs $5 per million BTU, resulting in a heating cost saving of $9,750 per year.

The filtering medium creates an additional air flow resistance of 2" of water gauge, which is imposed on the entire air flow of the air handling system. Fan power is approximately 0.25 horsepower per thousand CFM per inch of water gauge resistance. Thus, the total system fan power is increased by 50 horsepower, and the fans operate for a duration of 5,000 hours per year. At an electricity cost of $0.05 per kilowatt-hour, the additional fan power costs $9, 320 per year.

Thus, in this example, the saving in air heating energy is almost entirely cancelled by the cost of the additional fan power. In addition, the cost of the air cleaning equipment must be paid off. In this example, the cost of the air cleaning equipment might range from $6,000 to $100,000, depending on the equipment selected.

Available Methods of Air Cleaning

Each type of air treatment technique is effective only for certain classes of pollutants. Therefore, identify all the pollutants that need to be controlled. The current state of research and testing does not allow this to be done with complete certainty. Many substances found in air are suspected of causing discomfort and health problems. Many of these are lumped into broad categories, such as "volatile organic compounds."

In existing facilities, only a small fraction of identified pollutants will be a problem. To identify which, start by analyzing occupant complaints, or noting their absence. If occupant complaints suggest the presence of a specific pollutant, get an expert to check for it. The expert should check for noxious sources within the spaces and within the air handling systems. Test for chemical or biological pollutants, or both, depending on the nature of the symptoms.

In principle, any kind of air pollutant could be removed by some process. A practical criterion is that air cleaning should be less expensive than increasing outside air ventilation, while being reasonably reliable and free of serious disadvantages. For example, it is too expensive to remove carbon dioxide by chemical treatment in typical HVAC applications, even though this is done routinely aboard submarines.

The following discussion covers the application and limits of available air cleaning methods. These methods are presently used in HVAC applications, or they may become appropriate for HVAC applications within the foreseeable future. You may need to combine several pollutant removal techniques in your application.

■ **HEPA Filters**

HEPA filters are passive filters that pass essentially no particulate matter. (HEPA stands for "high efficiency particulate air.") They are used routinely in applications, such as surgery and microelectronics manufacture, which require removal of all biological and inorganic matter from the air. HEPA filtration is necessary to remove particulate matter below 10 micrometers (microns) in size. Particles of this size are particularly dangerous because they are retained in the lungs. Also, HEPA filtration is necessary to remove the small particles that are most responsible for soiling and dirt streaks.

The designation "HEPA" implies a specific filtering efficiency. There are filters that provide even higher filtering efficiency ("ultra" filtration), but these are unnecessary for dealing with ordinary particles. On the other hand, some purposes of this Measure can be served by filters that are less efficient than HEPA filters, such as "medium efficiency" filters. To keep the discussion simple, we will use the term "HEPA" for filters that clean air effectively enough to allow you to reduce outside air ventilation.

HEPA filters work by an entirely different principle than low-efficiency filters. The passages in HEPA filters are much smaller than those in conventional filters, but they are still much larger than the microscopic particles that they capture. Particles are trapped because they are attracted to the filter material by the localized electric charges on their surfaces.

(Conventional low-efficiency filters work by virtue of having passages that are smaller than the air flow passages in the coils they are designed to protect. Large particles straddle the fibers of filter material and are unable to pass through, while smaller particles pass through both the filter and coil. The principle of action is purely mechanical.)

The efficiency of a HEPA filter remains essentially constant throughout its service life. Photographs taken through the microscope show that the small particles collected by the filter grow into tree-like structures on the filter material, somewhat as a crystal grows. The electric charge on the surface of the dirt particles attract other particles, so that the captured particles in effect become part of the filter medium.

HEPA filters may perform satisfactorily for several years or longer, provided that they are isolated from large particles by prefilters. The open area of the filter is vast compared to the size of the tiny particles that the filters are designed to capture, so resistance to air flow increases only very slowly.

HEPA filters have virtually no ability to remove contaminants that are in the gaseous state. The electrons of gas molecules are too closely bound to serve as "hooks" to attach the molecules to the filter. Thus, HEPA filters are not effective against such major classes of indoor air pollutants as carbon monoxide, partially burned hydrocarbons, and the increasingly notorious class of "volatile organic compounds."

The resistance of HEPA filters to air flow may total several inches of water column or more. Therefore, adding HEPA filters may double or triple the fan load. This major increase in energy consumption seriously offsets the energy saving derived from reducing the intake of outside air.

While HEPA filters themselves do not require frequent replacement, they do require prefilters. The prefilters are similar to conventional coil protection filters. They must be changed or cleaned at regular intervals.

■ **Adsorbents**

Adsorption is a process of capturing atoms, molecules, or particles on a surface. The basic mechanism is one of electrical attraction. The electron distribution on the surfaces of particles and large molecules can distort easily, forming clumps of charge that induce "image charges" of opposite polarity on the surface of the adsorbent material. Attraction between the opposite charges causes the molecule or particle to stick.

Adsorbents require a vast surface area at the molecular level, which is achieved by using materials of extreme porosity, such as carbon, alumina, and silica gel that have been processed ("activated") to make them porous. The most common adsorbent is activated carbon. This material is often called "activated charcoal" because it is commonly made by partial combustion of coconut shells and other woody material.

The effectiveness of adsorbents in trapping smaller molecules varies widely. If the pollutant molecule has ionic bonds, i.e., if it has charges bulging out, it will stick to an adsorbent surface by electrical attraction. On the other hand, if a small molecule is formed by covalent bonds, it may easily pass through the filter.

Adsorbents have the advantage that they tend to trap all molecules and particles above a certain size, which corresponds to a molecular weight greater than about 80. Thus, adsorbents trap most incompletely burned hydrocarbons, and they trap many of the molecules included in the general class of "volatile organic compounds."

Fortunately, activated carbon is effective in trapping ozone, which has become a major indoor air polluter. Ozone is generated in copiers and laser printers, and to a lesser extent in the high voltage circuits of computer monitors and in cigarette smoke. Activated carbon not only traps ozone, but induces its chemical breakdown into ordinary oxygen. Activated carbon also traps radon.

A disadvantage of adsorption is that it is reversible, except where the trapped material (notably ozone) is chemically changed. When pollutants accumulate on the surface of the adsorber, they are less tightly bound. At the molecular level, they may escape the surface because of thermal agitation or impact from other molecules. At a visible level, the pollutants may be knocked loose by vibration or impact. Therefore, you need to replace adsorbent filters periodically. The interval depends on the concentration of contaminants in the air.

In order for adsorbent filters to be effective, they must be thick enough so that the air passes through a substantial amount of the filter medium. As a result, adsorbers have high resistance to air flow. Effective activated carbon filters have a flow resistance of several inches of water column. (The thin layer of carbon used in cheap carbon filters is not fully effective and does not last long.)

Because of their porous structure, adsorbents also capture particles by mechanical trapping. Adsorbents stop all particles that are bigger than the passages in the adsorbent. For this reason, adsorbents need prefilters to keep them from becoming clogged.

Adsorbent filters are often combined with HEPA filters. This is an effective combination because the two remove different types of pollutants, and because both require supplemental fan power and prefilters.

■ **Chemical Treatment**

Air cleaning by chemical treatment converts pollutants to harmless substances, or converts them to a form that can be trapped permanently. Chemical treatment has the advantage of being irreversible, unlike adsorption. A disadvantage is that chemical methods are specific to certain pollutants or classes of pollutants.

In principle, any chemically active pollutant material can be rendered benign by attacking it with an appropriate chemical reaction. However, there are too many pollutants to tailor a chemical antidote to each one. Instead, the most common approach is to use strong oxidizers against pollutants. Oxidizers used in air cleaning include potassium permanganate, potassium hydroxide, and sodium hydroxide. This approach assumes that the oxidation products of pollutants are harmless, or that they are trapped in the filter medium.

To be practical, a chemical treatment process should be inexpensive and it should not create a large amount of resistance to air flow. Mixing air and chemicals in a water spray is one method that satisfies these criteria, but it is messy and hence it is disliked by operating personnel.

An alternative is dry chemical treatment. One of the leading dry systems uses potassium permanganate, a strong oxidizing agent, contained in beads of activated alumina. Air is forced through a bed of these beads.

The effectiveness of chemical air cleaning depends on a number of factors that should be considered in selecting the equipment. These include:

- *temperature.* All chemical reactions are dependent on temperature, so install the chemical medium in a part of the system where the air temperature is appropriate to the chemicals being used.
- *contact effectiveness.* The effectiveness of pollutant removal varies radically with the length of time that the pollutant remains in the vicinity of the chemical medium. This can be done by making the filter larger in surface area or by making it thicker. The former reduces fan power, and the latter increases it. So, favor the approach of making the filter larger. Contact effectiveness also depends on the characteristics of the chemical bed. For example, in chemical processes that use media in the form of beads, the diameter of the beads is a major factor.
- *humidity.* Some dry treatment media, such as potassium permanganate, require a certain amount of moisture in the air. Normal humidity is usually sufficient, but it is conceivable that effectiveness could suffer in dry climates.

■ Electrostatic Filters

Electrostatic filters work by ionizing contaminant particles and then using electrostatic attraction to pull them out of the air stream. The process works with any kind of particle, although some types and sizes of particles are removed more effectively than others.

An electrostatic filter operates in two stages. At the upstream end of the unit, high electric fields are created by applying a positive potential of several thousand volts to electrodes located in the air stream. The strong positive field pulls electrons off air molecules and off small particles that pass close to the electrodes, giving them a positive charge. The charged particles are called "ions." Other particles in the air stream collect these ions and become charged also. The charged particles then pass between positive and negative (grounded) plates. The positively charged particles are attracted to the negative plates, where they stick. Adhesion of the particles may be enhanced by treating the surface of the plates with oils or adhesives. This type of device is often called a "precipitator."

"Ionizers" are less expensive electrostatic devices that ionize the particulates in air, but do not collect them on metal plates. Instead, the particles are allowed to collect on any nearby surface. The charged particle induces an opposite charge in the surface, so the two stick. In some designs, the charged particles are collected on filter media downstream of the electrode wires.

Electrostatic filters can create a serious problem of their own, namely, ozone. Ozone is a molecule consisting of three oxygen atoms, in contrast with normal atmospheric oxygen, which is a molecule consisting of two oxygen atoms. Ozone is an extremely strong oxidizer because it wants to get rid of the third oxygen atom. Ozone harms biological tissues and many other materials.

An electrostatic precipitator does not generate much ozone if it is functioning properly and if it is perfectly clean. A clean unit does not create electric fields strong enough to break down atmospheric oxygen to single atoms, which is the first stage in the formation of ozone. However, as the unit operates, fine dust fibers collect on the electrodes. These act as lightning rods, creating localized electric fields that are strong enough to generate ozone.

The rate of ozone generation rises rapidly as dust collects on the electrodes. This makes it important to keep electrostatic filters clean. Similarly, an apparently minor defect, such as a nicked electrode or a loose strand of wire, can create an area of high field strength in which ozone will form. Electrostatic filters should have effective prefilters to block the dust that causes ozone formation.

Fortunately, activated carbon is effective in neutralizing ozone, so a carbon filter may be installed downstream of the precipitator. The added resistance of the carbon filter increases the fan power requirement.

Dirt also reduces the filtering efficiency of an electrostatic filter, which is another reason why these units need regular maintenance. The maintenance consists of periodic cleaning of the electrodes and collecting plates. Maintenance is a major pitfall of electrostatic units, with the result that many are found abandoned. To ease maintenance, acquire electrostatic filters that have integral, automatic cleaning systems. These typically consist of detergent spray washers. Oil may be added to the plates as part of the cleaning process. The prefilters must be cleaned or replaced periodically.

Electrostatic filters require low air velocities, typically below one meter per second (200 feet per minute), to allow particles time to be trapped. Therefore, the unit must be large in cross section. By the same token, the air flow resistance of electrostatic filters is low. The prefilters serve to distribute air uniformly through the electrostatic unit. Prefilters are typically large in area, matching the size of the electrostatic filter, so they add only a modest amount of air resistance.

Electrostatic filters need energy to operate. They typically consume 20 to 40 watts for each 1,000 CFM of rated capacity.

■ **Ultraviolet Radiation**

Ultraviolet radiation is a method of killing microorganisms. It may prove to be increasingly important in air handling systems. In principle, ultraviolet radiation alone cannot reduce the outside air requirement, because biological contamination in an air handling system is not directly related to the rate of outside air ventilation. However, if a system is contaminated, a high rate of outside air ventilation may reduce the problem. By eliminating the growth of biological contamination, an ultraviolet system may make it possible to reduce outside air ventilation to a minimum level that is determined by other factors, such as the carbon dioxide level.

Ultraviolet radiation is a form of light that lies beyond the short wavelength end of the visible spectrum. The photons of ultraviolet light have enough energy to break the weak bonds that hold biological molecules together. Therefore, ultraviolet light will eventually kill any organism that it can penetrate. For example, bacteria and fungi are vulnerable. Larger microorganisms, such as amoebas, are not susceptible to ultraviolet at the levels of exposure that present equipment can produce.

The ultraviolet used for sterilizing applications lies in the "C" band of the ultraviolet spectrum, which is far from the visible spectrum. This radiation is produced by special lamps that are similar to fluorescent lamps, but without the phosphors. The lamps require special tubes to pass the radiation, which is absorbed by ordinary lamp glass.

(For those who are interested, the primary source of "C" band radiation in ultraviolet lamps is the 254-nanometer emission line of mercury. This is also the primary source of light in fluorescent lamps, before the phosphors convert it to visible light.)

Some ultraviolet lamps can operate in standard fluorescent fixtures. If the fixture needs a cover to protect the tube from moisture, dirt, or cold, the cover material must be made of a special material that is transparent to the ultraviolet radiation. As a practical matter, the output of present ultraviolet lamps is so weak that they should be installed bare wherever possible.

Extravagant claims of effectiveness have been made by advocates of ultraviolet sterilizing, including medical researchers. Nonetheless, ultraviolet has not yet been proven in HVAC applications. Research still needs to be conducted regarding the intensities and exposure times required, the appropriate wavelengths to be used, how different organisms respond, the best methods of installation, and other questions.

There are two separate problems in killing microorganisms. One is killing organisms that enter from outside the system, such as tuberculosis bacilli. The other is killing microorganisms that grow inside the system itself, or preventing them from growing.

A major limitation in both cases is the weak radiation output of the lamps. The killing ability of an ultraviolet lamp depends on the exposure time. In an air handling system, organisms in the air are carried past the lamp rapidly. For this reason, attention is being given to

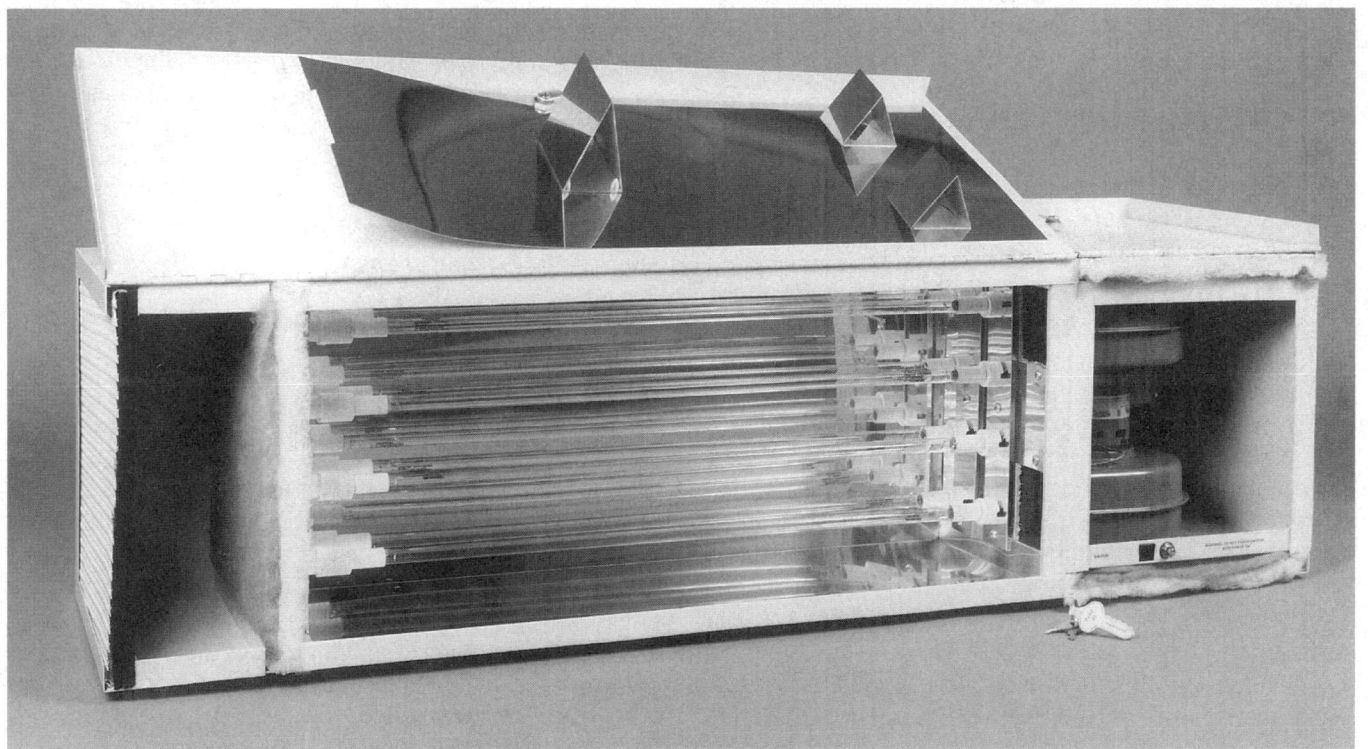

Pure Water / Clean Air Group

Fig. 1 Ultraviolet air cleaner This unit is designed to recirculate the air within a space. Note the large array of tubes that is required. A similar type of unit could be installed in areas of air handling systems where growth of microorganisms may occur.

increasing the output of the lamps. To kill transient organisms, you have to set up arrays of lamps in such a way that an organism anywhere in the air stream is exposed to the ultraviolet radiation for a sufficiently long time. Figure 1 shows a commercial ultraviolet air filtering unit that is used to recirculate air within a small space.

It is becoming increasingly clear that cooling coils, their drain pans, and related hardware, are the primary site for the growth of microorganisms within air handling systems. The organisms, which may be of many types, grow in the water that condenses from the cold surfaces of the cooling equipment. The primary solution to this problem is to locate the ultraviolet lamps where they can irradiate the wet areas continuously.

Irradiation is hampered by several factors. One is that the radiation is effectively line-of-sight, at least at the low intensities presently available. Therefore, organisms may continue to grow inside crevices of equipment. C-band ultraviolet radiation typically reflects from surfaces much less than visible light. This is especially true of rough surfaces, such as duct liner. It reflects moderately well from aluminum, and less well from stainless steel. Another limitation is that C-band ultraviolet does not penetrate deeply into water.

Ultraviolet air treatment may prove to be inexpensive, easy to install, noiseless, and undemanding of space. In contrast to other methods, it does not increase the power requirement or noise of the fans.

Do not install ultraviolet sources where people may be exposed to them. Moderate exposure to ultraviolet light causes eye inflammation. Extended exposure can cause cataracts, skin cancer, and other damage to exposed tissue. Ultraviolet radiation is energetic enough to break weak chemical bonds, so it harms all organic materials, such as wood, fabrics, and paper. The metal casing of an air handling system is immune to ultraviolet damage and it isolates the radiation completely, so it appears to be an ideal installation site.

■ Unproven Air Cleaning Methods

Various other types of air cleaning have been used, such as water baths, but none of these have gained the attention of building designers. Cleaning space air is likely to be a rapidly evolving field because of the intensity of concern about indoor air quality. When selecting an air treatment method, study the current status of developments. As in other aspects of energy conservation, be wary of any technology that is not well proven in installations similar to yours.

Trapping of Pollutants by Normal System Operation

Some air pollutants are reduced significantly just by passing through the air handling system. Contaminants such as cigarette smoke are trapped by sticking to the interior surfaces of the entire system. Highly reactive chemical pollutants, such as ozone, combine with the first thing they touch, so they tend to disappear in traveling through a long air distribution system. Large particulates settle out throughout the system. A look inside the ducts of a typical air handling system is enough to make you try to stop breathing, but this confirms that contaminants are trapped there.

Once pollutants settle out inside the air handling system, they tend to stay there. However, they are jarred loose by mechanical impact or vibration, causing a short burst of intense contamination in the space. This can be a potential health hazard.

Cooling coil surfaces that are wetted by condensation trap particulate material, such as pollen, that is too small to be removed efficiently by conventional filters. (This is unrelated to the growth of microorganisms in stagnant cooling coil condensate.)

The drying of the air that occurs during heating operation kills some microorganisms. On the other hand, alternating wetting and drying may hasten the growth of some microorganisms.

Ventilation standards make no allowance for removal of pollutants by normal system operation.

Where to Install Air Cleaning Equipment

You can install most types of air cleaners anywhere in the supply air stream that is convenient. You can install them in the air handling unit, in the ducts, or within the spaces themselves. The choice depends primarily on whether the air cleaner is supposed to serve all the spaces connected to the air handling system, or only certain areas. If you want to clean the air to all the spaces, it is usually least expensive to install a single filtration unit at the air handling unit. This approach also simplifies noise attenuation and maintenance.

On the other hand, if only a fraction of the space served by the air handling system requires special air treatment, it is usually more economical to install equipment only to serve that space. As we have seen, most types of air cleaners substantially increase fan power requirements. Therefore, try to avoid installing air cleaning equipment where it imposes an added fan power requirement on spaces that do not need air cleaning.

In most cases, the best way to compensate for the added resistance of the air cleaner is to install a supplemental fan at the unit, if the unit does not have its own fan. Otherwise, it would be necessary to increase the pressure of the air handling unit fan. This would increase duct leakage and might require heavier ductwork.

A disadvantage of integral fans is that they are closer to the occupants. The large dissipation of energy across the filter creates noise. Design the air cleaner to minimize the noise that enters the space.

Packaged in-space filtration units are available that recirculate the air within a space. All the types mentioned above, except wet processes, are available in in-space versions and even in portable versions. An advantage of localized units is that they can be turned on and off to conform to the occupancy of the space, minimizing fan energy consumption.

If a pollution source is localized, installing an air cleaner at the source is more efficient than letting the pollutants disperse before capturing them. Install the air cleaner as close to the pollution source as practical, and between the pollution source and the air return. If appropriate, install a hood or flexible duct to collect the emissions before they disperse into the space. (If the pollutant source is localized and intense, such as welding, exhausting the source to the outside is more economical than air cleaning.)

Another advantage of localized units is that you can control them individually. Control their operation automatically with a pollutant sensor in the space, as recommended by Measure 4.2.1.2, or with one of the control methods recommended by Measures 4.2.1 ff.

Increasing the Pressure of Air Handling System Fans

If you depend on the air handling system fans to overcome the resistance of the air cleaning equipment, expect to increase the pressure of the fans considerably. This will radically increase the pressure requirement of the fans. In retrofit applications, you may not be able to increase the pressure of the existing fans enough to do the job.

Ordinary coil protection filters add a resistance of 0.1 inch to 0.5 inch of water column, depending on air velocity, filter size, and the state of cleanliness of the filter. In comparison, HEPA and adsorbent filters offer much greater resistance to air flow, up to about 10 inches of water column in combination units.

In retrofits, you can usually increase fan pressure by increasing the fan speed. However, the potential increase is limited. Fan impellers have maximum rated speeds, and noise output rises rapidly with speed. Even a small increase in pressure may require a larger fan motor. As a result, you may have to replace the fan impeller or the entire fan assembly.

If a backward-inclined fan is presently installed, you can get a large increase in pressure at the same fan speed by substituting a forward-curved fan. However, the latter type is somewhat less efficient, and it is inherently noisier (for a given speed).

The air cleaning device will absorb some of the additional noise, but only in the direction of the device. Also, the noise level will increase in the space surrounding the air handling unit. Unless the air handling system was very quiet originally, you will probably have to redesign the air handling system to keep fan noise within acceptable limits.

Hazard to Occupants if Air Treatment Fails

Failure of the air treatment equipment will expose occupants to the pollutants that the equipment is supposed to remove. Consider the severity of this hazard in deciding whether to use air treatment as a substitute for higher ventilation rates.

The danger posed by an equipment failure depends on the nature of the pollutant and the rate at which its concentration builds up. With some pollutants, a brief increase in exposure is not harmful. For example, the U.S. Environmental Protection Agency specifies a long-term limit for sulfur oxides of 0.03 parts per million, but a short-term (24-hour) exposure limit of 0.14 parts per million. With other pollutants, even short-term exposure must be avoided.

In any environment where special air treatment is required, the application is serious enough to merit careful monitoring. The method of monitoring depends on the nature of the air cleaning methods. For example, mechanical filters can be monitored by sensing the pressure drop across them. On the other hand, adsorbent filters may become saturated with pollutants without exhibiting a marked rise in resistance. Monitor adsorbent filters on the basis of duration of usage, or by other means.

Where health is an issue, install alarms to indicate failure of the air cleaning equipment. In critical applications, make the alarms self-testing.

Alternatives to Air Cleaning

There are several ways to improve air quality, in addition to air cleaning. One is to minimize the generation of indoor pollutants. The most notable success of this type has been eliminating cigarette smoke from many environments. Increased attention is being given to reducing the hazard of pollutants emitted by structural materials and furnishings, especially when these are new. More needs to be done. It is becoming clear that chemicals brought into the building by people, such as cleaning materials and cosmetics, are significant pollutants. Unfortunately, there are presently no alternatives to ventilation in dealing with these pollutants.

Mechanical equipment that fosters biological hazards, such as condensate drain pans and duct insulation, can be redesigned to reduce these hazards. The HVAC industry is becoming more educated about this aspect of design, but no one has yet developed a comprehensive doctrine for the design of HVAC equipment to avoid these problems.

Isolating localized pollutant sources is another important technique. The most common way to do this is to install hoods over pollutant emitters. These should draw only enough air from the space to direct pollutants to the outside. Hoods are a good idea that can be used much more widely than they are at present. For example, they can be used with ordinary copy machines in spaces where the copy machine is the main source of pollution. Other forms of isolation, such as installing copying machines and laser printers in a separate room, are also effective. Integrate the design of hoods and isolation spaces with the overall design of the air handling system.

Another approach is to maintain conventional (i.e., high) outside air ventilation rates and recover the energy from the exhaust air. This approach is covered by Measure 4.2.8.

ECONOMICS

SAVINGS POTENTIAL: *A large fraction of the cost of conditioning outside air, in many cases. Deduct the cost of increased fan power from these savings.*

COST: *Varies widely. For example, HEPA filtration costs about $60 per thousand CFM. Oxidizing chemical filtration costs about $1,000 per thousand CFM, in large sizes, and much more in smaller sizes. Electrostatic air cleaners cost several hundred dollars per thousand CFM, the cost varying widely with size and features. Additional fan power needed to overcome the resistance of the filtering media is a major additional cost. Maintenance is a significant continuing cost with all air cleaning equipment. For example, replenishing chemical media may cost thousands of dollars per year.*

PAYBACK PERIOD: *One year to many years.*

TRAPS & TRICKS

CHOICE OF METHOD: *All air cleaning systems need unfailing maintenance. This is their biggest weakness. If you are not sure that the equipment will be maintained reliably for the life of the facility, pass up this Measure.*

DESIGN AND EQUIPMENT SELECTION: *The design of air cleaning systems is an extensive, specialized topic. Make sure that adequate talent is brought to the job. Selecting the best type of equipment is the biggest challenge. Some types of equipment, such as HEPA filters and adsorbents, are well proven. The quality of electrostatic equipment varies widely among models. The materials and maintenance costs of chemical systems are high. Any new or unusual type of equipment is suspect. Do your homework, and don't let vendors make this choice for you. Check with others who have used air cleaning systems of the types you are considering.*

INSTALLATION: *Make it easy to replace filters and filter materials, or the systems will be neglected. If potential pollutants may be dangerous in small concentrations, make sure that the installation does not allow the pollutants to leak around the cleaning units.*

EXPLAIN IT: *Establish and follow a rigid maintenance schedule. Keep the staff trained. Describe the system in the plant operating manual. Install placards on all parts of the system that require attention, stating what needs to be done.*

MONITOR PERFORMANCE: *Schedule periodic checks of equipment condition in your maintenance calendar. If the air cleaning system has observable characteristics that can indicate its state of maintenance, such as air pressure differential across filters, include appropriate columns in the plant operating log to record these conditions.*

MEASURE **4.2.5 Provide outside air economizer cycle operation of air handling units.**

SUMMARY

An important source of free cooling for buildings with high heat gains.

SELECTION SCORECARD

Savings Potential .... $ $ $ $

Rate of Return, New Facilities % % %

Rate of Return, Retrofit......... % %

Reliability .... ✓ ✓ ✓

Ease of Retrofit .... ☺ ☺

An economizer cycle is a method of providing free cooling during periods when the outside air is cooler than the indoor temperature, typically, when the outside air temperature is below about 60°F. The economizer cycle is a damper control sequence that increases outside air intake just enough to meet the cooling requirements of the air handling system. If outside air cannot provide sufficient cooling, mechanical cooling makes up the difference.

In order for an economizer cycle to work properly, the air handling system must be able to control the amount of outside air intake accurately. Also, the air intake and relief passage must be large enough to allow large volumes of air to be brought into the building. In new construction, these features can be included at modest cost.

In many existing facilities, the outside air intake capacity is ample and the necessary dampers are already installed in the system. In such cases, you can retrofit an economizer cycle without major mechanical modifications, and the additional controls are the only major expense. In other retrofit applications, you may have to make major modifications to the "front end" of the air handling system to handle the large volumes of outside air and relief air.

Energy Saving Potential

The savings potential of an economizer cycle is determined by three factors: (1) the climate at the time the air handling system is operating, (2) the net internal heat load in the spaces, and (3) the design of the air handling system.

■ The Weather Profile is a Primary Factor

An economizer cycle can save cooling energy only when the outside air temperature is less than the indoor temperature. To satisfy the entire cooling load, the outside air temperature must be as cool as the air distribution temperature that would be produced by mechanical cooling. Typically, this is about 55°F to 60°F. (This assumes that the system is capable of taking in 100% outside air. This point is discussed further below.)

Therefore, a critical issue is the number of hours each year that the ambient air is cool enough and dry enough to be used for cooling *while the facility is operating*. To calculate the saving potential with reasonable accuracy, you need temperature and humidity data grouped by time of day. It does little good for the outside air to be colder than inside while the air handling systems are shut down. For example, cold night temperatures are of little use to facilities that operate only during the day.

Most energy analysis computer programs include hourly weather data. If you want to make a manual estimate, a good source of weather data grouped by time of day is the publication *Engineering Weather Data*, available from the U.S. Government Printing Office. This covers many locations in the United States and major locations in other countries.

■ Internal Heat Gain Increases the Savings

The net heat load, i.e., the internal heat gain of a space minus the heat losses, determines the cooling requirement during cold weather. Interior zones have no envelope heat loss, and therefore have net positive heat loads under all weather conditions. Therefore, an economizer cycle can save energy in cooling interior zones any time the outside air temperature is lower than the indoor temperature.

Perimeter spaces may also have a considerable net heat gain, especially if they have large windows that are exposed to direct sunlight. For example, if windows are double- or triple-glazed, a perimeter space may have a net heat gain at an outside temperature of 40°F or 50°F. In typical mid-latitude climates, this results in many hours that an economizer cycle can save energy.

Heat gain may be increased substantially by user equipment, such as computers, photographic lighting, and sterilizers, increasing the value of an economizer cycle. On the other hand, other energy conservation measures reduce heat gains, making an economizer cycle less valuable.

■ Savings in Reheat Energy

In air handling systems that use reheat as a method of temperature control, the bulk of the cooling load that occurs during cool weather may be to chill air prior to reheating it. It may be economical to install an economizer cycle just to reduce the cost of reheat under this condition. But, a better alternative is to modify the system to eliminate or reduce reheat. In some cases, use both approaches. Refer to Subsections 4.4 through 4.8 for the methods that apply to specific types of systems.

■ Limitations with Multizone and Dual Duct Systems

Economizer cycles usually are not practical with conventional multizone or dual duct systems. In these systems, outside air is fed into both the heating and cooling coils of the system. Increasing the amount of cold outside air reduces the cooling load, but it increases the heating load. Therefore, an economizer cycle is valuable only if the cost of heating energy is always less than the cost of cooling energy.

This limitation does not apply if the dual duct or multizone system is designed so that outside air flows only into the cold air side of the system. This design feature is unconventional. It is likely to be too expensive for most retrofit applications.

Don't Use an Economizer Cycle with Polluted Outside Air

This Measure radically increases the quantity of outside air at certain times. This makes the Measure undesirable if the outside air is polluted or smelly. For example, an outside air economizer is not a wise choice near a paper mill or a fish processing plant.

Air Handling System Front End Configuration

An economizer cycle needs positive control of outside air intake. This is a function of the "front end" of the air handling system, as defined in the Section 4 Introduction. Not all air handling system designs are capable of controlling the amount of outside air intake over the wide range of flow rates that is needed by an economizer cycle. Refer to Figure 1 in Measure 4.2.1 for the different methods of controlling outside air intake. Systems B, C, and E in that Figure provide adequate control of outside air intake. There are several important issues that distinguish these three systems.

One issue is control of pressure inside the spaces. Only System C allows you to control space pressure independently of supply air flow and the outside air flow. This is important if the return air path has a significant amount of resistance to air flow. The pressure in the space with System B or E is always positive, and it is approximately equal to the flow resistance of the air return path. (See Measure 4.2.2 for details about this.) If the air return path has only minimal resistance, the space pressure will remain approximately neutral.

With an economizer cycle, the amount of air flow in the return air path changes over a wide range. Therefore, if the return path has high resistance, the pressure inside the spaces will vary with Systems B and E.

Another important issue is energy efficiency. In this regard, System E has a clear advantage because it does not require control dampers. It avoids the waste of fan power that is needed to move air past the resistance of the control dampers.

A third important issue is damper leakage, and the ability of the dampers to control flow accurately at low flow rates. Again, System E has a clear advantage in this regard because it has no control dampers. If you use System B or System C, the quality and characteristics of the air handling system dampers are important. The outside air damper should present only minimal resistance to air flow at maximum outside air intake. It should provide accurate control of outside air when the economizer cycle is not operating. And, it should close tightly when the facility is unoccupied. The return air damper should close tightly while the economizer cycle is in operation, and it should control air flow accurately at other times. These are stringent requirements that most existing air handling systems cannot meet without improvements. See Measure 4.2.2 for tips on improving damper performance.

System A cannot work with an economizer cycle because it has no means of modulating the intake of outside air. With such systems, there are two solutions: (1) convert the system to the configuration of System B, C, or E; or, (2) increase exhaust from the spaces using an exhaust fan that can handle full system flow. The second approach is rarely used. It would need a variable-output exhaust fan. It would also cause pressure variations inside the space, although these would usually be minor.

System D usually cannot work with an economizer cycle, because the relief fan limits the amount of relief air. The best way to adapt this system to economizer cycle operation is to convert it to the configuration of System C or System E. Converting to System E means installing a relief fan that is much larger, and providing it with a variable-speed drive and appropriate controls.

In summary, consider System E as your best choice, from the standpoint of energy efficiency and freedom from damper problems. However, if you need positive control of space air pressure, your only choice is System C. See Measure 4.2.2 for details of converting an air handling system to the configurations of System C or System E.

Air Intake and Relief Passages

In order for the system to fully exploit outside air cooling, the outside air intake ducting must be large enough to allow the entire system air flow to be drawn from the outside. Similarly, the relief air passages must able to exhaust the entire system air flow.

If an air handling unit does not presently have the capability of taking in a large fraction of outside air, consider enlarging the outside air intake and relief to accommodate an economizer cycle. If the air handling unit is installed next to an exterior wall, this may be relatively inexpensive. If the air handling unit is located far inside the building envelope, enlarging the ducting may be expensive. Sometimes, air flow is limited by duct features, such as a succession of sharp bends, that increase flow resistance.

Consider an economizer cycle even if the outside air intake is too small to handle the full system air flow, especially if there is a substantial amount of cold weather. Any amount of cold outside air reduces the amount of mechanical cooling that is required.

You may be able to increase outside air intake through the existing system by installing an intake fan. The fan is needed to overcome the resistance of the intake passage. It will consume additional energy. However, if it is controlled properly, it will consume much less energy than would be needed to provide a corresponding amount of mechanical cooling.

Control Sequences

Although the economizer cycle concept is simple, making it work efficiently requires thought. The major goal is to control the economizer so that mechanical cooling operates no more than necessary. The control sequence for an economizer cycle depends on the type of air handling system. These are the control sequences for different system types, in summary form:

- *single-zone systems with variable discharge temperature.* These systems typically use hydronic coils with throttling valves. The space thermostat initially opens the outside air damper instead of the cooling coil valve. After the outside air damper opens fully, the cooling coil valve starts to open to provide supplemental cooling. When the outside air temperature rises to equal the return air temperature (or space temperature), the outside air damper closes to minimum position.
- *single-zone systems with on-off or staged cooling.* These systems typically use direct-expansion refrigerant coils. The space thermostat initially opens the outside air damper instead of turning on the cooling compressor. After the outside air damper opens fully, the thermostatic control allows the compressor to operate to provide supplemental cooling. When the outside air temperature rises to equal the return air temperature (or space temperature), the outside air damper closes to minimum position.
- *single-duct reheat systems.* These systems maintain a constant cooling coil discharge temperature, which typically is 55°F. The coil discharge thermostat initially opens the outside air

damper instead of turning on the cooling compressor. If the outside air temperature is not able to maintain the coil discharge temperature, the coil discharge thermostat opens a hydronic cooling valve or starts a cooling compressor to provide supplemental cooling. When the outside air temperature rises to equal the return air temperature (or space temperature), the outside air damper closes to minimum position.

Dry-Bulb or Enthalpy Control?

The preceding control sequences are based on measuring ordinary (dry-bulb) temperature. The economizer cycle provides greater energy savings if the controls sense the enthalpy of the outside air instead of its temperature. Enthalpy controls can also be used to avoid discomfort from excessive intake of humid outside air. Enthalpy controls are explained in Measure 4.2.6, next.

Control Equipment Precautions

Many economizer cycle controls that are designed on a custom basis are installed incorrectly, or they fail to provide the full potential benefit. In addition, maintaining custom controls requires special training for plant operators.

The failure potential of economizer cycle controls is reduced by buying them as a dedicated control package, which is available commercially. Also, economizer software is available for energy management computer systems, although this is less desirable because of the reliability weaknesses of these systems. Even packaged economizer controls are vulnerable to failure from incorrect selection or installation.

Coordinate with Other "Front End" Measures

This Measure involves the same part of the air handling system as the other Measures of this Subsection. To avoid wasting money and opportunity, decide which of the Measures are appropriate for each air handling system, and accomplish all of them together.

Other Methods of Augmenting Mechanical Cooling

There are many ways of using outside air for cooling. This Measure is usually the best choice for sealed buildings of the modern type, provided that the mechanical installation is feasible and that the quality of the outside air is satisfactory.

An important class of alternatives consists of using the existing chilled water system, the condenser water system, or both for bringing outside cooling into the building. These methods are covered in Subsection 2.9. They exploit evaporative cooling to increase the cooling effect, so they can operate with somewhat warmer outside air than the outside air economizer cycle. This is a significant advantage in dry climates.

Here are some simpler methods of bringing outside air directly into spaces, or of augmenting mechanical cooling:

- *ventilate the space directly,* by installing openable windows, by using existing windows more effectively, or by installing ventilation openings and fans
- *use circulation fans inside the space.* They work well alone, and they increase the effectiveness of mechanical air conditioning.
- *in dry climates, use evaporative space coolers.*
- *in humid climates, use room dehumidifiers.* They reduce the need for mechanical cooling in humid climates. They can be used in combination with mechanical cooling to reduce overall energy consumption.

ECONOMICS

SAVINGS POTENTIAL: *An economizer cycle can eliminate the need for mechanical cooling during periods when the outside air temperature is below about 60°F. The savings depend on the cooling loads that exist during these periods.*

COST: *Varies widely. An economizer control package typically costs several thousand dollars. If modifications of the intake ducting or fans are required, the cost may be much higher.*

PAYBACK PERIOD: *Less than one year, to many years.*

TRAPS & TRICKS

CHOICE OF METHOD: *Compare this approach to all possible alternatives. In retrofit applications, consider an economizer cycle even if you do not have enough intake capacity to provide 100% of system air flow. Be careful in estimating the potential benefit, which varies widely with the type of air handling system.*

SELECTING THE CONTROLS: *You may be able to find a packaged control system for an economizer cycle, but make sure that it can control your type of system in an optimum manner. If the climate is humid, consider enthalpy control for the economizer cycle.*

EXPLAIN IT: *The system operates automatically, but air handling system controls are vulnerable to tampering, which may be prompted by comfort problems. The system will be reliable only if maintenance personnel are properly trained. Describe the system clearly in the plant operating manual.*

MONITOR PERFORMANCE: *Schedule periodic checks of the economizer cycle during periods of time when it is operating. If your plant has an operating log that covers air handling systems, include a column for the percentage of outside air. If you use enthalpy controls, schedule inspections and calibration at appropriate intervals.*

MEASURE **4.2.6 Install enthalpy control of economizer cycles.**

If humid outside air is used for cooling, the cooling coil may need to expend energy to reduce the air's humidity. The additional energy needed to dehumidify the outside air reduces the energy saving potential of an economizer cycle. For example, outside air at 65°F is useful for cooling if it is dry. But, if air at this temperature is humid, the cost of dehumidifying the air may exceed the saving in sensible cooling that it provides.

For this reason, an economizer control operating in a humid climate can be more efficient if it is controlled to sense humidity as well as temperature. (To emphasize that temperature indicates only sensible heat, the HVAC industry uses the redundant terms "sensible temperature" or "dry-bulb temperature.") The total energy content of air is the sum of its sensible heat and its latent heat. This total heat content is called "enthalpy." Economizer cycle controls that respond to both temperature and humidity are called "enthalpy controls."

Enthalpy controls are standard components. An enthalpy control consists of a temperature sensing element and a humidity sensing element calibrated so that the sum of their two signals provides an indication of the total enthalpy of the air. If an energy management computer system is being used to control economizer cycles, the computer can calculate enthalpy from the outside air temperature and humidity.

Where to Consider Enthalpy Control

If the outside air is often humid when the temperature is in the range of 40°F to 65°F, enthalpy controls can provide a significant improvement in economizer operation. For example, enthalpy controls are valuable in cities like New Orleans and London. On the other hand, humidity is almost irrelevant to economizer cycle operation in dry climates, as in the American Southwest.

Comfort Considerations

In conventional cooling design, humidity is controlled by adjusting the temperature of the cooling coil. In chilled water systems, this is done by controlling the chilled water temperature. However, economizer cycles shut down or reduce the output of the cooling coils, so they allow higher space humidity to occur. As a result, economizer cycles may be inappropriate for humid climates. Enthalpy controls partially avoid this problem, because they reduce the intake of outside air as it becomes more humid.

SUMMARY

A refinement of economizer control that improves efficiency in humid climates. Also avoids comfort problems caused by high humidity.

SELECTION SCORECARD

Savings Potential .... **$ $**

Rate of Return, New Facilities **% %**

Rate of Return, Retrofit.......... **% %**

Reliability .... ✓ ✓

Ease of Retrofit .... ☺ ☺

If you decide to use enthalpy control, be aware that comfort does not correlate closely to enthalpy. For example, most people consider a combination of 76°F and 40% relative humidity to be comfortable. Air has the same enthalpy at a combination of 60°F temperature and 100% relative humidity, but people feel clammy under those conditions. Going the other way, air has the same enthalpy at a condition of 90°F and 10% relative humidity, and this would feel arid.

Human comfort is related to temperature, humidity, air flow, interior heat radiation, clothing, the nature of the activity, and other factors. Enthalpy deals only with the first two of these. In humid climates, even using an enthalpy control to optimize efficiency may produce interior conditions that are considered too humid for comfort.

Within the narrow range of space temperatures that are considered acceptable with mechanical cooling, relative humidity is a reliable criterion for comfort. Therefore, when using an economizer cycle in humid climates, sense the relative humidity in the spaces. Use this signal as an override on the economizer cycle to limit the amount of outside air taken into the system when the outside air is humid.

Maintenance Requirements

Humidity sensing is the Achilles heel of enthalpy control. In the past, the humidity sensing elements used in HVAC controls were neither accurate nor stable. Inaccuracy in the enthalpy sensor causes the economizer cycle to operate inefficiently. An accurate dry-bulb temperature control is preferable to a miscalibrated enthalpy control. Do not use enthalpy controls unless you are reasonably sure that skilled personnel will

WESINC

Fig. 1 The trouble with enthalpy control All the control lines have been disconnected from this enthalpy control unit, disabling the economizer cycle. The staff could not understand it, so they abolished it. In this case, much more information is required than simply labeling the control.

maintain them throughout the life of the facility. Figure 1 shows what happens otherwise.

Factor skilled maintenance into the cost of using enthalpy controls. Contemporary enthalpy controls require periodic recalibration, although the best types of enthalpy sensors have much longer calibration intervals than earlier types.

Although enthalpy measurement technology is improving, the cheap unreliable varieties of sensors will probably remain in the market. When selecting enthalpy sensors, stress accuracy, reliability, and stability.

Don't expect the maintenance staff to do a good job with equipment that they do not understand. The concept of enthalpy is not widely understood. This makes it especially important for you to train maintenance personnel in calibrating enthalpy controls.

ECONOMICS

SAVINGS POTENTIAL: Depends on weather conditions. In humid climates, enthalpy controls may improve the efficiency of an economizer cycle from 10 to 30 percent. This may translate to approximately 2 to 10 percent of the total cooling load. In dry climates, enthalpy control provides little benefit.

COST: Several hundred dollars to several thousand dollars more than sensible temperature controls.

PAYBACK PERIOD: One year, to many years.

TRAPS & TRICKS

SELECTING THE EQUIPMENT: Your choice of the sensor type will determine the long-term reliability and accuracy of the enthalpy control. Investigate current offerings thoroughly. Some newer types of sensors may be able to operate for long periods without calibration. Older types are no longer acceptable. You may have to go outside the HVAC field, to "industrial controls," to get the best equipment.

INSTALLATION: Details are important. Install the enthalpy sensor near the center of the outside air intake duct, where it is not influenced by sunlight, motor heat, humidifiers, or other influences.

MAINTAIN IT: Provide maintenance training. Explain the control system in the plant operating manual. Schedule regular sensor inspections and calibration.

MEASURE 4.2.7 Install a purge cycle for overnight cooling.

A purge cycle is a ventilation control technique that applies to buildings that are unoccupied at night. The concept is to operate the air handling system fans at night to "sweep" heat out of the building, without operating the cooling coils. Typically, the system fans are operated for an hour or two just before sunrise, when the outside air is coolest. You might think of a purge cycle as an economizer cycle that operates during off-hours.

From a control standpoint, a purge cycle is fairly simple. It needs a time control, some temperature sensors, and damper control modifications.

Where to Consider a Purge Cycle

Purge cycles were widely advocated during the energy hysteria of the 1970's, when people were grasping for new ways to save energy. Like many energy conservation concepts that arose at that time, it promised greater savings than it can deliver. The main shortcoming of a purge cycle is that it works only within a narrow set conditions.

Most weather conditions do not allow a purge cycle to provide much benefit. The daytime temperature must be hot, but the outside air must cool to a temperature substantially below indoor temperature for a few hours during the night. In the United States, such conditions are typical of the Southwest, but they occur infrequently in the rest of the country.

A purge cycle is effective only with buildings that are occupied during normal daytime working hours, and are unoccupied the rest of the time. In this situation, the chillers are turned off at the end of normal occupancy, a few hours before sunset. After shutdown, the heat absorbed by the building exterior during the day soaks into the interior of the building. After sunset, the building exterior loses heat by conduction and radiation, but the interior may continue to warm up from the heat entering through the structure. Late at night, the outside air may become much cooler than the building interior. At this point, the air handling unit fans are operated to circulate outside air through the building.

This scenario describes the behavior of buildings that have a massive structure. Lightweight buildings, such as those using curtain walls, tend to cool off quickly as the outside air temperature drops, reducing the benefit of a purge cycle.

The building interior cannot be used to store much cooling for the daytime, because this would require the interior to be uncomfortably cool during the first few hours of occupancy.

SUMMARY

A control technique that may save a small amount of cooling. For a purge cycle to be useful, the building must be massive, it must be vacated at night, it must be located in a dry climate, and it must need cooling in the early morning.

SELECTION SCORECARD

Savings Potential .... $ $

Rate of Return, New Facilities % % % %

Rate of Return, Retrofit.......... % %

Reliability .... ✓ ✓ ✓

Ease of Retrofit .... ☺ ☺ ☺

The net effect of these factors is that a purge cycle provides only a limited ability to reduce cooling load. Where suitable weather conditions exist, an air purge cycle is useful mostly on the last night after an extended period of shutdown, such as Sunday night after a weekend.

Possible Humidity Problems

A purge cycle raises the humidity inside the building because the outside air is not dehumidified before it is brought into the building. This brief period of high humidity may cause problems if anything within the building is sensitive to humidity. However, condensation is unlikely to occur because the interior of the building is warmer than the air. The humidity level is reduced quickly once the cooling coils are turned on. In the dry climates that are most favorable for purge cycles, humidity is unlikely to cause problems.

If you are concerned that a purge cycle may cause humidity problems, you can avoid problems by sensing the relative humidity inside the space. Use this signal to turn off the purge cycle if the relative humidity becomes too high.

How to Control a Purge Cycle

When the purge cycle is operating, the outside air intake dampers and relief air dampers should open fully, and the recirculation dampers should close. The objective is to provide maximum exposure to the low outside air temperature.

When to turn on the fans, and for how long, depends on the outside air conditions and the inside temperature.

To minimize fan operation, the purge cycle should not start too early. Wait for the outside air to cool as much as possible. Then, the fans should operate long enough to reduce the inside temperature to the minimum acceptable daytime temperature. Purge cycle control is similar to economizer cycle control (Measure 4.2.5) in that both are controlled on the basis of the difference in temperature (and perhaps humidity) between the inside and outside.

■ Complications in VAV Systems

VAV systems with "shutoff" terminals have an additional complication. In these systems, the terminal units shut off air flow when the inside temperature falls below the normal daytime setting. You can work around this problem by modifying the thermostatic controls.

If the VAV terminals have pneumatically operated normally-open dampers, the controls can keep them open during the purge cycle by turning off control air pressure. You can do this from a central point, so the modification should be inexpensive. If the thermostats have a temperature setback feature that can be activated from a central point (e.g., by changing the pressure provided to pneumatic thermostats), you may be able to use this feature to open the terminals.

Some energy management control systems (EMCS's) have purge cycle control as a standard feature. The EMCS will open the VAV terminals as necessary. If you have an EMCS without this feature, you may be able to program it to provide a purge cycle. However, custom programming is difficult to do properly, and the cost may not be justified.

In VAV systems that use "minimum-CFM" terminals, the terminals cannot close beyond a certain fraction of maximum air flow, which is typically about 40% of maximum. The minimum-CFM position of the terminals may provide enough air flow to operate a purge cycle.

Outside Air Intake Capacity

The typical outside air ventilation capacity of most air handling systems is adequate for a purge cycle. The quantity of outside air that must be circulated in a purge cycle is relatively low because the quantity of heat being removed is much less than during maximum cooling conditions.

Coordinate with Other System Control Measures

Coordinate this Measure with other Measures involving control of air handling system dampers, including night shutdown control (Measure 4.2.1.1) and economizer cycle control (Measure 4.2.5).

ECONOMICS

SAVINGS POTENTIAL: *0.1 to 2 percent of cooling energy.*

COST: *In new construction, adding a purge cycle to the system controls costs little. Modifying existing controls may cost several thousand dollars.*

PAYBACK PERIOD: *Short, if you can use an existing EMCS. Several years or longer, otherwise.*

TRAPS & TRICKS

CHOICE OF MEASURE: *First, make sure that a purge cycle will provide a significant saving. Only a small fraction of buildings can benefit from it.*

DESIGN: *Although the concept seems simple, purge cycle control is an added complication in the "front end" controls of the air handling system. Make sure that it is done right.*

EXPLAIN IT: *This control feature is another source of confusion for maintenance personnel. Describe it clearly in the plant operating manual.*

MONITOR PERFORMANCE: *Failure of a purge cycle is easy to overlook, so schedule periodic checks. Somebody will have to get out of bed early to do this.*

MEASURE **4.2.8 Install an exhaust air heat recovery system.**

When air is exhausted from a building, it carries away a large fraction of the energy that was consumed to heat, cool, humidify, or dehumidify it. One method of reducing this energy waste is to recapture the useful energy from the exhausted air and to use this energy for conditioning the incoming air. In the most favorable applications, exhaust air heat recovery can save a large amount of energy.

This simple concept has been exploited for a long time. Still, it can be tricky to accomplish successfully. Several types of heat exchangers are used for exhaust air heat recovery. Each has its own limitations and quirks. All the types of systems are expensive, and estimating the savings potential can be subject to large uncertainties.

To avoid being tedious, we use the term "exhaust" in this Measure to mean both relief and exhaust, unless we need to make the distinction. "Relief air" is air that is vented from the air handling system to make room for incoming outside air. In the strictest sense, "exhaust air" is air that is vented directly from spaces, typically spaces that have major sources of contamination.

Consider Heat Recovery from Both Relief Air and Exhaust Air

Consider recovering energy from all the exhausts of the building, not just from the main air relief of the air handling system. A large amount of conditioned air may be dumped by toilet exhausts, hood exhausts, and other exhausts. Heat recovery from exhausts is more likely to be economical if the air streams are large, or if you can combine small streams before they enter the heat exchanger. Only one type of heat exchanger, the runaround loop, allows heat to be collected from several air streams.

Not all types of heat exchangers can be used safely with contaminated exhaust air.

Recovery of Sensible Heat

All common types of exhaust air heat exchangers are designed to recover sensible heat primarily. Sensible heat is the heat that changes the temperature of the air. Conversely, the flow of sensible heat is driven by a temperature difference. Therefore, the potential benefit of a sensible heat exchanger depends mainly on the average temperature difference between the exhaust air and the incoming air. In typical mid-latitude climates, much more heating energy can be recovered than cooling energy, because of the larger temperature differentials.

For example, a typical winter day may produce a 50°F temperature differential between exhaust air and

SUMMARY

A powerful method of using less energy to condition outside air. Expensive. Limited primarily to single-zone air handling systems and to harsher climates. Challenging to accomplish in the most efficient manner.

SELECTION SCORECARD

Savings Potential .... **$ $ $**

Rate of Return, New Facilities **% %**

Rate of Return, Retrofit .... **%** %

Reliability .... ✓ ✓

Ease of Retrofit .... ☺

incoming air. In the same location, a typical summer day may produce only a 15°F differential.

In some sensible heat exchangers, heat flows directly from the warmer air to the cooler air through the conducting material of the heat exchanger. In other types of heat exchangers, the heat from the warmer air stream flows into a heat transfer medium that moves between the warmer and cooler air streams.

Recovery of Latent Heat

Latent heat is the heat that is associated with changing water vapor to liquid, or vice versa. Latent heat is important in cooling, except in dry climates. This is because cooling systems typically provide a substantial amount of dehumidification as well as cooling. Latent heat is removed from incoming air when water vapor is condensed out of the air at the cooling coil. In typical mid-latitude climates, latent cooling (i.e., dehumidification) consumes less energy than sensible cooling on an annual basis, but it is still a large energy user. In a humid location, such as Houston or New Orleans, the cooling load during the winter months may consist mostly of dehumidification.

Water vapor is the carrier of latent heat in air. Therefore, recovery of latent heat consists of removing the water vapor from one air stream and putting into the other. The difference in moisture concentration is what drives the transfer of water vapor. Some equipment allows the moisture to migrate directly through heat transfer surfaces that are permeable to water vapor. Other equipment uses a hygroscopic material that shuttles between the incoming and outgoing air streams.

In principle, all sensible heat exchangers have the capability of dehumidifying incoming air by condensation, provided that the dew point of the incoming air is lower than the temperature of the exhaust air. For example, consider a muggy day in Houston when the outside air dew point is 85°F and the exhaust air temperature is 78°F. The surfaces of the heat exchanger, cooled by the exhaust air, would condense a certain amount of moisture out of the incoming air.

Weather data shows that this process can yield little dehumidification in most climates. Even in the most humid climates, this condensation process would salvage only a small fraction of the latent heat energy in the exhaust air stream. This is because the incoming air must be cooled to a temperature much lower than the temperature of the exhaust air. (Typically, the air is cooled to about 55°F). Thus, the cooling system must do most of the dehumidification. Furthermore, heat exchanger inefficiency reduces even the small amount of condensation that can occur.

Latent heat is not a factor in recovering energy for heating, because heating does not involve any evaporation or condensation of water vapor. A possible exception might be the recovery of humidity from humidified air, but this is not practical. If a latent heat exchanger were used for this purpose, it would load up with frost and stop the process (as well as creating other problems) whenever the outside air temperature drops below freezing temperature.

Heat Exchanger Types

The following are the air-to-air heat recovery methods that deserve your consideration at the present time. Most of these are well established. This is not to say that they are free of problems.

■ Plate Heat Exchangers

Plate heat exchangers consist of a stack of plates that are separated so that the entering and leaving air

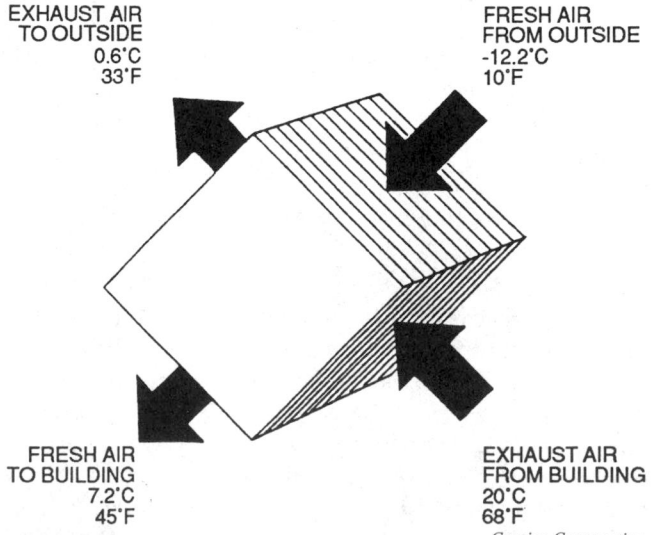

EXHAUST AIR TO OUTSIDE
0.6°C
33°F

FRESH AIR FROM OUTSIDE
-12.2°C
10°F

FRESH AIR TO BUILDING
7.2°C
45°F

EXHAUST AIR FROM BUILDING
20°C
68°F

Carrier Corporation

Fig. 1 Plate heat exchanger The core of a real unit is no more complicated than this. Accessories include washers and preheat coils.

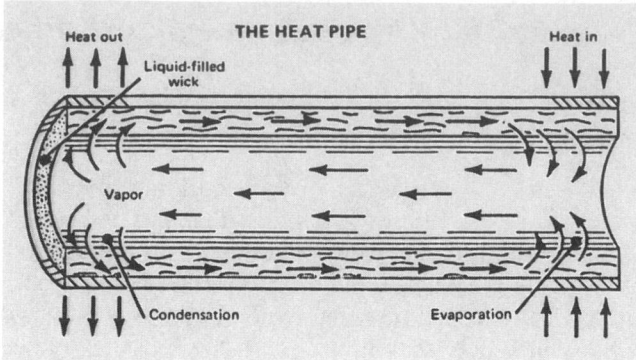

Trane Company

Fig. 2 Cross section of heat pipe The heat transfer rate is high because the vapor travels much faster than heat conducts through a solid.

streams pass between alternate plates. Heat is transferred between the air streams by conduction through the plates, which usually are metal. Figure 1 shows how they work.

Plate heat exchangers have been commercially available as packaged units for HVAC applications since the 1970's. Units are available with accessories, such as self-contained fans, washing systems, and preheaters for preventing freezing.

■ Permeable Medium Heat Exchangers

This type of heat exchanger is similar in construction to the previous one, except that a permeable material is used for the heat transfer plates that separate the intake and exhaust air streams. Molecules of water vapor are able to pass through the permeable material, driven by the difference in moisture concentration. Paper is one material that is used as a separator material.

Permeable medium heat exchangers are sold as packaged units for heat recovery in air handling systems. Originally commercialized in Japan, they have not been used extensively in the West. It seems reasonable to suspect that the permeable material is fragile, and that cleaning the material is a problem.

Because the heat exchange material is both permeable and fragile, it is not well suited to applications where it is important to prevent contamination of incoming air by the exhaust air.

■ Heat Pipe Heat Exchangers

A heat pipe is a sealed tube that contains a liquid that evaporates and condenses rapidly at the heat exchange temperatures. For example, halocarbon liquids, normally used as refrigerants, are commonly used as the fluid in heat pipes. When one end of the pipe is warmer than the other, the liquid at the warmer end evaporates and the vapor travels to the cooler end, where it condenses. Heat moves much faster in this way than it would by conduction through a solid. Figure 2 shows the interior of a typical heat pipe.

A typical heat exchanger consists of a bundle of heat pipes encased in fins and mounted in an enclosure, as shown in Figure 3. A partition in the middle of the

Trane Company

Fig. 3 Heat pipe heat exchanger One air stream flows through the near end, and the other air stream flows through the far end. The heat transfer capacity of the unit can be regulated by tilting the unit slightly, draining the liquid in the individual tubes toward one end or the other.

enclosure separates one end of the pipes from the other end. Exhaust air passes through one side of the enclosure, and entering air passes through the other side.

Heat transfer is sensitive to small angles of tilt of the tube bundle. If the cooler end is lower, the refrigerant all condenses there and heat transfer stops. If the warmer end is lower, pooling of the refrigerant at the warmer end reduces the evaporative surface, reducing heat transfer. Maximum heat transfer occurs when the warm end is slightly lower than the cold end.

Adjusting the tilt of the heat exchanger is the primary means of modulating heat transfer. Modulation is needed in exhaust air heat exchangers for reasons explained below. If the air handling system is used for both heating and cooling, the heat exchanger is mounted so that it can tilt in both directions. The range of tilt angle is typically only a few degrees.

To accommodate the changes in tilt, the heat exchanger needs flexible connections where it is connected to the intake and exhaust ductwork. These connections are prone to cracking, allowing air leakage to or from the space surrounding the heat exchanger. Careful installation can greatly extend the life of the flexible connections. Use an accordion-type material, and install enough material so that the connection is not stretched excessively.

Heat pipe heat exchangers are commercially available as packaged units for use in air handling systems. They have been sold for this purpose since the 1970's.

■ **Sensible Heat Wheels**

A heat wheel is a rotating cylinder of material that is honeycombed with air passages that run parallel to the axis. Exhaust air flows through one side of the cylinder, and incoming air flows through the other. The wheel is rotated slowly, transferring heat from one air stream to the other. The heat storage material of sensible heat wheels is usually thin sheetmetal. Figure 4 shows how the unit works.

The rotation rate of heat wheels must be matched to the velocity of the air streams. If the wheel rotates too slowly, heat is not transferred from one side to the other quickly enough. If the wheel rotates too rapidly, the warm portion of the wheel does not linger in the cool air stream long enough to lose its heat. By the same token, the rotation of the wheel can be slowed or stopped to modulate heat transfer.

A major shortcoming of heat wheels is transfer of contaminants from the exhaust air into the entering air. This occurs because each portion of the wheel alternately comes in contact with both air streams. Contaminants from the exhaust air may adhere to the heat wheel, and then be blown free into the entering air.

To minimize this problem, a purge section may be added to the heat wheel casing. This is a segment-shaped portion of the intake passage where the wheel first arrives

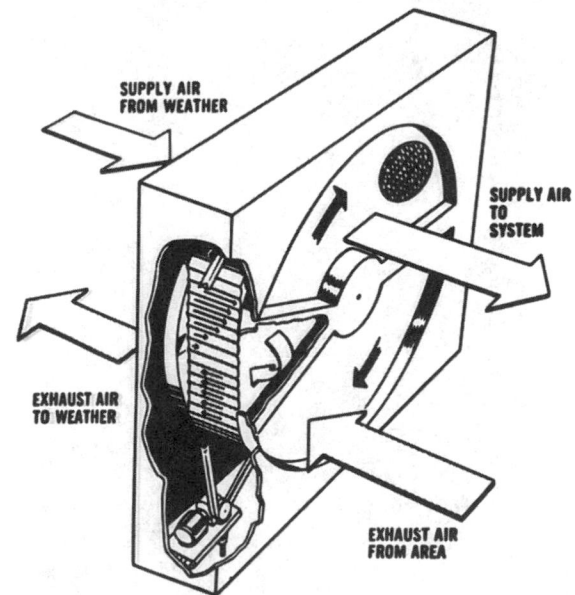

Trane Company

Fig. 4 Heat wheel One air stream heats half of the wheel. Slow rotation of the wheel moves the heated portion into the other air stream. The pie-shaped segment in front is the purge section, which reduces contamination from exhaust to intake, but it also wastes recovered heat.

after coming from the exhaust side. The passages in the casing are arranged so that the incoming air in this part of the wheel is diverted directly into the exhaust air stream. The reverse flow in this portion of the wheel dislodges contaminants. Unfortunately, the air passing through the purge section also carries away a significant portion of the heat being carried by the wheel.

Another source of cross contamination in heat wheels is the gap that exists between the wheel and the stationary casing. These gaps are closed by seals, but the seals wear out.

Heat wheels have a long history as combustion air preheaters in boilers. They were adapted for use in air handling systems before the other types of packaged heat exchangers were developed, and they experienced a great deal of popularity in the 1970's. Latent heat wheels, covered next, have always been more popular than sensible heat wheels. Sensible heat wheels have now faded away under competition from types that are simpler and less bulky.

■ Latent Heat Wheels

Latent heat wheels are similar to sensible heat wheels, except that the rotating cylinder is coated with a hygroscopic material (a material that has a strong attraction for water in the vapor state). In warm weather, moisture in the outside air is captured on the wheel and is released into the drier exhaust air. In cold weather, moisture is recovered from the exhaust air and added to the intake air.

(With a hygroscopic material, moisture transfers from one air stream to another because of the difference in moisture concentration, not because of a temperature difference. Condensation plays no role in this process.)

The problem of cross contamination between the exhaust air and the incoming air is even more severe with latent heat wheels than with sensible heat wheels. The hygroscopic material used in latent heat wheels has a rough texture, so it is more likely to catch contaminants. More seriously, it is difficult to clean the hygroscopic material because it cannot tolerate being wetted or scrubbed. As a result, latent heat wheels have the most serious cross contamination problem of any type of heat exchanger.

Latent heat wheels experienced a surge of popularity in the 1970's. Ironically, many were installed in medical and research facilities, applications where their tendency

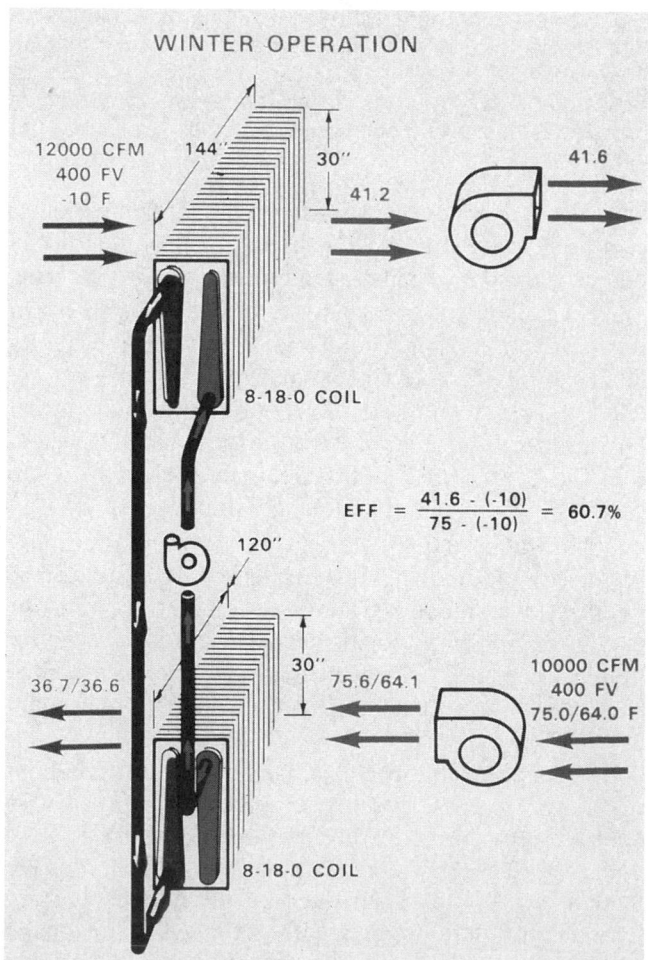

Trane Company

Fig. 5 Runaround loop Heat from one air stream is picked up in a coil, and the heat is transferred to another coil by a water loop. Heat can be transferred to and from any number of air streams in this way. There is no risk of contamination, and no fire hazard. However, the pump consumes energy, and the coils increase the fan power requirement.

toward cross contamination is most detrimental. This is a classic example of how fascination with new technology can overpower common sense.

■ Runaround Loops

A runaround loop is a system that uses water to transfer heat between air-to-water coils that are installed in the incoming and exhaust air passages. Figure 5 shows the operation of a typical system.

Runaround loops have several significant advantages over other types of heat exchangers:

• the intake and exhaust locations can be widely separated
• heat can be transferred from any number of exhaust locations to any number of intake locations. A single loop can even serve different air handling and ventilation systems.
• the intake air is completely isolated from the exhaust air, even in the event of equipment failure
• water coils are less bulky, and hence easier to install in tight spaces, than the other types of heat exchangers. However, installing the pipe may be difficult in retrofit applications.

On the negative side, the sensible heat recovery efficiency of runaround loops is somewhat lower than that of other sensible air-to-air heat recovery methods. This is because the two heat transfer steps absorb some of the temperature differential that drives the heat recovery.

Coil design is a major aspect of the design of runaround loops. The coils must provide efficient heat transfer, but they must also minimize resistance to the flow of air and water. You need to compromise between face size, fin configuration, and coil circuiting to achieve good efficiency at acceptable cost.

Modulate heat transfer in a runaround loop either by changing the pumping rate or by bypassing water flow around coils.

Freeze protection is a major factor in the design of runaround loops. A basic step is to use a heat transfer fluid having a low freezing temperature, such as glycol. Frosting of the exhaust air coil can be avoided by limiting the heat transfer.

Install the system pump(s) upstream of the coils that are installed in the cooler air stream. This recovers the heat that is added to the liquid by the pumps. If the air handling unit accomplishes both heating and cooling, the pump location will be wrong for one of the two applications. Therefore, locate the pump for the dominant usage.

Runaround loop systems are not available as standard packages. They are designed and assembled from individually selected coils, pumps, piping, and controls. Design is trickier than the simplicity of the concept may imply. A sloppily designed system may work, but it will be significantly less efficient than one that is designed by someone who takes the trouble to deal with all the issues. Certain manufacturers of HVAC equipment have developed expertise with these systems. They offer design support, including computer simulation programs.

Heat Recovery Efficiency

Table 1 gives nominal ranges of efficiencies for the common types of heat exchangers. Consider the following points when referring to this table.

■ Sensible Heat

All types of sensible heat recovery equipment claim efficiencies in the range of 50 to 70 percent, when used to recover heating. When used to recover cooling, claimed efficiencies are somewhat lower because of the lower temperature differentials. Efficiency increases with greater heat transfer surface, but heat exchanger size is a major cost factor. Therefore, efficiency is largely a matter of how much heat exchanger you can afford. For a given heat exchanger, efficiency is better in more severe climates.

■ Latent Heat

With units specifically designed to recover latent heat, calculated latent heat recovery efficiencies of 30 to 60 percent are claimed. Like sensible heat recovery, latent heat recovery is sensitive to weather conditions, being better in more humid climates. The features that allow latent heat recovery increase cost and maintenance problems. Because of these factors, many heat recovery systems do not include latent heat recovery.

■ Fan and Pump Energy Penalty

All types of air-to-air heat recovery incur a penalty of increased fan power, which is required to drive both the intake air and the exhaust air through the heat exchanger. In systems that use a liquid as a heat transfer medium, pump power is also required. If the control of the system is sloppy, or if the system is not well matched to the application, the additional fan and pump power may negate a large part of the savings from recovered heat energy.

Table 1. TYPICAL PERCENTAGES OF HEAT RECOVERY FOR EXHAUST AIR HEAT EXCHANGERS

Heat Exchanger Type	Sensible Heating	Sensible Cooling	Latent Cooling
Plate (Sensible)	50 - 70	40 - 60	minimal
Permeable Plate (Latent)	50 - 70	40 - 60	20 - 50
Heat Pipe	50 - 70	40 - 60	minimal
Sensible Heat Wheel	40*- 70	30*- 60	minimal
Latent Heat Wheel	40*- 70	30*- 60	20 - 50
Runaround Loop	40 - 60	30 - 50	minimal
Evaporative (Dry Air Cooler)	50 - 70	50 - 80	0 - 40**

* Recovery efficiency is lower in heat wheels that have purge sections.
** These figures are sensitive to climate.

■ Be Skeptical about Claims

You may have trouble finding objective data on the efficiency of specific types and models of heat recovery equipment. There are no independent organizations that rate these systems. Suspect that the efficiency claims of equipment manufacturers are optimistic, and that the claims of newer companies are based more on hope than on experience. It is prudent to purchase heat recovery equipment only from manufacturers with extensive experience. Request certified performance data.

How the Application Affects the Savings Potential

The amount of energy that you can save with exhaust air heat recovery depends on the application. The most important factors are:

• *the type of air handling system.* Single-zone air handling systems derive the greatest benefit from exhaust air heat recovery. Other types of systems are unable to exploit much of the energy saving potential of heat recovery.

In dual-duct and multizone systems, the benefit of heat recovery is reduced because one part of the incoming air is warmed and the other part is cooled. Thus, recovering heat saves heating cost but increases cooling cost. Likewise, recovering cooling energy reduces cooling cost but raises heating cost. This makes it unlikely that heat recovery would be economical in these types of systems.

In single-duct reheat and VAV systems, air is first cooled at the cooling coil regardless of the outside air temperature or the conditioning requirements of individual spaces. For example, if outside air enters at 60°F, it is first cooled to 55°F before being distributed to the spaces, where it is reheated to various temperatures. Increasing the temperature of the outside air with recovered heat would have the adverse effect of increasing the cooling load on the coil.

This behavior of reheat and VAV systems reduces the outside air temperature range in which heat recovery is useful. In the previous example, heat recovery is useful only when the outside air temperature is well below 55°F. Heat recovery cannot be economical with these types of systems unless the climate is very cold.

• *the required ventilation rate.* The savings potential is greatest in facilities that require a large amount of outside air ventilation, such as health care facilities.

• *the hours of operation of the facility.* The saving is roughly proportional to the number of hours during which outside air is used.

• *the weather profile.* Mild weather renders heat recovery uneconomical, both because conditioning energy requirements are reduced, and because the efficiency of heat recovery equipment declines.

• *heat gains.* If the facility has large heat gains, it is unable to benefit from heat recovery during the milder portions of the heating season. This is because the coolness of the incoming cold air is desirable to offset the heat gains, down to a certain temperature. In other words, heat gains reduce the outside air temperature range in which heat recovery is useful. There is no compensating improvement in the efficiency of cooling recovery.

Evaporative Cooling of Heat Exchangers

When the air handling system is providing cooling, the benefit of exhaust air heat recovery can be boosted by wetting the exhaust air side of the heat exchanger surfaces. This cools the heat exchanger surface by evaporation. During heating operation, the surfaces are not wetted.

This technique can be used with any type of heat exchanger that is not damaged by water and that confines the water to the exhaust passage. At least one manufacturer of plate heat exchangers offers a wetted version. In runaround loops, the exhaust air coil surfaces could be wetted.

A distinctive advantage of this method of evaporative cooling is that it works well even in humid climates. This is because the evaporation is caused by exhaust air, which has been dried by the action of the cooling system. Wetting the surfaces allows the heat exchanger temperature to approach the wet-bulb temperature of the exhaust air.

Furthermore, this lower temperature gives evaporative heat exchangers some ability to extract moisture from humid outside air, although not as much as the types of heat exchangers that provide direct vapor transfer between the air streams.

For example, consider air that has been cooled in the air handling unit to 55°F. By the time it becomes exhaust air at a temperature of 78°F, it has a dew point of about 55°F. (This assumes that the air is not saturated as it emerges from the cooling coil, but that it picks up some moisture in the building.) Air in this condition has a wet-bulb temperature of about 64°F. Evaporation may cool the incoming air surface of the heat exchanger to, say, 70°F. In a climate that is warm and humid, this process might accomplish as much as 50% of the dehumidification. (On the other hand, it will not accomplish any dehumidification as long as the incoming air has a dew point lower than 70°F.)

Evaporative cooling equipment needs frequent cleaning of the wetted surfaces, which accumulate the material dissolved or carried in the water. Anticipate problems with scale, dirt, and biological growth similar to those found in cooling towers and cooling coils. An automatic cleaning cycle is very desirable.

The cost of water for evaporative heat exchangers is significant. Design the equipment, and control the water flow, so that the minimum amount of water is needed to keep the exhaust passages wetted.

How to Select the Heat Exchanger

Here is a summary of the criteria to consider when selecting the general heat recovery method and the specific equipment:

- *efficiency* of heat recovery, as discussed previously
- *additional power* required for fans and auxiliary equipment, as discussed previously
- *ability to control* heat exchange. There may be times when heat exchange is not desirable. For example, during cool weather, it may be desirable to keep the incoming air at the outside temperature to offset high internal heat gains.

  In a single-duct reheat or VAV system, it is undesirable to increase the incoming air temperature as long as it is higher than the cooling coil discharge temperature. Lack of control, or sloppy control, of heat recovery in a single-duct reheat or VAV system will increase cooling energy requirements during mild weather.

  In extremely cold weather, you may need to limit heat transfer to keep ice from forming in the exhaust air passage of the heat exchanger. This is most likely to be a problem if the building is humidified. Plate heat exchangers (including permeable plate units) cannot modulate their heat exchange. They need external bypasses and dampers for this. Runaround loops and heat wheels can modulate heat exchange precisely. Heat pipe units can reduce heat transfer almost to zero.

- *need to bring the intake and exhaust air streams together* at the heat exchanger. This may require major modifications of the intake and exhaust duct systems. A runaround loop requires less sheetmetal work, but more piping.

- *ability to transfer heat between multiple air streams,* including the discharge from remote exhaust fans. Joining several air streams together at a heat exchanger is a clumsy proposition. In many cases, a runaround loop is the only practical way to recover heat from several exhaust air streams.

- *cleaning capability.* The heat transfer surfaces in all types of heat exchangers must be cleaned periodically. Some types of heat exchangers are easier to clean than others. Some manufacturers are more clever than others in providing cleaning capability for a particular type of heat exchanger. Automatic washers are desirable. They are mandatory if heat is recovered from especially dirty exhaust, such as kitchen or laundry exhaust. Cleaning is a major problem in all types of latent heat recovery equipment. Take a close look at the methods used by the equipment you are considering.

- *contaminant recirculation* between the exhaust and intake air in normal operation. Heat wheels inherently have this problem. Permeable plate heat exchangers may allow recirculation of microscopic contaminants.

- *risk of contaminant transfer in the event of failure* of a component, such as the heat transfer surfaces or seals. Heat wheels are worst in this regard, because of the continuous wear on seals. Permeable plate heat exchangers appear vulnerable to plate rupture. Only runaround loops are completely immune from the possibility of cross contamination.

- *spread of fire* from exhaust to intake. Runaround loops are immune from fire spread. Metal plate heat exchangers will heat the incoming air, but will not pass flame and smoke (unless they burn out). Heat wheels will leak smoke around the seals, especially if they burn out. Latent media will probably burn out quickly, leaving a large passage for recirculation of smoke.

- *freeze protection.* Any type of heat exchanger can form frost in the discharge air passages. Whether this will occur depends on the outside air temperature and on the humity of the exhaust air. Frosting can be prevented by reducing heat transfer, by bypassing the incoming air around the heat exchanger, or by preheating the incoming air.

- *maintenance requirements.* Some heat exchangers require considerably more maintenance effort and skill than others. Have a strong bias in favor of systems that require little maintenance, and that are easy to maintain.

- *design risk.* Consider how well proven the equipment is. Consider the amount of design support offered by the manufacturer. Favor equipment that is packaged to include auxiliary equipment, such as fans, washers, heaters, and controls.

- *installation difficulty.* This depends on the bulk of the equipment and the extent to which it is packaged.

First, Minimize Outside Air Intake

The cost and bulk of exhaust air heat recovery equipment is proportional to the volume of air that is handled. This makes it important to reduce the amount of outside air intake to the minimum that is required. Measure 4.2.1 covers ventilation requirements.

Exhaust air heat recovery is especially attractive in 100%-outside-air systems. But, consider the alternative of converting the system to recirculation, which reduces the outside air requirement. Measure 4.2.3 recommends such modifications.

Air cleaning may be used economically in some facilities to reduce outside air ventilation. This option is recommended by Measures 4.2.4 ff.

Accommodating an Outside Air Economizer Cycle

Outside air economizer cycles (see Measure 4.2.5) ingest large volumes of incoming air while they are operating. It is not economical to increase the size of an air-to-air heat exchanger to accommodate the intake requirements of the economizer cycle. Fortunately, this is not necessary, because the economizer cycle and the heat recovery system operate under different conditions. The heat recovery system operates when the weather is cold, and the conditioned space has a net loss of heat. In contrast, the economizer cycle operates during cool weather when the conditioning zone has a net heat gain. The economizer cycle never operates during warm weather.

Install a bypass around the heat exchanger to handle the large air flow during economizer operation. Either install controls to turn off the heat exchanger during economizer operation, or install an isolation damper for the heat exchanger that closes during economizer operation.

ECONOMICS

SAVINGS POTENTIAL: 10% to 50% of the energy used to condition outside air, depending on the climate, the type of air handling systems, the type of heat exchanger, and other factors. The savings are partially offset by higher fan energy consumption. Runaround loops also require additional pump energy. Systems that enhance cooling by evaporation in the exhaust air stream have a water cost.

COST: The heat exchange equipment itself costs several dollars per CFM of capacity. Connecting ducts, piping, other materials, and labor add additional cost, which may vary over a wide range.

PAYBACK PERIOD: Several years, to much longer.

TRAPS & TRICKS

ESTIMATING THE SAVINGS: Don't fool yourself. The energy saving potential is not the same as the heat exchanger efficiency. If the air handling systems have any kind of reheat, the potential saving is greatly reduced. The best applications are large, single-zone systems in extreme climates.

DESIGN AND EQUIPMENT SELECTION: The subject is more complex than it appears. Each type of heat exchanger has its own design lore. Talk with others who have years of experience with the types of equipment you are considering. Some types are well proven, while others require the owner to be a guinea pig.

MAINTAIN IT: Cleaning is a continuing requirement. It is easy to ignore, causing efficiency to decay invisibly. Schedule it in your maintenance calendar. At the same time, check the condition of the equipment, such as the integrity of the seals on heat wheels and heat pipe units.

EXPLAIN IT: Describe the system clearly in the plant operating manual. Install placards to identify the heat exchanger(s), controls, and pumps.

MONITOR PERFORMANCE: Schedule checks of the operation of the controls and systems temperatures to make sure that the system is really recovering as much energy as it should be. If you have a plant operating log, include columns to record all four air stream temperatures.

MEASURE 4.2.9 Improve the envelope penetrations of air handling systems to minimize air quality problems, wind problems, and energy requirements.

The openings in the building envelope where outside air is drawn into the air handling system and where air is exhausted are important parts of the air handling system. They deserve design attention, but they are commonly treated as mere holes in the wall. These neglected holes may cause significant problems, including:

- large variations in outside air intake
- building pressure fluctuations
- excessive preheat to prevent coil freezing
- intake of contaminated air from outside the building
- recirculation of building exhaust air back into the building.

This Measure suggests improvements to avoid these problems. In addition, by locating intakes to exploit local differences in ambient air temperature, you may be able to reduce heating and cooling costs by a modest amount. Consider all these factors together when planning modifications, because the solutions are interrelated. For example, beware of possible wind effects when relocating an air intake to avoid recirculation of exhaust air.

Design the Intakes to Avoid Pollutant Sources

A primary reason (in many cases, the only reason) for introducing outside air is to improve air quality inside the building. Therefore, it makes no sense to tolerate the intake of pollutants through the outside air. ASHRAE Standard 62-89, the primary ventilation standard in the United States, requires that "makeup air inlets and exhaust air outlets shall be located to avoid contamination of the makeup air." In some cases, problems arise from the location of the intakes, as implied by the Standard. However, problems may also be caused by the design of the intake and by conditions outside the building. The following are the most common sources of outdoor air pollution, and the ways to deal with them.

With respect to all the following pollutants, note that the location of the intake openings has a strong effect on the amount of pollutants that enter the air handling system. In addition, the velocity of the air intake has a strong influence on the tendency of the intake to ingest solid and particulate matter. High intake velocity creates a vacuum cleaner effect that can lift solid material. You can reduce intake velocity simply by increasing the free area of the intake.

SUMMARY

Corrects common installation problems that may waste a lot of energy, and cause health and comfort problems.

SELECTION SCORECARD

Savings Potential	$ $
Rate of Return, New Facilities	% % %
Rate of Return, Retrofit	%
Reliability	✓ ✓ ✓ ✓
Ease of Retrofit	☺ ☺ ☺

■ Vehicle Exhaust

Vehicle engine exhaust is the most common serious pollutant that is brought into a building along with outside air. Vehicle exhaust fumes are heavy, so an outside air intake that is located at street level or in a pit will ingest exhaust from nearby vehicles. Figure 1 shows a typical vulnerable location.

The problem of vehicle exhaust becomes much worse if the air intake is located in an area where vehicle exhaust is confined, as at a loading dock or garage. (Yes, there really are designers careless enough to locate outside air intakes in such areas.) For example, a diesel truck may idle for hours at a loading dock, poisoning the air throughout the area served by the air handling unit.

The only practical solution to this problem is to locate the outside air intake as high as possible. Vehicle exhaust is substantially warmer than the outside air at first, so it may tend to rise. For this reason, do not install the intake directly above vehicle traffic or loading areas. From the standpoint of minimizing pollution from ground-level sources, the roof is the best location for the air intake.

■ Vegetation and Botanical Chemicals

Decaying vegetation is a rich growth medium for molds and fungi. These are potential health hazards if they enter the air handling system in any concentration. This may occur if air intakes are located at or below grade level, especially if they are surrounded by soil or nearby vegetation. If trees are growing nearby, decaying leaves can be a problem. Figure 2 shows an example.

Consider these solutions:

- *relocate the intake* to a position where there is no rubbish to be ingested. This is generally to a position well above ground.

- *increase the surface area* of the screen. If necessary, install an extension of the intake opening that has a larger screen. This reduces the air velocity through the screen, along with the vacuum cleaner effect. This does not mean that you should increase the damper size. Dampers that are too large cause leakage and control problems.

- *mount the intake screen horizontally,* with air flow upward, so that gravity will keep the screen clear of material, such as leaves, that is large enough to block the screen.

A related problem is the use of herbicides, insecticides, fertilizers, and other chemicals that are applied to lawns, shrubbery, and other vegetation. Many such chemicals are volatile or dusty, and they are drawn into the facility, where they may cause health problems. One expedient is to ban the use of chemicals in the vicinity of air intakes, but this solution is unreliable in the long term.

■ **Animal Debris**

Serious health problems can be caused by microorganisms that grow in animal debris. For example, histoplasmosis is a serious fungal infection that originates in bird feces. Other debris includes feathers, nest material, and dead bodies. Birds can be a problem at any elevation. The solution is to install screens at the outermost intake point, blocking any sites where birds could perch between the screens and decorative louvers or grilles.

Pick the most effective place to install the screens. For example, one hospital had an attic containing all the air handling units for the surgical suites. All intake air for the spaces was drawn from the attic. The attic and everything in it was covered to a depth of several centimeters with pigeon excrement. The attic was cleaned out, but the problem immediately started to recur. It was not solved until all the entrances to the attic were effectively blocked.

■ **Evaporative Heat Rejection Equipment**

All equipment in which water is stagnant or recirculated is a potential breeding ground for microorganisms. Some of these organisms are lethal,

WESINC

Fig. 1 Outside air intakes below grade   Vehicle exhaust fumes from the adjacent street readily enter here, along with grass clippings, lawn chemicals, etc. Note the leaves accumulated behind the louver at right. On the positive side, easy access to the pit makes it easy to clean out. The large area keeps the entering air velocity low, which minimizes the tendency to pick up debris. Still, it would be better to install these intakes on the roof.

as in the case of Legionnaires' Disease. Others cause serious transient illness, such as Pontiac fever, and others cause long-term health problems, such as "hypersensitivity pneumonitis." Cooling towers, evaporative condensers, and evaporative space coolers convert contaminated water into a mist that may be breathed by occupants. (After the mist evaporates, most microorganisms that were carried by it die quickly.)

Mist is not the only potential danger. If the outside air intake ducting is in a vulnerable location, drippage from evaporative equipment may leak into the outside air intakes through leaks in the ductwork. This process was suspected of causing the original Legionnaires' Disease outbreak.

The only satisfactory protection against contamination by evaporative equipment is physical isolation. Install the equipment where cooling tower spray or water drippage cannot come in contact with people, either directly, through the air handling system, or by other paths.

Conventional water treatment chemicals kill most of the organisms found in evaporative equipment. However, do not depend on this. Expect that water treatment will be neglected at times.

■ **Radon**

Radon is a radioactive heavy element that exists in the form of an inert gas. Radon is a decay product of uranium, so it originates in the ground. The amount of uranium, and hence radon, varies radically from one geographic location to another. The concentration is very dangerous in some locations, and minimal in others. An expensive test can quickly measure the concentration of radon that is present in a building.

The primary danger of radon comes not from the gas itself, but from its radioactive decay products, which are radioactive isotopes of heavy metals. These metal atoms attach to small particles in the air that become lodged in the lungs. These radioactive materials decay further by emission of alpha particles, which are especially potent in creating cancerous changes in cells.

Excessive intake of radon is the result of certain construction features, such as an unvented area in contact with the soil.

Radon is also released from well water that originates in uranium-bearing soil. For example, showers act to release radon from the water, exposing bathers to radon. Cooling towers act in the same way. An outside air intake exposed to the discharge from a cooling tower would take in the radon.

■ **Emissions from Outlying Sites**

If the source of contamination is distant from the facility, or if it is large in size, such as a landfill, the contaminated air will envelope the facility. In such cases, local changes to the outside air intakes cannot solve the problem.

WESINC

Fig. 2 Air intakes for a public library The intakes are below grade, surrounded by small bays that accumulate debris. The bay at right is located behind bushes, where leaves, soil, and other matter fall directly into it.

Shield Intakes and Exhausts from Wind

Wind can cause significant changes in outside air intake and building pressurization. The kinetic energy of wind that strikes the upwind side of a building is partially converted into static pressure. If there is an opening in the building envelope, such as an air intake or relief, this increased static pressure is transmitted throughout the air handling system. Wind pressure can radically change the percentage of outside air that enters the air handling system.

The effect of wind on an outside air intake is minimized if the intake and relief dampers face in the same direction, because the wind pressure is balanced in the two. However, this same configuration maximizes building pressure effects caused by wind.

The positive pressure created by wind on the upwind side of the building is stronger than the negative pressure created on the downwind side. Buildings are poor airfoils, so the velocity of wind on the downwind side is mostly converted to turbulence rather than to negative static pressure.

Try to install air intakes, reliefs, and exhausts in shielded locations where the effects of wind pressure are averaged out. A shielded enclosure located on the roof approximates neutral pressure.

If you have to install an intake or discharge on the side of a building, reduce the effect of wind by installing a deflecting hood. A simple hood that faces downward or upward may not do the job. When air strikes a wall, it must turn and flow along the surface. The direction of the flow may be up, down, or sideways, depending on the wind direction and the location on the wall. Therefore, if you have to install an intake or exhaust in a wall, try to install a hood that shields in all directions.

The hoods used by restaurant exhaust fans are a good example. These have the shape of large mushrooms. They are not an elegant addition to the decor. If appearance matters, you can hide them with a porous decorative screen of some kind.

Design to Avoid Recirculation of Building Exhaust

A facility's own exhaust air is likely to be recirculated to some extent, perhaps from one air handling system to another. If the building exhaust does not contain dangerous contaminants, the small amount of recirculation that occurs in a typical facility is harmless.

The amount of recirculation depends on the air discharge and intake velocities, on the relative locations of the grilles, and on wind. The amount of recirculation may vary radically from moment to moment. Certain conditions can greatly increase recirculation. It is common practice to run both the intake and relief ducts to the closest point in the building envelope, often right next to each other. This sloppy practice allows exhaust air to short-circuit from the relief air grille to the adjacent air intake.

Approach a recirculation problem by studying the geometry of the situation. The best approach is to move either the outside air intake or the exhaust to a location where wind or convection cannot cause recirculation. In retrofit, this is likely to be expensive.

A less effective, but cheaper, approach is to install deflectors over the intake and relief that face in different directions. For example, if both are at the same level, the relief might have a deflector that directs the exhaust upward, while the intake is equipped with a hood that draws air from below. Design upward-facing deflectors to shed the rain water and debris that they will collect. For the reasons stated previously, deflectors that point in a particular direction are undesirable if wind can blast against the wall where the deflectors are installed.

If the exhaust air may be dangerous, a common solution is to discharge the exhaust from the roof with a high-velocity stack. Figure 3 shows examples. This requires exhaust fans of higher power to provide the necessary velocity. This solution is satisfactory only if the dangerous contaminants are decomposed or diluted sufficiently before they get back to earth. Wind and convection may defeat the protective action of high velocity exhaust.

For example, in one tall laboratory building where lethal microorganisms may have been discharged from the roof exhaust, it was found that the exhaust was being recirculated into the air intakes of the same building. The intakes were located below ground level. In summer, the cool exhaust air was heavier than the ambient air, and it fell to the ground on days when there was little wind.

If dangerous materials may be exhausted from the building, exhaust alone may not be the answer. Safety may require sterilizing the exhaust before releasing it into the environment. If you have a potentially dangerous situation, get an expert to help you solve it.

Exploit Local Differences in Ambient Air Temperature

The temperature of the air that surrounds a building is not uniform, especially during the daytime. For example, the air next to a wall exposed to sunlight may be 10°F warmer than the air next to a shaded wall. In a cool climate, it would be desirable to capture this thin layer of heated air. Conversely, in a warm climate, it would be desirable to avoid taking in this warmer air.

If you want to exploit this opportunity, understand how the air is heated and cooled. For all practical purposes, air itself does not gain or lose heat by radiation. Air cannot be warmed directly by the sun, and it cannot lose heat directly into the sky.

WESINC

Fig. 3 High-velocity exhaust stacks These exhaust stacks on a laboratory building reduce recirculation into the building. This method cannot completely eliminate recirculation. If the exhausted air is cooler than the surrounding air, it will fall back to the surface. Under such conditions, wind is needed to carry away the exhaust.

Instead, the differences in air temperature result from contact with the building or its surroundings. The solid surfaces of the building absorb solar radiation, which warms the surfaces, and the surfaces warm the adjacent air by conduction. Shaded portions of the buildings and its surroundings lose heat by radiation, become cooler, and cool the adjacent air by conduction.

The warmed or cooled air exists in a thin layer that is difficult to capture. For example, even though a black roof can become very hot, the heated air layer next to the roof is thin, and it escapes by convention and is blown away by wind.

To maximize useful air heating, you could cover the roof with glazing to channel the air on the way to the air intake. This exploits the greenhouse principle. Doing the same to a sunlighted wall creates a so-called Trombe wall. You could pass the air through a glazed box, which is called a solar air heater. All these methods are expensive because they require a lot of expensive glazing material. (This is one of the fundamental problems of solar energy. See Reference Note 23, Non-Fossil Energy Sources, for more about solar collectors.)

Specialized circumstances are needed to exploit the cooling of air by shaded surfaces. Air that is cooled by conduction becomes heavier and falls off the surface. The cooled air will not accumulate enough to be useful unless it is confined to an area where it can be collected, such as a shaded courtyard.

Exploiting natural cooling is limited by the volume of cooled air available. Typical ventilation guidelines call for the air in a building to be completely replaced one to three times per hour. Most facilities do not have an outside reservoir of cool air that is large enough to replenish the building air throughout the day.

In climates where both heating and cooling are significant loads, you would need one outside air intake for cooling and another for heating, along with foolproof controls to ensure that the proper intake is selected. With dual duct and multizone air handling units, an intake in a warm location might be used for the hot deck and an intake in a cool location for the cold deck.

In summary, if you have flexibility in locating your air intakes, try to find locations that exploit localized temperature variations. But, don't expect to save a lot of energy this way unless exceptional circumstances exist.

ECONOMICS

SAVINGS POTENTIAL: Varies widely. In the worst cases, wind effects may increase outside air intake severalfold, wasting the energy needed to condition this air. Effective shielding of intakes and exhausts can eliminate most of this waste, but not all of it.

Bad intake design may cause air quality problems. This can waste energy by motivating the staff to increase the amount of outside air in response to the bad air quality.

Conversely, ingenious methods of capturing air heat at the intake may save as much as 10% in outside air heating energy.

COST: In new construction, optimizing the design of envelope penetrations may not cost much. The cost of modifying existing penetrations varies widely.

PAYBACK PERIOD: Several months, to many years.

TRAPS & TRICKS

TURN PROBLEMS INTO OPPORTUNITIES: Look at your mechanical system drawings and catalog all the places where air enters and leaves the facility. Then, walk around and check them all for potential problems, such as intakes close to exhausts, openings facing into a prevailing wind, a truck dock adjacent to an air intake, heavy debris on screens, etc. Be aware of complaints about indoor air quality, but don't instigate them. Problems can often be fixed in a variety of ways. Look for ways to draw heating air off a hot roof, cooling air from a shaded courtyard, etc. Don't create eyesores.

MEASURE 4.2.10 Minimize the use of extra heat for freeze protection.

Air handling systems are vulnerable to freeze damage if they contain water (hydronic) or steam coils. Water coils used for either cooling or heating are vulnerable. When the system fans are running, the main hazard is entry of cold air through the outside air intake. When the fans are not running, cold air may be able to leak into the system through any of the entry or exit points.

When cold weather comes, facilities may waste energy in the methods they use to prevent coil freezing. The most wasteful practice is keeping the air handling system in operation just to prevent freezing. The freeze protection features of the air handling system may also waste energy throughout the rest of the year.

This Measure explains the methods that you can use to protect the system from freezing without wasting energy. These include changes to design, improvement in installation details, and better operating practices. Select the method, or combination of methods, that is most effective for your systems.

How to Protect Water (Hydronic) Coils from Freezing

■ Use Antifreeze

The most reliable method of protecting hydronic coils from freezing is using an antifreeze solution in the hydronic system. The only failure mode is neglecting to replace antifreeze that is lost through leakage or maintenance.

Antifreeze is relatively inexpensive. If it is removed from the system for maintenance, it can be reused.

Antifreeze solutions have two efficiency disadvantages. Both of them result from characteristics of the liquid that are shown in Figure 1. One characteristic is that ethylene glycol has less effective heat transfer than pure water. As a result, all the coils in the system must operate with higher temperature differentials. This effect is insignificant with heating systems. With cooling systems, it forces the chiller to operate at a somewhat lower evaporator temperature. This lowers the chiller efficiency. (Measure 2.2.1 explains why.)

The reduced heat transfer also reduces the heating or cooling capacity of the system.

Another efficiency disadvantage is that antifreeze solutions require more pump power to deliver a given amount of heat. This is primarily because glycol has a lower specific heat than water.

SUMMARY

The methods that you use to protect air handling systems from freezing can have a major effect on energy consumption, even during warm weather.

SELECTION SCORECARD

Savings Potential .... $ $ $

Rate of Return, New Facilities % % % %

Rate of Return, Retrofit .... % % % %

Reliability .... ✓ ✓ ✓

Ease of Retrofit .... ☺ ☺ ☺

In new construction, design the system to account for these factors. If an existing system was not designed for use with antifreeze, it is still likely to have enough reserve capacity to handle it.

The common antifreeze solution for hydronic systems is ethylene glycol, the same antifreeze used in automobiles. (The ethylene glycol used in HVAC systems has different additives than the ethylene glycol used in automobiles. Don't substitute these products for each other.) In unusual applications where the toxicity or environmental effects of ethylene glycol may be a problem, use propylene glycol. It is non-toxic. In fact, it is used as a food additive.

Antifreeze can be added to most water systems, including chilled water and heating water systems. An exception is systems where the water is exposed for evaporative cooling, such as "waterside economizer" systems that use an open cooling tower.

■ Drain the Coils

If the cooling coil is hydronic, the coil can be protected by draining it during periods of the year when freezing is a possibility. In most applications, it is practical to do this only on a seasonal basis. The main failure mode with this method is forgetting to drain the coil.

This method is desirable for applications where mechanical cooling is never needed during the period of the year when freezing may occur. Cooling can still be provided by an outside air economizer cycle (see Measure 4.2.5) even if the cooling coil is drained.

If the air handling system has water coils that are used for heating, other methods are needed to protect those coils.

How to Protect Steam Coils from Freezing

Protecting steam coils from freezing is a bigger challenge. If the flow of steam to a steam coil is throttled, condensation of the steam causes the pressure inside the coil to fall below atmospheric. The pressure of the atmosphere outside the coil then keeps condensate from draining. As a result, water may accumulate in the coil, the condensate piping, and steam traps. The condensate may freeze in parts of the equipment and rupture it, even though the coil is operating.

The most reliable way to protect the steam coil itself from freezing in extreme cold conditions is to eliminate the inlet valve, so the entire coil remains at full pressure and temperature. Heat output from the coil usually is controlled by face-and-bypass dampers.

This arrangement is susceptible to energy waste from air leakage because the dampers are large and the wide-open steam coil is very hot. Design the system controls to shut down the coil and drain it whenever it is not needed. Make sure that the dampers cannot leak enough hot air to exceed the heating requirements of the system under any operating condition. If the dampers leak excessive heat into the system, some form of cooling (either mechanical cooling or free cooling) is needed to cancel the excess heat.

If a throttling valve is used with a steam coil, freeze protection can be accomplished by installing vacuum breakers on the coil and on the steam traps to allow condensate to drain, as shown in Figure 2. This method is less reliable. Throttled steam coils must operate in cycles to allow condensate to drain. As a result, the temperature of the air through the coil fluctuates continually. For this reason, throttling a steam coil is inadvisable if the system cannot tolerate large temperature fluctuations.

You have to protect the condensate piping using different methods. The first step is to install the condensate piping so that all of it drains easily. In colder climates, keep the condensate lines warmed. Insulation may be sufficient for this purpose. However, insulation may not be completely reliable, because a low rate of condensate drainage may allow the condensate pipe to cool enough to freeze. If the incoming air is especially cold, or if the condensate pipe is long, install heat tracing on the condensate pipe.

Select steam traps with an eye toward their resistance to freeze damage. Unfortunately, all efficient types of traps are vulnerable to freezing to some extent, and other selection factors may conflict with freeze protection. See Measure 1.10.3 for the characteristics of steam traps.

How to Use Preheat Coils for Freeze Protection

Many air handling systems have a preheat coil installed in the outside air intake. Typically, the preheat coil is used only to bring the outside air up to a modest temperature, while the main heating coil in the air handling unit warms the return air and finishes the heating of the outside air in response to heating requirements. The usual reason for installing a preheat coil is to prevent freezing of the main air handling unit coils (both heating and cooling coils). Using a preheat coil for freeze protection provides a number of advantages:

- the preheat coil is sized only for the outside air flow, rather than the total system flow. Therefore, a preheat coil can be smaller and less expensive than a coil installed inside the air handling unit.

- the preheat coil offers resistance only to intake air, which reduces the fan power required to draw air through the coil

- system design is simplified by eliminating large variations in temperature at the main coils, and by eliminating the need to include freeze protection measures in the main coils

- by bringing the temperature of the outside air closer to the temperature of the return air, preheating minimizes the adverse effects of stratification. The preheat coil itself may reduce stratification by acting as a baffle to disperse the entering air stream. (Stratification is covered in Measure 4.2.11, next.)

Preheat coils are specially designed to resist freezing. They typically have short steam or hot water circuits consisting of straight tubes or U-bends, rather than long serpentine coils. This reduces the chance of overcooling the outlet ends of the circuits. With steam coils, the short tubes allow better condensate drainage. Steam or hot water is admitted to the air inlet side of the coil face, so that the air is warmed before it comes into contact with the outlet end of the coil. This is called "parallel flow," which is less efficient from a heat transfer standpoint than the "counterflow" pattern that is used

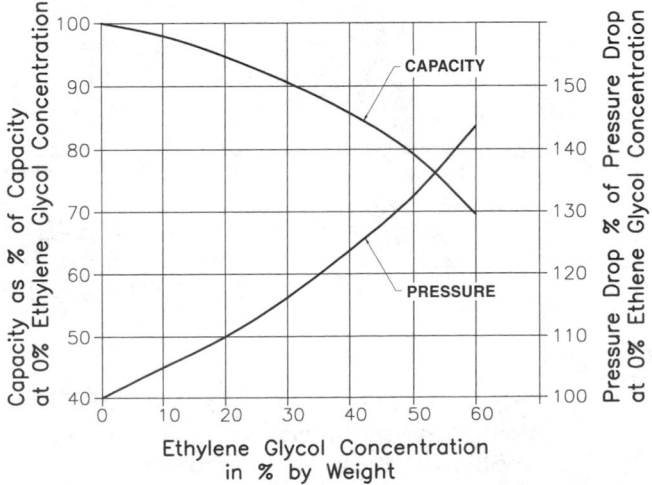

Blender Products, Inc.

Fig. 1 Adverse efficiency effects of antifreeze Antifreeze is a less effective heat transfer medium than water, so it increases system temperature differentials. Also, it requires more energy to pump.

in conditioning coils. Steam preheat coils may have steam distributor tubes installed inside the heating tubes to ensure uniform temperature across the entire coil surface.

Adding a preheat coil on a retrofit basis is too expensive, unless the system has a major freeze protection problem that cannot be solved by the methods given previously.

■ Optimize the Preheat Temperature

In systems where a preheat coil is installed, energy may be wasted if the preheat temperature is set too high. This possibility occurs when the coolness of the outside air is useful for cooling the building, whether or not the cooling coil is operating. In other words, you are trying to keep the preheat coil from interfering with the operation of an outside air economizer cycle. (See Measure 4.2.5 for the details of economizer cycles.)

In this case, energy waste may occur even if the preheat temperature is much lower than the supply air temperature. The reason is that a relatively small amount of outside air may be mixed with a relatively large amount of return air. The return air is warmer than the supply air, so the outside air would have to be much colder than the supply air to make the mixed air cooler than the supply air.

In single-zone systems and terminal reheat systems, you can avoid or minimize this energy waste by setting the preheat temperature as low as possible. See Figure 3. If possible, set the preheat coil thermostat to keep the entering air temperature just above freezing during normal system operation.

In dual-duct and multizone systems, follow the same practice of keeping the preheat temperature as low as possible. There is an exception if the preheat coil uses a cheaper fuel than the heating coil in the air handling unit. In that case, you have to calculate the preheat coil discharge temperature that yields the lowest overall operating cost. The answer depends on the magnitudes of the heating and cooling loads of the spaces served by the system.

The preheat coil may have to operate when the air handling unit is turned off in order to provide makeup air for exhaust fans. During that mode of operation, the controls must allow the preheat coils to produce an air temperature that is suitable for keeping the spaces warm.

■ Using Unit Heaters as Preheaters

Unit heaters are sometimes installed in the space between the outside air dampers and the cooling coil to provide freeze protection. In order for this method to work reliably, the unit heaters must be able to circulate warmed air throughout the space in front of the coil. Otherwise, a portion of the coil may freeze. If the fans are operating, the strong flow of air into the system may not allow adequate mixing to occur.

Unit heaters are fairly reliable as a method of preventing coil freezing when the fans are not running. If the dampers are leaky, the unit heaters will consume a lot of energy in attempting to heat the outdoors. A preheat coil that is installed intelligently will waste less energy in this way.

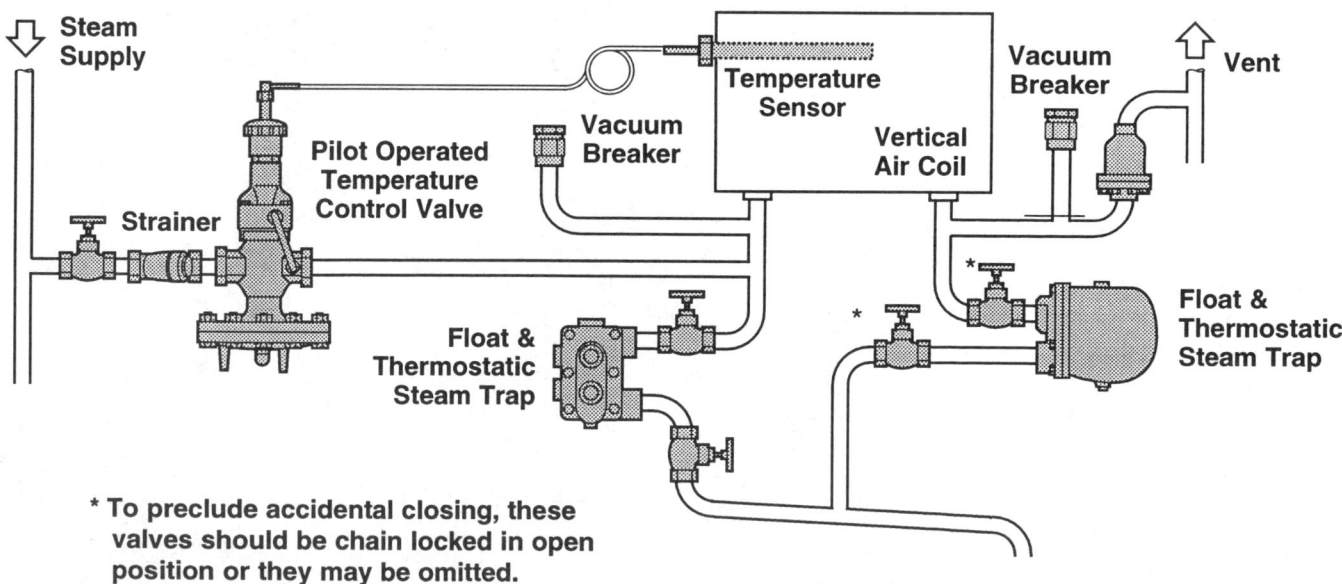

* To preclude accidental closing, these valves should be chain locked in open position or they may be omitted.

Spirax Sarco, Inc.

Fig. 2 Thermostatically controlled steam preheat coil In a throttled steam coil, the pressure may fall below atmospheric. This requires vacuum breakers. Note that a steam trap is needed here on the inlet side as well as the discharge side.

Don't Depend on Exhaust Air Heat Recovery for Freeze Protection

You may see exhaust air heat recovery promoted as a method of freeze protection. The idea is that heat removed from the exhaust air stream warms the incoming outside air. (See Measure 4.2.8 for details.) Unless the outside air is extremely cold, the heat recovered in this way is sufficient to prevent coil freezing. However, the heat recovery system provides freeze protection only as long as the fans are operating. Therefore, this benefit is useful only in applications that operate continuously, such as hospital wards and hotel lobbies.

Furthermore, the heat recovery system can fail, or it can be turned off. Therefore, if you plan to exploit the system for freeze protection, stress reliability in selecting it.

Efficient Freeze Protection When the System is Shut Down

When the main fans of the air handling system are running, the freeze hazard is limited to entry of cold air through the outside air intake. When the fans are not running, cold air may be able to leak into the system through any of the entry or exit points of the system. Every entry and exit point in the system, including exhaust fans throughout the building, should have dampers that close tightly.

Tightly closing dampers are an important energy conservation feature, but do not rely on dampers as a primary method of freeze protection. All dampers leak, and the leakage gets worse with time. Also, dampers are vulnerable to defects that keep them from closing fully. The best defense is to make the coils themselves freeze-proof.

If the air handling unit has a heating coil, keeping the coil warm protects all the coils on the downstream side of the heating coil. If there are exhaust fans that continue to operate, the heating coil warms the air entering the building.

One weakness of this method is that it does not protect the interior of the building from infiltration that occurs through the return portion of the air handling system. With the main fans turned off, the return duct is another avenue for air to enter the building. This makes it important to have exhaust air dampers and recirculation dampers that close tightly when the main fans turn off.

Keeping the heating coil active is not entirely reliable as a method of freeze protection. It fails if the source of steam or hot water fails, or if the coil itself is turned off. Also, in some configurations, enough cold air may bypass the heating coil to freeze the cooling coil. The heating coil itself may freeze at the discharge end if it has no antifreeze, because of the low flow rate through the coil.

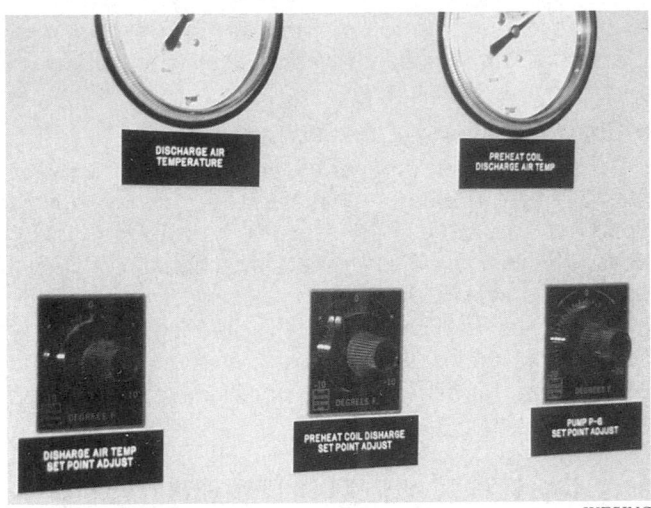

WESINC

Fig. 3 Air handling system temperature controls These settings can have a major effect on system energy consumption. In single-zone and reheat systems, do not raise the preheat coil discharge temperature higher than needed to avoid freezing and stratification.

■ Damper Improvements

See Measure 4.2.2 for the characteristics of good dampers. As suggested there, consider a combination of dampers. In addition to normal control dampers, which have a certain amount of leakage, another set of tightly closing dampers is used when the system is shut down. This arrangement is unusual and it involves extra cost. In new construction, it may pay for itself quickly. It may be difficult to retrofit.

■ Wind Protection

Wind is the main force that drives leakage through outside air dampers when the fans are turned off. As an adjunct to effective dampers, install wind deflecting hoods over the envelope penetrations where air enters and leaves the air handling system. Protection against wind is covered by Measure 4.2.9.

If the system is vulnerable to freezing, wind deflectors are important even if high winds occur only rarely. Wind pressure is proportional to the square of wind speed. A high wind velocity that occurs only once in several years may be able to force enough cold air past closed dampers to freeze vulnerable coils.

■ Effect of Duct Configuration on Air Leakage

The fact that a coil is installed far from the building envelope may provide less protection than your intuition may suggest. The duct connecting the coil to the outside provides virtually no resistance to air flow. Good dampers are the only effective barrier to air flow.

A difference in height between the coil and the envelope penetration may have some effect on air entry. If the penetration is above the coil, convection tends to force warm interior air through the damper to the outside. If the penetration is below the coil, convective force tends to oppose the flow of outside air into the system. Convective force is much weaker than wind, unless an air passage is very tall.

■ Control of Space Exhaust Fans

The outside air damper cannot be closed if exhaust fans are kept running anywhere within the portion of the building served by the air handling system. Whenever exhaust fans operate, the heating coils must operate to warm the outside air that replaces the exhausted air. Exhaust fans, such as toilet exhausts, may be operated unnecessarily during off hours. Refer to Measure 4.2.1.1 for methods of minimizing the operation of exhaust fans.

Avoid Unnecessary Energy Use for Freeze Protection

Don't expend energy for freeze protection when there is no need for it. During warm weather, unnecessary heating for freeze protection can waste cooling energy as well as heat.

■ Prevent Coil Valve Leakage

Leaking throttling valves on steam and hydronic heating coils cause them to lose heat continuously, even during periods when heating is not needed. This energy waste may remain undetected for years. Throttling valves on preheat coils tend to develop leakage because they operate for long periods at low output. This hastens erosion of seating surfaces ("wire drawing") that occurs when a valve is barely open.

You can detect valve leakage when the valve is supposed to be turned off by feeling the temperature of the pipe that connects the valve to the coil. This is a matter of good routine maintenance.

Most coils are fitted with isolation valves for maintenance purposes. Turn these valves off during periods when the coils are not needed. This reduces erosion of the control valves, and prevents energy loss if the control valves leak. This function is easy to forget, so install effective placards on the isolation valves to get the attention of the operating staff. See Reference Note 12, Placards, for placard design and installation procedures.

■ Minimize Stratification

Stratification of the air flow through the air handling unit may cause unnecessary preheat by confusing the freeze protection thermostats. If the cold portion of a stratified air stream passes over the thermostatic element of a freezestat or preheat control, this increases coil heat output.

The adverse effects of stratification are covered in Measure 4.2.11, along with remedies. Make a point of minimizing stratification whenever you modify the outside air intakes.

ECONOMICS

SAVINGS POTENTIAL: 0.5 to 5 percent of conditioning energy.

COST: Varies widely, depending on the nature of the repairs or modifications required.

PAYBACK PERIOD: Less than one year, to many years.

TRAPS & TRICKS

RECOGNIZE PROBLEMS: Check all the places where preheat is used. Don't be surprised if what you see does not match the drawings. Carry a thermometer to check that preheat is not being used excessively. Check for dampers that are poorly designed, poorly selected, or poorly installed. If you have steam coils and you are not a steam heating expert, get help.

EXPLAIN IT: Describe preheat controls in the plant operating manual. Install placards at controls that explain what they do, and state the correct settings.

MONITOR PERFORMANCE: Schedule periodic checks of your freeze protection equipment and controls during the heating season.

MEASURE **4.2.11 Eliminate air handling system stratification that increases energy consumption or reduces comfort.**

In air handling systems, "stratification" means that the air entering the coils of the air handling unit is not at a uniform temperature. Stratification is caused by failure of the incoming outside air to mix thoroughly with the recirculated air. As a result, air enters different parts of the coils at different temperatures. Stratification in an air handling unit is a simple phenomenon, although it may not be easy to fix. It is depicted in Figure 1.

In many air handling systems, stratification is serious enough to cause coil freezing, to reduce comfort, and/or to reduce system efficiency. Stratification is likely to occur unless the air handling unit, and the duct connections leading to the air handling unit, are designed specifically to prevent it.

Stratification occurs because air is a viscous fluid that does not mix easily. You can see this behavior in airplane condensation trails, which remain distinct from the surrounding air. There is a common misconception that stratification is caused by convection that results from the temperature differences. This is not true. The air moves through the system too rapidly for convection to be significant. (Convection does cause stratification of air within a space, which is an entirely different process.)

Stratification occurs most commonly at the coils in the air handling unit. However, in the worst cases, it may extend all the way to individual ducts in the conditioned spaces. This Measure tells you how to avoid stratification, and separately, how to keep it from causing problems.

Problems Caused by Stratification

How serious a problem is stratification? It depends on the climate, the type of air handling system, and whether a preheat coil is used in the outside air intake. In general, stratification causes damage only during freezing weather. In some cases, it can cause discomfort before it causes damage. It can waste energy in invisible ways whenever the weather is cold or hot by making supply air temperatures less than optimum.

Stratification is most troublesome in cold weather, when the temperature difference between outside air and return air is greatest. The most dramatic effect of stratification is coil freezing. The desire to avoid this hazard motivates excessive heating for freeze protection. (See Measure 4.2.10 about freeze protection methods in general.)

A related problem is nuisance tripping of freezestats, which are protective devices that turn off the fans if a temperature is sensed that is low enough to freeze the

SUMMARY

Stratification is a common problem that wastes energy indirectly, and may cause comfort problems. It can be tricky to eliminate.

SELECTION SCORECARD

Savings Potential .... **$ $**

Rate of Return, New Facilities **% %** % %

Rate of Return, Retrofit.......... **% %**

Reliability .... ✓ ✓ ✓ ✓

Ease of Retrofit .... ☺ ☺

coils. Freezestats err on the side of safety, causing unnecessary shutdowns if stratification exists during cold weather.

Another effect is inaccurate system temperature control. This results from the temperature sensor being located either on the warm or cold side of the stratified air stream. Depending on the type and layout of the air handling system, inaccurate temperature control can result in discomfort, energy waste, or both. For example, the cooling coil discharge temperature in a terminal reheat system has a major effect on the amount of energy wasted by reheat.

Discomfort results if stratification can extend to the spaces. The discomfort is bad in its own right, and it motivates desperate corrective actions by occupants and staff that typically increase the waste of energy.

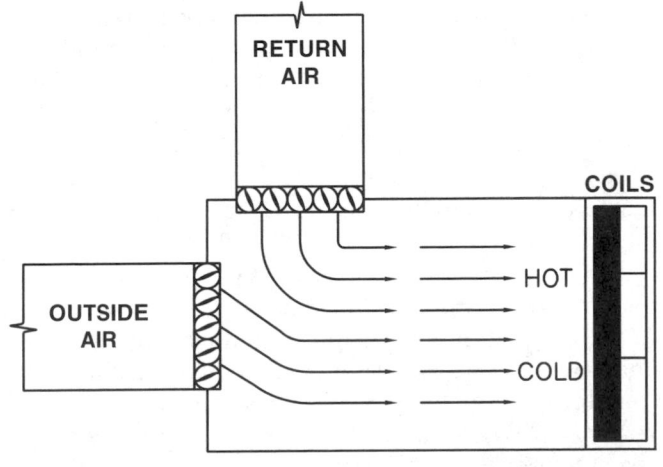

Blender Products, Inc.

Fig. 1 Stratification in an air handling system This is simply the tendency of air streams to maintain their individual identities. It may cause trouble when one air stream is colder than the other.

Stratification Problem or Thermostat Problem?

Stratification always occurs when two air streams join, as in Figure 1. If it causes trouble, the question is whether they are caused by the stratification itself, or by the way that the thermostatic controls in the air handling unit respond to the stratification. For example, a particular coil may not be in danger of freezing, even though a portion of the coil is exposed to freezing temperature by stratification. However, if the sensor of the freezestat is located where the incoming air is coldest, the air handling unit may be shut down by the freezestat.

There are two basic configurations of thermostat sensors used in air handling systems. One is a short probe that extends into the air stream. This type senses the air temperature only at the point where it is mounted. If a temperature sensing problem is caused primarily by the probe location, you can usually find a satisfactory place to relocate the probe.

Another common type of thermostat uses a long capillary that is stretched across the face of the coil. Contrary to appearances, this type of sensor does not average the temperature across the face of the coil. Instead, it responds preferentially to the lowest temperature on the capillary. This is because the capillary contains a liquid in equilibrium with a vapor. The coldest point on the capillary removes vapor by condensation, reducing the pressure that actuates the thermostat.

How to Visualize the Problem

You normally can't see directly how the air flows through an air handling system, but you need to have a clear picture of stratification problems to plan your corrective measures. In designing new facilities, you have to learn from experience with existing systems. The methods of avoiding stratification that are given below summarize this experience.

In existing facilities, you can visualize stratification by taking temperature measurements around the perimeter of the duct. You can take measurements from outside the duct by drilling holes to insert a thermometer. (Seal the holes afterwards.) The pattern of stratification is easier to visualize during colder weather, when the temperature differences in the stratified air stream are greatest.

Make temperature measurements upstream of the coils to determine whether there is a freeze problem. Make temperature measurements downstream of the coils to determine whether the coil discharge temperature is uniform. Make temperature measurements in different zone ducts to determine whether stratification is causing space temperature variations.

A smoke streamer can provide a dramatic visual clues to stratification. However, inserting the smoke and seeing it may take some effort. To see the smoke pattern, cover access panels or doors with transparent plastic. Make sure to announce the smoke tests to the occupants of affected spaces.

You can get a general impression of the potential for stratification by looking at the configuration of the system. In general, mixing occurs best when the geometry of the air handling system causes separately entering air streams to slam into each other. For example, in Figure 1, the return air enters the air handling unit perpendicular to the outside air stream, tending to mix the streams.

Stratification is greatly affected by the velocities of the air streams. Stratification tends to be severe when both air streams are moving at low velocity, regardless of the system configuration. If an air stream moves slowly, it hugs the side of the passage on which it enters.

Because of the effect of air velocity, check for stratification across the full range of damper settings and fan speeds. The stratification pattern may change considerably if the ratio of outside air to return air changes. Variations are greatest where an economizer cycle is used, because the outside air flow may vary from a small percentage of system flow to 100% of system flow.

How to Minimize Stratification

Because air is viscous, it takes energy to mix two air streams. To keep from wasting energy when you correct stratification, remember your objective. You are trying to avoid freezing, control, and comfort problems, not to achieve perfect mixing for its own sake. You can reduce stratification in a variety of ways, which we will discuss here. In most air handling systems, you will use more than one of these methods. Try to create no more turbulence in the air stream than necessary, because turbulence wastes fan energy. Furthermore, the wasted energy appears as heat in the air stream, which increases the cooling load. Choose a method, or a combination of methods, that will limit stratification sufficiently under all conditions, from minimum to maximum outside air intake, and from minimum to maximum supply air flow.

■ Exploit the Mixing Action of Fans

If a fan has only a single inlet, it acts as an effective mixer, virtually eliminating stratification in the discharge path. However, the supply fan is usually located downstream of the coils, so it cannot remedy troublesome stratification that occurs ahead of the coils, where outside air enters alongside the return air.

The main exception is multizone units, where the fan is almost always located ahead of the coils.

If a fan has a double inlet, it may fail to eliminate stratification. This occurs if colder air enters through one side of the fan wheel, and warmer air enters through the other. The flow may still be separated in the fan discharge. The stratification may continue through the supply duct system, with warmer air going to one side and cooler air to the other.

■ **Install Preheat Coils for Outside Air**

Preheat coils do not eliminate stratification, but they reduce its effects by reducing the temperature difference between the air streams. Preheat coils do not increase the fan power requirement, and they may even reduce it. See Measure 4.2.10 for more about them.

■ **Design the Front End Duct Routing to Reduce Stratification**

Experienced designers minimize stratification by using a number of tricks in laying out the front end of an air handling systems. Here they are:

• *make the passage leading to the outside air damper larger than the damper itself.* This causes the outside air to enter the damper at low velocity. Otherwise, the outside air flow could be concentrated through one part of the damper, creating a more concentrated stream of cold air in the air handling unit. (See Measure 4.2.9 about other improvements to outside air intakes.)

• *make the outside air damper as small as possible.* Injecting a small stream of cold outside air into a larger stream of warm return air reduces the chance that a large amount of undiluted cold air will reach the coils. Keeping the outside air damper small also minimizes air leakage into the system when it is turned off. Size the outside air damper for no more than the maximum amount of outside air that may be required. If the system uses an economizer cycle (see Measure 4.2.5 about this), the outside air damper must be large enough to provide the entire system air flow. Even in that case, the outside air damper should be no larger than the mixed air damper.

• *inject the outside air into the return air at right angles.* You don't want the two air streams to join smoothly at the front end of an air handling system. Instead, you want the outside air stream to crash into the return air stream, so that the inertia of the outside air carries it into the return air stream. Do not use parallel-blade dampers in the front end of the air handling system. They act as turning vanes, maintaining stratification around a bend. Opposed-blade dampers tend to aim the air flow into the crossing air stream, producing the desired result. In fact, they act as nozzles when they are in a partially closed position.

• *mix the outside air and return air, and then make them turn into the air handling unit together.* The basic idea is to avoid a situation in which one of the air streams can shoot directly into the air handling unit without mixing with the other air stream. Mixing occurs downstream of bends because the bends create turbulence. For most effective mixing, arrange for the outside air to merge with the return air in a direction perpendicular to the plane of turn that the mixed air makes into the air handling unit. For example, feed outside air downward into the return air, then have the mixed air turn horizontally into the air handling unit.

• *locate the last bend at some distance ahead of the filters.* Turbulence occurs downstream of a bend. Give the turbulence sufficient distance to work before the air enters the coil filters. This is because the filters damp out the turbulence and stop the mixing inside the air stream.

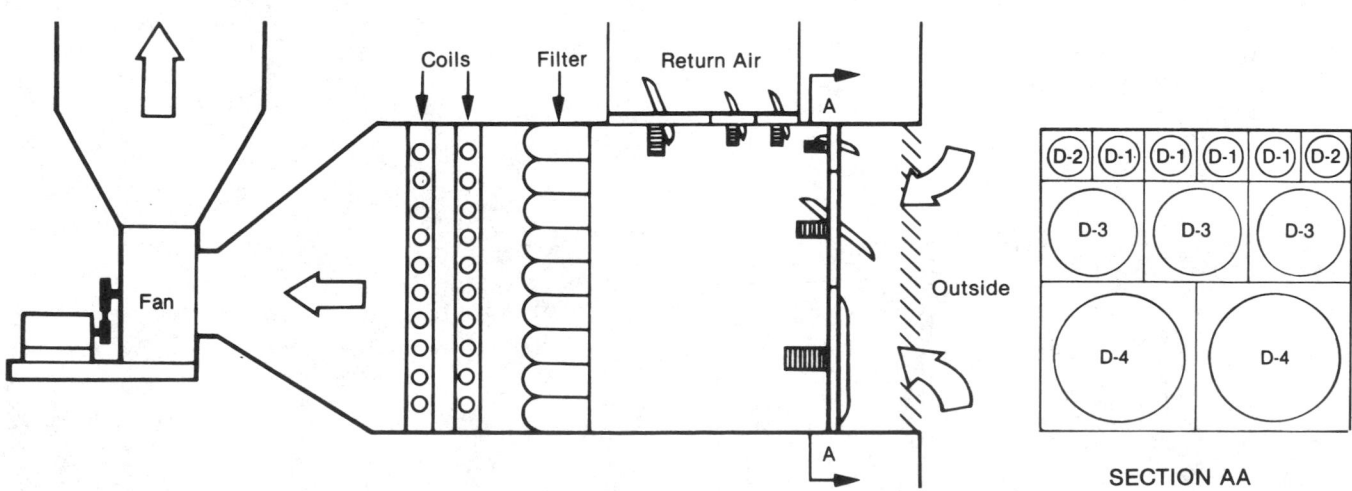

M&I Heat Transfer Products

Fig. 2 Reducing stratification by clever damper arrangement This air handling system has an economizer cycle, which requires a wide range of outside air flow rates. When the economizer is not operating, it uses tight dampers to block most of the outside air intake area. The dampers needed to satisfy the lowest outside air flow rates are located closest to the return air stream, and they are oriented to create the greatest mixing. See Figure 2 in Measure 4.2.2 for a photograph of this damper installation.

All this being said, don't expect the front end duct configuration to completely eliminate stratification, no matter how cleverly you design it. A stratified air stream can turn many corners without mixing entirely. If the outside air can become very cold, you will need to combine good front end duct design with one or more of the other methods given here.

■ Select Outside Air Dampers that Create Turbulence

In contemporary practice, most control dampers used in the front ends of air handling systems are of the opposed-blade type. These dampers do not tend to mix the air stream. As they close, they shed sheets of turbulence that flow downstream from each blade. These thin sheets of turbulence do not mix air across the air stream. Opposed-blade dampers also tend to be leaky, wasting energy and allowing freeze problems when the system is shut down during cold weather. You can avoid both of these limitations by using the type of tightly closing outside air dampers described in Measure 4.2.2. These dampers create larger, more penetrating vortexes than opposed-blade dampers.

Tight-shutoff dampers are effective for the outside air and relief air paths, where it is important to have tight closing. Use opposed-blade dampers in the mixed-air path, because their good control characteristics and lower turbulence are important there.

Tight-closing dampers are designed primarily to block leakage when they are closed. They are not designed to function well as control dampers, modulating flow. To use this type of damper for controlling outside air over a wide range of flow, you need to install an array of individual dampers, as we discuss next.

■ Group the Dampers to Minimize Stratification

If the amount of outside air varies over a wide range, arrange all three sets of dampers in groups so that individual damper blades are either fully closed or fully open. This technique is especially valuable with air handling systems that have economizer cycles. When the economizer cycle is operating, a large fraction of the outside air intake area may be open, whereas the outside air dampers are mostly closed the rest of the time.

The mixed air dampers are grouped in the same way. Arrange the opening sequence of the dampers so that the open blades of the outside air damper and the open blades of the mixed air damper are on the same side of the inlet passage. See Figure 2. In this way, the outside air is guided into collision with the return air.

■ Install Mixing Vanes

A compact, effective method of reducing stratification is to install mixing vanes inside the air handling unit, as shown in Figure 3. The basic idea is to create a swirl in the mixed air passage, so that a portion of one air stream is rotated into the other.

At least one company makes patented prefabricated swirling devices that are inserted into air handling systems, across the entire area of flow. Figure 4 shows one of these units. The vanes in this device create one swirling air stream inside another, with the two streams rotating in opposite directions. This creates a strong mixing action. Furthermore, any remaining temperature differences in the mixed air stream are distributed across the face of the coil.

A less effective method is to attach vanes to the sides of the duct. As with other mechanical methods of mixing

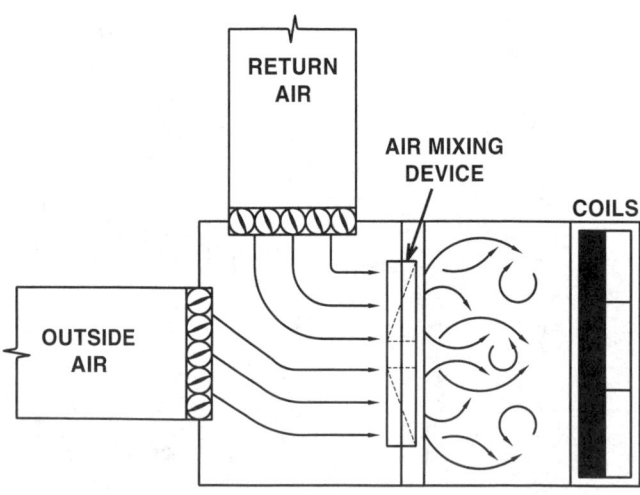

Blender Products, Inc.

Fig. 3 Using mixing vanes to minimize stratification An air mixing device is installed so that both air streams must pass through it. Compare this to Figure 1.

Blender Products, Inc.

Fig. 4 Air mixing device This set of vanes creates one swirling air stream inside another, the two rotating in opposite directions. The mixing action is strong, and the swirling motion distributes any remaining temperature differences across the face of the coil.

air streams, mixing vanes must be matched to the geometry of the duct and coil configuration.

Mixing vanes increase flow resistance. However, if a mixing vane is designed efficiently, it will create less flow resistance in relation to the degree of mixing than any other method of mixing. In most cases, the resistance of mixing vanes is lower than the resistance of a coil or a dirty filter. Still, make sure that you have enough fan power to compensate for this. Manufacturers of prefabricated air blending devices can probably tell you how much increased resistance to expect.

The most reliable approach is to use a mixing device that covers the entire area of flow. Especially at lower air velocities, a fraction of the air will not be affected by vanes that cover only part of the passage.

There is another tricky factor with air mixing devices. Try to use only one, as shown in Figure 5. If you use more than one, a large amount of stratification may remain. For example, consider an installation of two mixing devices of the type shown in Figure 4. If

they are installed so that cold outside air enters one of them, and warm return air enters the other, then you will still have two air streams of greatly different temperature.

Eliminate any rotary motion of the air stream before the air enters the fan inlet, to avoid loss of fan capacity and efficiency. In most air handling units, the coil is located ahead of the fan. (This is called a "draw-through" configuration.) In this configuration, the coil acts as an effective flow straightener, so you do not have to take any additional action.

■ **Avoid Baffle Plates**

Baffle plates are flat plates with holes. They are placed across an air stream. They are essentially turbulence generators, creating turbulence both in front of the plate and behind it. The turbulence in front of the plate is most effective in removing stratification because different portions of the air stream are mixed as flow backs up in front of the plate. The turbulence created downstream is not useful, because it is localized to the air streams coming out of individual holes in the plate.

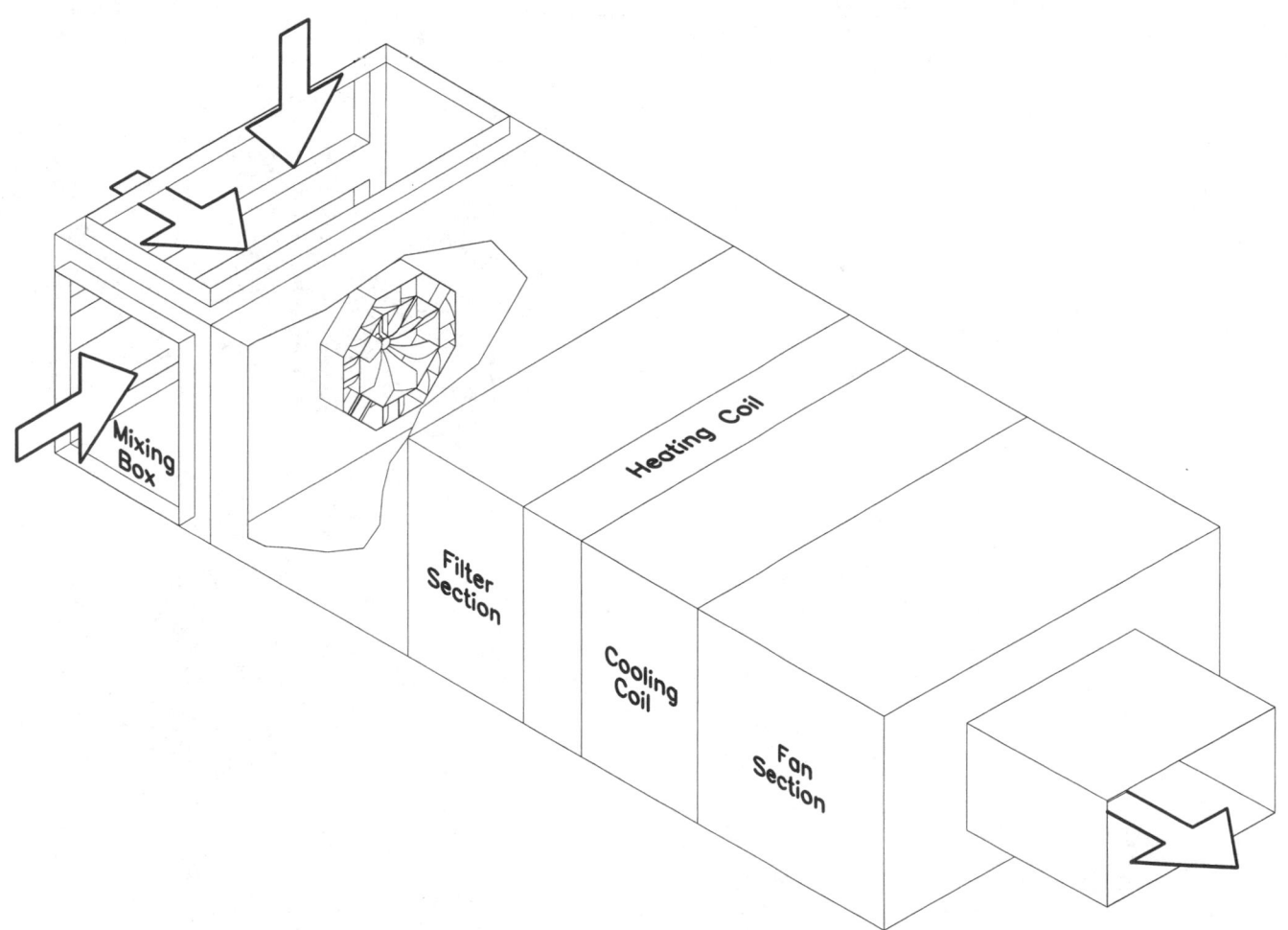

Fig. 5 Mixing device installation The mixing device should be as large as practical, to minimize the fan power penalty. Try to use only a single mixing device, so that all the air entering the air handling unit must pass through the mixing device together. Using an array of more than one mixing device allows stratification to remain.

Baffle plates are inefficient because they create too much flow resistance as the price of the mixing they provide. They are often a desperation measure employed in lieu of adequate thought.

(A coil acts as a baffle plate if the velocity of the air entering the coil is too high. Avoiding turbulence that results from high entry velocity is the main reason why coils are installed on the suction side of fans whenever possible.)

Do not confuse baffle plates with flow straighteners, which have the opposite effect. Flow straighteners are intended to make flow uniform. They typically are used in HVAC systems to provide greater accuracy of air flow measurements. They do not add much resistance, and they do not reduce stratification.

ECONOMICS

SAVINGS POTENTIAL: *Varies widely. Depends largely on how the system operators respond to problems caused by stratification.*

COST: *In new construction, extra sheetmetal and larger louvers for reducing stratification may add a few hundred dollars to a few thousand dollars per air handling unit. An array of tight-closing dampers costs about $50 to $100 per square foot ($500 to $1000 per square meter), for the equipment. Preheat coils cost thousands of dollars, depending on size and type. Prefabricated air mixing devices for air handling units cost about $0.25 per CFM in the smallest sizes, down to about $0.05 per CFM in large sizes.*

PAYBACK PERIOD: *Several years or longer.*

TRAPS & TRICKS

RECOGNIZE PROBLEMS: *In designing new facilities, take a careful look at the front end configurations of all the air handling systems. In existing facilities, visually inspect the air mixing sections to find potential stratification problems. For clues to stratification problems, recall previous incidents of coil freezing and complaints about comfort that might be related to stratification. Use a thermometer with a remote probe to scan for stratification at times when the outside air is cold.*

CLEVERNESS: *The big challenge is to avoid stratification at all damper positions and fan speeds. Try to get good mixing without wasting too much fan power in creating turbulence.*

Single-Zone Systems

The Measures in this Subsection improve the efficiency of single-zone air handling systems. Figure 1, on the next page, shows the components of a typical single-zone air handling unit. Figures 1 and 2, in the Section 4 Introduction, illustrate how single-zone systems compare to other types of air handling systems.

Single-zone systems are defined by the fact that they are controlled by a single space thermostat. They may heat or cool, or both. They may have separate heating and cooling coils, or a single coil may perform both functions. The coils may be electric, hydronic, or direct-expansion. Thermostatic control may be proportional or on-off. They may have any means of controlling outside air, humidification, and other functions.

Single-zone systems differ from room conditioning units primarily in size, in having more elaborate duct systems, and perhaps, in having more positive control of outside air intake. Their larger size often makes it economical to make design and component improvements that would not be practical with room conditioning units.

The basic single-zone system is an efficient type. The main opportunities for saving energy with single-zone systems are eliminating unnecessary conditioning and reducing fan energy consumption. The Measures in this Subsection exploit these opportunities.

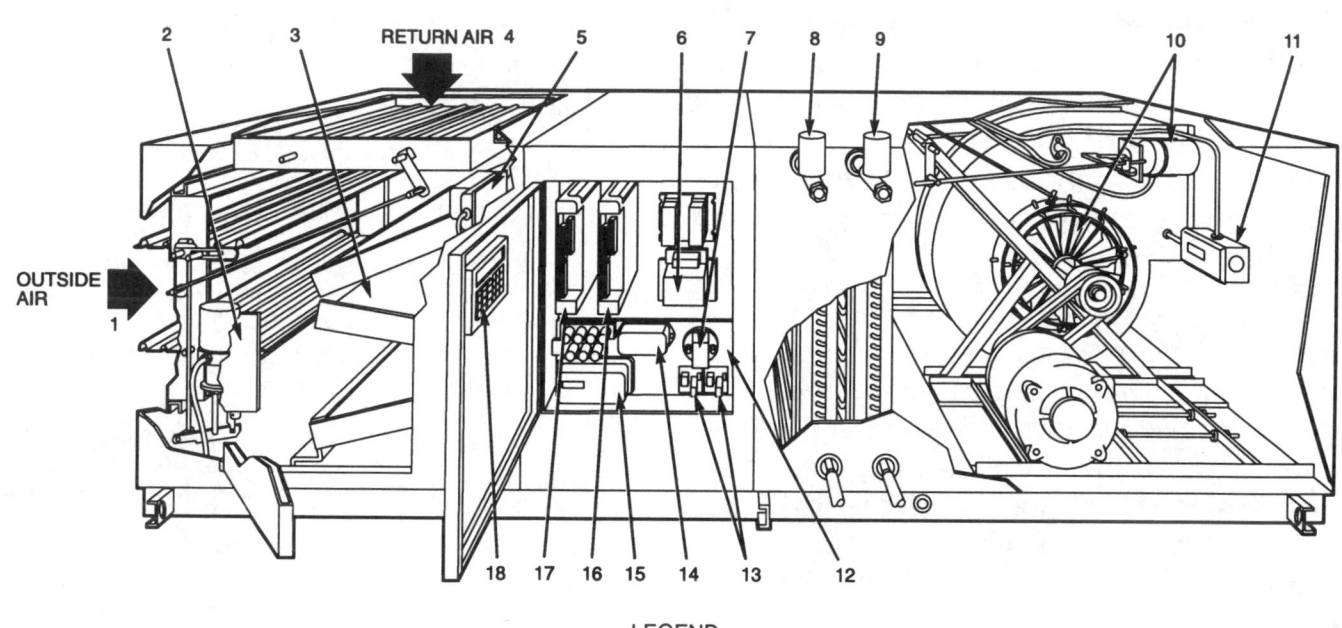

OUTSIDE
AIR

RETURN AIR

LEGEND

1 — Outside Temperature Sensor
2 — Modulating Mixed Air Damper
3 — Filter Maintenance
4 — Return Temperature Sensor
5 — Mixed Air Sensor
6 — Freeze Stat
7 — High Pressure Switch
8 — Heating Valve Assembly (2 or 3-way)
9 — Cooling Valve Assembly (2 or 3-way)

10 — Static Pressure Control
11 — Supply Air Temperature Sensor
12 — Integral Control Panel
13 — Fan Control
14 — Static Pressure Transmitter
15 — Air Flow Switch
16 — Option Module
17 — Processor Module
18 — Local Interface Device

Carrier Corporation

Fig. 1 Typical single-zone air handling unit

MEASURE 4.3.1 Install placards at user controls to encourage efficient operation.

See Reference Note 12, Placards, for guidance in using placards to promote efficiency. This Measure relates specifically to installing placards at the controls of single-zone air handling units.

Install placards at all thermostats. Explain the thermostat settings and any other features, including fan control switches and heat/cool selector switches.

Also, install effective placards at all timeclocks and other devices that control the running of the systems. (See Measures 4.1.1 ff for these controls.)

Applications and Limitations

Thermostat placards have the greatest potential for saving energy with systems that only heat or only cool, or with systems that have separate thermostat settings for heating and cooling.

Thermostat placards are less effective with systems that maintain a fixed space temperature by using both heating and cooling. (Improving the thermostatic control efficiency of such systems is covered in Measures 4.3.4 ff.)

Energy Saving Potential

■ Equipment Shutdown

Placards save the most energy by motivating the operators of air handling units to turn them off when they are not needed. Turning off the air handling units reduces energy usage for these functions:

- *space conditioning.* The unit consumes energy to change space temperature and humidity. If the spaces are unoccupied, it is usually harmless to let the temperature and humidity drift within wide limits. The amount of conditioning energy that is wasted by unnecessary operation depends on the heat loss (or gain) of the space.
- *conditioning of outside air.* Most air handling units take in a considerable amount of outside air, which typically accounts for a major fraction of conditioning energy. (If the unit must run during unoccupied periods to maintain space temperature, consider turning off outside air ventilation during these periods, as recommended by Measure 4.2.1.1. If this function is controlled manually, install appropriate placards at the controls.)
- *auxiliary energy consumption.* Turning off air handling units saves the energy consumed by their fans, along with other auxiliary equipment that may be turned off, such as hydronic coil booster pumps.

SUMMARY

A basic way to save a substantial amount of energy at minimal cost. But, not as simple as it looks.

SELECTION SCORECARD

Savings Potential	$ $ $
Rate of Return, New Facilities	% % % %
Rate of Return, Retrofit	% % % %
Reliability	✓ ✓ ✓
Ease of Retrofit	☺ ☺

■ Minimizing Heating and Cooling During Occupancy

The amount of energy required for heating or cooling is roughly proportional to the temperature differential between inside and outside. Placards save energy while the system is operating by motivating occupants to reduce this temperature differential. This benefit is greatest in climates where heating or cooling is needed for a larger fraction of the year.

If occupants set the space temperature to require less conditioning, this also reduces the number of hours that the system operates. This benefit is greatest in mild climates, where the temperature hovers for long periods in a range that does not require either heating or cooling. In addition, buildings in mild climates are often poorly insulated, so they derive larger savings from a given change in the temperature differential.

(As a point of interest, the energy saving is related only to the temperature difference, not to the outside temperature itself, as long as there is a heating or cooling load. For example, lowering the heating temperature by 3°F saves the same amount of energy if the outside temperature is 0°F or 50°F.)

What the Placards Should Say

■ Control Functions

Ask occupants to perform all of these actions for which manual controls are available:

- turn off the unit when the space is not occupied
- set the heating temperature for the lowest comfortable setting
- set the cooling temperature for the highest comfortable setting

- turn the fan to the lowest comfortable speed
- set the fan control so that the running of the fan is controlled by the thermostat (see Measure 4.3.3.2)
- turn off the outside air intake when it is not needed

■ Close Windows and Doors

If the space can lose conditioning through windows or doors, ask the occupants to keep these closed while the air handling unit is operating.

■ Identify the Space being Controlled

Each placard should say which spaces are covered by the controls, if this is not obvious. For example, the placard should say, "This thermostat controls the heating and cooling of the reception area." Otherwise, occupants cannot associate the controls with the conditioning of their space, or they may not know if other spaces are also controlled.

■ Inform Occupants about Automatic Controls

Placards should inform occupants and staff about the operation of any automatic controls, if this might create confusion. For example, the placard should advise occupants if the air handling system will shut down automatically after a certain time. Refer to Subsection 4.1 for automatic control Measures that should have placards.

■ Ask Occupants to Dress Appropriately

The way people dress has a major effect on the space temperatures they choose. Where appropriate, encourage occupants to dress warmly in cold weather, and to dress lightly in warm weather. In many environments, it is appropriate to say this on the thermostat placard.

Where to Install the Placards

Install placards adjacent to the controls that they explain. To encourage occupants to turn off conditioning at the end of the day, install another placard where occupants will see it upon leaving the space, requesting occupants to turn off the conditioning at the thermostat.

Let the Occupants Select the Space Temperatures

Don't bother telling occupants to set specific space temperatures. This is usually not effective or desirable. Accept the fact that occupants will not make themselves uncomfortable for the sake of energy conservation.

In practice, it is impossible to maintain a stated space temperature. Room temperature varies from one part of the space to another. Even at the thermostat itself, the calibration of thermostat dials is usually crude, so the numbers on the dial are not a reliable guide.

Installing a thermometer in the space, at the thermostat or elsewhere, can serve as a check on the space temperature. However, installing thermometers usually has little influence on the thermostat setting. They may be useful for keeping the peace in spaces where the occupants disagree about the thermostat setting.

Avoid Locked Thermostats

For the same reasons, don't use locked thermostats. Making thermostats inaccessible to the occupants of the spaces is a bad practice, except where it is necessary to protect controls from vandalism. If temperature limits are set too aggressively with a locked thermostat, occupants will be uncomfortable. On the other hand, if the locked thermostat is set to minimize complaints, energy efficiency is sacrificed.

It is bad policy to limit people's ability to remain comfortable. This reduces work efficiency, invites health problems, and creates antagonism toward energy conservation in general.

Locked thermostats that are visible within the space irritate people because they overtly prevent people from being able to control their environment. As a result, they are a magnet for tampering. If you have to keep thermostats out of reach of the occupants, then conceal them entirely in the air return from the space.

ECONOMICS

SAVINGS POTENTIAL: *1 to 40 percent of conditioning cost, depending on which control functions are manual, the operating schedule, etc.*

COST: *$5 to $200 per placard, depending on quantity, materials, and complexity. This includes design cost.*

PAYBACK PERIOD: *Less than one year, in most cases.*

TRAPS & TRICKS

EFFORT: *See Reference Note 12, Placards. The main hazard is failing to put enough effort into the design and installation of placards. Select materials that are durable and legible. Reject the temptation to use adhesives to attach placards where permanent methods are possible.*

MEASURE 4.3.2 If conditioning cannot be turned off during unoccupied hours, install temperature setback.

SUMMARY

A basic energy conservation measure that saves energy during unoccupied periods. A standard feature in most electronic control systems. With other types of controls, it requires special thermostats.

SELECTION SCORECARD

Savings Potential .... $ $ $

Rate of Return, New Facilities % % % %

Rate of Return, Retrofit.......... % % % %

Reliability .... ✓ ✓

Ease of Retrofit .... ☺ ☺

The most effective way to save conditioning energy after hours is to turn off the air handling system, as recommended by Measure 4.1.1. However, this is not possible in all cases. The air handling system may be kept in operation to maintain a minimum temperature, or to continue providing ventilation, or for other reasons. Even if the air handling system fans are turned off, the coils may be kept active to warm air that is drawn into the building by the action of exhaust fans.

In these situations, you can often use temperature setback as a way to save energy. Setback is a thermostatic control technique in which heating and cooling are turned off during unoccupied periods until the temperature in the space drifts to a lower (or higher) fixed temperature. This relaxed temperature setting is called the "setback temperature."

Energy Saving Potential

The conditioning load depends on the difference in temperature between the inside of the building and the outside. Temperature setback reduces this differential. For example, if the average space temperature can be reduced 10°F during the unoccupied period, and if the outside air temperature during this period averages 30°F below the occupied temperature, then heating energy consumption during this period can be reduced by about one third. (This example does not include the effect of heat gains, heat storage in the building structure, and other complications.)

This Measure also saves fan energy if the operation of the fans is controlled by the space thermostat. See Measure 4.3.3.2 for details.

You may be able to shut off outside air during the setback period, as recommended by Measure 4.2.1.1. This action typically saves a large part of the conditioning load.

How to Accomplish Temperature Setback

You can accomplish temperature setback in a variety of ways. The following are the main methods.

■ Individually Programmable Setback Thermostats

You can install individually programmable setback thermostats to control each air handling system. The time schedules and setback temperatures are set at each thermostat individually. In facilities that have only a small number of thermostats, this is the easiest and cheapest method. In facilities where thermostats are subject to vandalism, it takes disciplined maintenance to keep all the thermostats programmed and working properly.

A setback thermostat has two thermostatic elements in one enclosure, one of which is set for normal occupancy, and the other for unoccupied conditions. (Electronic units may have only a single sensor, but the effect is the same.) The unit has its own timeclock to switch between temperature settings.

Individually programmable setback thermostats are common items for on-off electrical control of heating and cooling and for pneumatic thermostatic control. They may be available for analog electrical control and for some electronic control systems.

Different models of setback thermostats offer different degrees of flexibility in scheduling. A variety of options are available, such as override control and a backup power supply. Ease of use is an important criterion. Some models are much easier to use than others. See Reference Note 10, Clock Controls and Programmable Thermostats, for equipment features and installation practices that are important for achieving satisfactory performance.

■ Electronic Control Systems

Temperature setback is a standard capability with almost all centralized electronic control systems. Most electronic control systems allow you to select the schedule and setback temperature of each thermostat individually.

Special thermostat models are needed to provide occupant override within the spaces.

■ Centralized Pneumatic Setback

In pneumatic control systems, you can accomplish setback by changing the control air pressure at the air source. This requires special thermostats. Also, you need a control station at the air source to change the

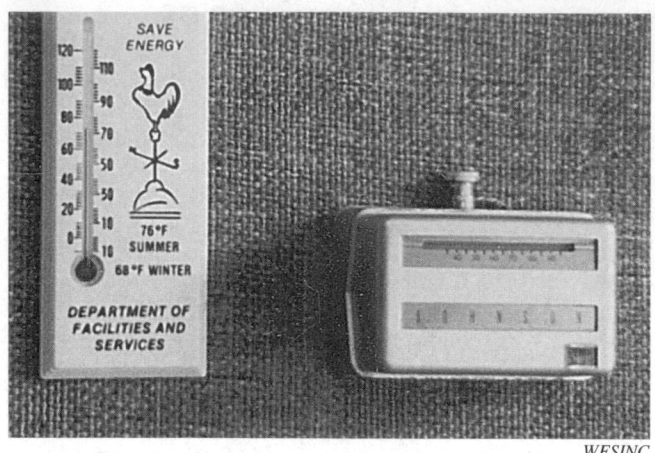

WESINC

Fig. 1 Pneumatic setback thermostat with override To restore normal space temperature during the setback period, depress the button on top of the thermostat. The button returns to its normal position during the next temperature cycle.

pressure at the appropriate times. A typical system sets the thermostats to daytime operation with 17 PSI supply pressure and nighttime operation with 22 PSI supply pressure.

The main disadvantage of this method is that all spaces served by the pneumatic system must operate on the same schedule.

Pneumatic setback thermostats with individual overrides are available. Figure 1 shows a typical unit.

Avoid Forcing the Setback Temperature

Be sure that conditioning energy is not used to drive the space to the setback temperature. This is a potential trap where a thermostat controls both heating and cooling equipment. This problem does not occur with true setback thermostats, but it can occur with simple changeover thermostats that are controlled automatically. It can also occur by programming a centralized electronic control system incorrectly.

Provide Efficient Overrides

A weakness of setback thermostats is that they operate on a fixed schedule, either individually or as part of an entire system. To overcome this problem, setback thermostats can be equipped with overrides. A person who wishes to work during the unoccupied period activates the override in his space, which causes the thermostat to call for normal temperature.

The override remains in effect until the next setback period, unless the occupant cancels the override. This is an efficiency weakness. It does not waste much energy

if the setback override is used only occasionally. However, if the override is used often by a user who forgets to cancel the override, much of the benefit of setback is lost.

One way to overcome this problem is to use a personnel sensor to activate the override. This is not yet a standard feature of setback controls, so you have to wire it on a custom basis.

Control Outside Air Appropriately

Many facilities do not need outside air ventilation during the unoccupied period. Use a control signal from a centralized setback system, or from an individually programmable setback thermostat, or from a timeclock to close the outside air dampers, if appropriate. See Measure 4.2.1.1 for details.

Placards are Essential

Setback thermostats are operated by occupants, so they need instruction placards. See Measure 4.3.1 for details. The most important part of the message is asking occupants to reset the temperature override after using it.

ECONOMICS

SAVINGS POTENTIAL: *5 to 15 percent of heating cost, if setback is applied only to heating. 20 to 60 percent of cooling cost. The saving are even greater if you can use setback to shut off outside air and to turn off the air handling system fans. See Measure 4.2.1.1 for these savings.*

COST: *An individually programmable setback thermostat for a low voltage thermostatic system typically costs about $100 to install. Setback typically adds about several thousand dollars to the cost of a centralized control system, plus about $100 for each thermostat.*

PAYBACK PERIOD: *Less than one year, to several years.*

TRAPS & TRICKS

SELECTING THE EQUIPMENT: *Select setback thermostats that are rugged and easy to use. Get sufficient scheduling flexibility. Always include an override and a backup power source.*

EXPLAIN IT: *Don't forget the instruction placards. These are essential to efficient operation and to minimize vandalism.*

MAINTENANCE: *Periodically, check the scheduling, setback operation, and override operation of each thermostat.*

MEASURE **4.3.3 Match fan output to the conditioning load.**

In constant-volume air handling systems, the fan output is selected to deliver heat or cooling during the most extreme load conditions. Most systems have an additional margin of capacity over this amount. Extreme heating or cooling conditions typically occur for only a small fraction of a system's running time. This makes it worthwhile to reduce fan output to meet the current load. You can do this in a variety of ways, which differ in efficiency, cost, and trouble. The four subsidiary Measures recommend these methods, respectively:

- *fan trimming* (Measure 4.3.3.1) is a one-time adjustment that gets rid of any excess margin above the capacity needed to meet peak loads
- *cycling the fans with the space thermostat* (Measure 4.3.3.2) stops the fans entirely when they are not needed to deliver heating or cooling
- *installing multi-speed fan motors* (Measure 4.3.3.3) reduces fan energy consumption for a large fraction of the time without sacrificing fan capacity
- *variable-air-volume (VAV) fan control* (Measure 4.3.3.4) keeps the output of the fan tuned precisely to the current load.

All these Measures can be used with hydronic cooling and heating coils. Only fan cycling (Measure 4.3.3.2) is advisable for direct-expansion (DX) cooling coils and staged electric heating coils. DX coils need a fixed amount of air flow to avoid freezing, and electric coils need a minimum air flow to keep from burning out.

You can combine any of the first three Measures, where they are appropriate. The last Measure eliminates the need for the others.

Energy Saving Potential

■ Fan Energy

A large saving in fan energy is often possible. The reason is that the power consumed by a fan discharging into an open duct system, such as a single-zone system, is approximately proportional to the third power of the air volume delivered. For example, reducing the air flow by half reduces the fan power requirement to approximately one eighth. Even a relatively small reduction in fan flow produces a significant reduction in fan power. For example, reducing air flow by 20% reduces the fan power requirement to approximately one half.

The key aspect of exploiting this is to reduce the output of the fan in an efficient manner. The subsidiary Measures reduce fan speed or turn off the fans, the most efficient methods of reducing air flow. In contrast, using dampers to reduce air flow would save little energy.

■ No Heating or Cooling Saving in Most Systems

Most single-zone systems do not offer a saving in heating or cooling energy when the fan output is reduced. The space conditioning load determines the heating and cooling requirement, except for conditioning of outside air. In systems with good control of outside air ventilation, outside air intake is fairly independent of the fan flow rate. If ventilation is affected by changes in fan output, this is usually an undesirable side effect.

■ Heating and Cooling Energy in 100%-Outside-Air Systems

Air handling systems that use 100% outside air are a special case where reducing fan output may save conditioning energy as well as fan energy. The reason is that the amount of outside air intake with such systems may greatly exceed the amount of outside air needed for ventilation. Heating, cooling, and dehumidifying outside air usually (but not always) requires more energy than conditioning air that is recirculated from inside the building.

■ Reheating Operation

In some single-zone systems, the cooling and heating coils are operated simultaneously (reheat) for the purpose of dehumidification. During periods of low conditioning load, reheat may consume much more energy than is actually delivered to the spaces. The sensible component of reheat energy loss is proportional to the air flow rate.

Reducing the air flow rate reduces reheat in these cases. However, this is not the best way of saving energy. In many cases, it would allow space humidity to rise, and it would not eliminate unnecessary reheat entirely. Instead, the efficient approach is to control reheat based on the need for dehumidification. You can do this by installing a humidistat in the space to control the heating coil, while the thermostat continues to control the operation of both coils to maintain the space temperature.

■ Auxiliary Equipment Energy

Some air handling systems have auxiliary equipment, such as hydronic coil booster pumps, that can be turned off if the fans are turned off. This opportunity for saving occurs only with fan cycling (Measure 4.3.3.2).

Factors to Consider Before Reducing Air Flow

Changing the fan output may have a surprisingly large number of effects. Consider the following factors when you decide whether, and how, to reduce fan output.

■ The Nature of the Application

The Measure does not apply in every facility that has an apparently extravagant air flow rate. Be sure that you understand all the reasons for the air flow settings. For example, surgical suites in hospitals typically require a high flow rate to provide "laminar" flow of air through the space. Laboratories typically require high air flow rates to prevent escape of contaminants from hoods. You may be able to save energy in such cases, but it will require more complex changes than these.

■ Heating and Cooling Capacity

Fan capacity trimming (Measure 4.3.3.1) reduces the system's heating and cooling capacity. All the other Measures vary fan output in response to load, so they do not reduce conditioning capacity.

■ Humidity

All the subsidiary Measures lower the space humidity to some extent. With hydronic cooling coils, this happens because the coils operate at lower average temperature. This effect is greatest with multi-speed fans and VAV operation.

Fan cycling may reduce humidity for a different reason in air handling systems that use a cycling cooling unit, such as a reciprocating compressor. The fins of cooling coils retain a large amount of moisture. When the cooling compressor is running, the coil is cold and removes moisture from the air. When the compressor turns off, the space air continues to circulate air through the wet coil. Until the coil dries out, the coil acts as a humidifier.

In extreme cases, the cooling coil never dries out, and the space is continually re-humidified whenever the compressor turns off. In that extreme situation, fan cycling may solve the problem.

Multi-speed fan operation also reduces the adverse effect of moisture trapping in cooling coils, but much less effectively than fan cycling. It does this by increasing the fraction of the time that the coils are cooling.

(In humid climates, humidity tends to become a problem during mild weather because the dehumidifying effect of the cooling system is reduced along with the cooling load. Humidity is generally not a problem during cold weather because cold outside air cannot carry much moisture, and air exchange keeps internal humidity tied to the external humidity. Humidity is not a problem when there is a high cooling load because the coils operate long enough to keep the air dry.)

■ Outside Air Ventilation

In recirculating air handling systems, the method of controlling outside air quantity depends on the "front end" configuration of the air handling system. See Measures 4.2.1 and 4.2.2 for details. A well designed air handling system maintains accurate control of outside air intake even as the fan output changes. However, this is an ideal rarely achieved. The amount of outside air generally varies with fan output, in a manner that differs from one system to another. Be sure to account for this when using multi-speed and VAV fan control.

When the conditioning load is low, fan cycling and VAV operation may reduce outside air ventilation below acceptable limits. Solutions for both problems are given in the respective Measures. Fan trimming usually does not reduce air intake enough to infringe on ventilation.

■ Air Cleaning

If the air handling system is used for air cleaning, by filtering or other processes, the rate of cleaning is proportional to the flow rate. All the methods reduce air flow, or interrupt it, so all reduce air cleaning. Air cleaning is covered in Measure 4.2.4.

■ Effect of Space Air Velocity on Comfort

The air velocity within a space may cause discomfort by being too high or too low. Air flow that is too high triggers complaints of "draft." These complaints usually occur during colder weather, when the fans need to be near full output. Thus, these Measures have little effect on perceived draft.

Low air velocity leads to complaints of "stagnation" or "stuffiness," which tend to occur during mild or warm weather. All these Measures either reduce or interrupt air flow in the spaces, so all may aggravate these complaints.

Research suggests that complaints of "draft" are due primarily to low space temperature, and complaints of "stagnation" are due primarily to high space temperature. Lowering the space temperature usually resolves these problems. However, this increases cooling energy, which offsets the advantage of reducing fan output. (Consider installing space circulation fans to supplement mechanical cooling, in applications where this is appropriate.)

■ Diffuser Performance

If fan output is reduced when the system is cooling, the cold air tends to dump out of the diffusers, rather than diffusing throughout the space. This causes discomfort to persons situated under the diffusers.

A diffusion problem also occurs with heating when air flow is reduced. In this case, the problem is that the warm air tends to stay at ceiling level, rather than flowing downward and circulating throughout the space.

VAV operation is notorious for causing these problems because of the great reduction in fan output. Air distribution problems may also occur with multi-speed fans. The other Measures do not affect diffuser velocity enough to cause air distribution problems.

■ Building Pressure

Operation of the air handling system fans may have a major effect on building pressure. This includes the outside-to-inside building pressure differential and the space-to-space pressure differentials.

All the Measures, except fan trimming, create large changes in fan output. Whether this will create pressure problems depends on the design of the air handling systems. Measure 4.2.1 explains how the system design affects space pressure.

■ **Fan Noise**

Reducing fan output by reducing fan speed substantially reduces fan noise. However, the sudden changes in fan noise that occur with fan cycling and multi-speed fans may be objectionable. The potential for annoyance depends on how much noise the fans project into the space, the amount of background noise in the space, how often the fans cycle, and other subjective factors, such as the formality of the space.

■ **Fan-Induced Resonances**

At certain specific speeds, a fan may trigger resonances in the air handling sheetmetal or in the building structure. This is most likely to occur with VAV fan operation, because the fan operates across a continuous range of speeds. A resonance may be triggered at a fixed fan speed after fan capacity trimming or installing multi-speed fan motors.

Air handling system vibration is rarely serious, if the system equipment is installed properly, because the vibrational frequency of centrifugal fans is higher than the resonant frequencies of typical structures and ductwork. If a serious resonance does occur, it is likely to be caused by an unbalanced fan wheel, which you can correct by balancing the wheel. If this does not solve the problem, stiffen the vibrating components, which increases their resonant frequency and reduces the intensity of vibration.

■ **Chiller Efficiency in Chilled Water Systems**

If the volume of air is reduced during cooling, the temperature of the supply air must be lower to compensate. At high cooling loads, achieving the lower air temperature may require lower chilled water temperature. Lowering the chilled water temperature reduces the efficiency of the chiller, as explained in Measure 2.2.1.

Forcing the chiller to operate at lower chiller water temperature for long periods is usually a bad bargain. To achieve the best overall efficiency, make sure that the system controls keep the chilled water temperature as high as possible, with fan output adjusted accordingly.

■ **Freeze Protection and Control of DX Refrigerant Coils**

If cooling is provided by a direct expansion (DX) coil, reducing the air flow through the coil may cause coil freezing. In addition, varying the air flow by a large amount may incur control difficulties. VAV operation is almost certain to cause trouble, and problems are likely if multi-speed fan motors are retrofitted. If you are not an expert in DX systems, don't reduce the air flow through a DX coil without the advice of an expert on this topic.

■ **Air Flow Requirements for Electric Heating Coils**

Reducing air flow through electric heating coils, if done to excess, shortens coil life and may create a fire hazard. Fan trimming typically does not reduce air flow enough to create a problem, but check the coil specifications. Operation with multi-speed fan motors may require modifications of the heating coils. VAV operation is unlikely to be feasible with electric heating coils unless the system has unusual controls for this purpose.

Don't Overlook Any Fans

Consider reducing the flow of all the fans in the air handling system, as appropriate. Do not overlook return fans, remote relief fans, toilet and kitchen hood exhaust fans, etc.

MEASURE 4.3.3.1 Trim fan output.

Fan trimming is a one-time adjustment that gets rid of fan capacity in excess of the amount needed to satisfy peak loads. Refer to Measure 10.3.1 for details.

This is the simplest method in this group of four Measures that save fan energy in single-zone air handling systems. It is cheap and quick. However, it does not tailor fan output to a changing load. It may be the only way to save fan energy that is practical in smaller single-zone systems.

MEASURE 4.3.3.2 Cycle the running of the fans and other air handling system equipment with the space thermostat.

Another way to improve the fan efficiency in single-zone air handling systems is to stop the fan(s) when conditioning is not needed. This method is common where cooling and heating are provided by equipment that cycles on and off. In such cases, simply use the space thermostat to cycle the fan as well as the conditioning equipment. You can also cycle the fans in hydronic heating and cooling systems, but the controls are more complex.

Where to Use Fan Cycling

This Measure is useful primarily for minimizing fan energy consumption during periods of normal operation. Other methods are appropriate for minimizing fan energy outside hours of normal operation, including turning off the equipment (Measures 4.1.1 ff) and temperature setback (Measure 4.3.2).

This Measure is used widely in smaller air handling systems and fan-coil units. However, this method has a number of limitations, especially in larger systems. The limitations are described below.

This Measure has a unique ability to solve humidity problems that occur during mild weather in some environments. See "Humidity," in Measure 4.3.3, for this special case.

Energy Saving Potential

■ Fan Energy

This Measure may reduce the operating time of fans to substantially less than half, especially in smaller systems. The relative saving may be substantially less in larger systems, because of steps needed to reduce adverse effects of fan cycling. The greatest savings in

fan energy occur during periods of low conditioning load. Therefore, this Measure is most effective in mild climates.

■ Heating and Cooling Energy in 100%-Outside-Air Systems

Air handling systems that use 100% outside air are a special case where fan cycling may save conditioning energy as well as fan energy. The reason is that the amount of outside air intake with such systems may greatly exceed the amount of outside air needed for ventilation. Heating, cooling, and dehumidifying outside air usually (but not always) requires more energy than conditioning air that is recirculated from inside the building.

Typical savings are 20 to 50 percent of conditioning energy. The saving in the excess energy needed to condition outside air is proportional to the outside air temperature differential and humidity, and to the time that the fans are turned off. These two factors oppose each other, because the fans run the longest during periods of high conditioning load.

For example, consider a 100%-outside-air system that is converted to on-off operation. During a period when the heating load is one third of the peak load, the fan operates only one third of the time, and only one third as much outside air is heated. In a typical case, the net saving is about one third of the heating energy requirement.

(In the previous example, the air is heated to a higher temperature during the periods when the fan is operating. This does not affect the energy saving. The saving results from reducing the volume of air that must be heated from outside air temperature to space temperature. The total amount of energy required to heat the outside air from space temperature to supply air temperature remains the same, regardless of whether the fan is cycled or not.)

■ Reheating Operation

If the system allows reheating, the amount of reheating energy is proportional to the air flow rate while reheat is being employed. (In a single-zone system, reheat means that the heating coil can operate while the cooling coil is operating. Some systems use this technique to control humidity.) If the system controls are designed for efficiency, they should limit reheat based on the need for dehumidification. Thus, cycling the fan capacity should have little net effect. Verify these aspects of system control.

■ Auxiliary Equipment Energy

Some air handling systems have auxiliary equipment, such as hydronic coil booster pumps, that can be turned off when the fans are turned off.

Limitations

■ Type of Thermostatic Control

This Measure is easiest to accomplish if the thermostat produces a discrete on/off signal. Such thermostats are associated with heating and cooling equipment that operates in an on-off manner, such as electric heating coils and direct-expansion cooling with reciprocating compressors. With on-off thermostats, the fan can be controlled simply with a relay that is activated by the "on" signal from the thermostat.

With systems that use proportional thermostats, such as hydronic heating and cooling, this Measure is more difficult to accomplish in a satisfactory fashion. With a proportional thermostatic system, you need to install

control logic that senses the zero-load condition and turns off the fan within a given range of this signal. For example, the zero-load output of a pneumatic thermostat may be 10 PSI, in which case the control logic would turn off the fan when the thermostat output is between 9 PSI and 11 PSI.

The control logic also needs to have a sensitivity range that keeps the fan from short cycling. In the previous example, the controls would be set so that the fan does not turn back on until the control pressure drops to 7 PSI or climbs to 13 PSI.

In analog control systems, drift and inaccuracy in the controls may make it impractical to accomplish this change in a satisfactory manner. Accuracy is not a problem in digital control systems. With digital systems, the question is whether the system provides an offset signal that you can use to control the fans.

■ Limitations on Fan Size

The amount of wear that occurs to equipment when fans are restarted increases rapidly as the fans get larger. This is because both the starting force and the inertial resistance increase with size. Slamming the torque of a large motor against the inertia of a large fan causes rapid wear of drive belts, bearings, and other components. Above a certain fan size, the saving in energy does not justify the increased maintenance.

■ Maintaining Adequate Outside Air Ventilation

Stopping the fan may excessively reduce outside air ventilation, especially under low-load conditions. Whether this will occur depends on the design of the air handling system. Measure 4.2.1 explains how the intake of outside air is affected by stopping the fans in different types of systems.

If necessary, you can solve this problem by adding a control feature that cycles the fans often enough to maintain adequate ventilation. If you do not have an electronic control system that you can program to perform this function, use two timer relays. One starts the fans if they have not run for a defined period of time, and the other keeps the fans running for the necessary duration.

■ Effect on Space Air Flow

This Measure stops air flow through the spaces when heating or cooling is not being provided. In some environments, this may lead to occupant complaints of stagnation or "stuffiness."

■ Noise Level Changes

An abrupt reduction in fan noise will be heard within the spaces as the fans cycle on and off. In some environments, this may be considered objectionable. On the other hand, occupants may welcome the intervals of silence.

The potential for annoyance depends on how much noise the fans project into the space, the amount of background noise in the space, how often the fans cycle, and other subjective factors, such as the elegance of the space.

■ Building Pressure Fluctuations

Cycling the fans causes fluctuations in building pressure. With smaller single-zone systems, the pressure changes may be too small to notice. Their magnitude also depends on the design of the air handling system, the duct layout, and the building layout. Refer to Measure 4.2.1 to understand how fan operation affects building pressure in different types of systems.

How to Reduce Short Cycling

Avoid frequent cycling of the fans, which causes most of the problems discussed previously.

One step in reducing the cycling rate is to widen the control range of the space thermostat. For example, a thermostat that keeps temperature within a range of 0.5°F cycles the fan approximately twice as often as a fan that controls temperature within 1.0°F. Many thermostats have an adjustable control range. If the installed thermostat is not adjustable, you may be able to substitute one with a larger control range.

Don't overdo this. If the control range is too great, there will be large swings in temperature, and occupants will increase conditioning to avoid discomfort.

The energy saving provided by this Measure occurs mostly at the low end of the load range, whereas cycling is a problem mostly at the higher end of the load range. You can exploit this fact by installing an override that keeps the fans running continuously when the conditioning load is high enough to cause the fans to cycle frequently. For example, you can use the outside air temperature to control the override.

Soft Starting

With larger fan motors, you can reduce the wear and noise that results from frequent starting by installing a motor starter that provides a "soft" start. Such units start the motor with reduced torque by limiting the starting current in one or more steps. Motors differ in their winding designs to respond to different starting torque requirements. Match the starter to the design of the motor windings and to the inertia of the fan.

Control of Outside Air Dampers

Air handling system controls often close the outside air dampers when the supply fan stops. This is to block infiltration during unoccupied periods. However, if the fans are cycled during occupied periods, control the dampers to stay open so that exhaust fans in the building are not starved for air. If you make this change, ensure that the outside air dampers still close normally during off hours. Coordinate this action with the Measures in Subsection 4.2 related to air handling system dampers.

Existing Thermostats with Integral "Fan" Switches

Existing systems that have on-off sources of heating and cooling, such as electric coils and reciprocating compressors, often have thermostats that allow occupants to control the running of the fans. In retrofitting such systems, the hard work is already done.

The remaining problem is that occupants are likely to set the fan switch haphazardly, defeating the purpose. The safest approach is to replace the thermostats or to re-wire them to keep the fan under the control of the thermostat.

Alternatively, if you want to give occupants control of fan operation, install a placard at the thermostat asking them to keep the fan under the control of the thermostat as much as possible. See Measure 4.3.1 about effective thermostat placards.

Increased Potential with Deadband

In applications where deadband is effective (see Measures 4.3.4 ff), it also increases the time that the fan can be stopped. This effect occurs during periods of low load. Deadband does not help to reduce short cycling, because short cycling occurs at higher loads.

ECONOMICS

SAVINGS POTENTIAL: *30 to 70 percent of the energy used by the air handling system fans and other cycled equipment. In 100%-outside-air systems, 20 to 50 percent of heating and cooling energy.*

COST: *Minimal to several hundred dollars, if only thermostatic modifications are required. Up to several thousand dollars, if motor starters and other equipment are installed.*

PAYBACK PERIOD: *Several months, to many years.*

TRAPS & TRICKS

CHOICE OF METHOD: *First, figure out whether this is a worthwhile way to save energy in your air handling system. Consider whether it may cause equipment or occupant problems. With analog control systems, is it worth the bother?*

MEASURE **4.3.3.3 Install multi-speed fan motors.**

SUMMARY

May be the best compromise between efficiency and cost for single-zone systems. Usually free of annoyances. Automatic fan speed switching requires custom control work.

SELECTION SCORECARD

Savings Potential .... $ $ $

Rate of Return, New Facilities % % %

Rate of Return, Retrofit .... % % %

Reliability .... ✓ ✓ ✓

Ease of Retrofit .... ☺ ☺ ☺

Installing multi-speed fan motors is a simple and efficient method of saving fan energy in a single-zone system. This method changes air flow in a stepwise manner that is appropriate for shifting between peak conditions and average conditions. It is well adapted to many single-zone systems, and it has few disadvantages. Unlike the previous two subsidiary Measures in this group, it requires more expensive equipment, but the additional cost is modest.

Energy Saving Potential

■ Fan Energy

In single-zone systems, fan power is proportional to approximately the third power of the air volume delivered. For example, if fan motors with a 2:1 speed ratio are installed, fan power is reduced to approximately one eighth at the lower fan speed. In typical applications, the fan motors operate at the lower speed for a large majority of the time. Thus, total fan energy consumption typically is reduced by well over half. Other speed ratios, and larger numbers of speeds, yield different results.

■ Heating and Cooling Energy in 100%-Outside-Air Systems

The excess energy consumption needed to condition outside air is proportional to the air flow rate, and to the temperature differential between inside and outside. For example, operating the fan at half speed reduces the conditioning of outside air by half. However, the fan must operate at high speed when the temperature differential is greatest, so the overall saving in heating and cooling energy is less than half.

(To be precise, the saving is half the amount of energy required to bring outside air to room temperature. Changing the fan speed has no effect on the amount of energy required to raise the air from room temperature to supply temperature, because this is determined by the space's heat losses.)

■ Reheating Operation

If the system allows reheating, the amount of reheating energy is proportional to the air flow rate while reheat is being employed. (In a single-zone system, reheat means that the heating coil can operate while the cooling coil is operating. Some systems use this technique to control humidity.)

If the system controls are designed for efficiency, they should limit reheat based on the need for dehumidification. Thus, fan speed should have little net effect. Verify these aspects of system control.

Fan Speeds and Motor Selection

See Reference Note 36, Variable-Speed Motors and Drives, for details of multi-speed motors. Select multi-speed motors for high efficiency, as with any motor. In retrofit applications, you can probably get a multi-speed motor that fits the mounting fixtures of the original motor, making installation simple. However, the motor will probably have a larger casing.

In most cases, a motor with two speeds is the reasonable choice. Motors with three or four speeds are available on a custom basis, but they are larger, more expensive, and suffer higher efficiency losses.

A 2:1 speed ratio is appropriate for most applications, although motors with other speed ratios are available. Operating at half speed reduces the system capacity to about 60 to 70 percent of maximum, which is enough for the load conditions that exist most of the time. (The loss of heating and cooling capacity is less than half because coil heat transfer improves at the lower air flow rate.)

How to Switch Fan Speeds

Manual switching of motor speeds is not reliable. After the fan is switched to high speed to satisfy a period of high load, switching the fan back to low speed is likely to be neglected. Therefore, automatic control is mandatory for success in most applications.

Automatic switching of motor speeds requires some custom control work of a simple nature. The speed control must sense when there is a high heating or cooling load in the space. For example, with a pneumatic control system, switching can be initiated with a pressure signal from the thermostat. If the zero-load output of the thermostat is 10 PSI, the motor might be switched to high speed by an output signal lower than 6 PSI (for heating) or higher than 14 PSI (for cooling). Electrical and electronic control systems may also provide a signal that is proportional to the magnitude of the space load. To keep the fan from short cycling between high and low speeds, there needs to be difference between the signal at which the fan switches up to high speed and back down to low speed.

If the thermostat operates on an on-off basis, you cannot use it to control the fan speed, so you need some other indication of load. This can be tricky. For example, if the motor speed is based on the outside air temperature alone, this would keep the motor at low speed under most weather conditions, preventing a fast warm-up or cool-down at the beginning of an occupancy period. In this situation, consider adding a time control that keeps the fan at high speed during the warm-up or cool-down period.

Units with Electric Heating Coils

Reducing the air flow through electric heating coils shortens their life, and it may create a fire hazard. Check the coil specifications to see if reducing the air flow may cause a problem.

You may be able to accommodate reduced air flow by switching the heating elements between parallel and series connections when the fan speed changes. The wiring modifications should be simple. However, there may be a mismatch between coil output and fan speed that reduces the potential for low-speed operation. For example, in a two-element electric coil, switching from parallel to series reduces the heat output of the coil by approximately a factor of four. (The actual reduction is somewhat less than a factor of four because the resistance of heating elements increases with temperature.) Since heating output is reduced so much at the lower fan speed, the fan has to spend more time at high speed.

How to Avoid Resonances

Check for noisy mechanical resonances at each fan speed. Resonances occur because of interaction between the fan and adjacent structure and ductwork. If a resonance occurs, correct any imbalance in the fan. Then, if necessary, try changing the fan speed. If this increases or reduces air flow too much, stiffen or preload the vibrating components to change their resonant frequencies.

Install Placards

Placards are important for both manual and automatic switching. See Reference Note 12, Placards, for details of effective placard design and installation.

ECONOMICS

SAVINGS POTENTIAL: *50 to 85 percent of fan energy, in typical HVAC applications. In 100%-outside-air systems, 25 to 45 percent of heating and cooling energy.*

COST: *Single-winding 2:1-ratio motors cost about $400 per horsepower in small sizes and about $100 per horsepower in large sizes. Motors with other speed ratios are more expensive. If a two-speed motor replaces an existing motor that is worn out, the cost differential is fairly small.*

Installation labor for the motor itself is usually minor. Installation of manual switching typically costs several hundred dollars and up. Design and installation of automatic switching may cost several thousand dollars.

PAYBACK PERIOD: *One year to many years, depending on the size of the motors, the duration of fan operation, and the complexity of controls.*

TRAPS & TRICKS

DESIGN: *Switch the motor speed automatically, if it is practical to do so. This is much more reliable than manual switching.*

SELECTING THE MOTORS: *Select the motors for high efficiency. This requires calls to motor manufacturers, because multi-speed motors are not yet rated for efficiency. See Measure 10.1.1 for other motor criteria.*

EXPLAIN IT: *Unusual switching controls confuse the maintenance staff. Describe the controls clearly in the plant operating manual.*

MEASURE **4.3.3.4 Convert the system to VAV operation.**

Variable-air-volume (VAV) is a mode of air handling system operation in which the temperature of the space is controlled by varying the rate of air delivery. It tailors the output of the fan precisely as the load changes. Therefore, VAV is the most efficient in this group of Measures. Unfortunately, a number of limitations keep this method of fan control from being practical in most single-zone air handling systems.

Feasibility and Alternatives

VAV is rare in single-zone systems. In most cases, the saving in fan energy does not justify the cost. Using a multi-speed fan (Measure 4.3.3.3) saves almost as much fan energy, and does it at much lower cost.

VAV saves conditioning energy in single-zone systems that use 100% outside air. VAV typically reduces conditioning energy in these systems by about a third. Multi-speed fans produce somewhat less energy saving than VAV in these cases.

VAV is more likely to be economical in large fan systems, because the cost per unit of fan power drops rapidly with larger fan sizes.

If cooling is provided by a direct expansion (DX) coil, reducing the air flow through the coil may cause coil freezing. If the DX coil is fed from a compressor that cycles, retrofitting VAV would require special design features that are not economical.

Similarly, if heating is provided by electric resistance coils, the switching of coils in stages generally causes control problems that make it impractical to retrofit VAV.

Energy Saving Potential

■ Fan Energy

A fan operating in VAV mode always delivers the minimum amount of air needed at any given moment. In a single-zone system, the fan power is proportional to the third power of the fan speed, so VAV operation usually saves well over half of the energy that would be used by a constant-volume fan.

All variable-speed fan drives have inefficiencies that partially offset the saving in fan power. The best of these devices have relatively low losses.

■ Heating and Cooling Energy in Once-Through Systems

Air handling systems that use 100% outside air are a special case where VAV may save a substantial amount of conditioning energy in a single-zone system. By reducing air intake at each instant to the amount that is needed for conditioning the spaces, VAV reduces the heating and cooling of outside air.

SUMMARY

The ultimate refinement in saving energy, but usually too expensive for single-zone systems. Usually not practical to retrofit with direction-expansion cooling coils or electric heating coils. May not provide adequate ventilation, dehumidification, and space air distribution at low loads.

SELECTION SCORECARD

Savings Potential .... **$ $** $

Rate of Return, New Facilities **%** **%**

Rate of Return, Retrofit.......... **%** %

Reliability .... ✓ ✓ ✓

Ease of Retrofit .... ☺ ☺

The energy consumption needed to condition outside air is proportional to the air flow rate, and to the temperature differential between inside and outside. Unfortunately, the flow rate required to deliver heating and cooling to the spaces is highest when the temperature differential is greatest. As a result of this interaction, the typical saving produced by VAV is limited to roughly a third of conditioning energy, in typical applications. Even so, this is a large amount of energy in big air handling systems.

(To go into greater detail, the energy saving results from reducing the volume of air that must be heated from outside air temperature to space temperature. The amount of heat required to heat the air from space temperature to supply air temperature depends on the space load. VAV does not save conditioning energy that is related to the space load.)

The potential saving is reduced if the application requires a high ventilation rate.

■ Reheating Operation

If the system allows reheating, the amount of reheating energy is proportional to the air flow rate while reheat is being employed. (In a single-zone system, reheat means that the heating coil can operate while the cooling coil is operating. Some systems use this technique to control humidity.)

If the system controls are designed for efficiency, they should limit reheat based on the need for dehumidification. Thus, fan speed should have little net effect. Verify these aspects of system control.

How to Use VAV in a Single-Zone System

Using VAV in a single-zone system is much easier than using it in a multiple-zone system. One reason is that no VAV terminal units are required. Another is that simultaneous heating and cooling does not occur, so you do not have to go through the contortions that are needed in typical VAV systems to minimize reheat energy waste. Nonetheless, VAV is always tricky. Consider the following aspects carefully.

■ Selecting the Fan Drive

Review Reference Note 36, Variable-Speed Motors and Drives. In the present state of technology, the best choice is usually a variable-speed electronic motor drive. In retrofit applications, consider replacing the motor as well.

■ Control Functions

Select a variable-speed drive package that includes an appropriate control package. The basic mode of operation is for the space thermostat to control the fan output. With hydronic coils, the coil discharge temperatures are held constant.

In most cases, you will want to establish a minimum fan speed. One reason is to maintain adequate ventilation. Another is to maintain proper operation of the diffusers, as explained below. Control packages for variable-speed drives typically include a minimum-speed setting.

When the fan speed falls to the minimum setting, the controls should switch to a mode of operation in which the space thermostat controls the coil discharge temperatures. You will probably need additional controls to perform this function.

If the system provides both heating and cooling, install a control changeover that selects heating or cooling as needed.

■ Outside Air Ventilation

In recirculating air handling systems, the method of controlling outside air quantity depends on the "front end" configuration of the air handling system. This aspect is covered by Measures 4.2.1 and 4.2.2. To maintain adequate ventilation under low loads, set the fan controls for an appropriate minimum air flow rate.

■ Air Distribution Velocity

Refer to "Effect of Space Air Velocity on Comfort" and "Diffuser Performance," in Measure 4.3.3.

Use diffusers that distribute air effectively at low air volumes. Some types of diffusers are designed to exploit Coanda effect to cause the air flow to adhere to the ceiling. This is effective for cooling, but creates a problem with heating. There is no perfect solution to this problem. Using the same diffusers for heating and cooling is a challenge in all VAV systems.

With any type of diffuser, you usually have to set a minimum fan output to prevent dumping of cold air during cooling and to force warm air downward during heating. You can improve diffuser performance at low air volumes by setting the temperature reset to commence before the fan reaches its minimum speed. In this way, the cold air is not as cold by the time the fan falls to minimum speed, and the warm air is not as warm at the minimum speed.

■ Flow and Control Requirements of Refrigerant (DX) Coils

Direct-expansion (DX) refrigerant coils, which are used with heat pumps and direct expansion cooling, usually have minimum flow requirements to avoid coil freezing.

If DX coils are fed by refrigerant units that vary output in stages (for example, by unloading individual cylinders in a reciprocating compressor), insurmountable control problems may occur. The controls that modulate air flow would be confused by the abrupt changes in coil temperature, causing unstable operation.

If the DX coils are fed refrigerant by a continuously modulating compressor, or if the coils draw refrigerant from an accumulator (refrigerant holding tank), the system may be able to operate satisfactorily in VAV mode.

■ Flow Requirements of Electric Coils

Electric coils require a specified air flow in order to avoid burning up. In VAV operation, air flow under low load conditions may not be sufficient for electric coils.

Switching coil elements to vary heat output is generally not satisfactory, because the abrupt changes in heat output would confuse the air flow controls, causing unstable operation. At the present time, there is no satisfactory solution to this problem.

■ Fan Noise and Resonances

VAV keeps the fans as quiet as possible. The fans always operate at the minimum possible speed, and there are no abrupt speed changes. On the other hand, you may have to take steps to correct vibration resonances in the fan system that may occur at certain speeds. See Measure 4.3.3 for more about these topics.

■ Building Pressure

VAV changes fan output over a large range, so it may have a significant effect on building pressure. Whether pressure problems will occur depends on the design of the air handling systems. Refer to Measure 4.2.1 for more about this. The changes in building pressure caused by VAV are gradual, unlike the changes with fan cycling (Measure 4.3.3.2) and multi-speed operation (Measure 4.3.3.3).

ECONOMICS

SAVINGS POTENTIAL: *50 to 80 percent of fan energy. In 100%-outside-air systems, 20 to 50 percent of heating and cooling energy.*

COST: *The cost of variable-flow fan drives is covered in Reference Note 36, Variable-Speed Motors and Drives. In retrofit applications, the control modifications, diffuser replacement, and design work cost at least a few thousand dollars.*

PAYBACK PERIOD: *Several years or longer.*

TRAPS & TRICKS

CHOICE OF METHOD: *Although this Measure provides the best fan efficiency, it is unusual and specialized. Choose another method unless you are sure that the resulting low air flow rates will not cause problems.*

DESIGN: *This is the simplest application for VAV. Still, design of VAV systems is tricky. Study successful VAV installations of this type, if you can find any.*

SELECTING THE EQUIPMENT: *Selecting the best fan drive involves many factors. To get it right, study Reference Note 36, Variable-Speed Motors and Drives.*

MEASURE 4.3.4 Install thermostatic controls that allow space temperature to drift within comfortable limits.

In most HVAC applications, comfort does not require a fixed space temperature. People remain comfortable within a range of temperatures. You can save energy by using thermostatic controls that turn off heating and cooling within this temperature range. The term "deadband," or "temperature deadband," means the range of space temperature in which the thermostatic controls do not call for heating or cooling.

The three subsidiary Measures offer different ways of achieving temperature deadband:

- Measure 4.3.4.1 recommends installing manual switches with the space thermostats that require occupants to select either heating or cooling. This method is commonly used with heating and cooling equipment that cycles on and off. It can be adapted to hydronic coils.

- Measure 4.3.4.2 recommends installing deadband thermostats, which turn off heating and cooling within the comfortable temperature range. This method is commonly used with hydronic coils, and it can be adapted to cycling equipment.

- Measure 4.3.4.3 recommends adjustments or modifications of hydronic coil control valves to create a temperature gap between heating and cooling. This method may be limited to a smaller deadband, but it requires no changes to the thermostats.

Energy Saving Potential

Adding a deadband temperature range in the thermostatic controls keeps the system from wasting energy in these ways:

- *unnecessary heating and cooling while the space temperature is within the comfort range.* The rate of energy waste is low because the conditioning load is small. However, in mild climates, the conditioning system may operate for many hours per year within the comfort range.

- *excessive heating or cooling whenever the system is operating.* Energy consumption for heating and cooling is proportional to the difference between inside and outside temperatures. If the temperature is set arbitrarily within the range of comfort, the system will maintain a larger temperature differential than necessary. For example, if the space temperature is fixed at 76°F during heating operation, instead of being allowed to drift down to 72°F, there is always an additional 4°F of temperature differential to force heat out of the building.

- *heating adds to cooling energy, and vice versa.* In mild climates, the space temperature may drift up and down within the comfort range. For example, it is usually cool in the morning, it gets warmer at midday, and it gets cooler again in the evening. If the space temperature is fixed, cooling is wasted in removing the heat that was added just a short while before, and vice versa.

- *unintentional reheat.* If the heating and cooling temperatures are set close together, small defects or incorrect adjustments in the control system can cause simultaneous heating and cooling. If this occurs, both heating and cooling operate continuously to cancel each other. This overlapping operation may result from incorrect thermostat adjustments, from sticky control components, or from aging of control components. This flaw can waste a lot of energy.

- *unnecessary operation of fans.* In some situations, the fans can be turned off when heating or cooling is not needed. (See Measure 4.3.3.2.) In those situations, failing to use deadband control increases the duration of fan operation.

How Wide Can You Make the Deadband?

People are comfortable indoors within a range of temperatures. For most people, the range extends from a low of 70°F to 72°F, to a high of 76°F to 80°F. This implies a possible deadband range of 4°F to 10°F. Experience confirms this. The highest and lowest acceptable temperatures are variables that are modified by humidity, air flow velocity inside the space, the temperature of walls, clothing, body weight, and other factors.

In many environments, you can increase the deadband range by encouraging occupants to dress appropriately for the weather.

On the other hand, the deadband range is limited by hot spots and cold spots within the space. For example, a person seated near a cold wall may require that the thermostat be set near the top of the comfort range, eliminating most opportunity to exploit deadband. This makes it important to eliminate hot spots and cold spots in the occupied portions of spaces.

Obviously, deadband is not acceptable in environments where close control of temperature is needed for reasons other than comfort.

Dealing with Reheat for Dehumidification

In some cases, a single-zone system may be designed so that the cooling and heating coils operate simultaneously ("reheat") to reduce space humidity. The cooling coil chills the air to remove moisture while the heating coil reheats the air to maintain the space temperature. Reheat is used for dehumidification primarily in humid climates during mild weather.

Reheat consumes large amounts of energy because the heating and cooling coils are continuously expending energy to cancel each other. The system controls should limit reheat operation to weather conditions that require dehumidification. At other times, deadband can operate in the usual way.

You can accomplish this fairly easily by using a humidistat in the space to energize the heating coil when the space humidity becomes too high. If the weather is warm and humid, the space humidistat will start the heating coil at the top of the deadband range. This increases the operation of the cooling coil, which acts to dehumidify the space.

If the weather is cool and humid, the space humidistat will turn on the heating coil, which raises the space temperature. The higher space temperature reduces the relative humidity. If the space temperature reaches the top of the deadband range, cooling will commence, holding the space temperature at the top of the deadband range.

MEASURE 4.3.4.1 Install thermostats that require manual switching between heating and cooling.

You can keep the conditioning system from shuttling between heating and cooling by installing a switch in the space that requires the occupants to select either heating or cooling exclusively. Thermostats that include such a switch are common items. They are intended primarily for use with cycling equipment that is controlled with electrical on-off signals, such as electric heating coils, forced-air furnaces, and direct-expansion cooling units. However, you can use the same principle to control any kind of heating and cooling equipment.

Standard Thermostat Features

Many control companies make thermostats that include a switch for selecting between heating and cooling. Figure 1 shows a typical unit. Look for these features:

- *a switch that selects "heating," "cooling," and "off."* In the "heating" position, the thermostatic can turn on only the heating equipment. In the "cooling" position, the switch can turn on only the cooling equipment. In the "off" position, the switch disables both the heating and cooling equipment. The action of this switch is simply to open or close an electric circuit activating the heating equipment, and another electric circuit activating the cooling equipment.
- *separate thermostats and thermostat settings for heating and cooling.* The thermostats can be any type (electric on-of, pneumatic, electronic, etc.) Not all changeover thermostats have dual settings, so

look for this feature. It allows the optimum heating and cooling temperatures to remain set. Without it, occupants must reset the temperature every time they change from heating to cooling, or vice versa.

- *a switch that controls the operation of the fan(s).* This switch typically has two positions, labeled "on" and "auto." In the "auto" position, the fan turns off when the heating or cooling equipment turns off. In the "on" position, the fan runs continuously. The usual reason to keep the fan running without heat or cooling is to provide outside air ventilation. This applies to systems where the main fans must run to bring in outside air. This feature is available with thermostats that control the heating and cooling equipment with an on-off electrical signal. You probably cannot find this feature with modulating (e.g., pneumatic) thermostats.

How to Switch with Modulating Controls

This Measure is most commonly used with heating and cooling equipment that is controlled with an on-off electrical signal. Connecting the thermostat to this kind of equipment is simple. If the heating or cooling equipment has modulating control (pneumatic, electric analog, etc.), it is easier to use a deadband thermostat, as recommended by Measure 4.3.4.2.

However, you can also install manual selection of heating and cooling with modulating temperature controls. Doing so has some advantages over Measure 4.3.4.2, even though it may be a bit awkward to install. It allows heating or cooling to be turned off individually. It provides for turning off the fan when conditioning is not needed. You can use the "off" position of the selector switch to turn off any other auxiliary equipment, such as hydronic booster pumps.

■ An Example

Figures 2 and 3 illustrate how to add heat-cool-off switching to a modulating thermostatic control. Figure 2 shows a typical air handling unit with hydronic heating and cooling coils and pneumatic coil valves. The heating valve is a "normally open" (NO) type, which means that it is held open by spring pressure. The cooling valve is a "normally closed" (NC) type, which means that it is held closed by spring pressure. The thermostat provides air pressure to the valves. Lower space temperatures lower the air pressure, and higher space temperatures

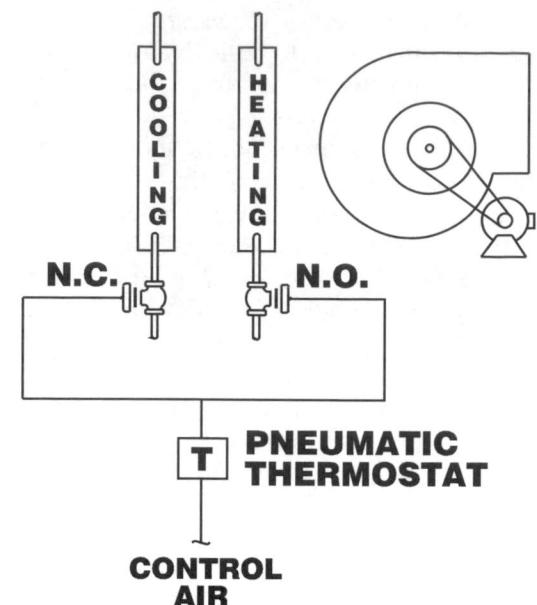

Fig. 2 Typical pneumatic temperature control This simple arrangement provides modulating temperature control, but not the ability to limit operation to either heating or cooling.

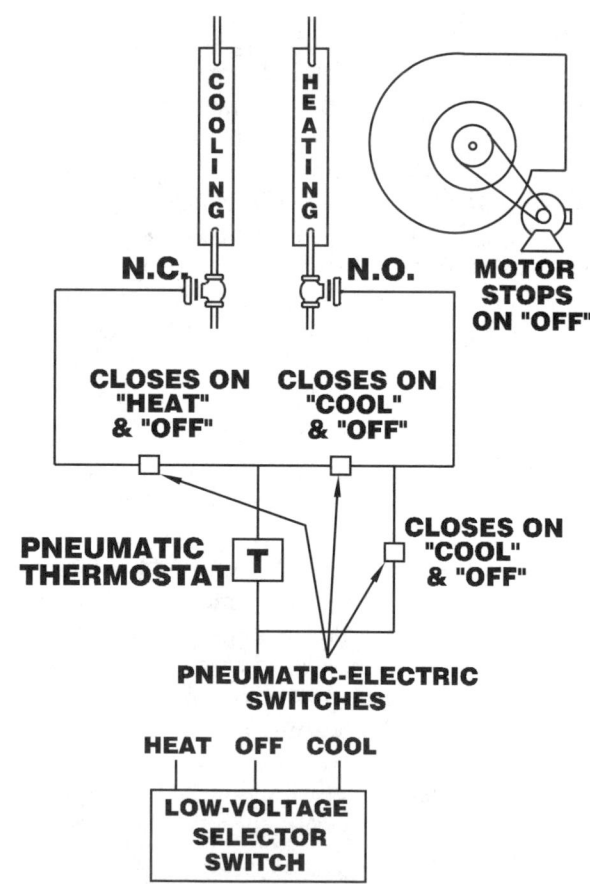

Fig. 3 How to add manual heat-cool-fan selection to modulating control This is the air handling system shown in Figure 2. The additional capability requires three P/E switches and a three-way manual selector switch.

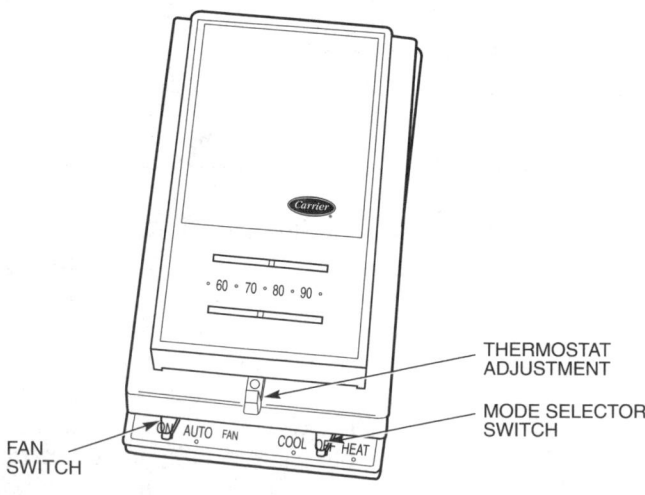

Carrier Corporation

Fig. 1 Thermostat with manual switching between heating and cooling The user can also select whether to operate the fan continuously or only during heating and cooling.

raise the air pressure. This type of control is the cheapest and simplest that is possible with pneumatic valves.

To provide separate control of heating and cooling, install pneumatic-electric switches in the control air lines to the coil valves, as shown in Figure 3. To turn off cooling, the switch in the air line to the cooling coil valve shuts off air pressure. To turn off heating, the switch in the air line to the heating coil valve applies full pressure to the valve. A check valve is installed in this air line to prevent the air pressure from flowing back toward the thermostat. The "heat/cool/off" switch sends the appropriate electric signals to the pneumatic-electric switches.

In retrofit applications, you can keep the original pneumatic thermostat. Add a three-position electric switch to control the pneumatic-electric valves in the control air lines. However, this retrofit does not provide separate heating and cooling temperature settings.

You can use similar methods with electric and electronic controls.

Install Effective Placards

This method of control depends on occupants to select the modes of operation and to set the space thermostat efficiently. Therefore, an effective placard is an integral part of the installation. See Measure 4.3.1 for details.

ECONOMICS

SAVINGS POTENTIAL: *1 to 10 percent of conditioning cost. The higher percentages occur in milder climates, where the overall energy consumption is lower.*

COST: *One hundred dollars to several hundred dollars per thermostat.*

PAYBACK PERIOD: *Less than one year, to many years.*

TRAPS & TRICKS

SELECTING THE EQUIPMENT: *Select controls that are easy to understand and easy to manipulate. In environments where vandalism is a risk, make them exceptionally rugged. If you can, use a standard thermostat with a built-in changeover switch and separate settings for heating and cooling. Let the thermostat control the fan, if the application allows it.*

EXPLAIN IT: *You are depending on the people in the space to select heating or cooling. Make it easy for them to understand. If necessary, install a big, conspicuous placard at the thermostat.*

MEASURE **4.3.4.2 Install deadband thermostats.**

Often the easiest way to achieve temperature deadband is to install a deadband thermostat. The purpose of a deadband thermostat is to insert a range of temperature between heating and cooling in which no conditioning occurs. In retrofit applications, a deadband thermostat is used to replace a single thermostat that controls both heating and cooling sequentially. Deadband thermostats are available for pneumatic, electric, and electronic control systems.

Deadband thermostats correct most of the causes of energy waste listed in Measure 4.3.4. Most deadband thermostats do not save fan energy, but some models are available with a fan switch and connections that allow the thermostat to turn off the fan when there is no heating or cooling.

Types of Deadband Thermostats

A deadband thermostat is essentially two independent thermostats in one housing. One thermostat controls heating and the other controls cooling. There

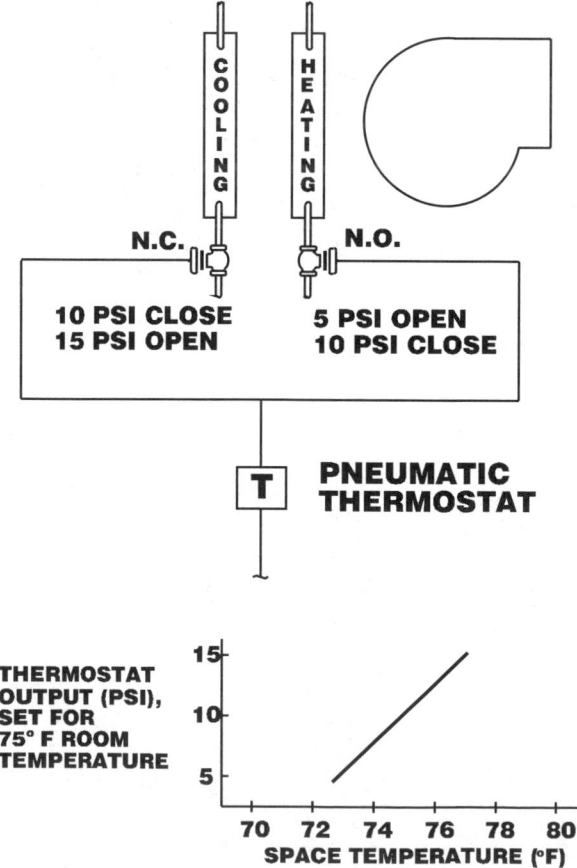

Fig. 1 Typical pneumatic temperature control The graph shows the strength of the output signal as the space temperature changes.

SUMMARY

Inexpensive. Usually easy to install. Different types provide different control characteristics.

SELECTION SCORECARD

Savings Potential .... $ $ $

Rate of Return, New Facilities % % % %

Rate of Return, Retrofit .... % % %

Reliability .... ✓ ✓ ✓

Ease of Retrofit .... ☺ ☺ ☺

---

are several different versions of dual-element thermostats. The differences between them are important, as follows.

■ **Fixed-Deadband, Dual Output Signal**

The most common type of deadband has a single temperature setting. The occupant cannot control the deadband interval.

The thermostat produces two separate control signals, one for the heating equipment and one for the cooling equipment. Some deadband thermostats allow you to adjust the sensitivity of each element individually. This is important if the heating and cooling equipment have different control characteristics. Also, some models allow one element to be connected direct-acting, while the other element is connected reverse-acting. ("Direct-acting" means that the output signal rises as the temperature rises. "Reverse-acting" means the opposite.)

Most deadband thermostats allow the staff to adjust the deadband interval by removing the thermostat cover. Get this important feature.

Figures 1 and 2 show how a deadband thermostat works. Figure 1 is the same type of system used as an example in Measure 4.3.4.1, without deadband. The graph in Figure 1 shows how the pressure output signal of the thermostat changes with space temperature, when the thermostat is set for a space temperature of 75°F.

Figure 2 shows how the control behavior changes when installing a deadband thermostat. One element in the deadband thermostat controls the heating valve, and the other controls the cooling valve. The graph in Figure 2 shows the pneumatic pressure output from the two elements when the thermostat is set to maintain a space temperature of 75°F. The graph shows that both valves are closed between 73°F and 77°F.

Electric and electronic (digital) deadband thermostats operate in a similar manner, except that they respond to electrical input signals rather than to air pressure.

■ Fixed-Deadband, Single Output Signal (Hesitation Thermostat)

A "hesitation" thermostat is a variation of a deadband thermostat that has a single output signal. The name derives from the fact that there is a plateau, or hesitation, in the output signal within the deadband temperature range. Figure 3 shows how a hesitation thermostat works in our example system.

In many applications, you can use either a conventional deadband thermostat or a hesitation thermostat. Hesitation thermostats do not have the control signal flexibility of the dual-output type. For example, use them where heating and cooling are both provided by hydronic coils having similar control valve characteristics.

■ Separately Adjustable Heating and Cooling Temperatures

Some thermostats have separate temperature settings for heating and cooling. The heating and cooling settings are interlocked so that the heating temperature is always lower than the cooling temperature. These are not true deadband thermostats, because there is no way to maintain a fixed deadband when either temperature setting is changed.

The amount of deadband is under the control of the occupants. The advantage of this type is that it gives occupants more explicit control of heating and cooling temperatures. The disadvantage is that occupants may not exploit the energy saving potential of the thermostat. Consider using this type in environments where occupants are permanently assigned to spaces, especially where a single responsible individual controls the thermostat.

How to Set the Deadband

Set the deadband as high as the thermostat allows. Reduce the deadband incrementally if there are legitimate comfort complaints. Expect the maximum acceptable deadband to differ between spaces, especially between interior and perimeter spaces.

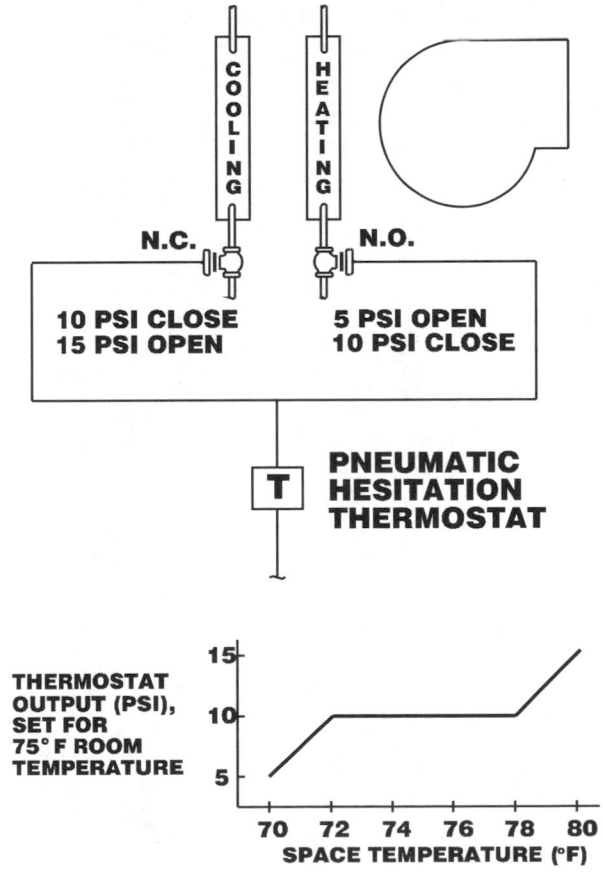

Fig. 2 Dual-output deadband thermostat This is the most common type of deadband thermostat. It has two separate sensing elements, each with its own output characteristics and adjustments. The cooling element controls only the cooling coil, and the heating element controls only the heating coil.

Fig. 3 Single-output (hesitation) deadband thermostat This variation has a single output that goes to both coils. It is less versatile than the dual-output type.

Install Thermostat Placards

This Measure depends on occupants to set the space thermostats efficiently. Therefore, an effective instruction placard is an integral part of each thermostat installation. See Measure 4.3.1 for details.

ECONOMICS

SAVINGS POTENTIAL: *1 to 10 percent of conditioning cost. The higher percentages occur in milder climates, where the overall energy consumption is lower.*

COST: *Deadband thermostats cost less than $100. In retrofit applications, they can replace the original thermostat in an hour or two.*

PAYBACK PERIOD: *Less than one year, to several years.*

TRAPS & TRICKS

SELECTING THE THERMOSTATS: *For most applications, select true deadband thermostats, the kind that have a fixed deadband setting that can be changed only by the maintenance staff. Select units that make it easy for occupants to set the space temperature.*

EXPLAIN THEM: *To forestall vandalism and comfort complaints, install instruction placards that tell occupants how the thermostats control the space temperature. See Measure 4.3.1.*

STAFF TRAINING: *The energy saving is determined by the person with the small screwdriver who sets the deadband in each thermostat. Make sure that the staff are trained to set the maximum acceptable deadband. Put clear instructions in the plant operating manual.*

MEASURE **4.3.4.3 Adjust or modify the coil controls to increase deadband.**

SUMMARY

Inexpensive ways to increase deadband in existing coil controls. With some types of controls, especially pneumatic, the available deadband range may be limited. Does not provide fan control.

SELECTION SCORECARD

Savings Potential .... $ $ $

Rate of Return .... % % % %

Reliability .... ✓ ✓ ✓

Ease of Retrofit .... ☺ ☺

Look for opportunities to create deadband in existing comfort control systems by adjusting or modifying the coil valves or valve actuators. In some installations, you can simply exploit adjustments that are available on the existing coils or actuators. Deadband adjustments may be available with any of the major types of controls (pneumatic, electric, and electronic). We will use pneumatic controls to illustrate the basic methods.

If the coil valves in a pneumatic system are not adjustable, you can achieve the same effect by adding a simple device that offsets the control signal to one of the coils.

These modifications affect the way the thermostat behaves, so expect to adjust the thermostat afterward.

Modifying Valve Actuator Response

Pneumatic controls work by using air pressure from the control system to force valves open or closed against spring pressure. The previous Measures, 4.3.4.1 and 4.3.4.2, increase deadband by modifying the control air pressure. The approach we use here is to change the response of the springs to the air pressure.

For example, in the typical pneumatic controls system shown in Measure 4.3.4.1, the heating coil closes fully at a pressure of 10 PSI and the cooling coil starts to open at the same pressure, so there is no deadband. However, if you increase the tension on the cooling coil spring so that it does not start to open until the air pressure is 15 PSI, this creates a deadband between heating and cooling operation.

Some pneumatic valve actuators have built-in spring tension adjustments. The adjustments may even be calibrated in terms of air pressure. Adjusting these units is simply a matter of turning a knob. Figure 3 in Measure 2.5.2 shows a valve like this. Other types of actuators provide for increasing the spring pressure by placing shims under springs.

Another possibility is replacing the existing actuators with units have different spring ranges. This approach is usually more expensive and less desirable than installing deadband thermostats.

With electric and electronic controls, you may be able to achieve similar results by adjusting the linkages between the electric motors and the valves. An electronic control system may provide a way for you to program it to provide deadband without making mechanical adjustments.

Signal Offset Relays

In pneumatic control systems, an offset relay is a device that changes the pressure in a line by a fixed amount. You can use offset relays in cases where the actuators themselves are not adjustable. In our previous example, the cooling coil valve starts to open at a pressure of 10 PSI. By inserting a offset relay in the line between the thermostat and the cooling valve to lower the air pressure, the opening of the cooling valve is delayed. For example, if you install a 5 PSI offset relay, the cooling coil valve will not open until the thermostat sends 15 PSI to the valve. This creates the same deadband range as in the previous example.

Offset relays that lower air pressure are inexpensive and easy to install. They are simply inserted into the pneumatic line, and require no other connections. Relays that increase air pressure are also available, but these require a connection to the system air supply, and they are more expensive.

Deadband Range Limitations with Pneumatic Controls

Depending on the type of control system, you may not be able to achieve the maximum comfortable deadband using these methods. With pneumatic controls, the deadband range is limited by the control air supply pressure. For example, if the maximum pressure that the thermostat can produce is 22 PSI and if the cooling valve operates over a 5 PSI range, then the offset relay should not increase the opening pressure above 17 PSI. Otherwise, the cooling valve will not be able to open fully.

With typical thermostat characteristics, supply air pressure limits the maximum deadband to about 2°F to 4°F. This is considerably less than comfort allows.

You can increase deadband somewhat by increasing the control air pressure. This is usually a simple matter of changing the setting of the pressure regulator at the air source. Do not exceed the maximum safe pressure of any component served by the pneumatic system. You may not be able to change the system pressure if the pneumatic system has devices, such as setback thermostats, that are activated by changes in the supply pressure.

Another way to increase deadband in pneumatic systems is to reduce the sensitivity of the space thermostats. The sensitivity of the thermostat is the difference in control output that the thermostat produces for a given change in temperature. For example, a typical pneumatic thermostat has an adjustable range of sensitivity from 1 to 4 PSI per degree Fahrenheit.

Here is an example that illustrates the effect of changing the thermostat sensitivity. Assume that the cooling valve in our example opens at a pressure 6 PSI greater than the heating valve closes. If the thermostat is set for highest sensitivity (4 PSI/°F), the deadband is only 1.5°F. On the other hand, if the thermostat is set for lowest sensitivity (1 PSI/°F), the deadband is 6°F.

If you push this method to far, it affects comfort. Reducing the thermostat sensitivity too much allows the space temperature to drift too far from the thermostat setting.

Thermostat Calibration

The temperature figures on the thermostat become meaningless when you create deadband. To minimize confusion, calibrate the thermostat dial so that the numbers correspond approximately to the center of the deadband range.

ECONOMICS

SAVINGS POTENTIAL: 1 to 10 percent of conditioning cost. The higher percentages occur in milder climates, where the overall energy consumption is lower.

COST: Equipment costs, if any, are typically minor. For example, an offset relay costs about $20. An experienced controls technician should be able to accomplish the work in an hour or two.

PAYBACK PERIOD: Less than one year, if you can create deadband with simple adjustments. Several years, if you need to make controls modifications.

TRAPS & TRICKS

EXPLAIN IT: Don't expect anyone to figure out what you did unless you explain it. Spell out the deadband setting procedure in the plant operating manual. Be sure to explain interactions between your method of setting deadband and other thermostat adjustments, such as sensitivity.

Single-Duct Reheat Systems

The Measures in this Subsection improve the efficiency of single-duct reheat systems. In these systems, the supply air is cooled at the air handling unit. The temperature of individual spaces is controlled by reheating the supply air in terminal units. The terminal units are controlled by the space thermostats. Figures 1 and 2, in the Section 4 Introduction, sketch a typical single-duct reheat system.

There are two common variations of single-duct reheat systems:

- *terminal reheat systems*, in which each space has one or more reheat terminal units controlled by the space thermostat
- *zone reheat systems*, in which one reheat terminal unit serves several spaces that make up a zone of thermostatic control

In physical features, there is no difference between these two types except for the number of terminal units. A larger number of terminal units provides more accurate control of temperature in localized areas. This is important when there are significant differences in heat gain or heat loss between areas.

Induction systems are a related type, but they have major differences in equipment. In these systems, the air handling unit delivers chilled air at exceptionally high pressure. This "primary" air is passed through a venturi terminal unit in each space that induces the flow of "secondary" space air through a heating or cooling coil. Because of their special hardware and operational requirements, induction systems are covered separately in Subsection 4.9.

By far the largest energy saving opportunity in single-duct reheat systems is minimizing reheat. Reheat is very wasteful. First, excess energy is expended to cool the supply air. Then, still more energy is expended to partially re-warm the same air before it enters the space. During low-load conditions, much more energy may be cancelled out

INDEX OF MEASURES

4.4.1 Set the cooling coil discharge at the highest temperature that maintains satisfactory cooling.

    4.4.1.1 Install automatic chilled air temperature reset control.

4.4.2 Turn off the air handling unit cooling coils when cooling is not needed.

4.4.3 Turn off reheat coils when practical.

4.4.4 In terminal units that blend supply air with reheated air, block the reheat passages during the cooling season.

4.4.5 Trim the fan output.

4.4.6 Install multi-speed fan motors.

4.4.7 Convert the system to variable-air-volume (VAV) operation.

4.4.8 Replace all reheat coils in a system with heating/cooling coils and minimize operation of the air handling unit cooling coil.

4.4.9 Install self-contained heating/cooling units, and use the air handling system only for ventilation.

RELATED MEASURES

- Subsection 4.7, for Measures that improve the efficiency of existing VAV single-duct systems
- Subsection 4.9, for Measures that improve the efficiency of induction systems

in this manner than actually enters the space to provide cooling or heating.

The main justification for the popularity of reheat systems is comfort. They provide a high level of comfort because air flow is constant, control of temperature can be maintained precisely as the load changes from heating to cooling, and humidity can be controlled at any desired level. In making changes for the sake of efficiency, try to preserve these comfort advantages as much as possible.

Perhaps another factor in the popularity of constant-volume reheat systems is that any fool can lay one out without spending much effort on calculations or control design. All that is required is to install an oversized cooling coil, oversized reheat coils, and let the sizing mistakes cancel when the air is mixed. The owner pays for this negligence with high energy bills throughout the life of the building.

The Measures in this Subsection are oriented primarily toward reducing the energy waste that occurs from reheat. Some of the Measures also save a substantial amount of fan energy. In new construction, none of these Measures adds a relatively large cost increase, unless the systems are small.

In retrofit applications, first decide whether or not to make major changes to the existing constant-volume system. This decision usually depends on two factors: how much money you can spend, and whether you can forego the tight control of humidity and ventilation that constant-volume reheat systems provide. If you cannot afford expensive changes, accomplish all of Measures 4.4.1 through 4.4.5 that are relevant to your systems and applications.

If you can afford major system modifications, consider Measures 4.4.7 through 4.4.9. These three Measures are alternatives to each other. Any of the three will provide the combined benefits of the less expensive Measures. Measure 4.4.6 is an intermediate approach that saves less energy than the latter three Measures, and also costs considerably less. You can combine it with the first five Measures.

MEASURE 4.4.1 Set the cooling coil discharge at the highest temperature that maintains satisfactory cooling.

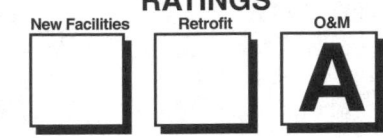

RATINGS
New Facilities   Retrofit   O&M

[ ]   [ ]   **A**

The amount of energy that is wasted by reheat depends on the cooling coil discharge temperature in the air handling unit. A lower chilled air temperature forces each terminal unit to expend more energy to re-warm the air to the appropriate supply temperature for the space. Therefore, the simple act of raising the chilled air temperature is a powerful energy conservation action. And, it is easy to accomplish. See Figure 1.

The optimum chilled air temperature changes continuously, unless the cooling load is constant. Don't expect the staff to keep resetting the chilled air manually as conditions change throughout the day. The best they can do manually is to reset the chilled air at intervals, perhaps on a seasonal basis. Subsidiary Measure 4.4.1.1 improves on this Measure with continuous automatic resetting of the chilled air temperature. Use manual adjustment as a stopgap method until you install automatic reset controls.

SUMMARY

Save a lot of energy just by turning a screw.

SELECTION SCORECARD

Savings Potential .... $ $ $ $

Rate of Return .... % % % %

Reliability .... ✓ ✓

Ease of Initiation .... ☺ ☺ ☺

Energy Saving

Let's calculate the energy saving at an individual terminal unit under typical conditions. Here are the conditions:

- the terminal unit delivers 1,000 CFM
- the space is cooled to 75°F
- the cooling coil maintains a supply air temperature of 55°F
- the air is dry enough so that all the cooling load is sensible
- the cooling load in the space is half its maximum, so that a supply air temperature of 65°F would suffice at this particular moment.

This table shows the effect of raising the chilled air temperature:

Energy Input, BTU/hr	Supply Air Temperature		
	55°F	60°F	65°F
cooling energy to space	10,800	10,800	10,800
cooling energy from coil	21,600	16,200	10,800
reheat energy	10,800	5,400	0
total input energy	32,400	21,600	10,800

This example shows how horribly wasteful reheat systems can be. At the original chilled air temperature of 55°F, the system consumes three times as much energy as is needed to cool the space. Experience with actual buildings confirms that reheat systems really are this wasteful.

Reheat systems are most wasteful at low loads. In fact, a reheat system consumes the most energy in absolute terms when the space load is zero, because the output of the cooling coil must be completely cancelled with reheat energy. In most climates, the cooling requirement is only a fraction of the design maximum

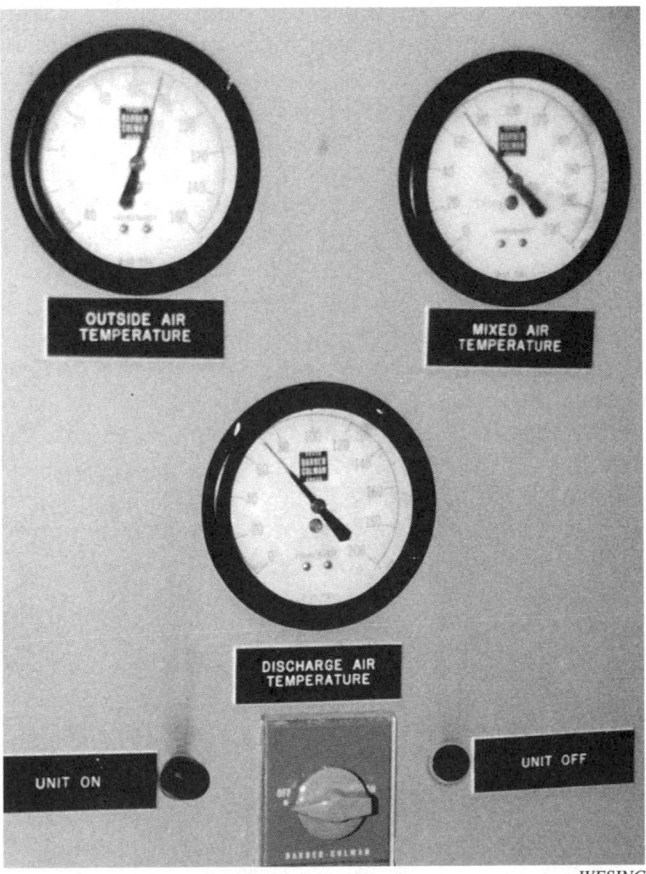

WESINC

Fig. 1 Chilled air temperature control Just by setting this knob the right way, you can save a large amount of energy in a reheat system.

load for much of the time. This fact keeps the reheat system in a very wasteful mode of operation.

How to Maximize the Chilled Air Temperature

Most reheat air handling systems have a chilled air thermostat that controls the cooling coil discharge temperature. You can reset this thermostat easily with a pocket screwdriver. If you have a centralized energy management system, you can achieve the same result by tapping on a keyboard.

Designers sometimes shave installation cost in hydronic reheat systems by omitting the control valve from the cooling coil. In such cases, the chilled air temperature is tied to the chilled water temperature. To raise the chilled air temperature independently, install a cooling coil valve that is controlled by a supply air thermostat. If you go to that expense, spend the additional money to install a reset controller, as recommended by subsidiary Measure 4.4.1.1.

Limitations Caused by Load Diversity

Your ability to raise the chilled air temperature may be seriously limited by diversity in the cooling loads of the different terminal units. The chilled air temperature can be set no higher than the temperature needed by the terminal unit having the largest cooling load.

A single space that has inadequate cooling can thwart your attempt to raise the chilled air temperature. The space may have an exceptionally high load, or the terminal unit serving the space may be undersized, or a control defect may limit the output of the terminal unit. As you raise the chilled air temperature, be aware of which spaces run out of cooling capacity first. Examine the cooling of these spaces to see if any of these situations exists. For example, if a space with high solar gain requires exceptionally low chilled air temperature, try shading the space. Or, replace the terminal unit with one having higher air flow capacity. (Of course, the former solution is more efficient.)

Load diversity depends on the way the building is zoned. For example, if a reheat system serves a zone with extensive glazing on both the east and west sides of the building, then the load remains high for most of the day. This illustrates a major reason to pay attention to zoning when laying out an air handling system. Seize this opportunity in new constructions and renovations, because changing zoning on a retrofit basis is expensive.

Effect on Humidity

Raising the supply air temperature setting causes the coil surface to be warmer, so the coil does not condense as much moisture out of the air. As a result, raising the chilled air temperature allows the humidity inside the building to rise, assuming that the air entering the coil is humid. The dew point of the supply air is always somewhat lower than the supply air temperature.

For example, if the supply air temperature is 65°F, the dew point may be 60°F. The difference depends on the surface temperature of the cooling coil surface, the bypass ratio of the coil, and the percentage of outside air.

Problems with excess humidity may arise when the weather is mild and humid. They may also occur if a large amount of moisture is released within the space, for example, if there is a lot of decorative foliage. Basements are a special humidity problem, because contact with the earth keeps basement walls cool enough to condense moisture out of the air. Lower humidity may be needed to satisfy comfort standards in posh environments. Some manufacturing and storage applications require exceptionally dry air. In such situations, be cautious about reducing the chilled air temperature.

Humidity is not a problem during cold weather because cold outside air cannot carry much moisture, and air exchange keeps internal humidity tied to the external humidity. Humidity is not a problem during warmer weather if mechanical cooling is used, because operation of the cooling equipment dries the air.

Two factors keep humidity from becoming a problem in most locations, even during mild weather. One is that the outside air temperature swings widely during the day in most locations, so that conditions that create high internal humidity persist only for a few hours, usually not long enough for humidity to become objectionable. The normal operation of cooling equipment during part of the day dries out the spaces.

The other helpful factor is heat gain. In a building with typical levels of heat gain, the heat gain alone usually keeps humidity from becoming excessive. As an extreme example, consider a rainy day in New Orleans with the outside temperature at 65°F. If heat gain (from solar gain, lighting, etc.) provides a temperature rise of 8°F, the relative humidity inside the building can be no higher than 77%, even if the air handling system is turned off. If the cooling system operates at all, even without reheat, the relative humidity in the space will fall even lower.

Explain It and Keep Reminding!

This activity is easily forgotten. Therefore, document the procedure in a manner that keeps it in the eye of the plant operators. Install an effective placard at the discharge air thermostat that explains how to set it. See Reference Note 12, Placards, for details of effective placard design and installation.

ECONOMICS

SAVINGS POTENTIAL: *10 to 40 percent of cooling and reheat energy.*

COST: *Minimal.*

PAYBACK PERIOD: *Immediate.*

TRAPS & TRICKS

TRAINING AND DILIGENCE: *This Measure depends on the diligence and skill of the plant operators to keep the chilled water temperature as high as possible. The cooling load changes widely over the course of a day, so almost continuous attention is needed to achieve the maximum saving. Make sure that plant operators are trained and understand the importance of the* adjustment. Put clear instructions in the plant operating manual. Install a placard at the thermostat that identifies it and explains its purpose. Schedule periodic checks of the procedure.

CHOICE OF METHOD: *If you cannot be sure of maintaining effective staff performance, install an automatic chilled water reset controller, as described next. In fact, there are few instances where manual control is more than a stopgap approach.*

MEASURE 4.4.1.1 Install automatic chilled air temperature reset control.

This is a refinement of Measure 4.4.1 that automatically keeps the chilled air temperature as high as possible under all load conditions.

How to Install Chilled Air Temperature Reset

Chilled air temperature reset is a standard control function. Usually, the reset controller determines the appropriate chilled air temperature by sensing the signal from a space thermostat. In pneumatic control systems, a receiver-controller is used to perform the function. In buildings that have an energy management computer system, you can usually accomplish this function through the computer system.

For maximum efficiency, connect the reset controller to all the space thermostats that may dictate the chilled air temperature at one time or another. You may need to sense the thermostat signals from a number of spaces, because the heat loads of different spaces do not vary in unison. The thermostat that is calling for the greatest cooling at any given moment is selected by installing a signal discriminator. For further detail, see Reference Note 14, Control Signal Polling.

For example, solar load is greatest in east-facing perimeter spaces in the morning, but in west-facing spaces in the afternoon. In addition, the reset controller should sense thermostats in any spaces that may experience exceptionally high heat gains, such as auditoriums, conference rooms, dining areas, and computer rooms.

■ Increase the Potential for Reset

The benefit of this Measure is limited by load diversity, or by any conditions that cause a space to require abnormally low chilled air temperature. Once the reset control has been installed, it becomes easier for you to find out which spaces are limiting your ability to raise the chilled water temperature.

If one space, or a small number of spaces, require substantially lower chilled water temperature, reduce the load on these spaces, or provide greater cooling air flow, as suggested by Measure 4.4.1. Repeat this process until further improvements no longer provide enough increase in chilled water temperature to be economical.

■ Humidity Override

In applications where space humidity may become excessive, install a space humidistat to override the chilled air temperature reset when necessary. See Measure 4.4.1 about conditions that may require a humidity override.

SUMMARY

Most reheat systems with hydronic cooling coils should have this control feature. It is the only way to keep chilled air near its optimum temperature.

SELECTION SCORECARD

Savings Potential	$ $ $ $
Rate of Return, New Facilities	% % % %
Rate of Return, Retrofit	% % % %
Reliability	✓ ✓
Ease of Retrofit	☺ ☺ ☺

■ Systems with Economizer Cycles

Economizer cycles are methods of exploiting low outside air temperatures to cool the interior of a building. In a reheat air handling system that is equipped with an economizer cycle, the controls should keep the chilled air temperature as high as possible while the economizer cycle is operating. This is the same as with the operation of the cooling coil. See Measure 4.2.5 about airside economizer cycles, and Measure 2.9.4 about waterside economizer cycles.

ECONOMICS

SAVINGS POTENTIAL: 20 to 60 percent of cooling and reheat energy, typically.

COST: The control components for a typical installation typically cost from several hundred dollars to several thousand dollars. Connection to thermostats at widely dispersed locations can incur high labor cost. If an energy management computer is installed, programming the computer to accomplish chilled air temperature reset typically costs several thousand dollars.

PAYBACK PERIOD: Less than one year, to several years.

TRAPS & TRICKS

PLANNING AND DESIGN: This is more than a control modification. To get the most benefit from this Measure, correct any conditions that cause some spaces to require lower chilled water temperature than the rest. In climates that are humid when temperatures are mild, you will need controls that respond to humidity as well as to temperature.

EXPLAIN IT: *Describe the reset control in the plant operating manual. If you use a separate control panel for this function, install a placard on the panel to identify it.*

MONITOR PERFORMANCE: *The reset control can fail, or it can be adjusted improperly, without anyone noticing.*

Therefore, check performance periodically. Include a column in the plant operating log to record the chilled air temperature, or use your energy management computer to log this information. Review the record periodically to verify that the chilled air temperature rises as the cooling load falls.

MEASURE 4.4.2 Turn off the air handling unit cooling coils when cooling is not needed.

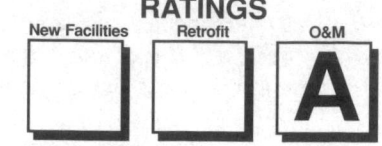

If you turn off the cooling coil in an air handling system, you eliminate all reheat. If your system lacks control features to eliminate reheat, turn off your cooling coils during periods when you expect that cooling will not be needed.

Almost all air handling systems with hydronic cooling have cooling coil isolation valves that you can shut. In systems with direct expansion cooling, you can turn off the power to the compressors.

If your systems do not have supply air temperature reset controls, this Measure is a stopgap until you install reset controls, as recommended by Measure 4.4.1.1. If the reset controls are adjusted properly, they will turn off the cooling coils when there is no cooling load.

Even if your systems have reset controls, it is still a good idea to turn off the coils on a seasonal basis to protect against inadvertent reheat resulting from coil leakage or a control failure. Put this on your schedule.

Energy Saving Potential and Limitations

The Measure eliminates reheat losses when they are greatest. This is because reheat systems consume the largest amount of energy when there is no cooling load. In a constant-volume reheat system, the cooling coil is always at maximum output (if the cooling coil discharge temperature is held constant). Therefore, operation of the system when there is no cooling load requires an amount of reheat energy equal to the cooling energy. All the cooling energy is wasted, along with the heating energy needed to cancel it.

Conversely, this Measure can provide an energy saving only during periods of time when the system has no cooling load. In practice, the staff cannot be standing by each air handling unit to turn off the cooling coils whenever the cooling load disappears. For this reason, this Measure is likely to be a seasonal activity.

The cooling coils may be turned off in different air handling units at different times. For example, air handling systems serving perimeter spaces with substantial solar gain must keep their cooling coils turned on longer than air handling units serving only shaded perimeter spaces. In many buildings, the core area of the building requires cooling all year. If a particular air handling system serves any spaces located in the core

---

SUMMARY

An easy method of eliminating reheat during periods when cooling is not needed. A stopgap measure to use until you install a chilled air temperature reset controller (Measure 4.4.1.1). Also a safety feature to use with a reset controller.

SELECTION SCORECARD

Savings Potential .... **$ $ $**

Rate of Return .... **% % % %**

Reliability .... **✓ ✓**

Ease of Initiation .... ☺ ☺ ☺

---

area, you may have to keep the cooling coil operating continuously for the benefit of these spaces.

Explain It and Keep Reminding!

Neglecting this procedure produces no discomfort or other apparent effects, so the action is likely to be neglected. Document the action in a manner that keeps the operating staff aware of it.

ECONOMICS

SAVINGS POTENTIAL: *10 to 60 percent of cooling and reheat energy, depending primarily on the weather and the system zoning.*

COST: *None, usually.*

PAYBACK PERIOD: *Immediate.*

TRAPS & TRICKS

EXPLAIN IT: *Success depends on training and supervision. Provide clear instructions that explain when to turn off the cooling coils, e.g., when the outside air temperature is below a certain temperature. Optimize the schedule for each air handling unit or group of units. Install placards at the cooling coil valves or at the controls of the compressors in self-contained air handling units.*

MONITOR PERFORMANCE: *This Measure is easy to forget. Schedule periodic checks of the procedure in your maintenance calendar.*

MEASURE 4.4.3 Turn off reheat coils when practical.

You can eliminate reheat by turning off the reheat coils. This method eliminates the ability of the air handling systems to control the different spaces individually. In effect, it turns a reheat system into a single-zone cooling system.

For most applications, this is a crude stopgap measure. It has been used widely during energy shortages because it eliminates energy waste due to reheat. The discomfort that results from eliminating individual space temperature control can be serious. When applied indiscriminately, this kind of desperate action gives energy conservation a bad name.

Applications, Limitations, and Alternatives

This method is used during warm weather, at times when all the spaces served by the air handling system require cooling. It can also be used during cooler weather with systems that exclusively serve the interior areas of a building.

This action is likely to create some discomfort except in the most favorable cases, such as a system that serves only similar core areas. If the system serves the perimeter area of the building, they must all have the same exposure to the sun, which is unlikely because of differences in window area, shading by adjacent buildings, etc.

Don't use this method unless you are desperate. If you need to reduce energy consumption immediately and you don't have much money for efficiency improvements, try Measures 4.4.1 ff and Measure 4.4.5 first.

Control Changes

With the reheat coils turned off, you need to change the way the cooling coil discharge temperature is controlled. Originally, the coil is controlled to maintain a constant discharge temperature. When you turn off the reheat coils, the discharge temperature needs to

SUMMARY

Saves a lot of energy cheaply, but is usually limited to situations where you need to save energy on an emergency basis. Likely to cause discomfort because it eliminates individual space temperature control.

SELECTION SCORECARD

Savings Potential .... $ $ $

Rate of Return .... % % % %

Reliability .... ✓ ✓

Ease of Initiation .... ☺ ☺ ☺

follow the space load. To do this with a hydronic cooling coil, use a space thermostat to control the coil, instead of the original coil discharge thermostat. With a refrigerant (DX) coil that is controlled by staging the compressor, use a space thermostat to control the compressor.

With the reheat coils in operation to provide heating, use the original thermostatic controls. Make it easy to switch between the original controls and the modified controls.

ECONOMICS

SAVINGS POTENTIAL: *10 to 50 percent of cooling energy. 30 to 80 percent of reheat energy.*

COST: *Minimal.*

PAYBACK PERIOD: *Short.*

TRAPS & TRICKS

CHOICE OF METHOD: *This is a desperation measure for systems that do not have the proper controls. If you need to do this, get your system fixed.*

MEASURE **4.4.4 In terminal units that blend supply air with reheated air, block the reheat passages during the cooling season.**

The Measure applies to a type of reheat terminal in which supply air is discharged through two parallel passages. One is straight through, and the other is through a heating coil. See Figure 1. Flow through the two passages is controlled by linked dampers that respond to the space thermostat. The volume of air flow is intended to remain approximately constant.

With this type of terminal, you can eliminate reheat during part of the year by blocking the reheat coil passage. This makes the terminal unit behave like a shutoff VAV terminal.

Applications and Limitations

If you have this unusual type of terminal unit, you can use this trick as long as no heating is required in the space served by the terminal. In a given air handling system, you may be able to use this method with some terminal units, but not with others. Or, you may have to use it on different schedules with different terminal units.

SUMMARY

An inexpensive but clumsy way to eliminate reheat seasonally in systems that use an uncommon type of terminal units. May cause noise and inadequate air flow.

SELECTION SCORECARD

Savings Potential .... $ $ $

Rate of Return .... % % % %

Reliability .... ✓ ✓ ✓

Ease of Initiation .... ☺

The activity is awkward and takes time, so you can do it only on a seasonal basis. You may be able to permanently modify terminal units that serve core areas, provided that the core areas need only cooling. This does not work if the core area needs a morning warm-up during cold weather.

In perimeter areas, the eligible period of the year depends on the climate and internal heat gains.

This Measure has all the limitations of a shutoff VAV system, which are explained in Measure 4.4.7. These include the possibilities of poor diffuser performance, inadequate air circulation, inadequate outside air ventilation, and diffuser noise at low loads.

This Measure is compatible with other inexpensive methods of reducing reheat losses, especially with chilled air temperature reset (Measures 4.4.1 ff) and fan trimming (Measure 4.4.5).

Energy Saving

This Measure eliminates all or most reheat and excess cooling in spaces where it is used, during the period of time when it can be used. This Measure functions only during the warmer part of the cooling season. Reheat losses are lowest under these conditions, but they are still serious.

The annual saving varies widely, depending mostly on the climate. In warm climates, the Measure may eliminate a majority of reheat energy, along with the cooling energy that it cancels. In colder climates, the Measure may be limited to only a couple of months per year.

The saving potential is greater in core spaces than in perimeter spaces, because the Measure can be used there for longer durations.

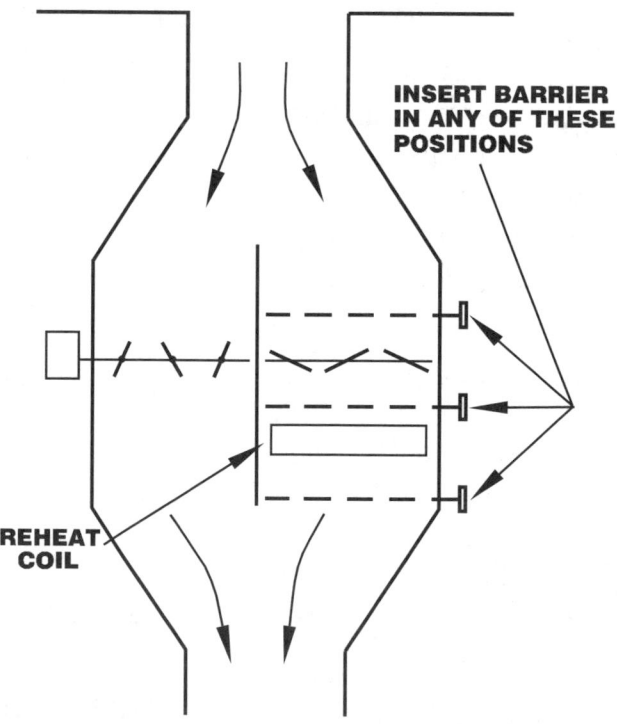

CHILLED AIR SUPPLY

INSERT BARRIER IN ANY OF THESE POSITIONS

REHEAT COIL

Fig. 1 Terminal unit with separate reheat path You can make this type of terminal unit act as a VAV terminal by blocking the reheat passage.

The Measure provides a small reduction in fan energy because the reduction in total system air flow causes the fan to draw less motor power.

How to Modify the Terminal Units

Take a look and see whether you have this type of terminal unit. It is uncommon because it is bulky and offers no significant advantages compared to simpler reheat terminals. (These units do not require reheat coil valves, but this cost saving is offset by additional dampers and sheetmetal.)

If you have this type of terminal, you need to devise a method of blocking the reheat passage of the terminal unit. You may have a choice of locations for doing this, as shown in Figure 1. With some models, you can do this easily by inserting a rectangular piece of sheetmetal through a joint in the casing. Unless you intend to block the reheat passage permanently, make the barrier easy to install and remove.

Expect it to be awkward to install and remove the inserts. You will have to bring a ladder into the space, find each terminal unit, and work through the ceiling.

ECONOMICS

SAVINGS POTENTIAL: *In predominantly warm climates, 50 to 80 percent of reheat energy and 30 to 60 percent of cooling energy. In predominantly cool climates, less than 30 percent of reheat and cooling energy. Also, 2 to 15 percent of fan energy.*

COST: *$30 to $100 per terminal unit for the initial modification, plus a fraction of an hour of labor per terminal unit each season to install and remove the blocking insert.*

PAYBACK PERIOD: *Several months, typically.*

TRAPS & TRICKS

SCHEDULING AND DILIGENCE: *This is a kind of job that the staff would rather avoid. Make sure that it gets done. Designate a reliable place to store the terminal unit inserts during the periods when they are not needed.*

MEASURE 4.4.5 Trim the fan output.

Fan trimming is a one-time adjustment that gets rid of fan capacity in excess of the amount needed to satisfy peak loads. Refer to Measure 10.3.1 for details.

This is a cheap and simple way to reduce reheat significantly while saving a modest amount of fan energy. In a reheat system, both excess cooling energy

and reheat energy are proportional to the air flow rate. This Measure reduces both "off the top." However, it does not tailor fan output to a changing load.

Consider this a stopgap action to save energy until you can accomplish one of Measures 4.4.6 through 4.4.9.

MEASURE 4.4.6 Install multi-speed fan motors.

In a reheat system, both excess cooling energy and reheat energy are proportional to the air flow rate. In a constant-volume reheat system, the fan output is determined by the need to deliver heat or cooling during extreme load conditions. However, extreme heating or cooling conditions typically occur for only a small fraction of a system's operating hours. This makes it worthwhile to adjust fan output to follow the cooling and heating loads.

One method of doing this is to install a multi-speed fan motor. This method changes air flow in a stepwise manner that is appropriate for shifting between peak conditions and average conditions. It is a compromise between efficiency and installation cost.

SUMMARY

Saves a large amount of reheat, cooling, and fan energy. Not as efficient as VAV, but much less expensive as a modification and less susceptible to problems.

SELECTION SCORECARD

Savings Potential	$ $ $ $
Rate of Return, New Facilities	% % % %
Rate of Return, Retrofit	% % %
Reliability	✓ ✓ ✓
Ease of Retrofit	☺ ☺

Comparison to VAV

The principle of this Measure can be extended to its extreme by converting the system to variable-air-volume (VAV) operation. This is recommended by Measure 4.4.7, next. Multi-speed fans do not provide all the savings that are possible with VAV. On the other hand, multi-speed fans maintain a substantial minimum air flow rate. This avoids much of the trouble that arises in VAV systems when they are operating at very low flow rates.

In retrofit, this Measure is much less expensive than VAV because it avoids the cost of replacing all the existing terminal units, and it avoids the higher cost of a VAV fan drive and fan controls.

Energy Saving Potential

▪ Reheat and Excess Cooling Energy

In a reheat system, both reheat energy and the excess cooling energy required to cancel it are proportional to the air flow rate. (This is not precise, because coil heat transfer improves somewhat as the flow rate is reduced.) For example, if you install a motor with a 2:1 speed ratio, air flow will be reduced by half for much more than half the time. Therefore, reheat energy and the corresponding excess cooling energy typically are reduced by more than half.

▪ Fan Energy

In constant-volume reheat systems, fan power is proportional to approximately the third power of the air volume delivered. For example, if fan motors with a 2:1 speed ratio are installed, fan power is reduced to approximately one eighth at the lower fan speed. In typical applications, the fan motors will operate at the lower speed for a large majority of the time. Thus, total fan energy consumption typically is reduced by well over half.

▪ Heating and Cooling Energy in 100%-Outside-Air Systems

In 100%-outside-air systems, the air handling systems must cool and/or dehumidify the air starting from outside conditions. In warm weather, this is more energy intensive than conditioning return air. This Measure reduces the excess energy consumption in proportion to the air volume reduction.

(To be precise, reducing the fan output reduces proportionately the amount of energy required to bring outside air to room temperature. It has no effect on the amount of energy required to raise the air from room temperature to supply temperature.)

In cool weather, reducing the amount of outside air may not save energy. It may require less energy to take in outside air than to cool recirculated air. This is the principle of the outside air economizer cycle, which is explained in Measure 4.2.5.

How to Select the Multi-Speed Motor

See Reference Note 36, Variable-Speed Motors and Drives, for details of multi-speed motors. Also, see Measure 10.1.1 about selecting efficient motors in general.

In retrofit applications, you can usually find a multi-speed motor that fits the mounting fixtures of the original motor, making installation simple. However, a multi-speed motor is larger in overall dimensions.

How to Switch Fan Speeds

For most applications, use automatic fan speed switching. Manual switching of motor speeds is not reliable. After the fan is switched to high speed to satisfy a period of high load, switching the fan back to low speed is likely to be neglected.

Automatic switching of motor speeds requires some custom control work that can sense when there is a high heating or cooling load in the space. For example, with a pneumatic control system, you can switch the motor speeds with a pressure signal from the thermostat. If the zero-load output of the thermostat is 10 PSI, the motor might be switched to high speed by an output signal lower than 6 PSI (for heating) or higher than 14 PSI (for cooling). Electrical and electronic controls systems may also provide an offset signal that indicates the magnitude of the space load. To keep the fan from short cycling between high and low speeds, there needs to be a difference between the signal at which the fan switches up to high speed and back down to low speed.

Control the fan speed with the space thermostat that has the greatest conditioning requirement. The fan switches to the lower speed only when the reduced air flow can satisfy the conditioning requirement of this thermostat. The maximum conditioning load may shift from one space to another, so you may have to install controls that identify the space with the greatest load. See Reference Note 14, Control Signal Polling, for the methods of doing this.

How to Avoid Resonances

Check for noisy mechanical resonances at each fan speed. Resonances occur because of interaction between the fan and adjacent structure and ductwork. If a resonance occurs, correct any imbalance in the fan. Then, if necessary, try an adjustment in fan speed to eliminate the noise. If this throws off the fan speed too much, stiffen or preload the vibrating components to change their resonant frequencies.

How to Avoid Problems at Reduced Air Flow

A number or problems may appear at the reduced air flow rate. You can probably find satisfactory solutions to these problems. See Measure 4.3.3 for details. For example, if changing the fan speed affects outside air intake, add control capability to readjust the outside air quantity when the fan speed changes.

ECONOMICS

SAVINGS POTENTIAL: 30 to 60 percent of cooling and reheat energy, typically. 50 to 80 percent of fan energy, typically.

COST: Single-winding 2:1-ratio motors cost about $400 per horsepower in small sizes and about $100 per horsepower in large sizes. Two-winding motors, which are needed to achieve other speed ratios and larger numbers of speeds, are considerably more expensive.

The cost of installing the motor itself is usually minor. Installing manual switching typically costs several hundred dollars and up. Designing and installing automatic switching may cost several thousand dollars.

PAYBACK PERIOD: One year to many years, depending on the size of the motors, the duration of fan operation, and the complexity of controls.

TRAPS & TRICKS

CHOICE OF METHOD: Consider this as a cheaper and less efficient alternative to VAV. In retrofit, it may be the only practical alternative, but consider Measures 4.4.8 and 4.4.9 also. When installing new systems, VAV or another efficient type of system is usually the best choice.

SELECTING THE EQUIPMENT: Select the motors for high efficiency. This requires calls to motor manufacturers, because multi-speed motors are not yet rated for efficiency. See Measure 10.1.1 for other motor criteria.

EXPLAIN IT: The operation of unusual switching controls is not obvious. Describe the controls clearly in the plant operating manual.

MEASURE 4.4.7 Convert the system to variable-air-volume (VAV) operation.

In principle, you can eliminate virtually all the energy waste of the constant-volume reheat design by converting the system to variable-air-volume (VAV) operation. The principle of VAV is simple. The cooling input to individual spaces is controlled by varying the flow rate of the supply air, rather than by reheating the supply air. This offers two major efficiency advantages: (1) the system provides only the amount of cooling that is needed by the space, which eliminates the need for reheat to control temperature, and (2) fan power is minimized at all times.

Although the VAV concept is simple, making it work well in practice has proven to be challenging. Many VAV systems infringe on comfort or they sacrifice much of the efficiency potential of the concept. The changing air flow rates that are an inherent part of the concept create a variety of problems. To achieve a satisfactory installation, you need to deal with each of these problems individually. As you do this, the simple concept may lead to complex control requirements.

Energy Saving

■ Reheat Energy and Excess Cooling Energy

The example at the beginning of Measure 4.4.1 illustrates the extreme level of energy waste that occurs in reheat systems. This example is confirmed by actual experience, which shows that reheat systems may consume several times more energy than the amount actually delivered to the spaces for cooling and heating.

To calculate the saving in reheat that you can expect from using VAV operation instead of constant-volume operation, you need to calculate the total energy consumption of each terminal unit for both types of systems. Do this as illustrated by the example in Measure 4.4.1.

In new construction, you can design a VAV system that eliminates all or most of this energy waste. A well designed VAV system typically has reheat losses less than 10% of those in a constant-volume reheat system. In retrofit applications, you may not be able to reduce reheat as well as you can in designing a new conditioning system.

For reasons explained below, you may be forced to use VAV terminals that have heating coils. This practice incurs considerable reheat. However, the reheat occurs only at the low end of the air flow range. As a result, the energy loss is much lower than that which occurs in constant-volume reheat systems, where reheat must overcome excess cooling at full air flow.

SUMMARY

The most efficient configuration for reheat systems. You have to work out a lot of details to achieve both comfort and high efficiency.

SELECTION SCORECARD

Savings Potential .... $ $ $ $

Rate of Return, New Facilities % % %

Rate of Return, Retrofit.......... % % %

Reliability .... ✓ ✓ ✓ ✓

Ease of Retrofit .... ☺ ☺

With VAV terminals that have heating coils, reheat energy waste occurs at each terminal unit individually. The amount of reheat that occurs at each terminal unit depends on these factors:

- *whether the space requires cooling or heating.* The cooling coil in the air handling unit produces chilled air. If the space needs cooling, and if the terminal unit can throttle the air flow sufficiently to match the amount of supply air to the cooling load in the space, then no reheating occurs.

  On the other hand, when the space requires heating, a heating unit must operate. The heating unit may be a coil inside the terminal unit, or it may be an entirely separate heating unit. The heating unit must satisfy the heating load in the space, and it must also cancel any cooling that enters the space through the air handling system.

- *the minimum-flow requirement.* For reasons explained below, it may be necessary to maintain a minimum flow through the terminal unit under all conditions. If the amount of chilled air that passes through the terminal provides more cooling than needed, the excess cooling must be cancelled by reheat. Using heat/cool changeover control of the terminal units, as discussed below, may eliminate or greatly reduce the minimum-flow requirement. If you use terminal units that can close to shutoff, leakage in the terminal units becomes an important factor. Any leakage behaves as a minimum-flow setting, and it wastes energy in the same way.

- *the chilled air temperature.* If the chilled air temperature is raised, the amount of cooling provided by a given amount of air flow through the terminal is reduced. Under conditions when reheat is required, raising the chilled air temperature reduces the amount of reheat that is necessary to cancel unwanted cooling. Chilled air temperature reset is a powerful energy conserving technique. However, it is limited by diversity of space cooling loads. This is because the chilled air temperature must be set low enough to satisfy the cooling load on the terminal unit with the greatest cooling requirement.

It is important to understand that the amount of reheat energy waste in a VAV system is not fixed and predictable at the time the system is designed. On the contrary, energy waste depends largely on system adjustments that are made at the time of installation and later throughout the life of the system. These adjustments are needed to deal with issues that we will discuss below. To a large extent, the adjustments are arbitrary, or they are a matter of preference. A terminal unit in one space may be adjusted to eliminate most reheat energy waste, whereas an identical terminal unit in the adjacent space may be adjusted in a way that causes a large amount of reheat energy waste. To minimize reheat energy waste in a VAV system, adjust each terminal unit properly. In addition, exploit chilled air temperature reset as much as possible.

■ Conditioning Outside Air in Once-Through Systems

In systems that require 100% outside air, such as those used in hospitals, VAV additionally provides a major reduction in the amount of energy required to condition outside air. This is because the constant-volume system must take in enough outside air to convey heating and cooling under peak load conditions. However, the average conditioning load is much less than the peak load. A VAV system takes in only enough outside air to convey the heat or cooling that is required at the moment.

This component of savings is greatest in warm, humid climates. In such locales, the constant-volume system must waste a large amount of energy in cooling and dehumidifying a large quantity of air. Conversely, in cool climates, bringing in outside air saves cooling energy by reducing the cooling load on the cooling coil. (This is the principle of the economizer cycle, which is covered by Measure 4.2.5.) When space heating becomes the dominant load, VAV substantially reduces the amount of energy needed to heat outside air.

■ Fan Energy

A fan operating in VAV mode delivers only enough air flow to carry the required amount of heating or cooling energy into the spaces.

The relationship between fan power and air flow varies considerably, depending on the type of terminal units, the zoning of the system, the duct layout, the method of fan control, and other factors. The power required by the fan to overcome the air flow resistance of the coils and ductwork is approximately proportional

to the third power of the air flow rate. The power required to overcome the resistance of the dampers in the terminal units is roughly proportional to the first power of the flow rate. (The latter is because the dampers act to keep the pressure opposing the fan constant, instead of letting the pressure drop at lower flow rates.) The actual power required by the fan varies between these two extremes.

(There is a widespread misconception that fan power in a VAV system is proportional to the third power of the air volume delivered. This error leads to excessively optimistic estimates of fan energy savings. This misconception results from failure to recognize the variable resistance of the terminal units.)

All the methods of varying fan output in VAV systems (variable-speed drives, inlet vortex dampers, discharge dampers, etc.) have inefficiencies that partially offset the saving in fan power. The best of the methods preserve most of the fan energy savings. The worst dissipate a large fraction of the potential savings.

VAV Design Challenges

Behind the simple concept of VAV lies a multitude of potential problems. These problems arise mainly because of two difficulties that are characteristic of VAV systems:

- *providing heating.* A single-duct VAV system is basically a cooling system, whereas a reheat system can provide either cooling or heating with equal ease. If some of the spaces served by a VAV system require heating, you need to add heating capability to the VAV system. Selecting the heating method involves serious compromises, especially in retrofits.

- *low air flow at low load.* VAV operation produces large variations in air delivery to the spaces. This variation may cause a number of problems. Dealing with these problems involves further compromises.

These characteristics tend to force VAV system design into a compromise between discomfort, energy waste, and control complexity. As you design a VAV installation, consider the following factors. Many of them are related, which makes VAV design a feat of mental juggling. If you deal with all these factors systematically, you will be able to predict the performance of your system with reasonable certainty and you will avoid unpleasant surprises. Further below, we discuss specific heating methods and specific methods of dealing with the problems of low space air flow.

■ Outside Air Ventilation

If you need to maintain uninterrupted outside air ventilation, this is likely to require some reheat. In a single-duct VAV system, the chilled air supply from the air handling unit is the only source of outside air. When cooling is needed, this presents no problem. However, when heating is needed, the outside air coming through the chilled air supply needs to be heated.

The amount of energy waste depends on the ventilation requirement, on the fraction of outside air in the supply air, and on the supply air temperature. For example, a typical U.S. ventilation requirement for commercial buildings is 15 CFM per person. If outside air comprises one third of the supply air, then 45 CFM of chilled air is delivered to the space for each person. The extra 30 CFM of supply air has to be cooled and then reheated to room temperature.

With a shutoff VAV system, no outside air ventilation can enter the space while the terminal unit damper is closed. In a shutoff system with deadband thermostats, this condition may persist for long periods during mild weather. This problem occurs only during mild weather, so the solution may be simply to open a window. However, this option is not available in many buildings.

A system may waste more energy by sloppy control of outside air quantity than by heating the air that is actually needed to satisfy ventilation standards. In recirculating air handling systems, the method of controlling outside air quantity depends on the "front end" configuration of the air handling system. A well designed air handling system maintains accurate control of outside air intake even as the fan output changes. Unfortunately, this ideal is rarely achieved. The amount of outside air intake tends to vary strongly as the fans change output. Overcoming this tendency may require improvements to the system's "front end," as recommended by Measure 4.2.2.

Even if the amount of outside air drawn into the total system is controlled well, the outside air is still distributed unevenly to individual spaces. For example, if a system serves both core and perimeter spaces, most of the air may be delivered to the core spaces on a cool day, and to the perimeter spaces on a hot day. You have to decide whether it is satisfactory to distribute ventilation air by letting it diffuse from one space to another. This may be acceptable if the occupant density is low and the air quality is benign.

The only way to eliminate energy waste in ventilation and to distribute it accurately is to install a separate outside air ventilation system, with its own ducts, fans, heating, and cooling elements. This approach is worth considering, although it is rare at present.

■ Dehumidification

Reducing the air flow rate into a space reduces the dehumidifying effect of the air handling system. This is most likely to be a problem in humid climates when mild temperatures keep the dampers in the terminal units closed. Measure 4.4.1 covers situations that may encounter humidity problems.

If there are times when normal cooling does not provide adequate control of humidity, some reheat may be needed. The cooling coil is operated to wring moisture out of the air, and the heating coil (or other heating equipment) restores the temperature. If the system uses reheat for dehumidification, design the controls to limit reheat to the amount needed.

The most efficient approach is to use a space humidistat to control reheating at low loads when necessary. A danger is that humidistats, especially cheaper types, are poorly calibrated and tend to drift. Thus, they could waste energy for years without the fact being noticed. To minimize this problem, use humidistats of the highest quality.

■ Diffuser Performance

When the system is providing cooling at low fan outputs, cold air tends to dump out of the diffusers, rather than diffusing throughout the space. This can cause discomfort to occupants located near the diffusers. VAV systems have become notorious for this problem.

Reduced air flow can also cause a problem when heating. In this case, the warm air tends to stay at ceiling level, rather than mixing with the rest of the air in the space.

There are large differences in the effectiveness of diffusers at low air velocity. In converting constant-volume systems, be prepared for the possibility that the present diffusers will not be satisfactory for VAV. This is especially likely if you install VAV boxes that can shut off completely. This issue is discussed further, below.

Chilled air temperature reset (explained in Measure 4.4.1) substantially reduces the tendency for chilled air to dump at low loads. Reset increases the volume of air at low cooling loads. This improves diffusion. It also decreases the temperature differential between the chilled air and the space air, which reduces the convective force that causes dumping.

Another solution is to use fan-powered VAV terminals, which maintain a minimum air flow through the diffusers by recirculating air within the space. This type of terminal unit is covered below.

■ Circulation of Air Within the Space

VAV operation greatly reduces the air delivery rate during periods of low cooling or heating load. It is not clear whether this factor contributes to discomfort, in itself. In early comfort theory, lack of air movement was considered to be uncomfortable. However, experience shows that many people are comfortable in spaces that have minimal air circulation.

Later research suggests that complaints of "stagnation" are due primarily to high space temperature or humidity. Lowering one or both usually satisfies a complaint of "stagnation," but at the cost of increased cooling energy.

Complaints of "draft" in a VAV system are usually not due to air flow that is too high, but to dumping of cold air from the diffusers, as discussed previously.

In many environments, you can provide air circulation with little energy cost by installing high-volume, low-velocity circulation fans, such as ceiling paddle fans. A bonus is that the cooling effect of circulation fans can significantly reduce the cost of mechanical cooling. However, each fan can serve only a limited amount of space. Ceiling fans also create an annoying stroboscopic effect if they are installed underneath ceiling light fixtures.

Fan-powered VAV terminals can maintain a minimum air velocity in the space, but they are not as effective as in-space circulation fans for this purpose. Both in-space fans and fan-powered VAV terminals have a noise penalty.

■ Space Pressure Differentials

The large variations of air delivery in a VAV system may cause troublesome changes in building pressure. Whether problems occur depends on the nature of the facility, the layout of the air handling systems, the layout of the spaces, and the "front end" design of the air handling systems. Measure 4.2.1 covers the design of systems to minimize pressure problems.

The amount of air delivered to each space is independent of the amount delivered to the others, so pressure differences can occur between spaces. In applications where isolation is important, such as hospitals and laboratory buildings, this characteristic may disqualify a VAV system. The pressure differences are usually not a problem in non-critical applications, such as office buildings.

■ Noise

Modulating the speed of the supply and return fans, which is the best method of modulating air flow, substantially reduces their average noise level. The changes in noise level that occur as the fan output changes are so gradual that they are not noticeable.

VAV terminal units may produce noise that is loud enough to be annoying, especially when they are operating near the fully closed position. Minimize this problem by selecting the terminal units for low noise, as discussed below. In addition, you can reduce terminal unit noise by lowering the duct pressure at low loads, as discussed below.

■ Fan-Induced Resonances

At certain specific speeds, a fan may trigger resonances in the air handling sheetmetal or in the building structure. This is a significant possibility in VAV systems because the fan operates across a continuous range of speeds.

Air handling system resonances are rarely serious, if the system equipment is installed properly. The vibrational frequency of centrifugal fans is higher than the resonant frequencies of typical structures and ductwork. If a serious low-frequency resonance does occur, it is likely to be caused by an unbalanced fan wheel, which you can correct by balancing it. If this does not solve the problem, stiffen the vibrating components, which increases their resonant frequency and reduces the intensity of vibration.

■ Air Cleaning

Air cleaning is covered by Measure 4.2.4. If the air handling system is used for air cleaning, by filtering or other processes, the rate of cleaning is proportional to the flow rate. VAV reduces air flow, so it also reduces air cleaning.

Design Outline for VAV

Follow these steps to convert a constant-volume air handling system to VAV, either in designing a new system or in converting an existing system:

- Decide how to provide heating in spaces that require heating. This decision affects most of the other aspects of the modification.
- Replace the reheat terminals with VAV terminals.
- If necessary, replace the air diffusers in the spaces to provide good performance at all air delivery rates.
- Install an efficient means of varying the output of the air handling system fans.
- Install a control system for the fan output.
- Install reset control for the supply air temperature, and perhaps, for the duct pressure.
- Provide controls for various auxiliary functions, such as morning warm-up, night setback, outside air economizer cycle operation, smoke relief, etc.

There may also be other modifications that are appropriate at the same time, such as changing the duct layout, even though these are not an inherent part of the VAV concept. The rest of this discussion covers these steps in greater detail.

How to Provide Heating

VAV systems are basically cooling systems. Converting a constant-volume system to VAV raises the question of how to provide heating. In principle, you could use any kind of heating equipment with a VAV systems. Practical considerations usually limit the choice to heating coils in the terminal units, or to convectors in the spaces, or to heating coils in the air handling unit. The difficulties of providing heating with a VAV system at reasonable cost tempt the designer to make compromises with comfort and health. If compromises are necessary, keep them reasonable. Here are some important factors to consider in selecting the heating equipment for a VAV system.

■ Morning Warm-Up

The need for morning warm-up may require heating equipment to be installed even where it is not needed for normal occupancy. This situation commonly occurs in core areas of the building. Even though these spaces may always have positive heat gain during normal occupancy, they can cool down during a period of temperature setback.

An alternative to installing heating equipment in all the spaces is installing a single warm-up coil in the air handling unit. This approach reduces initial cost if it avoids the need for a large number of heating units. However, in a system where some of the terminal units have heating coils, the fan must overcome the resistance of the coils in the terminal units in series with the resistance of the warm-up coil in the air handling unit.

■ Terminal Units with Heating Coils

The most common choice of heating method when converting a constant-volume system to VAV is to install terminal units that have heating coils. This decision usually is driven by cost constraints and by the need to minimize interference with the operation of the facility during conversion.

Unfortunately, this choice commits the system to a certain amount of reheat, because some chilled air is needed to carry heat from the coils into the space. The amount of energy that is wasted in this manner depends on the diversity of space loads. Especially during mild weather, some spaces may require heating while other spaces served by the same air handling system require cooling.

For example, if a system serves heavily glazed spaces on both the east and west sides of the building, the west side may require a large amount of cooling in the late afternoon of a cold day, while the east side needs heating. This example shows that zoning has a major effect on your ability to minimize reheat in a VAV system that uses reheating terminal units.

Reheat energy loss occurs even with shutoff VAV terminals, if they have heating coils. However, shutoff terminals have much less reheat loss than terminals with minimum-flow settings.

Fan-powered terminal units are an important special case, which is discussed below. Fan-powered terminals can avoid reheat entirely because they do not depend on the chilled air supply to carry heat into the space.

It may seem ironic to remove one set of terminal units with heating coils, only to replace them with other terminal units that also have heating coils. The key point of this swap is that the heating coils in VAV terminals waste energy only during periods of low cooling load, whereas the heating coils in a constant-volume reheat system waste energy during all cooling operation, and most severely during periods of high cooling load.

Selection of VAV terminal units is covered more extensively below.

■ Heating Equipment that is Separate from the VAV System

In principle, you can eliminate reheat entirely by installing a heating unit in each space that does not depend on air flow from the VAV system to distribute

heat. For example, you can combine a cooling-only VAV system with a baseboard heating system, or with fan-coil units, or with unit ventilators, etc.

(In practice, separate heating equipment may not entirely eliminate reheat. If the system is not controlled properly, or if perimeter spaces are not isolated from core spaces, heat from the perimeter system may move into the core and cancel cooling. If you are careful in laying out the system, this effect is minor.)

Baseboard convectors are by far the most common type of separate heating equipment that is used with VAV systems. Many people consider convectors to be the most comfortable type of heating. They are quiet and free of drafts, and they warm the interior surfaces of outside walls. They take up little space. They are less expensive than other types (but they are usually much more expensive to retrofit than terminal units with heating coils). Also, most other types of separate heating equipment, such as fan-coil units, could perform both heat and cooling, making VAV unnecessary.

If you install baseboard convectors, you have to work around their inability to provide outside air ventilation. You could install a small, separate ventilation system. Or, you could ventilate through the VAV system. The latter approach may not be possible without substantial reheat or control complexity.

A combination of in-space heating units and shutoff VAV terminals cannot provide dehumidification during mild, humid weather. Providing dehumidification requires reheat or a separate dehumidifier.

How to Select the Terminal Units

A VAV terminal unit, or "VAV box," is basically a damper that is controlled by the space temperature. The damper throttles the air supply as the cooling load drops. VAV terminals have many variations. Figures 1, 2, and 3 show typical examples. For a good introduction to VAV design, make a study of all the types available and the

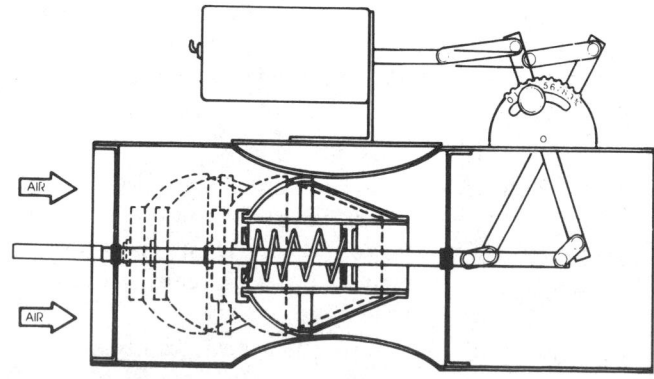

M&I Heat Transfer Products

Fig. 1 VAV terminal unit This externally powered unit has an aerodynamically shaped plug that provides accurate flow control, low noise, and tight shutoff. It wastes a minimum of fan power in creating turbulence.

reasons for them. The following are the major issues to consider.

■ Shutoff or Minimum-Flow?

VAV systems are designed in one of two basic ways. "Shutoff" systems have terminal units that can stop air flow completely. "Minimum-flow" systems have terminal units that do not close below a particular minimum flow rate. This flow rate is typically fairly high, around 40% of maximum flow.

The distinction between shutoff and minimum-flow is a matter of design intentions and terminal unit adjustments, not primarily a matter of hardware. Virtually all VAV boxes have mechanical stops that can be used to set minimum-flow limits. VAV boxes intended for shutoff applications need better construction to prevent air leakage at shutoff, and to minimize noise at low flow rates.

In minimum-flow systems, each terminal unit has reheat losses when the space load is low. This is because heating equipment must cancel the excess minimum-flow cooling from the terminal unit.

The amount of reheat loss that occurs in a minimum-flow system depends on the system zoning. If a system serves spaces with loads that vary in different patterns, this diversity limits the opportunity to raise the chilled air temperature. As a result, the reheat loss is higher.

Minimum-flow terminals are especially wasteful in systems where reheat depends on heating equipment installed in the space. (This configuration is rare.) To avoid discomfort from the cold supply air in such cases, the space temperature must be kept fairly high by the heating equipment, increasing the reheat loss.

■ Reheat Coils

Minimum-flow VAV terminals almost always have reheat coils. Hydronic reheat coils can be modulated to provide any percentage of heat output. This allows them to be controlled to minimize reheat losses. Hydronic heat may have much lower operating cost than electric heat, if a cheaper boiler fuel is used. The main disadvantage of hydronic reheat coils is the high cost of piping.

Reheat cannot be reduced as well with electric heating coils, because these are installed in fixed blocks of capacity. Furthermore, control of electric coils is more complex because of the need to avoid burn-out. The simple controls used for electric coils in many VAV systems are wasteful. See Measure 4.7.7 for details.

■ Power Source for Terminal Unit Dampers

VAV terminal units may be "externally powered" or "system powered." Externally powered units use pneumatic actuators or electric motors to move the dampers. System-powered terminal units use the pressure of the air in the duct system to drive the damper.

System-powered boxes require higher fan pressure than externally powered boxes because a minimum duct pressure is required to activate the damper mechanism. The minimum pressure at the box is typically about one inch of water column. The extra fan pressure required to operate system-powered VAV boxes costs a significant penalty in fan energy.

A third type of terminal unit is self-powered. It has a damper that is powered by the expansion of a fluid in a specialized space thermostat. The damper and the thermostat may be installed on the diffuser assembly, making a compact package that is inexpensive to install. The primary advantage of this type is low cost. The lack of external control can be a major disadvantage because spaces that are served by this type of terminal unit cannot be shut down individually.

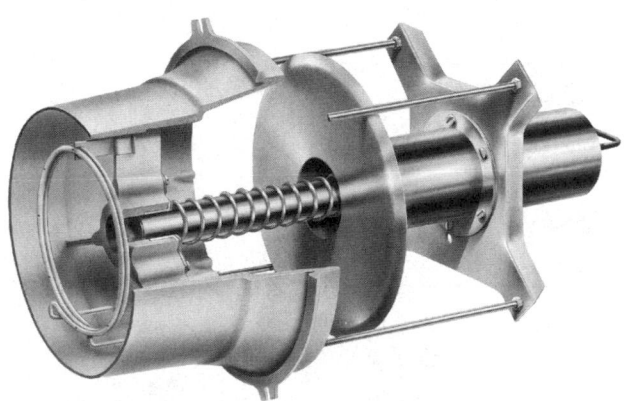

Trane Company

Fig. 2 VAV terminal valve with flow sensor Air flows from the left. The disc in the center moves along the shaft to throttle the flow. The narrow coiled tube at the entrance to the air valve has holes that face into the supply duct. This provides an indication of air flow and duct pressure that is used to optimize control of the unit.

Trane Company

Fig. 3 Simple VAV terminal This inexpensive model uses a single-blade damper that is driven by an electric motor. This type of damper has relatively poor control at the lowest and highest flow rates, and it is susceptible to leakage.

■ Sensitivity to Duct Pressure

For a given damper position, the amount of air flow through a terminal unit is strongly affected by the duct pressure. This is significant because large fluctuations can occur in the pressure at a given terminal. The fluctuations are greatest when a terminal unit is located at the end of a duct run that has other terminal units. The operation of the upstream terminal units bleeds air pressure out of the duct, reducing flow through the downstream terminals.

In principle, the space thermostat should compensate for differences in duct pressure. However, any thermostat requires a certain temperature offset to function, and the offset is greatest when the conditioning load is greatest. This resulting temperature drift can be large enough to cause discomfort.

To deal with system pressure fluctuations, "pressure-independent" VAV boxes were developed. These typically contain two dampers. One regulates flow in response to the thermostat. Another damper is added upstream to regulate the pressure that is felt by the flow regulating damper.

"Pressure-independent" boxes require more fan power than "pressure-dependent" units because of the additional resistance of the pressure regulating damper. For this reason, try to design the system so that pressure-independent boxes are not needed. Good duct layout helps to achieve this goal, by reducing pressure fluctuations in the duct system.

Another way to deal with duct pressure changes is to use thermostatic controls that are capable of maintaining close temperature control in the presence of large pressure fluctuations. See Reference Note 37, Control Characteristics, for methods.

■ Terminal Unit Noise

Terminal units can generate a substantial amount of flow noise when the damper approaches its minimum position. In the worst cases, the terminal unit emits a distinct whistle. Good terminal unit design can reduce this noise to levels that are acceptable for most applications.

Some types of terminal units have used internal silencers to reduce noise. Silencers increase flow resistance, so avoid units that use them.

■ Air Leakage at Shutoff

Shutoff terminal units should close tightly. Otherwise, cold air will leak into the space, causing discomfort and wasting energy. Tight shutoff is mainly a matter of sturdy construction. The units should be designed so that any seals needed for tight shutoff last a long time and can be replaced easily.

■ Retrofit Difficulty

Replacing the terminal units in an existing system can be a knuckle-bleeding chore. It helps to keep the replacement terminal units as small as possible. If the replacement units are larger than the original ones, there will probably be places where installing them is difficult.

Fan-Powered VAV Terminals

Fan-powered terminals are a newer development in the history of VAV. They avoid or minimize some of the problems of VAV, but they introduce some new problems of their own. They are important as an option to consider. We will discuss them separately because they are different from conventional VAV terminals in construction, operation, and control.

Fan-powered terminals are more than a substitute for conventional VAV terminals and diffusers. They

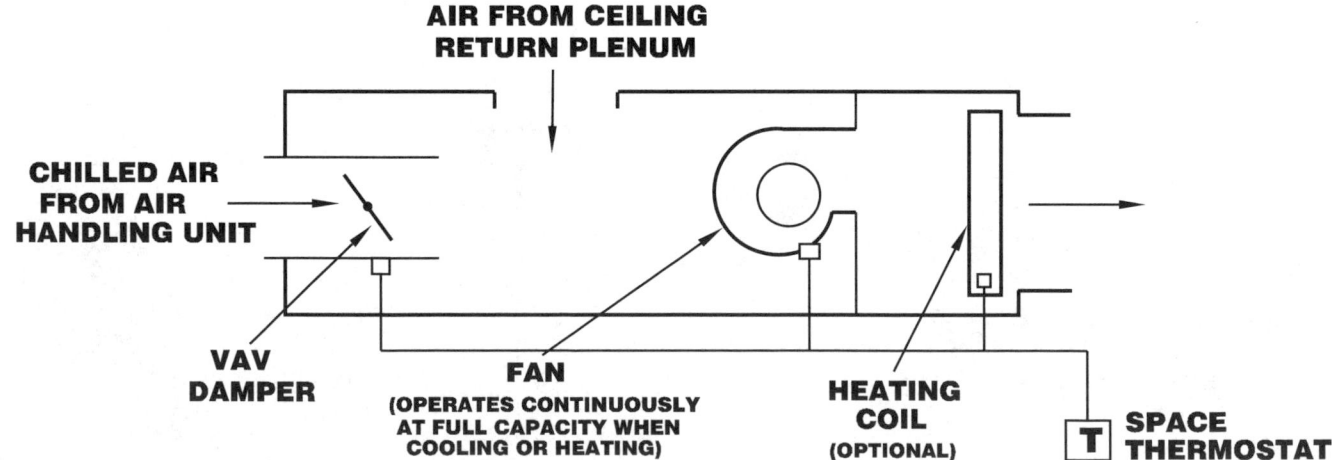

Fig. 4 Fan-powered VAV terminal, series flow   The fan operates continuously as long as conditioning is required. The difference between the fan's flow rate and the chilled air flow rate is made up by recirculation from the space or the ceiling plenum. The heating coil operates only when the space requires heating. During heating, the chilled air damper may close completely, or it may remain partially open to provide outside air ventilation. Recirculating air from the ceiling plenum recovers the heat of light fixtures, either for space heating or for reheat to reduce humidity. Most units have fans that operate at fixed output, but modulation is a possible option.

change the overall design of the system. A VAV system that uses fan-powered terminals is a hybrid between a central air handling system and a system that uses fan-coil units installed inside the spaces.

In terms of hardware, fan-powered VAV terminal units are fundamentally different from the terminal units discussed previously, which are simply dampers that restrict air flow. Fan-powered terminals contain a built-in fan, in addition to a chilled air damper, which helps to overcome problems related to low air flow rates into the spaces. They may also contain a heating coil, which can be controlled in a way that eliminates reheat losses.

Fan-powered terminals can make a VAV system much more reliable in terms of avoiding reheat energy waste over the life of the facility. This is because the control of the heating coil, if installed, is less subject to abuse than control of a reheat coil in a conventional terminal.

During cooling operation, the heating coil in a fan-powered unit turns off, and the flow of chilled air into the space is modulated by a simple damper. During heating operation, the chilled air damper closes, the heating coil turns on, and air from the return air plenum (usually, the ceiling) is circulated through the coil by the fan. Reheating is avoided because delivery of heated air does not depend on the flow of air from the air handling unit.

There are two distinctly different types of fan-powered terminals. One type, called "series flow," is sketched in Figure 4. The other type, "parallel flow," is shown in Figure 5.

In a series-flow unit, the fan operates continuously as long as the space requires conditioning, typically at a constant flow rate. If the chilled air damper closes completely, the air delivered to the space is entirely recirculated. At other times, the air delivered to the space is a mixture of chilled air and recirculated air, the ratio being determined only by the position of the chilled air damper.

In a parallel-flow unit, the fan operates only when space heating is required, either by the self-contained heating coil or by recirculating plenum air to the space. During cooling operation, air to the space comes only from the air handling system. During heating operation, the air delivered to the space may be entirely recirculated, or it may be a mixture of chilled air and recirculated air.

Fan-powered VAV terminals cannot solve the problem of inadequate outside air ventilation at low conditioning loads. If the units operate without any mixing of heated and cooled air, ventilation is limited by the position of the cooling damper. The terminal unit may have a minimum-flow setting, but this incurs the same reheat loss that would occur with any other type of minimum-flow terminal unit.

■ Advantages of Fan-Powered VAV Terminals

Both types of fan-powered VAV terminals offer these advantages, compared to other types of VAV terminal units:

- *reduced or eliminated reheat loss.* In principle, reheat energy waste can be eliminated entirely by not allowing heating coils to operate as long as the chilled air damper is open. Chilled air can be "tempered" by mixing it with plenum air, which requires no additional energy consumption for reheat. However, fan-powered units that have heating coils can waste energy through reheat. The amount of energy waste depends on the type of terminal unit and on the setting of the control that starts the heating coil.

- *light fixture heat recovery.* When the fan is recirculating air from the ceiling plenum, the heat

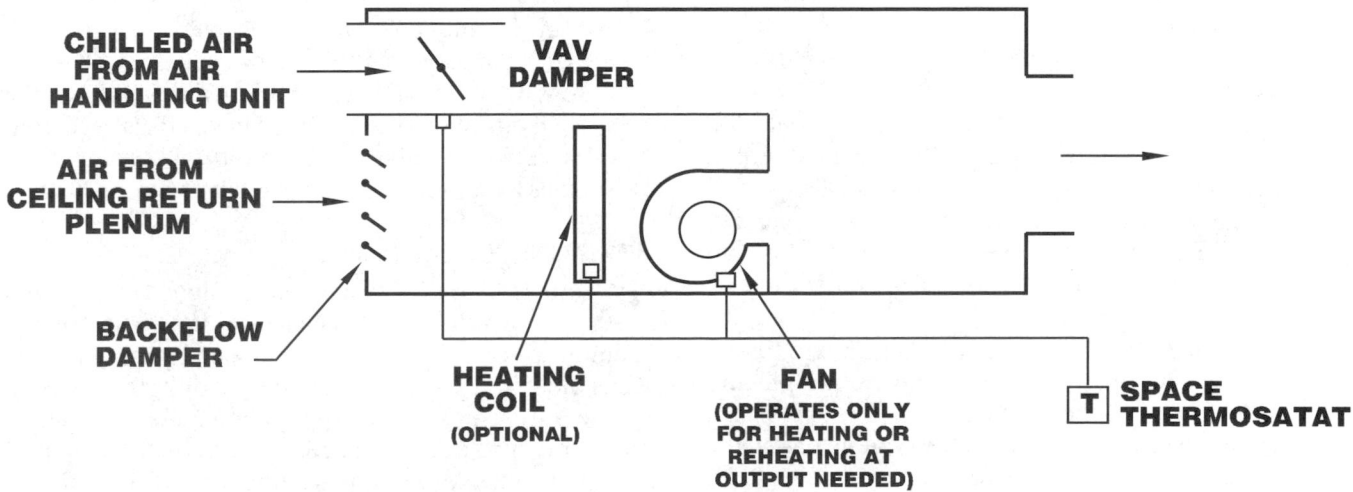

Fig. 5 Fan-powered VAV terminal, parallel flow During cooling operation, the fan is turned off, and the unit behaves like a simple shutoff VAV terminal. The fan operates only when the space requires heating, drawing air from the space or the ceiling plenum. In other respects, the control issues and options are similar to those for series-flow units.

of the light fixtures is captured to assist in space heating. This provides a double bonus because the heat of the lighting fixtures would otherwise be returned to the cooling coil, increasing the cooling load.

- *positive air circulation.* Fan-powered VAV terminals provide positive air circulation whenever the fan is running. The strong air flow also provides good diffuser performance.

■ Disadvantages of Fan-Powered VAV Terminals

Both types of fan-powered VAV terminals have these disadvantages, compared to other types of VAV terminals:

- *no outside air ventilation at low conditioning loads,* unless the terminal sacrifices its efficiency advantage by adding a reheat coil
- *noise.* The terminal unit fans are located close to the point of air discharge, so they transmit a significant amount of fan noise to the space.
- *large size.* The bulkiness of fan-powered terminals may be a major impediment to installation, especially if the ceiling space is already tightly packed. In some existing buildings, retrofit may be impractical.
- *higher cost.* The terminal units themselves are more expensive. However, using fan-powered terminals may reduce costs in other parts of the system, for example, by simplifying the system controls.

■ Comparison of Series and Parallel Units

There are significant differences between series and parallel fan-powered units. Neither is clearly better than the other for all applications. You have to decide which to use. Here are the major ways in which they differ:

- *space air circulation.* With series units, space air circulation is continuous and constant. With parallel units, circulation is constant only when heating. When cooling, space air circulation depends entirely on the position of the chilled air damper, unless the fan is started to provide supplementary circulation.
- *reheat loss.* As we said earlier, it is possible to control either a series or a parallel terminal unit to avoid reheat completely. However, the temptation to use a heating coil for reheat is greater in a parallel unit. This is because the fan does not run continuously, resulting in low air flow to the space as the chilled air damper closes.

On the other hand, in series terminals, light fixture heat that is recirculated from the ceiling plenum cancels some of the cooling energy. Some of this cooling load would return to the cooling coil in the air handling unit anyhow, but some of it would be discharged in the relief air. This is a factor only when cooling during cooler weather.

- *noise.* In series units, the fan operates continuously, creating noise continuously. On the other hand, the cycling operation of the fan in parallel units may also be objectionable.
- *bulk.* Parallel units are usually somewhat larger than series units. However, the fans in series units are larger than in parallel units.
- *overall fan power.* This is a tricky comparison. The fans in series units operate continuously, as long as the space requires conditioning, whereas the fans in parallel units may operate for only a small fraction of the time. Furthermore, the fans in series units must be sized for the larger of the maximum cooling load or the maximum heating load. In parallel units, the fan is sized only for the maximum heating load.

However, if all the terminal units are series units, they allow the air handling unit fan to operate at lower pressure differential. In the series unit, the chilled air is discharging into a box that is kept at pressure slightly lower than the space pressure. In a parallel unit, the chilled air must discharge at pressure comparable to the discharge pressure of the terminal unit fan.

(In any event, fan power is much less significant as a factor in overall system energy consumption than the ability of fan-powered terminals to minimize energy waste from reheat.)

- *control.* Control of series units is simple because the air flow rate is constant. The space thermostat controls the chilled air damper, along with the output of the heating coil (if installed). Control of parallel units is more complex. During cooling, the space thermostat controls the position of the chilled air damper. During heating, the space thermostat controls the running of the terminal unit fan, or the speed of the terminal unit, or the heating coil output, or some combination of these.
- *cost.* Parallel units are somewhat more expensive than series units. This is because they are bulkier, Also, they should have a backflow damper, and the fan motor should be designed to avoid damage from backward rotation of the fan at start-up. On the other hand, series units have larger fans, usually.

How to Select the Diffusers

Diffusers for VAV systems are designed to distribute air at the low velocities that can occur in VAV operation. Some types of diffusers are designed to exploit the Coanda effect, which causes the air flow to adhere to the ceiling. This is desirable for cooling, but it creates a problem with heating. There is no perfect solution to this problem. Reconciling heating air distribution with cooling air distribution is a challenge in all VAV systems. This problem was one of the major reasons for the introduction of the fan-powered VAV terminal, but only the series type of fan-powered VAV terminal can eliminate the dumping problem under all conditions.

With any type of diffuser, dumping of cold air will occur at some minimum velocity. The minimum velocity is higher if you use chilled air temperature reset, because this increases supply air velocity at low cooling loads. The higher supply air temperature also makes dumping more tolerable.

If you convert an existing constant-volume reheat system to VAV operation, you may have to replace the diffusers in the spaces. A diffuser selected for constant-volume operation is unlikely to work well in a VAV mode of operation.

Your Fan Modulation Options

As the VAV terminals throughout the system close, the system fans need to deliver less total air volume. The method of modulating the fans to accommodate changing air flow has major effects on efficiency, comfort, and cost.

See Measure 4.7.8 for an overview of your choices. In brief, these are the characteristics of the flow modulation methods that are used most commonly in contemporary VAV systems:

- *variable-frequency electronic drives* are the most efficient, the easiest to install, the most flexible, and the most expensive. Their major technical disadvantage is creating a high level of electrical noise in the electrical power system. See Reference Note 36, Variable-Speed Motors and Drives, for details.

- *variable-pulley drives* are efficient, reliable, and free of adverse side effects. They are limited in turndown range, and they cannot handle larger motors. See Reference Note 36 for details. They may be difficult to retrofit on existing fans.

- *inlet vortex dampers* are substantially less efficient than the previous two, especially at lower loads. They are simple, reliable, and inexpensive. Usually, they cannot be retrofitted, although this might be possible if the fan manufacturer offers a vortex damper for the particular fan model.

- *"riding the fan curve"* is a fancy way of saying that no fan modulation is used. Instead, the supply air is choked off by the terminal units. This is poor practice in all but the smallest systems. The only advantage is cheapness. The fan's power consumption declines relatively little as air flow is reduced. Riding the fan curve results in higher duct pressure as the VAV boxes close, which results in poor control and increased noise. To compensate, the designer may specify "pressure-independent" boxes. The higher cost of these boxes offsets the cost saved by omitting the fan modulation control.

- *multi-speed fan motors* are a compromise approach that may be appropriate in smaller systems. Multi-speed motors allow air flow to ride the fan curve at different fixed fan speeds. Fan power is greatly reduced at the lower speeds, along with the pressure at the terminal units. Multi-speed motors are covered by Reference Note 36.

How to Control the System

The efficiency and comfort of a VAV system depends on its terminal unit settings and its control settings. A VAV system may operate wastefully for its entire life because a control function was omitted from the design. Or, the system may waste energy for years because the controls are adjusted haphazardly. Optimize the following control functions as well as you can, and keep them optimized.

■ Setting the Minimum Flow of Terminal Units

Virtually all VAV boxes have mechanical stops that can be adjusted to prevent the damper from closing beyond a desired amount. The setting of minimum air flow at the VAV terminals has an enormous influence on the behavior of the system, to a greater extent than any other aspect of the system controls. The minimum-flow setting determines:

- the heating capacity of the terminal unit
- the amount of energy waste from reheat
- the amount of outside air ventilation
- the ability to dehumidify during periods of low cooling load
- diffuser performance and space air circulation
- if the heating coils are electric, the actions needed to prevent coil burn-out.

Minimum-flow is the Achilles heel of VAV system efficiency because it opens the door to reheat. Give priority to keeping the minimum-flow settings as low as possible, consistent with maintaining comfort and ventilation.

■ Heating/Cooling Changeover

If heating is provided by heating coils in the terminal units, there are two methods of changing from cooling to heating. The most efficient method is to let the damper close to zero, or to a much lower flow rate that is dictated by ventilation requirements. The heating coil turns on at the minimum-flow position of the damper. At the same point, the output signal of the thermostat reverses so that the damper opens as the space temperature drops. The thermostat may also control the output of the heating coil.

The other method is much less efficient, but it is more common because it is simpler. You should avoid it. In this method, the minimum flow of the terminal unit is set so that it is high enough to handle the peak heating load. Typically, this is about 40% of the maximum rated flow. (The air flow needed for heating is usually much less than the flow needed for cooling because the heating temperature differential is higher.)

As the cooling load drops, the damper closes until it reaches its minimum-flow limit, at which point the heating coil turns on. The heating coil's output is controlled by the space thermostat. As the cooling load drops further and becomes a heating load, the heat output of the coil continues to increase.

Energy waste from reheat begins as soon as the damper reaches the minimum-flow limit. Energy waste is high because the heating coil must heat the relatively large air flow dictated by the peak heating requirement. Energy waste continues as long as the cooling coil in the air handling unit produces chilled air.

The first method wastes less energy because reheat does not begin until the conditioning load is much lower, and the amount of reheat energy is greatly reduced within the range in which reheat occurs.

■ Control of Fan Modulation

The most common way of controlling the output of the supply fan is to sense the pressure in the duct system. The pressure in the duct system is set high enough to provide rated air flow through the terminals under maximum load conditions. This method is simple and fairly efficient. It provides predictable terminal unit performance.

If many terminals are installed along a supply duct, a large pressure drop occurs between the fan and the end of the duct run. Therefore, the pressure in the portion of the system near the fan must be higher than necessary for that part of the system. The location of the duct pressure sensor affects the response of the fan to the flow requirements of the terminal units. If many terminals are strung out along the duct runs, there is no optimum location for the duct pressure sensor. It is a common rule to place the pressure sensor two-thirds of the way toward the end of the longest duct run. As with all general rules, be cautious in applying this one.

The duct pressure you need depends on the type of terminal unit. Externally powered terminals with good thermostatic controls require the least pressure. System-powered terminals and "pressure-independent" terminals require additional pressure to overcome their increased resistance.

If a VAV system has a return fan or a relief fan, the return or relief fan must remain coordinated with the supply fan as the system flow changes. This is necessary to maintain the correct amount of outside air ventilation and to minimize the waste of fan energy that occurs in overcoming the resistance of the air handling system dampers. The only accurate control method is to control the return fan or relief fan by using air flow monitors. One flow monitor is always located in the outside air intake. The location of other air flow monitors, if needed, will depend on the front end configuration of the system. See Measure 4.2.2 for details about the different types of front end configurations that control outside air intake.

With a return fan, the return fan flow rate should always equal the supply fan flow rate minus the desired outside air intake rate. With a relief fan, the relief fan flow rate should always equal the outside air intake rate.

A less desirable method of controlling the supply fan output is to sense the return air volume. This method is based on an assumption that lower return air flow indicates that the zone dampers are more closed, which means that the duct pressure is higher. This method is not as accurate as measuring the fan discharge pressure directly. The only significant advantage of this method is that the air flow sensor can be packaged in the air handling unit, in contrast with the expense of installing a pressure sensor at some distant point in the duct system.

If the system uses multi-speed fans, sensing duct pressure is not the appropriate method of controlling fan speed, because this would cause the fan to oscillate between speeds. Usually, the best method of switching fan speed is to sense the conditioning load in the space that has the largest load. For details, see Reference Note 14, Control Signal Polling.

■ Chilled Air Temperature Reset

Chilled air temperature reset is covered by Measures 4.4.1 ff and 4.7.6 ff. The basic idea is that the reset control keeps the air temperature at a minimum until the VAV terminals close down to their minimum-flow settings. Then, the chilled air temperature is raised in response to decreasing cooling load.

Chilled air temperature reset lessens the problems that occur in VAV systems at low cooling load, especially reheat energy waste and poor diffuser performance. Your VAV systems should have this control feature.

The chilled air temperature must be kept low enough to serve the terminal with the highest cooling load. In a system with large load diversity, this limitation substantially reduces the benefit of chilled air temperature reset.

■ Duct Pressure Reset

An unusual technique is to reduce the duct pressure setting as fan output is reduced. This causes the terminal unit dampers to open more at low loads. The benefits are:

• reduced fan power
• reduced leakage through shutoff terminals when they are closed
• reduced leakage from the duct system
• reduced terminal unit noise
• more accurate and stable temperature control as terminals approach the fully closed position.

The cooling or heating output of the terminals is not affected, because the flow rate through the terminals remains unchanged. The only potential problem is failure to cool a space that has a higher load than the other spaces on the system. Prevent this problem with the same methods used to prevent excessive chilled air temperature reset.

■ **Temperature Setback**

During night setback operation, the system is used only for heating. At the beginning of the setback period, the cooling coil is turned off and the temperature is allowed to drift. (If the cooling coil were kept in operation, it would force the building to the setback temperature, wasting cooling energy.) The fan is usually turned off as well.

Setback operation can be controlled by a single thermostat at a critical location, saving control cost. In more sophisticated systems, setback operation may be controlled from all the space thermostats. If any space requires heating, the fan is started.

■ **Morning Warm-Up**

If the system has night setback, it should also have a special control sequence for morning warm-up. During warm-up operation, the cooling coil is turned off to avoid any reheat.

■ **Space Thermostat Deadband**

Converting a single-duct system to VAV opens the door to using deadband in the thermostatic control of the terminal units. Deadband is possible only if the terminal units can shut off completely. Deadband saves energy in several ways. Refer to Measure 4.7.2 for details of deadband in single-duct VAV systems.

■ **Control of Humidity**

As discussed previously, reheat may be necessary for dehumidification in climates that are mild and humid, and in spaces where a considerable amount of moisture is released. Reheat consumes a lot of energy, so limit it to the bare minimum needed to provide the desired level of dryness.

The simplest and most efficient way to control reheat for dehumidification is by using a reliable humidistat in the space to control the operation of the heating coil in the terminal unit. Connect the humidistat so that its signal overrides the signal from the thermostat to the heating coil.

Chilled air temperature reset is an important energy conserving feature, but it reduces dehumidification. The humidistat should override this control function when humidity is high.

■ **Control of Outside Air Intake**

Refer to the previous discussion of the problems of controlling outside air in a VAV system. Refer to Measures 4.2.1 ff for controls that tailor outside air intake to variable ventilation requirements.

■ **Control of Building Pressure**

Refer to the previous discussion of pressure fluctuation problems that may arise in VAV systems. Only specialized facilities, such as medical laboratories, have controls that are designed explicitly for controlling

space pressure. The appropriate design for the controls depends on the configuration and purposes of the spaces and the system, and on the pressure relationships that have to be maintained.

■ **Economizer Cycle Operation**

Economizer cycle control is not changed when converting from constant-volume to VAV operation. Refer to Measure 4.2.5 about economizer cycles.

■ **Purge Cycle Operation**

Refer to Measure 4.2.7 about purge cycles. If a VAV system has shutoff terminals, the purge cycle must include a control function that opens all the terminals during the purge period.

■ **Isolation of Space Conditioning**

If the terminal unit dampers can close to shutoff, you can shut off conditioning to the spaces they serve. This saves conditioning energy, reheat energy (if any), and fan energy in spaces that operate on different schedules from the rest. When you do this, treat each terminal unit as a separate conditioning unit. Use the methods recommended in Subsection 5.1 to turn off individual terminal units.

Improve Duct Zoning, If Practical

Diversity among the space loads increases the need for reheat, as explained previously. Resetting the chilled air temperature can reduce reheat. However, load diversity also limits the potential of chilled air temperature reset. (Measure 4.4.1 explains why.) To get around this basic limitation, lay out the different air handling systems so that each system serves spaces with similar loads.

In retrofit applications, you may be able to change the connections between ducts and air handling units to achieve this purpose. In order for this to be practical, the air handling units need to be close together, and you need clearance for relocating some of the ductwork.

Other Efficient Systems

Measures 4.4.8 and 4.4.9, which come next, are alternatives to VAV that avoid most of the energy waste of reheat systems and are less tricky to accomplish. However, they typically are more expensive.

ECONOMICS

SAVINGS POTENTIAL: All or most of the reheat energy used by the system; 20 to 60 percent of the cooling energy; 30 to 70 percent of the fan energy.

COST: In new construction, the additional cost elements are the VAV fan drive(s), controls for fan operation, the cost differential between reheating and VAV terminal units, the cost of heating equipment other than reheat coils, and the differential cost of space thermostatic controls. See Reference Note 36 for the cost of variable-

speed fan drives. VAV terminal units typically cost several hundred dollars apiece, although you can use units costing as little as $100 apiece in limited applications.

Retrofitting existing facilities is usually much more expensive than installing VAV in new facilities. Retrofitting heating capability that is separate from the VAV terminals greatly increases cost.

PAYBACK PERIOD: In new facilities, typically about one year. In existing facilities, typically several years.

TRAPS & TRICKS

DESIGN: VAV systems that are comfortable, highly efficient, and reliable are still rare. The large number of design issues makes it easy to overlook some of them. Do your homework. Find successful systems to study.

SELECTING THE EQUIPMENT: Your choice of terminal units is critical. Good equipment is available, but there is a lot of cheap junk on the market. Study Reference Note 36 about selecting variable-speed fan drives.

INSTALLATION: Details matter. Work with the installers of the terminal units and the fan drives. Ensure that minimum-flow limits and other settings are adjusted properly.

EXPLAIN IT: The way the facility staff adjust and maintain the system has a major effect on efficiency and comfort. These systems can have many important features that are not obvious. Describe the system clearly in the plant operating manual. Include a table of the settings for all terminal units. Install placards on all unfamiliar equipment. Buy an extra terminal unit to use as a training aid.

MEASURE **4.4.8 Replace all reheat coils in a system with heating/cooling coils and minimize operation of the air handling unit cooling coil.**

This Measure consists of replacing the reheat coils in the terminal units with coils that can provide both heating and cooling. In effect, each terminal unit becomes a fan-coil unit, with the air handling unit supply fan acting as a common fan for all the terminal units. No cooling coil is needed in the air handling unit.

Like VAV (Measure 4.4.7, previous), this design avoids most of the energy waste of reheat systems. The advantages of this Measure arise mostly from the fact that the heating and cooling of each space is independent of the central system. There is no common supply of chilled air that requires reheating to adapt to conditions in different spaces, so there are none of the associated problems involved in minimizing reheat.

This change to the system layout makes it behave like the "damperless multizone" system shown in Figure 2 of the Section 4 Introduction.

Comparison to VAV

Think of this Measure as a constant-volume alternative to VAV. In comparison to VAV, it has these advantages:

- *greater saving in reheat energy.* There is no need for reheat to compensate for differences in space conditioning loads, or to maintain a minimum air flow.
- *avoids comfort problems associated with low space air flow,* including space air stagnation and poor diffuser performance.
- *provides outside air ventilation* under all conditions
- *no fluctuations in space pressurization,* unless you modulate the fan output
- *simpler and less critical controls.* There is no possibility of reheat (in most installations), so no critical control adjustments are needed to minimize reheat.
- *simpler hardware*
- *low noise.* There are no dampers or fans in the terminal units to create noise.

These are the relative disadvantages:

- *difficult installation,* in both new construction and retrofit, because of the need to add cooling and condensate drainage at each terminal unit.
- *likely to be more expensive* than VAV, mostly for the same reason.
- *less fan energy saving.* In fact, fan energy may be increased by the resistance of the additional cooling coils in the terminal units.

SUMMARY

An unusual, major modification that eliminates reheat losses. Avoids most of the problems of VAV, but saves less fan energy and requires additional pump energy.

SELECTION SCORECARD

Savings Potential .... $ $ $ $

Rate of Return, New Facilities % % %

Rate of Return, Retrofit.......... % %

Reliability .... ✓ ✓ ✓

Ease of Retrofit .... ☺

- *requires pump energy* to move chilled water through the added cooling coils in the terminal units.

A VAV conversion with fan-powered terminals is similar to this Measure in some of its advantages. However, fan-powered terminals cannot provide steady ventilation without reheat, and they add noise to the space.

Comparison to Individual Fan-Coil Units

The terminal units are similar to fan-coil units. The main advantages of this method, compared to conditioning the spaces with individual fan-coil units, are:

- *no loss of floor space*
- *lower noise,* since there are no fans within the space
- *outside air ventilation,* without the need for wall penetrations

The main disadvantages of this method, compared to conventional fan-coil units, are:

- *higher cost,* because of the need for air handling units and ducts
- *additional building space* for air handling units and ducts
- *greater complication* to isolate spaces that operate on different schedules
- *draining condensate* from the cooling coils may be more difficult.

Where to Consider This Approach

This approach has no serious comfort or efficiency shortcomings. It is likely to be preferable to VAV in applications where the disadvantages of VAV pose serious limitations.

In particular, this approach is a strong contender for specialized applications that need a constant flow of air. For example, it might be used in a biological research facility, where the air handling system provides the air flow necessary to prevent escape of contaminants for fume hoods.

This approach is not a good choice in mild, humid climates where reheat is needed for dehumidification for long periods of time. By using tandem cooling and heating coils (an option discussed below), you can provide dehumidification in individual spaces. However, this is a significant complication. Furthermore, the waste of energy from reheat is proportional to the air flow through the terminal units. This Measure either does not reduce air flow, or it reduces air flow less effectively than VAV.

The difficulty of draining or pumping cooling coil condensate from each of the terminal units may make this Measure impractical in some facilities.

How to Install the Coils

■ Coil Arrangements

There are two practical ways to arrange the terminal unit coils for heating and cooling. If heating is hydronic, the most compact method is to install a single hydronic coil that is changed from heating to cooling by switching from heating water to chilled water. The weakness of this arrangement is the requirement for changeover valves. These may develop internal leakage, which wastes energy by mixing water from the heating and cooling supplies.

The other arrangement is to install separate heating and cooling coils in tandem, as is typical in small fan-coil units. The disadvantages of this arrangement are greater bulk, higher cost, and greater resistance to air flow. Also, a defect in the controls or leakage in the coil valves may allow some simultaneous operation of both coils.

■ Whether to Use Existing Reheat Coils

Existing reheat coils usually cannot be converted to serve as heating/cooling coils. Reheat coils have too little heat transfer surface for cooling, because they are designed to operate with higher temperature differentials. Also, they lack provision for condensate drainage.

■ Whether to Abandon the Air Handling Unit Cooling Coil

This modification eliminates the need for the cooling coil in the air handling unit. If practical, remove it to minimize air flow resistance and fan power.

It may be tempting to keep the cooling coil for pre-cooling, to reduce the size of the cooling coils in the terminal units, or as a means of dehumidification during mild weather. However, this will inevitably lead to waste of energy by reheat. In effect, this restores the original constant-volume reheat mode of operation.

Zoning

Unlike VAV systems, zoning is not critical with this type of system. This is because the air flow does not shift from one space to another, as it does in a VAV system that serves spaces with different load profiles. Even so, it pays to be clever about zoning. You can save money on terminal units, ductwork, and fans, and you can save fan energy in two ways.

You can save fan cost and fan energy by assigning the terminal units to air handling systems in such a way that the total system loads vary as little as possible. In particular, use one system for the core area alone, because the core load tends to remain constant. Similarly, use one system for the north perimeter alone, because the cooling load remains low on the north side.

If you modulate the fans to follow the system load, as recommended below, you can save fan energy by assigning terminal units with similar load profiles to the same air handling systems. For example, use one air handling system for the east perimeter, and another for the west perimeter.

Isolation of Unoccupied Spaces

If the system serves spaces that operate on different schedules, you can save more energy by turning off the conditioning to individual unoccupied spaces. To do this, you need a damper in the duct that leads to the space. To keep the cooling coil from sweating when the air supply is turned off, the cooling coil should also have a shutoff valve. If the heating coil is electric, it needs to be turned off to prevent burnout.

Use the methods recommended in Subsection 5.1 to turn conditioning on and off. This capability provides the greatest benefit if the unoccupied spaces have high heat gains or losses, and if they are physically isolated with walls and doors.

Chilled Water and Condensate Piping

Piping is a major consideration. Each terminal unit needs chilled water piping and a condensate drain. Installing new piping is especially troublesome in retrofit applications. Select the locations of the terminal units to minimize piping cost and difficulty.

Try to rely on gravity for condensate drainage. The alternative of many small, hidden condensate pumps would become a maintenance headache. However, gravity drainage requires a path that has a continuous downward pitch all the way to the sewer. In many cases, structural interferences make it impossible to install gravity drains.

Dehumidification at Low Loads

Refer to Measure 4.4.7 about humidity control during periods when the normal operation of the cooling coils does not provide adequate dehumidification. In summary, anticipate a problem if the climate is mild and humid, if there are major sources of moisture within the spaces, and in basement spaces.

In such cases, this Measure may not be the best choice. You could dehumidify during mild weather by overlapping the operation of the cooling and heating coils, if separate coils are installed. This reheating mode of control is covered in Measure 4.4.7.

Variable Fan Output

You can save fan energy in this type of system by modulating the fan output to follow the conditioning load. Fan modulation saves only fan energy, because this system usually has no reheat energy to be saved.

Once-through systems (systems that use 100% outside air) are an important exception. In such systems, reducing air flow may save a large fraction of conditioning energy.

In principle, you can use any type of fan modulation with this type of system. With small and medium sized fans, consider a two-speed fan motor. Variable-frequency electronic drives and variable-pulley drives provide even better turndown. However, low fan speeds incur ventilation and air flow problems, described in Measure 4.4.7. In large fan systems, electronic drives may be the only efficient method of fan modulation that is available, and the price per kilowatt drops sharply with increasing size. See Reference Note 36, Variable-Speed Motors and Drives, for details of fan drives.

Controls

This system is simple to control. The space thermostats control the coils in the terminal units, as with fan-coil units. For the best energy efficiency, install space thermostats that have heating/off/cooling selector switches. Add other control efficiency features, such as deadband and temperature setback, as appropriate. Refer to Subsections 4.1 and 4.3 for these features.

Install effective placards at all manual controls. See Measure 4.3.1 for tips in making the placards effective.

If the system has fan output modulation, the most efficient method of fan control is to follow the conditioning load in the space that has the largest load. See Reference Note 14, Control Signal Polling, for methods of doing this.

ECONOMICS

SAVINGS POTENTIAL: All or most of the reheat energy used by the system; 20 to 60 percent of the cooling energy. The savings are reduced by the need for greater chilled water pumping energy. Fan energy consumption may be more or less than in the original reheat system, depending on the nature of the coil installation.

COST: This Measure requires more expensive terminal units, widespread chilled water piping, cooling coil condensate drains, and more expensive thermostatic controls. These items are much more expensive in retrofit than in new construction. In new construction, this design avoids the cost of a cooling coil in the air handling unit.

PAYBACK PERIOD: Several years, or much longer.

TRAPS & TRICKS

DESIGN: Designing the system should be relatively simple, but it is unusual enough to make you a pioneer.

MEASURE **4.4.9 Install self-contained heating/cooling units, and use the air handling system only for ventilation.**

This Measure consists of replacing reheat terminal units with self-contained room conditioning units that provide both heating and cooling. The room conditioning units can be any type, depending on the needs of the space. The most likely replacements are fan-coil units. The air handling system is retained only to provide outside air ventilation. The cooling coil and the preheat coil (if any) are used only to condition outside air. The reheat coils in the terminal units are omitted or abandoned.

As with converting to a VAV configuration (Measure 4.4.7), this system layout avoids most of the energy waste of reheat systems. As with the layout of Measure 4.4.8, this layout has a major advantage over VAV, which is that the heating and cooling of each space is independent of the central system. There is no common supply of chilled air that requires reheating to adapt to conditions in different spaces, and none of the associated problems involved in minimizing reheat.

In new construction, this system layout approaches the ideal of efficiency that is suggested in the Section 4 Introduction, which is to completely separate the heating, cooling, and ventilation functions.

In retrofit applications, this approach uses even less of the original reheat system than Measure 4.4.8. It reduces the original system to the status of a ventilation system, delivering air at much lower flow rates than originally.

Comparison to VAV

This Measure may be considered as a more drastic alternative to VAV conversion. In comparison with VAV, the advantages of this Measure are:

- *no reheat,* except where reheat is used deliberately for dehumidification.
- *ability to localize conditioning* to occupied spaces
- *avoids problems of poor diffuser performance*
- *provides outside air ventilation* under all conditions
- *no fluctuations in space pressurization*
- *simpler hardware and controls*

This Measure has the following disadvantages, compared to VAV:

- *much higher cost to retrofit,* because most of the functions of the existing reheat system's equipment are abandoned. High conversion cost keeps this approach from being feasible as a retrofit except in specialized circumstances, such as a major renovation.

SUMMARY

Can be very efficient. Simpler than VAV, and free of its annoyances. As a retrofit, this is a drastic modification that involves abandoning most of the functions of the original system. Configurations vary with the types of heating and cooling equipment used.

SELECTION SCORECARD

Savings Potential	$ $ $ $
Rate of Return, New Facilities	% % % %
Rate of Return, Retrofit	% %
Reliability	✓ ✓ ✓
Ease of Retrofit	☺

- *increased maintenance workload.* Having a large number of individual space conditioning units increases the maintenance workload, especially for routine cleaning and changing filters.
- *higher noise level.* Fan-coil units and self-contained air conditioning units put more noise into the space than a well designed central system. However, this approach is no more noisy than fan-powered VAV terminals.

Aside from the saving in reheat energy, energy savings cannot be stated in general. The total fan power is greater than with VAV, but possibly less than with a reheat system. If hydronic heating and cooling equipment is used, additional pump energy is required.

Optimize Your Choice of Conditioning Equipment

This change in the system layout gives you the opportunity to exploit the advantages of any type of room conditioning equipment. By the same token, be prepared to minimize the disadvantages of the type you select. Each type of space conditioning equipment has its own art. Refer to Section 5 for the efficiency characteristics of the different types of room conditioning equipment.

Localize Conditioning As Much As Possible

A major advantage of this Measure is its ability to turn off conditioning in individual unoccupied spaces. If the air handling system serves spaces that operate on different schedules, take full advantage of this energy saving opportunity. Install dampers in the ducts that provide ventilation to the individual spaces, and control each damper to close when its space is vacated. See Subsection 5.1 for the appropriate control methods.

ECONOMICS

SAVINGS POTENTIAL: *All or most of the reheat energy used by the system; 20 to 60 percent of the cooling energy. The savings are reduced by additional chilled water pumping energy if you use hydronic space conditioning equipment. Fan energy consumption may be more or less than in the original system.*

COST: *In retrofit, typically several thousand dollars per heating/cooling unit, plus the cost of piping, pumps, modification of the original air handling system, and system controls. In new construction, the additional cost depends heavily on the design of the buildings. In the most favorable cases, this design approach may cost little more than a reheat system.*

PAYBACK PERIOD: *In retrofit, typically many years. This approach is too expensive to be retrofitted for the sake of energy conservation alone. Consider it as a design option in new construction and in major mechanical renovations, where appropriate.*

TRAPS & TRICKS

CHOICE OF METHOD: *This is close to starting with a blank slate. In retrofit, you have the additional challenge of exploiting pieces of an existing system. Lots of work and lots of fun for a clever designer. Study Section 5 to learn more about specific types of independent space conditioning equipment.*

Dual-Duct Reheat Systems

The Measures in this Subsection improve the efficiency of dual-duct air handling systems. This type of system is sketched in Figures 1 and 2, in the Section 4 Introduction.

In these systems, the air handling unit has a continuously operating cooling coil that feeds air into a cold air duct, and a continuously operating heating coil that feeds air into a hot air duct. The two ducts run in parallel throughout the facility. At each space, air from the two ducts is mixed in a terminal unit ("mixing box") that is controlled by the space thermostat. The terminal units mix the air in proportions that keep the volume of air flow into the space essentially constant.

In the early days of dual-duct systems, virtually all were designed to provide an essentially constant flow of air to the space. This requires mixing large quantities of heated and chilled air, which makes these "reheat" systems. The high energy cost of this mode of operation led to the later development of dual-duct VAV systems, which are covered in Subsection 4.8.

The primary justification for dual-duct reheat systems is close control of space conditions. They allow any number of spaces to be controlled independently. They provide a high level of comfort because air flow is constant, temperature control can be maintained precisely as the load changes from heating to cooling, humidity can be controlled at any desired level, and outside air ventilation is maintained under all load conditions. When you make changes for the sake of efficiency, preserve these qualities to the extent that the application requires them. Each of the Measures discusses the relevant comfort and health issues.

INDEX OF MEASURES

4.5.1 Keep the temperature of the cold duct as high as possible and the temperature of the hot duct as low as possible.

   4.5.1.1 Install temperature reset controllers for both the cold duct and the hot duct.

4.5.2 Turn off the heating coil and/or the cooling coil whenever practical.

4.5.3 Trim the output of the air handling system fans.

4.5.4 Install multi-speed fan motors.

4.5.5 Convert the system to variable-air-volume (VAV) operation.

RELATED MEASURES

- Subsection 4.8, for Measures that improve the efficiency of existing VAV dual-duct systems

The Measures in this Subsection are oriented primarily toward reducing the energy waste that occurs from reheat. Reheat systems are very wasteful because a large amount of heating and cooling energy is cancelled out in the mixing process. During low-load conditions, a dual-duct system may consume much more energy in reheat than it does in conditioning the space. Some of the Measures also save fan energy.

In new construction, none of the Measures adds much cost, unless the systems are small. Variable-output fan drives were once a major cost addition, but the relative cost of this equipment has fallen sharply, especially for larger systems. Furthermore, exploiting the VAV principle may allow some savings in initial cost, such as smaller ducts.

In retrofit applications, first decide whether or not to make major changes to the existing constant-volume system. This decision depends mainly on two factors: whether you can afford to make a major change, and whether you can sacrifice the tight control of humidity and the continuous ventilation that reheat systems provide. If you can afford only inexpensive changes, accomplish Measures 4.5.1 ff through 4.5.3. If you can afford a major system change, select Measure 4.5.5. Measure 4.5.4 is an intermediate approach that saves less energy than Measure 4.5.5, but it is less expensive.

Energy Behavior of Dual-Duct Systems

The energy consumption behavior of dual-duct systems is not obvious, so we will use a small dose of mathematics to help clarify it. Formula 1 tells us the amount of energy that is absorbed by an individual terminal unit in a dual-duct system. It is expressed in terms of the actual cooling load in the space, the air flow rate through the terminal unit, and the duct temperatures. The formula shows how important the temperature settings are to the efficiency of the system.

First of all, the formula reveals the wasteful behavior of dual-duct reheat systems under typical operating conditions. In the second term of the formula, input energy is directly proportional to the product of the cooling temperature differential times the heating temperature differential. If the temperature differentials are large, the second term of the formula becomes large in relation to the first. Therefore, at low loads, mixing losses consume much more energy than actually conditioning the spaces.

To see how dominant mixing losses can be, consider what happens when the heating and cooling temperature differentials happen to be equal. In that case, the first term in the formula becomes zero, and the amount of energy consumed by the terminal is completely independent of the space load! (What happens in this case is that mixing losses increase exactly as the space load declines, and vice versa.)

The way the two temperature differentials multiply in the second term of the formula makes it important to keep both the hot duct and cold duct temperature differentials as low as possible. (This is the basis of Measures 4.5.1 ff.) In dual-duct systems, the two temperature differentials are independent. This is different from single-duct reheat systems, where the reheat temperature differential is determined by the cooling temperature differential.

FORMULA 1

$$\text{Input Power} = \left(\frac{\triangle T(H) \ - \ \triangle T(C)}{\triangle T(H) \ + \ \triangle T(C)} \ \times \ \text{Space Conditioning Load} \right)$$

$$+ \ 2.16 \ \times \left(\frac{\triangle T(H) \ \times \ \triangle T(C)}{\triangle T(H) + \triangle T(C)} \ \times \ \text{Terminal Unit CFM} \right)$$

where:
- the units of power are BTU's per hour
- the units of temperature are degrees Fahrenheit
- the units of air flow are cubic feet per minute (CFM)
- $\triangle T(H)$ is the temperature difference between the hot duct and the space, always positive
- $\triangle T(C)$ is the temperature difference between the cold duct and the space, always positive
- "Space Conditioning Load" (in BTU/hr) is positive for heating, negative for cooling
- the elevation is sea level.

(The formula assumes that the air flow rate of the terminal unit is constant. The formula is valid only if the temperature differentials and flow rates are large enough to satisfy the space load.)

Going further, if conditions let you turn off either the cooling coil or the heating coil, the second term of the formula becomes zero. In this case, reheat loss disappears, and the input energy equals the conditioning load, as you would expect. (This is the basis of Measure 4.5.2.)

The second term of the formula also shows that energy waste from reheat is proportional to the supply air volume. You can reduce the supply air volume by trimming the fan output (Measure 4.5.3), by installing multi-speed fans (Measure 4.5.4), or by converting the system to variable-air-volume operation (Measure 4.5.5).

Note that this formula relates to the energy consumption of an individual terminal unit, not to the system as a whole. The flow rate in the formula relates to the terminal unit, but the hot and cold air temperature differentials are common to the entire system. Your ability to reduce the temperature differentials depends on the load diversity among the spaces.

Triple-Duct Systems

"Triple-duct" systems look similar to dual-duct systems. However, their energy behavior is entirely different. Triple-duct systems have a third duct that carries unconditioned air (a mixture of return air and outside air) for mixing with either the heated air or the chilled air. If properly installed, these systems have no mixing losses. Triple-duct systems are efficient, comfortable, and fairly simple. They are rare because they look expensive to designers and because they require a lot of duct space.

If you are dealing with triple-duct systems, refer to Subsection 4.8 for Measures that optimize their efficiency.

MEASURE **4.5.1 Keep the temperature of the cold duct as high as possible and the temperature of the hot duct as low as possible.**

The introduction to this Subsection makes the point that lowering the hot duct temperature and raising the cold duct temperature by even modest amounts can greatly reduce mixing losses. During the extended periods when conditioning loads are low, most of the system's energy consumption may consist of mixing losses.

Raising the chilled air temperature and lowering the hot air temperature is a powerful energy conservation technique. It is easy, and it costs nothing.

The optimum duct temperatures change continuously, along with the cooling load. Don't expect the staff to keep resetting the temperatures manually as conditions change throughout the day. The best they can do manually is to reset the temperatures at intervals, perhaps on a seasonal basis. Subsidiary Measure 4.5.1.1 improves on this Measure with continuous automatic resetting of the two duct temperatures. Use manual adjustment as a stopgap method until you install automatic reset controls.

Energy Saving Potential

■ Reduced Mixing Losses

The saving potential varies widely among facilities, typically ranging from 20 to 50 percent of total cooling and heating energy. Achieving this potential requires diligence in adjusting the duct temperatures to match changing load conditions.

■ Improved Chiller Efficiency

In hydronic systems, increasing the cold duct temperature may allow you to raise the chilled water temperature. This may improve chiller efficiency by several percent, as explained in Measure 2.2.1.

Try to Minimize Load Diversity

Your ability to reset duct temperatures is reduced if the air handling system serves spaces that have different loads. Some amount of reheat will occur if one space requires cooling at the same time that another space requires heating. The chilled air temperature must remain low enough to serve the terminal with the greatest cooling load, and the hot air temperature must remain high enough to serve the terminal with the greatest heating load.

Load diversity is determined by the way that spaces in the building are connected to the air handling systems. For example, if a dual-duct system serves a zone with extensive glazing on both the east and west sides of the building, there may be many hours per year when there

---

SUMMARY

Save a lot of energy just by turning two screws.

SELECTION SCORECARD

Savings Potential	$ $ $ $
Rate of Return	% % % %
Reliability	✓ ✓
Ease of Initiation	☺ ☺ ☺

---

is a large cooling load on one side of the system while there is a large heating load on the other side.

The load diversity is even worse if a system serves both core and perimeter spaces. This is because core areas may have a large cooling load throughout the year. If the same system serves perimeter spaces, reheat will occur any time a perimeter space has a heating load.

This makes it important to pay attention to zoning when designing air handling systems. Major renovations may also provide opportunities for improving zoning.

Duct temperature reset can be thwarted by a single space that has an inadequate air supply, because this space creates a demand for more extreme duct temperatures. Therefore, when planning the reset controls, make a survey of the cooling and heating capacities in all spaces, and correct any deficiencies in heating or cooling capacity. You can do this by increasing the capacity of the terminal unit, or by decreasing the load. For example, if a space has an exceptionally high solar load, consider shading the windows.

The diversity problem fades away when the weather becomes warm enough to allow heating in the hot duct to be turned off entirely. This eliminates all mixing losses. For this reason, be aggressive about minimizing the hot air temperature when the weather is mild.

Effect on Humidity

Raising the chilled air temperature may allow the humidity inside the building to rise excessively when the outside air is mild and humid. Refer to "Effect on Humidity," in Measure 4.4.1, to see if this may become a problem.

How to Adjust the Duct Temperatures

The discharge air temperatures of the hot and cold coils are controlled by thermostats in the air handling unit. You can easily reset these thermostats with your pocket screwdriver. If you have an electronic control system, you can change the discharge temperatures from the keyboard.

Adding Cooling Coil Control Valves

There is one unusual case in which you have to spend some money to optimize the cold duct temperature. Designers sometimes cut installation cost in chilled water coils by omitting a control valve. In such cases, the chilled water temperature directly controls the chilled air temperature. To get around this limitation, install a cooling coil valve in each air handling unit and control it with a chilled air discharge thermostat.

Don't Forget the Placards!

This activity is easy to forget. Therefore, install placards at the discharge air thermostats to explain how to set them. See Reference Note 12, Placards, for details of effective placard design and installation.

ECONOMICS

SAVINGS POTENTIAL: *20 to 50 percent of total cooling and heating energy, typically. The potential for savings is limited by space load diversity.*

COST: *Minimal, in most cases. If the cooling coil presently has no thermostatic control, installing a cooling coil valve and discharge thermostat typically costs about one thousand dollars.*

PAYBACK PERIOD: *Immediate, in most cases. One year to many years, if you have to install a cooling coil valve and discharge thermostat.*

TRAPS & TRICKS

CHOICE OF METHOD: *Consider manual control as a stopgap until you install automatic duct temperature reset controllers, as recommended next.*

DILIGENCE: *This Measure is completely dependent on the diligence and skill of the plant operators in adjusting the duct temperatures. Conditioning loads vary widely over the course of a day, so it would take almost continuous attention to achieve the maximum potential of this technique.*

EXPLAIN IT: *Make sure that plant operators are trained and understand the importance of the adjustments. Put clear instructions in the plant operating manual. Install placards at the thermostats that identify them and explain how to set them.*

MEASURE **4.5.1.1 Install temperature reset controllers for both the cold duct and the hot duct.**

The technical weakness of Measure 4.5.1 is that it cannot adapt to the continuous variations in conditioning load. This Measure consists of installing two automatic temperature reset controls, one to keep the cold duct temperature as high as possible and the other to keep the hot duct temperature as low as possible. The control components are identical for the two functions.

In many applications, this Measure can save large amounts of energy at relatively modest cost. Load diversity limits the potential savings, as explained in Measure 4.5.1. Take a good look at the system to see whether you can reduce the load diversity within each air handling system.

How to Control the Duct Temperatures

The air temperature in the cold duct is controlled by throttling the output of the cooling coil. The cooling coil is controlled by the space thermostat that has the greatest requirement for cooling. The reset control needs to sample the thermostat signals from a number of spaces because the cooling loads in spaces do not vary in unison. For example, solar load is greatest in east-facing perimeter spaces in the morning, but in west-facing spaces in the afternoon.

Therefore, connect the reset controller for the cold duct to all the space thermostats that may, at some time or other, dictate the minimum chilled air temperature. The cooling reset controller typically should sense at least one thermostat on the east, west, and south sides of the building and the core area. It should also sense thermostats in all spaces that may have exceptionally high heat gains, such as auditoriums, conference rooms, dining areas, and computer rooms.

Similarly, connect the reset controller for the hot duct to all the space thermostats that may dictate the maximum hot air temperature. The heating reset controller typically should sense at least one space on the north perimeter, and any other space that may get colder than the others.

Reference Note 14, Control Signal Polling, explains how to select a signal from a group of thermostats. If you have an energy management computer system, you can usually perform these functions through the computer system. In pneumatic control systems, an ordinary receiver-controller is used to accomplish the temperature reset function.

SUMMARY

This is the way to minimize reheat at all times. A basic control feature.

SELECTION SCORECARD

Savings Potential .... $ $ $ $

Rate of Return, New Facilities % % % %

Rate of Return, Retrofit .... % % % %

Reliability .... ✓ ✓ ✓

Ease of Retrofit .... ☺ ☺ ☺

ECONOMICS

SAVINGS POTENTIAL: *20 to 70 percent of cooling and heating energy.*

COST: *The control components for a typical installation typically costs from several hundred to several thousand dollars. Connection to thermostats at widely dispersed locations can incur extensive additional labor cost.*

If an energy management computer is installed, reprogramming the computer to accomplish chilled air temperature reset typically costs several thousand dollars.

PAYBACK PERIOD: *Less than one year, to several years.*

TRAPS & TRICKS

PLANNING AND DESIGN: *This is more than a control modification. To get most of the benefit from this Measure, you must fix any conditions that cause some spaces to require lower chilled air temperature or higher hot air temperature than the rest. In climates that commonly have high humidity during mild weather, you may have to add some control complications to maintain adequate dehumidification.*

EXPLAIN IT: *Describe the reset controls in the plant operating manual. Install placards on the controls to identify them.*

MONITOR PERFORMANCE: *These important controls can fail, or they can be adjusted improperly, without anyone noticing. Include a column in the plant operating log to record the duct temperatures, or use your energy management computer to log this information. Review the record periodically to verify that the cold duct temperature rises as the cooling load falls, and that the hot duct temperature falls as the heating load falls.*

MEASURE **4.5.2 Turn off the heating coil and/or the cooling coil whenever practical.**

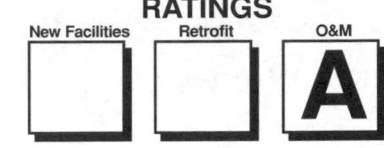

The formula in the introduction to this Subsection shows that mixing losses disappear in a dual-duct system if you turn off either the heating coil or the cooling coil. Once the weather has warmed up to the point that heating is no longer required, there is no reason why you cannot turn off the heating coil. If the coil is hydronic, close the isolation valve. If the coil is electric, disconnect the power to the coil.

Similarly, you can turn off the cooling coil whenever the system has no cooling load.

When to Turn Off the Heating and Cooling Coils

In principle, you can eliminate all mixing losses by turning off the cooling coil whenever the heating coil is turned on, and vice versa. In practice, this may not be acceptable. Even if the zoning of the building allows this, fluctuations in weather make it advisable to keep both coils in operation during the transition seasons. Turn off the cooling coil after cold weather has arrived to stay, and turn off the heating coil after cold weather has definitely gone for the year.

You can always turn off the heating coil as soon as the outside temperature becomes as warm as the inside temperature (except for warm-up periods). In contrast, you may not be able to turn off the cooling coil during cold weather. If the air handling system serves any spaces with high heat gains, you must keep the cooling coil operating for the benefit of those spaces.

If possible, try to arrange the air handling systems so that each serves spaces that have loads as similar as possible. This increases the fraction of time that you can turn off one coil or the other.

Relationship to Duct Temperature Reset

Duct temperature reset controls, which are recommended by Measure 4.5.1.1, perform the same function as this Measure. But, even with duct temperature reset, it is still a good idea to turn off the coils seasonally to protect against unnecessary reheat if a reset controller fails.

Conversely, you can use this Measure as a cheap substitute for installing a reset controller. However, it is effective only under certain conditions, whereas a reset controller works continuously.

Don't Forget

Neglecting this procedure produces no discomfort or other apparent effects, so it is easy to forget. Keep the procedure in mind by appropriate methods, especially a good reminder schedule and effective

placards. See Reference Note 12, Placards, for details of effective placard design and installation.

Alternative: Control Chiller and Boiler Operation

In cases where several air handling units are served by a central chiller, turning off the chiller has the same effect as turning off all the cooling coils. Minimizing the duration of chiller operation is covered by Measures 2.1.2 ff. However, control at the chiller usually cannot save as much energy, because some air handling units are likely to operate longer than others.

You can take the same approach with control at the heating plant, provided that the heating plant does not serve purposes, such as water heating, that keep it operating during warm weather. See Measures 1.1.1 ff. Again, this is usually less efficient than controlling the heating operation of the individual air handling systems.

RATINGS

New Facilities	Retrofit	O&M
		A

SUMMARY

A seasonal method of eliminating reheat that costs nothing.

SELECTION SCORECARD

Savings Potential	**$ $ $** $
Rate of Return	**% % %** %
Reliability	✓ ✓
Ease of Initiation	☺ ☺ ☺

ECONOMICS

SAVINGS POTENTIAL: *30 to 80 percent of mixing losses, depending on climate and system zoning. This typically corresponds to about 20 to 50 percent of total conditioning energy.*

COST: *Minimal.*

PAYBACK PERIOD: *Immediate.*

TRAPS & TRICKS

EXPLAIN IT: *Provide clear instructions that explain when to turn off the cooling coil and heating coil, e.g., when the morning outside air temperature is above or below a certain value. Optimize the schedule for each air handling unit or group of units. Install placards at the appropriate valves or at the controls of self-contained compressors.*

MONITOR PERFORMANCE: *This activity is easy to forget. Schedule periodic checks of the procedure in your maintenance calendar.*

MEASURE 4.5.3 Trim the output of the air handling system fans.

The formula at the beginning of this Subsection shows that the energy consumption of the terminal unit is partly proportional to their air flow rates. Fan trimming is a one-time adjustment that gets rid of fan capacity in excess of the amount needed to satisfy peak loads. Refer to Measure 10.3.1 for the details.

By reducing the constant air flow rate in the system, this action reduces reheat "off the top." However, it does not tailor fan output to a changing load. This is a cheap and simple way to reduce reheat significantly while saving a modest amount of fan energy. Consider this a stopgap action to save energy until you can accomplish either Measure 4.5.4 or Measure 4.5.5.

MEASURE 4.5.4 Install multi-speed fan motors.

The formula at the beginning of this Subsection shows that the energy requirement of the terminal units is partly proportional to their air flow rates. Installing multi-speed motors in the air handling units reduces the

air flow rate in steps that follow the system load. This Measure is virtually identical to Measure 4.4.6 in theory, installation, and economics. Refer to there for details.

MEASURE 4.5.5 Convert the system to variable-air-volume (VAV) operation.

The principle of a VAV dual-duct system is simple. Instead of controlling space temperature by mixing hot and cold air, the terminal unit selects either hot air or cold air, but not both. The terminal unit regulates the space temperature by throttling the amount of hot air, or cold air, that is delivered to the space.

Because of the throttling action of the terminal units, the total air flow rate delivered by the system fans varies over a wide range. Therefore, the fans should have an efficient method of modulating their output. This keeps fan power to a minimum under all load conditions, and maintains a constant pressure in the duct systems.

Although the VAV concept is simple, putting it into practice has proven to be tricky. You may need to make compromises that infringe on comfort or that give up some of the efficiency potential of the concept.

SUMMARY

This is the most efficient configuration for dual-duct systems. Achieving both comfort and high efficiency is challenging.

SELECTION SCORECARD

Savings Potential	$ $ $ $
Rate of Return, New Facilities	% % % %
Rate of Return, Retrofit..........	% % %
Reliability	✓ ✓ ✓
Ease of Retrofit	☺ ☺

To avoid repetition, some of the following topics refer you to the corresponding topics in Measure 4.4.7, which converts single-duct reheat systems to VAV. Converting a dual-duct reheat system is usually easier and cheaper than converting a single-duct system, but getting it right is still a challenge.

Less Expensive Alternatives

Measures 4.5.1 through 4.5.4 attack the same sources of energy waste as VAV conversion, namely, reducing reheat and fan power. In some applications, a combination of those Measures may provide a large part of the savings provided by VAV conversion, at much lower cost.

Energy Saving

■ Energy Waste from Mixing Heated Air and Chilled Air

The introduction to this Subsection points out that reheat systems may consume more energy in mixing losses than in actually conditioning the spaces. In principle, you could design a VAV dual-duct system that eliminates virtually all reheat. In actual practice, you may need to mix air at low flow rates to provide adequate ventilation or dehumidification.

Using VAV in dual-duct systems eliminates most reheat loss because it restricts reheat to the low end of the air flow range. Formula 1, in the introduction to this Subsection, shows that reheat loss is roughly proportional to air flow. Therefore, if you restrict reheat to low-flow conditions, the system wastes much less energy loss than a conventional reheat system, where the hot and cold air are mixed at maximum flow rate. Going further, you can add duct temperature reset to reduce reheat at the low end of the load range. By using all available techniques, you can typically eliminate over 90% of the reheat loss that would occur in a constant-volume dual-duct system.

To calculate the saving the saving in reheat that you can expect from using VAV operation instead of constant-volume operation, you need to calculate the total energy consumption of each terminal unit for both types of systems. Do this by using Formula 1, in the introduction to this Subsection.

Reheat energy waste occurs at each terminal unit individually. The amount of reheat that occurs at each terminal unit depends on these factors:

• *the minimum-flow requirement.* For reasons explained below, it may be necessary to maintain a minimum flow through the terminal unit. If the amount of chilled air that enters the terminal at the minimum-flow setting provides more cooling than needed, the excess cooling must be cancelled by mixing with heated air. Similarly, if the amount of heated air that enters the terminal provides more heating than needed, the excess heat must be cancelled by mixing with chilled air.

If you use terminal units that can close to shutoff, then leakage in the terminal units becomes an important factor. Any leakage behaves as a minimum-flow setting, and it wastes energy in the same way.

• *the duct temperatures.* If the chilled air temperature is raised, the amount of cooling provided by a given amount of air flow through the terminal is reduced. Under conditions when reheat is required, raising the chilled air temperature reduces the amount of reheat that is necessary to cancel unwanted cooling. The same applies to lowering the heated air temperature. Duct temperature reset is a powerful energy conserving technique.

Duct temperature reset is limited by diversity of conditioning loads. This is because the chilled air temperature must be set low enough to satisfy the cooling load on the terminal unit with the greatest cooling requirement, and the heated air temperature must be high enough to satisfy the terminal that has the highest heating load. This limitation occurs only during conditions when the system must provide both heating and cooling. Therefore, zoning of the system is an important factor in exploiting duct temperature reset.

It is important to understand that the amount of energy waste from air mixing in a dual-duct VAV system is not fixed and predictable. On the contrary, it depends largely on system adjustments that are made at the time of installation and later throughout the life of the system. These adjustments are needed to deal with issues that we will discuss below. To a large extent, the adjustments are arbitrary, or they are a matter of preference. A terminal unit in one space may be adjusted to eliminate most mixing losses, whereas an identical terminal unit in the adjacent space may be adjusted in a way that causes a large amount of energy waste. To minimize mixing loses, adjust each terminal unit to minimize overlap of the heated air and chilled air dampers. In addition, exploit duct temperature reset as much as possible.

■ Fan Energy

A fan operating in VAV mode delivers only enough air to carry the required amount of heating or cooling energy into the spaces.

The relationship between fan power and air flow in a dual-duct VAV system cannot be stated precisely. The power required by the fan to overcome the air flow resistance of the coils and ductwork is approximately proportional to the third power of the air flow rate. The power required to overcome the resistance of the dampers in the terminal units is roughly proportional to the first power of the flow rate. (The latter is because the dampers act to keep the pressure opposing the fan constant, instead of letting the pressure drop at lower flow rates.) The actual power required by the fan varies between these two extremes.

All the methods of varying fan output (variable-speed drives, inlet vortex dampers, discharge dampers, etc.) have inefficiencies that partially offset the saving in fan power. The best fan modulation methods have low losses, while the worst would dissipate a large fraction of the potential savings. A well designed VAV system should use substantially less than half the fan energy of a constant-volume system.

■ Energy to Condition Outside Air in Once-Through Systems

In systems that require 100% outside air, such as those used in hospitals, VAV can substantially reduce the amount of energy that is consumed to condition outside air.

A constant-volume reheat system is sized to provide enough air to convey heating and cooling under peak load conditions. This is wasteful because the average conditioning load is much lower than the peak load. A VAV system reduces this waste by taking in only enough outside air to convey the heat or cooling that is required at the moment.

The amount of energy saved by converting a 100%-outside-air dual-duct system to VAV depends primarily on the climate and the conditioning loads. In mid-latitude climates and typical applications, VAV conversion may reduce outside air heating and cooling by half.

How to Convert a Dual-Duct Reheat System to VAV

The steps in converting a dual-duct reheat system to VAV operation are:

- replace the constant-volume mixing terminals with VAV terminals, or modify the existing terminals to operate in a VAV mode
- replace the space thermostats, if necessary. If the terminals are able to close to shutoff, consider deadband thermostats.
- if necessary, replace the air diffusers in the spaces to provide good performance at all air delivery rates
- install an efficient means of varying the output of the air handling system fans
- install a control system for the fan output
- install temperature reset controls for both the hot and cold ducts
- provide controls for various auxiliary functions, such as morning warm-up, night setback, purge cycle operation, smoke relief, etc.

You may encounter a variety of problems, including unsatisfactory air distribution, inadequate outside air ventilation, and complaints of stagnation. The main source of these problems is the large reduction of space air flow that occurs during periods of low load. A tempting solution is to add reheat back into the design, but this wastes energy. The design challenge is to maintain comfort and health while minimizing reheat. Make sure you cover all the following points. Many of

WESINC

Fig. 1 Dual-duct terminal unit, partly disassembled The inlet end is on the right. The duct with foil faced insulation admits chilled air. The duct with unfaced insulation admits warm air. The entering air immediately encounters a baffle plate, which creates a large amount of resistance. The inside of the box contains a sound attenuator.

these points are related, which makes good VAV design a feat of mental juggling.

■ How to Select Terminal Units or Conversion Kits

Dual-duct reheat terminal units are often called "mixing boxes," which accurately describes their function. They mix air from the hot and cold ducts in a manner that provides approximately constant air flow to the space under all load conditions. The cheapest type of dual-duct mixing box has a single damper vane that swings between the chilled and hot air supplies inside the box, allowing more chilled air to flow as the hot air is blocked, and vice versa. Figures 1 and 2 show a mixing box of this type.

There are also mixing boxes with separate dampers for hot and cold air. The space thermostat controls the movement of the damper or dampers inside the terminal.

The VAV version of a dual-duct mixing box is similar to the constant-volume version. The difference is the way the dampers behave. In a VAV system, if the space has a cooling load, only the cooling damper opens, modulated by the space thermostat to satisfy the load. Similarly, if the space has a heating load, only the heating damper opens. There is no mixing of hot and cold air, except perhaps at low loads.

In retrofit applications, you may have a choice of two paths in converting to VAV. One is to install entirely new VAV terminal units in place of the original ones. The other is to convert the existing terminals to operate as VAV terminals.

Several manufacturers offer conversion kits for specific models of dual-duct terminals. Some constant-volume mixing boxes can be converted easily. This option provides a large saving in equipment cost. It may also save a large amount of labor cost because it is usually possible to convert the unit in place.

Before deciding to convert existing boxes, make sure that the converted boxes will operate efficiently. In particular, the converted box should not have an appreciable amount of leakage that allows air mixing. Compare the features that are available in new VAV terminals and in conversion kits.

■ Reduce Flow Resistance in Converted Terminal Units

In addition to the flow-regulating dampers, dual-duct terminal units may also contain volume dampers to limit the total flow through the box, baffles to mix the hot and cold air, and sound attenuators. These add a significant amount of resistance to air flow, increasing fan energy requirements. If you convert existing mixing boxes, take the opportunity to eliminate the flow resistance of these obstructions whenever possible. Figures 1 and 2 show some of the flow restricting hardware that you may encounter.

Remove any volume regulating dampers, because they serve no purpose in VAV operation.

If you intend to convert the terminal units to shutoff operation, remove any mixing baffles. They serve no

WESINC

Fig. 2 Mixing vane in dual-duct terminal unit This is the unit shown in Figure 1, looking into the ends of the chilled air and heated air ducts. The air is proportioned by a simple vane that swings between the two duct openings, so that the total flow remains approximately constant. The amount of energy wasted by air mixing is enormous, often much larger than the conditioning load of the space.

purpose if the converted terminal units do not allow air mixing.

If you intend to retain air mixing at low loads, you may need to retain the mixing baffles because low velocity air streams tend to remain stratified. In case of doubt, experiment to find out whether stratification continues into the space. Remove the mixing baffles from several units. Take a quick-response thermometer and see if the air comes out of the diffuser in all directions at a reasonably uniform temperature. The duct and the diffusers downstream of the mixing box may eliminate the stratification before the air enters the space.

Sound attenuators are installed in some mixing boxes to trap noise that originates in the air handling system. The attenuators may be intended to reduce noise that is generated upstream of the terminal unit, especially by the fan, or they may be intended to reduce noise generated inside the terminal unit itself. Fan noise may not be significant if the terminal is distant from the fan, but a noise attenuator may be necessary for terminals located near the fan.

The terminal unit itself may generate substantial noise. This noise is especially troublesome because of the short path to the space. Much of the noise generated inside the terminal unit is caused by mixing baffles and flow limiting dampers. If you remove those, the sound attenuator may become unnecessary.

The noise generated by any given air handling system rises sharply with increasing air flow. The VAV mode of operation reduces air flow significantly under most conditions, so converting the system to VAV should reduce the average system noise level considerably. The exception is noise generated by the terminal unit dampers as they approach the closed position. To minimize this whistling noise, control the fans to reduce their output to match system flow requirements.

■ Minimize Minimum-Flow Settings

VAV terminal units generally have some kind of adjustable mechanical stop that prevents each damper from closing fully. This feature allows you to establish a minimum flow from the cold and hot ducts. You may need to maintain a minimum air flow for several reasons, including:

- proper diffuser operation
- maintaining an adequate level of outside air ventilation
- maintaining in-space air velocity to prevent a feeling of "stuffiness" and to facilitate cooling by perspiration
- providing dehumidification during periods of low cooling load.

These requirements are covered below. Consider the requirements of each space to decide the appropriate minimum-flow settings for the terminals. Minimum-flow is the Achilles heel of VAV system efficiency

because it opens the door to reheat. This is so important that you should take all possible steps to minimize the minimum-flow setting of each terminal unit, consistent with maintaining comfort and ventilation.

A subtle difference in the manner of setting minimum flow can make a big difference in reheat energy waste. The least efficient method is to have a separate minimum-flow stop for the chilled air and hot air. With this method, there is some reheating under all load conditions. Some cheap retrofit kits may resort to this method.

In contrast, the proper way to deal with reheat (if it cannot be avoided) is to have mixing only if both the chilled air and hot air flows have fallen to a minimum value. For example, the chilled air damper remains fully closed during heating operation until the hot air flow falls to its minimum value, at which point the chilled air damper starts to open.

■ Diffuser Performance
See Measure 4.4.7.

■ Space Air Velocity
See Measure 4.4.7.

■ Night Temperature Setback and Morning Warm-Up

Night temperature setback is easy to accomplish in a dual-duct VAV system. The setback controls should perform these functions during the setback period:

- *control the heating coil with a setback thermostat.* The simplest way to do this is with a single thermostat located in the coldest or most critical space. If you have an energy management control system, you can probably program it to perform heating setback.
- *turn off the cooling coil.* Be sure that controls turn off the cooling coil at the beginning of the setback period, allowing the space temperature to drift. If the cooling coil is not turned off, it will force the space to the setback temperature, wasting cooling energy.
- usually, *turn off the air handling unit fans* as long as the spaces are within the deadband temperature range.
- for maximum efficiency, use *optimum-start* to extend the setback period as long as possible. (See Measure 4.1.1.2 for details.)

■ Space Temperature Deadband

In spaces where you use shutoff terminal units, you may be able to save additional energy by using deadband temperature control. Measure 4.7.2 explains deadband control in dual-duct systems.

You cannot use deadband control if you need to maintain a minimum flow through the terminals. Maintaining minimum air flow requires overlapping the operation of the heating and cooling dampers, whereas deadband involves inserting a temperature gap between the closing of one damper and the opening of the other. The way to minimize the energy waste resulting from minimum-flow is to use duct temperature reset (Measures 4.5.1 ff).

■ How to Select the Fan Modulation Method

See Measure 4.4.7 for your choices.

■ How to Control Fan Modulation

See Measure 4.4.7.

The hot and cold duct systems operate independently of each other. Sometimes one duct will have lower pressure, sometimes the other. To deal with this, install a pressure sensor in each. Control the fan to respond to the lower pressure.

Larger dual-duct systems may have two separate air handling units, one for the hot duct and the other for the cold duct. In this case, control the pressure in each duct independently.

■ Exploit Duct Temperature Reset

Duct temperature reset is explained in Measure 4.5.1.1. It is a primary method of improving efficiency and comfort at low space air flow rates. Virtually all VAV systems should have this control feature.

■ Dehumidification

See Measure 4.4.7.

■ Outside Air Ventilation

A dual-duct VAV system provides outside air ventilation through the air handling unit whenever a space is either heated or cooled. Outside air ventilation may be inadequate at low conditioning loads, when both the heating and cooling dampers are closed. If deadband thermostats are installed, both dampers may remain closed for long periods.

If a VAV system needs to provide full ventilation during low-load conditions, the only way to do it is to use reheat so that air flow can be increased enough to carry the ventilation air.

If you have recirculating air handling systems, your ability to control outside air intake depends on the "front end" configuration of the air handling system. Examine the design of your air handling systems to see if they allow accurate control. A well designed air handling system is able to maintain accurate control of outside air intake even as the fan output changes. See Measure 4.2.2 for the details.

Even if a VAV system is designed to maintain accurate control of total outside air intake, it cannot distribute ventilation air in proportional amounts to individual spaces. This is because the volume of air delivered to individual spaces varies with their conditioning loads. For example, if a system serves both core and perimeter spaces, most of the outside air may be delivered to the core spaces on a cool day, and to the perimeter spaces on a hot day. It may be satisfactory to

let ventilation air diffuse from the perimeter zone into the core zone, or vice versa. This is most likely to be acceptable if the occupant density is low and there are no major sources of pollution inside the spaces.

You can provide perfectly controlled outside air ventilation without energy waste by installing a separate outside air ventilation system that has its own ducts, fans, and heating elements. (The "triple-duct" system is one version of this arrangement.) A separate ventilation system also solves the problem of distributing outside air properly to individual spaces.

A simpler option is to open windows during mild weather, when the VAV boxes close down. This option is not available in buildings with fixed windows. Even in buildings with openable windows, it may not be acceptable. Controlling the use of windows to avoid wasting energy during hot and cold weather is difficult and unreliable.

Supply air temperature reset is important for reducing the reheat loss that occurs for the sake of ventilation. For example, if the outside air is warmer than the inside air, the hot duct reset control should completely turn off the heating coil, which eliminates reheat.

■ Zoning

Zoning affects energy consumption in dual-duct VAV systems, but its effects are less serious than they are with single-duct VAV systems. The main benefit of good zoning is that it maximizes the potential for using duct temperature reset. See Measure 4.5.1 for details. Exploit this in new construction. In retrofit applications, it is rarely economical to change the zoning of a dual-duct system.

■ Duct Layout

Converting an existing dual-duct system to VAV may provide a valuable opportunity for improving the basic duct layout. While you are making the conversion, the duct system is exposed and accessible. Use this opportunity to reduce pressure losses in the duct system. This allows you to reduce the fan pressure, saving fan energy. It may also improve temperature control in the spaces.

The upstream terminal units bleed air pressure out of the duct system, reducing flow through the downstream terminals. This may create flow problems in the downstream terminal units that motivate using "pressure-independent" terminals. This type of terminal requires higher fan pressure than the "pressure-dependent" type, and it is more expensive. Classic methods of "equal pressure" duct design cannot solve this problem, because the load relationship among VAV terminals varies continuously.

It is usually not necessary or economical to upgrade an entire duct system. You may be able to make some localized improvements that significantly improve air flow. For example, if two long pairs of ducts serve different zones, connecting them at the ends to form loops may considerably reduce the pressure differences in the system.

At the same time, ferret out localized duct defects that rob fan power and create excessive pressure differences. The common culprits are non-smooth bends, multiple bends within a short distance, and sharp-cornered takeoffs. Before making changes, spend some time studying how ducts can obstruct air flow. Air moves in strange ways in ducts, and you need to understand this behavior to get the most benefit from changes.

■ Shutting Off Conditioning of Unoccupied Spaces

If the terminal unit dampers can close to shutoff, you can stop conditioning individual spaces when they are not occupied. This saves conditioning energy and fan energy in spaces that operate on different schedules from the rest. Use Measures 4.1.1 ff to control conditioning based on occupancy.

■ Building Pressure Differentials

See Measure 4.4.7.

■ Economizer Cycle Operation

Installing an outside air economizer cycle (Measure 4.2.5) in a dual-duct system is not effective if the air handling unit has only a single supply fan. The single fan makes it impossible to route the cold outside air exclusively to the chilled air duct.

If the system has a separate fan for each duct, an outside air economizer can operate with full efficiency. See Measure 4.8.8 for details.

In existing systems, it is usually impractical to retrofit a two-fan configuration. As an alternative, consider installing a "waterside" economizer coil in the cooling side of the system. See Measure 2.9.4 about waterside economizers.

■ Purge Cycle Operation

See Measure 4.2.7 about purge cycles. A purge cycle requires an additional control function that opens all the terminal unit dampers during the purge period.

■ Fan-Induced Resonances

See Measure 4.4.7.

ECONOMICS

SAVINGS POTENTIAL: *30 to 70 percent of heating and cooling energy; 40 to 80 percent of fan energy. The savings depend on the climate, zoning, and other factors, including other energy conservation activities.*

COST: *In new construction, the cost elements are the VAV fan drive(s), perhaps a more complex air handling unit, controls for fan operation, the cost differential between reheat and VAV terminal units, and the differential cost of space thermostatic controls. See Reference Note 36 for the cost of variable-speed fan drives. The last two differential costs are small.*

Retrofitting existing facilities is much more expensive, primarily because of the need to replace or modify the terminal units. VAV terminal units typically cost several hundred dollars apiece.

PAYBACK PERIOD: *A few years or less, in new construction. Several years or longer, in retrofit.*

TRAPS & TRICKS

DESIGN: *VAV systems that are comfortable, highly efficient, and reliable are still rare. There are many design issues, so it is easy to overlook some of them. Do your homework. Try to find successful systems to study.*

SELECTING THE EQUIPMENT: *Make an effort to select the best terminal units or conversion kits. Good equipment is available, but there is a lot of flimsy junk on the market. Inspect sample units. Study Reference Note 36 about selecting variable-speed fan drives. Variable-speed fan drives are easier to design and have fewer practical problems than other methods of fan modulation.*

INSTALLATION: *Work with the installers of the terminal units to ensure that minimum-flow limits and other settings are adjusted properly.*

EXPLAIN IT: *Describe the system in the plant operating manual. Include a table of the settings for all terminal units. Buy an extra terminal unit to use as a training model.*

Multizone Systems

In multizone air handling systems, zone dampers mix heated air from a heating coil and chilled air from a cooling coil to regulate the temperature of the air supplied to a space, or zone. Each zone has a separate duct that extends all the way from the air handling unit. Each pair of zone dampers is controlled by a space thermostat. Figure 1, on the next page, shows the distinguishing features of a multizone unit. Figures 1 and 2, in the Section 4 Introduction, compare this type of system to other air handling systems. The energy behavior of multizone units is the same as that of dual-duct systems. See the introduction to Subsection 4.5 for details.

In conventional multizone units, the dampers mix hot and cold air in proportions that keep the flow of mixed air to each zone approximately constant. This can result in an enormous amount of mixing loss if both coils are operating. By far the largest opportunity for energy conservation in multizone reheat systems is minimizing this source of energy waste. During low-load conditions, the system may consume much more energy in mixing losses than in conditioning the spaces. All the Measures in this Subsection reduce mixing losses. Some of the Measures also save fan energy.

The primary justification for using multizone systems is comfort. They provide a high level of comfort because air flow is constant, temperature control can be maintained precisely as the load changes from heating to cooling, and humidity can be kept at any desired level. In making changes for the sake of efficiency, preserve these qualities to the extent that they are needed by the application.

INDEX OF MEASURES

4.6.1 Keep the temperature of the cold deck as high as possible and the temperature of the hot deck as low as possible.

4.6.1.1 Install temperature reset controllers for both the cold deck and the hot deck.

4.6.2 Turn off the cooling coil or the heating coil whenever practical.

4.6.3 Trim the fan output.

4.6.4 Install multi-speed fan motors.

4.6.5 Convert the system to variable-air-volume (VAV) operation.

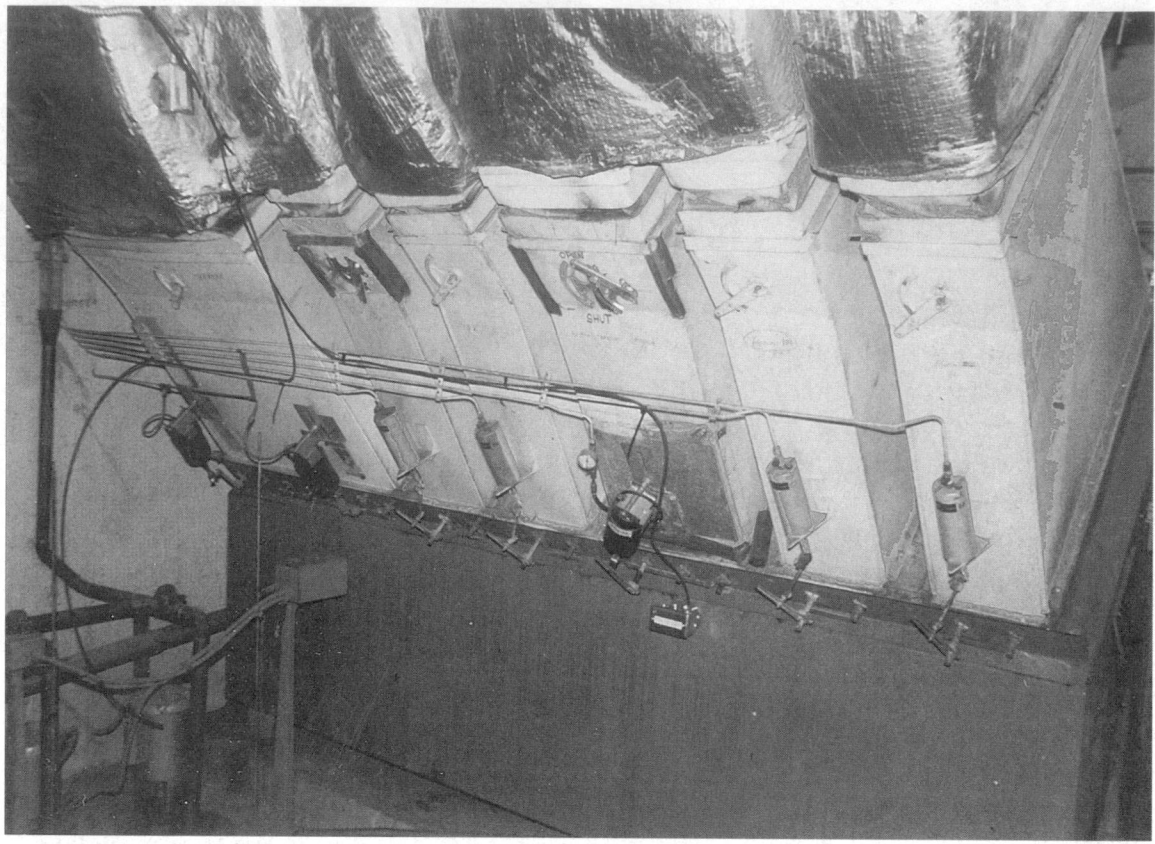

WESINC

Fig. 1 Large multizone air handling unit   Each of the ducts serves a zone with individual thermostatic control. The zone thermostat controls the damper actuator that is visible at the entrance to each zone duct. In this unit, the hot deck and cold deck dampers for each zone are installed rigidly on a common shaft, which is moved by the actuator. Large mixing losses are unavoidable with this arrangement.

A minor matter of terminology is the use of the terms "hot deck" and "cold deck" with multizone units, whereas the corresponding terms for dual-duct systems are "hot duct" and "cold duct." In multizone units, heated air is discharged from the hot coil into a chamber or "hot deck" in the air handling unit, from which it passes through mixing dampers into the individual zone ducts. Similarly, cooled air from the cooling coil is discharged into a "cold deck" before passing through mixing dampers into the zone ducts.

Three to eight zones is typically the maximum number that can be served by each air handling unit. The limitation is the amount of physical space needed to attach the zone ducts to the air handling unit. As a rule, larger air handling units can accommodate a larger number of zones.

In new construction, none of these modifications incurs a relatively large cost increase, unless the systems are small. Variable-output fan drives were once a major cost addition, but the relative cost of this equipment has fallen sharply, especially for larger systems.

In retrofit applications, first decide whether or not you will make major changes to the existing constant-volume system. This decision depends mainly on two issues: whether you can afford a major equipment change, and whether you can forego the tight control of humidity and the continuous ventilation that constant-flow operation provides. If you can afford only inexpensive changes, accomplish Measures 4.6.1 ff through 4.6.3. If you can afford a major modification and it is justified by the energy savings, accomplish Measure 4.6.5. Measure 4.6.4 is an intermediate approach that saves less energy than the latter, but it is much less expensive.

Three-Deck Multizone Systems

Three-deck multizone systems are similar in appearance to conventional multizone systems. However, their energy behavior is entirely different. Three-deck systems have a third deck that is not served by a coil. It recirculates unconditioned air (a mixture of return air and outside air) for mixing with either the heated air or the chilled air. If properly installed, these systems have greatly reduced losses. Three-deck systems are unusual. They are similar in energy behavior to the triple-duct systems described in the introduction to Subsection 4.5.

Damperless Multizone Systems

Another type of multizone system uses an air handling unit that has no dampers. Instead, each zone has its own heating and cooling coils. Each coil is controlled independently by a thermostat in its respective zone. This unit has no hot and cold decks. It is like a number of single-zone units in a single housing, all served by a common fan.

Figure 2, in the Section 4 Introduction, sketches this type of system.

Damperless multizone units do not mix heated and cooled air, so they avoid the most severe energy waste in conventional multizone units. This type does waste some fan energy because it uses a single blow-through fan to serve all the coils together, even though the flow requirements for different zones vary individually. Like all multizone units, this arrangement is not well adapted to an economizer cycle, because the zones differ in their supply temperature requirements.

If you are dealing with damperless multizone systems, refer to Subsection 4.3 for Measures that can optimize their efficiency.

MEASURE 4.6.1 Keep the temperature of the cold deck as high as possible and the temperature of the hot deck as low as possible.

This Measure is virtually identical to Measure 4.5.1 in theory, procedures, and economics. Refer to there for details.

MEASURE 4.6.1.1 Install temperature reset controllers for both the cold deck and the hot deck.

This Measure is similar to Measure 4.5.1.1 in theory, installation, and economics. Refer to there for details.

As a retrofit, the cost of this Measure is less than for Measure 4.5.1.1 because the signals from all the zone thermostats already go to the air handling unit, where they are available for the signal discriminators. This avoids the need to add extensive wiring or pneumatic lines.

MEASURE 4.6.2 Turn off the cooling coil or the heating coil whenever practical.

This Measure is virtually identical to Measure 4.5.2 in theory, procedures, and economics. Refer to there for details.

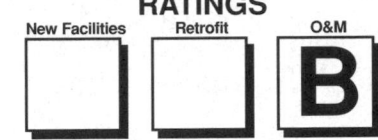

MEASURE **4.6.3 Trim the fan output.**

The formula at the beginning of Subsection 4.5 shows that the energy consumption of each zone is partly proportional to its air flow rate. Fan trimming is a one-time adjustment that gets rid of fan capacity in excess of the amount needed to satisfy peak system loads. Refer to Measure 10.3.1 for the details.

By reducing the constant air flow rate in the system, this action reduces reheat "off the top." However, it does not tailor fan output to a changing load. This is a cheap and simple way to reduce reheat significantly while saving a modest amount of fan energy. Consider this a stopgap action to save energy until you can accomplish either Measure 4.6.4 or Measure 4.6.5.

MEASURE **4.6.4 Install multi-speed fan motors.**

The formula at the beginning of Subsection 4.5 shows that the energy requirement of each zone is partly proportional to its air flow rate. Installing a multi-speed fan motor in the air handling unit reduces the air flow rate in steps that follow the system load. This Measure is virtually identical to Measure 4.4.6 in theory, installation, and economics. Refer to there for details.

MEASURE 4.6.5 Convert the system to variable-air-volume (VAV) operation.

The principle of a VAV multizone system is simple. The difference between VAV multizone operation and conventional multizone operation is primarily in the air handling unit dampers. Instead of controlling the temperature of each zone by mixing hot and cold air, the zone dampers operate independently to provide either hot air or cold air to each zone, but not both. If there is no overlap in the opening of hot and cold dampers, all mixing losses are eliminated.

Since the total amount of air delivered by a VAV system varies over a wide range, the system fan(s) should have variable output. This saves a certain amount of fan energy, but not as much as in other types of VAV systems.

Although the VAV concept is simple, putting it into practice has proven to be tricky. You may have to make some compromises that sacrifice efficiency to satisfy comfort and health requirements.

This Measure is primarily for improving the efficiency of existing multizone systems. In new construction, if you desire a multizone type of system, the damperless multizone system is a better choice. See Figure 2 in the Section 4 Introduction for a sketch of this type of system. Damperless multizone systems avoid the leakage and maintenance problems of dampers, and they avoid the problems of low air flow in the spaces. In addition, new construction affords you the opportunity of considering other types of efficient systems.

Is it Practical to Retrofit an Existing Multizone Unit?

The physical changes that you need to convert a conventional multizone system to VAV are concentrated at the damper section of the air handling unit. The way the mixing dampers were originally installed is the main factor that determines whether it is practical to convert the unit. Refer to "Zone Damper Modifications," below, for details.

The energy savings are proportional to the capacity of the air handling unit. In contrast, the cost of the modifications does not relate strongly to the size of the air handling unit. In fact, it may be easier to retrofit a larger unit. Therefore, this modification should pay for itself quickly in large systems, but it may not be economical in small systems.

When analyzing the potential savings, consider the benefit that this Measure can provide in excess of Measures 4.6.1 ff through 4.6.3, which you can accomplish at low cost.

SUMMARY

The most efficient configuration for multizone systems with dampers. Combining comfort and high efficiency is challenging. Lack of space at the damper end of the air handling unit can be a serious obstacle to retrofit.

SELECTION SCORECARD

Savings Potential .... $ $ $ $

Rate of Return, New Facilities % % % %

Rate of Return, Retrofit.......... % % %

Reliability .... ✓ ✓ ✓

Ease of Retrofit .... ☺ ☺ ☺

Energy Saving Potential

Refer to this heading in Measure 4.5.5. VAV conversion of multizone systems saves energy in the same ways as VAV conversion of dual-duct systems, and in similar amounts.

How to Convert a Multizone Unit to VAV

To convert a conventional mixing (reheat) multizone system to VAV operation:

- modify the zone dampers and their actuators so that the hot deck and cold deck dampers of each zone operate in a VAV mode
- replace the zone thermostats, if necessary. If the zone dampers can close to shutoff, consider deadband thermostats.
- if necessary, replace the air diffusers in the spaces to provide good performance at all air delivery rates
- install an efficient means of varying the output of the air handling system fans
- install a control system for the fan output
- install temperature reset controls for both the hot and cold decks
- provide controls for various auxiliary functions, such as morning warm-up, night setback, purge cycle operation, smoke relief, etc.

Be prepared to deal with a variety of problems in making this conversion, including unsatisfactory air distribution, inadequate outside air ventilation, and complaints of stagnation. The main source of these problems is the large reduction of space air flow that occurs for long periods of time. It is tempting to add reheat back into the design to increase the low-load air flow, but this wastes energy. The design challenge is to maintain comfort and health while minimizing reheat.

Consider the following points. You will see that many of them are related. To avoid repetition, some of the following topics refer you to the corresponding topics in Measure 4.4.7, which converts single-duct systems to VAV, and to Measure 4.5.5, which converts dual-duct systems to VAV.

■ How to Modify the Zone Dampers

The first step is to modify the hot and cold deck mixing dampers so that they (1) operate independently of each other, and (2) close tightly enough to restrict air flow to the minimum required by the application.

Many multizone units have crude damper arrangements in which the hot and cold damper for each zone are mounted on a common shaft. Figure 1 shows a typical arrangement of this type. In such cases, remove the existing dampers entirely and replace them with independent dampers and actuators. This change is not difficult in itself. However, the zone ducts and dampers are jammed together on one side of the air handling unit casing. If the air handling unit is crammed into a tight space, there may not be enough room to make the change.

In many multizone units, especially larger ones, the hot and cold dampers are mounted on the air handling unit casing as separate assemblies with separate actuators. In such cases, you may be able to retain the original dampers and actuators, and simply change the control signals to the actuators.

Don't let the apparent ease of this situation seduce you into keeping leaky dampers. The original dampers probably were not designed to close tightly. In most cases, you will want to replace the present dampers with dampers that are designed to close tightly. Damper leakage allows mixing of hot and cold air, which wastes energy.

Replace leaky dampers even if you design the system for minimum-flow operation. If the dampers cannot close tightly, they will cause mixing losses under all load conditions, not just minimum-flow conditions.

For example, consider a VAV multizone system in which the heating capacity equals the cooling capacity, and the dampers leak 5% of their full load capacity when closed. Assume that a zone has a cooling load that is 50% of maximum. The heating leakage, plus the additional cooling required to cancel it, represents another 10% of full load. Thus, apparently minor damper leakage adds 20% to energy consumption.

In reality, multizone unit dampers may leak much more than 5% when closed, because tight shutoff was not a requirement foreseen in the design of multizone units. Some retrofit damper improvements, such as blade seals, may substantially reduce leakage, but they are unlikely to provide tight closure. Furthermore, blade seals have a limited life. Therefore, expect to replace the dampers entirely.

■ Zone Damper Controls

The dampers in an ordinary, air mixing multizone system move together to maintain an approximately constant airflow into each zone duct. Both dampers are partially open at all times. In converting to VAV, the objective is to keep one damper closed, except perhaps for minimum-flow at low loads. You can achieve this mode of operation easily by using conventional damper actuators and space thermostats.

For example, if the system has pneumatic controls, select a zone thermostat that produces air pressure in the range of 5 to 15 PSI. Select the cooling damper actuator so that the cooling damper is wide open at 5 PSI, and closes at 10 PSI. Select the heating damper actuator so that the heating damper starts to open at 10 PSI, and is fully open at 15 PSI.

■ How to Select the Fan Modulation Method

See Measure 4.4.7 for your choices.

■ How to Control Fan Modulation

The device that modulates fan output must receive some signal that tells it how much air to deliver. The best method is to measure the pressure in the hot and cold decks. The pressures in the hot and cold decks are independent of each other, so install a pressure sensor in each deck, and design the controls so that the fan responds to the lower pressure.

There is another reason why it is a good idea to install the pressure sensors in the hot and cold decks.

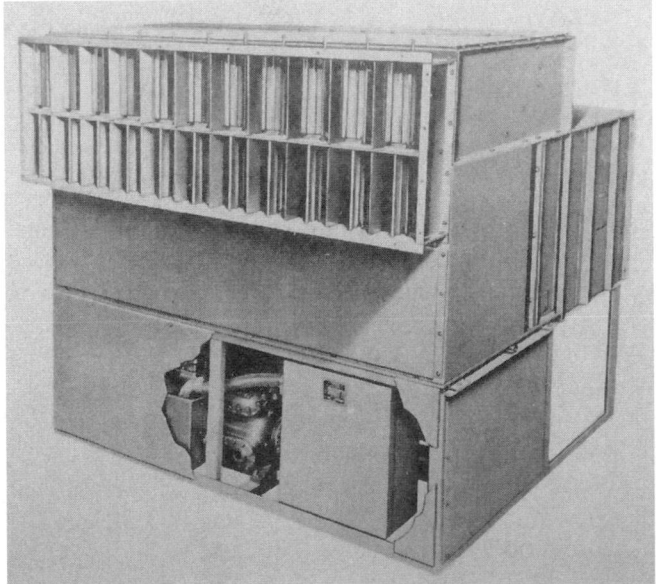

Trane Company

Fig. 1 Candidate for VAV retrofit This old multizone unit has pairs of hot and cold deck dampers installed on common shafts. There is no economical way to salvage the existing dampers. Instead, unbolt the old damper section and throw it away. Install separate hot and cold dampers of good quality for each zone.

The air that is discharged from the supply fan crashes into the front of the two coils, creating a region of great turbulence. For this reason, it would be difficult to get an accurate, stable pressure reading on the fan side of the coils. In contrast, the air flow on the discharge side of the coils is the most uniform anywhere inside the air handling unit. This is because the coils act as baffles, and straighten the air flow.

Even in the hot and cold decks, getting a good, stable signal from the pressure sensors may be tricky. Try to locate the pressure sensor where it is not excessively influenced by fan turbulence or by the drop in pressure at the zone dampers. You might install each sensor inside a pressure averaging tube that traverses the deck. The tube is perforated with tiny holes, so the pressure inside the tube is the average of all points on the outside of the tube.

The controls should keep the system pressure high enough to satisfy the air flow requirement of the zone with the greatest requirement. To avoid wasting fan energy, the pressure should be no higher than this.

(In this type of system, don't try to control fan output based on return air volume. Return air indicates only the average of the hot deck and cold deck flow volumes. The pressures in the hot deck and cold deck may differ radically for a given average flow volume.)

■ Satisfy Requirements for Minimum Air Flow

You may be forced to establish minimum-flow damper settings for a variety of reasons, including:

- proper diffuser operation
- maintaining an adequate level of outside air ventilation
- maintaining in-space air velocity to prevent a feeling of "stuffiness" and to aid cooling by perspiration
- providing dehumidification during periods of low cooling load.

These requirements are covered below. Assess each one to determine the minimum flow that each zone requires. Minimum-flow is the Achilles heel of VAV system efficiency because it causes reheat. Focus on keeping the minimum-flow requirement as low as possible, consistent with maintaining comfort and ventilation.

Minimum air flow is maintained by having both dampers open slightly at low load. For example, in the previous example of pneumatic control, the cold air damper might operate between 6 and 11 PSI, and the hot air damper might operate between 10 and 15 PSI.

■ Zone Temperature Deadband

Temperature deadband is a thermostatic control technique that creates a comfortable range of temperature in which there is no heating or cooling. If there are zones where you can shut off the air flow

completely, consider saving additional energy by using deadband temperature control. See Measure 4.7.2 for details.

You can't use deadband in zones where you set up the dampers to maintain minimum flow. Minimum-flow overlaps the opening of the heating and cooling dampers, whereas deadband involves inserting a temperature gap between the closing of one damper and the opening of the other. If you must use minimum-flow, conserve energy at low loads by using deck temperature reset (Measures 4.6.1 ff).

■ Space Air Velocity

See Measure 4.4.7.

■ Diffuser Performance

See Measure 4.4.7.

Multizone systems encounter an additional problem in diffuser performance because their duct systems are designed to withstand only minimal pressure. Diffusers designed to operate well at low flow rates may have a significant amount of resistance, thereby raising pressure in the zone ducts. The pressure is not great, but if the ducts are large, the additional pressure may deform the ducts or cause "oil can" noise.

■ Night Temperature Setback and Morning Warm-Up

See Measure 4.5.5.

■ Deck Temperature Reset

Automatic deck temperature reset is covered by Measure 4.6.1.1. It increases efficiency and reduces comfort problems at low space air flow rates. Expect to include this control feature.

■ Dehumidification

See Measure 4.4.7.

■ Outside Air Ventilation

See Measure 4.5.5.

■ Zoning

See Measure 4.5.5.

■ Isolation of Space Conditioning

See Measure 4.5.5.

■ Building Pressure Differentials

See Measure 4.4.7.

■ Economizer Cycle Operation

See Measure 4.5.5.

■ Purge Cycle Operation

See Measure 4.2.7 about purge cycles. If you want to have a purge cycle, include a function that opens the zone dampers fully during the purge period.

■ Fan-Induced Resonances

See Measure 4.4.7.

ECONOMICS

SAVINGS POTENTIAL: *30 to 70 percent of heating and cooling energy; 40 to 80 percent of fan energy. The savings depend on the climate, zoning, control features, and other factors.*

COST: *In new construction, the cost elements are the VAV fan drive(s), a more complex air handling unit, controls for fan operation, and the differential cost of space thermostatic controls, which is minor. See Reference Note 36 for the cost of variable-speed fan drives.*

In retrofit applications, the main costs are a VAV fan drive and controls, and damper and sheetmetal modifications. The latter cost varies widely. You may have to relocate nearby equipment or partitions to gain access for the modifications.

PAYBACK PERIOD: *One year or less, in new construction. One year or longer, in retrofit.*

TRAPS & TRICKS

DESIGN: *VAV multizone systems are uncommon. Furthermore, it is still difficult to find VAV systems of any type that are comfortable, efficient, and reliable. Therefore, you are pioneering. Make sure to deal with all the design issues covered above.*

SELECTING THE EQUIPMENT: *Select dampers for tight closing, long-term durability, and ease of maintenance. Install damper linkages that cannot slip. See Reference Note 36 about selecting variable-speed motors and drives.*

EXPLAIN IT: *Provide a clear explanation of how the system works in your plant operating manual. Include a table that specifies the correct settings of all dampers and controls, including minimum flows or deadband temperatures, the deck temperature reset schedule, and other control functions.*

Variable-Air-Volume Single-Duct Systems

The Measures in this Subsection improve the efficiency of variable-air-volume (VAV) single-duct systems. Figures 1 and 2, in the Section 4 Introduction, sketch this type of system.

In these systems, chilled air is distrib-uted to spaces from an air handling unit, and the temperature of individual spaces is controlled by throttling the quantity of air into each space. The throttling is accomplished by terminal units that are controlled by the space thermostats.

VAV systems were originally introduced as a more efficient alternative to constant-volume reheat systems (which are covered in Subsection 4.4). The VAV concept offers two major efficiency improvements: (1) it reduces or eliminates reheat and (2) it minimizes fan power. Unfortunately, the full efficiency potential of many VAV systems has not been achieved in practice. In many systems, the terminal units contain heating coils. These are used to provide space heating, or to reheat the chilled air to allow a minimum air flow to be maintained in the spaces, or for both purposes.

As VAV systems became widespread during the 1980's, it became apparent that they incur a number of conflicts with comfort and air quality. Indeed, VAV has become notorious for its comfort problems. Attempts to resolve the comfort problems and to reduce the installation cost of the systems often resulted in systems that squandered much of the efficiency potential of the VAV concept.

All the Measures in this Subsection, except the last, make system temperature control more efficient. The last Measure improves the efficiency of fan modulation.

INDEX OF MEASURES

4.7.1 In spaces with shutoff VAV terminals, install thermostat placards.

4.7.2 In spaces with shutoff VAV terminals, install deadband thermostats.

4.7.3 If the air handling systems cannot be turned off during unoccupied hours, install temperature setback.

4.7.4 Turn off the air handling unit cooling coil when cooling is not needed.

4.7.5 Minimize the minimum-flow settings of terminal units.

4.7.6 Set the cooling coil discharge at the highest temperature that maintains satisfactory cooling.

4.7.6.1 Install automatic chilled air temperature reset control.

4.7.7 With minimum-flow terminals, install heating/cooling changeover.

4.7.8 Improve the efficiency of fan modulation.

MEASURE **4.7.1 In spaces with shutoff VAV terminals, install thermostat placards.**

Placards are one of the most powerful tools of energy efficiency. See Reference Note 12, Placards, for general guidance in designing and installing placards. This Measure provides additional guidance for installing thermostat placards in VAV systems that have shutoff terminals.

Where to Use Thermostat Placards with VAV Systems

This Measure applies to individual spaces, not to the system as a whole. Thermostat placards are useful only in spaces where cooling and heating of the space can be shut off and controlled separately. To shut off cooling, the VAV system must have shutoff terminals. In addition, the system must have an explicit way to select either cooling or heating, such as a heating/cooling changeover control (Measure 4.7.7) or fan-powered VAV terminals.

Thermostat placards may be useless or counterproductive in spaces where the VAV system uses reheat. For example, consider what happens in a space where reheat terminals are installed. If the space has a small cooling load, the terminals in the space are held at their minimum-flow positions, and space temperature is controlled by using the reheat coils in the terminals. The normal instinct is to raise the thermostat setting to conserve cooling energy. However, if someone does this, the cooling output remains the same, and reheat energy increases.

Energy Saving

The amount of energy required for heating or cooling a space is roughly proportional to the temperature differential between the inside and the outside. Placards save energy by motivating occupants, in effect, to reduce this temperature differential. The benefit is greatest in climates where heating or cooling is needed for a larger fraction of the year.

During low-load conditions, resetting the space temperature may substantially reduce the number of hours that the space needs conditioning. This effect is more dramatic in mild climates, where the temperature hovers for long periods in a range that does not require either heating or cooling. In addition, buildings in mild climates are often poorly insulated, so they derive larger savings from a given temperature adjustment.

Thermostat placards can save the most energy in systems that have temperature deadband, or that have separate thermostat settings for heating and cooling. (Converting to deadband is covered by Measure 4.7.2.) Thermostat placards are less effective with systems that maintain a fixed space temperature by using both heating

and cooling. The advantage of deadband is greatest in mild climates.

(As a point of interest, the energy saving does not relate to the outside temperature itself, but only to the temperature difference between the inside and outside, as long as there is a heating or cooling load. For example, lowering the heating temperature by 3°F saves the same amount of energy whether the outside temperature is 0°F or 50°F.)

What the Placards Should Say

The placards should request occupants to:
- set the heating temperature as low as possible
- set the cooling temperature as high as possible
- close windows and exterior doors, if any, while the unit is operating.

If there may be doubt, the placard should state which spaces the thermostat controls. For example: "These controls operate the heating and cooling system for the reception area." Otherwise, occupants may not associate the controls with the conditioning of their space, or they may not know if other spaces are also controlled by them.

The placards should make occupants aware of the operation of any automatic controls, if this might create confusion. For example, if a timer control is installed to shut down the air handling system after a certain time interval, the placard should state this. Refer to Subsection 4.1 for automatic control Measures that require placards.

Don't Be Too Specific about Temperature

As a rule, don't try to tell people what temperatures to set in their spaces. The perception of comfort is what guides most occupants in deciding how far they will limit conditioning for the sake of energy conservation.

Furthermore, the calibration of thermostat dials may be inaccurate. Sometimes, thermometers are installed on or near thermostats as a guide to setting the thermostats. These little thermometers may not be accurate either, so the thermometers and the thermostats may disagree by several degrees.

Dress for the Environment

The temperatures that human beings prefer inside buildings depend largely on how they dress. Where appropriate, encourage occupants to dress warmly in cold weather, and to dress lightly in warm weather. Say this on the thermostat placard, if doing so is appropriate for the environment.

Make Thermostats Accessible, Usually

A basic principle of good energy conservation is to avoid infringing comfort. It is bad practice to try to conserve energy by making controls inaccessible. Locked thermostats irritate people because they overtly prevent people from being able to control their environment. Locked thermostats also create an adverse attitude toward energy conservation, resulting in loss of efficiency elsewhere.

If you put a locked enclosure around the thermostat and let the temperatures go to extremes, occupants will be uncomfortable. On the other hand, if you set the locked thermostat to minimize complaints, you waste energy. Only the occupants can optimize the temperature in each space. If the occupants disagree among themselves about the thermostat settings, at least it won't be your problem.

If you need to protect controls from vandalism, keep them entirely out of sight. Locked thermostats that are visible are a magnet for tampering.

ECONOMICS

SAVINGS POTENTIAL: *1 to 5 percent of conditioning cost, depending on which control functions are manual, the operating schedule, etc.*

COST: *$5 to $200 per placard, depending on quantity, materials, and complexity. Design accounts for a majority of placard cost, if you are doing it right.*

PAYBACK PERIOD: *Less than one year, in most cases.*

TRAPS & TRICKS

CHOICE OF METHOD: *Do not install placards if the system uses any reheat, unless you are sure that you can save energy this way.*

PLACARD DESIGN, MATERIALS, AND INSTALLATION: *The main pitfall is thinking that placards are simple. Study Reference Note 12, Placards.*

MEASURE 4.7.2 In spaces with shutoff VAV terminals, install deadband thermostats.

If the space thermostat maintains a fixed temperature, the system must consume energy continuously to maintain that temperature. Some of this energy consumption is unnecessary because human beings can be comfortable within a range of temperatures.

You can avoid unnecessary conditioning within the comfortable temperature range by using thermostatic "deadband." This is an optional feature of thermostatic controls in which both cooling and heating are turned off as long as the space temperature is within the comfort range. The way to achieve deadband in a single-duct VAV system is to install deadband thermostats to control the terminal units, instead of conventional thermostats.

How Wide Can You Make the Deadband?

The usual range of comfortable indoor temperature extends from a low of 70°F to 72°F, to a high of 76°F to 80°F. This implies a possible deadband range of 4°F to 10°F. Experience confirms that this is acceptable. These figures are for sedentary work. The deadband range is somewhat wider when people are physically active. You can increase the acceptable deadband range by encouraging occupants to dress appropriately for the weather.

On the other hand, the deadband range will be limited by any hot spots and cold spots within the space. For example, a person seated near a cold wall may require that the thermostat be set near the top of the comfort range, eliminating most opportunity to exploit deadband. This makes it important to eliminate hot spots and cold spots in the occupied portions of spaces.

Where to Use Deadband Thermostats

Deadband is possible only if the system can completely shut off delivery of heating and cooling to the space. To shut off cooling, the VAV system must have shutoff terminals.

Note that your ability to install deadband thermostats relates to the space, not to the system. A given VAV system may have some spaces that use shutoff terminals and other spaces that do not.

Obviously, deadband control is not appropriate in applications where close control of temperature is desired.

Ventilation and Dehumidification Limitations

In VAV systems with shutoff terminals, using deadband may greatly extend the periods of time during which the system does not provide any outside air

SUMMARY

A thermostatic control change that saves a few percent of system energy consumption at modest cost. Usually simple to accomplish. Not applicable where close control of temperature is needed.

SELECTION SCORECARD

Savings Potential .... $ $

Rate of Return, New Facilities % % %

Rate of Return, Retrofit.......... % %

Reliability .... ✓ ✓ ✓

Ease of Retrofit .... ☺ ☺ ☺

ventilation to the space. Therefore, use deadband only in spaces with limited ventilation requirements, or where ventilation air can be provided in other ways, such as by opening windows or by letting air circulate from other spaces. Outside air ventilation requirements are covered in Measure 4.2.1.

The system provides no control of humidity within the deadband temperature range. The climate determines whether this is likely to create a problem.

Energy Saving Potential

Deadband thermostatic control eliminates energy waste from these causes:

- *unnecessary operation while the space temperature is within the comfort range.* There is no need to provide conditioning when the space temperature is within the range of comfort.

- *excessive heating or cooling whenever the system is operating.* Conditioning load is largely proportional to the difference in temperature between the inside of the building and the outside. If the temperature is set arbitrarily within the range of comfort, the system maintains a larger temperature differential than necessary, with correspondingly greater energy consumption. For example, if the space thermostat is set at 75°F, this is several degrees higher than needed for heating, and several degrees lower than needed for cooling.

- *unnecessary shuttling between heating and cooling.* Especially in mild climates, the space temperature may tend to drift up and down within the comfort range. For example, it is cool in the morning, warmer at midday, and cooler again in the evening. If the space temperature is fixed, cooling energy will be expended to cancel the energy that was added by heating just a short while before, and vice versa.

- *unintentional reheat.* If the heating and cooling temperatures are set close together, it is easy for the temperature ranges to overlap inadvertently. This may result from a defect in the controls, from an incorrect adjustment, or from aging of control components. If the temperature ranges overlap, both heating and cooling will operate continuously to cancel each other.

- *unnecessary operation of equipment.* In some systems, fans or other equipment can be turned off within the deadband range. For example, this is possible with fan-powered VAV terminals.

Types of Deadband Thermostats

■ Manual Heat/Off/Cool Thermostats

This type of thermostat requires occupants to select either heat or cooling. Although this technique is not usually considered to be "deadband," it achieves the purposes of deadband. See Measure 4.3.4.1 for details about this type of thermostat.

This type of thermostat is well adapted to human temperature response, in that people usually do not want cooling until the temperature has started to rise considerably, and they do not want heating until the temperature starts a definite downward trend. In contrast with true deadband thermostats, a manual changeover thermostat keeps the system from alternating automatically between heating and cooling during mild weather.

The thermostat should have two elements that can be set independently for heating and cooling, respectively. If the thermostat has only one element, occupants are likely to leave the same temperature set when switching between heating and cooling.

■ Deadband Thermostats

A deadband thermostat is essentially two independent thermostats in one housing. One thermostat controls heating and the other controls cooling. Typically, a deadband thermostat is used to replace a single thermostat that controls both heating and cooling sequentially. The deadband thermostat inserts a range of temperatures between heating and cooling in which no conditioning is accomplished.

Some deadband thermostats allow the upper and lower temperature limits to be set by the occupants, and others do not. The former type has two dials, one for the low limit and one for the high limit. The latter type has a single dial, with the deadband set internally. The latter type is typically used in plusher environments, such as luxury hotels, which desire to maintain the illusion that the occupant has precise temperature control. The

deadband in these types is kept narrow to avoid occupant complaints, and hence they are less efficient.

A "hesitation" thermostat is a variation of a deadband thermostat that has a single output signal. The name derives from the fact that there is a plateau, or hesitation, in the output signal within the deadband temperature range.

In many cases, it does not matter whether you use a conventional deadband thermostat or a hesitation thermostat. The dual-output type is more flexible because it allows the sensitivity of each element to be adjusted individually. Furthermore, some models allow one element to be connected direct-acting, while the other element is connected reverse-acting. ("Direct-acting" means that the output signal rises as the temperature rises. "Reverse-acting" means the opposite.)

Don't Forget the Placards!

Getting the most benefit from deadband depends on the way that the occupants use the thermostats. Install effective thermostat placards. See Measure 4.7.1 for details.

ECONOMICS

SAVINGS POTENTIAL: *1 to 10 percent of conditioning cost. The larger percentages applying to mild climates, but the absolute saving is greater in buildings that have higher conditioning loads.*

COST: *Deadband thermostats cost less than $100. The cost of installation varies with the type of system, and the difficulty of wiring.*

PAYBACK PERIOD: *Several years, to many years. Depends on the conditioning load of the space.*

TRAPS & TRICKS

CHOICE OF METHOD: *Do not use deadband with terminal units that use reheat. Deadband is suitable for most comfort applications, but not for environments where close temperature control is needed.*

SELECTING THE THERMOSTATS: *Select thermostats that make it easy for occupants to set the space temperature. In environments where vandalism is a risk, select thermostats for ruggedness. If you select true deadband thermostats, buy the kind that have a fixed deadband setting that can be changed only by the maintenance staff. If you select "heat/cool/off" thermostats, buy the kind that has separate temperature setting levers for heating and cooling. Add setback features (see Measure 4.7.3), if appropriate.*

INSTALLATION: *When installing any thermostat, select a location where it senses the actual temperature of the space. Do not mount the thermostat on an exterior wall or other surface that is heated or cooled from behind. If* necessary, use a long standoff mounting bracket. Do not install the thermostat where sunlight can shine on it or on adjacent surfaces.

EXPLAIN THEM: All thermostats should have appropriate placards. With true deadband thermostats, the energy saving is determined by the person with the small screwdriver who sets the deadband in each thermostat, or in the case of electronic systems, by the person who sits at the keyboard. Make sure that the staff are trained to set the maximum acceptable deadband. Put clear instructions in the facility operating manual.

MEASURE 4.7.3 If the air handling systems cannot be turned off during unoccupied hours, install temperature setback.

SUMMARY

A basic energy conservation measure that saves energy during unoccupied periods. A standard feature in most electronic control systems. With other types of controls, requires special thermostats.

SELECTION SCORECARD

Savings Potential .... **$ $ $**

Rate of Return, New Facilities **% % % %**

Rate of Return, Retrofit.......... **% % %**

Reliability .... ✓ ✓

Ease of Retrofit .... ☺ ☺ ☺

The most effective way to save conditioning energy after hours is to turn off the air handling system, as recommended by Measures 4.1.1 ff. However, this is not possible in all cases. The air handling system may be kept in operation to maintain a minimum temperature, or to continue ventilation, or for other reasons. Even if the air handling system fans are turned off, the coils may be kept active to warm air that is drawn into the building by the action of exhaust fans.

In many of these situations, you can still save energy after hours by using temperature setback. This is a thermostatic control technique in which heating and cooling are turned off during unoccupied periods until the temperature in the space drifts to a lower (or higher) setback temperature.

Energy Saving Potential

Temperature setback reduces the temperature differential between the inside of the building and the outside during the setback period. The conditioning load is largely proportional to this temperature differential.

For example, if the average space temperature can be reduced 10°F during the setback period, and if the outside air temperature during the setback period averages 30°F below the occupied temperature, then approximately one third of heating energy consumption can be saved during the setback period. (This example does not include the effect of heat gains, heat storage in the building structure, and other complications, most of which would reduce the percentage of savings.)

Additional fan energy can be saved if fans or other equipment are controlled by the space thermostat.

In most cases, you can shut off outside air during the setback period, as recommended by Measure 4.2.1.1. This action typically saves a large part of the conditioning load.

Methods and Limitations

■ Setback That Shuts Down the Entire Air Handling System

If the entire air handling system is being controlled as a single unit, use a single setback thermostat to control the fans, the cooling coil, and all heating coils. Install the setback thermostat at the location that is most sensitive to temperature. If the space temperature falls to the setback level, the heating equipment alone is turned on.

This method is simple. However, if someone activates an override to provide conditioning for a single space outside of normal operating hours, the override will cause the system to provide conditioning to all the unoccupied spaces served by the system.

■ Setback that Shuts Down Individual Spaces

If the system serves some spaces that may be used during the setback period, install separate setback thermostats in each space. Each thermostat needs an override that can be used by after-hours occupants to obtain normal space temperature.

The air handling unit fan may continue to operate during the setback period, or it may be started if any one space calls for an override.

If all the spaces served by the system have shutoff terminals, each space can have both heating and cooling overrides during the setback period. During setback operation, the VAV boxes are closed by the setback thermostat, and the heating equipment is controlled to maintain the setback temperature. An override in any space restores normal conditioning to that space. This mode of operation requires the cooling coil in the air handling unit to operate.

If some of the spaces served by the system have minimum-flow VAV boxes, the central cooling coil must be turned off to save any energy. In this case, an override in an individual space can restore heating to the space, but it should not restore cooling.

Types of Setback Thermostats

■ Individually Programmable Setback Thermostats

A setback thermostat has two thermostatic elements in one enclosure, one of which is set for normal occupancy, and the other for unoccupied conditions. (Newer electronic units may have only a single sensor, but the effect is the same.) You can buy electronic thermostats that include setback along with a number of other functions, such as multiple setback schedules, an override, backup power supply, and perhaps other features. See Reference Note 10, Clock Controls and Programmable Thermostats, for the features, installation practices, and operating procedures that are important with time controls.

You can use a thermostat of this type to control an entire air handling system, or you can install them in individual spaces. In the latter case, the setback thermostat replaces the original thermostat.

■ Centralized Setback

In a facility with many thermostats, it may be preferable to use a centralized setback system. Such systems can send a signal from a central station to reset all thermostats to the setback temperature.

Temperature setback is a standard capability of centralized electronic control systems. These systems may allow different space thermostats to be set on individual schedules. If you have the capability, use it.

In pneumatic control systems, setback thermostats can be triggered by changing the control air pressure at the air source. This retrofit requires changing all the individual thermostats, plus adding a control station at the air source to change the pressure at the appropriate times. A typical system sets the thermostats to daytime operation with 17 PSI supply pressure and nighttime operation with 22 PSI supply pressure.

Control Issues

■ Avoid Forcing the Setback Temperature

Be sure that conditioning energy is not used to drive the space to the setback temperature. This is a potential trap where a thermostat controls both heating and cooling equipment. This problem does not occur with true setback thermostats. If you set space temperatures with a centralized control system, this problem may occur as a result of an incorrect input.

■ Setback Overrides

A weakness of centralized temperature setback systems is that all thermostats are controlled on the same schedule. To overcome this problem, thermostats used with centralized setback systems should include individual overrides. A person who wants to work during the unoccupied period activates the override in his space, which causes the thermostat to call for normal temperature. The override remains in effect until the next setback period, unless the occupant cancels the override.

Individually programmable setback thermostats usually have quick overrides of the setback temperature that work in a similar way.

Overrides have an efficiency weakness if they are used often. If a person who works late forgets to cancel the override when he leaves, the system will continue to operate at the daytime temperature. This does not waste much energy if it happens only occasionally. However, if the override is used often by someone who forgets to cancel the override, much of the benefit of the setback feature is lost.

■ Control of Outside Air

Many facilities do not need to continue outside air ventilation during the unoccupied period. In such cases, use a control signal from a centralized setback system, or from an individually programmable setback thermostat, or from a timeclock to close the outside air dampers. See Measure 4.2.1.1 for details.

Don't Forget the Placards!

Setback thermostats require occupant action to avoid wasting energy, so they need instruction placards. Make sure that the placards request occupants to reset the temperature override after using it. See Measure 4.7.1.

ECONOMICS

SAVINGS POTENTIAL: *5 to 15 percent of heating cost.*

COST: *In new construction, adding setback typically costs less than $100 per thermostat. It is a standard feature with some electronic control systems. In retrofit, adding setback typically costs several hundred dollars per thermostat.*

PAYBACK PERIOD: *Less than one year, to several years.*

TRAPS & TRICKS

CHOICE OF METHOD: *You may have a variety of ways to accomplish setback. Study your system to select the most effective.*

SELECTING THE EQUIPMENT: *Select thermostats for ease of use and for ruggedness. Add deadband features (see Measure 4.7.2), if appropriate.*

INSTALLATION: *When installing any thermostat, select a location where it senses the actual temperature of the space. Do not mount the thermostat on an exterior wall or other surface that is heated or cooled from behind. If necessary, use a long standoff mounting bracket. Do not install a thermostat where sunlight can shine on it or on adjacent surfaces.*

EXPLAIN IT: *Don't forget the instruction placards. Make sure that the staff are trained to set the proper setback temperature for each location. List these temperatures in the plant operating manual.*

MONITOR PERFORMANCE: *Verify proper operation of the setback feature at the start of the heating season. Put this in your tickler file.*

RATINGS

New Facilities | Retrofit | O&M

MEASURE **4.7.4 Turn off the air handling unit cooling coil when cooling is not needed.**

This is a method of eliminating reheat during the heating season. It applies to systems that have not been designed with a great deal of attention to efficiency. See Measure 4.4.2 for the procedures.

Energy Saving

The potential savings depend on how the VAV system is designed and operating, on the way systems are zoned, and on the diversity of heat gains within each system.

Limitations and Alternatives

This Measure cannot be used with systems that serve spaces with continuous cooling loads, such as spaces in core areas.

Chilled air temperature reset control (see Measures 4.7.6 ff) is more effective than this Measure. However, even if you have reset control, consider turning off the cooling coils seasonally as a safety feature.

ECONOMICS

SAVINGS POTENTIAL: *3 to 20 percent of cooling energy; 10 to 30 percent of reheat energy.*

COST: *Minimal.*

SUMMARY

An easy method of eliminating mixing losses when cooling is not needed.

SELECTION SCORECARD

Savings Potential .... **$ $** $

Rate of Return .... **% % %** %

Reliability .... ✓ ✓

Ease of Initiation .... ☺ ☺ ☺

PAYBACK PERIOD: Immediate.

TRAPS & TRICKS

EXPLAIN IT: Success depends on training and supervision. Provide clear instructions that explain when to turn off the cooling coils, e.g., when the outside air temperature is below a certain temperature. Optimize the schedule for each air handling unit or group of units. Install placards at the cooling coil valves or at the controls of the compressors in self-contained air handling units.

MONITOR PERFORMANCE: This Measure is easy to forget. Schedule periodic checks of the procedure in your maintenance calendar.

MEASURE **4.7.5 Minimize the minimum-flow settings of terminal units.**

SUMMARY

A simple adjustment in minimum-flow terminal units that reduces mixing losses. Much easier to accomplish if you have a central energy management control system. May cause discomfort if you overdo it.

SELECTION SCORECARD

Savings Potential .... $ $ $

Rate of Return .... % % % %

Reliability .... ✓ ✓ ✓

Ease of Initiation .... ☺ ☺

The main efficiency advantage of VAV systems over constant-volume reheat systems is eliminating the mixing of heated air with chilled air. However, many VAV systems are designed to retain a substantial amount of air mixing. This is done as a cheap way of providing heating, and as a way of avoiding comfort and ventilations problems that occur at low air flow rates.

The air mixing is accomplished by preventing the dampers in the terminal units from closing fully. This is usually done with mechanical stops inside the units. Such terminals are called "minimum-flow" units, in contrast with "shutoff" terminals. Minimum-flow settings can also be made through the control system, especially with terminal units that include their own air flow monitors.

The minimum-flow settings may be a large fraction of full flow, especially when the purpose is to provide space heating. It is not unusual for the minimum-flow settings to be as much as 40% of maximum flow. This wastes a large fraction of the savings potential of the VAV concept.

In many cases, you can reduce the minimum-flow setting by a large amount. This saves a substantial amount of energy at lower cooling loads, which typically occur for a large fraction of the time.

Minimum-Flow Always Causes Mixing Losses

Any VAV system that uses minimum-flow terminal units suffers mixing losses when the space load is less than the minimum-flow capacity of the terminal unit. Most minimum-flow terminals have reheat coils, but not all. Even if minimum-flow terminals do not have reheat coils, mixing losses must occur somewhere. Otherwise, the spaces would overcool if there were no means of compensating for the unnecessary cooling provided by the minimum supply air flow. In such cases, reheat is usually provided (whether or not the designer intended) by a separate heating system, such as a perimeter baseboard system.

Minimum-flow terminals without heating coils may be more wasteful than units that have integral heating coils. This is because they deliver colder supply air, so the space must be kept at a higher temperature to prevent discomfort. For a given chilled air temperature, a higher space temperature requires more reheat.

Energy Saving

Mixing losses occur whenever the terminal is operating at its minimum-flow setting. The loss increases as the cooling load declines, reaching a maximum at zero space load. The loss remains at this maximum whenever the space has a heating load.

To appreciate how large mixing losses can be, consider a typical VAV reheat terminal in which the minimum flow has been set at 40% of maximum flow. The cooling energy delivered by the terminal is proportional to the flow rate, so the cooling energy consumption is always at least 40% of maximum. If the space load is zero, an equal amount of heating energy is required to cancel the cooling energy. Thus, at zero space load, the mixing losses (in absolute energy units) are equal to 80% of the maximum cooling load! Furthermore, this same amount of mixing loss continues whenever the space has a net heating load.

Generalizing this example shows that the maximum mixing losses, expressed as a percentage of the maximum cooling load, are always equal to twice the minimum flow percentage. In many cases, you can reduce the minimum-flow settings by 10% to 50%, reducing the energy waste correspondingly.

How Much Can You Reduce Minimum-Flow Settings?

To decide how much you can reduce the minimum-flow settings, examine the system to find out which factors dictate the minimum-flow requirement of each terminal unit. Minimum-flow may be needed to:

- provide proper diffuser operation
- maintain an adequate level of outside air ventilation
- create in-space air velocity to prevent a feeling of "stuffiness" and to facilitate cooling by perspiration
- provide dehumidification during periods of low cooling load
- provide space heating

• if electric reheat coils are used, to ensure sufficient flow to prevent coil burn-out.

Review Measure 4.4.7 to see how each of these requirements may affect the minimum flow requirement.

The most wasteful minimum-flow requirements occur when the terminal unit is used for heating the space. This requirement exists only during heating weather, so consider adjusting the minimum-flow settings seasonally. For example, you might adjust the minimum-flow setting to a low level during warm weather, then increase it in steps as the weather becomes colder.

Where heating is the deciding factor in the minimum-flow requirements, adjust the minimum-flow settings on a space-by-space basis. The minimum-flow settings needed to provide adequate heating may vary radically from one terminal unit to another within a given VAV system. For example, the terminals in the core area of the building may be able to function with low minimum-flow settings, while terminals in the perimeter areas may need high minimum-flow settings.

The other factors that determine the minimum-flow setting vary considerably among different facilities, different air handling systems, and different terminal units. The minimum-flow requirement may be vague. In case of doubt, the best approach may be to reduce the minimum-CFM settings progressively until a problem becomes apparent, and then increase the flow slightly.

How to Adjust the Terminal Units

If you have a centralized electronic control system, you may be able to adjust the minimum-flow settings of the terminals from your keyboard. In this case, the effort is almost trivial, and you can easily experiment to find the optimum settings.

If you do not have a central control system, you can adjust each VAV box manually. Unfortunately, VAV terminals are often jammed tightly into the ceiling plenum, and reaching them can be time consuming. Terminal unit manufacturers have no shame about leaving sharp edges all over their equipment, so bring a box of bandages. Working with the boxes may disrupt the activities in the spaces, so expect the occupants to be annoyed. Because of these difficulties with manual

adjustments, you have to decide whether it is practical to keep adjusting the boxes on a periodic basis.

■ Record the Damper Settings

If you can reset the terminal units remotely, keep a list of the appropriate setting for each terminal, by season or load condition. If you adjust the boxes manually, mark the damper stop positions in a clear and permanent manner. If you change the damper stops at different seasons, label the appropriate positions for each season.

A Better Alternative

Measure 4.7.7 is a more efficient alternative to this Measure, and it is completely automatic. It keeps the dampers of each terminal unit at the minimum position needed to satisfy the current space conditioning requirements.

ECONOMICS

SAVINGS POTENTIAL: *20 to 60 percent of the energy expended for reheat; 5 to 30 percent of cooling energy. The saving is greatest in systems that have extended periods of heating load and low cooling load.*

COST: *The cost is entirely for labor, so it depends on the availability of manpower, and on the number and accessibility of terminals.*

PAYBACK PERIOD: *Less than one year, typically.*

TRAPS & TRICKS

PLANNING: *Before you start changing settings, figure out all the factors that determine the minimum-flow settings for each terminal or group of terminals. Then, decide whether to change the settings periodically, and when. If the system has electronic controls that allow you to adjust the settings remotely, plan to do so as often as needed.*

EXPLAIN IT: *Put a clear description of the procedures in the plant operating manual. Include a table of settings for each terminal unit, including different time periods. Do the adjustments yourself before delegating this job to others. It will be educational, and it will give you a feel for the drudgery involved.*

SCHEDULING: *This procedure can be forgotten with no visible effects. It is not the sort of work that the staff will remind you about. Schedule it in your maintenance calendar.*

MEASURE 4.7.6 Set the cooling coil discharge at the highest temperature that maintains satisfactory cooling.

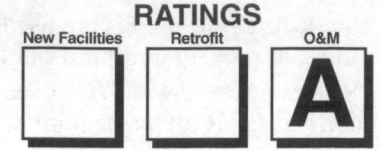

This is an easy way to save a lot of energy that is wasted by reheat. This Measure is similar to Measure 4.4.1 in theory, procedures, and limitations. Refer to there for details.

The energy saving is smaller than with Measure 4.4.1 (which deals with constant-flow reheat systems), but it is worth pursuing. This Measure has an additional benefit when used with VAV systems, namely, it lessens problems that are caused by low air flow, especially in spaces served by shutoff terminals.

Energy Saving

In any reheating VAV terminal, mixing losses occur whenever the reheat coil is operating. When the terminal is providing cooling, the reheat coil typically operates whenever the damper has closed to the minimum-flow limit. When the terminal is providing heating, the reheat coil is always turned on, and mixing losses are always at a maximum.

In the previous Measure 4.7.5, you saw the interesting fact that the maximum mixing losses, expressed as a percentage of the maximum cooling load, are equal to twice the minimum flow percentage. This assumed that the chilled air temperature is held constant. By raising the chilled air temperature, you can reduce mixing losses substantially below this amount.

■ An Example

To illustrate the benefit of raising the supply air temperature, let's examine the effect on an individual terminal unit, under these conditions:

- terminal unit maximum flow: 1,000 CFM
- terminal unit minimum-flow setting: 400 CFM
- space temperature: 75°F
- cooling load: none
- all cooling is sensible.

This table shows the energy savings that result from raising the chilled air temperature:

Energy Input, BTU/hr	Supply Air Temperature		
	55°F	65°F	75°F
cooling energy to space	0	0	0
cooling energy from coil	8,640	4,320	0
reheat energy	8,640	4,320	0
total input energy	17,280	8,640	0

To emphasize the magnitude of the energy waste, consider that the maximum cooling capacity of the VAV terminal is 21,600 BTU/HR (at a supply air temperature of 55°F). This shows that mixing losses can be severe even in a VAV system. For an even bigger thrill, compare this example with the similar one in Measure 4.4.1 for constant-volume reheat systems.

Comfort and Ventilation Advantages

VAV systems are notorious for problems that result from low air flow. These occur when the space has a low conditioning load. This Measure increases air flow at low loads. It may significantly improve performance in these areas:

- *diffuser operation.* The tendency of diffusers to "dump" cold supply air is reduced because the air is warmer and is delivered at higher velocity.
- *outside air ventilation.* If the terminals are able to close fully, they shut off outside air that is delivered through the air handling system. Increasing the chilled air flow to the spaces increases outside air ventilation proportionally.
- *in-space air velocity.* The increased air flow acts to prevent a feeling of "stuffiness" and it aids cooling by perspiration.

For more about these factors, see Measure 4.4.7.

ECONOMICS

SAVINGS POTENTIAL: *5 to 30 percent of cooling energy; 30 to 70 percent of reheat energy. Increases fan energy somewhat at low conditioning loads.*

COST: *Minimal.*

PAYBACK PERIOD: *Immediate.*

TRAPS & TRICKS

See Measure 4.4.1.

MEASURE **4.7.6.1 Install automatic chilled air temperature reset control.**

Most VAV systems should have this control feature to avoid energy waste from reheating. This Measure is similar to Measure 4.4.1.1 in theory, procedures, and limitations. Refer to there for details.

People tend to think of variable-air-volume systems as supplying chilled air at a constant temperature, in contrast with constant-volume systems, which provide air at variable temperature. However, you can achieve additional efficiency and comfort benefits by modulating the temperature of the chilled air supply as well as its volume. In a VAV system, temperature modulation is called "reset."

Where to Install Chilled Air Reset Controls

Chilled air reset controls are most valuable if the chilled air temperature is dictated by large, variable cooling loads, such as perimeter cooling loads in a glass curtain wall building. In such cases, a reset controller usually pays for itself quickly.

On the other hand, reset controls may not be economical in systems where the chilled air temperature is dictated by core zone cooling loads. Core areas always have a net cooling load (except perhaps for morning warm-up), and this load tends to remain constant. In this case, there is little need for an automatic reset control. Optimizing the chilled air temperature manually is almost as efficient.

How the Reset Control Should Work

To conserve fan energy, the reset control should keep the chilled air temperature low when the cooling load is high. Chilled air temperature reset should start to take effect just before any of the terminals shut down to their minimum-flow limit. Then, as the cooling load drops

SUMMARY

Most VAV systems with reheating terminal units should have this control feature. The only way to keep chilled air near its optimum temperature.

SELECTION SCORECARD

Savings Potential .... **$ $ $**

Rate of Return, New Facilities **% % %**

Rate of Return, Retrofit .... **% % %**

Reliability .... ✓ ✓ ✓

Ease of Retrofit .... ☺ ☺ ☺

further, the reset control should raise the supply air temperature as much as possible.

ECONOMICS

SAVINGS POTENTIAL: *30 to 80 percent of reheat energy, and 10 to 40 percent of cooling energy. The savings potential of this Measure is limited by system load diversity, and by spaces with large, constant cooling loads.*

COST: *The control components for a typical installation typically cost from several hundred to several thousand dollars. In addition, connections to thermostats at widely dispersed locations can add a large labor cost.*

PAYBACK PERIOD: *Less than one year, to many years.*

TRAPS & TRICKS

See Measure 4.4.1.1.

MEASURE 4.7.7 With minimum-flow terminals, install heating/cooling changeover.

Many VAV systems have a crude method of providing heating that wastes a lot of energy. In these systems, heating is provided by heating coils in the terminal units. The heating coils need a minimum amount of air flow to deliver peak heating capacity. To accommodate this, each terminal unit has a stop or limit that prevents the damper from closing below this amount of air flow. In typical systems, this may be about 40% of maximum flow.

The Problem

Energy waste occurs any time the cooling load falls below the level at which the minimum cooling air flow is required. This is because the heating coil must provide heat to cancel excess cooling that occurs because of the minimum-flow limit. As the cooling load of the space falls, even more heat is required to cancel the cooling.

The Solution

The key to reducing this energy waste is to allow the dampers in the terminal units to close fully. You can do this by using space thermostats that switch their terminal units between cooling and heating modes of operation. In cooling operation, space temperature is controlled in the normal manner, by the position of the chilled air damper in the terminal unit. The heating coils remain off during cooling operation. If the space temperature falls below the thermostat setting, the heating coil turns on and the thermostat reverses so that the damper opens as the space temperature falls. In some cases, the thermostat may also control the output of the heating coil.

The two new features are eliminating (or reducing) the minimum-flow limit, and adding a reversing feature to the space thermostats. There may be additional complications required by conditioning requirements, which are explained below.

At the same time, install chilled air temperature reset (see Measures 4.7.6 ff), if you don't already have it. This control feature raises the chilled air temperature as the cooling load falls, eventually turning off cooling completely. This reduces the reheat losses that occur during heating operation, while increasing the air flow to the space at low loads.

Energy Saving

We will use a typical example to illustrate the energy saving that can be achieved by using changeover between heating and cooling control. Assume that using changeover control allows the minimum-flow setting to be reduced from 40% of maximum flow to 10% of

SUMMARY

Greatly reduces the waste of energy from air mixing in systems where terminal units have minimum-flow limits. Requires new thermostatic controls, and in some cases, changes to terminal units, diffusers, and fan modulation. Can be tricky if the terminal units have electric heating elements.

SELECTION SCORECARD

Savings Potential .... $ $ $

Rate of Return, New Facilities % % % %

Rate of Return, Retrofit......... % % %

Reliability .... ✓ ✓ ✓

Ease of Retrofit .... ☺ ☺ ☺

maximum flow. The chilled air temperature reset schedule is the same for both configurations.

First, energy is saved because reheat does not commence until the cooling load falls to 10% of full load, rather than 40% of full load. Thus, reheating occurs for a much smaller fraction of the cooling season.

Second, energy is saved because reheating wastes much less energy when it does occur. When the conditioning load in the space is zero, the waste of cooling energy is reduced from 40% to 10%. In addition, there is an equal saving in the amount of heating energy needed to cancel the excess cooling.

If the space has a heating load that requires more than 10% of the maximum flow rate, the damper must open further to satisfy the heating load. This increases the amount of chilled air that enters the terminal, which further increases the amount of heat that is needed to cancel the unwanted cooling. As a result, chilled air temperature reset is very important for gaining the greatest benefit from changeover control.

Conditions That May Require Minimum-Flow Terminal Settings

The system is most efficient if the terminal unit dampers can close completely. Unfortunately, there may be factors that require minimum-flow setting in some or all of the terminal units. These are:

- *outside air ventilation.* Complete shutoff of terminal air flow blocks ventilation. In benign environments with a low occupant density, it may be acceptable to tolerate periods of shutoff. This is a judgement call that is usually driven more by code requirements than by occupant perception.

- *dehumidification,* especially in humid climates. Dehumidification requires some reheat. The terminal unit must provide a relatively low rate of dry air flow into the space to keep the humidity level down.

- *dumping of chilled air from diffusers* at low flow rates. This problem can be reduced or eliminated with chilled air temperature reset. Diffuser design also affects dumping, but any diffuser will dump at very low flow rates. You can also solve this problem by substituting fan-powered VAV terminals, but they are expensive and noisy.

- *"stagnation"* of air in the space. This problem is occasional and subjective. Research suggests that complaints of "stagnation" appear to be related more to temperature and humidity than to the air velocity in the space, although the latter is a factor. Other physical and psychological factors also affect the sensation of "stagnation." Many people are comfortable in spaces that have virtually no air velocity. You may be able to solve actual stagnation problems by using inexpensive circulation fans in the spaces. Substituting fan-powered VAV terminals is an expensive solution that may not solve the real problem.

- *terminal unit noise,* which increases as the damper approaches a closed position. This problem is reduced or eliminated by proper fan modulation, so that the air velocity through the partially closed damper is not excessive. A well designed terminal unit will not have this problem. You may be able to modify existing terminal units to reduce noise. For example, attach fuzzy material to the edges of damper blades.

Refer to "VAV Design Challenges," in Measure 4.4.7, for more about these problems.

Even if shutoff is not practical, try to minimize the minimum-flow setting. Start by identifying all the reasons why the system may need minimum-flow settings. Calculate the amount of air flow needed to satisfy each requirement. Then, figure out whether there is an economical way to reduce the air flow needed to satisfy the biggest requirement. Combat the energy waste further by using chilled air temperature reset aggressively.

Changeover control is simplest if the heating coil is fully modulating, i.e., if it is hydronic. After the damper reaches its minimum-flow position as the cooling load drops, the temperature in the space drops. This causes the thermostat to switch to heating mode, and heat is added by the coil to keep the space at proper temperature. On the other hand, if the heating coils are electric, get ready for more complications, which we now consider.

Dealing with Electric Heating Coils

The use of electric heating coils in terminal units, combined with inept design, is probably responsible for most of the energy waste in VAV systems. Electric heating coils need a minimum air flow rate to keep the coil itself from overheating and burning out. Designers who are careless about efficiency deal with this requirement by setting a minimum flow rate that is high enough to protect the coils. Not thinking beyond this, they simply specify a minimum-flow setting that is high enough to provide the peak heating load. While it avoids the need to think much about control design, this stupid approach is enormously wasteful.

To improve efficiency when using electric heating coils in terminal units, you need control methods that are presently unconventional, and therefore risky. If air flow to the space can be cut off completely, control cooling by letting the damper move all the way to shutoff, with the heating elements turned off. This completely eliminates reheat during cooling operation. When the thermostat calls for heating, the controls open the damper enough to protect the heating coil, while turning on the heating coil. When the space temperature is restored, the heating coil turns off and the damper closes.

This modification makes heating an on-off proposition, similar to the operation of a forced-air furnace. This mode of operation may surprise occupants who are accustomed to the continuous air flow usually provided by a VAV system. However, the cycling during heating operation is quiet, so it should not be objectionable.

When the damper first opens, the air entering the space is cool until the heating coil warms up slightly. This puff of cool air should last no longer than a few seconds, but it may be objectionable in some environments.

In all but the smallest terminal units, electric heating is provided in stages, with several heating elements. This may cause further complications. The air flow rate needed for the first stage may be less than the air flow needed for later stages. In this case, the damper position may have to be controlled in stages that correspond to the flow requirements of the heating stages.

■ Electric Heating Coils with Minimum-Flow Settings

If the terminal units require minimum-flow settings, there are still more complications. To minimize energy waste, satisfy two conditions:

- select the first-stage heating element to operate safely at the minimum-flow rate. This means that the element must operate at low temperature. You may be able to achieve this by connecting two of the elements in series as the first stage. The unit may already be connected this way. If not, you may be able to modify the switching of the elements.

- select the first-stage heat output to be no greater than necessary to cancel the cooling input at the minimum-flow rate. Bear in mind that this cooling output is determined by the chilled air temperature, which is a variable that is determined by the chilled air temperature reset control. You have to estimate the lowest chilled air temperature that will occur at minimum flow. You can achieve very low heat output by connecting three or four elements in series as the first stage. If the terminal unit does not have enough elements to allow this, install one coil of lower wattage to serve as the first stage, but make sure that you retain enough heating capacity for peak loads.

Load Diversity Reduces Changeover Efficiency

If the loads in all the spaces were to fall to zero at the same time, the chilled air temperature reset control could be adjusted so that cooling is turned off at zero load. This would eliminate reheat losses entirely. You can't achieve this ideal, because a VAV system serves spaces with differing loads. Instead, you have to keep the supply air temperature low enough to accommodate the space with the highest cooling load. This causes some reheat loss in the terminal units of the other spaces. To minimize these losses, optimize chilled air reset as much as possible. See Measures 4.7.6 ff for details.

How the Thermostats Should Work

When a space needs cooling, the thermostat closes the terminal unit dampers as the space temperature drops. When the space needs heating, the thermostat opens the dampers as the space temperature drops.

In retrofit applications where pneumatic or electric thermostats are installed, you can achieve this change in function by installing dual-element thermostats, such as a deadband thermostat. These are described in Measure 4.7.2.

If you have an electronic control system, you may be able to program the appropriate response using the existing control hardware.

Provide Effective Fan Modulation

In systems where the terminals are allowed to shut completely, or nearly so, inadequate turndown of fan output may cause excessive duct pressure. This aggravates terminal unit noise and leakage from the duct system. These problems are most severe in systems that "ride the fan curve." This is a fancy way of saying that the fan lacks any type of modulation. On the other hand, a VAV system that has good fan modulation probably will not suffer from these problems. See Measure 4.7.8, next, about improving fan modulation.

Consider Deadband with Shutoff Terminals

If the terminal units can close to shutoff, consider adding deadband to the space thermostats (Measure 4.7.2) at the same time. This saves additional energy during periods when the conditioning load is low. But, bear in mind that deadband may aggravate any of the problems discussed previously that are associated with low air flow rates.

ECONOMICS

SAVINGS POTENTIAL: *60 to 90 percent of the cooling and heating energy that is wasted by reheating.*

COST: *In new construction, the main cost element is additional control complexity, which is a small fraction of total system cost.*

In retrofit applications, the thermostatic control changes alone typically cost several thousand dollars. Changes to electric heating coils, if needed, may cost roughly $100 per terminal unit.

PAYBACK PERIOD: *In new construction, typically about one year. In retrofit, several years.*

TRAPS & TRICKS

PLANNING: *Before you can optimize your systems, you need to study how they work. Determine the actual minimum air flows that are acceptable for each terminal unit. Study the performance of other systems that have the features you intend to install. In retrofit, understand the present control sequences. Expect to be innovative in modifying terminal units having electric heating coils.*

INSTALLATION: *Don't expect the installers to understand your design or to adjust everything optimally by themselves. Be there to oversee the adjustments and to work out problems.*

EXPLAIN IT: *Describe how the system operates in your plant operating manual.*

MONITOR PERFORMANCE: *Wasteful operation is invisible, so it persists until you check for it. If your energy management control system has the ability to do trend logging, use it to record the chilled air temperature, the supply air temperature from each terminal unit, and the space temperature at each unit. Review the trend logs periodically to verify that the controls are operating properly.*

MEASURE 4.7.8 Improve the efficiency of fan modulation.

SUMMARY

Saves fan energy. May avoid system performance problems that occur at low flow rates. Usually expensive.

SELECTION SCORECARD

Savings Potential .... $ $ $

Rate of Return, New Facilities % % %

Rate of Return, Retrofit.......... % %

Reliability .... ✓ ✓ ✓

Ease of Retrofit .... ☺ ☺ ☺

The promise of reduced fan energy consumption was one of the main reasons for the initial popularity of VAV. Unfortunately, the methods that were available for modulating fan output in the early days of VAV had serious efficiency losses themselves. This is no longer true. Efficient fan modulation is now available, and it has become relatively inexpensive.

Efficient fan modulation also improves the heating and cooling efficiency of the system, and it may improve comfort. Also, you need fan modulation with a good turndown ratio to exploit the energy savings of Measures that reduce the minimum flow of terminal units (especially, Measures 4.7.5 and 4.7.7). This is to avoid problems that occur at low flow, including terminal unit noise, terminal and duct leakage, and unstable temperature control.

In new construction, use the most efficient method of fan modulation that is appropriate for the size of the air handling system. Efficient fan modulation simplifies the design of the system, and eliminates some potential problems.

In existing facilities, examine all your existing VAV systems to see whether you can improve fan modulation. Many VAV systems lack efficient fan modulation, or they may have none at all.

Energy Saving

The saving that you can achieve by improving fan modulation depends primarily on these factors:

- *the power of the fan motors.* The savings are directly proportional to the average power at which the fans operate.

- *the duration of system operation.* The savings are directly proportional to the number of hours that the system operates.

- *the load profile of the system.* All fan modulation methods provide the greatest benefit at reduced loads. VAV systems used for comfort conditioning in typical mid-latitude locations operate most of the time at low fan loads.

- *in retrofit, the efficiency improvement between the new and old modulation methods.* There are major differences in the efficiencies of the common fan modulation methods. Most incur a small efficiency penalty at peak load, but not all do. The largest differences occur at partial loads, where typical systems spend the most time operating.

If you are upgrading existing systems, don't expect to be able to make a precise estimate of savings. This is because the load changes continuously and the efficiency of the new and old modulation methods vary with load.

You get the most accurate estimate by using an energy analysis computer program. This takes a considerable amount of effort. (See Reference Note 17 about these programs.) If you use a computer program, you need to estimate the load profile with reasonable accuracy, and to get accurate efficiency profiles for both the new and existing fan modulation methods. However, you may not be able to get information about existing equipment that is accurate enough to justify the effort of a computer analysis.

In retrofit applications, if you cannot get accurate efficiency data on your existing fan modulation methods, you have to guess its efficiency based on the type of equipment. In principle, you could measure input energy and output flow. In practice, the accumulated error involved in measuring or estimating all the factors would make the result no better than a guess based on the type of modulation device.

You can make the most accurate estimate when the fans presently have no method of fan modulation, i.e., when they "ride the fan curve." The energy behavior of the system in this situation is shown in Figure 1. In this case, estimate the present fan efficiency by using a fan curve acquired from the fan manufacturer.

To make the savings estimates as accurate as possible, be resolute about getting reliable efficiency data. Even reputable manufacturers publish generalized performance figures that are sweetened for marketing purposes. Major manufacturers at least have the capability of measuring actual performance data, but they are likely to do so only if requested to certify the numbers.

Request performance figures that are based on the fan operating against the duct system pressure. Idealized performance curves issued by drive manufacturers may be based on fans discharging into an open duct (if they

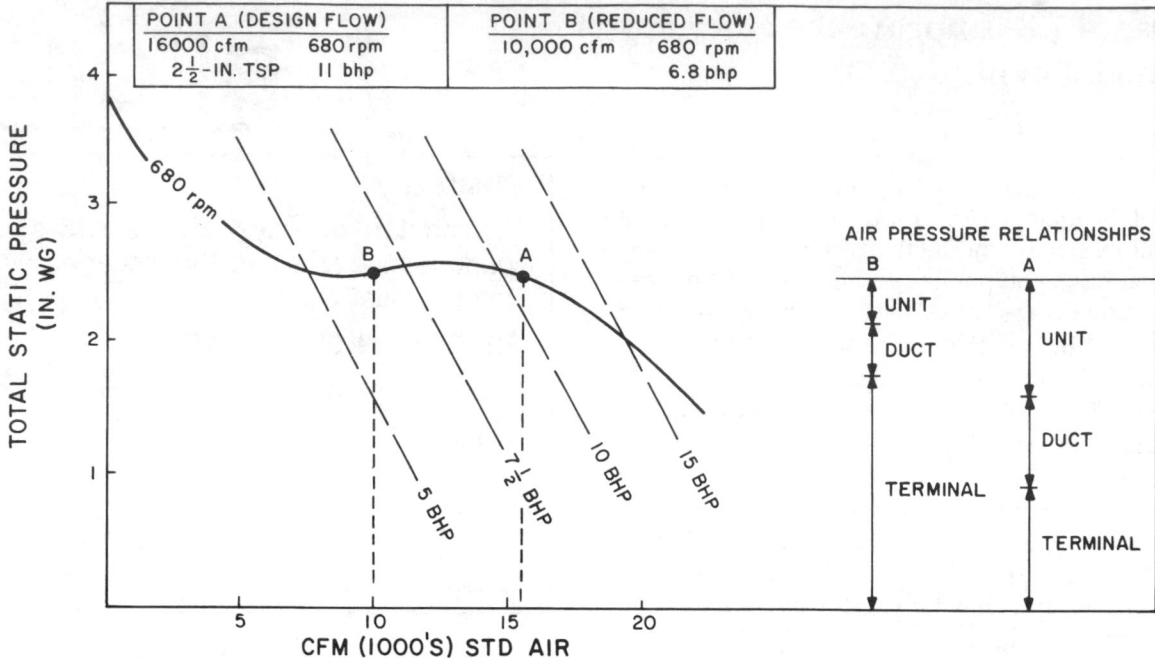

Carrier Corporation

Fig. 1 Riding the fan curve This is the typical behavior of the supply fan in a VAV system if it lacks any method of throttling output. As the flow of air is choked off by the closing of the terminal units, the fan power is reduced. However, in a typical range of throttling, as between points A and B, the power reduction is not great. The inset at lower right shows where the system pressure drops occur under the two conditions. Power consumption is proportional to the pressure drop.

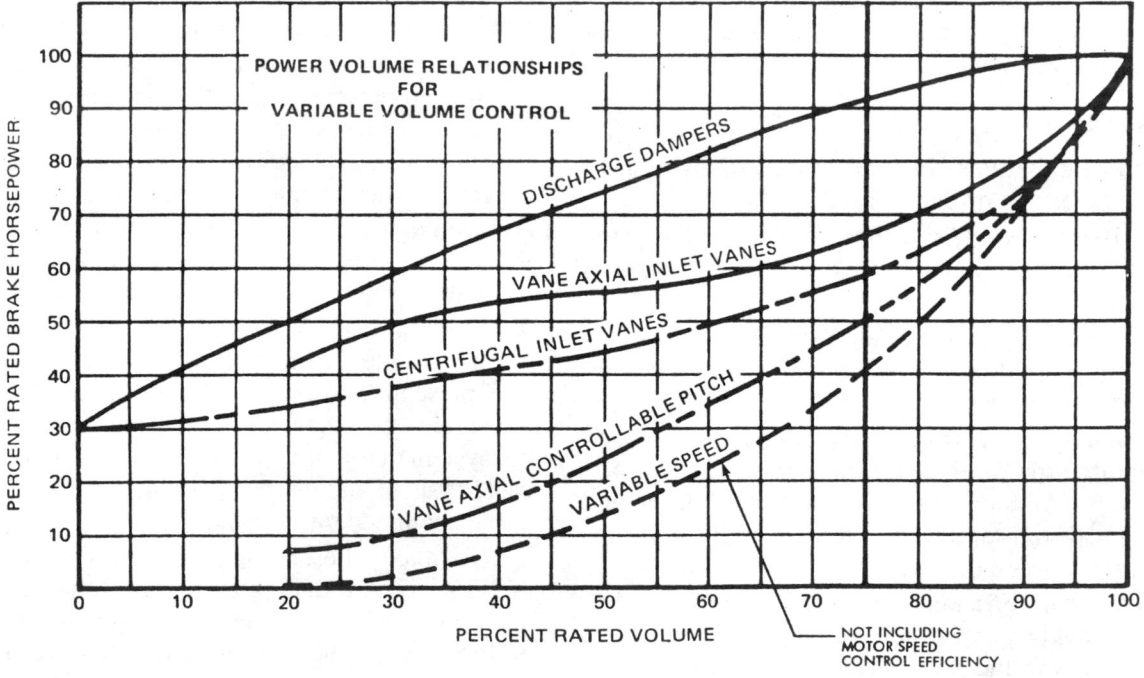

Trane Company

Fig. 2 Power saving of different fan throttling methods These curves are typical, and subject to wide variation. The top curve shows what happens to the power consumption of a fan with no throttling method as the terminal units close. Note that the curve for each modulation method is expressed as a fraction of the full-load power of that particular method. Compared to a fan with no modulation method, all these methods actually increase the power requirement near full load because they add drive losses and/or additional friction in the air stream. The curve for variable-speed drives does not include drive or motor losses, which vary widely among different types of equipment.

are based on anything). Such curves are not relevant to real VAV systems, and they are too optimistic. This is because VAV systems are not open duct systems. The terminal units must work against a certain amount of duct pressure in order to control air flow properly. Therefore, the action of the terminal units tends to maintain a constant duct pressure against which the fans must work.

Relative Performance of Fan Modulation Methods

Figure 2 shows the typical throttling performance of different methods of modulating fan output. In this figure, note that the curves show the power requirement as a fraction of the full-load power of each throttling method. Near full output, each throttling method actually increases power consumption, compared to a fan having no throttling. This is because of drive losses and/or additional friction in the air stream.

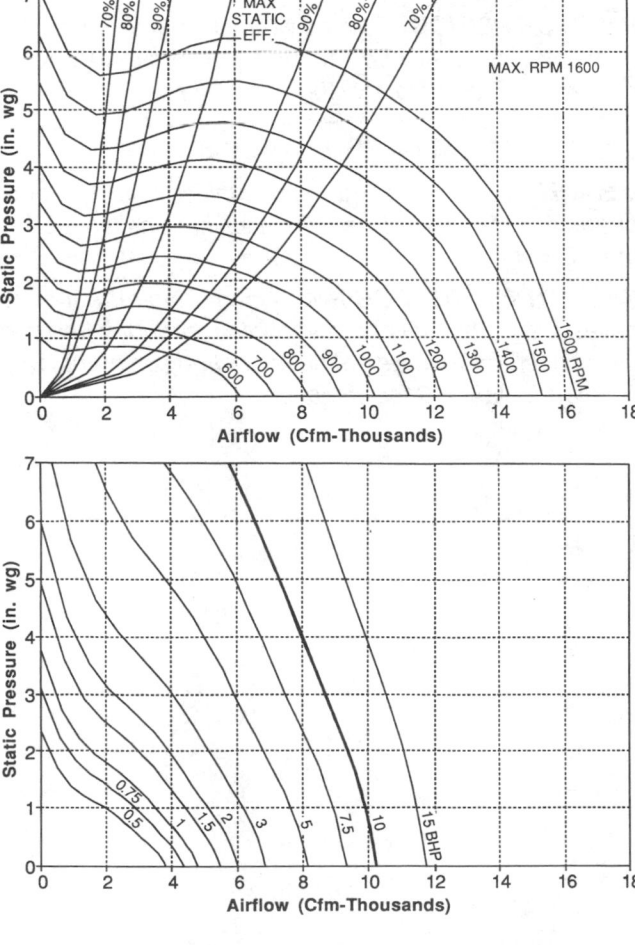

Carrier Corporation

Fig. 3 Effect of variable-speed drives on fan performance The upper curve shows the effect of changing speed on air flow and fan pressure. The lower curve shows how air flow and fan pressure affect fan power. These curves do not include losses in the fan drive or motor.

These are the most efficient fan modulation methods:
- *electronic variable-frequency drives (VFD's)*
- *variable-pulley drives*
- *direct current (DC) drives*
- *variable-pitch propeller fans*.

See Reference Note 36, Variable-Speed Motors and Drives, for a detailed explanation and comparison of the first three of these types. There are major differences within each of these types, mainly in characteristics other than efficiency. In existing facilities, it is not economical to upgrade drives of these types as long as they are working properly.

Figure 3 shows the effect on fan power of using any type of continuously modulating variable-speed drive. Note that this figure does not include drive losses.

These methods of modulating fan output are substantially less efficient than the previous methods:
- *multi-speed motors or multiple motors*
- *eddy current clutch drives*
- *inlet vortex dampers*.

Reference Note 36 covers the first two of these. Multi-speed motors may offer a good compromise between efficiency and cost in small fan systems. Eddy current clutches disappeared from the market under competition from VFD's, but they have recently returned. Inlet vortex dampers have a limited range of efficient turndown, and they are rarely a good choice, although they are relatively inexpensive. Figure 4 shows a fan with an inlet vortex damper.

In existing facilities, it is usually not economical to replace any of these types of modulation in smaller air

Trane Company

Fig. 4 Fan with inlet vortex damper This method of fan modulation is most efficient at high percentages of output. It is much less efficient than speed modulation at lower outputs. Its main attraction is low cost.

handling systems. Upgrading may be economical in systems that have larger fan motors and that operate for long durations at reduced output.

You can make the greatest improvement when a VAV system presently has no method of fan modulation, as shown in Figure 1. Installing efficient fan modulation in such systems may yield comfort improvements, such as reduced noise, in addition to major energy savings.

Some VAV systems that lack fan output modulation may have one-time output adjustments, including:

- *discharge dampers* (which may be installed at some distance from the fan)
- *scroll housing dampers*
- *inlet (not vortex) dampers*.

Get rid of these devices wherever you can. To make a one-time adjustment of fan output, trim the fan speed. This method is much more efficient, and it is usually cheap. See Measure 10.3.1 for details.

The Leading Contenders

For fan systems of medium and large size, variable-frequency drives are the leading contender. They are among the best of methods in terms of efficiency (except near peak load). They are easy to install, even in retrofit applications. They are a major cost item, but they are similar in price to other efficient variable-speed drives, or less expensive. Their cost per kilowatt falls rapidly with larger sizes. Adverse side effects, especially electrical interference, are being brought under control.

For small fan systems, multi-speed motors, two-motor drives, and inlet vortex dampers may offer the best compromise between efficiency and cost. Of these types, only multi-speed motors are easy to retrofit.

Coordinate with Other Measures

Upgrading the fan modulation method is expensive. If you are going to spend this much money, consider additional improvements to the system. The lower air flow rates that result from improved fan modulation may create problems. To improve system efficiency and minimize the problems, install supply air temperature reset (see Measure 4.7.6.1). If the system does not have heating/cooling changeover, think seriously about installing it (see Measure 4.7.7).

ECONOMICS

SAVINGS POTENTIAL: *30 to 70 percent of fan energy, depending on the present method of fan modulation, the load profile, the type of VAV terminals, and other factors. Several percent of system cooling energy may also be saved as a result of reduced system leakage.*

COST: *The costs of variable-speed drives are covered in Reference Note 36, Variable-Speed Motors and Drives.*

PAYBACK PERIOD: *Several years, to many years. The payback period is shorter in larger systems and in systems that operate for longer durations.*

TRAPS & TRICKS

DESIGN: *Consider all the opportunities and problems of this modification. See "VAV Design Challenges," in Measure 4.4.7.*

SELECTING THE EQUIPMENT: *Variable-frequency drives are probably your best choice, but don't assume this. Their popularity is not based entirely on merit. See Reference Note 36 for guidance.*

Variable-Air-Volume Dual-Duct Systems

The Measures in this Subsection improve the efficiency of VAV dual-duct systems. This type of system is sketched in Figures 1 and 2, in the Section 4 Introduction.

In these systems, the air handling unit has a continuously operating cooling coil that feeds air into a cold air duct, and a continuously operating heating coil that feeds air into a hot air duct. The two ducts run in parallel throughout the facility. At each space, air is tapped from the two ducts by a terminal unit. The terminal unit has a hot air damper and a cold air damper. When the space thermostat calls for heating, the hot air damper opens. When the thermostat calls for cooling, the cold air damper opens.

Dual-duct VAV systems are mechanically similar to dual-duct reheat systems, which are covered by Subsection 4.5. The crucial difference is that the VAV version throttles the air flow in the terminal units to conform to the space conditioning load, whereas the reheat version mixes hot and cold air in the terminal units to maintain a constant flow of air into the space.

VAV dual-duct systems have the potential of being efficient and comfortable. Efficiency suffers if chilled air is mixed with heated air in the terminal units. The system may be designed to do this deliberately under low conditioning loads to maintain a minimum air flow into the spaces.

The first three Measures, 4.8.1 through 4.8.3, reduce the conditioning load by improving the efficiency of thermostatic control. Measures 4.8.4 through 4.8.6 ff are inexpensive activities that reduce the energy waste that occurs from mixing heated and chilled air. Measure 4.8.7 is an expensive improvement in fan output modulation that saves fan energy, and may also improve system performance. Measure 4.8.8 is a major change to the fan

INDEX OF MEASURES

4.8.1 In spaces with shutoff VAV terminals, install temperature setting placards on thermostats.

4.8.2 In spaces with shutoff VAV terminals, modify thermostatic controls to maximize deadband.

4.8.3 If air handling equipment cannot be turned off during unoccupied hours, install temperature setback.

4.8.4 Turn off the heating coil and/or the cooling coil whenever practical.

4.8.5 Adjust minimum-flow terminals to minimize overlap of hot and cold air flow.

4.8.6 Keep the cold duct temperature as high as possible and the hot duct temperature as low as possible.

    4.8.6.1 Install duct temperature reset controllers for both the cold duct and the hot duct.

4.8.7 Improve the efficiency of fan modulation.

4.8.8 Install an outside air economizer cycle with separate hot and cold duct fans.

arrangement that allows a dual-duct system to exploit the principle of the outside air economizer cycle, saving both cooling and heating energy.

All the Measures in this Subsection may be used together.

Triple-Duct Systems

"Triple-duct" systems are similar in appearance to dual-duct systems. See the diagram in Figure 2 of the Section 4 Introduction. The main difference is a third duct that carries unconditioned air (a mixture of return air and outside air) for mixing with either the heated air or the chilled air. If properly installed, these systems have no mixing losses, except for leakage that occurs inside the terminal units. They can maintain any desired space air flow rate without mixing losses by recirculating unconditioned air through the third duct.

All the Measures of this Subsection apply to triple-duct systems as well as to dual-duct systems.

MEASURE **4.8.1 In spaces with shutoff VAV terminals, install temperature setting placards on thermostats.**

  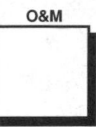

Placards are one of the most powerful tools of energy efficiency. Where appropriate, use them at thermostats that control dual-duct VAV terminal units. This activity is similar to Measure 4.7.1. Refer to there for details.

MEASURE **4.8.2 In spaces with shutoff VAV terminals, modify thermostatic controls to maximize deadband.**

 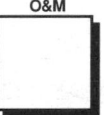

Deadband is a method of thermostatic control that eliminates unnecessary conditioning as long as the space temperature remains within comfortable limits. This Measure is similar to Measure 4.7.2. Refer to there for background.

The priority, methods, and economics of this Measure may differ from Measure 4.7.2. In particular, you can sometimes create deadband in dual-duct systems simply by making adjustments or minor modifications to the terminal units. This may require no changes to the thermostats or other control elements.

How to Create Deadband in Dual-Duct VAV Systems

Most dual-duct VAV terminals use a mechanical linkage to coordinate the operation of the hot and cold duct dampers. You increase deadband by adjusting this linkage to increase the range of motion in which both the chilled air damper and the hot air damper are closed.

If the linkage is not adjustable, you have to modify or replace the VAV terminals to increase deadband. Some manufacturers may offer retrofit kits for this purpose.

If the terminal unit has separate actuators for the hot and cold dampers, you can use other methods to achieve deadband. One is to use a deadband thermostat (see Measure 4.7.2). Another is to use a signal offset relay (see Measure 4.3.2.3).

ECONOMICS

SAVINGS POTENTIAL: *1 to 10 percent of conditioning cost. The larger percentages applying to mild climates, but the absolute saving is greater in buildings that have higher conditioning loads.*

COST: *If only adjustments are required, the cost is limited to a small amount of labor. Deadband thermostats, if needed, cost less than $100 apiece, plus*

SUMMARY

May save a few percent of system energy consumption at modest cost. Usually simple to accomplish. Not suitable where close control of temperature is needed.

SELECTION SCORECARD

Savings Potential	$	$	
Rate of Return, New Facilities	% % % %
Rate of Return, Retrofit	% % % %
Reliability	✓ ✓ ✓
Ease of Retrofit	☺ ☺

installation labor. Modifying or replacing terminal units may cost several hundred dollars per unit.

PAYBACK PERIOD: *Less than one year, if only adjustments are required. Several years, to many years, if deadband thermostats are installed.*

TRAPS & TRICKS

CHOICE OF METHOD: *This Measure applies to individual terminal units rather than to the entire system. It does not apply to terminal units that use reheat. Deadband is suitable for most comfort applications, but not for environments where close temperature control is needed. If your application can benefit from deadband, study your terminal units and their thermostats to select the most effective method of creating it. You may be able to design a simple modification that serves the purpose.*

SELECTING THERMOSTATS: *If you plan to achieve deadband by changing the thermostats, get all the* *efficiency features that apply. See Measures 4.7.2 and 4.8.3.*

EXPLAIN IT: *All thermostats should have placards, at least to state which spaces they control and that they allow a temperature deadband. With "heat/cool/off" thermostats, you are depending on the people in the spaces to select heating or cooling. With true deadband thermostats, the energy saving is determined by the person with the small screwdriver who sets the deadband*

in each thermostat. Make sure that the staff are trained to set the maximum acceptable deadband. Put clear instructions in the plant operating manual.

INSTALLATION: *When installing any thermostat, select a location where it senses the actual temperature of the space. Do not mount the thermostat on an exterior wall or other surface that is heated or cooled from behind. If necessary, use a long standoff mounting bracket. Do not install the thermostat where sunlight can shine on it or on adjacent surfaces.*

MEASURE 4.8.3 If air handling equipment cannot be turned off during unoccupied hours, install temperature setback.

RATINGS

New Facilities	Retrofit	O&M
A	**C**	

Temperature setback is a powerful and inexpensive method of saving conditioning energy during unoccupied periods. This Measure is similar to Measure 4.7.3. Refer to there for details.

MEASURE 4.8.4 Turn off the heating coil and/or the cooling coil whenever practical.

RATINGS

New Facilities	Retrofit	O&M
		A

This Measure can eliminate reheat losses for a large fraction of the system's operating hours. It costs nothing. It is identical to Measure 4.5.2 in methods and limitations. Refer to there for details.

The energy saving is substantially less than in Measure 4.5.2 because VAV systems have much lower mixing losses than constant-volume (reheat) systems. But, mixing losses are a major inefficiency even in some VAV systems. Even if the system has shutoff terminal units, this Measure eliminates mixing losses due to damper leakage inside the terminals.

ECONOMICS

SAVINGS POTENTIAL: *3 to 20 percent of cooling energy; 3 to 20 percent of reheat energy.*

COST: *Minimal.*

PAYBACK PERIOD: *Immediate.*

SUMMARY

A seasonal method of eliminating reheat that costs nothing.

SELECTION SCORECARD

Savings Potential	$ $
Rate of Return	% % % %
Reliability	✓ ✓
Ease of Initiation	☺ ☺ ☺

TRAPS & TRICKS

See Measure 4.5.2.

MEASURE **4.8.5 Adjust minimum-flow terminals to minimize overlap of hot and cold air flow.**

The main efficiency advantage of VAV systems over constant-volume reheat systems is eliminating the mixing of heated air with chilled air. However, many VAV systems use terminals that allow a reduced amount of mixing to maintain a minimum air flow into the spaces. Such terminals are called "minimum-flow" terminals, in contrast with "shutoff" terminals. In dual-duct VAV systems, the minimum flow typically is maintained by a linkage connecting the dampers for chilled air and heated air. The linkage prevents both dampers from closing simultaneously.

The air mixing that occurs with minimum-flow dampers wastes energy. This energy waste occurs only at low conditioning loads, so the energy waste is much less than in constant-volume dual-duct systems. Still, it is large enough to deserve aggressive action to minimize it.

Minimum-flow is used to avoid a number of comfort and ventilation problems that occur at low air flow rates, which are explained in Measure 4.5.5. Make sure that the minimum-flow settings are not higher than needed to satisfy these requirements. Check each terminal unit in the system and adjust it to satisfy the requirements of the space that it serves.

In some spaces, you may be able to eliminate damper overlap completely. In these cases, consider the possibility of saving more energy in these spaces by using "deadband" thermostats (Measure 4.8.2).

Energy Saving

The formula in the introduction to Subsection 4.5 applies to a VAV dual-duct terminal unit when it is operating in the minimum-flow region. (In this case, the term in the formula, "Terminal Unit CFM," is the value at the minimum-flow setting.) The formula shows that mixing losses are proportional to the minimum-flow rate.

Mixing losses peak at zero space load, because no useful conditioning is being done by the terminal air flow. As an example of how much energy waste occurs at zero load, consider a terminal in which the minimum flow has been set at 40% of the maximum cooling flow rate. If the heating temperature differential is three times the cooling temperature differential, some algebra will show that the amount of energy being consumed at zero load is equal to 60% of the maximum cooling load. That's a lot of wasted energy.

In this example, mixing losses occur as long as the cooling load remains below 40% of maximum. When the space has a heating load, mixing losses occur for an

SUMMARY

An important adjustment that reduces mixing losses and limits the conditions under which they occur. Affects comfort if done to excess.

SELECTION SCORECARD

Savings Potential	**$ $** $
Rate of Return	**% % % %**
Reliability	✓ ✓ ✓
Ease of Initiation	☺ ☺

even larger fraction of the time. This is because the high temperature of the heating air results in low air flow, so that chilled air must be mixed with the heated air to satisfy the minimum-flow limit. In this example, a typical load profile may result in mixing losses that consume as much energy as the space conditioning load itself.

Now, for example, if you reduce the overlap of heating and cooling so that the minimum flow is reduced to half its previous value, the mixing losses at zero space load are reduced to half. Even better, the total losses are reduced to much less than half. The reason is that mixing losses now disappear when the cooling load reaches 20% of maximum, and similarly, mixing losses stop sooner during the heating season. As a result, the system spends much less time operating with mixing losses.

What Determines the Minimum Acceptable Flow

A minimum rate of air flow through the terminals may be needed to:

- provide proper diffuser operation
- maintain an adequate level of outside air ventilation
- create in-space air velocity to prevent a feeling of "stuffiness" and to facilitate cooling by perspiration
- provide dehumidification during periods of low cooling load.

Review "VAV Design Challenges," in Measure 4.4.7, to figure out how each of these requirements affects the minimum-flow settings in your systems, and how you may be able to reduce these requirements.

In practice, you will probably determine the minimum air flow setting for each terminal unit by trial-and-error, keeping an eye on each of the previous requirements.

How to Adjust the Terminal Units

If you have a centralized electronic control system, you may be able to adjust the minimum-flow settings of the terminals from your keyboard. In this case, the effort is almost trivial, and you can easily experiment to find the optimum settings.

If you do not have a central control system, you can adjust each terminal unit manually. Unfortunately, they are often jammed tightly into the ceiling plenum, and reaching them can be time consuming. Terminal unit manufacturers have no shame about leaving sharp edges all over their equipment, so bring a box of bandages. Working with the boxes may disrupt the activities in the spaces, so expect the occupants to be annoyed. Because of these difficulties with manual adjustments, you have to decide whether it is practical to keep adjusting the boxes on a periodic basis.

■ Record the Settings

If you can reset the terminal units remotely, keep a list of the appropriate setting for each terminal, by season or load condition. If you adjust the boxes manually, mark the damper stop positions in a clear and permanent manner. If you change the damper stops at different seasons, label the appropriate positions for each season.

ECONOMICS

SAVINGS POTENTIAL: *10 to 50 percent of cooling energy; 10 to 50 percent of heating energy; a small fraction of fan energy.*

COST: *Perhaps one hour of labor per terminal, on average.*

PAYBACK PERIOD: *Short.*

TRAPS & TRICKS

OPTIMIZING THE SETTINGS: *The work is tedious, but not difficult. Before you start changing settings, figure out all the factors that determine the minimum-flow settings for each terminal or group of terminals. Expect some trial-and-error in paring down the reheat.*

EXPLAIN IT: *For maintenance and future reference, put a clear description of the procedures in the plant operating manual, including tables of settings for all the terminal units.*

MEASURE **4.8.6 Keep the cold duct temperature as high as possible and the hot duct temperature as low as possible.**

This Measure greatly reduces mixing loss. It is easy and it costs nothing. It is similar to Measure 4.5.1 in theory, procedures, and limitations.

The energy saving is substantially less than with Measure 4.5.1 because VAV systems have much lower mixing losses than constant-volume (reheat) systems. Still, minimum-flow terminals have substantial mixing losses. In shutoff terminals, this Measure reduces losses from unintended mixing that is caused by air leakage inside the terminals.

Energy Saving

The formula in the introduction to Subsection 4.5 applies to a VAV dual-duct terminal unit when it is operating in the minimum-flow region. The formula shows that mixing losses are approximately proportional to the cooling temperature differential and to the heating temperature differential. Thus, for example, reducing either temperature differential by half reduces the losses by half.

The two differentials are multiplied by each other in the formula, so reducing both differentials drastically reduces mixing losses. As explained in Measure 4.8.5, previous, the mixing losses in a dual-duct terminal peak at zero space load. Using the same example given there, if the heating and cooling temperature differentials are both reduced to half their previous values, the mixing losses at zero space load are reduced to one fourth of their previous level.

SUMMARY

Trim mixing losses just by turning two screws. Especially important in systems that have minimum-flow terminals.

SELECTION SCORECARD

Savings Potential .... **$ $** $

Rate of Return .... **% % %** %

Reliability .... ✓ ✓

Ease of Initiation .... ☺ ☺ ☺

The formula in Measure 4.5.1 applies to VAV systems only in the range where cooling and heating overlap inside the terminal. When the terminal air flow exceeds the minimum-flow setting, there are no mixing losses, other than losses caused by damper leakage inside the terminal.

ECONOMICS

SAVINGS POTENTIAL: *5 to 30 percent of cooling energy; 10 to 50 percent of heating energy.*

COST: *Minimal.*

PAYBACK PERIOD: *Immediate.*

TRAPS & TRICKS

See Measure 4.5.1.

MEASURE 4.8.6.1 Install duct temperature reset controllers for both the cold duct and the hot duct.

This Measure is similar to Measure 4.5.1.1 in theory, procedures, and limitations. The energy saving potential is substantially less, for the same reasons that apply to Measure 4.8.6, above.

ECONOMICS

SAVINGS POTENTIAL: *15 to 60 percent of cooling energy; 20 to 70 percent of heating energy. Savings are proportional to the minimum-flow settings of the terminals.*

COST: *The control components for a typical installation typically costs from several hundred to several thousand dollars. Connecting to thermostats at widely dispersed locations may cost considerably more.*

If an energy management computer is installed, reprogramming the computer to accomplish chilled air temperature reset typically costs several thousand dollars.

SUMMARY

This is the best way to control duct temperatures.

SELECTION SCORECARD

Savings Potential	$	$	$
Rate of Return, New Facilities	%	%	%	%
Rate of Return, Retrofit	%	%	%
Reliability	✓	✓	✓
Ease of Retrofit	☺	☺	☺

PAYBACK PERIOD: *Less than one year, to several years.*

TRAPS & TRICKS

See Measure 4.5.1.1.

MEASURE 4.8.7 Improve the efficiency of fan modulation.

This Measure is similar to Measure 4.7.8. Refer to there for details. Note the following difference in the method of controlling duct pressure, which is needed with dual-duct systems.

Sense Both Duct Pressures

In dual-duct systems, the hot and cold ducts operate independently. Therefore, design your system's controls to sense the pressure in both ducts, modulating the fans to maintain the minimum needed pressure in each duct.

MEASURE **4.8.8 Install an outside air economizer cycle with separate hot and cold duct fans.**

Conventional dual-duct systems that employ a single supply fan are poorly adapted to economizer cycle operation. (Measure 4.2.5 explains economizer cycles.) This is because the cold outside air used during economizer cycle operation would enter the heating coil as well as the cooling coil. The saving in cooling energy would be offset by increased heating energy. As a result, economizer cycles are usually not employed in dual-duct systems.

You can work around this obstacle by installing separate fans for the hot and cold ducts. In this configuration, outside air is taken in through the cold side of the system. Warm return air can be restricted to the hot side of the system. This eliminates heating of cold air, and vice versa. The two-fan configuration also saves energy in lesser ways that are not related to economizer operation.

Where to Apply This Measure

The primary benefit of the two-fan design is its ability to exploit economizer cycle operation. Consider it for facilities located in cooler climates where the air handling systems serve large spaces with high heat gains. Economic feasibility increases with the size of the system and the number of hours that an economizer cycle can provide an energy saving.

In new construction, the two-fan configuration increases cost by a relatively small percentage. In existing facilities, this is a specialized and expensive modification. It requires additional space in the vicinity of the present air handling unit. Typically, you would replace the existing air handling unit with two smaller air handling units, one for each duct.

Energy Saving

■ Cooling Energy

Refer to Measure 4.2.5 for the cooling energy savings of economizer cycles. In brief, an economizer cycle may save a large amount of energy if two conditions exist: (1) the facility has high internal heat gains, and (2) the climate has a large number of hours during which the outside air temperature is below about 55°F.

■ Heating Energy

If return air is heated as it returns to the air handling unit, using a separate fan for the hot duct allows the heat to be fully recovered, which saves heating energy and reduces cooling load. This advantage is significant in facilities where air is returned through ceiling plenums that are heated by lighting fixtures.

SUMMARY

The only way to make an economizer cycle work properly in a dual-duct system. Unusual.

SELECTION SCORECARD

Savings Potential .... $ $ $

Rate of Return, New Facilities % % %

Rate of Return, Retrofit.......... % %

Reliability .... ✓ ✓ ✓

Ease of Retrofit .... ☺

■ Fan Energy

Using two fans requires less total fan energy. With a single fan, the fan must deliver all the air at a pressure that is dictated by the hot or cold duct that has the highest pressure requirement. With two fans, each fan can modulate to satisfy the minimum pressure requirement of its respective duct.

Achieving this saving requires an efficient method of fan modulation. Without efficient fan modulation, fan energy consumption may be increased because of the larger total fan capacity.

In a two-fan installation, both fans can operate in a draw-through mode. Draw-through operation is more efficient than the blow-through configuration used in conventional dual-duct installations.

(In conventional dual-duct systems, the fan must operate in a blow-through mode so that a single fan can serve two coils. The disadvantage of this arrangement is that a fan discharges air at high velocity. Some of the kinetic energy that the fan adds to the air is dissipated as turbulence when the air strikes the coils. In a draw-through fan configuration, the high velocity of the fan discharge is converted to static pressure by proper design of the fan discharge duct. The static pressure does useful work in moving the air through the system.)

Outside Air Ventilation

Outside air ventilation requirements are covered in Measure 4.2.1. The cold duct system has a large outside air intake capability to allow economizer cycle operation. This does not satisfy ventilation requirements when the spaces are getting their air primarily from the hot duct, i.e., during cold weather. For this reason, the hot duct system also needs enough outside air intake capacity to satisfy ventilation requirements.

To control outside air intake efficiently, both the cold duct system and the hot duct system need good "front end" design. This is covered by Measure 4.2.1.

How to Design the System

The basic steps in adding an economizer cycle to a dual-duct VAV system are:

- Install two separate fans, one with the cooling coil and the other with the heating coil. Install the coils in a draw-through configuration. The cold duct fan is sized for the maximum cooling load, and the hot duct fan is sized for the maximum heating load. There is no need for the two air handling units to be located close together. For example, the cold-duct unit can be located near an outside wall to minimize economizer intake ductwork, and the hot-duct unit can be located close to the boiler to minimize the length of heating water pipe.

- Provide independent flow modulation controls for each fan. Use the most efficient method of fan modulation to minimize fan energy consumption. See Measure 4.8.7 about efficient fan modulation.

- Install appropriate dampers, ducting, and controls in the cold air system to achieve economizer cycle operation, as explained by Measure 4.2.5.

- If necessary, install a small set of dampers, ducting, and controls in the hot air system to provide outside air ventilation.

- Design the fan modulation controls of the system return air and exhaust fans to accommodate the fact that the hot air and cold air supply fans are modulating independently. You will probably need a fan modulation control system that monitors air flow and duct pressure.

In retrofit applications, the simplest approach is to replace the original air handling unit with two new ones, one for the cold air system and one for the hot air system. You won't be able to salvage much of the original equipment. In larger systems where the air handling units are field erected, you may be able to salvage the fan and coils.

Alternative: Waterside Economizer

An alternative way of adding an economizer cycle to a dual-duct system is to use a waterside economizer. This is an additional water coil for the cold duct that is cooled directly by a cooling tower, rather than by the chiller. Waterside economizers are described in Measure 2.9.4.

A waterside economizer avoids the extensive equipment modifications that are required for an outside air economizer, provided that the economizer coil can be added to the existing unit. This may be more difficult than it looks. Even if the additional coil can be fitted into the casing, the existing fan may not be able to overcome its additional resistance.

In comparison with outside air economizers, waterside economizers may be able to provide somewhat more free cooling in dry climates because of the evaporative cooling available from the cooling tower. However, in humid climates, a waterside economizer provides less cooling because of the additional resistance to heat flow in the coil. Waterside economizers require additional fan energy, as well as a small amount of energy for pumps. They do not require as much sheetmetal as an airside economizer, but they require additional piping to bring water from the cooling tower.

ECONOMICS

SAVINGS POTENTIAL: *30 to 80 percent of cooling energy. Up to 50 percent of heating energy (but limited to the amount of heat that is captured by the return air). A small fraction of fan energy, but only if the modulation method is efficient.*

COST: *In new construction, the increase in system cost may be small. Using separate air handling units for the hot and cold ducts provides design flexibility that may save some cost.*

In retrofit, the cost of this Measure may be prohibitive. The main cost components are the new air handling unit(s), economizer ductwork, relocation of equipment to provide space, and new controls.

PAYBACK PERIOD: *In new construction, typically a few years or less. In retrofit, typically several years, or much longer.*

TRAPS & TRICKS

PIONEERING: *This configuration is unusual, so be careful to resolve all the details. Start by studying Measure 4.2.5.*

INSTALLATION: *Make sure that the controls contractor understands what you are trying to achieve. Check out each mode of operation.*

EXPLAIN IT: *Since this involves dampers and controls, it is vulnerable to operating personnel trying to make quick fixes in response to comfort complaints. Explain the system to the staff. Describe it clearly in your plant operating manual.*

MONITOR PERFORMANCE: *Schedule checks of system performance during periods of the year when the economizer cycle should be operating. If your energy management control system has the ability to do trend logging, use it to record the temperature (or enthalpy) of the outside air, return air, and mixed air. These will tell you whether the economizer control is working properly.*

Induction Systems

The Measures in this Subsection improve the efficiency of induction air handling systems. Figure 2, in the Section 4 Introduction, sketches an induction system.

The most visible distinguishing feature of an induction system is the terminal unit. Figure 1, on the next page, shows a typical induction terminal unit. Induction terminal units have no fans. Air movement through coils in the terminal unit is induced by high-pressure air, called "primary" air, that comes from a central air handling unit. The primary air is passed through an array of nozzles in the terminal unit that create a venturi effect, or vacuum. The vacuum draws air from the space through the terminal unit coil. The space air, called "secondary air," mixes with the primary air. The mixed air is discharged into the space.

The primary air is cooled by a coil in the air handling unit. The temperature in individual spaces is regulated by the coil in the terminal unit, which is controlled by the space thermostat.

Induction terminal units commonly are packaged in enclosures that are similar in size and appearance to fan-coil units. They may also be installed in ceiling plenums. The coils are virtually always hydronic. The temperature of the coil may be controlled by throttling the water through the coil, or a damper may be installed in the terminal unit to bypass air around the coil.

Unlike other air handling systems, only a fraction of the air circulation in the space comes from the air handling unit. The volume of secondary air is approximately twice the volume of the primary air. As a result, an induction system can use smaller

INDEX OF MEASURES

4.9.1 Maximize the primary air temperature.

    4.9.1.1 Install automatic temperature reset control for the primary air.

4.9.2 Turn off the air handling unit cooling coil when cooling is not needed.

4.9.3 Install temperature setback.

4.9.4 Clean, adjust, and repair induction terminal units at appropriate intervals.

4.9.5 Avoid discharging conditioned air on exterior surfaces.

supply ducts, smaller return ducts, and a smaller air handling system than other systems. This saves space, and it may save material cost.

On the negative side, the high pressure requirement increases fan power, even though the volume of air delivered by the fan is less than in other types of air handling systems. Also, induction systems are designed to operate with low primary air temperatures. This may require lower chilled water temperatures, which reduces the efficiency of the chiller. The terminal units have a constant hissing sound, which is caused by the flow of high-velocity air through the venturis.

Induction systems cannot exploit economizer cycles to full effect, because the air from the air handling unit is only about a third of the air circulated in the space. For the same reason, induction systems are not well adapted to spaces that require high outside air ventilation rates.

Efficiency and System Variations

As with other central air handling systems that provide separate temperature control of individual

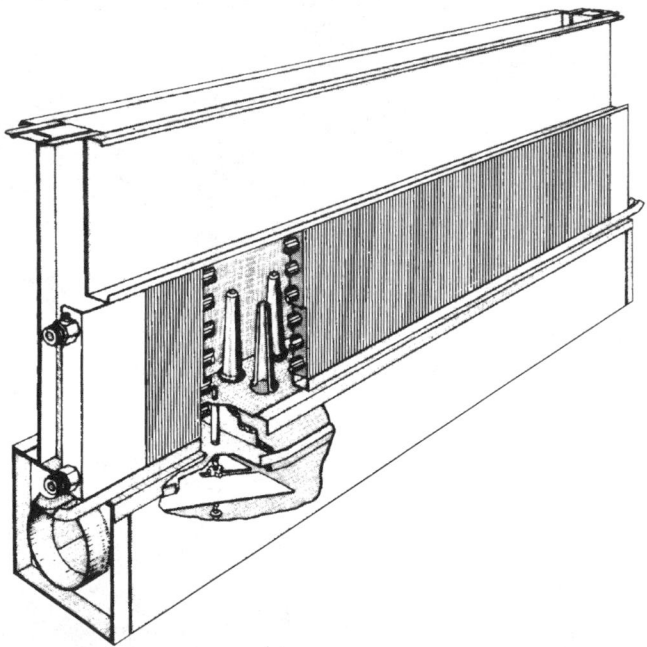

M&I Heat Transfer Products

Fig. 1 Induction terminal unit In this typical unit, high-pressure "primary" air enters through the duct at lower left. It passes through an array of nozzles, some of which are visible through the cutaway in the coil. The nozzles create a vacuum which draws air through the coil from the space. The conditioned space air mixes with the cooled primary air, and both are discharged into the space through the top of the unit.

spaces, reheat is a potential cause of energy waste. In reheat, excess energy is expended to cool the supply air, and then still more energy is then expended to partially re-warm the same air before it enters the space. During low-load conditions, much more energy may be cancelled out by reheat than actually enters the space to provide cooling or heating.

The amount of reheating that occurs in an induction system depends on the way that the terminal unit coils operate. If the terminal unit coils can only provide heat, then reheat will be a major cause of energy waste. In this case, the system operates similarly to other single-duct terminal reheat systems, which are covered in Subsection 4.4.

If the terminal unit coils are designed to change between heating and cooling, the system will have much less reheat energy waste than if the coils are designed only for heating. When the space requires cooling, both the coil in the air handling unit and the coils in the terminal units cool the air. Reheat does not occur under these circumstances. However, when a space requires heating, the coil in the terminal unit must cancel the cooling energy in the primary air. The amount of reheating that is unavoidable depends on the diversity of the space loads. If some spaces have a high cooling load at the same time that other spaces have a heating load, reheating of the primary air will occur in the spaces that require heating.

Terminal units that have changeover coils require effective drainage of the moisture that condenses on the coils during cooling operation. The design of an induction system with changeover terminal unit coils may assume that dehumidification is accomplished by the coil in the air handling unit, rather than by the coil in the terminal unit. However, conditions can still occur that cause moisture to condense on the coils. Furthermore, some of the Measures in this Subsection increase condensation on room coils.

Most induction systems operate at constant air volume. Some systems do attempt to achieve the benefits of VAV operation by throttling the primary air to the terminal units. However, the venturis in the terminal units rapidly lose their ability to induce flow of room air as the primary air flow is reduced. This minimizes any opportunity to save fan energy.

MEASURE 4.9.1 Maximize the primary air temperature.

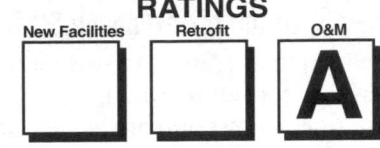

Reheat is the major cause of energy waste in induction systems. Reheat occurs when the primary air is being chilled and the secondary air is being heated. This is a problem that occurs in individual spaces. At any given time, reheating may be occurring in some spaces and not in others.

You can reduce reheat easily in an induction system by raising the temperature of the primary air. This Measure is similar to Measure 4.4.1 in theory, procedures, and limitations. Refer to there for details.

Subsidiary Measure 4.9.1.1 improves on this Measure with continuous automatic resetting of the chilled air temperature. Use manual adjustment as a stopgap method until you install automatic reset controls.

Energy Saving

Reheat wastes the most energy if the terminal unit coils provide only heating. By the same token, this activity saves the most energy in such units. There is less reheat, on average, if the terminal unit coils can switch between heating and cooling.

To illustrate the benefit of raising the primary air temperature, let's examine the effect on an individual terminal unit, under these conditions:

• primary air maximum flow: 200 CFM
• secondary flow is twice primary air flow
• space temperature: 75°F
• space conditioning load: none
• all cooling is sensible.

This table shows the energy savings that result from raising the chilled air temperature:

Energy Input, BTU/hr	Supply Air Temperature		
	55°F	65°F	75°F
conditioning energy to space	0	0	0
primary air cooling energy	4,320	2,160	0
reheat energy from terminal unit	4,320	2,160	0
total input energy	8,640	4,320	0

We have chosen these conditions to illustrate that an induction system can waste a large amount of energy even when there is no space conditioning load. The reheat energy waste is just as high any time the space requires heating, because the terminal unit must add enough energy to cancel the cooling of the primary air.

The induction system wastes energy even if the space has a low cooling load. This is because terminal unit must heat the secondary air enough to cancel the

SUMMARY

Reduce reheat losses just by turning a screw. May cause large amounts of condensate to form in the terminal unit, which makes cleaning important.

SELECTION SCORECARD

Savings Potential .... $ $ $

Rate of Return .... % % % %

Reliability .... ✓ ✓

Ease of Initiation .... ☺ ☺ ☺

excess cooling of the primary air. By reducing the energy content of the primary air, less energy is wasted in reheat.

In systems with changeover terminal units, the table shows that you can eliminate reheat entirely by setting the primary air temperature so high that the cooling coil in the air handling unit does not operate. In other words, the primary air temperature is set higher than the mixed air temperature returning to the air handling unit. (This is essentially the same as turning the cooling coil off, which is recommended by Measure 4.9.2.) In this situation, spaces that need cooling are cooled entirely by their terminal units, while spaces that need heating are heated by their terminal units. This works when the cooling load in all spaces is so low that the coils in the terminal units can carry the entire cooling load.

The potential saving is limited by diversity among the space conditioning loads. The worst case occurs when one or more spaces require a high cooling load continuously. See Measure 4.4.1 for details.

Potential Health Problems from Increased Condensate Drainage

Shifting the cooling load from the air handling system to the room units greatly increases the amount of dehumidification that is done by the room units, and correspondingly increases the amount of condensate that is formed within the room units. Make sure that this condensate be removed reliably by the drain pans inside the units. Regular cleaning of the condensate drains is essential, as recommended by Measure 4.9.4. Condensate that fails to drain is a breeding site for microorganisms, which enter the room air. This can cause serious health problems.

Explain It to the Maintenance Staff

Small temperature adjustments can have a big effect on energy consumption. Install a placard at the primary air temperature control that explains clearly how to adjust the temperature of the primary air.

ECONOMICS

SAVINGS POTENTIAL: *10 to 40 percent of cooling and reheat energy.*

COST: *Minimal.*

PAYBACK PERIOD: *Immediate.*

TRAPS & TRICKS

See Measure 4.4.1.

MEASURE 4.9.1.1 Install automatic temperature reset control for the primary air.

RATINGS

New Facilities	Retrofit	O&M
A	**B**	

Most induction systems should have this control feature to avoid energy waste from reheating. This Measure is similar to Measure 4.4.1.1 in theory, procedures, and limitations. Refer to there for details.

Where to Install Primary Air Temperature Reset Controls

Reset control for the primary air temperature is most valuable if all spaces have variable cooling loads, such as perimeter cooling loads in a glass curtain wall building. In such cases, a reset controller usually pays for itself quickly.

On the other hand, reset control may not be economical in systems where the primary air temperature is dictated by a constant high cooling load. This occurs in core zones, and in spaces with high internal heat gains, such as computer centers. In such cases, the reset control has little ability to raise the primary air temperature. Optimizing the primary air temperature manually is almost as efficient.

ECONOMICS

SAVINGS POTENTIAL: *30 to 80 percent of the energy expended for reheat; 10 to 40 percent of cooling energy. The savings potential of this Measure is limited by*

SUMMARY

Most induction systems should have this control feature. It is the only way to keep chilled air near its optimum temperature.

SELECTION SCORECARD

Savings Potential	$	$	$
Rate of Return, New Facilities	%	%	% %
Rate of Return, Retrofit	%	%	%
Reliability	✓	✓	✓
Ease of Retrofit	☺	☺	☺

system load diversity, and by spaces with large, constant cooling loads.

COST: *The control components for a typical installation typically cost from several hundred to several thousand dollars. In addition, connections to thermostats at widely dispersed locations can add a large labor cost.*

PAYBACK PERIOD: *Less than one year, to many years.*

TRAPS & TRICKS

See Measure 4.4.1.1.

MEASURE **4.9.2 Turn off the air handling unit cooling coil when cooling is not needed.**

You can eliminate all reheat losses in an induction system for extended periods of time by turning off the cooling coil in the air handling unit. The potential saving depends on the type of terminal unit, on the climate, and on the zoning of the systems.

Energy Saving Potential

The first question is whether the terminal units have heating-only coils or coils that can change between heating and cooling. If the terminal units have heating-only coils, this method can be used only when none of the spaces has a cooling requirement. If any space that is served by the system has a continuous cooling requirement, this method cannot be used.

If the terminal unit coils can provide both cooling and heating, you can turn off the air handling unit coil at all times, except when one or more of the spaces has a cooling load that cannot be satisfied by the cooling coil in the terminal unit. You may be able to eliminate reheat in this manner for more than half the year, depending on the climate and the type of facility.

Relationship to Primary Air Temperature Reset

Automatic reset for the primary air temperature (Measure 4.9.1.1) serves the same purpose as this Measure, and it is more effective. However, even if you have reset control, consider turning off the coils seasonally as additional security against unnecessary reheat.

SUMMARY

An easy method of eliminating mixing losses under a wide range of conditions.

SELECTION SCORECARD

Savings Potential .... **$ $** $

Rate of Return .... **% % %** %

Reliability .... ✓ ✓

Ease of Initiation .... ☺ ☺ ☺

ECONOMICS

SAVINGS POTENTIAL: *3 to 20 percent of cooling energy; 10 to 30 percent of reheat energy.*

COST: *Minimal.*

PAYBACK PERIOD: *Immediate.*

TRAPS & TRICKS

EXPLAIN IT: *Provide clear instructions that explain when to turn off the cooling coils, e.g., when the outside air is below a certain temperature. Optimize the schedule for each air handling unit or group of units. Install placards at the air handling unit cooling coil valves.*

MONITOR PERFORMANCE: *This Measure is easy to forget. Schedule periodic checks of the procedure in your maintenance calendar.*

MEASURE **4.9.3 Install temperature setback.**

Many facilities have unoccupied periods during which it is not necessary to condition the spaces. An effective way to exploit this opportunity for saving energy is to install setback temperature control. This is a control feature that maintains normal temperature during hours of occupancy, and allows the space temperature to drift to a lower (or higher) temperature during unoccupied hours.

With most types of air handling systems, you can save energy simply by turning off the air handling systems during unoccupied hours, as recommended by Measures 4.1.1 ff. However, this does not work well with induction systems, because the terminal units have a substantial amount of heating capacity even if the fan is turned off. Setback control avoids this problem by controlling the terminal units directly.

Compared to shutting the system down completely, temperature setback has the advantage that it maintains a minimum space temperature at all times. Typically, this is about 10°F less than the occupied temperature. This provides sufficient warmth for transient personnel, such as cleaning crews, and it provides for quick warm-up if the space is needed during unscheduled hours.

Also, setback control allows the air handling unit fan to keep running, if necessary, while allowing the space temperatures to drift within set limits.

Where to Use Temperature Setback

Temperature setback saves energy in any induction system, unless the temperature of all the spaces served by the system must remain constant. Ironically, induction systems are most popular in hospitals, where a desire to maintain constant space temperatures makes temperature setback irrelevant.

Energy Saving

■ Heating Load

The heating load is approximately proportional to the difference in temperature between the inside of the building and the outside, especially at night, when there are no solar effects. Temperature setback reduces this differential. Therefore, the greater the temperature setback, the greater is the energy saving. For example, if the average space temperature can be reduced 8°F during the setback period, and if the outside air temperature during the setback period averages 40°F below the occupied temperature, then heating energy consumption during the setback period can be reduced

by about 20%. (This example does not include the effect of heat gains, heat storage in the building structure, and other complications.)

■ Cooling and Reheat

Setback control turns off the cooling coil during the setback period. Therefore, it eliminates all cooling energy, and it also eliminates all reheat energy waste. The only exception would occur if one or more spaces need a high level of cooling during the setback period. This would be unusual. See Measure 4.9.1 for an analysis of the energy loss from reheat.

■ Conditioning Outside Air

In most cases, you can shut off outside air during the setback period. This action typically saves a large part of the conditioning load.

■ Fan Energy

The saving in fan energy is proportional to the length of time that the fan is turned off. Fan energy is especially important in induction systems because the need for high pressure at the terminal units increases the power of the fans.

How to Install Setback Temperature Control

Induction systems provide a great deal of flexibility in temperature setback because each terminal unit has individual heating capability, even with the air handling unit turned off. By the same token, exploiting the full efficiency potential of setback is more complicated in induction systems than in other types of reheating air handling systems. The following are the steps in achieving efficient setback temperature control.

■ Install a Setback Thermostat in Each Space

There are three common methods of installing setback temperature control:

- install individually programmable setback thermostats in each space
- if you have a central energy management system, program the setback schedule of each thermostat through the energy management system
- if you have pneumatic controls, program the setback schedule for the entire system collectively by using special thermostats that respond to the supply air pressure.

See Measure 4.3.2 for details of each approach.

■ Turn Off the Cooling Coil in the Air Handling Unit

The setback control should turn off the cooling coil in the air handling unit throughout the setback period. This keeps the cooling coil in the air handling unit from forcing the space temperature down to the setback temperature. It also avoids reheat if the air handling unit must operate to maintain the setback temperature.

This control feature involves tricky timing. The cooling coil in the air handling unit must be turned off before the beginning of the setback periods in the individual spaces, and it must be turned on after the end of the setback periods. This requires a separate timing function. A central timing system that controls all the space thermostats and the primary cooling coil is more reliable in this respect.

■ Turn Off the Air Handling Unit Fan, If Possible

If the air handling unit fan is not needed to provide ventilation during the setback period, turn it off. This saves fan energy.

Even if heat is needed during the setback period to maintain the minimum temperature, it may not be necessary to operate the fan. The space thermostat will turn on the heating coil in the terminal unit, and the convective heat output of the terminal unit may be sufficient to maintain the space temperature. Whether this is true depends on the heat loss from the space. (The terminal unit can provide a substantial amount of heat output only if it is installed near the floor. Units installed in the ceiling cannot provide convective heating.)

To exploit this possibility, you need to install two stages of setback. First, the setback thermostat turns on the heating coil in the terminal unit. Second, a separate thermostat starts the fan in the air handling unit if the convective heat output of the terminal unit is not sufficient. You can use a single thermostat in a critical space to provide the second stage.

An unusual complication occurs if you want to maintain a maximum allowable temperature as well as a minimum allowable temperature. In this case, you need an additional control function that turns on the fan if any space exceeds its maximum allowable temperature. This allows the cooling coil in the terminal unit to deliver cooling to the space. In extreme cases, it may also be necessary to restart the primary cooling coil (in the air handling unit) during the setback period.

■ Turn Off Outside Air Ventilation, If Possible

If the facility does not require ventilation during unoccupied hours, close the outside air dampers. In most cases, you can use the same control that turns off the primary cooling coil.

■ Provide Overrides in Each Space

In virtually all facilities, you should provide a setback override in each space that allows normal temperature to be restored during the setback period. A person who wishes to work during the unoccupied period activates the override in his space, which causes the thermostat to call for normal temperature.

The override should start the fan in the air handling unit, while leaving the cooling coil turned off. The fan is needed to provide full heating capacity from the terminal unit. Full heating capacity is needed to restore the space temperature quickly.

The override remains in effect until the next setback period, unless the occupant cancels the override. This is an efficiency weakness. It does not waste much energy if the setback override is used only occasionally. However, if the override is used often by a user who forgets to cancel the override, much of the benefit of setback is lost.

One way to overcome this problem is to use a personnel sensor to activate the override. This is not yet a standard feature of setback controls, so you have to wire it on a custom basis.

Don't Forget the Placards!

Setback thermostats require occupant action to avoid wasting energy, so they need instruction placards. Make sure that the placards request occupants to reset the temperature override after using it. See Reference Note 12, Placards, for general guidance in designing and installing placards.

ECONOMICS

SAVINGS POTENTIAL: *All cooling energy and all mixing losses are eliminated during the setback period (except in unusual applications). Approximately 5 to 30 percent of heating energy during the setback period is saved. Fan energy savings are approximately proportional to the length of the setback period.*

COST: *$100 to $300 per space, plus several thousand dollars for system connections.*

PAYBACK PERIOD: *Several years, or longer.*

MEASURE **4.9.4 Clean, adjust, and repair induction terminal units at appropriate intervals.**

RATINGS

New Facilities   Retrofit   O&M

B

Maintaining induction terminal units consists mostly of cleaning and filter replacement. Other maintenance requirements depend on the specific design of the room units. Give these items attention at regular intervals:

- ***filters, filter mountings, and lint screens.*** Cleaning filters is especially important in induction units. The venturi action that induces secondary air flow through the room unit coil is relatively weak. Therefore, dirty filters and lint screens can reduce air flow and conditioning capacity substantially.

  Induction units tend to be used in environments that generate a lot of lint, such as hospital rooms. This reduces the interval between filter changes. A common maintenance defect is improper fit of filters into the filter holders. Unless the fit is tight, dirt bypasses the filter and collects on the coils. Check the fit.

- ***coils.*** The items to check are cleanliness and fin spacing. You can clean coils with a vacuum cleaner or with compressed air (in the reverse direction). In extreme cases, you need to wash out a dirty coil with detergent and a brush. Straighten bent fins with a fin rake.

- ***coil condensate drain pans and lines.*** If the coils in the induction units are used for supplemental cooling, condensate forms on the coils. The condensate pans tend to become clogged by debris and by slime that grows in the condensate. When this happens, the stagnant water in the pans serves as a nutritious breeding site for microorganisms. Some of these organisms are health hazards. The

SUMMARY

Basic maintenance items needed to prevent energy waste, maintain comfort, and prevent health problems.

SELECTION SCORECARD

Savings Potential .... **$**

Rate of Return .... **% %** % %

Reliability .... ✓ ✓

Ease of Initiation .... ☺ ☺ ☺ ☺

air flow through the unit disperses these organisms into the space.

This hazard may be overlooked because most of the dehumidification in induction systems is supposed to be accomplished by the cooling coil in the air handling unit. However, high humidity in the space cause substantial condensation in the terminal unit. Also, increasing the temperature of the primary air, as recommended by Measures 4.9.1 ff, shifts more of the cooling load to the terminal unit, increasing condensation.

- ***thermostatic controls*** may be flimsy. Inspect them periodically and repair them as necessary.

- ***coil valves*** may leak, causing the coil to consume energy even when it is turned off. When the coil is in heating mode, valve leakage may increase the energy waste caused by mixing hot and cold air.

ECONOMICS

SAVINGS POTENTIAL: *A small fraction of conditioning energy. In induction units, maintenance is more important to comfort and health than to efficiency.*

COST: *$3 to $30 per unit per year, which is a routine maintenance cost.*

PAYBACK PERIOD: *Immediate. Maintenance costs less than neglect.*

TRAPS & TRICKS

SCHEDULING AND DILIGENCE: *This is simple, boring work that tends to be ignored. Schedule it in your maintenance calendar and make sure that it gets done.*

MEASURE **4.9.5 Avoid discharging conditioned air on exterior surfaces.**

RATINGS

New Facilities	Retrofit	O&M
A	**B**	

Keep induction units from blowing directly on exterior walls or windows. Also, keep draperies or other window treatments from trapping the discharge against exterior surfaces. Either condition increases heat loss through the window or wall. You can usually eliminate these problems with simple changes. See Measure 5.3.3 for the details.

Section 5. ROOM CONDITIONING UNITS & SELF-CONTAINED HVAC EQUIPMENT

INTRODUCTION

This Section covers a broad class of heating and cooling equipment that can be called "room conditioning units," "self-contained units," or "unitary equipment." Their common feature is that each unit serves a single space or a single thermostatic zone. Each unit has a single thermostat, a single air distribution fan, and minimal ducting. These units are distinguished from the air handling units in Section 4 by their lack of complexity in control and air distribution.

The approach to energy conservation with these units is largely governed by the fact that they are highly packaged and relatively inexpensive. If a unit is inefficient, the best course of action may be to replace it entirely. By the same token, it is generally impractical to change their basic design, although you may be able to upgrade components (such as motors) or add improvements (such as flue dampers). In general, the Measures of this Section involve much less engineering than the Measures in Section 4. A major exception is heat pump loop systems, which are covered in their own Subsection.

How to Use Section 5

Section 5 is organized primarily by type of conditioning equipment. Go to the Subsections that apply to your facility. The Measures in each Subsection are tailored to the type of equipment. But first, go to Subsection 5.1, which recommends control practices that minimize unnecessary operation of all types of unitary equipment.

As a general guide, each Subsection starts with the Measures that are least expensive and most widely applicable, and progresses to Measures that are more expensive and specialized. Measures providing exceptionally large savings are moved toward the front. Technically related Measures are grouped together.

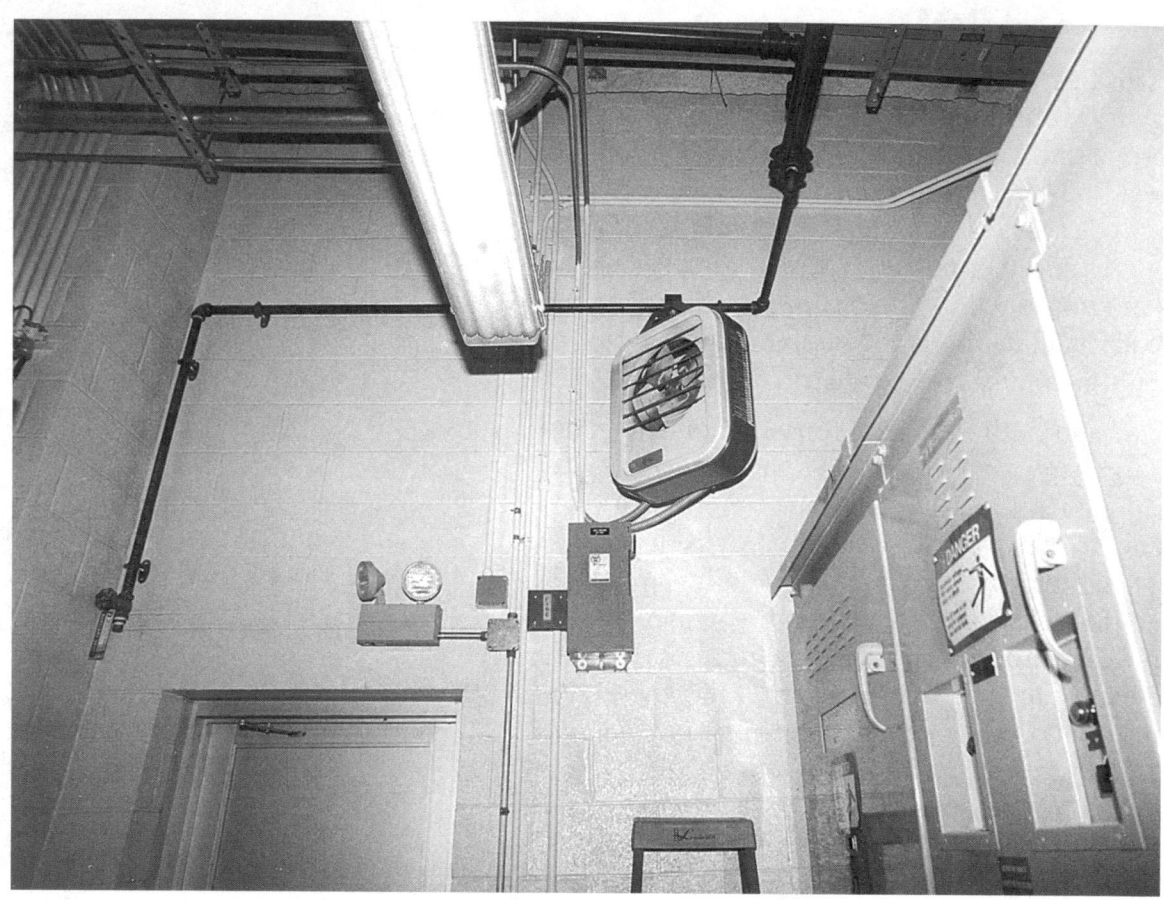

When should this heater operate?   The unit heater is located in a transformer room that is ventilated to the outside. There is probably nothing here that is vulnerable to freezing. No people work in the space, but people enter occasionally. What space temperature should be maintained? How would the heater be turned on to accommodate people in the space? What controls or combination of controls should be used?

Minimizing Equipment Operation

The first principle of energy conservation is turning off equipment when it isn't needed. The Measures in this Subsection tailor the output of unitary HVAC equipment to actual need. These Measures often provide large savings. They are relatively inexpensive. None of them requires special skills, but all of them require careful planning.

INDEX OF MEASURES

5.1.1 Install placards at the controls of conditioning units to motivate efficient operation.

5.1.2 Assign responsibility for turning conditioning units on and off to security personnel or to those who administer the spaces.

5.1.3 Install automatic controls to turn off conditioning units when they are not needed.

5.1.3.1 Where spaces operate on regular schedules, use setback thermostats or timeclocks to control the operation of conditioning units.

5.1.3.2 In spaces with irregular usage, install timed-turnoff switches to provide user control of conditioning units.

5.1.3.3 In spaces with irregular usage, install personnel sensors to control conditioning equipment.

5.1.3.4 Connect the power to conditioning units through an appropriate light switch.

5.1.4 Install proximity switches to turn off conditioning units when doors and windows are left open.

MEASURE **5.1.1 Install placards at the controls of conditioning units to motivate efficient operation.**

SUMMARY

A fundamental step in saving energy, but not as simple as it looks.

SELECTION SCORECARD

Savings Potential	$ $ $ $
Rate of Return, New Facilities	% % % %
Rate of Return, Retrofit	% % % %
Reliability	✓ ✓ ✓
Ease of Retrofit	☺ ☺ ☺

People cannot efficiently control what they do not understand. Therefore, install effective instruction placards on all control devices that are under the control of occupants. These include thermostats, fan control switches, heat/cool selectors switches, timeclocks, manual radiator valves, etc. Also, install placards to inform occupants about automatic controls that affect their comfort, such as timeclocks that turn conditioning on and off at particular times.

See Reference Note 12, Placards, for general guidance in exploiting placards. This Measure deals specifically with installing placards at the user controls of room conditioning units.

Where Placards are Effective

Placards provide the greatest energy savings in single-occupant spaces that have low occupant turnover and a high level of personal responsibility. If there is more than one occupant, each occupant is reluctant to turn off the conditioning unit or reset the thermostat for fear of annoying other occupants. Effective placard design can help to overcome this inhibition.

Placards are desirable in any space that has manual or automatic control of conditioning, even in spaces with many occupants. For example, Figure 1 shows a placard installed on a hotel room conditioning unit. At a minimum, they inform occupants how the conditioning in the space is controlled, which helps the occupants to deal with any comfort problems that may arise. Placards also help to create acceptance of efficient control.

Thermostat setting placards may not be useful with conditioning units that make heating and cooling available simultaneously. Such units are installed most commonly in posh environments, such a luxury hotels. With such equipment, the best thermostat setting may be ambiguous. Cheaper heating energy favors a higher thermostat setting, and cheaper cooling energy favors a lower thermostat setting. Therefore, the optimum temperature depends on the relative cost of heating and cooling energy. The occupants are not in a position to make this judgement.

If the equipment uses reheat for control of temperature or humidity, the effect of the thermostat setting on energy consumption is contrary to intuition. (Refer to Subsection 4.4 for an explanation of this behavior.) This is potentially so confusing to occupants that placards probably cannot help. Fortunately, reheat is rare in room conditioning units.

Energy Saving Potential

■ **Turning off the Equipment in Unoccupied Spaces**

The largest benefit of placards is motivating occupants to turn off conditioning units when they leave the space, which reduces these modes of energy usage:

• *space conditioning.* The energy delivered by a conditioning unit changes space temperature and humidity. If the spaces are unoccupied, you can usually allow temperature and humidity to drift within wide limits. The amount of conditioning energy that is wasted by unnecessary operation depends on the heat loss (or gain) of the space.

• *conditioning of outside air.* Some types of conditioning units take in outside air. Conditioning outside air requires a lot of conditioning energy. This energy expenditure usually is not needed when spaces are unoccupied. If a unit has to operate during unoccupied periods in order to maintain temperature, you may still be able to turn off ventilation.

• *auxiliary energy consumption.* Turning off space conditioning units that have fans saves the energy consumed by the fans. The fans typically are small, but the cumulative energy consumption of many small fans is expensive. For example, fans in typical room fan-coil units have a power consumption of 50 to 100 watts. The saving in fan energy depends on whether the fan runs continuously or intermittently.

■ **Minimizing Heating and Cooling in Occupied Spaces**

The amount of energy needed to heat a space, which includes conduction loss and the heating of outside air, is directly proportional to the temperature differential between inside and outside. Lowering the heating temperature saves energy by reducing this differential. For example, if the outside temperature is 30°F lower than the inside temperature, and the inside temperature is reduced 3°F, energy consumption is reduced about 10%. (This example ignores the effect of heat gains.)

Keeping the cooling temperature as high as possible is a major energy saver in most environments. The cooling energy requirement includes conduction heat gain, outside air cooling, dehumidification, and direct solar heat gain. The first two of these are proportional to the inside-to-outside temperature differential. (The dehumidification load is loosely related to the outside air temperature, and solar gain is independent of temperature.)

The overall energy saving depends on the length of time that the system operates. During low-load conditions, resetting the space temperature may substantially reduce the number of hours that the system operates. This effect is more dramatic in mild climates, where the temperature hovers for long periods in a range that does not require either heating or cooling.

Although the absolute saving is greater in more severe climates, the relative saving is greater with milder outside temperatures, and hence is more noticeable. In addition, buildings in mild climates are often poorly insulated, so they derive larger savings from a given temperature adjustment.

(As a point of interest, the energy saving does not relate to the outside temperature *per se*, as long as there is a heating or cooling load. For example, lowering the heating temperature by 3°F saves the same amount of energy if the outside temperature is 0°F or if it is 50°F.)

What the Placards Should Say

■ Control Functions

Ask occupants to perform all of these actions for which manual controls are available:

- turn off the unit when the space is not occupied
- set the heating temperature for the lowest comfortable setting
- set the cooling temperature for the highest comfortable setting
- turn the fan to the lowest comfortable speed
- set the fan control so that the running of the fan is controlled by the thermostat (see Measure 4.3.3.2)
- turn off the outside air intake when it is not needed

■ Close Windows and Doors

If the space can lose conditioning through windows or doors, ask the occupants to keep these closed while the air handling unit is operating.

■ Identify the Space being Controlled

If the controls are not mounted directly on the conditioning unit, state which spaces are covered by the controls, if this is not obvious. For example, the placard might say, "This thermostat controls the heating and cooling of the reception area." Otherwise, occupants may not associate the controls with the conditioning of their space, or they may not know if the controls affect conditioning in other spaces.

■ Inform Occupants about Automatic Controls

Inform the occupants about the operation of any automatic controls, if this might create confusion. If you use any of the automatic control methods recommended by Measures 5.1.3 ff and 5.1.4, let the occupants know. For example, the placard should advise occupants if a conditioning unit is controlled by a timeclock.

■ Ask Occupants to Dress Appropriately

The way people dress has a major effect on the space temperatures they choose. Where appropriate, encourage occupants to dress warmly in cold weather, and to dress lightly in warm weather. In many environments, it is appropriate to say this on the thermostat placard.

■ Whom to Call About Problems

If there is a problem with the conditioning controls, you want to know about it as soon as possible. Therefore, provide a contact for the occupants to call. Otherwise, the occupants may seek relief by vandalizing the control.

■ An Example

In an office that is conditioned by several four-pipe fan-coil units, an appropriate placard might say:

THIS THERMOSTAT CONTROLS THE THREE AIR CONDITIONING UNITS IN THIS ROOM.

To conserve energy, please adjust the temperature setting on this control unit as follows:

- **Select HEAT or COOL on the top selector switch.**
- **If HEAT is selected, set the HEAT lever to the lowest comfortable temperature.**
- **If COOL is selected, set the COOL lever to the highest comfortable temperature.**
- **Set the FAN switch to AUTOMATIC. If you leave the room for several hours or longer, please turn the FAN switch to OFF.**

Call Ext.123 if the heating or cooling does not work properly.

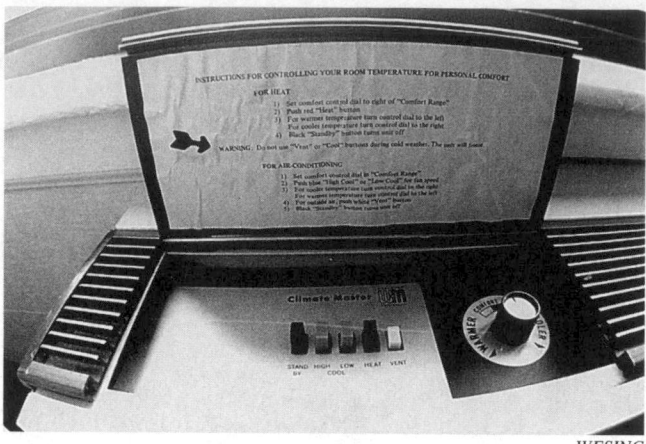

WESINC

Fig. 1 Hotel room conditioning unit placard These instructions are located in the right place, but they are difficult to read.

Where to Install the Placards

Install placards adjacent to the controls that they explain. If these locations are not plainly in view, consider installing additional placards at other locations. For example, if the purpose of the placard is to ask occupants to turn off equipment at the end of the day,

WESINC

Fig. 2 Bad practice Transparent thermostat covers irritate people by demonstrating that they cannot control their own comfort. To add insult to injury, the lower cover has a sticker that tells occupants what temperatures to select. If you cannot trust the occupants to set the space temperature, then hide the thermostats.

install it where the occupants are most likely to see it upon leaving the space. Typically, this is at the exit.

Let the Occupants Select the Space Temperatures

Don't bother telling occupants to set specific space temperatures. This is usually not effective or desirable. Accept the fact that occupants will not make themselves uncomfortable for the sake of energy conservation.

Furthermore, it is physically impossible to maintain a fixed space temperature. Room temperature varies from one part of the space to another. The thermostat controls the temperature only at the spot where it is installed. Furthermore, the calibration of thermostat dials is not precise.

Sometimes, thermometers are installed on or near thermostats as a guide to setting the thermostats. This is usually futile, because the thermometers themselves have limited accuracy, so the thermometers and the thermostats may disagree by several degrees.

A Poor Alternative: Locked Controls

It is bad practice to try to conserve energy by making conditioning unit controls inaccessible. Good energy conservation does not reduce comfort. If thermostat limits are set too aggressively with a locked thermostat, occupants will be uncomfortable. On the other hand, if the locked thermostat is set to minimize complaints, energy efficiency will be sacrificed.

Locked thermostats that are visible within the space irritate people because they overtly prevent people from being able to control their environment. Figure 2 shows a particularly silly example. As a result, they are a magnet for tampering. Locked thermostats create an adverse attitude toward energy conservation, resulting in loss of efficiency elsewhere.

If you have to keep conditioning controls out of reach of the occupants to prevent vandalism, conceal the controls, as in Figure 3. Putting a visible box around the controls will cause the occupants to break into the box.

Eliminate Hot Spots and Cold Spots

Some occupants may complain that they are too warm, while others may complain that they are too cold. In many cases, both are right. Efficient temperature control can work only if everyone is comfortable. You may have to correct localized areas of poor temperature regulation caused by inadequate air distribution, dumping diffusers, direct sunlight, poor wall insulation, or other problems.

ECONOMICS

SAVINGS POTENTIAL: *Turning off conditioning units during unoccupied hours may save from 10 to 50 percent of conditioning energy. Optimizing thermostat settings*

WESINC

Fig. 3 Opaque thermostat cover This is better than a transparent cover, but it still tells occupants that they cannot control their temperature. These are frequently knocked off the wall. Why does this cover have a sticker that tells occupants how to set the temperature?

may save from 1 to 10 percent of conditioning cost, with the larger percentages applying to mild climates and low conditioning loads.

COST: $1 to $50 per placard, depending on quantity, materials, and complexity. Expect design to account for a majority of the cost of effective placards.

PAYBACK PERIOD: Less than one year, usually.

TRAPS & TRICKS

PLANNING: Use Reference Note 12, Placards, as your guide. Make a tour of your entire facility and list all the conditioning units that should have placards. Design them as you make your list, then settle on a number of standard designs. Be sure to accommodate differences in equipment, occupants, or operating conditions.

INSTALLATION: Install the placards where they cannot be overlooked. Resist the temptation to use adhesives.

MONITOR PERFORMANCE: Schedule periodic checks of the placards in your maintenance calendar. Observe how well they are working. Improve the design, if necessary.

MEASURE 5.1.2 Assign responsibility for turning conditioning units on and off to security personnel or to those who administer the spaces.

RATINGS
New Facilities   Retrofit   O&M
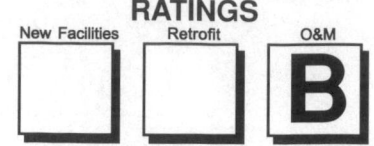

If the building has a security force, consider assigning them to turn off conditioning equipment at the end of the day. Similarly, if the spaces are under the administrative control of a particular group, that group can assign an individual to turn off the conditioning units at the end of the day.

Security patrols are also effective for turning off conditioning units that are turned on by people who work after normal occupancy hours. Similarly, they can be assigned to turn on conditioning units prior to occupancy, to allow time for warm-up.

This method may be the only practical way of turning off conditioning equipment in certain environments. Both occupant control and automatic control may be unreliable in spaces that have multiple occupants, especially if space usage is irregular.

Energy Saving Potential

See Measure 5.1.1.

Coordinate with Control of Air Handling Units

This Measure is identical in its administrative aspects with Measure 4.1.1.5, which relates to air handling units. Accomplish both Measures together if you have both types of conditioning equipment.

SUMMARY

A cheap and powerful way to save energy, but requires effective long-term management.

SELECTION SCORECARD

Savings Potential .... $ $ $ $

Rate of Return .... % % % %

Reliability .... ✓ ✓

Ease of Initiation .... ☺ ☺ ☺

ECONOMICS

SAVINGS POTENTIAL: *50 to 95 percent of unnecessary operation of the conditioning unit, depending on the effectiveness of management. The fraction of total system energy consumption that is saved varies widely, depending on operating schedule, conditioning load, and other factors.*

COST: *Minimal, in most cases.*

PAYBACK PERIOD: *Immediate.*

TRAPS & TRICKS

See Measure 4.1.1.5.

MEASURE 5.1.3 Install automatic controls to turn off conditioning units when they are not needed.

Human beings are not well adapted to starting and stopping equipment at precise times. Human control is especially unreliable if the starting and stopping times vary with changing conditions, such as weather. For these reasons, use automatic controls wherever possible to control the running of conditioning units. Figure 1 shows room conditioning units that are controlled automatically, and it also illustrates some of the issues that have to be satisfied to achieve efficient control.

The four subsidiary Measures provide a variety of automatic control methods:

- Measure 5.1.3.1 uses clock controls to make conditioning available on the basis of time, and also on the basis of minimum acceptable space temperature
- Measure 5.1.3.2 uses timed-turnoff switches to allow occupants to start conditioning at any time for a selected period of time
- Measure 5.1.3.3 uses personnel sensors to start and stop conditioning automatically, based on the presence of people in the space
- Measure 5.1.3.4 connects conditioning equipment through a light switch, to provide conditioning whenever the lights in the space are turned on.

Only the first of these can anticipate occupancy, allowing the space to be at normal temperature before anyone arrives in the space. On the other hand, the last three Measures allow an occupant to start conditioning at any time, and they automatically turn off conditioning after the occupants leave.

Consider combining the first Measure with one of the others for optimum control. For example, use a setback thermostat to start conditioning in the morning for the initial warm-up, and control conditioning for the rest of the day with an occupancy sensor.

Design Controls to Suit Occupants' Needs

Automatic controls for space conditioning have a poor record of reliability. They tend to be abandoned, or the occupants or staff disable them. There are two main reasons for this. One is that the occupants perceive the automatic controls as a threat to their comfort. The other is that occupants or staff cannot understand how the controls function. Therefore, to make automatic controls survive, you must make them user-friendly.

The first principle is to make conditioning available whenever people need it. Your challenge is to install controls that "know" when each conditioning unit should be running. In most applications, do not limit conditioning to fixed periods of time. Make conditioning available during unpredictable periods of space usage.

Think of regular occupants who work outside of normal working hours, and think of people who are not regular occupants, such as cleaning crews and maintenance staff.

Automatic controls are not obvious. You have to explain them so that an occupant understands how the conditioning is being controlled, as soon as he enters the space. An effective placard is your primary tool for doing this. This is especially important if the space is not at normal temperature when the occupant first arrives.

Limit Temperature Drift

Don't let the temperature wander too far when the equipment is turned off. The usual concern is letting temperature drop (or rise) to the point that it takes too long to restore comfortable temperature within the space.

WESINC

Fig. 1 Automatic control for classroom conditioning units
The classroom operates on an irregular schedule. However, the conditioning units are controlled from a central computer system located miles away. The arrow points to a computer interface. It would be more efficient and more comfortable to control the units with accessible local controls, and to make the school administration responsible for efficient operation.

In extreme cases, low temperature may cause freezing or other damage.

Temperature drift is caused by heat loss (or gain). Serious temperature drift is most likely to occur in perimeter spaces. In interior spaces that contain heat-producing equipment, the temperature may rise excessively.

Temperature drift is limited by the heat storage capacity of the building structure and contents, and by heat transfer from adjacent areas where the temperature is held constant. Even perimeter spaces may not experience large temperature drifts unless they are isolated from the adjacent spaces by closed doors.

Temperature drift usually is tolerable if the conditioning equipment can restore a comfortable temperature quickly. Temperature recovery is quick in smaller spaces because of the laws of geometry. The capacity of conditioning equipment is determined largely by the exterior surface area of the space. In contrast, the length of time that is required to bring the air temperature to a comfortable level depends on the air volume. Smaller spaces have a lower ratio of volume to exterior surface area, so the temperature of the air can be restored more quickly than in a larger space.

For example, a typical fan-coil unit in a small office may be able to restore comfortable air temperature within a few minutes, whereas a gymnasium heating unit may require an hour.

Restoring the air temperature in the space is the important part of recovering from setback, rather than reheating all the mass of the surroundings. As long as air is flowing from the conditioning unit, air temperature is the main factor in comfort. Restoring the temperature of the surrounding mass takes much longer than restoring the air temperature, but this typically does not cause a problem. The air temperature does need to be somewhat higher than normal to compensate for the cooler surfaces. Fortunately, this tends to occur naturally, because the thermostats are attached to the surfaces. As a result, the thermostat senses a temperature that lies between the air and surface temperatures.

To limit temperature drift, start the conditioning equipment when the space temperature drifts too far. A setback thermostat (see Measure 5.1.3.1) serves this purpose. As an alternative, you can install a separate thermostat to perform this function alone.

Coordinate with Control of the Central Plant

You can also use the control methods that are recommended here to control the running of the boiler plant (see Measures 1.1.1 ff), the chiller plant (see Measures 2.1.2 ff), and the air handling systems (see Measures 4.1.1 ff). Coordinate the controls for all these types of equipment to reduce cost and increase opportunities for conservation.

Shutting off the boilers and/or chillers that serve room conditioning units is less expensive than shutting off the conditioning units individually. However, this is acceptable only if the individual spaces all operate on approximately the same schedule, and if the room conditioning units operate on the same schedule as the boilers and/or chillers. Also, this method does not save fan energy in the room units.

MEASURE **5.1.3.1 Where spaces operate on regular schedules, use setback thermostats or timeclocks to control the operation of conditioning units.**

If a space is occupied on a predictable schedule, you can use either timeclocks or setback thermostats to turn the conditioning units on and off. A setback thermostat is a thermostat with an internal timeclock that changes the space temperature settings at different times.

A setback thermostat normally operates through the thermostatic control circuit of the conditioning unit. A timeclock is normally installed to control the power to the entire conditioning unit. However, you could install a setback thermostat so that it controls the power to the entire unit. Or, you could install a timeclock to interrupt the thermostatic circuit alone.

Where to Use Timeclocks and Setback Thermostats

If the space served by a conditioning unit operates on a regular schedule, a timeclock or setback thermostat is usually the best method of automatically starting and stopping conditioning. For greater control flexibility, you can combine timeclocks and setback thermostats with controls that respond to other conditions, such as outdoor temperature, occupancy, etc.

Timeclocks and setback thermostats are not appropriate for spaces that operate on irregular schedules, such as conference rooms, auditoriums, and surgical suites. Unfortunately, facility managers sometimes try to schedule specific times for conditioning such spaces, most commonly through an energy management system. This inevitably leads to dissatisfaction.

Don't use temperature setback if the thermostatic control system would use energy to force the setback temperature. This limitation exists in units that provide both heating and cooling, where neither can be turned off. For example, four-pipe fan-coil units with pneumatic controls are sometimes installed with a single thermostatic element controlling both the heating and cooling coils. These are installed in some luxury hotels so that guests do not have to select heating or cooling. If a setback thermostat were used with this type of system, switching to the lower nighttime temperature would consume energy to cool the space to the nighttime temperature, and to maintain that temperature until the setback thermostat switched back to the daytime temperature.

If the conditioning unit uses reheat, study the unit's controls before using temperature setback. Setback while reheating actually increases energy consumption. (Reheat is rarely used in room conditioning units. When it is used, the purpose is usually to control humidity. To

SUMMARY

Basic and powerful energy conservation tools with great flexibility. People will sabotage them if they are not user-friendly.

SELECTION SCORECARD

Savings Potential .... $ $ $ $

Rate of Return, New Facilities % % % %

Rate of Return, Retrofit.......... % % % %

Reliability .... ✓ ✓

Ease of Retrofit .... ☺ ☺ ☺

minimize energy consumption for dehumidification, the reheat should be controlled by a humidistat in the space.)

Comparison of Timeclocks and Setback Thermostats

In most cases, but not in all, a setback thermostat is preferable to a timeclock. Setback thermostats behave in the same way as timeclocks, until the temperature falls (or rises) to the temperature limit that is set. The main advantage of a setback thermostat is that it restarts the conditioning unit if the temperature drops to the lower setting. This feature keeps the space warm enough for a quick warm-up, or you can use a setback thermostat as a safety feature to protect against excessively low (or high) temperature.

If the conditioning unit has a continuously running fan, using a timeclock to control the power to the unit saves fan energy, whereas a setback thermostat allows the fan to keep running. In some cases, it is fairly easy to install a setback thermostat so that it controls the operation of the entire unit.

Setback thermostats cost more than timeclocks, but the labor cost of installing a setback thermostat in the thermostat circuit usually is less than the cost of installing a timeclock in the main power circuit. On the other hand, you could install a timeclock in the thermostat circuit.

Selecting the Setback Temperature

The setback temperature is a compromise. If the setback temperature is set much lower than the normal temperature, the space may get too cold for rapid warm-up. This is a problem if people may need to occupy the space at irregular times after the space has cooled down. The warm-up time is typically much longer in larger spaces, for the reason explained in Measure 5.1.3.

On the other hand, setting a higher setback temperature reduces the energy saving. This occurs if the space cool downs to the setback temperature, so that the conditioning unit operates during some of the unoccupied hours. More after-hours conditioning occurs if the space is vacated for long periods, such as a weekend. Not surprisingly, more after-hours conditioning occurs if the space has high heat loss.

Energy Saving Potential

The previous discussion shows that the potential savings vary over a wide range. The greatest savings occur when the conditioning can be turned off for a large fraction of the time and a large variation in temperature can be tolerated.

The difference between the inside and outside temperatures determines two major components of the conditioning load, the conduction load and the outside air load. Temperature setback reduces this difference. For example, if the space temperature is reduced an average of 5°F during the setback period, and if the outside air temperature during the setback period averages 25°F below the occupied temperature, then heating energy consumption during the setback period can be reduced by about 20%. (This example does not include the effect of heat gains, heat storage in the building structure, and other complications.)

How to Select the Controls

Many models of timeclocks and setback thermostats are available. Invest some effort in selecting the best one for each application. Specific features can make the difference between success and failure. The most important characteristics are ease of use, the scheduling options, and the method of override control. See Reference Note 10, Clock Controls and Programmable Thermostats, for equipment features, installation practices, and operating procedures that are needed for satisfactory performance.

Where older-style time controls are presently installed, consider replacing them to take advantage of the features of newer electronic models, including internal power backup, multiple schedules, and other features.

If your facility has an energy management computer system, you might use it for starting and stopping conditioning equipment. However, such large-scale control systems tend to be unreliable, and the cost of connecting each conditioning unit to the system is high. Simple localized controls are less expensive and individually more reliable. The main advantage offered by a centralized system is the ability to re-schedule conditioning units from a central location. See Reference Note 13, Energy Management Control Systems, for the complete story.

■ Occupant Override

In most applications, provide an easily accessible manual override. This allows a person to restore conditioning during hours when the timeclock or setback thermostat would normally have the conditioning turned off. This is essential for comfort. It also protects the time control. If an override is not easily available, people will eventually turn off the time control, or disable it.

Conventional overrides keep the equipment operating until the next "off" time. For example, if a conditioning unit is scheduled to turn off at 6 PM and is restarted manually at 8 PM, it will continue to operate until 6 PM of the following day. For this reason, overrides are inefficient if they are used often. Consider a conventional override to be an emergency feature rather than a routine control function.

If conditioning is often needed at odd hours, control the equipment with an interval timer, personnel sensor, or other control method that adapts to the duration of usage. These methods are covered by the Measures that follow.

Installation Options

A timeclock is usually installed to control all the power to the conditioning unit, including the fans. A setback thermostat is installed through the thermostatic control circuit. For example, a setback thermostat controlling a heat pump would control the compressor and the outside fan, but not the inside fan. With some equipment, such as convectors and hot air furnaces, there is no difference in the connections.

With small conditioning units that are powered through a receptacle, such as window air conditioners, you can plug an inexpensive timeclock into the receptacle, or use a timeclock as a replacement for the receptacle. Inexpensive timeclocks designed for such applications are common items.

Another option is to install a timeclock directly on the casing of the conditioning unit. Or, install a timeclock on a separate junction box.

To install a setback thermostat, remove the existing thermostat and replace it with the setback thermostat. Some setback thermostats are powered by the thermostat circuit itself, some are powered by a low-voltage transformer, and some are powered by an internal battery.

If the thermostatic controls are pneumatic, you can install a centralized setback system that controls all thermostats simultaneously. In this system, setback is triggered by changing the control air pressure at the air source. This retrofit requires changing all the individual thermostats, plus adding a control station at the air source to change the pressure at the appropriate times. A typical system sets the thermostats to daytime operation with 17 PSI supply pressure and to nighttime operation with 22 PSI supply pressure.

Install either timeclocks or setback thermostats where they are easily accessible to occupants, especially for override purposes.

Don't Forget the Placards!

People cannot use time controls intelligently unless you explain what they do. Refer to Measure 5.1.1.

ECONOMICS

SAVINGS POTENTIAL: *10 to 70 percent of the operating cost of the conditioning units, depending on the utilization of the spaces they serve. Setback thermostats that keep the space temperature within about 10°F of occupied temperature typically save about 10 to 30 percent of heating energy.*

COST: *Electronic setback thermostats with a variety of features cost about $100 to $200. Electrical and electronic timeclock units with overrides cost from $20 to $100. If the control uses existing thermostat wiring, installation cost is small. Installing a timeclock on a receptacle can be done in a few minutes. Installing a timeclock in the power circuit may cost $100 or more.*

PAYBACK PERIOD: *Less than one year, to several years.*

TRAPS & TRICKS

SELECTING THE EQUIPMENT: *Get all the appropriate features. Select units that are easy to use. A control will become junk if it is difficult to use or if it lacks a needed feature.*

INSTALLATION: *Install the timeclock or setback thermostat where it is obvious and accessible to those who should set it. Keep it out of sight of other people. Make sure that an accessible and well marked override is available for anyone who occupies the space at irregular times.*

EXPLAIN IT: *For the occupants of the space, install a well designed placard at each timeclock or setback thermostat that states what it controls, the normal schedule, and how to use the override. For the staff who are responsible for setting and maintaining the controls, provide clear instructions about these functions.*

MONITOR PERFORMANCE: *Time controls have a high failure rate, mainly from tampering. Schedule periodic inspections of all your time controls to check their condition and settings.*

MEASURE **5.1.3.2 In spaces with irregular usage, install timed-turnoff switches to provide user control of conditioning units.**

New Facilities | Retrofit | O&M

B **C**

This Measure is identical to Measure 4.1.1.3 in its procedures. The spaces affected are smaller, so the savings and the costs of each installation are lower.

MEASURE **5.1.3.3 In spaces with irregular usage, install personnel sensors to control conditioning equipment.**

RATINGS

 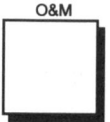

New Facilities | Retrofit | O&M

B **C**

This Measure is similar to Measure 4.1.1.4 in its procedures. Personnel sensors are especially appropriate for smaller spaces because the spaces warm up or cool down quickly, and because a single sensor can monitor the presence of people in the entire space. Because the spaces are smaller, the savings and costs of each installation are lower than for Measure 4.1.1.4.

MEASURE **5.1.3.4 Connect the power to conditioning units through an appropriate light switch.**

RATINGS

 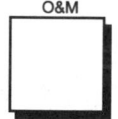

New Facilities | Retrofit | O&M

C **D**

You can control the conditioning of a space by activating the conditioning through a light switch. This method guarantees that conditioning is available whenever it is needed. It provides "automatic" control of conditioning without needing special controls. However, this method works well only under a specialized set of conditions.

Where to Use This Method

Consider this method for spaces that have all these characteristics:

- *occupancy that is independent of the schedule of the rest of the facility.* For example, this method can be effective for the offices of salesmen or field inspectors, and in a police station, for the detectives' and patrol officers' offices.
- *low personnel turnover and a high level of personal responsibility*. The occupants have to be the kind of people who respond to a request to turn off the lights when they leave the space.

SUMMARY

Simple. Easy to use. Unconventional. Depends on people to turn off the lights when spaces are vacated.

SELECTION SCORECARD

Savings Potential	$ $ $
Rate of Return, New Facilities	% % % %
Rate of Return, Retrofit..........	% % %
Reliability	✓ ✓ ✓ ✓
Ease of Retrofit	☺ ☺

- *single occupancy, or all occupants are visible to each other.* If more than one person is assigned to the space, each occupant may be reluctant to turn off the lights for fear of annoying the other occupants. Each person should be able to see that the space is entirely vacated at the time he leaves.

- *limited heat gain or heat loss.* Large temperature drift will occur in spaces with high heat losses or heat gains when the conditioning equipment is turned off. Therefore, if the climate can be extreme, limit this method to interior spaces. Occupants will tolerate an initial period of space temperature that is moderately cooler or moderately warmer than normal, especially if they are entering from the outside. However, if the temperature is too far from normal space temperature, the space will be perceived as uncomfortable. In that case, occupants will keep the lights turned on as a way of keeping the conditioning equipment running.

- *conditioning equipment that provides rapid warm-up and cool-down.* This Measure works best with fan-coil units, because these warm the air in the room quickly, making the space comfortable even before the structure and furnishings have time to warm up completely. It also works well with individual space air conditioners, because these generally have enough cooling and fan capacity to cool a small space quickly. On the other hand, you probably should not control baseboard heating with the light switch, because it warms the space too slowly.

- *at least one light fixture is always needed while the space is occupied, but not at other times.* Do not use this technique in spaces that can be lighted entirely by daylight. Also, do not use it in spaces where the lights may be turned off while the space is occupied, e.g., in sleeping quarters, or in rooms where audio-visual presentations are made.

The main competitor to this method is using personnel sensors to control conditioning (Measure 5.1.3.3). Personnel sensors are more reliable for conforming to actual occupancy. This Measure may be the best choice in spaces where personnel sensors cannot operate reliably, and in spaces where people enter and leave the space frequently during the period of time that the space is being used.

If you use security patrols to monitor the conditioning of spaces, as recommended by Measure 4.1.2, this Measure makes it easier for them to do the job.

Energy Saving Potential

See Measure 5.1.1.

How to Make the Connections

Select an easily accessible switch that controls a light fixture that is always required when the space is occupied. Replace it with a double-pole switch to isolate the conditioning unit power from the lighting power. Run power to the conditioning unit through the switch. If the switch is not in a convenient location, relocate it to a place where the occupant can turn it off easily when exiting.

Don't Forget the Placards!

This method is unusual. To make it acceptable, install a placard at the light switch to inform occupants that the switch also controls space conditioning. The placard should request occupants to turn off the switch whenever leaving the space for an extended period. See Reference Note 12, Placards, for guidance in effective placard design.

ECONOMICS

SAVINGS POTENTIAL: 50 to 95 percent of unnecessary operation of the conditioning unit. The fraction of total system energy consumption that is saved varies widely, depending on the occupancy of the space and its conditioning load.

COST: In new construction, the additional wiring cost typically is small. In retrofit, the cost may be several hundred dollars per space.

PAYBACK PERIOD: Short, in new construction. Several years or longer, in retrofit.

TRAPS & TRICKS

CHOICE OF METHOD: The space, the conditioning equipment, the type of lighting, and the nature of the activities must all be appropriate for this method. Review all these conditions for each candidate space, and be certain that they will continue for the life of the facility.

EXPLAIN IT: Let the occupants know how the conditioning is controlled. Install an effective placard that asks them to turn off the lights, and the conditioning, when they leave the space.

MEASURE 5.1.4 Install proximity switches to turn off conditioning units when doors and windows are left open.

If a space has exterior windows and doors, they should be closed while heating or cooling the space. Unfortunately, people may leave the windows or doors open. For example, this is a common problem in motels that have sliding glass doors to provide access to patios and balconies.

To solve this problem, install a proximity switch at each exterior window and door that turns off the conditioning equipment when the window or door is opened. This simple control connection is almost foolproof. If the switch breaks or the wiring is disconnected, the conditioning unit will not function.

This control feature has two desirable effects on occupant behavior. It motivates occupants to keep the doors and/or windows closed to avoid wasting conditioning. And, it highlights that fact that occupants have a choice between using outside air ventilation and using air conditioning.

Energy Saving Potential

Opening a door or a window allows so much outside air to enter the space that the conditioning unit is kept near maximum output if the weather is cold or hot. Even partially opening a window may increase the conditioning load of a space severalfold. This Measure saves the energy that is wasted in attempting to condition the entire outdoors.

How to Install the Switches

Install a magnetic proximity switch, a mechanical contactor, or other reliable switching device to sense the opening of each door and window. Wire the switch into the thermostat circuit of the conditioning unit (if the thermostat is electrical), or use the switch to control the power to the conditioning unit through a relay.

Select the sensing device for ruggedness. Protect the switch itself and its wiring. Keep both out of sight as much as possible.

Inform the Occupants!

An integral part of this method is informing occupants that the conditioning unit will stop automatically if a door or window is opened. Install an

---

SUMMARY

An important control feature in spaces that have openings to the outside. Especially desirable for lodging and rental facilities. Simple. Almost foolproof.

SELECTION SCORECARD

Savings Potential	$ $ $ $
Rate of Return, New Facilities	% % % %
Rate of Return, Retrofit	% % % %
Reliability	✓ ✓ ✓
Ease of Retrofit	☺ ☺ ☺

---

effective placard where occupants will notice it immediately, not in a location where it can be hidden by curtains or other objects. See Reference Note 12, Placards, for guidance in designing effective placards.

ECONOMICS

SAVINGS POTENTIAL: *5 to 50 percent of the conditioning energy of the space, depending on the pattern of space usage.*

COST: *$50 to $200, typically, mostly for labor.*

PAYBACK PERIOD: *Less than one year, to several years.*

TRAPS & TRICKS

EQUIPMENT SELECTION AND INSTALLATION: *Select the switch and the installation method to be rugged enough to survive the environment. Some people become aggressive if you try to keep them from heating or cooling the entire outdoors. Don't forget the placard in each space, or you will get a lot of calls complaining that the air conditioning is not working.*

MONITOR PERFORMANCE: *Schedule periodic checks of the equipment, at intervals that correspond to the likelihood of damage. If you observe frequent attempts to sabotage the switches, consider a different type of switch or a different approach to controlling the equipment.*

WESINC

Fig. 1 These windows need switches to turn off the conditioning   This luxury hotel is located in a climate that encourages guests to open the windows, and the ability to do so is an attraction of the hotel. The rooms are conditioned to maintain a precise temperature, so opening the windows can waste a large amount of energy.

Radiators and Convectors

The Measures in this Subsection improve the efficiency of heating with radiators and convectors. These are simple heating units in which heat is distributed solely by radiation and/or convection resulting from the surface temperature of the unit. They do not use fans, induction, or other means of distributing heat. The heating surfaces may be exposed or enclosed.

The distinction between radiators and convectors relates to their different modes of heat emission, as their names imply. Radiators have large, hot surfaces that are exposed to occupied areas of the space. This allows them to transmit a large fraction of their heat output by radiation. Radiators also release a large amount of heat by convection.

Convectors release heat primarily by convection. That is, they heat the surrounding air with fin coils, and the warm air circulates through the space by convection. Convectors typically consist of long horizontal heating elements that have sheetmetal guards. The surface temperature of the coils is too low for strong radiation of heat, and the guards block most radiation that does occur. The heating coils may be enclosed at the bottom of a tall enclosure. In this case, the enclosure acts as a chimney, increasing heat output somewhat.

The terminology is sloppy. For example, the term "baseboard radiator" is often used for convectors. A variety of names are given to different models of radiators and convectors. The distinction between radiators and convectors may be important to energy efficiency. Where this difference matters, it is discussed in the Measures.

Radiators and convectors operate with an efficiency of almost 100%, which means that virtually all energy absorbed by the units is delivered to the space as heat. Losses occur outside the space in the steam, hot water, or electrical systems that provide the energy.

Measure 5.2.1 recommends the appropriate maintenance for these units. Measure 5.2.2 recommends installing thermostatic radiator valves on units with manual valves, a "must do" improvement. Measures 5.2.3 through 5.2.6 recommend improvements to enclosures. Measures 5.2.7 through 5.2.10 recommend major, expensive system modifications.

INDEX OF MEASURES

5.2.1 Clean and repair radiators and convectors at appropriate intervals.

5.2.2 Install thermostatic control valves on units with manual valves.

5.2.3 On units with manual control valves, provide easy and safe access to the valves.

5.2.4 Ensure that convection and radiation are not obstructed by enclosures or other objects.

5.2.5 Avoid trapping of heat output against exterior walls.

5.2.6 Thermally isolate radiators and convectors from poorly insulated walls.

5.2.7 With steam heating, install a vacuum condensate system to improve temperature control.

5.2.8 Convert steam heating to hydronic heating.

5.2.9 Replace electric resistance convectors with heating units having lower energy cost.

5.2.10 Provide separate thermostatic control for each area with distinct heating requirements.

MEASURE 5.2.1 Clean and repair radiators and convectors at appropriate intervals.

The high efficiency of radiators and convectors does not depend on maintenance. However, inadequate maintenance may reduce the capacity of the units and it may cause conditions that motivate occupants to operate the equipment in a wasteful manner. To avoid these problems, clean or repair these components at regular intervals:

- *fins and coils.* Convectors have finned tubes that collect debris. Even cast iron radiators collect dust and dirt, especially when enclosed. The debris reduces capacity. At least once a year, arrange for your cleaning crews to suck the debris out the units with their vacuum cleaners. Supervise them when they do this.

- *manual control valves,* which are found mainly on older steam radiators. The chronic problem with these units is erosion of the valve disks and seats, which occurs rapidly when the valve is only slightly open. Typically, manual radiator valves should be overhauled every year.

  Failure of the control valves wastes energy by preventing the radiators from being turned off. This leads occupants to control temperature by leaving windows open, which vastly increases heating requirements.

- *thermostatic control valves* also develop leakage by erosion of the valve disks and seats, and they waste energy in the same way. In steam radiators, expect to overhaul the valves annually. In hydronic equipment, check valve closure at the beginning of the heating season.

- *steam traps* on steam radiators fail eventually and often. A steam trap is an automatic valve that prevents steam from blowing straight through the radiator. If the trap works properly, it passes only condensate. But, if the trap leaks, it can waste a large amount of energy by short-circuiting steam into the condensate system. If the trap sticks closed, the radiator will not function. If the trap sticks in a partially open position, the radiator will have reduced capacity and will not control properly. Inspect all your radiator steam traps annually, at the start of the heating season.

- *air eliminators* are used with steam radiators to vent air that accumulates inside the units. They are automatic devices similar in construction to steam traps. If they stick open, they waste steam. If they stick closed, the radiator fills with air, loses capacity, and becomes uncontrollable.

- *damper controls* are used to modulate output in some older types of convectors. The damper may

SUMMARY

Basic maintenance items needed to prevent energy waste and maintain comfort.

SELECTION SCORECARD

Savings Potential	$ $ $ $
Rate of Return	% % % %
Reliability	✓ ✓
Ease of Initiation	☺ ☺

be controlled by a manual knob, a bead chain, or a bimetallic element. Because of inexpensive construction, the dampers tend to jam. This tendency is aggravated by dirt and trash that falls into the cabinet. Maintenance consists of simple repairs, lubrication, and cleaning.

Differences Between Types

Electric units are easy to maintain, requiring only occasional coil cleaning and checking of thermostats.

Hot water radiators also require little maintenance. Be sure to check for control valve leakage. If the hot water supply system is not piped properly, air can accumulate in some heating units, mysteriously reducing capacity.

Diagnosing problems in steam radiators requires special skills related to steam equipment. Control valve operation is tricky for the reasons explained in Measure 5.2.7. Steam traps have a large amount of lore that is covered in Measures 1.10.3 and 1.10.4 ff.

ECONOMICS

SAVINGS POTENTIAL: *20 to 50 percent of heating energy, compared to systems in which maintenance is neglected.*

COST: *Usually modest.*

PAYBACK PERIOD: *Immediate.*

TRAPS & TRICKS

SKILLS AND TRAINING: *Electric and hydronic units require only conventional skills to check. Steam radiators require special knowledge that is disappearing as steam space heating fades into history. If you have steam radiators, study Measures 5.2.7, 1.10.3, and 1.10.4. Then, train the maintenance staff. Explain the diagnosis and repair procedures in the plant operating manual.*

MEASURE **5.2.2 Install thermostatic control valves on units with manual valves.**

Manual control valves on older radiators and convectors, such as those in Figures 1 and 2, can cause serious energy waste. The waste occurs because manual valves require continual adjustment as conditions change. Furthermore, the heat output of steam radiators tends to be unstable at low output, for the reasons explained in Measure 5.2.7. Frequent manipulation of the valves accelerates wear and leakage.

To compound these problems, manual valves are typically awkward and uncomfortable to operate, which makes occupants reluctant to grapple with them. As a result, occupants resort to leaving the radiator valves fully open and opening windows to vent the excess heat. Thus, the window becomes the control device.

You can solve these problems by installing thermostatic valves in place of the manual valves. The thermostatic valves follow the fluctuations in space load and radiator output, keeping space temperature stable.

Some types of thermostatic valves also provide the ability to turn off heat or lower space temperature automatically during unoccupied periods. Don't overlook this capability. You can't depend on people to turn off radiators manually. If radiators are turned off, they require a long time to get a cold space warmed up.

Energy Saving Potential

If occupants use the windows to control the temperature in the space, heating the space may require several times more energy than if the windows are kept closed. Installing thermostatic valves can eliminate all or most of this waste.

If you install thermostatic valves that allow shutdown or setback of the heating equipment during unoccupied hours, the saving in unnecessary conditioning can be major. See Measures 5.1.3 ff for details.

Types of Thermostatic Radiator Valves

You have a choice of three general types of thermostatic radiator valves:

SUMMARY

Eliminates the main source of energy waste and user dissatisfaction with manually controlled steam and hydronic heating units.

SELECTION SCORECARD

Savings Potential .... $ $ $ $

Rate of Return .... % % %

Reliability .... ✓ ✓ ✓

Ease of Retrofit .... ☺ ☺ ☺

• *remotely powered valves* may be pneumatic or electric. They are similar to valves used in other HVAC equipment. They require a source of compressed air or electricity, respectively, at each heating unit. This type provides the best temperature sensitivity and the quickest response. The thermostats can be installed anywhere in the space. The pneumatic or electric connections between the thermostats and the heating units can be hidden in the walls like other wiring. This is the only option that allows you to turn the radiators on and off, or to adjust the space temperatures, using a central control system. This option is the most expensive by far.

• *valves with internal self-powered thermostats* are the other extreme. They require no external control or power connections. What makes them work is a thermally expanding fluid that moves the valve disk. The temperature is set by turning a knob at the valve body. Figure 3 shows the insides of a typical self-powered thermostat.

Installation is easy, consisting of removing the manual valve and replacing it with the thermostatic unit. Self-powered valves are available as direct replacements for the manual valves on both hydronic and steam heating units. In the case of steam units, they are available for both two-pipe and one-pipe systems. This option is the least expensive.

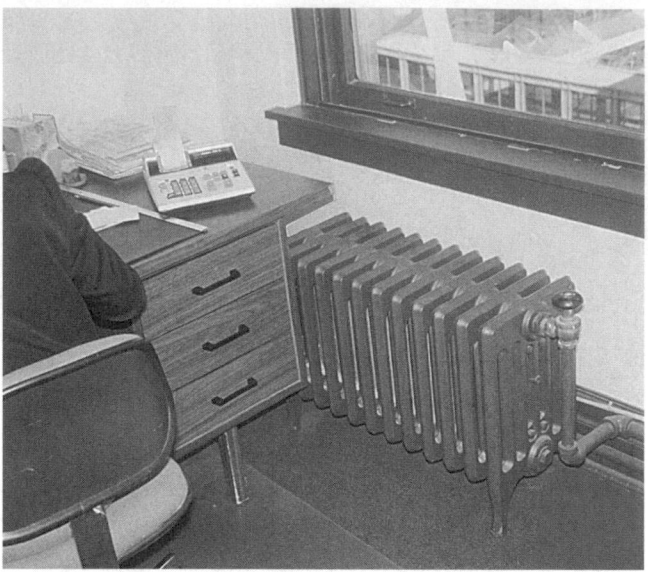

WESINC

Fig. 1 Radiator with manual control valve This radiator can be retrofitted easily with a self-contained thermostatic valve. The high position of the valve and the length of hot steam pipe under the valve make it advisable to install a valve that has a remote temperature sensor.

The major weakness of this type is the location of the thermostat, which is inside the valve. In this location, it is heavily influenced by the heat from the radiator, so it may respond poorly to the temperature of the space. Also, if the original manual valve was awkward or uncomfortable to reach, the same will be true of the thermostatic valve. This type of control has no provision for external control connections to provide automatic shutdown or temperature setback.

• *valves with external self-powered thermostats* operate on the same principle as the previous type. Figure 4 shows a typical unit. The difference is that the space thermostat is mounted remotely. The thermostat is connected to the valve on the radiator through a capillary. This type can provide good temperature control, and it is easy to use. The capillary has limited length, which may limit the thermostat locations. Also, the capillary is more difficult to get out of sight than electrical wiring or pneumatic tubing, because it cannot be snaked through walls. Retrofit is as simple as with self-contained units, and the cost is not much higher.

Provide for Turning Off Units During Unoccupied Periods

While you are planning the modification, provide for turning off the heat or lowering the space temperature when the space is unoccupied.

Measures 5.1.3 ff recommend automatic methods of turning off or reducing the heat in individual spaces.

These methods require externally controlled (pneumatic or electronic) valves.

If you use self-contained thermostatic valves, you will have to turn off the heat at the boiler, or at some point in the distribution system. For example, you could install a solenoid valve in the steam line to each radiator. However, the combined cost of the self-contained thermostatic valve and the solenoid valve, and its connections, would be as high as the cost of externally controlled electric or pneumatic control valves.

If you plan to turn off equipment during unoccupied hours, provide for overrides in individual spaces. See Reference Note 10, Clock Controls, about overrides.

Thermostat Location

If the valve has a remote thermostat, install it where it will sense only the space air temperature. This means, install it to the side of the radiator or convector, so that it is not blanketed by the warm air rising from the unit. Do not locate thermostats near hot or cold objects, such as exterior walls, coffee urns, etc. Install them where they will not be in direct sunlight coming through windows or skylights at any time of day. If necessary, install a simple sun shield around the thermostat.

Location is not an option with self-contained thermostatic valves that are mounted on the radiator. These units seem to work better than one has any reason to expect, but temperature control cannot be as efficient as with a remotely located thermostat.

WESINC

Fig. 2 Group of radiators For good space temperature control, all these radiators could be controlled collectively by installing conventional control valves and a single space thermostat.

Don't Forget the Placards!

See Measure 5.1.1.

Plan for Regular Maintenance

Maintenance is just as important for thermostatic valves as it is for manual valves. Automatic radiator valves do not avoid the burden of valve maintenance. The most frequent maintenance problem for both manual and automatic types is erosion of the valve seats. Both types require repair at similar intervals. See Measure 5.2.1 for the details.

Consider Your Alternatives

This Measure is by far the least expensive method of providing good control. In the case of steam systems, you may want to further improve the characteristics of the system by installing a vacuum condensate system, as recommended by Measure 5.2.7.

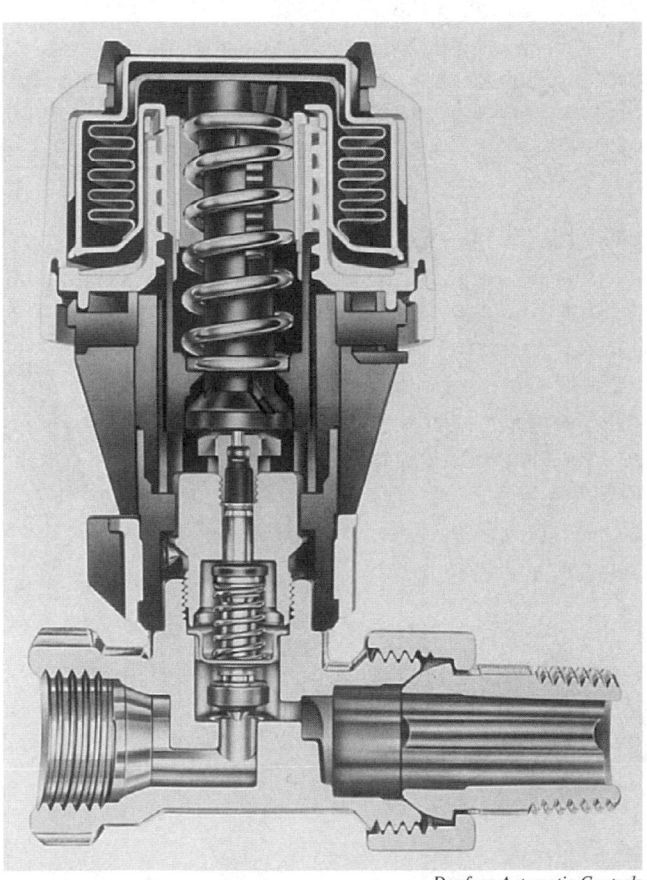

Danfoss Automatic Controls

Fig. 3 Self-contained radiator thermostat A liquid in the bellows assembly expands and contracts in response to space temperature. This motion is transmitted to a valve disk. Turning the knob changes the volume of the fluid, changing the temperature setting. The valve disk wears out and requires periodic maintenance, as with manual valves.

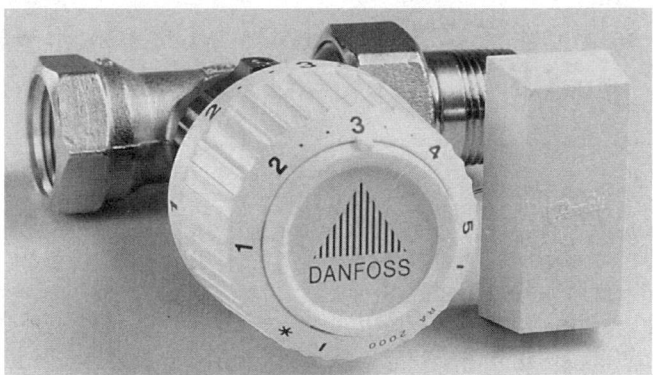

Danfoss Automatic Controls

Fig. 4 Self-contained radiator thermostat with remote temperature sensor The temperature is set by turning the knob, which is attached to the valve. However, the small rectangular sensor can be installed at some distance from the valve, so it avoids sensing the heat of the radiator itself. The sensor is connected to the valve through a capillary tube.

If you are planning a major renovation of the facility, consider converting a steam heating system to a hydronic system. Measure 5.2.8 recommends this approach, and explains the reasons why.

ECONOMICS

SAVINGS POTENTIAL: *20 to 80 percent of the energy used to heat the space, if the occupants are using wasteful practices to control temperature.*

COST: *Typically $100 to $200 per radiator, for valves with self contained thermostats, depending on the difficulty of installation. Up to $500 per radiator, with external thermostats.*

PAYBACK PERIOD: *One year to several years.*

TRAPS & TRICKS

SELECTING THE EQUIPMENT: *Your big decision is whether to install self-contained or externally powered units. The latter are more expensive, but they make it easy to shut down heating in individual spaces. If you decide to use self-contained units, shop the market thoroughly, as performance varies widely between models. Talk with others who have used the controls for at least several years. Select a model that uses a remotely mounted thermostat.*

EXPLAIN IT: *People who are accustomed to manual control valves on steam radiators need to learn the virtues of thermostatic control. Make sure that occupants understand that they do not have to leave the windows open all year. Tell them how to set the thermostats. Install effective placards on the thermostats, as recommended by Measure 5.1.1.*

MEASURE **5.2.3 On units with manual control valves, provide easy and safe access to the valves.**

RATINGS

New Facilities   Retrofit   O&M

C

Manual radiator valves are inherently a source of energy waste. You should get rid of them, as recommended by Measure 5.2.2. Until you do that, make the manual valves as easy to manipulate as possible.

The usual problem is that it is difficult or dangerous for occupants to reach radiator valves. The surface temperature of a steam radiator may exceed the boiling temperature of water, and occupants risk burns by trying to manipulate a control valve that is not out in the open. The problem is worsened by enclosures that have been added for the sake of appearance, which usually make the valves much more difficult to reach.

Improve Access or Add a Valve Extension

You have to work out solutions on an individual basis. There are two basic approaches: modify the enclosure to provide easy access to the valve, or add an extension to the valve stem to bring the handle outside the enclosure.

You probably will not be able to modify the enclosure to make it easy to reach the original valve handle. Even if you could, keeping the valve handle inside the enclosure makes it too hot for comfort.

The better solution in most cases is to install an accessible handle on an extension attached to the stem of the control valve. Get the handle out in free air where it can stay cool. This may not be pretty, but that is the price that must be paid for good control and safety.

If you add an extension to the valve stem, include a flexible joint at the valve end. You can buy inexpensive universal joints for this purpose from industrial supply houses. If you do not add a flexible joint, the valve stem and packing will be subjected to great stress, which will damage the valve and cause leakage.

SUMMARY

An improvement that may be necessary to allow people to use manual radiator valves efficiently.

SELECTION SCORECARD

Savings Potential .... $ $ $ $

Rate of Return .... % % %

Reliability .... ✓ ✓

Ease of Retrofit .... ☺ ☺

Related Enclosure Modifications

The same enclosures that cause difficulty with valve access often create other problems. Coordinate this Measure with Measures 5.2.4 through 5.2.6, which recommend improvements to radiator enclosures for other reasons.

Don't Forget the Placards!

Instruction placards are important with any manual control. Refer to Measure 5.1.1.

ECONOMICS

SAVINGS POTENTIAL: *10 to 50 percent of heating energy, if the occupants are opening windows to control temperature.*

COST: *$20 to $50 per unit, typically.*

PAYBACK PERIOD: *Less than one year.*

TRAPS & TRICKS

CHOICE OF METHOD: *Are you sure you want to do it this way? As long as you are going to all the trouble of reaching the valve, why not install a thermostatic valve? Go back to Measure 5.2.2 and think it over.*

MEASURE **5.2.4 Ensure that convection and radiation are not obstructed by enclosures or other objects.**

C

SUMMARY

Correcting a problem that is common in renovated buildings.

SELECTION SCORECARD

Savings Potential .... **$ $**

Rate of Return .... **% %**

Reliability .... ✓ ✓ ✓

Ease of Retrofit .... ☺ ☺

The heat flow from radiators and convectors may be obstructed by enclosures that were not originally intended. Figure 1 shows an example. This causes a loss of heating capacity, and in some cases, it increases heat loss through the building envelope. The loss of capacity may waste energy indirectly, such as overheating the remainder of the facility, causing occupants to turn on excess heating units, etc.

Heat flow may also be obstructed by objects that are placed around radiators and convectors. The problem may be obstruction of the heat output of the units or obstruction of the air flow returning to the units. Figure 2 shows both problems.

Recognize the difference between radiators and convectors when dealing with obstructed heat flow. As the name implies, convectors are designed to distribute heat almost solely by convection. On the other hand, radiators are designed to distribute heat both by radiation and by convection. Radiators have the ability to project heat far into a space, even into areas where convective

heat transfer does not occur. For example, radiators are installed high on the walls of some industrial buildings to project heat into interior space, where the heat is needed to offset losses through the roof and floor slab.

Loss of Convective Heat Output

Convection is a relatively weak force that is easily obstructed. A problem may occur when old style

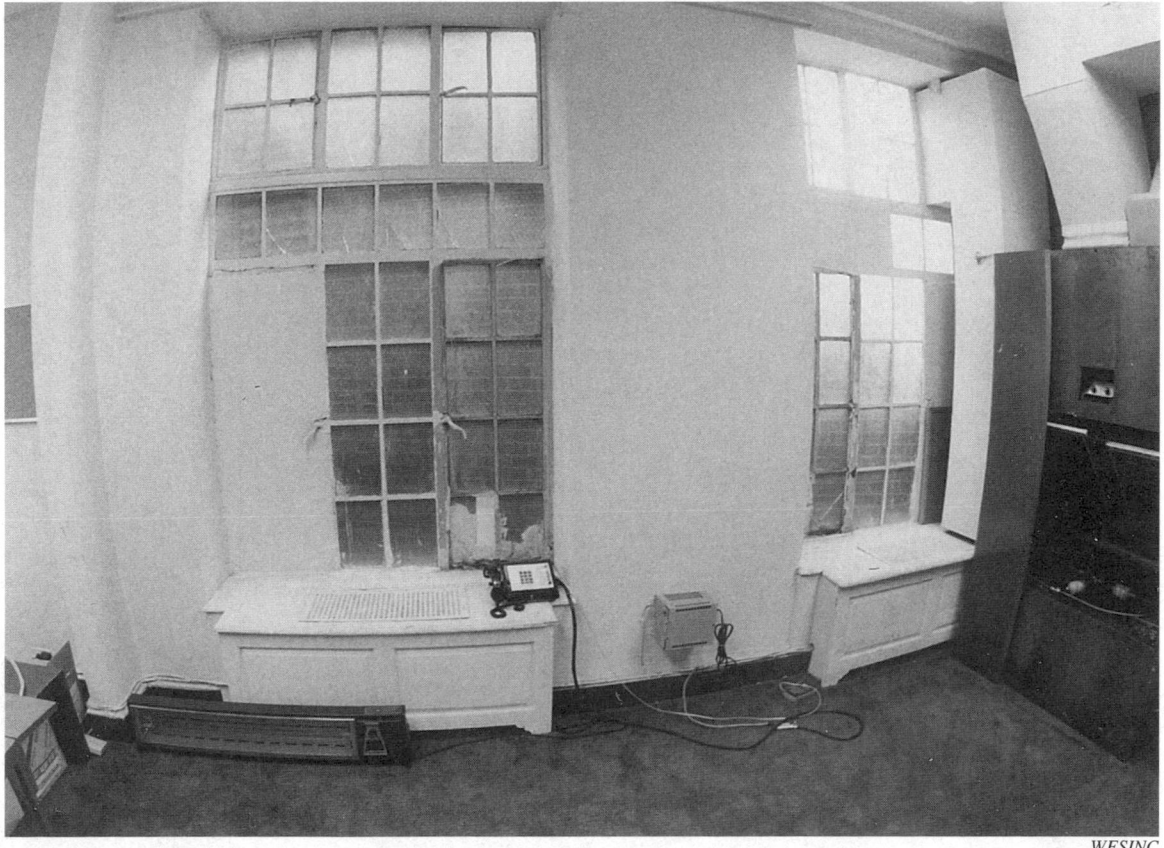

Fig. 1 Tightly enclosed radiators The mild force of convection is unable to overcome the resistance created by these decorative enclosures. Note the electric heater in front of the unit on the left. This building, located in Chicago, suffers great heat loss from the large, leaky, single-glazed windows. Upgrading the windows, and perhaps reducing their size, should be the first step in solving this problem.

radiators and convectors are enclosed to provide a more modern appearance or to keep people from being burned by the high surface temperatures. In many decorative enclosures, the convective discharge path is adequate, but a return path is lacking. In such cases, the lack of heating capacity may be puzzling.

The proper corrective action is to provide openings of ample area for both the outlet and inlet air flow. This is usually easy to accomplish.

Loss of Radiation Heat Output

Radiators become convectors if they are enclosed or obstructed. For example, if old cast iron radiators are enclosed, they lose all radiation ability. In many such cases, this does no harm. Convective heating may be satisfactory for the space, especially where it is needed mainly to offset heat loss through the walls where the radiators are installed. However, there are situations where the loss of radiation into the space causes a problem.

Heat radiation follows line of sight, so it can be blocked by almost any type of obstruction. For example,

radiation is obstructed by furniture or pallets of merchandise. Even glass stops heat radiation. (Heat radiation from radiators has much longer wavelength than the heat in sunlight, and it is almost totally absorbed by glass.)

If the radiation mode of heat output is required, the only corrective measure is to remove the enclosure. In some of these rare cases, you may have to relocate the radiators themselves or substitute a different type of heating unit.

Related Enclosure Modifications

Coordinate this modification with Measures 5.2.3, 5.2.5, and 5.2.6, as appropriate. Those Measures correct other problems related to heating unit enclosures.

ECONOMICS

SAVINGS POTENTIAL: *Up to 20 percent of the energy consumed by the affected heating units. In some cases, this activity may save a large amount of money by eliminating an apparent need for additional heating capacity.*

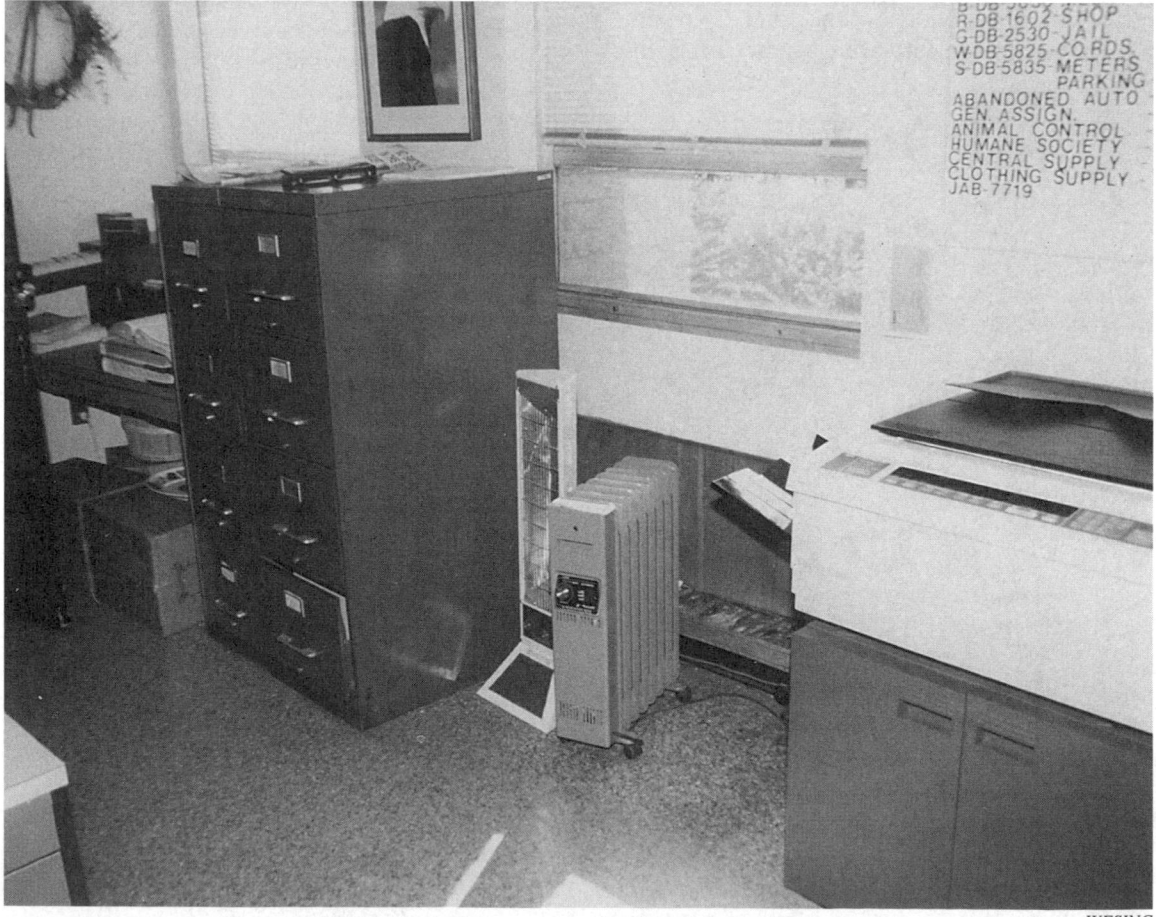

WESINC

Fig. 2 Obstructed convectors Air return to the convectors is blocked by the file cabinets and copy machine. Discharge from the convectors is obstructed by the table at left and by an accumulation of debris. The heat that does escape from the convectors is first confined against the uninsulated exterior wall, through which a fraction is lost. Note the electric heaters that compensate for these problems.

TRAPS & TRICKS

ANALYZE THE PROBLEM: *Before starting a lot of cabinetry work, be sure that the lack of heating is caused by the enclosure. Heating problems in steam radiators can be tricky to diagnose. However, if you see that an air return path is missing, create one.*

MEASURE 5.2.5 Avoid trapping heat output against exterior walls.

RATINGS

New Facilities	Retrofit	O&M
A	**C**	

Serious energy waste can occur if draperies or other window treatments are installed so they trap the heat output of the radiator or convector against an exterior wall. This is a common problem that occurs because HVAC designers and interior decorators do not communicate with each other.

The fact that the drapery or blind may be open at the top does not avoid the problem. Energy waste results from the greater temperature against the inner surface of the wall. This increase in temperature occurs near the heating unit even if air can flow freely behind the drapery.

The problem is usually caused by draperies, but other types of obstructions can cause the same problem. For example, in warehouses, tightly stacked goods may trap the heat from radiators against the wall.

SUMMARY

A common problem that may waste an unexpectedly large amount of energy. The worst cases involve draperies and large windows.

SELECTION SCORECARD

Savings Potential	**$**	$	$
Rate of Return, New Facilities	**%**	**%**	**%**	**%**
Rate of Return, Retrofit..........	**%**	%	%
Reliability	✓	✓	✓	✓
Ease of Retrofit	☺	☺	☺

Energy Saving Potential

The amount of heat loss added by trapping depends on the thermal resistance of the surface, the increase in temperature differential, the surface area affected, and the extent of heat trapping.

The thermal resistance of the wall is the dominant factor. Glass walls are the worst, with typical R-values ranging from 0.8 (single glazing) to 1.8 (double glazing with fancy coatings). At the other extreme, well insulated walls in newer commercial buildings have R-values between 10 and 20, but such buildings are exceptional. A typical uninsulated masonry wall with a furred interior surface has an R-value of about 4.

To illustrate the effect of the increased temperature differential, consider a situation where a heavy drapery hangs over a baseboard convector. If the outside temperature is 40°F and the room temperature is 70°F, the temperature of the space between the drapery and the wall could be 100°F. The greater temperature differential doubles the heat loss through the covered portion of the wall.

Also, the insulation value of the drapery ceases to be an asset and becomes a liability. A typical uninsulated masonry wall has an R-value of about 4, whereas a heavy drapery and its air layers has an R-value of about 1.0. Thus, about 20% of the insulation value is lost. In addition, the drapery now retards heat from entering the space.

Reroute the Heated Air

The solution is to relocate the drapery and/or the heating unit so that the heated air does not flow between the drapery and the wall. Or, install a fixture that deflects the output of the heating unit away from the drapery.

If the drapery must hang over the convector or radiator, consider installing a sill or deflector to divert the rising heated air toward the occupied side of the drapery. Figure 1 shows a simple device that is inexpensive, easy to install, reasonably attractive, and

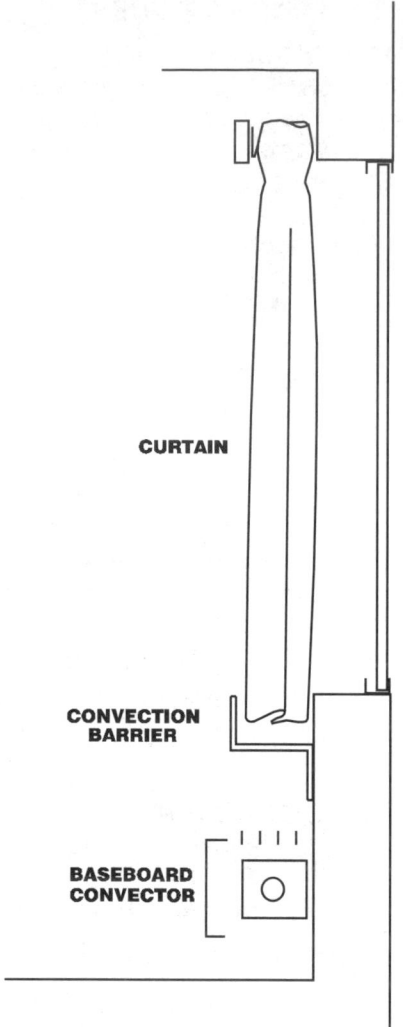

CURTAIN

CONVECTION
BARRIER

BASEBOARD
CONVECTOR

Fig. 1 Convection barrier This simple, inexpensive fixture eliminates the large heat loss that occurs if the heat output of the baseboard heater is able to rise between the curtain and the window.

completely effective. If the drapery goes all the way to the floor, you will have to shorten it.

If the drapery hangs over a window sill, install a retainer to keep the bottom of the drapery entirely over the sill. Go to a plastic fabrication shop and have them make attractive retainers from transparent plastic.

Related Enclosure Modifications

Coordinate this activity with Measures 5.2.3, 5.2.4, and 5.2.6, as appropriate. Those Measures correct other problems related to heating unit enclosures.

ECONOMICS

SAVINGS POTENTIAL: *3 to 50 percent of heating energy, depending on the configuration and on the thermal resistance of the wall.*

COST: *In new construction, this usually costs little, if anything. In retrofit applications, the cost varies greatly. It becomes expensive if window treatments have to be modified.*

PAYBACK PERIOD: *Less than one year, to many years.*

TRAPS & TRICKS

IMAGINATION: *This is an opportunity for creativity. Look for the simple solution that causes no new problems. For example, it is more satisfying to solve a problem with a simple deflector than to spend a fortune tearing out an expensive set of draperies and replacing them with something else.*

MEASURE 5.2.6 Thermally isolate radiators and convectors from poorly insulated walls.

Infrared photographs of the exteriors of buildings often show bright bands of heat loss through the portions of the walls adjacent to radiators or convectors. These localized areas of heavy heat loss are caused by the increased temperature differential that exists across the walls at these points. The increased heat loss is especially severe with radiators, and less severe with convectors.

It may be economical to reduce the heat loss in the vicinity of the heating units, even if it is impractical to improve the insulation of the entire wall. This Measure offers simple techniques for reducing wall heat loss in the vicinity of radiators and convectors.

Applications and Cost Saving Potential

This Measure is most valuable in buildings that do not have effective wall insulation adjacent to the radiators or convectors. A large fraction of commercial and industrial buildings fit this description. For an example, see Figure 1.

Most older buildings with masonry walls do not have wall insulation. Modern buildings with glass walls suffer from the high thermal conductivity of glass. Many modern buildings with lightweight curtain walls have little or no insulation.

As an example of the energy waste that occurs from this localized lack of insulation, consider an area of wall three feet wide and three feet high that is adjacent to a

SUMMARY

A small improvement that may be worthwhile in a facility with many radiators or convectors.

SELECTION SCORECARD

Savings Potential .... $ $

Rate of Return .... % %

Reliability .... ✓ ✓ ✓ ✓

Ease of Retrofit .... ☺ ☺

steam radiator. If the wall has an R-value of 4, and if the temperature differential across the wall is increased by 60°F during a heating season that lasts 5,000 hours, the extra heat loss is about 700,000 BTU per year. If the cost of heat delivered by the radiator is $4 per million BTU, the annual heat loss costs about $3. This is not a large amount, but in a facility that has several hundred radiators, it may be worthwhile to reduce this heat loss.

The increase in heat loss is lower with convectors than with radiators, and hence the potential saving is less. Convectors transfer less heat to the wall by radiation because they operate at lower temperature. They have fins that dissipate heat effectively to the surrounding air. The amount of wall heating may be strongly affected by whether a convector has a sheetmetal enclosure or backing plate. A metal surface between the convector element and the wall reflects heat radiation away from the wall. The design of the enclosure may also tend to keep the rising warm air current away from the wall.

Estimating the Extent of the Problem

You can calculate the amount of heat loss, as in the example above, provided that you know the characteristics of the wall and the interior surface temperature. You can determine the wall construction from the architectural drawings. Measure the interior surface temperature adjacent to the heating unit by holding a thermometer against the wall under a wad of insulating material. (Do this quickly, before the accumulated heat dissipates through the wall.) Or, use a remote reading (infrared) thermometer, if you have one.

An infrared scan of the building exterior can reveal whether there is aggravated heat loss near the heating units. Infrared thermographs are visually dramatic, but they do not provide an accurate indication of the rate of heat loss. See Reference Note 15, Infrared Thermal Scanning, for more about this diagnostic method.

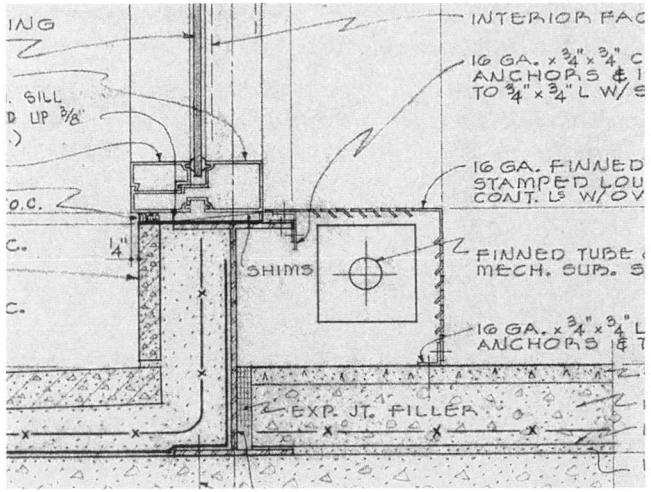

Fig. 1 Typical wall construction adjacent to convector There is no insulation between this baseboard and the short wall section. The uninsulated solid wall has little thermal resistance itself. This would have been easy to correct when the building was being designed, but it would be expensive to fix now. The single-glazed window also illustrates the architect's disregard for energy costs.

How to Reduce Heat Loss

Consider these methods of reducing wall heating by radiators and convectors:

- *insert insulation between the heating unit and the wall*
- *insert heat reflecting material between the heating unit and the wall*
- *move the heating unit away from the wall.* This eliminates conductive heat loss, and allows a convective air layer to form as a barrier between the heated air and the wall.

The first two methods are often inexpensive, but the third is usually too expensive to be practical. The easiest approach with freestanding radiators is to insert a board of rigid insulation behind the radiators. This insulation can have a foil face to provide heat reflection.

Select insulation that is safe for interior use and that will not break down from high temperature. This excludes all or most organic foam insulations. A good choice is glass or mineral fiber insulation, in the form of blankets or rigid boards. Unfortunately, the safest types of insulation are the bulkiest, so you usually cannot get much R-value installed. However, if the wall is uninsulated, even a modest amount of insulation may prevent most of the heat loss adjacent to the heating units. (See Reference Note 43, Insulation Selection, to find the best choices of insulation for this application.)

With enclosed units that are attached to the wall, you may be able to move the unit away from the wall and insert insulation behind it. On the other hand, if the unit has rigid piping connections, this approach is probably not economical.

The high surface temperature of radiators makes it possible to reduce heat loss to walls by installing heat (infrared) reflectors between them and the walls. Any polished metal plate or film is effective for this purpose, even aluminum foil. The performance of heat reflectors degrades seriously if they become dirty. In most environments, it is not practical to keep heat reflectors clean enough to keep them very effective. If the heating unit has a sheetmetal enclosure or backing plate, that metal surface serves to reflect heat as well as any other surface in the long term.

Another limitation of heat reflectors is that they require an air gap in front of the reflective surface. Thus, it does no go good to add a foil face to insulation that is butted against the back of a radiator enclosure.

Alternative: Wall Insulation

A more complete alternative to this Measure is improving the overall wall insulation. This is covered by the Measures in Subsection 7.2. In retrofit applications, improving overall wall insulation is not often economical unless the building is going to be gutted for a major renovation.

Related Enclosure Modifications

Coordinate this activity with Measures 5.2.3 through 5.2.5, as appropriate. Those Measures correct other problems related to heating unit enclosures.

ECONOMICS

SAVINGS POTENTIAL: *1 to 10 percent of heating cost. The greatest savings occur with freestanding radiators, and the lowest with convectors.*

COST: *Typically $5 to $20 per unit. The cost may be much higher if it is difficult to install the insulation.*

PAYBACK PERIOD: *Several years, or longer.*

TRAPS & TRICKS

ANALYZING THE PROBLEM: *This is not a big deal. While it is annoying to see inefficiently installed perimeter heating equipment, it may not be economical to improve it. Do the simple calculations first.*

MATERIALS: *Avoid the temptation to use foam board insulation. It is easy to use, but it creates toxic smoke in a fire.*

MEASURE 5.2.7 With steam heating, install a vacuum condensate system to improve temperature control.

SUMMARY

The only way to improve the inherently erratic control of steam heating units. Not cheap and not a panacea, but less expensive than system conversion.

SELECTION SCORECARD

Savings Potential	$ $ $
Rate of Return, New Facilities	% % %
Rate of Return, Retrofit	% %
Reliability	✓ ✓
Ease of Retrofit	☺ ☺ ☺

One reason why people tend to open windows that are located over steam radiators is that the heat output of a steam radiator is inherently erratic at low loads. Opening the window, and turning up the radiator to make up the heat loss, makes control of the radiator more stable. Also, the window is usually easier to manipulate than the radiator valve.

The erratic control of steam radiators relates to the properties of steam. Installing a vacuum condensate system allows radiators to operate in a stable manner at much lower heat output.

Other Ways to Improve Steam Heating Control

Adding a vacuum condensate system is expensive, especially in retrofit, and it does not solve other problems of manual radiator control. Installing thermostatic control valves (Measure 5.2.2) may provide adequate temperature control because they can follow the erratic behavior of radiators. If your steam radiators do not have thermostatic valves, install them before doing anything else. Then, decide whether to install a vacuum condensate system.

An alternative that is even more expensive is converting from steam to some other heat source. In retrofit applications, the most likely candidate is conversion to hydronic heating, which is recommended by Measure 5.2.8, next.

Energy Saving Potential

This Measure provides savings primarily by reducing wasteful occupant behavior in controlling steam radiators using manual valves. The savings may be large, but they cannot be calculated with precision. See Measure 5.2.2 for the range of savings.

A major deduction from the saving is the energy required to operate the vacuum pump motor. This motor may be several horsepower or larger. It must operate continuously as long as the heating system is in operation.

Why Steam Radiators are Difficult to Control

The steam in a radiator gives up most of its heat by condensing into water. The pressure at which the steam condenses depends on the temperature of the radiator surface. If the inlet valve is wide open and the radiator is at full output, the steam condenses at the supply pressure. The supply pressures for steam radiators is typically about 10 PSI, which gives a condensing temperature of about 240°F. The condensate collects in

the bottom of the radiator and is forced out through the steam trap by the pressure inside the radiator.

When less heat output is required, the occupant or the thermostatic control valve throttles the steam inlet valve, causing the pressure inside the radiator to fall. As the pressure falls, the condensing temperature falls, reducing heat output. When the temperature inside the radiator falls below the boiling temperature of water (212°F at sea level), the steam pressure inside the radiator falls below atmospheric pressure.

When the pressure inside the radiator is lower than the pressure in the condensate system, the condensate cannot exit freely. This is where stability problems begin. Condensate accumulates in the radiator. It discharges erratically as the gravity pressure of the accumulated water overcomes the pressure difference at the steam trap. When the steam trap opens to discharge condensate, air enters the radiator from the condensate system, flowing backward through the steam trap. You can see a similar situation when you turn a bottle of water upside down. The incoming air, trying to fill the vacuum above the water, interferes with the flow of water through the narrow neck.

The knocking of radiators is the result of air bursting through the pool of condensate inside the radiator. The air pressure combines with the steam pressure to produce an internal pressure that approaches atmospheric. The presence of air effectively reduces the area of the condensing surface that is exposed to steam, so it reduces the heat output. The exposed surface area is also reduced by the volume of water inside the radiator.

The situation is further complicated by the need to have an air vent on the radiator, which gets rid of air at some times and lets it enter at other times. For a given valve setting, the heat output of the radiator is continually altered by the unstable balance of air intake, air ejection, and condensate level. This makes it practically impossible to achieve satisfactory temperature control with the manual control valves that are found on old radiators.

Basics of Vacuum Condensate Systems

As the name implies, a vacuum condensate system is one in which the pressure inside the condensate piping is less than atmospheric. By keeping the pressure in the condensate system lower than the pressure inside the radiators, the condensate can drain freely at lower steam temperatures, eliminating the causes of instability.

The vacuum in the condensate system is created by using a vacuum pump to remove air from the condensate piping. The vacuum pump is located at the condensate receiver. Typically, it is part of an equipment package that includes a small condensate receiver, a separating chamber, and apparatus to minimize the loss of water vapor along with the air that is pumped out. The package may also include condensate or feedwater pumps. Figure 1 shows all these components packaged as a unit.

The vacuum that can be produced by the vacuum pump depends on the temperature of the condensate. The condensate is always cooler than the steam inside the radiator. Condensate is cooled by flashing as it leaves the steam trap, and it is cooled by conduction as it passes

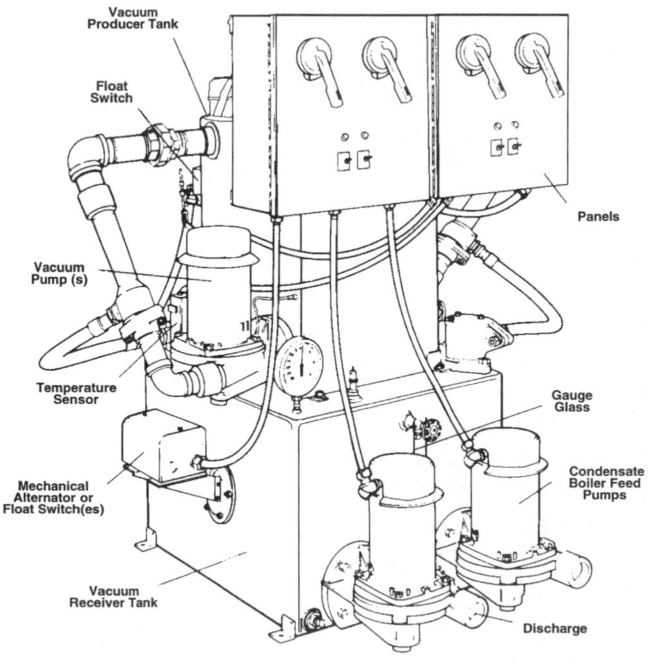

Spirax Sarco, Inc.

Fig. 1 Packaged vacuum condensate receiver This unit contains all the elements of a vacuum condensate system except the pipes. In retrofits, it can easily be substituted for the existing condensate receiver, or it can be installed at a different location in the system.

through the condensate piping. Therefore, the vacuum pump is able to maintain a lower pressure than exists inside the radiator.

The vacuum pump is limited in its ability to reduce condensate system pressure, but it can produce a pressure low enough to maintain stable operation even at greatly reduced radiator output. For example, most vacuum condensate systems can sustain a vacuum of about 9 PSI below atmospheric pressure, which corresponds to a condensing temperature of about 170°F.

Other Advantages of Vacuum Condensate Systems

The oxygen in air causes corrosion of the entire steam system, including radiators, piping, and boilers. A vacuum condensate system is totally closed, so it reduces entry of air into the steam system. Also, the rate of corrosion rises strongly with temperature, so vacuum condensate systems suffer less corrosion because they operate at lower average temperatures.

The air vents in the radiators relieve steam along with air. By reducing operation of the air vents, the vacuum condensate system reduces makeup water requirements. For this reason, a vacuum condensate system reduces water treatment costs.

Other benefits are improved steam distribution at low load and improved condensate return flow at high load. These benefits usually are not significant in retrofit applications, unless there is some problem with the piping that interferes with proper steam distribution.

How to Retrofit Vacuum Condensate

The first step is to survey the entire heating system to decide whether its condition is good enough to justify an investment in new equipment. The condition of the condensate piping is especially important, because a vacuum condensate system cannot tolerate much air leakage into the system. Test the condition of the pipe in the worst portions of the system.

Installing the vacuum equipment is usually straightforward. The receiver is small enough to be brought into the building through an existing entrance. The equipment package is usually installed near the return end of the condensate system, where the original condensate receiver is located. Location is not critical, but the vacuum receiver is heavy and it must be adequately supported.

Expect to Maintain It

If there is a significant amount of air leakage into the system, the vacuum is seriously degraded. This makes it important to maintain the entire condensate piping system to prevent air leaks. Maintaining vacuum condensate systems is covered by Measure 1.8.6.

Explain It!

Unless maintenance personnel are familiar with the operation of a vacuum condensate system, they are unlikely to monitor and maintain it properly. Explain the system clearly in the plant operating manual. Install an effective placard at the condensate receiver. (See Reference Note 12, Placards, for details.) The placard should state the appropriate vacuum gauge readings.

ECONOMICS

SAVINGS POTENTIAL: *10 to 50 percent of heating energy, if the occupants are opening windows to control temperature. The saving is reduced by the energy required to operate the vacuum pump.*

COST: *Vacuum condensate receiver packages for steam heating cost from $5,000 to $20,000, depending on size and features.*

PAYBACK PERIOD: *Several years, or longer.*

TRAPS & TRICKS

PLANNING: *Before deciding to install a vacuum condensate system, verify that the existing condensate system is tight and in good condition. Otherwise, you will not be able to maintain the system vacuum without continual repairs.*

EXPLAIN IT: *The purpose of a vacuum condensate system is not obvious. Make sure that plant operators understand how it works. Describe how the system operates in the plant operating manual. Install a placard at the vacuum pump and receiver that identifies these components and states their purpose.*

MONITOR PERFORMANCE: *Schedule periodic performance checks in your maintenance calendar. Include a column in the plant operating log to record the condensate system vacuum at a gauge that is installed well upstream of the vacuum pump.*

	RATINGS	
New Facilities	Retrofit	O&M
	C	

MEASURE 5.2.8 Convert steam heating to hydronic heating.

For the reasons explained in Measure 5.2.7, low-pressure steam heating systems are inherently unstable in output, making thermostatic control a problem. This problem and others have led to a preference in recent years for heated water (hydronic) heating instead of steam heating. Hydronic systems are simple, reliable, and easy to control.

Existing steam heating systems can be converted to hydronic heating with varying degrees of difficulty and expense. Conversion is a drastic step. It is economically more attractive if the existing steam system is approaching the end of its life. First, consider less expensive methods of improving the performance of the existing steam heating system. The most important of these are installing thermostatic control valves (Measure 5.2.2) and installing a vacuum condensate system (Measure 5.2.7).

Energy Saving Potential

In terms of the efficiency of the equipment itself, steam heating cannot be surpassed, because it requires no parasitic energy (pumps or fans). However, as explained in Measure 5.2.2, the unstable control characteristics of steam heating systems cause wasteful

SUMMARY

Not inherently more efficient than steam heating, but avoids control and comfort problems that tend to waste energy indirectly. Consider as part of a major renovation.

SELECTION SCORECARD

Savings Potential	**$ $** $
Rate of Return	**% %**
Reliability	✓ ✓ ✓ ✓
Ease of Retrofit	☺ ☺

behavior by occupants and maintenance staff. For this reason, getting rid of steam heating can provide savings as a result of more efficient occupant behavior.

This net saving is reduced by the hydronic system's need for pumps. These typically account for a few percent of the energy consumption of the system. If the operation of the pumps is not controlled well, excess pump energy consumption can become a major cause of energy waste.

If fan-powered space heating units are substituted for the present radiators, their fans typically have a rating of 50 to 100 watts per unit. This is small in relation to the amount of conditioning energy being delivered. Still, the system should be designed to minimize fan energy consumption.

Additional Benefits of Hydronic Systems

Hydronic systems are (almost) closed systems, and they operate at lower temperatures than steam systems. Therefore, if a hydronic system is operated and maintained properly, system corrosion is virtually eliminated. This provides long system life and freedom from maintenance.

Hydronic heating units do not have the knocking and whistling that are characteristic of steam radiators. However, fan-powered hydronic terminal units have fan noise.

It is easy to regulate the water temperature of hydronic systems to adapt to the heating load. During mild weather, this eliminates localized overheating, maintains stable control valve operation, and reduces valve wear.

Conversion Issues

The major components of a steam heating system are boilers, steam piping, condensate system (piping, receivers, and pumps), and radiators. You may be able to convert some of the existing steam system components to work with hot water. Your ability to do this is a major factor in economic feasibility. Here are some factors to consider.

■ Boilers

As a practical matter, don't attempt to convert a steam boiler to a hot water boiler, unless you are attracted to the technical challenge of doing it. The practical approach is to use the steam from the present boiler to make the heating water with a heat exchanger. (A heat exchanger used for this purpose is commonly called a "converter.") You can keep the original condensate receiver and condensate (or feedwater) pumps for this purpose. This is a relatively inexpensive and neat solution.

■ Piping

The pressure inside the piping of a hydronic system may be much higher than pressure inside a steam system. The maximum hydronic system pressure is the sum of the pressure due to gravity and the pump discharge pressure. For example, the gravity pressure in the basement of a 5-story building is about 30 PSI. The pump may add another 20 PSI to this to overcome system flow resistance, for a total of 50 PSI. Compare this to a typical maximum steam pressure of 15 PSI.

In most cases, even low-pressure steam piping can handle the higher hydronic pressure adequately, but make a thorough inspection of the existing piping to be sure. Water is much denser than steam, so the original pipe sizing will be adequate for the water flow. This minimizes the pump power that is required.

■ Alternative Heating Units

If you convert from steam heating to hydronic, and you need to get the same heat output at each unit, you will need to replace the radiators with units that are designed to operate with hot water. The usual choices are:

- *baseboard convectors.* These are a favored choice for many applications because they are inexpensive, quiet, dirt-free, and unobtrusive. They use finned tubes, which provide adequate heat output despite their lower operating temperature and lack of radiation heat transfer. Convectors are limited in their ability to emit heat, so you will probably have to replace the localized radiators with long runs of baseboard convectors.
- *fan-coil units,* which use a fan to circulate air. Using a fan and finned coils provides large heat in relation to the size of the unit. Also, the same unit can be used to provide both heating and cooling.
- *unit ventilators,* which are heating fan-coil units that have an outside air intake through the adjacent wall. These are appropriate if adequate outside air intake is not available from other sources. The wall penetrations are an invitation to excessive infiltration, and to freeze problems.

■ Converting Steam Radiators

There are two applications in which you may want to retain the old radiators. One is the renovation of historic buildings. The other is using the original radiators as an item of the decor. You may be able to retain the radiators as heating units for hydronic heating if the building envelope is well insulated. Older buildings that use steam radiators typically lose most of their heat through windows that have high heat loss, both from conduction and air leakage. Installing efficient windows in such buildings may reduce window heat loss to one half or one third of its original amount.

Consider additional steps to reduce envelope heat loss, such as reducing the window area by installing insulating panels to fill part of the window openings. Also, consider improving the wall insulation, although walls usually account for only a small fraction of heat loss in buildings with a large amount of glass area. Refer to Sections 6 and 7 for Measures to reduce envelope heat loss. It is a mistake to spend a lot of money upgrading the heating system without also taking economical steps to reduce heat loss.

With a certain amount of ingenuity, you can convert a steam radiator to operate with hot water. However, steam radiators lack the internal circuiting necessary to extract heat efficiently from water. Also, the heat output from the surface of a radiator increases exponentially as the surface temperature increases. Radiators typically are rated at 1 PSIG steam pressure, which corresponds to about 217°F at sea level. In contrast, hydronic systems typically operate at a temperature no higher than about 200°F. This factor alone represents a significant loss of capacity.

A radiator that is operated with a low surface temperature loses most of its radiating ability, so it tends to behave as a convector. This is acceptable if the unit is intended to offset conductive heat loss through the wall or window where it is mounted. However, it will be a problem if the radiator is intended to emit heat deep into the space, for example, to offset the roof heat loss.

■ Pumps

Since the original steam system does not use pumps, you need to add pumps to circulate the heating water. The pumps typically are small, both in power and physical size. Connecting them into the system is straightforward.

If different portions of the facility require heating on different schedules, consider subdividing the piping system into appropriate zones, with each zone having its own pumps. You can save energy by avoiding the heating of vacant areas, and by minimizing total pump power. The Measures in Subsection 5.1 recommend methods that you can use to control such arrangements.

■ Expansion Tank

The water in a hydronic system expands and contracts as its temperature changes, so a hydronic system requires an expansion tank. If you use an open, vented expansion tank, it must be located at a high point in the system. You can install a pressurized expansion tank at any level. The former approach is foolproof, the latter is not.

Air can enter the water through the expansion tank, if the air and water come into direct contact. The oxygen in the air causes system corrosion. With a vented tank, air dissolves into the water slowly. In a pressurized tank, air dissolves much more quickly. For this reason, a pressurized expansion tank should have an internal membrane to separate the air from the water.

■ Thermostatic Controls

You many be able to salvage the existing thermostatic controls, provided that they are not of the self-contained type. Make this a factor in selecting the type of heating units. It is expensive to replace thermostats and run connections between the thermostats and the heating units.

Install controls at the boiler or heat exchanger to reset the supply water temperature. This eliminates overheating during mild weather, provides stable operation of the throttling valves on the heating units, and extends the life of throttling valves by keeping them from operating in a barely-open position.

ECONOMICS

SAVINGS POTENTIAL: *10 to 50 percent of heating energy, if the existing system causes occupants to open windows to control temperature. Savings are reduced by the energy required to operate the system pumps. If fan-powered space heating units are installed, the fans add a small percentage of energy consumption.*

COST: *Varies widely, but expensive in all cases.*

PAYBACK PERIOD: *Several years or longer. The modification may be desirable largely for comfort and maintenance reasons.*

TRAPS & TRICKS

PLANNING: *The cost of conversion depends on the condition of the existing equipment. To be worth salvaging, the boilers and piping should have many years of useful life remaining. Be brutal in your inspection. Remove lengths of steam and condensate pipe from various parts of the system and cut them apart for inspection. Make a meticulous inspection of your existing steam boilers. If you are considering converting old steam radiators (a nice "deco" touch for some environments), calculate whether they are able to deliver sufficient heat. At the same time, consider improvements to the building envelope. It is far better to spend money to reduce heat loss than to spend the same amount for additional equipment to overcome the heat loss.*

MEASURE **5.2.9 Replace electric resistance convectors with heating units having lower energy cost.**

Convectors with electric resistance heating elements are widely installed because the equipment and the installation are cheap. The penalty paid for this cheapness is high operating cost, unless electricity prices are exceptionally low. In most areas, electric resistance heat is much more expensive than fuel heat, per unit of heat delivered to the space.

If other available energy sources are less expensive than electric resistance heating, consider substituting another heating energy source. There are many options. The choice has a major effect on equipment and space requirements, installation cost, occupant conditions, and other factors. The selection of heating method may also affect your choice of cooling method.

Energy and/or Cost Saving Potential

Making an estimate of the savings in energy and cost is not simple. The historical record shows that the relative prices of different energy sources fluctuate over a wide range. The best long-range estimate is obtained by understanding the underlying relationships in energy prices. This is a complex matter, and it involves great leaps of assumption by even the most knowledgeable experts. See Reference Note 21 for the basics of electricity pricing.

■ If You Convert to Fuel-Fired Heating

If you substitute fuel-fired heating for electric resistance heating, the saving results from the lower unit cost of energy. When making the cost comparison, you must factor in the efficiency of your fuel-burning equipment. Fuel-burning equipment and systems have efficiencies ranging from 55% to 90%. Losses are caused by boiler inefficiency, auxiliary energy requirements (pumps and fans, mostly), and heat leakage from distribution systems. In contrast, electric resistance heating is usually more than 95% efficient, the losses being limited to electrical distribution losses on the customer's side of the electric meter.

■ If You Convert to Heat Pumps

If you substitute heat pumps for electric resistance heating, the saving results primarily from reducing the amount of electricity you need to purchase. The cost of electricity is not entirely proportional to the amount used, as the price of electricity for commercial customers is usually tied to the consumption profile. Heat pumps have a higher ratio of peak consumption to average consumption than electric resistance heating, which raises the average price per kilowatt-hour.

The energy savings provided by heat pumps depend very much on climate and the type of heat pump. Air-

source heat pumps become less efficient as the outside air temperature drops, and at a certain temperature, the heat pump ceases to operate and heating is provided by a resistance heating element. In a relatively warm climate, such as the southern United States, air-source heat pumps may save substantially more than one half of the energy consumed by resistance heating. In cold climates, such as the northern United States, air-source heat pumps save substantially less than half.

A major new market for heat pumps is being opened up by units that use alternative heat sources and heat sinks, including the earth, ground water sources, and bodies of surface water. The anticipated benefit is to increase the potential efficiency of all heat pumps and to make heat pumps useful in much colder climates. See Reference Note 22 for details.

Fuel Availability

If there has not been a shortage recently, people tend to forget that fuel supplies can be cut off. In most locations, electricity is most likely to be available during an energy crisis, while oil or gas may be curtailed. Therefore, if a fuel-fired system is substituted for electric heating, give preference to systems that can burn several fuels. The combination of oil and gas is a good choice because it is unlikely that both fuels will become unavailable at the same time. The ability to burn multiple fuels also makes it possible to exploit changes in relative fuel prices.

Which fuels are available at the site to substitute for electric resistance heating? Oil is available almost everywhere. The major variable is whether natural gas is available at the site. This may not be a simple yes-or-no issue. Gas companies may be willing to install a pipeline where none exists, provided that the anticipated gas consumption is large enough. A request by a large potential customer may motivate a gas company to extend service into the locale to serve other customers as well.

Do not overlook propane as a fuel in locations where its pricing is favorable. Transportation limitations result in propane being available at relatively low cost in the vicinity of refineries. By the same token, the price of propane is less stable.

Alternative Heating Equipment

Eliminating electric resistance heating opens up a large field of possible replacements. The most important criteria are efficiency, fuel cost, installation cost, and comfort features. Here are some key points in selecting the best replacement.

■ Air-to-Air Heat Pumps

Heat pumps provide cooling as well as heating. This is a major factor in deciding whether to make the change, and how to make it. The economic payback of this Measure may be greatly improved if heat pumps also replace cooling units that are defective or near the end of their service lives.

Air-to-air heat pumps can be expensive to retrofit. They require openings through the exterior walls. They are much bulkier than electric resistance units, so rentable space is reduced somewhat. And, they make noise. You need to provide condensate drains for both the indoor and outdoor coils. As explained previously, the efficiency and cost savings of air-source heat pumps diminish seriously in colder climates.

On the positive side, there is no need to run piping or additional wiring to install air-to-air heat pumps. The existing wiring has sufficient capacity because the heat pumps use less electricity than the resistance heaters they replace.

Select heat pumps for high efficiency in both heating and cooling. A large range of efficiencies exists in the market. Measures 5.4.5 and 5.5.3 will help you select heat pumps.

■ Hydronic Heat Pumps

Hydronic heat pumps are similar to air-to-air heat pumps in their efficiency characteristics, but they are similar to fuel-fired systems in their installation requirements. There are two major variations in hydronic heat pumps. In one, each space has an individual heat pump, and heat is transported to or from the space units through water piping. In the other configuration, one or more large heat pumps serve the entire facility, and heat and cooling is transmitted through water piping.

The fact that hydronic heat pumps use water as a heat transfer medium allows a number of efficiency enhancements to be added to the system. These include:

• *cooling towers,* to improve cooling efficiency by exploiting evaporation

• *buried evaporators* ("earth-source heat pumps"), to improve heating effectiveness in cold climates. This feature may substantially improve efficiency, but it has serious practical difficulties. See Reference Note 22, Low-Temperature Heat Sources and Heat Sinks, for details.

• *ground water sources* to improve the efficiency of both cooling and heating. This feature is proven to be effective, but increasingly runs afoul of environmental and water conservation restrictions. See Reference Note 22 for details.

• *heat storage,* which is useful if a facility requires cooling during the daytime but heating at night

• *heat movement* from the core of the building for use in heating the perimeter. This is a difficult configuration to make efficient. See Subsection 5.6 for details.

One disadvantage of hydronic heat pumps is that they require additional energy to operate the pumps. Pump energy consumption tends to get out of hand in some of the more elaborate designs.

Using hydronic heat pumps efficiently requires a good deal of study and analysis for each application. Hydronic heat pumps provide large efficiency gains in some applications, but they provide little benefit in other applications. See Subsection 5.6 for details of these systems.

■ Hot Water Heating

Modern water boilers offer high fuel burning efficiency. Gas-fired boilers are especially efficient, if you select the condensing type. For details, see Reference Note 30, Boiler Types and Ratings, and the Measures related to boiler selection.

Hydronic heating is relatively easy to distribute through small water pipes. In retrofit, the cost and difficulty of installing the pipe may be the largest part of the cost.

Hot water heating is usually done by installing some type of terminal unit in each space, which may be a convector, a fan-coil unit, or a unit ventilator.

You can use a hydronic distribution system to distribute both heating and cooling, provided that the terminal units are suitable.

■ Forced-Air Furnaces

You can use an air distribution system for both heating and cooling. Unlike the other heating methods, forced air heating does not allow good thermostatic control of individual spaces unless you resort to unusual features. Retrofitting forced air distribution may be difficult. It is easiest in relatively open space layouts, such as manufacturing facilities and open-ceiling shopping malls.

■ Individual Fuel-Fired Heaters

Fuel-fired heaters for individual spaces take the form of overhead unit heaters, radiant heaters, freestanding furnaces, and wall-mounted furnaces. Don't assume that fuel-fired space heaters are limited to industrial applications. Particular models may be acceptable in many types of commercial space, including offices. The major installation issues are venting the combustion products, supplying combustion air, and supplying fuel. Subsection 5.7 deals with fuel-fired space conditioning units.

ECONOMICS

SAVINGS POTENTIAL: *If heat pumps are installed, 35 to 65 percent of electrical energy, with a somewhat smaller reduction in electricity cost. If fuel-fired heating is installed, the energy cost saving varies radically with fuel type and locale, and is typically from 30 to 70 percent.*

COST: *Depends on the type of equipment and the difficulty of installation. For example, gas furnaces cost $7 to $20 per MBH capacity, in typical sizes, plus installation. The price premium for condensing furnaces, compared to non-condensing furnaces of average efficiency, ranges from 50% to 80%. Smaller vented gas heaters for individual spaces cost from $15 to $30*

per MBH, plus installation. Small air-to-air heat pumps typically cost from $25 to $50 per MBH, plus installation, but these provide cooling also.

PAYBACK PERIOD: *Several years, or longer.*

TRAPS & TRICKS

GUESSING ENERGY COSTS: *This Measure is motivated entirely by energy cost, since nothing is more efficient than an electric resistance heater, at least when considered within the boundaries of the facility. This Measure takes at least a few years to pay off, and you want it to remain viable, so you must enter the Great Game of predicting relative energy costs and availability far into the future. This takes specialized knowledge, historical perspective, and chutzpah.*

SELECTING THE EQUIPMENT: *All conventional types of heating and cooling equipment are reliable if you select intelligently. However, unconventional equipment, such as earth-source heat pumps, is plagued by reliability problems and misrepresentation. Refer to the Measures or Reference Notes dealing with any unconventional energy sources that you are considering. By the same token, do not dismiss promising equipment just because it is unconventional. Decide how much pioneering you are willing to do.*

MEASURE **5.2.10 Provide separate thermostatic control for each area with distinct heating requirements.**

RATINGS

New Facilities	Retrofit	O&M
A	**D**	

Convectors typically are connected end to end, allowing an unlimited length. This tempts designers and contractors to control great lengths of convector with a single thermostat. Succumbing to this temptation can prevent separately occupied spaces from being shut down individually, and it keeps them from being controlled comfortably. For example, a common problem in office buildings is that one thermostat is installed to control a great length of perimeter, and later partitioning creates separate spaces that are under the control of the one thermostat.

The discomfort within localized areas motivates occupants to set a higher average temperature for the entire area served by the convectors. Some occupants may open windows to vent the heat. Even within a single space, it may not be satisfactory to control all the heating units together. For example, control a heater located near an exterior door separately from the other heaters.

SUMMARY

A control improvement for hydronic and electric convectors that typically saves a modest amount of energy, and may substantially improve comfort.

SELECTION SCORECARD

Savings Potential	$	$	
Rate of Return, New Facilities	%	%	%
Rate of Return, Retrofit..........	%	%	
Reliability	✓	✓	✓ ✓
Ease of Retrofit	☺	☺	

Hydronic heating units have an additional problem when a number of them are connected in series. The hot water cools as it flows from one unit to the next, so

WESINC

Fig. 1 An extraordinarily stupid thermostat installation
These thermostats control three sections of electric baseboard convectors in three areas. Installing them here, rather than in those areas, makes them useless for control. The thermostats are located next to a south-facing window, so they cut off heating when the sun shines on them. The window is single-glazed, so the thermostats overheat the spaces at other times. The locking cover makes it impossible for the occupants to adjust the thermostats to compensate.

that the downstream units produce less heat than the upstream units. This aggravates the problems.

The amount of energy waste from these causes varies widely, and it is often difficult to estimate. In general, heating units should have separate thermostatic

control if lack of separate control may cause a comfort problem or prevent shutting down an unoccupied space.

How to Provide Individual Control

The corrective measure is to subdivide thermostatic control of the convectors and radiators. Install as many individual thermostats as necessary to provide satisfactory local control of heating. Coordinate the thermostat installations with appropriate Measures of Subsection 5.1 to allow heating of individual spaces to be turned off separately.

The way that thermostats are installed is important. Each thermostat should be installed where it accurately senses the temperature within the space that is being heated by its corresponding heating unit. Space air should be able to circulate freely through the thermostat. The thermostat should not be under the influence of any source of heating or cooling, such as sunlight shining on a wall, the discharge from a conditioning unit, or nearby equipment. In addition, see Measure 5.1.1 about proper marking of thermostats. Figure 1 shows that some people in the heating business are thoughtless enough to violate all these requirements.

With hydronic convectors, you have to install additional piping to change the series flow to parallel flow. If a large number of heating units are served by the same supply pipe, install a reverse return to balance flow resistance without resorting to balancing valves. See Measure 5.2.2 about thermostatic control valves for hydronic units.

ECONOMICS

SAVINGS POTENTIAL: *Typically a small fraction of heating cost. Savings may be larger if comfort problems result in wasteful behavior, or if the Measure allows shutting down heating in vacant spaces.*

COST: *$100 to $1000 per heating unit, depending on the need for piping changes, trim changes, occupant displacement, etc.*

PAYBACK PERIOD: *Several years or longer.*

Fan-coil units are conditioning units for individual spaces that provide heating, cooling, or both. Units may be mounted in freestanding cabinets, inside walls, in ceiling plenums, or in other locations. Fan-coil units usually discharge air directly from their enclosures, although some may have rudimentary ducts for air distribution.

As the name implies, the main components of fan-coil units are a fan and one or two coils. Units may have separate heating and cooling coils, or a single water coil may be used for both functions. The coils may operate with hot water, chilled water, electric resistance, or rarely, steam. (Units with refrigerant coils are considered to be part of "split systems," and they are covered in Subsection 5.5).

The output of a fan-coil unit can be controlled by cycling the fan, by controlling the speed of the fan, by throttling the flow of water in the coil, or by turning electric coils on and off. Units typically have control panels to allow occupants to select heating or cooling, to select the fan speed, and to control outside air ventilation, if any is available. Automatic controls may shut off flow through hydronic coils when the fan stops, and they may perform other functions. The fan-coil unit may have thermostatic controls that are entirely self-contained, or the fan-coil unit may have actuators that are powered by external thermostats.

Fan-coil units that are designed to provide a large amount of outside air ventilation are called "unit ventilators." Unit ventilators are combined with relief air fans to provide positive control of outside air intake, maximize ventilation capacity, and direct the air flow. Consider the unit ventilator and its relief fan as an integral system.

The Measures in this Subsection improve the efficiency of fan-coil units. Measures 5.3.1 and 5.3.2 are maintenance activities that are important for maintaining energy efficiency. Measures 5.3.3 and 5.3.4 are improvements that apply to all kinds of fan-coil units in particular applications. Measures 5.3.5 through 5.3.7 recommend major system modifications.

MEASURE **5.3.1 Clean, adjust, lubricate, and repair fan-coil units at appropriate intervals.**

Fan-coil units are reliable if they are not abused. The maintenance they need consists mostly of cleaning and filter replacement. Other maintenance requirements depend on their specific design. Give these items attention at regular intervals:

- *filters and filter mountings.* Letting filters get clogged reduces conditioning capacity and causes fans be operate at higher power. In time, the filter will collapse because of the pressure difference across it, allowing dirt to accumulate on the coils.

  A common maintenance defect is improper fit of filters into the filter holders. Unless the fit is tight, dirt bypasses the filter and collects on the coils. The person who maintains the filters has to be careful to select the proper filter, and to insert it properly. See Figure 1.

- *coils.* The items to check are cleanliness and fin spacing. You can clean coils with a vacuum cleaner or with compressed air (in the reverse direction). In extreme cases, you need to wash out a dirty coil with detergent and a brush. Straighten bent fins with a fin rake.

- *coil condensate drain pans and lines.* Condensate pans, shown in Figure 2, are clogged by debris and by slime that grows in the condensate. Condensate

SUMMARY

Basic maintenance items that prevent energy waste, maintain comfort, prevent health problems, and extend equipment life.

SELECTION SCORECARD

Savings Potential .... $ $

Rate of Return .... % %

Reliability .... ✓ ✓

Ease of Initiation .... ☺ ☺ ☺

accumulation is a health hazard because it breeds microorganisms that are discharged into the surrounding air.

- *fan impellers* should be checked for balance, and for missing and bent blades. The cheap plastic impellers used in modern fan-coil units are easily broken by pens and other items that fall into the unit. Keep impellers clean to maintain their efficiency. Figure 3 shows a method of cleaning fan impellers. Be aware that cleaning an impeller may increase fan noise noticeably, because the dirt deadens noise.

Trane Company

Fig. 1 Filter maintenance The most important routine maintenance for fan-coil units. People hate to do it because it is awkward and messy. Failing to insert the filter properly is a common problem.

Trane Company

Fig. 2 Coil drain pan The pan must be cleaned out completely. A small amount of stagnant condensate allows colonies of organisms to grow, and these clog the drain and cause health problems.

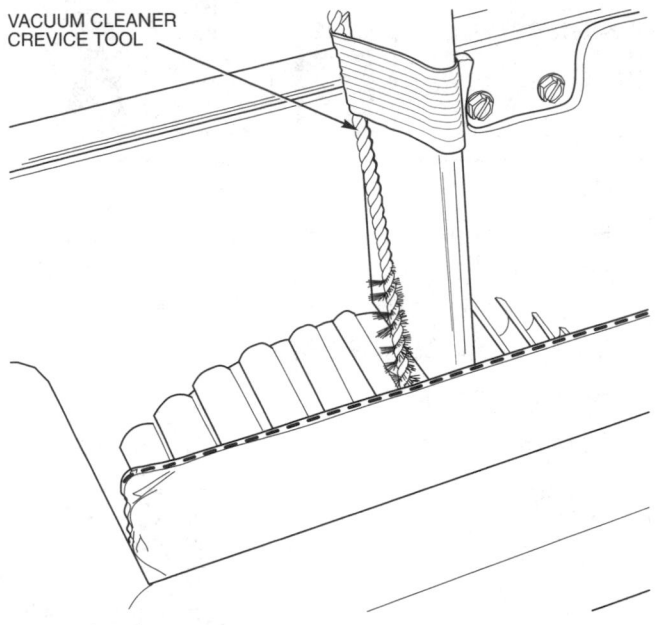

VACUUM CLEANER
CREVICE TOOL

Carrier Corporation

Fig. 3 Tool for cleaning fan impellers A brush taped to the end of a vacuum cleaner crevice tool works well. Be careful not to bend or break flimsy impeller blades.

Trane Company

Fig. 4 Outside air dampers in a unit ventilator The long, slender blades are vulnerable to damage. Be sure that dampers operate freely and that they close tightly.

- *fan motor lubrication.* If the fan motors are not permanently lubricated, oil them at appropriate intervals.

- *outside air ventilation dampers*, shown in Figure 4, have a major effect on energy consumption and comfort. Adjust them as well as possible. In many units, they are flimsy and they may fail to close properly. In some cases, the dampers may be jammed or limited in their motion by the way the fan-coil unit was installed. Correct such defects.

- *occupant controls,* such as fan speed switches and function buttons, are often flimsy. Inspect them periodically and repair them as necessary.

- *coil valves may leak*, causing the coil to consume energy even when it is turned off. In units that have both heating and cooling coils, valve leakage wastes energy by mixing hot and cold air.

- *steam traps* are used with steam coils, and they eventually fail. Inspect them annually, at the start of the heating season. Steam traps are covered in Measures 1.10.3 and 1.10.4 ff.

- *air eliminators* are used with steam coils to vent air that accumulates inside them. They are automatic devices similar in construction to steam traps. If they stick open, they waste steam. If they stick closed, the coil fills with air, loses capacity, and becomes uncontrollable.

ECONOMICS

SAVINGS POTENTIAL: *Usually a small fraction of conditioning energy. In fan-coil units, maintenance is more important to comfort and health than to efficiency.*

COST: *$3 to $30 per unit per year.*

PAYBACK PERIOD: *Short.*

TRAPS & TRICKS

DILIGENCE: *This kind of maintenance tends to be forgotten unless it is managed effectively. Schedule it in your maintenance calendar and make sure that it gets done.*

MEASURE **5.3.2 Select high-efficiency motors in new fan-coil units and when replacing failed motors.**

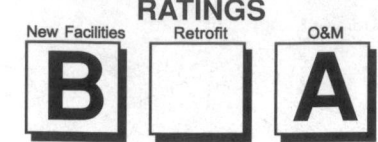

The motors used in fan-coil units typically are multi-speed shaded-pole or permanent-split-capacitor motors. The motors are small, typically drawing 50 to 200 watts, depending on the fan speed settings. Figure 1 shows a typical unit.

Small size and lack of efficiency standards make the efficiencies of these motors much lower than for larger motors. Efficiencies typically range between 25 and 65 percent. Significant improvements in motor efficiency in this size range began during the 1980's, so replacing older motors with high-efficiency units can offer improvements of 10 to 30 percent. Refer to Measure 10.1.1 for more about motor efficiency.

High-efficiency fan motors have become available as an option with many models of new fan-coil units. Exploit this option when installing new fan-coil units.

It is not economical to replace a small functioning motor for the sake of improving efficiency. This is because the cost of a small motor is high in relation to the amount of energy that it consumes. Instead, let existing motors serve out their lives, and establish a reliable procedure for replacing defective fan motors with high-efficiency models.

Be Prepared Ahead of Time

Replacing failed motors with the most efficient available replacement motors is a management

Trane Company

Fig. 1 Fan motor   The motor is small, but it operates for many hours. Upgrade efficiency wherever practical.

SUMMARY

High-efficiency motors for fan-coil units are an emerging option. Be ready to replace failed motors with high-efficiency units.

SELECTION SCORECARD

Savings Potential .... **$**

Rate of Return .... **% %** % %

Reliability .... ✓ ✓

Ease of Initiation .... ☺ ☺ ☺

---

challenge. The toughest part is ensuring that maintenance personnel remember to get the right motors at the unpredictable times when they will be needed. Otherwise, the failed motor will be replaced with any motor that fits. Don't expect the mechanic who picks up the motor to do a detailed analysis to find the most efficient model. You have to identify the appropriate replacement motor models in advance and find out where you can get them in a hurry.

If the facility has many similar fan-coil units, keep a number of high-efficiency motors as spares. This lets you order the motors for best efficiency and price, at your leisure. Also, it minimizes delay and waste of effort when making repairs.

ECONOMICS

SAVINGS POTENTIAL: *From 10 to 30 percent of the energy consumed by the fan motors.*

COST: *Small high-efficiency motors may cost from 10% to 50% more than less efficient models. The cost premium for smaller high-efficiency motors is changing as pricing policy and efficiency standards evolve.*

PAYBACK PERIOD: *Several years, typically. Depends on motor size and operating hours.*

TRAPS & TRICKS

TIMELY PREPARATION: *In existing facilities, the main challenge is remembering to install a high-efficiency replacement motor when the time comes, and being ready to do it. Identify vendors who can provide your selected motor models on short notice. If you have many motors of a particular size and type of service, keep at least one spare in stock. Make sure that all maintenance personnel, including those who work the late shift, are aware of the procedure. Designate the selected models and vendors in the plant operating manual. State where the spare units are stored.*

MEASURE **5.3.3 Keep conditioned air from discharging on windows and exterior walls.**

If a conditioning unit blows directly on the interior surface of a window or an exterior wall, heat loss through the window or wall is greatly increased. This is because conductive heat loss is proportional to temperature differential, and discharging directly on the wall raises the temperature differential across the wall. Also, the air blowing on the surface destroys the insulating air layer that attaches itself to the wall surface.

This problem is worst if draperies or other window treatments trap the discharge against windows. Figure 1 shows an example. Other types of obstructions can cause the same problem. For example, in warehouses, goods may be stacked tightly in front of a conditioning unit. This causes the unit to operate more intensely in order to heat or cool the space. While doing so, it increases heat loss through the wall.

If the windows allow air leakage, this may greatly increase the heat loss. The most extreme energy waste occurs if the fan-coil unit can discharge toward an open window, as in Figure 2.

SUMMARY

Corrects a common, unexpected cause of heat loss. Draperies are likely to cause the problem, or to aggravate it.

SELECTION SCORECARD

Savings Potential .... $ $ $

Rate of Return, New Facilities % % % %

Rate of Return, Retrofit.......... % % %

Reliability .... ✓ ✓ ✓ ✓

Ease of Retrofit .... ☺ ☺

In most cases, you can eliminate or greatly reduce these types of energy waste with simple changes.

Energy Saving Potential

■ Conductive Heat Loss

The additional heat loss depends on the thermal resistance of the surface, the discharge temperature of the fan-coil unit, and the amount of surface area that is affected.

If a drapery or other window treatment hangs above the fan-coil unit, it can greatly increase the area of the

WESINC

Fig. 1 Discharge trapped by drapery Almost all the heating output of this unit is discharged between the drapery and a large window. Much of the heat is lost by conduction through the window, and the small insulating value of the drapery is lost.

WESINC

Fig. 2 Discharge to an open window Much of the heating or cooling energy goes directly outdoors through this library window. And, books may block the discharge to the space. Another case where the architect did not talk to the engineer. There are several possible solutions.

wall or window that is affected. For example, if the outside temperature is 40°F and the room temperature is 70°F, the temperature differential is 30°F. Now, assume that a fan-coil unit discharges air at 140°F toward the wall. This air is trapped between the drapery and the wall, raising the surface temperature of the wall to 100°F. This doubles the temperature differential across the wall, and doubles the conductive heat loss.

In addition, the insulation value of the drapery itself is lost. A heavy drapery and its air layers has an R-value of about 1.0, which could account for a large fraction of the total thermal resistance of a window or a poorly insulated wall.

The thermal resistance of the wall determines the heat loss in absolute terms. Glass walls are the worst, with typical R-values ranging from 0.8 (single glazing) to 1.8 (double glazing with fancy coatings). At the other extreme, well insulated walls in newer commercial buildings have R-values between 10 and 20, but such buildings are exceptional. A typical uninsulated masonry wall with a furred interior surface has an R-value of about 4.

■ Loss of the Insulating Air Layer

Any vertical surface in relatively still air has an air layer attached to it that has an R-value in the range of 0.4 to 0.8. The direct discharge of a fan-coil unit is strong enough to scrub this air layer away, reducing the thermal resistance of the surface. The effect is especially significant with glazing, where the air layers on each side provide a large fraction of the total thermal resistance. With well insulated walls, the loss of the air layer matters much less than the increase in temperature differential.

WESINC

Fig. 3 Simple, effective baffle on a fan-coil unit This piece of flat clear plastic is bent to form a flange that is screwed to the top of the unit. It keeps the unit's air discharge from blowing behind the curtain and being trapped against the window.

■ Solar Heat Gain in Window Treatments

Curtains, blinds, and other interior shading devices absorb a significant amount of heat from the entering sunlight. Some of this heat is returned to the outside by a combination of radiation from the shading device and conduction through the window glass. If the fan-coil unit forces conditioned air through the space between the window and the shading device, the air picks up much of the absorbed heat. If the space requires heating, this effect saves energy. If the space requires cooling, this effect wastes energy.

■ Air Leakage

If the windows are leaky, blowing air against them increases leakage considerably because of the velocity pressure of the air. Furthermore, since the air from the fan-coil unit is warmed (or cooled), it carries away more energy than an equal amount of air from the space.

If a conditioning unit blows toward an open window, as in Figure 2, a large fraction of the heating or cooling energy is lost directly to the outside.

How to Avoid This Problem

In new construction, if the potential problem is draperies or window treatments, you can get models of fan-coil units that extend into the room and underneath curtains. Check with the interior designer. Bear in mind that the interior decor may change over the life of the building.

If the fan-coil unit is to be installed under a window, the window design can minimize the problem. If the wall is reasonably thick and the windows are installed flush with the outside surface of the wall, this leaves a space for installing curtains or other window treatments next to the window, without extending out over the fan-coil units. (Aside from this, the window design should exploit the Measures in Subsections 6.3, 7.3, 8.1, 8.3, and 8.4.)

In retrofit applications, the appropriate corrective action depends on the present installation. Most fan-coil units have louvers that control the direction of air discharge. These are usually adjustable. The trick is to fix them in position so that air does not discharge on the window or wall, and also does not cause an unwanted draft in the space. Once the louvers are adjusted properly, fasten them in place. You may be able to do this with a few sheetmetal screws.

In some cases, you may have to install an external baffle to direct the air discharge. Figure 3 shows an effective and inexpensive type of baffle that is widely applicable.

If there is a problem with heat trapping by draperies or other window treatments, use solutions similar to those recommended by Measure 5.2.5.

Comfort Dilemma with Poorly Insulated Walls

If a large fraction of the envelope surface is poorly insulated, e.g., if the building has a glass curtain wall, reducing the heating of the surface creates a comfort dilemma for occupants close to the wall. In order for a person to feel comfortable, heat loss from the person's body must be limited. Heat is lost both by conduction to the surrounding air and by radiation. If there is a large cold surface nearby, the person will feel chilly even though the air temperature would normally be comfortable. Therefore, designers locate heating units where they warm surfaces that have high heat loss.

To a certain extent, you can compensate for colder walls by increasing the space temperature. However, this makes people far inside the space too warm, and it increases heat loss elsewhere. Cold walls are an intractable comfort problem. The only solution that is entirely satisfactory is improving the insulation value of exterior surfaces. See Subsection 7.2 for the methods.

(Unfortunately, cold and leaky walls are a common problem in modern architecture, proliferated by a misconception that envelope integrity is unimportant in large buildings. Someone observed that larger buildings have a higher ratio of interior volume to surface area, so that the total heat gain of the building may be positive, even in cold weather. This may be true, but it provides no comfort for people in the perimeter zones.)

ECONOMICS

SAVINGS POTENTIAL: *3 to 50 percent of heating energy. Depends on the amount of wall or window area affected, the thermal resistance of the surface, the discharge temperature, and the extent of heat trapping.*

COST: *In new construction, this activity is mainly a matter of thoughtful equipment layout and selection, and it should add little cost. In retrofit, the cost is usually modest, unless you have to modify or replace window treatments.*

PAYBACK PERIOD: *Less than one year, to many years.*

TRAPS & TRICKS

INGENUITY: *A simple, neat solution is often possible. Make it permanent, and don't create a comfort problem.*

MEASURE **5.3.4 Install thermostatic controls that allow space temperature to drift within comfortable limits.**

RATINGS

New Facilities | Retrofit | O&M

A B B

In most HVAC applications, comfort does not require a fixed space temperature. People remain comfortable within a range of temperatures. You can save energy by using thermostatic controls that turn off heating and cooling within this temperature range. There are several ways to do this. See Measures 4.3.4 ff for the details.

MEASURE 5.3.5 Convert 3-pipe systems to 2-pipe operation.

SUMMARY

The least expensive way to eliminate the large energy waste that occurs in this type of system. Eliminates occupants' ability to switch between heating and cooling.

SELECTION SCORECARD

Savings Potential	$ $ $ $
Rate of Return	% % % %
Reliability	✓ ✓ ✓ ✓
Ease of Retrofit	☺ ☺ ☺

In "three-pipe" hydronic systems, each fan-coil unit is provided with a separate heating water and chilled water supply. This allows each unit individually to select either heating or cooling operation. However, the system uses a single common return line for both heating and cooling water, as shown in Figure 1. Using a common return line is intended to save the cost of separate return pipes for heating and cooling.

Three-pipe systems are an extreme example of false economy. The water returning from the heating coils retains a majority of its original heat energy. Similarly, the water returning from the cooling coils retains a large fraction of its original cooling energy. This does not waste energy if all the spaces in the building are being heated, or if they are all being cooled. However, if some spaces are being heated and others are being cooled, the return heating water and return cooling water are mixed in the common return line. This cancels out a large amount of energy.

The water that returns to the boiler may be cool, creating enormous thermal stresses in the boiler. Similarly, from the standpoint of the chiller, the return water is much warmer than normal. In effect, the boiler and chiller are feeding each other through the fan-coil units.

Energy Saving Potential

Three-pipe systems waste energy only when some fan-coil units are cooling while others are heating. Typically, this situation is limited to mild weather. However, if there are large heat load differences among spaces, simultaneous heating and cooling may occur during much of the year. This situation is especially common in buildings that have a large fraction of exterior glazing, and in buildings that have core zones. Let's look at several examples that illustrate the amount of energy waste that can be eliminated by getting rid of three-pipe operation.

First, consider a situation in which the heating and cooling loads are equal. The supply water temperature is 165°F and it returns at 145°F. At the same time, the chilled water is supplied at 45°F and returns at 55°F. A little calculation shows that the boiler load is four times higher than the actual space heating load, and the chiller load is four times higher than the actual space cooling load. In other words, three fourths of the energy produced by the boiler and chiller is wasted.

Next, consider a situation in which the heating load is ten times larger than the cooling load, with the same temperature differentials as before. In this case, the boiler is forced to produce 75% more output than the

space heating load, and the chiller is forced to produce 850% more output than the space cooling load. This energy waste might escape attention because the boiler and chiller are both operating at relatively small fractions of full load.

Similar energy waste results when the cooling load is much larger than the heating load. Obviously, it would take only a few months per year of overlapped heating and cooling operation to double the energy cost of the facility. That's a lot of energy wasted to save the cost of some pipe.

How to Eliminate the Energy Waste

■ **Method 1: Interlock Boiler and Chiller Operation**

The cheapest solution is simply to operate the boiler and chiller so that they do not feed the fan-coil units simultaneously. However, relying on an operating procedure alone is bound to fail. The only way to reliably avoid simultaneous heating and cooling is to connect the equipment in such a way that it cannot occur.

■ **Method 2: Abandon One Supply System**

The simplest reliable modification is to convert the three-pipe system to two-pipe operation. This is a relatively cheap solution in which one of the two supply lines is abandoned, so that all the fan-coil units are connected either to the boiler or to the chiller, but not to both simultaneously.

With luck, converting the fan-coil units should be relatively easy. The cost of the change is mostly for additional piping to connect both the boiler and chiller to the same supply pipes, plus valves to switch between the boilers and chiller.

■ **Method 3: Zone Changeover**

The disadvantage of Method 2 is that it eliminates the ability of spaces or groups of spaces to select heating or cooling individually. This disadvantage can be eased by using zone changeover valves, as shown in Figure 2.

As in Method 2, each riser is converted to two-pipe operation. But in addition, each riser is equipped with zone changeover valves so that all the units served by that riser can provide either heating or cooling. For example, the south side of a building may be switched to cooling, while the north side is switched to heating.

Zoning with changeover valves requires additional piping and valves. The cost depends on the original piping layout, the number of zones, and the way the zones are arranged.

How to Change Between Heating and Cooling

In a building with a three-pipe system, the occupants are used to being able to select heating or cooling at any time. A two-pipe system removes this desirable comfort feature. To retain comfort, the system must switch between heating and cooling appropriately.

Changing between heating and cooling manually invites complaints. It is best to install an automatic control to make the change.

There are several methods of sensing whether to provide heating or cooling. One is to sense the outside air temperature. The temperature setting should account for heat gains in the building. If the building has high heat gains, this method may not be satisfactory. For example, if the building has a large amount of glazing that creates high solar heat gain, cooling may be needed on a sunny day but not on a cloudy day, for a given outside air temperature.

To account for solar gain, you can install sensors that measure total heat gain from outside air and solar load. Figure 3 shows an array of such sensors oriented to control zones on each side of the building.

Another method is to control the changeover by using a thermostat that is installed in a typical space. This can be tricky. For example, if the system is providing heating because it is cold outside, allowing the space temperature to rise too much will cause the system to switch to cooling. Therefore, the thermostat must have a wide deadband range (see Measure 4.3.4.2) and it must not be adjustable by the occupants.

With Methods 1 and 2, the entire building must operate on either heating or cooling. A zone changeover system has much better comfort potential, but it needs a reliable method of selecting heating or cooling for each zone. For example, a zone facing toward the west may

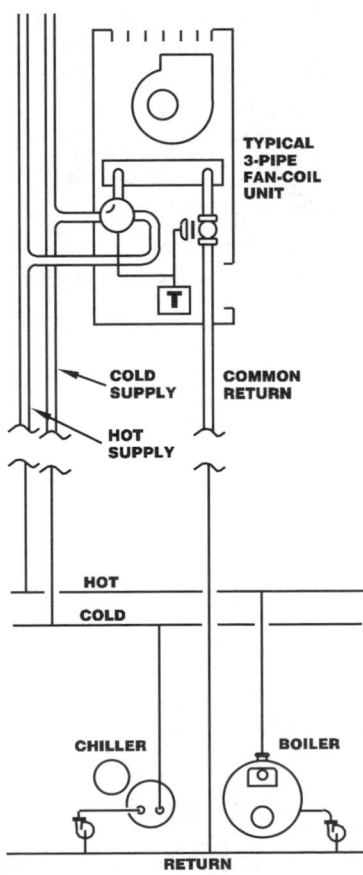

Fig. 1 Three-pipe system Each fan-coil unit may select heating or cooling. However, return water from the fan-coil units is mixed in common return pipes.

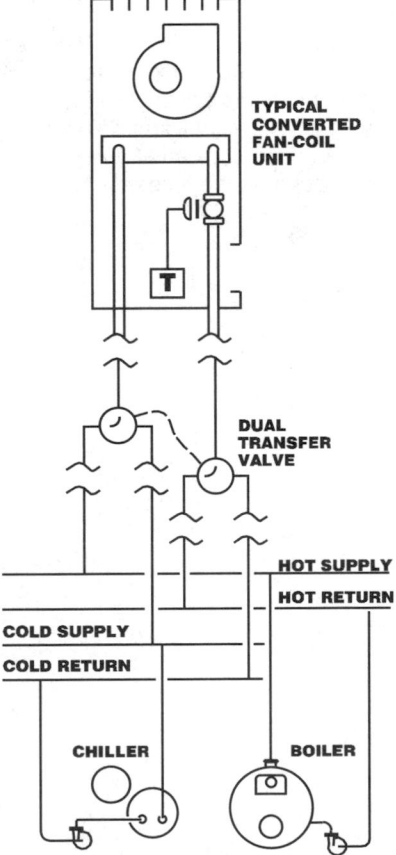

Fig. 2 Conversion to two-pipe changeover system Selection of heating or cooling by individual units is abandoned. Units grouped on individual risers, or zones, select heating or cooling collectively.

WESINC

Fig. 3 Zone cooling load sensors Each sensor simulates the combined load from outside air temperature and solar radiation. The four sensors are oriented to determine the load for each face of an office building. These provide only an approximation. A sensor located at one point cannot simulate the load at each point on the face of a building.

need heating in the morning and cooling in the afternoon. Therefore, you need sensors that determine the cooling load in each zone.

Alternative Modifications

Another possibility is to convert the system to four-pipe operation by adding a second return pipe. This approach is recommended by Measure 5.3.6, next.

Three-pipe systems are so wasteful that you should consider other major modifications. For example, replacing three-pipe fan-coil units with unitary heat pumps might be justified if it is important to provide for individual selection of heating or cooling in each space. Three-pipe systems are usually found in older buildings, so you may have the opportunity to replace the system as part of a major renovation.

ECONOMICS

SAVINGS POTENTIAL: *20 to 70 percent of heating and cooling costs.*

COST: *$10,000 to $100,000, in typical facilities.*

PAYBACK PERIOD: *One year, to many years.*

TRAPS & TRICKS

DESIGN: *The goal is to make the changeover as foolproof as possible while providing flexibility in zoning. The main point is to design the changeover valves so that the boiler and chiller cannot feed to any zone at the same time. For each zone, use special changeover valves that have dual passages to switch the supply and return connections together.*

EXPLAIN IT: *Install placards on the changeover valves to explain what they do. Describe the system in the plant operating manual.*

MEASURE 5.3.6 Convert 3-pipe systems to 4-pipe systems.

RATINGS

New Facilities | Retrofit | O&M
| **C** |

See the previous Measure 5.3.5 for an explanation of 3-pipe systems.

Another way of getting rid of 3-pipe operation is to convert to 4-pipe operation. This method preserves the ability of individual spaces to select heating or cooling independently. However, it is likely to be very expensive.

How to Convert to 4-Pipe Operation

The principle is simple: add a second set of return pipes to allow chilled water and hot water to be returned

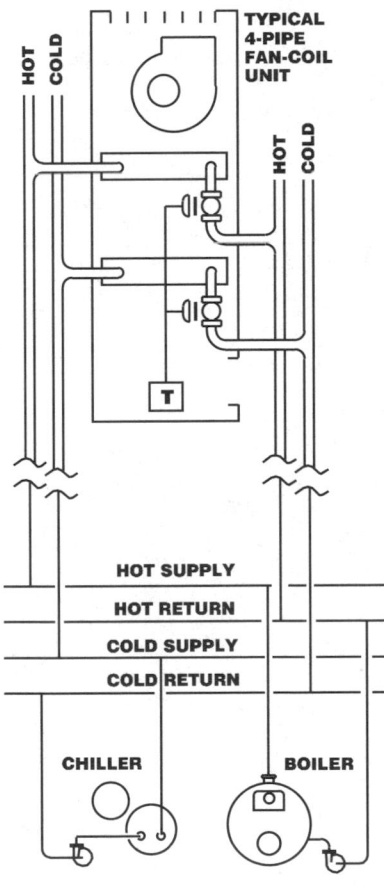

Fig. 1 Conversion to four-pipe system This system retains the ability of each fan-coil unit to select heating and cooling independently. It is expensive to retrofit.

SUMMARY

A much more expensive way of eliminating the large energy waste of 3-pipe systems. Retains the ability to switch between heating and cooling in individual spaces.

SELECTION SCORECARD

Savings Potential .... $ $ $ $

Rate of Return .... % % %

Reliability .... ✓ ✓ ✓ ✓

Ease of Retrofit .... ☺ ☺

separately. One return pipe goes to the chiller, and the other to the boiler. See Figure 1.

Attempting to convert existing fan-coil units is messy. The control valve and thermostatic control changes in the individual fan-coil units may be too expensive to be practical. In most cases, it makes more sense to replace the fan-coil units entirely.

Either way, piping work is needed at each fan-coil unit. Although holes already exist through floors and walls for the two supply pipes and the single return pipe, you may have to enlarge the holes to accommodate the second return pipe. Leave room to insulate both return pipes well because there is a large temperature difference between the chilled water return and the hot water return.

ECONOMICS

SAVINGS POTENTIAL: 20 to 70 percent of heating and cooling costs.

COST: The cost is mainly in installing extra pipe. Typical materials costs are $7 to $12 per foot. Labor costs depend on the difficulty of running the additional pipe.

PAYBACK PERIOD: Several years, to many years.

TRAPS & TRICKS

CHOICE OF METHOD: This conversion is foolproof if it is installed properly, but it is expensive. Consider all your options.

MEASURE 5.3.7 Replace electric resistance heating units with equipment having the lowest practical energy cost.

RATINGS

New Facilities   Retrofit   O&M

B | D | ☐

Fan-coil units with electric resistance heating elements are widely installed because the equipment and the installation are cheap. The penalty paid for this cheapness is high operating cost, unless electricity prices are exceptionally low. In most areas, electric resistance heat is much more expensive than fuel heat, per unit of heat delivered to the space.

If other available energy sources are less expensive than electric resistance heating, consider substituting another heating energy source. There are many options. The choice has a major effect on equipment and space requirements, installation cost, occupant conditions, and other factors. The selection of heating method may also affect your choice of cooling method.

This Measure is similar in economics and procedures to Measure 5.2.9. Refer to there for details.

One situation that is not covered explicitly in Measure 5.2.9 is the possibility of converting fan-coil units that have electric resistance heating and hydronic cooling. In this case, you may be able to convert the cooling coil to serve as a heating coil also. This creates a 2-pipe system.

Such fan-coil units are most likely to be found in facilities that do not have a boiler plant. Therefore, you have to install a boiler, if there is none. Also, you need to make thermostatic control changes to all the fan-coil units. This configuration sacrifices the ability of individual spaces to switch between heating and cooling.

Self-Contained Air Conditioners and Through-Wall Heat Pumps

The Measures in this Subsection improve the efficiency of self-contained cooling units and heat pumps. This category of equipment includes window air conditioners, packaged terminal air conditioners ("through the wall" units), and air-to-air heat pumps. Virtually all cooling units operate on the principle of refrigerant compression. (Absorption cooling is rare and specialized in this size range.) The units include all their components in a single package. Outside air is used for condenser cooling. In the case of heat pumps, outside air is the heat source. The units typically are installed in an opening in the building envelope. They often include heating coils, which are usually electric resistance elements.

INDEX OF MEASURES

5.4.1 Clean, adjust, lubricate, and repair through-wall air conditioners and heat pumps at appropriate intervals.

5.4.2 Install high-efficiency replacement fan motors.

5.4.3 Seal through-wall conditioning units to prevent outside air infiltration.

5.4.4 In units that provide outside air ventilation by exhausting conditioned air, abolish this feature.

5.4.5 Install air conditioning units and heat pumps having the highest practical efficiency.

5.4.6 If electric heating elements are installed with air conditioners, replace them with heating units having lower energy cost.

RELATED MEASURES

- Subsection 5.5, for Measures that improve the efficiency of larger air conditioning and heat pump systems
- Subsection 5.6, for Measures that improve the efficiency of heat pump loop systems

MEASURE **5.4.1 Clean, adjust, lubricate, and repair through-wall air conditioners and heat pumps at appropriate intervals.**

Modern self-contained air conditioning units are reliable throughout their service lives if they are not abused. They require little maintenance, but that maintenance is important. Give these items attention at appropriate intervals:

- *filters and filter mountings.* Clean filters at short intervals, usually every several months. In applications that generate a lot of lint, such as lodging facilities, clean the filters every several weeks. Letting filters get clogged reduces conditioning capacity substantially. If filters become too dirty, occupants are likely to remove them to restore cooling capacity, which leads to clogging of the coils.

  Have a stock of spare filters. They will wear out, especially if you clean them as often as you should. If you make replacement filters from bulk material, make sure that the replacements are installed properly. Unless the fit is tight, dirt bypasses the filter and collects on the coils.

- *inside coils.* Check for cleanliness and fin spacing. In almost all cases, fouled inside coils are caused by inadequate filter maintenance. Obstructed coils block air flow, forcing the evaporator coil to run colder and the condenser coil (in heat pumps) to run hotter. This reduces the cycle efficiency of the unit, which in turn forces the unit to operate longer, increasing fan energy consumption. In addition, restricting air flow causes the inside fan to draw more power and overheat.

  Fouled evaporator coils are a major potential source of indoor air quality problems. The condensate that forms on the coils keeps the dirt on the coils wet. The microorganisms that grow in the dirt are blown into the room by the fan.

  Clean dirty coils with a vacuum cleaner and a stiff brush. If necessary, wash out the coils with a detergent spray and compressed air. You may need a disinfectant to eliminate the health hazard from a badly fouled evaporator coil. Straighten bent fins with a fin rake.

- *coil condensate drain pans and lines.* Condensate pans are clogged by debris and by slime that grows in the condensate. Condensate accumulation is a health hazard because it breeds microorganisms that are discharged into the surrounding air. The condensate pan should drain freely. If it does not, check the installation for proper tilt.

- *outside coils.* Outside coils are most likely to become fouled by leaves and other debris, and possibly by mud. Dirty outside coils waste energy

---

SUMMARY

Basic maintenance items needed to prevent energy waste, maintain comfort, and prevent health problems.

SELECTION SCORECARD

Savings Potential .... **$ $** $ $

Rate of Return .... **% % % %**

Reliability .... ✓ ✓

Ease of Initiation .... ☺ ☺ ☺

---

in the same way as dirty inside coils. A brush and a water hose often suffice for cleaning. In extreme cases, use the methods given previously for inside coils.

- *fan blades* should be cleaned to keep air flow efficient. Be aware that cleaning a dirty centrifugal fan impeller makes it noisier, because dirt muffles fan noise.

- *fan motor lubrication.* If the fan motors are not permanently lubricated, oil them at appropriate intervals.

- *occupant controls,* such as fan speed switches and function buttons, are often flimsy. If occupants cannot operate the controls properly, the units will operate inefficiently and excessively. Inspect the controls often. Replace missing knobs. In environments where controls get rough treatment, as in motels, keep a stock of knobs and control panels on hand.

- *outside air ventilation dampers.* If outside air dampers are installed, check them for proper closure, connection to the occupants' control panel, and freedom of movement. A significant amount of conditioned air can be wasted if the dampers are inadvertently left open, or if they fail to close properly. See Measure 5.4.4 about modifying the outside air ventilation feature of room conditioning units.

- *refrigerant charge.* Improper refrigerant charge can seriously reduce efficiency and capacity. This should be a rare problem if units are repaired properly. Assign only qualified persons to check refrigerant charge, for the reasons explained next.

How to Check Refrigerant Charge

Most self-contained air conditioners and heat pumps are hermetically sealed. This makes refrigerant leakage an uncommon problem. Also, it is typically impossible to check the actual amount of refrigerant in the unit. For these reasons, you normally will not monitor the state of charge in air conditioning units. Your first indication of leakage is usually a loss of capacity that motivates an occupant to complain. Leakage usually occurs slowly at first, so a unit may operate at reduced efficiency for months before it loses enough cooling capacity for the defect to be noticed.

Specialized skill is needed to determine whether the refrigerant charge is low. Measure 2.7.2 explains the various methods. In summary, the most reliable method is to use refrigerant gauges to measure the suction and discharge pressures. This requires the unit to have service taps, which smaller units may lack. If no service taps are installed, an experienced technician can judge whether the charge is adequate from the external temperatures of the evaporator, condenser, liquid line, and suction line. A liquid line sight glass provides a qualitative indication of low charge that is less reliable.

If the unit is low on charge, find the leak using the methods explained in Measure 2.7.1.

Units can have too much refrigerant. This problem is almost always the result of attempted service by unskilled personnel. Depending on the design of the unit, excessive refrigerant may back up into the condenser, reducing condensing efficiency, or it may be sucked into the compressor as a liquid, causing compressor damage. Diagnosing excess refrigerant charge is even more subtle than diagnosing low charge.

Refrigerant-Side Repairs Require Skill!

Any repairs that require opening the refrigerant circuit, such as replacing a compressor, require special procedures to completely remove air from the unit prior to refilling with refrigerant. Even the small amount of the moisture that is contained in air combines with refrigerant to form an extremely corrosive product. The corrosion causes leaks, electrical insulation failure, compressor bearing damage, and reduced compression. The problem is insidious because the damage may not become apparent until months or years after the repair. Furthermore, mixing air with the refrigerant results in reduced cooling efficiency.

The only effective method of removing moisture, air, and other contaminants is to attach a vacuum pump to the unit and pump it down to a high vacuum for an appropriate period of time. If this procedure is not used, the unit will fail after a small fraction of its normal service life. See Figure 1.

Even trained refrigeration mechanics may neglect purging of air before recharging because the procedure

WESINC

Fig. 1 Where air conditioners go to die The air conditioners in this large hotel typically fail within a year or two after a compressor is replaced. We can see why. There is no equipment here for properly purging the units after the repair is made.

is tedious and time consuming. Some mechanics attempt to purge units by blowing refrigerant through them. This does not work, because moisture attaches to the materials and lubricants of the equipment. Also, the older types of refrigerants (containing fluorine) that are released into the atmosphere by this practice are believed to cause serious environmental harm.

Air conditioning units are as likely to suffer from too much maintenance as from too little, especially where the persons doing the work lack training. Most units are hermetically sealed, so getting inside them is comparable to surgery. Except for routine external cleaning and occasional repairs to external controls, leave cooling units alone unless there is a clear indication of internal trouble. If the unit does require opening the refrigerant circuit, make sure that the people doing the work are effectively trained and equipped.

Extra Maintenance for Units with Hydronic Heating

If the units include hydronic heating coils, refer to Measure 5.3.1 for additional maintenance items.

ECONOMICS

SAVINGS POTENTIAL: *5 to 20 percent of energy input. In addition, effective maintenance extends the service life of equipment.*

COST: *$3 to $20 per unit per year.*

PAYBACK PERIOD: *Short. The cumulative savings produced by proper maintenance greatly exceed the cost in the long term.*

TRAPS & TRICKS

MONITORING: *People hate to clean filters and coils, so stay on top of this. Filters need cleaning more often than commonly believed. Train maids and cleaning crews to clean filters, if possible. Check the condition of the user controls frequently, especially in transient environments. Be alert for comfort complaints that may indicate low refrigerant charge.*

SUPPLIES: *Keep an ample stock of spare filters, knobs, and control panels on hand. Treat these as expendable supplies. Getting spare parts on a piecemeal basis is a big obstacle to timely maintenance, and it wastes time and money.*

SKILLS AND SUPERVISION: *If you notice that your cooling units fail within a year or two after a repair that opens the refrigerant circuit, your maintenance people probably are not trained adequately, or they are being sloppy. Keep retraining and checking. Allow only designated technicians to do any service involving the refrigerant circuit.*

MEASURE **5.4.2 Install high-efficiency replacement fan motors.**

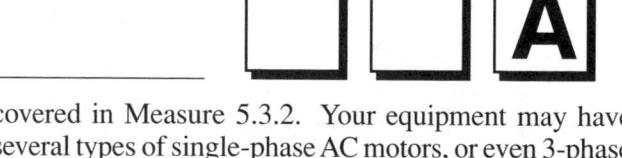

RATINGS		
New Facilities	Retrofit	O&M
		A

When you have to replace a failed fan motor, replace it with the most efficient motor available. This requires advance preparation. See Measure 5.3.2 for the details.

The motors used in self-contained air conditioners and heat pumps are larger than the fan-coil unit motors covered in Measure 5.3.2. Your equipment may have several types of single-phase AC motors, or even 3-phase motors. Look for the opportunity to substitute a more efficient type of motor, or a more efficient model of a particular type.

MEASURE **5.4.3 Seal through-wall conditioning units to prevent outside air infiltration.**

SUMMARY

A simple activity that prevents air leakage heat loss around the outside of the unit.

SELECTION SCORECARD

Savings Potential	$ $ $
Rate of Return	% % % %
Reliability	✓ ✓ ✓
Ease of Initiation	☺ ☺ ☺

With conditioning units that require an opening in the building envelope, there is a substantial gap between the unit and the window or wall opening. Seal these gaps effectively. Failing to do so wastes energy and may cause discomfort during cold weather. You can stop leakage using inexpensive materials that are available at most hardware stores.

Sealing Around the Unit

Seal units that are installed in windows by stuffing resilient foam weatherstripping into the gaps between the unit and the window.

If the conditioning unit is installed in a metal sleeve that is attached to the wall, use a durable caulking material between the sleeve and the wall. In addition, stuff resilient rubber or foam weatherstripping between the body of the air conditioner and the sleeve. The latter tends to be forgotten when the unit is removed for maintenance.

Seal wall openings on both the outside and inside. Sealing the gap on the outside prevents infiltration of outside air into the wall insulation or into the cavities of masonry blocks. Such infiltration may cool (or warm) a fairly large area of wall. Sealing the gap on the inside keeps air from leaking from the space into the wall, or vice versa.

Fill gaps with an impermeable material. Fill small gaps with a good grade of caulking compound. Fill large gaps with mortar, plastic foam, or some other solid material. Don't use fibrous insulation to fill gaps. Air passes through it easily.

Seasonal Covers for Air Conditioners

Another way to reduce leakage around air conditioners is to install an impermeable cover over the unit on a seasonal basis. Some people cover the inside of the unit, and others cover the outside. This approach works because air conditioners operate only during warm weather, and infiltration is a problem mostly during cold weather. To prevent infiltration into the wall from both sides, you would have to cover both the inside and the outside of the air conditioner. In most cases, this method is inferior to installing effective weatherstripping around the conditioning unit.

This method has the advantage that it stops leakage through the interior of the unit. However, leakage by this path is usually minimal unless internal seals have been omitted during careless repairs. If an air conditioner is installed in a leaky window, the cover can be extended over the entire window.

A cheap, but ugly, method of covering is to use a plastic sheet attached to the adjacent wall with tape. A more attractive but expensive approach is to make a rigid cover that fits against the adjacent wall with a soft gasket. Molded plastic covers for this purpose sometimes appear on the market.

Weather covers are sometimes installed on the outside of air conditioning units. You may not be able to achieve adequate sealing against an exterior wall.

If you decide to use air conditioner covers, be sure to disconnect the power to the unit at the same time. Otherwise, someone could turn on the unit with the cover in place. This may burn out the unit.

ECONOMICS

SAVINGS POTENTIAL: *3 to 30 percent of heating costs in the affected spaces.*

COST: *$10 to $100 per unit.*

PAYBACK PERIOD: *Less than one year, to several years.*

TRAPS & TRICKS

SELECTING THE METHOD: *Select the simplest and most reliable method that works. Permanent sealing is usually simple. Covers that require seasonal replacement are much less reliable than installing gaskets around the units. If you select removable covers, design them to be airtight, attractive, and easy to use.*

FOLLOWUP AND SCHEDULING: *Be sure to replace sealing materials after removing units for maintenance. If you select seasonal covers, schedule replacing the covers and disconnecting the units at the end of the cooling season.*

MEASURE 5.4.4 In units that provide outside air ventilation by exhausting conditioned air, abolish this feature.

Window-type air conditioners typically have a knob to provide outside air ventilation. Turning the knob opens a small door located in the discharge path of the fan, so that a certain fraction of the cooled air is vented directly to the outside. This creates a negative pressure within the space, causing air to leak into the space from elsewhere. The weaknesses of this feature are:

- the cooling or heating energy of the discharged air is completely wasted because the conditioned air never enters the space
- the air that enters the space to replace the discharged air is probably not fresh, because it comes from elsewhere in the building
- fresh air that does infiltrate into the space from the outside is unconditioned and may therefore create discomfort
- the volume of ventilation air is limited, because the ventilation opening inside the air conditioner typically is only a few square inches in size.

How to Disconnect the Ventilation Damper

You can easily disable this feature by disconnecting the link from the control knob to the air relief door. If you do this, remove the damper control knob and label the panel to state that the unit cannot provide outside air ventilation.

Provide Ventilation

Before accomplishing this Measure, determine the outside air ventilation requirement of the space and provide an alternative source, if necessary. Ventilation requirements are controversial. See Measure 4.2.1 for the details.

SUMMARY

Eliminates an inefficient feature of window-type air conditioners. Provide an alternative source of ventilation.

SELECTION SCORECARD

Savings Potential .... **$ $** $

Rate of Return .... **% % % %**

Reliability .... ✓ ✓ ✓ ✓

Ease of Initiation .... ☺ ☺

ECONOMICS

SAVINGS POTENTIAL: *5 to 20 percent of the cooling and heating energy delivered by the unit.*

COST: *Minimal, assuming that no alternative source of outside air must be installed.*

PAYBACK PERIOD: *Immediate, assuming that no alternative source of outside air must be installed.*

TRAPS & TRICKS

SELECTING THE METHOD: *Before doing this, make sure that the space will have enough ventilation air. If not, provide some other method of ventilation. This is rarely cheap or easy.*

EXPLAIN IT: *Don't forget to install the placard at each unit where you disconnect the ventilation control. Otherwise, the disconnected air dampers will be perceived as defects.*

MEASURE 5.4.5 Install air conditioning units and heat pumps having the highest practical efficiency.

There are large differences in the efficiencies of different models of air conditioners and heat pumps. In facilities where air conditioning is used a lot, or where heat pumps are used for heating for extended periods, you can save a lot of energy cost by selecting highly efficient units.

This applies to retrofit applications as well as to new construction. Packaged cooling units and heat pumps typically operate for 10 to 20 years before requiring major maintenance. When major maintenance is required, the cost may be so high that replacing the unit is a reasonable option. Consider this an opportunity to save energy by substituting a highly efficient unit. The most efficient models available today are much more efficient than older models.

Usually, it is not economical to replace a unit that is still operating for the sake of the energy savings. However, if the units are operated for a large number of hours per year and if they are approaching the end of their service lives, it may be economical to replace all the existing units as part of a single project. Replacing the units as a group is especially desirable if the replacement units require some custom fitting.

Efficiency Ratings

Many factors determine the efficiency of air conditioners and heat pumps. Better efficiency has been achieved by improving all the components, including the compressor, the compressor motor, the fan motors, the heat exchangers, and the refrigerant metering devices. Heat pump efficiency has also been improved by better control of evaporator defrosting and backup resistance heating.

Fortunately, you don't have to analyze all these factors to select an efficient unit. Many countries have laws that require efficiency ratings for air conditioners and heat pumps, along with minimum efficiency requirements. In the United States, you may find these ratings being used for packaged cooling units and heat pumps:

- *energy efficiency ratio (EER)* is expressed as BTU output per watt input. The EER of a particular unit varies widely depending on the temperature and humidity, so published values are based on continuous operation at stated conditions. This does not relate well to the way air conditioning equipment is actually used, so EER is no longer used in the United States as a consumer standard.
- *seasonal energy efficiency ratio (SEER)* is the average EER of a unit measured under a standardized variety of conditions that simulate

SUMMARY

An important design issue in new construction. In existing facilities, an important option to consider when units need major repair or approach the ends of their service lives.

SELECTION SCORECARD

Savings Potential	$	$	$	$
Rate of Return, New Facilities	%	%	%	%
Rate of Return, Retrofit	%	%		
Rate of Return, O&M	%	%	%	
Reliability of Equipment	✓	✓	✓	✓
Reliability of Procedure	✓	✓		
Ease of Retrofit	☺	☺	☺	☺
Ease of Initiation	☺	☺	☺	

typical operation. SEER is the consumer standard used in the United States for air conditioners and for the cooling performance of heat pumps.

- *heating seasonal performance factor (HSPF)* is defined in the same way as SEER. In the United States, it is used to rate the heating performance of heat pumps. The HSPF includes the energy consumption of the electric resistance heating elements that are turned on when the outside temperature is too low for the heat pump to operate alone.

Values of the SEER's for typical cooling units range from 8 to 15. Values of HSPF's for typical unitary heat pumps range from 6 to 10.

All other things remaining the same, the efficiency of newer units is reduced somewhat by a shift to new refrigerants intended to minimize environmental damage. However, this concern has stimulated a new round of innovation in refrigeration design, so equipment efficiencies and other characteristics may improve. Shop for high efficiency ratings, and don't worry too much about how the manufacturer achieves them. One caution, as always, is to avoid new technology. You are too busy to be a guinea pig.

Where to Get Comparative Efficiency Data

The Air Conditioning and Refrigeration Institute (ARI), an industry association located in Arlington, Virginia, publishes directories of most models of air conditioners and heat pumps sold in the United States, including their capacities and efficiency ratings. These directories are an invaluable aid in shopping for efficient units. Manufacturers' catalogs are not as useful for making your initial selection because they are limited to the offerings of a particular manufacturer or vendor. However, you need the manufacturers' catalogs to select other features.

Other Features to Consider

A variety of useful features have become available with newer space conditioning units, mostly by exploiting microprocessors. These features include accurately calibrated thermostats, time controls, temperature setback, automatic temperature reset during operation, and efficient control of fan operation. Some units allow these functions to be set with remote controls.

An especially valuable feature is turning off the evaporator (indoor) fan when the compressor is not operating. This prevents the fan from re-evaporating condensate from the coil and blowing it back into the room. This feature also saves fan energy and eliminates the noise of the fan between cooling cycles. Seasonal efficiency may be improved significantly. This function can be accomplished only with units that have their thermostats outside the unit and exposed directly to space temperature.

Cost and Size Penalties

Achieving greater efficiency requires greater weight, size, and cost. Compared to a unit of minimum efficiency standards, a highly efficient unit may be 50% larger in terms of weight and volume, and may cost 40% more.

The larger size may be the greatest impediment to replacing existing units. If you have to replace a unit in a space that cannot be enlarged, consider a unit of lower capacity, if possible. The original unit may have been oversized.

Consider Alternatives to Group Replacement

If all the conditioning units in a facility are approaching the end of their service lives, consider alternatives to replacing them. For example, consider a central conditioning system using fan-coil units.

ECONOMICS

SAVINGS POTENTIAL: *20 to 50 percent of the cost of operation of the units.*

COST: *Efficient units typically cost 20 to 50 percent more than the cheapest comparable units. This cost premium usually provides additional features that make the units more desirable, such as quieter operation and more accessible controls.*

PAYBACK PERIOD: *One year to several years, in new construction and when replacing failed units. Several years or longer, if you replace functioning units. Depends strongly on electricity costs and duration of operation.*

TRAPS & TRICKS

TIMELY PREPARATION: *Get a current copy of the ARI directory to help you select units for efficiency. To select for other features, you have to spend time with the catalogs. For replacing individual units when they fail, designate the selected models and identify vendors ahead of time. If you operate many units of a particular model, consider keeping a new replacement unit on hand.*

MEASURE 5.4.6 If electric heating elements are installed with air conditioners, replace them with heating units having lower energy cost.

This Measure is similar to Measure 5.2.9 in economics, options, and installation procedures. Refer to there for details.

Of the options presented in Measure 5.2.9, air-to-air heat pumps are the most likely replacement for air conditioners that have electric resistance heating elements. Heat pumps can use the envelope openings, power wiring, and condensate drain lines that already exist for the air conditioners. If the existing air conditioners are older units, substituting high-efficiency heat pumps substantially reduces the cooling cost as well as the heating cost.

Remotely Cooled Air Conditioning Units and Heat Pumps

The air conditioning units and heat pumps covered in this Subsection have outside heat exchangers that are separate from the inside equipment. This equipment is usually larger than the equipment in Subsection 5.4. It does not require wall or window openings to provide air flow from the outside. Virtually all equipment in this class operates on the principle of refrigerant compression (not absorption).

Most commonly, units reject heat directly to the atmosphere through air-cooled condensers. They may also exploit evaporative cooling by rejecting heat through an evaporative condenser or cooling tower. Or, for greater efficiency, they may reject heat to a cooler body of water, such as well water, a lake, or even service water. Such connections require heat exchangers.

The heat pumps in this group most commonly use the atmosphere as a heat source, by using an outside air coil. Such heat pumps usually require a supplemental purchased energy source during cold weather. For greater efficiency, and to avoid the need for a purchased energy back-up source, heat pumps may draw heat from water sources or from heat exchangers buried in the soil.

This category includes a variety of configurations. The four basic components of the vapor compression refrigeration cycle — compressor, condenser, refrigerant metering device, and evaporator — can be grouped in several ways. For example, the compressor may be packaged with the inside

INDEX OF MEASURES

5.5.1 Clean, adjust, lubricate, and repair air conditioners and heat pumps at appropriate intervals.

5.5.2 Replace fan, pump, and compressor motors with models having the highest economical efficiency.

5.5.3 Install compressors or new conditioning units having the highest economical efficiency.

5.5.4 Optimize the condenser temperature setting.

5.5.5 Turn off sump/crankcase heaters during extended shutdown intervals.

5.5.6 Drain cooling towers and evaporative condensers during the season when they are not required, and turn off water sump heaters.

5.5.7 Replace electric heating elements in cooling units with heating equipment that has lower energy cost.

5.5.8 With cooling towers and evaporative condensers, interlock the operation of the pumps and fans with the operation of the compressors.

5.5.9 Recover condenser heat for preheating service water or other low-temperature heating applications.

5.5.10 Improve cooling efficiency by using alternative heat sinks and heating efficiency by using alternative free heat sources.

RELATED MEASURES

- Section 2, for Measures related to central chiller plants
- Subsection 5.6, for Measures related to heat pump loop systems

equipment or with the outside condenser. The most common configuration is the "split system," which has an external package for the compressor and condenser, and an internal package for the refrigerant metering device and the evaporator. The grouping of components is based on practical considerations, such as available space, ease of installation, keeping noise outside occupied spaces, etc.

A system may have more than one of each basic element, and one element may serve several others.

For example, an external compressor/condenser unit may serve two or more inside evaporator units. Or, a single cooling tower may cool a water loop that serves any number of packaged room cooling units.

The inside equipment typically discharges conditioned air directly into the space, but some systems may have rudimentary ducts.

The inside equipment may include heating coils, which usually are electric resistance elements.

MEASURE 5.5.1 Clean, adjust, lubricate, and repair air conditioners and heat pumps at appropriate intervals.

Maintenance of remotely cooled air conditioners and heat pumps is generally similar to the maintenance of self-contained air conditioners and heat pumps, which is covered by Measure 5.4.1. The differences relate to the exterior location and larger size of the equipment. Figure 1 shows the exterior condensing unit of a small split system, and Figures 2 and 3 show air-cooled condensing units of larger systems.

Also, some systems employ wet condensing units, which have a separate set of maintenance requirements.

In addition to the maintenance items listed in Measure 5.4.1, make sure that these items get regular attention:

• *outside coils* have a strong tendency to collect organic debris. Clean this out regularly. Coils installed near the ground, as in Figure 2, collect leaves and grass clippings. Coils installed in locations that attract birds collect nests, feathers, the remains of prey animals, etc.

• *condenser temperature regulating valves* on water-cooled units strongly affect the efficiency of the unit. Adjust these appropriately. See Measure 5.5.4 for details.

• *cooling towers and evaporative coolers* have maintenance requirements that are almost identical

WESINC

Fig. 2 Air-cooled condensing unit This unit serves a medium size public library. Being low to the ground makes it vulnerable to debris. It should be inspected from underneath and from above. The concrete slab and the surrounding fence help to keep it clean.

to those of their larger counterparts used with central chiller systems. Refer to Subsection 2.3 for these procedures.

• *refrigerant lines* that connect the inside and outside equipment are sources of leaks. The major trouble spots are the connectors at the ends of the lines. Fortunately, these are localized and relatively accessible, making it easy to check them with leak detectors. Leaks can also occur where copper

Lennox Industries Inc.

Fig. 1 Air-cooled condensing unit for a residential-sized split system This is typical of units of all sizes. It contains the compressor and its accessories, a large condenser coil, and one or more fans to draw air through the coil. Refrigerant lines connect it to one or more evaporator units inside the building.

WESINC

Fig. 3 What is wrong with this condensing unit? A fan failed and was never replaced. This allows air to enter through the hole and bypass the condenser coils. Efficiency and capacity are suffering seriously. A thermostatically controlled damper is installed to maintain a minimum condensing temperature. It will create the same problems if it is not working properly, or if the damper blades do not close fully.

tubing is kinked and then straightened. Also, kinked lines can limit capacity and cause erratic operation. In addition to checking these specific items, take a slow walk around each unit and observe whether anything is wrong or suspicious. For example, the unit in Figure 3 has one serious defect and perhaps others.

MEASURE 5.5.2 Replace fan, pump, and compressor motors with models having the highest economical efficiency.

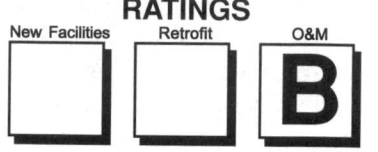

RATINGS

New Facilities	Retrofit	O&M
		B

Be prepared to replace failed motors with the most efficient available models. See Measure 10.1.1 for details. This Measure applies to replacing failed motors in existing cooling units. Motor efficiency in new units is covered by Measure 5.5.3.

MEASURE 5.5.3 Install compressors or new conditioning units having the highest economical efficiency.

A basic step in designing new facilities is selecting the most efficient conditioning equipment. The market offers a wide range of efficiencies, along with a wide range of other features, such as programmable thermostatic controls.

When the compressor of an existing split-system air conditioning unit or heat pump fails, you have several ways to improve efficiency. One is to replace the compressor with a more efficient unit. Compressors of improved efficiency are becoming available in smaller sizes. The efficiency of the compressors used in air conditioners ranges from 60 to 85 percent. The compressor consumes most of the energy used in air conditioners and heat pumps, so a 15% improvement in compressor efficiency, for example, produces almost a 15% reduction in energy consumption.

An efficient compressor alone cannot produce an efficient air conditioner or heat pump. The compressor is just one of several components that have a major effect on overall efficiency. If an air conditioning unit is old and inefficient, consider replacing the entire condensing unit, including the heat exchange coil, fan, and other apparatus. Also consider replacing the entire system. Newer systems have enhanced features, such as remote control and low noise level.

Normally, the right time to improve a system is when the compressor wears out. However, if you have many similar systems and they are all approaching the end of their service lives, it may be most economical to replace all the systems as a group.

Be Ready at Repair Time

Upgrading existing systems is a management and training challenge. Before the repair is needed, identify specific replacement models of compressors or conditioning units, along with sources. If this is not done, the need for quick repair will force maintenance personnel to use whatever replacement parts are readily available. Coordinate this activity with the acquisition of other high-efficiency replacement parts, such as motors (covered by Measure 5.5.2).

If the facility has a large number of similar conditioning systems, it helps to keep a number of high-efficiency units as spares. This practice gives you the leisure to order the equipment for best efficiency and price.

Finding Efficient Equipment

Refer to Measure 5.4.5 for efficiency ratings and sources of information on equipment efficiency. Ratings normally are available only for entire cooling units or

SUMMARY

An important design issue in new construction. In existing facilities, an important option to consider when replacing compressors or when conditioning units approach the ends of their service lives.

SELECTION SCORECARD

Savings Potential	$ $ $ $
Rate of Return, New Facilities	% % % %
Rate of Return, Retrofit	% %
Rate of Return, O&M	% % %
Reliability of Equipment	✓ ✓ ✓ ✓
Reliability of Procedure	✓ ✓
Ease of Retrofit	☺ ☺ ☺
Ease of Initiation	☺ ☺ ☺

heat pump systems. For efficiencies of individual compressors, refer to the manufacturers' literature.

ECONOMICS

SAVINGS POTENTIAL: *5 to 25 percent of the energy consumed by the conditioning unit.*

COST: *Widely variable. The cost premium for smaller high-efficiency units is changing as pricing policy and efficiency standards evolve.*

PAYBACK PERIOD: *One year to several years, in new construction and when replacing failed units. Several years or longer, when replacing old units that are still working. Varies widely depending on size and operating hours.*

TRAPS & TRICKS

SELECTING THE EQUIPMENT: *Call ARI (see Measure 5.4.5) to get a current copy of their efficiency directory for the type of equipment you are using. Use this to select entire conditioning units. For compressors alone, you have to search the catalogs.*

TIMELY PREPARATION: *For replacing failed units, identify vendors who can provide your selected models on short notice. If you operate many units of a particular model, consider keeping a spare replacement unit on hand. Make sure that all maintenance personnel involved with repairing air conditioning units are aware of the procedure.*

MEASURE **5.5.4 Optimize the condenser temperature setting.**

RATINGS

New Facilities   Retrofit   O&M

C

The efficiency of an air conditioning unit or heat pump improves one to two percent for each degree Fahrenheit that the condensing temperature is lowered. The reason is that lowering the condenser temperature reduces the pressure against which the compressor must operate.

Many types of cooling equipment need to maintain a minimum condensing temperature, typically in the range of 65°F to 85°F (18°C to 30°C). Such units have devices to regulate the flow of cooling water or cooling air so that condensing temperature remains high enough. It is not unusual to find that the condensing temperature is kept higher than necessary, which wastes energy.

You may be able to improve the efficiency of such units by lowering the condensing temperature. This is usually a simple change to the setting of a thermostatic control element. Just don't overdo it.

SUMMARY

Increase the efficiency of some cooling units by turning a screw. Be careful to avoid equipment damage.

SELECTION SCORECARD

Savings Potential	$ $ $
Rate of Return	% % % %
Reliability	✓ ✓
Ease of Initiation	☺ ☺ ☺

How to Lower the Condensing Temperature

The first step is to find out whether the cooling unit requires a minimum condensing temperature. In case

WESINC

Fig. 1 Condenser temperature control with city water cooling In this elegant installation, the arrow points to a thermostatic valve that regulates the flow of cooling water to the condenser. Using city water for cooling creates a serious compromise between the efficiency of the cooling unit, which affects electricity cost, and the cost of the water. The water cost is substantial, as witness the individual water meter installed for the unit. With air-cooled condensers and with condensers that are cooled by evaporative heat rejection units, it pays to keep the condensing temperature near the minimum that is possible.

of doubt, pick up the telephone and call the manufacturer's engineering department. When you make this call, find out which factors determine the lower temperature limit. See Measure 2.2.2 for the possible reasons. If a lower temperature setting does not pose a risk of equipment damage, but only a reduction of capacity, you may want to be more aggressive in lowering the temperature setting. Typically, lower condenser temperatures occur when the cooling load is low. A loss of capacity may not matter then.

The next step is to find all the methods of controlling condenser temperature that are used by the system. There may be several. For example, a cooling tower has a fan control and each of the individual room units served by the cooling tower may have a condenser water throttling valve. The adjustment procedure depends on the type of unit, as follows.

■ Water-Cooled Units

With water-cooled units, condenser temperature may be controlled by a thermostatic throttling valve that is installed in the condenser cooling water return line. Figure 1 shows an example. Measure the condenser water return temperature when the cooling unit is under load, and adjust the throttling valve to produce the minimum acceptable temperature. Refer to the manufacturer's literature for details.

If the cooling water is provided by a cooling tower, adjust the operation of the cooling tower fans to provide the most efficient compromise between compressor efficiency and fan energy consumption. See Measures 2.2.2 ff for details. Also, refer to Measure 5.5.8, which stops fan operation when the compressors are not running.

If a minimum cooling tower water temperature is maintained by means of a bypass around the cooling tower, set the bypass valve to the minimum safe temperature.

If the unit uses city water cooling, as in Figure 1, there is a serious compromise between the water cost and the cost of operating the cooling unit. Water costs have been rising dramatically in most locations. If you have such a unit, consider eliminating the city water cooling. If you have many such units, consider installing a cooling tower. If you have only a few smaller units,

consider installing air-cooled units having the best available efficiency.

■ Air-Cooled Units

With some air-cooled units, the minimum condensing temperature is controlled by a damper that restricts the flow of air through the condenser. The mechanism that controls the damper is exposed to weather, so it is likely to be defective or seriously out of calibration, especially in an older unit. Measure the condenser cooling air discharge temperature when the cooling unit is under load during cold weather, and adjust the damper control to produce the minimum acceptable temperature. Refer to the manufacturer's literature for details.

Some larger air-cooled condensers have multiple fans that operate in stages. Adjusting such units for maximum efficiency is tricky, because increasing condenser cooling also increases fan energy consumption. See Measures 2.2.2 ff for details.

Check the Adjustment Annually

This adjustment needs to be checked occasionally. In most cases, check it annually, at the beginning of the cooling season. Schedule this in your maintenance calendar. Install a placard at each thermostatic element that states the appropriate temperature setting. See Reference Note 12, Placards, for details of effective placard design and installation.

ECONOMICS

SAVINGS POTENTIAL: *Up to 10 percent of cooling energy.*

COST: *Minimal.*

PAYBACK PERIOD: *Immediate.*

TRAPS & TRICKS

PREPARATION: *Be certain that you know the minimum condensing temperature that your equipment can tolerate. Don't be shy about calling the factory to find out, if necessary.*

EXPLAIN IT: *Install an effective placard on the thermostat that controls the condensing temperature. Explain the settings in the plant operating manual.*

MEASURE **5.5.5 Turn off sump/crankcase heaters during extended shutdown intervals.**

This activity is identical with Measure 2.1.6, which applies to larger chiller systems. Refer to there for the details.

MEASURE **5.5.6 Drain cooling towers and evaporative condensers during the season when they are not required, and turn off water sump heaters.**

Operating cooling towers during cold weather may waste of a lot of energy if heaters are used to prevent freezing, as in Figure 1. If the cooling equipment is not needed, make it a practice to drain cooling towers during periods when freeze protection heaters may operate.

Turn off the power to any freeze protection heaters when you drain the water. Otherwise, the heaters will

turn on during cold weather even when the tower is dry. Lock open the electrical disconnects to air conditioning units and heat pumps served by the cooling tower. Tag them to say that the cooling water system must be refilled before the units can be operated.

It is good practice to drain the entire cooling water system, not just the portion exposed to freezing temperatures. Leaving water idle in the cooling system for long periods allows fouling to accumulate in the low points. Cooling tower water is dirty, so flush the system well when draining the tower.

SUMMARY

A routine seasonal procedure that avoids unnecessary sump heater operation.

SELECTION SCORECARD

Savings Potential .... $ $

Rate of Return .... % % % %

Reliability .... ✓ ✓ ✓

Ease of Initiation .... ☺ ☺ ☺

ECONOMICS

SAVINGS POTENTIAL: *All the energy used by the tower for freeze protection. In addition, water treatment and makeup costs are eliminated during the periods that the tower is shut down.*

COST: *Several hours of labor, mostly for flushing the system. The cost of the replacement water is small.*

PAYBACK PERIOD: *Less than one year.*

WESINC

Fig. 1 Cooling tower sump heater It must operate all winter to keep the water from freezing, unless the tower is drained.

MEASURE 5.5.7 Replace electric heating elements in cooling units with heating equipment that has lower energy cost.

RATINGS

New Facilities | Retrofit | O&M

B | **D** |

This Measure is similar to Measure 5.2.9 in economics, options, and procedures. Refer to there for details.

The modification that usually offers the quickest payback is installing heat pumps to replace air conditioning units with electric resistance heaters. Heat pumps can use the envelope openings, power wiring, and condensate drain lines that already exist for the air conditioners. If the existing air conditioners were manufactured to lower efficiency standards (which is true of most older units), replacement with high-efficiency heat pumps substantially reduces both cooling and heating costs.

Be aware that packaged air-to-air heat pumps generally rely on electric heating elements to carry the heating load during the coldest weather. For this reason, substituting heat pumps may not reduce the maximum electricity demand, although it will reduce the total amount of electricity consumed. (For an explanation of electricity demand and its effect on electricity pricing, see Reference Note 21, Electricity Pricing.)

To reduce both demand and consumption, consider heat pumps that use alternative heat sources and heat sinks. See Measure 5.5.10 about this.

MEASURE 5.5.8 With cooling towers and evaporative condensers, interlock the operation of the pumps and fans with the operation of the compressors.

SUMMARY

A simple control modification that saves a modest amount of energy.

SELECTION SCORECARD

Savings Potential	$ $
Rate of Return, New Facilities	% % % %
Rate of Return, Retrofit	% %
Reliability	✓ ✓ ✓ ✓
Ease of Retrofit	☺ ☺ ☺

Some systems use a cooling tower to provide condenser cooling water for a number of room cooling units. In such arrangements, the cooling tower fan and the cooling water pump typically run independently of the room units. You can save energy in these systems by keeping the pumps and fans from operating unless one or more of the room units is operating.

Some systems use an evaporative condenser. An evaporative condenser has a fan and a small pump to circulate water within the unit. You can save energy in these systems by keeping the fan and pump from running when the compressor turns off. (Evaporative condensers usually serve only a single compressor unit.)

Where to Do This

In a system with several room cooling units that are served by a single cooling tower, all the room units must turn off simultaneously in order for this Measure to save energy. If even a single room unit continues to operate, the pump and fans must keep running. For example, if one unit serves a computer room that requires continuous cooling, this Measure is ineffective. Similarly, if the climate is always warm, and a cooling tower serves many air conditioners, the cooling tower has to keep running most of the time.

Energy Saving Potential

The amount of energy that is saved depends on the facility's operating hours, the weather, the number of cooling units, and the size of the pumps and fans.

The energy saving also depends on the cooling water temperature that the fan is set to maintain. If the temperature is set relatively high, the fan will not operate unless the compressor is running. On the other hand, if the temperature is set low, the fan will run continuously any time the wet-bulb temperature of the air is higher than the temperature setting of the fan, regardless of the cooling load. See Measures 2.2.2 ff about optimizing fan operation.

How to Control the Equipment

A simple approach is to control power to the cooling tower or evaporative condensers through a relay. This relay is connected to relays in each of the room units, so that the operation of any room unit provides power to the pump and fans. Running the wiring is the main cost of this modification. Using relays allows low-voltage wiring to be used, which reduces cost.

If the facility has an energy management computer system, you may be able to program the computer to perform this function. In order to do this, the computer must be connected so that it senses a start signal from any cooling unit, and it must be connected so that it can control the power to the cooling tower or evaporative condenser.

Each room unit probably already has a flow switch to prove the availability of condenser water flow before the compressor is allowed to start. If this safety device is not installed, it is a good idea to install one. Virtually all compressor units are equipped with safety switches that turn off the compressors if head pressure is too high. However, these devices may allow the compressor to cycle on and off in the absence of flow, ultimately burning out the equipment.

Coordinate with Controls that Limit Equipment Running Time

To save effort, make this change at the same time that you install controls to limit the operation of the room cooling units. Refer to Subsection 5.1 for those controls.

ECONOMICS

SAVINGS POTENTIAL: *30 to 80 percent of the energy consumed by the pump. 5 to 30 percent of the energy consumed by the fans.*

COST: *Several hundred dollars to several thousand dollars. The main cost is running the control wiring between the cooling tower and the room units.*

PAYBACK PERIOD: *Less than one year, to several years.*

TRAPS & TRICKS

WHETHER TO DO IT: *In retrofit, make sure that this change is worth the bother. Check the running times of* all the air conditioning units connected to the cooling tower.

EXPLAIN IT: *Update the plant equipment records to record the change. Otherwise, no one will figure out the purpose of the control connections.*

MEASURE 5.5.9 Recover condenser heat for preheating service water or other low-temperature heating applications.

RATINGS

New Facilities	Retrofit	O&M
C	**C**	

Air conditioners and heat pumps remove heat from a space, raise its temperature above the outside air temperature, and throw it away. This heat may have substantial value. Try to find an application for it.

Heat recovery equipment is fairly simple, and it operates reliably. It is available from a number of sources, typically manufacturers of specialty refrigeration equipment.

Heat recovery from smaller cooling units is usually a simpler, scaled-down version of heat recovery from large cooling units. Refer to Subsection 2.10 for details about the various methods that are used in large systems.

Where to Consider Heat Recovery

Consider recovering heat from any equipment that has a condenser, including air conditioners, central chillers, refrigeration compressors, and heat pumps. (Heat recovery from central chillers, which involves a larger number of options and complications, is covered in Subsection 2.10.)

The main issue in feasibility is whether there is an application for the recovered heat. The main limitation is the low temperature of the heat, which is typically in the range of 100° to 120° (38°C to 50°C). By far, the most common application is preheating service (domestic) water. Raw water typically enters the facility at a temperature of 55°F to 65°F (13°C to 18°C). The difference between the incoming water temperature and the condensing temperature allows heat recovery to provide about half of the water heating energy, assuming that enough waste heat is available.

Effect on Cooling Unit Efficiency

The condensing temperature of a cooling unit has a major effect on its efficiency. (See Measure 2.2.2 for the details.) Heat recovery may improve the efficiency of the cooling unit if the liquid that is being heated (e.g., service water) is cooler than the normal condenser cooling medium (e.g., the outside atmosphere). However, the cooling unit must be able to tolerate the lower condensing temperature.

SUMMARY

A fairly simple and reliable method of recovering the heat rejected by cooling equipment.

SELECTION SCORECARD

Savings Potential	**$ $**
Rate of Return, New Facilities	**% %** % %
Rate of Return, Retrofit	**% %** %
Reliability	✓ ✓ ✓
Ease of Retrofit	☺ ☺ ☺

On the other hand, if you increase heat recovery by increasing the condensing temperature, the efficiency of the cooling unit is reduced.

Heat recovery reduces the energy consumption of the condenser fans, because they have less heat to reject.

How to Design the Heat Recovery Installation

In the simplest applications, just insert a heat exchanger between the compressor discharge pipe and the pipe carrying the liquid to be heated. This is economical, and it may work well in many applications. In other cases, there are complications. Consider the following issues.

■ Control

Heat recovery from cooling units may require no controls at all. The liquid being heated simply absorbs as much heat as it can, and the existing heat rejection equipment gets rid of the rest of the heat, if any. In some cases, you may have to install some simple controls, as explained next.

■ Maintain Minimum Condensing Temperature/Pressure

Many cooling units will operate improperly, or they will fail, if the condensing temperature is too low. (See Measure 2.2.2 for the reasons.) With such equipment, the manufacturer specifies a minimum condensing temperature. If the material you are heating with recovered condenser heat is colder than the minimum allowable condenser temperature, design the heat recovery installation to avoid trouble. A simple approach is installing a bypass valve on the liquid side of the heat exchanger that limits the liquid flow through the heat exchanger. Control the bypass valve by sensing the condensing temperature or pressure.

■ Consider Raising Condensing Temperature to Increase Heat Recovery

If the condensing temperature is a limiting factor in heat recovery, consider increasing the condensing temperature. You can do this by raising the temperature setting of the thermostat that controls the condenser cooling fans. Consult with the cooling unit manufacturer to learn the maximum allowable condensing temperature. Higher condensing temperature reduces the life of compressors and motors, but this effect should be minor if the equipment is designed for the higher condensing temperature.

Raising the condensing temperature reduces the efficiency of the cooling unit, so you should calculate the condensing temperature that provides the best compromise between heat recovery, cooling unit efficiency, and cooling unit maintenance cost.

■ Isolate from Refrigerant Contamination, if Necessary

In most heat recovery applications, the condensing pressure can be higher than the pressure of the liquid being heated. As a result, a leak in the heat exchanger could force refrigerant into the heated liquid. Common halocarbon refrigerants combine with water to form acids that are corrosive and toxic. (The properties of refrigerants are explained in Reference Note 34, Refrigerants.) For this reason, cooling unit heat recovery usually requires a double-wall heat exchanger, a standard item that is only slightly larger than a conventional single-wall heat exchanger. A more expensive alternative is using a separate heat transfer loop between the refrigerant and the heated liquid.

■ Can You Benefit from a Heat Storage Tank?

If the heat is needed at a different time than it is produced, you can increase heat recovery by adding a storage tank. The tank holds recovered heat until it is needed. You need a separate storage tank, even if the application already has a tank. This allows the heat exchanger to reject its heat at the low entering temperature of the heated liquid. For most systems, a simple stratified tank is the best choice.

A storage tank provides no benefit if (1) the supply of condenser heat always exceeds the amount of heat that can be recovered, or (2) the need for recovered heat always exceeds the amount available.

Refer to Measure 2.10.5 for more about heat recovery storage.

ECONOMICS

SAVINGS POTENTIAL: *Typically, a large fraction of the heat rejected by the cooling unit. The value of this heat depends on the heat source that is displaced.*

COST: *Several hundred dollars to several thousand dollars.*

PAYBACK PERIOD: *Several years, or longer.*

TRAPS & TRICKS

CHOICE OF METHOD: *Don't fool yourself about the amount of energy that you can save. Measure your hot water requirements, and calculate the fraction of heating energy that can be provided by heat recovery.*

DESIGN: *Provide for isolation from refrigerant contamination, if the application requires it. If you need a storage tank, err on the side of larger capacity. Keep the condensing temperature within acceptable limits.*

EXPLAIN IT: *If the installation is designed properly, operation is almost foolproof. Even so, make sure that the staff understand how the system is supposed to operate. Install a placard on the heat exchanger and/or storage tank. Describe the system in the plant operating manual.*

MEASURE 5.5.10 Improve cooling efficiency by using alternative heat sinks and heating efficiency by using alternative free heat sources.

The condensing temperature of an air conditioning system or heat pump determines its efficiency. (Measure 2.2.2 explains why.) With air-cooled equipment, the outside air temperature determines the condensing temperature. In comfort cooling applications, the cooling air is warmest when the cooling load is highest, so the system operates at lowest efficiency at the same time that it is consuming the most energy.

You may be able to maintain a substantially lower condensing temperature by using a heat sink other than the atmosphere. Your primary choices are the earth (soil), ground (well) water, and surface water. Figure 1 illustrates some of these heat sinks.

Similarly, the efficiency of a heat pump (while heating) depends on the temperature of the heat source. The main disadvantage of air as a heat source is that the air temperature is low when the heating load is high. In most climates, there are times when the air temperature is too low to provide satisfactory heat pump operation, and heating must be done by a separate heating unit. In most cases, the low-temperature heater is an electric resistance coil. Thus, energy cost is highest when the heating load is highest. (Reference Note 21, Electricity Pricing, explains why electricity prices are highest during peak periods.)

You may be able to improve heating efficiency by substituting a heat source that maintains a higher temperature. These heat sources are commonly called "free" or "renewable." As before, you can use soil, ground water, and surface water.

Also, you can use solar collectors and waste heat as heat sources, but not as heat sinks (with the exception of one unusual type of solar collector).

Heat pumps are especially well adapted to this technique because the heat sink for cooling can also act as the heat source for heating. Figure 2 shows how many heat pumps in a building can be connected in this way.

Where to Consider an Alternative Heat Source/Sink

Consider an alternative heat sink for cooling if the air conditioning equipment operates for a large number of hours each year. It is especially valuable in locations where the air temperature is high for much of the cooling season.

Consider an alternative heat source for heating if heating is needed for a large number of hours per year. It is especially valuable in locations where the air temperature is especially cold during the heating season. In fact, alternative heat sources, especially soil, have made heat pumps practical in much colder climates than are suitable for air-source heat pumps.

SUMMARY

An increasingly popular way to improve the efficiency of heat pumps. Also adaptable to cooling-only equipment. May be difficult to retrofit. Each type of alternative heat sink or free heat source is a separate field of study.

SELECTION SCORECARD

Savings Potential	$ $	$	$
Rate of Return, New Facilities	% %	%	
Rate of Return, Retrofit	%	%	
Reliability	✓ ✓	✓	✓
Ease of Retrofit	☺ ☺		

An alternative heat sink or heat source requires additional equipment, additional space, and connections to each conditioning unit. For these reasons, it is easier to install alternative heat sources/sinks in new construction than in existing facilities. However, there are many cases where it is practical to install an alternative heat source/sink on a retrofit basis.

Types of Alternative Heat Sources/Sinks

You may have a choice of possible heat sources and heat sinks. Start by reviewing Reference Note 22, Low-Temperature Heat Sources and Heat Sinks. You will see that each source or sink has significant advantages and disadvantages when compared to outside air and to the other sources. Each source or sink has a substantial body of knowledge that you need to learn to achieve an installation that is both efficient and reliable. This is not an activity to approach casually.

System Options in New Construction

All alternative heat sources and heat sinks are connected to the conditioning system by a water loop. Therefore, you need to select air conditioning equipment that is designed for water cooling, or select heat pumps that are designed to extract heat from a water loop and to reject heat to the water loop. The market offers a wide variety of such equipment. Any number of conditioning units can be connected to the water loop.

In selecting the equipment for efficiency, be sure that the equipment can exploit the lowest temperature of the heat sink. For reasons explained in Measure 2.2.2, some equipment maintains an artificially high condensing temperature. This negates the advantage of having a heat sink at lower temperature.

Generally speaking, using a water loop provides greater design flexibility than using air-cooled units. This is because water-cooled can be located anywhere in the facility. In contrast, self-contained air-cooled units must be installed next to a wall penetration, and split systems are somewhat limited by restrictions in refrigerant piping length.

In heat pump systems, each unit is limited to heating a small area, as with air-source units. The reason is that the heat produced by heat pumps is at too low a temperature for effective distribution over a large area. Each heat pump unit may discharge directly into the space, or it may serve a larger area through a short duct system.

In cooling systems, there is a broader choice of configurations. Individual spaces may have their own cooling units, or cooling may be distributed by air or water from larger centralized cooling units.

People tend to think of alternative heat sources and sinks as serving only smaller equipment. There is no fundamental reason for this. You can use an alternative heat sink for cooling equipment of any size. However, you may have trouble finding a heat sink of adequate capacity for a large cooling system.

Retrofit Possibilities

In some cases, it is practical to retrofit an alternative heat sink or heat source to an existing cooling or heat pump system. Again, the key point is that all alternative heat sources and sinks are connected to the conditioning equipment by a water loop. You have to figure out how to connect the new heat source/sink to the existing system. Here are some retrofit ideas for the common types of air conditioning units and heat pumps.

■ Water-Cooled Air Conditioners

The easiest case is a system in which one or more air conditioners is already water-cooled using a cooling tower. In this case, you can connect the alternative heat sink to the existing water circuit. Operate the alternative heat sink in parallel with the existing cooling tower. Install controls that select the cooler sources. For example, the cooling tower may provide lower temperatures than well water or a cooling pond at some times of the year. Also, the control should take into account the relative costs of operating the cooling tower (fan energy) and the heat sink loop (pump energy).

■ Cooling Units Cooled with Service Water

If an air conditioning system is cooled by domestic water, the alternative heat sink may be no cooler than the domestic water. However, using an alternative heat sink may save a substantial amount of water cost. Air conditioning that is cooled by domestic water is unusual. Most such units are too small to be worth converting.

■ Air-Cooled Cooling Units

Even if an air conditioning system is air-cooled, you can still use an alternative heat sink. Install a heat exchanger in the hot gas line that leads from the

WaterFurnace International, Inc.

Fig. 1 Alternative heat sources and heat sinks These improve the efficiency of cooling by lowering the average condensing temperature, especially during periods of high cooling load. They improve the average heating efficiency of heat pumps, especially in colder climates.

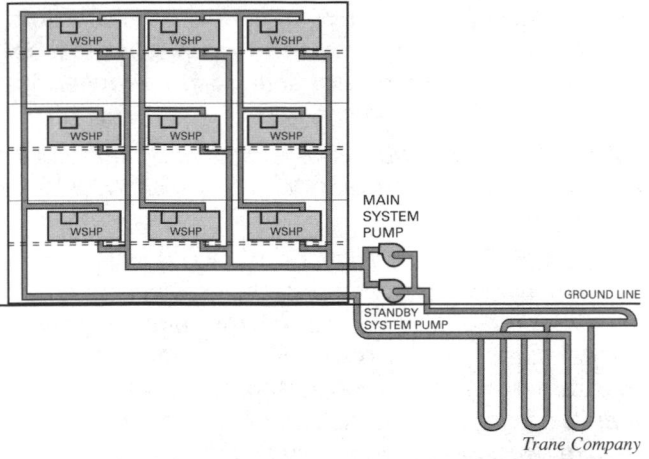

Trane Company

Fig. 2 Alternative heat sources and heat sinks serving many heat pumps Any number of heat pumps can be served easily by means of a water loop. The ultimate limitation is the capacity of the source and sink fields.

compressor to the existing condenser coil. This may require no extra controls, except for the water loop pump. This modification may be economical with larger systems, but not with smaller air conditioners.

If you have a large number of smaller air-cooled air conditioning units, consider scrapping them and replacing them with heat pump units that are connected to the alternative heat source/sink. This may be economical if electric resistance heat is presently used, if electricity is expensive, and if the existing air conditioners are nearing the end of their service lives.

■ Air-Source Heat Pumps

In principle, you can install a heat exchanger in the external circuit of an air-source heat pump. During cooling operation, the heat exchanger takes the heat out of the compressed gas line. During heating operation, the liquid refrigerant is evaporated in the heat exchanger. This would be a very unusual modification, and you would be fortunate to make it work well. It would not be economical with the smaller heat pumps used for comfort conditioning. You may find worthwhile applications in larger industrial heat pumps.

■ Heat Pump Loop Systems

Heat pump loop systems may benefit greatly from an alternative heat source/sink, and they are easily adaptable. Heat pump loops in general are complex. Measure 5.6.3 covers this modification in detail.

Common Design Issues

As we said before, each type of alternative heat source/sink involves its own body of knowledge. Use Reference Note 22 as your starting point. The following points apply to all the sources and sinks.

■ Minimize Pump Power

All the alternative sources and sinks connect to the rest of the system through a water loop. Therefore, you need pumps. Design the system to minimize the amount of pump energy required. Minimize the resistance of the heat exchangers and piping. Control the pumps to turn off when conditioning is not needed. In larger systems, consider variable-flow pumping.

■ Protect Against Refrigerant Contamination

Most cooling units and heat pumps operate with high refrigerant pressure. If ground water or surface water is used as the heat sink/source, you may have to protect the water source against refrigerant contamination that may occur if a heat exchanger leaks.

The easiest way to protect the water supply is by using a double-wall heat exchanger. In this type of heat exchanger, the heat transfer surface consists of two metal surfaces with a space between them that is vented to the outside. Thermal conduction occurs across closely spaced metal-to-metal contacts between the two surfaces.

If soil is used as the sink/source, a water loop is used to convey heat from or to the compressor, and the water loop acts to isolate the soil from a refrigerant leak.

■ Source/Sink Temperature Control

Using soil or ground water as the heat sink/source may require no additional temperature controls. Both act as essentially constant-temperature sources of heat. The refrigerant simply rejects or absorbs heat at the available temperature.

Using surface water as a heat sink may require a throttling valve in the heat exchanger to maintain a minimum condensing temperature when the water becomes very cold.

If you use solar collectors as a heat source, you will need controls to limit the temperature input to the heat pumps.

ECONOMICS

SAVINGS POTENTIAL: *Cooling energy consumption may be reduced from 20% to 40%. Heating energy consumption may be reduced from 10% to 50%, depending on the climate and on which alternative heat source you choose.*

COST: *Open-loop surface water systems are the least expensive, the cost determined largely by the distance to the body of water. Closed-loop surface water systems are next, if the distances are not too great. Water wells are usually more expensive than these. Well designed soil heat exchanger systems are expensive, typically costing several times more than a comparable air-source system. Solar collectors are expensive as heat sources, but they are more economical in combination with heat pumps than when used to provide heat directly. Waste heat sources may be fairly inexpensive.*

PAYBACK PERIOD: *Several years, or longer.*

TRAPS & TRICKS

SELECTING THE SOURCE: *Some sources/sinks are fairly easy to deal with, and some are tricky. Design is fairly conventional for water sources, but be sure to provide for easy cleaning of the heat exchangers. Soil heat exchangers and solar collectors involve difficult design choices. Study in detail each type of source that you are considering. Start with Reference Note 22.*

SELECTING THE EQUIPMENT: *Be relentless in checking references for every component of your system, including the plastic pipe. Don't buy unproven equipment. You don't want to add one more failed or uneconomical system to the dismal collection of failures that already litter the landscape.*

INSTALLATION: *Installing a water source is fairly routine, with few opportunities for major mistakes. On the other hand, heat exchange with soil involves a large temptation to take installation shortcuts. Soil heat exchangers are vulnerable to installation damage that*

causes later leaks that cannot be repaired. Solar collectors appear to have an evil genie that causes people to install them incorrectly or in a manner that requires continual maintenance.

EXPLAIN IT: *Describe the system clearly in the plant operating manual. Install placards to identify unusual components.*

MAINTAIN IT: *Schedule inspections and maintenance at appropriate intervals. Solar collector components tend to fail from weather exposure, from high stagnation temperatures, and in poorly designed systems, from thermal expansion and contraction. Open water sources require occasional heat exchanger cleaning. All systems require occasional pump maintenance.*

MONITOR PERFORMANCE: *If your energy management control system has the ability to do trend logging, use it to record heat exchanger supply and return temperatures. Or, do this manually in the plant operating log.*

Heat Pump Loop Systems (Additional Measures)

The Measures in this Subsection improve the efficiency of heat pump loop systems. Heat pumps in general are covered in Subsections 5.4 and 5.5. Heat pump loop systems are covered separately here because they are unique and difficult. Control is complex, selection of the heat sources and heat sinks takes you far away from the basic subject of heat pumps, and the piping can be tricky. If you deal with heat pump loop systems, budget a lot of time to work out the details and find a quiet place where you can concentrate. As you will see, it is a challenge to get these systems to provide the efficiency benefit that they promise.

GENERAL FEATURES OF HEAT PUMP LOOP SYSTEMS

In a heat pump loop system, individual air-to-water heat pumps are connected to a common water loop. Heat pumps that are cooling their spaces reject heat into the loop. Heat pumps that are heating their spaces draw heat from the loop.

Supplemental heat sources are connected to the loop to add heat at times when the total heating load of the system exceeds the total cooling load. Similarly, heat rejection equipment is connected to the loop to get rid of excess heat when the total cooling load of the system exceeds the total heating load.

In the simplest systems that are found, the supplemental heat source is a fuel-fired boiler, and the heat rejection equipment is some type of cooling tower. This type of system is shown in Figure 1.

The system in Figure 1 represents perhaps the most common concept of a heat pump loop system, and systems using this arrangement have been installed. Unfortunately, this arrangement is naive and wasteful. It throws away the primary advantage of a heat pump, which is to draw free energy from the environment for heating. To correct this basic flaw, a source of free energy is needed for all system operating conditions. More than one source of free energy can be included, to maximize system efficiency at different times. Such an improved system is shown in Figure 2.

INDEX OF MEASURES

5.6.1 Install controls to optimize the loop temperature under all operating conditions.

5.6.2 Install automatic valves and bypass lines to prevent the flow of loop water through heat sources and heat sinks when they are not needed.

5.6.3 Improve efficiency by substituting a heat source/sink with better temperature characteristics, or by adding a source of free heat.

5.6.4 Add thermal storage to the loop system.

RELATED MEASURES

- Section 2, for Measures related to chiller plants
- Subsection 2.10, for heat recovery from cooling equipment
- Subsection 5.6, for Measures related to centralized heat pumps in general

The overall system may have a net heating load that alternates with a net cooling load on a short cycle, usually daily. In such cases, installing a thermal storage system may improve average system efficiency. For simplicity, Figure 3 shows thermal storage added to a system that lacks the essential free heat source.

ADVANTAGES OF LOOP SYSTEMS

The water loop is the essential feature of the heat pump loop system. The water loop provides the following advantages:

• *space-to-space heat recovery.* A unique advantage of heat pump loop systems is that they allow the recovery of heat from spaces that are being cooled to provide heating for other spaces, thereby saving heating energy. The loop concept was popularized largely as a means of transferring heat from the core zones of buildings to perimeter spaces that require heating.

• *virtually any available type of heat source* can be connected to the system by means of the loop. An additional heat source is needed when the heat

pumps extract more heat from the loop than they reject to the loop. Some of these heat sources are "free." Even fuel-fired sources are typically less expensive than the electric resistance heat that is commonly used to supplement individual air-to-air heat pumps.

• *heat pump efficiency and capacity do not depend on the outside temperature,* unlike common air-to-air heat pumps. Therefore, the heat pumps in a loop system can be smaller. This advantage relates directly to the previous point. Fuel-fired heat sources and some free heat sources add heat to the loop in a way that is largely independent of the climate. (See Reference Note 22, Low-Temperature Heat Sources & Heat Sinks, for more about this.)

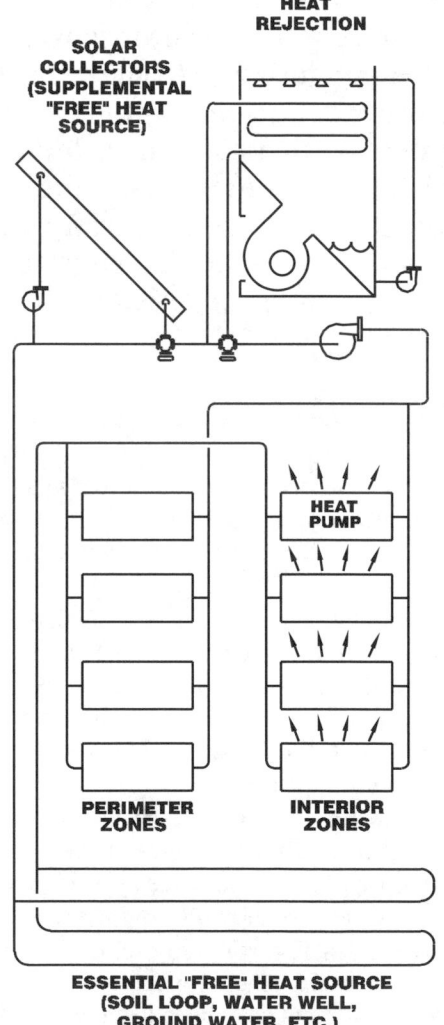

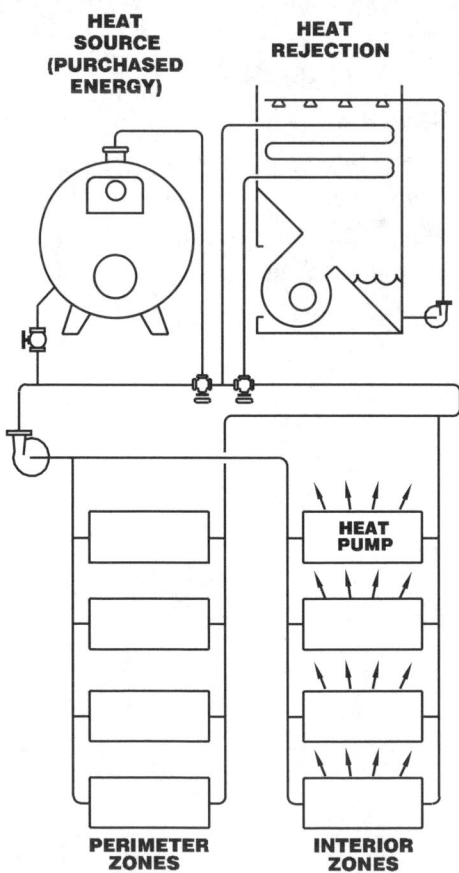

Fig. 1 Naive heat pump loop system   This illustrates the original concept of loop systems, which is to transfer heat from the core of the building to the perimeter. Although systems like this have actually been built, they throw away the primary advantage of heat pumps, which is their ability to draw free heat from the environment.

Fig. 2 Efficient heat pump loop system   This system corrects the flaw in Figure 1 by adding "free" heat sources. A free heat source that is available throughout the heating season is essential. Other free heat sources, such as solar collectors, can increase system efficiency by increasing the loop temperature for heating. The reduced energy consumption of this system is related mainly to the use of free heat sources. The core-to-perimeter heat transfer is usually less important.

• *solar collectors* can be connected to a loop system at substantially lower cost than for a freestanding solar system of equal heat contribution. This is because operation with heat pumps increases the heat collection capacity of solar collectors, and because the solar collectors do not require the other components of a freestanding solar heating system. (Reference Note 22 covers this point in greater detail.) As a result, solar heating may be economical in a loop system for applications where it would not be economical otherwise.

• *thermal storage* can be added to a loop system at relatively low cost. If there is an alternating requirement for heating and cooling, e.g., heating at night and cooling during the daytime, then thermal storage can save the daytime heat to make it available for nighttime heating. Thermal storage may also be used to reduce electricity demand charges.

Note that all of these advantages, except the first, can be exploited by any individual hydronic heat pump system. However, the economics of exploiting them tends to be better in loop systems.

EFFICIENCY PENALTIES OF LOOP SYSTEMS

Heat pump loop systems have several efficiency disadvantages compared to individual air-to-air heat pumps:

• the water loop adds *another stage of heat exchange*, which increases the temperature differential across the heat pump units and reduces their efficiency

• the various *loop circulation pump(s) require additional energy* to operate

• the central heat sources and heat rejection equipment, such as boilers, cooling towers, and solar collectors, require *additional energy for their own auxiliary equipment*, such as fans and pumps.

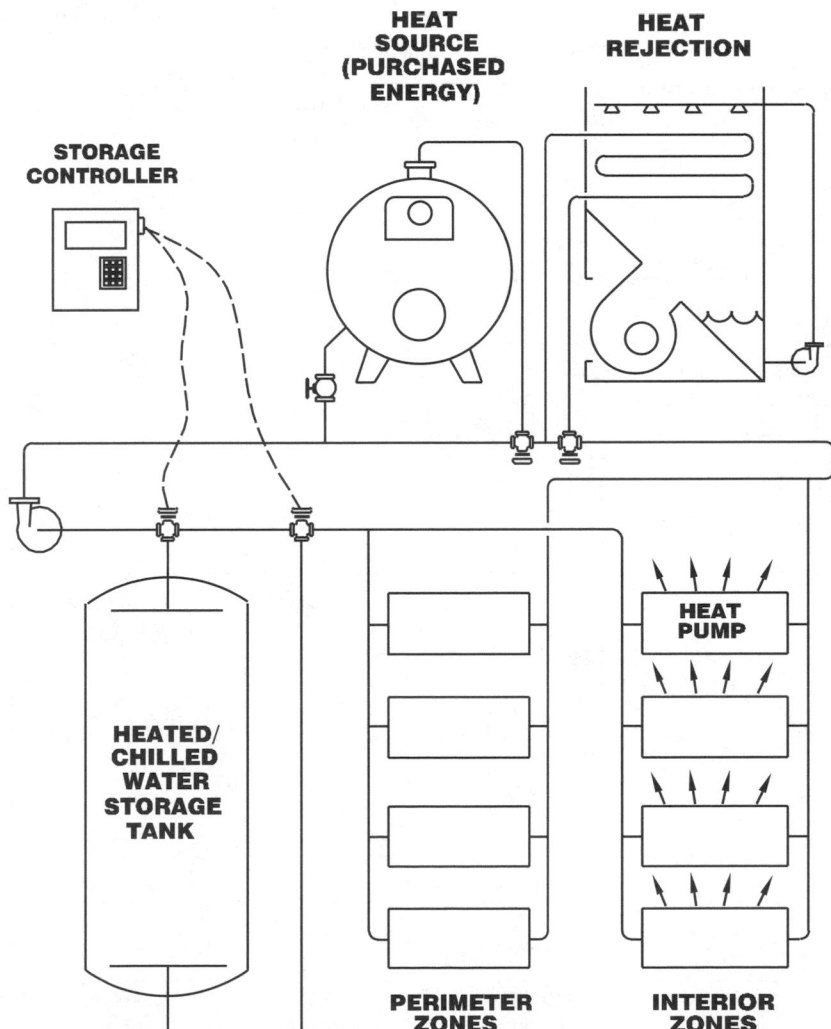

Fig. 3 Heat pump loop system with thermal storage Under some circumstances, thermal storage can improve the efficiency of a loop system. And, it allows buying electricity at times when rates are lower. This system does not show the essential free heat sources.

RELATIVE SYSTEM COST

The equipment cost for a heat pump loop system is higher than for individual air-to-air heat pumps. This is because of the extra piping and pumps, and because of the additional components that are required by the system. In all systems, these components include an additional heat source and a means of heat rejection. Optional equipment, especially thermal storage and solar collectors, raise the cost considerably. The total installation cost of all these items is typically much higher than for air-to-air heat pumps.

Using a free source of energy, such as a water well, eliminates or reduces the cost of conventional heat sources. Similarly, using the earth or a body of water as a heat sink eliminates or reduces the cost of a cooling tower or dry cooler. However, the equipment to exploit free sources and sinks is typically much more expensive to install than conventional heat sources and heat rejection equipment.

HOW VALUABLE IS CORE-TO-PERIMETER HEAT TRANSFER?

The original selling point of loop systems was their ability to transfer heat from interior spaces to perimeter spaces, and vice versa. The thermodynamic advantage of recovering heat from inside the building is that this heat is available at a higher temperature than is available when using outside air as a heat source. Likewise, units that are cooling the interior may be able to reject heat to a loop that is cooler than the outside air temperature.

Don't overestimate this potential benefit. It may be much less important than the connection possibilities provided by the water loop, discussed previously. The saving depends on the climate, on differences in load characteristics between different zones of the building, and on the heat storage characteristics of the building's structure. Calculating the saving with any degree of accuracy usually requires a sophisticated computer program. (See Reference Note 17, Energy Analysis Computer Programs, about these.)

The heat shifting potential of heat pump loops may be greatly reduced by other energy conservation measures. The interior or "core" zones of buildings require cooling year around, which provides a source of heat for perimeter areas. The amount of heat generated in the core of the building may be reduced dramatically by improvements in lighting efficiency, which are recommended in Section 9. Likewise, the heat requirement of perimeter spaces may be reduced substantially by envelope improvements, which are recommended in Sections 6 and 7.

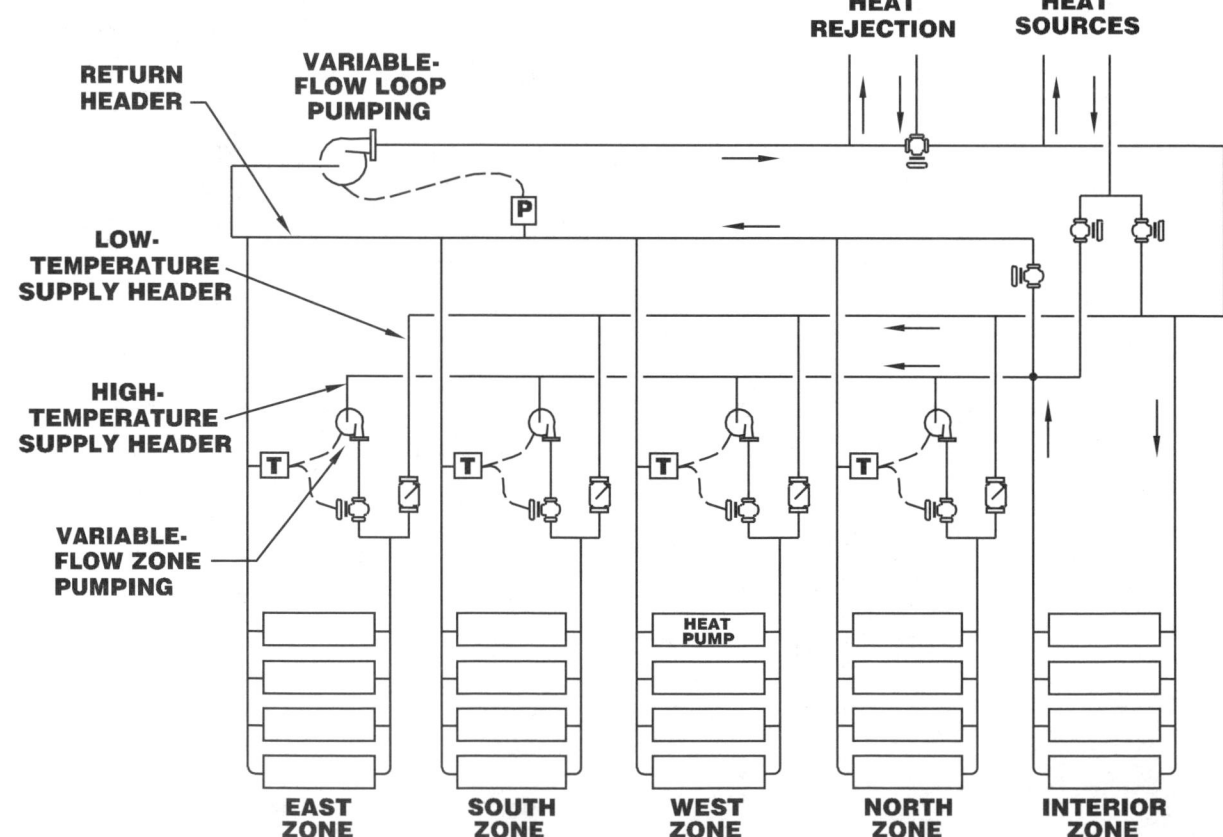

Fig. 4 Double-header heat pump loop system This system improves overall efficiency by routing warmer water to heat pumps that are heating, and cooler water to heat pumps that are cooling. All the details matter, including the sequence in which the heat pumps on different faces of the building are piped to the system.

COMMON FLAWS IN LOOP SYSTEMS

Control of heat pump loop systems is complicated by the many ways in which these systems can operate inefficiently. Overlooking any mode of operation can cause either the wrong loop temperature or improper operation of equipment units connected to the loop. Operating staffs find it difficult to understand the logic of system operation, so they cannot operate the systems efficiently. Optimizing system control is covered by Measure 5.6.1.

Extensive field investigations by one expert revealed that presently installed heat pump loop systems tend to be more expensive to operate than conventional fuel-fired systems. Many existing heat pump loop systems are defective in design, i.e., they lack elements necessary to achieve their potential efficiency. It appears that the designers of these systems fail to understand the underlying theory. Or, critical components are omitted to cut cost during construction. Or, the systems are installed in applications that cannot exploit their potential advantages.

The most serious design flaw in loop systems is failing to provide a free source of heat (outside air, solar panels, water wells, etc.) for heating operation. Measure 5.6.3 deals with this crucial issue.

SEPARATE HIGH- AND LOW-TEMPERATURE HEADERS

The simple loop circuit shown in Figures 1, 2, and 3 forfeits some potential efficiency because warm water (from units rejecting heat to the loop) is mixed with cool water (from units extracting heat from the loop). Greater efficiency is possible if the loop piping is arranged so that heat pumps that provide heating receive loop water that is warmed by the heat sources, while heat pumps that provide cooling receive loop water directly from the heat rejection equipment. An arrangement that does this is shown in Figure 4.

As far as is known, the arrangement shown in Figure 4 is entirely novel. If you attempt it, you will be executing an intricate concept with no experience to guide you. The details are tricky. Here are the main issues to consider:

- *the interior zone is treated as a heat source.* Note that the loop piping connects the interior zone in parallel with the other heat sources, and that the discharge water from the interior zone goes only to the high-temperature header (or to the return header).

- *the exterior zones are arranged in the sequence shown.* Each of the exterior zones may provide heating or cooling, at different times. Each is piped to the two supply headers identically. However, best overall efficiency is obtained when the exteriors zones that have the largest heating loads receive the warmest water. Therefore, the high-temperature header should be piped to the exterior zones as

shown. If the internal loads are the same in all the perimeter zones, then the north zone will have the largest heating load, because it receives no direct sunlight in winter. The west zone receives no sunlight until afternoon. The east zone receives the most sunlight early in the day, when the cooling load is highest, so it needs the least heating input. The south zone typically falls between the east and west zones in average heating requirement.

- *the main loop circulation pump is the only pump serving the interior zone.* The main loop circulation pump is designed to provide enough flow for all the heat pumps in the system. It is designed to provide enough pressure to overcome the resistance of the heat pump coils, but only when the heat pumps are being fed in parallel. This occurs in the exterior zones only when the individual zone is being fed from the low-temperature supply header, i.e., when cooling.

- *zone booster pumps are required for the exterior zones when heating.* When the heat pumps of an exterior zone are providing heating, the zone is fed from the high-temperature supply header. In this case, the exterior zone heat pumps are in series with the interior zone heat pumps. This increases the total flow resistance in the water loop. To overcome this additional resistance, each exterior zone has its own booster pump for heating operation. The booster pumps should be variable-flow, because the number of heats pumps operating in heating mode within each zone will change as the heating load changes.

- *control the operation of each zone booster pump to provide the highest overall efficiency.* In each exterior zone, there will be times when some of the individual heat pumps are providing heating, and others are providing cooling. This is inefficient. To minimize the waste of energy, supply the zone from the high-temperature header when most of the heat pumps are heating, and supply the zone from the low-temperature header when most of the heat pumps are cooling. A simple way to make the changeover is to control the zone booster pump in response to the return water temperature in its zone. The booster pumps draw only from the high-temperature header. When the booster pump is not operating, flow from the high-temperature header is valved off, and water feeds the zone from the low-temperature header.

- *balance the total system flow between the high-temperature and the low-temperature supply headers.* The total supply flow from the two supply headers must equal the return flow. This does not happen automatically, but depends on the control of the various valves in the water loop. Figure 4 does not show the control details, which can be executed in several ways.

Additional variations in this design can provide further small improvements in efficiency. However, they are a nightmare of complexity.

MEASURE 5.6.1 Install controls to optimize the loop temperature under all operating conditions.

SUMMARY

The controls needed to make loop systems efficient. Takes a lot of thought to get it right. Often missing in existing systems.

SELECTION SCORECARD

Savings Potential .... $ $ $ $

Rate of Return, New Facilities % % % %

Rate of Return, Retrofit.......... % % % %

Reliability .... ✓ ✓

Ease of Retrofit .... ☺ ☺

Even if a heat pump loop system has all the necessary components described in the introduction to this Subsection, it will still not provide a benefit unless it is controlled properly. Optimizing the control of a heat pump loop system is perhaps the most complex control challenge in HVAC. Stated simply, the objective of the loop temperature control is to maximize the overall system efficiency under all conditions.

To work out the controls properly, take a blank pad of paper, define all the operating conditions of the system, and calculate the optimum loop temperature for each set of conditions. To do this, you need to learn the efficiency characteristics and operating limitations of all the equipment in the system. If the system has any complications, such as thermal storage, the range of possible operating conditions expands radically.

Then, design the simplest set of controls that provide those loop temperatures. In all but the simplest cases, you need a computer to control the system efficiently. If an energy management control computer is installed, consider using that system. Or, consider using a programmable logic controller (PLC) exclusively for the loop system. Review Reference Note 13, Energy Management Computer Systems. Programming either type of computer is likely to be a daunting effort.

Consider the following issues in your control design.

The Loop Temperature Dilemma

Heat pump loop systems face a dilemma that apparently was not well understood in early systems. Namely, changing the loop temperature increases the efficiency of some components and lowers the efficiency of others. As a result, the temperature of the heat transfer loop has a complicated effect on the overall efficiency of the system. Specifically:

• *the individual heat pump units* become more efficient in the heating mode if the loop temperature is raised. On the other hand, they become more efficient in the cooling mode if the loop temperature is lowered. The fact that some heat pumps are heating while others are cooling creates a dilemma in setting the loop temperature. The efficiency (actually, the "coefficient of performance") of individual heat pumps may vary by a factor of two with changes in loop temperature.

• *solar and "renewable" heat sources* provide more heat to the loop as the loop temperature is lowered. This effect varies considerably among different types of sources. For example, solar collectors benefit greatly from reducing the loop temperature,

especially the types of collectors that have high heat loss. This is because solar heat that is not absorbed by the loop is lost to the atmosphere. In contrast, the amount of energy that can be extracted from natural water sources is not much affected by loop temperature, provided that the heat exchanger has ample capacity.

• *heat rejection equipment,* such as cooling towers and dry coolers, becomes less efficient as the loop temperature is lowered. The effect of heat rejection efficiency on overall system efficiency is much smaller than the effect of heat pump efficiency. Still, the effect is large enough that you should consider it in designing the loop temperature control. Cooling equipment that uses fans can be adjusted independently of the rest of the system to minimize waste of fan energy. Refer to Measures 2.2.2 ff for methods.

• *fuel-fired boilers* become somewhat less efficient as the loop temperature is raised. This is a minor factor within the temperature range in which heat pump loops operate. As a practical matter, you can ignore the effect on boiler efficiency. Electric boilers always operate with an efficiency near 100%, so their efficiency is not a factor in loop temperature.

The loop temperature that provides the best overall efficiency is a balance of all these factors. The balance changes continuously, depending on the relative number of units that are heating or cooling, the outside air conditions, the time of day, the amount of sunlight, the state of charge of the storage system, and other factors that depend on the specific design of the system. Another major factor is the temperature limits on heat pump operation, which we examine next.

Heat Pump Temperature Limits

Heat pump loop systems use air-to-water units that are designed to operate within a specified range of water temperatures. However, the stated temperature range may be nominal, and the manufacturer may not know how its units perform with water and air temperatures outside the stated range. Manufacturers typically design air-to-water heat pumps for a conservative range of temperatures, and may not anticipate the wide range of temperatures that may be necessary to optimize efficiency in loop systems. As a rule, heat pumps are designed with the assumption that the condenser temperature is always significantly higher than the evaporator temperature. In other words, manufacturers do not anticipate low or reversed temperature differentials across their units.

For example, consider operating a loop system in winter, with all the heat pumps providing heating. The loop heat is provided by a boiler, which can heat the loop to any temperature you choose. What should the loop temperature be? The efficiency of the heat pumps is increased by raising the loop temperature. However, as the condenser (room) temperature approaches the evaporator (loop) temperature, the pressure differential across the expansion valves in the heat pumps may drop to the point that they lose capacity. Or, the low pressure differential may allow pooling of oil away from the compressor, causing compressor failure.

If it appears desirable to operate the loop at temperatures higher or lower than those specified in the manufacturer's literature, or if no temperature limits are given, call the manufacturer's engineering department to discuss loop operation. Feel fortunate if you get a certain answer. Loop systems are still experimental, even though manufacturers and designers may not recognize the fact. Efficient operation requires pushing loop temperature control into the unknown.

While you are talking to the manufacturer, ask for verified data that shows how the heat pump efficiency relates to the water and air temperatures. You need this information to develop your schedule of loop temperatures.

Example: Loop Temperature Control in a Simple System

Consider a relatively simple loop system that has only heat pumps, a boiler, and a cooling tower. A hypothetical loop temperature schedule for this system might be:

- when the outside air temperature is less than 50°F, assume that most of the heat pumps are operating in heating mode. The loop temperature is maintained at 110°F by operation of the boiler.
- when the outside air temperature is greater than 70°F, assume that all heat pumps are in cooling mode. Keep the loop at the lowest temperature that

the cooling tower can provide without using excessive fan energy in the heat rejection equipment.
- in the range of outside air temperatures between 50°F and 70°F, the loop temperature is maintained between 110°F and 70°F using a linear reset schedule. The sliding temperature scale is based on an estimate of the relative numbers of heat pumps that are heating and cooling at each temperature.

This control schedule is purely for illustration. An evaporator temperature of 110°F for heating may not be possible with conventional heat pumps, but it illustrates an extreme that would be desirable from the standpoint of efficiency.

Adapt the Loop Temperature to the Heat Sources

Optimum loop temperature control depends on the type of heat sources installed in the system. Control with some heat sources is more complex than with others. As a rule, you can adjust the output temperature of conventional heat sources (i.e., boilers) to suit the needs of the loop system. However, the loop system temperature may have a major effect on your ability to exploit "free" energy sources.

In this Measure, we assume that your heat pump loop system includes a "free" source of energy. If this is not true, see Measure 5.6.3.

Each of the following "free" heat sources has its individual loop temperature control requirements. See Reference Note 22, Low-Temperature Heat Sources & Heat Sinks, for more about these heat sources.

■ Ambient Heat Sources and Heat Sinks

We define "ambient" sources as free heat available from the environment. Such sources include outside air, ground water, lakes, rivers, and deep soil. The quantity of heat is essentially unlimited, if you install enough equipment to tap it. Another major advantage of some ambient heat sources is that you can use them as heat sinks to get rid of heat when the system is primarily in a cooling mode. This is an advantage of ambient sources when compared to solar collectors or waste heat sources.

It is worth repeating that the lack of a free heat source for the loop system radically increases the consumption of purchased energy. Measure 5.6.3, below, describes how to correct this problem.

Ambient heat sources and heat sinks typically provide the simplest loop temperature control. This is because the temperature of the source determines the temperature of the loop directly. Specifically:

- *ground water* and *soil* control the loop temperature at all times, always acting as either a heat source or a heat sink. The temperature of ground water and deep soil does not vary much with the time of year. Almost everywhere on earth, the ground water and soil temperatures remain in a range that provides better efficiency than outside air for both heating and cooling. No boiler or cooling tower is required, provided that the source has sufficient capacity in relation to the system load.

- *river* and *lake* water behaves in generally the same way, except that there is a large seasonal change in the temperature of the source. When the river or lake approaches freezing temperature, it can no longer be used as a heat source, because the water freezes at the point where it gives up heat to the system. At that time, the boiler becomes the heat source and the loop is isolated from the ambient source. When the boiler is operating, control it to maintain the highest loop temperature at which the heat pumps can operate, for example, at 110°F.

- with *outside air* as the heat source and heat sink, the control schedule is about the same as with river and lake water. Most heat pump systems are able to draw heat from outside air at a substantially lower temperature than the freezing point of water. When the air temperature falls to the point that the heat pumps are no longer able to provide sufficient heating capacity, the boiler becomes the sole source of heat for the loop. At this point, the boiler is controlled to maintain the highest loop temperature at which the heat pumps can operate.

■ Waste Heat Sources

Waste heat typically is available at relatively low temperatures that keep it from being useful for direct recovery. However, most waste heat sources are warm enough to provide good recovery in heat pump systems.

The key issue in loop temperature control is whether or not the amount of available waste heat exceeds the heat requirement of the heat pump system. If the waste heat source can provide more heat than is needed by the heat pump system, then the waste heat source itself determines the loop temperature. On the other hand, if the amount of waste heat is insufficient to satisfy the loop heat requirement, the boiler operates and controls the loop temperature.

If the amount of waste heat is limited, you can increase recovery by lowering the loop temperature. For example, if a limited amount of cooling water discharge is available at 140°F, lowering the loop temperature from 110°F to 80°F would double the amount of heat that is recovered. However, this drop in loop temperature would also substantially reduce the heating efficiency of all heat pumps in the system.

To make best use of waste heat in such cases, you have to calculate a table of optimum loop temperatures based on the amount of waste heat available, the temperature of the waste heat, and the efficiency of the heat pumps at different loop temperatures. Then, you have to create automatic controls that maintain the temperature relationships that you calculated.

■ Solar Collectors

The economic limitation of solar collectors is the relatively small amount of energy that they can collect per unit of collector cost. A major collector design problem is reducing the loss of heat from the collectors themselves. For example, solar collectors used for direct space heating typically must deliver water at temperatures of 140°F or higher. At these temperatures, solar collectors suffer high losses by radiation and by conduction to the surrounding air.

These losses are greatly reduced when solar collectors are used as a heat source for a heat pump loop system. This is because the output of the collectors is useful at any temperature, down to the minimum evaporator temperature of the heat pump. As a result, the collectors can provide useful heat for a much larger number of hours per year than a conventional solar heating system. For example, on an overcast day in winter, a solar collector can provide only a minimal amount of heat at the typical flat-plate output temperature of 140°F, but it may be able to provide a substantial amount of heat at 70°F.

If the amount of solar heat is adequate to serve the heating load, the loop temperature is determined by the output of the solar collectors. Under this condition, keep the loop temperature as high as possible to maximize heat pump efficiency. If solar input starts to fall behind the heating load, the loop temperature drops as the heat pumps remove heat from the loop faster than the solar collectors add heat. If the loop temperature falls too far, the supplemental fuel-fired heat source is started to make up the difference.

Unlike other free energy sources, solar collectors can be used in parallel with fuel-fired heat sources at temperatures that provide good heat pump efficiency. This is because solar collectors can provide heat at a substantially higher temperature than other free sources.

Whenever the boiler is operating, i.e., when the solar collectors do not have sufficient output to carry the entire heating load, the boiler controls the loop temperature. In this case, the optimum loop temperature is a compromise between solar collector efficiency and heat pump efficiency. When the solar collectors are able to deliver a significant amount of heat, lower the loop temperature to maximize the output of the solar collectors. When the solar collectors cannot provide much energy (e.g., at night or during a heavy overcast on a cold day), increase the loop temperature to improve the efficiency of the heat pumps.

The output of the solar collectors can be measured directly with a BTU meter. To maximize the overall system efficiency, calculate a table of optimum loop temperatures based on the amount of solar heat available and the efficiency of the heat pumps at different loop temperatures. Then, design the system controls to maintain these temperatures.

How to Set the Loop Temperature in Systems with Thermal Storage

Control of loop temperature becomes still more complex if the system includes thermal storage. The simplest (but least efficient) case is a storage system that has no heat source other than a boiler. For this system, the basic problem is predicting how much thermal energy to store.

If the heat pumps are primarily providing cooling, they are adding heat to the loop. Should this heat be rejected to the cooling tower or saved in the thermal storage tank? The answer depends on the amount of heating operation that you anticipate. If not enough heat is stored, the boiler must operate longer to provide heating. On the other hand, if too much heat is stored, the higher temperature in the storage unit makes the heat pumps less efficient while they are still cooling.

Thermal storage in loop systems usually is limited to one-day cycles, so you can estimate the amount of heat to be stored from the outside air temperature and the cooling load. Delay turning off and isolating the cooling tower, which forces the loop and storage temperature to rise, until the latest possible moment. You can use an optimum-start controller (see Measure 4.1.1.2) to perform this control function. Alternatively, you can program an energy management computer for this.

Use the converse approach when storing cooling. Toward the end of a heating cycle, turn off the boiler at the latest possible moment that allows the heat pumps to pull down the loop temperature in anticipation of cooling operation.

In any event, don't expect to predict the load in the next cycle with much precision. Electric generating companies have full-time staffs who specialize in guessing what the cooling or heating load will be tomorrow.

If thermal storage is used in a system that also has a source of free energy, control can become even more complicated. The control sequence depends on the type of free energy source, as explained previously.

Avoid Efficiency Traps

To keep the system at maximum efficiency, optimize the loop temperature for every possible mode of system operation, at all seasons, all times of day, and with all possible combinations of heating and cooling loads. Here are some guidelines:

- *Avoid forcing a change of loop temperature with purchased energy.* Do not allow the boiler to raise the temperature of the loop as long as the temperature can be raised by the heat pumps or a free energy source. Similarly, do not use heat rejection equipment to blow away loop heat when entering a heating period.

- *Maintain a safe deadband between the operation of the boiler and the operation of the heat rejection equipment.* Otherwise, purchased energy will be expended alternately by the boiler and cooling tower to follow small fluctuations in the loop temperature.

- *Anticipate shutdowns periods when heating.* Turn off the boiler prior to shutting down the system and let the heat pumps extract the heat from the loop. Refer to the previous discussion of thermal storage. Coordinate with the controls recommended in Measure 5.1.3 that limit the duration of system operation.

- *Do not keep boilers warmed up.* Design the loop temperature controls to start the boiler only when it is needed. As soon as a boiler starts to fire, the heat pumps can extract the boiler's heat output from the loop. Letting the boiler stay cold eliminates standby losses. Furthermore, a boiler that is kept hot suffers from thermal shock when cold loop water is suddenly passed through it. (In heating systems that use the output of a boiler directly, the boiler is kept warm because a high temperature is needed to deliver the heat to the system. This is not necessary in heat pump loop systems.)

- *Install control interlocks to prevent simultaneous operation of the heat sources and the heat rejection equipment.* Defects may occur that cause the heat sources and heat rejection equipment to operate simultaneously. Interlocks can prevent this. The interlocks should include alarms to alert operators to the fact that a defect exists.

- *Be careful about the location of the temperature sensors.* The thermostatic controls for loop temperature regulate the temperature only at the point where the temperature sensor is located. The temperatures at other points in the system may be radically different. Figure out carefully what you are trying to measure, and install the control sensors accordingly. If possible, try to measure the temperature at the same point in the loop for all control functions. If this does not provide enough information for good control, add multiple sensors and program the controls accordingly.

ECONOMICS

SAVINGS POTENTIAL: *10 to 40 percent of the energy consumed by the system.*

COST: *A few thousand dollars, at least, assuming that the system presently does not have loop temperature optimization.*

PAYBACK PERIOD: *Less than one year, to several years.*

MEASURE 5.6.2 Install automatic valves and bypass lines to prevent the flow of loop water through heat sources and heat sinks when they are not needed.

RATINGS

New Facilities	Retrofit	O&M
A	**C**	

The controls recommended by Measure 5.6.1 should turn the heat sources and heat sinks on and off as necessary for optimum efficiency. However, controls can malfunction or they can be set incorrectly. Furthermore, heat may be able to leak into or out of the loop even if the sources or sinks are turned off.

These conditions may persist indefinitely, because they have no visible symptoms other than high energy bills. This makes it prudent to install a bypass line and automatic valves to isolate each heat source and heat sink from the loop. The loop system controls should open and close the isolation valves automatically. The system diagrams in the introduction to this Subsection show the boilers and the heat rejection equipment installed in this manner.

Isolate Boilers

Be able to physically isolate the boiler from the loop. Idle boilers act as heat exchangers. If the loop is kept warm during cold weather by other heat sources, such as solar collectors, heat can be lost from the loop by convection through the flue of an idle boiler. This problem does not arise with electric boilers, which have no flues. It is only a minor factor with boilers that have closed combustion chambers.

Plant operators may tend to keep the boiler running throughout the cold weather season, and perhaps all year. This is undesirable. The boiler should operate only when it is needed at the moment to feed heat into the loop. Measure 5.6.1 explains why. Installing an automatic isolation valve at least prevents leakage of hot water into the loop when it is not needed.

SUMMARY

Prevents loss of heat through idle equipment. Also acts as an additional safety feature against improper operation.

SELECTION SCORECARD

Savings Potential	**$**	$	$
Rate of Return, New Facilities	**% %**	**%**	%
Rate of Return, Retrofit..........	**% %**	%	
Reliability	✓ ✓		
Ease of Retrofit	☺	☺	☺

Isolate Heat Rejection Equipment

Bypass water around cooling towers and dry coolers when heat rejection from the loop is not required. These units are located outdoors, and they can reject a considerable amount of heat even when the cooling tower fans are not running.

The system controls should disable the fans in heat rejection equipment when they close the isolation valves. If this is neglected, the fans may continue to operate as a result of high outside air temperature or unintended operation of sump heaters.

Isolate Solar Collectors

When solar collectors are not providing heat to the loop, they can lose a large amount of heat. They are effective radiators because they are located outdoors and have large surface areas. For this reason, the controls should isolate solar collectors from the loop whenever the collectors are not feeding heat into the loop.

Isolate Ambient Heat Sources and Heat Sinks

Some ambient sources and sinks, especially water wells and earth coils, are connected to the loop continuously, and hence they do not require isolation. Others sources are not usable when freezing occurs, so these sources should be isolated. For example, air source coils suffer heat loss in the same way as dry coolers, and they should be isolated in the same way.

Select Isolation Valves for Longevity

Select appropriate types of isolation valves. The valves should stay free of leaks over the long term, have low maintenance requirements, and be easy to maintain. In small sizes, isolation typically is accomplished with solenoid valves. In larger sizes, motorized valves of various types may be used. Plug valves and ball valves can provide tight shut-off at low cost, and they are easy to operate automatically with standard actuators.

Install the Placards

Install effective placards on all bypass valves that explain their purpose and state when they should be opened or closed. See Reference Note 12, Placards, for details of effective placard design and installation.

ECONOMICS

SAVINGS POTENTIAL: *Perhaps 5 to 30 percent of system energy consumption, averaged over the life of the system. Specific savings cannot be predicted, because this is largely a safety measure to back up the operation of the system controls.*

COST: *Thousands of dollars.*

PAYBACK PERIOD: *Several years, typically.*

TRAPS & TRICKS

EXPLAIN IT: *Install an effective placard on each valve that states the conditions under which it should be open or closed. Describe the purpose of the valves in the plant operating manual.*

MONITOR PERFORMANCE: *Failure of this kind of auxiliary feature is easy to overlook. Schedule inspections of valve operation during periods when they should be closed.*

MEASURE 5.6.3 Improve efficiency by substituting a heat source/sink with better temperature characteristics, or by adding a source of free heat.

SUMMARY

A source of free energy is a fundamental feature needed to make heat pump loop systems efficient. If an existing system presently uses the atmosphere as a heat source/sink, you may be able to improve efficiency substantially by using a different heat source.

SELECTION SCORECARD

Savings Potential .... $ $ $ $

Rate of Return, New Facilities % % % %

Rate of Return, Retrofit.......... % % %

Reliability .... ✓ ✓ ✓

Ease of Retrofit .... ☺ ☺ ☺

The fundamental advantage of a heat pump over other types of heating equipment is that a heat pump can draw energy from the environment, so the actual source of heating energy is free. Make sure that your heat pump loop system exploits this advantage as effectively as possible. The key is to select the most efficient free source or sources that are available.

(A heat pump needs purchased energy only to force the heat to move against the inside-to-outside temperature gradient. This is the energy that goes into the motor of the compressor.)

Loop Systems Lacking a Free Heat Source

Individual heat pumps always have a free source of heat, and this is what makes them economically attractive. In contrast, loop systems sometimes lack a free heat source, as shown in Figure 1 in the introduction to this Subsection. These systems must rely on the system's "supplemental" heat source (i.e., a boiler) when there is a net heating requirement. It may seem inconceivable to undertake the trouble and expense of a heat pump loop system without installing a free energy source. Yet, such systems exist. Indeed, the basic design literature for engineers overlooked this fundamental feature for years after the concept was popularized. It appears that the engineering glamor of moving heat around a building distracted designers from the fact that the fundamental advantage of a heat pump is its ability to draw heat from the environment.

Without a source of free heat, a heat pump loop system consumes more energy than a conventional heating system using the same fuel. This is because the heat delivered to the space is derived mostly from the "supplemental" energy source, plus the electrical energy input to the compressor. Thus, the heat pump is not only extraneous, but adds to the energy cost. The only efficiency advantage provided by such a system is the core-to-perimeter heat transfer. Averaged over a typical operating schedule, this advantage cannot compensate for the lack of a free heat source.

With such a crippled heat pump loop system, the way to improve efficiency is to add a source of free heat, as shown in Figure 2 in the Subsection introduction. The free heat source is connected into the system through the water loop. The physical connection is typically easy to make.

Your Choice of Free Heat Sources

See Reference Note 22, Low-Temperature Heat Sources and Heat Sinks, for the main characteristics of soil, well water, ground water, solar collectors, and waste

heat as free energy sources. Each of these sources/sinks has its own advantages and disadvantages. Some of these heat sources also serve as heat sinks, providing better heat rejection efficiency than air-cooled heat rejection units. Your choice depends on the characteristics of the facility and its environment.

Also, see Measure 5.6.1 about the temperature and control characteristics of free heat sources.

Along with the heat sources covered in Reference Note 22, consider using outside air as the heat source and sink, even though it is less efficient in both roles than the other free heat sources/sinks. In many cases, outside air is the only available source/sink.

Outside Air as a Heat Source

The outside atmosphere is the most common heat source and heat sink for individual heat pumps. The main advantage of an air source is ready availability and low cost. You don't have to go far to find outside air, and connections to the water loop are simple. However, the atmosphere is rarely used as a heat source for heat pump loop systems. Approach the design of an air source with the caution appropriate to a complex, innovative activity.

The main disadvantage of air as the heat source is that the temperature of the air is usually falling as the heating load is increasing. Low air temperature substantially increases the compressor power and reduces heating capacity. This makes the atmosphere the least efficient of all free heat sources. Furthermore, the heating efficiency of a loop system with an air source, when the system is providing only heating, is less than the efficiency of individual air-to-air heat pumps. This is because the water loop needs pump energy, and because it adds an extra stage of heat transfer.

When operating at low temperatures, an air source heat pump expends energy for defrosting. At even lower temperatures, the heat pump may not be able to operate at all. For this reason, air source heat pumps often depend on electric resistance heating during cold weather.

Similarly, the main disadvantage of air as a heat sink for cooling is that the air temperature is usually high when the cooling load is high.

■ Equipment Options for Collecting Air Heat

If the system has a dry cooler to reject heat from the loop, consider using the dry cooler to collect heat from the air. The heat transfer capacity of dry coolers is about the same whether absorbing or rejecting heat. If the winter heating load is substantially higher than the summer cooling load, the dry cooler will be too small for the peak heating load, but it may still be used within its heat absorbing capacity. When the capacity of the dry cooler is exceeded, the controls should turn it off and isolate it. Then, additional heat is provided by the boiler or other existing heat source.

You can install a supplemental heat absorbing unit to provide full capacity. If this unit is used only as a heat intake for the heat pumps, try to install it where it will ingest air that is warmed by contact with a sunlighted surface, such as a large, dark roof.

If the system presently has a closed-loop cooling tower, you might consider using it as a heat source. However, this would be quite a novelty. It would work only in cases where the rated cooling capacity of the tower is much larger than the peak heat collection requirement. Closed loop cooling towers have bundles of smooth coils that are kept wetted during operation. Wetting provides high heat transfer capacity, and the heat is rejected mainly by evaporation. However, the tower would have to operate dry to absorb heat, which makes its heat absorption capacity much lower than its normal heat rejection capacity. Furthermore, it would be troublesome to continually switch between dry (heating) and wet (cooling) operation.

■ Defrosting Air Heat Sources

When the system is operating primarily in heating mode, the heat pumps are mostly extracting heat from the loop, which keeps the loop temperature lower than the outside air temperature. When the outside surface of the air coil gets below freezing temperature, ice can start to accumulate on the coil. When this occurs, the fins become choked and heat transfers falls off drastically. At this point, the air source can no longer be used. The system must have a means of defrosting the coil, or a supplemental heat source (typically, a boiler) must be energized.

Defrosting a heat pump loop is problematic. Unlike individual heat pumps, a loop system cannot briefly reverse its operating cycle to provide defrosting. Instead, the system needs a separate heat source to defrost the coil. It may be possible to defrost using an external heater. A more efficient method is to circulate warm water through the coils for defrosting. This requires a separate source of warm water, a separate defrost circuit, changeover valves, and control paraphernalia. The design work is tricky.

If the system does not operate for long periods of time when coil freezing can occur, it may be more cost effective to forget about defrosting, and use the boiler as the heat source during freezing conditions.

Cost Advantage for Solar Collectors

The cost of adding solar collectors as the heat source for a heat pump loop system may be much lower than the cost of installing a solar system from scratch. This is because the loop system includes most of the expensive distribution system, and it may also include the expensive storage equipment that an active solar system needs. Further, this application can use less expensive solar collectors that are designed to operate at lower temperature.

Consider Scrapping the Loop Design

Designers have tended to become infatuated with heat pump loop systems. Despite their theoretical appeal, consider other alternatives, even with an existing system. If your loop system is flawed, get a clean sheet of paper and consider all your options, including scrapping the loop system.

You can add a free heat source to heat pumps without using a loop system. See Measure 5.5.10 about this.

In many locations, using a free heat source may not be the cheapest or best way to improve the overall efficiency of heating and cooling. For example, in a mild climate, individual air-to-air heat pumps are likely to be the most efficient way to use heat pumps. In a cold climate, a system of hydronic fan-coils with fuel-fired heating and conventional cooling may be the most efficient approach.

ECONOMICS

SAVINGS POTENTIAL: *If the system presently lacks a source of free heat, this Measure saves all or most of the energy consumed by the heat sources in the original system. The saving may be reduced by increased heat pump energy consumption. However, this is true only if the free energy source operates at a temperature lower than the maximum-efficiency temperature of the heat pumps.*

If you substitute a different free source/sink for an existing outside air source/sink, expect savings in the range of 30 to 60 percent of the original energy consumption.

COST: Tens of thousands of dollars, typically.

PAYBACK PERIOD: One year, to many years.

TRAPS & TRICKS

SELECTING THE SYSTEM: If your loop system does not have a free source of energy, it is seriously flawed. In that case, review your heating and cooling options broadly. Review the introduction to this Subsection, as well as Measure 5.6.1, to appreciate what it takes to make a loop system work efficiently. If you decide to keep the loop system, start your selection of the free energy source by reviewing Reference Note 22. Examine your options thoroughly. Each type of free energy source requires distinct expertise.

SELECTING THE EQUIPMENT: If you select packaged equipment or systems, such as solar collectors or earth source heat pumps, be relentless in checking references. Do not buy equipment that is not proven by long service.

INSTALLATION: Installing water sources is fairly routine, with few opportunities for major mistakes. On the other hand, heat exchange with soil involves a large temptation to take installation shortcuts. Horizontal soil fields are vulnerable to installation damage that causes later leaks that cannot be repaired. Install solar collectors for ease of maintenance, because they will need it.

MAINTENANCE: Once installed, soil heat exchangers require no routine maintenance. Water sources require occasional heat exchanger cleaning. Solar collector components tend to fail from weather exposure, from high stagnation temperatures, and in poorly designed systems, from thermal expansion and contraction. Schedule periodic checks of condition and operating efficiency in your maintenance calendar.

MEASURE 5.6.4 Add thermal storage to the loop system.

Adding thermal storage to a heat pump loop system can substantially improve the system's efficiency and/or reduce electricity cost if certain specific conditions exist. The physical installation of the storage unit is straightforward, provided that you have sufficient space for it. In a loop system, adding thermal storage is less expensive than with other types of conditioning systems because the loop system already includes all the necessary connections.

The difficult part of thermal storage is getting the controls right. Failing to control a thermal storage system efficiently forfeits most of the savings, and may even increase energy cost. Like most aspects of heat pump loop systems, thermal storage is fraught with traps in design, installation, and operation.

A water tank is the appropriate storage unit for a heat pump loop system. Water storage is mechanically simple, it requires no additional heat exchangers, and it is suitable for both heating and cooling. However, storing heat in water requires very large quantities of water and big tanks.

Figure 3, in the introduction to this Subsection, shows a basic heat storage installation in a heat pump loop system. Be aware that this simplified diagram does not include the essential free heat sources, which are usually much more important than thermal storage.

Where to Consider Thermal Storage for a Loop System

Thermal storage may improve a heat pump system in these ways:

- *storing free energy that is not available continuously.* For example, thermal storage may increase the amount of heat that can be collected by solar collectors. This benefit occurs only if: (1) the source is not available continuously, and (2) the source provides more heat than is needed when the source is available. Thermal storage does not improve the availability of free energy if the source, such as well water, is always available with adequate capacity.

- *storing heat between alternating cycles of heating and cooling.* For example, the heat pumps may draw heat from the storage tank in the morning for heating the building, and add heat to the storage tank in the afternoon when cooling the building. In HVAC applications, this benefit occurs during the seasons of mild and variable weather. This benefit does not occur if there is a continuous net requirement for heating or for cooling.

- *reducing electricity cost.* Thermal storage allows heating or cooling energy to be produced at times

of day when the cost of electricity to operate the heat pumps is lowest. See Reference Note 21, Electricity Pricing, for more about this. This benefit occurs primarily during periods of peak cooling load and/or peak heating load.

Heat loss from the storage tank limits the duration of a storage cycle to a few days, e.g., a weekend or a short weather cycle.

The tank must be large because water has limited heat storage capacity. To store a useful amount of heat, the water volume can be enormous, typically thousands of gallons for a system of modest size. The tank itself is expensive, and structural support to carry the weight may cost even more than the tank. For this reason, it is usually not economical to make the tank large enough to cover peak conditions. Analyze your needs and calculate a storage capacity that provides a good compromise between equipment cost and system efficiency.

Energy and Cost Saving Potential

Thermal storage can completely eliminate the need for fuel heating as long as the facility's daily cooling load is larger than its daily heating load. In absolute terms, the amount of energy saved depends on the size of the heating load or the capacity of the storage tank, whichever is less. Assuming that the storage tank has enough capacity, the maximum amount of energy is saved when the daily heating load exactly equals the daily cooling load.

Using thermal storage probably reduces the average heat pump efficiency somewhat. When heat is being added to the system by a supplemental heat source, the temperature of the loop can be kept as high as desired, whereas drawing heat from thermal storage involves a significantly lower average loop temperature. On the other hand, the cooling efficiency of the heat pumps is increased during the beginning of a cooling period, when the temperature of the storage tank has been lowered by the preceding heating period.

The potential saving in electricity cost vary widely, depending on the load characteristics of the facility, the nature of the heat pump system, and electricity pricing.

How to Design the Storage Equipment

Refer to Measure 2.10.5 about storage of hot water. Refer to Measure 2.11.1 about storage of chilled water. A typical heat pump loop system stores both hot water and chilled water at different times.

ECONOMICS

SAVINGS POTENTIAL: *10 to 30 percent of purchased heating energy, but only in circumstances that favor installation of a thermal storage unit.*

COST: *Several thousand dollars, or more.*

PAYBACK PERIOD: *Several years, or longer.*

TRAPS & TRICKS

ESTIMATING THE BENEFIT: *In most applications, estimating the potential saving is tedious and somewhat uncertain. A lot of money is at stake, so consider using an energy analysis computer program for your calculations. See Reference Note 17 for details.*

DESIGN: *While the hardware is simple, the control design may have to be complex. Review Measure 5.6.1 about the interactions between storage temperature and loop temperature. It takes a great deal of thought and patience to design for all the operating conditions that may occur in the system. Design a sensor array to accurately indicate the state of charge of the storage tank.*

EXPLAIN IT: *The control logic is not easy for anyone but the designer to understand. As a result, control failures or inefficient control adjustment may persist. Describe the system clearly in the plant operating manual. Install placards on the controls.*

MONITOR PERFORMANCE: *If your energy management control system has the ability to do trend logging, use it to record parameters that indicate how the system is operating. Alternatively, make appropriate manual entries in the plant operating log. Review these records continually.*

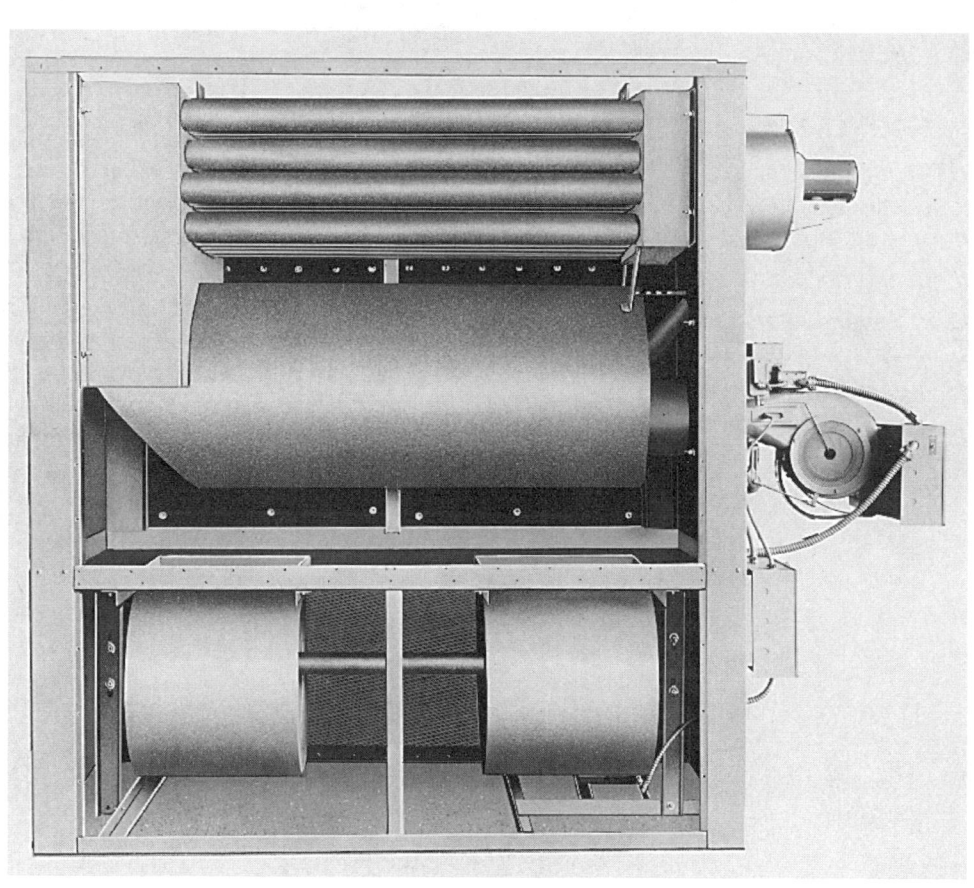

Direct-Fired Heating Units

The Measures in this Subsection improve the efficiency of space conditioning units that provide heating by firing of fuel within the unit itself. The types covered include direct-fired duct heaters, wall furnaces, residential-type forced air furnaces, and direct-fired heaters within air handling units.

INDEX OF MEASURES

5.7.1 Clean, adjust, lubricate, and repair direct-fired heating units at appropriate intervals.

5.7.2 Install high-efficiency heating units.

5.7.3 Install automatic flue dampers on existing heaters.

5.7.4 Set the anticipators of space thermostats to maximize the interval between firing cycles.

RELATED MEASURES

• Section 1, for Measures relating to larger heating plants and for all hot water heating systems

MEASURE 5.7.1 Clean, adjust, lubricate, and repair direct-fired heating units at appropriate intervals.

Direct-fired space conditioning units are generally reliable, and they require only a modest amount of maintenance. Give these items attention at regular intervals:

- *filters and filter mountings.* Clogged filters reduce the air flow through the furnace heat exchanger, causing the space air and the exhaust gases to be hotter. The higher exhaust gas temperature reduces furnace efficiency. Clogged filters also reduce heating capacity, requiring the furnace to operate for longer periods and to warm spaces more slowly. Dirty filters may be a breeding ground for microorganisms, especially if the furnace is located in an area of high humidity. The higher air temperature may cause some odors and air pollution, but only if the space air contains vapors or dusts that are changed chemically by the temperature in the heat exchanger.

- *burner cleanliness and air-fuel ratio.* Inadequate or improper burner maintenance wastes more energy than any other defect of maintenance. Routine burner maintenance consists of (1) cleaning the burner, followed by (2) readjusting the air-fuel ratio. Both of these functions require specialized equipment and techniques. See Measure 1.4.1 about burner cleaning. See Measure 1.3.1 about how to adjust the air-fuel ratio.

Gas-fired heating units typically require burner maintenance only at long intervals. Inspect the flame visually on a routine basis. Do a combustion gas analysis at the beginning of each heating season. If the test indicates the proper efficiency, leave the burner alone.

Oil-fired burners require cleaning and adjustment at regular intervals. Check the manufacturer's recommendations, and also inspect the units periodically.

- *fireside cleaning.* Combustion products that accumulate on the fireside surfaces of a furnace may cause serious energy waste. Adjust your cleaning interval based on the rate of soot accumulation. Typically, clean an oil-fired heater at the beginning of each heating season. See Measure 1.6.1 for details of fireside cleaning.

Gas-fired heaters that are adjusted properly may never require fireside cleaning. This is because soot burns off or spalls off faster than it can accumulate. If a heating unit accumulates soot or other fouling at an abnormal rate, the usual cause is a dirty or improperly adjusted burner.

SUMMARY

Basic maintenance items needed to maintain efficiency, comfort, and health.

SELECTION SCORECARD

Savings Potential .... $ $ $

Rate of Return .... % % % %

Reliability .... ✓ ✓

Ease of Initiation .... ☺ ☺ ☺

- *automatic flue dampers* are important to efficiency. Failure of a flue damper can be a serious safety hazard, so they are made to be highly reliable and to require no maintenance. Check the operation of flue dampers visually at the beginning of the heating season. Flue dampers are covered by Measure 5.7.3, below.

- *fan impellers* should be cleaned to keep air flow efficient. Be prepared for the fact that cleaning a centrifugal fan may make it noticeably louder.

- *fan motor lubrication.* If the fan motors are not permanently lubricated, oil them at the manufacturer's recommended intervals. If motor bearings are difficult to reach for lubrication, consider installing an access plate or a remote oiler tube.

- *outside air and combustion air dampers and actuators.* These are rarely found in direct-fired heating units. If a unit has a combustion air damper, check it for proper closure, freedom of movement, proper adjustment of linkages, and proper control. A malfunctioning or incorrectly adjusted combustion air damper reduces efficiency, and it may create a serious safety hazard.

ECONOMICS

SAVINGS POTENTIAL: *2 to 30 percent of energy input, the larger savings applying to oil-fired heaters. Effective maintenance extends the service life of equipment.*

COST: *$10 to $100 per unit per year, the larger costs applying to oil heaters.*

PAYBACK PERIOD: *Short. The long-term saving from proper maintenance is much higher than the cost.*

TRAPS & TRICKS

SCHEDULING: *Post a permanent inspection checklist at each heating unit. List the dates that each item is to* be performed. Make a duplicate of each checklist for the plant maintenance schedule.

SKILLS AND SUPERVISION: Burner maintenance and adjustment requires special equipment and skills.

Nobody likes to do burner and fireside cleaning because it is messy and awkward. Make sure that it is done correctly.

MEASURE **5.7.2 Install high-efficiency heating units.**

SUMMARY

The ultimate way to make heaters more efficient. Always do it when replacing old units. In some cases, it may even pay to replace a heater that is still functioning.

SELECTION SCORECARD

Savings Potential	$	$	$ $
Rate of Return, New Facilities	%	%	% %
Rate of Return, Retrofit..........	%	%	
Rate of Return, O&M	%	%	% %
Reliability of Equipment	✓	✓	✓
Reliability of Procedure	✓	✓	
Ease of Retrofit	☺	☺	☺
Ease of Initiation	☺	☺	☺

The efficiency of direct-fired heaters improved greatly in some models that became available during the 1980's. Some older heating units have seasonal average efficiencies of less than 60 percent. Modern non-condensing heaters may have efficiencies averaging around 80 percent. Condensing gas furnaces, such as the unit shown in Figure 1, typically have efficiencies better than 90 percent.

Select any new heating unit for the highest economical efficiency. The cost of the fuel burned by a heater in several years of operation may equal the cost of the heater itself. A wide range of efficiencies exists in the market, so a hasty replacement decision could waste a lot of money during the life of the new heater.

In existing facilities, consider replacing inefficient heaters even if they are still functioning. The economics are more favorable for this when:

- the existing heaters are very inefficient
- they are used for a large number of hours each year
- they are near the end of their service lives
- fuel costs are high.

How Efficient are Your Heaters?

To figure out whether it is worthwhile to replace an existing heater, you need to know the heater's efficiency. Older heaters generally were not rated for efficiency. The energy input stated on the nameplates of older heaters is nominal, so you cannot use it to calculate actual efficiency.

You can measure the combustion efficiency of a heater directly while it is operating. This is fairly easy. See Measure 1.2.1 for test procedures. A combustion efficiency test accounts for these deficiencies:

- **insufficient heat transfer surface.** Cheap heaters get their capacity by increasing the size of the burner and burning more fuel. Efficient heaters get their capacity by increasing the amount of heat exchange surface to extract as much heat as possible from the combustion gases. You can detect skimpy heat transfer surface easily by measuring the flue gas temperature. See Measure 1.2.1 for details. In older units, this factor typically wastes 10 to 20 percent of the fuel energy.

- **inability to capture the latent heat of flue gas moisture.** The latent heat of the water in flue gas represents about 10 percent of the combustion energy of natural gas, and less with oil. The heat loss is increased by the fact that non-condensing furnaces are designed to keep the combustion gases hot enough to eliminate any possibility of acid condensation on the heat exchanger. The gases are kept hot by reducing the heat transfer surface. Conversely, many modern heaters capture the latent heat of flue gas moisture by using a heat exchanger that is large enough to condense the water. These condensing furnaces are explained below.

- **incomplete fuel combustion.** This is a minor factor in all heaters as long as the burner is well maintained. The best modern heaters have slightly better fuel combustion than older heaters.

Combustion efficiency testing cannot reveal some causes of energy waste that may be important. You can only estimate these effects, which include:

- *standby losses,* which are explained in Measure 5.7.3. Efficient heaters reduce standby losses with an integral flue damper, or with a combustion gas path that has high resistance to convection.
- *continuous operation of a pilot light,* which wastes energy during periods of time when heating is not needed, and also increases standby losses during the heating season. Most modern heaters avoid this waste by using an igniter that operates only briefly to start the flame.
- *loss of conditioned space air through the flue.* This is important if the heater draws air from a conditioned space, either directly or indirectly. This

loss is increased somewhat in heaters that require a larger percentage of excess air for proper combustion. Many high-efficiency units make it easy to connect a ducted outside air intake for combustion, which eliminates this problem.

Condensing Furnaces

The most dramatic improvement in the efficiency of gas-fired heaters occurred with the introduction of condensing furnaces. Figure 1 shows a typical unit. These heaters capture the latent heat of the water in the gases by using so much heat transfer surface that the flue gas is cooled below the condensing temperature of water. The large heat exchanger also pushes sensible heat transfer to the limit. Another bonus is that the high-efficiency heat exchanger creates high resistance to combustion air flow, which minimizes the standby losses that occur between firing cycles.

This feature raises the price of the heater substantially. The heat exchanger is larger, and the low-temperature portion is made of stainless material. Also, there are additional components and more complex controls. The resistance of the heat exchanger requires a small fan to move the combustion gases. The energy consumption of this fan is a small fraction of the heater's total energy input.

Condensing units must be connected to a sewer drain to get rid of the condensate. The condensate may be somewhat acidic, even with natural gas as a fuel, so you may have to take steps to protect sewers from corrosion. (This is unlikely to be a problem with gas fuels.)

Another possible disadvantage is that condensing furnaces deliver air at lower temperatures than do non-condensing models. The temperature rise as space air passes through the furnace may be as low as 45°F. If a space is being warmed after a prolonged period of temperature setback, air delivered by the heating unit may create a drafty sensation until the room temperature rises to normal. (Even so, high-efficiency furnaces provide air at higher temperature than heat pumps. Low heating air temperature is a persistent complaint with heat pumps.)

At present, most condensing air heaters are fired with gaseous fuels, including natural gas, propane, and liquefied petroleum gas. Gas has a larger fraction of hydrogen than petroleum fuels, so the efficiency advantage is greater than with oil. This is because the hydrogen in fuel is the origin of the water in the flue gas.

There are various impediments to burning oil in condensing furnaces. The sulfur contained in petroleum forms enough sulfuric acid to be corrosive. Also, oil contains some quantity of dirt and it forms some amount of unburned hydrocarbons. Even light oil would eventually foul the cooler portions of the heat exchanger, which typically is difficult to clean.

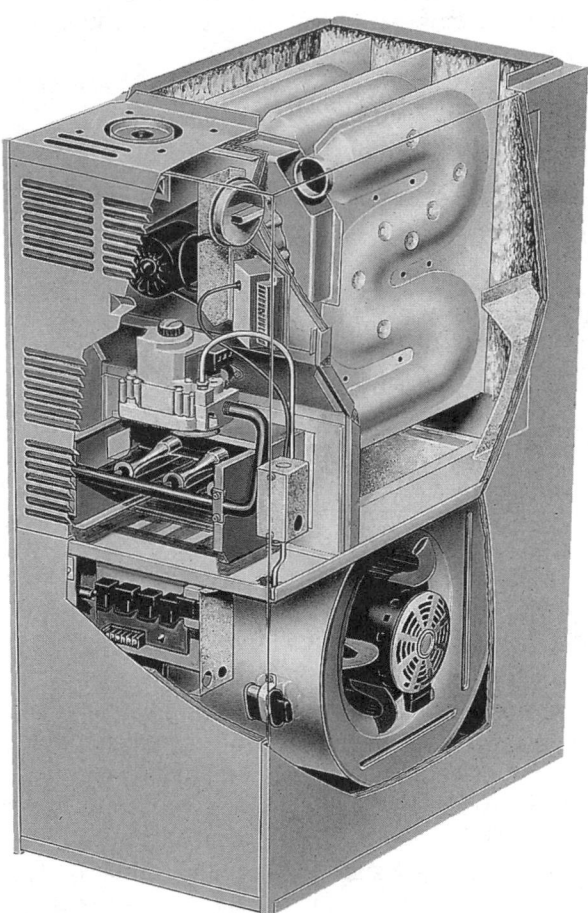

Lennox Industries Inc.

Fig. 1 High-efficiency condensing gas furnace Units of this type have two main features that enhance efficiency. A large heat exchanger made of non-corrosive material is able to cool the combustion gas sufficiently to extract the latent heat of the water in the flue gas. The resistance of the flue gas passage is so high that standby losses are eliminated when the furnace is not firing. A small fan is needed to draw the gas through the furnace.

Efficiency Ratings of Fuel-Fired Heaters

In the United States and some other countries, laws require furnaces to be rated for efficiency. In the U.S., these ratings show where each model's efficiency lies within the range of efficiencies for that type of heater. The efficiency ratings are displayed prominently in product literature and on the equipment itself. Efficiency standards have stimulated manufacturers to pursue efficiency as an end in itself, with the result that models are available that greatly exceed minimum efficiency requirements.

Some industry associations may be able to provide lists of efficiency ratings for units of particular types, such as gas-fired furnaces.

Multi-Fuel Units

The ability to burn multiple fuels is rare in direct-fired air heaters, but some dual-fuel furnaces have been marketed. Some units have dual burners to allow switching between gas and oil. Many gas-fired heaters can be converted to burn either natural gas or bottled gas. However, you cannot make this conversion at a moment's notice.

Consider dual-fuel heaters in situations where one of the available fuels may have the price advantage at one time, and a different fuel has the price advantage at

other times. Also, consider them if you fear that the favored type of fuel may become scarce.

ECONOMICS

SAVINGS POTENTIAL: *10 to 30 percent of the fuel consumed by the heating units.*

COST: *$7 to $20 per MBH capacity, in typical sizes. The price premium for condensing furnaces, compared to non-condensing furnaces of average efficiency, ranges from 50% to 80%.*

PAYBACK PERIOD: *Typically about one year, in new construction and when replacing a failed unit. Several years or longer, when replacing a functioning unit. Consider the remaining life of existing units.*

TRAPS & TRICKS

SELECTING THE EQUIPMENT: *High-efficiency heaters are more complex than conventional heaters, and are more prone to problems. Ask vendors for references, and talk with purchasers who have used the equipment for at least several years. Do not purchase a model that is entirely new. Purchase a popular model, so that you will be able to find spare parts in the future.*

INSTALLATION: *If you select a condensing heating unit, anticipate the need to drain the flue gas condensate into a sewer.*

MEASURE 5.7.3 Install automatic flue dampers on existing heaters.

The output of direct-fired heating units is usually controlled by turning them on and off. There are various ways in which heaters lose heat through the flue during the periods when heat is not being delivered to the space. These losses are called "standby losses." With an existing heater, you can prevent much of this energy waste by installing an automatic flue damper.

Which Heaters are Eligible for Flue Dampers

Flue dampers are appropriate for common types of heaters that have open combustion air paths and conventional flues. The heater must allow for installation of the damper at an appropriate point in the flue, as explained below.

Flue dampers are not needed with high-efficiency heating units that employ forced circulation through the combustion chamber. In such units, the resistance to convective flow is so high that heat loss is minimal when the fan is not running. Some high-efficiency furnaces include a flue gas damper.

Types of Standby Losses and Potential Savings

Flue dampers reduce standby loses. The potential for saving depends on the severity of these losses. To estimate the potential saving, examine your heating units to see whether they suffer from these types of standby losses:

- *loss of heat from the mass of the furnace.* This applies to all heaters. Combustion air continues to flow through the furnace by convection when the unit is not firing. This carries heat away from the furnace and the flue. This loss is small in direct-fired air heaters because they have lightweight construction. Also, the space air fan continues to operate after firing ceases, recovering furnace heat. The maximum loss from this cause is about four percent in air heaters. (It may be much more in hot water boilers, which have a large amount of heat stored in the mass of the water.)
- *loss of conditioned space air.* If the heater draws combustion air from a heated space, loss of space air through the flue continues by convection when the furnace is not operating. This may waste up to five percent of fuel energy. The amount is determined largely by factors unrelated to the heating unit itself, such as the leakiness of the building envelope, the outside temperature, and the effect of wind on the flue.
- *continuous operation of a pilot light* wastes most of the pilot light heat through the flue. (This

SUMMARY

A simple modification to improve the efficiency of heaters that have unrestricted flue gas flow. Not all heating units allow the retrofit.

SELECTION SCORECARD

Savings Potential .... $ $ $

Rate of Return .... % %

Reliability .... ✓ ✓ ✓ ✓

Ease of Retrofit .... ☺ ☺ ☺

assumes that the inside air fan does not operate between firing cycles.) A continuous pilot light may waste up to five percent of total fuel consumption, if the heater does not operate much. In absolute terms, the energy consumption of the pilot light is small.

Flue dampers differ in their effectiveness. Expect the flue damper to reduce losses by 50 to 80 percent.

Types of Flue Dampers

Self-powered thermostatically operated flue dampers are used in smaller heaters. This type opens in response to the high temperature of flue gases. Most thermostatic dampers have a blade made of bimetallic material that changes shape when heated. After the burner lights, hot gas reaching the damper blade causes it to warp to an open position, allowing the gas to pass up the chimney. No external power or external connections are required. The advantages of thermostatic dampers are low cost and reliability.

A limitation of self-powered thermostatic dampers is that they must remain partially open so that hot gas can flow quickly to the thermostatic element as soon as the burner starts firing. Partial opening is achieved by making a hole in the damper blade, or by undersizing the damper blade. This lack of tight closure undermines the effectiveness of the damper. As a result, thermostatic dampers are limited mainly to small furnaces and boilers, where their low cost is important.

Motorized dampers typically consist of a single blade damper that is moved by an electric actuator powered from the furnace's firing circuit. These dampers are more expensive and difficult to install than the self-powered variety, but they provide better closure of the flue. They may require additional safety controls to ensure that the fuel is shut off when the damper is not fully open.

How to Install a Flue Damper

The flue damper is installed in a short section of flue duct. A short piece of the existing duct is cut away, and it is replaced by the flue damper. Dampers are available with various diameters to match the flue diameter. Seal the connections well. The heater side of the damper is under positive pressure when the damper closes, so flue gases would be forced into the surrounding space through any leaks.

The flue damper must be installed between the heating unit and any draft hoods or draft regulating dampers ("vacuum breakers"). In other words, there can be no opening between the heater and the flue damper. Installing the flue damper on the chimney side of the draft regulator would defeat the purpose of both devices. Furthermore, it would create a safety hazard by forcing flue gases into the surrounding space. This requirement may make it impractical to install a flue damper on some existing heaters.

A flue damper should not restrict combustion air flow significantly when it is open. However, check the air-fuel ratio after installing a flue damper. Measure 1.2.1 explains how to do it.

Safety Issues

Flue dampers create a safety hazard if the damper remains in the closed position while the burner is firing. If this occurs, combustion gases will back up into the space surrounding the heating unit. Approved flue dampers include safety features that deal with this possibility. Therefore, select only approved flue dampers and install them in a safe manner. Standards for flue dampers are set by various agencies. For example, the American Gas Association acts as an approval agency for flue dampers that are used in gas-fired heating units. Make sure that the flue damper has the seal of the appropriate organization.

Alternative: Install Better Heating Units

Flue dampers can save only a fraction of the energy waste that occurs in less efficient heating units. If the existing heaters are much less efficient than the best modern units, consider replacing them. See Measure 5.7.2 for details.

ECONOMICS

SAVINGS POTENTIAL: 2 to 8 percent of the fuel consumption of the heating unit, depending on hether the unit has outside combustion air, a fixed pilot light, etc.

COST: Self-powered dampers cost from $20 to $50. Externally powered dampers cost $100 to $200. Installation requires one or two hours, typically.

PAYBACK PERIOD: At least several years. May be much longer.

TRAPS & TRICKS

CHOICE OF METHOD: Make sure that the heating unit allows you to do a good job of fitting the damper. Even if you can, consider whether you are trying to make a silk purse out of a sow's ear. Consider replacing an old, inefficient heater entirely, to exploit all the efficiency advantages of a new heater.

SELECTING THE DAMPER: Inspect the damper for quality before you buy it. Safety certification does not indicate reliability or efficient operation. Select a unit that closes as tightly as possible while satisfying safety requirements.

INSTALLATION: Some installations require ingenuity. Be sure that the entire flue gas passage on the heater side of the damper is well sealed. Be sure that the damper blade moves freely.

MEASURE **5.7.4 Set the anticipators of space thermostats to maximize the interval between firing cycles.**

Direct-fired heaters that control temperature by cycling on and off are subject to "standby losses," which are explained in Measure 5.7.3. The thermostats that control direct-fired heating units usually have a feature called an "anticipator." Anticipators cause the heating unit to cycle more frequently, so they increase standby losses. Reference Note 37, Control Characteristics, explains anticipators.

To reduce standby losses, set the anticipators inside the thermostat for the longest duration of heater operation that is consistent with comfort. You can adjust the anticipators in most thermostats quickly and easily. Typically, you move a small sliding contactor inside the thermostat that bypasses a portion of anticipator heating

Fig. 1 **Thermostat anticipators** In this two-stage heating thermostat, each stage has its own anticipator setting. To change the setting, slide the pointer over the numbered arc.

WESINC

SUMMARY

A simple adjustment that saves a small amount of energy. May affect comfort.

SELECTION SCORECARD

Savings Potential .... **$**

Rate of Return .... **% % % %**

Reliability .... ✓ ✓ ✓

Ease of Initiation .... ☺ ☺ ☺

element. Figure 1 shows the anticipator settings in a typical two-stage thermostat.

Don't make the firing interval so long that the space temperature can become uncomfortable, either because the space becomes too warm at the end of the firing cycle, or because it becomes too cold before the next firing cycle. In the latter case, occupants will raise the space temperature, increasing energy consumption.

Where This Works, and Where It Doesn't

This Measure is useful with conventional heating units that do not have a means of blocking the flow of combustion gases between firing cycles. It is not useful with heating units that are immune to standby losses. In particular, modern condensing furnaces (see Measure 5.7.2) have minimal standby losses. Heating units with automatic flue dampers also have reduced standby losses, so adding flue dampers (see Measure 5.7.3) eliminates most of the benefit of this Measure.

ECONOMICS

SAVINGS POTENTIAL: *0.5 to 2 percent of the fuel consumed by the heating units.*

COST: *Minimal.*

PAYBACK PERIOD: *Immediate.*

TRAPS & TRICKS

ADJUSTMENT: *Don't overdo it. Not much money is at stake, and energy conservation should never cause discomfort.*

Section 6. BUILDING AIR LEAKAGE

INTRODUCTION

New buildings can be designed and constructed to minimize air leakage at little additional cost. You can often reduce leakage in existing buildings by using methods that are inexpensive and simple. However, making existing buildings airtight is usually expensive or impractical. In any building, minimizing air leakage requires continuing maintenance.

Isn't Outside Air Needed Anyhow?

Air leakage through the building envelope may or may not be a significant source of energy waste. This depends largely on ventilation requirements. For example, it makes no sense to reduce the air leakage of a garage door if the garage has a fixed ventilation requirement that is higher than the leakage rate of the door. On the other hand, you should minimize leakage around a similar door that is installed in a heated warehouse where ventilation requirements are minimal.

Occupied spaces require a certain minimum amount of ventilation. In many countries, ventilation requirements are dictated by building codes, which require ventilation rates that are fairly high. (See Measure 4.2.1 for details.) In principle, air leakage can comprise part of the needed ventilation. If the leakage rate is less than the ventilation requirement, it does not "waste" energy, and there may be no economic merit in stopping it.

A building may have a high ventilation requirement when it is occupied, but little or no ventilation requirement when it is empty. This is true of most buildings intended primarily for human occupancy, such as schools, office buildings, stores, etc. Air leakage during the vacant period cools or heats the structure and contents, requiring additional conditioning energy when occupancy resumes.

Infiltration Causes Discomfort

Even if the rate of air leakage is much less than the required ventilation rate, you may still want to minimize air leakage for the sake of comfort. Unlike ventilation air that enters through the air conditioning system, air leakage enters the space at outside temperature and humidity. During cold weather, air leakage causes discomfort in the form of drafts that affect people located near walls, doors, and passages leading to the outside.

In addition to being objectionable from a comfort standpoint, cold drafts motivate occupants to increase heating and cooling energy consumption. If occupants bring in portable electric heaters, suspect serious infiltration.

Discomfort occurs only with infiltration (air leaking inward), not with exfiltration (air leaking outward). Many buildings maintain a positive pressure to avoid discomfort from infiltration. Positive pressure does not increase energy consumption if it is limited to occupied periods, and if exhaust fan capacity is reduced accordingly. In reality, such rational control is rare.

How to Use Section 6

The Section is organized by the envelope components where leakage occurs, including personnel doors, vehicle doors, windows, and other penetrations. The Measures in each Subsection provide specific improvements to these components. Go through all the Subsections, as some of the Measures in each Subsection apply to most buildings.

Before Getting Started ...

Read Reference Note 40, Building Air Leakage, to learn about the causes of air leakage, methods of finding air leaks, and methods of calculating leakage rates.

The Measures in this Subsection reduce air leakage through conventional personnel entrance doors. Air leakage wastes energy directly through the need to condition the entering air, and also indirectly by creating discomfort in the vicinity of the doors that motivates excessive heating of those areas. These effects vary widely, depending on the ventilation requirements and layout of the affected spaces.

INDEX OF MEASURES

6.1.1 Maintain the fit, closure, and sealing of exterior doors.

6.1.2 Install appropriate weatherstripping on exterior doors.

6.1.3 Install effective closers on exterior doors.

> **6.1.3.1 If manual opening of doors is acceptable, install spring-type door closers.**

> **6.1.3.2 If manual opening of doors is not acceptable, install automatic doors or door openers.**

6.1.4 Install high-efficiency doors.

6.1.5 If a pressure differential at the entrance is unavoidable, install a revolving door.

6.1.6 If a revolving door is installed, encourage people to use it instead of other doors.

6.1.7 Install storm doors.

6.1.8 Install vestibules for doors with frequent traffic.

6.1.9 Seal abandoned doors.

Related Measures:

- Measure 4.2.1, control of building pressurization
- Subsection 6.2, for Measures related to industrial doors
- Measure 6.4.2, sealing infiltration around door frames

MEASURE 6.1.1 Maintain the fit, closure, and sealing of exterior doors.

Exterior personnel doors allow a significant amount of air leakage. Leakage grows progressively worse as weatherstripping wears out, doors get out of alignment, and closers fail. When a point is reached that doors no longer close completely, the increase in air leakage becomes drastic.

To avoid this energy waste, perform these maintenance items at appropriate intervals:

• align doors for effective sealing and complete closure

• repair infiltration seals

• repair and adjust closing mechanisms.

Experience will tell you how often to schedule door maintenance. It will depend on how sturdy the doors are, and on the amount of traffic.

Related Measures

If the doors are not equipped with adequate infiltration seals, you may be able to install retrofit seals. See Measure 6.1.2, next. If the doors require excessive maintenance, consider replacing them with more efficient units, as recommended by Measure 6.1.4.

To save effort, coordinate this activity with maintenance of the industrial-type doors covered by Measure 6.2.1.

SUMMARY

Routine maintenance needed to prevent energy waste and maintain comfort.

SELECTION SCORECARD

Savings Potential .... $ $ $

Rate of Return .... % % % %

Reliability .... ✓ ✓ ✓ ✓

Ease of Initiation .... ☺ ☺ ☺

ECONOMICS

SAVINGS POTENTIAL: *$10 to $200 per year per door, depending on climate, energy cost, space layout, etc. Timely maintenance also avoids major repairs.*

COST: *$5 to $20 per year per door.*

PAYBACK PERIOD: *Short. The long-term saving is much greater than the cost.*

TRAPS & TRICKS

SCHEDULING: *Most of the challenge of door maintenance is remembering to do it. Make sure that you have an effective maintenance calendar.*

PREPARATION: *Keep an ample stock of spare parts on hand. Finding parts typically takes much longer than the work on the doors themselves.*

MEASURE **6.1.2 Install appropriate weatherstripping on exterior doors.**

Exterior personnel doors often have large gaps that you can seal using inexpensive weatherstripping. The trick is to install the weatherstripping so that it seals effectively, remains in place, and does not keep the door from closing completely.

First, Repair the Doors

It is futile to install weatherstripping on a door that is loose or out of alignment. If you do, the weatherstripping will fail quickly. Your first step is to make sure that the door is secure and moving properly. See Measure 6.1.1 for door repair procedures.

How to Install the Weatherstripping

Consider each of the four sides of a door separately when you select weatherstripping. Each type of door needs its own combination of weatherstripping materials. Make the effort to find the types of weatherstripping that will work effectively with your doors.

Some types of weatherstripping are designed to create a seal by sliding, and other types are designed to create a seal by compressing. Few are effective in both ways. If you install the wrong type of weatherstripping, it will be ineffective, or it will soon be damaged, or it will break loose. Installing the wrong kind of weatherstripping may keep the door from closing properly, which will make the air leakage even worse. Here are some hints for different types of doors.

■ Single Doors with Stop Strips

Single doors that open only in one direction are commonly used in residential facilities, and as secondary doors in commercial and industrial facilities. The top and two sides of the door close against a stop strip that is installed on the door frame. The door has butt hinges. This type of door can be sealed fairly tightly.

You can seal the top and both sides of the door by using a leaf-type flexible weatherstrip. Install it on the sides of the door frame, in a way that does not interfere with the contact between the door and the stop strip. On the top and latch side, the material works as a sliding seal. On the hinge side, it acts as a compression seal. On each side of the door, select the thickness of the weatherstripping to fill the gap between the door and the frame.

This material is available in plastic and spring bronze. The bronze version is much more durable. The bronze strip is fastened with nails, so it works only with wooden door frames. A plastic version is available with a self-adhesive backing.

SUMMARY

An easy way to eliminate most air leakage around doors. Requires continuing maintenance.

SELECTION SCORECARD

Savings Potential	$ $ $
Rate of Return, Retrofit	% % % %
Rate of Return, O&M	% % % %
Reliability of Equipment	✓ ✓
Reliability of Procedure	✓ ✓
Ease of Retrofit	☺ ☺ ☺
Ease of Initiation	☺ ☺ ☺

On the hinge side of the door only, if the gap is too large or too small for leaf-type weatherstripping, you can substitute a soft foam self-adhesive strip. This type is available in many thicknesses and widths. Install the strip near the hinge line, so it will create little resistance to closing the door.

Do not use compression strips between the door and the stop strip. No matter how soft the material is, it will tend to keep the door partially open, increasing leakage. A gap between the door and the stop strip does not matter, as long as the sides of the door are properly sealed to the sides of the frame. Reduce the gap by adjusting the hinges and the lock strike.

For the bottom of the door, you have a choice between installing a threshold or not. Thresholds cause people to stumble, so they are generally limited to private residences, where the residents are accustomed to stepping over them. The advantage of a threshold is that it provides a good seal without requiring a sweep on the door that contacts the floor. Door sweeps are incompatible with carpeting, they create a visible wear pattern on solid floors, they remove the varnish from wooden floors, etc.

You can buy thresholds that have an integral seal. Typically, the threshold is an aluminum extrusion, and a flexible plastic seal fits into a groove in the extrusion. These thresholds are satisfactory if the bottom of the door is parallel to the floor. However, you may not be able to find a replacement for the plastic seal when it wears out. A more reliable alternative is to install a sweep on the bottom of the door that contacts the threshold. You can easily adjust the sweep for any irregularity in the gap at the bottom of the door.

If you do not install a threshold, install a sweep strip on the bottom of the door. You can get these in the form of a plastic strip or in the form of a brush. Generally, the brush is better because is offers less resistance to opening and closing the door. A door sweep can interfere with opening the door fully if the bottom of the door gets closer to the floor as the door swings open. For this situation, try to find a brush-type sweep with very long bristles that can bend easily.

■ Double Doors with Stop Strips

Double doors that close against stop strips are found in many residential, commercial, and industrial buildings. Seal them in the same way as for single doors, discussed previously.

In addition, you have to seal the gap between the two doors. If both doors are intended to be used regularly, any seal that bridges the gap between the two doors tends to keep the doors from closing fully. The best choice in this case is a very soft brush seal, combined with a door closer that has sufficient force to ensure that the doors close fully. (See Measures 6.1.3 ff about door closers.)

If one door is intended to be opened preferentially, you can install a leaf-type seal on the door that usually remains closed. In effect, this door becomes part of the frame for the other door. (This door typically can be locked in place by a latch bolt that engages strike plates in the floor and the top of the frame.)

If the doors are closed in reverse order, a leaf-type seal will tear off. To prevent this, the door that opens first needs a strip of molding over the latch side that covers the gap between the two doors. This molding keeps the doors from being closed in reverse order. (In the door business, the strip that covers the gap between the two doors is called an "astragal.") This strip also provides protection from wind.

■ Single and Double Double-Acting Doors

A door that swings open in both directions is called a "double-acting" door. This type is difficult to seal. Any contact between the door and the frame tends to keep the door from closing fully. For this type of door, the best seal for all surfaces is a very soft brush that contacts the opposite surface very lightly. The door should have a closer the keeps the door centered with enough force to overcome the resistance of the seal.

Double-acting doors are usually made of metal. The door may have grooves to hold custom seals. If you cannot find these seals, or if they are unsatisfactory, you can attach brush seals to the door with sheetmetal screws. Self-adhesive seals would not survive the sliding contact very long.

Expect to Maintain the Weatherstripping

Weatherstripping wears out because of continual flexing, drying of adhesives, and abuse. Therefore, expect to replace it periodically. Schedule this as recommended by Measure 6.1.1.

Alternative: High-Efficiency Doors

Measure 6.1.4 recommends installing high-efficiency doors. The best models have integrated door and frame systems that include effective and durable seals. These doors offer much better sealing than you can achieve by weatherstripping a leaky door, and their weatherstripping requires less maintenance.

ECONOMICS

SAVINGS POTENTIAL: $10 to $200 per year per door, depending on climate, energy cost, space layout, etc.

COST: $10 to $30 per door, including labor.

PAYBACK PERIOD: Less than one year, usually.

TRAPS & TRICKS

SELECTING MATERIALS AND INSTALLATION METHODS: Select your weatherstripping materials for effective sealing, minimum interference with door closure, and durability. Expect to use two or three types of weatherstripping on each door. There are many sources. Keep looking until you find the right kinds.

MAINTENANCE: Weatherstripping wears out quickly compared to other maintenance items. Consider it an expendable commodity. See Measure 6.1.1.

MEASURE 6.1.3 Install effective closers on exterior doors.

Leaving an exterior door open or partially open allows vastly more air leakage than even the worst leakage around a closed door. The most effect way to keep doors closed is to install door closers. The first of the subsidiary Measures deals with environments where manual door opening is acceptable, and the second subsidiary Measure deals with environments where manual door operation is not acceptable.

Alternative: Revolving Doors

Revolving doors are an expensive alternative to doors with automatic closers. They are used primarily in applications where a large pressure differential exists across the door. See Measure 6.1.5 about them.

MEASURE 6.1.3.1 If manual opening of doors is acceptable, install spring-type door closers.

RATINGS

New Facilities	Retrofit	O&M
A	**B**	

Most exterior personnel doors in commercial and industrial buildings should have closers. Standard models are available that fit most personnel doors.

Features to Select

The door closer market offers reliable equipment and it offers junk that is intended to look like the real thing. The cheap units fail quickly, leaving you with problems you don't need. Make the effort to select good equipment. The best units are not much more expensive than the worst. Look for these features:

- *ruggedness.* Units of good quality last much longer than less durable units.
- *adjustable closing force.* The unit should be adjustable over a wide range because the optimum closing force may not be apparent until the unit is installed. A higher closing force makes the door harder to open, while a lower closing force allows the door to hang up in a partially open position.
- *smooth closing.* Good units have a hydraulic damper that closes the door at a uniform rate and keeps it from slamming shut.
- *back check.* A back check is a damper that operates as the door reaches the full open position. It keeps the door from slamming open. This is a standard feature of most units.

Reinforce the Attachment Points

A door closer exerts a great deal of force at the points where it is attached to the frame and to the door. Check

---

SUMMARY

The inexpensive and proven method of keeping doors closed.

SELECTION SCORECARD

Savings Potential .... **$ $** $

Rate of Return, New Facilities **% % %** %

Rate of Return, Retrofit.......... **% % %** %

Reliability .... ✓ ✓ ✓

Ease of Retrofit .... ☺ ☺ ☺

---

whether you need to install reinforcements at the attachment points. The leading door closer manufacturers offer reinforcement hardware that ease the job.

A door closer creates a strong reactive force at the top door hinge. Make sure that the top hinge is sturdy. Make sure that the hinge is well anchored to the door frame, and that the door frame is well anchored to the wall.

Install Handy Door Stops

All energy conservation measures should be user-friendly. Door closers may become a nuisance when the doors have to be kept open, for example, to move packages through the door or to provide ventilation. To make the door easy to use in these instances, install an effective door stop.

Otherwise, users will jam something into the hinge side of the door. When this happens, the leverage creates an enormous destructive force on the door, hinge, and frame.

When you install a door stop, also install a placard on the door requesting people to use the door stop properly, and to keep the door closed at other times. See Reference Note 12 about designing and installing placards.

ECONOMICS

SAVINGS POTENTIAL: $20 to $1,000 per year in heating and cooling costs, depending on climate, wind exposure, interior layout, energy costs, etc.

COST: Good door closers cost about $100 apiece. Installation may require one to several hours, depending on the door and frame configuration.

PAYBACK PERIOD: Less than one year, to several years.

TRAPS & TRICKS

SELECTING EQUIPMENT: Don't buy cheap door closers. They perform poorly, and fail early. Buy a model that is one grade tougher than you think you need.

INSTALLATION: Reinforce the attachment points, if necessary. Don't forget to install a door stop and placard, if the application needs it.

MAINTENANCE: See Measure 6.1.1.

MEASURE 6.1.3.2 If manual opening of doors is not acceptable, install automatic doors or door openers.

RATINGS

New Facilities	Retrofit	O&M
B	**C**	

Conventional door closers (recommended by Measure 6.1.3.1) require persons to overcome the spring force of the mechanism. This may be a serious impediment for handicapped persons or for persons using grocery carts, hand trucks, etc. In applications where conventional door closers are not advisable, consider installing automatic door openers. In most cases, expect to replace the entire door assembly, rather than retrofitting an automatic opener to an existing door.

Your Choices

You have two major decisions to make in selecting an automatic door, the type of door and the type of sensor to open and close it.

The two main types of automatic doors are swinging doors and sliding doors. Sliding doors are perhaps safer, because they cannot swing unexpectedly toward someone who is in the door's path. On the other hand, sliding doors require twice the width of the door opening. If you can afford separate doors for entering and leaving traffic, you can avoid the safety hazard of swinging doors by arranging the doors to open away from the traffic. In this case, use signs and railings to guide traffic through the doors in the proper directions.

The common methods of actuating automatic doors are tread switches, ultrasonic motion detectors, and photoelectric light beam sensors. See Reference Note 11, Personnel Sensors, about this equipment. For most

SUMMARY

An effective efficiency feature. Also a desirable convenience, especially for handicapped persons and for persons handling articles. Expensive.

SELECTION SCORECARD

Savings Potential	$	$	$
Rate of Return, New Facilities	%	%	%
Rate of Return, Retrofit	%	%	%
Reliability	✓	✓	✓
Ease of Retrofit			☺

applications, ultrasonic sensors have proven to be the best choice for opening the door. Figure 1 shows swing doors that are activated by ultrasonic sensors. In cases where a safety hazard may arise, consider installing a tread switch (weight sensor) to ensure that no one is within range of the door when it closes.

Decide Whether to Install a Vestibule

If you are willing to spend the money for an automatic door, consider making it part of an entry vestibule. A combination of an automatic door and a well designed vestibule is a powerful means of reducing infiltration. See Measure 6.1.8 about vestibules.

WESINC

Fig. 1 Automatic swing doors This pair of doors is the main entrance for a large grocery. They operate hundreds of times per day, so they must be rugged and reliable. For safety, each door must swing away from traffic. This requires two doors. The sensors that open each door should be triggered only by traffic from the correct direction. These doors are activated by ultrasonic sensors, which are small, inexpensive, and easy to install. The air curtain installed here plays no significant role in reducing infiltration. It discourages insects.

ECONOMICS

SAVINGS POTENTIAL: *$20 to $1,000 per year in heating and cooling costs, depending on climate, wind exposure, interior layout, energy costs, etc. Consider the value of increased convenience and comfort.*

COST: *Commercial automatic doors typically cost $5,000 to $10,000, including the operating hardware. There may be additional structural costs to accommodate the door.*

PAYBACK PERIOD: *Several years, in new construction, typically longer in retrofit.*

TRAPS & TRICKS

SELECTING EQUIPMENT: *Automatic doors are subject to severe use if traffic is heavy. Select units for ruggedness, ease of use, and absence of safety hazards.*

INSTALLATION: *Install the sensor where it will function reliably but will not open the door in response to false signals, such as people walking past the door. Install the sensor where the door has time to open fully without slowing traffic.*

MAINTENANCE: *Even an automatic door is an air leakage path. Keep it adjusted. See Measure 6.1.1.*

MEASURE **6.1.4 Install high-efficiency doors.**

You can buy standard doors that have minimal air leakage and good insulation. These are prefabricated assemblies that include the door itself, the frame, the threshold, and an effective sealing system.

For example, one common type uses a metal door skin and a magnetic gasket attached to a wooden frame to provide an almost perfect seal at the sides and top. The door is filled with foam insulation and has a wood thermal break at the edges, providing excellent thermal insulation.

Unfortunately, there is a compromise between effective sealing and the volume of traffic for which a door is appropriate. The most effective types of seals offer a certain amount of resistance to opening, and the seals themselves are subject to wear. Therefore, the doors with the most efficient seals are limited to applications that have relatively light traffic.

Install an Effective Closer

A high-efficiency door cannot save energy if it is left open. Consider installing an effective closer as part of the installation. See Measures 6.1.3 ff for details.

Minimize Infiltration Around the Door Frame

When you install the door, there will be a sizable gap between the door frame and the wall opening. The gap can act as a serious infiltration path. You have two ways to prevent this. One is to fill the gap with impermeable material. The other is to seal the moldings on each side of the door so that the moldings act as the infiltration barrier. See Measure 6.4.1 for effective methods of sealing against infiltration.

ECONOMICS

SAVINGS POTENTIAL: *$20 to $100 per year in heating and cooling costs, depending on climate, wind exposure, interior layout, energy costs, etc. This is in comparison to a mediocre door that has average weatherstripping.*

SUMMARY

The most efficient solution for air leakage in low-traffic applications. Also provides a small saving in conductive heat loss.

SELECTION SCORECARD

Savings Potential .... $ $ $

Rate of Return, New Facilities % % % %

Rate of Return, Retrofit.......... % % %

Reliability .... ✓ ✓ ✓

Ease of Retrofit .... ☺ ☺ ☺

COST: *In new construction, high-efficiency residential doors cost about $300 for single units, and about $600 for double units. Commercial doors cost from $15 to $30 per square foot, not including the cost of special materials and finishes. In retrofit, higher labor costs increase these prices, depending on the difficulty of retrofit.*

PAYBACK PERIOD: *One year to several years in new construction, several years or longer in retrofit. Consider the value of added comfort in the calculation.*

TRAPS & TRICKS

SELECTING EQUIPMENT: *Select doors for ruggedness, ease of use, sealing effectiveness, and durability of the seals. Inspect some installed units. Also, select an appropriate door closer, if appropriate, for installation at the same time.*

INSTALLATION: *Make sure that the door frame is attached securely to the wall opening, and that the leakage path between the frame and the wall opening is sealed tightly by filling gaps and caulking. Paint metal doors on arrival, as they are typically supplied with only a light primer coat.*

MEASURE **6.1.5 If a pressure differential at the entrance is unavoidable, install a revolving door.**

Some buildings may have a considerable differential pressure at entrances. (The reasons are explained by Reference Note 40, Building Air Leakage.) Even a relatively small pressure differential can keep a conventional door from operating properly. For example, a pressure differential of 0.5 inch of water column exerts a force of about 30 pounds on the handle of a door of average size. A swing door with an automatic closer cannot close fully against this pressure. If the pressure is in the direction that tends to keep the door closed, opening the door can be inconvenient or dangerously difficult.

Revolving doors were invented to avoid these problems. The door rotates on a central axis. The pressure differential acts symmetrically on both sides of the door axis, so operation is not affected by the building pressure. A revolving door also acts as an air lock, never allowing outside air to flow directly into the building.

Most revolving doors are moved manually, which requires considerable force to overcome the inertia of the door and the friction of infiltration seals. Revolving doors are built heavily to survive rough handling, and

SUMMARY

The most effective method of limiting air leakage at entrances with a large pressure differential. Expensive and inconvenient.

SELECTION SCORECARD

Savings Potential .... \$ \$ \$ \$

Rate of Return, New Facilities % % % %

Rate of Return, Retrofit.......... % %

Reliability .... ✓ ✓ ✓

Ease of Retrofit .... ☺ ☺

to resist the pressure differential on the door panels. Motorized revolving doors are becoming increasingly popular, but these may invite accidents. The revolving doors in Figure 1 are motorized.

Where to Consider a Revolving Door

Revolving doors are the most effective means of minimizing air leakage in applications with heavy traffic.

WESINC

Fig. 1 Powered revolving doors These doors are intended to minimize infiltration through the main entrance of a tall office building. They turn automatically in response to ultrasonic motions sensors. Users can choose to slow the rotation rate by pressing a button, visible here on the right-hand door. Even so, this pedestrian is using the manual door that is alongside the revolving door. Many of the people who use these doors are visitors. The installation lacks effective signs to tell them how to use the doors efficiently and safely.

Unfortunately, they are awkward and expensive. Their cost limits them to larger buildings where a pressure differential at the entrance is unavoidable.

You Still Need a Conventional Standby Door

Revolving doors should have a conventional door nearby for safety, for high volume traffic during an emergency, for delivery of items that cannot be carried through a revolving door, and for access by handicapped persons. Try to discourage people from using the conventional door. See Measure 6.1.6, next.

Alternative for Low Traffic

If traffic through the door is light, or if traffic is limited to short periods of time, consider using a powered sliding door. Even more effective is a long vestibule with a powered sliding door at each end. See Measure 6.1.8 for details. Neither of these options is as effective as a revolving door, but they may be easier to install in retrofit applications. Also, they provide easier access, which is important for handicapped persons and for moving goods through the door.

ECONOMICS

SAVINGS POTENTIAL: Thousands of dollar per year, typically, in applications where a high pressure differential exists at the entrance. Improved comfort in the vicinity of the entrance is a major benefit.

COST: The revolving door itself costs $15,000 to $30,000. This range includes powered units, but not exotic materials or finishes. In addition, a revolving door requires a separate conventional door, and the structure to support both.

PAYBACK PERIOD: In new construction, 1 to 10 years, depending on the climate, the height of the building, energy costs, etc. In retrofit, several years to much longer.

TRAPS & TRICKS

CHOICE OF METHOD: Balance the advantage against the disadvantages. Nothing else can stop air leakage as well as a revolving door. However, it is awkward, it needs a conventional standby door, and it is expensive. Do not try to cut costs by installing a lightweight unit, which will eventually leave you with a mess.

MEASURE **6.1.6 If a revolving door is installed, encourage people to use it instead of other doors.**

Where a revolving door is installed, a conventional door is usually installed nearby. The conventional door is an important safety feature, so it must not be locked or obstructed. Because the revolving door is more awkward to use, people are inclined to use the conventional door instead, which defeats the purpose of the revolving door. Figure 1 in the previous Measure shows this occurring.

To minimize this problem, install large, permanent, attractive signs on the conventional doors that say, "Please Use the Revolving Door," along with a large arrow pointing to the revolving door. This sign is a prominent part of the decor, so get someone with taste involved in designing it.

SUMMARY

A simple step to protect the benefit of revolving doors.

SELECTION SCORECARD

Savings Potential .... $ $ $

Rate of Return .... % % % %

Reliability .... ✓ ✓ ✓ ✓

Ease of Retrofit .... ☺ ☺ ☺

ECONOMICS

SAVINGS POTENTIAL: *Hundreds to thousands of dollar per year, typically.*

COST: *The placard may cost from $20 (engraved plastic) to $300 (cast bronze) per door.*

PAYBACK PERIOD: *Less than one year.*

MEASURE 6.1.7 Install storm doors.

If the existing doors are especially leaky, you can reduce infiltration significantly by installing storm doors. Storm doors are especially effective for doors exposed to wind. The storm door dissipates the velocity pressure of the wind, greatly reducing infiltration through the inner door, even if the storm door itself is leaky. A storm door also protects an expensive inner door from rain damage.

The main advantage of storm doors is that they are relatively easy to install. Typically, you buy a complete assembly that consists of the storm door installed in a lightweight frame, perhaps with a primitive closer. The assembly is screwed into place over the existing door frame. Figure 1 shows a unit at the high end of the quality range.

WESINC

Fig. 1 Storm door This is a storm door of much better than average quality. Still, it is fragile compared to a conventional door, and it costs more. Note the closers at top and bottom.

SUMMARY

Storm doors are clumsy and fragile, only cheap substitutes for a really good door. Usually limited to privately owned residences.

SELECTION SCORECARD

Savings Potential	**$ $**
Rate of Return	**% %**
Reliability	**✓ ✓**
Ease of Retrofit	☺ ☺ ☺

Storm doors have serious disadvantages. Fundamentally, they are cheap. They exist primarily as a cheaper alternative to installing a good primary door. They are lightweight, so they are vulnerable to damage from rough handling and from being slammed open by wind. They have cheap closers. Eventually, the storm door gets out of alignment and hangs open, which defeats its purpose. Also, they are awkward to manipulate, since the user must manipulate two doors at the same time.

Where to Consider Storm Doors

Storm doors remain open as long as the primary door, so they are effective only in applications where the doors remain closed most of the time.

Storm doors are weak, so they are suitable only for environments where everyone who uses the door has an interest in protecting it and closing it fully. This tends to limit storm doors to privately owned residences.

Features to Seek

The price of storm doors is proportional to their quality. The cheapest units cannot survive long. Expect to spend almost as much for a good storm door as for a good primary door. Remember, the only major advantage of the storm door is ease of retrofit. Look for these features:

- *rigidity.* The door must resist twisting and sagging to remain as airtight as possible. No storm door is really rigid enough, but some are better than others.
- *strong hinges and attachments.* The hinges should be strong enough to keep the door from shifting in its frame over time, which will prevent it from closing. The door and frame should be reinforced at the points where the hinges attach.
- *a durable latch.* The latch should be durable and easy to use. An aluminum frame needs a steel strike plate. The latch should allow for a certain amount

of shifting of the door in the frame. The latch should work with minimum closing force, or it will hold the door open.

• *effective weatherstripping.* The purpose of the storm door is to block infiltration, so it needs effective weatherstripping to do its job. The weatherstripping should be durable and easy to repair, and it should not require a large closing force. Get an adjustable sweep to minimize leakage at the bottom.

• *a good closer and back check.* No storm door comes with a really good closer, but get the best you can find. A storm door must have an effective back check or shock absorber, because wind tends to destroy storm doors by slamming them open.

Install Them Carefully

To achieve the most benefit, be careful to install the frame of the storm door in a way that minimizes leakage all around. This requires snug contact between the storm door frame and the frame of the main door. If necessary, insert weatherstripping or filler pieces between the two frames.

Better Alternatives

If you want to install a storm door because your main door is leaky, consider installing a better main door. This is more expensive, but it avoids the awkwardness of multiple doors. This alternative is also more reliable and more enduring, and it resists conductive heat loss as well as air leakage. See Measure 6.1.4.

If you want to create an enclosed space between entry doors, consider installing an entry vestibule. (In houses, this may be called a "mud room.") See Measure 6.1.8, next.

ECONOMICS

SAVINGS POTENTIAL: *$10 to $100 per year per door, depending on climate, energy cost, space layout, etc.*

COST: *Different models cost from $50 to $200. Installation typically takes less than one hour.*

PAYBACK PERIOD: *Several years, or longer.*

TRAPS & TRICKS

CHOICE OF METHOD: *Stop and reconsider before selecting this approach. Although storm doors are common, they are not very desirable. A storm door is just a cheap, flimsy door. It may not survive long enough to pay for itself, and it is a nuisance in the meantime.*

SELECTING THE EQUIPMENT: *Ordinary storm doors are too flimsy. Go out of your way to find sturdier units. Expect them to cost more. Check for all the important features.*

INSTALLATION: *Install weatherstripping between the storm door frame and the opening, to prevent leakage around the frame. Adjust the frame so the door closes crisply, with no tendency to hang partly open. Install the closer so it works effectively. Adjust all the sealing features.*

MAINTENANCE: *When the storm door gets out of alignment, fix it. A defective storm door saves little energy.*

MEASURE **6.1.8 Install vestibules for doors with frequent traffic.**

Entry vestibules are effective in reducing infiltration where wind is a major factor. The outer structure of the vestibule dissipates the velocity pressure of the wind, reducing infiltration through the inner door. Even when both doors of a vestibule are open, the vestibule may reduce infiltration significantly, provided that the outer door is not facing into the wind.

Where to Consider a Vestibule

Vestibules are more expensive than most other methods of reducing door infiltration. A vestibule may pay for itself in energy savings if the application has all these characteristics:

- *there is a lot of pedestrian traffic.* An open door is a big hole in the building envelope. If there is considerable traffic, a vestibule can reduce the effective length of time that the hole exists.

- *wind is a major factor in infiltration.* A vestibule may dissipate the velocity pressure of wind at the doors, greatly reducing the rate of infiltration. The orientation of the vestibule's openings with respect

SUMMARY

Most effective in reducing infiltration where wind is a major factor.

SELECTION SCORECARD

Savings Potential	$ $ $
Rate of Return, New Facilities	% % % %
Rate of Return, Retrofit	% %
Reliability	✓ ✓ ✓
Ease of Retrofit	☺ ☺ ☺

to the wind is a critical factor in the vestibule's effectiveness.

- *there is enough space to install the vestibule.* The vestibule must be wide enough to allow people to pass each other comfortably. Size is especially important if the vestibule may be used as a waiting area. In retrofit applications, consider installing a

WESINC

Fig. 1 Typical vestibule This vestibule is so short that the inner and outer doors are open simultaneously for a period of time when a person passes through. These doors are operated manually. The situation would be worse if the doors had automatic openers. The only solution is to make the vestibule much longer. This vestibule is heated, which attests to its lack of effectiveness in preventing infiltration.

vestibule either outside the existing doors or inside them, depending on the building layout and appearance considerations.

Even if a vestibule cannot meet the first two conditions enough to be economically desirable, it may still be desirable to minimize discomfort to occupants located near the door. For example, entry vestibules on restaurants prevent discomfort to patrons seated near the doors during cold weather, and the vestibule also acts as a protected area for patrons waiting to enter.

If people are transporting goods through the doors, make sure that a vestibule would not impede this traffic. For example, note how grocery stores have vestibules that allow customers to move shopping carts through them easily.

Distance Between the Vestibule Doors

If practical, make the vestibule long enough to allow the outer door to close before the inner one is opened, or vice versa. When both doors are open simultaneously, air flow into (or out of) the building can increase radically. Figure 1 shows a vestibule that is too short. Figure 2 shows how to install a vestibule of adequate length.

Automatic door openers keep doors open longer, increasing the separation required. Unless the vestibule is long, both doors will be open simultaneously for a large fraction of the door operating cycle.

Heating Vestibules

If the vestibule space itself is not heated, it always saves energy. However, heating the vestibule itself invites energy waste. The pressure inside the building makes a big difference.

■ Vestibules with Negative Indoor Pressure

If the area of the building adjacent to the vestibule has negative pressure, outside air flows through the vestibule toward the inside of the building. In this case, a vestibule heater improves comfort in the area of the building adjacent to the vestibule. To keep puffs of cold air from entering the building, the vestibule must be long enough, and the heater must be powerful enough, to heat the outside air before it reaches the inside door.

In this situation, a vestibule heater does not increase energy consumption very much. Most of the vestibule heating energy moves indoors, reducing the heating load on the interior heating systems.

■ Vestibules with Positive Indoor Pressure

If the area of the building adjacent to the vestibule has positive pressure, indoor air flows through the vestibule toward the outside. In this case, a vestibule heater is wasteful because the heat in the vestibule is contained only by the outer door. As a result, the

vestibule heater tends to run at full output during cold weather.

In this situation, a vestibule heater is not necessary for comfort inside the building, because outside air does not flow into the building. Furthermore, the flow of air from inside the building tends to keep the vestibule warm.

If the building has positive pressure, a vestibule heater is valuable only if the vestibule is used as a waiting area. To avoid the energy waste, try to design the area so that people can wait inside the building. If the people are waiting for rides, provide a window so that people waiting can see vehicles outside.

If you decide to install a vestibule heater, consider a radiant type. Radiant heaters heat the occupants and the mass of the space, rather than heating the air directly. As a result, they waste less heat to the outside, although they do not feel as cozy. If a radiant heater would require

WESINC

Fig. 2 Effective vestibule This vestibule for a large grocery is very long. It is built along the front face of the building. It accommodates people with shopping carts. The doors are operated automatically by ultrasonic sensors. Entering and leaving traffic are separated by a railing inside the vestibule, which expedites movement.

a more expensive energy source, calculate whether the saving in energy provides a saving in cost.

Alternative: Install a Revolving Door

In most cases, revolving doors are more effective than vestibules in preventing infiltration, but they have serious disadvantages. See Measure 6.1.5.

ECONOMICS

SAVINGS POTENTIAL: *Hundreds of dollars to thousands of dollars per year in typical applications, depending on climate, energy cost, building orientation, etc. Consider the value of improved comfort.*

COST: *Thousands of dollars.*

PAYBACK PERIOD: *Typically several years, in new construction. Several years or longer, in retrofit.*

TRAPS & TRICKS

DESIGN: *Invest some time in studying vestibules, noting their good and bad points. Remember that you are trying to mimic the performance of an air lock. Make the vestibule long, if possible. Using automatic doors affects the way you should lay out the vestibule. If the building has positive pressure, try to avoid installing a heater. If you must have a vestibule heater, consider a radiant unit.*

MEASURE 6.1.9 Seal abandoned doors.

B

Abandoned doors may be in poor condition, allowing a considerable amount of air leakage. If the door does not have to remain as an emergency exit, render it completely immobile and seal the air leaks. In effect, make it part of the wall. Mark the door plainly to state that it is no longer in use.

Since the door does not have to be opened, it is easy to select sealing materials. You can probably do the entire job with soft foam weatherstrips. With the door open, install self-adhesive foam strips on the hinge side of the door frame. After you close the door, stuff foam strips into the remaining gaps with a putty knife. Then, fasten the door rigidly on all sides.

If the door must continue to serve as an emergency exit, maintain it as you would any other door. See Measures 6.1.1 and 6.1.2. Mark the door boldly, stating that it is to be used only as an emergency exit.

ECONOMICS

SAVINGS POTENTIAL: *$10 to $200 per year per door, depending on climate, energy cost, space layout, etc.*

SUMMARY

Corrects a source of energy waste that is overlooked in older buildings.

SELECTION SCORECARD

Savings Potential .... **$ $ $**

Rate of Return .... **% % %**

Reliability .... ✓ ✓ ✓

Ease of Retrofit .... ☺ ☺ ☺

COST: *$20 to $100, typically.*

PAYBACK PERIOD: *Less than one year.*

TRAPS & TRICKS

MAKE IT STRONG: *Unless the door is blocked closed in an obviously strong way, some ignoramus will try to force it open. Make sure that the door is well secured. Mark it on both sides.*

Garage, Loading, and Equipment Doors

Industrial doors are immense holes in the building envelope, through which the heat and cooling inside a building easily spills into the outdoors. It is virtually impossible to maintain normal temperatures in a space that has a large door that is kept open. Even reduced levels of conditioning are very expensive to maintain. To avoid this energy waste, you need efficient doors, and you need to keep them closed. The first three Measures recommend maintenance and improvements for exterior doors. The fourth Measure recommends installing new high-efficiency exterior doors.

Large doors operate slowly, so they tend to be left open for long periods. This wastes the efficiency benefit of the door. The fifth Measure deals with this problem by using quick-acting doors that are installed inside the primary exterior doors. These can remain closed most of the time without interfering with traffic. The last Measure minimizes infiltration at loading docks by installing special seals between the building and the vehicle.

MEASURE 6.2.1 Maintain the fit, closure, and sealing of industrial-type doors.

Vehicle, loading dock, and other large exterior doors receive severe use. They require periodic maintenance to effectively minimize infiltration. You have to decide how often to inspect the doors, based on their type and usage. The adjustments and repairs you need to make are usually obvious when you inspect carefully. Check these conditions at appropriate intervals:

- repair and adjustment of infiltration seals, curtains, etc.
- operation of automatic closing mechanisms
- alignment of doors for effective sealing and complete closure.

Related Measures

Coordinate this activity with maintenance of personnel doors, which is covered by Measure 6.1.1.

ECONOMICS

SAVINGS POTENTIAL: *Hundreds of dollars to thousands of dollars per year per door, depending on climate, fuel cost, space layout, etc.*

COST: *$5 to $100 per year per door.*

SUMMARY

Routine maintenance needed to prevent energy waste and maintain comfort.

SELECTION SCORECARD

Savings Potential .... $ $ $ $

Rate of Return .... % % % %

Reliability .... ✓ ✓

Ease of Initiation .... ☺ ☺ ☺

PAYBACK PERIOD: *Short. The long-term saving greatly exceeds the maintenance cost.*

TRAPS & TRICKS

SCHEDULING: *The hard part is remembering to do it. Make sure that you have an effective maintenance calendar.*

PREPARATION: *Keep a full stock of spare parts on hand. Otherwise, searching for parts can be more bother than the maintenance work itself.*

MEASURE **6.2.2 Install infiltration seals on existing doors.**

It is especially important to keep industrial doors well sealed because they are so large. Unfortunately, industrial doors tend to be leaky because their large weight and size require mounting systems that are difficult to seal.

Sealing large garage and industrial-type doors is a challenge because sealing systems may make it more difficult to open and close the doors. Still, you may be able to substantially reduce the infiltration of existing doors by using methods that do not excessively interfere with the operation of the door. You can create effective seals by using standard weatherstripping materials and by creative carpentry. Also, you can buy specialized seals for industrial doors.

First, Repair the Doors

Don't bother trying to seal a door that is loose or out of alignment. Your sealing methods will be ineffective, or they will wear out quickly. Your first step is to make sure that the door is secure and aligned properly. See Measure 6.2.1.

How to Install the Weatherstripping

Make the effort to find the types of weatherstripping that will work effectively with your doors. The top, the sides, the bottom, and the gaps between individual panels each require different sealing methods. Each type of door needs its own combination of sealing materials. Study how your door operates before buying your materials. Measure 6.2.4 describes the characteristics of the main types of doors.

Select your methods for durability and for minimum interference with the operation of the door. Here are some guidelines for different types of doors.

■ Roll-Up Doors and Overhead Segmented Doors

These two types of doors may allow serious infiltration. They have large clearances at the top and sides that are difficult to seal effectively. The gaps between the segments provide many infiltration paths.

Both types slide vertically in tracks on each side of the door. If the segments are narrow, they may slide in the track directly. If the segments are wider, each segment may have wheels that roll in the track, as in Figure 1. The only difference between the two is that the roll-up door is stored on a spool at the top of the door, while the overhead door slides on its track to an overhead position. The roll-up door has narrower segments, and it takes up less ceiling space.

SUMMARY

An inexpensive method of reducing air leakage, but not effective with all doors. Expect to search for the best sealing devices.

SELECTION SCORECARD

Savings Potential .... $ $ $ $

Rate of Return .... % % % %

Reliability .... ✓ ✓ ✓

Ease of Retrofit .... ☺ ☺ ☺

There are two ways to deal with the large number of gaps between the door segments. If each segment butts against the next, try to install self-adhesive soft foam strips between the segments. The foam is pinched between the segments when the door is down, creating an effective seal. This is the most durable method. It may last several years.

If you cannot use this method between the segments, use a strong, flexible tape to cover the gaps on the inside of the door. The main weakness of this method is the life of the adhesive. Expect to replace the tape every year or two.

The sides are the most difficult to seal because of the need to maintain a clearance for the door to slide in its track. Various types of seals are offered to deal with this type of door. People keep inventing better types of seals, so make a search to find the best that are currently available. If the door is installed inside a jamb, you may be able to install brush seals on the outside of the door to contact the jamb. This method creates significant resistance, so limit it to doors that have motorized operators. You can reduce the resistance by arranging a brush seal on the door to mate with a "knife edge" installed on the jamb, or vice versa.

A less effective method is to install a rail on the wall, inside the door, that covers the gap. This still leaves a gap, but the rail greatly reduces the effect of wind.

To seal the top of the door, install a flexible or hinged flap that spans the space between the wall and the upper surface of the door. The door must slide underneath this seal, so make sure that it does not jam or create excessive friction. In the case of a roll-up door, the seal must able to move over an wide range as the door winds and unwinds on the spool.

The bottom of the door is usually easy to seal. Use a soft compression gasket. Several types are readily available.

■ Overhead Panel Doors

This type of door has a single large panel that swings down from the ceiling. It closes with a motion that is a combination of sliding and compression.

If the door fits within a frame, try to use a sliding seal for the sides. A plastic or rubber leaf seal, which is a larger version of the bronze seals used on older residential doors, works well and provides the least resistance.

If the door does not fit inside a frame, install brush seals on the wall to mate with the top and sides.

For the bottom of the door, use a soft, compressible gasket material that can tolerate sliding against the floor.

Don't rely on compression weatherstripping with this type of door. To be effective, compression weatherstripping requires a clamping pressure all around the perimeter of the door. This is not possible with a conventional operator. Conventional operators apply their closing force at one or two points, and the door would distort away from an all-around seal at other points.

■ Sliding Doors

Sliding doors have large surface areas in relation to their edge length. In principle, this provides for low infiltration. However, the edge gaps are large to provide clearance for the panels to move. Sealing a sliding door requires ingenuity and custom fitting.

To seal the gaps between the panels, install interlocking channels at the sides of each panel. Arrange the channels so that an edge of one channel seats in a very soft foam gasket in the channel of the adjacent door.

To seal between the end panels and the door frame, install a deep channel or pocket on the door frame that surrounds the edge of the end panel. Use a brush-type seal on the channel that seats against the inside surface of the door. Arrange the seal so that the panel does not contact the seal until the panel has entered the channel. To reduce the resistance of the seal, install it at a slight angle to the edge of the panel, from top to bottom. In this way, the panel pushes through only a portion of the seal at any point in its travel.

To seal the bottoms of the panels, install a limp curtain-type seal on each panel that drags across the floor with minimal resistance.

The top of a sliding door assembly is the most difficult to seal. One way is to install a horizontal shelf extending from the wall over the top of the panels. Install brush seals on top of each panel to contact the bottom of the shelf.

The total resistance of all these seals may make it impractical to operate the door by hand, forcing you to install an operator. See Measure 6.2.3 about door operators.

■ Vertically Folding (Bi-Fold) Doors

Bi-fold doors have the potential of sealing tightly. They have a low ratio of edge length to surface area. All sides of the panels may be relatively easy to seal.

There are two hinged joints, between the two panels and at the top of the top panel. You can seal both these joints tightly by installing a soft foam gasket in each joint so that it is pinched when the door is closed. Make the gaskets very soft so they do not interfere with closing the door.

Seal the bottom of the door using standard rubber weatherstripping that is widely available for installing on the bottoms of garage doors.

A sliding seal is best for the sides of the panels. If the door is installed inside a jamb, install the seals inside the jamb. If the door butts against the outside wall, install a channel along each side of the door, and install sliding seals between the channels and the sides of the panels. The side seals will increase the difficulty of closing the door fully. Make sure that the door operator is rigged to provide sufficient closing force.

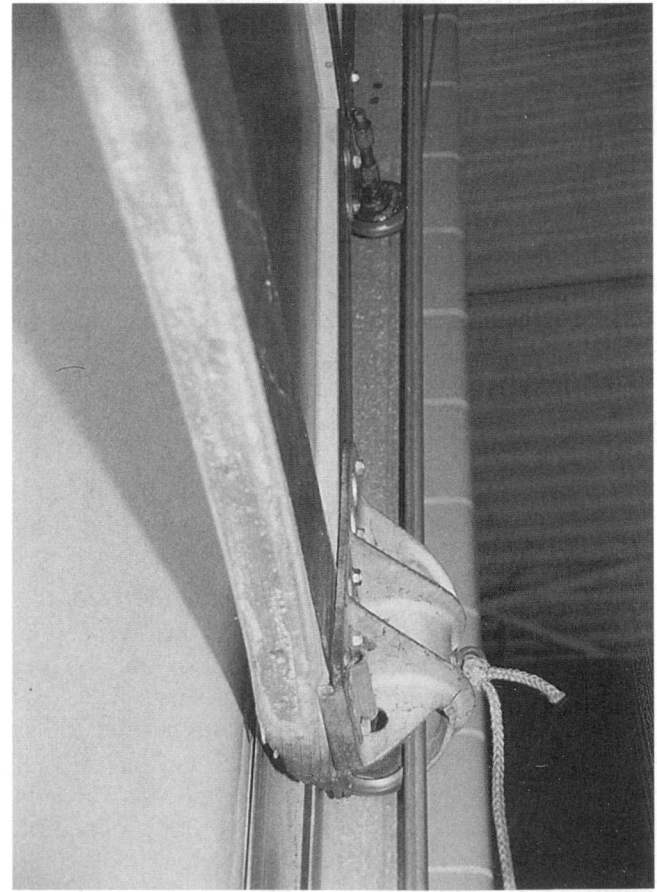

WESINC

Fig. 1 Track for an overhead door The rollers hold the edge of the door away from the wall. You can seal the sides of this door with a fixed seal that slides against the face of the door opposite the rollers. Sides seals must be very flexible to accommodate looseness in the track and to minimize resistance to opening and closing.

■ Horizontally Folding Doors

Horizontally folding doors consist of a series of hinged panels. The center of each panel is suspended from an overhead track. There may be one set of panels that closes across the entire door opening, or two sets of panels may close toward the center of the door.

Seal the hinged joints by installing a soft foam gasket in each joint so that it is pinched when the door is closed. Make the gaskets very soft so they do not resist closing the door.

If there are two sets of doors that close from opposite sides, install a channel at the end of one panel, such that the opposing panel enters the channel when the doors close. If the channel fits fair snugly, there is no need for additional sealing at this location. If there is only one set of panels, install the channel on the jamb.

Probably the best way to seal the top of the door is to install brush seals at the top of each panel. These contact the track as the door closes. The seals tend to keep the door from closing fully, so install them so they create as little resistance as possible.

To seal the bottoms of the panels, install a limp curtain-type seal on each panel that drags across the floor with minimal resistance. The door may have a bottom track to guide the bottom of the door and add strength. If the track is raised, most of the bottom seal will not contact the track until the door reaches a closed position. At that point, the bottom seal tends to keep the door from closing fully.

■ Swing Doors

Large swing doors are rare because they require a large clearance area, and because they place a great deal of stress on the hinge and wall.

Seal the hinged sides of the doors by installing soft foam gaskets that are pinched when the doors close. Make the gaskets very soft so they do not interfere with closing the door.

To seal the tops of the doors, install leaf seals, if possible. However, to get good results with a leaf seal, the gap needs to be uniform and stable in width. A leaf seal is destroyed quickly if the door can shift in its frame. If you cannot use a leaf seal, install a strip of molding on the outside of the door to cover the gap. This is effective for blocking wind, but does not provide much resistance to other infiltration.

If there are two doors, sealing the gap between them is tricky. The seal must offer little resistance to closing. It must bridge the gap and accommodate changes in the gap dimensions. It must work well regardless of the sequence in which the doors close. The best choice in most instances may be a very soft brush seal. Installing a strip of molding over the gap (an "astragal") requires users to close one door before the other.

To seal the bottoms of the doors, install sweeps. Select sweeps that are very limp, to minimize resistance.

Keep the Seals Maintained

All types of seals wear out. Consider them to be an expendable commodity that you expect to replace occasionally. Wear is aggravated by changes in the gap dimensions caused by shifting of the building, flexing of the doors, etc. See Measure 6.2.1 about scheduling door maintenance.

Alternative: High-Efficiency Doors

Some doors cannot be sealed reliably or economically. Consider installing an efficient new door in such cases. See Measure 6.2.4 for details.

ECONOMICS

SAVINGS POTENTIAL: *Typically hundreds of dollars per year per door, depending on climate, fuel cost, space layout, etc.*

COST: *Several hundred dollars, typically.*

PAYBACK PERIOD: *Less than one year, to several years.*

TRAPS & TRICKS

STUDY THE SITUATION: *Invest enough time to figure out a method of sealing your doors that is effective, reliable, and easy to maintain. Being clever at the beginning will save a lot of energy and effort later.*

MAINTENANCE: *Weatherstripping wears out quickly compared to most other maintenance items. Consider it an expendable material that you have to replace periodically.*

MEASURE **6.2.3 Install powered door operators.**

A powered door operator provides two benefits. One is providing the physical effort of opening and closing the door. The other is the option of installing an automatic control that keeps the door closed when it is not being used. Install an operator if the door is difficult to move manually or if people are likely to leave the door open unnecessarily. Most vehicle doors should have operators because a manually operated door would require the vehicle operator to dismount to operate the door.

Types of Door Operators

The typical door operator consists of a motor and a mechanism to convert the rotation of the motor into linear motion. Door operators use a variety of mechanisms for this purpose. The most common are:

- a chain traveling on a guide rod
- a threaded shaft driving a nut along a guide rod
- a pinion gear on the motor driving a rack (a rod with teeth)
- a cable wound on a drum or rod.

Also, you need a control device to start and stop the motor. Some operators include their own controls. For example, garage doors are offered with radio controls to open and close them. Other operators require you to design the control separately. A variety of operators and control devices are available as standard items.

How to Connect the Operator to the Door

Some doors have operators that are an integral part of the door assembly. With others, you have to figure out the best way of installing the operator. Here are some tips for different types of doors.

■ Overhead Panel and Segmented Doors

For these types of doors, the operator is usually one that provides a straight-line motion, and the operator is attached to the top of the door. The operator is usually installed on the ceiling, providing an easy connection.

Be especially careful in the way you attach this type of operator to the door. The operator pulls horizontally because it is mounted on the ceiling, while the top of the door initially moves vertically in its track. If the operator is attached to the center of the door, it applies a bending force to the top of the panel. Figure 1 shows the problem. This inward force may deform the upper door panel, making it impossible to seal. You can avoid this bending force by attaching the operator to both sides of the door with rigid rods. This requires installing the operator farther inside the building.

SUMMARY

All doors that handle traffic, or that are difficult to operate, should have powered operators. Standard units are reliable and modestly priced.

SELECTION SCORECARD

Savings Potential	$	$	$	$
Rate of Return, New Facilities	%	%	%	%
Rate of Return, Retrofit	%	%	%	%
Reliability	✓	✓	✓	
Ease of Retrofit	☺	☺	☺	

If you cannot install the operator at the proper orientation on the ceiling, you can use a system of cables to operate the door. You need one set of cables to open the door, and another to close it. An advantage of using cables is that you can operate them from a compact motorized winch.

■ Sliding Doors

If a sliding door has only one panel, or if it has two panels that open in opposite directions, installing an operator is fairly simple. The most compact solution is a simple cable system operated by a winch.

You can extend this method if the sliding door has more than one panel opening in each direction. For example, consider a set of three panels, all of which open in one direction. The cable opens one panel first. When the first panel completely overlaps the second panel, a stop on the second panel keeps the first panel from going past it. This causes the first and second panels to move as a unit until they overlap the third panel. Then, a stop on the third panel keeps the second from going past. From that point, the three overlapping panels move together to a fully open position.

The door closes in a similar manner. From the fully open position, the cable pulls the first panel toward the closed position. The first panel moves alone until its outer edge latches to the inner edge of the second panel, so the cable pulls the two panels in train. Then, the outer edge of the second panel latches to the inner edge of the third pane, and the three panels move in train until the door is fully closed.

Design the installation to provide effective sealing when the door is closed. You can do this by using interlocking channels to latch each panel to the next when it closes. See Measure 6.2.2 for details.

■ Horizontally Folding Doors

The panels in horizontally folding doors are connected. Therefore, opening and closing the door requires only a single connection to the end door. A simple cable system operated by a winch can do the job.

For sealing purposes, note that simply pulling the door closed does not force the panels tightly into alignment. With this type of door, the laws of trigonometry reduce the aligning force to zero as the door is pulled fully closed. You can overcome this problem with some fancy linkages, but you probably don't need to bother. As explained in Measure 6.2.2, you can effectively seal the gap in the panels and the gap underneath the door even when the panels are not aligned. However, sealing the top of the door requires the panels to be closed almost completely. By using a wide brush plate, as recommended in Measure 6.2.2, you can seal the top gap even if the door panels remain somewhat staggered.

■ Vertically Folding (Bi-Fold) Doors

Opening a bi-fold door requires lifting the entire weight of the door, so a bi-fold door usually has a powered operator. You can reduce the lifting force on the operator by using counterweights to balance the weight of the door. Bi-fold doors are almost always operated with a cable system. Cables are attached to the bottom of the door at several points, which keeps the lower door panel from bending excessively. The winch is typically installed on the door itself, which makes the installation neat and compact.

Bi-fold doors close by gravity. When the door approaches the fully closed position, the closing force on the sides of the door becomes zero. As a result, bi-fold doors tend to hang slightly open. This allows a substantial amount of infiltration if the door is installed outside the door opening, rather than inside a frame. Manual latches may be installed to complete closing the door, but manual latching tends to be ignored if the door is opened and closed frequently. Some manufacturers offer automatic latches. Get this feature.

■ Roll-Up Doors

Roll-up doors typically have an integral geared motor. They are difficult to retrofit with an operator. If you have a roll-up door that lacks an operator, see if you can attach a cable reel to the existing spool. As an alternative, consider replacing the door with another type that seals tightly.

■ Swing Doors

Probably the best way to operate a large swing door is to use a straight-line operator. Install the operator at a suitable location on the ceiling of the space. Attach a rod from the operator to the top edge of the door. Install a spring link on the rod so that the door is spring-loaded when it is closed.

How to Control the Operator

Select the most efficient type of control to open and close the doors. First, consider the way the door is used. Do vehicles pass entirely through the door, allowing the door to close behind them? Is the door at a loading dock, where it may have to be open for part of the time that a vehicle is present, but not all the time? Does the door have to opened in a way that does not relate clearly to the traffic through it, as in an aircraft maintenance hangar? Adapt the method of control to keep doors closed as much as possible, subject to usage requirements. These are your control choices:

- *manual switches.* These are desirable at loading docks, where a truck may linger for long periods without needing the door to be open. The truck driver or warehouse personnel can open the door when needed. However, this is an unreliable method of keeping doors closed. A loading dock door that seals tightly to the vehicle body is a better alternative in many cases. See Measure 6.2.6 for this approach.

- *driver-actuated remote controls.* These are commonly used for garages. The most popular type is a handheld radio transmitter that can be used from inside a vehicle. These are actually manual controls that operate over a distance. They suffer from the main disadvantage of manual control, which is that

WESINC

Fig. 1 Common door operator problem The arrow points to the link between the door operator and the top of the door. When the door starts to open, the link pulls horizontally, whereas the door can only move vertically. This creates a strong bending force on the center of the top panel. You can avoid this problem while using the same operator by using two long links between the operator and the sides of the door.

the driver can neglect to close the door. This type of control offers security because the signals are coded.

- *induction loops* that are buried under the pavement near the door. These are sensitive to the height of the vehicle chassis above the floor. For example, a sensor that is set to be triggered by a forklift truck may not be triggered by a tractor trailer.
- *motion detectors.* Ultrasonic motion detectors and photoelectric light beams are capable of restricting their sensing range to a localized area in front of the door. Other types of motion sensors, such as infrared units, are likely to be triggered by activities that are unrelated to the door.
- *card access controls* are used primarily for security against unauthorized entry. The driver inserts a card into a card reader to open the door, and a proximity sensor of some type closes the door after the vehicle passes through.
- *time delay controls* are used in combination with other controls to keep the door open for a minimum period of time. Time delays are used in applications that have heavy traffic, to reduce excessive cycling of doors.

If you install automatic or remote controls, consider installing an additional manual control at the door. This allows emergency exiting and provides backup if the automatic control fails.

Decide whether your application makes it desirable to stop the door at a partially open position. If so, provide controls that make this possible.

Placards

To avoid confusion and enhance safety, tell users how the doors work. Do this with a conspicuous sign. If the door is controlled manually, install a sign to remind users to keep the door closed. See Reference Note 12, Placards, for guidance in placard design and installation.

ECONOMICS

SAVINGS POTENTIAL: *Hundreds of dollars to thousands of dollars per year per door, depending on climate, fuel cost, space layout, etc.*

COST: *The equipment costs $200 to $500 for smaller standard systems, up to several thousand dollars for heavy or complex equipment. Installation costs vary widely, being higher in retrofit applications.*

PAYBACK PERIOD: *One year, to several years.*

TRAPS & TRICKS

SELECTING THE EQUIPMENT: *Powered doors and door openers are subject to severe use if traffic is heavy. Select units for ruggedness, ease of use, and absence of safety hazards. Select a control method, or a combination of methods, that is appropriate for the nature of the traffic.*

INSTALLATION: *If you use an automatic door opener, select a sensor appropriate for the type of traffic. Install it where it will function reliably, but will not open the door in response to false signals, such as people walking near the door. If doors are controlled manually, install conspicuous placards that tell users how to operate the controls.*

MAINTENANCE: *Even an automatic door is an air leakage path. Keep it adjusted. See Measure 6.1.1.*

MEASURE 6.2.4 Install efficient exterior doors.

In new construction, select the most efficient doors you can find. Much better doors have come on the market in recent years, but the old inefficient types still prevail, so you will have to select critically. You may have to adapt the design of the structure around the door opening to accommodate efficient doors. Doors that are well sealed and well insulated do not raise the overall price of the building very much, and they save a large amount of money.

In existing buildings, you have to decide whether to improve the existing leaky doors or replace them entirely. Some doors cannot be modified effectively to resist air leakage. Also, you may not be able to insulate some existing doors. These weaknesses are worst in doors that consist of many segments. The gaps between the segments cannot be sealed in a permanent way, and leakage between the door and the guide rails is difficult to seal at best. With such doors, the only possible remedy is to replace the door with a more efficient type.

This Measure deals with big doors that have good long-term resistance to outside weather conditions, including wind, precipitation, and sunlight. These doors are physically strong, so they also act as effective security barriers.

At present, all large doors that have these good durability characteristics are limited to slow opening and closing speeds. If you need doors that open and close quickly, and if you are willing to make compromises for speed, consider the types of doors covered by Measures 6.2.5 ff.

Types of Large Exterior Doors

The genius of humanity has devised many ways to cover a large entrance in a building. These are the main types of exterior doors that are used at present:

- *sliding panel doors.* This is the simplest type of large door. Individual panels hang from an overhead track. The panels overlap when the door opens. This type is used primarily in medium and large sizes, including the largest doors in the world. See Figure 1. Sliding doors are easy to insulate, but somewhat difficult to seal against infiltration.

- *horizontally folding doors.* This type is made of large panels that are suspended from an overhead track. The sides of the panels are hinged together, so the door folds open in accordion fashion. This type is similar to sliding doors in insulation and infiltration characteristics. It is typically used in medium sizes.

- *bi-fold (vertically folding) doors.* As the name implies, bi-fold doors consist of two panels. The

door folds vertically, usually toward the outside. This type has the potential to be well insulated and to provide good resistance to infiltration. Figures 2 and 4 show typical examples. The structure of bi-fold doors, shown in Figure 5, makes them expensive. Also, they require structural bracing of the top of the door opening. They are becoming increasingly popular in the medium size range.

- *roll-up doors.* This type consists of short segments that roll up vertically around a spindle. This provides a compact door installation. Figure 3 shows a typical unit. However, the short segments cannot be insulated effectively, and the design is difficult to seal against infiltration. It is typically used in sizes that accommodate individual vehicles.

- *overhead segmented doors.* This type opens by rising in a track that bends to a horizontal position underneath the ceiling. The door is made in hinged segments to accommodate the bend in the track. The number of segments keeps this type from being well insulated, and the track mounting makes it difficult to seal against infiltration. It is typically used in widths that accommodate one or two vehicles.

- *overhead single-panel doors.* This type consists of a single panel that swings upward to an overhead horizontal position. It can be well insulated, but sealing against infiltration is difficult. It has fallen out of popularity because the mounting hardware is awkward and the door sweeps through interior space when opening. It is typically used in single-vehicle sizes.

- *swinging doors.* This type is limited to smaller sizes because of the space that is swept when the door opens, the potential for slamming by wind, and the large loads on the hinges. Too bad, because swinging doors can be well insulated and they are easy to seal against infiltration.

How to Select Your Doors

In the past, you could expect that some types of doors would be better or worse than other types from an efficiency standpoint. This is true no longer. Interest in energy efficiency has spurred manufacturers of all types of doors to try to make their products more efficient. Therefore, do not limit your search to a particular type of door.

Although efficiency is important, it is not the only factor to consider. A large door is a major cost item, and it may have a major effect on the facility's activities. Make the effort to recognize every aspect of your application that affects your door selection. Use the following as a checklist.

■ Infiltration Sealing

The door should have an effective method of sealing each edge of each panel. Each edge is a separate design challenge for the manufacturer. Examine how the manufacturer deals with the top, the bottom, and the sides. With segmented doors, examine the sealing between segments. If the door closes from both sides, examine how the manufacturer closes the gap between the two halves. With sliding doors, look for a reliable method of sealing between the panels that does not make it awkward to close the door.

All seals eventually wear out from flexing and abrasion. Check that the seals are installed in a way

Kalwall Corporation

Fig. 1 Sliding doors At the top of the size range, sliding doors are about the only choice. In large sizes, it is easier to provide them with effective weather sealing and powered operators. Note how the structure at the top of the door is designed. This door is so big that it has two sizes of smaller doors at the bottom. The center panels in this door are box structures made of translucent material.

that makes them easy to replace. If the seals are specialized, you will be dependent on the manufacturer to provide replacement seals in the future.

The bottoms of vertically moving doors are usually sealed with a thick compression seal. These work well.

If the door moves horizontally, any type of seal that contacts the floor causes drag that makes the door difficult to close. Therefore, horizontally operating doors may lack bottom seals. Minimizing infiltration with such doors requires careful installation to minimize the size of the gap. If you plan to install a powered operator, select one that has enough force to overcome the drag of sweep seals, which you can add to the bottom of the door.

Another problem area in sliding and folding doors is the top of the door. These types are suspended from rollers that ride in a track or group of tracks. Examine how the door design minimizes leakage through the track.

Overhead segmented doors have a similar leakage problem at their side tracks. Some manufacturers of segmented doors offer side sealing that is much better than the products of others.

■ Adjustment and Reliability of Operators

If the door has a powered operator, examine how it affects infiltration sealing. With some types of doors, such as rolling and bi-fold types, the adjustment of the linkage between the operator and the door may not be critical.

Other types, such as sliding and folding doors, may need a complex linkage to provide effective sealing, and the adjustment of the linkage may have to be just right. To minimize air leakage in the long term, avoid linkages that may tend to get out of adjustment.

■ Insulation

Large doors have enough surface area to allow substantial heat loss by conduction. All types that have large panels can be insulated well, but this requires careful design by the manufacturer, or careful installation in the field. Effective insulation requires thickness, an unbroken surface area, and snug contact between the insulation and the door.

Doors with many small segments cannot be insulated well. This applies to rolling doors and overhead segmented doors. Insulation can be placed inside each segment, but the metal frames of segments comprise a large path for heat loss.

■ Dead Space for Open Door and Operating Hardware

Some types of doors take up a considerable amount of side space when they are open. Sliding doors are the worst in this regard. If the door must open to the full width of the building, the track must be extended beyond the sides of the building on an outrigger. This

Schweiss Distributing, Inc.

Fig. 2 Bi-fold door Almost all vertically opening panel doors consist of two panels, as shown here. A truss is needed at the bottom edge to resist bending from the weight of the panel. This model also has a truss at the center, which protects against wind loads when the door is closed. It is easy to install windows or a person door in the panels.

arrangement is very vulnerable to wind, and it impedes traffic around the sides of the building.

Folding doors waste relatively little end space when open.

Overhead segmented doors require ceiling space, which may interfere with lighting and ductwork.

Most bi-fold doors block a significant fraction of the upper part of the door opening. However, at least one manufacturer uses side braces that allow its bi-fold doors to open almost to the full height of the opening. This type is shown in Figure 4.

The operating hardware also takes up space. For example, overhead doors need tracks that extend into the space. Cable-operated doors need connections between the winch and the door. Models with self-contained actuators ease this problem. For example, some bi-fold doors have the motor, winch, and cables mounted on the door assembly itself. Figure 5 shows a compact operator installation.

■ Clearance While Operating

When the door opens, it sweeps a certain amount of space, which must be left clear. For example, horizontally folding doors require clear space inside and outside the door opening. Swinging doors require a large clear path, which is one reason why they are rarely found in large sizes.

The sweep area of the door reduces storage space somewhat, but this issue is usually minor. More important, it requires people to anticipate the opening of the door and to keep the space clear. By the same token, the sweep of the door is a safety issue, especially if the door has an automatic operator.

If you install a door that sweeps through an area that may be occupied by people or vehicles, install a prominent placard, guide lines, guard rails, or other means to avoid danger from the door.

■ Speed of Operation

All types of large exterior doors operate slowly, but some are faster than others. Higher speed creates less of an impediment to traffic. It also reduces the amount of heat that is lost while the door is open. If the door cannot open quickly, it is likely to be left open.

If there is frequent traffic, consider installing one or more quick-acting interior doors to supplement the exterior door. See Measure 6.2.5.4 for specialized doors that open quickly.

For loading docks, consider installing seals between the building and vehicles that are being loaded or unloaded. See Measure 6.2.6 for the details.

■ Ease of Operation

People are more likely to keep a door closed if it is easy to operate. This is a factor if the door is operated

WESINC

Fig. 3 Roll-up door The main virtue of this type of door is compactness. It can be motorized easily. The large number of segments make it impossible to insulate well. Some models seal against infiltration fairly well, and some do not.

Hi-Fold Door Corporation

Fig. 4 A bi-fold door that wastes a minimum of height This proprietary design allows the door to open almost completely by using integral jacks at each side of the door. Most bi-fold doors cannot open fully because the lower panel must act as a brace against the door frame.

manually. Any type of door can be easy to operate if you install a powered operator. Even then, select a method of controlling the door that is most efficient for the type of traffic. See Measure 6.2.3 for the details about operators.

■ Strength

Make sure that the door, the frame, and associated hardware are able to withstand all the forces that occur in the application.

Large, flat door surfaces are vulnerable to deformation by wind. The door is usually strong enough to withstand the force of strong winds, but slamming caused by wind may damage the door, the frame, or the building structure. Select and install the door to limit slamming forces. This means that the door should be held firmly while it is in any position.

In some cases, the closing and opening forces of the operator may be strong enough to deform the door. You can avoid this problem by the way you install the operator linkage. See Measure 6.2.3 for details.

Some types of doors, especially bi-fold doors, are subjected to large gravity loads that deform the doors when they are opened. Manufacturers of bi-fold doors offer additional bracing as an option. The bracing consists of horizontal trusses installed at the bottoms of the two panels. The trusses may be installed on the inner or outer sides of the door.

Conventional doors generally are not designed to resist impact, as from vehicles. If the application involves a risk of impact, consider supplementing the primary exterior doors with an inner door that is designed to tolerate impact. See Measures 6.2.4 ff for these doors.

■ Safety Features

If the door has a powered operator, you may wish to install safety devices to stop the door if something is in the door's path. Such features are unusual on large exterior doors, but you can probably find them if you search long enough. A common type of safety device is a light beam that stops the door if anything interrupts the beam. Another safety device is a "reversing edge," which stops the door if anything contacts the leading edge of the door. These devices are common with the types of quick-acting doors recommended by Measure 6.2.5.4. Refer to there for the details.

■ Ability to Install Windows and Personnel Doors

Some types of exterior doors offer windows as an option. These can provide natural lighting to a space that lacks it. Similarly, you can get a small personnel door installed in the larger door. This allows easy access without opening the large door itself. Windows and

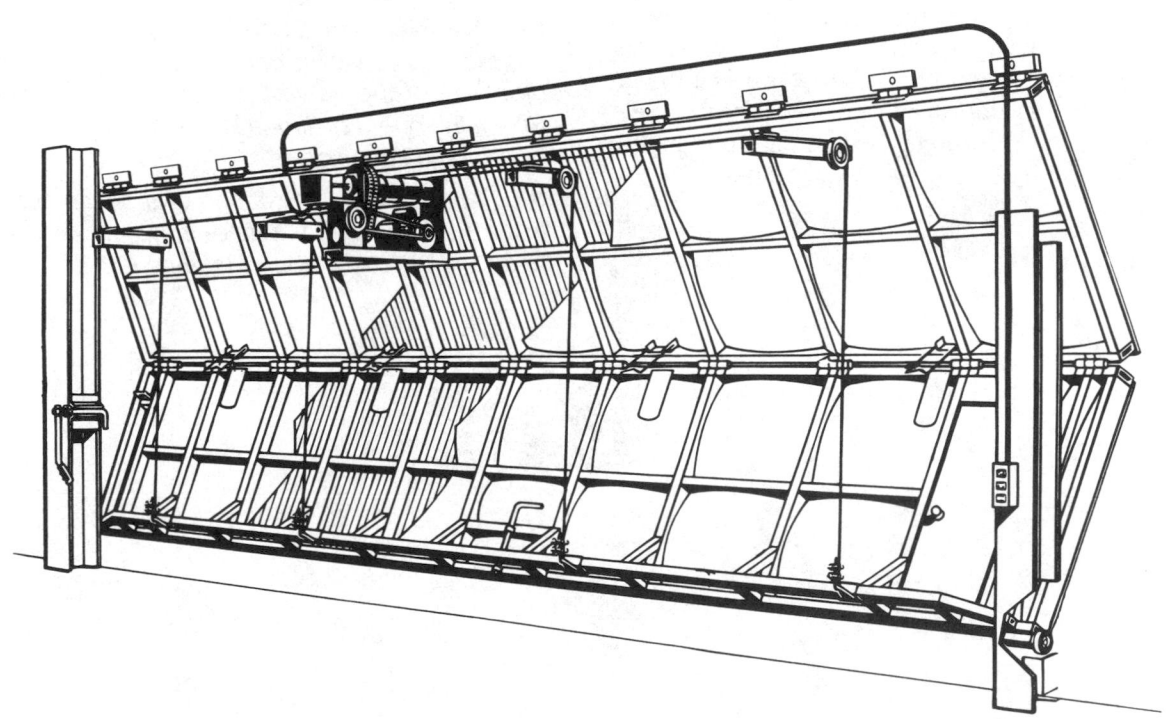

Fig. 5 Construction of a bi-fold door All bi-fold doors require a bottom truss to resist the strong bending forces that occur when the door is being opened. Similarly, a continuous line of hinges is needed to carry the top edge of the door, and the top of the door frame must be reinforced to carry this load. Bi-fold doors need operators, which can be installed in a compact manner. The edges of the door panels can be sealed against air leakage fairly easily.

personnel doors are offered in bi-fold, sliding, and horizontally folding doors. Windows are offered in overhead segmented and single-panel doors.

■ Security

The same factors that contribute to physical strength and to infiltration sealing also contribute to security. In addition, get appropriate locking devices.

■ Ease and Cost of Installation

Consider the price of the door itself, its accessories, installation labor, and any required strengthening of the building structure. For example, bi-fold doors typically require substantial bracing of the upper part of the door opening structure.

Which Type is Best?

Okay, now that we have gone through all that, which door is best for your application? For doors that are very wide or very tall, your choice is limited to doors that hang by gravity. Sliding doors are used in the largest sizes. Their main disadvantage is the difficulty of sealing the edges from air leakage. They also require a substantial amount of extra opening width. Folding doors probably deserve greater attention for large openings, because they are better in both respects.

In smaller sizes, swinging doors provide the best potential for insulation and sealing. Their major disadvantages is the swing clearance they require on opening. An overhead door may be the best choice for a relatively small opening. Infiltration sealing in overhead doors has improved considerably in recent years, but examine the sealing design carefully before you buy.

For the middle size range, from about 20 to 50 feet (7 to 16 meters) in width, bi-fold doors probably have the best efficiency potential. They can be made to seal tightly, and they can be well insulated. Bi-fold doors have other advantages. They fold outward, so they require no internal space, and they require no internal clearance to operate. This type initially came into widespread use for aircraft hangars, and they are gaining in popularity for other applications.

The main disadvantage of bi-fold doors is their relatively high cost. The door assembly itself is expensive, because it requires a lot of structure and an operator. And, it is custom-built at the site. Prefabricated bi-fold doors have not yet made an appearance on the market. The building structure must be braced to withstand the stress that the open door exerts on the top of the door opening.

ECONOMICS

SAVINGS POTENTIAL: *Hundreds of dollars to thousands of dollars per year per door, depending on climate, fuel cost, space layout, etc.*

COST: *$2,000 to $20,000 in typical sizes.*

PAYBACK PERIOD: *Several years, or longer.*

TRAPS & TRICKS

SELECTING THE EQUIPMENT: *Select doors for ruggedness, ease of use, sealing effectiveness, and durability of the seals. Inspect some installed units. Select the operator, if one is needed, for reliability, ease of use, and tight closing.*

INSTALLATION: *If the door has an automatic operator, install the sensor in a manner that is appropriate for the nature of the traffic. See Measure 6.2.3.*

MEASURE 6.2.5 Install lightweight quick-acting doors to supplement exterior doors and to separate interior spaces.

All the types of exterior doors covered by Measure 6.2.4 are inherently slow. That is, they cannot accommodate the rapid movement of vehicles and people. As a result, they are typically left open as long as frequent traffic may occur. You can solve this problem effectively by installing quick-acting interior doors to block infiltration.

You can choose from a variety of types, which differ in features and price. They are all made of light materials. The light weight allows users to push the door open easily, or it allows a powered actuator to operate the door quickly. You can find an infiltration door to fit virtually any size of door opening.

The subsidiary Measures cover the spectrum of contemporary quick-acting doors. In order of increasing price, the most common types are:

- *strip curtains,* covered by Measure 6.2.5.1. These are popular, primarily because they are cheap and simple. They are the least effective type, they are annoying if they touch people, and they soil clothing.

- *fabric doors,* covered by Measure 6.2.5.2. These are a compromise design that is moderately satisfactory for both people and vehicles, if the people can tolerate soiled clothing. They are simple, safe, and relatively inexpensive, but they do not seal well.

- *impact doors,* covered by Measure 6.2.5.3. These are effective for equipment handling, and less satisfactory for people. They are simple, reliable, and relatively inexpensive. They may pose a safety hazard.

- *powered quick-acting vehicle doors,* covered by Measure 6.2.5.4. These are the most effective type for resisting infiltration. Unlike the previous types, they require no contact with the vehicles or people passing through the door. Their main disadvantages are high initial cost and more expensive maintenance. Measure 6.2.5.4 covers the types of powered doors that are currently on the market.

There are other types that fit between these categories, and there are some relatively unusual types that combine features from several of these categories. If you study these four basic types, you will learn about the others also. The subsidiary Measures explain the advantages and disadvantages of each type. Often, several types are competitors for the same applications. It's your decision to make the compromise between efficiency, other features, and cost.

Comparison with Exterior Primary Doors

To date, nobody has invented a strong, durable type of exterior door that handles frequent traffic well. The doors covered by this Measure handle traffic well, but they cannot serve as primary exterior doors in most applications. In your planning, take into account these distinguishing characteristics of the quick-acting doors covered by this group of Measures:

- *little or no delay in passing through the door.* These doors are intended to keep traffic moving. Unpowered types are simply pushed aside by the user, either a vehicle or a person. Powered types have actuators that operate very rapidly.

- *designed to withstand impact.* These doors are designed to be hit. Unpowered types tolerate impact because they are light. Powered models are also relatively light, and they may have breakaway features that limit damage if the door is hit.

- *need for recurring maintenance.* All types of doors have components that wear out occasionally from repeated cycling, impact, and weather. The doors are (or should be) designed for quick repair.

- *minimal insulation value.* The primary benefit of these doors is minimizing infiltration. Most have little insulation value, because they are light and thin. Some types do have a limited amount of insulation value, which can be important for some applications.

- *minimal security.* The light weight of these doors keeps them from serving as an effective barrier to forced entry. A small vehicle could drive through any of them.

- *limited resistance to outdoor conditions.* Unpowered types have virtually no resistance to wind. Powered types can resist varying wind speeds, depending on the model. Strong sunlight would weaken the plastic materials within a few months or years and make them brittle. The mechanisms of powered types may be vulnerable to precipitation.

- *specific temperature ranges.* Temperature affects the flexibility of plastic materials. Flexibility is important for the operation of door types in which the fabric folds, rolls, or bends in operation. Flexibility is also important for impact resistance. Select a material that is suitable for the lowest temperature at which the door will operate.

Where to Install Quick-Acting Doors

In general, consider installing a quick-acting door if there is frequent traffic between a conditioned space and another space that is at different temperatures. These are common applications:

- **as a supplement for an exterior door.** This allows the primary door to remain open during hours of heavy traffic. Depending on the type of primary door (see Measure 6.2.4 for details), you may be able to install the infiltration door close to the primary door, taking up little additional space.
- **as the inner door of a vestibule.** If possible, try to create a vestibule between the primary door and the infiltration door. This blocks infiltration more effectively during the times when the primary door

is kept closed. For example, at a truck dock, install the infiltration door between the dock and the rest of the building. In this way, the entire dock acts as a vestibule. (See Measure 6.1.8 for vestibule design issues.)

- **for refrigerated spaces.** If practical, install an infiltration door between any interior spaces that differ substantially in temperature. For example, in a commissary building, install an infiltration door between the meat storage freezer and the warmer part of the refrigerated space. In addition, install an infiltration door at the truck docks.

MEASURE 6.2.5.1 Install strip curtains.

Strip curtains consist of overlapping strips of flexible plastic material that hang by gravity. They are used most commonly with loading vehicles, such as forklift trucks. Lighter versions are available for use by people. The strips are usually transparent, which is a valuable safety feature. Figure 1 shows how strip curtains work.

The main advantage of strip curtains is their low cost to install and to repair. The transparency of the plastic allows vehicles to approach the curtain safely and to drive through the strips without stopping. The curtain material is so light and supple that it does not pose a safety hazard when someone blasts through the door.

Strip curtains can be annoying in applications where people, rather than vehicles, push through the door. The transparent plastic cannot be reinforced, so the strips must be fairly heavy to survive. The plastic strips drag against exposed skin, which is especially unpleasant if the strips are stiff and cold. Also, the strips become soiled, so strip curtains are not appropriate for people wearing formal clothing. For this reason, don't use strip curtains as personnel doors except in rugged environments.

Strip curtains can be made in any width. All you need is a structural element or a suspended bar from which the strips can be hung. Strip curtains can also be fairly tall. However, long strips tend to twist, allowing more air leakage.

The transparent plastic is not crystal clear, and it becomes less clear with wear and soiling. To aid in guiding vehicles through the proper part of a strip

SUMMARY

A popular type that is inexpensive, simple, safe, and fairly effective. Annoying if the curtain touches people. Cannot resist wind.

SELECTION SCORECARD

Savings Potential	$ $ $
Rate of Return, New Facilities	% % % %
Rate of Return, Retrofit	% % % %
Reliability	✓ ✓ ✓
Ease of Retrofit	☺ ☺ ☺

curtain, you can install one or more colored strips among the transparent strips.

How to Install Strip Curtains

If the strip curtain is made of plastic, fabric, or any other organic material, try to install it out of sunlight. The ultraviolet component of sunlight causes plastic to become weak, stiff, and dark.

Select the plastic material for the temperature range in which it is used. Plastic becomes stiffer as it becomes colder. If the plastic is much colder than its normal temperature, it becomes brittle enough to break easily. If you have a choice, install the strip curtain in a location where the temperature varies least.

Try not to install the strip curtain where it is subject to strong wind. Wind flows easily through strip curtains if it is strong enough to blow the curtain partly open. If

Fig. 1 Strip curtains This is the least expensive kind of quick-acting door. It is foolproof in operation, and fairly safe. Because of the large amount of edge length, sealing against air leakage is relatively poor. They are annoying to walk through, especially with clean clothes. Different weights and widths of plastic can be used for vehicles and people. In this example, a colored strip of plastic is used to separate vehicle traffic from pedestrian traffic.

possible, install the strip curtain in combination with a vestibule (see Measure 6.1.8).

Keep Them Repaired

Strip curtains tempt some vehicle drivers to drive through them aggressively. Try to dampen their enthusiasm by installing a large sign on each side of the door that says, "CAUTION: Drive through the door SLOWLY."

The plastic strips eventually stiffen. This keeps them from closing properly, and pieces break off. Expect to replace strips as they break or lose their transparency.

Keep spare strips on hand to repair individual ones that become damaged. Keep the spares flat, in a dark, cool location. The strips are easy to replace, but they are also easy to overlook. Include maintenance of strip curtains as part of the regular door maintenance program recommended by Measure 6.2.1.

Roll-Up Strip Curtains

A hybrid type of door is available consisting of a strip curtain that rolls up on roller, which is usually motorized. This type avoids the annoyance of having to push through the strips. To keep the strips from jamming along the side of the door when the curtain rolls up, the strips are cut in a tapered shape. This type of door is not as effective as the tightly sealing roll-up doors recommended by Measure 6.2.5.4, but it is less expensive. The tapered shape of the strips sacrifices

sealing at the sides of the door, and the door may not be well sealed at the top.

ECONOMICS

SAVINGS POTENTIAL: *Hundreds of dollars to thousands of dollars per year per door, depending on climate, fuel cost, space layout, etc.*

COST: *Several hundred dollars, for a strip curtain large enough to pass a forklift truck. Roll-up strip curtains typically cost several thousand dollars.*

PAYBACK PERIOD: *Less than one year, to several years.*

TRAPS & TRICKS

CHOICE OF TYPE: *Install strip curtains if you can afford nothing else. Their low cost gives them the best cost-benefit ratio of all the door types, although they are less efficient. Use a better type of door if the location is windy, or if the door is used often by people on foot.*

INSTALLATION: *If you have a choice, install the strip curtain where it is protected from sunlight, wind, and large temperature variations. Install a highly visible placard that asks vehicle drivers to drive through the door slowly.*

MAINTENANCE: *Replace broken strips and strips that become too stiff to close properly. Schedule inspections in your maintenance calendar.*

MEASURE   **6.2.5.2 Install fabric swing doors.**

Fabric doors are soft swinging doors. The are usually installed as a double door, which makes them easier to use. The door surface is heavily reinforced fabric, which hangs from a pivoting bar at the top of the door. The lower part of the door is reinforced to withstand the impact. Typically, a window of clear plastic is installed in each door. Figure 1 shows one in action.

The door is opened by people or vehicles pushing through the door. The door is closed by a self-centering closer. Typically, the bar at the top of the door falls by gravity into a V-notch on the hinge, as shown in Figure 2 of Measure 6.2.5.3.

The characteristics of fabric doors lie between those of strip curtains (covered by the previous Measure) and rigid impact doors (covered by the next Measure). They are a close competitor to both types in many applications.

Fabric doors have several advantages over strip curtains. One is that they do not drag over persons passing through the door, although a person on foot must

SUMMARY

A compromise design that is effective for vehicles and people, especially for people moving goods on hand trucks. Inexpensive, simple, and safe. Does not seal well.

SELECTION SCORECARD

Savings Potential .... $ $ $

Rate of Return, New Facilities % % % %

Rate of Return, Retrofit.......... % % % %

Reliability .... ✓ ✓ ✓ ✓

Ease of Retrofit .... ☺ ☺ ☺

push his way past the door edges. Fabric doors are more durable than strip curtains because they can be made of stronger materials. Changes in the pliability of the material with temperature and age do not matter as much as they do with strip curtains.

Fabric doors also have several disadvantages. They are more expensive than strip curtains, although they are much less expensive than powered doors. Unlike strip curtains, they require a rigid frame, including a strong side frame for mounting the hinges. Fabric doors sweep open, which poses a safety hazard. However, the material is flexible and light, which minimizes the hazard. By the same token, if the object passing through the door is narrow or short, the door tends to stay closed, while the edges of the doors drag across the object or person.

Fabric doors are limited in width, but they are available in widths sufficient to pass all typical vehicles.

Fabric doors do not seal as well as rigid impact doors. The edges of the doors are thin and the material is pliable, so the edges typically do not meet closely at the center of the door. Also, fabric doors cannot resist significant wind or pressure differential.

You can order some models of fabric doors with enhanced insulation value, but this hardly matters in view of their relatively high air leakage.

Caution Drivers

Try to keep vehicle drivers from driving through fabric doors aggressively. This minimizes the safety hazard and extends the life of the door. Install a large sign on each side of the door that says, "CAUTION: Drive through the door SLOWLY."

Rite-Hite Corporation

Fig. 1 Fabric swing door This door is installed inside a rigid freezer door. It consists of heavy fabric suspended from two swinging arms. Use is somewhat awkward. Here, the cargo has already passed through the doors, but the upper arms have not yet opened against the centering force. This model has windows and reinforced bumper bottoms.

MEASURE 6.2.5.3 Install impact doors.

RATINGS

New Facilities	Retrofit	O&M
B	**B**	

Impact doors are lightweight rigid swing doors that are pushed open by a person or by a materials handling vehicle. Most impact doors are installed as double doors, which makes them easier to use. The doors themselves are typically made of plastic sheet materials over a light metal frame. The doors are installed in a strong frame that withstands the transmitted impact forces. Figure 1 shows an impact door in action.

To minimize air leakage, the door can be equipped with effective seals on all sides. They can also be insulated to a limited extent. Doors that seal tightly provide a significant amount of noise isolation.

Impact doors can be used by both people and vehicles. However, the door must have a fairly strong closing force to overcome the resistance of the seals. The closing force may be an impediment to some people on foot.

How to Select An Impact Door

Impact doors are simple devices. Still, there are significant differences between models. Consider these features:

- *closing and centering force.* The door must have a centering force that is strong enough to overcome the resistance of the seals, and to resist wind and pressure differential. Most impact doors are centered by gravity. In the closed position, a cam on the door side of the hinge rests in the bottom of a V-notch fitting that is installed on the frame. See Figure 2. When the door is opened, it rides up the side of the notch. When the door is released, gravity pulls the door back into the center position. A spring may be installed on a hinge rod to increase the centering force. Some manufacturers allow the purchaser to select the centering force by providing

SUMMARY

A popular type that is simple and fairly inexpensive. Good models provide the best resistance to air leakage and the best insulation of all types of quick-opening doors. They can also provide better security than other types. Can be used by powered vehicles and by people on foot. The swing of the door poses a safety hazard. Limited to smaller door sizes.

SELECTION SCORECARD

Savings Potential	$	$	$	$
Rate of Return, New Facilities	%	%	%	%
Rate of Return, Retrofit	%	%	%	%
Reliability	✓	✓	✓	✓
Ease of Retrofit	☺	☺	☺	

notch fittings of different depth or springs of different stiffness.

- *swing pattern.* By using a hinge installed on the corner of the frame, some models can open as much as 180° in one direction. In other words, the door folds flat against the wall. The door can open 90° in the other direction. See Figure 3. If the wider opening angle is in the direction of the vehicle's travel, the vehicle can make a turn even before it clears the door. This can be a valuable feature, especially in tight quarters.

- *sealing features.* The edge seals of the door are an important factor in stopping air leakage. Look at all four sides of the door. Manufacturers use several types of seals, all of which can block infiltration almost completely. The gap between the two doors is typically closed with soft hollow tube seals. Figure 4 shows typical seals for the door gap and the hinge side. The top seal is a special design because the door rises as it opens. This requires clearance between the top of the door and the door frame. To fill this gap and minimize resistance, many models have a soft seal on the top of the door that mates with a soft seal installed in the top of the frame. The bottom of the door can be sealed effectively with a rubber sweep. The door rises as it opens, so the sweep seal does not drag on the floor except when the door is near the closed position.

- *visibility.* Impact doors have enough weight and stiffness to be dangerous if they are slammed open. The primary safety feature of an impact door is a large window that allows vehicle drivers to see that the space on the other side of the door is clear. Unfortunately, the window cannot extend below the top of the bumper portion of the door, which is typically at about half the door height. This leaves the danger of hitting a low object on the other side

Rite-Hite Corporation

Fig. 1 Impact door   As the name implies, this type of door is designed to open by being pushed. The door survives by being light and strong. Plastic bumpers at the bottom of the door absorb much of the shock. The safety issues are obvious here.

WESINC

Fig. 2 Centering hinge   This is the most common type of centering device for impact doors. The weight of the door is carried by a roller fitting, which falls by gravity into a V-notch. This aligns the doors accurately enough to engage the infiltration seals. The hardware has virtually unlimited life. The infiltration seal on top of the door must accommodate the rising of the door as it opens.

of the door. Get the largest window that does not interfere with the strength of the door.

- *resistance to impact.* Impact doors use bumpers to absorb the shock of vehicles running into them. Select bumpers that are appropriate for the severity of the application. Select the height of the bumpers for the nature of the vehicles and goods passing through the door. For example, a stack of goods that is being carried on a forklift truck may strike the door before the truck itself. At the same time, do not make the bumper so tall that it interferes with visibility through the window.

- *insulation.* Impact doors are relatively small, so conductive heat loss through the door is not a major problem. Even so, it is worth insulating the door. You can get models that are filled with foam insulation. Typical door construction allows for somewhat more than one inch (about three centimeters) of insulation, which provides an overall insulation value of about R-4.

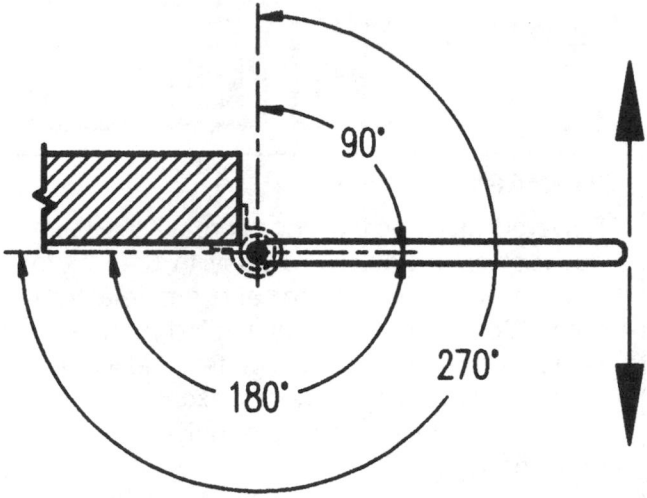

Fig. 3 Swing pattern An impact door mounted with the hinge on a corner of the wall opening can open flat against the wall in one direction. This allows materials handling vehicles to make a tight turn away from the door.

- *security features.* Impact doors can be made strong enough to provide a moderate degree of security. They cannot use conventional locks, because the thick seals between the doors create too much of a gap between the strong parts of the doors. The doors can be latched to the top of the frame and to the floor with sliding bolts. (These are commonly called "cane bolts" in the door trade.) The most common method of locking impact doors from the outside is to chain the doors together. To make this possible, order doors that have holes with bushings for the chain.

- *mounting features.* A vehicle slamming into an impact door transmits a large force to the adjacent wall. Therefore, the door needs a strong frame, and the frame needs to be attached strongly to the wall. In some cases, you may have to reinforce the wall itself. The bottom hinge absorbs more impact than the other hinges. For this reason, some models provide the option of bolting the bottom hinge to the floor.

- *ease of repair.* Select a door that is easy to repair. In particular, expect to replace the seals and bumpers occasionally. The windows should be easily replaceable, because they can be broken, and they will become obscured in time.

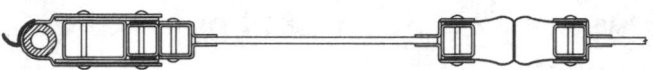

Fig. 4 Infiltration seals for door gap and hinge side
The center gap seal works by being big and soft. This door is made of a single sheet of rigid material. Many impact doors have a box construction, which maintains better alignment.

Caution Drivers

Safety is an issue with impact doors because they swing open. The best safety features are a large window and persuading vehicle drivers to go through them slowly. Less aggressive driving also extends the life of the door. Install a large sign on each side of the door that says, "CAUTION: Drive through the door SLOWLY."

ECONOMICS

SAVINGS POTENTIAL: *Hundreds of dollars to thousands of dollars per year per door, depending on climate, fuel cost, space layout, etc.*

COST: *$1,000 to $2,000.*

PAYBACK PERIOD: *Less than one year, to several years.*

TRAPS & TRICKS

CHOICE OF MEASURE: *Impact doors are a good choice from the standpoint of energy efficiency. They resist wind and pressure differentials better than the types in the previous subsidiary Measures, but not as well as powered doors. Powered doors, covered next, offer better safety and ease of operation, but at much higher cost.*

SELECTING THE EQUIPMENT: *You have lots of choices among impact doors. Make sure that you fit the door to the application.*

INSTALLATION: *Impact doors require a strong installation. Reinforce the wall, if necessary. Consider bolting the lower hinge to the floor. Install highly visible placards that ask vehicle drivers to drive through the door slowly.*

MEASURE 6.2.5.4 Install powered quick-acting vehicle doors.

The previous subsidiary Measures recommended doors that are pushed open by the people or vehicles who pass through them. In some applications, the requirement to push the door open is unacceptable, or at least undesirable. In such cases, you can minimize heat loss by installing powered quick-acting doors. Quick-acting doors are designed to open so rapidly that they do not delay traffic significantly. They are usually operated by sensors that detect approaching vehicles and open the doors automatically.

There are several types of quick-acting doors on the market. Each type has significant advantages and disadvantages. Within each type, different models may have important differences. This Measure describes the types that are available, and spells out the characteristics that distinguish them.

The primary advantage of quick-acting doors is that they avoid contact with people and vehicles. They resist wind and strong pressure differentials better than all types of push-through doors. However, they are probably no better than good impact doors for resisting small pressure differentials and ordinary convection.

Quick-acting doors are not foolproof, and they require maintenance that is more complex. They pose a safety hazard primarily when closing, in contrast with push-through doors, which pose a safety hazard when opening. Most types are designed to break loose when hit, so they do not provide strong security. All quick-acting doors are much more expensive than push-through doors.

Types of Powered Quick-Acting Doors

All quick-acting doors achieve high operating speed by using materials that are light in weight and by using a powerful motor. These are the main types:

- *vertical roll-up.* This type rolls up vertically on a spindle, like a roller blind. The material for this type of door is a fabric-reinforced plastic or rubber sheet. The door typically includes a plastic window. Figure 1 shows a typical unit in action.

- *horizontal roll-up.* This type typically consists of two doors, each of which rolls up on a spindle at the side of the door opening. The material is similar to that used in vertically rolling doors. The doors typically have plastic windows. Figure 2 shows a typical unit.

- *horizontally folding.* This type typically consists of two doors, each of which folds to the side in accordion fashion. The material can be thicker and tougher than that used in rolling doors. The doors typically have plastic windows. Models are

SUMMARY

Powered quick-acting doors create the least obstruction to traffic, are easy to use, have the best wind resistance, and are relatively safe. Some models offer limited insulation value. They do not offer a significant efficiency advantage over impact doors, except in windy locations. Much more expensive than the previous types.

SELECTION SCORECARD

Savings Potential .... $ $ $ $

Rate of Return, New Facilities % % % %

Rate of Return, Retrofit.......... % % %

Reliability .... ✓ ✓ ✓

Ease of Retrofit .... ☺ ☺ ☺

available that include a thin layer of insulation. These are sometimes called "freezer doors," because they are favored for cold storage applications. Figure 3 shows a unit, with particular features highlighted.

- *overhead sectional.* This type is a quick-acting version of overhead sectional doors that are used for exterior purposes (which are covered in Measure 6.2.4). The differences are faster operation, greatly reduced infiltration, and tolerance for impacts. Models are available with insulated panels, as shown in Figure 4.

How to Select a Powered Quick-Acting Door

The different types and models of quick-acting doors provide you with a broad choice of features. Their relative advantages and disadvantages may not be obvious, so pay attention to the details. Consider the following characteristics of quick-acting doors in relation to your particular applications.

■ Speed of Operation

Quick-acting doors typically operate at a speed of one to three meters per second (three to nine feet per second). Quick operation is important to minimize interference with traffic and to reduce the chance of hitting the door.

The door may close more slowly than it opens. Opening quickly is more important than closing quickly, so the door may have a spring to assist the motor in opening the door. When the door closes, the motor must overcome the force of the spring. Some vertically rolling doors close by gravity, whereas the other types require power to close.

Slower closing may be important for safety, since the closing cycle poses more of a danger than the opening cycle. Slower closing allows more heat loss, but this effect is minimal at the high speeds at which the door operates.

■ Infiltration Sealing

The door should have an effective method of sealing each edge of each panel or sheet. Each edge is a separate sealing problem. Examine how the manufacturer deals with the top, the bottom, and the sides. In the case of double doors, examine how the manufacturer closes the gap between the two halves. In the case of segmented doors, examine the sealing between segments.

All seals eventually wear out from flexing and abrasion. Check that the seals are installed in a way that makes them easy to replace. If the seals are specialized, you will be dependent on the manufacturer to provide replacement seals in the future.

The bottoms of vertically moving doors are usually sealed with a large, hollow cushion seal. These work well. The bottoms of folding doors usually have sweep seals. These work fairly well, but the seals wear on an uneven floor, and the drag of the seals tends to keep the door from closing fully near the bottom. Horizontally rolling doors may have no bottom seals. Minimizing infiltration with this type of door requires careful installation to minimize the size of the gap.

With double folding or horizontally rolling doors, look for an effective seal where the doors meet. Some doors rely on a large, soft edge seal. This method may not seal well, because the edges may not meet properly. Some models have magnetic seals installed in the edges to provide effective closure.

With rolling doors, pay attention to the manufacturer's method of minimizing air leakage where the edge of the sheet slides in the track. This part of the door design is a compromise between air leakage, reducing a tendency to jam, and breakaway during impact. Also, make sure that there is an effective seal between the door frame and the roller.

With folding doors, look for effective closure at the top and bottom. Some folding doors have a curtain of lighter fabric that seals the gap between the top of the door and the sliding track. The bottom of the door should have effective sweep seals.

With overhead segmented doors, examine the method of sealing the top and sides. It is a challenge

Rite-Hite Corporation

Fig. 1 Vertically rolling quick-acting door The motor that rolls up the door is on the upper right side of the frame. A large window is needed for safety. The driver has a view of the complete width of the floor on the other side as soon as the door starts to open.

for the manufacturer to combine good side sealing with effective breakaway.

■ Resistance to Wind and Pressure Differential

Wind and pressure differentials exert a strong force on large doors. In applications where wind or a strong pressure differential may act on the door, select a model designed to maintain effective sealing under these conditions.

Overhead segmented doors have the greatest ability to resist wind and air pressure. At the other extreme, horizontally rolling and folding doors have the least resistance, tending to blow apart where the two halves meet in the middle. Magnetic seals can substantially improve sealing by these types, up to a point where the magnets cannot hold the two halves of the door together. Vertically rolling doors resist wind and pressure up to a point, but they jam or blow open above their maximum pressure tolerance. Some models of vertically rolling doors offer limited reinforcement to resist pressure.

■ Insulation

Rolling doors have minimal insulation, because the door segments must be thin. Some horizontally rolling doors have double rollers and a double surface, so the air layers that adhere to the inside surfaces provide additional insulation. This may give the door an overall insulation value of about R-1.

With folding doors, the fabric can be padded with a thin layer of insulation. Insulated models are bulkier.

Segmented doors are much thicker than the other types, so a fairly large amount of insulation can be installed inside each panel. However, this advantage is partially cancelled by heat conduction through the many metal frames of the panels.

■ Durability

Durability is a primary factor in selecting a quick-acting door. All types have components that are subject to many cycles of flexing. The door should be designed to minimize stress on flexible components, and to make it easy to replace them when they fail.

Rite-Hite Corporation

Fig. 2 Horizontally rolling quick-acting door This type quickly provides clearance for tall, narrow loads. The bottom edges of the doors are not restrained, so the struts seen in front of the door are installed to hold the center edges in alignment.

The door is subject to moisture from condensation, and perhaps from blowing rain or snow. Moisture corrodes steel parts and it causes mildew on plastic and fabric components. The materials should resist this damage. (You can avoid condensation by heating the door, as described below.)

The ultraviolet radiation in sunlight harms all plastic and fabric materials. If the door is exposed to direct sunlight, these materials may fail prematurely. This is a factor with rolling and folding doors, where the plastic materials carry load and are subjected to flexing. Such applications may favor a segmented door, in which the loads are carried by metal parts.

Make sure that plastic components are rated for the temperature range of your application. Flexible materials become brittle when they are too cold.

■ Impact Resistance and Ease of Repair

Severe damage may occur if a vehicle strikes the door. Impacts occur because the door cannot open quickly enough, or because a sensor fails to detect the presence of the vehicle. To avoid or minimize damage,

most quick-acting doors are designed to be hit occasionally. The material of rolling and folding doors is intended to absorb light impacts.

To protect against impacts where the vehicle continues through the door opening, the door may be designed to come apart without breaking, as shown in Figure 5. Such breakaway features are offered with all types. However, the breakaway ability of horizontally rolling doors is limited. With folding doors, the individual folds may have magnetic or hook-and-loop tape attachments that separate. With vertically rolling doors, the fabric door material may pull out of the side guides. With segmented doors, the lower segments may pop out of the side frames. Or, the frame itself may separate or open on hinges, releasing the door.

When a breakaway door comes apart, the local staff must put it back together. If the door is well designed, this can be done quickly. If the door cannot be repaired quickly, it is likely to remain out of service for a long time, wasting energy.

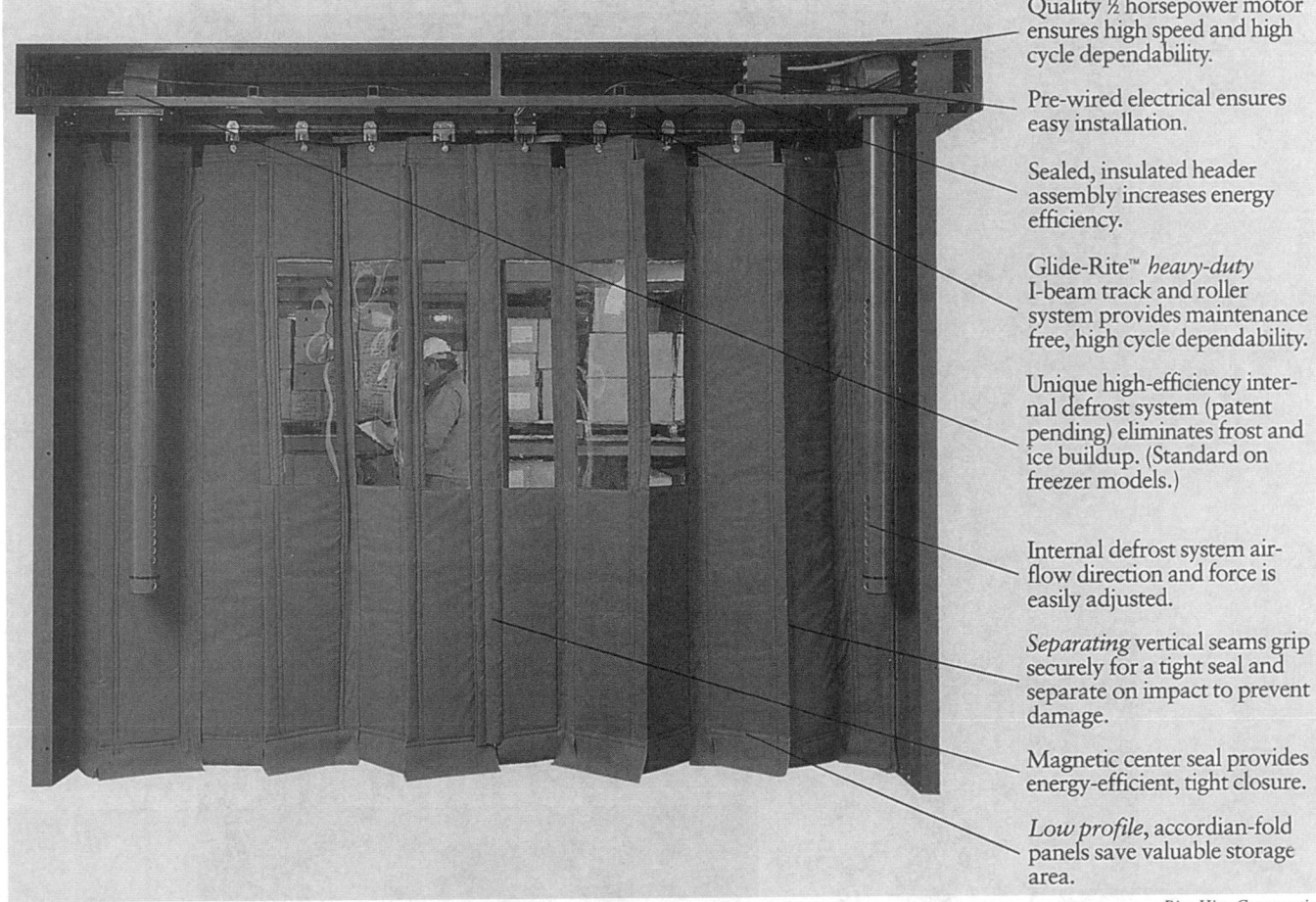

Quality ½ horsepower motor ensures high speed and high cycle dependability.

Pre-wired electrical ensures easy installation.

Sealed, insulated header assembly increases energy efficiency.

Glide-Rite™ *heavy-duty* I-beam track and roller system provides maintenance free, high cycle dependability.

Unique high-efficiency internal defrost system (patent pending) eliminates frost and ice buildup. (Standard on freezer models.)

Internal defrost system air-flow direction and force is easily adjusted.

Separating vertical seams grip securely for a tight seal and separate on impact to prevent damage.

Magnetic center seal provides energy-efficient, tight closure.

Low profile, accordian-fold panels save valuable storage area.

Rite-Hite Corporation

Fig. 3 Horizontally folding quick-acting door The door has two halves that fold to the sides. This model is designed for low-temperature cold storage. It highlights the features that one manufacturer offers for this application. The material of folding doors retains its flexibility better at low temperature than the material of rolling doors. Also, this type of door can be lightly insulated. Magnets provide a tight closure between the two halves of the door when it is closed. This door is made in vertical strips that are held together by hook-and-loop tape, which separate non-destructively in the event of a heavy impact.

Some doors make it easy to replace the window separately from the rest of the door. This is a valuable feature if the window is not expected to last as long as the door. However, the window attachment should not compromise the strength of the door.

■ Safety Features

The purpose of quick-acting doors is to allow traffic to keep moving. The motion of the traffic and the rapid action of the doors pose safety hazards. Different models of doors deal with safety in various ways. Investigate how the door deals with each of these three aspects of safety in your application:

- *visibility to the other side of the door.* A danger with any type of door is a collision between the moving vehicle and something on the other side of the door. To avoid this, the driver needs full visibility ahead of the vehicle. Rapid opening is the most effective way to provide safe visibility.

Doors that open horizontally provide full-height visibility through the door as soon as they start to open, but they do not provide a view of the entire door width until the door opens fully. Doors that open vertically immediately provide a full-width view of the floor area on the other side. However, the load in front of the driver obstructs the view of the center area until the door opens a large part of the way.

A large, clear window in the door is valuable in most applications. However, windows cannot cover the entire opening, and the plastic material becomes obscured over time.

- *danger when opening.* Rolling doors do not sweep through any additional space as they open, so they

Rite-Hite Corporation

Fig. 4 Overhead sectional quick-acting door   This is a variation of the common garage door. In addition to a powerful motor, the main differences are lighter weight, effective seals on all sides, and breakaway panels. The panels can be insulated.

Rite-Hite Corporation

Fig. 5 Breakaway feature Most quick-acting doors protect themselves against major impacts by breaking free of the frame. The design of the door should allow it to be replaced in the frame quickly when this occurs.

are relatively safe when opening. Segmented doors rise into space that is normally kept clear.

Folding doors pose some hazard when they open. You can minimize the hazard by selecting a folding door that has a larger number of folds, so that the folds do not stick out as far. Also, you can select a folding door that has soft folding edges. And, you can install guards to keep people away from the sides of the door.

• *danger when closing.* Quick-acting doors close rapidly, so their edges can be a serious hazard. A common solution is to install a "reversing edge" on the door. This is a lightweight strip that is installed in the extreme edge of the door. When the strip touches an object, it makes an electrical contact that stops or reverses the door. (You commonly find reversing edges on elevator doors.) Some vertically rolling doors rely on a yielding edge for safety. For example, one rolling door model that closes by gravity has an edge that consists of a pocket filled with small gravel. Figure 6 shows this feature.

Other doors use a breakaway edge. This is an edge that contains a weak link. Soft edges and breakaway edges protect the door itself from impact from any direction, as well as reducing the danger when closing.

Any type of door can be equipped with a sensor that holds the door open as long as anything is within the path of the door. Light beams are commonly used for this purpose. Some models

offer light beam switches as an integral feature. Figure 7 shows a typical installation.

■ Preventing Condensation and Frost

If the space on one side of the door is much colder than the space on the other side, condensation or frost is likely to form on the warmer side of the door. For example, this is a common problem in freezer applications, but it can occur in any application if the warmer space has high humidity. Condensation blocks visibility through the window, and it may harm the door materials. Frost may accumulate enough to interfere with the operation of the door.

Condensation occurs first on the window, because it is a single thickness of plastic that has minimal insulation value. You can keep the window clear by using a separately mounted heater to warm the side that faces the warmer space. You can use either a forced-air heater or a radiant heater. In either case, minimize the energy required by focussing the heater narrowly on the window portion of the door. Be careful to avoid overheating the plastic window or door material.

Some models offer a forced-air heater that is an integral part of the frame assembly. This is compact, and saves you the additional design and installation effort.

If the need for heating is not continuous, save energy by controlling the heater with a humidistat. Use a humidistat, not a thermostat, because condensation is related to the relative humidity in the warmer space. Install the humidistat in the warmer space at a distance from the door, so that the humidistat is not affected by the operation of the heater. Install a placard on the

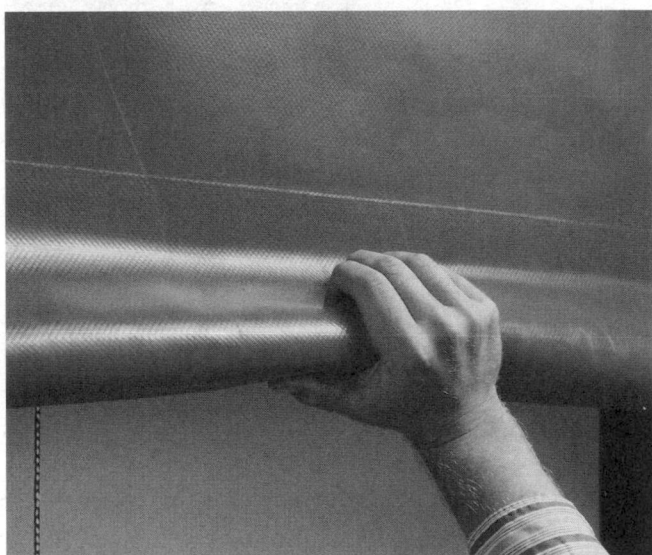

Rite-Hite Corporation

Fig. 6 Yielding edge This model of roll-up door uses a lower edge that is filled with gravel. The gravel also provides weight for closing the door by gravity. A common alternative is a "reversing edge," which has a contact switch that causes the motor to open the door. Some doors have a lower edge made in segments, with links that yield when the edge is struck.

humidistat to explain its operation. (Measure 5.1.1 tells how to design an effective placard for this purpose.)

You can minimize condensation on the rest of the door by selecting a door that has as much insulation as possible, and by avoiding a door that has metal conductive paths through the door. Fabric folding doors are perhaps the best in this regard. Selecting an insulated door minimizes the need to warm the entire door.

■ Avoiding Drippage

With doors that open vertically, debris and moisture picked up from the floor by the bottom edge can fall on people and goods as they pass through the door. This can be a hazard for sensitive goods, and it can be an annoyance to people using the door.

Horizontally opening doors avoid this problem completely. If you use a vertically opening door, you can reduce edge drippage by installing a threshold or by raising the floor under the edge. Make the threshold gradual enough to avoid jogging vehicles excessively as they pass through the door. This is most important with vehicles carrying tall loads.

■ Security

Plastic and fabric doors offer little security. An intruder can cut through them. Most rolling and folding doors have breakaway features, so an intruder can use a vehicle to push through the door.

Overhead sectional doors offer considerably more security, although not as much as a conventional exterior doors. However, the breakaway features make them vulnerable. In many models, the door can be locked to

WESINC

Fig. 7 Light beam safety switch This safety device holds the door fully open as long as the light beam is interrupted.

the frame with sliding bolts (cane bolts). The number of bolts increases with the number of panels that break away.

The staff must remember to insert the bolts during periods when the facility is not operating, and to remove them before traffic starts. Some manufacturers may offer doors with automatic bolts that unlock when the door is activated.

Installation Issues

Install the door so that it provides safety, efficient traffic flow, and reliability. Consider the following issues when deciding how to install the door.

■ Sensors for Opening and Closing

The type of sensor you install, and the way you install it, can make a big difference in the performance and the longevity of the door. You need one set of sensors to open the door reliably, and you may need another set of sensors to close the door safely.

For opening the door, the main point is to sense an approaching vehicle early enough so that the door is open enough to avoid damage by the time the vehicle reaches it. Calculate the appropriate distance between the sensor and the door by using the door opening speed and the maximum expected vehicle speed.

For example, consider a door that opens at the rate of eight feet per second. It is used by forklift trucks that require a vertical clearance of eight feet. Therefore, the door requires an opening time of one second to provide adequate clearance. If the maximum expected speed of the truck is ten feet per second (about seven miles per hour), then the sensor must start opening the door while the truck is at least ten feet from the door.

For closing the door safely, you need additional sensors that verify the absence of objects in the door safety zone. For example, an induction loop cannot sense a person standing underneath an open door. To keep the door from closing on the person, you need sensors that will detect anything within the door area itself.

In general, you can use any type or combination of sensors with any type of door. The sensor installation may be entirely separate from the door assembly. It's up to you to select a sensor that will work reliably in your application. The types used most commonly with quick-acting doors are:

- *induction loops.* An induction loop is a coil of wire buried in the floor that creates a magnetic field. It senses the presence of steel. The loop must be selected for the mass of steel in the vehicle, and for the height of the steel chassis above the floor. Induction loops are perhaps the most common type of sensor used to open quick-acting doors for vehicles.

- *ultrasonic motion detectors.* Ultrasonic sensors are fairly reliable as long as the person or vehicle keeps moving toward the door, but they fail if the vehicle has to stop near the door. They can "see" the entire path to the door. By the same token, they are susceptible to false triggering from activity near the door. See Reference Note 11, Personnel Sensors, for details about this type.

• *photoelectric cells (light beams, electric eyes).* These are commonly used for keeping the door open because the beam can be aligned closely with the operating path of the door. Some models offer them as an integral part of the door assembly for this purpose. Electric eyes can also be used to sense approaching vehicles or people.

Especially with materials handling equipment, you have to install the sensors so that they can accommodate differences in the vehicle configuration and loading. For example, an induction loop may be able to sense the fork of a forklift truck while it is lowered, but not when it is raised.

Even if vehicles are the primary traffic, expect that the door may be used by pedestrians. If the sensors necessary for vehicles cannot sense people reliably, install a separate manual control for pedestrian traffic. Install the manual control within the safety zone of the door, so the safety sensors keep the door open as long as people are nearby. Controls for pedestrian traffic are not needed if there is a separate personnel door that is located nearby. However, the door safety sensors should still protect against people attempting to scurry through the vehicle door.

Doors can also be operated by remote control. This is usually a security measure that is used outside of normal operating hours.

■ Mounting in the Wall Opening

Quick-acting doors are light, so they generally do not exert a strong reaction force on the wall when they operate. Typically, the open and closing forces are contained with the frame assembly.

As with any door, make sure that the gap between the door frame and the wall opening is sealed to prevent infiltration. Don't just stuff the gap with porous insulation. Fill the gap with solid material, such as wood and mortar, and finish it off with real caulking compound.

■ Door Guards

As we said before, the primary means of protecting the door from damage is to make sure that the door is fully open before a vehicle can reach it. In addition, it is prudent to install guards that protect the door from collision while it is in the open position. These should be heavy. Anchor them to the floor or to the wall, not to the door frame.

■ Traffic Pattern and Traffic Guides

For safety, and to avoid confusing the sensors, consider using guard rails or barriers to create an area on each side of the door that excludes people and vehicles that are not going to use the door.

Caution Drivers

Some vehicle drivers seem to view it as a challenge to ram the door before it opens. Try to dampen their enthusiasm by installing a large sign on each side of the door that says, "CAUTION: Drive through the door SLOWLY."

Keep Them Repaired

In the environments where quick-acting doors are used, some damage is almost inevitable. Establish a routine for getting the doors repaired as quickly as possible. Otherwise, energy will be wasted and refrigerated products may be spoiled.

ECONOMICS

SAVINGS POTENTIAL: *Hundreds of dollars to thousands of dollars per year per door, depending on climate, fuel cost, space layout, etc.*

COST: *In typical sizes, powered quick-acting doors cost from $10,000 to $20,000, installed.*

PAYBACK PERIOD: *One year to several years.*

TRAPS & TRICKS

CHOICE OF TYPE: *Do you really need a powered door? How important is ease of operation? Would a swinging door create a safety hazard? Impact doors provide comparable or better efficiency at much lower cost, unless there is significant wind or pressure differential. They also provide better security than most types of quick-acting doors.*

SELECTING THE EQUIPMENT: *The differences between the various types of powered doors may be important for your application. Also, the details of particular models may make a big difference in ease of maintenance.*

INSTALLATION: *Follow the manufacturer's instructions closely. Seal the gap between the frame and the wall. Install a highly visible placard on each side of the door that asks vehicle drivers to drive through the door carefully.*

MAINTENANCE AND SUPERVISION: *A powered door saves no energy while it is out of service. Teach the relevant people how to repair a door that has been knocked loose. Use your maintenance calendar to schedule periodic inspections.*

MEASURE **6.2.6 Install dock-to-truck seals.**

Most of the world's goods are transported at some time by trucks and tractor trailers. Many of these are loaded and unloaded at enclosed truck docks. When a truck pulls up to the dock, the gap between the vehicle and the dock is a major path for air leakage. In fact, the gaps along a line of vehicles in effect convert an enclosed truck dock to an outdoor space. This greatly increases heating requirements in the loading dock and adjacent spaces during cold weather. When loading to and from refrigerated trucks and spaces, the air leakage radically increases refrigeration cost.

You can prevent most of this loss by installing sealing systems on the outside of the truck dock wall that mate with the vehicles. These systems are commonly called "dock seals." Figure 1 shows a truck dock that uses dock seals effectively.

Dock seals have become popular. They continue to evolve, so get the latest catalogs.

SUMMARY

Greatly reduces infiltration into enclosed truck docks. Essential for cold weather, and when loading to and from refrigerated spaces.

SELECTION SCORECARD

Savings Potential	$ $ $ $
Rate of Return, New Facilities	% % % %
Rate of Return, Retrofit..........	% % %
Reliability	✓ ✓ ✓
Ease of Retrofit	☺ ☺ ☺

Types of Dock Seals

Contemporary dock seals come in two basic type. We will call these "butt seals" and "side seals," although

Rite-Hite Corporation

Fig. 1 Truck dock equipped to minimize infiltration Each entry has a tightly sealing door for times when it is not being used. To seal the gap between the trailer and the dock, butt-type seals are installed. Simple alignment marks are attached to the butt seals to help the driver mate the trailer to the seal properly. Dock levelers are installed to accommodate variations in trailer height. The dock leveler at left is in its vertical stowed position, where it acts as a safety feature to keep people from driving off the dock. A latching mechanism for the trailer's ICC bar is installed below the dock leveler. The trailer at right is being held in place by wheel chocks. The entire assembly is protected from precipitation by the roof overhang.

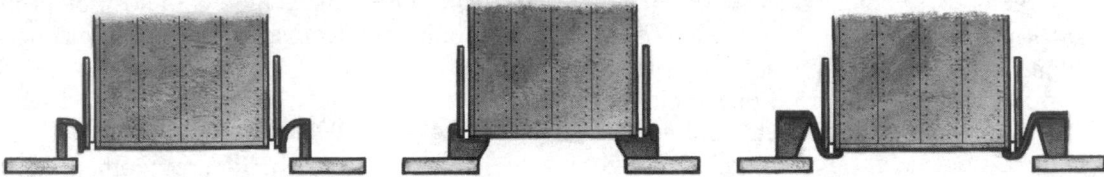

Rite-Hite Corporation

Fig. 2 Types of dock seals The type at left is a sliding seal, which contacts the sides of the trailer, seen here from above. The type in the center is a butt seal, which contacts the rear of the trailer. The type at right is a hybrid of the other two that seals the gap between the open door and the side of the trailer.

standard names for these types do not yet exist in the industry. Figure 2 shows both types. Each type has advantages and disadvantages in particular applications. There are variations within these two types.

With butt seals, the rear of the truck or trailer butts against thick pads on the outside of the wall opening, squeezing the pads between the vehicle and the wall. By far the most common seal of this type is made of a spongy rubber material, covered with a wear-resistant surface.

Less commonly, a butt seal may consist of an inflatable bladder. The bladder is inflated with a small high-pressure fan. This type is more adaptable, but more expensive. Figures 3 and 4 show this type.

Some docks seals may use an accordion seal made of rubberized fabric. This type is unusual.

The other major type, side seals, slide along the sides and roof of the vehicle. These consist of a more flexible rubberized material. The top seal may consist of a heavy curtain of adjustable height that drapes over the top of the truck.

Dock seals should be sheltered from precipitation. Otherwise, they will collect water, rot the wooden structure, leak into the wall, and suffer freeze damage. If the building does not provide adequate shelter, as in Figure 1, you can buy appropriate hoods from dock seal manufacturers, or from other sources. The hood may

WESINC

Fig. 3 Inflatable dock seal The upper part of the seal is inflated, the sides are not. Note the guide post, the bumpers to keep the trailer from backing too far into the seal, and the stoplight. Correct alignment is essential for sealing and for protecting the seals.

WESINC

Fig. 4 Inflatable seal deflated The small fan at the top of the unit provides all the pressure that is needed for effective sealing.

be separate from the dock seal, or it may be an integral part of some dock seal assemblies.

The seal along the floor of the vehicle is a special problem. The floor height varies from one vehicle to another, and the height varies as the vehicle is loaded and unloaded. If the floor height remains close enough to the dock height, a simple butt seal may suffice. This can be used in combination with butt seals or side seals for the sides and top of the vehicle.

An adjustable ramp, called a "dock leveler," is commonly installed in the floor of the loading dock. Figure 5 shows a typical unit. The dock leveler adjusts to changes in the height of the vehicle floor. Unfortunately, the space underneath the dock leveler is a large path for infiltration. Dock levelers may be equipped with side guards, as shown in Figure 5.

The side guards are intended as a safety feature, and to keep things from falling into the dock leveler pit. They may substantially reduce infiltration, especially from wind. One of these days, dock leveler manufacturers will realize that these guards can easily be augmented to make them effective infiltration seals. If you get dock levelers, make sure they have this feature.

Dock levelers are made of steel, so condensation and ice tend to form on the ramp when the outside air is cold. This is dangerous. To avoid this problem, you can order insulation as an option on the dock leveler. Get this feature. It also saves a small amount of heat loss.

An older method of minimizing infiltration at truck docks is to attach a short tunnel to the outside of the building. It is effective primarily against wind, rather than convective heat loss. The tunnel must fit the vehicle body closely to be effective. This requires that all vehicles have the same dimensions, and it is more

difficult to back the vehicle into position properly. The availability of effective dock seals has made this method obsolete.

How to Select a Dock Sealing System

Dock sealing systems are another example of an important energy conserving action that is simple in concept, but that requires careful planning. Consider these issues before you buy:

• *adaptability to different vehicle dimensions.* Truck or trailer dimensions will eventually change, even in fleet operations, where dimensions may be standardized. One day, your standard dimensions will change. Butt seals are more adaptable than side seals to differences in vehicle width. Major differences in vehicle height require a curtain-type seal to contact the roof of the vehicle.

• *adaptability to height change in loading.* The vehicle changes height at the dock as it is loaded and unloaded. Butt seals and sliding seals can accommodate these changes in height. However, the vertical movement tends to wear out the pads of butt seals. Different manufacturers address this problem in different ways. One method is to use pleats in the surface material, so the pleats slide as the vehicle height changes.

If you use a dock leveler to accommodate changes in vehicle floor height, be sure to include effective seals in the dock leveler.

• *adaptability to vehicle tilt.* At many loading docks, the vehicle ramps has an upward or downward slope. Most side seals can accommodate the tilt without special provisions. With butt seals, you have to order side pads that are tapered from top to bottom, as shown in Figure 6. Butt seals have only limited ability to deal with variations in the vehicle tilt. Inflatable seals are more adaptable than solid butt seals. If the ramp does not have a constant slope, vehicles may come to rest at different degrees of tilt, depending on the distance between the front and rear wheels.

• *adaptability to different types of vehicle doors.* Butt seals are efficient with all types of doors that swing completely out of the way of the rear face of the vehicle. Ordinary side seals have a problem with doors that swing open. The gap at the door hinge creates a channel between the open door and the side of the vehicle that is a path for infiltration. A variation of side seals, shown in Figure 2, has extensions that cover this gap at the rear of the vehicle. However, this type of side seal cannot accommodate much variation in vehicle tilt.

• *resistance of seal to damage from goods and loading equipment.* The opening between the pads of a butt seal must be narrower than the width of the vehicle. This makes the pads vulnerable to

WESINC

Fig. 5 Dock leveler This is a steel ramp that is moved by a hydraulic jack. The metal skirts on each side are intended as a safety feature. They also reduce infiltration from underneath the trailer into the space.

damage as goods are being loaded and unloaded. For example, a forklift truck carrying pallets into a trailer can sideswipe the seal. This problem does not exist with side seals.

• *resistance of wall to damage from vehicle pressure.* The seal should be designed so that the pressure of the vehicle is exerted primarily on the dock floor, not on the wall of the building. The force of a vehicle backing against a wall is enough to damage it. This is true of all wall types, even concrete block. (Concrete has great compression strength, but no significant tensile strength. The wall is held together by the weight above each block. A side load can buckle a concrete block wall fairly easily.) Side seals transmit much less force to the wall than butt seals. If you need a butt seal, consider an inflatable type. Alternatively, brace the wall or the door frame.

• *compatibility with dock doors.* Each opening in the truck dock needs a door, as shown in Figure 1. The door is normally installed on the inside of the wall, while the sealing system is normally installed on the outside of the wall. However, there may be interference between the two, especially if the sealing system must be bolted through the wall.

• *latching the vehicle to the dock.* Safety may require a system that latches the truck or trailer to the dock. The latch is below the level of the truck floor, as shown in Figure 1. In the United States, trucks and trailers must have an "ICC bar." This is a fender bar that is attached to the rear of the vehicle to keep following automobiles from riding under the truck or trailer in a collision. Some manufacturers offer systems that latch the ICC bar to the dock. Make sure that the sealing system does not interfere with the latch, or vice versa. An alternative is installing automatic chocks for the rear tires of the vehicle, which are also shown in Figure 1. These do not interfere with the dock seal.

• *durability and ease of repair.* The sealing system must withstand slamming by vehicles that weigh tons, sliding stresses from changing vehicle height, and uneven pressure on the pads. All dock seals have components that wear out and require

occasional replacement. Nobody likes maintenance, so repairs tend to be neglected. To minimize heat loss over the long term, select seals that last for a long time and are easy to repair.

• *ease of installation.* Labor is a large part of the total cost of a dock seal. Especially in retrofit applications, look for a sealing system that can be installed easily to the existing dock.

Align the Vehicles with the Dock

With butt seals, it is important to align the vehicle perpendicular to the dock. With side seals, it is important to position the vehicle laterally within the seal. Provide guides to aid drivers in aligning their vehicles.

An inexpensive method is to install painted stripes on the dock face to help drivers to align visually. For night operations, install reflective tape or position lights. Figure 1 shows simple alignment marks made of reflective tape that are attached to the face of a butt seal.

A more positive method is to install guide rails for the tires. With butt seals, make the rails long enough to guide both the front and rear tires. To prevent tire damage, make the rails rounded. Large steel pipe is effective for this purpose. Figure 3 shows guard posts that are installed near the dock seal.

Coordinate with Dock Doors and Interior Doors

Truck-to-dock seals are useless if the adjacent dock doors are left open to the outside. Make sure that you have good dock doors that close effectively and easily. See the Measure 6.2.4 about exterior doors. Also, see Measures 6.2.5 ff for interior doors that you can use to isolate the truck dock from the rest of the building.

ECONOMICS

SAVINGS POTENTIAL: *Thousands of dollars per year, typically. Depends on climate, fuel cost, space layout, etc.*

COST: *Butt seals and sliding seals typically cost $1,000 to $2,000, installed. Inflatable seals may cost twice this much. Additional sealing features for dock levelers may cost about $500.*

PAYBACK PERIOD: *Several years, or longer.*

TRAPS & TRICKS

SELECTING THE EQUIPMENT: *Shop the market to find all your options. If you have several dock doors, consider installing one sealing system on a trial basis. Talk to other operators about their systems.*

EXPLAIN IT: *Install large placards that tell drivers how to engage the sealing system.*

MAINTAIN IT: *Expect to replace parts from time to time. Keep spares on hand. Schedule inspections in your maintenance program.*

Rite-Hite Corporation

Fig. 6 Butt seals must be tapered to accommodate sloping ramps

Windows can be a major path for air leakage, especially in older buildings. Windows also affect energy consumption in several other ways. There are three Subsections that deal with different aspects of window efficiency. The Measures in this Subsection reduce air leakage through windows and window frames. Subsection 7.3 recommends improvements to windows that deal specifically with their insulation value. Subsection 8.1 deals with windows as a source of heat gain, and it provides the most detailed coverage of window characteristics. Become familiar with all the window-related Measures in these three Subsections before you make any major changes to your windows.

INDEX OF MEASURES

6.3.1 Maintain the fit, closure, and sealing of windows.

6.3.2 Install weatherstripping on openable windows.

6.3.3 Install supplemental ("storm") windows.

6.3.4 Install high-efficiency windows.

RELATED MEASURES

- Measure 4.2.1, control of building pressurization
- Measure 6.4.2, sealing infiltration around window frames
- Subsection 7.3, for window modifications to reduce conductive heat loss
- Measure 8.1.3, installation of high-efficiency windows

MEASURE **6.3.1 Maintain the fit, closure, and sealing of windows.**

All openable types of windows allow some air leakage. Modern versions of all types of windows have much less leakage than old models. However, even modern windows have serious infiltration when the weatherstripping wears out. With hinged windows, air leakage may become especially severe if the movable panes get out of alignment with the frame. Inspect all your windows periodically to check for air leakage, and maintain the sealing features.

Even fixed windows should be checked periodically. Thermal expansion causes relative motion between the glazing and the frame. This motion and the effects of weather and sunlight destroy sealing materials.

How often should you check the windows? It depends on the window type and the severity of usage. The main items to check are:

- *the integrity of the glazing.* Replace any panes that are broken, or that do not fit properly in the frame.

- *the integrity of glazing compounds and gaskets.* Glazing is held in the frames by rubber seals or by glazing compound. These need to be replaced over long intervals as they wear out.

- *the condition of infiltration seals.* All openable windows should have infiltration seals between the fixed and movable frame elements. These are made of flexible materials that deteriorate with time and usage. Replace them as necessary.

- *proper fit of movable elements.* Make sure that movable elements fit properly within the fixed frame, and that they move easily. Bad alignment defeats infiltration seals and wears them out.

- *operation and adjustment of latches.* Make sure that window latches operate easily and securely. A latch that is not adjusted properly can prevent the window from sealing properly.

SUMMARY

Routine maintenance needed to prevent energy waste and maintain comfort.

SELECTION SCORECARD

Savings Potential	$ $ $
Rate of Return	% % % %
Reliability	✓ ✓
Ease of Initiation	☺ ☺ ☺

What If the Windows are Inherently Leaky?

If windows lack adequate infiltration seals, you can usually improve them with weatherstripping. See Measure 6.3.2, next. If you cannot seal your existing windows adequately, consider replacing them with more efficient units, as recommended by Measure 6.3.4.

ECONOMICS

SAVINGS POTENTIAL: *Up to 50 percent of heating and cooling costs. The saving depends on the type and size of windows, space layout, etc.*

COST: *Usually minor.*

PAYBACK PERIOD: *Short. Serious window leakage can cost much more than routine maintenance.*

TRAPS & TRICKS

SCHEDULING: *Most of the challenge of window maintenance is remembering to do it. Schedule it in your maintenance calendar.*

PREPARATION: *Keep a full stock of spare parts and glazing materials on hand. Searching for specialized window hardware can be more bother than the actual maintenance work.*

MEASURE **6.3.2 Install weatherstripping on openable windows.**

Older openable windows tend to have serious air leakage. The worst leakage occurs with older hinged windows. These include casement (hinged on the side), hopper (hinged on the bottom), and awning (hinged on top) types. Old double-hung windows are also very leaky. Weatherstripping can reduce the air leakage of existing windows considerably, if you install it properly. However, it usually cannot provide the leakage resistance of modern efficient windows.

A given window may require several types of weatherstripping. The first step is to find out the types of weatherstripping that are available. Weatherstripping is available that you can use for most types of openable windows. Most weatherstripping is generic, so you have to figure out how to use it effectively on your windows. You will need a different combination of weatherstripping materials for each window type.

Weatherstripping is designed either to slide or to be compressed. Any particular type of weatherstripping will not work well unless it is used in the intended manner. Compression weatherstripping is the most important type for hinged windows. Select the thickness of compression weatherstripping carefully. If it is too thick, it will keep the window from closing fully, which may increase air leakage.

Weatherstripping for sliding windows is more specialized. Weatherstripping for old double-hung windows must be capable of sliding, and it must also fill the large gaps between the movable part and the frame.

Self-adhesive weatherstripping is convenient, but it has limited life because the adhesive eventually dries out. It tears off quickly if it is installed in a way that places stress on the adhesive.

Expect to Replace It Periodically

The major limitation of retrofit weatherstripping is that it does not last long, usually a couple of years at most. Weatherstripping is not an effective solution unless you can keep replacing it periodically. See Measure 6.3.1 about maintenance.

SUMMARY

May greatly reduce window air leakage. Requires careful installation and periodic replacement.

SELECTION SCORECARD

Savings Potential	$ $ $
Rate of Return	% % % %
Reliability	✓ ✓
Ease of Initiation	☺ ☺

Alternative: High-Efficiency Windows

To achieve the lowest air leakage and minimize maintenance requirements, as well as gaining other major benefits, install high-efficiency windows. The best modern windows have effective and durable sealing systems. Needless to say, this change is expensive. See Measure 6.3.4.

ECONOMICS

SAVINGS POTENTIAL: *3 to 60 percent of heating and cooling costs, depending on the type and size of windows, space layout, etc.*

COST: *$1.00 to $10.00 per movable element, most of which is for labor.*

PAYBACK PERIOD: *Typically less than one year.*

TRAPS & TRICKS

PLANNING: *This work requires thought, mostly in selecting the materials and in figuring out how to install them. Select your materials and methods to provide effective sealing, minimum interference with window closure, and durability.*

MAINTENANCE: *Consider weatherstripping to be an expendable material that you have to replace periodically. Check your installation methods by observing how long the weatherstripping lasts.*

MEASURE 6.3.3 Install supplemental ("storm") windows.

In existing facilities, installing supplemental glazing can be a powerful way of improving the efficiency of windows. (Common "storm windows" are supplemental glazing, but the concept can be extended much further.) Supplemental glazing can eliminate infiltration almost completely. If the existing windows are single-glazed, a single sheet of supplemental glazing can almost triple the insulation value. There are good ways and bad ways to install it. See Measure 7.3.2 for the details.

MEASURE 6.3.4 Install high-efficiency windows.

High-efficiency windows eliminate several major causes of energy waste. Installing high-efficiency windows is a fundamental, economical energy conservation measure in new construction. It is also an important improvement in existing facilities, although the economics are not as favorable. Selecting high-efficiency windows involves a variety of potential benefits, problems, and interactions with other aspects of the building design. See Measure 8.1.3 for the complete picture.

Other Envelope Leakage

The building envelope consists of the parts of the building that separate the inside from the outside. The envelope is a complex structure. In addition to walls and roof, it includes many surfaces that are hidden from view. Examples are elevator shafts, enclosed ceilings in loading docks, vented pipe chases, etc. All these surfaces may include paths for air leakage. In many cases, air leakage may occur along a long path inside the envelope structure. The envelope also includes specific components, such as hatches and smoke vents, that may become defective, allowing considerable air leakage. The Measures in this Subsection correct these leaks. Each Measure focuses on a specific type of envelope leakage and the particular methods used to correct it.

INDEX OF MEASURES

6.4.1 Seal gaps in the envelope structure.

6.4.2 Install gaskets at wall switches and receptacles that allow outside air leakage.

6.4.3 Install roof and attic hatches that close tightly and reliably.

6.4.4 Adjust and repair space air vents periodically.

6.4.5 Ventilate elevator shafts rationally.

RELATED MEASURES

• Subsection 7.2, for envelope insulation
• Measure 4.2.1, control of building pressurization

MEASURE **6.4.1 Seal gaps in the envelope structure.**

The building envelope is a complex system that is intended to provide structural strength, protection from the elements, and good appearance. Designers may leave the details of sealing the envelope structure against infiltration to the builders. However, builders are not in the business of devising reliable sealing methods in the midst of construction. Also, materials used for sealing the envelope may have shorter life than the building as a whole. As a result, most buildings have air leakage paths that should be sealed.

Where Envelope Air Leaks Occur

An air leak can occur anywhere that the skin of the building has an opening. In most buildings, the wall has an outer skin and an interior finish surface, with various layers of structure between the two. Outside air enters at an opening in the outer skin, travels through the envelope structure, and enters the interior of the building at a gap in the interior finish surface. Air may travel a long distance inside the envelope. Envelope air leakage typically occurs in certain areas of the structure. To organize your search, focus on the following sites.

■ Clearances Around Window and Door Frames

Windows and doors are installed after the wall is built. Therefore, the wall openings are made larger than the windows and doors. This leaves substantial gaps between the frames and the wall, as shown in Figure 1.

WESINC

Fig. 1 Typical window installation All window and door assemblies are surrounded by a gap. The gap should be filled with an effective sealant, but this is rarely done. In many cases, the molding around the door or window is the only barrier to air leakage.

SUMMARY

Typical buildings have many envelope leaks. You need one set of skills to find them, and another to fix them.

SELECTION SCORECARD

Savings Potential .... $ $

Rate of Return .... % % % %

Reliability .... ✓ ✓ ✓ ✓

Ease of Initiation .... ☺ ☺

If the building is well constructed, these gaps are filled when the windows and doors are installed. However, this rarely happens. Instead, the frames are shimmed into place, and the gaps are covered with molding.

If the molding is installed well, it acts as an effective infiltration seal. However, the molding may not be well attached, or it may loosen with age. If the molding is the infiltration seal, it should be tight on both the inside and outside. Otherwise, air can enter the wall structure through the molding.

■ Curtain Wall Structures

A "curtain wall" is a non-load-bearing exterior surface that is attached to a load-bearing steel building structure. The curtain wall is made of panels sized to be convenient for shipping and installation. Typically, each panel is attached to the steel frame with a small number of clips. There may be little or no obstruction to air movement inside the wall structure.

The primary barrier to air leakage in curtain walls is sealing material between the exterior panels. This may be done with pre-formed gaskets, or the gaps may be sealed with an elastomeric (rubbery) caulking material. Look for gaskets or caulking that may have been omitted. If the panels are not fitted closely, the sealing material may not close the gap between panels. In older buildings, the sealing material may have deteriorated to the point that it loses effectiveness.

A shoddy practice that is common in curtain wall construction is attempting to join the backing sheets of batt insulation to form a barrier to air leakage. Much of the tape fails immediately, and all of it fails eventually. The weight of the insulation causes it to collapse, as shown in Figure 2. Even if the exterior building surface is well sealed, air that is cooled by heat loss through the thin exterior surface is able to flow freely to the interior of the building, rendering the insulation useless.

■ Soffits

A soffit is an exposed underside of a structure. Soffits occur under the eaves of roofs and under overhanging floors. A soffit may form the roof of a loading dock or other entrance. The soffit is not exposed to precipitation and it carries no load, so it may be covered with a flimsy surface that allows considerable infiltration. In many buildings, the structure that is covered by the soffit has no provision to stop infiltration. Or, the air space inside the soffit may connect with the interior of the building, by way of ceiling plenums or curtain wall spaces. Figure 3 shows an example.

There are two general solutions to soffit infiltration. One is to install an infiltration barrier inside the soffit. The other is to use the soffit itself as the infiltration barrier. If you choose the latter approach, you may have to strip off the existing soffit material and replace it with a strong, impermeable surface.

■ Roof-to-Wall Joints

The heavy load of the roof is transmitted to the building structure by connecting the roof joists or rafters to the load-bearing columns. In this method of construction, the spaces between the roof joists are open to the spaces between the wall columns. To keep air leakage from traveling from the roof structure to the wall structure, or vice versa, the gaps between the joists or columns should be closed. However, this may not be

done, as there is no structural need. Infiltration at this point is especially likely if the roof overhangs the wall, forming a soffit.

In smaller buildings, the roof joists may rest on the top of the wall structure. In this case, the horizontal member that forms the top of the wall is called a "plate." The plate is an effective barrier to infiltration into the top of the wall if it is installed properly. However, there is still an infiltration path into the space between the joists. Failing to close the gap between the joists is a common oversight. In this type of construction, the roof or attic joists commonly overhang the wall, creating a soffit. Sealing this arrangement is complicated by the need to vent the attic space to prevent condensation.

■ Foundation Plates

The top of a concrete foundation is an irregular surface. A steel beam or a wooden board is attached to the top of the foundation to act as a connector to the wall structure above. This connector is called a "plate" or "sole plate." There is an irregular gap between the plate and the top of the foundation. There may also be a gap between the plate and the wall. Figures 4 and 5 illustrate the problem.

If the building is built well, the gap below the plate is filled by installing a strip of compressible material on top of the foundation before installing the plate. Similarly, if the wall rests on the plate, compressible material should be installed between the two. In many buildings, this gap sealing material is missing. This creates an infiltration path all along the top of the foundation, or along the top of the sole plate.

■ Expansion Joints

In larger buildings with rigid exterior surfaces (especially concrete and masonry), expansion joints must be installed in the exterior surfaces to allow for thermal expansion and contraction. These are potentially long infiltration paths. Expansions joints typically are filled

WESINC

Fig. 2 Air leakage superhighway This is the scene behind the opaque portions of the curtain wall of a new glass box office building. The insulation was joined by tape. The tape quickly fails, causing the insulation to collapse. Air cooled by the exterior surface of the curtain wall freely circulates to the interior of the building, rendering the insulation useless. Shoddy practice like this is common in glass box buildings.

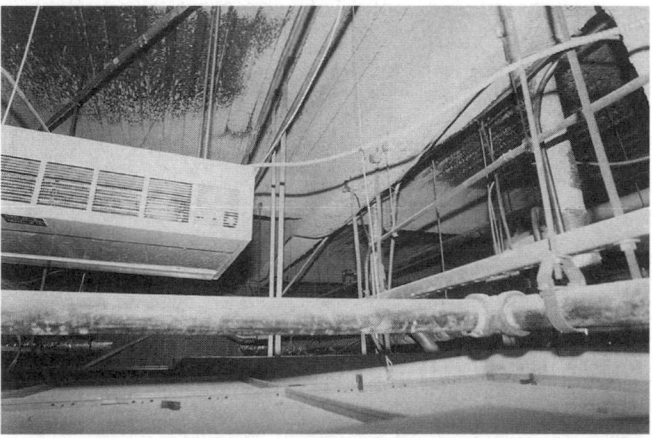

WESINC

Fig. 3 Inside a soffit over a loading dock The bottom surface, exposed to the outside, is loose fitting ceiling tile. The air space connects to the ceiling plenums and curtain wall spaces of the building. Several heaters are installed to keep water pipes in the space from freezing.

Fig. 4 **Foundation plate, from outside** A layer of insulation material is used to seal the gap between the concrete wall and the rim joist. Although sold commercially for the purpose, this is the wrong type of material for the job.

with gaskets or with very flexible caulking materials. All these materials have a life span that is long, but only a fraction of the life of the building.

■ **Parapet Flashing**

In buildings with flat roofs, the exterior wall structure typically extends above the roof level, forming a parapet. The wall structure typically is hollow, allowing a large chimney effect to move air inside the wall. To close the top of this potential chimney, a cover is installed over the top of the wall. In modern construction, this cover is typically sheetmetal called "flashing."

The flashing is not merely a decorative item. In most cases, it is an important infiltration barrier. To be effective in this role, it must be tightly sealed to the top of the wall. Commonly, the flashing is not installed in a manner that can keep it sealed over the life of the building.

■ **Gaps in Structures**

The people who build the structure may not have knowledge or interest about preventing infiltration. As

Fig. 5 **Foundation plate, from inside** Daylight shows where air can leak into the building. There are large gaps at the foundation plate and around the end of the center beam.

a result, air leaks are common in structures. For example, when concrete block is laid around structural steel, the bricklayers will not fill in the gaps that will occur. Figure 6 is a typical case.

Gaps may also occur after the building is built. For example, someone may remove a block to inspect hidden equipment, and neglect to replace the block. Figure 7 shows an example.

■ **Penetrations**

Any building envelope has a large number of penetrations. These include vent stacks, drain pipes, flues, exhaust ducts, power and telephone cables, etc. Most of these will be properly sealed. Some will not be. For example, if there is a flue pipe inside a chimney structure, air may be able to enter or leave the building through the space between the flue and the chimney.

How to Find Envelope Air Leaks

See Reference Note 40, Building Air Leakage, for the tools to use for finding air leaks. Most of these are simple and cheap.

To organize your search, focus on the areas discussed previously. Equip yourself with a set of as-built drawings to reveal concealed areas where leakage may occur. You can find some potential leak sites on the architectural drawings (e.g., expansion joints), some on the mechanical drawings (e.g., vent stacks and ducts), and some on the electrical drawings (e.g., power cables).

Don't overlook air leakage through the walls of internal spaces that are open to the outside, such as pipe chases and ventilation air shafts. Also, consider leakage

Fig. 6 **Gap around structural steel** Inside this elevator shaft, concrete block had to be fitted around steel beams. The masons filled the gap with clay bricks, the smallest material they had on hand, leaving a gap that is still large.

WESINC

Fig. 7 Missing block The block was removed by a plumber to gain access to a pipe fitting. Since the plumber is not a mason, he did not replace it. While this provides a home for birds, it also allows major air leakage into the building.

through the walls of spaces that are not heated to normal temperature, such as garages, loading docks, unheated storage spaces, unheated equipment rooms, crawl spaces, and voids.

If the facility is large, use a camera to help you make the survey. Film is cheaper than time, and photographs record conditions that you could not possibly describe verbally.

How to Seal the Leaks

Sealing an air leak is usually a simple procedure, but take the time to do it right. The trick is to seal the gap in a way that endures. These are the important points:

• *Seal the gap from the outside wherever possible,* to block leakage of outside air into the wall cavity.

Sometimes, this requires removing exterior sheathing to gain access to a gap that is hidden.

• *Close large gaps with structural material before caulking.* Fill large gaps with wood, masonry, mortar, sheetmetal panels, etc. Fasten this material securely to make it part of the structure. Then, caulk any small gaps that remain.

• *Use caulking materials,* rather than stuffing insulating materials into gaps. Air can flow easily through porous insulation.

• *Force the caulking material into the gap,* rather than merely over the gap. Caulking materials harden and shrink as they age. Install the caulking material so that the interior surfaces of the gap hold the caulking in place as it ages.

ECONOMICS

SAVINGS POTENTIAL: *Varies widely. See Measure 4.2.1.*

COST: *Low, unless you have to remove surfaces to gain access to gaps.*

PAYBACK PERIOD: *Less than one year, to several years.*

TRAPS & TRICKS

SELECTING METHODS: *Don't expect to fix all envelope leaks with a single method or material. Treat each leak as a separate engineering task, in which your goal is to find a sealing method that will last as long as the building. Combine caulking material with filler pieces for large gaps. Plastic foam applied with a spray can is versatile, but don't try to fill big holes with it. Large expanses of caulking material are ugly, and they will eventually crumble and fall away. Use silicone caulk and a small nozzle for most smaller gaps. Start with rubber strips and other flexible material in expansion joints and other gaps that may move over time.*

MEASURE **6.4.2 Install gaskets at wall switches and receptacles that allow outside air leakage.**

Among the many gadgets invented during the energy crisis of the 1970's are gaskets designed to be installed behind electrical switch plates and receptacle covers. These stop leakage between the inside of the wall and the space. The gaskets are precut to fit switches or receptacles snugly. They are available for common switch and receptacle configurations. They are cheap and easy to install.

Switch and receptacle gaskets provide a benefit if the outside of the wall is penetrated. For example a typical application is for a switch that is wired to an outside light fixture, where air can enter along the wiring. Gaskets are also helpful for switches and fixtures installed in an outside wall that has a large cavity. In this case, gaskets block the entry of convection currents from the inside of the wall.

The gaskets do not keep air from entering the wall on the outside. Air leakage into the wall can reduce the effectiveness of the wall insulation, even if the air cannot enter the interior spaces. Therefore, it is more efficient to seal any wall penetrations from the outside, as Measure 6.4.1 recommends.

The easiest way to check for incoming air leakage at the receptacle or switch is by feel. Use a smoke pencil to check for outward leakage.

SUMMARY

A cheap and easy way to stop minor air leaks along electrical wiring that feeds receptacles and wall switches.

SELECTION SCORECARD

Savings Potential .... **$**

Rate of Return .... **% % % %**

Reliability .... **✓ ✓ ✓**

Ease of Retrofit .... ☺ ☺ ☺ ☺

ECONOMICS

SAVINGS POTENTIAL: *Typically about one dollar per year per gasket.*

COST: *About $0.10 per gasket. Preparation and installation take a few minutes.*

PAYBACK PERIOD: *About one year.*

TRAPS & TRICKS

CHOICE OF METHOD: *Consider this an auxiliary activity of Measure 6.4.1. Used alone, this method is much less effective than sealing leaks from the outside.*

MEASURE **6.4.3 Install roof and attic hatches that close tightly and reliably.**

Roof hatches are potentially a serious path for air leakage because chimney effect is strong at the top of the building. During cold weather, chimney effect forces the warmer interior air through any defects in the hatch and through any gaps in installation. Hatches in older buildings are often damaged so badly that they cannot close effectively. These are the steps in avoiding this problem:

- purchase hatches of good quality
- seal them well
- make them easy to use
- ask users to keep the hatches closed
- keep the hatches maintained.

> **SUMMARY**
>
> Hatches may be a severe source of air leakage. Install them to stay airtight.
>
> **SELECTION SCORECARD**
>
> Savings Potential .... **$ $** $
>
> Rate of Return, New Facilities **% % %** %
>
> Rate of Return, Retrofit .... **% % %**
>
> Reliability .... ✓ ✓ ✓
>
> Ease of Retrofit .... ☺ ☺ ☺

Get Good Hatches

Strong roof hatches with effective gaskets are commercially available as prefabricated units. Many styles are available. Figure 1 shows a typical unit. Select a style for each application that is easy to operate by all users, including those burdened with tools and equipment.

Get an opening mechanism that operates easily. Unless the door is very light, get a model that is counterbalanced.

In new construction, good hatches add virtually nothing to the cost of the building. In older facilities, you can usually find a prefabricated unit that installs fairly easily as a replacement for a shabby older hatch.

Install Them Properly

When a hatch is installed, make sure that it is sealed tightly against air leakage between the hatch assembly and the roof structure.

Install a gasket of ample thickness between the hatch and the curb or other structure. To keep the gasket from slipping out of position during installation, staple it in place before mounting the hatch.

Do not warp the frame of the hatch during installation. This would keep it from closing easily and sealing properly.

Make Hatches Easy to Use

Hatches are most likely to be damaged if they are difficult to open or close. Put yourself in the position of a mechanic trying to get through a roof hatch while teetering on a vertical ladder and trying to hold equipment. If you are concerned about your safety, you are unlikely to be tender with the hatch.

Anything that eases passage through the hatch reduces the likelihood of damage. For example, if a hatch is installed at the top of a vertical ladder, consider installing a shelf at a convenient place just below the hatch. This allows users to place articles on the shelf while they open or close the hatch.

Even better, do not use awkward, dangerous vertical ladders to get to hatches. In new construction, design the access to the roof using a conventional stair and door. A hatch at the top of a tall vertical ladder indicates a lack of awareness by the architect.

WESINC

Fig. 1 Roof hatch A leaky roof hatch can waste a considerable amount of heat energy, especially in a tall building. For perspective, the hatch will waste much less energy than the adjacent fan, if the fan is left running unnecessarily.

Ask Users to Close Hatches

Install an effective placard on the inside that asks users to keep the hatch closed. Stencil a similar large sign on the outer surface of the hatch. See Reference Note 12 for guidance in effective placard installation.

Maintain Them

In existing facilities, check all accesses to roofs and attics for tight closure. Hatches are typically out of sight and in awkward locations, so they tend not be repaired. If you find that a hatch is damaged, question why. If necessary, make it easier to use.

If the hatch is leaking around the edges, you may be able to take it off, seal it, and reinstall it properly without too much effort.

ECONOMICS

SAVINGS POTENTIAL: *Hundreds of dollars to thousands of dollars per year in heating costs.*

COST: *Good hatch assemblies typically cost several hundred dollars. Retrofit installation cost, including*

features to make the hatch easy to use, may be several hundred dollars. In new construction, installing a hatch effectively adds little to cost.

PAYBACK PERIOD: *One year, to several years.*

TRAPS & TRICKS

SELECTING THE EQUIPMENT: *Select hatches for ruggedness and ease of use. If possible, select an unit in which the hatch is integral with the frame. A strong and user-friendly closing mechanism is essential. Get a balance spring to minimize the lifting force required to open the hatch.*

INSTALLATION: *Check the installation to ensure that the frame is well sealed at the time of installation. Don't forget the placards.*

MEASURE 6.4.4 Adjust and repair space air vents periodically.

A building may have a variety of dampers and louvers that vent air through the building envelope. Some have fans, and some do not. These may be intended for natural ventilation, for venting the warm

SUMMARY

Leaky vents can waste a lot of energy. You may have more than you realize.

SELECTION SCORECARD

Savings Potential	$ $ $
Rate of Return	% % % %
Reliability	✓ ✓
Ease of Initiation	☺ ☺ ☺

air that accumulates under skylights, or for venting smoke in the event of a fire. A vent that is open or leaky can waste a large amount of heating energy because of chimney effect.

Vents may be located in obscure places, such as the tops of stairwells or the perimeters of skylights. Sometimes, vents are completely hidden above suspended ceilings. Figure 1 shows a roof vent that is almost impossible to see from below. Make a careful search to find and inspect all envelope vents. If you find a leaky vent damper, repair it or replace it with a unit that is capable of sealing more tightly.

Fire-Related Dampers

As a rule, dampers installed as part of fire control systems should remain tightly closed under normal

WESINC

Fig. 1 Open roof vent Daylight reveals an open vent here, probably a ventilation fan with a damper that does not close properly. A permanent leak of this sort can waste a large amount of heating energy.

circumstances. The dampers should be held snugly in the closed position by springs in the release linkages. The dampers may be released in the event of a fire by fusible links, signals from a computer, or other means.

Theories of fire protection keep changing in fundamental respects. Figures 2 and 3 show vents that probably have something to do with fire protection, but their purposes are not clear. In any case, do not alter or adjust fire-related vents without expert knowledge. If in doubt, have a fire safety expert review all your smoke relief vents. You may want to update your fire control systems at the same time.

Label the Purpose of Vents

The purpose of a vent may not be obvious. The uncertainty keeps maintenance personnel from repairing leaky conditions. Correct this problem by stating the purpose of each vent on an adjacent placard. See Refer Note 12, Placards, for tips on doing this effectively.

WESINC

Fig. 3 Mystery vent This roof vent with a cheap, leaky gravity damper is installed at the top of a stairwell in a tall building. On a cold day, the damper opens, venting warm air from the building. Someone has placed a piece of lumber over the damper, apparently to hold it shut. It still wastes energy by leaking air through the bent blades, but it can no longer do whatever it was intended to do.

WESINC

Fig. 2 What are these vents? They have manual chains, apparently intended to release the dampers. They are probably part of a fire protection system. Only an expert could guess with any certainty. Since their appearance invites action, they need a clear placard to explain what they do, and how they should be used.

ECONOMICS

SAVINGS POTENTIAL: *Hundreds of dollars to thousands of dollars per year in heating costs, if leakage is severe.*

COST: *Variable, but usually low.*

PAYBACK PERIOD: *Usually short.*

TRAPS & TRICKS

SCHEDULE IT: *This kind of maintenance is always forgotten unless the facility has an effective maintenance scheduling system. Make sure you have one, and include vent maintenance in it.*

EXPLAIN IT: *Be especially detailed about adjusting vents that are part of safety systems.*

MEASURE 6.4.5 Ventilate elevator shafts rationally.

RATINGS

New Facilities   Retrofit   O&M

A   **B**   ☐

Elevator shafts are conditioned spaces. They are conditioned by air that enters when the elevator doors open, and they may have their own conditioning systems. The conditioned air can be lost to the outside through vents, as shown in Figure 1. They can also lose air through elevator cable openings between the top of the elevator shaft and the elevator machinery room.

Elevator shafts have maximum chimney effect, which tends to force air out of the elevator shaft into vents and penetrations at high points in the elevator shaft. The air may be lost directly to the outside, or to the elevator equipment room.

The motion of elevators has a piston effect which drives air out of the shaft ahead of the elevator rises, and sucks air into the shaft behind the elevator. The strength of this effect depends on how closely the elevator fits within the shaft. The effect is greatly reduced if more than one elevator rides in a shaft.

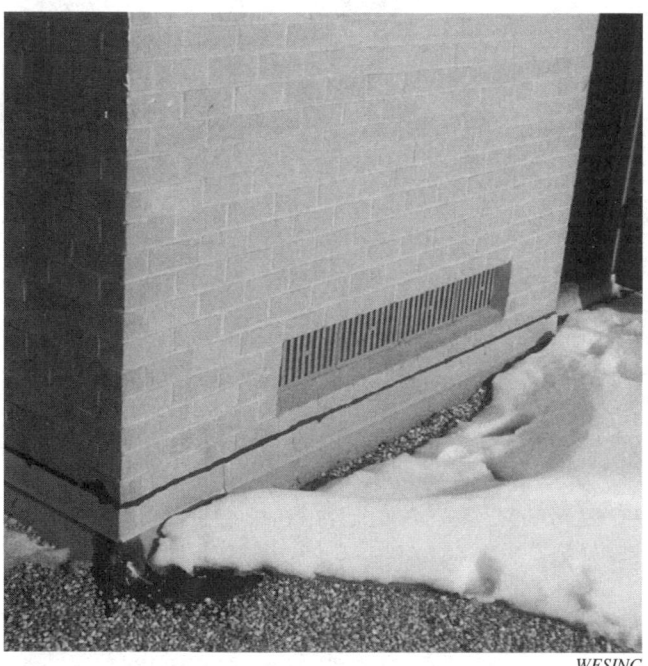

WESINC

Fig. 1 Elevator shaft ventilation The melting snow shows that a lot of heat is being lost from this elevator shaft, which has a number of vents this size.

SUMMARY
Use good sense with a serious leakage path.

SELECTION SCORECARD

Savings Potential	**$ $**
Rate of Return, New Facilities	**% % %** %
Rate of Return, Retrofit..........	**% % %**
Reliability	✓ ✓ ✓
Ease of Retrofit	☺ ☺ ☺

The question is, how much ventilation does an elevator shaft need? Ventilation may be required for respiration, in the event that the car becomes disabled. Or, it may be needed for some fire protection purpose. Or, it may be needed to minimize discomfort resulting from the piston effect. In any event, find out how much ventilation the shaft needs, and do not exceed a reasonable amount of ventilation.

Various types of seals are commercially available from elevator companies to seal around the cables. If the building lacks such seals, or if the present seals are not effective, ask your elevator contractor to install efficient seals. The seals should be designed to resist wear while allowing for sidewise motion of the cables. However, if the required ventilation rate is greater than the rate of leakage through the cable openings, then don't bother installing a seal.

ECONOMICS

SAVINGS POTENTIAL: *Varies widely, but several hundred dollars per year per elevator is a typical figure.*

COST: *Typically modest.*

PAYBACK PERIOD: *Less than one year, to several years.*

TRAPS & TRICKS

BE SURE OF THE REQUIREMENT: *If the shaft seems to have an excessive amount of ventilation, try to find out if there is a reason. Talk to your elevator expert and check the building codes.*

Section 7. BUILDING INSULATION

INTRODUCTION

In new construction, good insulation is probably the best bargain in energy conservation. Furthermore, improving insulation requires only conventional skills, and no unknown risks. There is enormous room for improvement in contemporary insulation practice. It is amazing how poorly buildings are still being insulated, given the ease and low cost of good insulation. You can do much better.

In existing buildings, improving insulation costs much more than in new construction. It also involves serious risks of damage to the structure that you have to be careful to avoid. Still, you can often greatly improve the insulation of existing buildings, making them more comfortable as well as more efficient.

How to Use Section 7

The Section is organized by the envelope components where you can improve insulation. The Subsections deal with roofs and attics, walls, and glazing, respectively. Each of these areas requires different insulation techniques. There are Measures in each of the Subsections that apply to most buildings.

This Section does not cover conductive heat loss through doors. Doors usually waste much more energy through air leakage than through heat conduction, so they are covered in Subsections 6.1 and 6.2.

Before Getting Started ...

For a quick refresher on insulation principles, see Reference Note 41, How Insulation Works, and Reference Note 42, Vapor Barriers.

For specific types of insulation and for insulation selection criteria, see Reference Note 43, Insulation Selection

For tips on installing insulation properly, see Reference Note 44, Insulation Integrity.

To decide how much insulation to install, see Reference Note 45, Insulation Economics.

Roofs and Attics

Roofs or attics make up a large fraction of the total surface area of most buildings. Their large size and their orientation make it important to insulate them well. Roof heat loss is especially severe during cold weather because the roof faces the night sky, allowing maximum radiation loss. The roof is exposed to precipitation and wind, which carry away any heat that makes its way to the outer surface. And, the interior temperature is highest inside the roof or attic.

Roof insulation is especially important for reducing solar heat gain during warm weather. Roofs have especially high solar gain because of their orientation and because roofing surfaces tend to be absorptive.

Roofs and attics have large, unbroken surfaces that offer relatively easy access for installing insulation. Therefore, they often provide the most economical opportunity for reducing heat loss and heat gain. In new construction, good roof insulation adds little to overall cost. In retrofit, improving roof insulation may be your only opportunity for improving the building's insulation on a large scale. You can usually find a way to improve roof or attic insulation without having to disrupt other building components.

The Measures in this Subsection cover the various methods that you can use to insulate roofs. Each has important advantages and drawbacks. They are alternatives to each other.

INDEX OF MEASURES

7.1.1 Increase the quantity of attic insulation.

7.1.2 Add rigid insulation to the top surface of roofs.

7.1.3 Apply sprayed foam insulation to the top surface of roofs.

7.1.4 Install insulation on the underside of roofs.

7.1.5 Install a suspended insulated ceiling.

RELATED MEASURES:

• Subsection 8.2, for methods of reducing cooling load on roofs and attics

MEASURE 7.1.1 Increase the quantity of attic insulation.

An attic is an unheated space between the roof and the highest occupied space. If a building has an attic, installing insulation on the floor of the attic is usually the best way to block heat conduction to and from the top of the building. Figure 1 shows a favorable application for attic insulation.

The attic space is almost always vented to the outside. In this way, the exterior roof acts as a sun shade, greatly reducing the cooling load. (In order for this to work properly, the attic needs good ventilation. See Measure 8.2.3 for details.) Since the attic is ventilated, the roof provides little reduction of heat loss during cold weather. Heat loss is blocked mostly by the attic insulation, which must be installed so that the air currents through the attic do not get under the insulation.

In new construction, a simple increase in the amount of attic insulation can make a major improvement in the thermal efficiency of the building at relatively modest additional cost.

Most existing buildings that are suitable for attic insulation already have some, but the amount may be scanty. Typically, you can easily add any amount of insulation that makes sense. However, the economic rate of return decreases sharply with the amount of existing insulation. Also, the cost of adding insulation may be raised considerably by the need to work around

SUMMARY

Usually the best way to minimize heat loss from the top of the building. Inadequate preparation or poor workmanship can degrade insulation performance and create safety hazards.

SELECTION SCORECARD

Savings Potential	$ $ $ $
Rate of Return, New Facilities	% % % %
Rate of Return, Retrofit..........	% %
Reliability	✓ ✓ ✓ ✓
Ease of Retrofit	☺ ☺ ☺

equipment and other obstructions. Figure 2 illustrates this.

Appropriate Types of Insulation

See Reference Note 43, Insulation Selection, for comparative insulation characteristics. For most attic insulation applications, the best materials for attic insulation are glass fiber or mineral fiber. Cellulose insulation is also used widely for attic insulation.

WESINC

Fig. 1 Uninsulated hospital attic with concrete deck   The hospital is located in a climate that has cold winters and hot summers, so attic insulation can reduce both the heating and cooling loads considerably. The concrete slab is a good base, and the equipment in the attic is installed above the slab, providing clearance for installation. But, why is this attic not already insulated? There may be concern about intake of fibers or microorganisms into air handling systems, or there may be insulation already installed under the slab. These questions need to be resolved first.

Cellulose insulation has been controversial because the raw material requires treatment to avoid certain problems that are explained in Reference Note 43. Loose fill, blankets, and batts are appropriate for attics. The choice depends on the configuration of the attic and the capabilities of the installer.

Loose fill is widely used in retrofit applications. It is favored by contractors because it is inexpensive to install. That does not mean that you should choose it. On the positive side, loose fill has much less tendency than batts to leave voids under wiring and other attic clutter, especially when it is installed properly. However, by the same token, it will easily fall into spaces where it does not belong, such as eaves and flue chases, and it is more likely to block proper venting of hot equipment.

How to Install Loose Fill

Loose fill insulation is commonly installed in large attics by blowing, because this method involves low labor costs. For smaller applications, the insulation can be poured from bags.

Preparation is important. Cover any deep cavities, such as the tops of interior walls or flue chases, into which the insulation may fall. Alternatively, install dams to hold back the insulation.

Similarly, keep loose fill away from any hot equipment, such as flues, incandescent light fixtures installed in the ceiling, and space heaters installed in the ceiling. If such equipment is covered with insulation, it may become hot enough to start a fire. Odd bits of loose fill may block vent holes in equipment, which may create a fire hazard or damage the equipment. For example, recessed light fixtures that are installed in ceilings must be vented.

Loose fill tends to settle, so add an extra amount of insulation to compensate. Rake the insulation after pouring it.

The Cellulose Insulation Manufacturers Association (Dayton, Ohio) publishes a pamphlet, "Standard Practice

WESINC

Fig. 2 Police station attic over drywall ceiling This attic is insulated, but with a meager amount of material. The arrows point to places where insulation is missing over gaps that lead to the interior of the building. The ductwork and other obstructions will make it difficult to add insulation properly.

for Installing Cellulose Building Insulation," that spells out practices for installing cellulose insulation. Much of this information applies to all types of loose fill insulation. This document is valuable both for building staffs doing their own insulation, and for owners to ensure that a proper job is done by insulation contractors.

How to Install Blankets and Batts

The major challenge with blanket and batt insulation is to avoid the formation of voids between the insulation and the ceiling. This problem is especially severe if the attic is filled with wiring, piping, and ductwork. The only solution is careful fitting of the insulation. This means cutting the blankets or batts to allow wiring to pass through it, or fitting the insulation under obstructions piecemeal.

If you use multiple layers of blankets or batts, crisscross successive layers to minimize convective heat loss.

Blankets and batts contain a small amount of combustible material to act as a binder. Also, dust that collects on the insulation over time is flammable. Therefore, even if the insulation is made from inert materials, avoid contact with hot equipment.

How to Install the Vapor Barrier

See Reference Note 42, Vapor Barriers, for an explanation of vapor barriers. With attic insulation, the primary purpose of a vapor barrier is to protect the insulation itself, and to a lesser extent, to protect the structure and equipment located above the insulation. In a cold climate, attic insulation that lacks a vapor barrier may become loaded with frost.

If possible, install an unbroken vapor barrier of plastic film under the insulation. In many attics, equipment clutter makes this impractical.

A poor second choice is to use insulation that has a kraft paper or foil vapor barrier attached. The weakness of this approach is that water vapor can escape between adjacent batts.

Vent Moisture from the Attic

Vent the space above the insulation to allow moisture to escape. Moisture can enter the attic either from below, as a result of the vapor barrier being imperfect, or from above, as a result of roof leaks and piping leaks.

It is tempting to close off the attic to reduce heat loss during cold weather. Resist this temptation. The insulation stops most of the heat loss, not the roof surface or attic space. Furthermore, if the attic space is closed during the long daylight hours of the cooling season, the attic will become an oven, perhaps doubling or tripling the temperature differential across the insulation.

Protect the Insulation from Physical Damage

If people need to enter the attic, such as for equipment maintenance, protect the insulation with walkways that are elevated above the top of the insulation.

Optimum R-Values and Economics

See Reference Note 45, Insulation Economics.

ECONOMICS

SAVINGS POTENTIAL: Insulating a bare attic typically reduces heat loss through the attic by 70 to 90 percent. It may reduce heat gain by 30 to 70 percent, depending on how well the attic is vented. This may be a large fraction of the total building heat gain or loss. In retrofit, the benefit depends on the amount of insulation previously installed.

COST: In new construction, a good quantity of blow-in insulation costs from $0.50 to $1.00 per square foot. Poured insulation averages about $2.00 per square foot. The cost of batt insulation varies widely, but averages perhaps $4.00 per square foot. In retrofit, blown-in insulation and poured insulation are not much more expensive than in new construction. Batt insulation may cost two or three times more.

PAYBACK PERIOD: One year, to many years, depending primarily on the original insulation value of the ceiling and attic structure.

TRAPS & TRICKS

PLANNING: Use Reference Note 43, Insulation Selection, as a guide in selecting the most appropriate materials. Figure out good ways of installing both the vapor barrier and the insulation. Refer to Reference Note 42 if you are unclear about vapor barriers. Make sure the installer knows how you want the job done. In retrofit, start by surveying the attic areas in detail, looking for voids, wiring, light fixtures penetrating the ceiling, and other problems.

INSTALLATION: Monitor the installation. The quality of the workmanship determines the effectiveness of the insulation. Also, you want to catch any potential safety problems.

MEASURE **7.1.2 Add rigid insulation to the top surface of roofs.**

Installing rigid insulation on the exterior roof surface is often the best method of insulating a roof, provided that the building has no attic. This method has several significant advantages:

• *ease of installation.* Roof deck insulation does not run afoul of obstructions from piping, ductwork, suspended ceilings, and other clutter that resides under the roof surface.

• *no physical restrictions on thickness.* The sky is the limit.

• *no separate vapor barrier is needed.* The insulation itself or its bonding layer form an unbroken impermeable surface.

• *does not interfere with a decorative ceiling surface.*

In retrofit, the major disadvantage of this method is that the existing roof covering must be stripped off, and a new roof covering installed over the insulation. This method gains appeal when the existing roof covering is due for renewal.

Each type of insulation, and each method of installation, involves its own set of details. These details are important. Follow the manufacturer's instructions

SUMMARY

An effective method for insulating roofs of buildings without attics. In retrofit, requires stripping the existing roof surface.

SELECTION SCORECARD

Savings Potential .... $ $ $ $

Rate of Return, New Facilities % % % %

Rate of Return, Retrofit .... % %

Reliability .... ✓ ✓ ✓ ✓

Ease of Retrofit .... ☺ ☺ ☺

carefully. This Measure covers the key issues for all methods of insulating the top surfaces of roofs by using rigid board insulation. Make sure that each of these issues is covered in adequate detail by the manufacturer's instructions, and that the installer pays attention to these issues.

Do not use this method if there is a vented attic space below the roof.

Fig. 1 Rigid insulation applied to a flat roof This insulation is applied in layers. Snug fit is essential, and it is achieved by using many fasteners. The top layer is tapered to provide drainage. However, the small amount of slope provided by tapered insulation does not provide reliable drainage unless the top surface membrane is applied carefully. Even then, the roof drains must collect all water reliably, or the water will work its way underneath the bottom layer of insulation.

Dow Chemical Company

Fig. 2 Rigid insulation applied to a sloped roof   Foam board insulation is applied under cedar shingles, allowing any desired amount of insulation value to be combined with a rustic appearance. With this common method of installation, any water that leaks through the shingles will be trapped by the horizontal battens, channeling the water between the joints of the insulation boards and through the nail holes into the underlayment.

Types of Insulation

Many manufacturers offer rigid roof insulation. Types range from simple blocks of insulation to interlocking composite panels. The material must be rigid enough to withstand compression from people walking on it. Plastic foam insulation is common. The types are covered in Reference Note 43, Insulation Selection. Even semi-rigid fiber can be used if it is bonded to a rigid surface. See Figures 1 through 4.

Plastic foam insulation may be acceptable from a fire safety standpoint in above-roof applications, especially if the insulation is installed on top of a concrete slab or steel pan. Make sure that occupants cannot be trapped by a roof fire. It is probably not a good idea to install plastic foam insulation on a wooden roof deck. Such a combination could burn fiercely.

Roof insulation is available that is bonded to a nail base, such as plywood. This is needed for applications where the outer roof surface consists of shingles, metal sheet, or other material that requires fasteners. The bonded nail base allows you to specify a less dense insulation, because the nail base distributes external compression loads.

Another composite insulation material provides a vented roof. In this type, a nailing surface is attached to the insulation board by spacers, providing a gap of about one inch (several centimeters) between the top surface and the insulation. This reduces the temperature differential across the insulation. This type works on a pitched roof, and it requires a continuous vent along the lower edge and the ridge of the roof.

Celotex Corporation

Fig. 3 A sample selection of rigid insulation and composites   Many types of rigid insulation are available. Many combinations with other structural materials are available as standard items, including insulation that is bonded to wood underlayment, to gypsum board, to interior and exterior finish surfaces, and to reflective foil. The composite on the bottom right allows circulation of air between the nailer surface and the insulation, for sloped roofs. Many variations of tapered insulation are available. All organic foam insulation is combustible in some manner, and it should be selected accordingly.

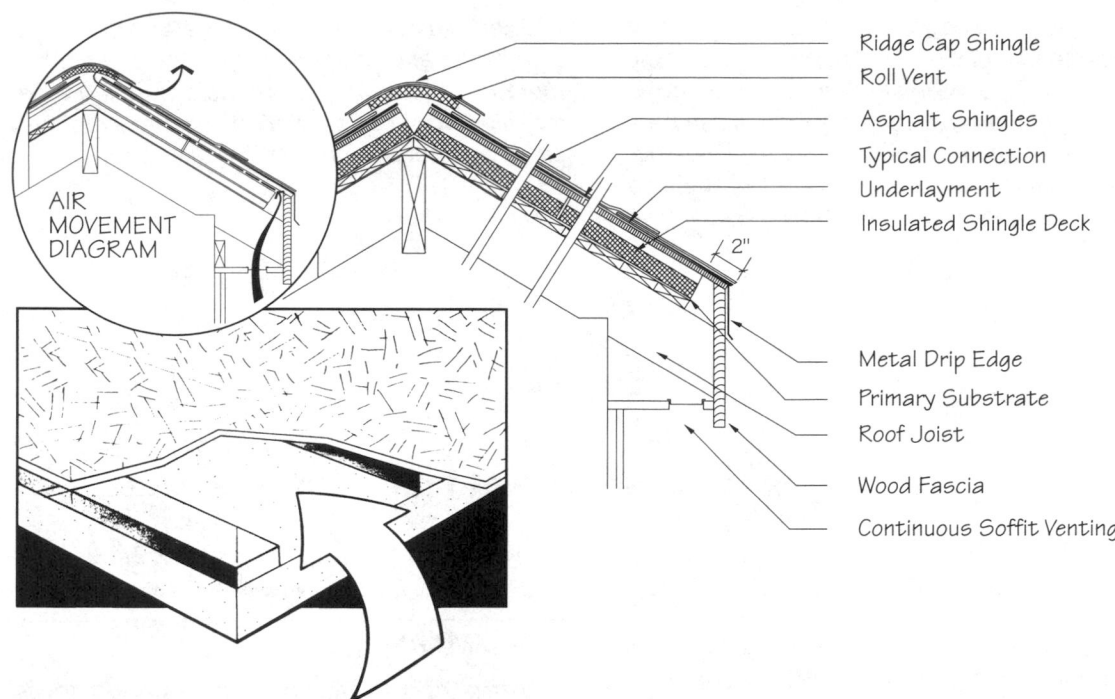

AIR MOVEMENT DIAGRAM

Ridge Cap Shingle
Roll Vent

Asphalt Shingles
Typical Connection
Underlayment
Insulated Shingle Deck

2"

Metal Drip Edge
Primary Substrate
Roof Joist

Wood Fascia
Continuous Soffit Venting

Celotex Corporation

Fig. 4 Insulation system that provides roof venting This system is available as prefabricated units that include the insulation, underlayment, and spacers. This configuration reduces the temperature differential across the insulation from solar heat gain, and it reduces the tendency to form ice dams in winter. In order to work well, it must be vented well at the top and bottom.

Reliable Drainage

Water is the enemy of roof insulation. If the insulation is permeable, water will soak it and destroy its thermal resistance. If the insulation is impermeable, water will conduct heat around it. Therefore, good drainage is a critical part of roof insulation.

The key to correct drainage is understanding a fact that has been forgotten in schools of architecture, namely, that water flows downhill. If the roof surface has enough slope and is smooth enough to prevent the formation of puddles, soaking will not occur as long as the top membrane remains intact. On the other hand, if the roof is flat, water will seek to go downhill through the roof structure itself, and it will eventually succeed.

With flat roofs, an easy way to achieve drainage is to install tapered insulation. Design the installation so that the insulation is thick enough at its thinnest point to provide good thermal resistance. Doing this may result in the insulation being thicker than is economical at the high points. On the other hand, tapered insulation is so easy to install that the saving in labor cost is likely to compensate for the additional material cost.

Other methods of protecting insulation on flat roofs are more expensive. About the only other reliable method is to construct a slanted structure on top of the roof to shed water.

Make sure that the deck or slab on which the insulation is installed has an effective drain system. All water that drains off the insulation, or through it, has to

be carried away before it can waterlog the insulation. Good drainage also avoids freeze damage.

Optimum R-Values and Economics

See Reference Note 45, Insulation Economics.

ECONOMICS

SAVINGS POTENTIAL: *Insulating a bare roof typically reduces heat loss and heat gain by 70 to 90 percent. This may be a large fraction of the total building heat gain or loss.*

COST: *In new construction, a good quantity of exterior roof deck insulation may cost from $8 to $20 per square foot. In typical retrofits, increase these figures by about 30%.*

PAYBACK PERIOD: *One year, to many years, depending primarily on the original insulation value of the ceiling and roof structure. The payback period in retrofit is improved if the roof surface is due for replacement.*

TRAPS & TRICKS

DESIGNING OR SELECTING THE INSULATION SYSTEM: *You can choose from a variety of established roof deck insulation systems. Start by studying all the available standard types. Most are reasonably good, but they are oriented heavily toward low cost and installation with unskilled labor. Even standard systems* *may fail to deal with drainage, but instead attempt to create an impermeable basin. This approach inevitably limits service life. Tapered insulation provides good drainage, if it is installed properly. In retrofit, it can be a problem to slope insulation toward the existing roof drains. A custom-engineered roof can solve all these problems, but the cost is usually much higher.*

INSTALLATION: With any type of roof, the quality of installation affects service life. This is especially true of membrane roofs intended to hold standing water. When installing tapered insulation, ensure continuity of the slope and effective runoff into drains.

MEASURE 7.1.3 Apply sprayed foam insulation to the top surface of roofs.

If the roof surface is flat or gently sloped, sprayed foam insulation can be applied to the upper surface. This type of roof insulation can be installed quickly by contractors who have specialized equipment for the work. In terms of installation, this method is almost the opposite of the multi-component roof systems covered previously by Measure 7.1.2.

There are various materials and application methods that are used for sprayed roof insulation. Each type of insulation, and each method of installation, involves its own set of details. These details are important. Make sure that the installer follows the insulation manufacturer's instructions carefully. This Measure covers the key issues for all methods of insulating the top surfaces of roofs with sprayed foam. Make sure that each of these issues is covered in adequate detail by the manufacturer's instructions, and that the installer pays attention to these issues.

As with all roof insulation, do not use this method if there is a vented attic space below the roof.

Advantages

Sprayed foam roof insulation is a radical departure from other types of external roof insulation. It has several significant advantages:

- *no separate outer membrane* is required. The foam itself, and an opaque coating, are the weather surface.
- *no separate vapor barrier* is needed. The insulation itself forms an unbroken, impermeable surface.
- *ease of installation.* Installation does not run afoul of the clutter of piping, ductwork, etc. that obstructs internal insulation.
- *no physical restrictions on thickness.*

SUMMARY

A method for flat or gently sloped roofs that combines major advantages with major disadvantages.

SELECTION SCORECARD

Savings Potential	$	$ $	$
Rate of Return	%	%	
Reliability	✓	✓ ✓	
Ease of Retrofit	☺	☺ ☺	

Disadvantages

Sprayed foam is not widely used, despite its important advantages. The reason is that it also has significant disadvantages:

- *uncertain longevity.* Foam is destroyed rapidly by sunlight, so a durable opaque coating is needed. Foam also shrinks and cracks. The long-term performance of external sprayed foam insulation is not well established.
- *vulnerability to traffic.* The foam does not have sufficient compression strength to resist footsteps. The foam surface is irregular, so it is not practical to apply flat sheathing to protect the foam. You may have to install independently supported walkways for protection.
- *puddling.* The method of application produces a lumpy surface that puddles water. At a minimum, this is messy. As the foam ages and cracks, puddled water penetrates the cracks, undermining the effectiveness of the insulation and destroying it with freeze and thaw cycles.

The disadvantages limit this type of roof insulation to retrofit applications where existing conditions make other approaches more difficult or much more expensive.

Installation Precautions

Some water will eventually penetrate the insulation. Therefore, install the insulation so that water penetrating the insulation is led to the roof drains.

Plastic foam is vulnerable to sunlight, so cover it with a completely opaque coating that will endure for the life of the foam. In fact, the coating will largely determine the life of the foam. In warm and moderate climates, make the coating as reflective as possible, to minimize heat gain and to prevent degradation of the foam from high temperature. (Do not use metallic paints for this purpose. Although they reflect light, they do not radiate heat well.)

Do everything you can to minimize puddling. Discuss this issue with potential contractors. If necessary, establish a slope on the surface of the foam, which requires additional material. Monitor the installation to make sure that this is handled properly.

Optimum R-Values and Economics

See Reference Note 45, Insulation Economics.

ECONOMICS

SAVINGS POTENTIAL: *Insulating a bare roof typically reduces the heat loss and heat gain through the roof by 70 to 90 percent. This may be a large fraction of the total building heat gain or loss.*

COST: *In typical retrofits, a good quantity of exterior sprayed foam roof deck insulation may cost from $5 to $10 per square foot.*

PAYBACK PERIOD: *Several years or longer, depending primarily on the original insulation value of the roof.*

TRAPS & TRICKS

SELECTING THE METHOD: *The appeal of this method is its speed and its apparent adaptability. Beware that the installer may deal with problems by burying them in foam, and the emphasis on speed causes important preparation to be ignored. If you select this method, have a clear agreement about how the job is to be done, and have an enforcer at the site.*

MATERIALS: *There is a common misconception that foam insulation is weatherproof. The opposite is true. Foam is destroyed by sunlight, and therefore requires a durable, opaque surface coating. It is also vulnerable to oxidation, and it is destroyed by freezing and thawing like any other material. A sprayed roof will probably have a shorter life than any other applicable type. A long-term warranty is likely to be meaningless, because a specialty applicator faced with many claims will disappear.*

RATINGS

New Facilities | Retrofit | O&M

D

MEASURE **7.1.4 Install insulation on the underside of roofs.**

In many existing buildings, the only economical way to improve the thermal resistance of the roof is to install insulation against the underside of the roof surface. This method has a number of major advantages, but one potentially disastrous disadvantage.

You can use various types of insulation underneath roof decks, including boards, batts, and spray-applied materials. Each type of material and each method of installation involves its own set of details. These details are important. Follow the manufacturer's instructions carefully. This Measure covers the key issues for all methods of insulating underneath roof surfaces. Make sure that each of these issues is covered in adequate detail by the manufacturer's instructions, and that the installer pays attention to all of them.

As with all roof deck insulation, this method should not be used if there is a vented attic space below the roof.

Advantages

This method has several significant advantages:

- *ease of installation.* In many buildings, the underside of the roof is the only reasonably accessible surface for installing insulation.
- *unrestricted thickness.* There is typically plenty of space to add as much insulation thickness as you want.
- *low cost.* Cost is relatively low because of the ease of installation, and because it is not necessary to remove and replace other roof components.
- *longevity.* The insulation is well protected from sunlight, weather, and impact.

Risk of Moisture Damage

The overwhelming disadvantage of this method is the risk of serious moisture damage if the building is located in a climate where the outside temperature can be much colder than the inside temperature. In such locations, the insulation must have an effective vapor barrier on its inner surface. (Reference Note 42, Vapor Barriers, explains the problems of moisture in insulation.) If the outer roof surface is also impermeable, a region is created between the outer roof surface and the vapor barrier that can trap and hold moisture. Persistent moisture will eventually destroy any type of roof structure, and may even cause structural collapse. You may not be able to provide drains and/or vents that are adequate to get rid of trapped moisture.

Because of this potential problem, this method of insulation is safe only if you can design a reliable

---

SUMMARY

May be the easiest way to add roof insulation to an existing building, but risks serious moisture damage. Safest in a warm, dry climate.

SELECTION SCORECARD

Savings Potential .... $ $ $ $

Rate of Return .... % %

Reliability .... ✓ ✓ ✓

Ease of Retrofit .... ☺ ☺ ☺

---

moisture venting for the space above the insulation, or if the outside air dew-point temperature is always higher than the inside temperature.

This method of insulation is least likely to have trouble in climates that are warm and dry, where the insulation is used primarily to reduce cooling load.

How to Avoid Moisture Damage

Insulation installed underneath the roof is much better protected from precipitation than insulation installed on top of the roof. However, even under-roof insulation is vulnerable to water damage from roof leaks. Expect any roof to leak occasionally during its life. Moisture inside the insulation may do more damage to the roof structure than to the insulation itself.

The best protection against roof leaks is to use insulation that is highly permeable, such as glass fiber board. This allows any moisture to evaporate freely and escape through the vents.

If you decide to install under-roof insulation in a colder climate, you face a serious challenge in avoiding moisture damage. In that case, two things are needed:

- *a very reliable vapor barrier.* You must keep water vapor from entering the insulation from the inside of the building. Use a very effective, reliable vapor barrier, such as thick plastic film with overlapped, taped joints. If you cannot be sure of the vapor barrier, do not use under-roof insulation. The exception is climates that are warm and dry, where it may be desirable to omit the vapor barrier.
- *a path to vent any accumulated moisture to the outside.* Aside from the issue of roof leaks, no vapor barrier can completely block the flow of water vapor into the insulated space. The installation must provide a path for water vapor to vent to the outside before it reaches a concentration that allows it to

condense. In a cold climate, the insulated space must vent to the outside of the building, not to the inside.

Adequate venting to the outside is not possible with typical commercial flat roofs that use an impermeable outer membrane. The best case occurs with a highly permeable outer roof surface, such as tiles on a sloped roof. However, standard practice involves laying an impermeable base under the tile. Even if an impermeable base does not presently exist, one may be installed later during a renovation.

Adequate venting may occur with a corrugated or channeled sheetmetal roof, in which the channels act as vapor vents. This requires keeping the channels open to the outside at the top and bottom. But again, an impermeable underlayment under the sheetmetal will block adequate venting.

Attach the Insulation Snugly

Make the insulation fit snugly against the roof surface. Otherwise, air will leak from the roof surface to the space, bypassing the insulation. Use a tight fastening that is reliable and will not work loose. Seal around the edges of each panel of insulation.

Optimum R-Values and Economics

See Reference Note 45, Insulation Economics.

ECONOMICS

SAVINGS POTENTIAL: *Insulating a bare roof typically reduces heat loss and heat gain by 70 to 90 percent. This may be a large fraction of the total building heat gain or loss.*

COST: *In typical retrofits, adding a good amount of under-roof insulation may cost from $4 to $10 per square foot, using conventional insulation materials. The unit cost varies with the size of the job and difficulty of access.*

PAYBACK PERIOD: *Several years or longer, depending primarily on the original insulation value of the ceiling and roof structure.*

TRAPS & TRICKS

CHOICE OF METHOD: *Approach this method with great caution. It is a challenge to design under-roof insulation in a manner that is thermally effective while avoiding moisture problems throughout the life of the building.*

MEASURE **7.1.5 Install a suspended insulated ceiling.**

Installing a suspended, insulated ceiling can reduce heat gain and heat loss from the ceiling and the upper portions of walls. Possible configurations range from a standard suspended grid ceiling tile system to a rigid, permanent ceiling. A suspended ceiling does not have to be horizontal.

This is generally a retrofit method. If you are designing a new building, the other Measures provide better ways of insulating the top surface of the building.

A suspended ceiling is most favorable in applications where cooling is the primary conditioning load, and there is little heating load. Unlike roof insulation, this method works even when the space below the roof is vented.

If heating is a major conditioning load, consider this a method of last resort. This is because a suspended ceiling is not a very reliable thermal barrier. For example, you might install an insulated suspended ceiling if the inner and outer surfaces of the roof are so cluttered that it is not practical to install insulation on them, or if you cannot insulate the inner roof surface because of the moisture problems explained in Measure 7.1.4.

Cooling-Only Applications

Suspended ceilings are most favorable in applications where cooling the space costs much more than heating it. They are especially effective where a hot roof radiates heat down into the occupied portion of the space, or where circulation fans move warm air from the upper portions of a tall space into the lower areas.

In cooling applications, the suspended ceiling performs two functions. One is creating a physical barrier to vertical air circulation. This enhances the natural tendency of cooler, denser air to stay in the lower part of the space, avoiding contact with the warmer surfaces in the upper part of the space. The other function is to block heat radiation from the upper part of the space into the lower part.

Both these functions could be performed by any thin, opaque material, even a sheet of paper. Insulation is needed only because the suspended ceiling itself is warmed by radiation and conduction from above. The insulation keeps the suspended ceiling itself from radiating heat into the space. The temperature rise in the ceiling space is relatively small, so only a small thickness of insulation is needed. Ordinary acoustical tile is satisfactory for most applications.

In these situations, the ceiling does not have to be tightly sealed. Stratification helps to keep the lower cool air separated from the upper warm air. If fans are

SUMMARY

A retrofit method of last resort. Better for reducing heat gain than heat loss.

SELECTION SCORECARD

Savings Potential	$ $ $ $
Rate of Return	% %
Reliability	✓ ✓ ✓
Ease of Retrofit	☺ ☺ ☺

used to circulate air within the cooled space, these are located below the suspended ceiling, so they do not mix air between the upper and lower spaces.

Applications involving cooling alone do not require a vapor barrier in the suspended ceiling. The lower, occupied space usually has the highest absolute humidity. The relative humidity is lower inside the ceiling insulation, and above the ceiling, because these areas are warmer.

For the suspended ceiling to be most effective, the space above the suspended ceiling should be well vented to the outside. This minimizes the temperature differential across the ceiling, and keeps the warm air above the ceiling from expanding into the occupied space.

However, if the space is heated at certain times, venting the space above the suspended ceiling poses the risk of serious heat loss, a problem that we will now explore.

Heating Applications

If heating is the primary need of the space, it is more difficult to make a suspended ceiling effective. The major challenge is creating an unbroken barrier to convective heat loss through the insulation. If the space is heated, the warmed space air tends to rise. Only an unbroken surface from wall to wall can prevent this. With lightweight ceilings, such as suspended grid systems, this is virtually impossible to achieve, so some convective heat loss occurs.

Similarly, it is difficult to create an effective vapor barrier, which is needed unless the space above the suspended ceiling is well vented to the outside. (See Reference Note 42, Vapor Barriers, for details.)

Both these considerations argue in favor of installing a rigid, permanent ceiling, instead of a suspended ceiling tile system. One way to install a rigid ceiling quickly is to use prefabricated panels that include insulation and a vapor barrier.

If there are air returns in the upper part of the space, a suspended ceiling cannot be used.

Reduce the Volume of the Space?

Suspended ceilings can be used to reduce the conditioned volume of a space. This was a popular energy conservation concept during the 1970's. However, space volume, by itself, does not consume energy. A suspended ceiling reduces the conditioning load only if it reduces the area of exterior surface, or if it reduces the rate of heat loss through exterior surfaces.

An example that illustrates the greatest advantage of installing a suspended ceiling is a poorly insulated metal building with a tall roof. The reduction in volume and the change in the shape of the space:

- *reduces the surface area for heat loss and heat gain*
- *minimizes stratification.* In a tall space, there may be a difference in air temperature between the top and bottom of the space of 10°F (6°C) or more. The higher air temperature at the top of the space increases conductive heat loss.
- *minimizes convective circulation.* Tall spaces have a great deal of convective heat loss if the insulation is poor. Buoyancy forces warmer air to the top of the space, where it is cooled by contact with the cold roof. The chilled air then falls to the bottom of the space, promoting air circulation and accelerating the heat loss. Reducing the height of the space greatly reduces this effect.
- *allows an attic space to be created above the conditioned space.* The space remaining above the suspended ceiling acts as an attic. The roof blocks solar radiation. If the space above the ceiling is well vented, the temperature differential across the ceiling is greatly reduced, compared to the differential that would exist if the insulation were installed against the roof.

These are important benefits. However, the suspended ceiling makes the space above the ceiling inaccessible. This space is now filled with suspension wires for the ceiling. Furthermore, all the lighting must be changed. In most cases, it would be more desirable to install a generous amount of insulation on the roof and walls, rather than install a suspended ceiling.

Persuade Occupants to Leave Ceilings Intact

A problem with suspended ceilings is the ease with which they can be rendered ineffective, especially as a barrier against heat loss. Suspended ceilings are vulnerable to openings created by the occupants and maintenance staff. For example, an occupant may prop up a ceiling tile in the hope of getting better ventilation.

Let everyone know that the insulated surface has to be kept unbroken. One possible method is installing placards in appropriate locations. See Reference Note 12, Placards, for details of effective placard design and installation.

Optimum R-Values and Economics

See Reference Note 45, Insulation Economics.

ECONOMICS

SAVINGS POTENTIAL: *Adding a suspended ceiling to a space with an uninsulated roof may reduce heat gain by 60 to 90 percent. Heat loss may be reduced 60 to 80 percent, but this saving is less reliable. In retrofit, the benefit depends on the amount of insulation previously installed in the roof and walls.*

COST: *In retrofit, installing a suspended ceiling may cost from $2 to $6 per square foot, depending on the difficulty of installation.*

PAYBACK PERIOD: *Several years, or longer, depending primarily on the original insulation value of the ceiling and roof structure.*

TRAPS & TRICKS

CHOICE OF METHOD: *Are you sure that this is the best approach for your application? The other Measures are more expensive, but they are also more reliable.*

SELECTING MATERIALS AND INSTALLATION METHODS: *There are many ways to install a suspended ceiling. Start by getting familiar with all the types of insulation materials you might use. Select a type that has a good fire safety rating.*

Heat loss through the opaque portions of walls contributes a large fraction of heating cost, except in buildings where large amounts of glazing dominate heating cost. Wall heat gain is typically a lesser factor in cooling load, but it can be important in many buildings.

The time to minimize this heat loss is while the building is still being designed. In new construction, improving wall insulation costs little. In fact, it may reduce overall construction cost by reducing the heating and cooling capacities required. Good wall insulation may even allow you to eliminate perimeter heating systems.

Improving wall insulation on a retrofit basis is practical primarily in buildings that have large areas of opaque wall. You can usually find a way of increasing the insulation value of walls, but the cost may be high in relation to the economic benefit.

The labor cost of retrofitting small or irregular wall areas is prohibitive in most cases. When improving the thermal performance of walls, consider where the heat loss is occurring. In buildings with a large fraction of window area, most of the heat loss and heat gain occurs through the glazing, so concentrate your attention there, using the Measures of Section 8.1.

The Measures in this Subsection cover the various methods that you can use to insulate walls. Each has important advantages and drawbacks. They are alternatives to each other.

INDEX OF MEASURES

7.2.1 Insulate wall cavities.

7.2.2 Insulate the inside surfaces of walls.

7.2.3 Insulate the outside surfaces of walls.

7.2.4 Increase the thermal resistance of the panels in curtain walls.

RELATED MEASURES:

- Measure 6.4.4, sealing of air leaks in wall structures
- Section 4, for modifications to central air handling systems that may be related to changes in envelope thermal characteristics
- Section 5, for modifications to perimeter heating systems

MEASURE **7.2.1 Insulate wall cavities.**

SUMMARY

Usually the preferred method of insulating walls in new construction, where it is relatively inexpensive. In retrofit, it is limited by the width of the wall cavity and by internal obstructions.

SELECTION SCORECARD

Savings Potential	$	$	$	$
Rate of Return, New Facilities	%	%	%	%
Rate of Return, Retrofit	%	%		
Reliability	✓	✓	✓	✓
Ease of Retrofit	☺	☺	☺	

In most new buildings, except shell buildings, wall insulation is an internal component of the wall structure. You can do a much better job of insulating the opaque portions of the walls than is conventional practice, and you can do it at modest additional cost. This does not require new techniques or materials. It requires only a modest increase in wall thickness to accommodate more insulation, along with attention to good installation practice.

Figure 1 shows a few common methods of installing insulation inside stud wall structures. All these methods are satisfactory, provided that enough insulation is installed to provide ample insulation value for the climate. Figures 2 and 3 show a common method of installing insulation inside masonry wall structures.

The main challenge is to install the insulation so that it will continue to be effective for the life of the building. This means avoiding convection paths around the insulation. If you use compressible insulation, such as glass fiber batt, the reliable installation method is to fill the wall cavity entirely and snugly. If you use rigid insulation, it is critical to fasten the insulation snugly to a strong wall component, and to permanently seal the gaps between the insulation boards. Figures 2 and 3 show the details.

With any type of wall insulation, you must provide a satisfactory vapor barrier and provide for venting any moisture that gets into the wall or insulation.

In many existing buildings, you can improve wall insulation by filling the cavities in the walls. Any type of wall that contains unfilled spaces is a candidate for cavity insulation. As a retrofit activity, filling the wall cavity has the advantage that there is less disturbance of the interior and exterior surfaces than with other methods. The practical question is whether you can gain access to the cavities to install the insulation at reasonable cost.

To help you select the best type of insulating material, see Reference Note 43, Insulation Selection. For retrofit installation, the type of insulation and the method of installing it are determined by the wall construction, especially by the obstacles that you can expect to encounter inside the wall. We discuss these issues for existing stud walls and masonry walls.

Energy Saving Potential

See Reference Note 45, Insulation Economics about the balance between energy savings and costs when increasing the amount of insulation.

In new construction, the cost of additional wall insulation is almost inconsequential. The major cost is changes to the wall structure to accommodate the greater thickness. A clever designer should be able to minimize this cost premium.

In retrofit applications, filling the cavity completely is the only practical approach. You can rarely overdo this. Typically, you can increase the R-value of a wall from 3 to 10 points by adding cavity insulation. The R-values of typical uninsulated walls range from 3 to 6, depending mostly on the number of layers in the wall section. Thus, filling wall cavities with insulation may double or triple their thermal resistance.

Retrofit Insulation for Stud Walls

In a stud wall, the inner and outer surfaces are attached to slender vertical members ("studs"). Studs may be made of wood or sheetmetal. In low buildings, the studs may carry the weight of the building above. In steel frame buildings, the studs only give thickness and rigidity to the wall surface.

The interior of a stud wall is mostly empty space. This space is oriented vertically, allowing a large amount of heat transfer by convection. Heat is also conducted through the studs themselves. However, the low thermal conductivity of wood studs and the thin cross section of steel studs limit this heat loss. In an uninsulated stud wall, most the of the heat transfer occurs in the space between the studs, as a result of convection.

In new construction, it is relatively easy to insulate a stud wall, although workmanship is a limiting factor. (See Reference Note 44 for details.) In existing buildings, the large gap inside stud walls is an attractive target for retrofit insulation. The problem in existing

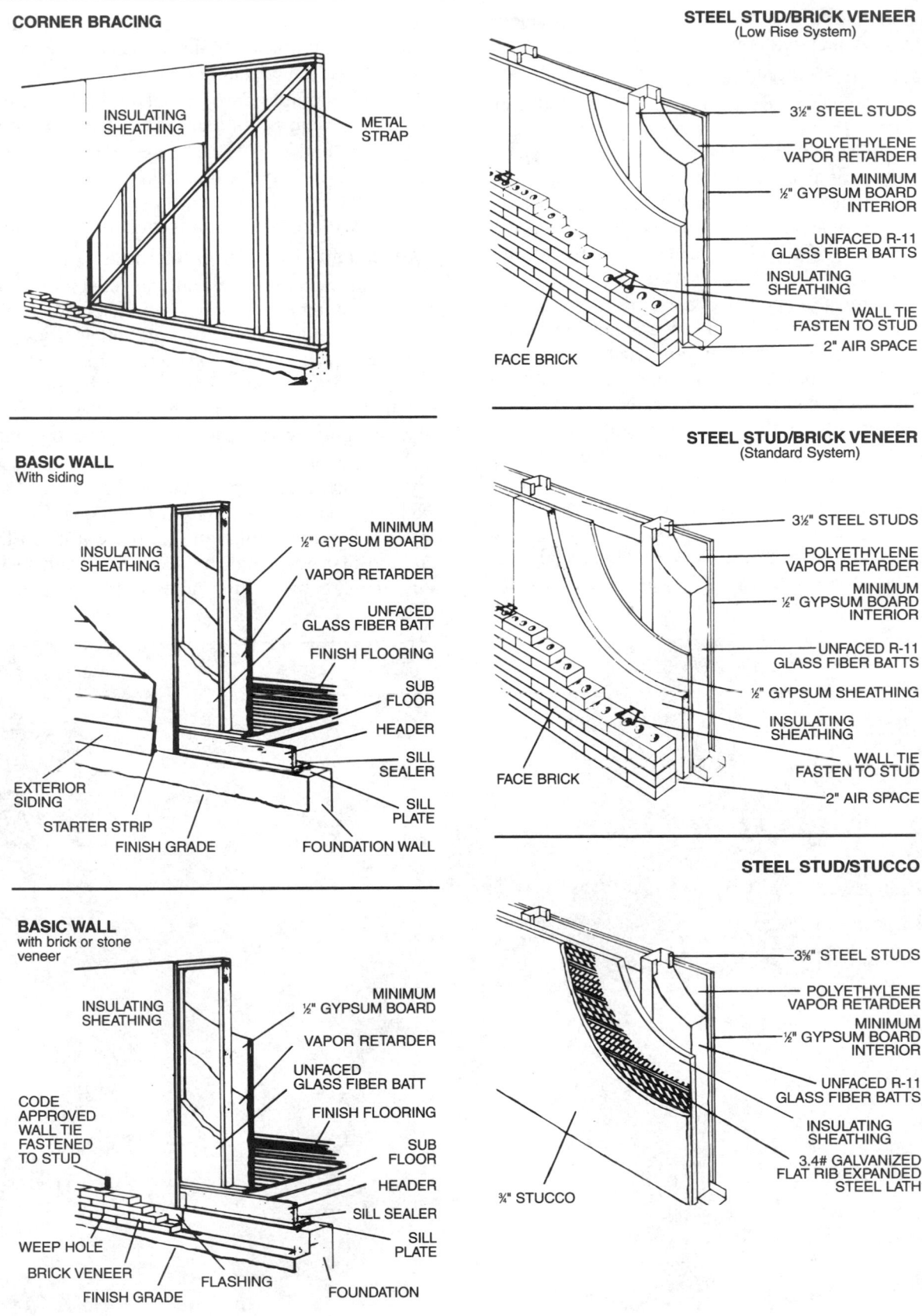

CORNER BRACING

INSULATING SHEATHING

METAL STRAP

BASIC WALL
With siding

INSULATING SHEATHING

MINIMUM ½" GYPSUM BOARD

VAPOR RETARDER

UNFACED GLASS FIBER BATT

FINISH FLOORING

SUB FLOOR

HEADER

SILL SEALER

EXTERIOR SIDING

SILL PLATE

STARTER STRIP

FINISH GRADE

FOUNDATION WALL

BASIC WALL
with brick or stone veneer

INSULATING SHEATHING

MINIMUM ½" GYPSUM BOARD

VAPOR RETARDER

UNFACED GLASS FIBER BATT

FINISH FLOORING

CODE APPROVED WALL TIE FASTENED TO STUD

SUB FLOOR

HEADER

SILL SEALER

SILL PLATE

WEEP HOLE

BRICK VENEER

FLASHING

FINISH GRADE

FOUNDATION

STEEL STUD/BRICK VENEER
(Low Rise System)

3½" STEEL STUDS

POLYETHYLENE VAPOR RETARDER

MINIMUM ½" GYPSUM BOARD INTERIOR

UNFACED R-11 GLASS FIBER BATTS

INSULATING SHEATHING

WALL TIE FASTEN TO STUD

2" AIR SPACE

FACE BRICK

STEEL STUD/BRICK VENEER
(Standard System)

3½" STEEL STUDS

POLYETHYLENE VAPOR RETARDER

MINIMUM ½" GYPSUM BOARD INTERIOR

UNFACED R-11 GLASS FIBER BATTS

½" GYPSUM SHEATHING

INSULATING SHEATHING

WALL TIE FASTEN TO STUD

2" AIR SPACE

FACE BRICK

STEEL STUD/STUCCO

3⅝" STEEL STUDS

POLYETHYLENE VAPOR RETARDER

MINIMUM ½" GYPSUM BOARD INTERIOR

UNFACED R-11 GLASS FIBER BATTS

INSULATING SHEATHING

3.4# GALVANIZED FLAT RIB EXPANDED STEEL LATH

¾" STUCCO

Celotex Corporation

Fig. 1 Various methods of insulating stud walls All these methods can provide good results, provided that you install enough insulation, avoid convective circulation around the insulation, and prevent moisture problems. Beware of methods that use impermeable sheathing on the cold side of the wall in combination with cavity insulation, as this invites moisture problems.

buildings is that the stud spaces may contain obstacles to insulation, which we now discuss.

■ Obstacles in Stud Walls

The main obstacle to retrofit insulation in stud walls is "fire stops." Fire stops are horizontal pieces that span from one stud to the next. Their purpose is to keep the empty stud space from acting as a chimney, hastening the spread of fire. In order to work, fire stops must block the stud space completely. They make it impossible to fill the wall with insulation by pouring into an opening at the top of the wall.

Your are most likely to encounter fire stops in types of construction where the stud space is not interrupted at each floor. These methods of construction are called "balloon frame," "post and beam," etc. In these types of construction, fire stops may be located anywhere. In construction where the stud starts at each floor ("platform" construction), fire stops typically are installed half way up the wall.

Not all stud walls have fire stops. They are not common in platform construction, or where non-flammable wall materials are used, such as metal studs and gypsum board. In case of doubt, check for the presence of fire stops by drilling access holes into some stud spaces and probe to determine whether the cavity is clear from top to bottom.

Window frames pose a problem similar to fire stops, except that there is no doubt where they are located.

Wiring in the walls can cause pouring insulation to fail to fall into the space below, especially if the insulation is fluffy. If the insulation is held up by wiring but later settles, a void will form at the top of the wall.

■ Access and Installation Methods

Loose insulation is usually injected into walls with blowing equipment that carries the insulation in a stream of air. This requires making holes of significant size in each stud space. Insulation contractors use specialized plugs to close access holes after the insulation is installed. The amount of refinishing work needed to cover the plugs varies widely, from none to prohibitive.

Some contractors claim that the holes can be located anywhere in the stud space, but be skeptical of how well the insulation fills the cavity if it is blown from below. Try to insert the insulation into the wall from the top. The top of a wall may be accessible from inside the attic, perhaps by drilling holes in the top plate of the wall.

Dow Chemical Company

Fig. 2 Rigid board insulation inside a wall cavity Using rigid board insulation may be the most reliable way to install insulation inside a masonry wall. The big challenge with rigid insulation is avoiding convection paths between the boards, and between the boards and the wall. Note the eyes installed in the mortar joints for holding the boards snugly against the wall. This requires good workmanship.

Any type of pouring insulation will leave an uninsulated void space at the top of the cavity because of settling. Also, the completeness of filling is dependent on the proficiency and honesty of the installer.

Another method of gaining access to the inside of stud walls is to remove either the exterior or interior skin. This is most likely to be economical when major renovation work is done. For example, if the siding of residential units needs to be replaced, it may be simple to remove the exterior sheathing, install batt insulation, and replace the sheathing. Figures 4 through 6 show an example. At the same time, seize the opportunity to eliminate air leak sites that become accessible.

■ Insulation Materials for Stud Walls

The choice of insulation materials is easiest when you insulate a wall by removing the skin. In this case, use glass or mineral fiber batts fitted between the studs.

If the insulation is to be installed by pouring or blowing, the choice involves compromises:

- *glass fiber* and *mineral fiber* have excellent thermal and safety characteristics. It tends to snag on anything inside the cavity, so a vigorous method of filling the cavity is required.

- *cellulose* insulation is vulnerable to moisture damage and settles more than other types. It also tends to snag.
- *perlite* and *vermiculite* pour well, but have a lower R-value than other types of insulation.
- *plastic foam beads* have excellent characteristics for pouring. However, their fire characteristics make them unsafe in most stud walls enclosing occupied spaces.

If the material consists of small particles, it will leak out of any crevices on either side of the wall, such as the gaps around electrical boxes.

Retrofit Insulation for Masonry Walls

Masonry walls may have several types of cavities. The concrete blocks themselves, and other types of masonry units, typically have large cavities that allow strong convective heat loss. Often, the cavities can be filled with insulation, which kills the convection. However, the insulation cannot reduce conductive heat loss, which occurs through the webs of the blocks. ASHRAE data show that filling the cores of masonry units typically doubles their effective thermal resistance.

Dow Chemical Company

Fig. 3 Details matter The long-term reliability of this insulation job depends on holding the boards snugly against the concrete block wall. Note the clips installed for this purpose. Also, note the sealant that is installed between the joints. We hope that the sealant was installed between the boards, not just skimmed over the top.

WESINC

Fig. 4 Retrofitting stud wall insulation from the outside, first step   The exterior siding and sheathing are removed. Porch light wiring penetrations in the wall are sealed.

WESINC

Fig. 5 Second step   The old single-glazed window is replaced with a new double-glazed window. The stud wall cavities are entirely filled with insulation, being careful to leave no voids.

WESINC

Fig. 6 Third step   Strong exterior sheathing is installed, each panel caulked and nailed to the studs. This method of installation does not allow a satisfactory vapor barrier to be installed inside the wall, so impermeable paint is applied inside the house.

This still results in R-values ranging only from 2 to 5, depending on the type of block.

If a masonry wall consists of a block core and an outer veneer of brick or other material, there is usually a gap of an inch or two (several centimeters) between the block the veneer. This gap is worth insulating, especially if the wall has no other insulation.

The space between the block and any interior finish surface is usually too narrow to insulate on a retrofit basis. Filling the narrow gaps would be difficult or impossible, and the economic benefit would be small.

■ Obstacles in Masonry Walls

The space between the inner and outer elements of a masonry wall is filled with wall ties. Wall ties hold the veneer in place, and improve the strength of the total wall structure. Wall ties can take many forms, such as heavy wire grids, steel rods, and bricks laid sideways across the gap.

The gap surface of block and brick walls is covered with hardened mortar squeezed from the joints. While not bridging the gap, the mortar snags insulation.

The wall cavity may also contain wiring, piping, and other equipment that causes the same problems as in stud walls.

■ **Insulation Materials for Masonry Walls**

Consider the same types of insulation as for stud walls. Foam beads are especially favorable in masonry walls, provided that they are installed outside a heavy, fireproof structural element, such as concrete block.

■ **Access Methods for Masonry Walls**

Try to fill the wall from the top. If the wall extends upward to an exposed parapet, removing the parapet flashing provides access to the interior of the wall. Or, chisel out blocks in the upper part of the wall, and cement them back into position after installing the insulation.

Fire Safety

Before selecting the type of insulation, survey the wall for potential fire hazards, such as wall-mounted heaters. The objectives are to keep from igniting the insulation and to keep from overheating the equipment itself. Insulation may catch fire even if it contains only a small amount of combustible material, such as the binder used in glass fiber blanket insulation. Even dust can fuel a fire. Isolate all electrical connections from contact with insulation. Electrical fires typically start when an electrical connection corrodes or loosens, creating resistance that acts as a heating element.

Vapor Barriers

Reference Note 42, Vapor Barriers, explains the need for a vapor barrier. In most cases, you can create a satisfactory vapor barrier with a reasonably impermeable surface inside the space, such as vinyl wall covering or a thick coat of latex paint. But, keep in mind that the interior finish may change during the life of the building. Leave an explanation in the building's permanent records that an impermeable interior finish is required to protect the wall structure, and hope that somebody reads it at the proper times. It is far safer to build a permanent vapor barrier into the wall structure.

ECONOMICS

SAVINGS POTENTIAL: 40 to 70 percent of the heat loss and heat gains through the insulated portions of the wall.

COST: In new construction, a good thickness of fibrous wall insulation costs from $0.50 to $1 per square foot. Plastic foam insulation costs $1 to $2 per square foot. In retrofit, filling wall cavities with fiber insulation may cost from $2 to $5 per square foot, depending on the difficulty of access.

PAYBACK PERIOD: Several years, or longer.

TRAPS & TRICKS

SELECTING THE MATERIALS AND INSTALLATION METHOD: Specific installation methods are used with specific types of materials as a system. Each system has its own contractors, who usually specialize in that system alone. It is your responsibility to understand all the advantages and disadvantages of each system, and to make the best choice for your facility.

MEASURE 7.2.2 Insulate the inside surfaces of walls.

If the wall structure itself does not provide enough space for adequate insulation, your remaining choices are to install the insulation outside the wall or inside the wall. Installing the insulation on the inside protects it from the elements. Also, the insulation may integrate well with an interior finish surface.

In many existing buildings, adding insulation to the inside surface of a wall is the only practical way to improve the wall's thermal resistance. Installation is easiest with industrial buildings that do not require an interior finish. However, the method works with almost any kind of wall by installing an appropriate cosmetic covering over the insulation.

Figure 1 shows a variety of methods for insulating the inside surface of a masonry wall using rigid board insulation. In retrofit, the same methods could be used with any kind of existing wall.

Figure 2 shows methods of insulating below-grade walls, both above and below floor slabs. Below-grade walls pose a much more difficult problem because of the possibility of soaking the wall from the outside. When insulating below-grade walls, the preparation of the outside of the wall matters as much as the insulation methods inside the wall.

Interior installation has one potentially serious disadvantage, which is a risk of serious moisture damage to the wall structure. Examine your situation closely to determine whether this can occur with your climate and your type of wall. If so, you need to take steps to prevent the problem.

Insulation Types to Consider

Glass or mineral fiber is best for most applications because of its fire safety and ease of attachment. For flat surfaces where no interior finish is needed, semi-rigid board works well. If a surface finish is needed, fiber insulation is available bonded to plywood, gypsum board, and other materials. Blankets or batts may be the best choice if the wall surface is irregular, or if there are many obstructions, such as structural elements, wiring, and piping. Review Reference Note 43, Insulation Selection, before making your final decision.

Each type of insulation, and each method of installation, involves its own set of details. These details are important. Follow the manufacturer's instructions carefully. The most important issues are preventing moisture damage and convective heat loss.

SUMMARY

Applicable to many building types. Becomes complicated if you need to add features to avoid moisture damage.

SELECTION SCORECARD

Savings Potential	$ $ $ $
Rate of Return, New Facilities	% % %
Rate of Return, Retrofit	% %
Reliability	✓ ✓ ✓ ✓
Ease of Retrofit	☺ ☺ ☺

How to Avoid Moisture Problems

To prevent moisture problems with any method of insulation, you have to provide two features: (1) a good vapor barrier on the warm side, and (2) a path on the cold side for venting any moisture that may get into the insulation. See Reference Note 42, Vapor Barriers, for more about vapor barriers. See Reference Note 44, Insulation Integrity, for an explanation of venting paths. You have to make a special effort to satisfy both of these requirements when installing insulation on the inside of the wall.

If the wall includes any elements that block the passage of water vapor, such as a sheetmetal skin, a region is created between that surface and the vapor barrier that traps moisture. To avoid this problem, create a gap on the outer (cold) side of the insulation for venting moisture. The existence of an impermeable trapping layer in the wall increases the importance of having an unbroken vapor barrier.

The climate and the permeability of the wall determine how critical the vapor barrier is. For example, in a building with concrete block walls that is located in a dry climate, it may be satisfactory to attach insulation board having an integral vapor barrier directly to the wall.

If the climate is fairly dry and the outer wall is corrugated aluminum, which does not corrode rapidly, the corrugations may serve as channels for venting moisture.

How to Avoid Convection Losses with Permeable Walls

Convection currents in walls can seriously reduce the effectiveness of wall insulation. This problem is

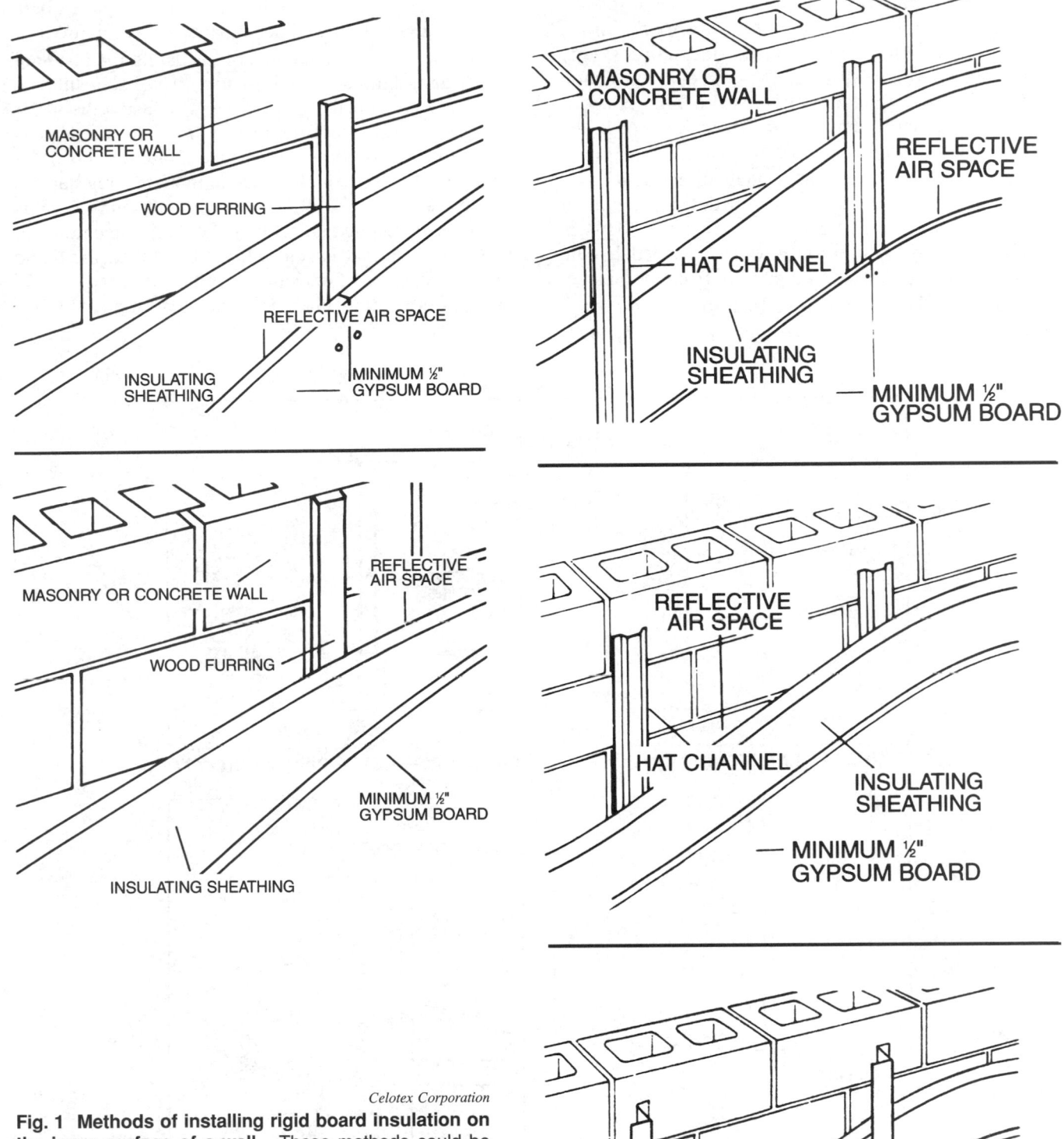

Celotex Corporation

Fig. 1 Methods of installing rigid board insulation on the inner surface of a wall These methods could be used inside a wall of any type, not just concrete block. The minimum gypsum board thickness is required for fire protection when using flammable insulation material.

most serious in walls because they are oriented vertically, creating a chimney effect. The way to minimize convection currents is to install the insulation snugly against an unbroken surface. This keeps convection currents from bypassing the insulation. Also, it minimizes the harm done by any air leaks through the insulation (from cable penetrations, etc.)

If the wall is permeable enough to act as its own vent path to the outside, you can install the insulation directly on the inside surface of the wall. This is satisfactory with most masonry walls, provided that the wall cannot be soaked by precipitation from the outside.

If the wall is flat enough so that you can use board insulation, attach it by using plenty of mechanical fasteners and by caulking the joints of each panel, both to the wall and to the adjacent panels. See Figure 3.

Figure 4 shows a method of insulating the inside of a wall using studs and batt insulation. Use batt insulation that has an attached vapor barrier, and fasten the vapor barrier to studs or furring strips. This method is adaptable to all types of walls. Batt insulation is generally fluffy enough to accommodate irregularities in the wall surface. To achieve the optimum R-value and to ensure snug contact with the wall, use a nominal batt thickness that is somewhat greater than the depth of the stud space. For example, if the stud space is 3.5" deep, install batt insulation that is rated for a stud space 5.5" deep.

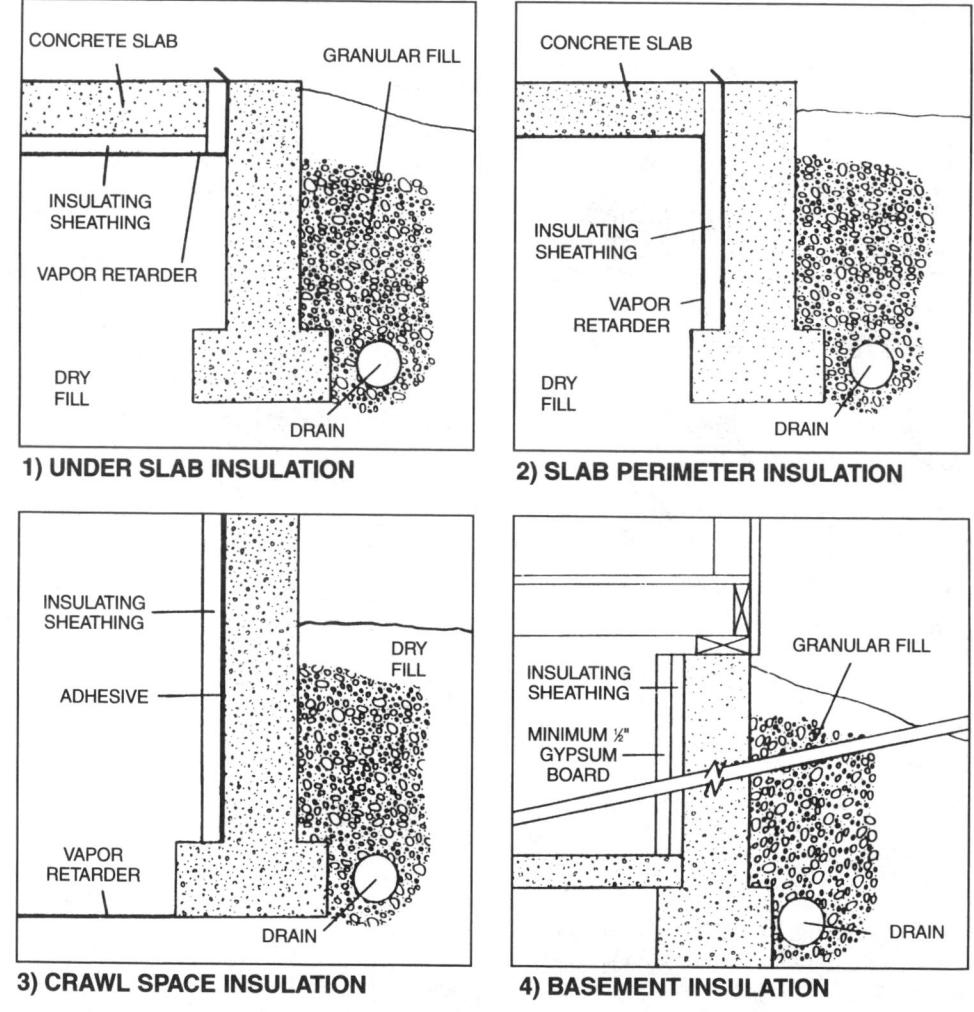

Celotex Corporation

Fig. 2 Methods of installing rigid board insulation on the inside surfaces of below-grade walls No common insulation material is able to function properly in a soaked condition. Therefore, the preparation outside the wall is as important as the installation methods inside the wall. The dry fill and drain tile shown here are necessary to deal with underground water sources, if any. The best protection against precipitation is a continuous downhill slope away from the wall. This can usually be accomplished easily and attractively by building a mild berm against the side of the building.

An important advantage of this method over rigid insulation is that the materials used are less dangerous in a fire. Fire safety should be primary issue when installing insulation inside a building.

Poor attachment is a common problem with insulation. Figure 5 shows an example. Tape and adhesives dry out with age. This releases the insulation and destroys the effectiveness of the vapor barrier. Therefore, use attachment methods that do not place a load on adhesives. If you must use taped joints to apply the insulation, use a method of installation that holds the joints together permanently. For example, install insulation between battens (furring strips) or cover the joints with battens. The battens then serve as an attachment for the interior finish surface.

Broad-head nails are commonly used to attach insulation board to walls, but this method is not always satisfactory. Unless the wall surface holds the nails well, the nails tend to come loose. This is a common problem when insulating masonry walls.

Do not use the wall as the mounting surface if it is so irregular that the insulation cannot contact the wall at all points, or if there is any equipment that prevents the insulation from lying snugly against the wall. In those cases, use the following general approach.

How to Avoid Convection Losses with Impermeable Walls

If the wall is not permeable enough to vent moisture under the climate conditions that exist at the site, avoiding convection becomes more complicated. Venting requires an air gap on the outer (cold) side of the insulation. To create the gap, you must attach the insulation to an interior surface that is strong enough to carry the insulation. The entire wall surface must be sealed from top to bottom, and from side to side, so that air cannot leak through the insulation.

One way to create an interior mounting surface is to use insulation that is bonded to a rigid backing, such as plywood or gypsum board. This material can also serve as an interior finish surface. Such composite material can be secured with screws, nails, or other permanent fasteners. If an air gap is needed on the cold side of the insulation, you have to mount the rigid boards on studs or other structural elements to separate them from the outer wall surface. Since the boards are rigid, you can achieve an unbroken interior surface with caulked or taped joints, provided that you support the joints properly.

Do not use batt insulation if it is not supported on both sides. If the side of batt insulation facing a vent space is unsupported, it will sag and tear loose, largely as a result of rough handling at the time of installation. This problem will be invisible, and it will be virtually impossible to fix. The torn insulation will also interfere with venting.

Minimize Air Leakage Into the Vent Space

If the insulation is installed in a way that leaves a vented air gap on the outer side of the insulation (whether deliberately or not), any penetration of the insulated inner surface will allow air leakage into the space. Therefore, if penetrations are needed for electrical wiring or other purposes, install stuffing boxes or take other steps to avoid air leakage. This is unconventional, so it is unlikely to succeed unless you work out the details and supervise the installation.

Optimum R-Values and Economics

See Reference Note 45, Insulation Economics.

Celotex Corporation

Fig. 3 Attaching rigid insulation inside a concrete block wall Rigid insulation works best when the wall surface is flat enough to provide a snug fit between the insulation board and the wall. Battens are an effective method of attachment, provided that plenty of fasteners are used and that the fasteners hold well in the wall. An important item missing from this picture is the caulking gun. Each insulation panel should be caulked to the wall and to adjacent panels.

WESINC

Fig. 4 Method of attaching batt insulation to the inside of a wall This works well with any type of wall, including the walls of old buildings. Note how the window openings are boxed, with flanges that attach to the wall to block infiltration. The batts should be somewhat thicker than the stud depth.

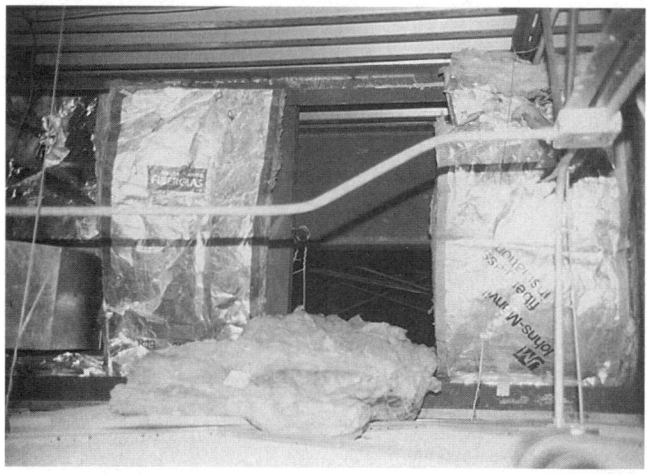

WESINC

Fig. 5 Irresponsible insulation job This batt insulation in a big, new office building was simply suspended by the paper backing sheet, and adjacent batts were taped together. The batt in the middle has fallen down already, and the rest will follow within a few years.

ECONOMICS

SAVINGS POTENTIAL: *60 to 95 percent of the heat loss and/or heat gain through the wall.*

COST: *In new construction, adding a good amount of insulation to the inside of the wall structure costs $2 to $5 per square foot. In retrofit, the cost is typically several times higher because of difficulty of access.*

PAYBACK PERIOD: *One year, to many years, depending mainly on climate and on the original insulation value of the wall.*

TRAPS & TRICKS

DESIGN: *The details are important. Work out a method of attachment appropriate for the insulation material. Do not depend on adhesives for attachment or sealing. Provide adequate moisture venting on the cold side. Select a vapor barrier material that will survive rough handling during installation.*

INSTALLATION: *Plan to supervise the installation. You will probably have to take some special steps to get a good job. Installers will resist any deviations from their usual procedures, no matter what it says in the contract.*

MEASURE **7.2.3 Insulate the outside surfaces of walls.**

Adding insulation to the outside of a wall is a method that is becoming increasingly popular. The development of plastic foam board insulation was a breakthrough for this method.

Exterior installation has these advantages:

- *ease of installation.* Installation from the outside is usually less hampered by obstructions. Foam board can be cut and fitted quickly to any wall configuration that consists of flat elements, even if the wall area is broken up by windows, ledges, etc. It can be attached to the structure directly. It is easy to protect from the elements.

- *unlimited insulation value,* in most applications. Plastic foam has high R-value per unit of thickness, and exterior application allows any desired thickness to be installed, within reason.

- *complete coverage.* Other methods of wall insulation cannot prevent heat loss through floor slabs, studs, and other structural elements that span the width of the wall. In contrast, exterior insulation covers the entire opaque area of the wall. The insulation can be extended below grade to any desired depth.

- *avoidance of moisture problems.* When the interior of the building is being heated, installing the insulation on the outside keeps the entire wall warmer than the dew point. When the interior is being cooled, the low permeability of plastic insulation (or of any insulation using a good vapor barrier) blocks moisture migration into the wall structure. If installed properly, exterior insulation also shields the wall from rain soaking.

Installing plastic insulation on the outside of the building minimizes the danger posed by its nasty fire characteristics. Of course, there may be cases where the fire hazard of petroleum-based foam insulation is unacceptable even for external applications.

Exterior insulation must be protected from sunlight and impact. If the insulation is located away from the possibility of contact with people and equipment, a coat of stucco is sufficient. Board insulation is available with various surfaces to attach the exterior finish surface. For example, board insulation is available that is bonded to a plywood nail base. You can achieve almost any desired exterior appearance using available materials.

Insulation Types to Consider

For most exterior applications, the appropriate type of insulation is some type of rigid, relatively impermeable foam board, such as extruded polystyrene or polyisocyanurate. In principle, any type of insulation

SUMMARY

Has significant advantages over other insulation methods, and is less prone to trouble. The insulation can be covered with almost any type of decor.

SELECTION SCORECARD

Savings Potential	$ $ $ $
Rate of Return, New Facilities	% % %
Rate of Return, Retrofit	% %
Reliability	✓ ✓ ✓ ✓
Ease of Retrofit	☺ ☺ ☺

can be installed externally. However, porous or permeable types are more difficult to install satisfactorily, especially from the standpoint of avoiding moisture problems.

Sprayed foam insulation is in development. Even if it comes to market, it probably will be a secondary choice, except for curved or irregular surfaces. Application of sprayed foam to vertical or slanted surfaces will be difficult to control, resulting in uneven thickness and roughness. On-site mixing of the materials will be subject to mistakes that may reduce the quality of the material.

Each type of insulation, and each method of installation, involves certain steps to achieve a good installation. These steps are important. Follow the manufacturer's instructions carefully. Make sure that all the important issues are covered in adequate detail by the manufacturer's instructions, and that the installer pays attention to them.

How to Avoid Convection Losses

Install the insulation snugly against the wall's exterior surface. Otherwise, outside air will bypass the insulation. If insulation boards are used, apply a bead of caulk around the entire perimeter of each board before sticking it to the wall. In addition, seal each board to the adjacent ones with caulk.

Try to select an exterior finish that holds the adjacent edges in position with respect to each other. For example, install the boards vertically, and install decorative battens over the joints.

Plastic insulation becomes weak and brittle with age, so install it in a manner that can keeps it physically intact over its useful life. Adhesives are unreliable over the long term because they dry out and separate from the foam and from the wall. In most cases, mechanical fasteners are best. However, the broad-head nails commonly used to attach insulation to masonry may perform poorly in certain materials. Test your method of installation beforehand to make sure that it will remain secure.

How to Avoid Moisture Problems

A major advantage of using plastic exterior insulation is that it usually does not require additional steps to avoid moisture problems, in either cold or hot climates. If you use board insulation, seal the joints between the boards to create an unbroken vapor barrier. This involves the same methods you use to prevent convection, which we just discussed. Keep both purposes in mind when you decide how to install the boards.

See Reference Note 42, Vapor Barriers, about the importance of blocking moisture movement through the envelope.

How to Protect the Insulation

In all external applications, plastic foam requires a completely opaque covering to block sunlight. Sunlight can destroy plastic insulation much more quickly than other environmental factors. Any lightweight surface can be used to cover the insulation for decorative

purposes and sunlight protection. The insulation material can be covered with a heavy paint or with a veneer, such as metal siding. A stucco appearance can be achieved by attaching a glass fiber mesh to foam boards with mechanical anchors and then applying a coating over the mesh and boards. Plastic foam can be bonded to a wide variety of surface materials, which speeds installation and aids attachment.

Lightweight plastic foam is easily dented and crushed. Protect vulnerable areas from impact, especially if appearance is important. If potential impacts are light and spread out, you may be able to get enough strength by using foam of higher density. To resist sharper impacts, you can use foam that is bonded to an impact resistant surface material. Manufacturers offer may types of resistant surfaces on insulation boards, including wooden nail base, rigid plastic skins, and metal.

Plastic insulation can be degraded by long exposure to moisture. However, if you protect the surface finish from excess exposure, the insulation underneath will be adequately protected. Moisture damage is a concern primarily with below-grade installation. Only certain types of insulation should be used underground, and even these types require special installation methods to protect them from waterlogging. See Figure 1.

Optimum R-Values and Economics

See Reference Note 45, Insulation Economics.

ECONOMICS

SAVINGS POTENTIAL: *60 to 90 percent of the heat loss and/or heat gain through the wall.*

COST: *In new construction, a good amount of exterior foam board insulation may cost $2 to $5 per square foot, including the surface finish. In retrofit, the cost may be 50% to 100% higher.*

PAYBACK PERIOD: *One year, to many years, depending mainly on climate and on the original insulation value of the wall.*

TRAPS & TRICKS

DESIGN: *Work out a secure method of attachment that will last as long as the building. Do not depend on adhesives for attachment or sealing. Make sure that the insulation is protected from sunlight and impact. Provide for drainage of moisture that may get behind the insulation. When insulating the outside of existing buildings, exploit the opportunity to give the building a better appearance.*

INSTALLATION: *Supervise the installation details, which can have a major effect on longevity and insulation effectiveness.*

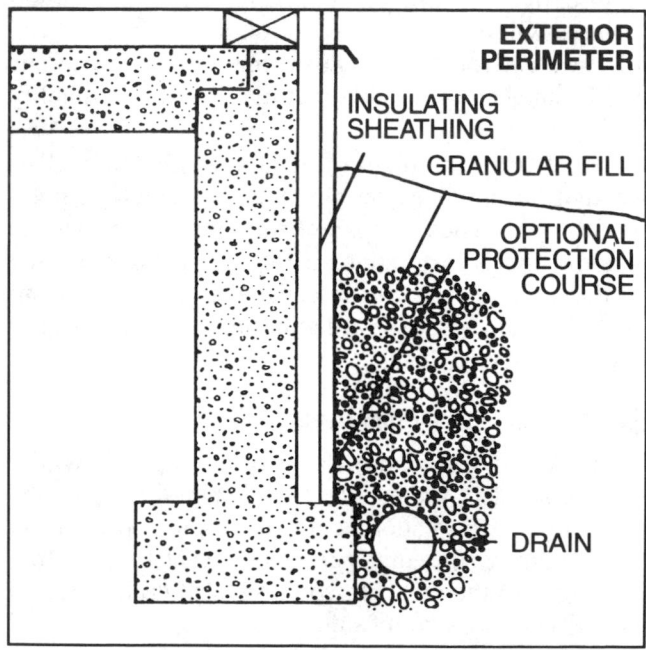

Celotex Corporation

Fig. 1 Exterior below-grade insulation This is a desirable method of insulating below-grade walls and foundations. However, no available material retains its insulation value when soaked, so protecting the insulation from moisture is critical.

MEASURE **7.2.4 Increase the thermal resistance of the panels in curtain walls.**

A large fraction of commercial buildings constructed since the 1950's, especially highrise buildings, use curtain wall construction. The structural strength of the building comes from a rectangular steel frame. Most of the wall surface consists of lightweight panels and/or glazing that are inserted in the spaces between the

SUMMARY

An easy, powerful improvement for new buildings. In retrofit, a radical method, but often the only way to substantially reduce heat loss. Lets you completely update the building's appearance. Keep this in mind for the day when an energy shortage makes your glass box energy hog uninhabitable.

SELECTION SCORECARD

Savings Potential .... $ $ $ $

Rate of Return, New Facilities % % %

Rate of Return, Retrofit .... % %

Reliability .... ✓ ✓ ✓ ✓

Ease of Retrofit .... ☺ ☺

WESINC

Fig. 1 Curtain wall construction This building uses several types of surface treatment over the same basic structural frame. This is the time to make the envelope efficient. However, in existing buildings, it may be possible to remove the exterior panels with surprising ease, providing an opportunity to improve the insulation, control sunlight, and update appearance.

exterior columns and beams. These panels are strictly fillers, with no structural purpose, even though they may be surfaced with materials such as brick or stone. Figure 1 shows all the phases of curtain wall construction in a large building.

In new construction, the designer has almost unlimited ability to increase the thermal resistance of the opaque portions of the curtain wall panels. (Improvements to glazing are covered by Measure 8.1.3.) Indeed, failing to exploit this opportunity is one of the worst oversights of modern architecture.

Another oversight is the potential for dramatically improving the thermal characteristics of existing curtain walls. Opaque curtain wall panels are usually attached with clips and bolts, so it may be fairly easy to remove them from existing buildings. This makes it practical to improve the thermal characteristics of the wall by exchanging the panels. Consider changes to the size and configuration of windows at the same time. With careful planning, the external appearance of an old building can be changed in ways that are limited only by the imagination of the designer. Figures 2, 3, and 4 show an example.

Improving wall panels involves two separate activities: (1) fabricating new panels or modifying the old ones, and (2) all the other work needed to remove and re-install the panels. The first activity is done by a fabricator who specializes in making curtain wall components. The second may involve a variety of contracting work. In the easiest cases, you may be able to upgrade the panels without seriously disturbing the interior finish and systems. In other cases, you may have to remove wallboard and move piping and equipment. The cost and feasibility of the modification are determined largely by whether you can remove and replace the panels without other expensive work.

Judging Whether Retrofit is Feasible

The first step is to examine the wall construction in detail to see how it is put together. Study the architectural details drawings (shop drawings) of the wall structure.

This will tell you how much insulation you can add, and how much effort is needed to take the wall apart.

Examine the mechanical and electrical drawings to find out if there is equipment that will interfere with modifying the wall panels.

If your examination of the drawings suggests that it may be feasible to upgrade the wall panels, select one or more representative areas of the wall as test subjects. Completely remove the curtain panels from these areas, and visualize the steps needed to make the changes. This

WESINC

Fig. 2 Curtain wall upgrade The original exterior, above, was replaced with the one below. Although they appear similar, the new curtain wall has much better insulation value and somewhat less glazing area. The improvement made it possible to abandon the original perimeter heating units, rather than replacing them.

will allow you to calculate the cost of the modification with some precision.

Insulation Types to Consider

In most applications, the appropriate type of insulation is glass or mineral fiber. Avoid plastic foam insulation because of the fire hazard. This is true even if the insulation is encapsulated within the panel. In a fire, plastic foam inside a sealed panel will melt, boil, and blow out the panel.

Panel Insulation Design

In most cases, you will want to fill the panels entirely with insulation to maximize the thermal benefit and to avoid convection.

Make sure to avoid moisture problems. See Reference Note 42 for the important issues. In summary, you need an impermeable surface on the warm side of the insulation and moisture vents on the cold side. In a humid climate where heating and cooling alternate, vent both sides of the insulation and use panel materials that resist moisture damage.

Seal the Panels to the Structure

Stop air leakage at the edges of the panels, using a method that will survive as long as the building itself. Relatively small air leaks can wipe out much of the benefit of the insulation. If necessary, use the modification to improve the design of the panel sealing system. Replace aged and defective sealing components.

Optimum R-Values and Economics

See Reference Note 45, Insulation Economics, about calculating the optimum amount of insulation. In most cases, the best approach is to install as much insulation as the existing structure allows.

Consider Other Improvements at the Same Time

In many curtain wall buildings, it makes sense not only to improve the R-value of the opaque wall components, but also to replace superfluous glazing with insulated panels. There is no type of glazing that has an insulation value even close the R-value of a well-insulated opaque panel. Therefore, reducing the glazing area may substantially reduce heat loss. Also, reducing the area of glazing that is exposed to sunlight greatly reduces solar heat gain, which is the strongest component of cooling load. See Measure 8.1.5 about this.

At the same time, consider other improvements. Consider installing windows that can be opened. Consider sun shades. And so forth. This is a major opportunity for creativity.

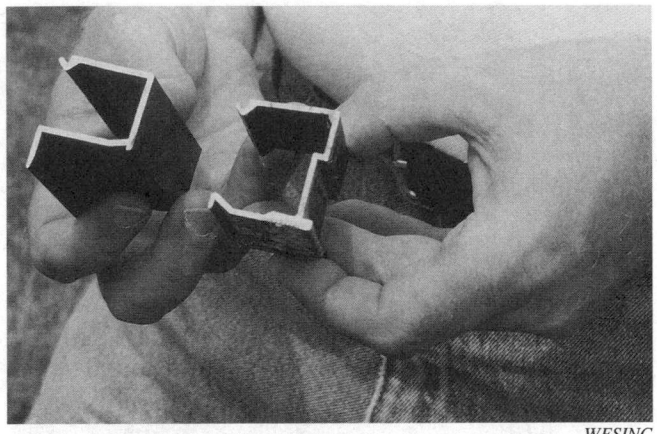

WESINC

Fig. 3 Curtain wall clips The clip on the right held the original curtain wall panels in Figure 2. The clip on the left is for the new panels. There is a mature industry that is capable of making all the necessary components for replacement curtain walls.

ECONOMICS

SAVINGS POTENTIAL: 60 to 90 percent of the heat loss and/or heat gain through the wall.

COST: In new construction, the cost for additional insulation may be $0.50 to $1 per square foot. In retrofit, the cost of replacing the curtain wall panels may range from $20 to $40 per square foot.

PAYBACK PERIOD: In new construction, the payback period for improved insulation may be relatively short. In retrofit, the payback period is many years. Consider the value added to the building by the improvement in appearance.

TRAPS & TRICKS

BE CREATIVE: Let your imagination roam free. This is the place to fix the flaws of modern envelope design. Even in retrofit, consider improving the windows, reducing their size, and adding sun shades. Design for effective sealing against air leakage and effective rain shedding. This improvement may radically reduce heat loss and heat gain through the walls, giving you an opportunity to eliminate or modify the perimeter conditioning systems. Make the building beautiful. Some inexpensive detailing on the exterior surfaces can make the building a showpiece.

INSTALLATION: Installation details are important for avoiding air leakage. Be actively involved in removing and installing the panels.

Glazing is responsible for much more heating load and much more cooling load, per unit of surface area, than other envelope components. Glazing includes windows, skylights, glass enclosures, glass doors, and any other transparent or translucent surfaces. Modern buildings tend to have large amounts of glazing, and therefore to have high energy loads from heat loss and heat gain.

The Measures in this Subsection reduce the heat loss of glazing during cold weather. The Measures of Subsection 8.1 reduce the heat gain of glazing. In many environments, it is important to reduce both heat gain and heat loss. For such cases, consider the Measures in this Subsection and in Subsection 8.1 together. Glazing improvements are expensive, so you want to optimize both aspects at the same time.

INDEX OF MEASURES

7.3.1 Install high-efficiency glazing.

7.3.2 Install storm windows or supplemental glazing.

7.3.3 Reduce the area of glazing.

7.3.4 Install thermal shutters.

7.3.5 Use window films that reflect heat back into the building.

RELATED MEASURES:

• Section 8, for glazing Measures that reduce cooling load

MEASURE 7.3.1 Install high-efficiency glazing.

RATINGS

New Facilities	Retrofit	O&M
B	C	

Installing high-efficiency windows, skylights, and other glazing is a powerful energy conservation measure because it can eliminate several major causes of energy waste. The heat loss of the best contemporary glazing is only a fraction of the heat loss from glazing that was conventional several decades ago. Before making any changes, consider all your options for improving the windows. See Measure 8.1.3 for the full story.

MEASURE 7.3.2 Install storm windows or supplemental glazing.

RATINGS

New Facilities	Retrofit	O&M
	C	

A common way of reducing both heat loss and air leakage through windows is installing supplemental windows, or "storm windows," outside the existing windows. A typical installation is shown in Figure 1.

A less common method is installing supplemental glazing inside the existing windows. This method is usually more efficient, and it may be easier to install in some cases. Figure 2 shows the features of a window with removable interior glazing.

ASHRAE

Fig. 1 Storm window A storm window will approximately double the insulation value of a single-glazed window, if it is installed properly. It is not as effective as installing new windows, but it is much less expensive. This version has a movable pane for ventilation, which is generally not a good idea in storm windows.

SUMMARY

A retrofit measure primarily for buildings with single-glazed windows. Substantially reduces heat loss and air leakage.

SELECTION SCORECARD

Savings Potential	$ $ $ $
Rate of Return	% %
Reliability	✓ ✓ ✓ ✓
Ease of Retrofit	☺ ☺ ☺

Energy Saving Potential

■ Conductive Heat Loss

The thermal resistance of a window depends mostly on the number of panes. Adding one additional pane to a single-pane window more than doubles the insulation value of the window, cutting heat loss to less than half.

Most of the thermal resistance of a pane of glass results from an air layer that attaches itself weakly to each surface of the glass. With a single-glazed window, both the inner and outer air layers are weakened by wind and air currents. If you add a second pane, you create two additional air layers that are well protected. A single-glazed window has an R-value that varies around 0.8. Adding a supplemental window increases the total R-value to about 2.0. You can make similar calculations for any other combination of original and supplemental glazing. Refer to Measure 8.1.3 for the details.

■ Air Leakage

Supplemental glazing eliminates all or most window air leakage, whether caused by wind, building pressurization, or chimney effect.

Exterior or Interior?

At present, the most common type of supplemental glazing is single-pane "storm windows" that are installed outside the existing windows. The prevalence of this type is due to the fact that they have the lowest installed price. Price is a dominant consideration in the residential market where storm windows are usually sold.

There is no inherent reason to limit supplemental glazing to residential applications or to single panes. In locations that have a significant amount of cold weather, consider double-pane supplemental glazing, even when the existing windows are double-glazed.

Also, there is no inherent reason why supplemental glazing must be installed outside the existing windows. On the contrary, installing supplemental glazing from the inside offers an efficiency advantage. The advantage stems from the fact that internal supplemental glazing does not have to be vented to prevent condensation. If the existing windows are very leaky, installing supplemental glazing inside the existing windows may be the only method that avoids serious condensation.

Both internal and external installation require careful fitting of the supplemental glazing to the existing window frame or to the adjacent wall.

How to Prevent Condensation

Condensation is a potential problem with any window that has more than one pane. If the air inside the building can leak past the inner pane, the moisture it carries may condense on the outer pane during cold weather. The condensation is unsightly and causes moisture damage. If the temperature is low enough, frost forms on the inside of the outer pane, obscuring the view.

To avoid condensation, be sure that the inner pane has substantially less air leakage than the outer panes. Stated differently, the outer panes must have significantly more leakage than the inner pane. If this condition is met, any moisture that makes its way between the panes will vent to the outside before it reaches a concentration high enough to condense. For this reason, externally mounted storm windows have small vent holes at the bottoms of their frames. This is also a reason why it is better to install supplemental glazing on the inside of existing windows that are leaky.

The same principle applies with any number of panes. Each pane must be sealed more tightly than the one outside of it. If the window ends up with more than

Fig. 2 Corner of window with interior removable glazing
The inner pane is held in place by the clips. The hole at left vents the space between the panes to the outside. These windows were installed as part of the original construction. Nowadays, in new construction, you would select a sealed double-glazed window. However, this arrangement shows the features that are important when retrofitting interior supplemental glazing.

two panes, you must vent each air space to the outside. As an alternative, you can install a sealed multiple glazing unit as the inner glazing. The second approach has greater efficiency potential, but it is heavier and more expensive.

The vent holes in external storm windows cause some loss of efficiency, but the penalty is fairly small. If the vent holes are kept small, air currents do not disrupt the air layer that is attached to the inside surface of the glass, which is the source of windows' insulation value. By the same token, the vent holes must be at the bottom of the frame so that convection currents will not arise in the space between the panes. Even if the original windows are leaky, the vent holes are so small that they do not allow much air leakage.

How to Provide Ventilation

Supplemental windows interfere with opening the original windows for ventilation. There are two ways to keep the ability to open the windows.

One way is to remove the supplemental windows seasonally. This requires labor and storage space. It also requires a removable frame system for the supplemental windows. Such systems invite a considerable amount of air leakage unless the supplemental windows are sealed carefully every time they are reinstalled.

The other way is to install supplemental windows that can be opened themselves. Such storm windows are usually unsatisfactory. Storm windows are light and flimsy in order to be competitive in cost with total window replacement. Openable versions soon become deformed to the point that they cannot close properly, which makes them worthless. Occupants find it clumsy to manipulate both the original window and the storm window. Hence, occupants become the agents of the demise of the storm windows.

Another possibility is abandoning the ability to open all the windows. For example, supplemental glazing may be installed on all the windows, except for two windows at opposite ends of the space that are left to provide cross ventilation.

Windows that open for ventilation require insect screens. Either install supplemental windows that do not interfere with the screens of the original windows, or install supplemental windows that have their own screens. Screens should be easy to remove and they should fit securely. Many commercial units lack these qualities.

How to Install Supplemental Glazing

The effectiveness of supplemental glazing depends on eliminating leakage all around the edges. This requires snug contact between the edges of the supplemental glazing and the frame of the original window. If necessary, insert weatherstripping or filler pieces between the two frames. But, make sure that the small vent holes at the bottom of the frames remain open.

If you install supplemental glazing inside the original windows, fit it to the existing window frame or to the adjacent area of the wall. If you expect to remove the supplemental windows seasonally (to allow opening the primary windows), spend some time designing a gasket surface that will remain effective as the supplemental glazing is reinstalled each season. Also, you need to design secure clips to hold the gaskets snugly. In most cases, the easiest approach is to install a jamb for the supplemental windows inside the existing window frame.

Comparison to Total Window Replacement

All the Measures of this Subsection are potential alternatives to supplemental glazing. In a majority of applications, the choice is between adding supplemental glazing and replacing the existing windows. (But, don't ignore the other possibilities.) Between these two alternatives, the main advantages of supplemental glazing are lower cost and relatively easy installation, without major disruption of normal activities.

Supplemental glazing is not able to provide the same efficiency as the best modern windows, which exploit tight sealing, insulating gases, infrared reflecting films, and other features. The efficiency advantage of primary windows has increased substantially in the past two decades as a result of new technology, and their efficiency advantage continues to grow. Refer to Measure 8.1.3 for details.

Total window replacement avoids the problem of seasonal removal of supplemental glazing. It also provides the opportunity for improving the appearance of the building.

ECONOMICS

SAVINGS POTENTIAL: About 35% to 65% of conductive heat loss, depending on the number of original panes and the number of supplemental panes. Supplemental glazing can eliminate most heat loss due to air leakage.

COST: $100 to $300 per window, for windows of typical size.

PAYBACK PERIOD: 5 to 15 years.

TRAPS & TRICKS

CHOICE OF METHOD: Supplemental windows are expensive, somewhat unattractive, and clumsy to open. Before deciding to use this approach, consider completely replacing the windows (Measure 7.3.1), replacing the windows with insulating panels (Measure 7.3.3), or if you are an innovator, installing thermal shutters (Measure 7.3.4). Whichever approach you choose, consider the need to open the windows and to install reliable insect screens. Decide in haste, and you may repent at leisure.

SELECTING THE EQUIPMENT: Commercial storm windows are rarely as rugged as they should be. Openable storm windows would have to be as rugged as regular windows in order to survive. With removable storm windows, examine how well the sealing system will survive 10 or 20 years of use. Use only aluminum alloy frame materials, as plastic will crack within a few years.

MEASURE 7.3.3 Reduce the area of glazing.

A powerful method of reducing the high heat loss of windows is to replace all or part of the window area with insulating panels. It is easy to fabricate panels that have much better insulation value than windows. This approach is widely applicable because of the excessive use of glazing in modern architecture. Obviously, this modification has a major effect on the appearance of the building, and it has other advantages and disadvantages.

Refer to Measure 8.1.5 for the full story.

MEASURE 7.3.4 Install thermal shutters.

"Thermal shutters" and "movable insulation" are general terms for methods of insulating windows and skylights during hours of darkness. The concept of thermal shutters first gained prominence in the context of passive solar heating, which requires large expanses of glazing. Thermal shutters prevent the serious heat loss that the passive solar collectors suffer at night.

Thermal shutters are especially important with non-vertical glazing, which has substantially lower thermal resistance than similar glazing would have if installed in a vertical orientation.

Thermal shutters are a novelty, still more of a concept than a practical type of equipment. Many configurations are possible, and none is yet free of major disadvantages. Use Reference Note 47, Passive Solar Heating Design, as a design guide.

At this point in time, start by considering thermal shutters in the form of rigid external panels, which may be moved in a variety of ways. This general type offers high insulation value, requires only simple materials and fabrication techniques, and is easiest to seal against air leakage. On the other hand, large panels are clumsy to manipulate. If a panel is designed to pivot away from the glazing, it must be constructed and mounted very strongly to resist wind.

Security Benefit

Thermal shutters can provide effective security for buildings that are left unoccupied, such as seasonal warehouses. With little additional effort, thermal shutters can be made an effective barrier to human entry.

SUMMARY

Still more of an innovative concept than a practical reality. Can radically reduce heat loss. May also be an effective security measure.

SELECTION SCORECARD

Savings Potential	$ $ $ $
Rate of Return, New Facilities	% %
Rate of Return, Retrofit	% %
Reliability	✓ ✓
Ease of Retrofit	☺

They also protect the glazing and building interior against hail damage and strong winds.

Comparison with Storm Windows

The most direct competition to thermal shutters is supplemental glazing (Measure 7.3.2). Thermal shutters have the advantage of much higher insulation value in the closed position, while supplemental glazing provides its insulation benefit at all times. As noted before, thermal shutters can be used to provide security if they are designed for this purpose.

The relative disadvantages of thermal shutters are higher cost, the need for custom engineering, the need to operate the thermal shutters on a daily basis, and odd appearance.

ECONOMICS

SAVINGS POTENTIAL: *60% to 80% of heat loss through the affected glazing.*

COST: *From several hundred dollars to several thousand dollar per window or skylight.*

PAYBACK PERIOD: *Expect the payback period to be longer than ten years, in both new construction and retrofit, until standardized shutter units are developed.*

TRAPS & TRICKS

DESIGN: *Thermal shutters are at the frontier of efficient building design. Anyone can come up with a variety of designs. The challenge is to design thermal shutters that are practical. They should be durable, they should continue to seal tightly over the years, they should open and close easily, they should not be too difficult to install, and they should not be too ugly.*

MEASURE 7.3.5 Use window films that reflect heat back into the building.

RATINGS

New Facilities	Retrofit	O&M
	C	

Retrofit window films are primarily a device for reducing cooling load. Some types of films offer a secondary advantage of reducing heat loss. They accomplish this with a low-emissivity ("low-E") coating that faces the interior of the building. The coating reflects the long-wavelength infrared radiation that is emitted by the warm interior of the building. Without the coating, this heat radiation would be absorbed in the glass, and a certain fraction of the heat would be lost to the outside by conduction.

In the past, low-E coatings were offered as an optional feature of films sold primarily to reduce solar heat gain. Now, some low-E films are being marketed primarily for reducing heat loss. Unfortunately, the benefit of such films is too limited to make them economical for this purpose alone. This is because heat radiation from the interior of the building represents only a small fraction of total heat loss. Other techniques, such as supplemental glazing, are much more effective for reducing heat loss. However, if you are considering window films for reducing the cooling load, consider the low-E option also.

See Measure 8.1.4 about the full range of window film types, including their heat reflecting properties.

Section 8. CONTROL AND USE OF SUNLIGHT

INTRODUCTION

This Section deals with sunlight, both as a source of unwanted cooling load and as a potential energy source for daylighting and heating.

The most glaring deficiency of modern architecture is failure to deal with the enormous amount of energy that falls on buildings in the form of sunlight. This amount of energy is enough to satisfy all or most of a typical building's energy requirement. At present, much of this potential is unavailable because of a lack of certain critical equipment, and because of the high cost of using available methods and materials. However, a large fraction of the solar energy that could be tapped economically with current technology is being overlooked, while careless architectural design allows too much uncontrolled sunlight into the building. The penalty paid by building owners for the first oversight is purchasing energy that could be obtained free. The penalty paid for the second oversight is purchasing still more energy to eject the unwanted solar energy.

Sunlight is a difficult energy source to harness. It is thinly distributed. It varies continuously in location and intensity. It is unavailable half the time, and it cannot be stored. Even during the daytime, it is unreliable. Using sunlight for one purpose, such as heating, may be difficult to reconcile with using sunlight for another purpose, such as lighting. As a result of these vexing characteristics, contemporary building design has thrown up its hands and ignored sunlight.

Nonetheless, a great deal of research and conceptual design relating to solar energy has been done, and there have been a few periods of intense commercial promotion. Specialized areas of interest have evolved from this experience, each with its own theory and methods.

The applications of solar energy that are summarized under the following headings deserve your attention in current and foreseeable building applications. This Section covers the application areas that are introduced under the first three headings. These three areas are grouped in this Section because of their heavy interaction with each other, and with the adjacent Section 7 (envelope insulation) and Section 9 (electric lighting). The remaining four application areas are summarized in Reference Note 23, where they are grouped with other "free" energy sources. These last four areas do not require difficult interactions with the building envelope design.

■ Reducing Solar Cooling Load

Ironically, the most important energy conservation aspect of solar energy for many buildings is keeping sunlight out. Since the advent of modern architecture, with its vast expanses of bare glazing, designers have failed to deal effectively with the enormous increase in cooling load that is caused by such design. As a result, reducing solar cooling load is a top priority for heavily glazed buildings, even those located in cooler climates. Consider almost any building as a candidate for the Measures that reduce solar cooling load, even if it has only a modest amount of glazing.

Reducing solar heat gain is covered by Subsections 8.1 and 8.2. Most of the methods given in these two Subsections are fairly well proven, although some of them need to be applied more intelligently than is common in contemporary architecture. In new construction, you can always design the building to control cooling load effectively at modest cost, although ideal control may be beyond reach. In retrofit, your choices are much more limited, and the economics are not as favorable.

All methods of controlling solar gain must be tailored closely to the structure and its surrounding conditions. Failure to provide this attention, rather than cost, has been the reason why this important area has been neglected.

The effect of cooling load on electric utilities has made this aspect of solar energy even more important to facility owners from an economic standpoint. Air conditioning causes generating loads that are large and of short duration. It is uneconomical and troublesome for electric utilities to satisfy such a demand profile. Utilities respond by creating strong financial incentives for reducing cooling loads. These include severe demand charges and providing subsidies for customer activities, such as installing thermal storage, that reduce cooling load. (See Reference Note 21, Electricity Pricing, for an explanation of electricity demand charges.)

■ Passive Solar Heating

In contrast with the previous, the objective of passive heating is to bring a large amount of solar heat into the building. The term "passive" solar heating contrasts with "active" solar systems, which collect solar heat outside the building and move it indoors. Passive solar heating is covered by Subsection 8.4.

The energy density of sunlight is sufficient to satisfy all or most of the heating requirements of typical buildings. However, passive solar heating has virtually never been successful in contemporary architecture. The main challenge is not collecting the heat, but avoiding adverse side effects.

Practical passive solar is probably just within reach for new construction, but it requires much more than the usual amount of brain work to design. In most existing buildings, you can exploit passive solar heating only to a limited extent.

■ Daylighting

The amount of sunlight falling on a building, even on a cloudy day, is theoretically sufficient to provide all lighting requirements. The practical challenge is distributing sunlight throughout the building. At present, daylighting is limited to perimeter areas and to interior areas that can be served by skylights. New methods and equipment are needed to approach the full potential of daylighting in all buildings.

Daylighting design is exceptionally complex, but this fact is not yet recognized. The result is that most contemporary attempts are failures. Daylighting must be integrated with control of solar gain, with passive solar heating, with control of artificial lighting, and with esthetic considerations. Each of these is complex in itself, and combining all of them effectively is a mind bending challenge.

Subsection 8.3 offers a limited number of daylighting Measures that you can accomplish without making a career of the subject. See

Subsection 9.5 for the methods that you need to control electric lighting in response to daylighting.

■ Active Solar Space Heating

From a design standpoint, collecting sunlight with an active solar system is relatively simple. This is because an active solar system has no effect on direct solar heat gain into the building, and it has no effect on daylighting. The main problem is equipment cost, which is high in relation to the value of the energy that is recovered. Governments heavily subsidized the development and installation of active solar space heating during the 1970's and 1980's.

Unfortunately, this epoch served mainly to demonstrate that active solar systems have a poor rate of return, in terms of both economics and energy recovery. Most active solar systems have not survived long enough to break even. This dismal record is not just the fault of the weak energy content of sunlight. There is something about solar energy that causes people to underestimate its practical requirements, leading to failures from sloppy design, installation, and maintenance.

■ Active Solar Heating of Domestic Water and Process Applications

Using solar collectors for low temperature process heat recovery may be much more economical than using solar collectors for space heating. Low-temperature process applications allow solar collectors to collect more heat, they use collector capacity more economically, and they do not need an expensive distribution system.

■ Solar Cooling

To date, solar cooling systems have consisted of active solar systems driving conventional absorption cooling machines. Thus, their economics are even worse than the economics of active solar heating. Solar cooling probably cannot become practical until someone invents a cheaper method of getting cooling from sunlight.

■ Photovoltaic Electricity Generation

Photovoltaic electricity generation is well proven technically. It is widely used in specialized remote applications. Also, many demonstration sites have proven that it can be technically successful in large arrays. It is simple and reliable. Unfortunately, it is not yet efficient enough or cheap enough to provide energy for typical building and plant applications. However, there is promise of major improvements in both areas.

How to Use Section 8

Go to Subsections 8.1 and 8.2 to reduce the cooling load of almost any building. Even if your building does not have mechanical cooling, the Measures in these Subsections will improve comfort during warm weather.

Subsections 8.3 and 8.4 exploit sunlight for illumination and for heating, respectively. Be prepared to get deeply involved with these subjects if you want to exploit them effectively. You are on the frontier of building design when you exploit daylighting and passive heating.

The Measures of this Section have a high degree of interaction with each other, even from one Subsection to another. Before you commit to any of them, identify all the Measures that apply to each situation. Then, apply them in a coordinated way.

Before Getting Started ...

Since this Subsection is all about sunlight, start by reading Reference Note 24, Characteristics of Sunlight.

Before you start with daylighting, review Reference Note 46, Daylighting Design, and Reference Note 51, Factors in Lighting Quality.

If you want to plunge into passive heating, read Reference Note 47, Passive Solar Heating Design.

NOTE FOR READERS IN THE SOUTHERN HEMISPHERE

To keep the discussion simple, compass directions in the text assume locations in the northern hemisphere. The corresponding situations for the southern hemisphere are obvious. Simply substitute "south" where the text says "north," and vice versa.

Reducing Cooling Load: Windows & Skylights

Control of solar heat gain has always been a blind spot of Western architecture, but architectural practices in recent decades have made this oversight a major cause of energy waste. In the past, limited window size, massive walls, and tall ceilings were able to limit overheating without the expense of cooling. However, the vast expanses of unshaded glazing in modern architecture make cooling a major building operating cost. By the same token, the popularity of mechanical space cooling allowed architects to remain oblivious to the problems caused by their extravagant use of glazing. As a result, a large fraction of modern buildings are candidates for improvement.

The heat content of direct sunlight is about 240 BTU per hour per square foot (about 0.7 kilowatts per square meter), measured perpendicular to the direction of the sunlight. Vertical glazing oriented east, west, or south receives roughly 1,000 BTU per square foot (about 3 kilowatt-hours per square meter) per day in clear weather. This figure applies to all middle latitudes, and it does not change much as the seasons change.

As a result, in warmer climates, reducing the cooling load is the most important energy conservation measure related to glazing. Even in northern climates, heat gain through glazing may cause a large fraction of annual energy costs, especially in heavily glazed buildings. In recent years, electricity demand charges have risen steeply as utilities seek to avoid building new power plants. Solar cooling load typically is the largest cause of high electricity demand charges in commercial buildings.

INDEX OF MEASURES

8.1.1 Install external shading devices appropriate for each exposure of the glazing.

8.1.2 Install internal shading devices.

8.1.3 Install high-efficiency glazing.

8.1.4 Install solar control films on existing glazing.

8.1.5 Reduce the area of glazing.

RELATED MEASURES:

- Measure 8.2.4, tree shading, which can be an effective method of shading windows
- Measure 8.3.2, light shelves, which must be integrated with exterior or interior shading devices
- Section 4, for modifications to air handling systems to make them respond more efficiently to reduced cooling load
- Subsection 6.3, for reduction of air leakage through windows
- Subsection 7.3, for reduction of conductive heat loss through glazing

From a practical standpoint, this is an important area of conservation because feasible methods exist to reduce solar heat gain in virtually all types of buildings. However, be careful. You may have to use unfamiliar methods, and unfamiliar methods may cause unexpected adverse effects.

Selecting a method of controlling sunlight is complicated by interactions with other important issues that involve glazing, including:

- *building heat loss.* If clear glazing is replaced with reflective glazing to reduce heat gain (Measure 8.1.3), the glazing can also be selected to reduce conductive heat loss during cold weather. Reducing the amount of glazing (Measure 8.1.5) may radically reduce both heat gain and heat loss. Other sunlight control methods offer only a minor improvement in heat loss.

- *passive heating.* Reducing solar heat gain during warm weather also reduces beneficial solar heating during cold weather. You can design shading (Measures 8.1.1 and 8.1.2) to reject sunlight during warm weather and to allow it into the building during cold weather. Other contemporary sun control methods do not provide this option.

- *daylighting.* All methods of reducing solar heat gain reduce the amount of visible sunlight. External shading (Measure 8.1.1) is the only method of reducing cooling load that can be combined with effective daylighting.

- *view.* Glazing treatment (Measures 8.1.3 and 8.1.4) and some types of external shading (Measure 8.1.1) may improve occupants' view. Other sun control methods interfere with view to some extent.

- *appearance.* Virtually anything you do to improve the efficiency of glazing affects the appearance of the building. If executed well, these Measures may substantially improve the appearance and value of the building. But, if you don't pay enough attention to appearance, you can make the building ugly.

As you can see, good sun control tries to kill several birds with one stone. This requires a great deal of thought and study. The responsible approach is to reduce the solar cooling load while exploiting other opportunities to conserve energy while minimizing adverse effects. Daylighting and passive heating are covered in Subsections 8.3 and 8.4, respectively. Consider all the Measures of Section 8 together if you want to push efficiency to the economical limit.

With all methods of reducing cooling load, an important guideline is to stop sunlight as far outside the space as possible. Once sunlight is absorbed, it becomes heat, which requires cooling energy to remove. External shading is the most efficient method of blocking sunlight, and internal shading is the least efficient. If you use coatings on glazing to reduce heat gain, install the coatings on the outer surfaces as much as possible. If you use internal shading devices, keep them close to the glazing.

Expect to customize your sunlight control methods to each orientation of the glazing, and also to localized shading conditions. Reference Note 24, Characteristics of Sunlight, explains how the sun moves through the sky. The way the sun moves with respect to each face of the building determines the techniques that you should consider for each face. This depends on the orientation of each face and the latitude of the site. The weather profile is also an important factor. If the sky is usually clear, you want different shading techniques than if the sky is usually cloudy. The yearly temperature profile is also a primary factor.

Big money is at stake with these Measures if a building has much glazing. Make accurate estimates of savings and costs. Sunlight is a complex energy source and shading interacts with other aspects of conservation, so consider using an energy analysis computer program to make your estimates. See Reference Note 17 if you are unfamiliar with these programs.

Finally, make sure that the building's HVAC equipment will respond efficiently to improvements in glazing. If the HVAC system uses reheat for temperature control, glazing improvements will fail to yield the anticipated benefits. Refer to Section 4 for Measures to improve the efficiency of such systems.

All the Measures of this Subsection are alternatives to each other, and each has major advantages and disadvantages. Tree shading, recommended by Measure 8.2.4, shades both glazed and opaque areas of the building. Consider it along with the Measures in this Subsection.

MEASURE 8.1.1 Install external shading devices appropriate for each exposure of the glazing.

SUMMARY

The most effective method of controlling solar heat gain. Each orientation must be designed separately. All new buildings should have it. In retrofit, it is expensive and often impractical.

SELECTION SCORECARD

Savings Potential	$	$	$	$
Rate of Return, New Facilities	%	%	%	%
Rate of Return, Retrofit..........	%	%		
Reliability	✓	✓	✓	✓
Ease of Retrofit	☺	☺	☺	

Well planned external shading is the most effective method of reducing solar heat gain. In addition, it offers possibilities for incorporating daylighting and passive heating. Some types of external shading interfere with view, while other types make it possible to exploit view that would otherwise be impossible because of solar glare.

External shading is much easier to integrate into the design of a new building than it is to retrofit. External shading has a major effect on the appearance of the building, and it must be anchored strongly to the building structure to resist wind loads. In retrofit, both of these factors are serious challenges.

Each face of a building requires a different shading treatment because sunlight strikes each side from different angles. A south face is best shaded with horizontal shading. East and west faces require shading that blocks sunlight entering at low angles. A north face can often be left unshaded.

Where to Consider External Shading

External shading is useful in almost all situations where direct sunlight through glazing increases the cooling energy requirement substantially. Shading can be adapted to virtually all sizes of windows and skylights. The most common limitations are high cost and the effect on the building's appearance.

If an existing building has glass that is treated to reduce solar heat gain, adding external shading is not likely to be economical from the standpoint of energy efficiency. External shading does reduce the temperature of the glass, improving comfort in some cases. (Exposed absorptive glass may become quite warm in the absence of wind.) See Measure 8.1.3 for details.

Energy Saving Potential

Effective external shading blocks all or most direct sunlight, although it admits indirect light from the sky. It typically reduces solar heat input by 80% to 90%. In buildings where solar load dominates the cooling requirement, shading may reduce a building's total cooling load by as much as half.

In new construction, this large reduction of cooling load may allow the capacity of the cooling equipment to be reduced by a similar amount. The saving in cooling equipment cost may pay for the shading, or a large part of it.

Shading Methods

External shading is a general technique that you can accomplish with many different types of hardware or architectural features. Shading may be fixed or movable. These are most of the types of external shading that you will encounter today:

- *projecting horizontal shelves.* These can be a primary method of shading south faces. They have little value elsewhere. They must be built into the building's structure, and hence they are limited to new construction. To be effective, they must be much wider than the windows, in the direction along the wall, to account for the sun's motion from east to west. Typically, they are installed above the level of the windows, as in Figure 1. If the windows are closely spaced, shelves typically are installed along the full width of the south face. They are vulnerable to strong wind forces, and in northern climates, to snow loads. Smaller shelves can also be built into tall windows at various levels, as in Figure 2.

- *balconies.* These have the same effect as horizontal shelves, but more so. They are deep enough to provide significant shading even if they do not face in a southerly direction. Figure 3 shows a variety of examples. They provide major additional value as usable space and as an ambiance feature. The shelves in Figure 1 also serve as balconies. They are limited to new construction.

- *eaves and overhangs*, which can provide effective shading for the floor level directly under the eave. They merit strong consideration as shading devices, and they can provide major additional value in protecting the wall finish and reducing below-grade moisture problems. Figure 4 shows a typical residential installation, and Figure 5 shows an installation for a public library. They are limited to new construction.

• *inset windows.* In effect, the entire wall acts as a shading device around the window, as shown in Figure 6. This is obviously a feature that is limited to new construction. It is usually done as a stylistic element, rather than as a rational approach to controlling sunlight. As a method of shading, it is very expensive and wasteful of occupiable space. However, it can be effective if it is used properly, namely, on southerly exposures at low geographic latitudes.

• *fixed louvers* may be useful on any exposure of a building, except north. The best orientation for the louver blades depends on the direction that the glazing faces. On the south side, the blades should be horizontal. For the north side, louvers are vertical. For other directions, they may be tilted.

Louvers can be arranged in a horizontal array, like a shelf. Figures 7, 8, and 9 show examples.

Or, they can be arranged in a vertical array, like a venetian blind. Figures 10 and 11 show examples.

Or, they can be installed at an angle, like an awning. Figures 12 and 13 show examples.

The choice of installation geometry depends on the issues discussed below, and perhaps on additional considerations, such as using the shading devices as storm shutters.

Louver blades can take many forms, including flat blades, airfoil shapes, and egg crates. They are easier to install than shelf-type shading because they have less wind resistance and they accumulate less snow load. Therefore, they can be much lighter and easier to attach. They can be attached to the wall, as in Figure 14, or they can be mounted on columns that carry their weight to the ground, as in Figures 9 and 15.

Louvers interfere with the view if they are installed in the line of sight, but they may not block the view entirely. For example, a vertical stack of horizontal louvers in front of a window interferes with the view upward and horizontally, but they are not too bad when viewing downward.

• *vertical fins* are useful for shading north faces from summer sunlight early and late in the day. Figure 16 shows a building with fins molded into the wall surface.

• *awnings,* which project downward over the windows. Figure 17 shows a typical installation. These may be fully effective on south faces, and provide partial shading of windows on east and west faces. A common mistake is making awnings to fit

WESINC

Fig. 1 Horizontal shelves These can be every effective for reducing cooling load on the south side of a building. As seen here, sunlight gets under the shelves on other orientations. These also serve as balconies, a nice touch that is rare in office buildings. They make window washing much easier, protect the glazing, and protect the surroundings from falling glass.

the width of the window. Such awnings are too narrow, allowing an excessive amount of sunlight to enter from the sides.

• *miniature fixed-louver materials* are supported in frames and installed directly over glazing. They are useful on all faces. One product is made from aluminum sheet punched to create tiny louvers. Another product consists of tiny bronze strips woven into a louver with wires. Within limits dictated by the manufacturing process, the manufacturer may offer a variety of louver spacings and tilt angles. Some fixed louver materials can be

oriented vertically for use in east-west shading, or to any other angle.

• *mesh materials* are loose weave fabrics made of materials such as glass fiber and plastics. They are largely non-directional, although some directional characteristics can be achieved by altering the pattern of the weave. The principal merit of these materials is low cost.

• *movable louvers,* which operate like venetian blinds. They are expensive because they must be rugged enough to survive the outside environment. They may be controlled manually or with sun tracking controls. Movable louvers can be installed in any orientation for shading of any face.

• *roll-up external blinds,* which typically are made of aluminum. These are widely used in Germany, for example. They provide security as well as sunlight control. Like internal roller shades, they are non-directional, and they do not offer good lighting quality.

Design and Selection Issues

Exterior shading requires more thought and innovation than most energy conservation techniques because you have many choices, but not a well established doctrine for using them. A wide variety of prefabricated shading devices have appeared on the market. For most applications, external shading is fabricated on a custom basis by a manufacturer who specializes in particular materials and fabrication techniques. Shading devices can be made from metal, wood, fabric, or any opaque material. There are well established companies that can fabricate almost anything you want, but they cannot tell you the best solution for your building. You have to design the installation. As you do, consider the following factors.

■ Shading Effectiveness

Simply hanging shading devices over windows may not provide much benefit, as shown in Figure 18. Overall shading effectiveness depends on the performance of the device at all sun positions. For example, a window awning on a south face may provide complete shading at noon, but poor shading in the morning and afternoon.

■ Effect on View

Shading always blocks a part of the view. As a minimum, it blocks the portion of the sky where the sun travels. On south faces, you can usually arrange window shading in a way that preserves the view of the surrounding landscape. On east and west faces, fixed shading may eliminate the view toward the south, or they may limit the view to a downward angle. Movable shading on the east and west can restore views during the portion of the day when the sun is on the other side of the building.

Vistawall Architectural Products

Fig. 2 Horizontal shelves integrated with windows These are effective cooling load control devices on a south face. Otherwise, they are mainly a decorative touch. This building must be at a southerly latitude for shelves this shallow to provide much benefit.

WESINC

Fig. 3 Balconies Buildings in this summer resort use many kinds of balconies to reduce the cooling load and to provide pleasant space for occupants. These buildings face the ocean toward the east, so the balconies are effective for controlling sunlight mainly around the middle of the day.

■ Daylighting Potential

External shading provides the potential of daylighting in perimeter areas, provided that the shading never allows direct sunlight to fall inside the space. If the shading method allows direct sunlight to enter the space even occasionally, occupants will resort to closing curtains or blinds. Daylighting is difficult to accomplish effectively, and it requires automatic light switching. See Subsections 8.3 and 9.5 for details.

WESINC

Fig. 4 Eaves and porches Roof overhangs have long been used to keep buildings cool in warm climates. The trees help, also.

WESINC

Fig. 5 Roof overhang The shadow pattern shows that the roof overhang of this library is effective in keeping direct sunlight out of the windows. The open windows show that natural ventilation is cooling the building, assuming that the air conditioning has been turned off.

Fig. 6 Inset windows The shadow patterns show that deeply insetting windows reduces solar heat gain. However, it cannot create satisfactory daylighting by itself.

■ Passive Heating Potential

Early in the project, read Reference Note 47, Passive Solar Heating Design. Effective shading kills passive heating. You can reconcile shading and passive heating by moving or removing the shading device when passive heating is desirable. For example, movable louvers and roller blinds provide shading when needed while allowing passive heating at other times.

On the south face of a building, you can achieve passive heating even with fixed shading. This is an important opportunity. The sun's path through the sky is much lower in winter. As a result, you can design horizontal shading over southerly windows so that it shades the windows in summer but allows sunlight to enter in winter. This fact has long been exploited in the architecture of various cultures, including pueblo Indians and the Zulu. For example, at a latitude of 40° (Philadelphia, Denver, Beijing, Madrid, Ankara, Wellington), the noonday sun is about 70° above the horizon in the middle of summer, while the noonday sun is about 30° above the horizon in the middle of winter. See Reference Note 24 for more about solar motion.

You can shade south faces either with a single shelf installed over the window, or with a set of louvers in front of the window. If you use a single shelf, it must project outward a distance that is proportional to the height of the windows. Therefore, this method is easiest to accomplish with windows that are not tall. In new construction, you can exploit this shading possibility by installing balconies along the south face.

Fixed shading used in this manner gives only crude control of solar heat input. It cannot adjust to changes in the intensity of sunlight or to internal heat gains. Also, the outside temperature lags behind seasonal solar motion, typically by four to eight weeks.

■ Appearance

External shading has a major effect on the appearance of a building. If the building is highly stylized (e.g., neoclassical or glass cube), it may be impossible to reconcile external shading with the original style. In such cases, the style of the building has to change.

A stylistic advantage of external shading is that you can make it have any color or surface finish without increasing heat gain. Very little of the heat that is absorbed by the shade is transmitted to the building interior by thermal conduction. Because of its exposed location, the shade is cooled by the atmosphere.

■ Longevity

Try to make external shading device last as long as the building. Generally, it is a mistake to use materials that have limited life outdoors, such as fabric and plastic.

Fig. 7 Horizontal louver shading This building is located at a latitude of 30 degrees. The louver arrays greatly reduce the cooling load. On the south side, they also provide effective daylighting by keeping direct sunlight out of the windows at all times.

Fig. 8 Shadow pattern of horizontal louvers   These are working well, except at the corner. We infer that the wall at right faces in a southerly direction. A double row of louvers is used to reduce the outward distance required. The louvers are attached to vertical stringers, which are bolted to the wall.

Fig. 9 Egg crate louvers   The egg crates block sunlight from all directions. They are supported by the columns, rather than being attached to the wall. This rather massive construction and the irregular placement of the louvers may have been motivated by an appearance concept as much as by sunlight control.

Why inflict the cost and effort of periodic replacement on future owners and operators? Also, such materials soon lose their sparkle and start looking shabby.

■ Attachment to the Building

Attaching external shading devices to the building can be a design challenge. Shading devices are subject to strong wind forces because they have a large surface area. Some shading is subject to snow loads. If the building envelope does not have easily accessible strong points for attachment, you have to create them.

In new construction, you can make some of the shading features described previously an integral part of the structure. This is generally very strong.

Separate shading devices can be bolted to the wall. However, do not assume that an existing wall is strong enough to withstand the weight and wind loads. You may have to reinforce the wall at the attachment points, and doing this may be awkward.

You can also attach shading devices to columns or vertical stringers. If the latter are attached to the wall, as in Figure 8, they distribute the weight of the devices along the wall, while minimizing bending loads that tend to tear attachments out of the wall.

If the attachment columns can extend to the ground, they can relieve the gravity load on the wall. In some installations, the columns are freestanding, except for a steadying attachment at the top.

■ Removable Shading

Try to avoid shading devices that have to be removed on a seasonal basis. Stowage is a major problem, and operating personnel hate to bother with this kind of chore. Such shading tends to be abandoned within a few years.

■ Method of Control

Movable shading may be operated manually or automatically. Automatic control is necessary to achieve efficient results in most applications. Controls based on some form of sun position sensor can be simple and reliable. Such controls are unconventional, so you have to make sure that they are installed properly.

Manual control is reliable only if people become uncomfortable in an obvious way when they fail to control the equipment efficiently. For example, manual control may work for a louver shade that allows glare when it is not adjusted to keep out sunlight. However, this method of control is not pleasant for the occupants, and it is not effective for exploiting passive heating.

Airolite Company

Fig. 10 Vertical array of louvers These provide effective shading of full-height windows across a narrow walkway. They are high enough to avoid obstructing the view.

WESINC

Fig. 11 Vertical louvers for a house These provide effective light and ventilation for the southeast side of a house located in southern Florida. Provides a limited downward view.

WESINC

Fig. 12 Louvers installed as awnings   These simple louver assemblies on the south side of a modest commercial building provide effective shading, along with a moderately good view of the street below. They have been lengthened to hide air conditioners installed below the windows.

Construction Specialties, Inc.

Fig. 13 Louvers arranged in a curved array   The curvature is strictly for appearance. The designer has to work a little harder to figure out the louver dimensions that keep direct sunlight from striking the window.

Airolite Company

Fig. 14 Shading louvers attached to wall by brackets

■ Envelope Penetrations

Avoid actuators for movable shading that require significant envelope penetrations. These become sources of air leakage.

■ Fire Egress

Windows may be potential escape routes in the event of fire, or points of entry for firefighters. Do not install shading devices in a way that interferes with emergency use of windows.

■ Property Lines and Setbacks

If a sun shading device would extend beyond a property line or beyond the limit of a construction setback, make arrangements beforehand to use the adjacent air space.

Orientation is Critical

Some shading methods are extremely specific to compass orientation (azimuth). For example, fixed horizontal shading may leak sunlight into the building during the morning or afternoon unless it is used on a face that is oriented almost exactly due south. The same is true of vertically oriented shading that is installed on east and west faces.

Do not assume that a north face does not require shading. See Figure 16. In summer, the sun rises and sets well to the north of due east and due west, respectively. Shading of the north side is simplified by the fact that the sun is low by the time it gets around to the north face, except at low latitudes. North faces that are oriented slightly toward the east or west experience a significant increase in solar gain during summer mornings or evenings, respectively.

■ Determine the Azimuth Accurately

If you are going to retrofit shading to an existing building, you need to know the actual orientation of the building faces within an accuracy of about 10°. The compass roses on most building drawings are not precise, and hence are worthless for shading design. Instead, work from a surveyor's plot plan that has an accurate north arrow. Use directions based on true north, rather than magnetic north. True north and magnetic north may differ by 30° or more. If you cannot find an accurate plot plan, use a surveyor's magnetic compass and set it up away from any ferrous objects (including buried pipe, pavement reinforcement steel, and the magnetized screwdriver in your shirt pocket). Correct the magnetic compass reading to find the true orientation.

WESINC

Fig. 16 Vertical fins Fins are cast into these concrete wall panels. They are too shallow to keep sunlight off the north windows during summer. They do keep the wall cool by shading most of the surface area.

Construction Specialties, Inc.

Fig. 15 Shading louvers supported by columns

Account for Reflection from Surrounding Features

Reference Note 24, Characteristics of Sunlight, points out that a substantial amount of solar radiation may arrive by reflection from features that surround the building. Some of these features may be too big to ignore, such as the lake on the south side, or the glass box building next door. Reflected sunlight that arrives from an unexpected direction is likely to get past your shading devices. So, don't assume that all sunlight comes directly from the sun. Take a careful look around your property.

Account for Shading by Surrounding Features

The lower portions of tall buildings may be shaded effectively by adjacent buildings and other features. Don't spend money unnecessarily to shade these portions of the building. However, external shading may still be desirable in such cases to eliminate glare for the purpose of preserving a good view.

WESINC

Fig. 17 Awnings Canvas awnings provide effective shading for the rooms of this hotel in a warm, sunny climate. The wall to the right faces east. As you can see from the shadows, the narrowness of the awnings allows some direct sunlight to enter the windows. This could be minimized by tapering the sides of the awnings outward. The awnings provide an important decorative accent for this plain rectangular building. However, fabric has a rather short service life.

Airolite Company

Fig. 18 Shading or deco? Look at the shadow patterns. The windows on the right face southeast, and most of the sunlight is getting into them now, at mid-morning. Fixed horizontal shading works well only on a south face. If horizontal shading wraps entirely around a building, only a part of it will be very useful. Given the limited outward reach of these shades, how far south must the building be located for the shading to be effective, even on the south side?

ECONOMICS

SAVINGS POTENTIAL: 70% to 95% of the cooling load caused by the shaded glazing.

COST: $2 to $20 per square foot of window area.

PAYBACK PERIOD: In new construction, may be immediate if the cost of shading reduces the cost of the cooling equipment by an equal amount. May be several years, in other cases. In retrofit, the payback period is several years or longer.

TRAPS & TRICKS

PIONEERING: If you want to attract attention, this will do it. For better or worse, sun shading radically changes the appearance of a building. A major decision is whether to use fixed or movable shading. Fixed shading cannot completely block direct sunlight while preserving an open view unless it faces directly south. Movable shading requires custom engineering, custom fabrication, and maintenance. Make sure that the structural attachments have sufficient strength to withstand the strongest wind loads. Consider the effects on fire safety, window cleaning, etc. Make a study of other facilities that use sun shading, but do not expect to find perfect shading performance on all sides of a building. If you achieve it, you will be making architectural history.

EXPLAIN IT: If you install movable shading devices, tell the staff how it is supposed to operate. Describe the system in the plant operating manual. Install effective placards at the controls.

MAINTENANCE: If you install movable shading devices, maintenance is critical. Schedule periodic checks of operation in your maintenance calendar. If you have an energy management control system, use it to monitor the operation of the movable shading devices.

MEASURE **8.1.2 Install internal shading devices.**

Internal shading comprises the category of shading techniques that architects call "window treatment." This includes draperies, curtains, and blinds. For most applications involving windows, you can use techniques that are commercially available. However, if you want to exploit daylighting and passive heating, be ready to innovate. Also, skylights require custom shading techniques.

In comparison with external shading (Measure 8.1.1), the main advantage of internal shading is low cost. The equipment is light because it is not subject to wind loads or precipitation. Some types of internal shading adapt to all sun positions, so you may be able to use a particular method throughout the building.

Internal shading cannot be as efficient or as visually pleasing as external shading. All types of internal shading interfere with view, and some eliminate view completely. See Figure 1.

Internal shading is primarily a retrofit technique. If you are designing a new building, attempt to accomplish as much shading as possible with external methods.

Where to Use Internal Shading

Internal shading can be used with almost all windows and skylights. Consider it for all glazing that lacks an effective method of limiting solar gain. For example, corridor windows are often unshaded. Large skylights in areas such as atriums and lobbies often admit excessive sunlight, even though they may be tinted.

On the other hand, if a building already has glass that is heavily treated to reduce solar heat gain, internal shading may not provide much additional reduction of heat gain. If the glass uses absorption to reduce solar gain, it will absorb most of the sunlight that is reflected by internal shading, making the glass warmer. The increased heating of the glass increases the risk of glass breakage. See Measure 8.1.3 for details.

A majority of windows that receive direct sunlight already have internal shading of some kind. Examine the existing shading and decide whether it is worthwhile to replace it with a more effective type. Existing shading that is made of light-absorbing materials, such as dark draperies, does little to reduce the cooling load. Replace it with a more efficient type of shading, if it is economical to do so.

Energy Saving Potential

■ Heat Gain

Internal shading reduces heat gain from 20% to 70%. The best performance is obtained from highly

SUMMARY

The least effective method of reducing solar heat gain, but often the only one that is economical in existing buildings. In a new building, a need for internal shading testifies to careless design.

SELECTION SCORECARD

Savings Potential .... $ $ $ $

Rate of Return, New Facilities % % %

Rate of Return, Retrofit.......... % %

Reliability .... ✔ ✔ ✔

Ease of Retrofit .... ☺ ☺ ☺

reflective shading devices that are installed close to the glazing.

■ Cold Weather Heat Loss

Do not expect interior shading devices to reduce conductive heat loss during cold weather. Convection between the shade and glazing virtually eliminates any benefit from the thermal resistance of the shade. In the past, people have tried to seal shading devices at the edges to serve as a form of supplemental window insulation. Doing this causes heavy condensation and moisture damage. See Measure 8.1.3 for the right way to reduce heat loss through new glazing. See Measure 7.3.2 for the right way to reduce heat loss through existing glazing.

■ Daylighting

Some interior shading devices can be part of a daylighting system. However, ordinary inexpensive internal shading methods do not provide significant opportunity for daylighting by themselves. Don't expect occupants to continually readjust their blinds to optimize daylighting. See Reference Note 46, Daylighting Design, for the main points to consider.

■ Passive Heating

Shading devices that reduce undesirable heat gain in summer also reduce desirable heat gain in winter. Most internal shading devices allow occupants to adjust the amount of sunlight admitted. However, don't expect occupants to optimize the use of their shading devices to exploit passive heating. Control of passive heating requires more elaborate techniques that are explained in Reference Note 47, Passive Solar Heating Design.

Available Methods

There is no limit to the methods of internal shading that might be devised. Some types of interior shading devices worth considering are:

- *horizontal louver (venetian) blinds* have remained popular over the years because they are the only type that controls solar gain while preserving the view and avoiding glare. For vertical windows that are accessible to occupants, standard venetian blinds are the best choice for a majority of cases.

- *vertical louver blinds* apparently became popular because interior decorators became bored with horizontal blinds. They are useless for sun control on south faces, and they require much more adjustment on east and west faces than horizontal blinds. Occupants typically keep them closed and sacrifice the view. They may be satisfactory for controlling glare through glazing that faces north, eliminating direct sunlight into the space during early morning and late afternoon in summer.

- *roller shades* are available in wide range of materials. If they are highly reflective, they can reduce solar gain more than any other type of internal shading, but only if they are kept closed. Roller shades made of ordinary materials must be kept partially open to preserve view, so they admit heat gain and glare. Roller shades can be made with translucent material to provide diffuse daylighting. However, the transparency of the material must be very low to avoid glare problems. (See Measure 8.3.1 for details.) Roller shades are also available with transparent material that has a reflective surface facing outward. The life of this material tends to be short.

- *draperies* have poor shading coefficients because the folds of the fabric trap sunlight. They cannot reconcile view with sunlight control, they have poor insulation properties, and they trap mechanical heating and cooling against the building envelope. Their merits are formal appearance and the ability to admit diffuse light.

- *miniature louver materials* (see Measure 8.1.1) could be used for internal shading. Their disadvantage when used indoors is that any sunlight they absorb adds to the cooling load. If they are used internally, their outward facing surfaces should be as reflective as possible.

- *diffusing materials* in general. These could include fabrics, mesh materials, milky plastics, or anything else that reflects most of the light while transmitting a part of it in a diffuse manner. Glare is a major issue with diffusing materials, as explained in Measure 8.3.1. When installed as fixed panels, they limit or prevent the use of windows for ventilation. Skylights and other inaccessible glazing can be shaded with a large variety of custom devices or fabric hangings. These can be very decorative.

- *reflective materials* in general. Reflective film can be applied directly to the glazing, as recommended by Measure 8.1.4. This is the most effective method of reducing cooling load that can be applied inside the building. This method cannot eliminate glare.

WESINC

Fig. 1 Typical interior window shading The first effect of most interior window shading is to abolish the window's original purpose, which is to provide a view and daylighting. Almost every occupied space on this side of the building has blocked out daylight completely. However, the windows still cause greatly increased heat loss and heat gain. The light color of the windows indicates that a considerable amount of sunlight is being reflected, but much is still trapped.

It has other disadvantages, such as increasing the possibility of glass breakage.

How to Select or Design Internal Shading

The following are the main points to consider when you select or design internal shading devices.

■ Shading Effectiveness

The term "shading coefficient" was defined to describe the effectiveness of various shading techniques for reducing cooling load. It is defined as the ratio of the heat gain that enters the space with the shading device installed, compared to the heat gain that would occur through a single pane of clear glass. The *Fundamentals* volume of the *ASHRAE Handbook* lists shading coefficients for common types of internal shading devices, such as venetian blinds and roller shades. Shading coefficients may also be provided by manufacturers of shading devices. These figures should not be treated as exact because the shading coefficient varies with factors such as the size of the window, the mounting position of the shading device, and the incoming angle of the sunlight.

For unconventional shading devices, base your estimates on the reflectance of the material and the geometry of the installation. It is easy to be too optimistic about the savings provided by custom shading. Unless the shading device is held snug to the glazing, a significant amount of sunlight will leak past it into the space. Also, any use of color on the outer surface of the material increases absorption.

■ View and Glare

All methods of internal shading, except for window films, significantly degrade the view through the windows. Some shading methods, such as draperies and roller shades, eliminate view entirely. Fixed or movable louver blinds may preserve most of the view while substantially reducing solar load. If a shading device allows glare, it harms the view.

Shading devices that allow direct sunlight to enter the space do not solve the problem of glare, even if they reduce the intensity of sunlight. (For an explanation of glare, see Reference Note 51, Factors in Lighting Quality.) Window films do little to reduce glare. A shading device that diffuses direct sunlight may create a serious glare problem, because it appears as a bright light source. Louver materials must have dark undersides to keep from becoming a glare source.

■ Daylighting Potential

Early in your planning, read Reference Note 46, Daylighting Design. On the positive side, achieving daylighting is compatible with control of solar heat gain. On the negative side, achieving good daylighting is more complicated than installing shading devices. In the realm of simple daylighting, louver blinds may make daylighting possible near windows. If you have glazing that is located well above normal sight lines, such as skylights and clerestory windows, diffusing materials may allow it to provide effective daylighting.

■ Adaptability to Different Orientations

Most of the common internal shading methods do not need to be customized to the azimuth of the window. This simplifies the interior decorator's task and reduces cost.

Expect to use one set of techniques for ordinary windows, and a different set of techniques for glazing that is located well above horizontal sight lines (e.g., skylights and clerestory windows).

■ External Appearance

In order to be reasonably effective, internal shading must have an outward facing surface that is highly reflective. This makes the shading devices quite visible from outside the building. Attempting to conceal internal shading devices by making them darker seriously reduces their effectiveness. If you want to maintain a uniform external appearance, consider fixed shading devices.

■ Longevity

The shading device needs a long service life to earn a good profit in energy savings. Select devices that are sturdy enough to survive normal handling and resist impact damage. (Recall how ugly venetian blinds become when their slats get bent.)

■ Installation

From an efficiency standpoint, the most important installation issue is mounting the shading device as close to the glazing as possible. If there is distance between the glazing and the shading device, sunlight will leak past the shading device, or it will be absorbed by the frame and adjacent wall structure.

Attaching internal shading is much less demanding than attaching external shading, which is subject to wind and weather. Still, expect to innovate and experiment to develop methods that are elegant and effective.

■ Control of Movable Shading

Flimsy or inaccessible controls undermine the effectiveness of movable shading, such as venetian blinds. Expect some difficulty in persuading the interior decorator that it is vital to have controls that are accessible and visible.

■ Fire Egress

Windows may be escape routes in the event of fire, or points of entry for firefighters. Do not install shading devices in a way that interferes with emergency use of windows.

Don't Forget the Placards

If a shading device needs occupant control, tell the occupants what to do. In most environments, you can do this best with placards. See Reference Note 12, Placards, for details of effective placard design and installation.

ECONOMICS

SAVINGS POTENTIAL: *20% to 50% of the cooling load caused by the shaded glazing.*

COST: *$1 to $5 per square foot of window area.*

PAYBACK PERIOD: *Several years. May be shorter if shading avoids the need to increase cooling capacity.*

TRAPS & TRICKS

SELECTING THE METHOD: *You can probably use a commercially available product. Shop around. The most popular product, horizontal venetian blinds, are the best compromise for most existing buildings. If you want to preserve the view, select blinds with reflective top surfaces and dark bottom surfaces. If you are selecting internal shading devices for a new building, you made a big mistake in the envelope design. Go back to Measure 8.1.1 and start over.*

EXPLAIN IT: *In most cases where internal shading devices are needed, the occupants simply close them and sacrifice the view. If occupant control is to have any chance of working, you need to install placards that ask occupants to adjust the shading appropriately.*

MEASURE **8.1.3 Install high-efficiency glazing.**

In typical buildings, the glazing accounts for most of the solar heat gain. Glazing also accounts for a large fraction of heat loss during cold weather. For these reasons, your choice of glazing is a major factor in energy costs.

This Measure deals primarily with the efficiency properties of glazing materials. It covers all types of glazing, including windows, skylights, and glazed enclosures. It covers all types of glazing materials, including glass and the various plastic glazing materials. However, one specialized type of glazing, translucent structural panel materials, is covered primarily by Measure 8.3.3.

This Measure does not deal with efficient layout of glazing. The layout aspects of glazing vary widely for windows, skylights, and glazed enclosures. These aspects are covered by other Measures throughout Section 8 that deal with specific situations.

Selecting glazing involves many more factors than it did a few decades ago. The key is to understand the theory of each aspect of glazing performance, and to combine these aspects in an optimum combination. On the positive side, both raw glass and fabricated glazing units are available with a broad range of properties that allow tailoring energy characteristics and appearance to the application. On the negative side, mistakes in glazing design or selection sacrifice energy efficiency, and they may cause major problems, such as glass breakage, overheated spaces, and appearance problems.

In new construction, select glazing to minimize heat loss during cold weather, but rely primarily on external shading to limit solar heat gain (see Measure 8.1.1). Using glazing to control heat gain is less efficient than using shading.

In existing buildings, where it may not be economical to add external shading, you can still make substantial improvements in controlling heat loss and heat gain by replacing the present glazing. This Measure explains the choices that you have to make to achieve both of these objectives.

In both new construction and retrofit, consider glazing as one element in a system of features that control heat loss and heat gain. Combine it efficiently with the techniques that are recommended by the other Measures of this Subsection.

Where to Retrofit High-Efficiency Glazing

From an efficiency standpoint, this Measure applies to any building that has a large cooling load caused by glazing or a large heating load caused by glazing, or both.

SUMMARY

A powerful method of reducing both cooling load and heating load. There are many factors to consider.

SELECTION SCORECARD

Savings Potential	$ $ $ $
Rate of Return, New Facilities	% % %
Rate of Return, Retrofit	% %
Reliability	✓ ✓ ✓ ✓
Ease of Retrofit	☺ ☺ ☺

Retrofitting high-efficiency glazing is expensive. The payback period is long, so this Measure is economical for existing buildings only if they have a long remaining life. These situations favor glazing retrofits:

- climate and building orientation that cause a large total amount of heat gain and/or heat loss through glazing. Glazing replacement may be important on some faces of a building, but not on others.
- large numbers of windows, especially if they are similar
- windows that are individually large
- frame systems that allow easy glazing replacement
- the building is due to receive an exterior renovation
- the HVAC system is due to be renovated, and improved glazing can reduce the need for new cooling equipment.

Energy Effects of Glazing

Glazing affects energy consumption in a number of ways. Tally all the following effects when you estimate the potential savings.

■ Solar Heat Gain

The most efficient available glazing can reduce solar radiation heat gain as much as 85%, compared to clear glass. Achieving this degree of improvement requires the glazing to be as reflective as possible, or to selectively absorb the infrared component of sunlight. It also requires low-E coatings and insulating gas fills (explained below) to prevent the inward flow of heat that is absorbed in the outer glazing.

Glazing treatment that relies solely on maximizing reflection can reduce direct radiation heat gain reduction by about 50% to 70%. If a high degree of reflection is not acceptable, and absorption is used instead, the reduction of heat gain may be limited to 40% to 60%.

These numbers are excessively optimistic because they use unshaded clear glass for comparison. When compared to a clear glass window with reflective interior curtains or blinds, the potential for reducing heat is perhaps half of the previous numbers.

■ Conductive Heat Loss During Cold Weather

In buildings with large amounts of glazing, heat loss through the glazing typically is responsible for about half the cost of heating.

The greatest relative improvement occurs when replacing single glazing with efficient triple glazing. This change reduces conductive heat loss by as much as 80%. On the other hand, replacing clear double glazing with "high-efficiency" double glazing reduces conductive heat loss by 10% to 50%.

■ Conductive Heat Gain During Warm Weather

Conductive heat gain during summer typically is less important than conductive heat loss in winter, because the temperature differentials are smaller. Still, conductive heat gain is important in a warm climate. It may account for 10% to 30% of the total heat gain through glazing.

The relative improvements that can be achieved are about the same as for conductive heat loss, just discussed.

■ Long-Wavelength Infrared Heat Loss During Cold Weather

Radiation heat loss from the interior of the building is much less important than conductive heat loss. Heat radiation from the interior of the building is absorbed in the glazing, where it warms the glazing. The heat moves by conduction and further radiation toward the outside of the building. Multiple glazing and low-emissivity coatings can reduce this component of heat loss by 50% to 80%.

■ Long-Wavelength Infrared Heat Gain During Warm Weather

This is radiation from the warm surfaces of the outside world (not including the sun) toward the building. The radiation is absorbed in the glazing, warming the glazing. This heat moves by conduction and further radiation toward the inside of the building. This is a minor component of heat gain. Wind and glazing treatment reduce it from 20% to 70%.

■ Passive Heating

All glazing that reduces solar heat gain also substantially reduces the potential for passive heating. This factor is unimportant in warm climates, but it may cancel much of the benefit of passive heating features in cold climates.

■ Daylighting

See Subsection 8.3 about daylighting. All types of treated glazing reduce the amount of daylight available. Whether this matters depends on whether the building has more glazing than it needs for daylighting.

Get the Most From Your Glazing Decisions

Consider all the available opportunities for conserving energy when you select glazing. Use this approach to sort through the possibilities:

• *First, consider conductive heat loss, for both cold weather and warm weather.* This defines the number of panes of glass (the industry calls them "lights") to be used. Make this decision before proceeding to the next step. If the conductive heat load is substantial, consider additional refinements, such as low-E coatings and low-conductivity fill gases.

• *Decide which types of glass and/or which coatings to use to reduce solar heat gain.* The number of possible combinations increases geometrically with the number of panes, but one combination usually stands out as the best. This decision has a major effect on the appearance of the building.

• *If you select openable windows, select a type of frame and weatherstripping that minimizes air leakage.* The configuration you select should keep out rain and minimize wind problems inside the space.

• *Consider whether to try to exploit daylighting or passive heating.* Either one commits you to a lot more brain work. Also, both usually require specialized equipment and/or building modifications in addition to the glazing replacement. Read Reference Note 46, Daylighting Design, and Reference Note 47, Passive Solar Heating Design, to help make this decision.

• *Consider the effect on the appearance of the building and the quality of the occupants' view.* If there are problems with either, repeat the selection process to achieve an acceptable compromise.

Unfortunately, you may not be able to resolve all these factors by dealing with a single vendor. As the glazing industry is presently structured, two different parties are involved in making a finished window. The glass manufacturer controls the basic physical properties of the glass, such as its thickness and whether it is heat treated. The glass manufacturer also determines the optical properties of the bulk glass, which have a major effect on energy efficiency. The manufacturer adjusts the optical properties by adding dyes to the glass and by applying reflective coatings to the glass surface.

The glass manufacturer sells glass sheets to window fabricators, who may create a wide variety of glazing characteristics by combining different types of glass.

The window fabricator makes the frame, which may have important efficiency features. The window manufacturer may increase efficiency by using insulating fill gas between the panes, suspending plastic films between the glass panes, and installing integral shading devices.

To get the best combination of window characteristics, become familiar with the offerings of both glass manufacturers and window fabricators. Not all window fabricators offer all the available techniques, and new features continue to emerge. Shop around to find the manufacturers who are leaders in technology.

Another problem is that the architect may be oblivious to efficiency, and the person responsible for energy efficiency may lack esthetic sense. You may have to become a diplomat to reconcile the appearance and efficiency issues.

As with all efficiency improvements that involve sunlight, customize your glazing selection for each orientation of the building. This is a major departure from conventional architectural practice, and you may have trouble with designers and vendors in doing it. Everyone is accustomed to using glass with a single set of characteristics on all surfaces of a building.

Now, let's look at the main design and selection issues in greater detail.

How to Select and Design Glazing to Reduce Conductive Heat Loss

There are four main methods that contemporary glazing can use to reduce conductive heat loss:

- increase the number of surfaces
- improve the insulating properties of the frame
- use "low-emissivity" coatings on the surfaces
- use insulating fill gases.

Let's look at your options in each of these areas, along with some other selection factors.

■ Number of Surfaces

Glass itself has only minimal insulating ability. Most of the resistance to heat flow comes from a thin, stagnant layer of air that attaches itself to any surface. If the air layer is well protected, as between two panes of double-glazed window, it has an R-value of about 0.5. The exposed air layer on the outside of the building has a much lower R-value because it is dissipated by wind. The inside air layer typically has a higher R-value than the outside layer, but it may be dissipated by air discharged from conditioning equipment. The *ASHRAE Handbook* lists R-values of 0.9 for single glazing, 2.0 for double glazing, and 3.0 for triple glazing. These figures are based on a wind speed of 15 MPH and other standardized conditions.

In principle, you could improve the thermal resistance of glazing by adding many more surfaces. In practice, three or four panes is presently the upper limit. Weight and cost increase with the number of panes, and clarity suffers.

In an attempt to reduce the weight and cost penalties of adding panes of glass, some manufacturers offer double-pane windows in which films of transparent plastic are suspended between the panes. These films can be coated in a manner similar to glass. However, using a plastic film inside glass does not seem to be a prudent approach, because the life of the film is only a fraction of the life of the rest of the window.

Under present conditions, these are typically appropriate numbers of panes:

- one, in climates where heating is minimal and air conditioning is not used
- two, in temperate climates
- three, in colder climates, such as the northern and mountain areas of the United States, and where energy costs are especially high
- four, in extremely cold climates, such as Alaska, Scandinavia, and northern Russia.

These recommendations may change if energy prices rise considerably, if multiple glazing becomes less expensive, if glazing with better thermal resistance becomes available, if movable insulation is used with the windows, etc.

■ Insulating Properties of Frames

Aluminum alloy is usually the best material for window and skylight frames. Don't bother with anything else unless you have a specialized application. Aluminum has much longer life than wood or plastic, and it does not change shape. It can be extruded into an unlimited number of shapes to ease installation, to accommodate effective thermal barriers, and to provide good sealing of movable panes. It can be anodized to hold any color, or to hold a smooth surface finish.

The main disadvantage of aluminum is high thermal conductivity. To prevent serious heat loss through the frame, window manufacturers include thermal barriers in the frame extrusions. Figure 1 shows how thermal barriers, or thermal breaks, are built into frames.

In effect, a frame with a thermal barrier consists of two frames, an outer unit and an inner unit. The two units are joined together by a plastic material that locks the two frame sections together. The plastic material has relatively low thermal conductivity, so it blocks heat flow between the inside and outside frame sections.

The plastic material is not a very good insulator, but it is much better than aluminum. The thermal barrier is fairly effective because it provides only a small cross section area for heat flow. All aluminum frames sweat somewhat during cold weather, especially the lower portion, unless the climate is dry. Deal with this by using trim materials around the window that are not harmed by occasional wetting.

Pay attention to the frame design. There are major differences in quality. Be sure that the thermal barrier does not weaken the frame excessively. If the thermal barrier is poorly designed, the frame could split apart eventually. A good frame design avoids this possibility completely by clever extrusion design, and by using a thermal barrier material that is reinforced with fibers of glass or other inorganic material.

You will not have trouble finding good frame design among the major window manufacturers. Unfortunately, skylights are a different story. Most large skylights are designed on a custom basis. You will have to work with the fabricator to develop a frame design that provides an adequate thermal barrier, as well as resisting leaking, allowing thermal expansion, etc.

Larger, undivided windows have less frame heat loss in relation to their size than smaller windows. This is because of the laws of geometry, not technology.

■ Low-Emissivity ("Low-E") Coatings

A lesser amount of heat loss from glazing occurs by radiation. (Most glazing heat loss occurs by conduction.) All objects radiate electromagnetic radiation. The intensity and the wavelength of the radiation depends on the temperature of the material. Objects that are near room temperature radiate in the portion of the electromagnetic spectrum called the "far infrared." (These wavelengths are much longer than the wavelengths of infrared light from the sun, which is called the "near infrared," i.e., near the visible spectrum).

The environment outside the building radiates heat toward the building in the same way as the interior of the building radiates heat outward. The net heat loss is proportional to the temperature difference between the inside and outside, just as with conductive heat loss.

Ordinary glass absorbs almost all far-infrared radiation that falls on it. The absorbed radiation warms the glass, increasing radiation by the glass itself. If the outside is colder than the inside, the net energy flow is outward. For this reason, ordinary window glass is only partially effective in blocking radiation heat loss, although multiple glazing is much more effective than single glazing.

Low-emissivity glass ("low-E" glass, for short) blocks radiation heat loss in two ways, depending on which surface has the low-E coating. If the coating is applied to an exterior surface, it inhibits the radiation of heat from the window toward the outside. This causes the glazing to become warmer and radiate heat back into the space. (But, the coating does nothing to prevent loss of heat to the outside by conduction.)

If a low-E coating is applied to an inside glazing surface, it reflects the internally generated heat radiation back into the building. This is because a low-emissivity surface inherently has high reflectance. (Reflectance is high because the sum of the emittance, reflectance, and

absorptance of a surface must equal one. If the emittance and absorptance are low, the reflectance must be high.)

Low-E coatings are metal films that are microscopically thin. They operate on the principle that smooth conducting materials have low emittance. For example, the emissivity of a polished aluminum surface is about 0.08. By comparison, the emittance of ordinary glass is 0.84.

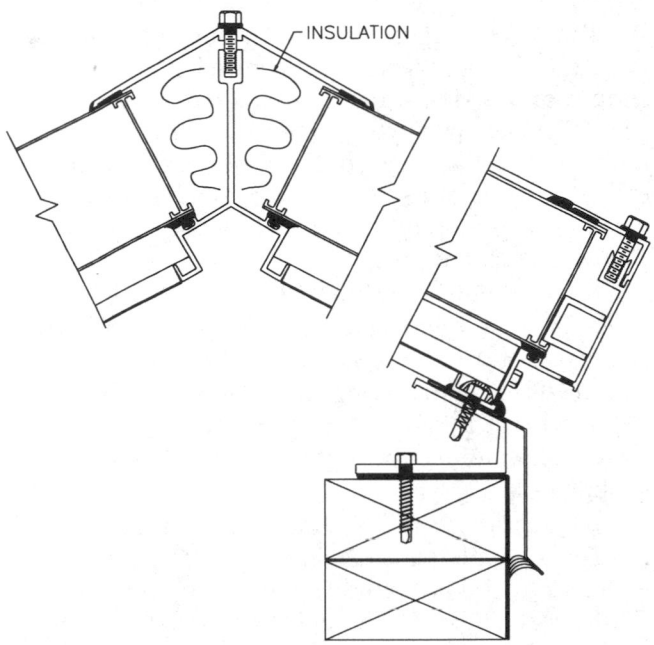

WITHOUT THERMAL BARRIERS

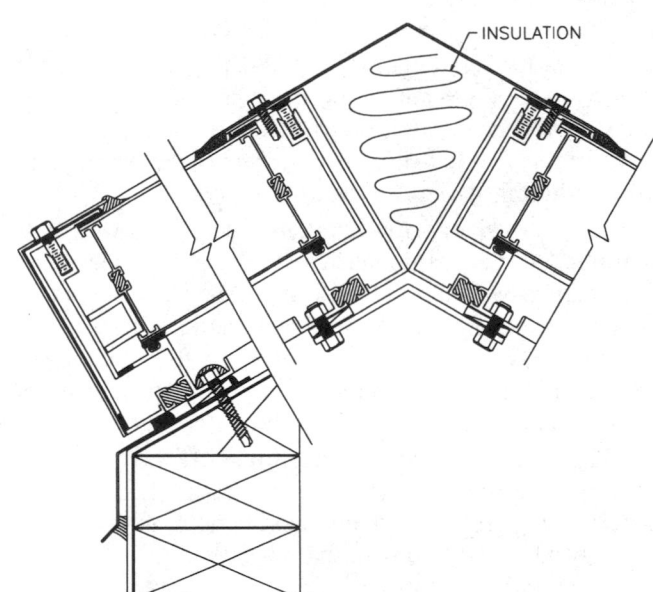

WITH THERMAL BARRIERS

Major Industries, Inc.

Fig. 1 Thermal breaks in glazing frames The difference between these two skylights is the thermal breaks installed in the lower frame. The thermal breaks are strips of reinforced plastic that separate the inner parts of the aluminum frame from the outer parts.

Two categories of low-E coatings are used on glass. One category includes coatings that are deposited on the surface of the solid glass by a variety of physical, chemical, and electrical processes. For example, "sputtered" coatings are metals evaporated on to glazing in a vacuum. These surface coatings are very smooth, which allows them to have minimum emissivity, in the range of 0.10. On the negative side, surface coatings are exposed to both abrasion and corrosion.

The other category of low-E coating is "pyrolytic." A chemical process implants the metal atoms under the surface of the glass while it is in a molten state. This process makes pyrolytic coatings much more durable than surface coatings, but it also makes them less efficient. The emissivities of commercial pyrolytic coatings are presently around 0.40, although emissivities as low as 0.20 are predicted.

Various metals are used for low-E coatings. For example, aluminum and silver are used for surface coatings, and tin oxide is used for pyrolytic coatings. One of the manufacturing tricks of low-E films is to deposit them in such a way that they reflect long-wavelength infrared, but not the shorter wavelengths of sunlight.

In practice, it makes little sense to use a low-E coating on single-pane glazing. It is cheaper and more efficient to install double-pane windows instead. Furthermore, the coating would be exposed, limiting the choice to the less fragile and less efficient pyrolytic type. Using two or more panes of glass solves these problems. The low-E coating is applied either to the outer side of the inner pane or the inner side of the outer pane.

Consider what happens when the low-E film is applied to the outer surface of the inner pane. The inner pane absorbs heat radiation from inside the building and is warmed. The low-E coating prevents the warm pane from radiating its heat outward.

Next, consider what happens when the low-E film is applied to the inside surface of the outer pane. The inner pane is warmed by absorbing radiation from inside the building. The warm inner pane radiates heat to the outer pane, but the low-E coating on the outer pane reflects that heat back to the inner pane, which absorbs it. Both approaches yield about the same heat retention.

This table shows the nominal effect of low-E coatings on the thermal resistance of glazing with different numbers of panes, assuming a temperature differential of 70°F across the window:

Coating	Approximate R-Value	
	Double Glazing	Triple Glazing
no coating	2.0	3.0
E = 0.40	2.5	4.0
E = 0.10	3.0	5.0

The double glazing uses a single low-E coating, which provides about the same performance if installed on either of the two interior surfaces. The triple glazing uses two low-E coatings, on the outward surfaces of the two inner panes. The figures include an allowance for edge losses.

These figures overstate the advantage of low-E coatings in most applications. They are based on ASHRAE data that assume a temperature differential of 70°F across the window. But, the average temperature differential is much smaller in most climates. (Other methods of improving window efficiency are not influenced strongly by the temperature differential.)

Solar reflective coatings that are used to reduce solar heat gain also act as low-E coatings, but not vice versa. Solar coatings reflect at all wavelengths, whereas low-E coatings reflect only at far-infrared wavelengths. Therefore, solar coatings can do double duty, reducing solar heat gain during warm weather as well as retaining building heat during cold weather. On the other hand, solar reflective coatings inhibit passive solar heating, whereas low-E coatings do not. Solar reflective coatings are discussed below.

■ Insulating Fill Gases

Many gases have better insulating properties than air. Some of these are used to fill the space between panes in high-efficiency windows. Gases used presently include carbon dioxide, argon, and krypton, all of which are relatively cheap because they are taken from the atmosphere. According to ASHRAE data, all three of these gases increase the R-value of clear double or triple glazing by about 7 percent. Chlorofluorocarbon gases offer better insulating properties, but these appear to be removed from consideration by concern about their effect on the ozone layer.

The benefit of insulating gases diminishes with age if the gas leak outs and is replaced by air. You might think that the gases would remain in the window because the pressure inside the window is the same as the pressure outside. However, the "partial pressure" of the insulating gas tries to drive it out of the window. (In a mixture of gases, each type of gas exerts its own pressure independently of the other gases. For example, the pressure of the atmosphere is the sum of the partial pressures of the oxygen and the nitrogen that make up air.)

The pure insulating gas has a partial pressure of one atmosphere inside the window, to balance atmospheric pressure. There is virtually no insulating gas in the atmosphere, so its partial pressure is almost zero outside the window. The difference in partial pressure tends to force the gas out of the window. Therefore, exploiting insulating gases requires reliable sealing, which we discuss next.

■ Seal Reliability

The Achilles heel of multiple-pane glazing has been the poor reliability of the seals between panes. If the seals leak, moisture enters the window and fogs it, leaving a permanent unsightly residue. If insulating gases are used in the window, failure of the seal allows the gases to escape. The best insurance against seal failure is to investigate the performance record of the types available, and to select a reliable window fabricator.

Ultimately, the best seal is provided by eliminating sealing compounds and fusing the panes of glass at the edges. This approach has been tried over the years, but glass breakage has been a problem. The search for better sealing methods continues.

One way of avoiding seal problems is to not use them. To avoid condensation and frosting in an unsealed window, the space between the panes must vent to the outside, i.e., the inner pane must seal more tightly than the outer pane. For this reason, old fashioned "storm windows" have small vent holes at the bottom of their frames.

Vented windows are less efficient than sealed units. Their primary market is retrofit applications. Adding a single pane of glass over existing windows may be much less expensive than replacing the existing windows with sealed units. This approach is antiquated, but it still merits consideration in rustic applications, such as country lodges, and in industrial applications. See Measure 7.3.2 for this approach.

■ Gap Between Panes

Research indicates that the optimum separation between panes is about one half inch (between one and two centimeters). Thermal resistance declines rapidly with smaller gaps, but only slowly with larger gaps. In case of doubt, err on the side of a larger gap.

How to Select and Design Glazing to Reduce Summer Heat Gain

Heat gain through windows occurs in three significant ways: direct solar radiation through the glass, conductive heat gain from the outside air, and long-wavelength infrared radiation from the surroundings of the building. If glazing is exposed to direct sunlight, the direct radiation heat gain is much greater than conductive heat gain. Solar gain through directly illuminated glass may average as much as 100 BTU per hour per square foot during daylight hours. Glazing that is illuminated by indirect light from they sky may average about 20 BTU per hour per square foot.

In contrast, conductive heat gain through double-pane glass may average about 10 BTU per hour per square foot in a hot climate. The heat transfer process is the same as for winter heat loss. If the window is designed to minimize conductive heat loss in winter, it also minimizes conductive heat gain in summer.

Long-wavelength infrared radiation from the outside is absorbed in the outer glass and becomes part of the conductive heat load. A low-E coating installed on the inner surface of the glazing keeps the warmed glazing from radiating heat into the space.

■ Retain Sufficient Visible Light Transmission

In order to keep the view through windows from appearing too dark, the glazing should transmit at least 20% of the visible light, and 30% is probably a better figure. Glass that is more opaque may be acceptable if the glazing does not provide a view of anything but sky, and if it is installed above normal sight lines.

■ Minimize Infrared Transmission

To minimize cooling load, the glazing should transmit as little of the solar infrared radiation as possible. This radiation brings in heat without providing illumination. You can now buy glass that absorbs a much larger percentage of the sun's infrared radiation than its visible light. The most effective glass of this type has a visible transmission greater than 60% and an infrared transmission less than 30%. This glass uses dyes to achieve its effect. The dyes absorb visible light more heavily at the red end of the spectrum, so the glass is green in color. This glass can be combined with low-E coating to minimize heat radiation from the absorptive glass to the interior of the building.

Because this glass blocks radiation by absorbing it, the energy of the radiation heats the glass. Therefore, the glass requires extra features to keep this heat from entering the space. At the present time, glass is not available that selectively blocks infrared radiation by reflection. This may change in the future.

■ Favor Reflection over Absorption, if Practical

Reflection is more effective than absorption in reducing solar input. This is because absorption increases the temperature of the glazing, causing the glazing itself to radiate heat into the building. The temperature increase is not trivial. For example, if a window has an absorption of 70%, it may absorb 150 BTU per hour on a bright day. This heat tends to be radiated toward the building interior, because the interior is cooler.

The high surface temperature of absorptive glass can create a serious comfort problem. The presence of a large warm surface on one side of a cooled space creates a strong heat flow gradient, like sitting around a camp fire on a cold night. The only comfort remedy available to occupants is to close the blinds to block the heat radiation. However, this eliminates the view and still allows the heat to enter the space by convection.

The heating of absorptive glass is strongly affected by wind. On a breezy day, wind may keep darkly tinted glazing near the outside air temperature. On a windless day, absorptive glass may become hot to the touch.

The inward radiation of heat from absorptive glass could be reduced by using a low-E film on the inner surface. However, this would increase the temperature of the glass, inviting glass breakage. For this reason, manufacturers advise against using low-E coatings with highly absorptive glass.

Reflective glass minimizes these problems. If you use reflective coatings without tints, the absorption of sunlight in the glass may be less than 20%. The actual absorption depends on the type of metal film used for the reflective coating. For example, aluminum and silver coatings have low absorption, while metals used to impart color to the glass have higher absorption.

There are two primary types of reflective coatings for glass, surface coatings and pyrolytic coatings, which are embedded under the surface. These two types are similar to the coatings used to create low-E glass, except that they are thicker. Pyrolytic coatings are tough enough to be used on the exposed outer surface of glazing, which minimizes absorption of light in the glass. Surface coatings are so fragile that they must be installed on an enclosed surface, which is the inside surface of the outer pane.

Reflective glazing has disadvantages, mostly related to its mirror-like appearance. The most acceptable compromise may be a combination of reflective and absorptive properties. We discuss this below, where we deal with the esthetic aspects of window design.

Note that this need to compromise between absorption and reflection arises *because designers use too much glazing area in the first place*. The most efficient solution is to limit the glazing area so that *clear* glazing admits no more sunlight than is appropriate, from the standpoints of cooling load, daylighting, view, etc. Even in existing buildings, consider reducing the glazing area. See Measure 8.1.5 for more about this.

■ Be Wary of Integral Shading Devices

Some window manufacturers offer prefabricated units that include movable louvers between the panes of glass. In principle, this arrangement is more efficient than installing a shading device inside the space. However, the internal mechanisms of the window will fail long before the rest of the window fails, leaving either an inoperative device or a large repair bill. For this reason, keep shading devices independent of the glazing.

How to Select Openable Windows to Minimize Air Leakage

In many locations, making windows openable to allow natural ventilation is an important efficiency feature during warmer weather. However, openable windows allow a certain amount of air leakage during times when ventilation is not desired. Well designed openable windows are able to limit leakage to a small quantity throughout the life of the building. Openable windows of poor quality develop serious leakage that cannot be corrected without major repairs.

Consider the effect of these factors on leakage when you select openable windows:

- *method of opening.* In the past, the worst leakage occurred with hinged windows, including casement (hinged on the side), hopper (hinged on the bottom), and awning (hinged on top) types, and with double-hung windows. Nowadays, hinged windows may be the most resistant to air leakage, because effective hinges and seals have been developed for them. Sliding windows have reasonably good sealing if they are well made. Double-hung windows continue to be vulnerable to leakage. (Vertical sliding windows, in which one element is fixed, have about the same leakage as horizontal sliding windows.)

  Louver windows have extremely high air leakage. Do not use them, except as wind barriers for unconditioned spaces, such as porches.

  Any window in which more than one element is moved by a single actuator cannot remain tight. This is another reason why louver windows are so bad.

- *sturdiness of construction.* Sturdiness equals longevity. Consider the sturdiness of frames, actuators, and locks individually.

- *weatherstripping type.* Examine the method of sealing. All parts of the weatherstripping should have positive contact in the closed position. The sealing system should be able to accommodate any looseness in the hinges and locks. The motion of opening and closing should not harm the weatherstripping. The weatherstripping should not be vulnerable to damage while the window is open.

- *ease of replacing weatherstripping.* All weatherstripping wears out, so select the weatherstripping system for ease of replacement. If the weatherstripping is specialized, bear in mind that you and your successors will need a source for replacement parts throughout the life of the building.

How to Design and Select Glazing for Passive Heating

Reducing light transmission blocks heat gain in winter as well as in summer. Therefore, the summer saving in cooling energy is partially cancelled by increased winter heating energy. The seriousness of the compromise between cooling load and heating load depends on the climate, the latitude, the operating schedule of the building, the amount of internal heat gain during daylight hours, and the relative costs of cooling and heating. Glazing is expensive, so make a detailed calculation to be sure of the net saving. Unless the situation is particularly simple, use an energy analysis capable computer program to make the calculation.

If passive heating capability is important, try to avoid treated glazing. Instead, use clear glazing and install separate external or internal shading devices that can be adjusted to meet shading and heating requirements. See Measure 8.1.1 and Subsection 8.4 about shading and passive solar heating, respectively.

Low-E coatings, which are used to reduce heat loss, are compatible with passive heating because they do not substantially reduce solar heat gain. In glazing with multiple panes, the location of the low-E coating is a compromise. Expert opinion says that low-E coatings provide the most heat gain if they are located on the outer surface of the inner pane.

In the future, glazing with variable characteristics may be able to reconcile passive heating with control of cooling load, but it does not appear likely that such glazing will become practical within the near future.

Daylighting Needs More than Special Glazing

Treated glazing can be used in combination with daylighting methods, which are covered in Subsection 8.3. However, reducing the transmission of glazing reduces the amount of sunlight available for daylighting.

Treated glazing alone cannot reduce the intensity of direct sunlight enough to make it acceptable for lighting. The light level outside on a clear day is about 6,000 footcandles, in contrast with a typical office space light level of 60 footcandles. Glazing would have to be limited to a transmission of 1 or 3 percent to allow reading by direct sunlight. Glass with such a low transmission would make the view of the outside appear almost black.

To combine daylighting with reduction of cooling load, select glazing that selectively blocks transmission of the infrared portion of sunlight. This feature was discussed previously.

Minimize Air leakage Around Frames

Where prefabricated window units are installed in wall openings, air leakage between the window unit and the wall can be a source of energy waste, discomfort, and moisture damage. As a building contractor once said, the only way to put a window into a hole in the wall is to make the hole bigger than the window. Typically, wall openings are about an inch wider and an inch taller than the window. Installers typically make no attempt to fill this gap, but depend on the interior finish to hide it. The gap allows significant heat loss around the window frame. Even worse, the gap may serve as an entry point for outside air to leak throughout the adjacent wall.

The gap is too large for caulking. If the installer does anything to fill the gap, it usually consists of stuffing some fibrous insulation into the gap. This is better than nothing, but fiber insulation does not block air leakage effectively. You need to specify a method of closing the gap that is tailored to the wall construction. If necessary, put detailed drawings in the glazing contract. Don't expect the glazing contractor to figure this out at the site.

The installation procedure should connect the vapor barrier in the wall to the window frame, so the gap cannot become a path for moisture in the building air to enter the wall structure. (For an explanation of vapor barriers, see Reference Note 42.)

How to Avoid Thermal Breakage

If you increase the amount of heat that glazing absorbs, this increases the risk of breakage from uneven thermal expansion. The problem is that the portion of the glazing that is exposed to sunlight expands, while the shaded edges of the glass do not expand. This creates tensile (pulling) stress in the edges of the panes. Glass does not resist this well. Because of the laws of geometry, edge stress increases with the size of the pane.

Thermal stress is increased by massive framing systems, such as concrete, that tend to remain cool. Partial shading of the glass, for example, by overhangs and large mullions, also increases thermal stress. However, partial shading is less of a problem, because it does not cause differential expansion between the center and the edges of the glass.

You can reduce the risk of thermal breakage by using reflective glass rather than absorptive glass. But, be careful about combining reflective and absorptive properties. For example, using a reflective film on the inside of colored glass increases the temperature of the glass, because heat is absorbed both before and after reflection. Low-E films may increase glass temperature by keeping the glass from cooling by radiation.

Heat treatment, which is done when the glass is made, reduces the risk of thermal breakage, but this adds to the cost.

Other causes of glass breakage are the weight of the glass itself, wind forces, and vibration (earthquake, sonic booms, etc.). Control these hazards primarily by selecting glass thickness, heat treatment, and frame design. These are all major cost factors. Competent designers and glazing contractors are aware of these issues, but they may fail to provide an adequate margin of safety.

Select Components for Longevity

An inherent advantage of glass is that it lasts as long as the building. It is folly to surrender this advantage by installing windows or skylights that have built-in components of limited life. Using plastic window components is poor economy. Plastic frames, internal plastic films, and internal shading devices that require cords and actuators made of synthetic materials are eventually destroyed by sunlight and high temperatures. Even plastic that is hidden from direct sunlight, such as thermal barrier strips, should not be used as a structural element, unless it is reinforced by strong inorganic materials, such as glass fiber. Gears and linkages made of aluminum and other soft metals also wear out prematurely.

Windows still rely on organic sealing and bedding materials. Recognize that these materials will eventually fail, and select the window design for easy repair. Keep an eye out for new windows coming on the market that seal by fusing the panes at the edges. This method avoids the need for organic sealants, but so far, it has been plagued by glass breakage.

All reflective and low-E coatings are vulnerable to damage, although some are much tougher than others. Coatings applied directly to the surface of the glass are fragile. They tolerate no mechanical abrasion, and even degrade from oxidation. Use them only on the interior surfaces of multiple-pane glazing units. The window assembly needs reliable seals to protect these films from oxidation and moisture corrosion.

Pyrolytic coatings are much tougher than surface coatings. Even so, these coatings are vulnerable to serious abrasion, for example, from contact with the metal parts of cleaning tools, scaffolds, etc. Also, pyrolytic coating gives glass a microscopically rough surface. As a result, materials that rub against the surface may create smudges that are difficult to remove. Oily substances create a visible change in surface appearance. Manufacturers who offer pyrolytic coatings claim that they can be used on exposed surfaces, but it seems prudent to restrict them to interior surfaces, as long as this does not reduce efficiency significantly or cause other problems.

Design for Good Appearance and View

Glazing is a dominant factor in the appearance of the building and the quality of the occupants' view. These factors demand careful consideration when selecting glazing. On the other hand, be careful not to let appearance considerations run away with the project. Architects tend to use glazing to make stylistic statements that are meaningful only to themselves, and they may be deficient in optical theory and practical experience. When you design glazing, make sure that technical and practical concerns (including appearance) take precedence over ethereal ones.

■ Color

Color can be created in a number of ways. One is by adding dyes to the glass that selectively absorb portions of the visible spectrum. The color imparted to the outside of the building is approximately the same as the color of light that is transmitted into the building. You have wide latitude in selecting color. However, dyes are absorptive, which increases heat gain.

A more efficient method of producing color is by using different metals for reflective coatings. For example, copper and bronze produce a reddish color. Reflective coatings with a noticeable color are somewhat absorptive, but less so than dyes that permeate the glass. The range of colors available is limited by the availability of suitable metals for the reflective films.

Another technique is to use coatings that reflect the visible spectrum selectively, transmitting most of the rest. With this method, the reflected color seen on the outside is the complement of the transmitted color seen on the inside. For example, glass that selectively reflects the red end of the visible spectrum makes the building look reddish from the outside, but the view from inside the building has a bluish appearance. Such coatings are still evolving.

■ Reflective Appearance

Glazing with even a modest amount of reflectance has a mirror-like appearance from the outside. Some architects use reflectance deliberately to create a specific appearance, but reflective buildings have become trite as an architectural "statement."

Reflective glazing can have adverse effects on those outside the building. People facing the building may find themselves staring at the sun. The reflected heat may increase the cooling load of adjacent buildings. As a result of these annoyances, some local ordinances have prohibited reflective glass.

Glass manufacturers are developing coatings that reflect in a more diffuse manner. This approach lessens mirroring of the surroundings, but does not entirely solve the annoyance of reflected sunlight. Instead of reflecting an image of the sun, such glass reflects a brilliant halo.

A compromise between reflective and absorptive properties may produce the best results. A common practice is to put a reflective film on the inside surface of a mildly colored outer pane, which mutes the reflection. However, this combination also increases the temperature of the glass, so the glass may require heat treatment.

■ View

Glazing was used originally to provide light and view. In modern architecture, glazing is used primarily because it is a cheap way of covering the building, and the original purposes have largely been forgotten. However, if there is a good view outside, preserve it for those inside.

Visible light transmission should be at least 20% to 30% to keep scenery from looking dark and the sky gloomy.

Within wide limits, the color of the glazing does not have a significant effect on the view, because the human eye compensates when looking through a colored medium. As a rule, glass companies do not make glass that imparts a strange color to the view, but beware of bizarre architectural uses of glazing color.

■ Uniformity of External Appearance

Current architectural fashion considers it desirable for buildings to have a uniform, featureless external appearance. With clear glazing, people outside the building can see curtains and blinds that are positioned irregularly, and other distressing sights.

Reflective glazing blocks the view of the inside almost completely during the daytime. Reflection of the sky in reflective glazing yields a uniform appearance, although reflection of surrounding buildings and other features gives a somewhat quirky appearance.

Absorptive glass does not shield the inside from view as much as reflective glazing. This is especially true of interior window treatment and other objects close to the windows that may be illuminated by direct sunlight.

At night, both types of window treatment reveal the lighted inside of the building to the outside, but irregularity of window blinds and other features is less conspicuous.

■ Privacy

For the reasons given above, reflective and absorptive glazing provide privacy during the daytime, but not at night.

Extra Considerations for Skylights

Refer to Measure 8.3.3 for special issues in the design of skylights.

United States Window Ratings

The National Fenestration Rating Council (NFRC), an organization funded by the United States Department of Energy, began to issue ratings of prefabricated windows in 1993. This work is evolving. By the time you read this, it may be worthwhile to refer to NFRC data when selecting windows. Some governmental units within the United States have already begun to use NFRC ratings as a basis for mandatory window standards in construction.

ECONOMICS

SAVINGS POTENTIAL: *40 to 80 percent of the solar cooling load related to glazing. 20 to 60 percent, if the building has interior curtains or shades.*

30 to 75 percent of the heating load related to glazing, depending mainly on the thermal resistance of the original glazing.

The net saving may be reduced by the loss of passive heating during cold weather. The importance of this varies widely. In facilities where effective daylighting is used, lighting costs may be increased, but this is likely to be a minor factor.

COST: *High-efficiency windows cost several hundred dollars apiece in typical sizes, in quantity. Costs in retrofit applications are typically 30% to 50% higher than in new construction.*

PAYBACK PERIOD: *In new construction, using high-efficiency glazing typically pays off within a few years. In retrofit, the payback period typically exceeds 10 years.*

TRAPS & TRICKS

SPECIFYING THE EQUIPMENT: *To select glazing systems properly, do your preparation in steps. First, study the glass characteristics that are offered by the glass manufacturers, including strength, color, types of coatings, etc. Second, decide how to combine the panes of glass to achieve the efficiency and the visual characteristics you want while protecting the coatings. Third, study the frame and support systems offered by window fabricators. Select a support system that minimizes the chance of breakage and fogging, and that minimizes leakage of insulating gases, if these are used. Effective seals are still evolving, so check on the latest developments. Make sure that metal window frames have a thermal barrier that is not vulnerable to long-term breakage. Make sure that the frame system is designed to make it easy to prevent air leakage around the window at the time of installation. Before signing the contract, check with sites where your proposed window system has been installed to see how well the installation worked out.*

INSTALLATION: *Window installers do not want to bother with sealing against air leakage when the windows are being installed, but that is the time to do it. Talk to the installer about this before the work commences. Have an enforcer at the site to make sure that sealing is done properly.*

MEASURE 8.1.4 Install solar control films on existing glazing.

A common method of reducing solar heat gain in existing buildings is to apply a film to the inside surface of the glazing that reflects and/or absorbs sunlight. These materials are commonly called "window film," but they can be used on any glazing, including skylights, greenhouses, etc. We will use the term "solar control films." This name does not tell the full story, because glazing films are used to reduce heat loss, to reduce breakage, and for other purposes. However, from the standpoint of energy efficiency, reducing solar heat gain is by far their most potent capability.

Solar control films reduce solar gain by about 35% to 65%. They are offered in a wide range of reflectances, absorptions, and colors. By the same token, you need to select films carefully to achieve a good balance of efficiency, external appearance, and occupant view, and to avoid potential problems.

Installing solar control film is strictly a retrofit way of getting some of the desirable characteristics of high-efficiency glazing. Retrofit films are a less expensive and less effective alternative to replacing the glazing itself, the approach which is recommended by Measure 8.1.3. Before going further with this Measure, review Measure 8.1.3 for an explanation of these characteristics. You need to understand them to select films intelligently.

The economics of solar control films are not very favorable, primarily because the films have limited service life. Therefore, films are profitable primarily when they are installed on glazing that experience a large amount of heat gain.

Application Limitations

Unfortunately, solar control films face obstacles to providing a good payback, and they may cause serious problems. Be aware of these limitations at the outset, as they may disqualify films from your applications.

■ Limited Service Life

The major functional limitation of retrofit films is their short service life. Warranties range from five to ten years, depending on the manufacturing process used and the seller's audacity. These warranties probably are not conservative, but may instead reflect a desire to make the life of the film appear as long as possible. The industry itself is reluctant to make definitive statements about film life. This undoubtedly reflects wide variations in film quality and operating conditions. The ultimate limitation is the nature of organic materials used for the films and the adhesives. All organic materials ultimately are destroyed by ultraviolet light and heat.

SUMMARY

A powerful retrofit method of reducing cooling load that is undermined by short life and other drawbacks.

SELECTION SCORECARD

Savings Potential	$ $ $ $
Rate of Return	% %
Reliability	✓ ✓
Ease of Retrofit	☺ ☺ ☺

■ Appearance Defects

Serious appearance defects may exist in the bulk material, or defective appearance may be caused by imperfect installation. Solar control films are made by a number of manufacturers. Some may not yet have mastered the ability to make films without appearance defects, such as stripes or blotches. These defects typically are not apparent until the film has been installed. The defense against this problem is purchasing the film from an experienced manufacturer who provides a reliable material warranty.

Another problem is peeling of the film, which is a failure of the adhesive. This may be the fault of the manufacturer, or may be due to poor installation. Any film will eventually peel. Once the film starts to peel, it must all be removed to keep the building from looking shabby. Selecting the best adhesive method and a good installer will delay peeling as long as possible.

Blotching of the film is usually caused by localized failure of the adhesive. This can occur early in the life of the film if the glass surface is not cleaned properly during installation.

Hazing of the film occurs over time as a result of cumulative scratching from cleaning and wiping. Hazing can be reduced by installing film with a scratch-resistant coating.

■ Are Shades or Curtains Installed?

For reasons explained in Subsection 8.3, people tend to keep any curtains or blinds closed if they can admit direct sunlight. Although this eliminates the view, it reduces heat gain considerably. In this case, installing solar control film may not offer enough additional benefit to pay for itself within its service life.

Internal shading devices that are absorptive, such as dark colored venetian blinds, are not effective for reducing heat gain. (See Measure 8.1.2 for details.) If the interior decorator insists on dark shades or curtains, solar control films may still be cost effective.

■ Risk of Thermal Breakage

All solar control films absorb a significant amount of heat. The heat absorbed by the film warms the glass, causing it to expand, while the shaded edges of the glass remain cooler. This creates tensile stress in the edges of the panes, which glass does not resist well. Because of the laws of geometry, this tensile stress increases with the size of the pane.

Thermal stress is increased by massive framing systems, such as concrete, that tend to remain cool. Partial shading of the glass, for example, by overhangs and large mullions, increases thermal stress. The risk of thermal breakage is reduced by using reflective films instead of absorptive films. But, bear in mind that all types of films have a significant amount of heat absorption (except for clear films used for security purposes).

Before proceeding, inspect the glazing for conditions that are likely to cause thermal breakage. A trade association of window film manufacturers suggests "extreme caution" in using absorptive or reflective films with:

• single pane glass larger than 100 square feet
• double pane glass larger than 40 square feet
• any triple pane glass
• clear glass thicker than 3/8 inch
• tinted glass thicker than 1/4 inch
• reflective glass
• laminated, wired, or textured glass
• frame systems that hold the glass tightly
• frame systems of concrete or solid metal.

Heat treating reduces the risk of thermal breakage. In case of doubt, examine the building's construction records to see whether heat treated glass is installed.

■ Whether to Install Film on Tinted or Reflective Glass

In principle, films can be used with glass that has already been treated to reduce solar gain. In practice, this is unfavorable for several reasons. From an energy saving standpoint, treated glazing already provides a large part of the potential benefit. Adding any type of film inside absorptive glass further increases the surface temperature of the glass, increasing discomfort and the chance of glass breakage. Film may undo the benefit of any low-E coatings on the original glass. And, further reducing light transmission may make the view dark and gloomy.

Energy Saving and Energy Penalties

■ Solar Heat Gain

When used with clear glazing, solar control films are typically selected to block 60% to 80% of visible sunlight. Films are usually not selected to block a larger percentage of visible sunlight, because this would make the view through the glazing appear too dark.

The reduction in solar heat gain is usually less than the reduction in visible light. This is because films work partly by absorption, so a part of the sunlight that is blocked is converted to heat within the glazing. Retrofit films are installed on the innermost surface of the glazing, so the glazing traps the heat that is absorbed in the film. A portion of this heat enters the space by conduction and radiation.

With single glazing that is cooled by wind, highly reflective film may reduce solar heat gain by as much as 65%. Multiple glazing traps the heat absorbed by the film more seriously. Therefore, the maximum reduction in heat gain is about 50% with multiple glazing. Films that are less reflective and rely more on absorption reduce solar heat gain even less, from about 35% to about 50%. Again, the benefit is reduced with multiple glazing.

Unfortunately, present films are not able to reflect the infrared component of sunlight selectively, while passing the desired amount of visible light. Films may be able to offer this capability in the future.

If reflective interior curtains or blinds were used with the original glazing, the reduction in heat gain may be half of these percentages.

■ Long-Wavelength Infrared Heat Loss

Low-emissivity coatings used with some films can reduce radiation heat loss from the interior of the building by about 40%. However, in films intended primarily to reflect solar heat gain, low-E coatings involve performance conflicts, which are explained below.

■ Passive Heating

All films that reduce solar heat gain also substantially reduce the potential for passive heating. This factor is unimportant in warm climates, but it may cancel much of the benefit in cold climates.

■ Daylighting

In buildings where effective daylighting exists, all types of films reduce the amount of daylight available. However, most contemporary buildings do not have effective daylighting, even if it was intended. See Reference Note 46, Daylighting Design, for the steps needed to make daylighting work.

Non-Energy Benefits

All films keep glass from fragmenting after breakage, and they may increase resistance to breakage. Some manufacturers of films for solar control also make clear films explicitly to provide shatter resistance against vandalism, storm damage, and other causes. Protective films are substantially thicker than solar control films, typically about 0.002 to 0.004 inches. Solar control films enhanced for breakage resistance are available.

Solar control films can also be used to enhance or modernize the appearance of a building. Films can be dyed in almost any color or shade. If films are intended for decorative purposes as well as for efficiency, make sure that the colors are compatible with the desired solar control effect, and that the inside view does not become bizarre. Also, make sure that the film characteristics do not invite breakage of the glass from thermal stress.

How Solar Control Films Work

Solar control films reduce the fraction of sunlight that passes through glazing by reflecting or absorbing the sunlight, or both. All films are similar in construction. The following are their main features.

■ Materials

The active ingredients in solar control films are reflective metal films, dyes, and ultraviolet absorbers. The carrier material of the film is one or more layers of plastic. Polyester is the most common carrier material. A layer of acrylic plastic may be added for scratch resistance. A polypropylene layer may be added to protect low-E coatings.

The metal layer, if one is used, is deposited on the surface of one of the plastic layers. The metal layer is transparent because it is very thin. A variety of metals and alloys are used for the reflective films, including aluminum, copper, bronze, silver, gold, titanium, stainless steel, and others.

Dyes and ultraviolet absorbers are added to the plastic material. Ultraviolet absorbers are added to virtually all films. These protect the film itself and the inside of the building.

■ Optical Properties

The optical properties of solar control films stem from the properties of their components:

- *the plastic carrier materials,* such as polyester and polypropylene. These materials are similar to glass in the visible and solar infrared spectrums, except that plastics absorb somewhat more energy in the solar infrared. Plastics pass much more ultraviolet than glass. They pass a small amount of long-wavelength infrared, which is almost totally absorbed by glass.
- *dyes* added to the plastic material, primarily to absorb sunlight and to give color to the film. Dyes are absorptive, and by increasing the quantity of the dye, any amount of absorption can be achieved. Some dyes used in films absorb visible light and solar infrared about equally, whereas other dyes may absorb infrared more strongly.
- *ultraviolet absorbers* added to the plastic material, are dyes selected specifically to absorb ultraviolet

wavelengths. They may be almost completely transparent to other light, or they may impart color to the film.

- *metal reflecting coatings* mostly have high reflectances in the near and far infrared spectrum. The reflectance of metals typically declines toward the short wavelength end of the visible spectrum. For example, the reflectances of aluminum and silver drop abruptly at the blue end of the visible spectrum, but both have high reflectance across the rest of the visible spectrum, which makes them efficient for rejecting solar radiation. The reflectance of copper and gold starts to fall off gradually near the blue end of the visible spectrum, which accounts for their reddish color. Even the thin metal coatings used in films absorb a large part of the light that they do not reflect.

■ Adhesive Methods

The film is attached to the glass with an adhesive. Deterioration of the adhesive is usually the first cause of film failure. Therefore, consider the adhesive when you select the film. Specific adhesives may be covered by patents, making those types available only from individual manufacturers. Check the performance of the films that are candidates for your application. Your best bet is to call users who have installed the film a long time ago, and ask how well it has behaved.

There are two classes of adhesives for solar control films, water-activated and pressure-sensitive. Water-activated adhesives bond chemically to the glass surface, keeping the film tightly attached to the surface and minimizing optical problems. By the same token, these adhesives make the film difficult to remove at the end of its service life.

Films with pressure-sensitive adhesives are applied by removing a backing sheet from the film and pressing it into place. The pressure-sensitive adhesive is quick to apply because no wetting of the adhesive is required. The disadvantage is that the bond is relatively weak.

■ Low-Emissivity ("Low-E") Coatings

The theory and composition of low-emissivity coatings are covered in Measure 8.1.3. Low-E films can be included on retrofit films, but they suffer efficiency limitations when applied in this manner. The low-E coating is fragile, so it must be protected by a layer of plastic film. None of the available plastic films are highly transparent to long-wavelength infrared, so a significant fraction of the infrared is absorbed in the film.

To reduce this problem, plastics with lower absorption may be used to cover the metal film. For example, polypropylene may be used to cover the metal film instead of polyester. Plastics with lower absorption tend to be soft, so there is a compromise between efficiency and scratch resistance.

At the present time, retrofit films with low-emissivity coatings are covered by patents, so one manufacturer makes most films having this feature.

How much can low-E coatings in glazing films reduce cold weather heat loss? Probably not much, except in very cold climates. Almost any other method of reducing the heat loss through glazing (installing storm windows, curtains, etc.) should yield results that are as good or better. The available performance claims still appear to lack substantiation, and reading between the lines leads one to suspect that performance is unremarkable. Given the short life of films and their other problems, it does not appear that low-E films are the best approach to reducing heat loss in most applications.

Unlike high-efficiency glazing, where low-E coatings are applied between the panes, existing glazing must have films applied to the inside surface. As a result, solar films cannot block heat gain from long-wavelength infrared entering from outside the building.

How to Select Solar Control Films

Solar control films are offered in a wide range of reflectances, absorptions, and colors. Some also offer a limited range of emissivities for long-wavelength heat radiation. In addition to these optical characteristics, films are available with options in adhesive methods, scratch resistant coatings, shatter strength, and perhaps other features.

The hazard in this wealth of choice is making a selection that is less than optimum for the application. Here are the most important selection issues.

■ Manufacturers

At the time this is written, there are fewer than a dozen manufacturers of solar control films. Perhaps two of these could be considered major companies. The performance history of glazing films emphasizes the important of dealing with manufacturers who have the capability of maintaining a high level of quality control, and who are likely to survive to honor any warranty claims. Check references. As a first filter, avoid any vendors who cannot provide serious technical product literature.

■ Visible Light Transmission

In order to keep the view through windows from appearing too dark, at least 20% of visible light should be transmitted, and 30% is probably a better figure. A lower visible transmission may be acceptable if the glazing does not provide a view of anything but sky, and the view is above normal sight lines.

■ Solar Infrared Transmission

The film's transmission in the solar infrared spectrum should be as low as possible, if the objective is reducing solar heat gain. Most films have somewhat lower transmission and higher absorption in the solar infrared band than in the visible band. Films with substantially lower infrared transmission than visible transmission may become available in the future.

■ Reflection vs. Absorption

This issue is essentially the same for retrofit films as for treated glazing. See Measure 8.1.3 about this issue.

■ Color

Color in glazing films is created in two ways. One is by adding dyes to the film that selectively absorb portions of the visible spectrum. The color imparted to the outside of the building is approximately the same as the color of light that is transmitted into the building. You can select from a wide range of colors. However, dyes are absorptive, which increases heat gain.

A more efficient method of producing color is by using different metals for reflective coatings. For example, copper and bronze produce a reddish color. Metals with a noticeable color are somewhat absorptive, but less so than dyes. The range of colors available is limited by the range of suitable metals.

■ Scratch Resistance

Some types of glazing films are much more resistant to scratching than others. A common scratch resistant coating consists of a thin outer layer of acrylic plastic applied over the main polyester film. All major manufacturers offer scratch resistant coatings as an option.

Unfortunately, the scratch resistant surfaces that are presently offered absorb long-wavelength infrared heat radiation. This makes them incompatible with low-E coatings.

■ Appearance and View

These issues are generally the same as for treated glass. See Measure 8.1.3 for the details.

The choice between the reflective and absorptive properties has a major effect on the exterior appearance of the building. Reflective films make the glazing look like a mirror. Absorptive films make the glazing appear colored. You can balance these properties by a combination of reflective and absorptive properties.

How to Install Retrofit Films

Selecting the right installer is as important as selecting the right film manufacturer. Installation does not require genius, but it does require experience and discipline. For this reason, have the film installed by specialists. Also, film warranties may require professional installation.

Labor is the largest part of the cost of solar control films, so the installer is faced with a large temptation to cut corners. The important aspects of installation are protecting the film from damage in handling, cleaning the glazing surface, eliminating bubbles and wrinkles, activating the adhesive uniformly, and trimming the film to fit the frame. If the glazing is wider than the film stock, the installer must know how to splice the film without making the seam too apparent.

Avoid Damage in Cleaning

Polyester film is relatively soft, and therefore develops hazing as a result of repeated cleaning. Scratch-resistant coatings resists hazing, and allow conventional window cleaning methods to be used. Some films have a silicone surface, which eases cleaning.

Relationship to Daylighting

Solar control films substantially reduce the amount of sunlight available for daylighting. However, sunlight is much brighter than is needed for daylighting, so films may be compatible with daylighting in some cases.

Films alone cannot produce acceptable daylighting. The light level outside on a clear day is about 6,000 footcandles, in contrast with a typical office space light level of 60 footcandles. Therefore, transmission would have to be limited to about 3 percent to allow reading by direct sunlight. Glass with such a low transmission would make the view of the outside appear almost black.

Relationship to Passive Heating

Solar control films block heat gain in winter as well as in summer. Therefore, the summer saving in cooling energy is partially lost as a result of increased winter heating cost. The compromise between cooling cost and heating cost depends on the climate, the latitude, the schedule of operation of the building, the amount of internal heat gain during daylight hours, and the relative costs of cooling and heating. If loss of passive heating may be an important issue, make a detailed calculation of the net saving.

If passive heating capability is important, solar control films are self defeating. Instead of using films, install separate external or internal shading devices that can be adjusted to meet shading or heating requirements. See Measure 8.1.1 and Subsection 8.4 about shading and passive solar heating, respectively.

ECONOMICS

SAVINGS POTENTIAL: 35 to 60 percent of the solar cooling load, if there is no other window treatment. 20 to 45 percent of the solar cooling load, if the building has interior curtains or shades.

The net saving is reduced by loss of passive heating during cold weather. The importance of this varies widely. In facilities where daylighting is used, a part of the saving may be offset by higher lighting costs.

COST: $2.50 to $5.00 per square foot, installed. Installation labor is the largest part of this cost.

PAYBACK PERIOD: 5 to 10 years, typically. Payback period is a misleading economic criterion for glazing films because they may not last much longer than the time needed to pay them off. A more effective indicator is the ratio of total lifetime savings to initial cost. With present films, this ratio typically ranges from about 4:1 in the best cases, to about 1.5:1 in the worst cases.

TRAPS & TRICKS

CHOICE OF METHOD: Do you have alternatives that may be better, such as installing sun shades, or planting trees, or eliminating large areas of glazing? Would installing film create a risk of glass breakage? Expect the film to start looking shabby during the second half of its service life. This is a method with serious drawbacks, but it may be your only choice.

SELECTING MATERIALS: All films have limited service life, but some films are better than others. Make sure that the film has a good reputation and has all the features you want.

INSTALLATION: Make sure that the installer follows the manufacturer's installation procedures meticulously. Watch how the installer performs on somebody else's job before you sign the contract. Deal with an installer who is likely to remain in business long enough to honor the warranty. Ask for references, and check them.

CLEANING: You need to exercise vigilance for the life of the film to ensure that cleaning practices do not cause mechanical damage or hazing. One careless cleaning can ruin the film.

MEASURE **8.1.5 Reduce the area of glazing.**

Glazing is responsible for two major energy costs, solar heat gain in summer and conductive heat loss in winter. You can reduce the heat gain and heat loss of glazing by replacing the glazing, or a portion of it, with insulating panels.

In new construction, this Measure is largely a matter of reversing the contemporary practice of using windows as walls. Deciding how much glazing to abolish is a judgement call. You may have to choose a different stylistic concept for the building. In fact, after you review all the Measures in this Section, your entire building envelope concept may change radically.

In existing buildings, reducing window area is a major change. However, it is more likely to be economical than you may expect. The challenge lies in taking the maximum advantage of the modification. Reducing glazing area may be a powerful opportunity to improve the appearance of the building. On the negative side, it may detract from the occupants' view. It also affects the potential for passive heating and daylighting. Reducing glazing area on a retrofit basis requires considerable thought, a sense of style, and attention to detail.

Where to Reduce Glazing Area

It pays to reduce glazing area in a wide variety of buildings. Obviously, the largest savings are achieved in buildings that have large areas of glazing. This Measure conserves both heating and cooling energy, so it is worth considering in most locations.

In the design of new buildings, reducing the glazing area requires rethinking the appearance concept of the building from the standpoint of energy efficiency. All building designs can benefit from such a review.

With existing buildings, retrofitting insulating panels tends to be less expensive if the contrast in appearance between an insulating panel and the adjacent wall and window is acceptable, or if the contrast can be used as an architectural accent. A requirement to match an existing decor can raise cost considerably. Figure 1 shows an example where the glazing area was reduced by a large amount without risk to the decor.

It might seem that glazing area cannot be reduced in "glass box" architecture because of the difficulty in maintaining uniform surface appearance. In fact, insulating panels can be made to match or complement virtually any style of glazing. Much of the "glazing" in many glass box buildings is actually the outer surface of an opaque curtain wall. However, using glass as the outer surface of an insulated panel requires careful

technique to prevent glass breakage and irregular appearance.

Energy Saving Potential

For a given window or skylight, solar heat gain and conductive heat loss are proportional to the glazing area. In buildings with a large amount of glazing, the glazing is responsible for most of the cooling load in warm weather, and most of the conductive heat loss in cold weather. You may be able to eliminate a large fraction of the glazing, thereby reducing your cooling and heating costs substantially.

In colder climates, reducing glazing area may reduce the benefit of passive solar heating, or it may not. This is discussed below.

Reducing the glazing area may reduce the potential for daylighting, or it may not. This is discussed below.

General Approach

Plan to eliminate glazing in a manner that provides an optimum balance of these factors:
- solar and conductive heat gain during warm weather
- conductive heat loss during cold weather
- passive heating during cold weather
- daylighting
- occupants' view
- exterior and interior appearance
- security.

Don't be afraid to reduce glazing area radically, if you can do this without causing problems. Much of the glazing that has been installed in buildings serves no purpose except to allow the architect to pass off the work of envelope design to the glazing fabricator.

SUMMARY

A simple and effective way of reducing both heat gain and heat loss. Applies to a large variety of buildings.

SELECTION SCORECARD

Savings Potential	$	$	$	$
Rate of Return, New Facilities	%	%	%	%
Rate of Return, Retrofit	%	%		
Reliability	✓	✓	✓	✓
Ease of Retrofit	☺	☺	☺	

■ Reduce Solar Heat Gain

The potential for reducing solar heat gain depends on the exposure of the glazing. Estimate the amount of solar heat, per unit area, falling on each face of the building where you may reduce the window area. Take into account any shading of the building exterior.

The *ASHRAE Handbook* and other references have tables that help you do this quickly. Also, review Reference Note 24, Characteristics of Sunlight, to see how the motion of the sun affects heat gain.

■ Reduce Conductive Heat Loss

The reduction of conductive heat loss is proportional to the reduction of glazing area. It does not matter much where the glazing is located in the building, assuming that it faces into a conditioned space.

Use materials with high insulation values to replace glazing. See Reference Note 43, Insulation Selection, for materials and methods. If the building has curtain walls, coordinate this Measure with Measure 7.2.4, which explains how to increase the insulation value of the opaque portions of curtain walls.

■ Optimize the Amount of Passive Solar Heating

Eliminating glazing sacrifices its ability to provide passive solar heating in cold weather. However, reducing

Kalwall Corporation

Fig. 1 Reduction of window area Most of the bottom window has been replaced with translucent insulated panels.

conductive heat loss may compensate for the loss of passive heating. You need a detailed calculation to estimate the net saving accurately.

In cold climates, you will usually achieve a net saving in winter heating if you replace single-pane glazing with well insulated panels. If the original windows have higher insulating value, passive heating may be providing a net benefit. However, if window treatments are installed that block sunlight, passive heating may not offset the conductive heat loss.

In milder climates, passive heating has a stronger effect than conductive heat loss. However, in mild climates, cooling load is likely to be a larger cost factor than heating load.

To exploit passive heating, retain more glazing on the building surfaces that are exposed to the sun for a large fraction of the day. Passive heating is tricky. To exploit it effectively, study Subsection 8.4 and Reference Note 47.

■ Exploit Daylighting

Whenever you do anything with windows, consider the possibility of exploiting daylighting. This takes some skill. See Subsection 8.3 and Reference Note 46.

Reducing glazing area reduces the amount of sunlight available for daylighting. This may not matter much, because the amount of sunlight falling on large windows is much greater than needed for perimeter lighting. Even reduced glazing area may leave plenty of daylighting potential, if the light is distributed effectively.

Keep in mind that daylighting is most effective when coming down from above. For example, in a building with tall windows, consider replacing the top portion of the glazing with light diffusing panels, and replacing the rest of the glazing with insulating panels.

■ Preserve Desirable Views

View is the only function that absolutely requires glazing. Examine the building skeptically to decide whether the existing windows are actually providing view. In a large fraction of buildings with extensive glazing, sunlight creates intolerable glare that causes occupants to block out all sunlight, along with the view. In such cases, decide whether to reduce glazing, to make changes to create acceptable view, or both.

With tall windows, you can eliminate glare by eliminating the upper portion of the windows, leaving the glazing at the bottom to provide view of surrounding landscape. If glare would still be a problem with the lower glazing, consider installing external or internal shading as part of the conversion. See Measures 8.1.1 and 8.1.2 for shading methods.

■ Improve the Interior and Exterior Appearance

This Measure strongly affects the appearance of a building. In many buildings, this is a major economic and esthetic issue. If the change is designed by a person with a good sense of style, it can enhance the appearance and value of the building. Conversely, a thoughtless job can make the building look derelict. Someone with good taste should have approval authority over the design.

You can achieve good insulating properties with any surface appearance. For example, you can use glass, polished metal, stone frieze, stucco, bronze bas relief, canvas awnings, etc. Let your imagination roam free.

■ **Enhance Security**

Unfortunately, the condition of our civilization makes it possible to claim increased security from burglary and vandalism as a benefit of eliminating glazing, especially for glazing near ground level. In locations where this factor is important, make the substitute material strong. The children of modern social policy are not easily denied.

Optimum R-Values and Economics

See Reference Note 45, Insulation Economics. In most climates, the optimum insulation thickness is the largest amount that you can install in the panels without causing appearance problems or increasing fabrication cost.

ECONOMICS

SAVINGS POTENTIAL: *Up to 50% of total cooling costs, and up to 50% of total heating costs. The saving depends on the glazing area, the climate, internal heat gains, building geometry, and other factors.*

COST: *In new construction, reducing the glazing area may increase or decrease the cost of the building by a small percentage, depending on the other options being considered. In retrofit, replacing glazing with insulating panels costs from $10 to $50 per square foot, depending on the size of the job, the difficulty of replacement, and the nature of the replacement panels.*

PAYBACK PERIOD: *In new construction, the payback period can be immediate or very short. In retrofit, the payback period is several years or longer.*

TRAPS & TRICKS

DESIGN: *The simplicity of the concept hides the fact that a lot of brain work is needed to develop an optimized design. This modification is unusual as a retrofit. When used with a sense of style and a sense of humor, it can turn an old building into a showpiece. On the other hand, a thoughtless design may fail to achieve potential savings, disfigure the building, and ruin views.*

Heat gain that enters through the opaque portions of a building's envelope is important. If the building has a large cooling load, consider steps to reduce this heat gain. The roofs of most buildings absorb enough heat to merit attention. Also, don't overlook the possibility of reducing heat gain through the walls.

The heat of sunlight passes through opaque surfaces in a sequence of steps. First, sunlight is absorbed in the outer skin, raising its temperature. Then, the absorbed heat is driven through the envelope by the temperature differential. Once inside the envelope, the heat is transferred to the cooled space by radiation, conduction, and perhaps by convection. You can reduce solar heat gain at each of these stages by using these methods:

• *prevent sunlight from reaching the outside surface,* by shading. Measure 8.2.4 recommends tree shading. Various methods of attaching mechanical shading to the building could be used, but this approach is not economical for opaque surfaces, except in rare cases.

• *reflect the sunlight from the outside surface.* Measure 8.2.2 recommends this approach.

• *cool the exterior surface.* Wind has an important cooling effect, but most locations do not have enough wind to keep the external surface near the air temperature. Evaporative cooling of exterior surfaces by keeping them wetted has been tried over the years, but this method no longer offers an advantage that makes it worth the bother.

• *minimize the conduction of heat through the envelope* by improving insulation. Measure 8.2.1 recommends better insulation.

• *vent heat from the inside of the envelope* before it becomes a part of the cooling load. Measure 8.2.3 recommends venting attics to reduce heat gain.

The Measures of this Subsection are presented in the approximate order of their importance in typical applications, considering energy saving potential and practical considerations. Try to combine the Measures where appropriate.

INDEX OF MEASURES

8.2.1 Improve the insulation of surfaces exposed to sunlight.

8.2.2 Apply paint, coating, or sheathing that minimizes absorption of sunlight.

8.2.3 Provide effective ventilation of attics.

8.2.4 Plant trees and other foliage to provide shading.

RELATED MEASURES:

• Section 4, for modifications to air handling systems to make them respond more efficiently to reduced cooling load

• Section 7, for building insulation

MEASURE 8.2.1 Improve the insulation of surfaces exposed to sunlight.

Insulation is a very effective way of stopping solar heat gain through opaque surfaces. The variety of available insulation methods makes it possible to add insulation to many buildings that lack insulation or have too little. Go to the Measures in Subsections 7.1 and 7.2 for the ways that you can improve insulation in different parts of the building envelope.

The value of envelope insulation for reducing heat gain in warmer climates tends to be underestimated. This occurs if a designer calculates heat gain based on the difference between outside air temperature and inside temperature. This is a mistake, because absorption of sunlight by exterior surfaces creates temperature differentials across the building shell that may be several times greater than the air temperature differential.

In retrofit applications, adding insulation is most cost effective when applied to large expanses, such as roofs. Whenever a roof requires major maintenance, consider improving its insulation value. The economics are much less favorable when installation requires a large amount of fitting or finish work, such as insulating interior walls between windows.

Even if heating is not required, install insulation below or inside any vented areas, such as attics. This reduces the temperature differential across the insulation, increasing its effectiveness.

Combine Insulation with Other Methods

Insulation does not eliminate the high temperature differentials that sunlight creates across surfaces. Wherever it is economical to do so, combine insulation with other Measures of this Subsection that reduce solar gain. For example, if insulation is installed on top of a roof deck, cover it with a reflective surface.

MEASURE 8.2.2 Apply paint, coating, or sheathing that minimizes absorption of sunlight.

The amount of solar heat that is absorbed by a building surface depends entirely on its finish. To minimize heat gain, the surface finish should do two things:

- *reflect sunlight* well. This includes both visible light and the sun's infrared heat radiation.
- *emit heat radiation* well. Surfaces stay cool by radiating their heat back into space.

Many ordinary finishing materials satisfy both these requirements. For example, a clean coat of white paint minimizes heat gain. However, many construction materials do not reject heat well, even though they have a light color. You need to do some simple testing to find the best finish materials.

A carefully selected surface finish is a powerful tool to reduce heat gain at low cost. In new construction or in refurbishing an old building, it adds nothing to the cost. Otherwise, it may cost no more than a new paint job.

SUMMARY

An important way to reduce cooling load, especially in sunny climates. Often inexpensive.

SELECTION SCORECARD

Savings Potential	$ $	$ $
Rate of Return, New Facilities	% %	% %
Rate of Return, Retrofit	% %	% %
Reliability	✓ ✓ ✓	
Ease of Retrofit	☺ ☺	☺ ☺

Where to Use Non-Absorbing Surface Finishes

This Measure is effective with roof and wall surfaces that receive a substantial amount of direct sunlight, regardless of orientation. Consider it for all new construction and for exterior renovations, including repainting.

As a retrofit activity, it merits attention with any building that has a high solar cooling load, provided that you are able to change the surface finish. For example, this Measure would not apply to a building with a decorative masonry surface. Some finishes, such as asphalt shingles, cannot be coated, so they must be replaced. Changing the surface materials is usually too expensive for the purpose of reducing cooling load alone.

Energy Saving Potential

In a fairly clear climate, each square foot of surface area absorbs about 300,000 BTU per year (about 900 kilowatt-hours per square meter per year). A fraction of this energy reaches the conditioned space. The fraction depends on the absorption of the surface, the insulation value of the envelope, the amount of wind, the nature of the surface, the length of the cooling season, and other factors. In the case of roofs, venting is an important factor (see Measure 8.2.3).

The reduction of cooling load is greatest with large, poorly insulated surfaces. For example, you can greatly reduce the cooling load in a lightly insulated metal building just by selecting a more reflective surface color for the roof and walls, as in Figure 1. (In such cases, also consider improving the insulation. See Subsections 7.1 and 7.2 for the Measures related to insulation.)

On the other hand, this Measure increases the heating load. If heating is a much larger cost than cooling, this activity may be self-defeating. However, the increase in heating load typically is much less than the reduction of cooling load. One reason is that the surface finish is important only when the sun is shining, and the sun shines longer in summer than in winter.

Another reason is that cooling temperature differentials are typically smaller than heating temperature differentials, so the effect of direct sunlight is more important for cooling.

The Tricky Part: Absorption and Emittance

You might think that reducing surface heat gain is simply a matter of selecting a surface finish that reflects light well. It is more complicated than that, for two reasons. First, visible light is only about 40% of solar radiation, and a surface that reflects visible light may not reflect the invisible portion. Second, the surface finish can stay cool only by radiating the heat that it absorbs back into space. The tendency of a surface to emit heat radiation is called "emittance." The emittance of different surfaces varies over a wide range.

(Emittance is defined formally as the amount of energy that is radiated by a warm surface compared to the amount that would be radiated by an ideal radiating surface. Paradoxically, such an ideal radiating surface is called a "black body." Surfaces at outdoor temperature radiate their heat at very long wavelengths, much longer than the wavelengths of the heat that reaches the earth from the sun.)

The visible color of a coating indicates the reflection, or absorption, of sunlight only at visible wavelengths. For example, bright white surfaces reflect about 90% of visible light. However, color is not a good indication of total heat absorption, because more than half of the heat in sunlight is at invisible infrared wavelengths. The surface may absorb infrared radiation much more strongly than it absorbs visible light. The energy of visible sunlight is concentrated in the narrow band of

WESINC

Fig. 1 Reflective paint provides most of the cooling of these buildings These buildings have metal roofs and walls, which makes it important to paint them with a fairly heavy coat of non-conducting paint. These buildings, located in a sunny climate, have good exhaust stacks, but they are too far apart. Continuous ridge vents would be better. The ventilation inlet louver on the end of the gable lacks sufficient open area.

wavelengths from 0.4 to 0.7 micrometers (microns), while the infrared energy of sunlight is spread out in the range of 0.7 to 2.5 micrometers. Thus, there is plenty of room for materials to have absorption bands within the infrared spectrum. For example, some white gravel used as ballast on flat roofs becomes hot in sunlight for this reason.

The emittance of a surface depends mostly on two factors: whether the surface is electrically conductive, and whether the surface is smooth. Almost all non-conductive construction materials have an emittance higher than 80%. This includes masonry, wood, gravel, glass, and non-metallic paint. However, as we noted earlier, some gravel appears to have low emittance, probably because it is made of crystals that have smooth surfaces.

Smooth metal surfaces have very low emittance, typically 0.1 or less. The emittance of metal increases if it becomes dirty or corroded, but it still remains down around 0.2. For this reason, you can observe that bare metal surfaces are usually warmer than other surfaces when exposed to sunlight.

Low emittance is a problem with metal surfaces even if they have high reflectance. For example, you can feel that the surface of a polished metal surface is much warmer in direct sunlight than a white painted surface, even though both reflect about 90% of the sunlight.

For this reason, metallic paint is not a good surface finish to protect against heat gain. For example, aluminum paint is sometimes used to reduce the heat gain of asphalt roofs, but you can observe that this does not work well. The high reflectance of the paint is offset by its low emittance.

If metal sheathing is used as a surface finish, it should have a thick coat of non-conducting paint. Emittance is a surface property, so the emittance of the sheathing is determined by the paint, rather than by the metal itself. (This is not true of anodized finishes, which are so thin that the emittance of the metal still matters.)

How to Select a Surface Finish

As a first step in selecting materials, apply the previous principles, i.e., start with light colors and avoid bare metals. Then, find out how well your candidate materials respond to actual sunlight.

The most reliable way to determine the heat rejection properties of surface finishes is to set up some test surfaces. This inexpensive test requires only a few hours in the sun, and it allows you to judge candidate materials before applying them on a large scale.

The procedure is simple. Use a portion of the existing surface to apply various candidate materials.

If possible, shield the test area and a portion of the original surface from wind. Measure the temperatures of the various surfaces under sunny conditions, and compare them to the air temperature.

You could try to get reflectance and emittance information from the manufacturer or from reference books, but this is usually not worth the bother. The manufacturer is unlikely to understand what you are asking. Testing the materials yourself is probably quicker.

Exploit Sheathing to Provide Shading

If you decide to install sheathing or siding over an exterior surface, try to use a method of attachment that allows ventilation between the sheathing and the envelope surface. In this way, the sheathing provides shading as well as reflection. For example, install sheetmetal siding over vertical furring strips. This is primarily a method for warm climates. If the weather spends much time being cold, a tight exterior surface is more valuable.

This method of installation makes the sheathing more vulnerable to wind and impact damage. Account for this by using more fasteners, and perhaps by installing battens over the surface material.

ECONOMICS

SAVINGS POTENTIAL: *Selecting the surface finish intelligently can reduce the cooling load through a sunlighted surface from 30% to 70%, compared to using a finish with high heat gain. Savings in cooling are offset by increased heating costs, so the net saving depends on the climate profile.*

COST: *In new construction and in refurbishing existing buildings, selecting surface finish materials to reduce heat gain usually costs little or nothing. A good exterior paint job typically costs $0.20 to $0.60 per square foot. Fancy surface materials may cost several dollars per square foot, but that expense is for appearance, not energy conservation.*

PAYBACK PERIOD: *Immediate, to several years.*

TRAPS & TRICKS

SELECTING THE COATING MATERIALS: *You get to be a scientist in doing this Measure. Don't go out of your way to find exotic materials. Many ordinary surface finishes perform well. Select materials that are durable, economical, and easily cleaned. Test them for low temperature rise in bright sunlight.*

MAINTENANCE: *Dirt increases the sunlight absorption of a reflective surface. If the surfaces tend to get dirty, clean them during the warm season. Put this in your maintenance calendar.*

MEASURE **8.2.3 Provide effective ventilation of attics.**

Many buildings have an attic space, which is a dead air space between the roof and the occupied space. When sun shines on the roof, the attic acts as an air heater. The heated air may spread to portions of the building that are cooled, or the heat may spread to occupied spaces by conduction. If the attic space contains cooling ductwork, the heating of the ducts increases cooling load.

To prevent these losses, make sure that the attic space is ventilated adequately. Warm air must be able to flow freely from every part of the attic to the outside. In some cases, air movement by convection is adequate. In others, you should install a ventilating fan. In some cases, you may be able to exploit wind to provide ventilation. Route the ventilation air to keep the roof surface as cool as possible. This reduces radiation of heat into the lower spaces.

Where to Improve Attic Ventilation

Most buildings that have an attic space can benefit from improved attic ventilation. There are some exceptions. If the attic space is used as an air return plenum, it cannot be ventilated, and the only practical way to reduce heat gain is to improve the roof insulation. If there is a partition between the attic and the

SUMMARY

Attics are often poorly vented. You may be able to substantially reduce the roof cooling load by improving attic ventilation.

SELECTION SCORECARD

Savings Potential	$ $ $
Rate of Return, New Facilities	% % % %
Rate of Return, Retrofit	% %
Reliability	✓ ✓ ✓
Ease of Retrofit	☺ ☺ ☺

conditioned space, such as a suspended ceiling, attic ventilation is less important.

The economic benefit of this Measure increases with the size of the roof. All large attics deserve attention, but adding ventilation equipment to smaller attics may be uneconomical. Attic ventilation is most important when there is equipment within the attic space that is vulnerable to heat gain, such as air conditioning ductwork.

WESINC

Fig. 1 Attic that provides convective ventilation This roof illustrates the main feature needed to exploit convection, which is a significant height difference between the ventilation air inlet and discharge points. This roof shape can distribute ventilation air exceptionally well. The air can flow upward at all points, parallel to the rafters. To cool the entire surface of the roof, air inlets must be installed all around the eaves. In this example, the air discharge louvers should be larger.

WESINC

Fig. 2 Wind ventilators   These are installed in pairs, as shown here. One unit has a vane that points it into the wind, and the other points downwind. They are placed at opposite ends of the attic. Dampers visible in the near unit blow shut in a high wind. They require a prevailing wind to be effective. In still air, they act as simple ventilation stacks, too far apart to be very effective.

Air conditioning and refrigeration condensers are sometimes installed in attics. This is a serious mistake that radically increases operating cost and reduces cooling capacity. You should correct this situation by moving the condensers to an external location. If that is not practical, ventilate the condensers aggressively with fans that draw air directly from the outside.

Energy Saving Potential

Improving attic ventilation rarely offers a dramatic saving. This is because an attic acts as a thermal buffer between the roof and the occupied space. Thermal isolation is increased by stratification, which creates a static air layer that has significant insulation value. Any partition between the roof and the occupied space enhances the isolating effect of the attic, and may have significant insulating value. For example, ordinary acoustical ceiling tile has an R-value of about 2.

If extensive cooling ductwork is installed in an attic, venting the attic may reduce heat losses through the ductwork by a factor of two or three. The importance of this depends on the amount of ductwork and how well it is insulated.

If you install fans for ventilation, their operating cost reduces the saving in cooling energy. However, if the ventilation layout is designed well, ventilation fans need relatively little power. Also, they should be equipped with controls that limit their operation to times when they provide a benefit.

How to Guide the Air Flow

You can observe that attic ventilation is typically a hit-or-miss proposition, with a few vents scattered around the perimeter of the attic, and maybe a couple of vents in the roof. You can do better than that. Follow these guidelines to provide the best possible ventilation.

(1) Create a defined path from inlet to outlet.

If the air does not know which way to go, it will just sit there in the attic and get hot. The air has to enter at one point and it has to leave at another. The layout depends on the geometry of the attic and the force (convection, wind, or fans) that moves the air.

If the roof is sloped, let air enter at the bottom and leave at the top. If the roof is flat, you may have the air move from one side of the roof to the other, or from the sides to center outlets.

(2) Circulate the entire air volume of the attic.

Don't leave any dead air space in the attic. If the roof is sloped, install outlets along the entire top of the ridge, and inlets along the entire eave or soffit.

If it is not practical to install an adequate number of inlets and outlets to guide the air flow, you can use lightweight ducts to route the air through the attic. Ducts can get ventilation air into all the nooks and crannies. Using ducts requires a fan or fans. Inexpensive ventilation ducts made of fabric or plastic film are widely available. They fit easily on inexpensive propeller fans designed for the purpose. You can get this equipment

WESINC

Fig. 3 Attic fan   The fan is needed to cool ductwork and other equipment, as well as to minimize attic heat gain. Its location in the gable wall is not good, because it must blow against the wind part of the time. Operation of the fan should be controlled by a thermostatic switch in the attic. A tall attic like this one could be ventilated effectively with a continuous ridge vent and large gable louvers.

from companies that supply greenhouses and poultry buildings, and from other sources.

(3) Ventilate the entire underside of the roof surface.

Keep the underside of the roof as cool as possible. This minimizes the heating of air in the attic, and it minimizes radiation of heat to the bottom of the attic. To do this, route the ventilation air so that it sweeps across the underside of the roof. In most cases, a layout that works well for the previous step also works well for cooling the roof surface. For example, with a gable roof, feed air along the eave and vent it out at the ridge.

Arrange the air flow parallel to rafters, joists, or purlins. Otherwise, these will block the flow, creating a dead space of hot air under the roof.

If you use ducts to distribute ventilation air, have the ducts discharge upward, toward the roof surface.

(4) Provide large inlet and outlet area.

Like any fluid, air has a certain amount of mass and viscosity. That means, it doesn't want to flow. If you expect to rely on convection or wind to move the air, make it easy for the air to flow. The main restrictions are the inlets and outlets. Make them big. It is virtually impossible to overdo this. For example, if you want to use natural ventilation for a gable attic, consider installing ventilation louvers that cover the entire area of the gable end walls.

If you use a fan or fans for ventilation, big openings and big ducts minimize the fan energy that you have to pay for.

How to Move the Air

Attic ventilation should use as little energy as possible, preferably none. The possible sources of air motion are convection, wind, and fans. The key to efficient ventilation with each of these is providing large inlet and outlet vents, so that a large volume of air can move easily at low velocity.

■ Ventilation by Convection

If possible, use convection to ventilate the attic. Convection is driven by the difference in temperature (and hence, density) between the air in the attic and the air outside. Convection does not depend on wind.

In order for convection to work, the warm air in the attic must flow upward toward the outside. The convection force is proportional to the temperature difference, and to the difference in height between the attic air outlet and the ventilation air inlet. The disadvantage of convection is that it is a weak force, except in very tall attics. Some styles of architecture enhance the height of the attic to increase attic ventilation by convection. Figure 1 is an example.

For effective ventilation coverage by convection, there must be outlet vents at all high points in the roof and there must inlet vents around the perimeter of the roof. On pitched roofs, full length ridge vents are effective outlets.

Ventilation stacks, such as those shown in Figure 1 of Measure 8.2.2, do not provide much convection beyond that which is provided by any ridge vent. However, they are effective for allowing the attic air to rise to an elevated position from which it can be blown away by wind.

■ Using Wind for Ventilation

Wind is effective for attic ventilation only in locations where there is a steady wind during warm weather. If you want to rely on wind, provide an effective means of funneling the wind into the attic. One method is to install vents all around the attic.

WESINC

Fig. 4 Turbine ventilators These are well placed. Static ventilators would probably be as effective. A continuous ridge vent would be even more effective, more attractive, and less expensive. The gable end louvers are large, and could be even larger.

If this is not practical, use large pivoting ventilators that point themselves using wind vanes. These were fairly common in the days before air conditioning, but they have since disappeared. They are arranged in pairs, with one unit facing into the wind and the other facing downwind. Figure 2 shows an example. You may have to get them made on a custom basis. Locate them at opposite end of the attic. Don't install these devices where they are shielded from wind, for example, below the level of parapets or behind decorative screens.

Using wind is tricky because its energy varies over an enormous range. Be prepared to deal with storm conditions that may occur only rarely during the life of the building. Wind inlets must prevent entry of precipitation during heavy winds. One solution is to use inlet dampers that blow closed above a certain wind speed. You can see this feature in Figure 2.

■ Ventilation Fans

If you use ventilation fans, install them so that they never discharge into the wind. For example, do not install fans in the vertical end walls of eaves, as is commonly done (see Figure 3). Fans that discharge vertically are not strongly affected by wind, unless there is a downdraft from an adjacent structure. Fans with hoods shield against wind, but even these should not be installed on vertical surfaces. In general, install the fans on the roof, at high points.

■ Turbine Ventilators

Turbine ventilators occasionally come into fashion among those who believe in perpetual motion machines. Figure 4 shows a typical installation. The idea is that convection causes the turbines to spin, and the spinning turbines draw air through the attic. In fact, turbine ventilators are usually spun by wind rather than by convection. They probably offer no significant advantage over static vents, as long as the vents are large enough. Turbine ventilators tend to squeak, annoying neighbors over a large distance.

How to Control Ventilation

■ Fan Control

If you use fans for attic ventilation, control them so that they do not operate unnecessarily. Simple, inexpensive thermostatic control is the way to do this. Install the sensing element of the thermostat in the stream of air that is being exhausted from the attic.

If there are days when the building is not cooled, such as weekends, consider time controls in addition to the thermostat. Control by the hour of the day, as well as the day of the week. For example, it may not be economical to run a ventilation fan at the end of the

day, because heat stored in the roof radiates away overnight. See Reference Note 10, Clock Controls and Programmable Thermostats, for guidance in selecting time controls.

■ Adapting to Cold Weather

Excessive attic ventilation during cold weather wastes heating energy. This should not be a problem if the attic insulation is installed properly (see Measure 7.1.1 for details). If you use an attic ventilation fan with a thermostat, the thermostat will prevent excessive ventilation.

Consider limiting air flow through the attic during cold weather if the attic contains heat distribution equipment. For example, steam heating systems typically feed downward, so the attic may be full of steam pipes. Excessive attic ventilation during cold weather may even create a freeze problem. In these specialized cases, it may be desirable to install dampers on the ventilation inlets and outlets. If so, the dampers need a reliable method of control. Again, thermostatic control is simple and relatively inexpensive. Manual operation is unreliable.

Do not seal the attic entirely during cold weather. A certain amount of venting is necessary to prevent condensation and ice formation in the insulation and in the attic. The appropriate amount of venting depends on weather conditions and on the effectiveness of the vapor barrier between the attic and the occupied space.

ECONOMICS

SAVINGS POTENTIAL: 30% to 70% of the cooling load entering through the attic. The amount of energy this represents varies widely depending on the roof and attic design.

COST: In new construction, inexpensive design changes may improve attic ventilation greatly. In existing facilities, the cost of improving ventilation varies widely. The cost per unit of roof area declines with roof size.

PAYBACK PERIOD: Less than one year, to many years.

TRAPS & TRICKS

DESIGN: The main point is to provide air flow throughout the attic. If you rely on vents alone, they must be large and numerous. If you use a ventilation fan, don't be shy about using lightweight duct to distribute the air.

MONITOR PERFORMANCE: Attic ventilation equipment is out of sight, and failure does not create alarms. Therefore, if the ventilation system relies on active devices, such as fans and motorized dampers, schedule periodic checks in your maintenance calendar.

MEASURE **8.2.4 Plant trees and other foliage to provide shading.**

SUMMARY

Works well for shading low structures and the lower portions of tall structures. Improves appearance. You have to wait several years to get results. Requires surrounding land. Requires maintenance.

SELECTION SCORECARD

Savings Potential .... $ $ $ $

Rate of Return, New Facilities % %

Rate of Return, Retrofit .... % %

Reliability .... ✓ ✓ ✓

Ease of Retrofit .... ☺ ☺ ☺

Tree shading is a technique as old as mankind, but it has not been used much in modern times to reduce mechanical cooling load. It deserves more attention for this purpose than it presently receives. Tree shading prevents direct entry of sunlight into the building through glazing, it reduces the surface temperature of the external opaque surfaces, and it lowers the temperature of the air surrounding the building. Figure 1 shows an application with a modern office building.

Trees may be the most economical and practical method of shading some buildings. Trees are inexpensive. They are fairly easy to plant. They do not have to be customized to the building. A special advantage is that they automatically adjust shading to the seasons. They may substantially improve the appearance of the property.

Tree shading has two major disadvantages. One is the long delay between the time the tree is planted and time it starts to produce savings. The other is a need for periodic maintenance.

Where to Consider Tree Shading

Consider tree shading for all types of buildings that receive a substantial amount of solar gain at an elevation below about 50 feet (15 meters). You can use trees to shade tall buildings, but they can shade only the lower portions of the buildings.

Trees also require soil, moisture, and sunlight conditions appropriate to the species planted. Trees can grow well even in areas that are heavily paved, provided that soil moisture is adequate.

Energy Saving Potential

Virtually no direct sunlight gets through the canopy of a healthy shade tree. A fully shaded surface has a solar gain of less than 20 BTU per hour per square foot. As a result, complete shading by trees eliminates over 90% of the solar energy falling on a surface. Shade trees are wide in relation to their height, so they continue to be effective when the sun is at low elevations.

Effect on Passive Heating

One of the major advantages of using deciduous trees for shading is that they do not seriously obstruct solar heat gain during cold weather. This makes tree shading compatible with passive solar heating.

Effect on Daylighting

Tree shading may help or hinder daylighting, depending on the geometry of the building and its glazing. Tree shading may enhance daylighting if it encourages occupants to open internal blinds to admit the pleasant sunlight of the shaded environment. On the other hand, tree shading is incompatible with daylighting techniques, such as "light shelves" (see Measure 8.3.2), which are intended to distribute the full intensity of sunlight.

Non-Energy Benefits

Tree shading has strong esthetic potential. If accomplished with a good sense of style, it can greatly enhance the appearance of a property.

Tree shading can also increase the usable area of space outside the building during warm weather for uses such as patio restaurants, cart vending, etc.

Tree shading has been promoted as a a means of purifying the atmosphere. This effect is minimal if trees are planted on a localized basis. Trees are more likely to succumb to air pollution than to correct it. Also, contrary to popular belief, trees do not reduce the amount of carbon dioxide in the atmosphere. Trees absorb a large amount of carbon dioxide as they grow, but they return carbon dioxide to the atmosphere as they decay.

Disadvantages

A serious disadvantage of tree shading is that it takes from three to eight years for trees to grow large enough to produce significant shading. This assumes that the trees are grown from small sizes. Trees that are purchased large enough to provide shading immediately are expensive. However, buying larger trees may be the most cost effective approach, especially if you need to shade large areas of glazing.

Trees require maintenance throughout their lives, although typically at long intervals. The life cycle cost of tree maintenance is not trivial, especially if you have to keep the trees manicured for decor.

How Tree Shading Works

Trees provide cooling in two principal ways:

- *blocking solar radiation.* Sunlight is either reflected from the tree, or it is absorbed by the foliage at a distance from the building. Heat absorbed by the tree is carried away by the surrounding air mass.

- *evaporative cooling through transpiration.* The surface of the foliage is cooled below ambient temperature by evaporation. Water drawn through the root system is evaporated from the leaves, cooling them. This process is called "transpiration." Its cooling effect does not appear to be well documented. Research by the Lawrence Berkeley Laboratory suggests that trees lower the air temperature by 3°F to 6°F (2°C to 4°C) in the vicinity. One could guess that the magnitude of the temperature drop is related to the evaporation rate (hence, to the species) and to the ambient wet-bulb temperature, as with any other form of evaporative cooling.

The beneficial effect of transpiration is limited if the climate is humid during warm weather.

WESINC

Fig. 1 Tree shading for an office building   Above, two trees planted on the south side of this four-story building substantially reduce the cooling load through its large expanses of glass. Below, the trees shed their leaves in winter, allowing the building to benefit from passive solar heating.   The trees form an effective visual divider between the office building and other buildings in this mixed-use neighborhood.

Evaporative cooling reduces sensible temperature by converting sensible heat in the air to latent heat, i.e., the heat in the air is used to evaporate water from the leaves. This increases humidity in the vicinity of the tree.

Some people have said that trees block "cooling breezes," reducing their overall cooling benefit. This is wrong. In warm weather, breezes do not cool, instead they warm. Furthermore, air motion has a cooling effect only on damp surfaces, notably human skin. This "wind chill" effect does not apply to the dry surface of a building.

How to Select Tree Species

A good way to start with selecting tree species is to observe which types of trees flourish in the vicinity. Take into account the soil type, the sun exposure, and the drainage at the planting site. Try to avoid trees that you cannot find growing in the area. No matter what the nursery says, foreign trees may grow more slowly than expected, or they may not survive at all. Even if a tree species is known to grow in a similar climate, it may not flourish locally because of conditions such as air pollution, soil acidity, etc.

Select deciduous trees, rather than evergreens, except for sites that need cooling all year. Evergreen trees would substantially increase heating requirements in cold weather. Most evergreen trees have a less desirable shape for shading, and they cannot be pruned as well to create a desirable shape.

From the standpoint of shading, you want trees with these characteristics:

- *rapid growth.* A lot of cooling energy may be wasted during the 3 to 8 years that the tree grows to a useful height.
- *adequate life.* Tree life ranges from 30 years to 1,000 years. The more rapidly growing trees tend to have shorter lives, but the correlation varies. The tree should have a life span as long as the remaining life of the building.
- *good shape,* not only for shading, but also for view and for traffic under the canopy.
- *proper timing of leaf growth and shedding.* The tree should grow a well developed canopy by the beginning of the cooling season, if possible, and it should shed all its leaves by the beginning of the heating system. Not all deciduous trees shed well. For example, some species of oaks keep their dead leaves through the winter, and shed them only when new leaves push the dead ones loose in spring.

The fact that a particular tree can provide good shading does not alone make it desirable. Other important characteristics are:

- *ability to flourish at the site* without watering, fertilizing, or other frequent tending.

- *absence of damage from root growth.* Roots can damage adjacent foundations, lift sidewalks, and destroy pavement. Roots that grow on the surface can interfere with lawn mowing, and can make the space under the tree unusable.
- *wind resistance.* Some trees survive high winds by sacrificing branches, which is dangerous. Some trees topple in high winds. Resistance to toppling depends on soil conditions and local geology as well as the tree's root pattern.
- *minimum shedding of annoying material,* such as twigs and seed pods that clog gutters, resin spattering that sticks to automobile finishes, etc.
- *compatibility with surrounding vegetation.* All trees have some effect on surrounding shrubbery, decorative plantings, and grass. Select both the trees and the surrounding vegetation to be compatible with each other.

Where to Plant

Consider these factors when deciding where to plant the trees:

- *the areas of greatest solar gain* in the building. Give top priority to shading large windows that are exposed to direct sunlight. Then, shade smaller windows. Then, shade dark roofs and walls.
- *how the sun moves* with respect to the areas to be shaded. Plant the trees so that they provide shade for the largest fraction of time, during the cooling season.
- *the tree's shape and dimensions* during its useful life
- *avoiding interferences,* as with power and telephone wires. Utility companies perceive trees as a problem, and may cut them back drastically, without regard to the energy saving intentions of the building owner.
- *appearance* of the property, including arrangements with other plantings.

A common mistake is planting the tree too close to the building. The small sapling that you plant today will become a huge structure weighing many tons. Planting trees too close to the building, or too close to each other, may force them to be removed just as they become most effective for shading. Professional tree planters make this mistake as well as amateurs. If a tree is planted at a proper distance from the building, it will seem too far away from the building when it is still young.

Planting Foliage on Buildings

The concept of tree shading can be extended to the planting of shading foliage on the building itself. For example, a building with a large flat roof may have the roof covered with shading foliage.

Although this approach may have merit, it surrenders many of the advantages of planting trees in the ground. The building must be reinforced to support the weight of the foliage. Automatic watering systems are required. If the water is purchased, the annual cost is substantial, in contrast to trees planted in the ground, which find their own water. And, maintenance requirements are much greater.

Keep the Benefit Apparent

It is important to keep the building's owners and operators aware of the fact that the trees are intended to improve their summertime comfort and reduce cooling costs. This may be forgotten as the years pass, allowing some future decorator to remove the trees on a whim. As with other energy conservation measures, effective placards are a principal means of preserving the benefit. Install decorative, permanent (e.g., cast bronze, embossed stainless steel) plaques that describe the shading and comfort purposes of the trees, along with the usual botanical information. It is easy to get people interested in tree shading.

Possible Financial Assistance

Tree planting has become an ideological issue, which has prompted various levels of government to subsidize tree planting. Governments may provide funds directly or they may mandate subsidies by others, especially by utility companies. If you are going to plant only a few trees, government subsidies are probably not worth the paperwork. If you are going to plant many trees, cash in on any subsidies that are available.

For Additional Information ...

There does not appear to be much scientific research on tree shading to reduce cooling load. Research probably is less important than common sense, but it pays to begin by learning from the experience of others. For example, some electric utilities have tree shading programs.

Don't overlook the importance of checking local growth conditions, which you can learn from county agricultural agents and public arboretums. Take the advice of commercial landscape contractors with a grain of salt. They have a bias toward exotic and expensive species that require maintenance.

ECONOMICS

SAVINGS POTENTIAL: *70 to 90 percent of the cooling load of the shaded portions of the building, but only after the trees have grown sufficiently. The net saving is reduced by a small increase in heating load.*

COST: *$50 to $1,000 per tree, planted, depending mostly on the initial size of the tree.*

PAYBACK PERIOD: *Payback period does not apply to this Measure in the usual way, because there is a long delay before it starts to produce benefits. If you start the clock when the tree becomes large enough to provide substantial shading, the payback period would be one or two years.*

TRAPS & TRICKS

CHOICE OF METHOD: *If tree shading is a possibility, the big question is whether it makes sense to wait several years for results. Tree shading is especially attractive for existing facilities where other methods of reducing solar gain are not economical. Consider the other advantages and disadvantages of tree shading, which can be major issues.*

SELECTING THE TREES: *Have a strong bias in favor of trees that you observe to flourish in your locale. Find out all the bad habits of each species you are considering.*

LAYOUT: *Little trees become great big trees, which cannot be moved. Plant the trees in locations suitable for their final size. To get shading quickly without planting trees too close to the building, select species that have rapid vertical growth and that are tolerant of heavy pruning to limit their horizontal spread.*

Enough sunlight falls on most buildings to provide ample illumination for the entire building. Daylighting is an important method of saving energy in large, open buildings. It can also save lighting energy in the perimeter spaces of other types of buildings.

A serious obstacle to daylighting is a lack of effective methods for distributing sunlight throughout the building. At present, daylighting is limited to spaces that are adjacent to roofs and exterior walls. Floors, interior walls, and opaque partitions remain impenetrable barriers. In many buildings, only a small fraction of the space can be illuminated with daylight.

Daylighting is a complicated source of illumination. Achieving lighting of good quality requires a great deal of attention to match the glazing to the space. Furthermore, daylighting has major interactions with heating, cooling, electric lighting, and the building structure. See Reference Note 46, Daylighting Design, for the issues that you need to consider.

The Measures of this Subsection provide daylighting to the extent that it is practical with current materials and equipment. Daylighting is much easier to accomplish well if it is designed into new buildings at the outset. These Measures show you how to exploit daylighting in new construction, how to retrofit it to existing buildings, and how to salvage defective daylighting installations in existing buildings.

INDEX OF MEASURES

8.3.1 Install skylights or light pipes.

8.3.2 Install diffusers for wexisting clear skylights.

8.3.3 Install translucent roof and wall sections for daylighting.

8.3.4 Install diffusers to make windows more effective for daylighting.

8.3.5 Install a system of light shelves and shading.

8.3.6 Use light interior colors or mirrored surfaces.

RELATED MEASURES:

- Subsection 6.3, for reducing air leakage through windows
- Subsection 8.1, for controlling solar heat gain, which has a major effect on the potential for daylighting
- Subsection 8.4, for passive solar heating, which should be coordinated with daylighting
- Subsection 9.5, for control of electric lighting to respond to daylighting

MEASURE   **8.3.1 Install skylights or light pipes.**

Skylights can provide satisfactory lighting for activities that can tolerate large variations in illumination level. Getting good performance from skylights is not as simple as it may appear. You have to satisfy a number of requirements, some of which may not be easily compatible with each other.

When considering skylights, also consider "light pipes." Light pipes perform the same function as skylights. They make it possible to transport daylight through thick roof structures and attics. They are easier to install in retrofit applications than skylights. For practical reasons, light pipes are limited to smaller light collection areas. They are still evolving.

In this Measure, we will use the term "skylights" to cover both skylights and light pipes, except when the distinction needs to be spelled out.

SUMMARY

Skylights are an old source of free lighting that needs to be improved in the way it is applied. Light pipes are a new way of achieving the same purpose with greater flexibility. Desirable for many large, open, non-office environments. Retrofit is usually difficult.

SELECTION SCORECARD

Savings Potential .... $ $ $ $

Rate of Return, New Facilities % % %

Rate of Return, Retrofit.......... % %

Reliability .... ✓ ✓ ✓

Ease of Retrofit .... ☺ ☺

WESINC

Fig. 1 Effective skylights in a shopping mall   The percentage of ceiling area is about right for the type of illumination needed. Light distribution is good. The diffusing skylights are installed at the tops of tapered recesses, minimizing glare.

Where to Use Skylights and Light Pipes

Many types of activities can be illuminated well by skylights, but many others cannot be. In terms of illumination quality, the major advantage of skylights is the ideal color rendition of daylight. Their major disadvantage is large fluctuations in illumination intensity caused by movement of clouds across the sun. This annoyance varies with location.

Experience indicates that skylights can be effective for retailing, even in posh environments, because sunlight has excellent color rendition and brilliance. See Figures 1, 2, and 3.

Skylights are also effective for many manufacturing and maintenance operations. Warehousing can be a favorable application. Skylights can be used to provide a sense of natural ambience, which is valuable in applications such as restaurants (Figure 4), transportation centers, and other public areas.

Skylights are less likely to be satisfactory where paperwork occurs, as in offices, drafting areas, and reading rooms. The wide range and fluctuations of sunlight intensity are more noticeable in applications that require concentration on text. Also, daylighting makes it more difficult to avoid veiling reflections, which are a problem especially with paperwork. See Reference Note 51, Factors in Visual Quality, for more about veiling reflections.

With current technology, skylights and light pipes are limited to illuminating the area directly underneath them. The roof structure must allow penetrations to be made without undue expense. Therefore, skylights are most likely to be worthwhile in industrial-type buildings and in large single-floor spaces, such as gymnasiums. Future types of light pipes may be able to transport sunlight far into the building interior.

Skylights must be located where the sun can shine on them directly. A skylight does not produce a useful amount of daylight if it is shaded by adjacent structures or foliage. Similarly, skylights are not worthwhile in areas that have heavy cloud cover for a large fraction of the time, unless the climate is mild and the structure can accommodate a large area of skylights. Clouds typically reduce solar illumination by a factor of five to ten.

When individual clouds pass in front of the sun, they cause abrupt changes in illumination level. The abruptness of the change is usually more objectionable than the reduction of light level. The large, quick variations of light level make skylights unacceptable for certain applications.

Retrofitting skylights in existing buildings is often impractical because of cost and structural interference. Even though skylights require only a relative small fraction of total roof area, installation usually requires structural changes, such as cutting through rafters and purlins.

If the building has an attic, installing skylights in the roof requires building a reflective enclosure to pass the light through the attic. Unless the attic is empty, this may be difficult. Light pipes are easier to pass through attics. In effect, a light pipe is a small skylight with an integral reflective enclosure.

Energy Saving Potential

In single-floor buildings, skylights may provide a large fraction of illumination requirements. Sunlight is so intense that skylights can provide virtually any illumination level that is required. Of course, artificial lighting is still needed at night.

Sunlight has a better ratio of light to heat than any type of electric lamp. Therefore, if the light from skylights is distributed efficiently and the skylights are not oversized, they may not substantially increase the cooling load. However, this ideal is difficult to achieve.

Skylights can provide significant passive heating during cold weather. This advantage is offset by conductive heat loss at night. In all but the coldest climates, there is a net heat gain if the skylights are located so that they collect the maximum amount of sunlight. On the other hand, skylights that face away from the sun may suffer a net heat loss even in relatively mild climates.

Surface Area Required for Daylighting

Outdoor sunlight from a clear sky produces an illumination of about 60,000 lux, most of which comes directly from the sun. Using this fact, you can easily calculate the fraction of the ceiling area that needs to be converted to skylight. (For a quick introduction to "lux" and other measures of light intensity, see Reference Note 50.)

For example, consider an application that needs an illumination level of 500 lux. To account for losses in reflection and diffusion within the skylight assembly, assume that 40% of the sunlight entering the skylight makes its way into the space. Thus, on a bright day, about 2% of the ceiling area needs to be skylights. To compensate for low sun angles, hazy conditions, dirty skylights, etc., double this to about 4%. To account for average cloudy conditions, increase this to 10% or 15%.

The installation in Figures 1 and 3 have skylights that are sized for approximately the latter percentages of ceiling area. The installation in Figure 2 has a much higher percentage, with the result that the glazing must be darkly tinted to avoid glare. The unusual installation in Figure 4 has a variable amount of skylight area.

These figures assume that the skylights are installed where they remain exposed to direct sun throughout most of the day. In some older industrial buildings, large skylights were installed facing north to avoid glare. This unnecessary practice is still followed in some new buildings, as shown in Figure 5. This greatly increases the glazing area required, which also increases heat loss in cold weather. To keep skylights as small as possible, install them so they face the sun as much as possible. Control glare with diffusion and careful space layout, as discussed below.

In predominantly warm climates, select the skylight area to give the best compromise between savings in lighting energy and extra cost for cooling energy. If the space is air conditioned, design the skylights for a clear sky. This means, keep them small.

In predominantly cold climates, the balance usually shifts toward larger skylights. This makes it more important to select the skylights for low conductive heat loss. If you want to exploit passive heating, the skylights must be much larger, and the whole arrangement

Super Sky Products, Inc.

Fig. 2 Large, attractive skylight This is a large amount of glazing area in relation to the illumination requirement.

becomes expensive and elaborate. For more about passive heating, see Reference Note 47, Passive Solar Heating Design.

Use the minimum total skylight area that you need to provide good illumination, and if appropriate, passive heating. This is because surface area increases heat loss, cost, and structural problems. Therefore, make all skylights as transparent as possible, subject to the need for diffusion, multiple glazing, reinforcing fibers in plastic material, etc. Do not use skylight materials with tints, and do not use skylight materials that reflect sunlight.

Efficient distribution of daylighting within the space is as important as the skylight area. Skylights should deliver their light where it is needed, and they should avoid creating visual problems. You achieve these characteristics by effective layout of the skylights, and by using diffusion. We will cover these two topics next.

WESINC

Fig. 3 Skylighting for a variety store The skylight consists of a small area of translucent panels surrounding the cupola. The lighting is effective for the colorful merchandise. The geometry of the skylight does not extend sufficient daylight into the ends of the space, so artificial lighting is needed there. Daylighting of this space was later abandoned, for unknown reasons.

Kalwall Corporation

Fig. 4 Adjustable skylight In this restaurant, insulated translucent panels with low light transmission can be retracted to provide a variable amount of skylighting. The layout must keep direct sunlight from entering the occupied area of the space, or diffusion must be used with the skylight glazing.

Skylight Layout

In general, it is better to use a larger number of smaller skylights, rather than one or a few large skylights. There are many examples of horrible daylighting in which someone attempted to illuminate a space with only one big skylight. Figure 6 shows a

good distribution of skylights for a space with a tall ceiling. Figure 1 in Measure 8.3.2 shows the interior of this space. Using smaller skylights has several important advantages:

- *you can tailor the light distribution within the space more accurately*. The skylights do not have to be installed in a regular pattern. In general, the size of skylights, and the spacing between them, should be proportional to the ceiling height. Stated differently, skylights should not have to throw light far to the side.
- *an array of smaller skylights provides illumination that is much more uniform* than the light from a single large skylight. Installations with large skylights commonly suffer from excess brightness directly below the skylight, accompanied by gloomy dark areas surrounding the skylight.
- *less modification of the roof structure is needed*. Large skylights require special roof design to carry the roof loads around the skylight. In existing buildings, it is often possible to retrofit small skylights, but not large ones.
- *it is easier to avoid leakage problems with small skylights*, for the reasons discussed below.

It helps to think of skylights as a class of light fixtures. They must obey the same rules of physics and lighting quality as electric light fixtures.

Compare the skylight installation in Figure 1 of Measure 8.3.2 with the skylight installation in Figure 7. The former provides better distribution of light throughout the space, with less glare. (However, it lacks proper diffusers, as we will discuss.)

Another useful guideline is that the dimensions of the skylight should be a small fraction of their height above the floor of the space. Thus, a large skylight over an atrium may be satisfactory, but not a large skylight over a dining room. The central skylight in Figure 3 is satisfactory because it is high above the floor. However,

WESINC

Fig. 5 Inefficient skylight orientation The skylights on this bus garage face north. This avoids glare problems and excessive solar heat gain, but it provides little light in relation to the heat loss from the glazing. Overhead equipment keeps much of the daylight from penetrating into the space.

WESINC

Fig. 6 Good skylight sizing and layout This shows the small fraction of roof area that is needed for typical illumination levels. These skylights are for a gymnasium, and the repetitive layout provides uniform illumination within the space. See Figure 1 of Measure 8.3.2 for the inside.

Vistawall Architectural Products

Fig. 7 Gymnasium daylighting with a single large skylight Compare this to Figure 1 of Measure 8.3.2. The barrel roof of this building makes it difficult to install distributed skylights.

the daylighting does not penetrate well to the perimeter of the space.

Skylights that are large with respect to the ceiling height may be decorative, but they are not optimum for daylighting, and they may be overwhelmingly bright. Figure 8 shows large clusters of skylights installed in a library. The brightness inside the space can be excessive. Figure 9 shows a huge skylight that covers an entire office.

Effective Diffusion is Essential

Skylights generally need diffusion. Direct sunlight through skylights is not suitable for illumination. It is

WESINC

Fig. 8 Large skylight arrays These illuminate areas of a public library with relatively low ceilings. The areas immediately beneath the skylights receive too much illumination and heat gain in bright daylight, while the light is unable to disperse to the rest of the space.

much too intense, it forms localized bright spots, and it shines on the wrong places. Diffusion corrects or reduces these problems by distributing sunlight in a fairly uniform pattern. It also minimizes changes in illumination caused the motion of the sun. Daylighting was abandoned in favor of electric lighting largely because of unsatisfactory illumination resulting from lack of diffusion.

For example, consider a space that has many small skylights. If the skylights are clear, they produce intense bright spots on the floor, surrounded by darkness. See Figure 1 in Measure 8.3.2. As a result of this oversight, the skylights produce no useful illumination, and the electric lights must be turned on to compensate for the glare. On the other hand, if the skylights diffuse the light, it is spread throughout the space.

You can make any surface of the skylight glazing a diffuser by selecting the material for this purpose. Or, you can install separate diffusers either above or underneath clear glazing. It is easier and usually preferable to make the diffuser an integral part of the skylight, unless you are also trying to accomplish passive solar heating.

Diffusion introduces its own set of issues to consider. The main ones are limiting glare, dealing with solar heat gain, and limiting light loss.

■ Limit Glare

The term "glare," as we use it here, means an area of intense brightness within the visual field. (Reference Note 51, Factors in Lighting Quality, explains glare in greater detail.) Diffusion has the potential of creating serious glare because it makes skylights look like bright light sources. For example, small skylights are similar to flat-faced fluorescent ceiling fixtures in appearance and light distribution pattern, although they can be significantly brighter.

Glare is a problem only when the bright surface is within the field of vision. Fortunately, people tolerate bright light sources that are overhead. As with other light sources, the solution to glare is to locate skylights well above the line of sight. In spaces with very tall ceilings, such as gymnasiums and manufacturing plants, the height of the ceiling alone may be sufficient to keep glare within acceptable limits.

If the skylight is installed at the top of a shaft or recess in the ceiling, this keeps the skylight out of normal lines of sight. For this to minimize glare in a space with a low ceiling, the shaft or recess should be at least as tall as the maximum dimension of the skylight. In other words, the recess needs to be taller with bigger skylights.

■ Locate Diffusers to Minimize or Exploit Heat Gain

The solar heat gain into the space is strongly affected by how the diffuser is installed, and by the characteristics of the diffuser. Heat gain is lowest if diffusion is limited to the outer surface of the glazing. In that case, a large fraction of the heat absorbed by the diffuser itself is carried away be the outside air. Heat gain is greatest if the innermost surface is used as the diffuser.

Heat gain is increased even more by locating the diffuser farther inside the space, so that less of the entering light is reflected back out. To capture solar heat for passive heating, install a separate diffuser inside the space that is made of absorptive material. Measure 8.4.2 explains this arrangement in detail.

■ Where to Locate the Diffuser if the Skylight is Installed Above a Ceiling Recess

The location of the diffuser matters most when there is a recess or shaft between the skylight and the interior of the space, typically to create a path for the daylight through the roof structure.

The location of the diffuser in a shaft radically affects the light distribution pattern. Installing the diffuser at the bottom of the shaft produces a broad pattern. Installing the diffuser high in a tall shaft produces illumination similar to that of a downlight.

Installing a diffuser higher in the shaft keeps people from seeing it, except when it is more nearly overhead. This reduces the possibility that glare will be a problem.

The surface of the shaft absorbs light. A diffuser installed high in the shaft deflects more of the light toward the shaft surface, making it especially important for the shaft surface to be highly reflective. A specular surface saves more light than a diffuse surface, because it reflects all the light downward. However, a diffuse surface may give better light distribution. It depends on the relative geometry of the skylight, the shaft, and the space.

If you install a separate diffuser, make sure that no sunlight leaks around the diffuser directly into the space. Direct sunlight is an intense source of glare, and it is useless for illumination.

Avoid Nasty Surprises: Heat Loss and Condensation

The heat loss of skylights may be much higher than you expect. Hidden away in the technical literature is the fact that heat loss through glazing is two to three times higher when the glazing is installed in a horizontal or steeply slanted orientation than when it is installed vertically. Thus, the double glazed skylight that you expected to have an R-value of 2 actually has an R-value less than 1.

One of the unpleasant surprises that results from this low thermal resistance is a tendency for skylights to sweat profusely. The condensation can damage or disfigure the surrounding structure. Poorly insulated skylights may drip heavily on the space below.

In cold climates, it is worth going to great lengths to limit the heat loss of skylights. These are your possible solutions:

Super Sky Products, Inc.

Fig. 9 An illustrative example   Analyze this skylight installation, which illustrates most of the issues of daylighting. How suitable is the skylight for the activities? How appropriate is the glazing area from the standpoints of glare, solar heat gain, and conductive heat loss? How effective are the controls for the electric lights? What should be the transparency of the glazing? Should diffusers be used? Internal shading? How would these issues be affected by the location of the building? Overall, is this effective daylighting or primarily an esthetic feature?

• *select multiple glazing.* Up to three or four sheets of glazing are practical. Light transmission is reduced somewhat, and weight and cost are increased.

• *use a glazing material that includes translucent insulation.* Translucent glazing systems are now available that offer R-values as high as 10. These systems combine a plastic or composite glazing material with a layer of translucent insulation, which may be glass fiber or foam. See Measure 8.3.3 for the details. There is a strong compromise between R-value and light transmission, so skylights using this material must be larger than skylights that use conventional glazing.

• *install movable insulation,* which can greatly reduce heat loss without reducing light transmission. Movable insulation is challenging to design and to install. For an introduction to movable insulation, see Reference Note 47, Passive Solar Heating Design.

All skylight frames should include gutters to catch condensation that flows off the interior surface of the glazing. This is important to keep the condensation from rotting or disfiguring the structure around the skylight. The gutters should be large enough to hold all the condensation until it can evaporate back into the space.

Skylight Materials

Skylights are commonly made from glass, glass composites, plastics, and plastic composites. All these materials can be treated to reduce light transmission and cooling load, either by adding dyes that absorb light or by adding a reflective surface. All glazing materials can be provided with diffusing properties. As discussed previously, the thermal insulation value of skylights can be increased by installing multiple sheets of glazing and by installing translucent insulation between the sheets.

All diffusing materials absorb a significant amount of the entering sunlight because there are multiple reflections within the material. Absorption is greatest with milky diffuser, and lowest with prismatic diffusers. Absorption is also increased by fibers, pigments, and other materials that are embedded in the material.

The advantages of glass include unlimited life, high light transmission, hardness, and rigidity. Glass can be treated to reduce cooling load by selectively absorbing the infrared portion of sunlight. At present, this capability is available only with glass, not with plastic. See Measure 8.1.3 for details. The infrared absorbing surface should be outermost.

The main disadvantage of glass is its vulnerability to breakage, along with the safety hazard that falling glass creates. Glass can be made more resistant to breakage by increasing its thickness, by heat treating it, and by combining it with reinforcing materials. All safety improvements for glass add cost, and they usually add weight.

Plastic materials are much lighter in weight, and they are resistant to shattering, so they pose only a minimal safety hazard. An entire skylight assembly can be molded from a single piece of non-reinforced plastic. Smaller plastic skylights can be molded so that they overlap a mounting curb, providing excellent resistance to water leakage and greatly reducing the cost of the frame. Plastic skylights can easily be fabricated with multiple layers of glazing to improve thermal resistance.

Plastics can be reinforced with fibers of various materials, including glass, to increase strength and service life. The fibers cause some light loss. They also diffuse light, which is useful in most applications. Reinforced plastic is more difficult to mold into compound shapes. It is normally made in flat sheets, which can be curved in one direction.

The plastics commonly used for glazing are acrylics and polycarbonates. Polycarbonates are stronger, but acrylics are more resistant to degradation by the ultraviolet component of sunlight. All plastics deteriorate in strength and light transmission over a number of years. The main causes of deterioration are ultraviolet light, heat, and oxidation, in that order.

The service life of plastic glazing can be extended greatly with additives. Unfortunately, you cannot judge the long-term performance of a plastic material except from manufacturers' claims, so purchase skylight material from a credible manufacturer. If you buy skylights as prefabricated assemblies, first examine the plastic manufacturer's data. If possible, investigate actual field experience with the particular products you are considering.

Flat plastic glazing material buckles as it ages. This can be quite noticeable when the material is observed from the outside, but not when looking at the skylight from the interior.

Glass and plastic can be combined in larger skylights to minimize their respective weaknesses. Glass is used for the outer sheet, where it can provide considerable protection to the plastic, while the inner plastic sheet protects against glass breakage. Ordinary window glass strongly absorbs the damaging ultraviolet portion of sunlight, so a plastic material will survive longer if it is installed inside glass. Design or select combination skylights so that the plastic elements can be replaced separately without a great deal of effort.

Skylight Configurations

Skylights are available in almost any configuration that you could want. Skylights made of glass and composite are usually built up from flat sections. Figure 10 shows a sampling. Plastic skylights can be molded into virtually any desired shape. Skylights made of reinforced plastic materials can easily be curved in

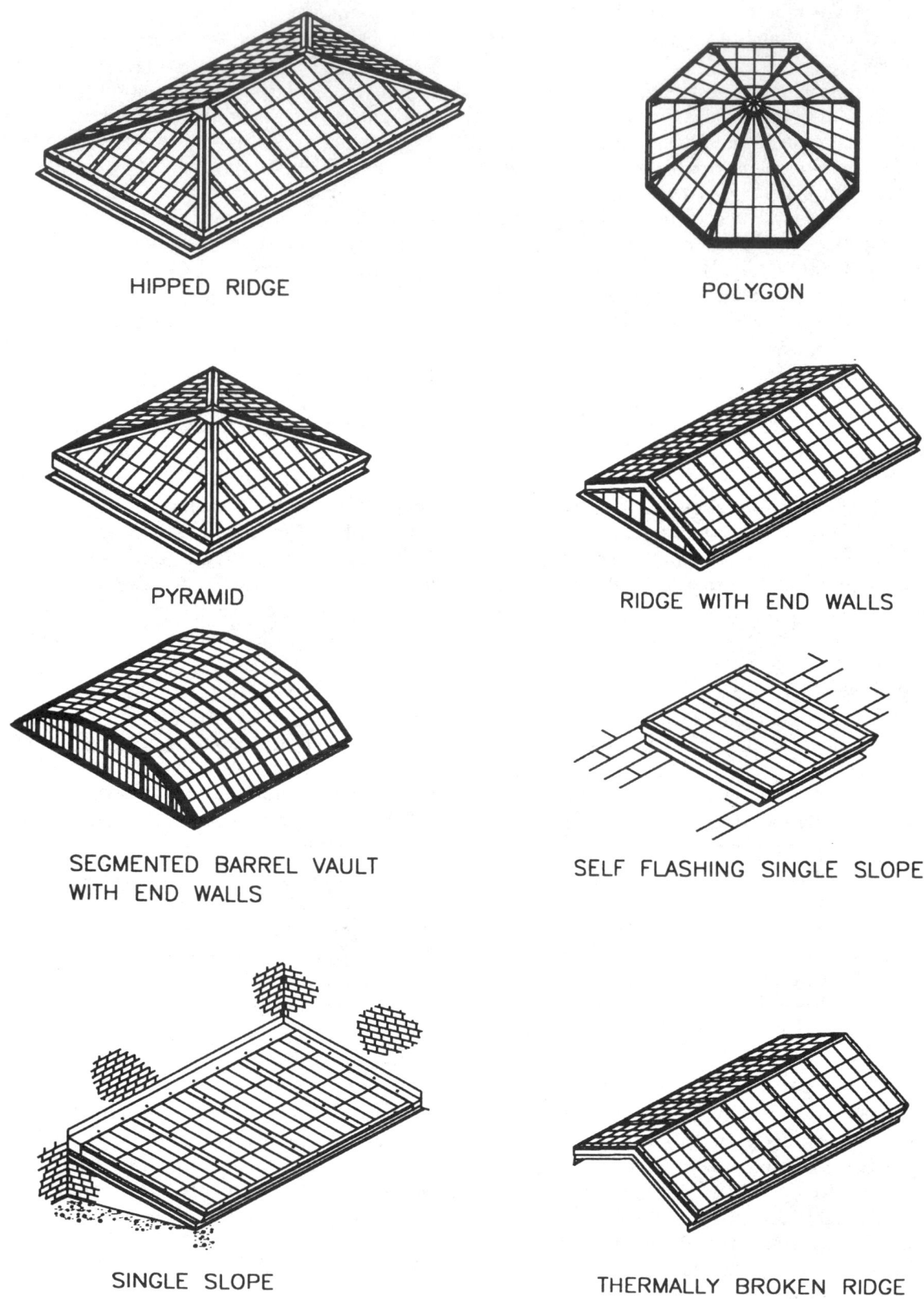

HIPPED RIDGE

POLYGON

PYRAMID

RIDGE WITH END WALLS

SEGMENTED BARREL VAULT
WITH END WALLS

SELF FLASHING SINGLE SLOPE

SINGLE SLOPE

THERMALLY BROKEN RIDGE

Major Industries, Inc.

Fig. 10 Built-up skylights These can be made in virtually any size. Materials that cannot be bent easily, such as glass and fiber-reinforced plastic, are usually made into skylights this way.

one direction. Figure 11 shows a sampling of the latter two types.

How to Prevent Water Leakage

Water leakage is a common problem with skylights. Do not select a skylight design that depends primarily on sealants to prevent leakage. The most reliable method of avoiding leaks is to use one-piece molded skylights that overlap the curb. Unfortunately, molded skylights are limited in size.

If a skylight must have joints, design it so that the joints are steeply sloped to shed water. Ideally, no joint or sealant should face uphill, but this condition cannot be met everywhere in built-up skylights. Instead, the frame extrusion should be designed so that it conveys any water that leaks through seals to the outside of the skylight.

Pay attention to the curb on which the skylight is mounted. The curb should be sealed to the roof as effectively as curbs used for installing scuttles or rooftop air handling units. If a curb is installed on a sloping roof, install an eave on the uphill side of the curb to shed water to the sides of the curb.

Control the Electric Lights to Exploit Daylighting

A skylight or light pipe is useless unless it reduces the energy consumed by electric lights. This is not just a matter of turning off the electric lights in daylighted areas. You also have to avoid the tendency to increase the electric lighting levels in adjacent parts of the space that are not daylighted to compete with the brightness of the daylighted areas.

See Measure 9.5.3 for methods of controlling electric lights in response to daylighting. Combine these

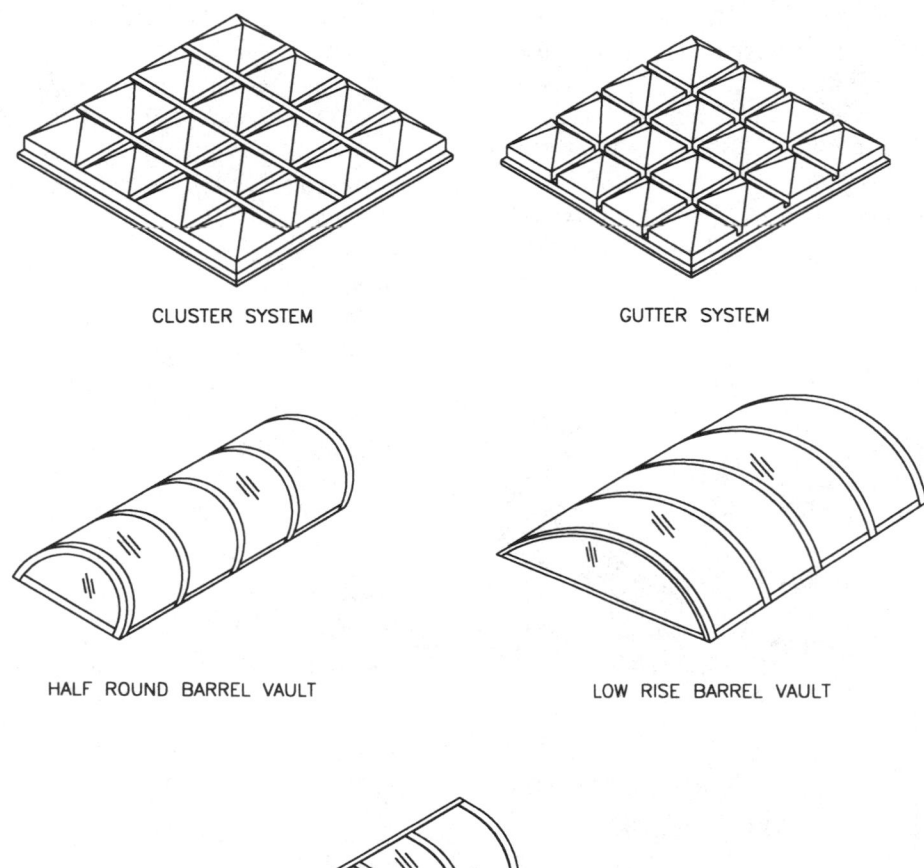

CLUSTER SYSTEM

GUTTER SYSTEM

HALF ROUND BARREL VAULT

LOW RISE BARREL VAULT

QUARTER-ROUND BARREL VAULT

Major Industries, Inc.

Fig. 11 Molded and curved skylights Molded units are limited in size, but they can be ganged together to create a skylight of any size. Individual units can be molded to overlap curbs, an excellent method of avoiding leaks. Large skylights can be made of sheets of reinforced plastic that are curved in one direction.

with the other automatic lighting controls of Subsection 9.5 that are appropriate for the activities that occur in the space.

Compatible Types of Electric Lighting

Fluorescent lighting is generally the best type to use in combination with skylights because its output can be adjusted efficiently to supplement reduced sunlight. Modulating dimmer systems are available for fluorescent lighting, and these work well with daylighting. Also, fluorescent lamps can be turned on and off repeatedly to respond to changes in daylight. However, frequent cycling reduces lamp life and annoys occupants.

It is common to use high-intensity discharge (HID) lighting in combination with skylights. (HID is the class of lamps that includes mercury vapor, high-pressure sodium, and metal halide.) This is because HID lamps have high output, so they can be installed at the height of the skylight in limited numbers, matching the light output of the skylight. Unfortunately, HID is usually a poor choice to combine with skylights. HID lamps take as long as ten minutes to reach full brightness, and they cannot restart for several minutes after being turned off. This makes them inappropriate for use with daylighting, where passage of clouds in front of the sun may require the output of the electric lights to change continually.

A newer type of HID lighting is available that turns on instantly. These operate by keeping the lamps hot continuously, which sacrifices efficiency. HID dimming does exists, but it is limited by problems that are described in Reference Note 56, HID and LPS Lighting.

There is a tendency to believe that fluorescent lighting cannot be used with tall ceilings. This is not true. However, individual fluorescent lamps are limited in light output. As a result, lighting a space with fluorescent lighting typically requires many more fixtures than with HID lighting. See Measure 9.3.3 for a comparison of HID and fluorescent lighting.

It makes no economic sense to use incandescent lighting as a complement to skylights, even though this is done often. The saving in lighting cost by daylighting is eliminated by the higher cost of incandescent lighting during the periods of time that it operates.

Special Features of Light Pipes

The desire to gain the benefits of skylights without suffering their disadvantages created an interest in "light pipes." As the name implies, light pipes convey light to locations within the building where it is needed. Many types of light pipes have been designed, but only a few are commercially available.

Some claims made for light pipes defy the laws of physics. A light pipe cannot deliver more light energy to the space than it collects on the outside of the building. At present, commercial light pipes do not concentrate sunlight from large exterior collectors into smaller pipes. Understand the principles of light pipes so that the installation will yield the results you expect.

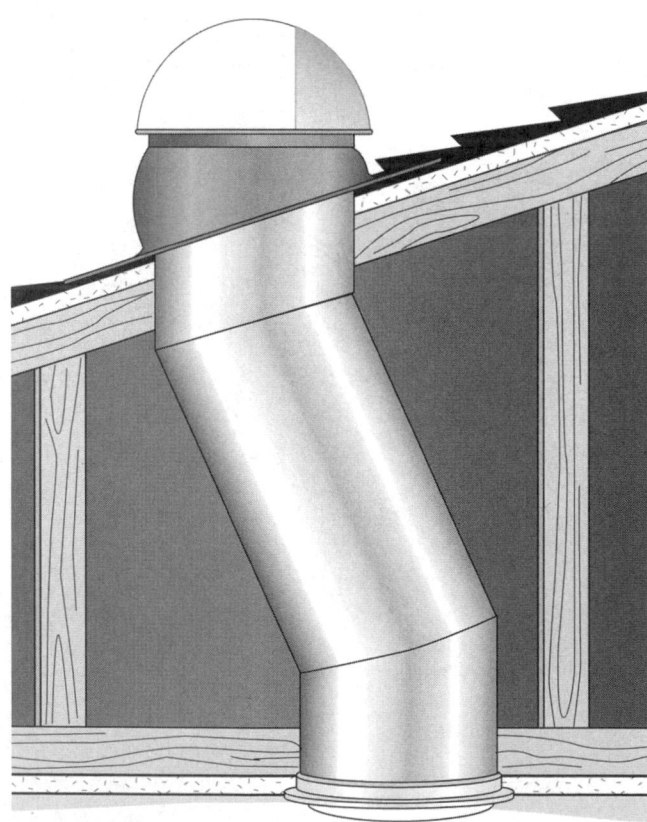

Huvco, LLC

Fig. 12 Rigid wall light pipe This unit has a reflector in the back of the dome to capture more daylight at low sun angles. A diffuser is installed at the ceiling.

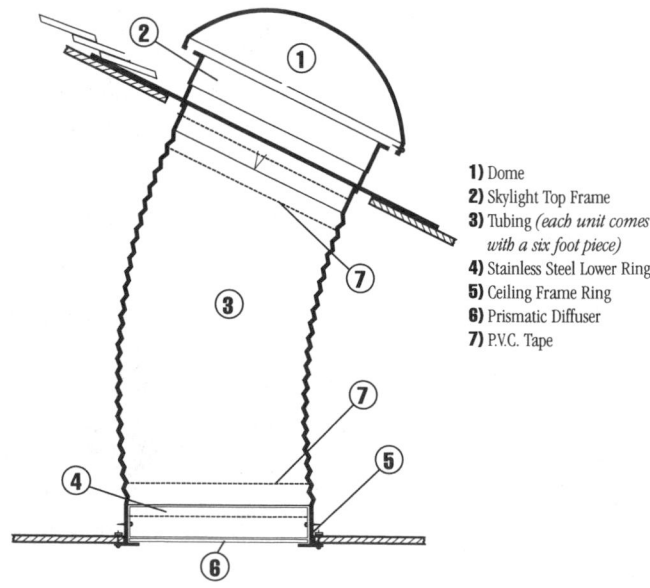

1) Dome
2) Skylight Top Frame
3) Tubing *(each unit comes with a six foot piece)*
4) Stainless Steel Lower Ring
5) Ceiling Frame Ring
6) Prismatic Diffuser
7) P.V.C. Tape

Sun Tunnel Skylight

Fig. 13 Flexible wall light pipe This type is easier to install than rigid wall light pipes. However, light losses are greater because of the corrugated surface of the tube.

■ Light Loss from Reflection Inside the Pipe

The efficiency of fixed light pipes suffers from absorption that occurs when light is reflected from the walls of the pipe. Unless the sun is lined up with the axis of the pipe, the light is reflected repeatedly as it travels through the pipe. Even if the surfaces of the pipe have high reflectance, say 90%, a large fraction of entering light is lost with a few reflections. The light loss is proportional to the length-to-width ratio of the pipe. Therefore, efficiency is sacrificed if the pipe is long in relation to its width.

■ Simple Light Pipes

Most light pipes that are available commercially consist of an exterior transparent dome, a reflecting metal pipe, and a diffuser for installation at the ceiling level of the space. The pipe may be rigid or flexible. Flexible pipes are easier to install, but they suffer more light loss from increased reflection and scatter inside the pipe. Figure 12 shows a rigid light pipe, and Figure 13 shows a flexible light pipe.

■ Sun Trackers

A movable mirror or refracting system can be used to align the incoming sunlight with the axis of the light pipe, minimizing reflection losses. A light pipe with this feature is called a "sun tracker."

Sun trackers have been built commercially. Figure 14 shows a number of installed units, and Figure 15 shows a cross section of the unit.

If mass produced, sun trackers could be relatively inexpensive. Their main limitation is that they lose effectiveness if the sky does not remain clear. The system is designed to collect light from the sun, which is a point source. The light reflecting apparatus gets in the way of the whole sky when the sun is obscured. Another disadvantage is the need for occasional maintenance.

■ Future Developments

An ideal light pipe would have a large exterior collecting surface, it would funnel the light into a narrow conduit, and it would deliver the light wherever it is needed.

A small conduit is desirable to minimize heat loss and to make the light pipe easy to install. Funneling the light from a large collector into a small pipe requires a tracking mirror and a lens system. These do not have to be precision components. For example, light can be concentrated with flat fresnel lenses made of molded plastic. Both the tracking mirror and the lens system should be able to adapt to changes in sky conditions, from direct sunlight to a diffuse sky.

Fig. 14 Sun tracking light pipes The rotating head contains mirrors that reflect direct sunlight straight down the pipe, minimizing losses. However, sun trackers may be less effective than simple light pipes for collecting sunlight from a diffuse sky.

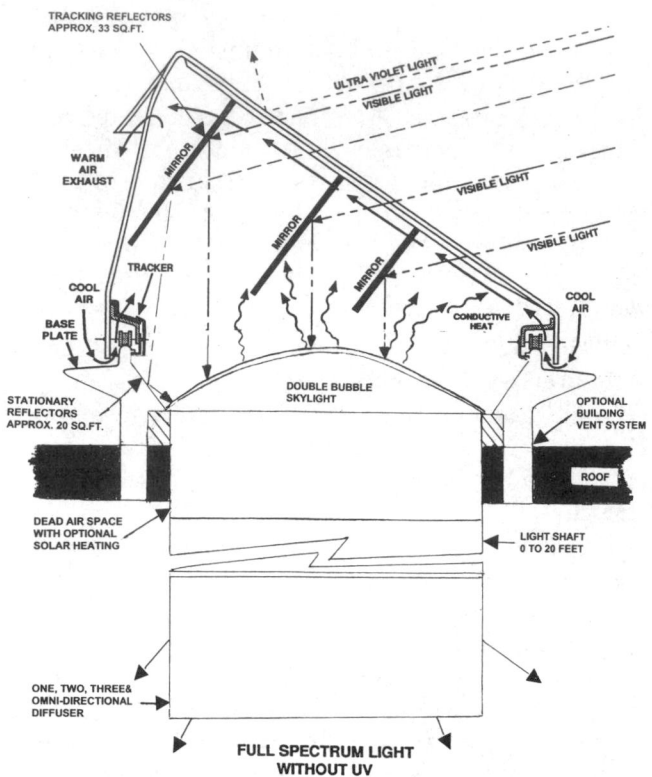

LightScience Corporation

Fig. 15 Cross section of sun tracker   The light shaft is sealed. The tracker head reflects sunlight through a double-glazed dome on top of the shaft.

Light pipes can avoid light loss by using the principle of fiber optics, which is an optical phenomenon called "total internal reflection." This requires the light pipe to be made of a solid transparent material, such as glass or plastic. The light pipe can be long, and it can have any number of bends.

To make this economical, all the light has to be squeezed into a light pipe of small diameter. The small diameter is a major advantage in itself, but it involves complication at each end of the light pipe. This concept is presently used with high-intensity electric lamps as the light source, for special effects. Present light pipes of this type are too expensive for daylighting.

ECONOMICS

SAVINGS POTENTIAL: *In the most favorable cases, skylights can provide a majority of lighting needed during the daytime. However, skylights increase heating cost. In mild climates, the heating cost penalty is not serious if the skylights are sized properly. Skylights may increase cooling cost, but this penalty should be minor if the skylights are sized and laid out properly.*

COST: *The major cost components are the skylights themselves, the roof penetrations, framing for the skylights, and electric lighting modifications. Molded plastic, double-glazed skylights cost $15 to $30 per square foot, depending on size. Large, self-supporting skylights with plastic composite glazing typically cost from $30 to $60 per square foot. Double-glazed glass skylights with safety glazing may cost up to $100 per square foot. Small, simple light pipes can be installed for less than one thousand dollars, even in retrofit. See Measure 9.5.3 for the cost of the electric lighting controls.*

PAYBACK PERIOD: *Up to several years, in new construction. Several years or longer, in retrofit.*

TRAPS & TRICKS

MAKING THE COMMITMENT: *Skylights do not disappear if you decide not to use them. They can create serious permanent problems, including glare, uncomfortable heat gain, heat loss, condensation, and water leakage. Do not use them unless you make a commitment to design and install them properly.*

DESIGN: *Skylight design is not something to do in a hurry. Consider the light distribution pattern, the effect on decor, the effect on cooling and heating, avoiding condensation problems, and avoiding water leakage. Provide effective controls to coordinate the electric lights with daylighting, or else the effort is wasted.*

SELECTING THE EQUIPMENT: *There are major differences between prefabricated units. Some have much better longevity and leakage resistance than others.*

INSTALLATION: *Even with the best design, avoiding water leakage requires good installation workmanship. See Measure 9.5.3 about installing the daylight-sensing lighting controls.*

MEASURE 8.3.2 Install diffusers for existing clear skylights.

RATINGS

New Facilities	Retrofit	O&M
	B	

SUMMARY

A fairly simple technique that converts useless skylights into effective sources of daylighting. May also reduce heat loss. Affects the appearance of the space.

SELECTION SCORECARD

Savings Potential .... $ $ $

Rate of Return, Retrofit.......... % % % %

Reliability .... ✓ ✓ ✓

Ease of Retrofit .... ☺ ☺ ☺

Direct sunlight through skylights is not suitable for illumination. It is much too intense, it forms localized bright spots, and it shines on the wrong places. For example, consider a space with a tall ceiling that has many small skylights. If the skylights are clear, they produce intense bright spots on the floor, surrounded by darkness. The skylights produce no useful illumination in either the areas that are directly daylighted or the areas that are not. Occupants typically turn on all the available electric lights to compensate for the glare of the bright spots. Figure 1 shows an example.

You can correct these problems and create useful illumination by installing diffusers under the skylights or above them. Diffusers distribute the daylight uniformly in a broad area underneath each skylight. If the skylights themselves are reasonably well arranged, effective and pleasant illumination results.

In new construction, diffusion can be built into the skylight itself at little or no additional cost. Diffusion in new skylights is covered by Measure 8.3.1. This Measure corrects the daylighting problems of clear skylights in existing buildings. This is often an easy and relatively inexpensive retrofit. It may also offer opportunities for decoration.

Fit the Solution to the Situation

Before you start making changes, figure out whether the skylights are oversized for daylighting, or not. If they are oversized, figure out what to do with the excess sunlight. In all cases, the location of the diffuser with respect to the skylight affects the light distribution pattern, the amount of light loss, the potential for glare problems, solar heat gain, and conductive heat loss. Here are guidelines for all the situations that can occur.

■ If the Skylights are Not Oversized

If the skylights are no larger than the optimum size for daylighting, as in Figure 1, you want to keep as much of the sunlight as possible, but distribute it better.

In this case, use diffuser materials that have high light transmission. You can use readily available diffuser materials made for fluorescent lighting fixtures. These have appropriate diffusing qualities, low light losses, and low cost. The most appropriate materials for most applications are clear prismatic diffusers and milky diffusers. The prismatic material has the highest light transmission, but a somewhat non-uniform light distribution pattern. It has a more intense, glittering appearance than milky diffusers, especially when viewed

from a short distance. Viewed from a distance, it is no brighter than any other type of diffuser that admits the same amount of light.

Locate the diffusers to avoid glare problems and to maximize or minimize heat gain. Glare is discussed later in this Measure. We will talk about heat gain next.

■ If the Skylights are Oversized for Daylighting, and You Want to Minimize the Cooling Load

In this case, you have to do two things. One is to admit enough of the sunlight to provide daylighting, and distribute it effectively. The other is to keep the unwanted fraction of sunlight outside the space. The solar heat gain into the space is strongly affected by the characteristics of the diffuser material and by the way it is installed.

To achieve the lowest possible heat gain in relation to the amount of daylighting, install the diffuser outside the skylight. In this way, almost none of the heat that is absorbed by the diffuser enters the building. The heat absorbed by the diffuser is carried away by the outside air. The disadvantage of this method is that the diffuser must be designed to have enough strength to withstand wind, snow loads, and hail. It is also exposed to the maximum solar ultraviolet radiation.

If you install the diffuser inside the skylight, try to reject the unwanted fraction of sunlight by reflecting it back out the skylight, rather than by absorbing it. Heat that is absorbed by the diffuser material becomes a cooling load inside the space. You may need two different materials to do this. For example, you may attach a reflective film to the glazing of the skylight (see Measure 8.1.4 about this), and use a separate diffusing material underneath the skylight. Keep the reflecting surface as close to the skylight as possible, to keep any reflected light from being absorbed by parts of the structure between the reflecting surface and the skylight.

All diffusing materials absorb a significant amount of the entering sunlight because there are multiple reflections within the material. For example, absorption is greater with milky diffusers than with prismatic diffusers. Absorption is also increased by fibers, pigments, and other materials that are embedded in the material.

■ If the Skylights are Oversized for Daylighting, and You Want to Exploit the Excess Heat Gain

If you want to absorb the excess daylight to warm the space, you can use any material for the diffuser that is both diffusing and absorbing. This case also offers opportunities for more decorative treatment. You can use many inexpensive translucent materials as diffusers, including fabrics, colored plastic sheets, and even paper. For example, you can create decorative effects by using patterned fabrics as diffusers.

For the most effective distribution of heat within the space, install the absorbing material as far inside the space as possible. This minimizes the amount of sunlight that is reflected back out the skylight, and it improves the distribution of heat within the space. Adjust the absorption of the diffuser material to achieve the desired balance between daylighting and heating. See Measure 8.4.2 for details of this arrangement.

In most geographic locations, you would like to exploit excess sunlight for heating during cold weather and reflect the excess sunlight during warm weather. In principle, you can design an arrangement that will do both. However, this requires the diffusers to be movable, and someone must control them. This involves a high degree of innovation, along with the risk that the installation will not be operated efficiently. See Reference Note 47, Passive Solar Heating Design, for the many issues that you need to consider when using sunlight for heating.

Skylights Installed Above Shafts or Recesses

Skylights are often installed above shafts or recesses in the ceiling. A shaft may be needed to carry the daylight through the roof structure, or a recess may be created as a decorative effect. In these cases, the location of the diffuser with respect to the shaft or recess has a

WESINC

Fig. 1 Potentially good daylighting rendered useless by lack of diffusers The skylights for this gymnasium are clear. Therefore, direct sunlight creates bright spots on the polished floor which are blinding to players and spectators. Note their reflections on the walls and ceiling. Because this daylighting is useless, the electric lights are turned on. The fixtures here are HID. They require a long time to brighten, so they are left on, despite the fact that the gymnasium is usually vacant.

major effect on the light distribution pattern and the potential for glare.

The surface of the shaft absorbs light. To minimize the loss, cover the surface of the shaft with a highly reflective material. A specular surface is better than a diffuse surface for preserving the light, because it reflects all the light downward.

The location of the diffuser in a shaft determines the breadth of light distribution. Installing the diffuser at the bottom of the shaft produces a broad light distribution pattern. Installing the diffuser high in a narrow shaft produces illumination similar to that of a downlight. A diffuser installed high in the shaft deflects more of the light toward the shaft surface, making it especially important for the shaft surface to be highly reflective. Installing a diffuser higher in the shaft keeps people from seeing it, except when it is more nearly overhead. This reduces the possibility that glare will be a problem. Make sure that no sunlight leaks around the diffuser directly into the space.

No matter where you install a diffuser inside a shaft or recess, you will have to deal with the condensation issue discussed below. Try to exploit the increased insulation benefit of sealing the diffuser to the shaft.

Avoid Glare

Diffusion may create serious glare, which is not a condition that can be tolerated. Refer to Measure 8.3.1 about avoiding glare with skylights.

Glare is a potential problem when the bright surface is within the field of vision. Fortunately, people tolerate bright light sources that are overhead, but the source cannot appear too bright. Diffusers are least likely to create a glare problem when they are far above the working level, or when they are installed in the top of shafts or ceiling recesses. Installing diffusers on skylights makes them similar to flat-faced fluorescent ceiling fixtures in appearance and light distribution pattern, although they may be much larger.

Select Materials for Longevity

Select diffuser materials for resistance to ultraviolet degradation. Solar ultraviolet light will darken or destroy all organic materials in time. Plastics, fabrics, and other materials are available with ultraviolet inhibitors to resist damage, but resistance to damage varies widely. Study the specifications of the materials you are considering. Ordinary window glass absorbs ultraviolet strongly, so organic materials will survive longer if the skylight is made of glass, rather than plastic.

Avoid Condensation and Exploit the Insulation Potential

In some cases, adding diffusers can create serious condensation problems. Or, adding diffusers may provide an opportunity to improve the insulation value of glazing.

If you add a diffuser inside an existing skylight, the diffuser will act as an insulating layer between the space and the skylight. The insulating effect lowers the inside surface temperature of the skylight. The lower temperature may radically increase condensation on the skylight glazing, its frame, and the surrounding structure. The problem is worst at night, when there is no sunlight to keep the glazing warm enough to prevent condensation.

There are several ways to avoid this problem. One is to install the diffuser far away from the glazing, allowing free convection around the diffuser so that it has minimal insulating effect. This solution is easy and cheap, but it wastes an important opportunity to reduce heat loss.

Another solution is to seal the edges of the diffuser tightly to the structure around the skylight, so that water vapor cannot migrate into the space between the diffuser and the glazing. The diffuser must be made of an impermeable material, such as plastic sheet. This method of installation makes the diffuser act as another pane of glazing, substantially reducing heat loss. For example, if the skylight has only single glazing, installing a diffuser may cut heat loss to less than half.

If you use this approach, the diffuser is acting as an interior storm window. See Measure 7.3.2 for details about how to do this without causing condensation problems. The main point is that you have to vent the space between the diffuser and the glazing to the outside. If you don't do this right, serious moisture damage will occur in the space between the diffuser and the skylight.

Still another solution is to install the diffuser outside the skylight. This actually reduces the tendency to condense moisture. If an exterior diffuser can be fitted tightly to the skylight, like a storm window, it will add a significant amount of insulation value.

Control the Electric Lighting to Exploit Daylighting

Daylighting is useless unless electric lighting energy is reduced when daylighting is available. This is not just a matter of turning off the electric lights in daylighted areas. You also have to avoid the tendency to increase the electric lighting levels in adjacent parts of the space that are not daylighted to compete with the brightness of the daylighted areas.

See Measure 9.5.3 for methods of controlling electric lights in response to daylighting. Combine these with the other automatic lighting controls of Subsection 9.5 that are appropriate for the activities that occur in the space.

ECONOMICS

SAVINGS POTENTIAL: *30% to 80% of lighting costs in eligible areas. The amount of potential daylighted* area varies widely, depending mostly on the building and glazing configurations.

COST: *Diffuser materials may cost from $1 to $10 per square foot, if you are not too fancy. Installation is often simple. The cost of the lighting controls is covered in Subsection 9.5.*

PAYBACK PERIOD: *Less than one year, to many years.*

TRAPS & TRICKS

DESIGN: *This is more complicated than it may seem. Consider light distribution pattern, decor, reducing heat loss, avoiding condensation problems, limiting heat gain, and avoiding potential glass breakage. Install effective controls for the electric lighting, or else the effort is wasted.*

MEASURE **8.3.3 Install translucent roof and wall sections for daylighting.**

SUMMARY

For generally the same applications as skylights, but very different to design. The material offers a wide range of compromise between optical and thermal properties. Requires particular attention to avoiding glare and minimizing electric lighting.

SELECTION SCORECARD

Savings Potential .... $ $ $ $

Rate of Return, New Facilities % % % %

Rate of Return, Retrofit.......... % %

Reliability .... ✓ ✓ ✓

Ease of Retrofit .... ☺ ☺

Measure 8.3.1 recommended installing skylights or light pipes to provide daylighting. A newer method is to bring in daylight through translucent panels that are an integral part of the building structure. "Translucent" means that the material passes light, but that the material is not clear.

Figures 1, 2, and 3 show some dramatic applications of this technique.

There is not a clear boundary between using translucent panels as construction components for a roof and using these same materials, or similar materials, in skylights. For roof applications, the distinction is largely one of size. However, for wall applications, there are much greater distinctions between translucent panels and conventional glazing.

The use of translucent panels as structural elements for daylighting has these distinctive and important characteristics:

- Because the translucent panels are an integral part of the building structure, they substitute for other roof or wall components, and they do not require additional structural support. As a result, the overall cost of building with translucent panels may be less than the overall cost of using skylights of similar area. The entire surface of a building could be made of such materials, and this is done with greenhouses, for example.

- Translucent panels can be installed in walls as well as roofs. This allows them to provide daylighting at any level of perimeter spaces. This may be valuable in spaces with tall walls.

- Translucent material is available that has much better insulating properties than conventional skylights and windows. On the other hand, the large surface areas that are typically used make conductive heat loss a more important issue than with skylights that are sized for daylighting alone.

- Glare is an even more serious design issue than with skylights. This is because large panels are more exposed to view, especially when they are used in walls.

- The large surface areas and relatively low cost favor passive solar heating more strongly than when using skylights.

- Presently available materials are all inherently diffusing. This avoids the need for separate diffusers. However, views are not possible through the material.

- All presently available translucent structural panels are made of plastic, which has limited life. In many

buildings, the panels will not last as long as the building, so they will need to replaced. (Do not be misled by the "economic life" of a building, which is a fiction that is meaningful only for accounting purposes.)

Where to Use Translucent Roof and Wall Panels

This method of daylighting applies to the same general categories of applications as skylights. To avoid duplication and to ease the comparison to skylights, refer to Measure 8.3.1 about applications. However, there are major differences between this method and using skylights. These differences may strongly favor one method over the other in particular applications.

Using translucent panels usually involves much larger surface areas than using skylights. None of the translucent materials can approach the thermal resistance of a well insulated roof or wall. This is a major negative factor in locations that have long, cold winters. On the other hand, the large surface areas can be effective for passive solar heating, which is valuable in locations that tend to have clear skies during cold weather.

This method is most widely applicable in new construction, where translucent panels offer the important economic advantage of substituting for other construction materials. Also, new construction can easily accommodate the installation requirements of the material. Retrofitting translucent panels in existing buildings is typically much more difficult than retrofitting small skylights. It may be practical to replace a lightweight roof in a building that does not have an attic. Retrofitting translucent wall panels may be practical if the building presently has large, contiguous areas of windows, or if the building has a curtain wall that needs replacement, or with metal buildings.

Energy Saving Potential

Translucent roofs and walls can provide virtually any illumination level that is required during the daytime, within the spaces that have the translucent surfaces. Of course, night operation still requires electric lighting.

Translucent panels may radically increase the cooling load if the panels are not designed carefully to avoid this problem, as discussed below. On the other hand, sunlight has a better ratio of light to heat than any type of electric lamp. Therefore, if the daylight is

Kalwall Corporation

Fig. 1 Extensive use of translucent panel construction in a manufacturing plant The translucent panels provide daylighting from overhead and from the sides.   Small conventional windows provide view.   The bottom photo hints at the degree of daylight penetration on a lower floor. Note the placement of the light fixtures.

distributed efficiently and the panels are selected properly, the increase in cooling load may not be serious.

By the same token, translucent panels can easily be designed to provide significant passive heating during cold weather. This advantage is offset by conductive heat loss at night. In all but the coldest climates, there is a net heat gain if the panels are located where they can collect the maximum amount of sunlight.

Characteristics of Fiber-Reinforced Composite Panel Materials

At the present time, two general types of translucent panels dominate the market. One is a composite material, which we will discuss first.

■ Construction Features and Structural Properties

This material is made of thin plastic sheets that are made by embedding glass fibers in polyester resin. This is similar to the construction of boat hulls and composite aircraft skins, except that less glass fiber is used in the resin. The glass fibers give the material high tensile strength, even as the plastic material itself loses strength with age.

To provide additional strength and to keep the outer surface intact as the resin decays, the manufacturer may embed a thin, dense mat of glass fibers into the outer surface of the outer sheet. As an alternative, the manufacturer may laminate a more durable plastic surface to the outer surface, but this method itself is limited in life.

To create a rigid panel and increase thermal resistance, the flat sheets are bonded to spacers, resulting in a box structure. Figure 4 shows a cross section detail of a typical installed unit. The spacers may be aluminum extrusions or composite materials. The spacers can be curved in either the length or the width, allowing the panels to be curved in one direction. Spacers that include thermal breaks to reduce conductive heat loss are an important option.

Kalwall Corporation

Fig. 2 Translucent panel construction for a library   Reading is a particularly challenging application for daylighting. The translucent roof section provides effective illumination for a relatively narrow central area that is limited by the tall joists that run underneath it. The translucent wall sections require careful selection to avoid glare problems. We hope that the lights are normally turned off with this much daylight.

The spacers act as beams that carry the loads that result from wind, snow, etc. The panels can be very large, each panel offering an unbroken surface that carries its own weight. The panels cannot carry loads from other parts of the building structure. Each panel is supported at its edges, so the frame is typically made of aluminum extrusions that include mounting flanges. A variety of mounting configurations are offered.

As with double-glazed windows, it is essential that the box structure be tightly sealed. Therefore, the manufacturer must install a frame that seals the edges reliably, even as the edge carries the weight on the panel and absorbs thermal expansion.

A radical improvement occurred with translucent glazing when manufacturers started to offer insulation inside the panels. At present, glass fiber batt insulation is the insulation used in composite panels.

Kalwall Corporation

Fig. 3 Translucent roof in airport waiting room The roof provides daylighting throughout the space, while conventional windows provide a view. The location is southern Florida, so roof heat loss is not a major issue. However, the translucent panels must be selected carefully to balance daylighting against solar heat gain.

■ Available Dimensions

Fiber-reinforced composite material is available in large, unbroken surfaces. For example, one major manufacturer offers widths up to 5 feet (1.5 meters) and lengths up to 20 feet (6 meters). The most common thickness (not including the frame) is somewhat less than 3 inches (about 70 millimeters).

■ Service Life

The glass fibers and the aluminum components of composite panels have virtually unlimited life. But, regardless of what the salesman says, plastic materials weaken, become brittle, and darken as they age. Also, the organic materials used for bonding and sealing plastic sheets to spacers and frames lose their effectiveness with time. The main cause of degradation is the ultraviolet component of sunlight. Oxidation, chemical instability of the material, and repeated flexing from thermal expansion are also factors.

The manufacturer of the glazing material can greatly extend the life of the resin by adding ultraviolet inhibitors and other agents to it, or less reliably, by adding protective surface coatings. However, in the long term, the glass fiber is what holds the material together as the plastic decays. Therefore, pay attention to the way that different manufacturers embed the glass fibers in the resin.

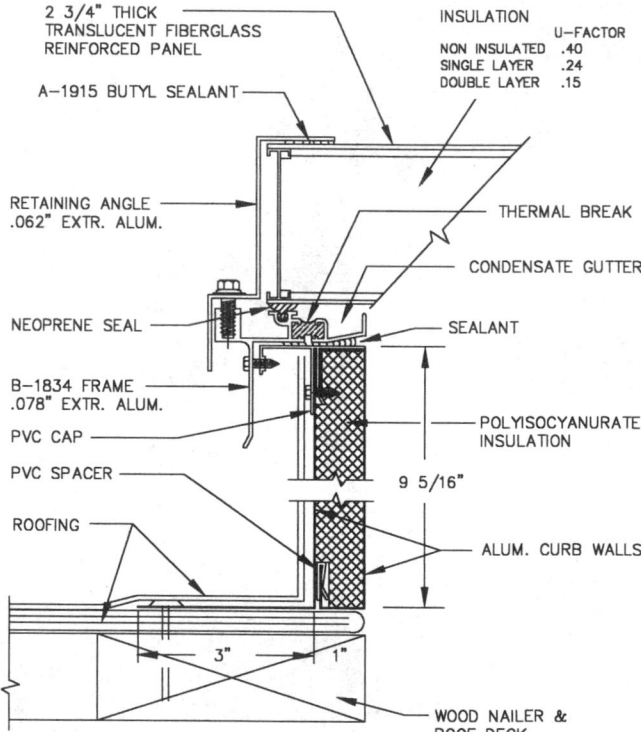

Vistawall Architectural Products

Fig. 4 Typical composite panel construction and installation This is for a roof installation. Notice the choice of insulation filler, the thermal breaks, and the condensation gutters.

Unfortunately, you cannot judge the long-term performance of the plastic material itself except from the manufacturers' claims, so purchase your panels from a manufacturer who can demonstrate that its material has long life in past installations.

■ Light Transmission vs. Insulation Value

In composite panels that use insulation inside the panels, the insulation interferes with light transmission. There is an inverse relationship between the maximum possible light transmission and the maximum possible insulation value.

For example, an uninsulated panel has a thermal resistance of approximately R-2 and a maximum light transmission of about 50%. Increasing the thermal resistance to R-4 reduces the maximum light transmission to about 30%. Panels that have the highest insulation value, presently about R-10, are limited to a maximum light transmission of about 10%. Increasing the strength of the material by adding additional glass fiber to the resin further reduces the light transmission that is possible for a given R-value. The overall average R-value of the panels is reduced by the high localized heat losses through the aluminum frames.

If desired, light transmission can be reduced by tinting the glazing material. However, it generally does not make sense to do this. Instead, use more insulation or glazing with more fibers to reduce the light transmission.

■ Diffusion and View

All composite panels diffuse light. All except the clearest versions appear white from the inside, which means that daylight enters the space in a broad, fairly uniform pattern. The clearest material, which has no insulation, has an opalescent or "crystal" appearance.

No translucent structural panel material is clear enough to provide a satisfactory view. To gain a view, you need to install regular windows. Windows are installed as with other types of wall construction, independently of the diffuser material.

■ Appearance

Composite panels have the appearance of a series of rectangles, created by the grid pattern of the internal spacers. From the inside, the material looks white when there is daylight, and neutral gray at night. From the outside, the material looks white to gray, depending primarily on the characteristics of the outer sheet.

The flat plastic glazing material buckles with thermal expansion and with age. This can be quite noticeable when the material is observed from the outside at a shallow angle, but it is usually not noticeable when looking at the panel from the interior.

The inner or outer sheets can be tinted, but this is rarely done, because it would waste light transmission or sacrifice insulation value.

■ Fire Characteristics

Plastic materials are made from petroleum, and they have considerable fuel value. All plastic materials will burn if they are exposed to temperatures that are hot enough. For this reason, manufacturers add fire retarding agents to the material. You have to decide whether installing plastic panels is acceptable in terms of fire safety.

Read the specifications carefully, including the temperatures at which the material is rated. A low rating temperature is meaningless in a hot fire. Bear in mind that buildings typically contain many types of combustible materials, and that the primary defense against fire is making it possible for all occupants to escape the building quickly.

Characteristics of Extruded Plastic Translucent Panels

The other common type of translucent glazing material consists of one-piece double-wall plastic extrusions. We will compare it to the composite material discussed previously. In summary, it has smaller maximum size and shorter life, but it may be less expensive to install, and it may be easier to retrofit.

■ Structural Properties

This type of glazing material is a continuous plastic extrusion. The plastic material is usually polycarbonate, which is preferred for its strength. The extrusion has two separate surfaces which are separated by ribs that run the length of the extrusion.

Extruded plastic is weaker than composite panels because the plastic is not reinforced, and because the ribs run only lengthwise. Therefore, the maximum panel dimensions are smaller. The ribs that join the inner and outer surfaces of the material act as beams to carry the loads of wind and snow. However, the ribs carry load only in the lengthwise dimension. The ability of the material to carry load and to resist deformation across the width of the panel depends primarily on the thickness of the inner and outer surfaces. Strength in both directions, and resistance to bending, depend strongly on the overall thickness of the extrusion.

The material is supported at its edges, typically by aluminum extrusions. The glazing extrusion is divided into lengthwise cells by the ribs, so in principle only the ends need to be sealed. However, it is common practice to leave the ends open inside the frame, and to seal the material by using gaskets on all four sides of the panels.

Leakage of water vapor into the cells of the plastic extrusion has the same effects as leakage of water vapor between the panes of multiple-glazed window, including permanent fogging of the material. (See Measure 8.1.3 about this.) The problem of fogging is much less noticeable than with clear windows if the panels are designed to diffuse light.

■ Insulation Value and Light Transmission

Extruded plastic glazing typically has two separate surfaces, which are separated by closely spaced ribs. Two glazing surfaces, without additional features to reduce heat loss, yield an insulation value of about R-2. This value is substantially reduced by heat conduction through the ribs and the frame. In addition, the R-value is greatly reduced if the material is installed in a tilted orientation. Thus, extruded plastic glazing, by itself, has poor thermal resistance.

To increase the thermal resistance of plastic glazing, at least one manufacturer offers a layer of translucent plastic foam insulation that is bonded to one of the surfaces or is sandwiched between two sheets of the extruded glazing. Figure 5 shows a detail of the latter. The foam insulation reduces light transmission and it strongly diffuses the light. The manufacturer presently offers a foam insulation thickness of 6 millimeters, combined with one or two sheets of extruded glazing, each of which is 8 millimeters thick. The manufacturer claims an insulation value of about R-6 for the sandwich configuration.

There is a compromise between insulation value and light transmission. The polycarbonate glazing material is clear, unless the customer requests tinted material. The foam insulation material has surprising high light transmission. The sandwich combination described previously has a light transmission of about 45% for light striking the material perpendicular to the surface, and about 25% for light striking the surface at an angle of about 60 degrees from the plane of the ribs.

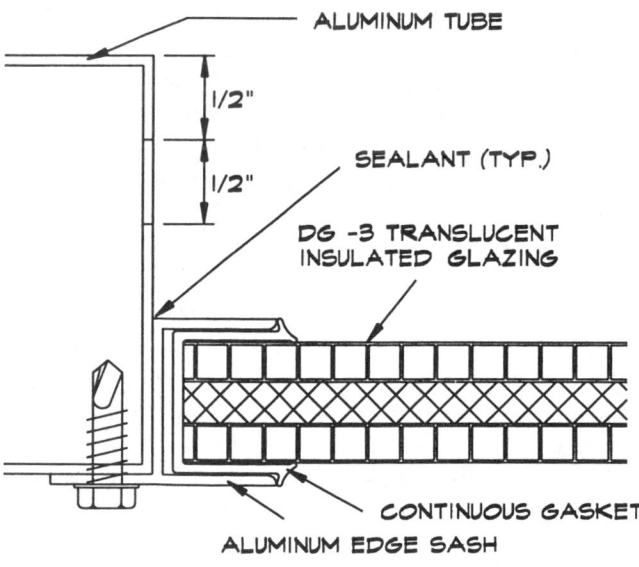

ALUMINUM TUBE

1/2"

1/2"

SEALANT (TYP.)

DG -3 TRANSLUCENT INSULATED GLAZING

CONTINUOUS GASKET

ALUMINUM EDGE SASH

Duo-Gard Industries, Inc.

Fig. 5 Extruded plastic panel assembly with insulation
This is an end view of two extruded panels with a layer of translucent plastic foam insulation between them. The rigidity of the assembly depends mainly on the thickness of the individual extruded panels.

At present, the compromise between light transmission and insulation value does not extend far enough in favor of insulation. The light transmission may be much higher than needed when glazing large areas. Manufacturers of extruded plastic glazing offer tints to reduce light transmission. However, this is an inefficient solution, because it does not offer any advantage in improved insulation value. Unfortunately, efficient solutions are more complex. See Reference Note 47, Passive Solar Heating Design, about dealing with more sunlight than you need.

All plastic materials darken as they age. The amount of darkening depends on the composition of the material and its exposure to ultraviolet light and other environmental factors. With the large surface areas being considered in this method of daylighting, it is usually easy to compensate for the darkening that can be expected during the life of the material.

■ Service Life

The life span of extruded plastic is shorter than the life of composite materials because the material lacks embedded inorganic fibers to hold the material together as it ages. Experience with polycarbonate plastics suggests that its maximum life in exterior applications is between ten and twenty years. This assumes that the manufacturer of the plastic material includes effective ultraviolet inhibitors and other agents to extend service life. In any event, the material weakens and becomes brittle as it ages, making it more vulnerable to damage from minor impacts.

Some manufacturers offer warranties as long as ten years. However, the duration of a warranty is a marketing decision as much as it is an indication of the reliability of the material. Again, purchase your panels from a manufacturer who can demonstrate that its material has long life in past installations.

■ Diffusion and View

Clear or tinted polycarbonate glazing is almost clear enough to provide a view, although the view is distorted by the molded ribs and by irregularities in the extrusion. However, in order to achieve diffusion for satisfactory daylighting, it is necessary to make the material diffusing, or to combine it with diffusing insulation. Tinting alone does not provide diffusion.

To provide a view, you need ordinary windows. Some manufacturers offer windows that are combined with diffusing panels in a single frame. However, this combination is usually limited to locations where the glazing is not in direct sunlight. Otherwise, direct sunlight entering the clear windows would cause excessive glare. In applications where the glazing is not in direct sunlight, the heat loss through the glazing during cold weather is not compensated by solar heating. Therefore, this combination should normally be limited to warmer climates.

■ Appearance

From the inside, the insulated material looks white when there is daylight, and neutral gray at night. If an inner sheet of extruded glazing is installed, the ribs add a striped pattern.

From the outside, the material looks white to gray, with a light striped pattern caused by the ribs of the extruded glazing. Tinting the glazing changes the grayness or color of the material.

Extruded glazing does not suffer from the surface buckling that occurs with the thin sheets of plastic used in composite panels. However, the exposed foam insulation has the esthetics of packaging material.

■ Fire Characteristics

The same cautions apply that were made previously with respect to composite panels.

Appropriate Surface Area

Translucent panels typically are used for all or a large fraction of the roof or wall area. In contrast, Measure 8.3.1 explains that the appropriate area for skylights ranges from 4% to about 15% of ceiling surface area. However, small areas of panels cannot economically substitute for other construction materials because they would have to be fitted into the structure as a separate item. (Translucent panel materials are sometimes used in skylights, but they are as expensive as any other kind of skylight when used in this manner.)

As a result, there is an economic attraction to using much larger areas, perhaps covering an entire roof or wall. This violates the guideline for skylights given in Measure 8.3.1, to keep the areas as small as possible. In order to avoid excess daylighting, the light transmission of the material must be reduced drastically. At the same time, this reduces the heat gain. Whether this is good or bad depends on the application.

A design based on large translucent areas suggests exploiting passive solar heating. In turn, this requires a separate set of techniques to maximize heat gain while controlling the illumination level. See Reference Note 47, Passive Solar Heating Design, for an introduction to this complex subject.

Avoiding Glare Should be a Dominant Design Issue

With daylighting, avoiding glare is always a major issue. Using translucent panels of large area raises the glare issue to potentially disastrous proportions. If the design of the installation does not avoid occupant discomfort from glare, it is virtually certain that the translucent panels will be covered with opaque material at some later time, eliminating all of their benefits. To understand the problem of glare, see Reference Note 51, Factors in Lighting Quality.

The problem of glare, and the methods of dealing with it, are closely related to the need for diffusion.

Daylighting requires diffusion, for the reasons explained in Measure 8.3.1. On the other hand, diffusion creates glare.

There are two ways to avoid glare from large surfaces. One is to limit the surface brightness. This technique is essential in most wall installations, as illustrated by Figure 6. The other way to avoid glare is to keep the bright surface out of the visual field of the occupants. This may be your primary approach with roof systems, as illustrated by Figure 7.

■ How to Limit Glare in Wall Installations

Using translucent panels in walls involves the greatest potential for glare problems. To understand the problem, consider an occupant looking toward a clear, south-facing window that is exposed to direct sunlight. If the window is clean, the occupant sees only the scenery outside, and there is no glare. Although direct sunlight is coming through the window, it disappears into the carpet. The occupant does not experience any visual problem as long as the bright spot on the floor is hidden from view by furniture, partitions, etc.

Now, consider what happens if the window is replaced by a diffusing panel that has high light transmission. The panel scatters the strong sunlight into the space. The occupant sees the diffuser as a bright, uniformly illuminated surface. The effect is like looking directly at a large fluorescent fixture. Now, visualize the effect if the entire wall is that bright. It would be unbearable. With this in mind, look at the various figures in this Measure that illustrate translucent walls. Decide how you would avoid glare problems with them.

In general, the way to avoid objectionable glare with translucent walls is to select panel material with light transmission that is low enough to allow people to look at the walls directly without discomfort, at least for the length of time that the occupants will be facing the walls. For example, if a large space has cubicles where people work, and the cubicles keep the translucent walls out of the visual field of people who are working, the translucent panels can be much brighter. However, even in this case, remember that the space layout and usage may change during the life of the building.

As a crude guideline, if the glazing will be within the occupants' field of view for extended periods, the glazing should be no brighter than a typical well illuminated white painted wall. At this point in time, we can offer nothing better than crude guidelines on this issue. Experience is the only reliable teacher when it comes to visual quality, and there are still relatively few translucent wall installations that serve as good examples.

If sunlight can strike the exterior surface of the panels directly, avoiding glare requires you to select the light transmission of the panels for times when this occurs. However, the panels will be illuminated by direct sunlight at some times, and not at others. The amount of useful daylight that you will get when the sun is not shining directly on the panels depends on the range of illumination levels that is acceptable within the space. Again, the nature of the activities in the space and the space layout determine this range. Of course, additional illumination can be provided by electric lighting, but then the translucent wall is not paying for itself.

Diffused glazing with low transparency may look gloomy when the sun is not shining on it directly. If appearance is an important factor, make sure that you inspect similar installations before committing to translucent wall panels.

If you intend to exploit daylighting alone, and not passive solar heating, you will probably want to select panel material that has much lower light transmission than you commonly find in translucent panel installations. Fortunately, as we have seen, you can now get panel material that reduces its light transmission by increasing its insulation value.

If the building is located in a warm climate, or if the space has high internal heat gain, you can use panels of high light transmission on walls that face north. However, remember that even north-facing surfaces are illuminated by grazing sunlight during the early morning and late afternoon in summer.

More generally, each face of the building should have panels with different amounts of light transmission. For example, in southern latitudes, sunlight always strikes the south wall of a building at much more grazing angles than it strikes the east and west walls.

As a final note, you could deal with glare, at least in principle, by facing all activities in the space away from the windows. However, this would rarely be practical. For example, in a classroom, all the students could face away from the translucent wall, but the teacher would have to face the bright panels. Furthermore, this layout would cause people to cast their own shadows on their activities.

Kalwall Corporation

Fig. 6 A challenging translucent wall application In this computer classroom, the students are required to face the translucent walls for long periods. Avoiding glare must be a major design issue here. The potential for glare, as well as the potential for saving energy by daylighting, is determined by the transparency of the wall. The dominant factor in selecting panel transparency is whether the exterior of the panels is exposed to direct sunlight. The relative brightness of the walls and the computer screens must also be considered. The ceiling is full of light fixtures. If the walls are too bright, the lights will be kept in operation to compensate.

■ How to Limit Glare in Roof Installations

The problem of glare with translucent roofs is similar to the glare problem with skylights that is discussed in Measure 8.3.1. However, there are significant differences. Skylights, if properly designed, are relatively small areas of high brightness. In contrast, translucent roofs have very large areas. However, if translucent roofs are selected for reasonable illumination levels, their surface brightness is much lower. As with skylights, exploit the fact that people tolerate areas of high brightness that are well above the line of sight.

To the extent that people must face toward translucent ceiling areas, follow the same guidelines given previously for walls. If necessary, consider making the lower portions of translucent ceilings less bright than the taller portions.

Select Panel Materials for the Solar Heat Gain Characteristics You Want

Solar heat gain through the panels is strongly affected by the optical characteristics of the sheets used to make the panels, and by their positions with respect to each other.

All diffusing materials absorb a significant amount of the entering sunlight because there are multiple reflections within the material. If the panels include a layer of insulation, most of the diffusion and absorption occur within the insulation. Absorption is also increased by fibers, pigments, and other materials that are embedded in the material. As with any glazing material, the manufacturer can reduce light transmission by adding dyes to make the material more absorptive. All light that is absorbed becomes heat within the panel. If you want to minimize heat gain, concentrate the absorbing and diffusing properties in the outer surface of the panel, to the extent that you have a choice. In this way, more of the absorbed heat is carried away by the outside air, rather than entering the space by radiation.

In principle, it is possible to reduce light transmission without absorbing so much heat in the glazing material by using a reflecting coating. This is a common technique with glass. However, reflecting surfaces are not yet offered as a feature of plastic panels by most manufacturers.

The best way to reduce heat gain is to use insulation as the method of reducing light transmission to the desired level, while minimizing other features that absorb light.

On the other hand, if you want to maximize heat gain to exploit passive solar heating, concentrate the absorbing and diffusing properties in the innermost surface, to the extent that you have a choice.

You can increase heat gain even more by making the panels as clear as possible, and by installing a separate diffuser inside the space that is made of absorptive material. Measure 8.4.2 explains this

arrangement. It minimizes the amount of sunlight that is reflected back outdoors, and it makes it easier to distribute the captured solar heat within the space. If you install a separate diffuser, make sure that no sunlight leaks around the diffuser directly into the space. Direct sunlight is an intense source of glare, and it is useless for illumination.

However, this may not be the best arrangement, on balance. Minimizing conductive heat loss through the glazing is a major factor in passive solar heating. Exploit the fact that insulation is now available as an integral part of translucent panels. With large panel areas, it is probably better to sacrifice some heat gain during hours of daylight by increasing the amount of insulation to reduce heat loss during hours of darkness.

Deal with the Higher Conductive Heat Loss

Using translucent panels involves much larger surface areas than using skylights. Furthermore, all translucent structural panels have much lower thermal

Major Industries, Inc.

Fig. 7 Translucent roof with relatively low glare potential
The translucent roof provides daylighting for several levels. Glare is a major issue in this building because the occupants are elderly. The potential for trouble is minimized by the elevation of the translucent surface well above the line of sight and by the nature of the activities, which allow the occupants to choose their direction of gaze.

insulation value than a typical well insulated wall. As a result, the insulation value of the material is a major issue if the climate can be cold for extended periods. Don't sweep this issue under the rug, or you may lose more in additional heating costs than you save in lighting costs.

Again, exploit the fact that translucent panels are now being offered that have much better insulation value than ordinary glazing. By the same token, using translucent panels that have integral insulation also minimizes the severe problem of sweating that occurs with ordinary glazing. (Measure 8.3.1 describes these problems.)

Another way to minimize heat loss is to use movable insulation, which provides high insulation value during hours of darkness. Movable insulation is challenging to design and to install. For an introduction to it, see Reference Note 47, Passive Solar Heating Design.

How to Prevent Water Leakage

Translucent panels have large unbroken surfaces. Established manufacturers offer frame systems that are reasonably reliable for preventing water leakage between the panels and the frames. However, you need to make sure that the frame design provides a watertight seal between the frame and the building structure. The glazing supplier has seen all the problems before, so work with the factory on this.

Design to Protect and Replace the Material

All types of plastic glazing deteriorate within a fraction of the normal life of a building, for the reasons discussed previously. Therefore, make it easy to replace the glazing several times during the life of the building. Your successors will thank you. Even you may still be around to replace the glazing.

Recognize the weaknesses of the material, and design the installation so that the glazing survives as long as possible. Both plastic and aluminum expand and contract by large amounts with changes in temperature. The frame system should be designed to avoid motion of the glazing with respect to the frame, which will wear out the glazing seals. Design the panel installation to allow for the difference between the expansion of the panel and the expansion of the structure, while maintaining an airtight seal where the two meet. The frame design is part of this, but additional features may be needed, such as sliding seals and elongated bolt holes. Mounting the panels too rigidly will create large stresses in the frames and in the glazing, accelerating failure of the seals between the glazing and the frame, and between the frame and the structure.

In locations where severe hail is likely to occur, select roof-mounted panel material for high surface strength. Composite materials are available with a dense layer of fibers embedded near the outer surface, and this may help resist impact damage. (Severe hail may not occur often in most locations, but only once is enough to destroy the glazing.)

Snow load is an issue with roof-mounted panels, as with any other roofing material. Fortunately, smooth glazing tends to shed snow. This is because heat that conducts through the glazing melts the snow adjacent to the panel surface, and slides it off. Of course, this effect is more dependable with steeper roof slopes.

Movable insulation can be employed to protect the glazing, if it is installed externally. However, movable insulation is a troublesome feature, and it does no good unless it closes reliably to provide protection when it is needed.

Control the Electric Lights to Exploit Daylighting

Daylighting is useless unless it reduces the amount of energy that is consumed by electric lights. The glare issue poses a particular problem with translucent glazing, especially when it is installed in walls. The reason is that occupants will tend to turn on all the interior lights in an attempt to match the electric lighting levels to the levels of lighting provided by the glazing. Thus, contrary to what you might expect, a high level of illumination by the translucent panels does not reduce the demand for electric lighting, but increases it. Reference Note 46, Daylighting Design, covers this issue in greater detail. In summary, human beings try to keep illumination levels uniform inside buildings.

You need automatic lighting controls to adjust the electric lighting in response to the amount of daylighting. See Measure 9.5.3 for the details. However, even automatic lighting control will not survive if glare makes occupants uncomfortable.

Compatible Types of Electric Lighting

Expect to use fluorescent lighting, rather than HID, in daylighted spaces. This is true even in the tall-ceilinged spaces where HID lighting is commonly used. See the discussion of this issue in Measure 8.3.1.

ECONOMICS

SAVINGS POTENTIAL: In the most favorable cases, translucent roofs and walls can provide a majority of lighting needed during the daytime. However, this method of daylighting may increase heating cost substantially, and it may also increase cooling cost.

COST: Plastic composite panels and plastic combination insulated glazing costs from $10 to $35 per square foot ($100 to $350 per square meter). Installation cost may be more or less than for the material that it replaces. See the Measures of Subsection 9.5 for the costs of automatic lighting controls.

PAYBACK PERIOD: *In new construction, several years or longer. In retrofit, the payback period is very long, unless the panels are a substitute for other materials that need to be replaced.*

TRAPS & TRICKS

MAKING THE COMMITMENT: *Do not use this feature unless you make a commitment to dealing with all the design and installation issues. Otherwise, you may create a big mess. Translucent roofs and walls are a major, permanent part of the building structure. They strongly affect lighting, heating, and cooling costs. They strongly affect the appearance of the building, both outside and inside. They have a major effect on occupant comfort. They may be incompatible with later uses of the building. The panels will probably require replacement during the normal life span of a building. To know what your building will look like twenty years from now, look at similar material that was installed twenty years ago.*

SELECTING THE MANUFACTURER: *The manufacturer is as important as the material. Select a manufacturer who knows how to avoid potential problems related to installation, including water leakage and panel replacement. The manufacturer should offer detailed design literature, and he should have the patience to work with you. However, don't expect the manufacturer to be candid with you about the inherent difficulties of the material, such as glare and heat loss. You have to take the initiative to resolve those issues.*

MEASURE 8.3.4 Install diffusers to make windows more effective for daylighting.

Direct sunlight through windows is not suitable for illumination. The illumination where the sunlight falls is much too intense, blinding the viewer. At high sun angles, daylighting is limited to a region close to the window, where most of the light falls on the floor. As a result, windows exposed to direct sunlight are equipped with curtains or shades that exclude almost all of the sunlight, wasting the potential of daylighting.

You may be able to overcome this problem and create useful illumination by installing diffusers on windows. Diffusers can be any translucent material that redirects the sunlight into the space in a fairly uniform pattern. This substantially increases the penetration of daylight into the space, eliminates the intense contrast between lighted and unlighted surfaces, and projects the light on the illuminated surfaces from a higher angle.

Diffusers are fairly easy to install, and they may not cost much. You can use a variety of inexpensive materials for them. However, the range of applications is severely limited by the potential for serious glare from the diffuser itself.

Where to Consider Diffusers for Windows

Diffusion is useful only with windows that are directly illuminated by the sun for a large fraction of the day. Light from the empty sky is already diffused. Also, for diffusion to provide a benefit, the glazing must be located properly with respect to the activities in the space.

Even where these conditions are met, diffusers are satisfactory only if they can be installed in a manner that does not cause discomfort to the occupants from glare. The discussion below will help you decide whether you can install diffusers without excessive glare. As a first criterion, do not expect to use window diffusers if occupants must face toward the windows for extended periods.

Aside from the issue of glare, daylighting should strike the illuminated surfaces at an angle that is high enough to avoid forming shadows on the surface. Tall windows are most favorable for this technique. If the windows are low, most of the sunlight will strike horizontal work surfaces at a grazing angle, which is generally unsatisfactory.

If you are constructing a new building, do not use this technique. There are much less troublesome ways of providing daylighting. See the other Measures in this Subsection.

SUMMARY

An unconventional technique that increases the penetration of daylight from windows. Uses inexpensive materials. Avoiding glare problems is the major challenge.

SELECTION SCORECARD

Savings Potential	$	$	$
Rate of Return, Retrofit	%	%	%
Reliability	✓	✓	✓
Ease of Retrofit	☺	☺	☺

Energy Saving Potential

The amount of space that can be daylighted effectively by windows depends on their area and on their shape. The window height is important because daylight must shine downward to be useful. This factor limits the useful penetration of daylight to approximately two to three times the height of the windows above the illuminated surfaces. For example, if the windows are ten feet tall and the illuminated surfaces are three feet above the floor, useful daylighting may penetrate fifteen to twenty feet into the space.

If highly transparent diffusers are potentially too bright, you may be able to use diffusers of lower transparency. In this case, the need to limit brightness may become the limiting factor.

Electric lighting typically requires one to four watts per square foot (10 to 40 watts per square meter) of floor space. This amount of lighting power is saved during periods when daylighting can substitute for electric lighting.

Avoiding Glare is Your Main Challenge

Using diffusers with windows involves a potential glare problem that is so bad that you must take steps to avoid it. (The term "glare," as we use it here, means an area of intense brightness within the visual field. Reference Note 51, Factors in Lighting Quality, explains glare.) To understand the problem, consider an occupant looking toward a south-facing window. If the window is clean, the occupant sees only the scenery outside, and there is no glare. Although direct sunlight is coming through the window, it disappears into the carpet. The occupant does not experience any visual problem as long as the bright spot on the floor is hidden from view by furniture, partitions, etc.

Figure 1 shows what happens if a highly transparent diffuser is installed in that window. The direct sunlight is captured by the diffuser and scattered into the space. The occupant sees the diffuser as a bright, uniformly illuminated surface. The effect is like looking directly at a large fluorescent fixture. Whether you can effectively deal with this glare depends on the geometry of the space and the nature of the activities in the space.

In general, glare is less of a problem if the light source is located well above the line of sight. This favors tall windows. Indeed, you may want to install diffusers only on the upper parts of tall windows, and cover the lower portions of the windows with conventional shades or replace them with insulating panels (as recommended by Measure 7.3.3).

Glare from the diffusers is not a problem if all the occupants face away from them. See Figure 2. However, there are few activities where this layout can be assured for the life of the building. For one thing, this orientation would cause people to cast their own shadows on their activities.

In most cases, expect to select diffuser material that has such low transmission that occupants can face the diffusers, at least for short periods, without discomfort. This reduces the amount of light entering the space, so the windows must be fairly large to provide a useful amount of illumination. (The extreme of this approach is installing translucent walls. See Measure 8.3.3 about this.)

The window will be illuminated by direct sunlight at some times, and not at others. You will probably not get a useful amount of daylight through diffused glazing when sunlight does not shine on it directly. Furthermore, diffused glazing with low transparency may look gloomy when the sun is not shining on it directly.

Not matter which approach you take, test your ideas with a mockup at a typical window and activity area before you commit to permanent installations.

WESINC

Fig. 1 How window diffusers behave In this demonstration, sunlight is shining on two clear windows of an office. A milky white plastic diffuser is installed inside the window at right. The light fixtures are turned on for comparison. The diffuser causes a considerable amount of glare, the intensity of which cannot be conveyed by the photograph. In contrast, the scene outside the clear window is not bright enough to create glare. From this point of view, the surface of the diffuser appears considerably brighter than the surfaces of the fluorescent light fixtures. Notice the amount of light scattered by the diffuser on the ceiling. This aids daylight penetration. The ceiling should be highly reflective to exploit this.

How to Install Window Diffusers

In existing buildings where you will not be replacing the windows, your only option is to install a separate diffuser material inside the windows. Consider readily available diffuser materials made for fluorescent lighting fixtures. These have appropriate diffusing qualities, low light losses, and low cost. The most appropriate materials for most applications are clear prismatic diffusers and milky diffusers. The prismatic material has the highest light transmission, but a somewhat non-uniform light distribution pattern. It has a more intense, glittering appearance than milky diffusers, especially when viewed from a short distance. Viewed from a distance, it is no brighter than any other type of diffuser that admits the same amount of light.

For applications where you have much more sunlight than you need, you can use colored plastic sheet, fabrics, and even paper. For example, you can create decorative effects by using patterned fabrics as diffusers.

You can buy roller shades with various kinds of diffusing material.

If you plan to replace the windows in a building, decide whether or not to make the windows themselves diffusing, or to install separate diffusers. If you install diffusing windows, the decision to exploit diffusion will be irrevocable, so you need to be sure that diffusion will not cause glare problems.

If you decide to replace windows with translucent panels because of their higher insulating value, diffusion will be part of the package. See Measure 8.3.3 if you decide to take this approach because it involves a number of choices and compromises, including the transparency and insulation value of the panels. Figures 3 and 4 show a typical window retrofit with diffusing panels.

Diffusing windows provide no view. If you want to retain some view, you must install regular windows along with the diffusing panels, as in Figure 3.

Select diffuser materials for resistance to ultraviolet degradation. Solar ultraviolet light will darken or destroy all organic materials in time. Plastics, fabrics, and other materials are available with ultraviolet inhibitors to resist damage, but resistance to damage varies widely. Study the specifications of the materials you are considering. Ordinary window glass absorbs ultraviolet strongly, but enough ultraviolet gets through glass to damage organic materials within a period of years.

Make the Ceiling Reflective

As we have seen, the ability of daylighting to penetrate into the space depends on the height of the windows. This is because light must shine downward to be useful. The window diffuser aims a large part of the incoming sunlight at the ceiling, as you can see in Figure 1. Don't waste this light. It is especially valuable

WESINC

Fig. 2 A lesson from the real world about window diffusion The orientation of activities with respect to the windows is a crucial factor in using window diffusers. In this office, the upper panes of the double-hung windows are made of diffusing glass. The diffuser to the side of the chair provides good illumination for the desk. However, the diffuser in front of the desk causes glare, so it has been covered with a venetian blind.

Kalwall Corporation

Fig. 3 Window retrofit with diffusing panels The top is the original configuration, with clear windows and roller shades. Notice that the outer row of light fixtures is turned on. The bottom shows the diffusing windows installed. These provide deeper daylight penetration and reduce heat loss. Some outside view is provided by clear hopper windows at the bottom, which can also be opened for ventilation. Notice that the children are all facing away from the diffusing windows, perhaps to avoid glare. However, the teacher must face the windows. See Figure 4 for the exterior appearance.

because it is shining downward from the highest angle. To exploit it, make sure that the ceiling is as reflective as possible. In fact, this is a good opportunity to use a specular (mirror-like) ceiling, if your decor can stand it. See Measure 8.3.6 for more about interior surfaces.

Avoid Condensation and Exploit Insulation Properties

In some cases, using diffusers can create serious condensation problems. Or, adding diffusers may provide an opportunity to improve the insulation value of glazing.

When you install a diffuser inside an existing window, the diffuser acts as an insulating layer between the space and the window. The insulating effect lowers the inside surface temperature of the window. The lower temperature may radically increase condensation on the glazing, its frame, and the surrounding structure. The problem is worst at night, when there is no sunlight to keep the glazing warm enough to prevent condensation.

There are two ways to avoid this problem. One is to install the diffuser far away from the glazing, allowing free convection around the diffuser so that it has minimal insulating effect. This solution is easy and cheap, but it wastes an important opportunity to reduce heat loss.

The other solution is to seal the edges of the diffuser tightly to the surrounding structure, so that water vapor cannot migrate into the space between the diffuser and the glazing. The diffuser must be made of an impermeable material, such as plastic sheet. This method of installation makes the diffuser act as another pane of glazing, substantially reducing heat loss. For example, if the window or skylight has only a single pane, installing a diffuser may cut heat loss to less than half.

If you use this approach, the diffuser is acting as an interior storm window. See Measure 7.3.2 for details about how to do this without causing condensation problems.

Minimize or Exploit Solar Heat Gain, as Appropriate

In many cases, the amount of sunlight falling on the window is much greater than you can use, even if the sunlight is diffused into the space. In such cases, you have to reduce the intensity of the sunlight as well as diffusing it. Even if you want to use all the daylight, all diffusing materials absorb a significant amount of the entering sunlight because there are multiple reflections within the material.

The sunlight that you do not use for illumination represents a considerable amount of heat energy. You may want to keep it outside the building to minimize the cooling load. Or, you may want to bring it inside the building to assist in warming the space. The characteristics of the diffusing material, and the way you install it, determine where this heat energy goes.

To minimize heat gain, minimize the amount of sunlight that gets into the space. For most effective results, consider installing diffusers outside the windows. In that way, almost none of the heat that is absorbed by the diffuser enters the building. This approach is highly visible from the outside, so the appearance of the diffusing material is important. Also, the diffusers must be installed so they resist wind, and the materials must resist exposure to the sun's ultraviolet radiation. Exterior diffusing screens made of glass fibers have been offered as a commercial product for reducing cooling load.

Kalwll Corporation

Fig. 4 The exterior of Figure 3 The windows on the first and second floors have been retrofitted. The windows on the third floor remain. The appearance is changed from aged to slightly industrial.

To minimize heat gain while keeping the diffusers indoors, use a material that reflects the unwanted portion of the light back out through the window. For example, you might use a plastic diffuser sheet that has an aluminized surface. Or, you could install a reflective window film (see Measure 8.1.4) and install a separate diffuser. Be aware that any reflective method noticeably changes the exterior appearance of the building.

Reflecting the unwanted sunlight from the inside of the space is much less effective if the glazing is tinted, because heat is absorbed in the glass in both directions. On the other hand, tinted glazing typically absorbs 60 to 80 percent of the incoming sunlight, so you may not have to eliminate more sunlight before it reaches the diffuser..

On the other hand, if you want to capture the excess sunlight for warming the space, install a diffuser inside the space that is made of absorptive material. See Measure 8.4.2 about this arrangement. Measure 8.4.2 also suggests ways to install diffusers and absorbers so that you can reject unwanted heat at some times and capture the heat at other times.

Control the Electric Lighting to Exploit Daylighting

Daylighting is useless unless it reduces the energy consumed by electric lights. This is not just a matter of turning off the electric lights in daylighted areas. You also have to avoid the tendency to increase the electric lighting levels in adjacent parts of the space that are not daylighted to compete with the brightness of the daylighted areas.

See Measure 9.5.3 for methods of controlling electric lights in response to daylighting. Combine these with the other automatic lighting controls of Subsection 9.5 that are appropriate for the activities that occur in the space.

Effect on View

Diffusers eliminate any view through the glazing. With windows that face the sun, this is usually not a practical disadvantage, because the window would be blocked otherwise to avoid the glare of direct sunlight. Think of diffusers on windows as an alternative to closing off windows completely to avoid the glare of direct sunlight. If you want to exploit a desirable view, select a different Measure.

You could install a movable diffuser, such as a translucent roller shade, but manually operated devices are unreliable for conserving energy.

ECONOMICS

SAVINGS POTENTIAL: *30% to 80% of lighting costs in eligible areas. The amount of potential daylighted area varies widely, depending mostly on the building and glazing configurations. Typical lighting power that is displaced by daylighting is 1 to 4 watts per square foot (10 to 40 watts per square meter).*

COST: *Diffuser materials cost from $1 to $10 per square foot ($10 to $100 per square meter), if you are not too fancy. Installation cost varies widely, depending on the design. The cost of the lighting controls is covered by Subsection 9.5.*

PAYBACK PERIOD: *Less than one year, to many years.*

TRAPS & TRICKS

CHOICE OF METHOD: *The potential for glare is a serious concern. Don't do this if it may create glare problems. Recognize that there is no view through a diffused window.*

DESIGN: *This is more complicated than it may seem. Consider the light distribution pattern, the decor, reducing heat loss, avoiding condensation problems, limiting heat gain, and avoiding potential glass breakage. Install effective controls for the electric lighting, or else the effort is wasted.*

MEASURE 8.3.5 Install a system of light shelves and shading.

You can make tall windows effective sources of daylighting by using "light shelves." Figure 1 shows how they are installed.

Light shelves avoid the glare problem that limits the use of diffusers to exploit daylighting through windows (see Measure 8.3.1). Light shelves also provide the unique advantage of shifting the light from the window so that it comes from a more overhead direction, improving the quality of illumination.

The light shelf itself is a simple device that is installed inside the window. In most applications, it must be combined with other devices to avoid glare from sunlight entering the lower portion of the window. The light shelf itself is not difficult to install. Light shelf systems, including exterior shading devices, are now available from manufacturers as prefabricated units.

A serious disadvantage of light shelves is that only the portion of windows above head height is usable for daylighting. Light shelves require periodic cleaning, which is easy to neglect.

Where to Install Light Shelves

Light shelves require direct sunlight. The windows should face toward the sun for a large fraction of the time that the space is occupied.

Tinted or reflective glazing may greatly reduce the potential benefit of light shelves, or make them uneconomical. These types of glazing typically block about 70 to 80 percent of incoming sunlight.

Light shelves are useful primarily with windows that have a large amount of glazing area at a height greater than about 6.5 feet (2.2 meters). Applications include tall conventional windows and clerestory windows. (A "clerestory" is an elevated vertical section in a wall or roof that is intended to have windows for illuminating the interior portion of a space.)

Light shelves may be used with glazing at lower heights in some specialized cases where people cannot get close to the glazing.

As with any kind of daylighting, the electric lighting must be arranged and controlled so that it can be turned off to exploit the daylight provided by the light shelf system.

Energy Saving Potential

Light shelf systems provide useful daylighting of a zone that lies along the exterior wall. The width of this zone depends on the height of the top of the window, the orientation of the window, the latitude of the site, the time of day, and the clarity of the sky. The width of the zone varies with sunlight conditions. With typical clear windows, the zone may extend inward a distance of 10 feet to 20 feet (3 meters to 7 meters). The light penetration may be deeper if the window is very tall.

With typical interior lighting levels, this translates to a saving in electricity ranging from 10 to 40 watts per foot (30 to 120 watts per meter) along the wall. This assumes that no daylighting is available without the light shelves.

If the windows are tall, light shelves can provide deeper penetration than daylighting that is achieved by shading windows. This is because light shelves can throw all the energy of direct sunlight into the space. In contrast, using shading to tame sunlight for daylighting leaves most of the potential daylighting energy outside the building.

As with any kind of daylighting, the daylight entering the space becomes heat energy. This increases the cooling load in warm weather and reduces the heating load in cold weather. Light shelves disperse sunlight fairly efficiently, so the amount of heat energy added to the space is not much greater than would be added by an equivalent amount of electric lighting. (This is true of electric area lighting. Electric task light is much more efficient in terms of localizing lighting energy.)

What is a Light Shelf?

Windows that face the sun receive an enormous amount of energy that could be used for daylighting. In principle, if a window faces anywhere between southeast and southwest and if it receives direct sunlight, each unit of window area could illuminate 20 to 100 units of interior area. However, this is possible only if the sunlight can be distributed efficiently.

The challenges in distributing this free lighting energy are lighting geometry and glare. In order for illumination to be useful, it must come from overhead. Raw sunlight coming through a window falls on the floor, so some method is needed to redirect the sunlight so that it comes from overhead.

The problem of glare is explained in Measure 8.3.1. In brief, if the light entering a window is simply deflected toward the interior of the space, occupants looking toward the window are blinded by glare.

Many concepts have arisen for taming the sunlight that enters through windows. A concept that appears promising at the present time is the "light shelf." A light shelf is essentially a mirror that is installed inside a window, facing upward. The mirror reflects incoming sunlight toward the ceiling. The ceiling then distributes the light into the working areas of the space.

An effective light shelf system needs four components:

- *the light shelf itself.* The light shelf is simply a reflector. It could be as simple as aluminum foil taped to a piece of cardboard. Commercial light shelves include features that ease cleaning and enhance safety. See Figures 2 and 3.

- *the window.* The light shelf distributes daylighting only from the portion of the window that extends above the light shelf . The bottom portion of the window contributes daylight only to the narrow zone underneath the light shelf. The window must face toward the sun for a large fraction of the time, and it cannot be shaded by outside objects. If the window glazing is tinted or reflective, the daylighting potential is reduced substantially.

- *the ceiling.* The light shelf aims sunlight at the ceiling. The ceiling then distributes the light to the

Construction Specialties, Inc.

Fig. 1 Light shelves These are installed directly against the windows, just high enough to avoid being a hazard. The top of each shelf is a mirror, and it must be kept clean. Only windows exposed to direct sunlight are eligible. Glare must be avoided from the portion of the window below the shelf. The windows at left have internal shades. The windows at right are shaded by an external fixture, shown in Figure 4.

occupants. The ceiling plays the same role as the fixtures in electric lighting. In most cases, the ceiling should be highly reflective to conserve as much light as possible. The height and orientation of the ceiling, and the diffusion characteristics of the ceiling surface, determine how the ceiling distributes the daylight.

• *shading device(s) to prevent glare from the bottom portion of the window.* A window must be exposed to direct sunlight to be a candidate for a light shelf. The portion of the window below the light shelf needs separate treatment to prevent glare. For example, you may install a shade or diffuser inside the lower portion of the window, as recommended by Measure 8.1.1. Or, you may install an external shade that increases the light collection of the upper part of the window, as we discuss below.

The rest of this Measure covers these components in greater detail. Think of light shelves as a system. Study Measures 8.1.1 (exterior shading devices), 8.1.2 (interior shading devices), and 9.5.3 (automatic light switching) before attempting to install light shelves.

Effective daylighting by any method is still a rarity. It is impossible to convey the visual effect of daylighting by words or pictures. If you are interested in using light shelves, visit sites where they are installed, and judge their performance over a range of sunlight conditions.

The Lighting Pattern of a Light Shelf System

The pattern of illumination depends on the reflection characteristics of the light shelf, and it depends on the geometry and reflection characteristics of the ceiling

surface. If the reflection by the light shelf is specular (like a mirror), the reflection is controlled by the law of optics that says the angle of reflection equals the angle of incidence. If the sun is low in the sky, sunlight penetrates deeply into the space. If the sun is high, the ceiling is illuminated close to the wall.

Pure specular reflection creates a sharply defined rectangular bright spot on the ceiling. This may not be considered attractive, and it may be bright enough to annoy occupants. You can reduce this effect by making the reflecting surface of the light shelf more diffuse. This spreads out the light on the ceiling. However, it also concentrates the ceiling illumination close to the

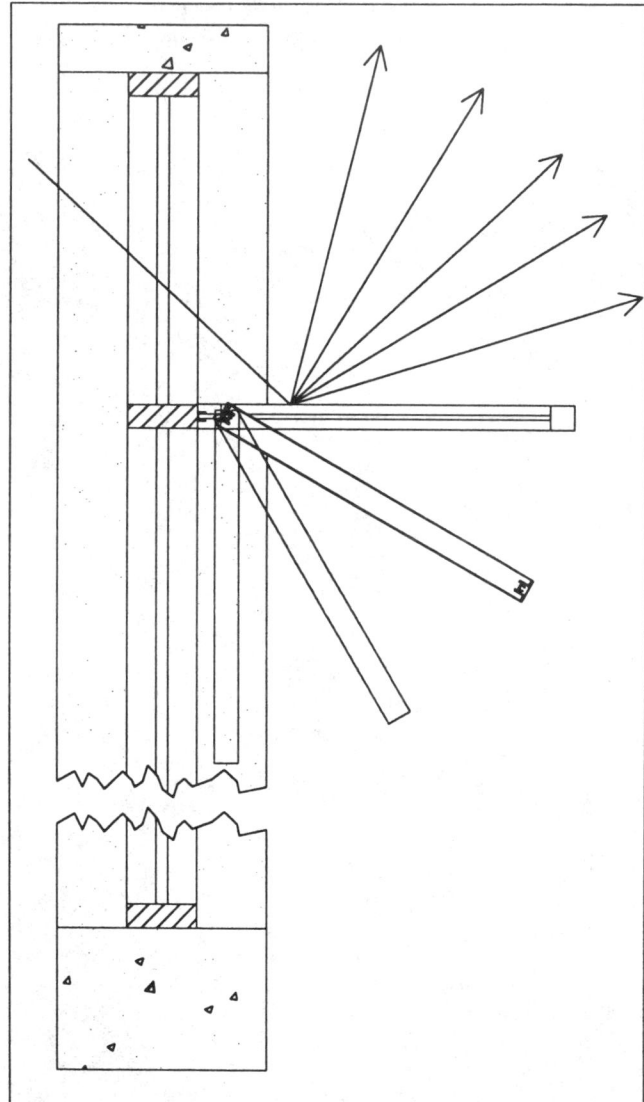

Construction Specialties, Inc.

Fig. 3 Tilting light shelf Ability to tilt is most valuable for making the top surface easy to clean. A small amount of tilt may also improve performance. Tilting the shelf downward increases the penetration of light into the space, but it also introduces the possibility of reflecting sunlight into occupants' eyes at low sun angles. By the same token, tilting the shelf upward may avoid glare and increase daylighting at low sun angles.

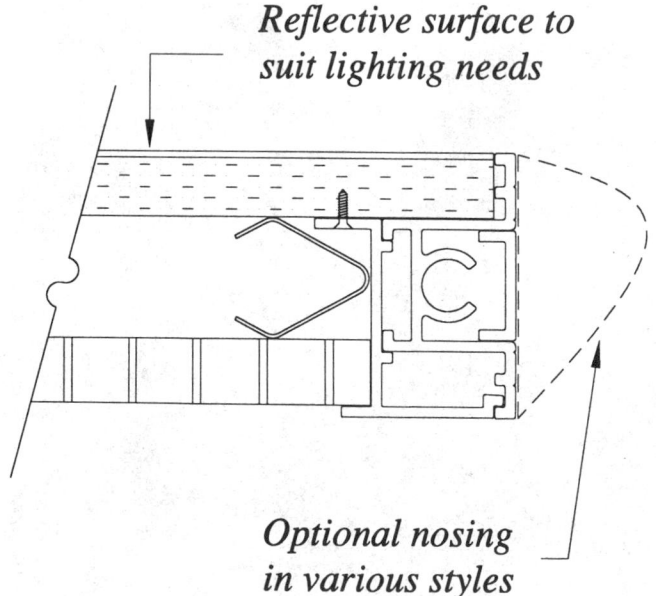

Reflective surface to suit lighting needs

Optional nosing in various styles

Construction Specialties, Inc.

Fig. 2 Detail of typical light shelf The main feature is the reflective upper surface. A soft edge is valuable for safety. A sign is needed somewhere that the shelf is not for storage and that it should be cleaned at stated intervals.

light shelf. If the reflection is very diffuse, penetration into the space is sacrificed.

A taller ceiling provides deeper light penetration, and it also distributes the light more widely within the space.

It helps to think of a light shelf system as having two zones of daylighting. The interior zone is illuminated by light that is thrown on the ceiling by the light shelf. The exterior zone, adjacent to the window, is illuminated by the portion of the window below the light shelf. The exterior zone is partially shaded from the ceiling light by the light shelf itself. (A light shelf could be made of partially transparent material to allow the exterior zone to be illuminated through the light shelf.)

The boundaries of these two zones change as the sun moves and sky conditions change. Both zones move toward the interior when the sun is low in the sky, and they move toward the exterior when the sun is high in the sky. Both zones move in and out together. However,

diffuse reflection by the light shelf reduces the movement of the interior zone as the sun angles change.

When the sun is higher, less sunlight is able to enter the window. At the same time, the sunlight is concentrated in a smaller area, closer to the exterior. Therefore, the illumination level within the zones tends to remain constant.

On the other hand, if the sun is obscured, there is less total daylight, but it is distributed more deeply into the space. As a result, the illumination level within the zones falls dramatically, perhaps to the point that artificial light is needed.

Equipment Choices and Layouts

You have two main design issues with a light shelf system: (1) where to locate the light shelf, and (2) how to avoid glare and excessive heat gain adjacent to the portion of the window below the shelf. With typical windows, the window area above the shelf cannot provide as much light as you want, and the window area below the shelf provides too much. We will look at that

Construction Specialties, Inc.

Fig. 4 Exterior shades that work with light shelves These are unusual because they are installed below the tops of the windows, at the same height as the light shelves. Their purpose is to keep direct sunlight from entering the portions of the windows below the light shelves. The shadow pattern shows that the shades would be more effective if they were installed as a continuous unit. Exterior shades can increase daylighting by reflecting more light into the top portions of the windows. This design fails to exploit that opportunity.

problem first, and then look at several practical equipment layouts.

■ Shelves are Above Head Height, Usually

A fundamental problem with light shelves is the need to locate the shelf too high for efficient use of the daylight entering the window. As a practical matter, the light shelf usually has to be installed above the level of people's heads. In modern buildings, the ceiling height is typically eight to ten feet, and the windows may not reach all the way to the ceiling. Therefore, only a small fraction of the total window area is available for daylighting.

There are two reasons why the light shelf has to be installed so high. One reason is safety. The light shelf extends into the space. People could walk into the shelf if it were low enough. Therefore, the shelf must be installed above the height of a tall person.

The other reason is glare. The top of the shelf is a reflecting surface. If the shelf were located below eye level, sunlight could reflect into people's eyes.

Light shelves can be installed at a lower level if people are not able to get close to the windows. Such situations are rare, but they do exist.

■ The Shape of the Light Shelf

As the name implies, a light shelf looks like a shelf. It is as wide as the window, or somewhat wider. The depth of the shelf is proportional to the height of the window above the shelf, but the ratio is a compromise. The depth required to reflect all entering sunlight depends on the elevation of the sun. If the sun facing the window is low in the sky, it would take a very deep shelf to reflect most of the sunlight to the ceiling. A shelf this deep might be difficult to mount, or it might interfere with lighting the space under the shelf, or it might look too strange.

The sides of the light shelf should be designed to keep sunlight from shining down the sides of the shelf when sunlight is entering the window sidewise. You can do this by making the shelf wider than the window, or by installing a side piece that extends upward from the shelf.

The shelf does not have to be exactly horizontal, although a horizontal orientation is near optimum if the shelf is fixed. You could make the shelf adjustable, as in Figure 3. Tilting the shelf downward would improve daylight penetration at high sun elevations, and tilting it slightly upward would reflect more light at low sun angles. However, don't expect the occupants to do this. You could automate the tilt with a fairly simple optical sensor, and win a design award for innovation.

There is another light shelf configuration to consider. It consists of a stack of light shelves, each fairly narrow. It looks something like a venetian blind with the blades adjusted to a horizontal position. However, it behaves differently, because the tops of the blades reflect. If the

shelves are narrow enough to fit within the depth of the window frame, they would not pose much of a safety hazard. Also, being horizontal, they would not interfere much with the view. The main disadvantage of this arrangement is that can reflect an image of the sun into the eyes of anyone standing nearby and looking out the window.

■ Relationship to Artificial Lighting

The spaces are illuminated with a combination of daylighting and artificial light. The light fixtures and their power circuits should be arranged so that they can dim or turn off when daylighting can provide sufficient illumination. The layout of the lighting should accommodate changes in sun position and cloud cover. This can be as simple as switching the fixtures in rows parallel to the exterior wall. Lighting switching and dimming should be automatic. See Measure 9.5.3 about automatic lighting controls and Measure 9.6.4 about the arrangement of the light fixtures and power circuits.

At night, the space must be illuminated with artificial lighting. The light shelves may interfere with illumination from ceiling-mounted fixtures. If necessary, modify the electric lighting to provide good nighttime illumination in the exterior zone.

■ Interior Shading for the Lower Window

In a typical light shelf installation, most of the window area is underneath the shelf. This provides an excess of sunlight to the exterior zone. You can use an interior shading device to block the excess sunlight. Refer to Measure 8.1.2 to find your options.

An ordinary venetian blind is probably your best choice. If you want to reduce heat gain in the space, make the outer surfaces of the blades highly reflective. If you want to capture heat gain, make the outer surfaces of the blades moderately absorptive.

■ Exterior Shading

If the windows face toward the south, you can shade the lower portions of the windows with an exterior horizontal sunshade. See Measure 8.1.1 for the issues to consider with exterior shading.

When exterior shading is used with a light shelf, it is installed in an unusual manner, as shown in Figure 4. It must be installed at exactly the same level as the light shelf. If it is higher, it will block the entry of useful sunlight to the space above the shelf. If it is lower, direct sunlight will enter the space through the gap between the light shelf and the shade. Both the shelf and the sunshade must butt against the glass closely to avoid leaving a gap through which sunlight can enter directly.

In this application, the exterior shade can serve double duty. If it has a reflective upper surface, it will reflect sunlight into the upper portion of the window. If the upper portion of the window is short, this will not help much, because the additional light will strike the ceiling close to the wall. However, if the window extends well above the exterior shade, the additional light will penetrate farther into the space. The external shade shown in Figure 4 fails to exploit this opportunity.

■ Layout with Clerestory Windows

Clerestory windows offer a major advantage for light shelves. They are well above head level, so the entire window can be used to capture daylighting to illuminate the interior area of the space.

With this layout, there are no longer two separate lighting zones, as for windows at lower levels. A space with clerestory windows has a tall ceiling. Light reflected from the ceiling can distribute light to the area underneath the light shelf.

An exception occurs if the clerestory is part of the roof structure, well inside the wall. In this case, the area adjacent to the wall may be shadowed by the light shelf.

Keep the Top Surface Clean

The light shelf is horizontal, or nearly so. This makes it an effective dust collector. Within a period of weeks or months, enough dust can settle on the shelf to seriously degrade illumination. Therefore, light shelves need to be cleaned on a regular basis.

The best approach is simply to make this part of the activity of the cleaning crew. However, contract cleaning crews will overlook this unfamiliar requirement. Therefore, a responsible manager must keep educating the cleaning crews about the need to dust the light shelves.

Some commercial light shelf units are designed to pivot downward to allow cleaning, as in Figure 3. This allows cleaning without a ladder, which is an important feature for increasing the likelihood that the shelf will actually be kept clean..

Cleaning shelves that are located at great heights, as with clerestory windows, can be a major task. Consider this before deciding to install light shelves in such locations.

Cleaning is so important that each light shelf should be marked in a durable, highly visible manner to tell cleaning crews to clean the top surface with each room cleaning.

If you use exterior shades with reflecting top surfaces, you will probably have to depend on precipitation and wind to keep them clean, at least between long intervals of manual cleaning. Select a finish for the top surface that sheds dirt easily. Give the top surface enough slope so that rain flows off the surface freely.

Safety

The height of the shelf is a compromise between lighting effectiveness and safety. It is usually not practical to install the shelf so high that the tallest

basketball player cannot run into it. Therefore, the shelf should be designed so that the edge is as yielding as possible. Also, the edge should be made very visible, as with a bright colored stripe. Some commercial units offer a soft rubber bumper around the edge of the shelf, as shown in Figure 2.

Explain It

Install a placard at each light shelf to explain its purpose. Ask the occupants to exploit the daylighting and minimize artificial lighting. (In most cases, there should be automatic lighting controls to do this.) Spell out that the shelf is not supposed to be used for storage. You can use the same placard for the cleaning instructions.

ECONOMICS

SAVINGS POTENTIAL: If light shelves make daylighting possible where it did not exist before, typical savings are 10 to 40 watts per foot (30 to 120 watts per meter) along the wall. Greater savings are possible with taller windows. This translates to a saving of about $1 to $10 per foot per year.

If external shading is used, the greatest saving may occur from the reduction of cooling load, especially in warm climates. See Measure 8.1.1 about this saving.

COST: The light shelf unit itself may cost about $100 per window, subject to wide variation. In addition, see Measure 8.1.1 for the cost of external shading devices, and Measure 8.1.2 for the cost of interior shading devices. See Measure 9.5.3 for the cost of automatic lighting controls. In existing buildings, you may have the cost of rearranging the light fixtures and power circuits to exploit daylighting. See Measure 9.6.4 about this.

PAYBACK PERIOD: Several years, in new construction. Ten years or longer, in retrofit.

TRAPS & TRICKS

CHOICE OF METHOD: Light shelves are still a novelty, so you are pioneering. Before you make any commitments, visit some sites where light shelves are installed, and judge their performance yourself. Remember that the efficiency objective is to minimize electric lighting and cooling load. See how well they do this.

SYSTEM FEATURES: The greatest design difficulties are not in the light shelf itself, but in the related shading. See Measures 8.1.1 and 8.1.2 about this. Include features that enhance cleaning and safety. Consider effective placards to be an integral part of the installation.

MONITOR PERFORMANCE: Light shelves are easy to neglect. If you see people storing objects on top of the light shelf, you know the system has failed. Inspect the shelves periodically to make sure that they are being cleaned.

MEASURE **8.3.6 Use light interior colors or mirrored surfaces.**

If a space has daylighting, and if it has appropriate controls for the electric lighting, you may be able to significantly improve the contribution of daylighting by using a highly reflective color scheme inside the space.

This Measure is similar to Measure 9.6.1. Refer to there for details.

Passive Solar Heating

The Measures of this Subsection exploit passive solar heating for a limited number of applications. These Measures tap only a small fraction of the theoretical potential of passive heating, but they are practical to try without a great deal of pioneering. Exploiting passive solar heating to a greater extent involves a major commitment to unproven methods. For the perspective that you need to be successful in this fascinating and risky area, see Reference Note 47, Passive Solar Heating Design.

INDEX OF MEASURES

8.4.1 Keep open the window shades of unoccupied spaces that need heating.

8.4.2 Install combinations of sunlight absorbers and reflectors inside windows and skylights.

8.4.3 Install solar enclosures over areas that can benefit from heating.

RELATED MEASURES:

- Subsection 5.6, for heat pump loop systems. These systems can transfer heat from sunlighted areas to other parts of the building

- Subsection 6.3, for reduction of air leakage through windows

- Subsection 7.3, for reduction of conductive heat loss through glazing

- Subsection 8.1, for glazing and shading principles

- Subsection 8.3, for daylighting issues

MEASURE 8.4.1 Keep open the window shades of unoccupied spaces that need heating.

If windows face the sun for at least part of the day, they receive enough sunlight to provide a significant amount of heating for the space. However, windows exposed to direct sunlight usually have shades or blinds to prevent glare. These shades may be left closed when the space is vacant, wasting the potential of passive heating.

If a space is vacant during long periods of daylight, open the shades during cold weather to admit solar heat. Accomplishing this requires people who can be assigned the task of opening the shades. For example, this responsibility can be assigned to maids for hotel rooms, to teachers for classrooms, to nursing staff in hospital rooms, and to the occupants of single-person spaces. Security patrols can be assigned this responsibility in many types of facilities.

Opening shades or curtains reduces their insulating value at night. This is not much of a penalty, because ordinary window shades and curtains provide little insulation. Convection bypasses them. The gain in solar heating during the daytime is much greater than the increase in heat loss overnight. This being said, try to close the curtains overnight, but only if you have a reliable procedure for opening them again before sunrise.

Energy Saving Potential

Throughout the middle latitudes, windows facing in any direction from southeast to southwest admit about 1,000 BTU per square foot of window area per day during winter, if the sky is clear. Conventional window treatments absorb 30% to 70% of this heat when they are closed. If they are kept open, all the heat gets into the space.

The total annual saving depends on the number of daylight hours during which the space is vacant, the size and orientation of the glazing, the clearness of the sky, the number of days requiring heating, and the heating requirements of the space.

This Measure may also save a significant amount of lighting energy, especially at the beginning of occupancy periods. For example, hotel guests arriving during the day typically rely on daylighting for most of their room lighting, provided that the windows provide satisfactory lighting.

SUMMARY

An easy way to save heating energy in spaces that are vacated for long periods.

SELECTION SCORECARD

Savings Potential .... $ $ $

Rate of Return .... % % % %

Reliability .... ✓ ✓

Ease of Initiation .... ☺ ☺ ☺

Explain It to the Staff and Occupants

This sort of procedure tends to be neglected. Use whatever administrative methods work in your facility to keep the procedure in effect.

Ask occupants to control the shading to conserve energy. Placards are usually the best way to do this. See Reference Note 12, Placards, for guidance in designing and installing placards.

ECONOMICS

SAVINGS POTENTIAL: *20% to 80% of heating costs in the sunlighted spaces, depending on the factors discussed above.*

COST: *Usually minimal. Customized placards typically cost about $1 to $5 apiece, in quantity.*

PAYBACK PERIOD: *Immediate.*

TRAPS & TRICKS

ADMINISTRATION AND COMMUNICATION: *This is primarily a matter of effective leadership and communication. Back up your request with a well written memorandum to the responsible managers. Show people what needs to be done. Say please and thank you.*

MONITOR PERFORMANCE: *This kind of procedure tends to fade away. Keep it refreshed by periodic contacts with the departments responsible. Put this in your tickler file.*

MEASURE 8.4.2 Install combinations of sunlight absorbers and reflectors inside windows and skylights.

SUMMARY

A simple but unconventional technique that can provide comfortable heating, and perhaps enhance daylighting. Uses conventional materials and equipment in a clever manner.

SELECTION SCORECARD

Savings Potential	$ $ $
Rate of Return	% % %
Reliability	✓ ✓ ✓
Ease of Retrofit	☺ ☺

Measure 8.4.1 points out that most glazing exposed to sunlight is equipped with shading devices that block out the sun completely, wasting the sun's heating potential. The inexpensive procedure recommended by Measure 8.4.1 is not effective for exploiting passive heating when spaces are occupied, because it does not correct glare problems that motivate people to close the shading.

A simple concept for exploiting solar heating while controlling glare is to install an absorbing surface inside the glazing that converts the sunlight to heat. There are two differences between the heat absorber and conventional window treatment:

- the absorber is intended to absorb heat, rather than reflecting it. It does this with an outward facing surface side that is dark.
- the absorber is installed far enough inside the space so that it acts as an efficient convector. The absorber does not have to intrude far into the space. However, you can install the absorber anywhere in the space where it will intercept direct sunlight and kill glare.

If absorbers are installed close to glazing, they can reduce the chill that occupants feel in cold weather when they are near glazing. When sun is shining through the glazing, the absorbing surfaces are warm, so they act as low-temperature radiators. When sun is not shining through the glazing, the surfaces limit radiation heat loss from the space to the windows.

You can make the heat absorbing surfaces from any opaque or translucent material, such as fabric. With a bit of creativity, you can use this Measure in most types of facilities, while adding a decorative touch.

In most cases, you still want to be able to minimize the cooling load during warm weather. You can do this by having a reflecting surface just inside the glazing when heating is not desired. In retrofit applications, you can keep the original window shading devices to serve this purpose.

An Example for Windows

You can put this concept into practice in many ways. For example, you can use two conventional venetian blinds. Install a highly reflective venetian blind as close to the window as possible, in the conventional manner. Install a second, highly absorptive venetian blind on a bracket that holds the blind about six or eight inches inside the wall. The inner blind can be a conventional blind that has a dark, fairly dull color. Suspending the dark blind inside the space keeps the convection from

the dark blind from scrubbing the stagnant air layer off the window surface.

Only one blind is used at a time. When the weather is warm enough that passive heating is not useful, the reflective blind is lowered, and the dark blind is raised out of the way. Occupants adjust the reflective blind in the normal way to avoid glare and exploit the view as much as possible.

When passive solar heating is desired, the reflective blind is raised and the dark blind is lowered. Again, occupants adjust the dark blind in the normal way to avoid glare and exploit the view as much as possible.

Sunlight comes through the window at various angles. Therefore, the absorbing blind may have to be much larger than the window to keep sunlight from entering the space around the sides of the blind. Alternatively, you can install side fins, projecting from the wall, to keep sunlight from getting around the dark blind.

An important advantage of this technique is that the effectiveness of passive heating does not depend on occupants adjusting the blinds in any particular way. Most of the sunlight that enters the window is captured, no matter how the dark blind is set. The purpose of the dark blind is to capture most of the solar heat even when occupants close the blind to avoid glare.

Heat circulates through the space by convection, as with a conventional convector. The effectiveness of the circulation depends on the geometry of the space. In a space with a horizontal ceiling, the warm air circulates throughout the space. Cooler air remains near the floor. This cooler air is heated only if the warm part of the heat absorber is low enough to reach it.

Therefore, this method alone may leave an uncomfortable cold air layer near the floor. This is true even if there is ample solar energy. For this reason, expect to use another method of heating the air near the floor. The most efficient solution is to let sunlight shine on a dark part of the floor near the window, if you can do this without creating glare.

Or, you might rely on your conventional heating equipment, such as a baseboard convector, but this uses purchased energy. (Any space needs a conventional heating system for periods when there is no solar energy available.)

Or, if the space has a forced-air heating system, keep the circulating fan running at low volume. The heating system must have a low-speed fan setting for this to be acceptable. Running a heating unit fan without heating the air will cause a wind chill effect on people. The air velocity must be kept low enough so that this effect is not objectionable.

An Example for Skylights

Skylights can produce oppressive heat and glare that motivates building owners to remove them or to paint over them. This wastes their potential for passive heating and daylighting.

A general approach that captures the heat of skylights while avoiding glare is to install a light absorbing screen under the skylight. The screen can as simple as a hanging of dark fabric. This solution avoids the need for any additional structure, and the fabric is easy to attach.

In a space that uses skylights, the heat absorber will be visually prominent. Use it to create pleasant daylighting. At the same time, exploit its potential as a decorative item.

There is a fundamental problem when using heat absorbers with skylights, or with other glazing that is located high in the space, above the level of the occupants. The heat absorber releases virtually of its heat by convection, so all the heated air rises to the space above the absorber. The absorber itself cannot project heat downward.

To solve this problem, try to feed the heat through the existing heating system, if it uses fan-forced circulation. This requires an air return located above the absorber, preferably near the top of the space. This is so important for the effectiveness of passive solar heating that you should consider adding an appropriate air return if none presently exists.

Even if the heat from the absorber cannot be distributed through existing heating equipment, installing the absorbers can still be a better alternative than abolishing the skylight. In a space with skylights, it is likely that most of the heat loss from the space occurs through the skylights themselves. The heat absorber allows the solar energy passing through the skylight to offset the heat loss from the skylights without creating visual problems inside the space. (During most daylight

hours, a skylight takes in much more energy than it loses by conduction.) The warm air bubble created at the ceiling level eliminates radiation cooling of the space below, improving comfort. And, the heat absorber can be designed to create effective daylighting.

Energy Saving Potential

Throughout the middle latitudes, windows facing from southeast to southwest admit about 1,000 BTU per square foot of window area per day (about three kilowatt-hours per square meter per day) during winter, if the sky is clear. Conventional shades typically absorb 30% to 70% of this heat. The heat collecting devices suggested here may capture 90% of the heat.

The total annual saving depends on the size and orientation of the glazing, the clearness of the sky, the number of days requiring heating, and the heating requirements of the space.

The potential saving depends on the transparency characteristics of the glazing. The devices recommended by this Measure capture only the solar heat energy that passes through the glazing. If the glazing is tinted or reflective, or if it has special features to block solar infrared radiation, the value of this Measure is greatly reduced.

Design Guidelines

This technique is unconventional, so you have to be an innovator. While you are being creative, follow these guidelines.

■ Select the glazing transparency for the application.

This Measure provides the greatest heating benefit if the glazing transmits a large fraction of the sunlight that falls on it, whether the glazing is clear or translucent. Glazing that is highly absorptive or highly reflective reduces the potential for passive heating so much that this Measure may not be worthwhile.

In new construction, you get to select the glazing. If you select clear glazing to maximize the benefit of this technique, the building will be married to it. This is a novel technique, so think hard about this decision.

■ Distribute the heat into the space effectively.

Provide enough clearance at the top and bottom of the absorber to allow free convective circulation. Install the absorber far enough from the wall so that the convective current from the screen does not scrub the surface of the wall. Leaving the wall's air layer intact reduces heat loss somewhat.

The absorbing device distributes its heat to the space mostly by convection. The absorber does not become warm enough to radiate strongly. The amount of heat that the absorber emits by radiation depends on the absorber's temperature and emissivity. Most non-metallic materials that you would use to make heat absorbers have high emissivity, which increases the amount of heat they transfer to the space by radiation.

Too much heat radiation may be uncomfortable for occupants who are near the absorbing surfaces, but this is unlikely. If you want to decrease radiation and increase convection, use surfaces of low emissivity. This means smooth metal surfaces, or metallic coatings. Another method is to install a second screen inside the absorber. This innermost screen will block heat radiation regardless of its other characteristics (color, weight, etc.), as long as it is opaque.

If the heat absorbing surfaces are located high in the space, as with skylights, design them to radiate as much heat as possible downward to the occupied space. Heat that is transferred by convection rises. If this heat accumulates above the occupied space and does not circulate, it has limited value.

■ Be able to reject excess solar heat.

In most buildings, there are times when sunlight could overheat the space. To deal with this, provide reflective shades or other methods of reducing cooling load (See Subsection 8.1). In retrofits, the original shades may serve this purpose adequately.

Some large skylights have vents or openable panes to get rid of heat gain during warm weather. Make sure that any vents are airtight when closed, or they may waste large amounts of heating energy. (See Measure 6.4.4 about this.)

■ Exploit daylighting.

This technique has potential for daylighting. The trick is not to absorb all the entering sunlight, but to distribute some of it within the space in a glare-free manner. See Reference Note 46, Daylighting Design, for the full story on daylighting.

■ Preserve the view, if there is one.

If the glazing offers a good view, try to make it available. For example, using a pair of venetian blinds, as in the previous example, accomplishes this.

■ Minimize waste of useful space.

Space inside a building is valuable. Also, devices that stick out are more likely to be a nuisance and get damaged. Design the absorbing devices to use a minimum of space that is useful for other purposes.

■ Have a sense of style.

Make this an opportunity to enhance the appearance of the space. For example, using fabric heat absorbers offers endless possibilities for decor.

■ Avoid condensation damage.

Skylights are prone to serious condensation. The absorbers will make the glazing colder at night because they shield the glazing from the warmth of the space. This aggravates the condensation problem. The condensation may damage your absorbers and cause other problems. See Reference Note 47, Passive Solar Heating Design, for ways of reducing condensation.

■ Don't interfere with fire egress.

Windows may be potential escape routes in the event of fire, or points of entry for firefighters. Do not install devices that conflict with safety.

■ Select materials for longevity.

Select materials that resist sun damage. If the absorbers can come in contact with people or moving equipment, design them to resist impact. Select movable components that resist wear. For example, pull cords should be designed so that they do not fray or jam. Select equipment that allows worn components to be replaced easily.

Explain It to the Occupants

The heat absorbing devices are vulnerable to being misused or abandoned because their purpose is not obvious, or because they appear to be a nuisance. Therefore, make people aware of their purpose. Also, if the shading devices require any human control, explain how to use them. Placards are an effective means of doing this. See Reference Note 12, Placards, for guidance in designing and installing placards.

ECONOMICS

SAVINGS POTENTIAL: 10% to 100% of heating costs in the sunlighted spaces, depending on the factors discussed above.

COST: Varies widely. You probably cannot invent anything that will cost less than $100 per window. Heat absorbers for large skylights will cost much more.

PAYBACK PERIOD: Several years or longer.

TRAPS & TRICKS

THE HAZARDS OF PIONEERING: This is an unusual concept. You have to develop it into a practical configuration. Don't try it unless you have the time and the discipline to work out all the details. Set up a test installation before extending your concept to many windows or skylights.

MEASURE 8.4.3 Install solar enclosures over areas that can benefit from heating.

The ultimate in passive solar heating is creating a space that is heated entirely by sunlight. Many facilities have opportunities for using a glazed enclosure to cover an outdoor space, providing free heating for at least part of the year. The enclosure does not have to be made of glass. It can be made of any material that admits enough solar heat to offset the heat losses. Materials include glass, plastic glazing materials, translucent panels, thin plastic sheet, and more unusual materials.

Where to Consider Glazed Enclosures

Greenhouses are by far the most common type of glazed enclosures. Figure 1 illustrates the range of types. Experience with greenhouses provides valuable lessons for employing glazed structures for any purpose. For example, Figure 1 shows the major heat venting and ventilation requirements.

One purpose of glazed enclosures is to make an open space usable for a more valuable purpose. For example, the front of a restaurant may be turned into a sidewalk cafe.

Another purpose is to extend that length of time that a facility can be used. For example, Figure 2 shows a patio restaurant in a mid-latitude location that is usable for three or four months per year. Installing a glazed enclosure would approximately double the usable period. Furthermore, it would make the space usable in rain and wind.

Enclosing an outdoor swimming pool, as in Figure 3, typically extends its season of usage by several months.

Another purpose is to reduce heat loss from adjacent space, as when enclosing the common area of a shopping center.

Large glazed enclosures have the ability to collect enormous amounts of passive heating, and they are able to keep themselves warm even in cool climates. Figure 4 shows a huge courtyard in a Tennessee tourist hotel that is enclosed with glazing, making it comfortable under almost all weather conditions.

Beware of Unexpected Heating and Cooling Costs

There is a fly in this ointment. The glazed enclosure is able to heat itself only during the daytime, and heating is substantially reduced when the sky is overcast. The nature of the activity may require the space to operate when passive heating is not adequate. Or, the owners may discover that the increased value of the space makes it worthwhile to operate the space at night. In this case, heating equipment is needed. Unfortunately, a simple

SUMMARY

An opportunity to make space more useful without increased heating costs. Not as simple as it looks.

SELECTION SCORECARD

Savings Potential .... $ $ $

Rate of Return, New Facilities % %

Rate of Return, Retrofit.......... % %

Reliability .... ✓ ✓ ✓

Ease of Retrofit .... ☺ ☺

glazed enclosure has only minimal insulation value. In the end, energy cost may be increased rather than decreased. If you examine glazed enclosures, you will see that many are heated in ways that require a large amount of energy.

Furthermore, a glazed enclosure may overheat badly during warm weather, and even during mildly cool weather. This sometimes prompts owners to install air conditioning. This can be very expensive, even in mild climates.

This leads us to an important point. Before installing a glazed enclosure, consider how the space is likely to be used afterward. If the space will be heated, design the enclosure to minimize heat loss. Also, try to prevent overheating without resorting to air conditioning.

Energy Saving and Comfort Benefits

Consider this Measure primarily as a way of improving comfort and/or space utilization. In some cases, you may reduce energy consumption at the same time. Or, you may achieve the improvement in comfort and/or space utilization at a relatively small additional energy cost.

If the newly enclosed space is not conditioned, there is no increase in energy cost. If it connects to conditioned spaces, it may substantially reduce energy cost. For example, enclosing the walkways of a shopping mall radically reduces heat loss from stores that are open to the walkway.

But, beware of using a glazed enclosure as a substitute for a properly designed building or building addition. Again, consider the ultimate mode of use.

Design Guidelines

A glazed enclosure is not as simple as it looks, if you want it to be comfortable and efficient. Concentrate on the following points to achieve efficiency and comfort.

■ Minimize the heating load.

The insulating value of all transparent and semi-transparent glazing materials is poor. Furthermore, if the glazing is slanted away from the vertical, its insulating value declines radically, typically by a factor of two to three, compared to vertical glazing. The poor insulation value does not matter as long as the enclosed space remains unheated, because solar heat gain will make the space much warmer during daylight than it would be if it were left open.

Some composite glazing materials, discussed in Measure 8.3.3, offer fairly good insulation value. These materials are bulky and expensive. They offer no sense of the outdoors, and they have low light transmission. However, they can be used effectively in combination with transparent materials.

If the enclosed area will be used extensively during hours of darkness, use insulating panels for a

WESINC

Super Sky Products, Inc.

Fig. 1 Plain and fancy greenhouses The upper one is made entirely of extruded double-wall plastic. The bottom one is made of glass. All glazed enclosures should apply the lessons that have been learned from a vast number of greenhouses. In both of these, notice the extensive vents that are located at high points. Venting is necessary for free temperature control in any application.

part of the enclosure surface area. Distribute the area of glazing and the area of insulated panels to save energy and create a pleasant environment. If practical, design the structure so that glazing and insulated panels can be swapped. This gives you a relatively inexpensive way of altering the daylighting and heat gain characteristics of the structure after it is built, if necessary. You may even be able to exchange glazing for insulated panels, or vice versa, on a seasonal basis.

■ Deal with excess heat gain efficiently.

A common mistake when adding glazed enclosures is overlooking the large cooling loads that they may create. If air conditioning is used to deal with the heat gain, the energy costs can be enormous in relation to the size of the space.

Deal with excess heat by venting it. Do this with a combination of vent dampers at high and low positions in the space. Locate the low vents where cold entering air will not cause discomfort. Provide dampers that seal tightly when venting is not required.

During warm weather, venting alone may not suffice. In such cases, consider using reflective glazing (see Measure 8.1.3) or opaque panels for part of the enclosure. If the weather can turn cold for significant periods, use insulating panels for as much of the enclosure as possible.

If the space will be air conditioned, avoid any vents at a low point in the structure, because these would waste cooled air. If you use vents at all when air conditioning,

limit them to locations at the highest points of the enclosure.

Even with adequate venting or with air conditioning, direct radiation of sunlight on occupants may be uncomfortable. This is most likely to be a problem if the occupants are stationary, e.g., at dining tables. In such cases, provide shade for the people.

■ Control condensation adequately for the application.

Glazing has low thermal resistance, so condensation forms if the outside temperature falls much below the dew point temperature of the inside air. This means that the glazing will sweat if there is any significant source of moisture inside the enclosed space. In many applications, sweating is transitory and causes no trouble, as long as the structure is made of materials that are not vulnerable to moisture.

Condensation is a major issue when enclosing moist environments, such as swimming pools and plant growing facilities. In such facilities, the glazing remains wet for long periods, and you need to use special techniques to prevent damage and unsightly conditions. Heavy condensation has a major effect on appearance and view. This may not matter with a swimming pool, but it is a problem for a restaurant that uses foliage for decor.

Swimming pools are a special challenge. The water temperature is typically around 80°F. Because of the large evaporating area of the water, the water temperature

WESINC

Fig. 2 Prime candidate for a glazed enclosure   This is the roof garden of a busy pizza restaurant at an oceanside resort. Although more dining space is needed, customers avoid this area because it is too exposed to chilly winds and frequent rains. Also, the still air temperature is somewhat too chilly for outdoor dining during much of the active season.

becomes the dew-point temperature of the enclosed space. The glazing will sweat heavily whenever the outside temperature is much below the water temperature. Provide effective drainage to carry condensate away from any areas where it does not belong, including the floor and wooden structures.

Lingering condensation grows mold. Sunlight stops mold growth, but only on parts of the structure that are directly illuminated. Make shaded surfaces accessible for frequent cleaning.

■ Provide good daylighting.

Psychologically, people may be more sensitive to bright sunlight when a space is enclosed, even if the function remains the same. You may want to compensate for this effect by using various types of diffusing and opaque panels.

As a rule, transparent glazing provides a more pleasing visual environment than glazing that diffuses light. You can tint the glazing or use reflective glazing to reduce heat gain. (At the same time, remember this rule: if you need to treat glazing to reduce the cooling load, there is too much glazing area. Few applications have a view that justifies excessive glazing.) To shade localized areas that cannot tolerate direct sunlight, provide localized shading inside the enclosure. For example, if you are enclosing an outdoor cafe, you could install sun umbrellas at the tables. Figure 5 shows an example.

ECONOMICS

SAVINGS POTENTIAL: *Various widely, depending on the application.*

COST: *Plain glazed enclosures start at about $15 per square foot of horizontal space. With heating, motorized venting, and few other accessories, the price may be about $30 per square foot in middle sizes. Very fancy enclosures can cost as much as several hundred dollars per square foot, but you are paying for glamor, not energy savings.*

PAYBACK PERIOD: *Many years, if you consider only energy cost. This is usually done to achieve purposes besides energy savings.*

Kalwall Corporation

Fig. 3 Swimming pool enclosure This swimming pool at an all-year campground uses an intelligent combination of translucent insulated panels, clear glazing, and venting. Notice that the upper portions of the roof can slide back to provide venting and daylight.

Vistawall Architectural Products

Fig. 4 A huge enclosed hotel courtyard This area the size of a village center is entirely under glass. Passive heating keeps it reasonably warm under most weather conditions, in a climate that can become fairly cold.

WESINC

Fig. 5 Localized shading This umbrella suspended inside a swimming pool enclosure shields people seated at the table from direct sunlight.

TRAPS & TRICKS

INVESTIGATE: *Spend time inside enclosures similar to the one that you wish to create, under all weather conditions. There is no other way to appreciate what the end result will be like.*

DETAILS, DETAILS: *Work out effective methods of ventilation, preventing glare, avoiding overheating, and dealing with condensation. Talk to people who are in the business of installing glass enclosures. Anticipate changes in usage that will occur as a result of enclosing the space, and design the enclosure to accommodate those changes efficiently.*

Section 9. ARTIFICIAL LIGHTING

INTRODUCTION

Lighting is an area of energy conservation that offers major savings in almost any facility. Lighting energy conservation offers a big payoff, but it is a big challenge for the lighting designer, the installer, and the facility operator. A typical facility has many lighting requirements. Each has its own set of opportunities for improving lighting efficiency. This makes it easy to overlook opportunities for savings. You must combine efficiency with good lighting quality. Doing this requires knowledge, imagination, and willingness to do things in a different way.

The Energy Saving Potential in Lighting

■ Lighting Energy Savings

In commercial buildings, lighting typically accounts for 20% to 50% of total energy consumption. In residential buildings, the figure typically is 10% or less. In energy-intensive industrial operations, lighting may account for no more than a few percent of total consumption, although representing a large amount of energy in absolute terms.

The percentage of lighting energy that can be saved typically is large, but impossible to define precisely. Centuries ago, a person might read by the light of a single candle. Today, a person in a typical office uses hundreds or even thousands of times more light. However, don't expect that you can reduce your lighting energy consumption by a factor of several hundred. Illumination standards have increased radically, modern activities require a larger illuminated area, and "background" lighting has come to be considered a necessity.

Experience suggests that aggressive lighting energy conservation can reduce average lighting energy consumption by a factor of three to ten compared to conventional practice, while providing good visual quality. In contrast, most contemporary lighting conservation programs reduce energy consumption by less than half. This says that there is plenty of room for improvement over present practice.

■ Contribution to Heating and Cooling Loads

In most non-industrial facilities, lighting is the largest source of internal heat gain, typically ranging from one watt to four watts per square foot. By comparison, the heat load from people in a typical office space is 0.3 to 1.0 watt per square foot. Even modern office equipment uses less energy than lighting, overall. For example, a desktop computer consumes only about as much energy as one or two typical fluorescent light fixtures.

Considered as heat sources, lighting fixtures are electric resistance heaters. Most of the energy consumed by lighting equipment remains inside the building. This energy input is beneficial if heating is required in the space. On the other hand, this heat adds to the cooling load if the space requires cooling. If electric cooling equipment is used, about one third watt of cooling energy is required for each watt of lighting energy.

Fixtures installed in a return air plenum ceiling typically deliver 30% to 50% of their heat into the return air stream. The amount of this heat that becomes a cooling load depends on the fraction of return air that is recirculated.

■ Daylighting

Lighting energy consumption can be reduced by using daylighting instead of artificial lighting. The energy saving potential varies widely. Many industrial facilities have a large potential for daylighting, while typical commercial buildings can displace only a small fraction of their lighting energy with daylighting. See Subsection 8.3 and Reference Note 46 for details.

The Cost of Lighting

Keep a sense of perspective when considering the cost of lighting and the cost of making changes. The major costs related to lighting, in approximate descending order, are:

- **human productivity.** The cost of labor may be hundreds of times higher than the cost of lighting for the worker. Hence, if a change in lighting causes even a small reduction in worker efficiency, any saving is lost. Properly executed lighting energy conservation does not degrade the quality of lighting, and may improve it. The key point is to invest the effort required to accomplish lighting conservation properly, and to avoid hasty, shotgun approaches.

- *energy* is by far the highest cost related to the lighting equipment itself, when tallied over the life of a facility. The most efficient lighting system is almost always the most economical overall.

- *lamps* typically cost an order of magnitude less than the energy they consume. The major classes of lamps — incandescent, fluorescent, and HID — all average about the same lamp cost per kilowatt-hour of service. However, there are large cost differences within each class.

- *lamp replacement labor* may cost more or less than the lamps themselves. Lamp life, lamp output (i.e., the number of lamps needed to illuminate the space), and accessibility of fixtures strongly influence the cost of lamp replacement. In the case of incandescent lighting, replacement labor usually costs much more than the lamps. Furthermore, incandescent lamps have short life, so the cost is repeated often. The labor cost for replacing fluorescent lamps is typically comparable to the cost of the lamps. The labor cost for replacing HID lamps is typically much less than the lamp cost, especially for lamps of larger size. However, the labor cost may be greatly inflated if the fixtures are difficult to reach. Make it as easy as possible to replace all types of lamps when they burn out. You can radically reduce the cost of lamp replacement by making fixtures accessible and by exploiting devices to aid lamp replacement, such as winches that allow fixtures installed in tall ceilings to be lowered to floor level.

- *fixtures* typically cost an order of magnitude less than the lamps inside them, when the lamps are totalled over the life of the facility. Lighting fixtures typically cost less than one percent of total construction cost. Fixture cost per lumen of light output does not differ much between lamp types, except for cheaper incandescent fixtures.

When you choose among different lighting options, include all these factors on your calculation pad or your computer spreadsheet. Compare lighting costs on a life-cycle basis.

How to Use Section 9

Ideal lighting provides the appropriate level of illumination at the activity with the minimum input of energy, and with good visual quality. Section 9 is organized to achieve ideal lighting as closely as current technology and your available budget will allow. The following "formula" breaks down the efficiency aspects of lighting into manageable parts:

Efficient Lighting = Efficient Lamps
+ Efficient Fixtures
+ Efficient Control
+ Efficient Light Path

The Subsections of Section 9 parallel this "formula." Each Subsection groups similar activities, including activities that are alternatives to each other. This grouping makes it easy for you to plan and execute your lighting improvements.

Most efficiency improvements involving lamps and their fixtures are done together. The methods of saving energy differ among the three major categories of lighting. Therefore Subsections 9.1, 9.2, and 9.3 deal with lamp and fixture improvements for each of the three major types of light sources individually.

Efficient control of lighting is subdivided into manual control and automatic control, each requiring different equipment and techniques. The two areas are covered by Subsections 9.4 and 9.5, respectively.

Establishing an efficient light path is essential to keep light from being lost on the way from the light source to the activity. The portion of the light path within the lamps and fixtures is covered implicitly by Subsections 9.1 through 9.3. The light path from the fixture to the activity is covered explicitly in Subsection 9.6.

The Measures in Subsection 9.7 maintain the long-term efficiency of lighting.

Efficient lighting also exploits daylighting as a source. Daylighting involves activities that are fundamentally different from those involved in improving artificial lighting. For this reason, daylighting is covered separately in Section 8.

The Measures in this Section are as specific as possible. However, the great diversity of lighting makes it impractical to define detailed improvements for each situation that may exist. To cover the range of possibilities most effectively, some Measures recommend specific actions that are related to clearly defined equipment or lighting situations, while other Measures are defined in a largely conceptual manner.

Use the Measures in this Section as components of an overall program of lighting efficiency. It is your job to knit the Measures together into an efficient lighting configuration for each application. You can achieve efficient and pleasing lighting in any environment. In retrofit applications, you will often find alternative ways to improve the lighting. You may have different Measures that you can apply to the same situation, or you may be able to combine Measures in different ways. Make your choices by analyzing the relative benefits and costs.

■ Deal with Each Activity Area and Each Fixture Individually

Each activity area has its own illumination requirements. Tailor the lighting to each activity. If a fixture is not providing illumination in the most efficient way possible, it does not belong.

In retrofit applications, start by inspecting each fixture individually. Ask yourself whether each fixture is located properly, whether it is oriented in the right direction, whether it is the right type for the job, whether it is using the appropriate type and number of lamps, and whether it is controlled properly. Lighting conservation tends to be time consuming and tedious, if you do it properly. There is no wholesale way to do it correctly.

■ Accommodate Future Changes in Activities and Space Layout

Try to make your lighting adaptable to future changes in the activities and space layout. Adaptable lighting equipment is still lacking, but the Measures in Subsection 9.6 will give you useful ideas.

■ Stress Visual Quality

Remember that visual quality is the dominant cost factor in lighting because of its effect on human activity. Never compromise visual quality. If you stress it, you will waste less light.

Why We Don't Use Conventional Lighting Design Methods

Every profession involved in the design of buildings and plants has its own set of design procedures, which are formalized and promoted by the profession's leading organizations. In the case of lighting, the formal doctrine includes design calculations that are based on factors, like "room cavity ratio," and idealized concepts, such as "equivalent sphere illumination." This design doctrine is fairly new, compared to the design methods used by other branches of engineering. It is an honest attempt to provide guidance in dealing with a subject that is more complex and subtle than it initially appears to be.

Unfortunately, some of the assumptions of this doctrine have turned out to be dubious or irrelevant, while some important factors are overlooked. Overall, the official lighting design doctrine has grown in a direction that is fundamentally in conflict with energy efficiency. It wastes energy to an extent that ranges from moderate to extreme, depending on the application. The present doctrine provides fairly good lighting quality, but it does this by expending energy to make lighting uniform over areas that are typically much larger than the areas where the illuminated activities are occurring. Even so, it does not provide the best possible lighting quality. For example, it largely ignores veiling reflections, a significant cause of visual fatigue as well as a waste of lighting energy. The doctrine cannot become a guide to efficient lighting without changing its nature entirely, so it is not worth salvaging as a general approach to lighting design. To the extent that the doctrine is useful, it is primarily a guide for applications that require uniform large-area lighting.

In the real world of construction, all this is largely academic. Perhaps a majority of lighting designers do not use the official methods of lighting design. Many are not aware of them. You don't have to examine too many lighting plans before you recognize that the basic tool of contemporary lighting design is a rectangular grid, and that lighting layout is usually guided by non-lighting constraints, such as the spacing of ceiling tiles. There is a gap between theory and practice in all professions and trades, but in the case of lighting design, the gap is a chasm that has only a tenuous bridge across it.

For all these reasons, the *Energy Efficiency Manual* does not use the present methods of lighting design, nor does it refer to them. Instead, it bases lighting selection, layout, and control on the principle of "task lighting" in its broadest conception. This principle is to design the lighting for each activity area as an individual entity, including the effects of lighting in adjacent areas, whether from artificial lighting or daylight.

This approach to lighting requires more care to design and to install than conventional lighting. Furthermore, it is hampered by the inability of presently available lighting fixtures to adapt easily to changes in the illuminated activities. Nonetheless, it is ultimately the path to best efficiency and best visual quality. We offer this approach to enable you to achieve the best available lighting efficiency. Also, we hope to reorient the lighting profession and lighting equipment manufacturers toward the full efficiency potential of lighting. Because this approach is still in its infancy, Section 9 subdivides the different aspects of efficient lighting, as described previously. This allows you to exploit those aspects that are presently practical for your situation, while leaving the rest to future developments.

Before Getting Started ...

For a practical explanation of how lamp output, illumination, and brightness are expressed, see Reference Note 50, Measuring Light Intensity.

Lighting quality is the most important aspect of lighting, so be familiar with the concepts in Reference Note 51, Factors in Lighting Quality.

For a comprehensive comparison of the major types of light sources, see Reference Note 52, Comparative Light Source Characteristics.

To help conform to standards of lighting efficiency, see Reference Note 53, Lighting Efficiency Standards.

For details of the three major categories of light sources – incandescent, fluorescent, and HID – see Reference Notes 54 through 56, respectively.

To understand the geometry of light distribution from fixtures, see Reference Note 57, Light Distribution Patterns of Fixtures.

Lamps and Fixtures, Incandescent

The "energy crisis" of the 1970's appeared to sound the death knell for incandescent lighting. It has long been recognized that incandescent lighting is much less efficient than fluorescent and HID lighting, and that it suffers from short service life. In recent years, further improvements in the efficiency and color characteristics of fluorescent and HID lighting have increased their advantage over incandescent lighting. All this makes it fair to say that eliminating incandescent lighting is a leading goal of energy conservation.

At the same time, recognize the advantages that have kept incandescent lighting popular. The color rendering of incandescent lamps is still the best available, although fluorescent and HID lamps have achieved radical color improvements in recent years. Incandescent lighting does not need ballasts, so it is free of the problems associated with them, including audible noise, electromagnetic interference, low power factor, and harmonic distortion of the power system. The small filaments of incandescent lamps allow tight focussing, which is important in some applications. Lamp and fixture costs are much lower than for other types. The lamps themselves are much less expensive, and they do not require the expensive auxiliary equipment of ballasts, capacitors, and ignitors that other light sources need.

The Measures of this Subsection improve the efficiency of incandescent lighting, in some cases by eliminating it. The Subsection starts with the Measures that are least expensive and most broadly applicable, and progresses to Measures that are more expensive or more narrowly applicable. Within this general sequence, similar Measures are grouped together.

Many of the Measures are defined broadly. The variety of possible lighting configurations is unlimited, so it is not practical for Measures to deal with each possible configuration. Combine the Measures in this Subsection and the following Subsections to create a lighting configuration that is optimum for each activity area.

INDEX OF MEASURES

9.1.1 Eliminate excessive lighting by reducing the total lamp wattage in each activity area.

9.1.2 Substitute higher-efficiency lamps in existing fixtures.

9.1.2.1 Substitute screw-in fluorescent lamps for incandescent lamps.

9.1.2.2 Substitute tungsten halogen lamps for conventional incandescent lamps.

9.1.3 Substitute lamps that minimize light trapping and/or improve light distribution.

9.1.4 Modify existing fixtures to reduce light trapping and/or improve light distribution.

9.1.4.1 In fixtures having shades that absorb light, modify or eliminate the shades.

9.1.4.2 Install reflective inserts in fixtures that have absorptive internal baffles or surfaces.

9.1.4.3 For task lighting, install focussing lamps on flexible extensions.

9.1.5 Replace incandescent fixtures with fluorescent or HID fixtures.

9.1.6 Modify or replace incandescent exit signs with fluorescent or LED light sources.

9.1.7 Install dimmers.

RELATED MEASURES:

• Measures 9.6.2 ff, fixture layout and replacement in general

• Measure 9.6.3, combinations of fixtures that adapt to different lighting requirements

Before getting started, review Reference Note 54, Incandescent Lighting. This explains how incandescent lighting works, the factors that affect efficiency and service life, and the ways that incandescent lamps are named. See Reference Note 52, Comparative Light Source Characteristics, for a comparison of incandescent lamps with other light sources.

MEASURE **9.1.1 Eliminate excessive lighting by reducing the total lamp wattage in each activity area.**

RATINGS

New Facilities	Retrofit	O&M
		B

The most common types of incandescent lamps all have the same type of socket, called a "medium" screw base. This allows a given incandescent fixture to accept lamps ranging from 10 watts to 1,000 watts in output. Similarly, most large incandescent lamps have "mogul" bases, and many sizes of smaller incandescent lamps have "miniature" bases. The fact that the common socket types accept many lamp sizes allows people to install incandescent lamps with excessive wattage.

You can save energy simply by substituting lamps that have the lowest wattage needed for the application. In facilities that use a lot of incandescent lighting, such as hotels and motels, this simple practice can yield large savings.

Of course, if the situation were really so simple, this opportunity for saving would not be so common. You have to expend the effort to determine the minimum acceptable wattage for each fixture. Then, you have to ensure that larger lamps are not substituted in the future. Incandescent lamps burn out often, so this is a continuing challenge.

Repeat with Later Improvements

Any time you make changes to incandescent fixtures, you have to repeat this Measure, even if the change is for the purpose of improving efficiency. For example, if you substitute screw-in fluorescent (Measure 9.1.2.1) or halogen lamps (Measure 9.1.2.2) in the existing fixtures, make sure that you use the lowest wattage of the new type that is satisfactory for the application.

First, Survey the Fixtures

Determine the appropriate wattage for each fixture individually. In a cafeteria, for example, the serving lines, cooking areas, cash registers, seating areas, etc., all have different illumination requirements.

Don't expect the people who are replacing the lamps to select the proper sizes as they do the work. Even with exceptional personnel, this would waste time and create confusion. If you have to deal with many fixtures, make a planning survey first. The easiest way to do this is to mark the lamp wattages on a copy of the lighting plan from the building drawings.

Use a good light meter to measure existing lighting levels. With a little experience, you will get the knack of estimating the lighting levels that result from changing lamp wattages.

Appropriate lighting levels are suggested in Reference Note 51, Factors in Lighting Quality.

SUMMARY

Saves a significant amount of energy in each fixture where it applies. Costs little or nothing. Repeat this activity as you accomplish other Measures. Fixture labeling is critical to preserve the benefit.

SELECTION SCORECARD

Savings Potential	**$** $ $
Rate of Return	**% % % %**
Reliability	✓ ✓
Ease of Initiation	☺ ☺

Label Fixtures to Deter Backsliding

When a lamp burns out, it is easy to replace it with the wrong wattage or type. The most effective way to prevent this is to label the fixtures with the proper lamp type. Where a socket is to be left without a lamp, label it to say so. Measure 9.7.3 provides guidance for fixture labeling.

Wattage Choices

The common incandescent lamp types are available in a wide range of wattages. In response to the market created by energy conservation, the older sizes were supplemented with intermediate sizes. These "energy saving" sizes appear to be based on the assumption that lighting designers add a margin of 20 to 35 percent to their lighting capacities. For example, one major manufacturer offers 65-watt reflector lamps to replace the standard 75-watt size, and 120-watt lamps to replace the standard 150-watt size.

Understand that these new "energy saving" lamp sizes are no more efficient than other lamps of the same type and similar wattage. The energy saving simply results from lower wattage, along with lower light output. To get greater lumens-per-watt efficacy, you have to substitute more efficient types of lamps.

■ Account for Lumen Degradation

Ordinary incandescent lamps suffer a lumen loss of 10% to 30% toward the end of their service lives. Halogen lamps suffer a lumen loss that is typically less than 10%. If lumen loss is high, provide enough margin in lamp sizing to provide adequate illumination at the end of lamp life. See Reference Note 54, Incandescent Lighting, for details of the lumen loss of different types of incandescent lamps.

Efficiency, Service Life, and Wattage

Reference Note 54 explains that incandescent lamps have an unavoidable conflict between efficiency and service life. All incandescent lamps have short lives, so the cost of lamp replacement labor may drive the compromise toward longer service life, at the expense of efficiency.

Common round ("A" style) bulbs larger than about 50 watts are mostly rated at 750 or 1,000 hours. Longer service lives are available in this wattage range, but at the expense of efficiency. Round bulbs of lower wattage have longer standard service lives, up to about 2,500 hours. Most reflector lamps are designed for a service life of 2,000 hours. These relationships are shown in Table 1 of Reference Note 54, Incandescent Lighting.

Substitute Smaller Numbers of Higher-Wattage Lamps

The efficacy of incandescent lamps increases somewhat with wattage, for a given type and service life. For example, a 75-watt round bulb has an efficacy of 16 lumens per watt, whereas a 150-watt bulb has an efficacy of 19 lumens per watt (both having a life of 750 hours). If you are willing to accept shorter bulb life, you can get even greater efficiencies. For example, a conventional 100-watt bulb is approximately 80 percent more efficient than a conventional 25-watt bulb, while a 200-watt bulb is about twice as efficient as the 25-watt bulb.

You may be able to exploit this fact by using a smaller number of higher-wattage lamps to replace lower-wattage lamps. The opportunities are limited, because there are not many cases where you can abandon a fraction of the sockets or fixtures without affecting lighting quality or creating a shabby appearance.

If you use this technique, be sure to disconnect and label the unused sockets or fixtures.

Lamps Operated by Dimmers

Incandescent lamps are much more efficient when operated at full power than when operated at reduced power. For example, a 50-watt bulb operated at full brightness uses much less energy than a 150-watt bulb that is dimmed to produce the same lumen output.

Installing lamps with unnecessarily high wattage is commonplace in fixtures that have dimmers. Therefore, check all the lamps on dimmer circuits to ensure that the lamps have the lowest wattages that are appropriate for the application.

This problem occurs commonly in facilities such as restaurants and lounges. Low lighting levels are desired during service hours, but high lighting levels are needed for preparation and cleaning. The only efficient solution in this situation is to install separate fixtures for each function. See Measure 9.6.3.

Light Distribution Patterns of Lamps

Incandescent lamps are available with a wide variety of light distribution patterns. Select the proper types for each application. See Measure 9.1.3.

Avoid Tinted or Color-Corrected Lamps

Even mild tints absorb a large fraction of a lamp's light output. For example, about one third of the light output of a "daylight" incandescent bulb is absorbed in a blue tint that is used to modify the color. Do not use such lamps unless the application demands them.

ECONOMICS

SAVINGS POTENTIAL: *20 to 40 percent of the energy used by incandescent fixtures, typically.*

COST: *Minimal, if the changes are made as lamps burn out.*

PAYBACK PERIOD: *Less than one year.*

TRAPS & TRICKS

SURVEY: *It is tedious to figure out the optimum wattage for each fixture and to label it accordingly. But, if you skip this preparation, you will miss energy savings, and the savings that you do achieve will disappear within a few lamp replacements.*

EXPLAIN IT: *Explain the importance of lamp selection to the staff. Make a list of the lamp types and wattages to be used in each location or type of fixture. Post the list in the lamp storage room.*

REPEAT AS NECESSARY: *Repeat this Measure any time light fixtures or lighting requirements change. Make this a long-term repetitve item in your maintenance calendar.*

MEASURE 9.1.2 Substitute higher-efficiency lamps in existing fixtures.

The low efficacy of incandescent lamps has created a market for more efficient lamps that can be used in the same fixtures. The subsidiary Measures recommend two different types of higher-efficiency lamps:

- *screw-in fluorescent lamps* (Measure 9.1.2.1) provide major improvements in efficacy and service life
- *tungsten halogen lamps* (Measure 9.1.2.2) provide only modest improvements in efficacy and service life, but they are less expensive. They may be the only option in some applications.

Other Replacements for Incandescent Lamps

Two other types of direct replacements for conventional incandescent lamps are worth mentioning because they have been promoted under the banner of energy conservation. Both of these are probably dead ends in terms of energy conservation:

- *self-ballasted mercury vapor lamps* use an internal tungsten filament to serve as a ballast. This type of lamp is no more efficient than incandescent lamps

of comparable wattage, and it suffers serious lumen degradation. Its advantage is longer life, typically about 12,000 hours. It reduces labor cost for lamp replacement, but not energy cost.

- *krypton-filled conventional incandescent lamps* may improve energy efficacy by 5 to 15 percent, depending on wattage. The heavy krypton gas retards filament evaporation, allowing the filament to be operated at a somewhat higher temperature. The cost premium is high in relation to the small efficiency improvement. For applications where small lamps are needed, screw-in fluorescent lamps are a much better choice, from the standpoints of efficacy and service life. For applications where large lamps are needed, halogen or HID lamps are a much better choice.

(Krypton is also used in the small filament capsule of halogen lamps, where it serves an important purpose. It is more economical in halogen lamps because of the small amount used and the greater relative efficiency improvement.)

MEASURE 9.1.2.1 Substitute screw-in fluorescent lamps for incandescent lamps.

RATINGS

New Facilities | Retrofit **B** | O&M

Screw-in fluorescent "lamps" are actually complete fixtures packaged on a standard lamp base. They include a fluorescent tube, a starter, and a ballast. The fluorescent tube itself may be permanently attached to the ballast or it may be separately replaceable. A large variety of configurations are available, including units with integral reflectors and various types of globes or diffusers. See Figure 1.

The best screw-in fluorescent lamps are somewhat less efficient than the best conventional fluorescent fixtures, but they are still three to four times more efficient than ordinary incandescent bulbs of similar light output. In applications such as the example shown in Figure 2, they can significantly reduce electricity cost. They last about 10 times longer, but they cost about 20 times more.

No special steps are needed to replace incandescent lamps with screw-in fluorescent units in most situations.

SUMMARY

Radically improves lighting efficiency. Easy, quick, and fairly cheap, but not foolproof. Be careful to avoid junk brands.

SELECTION SCORECARD

Savings Potential .... $ $ $ $

Rate of Return .... % % %

Reliability .... ✓ ✓ ✓

Ease of Retrofit .... ☺ ☺ ☺

In some cases, a minor modification of the fixture is needed. However, there are many situations where screw-in fluorescent replacements cannot be used.

Some screw-in fluorescent lamps have magnetic ballasts and some have electronic ballasts. In units made

Osram Sylvania Inc.

Fig. 1 Three basic types You can select a bare lamp, or a lamp with a reflector, or a lamp with a diffuser.

by reputable manufacturers, electronic ballasts are more efficient. They add some problems, and they eliminate others. See Reference Note 55, Fluorescent Lighting, for details about fluorescent lamps and ballasts.

The newest types of screw-in fluorescent fixtures use tubes of small diameter and high brightness. These were originally called "PL" fixtures (after Philips Lighting, who introduced them), and they are now called "compact fluorescent" fixtures. Older types of screw-in fluorescent fixtures, still being offered, use a conventional circular tube of large diameter that is mounted on a screw-in ballast.

Limitations

Screw-in compact fluorescent lamps were designed as replacements for incandescent lamps. However, they cannot be substituted in these common situations:

- *fixtures too small for the lamps.* The need for a ballast, which is mounted at the screw base, makes the units too fat for many existing fixtures. Also, screw-in units are too long for some fixtures.

  Some table lamps that use harps to hold the shades may be adapted to accommodate the lamps by using harp extenders.

- *need for high light output.* The largest compact fluorescent lamps cannot match the light output of the largest incandescent lamps.

- *need for dimming.* Compact fluorescent lamps should not be used with conventional lighting dimmers. Doing so shortens their life, reduces their efficiency, and may create a fire in the fixture.

- *frequent switching.* The filaments and starters of fluorescent lamps wear out with frequent on/off switching.

- *need for instant light.* Most compact fluorescent fixtures have preheat starters, which take a second or two to work. After starting, the lamps require a few seconds to reach full brightness. Lamps that use mercury amalgams (explained below) require almost a minute to reach full brightness.

There are some other, less common situations where you should not use compact fluorescent lamps. These include extremely low temperatures, a need for very good color rendering, sensitivity to acoustical noise, and sensitivity to electrical interference. See "How to Select Compact Fluorescent Lamps," below, for more about these factors.

Energy Saving Potential

The following table compares the energy consumption of ordinary incandescent bulbs to the energy consumption of some typical compact fluorescent lamps that might be used to replace them. The figures indicate a typical saving of about 70%. Few other energy conservation measures offer this much potential for savings.

Incandescent		Fluorescent			Wattage
Watts	Lumens	Watts	Lumens	Type	Reduction (%)
25	230	9	400	bare lamp	64
40	460	11	600	bare lamp	72
60	890	15	900	bare lamp	75
60	890	17	950	with diffuser	72
75	1,200	20	1,200	bare lamp	73
75	1,200	23	1,550	bare lamp	69

However, this simple comparison does not tell the whole story. You need to make an energy savings analysis for each potential lamp substitution. For example, much incandescent lighting in commercial facilities is done with downlights and other shrouded fixtures. The lateral light distribution pattern of bare compact fluorescent lamps is inappropriate in such fixtures, so you may have to use fluorescent reflector lamps that are less efficient.

How to Select Compact Fluorescent Lamps

Replacing incandescent lamps with compact fluorescent lamps is a simple operation, but you need to be careful in selecting the substitute lamps. The following are the important criteria.

■ Lumen Output

Start by finding out whether the illumination levels provided by the present lighting match the requirements of the activities. Calculate any increases or reductions in the lumen output that are appropriate. Then, search the catalogs to find lamps that provide the output you need.

Screw-in fluorescent lamps are available in a range of sizes to replace the most common incandescent bulb sizes. See Figure 3. The output of the fluorescent units was somewhat limited at first, but output continues to be increased.

Experience indicates that manufacturers exaggerate the effective light output of compact fluorescent fixtures. For example, if a compact fluorescent lamp promises "as much light as a 75-watt bulb," a person probably will not be able to see as well as with the original 75-watt bulb. This may be due to differences in color rendering, or to exaggerating the lumen output.

On the other hand, where fixtures are largely decorative, as with downlights in lobbies, compact fluorescent fixtures tend to produce an illusion of greater brightness than their lumen output suggests.

Light output deteriorates with age. Provide enough reserve capacity to provide adequate lighting at the end of the replacement cycle. The best compact fluorescent lamps lose about 20% of their light output at the end of normal service life.

■ Wattage Input and Efficacy

After you find the lamps that can provide the desired lumen levels, select among them for the lowest input wattage. In other words, select the highest efficacy.

Be skeptical of the efficacy figures published for compact fluorescent fixtures. Experience indicates that the actual efficacies of cheap compact fluorescent fixtures may be less than half that of better units. Reliable comparisons of the products of different manufacturers are not available, so it is reckless to acquire units from any but the best manufacturers. (We said "best," not "most popular." One of the largest vendors of compact fluorescent fixtures produces cheap junk.)

These are efficacies for compact fluorescent fixtures derived from the published figures of one major manufacturer:

Watts	Lumens/Watt	Type
9	44	bare lamp
11	54	"
15	60	"
20	60	"
23	67	"
17	55	diffuser lamp
18	61	"
18	44	reflector lamp

Note that *these efficacy figures include the effects of the ballast, whereas this is not true of ratings for conventional fluorescent lamps.*

■ Reliability and Service Life

Some units presently on the market have poor reliability. Screw-in fluorescent lamps can be produced by marginal manufacturers, and units of poor quality can be made substantially cheaper than units of good quality. The market has not shaken out the poor

WESINC

Fig. 2 A good place to retrofit screw-in fluorescent lamps The ceiling of this lobby is a forest of downlights. Screw-in fluorescent lamps provide adequate illumination. Bare lamps provide the best efficiency in the existing reflector fixtures. Their appearance is satisfactory at this distance. Their longer service life greatly reduces the amount of labor required for maintenance.

products. On the contrary, because compact fluorescent lamps are perceived as expensive (compared to incandescent lamps), the lower cost of the poor quality units has allowed them to flood the market. The rated life of most compact fluorescent lamps is 10,000 hours. However, the rating does not guarantee quality. Some units of low quality last only a few hours or a few weeks.

Purchase compact fluorescent lamps only from manufacturers of good reputation. If you are going to install large numbers of lamps, make some calls to experienced users of compact fluorescent lamps to find which units are best.

If the unit allows the fluorescent tube to be replaced, the life of the ballast is rated separately. A life of 50,000 hours is commonly claimed for ballasts. Take such numbers with a grain of salt. Not enough time has gone by to allow such figures to be verified by field experience.

The life of a magnetic ballast declines with higher operating temperature. The operating temperature is increased if the unit is installed with the ballast above the lamp, and if the fixture is poorly ventilated. The life of an electronic ballast declines as the number of switching cycles increases.

■ Lamp Dimensions

You may have trouble finding models of compact fluorescent fixtures that fit many incandescent fixtures. Figures 1 and 3 show the problem. Compact fluorescent fixtures are all much wider at the base than incandescent lamps. Most are substantially longer, although shorter compact fluorescent tubes are entering the market. Manufacturers are achieving shorter lengths by bending the tubes into hairpin and corkscrew shapes.

■ Light Distribution Pattern

The light distribution pattern of the lamps affects the overall efficiency of lighting. For example, a bare compact fluorescent lamp radiates most of its energy in a direction perpendicular to the tube. If such a lamp is installed in a recessed downlight with an absorptive interior, most of the light is absorbed inside the fixture.

Compact fluorescent lamps are available with integral reflectors or diffusers to improve light distribution, as shown in Figures 1 and 4. For example, they are available with parabolic reflectors to minimize light loss when installed in downlights. The combination of greater light source efficiency and lower light losses can increase the overall efficiency of the fixture by a factor of 10 to 30 (that's not a misprint) when these lamps are installed in highly absorptive fixtures.

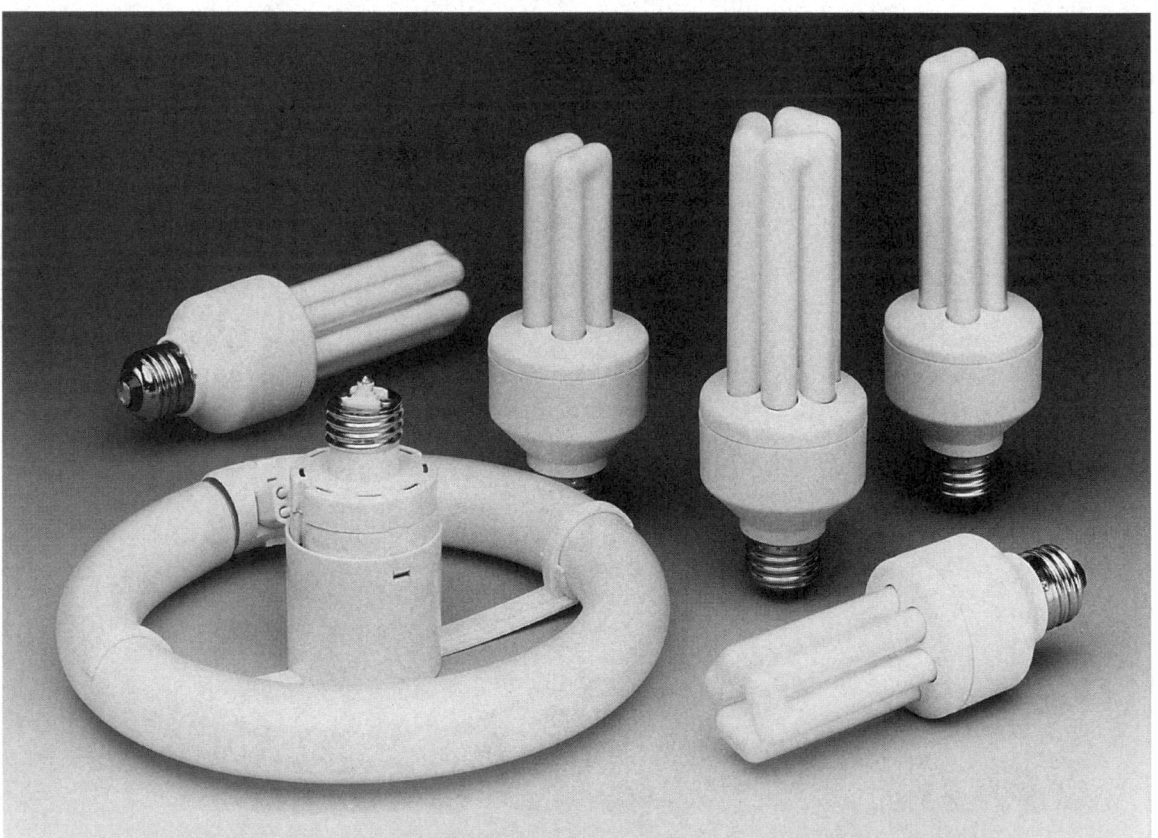

Osram Sylvania Inc.

Fig. 3 A range of light outputs The maximum size of screw-in units with bare lamps continues to grow. Units are available that substitute for the most common incandescent lamp sizes. However, beware of exagerrated claims about "equivalent" light output.

■ Ability to Replace Fluorescent Tube Separately

The ballasts in screw-in fluorescent lamps may last several times longer than the tubes. Some units allow the lamp to be replaced separately from the ballast. For example, see Figure 4. Units that allow the tube to be replaced separately may have better long-term economics. This is illustrated by the economic example given below.

■ Color Characteristics

The color rendering of fluorescent lamps is different from the color rendering of incandescent lamps, and this may be important in some applications. Fluorescent lamps emit light strongly in a few narrow bands of color that are created by the mercury vapor in the lamps. These color spikes are superimposed on a broader light spectrum produced by the phosphors.

The perceived white color of the fluorescent lamp is not the result of a continuous light distribution spectrum. Instead, it is the result of selecting the phosphors to produce the illusion of white light. This causes the lamp to distort the colors of illuminated objects. Human vision corrects for this, so people do not notice it, except in certain color-critical applications.

For more about lamp color in general, see "Color Rendering Index" and "Lamp Color" in Reference Note 52, Comparative Light Source Characteristics. For more details of fluorescent lamp color, see Reference Note 55, Fluorescent Lighting.

■ Starting and Operating Temperatures

All fluorescent lamps, including compact units, operate at peak efficiency only if the glass tube is near a particular temperature. The tube temperature is determined by the environment of the lamp and by the lamp's heat output. The optimum temperature is about 105°F (40°C). The lamp usually starts at a lower temperature than this, and then warms up to a temperature that may be well below or well above the optimum temperature. If the space is kept at normal indoor temperature and the fixture is well ventilated, the lamp will stabilize near its optimum temperature. If the lamp is used outdoors, or if it is installed in a poorly vented ceiling fixture, its efficiency suffers seriously.

Fluorescent lamps require a minimum ambient temperature to start reliably. Compact fluorescent lamps have higher current densities, shorter arcs, preheat starting, and other features that allow them to operate at lower temperatures than conventional fluorescent lamps. Different models of compact fluorescent fixtures claim starting temperatures from -20°F to 32°F (-29°C to 0°C).

The product literature does not make it clear how much efficacy suffers at lower operating temperatures. Some manufacturers use an amalgam of mercury with another metal to stabilize the mercury vapor pressure inside the lamp. This keeps the lamp efficacy high over a wider range of temperature. A disadvantage of this method is that the lamp may require as much as a minute to reach full brightness as the mercury separates from the amalgam.

Screw-in fluorescent fixtures that use conventional circular lamps should be used only at normal indoor temperatures.

■ Acoustical Noise

Units of good quality emit little noise. Cheaper units may be noticeably noisy. Refer to Measure 9.2.4 for more about this.

■ Power Factor

Low power factor is a potential problem with magnetic ballasts, but generally not with electronic ballasts. Measure 9.2.4 gives the details.

■ Radiated Electromagnetic Noise

This is a potential problem with units that have electronic ballasts. For example, problems have been noted with remote control units and security scanners. Refer to Measure 9.2.4 for details.

Compact lamps are smaller and they draw less current than conventional fluorescent fixtures, so they have the potential of emitting less electromagnetic radiation. This does not mean that they actually make less noise. Cheaper units that lack adequate input filtering may cause the power wiring to act as an antenna, causing trouble at some distance from the lamps.

■ Harmonic Distortion

Harmonic distortion in compact fluorescent fixtures worries electric utilities, who have been using incentives and penalties to promote lower harmonic distortion. Units are commonly rated in terms of "total harmonic distortion" (THD). At present, a THD of 20% is considered acceptable. Some cheaper units may have a

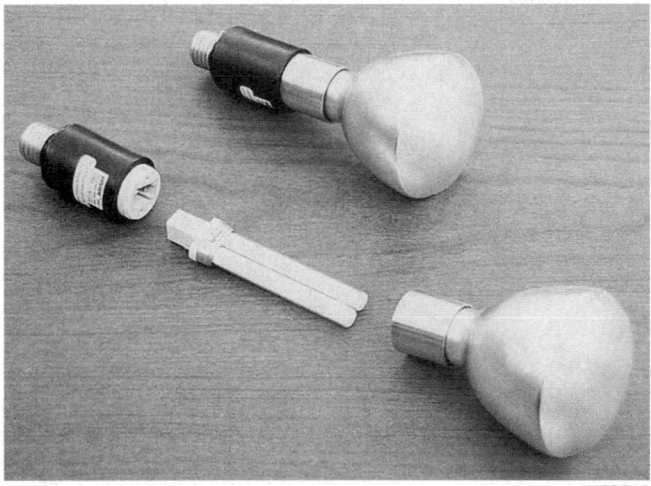

WESINC

Fig. 4 Reflector fixture with separately replaceable lamp
The ballast may last several times longer than the lamp. A unit like this should be made of glass, rather than plastic, because the ultraviolet radiation from a fluorescent lamp will darken plastic.

THD much worse than this. See Measure 9.2.4 for details.

Label Fixtures to Deter Backsliding

When a lamp burns out, it is easy to replace it with the wrong type. The most effective way to prevent this is to label the original fixtures with the lamp types to be used. Measure 9.7.3 explains how to label fixtures to minimize these failure modes.

An Economic Example

Compact fluorescent fixtures are good devices in their own right, but they are also a fad, which causes people to use them uncritically. The economic benefit they provide may not be as great as widely assumed, so it is worthwhile to examine a typical case.

Assume that 23-watt compact fluorescent lamps are used to replace 75-watt incandescent lamps, resulting in a saving of 52 watts per lamp. Assuming a 10,000-hour life and an electricity cost of 8 cents per kilowatt-hour, each lamp replacement saves $42 in electricity cost during its life.

During the unit's lifetime, it saves replacement of ten incandescent lamps costing $0.50 apiece, for an additional saving of $5.00. The compact fluorescent lamp costs $18. Thus, the facility saves $42 for a net investment of $13.

This is a good ratio, but variations in the assumed factors could make the economics much worse or much better. There is great variation in electricity rates. The example did not include demand charges, which are a major factor is some locations, but not in others. The example did not include labor cost for lamp replacement, which is much lower for compact fluorescent fixtures.

The rate of return, as opposed to total savings, depends on the number of hours that the lamps operate each year.

If the fluorescent tube can be replaced separately, the typical replacement tube cost is about $5. Therefore, lamps with replaceable tubes may have considerably better long-term economics.

It is worth stressing that cheap junk units, which infest the market, will not last long enough to pay off. Buy for quality, not for price.

ECONOMICS

SAVINGS POTENTIAL: *50 to 75 percent of lighting energy, depending on how closely the needed lumen levels can be matched by the fluorescent lamps.*

COST: *Good screw-in fluorescent lamp assemblies cost from $15 to $30, depending on wattage, features, and quantity purchased. Replacement tubes alone cost about $5, for the more common types. The cost is reduced by the cost of all the incandescent lamps that would have burned out during the life of the fluorescent lamp. This is typically $5 to $10.*

PAYBACK PERIOD: *Less than one year, to many years. Refer to the economic example above.*

TRAPS & TRICKS

SELECTING THE LAMPS: *The continued popularity of junk units shows that buyers are not selecting wisely. Even the best equipment is too new to have a reliable performance record in real applications. Take care to select the proper lamp wattage and configuration for each application. Check with previous users about service life and problems.*

MAINTAIN THE BENEFIT: *Don't forget to label the fixtures.*

MEASURE 9.1.2.2 Substitute tungsten halogen lamps for conventional incandescent lamps.

Tungsten halogen lamps are a variation of ordinary incandescent lamps that have a modest efficiency advantage. Reference Note 54 explains how they work and shows a range of available types.

Remember that halogen lamps must compromise between improved efficacy and extended service life. Look at the performance figures to find out whether you are getting an advantage in efficiency, or an advantage in service life. You won't get a big improvement in both.

Halogen lamps are much less efficient than fluorescent or HID lamps. Select them only for applications where their advantages in color rendering, tighter focussing, and lower cost are more important than efficiency.

There is only a limited selection of halogen lamps designed to replace incandescent lamps directly, i.e., that have medium screw bases. At present, halogen lamps that replace ordinary bulbs are available in two configurations, bulbs and PAR-type reflectors. Wattages are limited to small and medium sizes.

Applications and Limitations

You can substitute halogen lamps for conventional incandescent lamps in most applications. Halogen lamps designed to fit conventional incandescent lamp screw sockets do not suffer the same restrictions on mounting position that apply to some other types of halogen lamps. Also, they do not emit amounts of ultraviolet light that require protective measures.

As a rule, do not use tungsten halogen lamps with dimmers. Some halogen lamps have diodes, and these lamps do not work with conventional dimmers. Most halogen lamps can be dimmed, but this is undesirable. Halogen lamps must operate at or near full power for the halogen cycle to work. Manufacturers state that operating the lamps at full power for a fraction of the time is sufficient to keep the glass clean, but forget about trying to do this on a scheduled basis.

Tungsten halogen lamps are much more dangerous than conventional incandescent lamps if they shatter. The inner quartz capsule operates at high temperature. It has enough mass to store considerable heat, unlike the naked filament of a conventional bulb. The fill gas inside the capsule operates at positive pressure. If the lamp shatters, the fragments may cause injury or fire. Halogen lamps that replace conventional incandescent lamps have heavy outer glass shells to contain failures of the filament capsule. However, keep the danger in mind, especially near combustible materials. Manufacturers caution against using halogen lamps in

wet locations because water droplets may shatter the hot outer envelope.

More Efficient Alternatives

Screw-in fluorescent lamps (Measure 9.1.2.1) are available in the same lumen output range as the halogen lamps that are designed to replace incandescent round bulbs. However, screw-in fluorescent reflector lamps cannot match the lumen outputs of larger halogen reflector lamps.

Compact HID lamps in reflector fixtures can substitute for larger incandescent lamps, but these require expensive fixture replacement. Refer to Measure 9.1.5 for fluorescent and HID fixture retrofits.

How to Select Tungsten Halogen Lamps

Consider these factors in selecting halogen lamps to replace conventional incandescent lamps.

■ Lumen Output

Start by finding out whether the illumination levels provided by the present lighting match the requirements of the activities. Calculate any increases or reductions in the lumen output, and then scan the catalogs to find halogen lamps that provide this output.

Halogen lamps suffer less light loss with age than conventional incandescent lamps. Typically, this accounts for about 15% to 20% more light output at the time of lamp replacement. In marginal cases, this may allow you to reduce the initial lumens and wattage.

■ Lamp Dimensions

Check that the halogen lamp will fit the existing fixture. This is most likely to be a problem when replacing miniature styles of incandescent reflector lamps.

■ Light Distribution Pattern

Match the light distribution pattern of reflector lamps to the application. Check the catalog ratings for beam spread and for the candlepower at the center of the beam.

Halogen lamps that replace round bulbs distribute somewhat more of their light perpendicular to the lamp axis.

■ Wattage Input, Efficacy, and Service Life

Reference Note 54 explains that there is an inherent trade-off between efficacy and service life in halogen lamps. By the same token, there is a compromise between energy cost and lamp replacement labor cost. Compare the lumens-per-watt efficacy of the candidate lamps with the efficacy of the lamps you are presently using, to ensure that the improvement is large enough to be cost effective.

At present, halogen lamps designed to replace conventional round bulbs put most of their advantage into service life, rather than efficacy. These lamps have a nominal life of 3,500 hours, which is three to four times longer than the lamps they replace. They offer little or no efficiency improvement.

Halogen reflector lamps offer a broader range of choice between efficacy and service life. Some units put all their advantage into service life, which can be as high as 6,000 hours, three times longer than with conventional lamps. These lamps offer no efficiency improvement.

A newer type of halogen reflector lamp, introduced commercially in 1990, uses special coatings to reflect infrared heat emission back to the filament, reducing the amount of electricity required to heat the filament. This type of lamp offers efficacy improvements up to 80%. Some older types of halogen reflector lamps offer efficacy improvements as high as 40%, while maintaining a service life of 2,000 hours.

■ Color Temperature

Where color is a factor, select the appropriate "color temperature" rating. For an explanation of this rating, see Reference Note 52, Comparative Light Source Characteristics.

Halogen lamps are often used because they offer light that is "whiter" and "brighter," in the sense that its spectrum is closer to sunlight (has higher color temperature) than conventional incandescent lamps. This improvement results from higher filament temperature, which also provides higher efficacy. Be aware that halogen lamps optimized for long service life operate at lower filament temperature, and therefore they do not provide the "brighter," bluer light.

Label Fixtures to Deter Backsliding

When a lamp burns out, it is easy to replace it with the wrong type. The most effective way to prevent this is to label the original fixtures with the lamp types to be used. Measure 9.7.3 explains how to label fixtures.

***

ECONOMICS

SAVINGS POTENTIAL: *Halogen lamps that are optimized for efficiency (rather than extended life) save from 30 to 40 percent of the energy of the conventional incandescent lamps they replace.*

COST: *Halogen lamps typically cost several dollars more than their conventional counterparts. The higher cost is partially compensated by longer life.*

PAYBACK PERIOD: *One year, with typical operating schedules. However, this is misleading, because halogen lamps may burn out soon after paying off. The ratio of savings to investment may range from slightly more than 1 (not worth the effort) to 3 (a fairly good investment).*

TRAPS & TRICKS

SELECTING THE LAMPS: *Check the latest offerings of the reputable lamp manufacturers. The technology is still evolving. Decide whether you want higher efficiency or longer life.*

MAINTAIN THE BENEFIT: *Don't forget to label the fixtures.*

MEASURE 9.1.3 Substitute lamps that minimize light trapping and/or improve light distribution.

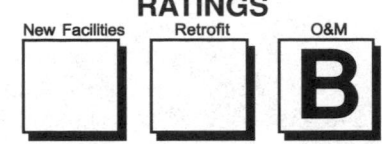

SUMMARY

Sometimes provides major savings in lamp wattage. Lamp selection and fixture labeling are the important steps.

SELECTION SCORECARD

Savings Potential .... $ $ $

Rate of Return .... % % % %

Reliability .... ✓ ✓ ✓

Ease of Initiation .... ☺ ☺ ☺

Many incandescent fixtures absorb a substantial fraction of the lamps' light output. These appear to be designed with a blatant disregard for the amount of energy they waste. You can reduce this energy waste, sometimes substantially, by exploiting the fact that incandescent lamps are available with many light distribution patterns. In addition to reducing light losses within the fixtures, you may also direct a larger fraction of the light in a useful direction.

Conversely, the variety of light distribution patterns makes it easy to install the wrong types of lamps in fixtures. Incandescent lamps burn out often, so it is a continuing challenge to keep the most efficient lamps in fixtures.

An Extreme Example: Downlights

An extreme example, but a common one, is recessed downlights with absorptive interior surfaces. The amount of light absorbed by this type of fixture varies enormously depending on the light distribution pattern of the lamp that is installed. For different types of lamps, these are typical percentages of the lamp's light output that manages to emerge from the fixture:

- 2%, with a bottom silvered bulb
- 5%, with a clear vertical-filament bulb
- 10%, with an ordinary frosted bulb
- 40%, with an R-series reflector lamp
- 60%, with an ellipsoidal reflector lamp
- 85%, with a screw-in fluorescent lamp having an integral reflector that extends to the rim of the fixture
- 70%, with a screw-in fluorescent diffuser lamp having an extender that allows surface mounting. (This lamp, unlike the others, greatly broadens the light distribution.)

These percentages show that you may be able to improve the efficiency of the fixture by a factor of 5 to 40 just by changing the type of lamp. This range of savings may seem too good to be true, but it is real. With many other types of existing fixtures, the potential savings are less dramatic, but well worth pursuing.

Better Ways of Improving Light Distribution

This Measure does not require fixture modifications, so it is relatively inexpensive in the short term. However, it may not be your best choice in the long term. The lamps are still incandescent, and hence inefficient. Furthermore, you still have to keep replacing them at short intervals. You can often achieve better long-term savings by replacing the fixtures entirely. So, review Measure 9.1.5 before you get started.

Even if you cannot afford major fixture changes at present, consider whether you can improve the existing fixtures inexpensively. See Measures 9.1.4 ff for ideas.

Your Lamp Choices

Incandescent lamps are available with a seemingly unlimited variety of characteristics. Consider these lamp choices for improving light distribution:

- **round bulbs with more efficient filament or lamp configurations.** Not all round bulbs ("A" shape) have the same distribution pattern. Select the right type for each application. A frosted round bulb emits light fairly uniformly in all directions, except toward the base. In clear incandescent bulbs, the filament configuration affects the light distribution pattern. The filament configuration also affects lumen degradation, as described in Reference Note 54.

 For example, a straight filament emits most of its light perpendicular to its axis. This makes it waste a lot of energy if the lamp is installed inside a downlight with an absorbing surface, or inside a shade. A circular filament is better in this case. (Ironically, straight-filament lamps have been favored for use in downlights, because this type of lamp darkens less when the base is upward.)

 Screw-in fluorescent lamps behave in a similar way. They have little light emission along the axis of the tube(s). Install these with reflectors if you want the light to shine away from the base. Install them with diffusers if you want a broader distribution pattern.

- **R-series reflector lamps.** These have broad (floodlamp) light distribution. They are relatively fragile. Use them in fixtures that protect the lamps well.

- **PAR (parabolic reflector) lamps.** These are available in a variety of specified beam widths, ranging from "spot" to "flood" patterns. They focus more sharply, with more abrupt cutoff at the edge of the beam. They are relatively rugged, so they are often used without an enclosure. Figure 1 shows a selection of rugged reflector lamps.

- *ellipsoidal reflector (ER) lamps.* These were developed specifically to overcome the trapping of light inside deeply recessed downlights. Figure 2 shows how they work. As the name implies, the glass bulb has an ellipsoidal shape. The inside of the bulb has a reflective coating, except for a flat end that allows the light to escape. The filament is at one focus of the ellipsoid. The other focal point is in front of the lamp, outside the fixture. From this point, the light can spread without being trapped.

(An ellipsoid is a three-dimensional shape created by rotating an ellipse around its long axis. An ellipse, or an ellipsoid, has two focal points. Light emitted from one focal point reflects from the surface of an ellipsoid in such a way that it focuses on the other focal point.)

- *screw-in fluorescent reflector lamps.* Some are designed specifically to aim light out of recessed fixtures. The reflector typically covers all or most of the absorptive inner surface of the fixtures in which they are used.

- *screw-in fluorescent extension lamps.* Another type of retrofit lamp intended for deep fixtures is mounted on an extension, so that the lamp is actually outside the original fixture. The extension lamp may have its own diffuser, creating the appearance of a surface mounted fixture. The extension fixture may have a much broader distribution pattern than the recessed fixture. Recessed fixtures typically are installed close together, and in large numbers, because of their

narrow light distribution patterns. Using screw-in extension fixtures may allow you to disconnect some of the original fixtures.

How to Select Replacement Lamps

Consider the following factors when you select retrofit lamps to improve light distribution.

■ Lumen Output and Wattage

Start by finding out whether the illumination levels provided by the present lighting match the requirements of the activities. Using a light meter, experiment by changing lamps to find the lowest wattage with any type of lamp that provides the desired illumination levels. If the fixtures are especially wasteful with the present types of lamps, you may be able to radically improve their light output.

■ Efficacy and Service Life

There is a big spread in efficacy and service life between the different types of lamps that can fit an incandescent fixture. Recognize that you have to compromise between efficacy and service life. See Reference Note 54, Incandescent Lighting, for typical efficacy and service life ratings for incandescent lamps.

The lighting industry has effectively standardized on a service life of 750 or 1,000 hours for common bulbs of medium and large wattage, and a life of 2,000 hours for reflector lamps. See Measure 9.1.2.2 for the service lives of tungsten halogen lamps. Manufacturers offer lamps with longer lives to reduce lamp replacement

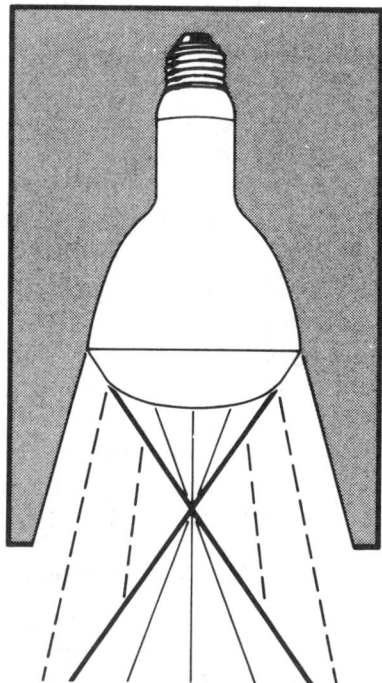

Philips Lighting

Osram Sylvania Inc.

Fig. 1 A selection of reflector lamps These may reduce light trapping greatly. However, the light still needs to go in the proper direction, and these lamps are still incandescent.

Fig. 2 Ellipsoidal reflector lamp This is a special type of reflector lamp that focuses through a point well in front of the lamp. These are intended specifically to reduce light loss in deep fixtures.

labor, at the expense of efficacy. Manufacturers also offer lamps with shorter lives to achieve brighter color.

Screw-fluorescent lamps have radically different characteristics. Their efficacy is much higher, ranging from 40 to 60 lumens per watt, for good units. In comparison, incandescent lamps of similar light output have efficacies ranging from 10 to 20 lumens per watt. There is no conflict between efficacy and service life in screw-in fluorescent lamps. In fact, longer service life tends to go with higher efficacy, for reasons that are practical rather than theoretical. Most screw-in fluorescent lamps have a rated life around 10,000 hours, which is another major advantage they offer. The great efficacy and service life advantages of screw-in fluorescent lamps merit your attention. See Measure 9.1.2.1 for more about them.

You may not be able to gain a reduction of energy input that is proportional to the greater efficiency of the substitute lamp. This is because of limitations in available wattages. For example, you might replace a 40-watt frosted bulb in a downlight with either a 17-watt R-20 incandescent reflector bulb or with an 18-watt screw-in fluorescent reflector fixture. These are the smallest sizes of each type. The fluorescent lamp produces more light than the incandescent reflector, but it cannot save more energy in this situation.

■ **Light Distribution Pattern**

Select the light distribution pattern of the lamp to serve two purposes. First, the light has to get out of the fixture with a minimum of loss. Second, the light has to go where it is needed within the space.

To satisfy the first requirement, match the light distribution pattern of the lamp to the characteristics of the fixture. For example, in a deeply recessed downlight, use an ellipsoidal reflector (ER) lamp to avoid light loss. On the other hand, if a downlight is relatively shallow, so that the rim of an R-lamp can rest at the rim of the downlight, the R-lamp provides a broader light distribution pattern, without light loss. To get an even broader distribution pattern, install extension lamps.

■ **Cost**

Try to avoid specialty types of incandescent lamps, such as bottom-silvered bulbs, tubular bulbs, and others. All types of specialty incandescent lamps are much more expensive than ordinary bulbs ("A" or "P" shapes). They do not last long enough to pay for themselves, unless they are replacing a type of lamp that is even more expensive.

Incandescent reflector lamps can provide a significant cost saving only if they provide a large wattage saving, or if electricity costs are high. You need to make some calculations to find the replacement that provides the best economics. The longer life of reflector lamps partially compensates for the higher cost. For example, long-life halogen reflector lamps merit special attention in cases where their improved light distribution allows wattage to be reduced significantly.

■ **Color Characteristics**

For a general explanation of lamp color, refer to "Color Rendering Index" and "Lamp Color" in Reference Note 52, Comparative Light Source Characteristics. For specifics about the color characteristics of incandescent lamps, see Reference Note 54, Incandescent Lighting.

■ **Lamp Appearance**

The appearance of the lamp itself may be a factor in the decor. For example, compact fluorescent lamps are not considered as beautiful as some incandescent lamps. Some designers make a fetish of keeping lamps completely out of sight, which invites extreme energy waste. Be prepared to negotiate with the decorator.

■ **Additional Criteria for Compact Fluorescent Lamps**

Refer to Measure 9.1.2.1 for additional selection considerations that apply specifically to screw-in fluorescent lamps.

Label the Fixtures to Deter Backsliding

When a lamp burns out, it is easy to replace it with the wrong type. The most effective way to prevent this is to label the original fixtures with the lamp types to be used.

Be very specific in defining the lamp type. For example, an important feature of the filament may be buried in the lamp's catalog number. See Measure 9.7.3 about fixture labeling.

ECONOMICS

SAVINGS POTENTIAL: 20 to 90 percent of lamp energy.

COST: The price of incandescent lamps varies over a wide range among the different types. For example, common "A" lamps cost about $0.25 apiece, in quantity. Reflector lamps cost $3 to $12. Screw-in fluorescent lamps with reflectors or extensions cost $20 to $30.

PAYBACK PERIOD: Less than one year, typically. However, this is misleading, because incandescent lamps burn out quickly and do not continue to generate savings. The ratio of savings to investment may range from slightly more than 1 (not worth the effort) to 5 (a good investment).

TRAPS & TRICKS

SELECTING THE LAMPS: Get into the details of the lamp manufacturers' catalogs. Select the proper lamp type and wattage on a fixture-by-fixture basis. Make an economic calculation if the candidate lamp is more expensive than the original lamp.

LABEL THE FIXTURES: Don't forget to label the fixtures with the correct lamp type. See Measure 9.7.3.

MEASURE **9.1.4 Modify existing fixtures to reduce light trapping and/or improve light distribution.**

Many incandescent fixtures trap a large fraction of the light produced by the lamps. Also, they may distribute light inefficiently, i.e., they do not put the light where it is needed. In some cases, you can improve the fixtures inexpensively. The subsidiary Measures deal with several specific opportunities. As part of the change, select the appropriate lamp type and wattage. See Measure 9.1.3 about the lamps.

Alternative: Replace the Fixtures

It may be more economical to replace incandescent fixtures entirely than to modify them. Total replacement provides a clean slate for designing more efficient lighting. See Measure 9.1.5 for this most general approach to improving lighting. Replacing fixtures provides the greatest benefit when switching to fluorescent or HID lighting.

MEASURE **9.1.4.1 In fixtures having shades that absorb light, modify or eliminate the shades.**

RATINGS

New Facilities	Retrofit	O&M
	B	

Many incandescent fixtures use shades. Shades serve a useful purpose in preventing glare, but they often are chosen without regard to efficiency or light distribution pattern. For example, table lamps in hotel rooms may have shades that are virtually opaque. For example, see Figure 1. Making shades more translucent reduces light absorption, improves lighting throughout the room, and improves appearance.

> **SUMMARY**
> Simple and easy. Often inexpensive.
> **SELECTION SCORECARD**
> Savings Potential .... **\$ \$ \$**
> Rate of Return .... **% % %** %
> Reliability .... ✓ ✓ ✓
> Ease of Retrofit .... ☺ ☺ ☺

This change is often simple and inexpensive, and it may reduce the need for lamp wattage by half. In other cases, it may take a certain amount of effort to improve a fixture. See Figure 2, for example. However, if a fixture has a large amount of incandescent lamp wattage, the effort is likely to pay off quickly.

With some fixtures and applications, such as a chandelier in a lobby, you may be able to remove shades completely and install decorative lamps.

Guidelines

There are innumerable styles of incandescent fixtures, so you have to figure out your own improvements. Here are some guidelines:

- *Make shades as transparent as possible.* It is virtually impossible to make shades too transparent. If a lot of wattage is at stake, it is worth buying some candidate types of shades and testing them with a light meter to find the most transparent models.

WESINC

Fig. 1 Light shade, dark shade The fixture with the dark shade is located where it could be useful for reading in the chair, for writing at the desk, or for overall room light. However, the opaque shade makes it useless for all these purposes. This hotel knows where to find translucent shades, because one is installed at left.

WESINC

Fig. 2 Quirky chandelier This unattractive chandelier has the energy waste of 18 downlights, and equally bad light distribution. Replacing the cylinders with globes or some other open type of lamp mounting would make it much more effective.

• *If you eliminate shades, limit glare appropriately for the environment.* The acceptable level of glare varies widely. For example, small bright bulbs may be a desirable accent in a lobby chandelier. On the other hand, keep lamp glare muted in locations where people linger for long periods, as in a dining room. Glare is less of a problem when the lamps are located well above the line of sight. Globe lamps and other types with large surface areas reduce glare. Read Reference Note 51, Factors in Lighting Quality, to understand how people respond to glare.

• *Match the lamps to the fixture.* With shaded fixtures, the light distribution pattern from the fixture and the amount of light that is trapped are strongly affected by the light distribution pattern of the lamp. Measure 9.1.3 deals with this aspect of lamp selection.

• *Keep the fixtures and lighting attractive.* Appearance is important with most shaded incandescent fixtures. More efficient shades are brighter, so they draw more attention to themselves. Better light distribution tends to provide lighting that is more pleasing.

ECONOMICS

SAVINGS POTENTIAL: *Fixture wattages typically can be reduced from 30 to 60 percent.*

COST: *Varies widely, but usually modest.*

PAYBACK PERIOD: *Immediate, to several years.*

TRAPS & TRICKS

DESIGN: *Be creative. Find the most appropriate lamps. Use light measurements to find the minimum acceptable lamp wattage.*

PRESERVE THE BENEFIT: *Label each fixture to state the proper lamp type and wattage.*

MEASURE **9.1.4.2 Install reflective inserts in fixtures that have absorptive internal baffles or surfaces.**

Reflective inserts are a product designed specifically to correct the energy waste of downlights with light absorbing interiors. Most inserts have a shape that is approximately parabolic, so the light is projected outward through the fixture opening. These inserts can also be used anywhere else they fit, for example, in track lighting fixtures.

It may take some searching to find an insert that fits properly in the specific fixture. Exact fit is not essential to efficiency, but it may be important for appearance and ease of installation.

Some reflective inserts are offered with tints. Avoid them. The tints absorb a significant amount of light. If the decor needs a particular light source color, select the lamp appropriately.

ECONOMICS

SAVINGS POTENTIAL: *60 to 90 percent of lamp power.*

COST: *$10 to $20.*

PAYBACK PERIOD: *One year, typically.*

SUMMARY

Simple. Foolproof. Fairly cheap.

SELECTION SCORECARD

Savings Potential	$ $ $
Rate of Return	% % % %
Reliability	✓ ✓ ✓ ✓
Ease of Retrofit	☺ ☺ ☺

TRAPS & TRICKS

SELECTING THE EQUIPMENT: *Shop around. There are many manufacturers of this equipment. Select for efficient light distribution, ease of installation, safety, and low cost.*

INSTALLATION: *Use light measurements to determine the lowest lamp wattage you can use. Label each fixture to state the proper lamp type and wattage.*

RATINGS

MEASURE **9.1.4.3 For task lighting, install focussing lamps on flexible extensions.**

RATINGS

New Facilities | Retrofit | O&M

B

Flexible extension lamps are available that you can screw into existing fixtures, typically ceiling mounted downlights. You can use these to achieve task lighting, provided that the relative positions of the fixture and the activity are satisfactory. See Reference Note 51, Factors in Lighting Quality, about glare and veiling reflections.

This is a specialized improvement that you may be able to use in a small number of locations. Extension lamps are not pretty, so this technique may be unacceptable where decor is a factor.

SUMMARY

An easy way to achieve efficient task lighting. Applications are infrequent. Not elegant.

SELECTION SCORECARD

Savings Potential .... **$**

Rate of Return .... **% % %** %

Reliability .... ✓ ✓ ✓ ✓

Ease of Retrofit .... ☺ ☺ ☺

ECONOMICS

SAVINGS POTENTIAL: *50 to 90 percent of the energy needed to illuminate the activity.*

COST: *Several dollars.*

PAYBACK PERIOD: *Less than one year.*

TRAPS & TRICKS

CHOICE OF METHOD: *Be sure that this method avoids glare and veiling reflections. If not, or if appearance is important, this is probably not the right approach. Instead, install a surface mounted aiming fixture.*

MEASURE **9.1.5 Replace incandescent fixtures with fluorescent or HID fixtures.**

Wherever incandescent lighting is installed or is part of the design of a new facility, consider whether you can replace it with fluorescent or HID lighting.

Design the lighting for each area from scratch. Consider all the factors that affect efficiency, visual quality, cost, decor, etc., for the particular activities and the space geometry. For overall lighting design issues, see Reference Note 51, Factors in Lighting Quality. Refer to Subsection 9.2 for Measures that optimize the efficiency of fluorescent lighting. Refer to Subsection 9.3 for Measures about HID lighting.

Retrofit Alternative: Screw-In Fluorescent Lamps

You can replace many incandescent lamps with screw-in fluorescent lamps (Measure 9.1.2.1). This is by far the easiest and least expensive way to install fluorescent lighting. However, screw-in fluorescent lamps are not as efficient as the best fluorescent or HID lighting, they are limited in light output, and the existing fixture locations may not provide efficient light distribution.

Fluorescent or HID?

In the past, large differences in efficacy and color rendering created a natural division between applications for fluorescent lighting and applications for HID lighting. This division has largely disappeared. Fluorescent lighting now approaches the efficacy of the best HID lamps, and HID lighting has achieved color rendering almost as good as the best fluorescent lamps. As a result, many applications could use either type of lighting satisfactorily. See Measure 9.3.3 for guidance in choosing between the two.

Don't Forget Efficient Control

Efficient control is a major part of lighting efficiency. Replacing incandescent lighting with another type will provide you with a number of opportunities to optimize control. Make this an integral part of your project. Refer to Subsections 9.4, 9.5, and 9.6 for lighting control Measures.

SUMMARY

The most efficient and most expensive approach. Widely applicable. Requires a lot of thought to yield the most benefit.

SELECTION SCORECARD

Savings Potential .... $ $ $ $

Rate of Return, New Facilities % % % %

Rate of Return, Retrofit .... % % %

Reliability .... ✓ ✓ ✓ ✓

Ease of Retrofit .... ☺ ☺ ☺

ECONOMICS

SAVINGS POTENTIAL: *60 to 90 percent of lighting energy consumption in the affected areas.*

COST: *Fluorescent fixtures cost $0.20 to $1.00 per watt. The price depends largely on appearance and diffuser type. HID fixtures cost $0.20 to $3.00 per watt, the price depending largely on wattage. Installation costs are $0.20 to $2.00 per watt, depending on wattage, fixture accessibility, and labor rates. In new construction, substituting fluorescent or HID lighting for incandescent lighting adds only a fraction of these costs.*

PAYBACK PERIOD: *1 to 5 years, in new construction. 3 to 10 years, in retrofit.*

TRAPS & TRICKS

PLANNING AND EQUIPMENT SELECTION: *This change gives you a chance to design lighting design from scratch. Start with a clean sheet of paper and consider all your options. See Subsection 9.6 for general layout. See Subsections 9.2 and 9.3 for fluorescent and HID lighting, respectively.*

MEASURE 9.1.6 Modify or replace incandescent exit signs with fluorescent or LED light sources.

RATINGS
New Facilities | Retrofit | O&M
A | **B** |

Commercial buildings are required by fire codes to have lighted exit signs. Until the 1980's, most units were illuminated with internal incandescent lamps, typically with two bulbs ranging in size from 15 to 40 watts each. The wattage of each fixture is small, but the units operate continuously, so it is usually worthwhile to replace the incandescent lamps with more efficient light sources.

Exit signs are visible everywhere and they are accessible, so upgrading them has become a popular energy conservation measure. In fact, this opportunity to save energy has fascinated some jurisdictions so much that they have passed laws requiring efficient exit signs.

Choices

You presently have these choices for improving exit sign efficiency:

- *retrofit existing fixtures with compact fluorescent lamps and ballasts.* Conversion kits are available with ballasts, lamps, and mounting brackets. The compact fluorescent lamps typically are from 5 to 9 watts. There may be one lamp and one ballast, two lamps and two ballasts, or two lamps and one ballast. Some of the ballasts are designed to operate with a pair of lamps, but they can operate only one lamp at a time. If one lamp burns out, the other lamp lights.

Compact fluorescent lamps of good quality are rated to last 10,000 hours, and the ballasts perhaps 50,000 hours. This greatly reduces lamp replacement labor.

WESINC

Fig. 1 Incandescent exit light The lamps in these are small, but they operate all the time. At your first convenient opportunity, replace incandescent units with one of the much more efficient models that are now available.

SUMMARY

A specialized activity that saves a modest amount of energy. Retrofit equipment is readily available.

SELECTION SCORECARD

Savings Potential .... **$ $**

Rate of Return .... **% %** %

Reliability .... ✓ ✓ ✓ ✓

Ease of Retrofit .... ☺ ☺ ☺

- *replace the fixtures with new fluorescent fixtures.* This approach achieves the same result as the previous. It is preferable if the existing fixture is difficult to convert or is unattractive. The equipment cost is higher, but the labor cost is usually lower.

- *retrofit existing fixtures with LED light strips.* Conversion kits are available that replace the incandescent bulbs with strips of light-emitting diode (LED) lamps. LED's use very little energy, typically 2 watts for each side of the exit sign. Also, LED's are claimed to have virtually unlimited life. LED retrofits are novel, so it is too early to believe these claims. The light strip is mounted at one side of the fixture, so the lighting may be uneven. At present, LED's are considerably more expensive than compact fluorescent lamps.

- *replace the fixtures with new LED fixtures.* This achieves the same result as the previous. In some fixtures, the LED's are used to backlight transparent letters. In others, the LED's themselves form the letters, which produces light that is stronger and more uniform.

Reliability

Reliability is a major requirement for exit signs. Compact fluorescent lamps of good quality have proven reliability. Cheap compact fluorescent lamps, which are common, have no reliability. See Measure 9.1.2.1 for guidance in selecting compact fluorescent lamps.

If you select fluorescent lamps, consider installing double lamps and ballasts for adequate reliability. Check the codes in your jurisdiction for any specific requirements.

Fluorescent lamps do not function properly when the ambient temperature is too low. Compact fluorescent units typically start and operate reliably down to 32°F (0°C) and below. Check the unit's temperature specifications if it will be located where it can get much colder than room temperature.

Fluorescent and LED exit signs are available with all the usual backup power arrangements. They draw little power, so they are well adapted to battery backup.

ECONOMICS

SAVINGS POTENTIAL: *70 to 90 percent of exit sign lighting energy. It isn't much energy overall, but everybody's doing it.*

COST: *Fluorescent retrofit kits cost $8 to $20. New fluorescent exit signs cost about $15 for cheap plastic units, to much higher for decorative units. LED retrofit kits cost $40 to $60 at this time. New LED exit signs* presently cost $80 to $120, depending on features. Installation takes from a fraction of an hour to one hour per unit, depending largely on the number of units to be installed.

PAYBACK PERIOD: *About one or two years, for fluorescent retrofits. Several years, for new fixtures or LED retrofits.*

TRAPS & TRICKS

CHOICE OF EQUIPMENT: *LED's are more expensive than compact fluorescent lamps, but they use somewhat less energy and they probably will last much longer. Considering that not much money is at stake, they probably are the best choice in most cases. Check with others who have experience with the equipment you are considering.*

MEASURE **9.1.7 Install dimmers.**

Dimmers are useful for saving energy with incandescent lighting in situations where the desired illumination requirement varies. Dimming may also be useful for decorative or esthetic purposes, with energy savings as a bonus. In retrofit applications, dimmers for incandescent lighting can be installed easily in place of the original switches.

If you want to exploit daylighting, don't use manual dimmers. Instead, use automatic dimming that senses the amount of daylight in the space. See Measure 9.5.3 about daylighting controls.

Efficiency Limitations

Unfortunately, it is difficult to achieve the savings potential of manual dimmers. The response characteristics of the eye make it difficult for the person setting the dimmer to quickly judge an acceptable minimum position, even if he is willing to make the effort. In the real world, dimmers tend to be set willy-nilly.

Another limitation is that the efficiency of incandescent lamps declines rapidly as they are dimmed. Therefore, dimming is partially self-defeating.

Even more self-defeating, the availability of dimmers motivates people to install larger lamps, so that more power is consumed to achieve a given lighting level than before the dimmers were installed. This makes

SUMMARY

For applications where the desired illumination level changes. Dimmers are inexpensive and easy to install, but they are usually not the most efficient approach. It is difficult to get people to use dimmers efficiently.

SELECTION SCORECARD

Savings Potential	$ $
Rate of Return, New Facilities	% % % %
Rate of Return, Retrofit..........	% % % %
Reliability of Equipment	✓ ✓ ✓
Reliability of Usage	✓ ✓
Ease of Retrofit	☺ ☺ ☺

it essential to limit lamp wattage, as recommended by Measure 9.1.1.

Types of Dimmers

Dimmers are available in a variety of styles, such as sliders, rotary knobs, and rotary knobs with push switches. Dimmers are available for 3-way operation.

Modern dimmers do not absorb much energy, but cheaper units may cause electrical interference.

Dimmers are easy to retrofit. Simply insert them in place of the original toggle switch. However, some dimmers are too wide or too deep to fit into some junction boxes. With deeper dimmer units, you may have to cut out excess wiring inside the box in order to install the dimmer. Check the dimensions of the dimmer before you buy.

Install Placards at the Dimmers and Fixtures

People tend to use dimmers inefficiently, so provide effective instructions. See Measure 9.4.1 for the details.

Label fixtures that are controlled by dimmers, as recommended by Measure 9.7.3. This tells the staff not to install lamps that are too large.

Electrical Interference

Modern incandescent dimmers operate by electronically interrupting (chopping) the line current at a frequency that is much higher than the line frequency. This higher frequency spreads through the power wiring. It can cause interference with radios, public address systems, computers, and other electronic equipment. Dimmers now have filters to minimize these problems, but the effectiveness of the filtering varies among models. In environments where electronic interference may be a problem, install a few dimmers on a test basis before proceeding with wholesale installation.

Audible Noise

Some dimmer models may make an audible buzzing noise at certain light levels. Modern units of good quality are generally free of this problem.

More Efficient Ways to Achieve Variable Lighting Levels

If the application requires incandescent light, consider multi-level switching of lamps or fixtures (Measures 9.6.4 ff). This may be satisfactory if fixtures have more than one lamp, or if a number of lamps are used to illuminate a particular area. Switching lamps,

rather than dimming them, keeps the lamps operating at maximum efficiency and provides positive control of the illumination level.

In new construction, you no longer have to install incandescent lighting to get dimming capability. Fluorescent dimming (Measure 9.2.6) is now an efficient and reasonably economical option, and most its previous weaknesses have been eliminated.

In retrofit applications, if a space has both incandescent and fluorescent lighting, consider installing dimmers for the fluorescent lighting and eliminating the incandescent lighting. Retrofitting fluorescent dimming costs much more than retrofitting incandescent dimmers, but dimming fluorescent lighting is much more efficient than dimming incandescent lighting.

HID lighting (i.e., mercury vapor, metal halide, and high-pressure sodium) does not yet have the dimming flexibility of fluorescent lighting. This may change in the future. See Measure 9.3.2 about HID dimming.

ECONOMICS

SAVINGS POTENTIAL: 20 to 60 percent of the energy of controlled fixtures.

COST: Less than $1 per ampere in common styles. Large dimming systems may be much more expensive.

PAYBACK PERIOD: Less than one year, typically, if it is properly applied.

TRAPS & TRICKS

CHOICE OF METHOD: This Measure will not save much energy unless the staff or occupants are motivated to use the dimmers efficiently. Even in the most favorable cases, dimming is usually less efficient than other techniques you might use, such as multi-level switching and automatic daylighting control. Consider all your options.

SELECTING THE EQUIPMENT: Avoid cheap junk models, which may fail quickly or make electrical noise. Test units in a sensitive location before buying in quantity.

INSTRUCTIONS: Don't forget to install effective instruction placards, which are just as important as the dimmers themselves.

Lamps and Fixtures, Fluorescent

Fluorescent lighting provides the best combination of efficiency, good lighting qualities, and other characteristics for most applications. Fluorescent lighting is by far the most common type used in commercial and industrial applications, the main exceptions being outdoor lighting, lighting for spaces with very tall ceilings, and decorative applications. Fluorescent lighting is even spreading rapidly in residential applications, largely in the form of compact screw-in lamps.

It appears likely that fluorescent lighting will continue to dominate. The technology is presently undergoing major changes, oriented primarily toward improved efficiency and color rendering, but also driven by style and marketing. A dark cloud on the horizon is concern about the mercury content of fluorescent lamps, which probably will lead to major changes in design, and perhaps to compromises with performance and economy.

The Measures of this Subsection reduce the energy consumption of fluorescent lighting. The Subsection starts with the least expensive and most broadly applicable Measures, and progresses to Measures that are more expensive and more narrowly applicable. Within this general sequence, similar Measures are kept next to each other.

Use the Measures in this and other Subsections as building blocks to create a lighting configuration that is optimum for each area. Tailor the lighting to each activity area individually. In retrofit applications, start by examining the environment of each fixture, and then select the appropriate Measures that apply to that fixture. For example, delamping (Measure 9.2.1) is appropriate if the environment is grossly overlighted, but replacement with high-efficiency tubes (Measure 9.2.3) is appropriate if the present lighting level is about right.

Before getting started, review Reference Note 55, Fluorescent Lighting, for an overview of the way that fluorescent lighting works. This will give you a basis for selecting from among the many types and models of fluorescent equipment. For a comprehensive comparison of fluorescent lamps with other lamp types, see Reference Note 52, Comparative Light Source Characteristics.

INDEX OF MEASURES

9.2.1 Eliminate excessive lighting by removing lamps and disconnecting or removing their ballasts.

    9.2.1.1 To remove single tubes from 2-tube ballasts, substitute dummy lamps.

    9.2.1.2 To remove single tubes where 2-tube ballasts are installed, substitute single-tube ballasts.

    9.2.1.3 To remove single tubes from groups of fixtures, rewire the ballasts between fixtures.

9.2.2 Where fixtures have been delamped, disconnect or remove the ballasts.

9.2.3 Replace fluorescent lamps with high-efficiency or reduced-wattage types.

9.2.4 Replace ballasts with high-efficiency or reduced-wattage types, or upgrade ballasts and lamps together.

9.2.5 Install current limiters.

9.2.6 Install fluorescent dimming equipment.

9.2.7 Consider retrofit "reflectors" for fluorescent fixtures.

RELATED MEASURES:

- Measures 9.6.2 ff, fixture layout and replacement in general
- Measure 9.6.3, combinations of fixtures
- Measure 9.6.5.1, installation of pullcord switches on fluorescent fixtures

MEASURE **9.2.1 Eliminate excessive lighting by removing lamps and disconnecting or removing their ballasts.**

SUMMARY

Widely applicable, saves lots of energy, cheap to accomplish. Needs a detailed survey before making the changes. Needs public relations and fixture labeling to survive.

SELECTION SCORECARD

Savings Potential	$ $ $ $
Rate of Return	% % % %
Reliability	✓ ✓
Ease of Retrofit	☺ ☺

The two most common causes of lighting energy waste are excessive illumination levels overall and failure to locate fixtures in relation to the activity areas. Typically, the entire space is illuminated at a uniform level that is sufficient for the most demanding work, with an additional margin thrown in. This pattern is especially common with fluorescent lighting.

Where this kind of lighting exists, you can save a large amount of energy quickly and cheaply by removing lamps from fixtures that are producing more light than needed. To maintain the savings in the long term, disconnect or remove the ballasts serving the lamps that are removed.

Removing lamps is often called "delamping", and disconnecting or removing ballasts is often called "deballasting." We will use these terms because they are handy.

Benefits of Delamping

Energy saving is the main benefit of delamping. Delamping eliminates a major fraction of the energy consumption of each fixture that is delamped. In many facilities, delamping can eliminate a major fraction of overall lighting cost.

Delamping can also improve the quality of illumination, primarily by eliminating lamps that cause veiling reflections and glare. In facilities that have uncomfortably high illumination levels, delamping eliminates discomfort from this cause.

Delamping saves the cost of the lamps and ballasts that do not have to be replaced. It may also save some replacement labor cost, but this saving may be less than expected, because a fixture with half its lamps operating still requires lamp replacement at the same intervals as if all the lamps were operating. (See "The Cost of Lighting", in the Section 9 Introduction.)

Drawbacks of Delamping

Delamping produces an irregular pattern of fixture brightness and the appearance of having a lot of burned-out lamps. Whether this is important depends on the nature of the space.

Immediately after delamping, occupants may complain about reduced lighting levels, for the reasons explained below. Good public relations eases this transition. Complaints should cease after an initial period, provided that the delamping is accomplished properly.

Where to Delamp

Consider delamping where a large reduction in the light output of an individual fixture is acceptable. Such cases are common. For example, offices with multiple fixtures usually do not require full output from all the fixtures. Storage and transit areas do not need illumination at the same levels as office work.

Once you get started, don't be surprised if you find some fixtures that you can delamp completely. For example, a kitchen fixture that illuminates only the top of a hood or upright freezer is useless. Refer to Measure 9.6.2.1 if it makes sense to remove these fixtures.

Ballast Connection Limitations

Most fluorescent ballasts are designed to operate two tubes. Ballasts are also available to operate a single tube. In fixtures having more than one tube, the usual practice is to use 2-tube ballasts, except for odd tubes (e.g., in 3-tube fixtures). Some electronic ballasts can drive a variable number of tubes.

With most 2-tube ballasts, the lamp current flows in series through the two tubes. This makes it easy to tell which tubes are connected to a common ballast, because removing one of the tubes extinguishes the other. An exception occurs with some electronic ballasts that power the tubes in parallel. These ballasts may operate efficiently with one tube disconnected, or they may not.

The minimum reduction of light output that is possible by removing lamps depends on the number of ballasts in the fixture. If a fixture has only one ballast, delamping disables the fixture completely. In a 4-tube fixture with 2-tube ballasts, the choices are to reduce the light level to one half or to disable the fixture completely. In a 6-tube fixture with 2-tube ballasts, the choices are two thirds of full output, one third, or none. And so forth.

The subsidiary Measures provide ways to work around these limitations, but they require lamp or fixture modifications.

Survey Before You Delamp

Delamping is not dummy work. You need to make a detailed survey to determine the number of lamps that need to be operated in each fixture. Delamping without adequate planning creates vision problems, misses opportunities for savings, and gives energy conservation a bad reputation. Figure 1 shows an example.

Do not expect to get satisfactory results just by giving general guidelines to the people who remove the lamps, unless they are exceptionally well trained. Even with exceptional personnel, it probably wastes time and creates confusion to plan the delamping at the same time as the lamps and ballasts are being removed.

Instead, make a detailed, fixture-by-fixture plan that shows the number and location of the tubes to be removed from each fixture. An easy way to do this is to mark the delamping instructions on a copy of the lighting plan from the building's architectural drawings.

Assign one individual on each work team exclusively to call out the modifications for each fixture and to check the work.

■ Select the Appropriate Illumination Levels

Use a good light meter to measure lighting levels at the activity areas. You will quickly become adept at estimating the lighting levels that will result from delamping by measuring the existing lighting levels.

Appropriate lighting levels are suggested in Reference Note 51, Factors in Lighting Quality. Illumination levels are controversial, and widely varying standards have been established in the United States and other countries. Lighting efficiency standards are covered in Reference Note 53.

As an example of the need to tailor delamping within a space, consider a typical office area illuminated by fixtures using tubes of 32 to 40 watts. If fixtures and ceiling heights are typical, you might use these guidelines for the numbers of tubes at each activity area:

- for routine paperwork, two tubes
- for detail work, four tubes
- for work at computer terminals with bright displays, and no other detail work, two tubes per 150 square feet
- for filing and storage areas, two tubes per 100-200 square feet, depending on the level of illumination required to find objects in storage
- for transit areas, one or two tubes for each 15 to 20 feet of length, but separate lighting is needed only if sufficient illumination is not available reliably from adjacent areas.

For other types of activities, refer to standard tables of illumination levels, such as those in the IESNA *Lighting Handbook*. Temper these guidelines with good judgement. The lighting trade is still governed by the "more is better" philosophy of lighting.

On the other hand, youthful, sharp-eyed lighting planners should leave a margin of lighting intensity because:

- *elderly persons need more light than young persons.* The illumination levels that people need to see well start to increase noticeably at age 40, and escalate rapidly thereafter.

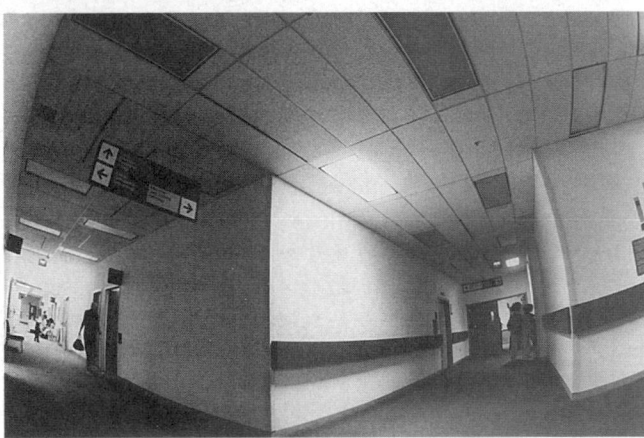

WESINC

Fig. 1 Stupid delamping   Some of these hospital corridors are much brighter than necessary, while others have long unlighted areas. Nobody planned this work, and nobody checked it.

• *all types of lamps lose light output as they age and get dirty.* Fluorescent lamps may lose 10 to 40 percent of their light output as they age, depending on the type.

■ Delamp to Improve Visual Quality

A task area within a space may be illuminated by several fixtures, presenting several different ways of delamping to satisfy the illumination level at the task area. Whenever possible, delamp fixtures in a way that improves visual quality. Try to minimize:

• *veiling reflections,* which occur when the fixture is in front of the viewer
• *source glare,* which can occur with a bright fixture located anywhere in the viewer's visual field
• *shadows on the task,* which occur when the fixture is behind the viewer or behind a shadowing object
• *excessively uneven illumination,* which results from the fixture being too far off to the side.

Veiling reflections is the most common problem with fluorescent lighting. This is because paperwork and other common activities that are vulnerable to veiling reflections commonly use fluorescent lighting. Source glare may sometimes be a problem, but delamping is less effective as a cure for glare.

Your ability to solve visual problems by delamping is limited by the present locations of the fixtures and the tasks. In open spaces with a number of occupants, fixtures may contribute collectively to the lighting of several task areas. This requires you to work out delamping patterns that provide satisfactory lighting at all the task areas. In many cases, delamping alone cannot correct poor lighting geometry. You may have to move fixtures (Measure 9.6.2.2) or use other methods.

As part of your planning, review Reference Note 51, Factors in Lighting Quality. Remember that the best position for a light source usually is to the side of the task area, and slightly behind the viewer.

If the space has daylighting, it is even more difficult to eliminate visual problems. Try to arrange the tasks so that both daylighting and artificial lighting come from the same or opposite sides. See Reference Note 46, Daylighting Design, for the special steps needed to make daylighting satisfactory.

■ Which Tubes to Remove?

Removing tubes changes the light distribution pattern of the fixture as well as reducing its total light output. For example, in most 4-tube fixtures, the outer tubes are connected to one ballast, and the inner tubes to another. Delamping the outer tubes may narrow the light distribution pattern much more than delamping the inner tubes. The difference in light distribution pattern is probably important. Use your light meter to study how your fixtures will behave when they are delamped.

Socket adapters are available that relocate the remaining tubes to a position halfway between the original sockets. These adapters simply plug into the existing sockets. You probably will not need these, but keep them in mind. They would be useful if they provide the only way to achieve a satisfactory light distribution pattern.

Another possibility is changing the connections of the ballasts to the sockets. For example, in a 6-lamp fixture with 2-tube ballasts, you could rewire the ballasts to deactivate any two sockets in the fixture, leaving the rest in operation.

Public Relations is Essential

Objections by the occupants can be a major obstacle to delamping. To avoid this problem, explain the purpose of the activity ahead of time, and assure occupants that lighting will be maintained at a level that is appropriate for each area. Stress that the changes will maintain or even improve lighting quality.

An odd phenomenon makes people averse to reasonably reduced lighting levels in the short term. It has been observed that some people complain of insufficient lighting when they work in an inside environment where the lighting level is maintained higher (yes, higher) than approximately 70 footcandles. The reason for this has not been discovered, but one could guess that this behavior is related to visual fatigue.

Similarly, people tend to assume that visual problems are due to insufficient light, rather than too much. Thus, occupants get the wrong impression that delamping will make any discomfort worse. (Among older persons, this may be a residue of an intense public relations campaign waged by electric utilities before the 1970's that tried to convince the public to increase lighting levels.)

Veiling reflections appear to be a factor in this phenomenon, but they do not explain all of it. Daylighting worsens the problem by making the non-daylighted portions of the space appear darker. This is one of the main factors that make daylighting tricky. See Reference Note 46, Daylighting Design, for methods of avoiding visual problems.

People will eventually accommodate to reasonably reduced lighting levels, but there appears to be a delay of one or more weeks before individuals become comfortable with it.

How to Delamp and Deballast

■ Remove the Lamps

Remove the lamps entirely and place them in storage for use elsewhere. Do not leave them to dangle in the lampholders. This is insecure, and invites putting the lamps back into operation.

Remove all the tubes connected to a ballast. Don't leave one tube connected to a 2-tube ballast, as this may damage the tube or ballast. In tubes that have filaments, the filaments will continue to operate until they burn out. These problems will not occur if you disconnect the ballast, but there is no sense in leaving tubes connected to a dead ballast.

■ **Disable the Ballasts**

After you remove the tubes and verify that the remaining light level is adequate, disconnect the ballast or ballasts. Disabling the ballast is important because:

• ballasts that are left connected continue to draw a significant amount of power even though the tubes are removed

• if the ballasts are not disconnected or removed, the removed tubes will be replaced sooner or later. The absence of a ballast is an effective signal that the tubes are supposed to be absent. (Delamped fixtures should be labeled, but people may ignore the labels.)

Ballasts are almost always installed inside the fixtures. When you are delamping, the ballast is easy to reach because the diffuser and the lamps in front of the ballast have already been removed. The ballast is usually installed behind a cover plate that usually comes off quickly. To disconnect the ballast, simply remove the power wiring from the primary. Wire nuts are commonly used for ballast installation, and they can be removed in a few seconds.

Occasionally, ballasts may be installed remotely from fixtures. This is done to isolate the noise emitted by ballasts. The ballasts are usually accessible inside an electrical panel.

Of course, there is an electrical shock hazard in doing this work, so the people doing the electrical work should be properly qualified.

Removing the ballast takes only a few more seconds. Remove the wire nuts that connect the secondary side of the ballast to the lamp sockets. The ballast is typically held in place by two easily accessible nuts. Remove these, and the ballast drops out of the fixture.

■ **Disconnect or Remove Ballasts?**

You can leave the disconnected ballasts in the fixtures or remove them. Removing the ballasts is better. This makes it much less likely that excess lamps will be put back into service. Use the removed ballasts as spare parts, or sell them.

■ **How to Deactivate a Single Tube**

Sometimes, you want to eliminate only one tube in a fixture. There are ways to do this, even if the fixture has multi-tube ballasts. However, it takes more effort and expense. The subsidiary Measures tell you how.

■ **Delamping Screw-In Fluorescent Lamps**

Screw-in fluorescent lamps usually include an integral ballast, so disconnecting ballasts is not a separate requirement with these lamps. You still need to electrically disconnect the fixture that held the lamp.

Label the Delamped Fixtures

Label each fixture that has been delamped. Otherwise, people will assume that the fixtures are defective, and the lamps will be replaced. This step is as important as the rest of the work. See Measure 9.7.3 for the details of labeling fixtures.

Repeat the Process When Changing Space Layout

If delamping is tailored to individual areas, future changes in the location or orientation of the activities will require corresponding modifications of the delamping pattern. Neglecting to keep up with changes will result in unsatisfactory lighting.

ECONOMICS

SAVINGS POTENTIAL: 30 to 70 percent of lighting energy in the delamped spaces. Delamping one pair of 40-watt tubes saves about 300 KWH per year, on a typical workday schedule. In addition, this saves several dollars per year in lamp replacement cost.

COST: $1 to $10 per fixture, depending on labor rates, the number of fixtures, and accessibility.

PAYBACK PERIOD: Less than one year.

TRAPS & TRICKS

PREPARATION: The big risk is the temptation to start delamping without doing a thorough planning survey first. Don't expect people to select lamps for delamping on the fly. Even if they could, it would be an inefficient use of labor. Don't neglect an effective public relations program immediately before you start delamping.

FIXTURE LABELING: Train the supervisor of the delamping crew to label the delamped fixtures as the work is accomplished.

MEASURE 9.2.1.1 To remove single tubes from 2-tube ballasts, substitute dummy lamps.

A dummy lamp is simply a wire that runs from one socket of a fluorescent fixture to the opposite one, in place of a real tube. The wire has minimal electrical resistance. It mimics the ionized gas of an operating fluorescent tube, which also has minimal resistance. The wire dissipates no power, and there is no filament. Some models enclose the wire in a glass envelope that looks like a lamp, but the glass plays no role except to protect the wire and look official. A capacitor may be included with the dummy lamp to improve power factor.

All magnetic 2-tube ballasts operate the lamps in series. Fluorescent lamps are essentially short circuits while they are operating, so they do not affect each other's operation. Therefore, substituting a wire for one of the lamps makes no difference to the other lamp or to the ballast, at least during operation.

Potential Problems

Dummy lamps have been used successfully, but they never became popular, probably because people suspect that they may damage the ballast or the remaining lamp. In fact, some technical representatives of ballast manufacturers have stated that dummy lamps are harmless to conventional ballasts. Even so, using them may void ballast warranties.

There is at least one real problem. With rapid-start tubes, which have filaments, using dummy lamps may shorten the life of the lamp by increasing the voltage across the tube when starting. The problem occurs because the tubes are connected to the ballast in series. When starting, the initial high resistance of both tubes is additive. However, a dummy lamp has zero resistance even when the fixture is starting, so the entire starting voltage is applied to the real tube.

This suggests that you should use dummy lamps only if the fixtures are not turned on and off frequently. For example, corridor lighting is a good application. In appropriate applications, dummy lamps probably have little effect on lamp replacement cost. On one hand, there is only one lamp to wear out, instead of the original two. On the other hand, the lamps may have reduced lamp life.

SUMMARY

Efficient, easy, and cheap. Check for possible lamp or ballast problems.

SELECTION SCORECARD

Savings Potential .... **$ $ $ $**

Rate of Return .... **% % %** %

Reliability .... ✓ ✓ ✓ ✓

Ease of Retrofit .... ☺ ☺ ☺

There is always the possibility that a dummy lamp can cause trouble with some ballast models. For example, some electronic ballasts may not be able to cope with the absence of starting resistance. So, check with the engineering departments of both the lamp and ballast manufacturers before using dummy lamps.

How to Install Dummy Lamps

Installing dummy lamps is simplicity itself. Just remove one of the fluorescent tubes and insert the dummy lamp in its place. Nothing needs to be done with the ballast.

Most of the work is in the supporting steps needed to make any delamping program successful, which are spelled out in Measure 9.2.1, above. Label the fixtures and do your public relations.

ECONOMICS

SAVINGS POTENTIAL: *All the energy of the removed tubes. For example, eliminating one 40-watt tube save about 150 KWH per year on a typical workday schedule.*

COST: *Dummy lamps cost $2 to $10 apiece, depending on style. Labor for installation and supporting activities may cost from $0.30 to $3 per fixture, depending on the environment.*

PAYBACK PERIOD: *One year, typically.*

TRAPS & TRICKS

See Measure 9.2.1. In addition, be sure to check that the dummy tubes do not have significant adverse effects on the ballasts or the remaining tubes.

MEASURE **9.2.1.2 To remove single tubes where 2-tube ballasts are installed, substitute single-tube ballasts.**

C

In fluorescent fixtures that have even numbers of tubes, the tubes are usually powered in pairs by 2-tube ballasts. Magnetic 2-tube ballasts cannot operate a single tube, and the same is true of some electronic 2-tube ballasts. You can get around this limitation by replacing the 2-tube ballasts with single-tube ballasts.

How to Select the Ballasts

Refer to Measure 9.2.4 for the factors to consider in selecting ballasts.

First, Survey and Plan

To be successful, this activity needs the kind of planning that is spelled out in Measure 9.2.1, above.

Opportunity to Add Switching Flexibility

Installing a single-tube ballast for one of the lamps in a fixture allows light output to be adjusted with great flexibility, provided that the appropriate switching circuits are installed (see Measure 9.6.4). This is valuable in spaces that have widely varying lighting requirements.

For example, in a function room that has 4-tube fixtures, one of the pairs of tubes in each fixture can be replaced by a single tube. This gives occupants the option of turning on one, two, or three tubes in each fixture.

If you provide manual switching options, be sure to label the switches effectively. See Measure 9.4.1 for tips.

SUMMARY

Efficient and foolproof. Moderately expensive.

SELECTION SCORECARD

Savings Potential .... **$ $** $

Rate of Return .... **% %** %

Reliability .... ✓ ✓ ✓ ✓

Ease of Retrofit .... ☺ ☺ ☺

ECONOMICS

SAVINGS POTENTIAL: *All the energy of the removed tubes. For example, eliminating one 40-watt tube saves about 150 KWH per year, on a typical workday schedule.*

COST: *Efficient single-tube ballasts cost $10 to $40, depending on wattage and type. Labor for installation and supporting activities adds $2 to $10 per fixture, depending on the environment. The removed ballasts have salvage value, at least as spare parts.*

PAYBACK PERIOD: *One to five years.*

TRAPS & TRICKS

See Measure 9.2.1. In addition, select the replacement ballasts for efficiency and reliability.

MEASURE 9.2.1.3 To remove single tubes from groups of fixtures, rewire the ballasts between fixtures.

If adjacent fixtures have 2-tube ballasts, you can rewire the ballast connections to operate a single tube in that fixture and a single tube in another fixture. This technique can be applied in a variety of ways. It applies to fixtures with any number of ballasts.

For example, consider a row of 2-tube fixtures connected end-to-end. A single tube is left in each fixture, and the ballast in every other fixture is used to power the lamps.

Or, in this same situation, you can create two levels of lighting. To do this, retain both tubes in each fixture, rewire the ballasts, and install two light switches. The two switches control alternate ballasts, so that one switch controls one tube in each fixture, and the other switch controls the other tube in each fixture. In this way, a single continuous row of tubes can be turned on, or two continuous rows.

The wiring modifications are awkward, especially with rapid-start lamps. Rapid-start ballasts provide filament heating for each socket, so you have to feed several socket wires from one fixture to the next.

Limitations on Wiring Length

If the present ballasts are an electronic type that operates at high frequency, there may be problems if the ballast wiring is extended to great lengths. The ballast may not be able to drive long connections, or there may be noise propagation problems. Check with the ballast manufacturer to find out about wiring length limitations.

SUMMARY

Efficient. Labor intensive. Limited to closely spaced groups of fixtures.

SELECTION SCORECARD

Savings Potential .... $ $ $

Rate of Return .... % % %

Reliability .... ✓ ✓ ✓ ✓

Ease of Retrofit .... ☺ ☺ ☺

Supporting Activities

As with any method of delamping, the planning and public relations steps recommended in Measure 9.2.1 are needed to make the activity successful.

ECONOMICS

SAVINGS POTENTIAL: *All the energy of the removed tubes. For example, eliminating one 40-watt tube in typical daytime service saves about 150 KWH per year.*

COST: *$5 to $40 per fixture. The cost is mostly for labor, and depends on labor rates and the ease of wiring between fixtures. Leftover ballasts may have substantial salvage value.*

PAYBACK PERIOD: *Less than one year, to several years.*

TRAPS & TRICKS

See Measure 9.2.1. In addition, the awkwardness of the wiring work invites mistakes.

MEASURE **9.2.2** **Where fixtures have been delamped, disconnect or remove the ballasts.**

RATINGS

New Facilities | Retrofit **B** | O&M

SUMMARY

An essential step in delamping that is often neglected.

SELECTION SCORECARD

Savings Potential .... **$ $**

Rate of Return .... **% % %**

Reliability .... ✓ ✓ ✓ ✓

Ease of Retrofit .... ☺ ☺ ☺

Delamping fluorescent fixtures (see Measure 9.2.1) is a quick and easy method of saving energy. However, in facilities where delamping was accomplished in the past, it is common to find that disconnecting the ballasts was neglected. If you have this situation, correct it.

Disconnecting the ballasts is important for two reasons. One is that a ballast continues to dissipate a certain amount of energy even with no lamps connected to it. The other reason is that disconnecting the ballast is needed to prevent lamps from being replaced and put back into operation.

Amount of Energy Waste

Energy waste by the ballast itself is primarily a problem with magnetic ballasts. These are similar to transformers in construction. The primary winding continues to draw a certain amount of power even when the secondary is disconnected. This power is dissipated as heat. The energy waste typically is a few watts per lamp.

Electronic ballasts may not draw much current when the lamps are removed, but it is still important to disconnect them to prevent lamps from being reinstalled.

How to Disconnect Ballasts

Disconnecting ballasts is simple. See Measure 9.2.1 for the details.

Label the Fixtures

Label the fixture when you disconnect the ballast. See Measure 9.2.1 for the details.

ECONOMICS

SAVINGS POTENTIAL: *The power consumption of a magnetic ballast with no lamps is typically several watts per lamp connection. Disconnecting ballasts may save much more energy by preventing lamps from being re-installed in delamped fixtures.*

COST: *$0.50 to $5 per fixture, depending on labor rates, the number of fixtures, and fixture accessibility.*

PAYBACK PERIOD: *Less than one year, to several years.*

TRAPS & TRICKS

CONSIDER STARTING OVER: *If the ballasts were not disconnected when delamping was first accomplished, the delamping was probably not planned properly. Start all over again with a thorough delamping survey. Go back to Measure 9.2.1.*

MEASURE **9.2.3 Replace fluorescent lamps with high-efficiency or reduced-wattage types.**

SUMMARY

Reduces lamp wattage up to 20 percent, and may improve efficiency. There are many selection issues, and some pitfalls. Backsliding is a challenge.

SELECTION SCORECARD

Savings Potential	$ $ $ $
Rate of Return, New Facilities	% % % %
Rate of Return, Retrofit..........	% % %
Reliability	✓ ✓ ✓
Ease of Retrofit	☺ ☺ ☺

The efficiency of fluorescent lamps improved considerably between the 1970's and the 1990's. In new construction, where you can design the lighting with a clean sheet of paper, it is virtually always desirable to used lamps of the highest efficiency available. Efficient lamps cost more, but the saving in energy cost is usually greater than the cost premium for the lamps.

In retrofit applications, one of the more popular energy conservation measures is replacing older types of fluorescent tubes with newer types that operate at lower wattage. Reduced-wattage and high-efficiency lamps are now available in all the common sizes and shapes. For example, the venerable 4-foot 40-watt tubes are widely being replaced with 32-watt and 34-watt tubes.

Some lower-wattage substitute lamps are more efficient than older types, and some are not. We will use the term "reduced-wattage lamps" to refer to both types collectively. When referring specifically to lamps with improved efficacy, we will use the term "high-efficiency lamps."

Replacing one type of lamp with another requires no physical effort. However, you have to be careful to select the replacement lamps properly. This Measure explains the factors that you should consider.

Applications and Alternatives

In retrofit applications, reduced-wattage lamps are most likely to be the best technique in cases where illumination is slightly excessive and lighting requirements are uniform. This situation is found in public spaces, such as corridors and cafeterias, and in large, open spaces with tall ceilings, such as auditoriums.

Reduced-wattage fluorescent lamps typically reduce power consumption by 15 to 20 percent. They may also reduce lighting levels up to 20 percent. More efficient models may retain original lighting levels while reducing power consumption by a lesser amount.

If the spaces have a substantial excess of illumination, reduced-wattage lamps are not the appropriate way to reduce lighting levels. You can easily achieve bigger reductions in energy consumption by delamping (Measures 9.2.1 ff), while improving lighting patterns. Other Measures, such as dimming (Measure 9.2.6) and overall layout changes (Subsection 9.6), may provide even larger reductions in lighting energy.

The next stage in improving efficiency is operating high-efficiency lamps with high-efficiency ballasts. This may require different lamps than the higher-efficiency lamps used with older ballasts. See Measure 9.2.4 about upgrading ballasts. You should decide at the outset whether to upgrade the lamps alone, or to replace the lamps and ballasts as a matched system.

Higher Efficiency or Just Lower Wattage?

Some replacement lamps are more efficient, so they can maintain the original light output at the lower wattage. Other replacement lamps have about the same efficacy as the older types, or less, so they reduce light output in proportion to the power reduction.

High-efficiency types are considerably more expensive than other reduced-wattage lamps. In retrofit applications, you need true "high-efficiency" lamps only in the relatively uncommon cases where present light levels are just barely adequate. Or, you may want them in some specialized applications for their superior color characteristics.

All or Nothing

Try to make an all-or-nothing decision about using reduced-wattage lamps. Reduced-wattage lamps have the same appearance as conventional lamps, so keeping both types in stock invites putting the wrong lamps in fixtures. It also increases the bother of ordering and storage.

Furthermore, reduced-wattage lamps are offered in different colors than conventional lamps. Occupants will notice the difference in lamp color if fixtures with different lamp types are within sight of each other.

How to Select Replacement Lamps

See Reference Note 52, Comparative Light Source Characteristics, for the criteria that may be important in selecting lamps for a given application. Also, review Reference Note 55, Fluorescent Lighting, which explains the factors that determine fluorescent lamp efficiency. Then, for each type of lamp that you need, study the catalogs for the following selection factors.

■ Ballast Compatibility

Check the ballast's specifications for the types of lamps that it is designed to drive. This information is found in lighting supply catalogs. If a ballast is fairly new, the types of lamps it drives may be printed on the ballast.

There are many types of lamps and ballasts, and the number is increasing rapidly. If a lamp is connected to the wrong ballast, efficiency always suffers. Lamp life and light output may also be reduced. If your existing ballasts are not appropriate for the kind of lamps you want to install, you may be able to find another type of lamp that achieves your purpose. Alternatively, you could change the ballasts, which gives you an unlimited choice of lamps. This is an expensive step, but it is often profitable. See Measure 9.2.4 if you take that path.

Make sure that the lamps you select are compatible with these characteristics of the ballasts:

• *current rating*

• *number of tubes*

• *filament connections.* This is usually determined by the socket type, so it is usually not expressed.

• *starting method.* This may be expressed by the terms "instant-start," "rapid-start." "preheat," or "trigger-start."

• *starting voltage.* This may be higher than the operating voltage. For example, a lamp may need a higher starting voltage if it is located in a cold location. The higher starting voltage may be expressed as a temperature rating, or in other ways.

If the ballasts are electronic or a hybrid type, you need to check additional compatibility issues, including:

• *the frequency of power supplied by the ballast.* Electronic ballasts operate the tube at a frequency that is much higher than line frequency. Most tubes can tolerate the higher frequency, even though this is not stated.

• *dimming provisions,* if the ballast has dimming capability. Not all fluorescent tubes operate properly with all dimming ballasts.

• *filament switching.* Electronic and hybrid ballasts usually turn off the filament after the lamp has started.

For example, most high-efficiency tubes require a specialized ballast to operate properly. Also, the new T8 series of lamps are not suited to any ballasts for T12 lamps, even though they fit in the original sockets.

If the ballasts were installed before 1979, it is not prudent to install any kind of reduced-wattage lamp with them. Experience has shown that reduced-wattage lamps hasten the failure of older ballasts, and that they use an increased amount of power while doing so.

Unless the ballasts have a high power factor, reduced-wattage lamps may be difficult to start, or they may operate in an unstable manner.

■ Wattage

For each lamp configuration, there may be a range of wattages. For example, 4-foot tubes typically are available in 40-watt, 34-watt, and 32-watt families.

■ Lumen Output

Lumen output can vary widely for a given wattage. For example, one manufacturer offers 34-watt, 4-foot rapid-start lamps with lumen outputs ranging from 1,900 lumens to 2,900 lumens. With a range this wide, the wrong lamp selection will make a space too dark if the original light level is marginal.

Take into account the normal degradation of light output as the lamp ages. Select the lamps to provide satisfactory illumination at the end of the service life. Fluorescent lamps may lose from 10 to 40 percent of their light output with age. High-efficiency lamps with the newer types of phosphors have substantially less lumen loss than older types.

■ Combined Lamp-and-Ballast Efficiency

The efficiency and light output of lamps are affected by the ballasts that drive them, so try to look up the efficiencies of the particular lamp and ballast combinations you are considering. If this information is not readily available, it is worth making a call to the lamp and ballast manufacturers to see if they have the figures. For any given ballast, the "lumens per watt" rating of the lamp provides an approximate indication of relative performance. If only the lumen rating of the lamp is given, you can estimate its efficacy by dividing the lumen output rating by the wattage rating.

■ Filament Operation

Some rapid-start lamps save energy with internal switches that turn off the filaments after the lamp starts. The switch is a simple bimetal contact that is kept open by the warmth of lamp operation. A disadvantage of this feature is that the bimetal contact may take up to a minute to cool after the lamp is turned off, preventing restarting for that period of time.

Turning off the filaments of rapid-start lamps can be done externally by hybrid ballasts and by some electronic ballasts. If you are going to upgrade ballasts (see Measure 9.2.4), it is generally better to let the new ballasts perform this function.

■ Color Characteristics

In fluorescent lamps, efficacy and cost increase along with color rendering because all three are related to the composition of the phosphors.

The color of older fluorescent lamps was commonly described by colloquial terms, such as "cool white," "warm white," and "daylight." Newer models of fluorescent lamps for general lighting designate lamp color by "color temperature." The current versions of older types still carry the old color descriptions, and any lamp may have a promotional name that implies a color.

For an explanation of lamp color in general, refer to "Color Rendering Index" and "Lamp Color" in Reference Note 52, Comparative Light Source Characteristics. For specifics about fluorescent lamp color, see Reference Note 55.

■ **Lamp Life**

Lamp life is not a factor in efficiency, but it is a major factor in labor and lamp cost.

Catalog specifications typically show little or no difference between the lives of different lamps within a family. For example, one major manufacturer rates virtually all of its 4-foot rapid-start lamps at 20,000 hours. However, experience has shown that some reduced-wattage lamps fail well short of their rated service lives. This fact makes it worthwhile to ask major lamp users about their experience with the types of lamps you are considering.

■ **Voltage Sensitivity**

Fluorescent lamps turn on and off with each cycle of the alternating current. Reduced-wattage lamps contain less of the gases needed to aid starting, so they require higher voltage to start. This means that the lamp turns on close to the voltage peak of each cycle of alternating current. If the supply voltage is reduced, the lamp may flicker or fail to operate. Therefore, avoid reduced-wattage lamps in areas where brownouts or voltage fluctuations are common.

There is an exception. Some electronic ballasts compensate for voltage fluctuations. However, if you are upgrading older types of lamps, it is unlikely that these ballasts are installed. By the same token, this would be an argument for upgrading the ballasts.

■ **Starting and Operating Temperatures**

Conventional indoor fluorescent lamps usually are rated with a minimum temperature of 50°F, but high-efficiency types usually are rated at 60°F. Again, this is related to the lower gas density inside the lamps. Temperature ratings are not sharp cutoff limits, but stability, light output, and efficacy diminish rapidly below the rated temperatures. You are asking for trouble if you install reduced-wattage lamps in spaces that operate below normal room temperature.

■ **Cost**

There is a large range of prices among fluorescent lamps of a given shape, and some of the newer high-efficiency lamps are quite expensive. It is pointless to pay high prices for high light output or outstanding color rendering if you do not need these characteristics.

Factors other than performance affect lamp cost. For example, curved tubes are more expensive than comparable straight tubes. As a rule, the most common lamp sizes are the least expensive, usually by a wide margin.

Label Your Fixtures to Prevent Backsliding

When a reduced-wattage tube burns out, it may be replaced incorrectly with any lamp that fits the same socket, unless you use effective measures to prevent this. The most effective method is to label each fixture with the correct tube type. See Measure 9.7.3 for guidance in fixture labeling.

It helps if you can standardize on a particular type of tube for the entire facility. This lets you establish a purchasing policy that restricts tube purchases to the proper type. However, a fluorescent lamp may last ten years, and administrative policies rarely survive so long.

ECONOMICS

SAVINGS POTENTIAL: 10 to 20 percent of lamp energy.

COST: Reduced-wattage lamps have a cost premium of $1 to $5 apiece, depending on efficacy and color quality. In retrofit installations, planning and installation costs $0.50 to $5.00 per fixture, depending on labor rates and accessibility.

PAYBACK PERIOD: One year, to several years, for typical workday operating schedules.

TRAPS & TRICKS

CHOICE OF METHOD: Use this approach only if it is appropriate for the environment, namely, if lighting levels are uniformly and only moderately excessive. This Measure is not a substitute for careful delamping, or for careful fixture selection and placement.

SELECTING THE LAMPS: Take the trouble to find the proper types of lamps for the existing ballasts. A mismatch will save little or no energy, while reducing light output excessively and possibly shortening equipment life. Check the ballasts in all the fixtures where you plan to change lamp types. After you have identified your candidate lamp type, check with other users who have operated the lamps with your type of ballasts for at least one life cycle.

FIXTURE LABELING: As sure as the sun rises, if you fail to label the fixtures with the types of lamps to be installed, the wrong type will be installed within a few years.

MEASURE **9.2.4 Replace ballasts with high-efficiency or reduced-wattage types, or upgrade ballasts and lamps together.**

SUMMARY

A major equipment change involving many selection issues. Select ballasts and lamps for compatibility. Beware of ballast and lamp reliability problems.

SELECTION SCORECARD

Savings Potential .... $ $ $ $

Rate of Return, New Facilities % % %

Rate of Return, Retrofit.......... % % %

Reliability .... ✓ ✓ ✓

Ease of Retrofit .... ☺ ☺ ☺

The efficiency of fluorescent ballasts improved considerably between the 1970's and the 1990's. In new construction, it usually pays to install high-efficiency ballasts. Ballast efficiency has a large price premium, but the energy consumed by the ballasts and the lamps costs much more than the ballast itself. You need high-efficiency ballasts to get the most efficiency from lamps, and to be able to use the most efficient types of lamps. Also, high-efficiency ballasts offer a variety of other advantages.

Ballasts have several functions in fluorescent lighting. They regulate lamp current, which is necessary because an operating fluorescent lamp has minimal electrical resistance. They increase the supplied voltage for starting. And, some types of electronic ballasts provide dimming capability.

Electronic ballasts are the most efficient type. They have only about 25% of the internal losses of older magnetic ballasts. In addition, electronic ballasts improve the efficiency of the lamps themselves by about 10%. An optimum combination of ballast and lamp types may improve lighting efficiency by as much as 30%, compared to older fluorescent lighting.

Some reduced-wattage ballasts do not have particularly high efficiency. They save energy by reducing the power input to the lamps. Such ballasts reduce light output roughly in proportion to the reduction in power. These ballasts are limited to retrofit applications as a means of reducing excess lighting.

Upgrade Ballasts, Lamps, or Both?

If you are upgrading existing fluorescent lighting, you have to make a big decision at the outset. The ballast type you select will limit your ability to make further improvements in the future. The limitation is compatibility between ballasts and lamps. You have to decide whether to install high-efficiency ballasts that are compatible with your existing types of lamps, or to install a matched combination of new ballasts and lamps.

As a matter of comparison, upgrading ballasts alone provides roughly the same magnitude of savings as upgrading lamps alone, as recommended by Measure 9.2.3. As with lamp replacement, ballast replacement can provide a simple reduction of wattage, or an improvement in lumens-per-watt efficiency, or both. However, upgrading ballasts is much more expensive than upgrading lamps.

Then, why would you want to install upgraded ballasts, while continuing to use an old type of lamp? Two major reasons are reliability and long-term lamp cost. The older lamps are less sensitive to voltage fluctuations and to low temperature, and they seem to last longer in typical usage. Also, the older lamps are much less expensive, primarily because the old phosphors are less expensive than the tri-phosphors of high-efficiency lamps. The cost difference is a significant factor over the life of the facility. Thus, using high-efficiency ballasts with older lamps may be a rational choice.

This approach lets you phase in high-efficiency ballasts as replacements for older ballasts that fail. However, ballast life is so long that this approach produces savings at a snail's pace.

If you decide to take this course, you may not be able to upgrade later to lamps of higher efficiency. This is because high-efficiency ballasts that are optimized for older tubes are not optimized for newer high-efficiency tubes. For example, ballasts designed to drive T12 tubes will not drive T8 tubes at all. Even if the ballast works with a high-efficiency T12 tube, there may be a serious loss of performance.

If you want to upgrade both ballasts and tubes, the proper course is to select high-efficiency ballasts and high-efficiency tubes that are specifically matched to each other. Until recently, ensuring compatibility between high-efficiency lamps and ballasts was an uncertain matter because the development of lamps and ballasts proceeded almost independently of each other.

At the present time, the new T8 lamp system appears to the be only system rationally designed for maximum overall efficiency in fluorescent fixtures of conventional type. You can get T8 lamps in a variety of lamp sizes that correspond to the most common types of T12 lamps. T8 lamps fit conventional T12 lampholders, so you can retrofit lamps and ballasts in existing fixtures without too much trouble. See Figure 1.

It also appears that T8 lamps, when operated with ballasts designed for them, are less prone to the temperamental performance that afflicts other high-efficiency lamps. However, it is still too early to judge the long-term performance of this system.

If you consider replacing both lamps and ballasts, do not limit your choices to the old fashioned lamp shapes. Consider the new possibilities that compact fluorescent lamps provide. These lamps are available in smaller wattages that allow you to tailor lighting energy to small task areas. Also, the small physical size of the lamps may allow you to provide more efficient light distribution. See Figure 2.

Compact fluorescent lamps have major potential for task lighting, which cannot be done well with large fluorescent lamps. See Measure 9.6.2 about this.

However, compact fluorescent lamps are not as efficient, in terms of lumens per watt, as the larger conventional lamps. We do not know how good they really are. Manufacturers remain coy about their actual efficiency, resorting to the word "nominal" in the specifications.

They are an evolving type. They have not been in the field long enough to establish a reliable record of actual service life. Too many new shapes and base types

are coming on the market for all of them to remain in service over the long term.

Retrofit Applications and Alternatives

High-efficiency ballasts, or combinations of high-efficiency ballasts and lamps, provide approximately the same lighting levels as older equipment, but use less energy. Therefore, this Measure is appropriate for retrofit if present lighting levels are about right.

If the spaces have a substantial excess of illumination, consider other Measures, perhaps in combination with this one. For reductions in wattage and light output, but without improved efficiency, consider current limiters (Measure 9.2.5, next). Note that current limiters are compatible only with magnetic ballasts. Bigger reductions in energy consumption can be made at lower cost by delamping (Measures 9.2.1 ff), while providing an opportunity to optimize lighting patterns. Other Measures, such as dimming (Measure 9.2.6) and overall layout changes (Subsection 9.6) also provide large reductions in lighting energy.

Types of Reduced-Wattage Ballasts

Reduced-wattage ballasts and "high-efficiency" ballasts are available both in the original magnetic type

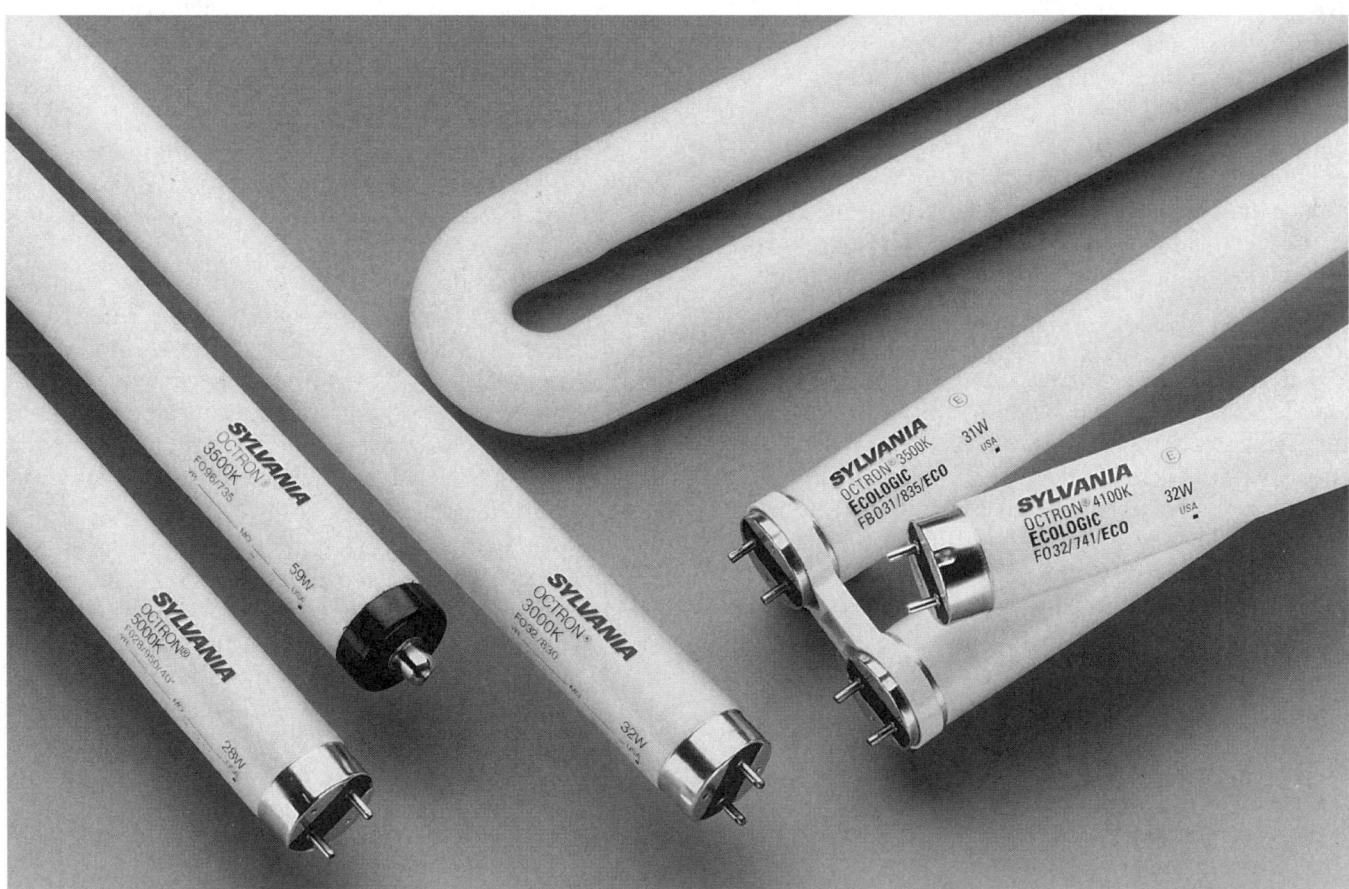

Osram Sylvania Inc.

Fig. 1 T8 lamps These lamps have high efficiency and very good color rendering. They fit in the same fixture dimensions as the most common older lamps, but they are not interchangeable with the older lamps. Consider these along with high-efficiency ballasts.

and in the newer electronic types. In addition, a hybrid type has been developed that combines features of the two. The following are the main features of reduced-wattage and high-efficiency ballasts.

■ Reduced-Wattage Magnetic Ballasts

The energy consumption of a conventional magnetic ballast can be reduced simply by designing it to deliver less current to the lamp. This does not by itself change the efficiency of the ballast. In fact, this approach is less efficient overall because the ballast and lamp losses do not decline with lamp current. In the U.S., reduced-wattage ballasts that do not offer an efficiency improvement are being forced off the market by recently enacted efficiency standards.

The efficiency of magnetic ballasts has been increased by a number of improvements that are explained in Reference Note 55, Fluorescent Lighting. High-efficiency versions of conventional ballasts eliminate about 40-50 percent of the losses of older magnetic ballasts. In a typical ballast for two 40-watt tubes, this amounts to a saving of about 8 watts.

The description "high efficiency" does not have a quantitative meaning as it applies to ballasts. The efficiency advantage of improved conventional ballasts may differ substantially from one manufacturer to another, or from one model to another.

High-efficiency magnetic ballasts generate less internal heat, so they operate at a lower temperature and they promise longer life. It remains to be seen whether they will deliver on this promise.

■ Electronic Ballasts

The efficiency of electronic components gives electronic ballasts inherently higher efficiencies than magnetic ballasts. There are significant differences in efficiency among electronic ballasts, but the less efficient units have already demonstrated low reliability and have largely died out.

(The ballasts in compact fluorescent fixtures are an exception, with junk units surviving on the basis of cheapness, largely in residential sales. Subsidies by electric utility companies keep this junk on the market because the homeowner whose unit fails after a few hours of operation has little incentive to take it back to the store. This is a lesson in why centrally planned, third-party programs of any kind tend to fail.)

The best electronic ballasts consume about one fourth as much energy as older magnetic ballasts. For example, an electronic ballast driving two 40-watt tubes dissipates about 5 watts, in contrast with losses of about 20 watts for the older ballast.

Virtually all electronic ballasts create alternating current at high frequency (over 20 KHz) to drive the

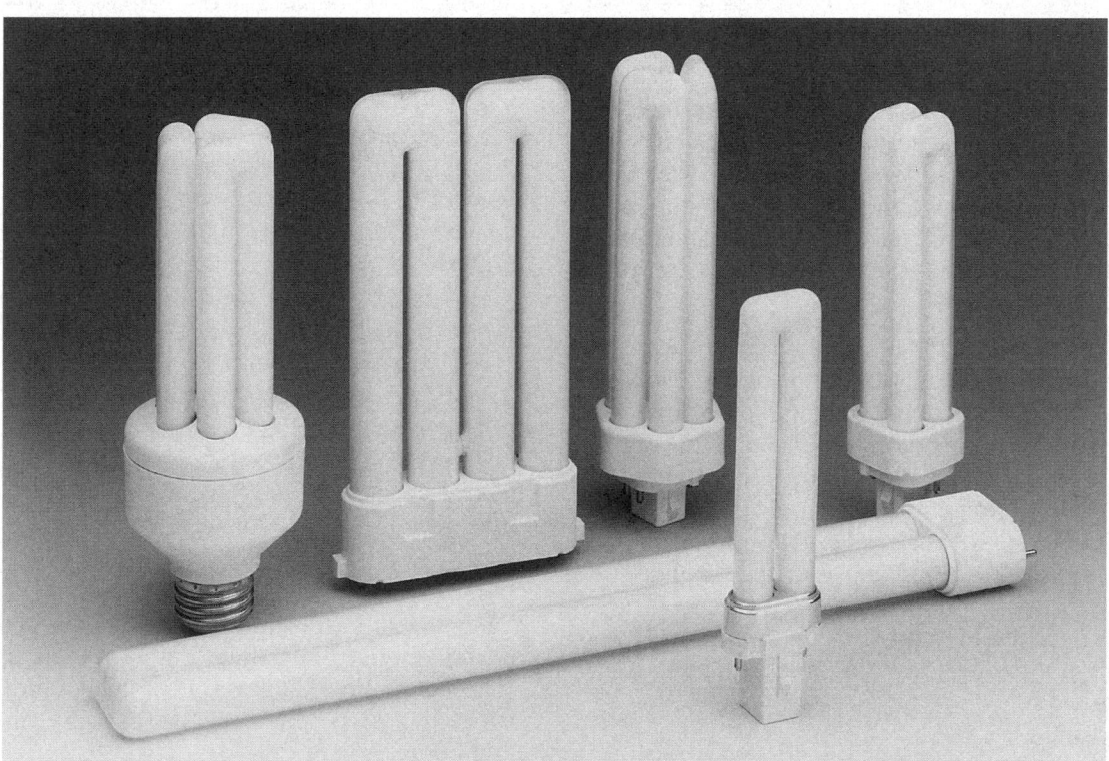

Osram Sylvania Inc.

Fig. 2 Compact fluorescent lamps These are available in smaller wattages than common conventional lamps. Their small physical size allows them to concentrate light on small task areas, and they can be mounted in aiming fixtures. For these reasons, they hold the key to using fluorescent lighting in task lighting. However, they are less efficient, per watt, than the larger lamp styles. The emerging multitude of lamp wattages and base types is creating a future standardization problem that will probably cause many of the present types to become obsolete.

lamps. This improves the efficiency of the lamps themselves by about 10 percent, for reasons explained in Reference Note 55.

Electronic ballasts usually are designed to reduce their input power to maintain approximately the same light output as with a conventional magnetic ballast. Thus, for example, retrofitting electronic ballasts to an older fixture with two 40-watt tubes may save about 23 watts, while retaining the same light level.

■ Hybrid Ballasts

"Hybrid" ballasts are magnetic ballasts that include an automatic switch to turn off the filaments of rapid-start lamps after the lamp starts. The switching feature saves about 2 watts per lamp.

Major ballast manufacturers base their hybrid units on magnetic ballasts whose efficiency has been substantially improved over earlier models. Therefore, hybrid ballasts typically reduce overall energy consumption by about 10 percent compared to older magnetic ballasts.

Like standard magnetic ballasts, hybrid ballasts can operate reduced-wattage lamps as well as standard lamps. Thus, hybrid ballasts combined with reduced-wattage lamps may lower energy input to almost the same levels as some electronic ballasts operating with standard lamps. However, the lumen-per-watt efficacy is not as high.

How to Select Ballasts

Once you decide whether to seek high efficiency, reduced wattage, or both, you need to select the specific equipment. There is a fairly large number of factors to consider, although some of them may not be important to a particular application. You can get information about the factors that matter in most applications from the ballast catalogs.

■ Compatible Lamp Types

As we discussed before, the need for compatibility forces you to decide what kind of lamps you plan to use in the future. Some of the more sophisticated electronic ballasts adjust themselves to drive a broad range of lamps, including conventional, high-efficiency, and reduced-wattage types. Such ballasts allow you to keep using your original type of lamps with the option of possibly upgrading lamp efficiency in the future. However, the range of lamp options is still limited. For example, do not expect to find a ballast that will drive both T12 and T8 lamps.

■ Wattage Input

The wattage input of the ballast and lamp combination is the bottom line in terms of energy consumption, but it does not indicate efficiency. A ballast may have different wattage input when driving different types of lamps.

Reduced-wattage ballasts may reduce input power by as much as 20%. In principle, ballasts could be designed for much larger reductions in input power, by reducing light output also. However, the major manufacturers do not produce such ballasts. They would make no sense in new construction, and would be too expensive as retrofit devices. If you want larger reductions in wattage and light output than reduced-wattage ballasts provide, you can add current limiters to the present ballasts (see Measure 9.2.5).

In the past, the wattage input was the criterion that you would select to tailor the light output of the lamp and ballast combination to the application. Now, a better approach is to use "ballast factor," which we discuss next.

■ Ballast Factor

"Ballast factor" is a measure of light output. Ballast factor is a comparison of the light output of a particular ballast when it is driving a standard lamp, compared to the light output of the same lamp when driven by an ANSI standard ballast. The ballast factor is expressed as a ratio. When a ballast drives a lamp to the "standard" output, the ballast factor is 1.0.

Ballast factor is now routinely specified in ballast catalogs. Use it as your starting point in "fine tuning" the light output to the application, after you have selected the proper number and size of lamps. If you don't pay attention to ballast factor, you may get a nasty surprise when you check light levels. You may get much more light than you expect, or much less.

Most ballasts have a ballast factor less than 1.0. "Normal" ballasts have ballast factors in the range of 0.85 to 1.0. In new construction, the ballast factor should fall within this range if the lighting has been designed properly.

Some ballasts are available with exceptionally low ballast factors, as low as 0.7. These are used specifically to reduce power consumption. Ballasts that have an exceptionally low ballast factor may be labeled as "energy saving." However, note that the ballast factor does not indicate efficiency, only light output. To know relative efficiency, you have to divide the light output (measured in lumens) by the energy input (measured in watts).

For an efficient installation, you need to do two things together. Select the ballast factor and the lamp to provide the proper light output for the application. At the same time, look for a combination of lamp and ballast that gives you the lowest energy input for the light output you desire.

You can also find ballasts that have ballast factors higher than 1.0, even as high as 1.3. These are intended to increase light output. Exceptionally low and exceptionally high ballast factors are likely to reduce overall efficiency. They are used primarily in retrofit applications, where changing the ballast factor is the most efficient and economical way to adjust the illumination level of existing fixtures.

If you are upgrading both lamp and ballast efficiency in an existing facility where the illumination level is correct for the application, you will want a ballast with a lower ballast factor. This is to take advantage of the higher efficiency of the new lamps. For example, if you are upgrading from old T12 lamps to the newer T8 lamps, the new ballasts will have a lower ballast factor to take advantage of the higher efficiency of the T8 lamps. Calculate the new ballast factor that you need from the relative efficiency of the old lamps and the new lamps.

■ Efficiency

Strictly speaking, the efficiency of a ballast alone cannot be stated, because the characteristics of a ballast depend on the lamp to which it is connected, and vice versa. Therefore, look for the efficiencies of the lamp-and-ballast combinations that you are considering. This information may be easy to find for common combinations. If you cannot find the information in the catalogs, contact the ballast manufacturer's engineering department.

Lamps and ballasts are not made by the same companies. It is usually better to call the ballast manufacturer than the lamp manufacturer, because the ballast is subject to more variations in efficiency, compatibility, and other features than the lamp.

"Ballast efficiency factor" (BEF) is a rating that was developed to provide a standard baseline for ballast efficiency. It is defined as the ballast factor divided by the ballast input wattage. Ballast efficiency factor is specified in some laws that require minimum ballast efficiency. It is intended primarily for ballast manufacturers. Unfortunately, it is awkward for ballast purchasers to use as a criterion because the rating numbers indicate relative performance only within individual lamp families corresponding to the ANSI standard ballasts. You would have to do some arithmetic to compare BEF's between lamp families. Furthermore, it does not tell you the efficiency of an existing ballast when retrofitted with newer types of lamps.

■ How the Lamps are Connected to the Ballast

Magnetic ballasts are available to drive either one or two lamps. At present, electronic ballasts are available to drive from one to four lamps. In fixtures with more than two lamps, using a single electronic ballast to drive all the lamps partially compensates for the higher cost of electronic ballasts.

If a magnetic ballast is designed to drive two lamps, it drives the lamps in series. Therefore, if one lamp fails, the other lamp cannot operate properly, either. In contrast, electronic ballasts may (or may not) drive lamps in parallel. With parallel ballasts, if one lamp fails, the other lamp or lamps continue to operate normally.

■ Lamp Current Frequency

This is basically a choice between magnetic and electronic ballasts. Magnetic and hybrid ballasts drive lamps at line frequency. Electronic ballasts drive lamps at high frequencies, above 20 KHz. High frequency improves lamp efficacy about 10%, but it may cause electronic noise problems, as explained below.

■ Filament Operation

The filaments of rapid-start and preheat lamps may be operated in different ways by different ballasts, affecting efficiency, lamp life, and starting characteristics. The three basic modes of operation are:

* *leaving the filaments heated continuously,* which is the conventional mode of operation for "rapid-start" lamps

* *providing no heating power for the filaments,* and starting the lamps solely by using higher voltage. This is the normal mode of operation for "instant-start" lamps. Some magnetic and electronic ballasts operate "rapid-start" and "preheat" lamps in this manner. This saves filament operating energy.

  Ironically, this mode of operation reduces filament life, sometimes drastically. The reason is that most loss of filament material is caused by the high electric field that exists during starting. According to industry figures, about 25% of service life is lost with 3-hour starting cycles. The loss in lamp life is proportional to the number of starting cycles. Ballasts that operate this way are most suitable for applications where the fixtures operate for long durations, as in corridor lighting.

* *operating the filaments only for starting.* This mode of operation eliminates most filament energy consumption and preserves lamp life. It is available with hybrid ballasts and some electronic ballasts.

The bare catalog data may not be clear about the specific mode of filament operation. In case of doubt, call the manufacturer.

■ Dimming Capability

Some types of electronic ballasts are capable of dimming the lamps by varying the lamp current. They have input connections that allow them to be controlled by a variety of lighting level controllers. Refer to Measure 9.2.6 about fluorescent dimming.

Various types of add-on controls provide limited dimming capability with non-dimming magnetic and electronic ballasts by varying the primary voltage to the ballast. Most such arrangements create unsatisfactory lighting, and all are now outdated.

■ Crest Factor

"Crest factor" is the ratio of the peak lamp current produced by the ballast to the average lamp current. High crest factor implies high peak current, which hastens the loss of the emissive material on the lamp electrodes. This loss reduces lamp efficiency and service life. Crest factor is not a commonly published rating for fluorescent ballasts, but if you find it, give preference to low numbers.

■ Lamp Starting Temperature

All fluorescent lamps suffer serious loss of light output and efficacy when they operate much below rated temperature (about 100°F at the lamp surface), although they may continue to operate weakly at much lower temperatures. If lamps in cold environments are well enclosed, they may operate adequately. However, getting them started may require special low-temperature ballasts.

The nominal starting temperature is 50°F for most conventional lamps, and 60°F for reduced-wattage lamps. The newest types of compact fluorescent lamps are designed to start at lower temperatures, some as low as 32°F (0°C).

■ Acoustical Noise

The cores of magnetic ballasts are made of thin metal laminations to minimize eddy currents. These laminations want to respond to the 60 Hz magnetic field like the voice coil of a loudspeaker. Magnetic ballasts all have a certain amount of hum, and it is easy for them to become abnormally noisy.

Some electronic ballasts have large filtering inductors that generate noise in the same way. Furthermore, any device driven by alternating current may find ways to vibrate.

Good ballasts typically have an "A" noise rating. Sound covers a broad frequency spectrum, and human hearing response varies widely with frequency, so it is not possible to characterize noise output with a single rating. Therefore, noise ratings are defined in several ways. If ballast noise could be important in your facility, check the manufacturers' technical literature for details.

■ Power Factor

See Reference Note 21 for an explanation of power factor. Many electric utilities impose a penalty for low power factor. Therefore, power factor may be an issue if lighting comprises a large part of the facility's electrical load. Low power factor can also cause overloading of circuits.

Low power factor is a problem primarily with magnetic ballasts, since they are essentially big inductors. The power factor in magnetic ballasts can be improved by adding a large capacitor in series with the coil. Power factor is generally not a problem with electronic ballasts, because they do not use the inductive principle to control current. Both types of ballasts are available with power factors better than 90%.

■ Radiated Electromagnetic Noise

Electronic ballasts radiate electromagnetic noise that may interfere with communications equipment and other electronic apparatus. The reason why they have this bad habit is that electromagnetic radiation is emitted more easily at the high frequencies at which electronic ballasts operate. Another reason is that electronic equipment is more susceptible to interference at these frequencies.

Both the fixture wiring and the lamps themselves act as transmitting antennas. If the metal case of the fixture is properly designed, not much radiation escapes from it. However, the arc in the lamp acts as an antenna, so the lamp can radiate through the plastic diffuser. In applications where electrical noise radiation is a potential problem, the radiation can be blocked by using lamps that have a transparent conductive coating, or by using fixtures that have a conductive, grounded coating on the diffusers. These solutions are expense.

Even the best electronic ballasts may interfere with some electronic equipment. One problem is interference with some theft detection equipment made before 1992. The problem occurs in the vicinity of the checkout locations where scanners that operate on the same frequency as electronic ballasts search for tags on stolen merchandise. Electronic ballasts may also interfere with consumer electronics, such as audio and video equipment that has circuitry operating at similar frequencies. There may also be interference with power line carrier signals.

The problem of electromagnetic noise radiation from the wiring usually makes it impractical to install electronic ballasts remotely from the fixtures.

In the United States, electromagnetic interference from the ballast itself is governed by Federal Communications Commission Regulations, Part 18, Subpart J. "Class A" criteria apply to equipment intended for use in commercial environments. "Class B" criteria, which are more restrictive, apply to equipment intended for use in residential environments. In other words, Class B is better.

■ Harmonic Distortion

Ballasts do not absorb power evenly during each part of the input cycle of alternating current. As a result, they distort the waveform in the power wiring. This distortion is expressed as "harmonic distortion" because it can be described as multiples, or harmonics, of the input power frequency. Harmonic distortion can be viewed as low-frequency electrical noise that enters the power distribution system from the input end of the ballast.

Like power factor, harmonic distortion is a problem that concerns electric utilities. Third-harmonic distortion is especially troublesome in 3-phase electrical power distribution because it creates large return currents in the neutral leg, which normally carries little current and is smaller in capacity than the main phase wires.

Harmonic distortion also affects the internal wiring of the building. It causes inefficiency and overheating in motors, it interferes with power line carrier communication, and it may interfere with electronic equipment that is connected to the same power system.

Electronic ballasts can cause the worst harmonic distortion. For this reason, electronic ballasts have input filters. However, low-frequency filtering is expensive and bulky, so lower-frequency distortion of the supply waveform still occurs.

Ballasts are now rated in terms of total harmonic distortion (THD). THD of 10% to 20% is presently typical. Some units have THD higher or lower than these figures. Magnetic ballasts typically have high THD, while electronic ballasts cover the range. The type of distortion differs between magnetic and electronic ballasts, and this difference is not revealed by the THD figure.

■ Reliability and Service Life

A magnetic ballast consists of little more than a coil of wire wrapped around a steel core and embedded in tar. If reasonably well made, a magnetic ballast can last forever. On the other hand, an electronic ballast is an active semiconductor device. While semiconductors may last a long time, they have a finite service life. Thermal cycling, when the lamp turns on and off, causes failures. Also, semiconductors are vulnerable to voltage spikes in the electrical power system that may occur for a variety of reasons.

High-efficiency ballasts have not enjoyed a good reliability history overall. The design of electronic power handling equipment is a compromise between reliability and cost, and many models err on the side of cheapness. Therefore, check with other users who have experience with a particular model before selecting it. In any event, expect to replace electronic ballasts on a continual but unpredictable basis.

Ballast Standards and Certification

In the United States, standards for ballasts are set by the American National Standards Institute (ANSI), but ballasts can be sold without meeting their standards. The Certified Ballast Manufacturers Association (CBM), an association of ballast manufacturers, has a program of certifying individual ballast models for compliance with ANSI standards. ANSI ballast standards specify various performance factors, including lamp output, reliable starting, ballast life, and preservation of lamp life.

An inherent problem in ballast certification is that the performance of a ballast depends on the lamp which it drives. For this reason, ANSI ballast testing is based on standard lamp types. ANSI or CBM certification may not provide useful guidance about performance with newer lamp types.

In the United States, a 1988 federal law established minimum ballast efficiencies for the most common types of ballasts, namely, those that drive:

- one or two F40/T12 lamps
- two F96/T12 lamps
- two F96/T12/HO lamps.

The efficiency requirements are expressed in terms of ballast efficacy factor (BEF), which is defined above.

Manufacturers may choose to certify other characteristics, such as electronic and acoustical noise output. Or, they may do so if requested by a potential purchaser.

Label the Fixtures

Label the fixtures with the lamp type, as explained in Measure 9.7.3. This is especially important if the new ballasts are intended to be used with new types of lamps.

ECONOMICS

SAVINGS POTENTIAL: Improved efficiency can save up to 4 watts per lamp with magnetic ballasts, up to 6 watts per lamp with hybrid ballasts, and up to 8 watts per lamp with electronic ballasts. Electronic ballasts also improve the efficacy of lamps by about 10%, usually reducing lamp power to maintain the same light output.

Ballasts may be designed to reduce wattage input by reducing lamp current, without efficiency improvement, but major U.S. ballast manufacturers do not use this approach.

COST: High-efficiency magnetic ballasts average about $10 apiece, for two F40 tubes. Hybrid ballasts cost about $15 apiece, for two F40 tubes. Good electronic ballasts cost from $25 to $45 apiece, for F40 tubes, depending on features. Installation labor costs $3 to $15 per fixture, depending on labor rates and accessibility. Refer to Measure 9.2.3 for lamp replacement cost.

PAYBACK PERIOD: Several years, or longer.

TRAPS & TRICKS

CHOOSING THIS APPROACH: This is one of the most expensive lighting modifications you can make. Don't do it until you have planned your lighting configuration to eliminate excess fixtures and lamps.

SELECTING THE EQUIPMENT: At the present time, you need to decide whether to use T8 lamps, because these require ballasts that are designed especially for them. If you want to continue using the less expensive T12 lamps, you can protect your options by selecting an electronic ballast that adapts to different lamp types. However, you will pay a premium for this flexibility. Check the performance of any type of ballast or lamp that you are considering. Reliability is still evolving, even among the major manufacturers, and there is a lot of cheap junk on the market.

FIXTURE LABELING: Be sure to label each fixture with the type of lamps that should be installed in it.

MEASURE **9.2.5 Install current limiters.**

A fluorescent current limiter is a coil that is installed in the secondary circuit of an existing magnetic ballast to reduce the lamp current. In effect, it is an extension of the coil in the ballast. Current limiters are smaller than ballasts, and there is usually plenty of room for them inside the fixture, but not always.

Current limiters reduce lamp wattage and light output by approximately equal amounts. Ballast losses are not affected significantly, so the overall efficiency of the lighting system is somewhat reduced. The current limiter itself incurs a small loss.

In principle, a current reducer can be designed to limit lamp current to any value up to the limit of stable lamp operation, which is about a 50% reduction for conventional lamps.

Applications and Alternatives

Current limiters fit into a narrow range of retrofit applications between delamping and reduced-wattage lamps. Current limiters have the field almost to themselves for providing power and light reductions in the range of 30% to 50%. Most units provide reductions in this range.

Delamping (Measure 9.2.1) may be an alternative to current limiters, depending on the number of tubes in each fixture, and the excess in the number of fixtures. Delamping is more efficient, much less expensive, and much simpler. For example, delamping a fixture with two ballasts provides a 50% power reduction, with a corresponding reduction of light output. This is about the same as the most aggressive current limiters. In fixtures that have more than four tubes, delamping can provide smaller percentage reductions in light output, e.g., by removing two tubes in a 6-tube fixture.

At the other extreme are reduced-wattage lamps (Measure 9.2.3) and reduced-wattage ballasts (Measure 9.2.4). These provide power reductions up to 20 percent, and some models improve lighting efficiency up to 20 percent. Reduced-wattage lamps are much cheaper and much easier to install than current limiters. Current limiters operate with lamps that are less expensive, but the break-even point is not reached until several sets of lamps have lived out their lives.

Reduced-wattage ballasts are more expensive, but the better models offer a significant improvement in efficiency, which current limiters do not. The labor cost for replacing ballasts is about the same as for installing current limiters, or somewhat less.

SUMMARY

Reduces wattage and light output 20 to 50 percent. Labor intensive. Use only with magnetic ballasts. Becoming an obsolete method.

SELECTION SCORECARD

Savings Potential .... $ $ $ $

Rate of Return .... % % %

Reliability .... ✓ ✓ ✓

Ease of Retrofit .... ☺ ☺ ☺

Ballast and Lamp Compatibility

Current limiters can be used only with magnetic ballasts, including high-efficiency models.

Current limiters should not be used with reduced-wattage lamps, which are too sensitive to reduced current.

Different models of current limiters are required for different families of lamps, depending on their lamp current and whether they have filaments to heat the electrodes.

Effect on Lamp and Ballast Life

Manufacturers of current limiters claim that the devices increase lamp and ballast life, and reduce lumen degradation. These claims seem credible. It is true that lumen degradation is proportional to lamp current. Also, ballast temperature is reduced by reducing the lamp current, and ballast temperature is the primary factor in ballast life.

On the other hand, there is concern that some current limiters may reduce filament life in rapid-start and preheat lamps. The current limiter is installed in a branch of the ballast wiring that powers one of the lamp filaments in addition to carrying the lamp current. The current limiter must include a feature, typically a transformer, to maintain full filament voltage in this circuit while reducing the lamp current. If this feature is not properly designed, the filament may burn out early.

How to Install Current Limiters

The current limiter is screwed to the sheetmetal of the light fixture, near the ballast. Then, the current limiter is wired into the secondary circuit of the ballast as explained in the instructions.

A potential problem is finding enough space to install the device inside the fixture. This is most likely in a fixture that has several ballasts. Install a unit on a trial basis before purchasing the units in quantity.

Performance Claims

No certification standards presently apply to current limiters, so be skeptical about manufacturers' performance claims. The devices are simple and legitimate, but not foolproof. Purchase them from a company that can refer you to customers who have had a long history of satisfactory experience.

ECONOMICS

SAVINGS POTENTIAL: *Typically 30 percent of lamp energy.*

COST: *Current reducers cost about $10 each. Installation costs $3 to $15 per fixture, depending on labor rates and accessibility.*

PAYBACK PERIOD: *One to five years.*

TRAPS & TRICKS

CHOICE OF METHOD: *Make sure that this Measure is the best approach for your situation. Remember that current limiters do not improve efficiency, but only reduce energy consumption at the expense of light output. They work only with magnetic ballasts. You may want to combine this Measure with delamping for tailoring light output in big increments.*

PLANNING: *Approach this Measure with the same thorough planning described in Measure 9.2.1.*

SELECTING THE EQUIPMENT: *Current limiters are not a mainstream product, so you will have trouble getting credible information about how they perform. First, ask vendors for references, and talk with purchasers who have used the units for several years. Before buying quantity, test a few units in typical fixtures. Use a light meter to measure the effect on light output, and use a wattmeter to measure the reduction of input power.*

EXPLAIN THEM: *The purpose of the current limiters will not be obvious to maintenance personnel who have to work on the fixtures later. Therefore, label the fixtures where current limiters are installed, explaining their purpose.*

MEASURE 9.2.6 Install fluorescent dimming equipment.

SUMMARY

Adjusts wattage and lighting levels over a wide range. Makes daylighting practical, but is expensive overkill in many other applications. Does not eliminate the need to customize control for each space. The best equipment is fairly efficient. Reliability is still in doubt.

SELECTION SCORECARD

Savings Potential	$ $ $
Rate of Return, New Facilities	% % % %
Rate of Return, Retrofit	% % %
Reliability	✓ ✓ ✓
Ease of Retrofit	☺ ☺ ☺

Dimming of fluorescent lighting without serious performance penalties became a practical reality during the 1990's. The price of satisfactory dimming equipment is still high, but it continues to fall as the equipment becomes more common and competition increases.

The market is still evolving. Many types have been put on the market. Manufacturers have failed to survive, leaving owners with orphaned equipment. Fluorescent dimming remains a pioneering activity that demands study to develop an optimum system and to minimize the chance of failure.

Where to Consider Fluorescent Dimming

There are several broad categories of applications where fluorescent dimming offers important capability:

- *exploiting daylighting.* Dimming is a major breakthrough for daylighting because it adjusts the fluorescent lighting level smoothly and continuously to complement daylighting. A less expensive alternative is using ordinary photoelectric switching to control the number of lighted tubes in response to daylight (see Measure 9.5.3). However, switching lamps creates abrupt changes in lighting that may annoy occupants.

- *other applications that benefit from a gradual change of lighting level.* In lavish environments, smooth transitions in lighting level are considered more elegant than stepped switching. In dark environments, such as movie theaters, dimming the lights gradually avoids the sensation of being plunged into darkness. In the past, such environments would have installed incandescent lighting and older, less efficient types of dimmers.

- *applications that benefit from fine adjustment of lighting level.* An example is changing the lighting level in restaurants and bars from day to night. Alternatives are stepped switching and installing separate fixtures for each lighting level.

- *where a high turndown ratio is needed.* The best fluorescent dimming equipment can reduce lighting levels to a few percent of maximum levels. Dimming is the only way to achieve such a large range without installing separate types of fixtures for low-level lighting. However, installing separate fixtures to handle low lighting levels is usually the best approach.

- *where dimming is needed with color stability.* In contrast with dimming incandescent lamps, dimming fluorescent lamps produces little color change. This is because the color of fluorescent light is determined primarily by the composition of the lamp's phosphors and gases, which are constant factors. Color stability is most likely to be important in technical applications involving photography or video recording.

Other Ways of Changing Lighting Level

In principle, you could use dimming equipment for most applications that need variable lighting level. However, dimming is not always the best approach. Fluorescent dimming is expensive. It also reduces lamp efficiency, especially at low settings. The least expensive method of changing fluorescent lighting level is switching individual lamps. Measures in Subsections 9.4 through 9.6 recommend various ways of accomplishing this.

Turndown Ratio and Efficiency

The turndown ratio of a dimming system is the ratio of the maximum lighting level to the minimum lighting level. A high turndown ratio is essential to economic success, because the high initial cost of the dimming equipment can be justified only in applications where the average lighting power is much lower than the maximum.

Typical electronic fluorescent dimmers provide turndown to about 20% of full light output with good stability. Some dimmers claim to turn down as low as 1% of full light output. However, the fine print for these units states outlandish requirements to achieve this extreme turndown, such as aging new lamps for 100 hours at full output.

Manufacturers claim that electronic ballasts maintain full lumens-per-watt efficiency down to about 50% of full output, and that efficiency suffers only modestly below that point. However, the technology is new, standards are not established, and there is little independent testing to deter manufacturers from fudging their claims.

How Fluorescent Dimming Works

Fluorescent dimming works by varying the lamp current. Doing this while maintaining stable lamp operation and good energy efficiency is the technical challenge that obstructed fluorescent dimming until electronic ballasts appeared.

To understand the problem, recognize that fluorescent lamps turn on and off with each half cycle of the alternating supply current. At the beginning of each half cycle, the ballast must provide sufficient voltage to restart the lamp.

With magnetic ballasts, the voltage supplied to the lamp has a fixed relationship to the current limiting characteristics of the ballast. The ballast voltage is designed for cold starting, so it has a certain amount of excess voltage for restarting the lamp each half cycle when it is warm. The first types of magnetic ballast dimmers exploited this voltage margin by reducing the ballast voltage once the lamp started. However, larger reductions of voltage and current pushed beyond the stability range of the lamps, making them operate inefficiently and giving the lamp a wavering, defective appearance.

The development of electronic ballasts was a breakthrough for fluorescent dimming because electronic ballasts can be designed to adjust voltage and current separately. The ballast feeds current to the lamps in controlled pulses. The circuitry of electronic ballasts can be designed so that the appropriate starting voltage is applied with each cycle, or pulse, of lamp current.

Electronic ballasts also keep the mercury arc stable because they provide power to the lamps at high frequencies, above 20 KHz. The mercury vapor in the lamps remains partially ionized during the much shorter off cycles, so the arc does not wander and the flow of current resumes easily.

Modern Dimming Equipment

The first fully satisfactory types of fluorescent dimming equipment started to appear on the market in the early 1990's. Earlier types of fluorescent dimming equipment, based on either magnetic or electronic ballasts, are now outmoded.

Good dimming requires control of the lamp current from within the ballast. Some early types of dimming systems attempted to minimize cost and simplify installation by controlling the primary power to existing ballasts, both magnetic and electronic, but this method did not yield a good turndown ratio.

A modern fluorescent dimming system has two basic elements. The heart of the system is an electronic ballast that is designed specifically for dimming control. The other element is control of the light level by one or more control devices that are located outside the fixture where the need for lighting can be sensed or adjusted.

■ Light Level Controllers

Dimming systems that are based on controlled electronic ballasts have essentially unlimited control flexibility. Any device that generates a compatible control signal (e.g., a standard 4-to-20 milliamp analog signal) can be used to control the ballast output. Standard lighting control devices for use with controllable ballasts are becoming increasingly available.

A variety of specialized lighting control devices are presently available from established manufacturers of lighting equipment. These include manual dimmers (that you can get to coordinate in appearance with other light switches), light level sensors, motion sensors, time controls for programming lighting levels, etc.

Equipment Selection Factors

The dimming electronic ballast is the critical element in a fluorescent dimming system. The selection factors for dimming ballasts are the same as for other electronic ballasts. Refer to Measure 9.2.4 for these factors. In addition, turndown ratio is an important characteristic. The ballast should produce its minimum light output without discernible flickering, sensitivity to voltage fluctuations, or other instability.

Install Effective Light Level Controls

A dimming system will not provide its potential energy savings unless it is controlled to keep lamp output at the minimum needed under all circumstances. Without effective control, a dimming system is just an expensive frill. For example, if the light level is set with manual slider switches in a space where people have no reason to minimize lighting levels, the sliders may be left in any position, and most likely will tend to stay near maximum output.

As with other aspects of lighting, lay out the dimming control on a fixture-by-fixture basis. For example, control fixtures near windows with daylight sensors, control fixtures in spaces with sporadic occupancy with personnel sensors, and so forth. Refer to Subsections 9.4 and 9.5 for manual and automatic lighting control techniques, respectively.

Lamp Compatibility

Lamp compatibility varies widely among dimming systems. The more sophisticated and expensive dimming ballasts adjust themselves to drive a broad range of lamp types. Other dimming systems are specific as to lamp type. In particular, some dimming ballasts are not compatible with reduced-wattage lamps.

Compact fluorescent fixtures that have integral preheat starters cannot be dimmed. Compact fluorescent lamps without integral ballasts can be dimmed effectively with compatible dimming equipment.

Reliability

Electronic lighting equipment has had protracted teething problems. Electronic ballasts needed about ten years to become reasonably reliable after they entered that market, and that was for major manufacturers. Fluorescent dimming equipment is an outgrowth of electronic ballasts, and is newer. Check the performance record of the specific equipment being considered.

Don't Expect Daylighting to be Easy

If you want to use fluorescent dimming to exploit daylighting, understand that installing the dimming equipment is only part of the job. Daylighting is still a pioneering activity that is characterized by more failure than success. Study Subsection 8.3 before venturing into this challenging area.

ECONOMICS

SAVINGS POTENTIAL: *In applications that are appropriate for dimming, savings should average 30 to 60 percent of maximum wattage.*

COST: *Good dimming ballasts cost $50 to $100 apiece, and may control up to four lamps. The cost of control devices varies widely. Installation labor costs $10 to $200 per fixture, depending on labor rates, accessibility, and the number fixtures per control device.*

PAYBACK PERIOD: *Several years, or longer.*

TRAPS & TRICKS

CHOICE OF METHOD: *Fluorescent dimming is legitimate technology with limited application. Daylighting and theater dimming comprise a large fraction of the appropriate applications. At present, fluorescent dimming is largely a fad, giving bored lighting designers a chance to make a splash with something expensive and technical. It is irresponsible to use dimming where cheaper, more reliable methods will serve as well.*

PLANNING: *Daylighting control is tricky. Study Subsection 8.3 and Measure 9.5.3. For manual dimming applications, you have the challenge of getting people to use the controls efficiently. Make a complete installation plan. Optimize the controls separately for each location.*

SELECTING THE EQUIPMENT: *Fluorescent dimming equipment may still be evolving, and the most efficient types have not been in service long enough to demonstrate long-term reliability. Check with others who have used the equipment you are considering. Avoid any equipment that has been in service for less than a few years.*

EXPLAIN IT: *All lighting controls require placards, even if the controls are automatic. See Measure 9.4.1.*

MEASURE **9.2.7 Consider retrofit "reflectors" for fluorescent fixtures.**

SUMMARY

Widely promoted with false claims, "reflectors" for straight-tube fixtures typically provide little real benefit, and may cause unsatisfactory light distribution. They may be useful in fixtures with inefficient diffusers.

SELECTION SCORECARD

Savings Potential .... **$**

Rate of Return .... **%**

Reliability .... ✓ ✓ ✓

Ease of Retrofit .... ☺ ☺ ☺

"Reflector" retrofits for straight-tube fluorescent fixtures are thin reflective overlays inserted into fixtures behind the tubes. The common feature of these overlays is that they have mirror-like ("specular") surfaces instead of the diffuse white surfaces that are commonly used in fluorescent fixtures. The overlays may also have curvatures to concentrate light downward more narrowly than the fixture does.

The most common claim for specular reflectors is that installing them in 4-tube fluorescent fixtures allows two of the tubes to be removed with minimal loss of light. This claim is not credible. Although the devices do indeed reflect light from the fixture, they do not provide the improvement promised. When promoted with such claims, reflectors fall into the category of snake oil.

Bogus Claims

The explanations usually given for this miracle of free light are that the specular reflectors have better reflectivity than the original fixture, and that the light distribution pattern is improved. Both of these claims are false.

With respect to the issue of reflectivity, the original surface of the fixture typically is a glossy white enamel having a reflectivity in excess of 90%. The retrofit reflector typically has an aluminum or silver surface with about the same reflectivity. The significant difference is that the specular retrofit reflector yields the glinty brilliance of sharply reflected light. This glinting effect leads the naive purchaser to believe that the light output is higher.

Still, tests have been conducted that seem to demonstrate major increases in light output with specular reflectors! Why? In some of the tests, dirty fixtures with old tubes were used as a basis of comparison. The new reflectors are clean, and two new tubes are installed to replace the four old ones.

Bogus "tests" of specular reflectors also exploit the fact that they may concentrate light more narrowly, so they show a higher footcandle level in the brightest part of the distribution pattern. Of course, this robs light from the rest of the distribution pattern.

Presumably, the light distribution pattern of the original fluorescent fixture was selected by the building designer to suit the illumination requirements. Therefore, changing the pattern is likely to do more harm than good. In addition, the light distribution pattern of the modified fixture is likely to be irregular, in contrast with the carefully shaped light distribution pattern of the fixture originally. "Reflector" retrofits are like converting a van to a pickup truck when the original van is the proper vehicle for the job. As P. T. Barnum once said, . . .

Actual Efficiency Improvement

The only way that reflectors can provide a real benefit is to reduce trapping of light within the fixture. Some common types of fluorescent fixtures do trap a significant amount of light. To judge whether specular reflectors may provide a benefit, consider the three ways that light may be trapped in a fluorescent fixture:

- *shadowing by the fluorescent tubes.* In fixtures where tubes are close together, as in most fixtures with four or more tubes, each tube blocks light output from its neighbors. Tubes are fairly reflective, but they do absorb some light. Also, they reflect some light back into the fixture. Removing alternate tubes from a fixture eliminates most of this loss. Installing a specular reflector does not contribute to this improvement

- *internal reflections.* For a given reflector shape, a diffuse reflector surface may direct more light back into the fixture than specular reflectors, but this is not necessarily true. Internal reflection increases light absorption by the reflector itself and by the tubes. However, most fluorescent fixtures have shallow reflectors, so there is little reflection from one part of the fixture to another.

- *absorption in the diffuser.* Diffusers are responsible for most of the light loss in fluorescent fixtures that have them. Clear prismatic diffusers, which are the most efficient common type, absorb about 30% of lamp output. Colored egg crate diffusers may absorb as much as 75% of lamp output, but these are a rarity. Installing a specular reflector would probably have little effect on absorption by the diffuser. There is one possible exception. A retrofit reflector that is shaped to concentrate light in a downward direction may increase the light output from fixtures with absorptive diffusers, especially diffusers having some depth, such as egg crate styles and deep parabolic louvers.

In summary, reflectors that concentrate light downward may increase light output somewhat in rare cases involving wasteful diffusers. Even in these cases, the narrower light distribution patterns of the fixtures are likely to produce spotty coverage. Where fixtures are designed to provide lighting on an area basis, the combination of specular reflectors and delamping is likely to deprive some task areas of adequate lighting.

Energy Saving Potential

Specular reflectors do not save energy unless they are necessary to make some other modification to the fixture that reduces energy input. The small increase in light output that reflectors may provide even in unusual cases usually does not allow such modifications.

A Better Alternative: Improve the Diffusers

In grasping for a legitimate application for reflectors, we suggested that specular reflectors might be useful in fixtures with diffusers that absorb a large amount of light. However, a more direct approach is to improve the diffusers. See Measure 9.7.2 about this.

Types of Specular Reflectors

Retrofit reflectors are formed from thin sheets of aluminum. Differences relate to the type of reflective surface and to the shape.

The cheapest reflectors simply have a polished surface, with a coat of varnish to delay corrosion. More expensive reflectors improve reflectivity by using sputtered (vacuum deposited) coatings. These coatings need to be protected, and a tough polyester film is added for this purpose. Silver is sometimes used as a reflective material because its reflectivity is a few percent higher than that of aluminum. However, silver is even more vulnerable to oxidation, so the protective coating is even more important with silver.

Reflector shapes are usually generic, rather than being customized to particular fixture models. Some reflectors are shaped to follow the general contours of the fixture. Others have pseudo-parabolic shapes to concentrate light downward. Some reflectors cover the entire inside surface of the fixture, and some cover just the ballast housing. Some surround individual tubes.

Degradation of Reflectivity

The original enamel diffuser of the fixture suffers little loss of reflectivity over the life of the fixture, but retrofit reflectors suffer some degradation of reflectivity and specular quality as the aluminum or silver oxidizes. Plastic coatings can be effective for blocking oxidation, but the film itself is subject to long-term degradation, primarily from the ultraviolet light emitted by the lamps. Therefore, over the life of the fixture, the average reflectivity of the retrofit reflector is likely to be lower than the reflectivity of the original surface that it covers.

Socket Modifications

Reflectors are installed with the assumption that pairs of tubes will be removed. Removing the tubes alters the light distribution pattern of the fixture. To compensate for this, some reflector kits include socket adapters that relocate the remaining tubes to a position halfway between the original sockets. This is done to keep the light distribution pattern from becoming too lopsided, and it may also reduce light trapping within the fixture.

Socket adapters are available separately. You can use them when delamping fixtures even if you do not use reflectors.

Reflectors for Screw-in Fluorescent Fixtures

Many fixtures that use screw-in fluorescent fixtures, especially downlights, can benefit from reflective inserts to reduce absorption inside the fixtures. These inserts are covered in Measure 9.1.4.2.

ECONOMICS

SAVINGS POTENTIAL: *There are probably no mainstream applications where reflectors can save money. It is possible to imagine some oddball applications where they might, but these cases are too speculative to estimate.*

COST: *Reflectors cost $5 to $30 per fixture, depending on size, shape, and other features. Installation costs $5 to $30 per fixture, depending on labor rates, accessibility, and whether other modifications and cleaning are performed at the same time.*

PAYBACK PERIOD: *Reflector retrofits usually do not pay back.*

TRAPS & TRICKS

CHOICE OF METHOD: *Treat reflectors as snake oil. They rarely provide significant benefit, so they almost never pay off. There is a common misconception that you need reflectors in order to delamp fixtures. You don't.*

SELECTING THE EQUIPMENT: *In those rare cases where reflectors may be useful, select the reflector shape to give you the light distribution pattern you want. Select reflector materials to resist oxidation, ultraviolet degradation, and abrasion from cleaning.*

High-intensity discharge ("HID") lighting has long been the dominant type for industrial facilities and outdoor lighting. Major improvements in color rendering have allowed HID lighting to move into the commercial sector, where it competes with incandescent and fluorescent lighting for many applications.

The first type of HID lighting was mercury vapor (which is abbreviated "MV" in this book). In recent decades, mercury vapor lighting was joined by metal halide ("MH") and high-pressure sodium ("HPS") lighting. These two newer types offer much better color rendering and higher efficiency, leaving lower cost as the only remaining advantage of mercury vapor.

This Subsection also covers low-pressure sodium ("LPS") lighting. LPS lighting does not fit within the strict definition of HID lighting, but it is similar in operation and installation. It has exceptionally high electrical efficiency. Unfortunately, it is limited in application by poor color rendering, mainly to outdoor lighting.

At present, there are fewer opportunities for improving HID lighting than for fluorescent and incandescent lighting. Unlike the other two types of electric lighting, HID lighting forces you to compromise between efficiency and color rendering. HID provides fewer ways to make small reductions in wattage. Delamping opportunities are limited because the high intensity and focussed light distribution of HID lamps make each lamp essential. Dimming options are presently limited, and not well proven. Opportunities for improving control are limited by long starting and restarting times.

Even so, you may be able to make major improvements in the efficiency of your HID lighting. At present, your main opportunities are selecting the best available equipment and improving lighting layout. For the latter, combine the Measures of this Subsection with the Measures of Subsection 9.6.

HID lamps with much better efficiency have become available in recent years. Unfortunately, the efficiency of HID ballasts continues to be poor. Electronic ballasts may create the same opportunities for conservation in HID lighting that they did in fluorescent lighting, and they may lessen or overcome the starting and restarting problems of HID. Watch for developments, but beware of the kinds of troubles that first beset fluorescent electronic ballasts.

Before getting started, review Reference Note 56, HID and LPS Lighting. It tells you how HID lighting works, and gives details of the main types of HID lamps. Armed with this knowledge, you can select the best types and models of HID equipment. Reference Note 56 also provides an overview of the present status of HID lamp and ballast efficiency. For a comprehensive comparison of HID lighting with other light sources, see Reference Note 52, Comparative Light Source Characteristics.

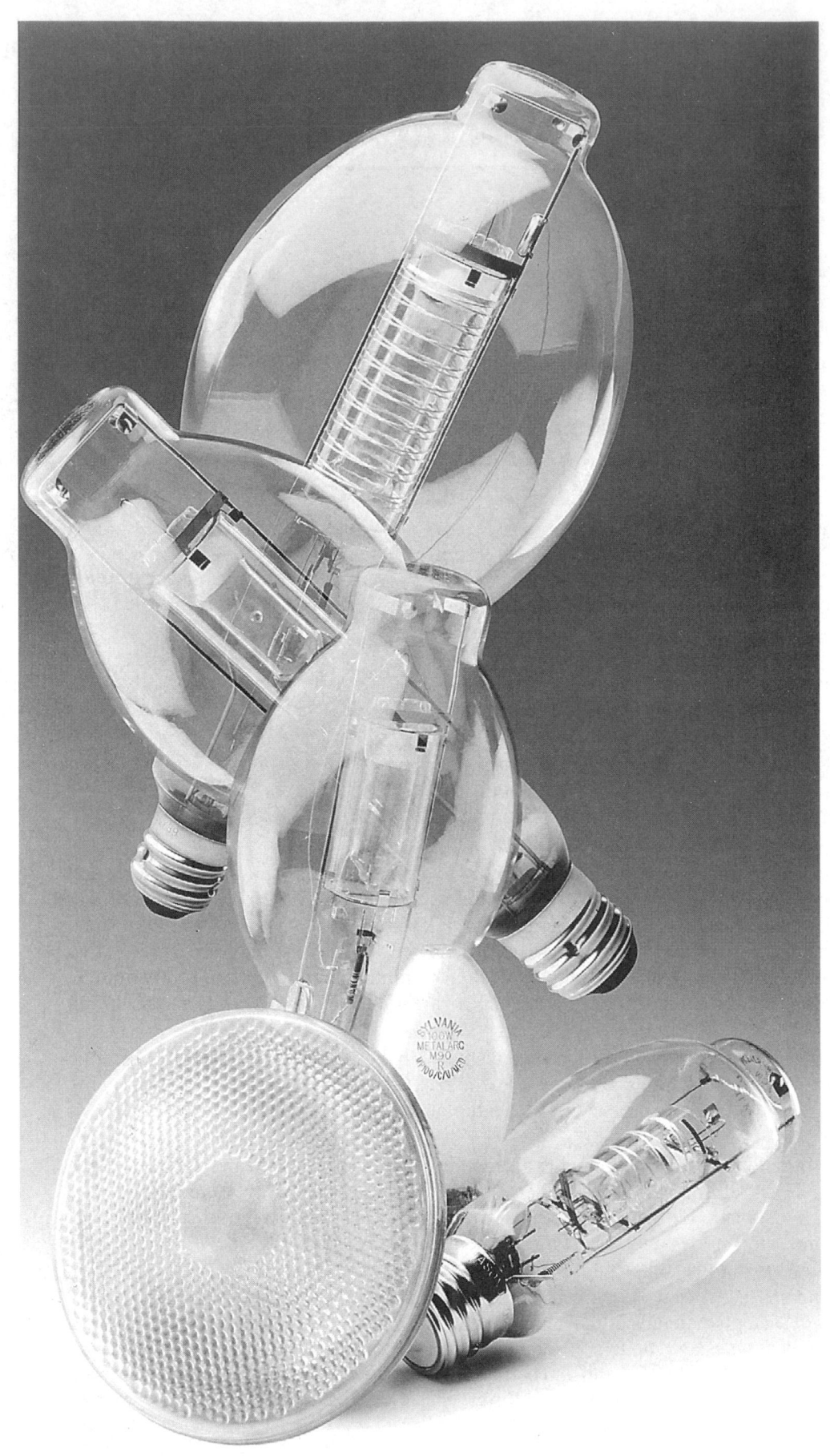

MEASURE **9.3.1 Install the most efficient HID lamps, ballasts, and fixtures.**

A	B	

A flurry of new HID lamp models have appeared over the past several decades. These provide a broad selection of characteristics among all three types of HID lamps. You can now get a combination of energy efficiency and good color rendering in HID lighting. However, you still have to select a compromise between these two fundamentally important characteristics.

It is relatively easy to upgrade many existing HID fixtures. In most cases, this requires changing the ballasts, along with capacitors and ignitors, if they are needed. Ballast conversion kits are available for most modern types of HID lamps. For example, see Figure 1.

However, even in retrofit applications, consider replacing the entire fixture. HID fixtures, including ballasts, may not be much more expensive than installing a ballast conversion kit. This approach becomes more desirable if a new fixture would provide better light distribution.

Energy Saving Potential

See Table 1 of Reference Note 56, HID and LPS Lighting. It shows that there are large differences among different types of HID lighting. Careful selection can reduce lamp energy consumption from 40 to 65 percent in typical applications.

Where to Upgrade Existing HID Lighting

You can improve most HID lighting, except for the most efficient modern types. Upgrading is fairly expensive because you have to change both the lamps and ballasts.

Retrofitting HID lighting for greater efficiency is most likely to be economical where you can replace existing mercury vapor lighting with metal halide or sodium vapor lighting. Some metal halide and high-pressure sodium lamps have been designed as direct replacements for mercury vapor lamps, without the need to change ballasts. This upgrade is covered by Measure 9.3.1.1. Unfortunately, it improves color more than it saves energy. Replacing HID lamps without replacing ballasts is not yet possible without paying a high price in efficiency.

Upgrading becomes more valuable in applications where the color characteristics of the lighting are important. The greatest improvements to HID lighting in recent years have been in color rendering. The muddy color seen under older mercury vapor lamps is no longer a penalty that must be paid. Some models of metal halide and sodium vapor lamps approach the color rendering of the best fluorescent lamps.

SUMMARY

A reliable source of energy savings. Fairly easy to retrofit. The biggest improvement is replacing mercury vapor lighting with metal halide or high-pressure sodium. Select carefully for efficiency and color rendering.

SELECTION SCORECARD

Savings Potential	$ $ $ $
Rate of Return, New Facilities	% % % %
Rate of Return, Retrofit..........	% % %
Reliability	✓ ✓ ✓ ✓
Ease of Retrofit	☺ ☺ ☺

Low-pressure sodium lighting is rarely a suitable replacement for other lamp types, despite its exceptionally high lumens-per-watt output. It may be the best choice in uncommon applications where color is unimportant and people are exposed to the light only for short durations.

■ Ballast Conversion Kits

Conversion kits are available for modifying existing mercury vapor fixtures to use other types of HID lamps. These typically consist of a new ballast, capacitor, ignitor (if needed), and mounting hardware.

Before committing to conversion on a large scale, try a conversion kit in one lamp to make sure that it does not require too much troublesome fitting.

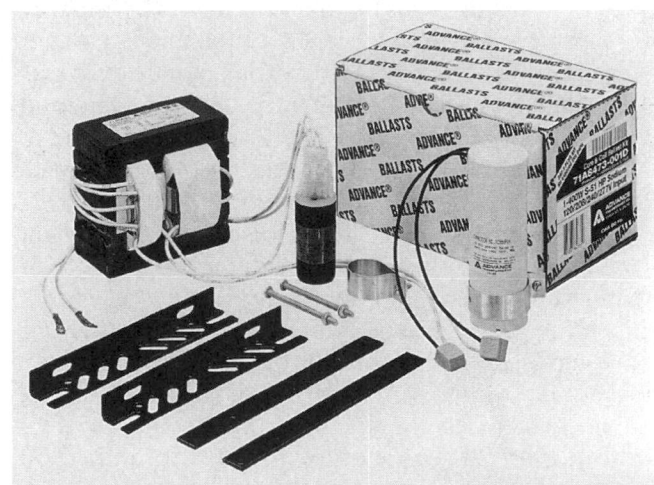

Advance Transformer Co.

Fig. 1 HID conversion kit This kit converts a fixture to use a high-pressure sodium lamp. From left to right are the ballast, ignitor, and capacitor, along with some universal mounting hardware.

■ Consider Improving the Entire Fixture

Replacing fixtures entirely may be a better alternative than retrofitting new ballasts in the old fixtures, especially if new fixtures can provide more efficient light distribution and more efficient lamp layout. All other factors being equal, efficacy increases sharply with lamp size. Where you have a choice, it is usually better to use a smaller number of larger lamps. Refer to Measures 9.6.2 ff for guidance on fixture layout.

In some applications, replacing the present HID lighting with fluorescent lighting provides better efficacy than replacing it with improved HID lighting. See Measure 9.3.3 for this approach.

■ Do Not Substitute Smaller Lamps on the Same Ballasts

HID lamps are not incandescent lamps. You can't just install a lower-wattage lamp without changing the ballast, unless the lamp is explicitly designed for such service. Using a smaller lamp will not reduce wattage much. This is because the lamp current is controlled by the ballast, not by the lamp itself. The smaller lamp is driven at greater power than it was designed to handle, and it burns out quickly.

Consolidate Lamp Types

Try to make an all-or-nothing decision about the types of HID lamps and ballasts that you plan to use, for these reasons:

- having several types of lamps and ballasts makes it easy to install the wrong type of lamp in a fixture
- keeping similar-looking lamps in stock invites ordering and warehousing mistakes
- occupants will notice differences in lamp color if fixtures with different lamp types are within sight of each other.

There is an interesting exception to this guideline. You can mix lamp types to improve color. This is a way of getting around the fact that HID lamps with better color characteristics have lower efficacy and cost more. This method works only if there is a lot of overlap in the light from different fixtures. For example, in a tall ceiling with closely spaced fixtures, you could alternate metal halide lamps with high-pressure sodium lamps in a checkerboard pattern. The bluish light of efficient metal halide lamps blends well with the reddish light of efficient high-pressure sodium lamps. But, be careful. This combination provides light that is attractive for many applications, but the color rendering is irregular. Furthermore, with recent advances in HID lamp design, it is no longer necessary to go to this bother to get a combination of good color rendering and good efficiency.

How to Select HID Lamps

Reference Notes 52 and 56 give you the general background you need to select HID lamps. When you get ready to buy, select for the following specific characteristics in the lamp catalogs.

■ Initial Lumen Output

In retrofit applications, start by finding out whether the illumination levels provided by the present lighting match the requirements of the activities. Calculate any increases or reductions in the lumen output, and then scan the catalogs to find lamps that provide this output.

Take into account the effect of color rendering on lumen requirements. Poorer color rendering may require higher lumen output, but only if the work involves colors. Color rendering does not strongly affect lumen levels with activities that do not rely on colors, such as routine paperwork, reading stencils on crates, and work at computer terminals (which are their own light sources).

■ Degradation of Lumen Output

HID lamps suffer serious loss of light output as they age. Ensure that enough light is provided at the end of the lamp replacement cycle (which is not necessarily at the end of rated service life). The lamp specifications usually do not provide this information. Instead, they state the initial lumen output, and perhaps some sort of average figure. The lumen output at the end of the lamp's rated life typically is 60 to 75 percent of its initial lumens.

High-pressure sodium lamps have somewhat lower lumen loss than mercury vapor and metal halide. Also, the most lumen loss in high-pressure sodium lamps occurs toward the end of lamp life, whereas the other two types have their highest rate of lumen loss at the beginning.

■ Color Characteristics

All three types of HID lamps have improved their color rendering in recent years. Both metal halide and high-pressure sodium models are available with "color rendering index" (CRI) ratings that are almost as good as the best fluorescent lamps. Older types with low CRI still remain, leaving a wide range of color rendering and efficacy for each type of HID lamp.

Unlike fluorescent lighting, HID lamps have a conflict between color rendering and lumen-per-watt efficacy, so do not select a higher CRI than you need. Also, lamps with higher CRI usually are more expensive.

Low-pressure sodium lamps stand apart from the other types because they have almost no ability to bring out colors. This serious deficiency is covered below.

■ Mounting Position

Some HID lamps need specific mounting positions to keep the arc from touching the wall of the arc tube, or to protect the lamp from overheating. These lamps are specified for mounting "base up," "base down," and "horizontal." Lamps that can be operated in any position are listed as "universal."

If a lamp is specified for vertical mounting (base-up or base-down), it is usually limited to a variation of about 15 degrees from the vertical. If a lamp is specified for horizontal mounting, it is usually limited to within 45 degrees of horizontal.

■ Starting and Operating Temperatures

All mainstream HID lamps operate well down to a temperature of about -20°F (about -30°C), and they may operate at lower temperatures if enclosed. The minimum starting temperature is determined by the ballast, as explained below.

■ Wattage and Efficacy

Select lamps that satisfy the previous requirements with the best efficiency. Lamp efficacy may not be stated outright in the catalogs. It is easy to calculate. Simply divide the lumen rating by the lamp wattage. Table 1 of Reference Note 56, HID and LPS Lighting, illustrates the range of efficacies you can expect to find.

■ Service Life

Lamp life does not relate directly to efficacy, but it is a major factor in labor and lamp cost. These are the ranges of lamp life for mainstream lamps of each type:

mercury vapor	24,000 hours
metal halide	10,000 - 20,000 hours
high-pressure sodium	15,000 - 24,000 hours
low-pressure sodium	18,000 hours

Specialized lamps designed for exceptional color rendering, compact size, or other needs may have much shorter life.

How to Select HID Ballasts

Select the ballast to match the lamp. Get the following information from the ballast catalogs.

■ Dimming Capability

Decide first whether you want dimming capability. If so, your choice of ballasts will be limited, perhaps to relatively unknown equipment. Refer to Measure 9.3.2 for the details of HID dimming.

■ Lamp Compatibility

The lamp and the ballast must be matched. Otherwise, efficiency and service life suffer, if the unit operates at all.

The ANSI ballast code system simplifies ballast selection by assigning a standard ballast code to each lamp model. The ANSI lamp code consists of a single letter that indicates the type of lamp, followed by a number that indicates the ballast type. The lamp type codes are:

H	mercury vapor
M	metal halide
S	high-pressure sodium
L	low-pressure sodium

A particular ballast may have more than one ANSI code. For example, a particular 400-watt ballast may have ANSI codes of H-33 and M-59. This means that it can operate some types of mercury vapor lamps and some times of metal halide lamps.

The ANSI code is not a rating system for either the lamp or the ballast. It tells you only the electrical power characteristics that the lamp requires from the ballast. Different ballast models for a given lamp may have major differences in efficiency, power factor, etc. You still have to select the specific ballast model.

Some lamps may not have ANSI ballast codes. For these, the manufacturer provides a list of specific ballasts that can be used.

■ Number of Lamps

HID ballasts are available to drive either one or two lamps. Using a single ballast to drive two lamps saves money in purchasing the ballast, but it eliminates the possibility of operating one of the lamps alone.

■ Sensitivity to Voltage Fluctuations

"Reactor" ballasts allow large changes in lamp wattage if the line voltage varies. To avoid this problem, use "constant wattage" (CW) or "constant wattage autotransformer" (CWA) ballasts.

■ Number of Fixtures per Circuit

The low power factor of "reactor" ballasts is improved by installing a capacitor in parallel with the ballast coil. Unfortunately, this simple technique has the disadvantage that it limits the number of fixtures that can be installed on a circuit. If the lamp stops drawing current, because it fails or is cooling down to restart, the current drawn by the ballast actually increases because the capacitor becomes a short circuit for the alternating current. For this reason, the number of power circuits required may be more than doubled by using high-power-factor reactor ballasts.

One way around this problem is to use reactor ballasts that are designed to disconnect the capacitor when the lamp circuit opens. Another solution is to use a different type of ballast, such as the common CWA type.

■ Crest Factor

"Crest factor" is the ratio of the peak lamp current to the average lamp current. High crest factors cause more rapid loss of the emissive material on the lamp electrodes, reducing lamp efficacy and service life. Crest factor varies widely among different types of HID ballasts. Look for the lowest numbers.

■ Acoustical Noise

This issue is the same as for fluorescent ballasts. Refer to Measure 9.2.4.

■ Power Factor

Power factor measures the phase relationship between the current and voltage of an electrical device. A lower power factor increases the current required to deliver a given amount of power. This increases the size of wiring, transformers, and switching in the power system. Many electric utilities impose a penalty for low power factor. Therefore, power factor may be a major issue if lighting comprises a large part of the facility's electrical load. Low power factor can also cause overloading of circuits.

Reactor ballasts without capacitors have very low power factors. Aside from this case, all types of HID ballasts are available in models that provide a power factor of 90% or better.

■ Starting Temperature

HID ballasts start lamps down to their normal minimum operating temperature, which is about -20°F (about -30°C). Lamps can operate at lower ambient temperatures if they are enclosed in suitable fixtures. However, to start the lamps in extremely cold weather, you may need special low-temperature ballasts.

■ Lamp-to-Ballast Distance

Ballasts for large lamps may be mounted far from the lamps, for example, on the side of a light pole. Special ballasts and ignitors may be needed to cope with separations greater than about 30 feet (10 meters).

■ Reliability and Service Life

The main areas of failure in conventional magnetic HID ballasts are the coil insulation, the capacitor, and the ignitor (if used).

Ballast insulation failure is usually caused by long exposure to high temperature. Ballasts that are designed for improved efficiency offer the potential for extended coil life because they operate at lower temperatures.

"Dry" capacitors are preferable to electrolytic capacitors, which fail by drying out. Also, dry capacitors do not require attention to grounding.

■ Integral or Separate Ignitors and Capacitors

Capacitors and ignitors may be separate or they may be an integral part of the ballast assembly. The choice is usually dictated by the fixture. If there is an option, installing separate components makes it less expensive to replace failed components in the future, but initial installation takes more time.

In HID lighting, the term "core-and-coil" ballast usually means the bare ballast without the capacitor or ignitor.

■ Wattage Input and Efficiency

Once all these mandatory features have been defined, select among the candidate models for the lowest input wattage. This is important because contemporary HID ballast losses are high, especially for small sizes.

■ Wiring Convenience Features

Some models of HID ballasts offer features that simplify installation. Take advantage of these features, if they are useful in your installations:

- *power leads for external control.* HID is commonly used for exterior lighting, so ballasts may include leads intended for easy connection of photocell control. The same leads could be used for any other mode of on-off control.
- *multiple line voltages.* This feature simplifies spare parts inventory if a particular type of lamp is installed in fixtures that are supplied with different voltages.
- *power leads for emergency lighting.* If there is momentary loss of power, all the HID lighting may be lost for several minutes while the lamps cool down to restart. For this reason, incandescent lamps may be installed as emergency lighting. Some ballasts include leads for powering emergency lamps.

If a ballast has taps for multiple input voltages, you may be able to use one of the taps to power emergency lighting. For example, you could use the 120-volt tap for emergency lighting if the ballast is operated at higher voltage. This avoids the need to run a separate 120-volt circuit to the emergency lamp. However, the emergency lamp current is limited by the current capacity of the ballast coil. Therefore, the amount of light available from the emergency lamp is substantially less than the light from the HID lamp, because incandescent lighting is less efficient than HID.

■ Additional Concerns for Electronic Ballasts

It seems inevitable that electronic ballasts will soon be widely available for HID lighting. When this occurs, they will bring the same additional selections concerns that presently apply to fluorescent electronic ballasts. These include harmonic distortion, radiated electromagnetic noise, power line noise, and initially low reliability. For a preview of these problems, refer to Measure 9.2.4

How to Select New Fixtures

In selecting new fixtures, use the same ballast criteria given previously. A new fixture will not provide efficiency or good color rendering unless you shop for these characteristics.

Be sure that you know the light distribution patterns of fixtures you select. If half of the fixture's light goes where you don't need it, overall lighting efficiency has been cut in half, regardless of what else you do.

Select fixtures for low light loss. Fixture specifications generally do not state the light loss, so estimate this by examining the fixture design. Most light loss in HID fixtures occurs in diffusers. Therefore, fixtures without diffusers are likely to be most efficient. Diffusers that create a great deal of diffusion are likely to be least efficient. Plastic diffusers absorb more light than glass, and they darken with age.

Don't Misuse Low-Pressure Sodium Lamps

Low-pressure sodium (LPS) lamps have exceptionally high lumens-per-watt ratings. This seduces people into using them where they do not belong. Remember that LPS lamps produce very poor visual quality. (Indeed, low-pressure sodium lighting provides an extreme example of the weakness of the lumens-per-watt method of judging lighting efficiency.)

What really matters to vision is the amount of light reflected from the scene, not the amount of light that falls on the scene. If an object does not contain the exact color emitted by the LPS lamp, the object appears black, i.e., it is not illuminated. This is why scenes lighted solely by LPS lamps appear black-and-white, and dim.

Another defect of the lumens/watt rating of efficiency is that vision is largely a matter of discriminating colors. LPS lighting fails completely in this regard because all its light is emitted at (essentially) one color. Even a colorblind person sees less well in LPS light. This is because reflection of light from an object is only at the wavelength of the sodium light, so there is no gradation of intensity provided by other colors.

Low-pressure sodium lighting is so poor in color rendering that it has virtually no applications as a replacement for other types of lamps. (The CRI of LPS lamps is typically listed as zero!) There are certain black-and-white applications, such as reading labels on cartons, where the superior lumens-per-watt efficiency of LPS can reduce lighting power. However, the monochromatic light is distressing to people.

In terms of overall lighting effectiveness, there are probably more cases where it is desirable to get rid of LPS lighting than to install it. This is even true for road and parking lot lighting, the most common applications for LPS lamps. The most likely replacement for LPS lamps is high-pressure sodium lamps of moderate CRI

rating. With the improved vision that results, it probably would not be necessary to increase the wattage of the lamps.

Label Your Fixtures to Deter Backsliding

When a lamp burns out, it is easy to replace it with the wrong type. The most effective way to prevent this is to label the fixtures with the lamp type. See Measure 9.7.3 for guidance in fixture labeling.

ECONOMICS

SAVINGS POTENTIAL: 40 to 65 percent of lamp power in typical applications.

COST: Metal halide and high-pressure sodium lamps cost 20% to 80% more per watt than mercury vapor lamps. The cost per lumen may be higher or lower than for mercury vapor. Metal halide lamps must be replaced more often than mercury vapor, increasing the lifetime lamp cost.

Ballast conversion kits for metal halide and high-pressure sodium lamps cost $100 to $300, depending on wattage. Entirely new fixtures are more expensive in retrofit, but perhaps not much more. Labor costs depends on local rates and the accessibility of the fixtures.

PAYBACK PERIOD: On or two years, in new construction. Several years, or longer, in retrofit. The payback period may be very long for smaller lamps. On the other hand, the value of improved color rendering may be important.

TRAPS & TRICKS

PLANNING: Make a thorough planning survey to determine the appropriate light levels for each area. Decide on the CRI that you really need, which may differ from one area to another.

SELECTING THE EQUIPMENT: Fortunately, you have many choices. Search the catalogs. Different manufacturers offer different features. Buy from major manufacturers, and check with other users before buying. There are wide variations in lamp and ballast quality. Be wary of new lamp types.

SPELL IT OUT: Label each fixture with the type of lamp that should be installed. Put a list of ballast and lamp types for each fixture or group of fixtures in the lamp stock room.

MEASURE 9.3.1.1 For lowest retrofit cost, replace mercury vapor lamps with metal halide or high-pressure sodium lamps that do not require ballast replacement.

Normally, each type and size of HID lamp needs a different ballast. This makes it expensive to replace mercury vapor lighting with metal halide or high-pressure sodium. To avoid this cost barrier, some models of metal halide and high-pressure sodium lamps have been developed specifically to replace mercury vapor lamps without the need to replace the ballasts.

These lamps are intended primarily to improve color rendering and light output. They are substantially more efficient than mercury vapor lamps, but not as efficient as mainstream metal halide and high-pressure sodium lamp models. As a price for ease of replacement, they compromise some other characteristics, including service life.

Energy Saving Potential

Generally, this substitution is not economical on the basis of energy savings alone. Although the replacement lamps have higher efficacy, most of this advantage appears as higher light output rather than lower input energy. Depending on the wattage of the original lamp, the maximum energy saving ranges from 5 to 20 percent.

The energy saving may be greater in uncommon cases where the improved light output of the replacement lamps allows less lighting wattage to be used elsewhere.

Where to Consider this Method

Consider this approach where you need greater light output and better color rendering. These lamps provide 40% to 75% more light output than the original mercury vapor lamps. Their color rendering is considerably better than that of mercury vapor lamps, but not as good as the color rendering that is provided by HID lamps that are designed for high color rendering.

Not many lamps of this type are presently on the market, so you may not be able to find the right wattage for upgrading your existing lamps.

How to Select the Lamps

The selection factors for this type of lamp are generally the same as for other HID lamps, as listed in Measure 9.3.1. The lamp manufacturer usually provides a list of the specific ballast models with which the lamps operate. The manufacturer may also provide a list of ballasts to avoid. However, manufacturers do not claim that these lists are complete, so you may have to call either the ballast manufacturer or the lamp manufacturer for further information.

SUMMARY

Provides more light output and better color rendering than the original lamp, but does not reduce energy consumption as much as lamps using metal halide or high-pressure sodium ballasts.

SELECTION SCORECARD

Savings Potential .... $ $ $

Rate of Return .... % % %

Reliability .... ✓ ✓ ✓ ✓

Ease of Retrofit .... ☺ ☺ ☺

The ANSI lamp designation system (described in Measure 9.3.1) does not help in selecting this type of lamp. The ANSI system defines compatibility narrowly for optimum efficiency and performance, whereas direct-replacement lamps involve serious performance compromises.

Label Fixtures to Prevent Backsliding

When a lamp burns out, someone may replace it with any type of lamp that fits the socket. To ensure that the right kind of lamp is used, label the fixture with the exact lamp model to be used. Measure 9.7.3 provides details about this important step.

ECONOMICS

SAVINGS POTENTIAL: 5 to 20 percent of the original lamp wattage. The increased lumen output and improved color rendering add substantial value in some applications.

COST: The replacement lamps cost 20 to 50 percent more than the original mercury vapor lamps. In addition, they have shorter service lives, which increases the lifetime cost of lamps.

PAYBACK PERIOD: Several years, or longer.

TRAPS & TRICKS

CHOICE OF METHOD: There are few applications for using this type of lamp purely as an energy conservation measure. Other Measures produce higher efficiency. Consider this approach primarily where you need more light, and light of better quality, at modest cost.

MEASURE 9.3.2 Install HID dimming equipment.

Dimming works by reducing the lamp current. The only HID dimming equipment with much application history is single-stage dimming. The first generation of this equipment has become notorious for failing after a relatively short period of service. This equipment still needs to mature. Electronic dimming, which promises a large turndown ratio while maintaining stable lamp operation and good energy efficiency, is not yet a proven technology for HID lighting.

Step Dimming

Step dimming works by inserting a capacitor in series with the lamp. The reactance of the capacitor reduces the lamp current by a fixed amount. Retrofit equipment is available that consists of the capacitor plus switching equipment to switch the capacitor into and out of the lamp circuit as dimming is needed.

Step dimming works only with transformer or autotransformer ballasts, especially CWA types. If the ballast is incompatible, one option is to replace the ballast to allow dimming. Ballasts with built-in switching are available for some lamp models. A more expensive option is to buy fixtures with built-in dimming capability.

Electronic Dimming

Electronic dimming of HID lamps is similar in concept to electronic fluorescent dimming. Electronic ballasts power the lamps with high-frequency pulses. This allows the lamp current and voltage to be controlled independently of each other, so that the ratio of voltage to current can be optimized for all operating conditions.

As of the mid-1990's, major U.S. ballast manufacturers were not offering electronic dimming, but some smaller companies were introducing units. The present situation is reminiscent of the early years of fluorescent electronic dimming, which had a long and troubled period of introduction to the market. While that history of troubles may not repeat itself, it seems

SUMMARY

Useful mainly for responding to occupancy, rather than exploiting daylighting. Lack of long-term reliability remains a problem.

SELECTION SCORECARD

Savings Potential	$ $ $
Rate of Return, New Facilities	% % %
Rate of Return, Retrofit	% % %
Reliability	✓
Ease of Retrofit	☺ ☺ ☺

prudent to wait until the technology of HID dimming has become more mature.

Applications and Alternatives

The most common application for step dimming is reducing lighting power when a space is unoccupied. Dimmers are available with matching occupancy sensors for this purpose.

Dimming is also a necessary part of daylighting. However, step dimmers cause abrupt changes in lighting, which annoys occupants. If and when electronic dimmers become proven, refer to Measure 9.2.6 for the selection factors. (They will probably be the same as for fluorescent dimmers.)

If the application offers the potential of major savings by exploiting dimming, consider replacing the HID lighting with fluorescent lighting (Measure 9.3.3). This change also allows the lights to be turned on and off at short intervals. It is especially worth considering if the existing HID lighting is inefficient.

Turndown Ratio

The maximum turndown ratio depends on the type of lamp and the type of dimming equipment. Simple step dimming, using a capacitor to limit lamp current, is claimed to provide a maximum turndown ratio of 70% with high-pressure sodium lamps, and a maximum turndown of 50% with metal halide lamps. Electronic dimming may allow higher turndown ratios, but it is too early to tell what the ultimate limits will be.

Efficiency

Dimming is a fairly efficient process in HID lamps, within limits. Reliable numbers are not available, but it appears that an efficacy loss of 10 to 20 percent can be expected near maximum turndown. This does not include any losses in the dimming equipment itself. Capacitor dimmers do not have much loss in themselves. Electronic dimmers may add about 10% loss, judging from experience with fluorescent dimmers.

Adverse Effects on Lamp Operation

Dimming is possible with all three types of HID lamps, but it imposes some constraints on lamp operation, including:

- *change in color and color rendering.* Metal halide lamps contain a variety of light emitting metals that differ in their response to changing lamp current. Therefore, metal halide lamps change color as they are dimmed. The severity of the change varies with lamp model and wattage.

  High-pressure sodium lamps lose their broad color spectrum at low current because more of the lamp energy goes to generating the characteristic monochromatic color of low-pressure sodium lamps. The lamp color changes from white to yellow, and color rendering suffers.

  Mercury vapor lamps do not experience much color change with dimming. However, their color rendering is poor in any event.

- *operating temperature.* Reducing the lamp current slows the processes that create and maintain the arc. These processes are also proportional to temperature, so lowering the current requires a higher minimum operating temperature. Electronic dimming may be able to avoid this problem by using higher voltage to compensate for lower temperature.

- *mounting position.* Reduced lamp current may allow the arc to wander closer to the lamp envelope, which quenches the arc. Therefore, there may be stricter limitations in lamp position.

Dimming Controllers

HID lamps must be started at full output so that the light-emitting metals will vaporize. Therefore, any controls used with dimmers should include a feature that starts them at full output.

HID lamps normally must cool down for an extended period before restarting. Therefore, the controller should have a feature that resets the lamp to full output after a shutdown, including shutdowns from brief power interruptions.

In the future, electronic modulating dimmers may be able to start lamps at any level of output. However, starting at low output extends the warm-up time, during which the lamp is operating inefficiently.

Dimmers can be controlled by signals from any appropriate source, such as manual controls, light level sensors, motion sensors, time controls, etc. Controllers for electronic dimmers usually provide an analog input signal to the ballast.

Daylighting is a Complex Application

If you really want to install HID dimming equipment to exploit daylighting, study Subsection 8.3 before venturing into this complex area. Dimming is only one part of daylighting.

ECONOMICS

SAVINGS POTENTIAL: *The maximum wattage reduction is 50 to 70 percent, depending on lamp type. The average reduction depends on the application.*

COST: *Step dimmers and appropriate controllers cost about $100 per lamp, typically. Installation may cost about the same amount. The cost of electronic dimming systems remains to be seen.*

PAYBACK PERIOD: *Several years, typically.*

TRAPS & TRICKS

CHOICE OF METHOD: *It's probably too soon for this. Try something else, or wait a while. If you want to gamble, try step controllers for the time being.*

MEASURE 9.3.3 In appropriate applications, substitute fluorescent lighting for HID lighting.

SUMMARY

Don't install HID lighting by rote. Some "typical" HID lighting applications can be served better with fluorescent lighting.

SELECTION SCORECARD

Savings Potential	$ $ $ $
Rate of Return, New Facilities	% % % %
Rate of Return, Retrofit..........	% % %
Reliability	✓ ✓ ✓ ✓
Ease of Retrofit	☺ ☺ ☺

In the past, HID lighting tended to be limited to applications that were clearly inappropriate for fluorescent lighting. However, there are now large areas of overlap in applications for fluorescent and HID lighting. This has led to one type being used in applications where the other type would serve more efficiently. Examine all your HID lighting to see whether you can save energy and cost by shifting to fluorescent lighting.

Reasons to Substitute Fluorescent Lighting

The following situations usually favor fluorescent lighting enough to provide a significant saving in energy consumption.

- *The space, or areas within the space, are occupied irregularly.* HID lamps take a long time to warm up from a cold state, and they may take even longer to restart if they are turned off when hot. This makes it impossible to save energy by turning off HID lighting to follow unscheduled or irregular occupancy. In contrast, fluorescent lighting can respond instantly.

  Applications where this is important include gymnasiums, loading docks, parking garages, localized areas within warehouses, maintenance shops, etc.

  One way to improve the efficiency of HID lighting in these situations is to install dimming equipment, as recommended by Measure 9.3.2. However, the energy saving is limited to about half of full power.

- *The application would benefit from dimming or variable lighting levels.* Fluorescent lighting offers two efficient method of adjusting light output, switching individual lamps in fixtures and electronic dimming. In comparison, present HID dimming technology has limited range, it causes color changes in metal halide and sodium vapor lamps, and it still has a poor reliability record.

- *Color rendering is important.* HID lamps can now provide CRI's that are almost as high as those of the best fluorescent lamps. However, HID lighting pays a large efficiency penalty to get good color rendering. In contrast, the fluorescent lamps with the best color rendering also have the highest efficiency. Color rendering is important in a large fraction of all lighting applications, including applications that traditionally use HID lighting.

- *HID lamps have too much light output for efficient light distribution.* HID lamps must be large to be efficient. This means that they illuminate large areas. An individual HID lamp with enough wattage to be efficient typically illuminates an area of 200 square feet (20 square meters) or more. In contrast, an individual fluorescent lamp may efficiently illuminate an area of 10 square feet (one square meter).

  If the layout of spaces is irregular, or if illumination requirements vary between areas, efficient light distribution requires a larger number of smaller lamps. However, smaller HID lamps and ballasts are substantially less efficient than good fluorescent lighting. Also, if the fixture must be close to the viewer, fluorescent lighting is more pleasant.

- *The installed HID lamps are inefficient.* In retrofit applications, consider fluorescent lighting as a replacement for inefficient HID lighting. Of course, you can also replace inefficient HID with more efficient HID lighting, as recommended by Measure 9.3.1.

Reasons to Keep HID Lighting

It is not always feasible to substitute fluorescent lighting for HID, even in cases where it might provide a large energy saving. The following situations make it difficult or impractical to substitute fluorescent lighting.

- *The space operates at low temperature,* either continuously or occasionally. The optimum lamp temperature for fluorescent lamps is about 100°F (38°C). Efficacy and light output decline rapidly below this temperature. Fluorescent lamps can be kept warm by using tight enclosures, but this creates temperatures that are too high when the ambient temperature rises. Also, special ballasts are required to start fluorescent lamps in cold conditions. In contrast, HID lighting is tolerant of wide temperature variations.

- **The fixture environment is dirty.** Incandescent lamps and fixtures have large surface areas in relation to the amount of light they produce. If the environment is dirty, the cost of cleaning the lamps and fixtures to maintain efficiency may be prohibitive.
- **Fixture installation cost is high.** It takes a number of fluorescent fixtures to provide as much light as a single large HID fixture. If installation costs are high, retrofitting fluorescent lighting is unlikely to be economical.
- **Lamp maintenance is difficult.** The labor cost for lamp replacement may be prohibitive if fixtures are hard to reach. Consider whether you can install fluorescent fixtures at lower heights, making lamp replacement easier. This possibility is explored further, below.
- **It is difficult to locate the fixtures to provide efficient light distribution.** HID fixtures are typically installed in tall ceilings. Their ability to focus sharply makes them well adapted to this geometry. In contrast, conventional fluorescent fixtures have broad light distribution patterns. As a result, they cannot exploit their ability to illuminate small areas efficiently if they are installed high above the task areas. Fluorescent lighting in tall ceilings can be efficient only for illuminating large areas, such as gymnasiums. See Figure 1, for example. In spaces with tall ceilings, consider whether you can install the fluorescent fixtures closer to the task areas, as discussed below.

Comparative Lamp and Ballast Efficiencies

If none of the previous factors tilt your decision strongly toward either fluorescent or HID lighting, make your decision largely on the basis of comparative lamp and ballast efficiencies. The efficiency of HID lamps is inversely related to their CRI, and higher CRI increases the cost of both HID and fluorescent lamps. Bearing this in mind, shorten your analysis at the outset by deciding the lowest CRI that is acceptable for the application.

It is widely believed that HID lighting is more efficient than fluorescent lighting. Table 1 of Reference Note 56, HID and LPS Lighting, shows that this is not true in general. The table shows that HID lamps with good color rendering may be much less efficient than the best fluorescent lamps. Only a few HID models combine efficiency and color rendering as well as the best fluorescent lamps. Only certain models of high-pressure sodium lamps have efficacies that are much higher than the best fluorescent efficacies, and they achieve this at the expense of color rendering.

The common, inexpensive 40-watt cool white fluorescent lamp, which has a CRI of 62, has an efficacy of 78 lumens/watt when driven by a magnetic ballast, and about 85 lumens/watt when driven by an electronic ballast. Comparable "high-efficiency" fluorescent lamps achieve efficacies near 100 lumens/watt with electronic ballasts, and they have CRI's as high as 85.

Consider the effects of ballasts when making your efficiency comparison. (See Table 2 of Reference Note 56, HID and LPS Lighting.) HID ballasts consume a

WESINC

Fig. 1 Fluorescent lighting in a gymnasium It is possible, and it produces good illumination on short notice. The activity here is wrestling, which poses no hazard to the fixtures.

much larger fraction of the input energy than the best fluorescent ballasts, except with very large HID lamps. In the middle range of HID lamp wattages, ballast losses largely negate the efficiency advantage that some HID lamps have over fluorescent lamps.

Fluorescent lighting still has another important efficiency advantage related to ballasts. The efficacy ratings of both fluorescent and HID lamps assume that they are operated from magnetic ballasts. However, the electronic ballasts that are available for fluorescent lighting improve fluorescent lamp efficacy by about 10% over their published lumens-per-watt figures. In fact, some high-efficiency electronic ballasts may save more energy in the lamp than they consume themselves.

A Special Case: Compact Fluorescent Lighting

Compact fluorescent fixtures are still evolving in efficiency and variety. At present, there are not many applications where you would consider compact fluorescent lamps as a replacement for HID lighting. One reason is that their efficiency is lower than the efficiency of conventional fluorescent lighting. Another is that they produce less light output per fixture, so it would take many compact fluorescent fixtures to substitute for one large HID fixture.

On the positive side, compact fluorescent fixtures are not as sensitive to space temperature, and they can be focussed more tightly. Because they are smaller, they allow even more localized control of lighting.

How to Select Fluorescent Lighting

The main selection factors for fluorescent lighting equipment are:

- *ballast efficiency.* See Measure 9.2.4 for the full story on fluorescent ballasts.
- *ballast dimming capability.* Decide whether you want dimming capability. If so, the only satisfactory method is using electronic ballasts that are specialized for this purpose. Refer to Measure 9.2.6 about fluorescent dimming.
- *lamp efficiency.* Coordinate lamp selection with ballast selection. Refer to Measure 9.2.3 for lamp characteristics. Unlike HID lamps, fluorescent lamps do not sacrifice other performance characteristics to achieve efficiency (except for needing slightly higher minimum operating temperature).
- *diffuser efficiency.* Fluorescent fixtures typically have larger diffuser losses than HID fixtures. If appropriate, select fluorescent fixtures that do not have diffusers. Otherwise, refer to Measure 9.7.2 about diffuser selection.
- *light distribution pattern.* The light distribution geometry of fluorescent lighting is radically different from the geometry used with HID lighting. Design your layout from scratch, rather than trying

to mimic the light distribution pattern of the HID lighting. The localized light of fluorescent lamps provides an important opportunity for task lighting. Refer to Subsection 9.6 for guidance in lighting layout.

Make it Easy to Clean and Replace Lamps

Design your fluorescent lighting to ease lamp maintenance. Try to compensate for the larger number of fluorescent fixtures by installing them at a height that is convenient for maintenance. The fact that HID fixtures are installed at great heights does not necessarily mean that fluorescent fixtures must be installed at the same height. In fact, one of the advantages of fluorescent fixtures is that they can be installed close to people.

In many facilities, you can suspend fluorescent fixtures from the ceiling at any desired height. If you install a suspended ceiling to save heating energy, you can lay fluorescent fixtures into the suspended ceiling.

If fork lift trucks, traveling cranes, or other interferences make it impractical to suspend the fixtures from overhead, consider mounting them on equipment, on wall brackets, or even on floor stands. This is an opportunity to innovate.

Aiming Fluorescent Fixtures

Fluorescent fixtures do not have to aim straight down. If you tilt a fixture that uses long tubes, try to keep the tubes nearly horizontal. Too much tilt in long tubes may cause the mercury to accumulate at the lower end under some circumstances, making it difficult to start.

Be aware that non-vertical orientation can cause serious glare for people facing the fixture. See Measures 9.6.2 ff, which cover task lighting, for guidance in dealing with this problem.

Lighting Controls

Installing controls to minimize lighting operation is an integral part of this Measure. Refer to Subsections 9.4, 9.5, and 9.6 for control techniques.

ECONOMICS

SAVINGS POTENTIAL: *30 to 60 percent of lighting energy consumption, in appropriate applications.*

COST: *Fluorescent fixtures typically cost $0.20 to $1.00 per watt of capacity. Installation costs are similar. Lamp replacement cost, per lumen, is much lower than for HID lamps, except for the largest HID lamps. An important difference is the larger numbers of fluorescent fixtures that are required. This increases the costs of fixtures, fixture mounting, and wiring.*

PAYBACK PERIOD: *One year to several years, in new construction. Several years or longer, in retrofit.*

TRAPS & TRICKS

CHOICE OF METHOD: *This Measure is mostly a matter of avoiding or correcting inappropriate HID installations. These are most likely to occur where HID lighting cannot respond efficiently to irregular schedules or to variable illumination requirements.*

SELECTING THE EQUIPMENT: *Take full advantage of the opportunity to maximize efficiency, reliability, color rendering, and other features. Review Reference Note 55, Fluorescent Lighting. Plan to exploit the Measures in Subsection 9.2.*

Lighting Controls, Manual

The Measures in this Subsection recommend manual lighting controls that enable people to use lighting efficiently, and encourage them to do so. These Measures can be accomplished easily, with little or no change to the connections between the controls and the fixtures. However, in existing facilities, you may need to relocate or rewire lighting controls to make it possible for people to use them effectively. See Measures 9.6.4 ff and Measure 9.6.5 for these aspects of your improvements.

Understand that manual control is unreliable for saving energy, especially in commercial and industrial environments. Advances in technology and lower equipment prices have greatly expanded the opportunities for automatic lighting control, which is covered in Subsection 9.5. Examine all your manual lighting controls to see whether you can replace or supplement them with automatic controls.

A NOTE ON TERMINOLOGY

To avoid confusion, be aware that lighting specialists use the word "control" to mean shaping the light distribution pattern. In this context, a parabolic reflector is a "control" device. However, "control" is used in this Subsection in the same way that it is used for other equipment, namely, turning things on and off, changing levels of output, etc.

INDEX OF MEASURES

9.4.1 Install effective placards at lighting controls.

9.4.2 Use security forces, watch engineers, or other regularly assigned personnel to keep unnecessary lights turned off.

9.4.3 Install all single-pole toggle switches so that the toggle is down when the switch is off.

9.4.4 Replace rheostat dimmers with efficient electronic dimmers.

9.4.5 Where fixtures are not easily visible from the switch locations, install telltale lights.

9.4.6 Draw attention to switches that should be used in preference to others.

9.4.7 In applications where fixtures may be operated improperly by unauthorized personnel, use key switches.

RELATED MEASURES:

- Measure 9.1.7, incandescent dimmers
- Measure 9.2.6, fluorescent dimmers
- Measure 9.3.2, HID dimmers
- Subsection 9.5, for automatic controls that can be used in combination with manual controls
- Measure 9.6.4, for circuit changes that improve control layout
- Measure 9.6.4.1, programmable lighting controllers
- Measure 9.6.5.1, pullcord light switches

MEASURE **9.4.1 Install effective placards at lighting controls.**

SUMMARY

A fundamental part of lighting energy conservation. It takes effort to earn results.

SELECTION SCORECARD

Savings Potential	$ $ $ $
Rate of Return, New Facilities	% % % %
Rate of Return, Retrofit.........	% % % %
Reliability	✓ ✓ ✓
Ease of Retrofit	☺ ☺

People need information to control lighting efficiently. Placards are an effective and inexpensive method of providing this information. Using placards is a powerful conservation measure that remains to be exploited in most facilities. However, even where placards have been used, they have not yet enjoyed great success. They require more effort to design and install than you probably expect. To be effective, placards must attract attention, they must provide enough information, and they must be easy to comprehend. This Measure tells you how to get the most benefit from lighting control placards.

The effectiveness of placards varies considerably in different environments. You get the best results in facilities where there a high level of personal responsibility, where occupant turnover is low, and where the space layout allows a person at the switch locations to see the other people in the space.

In most environments, placards can be effective in keeping excess lights from being turned on. Placards are less successful in getting people to turn lights off, especially in spaces with multiple occupants.

The only effective alternative to installing placards for manual controls is installing automatic controls, which is covered in Subsection 9.5. However, even automatic lighting controls may require placards.

WESINC

Fig. 1 An effective placard for the environment This placard works because it is big and because it is located in the right place. The light switches are next to the door handle, where they belong, so the user has all the needed information. In contrast, see Figure 3, where the amount of information given by this placard is inadequate.

Where to Install Placards

Install instruction placards at virtually all manual switches and dimmers.

Why Placards Are Important

Lighting control placards serve two distinct functions: (1) when a person enters the space, they provide information about how to get the desired lighting without turning on unnecessary lamps, and (2) when a person leaves the space, they motivate and grant permission to turn off the lights.

When a person enters a space, he needs to know which switches and dimmers control the lighting in the part of the space that he will be occupying. If this information is not immediately apparent, the person is likely to turn on more lighting equipment than he needs.

The second function is primarily a matter of overcoming inhibition. Most people realize that leaving lights turned on in a vacant space wastes energy, and they would like to save energy. However, people may not be certain which parts of the space are controlled by which switches. They may fear criticism for changing the lighting. And, people wish to avoid causing inconvenience to others. Except in their own homes or in personal spaces, people feel that they lack permission to turn off unnecessary lighting. Placards provide both the needed information and the permission to act.

How to Design Placards

Expect to spend money and effort to create effective lighting control placards. The key to success is customizing each placard to the particular space and lighting equipment. Generic "turn off the lights" signs are of little value, except in single-person spaces.

■ Make Placards Easy to Use

Make placards conspicuous. Make them easy to understand. Locate each placard where it has an obvious relationship to the location of the controls. Figure 1 shows a crude but effective placard that illustrates these principles.

(By the same token, the controls themselves must be accessible. See Measures 9.6.5 ff about this.)

Words are cheap, so don't be cryptic. For example, instead of labeling the light switch for a kitchen hood as "hood," say "This switch controls the lights inside the hood." Just saying "Hood" might mean that the switch starts a fan in the hood, or that it starts the stove, or that it activates a hood fire alarm, etc.

■ Tell Users What They Need to Know

Be complete and specific. The placard should:

• *identify the controls,* if this is not readily apparent ("This switch controls the lights in the men's restroom.") Do not assume any prior knowledge. For example, a person cannot tell that an unmarked toggle switch is even a light switch (it might start a fan, trigger an alarm, etc.).

• *identify the fixtures that are controlled* (for example, "chandeliers," "fluorescent fixtures," "perimeter wall light dimmers," "the row of downlights in front of the projection screen," etc.) Figure 2 illustrates this aspect of placards. Without this, a person cannot guess which light fixtures are controlled by each switch.

• *tell what to do* ("Please turn off the lights ...") However, saying this alone is usually useless. See Figure 3 as an example.

• *say when to do it* ("... when leaving the room," "... when the blinds are open," etc.)

• *alert occupants to automatic lighting controls* ("The row of fixtures adjacent to the windows is controlled by an automatic dimmer that responds to daylight. This switch allows the fixtures to be turned off when the room is vacant.")

• *tell occupants whom to contact if the controls fail.* ("Call Extension 329 if ...") Otherwise, an inefficient condition may continue indefinitely.

Specific Situations

The following are guidelines for some common lighting control situations. Combine them as appropriate.

■ All Lighting is Controlled by a Single Switch

This is the easiest case. Install a placard asking the last person leaving the space to turn off the lights.

■ Lights are Controlled from Multiple Locations

If lighting is controlled from more than one switch location, install the appropriate placards at each switch.

■ There is a Group of Switches

When a person encounters an unmarked group of light switches, the common response is to turn them all on. Users also are inhibited from turning off lights

selectively because of the effort or embarrassment that is involved in experimenting to determine which switches to use.

The solution is to label each switch to indicate the fixtures that it controls. Label the switches in a clear, permanent manner. Make the lettering big enough for people to see without their reading glasses. The placard should also request occupants to turn on a minimum number of fixtures ("Use only as much light as you need.").

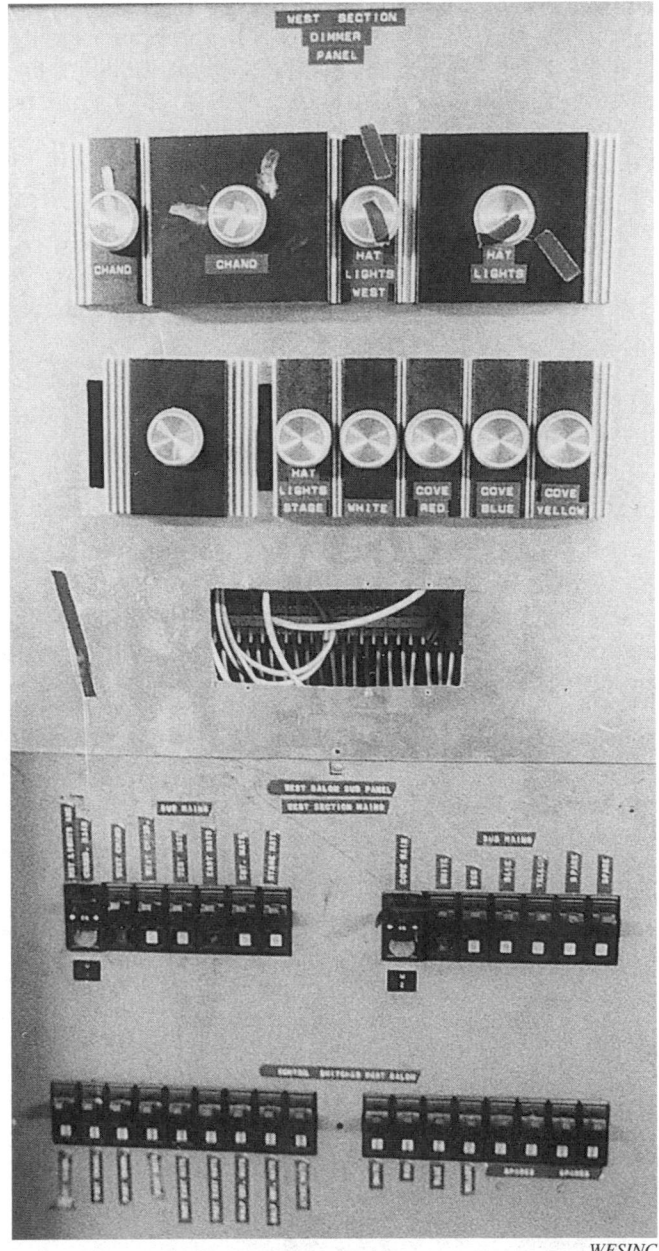

WESINC

Fig. 2 Lighting control panel for a large ballroom
This marking performs the basic function of identifying each control. It is fairly permanent. Pieces of tape are used on the top three dimmers to suggest brightness settings, but this is not explained. The lower left master dimmer controls a lot of electricity, but it is not marked. There is no guidance about how to set the lighting so that it is most pleasing and efficient for all the functions that may occur in the space.

■ The Space has Daylighted Areas

Many spaces have windows that provide effective daylighting for a portion of the space. Often, switches are installed in such spaces to provide separate control of fixtures near the windows. In addition to labeling each switch, ask occupants not to turn on the lights in daylighted areas when daylight is available.

■ Light Switches are Not Located Near Exits

The common and expected location for light switches is on the wall adjacent to the knob side of the door. If the light switch is not located there, users will not search for it. In such cases, install an appropriate placard at the switch and also at the exit, to direct the user to the light control. ("Please turn off the bathroom lights when you leave. The switch is located outside on the right.")

■ Fixtures are Not Visible from the Switches

It is poor practice to install switches and dimmers where users cannot see the fixtures they control. In such cases, the preferred solution is to relocate the controls (see Measure 9.6.5).

If this is not practical, label the switches to indicate which fixtures they control. ("This switch controls the parking lot lights.")

In this situation, it is also valuable to install telltale lights at the switches (see Measure 9.4.5).

■ The Spaces have Emergency Lighting

People are inhibited from turning off lights in large spaces because they are concerned that others may be trapped in the dark. If spaces have continuous emergency lighting, the placard should say so. ("Please

WESINC

Fig. 3 Useless placard This sticker tells us to turn off the lights. So, now what? These switches are in an obscure location. Which spaces do they control? Which fixtures does each switch control? How do we know if there are other people still in the space?

turn off all the light switches when you leave. The emergency lights will remain lighted.")

■ Some Lighting has Automatic Controls

If some of the lighting has automatic controls (e.g., timeclocks, photocontrols, motion detectors, interval timers), describe the operation of the automatic controls at the location of the manual switches. The purpose is to keep the automatic controls from confusing people, as this discourages them from taking initiative to use manual switches.

For example, in a room where daylighting is available, you might install a photoelectric switch to override the manual switch for the fixtures adjacent to the windows. However, this has the effect of making the manual controls incomprehensible. To correct this problem, explain what the photocontrol is doing. On the placard for the switch group, highlight the switches for the daylighted area and say that they alone have automatic control.

■ The Lighting has Multi-Level Control

Where separate switches are used to control different numbers of lamps within fixtures, this capability is often ignored by the occupants. The solution is to label each switch to state the brightness level ("normal brightness", "extra bright", etc.) The placard should also request occupants to turn on the minimum number of lamps in each fixture ("Use only as much light as you need.").

■ Programmable Lighting Controllers are Installed

Refer to Measure 9.6.4.1 about programmable lighting controllers. These are used with low-voltage relay control systems. The controller should have clear instructions stating, as a minimum:

- what the available lighting patterns are, using diagrams if necessary
- when each lighting pattern should be used
- how to press the buttons or turn the dials to select the desired pattern. This is important because some programmable controls are user-hostile.

■ Dimmers are Installed

People habitually turn up dimmers to about two thirds of full power, regardless of the actual illumination requirement. This may result from the insensitivity of human vision to relative differences in lighting levels, and from the long time required for humans to adapt to lower light levels. It may also reflect a desire by people to make use of available lighting capacity. In any event, install dimmer placards that say something like, "To save energy, set the dimmer to the lowest level you need."

People tend to be careless in turning off dimmers. The lights may seem to be turned off, but they continue to operate at low power during unoccupied hours. Manufacturers contribute to this problem by using click stops in the off position of many dimmers. Users tend not to move the dimmers forcefully enough to overcome these stops unless they are told to do so. This makes it important to customize the placard to the particular dimmer configuration (slider, rotary knob, etc.). For example, "To turn off the lights, push the knob down past the click."

Wherever it is practical to do so, label the faceplates of the dimmers to tell occupants the recommended settings for each type of activity. For example, in a hotel ballroom, label different levels as "banquet", "dance", "slide projector", etc.

How to Make Effective Placards

See Reference Note 12, Placards, for the details of effective placard design, materials, and installation.

Make placards from materials appropriate for each application, considering factors such as durability, visibility, and cost. For example, inexpensive adhesive labels applied over switch plates are effective and economical in applications where the only function of the placard is to remind occupants to turn off the lights. At the other extreme, instructing people how to use complex lighting controls usually requires custom placards and special materials, such as photographically printed metal. In all cases, make placards durable enough to survive the environments where they are installed.

Appearance is Important

Facility managers may resist installing placards in visible locations because of their previous experience with unattractive energy conservation stickers, such as the ones given away by utility companies and government agencies. The solution is to design the placards to fit the decor.

For example, in designing placards for the guest rooms of a luxury hotel, use a complementary color scheme and attractive lettering. Invite the hotel's interior decorator to participate in the design.

Coordinate with Other Energy Conservation Placards

Energy consuming equipment throughout the facility should have instruction placards to motivate efficient operation. Placards for different types of equipment are covered in the Measures dealing with those types of equipment. Placard design is an intensive activity, so try to do it throughout the facility at one time. This saves effort and gives you an opportunity to combine information. For example, use a single placard to ask occupants to turn off both the lights and the air conditioning when departing.

Introduce the Placards with Fanfare

Placards are effective for providing information. However, passive objects such as placards are not very effective in getting people to change their behavior. Personal contact is the best way to do this. When you install placards, ask managers to explain them to occupants and to ask everyone to take the indicated actions to save energy.

To keep placards effective, make them part of a continuing energy conservation motivational program that includes posters, newsletters, talks by managers, etc.

ECONOMICS

SAVINGS POTENTIAL: *10 to 60 percent of the lighting energy under manual control. Lamp cost and replacement labor may be reduced by a similar amount.*

COST: *$0.10 to $50 per placard, depending on quantity, materials, and complexity. Expect high placard design costs for special situations.*

PAYBACK PERIOD: *Less than one year, usually.*

TRAPS & TRICKS

DESIGN: *Designing effective placards takes a lot of time and effort. Don't take shortcuts. Read Reference Note 12.*

MATERIALS: *Your choice of placard material matters. See Reference Note 12.*

INSTALLATION: *Your choice of installation methods matters. See Reference Note 12.*

KEEP IT UP: *Placards are part of a system of communication. Keep the whole system alive, in a manner appropriate to your environment.*

MEASURE **9.4.2 Use security forces, watch engineers, or other regularly assigned personnel to keep unnecessary lights turned off.**

A powerful way of reducing energy costs is to use the facility staff to turn off excess lighting, as well as to perform other energy management functions. Eligible personnel may include security police, watch engineers, or anyone else who is assigned to the facility on a regular basis.

If the facility has security police, these may be the best people to carry out this responsibility. They usually are present for the largest number of hours per day, and they have access to most spaces. In fact, making police responsible for monitoring lighting makes them more alert to security conditions.

Also, consider assigning engineering personnel to control lighting. Engineers typically have access to most of the facility. However, they are less inclined toward patrolling.

Alternatively, if spaces are under the administrative control of a particular group, those groups may have individuals who can be assigned to turn off unnecessary lighting. For example, ask the executive chef to be responsible for turning off lights in the kitchen and dining areas.

In some facilities, the savings potential may be large enough to justify paying overtime, or hiring an additional person, to monitor the after-hours operation of lighting and other energy consuming equipment.

Where to Use This Method

This method of turning off lights is desirable in facilities that have some degree of irregular occupancy. Examples are office buildings, laboratories, recreational facilities, etc.

Where suitable personnel are available for lighting patrols, this Measure may cost little or nothing, making it the most economical method of lighting control. In some environments, it may also be the most efficient method. For example, in large spaces occupied by many people, both occupant control and automatic control may be impractical.

Using facility personnel to control lighting may be valuable in combination with other control methods. It also serves as a check on other methods. No manual method of lighting control is completely effective, and automatic control methods should be checked periodically for proper operation.

Provide Effective Instructions

Achieving success with this Measure is largely a matter of effective personnel management. This starts with effective instruction at the outset, and continues with effective supervision. To provide motivation, explain the high cost of lighting.

Monitor all spaces with manual lighting control, except where special circumstances exist. There is rarely a reason to keep lights turned on in a vacant space.

Look for unnecessary operation of lighting both during and outside of normal hours of occupancy. After hours, lights are turned on by late workers and by cleaning crews. During normal working hours, lights may be left on by people who leave for activities elsewhere.

The placards recommended by Measure 9.4.1 are a valuable way of instructing lighting patrols as well as the regular occupants. Design them accordingly.

Combine with Other Energy Conservation Activities

Don't limit this activity to lighting. The same people who monitor lighting can also be assigned to other energy conservation functions, such as turning off air handling units (Measure 4.1.1.5) and room conditioning units (Measure 5.1.2), closing windows and doors, etc.

ECONOMICS

SAVINGS POTENTIAL: *A significant fraction of all facility lighting energy. Lamp cost and replacement labor may be reduced by a similar amount.*

COST: *Minimal, if regularly assigned personnel are used.*

BENEFIT-COST RATIO: *Usually high.*

TRAPS & TRICKS

See Measure 4.1.1.5, which uses the same technique for control of air conditioning equipment.

MEASURE 9.4.3 Install all single-pole toggle switches so that the toggle is down when the switch is off.

People in the United States expect lights to turn off when the switch is pushed down. In some other countries, it is customary for the up position to be off. If a switch is installed the wrong way, people tend to leave those lights turned on. Furthermore, reversed switches interfere with the habit of turning lights off.

If a switch is installed upside down, simply remove it from the junction box and turn it over. This takes only a few minutes.

This Measure applies only to single-pole switches, i.e., switches that control lights from one location only. Switches that are designed to control fixtures from multiple locations do not have defined on and off positions. These switches are either "3-way" or "4-way" switches. To preserve energy conserving habits, install 3-way and 4-way switches separately from single-pole switches, and mark them with distinctive placards.

SUMMARY

An easy fix for a common flaw that promotes energy waste.

SELECTION SCORECARD

Savings Potential .... $ $

Rate of Return .... % % % %

Reliability .... ✓ ✓ ✓ ✓

Ease of Retrofit .... ☺ ☺ ☺ ☺

ECONOMICS

SAVINGS POTENTIAL: *Small and unpredictable.*

COST: *Minor.*

PAYBACK PERIOD: *Short.*

MEASURE 9.4.4 Replace rheostat dimmers with efficient electronic dimmers.

Rheostat dimmers are found with incandescent lighting in older facilities. Rheostats are large variable resistors in series with the lamps. The rheostats absorb power in the same way as the lamps themselves.

You can eliminate most of this energy waste by replacing the rheostats with modern electronic dimmers. These absorb little energy. See Measure 9.1.7 for details about electronic dimmers. They do have some limitations.

Energy Saving Potential

The greatest amount of energy is wasted by rheostats when they reduce the lamp power to half, at which point they absorb as much energy as the lamps themselves. This is an amount equal to one fourth of the rated lamp capacity. At lower light output, rheostats absorb an even larger fraction of the total energy consumed, although they consume less energy in absolute terms.

Where to Upgrade Dimmers

It is almost always economical to get rid of rheostats if the lights are dimmed for long periods of time. On

SUMMARY

Usually a simple and inexpensive change. However, there are some applications that still require rheostats.

SELECTION SCORECARD

Savings Potential .... $ $

Rate of Return .... % % % %

Reliability .... ✓ ✓ ✓ ✓

Ease of Retrofit .... ☺ ☺ ☺ ☺

the other hand, rheostats waste little energy in applications where they quickly change the lighting from full on to full off, or vice versa. For example, if rheostats are used in a theater to dim out the house lights, they waste energy for only a few minutes per year. In such cases, keep them.

Electronic dimmers can replace rheostats in most cases. However, they create some electrical interference and faint audible noise. Therefore, you may have to retain rheostats in sensitive applications, such as laboratories with sensitive electronic equipment.

Alternative: Fluorescent Dimming

If incandescent lighting is still being used because the application requires dimming, consider substituting fluorescent lighting that has dimming capability. See Measure 9.2.6 for this option.

ECONOMICS

SAVINGS POTENTIAL: Varies widely. Don't expect to save much money in applications where dimmers are used only for brief periods.

COST: Less than $1 per ampere in common styles. Large dimming systems may be much more expensive.

PAYBACK PERIOD: Varies widely, depending mainly on the length of time that lights operate in a dimmed state.

TRAPS & TRICKS

SELECTING THE EQUIPMENT: Be wary of cheap dimmers. Make sure that the dimmers you buy do not cause radio interference or audible noise. Good dimmers have become inexpensive, if you know how to shop.

MEASURE **9.4.5 Where fixtures are not easily visible from the switch locations, install telltale lights.**

Lighting is likely to operate unnecessarily if it is not visible from the switch location. For example, this is a common problem with outside lighting.

To reduce this energy waste, install a telltale light at a location that is visible at the switch. The telltale light is simply a small light that is wired in parallel with the fixtures being controlled.

The easiest way to install a telltale light is to install a switch that is combined with a telltale light. These are available in different styles. A common type consists of a toggle switch and a separate neon light. It is installed in a standard double receptacle box. Figure 1 shows this type of telltale light.

Fig. 1 Telltale light The 3-way switch on top controls an outdoor light from several locations. The light is not visible from this location. The telltale reveals whether the light is on. In the bottom group of switches, the dark toggle is for one 3-way switch among a group of single-pole switches. All these switches need to be identified individually.

WESINC

SUMMARY

An important switch feature for fixtures that are out of sight. Cheap with single-location switches, more expensive with multiple locations.

SELECTION SCORECARD

Savings Potential	$	$	$
Rate of Return, New Facilities	% % %	%	
Rate of Return, Retrofit	% %	%	
Reliability	✓	✓	✓
Ease of Retrofit	☺	☺	☺

An even more compact arrangement is a switch that has a lighted toggle. This type fits in a single-switch connection box.

Installation is simple and cheap if the lights are controlled from a single location. No change in wiring is needed, outside the connection box.

You can install any number of telltale lights for a given fixture. If lighting is controlled from multiple locations, you need a telltale light at each switch. This is especially important because the position of 3-way and 4-way switches (up or down) does not indicate whether the fixture is on or off. Installing telltale lights at multiple locations involves a significant wiring cost, because each telltale light needs its own wiring run.

If the switch itself is installed out of sight, install an extra telltale light where it can attract attention to the switch. In such cases, install an appropriate placard at the remote telltale (see Measure 9.4.1).

ECONOMICS

SAVINGS POTENTIAL: A substantial fraction of the energy used by the affected fixtures. Lamp cost and replacement labor may be reduced by a similar amount.

COST: $10 to $30, if no additional wiring is required. With fixtures controlled from multiple locations, the additional wiring for the telltale lights may be expensive.

PAYBACK PERIOD: Less than one year, to several years.

TRAPS & TRICKS

ANALYZE THE PROBLEM: Is the main problem that the switch is inaccessible or out of sight? In such cases, relocate the switch, but still use a telltale light if appropriate.

MEASURE 9.4.6 Draw attention to switches that should be used in preference to others.

Groups of switches are a problem for efficient lighting control because they make it difficult for occupants to associate an individual switch with the lights that the switch controls. The primary method of getting people to use groups of switches properly is installing an effective placard that points to each switch and identifies it. See Measure 9.4.1 for this step.

An additional step is important if one switch should be used in a manner different from the rest. For example, draw attention to a switch that controls night transit lighting, to keep people from turning on the other lights in the space.

Here are some ways of drawing attention to a particular switch in a group:

- *use a contrasting toggle color.* For example, install a dark brown toggle in a group of ivory-colored switches. Figure 1 in Measure 9.4.5 shows an example.

- *use a lighted toggle.* Switches with lighted toggles are a direct replacement for conventional toggle switches. You can also get lighted toggles on some models of personnel sensor switches (Measure 9.5.4) and rundown timer switches (Measure 9.5.5).

- *install a decal around the toggle.* Design a sticker to attach to the switch plate, with a hole for the toggle. You can print a message on the sticker. Be aware that a sticker will eventually come off when the adhesive dries out, and that rubbing from using the switch will wear through the sticker.

- *install the switch alone or in a separate group.* This is more expensive, especially to retrofit. However, it is a powerful way of setting a switch apart. See Figure 1 in Measure 9.4.5 for a typical example. This method is particularly valuable for separating 3-way and 4-way switches from single-pole toggle switches. This is because the position of a single-pole switch tells whether it is on or off,

SUMMARY

An important ergonomic feature that works in combination with an effective placard. Inexpensive and reliable.

SELECTION SCORECARD

Savings Potential	**$ $**
Rate of Return, New Facilities	**% % % %**
Rate of Return, Retrofit	**% % % %**
Reliability	✓ ✓ ✓
Ease of Retrofit	☺ ☺ ☺ ☺

but the same is not true for 3-way and 4-way switches. (See Measure 9.4.3 about this.)

You may be able to think of other effective methods. Use your imagination.

ECONOMICS

SAVINGS POTENTIAL: *A significant fraction of the energy controlled by switches other than the lighted switch. This also reduces the cost of lamps and replacement labor.*

COST: *Lighted toggle switches cost several dollars apiece. In retrofit, replacement labor adds $10 to $30 per switch.*

PAYBACK PERIOD: *Less than one year, typically.*

TRAPS & TRICKS

BE BOLD: *Installing the lighted toggle switch is easy. The part that takes effort is designing an effective placard for the whole group of switches. Don't be inhibited. For example, maybe install a big, bold arrow pointing to the lighted toggle, with a sign that says, "Use only the lighted switch when passing through the space."*

MEASURE **9.4.7 In applications where fixtures may be operated improperly by unauthorized personnel, use key switches.**

Using key switches is an absolute method of controlling lighting. Its major advantage is that it avoids the need to depend on occupants to control lighting. Low cost is another advantage, compared to other control methods. Figure 1 shows a typical installation.

A serious limitation is that an individual with the proper key always needs to be available to turn the lights on and off. Key switches make it awkward to use spaces outside normal hours of occupancy. As a result of the difficulty that they cause, key switches tend to be abandoned. Use them only where you cannot find a better method.

In most applications where the purpose of lighting controls is to keep unauthorized persons from turning on lights, consider automatic control (see Subsection 9.5). You can combine these with key switches. For

SUMMARY

Avoids the uncertainty of occupant behavior. Inexpensive. Easy to retrofit. But, too inflexible for most applications.

SELECTION SCORECARD

Savings Potential .... **$ $ $**

Rate of Return, New Facilities **% % % %**

Rate of Return, Retrofit.......... **% % % %**

Reliability .... ✓ ✓

Ease of Retrofit .... ☺ ☺ ☺

example, use key switches to turn off the overall lighting in a large space, and use motions sensors to provide localized lighting within the space.

ECONOMICS

SAVINGS POTENTIAL: *All the lighting energy controlled by the switches during off periods. Lamp cost and replacement labor may be reduced by a similar amount.*

COST: *Key switches typically cost several dollars apiece. In retrofit, labor for replacing existing switches typically costs $10 to $30 per switch.*

PAYBACK PERIOD: *Less than one year, typically.*

TRAPS & TRICKS

CHOICE OF METHOD: *This method is not as simple as it looks. Key switches that are used alone tend to be abandoned, because some light must always be made available. Either limit key switches to a fraction of the lighting, or combine them with other methods of control.*

WESINC

Fig. 1 Key switches The "keys" are inexpensive, non-custom items. However, anyone who has a need to turn the lights on or off must have one. As a result, the lights tend to be left on, and key switches tend to be abandoned.

Lighting Controls, Automatic

In many existing facilities, installing automatic lighting control can save more energy than any other lighting improvements. The Measures in this Subsection automatically turn off lighting when it is not needed. Many of these Measures can be accomplished easily, with little or no rewiring of the connections between the controls and the fixtures. In other cases, optimum performance requires changes to the lighting circuits, which are covered by the Measures in Subsection 9.6.

Safety is a Major Issue

Any automatic method of lighting control involves some possibility of turning off the controlled lights unexpectedly while people are still in the lighted area. Therefore, with any automatic lighting control, ensure safe egress from the space if the controls turn off the lights. Generally, you can do this by providing a minimum amount of light that is not under automatic control.

Tailor Lighting Control to Each Space

Don't install automatic lighting controls in a shotgun fashion. This will waste money and conservation potential, and create user problems. Review all the Measures in this Subsection to find the most appropriate automatic lighting control methods for each space. Expect to use a combination of control methods to achieve optimum performance.

A NOTE ON TERMINOLOGY

To avoid confusion, be aware that lighting specialists use the word "control" to mean shaping the light distribution pattern. In this context, a parabolic reflector is a "control" device. However, "control" is used in this Subsection in the same way that it is used for other equipment, namely, turning things on and off, changing levels of output, etc.

INDEX OF MEASURES

9.5.1 Where lighting is needed on a repetitive schedule, use timeclock control.

    9.5.1.1 To combine time switching with daylighting, use astronomical timeclocks.

9.5.2 Control exterior lighting with photocontrols.

9.5.3 Install interior photocontrols to exploit daylighting.

9.5.4 Where the need for lighting is determined by the presence of people, use personnel sensor switching.

9.5.5 Where lighting can be turned off after a fixed interval, install timed-turnoff switches.

9.5.6 If a door remains open when lighting is needed, use door switches.

RELATED MEASURES:

- Subsection 9.4, for manual controls to be used in combination with automatic controls

- Measure 9.2.6, daylight sensing dimmers for fluorescent fixtures

- Measure 9.3.2, daylight sensing dimmers for HID fixtures

- Measure 9.6.2.1, pullcord light switches, a control method that competes with automatic controls

- Measure 9.6.4, for circuit changes needed to maximize the effectiveness of some control improvements

MEASURE **9.5.1 Where lighting is needed on a repetitive schedule, use timeclock control.**

There are many lighting applications where occupancy occurs on a repetitive schedule. Safety is the main factor that determines whether you should use timeclocks to control the lighting in such applications. The controls should not trap people in unexpectedly darkened spaces.

Where to Use Time Control for Lighting

You can use time control by itself for these types of applications:

- *lighting that is not related to safety.* Examples include advertising sign lighting and decorative lighting.
- *lighting that exceeds levels needed for safety.* For example, the lighting in the dining area of a cafeteria may be turned off on a time schedule if the space has continuous emergency lighting. (Motion sensors would not be effective in this application because people walking through the space would keep turning on the lights unnecessarily.)

Time controls can be used more broadly in combination with other types of automatic or manual control. For example, in a large, open office area, use time controls to turn off all but the emergency lighting outside normal working hours. For late workers, use personnel sensors (Measure 9.5.4) to control lighting of smaller areas within the larger space.

SUMMARY

May save lots of energy at relatively low cost. Usually needs to be combined with other types of control.

SELECTION SCORECARD

Savings Potential .... $ $ $ $

Rate of Return, New Facilities % % % %

Rate of Return, Retrofit .... % % % %

Reliability .... ✓ ✓

Ease of Retrofit .... ☺ ☺ ☺

Time control is often useful in combination with daylight sensing control. For example, photocontrols are used to turn on parking lot lights at sunset. If there is no need to continue lighting the parking lot after midnight, install a time control to turn off the lights after midnight. For example, see Figure 1.

Another way to achieve the same effect is to use a time control called an "astronomical" timeclock, which is recommended by the subsidiary Measure 9.5.1.1.

You can use a timeclock to change the patterns of a programmable lighting controller (Measure 9.6.4.1) at different times of day. For example, in a restaurant, a programmable controller can change from incandescent

WESINC

Fig. 1 Subway station parking lot After the subway stops running, why is it necessary to keep all these lights on? A timeclock is the appropriate method of turning these lights off. A photocontrol is the appropriate method of turning them on.

mood lighting during meal hours to more efficient incandescent lighting during set-up and cleaning hours.

How to Select and Install Time Controls

A great variety of timeclocks are available, with virtually any degree of scheduling flexibility that you could want. This includes separate schedules for weekends and holidays, automatic changeover to daylight saving time, etc.

See Reference Note 10, Clock Controls and Programmable Thermostats, for the equipment features, installation practices, operating procedures, and overrides that you need to achieve satisfactory performance with time controls.

If the facility has an energy management computer system, you can use it for time control instead of installing individual timeclocks, but this approach is usually much more expensive. For more about this, see Reference Note 13, Energy Management Control Systems.

ECONOMICS

SAVINGS POTENTIAL: 20 to 70 percent of the energy used by controlled fixtures. Lamp cost and replacement labor may be reduced by a similar percentage.

COST: Appropriate modern electronic time controls cost from $80 to $500, depending on schedule options, number of circuits, power backup features, etc. Installation labor varies widely.

PAYBACK PERIOD: Less than one year, to several years. Depends on the lighting wattage, occupancy patterns, labor costs, etc.

TRAPS & TRICKS

CHOICE OF METHODS: In many applications, this is the best method to turn lights on, or to turn them off, but not both. Think in terms of combining time control with other methods.

SELECTING THE EQUIPMENT: Time controls are inexpensive and readily available, but don't grab the first unit you find. Shop to find the features that your application needs. Use Reference Note 10 as your guide.

INSTALLATION: Be careful to avoid safety problems. Install the timeclock itself so that it is obvious and accessible only to those who should set it. Install a placard at the timeclock that says which lighting it controls and the schedules that should be set.

MONITOR PERFORMANCE: Sooner or later, all time controls fail or get out of time. Schedule periodic checks of all time controls to verify proper operation and settings. Keep a list of all timeclocks and the functions they control in your plant maintenance manual. (You do have one, right?)

MEASURE 9.5.1.1 To combine time switching with daylighting, use astronomical timeclocks.

Most outdoor lighting, such as security, parking lot, decorative, and advertising lighting, is turned on at sunset. The same is true of heavily daylighted interior spaces, such as atriums and greenhouses. However, most of these applications do not need lighting all night. For such applications, an "astronomical" timeclock is a single piece of equipment that can turn the lights on and off at the proper times.

SUMMARY

An easy way to combine control by time and by daylight.

SELECTION SCORECARD

Savings Potential .... $ $ $ $

Rate of Return, New Facilities % % % %

Rate of Return, Retrofit.......... % % % %

Reliability .... ✓ ✓

Ease of Retrofit .... ☺ ☺ ☺

The time of sunrise and sunset varies considerably with the seasons. An astronomical timeclock calculates the times of sunset and sunrise, based on the date and the facility's geographic location. You can install an astronomical timeclock anywhere that is convenient, minimizing wiring cost.

How to Select and Install Time Controls

Select and install an astronomical timeclock like any other timeclock, except for the additional "astronomical" feature. See Measure 9.5.1 for guidelines. Earlier astronomical timeclocks were mechanical units, but these are now obsolete. Buy an electronic unit made by an established manufacturer.

Alternative: Photocontrol

An alternative is using a photosensitive control (Measures 9.5.2 and 9.5.3) in combination with a conventional timeclock. Photocontrols provide a direct indication of available daylight, so they never get out of time. Also, they can account for overcasts, eclipses, etc. On the other hand, you have to install them where they can "see" daylight. The wiring may cost much more than the controls themselves.

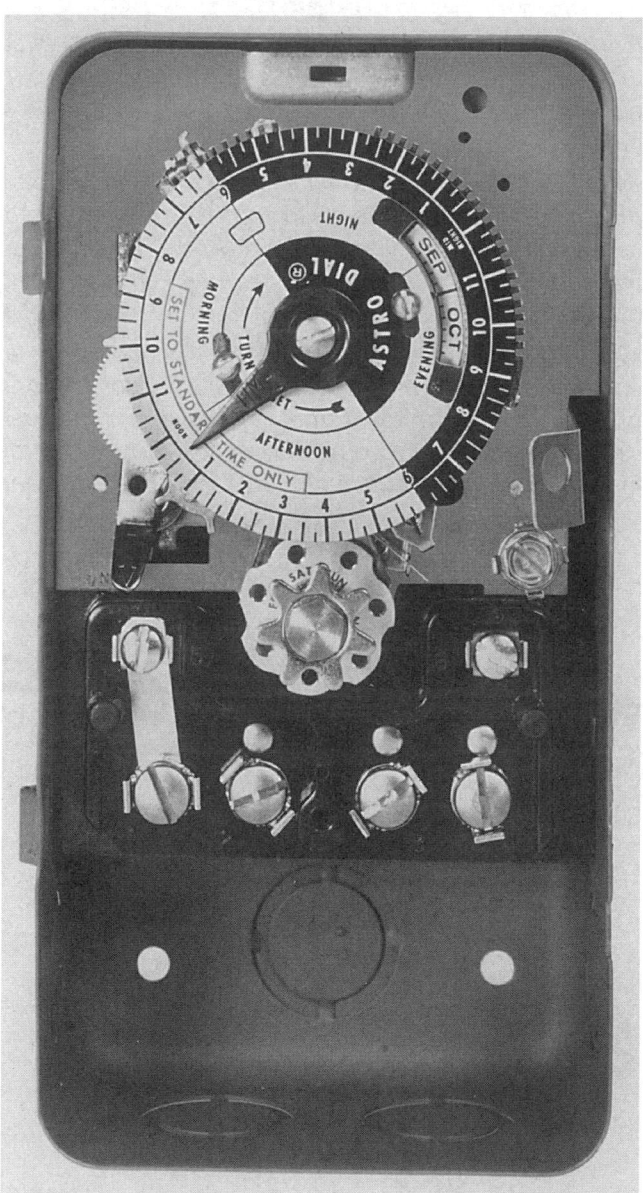

Paragon Electric Company, Inc.

Fig. 1 Old fashioned astronomical timeclock This type adjusts for seasonal changes in the length of daylight. The latitude of the location has a major effect on seasonal daylight variation, so make sure that it is set properly. In most cases, you would prefer to use an electronic version, primarily for benefits such as battery backup, flexible scheduling, etc.

ECONOMICS

SAVINGS POTENTIAL: *40 to 70 percent of the energy used by controlled fixtures, compared to operating all night. Lamp cost and replacement labor may be reduced by a similar percentage.*

COST: *A typical single-circuit electronic astronomical time control with battery backup and other features costs about $200. Installation typically costs no more than one or two hundred dollars, because the unit can be installed anywhere that is convenient.*

PAYBACK PERIOD: Less than one year, to several years. Depends on the lighting wattage, operating schedule, wiring costs, etc.

TRAPS & TRICKS

CHOICE OF METHOD: An astronomical timeclock can get off schedule, and it cannot respond to unexpected periods of darkness, which makes it less reliable than photocontrol. In new construction, photocontrol is usually preferable. In retrofit applications, the lower wiring cost of an astronomical timeclock may be the deciding factor.

SELECTING THE EQUIPMENT: An astronomical timeclock is basically a timeclock. Get all the timeclock features that would benefit the application. Use Reference Note 10 as a guide.

INSTALLATION: See Measure 9.5.1.

MONITOR PERFORMANCE: See Measure 9.5.1.

MEASURE **9.5.2 Control exterior lighting with photocontrols.**

RATINGS

New Facilities	Retrofit	O&M
A	B	

Leaving exterior lights turned on during daylight, as in Figure 1, is a useless waste of energy. On the other hand, exterior lighting applications are often critical, so they need to be turned on as soon as it becomes dark. The most precise way to tailor exterior lighting to the availability of daylight is to control the lighting with a daylight sensor. Such devices are called by a variety of names, including "photoelectric," "photocontrol," and "photocell."

This is also the most reliable method of controlling exterior lighting. The equipment is highly reliable, and it is much less vulnerable to tinkering than other methods, such as timeclocks.

A photocontrol is essentially an on-off switch that is activated by a light sensor. Most photocontrols are small units that typically weigh a few ounces. Photocontrol is so popular for exterior lighting that inexpensive units are available in many configurations, with standardized features. Figures 2 and 3 show two common types.

You can control any type of lighting with photocontrols. However, if you are controlling HID lamps (mercury vapor, metal halide, or sodium vapor) or low-pressure sodium lamps, it is important to select a photocontrol that avoids turning off the lamps for brief periods.

How to Select Photocontrols

Consider these features in selecting a photocontrol for exterior lighting:

- *mounting method.* Photocontrols are available for convenient installation using all common electrical mounting methods. Many types are available that fit standard knockout holes in electrical boxes and lamp posts. Units are available that screw into

SUMMARY

Simple and reliable. Often inexpensive. Usually combined with time control.

SELECTION SCORECARD

Savings Potential	$	$	$	
Rate of Return, New Facilities	%	%	%	%
Rate of Return, Retrofit	%	%	%	%
Reliability	✓	✓	✓	
Ease of Retrofit	☺	☺	☺	

medium lamp bases, and require no wiring. Figure 3 shows a common type that is installed on a bayonet base.

- *orientation flexibility.* Some units have a swivel base that provides extra flexibility in pointing the unit. This may be necessary to keep the sensor out of direct sunlight or to receive daylight from the right direction for proper sensing.

- *ruggedness.* Select units that can withstand abuse. This is primarily a matter of selecting metal over plastic. Plastic units are generally flimsy, and they deteriorate from sunlight and lamp heat.

- *light level adjustment.* Some photocontrols have an adjustable shutter over the sensor window that lets you adjust the light levels at which the lights turn on and off. Figure 2 shows an example. With these, the ratio of on and off light levels is fixed at about 1:3. Newer, more expensive photocontrols may provide more flexibility.

- *response delay.* To keep photocontrols from responding to light flashes from vehicle headlights, lightning, and other sources, most photocontrols have a built-in delay. The delay is usually not adjustable.

- *amperage.* For cost reasons, photocontrols typically do not have enough current capacity to control all the amperage of a single power circuit. This is easy to handle. Either buy a photocontrol that has enough capacity for the lighting load, or use the photocontrol to provide power to the lighting through a relay of sufficient capacity.

How to Install Photocontrols

Installing photocontrols for exterior lighting is usually simple. The unit can be inserted anywhere in the lighting power circuit. If the wiring is accessible, installation may take only a few minutes. Some exterior fixtures have standardized receptacles for photocontrols.

Keep the photocontrol out of precipitation and direct sunlight, unless it is designed specifically for exposed locations. If necessary, make a small hood for the unit.

A unit that is properly selected and installed may last for the life of the facility.

Keep the sensor from seeing direct sunlight, as this shortens sensor life. At the same time, install the sensor so that it sees enough daylight to keep the lights turned off during cloudy days. Typical units can be set to remain off until the light level falls to about one footcandle (about ten lux). The setting may drift from the effects of aging and from dirt on the sensor window.

Combine Photocontrol with Other Methods

Unless the application needs continuous lighting all night, combine photocontrols with other controls that turn off the lights during selected hours of darkness. To turn the lights on or off at specific times, combine with time controls, as recommended by Measure 9.5.1.

If you provide any manual control, be sure to install effective control placards. See Measure 9.4.1 for guidance. Also, telltales lights (Measure 9.4.5) are valuable with manually controlled exterior lighting.

WESINC

Fig. 1 Lights turned on in the middle of the day These lights are turned on because the switch is installed at some remote location where the lamps cannot be seen. Photocontrol is the most reliable and least expensive control method for this situation.

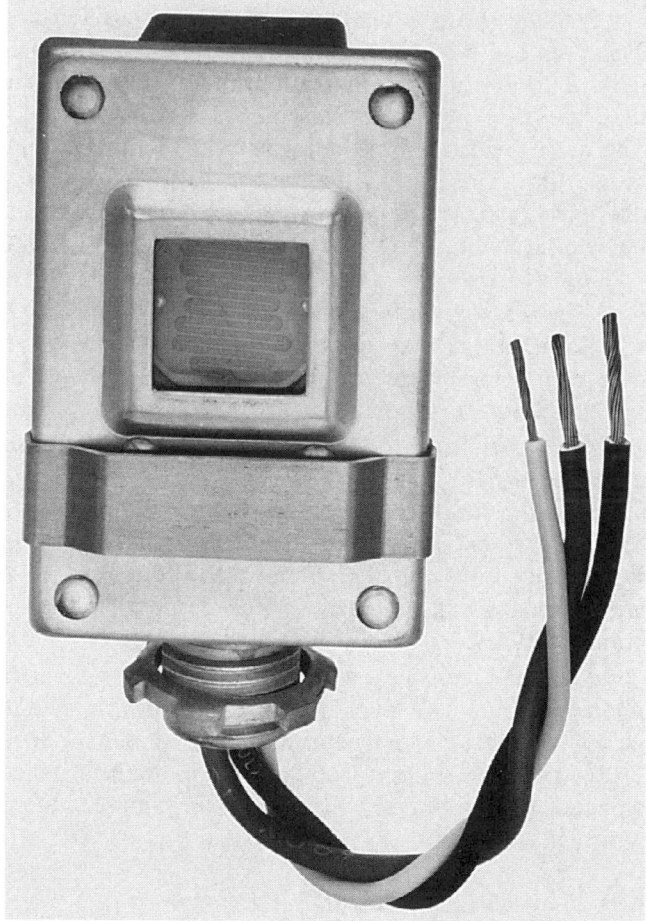

Paragon Electric Company, Inc.

Fig. 2 Exterior photocontrol This unit, shown approximately actual size, is inexpensive and rugged. The sliding metal strip on the cover acts as a mask to reduce the sensitivity of the cell behind the window. The on-off light level ratio is fixed.

Paragon Electric Company, Inc.

Fig. 3 Bayonet mounted photocontrol This style allows quick replacement, which is useful in awkward locations, such as the tops of light poles. Some outdoor light fixtures have an integral socket to accept sensors of this type.

Alternative: Astronomical Timeclock

An astronomical timeclock (Measure 9.5.1.1) calculates the times of sunset and sunrise, and can turn lighting on and off accordingly, as well as providing other timeclock functions. They are more expensive than photocontrols, and they are more vulnerable to tampering and setting errors. The main advantage of astronomical timeclocks is that they do not have to be installed where there is daylight.

ECONOMICS

SAVINGS POTENTIAL: *10 to 70 percent of exterior lighting energy. Lamp cost and replacement labor may be reduced by a similar percentage.*

COST: *Typical photocontrols for exterior applications cost less than $10. New fixtures can be selected with integral daylight sensors. In retrofit, sensors can be mounted at the fixtures, minimizing wiring cost.*

PAYBACK PERIOD: *Usually less than one year.*

TRAPS & TRICKS

CHOICE OF METHODS: *Unless the lights are needed during all hours of darkness, combine this method with timeclocks or other appropriate methods of control.*

MEASURE **9.5.3 Install interior photocontrols to exploit daylighting.**

Turning interior lighting on and off with photocontrols is a possibility that deserves more attention because it may save substantial amounts of energy in spaces that have glazing. Photocontrol of interior lighting is more difficult to accomplish in a satisfactory manner than photocontrol of exterior lighting, which was recommended by the previous Measure. Here, we cover the additional complications of indoor photocontrol.

Where to Consider Photocontrol for Inside Lighting

Consider photocontrols for any space that has lighting from windows or skylights, provided that the space has fluorescent or incandescent lighting. Figures 1 through 4 show favorable examples.

Don't overlook opportunities in spaces that are only partially daylighted. A common example is using photocontrol for the row of fixtures adjacent to the windows of a room, as in Figure 2. Figure 4 shows where photocontrols should be installed in the area adjacent to a daylighted atrium.

To find out whether there is enough daylight to make photocontrol worthwhile, survey candidate areas using a light meter. There should be a many hours during

SUMMARY

A technique with great untapped potential. Don't overlook any lighting that would benefit, such as emergency lighting. May be tricky to design and install. Expect to combine photocontrol with other types of control.

SELECTION SCORECARD

Savings Potential	$ $ $
Rate of Return, New Facilities	% % % %
Rate of Return, Retrofit..........	% % %
Reliability	✓ ✓
Ease of Retrofit	☺ ☺

which the daylight level is about three times higher than the minimum lighting level required for the activities. The reason for this ratio is explained below.

If there is adequate daylight, consider whether the abrupt changes in lighting level that occur as fixtures turn on and off will be an annoyance. The worst problem is created by thick, localized clouds of the kind shown in Figure 5. However, don't assume that clouds are an insurmountable obstacle. When light fixtures are turned on and off in a brightly daylighted space, the change is much less noticeable than turning lights on and off in a dark space. Furthermore, if the photocontrols are installed properly, the lights will not switch often.

Interior photocontrol is most acceptable in transient areas, such as daylighted corridors (Figure 1), common

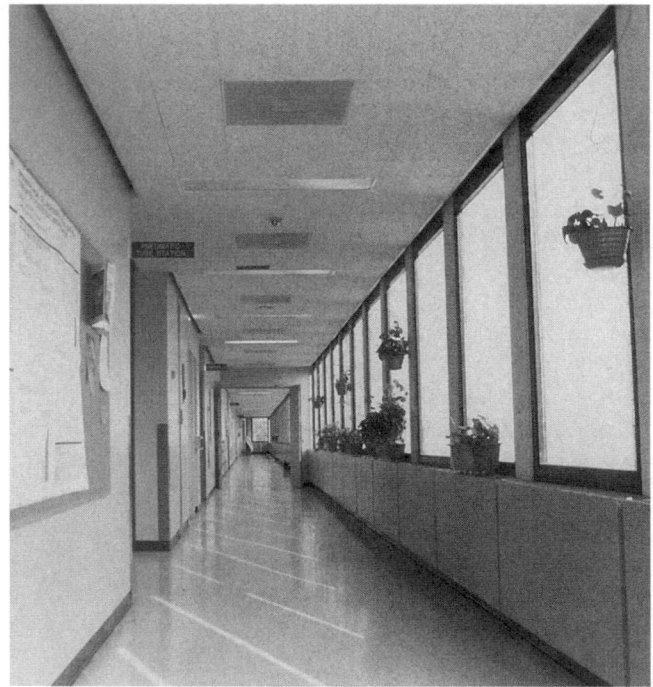

WESINC

Fig. 1 Daylighted corridor This is an ideal application for photocontrol of lighting. Daylight provides ample illumination under most daytime conditions. The light fixtures are fluorescent, with multiple lamps, so they can even be switched in stages for a more gradual response.

WESINC

Fig. 2 Daylighted reading area in a library The row of light fixtures adjacent to the windows is a prime candidate for photocontrol.

areas in shopping malls that are illuminated by skylights, and daylighted atriums (Figure 3). Photocontrols are less likely to be acceptable in offices and classrooms. However, a greater awareness of the need for energy conservation may make photocontrol acceptable even in such locations.

Security lighting is an especially fruitful application for photocontrol. Only low levels of light are needed for emergency egress, so daylighting may serve as emergency lighting for a much larger portion of the building than for conventional lighting. For example, photocontrol may eliminate the need for most emergency lighting on weekend days, when no people are present to require high lighting levels. Also, photocontrol of emergency lighting may not require the complexity of other applications. Check that local codes allow turning off electric emergency lighting when daylight is available. This is an unusual way of dealing with emergency lighting, so you may have to request a variance.

Limitations with HID Lighting

All types of HID lamps (mercury vapor, metal halide, and high-pressure sodium) require a time of several minutes to start, and most require an even longer time to restart from a hot condition. Therefore, do not use photocontrols with HID lamps in applications where fluctuations in daylight during the day, typically caused by clouds, might require starting or restarting during occupied periods.

Combine with Other Types of Lighting Controls

Expect to combine photocontrol of interior lighting with other methods of control. Photocontrols alone would keep the lights turned on during all hours of darkness. Provide additional controls to turn the lights on and off in response to the presence of people. Photocontrol alone is useful only for lighting that is required throughout hours of darkness, such as security lighting.

A common pitfall is using manual switches in combination with photocontrols. In such cases, the manual switch may inadvertently be left in the on position even though the photocontrol has turned off the lights. At sunset, the photocontrol will turn the lights on even though the facility is vacant.

Solving this problem requires a combination of controls. Use a timeclock to keep the lights turned off during hours of darkness when the space is normally unoccupied. In addition, install another type of automatic control, such as a motion sensor, to override the timeclock if people are present during these hours.

Types of Photocontrols

Most photocontrols are designed for exterior lighting. They are rugged and inexpensive, but they

Fig. 3 Why are all these lights turned on? This atrium in a government building is excessively illuminated by a huge skylight. The illumination provided by the light fixtures is not noticeable during the daytime. A single inexpensive photocontrol could save a lot of money and many light bulbs in this area alone.

lack setting flexibility. They are described in Measure 9.5.2. Interior photocontrol is uncommon, so the market does not offer a wide selection of photocontrols intended specifically for indoor applications, although some are entering the market. You can use exterior photocontrols for interior applications, but you may have to be creative in how you install them.

How to Install Interior Photocontrols

Interior photocontrol usually is part of a system consisting of several types of lighting control. Designing the system requires some thought, and installation details are important. Work out the following issues carefully.

■ Sensor Location and Orientation

Choosing the location of the light sensors is tricky. Remember that the purpose of the sensor is to detect how much daylighting is available within the space, not to sense the total lighting level. Therefore, aim the sensor somewhat toward the outside, so that it senses the available daylighting without being influenced too much by the electric lights inside the space.

On the other hand, if you aim the sensor too far outside the space, it will see a scene whose brightness does not correspond to the amount of sunlight entering the space. Also, this can lead to situations that trick the sensor into turning the lights off. For example, if the sensor looks toward a parking lot, reflections of the sun from the windshields of parked cars may blind the sensor. Or, reflections from the windows of adjacent buildings may do the same thing.

To minimize these difficulties, locate the sensor as close to the perimeter as possible. Ideally, you would like to install the sensor outside the window, looking inward. If there is a soffit over the window, consider this seriously. If you choose this arrangement, be careful to install the sensor so that it cannot be fooled by reflections from the glass.

Sensors mounted on swivels are worth the small additional cost. They provide flexibility in mounting, and they make it easy to experiment to find the best orientation. Figure 6 shows a unit that has this feature.

■ Number of Sensors

Use separate sensors for each area that has a different response to exterior changes in sunlight. For example, use separate sensors for spaces that face in different directions, and for spaces that fall into the shadow of adjacent buildings at different times of day.

One sensor can control any number of fixtures, either directly or through a power relay. However, sensors are cheap in comparison with the cost of wiring labor. This makes it more economical to install a sensor in each lighting circuit, rather than extending the wiring to allow a single sensor to control a larger area.

■ Light Sensitivity Settings

Most photocontrols allow you to adjust the light levels at which the fixtures turn on and off. Most inexpensive photocontrols achieve this with a simple shutter that changes the amount of light entering the light sensor. However, even with the shutter set to admit minimum light, exterior photocontrols typically turn lights off when the illumination level rises to about 15 footcandles (150 lux). This may be adequate for corridors and other peripheral areas, but it is too dim for most work activities. For example, offices and retailing typically require about 50 footcandles (500 lux).

WESINC

Fig. 4 Area surrounding a large, brightly lighted atrium   This area deserves several photocontrols. The inner and outer rows of light fixtures should have separate control. And, each side of the atrium should have separate control.

You can solve this problem by installing a simple light-absorbing filter in front of the sensor. For example, make the filter from translucent plastic diffuser material. Each sheet typically has a transparency of about 60%, so three layers reduce the light to the sensor by a factor of about five. Keep plastic diffuser material out of direct sunlight and away from light fixtures, as heat and ultraviolet light darken it.

Ordinary exterior photocontrols typically turn lights off at about three times the lighting level at which they turn lights on. In inexpensive photocontrols, this ratio is fixed. The wide ratio keeps small fluctuations in the light level from cycling the lights on and off. Clouds passing in front of the sun reduce sunlight by a factor of three to five. This is not likely to cause much cycling, unless maximum and minimum light levels happen to coincide with the upper and lower limits of the photocontrol.

The wide on-off ratio also makes the orientation of the sensor less critical. It allows the sensor to be aimed more into the space, where it senses artificial light along with daylight, without the risk of short-cycling the lights.

To save as much energy as possible, adjust the photocontrol to turn on the lights at the lowest acceptable lighting level. Unfortunately, the wide on-off ratio keeps the lights on until the light level in the space has risen considerably. How much energy saving is lost by this factor depends on the geometry of the space, the effectiveness of daylight distribution, and other factors that are covered in Reference Note 46, Daylighting Design.

Explain the Controls to the Occupants

Install effective placards to inform occupants how the lighting is controlled. This will get them interested, and minimize annoyance. See Measure 9.4.1 about placards.

Arrange Circuits to Maximize Daylighting Potential

Lighting control cannot exploit daylighting unless the fixtures are grouped in a way that corresponds to the availability of daylight in the space. For example, if the space has a continuous row of windows, the fixtures should be wired in rows parallel to the windows, with each row switched separately. The outer row of fixtures is turned off during most daylight hours, the next row is turned off for a lesser period of time, and so forth.

In existing buildings, you may have to rearrange the light fixture circuits to make daylighting possible. See Measure 9.6.4 for details.

There are various techniques that you can use to increase the raw amount of daylight that is available to exploit. See Reference Note 46 and Subsection 8.3 for the full story.

Alternative Method: Dimming Controls

If switching lamps on and off is not acceptable as a method of exploiting daylight, consider automatic dimming controls. These change the output of fixtures smoothly in response to the light level in the space.

Automatic dimming has become a fairly reliable technique with fluorescent lighting. Fluorescent dimming requires special electronic dimming ballasts

WESINC

Fig. 5 Puffy clouds Cumulus clouds like this are the worst problem for daylighting control because the clouds are small and thick. Thus, the light level changes are large, abrupt, and frequent. Think of this situation when you decide whether to use photocontrol for interior lighting.

Paragon Electric Company, Inc.

Fig. 6 Photocontrol with swivel base   The ability to point the sensor accurately is important. Aiming the sensor for proper response can be tricky, and it is likely to require trial-and-error.

and controls that respond to daylight. The best equipment can control the output of fluorescent lamps over a wide range. However, fluorescent dimming is much more expensive than on-off photocontrol. See Measure 9.2.6 for the full story.

HID lighting can be dimmed over a limited range. It is not yet well established. See Measure 9.3.2 for the details.

Automatic dimming is not commercially available for incandescent lighting. The best way to save energy with incandescent lighting is to get rid of it.

ECONOMICS

SAVINGS POTENTIAL: *30 to 70 percent of the energy used by controlled fixtures. Lamp cost and replacement labor may be reduced by a similar percentage.*

COST: *Common photocontrols designed for exterior applications cost less than $10 apiece. The cost of wiring and finish work varies widely, averaging perhaps several hundred dollars per sensor.*

PAYBACK PERIOD: *Less than one year, to several years.*

TRAPS & TRICKS

PLANNING: *Wherever daylight enters a building, consider using it instead of electric lighting. Make sure that occasional abrupt changes of lighting level are acceptable to occupants. Lay out the fixture wiring to exploit daylight as much as possible. Be sure to satisfy all lighting needs, including emergency and off-hours lighting.*

INSTALLATION: *Expect to spend some time finding the best sensor location for each space. Be sure to install informative placards.*

MONITOR PERFORMANCE: *This is still an experimental activity. It affects occupant comfort and productivity. Monitor it carefully, until you are sure that lighting needs are satisfied under all conditions. Expect to make adjustments. Schedule periodic checks in your maintenance calendar.*

MEASURE **9.5.4 Where the need for lighting is determined by the presence of people, use personnel sensor switching.**

In most spaces, lighting is needed only when people are present. Personnel sensors are devices that sense the actual presence of people. Therefore, personnel sensors can match lighting to space occupancy more accurately than any other method. They are not satisfactory for all spaces, but they are widely applicable.

Where to Use Personnel Sensors

Personnel sensors are desirable for lighting control in an exceptionally broad range of applications. For example, a lighting retrofit program for a large, diverse research hospital found personnel sensors to be the most efficient method of lighting control for a large fraction of the spaces in the hospital, including offices, laboratories, stock rooms, loading docks, etc. Figure 1 shows yet another possible application.

See Reference Note 11, Personnel Sensors, about the conditions that are required for reliable, safe lighting control using personnel sensors. Personnel sensors are not a panacea in the realm of lighting control. In many applications, personnel sensors are simply the wrong tool. Some applications have quirks that make it inadvisable to use them, for reasons that may not be obvious.

In brief, there are three conditions that must be met for a successful installation. One is that the sensor should be able to reliably detect the presence of people anywhere within the controlled area. This requires that people move continually, but only small motions are needed. The second condition is that the sensor should not be subject to an excessive amount of false triggering by things other than people. The third condition is that the lighting must be able to respond quickly enough to the presence of people. This excludes HID lighting, and it excludes some types of fluorescent lamps in some applications.

Space-by-Space Planning is Essential

You cannot successfully install personnel sensors in a broadcast manner, as you might install high-efficiency lamps or ballasts. Analyze each activity area to decide whether personnel sensors are satisfactory for that environment. In locations where they are appropriate, figure out the best places to install them. If necessary, combine personnel sensor control with other methods of control.

Expect to spend a lot of time and effort on this. For example, in a complex facility, such as a hospital or hotel, you may need three or four different types of personnel sensors to deal properly with all the conditions that exist, along with many mounting configurations.

SUMMARY

A powerful lighting control option. Applicable widely, but not everywhere. Often inexpensive. Takes thought and effort to get it right.

SELECTION SCORECARD

Savings Potential	$ $ $ $
Rate of Return, New Facilities	% % % %
Rate of Return, Retrofit..........	% % % %
Reliability	✓ ✓ ✓
Ease of Retrofit	☺ ☺ ☺

You need to select the location of each sensor carefully. Installing personnel sensors carelessly or in the wrong applications can waste energy, annoy occupants, and create safety hazards.

Types of Personnel Sensors

The main types of personnel sensors are explained in Reference Note 11, Personnel Sensors. Most lighting applications favor passive infrared detectors. Many inexpensive models are available that replace existing light switches, without further wiring or installation work. These units are a major breakthrough in lower cost and ease of installation. However, these advantages tempt people to use them in locations where they cannot provide satisfactory performance. In many cases, only ceiling mounted sensors are satisfactory, and these are expensive to install.

Passive infrared sensors are now commonly built into floodlamp fixtures, intended primarily for exterior applications. Integral motion sensors are starting to appear in other types of light fixtures. Check the latest catalogs.

Avoid "On" Overrides

As a rule, do not select units that have an "on" switch to override the automatic turn-off. Occupants will eventually leave the control in the "on" option, effectively eliminating its efficiency benefit. Unfortunately, it is getting difficult to find personnel sensors that are free of this bad feature.

If activities occur in the space that require an "on" override, then a personnel sensor is not the appropriate type of lighting control.

On the other hand, all personnel sensors should have an "off" override. This allows the lights to be turned off while people are in the space. This is useful for using slide projectors, watching television, sleeping, etc.

ECONOMICS

SAVINGS POTENTIAL: *20 to 70 percent of the lighting energy in controlled areas, depending primarily on occupancy patterns. Lamp cost and replacement labor may be reduced by a similar percentage.*

COST: *$15 to $50 apiece for infrared sensors that replace wall switches, which can usually be installed in minutes. Ceiling mounted sensors cost $100 to $200 apiece, with installation costs depending on location, labor rates, etc.*

PAYBACK PERIOD: *Less than one year, to several years.*

TRAPS & TRICKS

PLANNING: *Expect to spend time and effort to select the best model of sensor and the best location for each space.*

SELECTING THE EQUIPMENT: *See Reference Note 11. You can get good equipment at modest cost if you shop carefully and test samples. Cheap units may fail quickly, break easily, and create radio interference. Units should have an "off" switch but no "on" position. Try to get a lighted toggle.*

INFORM THE OCCUPANTS: *Don't forget the placards. They get people involved, and keep occupants from being annoyed by seemingly erratic lighting operation.*

WESINC

Fig. 1 An ideal application for personnel sensor lighting control   Most of the lighting in this reference library is wasted. Personnel sensors could be installed to control the lighting in each of the aisles between the shelves. Control would be certain. False triggering is easily avoided. People in adjacent areas would not be bothered. There is no safety hazard.

MEASURE **9.5.5 Where lighting can be turned off after a fixed interval, install timed-turnoff switches.**

Timed-turnoff switches are the least expensive type of automatic lighting control. In some cases, their low cost and ease of installation makes it desirable to use them where more efficient controls would be too expensive.

Types and Features

The oldest and most common type of timed-turnoff switch is the "dial timer," a spring-wound mechanical timer that is set by twisting the knob to the desired time. Typical units of this type are vulnerable to damage because the shaft is weak and the knob is not securely attached to the shaft. Some spring-wound units make an annoying ticking sound as they operate.

Newer types of timed-turnoff switches are completely electronic and silent. Electronic switches can be made much more rugged than the spring-wound dial timer. These units typically have a spring-loaded toggle switch that turns on the circuit for a preset time interval. Some electronic models provide a choice of time intervals, which you select by adjusting a knob located behind the faceplate.

Most models allow occupants to turn off the lights manually. Some models allow occupants to keep the lights on, overriding the timer.

Timed-turnoff switches are available with a wide range of time spans. The choice of time span is a compromise. Shorter time spans waste less energy but increase the probability that the lights will turn off while someone is in the space. Dial timers allow the occupant to set the time span, but this is not likely to be done with a view toward optimizing efficiency. For most applications, the best choice is an electronic unit that allows the engineering staff to set a fixed time interval behind the cover plate.

Avoid "On" Overrides

As a rule, do not select units that have an "on" switch to override the automatic turn-off. Occupants will eventually select the "on" option, effectively eliminating the efficiency benefit of the control. If activities occur in the space that require an "on" override, then a timed-turnoff switch is not the appropriate control method.

Safety Limits Applications

Time switches turn off lights without regard to whether anyone is in the space. If they are used inappropriately, they may create a safety hazard, or at least annoyance. The most dangerous situation occurs when one person turns on the lights and then leaves. A

SUMMARY

An inexpensive and reliable way to provide brief periods of lighting. Limited to spaces that cannot trap people in the dark.

SELECTION SCORECARD

Savings Potential .... $ $

Rate of Return, New Facilities % % % %

Rate of Return, Retrofit .... % % % %

Reliability .... ✓ ✓ ✓

Ease of Retrofit .... ☺ ☺ ☺

second person enters the space while the lights are still turned on, but the lights turn off unexpectedly while the second person is still in the space. For this reason, time switches are desirable primarily for spaces that are used sporadically, especially by one person or one group at a time.

Because of the hazard of unexpected darkness, do not use time switches for lighting unless:

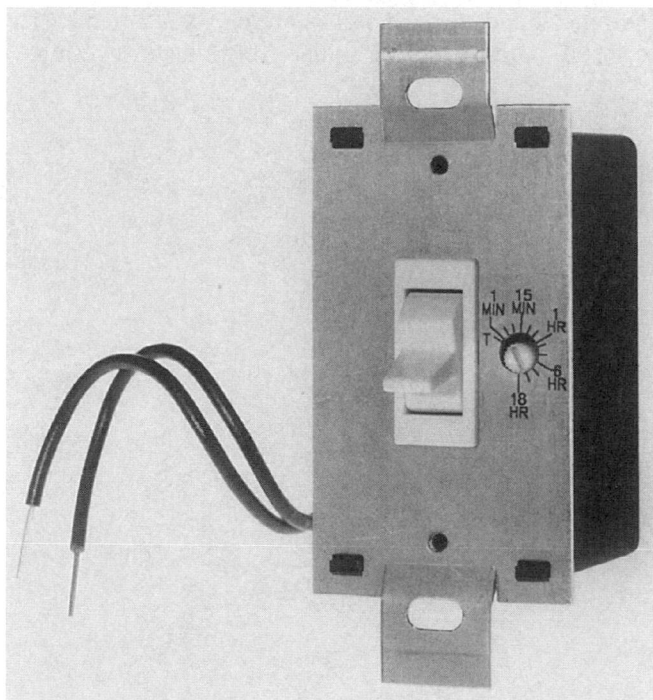

Paragon Electric Company, Inc.

Fig. 1 Timed-turnoff switch Pressing the toggle upward starts a timer that keeps the lights turned on for a fixed period. Pressing the toggle downward turns off the lights if they are on. On this unit, the time interval is adjustable from one minute to 18 hours. This wide range makes it difficult to tune the time interval for greatest efficiency.

- *the space is so small that the exit is easily within reach.* Storage rooms and janitor closets are typical applications.
- *the space has adequate emergency lighting.* For example, you can use timer switches to provide lighting for cleaning crews who work in large office spaces late at night. The timer switch overrides a timeclock or key switch that is used to turn off the lights after hours. The emergency lighting is bright enough to provide safe access to the timer switch.
- *a minimum lighting level is provided by adjacent lighted spaces.* For example, you could safely use timed-turnoff switches for individual stacks in a library, or for individual stock shelves in a warehouse, where individual areas are used only occasionally and for short periods. Even here, timer switches would be unsatisfactory if the lighted spaces are visited frequently. Exploiting this kind of opportunity is likely to need light circuit modifications and installing the switches in unconventional locations, such as from overhead fixtures.

Timed-Turnoff Switch or Occupancy Sensor?

Personnel sensors (Measure 9.5.4) are effective in many of the same applications where time switches are appropriate. Personnel sensors can match lighting to occupancy more accurately than time switches, and they avoid the problem of unexpected loss of lighting. On the other hand, time switches are cheaper, often easier to install, and they are immune from false triggering.

Warn Occupants

Timed-turnoff switches may turn off lights unexpectedly. This makes it especially important to install conspicuous placards at the entrances to tell occupants how the lights are controlled. See Reference Note 12, Placards, for details of effective placard design, materials, and installation.

ECONOMICS

SAVINGS POTENTIAL: Usually more than 50 percent of controlled lighting. Lamp cost and replacement labor may be reduced by a similar percentage.

COST: Spring-wound units cost less than $20, electronic units less than $30. Installation usually takes only a few minutes if replacing an existing light switch. Subdividing light circuits to provide localized control is much more expensive.

PAYBACK PERIOD: Less than one year, if replacing existing switches. Several years, typically, if rewiring is needed.

TRAPS & TRICKS

CHOICE OF METHOD: Use this type of switch only where it cannot cause safety or convenience problems.

SELECTING THE EQUIPMENT: These switches are subject to abuse, so select them for ruggedness and ease of use. Select the appropriate time interval for each location. Do not install units with an "on" override.

INSTALLATION: An effective placard is essential for this type of switch.

MEASURE **9.5.6 If a door remains open when lighting is needed, use door switches.**

Janitor closets, linen rooms, small stockrooms, and similar spaces are entered by a single door that is kept open while the room is in use. In such spaces, lighting can be controlled efficiently by a door switch. The switch mounts easily on the door jamb and is connected into the fixture wiring. The switch is inexpensive. The main cost is routing the wiring from the switch to the rest of the circuit.

Unless you are sure that the door is kept closed whenever the space is vacated, install a placard that says, "The lights in this room turn off automatically when the door is closed. Please keep the door closed when you are not in the room."

Alternative Method: Timed-Turnoff Switch

An alternative to door switches is using a timed-turnoff switch. See Measure 9.5.5, previous. A door switch is more convenient, but it depends on people keeping the door closed. In retrofit applications where there is an existing toggle switch, a timed-turnoff switch is less expensive to install.

SUMMARY

Simple and reliable. Usually limited to small, unoccupied spaces.

SELECTION SCORECARD

Savings Potential .... **$**

Rate of Return, New Facilities **% %** % %

Rate of Return, Retrofit .... **% %** %

Reliability .... ✓ ✓ ✓

Ease of Retrofit .... ☺ ☺ ☺

ECONOMICS

SAVINGS POTENTIAL: *A large fraction of the energy of the controlled lighting. The total amount of energy at stake is usually small.*

COST: *The hardware costs less than $20. Labor cost varies widely, but is typically less than $100.*

PAYBACK PERIOD: *Several years, typically.*

This Subsection deals with the spatial layout of fixtures and the wiring that controls them. Efficient lighting layout involves getting the light from the lamps to the task with a minimum of loss on the way, and delivering the light by paths that avoid visual problems. The light fixture controls light distribution after the light leaves the lamp. The surfaces within the space may be important to light distribution after light leaves the fixture. The first Measure deals with the surfaces of the space. The other Measures deal with fixtures and their locations in the space.

In new construction, review the Measures of this Subsection as part of the lighting design. Typically, you do not know the arrangement of activities in the spaces at the time the building or space is designed. This is a serious challenge to efficiency, which makes it especially important to design lighting that is adaptable.

In existing facilities, prioritize the Measures in the usual way. Try to combine your improvements. Most of the Measures require a detailed survey and a heavy dose of planning. Get it all done at one time.

Coordinate these Measures with the delamping, lamp improvement, and fixture improvement Measures of Subsections 9.1 through 9.3. Also, coordinate fixture placement with the appropriate control methods from Subsections 9.4 and 9.5. Remember that efficient lighting is accomplished one fixture at a time, matching each fixture carefully to the lighting requirements of the task.

For these activities, you need to know about the light distribution patterns of lamps and fixtures. This topic is covered in Reference Note 57, Light Distribution Patterns of Fixtures.

INDEX OF MEASURES

9.6.1 Make the surfaces of spaces highly reflective.

9.6.2 Lay out lighting using the task lighting principle.

> **9.6.2.1 Disconnect or remove fixtures where they are not needed.**

> **9.6.2.2 Relocate and reorient fixtures to improve energy efficiency and visual quality.**

> **9.6.2.3 Replace fixtures and improve fixture installations that waste light.**

9.6.3 Install fixtures or combinations of fixtures that provide efficient lighting for all modes of space usage.

9.6.4 Install a separate control circuit for each lighting element that operates on a distinct schedule.

> **9.6.4.1 Where light fixtures are needed in a predictable variety of patterns, install programmable switches.**

9.6.5 Install lighting controls at visible, accessible locations.

> **9.6.5.1 Provide localized control of ceiling fixtures by installing pullcord switches.**

RELATED MEASURES:

- Subsection 9.1, for improvements to incandescent lamps and fixtures

- Subsection 9.2, for improvements to fluorescent lamps and fixtures

- Subsection 9.3, for improvements to HID lamps and fixtures

- Subsections 9.4 and 9.5, for improved control of lamps and fixtures

MEASURE 9.6.1 Make the surfaces of spaces highly reflective.

You may be able to save a substantial amount of lighting energy in some spaces by using highly reflective surfaces for the walls and ceilings, and even for the floor. In most cases, this simply means using light colors. In some applications, mirrored surfaces are appropriate. Highly reflective surface colors are already accepted in offices and other "clean" spaces. Extend them to environments where dark or drab colors are commonplace, such as factories, hotel corridors, etc.

Using light colors and mirrored surfaces is relatively forgiving as a technique, i.e., it is unlikely to create any lighting problems. Improving surface reflection is usually simple, consisting of a new paint job or applying surface coverings.

However, make sure that the light surface treatment will continue to be acceptable in the future. Once you minimize the lighting equipment to exploit highly reflective interior surfaces, it can be expensive to increase lighting at a later date. For example, an interior decorator may get a notion to redecorate with darker colors, which would leave the space gloomy unless the lighting power is increased.

Energy Saving Potential

This Measure does not save energy by itself. Instead, it is a method of reducing lighting requirements. To save energy, you have to reduce the wattage of lamps and fixtures, and disconnect excess fixtures. Otherwise, this Measure will increase illumination levels, but it will not save energy.

■ **Relative Light Absorption**

The savings potential depends largely on the darkness of the existing color scheme. These are the light absorption percentages for some typical surfaces:

clean, bright white paint	about 10%
pastel colors of paint, stone, etc.	20% to 40%
typical "dark" colors	70% to 90%
clear mirror surfaces	about 10%
typical "smoky" mirrors	30% to 50%

■ **The Path of Light from Fixture to Task**

The savings potential depends on the fraction of light that is reflected from surfaces as it travels through the space. The reflectivity of surfaces is most important if light must reflect repeatedly before reaching the task.

For example, consider the amount of light that is lost if light is reflected three times on the way to the task. With a surface reflectance of 90%, about 72% (multiplying by 90% three times) of the light reaches

the task. In contrast, with a surface reflectance of 50%, only about 12% (multiplying by 50% three times) of the light reaches the task. Thus, even "medium" colors can absorb a large fraction of light if the path is indirect.

Reflective surfaces are important in spaces, such as corridors, guest rooms, and function rooms, where a large fraction of the light reflects from the surfaces. Such spaces may not have specific task areas, and the surfaces may be a primary means of distributing light in the space.

High reflectivity is also important when using area illumination, indirect lighting, and shaded light fixtures. Light loss is severe with cove lighting and valance lighting, where the light must bounce off several surfaces just to get into the space. Figure 1 shows a large area that is lighted entirely by reflection from ceiling and wall surfaces.

The surfaces of the space do not affect light that travels directly from a fixture to the task. Therefore, reflective surfaces cannot save much energy if the activities are illuminated mostly by direct lighting. For example, this is typical of work in graphics studios.

Effects on Lighting Equipment and Layout

Reflective surfaces make the layout of fixtures less critical. The fact that light can survive several reflections on the way to the activity makes it less necessary to get the fixtures close to the task.

You have to get rid of some lighting capacity to reap the energy saving benefit. If you have fixtures that are less efficient, such as incandescent wall washers, exploit the opportunity to get rid of them. The reduced need for lighting equipment is an important cost saving feature in new construction and renovations.

Effects on Space Appearance and Lighting Quality

Making surfaces brighter creates a major change in the esthetics of the space. For one thing, it gives the space a more "open" feel. This is desirable in many applications, but it may be undesirable in intimate settings.

Making the surfaces of the space more reflective increases the indirect component of lighting. This usually improves visual quality. In particular, this can improve the appearance of a space that uses task lighting, which tends to create islands of light in darker surroundings. This is unpleasant psychologically, if not visually. Making the surfaces brighter also reduces the intensity of shadows at the task.

■ Can You Make Surfaces Too Bright?

Some people are concerned that making the surfaces of spaces too reflective may cause glare. Experience indicates that this is generally not a problem. The world is full of buildings that have white walls and ceilings, and they do not appear to cause discomfort. Light interior surface colors are comfortable because the surface brightness is low and it is fairly uniform over a large area. If lighting is laid out for efficiency, the task area is brighter than the surrounding surfaces, so the eyes are not overloaded by brightness from outside the task area.

Usually, the only place you need to worry about excessive surface brightness is within the "working area" of the visual field, such as the tops of desks and tables. Even in this area, highly reflective surfaces, such as white paper, do not cause comfort problems if the lighting is laid out properly. Try to make large surfaces within the task area highly diffuse. Specular surfaces invite trouble with veiling reflections, which are explained in Reference Note 51, Factors in Lighting Quality.

Not Just Walls and Ceilings

All surfaces within the space are important if they have a substantial amount of exposed surface. This includes floors, draperies, blinds, furniture, decoration, etc.

Floor color is important because a large fraction of the light emitted from typical lighting systems first strikes the floor. It is commonly believed that floors should have dark colors to hide dirt and scuff marks. This is not true in general. Many facilities maintain a satisfactory appearance with light-colored floors. Smooth-surfaced flooring, such as vinyl, can be light in color in most commercial environments.

Whether to Use Diffuse or Specular Reflection

The directional characteristics of reflecting surfaces range from "diffuse" to "specular." "Diffuse" surfaces reflect light equally in all directions. A surface reflects diffusely if it is rough at a scale corresponding to the wavelengths of light. For example, rough masonry and flat paint are very diffuse. A diffuse surface that is reflective at all visible wavelengths looks "white."

In contrast, "specular" surfaces reflect light at exactly the same angle as it arrives. For this reason, completely specular reflection preserves images. (The word "specular" is derived from the Latin word for mirror.) Any surface that is smooth at the microscopic level produces specular reflection. For example, plate glass and polished metal are highly specular.

WESINC

Fig. 1 Lighting power is highly dependent on surface colors here Before light gets to any book in this library, it has to reflect from a ceiling surface at least once, and probably several times from ceiling, wall, floor and furniture surfaces. A slight reduction of surface reflectivity greatly reduces light loss.

In most cases, you want the surfaces of the space to be diffuse. If walls are painted flat white, they reflect a large fraction the light from ceiling fixtures horizontally into the space. In contrast, a mirror wall reflects the light from ceiling fixtures directly to the floor, where most of the light is lost.

Specular ceilings are about as efficient as diffuse ceilings, if they are not tinted. Specular floors are used only in discotheques, where they are appreciated by the boys.

Specular surface finishes are more expensive than diffuse finishes. Mirrors are used primarily as an item of decor, often to make a space appear larger.

Most smooth materials used for interior surfaces have a combination of specular and diffuse reflection. This is because the outer surface is both smooth and somewhat transparent. A portion of light is reflected from the smooth surface in a specular manner, but the rest of the light penetrates to a rough surface below, where it may be reflected diffusely. Examples are glossy paint, polished stone, and varnished wood.

Whether a surface is diffuse or specular is a separate issue from whether it is reflective or absorptive. For example, white wool and black wool fabrics are both diffuse, but the white wool is much more reflective. Black glass reflects in a specular manner, but it also absorbs a large fraction of the incoming light.

Unusual Color Effects

When using fluorescent or HID lighting, any color used in the surface may absorb an unexpectedly large amount of light. The light output of these types of lamps is concentrated in a few narrow bands of wavelengths. Color is created by pigments in the surface material. If any of the pigments has a high absorption at one of the major wavelengths emitted by the lamp, the surface will absorb more light than expected. For example, a pastel paint may have a reflectance of 80% in daylight, but it may reflect only 60% of the light from a particular HID lamp.

In most cases, you don't need to worry about this effect. Just avoid pigmented surfaces. A highly

reflective "white" is the most efficient color for general decor. Titanium dioxide, a common pigment for white paint, reflects all colors well. The same is true of most white surface materials. From the standpoints of esthetics and visual comfort, white has proven to be satisfactory for walls and ceilings in most applications.

Keep the Surfaces Clean

Keep the surfaces clean to maintain their high reflectivity. In most environments, using light colors does not increase the cleaning requirement. In industrial environments where surface cleaning may not be routine, initiate period cleaning. Dirtier environments also need a regular program of fixture cleaning, which is recommended by Measure 9.7.1.

ECONOMICS

SAVINGS POTENTIAL: *30 to 60 percent of lighting energy, typically. Depends primarily on the original color scheme and on the amount of indirect lighting.*

COST: *In new construction, it usually costs nothing to make interior surfaces reflective. In retrofit, cost varies widely. Typical painting costs about $0.15 to $0.50 per square foot. Other types of surface finish may be much more expensive. If you do the work as part of repainting or redecoration that is needed anyway, there may be little additional cost. The total cost should include the cost of changes to the lighting fixtures.*

PAYBACK PERIOD: *Usually immediate, in new construction. Up to several years, in retrofit.*

TRAPS & TRICKS

MAINTENANCE: *Light surfaces become dirty, especially in industrial environments. To preserve the benefit, schedule cleaning of the surfaces at appropriate intervals.*

EXPLAIN IT: *Future managers of the facility are unlikely to realize that the color scheme was selected to maximize lighting efficiency. Leave a record that is not likely to get lost, explaining why a light color scheme should be retained.*

MEASURE 9.6.2 Lay out lighting using the task lighting principle.

SUMMARY

Task lighting is the most efficient method of lighting layout. Its essence is close matching of the lighting geometry to individual activity areas. Try to exploit task lighting principles in all lighting. Appropriate fixtures, wiring methods, and other accessories are still lacking, so expect to improvise. Changes the appearance of spaces from "open" to "localized."

SELECTION SCORECARD

Savings Potential	$ $ $ $
Rate of Return, New Facilities	% % % %
Rate of Return, Retrofit	% % % %
Reliability	✓ ✓ ✓
Ease of Retrofit	☺ ☺

This Measure covers the main points of task lighting in its broadest sense. It also offers some suggestions for equipment used in task lighting. Task lighting is still embryonic, so it is not yet possible to define a complete set of well proven task lighting techniques. To a greater extent than in any other aspect of lighting, it is your challenge to apply general principles to develop lighting configurations that are effective for your facilities.

The most important point in this Measure is a "standard" task lighting layout that provides illumination that is both efficient and visually pleasing. This is Figure 1.

Since task lighting has not yet come into existence as a formal method of lighting design, we cannot show rationally designed examples of it in actual installations. Instead, we will show a number of typical spaces with poor lighting efficiency, and we will discuss how these spaces could be retrofitted with task lighting in a way that would radically reduce lighting power. These spaces are shown in Figures 2 through 5.

Subsidiary Measures 9.6.2.1 through 9.6.2.3 are important lighting improvements that exploit the principles of task lighting without requiring any unconventional equipment or methods. Use this Measure and the subsidiary Measures as a checklist in your new designs and in reviewing your existing lighting.

What is Task Lighting?

Task lighting can be defined as lighting that is optimized to individual activities. Task lighting minimizes the energy waste that results from spreading light around where it is not needed.

Task lighting is partly a general concept, and partly a collection of lighting techniques. Task lighting has been discussed for years, but it has not yet been applied as a mainstream approach to lighting design. It remains a sleeping giant of lighting conservation.

Task lighting was popularized during the 1970's, largely based on the notion of installing fixtures close to the task on furniture or equipment. That approach died on the vine, largely because the needed light fixtures were never developed. If that type of task lighting had been developed, the lighting quality probably would have been considered unacceptable, especially in comparison with the area lighting approach that has become almost universal today. To achieve good visual quality along with energy efficiency, a more general approach is needed, which we introduce here.

A "Standard" Task Lighting Layout

The present wasteful method of broadcast lighting is forgiving in terms of visual quality. This is because light comes from many directions. If the light coming from a particular direction causes a vision problem, such as veiling reflections, humans can still see effectively using light that is coming from the other directions.

In contrast, experience suggests that task lighting is much less forgiving of poor layout. Task lighting provides illumination from a limited range of directions, so the lighting is either mostly good or mostly bad. This makes it especially important for the layout of task lighting to avoid all the causes of lighting discomfort that are related to fixture positioning. These are the problems to be avoided, summarized from Reference Note 51:

- *source glare.* Source glare is avoided by keeping the fixture, and any localized bright spots caused by it, out of the visual field of the viewer.
- *veiling reflections.* Veiling reflections are usually caused by fixtures that are located in the direction that the viewer is facing.
- *uneven illumination.* If a single fixture is to provide reasonably uniform illumination over the task area, the distance of the fixture from the task needs to be several times greater than the width of the task area. Low ceiling heights make it impossible to satisfy this requirement. Therefore, you typically have to

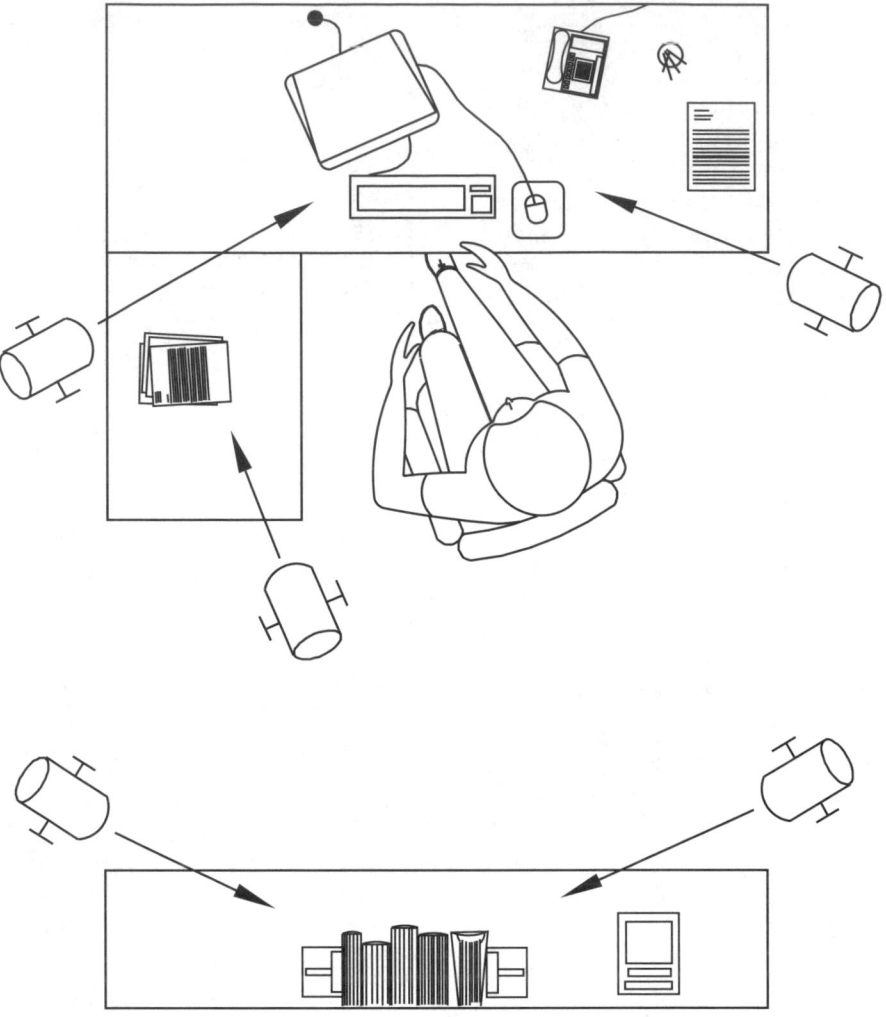

Fig. 1 A "standard" layout for task lighting To make lighting energy efficient, use efficient light sources, shine the light directly on the task area, and limit light distribution to the task area. To make lighting visually pleasing, avoid glare from any light source, avoid veiling reflections, make the illumination within the task area fairly uniform, and minimize shadowing at the task. A lighting layout that satisfies all these conditions limits the fixtures to a narrow range of positions, which are illustrated by this example of office work.

The essential features are mounting the fixtures well above the viewer, pointing them steeply downward, and placing them alongside or slightly behind the viewer. In most cases, two fixtures are needed for each task area. The fixtures can be mounted on the ceiling, on furniture extensions, or on anything else. The fixtures must be able to be aimed.

At present, compact fluorescent lamps are the only type that provides the necessary combination of energy efficiency, color rendering, and small size (for focussing). With two of these lamps, the power required to illuminate each task area is about 40 watts.

To avoid "harsh" lighting, and to minimize shadowing at the task, the fixtures should have a radiating surface area much larger than shown here. A target figure is about 20% of the distance from the fixture to the task. This can be achieved with reflector fixtures.

Task lighting allows efficient localized switching. The three fixtures for the desk might be controlled by one motion sensor, and the two fixtures for the credenza by another.

Maintaining high visual quality is challenging when the viewer changes position. In this example, if the person turns to the left to work at the desk extension, the left-hand desk fixture may cause glare. Precise adjustment of the fixture position is needed to avoid this.

install two fixtures, on opposite sides of the viewer and with partially overlapping illumination patterns.

- *self shadowing.* If a fixture is located behind the viewer, the viewer's body casts a shadow on the task. If a fixture is located to the side, the viewer's hands may cast a shadow on the task. The way to avoid this shadowing is to install fixtures on both sides of the viewer.

Fixture locations that avoid all these problems are limited. In fact, acceptable locations are so limited that they become, in effect, a "standard" fixture geometry for task lighting.

This "standard" task lighting layout is shown in Figure 1. In this layout, any fixture is placed alongside the viewer, so that the beam makes an angle of about 70 degrees with the ceiling. If two fixtures are used, one on each side, both can be moved slightly to the rear of the viewer.

You have to establish this layout separately for each task area. From the standpoints of both efficiency and visual comfort, a task area can be defined as the area within a stationary person's visual field where illumination is needed. If the person moves to look in a different direction, that defines a different task area.

For example, Figure 1 is an office layout with a desk and a credenza. Each is a separate task area, and each needs its own pair of fixtures.

In principle, the "standard" layout geometry applies to any light source, whether it is an actual fixture, a reflector, a skylight, bounce light, etc. However, achieving optimum energy efficiency with electric lighting (as opposed to daylighting) generally requires fixtures that can be aimed at the task area. The "standard" layout of Figure 1 uses swiveling surface-mounted compact fluorescent aiming fixtures.

Advantages of Task Lighting

As you can see, task lighting is a big departure from conventional lighting design. To persuade yourself to try it, consider these potential advantages:

- *good visual quality.* There is no source glare, because the fixtures are entirely outside the visual field. There are no veiling reflections. It is easy to make the illumination level for each task area adjustable individually. It is easy to select lamps with different color rendering characteristics for each task area. Shadowing in the task area can be minimized by using fixtures with large surface area.

- *maximum energy efficiency.* Light is placed precisely *where* it is needed. It is easy to control the lighting of each task area to the times *when* it is needed.

- *easy maintenance.* The location and types of fixtures make lamps easy to replace. In most facilities, only a minimum variety of lamp types are required.

- *adaptability.* Because the fixtures are surface-mounted, ceiling grid patterns and equipment above the ceiling do not constrain the location of fixtures. The fixtures are much easier to relocate when the activities in the space change. (However, elegant and inexpensive electrical connections that simplify moving the fixtures are not yet available as a standard product.)

- *low initial cost.* The cost of individual fixtures is likely to be lower. Fewer fixtures are likely to be needed. Installation cost may be much lower. Far less difficulty may be encountered in retrofit.

- *independence from adjacent lighting.* Each task area has totally independent lighting, so it does not matter whether the lighting in adjacent areas is turned on or off. If necessary, glare shielding for the benefit of adjacent areas (discussed below) is easy to install.

Task Lighting Challenges

Tapping the full potential of task lighting requires innovation. Innovation involves cost, risk, and delay. Most organizations with modest resources have to wait for industry-wide improvements before they can apply task lighting in its most efficient form. On the other hand, an organization that is able to assign a talented individual to work on lighting efficiency may be able to achieve great improvements after a period of innovation and experimentation. When you set out to apply task lighting, be prepared for these differences between task lighting and conventional lighting:

- *still emerging equipment.* Until a few years ago, the right combination of fixtures and lamps for task lighting was not available. Conventional fluorescent and HID light fixtures are poorly adapted to illuminating individual task areas. Lack of ability to aim these types of fixtures is a critical weakness. Some incandescent lamps and fixtures are well adapted to aiming and localizing lighting distribution, but incandescent lighting is inefficient. However, the compact fluorescent unit is probably the breakthrough that will bring all the essential elements together. It has reasonably high efficiency, the appropriate range of wattage and light output, and fairly good color rendering. The small size of the unit makes it usable in fixtures that can be aimed. The fixtures themselves are not yet available as standard items, but they could be produced for a low price with only modest equipment and investment. The discussion below reports some successful task lighting that was set up on an experimental basis.

- *incomplete understanding of lighting comfort.* The initial failure of task lighting was caused partly by dissatisfaction with the quality of the light. In task lighting, the factors involved in visual comfort are less forgiving than in general lighting. However, all the essentials of visual comfort are covered in Reference Note 51, and it appears that task lighting can be created that will satisfy all the requirements.

- *need for greater design effort.* Contemporary lighting design achieves comfortable, attractive lighting with rote layout and conventional equipment. In contrast, task lighting requires a great deal of study, experimentation, innovation, and risk. These involve liability, and they add time to the design phase of the project.

- *radically different appearance.* Task lighting may dramatically change the appearance of the space. Task lighting does not have the regular checkerboard appearance of contemporary lighting. Light levels are not uniform throughout the space. If fixtures are ceiling-mounted, they may point in different directions and may extend different distances below the ceiling. Fixtures need large surface areas to avoid glare, but task lighting fixtures cannot be hidden in the ceiling, as in contemporary lighting. Therefore, task lighting fixtures are prominently visible as an element of the decor. Reflectors and glare shields, essential elements of task lighting, also project below the ceiling. The present stylistic prejudice toward making light sources invisible must be abandoned.

The Cost of Task Lighting

It is too early to make generalizations about the cost of task lighting. Efficient lighting in the future may require far fewer fixtures and lamps, but individual fixtures may (or may not) be more expensive. Wiring labor, which is a major component of the cost of installing lighting, will be reduced by developments that allow fixtures to be relocated more easily. In the long run, it seems likely that task lighting may reduce the initial cost of lighting. In the near term, task lighting is burdened with the high cost of innovation.

Relationship to Lighting of Surrounding Area

Task lighting can provide good illumination within the boundaries of a localized activity area. But, by definition, task lighting does not illuminate the surroundings. People do not like the sensation of being stranded in an island of light. This problem is largely subjective, as is the amount of background lighting that is considered desirable. A person who expects dark surroundings to be unpleasant is likely to want more background lighting. On the other hand, people who are interested in conserving energy may not feel a need for much background lighting. For more about background lighting, see Reference Note 51.

A task-lighted area that is surrounded by other task-lighted areas may not need background or transition lighting. For example, individual desks in an open office area typically do not require additional lighting between

them. Whether background lighting is needed depends on the nature of the activities, the color scheme, the amount of light spilled out of the task lighted areas, the amount of traffic through the unlighted areas, etc. In many spaces that have windows, daylight can provide ample background lighting for daytime activities.

How to Avoid Excessive Shadowing at the Task

Light sources with small surface area create "harsh" lighting, which is primarily the result of sharp shadows cast by objects in the task area. For details, see Reference Note 51. This factor is especially important in task lighting because the fixtures tend to be smaller and less widely distributed.

To avoid excessive shadowing, you need fixtures that have a large surface area. This is perhaps the most radical innovation that is needed to make task lighting suitable for general use. Specific methods of achieving this are given below.

How Important is Glare from Reflections of Light Sources?

Bright light sources may create glare by reflection from objects in the viewer's field of vision. Most task areas include objects that can pick up reflections of light sources, for example, computer screens, the polished surfaces of machine tools, crystal paperweights, etc. Whether reflections of the light sources will cause an objectionable amount of glare depends on these factors:

- *the surface texture of the illuminated objects.* Smooth surfaces produce specular (mirror-like) reflections that concentrate glare. On the other hand, diffuse surfaces cannot create reflected glare. For example, reflections of the light fixtures are more troublesome for a machinist handling metal objects than for a sales clerk handling stuffed toys.

- *the curvature of illuminated objects.* If smooth surfaces within the visual field are highly curved, they reflect little of the light from the source into the eyes of the viewer. For example, reflections from the curved handle of a coffee cup are insignificant. On the other hand, fairly flat surfaces focus much more light into the viewer's eyes. They can cause extreme glare, as in sheetmetal work.

- *the orientation of illuminated surfaces.* In order for glare to occur, the surface must be oriented to reflect the image of the fixture into the eyes of the viewer. If the surfaces are fixed, you can avoid glare by laying out the lighting geometry to avoid it. Trouble arises when the surfaces are movable, for example, when reading magazines.

- *the steadiness of the viewer's gaze.* The importance of reflected glare depends on how much the viewer's gaze is moving. For example, reflections from a computer monitor are serious because the user stares at the screen. In contrast, reflections from the bright stainless steel surfaces in a kitchen are tolerable to the cooks because they shift their gaze continually.

• *the brightness of the light source.* Annoyance increases with the surface brightness of the light source. For a given amount of light delivered to the task, average surface brightness is inversely proportional to the surface area of the light source. A naked incandescent bulb is much more annoying than a fluorescent fixture with a diffuser.

Of these factors, the only one that you can control in the lighting design is the last one. If the task area contains objects that can reflect an image of the light sources into the eyes of the viewer, reduce the brightness of the light sources by increasing the surface area of the fixtures.

How to Achieve a Large Light Source Area

Previously, we made the point that task lighting fixtures should have a large surface area of uniform brightness. These characteristics reduce shadowing and reduce glare from reflections of the light source. Because task lighting is a pioneering area, there are no established guidelines for fixture surface area. As explained in Reference Note 51, experiments suggest that there should be a ratio of at least 1:5 between the average width of the fixture's radiating surface and the distance between the fixture and the task.

There are two approaches for reducing the brightness of lamps, using diffusers and using reflectors. The two differ in light loss, control of the light distribution pattern, ability to shield the lamp from view, and ease of installation. We'll talk about diffusers first, then reflectors.

■ Lamp and Fixture Diffusers

Most general commercial lighting uses diffusers. These reduce surface brightness and disperse the light. In task lighting, the first function is useful, but the second is usually a disadvantage. Conventional diffuser material disperses light too widely for efficient task lighting, unless the task area is large or the fixtures are very close to the task.

In principle, surface brightness can be reduced to any desired level by increasing the diffuser area. The diffusers of typical fluorescent fixtures increase surface area and reduce the surface brightness by a factor of two to five, compared to the surface area of the bare tubes. This is usually adequate for most task lighting. Diffusing incandescent or HID task lighting requires some creativity. At present, there are no standard types of large-area diffusers for these sources. Custom fixtures can be fabricated using standard fixture components and diffuser materials.

Conventional diffuser materials used with fluorescent lighting absorb 30% (e.g., clear prismatic) to 50% (e.g., milky white) of the light produced by the lamps. Some diffuser materials, such as the typical shades of table lamps, are too absorptive for efficient lighting.

■ Reflector Fixtures

A new concept for achieving large fixture surface areas in task lighting is using reflector fixtures. Reflector fixtures not only provide large surface areas, they also provide great flexibility. They can easily be aimed and they are less constrained in mounting location than light fixtures. Thus, reflector fixtures can adapt to changing conditions, which are an essential feature of efficient lighting.

Shaped specular reflectors, such as the photoflood fixtures discussed below, can provide good control of the light distribution pattern when used with incandescent lamps, compact fluorescent lamps, and HID lamps. In fact, reflector fixtures can limit the illuminated area so sharply that it may be desirable to broaden and soften the edges of their distribution patterns. For example, some photoflood fixtures allow you to modify the breadth of the pattern by adjusting the depth of the lamp within the fixture.

Diffuse reflector fixtures that use conventional fluorescent lamps, also discussed below, can provide good task lighting for larger task areas.

A problem with reflector fixtures is that they leave the bright lamp visible from the side, causing glare in adjacent areas. Even conventional fluorescent lamps may be too bright for some applications. A solution to this problem is to install a light shield around the lamp, as described below.

Reflectors generally absorb less light than diffusers. The reflectivity of most eligible reflector types is 85% to 90% when clean, whether the reflector is diffuse (e.g., white enamel) or specular (e.g., spun aluminum). As with diffusing fixtures, the total light loss is increased by multiple reflections and light trapping in the fixture.

■ Bounce Lighting

Bounce lighting achieves a large light source surface area by using a surface of the space as a diffuse reflector. A common example is using a floor-mounted torchere lamp to create a bright spot on the ceiling adjacent to a task area. Wall-mounted sconces bounce light from both walls and ceilings, which can be efficient for some task layouts.

At present, bounce lighting is used primarily for area lighting, which can be done on any scale. For example, the subway stations in Washington, D.C. are lighted by upward-aimed HID fixtures installed on the subway platforms that bounce light off the concrete ceiling. Bounce lighting has rarely been used for task lighting.

Bounce lighting has several advantages. Many common types of fixtures can be used to project light on a surface, so equipment is inexpensive and installation is simplified. Also, you can create a virtual light source of any desired size. Bounce lighting makes it easy to keep the lamp hidden from view, eliminating light source glare.

The main disadvantage of bounce lighting is inefficient aiming. Bounce lighting provides no control of light distribution other than by proximity to the task. The "standard" task lighting layout defined previously requires the lights source to be above, alongside, and somewhat behind the viewer. If a fixture is aimed at this location on the ceiling to create bounce light, a large fraction of the light is lost outside the visual field.

You can use reflectors to reduce this light loss, as discussed below. For example, aim a directional lamp into a reflector to concentrate the reflected light toward the task. However, aiming is crude even with such arrangements. Experiments conducted by Wulfinghoff Energy Services, Inc. showed that the energy efficiency of bounce lighting cannot approach the efficiency of other methods of task lighting. Diffuse reflection simply does not provide sufficient control of the light distribution pattern.

Absorption losses for bounce lighting are about the same as for diffuser and reflector fixtures. The absorption of typical clean white and light pastel surfaces ranges from 15% to 30%.

Aiming Techniques

One of the basic elements of task lighting is careful aiming of the light at the task area. At present, there is a lack of commercially available fixtures that combine good aiming with efficient (i.e., fluorescent) lamps. Therefore, this aspect of task lighting requires innovation. The following are two general ways of aiming light toward the task area.

■ Fixtures That Can be Aimed

At present, aiming fixtures are available as standard equipment only for incandescent lamps. Compact fluorescent lamps can be used in many incandescent fixtures in a way that provides adequate illumination levels and beam control for task lighting. However, compact fluorescent lamps are too small in frontal area to avoid shadowing.

To explore task lighting with fixtures of large surface area, Wulfinghoff Energy Services, Inc. conducted experiments with a variety of fixtures. By far the best results were obtained by using two large parabolic aluminum photoflood fixtures for each task area. These reflectors have some surface etching to soften the beam. The fixtures were installed on opposite sides of the viewer, just below the ceiling, in the "standard" task lighting configuration described previously. Each

WESINC

Fig. 2 Task lighting challenge No. 1 This is a large reading room in a public library. The present light fixtures shine mostly on the carpet and on empty tables. Daylighting is not being exploited. All reading occurs at individual furniture units, which can be moved and reoriented. There are many windows, and they extend high on the wall. What would you do to make this lighting efficient and pleasing?

photoflood fixture used one 23-watt screw-in fluorescent lamp. The combination of a compact fluorescent lamp inside a deep reflector fixture works well because the lamp emits most of its light perpendicular to its axis, so there is little direct light from the small lamp in comparison with the light coming from the reflector.

In experiments conducted in a typical office, the beam of each reflector was aimed at its respective side of the desk. The beams overlapped, providing illumination that was essentially uniform within the task area. The average illumination level was about 25 footcandles (250 lux), which was more than adequate for typical office paperwork and word processing. (25 footcandles is considered a low illumination level for office work in the United States, but not in most other countries. This illustrates that illumination levels can be reduced substantially if glare and veiling reflections are eliminated.) Shadows were light and diffuse. There were no bright reflections in the computer screen or elsewhere, but the screen had a light haze from being directly illuminated by the fixtures. Visual comfort remained good over a test period of several weeks. There was no discomfort when facing either side of the desk.

In summary, this arrangement provided illumination of high quality with an energy expenditure of only 50 watts per work station.

Good results were also achieved using a pair of industrial-type skirted fluorescent fixtures for each work station. The fixtures were fitted with brackets to allow aiming, and were installed just below ceiling level. Each fixture has two 24-inch, 20-watt tubes, for a total power of about 100 watts per work station. The illumination level was somewhat lower, about 20 footcandles (200 lux), which was satisfactory for routine word processing. The fixtures spread light well outside the task area, which was a desk of normal size. This factor accounts for the lower illumination level. The illumination was uniform and pleasant.

These experiments demonstrated the feasibility of task lighting for general office work using aiming fixtures. They also showed that fluorescent task lighting fixtures can be produced at low cost. If appearance is not a major factor, task lighting fixtures can be adapted from fixtures that are readily available.

Note that fluorescent and HID lamps may have limitations in mounting position. These are explained in Subsections 9.2 and 9.3, respectively.

WESINC

Fig. 3 Task lighting challenge No. 2 This is an office area with very tall ceilings in a public administration building. There are several desks, only one of which is being used at the moment. There is a conference table, some visitors' chairs, and a lot of empty space. It appears likely that the arrangement of the space will change repeatedly. Illumination is provided by uniformly spaced downlights that bear no relationship to the work in the space. There is a large amount of daylighting from tall windows, which is not being exploited. How would you make the lighting in this space efficient?

■ Aiming with Reflector Panels

Reflector panels are a way of improving the aiming of conventional fixtures. For example, if a reflecting panel is installed vertically alongside a fluorescent fixture that has a "batwing" distribution pattern, the distribution pattern becomes a single-lobed beam centered about an angle of 70 degrees from the ceiling, a geometry that is desirable for avoiding veiling reflections. The additional reflected light on the task may allow partial delamping of the fixture. In experiments with office task lighting, reflector panels increased the illumination of the task from 40 to 80 percent when installed with existing fluorescent ceiling fixtures.

This is primarily a retrofit technique. For example, it may be the most practical way to improve the light distribution of existing fluorescent fixtures. In new construction, consider other methods of controlling light distribution.

The surfaces of the reflector may be specular or diffuse, depending primarily on the existing fixtures. With fixtures that have broad light distribution, such as conventional fluorescent fixtures, use specular reflectors to avoid further scattering of the light.

Reflector panels can be fabricated easily from a wide variety of inexpensive materials, including paper, fabric, aluminized paper and plastic, and polished aluminum with an anti-oxidant coating. For example, specular reflectors can be made from various types of aluminum stock used to make the fluorescent fixture "reflectors" covered by Measure 9.2.7.

Panels may be flat, bent, or curved as needed to improve light distribution, to provide more effective glare shielding, and to provide rigidity. For example, a reflector may be installed around two of the four sides of a fluorescent ceiling fixture.

Coordinate reflectors with glare shielding, as described under "Glare Shielding," below. With diffuse reflector panels, the illuminated side will appear excessively bright to persons outside the task area. On the other hand, a specular reflector will not allow a distant viewer to see an image of the bright fixture, provided that the reflector is tilted backward somewhat from the vertical. Any type of downward projecting reflector shields against glare from the rear of the panel. The rear sides of the panels can be decorated.

Fixture Mounting

Fixture mounting is another area that needs development to make task lighting practical on a large scale. The fixture mounting method should provide:

- *ability to locate the fixtures precisely.* The "standard" geometry for task lighting requires fairly precise fixture placement.

WESINC

Fig. 4 Task lighting challenge No. 3 This credit union office has fluorescent lighting in a tall ceiling, which is illuminating nothing at this moment. There are a few desks and work tables, some shelves, and a counter for the public. The arrangement of the space, and even its basic use, will probably keep changing. How would you provide efficient and pleasing lighting for all the activities, now and in the future?

• *ability to aim the fixtures precisely.* The mounting should allow the fixture to be aimed horizontally and vertically.

• *adaptability to changing conditions.* It should be easy, inexpensive, and safe to move fixtures from one location to another as activities are moved.

■ Distance from Fixture to Task

Contrary to intuition, the distance of a fixture from the task has no effect on task lighting efficiency, if the light distribution pattern of the fixture can be shaped properly. Light is not absorbed in the space between the fixture and the task. However, fixtures with poor control of the light distribution pattern must be mounted close to the task to minimize light loss outside the task area.

The distance between the fixture and the task matters primarily because it affects the ratio of the fixture surface area to the distance from the task. (To review, we want this ratio to be large to prevent excessive shadowing and to reduce surface brightness.) For example, if the distance of a fixture from the task is doubled, the surface area of the fixture must be increased by a factor of four to maintain the same ratio. Larger fixtures are more expensive, more difficult to install, and more difficult to make attractive. So, try to keep task lighting fixtures

reasonably close to the activities. The limitation on getting the fixture close to the task is maintaining sufficiently uniform illumination.

■ Ceiling Mounting

Expect task lighting fixtures to be surface-mounted on ceilings, rather than recessed into the ceilings. Surface mounting allows fixtures to be moved laterally by any desired amount. It also allows fixtures to be rotated, which is essential for aiming the light. Suspended ceiling systems with lay-in fixtures are not suitable for task lighting, because the grid system does not allow the fixture to be positioned properly, and it does not allow fixtures to be rotated.

■ Task Mounting

Task lighting fixtures may be installed on furniture, equipment, or partitions. Task mounting offers a number of significant advantages. It fixes the lighting geometry, it eliminates the problem of relocating the lighting when the task is relocated, and it simplifies the installation of controls.

However, task mounting produced unsatisfactory results during the 1970's wave of task lighting. The problems were basic, namely, source glare, veiling reflections, and non-uniform illumination within the visual field. The underlying cause was that the fixtures

WESINC

Fig. 5 Task lighting challenge No. 4 Lighting is a fundamental part of merchandising in this department store, and it is a major part of the store's operating cost. The present lighting is a combination of fluorescent fixtures and incandescent downlights located in a repetitive pattern, without regard to the location of the merchandise or the traffic flow. How would you save lighting energy here while improving the merchandising appeal of the lighting?

were too close to the task. A typical example was mounting a fluorescent lamp underneath a shelf in an office module.

These visual problems can be alleviated by installing the fixtures at some distance from the tasks. With task mounting, this can be done by installing outriggers on desks for holding the fixtures. This may be awkward. However, keep an open mind. There are no fixed rules in this game, yet.

■ Mounting Bounce Light Fixtures

Fixtures that bounce light off the ceiling or other surfaces can be installed in many ways. The reflecting surface usually is diffuse, so the location of the fixture does not materially affect the direction that light is reflected. You can use many types of fixtures to create bounce lighting. Torchere lamps are handy, but they are limited to creating the lighted spot directly overhead. Pole lamps with one or more movable sockets overcome this problem. Table-mounted lamps may be suitable, but they take up space. Attractive reflector lamps can be installed on any surface or object, and they have great flexibility in aiming.

The main guideline is to install the fixtures so that the lamps and the bright surfaces of the fixtures are not visible from occupied portions of the space. One way to do this is to make the fixtures tall enough so that the bright portions of the fixtures are above eye level.

Glare Shielding

Task lighting for one task area may create serious glare at other task areas in the same space. This occurs if the fixtures are installed so that their lamps or bright surfaces are visible from adjacent areas. Glare from adjacent task lighting may be especially problematical if the fixture is aimed away from the vertical. Even though a fixture is tilted only slightly in the direction of an adjacent viewer, it will appear much brighter to a person in the distance than if the fixture were facing straight down. Furthermore, task lighting usually has to be below the ceiling level, so it intrudes more into the field of view of adjacent viewers.

The solution to this problem is to shield the fixtures from view from other occupied areas. This is another novel element of successful task lighting. Glare shielding is limited only by your creativity. Fortunately, it can be easy and inexpensive.

The shape and location of glare shields are not critical. There are only two conditions: (1) the glare shields must lie in the line of sight between the bright parts of the fixture and the eyes of persons in adjacent areas, and (2) they must lie outside the main beam of the fixtures, so that they do not become sources of glare themselves. You can use any opaque or translucent material for shielding, such as fabric, colored paper, magazine centerfolds, etc.

With aimed fixtures, glare shielding can take the form of skirts or blinders attached to the fixtures themselves. The sides of the skirts facing the lamps are flat black. This technique is commonly used with theatrical lighting fixtures and with some display lighting. To avoid wasting energy, install the skirts well outside the main part of the light beam. You can shield a group of fixtures by suspending a curtain or a series of panels around the fixtures.

If you use reflector panels for aiming, as suggested above, integrate glare shielding with the reflector panels.

Get people used to the fact that glare shielding will stand out in the decor. With a bit of imagination, you can accomplish the shielding in ways that are quite decorative. For example, one office uses marine signal pennants suspended from the ceiling alongside fixtures.

Glare in adjacent areas is generally not a problem with task lighting that is installed below the tops of furniture or partitions. However, it is difficult to achieve good visual quality with low-mounted task lighting, for the reasons given previously.

Lamp Efficiency

You want task lighting for its energy efficiency. Therefore, use efficient lamps. Until recently, there was a strong temptation to use incandescent lamps and fixtures for task lighting because these were the only kind that allowed precise aiming at short range. However, the inefficiency of incandescent lamps cancels the advantage of accurate aiming.

For most applications, only fluorescent lighting offers good lamp efficiency. Much of the bother and innovation involved in the preceding discussion stems from the need to make the light distribution of conventional fluorescent lighting more controllable.

Compact fluorescent lamps may prove to be a breakthrough. They combine most of the efficiency potential of fluorescent lighting with the aiming capability of incandescent lamps. The best compact fluorescent lamps are perhaps 70% to 80% as efficient as the best conventional fluorescent lighting, and they have all the other advantages of fluorescent lighting.

For these reasons, base your initial task lighting designs on compact fluorescent lamps. Consider installing them in large reflector fixtures, as described previously.

HID lamps are almost point sources, so you can combine them with reflector fixtures to tailor the light distribution pattern accurately. Consider HID lighting for task lighting of large areas, especially where the fixtures cannot be installed close to the tasks. Color rendering, previously a serious weakness of HID lighting, has been greatly improved in recent years.

HID lighting is not suitable for small task areas, because small HID lamps are substantially less efficient than fluorescent lamps in small wattages. Also, remember that HID lamps cannot be turned on and off at short intervals.

Control Fixtures Efficiently

The ability to turn lights off at individual activity areas is one of the major efficiency advantages of task lighting. Be sure to exploit this advantage by using controls that minimize the duration of operation within each task area. Refer to Subsections 9.4 and 9.5 for control devices. Also, see Measure 9.6.4, in this Subsection, for efficient control circuiting.

Do Not Duplicate Lighting!

Task lighting saves energy only if it displaces other lighting. It makes no sense to add task lighting to area lighting. Doing so is likely to increase energy consumption, rather than reduce it. Eliminate area lighting when you install task lighting.

Accommodate Each Mode of Space Usage

If the activities in the space change on a predictable basis, provide lighting that is efficient for each mode of space usage, along with a separate control for each lighting function. This is covered by Measure 9.6.3. Task lighting is likely to be inadequate for auxiliary functions, such as transit and cleaning. If you are retrofitting task lighting in an existing facility, it may be appropriate to leave some of the original area lighting to serve these functions.

Rehearse Before You Perform

As we said at the beginning, the methods of effective task lighting have not yet found their way into the mainstream of contemporary lighting design. You will have to be a pioneer to exploit this powerful method. Before you commit to actual installations, design task lighting for a variety of sample situations in detail. Practice dealing with all possible considerations, such as the presence of daylight and the glare from it, uncertainly in work station locations, and so forth. Because task lighting may be affected by virtually every other aspect of lighting design, review all the Measures of Section 9 in terms of their potential significance to task lighting.

As a start, think how you would redesign the lighting in the spaces shown in Figures 2 through 5. Then, go out and practice redesigning as many inefficient lighting installations as you can find. Look for an opportunity to actually test the concepts we have discussed here. Start small. Build on your successes. If you are successful, tell others how you did it.

ECONOMICS

SAVINGS POTENTIAL: *50 to 80 percent of lighting energy, compared to conventional area lighting. This Measure also provides large savings in lamp and lamp replacement costs.*

COST: *$1 to $10 per watt of task lighting capacity, in retrofit applications. Eventually, task lighting will reduce equipment cost in new construction. In the meantime, the cost for task lighting in new construction depends largely on the imagination of the lighting designer and the scale of the project. Design and development costs will be high until task lighting becomes common.*

PAYBACK PERIOD: *One year, to many years.*

TRAPS & TRICKS

INNOVATION: *Want to become famous? Have either a major success or a major failure with task lighting. Don't tackle task lighting, in the most innovative sense, unless you have a lot of time, a modest budget for equipment, and some spaces with occupants who are willing to be test subjects. If you decide to become a pioneer in task lighting, check everything you do against the lighting quality factors explained in Reference Note 51. Let your imagination roam free, but start with compact fluorescent lamps as your lamp type. If appearance matters and your facility has an interior decorator, get her involved as soon as you are ready to go beyond the prototype stage.*

EQUIPMENT: *Don't expect to find ideal task lighting fixtures in the standard catalogs. Specialty manufacturers will produce custom fixtures for you. But, don't place a large order until you have thoroughly tested your concepts with prototypes.*

INSTALLATION: *Install every fixture on a custom basis, including working out sight lines, installing glare shields, etc. If you contract for the labor, make sure that everybody knows this beforehand. Demand the time and skill required.*

MEASURE 9.6.2.1 Disconnect or remove fixtures where they are not needed.

Contemporary lighting practice uses many more fixtures than required for efficient and effective illumination. Typically, a uniformly high level of lighting is installed, avoiding the need to customize lighting to the activities. This creates an opportunity to save energy by eliminating the excess fixtures. This is quick and inexpensive, but it requires thought. You need to examine the environment of each fixture individually to decide whether the fixture can be disconnected.

In the case of fluorescent lighting, consider this Measure to be an extension of Measure 9.2.1, which recommends "delamping" and "deballasting" fluorescent fixtures.

Figures 1 and 2 provide two examples of fluorescent lighting where both delamping and removing excess fixtures are appropriate.

Remove the Lamps, Disconnect the Fixture, or Remove the Fixture?

You can take a fixture out of service by removing the lamps, by disconnecting the power to the fixture, or by physically removing the fixture. The last method is usually best, because the presence of an idle fixture motivates putting it back into operation. Also, it looks better to remove the fixture completely and refinish the previous mounting area.

WESINC

Fig. 1 One desk does not need twelve fluorescent lamps
This small office has three 4-lamp fixtures. The right front fixture creates serious veiling reflections at the desk. The right rear fixture cannot illuminate the desk properly because it is behind the viewer's body. Both could be eliminated. The fixture on the left will provide ample illumination. The office has a large window, which could provide daylighting. This would require reorienting the desk and improving the window treatment.

SUMMARY

Typically saves a modest amount of lighting energy. Quick and cheap.

SELECTION SCORECARD

Savings Potential .... $ $ $

Rate of Return, New Facilities % % % %

Rate of Return, Retrofit.......... % % % %

Reliability .... ✓ ✓ ✓

Ease of Retrofit .... ☺ ☺ ☺

If you do not remove idle fixtures, at least disconnect the wiring to them. Disconnecting power at a circuit breaker panel makes it too easy to restore power. Similarly, removing the lamps is ineffective by itself because the lamps will be replaced sooner or later. Furthermore, this practice involves some safety hazard because the electrodes are left exposed.

All fluorescent and HID fixtures have ballasts. A ballast continues to draw current even if the lamp is removed. Therefore, disconnect the wiring ahead of the ballasts, rather than just removing the lamps. With fluorescent fixtures, you have the choice of leaving some of the lamps in operation. See Measures 9.2.1 ff for the details of delamping fluorescent fixtures.

Label Disconnected Fixtures

If you leave disconnected fixtures in place, label them to point out they have been disconnected for the sake of energy conservation. Otherwise, occupants and staff will be in doubt as to whether the fixture should be operating. Use the labeling techniques recommended by Measure 9.7.3.

Maintaining or Improving Visual Effectiveness

At first glance, it might seem that deactivating fixtures must reduce visual effectiveness. On the contrary, careful planning can often improve visual effectiveness in facilities that have excessive lighting. Make this a primary goal. Review the lighting quality issues in Reference Note 51, Factors in Lighting Quality. When eliminating fixtures, pay special attention to:

- *maintaining appropriate illumination intensity.* Carry a light meter when identifying fixtures. Measure illumination levels at the task. Account for decline in light output as lamps age, and for greater light requirements by older people.

• *eliminating veiling reflections and source glare.* In many cases, deactivating fixtures can reduce veiling reflections and glare. The improvement in visual quality partially compensates for the reduced illumination level.

• *color rendering.* Consider whether the fixtures to be deactivated may be necessary for good color rendering, for example, incandescent fixtures used to highlight paintings.

• *background lighting.* All environments require some amount of background lighting, so do not disconnect fixtures to the extent that the surroundings appear excessively dark.

Conflict with Lighting Decor

The most serious resistance to deactivating lighting is likely to occur in environments where a regular pattern of fixtures is considered to be part of the decor. Disconnecting fixtures makes them appear defective. In many cases, you can avoid this problem by removing the fixture completely, leaving no evidence that a fixture was once located there. For example, remove lay-in fluorescent fixtures and replace them with ceiling tiles.

The esthetic problem tends to disappear if enough fixtures are removed to eliminate the appearance of a repetitive pattern of fixtures. In many cases, removing about a third of the fixtures accomplishes this.

Common Opportunities

The following are some common situations where fixtures can be taken out of service. Many other situations may exist.

■ Area Lighting

Almost anywhere fixtures are installed without regard to individual tasks, as in the common checkerboard pattern, there is an opportunity to deactivate a major fraction of the fixtures. Indiscriminate area lighting is sometimes justified on the grounds that it provides "background lighting." In most cases, area lighting is much brighter than needed for this purpose.

■ Fixtures that Illuminate Dead Space

Examine where each fixture is throwing its light. If most of the light is falling on places that people cannot see, or have no need to see, disconnect the fixture. For example, disconnect fixtures installed over the tops of shelving, over walk-in freezers, etc.

■ Overlapping Fixtures

Facilities such as ballrooms and upscale restaurants often have various types of lighting fixtures that illuminate the same area. This is done to make the lighting adaptable to different types of functions. Such configurations are an invitation to use more light fixtures than are needed for any given purpose. You can often eliminate the least efficient fixtures. Retain lighting

WESINC

Fig. 2 Easy savings This government office has a high level of illumination for a few desks and work tables, some filing cabinets, and a lot of empty space. Perhaps 80% of the energy cost of lighting this space could be eliminated with easy changes. Most of the fixtures could be eliminated or delamped. Adding some inexpensive controls would adapt the space to occupancy and exploit the ample daylighting that is available. Task lighting would further reduce the remaining energy consumption.

flexibility by improving the controls (Subsections 9.4 and 9.5) and switching layout (Subsection 9.6). Using dimming with fluorescent fixtures (Measure 9.2.6) is another approach that is now practical.

■ Downlights

Installing downlights indiscriminately has become a careless habit of architects. As usually installed, downlights are very inefficient, so that other fixtures in the space do a majority of the actual lighting. In such cases, you can remove the downlights. Even in cases where downlights provide useful illumination, it may be desirable to replace them with more efficient types. See Measure 9.6.2.3.

■ Wall Washers

In most cases, wall washers are a decorative extravagance. In interior spaces, they may contribute to overall illumination. As an alternative, consider making the walls more reflective. See Measure 9.6.1 for the details.

■ Cove Lighting

Cove lighting is inefficient because light must be reflected many times to reach the area to be illuminated. It is so inefficient that other fixtures may be installed to do most of the actual space lighting. In many facilities, cove lighting is primarily a decoration. See Measure 9.6.2.3 for tips on eliminating cove lighting.

■ Outside Lighting

Take a walk around the facility at night. You may find some fixtures that no longer serve a useful purpose.

ECONOMICS

SAVINGS POTENTIAL: *10 to 70 percent of lighting energy. This Measure may also provide large savings in lamp and lamp replacement costs.*

COST: *In new construction, this activity may save money. In retrofit, the cost is usually minimal.*

PAYBACK PERIOD: *Less than one year.*

TRAPS & TRICKS

PLANNING: *In existing facilities, survey the facility fixture by fixture. Eliminate any that are providing only useless illumination. Eliminate any that cause veiling reflections or glare, and see whether the illumination level remains high enough. If it does not, go to Measure 9.6.2.2.*

MEASURE **9.6.2.2 Relocate and reorient fixtures to improve energy efficiency and visual quality.**

RATINGS

New Facilities   Retrofit   O&M

A   **C**   ☐

This Measure goes one step beyond Measure 9.6.2.1 by relocating the fixtures to make them more effective. This Measure requires much more labor. In spaces that have more than one type of lighting, combine this Measure with Measure 9.6.3.

Relocation Strategy

Your objective is to provide good illumination while saving energy. To provide good illumination, observe the principles summarized in Reference Note 51, Factors in Lighting Quality. To save energy, try to achieve these objectives as you move fixtures:

- *reduce lamp wattage.* For example, moving a 4-tube fluorescent fixture to an optimum location over a work station may allow two of the four tubes to be removed.

  You can reduce fixture wattage in a variety of ways. Refer to Subsection 9.1 for incandescent lighting, Subsection 9.2 for fluorescent lighting, and Subsection 9.3 for HID lighting.

SUMMARY

After eliminating unnecessary fixtures, this step makes better use of the fixtures that remain. Can often improve visual quality. May be difficult in retrofit because of interference in the ceiling.

SELECTION SCORECARD

Savings Potential .... **$** $ $

Rate of Return, New Facilities **% % % %**

Rate of Return, Retrofit.......... **% %** % %

Reliability .... ✓ ✓ ✓ ✓

Ease of Retrofit .... ☺ ☺

- *eliminate fixtures.* For example, in a restaurant that uses downlights for general lighting, relocating the downlights directly over the tables may allow you to remove half of them.
- *increase the subdivision of control.* For example, in a large office area, moving the fixtures over individual desks allows them to be turned off when the individual user leaves the space. Exploiting this opportunity usually requires circuit modifications and additional controls, as explained in Measures 9.6.4 ff.

How to Relocate Fixtures

If fixtures are laid in suspended ceilings, it may be fairly easy to move the fixtures, but the available positions are limited by the dimensions of the ceiling grid. The lateral grid bars can be moved in small increments. This results in odd-sized ceiling tiles between fixtures, which may present a strange appearance.

Expect interference from ductwork, sprinkler piping, etc., above the ceiling. See Figure 1. If there is sufficient space above the ceiling, it may be worth elevating these obstructions to allow fixtures to be moved.

Surface-mounted fixtures need more effort and cost to relocate because new mounting boxes need to be installed. If appearance is not a prohibitive factor, surface wiring may reduce this cost. Touch-up work is needed where the fixtures were previously located.

If the fixture is not symmetrical, its orientation matters as well as its lateral location. See Reference Note 57, Light Distribution Pattern of Fixtures, about this. Figure 2 shows a typical example of long, thin

WESINC

Fig. 1 Why grid-mounted light fixtures cannot be located efficiently The grid pattern limits the fixture location to increments of spacing that are too large for efficient placement. Also, duct and pipe installed above the grid may make it difficult or impossible to move the fixtures.

fixtures that should be turned 90 degrees to provide better light distribution.

ECONOMICS

SAVINGS POTENTIAL: 20 to 70 percent of the energy used by the affected fixtures. This Measure may also provide large savings in lamp and lamp replacement costs.

COST: In new construction, this activity typically costs nothing. In retrofit, the cost varies widely, depending on the existing mounting configurations. Most of the cost is for labor.

PAYBACK PERIOD: Immediate, in new construction. Less than one year, to several years, in retrofit.

TRAPS & TRICKS

PLANNING AND LEARNING: The key to results is effective planning. Expect to spend a lot of time on this. Build experience by doing this first in a few spaces of different types, and seeing how well the changes work. If the fixtures are installed in suspended ceilings, decide how much effort you are willing to invest to work around interferences with duct, piping, and other equipment located above the suspended ceilings.

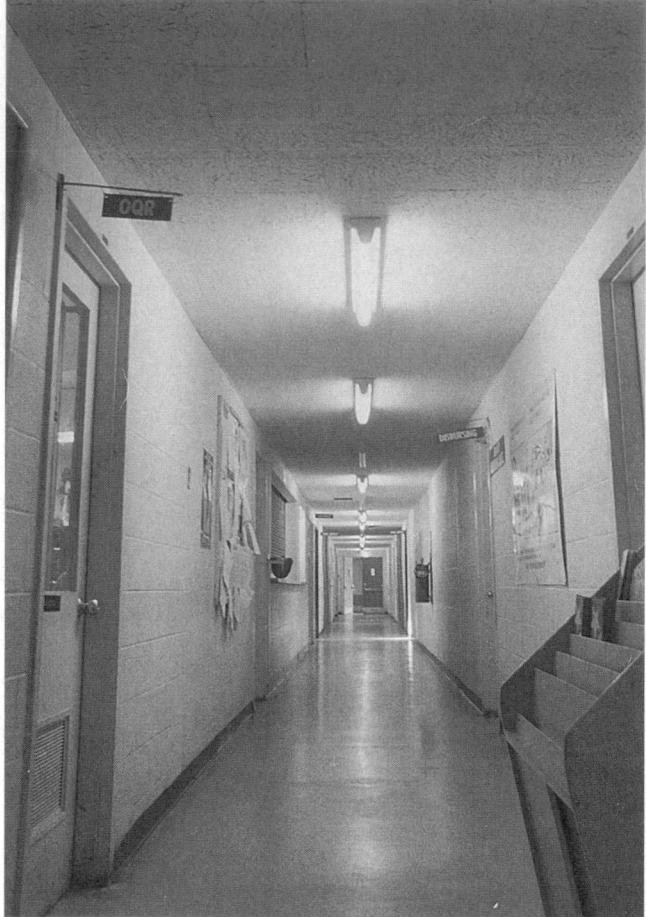

WESINC

Fig. 2 Fixture orientation matters as well as location
This scene shows clearly that a long fixture emits light mostly perpendicular to its axis. Rotating the fixtures 90 degrees will distribute light along the corridor more uniformly. To avoid the glare that would result, it is also necessary to suspend short glare shields from the ceiling, halfway between the fixtures. The same principle applies to tunnels, walkways, etc.

MEASURE **9.6.2.3 Replace fixtures and improve fixture installations that waste light.**

The previous Measures eliminated unnecessary fixtures or moved fixtures to more efficient locations. In many cases, fixtures are so inefficient that they are not worth moving or salvaging. Discard such fixtures and, if necessary, replace them with more efficient fixtures.

How efficient are your fixtures? Survey your facility to find out. You may find that some are inherently inefficient. You may find that others waste energy because they are the wrong type for the lighting application. These characteristics may make a fixture too wasteful to keep:

- *the fixture creates light inefficiently.* Incandescent fixtures are the main offenders. Some HID lamps, especially mercury vapor, are much less efficient than the best HID lamps that are presently available.

- *the fixture absorbs light internally.* Fixtures with dark interior surfaces or tight baffles have this problem.

- *the fixture distributes light inefficiently.* Downlights are a common example. HID fixtures and other high-output fixtures may put much of their light in places where it is not needed. (But, note that some HID fixtures let you adjust the light distribution pattern.)

- *the fixture cannot be controlled efficiently.* HID fixtures are the worst in this regard, because they cannot be turned on and off repeatedly.

Several types of lighting are especially common energy wasters, so we will cover them in greater detail. These culprits are downlights, indirect lighting in general, and fixtures installed in ceiling plenums.

Downlights

Downlights are a fad that has created an epidemic of energy waste. Figures 1 through 4 illustrate how they are abused. Downlights waste light in several ways:

- *absorption inside the fixture.* Many downlights are designed as light traps that absorb all light from the lamp except the light that is aimed directly at the opening. Such downlights may trap from 30% to 98% of the light emitted by the lamp, depending on the type of lamp (see Measure 9.1.3). You may be able to improve the light emission of downlights by installing reflective inserts in them (Measure 9.1.4.2).

- *absorption outside the fixture.* Downlights do not illuminate activities directly, except by accident or in cases where they are deliberately installed over the activity. They typically project light on the floor,

SUMMARY

Corrections for fixture installations that trap light or send it to the wrong places.

SELECTION SCORECARD

Savings Potential	$ $ $
Rate of Return, New Facilities	% % % %
Rate of Return, Retrofit..........	% % % %
Reliability	✓ ✓ ✓ ✓
Ease of Retrofit	☺ ☺ ☺

WESINC

Fig. 1 Wrong fixtures for the application The dark interiors of these deeply recessed downlights absorb most of the light of the lamps. The light that escapes from the fixtures mostly goes into the carpet. Single-tube fluorescent fixtures, mounted on the ceiling surface, would require only a few percent of the energy of these fixtures, for a given amount of illumination on the books.

which is darker than other surfaces and is too low to provide efficient lighting by reflection. As a result, most of the light emitted by a downlight is absorbed in the course of multiple reflections before it reaches the area intended for illumination.

- *inefficient lamps.* Most downlights use incandescent lamps, which are about one fourth as efficient as fluorescent or HID lighting. You can substitute screw-in fluorescent lamps for incandescent lamps in most cases (Measure 9.1.2.1), although the maximum light output presently available is not as high as with incandescent lamps.

In violation of all logic, downlights are often used for general lighting. Where this situation exists, take inventory of the areas that need to be illuminated, and select replacement fixtures that will illuminate them most efficiently. In most cases, the number of replacement fixtures required is a fraction of the original number of downlights.

Legitimate applications for downlights are rare, the most common being applications where an atmosphere of intimacy is desired, as in fine dining, or where it is necessary to eliminate all glare, as in museums. Even in such cases, the manner of installing the downlight often wastes energy.

■ Concern About Glare

The popularity of downlights appears to stem from a notion that light fixtures should not be visible. In turn, this notion may be based on a concern about glare. However, visible fixtures do not cause visual discomfort if their surface brightness is not excessive, if they are installed outside of the viewers' usual lines of sight, and if they are not installed where they cause veiling reflections. Reference Note 51, Factors in Lighting Quality, explains how to avoid glare.

■ Close Ceiling Penetrations

Downlights installed inside ceilings usually are vented to the space above the ceiling to reduce the temperature inside the fixture. If the space above the ceiling is unconditioned, such as an attic, a downlight serves as a path for wasting conditioned air. For this reason, if you deactivate a vented downlight that is installed into an unconditioned space, remove the fixture. Close and refinish the penetration. Dead downlights are not attractive, so this also improves appearance.

Indirect Lighting

Indirect lighting includes cove lighting, valance lighting, wall washers, and bounce lighting. Indirect lighting wastes energy by requiring light to be reflected repeatedly on the way from the lamp to the activity. A fraction of the light is lost at each reflection.

Even the most reflective surfaces absorb some light. For example, clean white paint has a reflectivity of about 90%. If light must reflect three times from such a surface, somewhat less than one third of the light is lost. If a the surface is dirty or has a pastel color, its reflectivity

WESINC

Fig. 2 Downlight abuse in a courthouse Downlights are used as the only source of illumination in this large courthouse lobby. Most of the light is absorbed in the carpet, making this large space gloomy. Note the dark walls, including the portraits of the learned judges. Compare the brightness of the wall in the office at left, which is illuminated with conventional fluorescent ceiling fixtures.

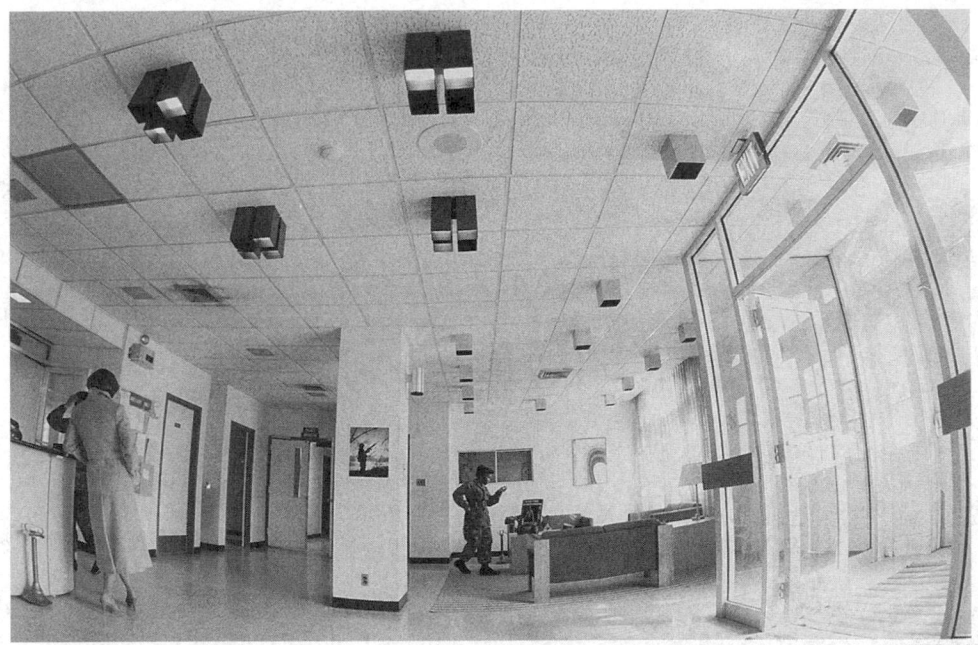

WESINC

Fig. 3 Downlight abuse in a military barracks Downlights are utterly inappropriate for this environment. At least the floors are polished, so some light gets into the space. These fixtures are so ineffectual that nobody notices that many of the bulbs are burned out. The remaining ones are just wasting energy, because the glass front is providing plenty of daylight. A few fluorescent fixtures and some basic controls would illuminate this space for less than ten percent of the present cost.

WESINC

Fig. 4 Downlight abuse in a university building entrance This forest of downlights is intended to provide security for the entrance to an urban classroom building. Most of the light that escapes from the highly absorbing fixtures is lost in the dark stone floor. About three small surface-mounted HID lamps would provide vastly better illumination, at a small fraction of the energy and lamp replacement cost.

is reduced to about 60% to 80%. In this case, three reflections will result in well over half of the light being lost.

Eliminating cove and valance lighting may leave the walls too dark, especially if they have decor. The solution is to install ceiling fixtures near the wall that have enough lateral distribution to highlight the walls.

WESINC

Fig. 5 Ceiling plenum lighting This escalator in a department store is illuminated by fluorescent strip fixtures located in the ceiling plenum. Grilles let some of the light escape, but this arrangement is an expensive light trap. It has no advantage over conventional fluorescent fixtures.

Eliminating cove and valance lighting leaves a dark slot. You can cover this with a decorative frieze. If you want to create the appearance of illumination coming from the slot, you can create this effect with much less power by installing low-wattage lamps behind a translucent panel that covers the slot. Or, if you have budget for deco, install a neon tube along the slot.

Bounce lighting can be reasonably efficient if it involves only a single reflection. For example, you can use upward pointing HID lamps to reflect light off a white ceiling to provide area lighting. Make the surface very reflective. See Measure 9.6.2 about using bounce light for task lighting.

Fixtures Installed in Ceiling Plenums

In some commercial buildings, you may find fluorescent strip fixtures installed in the ceiling plenum, illuminating the space below through grilles installed in a suspended ceiling. This arrangement is especially inefficient. It traps most of the light that does not pass directly from the lamps through the opening. Such arrangements are usually installed for appearance, but they are not particularly attractive. Figure 5 shows an example.

To eliminate the energy waste while maintaining a flush ceiling, simply install flush-mounted fluorescent fixtures that have efficient reflectors. Surface-mounted fixtures also provide greater flexibility in fixture location and light distribution.

ECONOMICS

SAVINGS POTENTIAL: 20 to 70 percent of lighting energy in the affected spaces. Lamp and replacement labor costs may be reduced significantly.

COST: In new construction, this modification is typically inexpensive, and in some cases, it may have a lower initial cost. In retrofit, replacement lighting typically costs $0.50 to $3.00 per watt of capacity.

PAYBACK PERIOD: Several years, or longer.

TRAPS & TRICKS

PLANNING: Start with a survey of all your fixtures. Since you are entirely eliminating certain types of fixtures or lighting arrangements, you have a wide range of options in replacing the lighting. Make the most of this opportunity. Consider layout and fixture type together.

MEASURE **9.6.3 Install fixtures or combinations of fixtures that provide efficient lighting for all modes of space usage.**

SUMMARY

Applies to many environments where different types of activities occur on a regular basis. Needs reliable controls to produce results.

SELECTION SCORECARD

Savings Potential .... $ $ $ $

Rate of Return, New Facilities % % % %

Rate of Return, Retrofit......... % % %

Reliability .... ✓ ✓ ✓

Ease of Retrofit .... ☺ ☺ ☺

In many cases, the lighting of a space is designed for one type of activity, and it is inefficient for other usage. For example, nightclubs use incandescent lighting operated by dimmers because the low intensity and warm color of incandescent lighting creates the desired mood. While, this is appropriate during evening business hours, the nightclub needs lighting at much higher intensity for many hours during the day, for activities such as cleaning, rehearsals, and serving the lunch trade. The incandescent lighting is inefficient for such activities.

The most efficient way to provide lighting for such facilities is to install the types of fixtures, or a combination of fixture types, that can accommodate each mode of space usage as efficiently as possible. How well this pays off depends on the total number of hours each year that lighting is needed for each function.

Common Opportunities

Consider all the times that lighting is required in each space, and provide appropriate lighting for each requirement. The following are common situations where different modes of lighting are needed for efficiency.

■ Spaces that Need Incandescent Lighting for Appearance

Facilities such as restaurants, nightclubs, theaters, and retail stores commonly use incandescent lighting for esthetic reasons during hours when they are open to the public. However, such facilities commonly operate for many hours when esthetics are not a factor. For example, cleaning and restocking require illumination that is bright and widespread, rather than subtle. Consider installing more efficient fixtures to provide illumination for these functions. Fluorescent and HID lighting is about four times as efficient as incandescent lighting, and the lamps last about ten times longer.

Incandescent fixtures often distribute light inefficiently. Downlights, commonly installed without regard to the location of activities, are the most common example. For times when incandescent lighting is needed, try to install incandescent fixtures that have efficient light distribution.

■ Spaces with HID Lighting

HID lighting tends to be left on in empty spaces because it cannot be restarted quickly. To eliminate this problem, install additional fluorescent fixtures for emergency lighting, transit, and activities that occur outside normal occupancy hours. An alternative is replacing all the existing HID lighting with fluorescent lighting, as recommended by Measure 9.3.3.

■ Emergency Lighting

It is common practice to provide emergency lighting with some of the conventional fixtures. Often, the fixtures are much too powerful for emergency lighting, even though only a fraction of them are used. For example, the 4-tube fluorescent fixtures commonly used in office space each consume about 200 watts. This is vastly more power than needed for emergency egress.

Another disadvantage of this practice is that the lamps in the emergency fixtures burn out more often than the lamps in the other fixtures. This upsets the schedule for group relamping, if this practice is followed.

There are a number of economical solutions. The easiest, most efficient, and cheapest may be to install specialized emergency fixtures in locations that are appropriate for emergency egress. Instead of expending 200 watts per fixture, as in the previous example, each emergency fixture might have a power consumption of about 20 watts. Another improvement is to rewire some of the primary fixtures so that only a reduced number of lamps operate during unoccupied hours (see Measure 9.6.4 for this). However, this solution still burns out one set of lamps at a different rate than the others.

■ Transit Lighting

The worst situation is turning on all the lights in a space only to pass through it. You can keep this from happening with the circuiting and control methods recommended by Measure 9.6.4. If the lights in the space need to be turned on for long durations to allow transit, install appropriate fixtures for this purpose.

■ Maintenance Lighting

Industrial facilities typically have equipment that requires frequent maintenance, inspection, and cleaning outside of normal production hours. It may be practical to install specialized lighting for these functions, avoiding the need to turn on the lights in the entire plant.

Control to Prevent Duplication of Lighting

Merely installing a variety of fixtures does not make the lighting efficient. In fact, it may make the lighting even less efficient, and also annoying. Figure 1 illustrates this.

It is a big challenge to make sure that only the lighting needed for each mode of space usage is turned on. For example, if a nightclub has fluorescent lighting for cleaning, make sure that the incandescent lights are turned off when the fluorescent lights are turned on. Otherwise, the money spent on the extra fixtures is wasted and overall consumption may even increase.

An important requirement is to have efficient switching for the fixtures. To achieve this, the fixtures must be arranged properly on the power circuits, as prescribed by Measure 9.6.4. In order for the switches to be used properly, they must be located where they are accessible and obvious, as prescribed by Measure 9.6.4. And, the controls must be marked in a manner

that is effective for its environment, as prescribed by Measure 9.4.1.

One way to keep the wrong lights from being turned on is to install a programmable light switch that turns on preset patterns of fixtures tailored to the different activities. See Measure 9.6.4.1 for the details.

A foolproof method is to install a transfer switch between lighting circuits that feed different types of lighting. For example, you could put all the incandescent lighting on one circuit, and all the fluorescent lighting on the other. See Measure 10.4.1 for details. The disadvantage of this method is that you can never operate both groups of fixtures at once.

See Subsections 9.4 and 9.5 for other control techniques that customize lighting to the nature of the activity. Methods that involve manual switching are inherently unreliable. They depend on effective placards (see Measure 9.4.1) to produce a significant benefit.

While you are planning this Measure, review Measure 9.6.4 about control circuits. As you install the wiring for the new fixtures, improve the control wiring of the existing lighting, if appropriate.

ECONOMICS

SAVINGS POTENTIAL: *20 to 70 percent of lighting energy in the affected spaces. Lamp and replacement*

WESINC

Fig. 1 Just installing a variety of fixtures does not make efficient lighting
This conference room has four different types of light fixtures, intended to serve a variety of functions. However, this feature alone has increased energy consumption, and it has created the potential of being extremely annoying. Note that bright reflector lamps, installed adjacent to the suspended fixtures, are shining directly into the eyes of the two persons. A dimmer for one type of fixture is hidden behind a drapery. The switches for the other fixtures are located in a different room, with no markings. The fixtures are not circuited to make it possible to darken the projection screen area without darkening the entire room.

labor costs may be reduced significantly, especially where incandescent lighting is reduced.

COST: *$0.50 to $3.00 per watt of additional lighting capacity.*

PAYBACK PERIOD: *One year to several years, in new facilities. Several years or longer, in retrofit.*

MEASURE 9.6.4 Install a separate control circuit for each lighting element that operates on a distinct schedule.

RATINGS

New Facilities	Retrofit	O&M
B	**C**	

A common cause of lighting energy waste is controlling too many fixtures with a single circuit. For example, if one lighting circuit serves two desks, energy is wasted whenever one of the desks is vacated.

Try to provide a separate circuit for each lamp, fixture, or group of fixtures that serves a separate activity or function. This may not be economical in all cases, but it is an important general principle. In deciding whether to provide separate circuits for the lighting that serves adjacent activity areas, a key criterion is the number of hours per year that the areas are occupied on different schedules. Figure 1 shows an example where improved circuiting would pay off quickly.

Lay out the lighting circuits so that lighting can adapt efficiently to changing conditions. Planning circuit layout does not require a crystal ball to foresee all possible future layouts. In case of doubt, more partitioning of the lighting circuits is better than too little.

The final step in providing appropriate lighting circuits is installing the switches and other controls in appropriate locations. This aspect is covered in Measures 9.6.5 ff, below.

The following are common situations where you should design the lighting circuits to provide flexibility in selecting fixtures and lamps.

■ Separate Activity Areas

Take stock of all the different lighting needs that may occur within a space. For example, in a conference room, a podium may be installed while showing slides. To accommodate this, provide separate circuits to provide bright lighting on the podium area while allowing the lighting over the projection screen to be turned off.

Do not assume that walls or partitions are the only appropriate boundaries for lighting circuits. For

SUMMARY

Electrical wiring that is needed to minimize unnecessary lighting. Applies to individual fixtures and to groups of fixtures.

SELECTION SCORECARD

Savings Potential	$ $ $ $
Rate of Return, New Facilities	% % % %
Rate of Return, Retrofit	% % %
Reliability	✓ ✓ ✓ ✓
Ease of Retrofit	☺ ☺ ☺

example, consider whether you can provide separate lighting circuits for individual desks in a large, open office space.

Even areas that are occupied by a single person may have several areas that merit separate lighting control. For example, a large one-person office may have different fixtures that illuminate the desk, the credenza, bookshelves, file cabinets, and a supply cabinet. Separate switching may be appropriate for each of these. Provide easy access to the controls (Measure 9.6.5) or install automatic controls (Subsection 9.5) to exploit such opportunities.

■ Movable Locations of Activities Within a Space

In many spaces, such as hotel ballrooms and school gymnasiums, different activities occur in different parts of the space. Provide separate circuits for each area in such spaces.

■ Separate Lighting Functions

Provide separate circuits for separate lighting functions. For, example, put the display lighting in a store on different circuits than the general lighting. When the store is restocking or taking inventory, the display lighting can be turned off.

Look for separate functions even where there are no separate types of lighting. For example, on an assembly line, individual machines receive maintenance outside of normal working hours. In this example, provide separate circuits to allow maintenance of each machine without turning on the lights for the entire assembly line.

■ Transit Lighting

The situation to be avoided is turning on all the lights in a space only to pass through it. A common example is turning on many lights in a large hotel conference room so that waiters can carry trays to an adjacent conference room. In such cases, connect the minimum number of lamps needed for safe transit to a separate circuit. Effective switch location and placards (Measure 9.6.5) are especially important for transit lighting.

■ Multiple Lighting Levels

If fixtures have multiple lamps, you can provide a choice of light levels by wiring separate circuits for individual lamps within the fixtures. For example, with 3-tube fluorescent fixtures, install an additional circuit to power the single-tube ballasts in each fixture. This provides three levels of illumination: a single tube on the one circuit, two tubes on the other circuit, or three tubes with both circuits.

In fluorescent fixtures where a single ballasts serves two tubes, you can control individual tubes by wiring ballasts between fixtures. Refer to Measure 9.2.1.3 about this arrangement.

In larger spaces, you can adjust illumination levels by circuiting the fixtures in interlaced patterns. Typical applications are banquet halls and auditoriums. To achieve sufficiently uniform illumination, make the distance between fixtures small in relation to the ceiling height.

With fluorescent lighting, dimming (Measure 9.2.6) is an alternative to switching individual lamps. Dimming incandescent lighting is undesirable because it further reduces the low efficiency of incandescent lamps.

■ Daylighted Areas

Exploit daylighting by providing separate circuits for each area of a space that is daylighted at equal intensity. For example, install one circuit for the row of fixtures closest to the windows, and another circuit for the second row of fixtures. Daylighting requires automatic controls (Measure 9.5.3) to be successful. You may also have to install window treatment (see Subsection 8.3) to exploit daylighting.

Wiring Cost and Methods

Most of the cost of this Measure is for labor, with a lesser amount for materials. Rewiring is easiest if the wiring is accessible and flexible, conditions commonly encountered in fixture wiring above suspended ceilings.

WESINC

Fig. 1 Inadequate light fixture circuiting All the lighting in this office is controlled by a single switch. The row of fixtures adjacent to the windows should have a separate circuit, to exploit daylighting. The fixtures in all three rows on the same side as the door should have separate switches. Even better, they should be removed. 36 fluorescent tubes is far too many for two desks. This room needs extensive delamping, or even better, task lighting.

In other cases, for example, where conduit is embedded in concrete, the cost of changing circuit layout may be prohibitive. Be sure to consider all your wiring options. For example, raceways may be acceptable for surface wiring if you can't rewire behind the surface.

■ Switching Fixtures from Multiple Locations

A fixture or group of fixtures can be controlled from any number of locations. The most common method is using inexpensive 3-way and 4-way switches. All the switches are connected in a chain between the power source and the fixtures. A 3-way switch is installed at each end of the chain, and 4-way switches are used in between.

Another method of switching fixtures from several locations is to use relay lighting control, which offers a number of other capabilities.

■ Relay Lighting Control

In most buildings, the power circuits for lighting also act as the control circuits, which limits control flexibility. You can overcome this limitation by adding low-voltage relay lighting control. A relay is installed at each controlled point, which may be an individual fixture, a junction box feeding a group of fixtures, or even an individual lamp within a fixture. Each relay is actuated by a separate low-voltage control wire from the switch or other control device. Consider relay lighting control for these situations:

- *to connect lighting in variable patterns.* Relay control is mandatory for switching fixtures in variable combinations. It allows a particular fixture to be controlled by more than one switch. Ordinary power switching does not allow this, because current would flow from the live circuits into the dead circuits. (If you need to be convinced of this, try sketching a variable lighting pattern using ordinary switches.) Programmable lighting controls use relay control. See Measure 9.6.4.1, next, for details.

- *where the controlled lighting exceeds the amperage of individual circuits.* Individual power circuits typically are limited to 20 amperes. Using relays allows a single switch or controller to control any lighting wattage while preserving the protection of individual circuits.

- *to control fixtures from multiple locations.* It may be less expensive to run control wires from each switch location directly to the fixture than to connect the existing switches together using 3-way and 4-way switches.

- *where modifying the power wiring would be more expensive.* Low-voltage wiring may be less expensive to install than power wiring, mainly because it is smaller and easier to handle, and because code requirements for protecting low-voltage control wiring are less stringent than for power wiring. However, the cost of relays, special low-voltage switches, transformers, and their installation subtract from the saving in wiring cost.

- *where it is not practical to move the existing power wiring.* Relay control wiring can be routed independently of the power wiring, so installation is not limited by the present location of the power wiring. For example, relay control can be retrofitted even if the power wiring is buried in concrete.

Relay control involves a compromise between flexibility and installation cost. To get the most control flexibility, you need to install a relay on each light fixture, but this may be unnecessary. Use a single relay to control several fixtures if all of them are needed on the same schedule.

Alternative: Individual Control

An alternative to installing new control circuits is to provide localized control at the fixtures themselves, for example, by installing pullcord switches (Measure 9.6.5.1). This method may be much less expensive, and it may make switching more accessible to users.

ECONOMICS

SAVINGS POTENTIAL: *20 to 70 percent of the energy of controlled lighting, depending on fixture and activity layout. Lamp and replacement labor costs may be reduced by similar amounts.*

COST: *$20 to $200 per fixture, most of which is for labor. Depends on space layout, number of fixtures, labor costs, etc.*

PAYBACK PERIOD: *One year, to many years.*

TRAPS & TRICKS

PLANNING: *The location of activities and partitions changes throughout the life of the facility. Design the switch circuits to accommodate foreseeable changes.*

LEAVE A RECORD: *Don't forget to record the changes in the facility's electrical drawings. Future generations of electricians will thank you.*

MEASURE **9.6.4.1 Where light fixtures are needed in a predictable variety of patterns, install programmable switches.**

SUMMARY

A convenient and accurate method of matching lighting to changing requirements. Vulnerable to poor user instructions.

SELECTION SCORECARD

Savings Potential	$	$	$
Rate of Return, New Facilities	%	%	%
Rate of Return, Retrofit	%	%	
Reliability	✓	✓	✓
Ease of Retrofit	☺	☺	☺

Programmable lighting controls allow you to change instantly from one pattern of lighting to another by selecting different groups of fixtures. A programmable lighting controller is simply a multi-pole, multi-position switch that activates relays in patterns. The controller can store a variety of patterns for instant recall. The patterns are selected by the installer or by the facility staff. Modern programmable switches are solid state devices.

Programmable switching requires relay lighting control, which is explained in Measure 9.6.4, above. As with all relay-controlled lighting, the degree of control flexibility depends of the number and arrangement of lighting relays.

A typical application for programmable switching is a large multipurpose space, such as a school cafeteria. At meal times, the controller turns on all the lights except those adjacent to the windows. For evening functions, the controller turns on all the lights except those in the serving area. During meal preparation hours, only the lights in the serving area are turned on. And so forth.

Typically, an occupant will select the desired lighting pattern by pressing a button on the controller. However, you can use other control devices to change lighting patterns, provided that the controller can accept electrical input signals. In the previous example, the changeover in lighting patterns occurs at predictable times, so you could use a timeclock to tell the controller to change the lighting patterns.

Programmable lighting controllers are a convenience, rather than a necessity. You could create any lighting pattern by installing individual switches for each fixture, although this approach would be awkward to use in a space with many fixtures. Programmable controllers improve efficiency only to the extent that they increase the likelihood that users will tailor lighting to the activities.

The Achilles Heel: User Instructions

Many programmable lighting controllers have been installed. Unfortunately, these are often installed in a way that only serves to confuse occupants and waste energy. For some peculiar reason, programmable controllers never seem to be labeled with adequate instructions for those who set the lighting patterns or for space occupants who select the patterns. To be effective, the controller needs clear instructions that tell:

- what the available lighting patterns are, using diagrams if necessary
- when to use each lighting pattern
- how to press the buttons or turn the dials to make the desired selection.

If the controller does not come with clear instructions, you have to create them. Refer to Measure 9.4.1 for guidance on how to do this.

ECONOMICS

SAVINGS POTENTIAL: *10 to 70 percent of the energy of controlled lighting, depending on fixture and activity layout. Lamp and replacement labor costs may be reduced by similar amounts.*

COST: *Several hundred to several thousand dollars, for the programmable controller itself. $50 to $200 per fixture to install relay lighting control.*

PAYBACK PERIOD: *One year, to many years.*

TRAPS & TRICKS

SELECTING THE EQUIPMENT: *All programmable switches are probably reliable. Select a model that is as user-friendly as possible. Try out the models you are considering before you buy.*

INSTALLATION: *Install programmable switches in an obvious, easily accessible location. See Measure 9.6.5, next. Invest the effort to produce clear instructions.*

MONITOR PERFORMANCE: *Check periodically to see whether the programmable switches are being used efficiently. If not, figure out why and make appropriate changes. Keep checking.*

MEASURE 9.6.5 Install lighting controls at visible, accessible locations.

RATINGS

New Facilities — Retrofit — O&M

A | C | ☐

It should be obvious. If lighting controls are not convenient, they will not be used efficiently. However, this important point is commonly overlooked, and a lot of energy is wasted as a result. Figure 1 shows an extreme example.

The most important point is to locate lighting controls where people would expect to find them. But, there is more. In particular, locate lighting controls so that:

- they can easily be seen and identified
- people do not have to go out of their way to use them
- the fixtures controlled by the switches can be seen from the control locations
- the occupied areas can be seen, to avoid inadvertently turning off lights needed by persons remaining within the space
- there is no risk of persons being trapped within a darkened space.

WESINC

Fig. 1 Light switch folly These are all the light switches for a large public library. It must have been expensive to bring all the lighting controls to this one point, in an obscure location from which most of the fixtures are invisible. This situation is so hopeless in terms of efficient control that nobody has bothered to label the switches. They are all on, or all off. The whole building might as well be controlled by a single switch.

> **SUMMARY**
>
> A basic step in making lighting controls user-friendly.
>
> **SELECTION SCORECARD**
>
> Savings Potential .... $ $ $ $
>
> Rate of Return, New Facilities % % % %
>
> Rate of Return, Retrofit......... % % % %
>
> Reliability .... ✓ ✓ ✓ ✓
>
> Ease of Retrofit .... ☺ ☺ ☺

These guidelines apply to automatic lighting controls as well as to manual switches, if they may require any action by occupants. For example, automatic controls often have manual override.

How to Install Lighting Controls at Exits

People expect to find light switches at exits. If the switches are presently located in some obscure or inaccessible location, relocate them to the primary exits, unless it is more effective to move them to individual activity areas, as explained below. If it is appropriate to install lighting controls at doors, install them on the latch side.

Use the same locations for manual overrides that are combined with automatic controls, such as personnel sensors.

Where there is more than one usual exit, provide switches at each exit. People will not cross a space to turn off lights at an exit that they do not intend to use. By the same token, lack of switches at each exit may require entering people to cross a darkened space in order to turn on the lights. A fixture can be controlled from any number of locations by using inexpensive 3-way and 4-way switches, or by using relay control, as explained in Measure 9.6.4.

If lighting within the space is subdivided into activity areas, as discussed next, consider installing controls at both the exits and the individual activity areas.

How to Install Lighting Controls at Activity Areas

Provide a separate control circuit for each area that has distinct lighting requirements, as recommended by Measure 9.6.4. Install the controls for each task area where they are most accessible to the individuals who should operate them. You may want to retain the original wall switches for collective control of the lighting. This simplifies the task of security personnel or others who may be instructed to turn off lights after hours (see Measure 9.4.2).

You may have to use your imagination to find satisfactory mounting locations for the controls. Consider installing switches on furniture or equipment, but do not violate any safety codes. Consider installing controls on stalks suspended from the ceiling at a safe height. Anticipate changes to the space layout.

Pullcord switches are a technique for providing localized control in many environments, including offices and other activities where they are not presently used. Because of their potential, pullcord switches get separate treatment in Measure 9.6.5.1, the subsidiary Measure.

How to Control Transit Lighting

If a space is used as a passage between other spaces, especially outside normal occupancy hours, people tend to turn on more light in the space than they need to pass through safely. If you can limit transit lighting to certain fixtures (as recommended in Measure 9.6.3), consider removing all switches from the entrances except for the transit lighting switches. Install the switches for individual activity areas within those areas. If you do this, arrange the transit lighting so that regular occupants of the space can safely reach their work areas.

ECONOMICS

SAVINGS POTENTIAL: *A large fraction of excess lighting usage. Lamp and replacement labor costs may be reduced by similar amounts.*

COST: *In new construction, intelligent control layout costs little. In retrofit, modifications may cost hundreds of dollars per circuit.*

PAYBACK PERIOD: *Short, in new construction. Typically several years, in retrofit.*

TRAPS & TRICKS

INNOVATION: *This Measure is mostly about taking more initiative than the usual lighting designer in finding ways to make controls user-friendly. Let your imagination roam free.*

MEASURE 9.6.5.1 Provide localized control of ceiling fixtures by installing pullcord switches.

SUMMARY

An unusual, relatively inexpensive way to achieve localized lighting control. Widely applicable to environments where people are well behaved. Requires some innovation in mounting.

SELECTION SCORECARD

Savings Potential .... $ $ $

Rate of Return, New Facilities % % % %

Rate of Return, Retrofit .... % % %

Reliability .... ✓ ✓

Ease of Retrofit .... ☺ ☺ ☺

Installing pullcord switches may be the most practical way to provide individual control of fixtures in many applications. Consider pullcord switches for both new construction and retrofit applications. They have a lowbrow appearance, but they offer a number of potent advantages:

- they are much less expensive than switches installed remotely from the fixture
- they are easily accessible at the task location itself
- it is obvious how to use them
- they move along with the fixtures when fixtures are moved to accommodate new space layouts.

The last of these advantages is an important aspect of maintaining long-term efficiency that is not available with other types of switching.

Other methods are available for controlling individual fixtures locally. For example, some manufacturers offer switching that uses an infrared transmitter that works on the same principle as a television remote control. This method is more technically glamorous, but it is more expensive, and it requires a transmitter that can get lost. A piece of string is hard to beat for simplicity and reliability.

Where to Use Pullcord Switches

This method is appropriate in a variety of environments. In retrofit applications, it is an effective way of providing individual control of light fixtures that are connected in groups. Typical applications are individual desks in large office areas and individual work stations in factories. Figure 1 shows a typical environment where they might be used.

The pullcord switch does not have to control all the lamps in an individual fixture. You can connect pullcord switches to turn off some of the lamps in a fixture. Or, you can install more than one pullcord to control different lamps in a fixture. Going the other way, you can install a single pullcord switch to control several fixtures. (This may involve some rewiring between fixtures.) The amount of lighting that an individual switch can control is limited by its current rating, which typically is a few amperes. If necessary, you can use a relay with a pullcord switch to control any amount of wattage.

Be aware that a light fixture may provide lighting for more than one work area. Do not install pullcord switches on such fixtures if the adjacent work areas may operate on different schedules.

Perhaps the main reason that this simple, powerful technique is not popular is the concern that a jungle of pullcords dangling from fixtures would look messy. Pullcords do not have to be unattractive. By using decorative tassels or other ornaments, an appearance can be created that is attractive, or at least amusing.

How to Install Pullcord Switches

Success depends on details. Work out the following aspects of the installation.

■ How to Install the Switch on the Fixture

The Achilles heel of common pullcord switches is the flimsy bead chain and string. Replace the cotton string with strong nylon cord.

Also, figure out a way to install the switch so that the chain and cord run free without binding or chafing. Common pullcord switches are designed to be installed on the sides of shop lights, which allow the chain and cord to hang freely. These switches typically extend the chain slightly more than a half inch (about 15 millimeters) beyond the side of the fixture. This is fortunate for fixtures that are installed in suspended ceilings, because it may allow the chain to clear the T-bars that support the ceiling.

Even so, the chains may not last long if they have to pass through ceiling tile. A better method is to install a small mounting plate for the switch anywhere on the ceiling tile that is convenient. A variation of this technique is to install a common junction box on the ceiling tile. You need to do this if you use one switch to control several fixtures. Do all wiring in accordance with your local electrical code.

This method of switching overcomes one of the main problems of lighting layout, namely, difficulty in adapting lighting control to changes in space configuration. Work out the details of your switch installations to exploit this advantage. For example, mount the switches on ceiling tiles. This allows you to relocate the switch to almost any position by shifting the tile to a different point in the suspension grid and by rotating it to different positions.

■ Length of Cord

Adjust the length of the pullcord so that it is easy to reach, but not so long that it strikes the heads of occupants.

■ Tagging and Placards

Even though the operation of pullcord switches is obvious, instructions are important to provide permission and motivation. A handy method is to attach a small, durable plastic tag to the end of each pullcord that says, "Turn me off when not in use". In addition, install placards at the original wall switches, if you decide to keep them. See Measure 9.4.1 for details.

■ Maintain Emergency Lighting

Do not attach switches to fixtures that are wired as emergency fixtures. It is common practice to use a fraction of the conventional fixtures for emergency lighting. You can identify emergency fixtures by turning off all the light switches and observing which fixtures keep operating.

(If you observe that a lot of wattage is being wasted by using large fixtures for emergency lighting, change the emergency lighting as recommended by Measure 9.6.3.)

■ Overall Space Lighting Override

In retrofits, retain the original switches to provide a quick means of turning off all lights when the space is unoccupied. Typically, the pullcord switches are used by regular occupants for their own work areas, while the original switches are used by security personnel or others to turn off all the lights at the end of the day. Install a large placard at the original switches that explains how the lighting in the space is controlled.

In new construction, design efficient transit light to get occupants from the entrances to their activity areas.

ECONOMICS

SAVINGS POTENTIAL: *A large fraction of the lighting energy in controlled areas. Lamp and replacement labor costs may be reduced by similar amounts.*

COST: *Pullcord switches cost about $2. Installation costs $2 to $20 per fixture.*

PAYBACK PERIOD: *In new construction, this method may save a large amount of money compared to other methods of switching individual fixtures. In retrofit, the payback period is typically one year to several years.*

TRAPS & TRICKS

CHOICE OF METHOD: *Consider all the ways that you can provide localized light switching. Go back to subsections 9.4 and 9.5. This method is usually the easiest and cheapest, but it is less reliable than*

WESINC

Fig. 1 An environment for pullcord switches It takes the right kind of people to exploit pullcord switches. The main issue in installing them here is how to deal with fixtures that provide light for more than one cubicle.

automatic control. Restrict pullcords to environments where the occupants are genteel.

PLANNING AND INSTALLATION: This is an opportunity for innovation. Make the pullcords convenient for users. Install them to resist damage. Tie something attractive to the bottom of the cords, to attract attention and to make them easier to use. Don't forget to install placards at the original switches.

SELECTING THE EQUIPMENT: Search for the most rugged pullcord switches available. The cost of the switch itself is minor in comparison with the cost of installing and repairing the switch.

MAINTAIN THEM: Unless you find very rugged pullcord switches, they will break regularly. Repair them quickly, so that occupants do not revert to using the original switches.

Fixture Maintenance and Marking

The Measures in this Subsection maintain the efficiency of lighting systems over the long term. They are important for most kinds of light fixtures. Accomplish each Measure as a group activity throughout your facility.

INDEX OF MEASURES

9.7.1 Clean fixtures and lamps at appropriate intervals.

9.7.2 Replace darkened diffusers.

9.7.3 In fixtures where the type or number of lamps may vary, mark the fixtures to indicate the proper type of lamp.

MEASURE 9.7.1 Clean fixtures and lamps at appropriate intervals.

Dirt that collects on lamps and fixture surfaces may substantially reduce light output. Dirt also traps heat, and thereby reduces lamp and ballast life.

Energy Saving Potential

The amount of light that is lost from dirt depends on the cleanliness of the environment and the type of enclosure. The problem is minimal in clean environments, but it can be a major source of light loss in dirty environments. Lamps and reflectors become dirty in open fixtures, and stay much cleaner in enclosed fixtures. Diffusers collect dirt whether they enclose the lamps or not.

The amount of dirt that typically collects during the life of a lamp would absorb 10 to 50 percent of the light output, if the lamp and fixture are not cleaned.

Cleaning the lamp and fixture does not reduce the energy consumption of the lamps. Instead, it allows you to use other methods of reducing lighting power, such as using lamps of lower wattage.

What to Clean

Clean all the surfaces of lamps and fixtures that transmit or reflect light. This includes the lamps themselves, the fixture's reflecting surfaces, and both sides of diffusers.

How Often to Clean

Fixtures in almost any environment accumulate enough dirt to substantially reduce their efficiency. The rate of dirt accumulation depends on the nature of the environment and the type of fixture. Do a "white glove inspection" to determine the need for cleaning. If enough dirt wipes off a fixture or lamp to smudge a clean cloth, dirt is absorbing a significant amount of light. Schedule fixture cleaning based on your experience with the rate of dirt accumulation.

Safety

Lamps have to be removed to clean them and the fixtures properly. This involves a potential for electric shock. If wet or electrically conductive cleaners are used, make sure to remove any residue from the electrical contact surfaces. Lock out the electrical circuits while cleaning the equipment.

SUMMARY

Routine maintenance needed to prevent energy waste and maintain lighting levels.

SELECTION SCORECARD

Savings Potential .... $ $ $

Rate of Return .... % % % %

Reliability .... ✓ ✓ ✓ ✓

Ease of Initiation .... ☺ ☺ ☺

Let lamps cool before handling them. This is to protect both the lamp and the worker.

Select Appropriate Cleaning Materials

Typical glass cleaners are harmful to plastics, especially to the acrylic and polycarbonate plastics used for diffusers. Using improper cleaners on plastics crazes the surface of the plastic, which reduces its light transmission and shortens its life. When cleaning plastic components, use cleaners that are intended for the specific plastics being cleaned.

Residue of cleaning materials absorbs light. This is a problem with diffusers that are not perfectly flat. If the diffuser has a textured surface, such as a prismatic plastic sheet, remove the diffuser, clean it with a method that gets into the entire surface, and then rinse it.

Special Care for Lamps with High Surface Temperatures

Quartz lamps and some other types have very high surface temperatures. These are vulnerable to breakage caused by localized accumulations of dirt that concentrate heat. Even fingerprints can break these lamps. The problem is limited to lamps that do not have an outer glass envelope. Train the cleaning personnel to make sure that the entire transparent surface of the lamp is free of specks and smudges.

Coordinate with Group Relamping

Group relamping is a matter of economics more than a matter of efficiency. If you do group relamping, you can save some money by coordinating one of the cleaning cycles with a cycle of lamp replacement, or vice versa.

ECONOMICS

SAVINGS POTENTIAL: *Up to 30 percent of lighting energy. The saving depends on how you can exploit the greater light output.*

COST: *Varies widely, depending on the accessibility of the lamps, the cleanliness of the environment, and lamp life. Refer to "The Cost of Lighting" in the Section 9 Introduction.*

PAYBACK PERIOD: *Less than one year, to several years.*

TRAPS & TRICKS

PREPARATION: *Write explicit instructions for cleaning personnel, including the materials and procedures to be used for each type of lamp and fixture. Include training of the cleaning crews as part of the cleaning schedule.*

MEASURE **9.7.2 Replace darkened diffusers.**

RATINGS
New Facilities | Retrofit | O&M
A

Plastic diffusers need to be replaced at long intervals because plastic diffuser materials lose their transparency with age. This is caused primarily by the ultraviolet light produced in fluorescent and HID lamps. Heat is also a factor, and it may cause darkening of diffusers in incandescent fixtures.

Energy Saving Potential

By the time a diffuser becomes noticeably darkened, it is probably absorbing more than 10% of the lamp's light output. Very darkened diffusers may be absorbing 50% of light output.

Replacing the diffusers does not reduce the energy consumption of the lamps. Instead, it allows other methods of reducing lighting power to be used, such as using lamps of lower wattage.

Diffuser Materials

Some types of plastic darken more slowly than others, and inhibitors can be added to plastics to slow the rate of darkening. You can't judge this from the appearance of the material. Study the manufacturers' data to determine the rate of darkening of each available material. Acrylic plastic retains its clarity best, but polycarbonate is more popular because it is cheaper and stronger. Styrene is used for diffusers because it is cheapest of all, but it darkens quickly, and is a bad bargain.

The method of diffusion has a large effect on light transmission. For example, milky white diffusers absorb considerably more light than clear prismatic diffusers. All other things being equal, use the most transparent material.

SUMMARY

Routine maintenance needed with plastic diffusers to prevent energy waste and maintain lighting levels.

SELECTION SCORECARD

Savings Potential .... **$ $ $**

Rate of Return .... **% % %**

Reliability .... ✓ ✓ ✓ ✓

Ease of Initiation .... ☺ ☺ ☺

Changing the Light Distribution Pattern

Your choice of diffuser material has a major effect on the light distribution pattern of the fixture. For example, the small prism patterns that are molded into clear plastic diffusers disperse light widely, but in particular directions. As another example, you can increase lateral light distribution, if this is appropriate for the activities, by replacing flat sheet diffusers with molded diffusers that are domed downward. Some manufacturers of replacement diffusers offer this as a standard feature.

You may want to replace transparent diffusers with open types, such as egg crates and parabolic cell units. These typically narrow the light distribution pattern, and the best of them provide excellent glare isolation. They may absorb more light than transparent types, especially when they get dirty. They are difficult to clean.

However, be cautious about making changes to the diffuser design. Test a few fixtures in a typical environment before changing all your fixtures. For example, increasing lateral light distribution may cause glare problems by making fixtures appear brighter to people in adjacent areas. Broadening the light distribution pattern also reduces the light level below the fixture. See Reference Note 57, Light Distribution Pattern of Fixtures, for tips.

Where to Find Replacement Diffusers

Flat sheet diffusers are widely available in standard lamp dimensions from lighting, plastics, and general hardware supply companies.

Replacement molded diffusers are offered by companies that specialize in making replacement lighting diffusers. If the fixture model number is not available, these specialty companies can replicate the diffusers using one of the original diffusers as a pattern. Compare the prices and materials offered by such companies to the prices and materials offered by the original fixture manufacturer.

Coordinate with Cleaning and Fixture Improvements

Plan to clean the fixtures when you replace the diffusers. See Measure 9.7.1 about fixture cleaning.

Also, consider whether you should delamp the fixtures or make other changes as part of the same effort. Refer to the other Measures in this Section for appropriate fixture modifications.

ECONOMICS

SAVINGS POTENTIAL: *Up to 30 percent of the energy consumed by the fixtures. The saving depends on how you can exploit the greater light output.*

COST: *$2 to $40 per fixture, depending mostly on whether the diffusers are flat or molded, and on the accessibility of the fixtures.*

PAYBACK PERIOD: *Less than one year, to several years.*

TRAPS & TRICKS

PLANNING: *If you want to change the light distribution characteristics of the diffusers, study Reference Note 51, Factors in Lighting Quality. Then, test your ideas on a few fixtures before changing them all.*

SELECTING THE MATERIALS: *Plastic is vulnerable to degradation by the ultraviolet light that leaks out of fluorescent lamps. Find out all you can about the materials that are offered. Ask vendors for references, and talk with purchasers who have used the material for several years or longer. Select the specific diffuser pattern for high transparency. Avoid milky diffusers.*

MEASURE **9.7.3 In fixtures where the type or number of lamps may vary, mark the fixtures to indicate the proper type of lamp.**

SUMMARY

An innovative step needed to maintain the long-term efficiency of lighting.

SELECTION SCORECARD

Savings Potential .... $ $ $

Rate of Return, New Facilities % % % %

Rate of Return, Retrofit.......... % % % %

Reliability .... ✓ ✓ ✓

Ease of Retrofit .... ☺ ☺ ☺

Only a few styles of sockets are used for the majority of lamps in each of the major lamp categories — incandescent, HID, and fluorescent. This situation allows different types and wattages of lamps to be installed in a given fixture. This provides flexibility, but it also invites energy waste.

With incandescent lighting, it is easy to install lamps that are too big for the application. With fluorescent and HID lighting, the main problem is installing lamps that do not operate efficiently with the ballasts. This has become a major problem in recent years because of a proliferation of new lamp and ballast types.

Maintenance personnel who must replace burned-out lamps do not know the correct type of lamp to put into the socket unless you tell them. If you don't do this in an effective manner, any lamp that is on hand may find its way into the fixture. Figure 1 illustrates the problem that the maintenance staff faces.

The most convenient and foolproof way of ensuring that the lamp is matched to the fixture is to mark each fixture with the specific type of lamp required. This eliminates the need for detailed verbal instructions, and it avoids the loss of information that occurs as people change jobs and records get lost.

What Can Go Wrong If You Don't Mark the Fixtures

Putting the wrong lamps in fixtures can cause trouble in a number of ways, including these:

• *loss of delamping.* Experience shows that delamping programs tend to regress quickly. The first time a senior manager looks up at the ceiling and observes that some lamps are not operating, the maintenance staff are called to replace the lamps. Nobody tells the manager that the lamps have been removed for the sake of energy conservation, and after a number of years, nobody remembers that fact.

• *excessive lamp wattage.* It is easy to install a lamp that draws too much power, sometimes by a large margin. The wattage of incandescent lamps with medium bases can vary by a factor of 100. The wattage of HID lamps may vary by a factor of ten or more within a given socket type. The wattage of fluorescent lamps may vary by as much as 20% in a given fixture.

• *low lamp efficiency.* Lamps of lower efficiency may be installed. For example, an conventional incandescent bulb may be used to replace a halogen bulb or a compact fluorescent lamp. The efficiency of fluorescent lamps may vary by 30% for a given wattage. The efficiency of HID lamps varies over an even wider range.

• *poor color rendering.* HID and fluorescent lamps can have radically different color rendering within a given socket type, or wattage. Incandescent lamps all have good color rendering, but the "color temperature" (the peak wavelength of the color spectrum, roughly speaking) can vary widely.

• *inefficient light distribution pattern.* In many fixtures, the lamps play an important role in the light distribution pattern. For example, elliptical-focus lamps get light out of recessed fixtures much more efficiently than other types of incandescent bulbs.

• *incompatibility with the ballast.* Fluorescent and HID lamps should be matched to their ballasts. Mismatches reduce efficiency, light output, and equipment life.

• *inefficient or unreliable dimming.* Dimming systems for fluorescent and HID lighting may require specific lamp types. For example, the most efficient type of fluorescent dimming systems does not perform properly with some "high-efficiency" fluorescent lamps, as odd as this may seem.

• *sensitivity to ambient temperature.* Fluorescent and HID lamps vary in their minimum operating temperatures. This is why you see dim and flickering fluorescent lamps outdoors during cold weather. "High-efficiency" fluorescent lamps are even more sensitive to temperature, even to cool indoor temperatures.

Refer to the Measures and Reference Notes dealing with specific lamp types for details of these aspects. In addition to causing these performance problems, using the wrong lamps wastes money and may increase workload for more frequent lamp replacement.

What the Fixture Labels Should Say

For all types of fixtures, provide this information about the lamps:

- *the specific lamp type.* This may be a verbal description ("frosted," "elliptical focus"), an industry standard designation ("F40/T12/CW"), or a combination of the two ("M58-250/U coated")
- *the wattage,* if this is not included in the lamp designation.

Also, if any fixtures have been delamped, put the following information on all the fixtures of the same type, whether or not they have been delamped:

- *whether any sockets are to be left empty*
- *whether the ballasts of the delamped sockets have been disconnected or removed.*

Spell out this information. Color-coded pieces of tape or general energy conservation stickers are unsatisfactory because people will forget what they are supposed to mean.

In some cases, you may want to use external labels for motivational purposes. For example, you may do this to make delamping more acceptable to the occupants, as recommended in Measure 9.2.1.

Outside or Inside the Fixture?

Marking fixtures externally is more convenient for maintenance personnel. If the markings are not visible from floor level, the maintenance person must make an extra trip up the ladder to find out what kind of lamp is

required. Furthermore, if a fixture is delamped, external marking reassures anyone who notices that the lamps are supposed to be removed. A bonus is that such markings improve general awareness of energy conservation.

In posh environments, appearance may be an objection to external labeling. You may be able to overcome this objection by making the labels attractive and/or unobtrusive.

If fixtures are partially delamped, install labels over the empty sockets that say that the lamps should not be replaced.

Make the Information Survive

Use a marking method that can survive repeated cleaning and the heat of the lamps. If you mark the fixtures internally, you can write the lamp type on the reflector with an indelible pen having a fine point.

If you really must use adhesive labels, select label materials and adhesives that can tolerate the fixture temperatures as long as possible. Install the labels on a cool, protected surface, if possible. If you must install the labels inside the fixtures, install them at the coolest part of the fixture where they are easy to read. Using a reflective color helps to keep labels cool.

Coordinate with Cleaning

Clean the fixtures before marking them. Consider reducing cost by combining the marking with one of the cleaning cycles recommended by Measure 9.7.1.

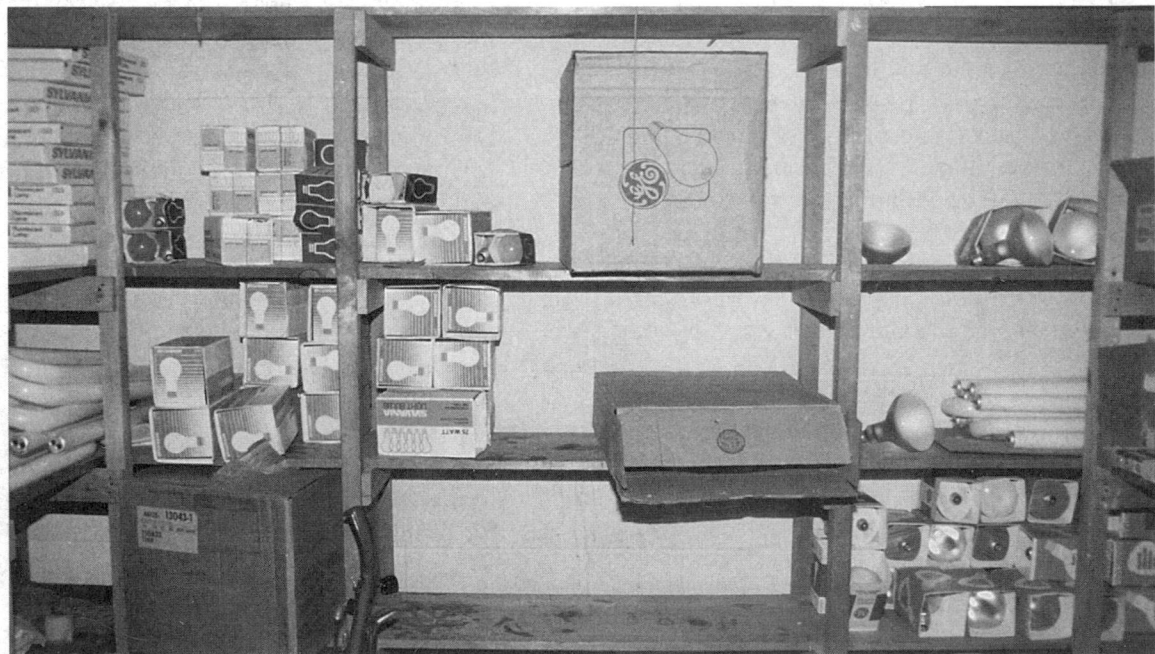

WESINC

Fig. 1 Lamp stockroom This is what the maintenance staff faces. If there are several types of lamps that will fit into the same socket, the wrong lamps will eventually be installed unless you mark all fixtures with the types of lamps they use.

ECONOMICS

SAVINGS POTENTIAL: *For incandescent and HID fixtures, up to 70 percent of individual fixture energy consumption. For fluorescent fixtures, up to 20 percent for individual lamps, up to 60 percent if the fixtures are delamped, and up to 50 percent if the fixtures have dimmers.*

COST: *$0.05 to $1.00 per fixture. Most of the cost is for labor to fill out the labels, clean the fixtures, and install the labels. You can purchase custom labels in large quantities for a few cents apiece.*

PAYBACK PERIOD: *Less than one year, typically.*

TRAPS & TRICKS

METHOD AND MATERIALS: *Your choice of marking method is important. A label that falls off in a year or two is useless. See Reference Note 12.*

STAY CURRENT: *Put an entry in your long-term maintenance schedule to check the marking of fixtures every few years. Update the markings any time you change the lamp or ballast type.*

Section 10. INDEPENDENT ENERGY-USING COMPONENTS

INTRODUCTION

This Section covers common energy-consuming components of buildings and plants that you can deal with in a similar manner throughout the facility. These components – motors, pumps, and fans – are responsible for a large fraction of the total energy consumption in most facilities. You can often reduce their energy consumption without needing to change the systems in which they are installed.

How to Use Section 10

To save effort and get the maximum benefit, take inventory of all the equipment in your facility that is covered by this Section. Then apply each Measure to all the applicable equipment as a group.

A. C. Induction Motors

Motors consume more electricity than any other type of equipment, perhaps more than half the electricity produced in the world. By far the largest portion of motor energy is consumed by large motors. Almost all of these are 3-phase induction motors.

In the industrial sector, large, replaceable motors account for a majority of total energy consumption. In the commercial sector, motors are smaller in size and they account for a smaller fraction of total energy consumption, but they are still major candidates for efficiency improvements. In the residential sector, separately replaceable motors represent only a small fraction of total energy consumption, and opportunities to upgrade them are infrequent.

This Subsection contains a single Measure that recommends installing the most efficient models of motors. It is important in new construction and in existing facilities.

Variable-frequency motor drives can have adverse effects on motor efficiency and reliability. These factors are covered in Reference Note 36, Variable-Speed Motors and Drives. Refer to there before making changes to motors used in variable-speed drives. Also, see Reference Note 36 about types of motors that are designed to operate at variable speed, including multi-winding motors, DC motors, and wound-rotor motors.

MEASURE **10.1.1 Install motors having the highest economical efficiency.**

SUMMARY

This is one of the most important and most widely applicable energy conservation measures. Needs more attention in new construction. In existing facilities, replace failed motors with more efficient models. In some cases, it is economical to replace motors that are still operating.

SELECTION SCORECARD

Savings Potential .... $ $ $

Rate of Return, New Facilities % % % %

Rate of Return, Retrofit.......... % % %

Rate of Return, O&M .... % % % %

Reliability of Equipment .... ✓ ✓ ✓ ✓

Reliability of Procedure .... ✓ ✓

Ease of Retrofit .... ☺ ☺ ☺

Ease of Initiation .... ☺ ☺

The efficiency of motors has increased in recent decades because of improvements in motor design and manufacturing processes. These improvements were spurred by market pressures and regulations. The percentage increase in efficiency has been modest because all motors, especially larger ones, have always been fairly efficient. However, the total amount of energy that can be saved by using more efficient motors is large.

Upgrading motor efficiency is economical because motors, except for smaller units, are relatively inexpensive in relation to the cost of the energy that they consume. Stated differently, larger electric motors in typical applications annually consume electricity that costs several times the price of the motor.

In the past, the biggest obstacle to optimizing motor efficiency was a lack of objective information about the efficiency characteristics of individual models. This is no longer a problem when selecting the most common types of 3-phase motors. These motors are now rated on the basis of efficiency. Also, efficiency standards now exist that allow you to judge the efficiency of a motor in relation to an objective baseline. A simple computer program, described below, provides comprehensive efficiency data on the most common types of 3-phase motors. This program lets you accomplish savings calculations easily.

Improving efficiency has received the most attention in the manufacture of 3-phase motors. Efficiency improvements in single-phase motors, which are smaller, are presently limited to the offering of "high-efficiency" models by some manufacturers in a limited range of sizes and types. There are presently no standards or objective guidelines for smaller motors, and little reliable efficiency data.

This Measure relates to conventional alternating-current (AC) induction motors, both 3-phase and single-phase. Other types of motors, such as DC motors and wound-rotor motors, are designed to provide variable speed. See Reference Note 36, Variable-Speed Motors and Drives, for these types. They are not presently covered by efficiency standards.

Motor Efficiency in New Construction

The best time to optimize motor efficiency is before the facility is built. Doing this is as simple as specifying the motor model or the efficiency rating that you want. Motors last a long time, so missing this opportunity condemns the facility to inefficient operation for all of its early life.

The additional effort required is minimal. In additional to specifying the usual necessary characteristics (speed, voltage, torque, etc.), also specify the motor efficiency, using the prevailing rating method. (Rating methods are described below.) Using a readily available database, select the highest efficiency rating that is appropriate for the motor size and application. Then, write it into the drawings and specifications.

There is no excuse for neglecting this simple, powerful efficiency improvement.

When to Upgrade Motor Efficiency in Existing Facilities

If a motor fails, the cost premium to replace it with a high-efficiency motor is small in comparison with the energy cost that is saved, provided that the motor runs for a reasonably large number of hours per year. Make it a policy to replace motors in such applications with high-efficiency models, regardless of size.

What if an inefficient motor is still running well? Does it make sense to replace it with a high-efficiency model? You have to answer this question on an individual basis. For motors larger than a few horsepower, the answer depends primarily on the potential efficiency improvement, the number of operating hours, electricity costs, and the economic rate of return that is required. Under present conditions, replacing an existing low-efficiency motor is likely to be economical if it runs continuously, or if electricity costs are higher than average.

Smaller motors generally do not warrant replacement while they are still operating. The break point in size is about one horsepower (or one kilowatt). There are for two reasons for this. One is that smaller motors are more expensive in relation to the cost of the energy they consume. The other is that motor efficiency ratings presently do not apply to motors smaller than one horsepower. Both conditions may change in the future. The market for efficient small motors is still evolving.

To avoid duplicating effort, identify and analyze all your motors as a batch. Immediately replace those motors where the economics are favorable for doing so. For the rest, be sure to use the most efficient replacement model when a motor fails.

■ Be Ready at Replacement Time

If a motor fails, it usually has to be replaced quickly. The delay involved in shopping for a more efficient replacement may be unacceptable if an important piece of equipment sitting idle. Therefore, identify your sources of efficient motors before the need arises. The MotorMaster computer program, described below, provides a listing of most conventional motors of one horsepower and larger that are sold in the United States.

To avoid interruptions of equipment operation, consider keeping high-efficiency motors on hand as spares, especially if the application is critical or if the facility uses many motors of a particular type. This practice allows the motors to be ordered for best efficiency and price, rather than in haste.

■ What About Rewinding?

It is common practice to rewind larger motors when they fail. The cost of rewinding a burned-out motor is roughly half the cost of a new motor, in large sizes. Therefore, facility operators have developed the habit of rewinding large motors instead of replacing them. Nowadays, this viewpoint needs to be broadened to include the possibility of replacing the old motor with a new, higher-efficiency motor. Eliminating the cost of rewinding makes it economically attractive to replace most failed motors of lower efficiency.

Can motors be rewound in a way that increases their efficiency? Generally, no. The features of a motor that affect efficiency are explained below. Most of these features cannot be improved in the repair process. On the contrary, a rewound motor is likely to be less efficient than it was originally. The typical efficiency loss in a rewound motor, including normal deterioration of other parts of the motor, is a few percent.

If the motor has to be replaced quickly, you can usually get a new high-efficiency motor in less time than it takes to rewind a failed motor.

The MotorMaster computer program provides a quick analysis of the savings and costs of higher-efficiency motor replacement in comparison with rewinding.

Overview of Motor Efficiency Characteristics

There are substantial differences in the efficiencies of motors of a given type and size. Figure 1 depicts the range of full-load efficiencies of common types of 3-phase motors. The figure also shows the U.S. minimum efficiency standard for motors, which was enacted in 1997. These curves do not include the lower efficiencies of some special-purpose motors.

Part-load efficiency is important in applications that involve extended operation at reduced loads. Size is a major factor in part-load efficiency, as shown in Figure 2. This is usually academic, because the application dictates the size of the motor. However, there are also major differences in part-load efficiency among motors of a given size.

Fractional-horsepower motors are less efficient than larger motors for reasons that are unavoidable. Also, small motors have been less efficient for practical and economic reasons.

Small motors can tolerate greater losses because they can dissipate heat better than larger motors. This is because the ratio of surface area to volume increases as an object becomes smaller. Also, they tend to be used in applications that allow open, ventilated construction.

In addition to these physical limitations, the present lack of efficiency standards for fractional-horsepower motors causes the compromise between cost and

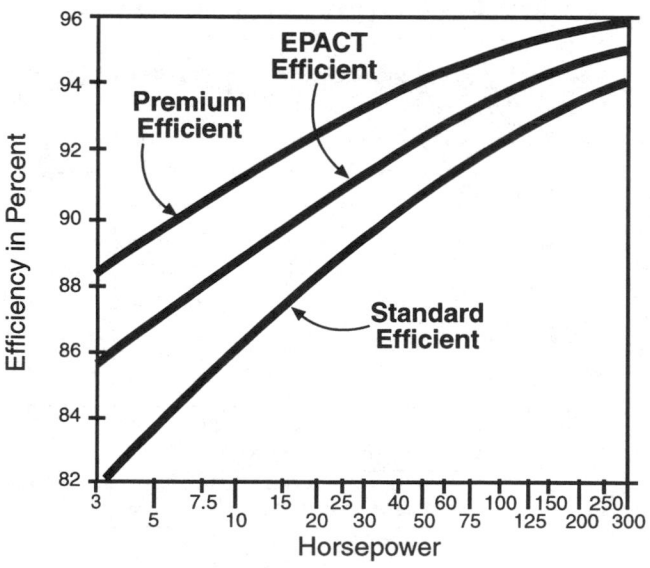

Reliance Electric

Fig. 1 Efficiencies of AC induction motors The "EPACT" curve is the minimum allowable for new motors under current U.S. law. The "standard" curve is typical of older motors. The "premium" curve is near the highest efficiency available.

efficiency to lean heavily toward cheapness. As a result, there is still plenty of opportunity for improving the efficiency of fractional-horsepower motors. Still, the overall savings potential is much smaller than for larger motors.

The full-load efficiencies of single-phase, fractional-horsepower motors range from 25 to 65 percent, depending mainly on size and type. Significant efficiency improvements in this size range occurred in the 1980's, so that newer "high-efficiency" motors in this category offer efficiency improvements in the range of 10 to 30 percent. The "high-efficiency" designation does not have a common definition. You can depend only on explicit efficiency ratings.

Construction Features of High-Efficiency Motors

The recent improvements in the efficiency of electric motors were not achieved by new features, but by improving the basic components that all motors have. Figure 3 shows a typical high-efficiency motor. The efficiency improvements include:

- *wire with lower resistance,* which reduces heat generation in the stator coils. Lower resistance is achieved mainly by making the wire thicker. This requires reducing the insulation thickness, which requires improved insulation materials. (There have been excessive failures of high-efficiency motors that appear to be related to this issue.) Also, the space for the stator coils may be increased by greater sophistication in the design of the steel core that holds the coils.

- *improved design of the rotor electric circuit.* The rotor current in an AC induction motor is induced by the magnetic field of the stator coils. The strength of these currents determines the power and

torque of the motor. The induced voltage is low, so the rotor conductors must have very low resistance to allow the high currents that are required. Reducing the resistance of the rotor conductors reduces heat losses and increases the power output in relation to the input energy.

- *higher permeability in the magnetic circuits of the stator and rotor.* The "permeability" of the steel cores affects the strength of the magnetic field that is induced in the steel for a given amount of excitation.

- *thinner steel laminations in the magnetic circuits.* This reduces eddy currents, which are circular electric currents induced in the steel of the magnetic circuits. These currents dissipate energy by heating the steel. They are prevented from forming by making the magnetic circuits (the "core") from steel laminations that are thinner than the diameter of the eddies.

- *improved shape of the steel stator core and rotor magnetic circuits,* to increase the force produced in relation to the losses

- *smaller gaps between stator and rotor,* to reduce magnetic flux losses

- *design of the internal fan, cooling fins, and cooling air passages to reduce the cooling power requirement.* Higher-efficiency motors produce less heat, so they need less cooling. They may also be designed to survive at higher temperatures.

- *bearings with lower friction.*

■ Additional Efficiency Factors in Single-Phase Motors

Improving the efficiency of single-phase motors involves the same techniques listed above. In addition, the design of starting features for single-phase motors may be improved.

The stator magnetic field of a single-phase motor does not rotate, unlike the stator field of a 3-phase motor. Therefore, any single-phase motor requires a separate method of getting it started. The names of the different types of single-phase motors are based on the starting methods.

All starting methods in single-phase motors cause some efficiency loss, but they differ significantly in this regard. In decreasing order of efficiency, the main types of single-phase motors are:

- capacitor start
- split phase
- permanent split capacitor
- shaded pole.

This same ranking also applies to starting torque, and to cost. In other words, you pay more for higher efficiency and higher torque.

"Energy-saving" versions are offered in each of these types, despite the large efficiency differences between the types. This is because economics favors

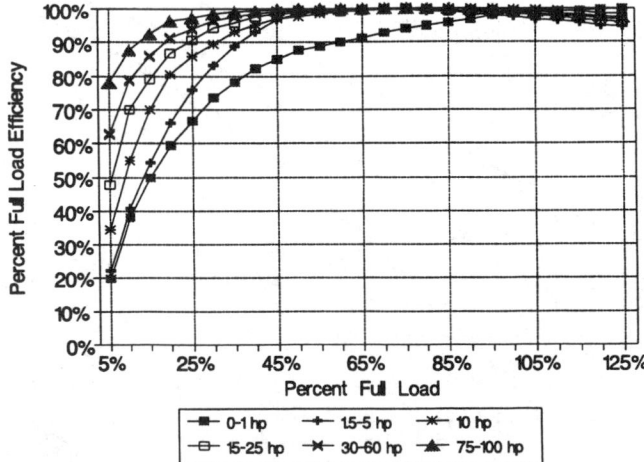

U.S. Department of Energy

Fig. 2 Part-load efficiency of AC induction motors   In many applications, part-load efficiency is even more important than peak efficiency. This graph shows that part-load efficiency is better in larger motors. All motors retain good efficiency above half load, except for the smallest.

using less efficient types in certain applications. The relative price differential between the types increases in smaller sizes.

Energy and Cost Saving Potential

Better motor efficiency saves energy, both directly and indirectly. Most of the cost savings result from the energy savings, but some cost saving also results from the way electricity is priced.

■ Reduced Motor Energy Consumption

Larger electric motors annually consume electricity that costs several times the price of the motor. The energy savings from upgrading motor efficiency depend primarily on the size of the motor, its hours of operation, its load profile, its efficiency, and the efficiencies of potential replacement motors. In addition, the cost saving depends on the unit price of electricity. Most of the economic savings from improving motor efficiency typically results from a reduction in kilowatt-hour consumption.

■ Reduced Electrical Demand

In facilities where larger motors are used, billings for electricity usually include a substantial demand charge. Reference Note 21, Electricity Pricing, explains demand charges.

With the rate schedules presently used by most electric utilities, improving motor efficiency may provide significant savings in demand charges. Even so, these savings typically are small in comparison with the savings in kilowatt-hour consumption.

■ Space Heating and Cooling Load

Losses from electric motors are released into the surrounding space as heat. Whether this affects the facility's energy cost depends on whether the surrounding space is heated or cooled. If the space is unconditioned, savings are not affected.

If the space is cooled with a cooling system of typical efficiency, it requires about one additional unit of electricity to remove three units of electricity lost into the space. This increases the savings by about one third, during the period when the space is cooled.

If the space is heated, reducing motor losses increases the space heating requirement. The additional heating cost depends on the energy source used for heating, and on the efficiency of the system. The motor acts as an electric resistance heater, with respect to the

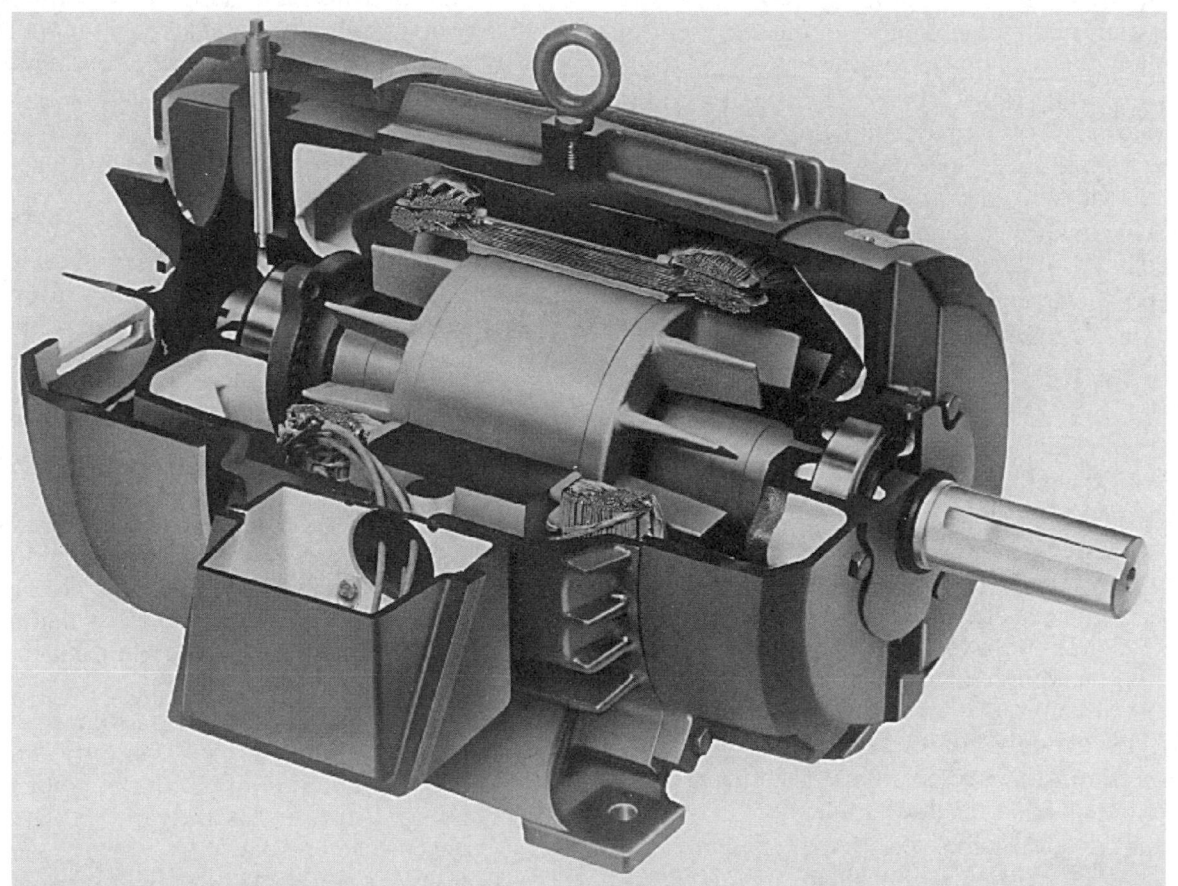

Reliance Electric

Fig. 3 Construction of AC induction motor This is a totally enclosed, fan cooled (TEFC) model. Note the internal and external fan impellers. The internal fans are needed to move heat from the rotor and windings to the motor case. The external fan is needed to blow air over the motor case. The fans are a factor in motor efficiency and insulation life.

heat that it releases into the space. If the space were continuously heated by electric resistance heat, there would be no advantage in improving the efficiency of the motor. Shorter durations of heating, and less expensive heat sources, offset efficiency savings to a lesser extent.

Cost Premium for High-Efficiency Motors

The cost premium for a motor near the maximum available efficiency averages about 15 to 30 percent above the cost of an average motor. The relationship between efficiency and price is not rigid. Careful shopping and effective price negotiation appear to affect price as much as the efficiency characteristics of the motor.

The MotorMaster Computer Program

The arithmetic of calculating the savings from upgrading motor efficiency is fairly simple. However, finding data on the efficiency of potential replacement motors used to be a major chore. Both these tasks have been made effortless by the MotorMaster computer program, which is an elegant, simple program for motor selection. It is sponsored by the U.S. Department of Energy under its Motor Challenge program. The program is updated semiannually, so you can use it to find recent additions to the motor market.

With a few keystrokes, the program prepares a list of all motors matching your input criteria, which include size, speed, enclosure type, etc. Another keystroke presents all available data on any individual motor.

The MotorMaster program covers motors sold in the United States that are within these characteristics:
- 1 to 600 horsepower
- 3-phase
- 900 to 3,600 RPM
- 200 to 4,160 volts
- all major enclosure types
- NEMA Designs A and B (motors with conventional torque characteristics)

MotorMaster provides all the usual nameplate information for each motor, including temperature ratings, service factor, etc. It includes price and warranty information. It includes all available efficiency information, including part-load efficiency. This is a significant advantage over catalog ratings, which normally indicate only full-load efficiencies.

The program quickly calculates the savings and simple payback period for three situations:
- comparing new motors of different efficiency
- replacing an existing motor with a more efficient motor
- comparing rewinding the original motor with installing a new motor.

You have to input the hours of use, the load percentage, installation cost, the local price of electricity, the demand charge, and the demand schedule. The program can accommodate dealer discounts and utility rebates, but cannot calculate based on a load profile. (You can approximate a load profile by doing a series of calculations with various percentages of load.) The program does not account for space heating or cooling load.

How to Estimate the Efficiency of Existing Motors

When you consider replacing a functioning motor with a high-efficiency motor, you need to know the efficiency of your present motor to calculate the potential savings. Unfortunately, you may not be able to get this information precisely for the present crop of installed motors. The best you can do is to estimate the efficiency of your existing motors accurately enough to make a replacement decision. Try one or more of these methods:

- *check the motor nameplate.* Older motors typically do not have efficiency information on their nameplates, but they may. It takes only a moment to check. However, be skeptical. Any efficiency ratings on motors that date prior to the present system of efficiency ratings may be nominal, or they may not be comparable to present ratings. As time goes on, a larger fraction of installed motors will carry credible efficiency ratings.

- *contact the manufacturer.* Start with the manufacturer's engineering department. Provide all the nameplate information.

- *check the original installation file.* The information submitted with the motor when it was installed may contain efficiency information. Beware of efficiency figures in promotional literature, especially if the test method is not stated. The fact that a motor efficiency may have been specified in the designer's specifications does not guarantee that the motor actually installed has this efficiency.

- *make a guess based on other characteristics.* Even if the motor efficiency is not stated explicitly, you may be able to judge the relative efficiency of the motor, compared to new motors of the same size and type. You can get clues from the nameplate or elsewhere. Consider:

  - the *age* of the motor. The maximum efficiency of motors increased substantially during the 1980's. Therefore, replacing a motor that was installed before then is likely to yield a substantial efficiency improvement.

  - the *full-load RPM*, which is shown on the nameplate, relates to efficiency. In general, a higher full-load RPM means higher efficiency. However, a comparison is valid only within a particular torque classification. High-torque motors are designed for lower full-load RPM. Especially in older motors, the full-load RPM may be nominal.

- the *service factor* of the motor relates indirectly to its efficiency. A service factor of 1.0 suggests mediocre efficiency, but this is not always true.

- the *temperature rating* of a motor relates indirectly to its efficiency. More efficient motors usually have higher temperature ratings. Insulation technology has changed over the years, so compare the temperature rating to the temperature ratings of other motors of the same age. These ratings are explained below.

• *send the motor to a testing laboratory.* You usually cannot measure motor efficiency on site, because you typically have no accurate way of measuring output power or correcting for variations in input power. An alternative is to send out a motor to be tested. This may be an economical approach for larger motors, provided that you can remove the motor without expensive rigging, and if you can afford to have the motor missing for a week or two.

In response to the growing emphasis on motor efficiency, a number of independent testing laboratories for motors have been established. You can find motor testing laboratories in the United States and Canada through the Motor Challenge program of the U.S. Department of Energy. Each laboratory has its own limitations in the types and sizes of motors that it tests.

Motor Efficiency Guidelines and Standards

Many motor manufacturers call their motors "high efficiency," or they may have separate "high-efficiency" models. Such claims have no specific meaning. For example, one manufacturer claims that all its standard motors have efficiencies as good as the average "high-efficiency" models of other manufacturers.

Fortunately, you no longer have to select motors based on vaguely worded claims. In recent years, official ratings and standards have appeared to guide you in selecting motors for efficiency. Use the hard numbers that are now available as your primary selection tool. Here are the details.

■ NEMA Efficiency Ratings

In the United States, the National Electrical Manufacturers Association (NEMA) has established a voluntary system for stating the efficiency of the most common types and sizes of 3-phase motors.

Motor manufacturers who use this rating system now stamp the NEMA ratings on the motor nameplates, as shown in Figure 4. The ratings are usually stated in catalogs.

The NEMA system establishes a ladder of standardized efficiency levels. Each rung of the ladder has two ratings, "nominal" and "minimum." The "minimum" rating may also be called "guaranteed." An individual "minimum" or "guaranteed" efficiency rating may be substantially lower than the "nominal" rating. For example, if the "nominal" efficiency is 84.0%, the "minimum" efficiency is 81.5%. The differences

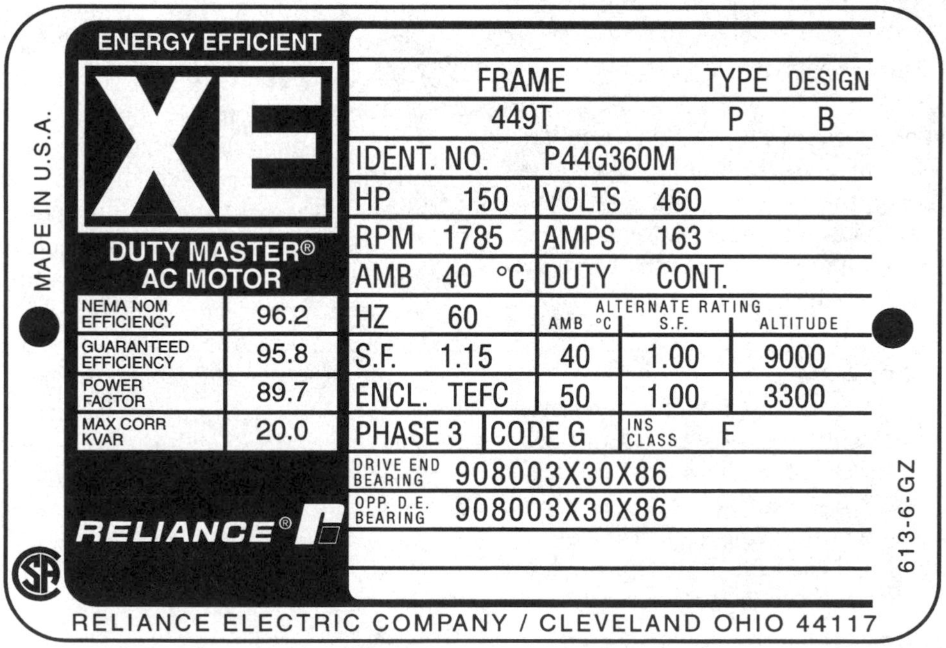

Reliance Electric

Fig. 4 Nameplate for high-efficiency motor This nameplate offers a rich selection of the most important information for typical motor applications. However, you have to know what all those things mean. The key efficiency ratings are on the left.

between "nominal" and "minimum" efficiencies are substantially smaller at the higher end of the efficiency range, which applies to larger motors.

Under the NEMA system, the term "energy-efficient" means that a motor has a full-load efficiency that meets or exceeds the NEMA ratings for its horsepower. The NEMA table has changed with time, so the current ratings do not apply to older "energy-efficient" motors.

Note that the similar sounding term "high efficiency" has no official meaning. This term is typically used by manufacturers to designate their own more efficient motors.

The complete listing of efficiency ratings is given in Table 10 of NEMA Publication MG1-1993. The efficiency levels differ by as little as 0.4 percent at the top of the efficiency scale, and by as much as 1.5 percent at the bottom. In order for a manufacturer to claim a particular efficiency level for a motor, a typical batch of these motors must have an average measured efficiency as high as the claimed level. The motors must be tested in accordance with the NEMA standard test method, which we examine below.

The ratings presently apply to the more common 3-phase motor designs, including NEMA Designs A and B. In 1994, NEMA published a table of efficiency guidelines for a new standard motor type, called Design E. Design E was developed primarily as a counterpart to the international (IEC) standard Design N motor. The efficiency standards for Design E motors are higher than for Designs A and B.

The ratings are updated and expanded from time to time. They have been extended progressively to larger motor sizes. They now cover the largest motors that you are likely to order.

There are presently no NEMA efficiency standards for single-phase motors of any size. Single-phase motors typically are one horsepower or less, but some are as large as 20 horsepower. Larger single-phase motors are used in applications that may be served only by single-phase electrical power. For this reason, they are common on farms.

■ **Part-Load Efficiency Data**

The database in the MotorMaster computer program includes part-load efficiencies as well as full-load efficiencies. The MotorMaster database covers all the motors covered by the NEMA guidelines, plus motors of larger horsepower and higher voltage. The efficiency data in MotorMaster are based on the test procedures that are used for NEMA Nominal Efficiency ratings.

■ **United States Motor Efficiency Laws**

In the United States, the Energy Policy Act of 1992 (EPACT) established the NEMA efficiency guidelines for Designs A and B motors as the minimum efficiency for motors manufactured for sale in the United States. The Act went into effect in 1997.

The Act applies only to NEMA Designs A and B, 3-phase motors between 1 and 200 horsepower, driven by 60 Hertz power, that operate at 1200, 1800, or 3600 RPM. The Act does not apply to motors designed for special applications.

In addition, the Act offers the possibility of establishing standards for smaller motors, but it is not clear that any standards for small motors will actually result from the 1992 Act.

ASHRAE Standard 90, which has been revised repeatedly, is a comprehensive standard for energy efficiency in new building construction. Versions of this standard have been incorporated into building codes throughout the United States. The latest version, Standard 90.1-1989, specifies efficiency requirements for electric motors by using a table that is based on horsepower. The ASHRAE standard applies to generally the same motors as NEMA standards, and it specifies the same test method. However, the ASHRAE standard is limited to T-frame motors that are expected to operate more than 500 hours per year. The required efficiency levels are lower than those presently specified for NEMA "energy-efficient" motors.

■ **Be Aware of the Efficiency Test Method**

When you get efficiency data from motor manufacturers, be sure to check the test method used. There may be large differences in the results produced by different test methods.

The standard now used almost exclusively in the United States is NEMA Standard MG1-20.52. This is derived from Standard 112, Method B, of the Institute of Electrical and Electronic Engineers (IEEE).

The international (IEC) test method and test methods of other major motor manufacturing nations are also derived from IEEE Method 112-B, but these variations may produce significantly different efficiency figures.

U.S. and Canadian test methods are the most realistic. They are based on actual measured output and measured input. The standards of other nations may estimate some of the motor losses, or may ignore them.

For example, a particular 7.5 horsepower motor has a nominal efficiency of 80.3% using the U.S. and the Canadian (CSA C390-M1985) standards. The same motor has a nominal efficiency of 82.3% using the British and international (IEC 34-2) standards. It has a nominal efficiency of 85.0% using the Japanese (JEC 37) standard. In this way, the Japanese or British manufacturer may claim a much higher efficiency for a motor than an American manufacturer may claim for an identical motor.

A remaining weakness of some standards, including the U.S. and Canadian standards, is that the tests are performed by the manufacturers themselves, without independent verification. Steps are presently being taken in the United States to certify the testing laboratories of motor manufacturers.

What to Consider When Selecting a Motor

You need to consider all the following motor characteristics to select the best motor for each application. Most of these characteristics are expressed by ratings or specifications that are available in the catalogs. The MotorMaster computer program includes related data for the motors in its database. As a point of interest, all these issues relate to efficiency directly or indirectly. For specialized applications, you may need to consider additional characteristics, such as low sound level.

■ Full-Load Efficiency

Figure 1 shows the range of efficiencies that presently exists among 3-phase motors. You can usually find an economical choice with an efficiency near the highest available. It is worth repeating that careful shopping may have a greater effect on price than the efficiency rating.

For single-phase motors, research the catalogs to find the maximum efficiencies that are presently available. Most major motor manufacturers have a "high-efficiency" line of single-phase motors, so you can narrow your search to these models. The highest efficiency is found in "capacitor start" motors, followed in declining order by "split phase," "permanent split capacitor," and "shaded pole" motors.

■ Part-Load Efficiency

The part-load efficiency characteristics of motors were covered previously. See Figure 2. If motors operate for long periods at reduced loads, be sure to study the part-load efficiency figures. Some manufacturers offer special models that are claimed to provide high efficiency at low loads.

Unfortunately, motor catalogs often omit part-load efficiency information. At the present time, the MotorMaster program is the only comprehensive source of information on the part-load performance of 3-phase motors. The part-load performance of single-phase motors is still a blank area.

■ Power Rating

When replacing a motor, consider whether you can substitute a motor with a lower power rating. To do this, you usually need to reduce the load. This opportunity arises commonly as a result of energy conservation measures. For example, you can trim the impeller of a pump to eliminate excess capacity, which may greatly reduce the pump's power requirement.

Serious oversizing of the motor causes it to operate at reduced efficiency. However, it rarely pays to replace a motor simply because it is oversized. Peak motor efficiency typically occurs around 80% of full load, but

efficiency remains near maximum until the load falls well below half, except in smaller sizes. Furthermore, the larger motor is more efficient overall, and its efficiency drops less at low loads.

When operating a motor with a variable-frequency drive, it may be desirable to oversize the motor to avoid overheating. See Reference Note 36, Variable-Speed Motors and Drives, for details.

To measure the present motor load accurately, use a wattmeter. A wattmeter accounts for the difference in motor power factor at different loads. If you need only a rough idea of the load, you can use an ammeter. Check the motor nameplate for the Full-Load Amps (FLA), and calculate the motor load from the ratio of the measured amperage to this value. Power factor remains fairly constant down to about 65% load, so this method is reasonably accurate at loads higher than 65%. Below this value, power factor drops rapidly. Measuring the amperage of a lightly loaded motor yields a false value of power output that is higher than actual.

■ Service Factor

"Service factor" is the highest percentage of full load at which a motor can operate continuously under standard test conditions without overheating. Service factor is related to motor temperature ratings, which are explained below. Service factor is always displayed on motor nameplates, as in Figure 4.

Service factor provides a margin for error in calculating the size of motor that is needed for an application. The lowest service factor is 1.0. High-efficiency motors usually have a service factor of at least 1.15, but this is not always true. Motors designed for high-torque applications typically have higher service factors, perhaps as high as 1.40 in smaller sizes.

A service factor higher than 1.0 is useful in applications where the motor operates for a large fraction of the time at low loads. This allows you to select the smallest possible motor, thereby achieving greater efficiency at low loads, without fear of overheating the motor during occasional periods of peak load. A higher service factor also allows operation at higher ambient temperatures.

Even if a motor has a service factor higher than 1.0, operating the motor above its rated load reduces its service life, and usually reduces its efficiency.

■ Torque Characteristics

Torque is rotational force. The torque that a motor produces changes radically with the speed. Motor applications vary widely in the torque they require at different fractions of full speed. Therefore, special names are given to the torque that develops in different parts of a motor's speed range. Figure 5 illustrates these aspects of motor torque for a typical AC induction motor. These are the important torque characteristics:

- *full-load torque* is the torque measured at the motor's rated load. This load occurs at the motor's rated speed. This rating is especially important for loads that develop more resistance as the speed increases. Pumps and fans are the most common loads of this type.

- *locked-rotor torque* is the motor's torque when the shaft is held stationary. This is the motor's available *starting torque*. It is important when selecting a motor to drive a load that has large starting resistance, such as a loaded conveyor belt. It also indicates how rapidly a motor can accelerate a load that has high inertia, such as a centrifugal fan or a balanced elevator. Starting torque is less important with loads that move freely and have little rotational inertia, such as centrifugal pumps.

- *pull-up torque* is the minimum torque that the motor can produce. This typically occurs at a fraction of full speed. The motor will stall and burn out after starting if the load requires more torque than this amount, even if the motor is able to drive the load at normal speed. This rating is important when driving loads that quickly develop resistance when starting, such as reciprocating compressors (see Figure 6).

- *breakdown torque* is the maximum torque that the motor can produce. The motor passes through breakdown torque on the way to stalling from overload. Breakdown torque is developed at a speed slower than the rated speed, with the motor in an overloaded condition. Breakdown torque is most important in applications that have transient

requirements for extreme torque, such as a rock crusher.

To deal with different torque requirements, motor manufacturers produce motors with several standard torque characteristics. NEMA Designs "A" and "B" are motors with conventional torque. NEMA Designs "C" and "D" produce high starting torque. Torque would have little effect on efficiency if all other factors remained the same. However, achieving higher torque requires major changes in the motor geometry, and the motor geometry affects efficiency. Torque is the product of force times moment arm, so high-torque motors typically are larger in diameter.

Some motors list the NEMA Design on the nameplate, as in Figure 4.

Among single-phase motors, the highest torque is found in "capacitor start" motors, followed in declining order by "split phase", "permanent split capacitor", and "shaded pole" motors. This sequence is the same as the ranking of single-phase motors in terms of efficiency. In other words, among single-phase motors, higher efficiency correlates to higher torque.

■ Full-Load RPM

The magnetic field of the stator in a 3-phase motor rotates at a rate that is determined by the line frequency and by the number of poles. This rate is called the "synchronous speed." With 60-Hertz power, this speed is 3,600 RPM for a 2-pole motor, 1,800 RPM for a 4-pole motor, 1,200 RPM for a 6-pole motor, etc. In single-phase motors, the synchronous speed is calculated the same way, even though the field does not actually rotate.

The actual speed of an induction (ordinary) motor at full rated load is a few percent slower than its synchronous speed. This speed differential is necessary for the operation of an induction motor. The difference in speed is called "slip." In an induction motor, the speed difference between the rotor and the magnetic field of the stator is what induces the magnetic force that turns the motor.

The actual speed of the motor when operating at full-load torque is listed on the motor nameplate, as

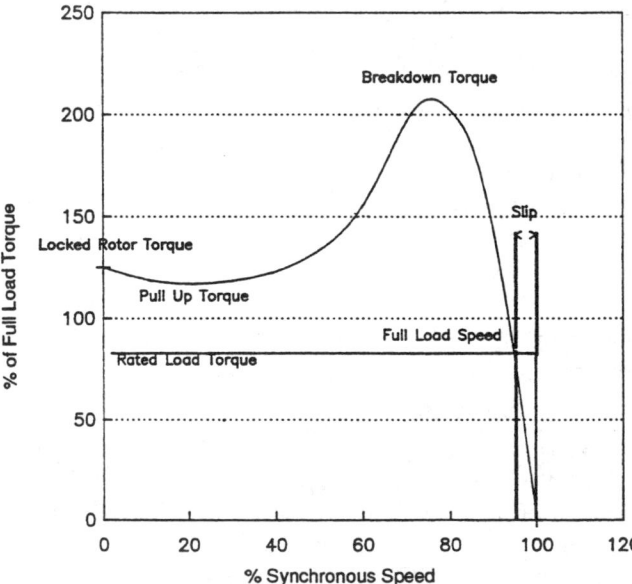

U.S. Department of Energy

Fig. 5 Torque characteristics of a typical AC induction motor To select a motor's torque characteristics, start by knowing the torque requirements of the load. Then, select a motor in the appropriate NEMA Design category.

WESINC

Fig. 6 High-torque application This reciprocating air compressor requires full torque almost immediately after it starts. This is because the force needed to compress the air in the cylinders is almost independent of the speed.

shown in Figure 4. All other things being equal, a more efficient motor has less slip at a given load, so its speed is slightly higher.

The amount of slip has a very large effect on torque, even though it represents only a small fractional change in speed. See Figure 5.

You might think that this difference in speed is inconsequential in terms of energy consumption, but it may be significant if the motor is used to drive a pump or fan. With such loads, the power requirement rises sharply with speed. The extreme case exists where a fan or pump discharges into a system where the flow is not regulated. In this case, the energy requirement is proportional to the third power of the speed.

For example, this is true of a fan discharging into an open duct, or of a pump discharging sewage into a basin. In these examples, if the speed is increased by one percent, the power input is increased by about three percent. In such cases, the increased power requirement may cancel a large part of the efficiency advantage of the higher-efficiency motor.

This well-known third-power relationship does not occur in all applications. If the resistance facing the fan or pump is held constant, the increase in power is proportion to the first power of speed. For example, VAV air handling systems and hydronic systems with control valves tend to maintain a constant pressure. In such systems, a one percent increase in speed might increase the power requirement between one and two percent.

One way to deal with this problem is to select the replacement high-efficiency motor to have the same speed as the original motor. However, you may not be able to find such a motor without compromising efficiency or cost. In practice, you cannot adjust the power input of pumps and fans with accuracies of one percent. Instead of trying to split hairs in this manner, adjust the fan or pump to match its load requirement at the same time you upgrade motor efficiency.

With fans, you have a variety of methods to reduce output efficiently. See Measure 10.3.1 for the details. With direct-drive pumps, trimming the pump impeller is the only way of reducing excessive pump output efficiently. See Measure 10.2.1 for the details. These methods typically yield power reductions of 10% to 40%. With any kind of equipment that is driven by belts, you can adjust the speed by changing the size ratio of the drive pulleys.

A motor's full-load speed rating is related to its torque characteristics. Higher torque is achieved largely by designing the motor to have greater slip. For this reason, NEMA Designs A and B (conventional torque characteristics) have full-load slips well under 5%, while NEMA Designs C and D (high torque) may have full-load slips exceeding 5%. All other things remaining equal (which never occurs), achieving high torque requires some sacrifice of efficiency.

■ Operating Temperature

The most common motor failures are insulation failure and bearing failure. Both are related to operating temperature. For a given type of insulation, the service life of a motor declines exponentially with increasing operating temperature. Experience suggests that half of all motor failures occur from the long-term effect of high temperature on the windings. It is estimated that a 10°C (18°F) increase in winding temperature reduces insulation life by about half, and bearing life by about one quarter.

In the United States, NEMA has established four classes of insulation, which are based on the operating temperature of the motor. In order of increasing temperature tolerance, these are Classes A, B, F, and H. Class B is the common class in conventional motors, and Class F is the most common in high-efficiency motors.

The temperature limitations of a motor may be indicated by stating the insulation class. Or, a specific ambient temperature may be stated. If an ambient temperature is given, the assumed service factor and the altitude may also be stated. Altitude is important because air is less dense at higher altitude, so it has less cooling ability. The nameplate in Figure 4 uses both rating methods.

In most applications, the operating temperature of a motor depends more on its load than it depends on the ambient temperature. Totally enclosed motors cannot get rid of heat by direct ventilation, so they rely on conduction through the case. Their internal temperature largely depends on the rate of heat generation inside the motor. Near full load, a motor becomes quite hot. For example, a Class F motor with a service factor of 1.15 has an allowable temperature rise of 115°C (207°F) above an ambient temperature of 40°C (104°F), for a maximum temperature of 155°C (311°F).

If a motor must operate in a high-temperature environment, there are two ways to achieve a normal life span. One is to get a motor with insulation that is rated for higher temperatures. The other is to get a motor with a larger power rating, so that the motor operates at a lower percentage of full load.

Make allowances for other factors that may raise motor temperature, including:

• *frequent starting.* The starting current of a motor is several times its running current. The heat generated by a starting cycle is dissipated slowly, so frequent starting raises temperature drastically.

• *using the motor with variable-frequency drives.* Some electronic variable-frequency drives produce a substantial amount of distorted input power that the motor converts to heat. In "constant-torque" applications, motors overheat because the current remains high as the rotor and the cooling fan slow. Using motors with electronic drives is discussed below.

- ***poor power quality.*** A motor cannot efficiently use electrical power that is in the form of distortion or electrical noise. The motor converts much of this erratic power into heat. If the power supply contains a substantial amount of distortion, the motor runs hotter.
- ***phase balance.*** Small phase imbalances can seriously increase motor temperature. This is explained below.
- ***dirty environment.*** If the motor uses ambient air for internal ventilation, dirt in the environment quickly fouls the motor interior and reduces heat transfer. A coating of dirt also reduces the heat rejection from a totally-enclosed motor.
- ***high altitude.*** The air inside and outside a motor is what carries away most of the motor's heat. Air is thinner at higher altitudes, causing the motor to run hotter.

High-efficiency motors do not necessarily run cooler than conventional motors. Although high-efficiency motors generate less internal heat, the size of the internal cooling fan may be reduced as one way of reducing energy consumption. Base your temperature-related selection decisions on temperature ratings, not on efficiency ratings.

■ High-Voltage vs. Low-Voltage

Lower operating voltages allow higher motor efficiencies. We are talking about big differences in voltage ratings here, for example, the difference between a 460-volt motor and a 2,300-volt motor.

Lower voltage allows thinner insulation in the windings, which allows the windings to be larger in diameter. This reduces resistance losses and increases magnetic flux. The efficiency difference related to this aspect of motor design ranges from a fraction of one percent to as much as two percent.

Before you make this decision, step back and look at the bigger picture. Let's say that your facility receives power from the electric utility at 2,300 volts. For lower voltages, the facility owns a transformer that reduces the incoming voltage to 460 volts. If you select a 460-volt motor because of its higher efficiency rating, you may not save energy overall. This is because the electricity for the 460-volt motor must come through the transformer, where it suffers additional losses. In this case, you may save more energy by installing a 2,300-volt motor. In new construction, this will also reduce the cost of your transformer.

Thus, when you select your motor voltage, look at your incoming voltage, and consider who owns the transformers on site.

■ Small Voltage Differences

A motor operates at highest efficiency when it is supplied with its precise rated voltage. Lower voltage reduces efficiency, reduces starting torque, and causes the motor to run hotter. Higher voltage also reduces efficiency, and it increases starting current. If possible, adjust the taps in the power supply transformers to match the supply voltage to the motor rating.

In order to reduce inventory requirements, many motors are designed to operate at a number of standard voltages, such as 208 volts and 230 volts. Such motors tend to be less efficient than motors designed to operate at a single voltage. Select your motors for efficiency based on the actual voltage that is used.

Motors that are designed to operate at several voltages operate most efficiently at the highest rated voltage. The difference in efficiency at the different voltages may be as high as several percent. Therefore, check whether the transformer that supplies power to the motors has taps that you can reset to the higher voltage. This is often the case. However, this may not be practical if the transformer serves loads other than the motors. See whether the other equipment can operate satisfactorily at the higher voltage. Or, if possible, rearrange the power circuits to allow the motors to be operated by higher-voltage transformers.

■ Power Factor

(If you are unsure about the meaning of power factor, see Reference Note 21, Electricity Pricing, for an explanation.)

Electric motors are the major cause of low power factor (along with magnetic fluorescent lighting ballasts). The power factor of motors is fairly high at high load, but drops radically at loads below about half. Larger motors have somewhat higher power factor at all percentages of load. Motor efficiency, by itself, does not have a significant effect on power factor.

Power factor may become a significant economic issue if the utility company has a large power factor penalty, and if motors operate at low loads. However, there is not much difference in the power factors of similar motors. The effective way to deal with power factor is to increase the power factor of the facility's electrical system by using power factor correction capacitors or other devices intended for this purpose.

The full-load power factor of a motor may be stated on the nameplate, as in Figure 4. This makes it easy to select power factor correction capacitors.

■ Locked-Rotor and Starting Amperage

Higher-efficiency motors may have higher starting currents than ordinary motors. This is because of the lower resistance in the windings. In retrofits, the starting current of a high-efficiency replacement motor may be too high for the existing fuses, circuit breakers, or wiring.

If this problem occurs, a fairly inexpensive solution is to use a staged ("soft") motor starter. The starter limits the motor current until the motor is running fast enough to limit its current draw. A larger motor may already have a soft starter, except in applications requiring high starting torque.

NEMA Design B motors have the same torque characteristics as Design A motors, but they have lower starting current. However, Design B motors usually pay for this advantage with an efficiency penalty.

The starting amperage is commonly abbreviated on motor nameplates as "SLA." This distinguishes from full-load amperage, which is commonly abbreviated "FLA." If the nameplate lists only a single amperage, it is the starting amperage.

■ Enclosure Type

Select the type of motor casing, or enclosure, on the basis of the motor's environment. The type of casing has little effect on efficiency.

Motor enclosures differ primarily in the degree of isolation that they provide between the inside of the motor and the outside environment. This is reflected in the names of enclosure types, such as "open," "splash-proof," etc.

A fully enclosed motor cannot be cooled by direct ventilation, so selecting a fully enclosed motor requires careful attention to the way that the motor is cooled. Fully enclosed motors are designated by their cooling characteristics, such as "totally enclosed nonventilated" (TENV) and "totally enclosed fan cooled" (TEFC).

The enclosure type is usually shown on the motor nameplate, as in Figure 4.

■ Frame and Face Type

High-efficiency models have become available in all the common frame and face configurations of motors larger than about one horsepower. The frame and face configuration do not strongly affect efficiency, except perhaps with some specialized motors. The frame type may be shown on the motor nameplate, as in Figure 4.

Phase Balance

A three-phase motor is served in effect by three separate power circuits, or "phases." Voltage differences between the phases cause much more efficiency loss than fluctuations in the overall voltage. Typical phase voltage imbalances may reduce motor efficiency by several percent.

Voltage imbalance between the phases can also cause vibration and overheating. The overheating effect is significant enough that NEMA requires motors operating with an imbalance of more than 1% to be derated. Figure 7 shows the amount of derating that is required for different amounts of voltage imbalance.

If you can't afford alarms to monitor phase balance at the motors, check it regularly with a voltmeter of good quality. Preferably, use a digital display that makes it easy to read small differences in voltage.

Correcting phase imbalance is entirely a matter of the power supply system, rather than the motor itself. The appropriate methods of correcting phase imbalance depend on the causes, which are:

- **uneven distribution of load among phases within the facility.** The single-phase loads in a facility, such as lighting, are divided among the phases of the incoming 3-phase power supply. The load is rarely divided precisely. The differences in phase current produce differences in phase voltage, especially if the conductors are heavily loaded.

  The appropriate solution may be to redistribute the single-phase loads among the phases, to increase the capacity of heavily loaded conductors, or to install separate transformers for 3-phase loads.

  It is usually not a good idea to balance phase voltage by using the transformer taps. If the phase imbalance is so bad that transformer taps would make a difference, the problem is bad enough that you should correct the underlying problem.

- **power wiring phase defects.** A major current leak from a phase to ground, or less commonly, from one phase to another, can cause a voltage imbalance. A 3-phase motor may continue to run even if one of the phases is completely dead, but the motor may also fail from overheating.

  The solution is to find and correct phase defects immediately. If a single phase fails, the single-phase equipment on that circuit will stop. If you see this happening, recognize the implication for the 3-phase motors and stop them immediately.

  A reliable way to protect motors from phase imbalance is to install a phase monitor on each motor power circuit. This device will either sound

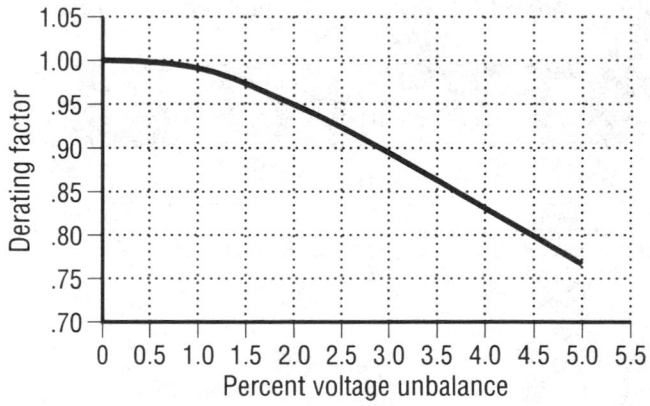

U.S. DOE Motor Challenge Turning Point

Fig. 7 Motor derating required to compensate for unbalanced phase voltage The message here is that very small differences in phase voltage create a serious danger to the motor.

an alarm or trip out the motors if a phase is lost or if the phases become seriously unbalanced. The device itself is inexpensive. Most of the cost is for the wiring.

- *failure of power factor correction capacitors.* Separate power factor correction capacitors are installed on each phase of a 3-phase system. Capacitors fail with age, in effect creating a ground fault. You can detect failed capacitors by testing for ground faults or by checking the power factor of each phase.

- *unbalanced power from the electric utility.* Phase imbalance in the power supplied by the electric utility occurs because the single-phase power provided to its other customers is not evenly divided among the phases. Try to solve this problem by asking the electric utility to improve its phase balance. Phase imbalance in the utility feeders may vary greatly with the season and time of day, as the utility's loads change. The phase balance of the electricity supply tends to fluctuate more when the facility is farther away from the generating plant or the substation.

Motors Used with Electronic Variable-Speed Drives

Electronic variable-speed motor drives are becoming popular for many applications. Reference Note 36 describes this device in detail. In summary, a variable-frequency drive is an electronic device that changes the driving frequency of the electrical power provided to the motor. Unfortunately, electronic drives may reduce the efficiency of motors and increase their internal heating. This may make it prudent to install a special motor that is designed to operate with electronic drives.

The basis of the problem is that the simulated alternating current produced by the electronic drive is not a smooth sine wave. In fact, the output of an

electronic drive may be only a rough approximation of a sine wave. It may be a square wave, a sequence of square waves, or a series of pulses. As a result, the power delivered to the motor has a large fraction of high-frequency components. Conventional motors are designed to use pure sine waves at line frequency, so the high-frequency components are not efficiently converted to mechanical energy. The unusable high-frequency energy is converted to heat, which shortens the life of the motor. In addition, the crude waveform produced by the drive may have voltage spikes that break down the motor insulation.

Certain features of higher-efficiency motors make better use of the high-frequency components. For example, thinner core laminations reduce the eddy currents resulting from the high-frequency components. As a result, high-efficiency motors offer an even greater efficiency premium when used with variable-frequency drives than when used in other applications.

A high-efficiency motor does not necessarily deal better with all the problems created by electronic drives. In particular, a high-efficiency motor may be more vulnerable to insulation breakdown (because of its thinner insulation). For this reason, some manufacturers offer motors that are claimed to be designed specifically for use with variable-frequency drives. These models usually have a large price premium. At least one manufacturer offers different motors for different turndown ratios. A motor advertised with a 10:1 turndown ratio may cost much more than a motor advertised with a 4:1 turndown ratio. Claims of improved performance with electronic drives cannot be verified by the purchaser, because there are no standards specifically for motors used with electronic drives.

Selecting a motor for variable-frequency operation requires even more information than is needed for ordering a single-speed motor, such as the maximum allowable speed, torque limitations, etc. Figure 8 shows a typical nameplate of an "inverter duty" motor, which provides only a fraction of the information needed. See Reference Note 36 for more details.

Motors advertised specifically for use with electronic drives are identified as such in the MotorMaster computer program. The program can search for such motors as a category.

Multi-Speed Motors

A multi-speed 3-phase motor is like a single-speed 3-phase motor, except that it has extra stator coils and poles. Speed is changed in steps by energizing different numbers of poles.

Multi-speed motors are fairly efficient. They are economical, and they do not require any complex controls. Reference Note 36, Variable-Speed Motors and Drives, describes multi-speed motors in detail, and

RELIANCE ELECTRIC	INVERTER DUTY	V-XS VARIABLE SPEED MOTOR				
I.D. NO.	G	MODEL NO. P25G3371		FRAME SIZE. 254 T		
HP	RPM	VOLTS	AMPS	HZ	TYPE P	
1.5	1765	460	18.3	60	S.F. 1.00	
1.5	2645	460	18.2	90	INSUL CLASS F	
1.32	155	64	18.3	6	CODE G	
DRIVE END BEARING	45 BC 03J30X			ENCL TEFC	DESIGN B	
OPP D.E. BEARING	45 BC 03J30X			DUTY CONT	PHASE 3	
					AMB 40°C	
MAX SAFE R.P.M. 5400	OVERTEMP PROT 2	wk2 1.9 lb•ft2	POLES 4	MOTOR WEIGHT 325 LBS.		

MFD. BY RELIANCE ELECTRIC INDUSTRIAL COMPANY   PLANT 010 (G)
CLEVELAND, OHIO 44117   MADE IN U.S.A.

Reliance Electric

Fig. 8 Typical nameplate for an inverter-duty motor
What is unusual here is the information for a wide range of speeds and power frequencies. You need much more information than this to properly select a motor for your application.

compares them with other methods of driving motor loads at variable speed.

ECONOMICS

SAVINGS POTENTIAL: *2% to 15% of energy consumption, with 3-phase motors. Up to 30% of energy consumption, with small single-phase motors. Larger motors have smaller percentages of savings, but save much more energy in absolute terms.*

COST: *High-efficiency motors of common types cost $50 to $100 per horsepower in sizes larger than 10 horsepower. In sizes near one horsepower, the cost is about $200 per horsepower. The cost premium for high-efficiency motors averages about 20%, but this varies widely.*

PAYBACK PERIOD: *Usually less than one year, in new construction and for replacing failed motors. Several years or much longer, for upgrading motors that are still working.*

TRAPS & TRICKS

SELECTING MODELS: *Get a current version of the MotorMaster computer program, and use it. It will make your life easier, save you money in buying motors, and maximize your energy savings. Go to the motor catalogs for additional information.*

ADVANCE PREPARATION: *For upgrading failed motors, doing your homework beforehand is the key to success. Identify vendors who can provide your selected motor models on short notice. If you have many motors of a particular size and type of service, keep at least one spare in stock.*

SPELL IT OUT: *To guide maintenance personnel in replacing failed motors, designate the selected models and vendors in the plant operating manual. Make sure that all maintenance personnel, including those who work the late shift, know what to do. Tell them where the spare units are stored.*

Centrifugal pumps are used in virtually all buildings and plants. In many, pumps are major energy consumers. This Subsection consists of two broadly applicable Measures for energy conservation in simple pump systems. Both are methods for matching the output of a pump to the system requirement. In many cases, the two Measures are alternatives to each other. The first Measure, pump impeller trimming, is much more efficient than the second Measure, pump throttling.

Also, Consider Variable-Flow Pumping

In systems with larger pumps, installing variable-flow pumping may provide large savings in pump energy. Consider variable-flow pumping for applications where the flow requirement varies widely and pump pressure can be allowed to fall along with the flow. Unlike the Measures in this Subsection, variable-flow pumping must be designed to fit the particular application. For this reason, it is covered in the Subsections dealing with specific applications. For example, refer to Measure 2.5.2 for variable-flow pumping in chilled water systems.

Before Getting Started ...

These Measures require an understanding of pump performance. If you need a refresher on this topic, see Reference Note 35, Centrifugal Pumps, which also explains pump construction and selection.

INDEX OF MEASURES

10.2.1 Trim pump impellers to match pump output to system pressure and/or flow requirements.

10.2.2 Throttle the discharge of pumps to match system pressure and/or flow requirements.

MEASURE 10.2.1 Trim pump impellers to match pump output to system pressure and/or flow requirements.

If a pump is producing more pressure than the application requires, it is also consuming more energy than necessary. You can eliminate the excess energy consumption in many cases by reducing the diameter of the pump impeller to the minimum needed to satisfy the pressure or flow requirement. This is a standard procedure. It is called "trimming" in the lingo of the pump business.

Trimming is a simple operation that is done on a lathe. The greatest part of the work is removing the impeller from the pump. The most complicated aspect is calculating how much to reduce the impeller diameter.

Trimming the impeller of a centrifugal pump reduces its pressure output, rather than its flow capacity. However, the resistance of the system interacts with the pump pressure to determine the actual system flow. Thus, in many applications, you can use impeller trimming to reduce excess flow as well as excess pressure.

Where to Trim Pump Impellers

In many pump installations, the pressure and/or flow delivered by the pump substantially exceeds the system requirement. This occurs because the designer who selects the pump adds a margin for uncertainties, and sometimes because the pump is not delivered in accordance with the engineer's specifications.

If a pumping system has a partially closed valve in the discharge pipe that is acting as a "balancing" or throttling valve, this indicates that the pump should probably have its impeller trimmed. Similarly, if the pumping system has a pressure-regulating bypass valve, and the valve is always open, this indicates an opportunity to reduce pump pressure. Even if such clues are not visible, an opportunity for reducing pressure often exists.

You may be able to create opportunities for reducing pump power by modifying the system that the pump serves. An important example is converting bypass control valves on heating and cooling equipment to throttling valves.

In new construction, if you design a constant-flow pump system, write pump trimming into the construction contract as part of the system installation. Also, select pumps that will not lose efficiency badly when the impeller is trimmed.

Graphical Analysis of Impeller Trimming

As an example, we will trace the effects of impeller trimming using the typical pump curves shown in

SUMMARY

Matches pump pressure efficiently to the maximum system pressure requirement. Reduces flow capacity somewhat. Calculating the diameter reduction may be tricky.

SELECTION SCORECARD

Savings Potential .... $ $ $
Rate of Return, New Facilities % % % %
Rate of Return, Retrofit .... % % % %
Reliability .... ✓ ✓ ✓ ✓
Ease of Retrofit .... ☺ ☺

Figure 1. (These same curves are used in the example in Measure 10.2.2, allowing you to compare the two methods.) Let's take this step by step:

• In designing the pumping system, the engineer calculates that the system needs a flow rate of 800 GPM, and that the pump must produce a pressure of 60 feet of head at this flow rate. These pressure and flow rate requirements are marked by point "A."

• Based on these requirements, the engineer selects a pump that has pump curve "B."

• When the system is built, the actual system curve turns out to be "C," so the pump actually operates at point "D." Although the pump pressure is approximately the same as originally estimated, the flow rate is much higher than the system requires. Pump power is proportional to the third power of flow rate, so a large amount of energy is wasted.

• To eliminate this energy waste, trim the pump impeller so that the pump operates at point "E." This produces the originally required flow rate of 800 GPM, but the pump is able to operate at a lower pressure.

• The pump curves show that the pumping power has been reduced from about 18 horsepower to about 8 horsepower. Taking into account some efficiency loss in the pump motor, the net energy input is reduced to about half.

The Effect on Pump Efficiency

Trimming the impeller usually reduces the efficiency of a pump somewhat. As long as the diameter reduction is not excessive, the efficiency loss is moderate and the reduction in pump power is substantial. In the previous example, the pump efficiency was reduced from about 82% to about 77%, but the overall energy consumption was still reduced by half.

How much you can trim a pump impeller may depend on the application or it may depend on the pump. If the application needs only a small reduction of impeller diameter, the pump is not likely to be a limitation. However, if the application allows a large reduction of flow rate, the characteristics of your particular pump may not allow you to reduce the impeller diameter enough to eliminate all the excess pressure.

It depends on the original relationship between the impeller and the pump casing. Manufacturers make only a limited number of casing sizes. They satisfy individual customer requirements by adjusting the impeller diameter. If the impeller is originally at the small end of the manufacturer's diameter range, reducing the diameter much more would cause a drop in efficiency that cancels part of the saving in power. For this reason, it is a good idea to call the pump manufacturer's engineering department to discuss the potential effect of trimming your pump, and to find out the maximum reduction that is advisable.

You may be wondering whether trimming the impeller of an existing pump makes it less efficient than it would have been if the manufacturer had supplied the optimum diameter. Usually, the answer is no, provided that you do not trim the impeller diameter excessively. The manufacturer trims the pump impeller in the same way that you do to satisfy an order for particular pump characteristics. In some cases, particularly with exceptionally large pumps, the manufacturer may file

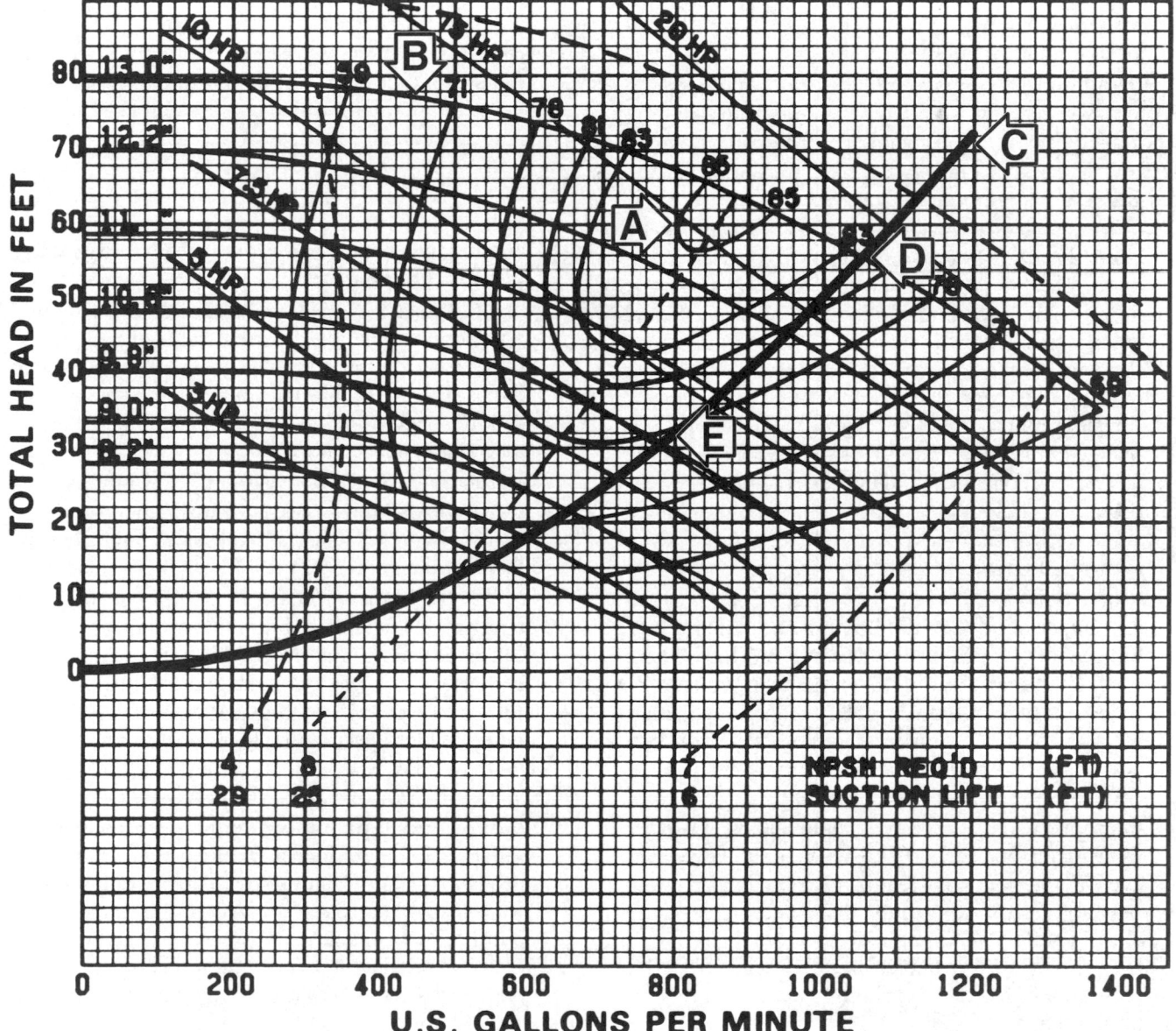

Fig. 1 Analysis of pump impeller trimming Impeller trimming may save a large fraction of pump energy because it efficiently eliminates excess pressure and flow capacity.

down the impeller vanes to adjust the pump's performance characteristics. If this refinement is appropriate, you can get it done yourself.

Pumps with higher "specific speed" suffer greater efficiency loss from impeller trimming. This may be something to consider if you are designing a large pump system that will operate at constant flow. With smaller pumps, you have little or no choice of specific speed. See Reference Note 35, Centrifugal Pumps, for an explanation of specific speed.

How Much Can You Reduce Pressure and Flow?

Do not reduce either pressure or flow to an extent that harms system performance. In most cases, the optimum operating point is not clearly defined. At some point, system performance will begin to suffer. In some systems, reducing pump output will increase energy consumption elsewhere in the system. Study the system to determine what the limits are. In case of doubt, leave a modest amount of excess pressure.

Trimming the impeller reduces both the pressure and the flow rate. In some systems, pump pressure is important primarily as a means of producing flow, so the flow rate is the main issue. In other systems, the pressure itself is important. These are the pressure and flow characteristics of common devices and applications:

- *control valves.* For a control valve to provide stable control of flow into a device, the control valve itself needs to create a certain amount of flow resistance, even when it is wide open. (Without creating resistance of its own, a control valve would have nothing to change to control the flow.) A common rule of thumb is that the control valve, when fully open, should have about the same resistance as the device it is controlling. Therefore, the pump should provide sufficient pressure to overcome the control valve resistance, in addition to the resistance of the controlled device.

- *heat exchangers.* The flow rate through a heat exchanger or coil should be high enough to create turbulent flow, which is necessary for effective heat transfer. As long as the flow remains turbulent, the reduction of heat transfer is less than the reduction of flow rate. This is because a slower flow rate allows more time for heat transfer.

  Minimizing heat exchanger fouling requires a minimum water flow rate of about 3 feet per second (one meter per second). This velocity is needed even if the water is treated. Increasing the flow rate above this minimum does not increase the cleaning effect very much. A flow rate high enough for good heat transfer usually satisfies the requirement for tube scrubbing, since both rely on turbulence.

- *water-to-air coils.* Heating and cooling coils are heat exchangers, so they involve the same considerations in reducing flow as stated previously. These coils have narrow passages, so flow is always turbulent. Typical water-to-air coils suffer little loss of capacity with a modest reduction of water flow, because of the circuiting of the coil and the longer time that the water spends in the coil.

- *chilled water systems.* Reducing the chilled water flow rate lowers the chilled water temperature, which reduces chiller efficiency and capacity. (See Measure 2.2.1 for details.) If the chilled water system serves variable-air-volume (VAV) air handling systems, higher chilled water temperature causes the air handling unit fans to operate at higher power. In addition to overcoming the resistance of the evaporator and piping, the pump must produce enough pressure for proper operation of the cooling coil control valves.

- *condenser water systems.* Reducing the condenser cooling water flow rate increases the condenser water temperature, which reduces chiller efficiency and capacity. (See Measure 2.2.2 for details.) In addition to overcoming the resistance of the condenser and piping, the pump must produce pressure to lift the water from the level of the cooling tower sump to the level of the cooling water distributor pipe on top.

- *cooling towers.* Open cooling towers may require a minimum flow rate in order to avoid gaps in the water distribution pattern that would allow air to bypass the water. The minimum percentage of flow depends on the cooling tower's design, installation, and maintenance.

- *hot water space heating systems.* The pumps must provide enough pressure for satisfactory operation of the control valves, as well as delivering flow through the coils.

- *boiler feedwater pumping.* Insufficient pressure may cause a boiler to run dry under high load. This causes a sudden shutdown of the boiler by the safety controls. If the safety controls fail, the boiler is likely to explode. Don't mess with feedwater pumps unless you are really sure you know what you are doing.

- *hot and cold service water pressurization.* The pumps must provide adequate pressure for satisfactory operation of fixtures, such as shower heads and toilet flush valves. The pressure falls as water usage increases, because of the shape of the pump curve. Inadequate pressure first becomes apparent when water consumption is high in those parts of the system that need the greatest pressure.

How to Calculate the Diameter Reduction

Calculate the diameter reduction precisely, because the diameter affects the flow, pressure, and energy consumption. Pump power is especially sensitive to impeller diameter, since it is proportional to the third power of the diameter.

For example, in Figure 1, a radius cut of little more than two inches reduces the pressure from 56 feet of head to 32 feet of head, and it reduces the power requirement from 18 HP to less than 8 HP.

Use this procedure to determine the minimum acceptable pump pressure:

• If a particular flow requirement is needed to satisfy an equipment or system requirement, find out what the requirement is. For example, a cooling tower may need a certain minimum flow rate to maintain full efficiency.

• Fully open any "balancing" or throttling valves in the system, and close any bypass valves used for regulating pressure in the system.

• If system flow is variable, create the conditions for maximum flow. For example, in a heating water system, open all the control valves on heating equipment.

• Measure the flow rate. (See below for the methods.)

• Record the pressure differential across the pump, i.e., the difference in pressure between the pump suction and the pump discharge. Correct for static pressure due to differences in the height of the gauges. Call the corrected pressure difference "P1." (See below for methods of making pressure readings accurately.)

• Use any available valve in the pump discharge to throttle the pump discharge down to the minimum acceptable system pressure or flow rate. If flow rate is the critical factor, rather than pressure, record the difference in pressure between the pump suction and the downstream side of the throttling valve. Correct for static pressure due to differences in the height of the gauges. Call the corrected pressure difference "P2." If the system is being set to achieve a given pressure, measure the flow rate.

With this information in hand, use one of these methods to determine the new impeller diameter:

• *call the pump manufacturer's engineering department.* In addition to the before-and-after pressures and the before-and-after flow rates, be prepared to provide all the pump identification information from the pump nameplate. Do not rely on pump information contained in engineering documents, as the actual pump may be different.

• *use the pump curves.* If the pump curves for the present pump model are available, and if they show the pump's characteristics for different impeller diameters, use the information you collected to read

the new impeller diameter from the curves, as discussed above. Remember that performance with a trimmed impeller may be worse than performance with a new pump having an impeller of the same diameter.

• *calculate from the pressure measurements.* You can use the fact that the pressure developed by a pump is proportional to the square of the impeller diameter. Use this formula:

$$\textbf{Final Diameter} \ = \ \textbf{Initial Diameter} \ \times \ \sqrt{\frac{\textbf{P2}}{\textbf{P1}}}$$

The formula is not precise, because it fails to account for the reduction of pump efficiency that occurs with trimming. To account for this, make the final diameter somewhat greater than indicated by the formula.

• *trial-and-error.* In case of doubt, make an initial cut on the impeller that is conservative. Reinstall the impeller, and observe the changes in pressure and flow rate. Repeat as necessary.

■ How to Measure Flow Rate

To measure flow rate in a closed system, you need a flowmeter. The variety of flowmeter types has mushroomed in recent years. Externally mounted flowmeters are now available that avoid the cost and trouble of tapping into the pipe. See Reference Note 16, Measurement of Liquid, Gas, and Heat Flow, for details.

With open pumping systems, such as cooling towers and service water systems, you can measure flow rate

WESINC

Fig. 2 Now you have a chance to do something useful with these gauges Supply and return pressure gauges in constant-flow systems rarely serve any purpose, except perhaps to tell whether the pumps are running. You can use them to calculate impeller trimming, but you have to do some tricks to get the information you need.

without a flowmeter. To do this, record the length of time it takes the system to fill a container of known volume. For example, you can measure the flow rate of a cooling tower water pump by recording the time required to pump water from one cooling tower sump to an adjacent sump. However, even with an open system, it is more convenient to use an external flowmeter.

■ How to Measure Pressure

You can make the necessary pressure measurements by using the existing pressure gauges that are probably installed in the pump suction and discharge lines. (See Figure 2.) You may have to install at least one new pressure gauge tap on the downstream side of the valve that you use for throttling flow.

Gauge accuracy is important in making these measurements. You can use some tricks to compensate for the fact that the pressure gauges installed in mechanical systems are typically not very accurate. First, set all gauges with the pumps turned off. With the pumps turned off, each gauge should read the static pressure at its location in the system. You can calculate the static pressure from the density of the liquid and the difference in height between the top of the system and the gauge location. (Fresh water, without antifreeze or other additives, exerts a static pressure of 0.43 PSI per foot of height.) The architectural drawings provide these heights. The pressure difference is more important than the absolute pressures. You can figure this by accurately measuring the difference in height between the gauges.

With the pumps running, read the pressure differentials. Then, swap the gauges to average out their errors. I.e., make a set of measurements, swap the gauges in pairs, repeat the measurements, and average the differences of each pair of measurements.

How to Trim the Impeller

Any competent machine shop should be able to trim the impeller using a lathe. Send the impeller to the shop on its shaft assembly, if this is practical, so that the shaft can be used to hold the impeller while it is being turned on the lathe.

Trimming is irreversible. If you reduce the impeller diameter too much, the only solution is to replace the impeller. This can be expensive, especially if the impeller is integral with the shaft assembly.

Make System Modifications First

Before modifying pumps, accomplish any energy conservation measures that reduce the flow and pressure requirements on the pumps. For example, if there is an opportunity to convert bypass control valves to throttling valves, do this before trimming the impellers.

ECONOMICS

SAVINGS POTENTIAL: *Typically 20 to 60 percent of pump power.*

COST: *The cost to remove, machine, and replace an impeller ranges from about one hundred dollars for a small pump to several thousand dollars for a large pump. In addition, the cost of making flow and pressure measurements may range from several hundred to several thousand dollars.*

PAYBACK PERIOD: *Less than one year, to several years.*

TRAPS & TRICKS

CALCULATIONS: *Calculating the diameter reduction requires a combination of theory and practical technique that may be unfamiliar. Proceed carefully.*

REMOVAL, REPLACEMENT, AND HANDLING: *This does not require special skill, but it does require care. Avoid damaging the shaft seals, wearing rings, and bearings. Protect the sealing surfaces on the shaft. Make a wooden cradle to carry the impeller assembly to the machine shop and back.*

MACHINE WORK: *Impeller trimming is routine lathe work. Even so, try to find a shop that has experience with impeller trimming.*

LEAVE A RECORD: *Don't forget to record the changes in the equipment records, because the pump no longer has the characteristics stated on the manufacturer's nameplate. Install a durable metal tag on the pump that indicates its new impeller diameter, and the new pressure and flow ratings.*

MEASURE **10.2.2 Throttle the discharge of pumps to match system pressure and/or flow requirements.**

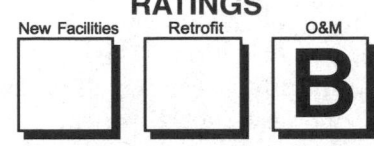

New Facilities   Retrofit   O&M

If a pump delivers more flow than the system needs, you can usually save some energy with little effort by throttling the pump discharge. You can do this with any suitable valve in the pump discharge line, or by placing a restriction in the discharge line.

This may seem contrary to intuition. A clever individual observes that adding resistance to the system dissipates power across the throttling device. This is true. However, this loss is more than compensated by the power saving that results from throttling the pump.

A similar situation exists when a farmer reduces the speed of a tractor by applying the brakes, instead of reducing the throttle. The engine is forced to operate at lower power output, but part of this reduced power is dissipated in the brakes. Of course, it would be more efficient to reduce the throttle setting.

Likewise, it would be more efficient to reduce the pump's output by reducing the diameter of the impeller or by reducing the impeller speed. Compared to those methods of reducing excess flow, the only advantage of this Measure is its low cost. The major disadvantage of this method is that it offers relatively little energy saving. In a given situation, impeller trimming (Measure 10.2.1) saves much more energy.

Graphical Analysis of Pump Throttling

You don't need any fancy analysis to do pump throttling, so you can skip this part if you are not interested in the theory. However, this analysis will show you the extent to which throttling saves pump energy. It makes an interesting comparison with impeller trimming. As an example, we will trace the effects of pump throttling on the same pump curves that we used in the example in Measure 10.2.1. Let's take it step by step:

- In designing the system, the engineer calculates that it needs a flow rate of 800 GPM, and that the pump needs to produce a pressure of 60 feet of head. These pressure and flow rate requirements are marked by point "A."
- To meet these requirements, the engineer specifies a pump having pump curve "B."
- When the system is built, it turns out to have system curve "C," so the pump operates at point "D." Although the pump pressure is approximately the same as originally estimated, the flow rate is much higher than the system requires. Pump power is proportional to the third power of flow rate, so a large amount of energy is wasted.

SUMMARY

An easy adjustment that provides a modest energy saving. Costs nothing, in most cases. Probably not economical if you have to install a throttling valve or orifice. Impeller trimming is usually a much better alternative.

SELECTION SCORECARD

Savings Potential .... **$**

Rate of Return .... **% % % %**

Reliability .... ✓ ✓ ✓

Ease of Initiation .... ☺ ☺ ☺

- To eliminate the excess flow, a valve in the pump discharge line is partially closed to make the system flow rate equal to the original system specification. From the standpoint of the pump, this creates a new system curve "C," causing the pump to operate at point "E." The pressure at the pump discharge is now higher than the original specification, but the pressure in the system is lower.
- The pump curves show that the pumping power has been reduced by about one horsepower. The saving is only a small fraction of the original pump power.

This example illustrates the weakness of the throttling method, which is the waste of energy that occurs in forcing water past the throttling device. In comparison, impeller trimming would save more than half the original pump power, as we saw in Measure 10.2.1.

Do You Have a Suitable Throttling Valve?

Throttling does not save much energy, so you cannot afford to spend much money to do it. Whether it is practical depends largely on whether a valve suitable for throttling is already installed.

You need a valve somewhere on the discharge side of the pump, i.e., between the pump and the portions of the system where the liquid is being used. Throttling the suction side of the pump could result in pump cavitation, which is destructive to the pump.

You are likely to find a valve installed in a suitable part of the system, because valves are installed to isolate system components for maintenance. The question is whether the valve is satisfactory for throttling. Two types of valves are commonly used in pump systems, butterfly valves and gate valves. Butterfly valves are usually suitable for throttling, while gate valves are not.

Butterfly valves are chosen for their ease and quickness of shutoff. They require only a quarter turn to go from full open to full closed. This makes them desirable where a component may be turned on or off frequently, e.g., with individual chillers in a multi-chiller installation. You will also find them in other applications, because they are relatively inexpensive. With butterfly valves, you can control flow accurately at a relatively high percentage of maximum flow, because a large valve movement is required to achieve the relatively small reduction of flow that is typically appropriate.

If your luck is bad, your system may have gate valves instead of butterfly valves. Gate valves are used for isolation because they provide tight shutoff.

Unfortunately, gate valves are not recommended for throttling. As the name implies, gate valves have a disk that slides across the fluid stream like a gate. Tight closure requires preserving the close tolerances between the metal sealing surfaces of the disk and the seat. (Elastic seats cannot be used because of the sliding contact between disk and seat.) If the valve is left in a partially open position for extended periods, the turbulence that occurs at the sealing surfaces of the disk causes erosion that allows leakage. For this reason, gate valves should be kept either fully open or fully closed.

If a valve suitable for throttling is not presently installed, you could install a butterfly valve for this purpose. However, you might be able to trim the pump impeller for about the same amount of money, and save much more energy.

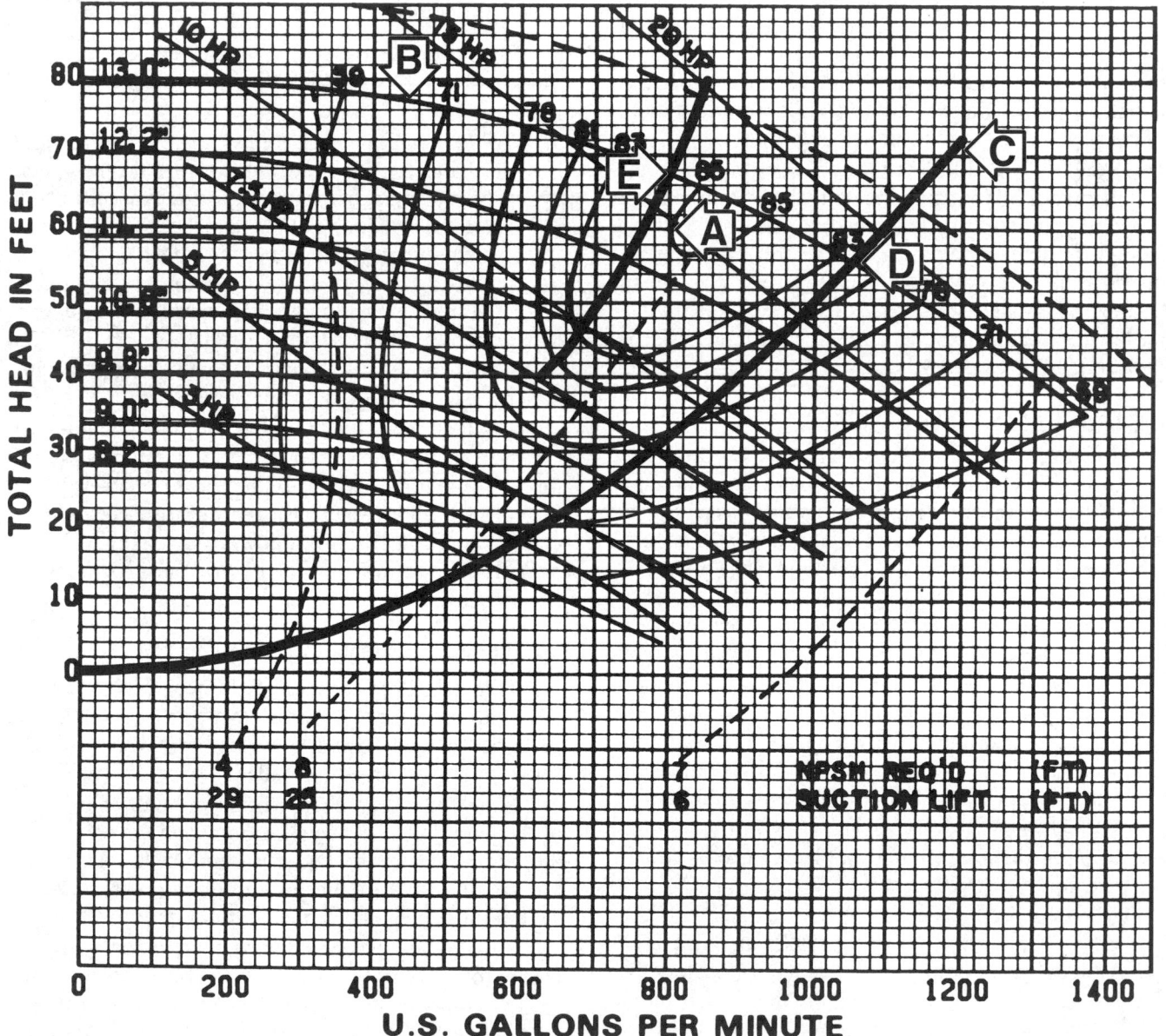

Fig. 1 Analysis of pump throttling This method of reducing excess pump capacity saves a relatively small amount of energy. The problem is that throttling forces the pump to operate at higher pressure, which largely cancels the saving from the reduced flow rate. Compare this analysis to Figure 1 in Measure 10.2.1.

Another possibility is installing a simple orifice in the pump discharge line, instead of a valve. Orifice plates intended to provide a pressure differential for flow metering are available as a commercial item, and might be used for this purpose. This might be economical if there is a flanged joint where you can install the orifice plate. If you take this unusual approach, be sure to tag the flange to indicate that an orifice is installed, and state that its purpose is to optimize pump flow.

How Much Can You Reduce Pressure and Flow?

Refer to this heading in Measure 10.2.1. Reducing system flow excessively degrades system performance, and may waste energy in other parts of the system.

How to Set the Throttling Valve

An advantage of this method over impeller trimming is that you can experiment until you find the optimum setting. The question is, how do you know when the setting is correct? You don't have to do any calculations. Just change the valve setting until you get the desired result.

The simplest case is when you can observe the desired flow rate directly. For example, in a service water system serving showers and toilets, observe how well the fixtures work in the highest part of the building under conditions of maximum water consumption.

If the requirement is based on flow rate, use a flowmeter while you are adjusting the valve. Strap-on, externally mounted flowmeters are available that avoid the cost and trouble of tapping into the pipe. (See Reference Note 16, Measurement of Liquid, Gas, and Heat Flow.)

Another method of measuring flow in open pumping systems, such as cooling towers and service water systems, is to record the length of time it takes the system to fill a container of known volume. For example, record the time required to pump water from one cooling tower sump to an adjacent sump.

If the requirement is for a particular pressure, monitor a pressure gauge installed at the point in the system where the pressure needs to be set. If no gauge is available there, observe a gauge at a nearby location in the same run of pipe, and correct for the difference in height. The other gauge should be close enough so that pipe friction does not create a significant pressure difference between the two locations.

How to Measure the Energy Saving

You are doing this to reduce the amount of energy going into the pump motor. So, you can determine the energy saving just by measuring the motor power before and after setting the valve. For absolute accuracy, use a wattmeter. If all you have is an ammeter, it will work reasonably well. This technique does not reduce the motor load very much, so the motor power factor will not change enough to cause a big error.

Mark and Lock the Throttling Valve

The position of the throttling valve is critical. Too little throttling wastes energy, and too much throttling may cause poor performance, inefficiency, and/or damage at the end-use equipment. Once you find the optimum throttling valve position, mark it clearly and install a lock on the valve handle.

Explain It!

The purpose of throttling the valve is not obvious. It may look wrong to anyone who does not understand the theory. Therefore, install a permanent placard at the valve that explains the valve position. See Reference Note 12, Placards, for tips on effective placard design.

ECONOMICS

SAVINGS POTENTIAL: *Typically, 5 to 25 percent of pump power.*

COST: *If a valve suitable for throttling is already installed, cost is limited to measuring and adjusting the flow, and to installing a lock and placard. If you have to install a throttling valve, the additional cost ranges from several hundred dollars in a small system to thousands of dollars in a large system.*

PAYBACK PERIOD: *Less than one year, to several years.*

TRAPS & TRICKS

MAKE IT LAST: *Lock the valves in position. Mark the valve position clearly and install an effective placard. Explain the valve settings in the plant operating manual.*

Fans consume a substantial portion of the energy input in most commercial facilities, primarily for HVAC (heating, ventilating, and air conditioning) applications. Fans serve many purposes in industrial facilities, where their importance as energy consumers varies widely. In residential facilities, fans usually do not offer much opportunity for improvement, but consider the possibility.

This Subsection contains a single Measure that recommends eliminating excess output in constant-flow fans. This is a common opportunity for savings. Save effort by applying this Measure to all your constant-flow fan systems as a group.

In many applications, you can save much more fan energy by using the variable-air-volume (VAV) concept. This is usually a complex system feature that has to be customized to the installation. See Section 4 to learn how to apply it in different types of air handling systems.

MEASURE **10.3.1 Adjust the output of constant-flow fans to the minimum needed.**

Fans are often installed with hefty margins of excess capacity to account for inaccuracy in calculations and for unforeseen circumstances. This extra capacity wastes energy in a variety of ways, depending on the type of system.

You can reduce this energy waste with a simple adjustment of fan output. It is usually safe to do this on a trial basis. It costs little or nothing, and it is easily reversible.

There may be two steps in trimming fan capacity. One is adjusting the fan output to the minimum that is acceptable, using one or more of the methods suggested below. The other is eliminating any dampers or other inefficient methods that were previously used to limit fan output.

This Measure applies to fans that do not have output modulation devices. Fans with efficient modulation devices, typically found in VAV systems, do not need this action. In HVAC fan systems, this Measure is one of several that apply to the system. Refer to the Subsections covering your types of fan systems for all relevant Measures.

Energy Saving Potential

■ Fan Energy

The power required to drive a fan is approximately proportional to the third power of the air volume delivered. Therefore, a relatively small reduction in fan output produces a significant reduction in fan power. For example, reducing the flow rate by 20% reduces the fan power requirement by approximately one half.

There is a significant exception to the third-power relationship of fan power to air flow rate. This is variable-air-volume (VAV) systems that "ride the fan curve." In these systems, if you reduce the fan output pressure, the terminal units open wider to keep the flow rates essentially the same as before. As a result, the energy saving is approximately proportional to the reduction in fan output pressure.

In air conditioning systems, the potential for reducing fan output is increased somewhat by the fact that a heating or cooling coil transfers more energy to the air when the air velocity through the coil is reduced. For example, a 40% reduction in air flow rate might produce only a 20% reduction in maximum coil output.

Reducing the fan output saves energy even if the fan operates in an on-off mode. This is because the reduction in fan power more than offsets the longer duration of fan operation.

SUMMARY

A simple and inexpensive adjustment that may save a significant amount of fan energy. It may also save substantial amounts of cooling, heating, and/or reheat energy in certain types of air handling systems. Overdoing it causes problems, but the adjustment is easily reversible.

SELECTION SCORECARD

Savings Potential .... $ $ $

Rate of Return .... % % % %

Reliability .... ✓ ✓ ✓

Ease of Initiation .... ☺ ☺ ☺

■ Heating and Cooling Energy in 100%-Outside-Air Systems

In 100%-outside-air systems, the air handling systems must cool and/or dehumidify the air starting from outside conditions. In warm weather, this is more energy intensive than conditioning return air. Fan trimming reduces the excess energy consumption in proportion to the air volume reduction.

(To be precise, trimming the fan output reduces proportionately the amount of energy required to bring outside air to room temperature. It has no effect on the amount of energy required to raise the air from room temperature to supply temperature.)

In cool weather, reducing the amount of outside air may or may not save energy, depending on the weather and cooling load profiles. When the spaces need cooling, it sometimes requires less energy to take in outside air than to cool recirculated air. (This is the principle of the outside air economizer cycle, which is covered by Measure 4.2.5.)

■ Reheating Operation

In a system that uses reheat, both reheat energy and the excess cooling energy required to cancel it are proportional to the air flow rate. For example, reducing the fan output by 20% reduces both the reheat energy and the excess cooling energy by 20%. Reheat operation can waste a large fraction of a system's energy input. Consider any reheat system as a candidate for modification to VAV or some other type. See Section 4 for these improvements.

Other Benefits

Trimming fan output may provide other benefits, including:

- *noise reduction.* The loudness of noise in an air handling system increases rapidly with increasing air velocity. Therefore, a small reduction in fan output substantially reduces the physical noise level. However, human hearing is relatively insensitive to sound level changes, so people may not perceive much reduction in the noise level.

- *improved control in VAV systems that "ride the fan curve."* In this cheap type of VAV system, the fans themselves have no method of modulation, and air flow into the spaces is controlled only by the dampers in the terminal units. This causes pressure to rise in the system at low flow, which may cause control instability, increased terminal noise, and leakage of cold air into the space. Reducing the maximum fan output eases these conditions, although it is not a proper cure. (See Subsection 4.4 for more effective methods.)

Limitations

The major question is, how much should you reduce the air flow? In most applications, the reduction will not be a large percentage. In HVAC applications, these factors limit the reduction that is possible, or desirable:

- *increased time to warm up or cool down* the space after a period of temperature setback. As you reduce the flow, the increased operating time of the heating and cooling equipment increasingly offsets the saving in fan energy.

- *discomfort during severe load conditions*, if the fan capacity is reduced too aggressively

- *conflict with chiller efficiency.* In cooling systems where the supply air temperature is held constant, reducing the air flow rate requires lower chilled air temperature near full load. (This applies to dual-duct and multizone systems, and to VAV systems that "ride the fan curve.") Lower chilled air temperature requires lower evaporator temperature, which reduces chiller efficiency. Using chilled air temperature reset reduces this conflict. For details, refer to the Measures for the respective types of air handling systems.

- *other limitations* on reducing air flow are covered in Measure 4.3.3, under "Factors to Consider Before Reducing Air Flow." The limited flow reduction recommended by this Measure is usually too small to cause any of the problems covered there.

This Measure eliminates only that portion of fan power in excess of the maximum power output needed. It does not reduce fan power to follow the moment-to-moment load, which may be much less than the maximum. To follow the load, the fan must be cycled, or it must be driven by a variable-output drive. The latter methods are recommended in Section 4, in the Subsections dealing with particular types of air handling systems. You can use this Measure in combination with fan cycling, multi-speed fan motors, and damper control.

How to Adjust Belt-Driven Fans

The most efficient method of reducing the output of a centrifugal fan is to reduce fan speed. Fortunately, most larger centrifugal fans are driven by belts, as shown in Figure 1. You can easily change the fan speed by changing one or both of the sheaves. Sheaves are available in small increments of diameter, so you can adjust the speed by virtually any amount.

The fan may have an adjustable sheave. If its adjustment range is too limited to trim the fan speed as much as possible, simply change the size of the fixed sheave. Use the adjustable sheave for finer adjustment.

If the fan does not have an adjustable sheave, consider installing one. Adjusting fan output involves trial-and-error, and an adjustable sheave simplifies the process.

If the fan also has a vortex damper, inlet damper, or discharge damper, remove it or block it fully open. Reducing fan output by reducing fan speed is much more efficient than any of those methods.

Drive belts have moderate losses, typically in the range of 3% to 10% of the energy transmitted. Drive

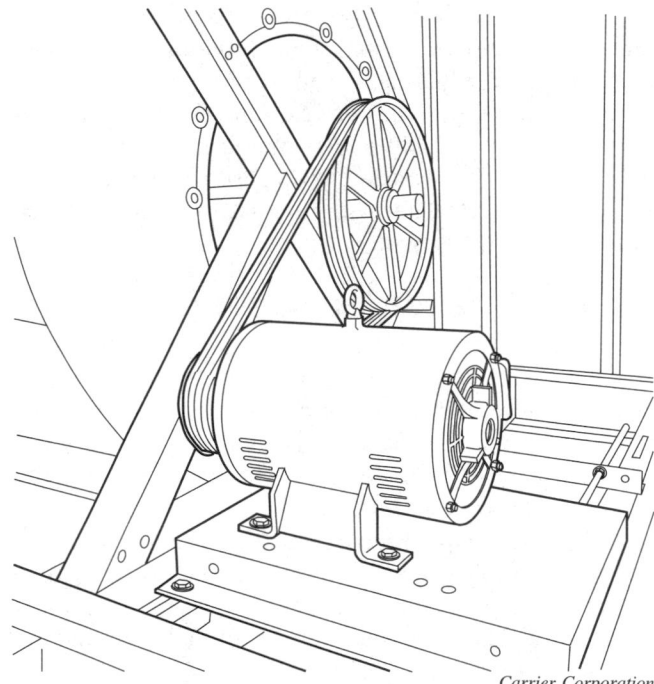

Carrier Corporation

Fig. 1 Belt drive Changing the diameter of one or both of the sheaves allows you to reduce fan speed by any amount. This is precise, easy, cheap, and almost foolproof. Don't make either sheave too small. Installing an adjustable sheave makes it easier to find the optimum fan output.

belt losses are affected by sheave diameter in a manner that is somewhat complex. The main point is to avoid sheaves that are too small, which force the belt to make a tight turn. If the fan has a sheave that is too small, you can get rid of it by increasing the diameter of the sheaves at each end of the belt. Excessively large sheaves also waste energy.

How to Adjust Fans with Inlet Vortex Dampers

Vortex "dampers" are aerodynamic devices that alter the performance of the fan itself. The output of a centrifugal fan is proportional to the difference between the speed of air flow at the outer tips of the vanes and the speed of air flow at the inlet end of the vanes. Inlet vortex dampers provide a rotation of the inlet air flow in the direction of fan rotation. This increases the speed of the air entering the vanes, thereby reducing the speed difference between the discharge and inlet ends of the vanes.

(In contrast, conventional inlet and outlet dampers function largely by dissipating fan energy in the form of turbulence, and they force fans to operate at less efficient points on their operating curves.)

Reducing fan output by giving the air a swirl at the fan inlet is an efficient method in principle, but not as efficient as reducing the fan speed. Furthermore, most vortex dampers are aerodynamically crude. If the fan is belt-driven, use the option of slowing the fan speed, and leave the vortex damper wide open.

If the fan is coupled directly to the motor, adjusting the vortex dampers is usually the only practical method of reducing the flow rate. You would not consider any other method unless the vortex damper approaches its closed position, where it becomes inefficient. In that case, it might be worthwhile to insert a belt drive, or even to replace the fan with one that is better matched to the flow requirement.

How to Adjust Variable-Pitch Propeller Fans

Changing the blade pitch of a propeller fan is an efficient method of reducing its output. The method of changing the pitch depends on the design of the fan. Follow the manufacturer's instructions carefully. A propeller fan with incorrectly adjusted blades may fail with catastrophic consequences.

If a variable-pitch propeller fan is belt-driven, you can reduce the flow rate by changing the propeller's pitch or its speed. A combination of both may be most efficient. Study the manufacturer's literature to find out how to set the fan most efficiently.

Propeller fans are especially susceptible to resonant vibrations. These may occur in the propeller itself, or the rotation of the propeller may drive resonant vibrations in the surrounding structure. Resonances can build to large amplitudes, possibly leading to catastrophic failure. Even low-amplitude resonances can create noise problems. Resonances may be masked by the propeller noise.

Resonance depends on the speed of the fan, rather than the pitch. The speed at which resonances can occur is not predictable. In practice, you find them from observation. Set the fan speed well away from any observed resonance frequency.

Do not use variable sheaves with propeller fans. Variable sheaves allow people to set fan speeds that might inadvertently fall within a resonance band.

What to Do About Wound-Rotor Motors

Wound-rotor motors are described in Reference Note 36, Variable-Speed Motors and Drives. The speed of a wound-rotor motor is usually reduced by inserting resistance into the rotor winding circuits. The current flow through these resistors wastes a considerable amount of energy. Indeed, the presence of an external resistor bank is the most obvious clue that a wound-rotor motor is installed. The resistors may be installed in the chilled air stream, compounding the energy waste. Wound-rotor motors are most likely to be found with direct-drive fans, which lack other means of adjusting speed.

Wherever possible, get rid of a wound-rotor motor and replace it with a more efficient method of speed control. For constant-speed adjustment, consider installing a conventional high-efficiency motor with a belt drive. See Measure 10.1.1 about selecting conventional motors. In applications where energy can be saved by modulating the fan output, as in typical air handling systems, consider a variable-speed drive. These are expensive modifications.

Should You Install a Smaller Motor?

If you can reduce fan output substantially, it may appear desirable to substitute a smaller motor. In reality, this is rarely economical. Although motors lose efficiency at lower loads, smaller motors are less efficient than large motors. The two effects offset each other.

Installing a smaller motor may be economical in cases where the existing motor is very oversized or exceptionally inefficient. In the latter case, a large part of the saving will come from greater efficiency in the replacement motor. See Measure 10.1.1 about selecting motors for efficiency.

Eliminate Inefficient Fan Throttling

As part of the fan output adjustment, eliminate any inefficient methods of fan throttling that may have been installed originally, including these:

- *discharge dampers,* which are inefficient because they require the fan to operate at a higher pressure for a given air flow. The existence of discharge dampers may not be obvious. They may be installed in the discharge duct at some distance from the air handling unit.

- *scroll housing dampers,* which are devices that modify the discharge portion of a centrifugal fan housing. The idea is to reduce fan output by changing the fan characteristics, rather than by the brute resistance of discharge dampers. (The casing of a straight-line discharge centrifugal fan is called a scroll housing because of its shape.) Scroll dampers are somewhat more efficient than discharge dampers, but not much more.

- *inlet (not vortex) dampers,* which simply choke off the air flow at the inlet to the fan. This method is rarely used, because it may cause fan surge.

Try to physically remove these types of dampers, if the application permits. If this would require too much effort, lock the damper open by a secure method. As long as the damper remains, the possibility exists that it will work its way into a partially closed position without anyone being aware of the fact.

Before making this change, find out why the dampers were installed in the first place. In crude VAV systems, discharge dampers may be used to adjust the input pressure to the terminal units. Reducing the fan output by one of the efficient methods described previously limits pressure just as well as a damper.

Discharge dampers are sometimes installed to limit the pressure in the duct system, to keep from blowing out ductwork. If the fan discharge pressure is limited for reasons of safety, it is prudent to install a means of limiting pressure that is independent of the fan output adjustment. For example, install a spring-loaded pressure relief door in the duct, along with an alarm that indicates when the door blows open.

Don't Overlook Any Fans

Trim the capacity of all the fans in the air handling system as appropriate. Do not overlook return fans, remote relief fans, toilet and kitchen hood exhaust fans, etc.

ECONOMICS

SAVINGS POTENTIAL: *30 to 80 percent of fan energy. In reheat systems, 10 to 30 percent of both cooling and reheat energy. In 100%-outside-air systems, 10 to 40 percent of conditioning energy.*

COST: *Minimal, if only adjustments of existing equipment are required. Adjustable sheaves cost from $20 to $200 in common sizes.*

PAYBACK PERIOD: *Less than one year, in most cases.*

TRAPS & TRICKS

EXPLAIN IT: *Explain the energy conservation purpose of the fan settings in the plant operating manual.*

MONITOR PERFORMANCE: *Check the performance of the fan systems at times of peak loads. Schedule this in your maintenance calendar, perhaps annually,*

Standby Equipment

This Subsection contains a single Measure that recommends simple methods of avoiding unnecessary operation of standby equipment. To save effort, apply this Measure to all your standby equipment as a group.

MEASURE **10.4.1 Install power switching that prevents unnecessary operation of standby equipment.**

RATINGS
New Facilities   Retrofit   O&M

In many systems, standby components are installed for the sake of reliability. This practice can waste energy if it allows duplicate units to be operated unnecessarily. For example, this often occurs with spare pumps because operators are not sure of how many pumps to operate.

There is a simple and reliable way to prevent this problem: connect power to the equipment so that only the proper number of units can receive power simultaneously. Where only two units of equipment are installed, a transfer switch is the simplest way of achieving this. Where more than two units are installed, a rotary switch is the simplest approach.

The only caution is to be sure that no conditions may arise that require the standby unit to operate along with the other units.

Method for Two Units: Transfer Switch

Where equipment is installed in pairs, you prevent operation of both units by providing power to the pair through a transfer switch. A transfer switch routes power to one unit or the other, but not to both.

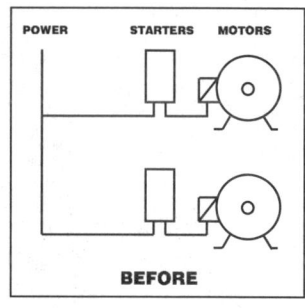

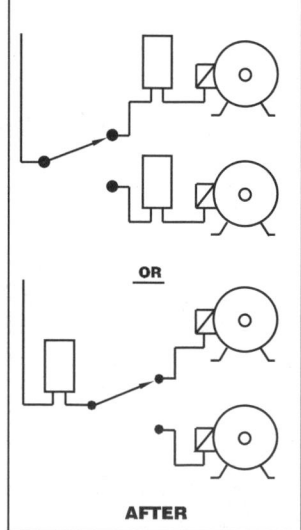

SUMMARY

Simple, foolproof methods of preventing unnecessary operation of standby equipment.

SELECTION SCORECARD

Savings Potential .... $ $ $

Rate of Return, New Facilities % % % %

Rate of Return, Retrofit.......... % % %

Reliability .... ✓ ✓ ✓ ✓

Ease of Retrofit .... ☺ ☺ ☺

Figure 1 shows how a transfer switch is wired. Typically, an ordinary double-throw knife switch suffices for this purpose. The switch is inexpensive, and installation is routine.

In new construction, this method can save the cost of one starter and/or one set of circuit protection equipment. Only one motor can operate at a time, so you need only one set of this equipment.

Method for More Than Two Units: Rotary Switch

If more than two identical units are installed, and one is intended to be a spare, you can use a rotary switch to deny power to any one of the units. The rotary switch feeds power to all the units except one, the unpowered unit being selected by the position of the switch. Figure 2 shows how to install the rotary switch.

You can use a rotary switch even in applications where the controlled units are not operating in parallel. For example, a chiller plant has two chillers and three chilled water pumps. One of the pumps is connected to the first chiller, while another pump is connected to the second chiller. An identical third pump is connected between the first two pumps for use as a spare. The rotary switch keeps this pump from running. If one of the first two pumps fails, valve positions are changed to allow the third pump to substitute for it. Then, the rotary switch is simply turned to remove power from the defective pump.

If you cannot find an appropriate rotary switch in stock, you can order a custom switch at modest cost from a number of manufacturers. If the equipment requires too much current for an available rotary switch to handle directly, use a rotary switch to control each unit through a relay.

Fig. 1 How to install a transfer switch

■ Rotary Switch Limitations

The rotary switch removes only one unit from operation. It cannot prevent waste of energy from unnecessary operation of the remaining units. For example, consider a chiller system having four chilled water pumps, one of which is a spare. The rotary switch ensures that only three of the pumps are operated. However, the system may need only one or two pumps when the cooling load is low. In this situation, you need different methods to avoid wasting pump power. Refer to the Measures dealing with the specific type of system.

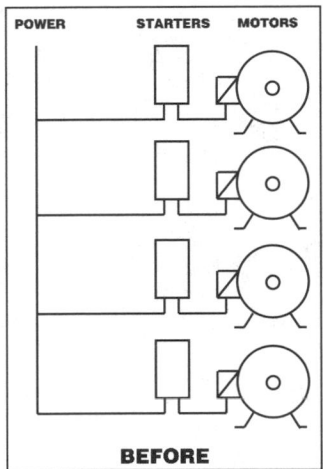

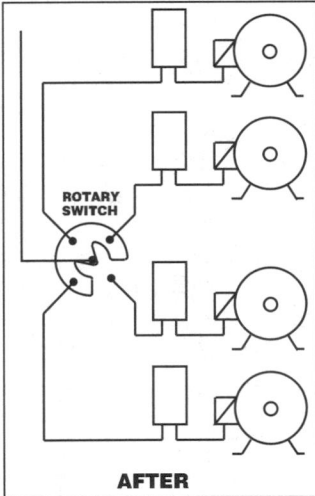

Fig. 2 How to install a rotary switch

Do not install a rotary switch in addition to automatic controls that control the running of equipment. For example, a battery of pumps may be installed to provide service water pressure, and the number operating may be controlled automatically in response to the water demand. In such cases, a rotary switch would complicate the control installation without doing much to reduce the possibility of energy waste.

Automatic Switching

These switching functions do not have to be manual. They can be performed automatically if you install appropriate control devices. Automatic switching is desirable in high-reliability and unmanned installations. For switching power between two units, you can use a simple solenoid-actuated transfer switch. For switching power between more than two units, you need a more complex set of relays. Or, you can program an energy management computer for this purpose.

ECONOMICS

SAVINGS POTENTIAL: *All the cost of operating standby equipment unnecessarily, including excess energy and maintenance.*

COST: *The switches are relatively inexpensive, costing a few dollars to several hundred dollars, depending mainly on their current capacity. In designing new facilities, this modification costs little or nothing. In retrofit, the installation cost may range from several hundred to several thousand dollars, depending on the amount of rewiring needed.*

PAYBACK PERIOD: *Less than one year, to many years, depending on the power consumption of the equipment and the likelihood that someone will operate the units unnecessarily.*

TRAPS & TRICKS

EXPLAIN IT: *Install a placard on the power switch to tell the staff how to use it. Do this for both manual and automatic switches. See Reference Note 12, Placards, for tips on installing effective placards.*

Section 11. REFERENCE NOTES

INTRODUCTION

The Reference Notes are self-contained explanations of specific topics. You will use them in different ways, depending on how you are using the Energy Efficiency Manual.

The Reference Notes Enhance the Measures

The Reference Notes support the Measures, which are in the first ten Sections of this Manual. As you accomplish an energy conservation program, each Measure that you select refers you to relevant Reference Notes. These give you additional in-depth information about equipment, principles of operation, installation and operating practices, calculation tools, and energy sources. Most of the Reference Notes support a number of Measures. Keeping this information separate from the Measures avoids repetition and keeps each Measure focussed on its particular applications.

A word of caution. Do not use the Reference Notes as a starting point for your energy conservation actions. Your starting point should always be the individual Measure. The Measure provides the details that are important for specific applications, along with ratings, economic analysis, and practical guidance.

Use the Reference Notes for Study

If you are using the Energy Efficiency Manual as a textbook, the Reference Notes are the place to start. As you can see from the index to the Reference Notes on the right side of this page, they are arranged in Groups. For formal study, start with any Group that is appropriate for your study plans, and read all the Notes in that Group.

As you study the Reference Notes, refer to related Measures as examples of practical applications. Find facilities, construction sites, and equipment where you can see the principles in action. Take this book with you. As its pages get dirty and full of highlighting pen, your knowledge will be growing.

This approach applies to anyone who is serious about learning energy efficiency. Use it if you are enrolled as a student, or if you are a conscientious architect or engineer seeking to improve the efficiency of your

1198 11. REFERENCE NOTES

designs, or if you are an advocate seeking to learn about the most effective methods of protecting the environment.

Read the Reference Notes for General Interest

The Reference Notes are written to be clear to anyone who has a general interest in energy efficiency. If you want to browse, you can read them in any order that you find interesting. The shorter Reference Notes give you concise, simple explanations of individual efficiency concepts that you cannot find in literature that is written for specialists. The longer Reference Notes will provide you with information that is clearer, more complete, and better balanced than you will be able to find elsewhere at a practical level.

Clock Controls and Programmable Thermostats

Upgrade to Electronic Time Controls
Scheduling Options
- **Number of Circuits**
- **Variety of Schedules**
- **Daylight Saving Time and Leap Year Reset**
- **Cycling**
- **Astronomical Settings**
- **Advance Warning**
- **Optimum Start**

Electrical Options
- **Types of Switching Output**
- **Current, Power, and Voltage Ratings**
- **Backup Against Loss of Power**

Important Physical Characteristics
- **Ease of Use**
- **Ruggedness**

Programmable Thermostats
Overrides
Installation Issues

The most powerful technique of energy conservation is turning off equipment when it is not needed. Timeclock controls are a primary tool for this function, because most equipment is needed on the basis of specific time schedules. This Note gives you the main points to consider when selecting and installing clock controls.

Achieving effective and long-lasting clock control involves three main steps: (1) select clock controls that have sufficient flexibility to adapt to all the predictable variations in schedule, (2) provide very clear instructions for setting and adjusting them, and (3) install overrides that allow users to easily bypass the timeclock when equipment is needed during unscheduled periods.

A programmable thermostat is a combination of a timeclock and a thermostat. It provides ease and flexibility in tailoring thermostatic control to requirements. However, you have to be as careful in selecting one as in selecting a timeclock, plus you have to select the best thermostatic features.

Upgrade to Electronic Time Controls

Until recently, lack of flexibility was a serious limitation of clock controls. For most of the modern era, time switches were mechanical devices driven by little electric motors. Figure 1 shows an example. Consider these units obsolete for most applications, even though many are still sold. These clocks are inherently limited in their ability to adapt to different schedules of operation. They lose time with every power outage. They are clumsy to set. Most cannot be set accurately. The method of override, if any, is usually obscure.

Electronic timeclocks can avoid all these shortcomings. Electronic time controls are now available that have most of the features that anyone could desire. Figure 2 shows a typical selection of models. Very capable units are available at modest cost. Even in existing facilities, it may be economical to replace old electro-mechanical timeclocks where they are presently installed.

The more sophisticated time controls approach the capabilities of the lower end of the range for energy management control systems, which are covered by Reference Note 13. In fact, the boundary between the two classes of controls is not distinct. If you have a complex control application, read Reference Note 13 along with this Note.

The wide variety of features available in electronic time controls makes it important to invest some study before selecting time controls for your applications. The following are the important selection considerations.

Scheduling Options

■ Number of Circuits

Some electronic timeclocks have a number of separate circuits, each of which can be controlled on a different schedule. This allows a single timeclock to

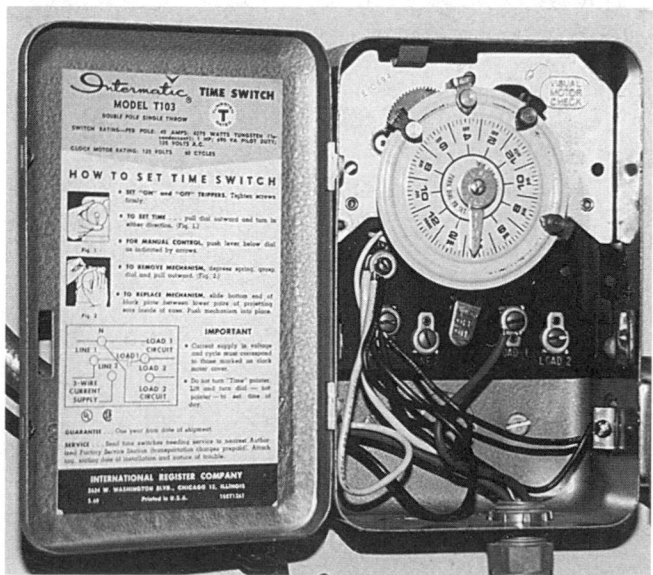

WESINC

Fig. 1 Classic electro-mechanical timeclock This type is now obsolete for most applications. It is too inflexible, and it loses time with each power interruption. However, it was simple to understand and set. This unit has good instructions.

act as several separate timeclocks. In most applications, this feature does not provide a major advantage over using separate timeclocks, and it has some disadvantages.

Using separate single-circuit timeclocks for different equipment is intuitively easier for the operating staff, it makes programming the timeclock less confusing, and it simplifies troubleshooting. Furthermore, failure of a multi-circuit timeclock can disable more equipment than failure of a single-circuit unit.

Several individual timeclocks may be cheaper than a multi-circuit timeclock. Furthermore, the ability to install each single-circuit timeclock closer to the equipment it controls may save wiring cost.

As a rule, multi-circuit timeclocks offer more features than single-circuit units, simply because they are in a higher price class. You may have to buy a multi-circuit unit to get all the features you want.

■ Variety of Schedules

Electronic timeclocks generally allow "on" and "off" switching to be scheduled to the nearest minute. Models differ in the number of on and off actions that can be programmed per day, or per week. Most units provide an ample number of on/cycles per day. The cheapest timers provide only one daily sequence of on/off actions, while more sophisticated units provide a number of sequences for different day types.

Paragon Electric Company

Fig. 2 A selection of electronic timeclocks Almost any time control features you could want are available in inexpensive models. Be sure that the control you select is the best for your application.

Models differ in the way that they account for different schedules on different days. Some allow separate schedules for a few different day types, such as "weekday", "weekend", and "holiday." Others allow an entire week to be scheduled, with separate settings available for each day of the week, plus a separate "holiday" schedule. All but the simplest models have sufficient flexibility for conventional applications, for example, heating and air conditioning a rental office building. Pay attention to the details of the scheduling features, especially if your schedules are more complex.

■ Daylight Saving Time and Leap Year Reset

An increasingly common feature is automatic resetting of the time when daylight saving time begins and ends. Be sure to get this.

Another feature is automatically adding February 29th on leap years. Get this feature if you plan to operate equipment on more than one daily schedule, e.g., weekday and weekend.

■ Cycling

Some timers simply repeat the same on/off switching sequence endlessly. The user can select the on and off times, and usually the duration of the cycle. For specialized cycling applications, such as control of poultry house ventilation, it is best to install a small timer that is specialized for the type of cycling required. Cycling is also included as a feature of some of the more complex models of timeclocks.

The most common energy management application for cycling is minimizing electricity demand charges (see Reference Note 21, Electricity Pricing). In this application, a timeclock is used to provide power to different equipment on alternating time cycles, so that all the equipment cannot operate simultaneously. For example, a timeclock can be used to provide power alternately to two groups of air conditioners.

A single-circuit timer can be used to alternate power between two groups of equipment if it has "double-throw" output. A multi-circuit timer is needed for cycling between equipment in a non-alternating manner, or for cycling more than two groups of equipment.

■ Astronomical Settings

"Astronomical" timeclocks automatically calculate the time of sunset and sunrise each day for switching purposes. Astronomical switching is commonly used to control lighting fixtures that are installed outdoors or under skylights. It could also be used for other functions related to the position of the sun, such as air conditioning in dry climates.

Astronomical switching is provided in some specialized single-circuit timers. It is also included as a capability in some of the more expensive general-purpose models. You need to input your latitude when installing an astronomical clock.

Check for the ability to combine astronomical switching with switching at fixed times. For example, it may be desirable to turn on parking lot lights at sunset, and to turn them off at midnight.

Photoelectric switches are an alternative to astronomical time switches. Photocontrols are cheaper, and they respond to unexpected periods of darkness, such as eclipses and heavy overcasts. But, you have to install them where they can "see" daylight. You can install an astronomical timeclock anywhere. Also, the astronomical timeclock can perform both functions, turning the lights off at sunset and turning them back on at any time.

■ Advance Warning

A timeclock with an advance-warning feature has a separate circuit that is switched on at a fixed interval ahead of another circuit. Typically, the first circuit is used to activate an alarm before the second circuit takes some action, such as starting a large piece of equipment.

■ Optimum Start

Optimum-start control is a feature of some timeclocks that automatically adjusts the equipment start time in response to weather conditions. Some models provide optimum start for both heating and cooling equipment, while others control only for heating equipment. Refer to the Measures that recommend optimum-start control for more details.

Electrical Options

■ Types of Switching Output

Timeclocks have different types of electrical output. Depending on the model, the timeclock may:

- act as a switch, simply opening or closing a circuit. This is called "single-throw" operation. If the timer switches only one wire of a power circuit, the action is called "single-pole, single-throw" (SPST). If the timer switches both wires of the circuit, the action is called "double-pole, single-throw" (DPST).

- act as a switch, transferring from one contact to another. This is called "double-throw" operation. If the timer switches only one wire of a power circuit, the action is called "single-pole, double-throw" (SPDT). If the timer switches both wires of the circuit, the action is called "double-pole, double-throw" (DPDT). Double-throw output can be used for single-throw applications by leaving one of the terminals disconnected.

- connect the equipment to power ("on") or ground ("off")

- provide pulses. The duration of pulses may be fixed, or they may be adjustable, typically from one second to one hour. This feature can be used directly for applications such as ringing bells, or the pulses can be used as input signals for relays or digital control systems.

Be sure to select the right kind of output for your application. For example, if you want the timer to close a switch in a low-voltage control circuit, you do not want to discover that your timer applies full line voltage to the circuit.

Some multi-circuit timeclocks can provide pulses on one circuit and switching on another.

■ Current, Power, and Voltage Ratings

Timers are rated by the amperage that they can control. This merely indicates the amperage that the timer can control directly. You can use any timer to control any amount of current by using a suitable relay. A high amperage rating may save the cost and effort of installing a relay.

The amperage rating of a given model may be different for incandescent lighting than for other types of loads. This is because incandescent lamp filaments have much lower resistance when cold than at operating temperature, so they have high starting current.

If the timer is rated by the power that it can control, rather than by amperage, the rating may be lower for

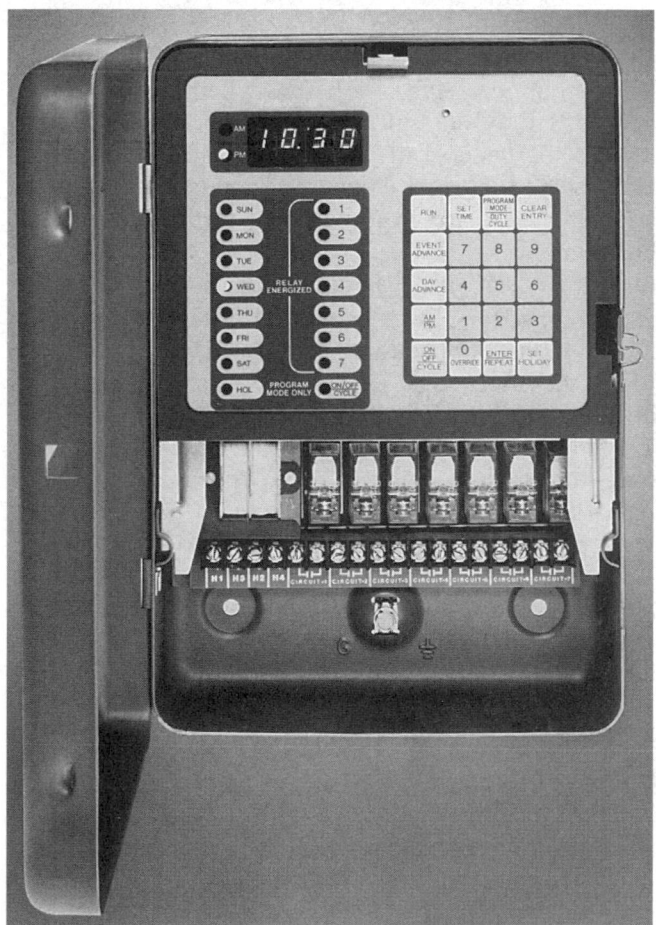

Paragon Electric Company

Fig. 3 Could you program this time control properly?
This time control has a good range of features. But, what are they? Where are the instructions? Can you read the buttons easily? Will the letters rub off in a few years? A time control is useless unless everyone involved with it understands it.

motors and fluorescent lighting. This is because motors and magnetic lamp ballasts have low power factor, which requires them to draw extra current for a given amount of power.

The timeclock itself usually operates from normal line voltage. Timeclocks need little power.

■ Backup Against Loss of Power

Most electronic timeclocks avoid the problem of power outages by having a backup power source to protect against loss of external power. Continuous power is needed to keep the clock on time and to preserve the switching schedules.

The backup power source can be a battery or a capacitor. A battery typically can keep the timer in operation for periods up to several months. A capacitor may provide power for periods up to several days.

Some time controls use conventional batteries, and some use rechargeable batteries. The most convenient and economical choice is conventional, non-rechargeable batteries. Rechargeable batteries are desirable only if the facility is subject to frequent power outages, or if equipment operation is critical.

Battery failure from old age is a long-term failure mode. If the battery fails, the unit will probably lose all its programming during a period when external power is lost. Try to find timeclocks that warn of the need to replace the battery. If you cannot find this feature in the type of timeclock you want, put specific dates for replacing the battery on the unit's instruction placard. For example, the placard might say, "Replace the battery in June of each year that ends with a '0' or '5'."

A capacitor power backup almost eliminates this failure mode. Capacitors generally should last for the life of the time control, although electrolytic capacitors do fail more often than other electronic components. The main disadvantage of capacitors is a shorter continuous backup period. However, the capacitor recharges almost immediately if power is restored for a short period. For frequent interruptions of power that last only for a short period, the capacitor is probably a better choice. Look at your application to decide whether a battery or a capacitor provides greater security in your environment.

Important Physical Characteristics

■ Ease of Use

Ease of use is vital to ensure that a timeclock is used properly, and is not abused. Some operators are likely to ignore or abuse a timeclock if it appears to require "programming." The timer should have simple and obvious procedures for setting, for changing settings, and for override. Also, it should easily display the schedules that are currently programmed into the unit. Figure 3 shows the programming panel of a typical time control of medium capability.

Unfortunately, most electronic timeclocks are not yet user-friendly. Find the models that are best in this regard. Ideally, the setting procedures should be almost as easy as using a pushbutton telephone. If you cannot find such a unit, at least try to find models that have large numbers in the displays, large buttons for programming, and large text for the instructions. This is important to middle-aged operators who leave their glasses at home.

Obscure and poorly illustrated instructions are a major failing of present time controls. If you find a unit that has all the control features you want, be prepared to design a good set of instructions to install along with the hardware. See Reference Note 12, Placards, about how to do this.

■ Ruggedness

A weakness of most electronic timeclocks is that they are built more for the environment of a library or a laboratory than for the rugged environment of equipment rooms. Not all people who operate facilities are dainty in their handling of equipment. It takes only one ham-fisted or harried individual to destroy a time control that is too delicate for the environment. Damage becomes almost a certainty if the timer is not easy to use, which incites operators to intemperate behavior.

Many models are available with sturdy steel enclosures. The enclosures are adequate to protect against accidental impact damage, but they do not protect against the greatest hazard, which is the operator in a hurry to get equipment started. In severe environments, the best approach is to make the timeclock itself inaccessible to all but a chosen few, and to install readily accessible overrides, as explained below.

Programmable Thermostats

A programmable thermostat is a combination of a timeclock and a thermostat. The timeclock switches the thermostat from one temperature during the occupied periods to a lower temperature during unoccupied periods. Figure 4 shows the layout and features of a typical programmable thermostat.

In principle, programmable thermostats provide ease and flexibility in controlling heating and cooling equipment. However, most programmable timeclocks are fairly cheap, so they may lack important features. In particular, many models may not be rugged enough for some environments and they may not be easy to program.

When you select a programmable thermostat, consider all the criteria that we have discussed for clock controls. In addition, be sure that the unit has all the thermostatic features needed to provide maximum efficiency for the application. For example, if the space can benefit from deadband temperature control during occupied periods, make sure that the unit includes this capability.

Overrides

The main weakness of clock controls, when used alone, is that they do not provide an easy method of adapting the operation of equipment to irregular times of usage. As a result, clock control is often sabotaged. Figure 5 shows a common example. In many cases, the timeclock is disabled the first time that someone needs the controlled equipment at an unexpected time.

The problem is most severe with time controls that are set remotely, or that are designed to keep people from making changes to them. Figure 6 shows an example.

The solution to this problem is to provide an override feature that allows the equipment to be operated quickly and easily without the need to re-program the timeclock. The override may be part of the clock control itself. Or,

it may be an entirely separate switching device that provides power to the equipment separately from the time control.

Many models of timeclocks have some form of override. These may be satisfactory if the override is easy to find, easy to use, and well marked. Unfortunately, timeclock manufacturers still fail to understand the importance of overrides that are rugged and well marked. Expect that an override feature on the timer may not satisfy the needs of a person who is in a hurry and unfamiliar with the unit.

Fortunately, it is easy to provide an external override. In many applications, the best override is a simple rundown timer (such as a spring-wound dial timer) that is wired in parallel with the timeclock. The rundown timer can be installed where it is most obvious to users

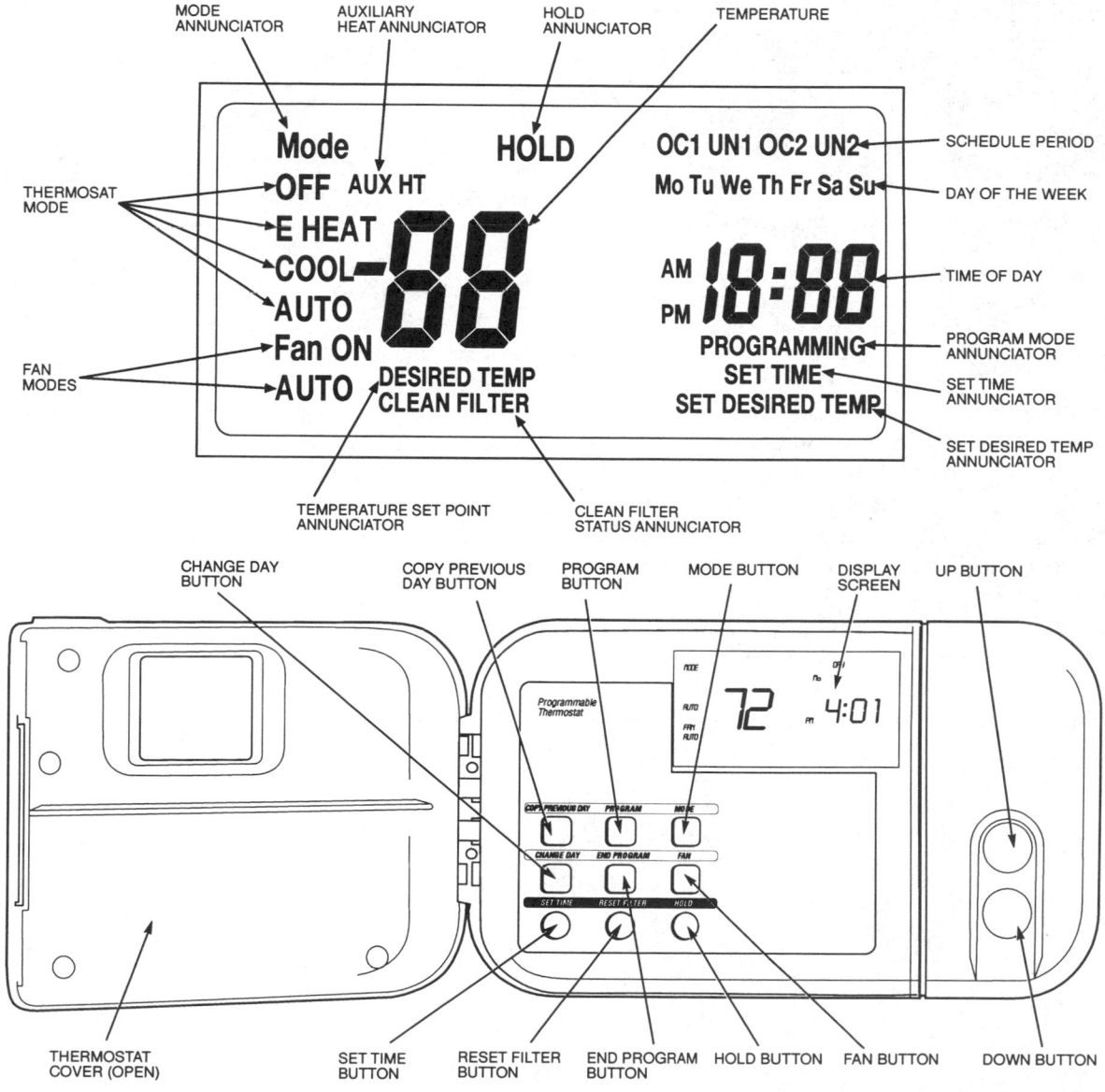

Carrier Corporation

Fig. 4 Programmable thermostat It combines the features of a thermostat with a simple electronic timeclock to change the space temperature settings at different times.

WESINC

Fig. 5 The fate of any timeclock that lacks an effective override This old timeclock has been sabotaged in the usual way, by removing the "off" trippers, which are lying below the timeclock. An electronic timeclock without an easily accessible override would probably have been put out of commission in a more destructive way.

of the equipment, and the timeclock itself can be kept isolated in a place that is accessible only to persons authorized to program the timeclock.

Overrides that are provided as features of electronic timeclocks typically self-cancel after one on/off cycle. This feature is important. Without it, an override effectively disables the timeclock.

However, even a self-cancelling override can be inefficient. For example, a person working late in an office may use the override to keep a heater running until he leaves, but the override will keep the heater

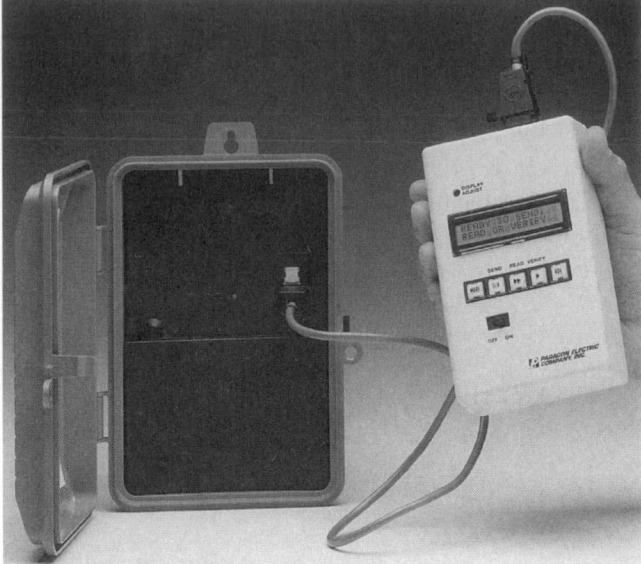

Paragon Electric Company

Fig. 6 Remote-setting time control This is a handy feature for some applications. However, an override is needed at the point where this control affects people.

running all night. In such cases, it is more efficient to use a separate rundown timer as an override, as this limits the override period.

Installation Issues

Like all control equipment, clock controls should be accessible, well protected, and well explained. The timeclock itself should be installed where it is easily accessible to the people assigned to program it. On the other hand, the override should be installed at a location that is highly visible and accessible to the people who need the controlled equipment. In facilities, such as office buildings, where occupants are not members of the operating staff, expect to install the timeclock in one location, and to install one or more overrides in other locations.

See Reference Note 12 for details of installing the placard, which is an integral part of an effective control installation.

Where to Use Personnel Sensors
Types of Personnel Sensors
- **Passive Infrared Sensors**
- **Ultrasonic Sensors**
- **Audible Sound Sensors**
- **Microwave Sensors**
- **Tread Switches**
- **Photoelectric (Light Beam) Switches**
- **Interlocks with Other Controls**

How to Select and Install Personnel Sensors
- **Safety**
- **Minimize the Perceived Delay in Turning On Lights**
- **Range, Coverage Pattern, and Sensitivity**
- **Mounting Location**

- **Daylight Override**
- **Turn-Off Delay**
- **Minimize False Triggering**
- **How to Confine the Area of Control**
- **How to Control Large Areas with Few Sensors**
- **Accommodate Diverse Control Requirements**
- **Anticipate Changes in Space Configurations**
- **On and Off Override Switches**
- **Sensors with Three-Way Switches**
- **Maximum Amperage or Wattage**
- **Minimum Wattage**
- **Lamp Flicker**
- **Grounding**
- **Placards**

Test Unfamiliar Situations
Monitor Performance

Personnel sensors, also called "motion detectors" and "occupancy sensors," are devices that sense the presence of people within an area. Originally developed as security devices, personnel sensors have become an efficient method of controlling equipment that is needed only when people are present.

In appropriate applications, personnel sensors match the operation of equipment to space occupancy more accurately than any other method. At the same time, using personnel sensors can be tricky. Inappropriate or careless installation can waste energy, annoy occupants, and create safety hazards. This Note tells you how to select and install personnel sensors successfully.

Where to Use Personnel Sensors

The most common energy conservation application for personnel sensors is controlling lighting. For example, a lighting retrofit program for a large, diverse research hospital found personnel sensors to be the most efficient method of lighting control for a large fraction of the spaces in the hospital, including offices, laboratories, restrooms, etc. Lighting is well suited to control by people sensors because it is needed only when people are present. However, even with lighting, it is sometimes better to control with other methods. For example, sensor control is not appropriate for patient rooms in hospitals.

Personnel sensors can also be used to control other types of equipment that are related to occupancy, such as ventilation fans. Personnel sensors cannot be used to start equipment that needs a warm-up or cool-down period, but they may be an effective means of stopping such equipment in applications where occupancy ends at varying times. For example, timeclocks can be used

to start heating and cooling equipment at the beginning of the day, while personnel sensors are used to control this equipment later in the day.

Do not expect to use one particular model of sensor or one particular mounting method throughout a facility. Typically, you will need three or four different models and many mounting configurations to cover a building of average size and complexity. Expect to tailor each personnel sensor installation to the space that it covers. Expect to spend a lot of time and effort on this.

Personnel sensor control has the potential of annoying people, and even of creating safety hazards. Experience shows that occupants do not object to personnel sensors that are installed with adequate attention to the considerations discussed below. In proper installations, occupants typically are amused by personnel sensors until they no longer notice them.

Types of Personnel Sensors

Various methods have been developed to detect the presence of people in a space. None of them can absolutely distinguish a person from another object, and none can absolutely detect the mode of behavior, such as remaining in a space or just passing through. You have to select the device, or combination of devices, that most reliably detects the person, object, or mode of behavior that is appropriate for your control application. The following are the main types being used to detect people, and objects associated with people.

■ Passive Infrared Sensors

Infrared sensors detect the long-wavelength heat radiation that is emitted by people or other warm objects. The sensor does not emit any radiation itself, so it is "passive." For this reason, infrared personnel sensors are commonly called "passive infrared" sensors.

The world is full of warm objects that we do not want to detect, such as heating vents and coffee pots. To avoid being triggered by these objects, infrared personnel sensors exploit the fact that most of these objects are stationary. The device uses a grid and an optical system to sense the motion of people. Only a small amount of motion is needed by a person to trigger the sensor.

Infrared sensors have become the most common type of personnel sensor for energy conservation applications. Figure 1 shows a small ceiling-mounted model that could be used to control lighting, air conditioning, a security system, or other applications.

Infrared sensor models that are designed to replace existing light switches have become popular because of their low cost and ease of installation. Figure 2 shows a good example.

The number of configurations is growing, especially for lighting control. Passive infrared sensors are now commonly available as an accessory of exterior floodlight fixtures, and they are also installed in other types of light fixtures. For controlling existing lamps, an inexpensive infrared lighting control can be inserted between a light bulb and its socket. Portable infrared sensors are available with a power cord and a male/female plug that allow them to control any appliance.

■ Ultrasonic Sensors

Ultrasonic sensors function by radiating high-frequency sound waves and sensing the frequency shifts (Doppler effect) in the sound that is reflected back to the sensor by moving objects. They cannot distinguish between people and other objects, except by size and location. Ultrasonic sensors are commonly used to open automatic doors, for example. They function well in this application because they open the door for any moving object, whether living or not.

Ultrasonic sensors are used much less commonly than infrared sensors for detecting people inside a space. However, ultrasonic sensors designed for this purpose are available. Figure 3 shows an example.

■ Audible Sound Sensors

Sensitive microphones can be installed to sense the presence of people by the sounds they make. Sound detectors are rarely used for control of energy systems, primarily because there is too much background noise in most applications. They are used most commonly in security systems. However, audible sound detectors may be a good choice for controlling interior security lighting, for controlling lighting in very large spaces, and for other specialized applications.

■ Microwave Sensors

Microwave sensors fill a space with microwave radiation. They detect movement from the distortion of the reflected radiation. They are like the early type of non-directional radar. That is why the radar detector of your car may sound an alarm when you drive past a large building. Microwave sensors are common in security applications, but they are not generally used for control applications.

■ Tread Switches

A tread switch makes or breaks an electrical contact when someone walks on it or rolls equipment over it. For example, tread switches are commonly used to open automatic doors. They are not commonly used for sensing people in a space, because they cannot cover a large area. In principle, you could use a tread switch at the entrance to a space to turn on the lights in the space. However, the switch would not know when the space has been completely vacated. Thus, they are generally not useful for turning equipment off.

Unenco Electronics, Inc.

Fig. 1 Passive infrared sensor for ceiling mounting
This unit sees in all directions. Installing it on the ceiling provides the most reliable coverage of the space. Settings are made with the tiny DIP switches, at the time of installation.

A tread switch may be a good way of conserving energy in a very localized application. For example, a tread pad can be used to turn off a machine tool when the operator is not standing on it.

■ Photoelectric (Light Beam) Switches

A light beam switch senses a person or object interrupting the light beam. A lamp is placed on one side of a path and shines a narrow beam into a photoelectric cell on the other side of the path. For example, light beam switches are commonly used as safety devices to keep automatic doors open when a person or object may be passing through the door.

This device is similar in application to tread switches, and it has similar weaknesses. Light beam switches are reliable and they are fairly inexpensive. In applications involving safety, they have the advantage of failing in a safe mode. For example, if the bulb burns out, the effect is the same as if a person crosses the light beam.

■ Interlocks with Other Controls

Any control that relates to the presence of people can be used as a personnel sensor. For example, the heating of a space may be controlled by the light switch for the space. Or, a contact switch can be installed on a door to turn on lights when the door is opened. Various Measures recommend these techniques where they are appropriate.

How to Select and Install Personnel Sensors

■ Safety

If you use personnel sensors to control lighting, take care to design the lighting layout so that a person will not be trapped in a darkened space if a personnel sensor fails to detect the person. For example, a personnel sensor controlling the light in the restroom may lose sight of a person who enters an enclosed stall. If the

WESINC photo

Fig. 2 Passive infrared sensor for light switch location
This type is inexpensive, easy to install, and easy to use. Its main weakness is the mounting location, which may not provide good coverage and may be easily obstructed. The details of this unit are shown in Figures 6, 7, and 8.

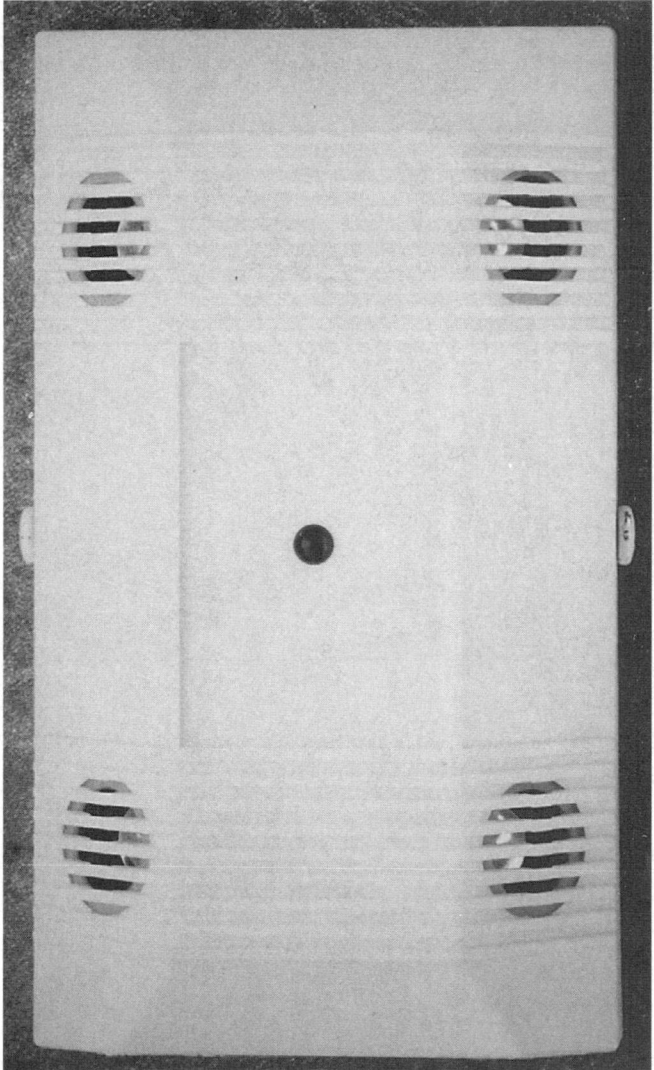

Unenco Electronics, Inc.

Fig. 3 Ultrasonic motion sensor for ceiling mounting
Ultrasound is directional, so this unit has four emitters to provide reasonably complete coverage. It is somewhat larger than a passive infrared unit.

person lingers there, the lights in the restroom may turn off, stranding the person.

The general solution to this problem is to provide enough continuous lighting to allow occupants to find their way safely to an exit. In most commercial buildings, this purpose is served by emergency lighting. If the facility operates only in the daytime, letting in some daylight may suffice.

Safety is also a major consideration where personnel sensors are used to control moving equipment, such as a conveyor belt in a factory. Design the controls to protect both equipment users and maintenance personnel from unexpected starts. The appropriate safeguards depend on the application, and may include warning signs, guards, interlocks, dead man switches, etc.

Ultrasonic sensors may be ineligible for some applications because of concern about their long-term effects on occupants. The fact that ultrasonic sensors remain on the market after years of use for opening doors indicates that no serious problems have been proven. However, it remains questionable whether ultrasonic

controls should be installed where people will be exposed to them for long periods.

Do not use ultrasonic detectors in spaces containing animals that are sensitive to high-frequency sound, unless the animals are unwelcome.

■ Minimize the Perceived Delay in Turning On Lights

Locate sensors where they can immediately see a person entering a space, as when a person abruptly opens the door to a restroom with sensor lighting control. Otherwise, the person will have the unpleasant sensation of walking into a dark space. This problem is especially severe if persons are not familiar with the sensor control, for example, in public restrooms.

If the lighting is fluorescent, the problem is worsened by the starting delay of the lamp. The delay ranges from a fraction of a second for rapid-start lamps, to several seconds for preheat lamps. Most compact fluorescent lamps are preheat types, and many have a noticeable starting interval. Some compact fluorescent lamps take almost a minute to reach full brightness, and these lamps should not be used where instant light is required.

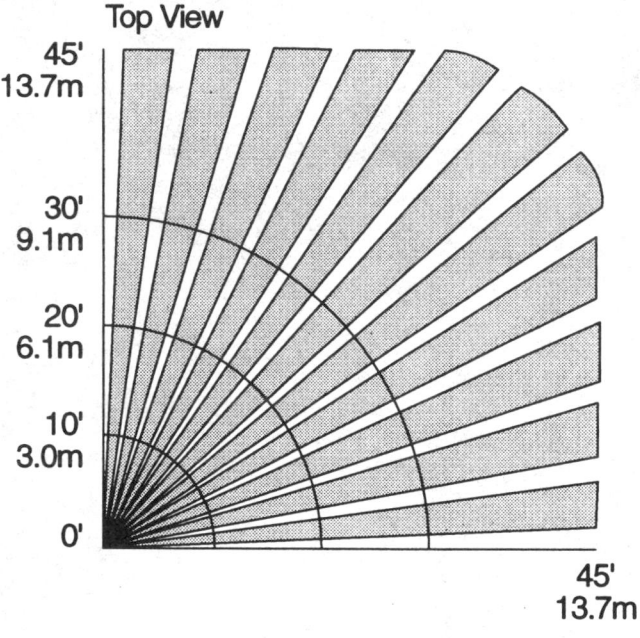

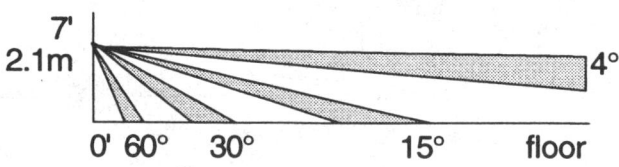

Unenco Electronics, Inc.

Fig. 4 Coverage pattern of an infrared sensor for corner mounting Infrared sensors use a lens system, so they can be made with any coverage pattern. Coverage can be uniform across the field of view, but sensitivity varies with background temperature.

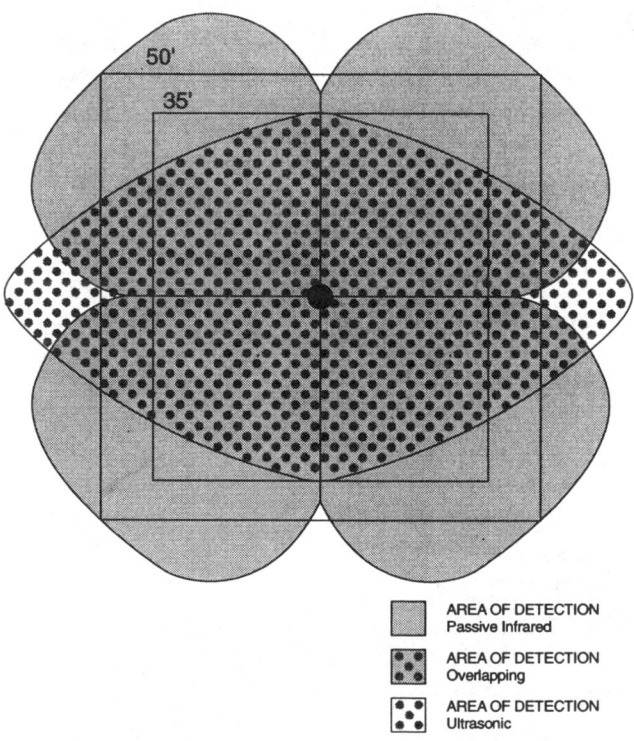

Unenco Electronics, Inc.

Fig. 5 Coverage pattern of combined infrared and ultrasonic sensor This diagram does not tell the whole story, because each type of sensor is more sensitive in some ways than the other. For example, the IR sensor is most sensitive to motion across the field of view, but the ultrasonic sensor is most sensitive to approaching or departing motion.

Do not use personnel sensors to control high-intensity discharge (HID) lighting. It has a very long delay in starting and restarting, of several minutes or longer.

■ Range, Coverage Pattern, and Sensitivity

Infrared personnel detectors typically have a range from 20 to 70 feet (6 to 20 meters). Detection is strictly line-of-sight. The coverage pattern, or field of view, depends on the geometry of the lenses that focus the heat radiation on the sensor. Ceiling-mounted sensors may have 360° coverage. Infrared sensors that replace existing wall switches may achieve almost 180° coverage against a flat wall. Figure 4 shows the coverage pattern of a model designed to be installed in a corner.

The range of infrared sensors varies with the magnitude of the temperature differential between the person's exposed skin and the background. Infrared sensors can detect remarkably small temperature differences, but it is conceivable that persons could become invisible to the sensors if the background temperature is close to skin temperature.

Infrared detectors sense motion across the sensor's field of view. Hence, the amount of motion required to trigger the unit depends on distance from the sensor. A unit installed close to a desk can sense a person nodding his head or moving his hand across a page, provided that this motion is across the detector's field of view.

Ultrasonic detectors have about the same range as infrared units, or somewhat less. Their range depends on the person's size and direction of motion. Ultrasonic detectors sense motion primarily toward or away from the detector. They are generally less sensitive to motion than infrared detectors. As a result, they are limited to sensing large motions, such as people walking toward a door.

In order to get good sensing reliability, it may be desirable to combine the features of infrared sensing and ultrasonic sensing. Figure 5 shows the coverage pattern of a sensor that uses both types of technology in a single small unit.

Audible sound detectors have unlimited sensitivity potential. Their effective range is limited by the need to avoid false triggering from background noise. Occasional outside noises may be louder than typical human activity inside the space. Furthermore, human activity does not tend to be noisy on a continuous basis. Therefore, audible sound detectors are limited to applications where noise-producing activity is associated with the presence of people.

■ Mounting Location

Sensors must be mounted where they can reliably "see" people in the controlled area.

Infrared radiation is a form of electromagnetic radiation, like visible light. It has about the same penetration. If visible light cannot pass through something, infrared radiation cannot pass through it either. Furthermore, glass and other materials block heat radiation from people.

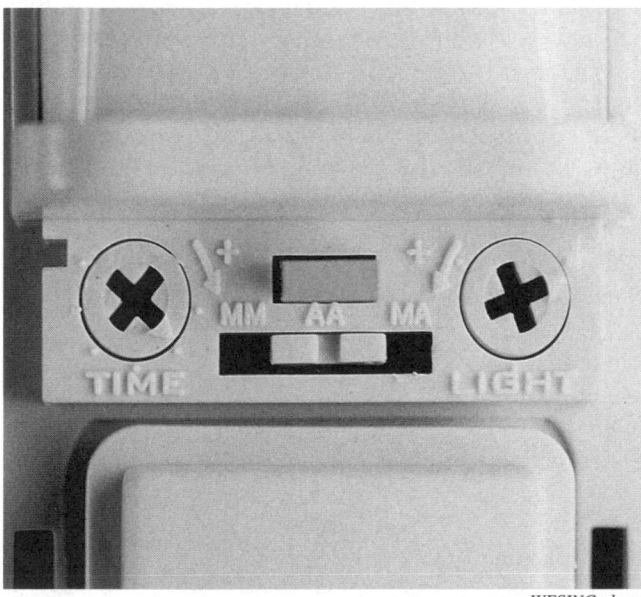

WESINC photo

Fig. 6 A good set of features in an infrared sensor
The screw on the right adjusts the daylight level that keeps lights turned off. The screw on the left adjusts the turn-off delay. Both require trial-and-error to set. The slide switch in the middle selects the override mode, explained in Figure 7. The switch plate must be removed to change any of these settings, which is an important safeguard. The little rectangular light amuses occupants by blinking when the sensor detects motion.

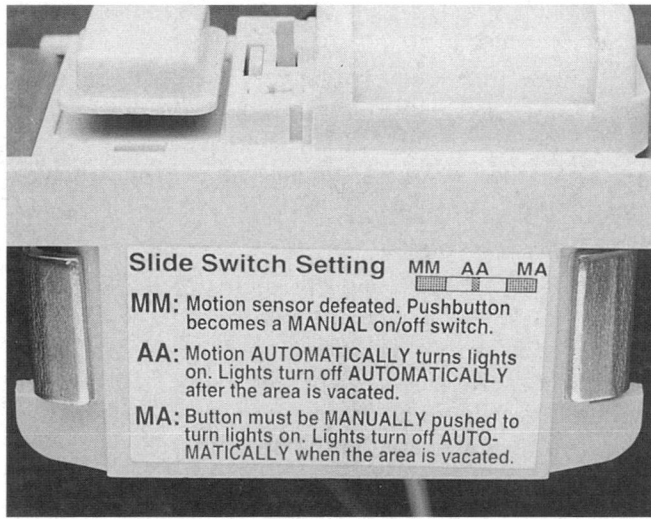

WESINC photo

Fig. 7 Override options This sticker explains the override modes that are selected by the switch in Figure 6. Option AA is the one to use in most applications. Option MA is efficient, but has few applications. Option MM abolishes the benefit of the sensor. The sticker becomes inaccessible, and hence useless, when the switch is installed.

Infrared sensors that replace wall switches are tempting with their low cost and ease of installation, but make sure that they provide adequate line-of-sight coverage of the space. This may be a problem, because light switches are mounted low on the wall, where the line of sight into the space may be obstructed by partitions, coat racks, etc.

This type of infrared sensor may be difficult to install in place of an existing toggle switch if the present switches are ganged, because the face plate of the sensor unit may be too wide for mounting alongside toggle switches. At least one manufacturer includes a replacement face plate to be used with double switch boxes. With larger numbers of ganged switches, you may need to install a separate junction box for the sensor.

Ultrasonic personnel detectors emit sound at very high frequency, which makes the sound directional. Ultrasound does not bend around objects very much. It does pass through lightweight objects, such as curtains and plants, and it reflects from the surfaces of the space, especially from hard surfaces. Thus, ultrasonic sensors are not limited entirely to line-of-sight detection. However, don't depend on this. Most ultrasonic detectors emit sound in a lobe pattern, with an effective pattern width around 50°. They are usually mounted high on a wall.

Audible sound detectors are non-directional, especially for sounds of lower frequency. This makes them worth considering where the other two methods cannot provide adequate coverage, provided that the environment is fairly quiet. Audible sound detectors can be installed almost anywhere, except that they should be installed away from any sources of noise that are not related to occupancy.

Wall-mounted sensors typically have self-contained override switches, sensitivity and delay adjustments, etc. Ceiling-mounted units either sacrifice these control options or they require separate control panels, which adds wiring cost.

■ Daylight Override

Sensors that are designed to control lighting may have a feature that detects the amount of daylight in the space. It keeps the switch from turning on the electric lights when sufficient daylight is available. This feature requires a setting to adjust the amount of daylight that is needed to deactivate the switch. Figure 6 shows the setting on a typical unit.

■ Turn-Off Delay

Virtually all personnel sensors have delay mechanisms to keep the equipment turned on for a period of time after the last motion or sound is detected. This delay feature compensates for lack of sensitivity. Extending the time delay makes it more probable that a person within the space will move enough to keep the equipment turned on. For example, even a person

reading a book moves enough every few minutes to reactivate an infrared sensor.

The energy penalty of extending the turn-off delay is usually small. For example, if the equipment is kept running for ten minutes after the last person leaves the space, this does not waste much of the energy-saving potential in most applications.

In applications where people frequently enter and leave a space, the turn-off delay minimizes short-cycling of equipment, such as lamps, motors, compressors, etc.

Many delay options are available on different models. Some units have a single fixed delay period. Others have a choice of fixed delay periods. Many units provide a continuous range of delays, as in Figure 6. The latter type can be a nuisance to set, because the setting feature does not accurately indicate the delay. This requires wasting a lot time with trial and error.

■ Minimize False Triggering

Personnel sensors can be triggered by stimuli other than people, or by the wrong people. False triggering wastes energy and wears out equipment. To determine whether false triggering is occurring, observe the operation of the equipment during unoccupied periods. You can temporarily connect an inexpensive data logger to the controlled equipment for this purpose.

Infrared detectors may be triggered by equipment that turns on and off while the space is vacant. Examples are heating and cooling equipment, refrigerator motors and thermostatically controlled coffee pots. Infrared sensors are not triggered by the air from the conditioning units, but by surfaces that are heated or cooled rapidly by the air, including the surfaces of the conditioning units themselves. The same problem can be created by sunlight that enters through windows and skylights. Fortunately, window glass is opaque to heat radiation, which minimizes (but does not eliminate) false triggering caused by moving heat sources seen through windows.

Experience will teach you where to expect false triggering. Infrared control is line-of-sight, so you can usually prevent false triggering by installing shutters or blinders to keep the sensor from seeing the heated areas. Some infrared units include adjustable shutters to limit their field of view. Some infrared sensors have sensitivity controls that may avoid false triggering.

Ultrasonic personnel sensors may be triggered by moving objects within vacant spaces, such as machinery, exposed fans, blowing curtains, kinetic artworks, etc. They may also be triggered by transient sounds, such as thunder or an air conditioner turning on. Ultrasonic sensors may have sensitivity controls to limit unwanted triggering.

With audible sound detectors, it is virtually impossible to prevent some false triggering by sounds from outside the controlled space. Also, false triggering may be caused by equipment sounds that occur within the space while it is vacant. Selecting the best mounting location and adjusting the sensitivity control are the ways to minimize false triggering.

■ How to Confine the Area of Control

Arrange the control layout so that people located in one area do not trigger sensors in other areas. You can do this by limiting the sensitive area of each sensor. For example, to provide individual lighting for a number of indoor tennis courts, install a sensor for each court and aim it at the center of its court. This may require adjusting shutters or installing blinders on the sensors.

The weakness of this technique is that people who are near the edges of controlled areas may not trigger sensors reliably. You can solve this problem, at increased cost, by installing a larger number of short-range sensors. For example, the lighting for each desk in an open office bay can be controlled by its own sensor.

Another approach is to install partitions. This may not be as expensive as it sounds. With infrared sensors, only lightweight visual screens are needed.

■ How to Control Large Areas with Few Sensors

Use clever sensor layout to minimize the number and cost of the sensors needed to provide coverage of a large area. For example, you may provide adequate control for lighting a basketball court by installing sensors near the basket, at the entrances, and at the front of the seating area.

If you use this "sampling" technique to control lighting, it is especially important to design the lighting so that safe egress is possible under all conditions.

Consider audible sound detectors for large areas that are sparsely or irregularly occupied, provided that the background noise is not too high. Audible detectors provide the advantage of complete coverage, avoiding the uncertainties of control on a sampling basis.

■ Accommodate Diverse Control Requirements

It may be desirable for equipment to respond differently to different types of occupancy. For example, the air conditioning unit in a room should start when the first occupant enters, but not when a security guard looks inside. One way to make this distinction is to mask out the portion of the space where unwanted triggering can occur. If this is not practical, consider subdividing the space into smaller controlled areas by using a number of sensors of limited range.

Cleaning crews deserve special attention because they go almost everywhere and they work at hours when the regular occupants are gone. For control of lighting, personnel sensors respond almost perfectly to cleaning crews. However, for control of conditioning equipment, they have just the wrong response. You may need to combine personnel sensors with other controls to keep equipment from starting as a result of the transient presence of people in an area.

(Permit a small sermon. Designers fail to account for operational factors like cleaning crews, window washing, light bulb changing, and equipment maintenance because they do not acquire experience in facility operations during the years they are in training. This is a glaring defect in our present way of training architects and engineers.)

■ Anticipate Changes in Space Configurations

Try to anticipate future changes in space configuration that may interfere with the operation of personnel sensors, such as relocation of furniture, addition of privacy screens, etc. Even small changes may mask the coverage of a sensor, such as placing a coat rack next to a wall sensor. Retrofit sensors that replace toggle switches are most vulnerable to changes because they are installed at a low height.

■ On and Off Override Switches

Personnel sensors cannot respond to all the conditions that may occur, so they may include a manual "off" switch, a manual "on" switch, or both. Figures 6 and 7 show an interesting combination of choices for selecting manual or automatic operation.

It is usually desirable to provide a manual "off" override, either on the sensor itself or separately. For example, in lighting applications, there is usually an occasional need to darken a space while it is occupied, e.g., to show slides in a conference room. Make the override switch easily accessible.

On the other hand, avoid sensors that include an "on" override. If the sensor has this feature, it is likely that someone will leave the switch in the "on" position, disabling its ability to turn equipment off. There should be no circumstances that require an "on" override. If a personnel sensor fails to turn on equipment automatically

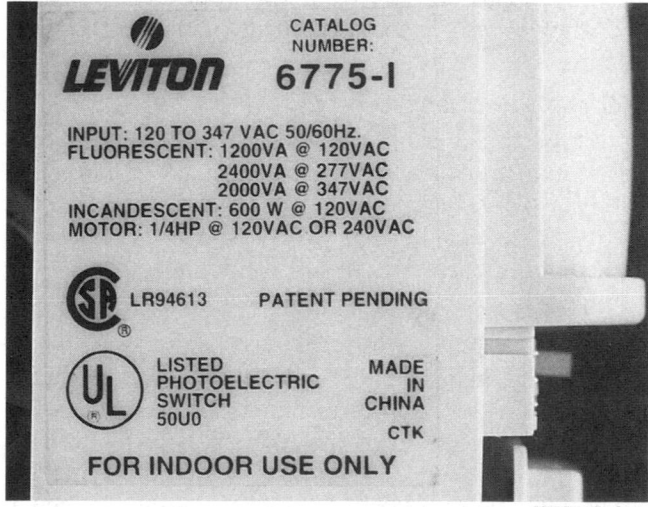

WESINC photo

Fig. 8 Critical electrical information This infrared switch can control different kinds of electrical equipment, including motors, but the capacity ratings are different for each kind.

when it is needed, it is either installed improperly or it is not appropriate for the application.

■ Sensors with Three-Way Switches

Some wall-mounted sensors can act as a 3-way switch. These units are intended as replacements for existing 3-way toggle switches. These can be used in two ways. If all the manual switches are replaced with sensor switches, sensors in different parts of the space can control a single set of equipment. Or, if a single sensor switch is used in combination with manual switches, the manual switches can act as overrides to force the controlled equipment off.

■ Maximum Amperage or Wattage

The cost of electronic devices is related to the current they handle, so many personnel sensors are not designed to handle full circuit capacity. Be sure to check that the amperage or wattage rating of the sensor is adequate for the controlled load.

The amperage rating may be lower for incandescent lighting than for other types of loads. This is because incandescent lamp filaments have lower resistance when cold than at operating temperature, so they have high starting current.

If the sensor is rated by power, rather than by amperage, the rating may be lower for motors and fluorescent lighting. This is because motors and magnetic lamp ballasts have low power factor, which requires them to draw extra current for a given amount of power.

Figure 8 illustrates how a single sensor may have a number of different wattage and volt-amp ratings for different types of loads.

If you need to control a large amount of amperage or power, consider using a power relay instead of a more expensive sensor. Using multiple relays allows a single sensor to control as many circuits as desired. A relay also allows any sensor to control equipment of any voltage.

■ Minimum Wattage

An odd feature of some personnel sensors is that they require the controlled equipment to have a minimum wattage. For example, some lighting sensors cannot control lamps smaller than 40 watts. This is related to the switching characteristics of semiconductors.

■ Lamp Flicker

Some personnel sensors cause fluorescent lamps to flicker. This appears to be a problem associated with aged lamps. Also, high-efficiency fluorescent tubes are more susceptible to flicker than conventional tubes. The problem is most serious immediately after the lights are turned on. These problems suggest that the personnel sensors are lowering the peak voltage to the lamps.

For fluorescent lighting and other applications where the voltage and waveform of the power are critical, test candidate units to ensure that they do not cause problems.

Some electronic fluorescent lighting ballasts compensate for variations in input voltage, including variations caused by personnel detectors.

■ Grounding

Good grounding is especially important for reliable operation of semiconductor switching equipment, as well as for safety. It is good practice to ground the sensor units using separate ground wires attached to the boxes, rather than relying on the mounting screws for grounding. Trouble is likely to occur if the existing electrical system is not well grounded.

■ Placards

Even though personnel sensors are fully automatic, expect to install placards to inform occupants about them. Incomprehensible control operation is annoying, and people may respond in a manner that is wasteful or unsafe. See Reference Note 12, Placards, for details of effective placard design, materials, and installation. Placards about personnel sensors should explain the following:

- which equipment is controlled
- that the equipment responds to the presence of people in the space
- that the control will delay turning off the equipment for a short period after it ceases to sense people in the space
- how to operate the manual features of the control (if any)
- that the user should set the switch back to "automatic" after using any manual settings
- whom to contact in the event of malfunction.

Test Unfamiliar Situations

It is difficult to anticipate exactly how particular models of personnel sensors will work in any particular environment. In case of doubt, test a few candidate sensors in some typical spaces before purchasing large numbers.

Monitor Performance

For a period of several days after a sensor is installed, check with the occupants of the space to see whether the controls are causing any problems. Make any appropriate corrections.

Don't forget this step. If a dangerous or inconvenient situation is created by the sensor control, you want to know about it before trouble occurs.

Placards

Why Placards are Important
Where Placards are Most Effective
Where to Install Placards
What Placards Should Say
Examples
Placard Materials

How to Attach Placards
Make Them Attractive
Update Them
Coordinate Your Placard Installations
Coordinate with Other Communications Methods

A large fraction of the energy consumption of a facility is under the control of people. These people include managers, operating staff, employees, occupants, tenants, guests, etc. Where people control energy consumption, much energy waste is caused by failing to effectively tell those people how to conserve energy.

In many environments, placards are potentially the best method of communicating with people about the energy conservation actions that they should take. Placards themselves are cheap, but designing and installing them properly requires perception, effort, and a certain gift for communications. This Note will guide you in using these powerful energy saving tools.

Why Placards are Important

When deciding where to use placards and how to design them, put yourself in the position of the people you are trying to reach. The common communication problems that lead to missed conservation opportunities are:

- people are not aware that particular equipment, such as lighting or air conditioning or machinery, is using energy, or how much
- people are not aware of energy-conserving actions that they can take
- even if they are aware, they are not sure about what to do, or when to do it
- the purpose of controls is not clear
- controls are inaccessible, inconvenient, or out of sight
- people are concerned that taking action will offend others, draw criticism, or be ridiculed
- people lack motivation to act
- people lack the skill required.

Placards attack each of these problems. Placards do more than provide information. They have important psychological functions. They request action, and equally important, they grant permission to act. They assure the person that actions taken to conserve energy are appropriate.

Placards can be a powerful means of overcoming social inhibition that deters energy conservation. It is not generally true that "people don't care" about energy conservation, or that people are irresponsible. Instead, people are hesitant to act if doing so may offend others, or if they may be subjected to ridicule for taking unusual initiative. By the same token, people will take appropriate action if two conditions are satisfied: (1) the desired action is stated clearly, and (2) the desired action appears to be appropriate and expected. Design your placards to satisfy both of these requirements.

The presence of an official-looking placard says, in effect, "It is polite and conscientious to conserve energy." This is especially important for persons who are not employed in the operation of the facility.

Expect to break new ground and invest effort. In the past, people responsible for energy efficiency failed to recognize the potential benefit of placards. When they have used placards, they failed to understand the amount of effort needed to produce good ones. Most placards have been overly general, poorly presented, and impermanent. For example, a sign scrawled on a piece of cardboard with pencil is illegible, it lacks authority, and it will not survive.

Generic "conserve energy" stickers and posters are almost worthless. To be effective, each placard must be customized to its application and environment.

Where Placards are Most Effective

Placards are an economical way of motivating efficient operation of almost any kind of equipment. However, the potential effectiveness of placards varies considerably in different situations. Placards are most effective if they are related to individual responsibilities. For example, a placard on a chilled water temperature reset control is most effective if specific persons are assigned to operate the chiller plant.

If responsibility cannot be assigned to a specific individual, placards work best in single-occupant spaces where there is low personnel turnover and a high level of personal responsibility. If there is more than one occupant, each occupant is reluctant to act for fear of annoying other occupants. In such situations, you can improve the effectiveness of placards with additional publicity, as discussed below.

By the same token, you can enhance energy conservation by making designated individuals responsible for specific energy conservation activities. When you do this, the placard acts as a "homing device," guiding the designated individual to the activity.

Where to Install Placards

Install placards on all devices affecting energy consumption that require some human action. This includes windows, light fixtures, thermostats, fan control switches, heat/cool selector switches, timeclocks, manual radiator valves, etc.

Also, install placards to inform people of the operation of automatic controls that have overt effects. For example, if lighting is controlled with motion sensors, install a placard at the entrance to the space that states this fact. This prevents unpleasant surprise, avoids safety problems, and limits resentment toward energy conservation in general.

The location of placards is critical. Locate placards where they are conspicuous, where they will be noticed at the time they are needed, and where their relationship to the equipment is obvious. The best location depends on the information that is being conveyed. For example, if the purpose of the placard is to encourage efficient temperature setting, install the placard at the thermostat. If the purpose of the placard is to get people to turn off their equipment at the end of the day, you might install the placard at the switch controlling the equipment, or at the exit.

Do not be reluctant to install more than one placard if there is no single best location. For example, you might install one placard at the thermostat for temperature control, and another at the exit to suggest turning off the conditioning unit when the last occupant leaves the space.

What Placards Should Say

In order for a placard to be effective, it should state clearly:
- the purpose or identity of the item
- what to do (or avoid doing)
- when to do it
- how to do it
- what response to expect
- how to get help, if needed,

Some of these items can be omitted if they are obvious.

When designing a placard, put yourself in the position of the intended user. Do not assume any prior knowledge, and do not ignore any steps.

Current technology lets you make durable placards that include diagrams and photographs. Use illustrations where they can communicate more effectively than words alone.

Examples

■ Thermostat Placard

In an office that is conditioned by several four-pipe fan-coil units, install a placard at the room thermostat that says:

> **THIS THERMOSTAT CONTROLS THE THREE HEATING AND COOLING UNITS IN THIS ROOM.**
> **To conserve energy, please adjust the temperature setting on this control unit as follows:**
> - **Select HEAT or COOL on the top selector switch.**
> - **If HEAT is selected, set the HEAT lever to the lowest comfortable temperature.**
> - **If COOL is selected, set the COOL lever to the highest comfortable temperature.**
> - **Set the FAN switch to AUTOMATIC.**
>
> - **If you leave the room for several hours or longer, please turn the FAN switch to OFF.**
> - **Call Extension 89 if there is a problem with heating or cooling.**

■ Local Control of Heating and Cooling

In a conference room where conditioning is controlled by rundown timer, install a placard at the timer that says:

> **THIS TIME SWITCH CONTROLS THE HEATING AND COOLING IN THIS ROOM.**
> **The room temperature will become comfortable within a few minutes.**
> - **Please set the timer for the minimum length of time that you will be using the room. You can add more time as needed.**
> - **Call Extension 89 if there is a problem with heating or cooling.**

■ Kitchen Range Hood

On the range hood in the kitchen of a hotel located in a cold climate, install a placard that says:

> **PLEASE TURN OFF THE HOOD FAN WHEN THE COOKING EQUIPMENT IS TURNED OFF.**
> **The fan draws a large amount of heat out of the building when it is running. The fan switch is located inside the door of the dry storage room.**

■ Openable Windows

In a room that has user-controlled heating and cooling, and openable windows, install a placard at a visible location near the windows that says:

> **PLEASE CONSERVE ENERGY.**
> **Close the windows**
> **when heating or cooling.**

■ Lighting Controlled by Motion Sensors

In a small room that has lights controlled by a motion sensor, install a placard at the entrance that says:

> **THE LIGHTS IN THIS ROOM ARE CONTROLLED BY AN AUTOMATIC SWITCH THAT SENSES MOTION.**
> - **The lights will turn on as soon as you enter the room.**
> - **To turn the lights off while you are in the room, use the "lights off" switch beside the door.**
> - **Call Extension 89 if there is a problem with the lights.**

■ Marking Delamped Light Fixtures

To mark fluorescent fixtures that have been delamped, attach stickers to the outside of the fixtures that say:

> - **Use only (wattage), (color) lamps in this fixture.**
> - **(number) lamps have been removed from this fixture for energy conservation.**

■ System Diagram

In the central monitoring station of the heating and cooling plant, install a large diagram that indicates the location of all the controls that should be adjusted to optimize energy efficiency, with inset photographs of each control.

■ Some Real Examples

Figures 1 through 4 illustrate the present state of placard usage. These actual examples range from fairly good to useless. Clearly, there is a long way to go. You may find it useful to score each of these examples in terms of how well it satisfies the criteria that we stated previously.

Placard Materials

Make placards out of materials that are durable enough to survive as long as the equipment to which they relate. Consider these materials:

- *photographically etched metal.* Virtually any kind of text or graphics can be imprinted on aluminum placards with a photographic etching process. At present, this process is limited to a single color, and is somewhat limited in contrast. This product is available from trophy shops, some sign companies, and miscellaneous specialty companies.
- *painted metal.* Text and bold graphics can be made on painted metal signs. Several colors may be used, and silk screening can be used to reproduce virtually any type of copy. Aluminum or enameled steel signs can survive harsher environments. Simple signs with bold lettering are produced by sign shops. More complex signs are produced by specialty silk screening shops.

- *engraved laminated plastic.* Engraved plastic signs are bold, but limited to plain text and single colors. Laminated plastic cannot survive mechanical stress, strong sunlight, or outdoor conditions. Engraved plastic signs are available from office supply stores and specialty marking companies.
- *printed plastic.* Placards can be printed on many types of plastic materials, including the reverse sides of transparent plastic. Any type of text or graphics can be produced using silk screening. Plastic cannot survive mechanical stress or strong sunlight. Printed plastic signs are produced by specialty silk screening shops, and by specialty sign shops.
- *plain paper.* Plain paper is rarely durable enough for placards. Its merit is cheapness and the ability to be printed with any type of material. Paper may be the only choice in highly protected applications where low cost is important, as in marking delamped light fixtures. Self-adhesive stickers are produced in large quantities at low cost by label manufacturing companies. Also, you can make them yourself using a desktop computer.
- *plastic-coated paper.* Anything that can be printed on paper can be protected with a plastic film. The film greatly increases the durability of the placard. Still, plastic-covered paper is suitable only for dry indoor environments. Many printing shops can produce placards on plastic-coated paper. If you make paper placards yourself, you can easily laminate them by using self-adhesive laminating sheets or an inexpensive laminating machine.

WESINC

Fig. 1 Not pretty, but pretty good This placard is on the door of a motel room. It is big, and easy to read. It spells out what is wanted, and when. Guests see it at the proper time. It will last a long time. It could be presented more elegantly in an upscale setting.

How to Attach Placards

Make the method of attachment as durable as the placard itself. In most cases, install placards with screws or bolts having ample grip in the material to which the placard is attached.

Adhesives are usually unsatisfactory because they dry out and lose their stickiness within a few years. Adhesives may be the only practical choice for some applications, such as light fixture delamping stickers. However, expect that any placards attached with adhesives will need periodic replacement, which is unlikely to occur when it is needed.

Select the attachment method to survive the environment and future actions to the related equipment. For example, if you need a placard to tell how to set a manual valve in a machinery room, make the placard out of non-corroding metal and attach it to the adjacent pipe with metal straps. Don't bolt the placard to a flange on the valve itself, because the placard is likely to be lost when the valve is removed for maintenance.

Make Them Attractive

Facility managers may resist installing placards in visible locations because they feel that placards are unsightly. The solution is to design the placards to fit the decor. Generic "conserve energy" stickers, such as the ones given away by utility companies and government agencies, typically clash with a well planned decor.

For example, when designing placards for the guest rooms of a luxury hotel, select a complementary color scheme and attractive lettering, and perhaps the logo of the hotel. Invite the hotel's interior decorator to participate in the design.

Update Them

Any sign becomes invisible to a person who sees it repeatedly. That is why Coca-Cola keeps changing its advertisements. The same is true of placards. In particular, placards that request people to take optional action, such as turning off lights, tend to fade from consciousness. The first way to reduce this effect is to be specific about the action requested, as previously discussed.

Another way is to periodically redesign and replace the placard. Present the same information in a different style. This gives you a chance to discover your hidden artistic talents. You can improve the information provided by the placards using experience gained from the response to the earlier placards.

When you design a placard, decide whether to make it permanent or to replace it periodically. This affects the selection of materials and installation methods. For example, if you plan to replace an indoor placard every year, you may print it on plastic-coated paper instead of more expensive material. Install replaceable placards with screws, clips, or other easily detachable methods, rather than with adhesive.

Placards that provide needed information generally do not have to be refreshed. For example, people do not need novelty in a placard that indicates where a timeclock is located. The main guideline for such placards is to install them where they are conspicuous.

Keep an updated list of all placards. Use it periodically to survey your facility. Check to see if placards need to be replaced or modified. Find new opportunities for exploiting placards. The list also serves as a guide to all your energy conservation activities that need action by people.

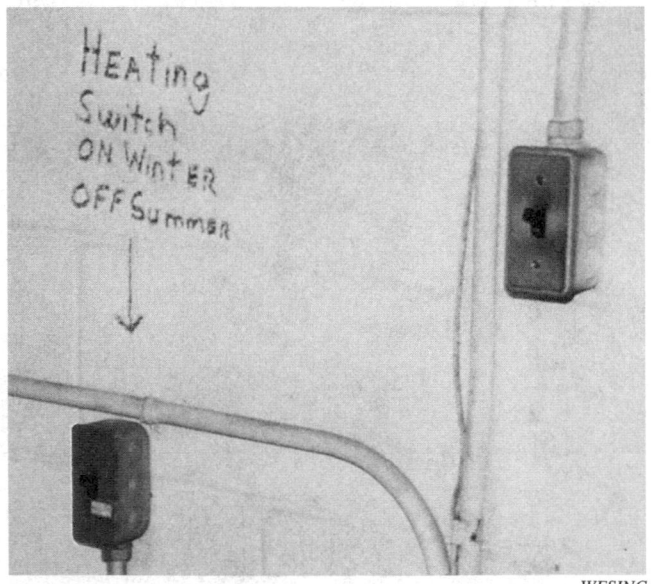

WESINC

Fig. 2 Much better than nothing This sign is probably well attuned to the staff. It identifies the switch, and spells out the essential action. It should be more specific about time. It will survive until the next paint job.

WESINC

Fig. 3 Useless placard This sticker on a light fixture is just visual clutter. It gives no guidance about what is expected, if anything. It numbs people toward conserving energy.

Coordinate Your Placard Installations

You will be amazed at how many places you can improve efficiency and occupant satisfaction with effective placards. Save time, effort, and money by planning and producing all your placards as a batch. You will learn a lot about your facility in the process.

Where it is more effective to do so, use a single placard for several energy conserving actions. For example, use one placard to request turning off the lights and turning off the conditioning equipment at the end of the day.

The material cost of placards drops dramatically with quantity, which can save money with placards used on common items, such as thermostats and light switches. However, do not allow this to lure you into using general instructions where more customized instructions are appropriate.

Coordinate with Other Communications Methods

Placards are a means of communication. They are usually not fully effective by themselves. Make each placard part of an overall program of communicating with each intended audience. For example, coordinate placards for maintenance personnel with maintenance training programs. Publicize your use of placards in company newsletters, posters, incentive programs, etc.

WESINC

Fig. 4 Ridiculous placard The tiny sticker on this thermostat cover tells people what temperatures to set. The cover is locked, so they can't reach the thermostat. Anyway, the print is too small to read.

Energy Management Control Systems

An Alternative to Specialized Controls
Unique or Advantageous EMCS Capabilities
  ■ **Monitoring**
  ■ **Reports**
  ■ **Controlling Many Items**
  ■ **Coordinated Control**
  ■ **Complex Control Functions**
Inappropriate Applications
System Configurations
	True System Cost
	Continuing Problems
	  ■ **Management Failure to Use the System**
	  ■ **Operator Confusion**
	  ■ **Staffing**
	  ■ **Inadequate Software**
	  ■ **Inadequate Programming Support**
	  ■ **Hardware Defects**
	  ■ **Rapid Obsolescence**
	  ■ **Lack of Standardization**

An energy management control system (EMCS) is a centralized control system intended to operate a facility's equipment efficiently. These systems are also known by a variety of other names, including "energy management systems" (EMS), "smart building controls," etc. A system typically has a central computer, distributed programmable controllers, and a digital communication system. The communication system may carry signals directly between the computer and the controlled equipment, or there may be tiers of communications.

These systems are still evolving rapidly, and they are controversial. This Note explains the basic elements of an EMCS, the major alternative configurations, and the advantages and disadvantages of EMCS's.

An Alternative to Specialized Controls

As a method of control, an EMCS is an alternative to using specialized controllers for individual items of equipment. In general, any control function that can be performed with an EMCS can also be performed with a less expensive single-purpose controller.

An EMCS has two inherent advantages over local controllers. One is two-way communication with the equipment, which makes it possible to monitor conditions as well as to exercise control. The other advantage is the versatility provided by the general-purpose computer, which in principle can be programmed at any time to perform any function.

On the other hand, dedicated controllers are more reliable and easier to maintain. In most facilities, the total cost of individual controllers would be much less than the cost of an EMCS. Install an EMCS only if its unique advantages outweigh its inherent disadvantages.

Unique or Advantageous EMCS Capabilities

In deciding whether to install an EMCS, be clear about what you expect it to do. Some of the following are functions that only an EMCS can accomplish. Others are functions that an EMCS may be able to accomplish better than local controls.

■ Monitoring

A computer has input as well as output, so it can monitor equipment as well as control it. Part of the original EMCS concept was the ideal of having perfect knowledge of the facility's status from the splendid isolation of the air conditioned office. As a real example, an operator can call up the supply air temperature of a particular air handling unit at the keyboard, and the temperature may be presented on a multi-color diagram of the system that appears on the computer screen. Figure 1 shows the general idea.

In reality, the monitoring capability of an EMCS is presently far from the ideal of universal knowledge of facility conditions. An EMCS can monitor only those functions for which a sensor is installed. The cost of installing sensors and wiring limits monitoring to conditions that are easily measurable and that are likely to be monitored repeatedly.

The monitoring done by most EMCS's consists of little more than telling whether equipment is running, and reporting system and space temperatures. EMCS's have not yet fulfilled an early promise of universal diagnostics. For example, it is not economical to use an EMCS to report pump packing leakage or strange bearing noises, even though such conditions could be monitored in theory.

At present, facilities typically make little use of the monitoring capability of their EMCS's. The most common monitoring function is creating an alarm if some variable (space temperature, air flow, the operation of a pump, etc.) is outside limits that are specified at the computer. EMCS's are not yet ready to monitor system efficiency at a sophisticated level, because the software for this function is not available.

■ Reports

By combining monitoring capability with the analysis and graphical capabilities of a computer, an

EMCS can be produce reports on the data that it collects. The usefulness of reports and their ease of generation varies between systems. This valuable function adds little to the cost of an EMCS because it requires only software.

"Trend logging" is the elementary level of reporting. This is simply a printout of selected variables at short intervals. Trend logging uses a lot of paper, and is tedious to use. Trend logging can be made more useful by presenting the data graphically, if the software has this capability.

The report generation capability of EMCS's makes them potentially an important tool of energy management at the facility management level. To fulfill this potential, the computer should be able to produce easily understandable reports that show the energy consumption patterns of each item of equipment in the facility, and the energy consumption and cost patterns of the facility as a whole.

To prepare such reports, the computer needs input data about energy consumption. This means you have to provide metering and signal interfaces for each mode of energy consumption that you wish to monitor separately. For example, to develop a profile of boiler energy consumption, you need to install a boiler fuel flowmeter and interface it to the EMCS. To monitor chiller energy consumption, you need to install a wattmeter on the chiller input, and interface it to the computer. To monitor lighting energy consumption, you need to install an ammeter on each lighting circuit that you want to monitor separately.

■ Controlling Many Items

An EMCS can control a virtually unlimited number of devices by switching between them. Is it better to have many individual controllers, or only one EMCS? The EMCS may have a reliability advantage because of its monitoring ability, especially in facilities with a large number of similar devices to be controlled. The occasional failure of a local controller could escape attention without the performance monitoring provided by an EMCS. From a statistical standpoint, a single EMCS may be more reliable than a large number of local controllers, provided that the EMCS is well maintained. However, EMCS's generally increase reliability problems rather than reducing them, for reasons covered below.

At present, the ability of an EMCS to control many units does not save money compared to local control. For simpler control functions, such as time control and optimum start, installing an individual controller located at the equipment is typically less expensive than connecting the equipment to an EMCS, even if the facility already has an EMCS.

■ Coordinated Control

Since an EMCS has two-way communication, you can use it to coordinate the control of different items of

WESINC

Fig. 1 The person is important, the computer is not For an effective energy manager, an EMCS is a convenience. In the absence of an effective energy manager, an EMCS is electronic junk that increases construction cost, increases staffing requirements, and confuses the maintenance staff.

equipment to optimize overall system efficiency. Such applications are few, but they may save a significant amount of energy or money.

For example, this coordinated control capability makes an EMCS useful for demand limiting, in which the EMCS turns off selected items of equipment as the overall facility approaches a demand limit. (See Reference Note 21, Electricity Pricing, about demand charges.) The EMCS can be programmed to turn off different equipment on a rotating basis, or to turn off equipment based on certain criteria, such as relative importance, the time of day, operation of other equipment, etc. However, specialized demand controllers are available to perform these same functions.

Another example is using an EMCS to control the running of a cooling tower that serves several self-contained air handling systems. The EMCS can sense when all the air conditioning units are turned off, and can then turn off the cooling tower fans. However, the same function can be performed cheaply by using a few relays and some low-voltage wire.

■ Complex Control Functions

Because an EMCS includes a powerful computer, it can perform any control function for which you program it, no matter how complex. Further, it can perform any number of different control functions.

For example, in a chiller system serving VAV air handling units, an EMCS could program the computer to continuously optimize the chilled water temperature, the supply air temperature, and the condenser water temperature settings to minimize the overall system energy consumption under all operating conditions. This assumes that the appropriate software is available, or that a highly skilled individual is available to program the system for this purpose.

The unlimited programming capability of EMCS's makes it possible to specify system performance in much greater detail than was possible with conventional controls. For example, some EMCS's offer the ability to tune the response characteristics of controls by specifying their proportional, integral, and derivative (PID) control characteristics. For some applications, such advanced features may provide a significant benefit. For other applications, they add an unnecessary burden of complexity.

The fact that you want to perform a complex control function does not necessarily mean you need an EMCS. Cheap microprocessors provide complex control at the local level for many common functions, such as supply air temperature reset, enthalpy control, and optimum start. These controllers cannot be programmed in the general sense, although they may offer ranges of settings. The additional capabilities of the EMCS, such as the ability to monitor and to change settings remotely, may justify the additional cost and complexity of the EMCS.

Inappropriate Applications

Many EMCS's have been purchased on the basis of vague expectations, rather than a clear idea of functions that an EMCS is expected to perform. As a result, most EMCS's operate at a rudimentary level. Even in cases where a permanent EMCS staff has been established to keep systems working, the EMCS typically does little more than serve as a remote timeclock.

EMCS's became a fad item toward the end of the 1970's as the public, already aroused to energy conservation, started to take an interest in computers. The sales pitch given to enthralled facility owners implied that installing a computer would in obscure ways make the facility more efficient. This claim could not be fulfilled, because energy waste is caused by many factors unrelated to control, including sloppy design, poor installation, and improper operation. These factors cause hundreds or thousands of inefficiencies to be scattered throughout a facility. Overlaying an electronic control system on a wasteful facility does not correct these inefficiencies.

Undoubtedly, a primary reason for the initial popularity of EMCS's was their expected entertainment and status value. Managers could play with the EMCS and show it off to visitors. Mechanics could pretend to be managers by sitting at the EMCS in an air conditioned room, and avoid getting dirty. This appeal soon wears out, and the staff has to get back to running the facility. Many EMCS's purchased as toys have been abandoned.

Like computers in general, EMCS's are often used in ways that produce more trouble than benefit. To justify the large investment, it is common to use the EMCS to control inappropriate functions, sometimes to the point of absurdity. For example, some facilities use an EMCS to control the conditioning of conference rooms, ballrooms, and other irregularly used spaces. This requires the EMCS operators to become involved in the scheduling of the spaces, and it requires all potential users of the conference rooms to know how to contact the EMCS staff. In addition to the increased workload and confusion, it becomes virtually impossible to reschedule the spaces on short notice.

A common abuse of EMCS's is remote setting of space thermostats. In most cases, occupants should be allowed to control space temperature locally. A central computer cannot compensate for variations in personal preference, cold spots, changing solar gain, etc. Connecting an EMCS to every thermostat in a building typically costs a lot of money that produces no benefit and reduces comfort.

(In general, equipment that is used for local functions should be started and stopped locally. For example, the air conditioning for a conference room is best controlled by the users of the space. Each Section of this book includes appropriate Measures for minimizing unnecessary equipment operation while maintaining high comfort levels. Most of these Measures recommend local control by the users or occupants.)

Ability to control from a remote location is not an inherent advantage of EMCS's. Any type of control, such as a simple timeclock, could control equipment from any location, if remote control is desirable. In the case of simple controls, this may seem silly because of the high wiring cost. However, the wiring cost does not seem silly with an EMCS. The difference is psychological, not technical.

System Configurations

The original configuration of EMCS's was the simplest. A central computer was connected by wires to sensors at locations being monitored, and to actuators on equipment being controlled. Figure 2 shows this arrangement.

The early systems could do little more than turn equipment on and off. Additional electronic devices were then developed to allow the computer to interact at a more "intelligent" level with controlled equipment, such as making changes to thermostat settings. This required the development of complex digital thermostats and other local control devices.

The high end of clock controls, covered in Reference Note 10, have evolved to the point that they have as much capability as the early EMCS's. These still depend on dedicated connections between the controller and each item of controlled equipment. Figure 6 shows a unit that straddles the boundary between clock controls and EMCS's.

Soon after EMCS's were introduced, it became apparent that failure of the computer would disable all the equipment connected to the computer, or make it impossible to control the equipment. Even failure of a cable, a telephone line, or other communication device would disable a large part of the EMCS. To reduce vulnerability to these failures, EMCS's evolved a configuration that uses distributed microprocessors, called "slave panels," "local panels," "standalone control units," "terminal equipment controllers," and other names, in addition to the central computer. For example, a university campus might have a single central computer, and a local panel in each building. Figure 3 shows this variation.

If the central computer fails, the local panels keep systems operating normally. In this arrangement, the central computer still performs the monitoring function, and it is used to reset or reprogram the local panels as needed. This configuration is presently the most common, and there is a trend for more of the "intelligence" to be put into the local panels, while the central computer is reduced to the role of a "dumb terminal."

The original system layout, with wiring that spreads out from a central computer, is being replaced by loop systems. Figure 4 shows this configuration. These are similar to the local area networks (LAN's) used in office computer systems.

Local panels, equipment controllers, sensors, and operator terminals can be connected to any convenient point in the loop. This makes wiring easier and increases reliability, at least in principle. It also allows information to be introduced at any point. For example, for load shedding, each item of equipment reports its electric load to the loop, which allows any local panel to calculate the total facility load. This allows each local panel to decide which equipment to shed as the facility approaches its maximum allowable demand.

The local panel itself may soon become obsolete by losing its role as middleman. At the center of the system, most EMCS's now use an inexpensive personal computer as the central computer. This radically reduces the cost of the most expensive component of the EMCS, and makes it a standardized item. At the remote ends of the system, the low cost of small microprocessors now makes it practical for equipment manufacturers to install

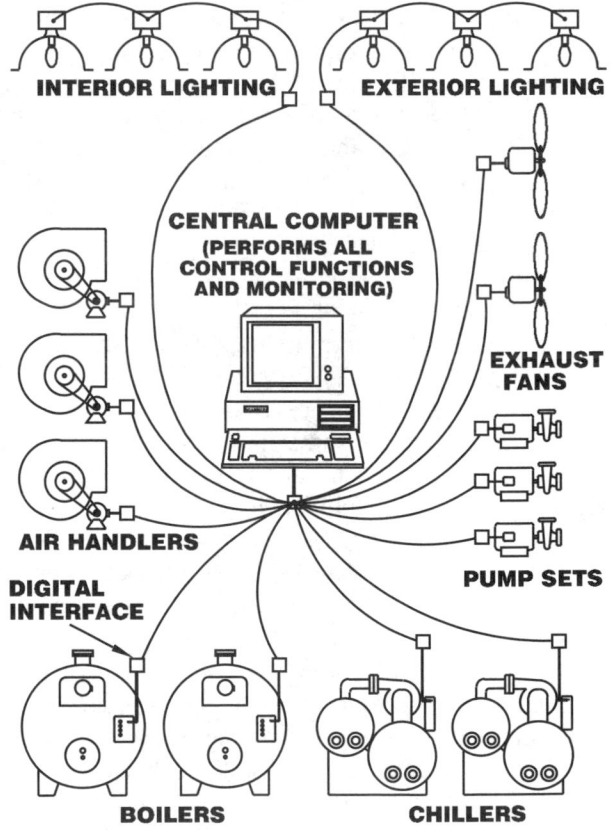

Fig. 2 Early EMCS layout A single computer performs all control functions, which makes the entire system vulnerable. The computer is connected to each item of equipment individually, which makes the equipment vulnerable to fragile wiring.

specialized, low-cost "smart controls" on each controlled item of equipment.

Equipment-mounted controls are being produced by both controls companies and equipment manufacturers. This allows anyone to build an EMCS around a standard personal computer. Furthermore, anyone can now produce software that would be usable by such a generic system. This squeezes out the need for local panels, which remain the most non-standard items of EMCS hardware, and collectively the most expensive.

True System Cost

If you compare the true cost of an EMCS to a realistic estimate of the savings that it can provide, you are likely to find a low rate of return. Remember, an EMCS is usually an alternative way of performing a function. It rarely lets you do something that would be impossible otherwise. The overall cost of control is usually much higher with an EMCS than it is with local controllers, although this comparison does not account for the unique capabilities of the EMCS.

The cost of the installed EMCS may be much higher than anticipated because the cost of the equipment quoted by the vendor is typically only a small fraction of the installed system cost. Most of the system cost is in the wiring and interfaces needed to connect the central computer, local panels, equipment controllers, and sensors throughout the facility. Also, vendors may not be candid about the programming costs needed to make the system useful once it is installed.

If EMCS's follow the pattern of other electronic devices, costs may decline substantially in the future. Newer systems use common personal computers in place of expensive specialized computers. Eliminating local panels will save a substantial amount of money, if this development occurs. However, wiring costs, which are mostly labor, are unlikely to decline substantially unless new methods of low-cost communication are developed.

Continuing Problems

All new, complex equipment goes through a period when it is difficult to use, it is difficult to maintain, its reliability is low, and its capabilities do not match its potential. EMCS's are still in this phase. Be aware of the following sources of trouble that will continue to bedevil these systems for the foreseeable future. If you

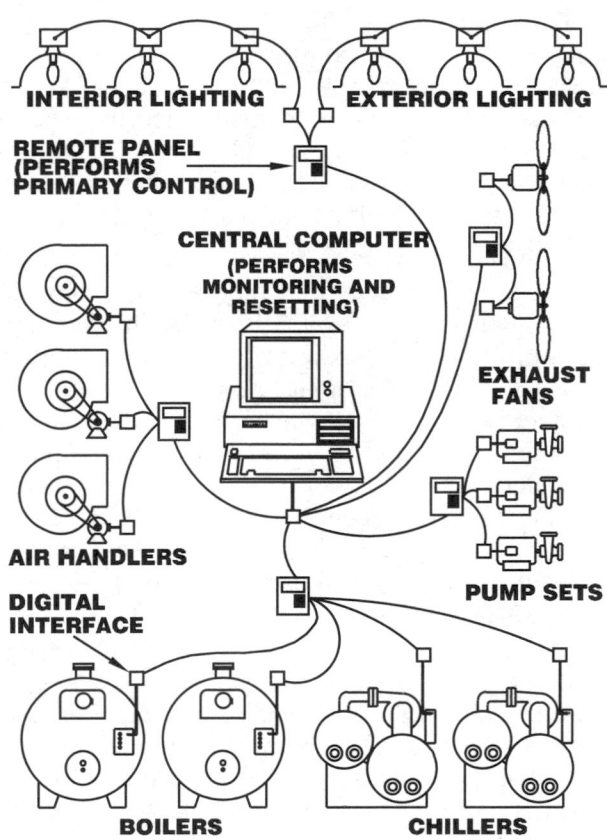

Fig. 3 Later EMCS layout Moment-to-moment control functions are performed by local panels, each of which controls a fraction of the equipment. The central computer monitors equipment performance, resets the local panels remotely, and prepares reports. The system continues to function if the central computer fails, or if a communication link is broken.

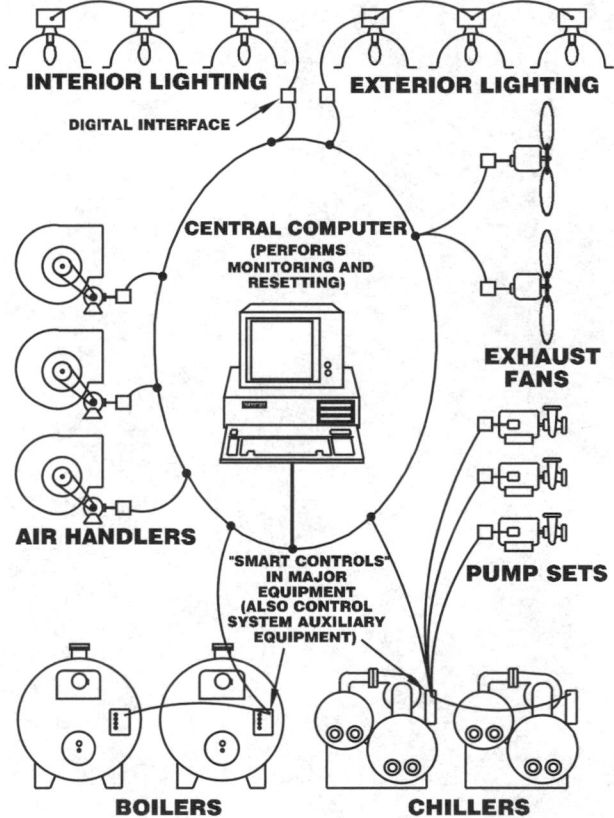

Fig. 4 Recent EMCS layout Communications occur on a loop, which increases reliability. A large part of the control functions are accomplished by microcomputers installed on major items of equipment. This allows optimum control of each system, without requiring the programmer of the central computer to know the details of individual equipment.

do not get all these potential problems under control, your EMCS will inevitably fail, either physically or by failing to deliver its promised benefits.

■ Management Failure to Use the System

Perhaps the greatest failure of EMCS's has been the failure of facility managers to effectively use the analysis and reporting capabilities of these systems as a tool of energy management. Even though managers may hire bright, highly paid individuals to operate and maintain the system, they themselves do not use it. This reflects a lack of appreciation by senior managers of the economic potential of energy management, and of their own roles in it. A practical problem is lack of software for producing reports that are easily usable by high-level managers.

■ Operator Confusion

An EMCS greatly increases the confusion and difficulty of plant operators. Indeed, from the operators' standpoint, the EMCS is much more difficult to deal with than the familiar equipment being controlled. Figure 5 shows the problem. An operator who is under pressure to keep an air conditioning system running is

WESINC

Fig. 5 Can you tell what this does? Neither can the maintenance staff. This is a local panel of an EMCS. It is much more difficult to understand than the facility's primary equipment. The EMCS requires its own specialized skills and additional manpower. Without these, the EMCS will fail and be abandoned.

tempted to deal with an unfamiliar EMCS by taking the wire cutters from his tool pouch, lobotomizing the EMCS, and operating the air conditioning system in the old way.

The capability of EMCS's to customize control performance is a source of confusion. For example, if an EMCS is used to control a VAV air handling system, each VAV terminal may have a hundred control parameters. Some of these are obvious, such as temperature setpoints. However, others are incomprehensible to the building staff, such as the time constants for control response characteristics. In theory, most of these parameters are set by the installing contractor, but it is unlikely that all settings will be proper, even with the most conscientious installation. This glut of control parameters inevitably creates problems that are difficult to diagnose.

As with computers in general, the documentation provided with EMCS's tends to be obscure and incomplete. The problem is especially severe with EMCS's because they are produced in small quantities and are still evolving. Manufacturers presently try to deal with this problem by offering classes to operators. However, presenting an overwhelming subject in a class of two or three days is not the answer.

This situation should improve in time. Manufacturers will learn to make their hardware and software more user-friendly, and will document it better. The adoption of personal computers and the possible elimination of local panels will reduce the initial mystery of the system. However, EMCS's will never be as simple as hard-wired single-function local controllers.

■ Staffing

Labor saving was one of the major benefits originally claimed for EMCS's. In facilities of small and medium size, the opposite has been true. Systems that are working usually have one or more full-time operators who work exclusively on the EMCS, without providing a reduction of other staff. Talented people are needed continuously to monitor what the system is reporting, to analyze reports, to respond to trouble calls, and to keep the system repaired. Without a permanent staff of its own, an EMCS usually collapses.

In a dispersed group of facilities, such as a school system, the EMCS may save labor compared to installing local control systems and checking their operation with traveling maintenance personnel. However, even this is not certain. A large EMCS may generate so many faults of its own that keeping the EMCS in operation may require as much labor as it would take to monitor the facilities in the old way.

Maintaining an effective EMCS staff has proven to be difficult. The work requires individuals with exceptional talent and initiative, along with detailed knowledge of the specific system. If such a person leaves, he is not easily replaced. Another problem is that talented EMCS operators tend to be diverted to solving other problems, leaving the EMCS neglected. The EMCS staff must be protected and supported by the facility's management to remain effective.

■ Inadequate Software

Software is what makes computers powerful. As of the late 1990's, EMCS's still lacked the software needed to make them an effective tool for energy administration or for system optimization. The software available is largely limited to graphics displays and static changes to system settings.

To generate reports for managing energy costs, you need to meter each item of energy consuming equipment. In addition, you need software to process the energy consumption information and present it in a useful manner. Such software is still inadequate, probably because of inadequate market demand from facility managers who understand its value as a cost saving tool.

Energy optimization software is an even greater development challenge. This function requires the software to perform systems simulations similar to those used in energy consumption modeling programs (which are explained in Reference Note 17). To perform system optimization, the software must be able to simulate the operation of the types of systems installed in the facility, including the part-load performance of the equipment. Then, the software must be able to calculate the combination of equipment operating conditions that yields the lowest overall energy consumption, and to send the appropriate signals to the equipment.

■ Inadequate Programming Support

Experience shows that EMCS vendors' dedication to service generally does not match their zeal to sell systems. The worst aspects of service have been programming support. Comprehensive, easy-to-use software is not yet available. This makes you dependent on the EMCS vendor to reprogram the EMCS when you make systems changes, such as adding sensors or new reporting functions. This service has not been dependable.

A common situation is that the vendor's local office has one individual who is competent to program the equipment at the time the system is installed. When that individual moves to a different job, no one is left who knows how to program the system. This can make it virtually impossible to add a new control function, to adapt to changes in the controlled equipment, or to correct defects in the EMCS.

This problem will remain as long as EMCS's are programmed by local vendor offices. Programming may become less of a problem if software is developed that allows the facility operators to do more of the programming, or to take over this function completely. The ultimate goal is to have a situation similar to the emerging "plug and play" installation of accessories in personal computers, whereby the EMCS automatically integrates any new component without the need for additional programming.

■ Hardware Defects

The EMCS itself adds serious reliability problems to the facility. The processors, sensors, actuators, and communications links are complex and fragile. Historically, EMCS's have been more likely to fail than the equipment they control. Similarly, if the EMCS indicates an alarm, the problem is more likely to be in the EMCS than in the equipment being monitored. This tendency to indicate false alarms makes EMCS's almost useless for critical monitoring. For these reasons, some building code jurisdictions forbid the use of an EMCS for safety-related monitoring.

Reliability may improve dramatically in the future, but EMCS's will never be as reliable as dedicated local

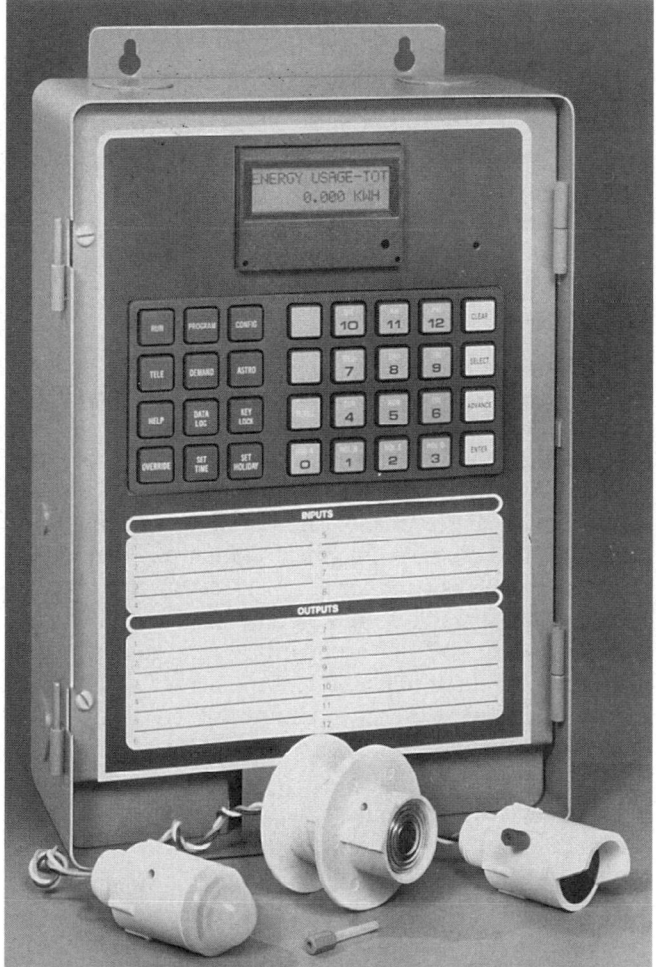

Paragon Electric Company

Fig. 6 Consider going back to basics This is a high-end multi-channel controller that evolved from a timeclock. It has about the same control capability as an early EMCS, including demand limiting, outside air temperature reset, and daylight sensing. It can log performance data. It requires no special programming skill. It controls a limited number of items, so it is installed close to the equipment, limiting communication problems. One or more units like this may be all the central control you need.

controllers. The standardization provided by the use of personal computers improves reliability in the central computer and its peripherals. Custom components, especially local panels and signal converters on equipment, will continue to be a problem. Experience shows that new electronic components tend to be unreliable for the first few years of manufacture. The large number of sensors and actuators used in EMCS's will continue to contribute a certain level of defects. Each new edition of software will have bugs. Poorly protected sensors and wiring will continue to be damaged.

■ Rapid Obsolescence

EMCS's suffer from the rapid obsolescence that is characteristic of all evolving computer equipment. The effects are short service life, bugs that never get fixed before the next model appears, and inability to obtain maintenance and programming support. Continually improving performance and declining prices make it uneconomical for the manufacturer to support old models. Furthermore, the manufacturer wants to keep selling you newer equipment.

Expect to upgrade major components of the system at relatively short intervals, and perhaps to replace the entire system. EMCS models will probably continue to have short service lives well into the 21st century, especially if standardization of components and signals does not occur soon.

■ Lack of Standardization

Until the mid-1990's, manufacturers of EMCS's tried to keep their systems proprietary, so that the purchaser had to remain married to the manufacturer for programming, repair, and upgrading. The signals produced by sensors and the signal requirements of controlled devices have long been standardized in HVAC and process applications. However, the system components and the digital signals used with EMCS's themselves are still not compatible on a general basis. The components of one EMCS typically do not produce signals usable by the components of another. Even where standardized signals are used, components may not be compatible physically, meaning that one manufacturer's components will not plug into another manufacturer's sockets.

In the past, this situation made it desirable to purchase an EMCS from one of the major manu-facturers, in the hope that the manufacturer will survive to support the system. However, any manufacturer may abandon support for a system as it becomes obsolete.

In 1996, the American Society of Heating, Refrigerating, and Air Conditioning Engineers (ASHRAE) published an enormous document, Standard 135-1995, that specifies communications standards between building control components. The full name of the Standard is "A Communications Protocol for Automation and Control Networks." The common name

of the Standard is "BACnet," which is an ASHRAE trade mark. BACnet is intended specifically to solve the problem of compatibility among digital control components, especially in EMCS's.

The problem of incompatibility has long created pressure from users for standardization. However, early attempts at defining standards for EMCS's failed. Industry groups have often been successful in establishing standards, but not for an entire complex and evolving system. In the arena of personal computers, standardization arose primarily as a result of one manufacturer gaining a dominant sales position with a particular item of equipment. The characteristics of that manufacturer's equipment become a *de facto* standard, especially if the manufacturer takes a leadership role in establishing an "open architecture," that is, a set of characteristics that can be copied by others.

At the present time, a pattern similar to this appears to be evolving around a control technology called LonWorks, which is a trademark of the Echelon Corporation. LonWorks, introduced in 1991, is a distributed control system, which means that it does not require a centralized controller or computer. In 1994, 36 companies formed the LonMark Interoperability Association to develop design standards based on LonWorks. The Association certifies that particular items of control equipment conform to the LonMark requirements.

BACnet is oriented toward centralized control with high data transfer rates. LonMark is oriented toward localized control with lower data transfer rates. Some systems use both BACnet and LonMark. In these configurations, the BACnet system serves as a high-speed communication system and LonMark controls equipment locally. As of the late 1990's, it is too early to tell how either system, or a possible combination, will fare.

BACnet and LonMark illustrate how two different approaches to standardization work in a free market. BACnet, as a "standard," is established by consensus within an industry. Standards are immune from proprietary limitations that keep competing manufacturers from entering the field, but the consensus process that guides them is slow and is subject to resistance by individual members of the industry. In contrast, LonMark is an "open architecture" that is the product of a private entity, so it can evolve rapidly. However, open architecture, at least in principle, favors the company that produces it.

If any protocol becomes a successful standard, it is likely to make existing EMCS's obsolete. Any manufacturer can produce interface devices (called "gateways") to connect older equipment to a network based on BACnet or LonMark, but the cost and uncertainty of retrofitting older EMCS's may doom this prospect.

Control Signal Discriminators
Using Energy Management Computer Systems
Minimize the Number of Signals to be Polled

A common control function in optimizing the efficiency of a system is selecting the highest or lowest signal from a number of inputs. There is no standard name for this function, so we will call it "signal polling." This Note briefly explains the methods used to perform this control function.

A typical application requiring signal polling is chilled air temperature reset. Signals from all the thermostats served by an air handling system are routed to a control device that picks out the signal calling for the greatest cooling load. This signal is used by the control system to adjust the supply air temperature of the air handling unit.

Signal polling can be done by standard control components designed for this purpose. There is no standard name for these components, so we will call them "signal discriminators." Figure 1 is a typical unit. Signal polling can also be done with an energy management control system.

Control Signal Discriminators

A signal discriminator is a standard control component that senses the signals coming from a number of sources and selects the highest or lowest, as appropriate. This device is given many names, depending on the manufacturer. Variations of this

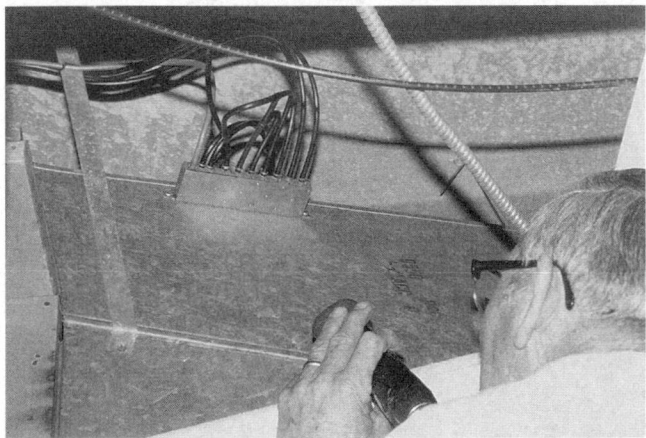

WESINC

Fig. 1 Pneumatic signal discriminator Attached to a piece of duct, this small device accepts eight pneumatic input signals. It sends out the highest and the lowest of the input pressures.

component are available for pneumatic, electrical analog, and electronic digital control systems.

Each discriminator typically accepts from four to eight inputs. Any number of signal sources can be sampled by cascading discriminators together. For example, if four-input units are being used, the outputs of four units can be fed to a fifth, so that the highest or lowest signal from sixteen sampling points can be selected. This method can be extended indefinitely.

Discriminators themselves are inexpensive. The main cost of signal polling is routing wiring or pneumatic tubing from dispersed locations to a central point. Labor is the largest cost in a discriminator system.

Discriminator systems are somewhat fragile because they require an extended network of pneumatic or electrical lines that are vulnerable to damage. Failure of any one of the sensed elements (e.g., a thermostat), or of any discriminator, or of any connection can disable the entire system. For example, if there are several pneumatic signal lines leading from thermostats to a discriminator, and there is a leak in one of the lines, the false low pressure signal in the leaky line may be selected as the controlling signal.

Using Energy Management Computer Systems

An energy management computer system can perform the same function as a discriminator network, if all the necessary signal sources are connected to the computer. A computer system is subject to the same requirements and weaknesses as a discriminator network that is made from discrete components, as well as to the weaknesses that are characteristic of computer systems. See Reference Note 13, Energy Management Control Systems, about these systems.

Minimize the Number of Signals to be Polled

You can reduce the number of signal discriminators or signal connections if the maximum or minimum signal is always produced by one of a small number of sources. For example, if you use space temperature polling to start a heating plant, exploit the fact that the lowest temperatures always occurs in certain spaces on the north side of the building. For this application, there is no need to connect the rest of the thermostats in the building to the polling network.

Infrared Thermal Scanning

Capabilities and Limitations
What a Thermal Scanner Sees
A Quick Tour of the Infrared Neighborhood
How Thermal Scanners Work
Wait for Good Survey Conditions

How to Interpret Thermal Surveys
- Estimating Conductive Heat Loss
- Estimating Heat Loss Caused by Air Flow
Where to Get Scanning Equipment or Services

Thermal scanning equipment is used to detect heat radiation emitted by surfaces. For energy conservation, a thermal scanner is a powerful tool for quickly finding and interpreting heat loss through walls, roofs, equipment insulation, etc. This Note explains the basics, and helps you to decide when to use thermal scanning.

Capabilities and Limitations

Compared to other methods of measuring or estimating heat loss, thermal scanners have several unique capabilities:

- they can read temperatures from virtually any distance, even from a flying aircraft
- the temperature readings are practically instantaneous
- their sensitivity to temperature differences is ample for most energy survey work
- imaging-type scanners can provide a photographic depiction of the surface temperatures.

However, thermal scanning is not a casual activity, and it has significant limitations. These are its main drawbacks:

- the need for skill and experience to translate a surface temperature survey into reasonably accurate estimates of heat loss
- the need to survey at night and in cold weather when surveying for low-temperature heat losses (for example, building insulation defects)
- high cost for imaging-type scanners, although the equipment can be rented
- awkwardness in handling the equipment, although it is becoming much smaller and much lighter.

For a direct, visual indication of heat loss over a large area, there is no practical alternative to thermal scanning equipment. If you want to survey heat loss over any large area, this is the method you must use.

Contact thermometers can also provide surface temperature measurements, but only on a point basis. Furthermore, they are subject to error, measurements are too slow for survey work, and it may be difficult or dangerous to reach the surfaces being measured.

What a Thermal Scanner Sees

Any solid object is made of atoms. The atoms have kinetic energy but no place to go, so they jiggle around nervously. This nervous energy collectively is called the "heat" energy of the solid. The atoms are packed against each other, so they keep each other agitated with their jiggling. This continually excites the atoms' electrons to higher energy states. The electrons continually get rid of this excess energy by emitting radiation. Most of this radiation is absorbed back into the solid, but some of it escapes from the surface of the solid. This escaping radiation is called "heat" radiation. (Note that the word "heat" is used in two entirely different ways.) Thermal scanners see this radiated energy.

The "temperature" of an object is a measure of the average amount of kinetic energy in the object's atoms. Temperature has two major effects on the heat radiation that an object emits. One is obvious: hotter objects emit more radiation than cooler objects. This effect is very strong. In fact, the rate of heat radiation is proportional to the fourth power of the absolute temperature.

The second effect is that the object's temperature determines the wavelengths (or frequencies) of the radiation. The jiggling of the atoms in the object is chaotic, so the radiation is emitted in a broad pattern of wavelengths. This pattern is a lopsided bell curve, with the short wavelengths dropping off abruptly, and the long wavelengths tailing off over a long range. The peak of the curve shifts toward shorter wavelengths as the object gets hotter. More precisely, the frequency of strongest radiation is directly proportional to the temperature, or the wavelength is inversely proportional to the temperature. (This relationship lets astronomers know the temperatures of stars located on the other side of the universe.)

Using this relationship, thermal scanners calculate temperature by determining the wavelength at the peak of the curve. Thermal scanners used for energy conservation work can discriminate temperature differences of a fraction of one degree (Celsius or Fahrenheit, take your pick).

A Quick Tour of the Infrared Neighborhood

Like light, heat radiation is electromagnetic radiation. The two forms of radiation differ only in wavelength. Heat radiation is next to light in the electromagnetic spectrum, in the range of wavelengths called "infrared." The name means "below red," in terms of frequency. For this reason, thermal scanners are commonly called infrared scanners.

Objects at room temperature radiate heat at wavelengths that are about 20 times longer than the wavelengths of visible light. If an object is very hot, it emits enough radiation at visible wavelengths for you to see it. "Incandescent" light is heat radiation that has wavelengths short enough to see. (For more about incandescent light, see Reference Note 54, Incandescent Lighting.)

You can easily observe the transition from the infrared spectrum into the visible spectrum. Go into a kitchen with an electric stove, turn on a burner and turn off the lights, and watch the increasing light emission as the burner gets hotter. You will also see that the color changes as the burner gets hotter, showing that the average wavelength is getting shorter. If you hold your hand at a fixed distance from the burner, you can sense that the rate of heat emission increases radically as the burner gets hotter.

A thermal scanner discriminates temperature in the same way that the human eye discriminates color, by sensing the differences in the wavelengths. The human eye works in the region of 0.3 to 0.7 microns (millionths of a meter). Most imaging-type thermal scanners work in one of two wavelength bands, 3-to-5 microns or 8-to-12 microns. These wavelength bands are determined by the available detector and lens materials. Also, for scanning at long range, they are determined by the transmission spectrum of the atmosphere. Water vapor in the air strongly absorbs some infrared wavelengths.

For surveying buildings to find heat leakage, the 8-to-12 micron band is preferable, because objects near ambient temperature emit more heat in this band. Radiation from objects at ambient temperature effectively ceases at wavelengths below 3 microns. The 3-to-5 micron band is where objects at temperatures of 300°C to 700°C (approximately 600°F to 1,300°F) emit most of their energy. It is important to know this when selecting a scanner. Most scanners are sold for industrial monitoring applications, where the temperatures involved are much higher than ambient.

Incidentally, you cannot do thermal scanning by going to your local photo supplier and buying "infrared" film to use in a regular camera. This film senses only so-called "near infrared," which means, near the visible spectrum in wavelength. This infrared radiation is reflected light from the sun, like visible light, and it can pass through the glass lenses of conventional camera lenses. In contrast, the heat radiation sensed by scanners is classified as "middle" or "far" infrared, and this radiation cannot pass through ordinary lenses.

How Thermal Scanners Work

There are two broad types of thermal scanners. The simplest type is called a "pyrometer." You point it at a surface, the heat energy is focussed on a detector, and the wavelength distribution is analyzed to determine the temperature. The process takes about one second. The field of view of pyrometers is narrow, because the temperature reading is averaged from the characteristics of all the radiation in the field of view. Pyrometers are small and light. They are usually housed in a pistol-grip enclosure that looks like an automotive timing light. They generally cost less than $1,000.

When using a pyrometer, be aware of the field of view that it is scanning. In more expensive units, the field of view is variable, and the unit is equipped with an optical sight so you can see the field that is covered. This is important for measuring the temperature of a small object against a background of different temperature, such as an elevated steam line.

The other type of thermal scanning equipment is an imaging, or camera-type, device. Modern handheld units are similar in appearance and size to a video camera. In the most expensive units, a lens system focuses the heat radiation on a plate consisting of many separate sensors, or pixels. Temperature and intensity are calculated separately for each sensor, producing a photographic image of the scene. Less expensive units achieve the same effect by scanning across a small array of sensors.

Older scanners produce a "black and white" image, in which surfaces with higher temperatures appear brighter. Most modern units produce the image in color. Color is used to represent temperature, and brightness is used to represent the intensity of the radiation.

Imaging scanners allow rapid survey of large areas, and make it possible to detect problems that would not be apparent with a pyrometer. For example, an infrared camera can reveal waterlogged insulation on buried steam pipe from a subtle pattern of heating on the surface of the soil. Most camera-type units allow images to be preserved on film or using a video recorder. More expense units can store images in digital form, allowing the images to be printed on an ordinary computer printer. Some vendors offer software that analyzes the digital data, although the direct photographic representation of the data is the most useful for energy conservation work.

A basic advantage of camera-type scanners over pyrometers is that they can depict differences in intensity as well as temperature. Because the amount of heat energy emitted by a surface rises rapidly as the surface temperature increases, even small differences in surface temperature result in easily perceptible differences in the amount of radiation emitted. Camera-type units can discriminate between much smaller temperature differences than pyrometers because each pixel in the image covers a very small angle of view, eliminating errors due to variations of temperature on a surface.

Imaging scanners have unusual construction features. The lenses must be made of exotic materials to pass heat radiation. The interior of the camera is cooled to a very low temperature, so that heat emitted by the structure of the camera itself does not interfere with the image. Older units have an interior liner that must be continually refilled with a liquified gas coolant, typically nitrogen or argon. Contemporary units have miraculously small cooling units that require no external input, except for a small amount of power that can be provided by an easily portable battery pack.

Prices for imaging scanners start above $20,000, and range to several times that amount.

Wait for Good Survey Conditions

To find heat loss through the building envelope, survey the building during cold weather. This makes the areas of heat loss show up with the greatest contrast.

To find conduction losses in cooling equipment, such as chilled water piping, make your survey during the warmest weather.

You can make heat loss surveys of hot equipment, such as steam lines, during any season. Be sure to keep all the heating systems in operation during the survey.

Regardless of the season, make outdoor surveys at night, starting several hours after sunset. This allows the exterior surfaces to cool off, and prevents false indications of heat loss from surfaces that were actually heated by the sun. Make the survey when there is little wind, as wind tends to obliterate variations in surface temperature.

If you survey for envelope heat loss from the inside of the building, start several hours after sunset. Solar heat gain on external surfaces may reduce internal cold spots, making them harder to find. Keep the HVAC equipment running when you do an indoor survey. This increases the temperature differential, and helps to reveal air leakage inside the envelope structure that may be induced by the operation of the HVAC equipment. It also helps to make indoor surveys during cold, windy conditions, with the wind coming from the prevailing winter direction.

You can survey indoor equipment for heat loss during the daytime, provided that it is not heated by direct sunlight through windows or skylights.

How to Interpret Thermal Surveys

Thermal scanners sense heat radiation, but heat radiation itself accounts for a relatively small fraction of heat loss at normal ambient temperatures. Most heat loss is by conduction, by air leakage, or by excessive ventilation. Thermal scanning can help to find all these types of heat loss, but it cannot directly indicate the rate

of heat loss due to each. The person doing the thermal survey must interpret the infrared scan to estimate these modes of heat loss.

■ Estimating Conductive Heat Loss

Surface temperature is the key to estimating conductive heat loss, because temperature differential is what drives conductive heat loss. If an external surface is warmer than the surroundings, then heat is coming through that surface. There are two ways to estimate the heat loss rate:

• calculate the heat loss from the surface to the outside air based on the measured surface temperature, the ambient temperature, and the surface characteristics. This method, described in the *ASHRAE Handbook*, is vulnerable to wind, which lowers the surface temperature and increases the heat loss rate. If you plan to use this method of calculation, make the survey under windless conditions.

• use the scanner to find the areas that need to be analyzed. Then, visually inspect the wall, insulation, or other component to determine its construction and condition. From this, perform a conventional heat flow calculation. This method works only if you can inspect the component. For example, you may not be able to inspect the structure and insulation inside the walls of an old building.

Infrared measurements of surface temperature are affected by differences in the "emissivity" of surfaces. Emissivity is the ratio of the actual amount of heat radiation to the amount of radiation that would be emitted by an ideal ("black body") surface at the same temperature. The emissivity of most non-conducting materials is about 0.9. The emissivity of conducting surfaces ranges from 0.1 for polished metal to about 0.3 for dirty, heavily oxidized metal. Low emissivity keeps heat in the surface, which makes the surface warmer. The scanner indicates the correct surface temperature, but since less heat is being radiated, the user may be fooled into thinking that the rate of heat loss is lower than it actually is.

For example, when surveying a steam line that has an aluminum weather shield on the top surface, the scanner shows a different appearance for the shielded and unshielded surfaces, even though the heat loss is about the same for both.

Some pyrometers are dressed up as "R-value meters." They are used mainly for surveying building envelopes from the inside. To use such a device, you first set up the device by measuring the outside and inside air temperatures. From this, the device calculates an R-value by measuring the inside surface heat radiation.

This method is subject to error because of draft effects on the internal surfaces, thermal lag in the structure, and variations in emissivity. To reduce this error, "R-value" or "heat loss" meters usually have a manual setting for the emissivity. You have to estimate the emissivity of the surface you are scanning, and adjust the meter accordingly.

Even an imaging scanner cannot provide reliable information about conductive heat loss if the insulated surface is hidden from view by another surface. One of the worst cases is heat loss through an attic. The only external indication is the temperature of the roof. This provides no indication of the nature of the heat leaks within the attic, and only a vague indication of their location. In such buildings, you need to make a second survey within the attic to find the locations and intensities of the heat leaks.

■ Estimating Heat Loss Caused by Air Flow

Air emits virtually no thermal radiation. Therefore, thermal scanning cannot directly detect the escape of heated air or cooled air. Loss of heat by air movement is revealed only by the warming or cooling of solid surfaces in the path of the escaping air. For example, air leakage around windows and doors during cold weather appears as localized hot spots, which might be mistaken for areas that are poorly insulated.

If you can find the point of leakage by scanning, investigate to determine the amount of air flow and the reasons for it. For example, the heated air exhausted from a building for normal ventilation may carry away much more energy than the building walls and roof lose by conduction.

To find air leakage into a building, survey from inside the building. Air leaks may follow long, indirect paths through walls and ceilings. A thermal scanner can reveal such a path by sensing the reduced inside surface temperature along the path.

Where to Get Scanning Equipment or Services

It is not economical to purchase thermal scanning equipment unless you plan to use it often in facilities that lose a large amount of energy by heat leakage. The alternatives are to hire a specialist to perform a survey of your facility, or to rent the equipment and do the survey yourself.

There are companies that specialize in thermal surveys. If you decide to hire the service, be sure to select a company that knows how to do surveys for the purpose of energy conservation. A thermal survey company specializing in different applications, such as searching for defects in electrical distribution systems or checking coal piles for spontaneous combustion, may perform a poor energy conservation survey. Some utility companies offer thermal scans as an energy conservation service for their customers.

The person who does the survey may not be the best one to calculate heat losses from the survey results. Make sure that the person who does the calculations gets good images from the survey, along with color codes and other information needed to interpret the images.

Expect to make later localized scans to follow up your initial overall survey. For example, if the first scan shows serious heat loss through the roof, you will want to make a second survey of the attic and the inside surfaces to localize the problems.

If you decide to do the survey yourself (lots of fun, especially the first few times), you can rent the equipment from a number of places. Get adequate training, which is provided by some manufacturers and others.

Measurement of Liquid, Gas, and Heat Flow

The Elements of a Flow Metering System
Common Types of Flowmeters
- Propeller Flowmeters
- Vortex Shedding Flowmeters
- Ultrasonic Flowmeters
- Pitot Tube Flowmeters
- Venturi Flowmeters
- Orifice Flowmeters
- Variable-Area Flowmeters
- Spring-Loaded Variable-Area Flowmeters

Totalizers
Heat Meters
Equipment Quality
Piping Issues
Calibration
Displays

Measuring the flow of water, steam, and other liquids and gases plays an important role in keeping systems at peak efficiency. For example, it is necessary for adjusting pump output in constant-flow systems and for controlling variable-flow systems.

Flow measurement is the most challenging part of measuring the energy content of hot water, chilled water, and steam. Measuring energy flow directly is essential in sophisticated control systems that optimize energy efficiency. It is also necessary for accurate billing of energy production.

Flow and energy measurement are important tools for monitoring system performance and for troubleshooting.

Flow and heat metering continue to be areas of difficulty. There are many types of flowmeters because no single type is entirely satisfactory in all respects. A given type of flowmeter may be best for some applications, but inappropriate for others. With all types, installation details are important. This Note reviews the basics, and helps you to select the best flow measuring device for your applications.

The Elements of a Flow Metering System

A flow metering system consists of three elements: (1) a sensor that responds to flow, (2) a device that converts the output of the sensor to a form that can be displayed or used for other purposes, and (3) a display device. A flow metering system may also include devices to compute information, such as total flow or heat flow, from the flow measurement.

Common Types of Flowmeters

The following are the most common types of devices used to measure flow of liquids and vapors for energy measurement purposes. This Note does not cover flowmeters used primarily in other applications, such as handling fluids that are caustic or of high viscosity. Some of the following types are fading out, and others are still developmental. Other types not listed here may become prominent by the time you read this. When selecting a flowmeter, search the market for the best type currently available for the application.

The main distinction between the different types of flowmeters is the type of sensor. Any type of display, control, or calculation device can be attached to any type of sensor by using the appropriate interface.

■ Propeller Flowmeters

Propeller flowmeters, sometimes called "turbine" meters, use a propeller that spins freely in the fluid stream. Flow is measured by counting the rotation rate of the propeller, typically with an inductive pickup that offers no resistance to rotation.

This type works well with liquid, steam, and air, and in general, with low-viscosity fluids that are free of large particles.

Propeller flowmeters can be accurate and reliable, they can sense low flow rates accurately, and they can provide large turndown ratios. If the propeller diameter is a large fraction of the pipe diameter, the propeller tends to average the differences in flow velocity that exist inside the pipe because of pipe wall friction. This type is less sensitive to turbulence than most other types, and it is relatively insensitive to dirt.

Propeller flowmeters have been afflicted with bearing friction problems. There appears to be no reason why this problem cannot be eliminated by protecting the bearing from fouling. Check the performance record of any unit you are considering. Larger propeller diameters produce more torque to overcome any bearing friction.

Manufacturers recommend periodic maintenance of the propeller bearings, which is done by the manufacturer. The maintenance interval depends on the design, the propeller size, and the nature of the fluid. In some units, the propeller assembly is simply replaced.

Before the introduction of electronic revolution counters, the propeller was required to drive an indicator through a gear train. This caused the propeller to run slow at low flow rates. This problem does not exist in modern units.

■ Vortex Shedding Flowmeters

When a non-streamlined object is placed in a moving stream, it create vortexes in the stream that are generated alternately from both sides of the object. As each vortex forms, it falls away from the object and continues downstream. The rate of vortex formation is related to the flow velocity.

(You can see these vortexes easily by moving a paddle through calm water. They are named "von Karman" vortexes after the aeronautical engineer who studied them.)

Vortex shedding flowmeters exploit this principle by inserting a small flat surface perpendicular to the flow in the pipe. The vortexes shed by the flat surface strike a flexible vane located behind the surface. The resulting vibration of the vane is detected either by a piezoelectric sensor or by a strain gauge.

This type of sensor can be used with low-viscosity liquids at all normal velocities, and with air and steam at higher velocities.

Vortex sensors are sensitive to turbulence in the fluid stream, so installation is critical. Manufacturers typically require the sensor to be installed in a straight section of pipe, typically ten diameters in length upstream and five diameters downstream.

The turndown ratio is high in liquids, which shed vortexes down to low velocities. With steam and other gases, the turndown ratio is narrower because the

vortexes are too weak at low velocities to be sensed reliably.

Manufacturers recommend replacing the sensors periodically, typically every year or two, because the vibration sensing components wear out. This type of sensor is still new, so check with other users of the models you are considering.

■ Ultrasonic Flowmeters

Ultrasonic flowmeters are used with low-viscosity liquids. They sense the Doppler shift of particulate material carried in the stream, or they may sense turbulence carried along in the liquid. The technology is new, so check the current status of this type of flowmeter. In principle, the device should be accurate at any flow rate, providing an effectively unlimited turndown ratio. You should question the ability of ultrasonic flowmeters to work with clean water.

The major attraction of ultrasonic flowmeters is that they can be used from outside the pipe, needing no taps or other devices to be installed. This eliminates a major cost factor in flow metering for one-time measurements, such as are needed for pump throttling and pump impeller trimming.

■ Pitot Tube Flowmeters

The sensor of a pitot flowmeter is the simplest of devices, consisting of nothing more than a tube mounted in a holder. Figure 1 shows a typical unit for use in air handling systems.

The tube faces into the flow. Flowing gas or liquid has inertia. When the fluid is stopped inside the tube, the inertia is converted into pressure, which is called "velocity pressure." The velocity pressure is proportional to the square of the velocity. A device called a "square root extractor" is used to derive the flow rate from the pressure signal.

Pitot tubes provide strong output pressures when they are used with liquids, but the output pressures are weak when measuring air or steam flow, especially at the lowest velocities that may occur. This invites error in the pressure transducer that converts the pressure to an output signal. (Pitot tubes are the primary type of sensor for air speed in aircraft. However, they function reliably only at the higher speeds needed to keep the aircraft flying.)

Liquid flow in a pipe is faster in the center than at the perimeter because of friction at the pipe wall. The resulting error in flow rate depends on the diameter and roughness of the pipe, and on the properties of the liquid. You can get pitot systems that have adjustments for these factors. Pitot tubes are sensitive to turbulence, making installation critical.

The accuracy of pitot tubes is not very sensitive to dirt in the tube, because there is no flow through the tube. Of course, the tube can be plugged by dirt, making the unit inoperative.

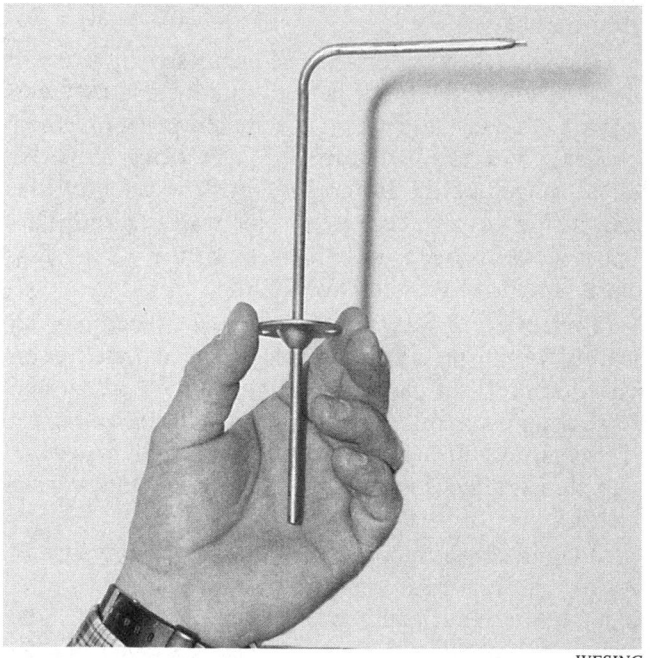

WESINC

Fig. 1 Pitot tube This pitot tube is easily installed through a small hole in a duct. Additional hardware is needed to convert the pressure to a control signal. The pressure generated in the tube is very small at low air flow rates, limiting accuracy.

Pitot tube systems are proven by long experience. Consider them for applications where flow rates are always fairly high. Do not use them in applications, such as variable-flow pumping systems, where velocities can fall into the range of poor accuracy.

■ Venturi Flowmeters

A venturi is a smooth contraction in a section of pipe. The fluid speeds up to get through the contracted area. The faster fluid has higher kinetic energy, so its pressure energy drops. The flow rate is determined by sensing the pressure difference between the throat of the venturi and the wide open pipe.

(The pressure drops in a venturi because of conservation of energy. The energy of a fluid stream consists of its kinetic energy and its internal pressure. If one of these increases at a particular point in the system, the other must fall to keep the fluid's total energy constant.)

Like pitot tubes, venturis are simple, they have no moving parts, and they are insensitive to dirt. The venturi restriction is smooth, so it dissipates very little energy. Venturis are proven by long experience.

Venturis are used with liquids, gases, and vapors. Unfortunately, venturis require relatively high velocities to provide a signal of adequate strength. For this reason, they have poor turndown ratio. This makes them undesirable in typical energy measurement applications.

■ Orifice Flowmeters

Flow can be measured by sensing the small pressure differential that is created between the upstream and downstream sides of an orifice in a plate that is inserted inside the pipe. The sensing element itself is nothing more than a hole. The cost and complexity come from the additional apparatus that is needed to convert the small pressure difference into a useful indication of flow rate. Figure 2 shows a typical orifice metering system.

Orifice flowmeters are used with liquids and vapors. They have the same advantages and disadvantages mentioned previously for pitot tubes and venturis. Of these, the main disadvantage is a limited turndown range. They have the additional disadvantage that the orifice dissipates pump energy in the form of turbulence. The energy loss is reduced somewhat if a specialized nozzle is used in place of a simple orifice.

Orifice plates can be installed in very little space. You may be able to slip an orifice plate into an existing flange joint. However, turbulence in the fluid makes an orifice plate inaccurate, so you have to install the orifice in a straight section of pipe. Orifice plates cannot be calibrated accurately in the field unless the installation conditions are almost ideal. There are few applications where orifice plates are the best choice. Better methods are making them obsolete.

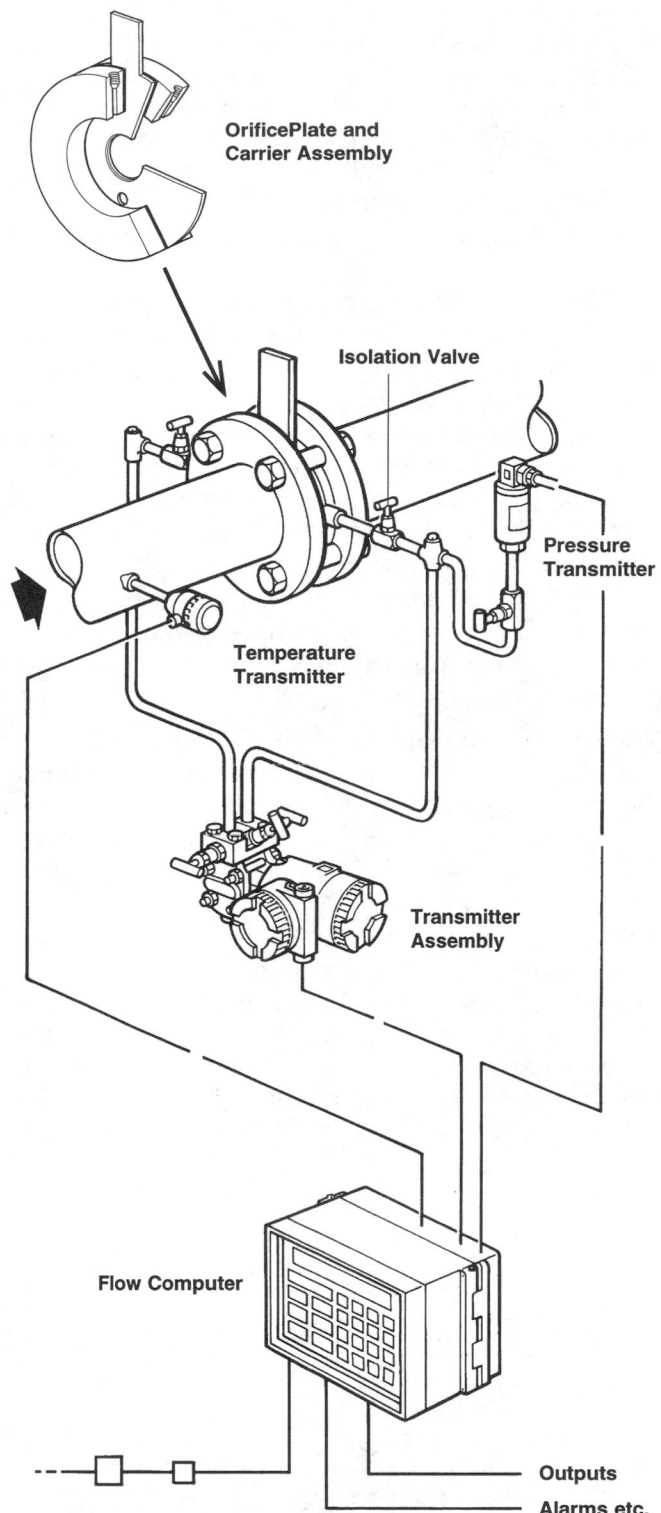

Spirax Sarco, Inc.

Fig. 2 Orifice flow metering system The sensor itself is simply a hole. The hole is too big to be vulnerable to fouling. The main weakness of this method is poor accuracy at low flow rates. The orifice itself requires only a minimal length of pipe, but the pipe must be straight upstream and downstream to ensure accuracy.

■ Variable-Area Flowmeters

A variable-area flowmeter is a simple device that measures flow with reasonable accuracy. (This type is sometimes called a "rotameter," although no rotation is involved.) It consists of a loose plug inside a vertical, slightly tapered, transparent body. Liquid flows upward through the body, lifting the plug. Because the body is tapered, greater flow is required to lift the plug higher. Flow can be read directly by marking the body in units of flow rate.

Variable-area flowmeters are used with clean liquids and gases. For energy measurement, they are generally limited to liquids. With gases, they require high velocity and they tend to stick.

Variable-area flowmeters are made in small sizes. They must be installed in a vertical orientation. Installation is not critical. Only a minimal amount of energy is dissipated in the device. They are relatively insensitive to turbulence.

The basic variable-area flowmeter is read directly by sight. With this type, no interface or other display is required. Some manufacturers may offer models that produce a signal for remote reading.

■ Spring-Loaded Variable-Area Flowmeters

There is a hybrid type of flowmeter that combines the characteristics of an orifice meter and a variable-area meter. In this type, a spring-loaded conical plug fits into a large orifice. Figure 3 is a cutaway of a typical unit.

Flow moves the plug out of the orifice. Pressure taps at each end of the assembly sense the pressure differential across the orifice, which is used to calculate flow rate. The relationship between the pressure differential and the flow rate is linear. This gives the unit a high turndown ratio. (In a simple orifice, the pressure differential is proportional to the square of the flow rate.)

This type of flowmeter can be installed in any position. It is adaptable to any flow rate. It can be used with liquids and steam. It is relative insensitive to turbulence.

The main disadvantage of this flowmeter is that it dissipates a significant amount of energy by creating resistance to flow in the entire stream.

Totalizers

A totalizer is a device that uses the flow rate signal to calculate a running tally of the amount of substance that has passed through the flowmeter. It is commonly used as part of a heat metering system, for example.

Totalizers eliminate the need for separate consumption meters, such as gas, water, and fuel meters. Consumption meters measure the amount of material directly, usually by passing it through a positive displacement device.

Heat Meters

Heat meters are used to measure the amount of energy carried by a fluid, such as chilled water. (Heat meters are commonly called "BTU meters" in countries where that archaic unit is still used.) A heat meter is a combination of devices that:

- measure the flow rate
- measure the temperature difference between the supply and return
- calculate the rate of heat delivery by multiplying the temperature difference times the flow rate
- calculate the total amount of heat delivered.

The calculations are done by an electronic module. Any kind of flowmeter and temperature sensor can be used with the module, provided that they produce compatible signals. Figure 4 shows a basic system.

Equipment Quality

Instrument quality varies widely. In the HVAC industry, cheap, inaccurate instruments are the norm. There seems to be an assumption that instruments installed on HVAC equipment will not be used anyhow. To get adequate stability and ruggedness, consider so-called "industrial grade" controls. You may have to go to a different set of suppliers to get these.

Piping Issues

Accuracy requires smooth flow at the flow sensor. For this reason, the manufacturer usually specifies that the sensor should be installed in a long run of straight pipe. Typically, the manufacturer specifies a certain number of pipe diameters of straight pipe ahead of the sensor, and a smaller number of pipe diameters downstream.

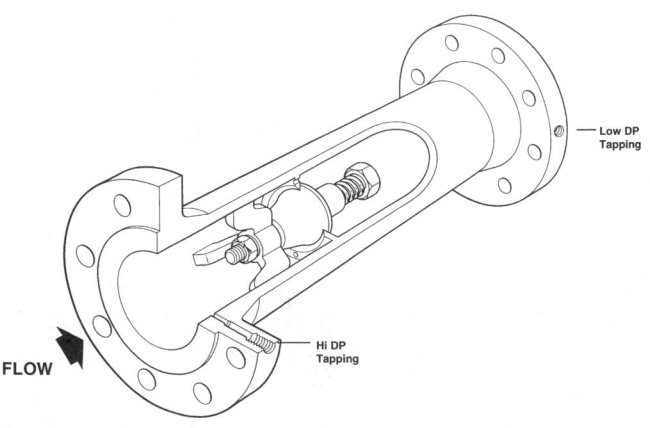

Spirax Sarco, Inc.

Fig. 3 Spring-loaded variable-area flowmeter This device is compact, simple, and relatively insensitive to turbulence. It provides a linear output signal, which improves the range of accuracy. Its main drawback is significant resistance to flow, which dissipates energy.

Some types of sensors are much more sensitive to irregular flow than others. Orifice meters and pitot tubes require smooth flow. Propeller, ultrasonic, and variable-area sensors are most tolerant of turbulence, so they can be installed in sections of pipe that are less than ideal. Venturi and vortex shedding sensors are somewhere between these extremes.

Calibration

Calibration is the Achilles heel of accuracy in flow measurement. Installations in the field provide no opportunity for comparison with standards. Each type of sensor has its own calibration instructions, but these may be difficult or impractical. The installing contractor may not understand the calibration techniques, or may not bother with them. Accuracy of calibration depends on being able to fulfill all the installation conditions properly, especially on providing the specified lengths of straight pipe at the inlet and discharge.

The solution to this problem is to acquire a type of flowmeter that does not require field calibration. The newer types of flowmeters, including propeller, vortex shedding, and ultrasonic types, do not require field calibration, but they must be matched to the dimensions of the pipe in which they are used.

Displays

Select a display that is appropriate for the application. A good digital display is usually easier to read than an analog display. This is important for displays that are read often. On the other hand, any accurate display is satisfactory for one-time readings, as in adjusting pump throttling.

WESINC

Fig. 5 Chart recorders Charts display important aspects of system operation in a very obvious way. These recorders were installed to show system loads. They are also revealing serious oscillations in system controls. The old pen-and-ink versions are obsolete. Now, digital data logging and computer report processing lets you print out charts on a printer with perfect precision. And, no more mess with cleaning ink pens.

WESINC

Fig. 4 Chilled water energy meter The supply and return pipes are on the right. Each has a thermometer, and one has a flowmeter. The box at left calculates the heat flow rate from the chilled water flow and temperature signals, and displays it on the chart recorder. This is a pre-digital installation that allows considerable error in flow sensing, calculation, display, and reading the display. A digital system that is based on a reliable flowmeter would give much more accurate results, and in many more useful forms.

Any measurement system that uses analog components must be zeroed and calibrated periodically. This includes digital displays that are driven by analog-to-digital converters.

Digital displays may or may not lend accuracy to the measurement. For example, propeller and vortex shedding flowmeters have digital inputs from the sensors, so a digital display preserves the accuracy of the input. On the other hand, a digital display of an initially analog signal, such as that from orifice meters or pitot tubes, adds no accuracy. In fact, it may produce a false sense of accuracy.

Charts are a very effective method of displaying system information of many kinds. Charts of loads and usage patterns are an important tool for routine energy management. Charts can also indicate trouble in the system. For example, a wandering display may indicate an unreliable flow sensor. An oscillating readout may indicate instability in control valves or in variable-flow pump controls. Figure 5 shows a real example.

Old-style chart recorders are now obsolete. No more ink pens! You get much better results by preparing charts digitally. Any good energy management computer system can do this. (See Reference Note 13 for details.) You can also get good charts from digital data logging systems, without the computer hassles. Digital data reporting gives you a wide choice of formats, the ability to perform calculations, and perfect accuracy in computation and display. You do not need special display equipment, only an ordinary computer printer.

Use a totalizer to calculate consumption. In the past, estimates of consumption were often made by visually scanning chart recordings, or from hourly readings. These practices are inaccurate and misleading.

Energy Analysis Computer Programs

The Structure of Energy Analysis Programs
- Space Loads
- Systems Loads
- Central Plant Loads
- Economic Calculations

The Difference Between Load Calculations and Energy Analysis

Typical Program Output

How to Use the Program
- Input Data Libraries
- Sensitivity Analysis

How Accurate are Computer Simulations?
- Range of Simulation Capabilities
- Accuracy of Component Simulation

- Number of Calculation Intervals
- Weather Data
- Number of Zones
- Equipment Defects and Unpredictable Behavior
- Program Errors
- Input Errors and Misunderstandings

How to Catch Errors
- Read the Output in Detail
- Check with Pencil Calculations
- Use Sensitivity Analysis
- Analyze the Relative Performance of Dummy Configurations
- Compare to Measured Energy Consumption
- Check the Computer Code

How to Select Your Energy Analysis Program

Energy analysis programs are used to analyze the energy efficiency of energy systems, building design features, entire buildings, and in some cases, whole facilities. Examples of these programs are DOE2, TRACE, HAP, BLAST, and AXCESS.

These complex, sophisticated programs require a great deal of input effort by the user. If used properly, they provide the best possible accuracy for making design and investment decisions. However, computer programs can also give a false veneer of authority to bad analysis. This Note gives you an overview of energy analysis programs, and tells you how to get results you can trust.

Originally, energy analysis programs had to be run on mainframe computers. Today, all the major energy analysis programs have been adapted to run on personal computers. You need a fast personal computer equipped with math coprocessing to achieve reasonably short calculation times.

The Structure of Energy Analysis Programs

All the sophisticated energy analysis programs perform four basic groups of calculations, which are described here. Different programs link these calculations in various ways. It is worth your effort to understand the general flow of calculations in the program you are using.

The facility's energy requirements change continuously, so the sequence of calculations is repeated many times to simulate a full yearly cycle of operation under different conditions of weather, occupancy, etc. At the end, the results of all the repeated calculations are summed to produce the total yearly energy consumption and costs.

For input and calculation purposes, the building is divided into "zones." Each zone is an area of the building that has particular load characteristics, and is served by specific types of conditioning, lighting, and other energy-consuming systems. The program does most of its calculations separately for each zone.

■ Space Loads

The program starts by calculating the end-use energy requirements of the spaces. It calculates heating and cooling loads in the usual way, by adding conduction gains and losses, solar gain, heat gained or lost from outside air, humidification or dehumidification, and internal heat gains.

Weather-related data needed for these calculations are usually taken automatically from a weather data library. Solar inputs are calculated by the computer based on the geographic location and building envelope characteristics, as well as weather data. The program requires manual input of the physical characteristics of the structure, the sources of internal heat gain (people and equipment), and temperature settings.

The program requires manual input of non-weather-related loads. These include lighting, electrical equipment (computers, etc.), domestic water heating, specialized process loads, etc. Typically, the user inputs the peak load of each type, along with an hourly schedule of the percentages of peak load.

Most energy analysis programs provide the option of performing these load calculations alone, without the need to perform the calculations described next. For example, in designing a new building, the space load calculations allow the designer to refine the building exterior to reduce solar cooling loads, or to find the most economical amount of wall insulation.

■ Systems Loads

The "systems" loads include the space loads, calculated previously, plus the extra energy needed to run the conditioning equipment, or "systems." All or most of this extra energy is for fans, in typical systems. The "systems" include air handling systems, fan-coil units, air conditioners, radiators, etc., but usually not the central plant equipment.

The program also calculates energy losses that may be involved in the operation of the systems. The worst of these losses is reheat. For example, if a particular zone is conditioned by an air handling system with reheating terminal units, the program calculates the fan load, the reheat energy, and the extra cooling energy needed to offset the reheat.

This part of the program requires manual input to describe the efficiency characteristics of the equipment. The equipment operating schedules may also be entered manually, to account for schedules of occupancy. In some cases, the systems' characteristics can be selected from the program's equipment data library.

The program will pick out the maximum energy consumption that occurs during the yearly cycle, for each system. The designer uses these maximum loads to determine the sizing of the conditioning equipment.

■ Central Plant Loads

Up to this point, the program has calculated all loads on a zone-by-zone basis. Next, the program adds all the zone loads as a total load on the central plant equipment, which includes boilers, chillers, electric generators, cogeneration plants, thermal storage systems, solar collectors, etc. The program further adds the energy consumption of central plant auxiliary equipment, such as hydronic system pumps and cooling tower fans.

For this part of the calculation, the program requires manual input to describe the plant equipment and its operating schedules. In some cases, plant equipment characteristics can be selected from the program's equipment data library.

This part of the program yields the "bottom line" energy input to the facility as a whole. This includes energy that is used directly, such as electricity consumption for lighting and receptacle loads.

■ Economic Calculations

All the programs can provide energy cost estimates, as well as raw energy consumptions. To do this, the program requires manual input of energy costs and rate schedules.

Some programs can also calculate the life-cycle costs of alternatives. This requires manual input of equipment and construction costs, at least for the features or equipment being compared. Most programs can incorporate desired interest rates, inflation factors, and other economic variables in the calculation.

The Difference Between Load Calculations and Energy Analysis

To avoid confusion, we will clarify the distinction between energy analysis programs and load calculations. A load calculation is used primarily to select equipment size based on maximum-load conditions. It includes only the space loads, as discussed above, and it is done only for the single instance of peak load conditions.

For example, when selecting an air conditioning system for a house, a contractor makes a load calculation to estimate the peak cooling requirement. This is done by calculating the individual load components, and then adding them. For cooling, the load components are solar gain, conduction gain through the walls and roof, the heat gained by air leakage, and dehumidification. The contractor typically uses a one-page worksheet to perform this calculation.

Basic peak-load calculations are simple. They can be done manually, and they are adequate for many applications. However, manual calculations are not very accurate. One cause of error is that the components of load do not all peak at the same time. Another limitation is inability to deal with thermal storage effects. The degree of detail is limited. For example, it is not practical to include the precise orientation of the surfaces in a manual calculation.

As a result, a number of computerized load calculation programs are now available. They provide greater accuracy, they eliminate the drudgery of arithmetic, and they provide a checklist to make sure that all components of load are included. There is a range of complexity among computerized programs. Some are relatively simple. The most detailed load calculations are provided by using the space loads portion of an energy analysis program, as discussed previously.

(Note that the capacity of heating and cooling equipment may be much greater than the peak conditioning load caused by weather conditions. Heating and cooling equipment is usually sized to provide a reasonably short warm-up or cool-down interval after starting. For this reason, many designers and contractors do not use load calculations. Instead, they select equipment capacity using simple rules of thumb that are based on the size and mass of the building.)

Typical Program Output

All the major programs offer the same types of output information. In order of calculation, the available output includes:

- *the input data.* The report usually repeats the input data for ease of review. This includes data drawn from the program's data libraries, which are discussed below. For example, the output may indicate the outside air temperature and humidity that were assumed for each hour.

- *building loads.* Loads are divided into heating, cooling, lighting, process, etc. Some programs may report the components of these loads. For example, cooling load may be divided into solar gain, conduction load, internal heat gain, and latent load. The loads for individual hours may be displayed. Most programs report the time of occurrence of each of the peak loads.

- *equipment sizing data.* You select equipment capacities by using the calculations of peak equipment load. For example, the program may report the peak air flow of air handling units, the peak steam flow from boilers, the peak energy input to individual chillers, etc.

- *energy consumption.* This can be reported in many ways, including totals by energy type; by types of loads (e.g., heating, cooling, lighting, process loads, etc.); by different intervals, including hourly, by day type (i.e., weekday, weekend, holiday), monthly, and yearly; and by system (e.g., chiller plant, air handling units, etc.).

- *energy costs.* These are derived directly from the consumption calculation, making corrections for variations in price at different times. The costs can be reported separately by energy type, e.g., electricity, fuel oil, natural gas, etc.

- *life-cycle cost.* This is the total cost for energy over a facility's life cycle, or other long period of time. This output includes the effects of changes in fuel price, inflation, and interest rates.

The major programs allow you to specify the type of output information you want, and the degree of detail. The program usually allows you to get output in the same subdivisions as input. The loads can usually be displayed by individual zones. Similarly, the characteristics of each system can be displayed separately. All the major programs can report in a variety of tabular and graphical formats.

How to Use the Program

You use an energy analysis program in the same way whether you are designing a new facility or making improvements to an existing facility. In both cases, the first step is to input the characteristics of a "baseline" facility configuration. In retrofits, the baseline is the existing facility. The baseline input is a lot of work. Once you make this investment of effort, you can easily modify the input and rerun the program to simulate changes to individual components.

To develop your design into an optimum configuration, you change the input characteristics of individual components or systems (windows, chillers, lights, etc.), and repeat the computer run to see the effect on overall energy consumption and cost. This is a cut-

and-try process. For example, analyzing the retrofit of a single system typically requires a few computer runs. Designing a big, complex facility may require hundreds of runs to achieve high efficiency.

Input is the main difficulty of using energy analysis programs. It is a slow, tedious, and often frustrating process. In order for the program to calculate loads and energy consumption with a useful degree of accuracy, you need to "describe" the facility and its equipment to the computer in detail. This typically requires many hours of sitting at a keyboard. It also requires knowing the energy characteristics of the structures and equipment.

For example, you need to specify the insulation value of each type of wall, window, roof element, and floor element; an occupancy schedule for each day type; a usage schedule for each type of light fixture; information about each type of heating and air conditioning unit; temperature reset schedules; the times that fuel types are changed; etc.

Each zone requires its own input. A small building usually has at least five zones. A large, complex building typically has dozens of zones. The more sophisticated programs allow you to copy input from one zone to another similar zone. For example, a building may have zones that are identical, except for the fact that they face in different directions.

Input requires a great deal of poring over drawings. When simulating existing facilities, it also requires a lot of legwork to inspect actual conditions. In existing facilities, you may need to measure some energy loads that are difficult to estimate, such as hot water consumption.

To calculate energy cost, you need to input energy prices, by type. This requires learning how to read rate schedules and fuel contracts. (Electricity rates are the most complex. See Reference Note 21, Electricity Pricing, for details.)

The temptation is great to fudge the input data if it is not readily available. The person doing the input needs an intimate knowledge of the facility, so it is generally not advisable to get the analysis performed by a service bureau that is remote from the existing facility. For retrofit projects, input should be done at the site. In designing new facilities, the designer should do the analysis.

■ Input Data Libraries

An important advantage of energy analysis programs is that they include libraries of input data. Most importantly, they include libraries of weather data for many locations, including all major cities of the world and many outlying areas. The weather data are compiled from hourly observations, allowing a high level of detail in calculation. The weather in almost any location can be simulated by using data from some location that is available in the weather library.

Some programs have libraries of heat conduction characteristics for wall and roof components, which eliminates the tedium of defining a wall or roof as many thermal layers. For example, the program may have a pre-defined wall section that consists of common brick-and-block construction with one inch of foam insulation and an internal finish on furring strips.

There are major differences in the structural component data libraries of different programs. The programs also differ in their ability to accurately simulate the thermal characteristics of structures that you define in terms of raw components. However, the capabilities of most programs are good enough for typical applications.

Some programs have libraries of characteristics for mechanical and electrical equipment, including chillers, boilers, pumps, fans, engines, and other types of equipment. Again, there are major differences between programs in this regard. For example, if the program can simulate cogeneration systems, it may include efficiency characteristics of prime movers, such as steam turbines and diesel engines, that are not found in other programs. However, be wary of using these generic data, because of the accuracy limitations discussed below.

■ Sensitivity Analysis

"Sensitivity analysis" means changing a single input over a range of possible values to determine the effect of the variation on the overall output. (This is sometimes called "what-if" analysis.) Use sensitivity analysis when you are uncertain about the actual value of the variable. For example, you cannot predict future fuel prices. To determine the range of economic benefit of installing a dual-fuel boiler, repeat the computer run several times with different combinations of prices for the two fuels.

You should also use sensitivity analysis to optimize equipment characteristics. For example, it is difficult to size the tank capacity in a heat recovery storage system because it is difficult to predict the amount of recovered heat that may be usable. To analyze such a system, repeat the program for a range of tank sizes, and for a reasonable variety of heat recovery and heat usage profiles.

Sensitivity analysis tends to be neglected, resulting in projects that fail because actual conditions differ from those expected. Make a sensitivity analysis for each input that is in doubt, if it may have a significant effect on performance. Sensitivity analysis multiplies the number of computer runs, but you can do it quickly. You simply change the variable that is in doubt, and repeat the run.

How Accurate are Computer Simulations?

Computers do not make arithmetic mistakes. Even so, accuracy in simulating actual energy consumption is not a virtue of energy analysis programs. The problem is that the actual characteristics of the building and its equipment will differ from the input to some extent. Furthermore, the characteristics of the building will change throughout its life.

Computer energy analysis is an example of measuring something with a micrometer that can only be cut with an axe. In other words, all those decimal places in the computer output don't mean anything. Remember what your grade school teacher told you about "significant digits."

Reasonable accuracy is possible only when comparing the difference between two alternatives, where only one or two variables differ and all other factors are the same. To keep from fooling yourself, understand the following limitations in the accuracy of energy analysis programs.

■ Range of Simulation Capabilities

An energy analysis program cannot precisely simulate the wide variety of building shapes, equipment, controls, and conditions that exist. The simulation capabilities of energy analysis programs are being expanded continually, but they always lag behind the latest developments. For example, the major programs only gradually acquired the ability to simulate window shading features, tilted glazing, VAV systems, temperature reset controls, variable-flow pumping, thermal storage, etc. If you are an innovator, you will not find a program that easily simulates all the configurations that you want to investigate.

Energy analysis programs differ in their strengths and weaknesses. For example, one program may be able to simulate the greatest range of equipment types, while another provides the most accurate description of wall thermal characteristics, and yet another provides the best simulation of complex glazing configurations. Be sure that the program you choose is able to simulate the engineering and architectural features that are important for your project.

Veteran users of energy analysis programs become adept at simulating equipment operation that is not explicitly available in the program. For example, heat recovery from a cogeneration plant can be simulated as a "boiler," and a solar collector can be simulated as heat gain through a skylight. The effect of a temperature reset control can be simulated by running the program for different temperature settings, and averaging the results. These tricks can be tedious, and they usually involve inaccuracies.

■ Accuracy of Component Simulation

The program must be told the efficiency (input-output) characteristics of each component that uses energy, including part-load performance. Input this information by specifying percentages of maximum output for various percentages of energy input. You may have to call the manufacturer's engineering department to get part-load information for the equipment you are simulating.

Some energy analysis programs save you this work with libraries of efficiency curves for different types of mechanical equipment, such as fans, pumps, chillers, and boilers. But beware. It is a mistake to rely on generic efficiency curves for major items of equipment, because there are major differences among models.

For example, different models of centrifugal chillers differ widely in their part-load behavior and in their minimum loads, which are important factors in their overall energy consumption. Programs developed by equipment manufacturers may have accurate performance data for their own specific models of equipment, but not for the equipment of other manufacturers.

Data for pumps and fans may be used if they are sufficiently specific as to type. For example, a generic curve for "forward-curved fan" may be satisfactory for most applications, but a general curve for "fan" is not specific enough, unless the fan is a minor item.

You may have to manually calculate the efficiency curve for a complex combination, such as a pump driven by a variable-speed drive. In this example, the program may not be able to account for the efficiencies of both the pump and the drive at each percentage of load.

■ Number of Calculation Intervals

There would be no need for computer energy analysis if the energy loads were constant, or if the loads changed in regular, repeating patterns. In HVAC applications, changing weather conditions make it necessary to use a computer to achieve accurate estimates.

The computer achieves accuracy by repeating the entire sequence of calculations for many short time periods. Some energy analysis programs calculate for each hour of the year, assuming a constant load during each hour. This method repeats each calculation 8,760 times, the number of hours per year. This degree of refinement requires relatively long computing times. The newest personal computers may be able to run such programs in less than one hour per run, and perhaps much less.

At the other extreme are programs that use "bin method" calculations. The bin method was originally developed to allow energy consumption calculations to be done with hand calculators, and the method was later refined for use with computers of limited capability.

The bin method calculates energy consumption only for a series of discrete temperatures, typically in 5°F increments. The increments are called "bins." The results are weighted by the number of hours that the ambient temperature falls within each bin. There are only twenty 5°F increments in a yearly ambient temperature range of about 100°F, so the bin method requires far fewer calculations than the hourly method.

To use the bin method, you need weather data for the total number of hours that the temperature falls within each increment. The U.S. Government publication, Engineering Weather Data, provides temperature and humidity data arranged by bins for many geographic locations. This publication subdivides the bins by month. This allows you to assign different cloud cover and solar characteristics for each month. Of course, breaking down the calculations by month increases the number of calculations by twelve.

The main weakness of the bin method is that not all energy consumption is based on temperature differential. For example, cooling load is also determined by sun position, sky clearness, and internal loads, which vary over wide ranges. The bin method can be extended to deal with these factors by repeating the calculations for different ranges of these conditions, but this requires weather data in a form that is not published, so the computer itself must create bin weather data from actual weather data. Attempting to refine the bin method increases the complexity of the program and the computer running time. The widespread availability of fast personal computers has made the bin method an unnecessary compromise. It is still used in some energy analysis programs, but it is probably a dead end for future development.

Some programs take a path between these two extremes of detail by creating a single day of synthetic average weather for each month of the year. This makes weather the only compromise. Sun angles, usage schedules, and other variables that change on an hourly basis can be calculated properly for each hour of the day. If there are several day types (e.g., weekday and holiday), these can be calculated separately for each month.

■ Weather Data

Even an 8,760-hours-per-year simulation does not account for variations in weather from one year to the next. If the program uses a true hourly calculation interval, the year is probably an actual calendar year of weather data that is selected to be typical. If the program does not use an hourly calculation, it generates synthetic weather input data from actual weather data or from a bin method.

It is a good idea to repeat the program using different years of weather input to represent extremes of weather conditions. Some programs allow different years to be selected. If the program does not provide this option, you can work around this limitation by repeating the calculation with weather data from other locations that are similar to the extremes of weather at the actual location. For example, for a building in Cincinnati, run the analysis with weather data from Atlanta and Chicago. If this produces big differences in energy consumption, there are aspects of the design that are too sensitive to weather conditions.

■ **Number of Zones**

All energy analysis programs are designed to divide the building into zones. They do this as an easy way of dealing with differences in load characteristics. The program user defines the number of zones, and the way that the building is divided into zones. This is a major decision that the user has to make at the beginning of the input process. A larger number of zones provides greater accuracy, but requires more input work. The computer program calculates the energy requirements of each zone separately, so more zones require longer computer runs.

It is conventional to create separate zones for each exterior orientation of the building, plus interior zones. For example, the user may divide a simple, single-story building into five zones: north, south, east, west, and interior. All five zones in this example differ in solar gain, daylighting, and heat loss, and they may differ in internal heat gain. Users typically define major spaces with distinct energy characteristics as separate zones. Examples of such spaces are kitchens, computer rooms, large conference rooms, and atriums.

More sophisticated programs may require fewer zones, because they allow the user to specify a greater variety of features per zone. For example, the program may be able to deal with several window types in each zone. Such capability does not reduce the input work by much, because the additional description detail requires more input work for each zone.

■ **Equipment Defects and Unpredictable Behavior**

The program's simulation may differ substantially from real performance because components are not ideal. For example, some energy analysis programs calculate theoretical thermal characteristics of walls in excruciating detail. Real wall construction includes sloppy insulation, thermal short circuits through studs, air infiltration, and other factors that substantially degrade thermal performance. Sophisticated program users try to compensate for such predictable flaws by adjusting the input, but it is not possible to predict the performance of bricklayers and carpenters with a great deal of accuracy.

The program cannot simulate factors that change during the life of the facility. For example, you have to assume a particular chilled water temperature in the input, as this affects chiller efficiency. However, this setting is likely to be changed repeatedly. Use sensitivity analysis to estimate the effect of unpredictable changes.

■ **Program Errors**

There are only a few energy analysis programs in existence that have proven to be reasonably reliable, and all have limitations and flaws. These programs have bugs for the same reasons that all large computer programs have bugs. A common bug in energy analysis programs is failure of the program to accurately simulate the way equipment behaves. For example, a leading energy analysis program of the 1970's was abandoned after years of use when it was discovered that some of its equipment simulations were written incorrectly.

This is a continuing problem. Few people have a thorough understanding of the energy behavior of systems, not many of these write energy analysis programs, and those who do may not remain involved with the program to upgrade and debug it.

■ **Input Errors and Misunderstandings**

The computer cannot read your mind to determine what you really mean, and it will not fix your input errors. The computer is completely literal in reading the input. A typically analysis requires hundreds or thousands of input numbers. An error in any single number will cause an error in the output. Therefore, your input needs to be meticulous.

Some programs can flag input data that falls outside a normal range. This is helpful in catching gross input errors, but errors can still slip through this screen.

How to Catch Errors

Your first computer run usually contains mistakes. Serious mistakes keep the program from running, while subtle mistakes produce wrong answers that may look right. Anticipate the problems discussed previously, and make a careful effort to debug the run and verify the validity of the results. Use the following methods to do this.

■ **Read the Output in Detail**

The first step is to print all the summary reports. Examine them in detail to see if they make sense.

If the summary reports appear credible, print the entire output. This produces a large stack of paper. If the stack would be too large, at least print the months of January, April, and August. Track through the calculations to see if the trends make sense. Expect to spend hours in checking the output for a building of average complexity.

■ **Check with Pencil Calculations**

Use pencil calculation to check the computer analysis for several different hours, at different times of the day and different times of the year. To do this, get a complete printout for a particular hour, showing all the inputs. Based on the inputs for that hour, manually calculate all the loads and equipment energy inputs. What you calculate should match what the computer calculates.

A difference may be caused by an input error. Or, you may not be understanding the output correctly. Or there may be a difference between your assumptions and the computer's assumptions. You may find that the program is not calculating what you think it is calculating. For example, the computer may model the behavior of a system differently from the way you intend the system to work. Only a pencil check can reveal such a misunderstanding.

■ Use Sensitivity Analysis

When running a sensitivity analysis, see whether the changes in output make sense. For example, separate runs for progressively larger tank sizes in a thermal storage system should yield progressively greater energy recovery, but in progressively smaller increments.

■ Analyze the Relative Performance of Dummy Configurations

Calculate the performance of several dummy configurations in addition to the desired configuration, where you know the relative performance of the different configurations.

For example, you may wish to analyze the benefit of converting a constant-volume air handling system to a variable-volume system that uses supply air temperature reset. As a dummy configuration, also analyze a variable-volume system without supply air temperature rest, with everything else remaining the same. You expect the dummy configuration to have higher overall energy consumption than the proposed system, but lower fan load. If the computer does not produce these results, there is probably an error or misunderstanding somewhere.

■ Compare to Measured Energy Consumption

In retrofit applications, check the program's baseline calculation (i.e., the existing configuration) against actual consumption. Use past utilities bills and measurements to determine actual consumption.

■ Check the Computer Code

If you know how to read the computer language in which the program is written, you can check the code itself. Energy analysis programs are huge, so finding errors this way is like searching for a needle in a haystack. Still, the portion of the program that applies to a particular problem may be fairly small. If the program vendor will not supply the computer code, you can use various computer utilities that are available for making computer code visible. If you achieve nothing else, you can satisfy yourself that the program is calculating what you think it is calculating.

How to Select Your Energy Analysis Program

When you select an energy analysis program, consider both the features of the program and the support that will be available for it. In summary, look for:

• *simulation capabilities.* The program should be able to make accurate distinctions between options you are considering. Accuracy is less important

for features of your project that are common to all the options. For example, if you are doing a sophisticated envelope design to minimize solar cooling loads, you need a program that can accurately calculate complex glazing and shading configurations. On the other hand, you may not need an elaborate calculation of wall thermal characteristics. All the major energy analysis programs can simulate common building features and systems. The programs differ widely in their ability to simulate unusual features, such as complex envelope shapes, cogeneration, solar energy, thermal storage, and sophisticated control systems.

• *library data.* Consider the types of weather data, equipment data, envelope component data, and other library data that are offered with the program. Consider whether they contain sufficient detail and specificity to distinguish between alternative designs that you are considering.

• *output options.* All programs offer the same general types of output, but some offer considerably more detail than others. Some output formats are more user-friendly.

• *ease of input.* Some programs, such as the raw government version of DOE2, require you to write elaborate input text. Other programs have a simple, orderly fill-in-the-blanks format.

• *technical support.* You will have trouble, no matter which program you choose. This makes it important to acquire the program from a vendor who provides competent technical support on short notice. Energy analysis programs developed by private companies usually include technical support as part of the purchase price. Check with other customers about the quality and availability of the support provided. Typically, only one or two individuals at each company are well versed in the details of the program. If a critical person departs the organization, the program may become orphaned.

Government-sponsored programs are not copyrighted, so there is no single party with an ownership interest in supporting them. For these programs, there may be several private parties that offer program interfaces for personal computers, which are mainly a vehicle for charging for technical support. The quality of support from such vendors can be poor or spotty, because the interface vendor may not have a good understanding of the main program or of the building systems you want to simulate.

Efficiency Characteristics
- **Chemical Composition of Fossil Fuels**
- **Combustion End Products**
- **Energy Content of Fuels**
- **Excess Air Requirements**
- **Minimum Flue Gas Temperature**

Fuel-Related Costs

The name "fossil fuel" applies to solid, liquid, and gas fuels that are believed to be products of the decomposition of ancient plants and animals. In some coal deposits, the outlines of the original prehistoric plants are plainly visible. Petroleum appears to contain components of biological origin. The simple molecules of natural gas cannot be traced to their origins, but natural gas also seems to be of biological origin.

Fossil fuels have been the primary energy source of civilization for about one century, and they will continue to be the primary energy source for the foreseeable future. No other energy source has their combination of desirable characteristics, including high energy density, ease of use, ease of storage, portability, and relative safety. However, fossil fuels are a limited resource that is being rapidly depleted.

Selecting fuel sources is more complicated than simply selecting the source with the lowest price. Nowadays, fuel selection involves environmental considerations, which may limit your choices of fuels, or make it more expensive and difficult to burn them. It involves equipment for transportation and storage, and perhaps for disposal of residue. It affects the efficiency

of the fuel burning equipment. It affects maintenance requirements. This Note will give you a basic knowledge of fuels that will help to deal with these considerations rationally.

Efficiency Characteristics

The purpose of a boiler, a furnace, or an engine is to release the energy of fuel in a useful manner. The following characteristics of fuels affect the efficiency of this process. Table 1 summarizes some of the key features of the common fossil fuels.

■ Chemical Composition of Fossil Fuels

In their "pure" form, all fossil fuels are hydrocarbons, which means that they are compounds of only two elements, hydrogen and carbon. The main difference between fossil fuels is the ratio of the two. Table 1 gives the carbon and hydrogen contents of typical fuels by weight.

Methane, the main constituent of natural gas, consists of one carbon atom and four hydrogen atoms, and it has the highest percentage of hydrogen. At the other extreme, coal has mostly carbon as its combustible component, and contains little hydrogen. A hydrogen

Table 1. EFFICIENCY-RELATED CHARACTERISTICS OF FUELS

Fuel	Heat Content, Total BTU/lb (kJ/kg)	Volumetric Heat Content *BTU/cu.ft. (kJ/cu. m) **BTU/gal. (mJ/cu. m)	Components Percent by Weight		
			H	C	S
Hydrogen, pure	61,100 (142,000)	325 (12,000)*	100	--	--
Carbon, pure	14,100 (32,800)		--	100	--
Sulfur, pure	4,000 (9,250)		--	--	100
Natural gas (mostly methane)	23,800 (55,300)	1,000 (37,000)*	24	76	--
Propane	21,600 (50,200)	2,600 (96,800)* 91,600** as liquid	18	82	--
No. 2 Oil (average)	19,600 (45,600)	138,000 (37,300)**	13	86	0.0 to 0.5
No. 4 Oil (average)	19,200 (44,700)	145,000 (40,500)**	12	87	0.5 to 1.5
No. 6 Oil (average)	18,800 (43,700)	150,500 (41,900)**	12	86	0.3 to 4.0
Coal, bituminous (typical)	14,000 (32,600)		4	84	0.5 to 4.0
Wood (typical)	8,600 (20,000)		6	49	--
Bagasse (typical)	8,200 (19,000)		6	45	--

atom has an atomic weight of 1, while a carbon atom has an atomic weight of 12, so carbon dominates the weight of fossil fuels.

Some fossil fuels are almost perfectly pure, in the sense that they consist almost entirely of hydrocarbons that have high fuel value. Other fossil fuels stray far from this ideal. Fossil fuels contain various contaminants. These may be a major factor in the design of the equipment that burns it. We can divide fossil fuel contaminants into these groups:

- *other types of hydrocarbons.* For example, fuel oil is contaminated by waxes. The waxes have high fuel value, but they cause equipment problems in storage and burning.

- *non-hydrocarbon materials that burn.* An common contaminant of this type is sulfur. It burns, but its fuel value is lower than that of the hydrocarbons. In the process of burning, it causes compounds that are serious pollution sources.

- *non-combustible materials.* Minerals, metals, and water are common contaminants of solid fuels and heavy oil. These cause a variety of problems in storage and combustion.

Natural gas is mostly methane, but it also contains small amounts of ethane, propane, and other light hydrocarbon gases. Contaminants include free nitrogen, oxygen, and carbon dioxide, along with trace amounts of inert gas and sulfur oxides. The foul smell of natural gas is a chemical added in retail distribution to reveal leaks.

Fuel oil is a mixture of many different liquid hydrocarbons. The characteristics of fuel oil represent an average of the characteristics of its components. The energy content, viscosity, and other characteristics of fuel oil vary widely. Fuel oils are commonly classified by numbered grades, from Number 1 (light, least carbon) to Number 6 (heavy, highest carbon content).

The properties of fuel oil can vary significantly, even within a grade. Properties depend on the location of origin, the refining and "cracking" processes used, and blending that is done by the seller to achieve certain properties. Common contaminants of fuel oil include water, sulfur, dirt, and trace amounts of metals.

Coal varies widely in carbon content, being the residue of the decay and cooking of ancient vegetation and animals, along with a substantial amount of inorganic matter. Coal is categorized by grades that have different names. The grades, in order of diminishing carbon and energy content, are bituminous, anthracite, sub-bituminous, and lignite. There are sub-classifications within these grades.

The energy content of coal is affected more by its water and dirt content than by the nature of its hydrocarbon content. Contaminants include anything that is in the ground where the coal is mined. Sulfur is a major contaminant, and metals may also cause trouble.

Sulfur deserves special mention because it occurs in large quantities in oil and coil, and because it has major detrimental effects on efficiency, air quality, and maintenance. Ranges of sulfur content for different fuels are given in Table 1.

The efficiency effects of contaminants are to reduce the energy content of the fuel, to absorb some heat from the combustion gases, and to dilute the hot combustion gases. Only coal has enough contaminants to have a major effect on energy content.

Coal and other solid fuels typically contain a large amount of non-combustible material that becomes slag and fly ash. This inert matter requires special boiler design and auxiliary equipment for disposal. Solid contaminants also create particulate air pollutants. The types and amounts of contaminants vary widely, depending on the geographic location of the source, the type of mine or well, the refining process, the method of transportation and storage, and other factors.

■ Combustion End Products

In combustion, fuel gives up its chemical energy by a chain of reactions that convert the fuel into end products that have no energy left to release. With hydrocarbon fuels, these end products are always carbon dioxide and water vapor. Carbon dioxide is produced by the carbon in fuel. Water vapor is produced by the hydrogen. Thus, for example, natural gas produces a large amount of water vapor, and coal produces little. (However, water that contaminates oil and coal, often in large amounts, ends up as water vapor in the flue gases.)

In perfect combustion, no oxygen would be left in flue gases, because any excess oxygen is deadweight and would absorb combustion heat. By the same logic, the presence of oxygen is an excellent indicator of boiler or engine efficiency. Similarly, the quantity of carbon dioxide in the fuel is also used to measure boiler efficiency. Subsections 1.2 and 1.3 cover this in detail.

The sulfur in fuels burns to form sulfur dioxide. This produces a relatively small amount of energy, but it causes major problems.

■ Energy Content of Fuels

The energy content of a hydrocarbon fuel depends primarily on the relative amounts of hydrogen and carbon in the fuel. Burning pure hydrogen produces about 61,000 BTU per pound, while burning carbon produces only about 14,000 BTU per pound. Thus, fuels with a high hydrogen content have the most energy per unit of weight. Table 1 gives the total heat content of common fuels.

Sulfur plays a minor role in the energy content of fuel, mostly by diluting the more energetic elements. Burning sulfur produces about 4,000 BTU per pound.

When purchasing fuel or calculating the efficiency of a boiler or engine from the fuel energy content, distinguish between "high heat value" and "low heat value." (These are sometimes called "gross" and "net" heat contents, respectively.) The high heat value is the total energy content of the fuel, with the flue gas reduced to room temperature.

The high heat value is generally unrealistic because the water that is produced escapes from most boilers and engines in the vapor state, so its latent heat is not captured. To account for this, the low heat value was defined as the high heat value minus the latent heat of the water vapor. Water vapor is produced only by the hydrogen in fuels, so the difference between high and low heat values is greatest in fuels with a high hydrogen content.

Even the low heat value overestimates the amount of heat that may be extracted from fuel, even in a conventional boiler. Conventional boilers extract less than the low heat value of a fuel because they need to maintain a margin of temperature in the flue gas, to prevent corrosion. The exhaust temperature of an internal combustion engine is limited by its compression ratio.

High heat value is generally used in calculating the efficiency of boilers, which provides a consistent baseline for comparing all types of boilers, including boilers that condense the water in the flue gas. Low heat value is commonly used in rating the efficiency of gas turbines, probably because it gives an apparent boost to the poor efficiency of these units. Be aware that low heat value may be used to embellish the efficiency of other equipment as well.

The energy content per unit of weight increases with relative hydrogen content for all fuels, because hydrogen has the highest energy content in relation to its weight. On the other hand, the energy content per unit volume decreases with hydrogen content. In the case of gas fuels, this is simply because the number of molecules per unit volume of gas is constant (Avogadro's Law), and the heavier gases have more hydrogen along with a lower ratio of hydrogen to carbon. In the case of oil, the increased density of heavier oils more than compensates for their lower percentage of hydrogen. Table 1 gives heat content of gas and liquid fuels in terms of both volume and weight. The volumetric heat content is useful because gas and liquid fuels are usually purchased on the basis of volume.

■ Excess Air Requirements

The choice of fuel affects efficiency indirectly by dictating the amount of excess air needed to burn the fuel completely. Excess air wastes energy primarily by lowering the combustion gas temperature, which reduces heat transfer.

Little excess air is needed to burn gas fuels. More excess air is needed for fuels that take longer to be vaporized in the combustion chamber. The quantity of excess air that is typically needed to burn oil and solid fuels has a significant effect on combustion efficiency.

The best modern burners greatly reduce the excess air requirement compared to earlier burners. Therefore, excess air no longer enters as a major factor in efficiency differences among fuels, except perhaps with solid fuel that is fired on stokers. Excess air requirements are covered in Subsection 1.3, and burner improvements are covered in Subsection 1.4.

■ Minimum Flue Gas Temperature

The flue gas temperature is the primary indication of the amount of heat that has been extracted from the combustion gases. The lower the flue gas temperature, the higher the boiler or engine efficiency, all other things being equal. In a boiler, the tube surface area is the main factor that determines the amount of heat that can be extracted from the flue gas, so it determines the minimum flue gas temperature that can be achieved.

Conventional boilers must keep the flue gas temperature higher than a certain temperature, which is called the "acid dew point." The acid dew point is the temperature at which corrosive acids can condense out of the flue gas and corrode the boiler. The acid dew point of some fuels is high enough to impose an efficiency penalty on the boilers. The acid dew point is determined primarily by the sulfur content of the fuel, because it is sulfur that forms the corrosive acids.

Conventional boilers remove only sensible heat from the flue gases. In contrast, a special class of boilers lowers the flue gas temperature enough to recover the latent heat of the water vapor in the flue gas. Recovering this latent heat may provide a substantial increment of additional efficiency. This general class of boilers is called condensing boilers.

Subsection 1.7 explains acid dew point limitations in greater detail, and presents methods of dealing with acid in flue gases.

Fuel-Related Costs

Everyone understands that there are large price differences between fuel types. Consider these other cost factors as well:

- *future changes in relative fuel prices.* The relationship between the prices of different fuels can vary radically over time. These changes are caused by a number of factors, such as exhaustion of a nearby source, changes in political policy, developments in energy technology, and other factors. For example, political policy in the U.S. once made natural gas much cheaper than other fuels, but gas may now be more expensive than oil. Most of these factors cannot be predicted with any accuracy.
- *effect on combustion efficiency.* A boiler that is designed to burn several types of hydrocarbon fuels

will burn all the them with about the same efficiency. All oils will burn within one or two percent of the same efficiency. Natural gas burns at a few percent lower efficiency than oil, in conventional boilers. However, only gas can burn in condensing boilers, which have the highest efficiency. Liquid and solid fuels with higher sulfur content require higher flue gas temperatures, which lowers efficiency.

• *environmental protection costs.* Burning some fuels requires expensive equipment to minimize the adverse environmental effects of the combustion products. This is especially true with coal, which requires equipment and special processes to reduce sulfur emission, to minimize creation of nitrogen oxides, and to prevent emission of airborne particulates. These costs depend on the nature of the fuel, the location of the boiler, and the technologies used. For example, sulfur emissions can be reduced by burning low-sulfur coal (which is more expensive), by using a fluidized bed boiler, or by using stack gas scrubbers. Fuels may be penalized for their contribution to atmospheric carbon dioxide by a "carbon tax" that is based on the carbon content of the fuel.

• *effect on maintenance cost.* Cleaning firesides is a major maintenance cost related to the choice of fuel. Boilers may fire indefinitely with gas fuels without the need for fireside cleaning, whereas burning heavy oil may require fireside cleaning several times per year. Solid fuels may require considerable maintenance of fuel handling and preparation equipment. Oil burning requires a lesser degree of maintenance for fuel strainers, heaters, etc. The choice of fuel also affects long-term maintenance. For example, burning oil instead of gas in a boiler hastens deterioration due to fireside corrosion, such as tube sheet leakage.

• *storage and handling equipment.* Natural gas requires no storage facilities and delivery costs are low, assuming that a pipeline is nearby. Oil requires storage tanks, pumps, and heaters. Coal requires expensive facilities for unloading, storage, transfer to the boiler, crushing or pulverizing, and removal of ash. Burning coal may also require storage and handling of limestone for removing sulfur, and for disposal of the calcium sulfate that results from this treatment.

• *availability options.* Large variability of price, and possible interruptions of supply, make a strong argument for having the capability of burning different fuels in a boiler or engine. This option requires specialized burners, perhaps some compromises with efficiency, and the purchase of storage and handling equipment for each type of fuel. If availability is a significant concern, select your combination of fuels so that they are least likely to become scarce at the same time. For example, if you feel vulnerable to imported oil, install burners to burn natural gas that is available domestically. In some cases, it may even be desirable to have separate boilers for different fuels.

Rate Schedules
Commodity Charges
Fuel Surcharges
Demand Charges
- How the Demand Peak is Measured and Calculated
- Variations of Demand Charges
- Time-Related Aspects of Demand Charges
- Ratchet Provisions
- "Hour Charges"and Load Factor
- Effect of Energy Conservation on Demand

Electricity Charges Based on Customer Equipment
- Power Factor Penalties
- Delivery Voltage
- Ownership of On-Site Distribution Equipment
- Incentive Rates
- Effect of Multiple Meters

Administrative Charges
Excess-Consumption Charges
Competitive Purchasing of Electricity
Negotiating Rates

You can reduce your electricity cost in three general ways:
- use less electricity
- obtain electricity under the most favorable rates
- modify your facility's pattern of electricity consumption to exploit the utility's pricing structure.

The first of these is classical energy conservation. The second is a matter of intelligent shopping. The last allows the utility to generate more efficiently, or to generate with a less critical energy source, for which the utility rewards the customer with a lower rate.

To exploit the last two possibilities for reducing electricity cost, you need to understand the factors that an electric utility considers in setting its prices. By understanding electricity pricing, you may be able to reduce your electricity cost, often with little effort and sometimes by a large amount. As with income taxes, those who understand the rules may pay much less than those who do not. This Note explains the basics of electricity pricing and suggests ways that you can exploit the pricing structure to reduce your costs.

The price of electricity reflects its origin. Electricity from large hydroelectric generators is cheap. The price marches upward as the source progresses from nuclear reactors to lignite, coal, heavy oil, light oil, and natural gas. At this time, electricity generated from "renewable" sources of energy, such as wind and sunlight, is the most expensive.

There are many other factors that affect electricity prices. Electricity pricing varies from one locale to another, from one utility to another, and from one customer category to another. Negotiation is an important factor in pricing for larger customers.

Rate Schedules

Unlike the prices of other energy sources, the price of electricity is regulated. This occurs even in countries with free markets because a public utility is granted a monopoly to provide electrical service within a specified geographical area, so its prices are not restrained by competition. In the United States and in some other countries where electric utilities are privately owned, each utility publishes its own prices and method of pricing in the form of "rate schedules." Figure 1 shows a key extract from a typical rate schedule.

In the United States, utility regulation is done by each state. The state regulatory agency audits the rate schedules, negotiates changes requested by utilities, and sometimes initiates changes on its own. Each electric utility operating within a state establishes its own rate schedules subject to regulation by that state. An electric utility that operates in more than one state has a different set of rate schedules for each state.

The central principle in utility regulation has been to make the utility's charges reflect its actual costs. The utility's main cost categories are:

- **generating plant,** the cost of which varies greatly with the type of equipment. For example, gas turbines are much cheaper to install than nuclear or coal-fired power plants.
- **distribution equipment,** such as that shown in Figure 2. Distribution cost depends mostly on the distance covered, the nature of the terrain, and the power capacity. For example, it costs much less to serve a single large customer located near the power plant than to serve many residential customers distributed over a large area.
- **fuel or energy,** which differs greatly in cost for different types of generating equipment. For example, gas turbines need more fuel than boilers that drive steam turbines, and gas turbine fuel is more expensive. On the other hand, dirty boiler fuels incur a large equipment cost for reducing air pollution. Generation with water power, wind, solar, geothermal heat, etc., involves little fuel cost, but it may have a high initial cost and low reliability.
- **administration,** which is distributed over all aspects of utility operation. Administration is a larger fraction of the electric bill for smaller customers.

• *ability to buy electricity from, and sell to, other utility systems.* Utilities use their excess generating capacity to sell electricity to other utilities. Likewise, a utility may purchase electricity at times when it lacks capacity. The price of electricity transferred between utilities fluctuates from moment to moment, depending on the load characteristics of all the utilities that are connected together.

These cost elements are reflected in the rate schedule by a variety of charges. Most rate schedules have three major categories of charges that customers pay:

• *commodity charge,* which is based on the number of kilowatt-hours purchased

• *demand charge,* which is based on the maximum rate of consumption, expressed in kilowatts, during certain time periods defined by the utility

• *fuel surcharge,* which accounts for fluctuations in the utility's fuel cost.

It is essential to understand the difference between the charges that are based on kilowatt-hours, and those that are based on kilowatts. See Figure 3.

The names and details of these charges differ between utilities and jurisdictions. Rate schedules also include a somewhat random sprinkling of charges that reflect less important cost factors, such as power factor and supply voltage. Each of these types of charges is discussed in greater detail, below.

It is impractical to match the utility's actual cost of serving each customer to its charges to that customer. For each customer, there are differences in the cost of distribution, billing, and customer service. Different customers use electricity in different ways. Political considerations create pressures to shift costs between classes of customers. To deal with these variations in cost of service, the utility creates different rate schedules for different classes of customers, and for different consumption patterns.

As a rule, the actual cost of electricity is tailored more closely for larger customers with more complex rate schedules. Smaller customers, which include individual residential accounts, are served under simple, standardized rate schedules.

"Riders" are supplements or variations of a basic rate schedule. The number of schedules and riders generally increases with the size of the utility and the diversity of its customers.

For example, one typical large utility has separate rate schedules for general residential service, time-metered residential service, commercial facilities in several size ranges, time-metered commercial service in several size ranges, several types of interruptible service, electric heating, outdoor lighting, street lighting, traffic signals, electric vehicles, electric mass transit, and cogeneration.

MONTHLY RATE -

		Billing Months of June - October (Summer)	Billing Months of November - May (Winter)
A.	Customer Charge	$292.20 per month	$292.20 per month
B.	Energy Charge		
	On-Peak Period	3.409¢ per kwhr	2.518¢ per kwhr
	Intermediate Period	2.714¢ per kwhr	1.941¢ per kwhr
	Off-Peak Period	1.080¢ per kwhr	.625¢ per kwhr
C.	On-Peak Demand Charge	$ 10.30 per kw	
D.	Maximum Demand Charge	$ 4.25 per kw	$ 4.25 per kw
E.	Minimum Charge - The Customer Charge		

Fig. 1 Key portion of an electric rate schedule This is the core of one rate schedule created by a large electric utility. This rate applies to medium and large commercial customers. The charges vary by the time of year. The charges also vary by the time of day, and by the day of the week. The hourly and daily variations are covered by the three periods under the "energy charge," which are explained elsewhere in the rate. There is a "demand charge," which is made only during the utility's peak generating hours. The charges are calculated based on the pattern of consumption during individual months. The full rate schedule is several pages long, defining the terms used in this table, establishing various conditions, and specifying variations, or "riders." The same utility has about a dozen other rates for different classes of customers.

Electric utilities structure their rates to encourage actions by their customers that are advantageous to the utilities. For example, utilities faced with the cost and difficulty of constructing new generating facilities increase their demand charges to encourage their customers to find ways of reducing peak power consumption.

In most cases, the utility selects the rate schedule and riders that apply to a particular customer. However, a customer may be eligible for several rates or riders. The difference in electricity cost between alternative rate schedules is not always obvious. For example, a time-of-day rate may reduce electricity costs if the facility has a low daytime peak, while it may increase costs otherwise.

Rate schedules are often written in a manner that is inscrutable, incomplete, or illogical. A call to the utility's rate department may be necessary for clarification. Some utilities cannot interpret their own rate schedules, making it necessary to divine from actual billings how the utility is charging for electricity.

In recent decades, new regulatory principles have influenced rate schedules. Some regulatory bodies use rates to transfer costs from one class of customer to another. Many regulatory bodies have been requiring utilities to promote "demand-side management" (DSM) and energy conservation. This skews existing rate schedules and creates entirely new rate schedules for certain activities, such as interruptible service and thermal storage.

Commodity Charges

A commodity charge is an electric rate that is based on the amount of electrical energy delivered. The billing units are kilowatt-hours (KWH). A commodity charge may be called an "energy charge" or some other name. A commodity charge is similar in form to the prices of other energy sources, which are sold per-barrel, per-ton, etc. A rate schedule may have different commodity rates for different months or seasons.

For small customers, including most residential customers, a commodity charge and a fuel surcharge are the only types of charge that are found on the bill. For such customers, the commodity charge reflects all the utility's costs, except fuel.

For larger customers, a commodity charge is usually one of several charges. In this case, the commodity charge tends to reflect the utility's major fixed costs, especially the generating plant and long-range distribution equipment. For this reason, the commodity rate tends to remain unchanged for long periods of time.

Fuel Surcharges

The utility's fuel costs change continually, sometimes over a wide range. Rate negotiations with the regulatory agency take too long to reflect changes in fuel cost, so the utility is allowed to adjust for its fuel cost with each billing by using an approved formula. This is simplified by the fact that a utility is generally not allowed to make a profit on its fuel costs. (In effect, a utility is allowed to make a profit only by operating and renting out its equipment.)

Fuel surcharges usually do not discriminate by time, even though the utility's fuel cost may vary widely from on-peak to off-peak periods because of differences in generating equipment. This is another compromise to simplify billing.

WESINC

Fig. 2 Utility distribution equipment Customers pay for this largely by demand charges, because the size of the equipment and wires is determined by the customers' peak rate of consumption. Demand charges are higher if there are fewer customers served by this substation to average the peaks in consumption.

Demand Charges

"Demand" is defined as the highest rate of consumption, expressed in kilowatts, during a defined time period. The time period is usually a calendar month. Note that demand is measured in kilowatts (KW), not in kilowatt-hours (KWH). Demand is a peak rate of consumption, not an amount of consumption.

A demand charge can be considered a surcharge on customers who use electricity in large amounts during the utility's peak generating periods, thereby forcing the utility to install extra generation and distribution capacity. Demand charges are highest during time periods when the utility is forced to use supplemental ("peaking") generators that are inefficient and use expensive fuels. Demand charges may differ by the hour, the day, or the season, as discussed below. Most larger customers are subject to a demand charge, which may be given a variety of names.

The demand charge is often a large fraction of the total electric bill, and in many cases, it may account for more than half of the bill. The distinction between the demand charge and other charges is important, because the facility often has the ability to reduce its demand separately from its total consumption. Utilities base demand charges on their own load, not on the individual customer's load. By shifting your peak consumption to a time that is more favorable for the utility, you may reduce your demand charges substantially.

The demand charge becomes a larger fraction of the total electric bill if the customer's electric load profile is "peaky," for example, if there is a large, brief air conditioning load on an unusually hot day. In effect, a demand charge is a penalty for having high peak loads. By the same token, a customer can reduce its demand charges by reducing its load peaks, for example, by "peak shaving" or using thermal storage.

Demand billing requires special metering equipment and extra billing work, so demand charges are usually not levied on small users. Instead, peak-related costs for a given class of customers, such as residential or small commercial, are averaged into a rate that is expressed in terms of a commodity charge. (There has long been pressure to include demand charges in residential billing. Improvements in metering technology may make this practical in the near future.)

■ How the Demand Peak is Measured and Calculated

The "demand" that appears on the monthly electric bill is not an instantaneous value. In most cases, billing demand is the highest average consumption during a short period of time, which is usually 15 to 30 minutes. Therefore, you don't need to worry about surges of motor starting current that last only a few seconds.

The utility typically measures demand by installing special metering equipment that totals the number of kilowatt-hours used during each interval of this duration, and uses a computer to find the highest total. In some cases, the measurement intervals start exactly on the hour, and at intervals of 15, 20, or 30 minutes thereafter. This is called a "fixed window." In some cases, a "sliding window" is used by the utility company, which means that the demand measurement interval can be selected to cover the customer's actual highest peak period.

Utilities with old equipment may still measure demand visually by scanning a continuous chart recording of power consumption.

Almost all utilities start the demand measurement afresh on the first day of each billing period, which is typically a calendar month. However, "ratchet charges," discussed below, may cause the demand recorded in one month to affect demand charges in succeeding months.

■ Variations of Demand Charges

Demand charges may occur in the electricity bill under a variety of names other than "demand." There may be a single demand charge designed to recover all the fixed costs of the utility, or there may be different types of demand charges that recover different types of costs. The different rate schedules of a given utility typically express demand charges in different ways.

For example, some utilities define a "generation charge" covering the generation equipment at the power plant, and a separate "distribution charge" that compensates the utility for its equipment at the user site and in the adjacent geographic area. The "distribution charge" typically is charged at the same rate per kilowatt all year, while the "generation charge" typically is based on the demand measured during peak generation periods.

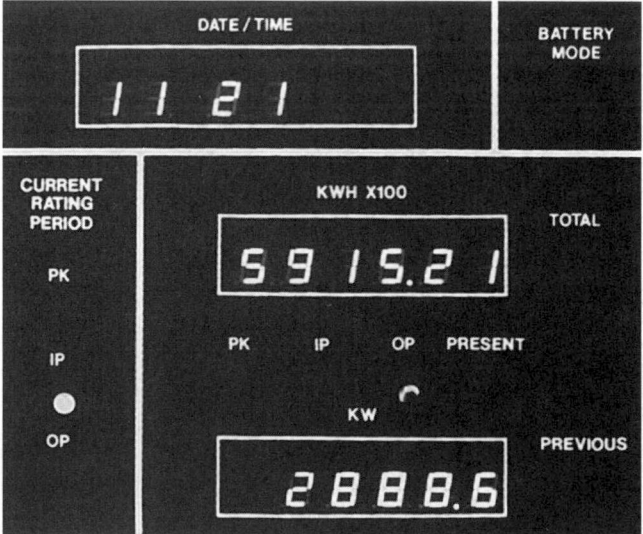

Fig. 3 Be clear about the difference between kilowatts and kilowatt-hours This electric power monitor in a commercial building shows the highest demand peak of the current billing month, expressed in kilowatts. It also shows the current cumulative total of electricity consumption for the month, expressed in kilowatt-hours. You need to know how these numbers relate to your utility's rate schedule.

There is logic in this. The distribution charge is simply a way for the utility to recover its cost of installing equipment to deliver electricity to an individual customer. This includes the cost of transformers, meters, and local power lines. The distribution charge is like a mortgage payment for equipment that has already been bought.

In contrast, the generation charge largely relates to central generating equipment. It applies to future costs as well as past costs. It is structured to discourage customers from using electricity during time periods when the utility is generating near maximum capacity. Consumption during the utility's peak periods forces the utility to build more plants. It also forces the utility to generate electricity using equipment, such as gas turbines, that have a higher unit energy cost. (This is because the gas turbine is less efficient, and because it uses fuel that is more expensive.) For these reasons, demand charges depend on the utility's present reserve capacity. It also depends on the types of generating equipment that the utility owns.

This is not to say that all electricity rate schedules are logical or consistent. Many utilities lack the capability to develop rate schedules that reflect their costs in a proportional manner. Any rate schedule is a compromise between complexity and accuracy in proportioning costs. This is one reason why utilities reserve their most complex rate schedules for their largest customers.

■ Time-Related Aspects of Demand Charges

The price of any commodity changes with time. Some commodities have price fluctuations that are cyclic, such as strawberries (more expensive in winter), air fares (more expensive in summer), swimsuits (more expensive in spring), newspapers (more expensive on Sunday), and telephone calls (more expensive during business hours).

The price of electricity may vary in several sets of time cycles. This is because the cost of producing electricity varies with the hour of the day, the day of the week, and the month of the year. The term "time-of-day rate" is used for any rate schedule that varies the price based on the time that electricity is delivered. In most cases, time variations in electricity price are included in demand charges.

If the utility experiences its highest loads in summer, it is called "summer-peaking." Summer demand peaks are usually created by air conditioning loads on exceptionally hot days during the mid-afternoon. Therefore, summer-peaking utilities typically have high demand rates that apply during the daytime hours of the summer months.

Similarly, winter demand peaks are usually created by electric heating equipment on exceptionally cold days. Heating peaks may occur during the night, if the utility serves primarily residential customers. Or, a heating peak may occur during the daytime, if the utility serves a large commercial heating load.

Some utilities have both summer and winter peaks, which are factored into demand charges for both seasons.

Some utilities levy demand charges without regard to the time of consumption. Usually, such rate schedules are a mistake, and they may motivate customers to make changes to consumption patterns that are pointless, or even contrary to the interests of the utility.

■ Ratchet Provisions

A "ratchet charge" is a demand charge that carries over from one month to later months. For example, a summer-peaking utility may have a three-month period of high demand charges during the summer. Its demand rate may have a ratchet provision that stipulates that the highest demand measured during this three-month period will be used to calculate demand charges for each of the following nine months, although the demand charge may be lower outside the summer months. In some cases, a ratchet provision stipulates that the demand for months outside the peak period is considered to be a certain fraction of the demand that occurs within the peak period.

The logic of a ratchet charge is that the utility must pay for equipment that it owns all year, even though the equipment is used to its full capacity only during a small part of the year. A customer who contributes to the utility's peak load is penalized by a ratchet charge, whereas a customer whose demand peak occurs outside the utility's peak load period receives a lower demand rate.

■ "Hour Charges" and Load Factor

In constructing its rate schedules, a utility faces the dilemma that it wants to encourage sales of energy while discouraging demand. Some utilities respond to this dilemma by using "hour charges," which are demand charges discounted for high consumption. (The term "hour charge" stems from the fact that it is calculated on the basis of kilowatt-hours per kilowatt, which results in the dimension of hours.)

Hour charges benefit customers who have a high "load factor," which is the ratio of the average load to the peak demand.

■ Effect of Energy Conservation on Demand

An energy conservation measure will reduce demand charges if it operates continuously, at least during the period when demand charges are levied. For example, installing light fixtures of lower wattage reduces demand as well as consumption. On the other hand, an energy conservation measure may not reduce demand if it fails to operate, even one time, during the demand charge period.

For example, an expensive thermal storage system will not reduce demand if it experiences even one failure during the demand measurement period. Likewise, the "demand limiter" controls commonly installed on chillers do not reduce demand charges if the chillers are allowed to operate at full output during periods of high cooling load.

Electricity Charges Based on Customer Equipment

Some aspects of electricity pricing are based on the characteristics of the customer's electrical system or equipment. The customer may be able to change his equipment to reduce electricity costs.

■ Power Factor Penalties

Power factor is a measure of the phase relationship between the voltage of the power system and the current that is drawn by a device in response to this voltage. Power factor is 1.00 if the current is perfectly in phase with the voltage, and is less than 1.00 if the current is out of phase with the voltage. It is also common to express power factor as a percentage, with 100% indicating that the current is perfectly in phase with the voltage.

(Mathematically, power factor is the cosine of the phase angle between the voltage and the current waveforms. This assumes that both waveforms are pure sine waves. Some modern electrical equipment, such as variable-frequency motor drives and electronic transformer ballasts, distort the waveform in the electrical system. This may create an effective reduction of power factor.)

If a device is essentially a resistor, such as an incandescent light bulb filament, the current flows in phase with the voltage, and power factor is high.

Some major electrical equipment, especially motors and magnetic lighting ballasts, are inductors. The current flow through an inductor rises slowly (lags) after the voltage is applied. The current through a pure inductor lags the voltage across the inductor by 90°.

Conversely, the current flow through a capacitor leads the voltage by 90°. (This does not mean that the current in a capacitor has a magical ability to anticipate what the voltage will be. It simply means that voltage cannot develop across a capacitor until it accumulates electric charge. Perhaps it is more realistic to say the current flow in a capacitor lags the voltage by 270°.) Hardly any modern electrical equipment is capacitive.

Utilities are concerned about power factor because low power factor raises their distribution cost. If the current is significantly out of phase with the voltage, more current is required to deliver a given amount of power. (Power is delivered only by current that is in phase with the voltage.) The additional useless current is called "reactive current." The reactive current increases the utility's costs for wiring, transformers, and switchgear. To compensate for this, utility rate schedules often include a penalty for low power factor. This is sometimes expressed as a discount for high power factor.

Power factor penalties typically commence when the power factor falls below about 85%.

Power factor penalties vary widely from one utility to another. Some utilities do not include power factor in their rate schedules, while others charge a fairly severe penalty for low power factor. Low power factor is commonly caused by inductive loads, especially by electric motors and old-style magnetic lamp ballasts. The customer can improve power factor by installing capacitors in his power circuits.

The manner in which the utility measures power factor may be important. The utility may measure a facility's power factor during the facility's peak demand period, or at other times. The measured power factor is lowest when equipment that operates with low power factor comprises the largest fraction of the overall load. For example, if the facility has electric resistance heating, the power factor will be relatively high during the peak heating period. On the other hand, the power factor may be low during a cooling period, when the load is dominated by electric motors and fluorescent lights.

An odd case occurs with variable-frequency motor drives of a type whose power factor declines with motor speed. These drives may reduce the facility's overall power factor during periods of low load, but not during periods of high load. (For details, see Reference Note 36, Variable-Speed Motors and Drives.)

■ Delivery Voltage

Rate schedules for large customers may include a discount if the electricity can be provided at a specified high voltage. Delivery at the utility's distribution voltage saves the utility the cost of a step-down transformer.

■ Ownership of On-Site Distribution Equipment

The customer may have the option of owning the electrical distribution equipment (transformers, poles, wires, etc.) on its own property. This may reduce the electricity rate, since the utility does not have to purchase this equipment.

■ Incentive Rates

Many utilities offer favorable rates to motivate customers to install equipment that benefits the utility. Some rates are intended to increase electricity sales, such as discounted rates for heat pumps and electric heating. Other rates are intended to reduce the utility's peak load, such as discounted rates if thermal storage is installed. Some rates promote both purposes, as for electric storage heating.

Many utilities offer a one-time rebate or subsidy to promote the installation of such equipment.

■ Effect of Multiple Meters

It is generally cheaper to purchase electricity in larger quantities. This quantity discount often takes the form of a commodity charge that declines in several steps as more kilowatt-hours are used. The rate of decline between steps may be steep, with the last kilowatt-hours purchased during a month costing only a fraction of the first.

In many facilities, electricity is provided through multiple utility meters, each of which has a separate billing account. Multiple meters are often the residue of successive additions to a facility. In each account, the charges for electricity start at the highest rate. This often makes it advantageous to combine billings on a single account.

But, be aware that this may not be true, especially if there are differences in the demand rates. For example, combining metering might require a different rate schedule having a lower commodity charge but higher demand charges. Combining metering may reduce total demand charges if the different users operate on different schedules, so that the demand peaks are not superimposed. Conversely, if the demand peaks of all users occur simultaneously, the net demand charge may be increased.

Administrative Charges

A common feature of rate schedules is a fixed monthly charge that covers the administrative costs of servicing the account. This can take a number of forms. A "customer charge" is a fixed monthly charge for each customer. A "meter charge" is a fixed monthly charge for each meter. Another type of administrative charge is a minimum charge for KWH that applies even if a smaller quantity of electricity is actually used.

Excess-Consumption Charges

An excess-consumption charge is a rate that increases with the amount consumed above a certain amount. This is opposite the usual practice, which is to give a discount for increased consumption. Excess-consumption charges usually are limited to periods of energy shortage, when they may be initiated by the regulatory authority.

Competitive Purchasing of Electricity

Ordinarily, a customer deals with the one electric utility that serves his geographic area. However, new utility regulation may allow the customer to negotiate with electric utilities serving adjacent areas. In effect, the customer may shop around for electricity among various sources. The local utility may be required to serve as a conduit for electricity purchased from another utility, although it may charge a fee for the use of its lines.

This procedure is called "wheeling." This area of utility regulation is evolving and controversial. Larger customers should look into competitive purchasing to determine whether it offers any potential for reducing electricity costs.

Negotiating Rates

As with other commodities, the price of electricity is negotiable, especially with large customers. As in any negotiation, know when to negotiate, know what your alternatives are, and understand the desires of the other party. The best time to negotiate is before building a facility, when you have the greatest variety of options. However, conditions in the utility market change continually, providing new opportunities for negotiating rates.

You may negotiate on the basis of competition from other suppliers, including the possibility of generating your own power, perhaps with a cogeneration plant. You may offer to change consumption in a way that benefits the utility, such as by installing a thermal storage system. Changes in utility industry regulation that allow competitive purchasing provide an opportunity for negotiating with your present supplier.

Before negotiating, develop an understanding of the utility's situation, including its load profiles, its reserve generating capacity, the combination of generating equipment it owns, the type and availability of fuels that it uses, its ability to buy and sell electricity in the adjacent power pool, and its future plans and projections. Changes in the utility's load characteristics, e.g., whether it is summer-peaking or winter-peaking, can make the utility more or less willing to negotiate rates, depending on how the customer's load affects the utility.

Also, assess external factors, such as political pressure on utilities to promote demand-side management, the fuels that are presently considered to be the most environmentally acceptable, and whims of regulatory agencies. Such factors tend to change more quickly and erratically than changes to the situation of the utility itself.

Low-Temperature Heat Sources & Heat Sinks
for Heat Pumps and Cooling Equipment

Soil
Ground Water
Surface Water

Solar Collectors
Waste Heat

You can increase the efficiency of cooling equipment by lowering the temperature at which the equipment rejects heat. The most common heat sink for cooling is the atmosphere. The disadvantage of the atmosphere as a heat sink is that it gets warmer at the same time you need cooling. In many cases, you can reject heat at lower temperature by using the soil, well water, or surface water as a heat sink. These alternative heat sinks can typically lower the average condensing temperature by 20°F to 30°F (11°C to 17°C), substantially increasing cooling efficiency.

Similarly, you can increase the efficiency of a heat pump, when it is heating, by increasing the temperature of the heat source. Again, if the atmosphere is the heat source, its main disadvantage is that it gets colder when you need heating. Soil, well water, and sometimes surface water are warmer heat sources during cold weather. Depending on climate and the nature of the heat source, the average heat source temperature may be 10°F to 30°F (6°C to 17°C) higher than the air temperature.

Using an alternative heat source for heat pumps is especially important because the capacity and efficiency of heat pumps drop radically as the temperature of the heat source drops. In fact, heat pumps must cease operating and rely on a secondary heat source (typically electric resistance heat) when the outside air temperature becomes too cold. Using a higher-temperature heat source eliminates this problem, and greatly extends the range of climates in which heat pumps can be used for heating.

Heat pumps can also use solar collectors and waste heat as low-temperature heat sources. These sources are available at all times of the year. Certain types of solar collectors offer the possibility of using the collector as a heat radiator at night, so this type of collector can operate as both heat source and heat sink.

Use this Note to help you screen the possibilities for heat sources and heat sinks other than the atmosphere. If you find a likely candidate, study it further. Each of the sources has an ample literature, so we won't repeat a lot of detailed information here. A good starting point for more detail is the *ASHRAE Handbook*, which in turn offers additional references.

But, beware of the optimistic claims of promoters, who write much of the literature on this topic.

Reference Note 23, Non-Fossil Energy Sources, covers an entirely different group of free energy sources, which provide heat at much higher temperatures than is available from the heat sources covered in this Reference Note.

Soil

The temperature of the earth's surface tends to remain constant below a certain depth, which is typically about 12 feet (4 meters). This makes deep soil a potential heat source and heat sink. Heat pump systems exploiting this principle were first promoted during the 1980's as a way of using heat pumps in colder climates where air-to-air heat pumps are considered uneconomical. Such systems have been called "ground source" or "earth-coupled." Recently, promoters have started calling them "geothermal" systems, which is a misapplication of a term that describes an entirely different class of heat sources. (Geothermal sources typically are hot enough for direct use, without needing heat pumps to raise their temperature. See "Non-Fossil Energy Sources" for details.)

The distinguishing feature of an earth-source heat pump installation is its buried heat exchanger, which consists of a great length of pipe in which the loop water is circulated. The pipe may be buried in a horizontal grid pattern or in deep vertical hairpin loops. The pipe may be standard plastic water or gas pipe, or specialized copper alloy tubing. The pipe must be able to resist soil conditions, such as acidity, for many years. It must also be compatible with any antifreeze material that is used.

With horizontal fields, the pipe is laid in trenches spaced one or two meters (three to six feet) apart. This requires a large amount of trenching and many pipe joints. Figure 1 shows a horizontal field being installed.

An alternative method of installing a horizontal field is to excavate the entire field, lay down the pipe, and backfill over the pipe.

With vertical installation, one or more hairpin loops may be buried in holes that typically range in depth from 100 feet (30 meters) to 400 feet (120 meters). Figure 2 diagrams this method. The holes are filled with a grout, such as bentonite clay, that must be kept in a slurry state. The grout provides heat transfer, allows for thermal expansion and contraction, and prevents penetration of surface contaminants into the ground through the bore holes. The need for grout is a potential problem in locations where the grout can dry out.

Earth-source heat pumps are still in the experimental stage of development, notwithstanding the fact that some promoters have been pushing them aggressively. By the 1990's, many earth-coupled heat pump systems were being installed, including systems for large facilities. However, major problems were continuing to come to the surface. Anyone contemplating designing or using soil as a heat source should study the latest experience, and be prepared to take risks.

Burying the heat exchanger is the largest part of the system's cost, and the pipe itself is not cheap. Engineering cost may be relatively high because these systems are novel. Certain contractors specialize in earth-source heat pumps, and they do the engineering as well as the installation. The disadvantage of this arrangement is that the contractor will lean toward minimizing his own costs rather than toward maximizing the efficiency of the system.

The installation cost of an earth source is comprised mostly of the labor cost and equipment rental required to dig holes in the ground, to bury the heat transfer pipe in a suitable bedding material, and to make piping connections. These costs are subject to great variation, but generally are high.

Trenching becomes much more expensive as the depth increases below about one meter, which tempts installers to bury horizontal fields too shallow for good efficiency. It has been claimed that safety regulations requiring shoring of deep trenches makes it uneconomical to bury horizontal fields at an efficient depth.

The main advantages of using soil as a heat source and heat sink are:

• *good compromise temperature.* In most locations, the temperature of deep soil is a good compromise for use in both heating and cooling. Vertical heat exchangers are buried at depths where the soil temperature is essentially constant. With horizontal fields, it is usually not economical to bury the heat exchanger deep enough to reach a constant soil temperature.

To date, most heat exchangers are horizontal fields that are buried at depths ranging from just below the frost line down to about 6 feet (2 meters). At these depths, the temperature of soil varies seasonally. A redeeming factor is that soil temperature lags several months behind air temperature. This provides a useful storage effect that cools the soil in advance of the cooling season, and warms the soil in advance of the heating season.

The heat exchanger itself changes the temperature of the soil. This may substantially reduce the efficiency of system operation. The importance of this effect depends on soil type and water content, as well as the heat exchanger size and system operating schedule. Engineering data on this issue is still sketchy.

WESINC

Fig. 1 Horizontal soil heat exchanger field This field is being installed to serve as the heat source and heat sink for a church in Minnesota. The climate here would not allow efficient operation of an air-source heat pump. Obviously, a large amount of open space is required. Loose soil makes it too expensive to bury the pipe at a depth that would provide the best efficiency.

• *no need for supplemental heat sources or heat rejection equipment.* Soil temperature remains fairly constant below a certain depth, so there is no need for either a supplemental heat source (boiler) or heat rejection equipment, provided that the buried heat exchanger is made large enough. On the other hand, economics or limitations in land area may dictate a hybrid system, most likely involving a cooling tower to supplement cooling operation.

• **simple control.** Complex loop temperature control is avoided because the soil temperature controls the loop temperature directly.

The main disadvantages of using soil as a heat source and heat sink are:

• *land area requirement.* If horizontal fields are used, a large amount of bare ground is required for installation of the piping field. The entire area must be trenched to lay down the piping, and the ground cannot be used for other construction. Vertical heat exchanger coils need much less land area.

• *uncertain heat exchanger reliability.* The long-term reliability of the heat exchanger piping is unknown. Although the pipe appears to be well protected, it is subject to stress from thermal expansion and contraction, and from soil conditions. One type of plastic originally used

Fig. 2 Vertical soil heat exchanger field This requires much less surface area than a horizontal field. The cost of trenching is replaced by the cost of drilling. Leakage is less of a potential problem. However, tricky methods are needed to maintain good contact between the pipe and the soil while allowing for thermal expansion.

(polybutylene) has experienced serious failures. Polyethylene pipe is now being used, but is subject to softening from some antifreeze impurities. The long-term performance of buried copper heat exchangers is untested.

• *shortcuts in design and installation.* System efficiency and reliability are in conflict with the desire of the owner and the contractor to limit costs. The efficiency of the system may be seriously degraded by inadequate heat exchanger size, or failure to bury horizontal fields deeply enough.

• *failure-prone installation.* Damage in installation and defective joints are serious problems. Especially in horizontal fields, rough handling of the heat exchanger pipe is virtually unavoidable. These problems cause long-term failures that may occur long after the warranty has expired and the original occupants have departed.

• *repair difficulties.* If a leak occurs in a horizontal field, it may be extremely difficult and expensive to find the leak. Leaks are less likely to occur in a vertical heat exchanger, but they may be virtually impossible to repair. Leakage is usually handled by plugging and abandoning the leaking loop. This reduces heat exchanger size and system efficiency.

In terms of thermal characteristics, soil is similar to ground water. A ground water system is usually less expensive to exploit, but the water must be pumped out of the ground, and often, pumped back.

Ground Water

Ground water is water that comes from wells. Water may be taken out of the ground, passed through the heat pump system, and then used on the surface for other purposes, such as irrigation. Figure 3 shows this arrangement.

Or, it may be necessary to inject the water back into the aquifer. This requires two wells. One draws water out of an aquifer at one depth, and the other returns the water to a different aquifer at a different depth. It may be possible to use a single bore hole to reach both aquifers with a coaxial pipe.

The cost of tapping ground water varies widely, depending on the depth, the nature of the ground, the need for reinjection, and other factors.

The main advantages of ground water as a heat source and heat sink are:

• *good compromise temperature.* Wells are deep, so the temperature of ground water remains essentially constant. In most locations, the temperature is a good compromise for use in both heating and cooling.

• *no need for supplemental heat sources or heat rejection equipment.* Ground water sources usually have enough capacity to eliminate the need for either a supplemental heat source (boiler) or heat rejection equipment. This is a major factor in reducing the system's initial cost.

- *simple control.* Complex loop temperature control is avoided because the ground water temperature dictates the loop temperature.

- *ability to use water extracted for other purposes.* If well water is used for other purposes, such as service water, irrigation, lawn watering, and industrial processes, it can first be used as a heat source or heat sink. If the water requires heating for the application, using it as cooling water warms it, providing a double saving in energy.

The main disadvantages of ground water are:

- *limited availability.* Ground water is not available in all locations, at least not at an economical depth.

- *restrictions on use.* Even in areas where ground water is available, environmental restrictions may limit its use in heat pump systems. Environmental restrictions may be based on several objections. One is wasting of the resource. This objection can be overcome by returning the water to the aquifer ("reinjection") after it is used.

Another objection is that an internal leak in the heat pumps may cause refrigerant to be injected into the aquifer. It may be possible to allay this concern by isolating the ground water from the refrigerant with a double-wall heat exchanger.

Another objection is that altering the temperature of the water may cause some harm. There appears to be no strong scientific basis for this concern, but projects may be halted during the public debate about it.

Fig. 3 Ground water (well water) heat source and heat sink This is an open loop, in which the well water is discharged to the surface after passing through the heat pump system. In some locations, the water would have to be injected back into the ground, at a different depth.

- *heat exchanger maintenance.* Ground water usually is saturated with minerals. When ground water is used as a heat source, it is cooled in the heat exchanger, and becomes less able to dissolve minerals. As a result, the minerals are deposited in the heat exchanger, which requires periodic cleaning.

Surface Water

Surface water is water that originates on the earth's surface from rain. Large bodies of surface water, such as oceans, lakes, and rivers can serve as heat sinks. Inland bodies of water change in temperature with the seasons, but they are generally cooler than the atmosphere throughout the cooling season. Exposed bodies of water are cooled by evaporation and by radiation into the night sky. On the other hand, water that is darkened by contaminants absorbs solar heat during the day.

Surface water may also serve as a heat source. However, the air temperature may often be warmer than the water temperature during the heating season, especially in spring. For this reason, surface water is usually not a good alternative to an air source for heat pumps.

Surface water is dirty and corrosive, so it is generally not used directly in the system loop. (Surface water is corrosive because exposure to the atmosphere keeps it saturated with oxygen.) One way of dealing with this problem is to use a heat exchanger located on dry land. Figure 4 shows this arrangement.

Another approach is to keep the entire water loop closed, and circulate the water through heat exchange loops that are submerged in the body of water. Figure 5 illustrates this approach. The pipe may be made of inexpensive material, such as plastic. This arrangement normally requires no cleaning, but it is more vulnerable to leakage. The submerged pipe should be kept under positive pressure to ensure that no dirt from the body of water can enter the system.

Piping is usually the major cost component in exploiting surface water. This cost varies widely from one location to another, depending mainly on the accessibility of the body of water.

The main advantages of surface water as a heat sink are:

- *lower temperature.* In many locations, surface water is cooler than ground water.

- *no need for heat rejection equipment.* Almost any permanent body of surface water has enough heat rejection capacity to eliminate the need for other heat rejection equipment.

The main advantage of surface water as a heat source is:

- *in warmer climates,* there is no need for supplemental heat sources. This requires the body of water to be warm enough and large enough so that the heat exchanger does not become encrusted with ice at any time.

A disadvantage of surface water as a heat source and heat sink is:

- *uncertain future availability.* Environmental restrictions already exist against the use of surface water for cooling large power plants. The objection is that altering the temperature of the water may harm fish, and this assertion might be extended to other organisms. These restrictions may be extended to smaller applications.

Another concern is that an internal leak in the heat pumps may cause refrigerant to be released into the body of water. It may be possible to allay this concern by isolating the water from refrigerant contamination with a double-wall heat exchanger.

Another disadvantage of surface water as a heat source is:

- *lower temperature.* The water temperature falls below the air temperature much of the time, especially in spring.

Solar Collectors

Using solar collectors as a low-temperature heat source for use with heat pumps is one of the more promising applications of solar energy. Heat is provided at a temperature much higher than is available from other low-temperature sources, maximizing the heating efficiency of the heat pumps.

Using solar collectors with a heat pump system improves the efficiency of the collectors themselves. When solar collector are used alone, they provide no benefit until the temperature of the collectors rises enough to drive heat into the facility's distribution system. On the other hand, heat pumps can exploit any heat that is collected. By the same token, using solar collectors as input to heat pumps allows the collectors to operate at lower temperature, which reduces their losses.

For the same reason, you can use less expensive types of solar collectors in which the collector plates are exposed. These units avoid the cost of glazing, rigid frames, insulation, and thermal seals. The lack of an insulating jacket increases the heat loss of these collectors, but this disadvantage is lessened when they are used at low temperature with heat pump systems.

The main advantage of solar collectors as a heat source for heat pumps is:

- *high temperature.* The average operating temperature of solar collectors is high enough to keep heat pumps at peak efficiency.

WaterFurnace International, Inc.

Fig. 4 Surface water heat sink, and possibly, heat source Surface water is almost always cool enough to be a good heat sink, if there is enough of it. It can also be a heat source, if no ice forms in the system. This system uses a heat exchanger to keep dirt out of the building water loop.

WaterFurnace International, Inc.

Fig. 5 Surface water heat source/sink with enclosed water loop This arrangement is similar in performance to the one shown in Figure 4. It is more vulnerable to leakage, but it may be less troubled by ice when the water is being used as a heat source.

The main disadvantages of solar collectors are:

- *lack of heat input at night.* If the facility requires heating at night, the system must have another source of heat.
- *high cost.* Combining solar energy with heat pumps may reduce the cost of solar heating, but it raises the cost of the heat pump system considerably.
- *complexity.* The number of factors that need to be addressed can be overwhelming, especially if the system has thermal storage and/or other heat sources.

If you use unshielded solar collectors that are non-metallic, these can also function as heat rejection devices at night. In addition to rejecting heat to the night air by conduction, they also radiate heat into the night sky. The night sky has an effective temperature that is colder than the air temperature. The difference between the night air temperature and the night sky temperature ranges from less than 10°F (5°C) to almost 50°F (28°C). The difference depends primarily on the cloud cover and the amount of moisture in the air. For example, over the entire western United States, the average sky temperature in summer is more than 20°F lower than the air temperature.

(You can't use conventional solar collectors with glass covers as heat radiators, because heat radiation does not pass through glass. Also, unshielded metal surfaces are not good radiators because they have low emissivity.)

Even so, if you want to use this unusual method of getting rid of heat, be aware that the heat rejection capacity at night is much less than the heat collection capacity during the day.

Waste Heat

Potential waste heat sources include any gas, liquid, or solid that is discharged from the facility in sufficient quantity to satisfy at least a substantial fraction of the heat input requirements of the heat pump system. Exhaust air, sewage, and process effluent have all been exploited as heat sources.

Figure 6 shows a system for recovering heat from industrial waste water. If the water is dirty, the system needs a flushing system to clean out the heat exchanger periodically.

Figure 7 shows an interesting method of recovering the heat from domestic water drains. This figure shows a special type of heat exchanger, but the same function could be performed by a conventional heat exchanger.

The typical advantages of waste heat sources are:

- *higher temperature.* The temperature of most waste heat sources is higher than that of soil, well water, or surface water.
- *low cost.* With many waste heat sources, only rudimentary heat exchange equipment is needed to convey the heat into the loop system.

The typical disadvantage of waste heat sources is:

- *limited and/or sporadic availability.* Many waste heat sources can provide only a fraction of the heating requirement, so other heat sources are required.

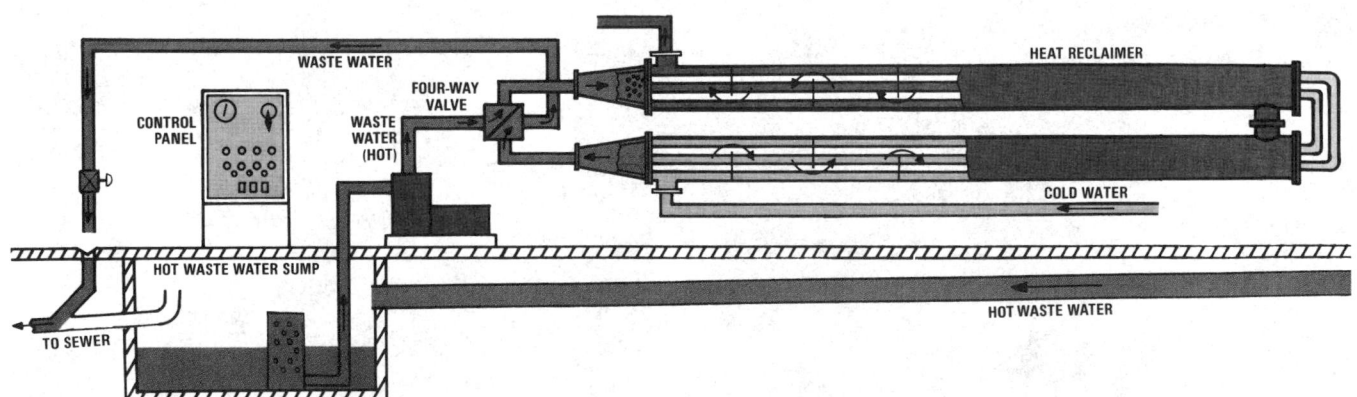

Ludell Manufacturing Company

Fig. 6 System for recovering heat from process waste or sewage   This is a conventional tube-and-shell heat exchanger combined with an automatic backflush system. Flushing is needed to keep the heat exchanger effective if the waste fluid is dirty.

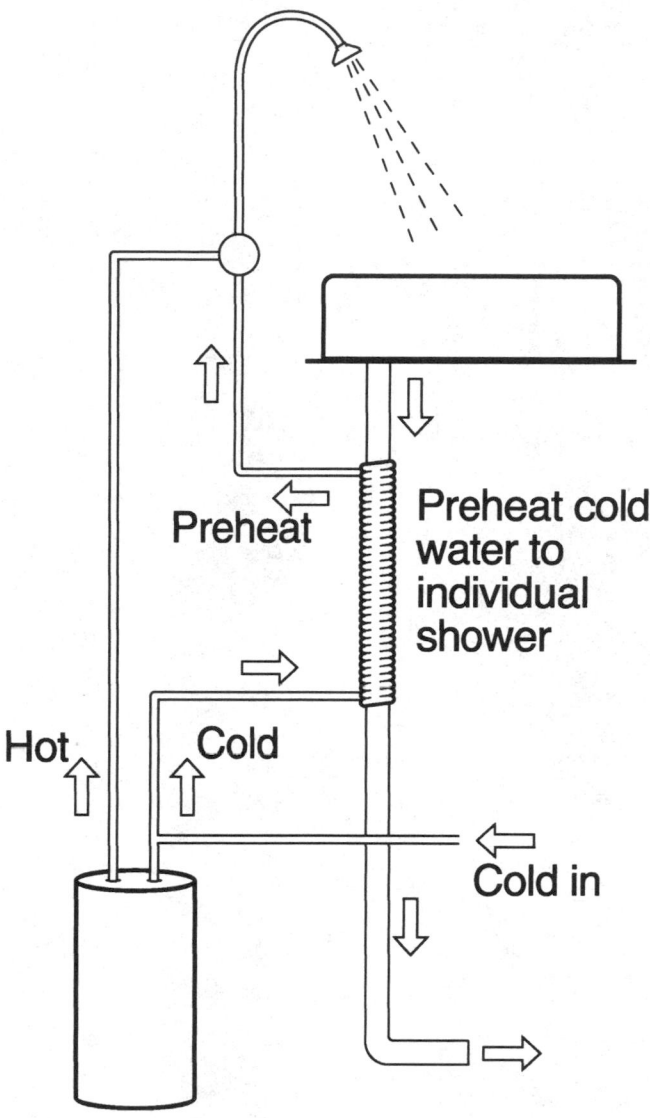

Vaughn Manufacturing Company

Fig. 7 System for recovering heat from domestic water drains This could be scaled up for major hot water users, such as gymnasiums and swimming centers. The heat exchanger here is a special type that replaces a section of the drain pipe. It must be almost perfectly vertical, so that waste water will drain uniformly down the sides of the drain pipe. A conventional heat exchanger would also work in this application.

Non-Fossil Energy Sources

Active Solar Heating and Cooling
- **Equipment**
- **Applications**
- **Development Status and Economics**
- **Energy Density and Availability**
- **Environmental Considerations**
- **Need for Conventional Energy Back-up**

Photovoltaic Electricity Generation
- **Photovoltaic Cells and Their Development Status**
- **Photovoltaic Systems and Applications**
- **Energy Density**
- **Availability**
- **Environmental Considerations**
- **Economics**
- **Relationship to Conventional Energy Sources**

Solar Thermal Power Generation
- **Equipment and Development Status**
- **Energy Density and Availability**
- **Environmental Considerations**
- **Economics and Feasibility**
- **Relationship to Conventional Energy Sources**

Wind Power
- **Equipment and Applications**
- **State of Development**
- **Energy Density and Availability**
- **Environmental Considerations**
- **Economics**
- **Relationship to Conventional Energy Sources**

Local Hydropower
- **Equipment and Stream Preparation**
- **Applications**
- **State of Development**
- **Energy Density and Availability**
- **Environmental Considerations**
- **Economics**
- **Relationship to Conventional Energy Sources**

Geothermal Hot Water and Steam
- **Origin, Availability, and Energy Characteristics**
- **Extraction Techniques and Development Status**
- **Applications**
- **Economics**
- **Environmental Considerations**
- **Relationship to Conventional Energy Sources**

Waste Product Combustion
- **Types of Combustible Waste**
- **Equipment and Applications**
- **Energy Content**
- **Availability**
- **Environmental Considerations**
- **Economics and Feasibility**
- **Relationship to Conventional Energy Sources**

Wood and Biomass Combustion
- **Energy Content**
- **Equipment**
- **Applications and Limitations**
- **Availability and Economics**
- **Environmental Considerations**
- **Relationship to Conventional Energy Sources**

This Note is a quick review of energy sources that are alternatives to conventional fossil fuels. These energy sources are commonly called "renewable energy" or "free energy." We use the somewhat clumsy term "non-fossil" for this category of energy sources because none of them is literally free, either in terms of money or natural resource consumption. Some of them are not actually renewable or non-depleting.

These energy sources are capable of producing temperatures that are high enough for direct use, or of producing electricity or mechanical work directly. Reference Note 22, Low-Temperature Heat Sources and Heat Sinks for Heat Pumps and Cooling Equipment, covers a different category of "free" energy sources that have temperatures so low that they are generally usable only as heat sources for heat pumps.

This Note is limited to energy sources that may be exploited at the present time, even though some of them are still in the experimental stage. It is also limited to sources that are appropriate for exploitation by individual sites, although the sites may have to be large. This Note does not cover sources that require action at a regional or governmental level, such as nuclear energy, ocean thermal energy, tidal energy, and solar arrays in space.

Solar energy is potentially so important that three of its major applications are discussed separately here. Daylighting and passive heating applications of solar energy are covered individually in Subsections 8.3 and 8.4, respectively. For the general characteristics of sunlight as an energy source, see Reference Note 24, Characteristics of Sunlight.

This Note will help you make an initial screening of the non-fossil energy sources that may be available to you. If you decide to go further with any of these sources, you have more study ahead. Each non-fossil source involves many issues, and the details are important. You will also want to learn the current status, as some of these sources are still evolving. Each source has an ample literature. Two good starting points for further reading are the *ASHRAE Handbook* and Marks' *Standard Handbook for Mechanical Engineers*, both of which include references to leading books on each topic.

As you study "renewable" energy sources, beware of the widespread nonsense and unrealistic assumptions that confuse the discussion. Much of the literature, and perhaps most of the public comment, is by enthusiasts who either know little about the subject or are blinded by their fervor. As a rule, the most credible reference sources tend to be dry and technical. Sorry.

Active Solar Heating and Cooling

An active solar heating system uses collectors outside the structure to capture the energy of sunlight in a heat transfer medium, which usually is water. Air is occasionally used as the heat transfer medium. The system usually includes a tank to store excess heat. The flow of heat from the collectors or the storage tank can be regulated precisely, as long as the available supply of energy exceeds the requirement.

The name "active" distinguishes these systems from "passive" systems, in which sunlight is admitted directly for space heating. Passive solar heating is simpler in concept, but it is proving to be difficult to accomplish well. It is covered in Subsection 8.4.

■ Equipment

An active solar system consists of collectors, structural supports for the collectors, a storage tank, structural support for the storage tank, piping, valves, and thermostatic controls. In addition, there is the conventional heating equipment within the space. Solar systems can be piped in a variety of ways, which differ in their ability to avoid freeze damage and deal with excess solar input. Solar collectors come in two general types, flat plate and focussing.

The typical flat collector is a shallow sealed box. The sunlight is captured by a metal plate that contains water passages. Insulation is installed behind the metal plate to reduce conductive heat loss. Heat loss from the front of the collector is minimized by one or two transparent cover sheets. Figure 1 shows an array of flat plate collectors.

Flat collectors usually are fixed, but some are installed on pivoting bases to improve collection efficiency.

Flat plate collectors are limited in the temperature they can produce. At flow rates that provide reasonable output, the best units produce temperatures that range from 120°F to 170°F (50°C to 80°C) on sunny days.

The collector is a major part of the system cost, so various types of cheaper flat collectors have been developed. The cheapest consists of continuous rolls of black plastic with extruded water passages. The material is cut to any desired length and fitted with snap-in headers. Such collectors have higher losses and produce lower temperatures, but they may be satisfactory in warmer climates for preheating domestic water and boosting heat pumps.

Focussing collectors use reflectors to focus sunlight on small water tubes. Most systems use parabolic reflectors that rotate to track the sun. Figure 2 shows an array of such collectors. These can provide temperatures of 200°F to 300°F (90°C to 150°C), depending on their design and the flow rate.

Focussing collectors are disabled by any cloud cover, haze, or fog that renders the sun indistinct, because the sunlight cannot be focussed under those conditions.

Some focussing reflectors are designed to remain stationary. These focus less accurately, so their efficiency is lower. Some hybrid arrangements supplement non-directional collectors with reflecting skirts to increase the sunlight collection area. These provide somewhat higher temperatures, and they operate well with diffused sunlight.

WESINC

Fig. 1 An array of flat-plate solar collectors This installation shows typical dimensions and mounting of flat-plate collectors. Sadly, it also illustrates the poor design and maintenance that is given to this basically simple technology. Individual collectors had to be removed because no access was provided for maintenance of the connecting joints. These are subject to rapid wear because of the large thermal expansion coefficient of aluminum and the large temperature cycles in daily operation. The system was eventually abandoned because of avoidable freeze damage.

At times when the solar system collects more heat than is needed, where does the excess heat go? This design issue is a major factor in system failure and reduced collector life. When there is no place to reject the heat, the collector temperature rises until the heat radiation from the collector equals the heat input, which occurs at the "stagnation temperature." The stagnation temperature may be well above the boiling point of water.

The problem is easiest to solve with movable focussing collectors. Heat input can be halted by aiming the collectors away from the sun, as shown in Figure 2. Some systems are controlled in a fail-safe manner, so that the collector points downward if the controls fail or if safe temperatures are exceeded.

With flat plate collectors, various methods are used to keep collector temperature within limits. None of them are entirely satisfactory. For example, see Figure 3. A common method is installing an external heat rejection coil to vent excess heat from the solar system. Movable shading would be more efficient, and would reduce weathering of the collectors during idle periods, but this approach has not yet been brought to market.

Avoiding freezing is easier. If the system uses a heat transfer loop, an antifreeze solution can be used in the collector circuit. Alternatively, the collectors can be designed to drain into a holding tank when the outside temperature approaches freezing.

Solar cooling is not a distinct technology. Instead, the heat from an active solar system is used to operate a conventional absorption chiller. The cooling output of an absorption chiller is degraded severely if the input temperature of the input heat is below the design temperature. For conventional lithium bromide absorption chillers, the rated minimum input temperature is about 250°F (120°C). Therefore, the absorption chiller needs high-temperature solar collectors, which typically are parabolic tracking units. These are more expensive than flat plate collectors, for a given heat input, so adding the absorption machine increases the cost of the solar system that feeds it, making the overall system very expensive. (Some solar cooling systems have been built with flat plate collectors, but they required the chiller to be horrendously oversized.)

"Solar cooling" has also been accomplished by using a photovoltaic source to power a conventional electrical air conditioner, but there is nothing inherently "solar" about either method of cooling.

■ **Applications**

The usual applications for active solar systems are space heating and service water heating. Figure 4 shows a typical installation for these purposes. Solar systems can be used to provide thermal energy for any application if the temperatures are compatible. The same is true of solar cooling systems.

WESINC

Fig. 2 Focussing solar collectors The parabolic reflectors track the sun, focussing the heat on the small tube containing water. This array, in a sunny climate, uses the high-temperature water to drive absorption cooling equipment for a government hospital. The individual rows of collectors are fail-safe. If there is a defect or excessive temperature, the row tips upside down, as on the right. This avoids stagnation damage and protects the reflecting surface. Electrical power for tracking is provided by batteries. Maintenance is simple, but when a key individual goes on vacation, the arrays start to fail within days. Several abandoned batteries have been left in the foreground.

Solar water heating is simpler, more efficient, and less expensive than solar space heating. The fact that the system heats a liquid from a cold state substantially improves the efficiency and output of the collectors. Storage costs are lower because a water heating system requires some storage anyhow.

Solar collectors can be used to boost the efficiency of heat pumps, without much additional equipment. This application is covered in Reference Note 22, Low-Temperature Heat Sources and Heat Sinks.

■ Development Status and Economics

As a technology, active solar heating systems are mature. They do not require exotic technology, and the principles are well known. Maintenance requirements are modest, but many have failed from lack of maintenance.

Simple solar water heaters proliferated in Florida before World War II, and more recently in Israel. The American systems were abandoned as energy prices dropped, and even Israelis are reported to be losing enthusiasm for their solar heaters.

As a result of the energy crises of the 1970's, active solar space heating went through two periods of intense testing in the United States. The first was the installation of many solar systems as research projects sponsored by the U.S. Department of Energy and its predecessor agencies. The second was the installation of many commercial solar energy systems by private parties in response to a Federal solar energy tax credit that was in effect during the 1970's and 1980's.

The resulting experience confirmed calculations showing that active solar heating has a long payback period. Unfortunately, this fact was deliberately obscured in Federal demonstrations by mandating cost factors that were absurdly understated, and by including tax credits and other adjustments into the calculations for private sector installations.

The experience of the 1970's and 1980's was sufficient to indicate where the bottom of the cost curve lies. The cost of collectors is proportional to collector capacity, while the cost of other system components is subject to a substantial economy of scale.

The systems installed during this period suffered a high failure rate, and many of them are now abandoned. Figures 1 and 3 are sad example. The high failure rate of solar energy is not caused by a fundamental flaw in the concept, but by incompetence and sloppiness in design and installation. These factors were aggravated in demonstration installations by the easy availability of government money, and by lack of continuing interest. Also, many systems failed because of a blatant disregard for their modest maintenance requirements. In many cases, a solar system is abandoned before it collects as much energy as was invested to build it.

Better design, installation, and maintenance can keep solar systems operating, but they can never have a good economic rate of return. The diffuseness of sunlight gives solar heating inherently poor economics.

WESINC

Fig. 3 Pressure relief valves to protect against stagnation
This abandoned solar heating system failed to provide adequate protection against stagnation. The pressure became high enough to repeatedly vent the liquid, which contained antifreeze. The replacement water did not contain antifreeze. Eventually, the system was destroyed by freeze damage.

WESINC

Fig. 4 Typical space heating and domestic water heating application for solar energy This array is nicely integrated into the architecture of the building. It even faces in the correct direction, unlike many solar installations.

Solar energy is what heats the great outdoors. People want solar heat where it is cold, but the reason it is cold is that there is not much solar heat available. Concentration of solar energy costs money and takes up space. Even cheap methods are expensive in relation to the value of the energy recovered.

A number of solar cooling installations have been built, all as demonstration projects. As of this writing, several larger installations powered by focussing solar collectors continue to operate. Other installations have been abandoned because of poor economics and excessive maintenance requirements.

■ Energy Density and Availability

The energy density of sunlight is low. At the earth's surface, the maximum solar energy density is somewhat less than one kilowatt per square meter of collecting area. Night, clouds, haze, collector losses, and failure to point collectors directly at the sun greatly reduce the average amount of energy that can be collected.

Another limiting factor of active solar systems is that the collector temperature must be higher than the temperature of the space or the medium being heated. This reduces the operating time of active systems during cold weather, especially with flat plate collectors. (Passive solar energy collection does not have this limitation.)

■ Environmental Considerations

Solar heating is almost ideal in its lack of direct environmental hazards or nuisances. It is quiet, it emits nothing into the surroundings, and it requires nothing from the surroundings, such as cooling water. The usual antifreeze solution, propylene glycol, is harmless.

The main environmental cost of solar systems is the energy and materials required to produce them. Unfortunately, this is a large amount of energy in relation to the amount of energy that they can collect over their service lives.

Solar systems can have a major effect on architectural appearance and/or space requirements, because collector area typically is large in relation to the size of the facility. The faces of efficient solar collectors are dark and blotchy, which makes it desirable to hide them. They are often installed on the flat roofs of commercial buildings.

■ Need for Conventional Energy Back-up

Almost any location has occasional long periods of low sunshine that will deplete the economical storage capacity of a solar system. Therefore, most solar heating and cooling applications must be backed up completely with conventional energy sources.

Photovoltaic Electricity Generation

Photovoltaic electricity generation is the direct production of electricity from sunlight, without any thermal processes. Photovoltaic generation is potentially an ideal energy source. It has no moving parts, requires no maintenance, makes no noise, requires no input of water or other substances from the environment, and produces no pollution. The main problem is high cost, which may be reduced in the future by technological developments.

WESINC

Fig. 5 Large photovoltaic system This is part of an array of many similar rows of photovoltaic cells that cover the roof of a large classroom building. The array provides a peak capacity of about 100 kilowatts, which is one fourth of the peak load on the building. The system supplements power provided by a public utility. The panels are light, thin, easy to mount, and easy to connect together.

Photovoltaic Cells and Their Development Status

At the present time, photovoltaic electricity generation is accomplished primarily using materials that have semiconductor properties. In these materials, electrons are held fast, preventing conduction. Photons of sunlight are sufficiently energetic to release the electrons in these materials. If a thin conducting film is overlaid on such a material, electrons released by light are collected in the conducting layer, where they can flow as an electric current. An assembly of this type is called a solar cell. To achieve higher voltages and currents, cells are connected in arrays.

The first solar cells were made from thin slices of silicon, germanium, or other semiconductors cut from pure crystals of the material. This part of the process is essentially the same as used in the semiconductor electronics industry. Cells made in this manner are expensive. Achieving lower cost is helped by the fact that the purity of photovoltaic semiconductors is less critical than in electronic semiconductors. At present, crystalline semiconductors have a peak efficiency of about 20%.

Amorphous (non-crystalline) semiconductors have been developed. These are less expensive than crystalline semiconductors, but they are also substantially less efficient. They also lose efficiency rapidly in service. At present, amorphous semiconductors have a peak efficiency of about 9%, but efficiency soon drops to a fraction of that value.

The ultimate goal is to produce fairly efficient photovoltaic material that can be produced in large sheets, without the need to connect arrays of discrete small cells. Effort in this direction continues, but a fundamental breakthrough may be needed to achieve an acceptable combination of cost and efficiency for widespread application.

Solarex

Fig. 6 Residential photovoltaic system Photovoltaic cells cover one side of the roof of this house in Japan. The system provides enough power for the average needs of the house, but batteries are needed to match the generation of power to the time of use.

Photovoltaic Systems and Applications

The most common photovoltaic systems serve small direct-current applications, such as navigation lights and remote radio transmitters. These simple systems consist of an array of photovoltaic cells that charge a battery. The battery stores power for periods when sunlight is inadequate and it helps to regulate the voltage.

Batteries are a major part of the cost of such systems. They are typically responsible for most of the maintenance cost and trouble. Also, they may create a serious environmental problem.

If a photovoltaic system is going to operate equipment that requires alternating current, the system needs additional equipment to convert the direct current of the cells to the desired voltage and frequency. This equipment is now well proven, and it is fairly inexpensive.

Many photovoltaic systems have been installed in recent years to provide power to buildings and facilities, mostly as demonstration projects. Figures 5 and 6 show two of these. Most of these systems provide power as a supplement to a public utility system.

Photovoltaic systems that supplement other power sources avoid the need for batteries. They require extra equipment for producing alternating current, and for keeping their output in phase with the public utility's power. Such applications have most promise for large photovoltaic arrays. Large installations provide economy of scale, provided that the cell arrays can be installed on space that has little other value. Such systems may be owned by an electric utility or by an individual customer who has available space, such as the roof of a large building.

Another possibility is that a customer may have an independent photovoltaic system, and switch between it and the public utility. This arrangement might be desirable for "load shedding," especially for daytime applications. The battery capacity can be reduced to the minimum necessary to carry the load during interruptions of sunlight by clouds. The battery capacity depends on the pattern of sunlight and how often it is acceptable to switch between solar and conventional power. If the system operates at night, the cost of purchased electricity must be balanced against the greatly increased battery capacity that is needed to carry the load through the night.

Energy Density

The energy density of sunlight is low. At the earth's surface, the maximum solar energy density is somewhat less than one kilowatt per square meter of collecting area. Using an optimistic cell efficiency of 20%, surface coverage of 50%, and a clear climate, the annual average power output of a photovoltaic array might be about 30 watts per square meter.

If photovoltaic generation is concentrated in large arrays, this translates to an average power output of about

30 megawatts per square kilometer. This is lower by a factor of about 100 than the energy density of conventional generating plants and their surrounding space.

Focussing mirrors can be used to concentrate the sunlight on the photovoltaic material. This does not increase the energy density in terms of surface area, although it reduces the cost of the photovoltaic material.

The energy density in terms of weight is much more favorable. The active area of a photovoltaic cell consists of a thin film of semiconductor overlaid with an almost weightless film of conducting material. Almost all the weight in a photovoltaic system is in the mechanical support of the active material. With typical cell weights, photovoltaic generation has the highest energy density of any energy conversion device, and the fuel weight is zero.

■ Availability

The availability of sunlight varies widely with location and time. Detailed data are available on the average daily, monthly, and yearly amounts of sunlight for many locations. The yearly figures tell how much energy can be collected. Design the collector capacity and the storage capacity to provide for several successive days of low sunlight. Weather data is available to indicate the length of overcast spells.

■ Environmental Considerations

Photovoltaic electricity generation is almost ideal in its lack of direct environmental hazards or nuisances. It is quiet, it emits nothing into the surroundings, and it requires nothing from the surroundings, such as cooling water. It is not especially ugly. Unlike wind machines, photovoltaic arrays do not interfere with communications or endanger wildlife.

Even disposal and recycling of photovoltaic equipment is free of major problems. Most of the weight of the apparatus consists of steel and aluminum, which is common, benign, and easily recyclable. The small weight of plastic materials can be incinerated easily. Any exotic materials that may be used in the cells, such as selenium or gallium, is sufficiently localized in the cells to be easily recovered, if desired.

In a large photovoltaic system, the required land area is a major environmental consideration. Land requirements are large in relation to power output because of the low average energy density of sunlight and the low (at present) efficiency of photovoltaic cells. This consideration does not apply to many smaller systems, especially if the cells can be attached to buildings or other existing structures. Even if the cells are mounted on the ground in large arrays, it may be possible to use the land for other purposes, such as grazing.

Batteries are the malign intruder to this idyllic scene. To date, lead-acid batteries are the preferred storage medium for photovoltaic systems. Batteries are limited to a number of charging cycles, after which they must be replaced. Batteries are difficult to move out of remote locations, and recycling is an environmentally dirty process. In some less developed areas, photovoltaic systems are surrounded by huge piles of discarded batteries that drip acid and leak lead into the environment.

Many other types of energy storage exist, and some have been studied extensively. For example, excess electricity can be used to pump water to an elevated reservoir. The water is later released through a turbine to recover the stored energy. Most types of storage other than batteries have promise only on a large scale.

■ Economics

The cost elements in a photovoltaic system are the cells, their supporting structures, electricity storage, power conversion equipment, switchgear, and land. The cells, batteries, and mechanical supports account for most of the cost.

As noted previously, there is presently an inverse relationship between cell efficiency and cost. With lower efficiency, the size of the collecting array must increase, which increases structure and land costs.

At present, electricity from photovoltaic systems costs roughly ten times more than electricity from conventional sources. Of course, this ratio varies widely with location. In order for photovoltaic systems to become competitive, the cost of cells must be reduced by approximately a factor of ten, and systems must be designed to minimize the need for storage.

■ Relationship to Conventional Energy Sources

At the present time, most photovoltaic systems are installed where conventional sources of electricity are not available, or where they would be prohibitively expensive.

In principle, a photovoltaic system can be operated along with a conventional source of electricity. This would eliminate most of the storage requirement, but the present cost of photovoltaic cells still keeps the system from being economically competitive in areas served by conventional utilities.

In the U.S., such systems receive various advantages from the federal Public Utilities Regulatory Policy Act (PURPA). The major advantage is the privilege of buying supplemental power from the public utility when needed and selling excess power to the utility. Public utilities are not fond of such projects, because the sporadic excess electricity produced by a photovoltaic system is of little value to the utility. Public utilities are allowed to levy connection charges that reflect their cost of protecting their systems against faults and power quality problems caused by private generators. These charges can be prohibitive for small installations.

Solar Thermal Power Generation

By using mirrors, solar energy can be concentrated to produce high temperatures. The concentrated solar energy can be used to produce steam or hot water. It is possible to sustain temperatures that are high enough for efficient operation of steam turbines. The name "solar thermal" is used to distinguish this use of solar energy from photovoltaic power generation, which produces electricity directly, without going through a thermal process.

■ Equipment and Development Status

The highest operating pressure in a solar thermal power plant is probably produced by a U.S. Department of Energy demonstration project that is rated at 10 megawatts. This installation uses mirrors to focus sunlight on a boiler installed in a tower, producing exceptionally high temperatures.

Another approach is to use tracking parabolic reflectors to concentrate sunlight on water-filled tubes. The water is routed to a drum, where steam is produced. This design is used in the world's largest solar thermal electric generating plant, a 300 MW installation located in a sunny area of California. Reliable figures on the operating cost of this plant are not available, but it does not seem that this facility is competitive in the absence of subsidies.

Direct storage of high-temperature hot water or steam is not practical, because the storage vessels would have to be massive to contain the pressure. The only means of storage is to convert the energy to some other form that can be stored. All such possibilities involve significant losses, which degrade the economics.

■ Energy Density and Availability

The basic limitation of this technique is the low energy density and the irregular availability of sunlight, which were discussed previously.

A major limitation is that the sky must be clear in order for this technology to work. If the sun is not visible as a distinct point, the mirrors cannot focus, and energy collection virtually stops. Therefore, this technology is limited to areas of continuously clear skies, which are usually deserts.

■ Environmental Considerations

From an environmental standpoint, solar power production is like conventional power production, but without any of the pollution that is related to fuels. The only effluent is a small quantity of water treatment chemicals. If turbines are used, cooling water is required. On a relative scale, this is a very clean technology for power production.

■ Economics and Feasibility

At first glance, it appears that the cost of solar-powered boilers ought to be roughly comparable to the cost of fuel-fired boilers, if they are designed for economy and produced in sufficient quantity. Solar-powered boilers require mirrors and tracking systems, but they are much simpler in construction, and do not require fuel handling and storage facilities.

The big barrier to economic feasibility is the low density and the intermittent availability of sunlight, which limits the amount of energy that can be collected by the equipment. An analogy is operating a boiler with a burner that has only a small fraction of the boiler's capacity, and that fails to function over half the time.

■ Relationship to Conventional Energy Sources

Because storage is not feasible with this technology, it can be used only if conventional power is also available with sufficient capacity to carry the full load. In this regard, it is similar to photovoltaic and wind installations.

Wind Power

Windmills are perhaps the oldest power producing technology that is not dependent on human or animal muscle. Modern wind generators are direct descendants of the older technology, but problems of an elementary nature are still being worked out. The major inherent problem is the sporadic nature of wind.

■ Equipment and Applications

Modern wind turbines are used almost exclusively for generating electricity. By far the most common configuration consists of a propeller, a speed-increasing gear, and an electric generator. The assembly pivots on top of a tower to face into the wind. Various methods are used to control the speed of the propeller and to protect the equipment from exceptionally high winds. Figure 7 shows a large wind generator that uses variable-pitch blades for these purposes.

Vertical-axis wind turbines, typically with one end resting on or near the ground, have attracted a great deal of attention. These avoid the expense and complication of the tower and the mechanism to aim a propeller into the wind, but they have not yet provided strong competition to propeller-type wind machines.

If the wind generator operates in parallel with the public utility system, it requires expensive equipment for paralleling the two power systems, and for protecting each system from faults in the other.

If a wind generator operates independently of the public utility system, it usually requires energy storage. In smaller systems, storage usually takes the form of batteries. In larger systems, it may take many forms, such as pumping water into a tall tower. Storage is expensive in relation to the overall cost of the system.

About the only direct application of mechanical energy is the classic one of pumping water into storage tanks. For example, windmills are used for cattle watering, irrigation, and other purposes where sporadic pumping is acceptable. Such applications still use old-style, high-torque, multi-blade windmills. Although these machines look antique, their efficiency is fairly good.

■ State of Development

Wind machines have been used for two thousand years, the earliest known machine being used in Persia about 200 BC for grinding grain. The famous Dutch windmills were pumping water reliably and in large numbers by the fifteenth century. These large structures produced from 20 to 80 kilowatts. In the United States, several million windmills were installed for water well pumping since the middle of the nineteenth century. These machines produce less than one kilowatt.

The largest wind generator built to date was constructed on a mountain in Vermont and operated during the years of World War II. It could produce over one megawatt in winds of 50 kilometers per hour (30 miles per hour). It was abandoned because of blade failure and lack of economic advantage. During the next two decades, the Danes built on earlier experience to construct wind generators that served their public utility network. These were abandoned because they could not compete economically with steam turbine plants. Britain, France, and Germany developed wind generators larger than 100 KW during the 1950's and 1960's. The U.S. government responded to the energy crisis of 1973 by funding the development of wind generators. This work, accomplished largely by the aerospace industry, produced highly refined designs with many blade failures.

At the present time, a large fraction of the world's total wind generation capacity is located in a windy valley area near Palm Desert, California. Several thousand wind generators there, ranging in size from about 30 KW to several hundred kilowatts, are connected to the public utility network. These machines, especially the newer ones, have demonstrated reasonably good reliability. Much of this effort was subsidized. Wind generators are not yet quite competitive with other sources of electricity, even in this favorable environment.

Many wind generators, even very large ones, have been installed in locations where there is little wind. Stationary windmills decorate the landscape across the United States. This is the result of government grants, government pressure on electric utilities to use "renewable" energy, excessive optimism about wind availability, and scratching an itch to build windmills.

■ Energy Density and Availability

When wind is blowing, it is a very desirable energy source. Wind is capable of producing about one kilowatt per square meter of rotor disc area in windy locations. Large wind generators are typically designed with a capacity somewhat less than one kilowatt per square meter.

The output of a wind generator is approximately proportional to the square of the wind speed. The fraction of the wind's energy that can be extracted by a wind turbine has a theoretical upper limit of 0.59, which is called the Betz Coefficient. The best wind machines extract about 40% of the wind's energy.

The main shortcoming of wind energy is that high wind speeds occur for only a small fraction of the time. As a result, the average output of a wind machine is much less than its power rating. The average power density of wind over most of the continental United States ranges from less than 50 watts per square meter to over 400 watts per square meter, with the higher values being limited to a small fraction of the total land area.

To get good power output from a wind machine, the location must have wind speeds consistently higher than about 35 kilometers per hour (22 miles per hour). There are few such areas, and the high winds make them

Fig. 7 Large wind generator This unit has a maximum capacity of several hundred kilowatts. The pitch of the blades is varied to maintain a fixed rotation speed when operating. The average output of this expensive machine is only a small fraction of its capacity. The placid water of the lake reveals a human weakness of wind power, which is a tendency to build windmills where there is insufficient wind to justify them.

inhospitable for residential or commercial purposes. Such sites may be acceptable for remote wind farms that serve utility networks.

In order for wind to be a reliable energy source, it must be anchored to a particular topographical feature that creates high wind speed. The shoreline of oceans, seas, and large lakes may be windy on a daily cycle if the temperature of the land and water reverse on a daily basis. Mountain gaps funnel wind, especially if the mountain range is perpendicular to a prevailing wind direction. This is true of the major U.S. mountain ranges. Wind speed may increase at the top of a hill that lies perpendicular to a prevailing wind, but only if the hill is well rounded.

High winds are also found sporadically on open plains. These winds wander too much to be useful. A wind survey conducted in one year may show that an area is favorable for wind energy, whereas the winds may fail to occur the next year.

At altitudes of 1,000 meters to 3,000 meters (3,000 feet to 10,000 feet) above ground level in the mid-latitudes, there are prevailing winds over a large fraction of the earth's surface that would be fast enough to sustain good energy output for a large fraction of the time. The trouble is that wind speed declines rapidly toward the surface, because of ground friction. The rate of decline varies widely with the nature of the terrain and ground cover. Trees and buildings seriously reduce wind speed. Open water and flat land (desert) are best, which is one advantage of shoreline locations. In any event, it is important to install a wind generator as high above the ground as possible, which adds the cost of an expensive tower. The fact that larger wind turbines require taller towers anyhow contributes to their better economics.

The limitation imposed by ground friction suggests installing wind turbines on tethered blimps. This idea has not yet received serious attention. It probably should. Balloons used as radar platforms are already tethered at altitudes up to 10,000 feet (3,000 meters), and they approach the size needed to carry wind machines.

■ Environmental Considerations

It comes as a surprise to those unfamiliar with wind machines that noise is a major problem with them. Wind machines have an insistent, choppy whooshing sound. The sound has a low frequency that travels far. This problem may contribute to the fact that so many wind machines are shut down. Frederick the Great complained about the noise of a grain mill adjacent to his hideaway palace, Sans Souci, but the miller refused to shut down. Modern neighbors may not be so tolerant.

Wind machines have also been found to cause communications interference, such as with television reception. This problem occurs with large machines that are installed where their spinning blades can reflect signals.

There are some wildlife concerns. For example, in California, there is a concern that wind machines pose a hazard to the giant condor.

Installing wind machines on balloons would avoid the previous problems. However, this approach would pose a hazard to aircraft, and would make the machines visible over a much larger area.

Wind machines are presently sanctified in the eyes of environmentalists, and hence to legislators, so installing them may face little in the way of permit hurdles. The neighboring community may raise bigger objections.

■ Economics

The experience of the past two thousand years indicates that wind machines can produce enough energy to pay for themselves. However, at the present time, wind machines cannot compete with the contemporary large-scale energy sources.

By the late 1970's, it became well understood that the economics of wind machines improve with size, up to an optimum size that is somewhere between several hundred kilowatts to one megawatt per machine.

■ Relationship to Conventional Energy Sources

In the U.S., private operators of wind generators are given various advantages by the federal Public Utilities Regulatory Policy Act (PURPA). Among these are the privilege of buying supplemental power from the public utility when needed, and selling excess power to the utility. This option is crucial because of the sporadic nature of wind.

Public utilities are not fond of interconnected wind generators because they produce power in irregular quantities that bear no relationship to the load on the utility network. Unlike other types of private power production, wind generation cannot help the utility to shave its peak demand, which is the utility's major concern in acquiring generating capacity.

Public utilities are allowed to levy connection charges that reflect their cost of protecting their systems against faults and power quality problems caused by wind generators. These charges can be prohibitive for small installations.

Interconnection charges and other problems can be avoided if the wind power system can operate independently of the public power system. For example, the wind power system can operate through storage batteries to provide a separate part of the facility. If the storage batteries are depleted, that part of the system can be switched over to public power entirely while the wind generator recharges the batteries.

Local Hydropower

Hydropower is presently second only to nuclear power as a non-fossil source of energy. Favorable sites can be very economical. Adverse environmental effects are related to damming, which costs land area and alters ecosystems. Exploiting hydropower on a local basis has become uncommon. It probably deserves revived attention in the relatively small number of sites where it is a potential source.

Opportunities for hydropower can be divided into high-head and low-head, although there is no clear dividing line between the two. High-head hydropower requires a large fall in height for the water source, but it produces a given amount of power with a much smaller volume of water. Low-head hydropower operates with relatively small height differences, typically a few meters, but it requires large streams to produce much power.

■ Equipment and Stream Preparation

The basic elements of hydropower have remained the same for the past thousand years:

- *a turbine or wheel,* which converts the potential and/or kinetic energy of the water to mechanical energy
- *a watershed,* which is a geographic drainage area that provides a sufficient volume of water
- *a watercourse,* which is a stream or channel that has an adequate drop in elevation between the points that water is taken from the stream and returned to it
- *a flume,* which is a conduit from the upper portion of the watercourse to the water turbine.

Low-head hydro typically is installed at a river or stream, where a dam is constructed to provide the vertical fall through the turbine. With the old-fashioned overshot water wheel, all the vertical fall occurs in the wheel itself. If a turbine is used, it is installed in the bottom of the dam, so that the effective water column is from the surface of the stream to the turbine outlet. With low-head hydro, damming of the stream usually is necessary to produce an adequate fall, and this may flood upstream areas adjacent to the original stream. Alternatively, it may be necessary to build up the banks upstream of the dam.

High-head hydro is simpler and much more compact. A closed flume pipe is extended to a watercourse as high above the facility as feasible. High-head hydropower usually requires much less damming of streams than low-head hydro. Only a small dam is needed to feed water into the pipe, but storage requirements may dictate the need for a larger storage basin. The turbine can be installed on dry ground, even at a distance from the stream. This avoids the cost of a structure to isolate the turbine from the stream.

The hydrostatic pressure that develops in the pipe is delivered to the turbine, which is located as low as possible. The flume pipe is inexpensive. It must only be strong enough to withstand the hydrostatic pressure of the water column, and it must be large enough and smooth enough to avoid dissipating the energy of the water by friction.

High-head hydropower is generally limited to mountainous terrain, which provides the needed difference in height over a short distance. The stream at the high end must be fed by a watershed of adequate area.

■ Applications

In most hydropower applications, the turbine or wheel is used to drive an electric generator. This provides a great deal of versatility, and eliminates the need for expensive mechanical transmissions. When used to generate electricity, hydropower can serve virtually any type of facility. Also, this provides the hydropower installation with the option of purchasing additional power or selling excess power to the utility. Hydropower is no longer used to produce mechanical work directly, although it could be.

Hydroelectric generators require little tending. However, generating electricity is a dangerous activity, so you probably would not install a system unless it is large enough to justify the cost of continuous monitoring.

■ State of Development

Hydropower is an old and stable technology. There probably will be no further technical developments that significantly affect its economics or availability.

■ Energy Density and Availability

The available amount of energy is determined by the water supply and the vertical fall that acts on the water wheel or water turbine. For example, one inch of rainfall that accumulates on one acre of watershed has an energy potential of 55 kilowatt-hours per foot of fall. Information on stream flow may be available from government sources, or you may have to measure the stream flow. Stream flow usually varies widely over the course of a year, so it is best to collect a year of data. In many locations, stream flow also varies widely from year to year. You can estimate this variation from historical precipitation data.

Topographical maps are available for most areas that indicate the potential drop in elevation that is available, as well as the area of the watershed.

Water wheels and water turbines are remarkably efficient, given their simplicity. If properly applied, either will yield efficiencies greater than 50%.

The basic limitation of hydropower is that the facility must have access to a watercourse that is served by a watershed with sufficient drainage to make the effort worthwhile. Low-head installations are limited to rivers and streams with substantial flow.

High-head installations require steep, mountainous terrain with an overlying watershed. The high pressure developed in high-head hydro installations provides substantial power from a modest stream flow. For example, water turbines with flumes several hundred feet or more in elevation were used to power mining camps in the Rocky Mountains early in the twentieth century.

■ Environmental Considerations

Hydropower has minimal effect on the water resource itself. It does not change its quantity, content, or temperature. However, it may have a substantial effect on the watercourse and its surroundings, if only esthetically. Aside from any objective effects on the environment, anything related to bodies of water evokes strong and often irrational responses that make their way into law. Therefore, it may be difficult to get permission to build a hydropower installation, even on one's own property. Once an installation is built, it may appear sufficiently benign to avoid further attack. However, environmental regulation is likely to remain uncertain and quirky.

■ Economics

The cost of hydropower varies widely. The ubiquity of hydropower demonstrates its economy. The fact that remaining hydropower is mostly in large installations does not mean that smaller installations are uneconomical.

Low-head installations have a high initial cost for damming and conditioning the watercourse, and for building a heavy structure to contain the turbine or wheel. Equipment cost is increased by the low power density.

High-head hydro may be much less expensive. It requires only a few thousand feet of pipe, a compact turbine and generator, and controls.

In large installations, operating and maintenance costs are lower than for fuel-fired generators, aside from the fact that there is no fuel cost. The relative maintenance costs of small systems depend on local details.

■ Relationship to Conventional Energy Sources

In some low-head installations, stream flow is continuous, allowing hydropower to be a primary source. In most cases, hydropower should be backed up by purchased electricity. High-head hydropower is likely to be seasonal, depending on rain patterns and snow melting at high elevations. Also, the equipment must be shut down occasionally for maintenance.

In the U.S., private hydropower producers are given various advantages by the federal Public Utilities Regulatory Policy Act (PURPA). Among these are the privilege of buying supplemental power from the public utility when needed, and selling excess power to the utility. The attractiveness of a hydropower facility to the local utility, and hence the rates, will depend primarily on the extent to which the utility can use the hydropower as a supplement during the utility's peak load periods.

Geothermal Hot Water and Steam

Geothermal energy consists of hot water, steam, or a mixture of hot water and steam that is created by hot geological features below the surface of the earth. The energy is tapped by drilling wells down to a source.

(The term "geothermal" is now being used by promoters of earth-source heat pump systems, which are unrelated to this topic. For more about those systems, see Reference Note 22, Low-Temperature Heat Sources and Heat Sinks. Also, see Measure 5.5.10.)

■ Origin, Availability, and Energy Characteristics

The source of geothermal energy is the molten core of the earth, which is kept hot by decay of radioactive isotopes. If the solid outer crust of the planet were uniform and unbroken, the temperature of the earth would increase by about 14°F for each 1,000 feet (25°C per 1,000 meters) increase in depth. In fact, variations in the conductivity of rock formations cause the temperature gradient to vary from 5°F to 27°F per 1,000 feet (9°C to 50°C per 1,000 meters).

Most geothermal sources do not depend solely on these temperature gradients, but on localized irregularities in the earth's crust that provide higher temperatures. The most dramatic of these are geologic faults that allow the molten core of the earth to penetrate to the surface to become volcanoes. There are also numerous underground faults that bring the molten core near to the surface. These faults are concentrated along the boundaries of tectonic plates, which accounts for the prevalence of geothermal sources in the western United States.

Geothermal sources are also created by subsurface radioactive rock formations, which generally have high thermal conductivity. At some locations in the eastern United States, such deposits are covered by sediments of low thermal conductivity, trapping the heat and increasing the temperature of the resource.

The temperature of any particular geothermal resource is fairly constant. To date, most geothermal wells produce water at temperatures ranging from tepid to about 300°F (150°C). A small number of wells produce much hotter water and steam, up to about 600°F (320°C).

The life of a typical geothermal resource is fairly long. Some geothermal sources are continuously replenished by heat from the earth's core, by convection, or by radioactive decay. If heat is extracted at a rate less than the rate of replenishment, the source should produce indefinitely. On the other hand, if the source is an isolated hot rock formation, its heat will become depleted. Useful lives over 30 years are reported, but such numbers are dependent on the rate at which the resource is tapped. Overdevelopment of a particular site could result in a much shorter life. In this regard, geothermal energy is similar to extracting oil from an oil well, or water from an aquifer.

The required well depth varies considerably from one site to another. Hot sources in the western United States can be tapped with relatively shallow wells, typically 500 to 2,000 feet (200 to 700 meters) in depth. Other geothermal sources may require much deeper wells.

The government energy agency is the place to start in determining whether there are useful geothermal resources in your area. Geothermal resources have not been surveyed thoroughly, but they are most likely to be documented in areas of greatest promise.

■ **Extraction Techniques and Development Status**

Hot water breaking the surface near volcanic activity has been used for heating for centuries. Deeper drilling for geothermal energy received an impetus from the energy crises of the 1970's. Geothermal energy is less visible than other non-fossil sources, so it has received less attention, even though its economics are often comparable or better. Most modern geothermal applications can be classed as demonstration projects. Major commercial development of geothermal energy is limited to sources near the surface, as in Iceland.

In all geothermal developments to date, water acts as a heat transfer medium between the hot rock and the heating application. Natural convection of water from a hot source can bring water near the surface at a much higher temperature than is produced by the normal temperature gradient. This requires geological faults or fractures that are oriented vertically.

Extraction technology is well developed, consisting mostly of conventional drilling techniques and equipment. Hot water or steam may emerge from a geothermal well under its own pressure. More commonly, hot water must be pumped from the well. In some areas, such as the Gulf coast of the United States, water is highly pressurized by overlying formations and bodies of water. These sources produce very hot water from very deep wells.

The water comes from aquifers near the heat source. An exception is hot dry rock, which requires circulating water from the surface through a naturally occurring fracture system. If the aquifer does not provide sufficient water for the application, there must be a supply of surface water to feed or replenish the wells.

For most applications, a heat exchanger is needed because the raw water from the well typically has a high concentration of dissolved minerals. As the water is cooled, the solubility of the minerals is reduced, and

they come out of solution. The heat exchanger should be designed for ease of cleaning.

■ **Applications**

The applications for geothermal heat are determined by the available temperature and flow rate. Geothermal wells producing lower temperatures have been used for greenhouse heating and aquaculture. Warmer sources are used for heating and food processing. Even if the geothermal source is not as hot as the application, it can be used for preheating cold water and other liquids. If the water is free of noxious dissolved materials, it can be used directly. Sources hotter than about 300°F are being used for generating electricity. Many other applications are possible.

Individual geothermal wells typically produce about 200 to 1,000 gallons per minute (800 to 4,000 liters per minute). Systems are designed to operate at a high temperature differential, from 30°F to 100°F (17°C to 55°C), to exploit as much of the energy as possible. This translates to heat outputs for individual wells ranging from 3 million to 50 million BTU per hour (1 to 15 megawatts, thermal). The cost of well drilling makes geothermal energy inappropriate for smaller applications.

■ **Economics**

Geothermal energy has a high initial cost, and relatively low operating cost. The drilling and equipment for the well account for most of the initial cost. A heat exchanger adds a modest cost. Geothermal wells are usually located at or near the site, so surface piping costs are low. Increasing the distance between the geothermal well and the user adds major piping cost.

Unlike most other unconventional energy sources, geothermal sources are usually constant and reliable once they are developed. On the other hand, there is a big gamble at the beginning, when the depth of the well, the amount of flow, and the temperature are uncertain.

■ **Environmental Considerations**

In many cases, the water from geothermal wells is similar to the water from shallow wells. After the heat is extracted, it can be used for irrigation or other purposes, or it can be discharged to the surface.

Using the water at the surface depletes the aquifer. For this reason, local regulations may require the water to be reinjected into the aquifer. It may be possible to use a single coaxial well if the water can be reinjected at a level higher than the source level.

The water from a geothermal source may contain noxious or dangerous materials that are not acceptable in the surface environment. These include sulfur dioxide, fluorine, boron, and radioactive isotopes. Water containing excessive concentrations of undesirable materials must be reinjected into the same aquifer. This requires an additional well located at a distance from the source well.

■ Relationship to Conventional Energy Sources

The capacity of a geothermal source is constant, so it can be used as a substitute for conventional energy sources. The useful life of geothermal sources, if not overused, is comparable to the life of conventional heating equipment.

Geothermal sources are expensive to develop, so they are most economical when used near full capacity. This may make it worthwhile to use geothermal energy to carry a base load, while other energy sources are used for peaking.

Waste Product Combustion

Consider burning waste products as an energy source at sites where a significant amount of combustible waste is available. This usually requires a specialized boiler. The adverse environmental effects vary widely with the nature of the waste.

■ Types of Combustible Waste

Almost any kind of waste that has a significant energy content can be burned to produce energy. The waste may be gaseous, liquid, or solid. For example, the trademark flares of the petroleum industry were long recognized as a wasted energy source, and steel smelting produces a gas rich in combustible carbon monoxide. The petroleum industry also produces heavy liquid end products of distillation that can be used as fuel, and paper processing produces "black liquor," a liquid that contains dissolved combustible components of wood. A great variety of combustible waste occurs in solid form. A common example is bagasse, the solid residue of sugar cane refining.

There is usually a well developed technology for recovering heat from the combustion of waste that is specific to a particular industry, such as the burning of black liquor in the paper industry. Such applications tend to be integrated into the overall production cycle, optimizing efficiency and minimizing cost. If you are in the industry, you will know about these specialized waste recovery opportunities.

Untapped sources of waste energy are most likely to be found in smaller, less common, or evolving operations. There remains considerable opportunity for economic exploitation of waste that is not an inherent part of the process, for example, packaging discarded from sales and warehousing facilities. Take inventory of the materials you are discarding, and see whether it makes sense to use them as a fuel.

■ Equipment and Applications

Waste is almost always burned in a boiler to produce hot water or steam, which can be used in the same way as hot water or steam produced by any other boiler. Waste burning is a branch of boiler technology that is mature. Advances are related primarily to reducing environmental pollution. Waste burning boilers typically are different from those that burn fossil fuels exclusively, but some boilers are designed to burn both specific wastes and conventional fuels efficiently. In new installations, this allows waste to be used as a supplemental fuel without a large cost penalty for separate equipment.

Waste can also be used for purposes other than combustion, such as fertilizer, construction materials, etc. Consider the broad range of possibilities.

■ Energy Content

Waste fuels have a wide range of energy content, from 20% to 100% of the energy content of conventional fuels, on a mass basis. Waste with the lowest energy content requires boilers of the largest size and most specialized design.

Low energy content does not argue against waste burning as strongly as might be expected. The waste has to be discarded in any event, and the distance to an on-site boiler is shorter than the distance to a landfill. Even though it is still necessary to dispose of the products of waste burning, the reduction in volume may have substantial value in terms of reduced disposal costs.

■ Availability

Obviously, the availability of waste suitable for burning variously widely among different activities. In most processes, the amount of waste is roughly proportional to the amount of product that is manufactured or handled. Liquid and solid waste can be stored until the heat is needed, or until there is enough of it to be worth burning as a batch.

■ Environmental Considerations

Waste burning reduces the solid waste disposal problem, and adds a local air pollution problem. From the standpoints of permits and public relations, waste burning tends to be viewed as the lesser evil in lightly populated areas, but is likely to encounter strong opposition in more heavily populated areas. Well designed waste burners can operate with stack emissions that are almost invisible, without using technology that is particularly exotic. However, minimizing air pollution requires careful control and a well trained staff. Along with equipment cost, these factors tend to limit waste burning to larger installations.

■ Economics and Feasibility

There are many applications where waste burning can be profitable, even to the point that it makes a major contribution to the economics of the overall process. Such applications tend to include waste burning as an integral part of the process, but this is not always true.

In applications where waste burning is not closely related to the activity of the facility, the chief obstacle may be the need to train the staff for an unfamiliar and unrelated function. While this may not add much to staffing cost, it is a management challenge. An abandoned waste burning boiler makes no money.

■ Relationship to Conventional Energy Sources

In most cases, waste burning is not able to satisfy all the energy requirements of the facility. Therefore, waste is a supplemental energy source that must be backed up with purchased energy.

Wood and Biomass Combustion

Organic matter that is used as a fuel is called "biomass." Almost all biomass that is burned directly consists of wood or woody plants. Wood was once the primary fuel of mankind, and it is still a primary fuel in less developed countries. In developed countries, wood that is used primarily as a fuel is presently limited to smaller applications, and it has special requirements and problems. The burning of wood waste in industrial applications involves the same general considerations as burning other waste fuels, discussed previously.

■ Energy Content

Wood has an energy content of about 5,000 to 7,000 BTU per pound. Pine bark, which is exceptionally resinous, has an energy content of about 9,000 BTU per pound. Therefore, wood and wood waste have about half the energy content of petroleum, on a mass basis. This qualifies it as an exceptionally good non-fossil fuel.

■ Equipment

In larger applications, wood is burned in specially designed boilers. The wood is usually a manufacturing byproduct, such as sawdust, chips, small pieces, and bark. These forms have large surface areas and burn to completion readily. Wood that is to be burned in industrial boilers may be further processed for ease of handling and burning, if necessary.

At the other end of the spectrum are wood stoves that burn whole or split logs. Combustion is incomplete, air pollution is a problem, and a great deal of labor is required for fuel preparation and tending the fire.

In response to the problems of safety and soot that are created by wood stoves, small externally installed wood heating boilers have been introduced. Heat is conveyed into the building with a water loop, which requires the installation of hydronic heating equipment. The long-term performance of external wood burning systems is not yet well documented. Smaller contemporary units are typically crude contraptions that lack adequate access for fireside and waterside maintenance, which is needed for efficiency and longevity. The fact that wood burning fell from favor during the 1980's has halted the refinement of this type of equipment.

■ Applications and Limitations

Wood stoves are widely used for direct radiant space heating. Wood-fired boilers can produce hot water or low-pressure steam for any application.

A serious limitation of small-scale wood burning is the need for tending the fire. Wood stoves require repeated dampering and restarting. Some newer, larger types can burn unattended for many hours at reduced load, but require periodic refueling.

Where wood is grown or recovered whole as a fuel, another serious limitation is the work required for preparation and storage. Many wood burning enthusiasts have abandoned wood because of the annoyances of starting the fires, cutting and storing the wood, and removing ashes.

Serious injuries are related to the harvesting, storage, and handling of wood, including chain saw accidents, hernias, trees falling on people and equipment, heart attacks from unaccustomed exertion during cold weather, and other causes.

Wood burning has been a major cause of house fires. As a result, some jurisdictions are banning wood stoves, and insurance companies are imposing limitations and penalties on the use of wood stoves. In addition, overuse of portable electric heaters that may occur in conjunction with wood heating is another fire hazard.

■ Availability and Economics

At the present time, the growing scarcity of timber makes it unusual to grow wood specifically for burning, except in lightly populated rural areas. Most log wood that is burned in developed areas is the residue of land clearing and tree trimming. Wood is cheap and available when such supplies exceed the demand, the cost being determined largely by the need for cutting and transportation. However, if the demand exceeds this incidental supply, the cost of log wood becomes prohibitive in relation to the cost of other energy sources.

■ Environmental Considerations

Wood contains far less of environmentally troublesome elements, such as sulfur or mercury, than fossil fuels. It has a somewhat lower ratio of carbon to hydrogen than fossil fuels. Therefore, wood burning has the potential of being relatively clean from an air pollution standpoint. However, thorough burning is necessary to achieve this. Specialized industrial boilers designed to burn wood refuse are capable of thorough burning, but smaller boilers and wood stoves that burn logs are not.

In small furnaces or boilers where only a few logs are burned, the temperature in the firebox typically is not hot enough to completely burn the volatile compounds that are distilled out of the logs. This is especially true at the beginning of the burn, and when output is dampered back. The result is the formation of a great variety of intermediate organic compounds, which contribute to air pollution. In addition, small furnaces and boilers lack provisions to remove particulate matter, which is a major element in wood smoke.

Newer models of small, log burning wood furnaces are being equipped with catalytic afterburners that are claimed to substantially increase the efficiency of combustion. However, catalytic burners appear to be difficult to keep in operation because they are fouled by the particulates of wood smoke. Interest in small-scale wood burning declined so rapidly in the 1980's that development of clean-burning wood stoves petered out, along with research into their effectiveness.

Attitudes toward wood burning illustrate the mood swings that can occur in environmental issues. As late as the 1970's, wood burning was one of the darlings of energy conservation. Now, it is asserted that the products of incomplete wood combustion are seriously carcinogenic, and small-scale wood burning is accused of being a major atmospheric polluter.

■ Relationship to Conventional Energy Sources

Wood burning can be a primary source of heating energy if the supply of wood is adequate and dependable. However, even in such cases, the inconvenience of burning wood causes it to be used as a secondary or occasional fuel.

Some rural electric utilities have promoted wood burning as a means of demand reduction during periods of high electric heating demand. However, this policy backfired because dependence on wood heating prompts customers to use portable electric heaters at times when it is inconvenient to operate a wood fire, and this is especially likely to occur during the peak cold period. A related problem is that the typical wood heater does not effectively heat the entire house, so portable electric heaters are likely to be used in areas of the house remote from the location of the wood heater.

Amount of Energy Available
Wavelength Characteristics
Visible and Invisible Components
The Ultraviolet Component
How the Sun Moves Through the Sky

Weather and Atmospheric
Effects Effect of Glazing Orientation
Effect of Glazing Materials on Sunlight
Smooth-Surface Reflectance
Sunlight Reflected from Exterior Surfaces

> **NOTE FOR READERS IN THE SOUTHERN HEMISPHERE**
>
> To keep the discussion simple, the compass directions given in this Note are based on locations in the northern hemisphere. The corresponding situations for the southern hemisphere are obvious. Simply substitute "south" where the text says "north," and vice versa.

To use sunlight effectively, or to limit its effects, you need to know its characteristics. This Note covers the energy content of sunlight; how light is emitted by the sun; how the sun moves through the sky; and, how sunlight is affected by the atmosphere, by glazing materials, and by surrounding surfaces.

More than with any other important energy source, sunlight is variable. Some of the variations are predictable, especially those caused by the earth's motion and by the location and orientation of the facility. Some of the variations are unpredictable, especially those caused by weather and air pollution. More unusual occurrences, such as the construction of a tall building next door, may also have a major effect on sunlight.

Amount of Energy Available

Here are some handy facts about the amount of solar energy in its different forms:

- A beam of bright sunlight one square foot in cross section delivers energy at the rate of about 240 BTU per hour in clear air. This number varies somewhat with latitude, altitude, and time of day.
- In middle latitudes, vertical glass collects about 300,000 BTU per square foot per year, assuming a continuously clear sky. This number is substantially reduced by cloudy weather, smog, glazing orientations that face toward the north, and shading by adjacent features.
- A significant amount of solar energy comes from the sky indirectly by scattering. Half of a clear sky, with the direct sun blocked (for example, the amount of energy falling on the east side of a building when the sun is in the west) delivers about 20 to 40 BTU per hour per square foot of building surface.
- The amount of solar energy falling on a horizontal surface from an overcast sky typically ranges from 20 to 60 BTU per hour per square foot.
- As a light source, the sun provides an illumination of about 6,000 footcandles (60,000 lux) on a clear day. An overcast sky typically provides an illumination of 500 to 1,500 footcandles (5,000 to 15,000 lux). By comparison, typical office lighting ranges from 50 to 100 footcandles (500 to 1,000 lux).
- At night, the amount of solar energy is essentially nil.

Various ASHRAE publications and other references provide data on solar heat input for different latitudes, dates, hours, and glazing orientations. The more sophisticated energy analysis computer programs have the ability to calculate these data, as well as programs designed explicitly for glass shading and daylighting.

Wavelength Characteristics

The sun is an incandescent light source. This means that the sun's wavelength spectrum is determined by its surface temperature, which is about 10,000°F (5,500°C). By comparison, an ordinary incandescent light bulb has a filament temperature of about 3,800°F (2,100°C), and an arc lamp has a temperature of about 6,000°F (3,500°C). The characteristics of incandescent light sources are covered in Reference Note 54, Incandescent Lighting.

A small fraction of sunlight is absorbed at specific wavelengths as it passes through the earth's atmosphere. Most of this absorption is by water vapor. An even smaller fraction is absorbed by carbon dioxide, mostly at the longest wavelengths of sunlight. (Carbon dioxide absorbs a much larger fraction of the heat radiation emitted by the earth at very long infrared wavelengths. This is the basis of the "greenhouse effect.")

Visible and Invisible Components

All sunlight that is absorbed by a building becomes heat, but only a portion of sunlight is useful for vision. The light sensitivity of human vision falls within the peak portion of the solar spectrum, which is probably not an evolutionary coincidence. About 35% of solar energy that penetrates to sea level is in the visible spectrum. Most of the remainder consists of invisible "infrared" radiation, i.e., light of wavelengths longer than the red end of the visible spectrum. Only about 3% of sunlight arriving on earth is "ultraviolet" radiation, i.e., light of wavelengths shorter than the violet end of the visible spectrum.

The percentage of sunlight in the visible spectrum depends on how broadly the visible spectrum is defined. People see little at the extreme red and blue ends of the vision spectrum. If this sunlight is filtered out, very little vision is lost, but a lot of heat energy is stopped. Blocking all sunlight except the 20% to 30% in the middle of the visible spectrum would provide almost unimpaired vision while minimizing heat gain.

A major challenge in juggling cooling load, passive heating, and daylighting is balancing solar input from the visible and invisible portions of the solar spectrum. Methods exist to filter sunlight by changing the characteristics of the glazing materials. Current methods are not variable. Once the glass is made and coated, its light transmission characteristics are fixed. This may change in the future. The only exception at present is retrofit plastic films, a compromise solution with major drawbacks.

The Ultraviolet Component

Unlike incandescent light sources on earth, the sun is hot enough to produce a significant amount of ultraviolet energy. The amount reaching the earth, 3% of total sunlight, is not important in terms of its heat contents. However, at approximately the boundary between visible light and ultraviolet light, photons become energetic enough to break chemical bonds in organic materials. (The energy of individual photons

depends only on their wavelength.) As a result, ultraviolet light can cause skin cancer, kill microorganisms, weaken fibers, bleach dyes, and create other chemical effects. Most of these effects are undesirable.

A fortunate coincidence is that ordinary window glass blocks the most energetic ultraviolet component of sunlight. With typical windows, ultraviolet damage usually is limited to bleaching of curtains that are exposed to direct sunlight for years. However, if you increase the amount of sunlight enough to provide passive heating or daylighting, expect to take steps to limit ultraviolet damage.

How the Sun Moves Through the Sky

Solar motion is completely predictable. The sun moves with respect to a point on the earth for three reasons, shown in Figure 1:

- the earth rotates about its own axis once per day
- the earth revolves around the sun once per year
- the axis of the earth's rotation is tilted 23.5° with respect to the plane of the earth's revolution around the sun.

The earth is farther away from the sun during the northern hemisphere summer than during the southern hemisphere summer. You might suspect that this would make the seasonal change from summer to winter more severe in the southern hemisphere. However, this factor is obscured by other factors, such as differences in land mass, ocean currents, jet streams, etc.

Figure 2 shows how the sun appears to move with respect to the earth's surface. The tilt of the earth's axis is what causes the change of seasons and what complicates solar motion. Here are a few key aspects of the solar motion we observe:

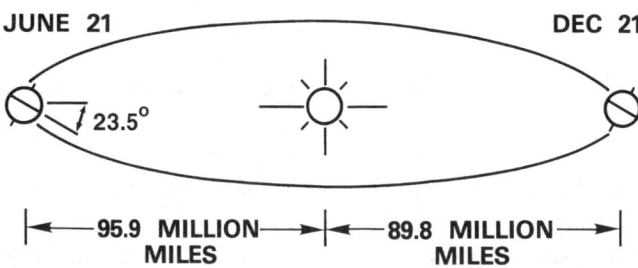

Fig. 1 How the earth moves around the sun The most important directional aspects of sunlight result from the daily rotation of the earth about its own axis, and from the tilt of the earth's axis with respect to the earth's orbit. The seasonal differences in distance from the sun have relatively little effect on solar energy use.

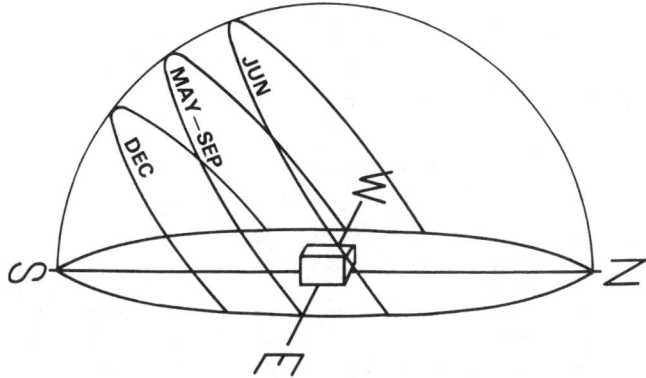

Fig. 2 How the sun moves through the sky More solar energy is available in summer than in winter because the sun's path is longer and because sunlight strikes the earth at a higher angle.

• **The path of the sun's motion through the sky moves north in summer, south in winter.** The sun literally "rises in the east" and "sets in the west" only on two days each year, the spring equinox (March 21) and the autumn equinox (September 21). The sun's path reaches its extreme northern limit at the summer solstice (June 21), and its extreme southern limit at the winter solstice (December 21). The northward movement of the sun's path is what causes the sun to be higher in the sky in summer. This is one of two reasons why it gets hotter in summer.

• **The length of the sun's path through the sky increases as the sun's path moves north.** The sun moves through the sky at a constant angular rate, so the sun remains in the sky longer as its path moves north. This is the second reason why it gets hotter in summer.

• **The azimuth of the sun at sunrise and sunset varies more radically at higher latitudes.** (The sun's azimuth is its horizontal position with respect to the horizon, i.e., its compass direction.) At latitudes closer to the poles, the sun rises and sets farther north in summer, and farther south in winter. (Above the Arctic Circle, in summer, the sun may move all the way around the horizon without setting at all.)

• **The tilt of the sun's path never changes at any given location,** but the path itself shifts north and south with the seasons. The tilt of the sun's path across the sky is equal to 90° minus the latitude. For example, at a latitude of 30°, the sun's path is tilted at angle of 60° to the horizon.

You need a detailed knowledge of how the sun moves through the sky in order to exploit solar energy effectively, or to minimize solar heat gain. You need the most detailed knowledge of the sun's motion to design shading, either to limit heat gain or to provide effective daylighting. For example, the noontime sun at the summer solstice is 47° degree higher than the noontime sun at the winter solstice. You can exploit this difference easily to provide shading for south-facing surfaces that blocks sunlight in summer but admits it in winter.

(In the northern hemisphere, the difference of 47° is constant anywhere at latitudes between the Tropic of Cancer and the Arctic Circle. Below the Tropic of Cancer, the sun crosses into the northern sky at noon during the summer. Above the Arctic Circle, the sun may remain below the horizon at noon during the winter. In the southern hemisphere, the same situation applies to the Tropic of Capricorn and the Antarctic Circle.)

The best tool for laying out shading is a set of solar position charts. Each chart depicts the position of the sun for all seasons and times of day, for a given latitude.

Such charts are available from the Smithsonian Institution and in some books on solar energy.

Weather and Atmospheric Effects

Weather changes make sunlight unpredictable. The largest changes come from clouds, which result from both local and long range atmospheric behavior. For sites in the United States, information on cloud cover is given in the *Climatic Atlas of the United States*, published by the U.S. Government Printing Office.

Atmospheric moisture affects solar input, because water vapor is the most potent absorber of sunlight in a clean, cloudless sky. Most absorption by water vapor occurs in the infrared spectrum. Arid climates get about 10% more total solar energy than average, and humid coastal climates get about 10% less than average. The effect of atmospheric moisture is accounted for by a calculation factor called "clearness number." This is given for different U.S. locations in the *Fundamentals* volume of the *ASHRAE Handbook*.

Air pollution is a major factor in some urban areas. The effect of pollution on solar energy is localized and highly variable.

For a given elevation of the sun above the horizon, heat gain is usually somewhat higher in winter than in summer because the atmosphere is clearer. Of course, total heat gain is less in winter because the sun spends less time above the horizon.

Water vapor, dust, and pollutants not only reduce the intensity of sunlight, but also diffuse it. The increased diffusion usually is not troublesome for exploiting daylighting and passive heating, or for reducing cooling load. However, it reduces the efficiency of directional devices, such as concentrating solar collectors and sun tracking light pipes.

Effect of Glazing Orientation

People tend to assume that the greatest amount of solar gain occurs when the sun is high in the sky. This is not true for most buildings, because solar energy typically enters a building through windows that are vertical. Figure 3 shows how the height of the sun above the horizon affects solar heat gain through vertical glazing. Solar heat gain is greatest through vertical glazing when the sun is about 30° above the horizon. When the sun is close to the horizon, its energy is absorbed heavily by the atmosphere. This effect varies with the clarity of the atmosphere.

At low and middle latitudes, the greatest total solar gain occurs through windows and other vertical glazing that face east and west. At higher latitudes, the greatest total heat gain occurs through windows that face south. Figure 4 shows hourly heat gain for vertical glazing that faces east, west, and south, at a latitude of 40°.

The compass orientation of a vertical window obviously has a major effect on its heat gain. This is illustrated in Figure 5. What this figure does not reveal is that relatively small changes in orientation may have a major effect on shading. This is discussed in Subsection 8.1.

Virtually all methods of controlling sunlight must be tailored to the orientation of the glazing. A building that has been designed to control sunlight effectively will have different sun control techniques on different sides. If a building looks the same from all directions, the designer was not serious about using or controlling sunlight.

Effect of Glazing Materials on Sunlight

Three things may happen to light that strikes a window or skylight. It may pass through the window, or it may be reflected, or it may be absorbed by the glazing material. There are three corresponding characteristics that describe a window or other glazing element: its "transmittance" (the fraction of light that passes through), its "reflectance" (the fraction that is reflected), and its "absorptance" (the fraction that is absorbed). Each of these characteristics is expressed as a fraction or percentage. For any given wavelength and incident angle, the sum of these three numbers equals one, or one hundred percent. Controlling sunlight largely consists of controlling these three characteristics of glazing.

(A note on terminology. You will see variations of each of these terms. For example, you may see "absortance," "absorptivity," or simply, "absorption." In most cases, there are no differences in meaning that you need to worry about.)

Almost all glazing in buildings uses a single basic type of glass. Clear glass has transmission characteristics that are easy to remember. By a remarkable coincidence, window glass has a high transmittance across the spectrum of sunlight, including both visible and infrared portions, and it has almost zero transmittance at other wavelengths. For typical thicknesses used in glazing, the transmittance of clear glass is about 90% in the visible spectrum, and is slightly less in the solar infrared spectrum.

At wavelengths longer than solar infrared radiation, window glass absorbs the energy almost completely. This characteristic reduces loss of heat from the interior of the building by radiation, especially if the glazing has multiple panes. It also reduces heat gain during warm weather from outside objects, such as pavement and adjacent buildings. The latter mode of heat gain is minor compared to heat gain from the wavelengths in sunlight.

By another happy coincidence, the light transmission of glass cuts off just above the blue end of the visible spectrum. Thus, glass absorbs the shorter, more damaging ultraviolet wavelengths at the extreme end of the solar spectrum. The filtering of ultraviolet light by glass has only a minimal effect on plant growth.

Plastic glazing materials, such as acrylic and polycarbonate plastics, are similar to glass in their transmittance for visible light. They have lower transmittance in the solar infrared region, and they may allow more ultraviolet to pass. Plastics are diverse in their optical characteristics.

In principle, you can block entry of sunlight through glazing as much as you want by increasing its reflection

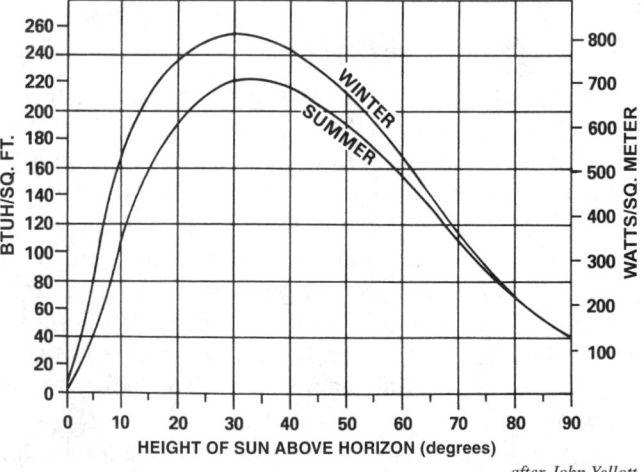

after John Yellott

Fig. 3 The effect of the sun's height on heat gain through vertical glazing Maximum solar gain through windows occurs when the sun has an elevation of about 30 degrees. Sunlight is absorbed heavily by the atmosphere at lower elevations. Also, the sun is more likely to be obstructed by foliage, other buildings, and terrain at lower elevations. At high sun elevations, much of the sunlight reflects off the outer surface of the glazing.

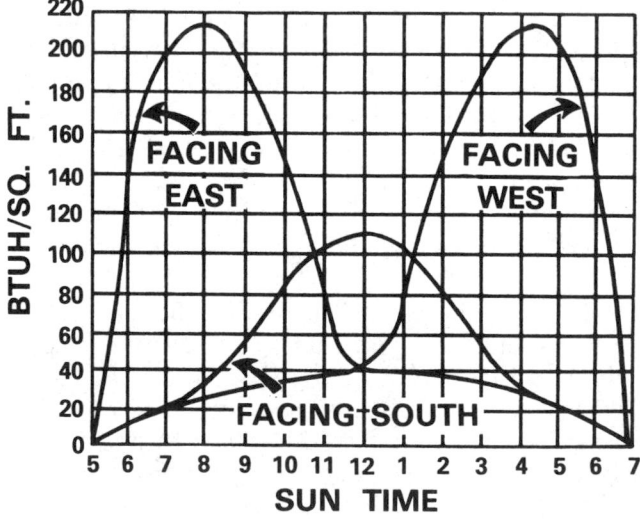

after John Yellott

Fig. 4 The effect of orientation on solar heat gain through vertical glazing, at 40 degrees latitude At this middle latitude, solar gain is much more intense through windows that face east and west than through windows that face south. The difference is even greater at more southerly latitudes.

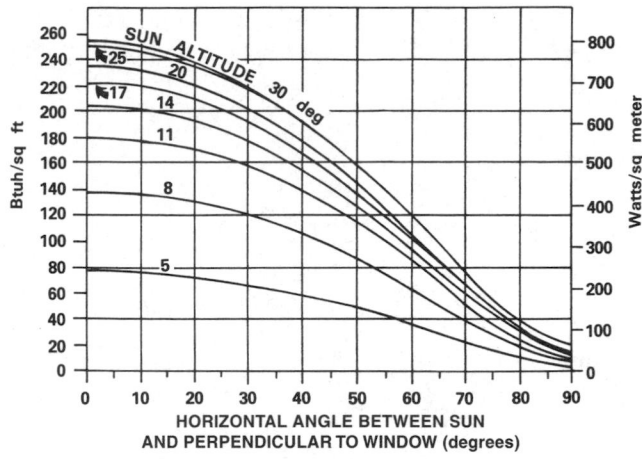

after John Yellott

Fig. 5 The effect of azimuth on solar heat gain through vertical glazing This shows that the sun's altitude is as important as its azimuth. Also, the heat gain from clear daylight sky is about 10% of maximum direct solar radiation, for a given sun elevation.

or absorption, or both. Reflection is increased by applying a reflective coating to the surface of the glass, either as part of the glass manufacturing process or by adding a retrofit reflective film. Absorption is increased by adding dyes during manufacture, or by using a retrofit absorptive film. In practice, you cannot block more than about 80% of visible sunlight without making the view of the outside look too dark.

Controlling the characteristics of glazing is one of the most important tools in using and controlling sunlight. Further details of glazing characteristics are provided in Subsections 8.1 and 8.3.

Smooth-Surface Reflectance

Glass, plastic, and other non-conducting materials with a smooth surface inherently have a certain amount of reflectance, which is independent of the absorptance or transmittance of the material. (This is why you see reflections in store windows and polished table tops.) This smooth-surface reflectance depends only on a physical property called the index of refraction, which is loosely related to the density of the material.

If light strikes glazing from directions near perpendicular, the smooth-surface reflectance is only a few percent. At grazing angles less than about 30° from the glass surface, the reflectance starts to increase rapidly, becoming about 50% at an angle of 10°, and 100% as the angle becomes zero.

Smooth-surface reflectance can be a desirable design feature. For example, in passive heating with vertical windows, smooth-surface reflectance reduces heat gain during warm weather, when sun angles are high, yet it does not reduce heat gain much during cold weather, when sun angles remain low.

Sunlight Reflected from Exterior Surfaces

A substantial amount of solar input may occur by reflection from surrounding terrain features or from other structures. The *ASHRAE Handbook* lists reflectances of about 20% for most common ground surfaces, such as weathered concrete, gravel, and grass, and about 10% for asphalt. Reflection from an adjacent body of water may increase solar energy input by 50% or more. Adjacent highrise buildings with a large fraction of glazing act as huge mirrors, especially if they have glazing with reflective coatings.

Types of Boilers
- **Firetube Boilers**
- **Watertube Boilers**
- **Cast Iron Sectional Boilers**
- **Condensing Boilers**
- **Electric Boilers**

Boiler Capacity Ratings

This Note gives you an overview of boiler types. It also defines terms that you may encounter in accomplishing the Measures related to boilers. The rating methods that you may encounter are explained at the end of this Note.

Boilers are old technology, and they have been used for many applications. As a result, they are rated in a variety of ways. The rating methods that you may encounter are explained at the end of this Note.

The best source of information about boilers that you already own is the manufacturer's literature. A good selection of manufacturers' literature is also the best source of detailed information about boilers that you may be interested in purchasing.

Types of Boilers

Someone has described a boiler as "a bunch of tubes and a furnace." This is a reasonably accurate picture of the construction of most boilers.

In terms of energy efficiency characteristics, it helps to think of a boiler as a heat exchanger that has its own heat source. The boiler exists to transfer heat efficiently from a heat source to water. The heat source is usually a fuel burner or an electric heating element. The type of burner is determined primarily by the type of fuel being burned.

The heat transfer layout is what gives each boiler type its name. The two main layouts are "firetube" and "watertube." In both types, the name derives from the fact that tubes are used as the heat transfer surfaces. If the combustion gases are on the inside of the tubes and the water is on the outside, the boiler is called a "firetube" boiler. If the water is on the inside of the tubes and the flame on the outside, the boiler is called a "watertube" boiler. Both types have been used since the beginning of the industrial age, and each continues to dominate its own size and pressure range. Other types of boilers are significant in narrower ranges of applications.

■ Firetube Boilers

Firetube boilers are the most common type in small and medium commercial sizes, at lower pressures. Firetube boilers are most economical in this range because their construction is simple. Figure 1 diagrams of number of different variations of firetube boilers. In all these, the construction consists of a large number of straight tubes held in place by "tube sheets" at each end.

Modern firetube boilers almost all have a cylindrical outer shell, because this provides the greatest strength in relation to weight, and it is easy to fabricate. For the same reasons, the combustion chamber is a smaller round tube installed inside the bottom of the shell. Figure 2 is a cutaway drawing of an efficient modern firetube boiler.

Firetube boilers usually have a number of "passes" to increase the heat transfer area and to accommodate the shrinking of the combustion gases as they cool. Each pass is a bundle of tubes through which the gases travel. At the end of each pass, the gas reverses and goes into the next pass, and finally into the flue. Figures 1 and 2 show the passes clearly. The most efficient firetube boilers have three or four passes.

Firetube boilers cannot be used at high pressures, because they would be too heavy. The wall thickness of any pressure vessel is proportional to its own diameter and to the pressure. Therefore, a firetube boiler would require an enormously thick shell at larger sizes and higher pressures. Also, the flat tube sheets of a firetube boiler cannot be supported properly at high pressures.

Firetube boilers can make either hot water or steam. The shell of a hot water boiler is filled with tubes. The shell of a steam boiler has a steam space without tubes inside the top, where a steam separator typically is installed. Steam boilers also have external equipment to maintain the water level.

■ Watertube Boilers

Watertube boilers come in many configurations. Most have a large number of individual tubes that extend between a single upper header, called the "steam drum," and one or more lower headers. The lower header, or the lowest of several lower headers, is often called the "mud drum." This is because dirt and the residue of water treatment settles here.

Figure 3 diagrams a variety of older types of watertube boilers, many of which are still in service. These were designed to use straight tubes, or tubes with only simple bends.

Figure 4 shows two watertube boilers of modern design. These are more compact because they are made with tubes that have complex and diverse bends, made economical by modern tube bending machinery. These boilers also make more efficient use of materials, so they are lighter.

Modern watertube boilers are better adapted to including superheaters than older types. A superheater is a section of boiler tubes that continue to heat the steam after it has been vaporized. Superheated steam may be hundreds of degrees (Celsius or Fahrenheit) hotter than saturated steam at the same pressure. Superheated steam operates steam turbines more efficiently than saturated steam, and it minimizes turbine damage that is caused by droplets of condensed steam passing through the blades. Steam may also be superheated to keep it from condensing in long distribution pipes.

The design of superheaters is critical because the steam inside them has little ability to cool the tubes to prevent burnout by the hot combustion gases. Therefore, the superheater tubes reside in a portion of the boiler where the gases have already been cooled sufficiently by evaporating water in the primary boiler tubes, which are called "generating" tubes.

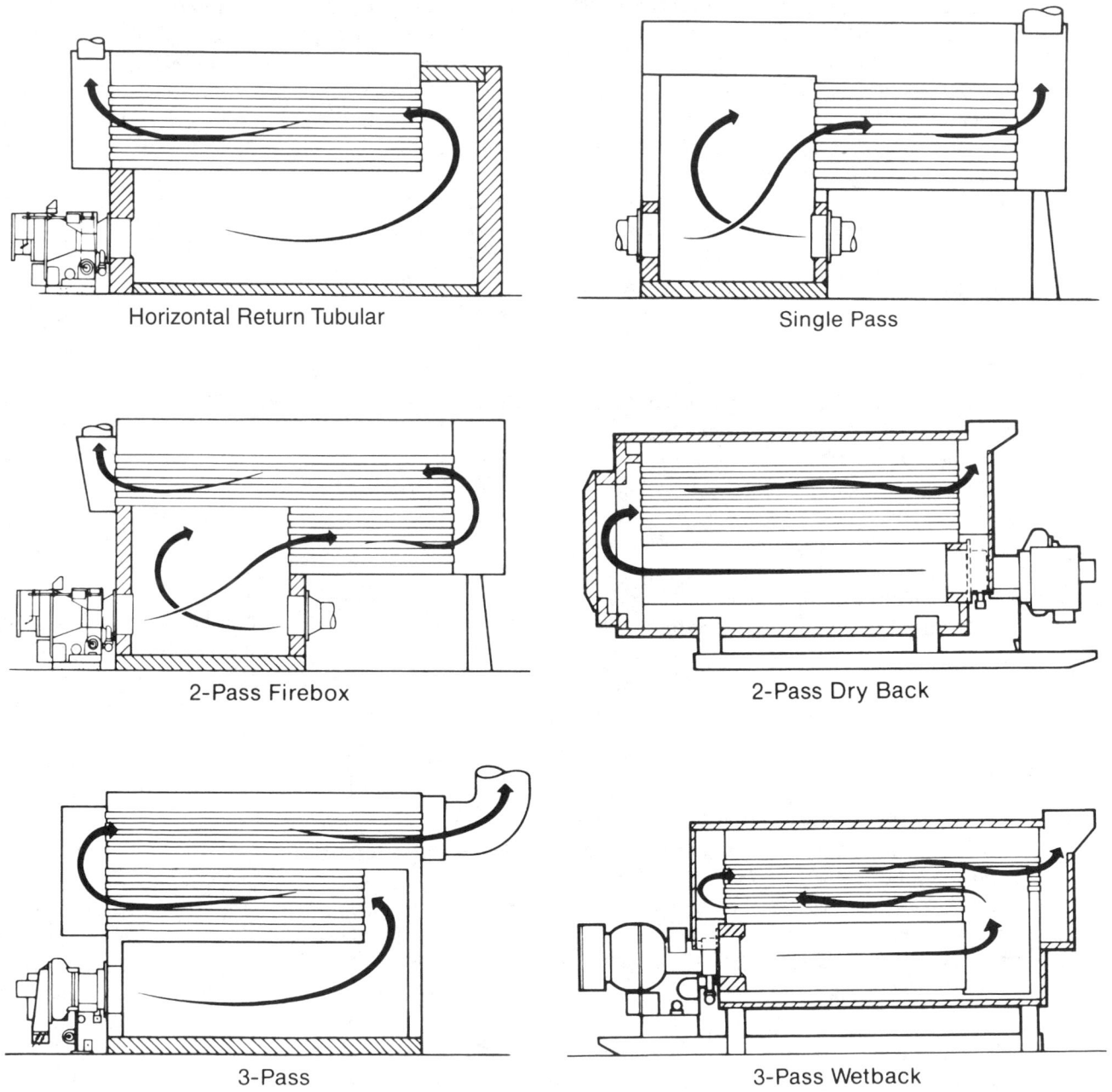

Horizontal Return Tubular

Single Pass

2-Pass Firebox

2-Pass Dry Back

3-Pass

3-Pass Wetback

Industrial Combustion

Fig. 1 Firetube boilers   The basic design consists of many straight tubes held at the ends by "tube sheets." Water surrounds the tubes. The combustion gases work their way through the tubes in a series of "passes." Most modern firetube boilers are enclosed in a cylindrical steel shell. Many older boilers were set on brickwork, or were enclosed in iron casings of many shapes.

Watertube boilers are used in all high-pressure applications and in most larger sizes. For the reasons stated previously, the tubes can be relatively thin, although the headers become quite thick at high pressures. The capacity of a watertube boiler is determined by the number and length of the tubes.

The relatively small water volume of watertube boilers allows them to heat up more quickly than firetube boilers. This is an advantage for producing hot water as well as for producing steam. Also, watertube boilers may have more desirable dimensions than firetube boilers because the tubes can be bent to provide the desired overall shape. Figure 5 shows a relatively simple watertube boiler for producing hot water. Figure 6 shows a larger, more complex unit.

Watertube boilers can be designed to produce hot water or to produce steam. As a practical matter, only watertube boilers can produce superheated steam. The layout of watertube boilers can be adapted readily to accommodate superheater tubes, whereas there is no space for superheater tubes in a firetube boiler. For the

same reason, watertube boilers can easily be designed with integral economizers. (In boilers, the overworked term "economizer" means a heat exchanger that captures the heat of the leaving flue gases to heat the feedwater entering the boiler.)

Maintaining water circulation through all the tubes is a critical aspect of design and operation with watertube boilers, and especially with steam boilers. The water in the tubes protects the steel from the heat of the gases by acting as a coolant. Inadequate circulation allows the tube space to fill with steam, which is much less effective as a coolant. This is a delicate balance, because circulation in most small and medium watertube boilers is by convection, which requires the formation of small steam bubbles inside the tubes to lighten the water column.

Some large, complex boilers use pumped circulation. This becomes necessary at very high pressures, because the density difference between steam and hot water becomes smaller, and the temperature of the steam is higher. (The extreme case occurs in

Cleaver-Brooks

Fig. 2 Modern firetube boiler This boiler has four passes, which includes the cylindrical combustion chamber. The only exterior surface contacted by the combustion gases is the rear end of the casing. In some boilers, even this end contains boiler water, in which case it is called a "wetback" boiler.

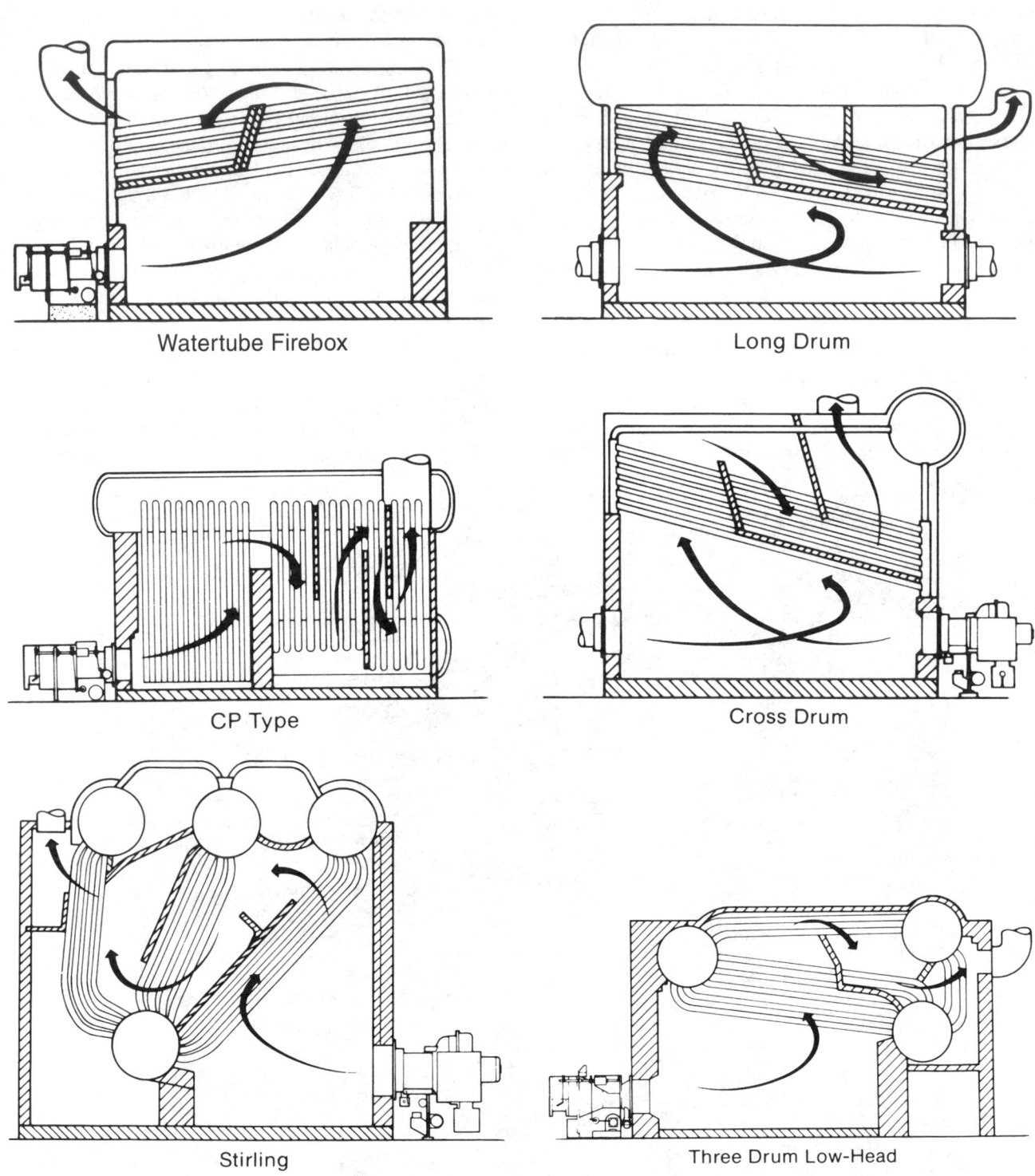

Industrial Combustion

Fig. 3 Older types of watertube boilers   Watertube boilers are complicated by the need to provide effective water circulation to prevent tube burnout, which is much easier to avoid in firetube boilers. Also, the combustion gas must circulate around the tubes by a path that efficiently removes the heat from the gas over a wide range of firing rates. These boiler designs were further complicated by the inability of early manufacturers to bend tubes into complex shapes. Many boilers of these types are still in service.

"supercritical" boilers, where the pressure is so high that the physical differences between liquid water and steam disappear.)

All hot water boilers depend on pumps for circulation, because no steam forms inside them to provide convective circulation. In simple systems, boiler circulation may be provided by the hot water system distribution pumps.

In watertube steam boilers, the upper drum (called the "steam drum") serves mainly as a steam separator. Steam-separating baffles are usually installed in the steam space of the drum. Water level is critical because the volume of the steam drum is small. If the water level falls to the level of the tubes, the dry portions of the tubes will burn out, causing an explosion. If the water level rises too high, the baffles are flooded, and water carries over into the steam system.

Conventional watertube designs tend to be uneconomical in smaller sizes. They are expensive to design and manufacture because each row of tubes must be bent at different angles. The boiler needs an outer casing lined with heavy refractory material because the hot combustion gases are not contained inside a water vessel. These cost factors become less of a disadvantage in larger boilers.

Coil boilers are specialized watertube boilers that are usually found in smaller sizes. As the name implies, they contain only one or a few long coils wound in a tight helix around the combustion chamber. Their major advantage is rapid warm-up, since they have an exceptionally high ratio of surface area to water volume. The coil configuration accommodates rapid thermal expansion. These boilers may require forced circulation to avoid coil burnout. Figure 7 shows an example.

■ **Cast Iron Sectional Boilers**

The "cast iron sectional" boiler does not use tubes. Instead, it is built up from cast iron sections that have water and combustion gas passages. Figure 8 shows a cross section of this boiler type. The iron castings are bolted together, much like an old steam radiator. This type of boiler usually does not have separate headers, but it achieves the same effect with openings that line up between adjacent sections. Although this type of boiler looks crude, it is capable of high efficiency. Cast iron boilers are available for producing either steam or hot water.

The main advantage of cast iron sectional boilers is that they can be assembled inside the boiler room, requiring only a small door for the sections. For this reason, they are often found as replacement units where no other type could be brought into the boiler room.

The main disadvantage of this type of boiler is that the sections are sealed together with gaskets. This invites leakage as the gaskets age and as they are attacked by boiler treatment chemicals. By the same token, the

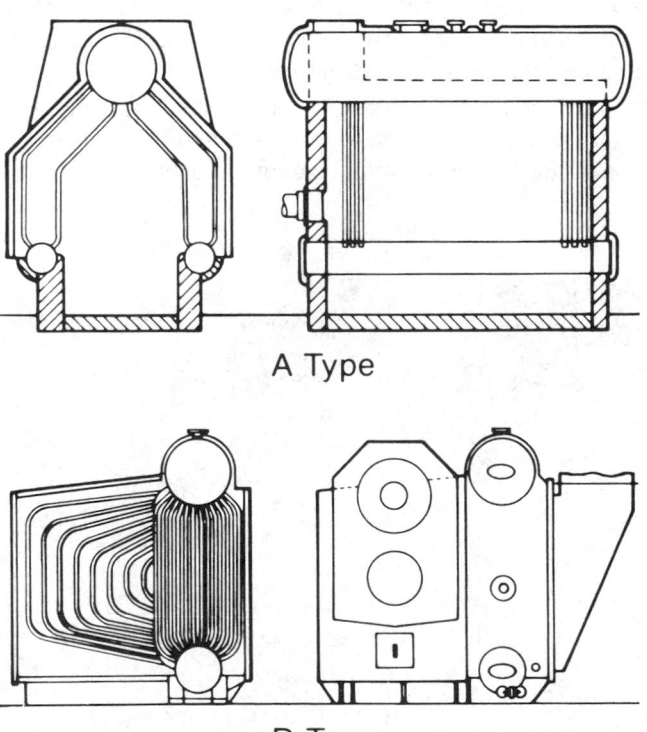

A Type

D Type

Industrial Combustion

Fig. 4 Modern watertube boilers These are more compact and simpler in construction than older types because they can exploit an unlimited combination of complex tube shapes to optimize water circulation and heat transfer. These types can also include superheater sections that safely heat the steam to high temperatures.

Parker Boiler Co.

Fig. 5 The inside of a watertube hot water boiler designed for atmospheric gas burners This simple design is compact, and it heats up quickly. However, the atmospheric burner design causes large standby losses.

presence of the gaskets is a factor to consider in water treatment. Cast iron boilers are limited to smaller sizes. They are also limited to low pressure by the presence of the gaskets, and by limitations in the strength of the tension rods ("stay rods") that hold the sections together.

■ Condensing Boilers

Condensing boilers are the newest major type. They are important because they have higher efficiencies than conventional boilers. Their defining characteristic is a large, corrosion-resistant heat exchange surface, which allows the water vapor in the flue gases to cool enough to condense while still inside the boiler. Condensing recovers the latent heat of the water vapor in the flue gas. Another major advantage is that these boilers require minimal flues, because the flue gases are cool and reduced in volume. In many installations, the flue gases are expelled directly through the wall of the boiler room.

(Condensing furnaces have also become popular. They are based on the same principle.)

Condensing boilers are different in their internal configuration from conventional firetube and watertube boilers. A major design consideration is minimizing the amount of expensive stainless steel that is required.

Condensing boilers and furnaces are generally limited to clean burning fuels, especially gas fuels. At present, they are made only in smaller sizes, but you can easily achieve greater capacity by installing many units in parallel.

■ Electric Boilers

Electric boilers come in two types. One uses an electric resistance element to heat the water. The other, called an "electrode boiler," has electrodes that use the water itself as the resistance element.

Electric boilers are exceptionally efficient as energy conversion devices, suffering only surface heat loss and losses in the electricity supply system. They also have the major advantages of requiring no flue and little maintenance.

A severe disadvantage is the high cost of electricity, at least in most locations. If the electricity is produced in a thermal power plant, electric boilers can be considered as having large losses, which occur at the generating plant rather than in the boiler itself.

Boiler Capacity Ratings

Boiler capacity is expressed by a jumble of different rating methods. Some of these methods are obsolete, but it is useful to know about them when trying to figure out the capacities of older boilers.

Boilers may be rated either by fuel input or by energy output. Input is usually expressed in power units, typically BTU per hour. In many cases, the output is not stated and is assumed to be a standard percentage of the fuel energy input. In particular, smaller gas boilers and furnaces made in the U.S. until recently assumed that the output is 80% of the input.

Boiler output is stated on the nameplates of most larger boilers, and on newer boilers and furnaces of all sizes. Boiler output may be expressed as:

Cleaver-Brooks

Fig. 6 A larger, more complex watertube hot water boiler Watertube boilers are commonly used for producing hot water as well as for producing steam. Water circulation is similar in both types. Water enters the tubes at the lower header, and the tubes discharge into the upper header.

WESINC

Fig. 7 A coil boiler This is a watertube boiler in which a coil of tubes surrounds the combustion chamber. The arrow points to a 50-horsepower pump that must circulate water through the tubes continuously to keep them from burning out. This boiler is appropriate only for applications requiring fast warm-up and short durations of operation. It was a very bad bargain for this building heating application.

• *power units.* Metric practice is to express boiler output in watts. Where English units are still used, the usual units are BTU per hour.

• *boiler horsepower,* which is an archaic rating that still persists needlessly. One boiler horsepower equals 33,475 BTU per hour. This method of rating originated as a means of sizing boilers to operate old non-condensing steam engines.

• *heating surface,* found mainly on older boilers. By convention, one square foot of heating surface is considered to be one fifth of a boiler horsepower, or about 6,700 BTU per hour. This rating refers to the area of the heat transfer surface within the boiler itself. If the heating surface is given on the boiler nameplate, be sure that the figure is intended as an output rating. If the figure refers to actual heat transfer surface, the actual boiler output may be lower or higher than indicated by the nominal heating surface.

• *square feet of equivalent direct radiation (EDR)* is used with older boilers intended to make steam for radiators. The rating refers to the surface area of the radiators, not to the heating surface of the boiler. One square foot EDR equals 240 BTU per hour if the boiler is intended to serve steam radiators, or 150 BTU per hour if the boiler is intended to serve hot water radiators.

• *"net output"* is a rating used on some boilers intended for space heating applications. The net output is equal to the actual output of the boiler divided by a capacity safety factor. Institute of Boiler and Radiator Manufacturers (IBR) ratings

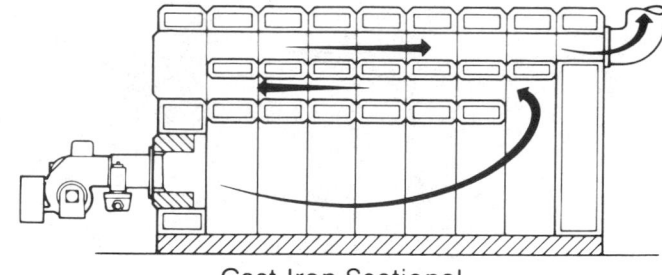

Cast Iron Sectional

Industrial Combustion

Fig. 8 Cast iron sectional boiler The boiler can easily be assembled at the site from many individual sections. The sections are sealed by gaskets and held together by compression rods. The main advantage of this type is that the individual pieces can be transported into the boiler room through a small entrance.

assume a factor of 1.13, and Mechanical Contractors Association (MCA) ratings assume a factor of 1.25. To calculate the actual output of the boiler, check the nameplate to see whether the IBR or MCA rating is used, and multiply by the appropriate factor.

Many older boilers have been upgraded with retrofit burners. If the burner capacity is fixed, check the capacity shown on the burner nameplate. Burner capacity is commonly expressed as fuel input rate, such as gallons per hour of fuel oil. In this case, you need to calculate the energy input by looking up the energy content of the fuel. To convert this to an accurate estimate of output, do an efficiency test of the boiler, as explained in Subsection 1.2.

How Cooling Efficiency is Expressed

Coefficient of Performance (COP)
Other Performance Ratings
Rating Temperatures

COP's of Direct-Fired Absorption Chillers
Auxiliary Equipment Energy Consumption
Consider Part-Load Efficiency

Energy flows through a cooling machine (an air conditioner or water chiller) in this way:

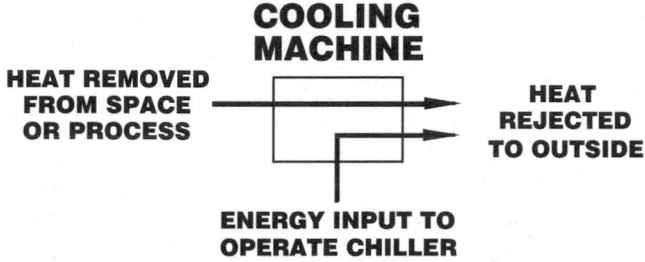

Energy enters the machine at two points (the evaporator and the drive motor) and leaves at one point (the condenser). For this reason, the term "efficiency" does not apply to chillers in the usual sense. Efficiency is defined as useful energy output divided by total energy input. However, the energy output of a chiller is the amount of heat rejected to the outside, which is not of primary concern.

What really matters is the amount of heat withdrawn from the space, and the energy required to operate the chiller, which are two different modes of energy input. For this reason, the performance of chillers and air conditioners is defined in a special way, in terms of "coefficient of performance."

Coefficient of Performance (COP)

Chiller performance is expressed as the ratio of the amount of heat removed from the space to the amount of energy required to operate the chiller. This ratio is called the "coefficient of performance," commonly abbreviated as "COP," so

$$COP = \frac{\text{heat removed from load}}{\text{energy to operate chiller}}$$

COP is defined using common units, i.e., BTU's removed per BTU input, or watts per watt, etc. Therefore, COP is "dimensionless." Since the COP has no units, it applies equally to metric and non-metric engineering.

Other Performance Ratings

In the U.S., the performance of smaller cooling equipment may be rated with a bastard measure called the Energy Efficiency Ratio (EER), which has units of BTU cooling per KWH input. Therefore, EER equals the COP times 3.413. A variation of this is the Seasonal Energy Efficiency Ratio (SEER), which is the average EER of the unit when operated over a standard load profile.

A common way of expressing the performance of large chillers in the U.S. is "kilowatts per ton." Unlike COP, a smaller number is better. The "ton" is an archaic but common term used to express the rate of cooling. (A ton equals 12,000 BTU's per hour. The name originated in icehouse refrigeration, and derives from the fact that a ton is the rate of heat removal needed to freeze one ton of ice per day.) At the present time, the highest efficiency achieved by chillers is about 0.55 kilowatts per ton, at standard rating conditions.

Rating Temperatures

The COP rating is meaningless unless the evaporating and condensing temperatures are specified. This is because the COP and cooling capacity of a chiller are affected strongly by the evaporating and condensing temperatures. In the U.S., the Air Conditioning and Refrigeration Institute (ARI) has established standard rating temperatures for different types of cooling equipment.

For example, water chillers are rated on the basis of the chilled water temperature as it leaves the evaporator and the cooling water temperature as it enters the condenser. The specific rating temperatures differ among manufacturers, equipment types, and applications. For example, one chiller may be rated at 44°F chilled water temperature and 95°F condenser water temperature, while other chillers may be rated at different temperatures. Be sure to make chiller comparisons using the same temperatures.

COP's of Direct-Fired Absorption Chillers

The purchased energy that operates direct-fired absorption chillers is in the form of fuel. Therefore, the COP rating depends on the energy content of the fuel. However, the energy content of fuels may be expressed in two different ways, "high heat value" and "low heat value." The COP's of direct-fired chillers are usually based on the "high heat value" of the fuel. Be aware that a manufacturer may artificially boost the COP figure by using the "low heat value" of the fuel. Reference Note 20, Fossil Fuels, explains this difference in fuel ratings.

Consider the Energy Consumption of Auxiliary Equipment

Even when using standard rating conditions, the COP or EER does not necessarily provide a valid comparison between cooling systems. For example, the COP quoted for a large packaged chiller includes only the energy input to the compressor. In contrast, the EER or SEER of a window air conditioner includes the energy input to all the system equipment, including the evaporator and condenser fans. The SEER of a packaged split system may include the energy input to the condenser and evaporator fans, or if the evaporator is specified separately, for just the condenser fan.

COP's quoted for absorption chillers typically include only the heat input, and not the electrical input for pumps, which is substantial.

To make the most valid efficiency comparison between cooling systems, include all energy consuming devices in the chiller system as part of your efficiency calculation. This means taking into account the energy consumption of auxiliaries, especially pumps and fans.

The energy consumption of other auxiliaries, such as purge units, is usually minor, but it is easy to include.

When you do this, account for efficiency effects caused by the auxiliary equipment. For example, reducing the energy consumption of condenser or evaporator fans increases the compressor power requirement.

Consider Part-Load Efficiency

Manufacturers rate their chillers at their most efficient operating points, which typically are near full load. However, most chiller systems operate at low loads for a large fraction of the time. Therefore, the average COP of a chiller may be substantially lower than its rated COP. There are large differences in part-load performance between different types of chillers.

Auxiliary equipment usually is even worse in this regard. For example, the pump power needed for chilled water distribution may remain fixed at all cooling loads.

To compare the efficiencies of different systems, you need to compute their energy consumptions when operated over a typical range of loads. You may need help from a computer program to do this accurately. See Reference Note 17, Energy Analysis Computer Programs, for details.

How Compression Cooling Works
Refrigerants
The Components of a Compression Cooling
  System
System Layouts
The Effect of Operating Temperatures on COP
  - Maximum Theoretical COP
  - The COP of an Ideal Machine with Real Refrigerants
  - The COP of Real Systems with Real Refrigerants
System Temperature Differential is Larger than
  Cooling Load Temperature Differential
The Effect of Operating Temperatures on Capacity
The Compressor and System Efficiency
  - The Refrigerant
  - Efficiency of Gas Flow in the Compressor
  - Interstage Heat Transfer
  - Fixed or Variable Compression Ratio
  - Efficiency of Output Modulation
  - Limitations in Reducing Condenser Temperature

Types of Compressors and Throttling Methods
  - Centrifugal Compressors
  - Reciprocating Compressors
  - Screw Compressors
  - Scroll Compressors
  - Other Compressor Types
Compressor Drivers
  - Internal vs. External Motors
  - Motor Efficiency
  - Other Types of Drivers
Evaporators
  - Flooded Evaporators
  - Liquid Overfeed Evaporators
  - Evaporators with Expansion Valves
  - How Evaporator Superheat Reduces COP
Energy Recovery from Refrigerant Expansion
Evaporator and Condenser Heat Transfer
  - Fouling Factor
Parallel and Series Evaporator Circuits
Condensate Subcooling

The great majority of cooling equipment worldwide operates by compression. This Note is an overview of compression cooling systems, emphasizing the aspects of design and equipment selection that affect system efficiency.

In principle, you could select a system with the highest efficiency simply by using manufacturers' efficiency quotations. However, this is not a prudent approach. The efficiency of a cooling system is determined by a variety of factors involving cost, safety, space requirements, maintenance, and other issues. You need to consider all of these to achieve a good system.

There is fierce competition among manufacturers with different designs, and each touts its own advantages while being silent about its own disadvantages. This Note will help you to navigate through the competing claims by understanding the principles of compression cooling. It will enable you to ask the right questions when selecting new equipment, and it will give you a solid foundation for making efficiency improvements to existing cooling systems.

How Compression Cooling Works

All contemporary space cooling and process cooling equipment exploits the fact that a liquid absorbs heat when it evaporates. (There are a few exceptions, which are limited to unusual applications.) A liquid used for cooling is called a "refrigerant." The energy absorbed by the refrigerant in changing from a liquid to a vapor is called its "latent heat." When this heat is drawn from a body, the body is cooled. For a good demonstration

of this, swab some alcohol on your arm. The wet spot will chill your skin until it all evaporates.

You can create cooling without any machinery if you have enough liquid to evaporate. Water has been used for cooling since before the dawn of humanity. However, it is not very satisfactory as a coolant under ambient conditions because it evaporates too slowly. There are many liquids that provide a large evaporative cooling effect, but they are not cheap enough to be used on a once-through basis.

Mechanical cooling can use more effective refrigerants and recycle them indefinitely. The process of doing this is called a "cooling cycle." The outer shell or casing of the cooling equipment serves as a pressure vessel, isolating the refrigerant from air and atmospheric pressure. This allows the cooling cycle to operate under conditions that provide the greatest cooling capacity and efficiency from the refrigerant.

The most common kind of cooling equipment uses a compression cooling cycle. After the refrigerant has been evaporated by heat from the cooling load, the vapor is compressed. This raises the temperature of the gas well above ambient temperature, so that the heat in the gas can be removed by cooling it with air or water at ambient temperature. Removing the heat causes the compressed gas to condense back to a warm liquid.

The warm refrigerant liquid from the condenser is metered into the cooling area (evaporator) by a flow control device of some kind. The pressure in the evaporator is determined by the suction of the compressor and by the rate of evaporation. Because the evaporator pressure is lower than the condenser pressure, a small portion of the liquid refrigerant "flashes" into vapor when it passes through the control device. This flashing cools the remaining liquid to the temperature of the evaporator. The liquid refrigerant is now ready to absorb heat from the cooling load, repeating the cycle.

Refrigerants

Nowadays, selecting the refrigerant may be the first choice you make in selecting or designing a cooling system. The choice of refrigerant limits the system's efficiency and affects the design of all the system hardware. Safety concerns with some refrigerants dictate where the equipment can be located, and how the surrounding space must be designed. Refrigerant selection has recently become more important and complex because of environmental concerns. To learn more about refrigerants and how to select them, see Reference Note 34, Refrigerants.

The Components of a Compression Cooling System

All compression cooling systems have five basic elements. As illustrated by the typical system in Figure 1, they are:

- the *compressor.* On its discharge side, the compressor increases the pressure of the evaporated refrigerant vapor to raise its temperature. On the suction side, the compressor lowers the pressure of the liquid refrigerant, causing the refrigerant to evaporate more rapidly, and at a lower temperature. In most systems, cooling capacity is regulated by varying the output of the compressor. There are several major types of compressors, discussed below.

- the *evaporator,* which is a heat exchanger on the suction side of the compressor. This is where the liquid refrigerant evaporates to do its work. If the machine is designed to cool air directly, as in Figure 1, the evaporator is an air coil. If the machine is designed to cool liquid, the evaporator typically is a tube-and-shell heat exchanger. (A cooling machine that is designed to deliver cooling by means of a liquid is commonly called a "liquid chiller," or simply a "chiller.")

- the *condenser,* which is a heat exchanger on the discharge side of the compressor. The high-pressure refrigerant gas from the compressor is cooled in the condenser so that it returns to a liquid state. If

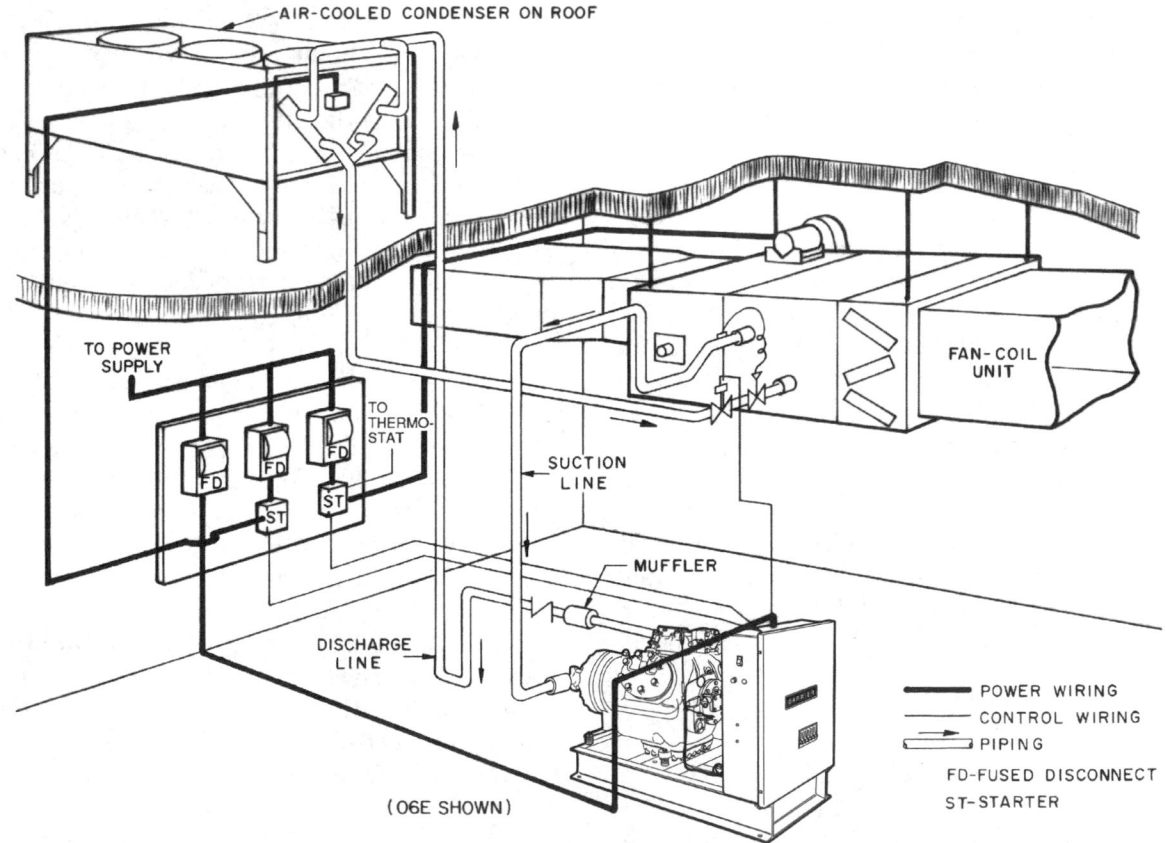

Carrier Corporation

Fig. 1 The elements of a compression cooling system   In this system, the evaporator is the coil in the fan-coil unit. An expansion valve, installed on the side of the fan-coil unit, serves as a refrigerant metering device. The condenser is an air coil on the outside of the building. The system does not have a distinct refrigerant accumulator, but the bottom of the condenser coil and the refrigerant liquid pipe serve this function.

the cooling medium is air, the condenser is an air coil. If the cooling medium is water, the condenser is usually a tube-and-shell heat exchanger, or sometimes a coaxial tube heat exchanger.

- an *accumulator* or *receiver*, which is a reservoir that holds the liquid refrigerant until it is needed. Many cooling systems do not have a separate accumulator, instead depending on the volume of the condenser, evaporator, and/or refrigerant lines to hold the liquid refrigerant. For example, packaged water chillers typically use the evaporator shell as the liquid storage vessel. In Figure 1, the bottom of the condenser coil and the liquid discharge line act as the accumulator.

- a device or combination of devices for *separating the high-pressure side of the system (condenser) from the low-pressure side (evaporator)* and, in some systems, for *metering the flow of refrigerant into the evaporator*. The device keeps the hot refrigerant gas from blasting through the condenser

coil, and it maintains the discharge pressure that is necessary for condensing the refrigerant. In most liquid chillers, a simple *orifice* or *float valve* between the condenser and the evaporator serves this purpose. If the evaporator is partially wetted, rather than flooded, the flow of refrigerant into the evaporator is controlled by a metering device called an *expansion valve*. Most air coils and some water-cooled evaporators use expansion valves. A recent development is the use of a small *turbine* to act as a metering device. This design recovers energy from the pressure difference that exists between the condenser and evaporator.

A compression cooling system may also have accessory devices or specialized features, such as crankcase heaters, purge units, hot gas bypass circuits, valves for controlling the flow of refrigerant to different parts of a coil, etc. Most of these accessories are specific to particular types, models, or system designs.

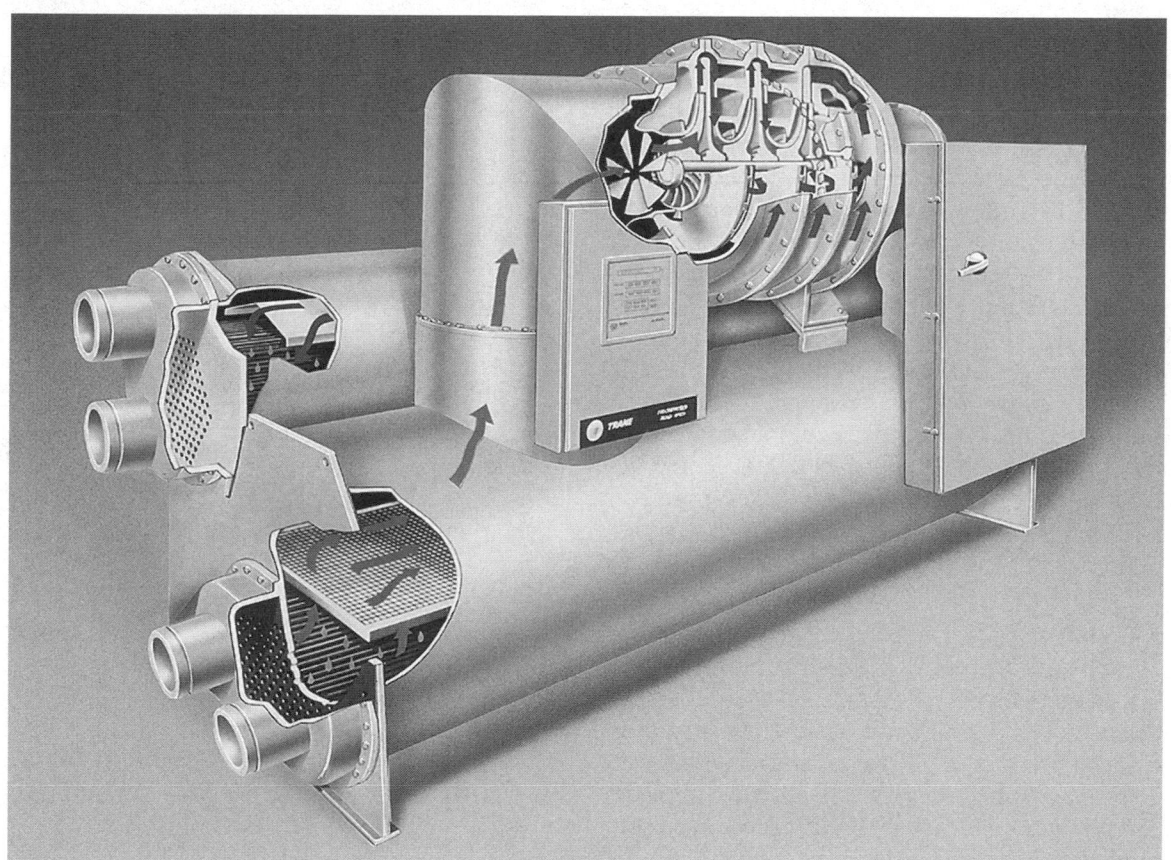

Trane Company

Fig. 2 Packaged water chiller The machine packages all the primary elements of a cooling system together. This model has a three-stage centrifugal compressor. The cooling output of the machine is controlled by throttling the flow of gas into the compressor with an inlet vortex damper. The evaporator, at bottom, is flooded with refrigerant. Tubes for chilling water are immersed in the refrigerant. The refrigerant evaporates under the reduced pressure of the compressor suction, chilling the water. The vapor is compressed to a hot gas by the compressor, and is discharged to the condenser in the rear. Cool water from a cooling tower flows through the tubes in the condenser, absorbing heat from the hot gas and condensing it. The warm refrigerant liquid drops down into the evaporator through an orifice. Upon exposure to the lower pressure in the evaporator, a portion of the warm refrigerant liquid evaporates, cooling the remaining liquid to the saturation temperature that corresponds to the evaporator pressure. The cycle continues as long as the compressor runs.

System Layouts

The five elements of a compression cooling system can be grouped in a variety of ways. You can find virtually all possible combinations in practice.

For example, Figure 1 shows a common arrangement where the compressor is located in the basement, an air-cooled condenser and accumulator is installed on the roof, and air coils with their expansion valves are installed on each floor.

In the popular "split system," the compressor is packaged with the condenser outside the building, and liquid refrigerant is piped to air coils or to a water chilling evaporator inside the building. This keeps the noisiest equipment outside the building.

Large water chillers generally include all their components in a compact package that is assembled at the factory. Figure 2 shows a typical packaged water chiller. Such packaging is becoming increasingly common in smaller systems as well. On the other hand, a cooling system may distribute its components throughout the facility.

If a system distributes its cooling by means of chilled water, this is called "hydronic" distribution. If liquid refrigerant is evaporated inside an air cooling coil, this is called "direct expansion" or "DX."

Larger systems generally use water for condenser cooling, because this allows the condensing temperature to be reduced by evaporative cooling of the water in a cooling tower. Smaller systems, and systems used in locations where the supply of water is limited, may reject heat directly to the outside air through an air coil that is called an "air-cooled condenser." A compressor that is packaged with an air-cooled condenser is called an "air-cooled condensing unit." An intermediate approach is spraying water over a condensing coil that is cooled by air flow. This arrangement is called an "evaporative condenser."

The Effect of Operating Temperatures on COP

The amount of power needed by the compressor depends on the difference in pressure between the compressor inlet (from the evaporator) and outlet (to the condenser). You know from basic chemistry that the temperature of a gas is proportional to its pressure. Therefore, we can express the power needed to compress the refrigerant vapor in terms of the difference between the evaporator and condenser temperatures.

This is a useful way of looking at compressor power, because the evaporator and condenser temperatures are determined largely by factors that are under the control of the equipment manufacturer, the system designer, and the plant operator. These factors are discussed below. But first, we need to understand the relationship between system temperatures and COP. If you need an explanation of COP, see Reference Note 31, How Cooling Efficiency is Expressed.

■ Maximum Theoretical COP

In 1824, the French engineer Sadi Carnot developed an analysis of compression cycles that allows their theoretical maximum efficiency to be expressed in terms of temperature alone. Carnot found that the maximum possible COP of a compression cooling machine is determined by this remarkably simple formula:

$$\text{Ideal COP} = \frac{\text{lowest temperature}}{(\text{highest temperature}) - (\text{lowest temperature})}$$

In this formula, the temperatures are absolute. You can use any system of absolute temperature units in this formula because the units cancel. (The Rankine absolute scale equals Fahrenheit plus 460°F. The Kelvin absolute scale equals Celsius plus 273°C.)

The lowest temperature in the system occurs at the compressor inlet. The highest temperature occurs at the compressor discharge. These temperatures are not fixed by the compressor. Instead, they are imposed on the compressor by the cooling application, by the

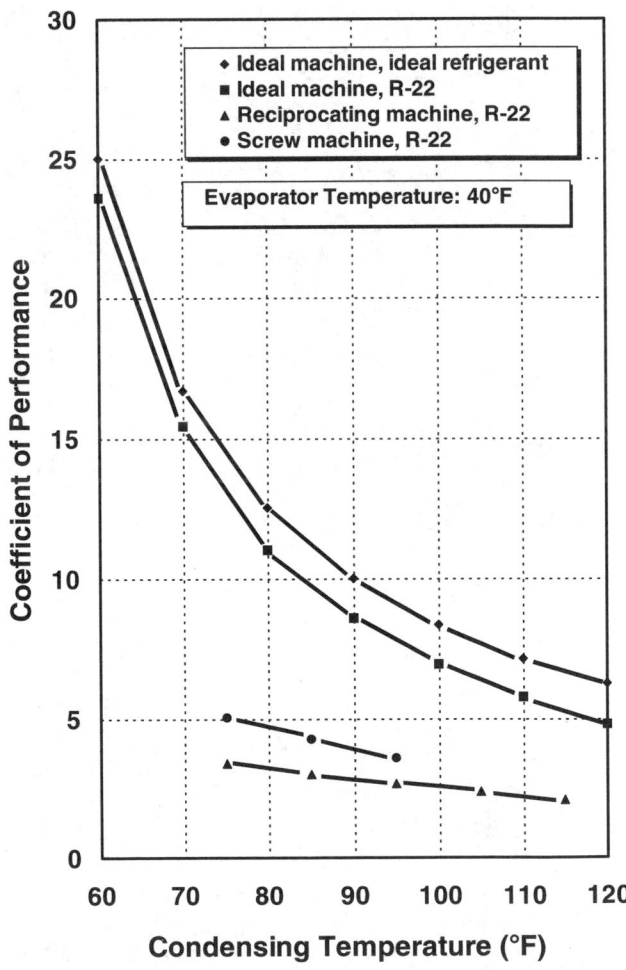

Fig. 3 The effect of system temperatures on COP The top curve shows the theoretical maximum COP that is possible, which is determined solely by the system temperatures. The second curve shows the efficiency loss related to a highly efficient real refrigerant. The bottom two curves show the COP's of real cooling machines with the same refrigerant.

temperature of the environment to which the heat is rejected, and by many system design factors.

This formula reveals the crucial role of the evaporator-to-condenser temperature differential in system efficiency. Cooling equipment moves heat from one place to another. The energy input to a cooling system is needed to overcome the "resistance" of the temperature differential. This viewpoint is evoked by the term "heat pump," which is used when a cooling machine is turned around to make it a heating device rather than a cooling device.

In Figure 3, the top curve shows the maximum theoretical COP's that can be achieved with an ideal cooling machine and an ideal refrigerant. This curve is calculated using Carnot's formula, assuming a cooling temperature of 40°F and a typical range of condensing temperatures.

■ The COP of an Ideal Machine with Real Refrigerants

In Figure 3, the second curve shows the COP's that would be achieved by an ideal cooling machine using HCFC-22 as a refrigerant. The curve assumes a cooling temperature of 40°F and a typical range of condensing temperatures. HCFC-22 (originally called Freon-22) is the most common refrigerant at present, and it is one of the most efficient.

This curve shows that some real refrigerants can approach the theoretical maximum COP. However, there is wide variation in efficiency among different refrigerants. The theoretical COP's of the most popular refrigerants are tabulated in Reference Note 34, Refrigerants.

■ The COP of Real Systems with Real Refrigerants

In Figure 3, the bottom two curves shows the manufacturer's COP ratings for two typical chillers, both using HCFC-22 refrigerant. As with the other curves, the cooling temperature is 40°F.

These two curves show that the COP's of real machines are considerably lower than the theoretical maximum, even accounting for limitations in the refrigerants. The losses are caused by mechanical inefficiencies in the compressor, by inefficiencies in the compression process, by energy waste in the expansion of refrigerant when it returns to the evaporator, and by heat transfer losses within the system. There is potential for improvement in all these areas.

The two curves also show that there is a considerable difference in COP between screw compressor machines and reciprocating machines. The COP of a good centrifugal machine would be somewhat higher than the COP of the screw machine.

Condensing temperatures below 75°F are not shown for the two real machines, because various design features keep them from operating with lower condensing temperatures.

The System Temperature Differential is Much Larger than the Cooling Load Temperature Differential

The temperature differential between evaporator and condenser is generally much higher than the temperature differential of the actual cooling load. This is a matter of great concern, because the temperature differential is the underlying theoretical factor that limits COP. As a realistic example of how the temperature differential builds up, consider the example of an air-cooled water chiller used for air conditioning. The temperatures throughout the system are depicted in Figure 4.

An air-cooled system was selected for this example to avoid the complications of temperature changes in cooling towers, which are really supplementary cooling systems. Other thermal complications, especially evaporator superheat, condensate subcooling, and interstage heat exchange, are discussed below. The illustration ignores small losses, such as the rise in chilled water temperature in the distribution system.

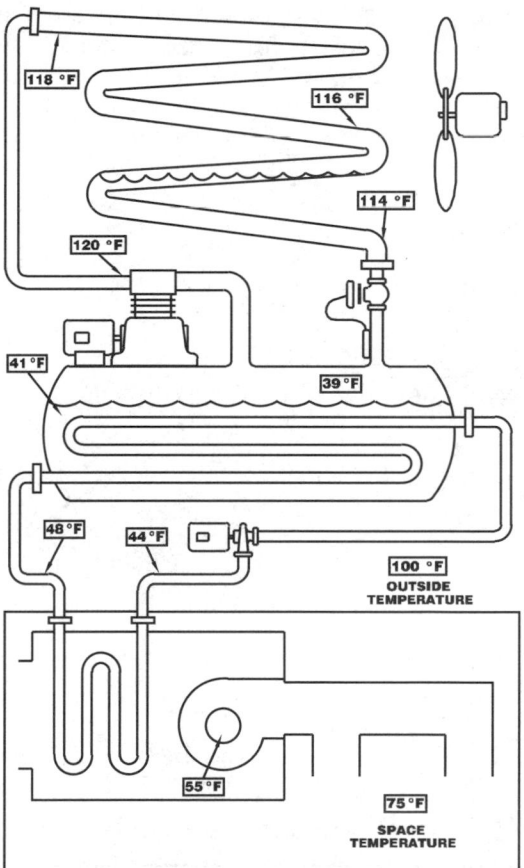

Fig. 4 The compressor temperature differential is much higher than the cooling load temperature differential
In this example, the outside-to-inside temperature differential is only 25 degrees. Yet, the temperature differential on the compressor, which determines the system energy requirement, is 81 degrees! Of the excess, the evaporator side, including hydronics, accounts for 9 degrees; the condenser side accounts for 20 degrees; and, the air distribution system accounts for 27 degrees.

In this example, the original 25°F temperature differential created by the weather becomes an 81°F differential at the compressor! This has the effect of reducing the theoretical COP from 21.4 to 6.2, an enormous loss of efficiency. The COP of the actual chiller is reduced by roughly the same ratio.

Figure 4 shows that the compressor discharge temperature is determined primarily by the outside air temperature. However, the compressor suction temperature is much lower than the inside air temperature. This is mostly because of heat transfer losses in the cooling system and in the air handling system. Indeed, most of the temperature loss in this example occurs in the air handling systems.

This example illustrates an important point, namely, that *cooling system efficiency is determined largely by factors outside the design of the cooling equipment. Much of the waste in a typical cooling system occurs because the compressor must operate at a higher temperature (or pressure) differential than necessary.* The struggle to achieve better cooling efficiency is largely a war against temperature differential that is fought in many parts of the system.

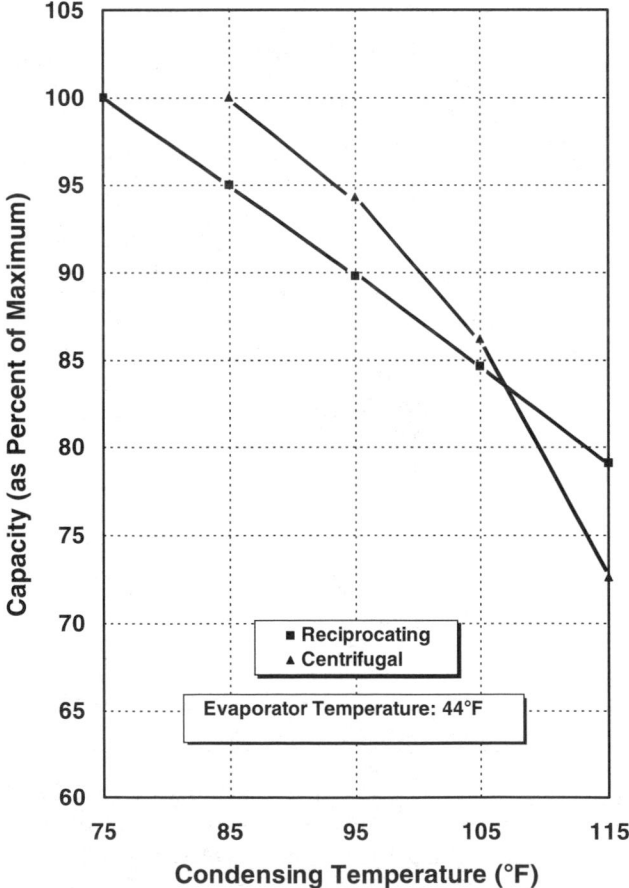

Fig. 5 The effect of system temperatures on capacity
Increasing the temperature differential across a cooling machine reduces its capacity. The effect is important in typical applications. It varies with the type of compressor.

Reducing the temperature differential in air handling systems is accomplished by various Measures in Section 4. Increasing chilled water temperature and lowering condenser temperature is accomplished in Subsection 2.2. Reducing the temperature differentials inside the cooling system itself is a matter of selecting the best design options, which are explained below.

The greatest potential for reducing the system temperature differentials occurs during off-peak periods, when the system temperature differentials may be much larger than needed to satisfy the load.

The Effect of Operating Temperatures on System Capacity

The cooling system must overcome a temperature differential that resists the flow of heat. Reducing the temperature differential increases the cooling capacity of the system. This effect is dramatic. Figure 5 shows the loss of capacity for two typical machines, one with a reciprocating compressor and the other with a centrifugal compressor.

Conversely, by reducing the system temperature differentials that occur under peak load conditions, you can reduce the nominal capacity of the equipment that you need to buy for new installations. You do this by spending more money for larger heat exchangers, and by designing air distribution systems for more efficient flow. This extra cost is partially offset because the increased efficiency allows smaller compressors to satisfy the load.

The change in cooling capacity with temperature differential varies substantially among different types of compressors. Centrifugal compressors suffer more capacity loss than other types. This is because the gas is driven through the compressor only by its own centrifugal force. As the condenser temperature increases, this force is counteracted by the greater pressure in the condenser.

In contrast, positive displacement compressors push the gas through the machine without regard to the temperature differential. The cooling capacity of compressors with a fixed compression ratio, such as screw compressors, is affected least. This is because the refrigerant gas is isolated from the discharge pressure during the compression process.

The Compressor and System Efficiency

In a compression cooling machine, most of the input energy goes to compressing the refrigerant gas to make it liquify. Therefore, the design of the compressor and other aspects of the compression process play a large role in the system's efficiency. The following are the major factors that determine the efficiency of the compression process.

■ The Refrigerant

Until about 1990, it was possible to read a cooling equipment catalog without finding any mention of the refrigerant used by the equipment. That situation has reversed because of environmental concerns about refrigerants, so that the choice of refrigerant may be the first consideration in selecting a compressor. As discussed previously, the choice of refrigerant limits the potential system efficiency.

With respect to the system hardware, the refrigerant affects the system pressures, the types of metals that can be used, the handling of lubricants inside and outside the compressor, safety features, and other equipment selection considerations. See Reference Note 34 for more about these issues.

■ Efficiency of Gas Flow in the Compressor

Refrigerant vapor has mass, so it has kinetic energy. If the flow through the compressor is turbulent, a portion of this energy is wasted by converting it to heat energy. Reciprocating compressors unavoidably generate a large amount of turbulence as a result of the oscillating piston motion. In contrast, the other major types of compressors have smoother gas flow.

In order to achieve ideal thermodynamic efficiency, a compressor would have to provide the minimum power at each incremental stage of compression. No type of compressor is perfect in this regard. For example, efficient compression is achieved in a centrifugal compressor by careful attention to the shape of the impeller and diffuser, but this shape achieves maximum efficiency only at one refrigerant flow rate.

Any leakage within a compressor wastes compressor energy by allowing compressed gas to re-expand uselessly. Leakage occurs around the piston rings of a reciprocating compressor, around the impeller of a centrifugal compressor, between the scrolls of a scroll compressor, and between the rotors and casing of a screw compressor. Minimizing leakage is a challenge in all types of compressors.

All types of positive-displacement compressors suffer re-expansion as a result of "clearance volume", which is the volume of gas left to re-expand at the end of the compression process. This problem is most severe with reciprocating compressors, which leave a significant amount of compressed gas in the cylinder at the top of the piston stroke. After the exhaust valve closes, the remaining gas expands in the first part of the suction stroke. The energy consumed to compress this gas is wasted.

■ Interstage Heat Transfer

When a gas is compressed, the heat of compression creates an expansion force that opposes the compression and increases compressor power. In an ideal compressor, the heat of compression would be removed as the gas is compressed. (If you are interested in the theory, this corresponds to the isothermal compression phase of the Carnot cycle.) Unfortunately, no conventional compressor design allows the gas to be cooled continuously during compression.

If the compressor has more than one stage, heat can be removed between the stages. Unfortunately, the interstage cooling cannot be done with ambient air or domestic water, because the heat exchange process occurs at low temperature. The compression process starts with a cold refrigerant gas and never gets very hot. The cooling is usually done by taking some liquid refrigerant from the condenser at high pressure, flashing it at a reduced pressure to get cool vapor, and injecting this vapor into the compressor between the stages. The mass of the added refrigerant also increases the cooling capacity.

The general technique of lowering gas temperature and/or increasing mass flow between stages is called an "economizer." (The same name is given to efficiency measures in other types of equipment, including absorption chillers, boilers, and air handling units). The

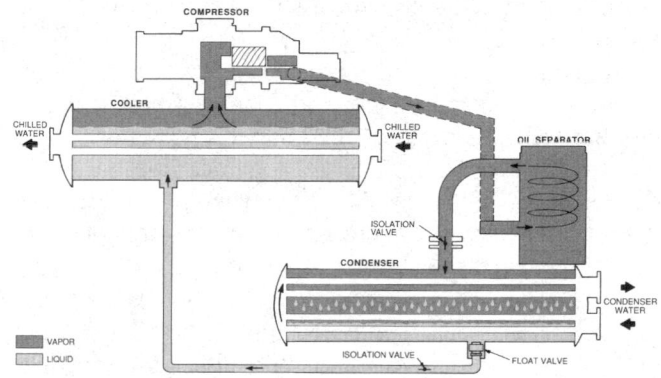

WITHOUT ECONOMIZER

WITH ECONOMIZER

Carrier Corporation

Fig. 6 Interstage heat transfer in a screw compressor system A portion of the liquid refrigerant from the condenser is flashed into gas, which is introduced part way through the compression process. This lowers the average temperature rise in relation to the mass of gas that is compressed, which lowers the average work of compression.

specific technique of using refrigerant flash gas for this purpose is called "flash intercooling."

With compressors that convey an isolated volume of gas through the compression process, as occurs in screw and scroll compressors, ports can be built into the compressor to inject gas for flash intercooling. Figure 6 shows how an economizer operates with a screw compressor chiller.

Interstage heat transfer produces a significant efficiency gain only if the compressor has a fairly high compression ratio. In this case, the heat of compression adds substantially to the gas pressure, increasing the work required from the compressor.

■ Fixed or Variable Compression Ratio

If the compressor discharge is open to the condenser as the compression is taking place, then the discharge pressure never exceeds the pressure needed to condense the refrigerant (except for a small expansion loss between the compressor and condenser). The leading examples of such compressors are the centrifugal and reciprocating types. In such compressors, the compression ratio adjusts itself to the needs of the system.

On the other hand, some compressors work by trapping a volume of gas and then compressing this amount independently of the condenser pressure. The gas is released to the condenser only near the end of the compression process. This is true of some screw compressors and most scroll compressors, and of some other types, such as vane compressors.

If the compressor discharge pressure exceeds the condenser pressure, then a fraction of the compression energy is wasted in compressing the gas more than necessary. On the other hand, if the compressor discharge pressure is lower than the desired condenser pressure, the condenser is not able to condense the gas at higher coolant temperatures. Therefore, systems using compressors with fixed compression ratios always err on the side of excess discharge pressure, which wastes energy at reduced loads.

The compression ratio corresponds to a ratio of temperatures between the evaporator and the condenser. Therefore, energy conservation measures that attempt to improve efficiency by raising the evaporator temperature or lowering the condenser temperature are less effective with compressors having fixed compression ratios.

■ Efficiency of Output Modulation

In all common types of cooling machines, the method used to throttle output is an integral part of the compressor, and may account for much of its efficiency limitations. Some compressors that are efficient at high load become inefficient at low loads. Some compressors cannot operate at all below some minimum load.

In typical applications, cooling systems operate over a wide range of loads, with full-load operation being an exception that occurs mostly on start-up. Many older cooling plants were designed at a time when manufacturers did not stress part-load efficiency, and these commonly provide major opportunities for improving efficiency by substituting machines that respond more efficiently to partial loads.

Losses in throttling output are most severe in centrifugal compressors. In the past, economies of scale led to decisions to cool a facility with one or two large centrifugal machines, but this configuration is wasteful at low loads. To increase part-load efficiency, you need to use a larger number of smaller machines, or to use different types of compressors, perhaps in combination with centrifugal machines. Improving part-load efficiency in an existing plant may require installing an additional, smaller machine to serve low loads. (Measure 2.8.1 recommends this technique.)

Reciprocating compressors retain their efficiency fairly well at lower loads, although part-load efficiency varies among different models. Smaller reciprocating compressors usually operate in an on-off mode, so they suffer efficiency loss only during the start of each cycle, while refrigerant flow in the system is stabilizing.

Scroll compressors usually operate in an on-off mode, so the efficiency of capacity reduction is not a factor. Other methods of modulation are being developed for scroll compressors, as discussed below. The efficiency performance of modulating scroll compressors is not yet well documented.

■ Limitations in Reducing Condenser Temperature

The previous discussion emphasized the importance of keeping condenser temperature as low as possible. With some compressors, the condensing temperature must be kept above a certain level. In such cases, the condenser temperature is kept artificially high by choking off the air or water that cools the condenser. This wastes energy in applications where the system may operate during cooler weather.

There are several reasons why a system may require an elevated condenser temperature. In some cases, the compressor is the limiting factor, because high condensing temperature may be needed to prevent migration of lubricants away from the compressor. More commonly, high condensing temperature may be needed for reasons not related to the compressor, such as maintaining adequate pressure for operation of an expansion valve or capillary refrigerant distributors. Other reasons are given in Measure 2.2.2.

Types of Compressors and Throttling Methods

The following are the efficiency characteristics of the most common types of cooling compressors. One of the most important efficiency characteristics is the method used by the compressor to reduce its output.

■ Centrifugal Compressors

Centrifugal compressors are the most efficient type when they are operating near full load. Their efficiency advantage is greatest in large sizes, and they offer considerable economy of scale, so they dominate the market for large chillers. They are able to use a wide range of refrigerants efficiently, so they will probably continue to be the dominant type in large sizes. The peak efficiency of centrifugal compressors has improved dramatically and progressively since about 1980, and several major improvements to part-load efficiency have been made in that time.

Centrifugal compressors have a single major moving part, an impeller that compresses the refrigerant gas by centrifugal force. The gas is given kinetic energy as it flows through the impeller. This kinetic energy is not useful in itself, so it must be converted to pressure energy. This is done by allowing the gas to slow down smoothly in a stationary diffuser surrounding the impeller.

Figure 7 is a detailed drawing of a centrifugal compressor that has two impellers on the same shaft, for two stages of compression. The impeller speed is increased by using a gear train between the motor and the impellers.

Figure 2 shows a centrifugal compressor that has three impellers, for three stages of compression. The impellers are driven directly by an electric motor, which limits the rotation speed. Three stages are needed to achieve the needed condensing pressure with the particular refrigerant that is used in this machine. Using a large impeller diameter also compensates for the low rotation speed.

Figure 8 shows a centrifugal chiller in which the compressor has only a single, relatively small impeller. The motor speed is limited by the line frequency, so this machine uses gears to increase the impeller speed.

High energy efficiency in a centrifugal compressor requires efficient gas flow through the impeller and the diffuser. Unfortunately, the design of the impeller and

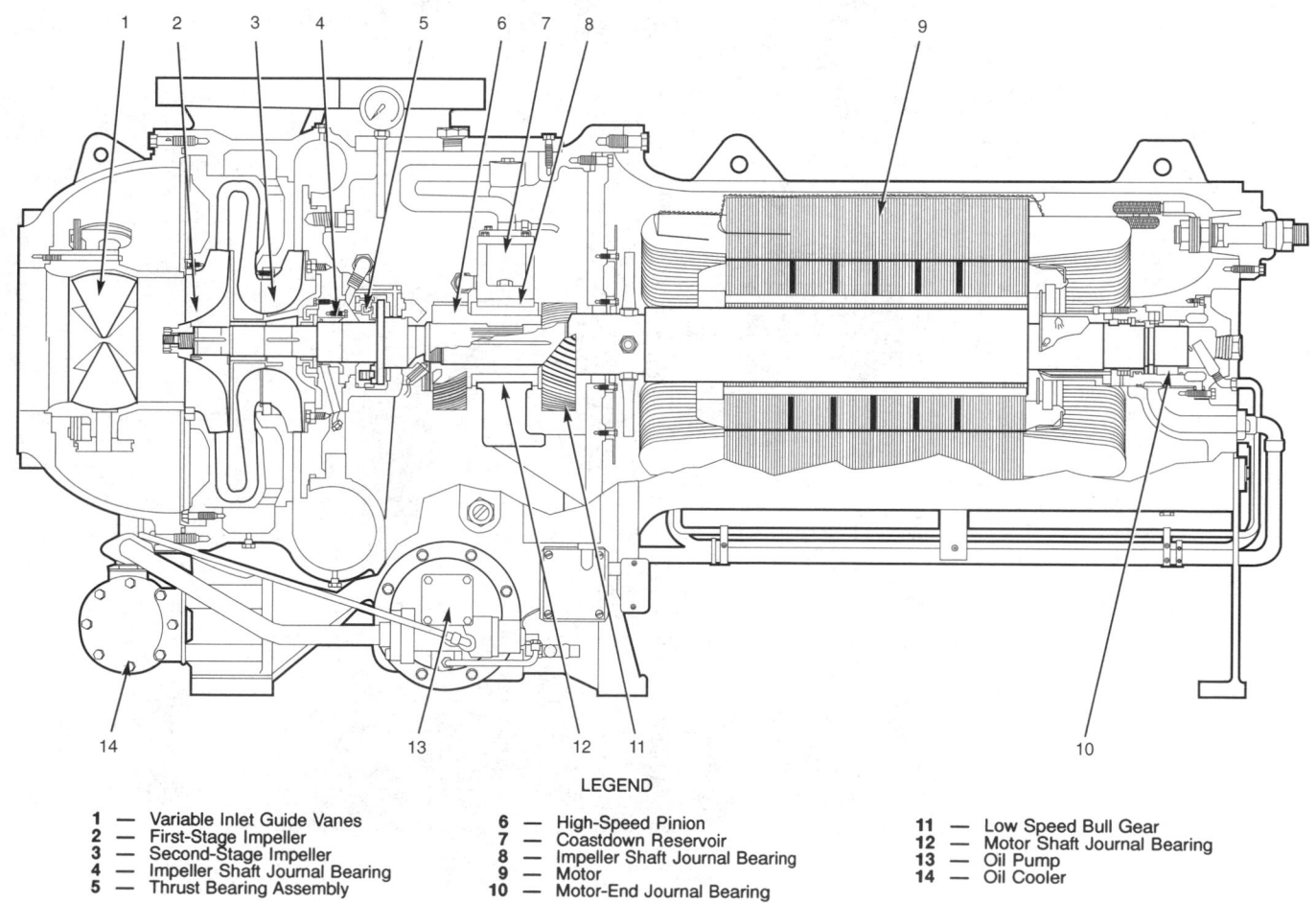

LEGEND

1	—	Variable Inlet Guide Vanes	6	—	High-Speed Pinion	11	— Low Speed Bull Gear
2	—	First-Stage Impeller	7	—	Coastdown Reservoir	12	— Motor Shaft Journal Bearing
3	—	Second-Stage Impeller	8	—	Impeller Shaft Journal Bearing	13	— Oil Pump
4	—	Impeller Shaft Journal Bearing	9	—	Motor	14	— Oil Cooler
5	—	Thrust Bearing Assembly	10	—	Motor-End Journal Bearing		

Carrier Corporation

Fig. 7 Centrifugal compressor with internal motor This model has two impellers on a common shaft, providing two stages of compression. To further increase compression, the impeller speed is increased by gears. Typical of most centrifugal compressors, the output of the machine is adjusted by throttling the flow of gas into the first stage using pre-rotation inlet vanes.

diffuser can be optimized for only one gas flow rate. To minimize efficiency loss at reduced loads, centrifugal compressors typically throttle output with pre-rotation vanes (inlet guide vanes) located at the inlet to the impeller(s). This method is efficient down to about half load, but the efficiency of this method decays rapidly below half load.

Older centrifugal machines are not able to reduce load much below 50%. This is because of "surge" in the impeller. As the flow through the impeller is choked off, the gas does not acquire enough energy to overcome the discharge pressure. Flow drops abruptly at this point, and an oscillation begins as the gas flutters back and forth in the impeller. Efficiency drops abruptly, and the resulting vibration can damage the machine.

Many older centrifugal machines deal with low loads by creating a false load on the system, such as by using hot gas bypass. This wastes the portion of the cooling output that is not required.

To improve both turndown range and low-load efficiency, some machines use two separate centrifugal compressors. The two compressors share a common condenser shell and a common evaporator shell. Figure 9 shows this approach used with a large chiller.

Another approach is to use variable-speed drives in combination with inlet guide vanes. This may allow the compressor to throttle down to about 20% of full load, or less, without false loading. Changing the impeller speed causes a departure from optimum performance, so efficiency still declines badly at low loads.

A compressor that uses a variable-speed drive reduces its output in the range between full load and approximately half load by slowing the impeller speed. At lower loads, the impeller cannot be slowed further, because the discharge pressure would become too low to condense the refrigerant. Below the minimum load provided by the variable-speed drive, inlet guide vanes are used to provide further capacity reduction.

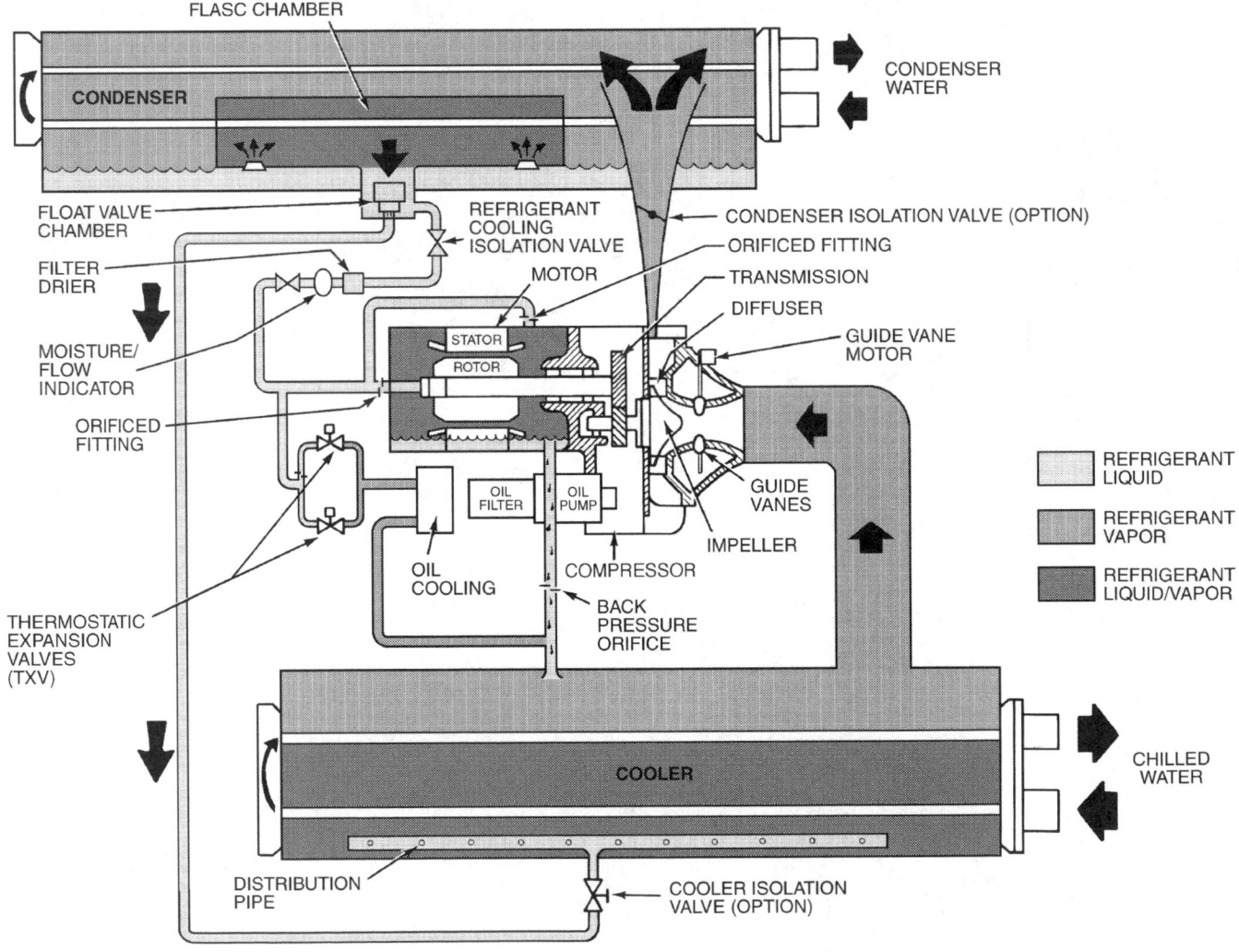

Carrier Corporation

Fig. 8 Packaged centrifugal water chiller Details and accessories vary from one model to another, and from one manufacturer to another. Here, the flash chamber provides cool gas for motor cooling and oil cooling. This compressor has only a single stage, so an economizer is not possible.

Fig. 9 A large water chiller with two centrifugal compressors This is one way of maintaining efficiency at reduced cooling loads.

Unlike positive-displacement compressors, which are not subject to surge, the minimum load of a centrifugal compressor is sensitive to the evaporator and condensing temperatures. Raising the evaporator temperature and lowering the condenser temperature reduce the tendency to surge, and hence reduce the minimum load. The minimum load of an individual centrifugal machine varies widely, depending on these temperatures.

It is not advisable to turn a centrifugal compressor on and off to deal with low loads. Most centrifugal compressors use journal bearings, which are not positively lubricated until the shaft reaches a certain speed. Refrigerant flow within the machine may require several minutes to stabilize after each start-up. The types of motors used to drive centrifugal machines are more likely to overheat with frequent start cycles. And, the gears in gear-driven compressors are subjected to high stress on start-up.

The pressure output of a centrifugal compressor depends on the diameter and speed of the impeller(s). Refrigerants that have low condensing pressures can be compressed with a centrifugal machine that is driven directly from a 50 Hz or 60 Hz motor. Gases that require higher pressures require the compressor to have speed increasing gears, or several stages of compression (several impellers), or both.

For a given refrigerant, the equipment designer can choose to increase the number of stages of compression, while reducing the impeller speed and diameter. Using several stages allows an economizer (described previously) to be used between each stage. The theoretical efficiency improvement becomes smaller with each additional stage, and flow losses accumulate at the inlet and outlet of each impeller, so there is an optimum number of stages for each refrigerant. Increasing the number of stages increases cost and complexity.

Increasing the impeller speed by using gears incurs losses in the gears and in the additional bearings. It also eliminates the opportunity of exploiting interstage heat transfer.

■ Reciprocating Compressors

Reciprocating compressors are the oldest type, and they continue to be the dominant type in the small and medium size range. Figure 10 shows a typical unit.

In principle, reciprocating compressors can be designed for virtually any refrigerant. In practice, they are most economical for refrigerants that operate at medium and high pressures, because the higher gas densities provide more refrigeration effect for a machine of a given size.

The maximum efficiency of reciprocating compressors is lower than that of centrifugal and screw compressors. Efficiency is reduced by clearance volume (the compressed gas volume that is left at the top of the piston stroke), throttling losses at the intake and discharge valves, abrupt changes in gas flow, and friction. Also, reciprocating compressors tend to be used with systems having expansion valves, which are somewhat less efficient than the fluid metering methods used in larger systems. Lower efficiency also results from the smaller sizes of reciprocating units, because motor losses and friction account for a larger fraction of energy input in smaller systems.

For most air conditioning applications, reciprocating compressors can easily compress the refrigerant gas in a single stage, so there is no opportunity for improving the theoretical efficiency of compression by using an economizer (described previously). However, multiple stages of compression are commonly used in low-temperature refrigeration, and compression efficiency can be improved by using economizers in those applications.

Reciprocating compressors suffer less efficiency loss at partial loads than other types, and they may actually have a higher absolute efficiency at low loads

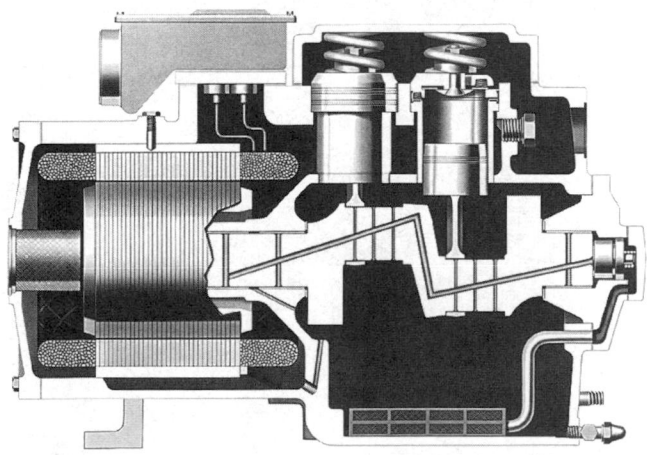

Fig. 10 Reciprocating compressor Capacity is controlled by disabling individual cylinders, which can be done in several ways, or by varying the speed. The large springs allow the cylinder heads to lift if liquid refrigerant enters the cylinders, avoiding destruction. This type of enclosure, with an internal motor and bolted access plates, is called "semi-hermetic."

than the other types. Smaller reciprocating compressors control output by turning on and off. This eliminates all part-load losses, except for a short period of inefficient operation when the machine starts.

Larger multi-cylinder reciprocating compressors commonly reduce output by disabling ("unloading") individual cylinders. When the load falls to the point that even one cylinder provides too much capacity, the machine turns off.

Several methods of cylinder unloading are used, and they differ in efficiency. The most common is holding open the intake valves of the unloaded cylinders. This eliminates most of the work of compression, but a small amount of power is still wasted in pumping refrigerant gas to-and-fro through the unloaded cylinders. Another method is blocking gas flow to the unloaded cylinders, which is called "suction cutoff."

If a multi-cylinder reciprocating compressor can reduce its capacity by more than 50%, a significant amount of additional complication is required to provide stable refrigerant flow through expansion valves, to avoid coil frosting, to achieve proper return of lubricating oil to the compressor, etc. For this reason, the trend in larger reciprocating systems is to use multiple compressors with independent refrigerant circuits, with each compressor unloading no more than 50% (i.e., in two steps of capacity).

Variable-speed drives can be used with reciprocating compressors, eliminating the complications of cylinder unloading. This method has not become popular yet, probably because the attention of designers is now focussed on newer types of compressors. Automotive air conditioning compressors, which are driven directly by the engine, operate successfully over a wide range of speeds.

■ Screw Compressors

Screw compressors, sometimes called "helical rotary" compressors, compress refrigerant by trapping it in the "threads" of a rotating screw-shaped rotor. There are two types of screw compressors, twin-screw and single-screw. At present, the twin-screw version is the most common.

In the twin-screw compressor, the refrigerant is pinched between the lobes (raised portions) and flutes (recessed portions) of two parallel mating screw rotors. Figure 11 is a cutaway drawing of a typical twin-screw compressor.

Figure 12 is a cutaway drawing of a chiller that uses a twin-screw compressor.

In single-screw compressors, the refrigerant is pinched between the threads of a single helical rotor and one or two star-shaped rotors (called "gaterotors") that rotate at right angles to the main rotor and mesh

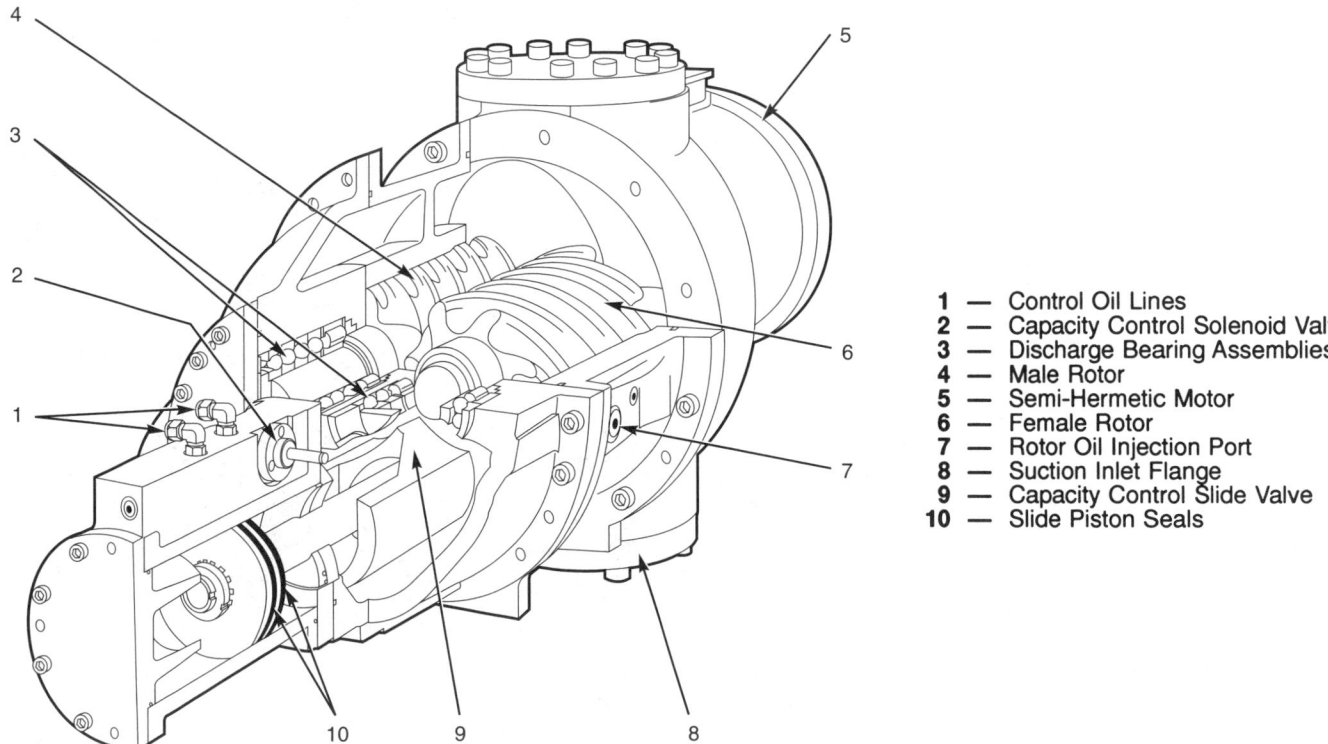

1 — Control Oil Lines
2 — Capacity Control Solenoid Valve
3 — Discharge Bearing Assemblies
4 — Male Rotor
5 — Semi-Hermetic Motor
6 — Female Rotor
7 — Rotor Oil Injection Port
8 — Suction Inlet Flange
9 — Capacity Control Slide Valve
10 — Slide Piston Seals

Carrier Corporation

Fig. 11 Twin-screw compressor   The two helical rotors mesh, squeezing gas from the rear toward the front. The capacity control slide valve, 9, is a movable part of the housing that surrounds the rotors. By sliding away from the rest of the housing, it creates a gap of variable size that disables the adjacent portion of the rotors.

with the helix, like pinions being driven by a worm gear. The only purpose of the gaterotors is to block flow through the threads. They are idlers, and play no part in moving the gas.

Screw compressors have been used for a long time as low-pressure air compressors and engine superchargers, but leakage between the screws has made these units too inefficient for refrigeration work until the 1980's. At that time, improved machining technology made it possible to achieve the close tolerances that are necessary for good efficiency, and improved designs reduced other efficiency losses. Now, the best screw compressors are significantly more efficient than reciprocating compressors at all loads, and they may be more efficient than centrifugal compressors at low loads.

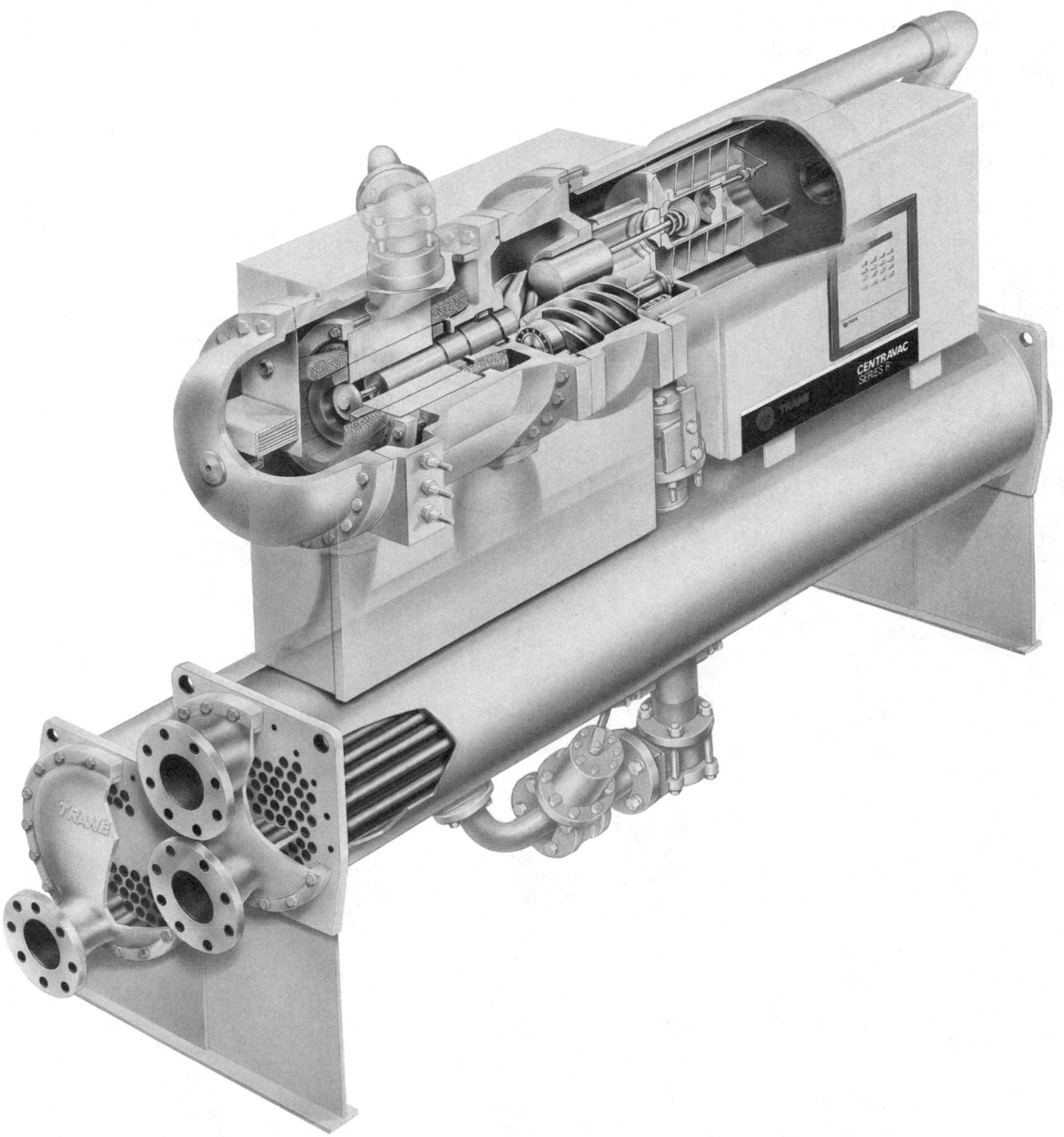

Fig. 12 Chiller with twin-screw compressor Refrigerant gas enters the near end of the screw from the evaporator, in the lower rear. After a short passage through the rotors, the compressed gas is discharged into the condenser. The valves under the condenser flash liquid refrigerant for motor cooling, and perhaps, for an economizer.

Screw compressors have increasingly taken over from reciprocating compressors of medium sizes and large sizes, and they have even entered the size domain of centrifugal machines.

Screw compressors are applicable to refrigerants that have higher condensing pressures, such as HCFC-22 and ammonia. They are especially compact. The practical upper limit of screw compressor size is determined by competition from centrifugal machines, which offer greater efficiency in larger sizes. However, it may turn out that screw compressors are better able to exploit the efficiency potential of their refrigerants, especially in applications with high temperature differentials (such as making ice), so screw compressors may prove to have no economical upper size limit. The lower limit of size is determined by price competition from reciprocating compressors and other types.

A major distinction between screw compressors is whether the unit operates with or without oil injection into the rotors. Oil is used to cool the refrigerant as it is compressed, reducing the work of compression and improving cycle efficiency. It also seals the leakage paths, allowing the compressor to operate efficiently at lower speeds. The oil typically takes up less than one percent of the compression volume.

In twin-screw compressors, the oil serves as a lubricant between the rotors, allowing one rotor to drive the other by direct contact. Rotor wear is small in this drive arrangement because the shape of the rotors is designed to mesh them with a rolling contact.

The oil is cooled in an oil cooler to get rid of the heat of compression. Cooling may be provided by any available cooling medium, including condenser water,

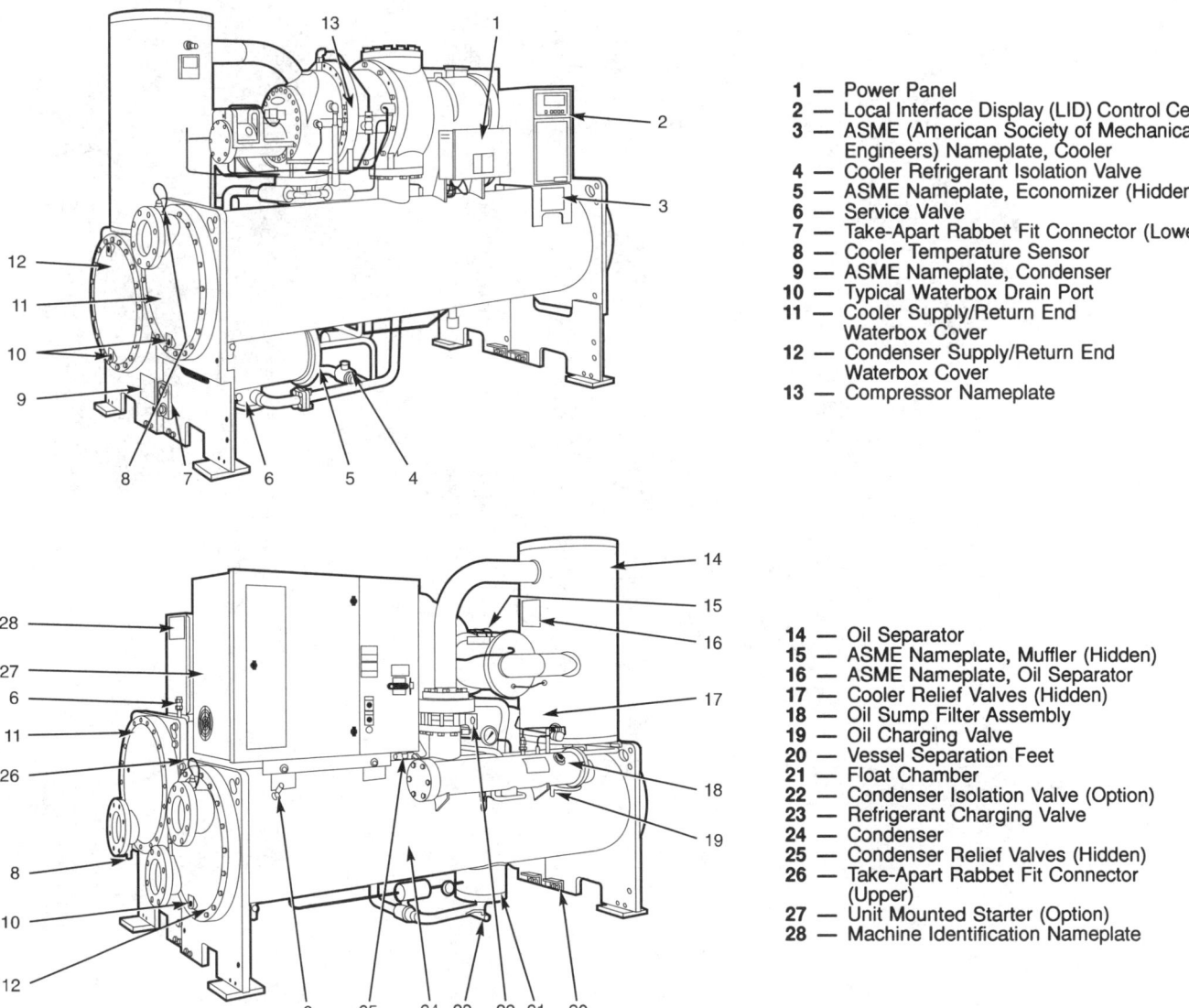

1 — Power Panel
2 — Local Interface Display (LID) Control Center
3 — ASME (American Society of Mechanical Engineers) Nameplate, Cooler
4 — Cooler Refrigerant Isolation Valve
5 — ASME Nameplate, Economizer (Hidden)
6 — Service Valve
7 — Take-Apart Rabbet Fit Connector (Lower)
8 — Cooler Temperature Sensor
9 — ASME Nameplate, Condenser
10 — Typical Waterbox Drain Port
11 — Cooler Supply/Return End Waterbox Cover
12 — Condenser Supply/Return End Waterbox Cover
13 — Compressor Nameplate

14 — Oil Separator
15 — ASME Nameplate, Muffler (Hidden)
16 — ASME Nameplate, Oil Separator
17 — Cooler Relief Valves (Hidden)
18 — Oil Sump Filter Assembly
19 — Oil Charging Valve
20 — Vessel Separation Feet
21 — Float Chamber
22 — Condenser Isolation Valve (Option)
23 — Refrigerant Charging Valve
24 — Condenser
25 — Condenser Relief Valves (Hidden)
26 — Take-Apart Rabbet Fit Connector (Upper)
27 — Unit Mounted Starter (Option)
28 — Machine Identification Nameplate

Carrier Corporation

Fig. 13 What is all that stuff? Screw compressor cooling machines have some large accessories. Note the size of the oil separator, 14. Other models have equally large suction filters to protect the narrow clearances of the rotors from debris in the refrigerant system.

ambient air, refrigerant, or chilled water. Cooling the oil with refrigerant or chilled water reduces the overall system efficiency, so avoid these methods of oil cooling.

Oil-injected compressors also require an oil separator to remove oil from refrigerant after it leaves the compressor. The oil separator presents some resistance to the flow of refrigerant gas on the discharge side, so it adds an efficiency penalty. Also, it typically is larger than the compressor itself. Figure 13 shows a screw compressor chiller with all its accessories, including the large oil separator.

Some screw compressors have another large accessory, a suction filter. This protects the compressor, which has very close tolerances, from any debris that travels through the refrigerant system, such as pieces of solder, motor insulation, etc.

In twin-screw compressors without oil injection, clearance between the two rotors is maintained by driving them through meshed gears. This increases cost, size, and maintenance requirements. Dry compressors must operate at high speed to minimize leakage of gas between the rotors, which increases noise. Since there is no oil to seal the gaps, more elaborate mechanical seals are required. On the positive side, no oil separator or oil cooler is needed.

Single-screw compressors may also operate with or without oil injection. In machines that do not inject oil, liquid refrigerant may be injected into the rotor, achieving both sealing and improved mass flow. These compressors can operate with minimal lubrication because the gaterotors are made of a compliant plastic material that provides tight sealing against the main rotor.

A variety of methods are used to control the output of screw compressors. There are major efficiency differences among the different methods. The most common is a slide valve that forms a portion of the housing that surrounds the screws. When the valve slides away from the rest of the housing, it creates a gap that deactivates the adjacent portion of the rotors.

Using a variable-speed drive is another method of capacity control. It is limited to oil-injected compressors, because slowing the speed of a dry compressor would allow excessive internal leakage.

There are other methods of reducing capacity, such as suction throttling, that are inherently less efficient than the previous two.

Screw compressors can achieve a much better turndown ratio than centrifugal compressors, typically down to about 10% of full load. Screw machines tolerate frequent on-off cycling for low-load operation, as with reciprocating machines. This avoids the temptation to create false loads to keep the machine running.

The basic screw compressor has a fixed compression ratio, which has efficiency limitations discussed

previously. To minimize this problem, some screw compressors have a variable compression ratio. This is achieved with a variable discharge port that taps the compressed gas from different areas near the discharge end of the rotor(s). The most common form is a slide valve. The mechanism for this slide valve is sometimes combined with the inlet slide valve that controls compressor capacity.

The efficiency of single-screw and twin-screw compressors can be improved by flashing some liquid refrigerant and injecting the cool refrigerant gas into the rotor at a point downstream of the primary suction port. This feature is called an "economizer," and it is described above.

Don't confuse the economizer process with injecting liquid refrigerant into oil-injected compressors to cool the oil. The latter function has an efficiency penalty, because the oil heat adds to the cooling load.

Trane Company

Fig. 14 Scroll compressor The compressor itself occupies the upper third of the case. The rest is the motor.

■ Scroll Compressors

The scroll compressor is an old invention that has finally come to market. Figure 14 shows a typical unit.

The gas is compressed between two scroll-shaped vanes, shown in Figure 15. One of the vanes is fixed, and the other moves within it. The moving vane does not rotate, but its center revolves with respect to the center of the fixed vane, as shown in Figure 16. This motion squeezes the refrigerant gas along a spiral path, from the outside of the vanes toward the center, where the discharge port is located. The compressor has only two moving parts, the moving vane and a shaft with an off-center crank to drive the moving vane.

Scroll compressors have only recently become practical, because close machining tolerances are needed to prevent leakage between the vanes, and between the vanes and the casing. A variety of designs are used to achieve good sealing, including oil flooding and pressure-loaded sliding contacts between the scrolls.

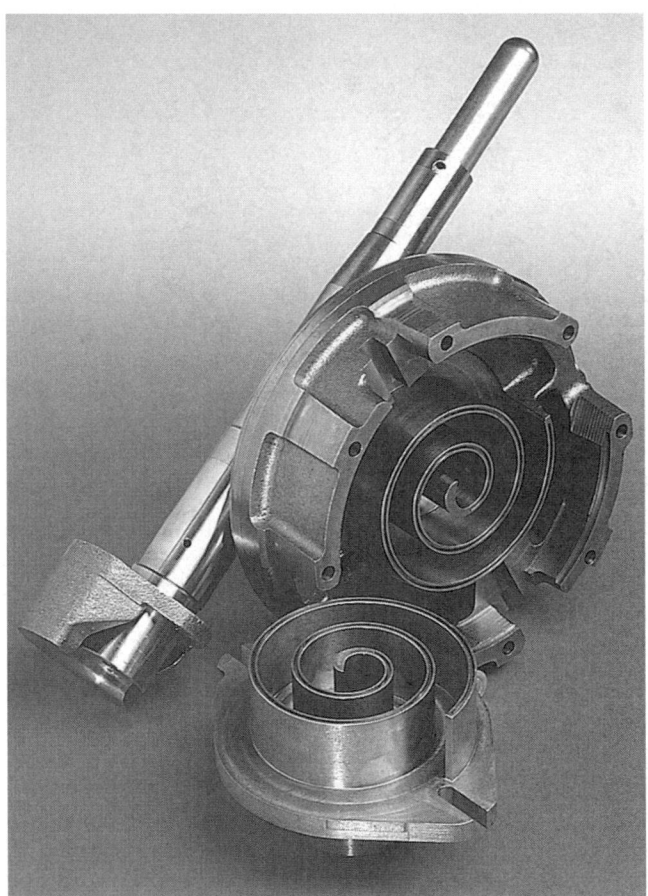

Trane Company

Fig. 15 The parts of a scroll compressor The top scroll is fixed. The bottom scroll moves within the fixed scroll. It does not rotate, but moves with an eccentric motion that squeezes the refrigerant gas from the outside of the spiral toward the center, as shown in Figure 16. The slot on the side of the lower scroll engages a fixed pin that keeps it from rotating. In back is the drive shaft for the moving scroll, with a counterbalanced eccentric drive at the bottom. The performance of the compressor is dependent on very accurate machining of the scrolls.

Scroll compressors promise good efficiency because of their smooth gas flow. Unlike the gas flow in a reciprocating compressor, the gas moves through a screw compressor in only one direction, from the inlet port toward the outlet port. The absence of reversing gas flow greatly reduces the variation of motor torque during the compression cycle, as shown in Figure 17. This promises reduced vibration and noise.

Scroll compressors have a fixed compression ratio, which has efficiency limitations discussed previously. If they prove to be satisfactory, they seem likely to become prominent at the low end of the system size range.

Like other small compressors, scroll compressors usually control output by turning on and off. Variable capacity can be achieved by using several compressors. Electronic variable-speed drives have been used on some scroll compressors, but this method of capacity control is expensive in small sizes. Another method is to open ports that bleed gas from the early (outer) portion of the compression cycle, but this method wastes some energy.

■ Other Compressor Types

Mankind has devoted a great deal of ingenuity to finding ways of compressing gases, resulting in many compressor designs. The most efficient types approach the maximum achievable efficiency, but still leave room for significant improvement. Greater efficiency, reduced cost, reduced maintenance, and lower noise are all factors that may cause a new type of compressor to burst on the scene.

At present, there are no major contenders to displace the previous types, but this could change quickly, as witness the rapid commercialization of screw compressors and scroll compressors after years of dormancy. The best advice is to check the latest developments before selecting a compressor, but make sure that a model is well proven before buying it. The potential advantages of novel types are too small to justify reliability risks.

Vane compressors and trochoid compressors are types that have been used in various smaller refrigeration applications, such as residential refrigerators and air conditioners, but not yet in larger cooling systems. They are appealing because of their apparent simplicity, their compactness, and their ability to operate quietly. The major practical problem with both has been sealing.

Vane compressors use a rotor that is located eccentrically in a chamber. Sliding vanes in the rotor create spaces between the rotor and the chamber wall. The volume of these spaces changes as the rotor rotates. Refrigerant gas is drawn into the spaces in the half of the rotation where the size of the spaces is increasing. The gas is compressed as the rotor turns, and is discharged at the point where the spaces are smallest.

Trochoidal compressors are a general type in which a rotor rolls along the inside of a chamber, compressing the gas into pockets formed between the rotor and the chamber wall. ("Trochoid" is the name of a curve that is traced by the perimeter of a rolling circle.) The best known example of a trochoidal device is the Wankel rotary engine, a variant of which has been made into a commercial compressor.

Compressor Drivers

All the energy input to a compression cooling system goes into the compressor driver. This may be an electric motor, a reciprocating engine, a gas turbine, or other machine. Your selection of the compressor driver is a critical factor in efficiency and operation. The following are the most common choices to be made.

■ Internal vs. External Motors

If the compressor is driven by an electric motor that is located outside the compressor housing, a shaft must penetrate the compressor housing. This type of installation is called an "open drive." Figure 18 shows a typical open drive.

The motor may also be installed inside the compressor casing, along with all the other moving parts.

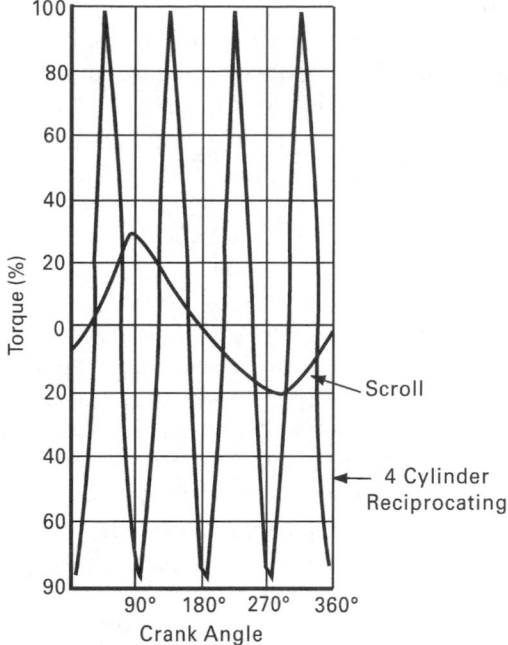

Fig. 17 Torque characteristics of scroll and reciprocating compressors The scroll compressor has less torque vibration, largely because the gas does not have to reverse direction in going through compression.

Fig. 16 How a scroll compressor moves The dark scroll is fixed. The light scroll moves within it. The moving scroll does not rotate, but its center moves in a small circle. The effect is to squeeze gas from the outside toward the inside along a spiral path.

If the compressor and motor is totally encapsulated within a welded housing, the compressor is called "hermetic." True hermetic compressors are found only in smaller sizes. Figure 19 shows a typical example.

If the motor or other components are accessible through bolted access plates, the compressor is properly called "semi-hermetic," although it is common to call such compressors "hermetic" also. The compressor shown in Figure 10 is semi-hermetic.

Are open or hermetic compressors better? There is no definite answer. Both types work well within their appropriate applications. However, increasing concern about the environmental hazard of certain refrigerants has shifted the balance toward hermetic machines, at least for machines that use harmful refrigerants. The advantages of hermetic systems are:

- *reduced refrigerant leakage.* Installing the motor inside the compressor housing eliminates the shaft penetration, which is the main source of refrigerant loss with open drive compressors. Low leakage is an important advantage if the refrigerant has objectionable environmental properties.

- *motor longevity.* Refrigerant cooling promises long motor life because temperatures are lower, and the coolant is clean. However, all compression refrigerants have some solvent behavior, and they tend to be absorbed by the organic materials used for motor insulation. This may create unexpected problems with newer refrigerants.

Hermetic motors have these disadvantages:

- *reduced cycle efficiency and/or increased condenser size.* The motor of a hermetic compressor may be cooled by cold suction gas from the evaporator, or by liquid refrigerant from the condenser. In both cases, the motor heat goes into the compression cycle.

For example, the efficiency of the large motors used with centrifugal chillers ranges from 92% to 96%. The remaining percentage becomes motor heat, which reduces the COP of the chiller by about one or two percent.

- *difficult access to the motor for maintenance.* Getting to the motor requires pumping the refrigerant out of the system, and removing the access covers.

- *need to clean out the entire refrigerant system in the event of a motor burnout.* The electric arcing that occurs in an insulation burnout creates corrosive materials from most refrigerants. This requires cleaning out the entire refrigerant side of the cooling system.

- *possible inability to retrofit variable-speed drives.* If a hermetic compressor was originally designed to operate at a constant speed, it may not be possible to retrofit an electronic variable-speed motor drive. This is because the full motor speed may be needed for proper bearing lubrication and motor cooling.

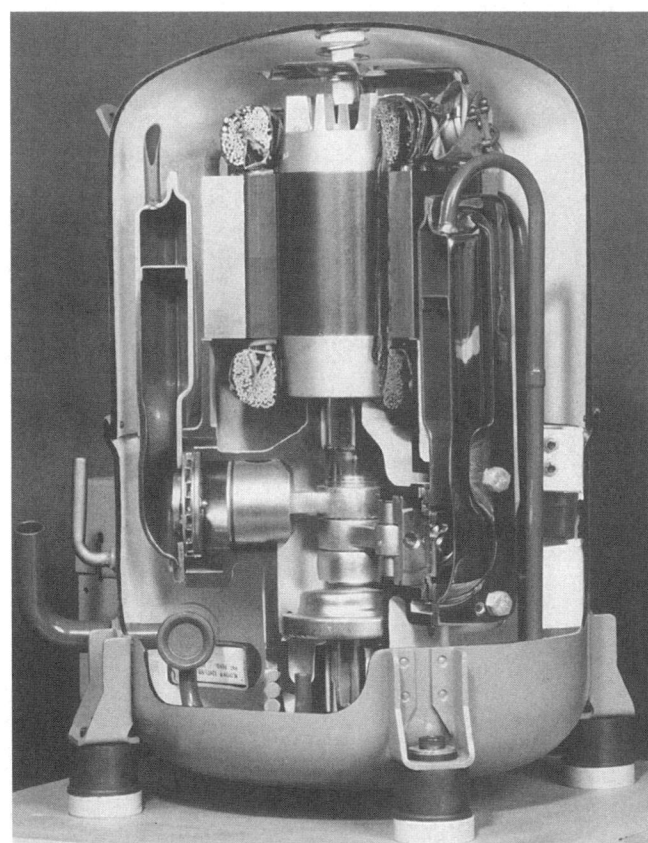

Trane Company

Fig. 19 Hermetic compressor   This arrangement virtually eliminates refrigerant leakage. However, it is not practical to repair this type of unit at the site.

WESINC

Fig. 18 Compressor with open drive   An important disadvantage of this arrangement is leakage of refrigerant around the drive shaft. On the other hand, the arrangement has several advantages.

■ Motor Efficiency

Motor efficiency is generally not an optional factor with hermetic compressors, where the motors are custom made for the machine. Motors of higher efficiency may be an option with open-drive compressors.

Controlling the compressor output with an electronic variable-frequency motor drive keeps the motor operating more efficiently at partial loads, although there is a small efficiency penalty at high loads. For more about this, see Reference Note 36, Variable-Speed Motors and Drives.

■ Other Types of Drivers

The great majority of compression cooling machines are driven by electric motors, but there are circumstances in which a different source of shaft power may be appropriate. The first refrigeration compressors were driven by reciprocating steam engines. Modern compressors are sometimes driven by diesel engines, natural gas reciprocating engines, gas turbines, and steam turbines.

A decision whether to choose such a drive requires detailed knowledge of the engineering, operating practices, and economics of the drive in question. There may be a temptation to use non-motor drives for "obvious" but illogical reasons. For example, a large urban hospital uses a condensing steam turbine to drive a chiller, apparently because the facility was designed with a high-pressure steam supply for other reasons. The steam turbine drive in this facility is much more expensive to operate than an electric motor would have been.

In general, electric motors combined with variable-frequency drives provide better part-load performance than other types of drives, along with lowest first cost and minimal maintenance requirements. If you are interested in a reciprocating engine drive, you have to find out whether the compressor is vulnerable to the torsional vibrations of the engine drive. This requires a sophisticated analysis of the rotating components by the compressor manufacturer.

Packaged engine-driven cooling machines are available. These are sold primarily as a means of reducing electrical demand during peak load periods.

Evaporators

The design of the evaporator can have a significant effect on system efficiency. We will cover the most common types of evaporators. Then, we will discuss evaporator superheat, which is the most important factor in evaporator efficiency.

■ Flooded Evaporators

Flooded evaporators are used in a large fraction of packaged water chillers, as shown in Figure 2. Chilled water is cooled in tubes that are completely submerged in liquid refrigerant. This provides complete utilization of the tube surface. The large free surface of the liquid refrigerant is directly exposed to the compressor suction so there is virtually no pressure loss between the evaporation area and the compressor suction.

This arrangement is simple as well as efficient. Liquid refrigerant from the condenser is simply drained back to the evaporator, typically using an orifice or float valve to isolate the evaporator from the pressure of the condenser.

The only significant disadvantage of flooded evaporators is that they require a larger quantity of refrigerant than evaporators where the refrigerant evaporates inside the tubes.

■ Liquid Overfeed Evaporators

In liquid overfeed systems, refrigerant is continuously pumped, or flows by gravity, through any number of evaporators. The evaporators can take any form needed by the application, such as an air coil or a plate ice maker. The name "liquid overfeed" derives from the fact that more liquid refrigerant is fed to each evaporator than evaporates, the excess liquid flowing back to a receiver. The receiver is open to the compressor suction, so the refrigerant liquid is kept at the temperature corresponding to the compressor suction pressure. Thus, each evaporator behaves almost like a flooded evaporator.

The system is flexible, in that it can serve any number and variety of evaporators. The only limitation is that all evaporators must operate at the same pressure, and hence, at the same temperature. Liquid overfeed has generally the same efficiency advantages as a flooded evaporator, except for a minimal energy loss in refrigerant pumping and piping pressure loss. There is no pressure difference between the evaporator receivers and the evaporators, except for any gravity head that may exist as a result of differences between the heights of the evaporators. As a result, the pumping power is minimal.

In addition to the low-pressure receiver for each evaporator, the system generally has a single high-pressure accumulator to allow for fluctuations in the quantity of refrigerant in the evaporators.

Since each evaporator in a liquid overfeed system is kept flooded with refrigerant, there is no way of controlling the evaporating temperature individually. To control the cooling output of each evaporator, it is necessary to throttle the flow of the cooled medium through the evaporator, or to bypass a portion of the cooled medium around the evaporator.

If only a single evaporator is needed, as in a central chiller system, liquid overfeed offers no significant advantage over a flooded evaporator, and it would be unnecessarily complex.

■ Evaporators with Expansion Valves

An expansion valve is an automatic valve for metering the flow of refrigerant into the evaporator. Figure 1 shows where an expansion valve is installed in the system.

Unlike flooded evaporators and liquid overfeed evaporators, the flow of refrigerant into evaporators that use expansion valves is restricted.

The name "expansion valve" is a misnomer. An expansion valve is actually a liquid refrigerant metering device. The only expansion that occurs at the valve itself is a relatively small fraction of the refrigerant that flashes because of the drop in pressure upon entering the evaporator. Most of refrigerant evaporates as a result of absorbing heat inside the evaporator.

Most air coil (DX) evaporators use expansion valves. Many smaller water chilling evaporators use them. Expansion valves limit the amount of refrigerant used in the evaporator to the minimum needed to satisfy the cooling load. This reduces the evaporator size and the refrigerant charge. Expansion valves are the most economical way of serving multiple evaporators in a single system. Compared to liquid overfeed, the piping is smaller and simpler. The main disadvantage of expansion valves is that they reduce the system COP, for reasons discussed below.

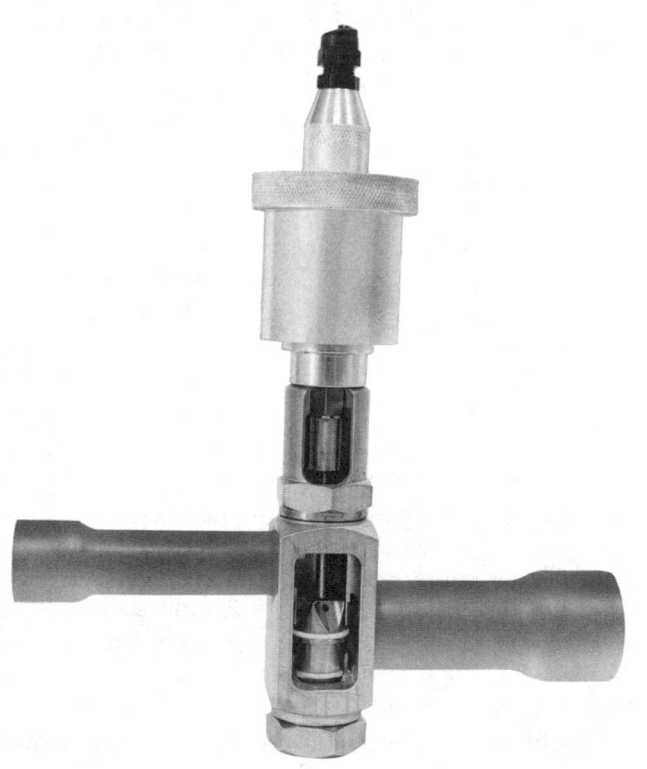

Fig. 20 An "electronic" expansion valve This expansion valve minimizes superheat under all load conditions by changing its throttling behavior. The valve receives input signals from a microcomputer that senses system conditions.

Until recently, virtually all expansion valves were set to maintain a constant difference between the temperature of the vapor leaving the evaporator and the temperature of the liquid entering the evaporator. This temperature difference is called "superheat" because it represents sensible heating of the vapor above its saturation temperature after all the liquid is evaporated. Typical superheat ranges from 5°F to 10°F. Using an expansion valve forces the vapor outlet end of the coil to be dry, so this portion of the coil provides only minimal cooling capacity.

Direct-expansion heat exchangers, such as air coils, are commonly divided into many separate refrigerant circuits to increase the effective cooling area. In order to use all the surface most effectively, the refrigerant must be distributed evenly to all the circuits. To do this, thin distributor tubes are installed between the expansion valve and the individual circuits. The distributor tubes all have the same resistance to flow, so refrigerant is distributed uniformly.

An important energy conservation principle is keeping the condenser temperature as low as possible. Unfortunately, using an expansion valve limits this technique during cooler weather, when it is most valuable. The reason is that the condenser pressure must be kept high enough to drive refrigerant through the expansion valve and distributor tubes.

In small cooling equipment, the function of the expansion valve may be performed solely by the distributor tubes, in which case the internal diameter of the distributor tubes is carefully selected so that they provide the proper amount of resistance to isolate the evaporator from the condenser pressure. In such applications, the distributor tubes are called "capillary tubes." This method cannot accurately control superheat, or prevent liquid refrigerant from getting all the way through the evaporator. Therefore, it may require additional features to protect against liquid entering the compressor, such as using an accumulator at the compressor suction.

"Electronic" expansion valves are an improvement that became widespread during the 1990's. An electronic expansion valve adapts to changing load conditions, minimizing the superheat at all times. Figure 20 shows an electronic expansion valve.

■ How Evaporator Superheat Reduces COP

In an ideal cooling machine, the liquid refrigerant in the evaporator would be cooled down to the compressor suction temperature. This condition is nearly met in flooded evaporators and in liquid overfeed evaporators. This is because the liquid refrigerant is exposed directly to the compressor suction.

However, if an expansion valve is used to control the flow of refrigerant through an evaporator, all the refrigerant must evaporate before it leaves the evaporator. To ensure that this happens, the evaporator is oversized in relation to the refrigerant flow. As the refrigerant continues to travel through the evaporator after it has evaporated, it picks up sensible heat, increasing its temperature. The rise in vapor temperature above the evaporation (saturation) temperature is called "superheat."

Superheat reduces the system COP. To understand this, consider the example of a direct-expansion air coil evaporator that is operating with a constant cooling load. Assume that the superheat is adjustable. Initially, the superheat is set at zero, so the coil is completely filled with evaporating refrigerant. Then, increase the superheat setting. The immediate effect of this is to restrict the flow of refrigerant into the coil. Since the quantity of refrigerant must be constant to satisfy the cooling load, there is a rise in the air temperature. The system controls sense the rise in temperature, and increases the compressor power input to compensate. The greater compressor suction restores the flow of refrigerant through the expansion valve to its former value (less a small amount that accounts for the sensible cooling of the air by the superheated refrigerant vapor).

The increase in compressor power can be expressed in terms of the temperature difference across the compressor. In the previous example, the compressor power increased as the suction pressure dropped, requiring evaporation to occur at a lower temperature. As we discussed earlier, the COP of a cooling system suffers if the temperature differential between the evaporator and the condenser increases.

It is a common belief that superheat improves the efficiency of the machine because the absorption of sensible heat by the refrigerant provides additional cooling effect. This is not true. The sensible cooling is minor in relation to the increased compressor power required by the lower suction pressure.

Energy Recovery from Refrigerant Expansion

The refrigerant in the condenser is at high pressure and the refrigerant in the evaporator is at low pressure. When refrigerant is released into the evaporator, the reduction of pressure causes a fraction of the refrigerant to flash into vapor. In most large cooling machines, flow from the condenser to the evaporator is regulated by a float valve, a simple orifice, or an expansion valve. With any of these devices, refrigerant flashing is a free expansion process.

Free expansion wastes most of the energy that was required to compress the flashed refrigerant initially, because the expansion does no useful work. The flow of liquid refrigerant across the pressure difference also comprises a waste of energy, but this is a minor factor because the volume of the liquid is small.

Some new chiller designs recover this lost energy. The most direct approach is to pass the refrigerant from condenser to evaporator through a turbine. Figure 21 shows a machine that uses this approach.

The liquid refrigerant enters the turbine from the condenser. Inside the turbine, the liquid tends to flash into vapor, but it also tends to re-condense because it is losing energy. The turbine is designed to minimize the energy left in the refrigerant. The turbine, which is driven by the pressure differential and the expansion of the flash gas, helps to run the compressor.

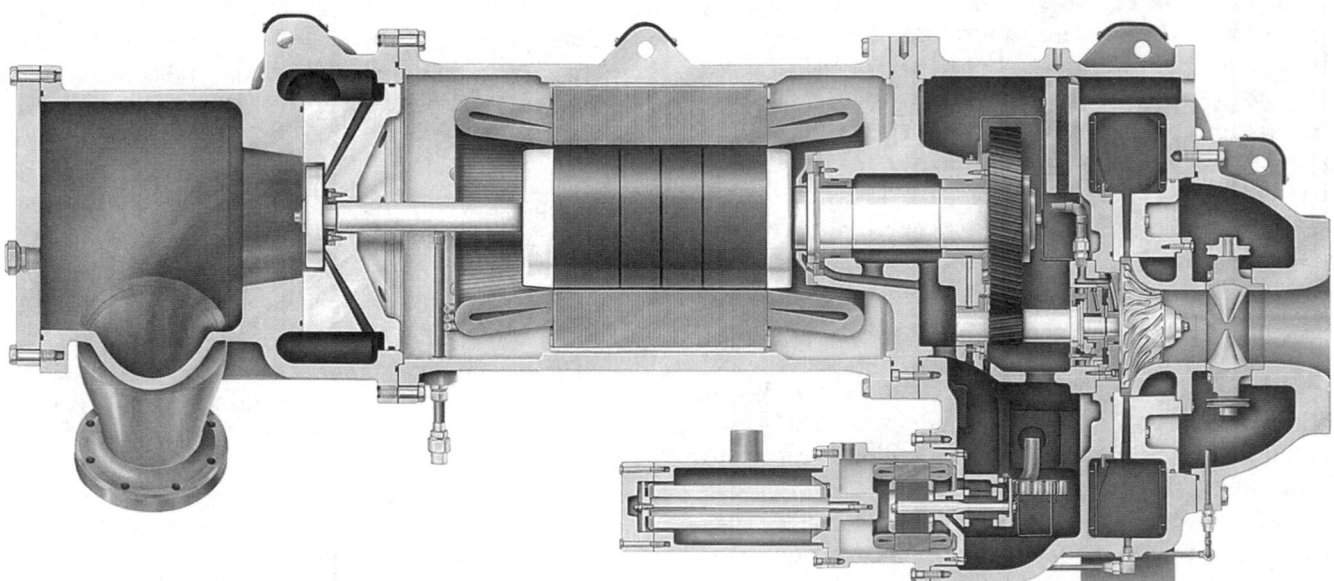

Carrier Corporation

Fig. 21 Energy recovery from refrigerant expansion The condenser is at much higher pressure than the evaporator. Energy is recovered by passing refrigerant through a turbine on its way from the condenser to the evaporator. The turbine helps the motor to drive the compressor. In this machine, the turbine wheel is on the left end of the motor shaft, and the compressor is on the right. The turbine refrigerant flow is entirely independent of the compressor flow.

(A similar idea is used in steam plants, where pressure reducing turbines are used in place of pressure reducing valves to get useful work.)

Since energy has been removed from the refrigerant before it enters the evaporator, it has more ability to absorb heat from the load. It also has less superheat, which reduces the power needed by the compressor to compress the gas.

This method can be used with flooded evaporators and liquid overfeed systems. It has little or no potential with expansion valve systems and with air coils that use capillary tubes or other types of flow restrictors to distribute refrigerant within the coils. This is because these devices require a significant pressure differential to operate properly.

Evaporator and Condenser Heat Transfer

Temperature differentials that occur in the evaporator and the condenser are important factors in the amount of work that the compressor must do, because these temperature differentials translate to an increased pressure difference across the compressor. Typical evaporator and condenser temperature drops are shown in Figure 4. The temperature differential is proportional to three factors: the heat transfer coefficient of the surfaces, the area of the heat transfer surfaces, and the rate of heat flow, or cooling load.

The evaporator and condenser temperature differentials are determined largely when the cooling machine is selected. They are a compromise between efficiency and first cost that the engineer makes in the selection process. Larger heat exchangers improve heat transfer, but heat exchangers are a major part of the system cost. The efficiency benefit is subject to a severe limitation of diminishing returns as the temperature differentials become small. For example, reducing the temperature differential from 4°F to 2°F requires doubling the heat transfer surface. This makes it very expensive to reduce the last few degrees of temperature differential.

Heat transfer is not limited primarily by the conductivity of the metal, but by heat transfer at the interfaces between the metal, liquid, and gas. The refrigerant side of evaporators involves two heat transfer processes, from the gas to the liquid film on the tube surface, and from the liquid film to the metal. Where water is used in evaporators and condensers, heat transfer is reduced by fouling from the water. This is especially severe with open cooling towers.

With refrigerant-to-water evaporators and condensers, minimum water flow rates are required to maintain adequate turbulence for good heat transfer. In variable-flow systems designed to reduce pump power, a separate loop must be provided to maintain a minimum water velocity through the heat exchanger.

Major improvements in the heat transfer coefficients of evaporators and condensers have been made since the 1970's. These improvements are achieved primarily by giving the tube surfaces a texture that increases turbulence, enhancing contact between the different mediums.

Heat transfer in air coils involves air velocity, fin spacing, number of rows of coils, and refrigerant circuiting. Air-cooled (DX) evaporators may require separate coil sections to allow operation at partial loads. Distribution of refrigerant within a DX evaporator requires elaborate distributors, and is difficult to optimize.

Heat transfer efficiency improves at partial loads because the heat transfer surface is then larger in relation to the amount of heat to be transferred.

■ Fouling Factor

When a system uses water for heat transfer in either the evaporator or condenser, a "fouling factor" is used in the rating conditions to account for fouling of the tube surfaces by dirt in the water. The fouling factor is a measure of thermal resistance, like the R-value used to rate insulation. The units are the same, for example, square feet-°F-hour per BTU in U.S. practice.

In the U.S., a fouling factor of 0.00025 is commonly assumed in rating large cooling machines. Experience has shown that this figure is appropriate for both evaporators and condensers. It is assumed that the fouling factor is kept from rising above this value by the scrubbing action of the water flow. For this assumption to be valid, the equipment must be designed with appropriate water velocities to maintain turbulent flow at the tube surfaces. Also, if the condenser uses water from an open cooling tower, you need to maintain an appropriate water treatment program for the tower water. The rating also assumes that tube surfaces are cleaned at appropriate intervals.

When comparing cooling equipment specifications, be careful to check the fouling factor that is actually being quoted by the manufacturer in its efficiency and capacity ratings. Although a fouling factor of 0.00025 is a standard value, a manufacturer may fudge comparisons with competing equipment by using different fouling factors.

The refrigerant sides of evaporators and condensers are kept clean by the solvent and detergent properties of the refrigerant materials. (Halocarbon refrigerants were widely used for cleaning purposes, before they were banned for environmental reasons. Ammonia is still a common cleaning agent.) Therefore, no fouling factor is needed for the refrigerant sides.

Parallel and Series Evaporator Circuits

In most facilities that have multiple water chillers, the evaporators are piped in parallel. This allows the chillers to be operated in any combination. However, it is possible to improve efficiency at high loads, at least in principle, by connecting the evaporators so that the chilled water flows through them in series.

To understand this, let's look at an example. Consider a facility that has two chillers of equal capacity, each operating at full load on a peak-load day. There is a large spread between the chilled water supply and return temperatures, for example, supply at 40°F and return at 52°F. If the chillers were operating in parallel, each would have to chill the water to 40°F. However, if the evaporators are connected in series, one of the machines can operate with a higher evaporator temperature. In this example, all the return chilled water is cooled in the first machine from 52°F to 46°F, and in the second machine from 46°F to 40°F. The COP of the first machine is increased substantially, while the COP of the second machine is reduced by a small amount. The average COP of the two machines might be increased by perhaps five percent under full load.

The efficiency advantage disappears below about half load, because one of the chillers is turned off in either the parallel or series arrangement. In most comfort cooling applications, the cooling load is less than half the peak design load for a majority of the time. Thus, the opportunity of exploiting a series connection is limited. However, if there are more than two chillers in the plant, the number of efficient operating hours may be extended by combining series and parallel connections.

The efficiency advantage of a series connection also drops rapidly below full load because the evaporator temperature of the chillers in a parallel connection can be raised to follow the load.

The main disadvantage of a series connection is that the chilled water must always flow through the evaporators of both chillers, even if one chiller is turned off. This wastes pump power. To minimize the pump power, the evaporators of series-connected chillers are usually designed for lower resistance, which wastes some efficiency and capacity in the individual chillers. For a given plant cost, these factors may eliminate any efficiency advantage that the series connection offers.

In principle, any number of chillers could be connected in series, but practical considerations limit the number to two.

Series-connected chillers should be similar in size, because the temperature drop across the evaporator of each machine will be proportional to its relative capacity. Some simple mathematics shows that the efficiency advantage diminishes if the temperature drops across the chillers are not divided almost equally.

To maximize the benefit of this configuration, the chillers should be controlled so that they share the load while both are running, and that one of the chillers is turned off at the appropriate load, typically at half of the cooling load or somewhat less. Inefficient load sharing between series-connected chillers wastes more energy than inefficient load sharing among parallel-connected chillers, which is another reason to be cautious about this configuration.

Condensate Subcooling

In an ideal cooling machine, the liquid refrigerant in the condenser would fall to the same temperature as the cooling water or air. Because heat transfer is imperfect, the condensed refrigerant is actually somewhat warmer than the cooling medium. This reduces efficiency by holding the compressor discharge temperature higher, increasing compressor power.

A condenser may cool the refrigerant below the temperature at which it condenses on the condenser surface. This is called "condenser subcooling." It is a cheaper alternative to more efficient condenser heat exchanger design. Subcooling provides some additional sensible cooling of the refrigerant, but sensible cooling is a minor part of the cooling potential of the refrigerant. Note that condenser subcooling does not significantly reduce the compressor discharge pressure.

Subcooling can be done in several ways. Most commonly, the refrigerant is routed through the coldest part of the condenser on the way out to the evaporator. Most condensers are tube-and-shell units, with the cooling water running in the tubes. To achieve subcooling of the condensed refrigerant, the water is brought into the condenser through the bottom tubes, and the condensate is allowed to pool around these lower tubes before flowing out of the condenser.

Condensate subcooling can be combined with an economizer. To do this, a portion of the condensate is routed through a separate vessel that is intermediate in pressure between the condenser and the evaporator. A portion of the condensate is flashed into vapor, cooling the condensate, and the cool vapor is injected into an intermediate stage of the compressor, as described previously.

The term "condensate subcooling" is also used for methods of improving the overall cycle efficiency of multi-stage compressors. This involves more theory than we need to discuss here.

How Absorption Cooling Works
Absorption Chiller Components
Refrigerant Materials and Absorber Materials
Number of Stages
Externally-Fired and Direct-Fired Chillers
The Problem of Crystallization

The Effect of Cooling Water and Chilled Water
 Temperatures on COP
Part-Load Efficiency
The Effect of System Temperatures on Capacity
Heating with Direct-Fired Absorption Chillers
Compression Chillers vs. Absorption Chillers
Combination Chiller Systems

The distinguishing characteristic of absorption cooling equipment is that it produces cooling by using heat energy as an input, rather than by using mechanical energy. For this reason, absorption chillers were once common in facilities that had large boiler plants with excess capacity during the cooling season. Circumstances later changed, and absorption chillers appeared destined for extinction. However, absorption chillers are now experiencing a revival, for reasons explained below.

The components of an absorption chiller must be integrated much more closely than the components of a compression cooling system. As a result, all absorption chillers are contained within a single compact package. For the same reason, absorption chillers have few variations. The main differences between models are in the heat source and in the number of stages.

In all large absorption systems, cooling is distributed by chilled water. Similarly, all condensers are cooled by water, usually from a cooling tower. For a variety of reasons, large absorption systems do not use air coils for either evaporators or condensers.

How Absorption Cooling Works

As with compression chillers, absorption chillers operate by absorbing heat to evaporate a liquid. In all larger absorption chillers, the evaporating liquid is water.

In dry climates, water can be used to provide cooling without a chiller. Unfortunately, the evaporation rate of water approaches zero at high humidity and lower temperatures. Also, water does not evaporate rapidly under atmospheric pressure and normal temperatures, so its cooling capacity is limited.

To achieve a large cooling capacity using water as the refrigerant, an absorption chiller keeps the water at a pressure so low that the water boils at the cooling temperature. This pressure is about 0.1 PSIA, which is less than one hundredth of atmospheric pressure. An absorption machine is big because it must handle large volumes of water vapor. This is because the density of the water vapor is very low at the extremely low pressure needed to achieve rapid evaporation.

The absorption machine maintains continuous rapid evaporation of water under the very low pressure. This is done by absorbing the evaporated water into a highly hygroscopic (water absorbing) material. In absorption chillers large enough for commercial applications, the absorber is usually lithium bromide, a salt.

A more technical way of explaining absorption is that the vapor pressure of a salt solution is lower than the vapor pressure of pure water. This causes the vapor to move out of the evaporator toward the salt solution. For example, pure water has the same vapor pressure at 38°F as a strong lithium bromide solution at 110°F. Therefore, if a cooling tower can keep the salt solution cooled to 110°F, water evaporating at any temperature higher than 38°F will produce vapor at a higher pressure than the salt solution.

In order to keep the absorption process going, the concentration of the absorbing salt solution must be restored continuously. This is done by circulating the salt solution through a distiller, which is located in a different part of the machine. The distiller returns the pure water to the evaporating chamber, and returns the concentrated salt solution to the absorbing chamber.

Absorption Chiller Components

The simplest type of large commercial absorption chiller consists of these major components, which are sketched in Figure 1:

- the *evaporator,* which is a chamber that contains coils to transfer heat to an external water ("chilled water") circuit. The water refrigerant inside the chiller enters the evaporator from the distiller in a warm state. A pump and an array of nozzles spray the refrigerant water inside the evaporator chamber, hastening evaporation and cooling it before it falls on the chilled water coils.

- the *absorber,* which is a chamber that contains the concentrated absorbent. The absorber shares a common shell with the evaporator, and water vapor flows freely from the evaporator to the absorber. The absorbing salt solution is cooled by distributing it over a bundle of tubes that are cooled by the entering cooling water. This minimizes the vapor pressure of the absorbing solution, and most of the vapor is absorbed by the film of solution on the tube bundle. The tube bundles carries away the latent heat that the water vapor had acquired when it evaporated, as well as heat that is released by the dilution of the salt solution.

• the *distiller section,* which removes water from the salt solution to restore its concentration. The distiller consists of two chambers, a *concentrator* (sometimes called a *generator*) where the water is boiled out of the salt solution, and a *condenser*

where the water vapor is condensed for further use. The design of the distiller section is where major differences occur between different types of absorption chillers, as discussed below.

• a *heat exchanger* between the absorber and the concentrator. The salt solution in the concentrator should be as hot as possible to boil off water, while the salt solution in the absorber should be as cool as possible to absorb water. The concentrated salt solution flowing from the concentrator to the absorber is hot, and the diluted salt solution flowing in the opposite direction is cool. The heat exchanger

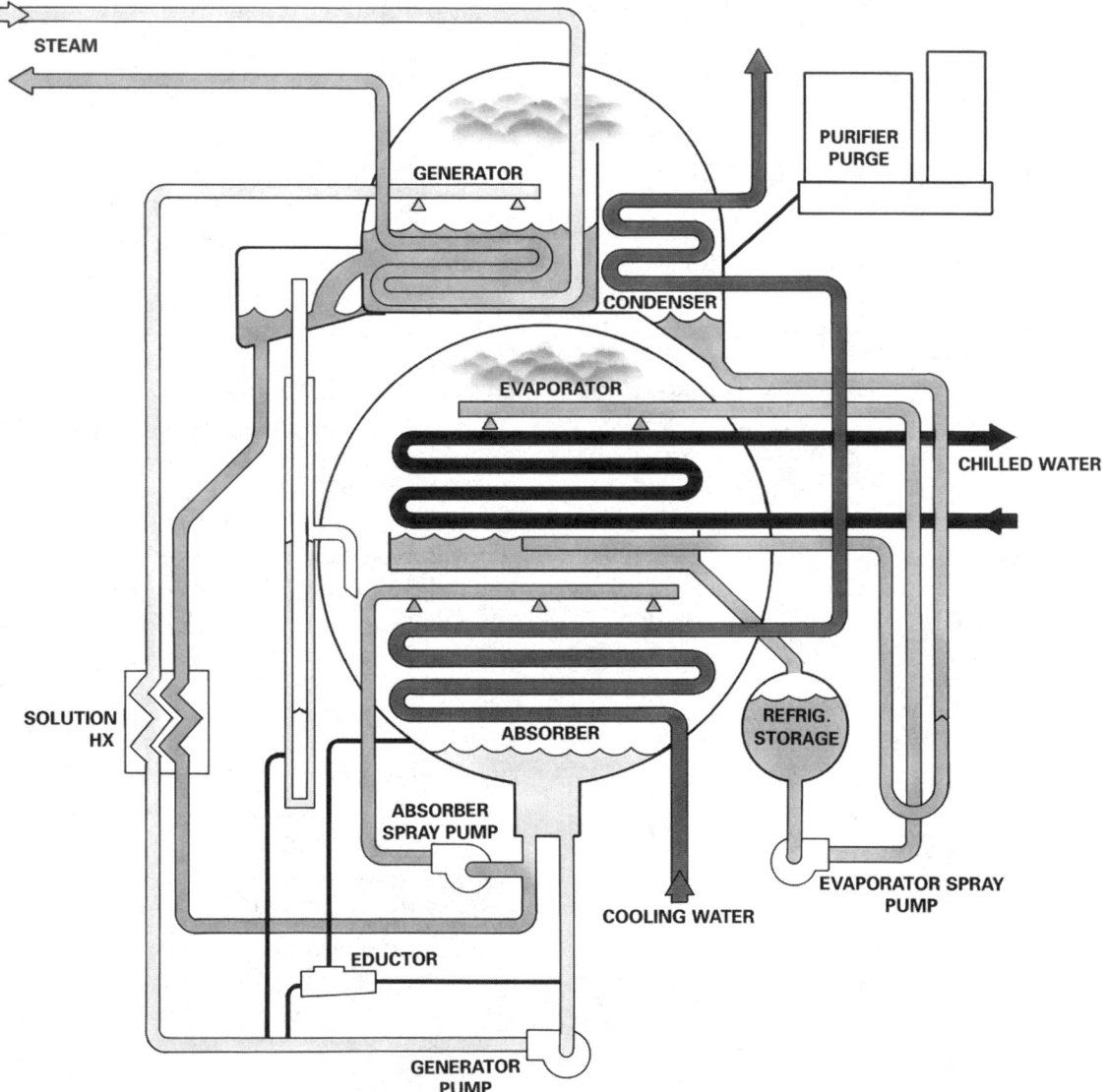

Trane Company

Fig. 1 How an absorption cooling machine works   The "refrigerant" is water. The "absorber" is a strong solution of salt in water. The part of the machine on top is a distiller, which continuously separates pure water from the salt solution, while making the salt solution stronger. The pure water is pumped over cooling coils, where it evaporates rapidly in the extremely low pressure of the machine. The evaporated vapor is drawn to the salt solution in the chamber below, where the pressure is even lower, and it is absorbed in the salt solution. We want the salt solution to be hot in the distiller and cool in the absorber chamber, so the heat exchanger on the left keeps it that way. The purge unit on the upper right is a separate device that removes any foreign gases from the machine, as these would destroy the low vacuum needed for operation.

extracts the heat from the concentrated solution and transfers it to the diluted solution.

- the *solution pump,* which pumps the diluted salt solution to the distiller.
- the *refrigerant pump,* which circulates the water in the evaporator over the chilled water coils.
- a device for regulating the return of distilled water from the high pressure condenser chamber to the low pressure evaporator chamber. This is usually a simple *orifice* or *float valve.*
- the *purge unit.* The portion of the chiller where evaporation occurs must be kept at an absolute pressure around 0.1 PSI. This requires keeping all gases but water vapor out of the chiller. Gases that leak into the chiller are removed by the purge unit, which is a vacuum pump combined with a condenser. The purge unit's condenser captures water vapor that is drawn out of the chiller along with the contaminant gases, condenses it, and returns it to the chiller.

Figure 2 shows an actual absorption machine that has these components. Machines of this type have capacities ranging from several hundred tons to more than one thousand tons.

Refrigerant Materials and Absorber Materials

Water has three roles in a large absorption machine. Don't get confused about how the machine works because there is water in it everywhere.

As we said before, water is the actual refrigerant. In the part of the machine where water acts as a refrigerant, it is very pure. In fact, it is distilled water.

Most large absorption machines are water chillers. Water is used to distribute cooling energy to the facility. This water travels in a separate loop. It enters the machine only in the evaporator section.

Water also acts as the carrier for the absorber salt. It is not desirable to use the salt in the solid state because it would require too much energy to dry it out. Also, it would be harder to handle. A salt solution is easy to move around inside the machine by pumping. Also, the absorbing surface area of the salt solution can be increased greatly by spraying it inside the absorbing chamber.

The absorber material in machines that are large enough for commercial applications is lithium bromide. Lithium bromide has a very strong tendency to absorb water. It is chemically a close relative of table salt, which is sodium chloride. You can easily see the tendency of table salt to absorb water on humid days.

The lithium bromide absorbent is extremely corrosive. To be practical in chillers, lithium bromide needs to be used in combination with special corrosion inhibitors. Other materials are added to the absorbing solution to increase the rate of vapor absorption.

In very small chillers and refrigerators (up to several tons in capacity), ammonia is used as the refrigerant and water is used as the absorbent. These machines have lower COP's than lithium bromide machines, so they are unlikely to evolve into larger sizes. Their advantages are smaller size, less critical control, and freedom from the corrosion problems of salt solutions.

Number of Stages

A major variant among absorption chillers is the two-stage machine. The difference is in the distiller section. A theoretical analysis of the efficiency of absorption chillers shows that a large part of the energy input into the chiller is related to the operation of the distiller section. The efficiency of the chiller can be improved considerably by installing a distiller that works in a series of cascading temperature stages.

In any distiller, heat must be added to the solution to boil off the pure water. Then, this heat must be removed from the water vapor to condense it. In a two-stage chiller, water evaporated in the first (high temperature) stage of distillation is condensed by using the hot vapor to heat a second (low temperature) stage of distillation. Only the second stage water vapor gives up its heat to the condenser cooling water.

Figure 3 shows how the additional components of a two-stage machine are arranged. Compare this to Figure 1.

Figure 4 is a cutaway drawing of an actual machine. As you can see, absorption chillers consist mainly of huge numbers of tubes inside steel pressure vessels.

Two-stage absorption machines are approximately 50% more efficient than single-stage machines. The limitation of two-stage machines is that they require much higher input temperature for the first stage of

Trane Company

Fig. 2 Single-stage steam-powered absorption chiller This machine is constructed as shown in Figure 1. The three pumps shown there are combined in a common housing underneath the main shell.

distillation. Single-stage steam absorption chillers are usually designed to operate with 10 PSI steam or 220°F hot water, whereas two-stage chillers require approximately 100 PSI steam or 350°F hot water to achieve their full efficiency potential.

The present limitation to two stages is not a matter of technology. For example, distillers used aboard ships to make fresh water from seawater commonly operate with three or more stages. The major practical limitation is increased complexity and cost.

Externally-Fired and Direct-Fired Chillers

Until the 1980's, all large absorption machines used steam or high-temperature hot water as the heat source. Unless a facility could purchase steam, using an absorption chiller required the expense, the space, and the efficiency losses of a boiler plant. As stated previously, two-stage machines require much higher pressures and temperatures than single-stage machines.

Since that time, absorption chillers with internal fuel firing have created a resurgence of interest in absorption chillers. These machines are capable of burning natural gas and/or oil directly. Since flame has a very high temperature, direct firing allows multi-stage distillers

to be used. At present, all large direct-fired absorption chillers have two-stage distillers.

Figure 5 shows the components of a direct-fired absorption machine. As you can see, it is not much different from an externally fired two-stage unit.

Figure 6 is a cutaway drawing of a direct-fired chiller. It looks entirely similar to an externally-fired two-stage chiller, except that it has its own little boiler alongside.

Absorption chillers designed for direct fuel firing can be adapted for operation with hot exhaust gases from gas turbines, reciprocating engines, industrial processes, etc. These sources must be predictable, because sudden reduction of the heat input can cause crystallization of the highly concentrated salt solution.

The Problem of Crystallization

To achieve maximum capacity, the concentration of salt in the absorber solution is kept as high as possible. The salt solution is concentrated in the distiller. If the heat source is suddenly removed from the distiller and the solution is allowed to cool, the solution becomes supersaturated. The salt comes out of solution and encrusts the distiller heat exchanger and other surfaces

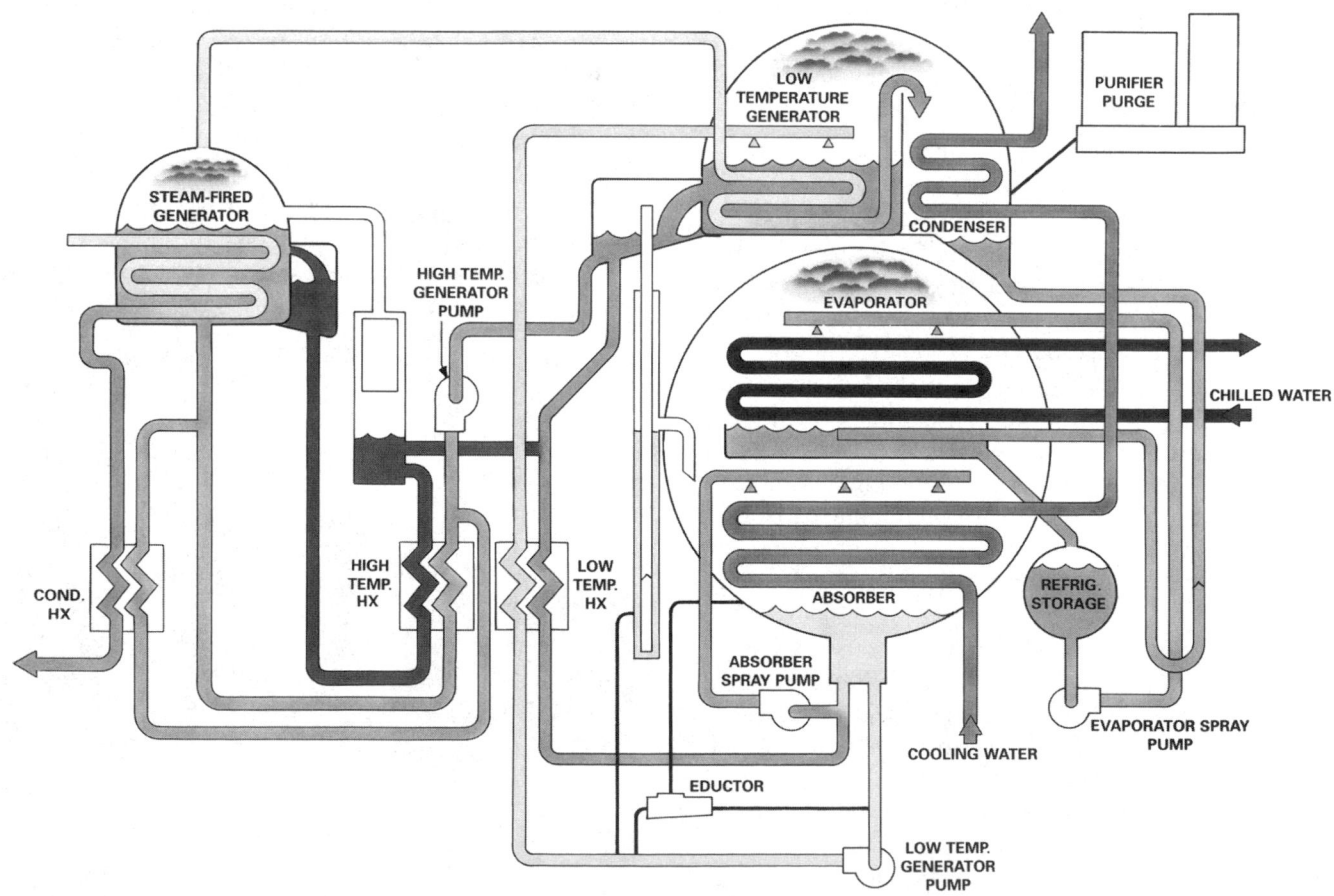

Trane Company

Fig. 3 Two-stage steam-powered absorption chiller This machine is mostly identical to the one shown in Figure 1. The main difference is having a two-stage distiller. The new equipment on the left is the first, high-temperature stage of the distiller. This machine is more efficient than a single-stage machine, but it requires steam at much higher pressure.

as a layer of crystals. The chiller cannot operate in this state, or it operates inefficiently.

If crystallization occurs, the chiller must be shut down. The crystals must be dissolved by circulating water from the evaporator through the distiller. After this is done, a period of operation is needed to get the water and salt distributed properly within the chiller.

The Effect of Cooling Water and Chilled Water Temperatures on COP

In addition to the temperature of the heat source, the temperatures of the cooling water and chilled water have a major effect on the COP of an absorption chiller. This is analogous to the effect of condenser and evaporator temperatures in compression chillers, although the physical processes are entirely different.

In an absorption chiller, the cooling water is first routed to the absorber section, where it determines the vapor pressure of the absorbing salt solution. Lowering the cooling water temperature lowers the vapor pressure in the absorber, which allows the concentration of the salt solution to be reduced. Since an absorption chiller uses its energy input to concentrate the salt solution, lowering the cooling water temperature reduces the energy requirement, provided that the chiller is able to reduce the salt concentration to match the load.

The cooling water also travels to the condenser section of the distiller. Lowering the condensing temperature in the distiller also reduces the amount of

energy needed to distill the water out of the salt solution, for a given concentration. Thus, the cooling water temperature affects efficiency in two separate ways.

The chilled water temperature determines the vapor pressure in the evaporator section of the chiller. A higher chilled water temperature increases the rate of evaporation, which allows the salt concentration to be reduced correspondingly, increasing the COP as before.

Part-Load Efficiency

The part-load efficiency of an absorption chiller depends on how the concentration of the salt solution is controlled as the load changes. The heat input to an absorption chiller is used to distill water out of the absorbing salt solution. The amount of heat required increases with the salt concentration. Therefore, efficiency can be improved by keeping the salt solution as dilute as possible. This is done by slowing the flow of liquid from the absorber to the concentrator, so that the liquid absorbs more water in the absorber. This feature has been called an "economizer" (that word again!). It is becoming a common feature of new machines, but be aware that you may still have to order it as an option.

On older machines, the concentration was changed at one specific load point. Newer designs adjust the concentration continuously as the load changes. The solution pump absorbs a substantial amount of energy, so it is worth using a variable-speed pump drive to

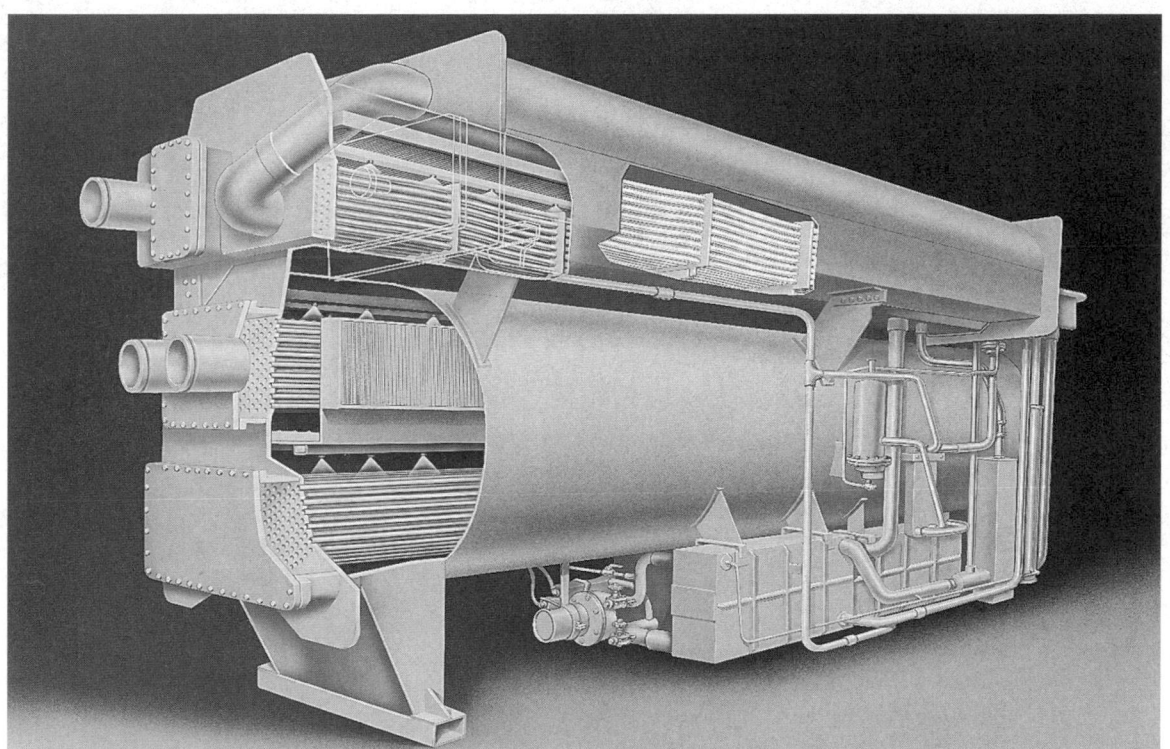

Trane Company

Fig. 4 Two-stage steam-powered absorption chiller This is the same machine diagrammed in Figure 3. Most of the material in the machine consists of huge bundles of copper alloy heat exchanger tubes that are enclosed in steel cylinders.

control the flow of salt solution, rather than simply choking the flow with a valve.

With present absorption machines, the purchaser has no options to improve the efficiency features of a particular chiller model, except perhaps for the concentration control economizer. Pay particular attention to the part-load efficiency specifications, where significant differences among models may appear. Ask the manufacturer to certify all performance figures.

Part-load efficiency has been improved in newer machines by microprocessor controls and improved sensors, which keep the machines' operating variables close to optimum at each combination of loads and temperatures. The efficiency features of newer machines generally are not practical to retrofit to older machines.

The Effect of System Temperatures on Capacity

The temperature in the concentrator section of the distiller determines the rate at which the salt solution is concentrated, and hence the capacity of the chiller. The rate of evaporation in the distiller is strongly affected by the input temperature.

Capacity is also determined by the maximum rate of vapor flow between the evaporator and condenser. Vapor flows in response to a pressure differential, so

the flow rate is increased by increasing the pressure in the evaporator and by lowering the pressure in the absorber. Raising the chilled water temperature increases the evaporator pressure, and lowering the cooling water temperature lowers the absorber pressure. Both have a strong effect on capacity.

Heating with Direct-Fired Absorption Chillers

Since direct-fired absorption chillers include their own boilers, they may offer the option of serving as a heat source. They can be used for space heating, domestic water heating, or other purposes. In new construction, this may save the cost of a separate boiler.

Figure 5 shows the simplest arrangement. Steam is simply tapped off the boiler and is used to heat a water loop. The heating function operates independently of the operation of the rest of the machine.

In some designs, the heating option may offer a net efficiency advantage, compared to heating with a separate boiler. If the unit is cooling and heating simultaneously, and if it is heating a cold water source, the cold water source may be heated in the condenser section of the chiller. This reduces the amount of cooling water required to operate the chiller, and it avoids the extra fuel needed to heat the water. However, this

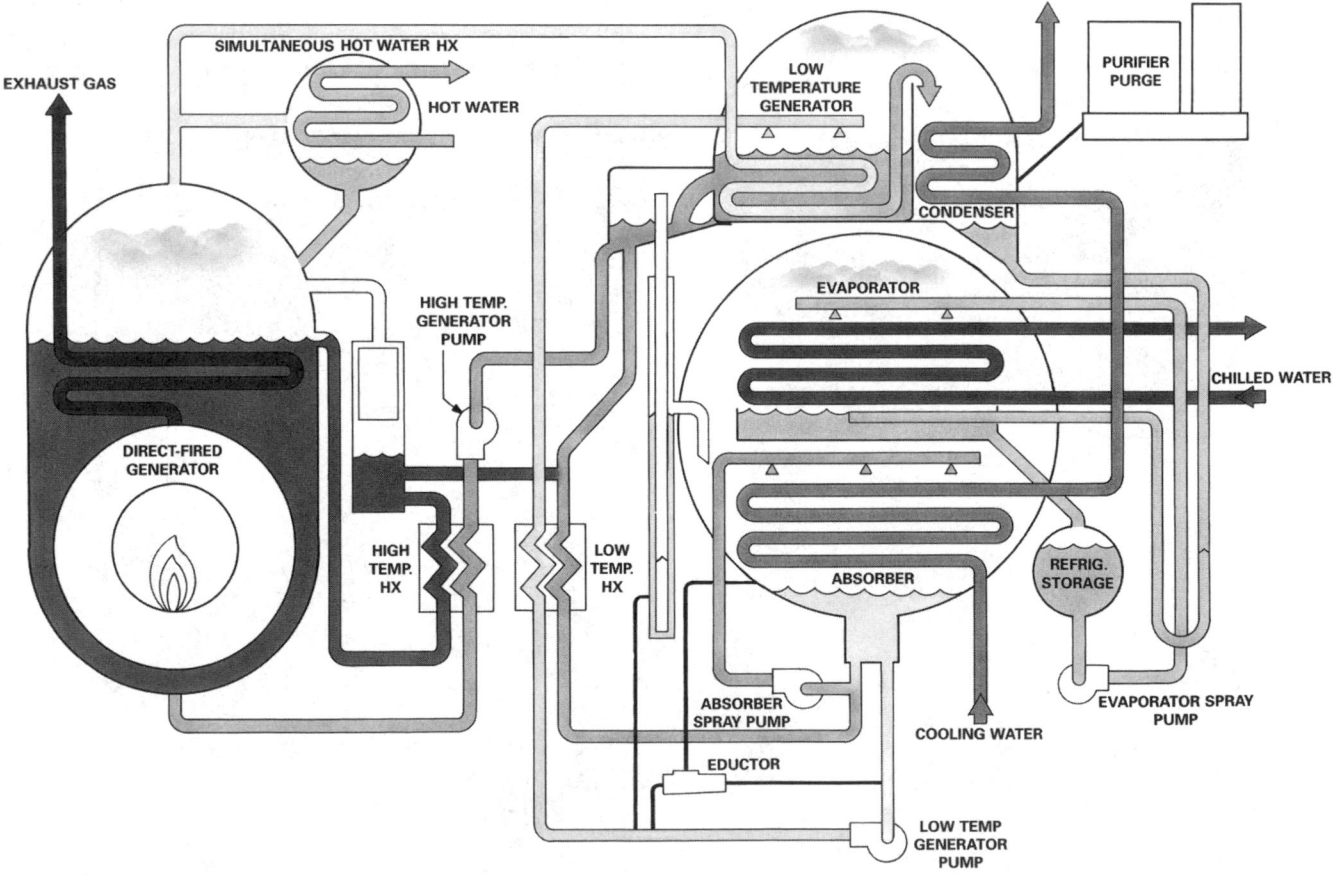

Trane Company

Fig. 5 Direct-fired two-stage absorption chiller Compare this to Figure 3. The only major difference is that a fuel-fired boiler has been substituted for the steam-powered first stage of the distiller. As an accessory feature, the boiler can provide water heating independently of cooling operation.

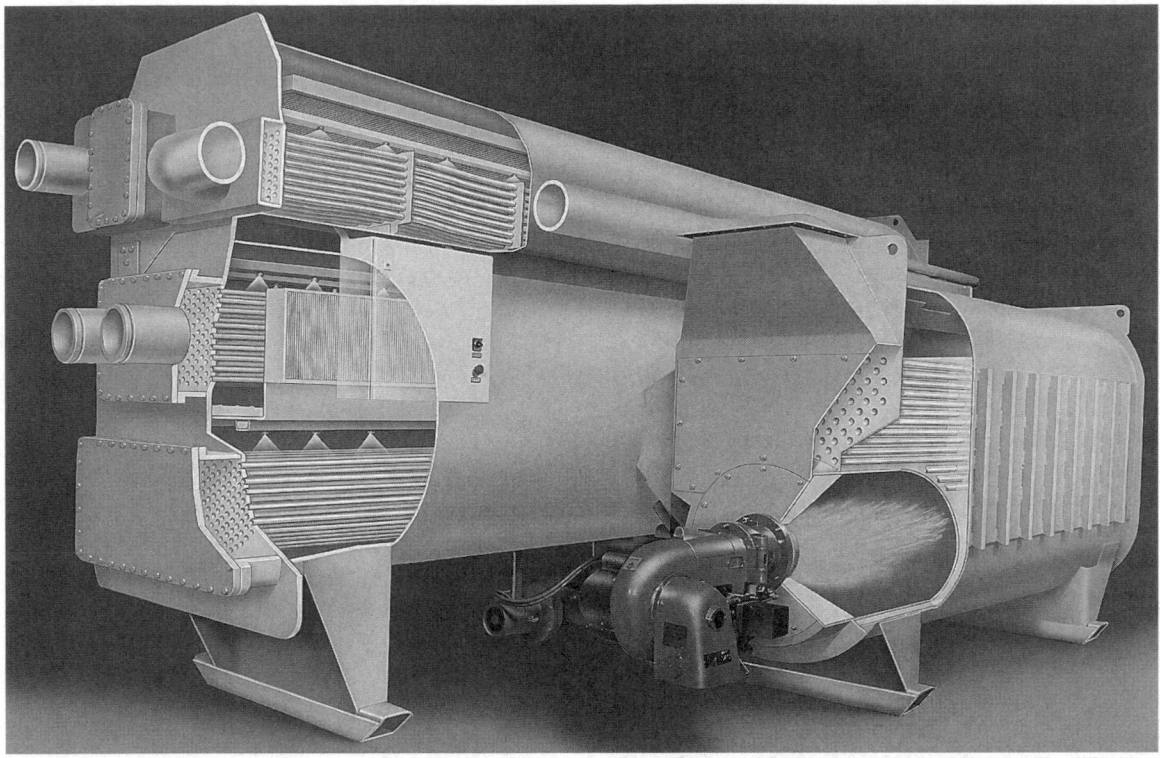

Trane Company

Fig. 6 Direct-fired two-stage absorption chiller This is the machine diagrammed in Figure 5. Compare this to the steam-powered machine in Figure 4. This is essentially the same machine, except that it has its own small boiler.

capability complicates the internal plumbing of the machine. Also, no saving is possible when heating a hot-water system, such as a hydronic heating system.

Some models can provide heating simultaneously with cooling, although the capacity of the boiler unit will limit total heating and cooling capacity. With other models, you have to choose between cooling or heating operation.

Compression Chillers vs. Absorption Chillers

Of the two major types of chillers, compression chillers are the more efficient type, in terms of COP, by a considerable margin. For this reason, absorption chillers appeared to be headed for obsolescence, except in specialized applications. However, there has been a resurgence of interest in absorption chillers, for these reasons:

• utility regulatory agencies are making it difficult for utilities to build new generating plants. As a result, utilities are running out of generating capacity for periods of peak consumption, which are caused mainly by cooling load. Since absorption chillers have only a small requirement for electrical power, electric utilities have promoted absorption chillers as a means of reducing utility demand peaks. The incentives include demand-penalizing rate structures and direct subsidies to customers. (This costs the utility electricity sales.

Utilities prefer to promote electric storage cooling as a means of reducing demand without surrendering electricity sales.)

• gas utilities are marketing absorption chillers as a major market for natural gas, and a means of leveling annual gas sales.

• the refrigerants and absorbents used in absorption chillers are not considered harmful to the environment at a global level, unlike the CFC's and HCFC's used in compression chillers. However, environmentally safe replacement refrigerants are available for compression chillers, so this is not likely to continue as a significant factor.

• the development of direct-fired absorption chillers made it possible to achieve higher COP's while avoiding the need for a separate boiler system to drive the absorption machines.

• the direct-fired type of absorption chillers can be powered by waste heat, creating opportunities for improving overall efficiency in local power generation and industrial processes.

Combination Chiller Systems

Efficiency favors compression chillers, while other considerations may favor absorption chillers. In some applications, the optimum choice is a combination of the two types. This combination has lower electric demand than a compression chiller alone, higher efficiency than an absorption chiller alone, and perhaps the ability to exploit waste heat and cheaper fuels. Typical combinations of the two types are:

- a steam turbine drives a compression chiller, and the turbine exhaust steam is used to drive an absorption chiller and perhaps other thermal loads
- a diesel engine, a gas reciprocating engine, or a gas turbine drives a compression chiller, and heat recovered from the engine is used to drive an absorption chiller and perhaps other thermal loads
- an engine is used to drive an electric generator, the electricity is used to drive electric compression chillers and perhaps other electrical loads, and heat recovered from the engine is used to drive an

absorption chiller and perhaps other thermal loads. This configuration is a common form of "cogeneration."

The disadvantages of all combination chiller systems are greater complexity, and hence reduced reliability. Combination chiller systems are tricky to design, especially because their performance is sensitive to the load scenarios under which they may operate. Also, combination systems require enhanced operating skills, so they should not be installed unless the owner is certain that the necessary operating skills can be maintained. There have been many failures of combination systems because of insufficient attention to these issues.

Refrigerants

What is a Refrigerant?
Halocarbon Refrigerants
The Naming System for Refrigerants
The Phase-Out of Chlorine-Containing Refrigerants
- Phase-Out Schedule
- Substituting Refrigerant Types in Existing Cooling Units

How to Select the Refrigerant for Your Cooling Equipment
- Availability
- Won't New Refrigerants be Developed?
- Refrigerant Mixtures and Azeotropes
- Ozone Depletion Potential (ODP)
- Global Warming Potential (GWP)
- Energy Efficiency
- Toxicity
- Flammability and Explosion Potential
- Condensing Pressure
- Evaporating Pressure

This Reference Note covers the main points of refrigerant selection from the standpoint of the system purchaser and the system designer. You cannot select a refrigerant for efficiency alone. The selection of the refrigerant has other major consequences, such as environmental protection, safety, and equipment size. Refrigerant selection is presently in a state of uncertainty because of environmental concerns about the most popular contemporary refrigerants. This Note covers enough of the theory to enable you to understand how refrigerant selection issues may evolve in the future.

Table 1 lists the efficiency, environmental, and safety properties of the refrigerants that are commonly used today, along with the properties of newer refrigerants that are considered leading candidates for future use. For the person who must select a cooling machine, these are the refrigerant characteristics that matter most urgently. The table does not include refrigerants for specialized applications, such as low-temperature refrigeration and high-temperature heat pumps. For many refrigerants, the characteristics given in the table have not yet been determined.

Table 1. SELECTION CHARACTERISTICS OF COMMON REFRIGERANTS

Refrigerant	Theoretical COP (Note 1)	Ozone Depletion Potential (Note 2)	Global Warming Potential (Note 3)	Exposure Limit (PPM)	Safety (Note 4)
CFC-113	4.3	0.8			A1
CFC-11	5.0	1.0	4,400	1,000	A1
HCFC-123	4.8	0.02	85	100	B1
CFC-12	4.7	1.0	13,700	1,000	A1
HFC-134a	4.4	0	1,200	1,000	A1
HCFC-22	4.7	0.05	1,500	1,000	A1
HCFC-502	4.4				A1
Ammonia	4.8	0	0	25	B2
Ideal refrigerant (Carnot cycle)	6.7	0	0	unlimited	A1
Water		0	(Note 5)		
Carbon dioxide	2.8	0	1.0	5,000	

Notes:
1. These figures assume a condensing temperature of 86°F and an evaporating temperature of 5°F.
2. The figures are based on a value of 1.0 for CFC-11.
3. The figures are based on a value of 1.0 for carbon dioxide.
4. The letter is the toxicity class. The number is the flammability class. See ASHRAE Standard 34 for details.
5. Water is the most powerful atmospheric regulator of global temperature, by far.

The *ASHRAE Handbook* is the primary reference for the detailed engineering properties of refrigerants. Because it is published on a four-year cycle, it may not have the latest information about refrigerants that are currently exciting interest. Environmental concerns have triggered a frantic search for new refrigerants, and the leading contenders are still changing from moment to moment.

What is a Refrigerant?

The liquid that is evaporated in cooling equipment to produce a cooling effect is called a "refrigerant." In the process of evaporating, the liquid absorbs heat. The heat needed to change a refrigerant from its liquid state to a vapor state is called "latent heat of evaporation." For example, if a pound of water evaporates at room temperature, it absorbs about 1050 BTU of heat in the process. The change in state does not, in itself, involve a change in temperature,

How do we get a temperature reduction ("sensible cooling") from this? You can think of it as a two-step process. First, the portion of the liquid that evaporates absorbs heat from the surrounding mass, which includes the remaining liquid, refrigerant tubes, etc. The withdrawal of heat from this mass lowers its temperature. The actual temperature drop depends on the amount of mass that is giving up its heat, and on its specific heat. If the material has a low specific heat, its temperature is reduced more for a given amount of heat that is removed from it. Therefore, the greatest cooling effect occurs if the refrigerant has a high latent heat, and a low specific heat.

The cooling effect of a liquid is determined by its tendency to evaporate, as well as by its latent heat. This tendency is related to the "vapor pressure" of the material. For example, at room temperature, alcohol has a greater cooling effect than water because it has a higher vapor pressure, even though its latent heat is only about half that of water.

(Any liquid has a tendency to eject some of its molecules from its surface. These molecules form a gas, which is called a "vapor" as long as it remains in equilibrium with the liquid. Like any gas, the vapor exerts a pressure, which is called the "vapor pressure." The tendency to form a vapor varies greatly among different materials. For any given material, it increases greatly with temperature.)

Many substances that occur in nature have been used as refrigerants, along with many manufactured materials. Older refrigerants include ammonia, sulfur dioxide, carbon dioxide, propane, butane, and methyl chloride. Prior to the 1930's, all practical refrigerants were flammable or toxic, or both. These were largely displaced by the modern "halocarbon" refrigerants, which will occupy most of our attention for the rest of this Note.

Halocarbon Refrigerants

The halocarbon refrigerants were developed by Thomas Midgely and his associates at the Frigidaire Division of General Motors. They were first marketed under the trade name of Freon by Kinetic Chemicals, Inc., which was formed in 1930 by GM and DuPont.

As their name implies, the halocarbon refrigerants are compounds of the halogens and carbon. (The halogens are the group of elements consisting of fluorine, chlorine, bromine, and iodine.) In all these refrigerants, the basic chemistry consists of mimicking the structure of simple hydrocarbon gases (methane, ethane, etc.).

To retain the good refrigerant qualities of the original hydrocarbons while avoiding their safety hazards, halogen atoms are substituted for some or all of the hydrogen atoms. The halogen compounds bind tightly to carbon, producing compounds that are relatively stable and non-reactive. The halocarbon refrigerants were a breakthrough because they combine good cooling properties with safety. In fact, the halocarbons are better refrigerants in general than the hydrocarbons of similar structure.

The most popular of the original halocarbon refrigerants, numbered 11, 12, and 22, are all variations of methane, the simplest hydrocarbon gas. Methane consists of one carbon atom surrounded by four hydrogen atoms. In the halocarbons, the hydrogen atoms of the methane structure are replaced with differing numbers of chlorine and fluorine atoms. Other refrigerants were developed that are based on the chemical structures of the other simple hydrocarbon gases.

The Naming System for Refrigerants

ASHRAE has developed a numbering system that covers all refrigerants. The original Freon numbering system has been carried forward for those refrigerants to which it applies. ASHRAE uses the term "Refrigerant" or "R." For example, R-12 has the same composition as Freon-12.

For hydrocarbons and for the halocarbon refrigerants, the refrigerant number reveals the number of atoms of each type. The trick is to add 90 to the refrigerant number. For example, adding 90 to the number in R-22 gives 112. Then,

- the first digit is the number of carbon atoms
- the second digit is the number of hydrogen atoms
- the third digit is the number of fluorine atoms
- any remaining atoms are chlorine.

For example, R-22 has one carbon atom, one hydrogen atom, and two fluorine atoms. Because R-22 has the structure of methane (one carbon atom surrounded by four other atoms), there is one site remaining. The remaining atom is chlorine.

This numbering system turns out to be handy when considering environmental effects of refrigerants, because carbon, hydrogen, fluorine, and chlorine each has a distinct role in these effects. Unfortunately, the numbering system does not explicitly state the number of chlorine atoms, which is the main source of concern about ozone depletion among the common refrigerants.

Inorganic refrigerants (e.g., ammonia, carbon dioxide, water, etc.) have numbers starting with "7", and the last two digits are the molecular weight. For example, ammonia is named R-717 because it has a molecular weight of 17.

In response to the concern about environmental effects of halocarbon refrigerants, the simple designation "R" for these materials is now replaced with identifiers that highlight their chemical compositions, as follows:

- CFC ("chlorofluorocarbon") indicates that the carbon atom is surrounded with chlorine and fluorine. CFC-11 and CFC-12 are the most common CFC's.
- HCFC ("hydrochlorofluorocarbon") indicates that the refrigerant contains hydrogen as well as halogens. HCFC-22 is the most common HCFC.
- HFC ("hydrofluorocarbon") indicates that the refrigerant has hydrogen and fluorine, but no chlorine. HFC-134a is an important example because it is presently a leading candidate as a replacement for CFC-12.

The Phase-Out of Chlorine-Containing Refrigerants

The greatest upheaval in the history of compression cooling is presently occurring as a result of concerns about the environmental effects of the halocarbon refrigerants. Some of the most commonly used halocarbon refrigerants must sooner or later be replaced with other types. International agreements and various national laws limit present and future choices of refrigerants. In turn, this affects the design and selection of cooling equipment.

The popular halocarbon refrigerants provide an almost ideal combination of efficiency, safety, and moderate cost. Alternative refrigerants with equally good attributes are not presently available, so replacing the present refrigerants involves serious compromises. At this time, it appears unlikely that the compromise will focus neatly on a few refrigerants. This because a larger number of competing selection factors (discussed below) must now be balanced against each other, and the search will continue for new refrigerants that can match the old refrigerants in performance.

■ Phase-Out Schedule

Curtailment of CFC's in developed countries was initiated by the Montreal Protocol, an international agreement signed in 1987. Meetings in subsequent years expanded the curtailment to call for complete phase-out of CFC's, to limit other gases, and to include developing countries in the agreement. As of the Montreal meeting of September 1997, ten years after the original agreement, the status of refrigerants is:

- CFC consumption for new equipment, in developed countries, was phased out at the beginning of 1996. CFC's for servicing existing equipment are available only from stockpiles of virgin material produced before 1996, or from recovery of refrigerant from abandoned or converted equipment. Phase-out of CFC's in developing countries is deferred until 2010.
- HCFC consumption, in developed countries, is reduced in a series of steps starting in 2004 and ending in 2030. The steps are percentages of HCFC consumption in 1989, plus an allowance based on 1989 CFC consumption. A small quantity is reserved in the period between 2020 and 2030 for servicing refrigeration and air conditioning equipment. In developing countries, phase-out is scheduled to start in 2016 and to be complete by 2040.

Individual nations may have additional requirements. In the United States, Section 608 of the Clean Air Act phases out individual HCFC's on different schedules, based on their ozone depletion potentials (discussed below). Under this Act, HCFC-22 is available for new equipment until 2010 and for service until 2020. HCFC-123 is available for new equipment until 2020 and for service until 2030.

HFC's are not subject to the Montreal Protocol. However, they are likely to come under some form of control under the Framework Convention on Climate Change.

■ Substituting Refrigerant Types in Existing Cooling Units

A question that looms in the minds of those who operate cooling equipment with banned refrigerants is whether acceptable refrigerants can be substituted in the existing equipment. The refrigerant must be closely matched to the equipment. Even if it is possible to substitute refrigerants, efficiency may suffer. Also, capacity may be lost. Here are the substitutions that are presently attracting the most attention:

- CFC-11 is being replaced by HCFC-123 in some machines. In older machines, this involves a capacity loss up to about 18%, and an efficiency loss up to 5%. These losses may be largely restored by changes to the impellers and/or gears. This is expensive, even assuming that the chiller manufacturer offers the conversion hardware. Chillers made in the era of the Montreal Protocol were designed with conversion in mind. Conversion of these chillers is easier, and provides better performance.

Expect to replace sealing materials in most machines, because HCFC-123 attacks previously used sealing materials. With hermetic chillers, expect to replace the motor, because HCFC-123 attacks the original motor insulation. Replacement motors and seals adapted to HCFC-123 are presently available for most hermetic chillers.

Note that HCFC-123, the replacement refrigerant, is subject to phase-out itself, as discussed previously. HCFC-123 is not highly toxic, but less exposure is allowed than with CFC-11.

HFC-254ca has also been considered as a replacement for CFC-11. It contains no chlorine, so it is not subject to phase-out under the Montreal Protocol. However, it is flammable, which is a major disadvantage. So far, its other properties have received less study.

• CFC-12 is being replaced by HFC-134a in some machines. When making this swap, it is critical to leave no trace of the original CFC-12 or the original oil. HFC's must be used with synthetic lubricants. These react with any residual chlorine in the system to produce destructive substances.

HFC-134a attacks the insulation of some hermetic motors. Be sure that the motor is compatible with the refrigerant before making the change.

• HCFC-22, which is used in more than half of all commercial and residential cooling, does not yet have a fully satisfactory replacement. Various HFC's and HFC mixtures are being considered. At present, Refrigerant 407C is being considered as a near-term substitute, while Refrigerant 410A may take over from HCFC-22 in the long term.

• HCFC-502, which is a common refrigerant for refrigeration applications, also does not yet have a clear successor. Refrigerants 404A, 410A, and 410B are being considered.

Many other substitutions are being considered. The greater the difference in characteristics between the original refrigerant and its replacement, the less effective is the original equipment for the new refrigerant. Mismatches cause significant losses of efficiency and capacity beyond those related to the inherent characteristics of the replacement refrigerant. If the loss of efficiency with the new refrigerant is greater than the loss of capacity, the original motor may not be powerful enough to drive the compressor with the new refrigerant.

How to Select the Refrigerant for Your Cooling Equipment

In the past, purchasers of cooling equipment did not concern themselves much with the refrigerant used. Machines were selected on the basis of their capacity, efficiency, size, etc., and the refrigerant choice was a matter for the factory to decide. These days, environmental concerns make it prudent to start your selection by considering the long-term availability prospects of specific refrigerants.

After considering availability, place greatest weight on efficiency. The efficiency potential of the refrigerant is a dominant factor in the overall efficiency of the cooling system. The refrigerants designed to replace CFC's and HCFC's typically have lower efficiency. At the same time, be aware that equipment designers have been working hard to compensate. In fact, the best contemporary cooling equipment is more efficient than the best that was available before the Montreal Protocol.

The banning of chlorine-containing refrigerants forces you to consider refrigerants with varying degrees of toxicity and flammability. Keeping these hazards under control may add cost and complexity to the overall plant. These hazards make it impractical to use some refrigerants near enclosed concentrations of people. At the same time, recognition of these hazards by manufacturers and professional organizations is resulting in new codes, standards, and publications that you can use for guidance in dealing with the hazards.

After considering availability, efficiency, and safety, study the characteristics that determine the type of compressor, the size of the cooling unit, noise, and maintenance. The following discussion covers all these selection considerations in greater detail.

■ Availability

The future availability of any particular refrigerant is a gamble because of several uncertainties. The first uncertainty is future opinion regarding the environmental dangers posed by particular refrigerants. There is a wide spectrum of opinion, and few consider the issue settled. There is much uncertainty in the science. The perception of environmental danger in the coming years will be determined largely by the weather, by politics, and by commercial opportunities to sell new equipment and refrigerants.

The underlying assumptions about the environmental dangers of refrigerants may change. At this moment, attention is focussed on two types of environmental harm that may be caused by chlorine-containing refrigerants: depletion of the ozone layer and global warming. Of the two, concern about ozone depletion is presently having a stronger effect on the selection of refrigerants. Indeed, the phase-out of CFC's and HCFC's under the Montreal Protocol is related only to the ozone depletion issue.

Subsequently, global warming entered the discussion of refrigerants. Concern about global warming is partly the result of scientific findings and computer modeling of atmospheric behavior. It has also been used for marketing of environmental hysteria, and it has become part of the competition between cooling equipment manufacturers who have made major commitments to particular refrigerants. For this reason, it is important to understand the specific ozone depletion and global warming characteristics of refrigerants, which we discuss below.

It seems likely that the present popular refrigerants will remain in production until their scheduled phase-out dates because of the demands of existing equipment. It is presently assumed that HCFC 123 will be available because it is similar in refrigeration properties to the phased-out CFC-11, and that HFC-134a will be available because it is similar to the phased-out CFC-12. The production of these refrigerants as substitutes in old equipment guarantees their availability in new equipment for the short term, but not for the long term.

■ Won't New Refrigerants be Developed?

Another possibility is that new, more desirable refrigerants will be developed. However, there is a viewpoint in the industry that all possible refrigerants have already been identified, and that uncertainties are limited to exploring their properties further.

There is actually a fairly solid basis for this bold assumption. As we said earlier, a good refrigerant will have a high latent heat and a low specific heat. However, as molecules become larger, their latent heat diminishes and their specific heat increases. Thus, only small molecules can be good refrigerants.

Furthermore, a refrigerant must be a liquid that changes to a vapor, and back, at useful temperatures and feasible pressures. Also, it must not be highly toxic. Of the 92 naturally occurring elements, only eight produce compounds that satisfy these requirements. The eight are carbon, hydrogen, oxygen, nitrogen, sulfur, fluorine, chlorine, and bromine. Of these, the last two are being banned from future refrigerants for environmental reasons.

These requirements narrow the field so much that it is possible to list all possible materials that would have good refrigerant properties. In fact, Thomas Midgely did this in 1928, and he asserted that no other satisfactory refrigerants would be possible. However, history illustrates that it is risky to predict that radical new developments will not occur.

■ Refrigerant Mixtures and Azeotropes

By this time, the properties of individual refrigerants have been studied extensively. The search for substitute refrigerants is now extending to mixtures of refrigerants. For example, Refrigerant 407C is a mixture of Refrigerant 32 (23%), Refrigerant 125 (25%), and Refrigerant 134a (52%). Refrigerant 407C is a possible substitute for Refrigerant 22.

A fundamental problem with refrigerant mixtures is that they tend to separate. This occurs because the individual components have different evaporation and condensing temperatures, and different leakage rates. Thus, if a cooling system develops a leak, one component of the mixture will leak out faster than the other, resulting in a different mixture with different refrigerant characteristics. Refrigerant mixtures also tend to separate in storage and transit, so that the wrong mixture may be put into the system when it is charged.

For this reason, there is great interest in "azeotropes." An azeotrope is a combination of two or more refrigerants that behaves as a single material in all relevant characteristics. In this respect, an azeotrope is more like a compound than a mixture. Unlike a compound, the physical properties of the azeotrope are an average of the properties of the original materials, as in a mixture. A common azeotrope is Refrigerant 502, which is a blend of Refrigerant 22 (49%) and Refrigerant 115 (51%).

A combination of refrigerants may behave as an azeotrope only within a certain range of mixing ratios. This range may change with the temperature. Thus, azeotropes are not "foolproof."

■ Ozone Depletion Potential (ODP)

Ordinary atmospheric oxygen is a molecule consisting of two oxygen atoms. Ozone is a molecule consisting of three oxygen atoms. At the level of the stratosphere, the concentration of ozone is increased by the action of sunlight, where intense solar ultraviolet radiation breaks atmospheric oxygen into two single oxygen atoms. Individual oxygen atoms are extremely reactive, so the two individual atoms combine with two other molecules of oxygen to form two molecules of ozone. The high-altitude stratum of ozone concentration is called the "ozone layer."

Ozone strongly absorbs ultraviolet light. The main significance of the ozone layer is that it shields the surface of the earth from the sun's intense ultraviolet radiation. Depletion of the ozone layer is expected to increase the harmful effects of ultraviolet radiation. Present health concerns include an increase in skin cancers and cataracts, and harm to the immune system. More generally, ultraviolet light can damage any organic matter that it can penetrate. In addition to human skin and eyes, this includes microorganisms, insects, and even plants. There may be other adverse effects that are not yet recognized.

(Ozone has become notorious as an air pollutant at ground level. Refrigerants are not relevant to this issue. Ozone at ground level is harmful because of its powerful oxidizing effect. Ozone is produced by the action of sunlight on hydrocarbon vapors, such as gasoline fumes, and it probably promotes the formation of other harmful air pollutants.)

The ozone depletion theory asserts that the damage to the ozone layer is largely being done by the chlorine in refrigerants. (Many other natural and synthetic materials are suspected of causing ozone depletion. Some of these, such as methyl bromide and carbon tetrachloride, are being regulated under the Montreal Protocol.) Chlorine-containing refrigerants are the most important suspect because they exist in large quantities and are relatively stable. Their stability allows them to survive long enough in the open atmosphere to reach the stratosphere. Once there, the strong ultraviolet breaks down the refrigerant molecules, releasing chlorine to do its damage. The atmospheric lifetime of halocarbon refrigerants ranges from about one year to about one century.

(There is a large amount of chlorine in the surface environment, such as in laundry bleach, but most of this is not a concern. Free chlorine is highly reactive, forming compounds that remain near the surface. Furthermore, chlorine is very soluble in water, so rain keeps bringing the chlorine back to earth.)

In the framework of this theory, the potential harm that a refrigerant may cause the ozone layer is defined by a relative number called the Ozone Depletion Potential (ODP), which is defined on a mass basis. The baseline of this ranking system is the ODP of CFC-11, which is defined as 1.0. Table 1 gives ODP ratings for the most common refrigerants and leading candidates.

From the standpoint of refrigerant selection, the most important aspect of this theory is that chlorine is the only component of common refrigerants that is responsible for ozone depletion. (Bromine, much less commonly used in refrigerants, is up to 50 times more damaging than chlorine.) Thus, refrigerants without chlorine (or bromine) pose no danger to the ozone layer. By definition, none of the HFC's contain chlorine, so they all have an ODP of zero.

One obvious approach to reducing ozone depletion is to substitute refrigerants that contain no chlorine. The leading new refrigerant of this type is HFC-134a. For the same reason, there is a resurgence of interest in ammonia, including greater consideration of ammonia in air conditioning applications.

Unfortunately, there are not enough chlorine-free refrigerants to serve the present spectrum of applications without serious compromises involving efficiency, safety, or machine design. Therefore, a second approach is being used, which is to use refrigerants that break down before they can reach the stratosphere. According to theory, it takes many years for CFC's to diffuse upward into the ozone layer. If the refrigerant molecule decomposes before it arrives there, its chlorine will be washed out of the atmosphere.

Refrigerants will decompose in the open atmosphere if they contain a chemical weak link. Hydrogen serves as the weak link. Halocarbon refrigerants that contain hydrogen are called HCFC's to distinguish them from CFC's.

For example, the estimated atmospheric lifetime of HCFC-123 is 1.4 years, so short that the actual ozone depletion caused by this refrigerant may be much less than the present ODP rating indicates. In contrast, the other major HCFC, HCFC-22, has an atmospheric lifetime estimated at 19 years.

(Some people suspect that the short atmospheric life of HCFC-123 may earn it a reprieve from the Montreal Protocol phase-out. However, HCFC-123 is used only in large centrifugal chillers. These are less likely to be found in the developing countries from which most of the political pressure for a reprieve are likely to come. Anyway, the phase-out date for HCFC's is still an equipment lifetime away.)

For all CFC's and HCFC's, leakage of refrigerant from the cooling equipment is a dominant factor in ozone depletion. If refrigerant does not leak out, it can do no harm. The most important factors in leakage are:

- *the pressure of the refrigerant.* All other things being equal, leakage increases with pressure. Some refrigerants operate at much higher pressures than others. For example, HCFC-22 condenses and evaporates at relatively high pressure, whereas HCFC-123 condenses at about atmospheric pressure, and evaporates at sub-atmospheric pressure.

- *for low-pressure gases, the efficiency of the purge unit.* If a refrigerant operates at a pressure below atmospheric, some air leaks into the system and it must be removed. This is done with a purge unit. Unfortunately, the purge unit ejects some refrigerant along with the air. Operation of the purge unit, especially of older models, can cause as much refrigerant loss as using refrigerants that operate at higher pressures. Modern purge units cause far less refrigerant loss.

- *whether the motor is open or hermetic.* A compressor with an open drive requires a shaft penetration into the system. This is a significant leakage path that cannot be sealed completely throughout the life of the system.

- *maintenance practices.* Maintenance practices have a major effect on the amount of refrigerant leakage. Failure to repair seals, venting of refrigerant to clear air out of the unit, failure to recover refrigerant, and sloppy charging practices may all cause more loss of refrigerant than the design of the machine itself. These bad practices are actually outlawed in the United States, but such laws are unenforceable in practice.

■ Global Warming Potential (GWP)

The global warming theory asserts that the surface temperature of the earth is being raised as a result of "greenhouse effect," in which certain gases in the atmosphere act like the glass in a greenhouse. Incoming short-wavelength solar radiation passes through the atmosphere and heats the surface of the earth, without being affected by the gases. The warmed surface of the earth emits heat into space at much longer wavelength. This long-wavelength infrared radiation is absorbed by the "greenhouse gases," preventing its escape from the earth and heating the atmosphere.

In the framework of this theory, the potential of a refrigerant gas to cause global warming is defined by a relative number called the Global Warming Potential

(GWP), which is defined on a mass basis. This rating system assigns a baseline GWP of 1.0 to carbon dioxide. Carbon dioxide is used as the basis of comparison because it is the second most significant greenhouse gas in the atmosphere, after water vapor. Also, carbon dioxide is the greenhouse gas most abundantly produced by human activity. Table 1 gives GWP ratings for the most common refrigerants and leading candidates.

Global warming differs from ozone depletion in several fundamental ways that relate to refrigerants. The global warming potential of refrigerants is not related to the chlorine content in the refrigerant. Global warming may be caused by many types of gases, including CFC's, HCFC's, and some hydrocarbons. The heating effect occurs at all levels of the atmosphere. It commences as soon as the refrigerant is released into the atmosphere.

The contribution of refrigerants to global warming is much less important on a relative scale than their presumed contribution to ozone depletion. By far the most important gas regulating the earth's heat balance is water vapor. It does this in at least two important ways. One is that it absorbs a large part of the infrared spectrum of incoming solar radiation. Another is that it forms clouds that reflect both incoming and outgoing heat.

Of the other gases, the most important is carbon dioxide, because of the large amount of it in the atmosphere. Next in importance is methane, which comes from many natural sources, from landfills, and from domestic animals, especially cattle.

As of the early 1990's, halocarbon refrigerants were assumed to be next in importance. This included the used of refrigerant-type gases in non-refrigeration applications, such as blowing the bubbles in plastic foam insulation.

Next in importance are nitrogen oxides, which come primarily from combustion emissions. These are followed in importance by ozone.

The selection of refrigerant also influences global warming through its effect on the efficiency of the equipment. If the machine is more efficient, it requires less energy. Most of the energy for cooling is produced by the burning of fossil fuels, which produces carbon dioxide. The efficiency of the refrigerant may be more important with respect to global warming than its GWP.

Similarly, control of refrigerant leakage from the system is as important as the selection of refrigerant. Refrigerant leakage is determined by the factors listed previously in regard to ozone depletion.

The atmospheric lifetime of refrigerants is important in global warming, because the total warming effect is proportional to the length of time that the refrigerant survives in the atmosphere. The atmospheric lifetime is factored into the GWP rating. Of the major HCFC and HFC refrigerants, the atmospheric lifetimes are 1.4 years for HCFC-123, 16 years for HFC-134a, and 19 years for HCFC-22. For CFC's, which are very stable, the atmospheric lifetimes are much longer.

■ Energy Efficiency

Refrigerants differ in the amount of cooling they produce in relation to the energy required to compress them. As with the hardware, the efficiency of a refrigerant can be expressed in terms of COP, which is the COP of the refrigerant in an idealized cooling machine.

The refrigerants presently used for conventional applications all have fairly high COP's, within the range of 75% to 90% of the theoretical maximum. However, the differences between currently available combinations of refrigerants and cooling equipment have a significant effect on the system COP. This makes refrigerant selection a primary decision affecting efficiency. See Table 1 for the COP's of the most common refrigerants and leading candidates.

■ Toxicity

ASHRAE Standard 34 rates refrigerants according to their toxicity on a simple two-level scale, Class A and Class B. The wording of the Standard is technical. Both toxicity classes are based on exposure at a refrigerant concentration of 400 parts per million (PPM). In essence, the Standard says that a Class A refrigerant has a low degree of toxicity at this exposure level, while a Class B refrigerant has a higher degree of toxicity at this exposure level.

The toxicity rating of the refrigerant is only one of several safety considerations. Others are the average duration of exposure to the refrigerant, the ability of persons to escape rapidly from a massive leak, ventilation of the cooling equipment, using refrigerant detectors to warn of leaks, etc. For example, ammonia should not be used in a cooling system that is located where a leak in the system could allow the ammonia to leak into a crowded space. However, ammonia may be acceptable in a central chiller plant that distributes cooling via water.

ASHRAE Standard 15, Safety Code for Mechanical Refrigeration, sets forth the safety features that should be included in the design of cooling plants, based on safety considerations.

■ Flammability and Explosion Potential

ASHRAE Standard 34 also rates refrigerants according to their flammability. The flammability rating has three levels, Classes 1, 2, and 3. Class 1 refrigerants are not flammable in air under particular test conditions. The currently popular CFC's, HCFC's, and some of the HFC's are in this class. Class 2 refrigerants are moderately flammable and/or explosive. For example, ammonia is in this class. Class 3 refrigerants are very flammable and/or explosive. It will come as no surprise that propane and butane are in this class, for example.

Table 1 gives flammability ratings for the most common refrigerants and leading candidates.

ASHRAE Standard 15, Safety Code for Mechanical Refrigeration deals with the safety aspects of flammable refrigerants, as it does for toxicity. The Standard does not make a clear separation between design features for dealing with the two hazards. Instead, it suggests common techniques of isolation, ventilation, etc., to deal with both. This simplifies the design effort.

■ **Condensing Pressure**

The condensing pressure is (almost) the highest pressure that occurs inside the system. A large chiller is a major pressure vessel, and the condensing pressure determines the weight of the shell, piping, and accessories.

On the other hand, operation at higher pressure reduces the volume of the gas, all other things being equal (which they are not). This results in smaller machinery, which reduces cost. For example, HCFC-22 condenses at much higher pressure than HCFC-123, but it has about nine times the refrigerating effect per volume of gas.

The condensing pressure is a major factor in the rate of refrigerant leakage. This is important if the refrigerant has environmental hazards. Leakage also contributes to maintenance cost, but this is a relatively small factor in the overall operating cost of a plant.

■ **Evaporating Pressure**

The evaporating pressure is (almost) the lowest pressure that occurs inside the system. The major consideration is whether the evaporating pressure is below atmospheric, which is true of CFC-11 and HCFC-123. If so, there will be some leakage of air into the system. Air contaminates the refrigerant and reduces performance. For this reason, systems with evaporator pressures below atmospheric require purge units to remove the air. The operation of the purge unit expels some refrigerant into the atmosphere, as discussed previously.

How Centrifugal Pumps Work
Names for Centrifugal Pumps
The Impeller Determines the Pump Characteristics
Pump Efficiency and Operating Points
Pump Curves
- **Flatness of Pump Curves**

The Formula for Pump Power Input

System Curves
Pump Selection Issues that Affect Efficiency
- **Specific Speed**
- **Pump Size**
- **Liquid Viscosity**
- **Impeller Compromises to Handle Solids**
- **Smoothness of Pump Passages**
- **Internal Leakage**
- **Mechanical Drag**

This Note provides a quick introduction to the efficiency aspects of centrifugal pumps. It gives you a basis for selecting pumps for maximum efficiency in typical applications, and for improving the efficiency of existing pump installations.

How Centrifugal Pumps Work

Centrifugal pumps are one of the quiet mainstays of modern civilization, notable for their simplicity. Like those other technological marvels, the centrifugal fan and the electric motor, centrifugal pumps have only one moving part. That part is the impeller, which discharges liquid from its periphery by centrifugal force and draws in liquid at its center. The pump casing that surrounds the impeller is designed to feed the liquid into the impeller efficiently, and to discharge the liquid into the system efficiently.

The liquid velocity leaving the impeller is usually higher than needed. To slow it down, the pump has passages of increasing cross sectional area through which the liquid flows after leaving the impeller. A common configuration is an expanding passage that is wrapped around the perimeter of the pump. This gives the pump casing the classic spiral or "volute" shape, which is shown in Figure 1. The output velocity can be reduced further by installing an expanding discharge throat.

An alternative way of reducing the impeller discharge velocity is to use a set of diffuser vanes surrounding the impeller. Pumps of this design have a casing that is shaped like a pillbox.

Multi-stage pumps are used to achieve higher discharge pressures without resorting to large impeller diameters. In multi-stage pumps, a number of impellers are mounted on a single shaft, with the discharge from one stage being routed to the inlet of the next.

Names for Centrifugal Pumps

There is no consistent naming system for centrifugal pumps. A pump may be named for its number of stages, orientation ("horizontal" or "vertical"), type of case ("end suction," "double-volute," etc.), method of disassembly ("split case," etc.), impeller type ("propeller," "mixed flow," etc.), or other characteristics.

The Impeller Determines the Pump's Characteristics

The main performance characteristics of a pump are its flow rate, pressure, and power input. These characteristics are determined primarily by the diameter, the width, and the rotational speed of the impeller.

From simple geometry, the volume of flow produced by a centrifugal pump is the product of the cross sectional area of the impeller passages multiplied by velocity of the liquid as it exits the impeller. The exit velocity is related to the peripheral velocity of the impeller by a constant relationship that depends on the curvature of the impeller blades.

Since the flow volume is directly proportional to the peripheral velocity of the impeller, it is proportional to the impeller diameter and to the rotational speed.

The pressure output of a centrifugal pump is derived from the kinetic energy that it imparts to the liquid. Kinetic energy is proportional to the square of the liquid's speed. Since the speed of the liquid is related to the peripheral velocity of the impeller, the pressure developed by the pump is proportional to the square of the impeller diameter, and also to the square of the impeller speed.

The power consumed by a pump is proportional to the product of the pressure and the flow rate, so the power is proportional to the third power of the impeller diameter, and to the third power of the impeller speed.

Since the pressure and flow of a given pump are both related to the characteristics of its impeller, you cannot change one without changing the other. Changing the impeller diameter or speed has a larger effect on the pressure than it does on the flow rate. This is because pressure is proportional to the second power of the impeller diameter or speed, while flow is proportional to the first power.

The flow rate is not affected by the specific gravity, or mass, of the liquid being pumped. However, the

WESINC

Fig. 1 Centrifugal pump A simple impeller inside the case is spun by the motor. Water enters from the left through an opening in the center of the impeller. Centrifugal force flings the water toward the perimeter of the case. The spiral shape of the case collects the water as it follows the rotation of the impeller, and discharges the water into the vertical discharge pipe. The width of the impeller determines the flow rate through the pump. The speed at the outer rim of the impeller, which is determined by its diameter and rotational speed, determines the discharge pressure.

Table 1. PUMP PERFORMANCE RELATIONSHIPS		
Effects of Speed		
flow rate	~	speed
pressure	~	speed^2
power	~	speed^3
Effects of Diameter		
flow rate	~	diameter
pressure	~	diameter^2
power	~	diameter^3
Effects of Specific Gravity		
flow	*is independent of*	specific gravity
pressure	~	specific gravity
power	~	specific gravity

NOTE:
• "~" means "proportional to".
• In each of these relationships, all other variables are held constant.
• These relationships are approximate, because any changes affect the efficiency of the pump.

kinetic energy of the liquid is proportional to the first power of the specific gravity. Therefore, the pressure and the input power are also proportional to the first power of specific gravity. The nature of the liquid usually does not change in a given pumping application, so this is usually not a factor in energy conservation modifications.

These relationships, which are sometimes called "affinity laws," are summarized in Table 1.

The fact that the impeller determines the pump characteristics is especially useful if you have an opportunity to modify existing pumps to reduce energy consumption. You can reduce pressure, flow, and power input on a one-time basis by trimming the diameter of the impeller. Or, you can adjust the speed of the impeller, usually with a variable-speed drive. The latter approach provides a great deal of flexibility in adapting pump power to changing system requirements.

Pump Efficiency and Operating Points

A pump has its best efficiency at a single combination of flow rate and pressure. This point

MAXIMUM EFFICIENCY
TYPICAL OPERATING LIMITS
Pressure Produced by Pump
Flow Rate

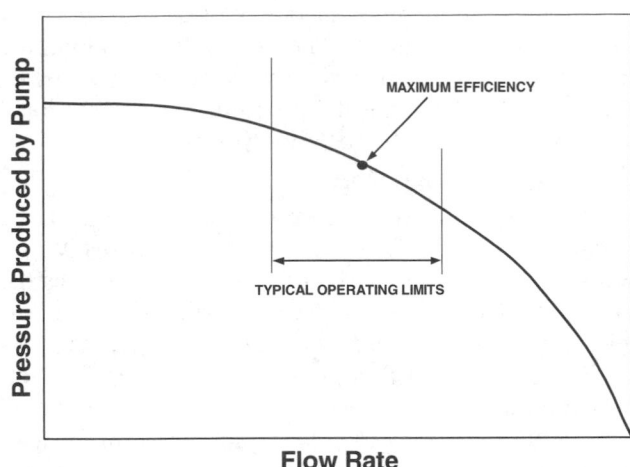

Fig. 2 Basic pump curve This relationship between the flow rate and the pressure that the pump produces is typical of all centrifugal pumps. Also, the point of maximum efficiency is usually near the point shown here. Operating far from this point makes the pump very inefficient, and may damage the pump.

typically is near the maximum pressure, and at about one third of maximum flow. At lower pressures, efficiency declines because the impeller is producing pressure internally that is not utilized. Similarly, at higher flow rates, the flow is occurring without the need for all the pressure that is being produced. At lower flow rates, efficiency declines because the impeller produces turbulence in trying to move the liquid into a system that will not take it.

A centrifugal pump is able to operate well away from its point of maximum efficiency, which simplifies system design and allows a margin for error in estimating system characteristics. The efficiency loss is considered moderate as long as the actual operating point is reasonably close to the point of maximum efficiency. On the other hand, if the pump is not well matched to system flow and pressure requirements, or if the system flow requirement varies widely, the pump may spend a large fraction of the time operating at substantially reduced efficiency.

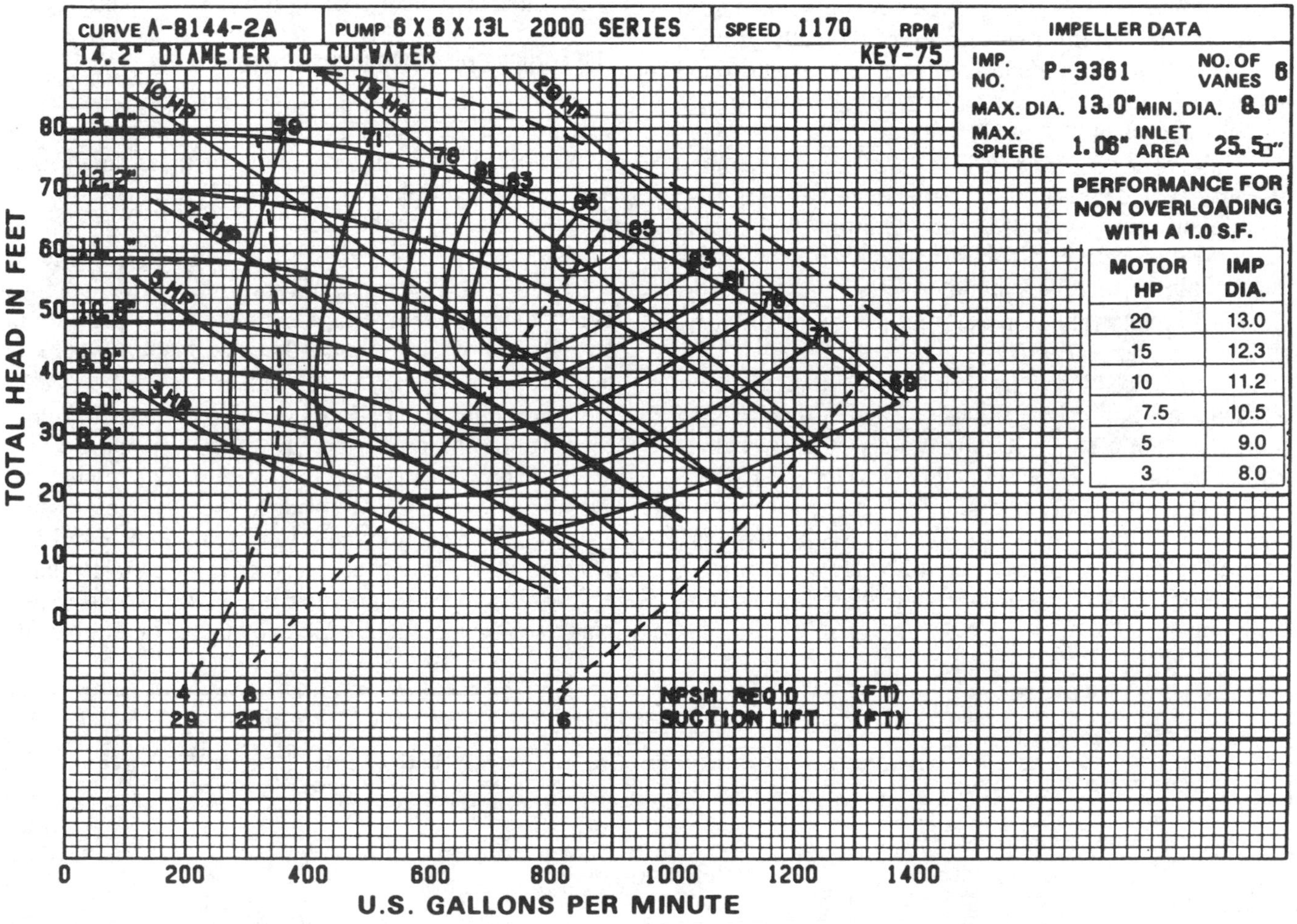

CURVE A-8144-2A
PUMP 6 X 6 X 13L 2000 SERIES
SPEED 1170 RPM
14.2" DIAMETER TO CUTWATER
KEY-75
IMPELLER DATA
IMP. NO. P-3361
NO. OF VANES 6
MAX. DIA. 13.0" MIN. DIA. 8.0"
MAX. SPHERE 1.06" INLET AREA 25.5□"
PERFORMANCE FOR NON OVERLOADING WITH A 1.0 S.F.
MOTOR HP | IMP DIA.
20 | 13.0
15 | 12.3
10 | 11.2
7.5 | 10.5
5 | 9.0
3 | 8.0
20 HP
15 HP
10 HP
7.5 HP
5 HP
3 HP
13.0"
12.2"
11."
10.6"
9.8"
9.0"
8.2"
NPSH REQ'D (FT)
SUCTION LIFT (FT)
TOTAL HEAD IN FEET
U.S. GALLONS PER MINUTE
0
200
400
600
800
1000
1200
1400

Allis-Chalmers

Fig. 3 Typical catalog pump curves This important engineering tool displays all the information that you need for selecting or modifying a pump for most applications. Overlaid on the pressure-vs.-flow curves are curves of efficiency and input power. "NPSH" means "net positive suction head," which is the amount of pressure needed at the pump inlet to avoid cavitation.

Pump Curves

The relationship between a particular pump's pressure and flow is called its "pump curve." A typical pump curve is shown in Figure 2. All centrifugal pumps have pump curves similar to this one.

One characteristic of the pump curve is that the pressure produced by the pump decreases as the flow rate through the pump increases. For a given pump, the flow rate through the pump is determined by the resistance of the system, and the pump pressure is determined accordingly.

Another characteristic of the pump curve is that the point of maximum efficiency is part way down the curve, usually near the point shown in Figure 2. Efficiency falls rapidly above and below this point.

A pump that is properly selected and applied will operate within a limited range of flow rates, as shown in Figure 2. This keeps the pump operating at high efficiency. Also, throttling the flow excessively creates turbulence and cavitation, which damages the pump.

This basic curve of flow-vs.-pressure is commonly used as a platform to display other pump characteristics, such as power consumption, efficiency, and suction head requirements. Pump manufacturers typically provide combined sets of curves in their catalogs, as illustrated in Figure 3. Such sets of curves are your basic reference for selecting or modifying a pump.

The typical set of curves in Figure 3 includes a range of impeller diameters. This is because pump manufacturers make only a limited number of casing sizes, so they satisfy different customer requirements by varying the impeller diameter.

An increasingly popular method of modulating pump output is varying the speed of the pump. As discussed previously, impeller speed and impeller diameter have similar effects on flow, pressure, and energy consumption. Therefore, a set of pump curves similar to those in Figure 3 can be developed in which the curves for different diameters are replaced with curves for different impeller speeds. Such curves are needed for designing variable-flow pumping applications, where the flow is modulated by varying the pump speed.

Reducing either impeller diameter or impeller speed shrinks the pump curve. The curve shrinks along the flow axis proportional to the diameter or speed, while the curve shrinks along the pressure axis proportional to the square of the diameter or speed.

■ Flatness of Pump Curves

With all centrifugal pumps, the pump discharge pressure falls as the flow increases. (Alternatively, you can visualize that the pump pressure "backs up" as the discharge from the pump is obstructed by higher discharge pressure.) This tendency varies among pumps of different design. Pumps that tend to maintain their pressure as flow increases are said to have "flat" pump curves, while pumps that lose pressure rapidly with increasing flow are said to have "steep" pump curves.

Select a pump with a flat curve for applications that require a fairly constant pressure over a wide range of flow rates. Such applications include boiler feedwater pumping, pumping of service water in tall buildings, and pumping of chilled water in systems where the cooling coils have throttling valves. A flat pump curve is not needed in applications involving a single operating point, such as typical cooling tower water pumping.

The Formula for Pump Power Input

In the most general sense, power is consumed when some form of force is exerted at some rate of displacement. With pumps, the force takes the form of pressure. The displacement takes the form of flow rate times the density of the fluid.

Formula 1 expresses this relationship mathematically. This formula appears with a variety of units. For example, some metric versions express pressure in terms of kiloPascals and others express pressure as meters of water.

In existing installations, you can often save energy by reducing the pressure or flow rate, or both. For a given pump, changes to the system that reduce the average flow rate usually reduce the average pressure as well. Since pressure and flow are multiplied by each other in the formula, this results in a relatively large saving. Changes to pressure and flow also affect the pump's efficiency. The efficiency factor in the formula

---

FORMULA 1

$$\text{Pump power input} = \frac{(\text{discharge pressure - suction pressure}) \times \text{flow rate} \times \text{specific gravity}}{\text{constant} \times \text{pump efficiency}}$$

where the units are as follows:

	metric	inch-pound
power	kilowatts	horsepower
pressures	kiloPascals	feet of head
flow rate	cubic meters per minute	gallons per minute
constant	60	3960

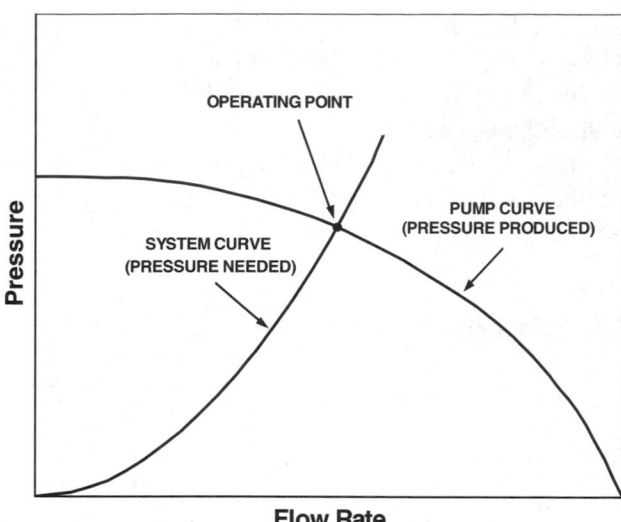

Fig. 4 System curve and pump curve   The pump must operate somewhere along the pump curve. The system must operate somewhere along the system curve. The point where the two curves cross is where the pump and the system will operate.

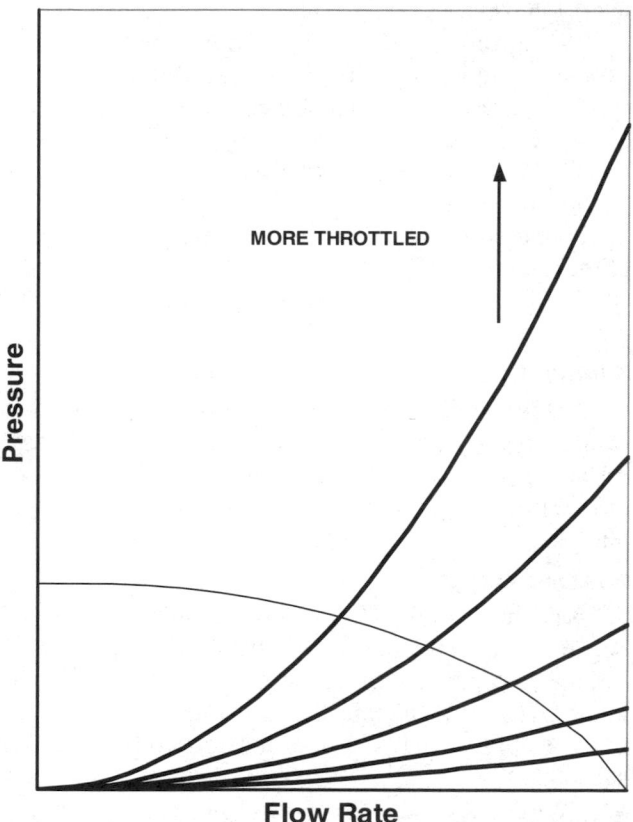

Fig. 5 Family of system curves for a system that has variable resistance   These curves might represent a hot water heating system or a chilled water system that uses throttling valves on the coils. The operating point moves over a wide range on the pump curve.

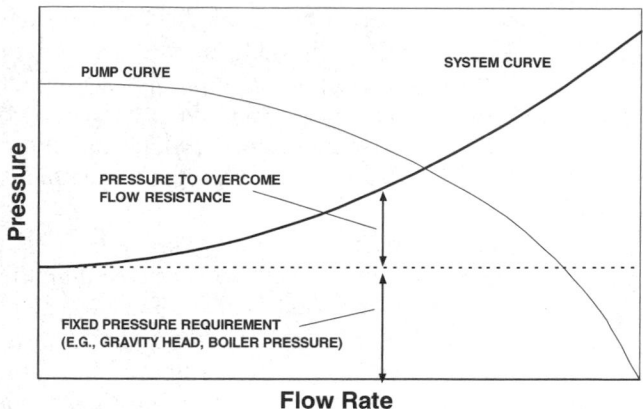

Fig. 6 System that has constant pressure requirement This type of system curve is typical of open systems in which the pump must overcome gravity head in addition to piping friction.

depends on how close the pump's actual operating point comes to the point of maximum efficiency, as discussed previously.

In designing a new pump system, you have the additional option of selecting the pumps for higher inherent efficiency. The factors that determine inherent efficiency are covered below.

System Curves

To achieve efficient pump operation, you have to match the pump's pressure and flow characteristics to the pressure and flow requirements of the system. The characteristics of the system can be depicted by a

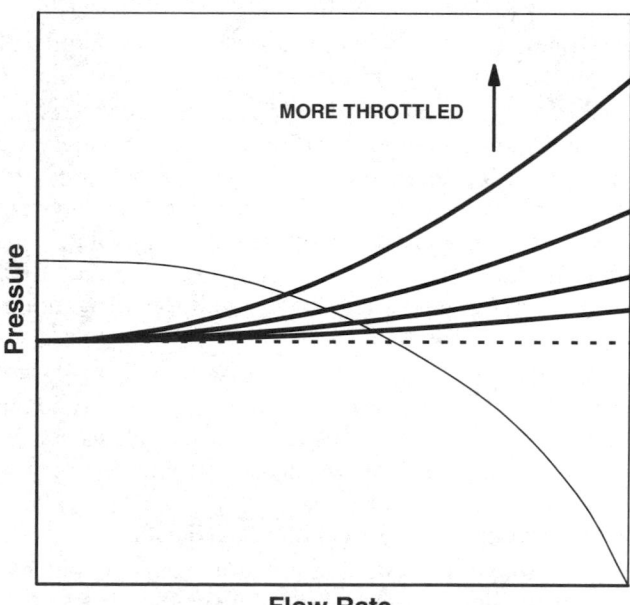

Fig. 7 System curves for throttling boiler feedwater system In this system, the pump must overcome the high pressure of the boiler and also the variable resistance of the feedwater throttling valve.

"system curve." This is a graph of the pressure that would be required to drive liquid through the system at various flow rates. When a particular pump is used with a particular system, the flow rate and the pressure at the pump are determined by the point at which the pump curve intersects the system curve. A typical example is shown in Figure 4.

The system curve shown in Figure 4 is typical of a system in which the only pump load is overcoming the friction of pipe and other components. This is typical of closed systems. Gravity can be disregarded in most closed systems, because the weight of liquid in the return leg balances the weight of liquid in the supply leg. Such a system has a zero pressure requirement at zero flow, and the pressure requirement rises proportional to the square of the flow rate.

If the system has valves that throttle the flow, the position of the valves affects the shape of the system curve. Figure 5 shows how the system curve changes as the flow resistance increases. For example, Figure 5 might represent a hot water heating system or a chilled water system with throttling valves on the coils.

In open systems, such as domestic water systems and cooling tower systems, the pump must overcome gravity in addition to the friction of the pipe and fittings. The gravity head is constant, while the system friction varies with the flow rate. Figure 6 shows a typical system curve for such a system.

Boiler feedwater systems are different from the previous. These are closed systems, but they have a constant pressure requirement, which is the boiler pressure. In addition, the pump must overcome the resistance of the boiler inlet valve. If the boiler has a throttling feedwater valve, the system has a family of curves, such as those in Figure 7.

The system curve is useful for visualizing the effect of changes in operating conditions, and for visualizing the effect of changes to the system. In most cases, the system designer does not select a pump with a system curve, but instead selects a single operating point. This point is based on the calculated pressure and flow requirements of the system at the condition of maximum pump load.

When making changes to an existing system, you cannot determine the actual system curve with enough accuracy to be useful. Instead, make changes on the basis of pressure and flow measurements.

Pump Selection Issues that Affect Efficiency

The maximum efficiency that a pump can achieve depends on certain inherent design characteristics. You usually determine the efficiency of a particular pump from the catalog, but it helps to understand the underlying factors. In larger pump installations, understanding these factors is essential for selecting the most efficient pump, and for designing the system to take advantage of pumps' characteristics. The following are the pump characteristics that determine efficiency.

■ Specific Speed

The "specific speed" of a pump is an abstract number used in designing pumps. It is defined by this formula:

$$\text{specific speed} = \frac{\text{RPM} \times (\text{flow rate})^{1/2}}{(\text{pressure})^{3/4}}$$

where the pump RPM, flow rate, and pressure are the values at the pump's point of maximum efficiency.

If the flow rate is expressed in cubic meters per second, and the pressure is expressed in meters of water column, the highest efficiency is achievable when the pump is designed for a specific speed in the range of about 30 to 100. If the flow rate is expressed in gallons per minute, and pressure is expressed in feet of water column, the corresponding specific speeds are in the range of about 1,500 to 5,000. Maximum achievable efficiency drops steadily at values above and below these ranges.

The specific speed is determined by the pump's geometry, but is not related to its size. Pumps having a similar shape have a similar specific speed, regardless of size. Low specific speeds are characterized by radial flow and narrow impeller passages. High specific speeds are characterized by increasingly axial flow, and large, open impellers that start to resemble propellers. Larger pumps tend to be designed for higher specific speeds because this reduces the size and cost of the pump.

■ Pump Size

Larger pumps tend to be more efficient, for a given shape or specific speed. The main reason is that mechanical imperfections and design compromises that reduce efficiency are relatively less important in larger pumps.

■ Liquid Viscosity

Viscous (thick) fluids reduce pump efficiency because the deformation of the liquid as it passes through the pump requires shearing force within the liquid. This is not a major factor with water and water-based solutions, but is important in pumping thicker substances.

■ Impeller Compromises to Handle Solids

If the pump is used for liquid that carries a content of solids, such as sewage and slurries, the impeller must be designed to pass the solids. This design compromise usually reduces efficiency.

■ Smoothness of Pump Passages

Surface roughness in the liquid passages reduces pump efficiency, especially in pumps where the passages are narrow. The efficiency loss declines as the impeller passages become larger. The loss also depends on the flow properties of the liquid.

Pump models with smoother liquid passages are an option in larger sizes. These models may use different materials and/or more expensive manufacturing techniques.

■ Internal Leakage

The high-pressure liquid in the discharge of the pump tends to leak back to the low-pressure inlet, wasting pump energy and reducing capacity. Leakage is worst in pumps with low specific speeds, i.e., in pumps where the pressure is high with respect to the flow rate. For a given specific speed, leakage loss is lower in larger pumps.

Leakage is minimized by maintaining small clearances between the inner and outer rims of the impeller and the pump casing, and in larger pumps, by using seals called "wearing rings." Close manufacturing tolerances are needed to keep the impeller from changing its position with respect to the casing. This aspect of efficiency is a quality and price choice to be made by the system designer and the purchaser.

■ Mechanical Drag

Pumps require seals and packings to prevent the escape of liquid from the pump to the outside, to prevent internal leakage, and perhaps to prevent the entry of air into the pump. These items exert mechanical drag that must be overcome by the motor. The pump and motor bearings also exert drag. This drag absorbs a substantial fraction of motor power in small pumps, but relatively little in large pumps. In most pump sizes, this is a selection option for the system designer and purchaser.

Variable-Speed Motors and Drives

This Note covers all the contemporary methods of getting variable-speed output from electric motors. It will guide you to the best choice of a variable-speed motor drive for your application.

Types of Variable-Speed Motors and Drives

One broad class of motors, direct-current (DC) motors, are inherently variable-speed devices. These motors are historically the oldest type, and they are well proven. However, since the end of the nineteenth century, they have been relegated to the status of specialty devices by the development of the alternating-current (AC) induction motor and the spread of AC power.

The three-phase AC induction motor is the workhorse of modern civilization, providing most of the mechanical power used in stationary applications. It is simple, compact, reliable, and cheap. About 60% of all electricity is used in these motors. Most of that is used in motors of larger sizes. Single-phase induction motors are more complex, and they are limited to smaller sizes.

The only major disadvantage of induction motors is that they are single-speed devices. Their speed is determined by the frequency of the alternating current that drives them. If the power requirement of the driven equipment depends on speed, and if the load varies, the excess input energy at partial loads is wasted. Many applications for fans and pumps waste energy in this way.

Now, efficient means are available for varying the speed at which induction motors can drive equipment. This opens up many opportunities for energy conservation. It also offers significant secondary benefits, such as reducing the average noise level of driven equipment, and extending equipment life.

The most direct method of varying the speed of an AC motor is to change the frequency of the power that drives it. The electronic variable-frequency drive (VFD) has become the primary method for applications requiring smooth speed modulation over a wide range.

An AC motor can change its speed in steps by having several distinct sets of windings. Speed is changed by switching between the windings. A similar result is achieved by driving the load with two different motors, each of which has a different number of poles.

One type of AC motor, the wound-rotor motor, attempts to mimic the behavior of DC motors. It controls speed by varying the rotor current. This type is inefficient and limited in speed range. It has become obsolete, except in combination with special equipment that is described below.

Older methods let the motor run at constant speed, and use a speed changing transmission. The transmission may be mechanical, hydraulic, or electromagnetic. Some of these methods are still worth considering,

This Note covers most of these types of variable-speed motor drives, except for ones that have virtually disappeared.

Consider Overall Efficiency

When selecting a method of achieving variable speed, calculate the total losses for each method you are considering. The total losses consist of these components:

- *motor losses.* Multi-speed motors, wound-rotor motors, and variable-frequency drives incur some energy losses within the motor itself.
- *drive losses.* Any type of drive that requires additional equipment, such as a variable-frequency drive or a variable pulley, incurs losses in that equipment.
- *losses in the power supply system.* Any drive that causes distortion of the input power incurs losses in transformers and other parts of the power supply system. This is a problem primarily with variable-frequency drives.
- *losses in the application.* Any drive that does not modulate speed precisely to minimize energy consumption in the drive equipment wastes energy. The amount of waste depends on the nature of the application. For example, a two-speed drive may capture most of the potential savings when driving a cooling tower fan, but it may be much less efficient than a fully modulating fan drive when used in a reheat air handling system.

Other Selection Factors

Efficiency is not the only important selection criterion for a variable-speed drive. The reason that different approaches remain in the market is that each has a number of advantages and disadvantages. The most popular type of drive is not necessarily the best for your application. Be prepared to research the market, and keep an open mind.

For example, multi-speed motors and multiple-motor drives are inexpensive, reliable, and free of electrical problems, but they cannot provide continuous modulation of speed. Variable-pulley drives are also simple, reliable, and free of electrical problems, but they are limited in capacity and speed range. Variable-frequency drives provide good efficiency and almost unlimited speed and torque range, but they may cause a variety of serious electrical problems. Direct-current (DC) drives are efficient and provide the best speed control, but they are expensive and require more maintenance than AC motors. Eddy current clutches and wound-rotor motors are simple and inexpensive, but they are less efficient than the other methods.

MULTI-SPEED A.C. MOTORS

Using 3-phase multi-speed motors is one of the most efficient methods of minimizing motor energy requirements, from the standpoint of the motor itself. Multi-speed motors are relatively inexpensive, simple to install, and free of the problems that plague some modulating types of variable-speed drives. Their main limitation is the stepwise nature of the speed change, which usually imposes an efficiency penalty in the overall application. The size of this penalty depends on the speed and load characteristics of the application.

The discussion here relates to 3-phase motors that change speed by switching poles. Switching poles is the only efficient method of changing the speed of a motor. Pole-switching motors are made in sizes ranging from fractional horsepower to several hundred horsepower (or kilowatts).

Speed and Number of Poles

The highest possible motor speed with 60-Hz power is 3,600 RPM, where the rotor turns 360° during a single cycle of alternating current. Doubling the number of poles produces a nominal speed of 1,800 RPM because the rotor makes only half a turn during the cycle. Tripling the number of poles produces a nominal speed of 1,200 RPM, and so forth. Thus, motor speed is inversely proportional to the number of poles.

(The actual speed of ordinary induction motors is somewhat lower than the nominal speed. This is because of the "slip" needed to induce the magnetic fields between stator and rotor. Slip increases with load. The motor speeds given in catalogs are usually the actual speeds at full load.)

Types and Speed Ratios of Two-Speed Motors

Most multi-speed motors have two speeds. There are two common types of 3-phase two-speed motors, "one-winding" and "two-winding."

As the name implies, a one-winding motor has a single set of windings that drive the motor at both of its speeds. The speed is changed by switching the connections of the windings so that the number of

electrical poles is doubled to achieve the lower speed. One-winding motors are limited to a speed ratio of 2:1.

Figure 1 shows all the important characteristics of a one-winding, two-speed motor. Notice that all these characteristics are related to the amount of torque that the motor is producing.

Two-winding motors have two entirely independent sets of windings. This allows any combination of poles that is a ratio of whole numbers. Speed ratios of 4:3, 3:2, and 2:1 are common. Motors with higher speed ratios, such as 3:1 and 4:1, can be produced on a custom basis.

The two-winding design also allows the motor torque characteristics at each speed to be matched to the requirements of the application.

Two-winding motors are larger, heavier, and more expensive than one-winding motors. The inactive winding takes up space, which requires a larger core. This increases magnetization losses, which makes two-winding motors less efficient than one-winding motors.

Motors With More Than Two Speeds

It is possible to make 3-phase motors with three or four speeds. Such motors are made on a custom basis. Each speed is achieved with a separate set of windings. Alternatively, a four-speed motor can be made with two 2-speed windings.

These motors are even larger, heavier, and more expensive than two-speed motors. Efficiency is further reduced by increased losses in the magnetic circuit. These motors are worthwhile if the efficiency

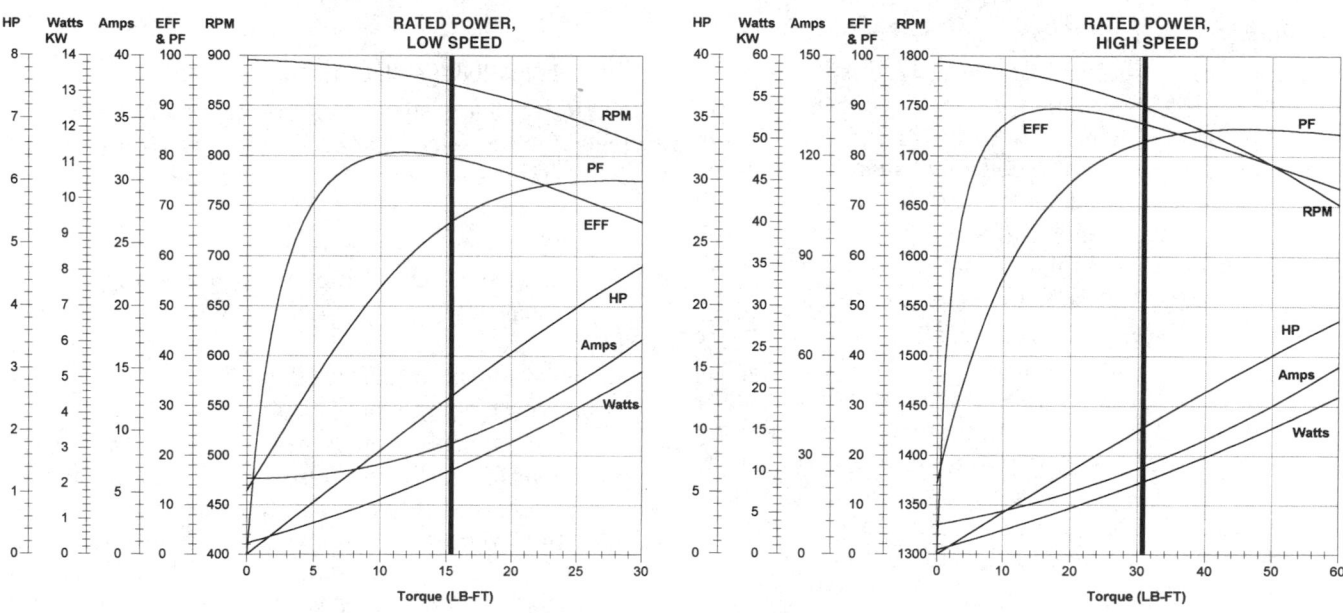

Leeson Electric Corporation

Fig. 1 Characteristics of a two-speed, one-winding motor All the important operating characteristics of the motor are presented here. The curve on the left is for low-speed operation, at a rated output of 2.5 horsepower. The curve on the right is for high-speed operation, at a rated output of 10 horsepower. In this manner of presentation, the motor's performance is displayed in relation to the motor's torque load.

improvement in the application offsets the greater losses in the motor.

Efficiency Characteristics

At the present time, multi-speed motors are not covered by efficiency standards, and efficiency is usually not part of the catalog information. To get efficiency data, you must request it from the manufacturer for a particular model. This information will probably be a computer-generated estimate, since most multi-speed motors are built to custom order even though they may have standard catalog listings.

One-winding motors are somewhat less efficient than conventional single-speed motors when operating at high speed. Their efficiency drops further at the lower speed. For example, a typical 10-horsepower variable-torque motor has an efficiency of about 88% at high speed, and about 79% at low speed. The high-speed efficiency is slightly less than the current U.S. efficiency standard for single-speed motors.

As stated previously, two-winding motors have somewhat lower efficiency than one-winding motors, at least at the high speed. Motors with more than two speeds are even less efficient.

The overall drive efficiency of multi-speed motors compares favorably with that of variable-speed drives. Multi-speed motors require no additional speed control devices and they cause no degradation of power system efficiency. From an efficiency standpoint, the key question is whether the stepwise nature of the speed control is able to exploit most of the energy-saving potential in the application.

Torque and Power Ratings

The torque characteristics of single-speed motors are covered in Measure 10.1.1. Multi-speed motors have the additional complication that the torque characteristics at each speed must match the torque requirements of the load at that speed. The relationship of the torque requirement to speed varies widely with the application. For this reason, it is conventional to manufacture multi-speed motors in three torque classes:

- *"variable-torque"* motors, in which the torque requirement increases with speed. The low-speed motor windings produce much less torque than the high-speed windings. The motor horsepower rating at each speed is proportional to the square of the speed. The most common variable-torque applications are centrifugal fans and centrifugal pumps, in which the load increases as the square of the speed.
- *"constant-torque"* motors, in which the maximum torque may be required at any speed. For these applications, the low-speed motor windings must produce as much torque as the high-speed windings.

The motor horsepower rating at each speed is directly proportional to the speed. Typical applications are conveyor belts, tumblers, and reciprocating compressors.

- *"constant-power"* or *"constant-horsepower"* motors, in which the torque requirement is related inversely to speed. For these applications, the low-speed motor windings must produce much higher torque than the high-speed windings. The motor horsepower rating at all speeds is the same. Typical applications are drill presses, milling machines, lathes, and some grinders and crushers.

Relative Prices

A two-speed motor is the least expensive way of achieving variable speed. A two-speed motor is considerably less expensive than any other method of changing speed, except for multiple-motor drives, which are discussed next.

A one-winding, variable-torque, two-speed motor may cost almost twice as much as a comparable single-speed motor. However, by comparison, it may still cost less than some single-speed motors designed to operate with variable-frequency drives.

Two-winding motors cost considerably more than one-winding motors. However, if a motor starter is needed, it is less expensive for two-winding motors.

Increasing the torque rating of a multi-speed motor increases the price considerably, but the same is true of any other type of speed control.

Three- and four-speed motors are custom items. They are more expensive than two-speed, two-winding motors.

Other Selection Considerations

Refer to Measure 10.1.1 for general considerations related to the selection and installation of motors, such as service factor, power factor, etc.

Single-Phase Multi-Speed Motors

The only efficient method of changing speed in single-phase motors is switching the number of poles, as in three-phase motors. Old ceiling fan motors demonstrate the large speed ratios that are possible, along with the large number of poles required.

Single-phase motors require separate winding provisions for starting because the field of a single-phase motor does not rotate. This complicates two-speed operation, requiring compromises in efficiency and starting torque. At present, there is little interest in improving the efficiency in single-phase, multi-speed motors.

There are several other types of single-phase motors that provide speed variation. All of them are inefficient. Their size is limited to a small fraction of one horsepower (or kilowatt) because their losses become heat that would burn up the motor in larger sizes.

MULTIPLE-MOTOR DRIVES

Another method of achieving multi-speed motor operation is to use two or more motors to drive a common piece of equipment. One motor runs while the other motor or motors idle. For example, two-motor arrangements are commonly used to drive fans in air handling units and cooling towers. Figure 2 shows a typical application.

You can achieve virtually any combination of speeds by combining single-speed motors. Motor speeds of 3600, 1800, 1200, and 900 RPM are readily available. (These are the nominal, or no-slip speeds. Also, we are assuming 60-Hertz power here.) Therefore, by using two motors, you can have speed ratios of 4:1, 3:1, 2:1, 3:2, or 4:3 in direct-drive applications. By using a belt drive for at least one of the motors, you can create any combination of two speeds.

This method of achieving multi-speed operation offers the highest motor efficiency. This is because each motor is conventional, so you can select it for high efficiency without making other compromises. Additional drive losses, compared to a single motor, are minimal. Also, this method provides increased reliability, since any motor can operate the equipment independently.

Equipment cost is modest, because the motors are conventional, and because the smaller motor is usually a fraction of the size of the larger motor.

This method is often impractical to retrofit, especially in smaller sizes. In larger sizes, where a lot of energy can be saved, it may be worthwhile to modify the equipment to take a second motor. This may be easier than it looks. There is often room to install a pulley somewhere on a shaft, allowing the second motor to be installed at some distance from the equipment by using a belt.

Baltimore Aircoil Company

Fig. 2 Two-motor fan drive for a cooling tower The two centrifugal fans are driven by a common shaft. The larger motor drives the shaft from the left side, for high speed operation. The smaller motor drives the shaft from the right side, for low speed operation. Each motor is connected to the shaft by drive belts, allowing any possible combination of two speeds to be selected. When each motor operates, the other idles. Multiple-motor drives can be used for many applications. The motors are entirely standard, and no special drive equipment is required.

VARIABLE-FREQUENCY MOTOR DRIVES

During the 1980's, the variable-frequency electronic drive (VFD) became the preferred method of driving many types of loads at continuously variable speeds. VFD's are used with ordinary types of AC induction motors or synchronous motors, although it is prudent to use special models designed to operate with VFD's. Figure 3 shows a typical installation.

Rather than using a speed changing mechanism between the motor and the load, the VFD changes the speed of the motor itself, a capability that previously was limited to DC motors. The VFD, which is totally electronic, changes motor speed by changing the frequency of the current that is supplied to the motor. The rotational speed of AC motors is tied to the frequency of the input power, so changing the frequency changes the speed.

VFD's can be reliable and trouble-free if they are selected and installed properly. There are many things you ought to know before selecting the equipment, including problems that VFD's can cause. At first, manufacturers kept silent about the problems, but the development of adequate solutions to these problems now stimulates vigorous discussion, largely in the form of competing claims. Other selection considerations are important for matching the drive to the application. VFD's are still evolving. The following tells you what to look for, and gives you the background to understand future developments.

VFD's are a major breakthrough in energy efficiency because they make it practical to use precise motor speed control in a wide variety of applications. Figure 4 shows a realistic number of applications in a typical building HVAC system.

Advantages of Variable-Frequency Drives

■ High Efficiency

Variable-frequency drives are an efficient means of modulating the output of conventional induction motors and synchronous motors. For the drives themselves, full-load efficiencies above 95% are claimed. Efficiency falls with speed, with wide variation of part-load efficiency between drive types and models. Drive efficiency is discussed below.

When a motor is driven by a VFD, it wastes somewhat more energy as heat because the waveform

Fig. 3 Variable-frequency drives for large pumps   The cabinets are full of solid state electronics and electrical filters. There are no moving parts, and little audible noise. The drives can be installed anywhere with respect to the location of the equipment. The size of the cabinets is roughly proportional to the power of the motors being driven.

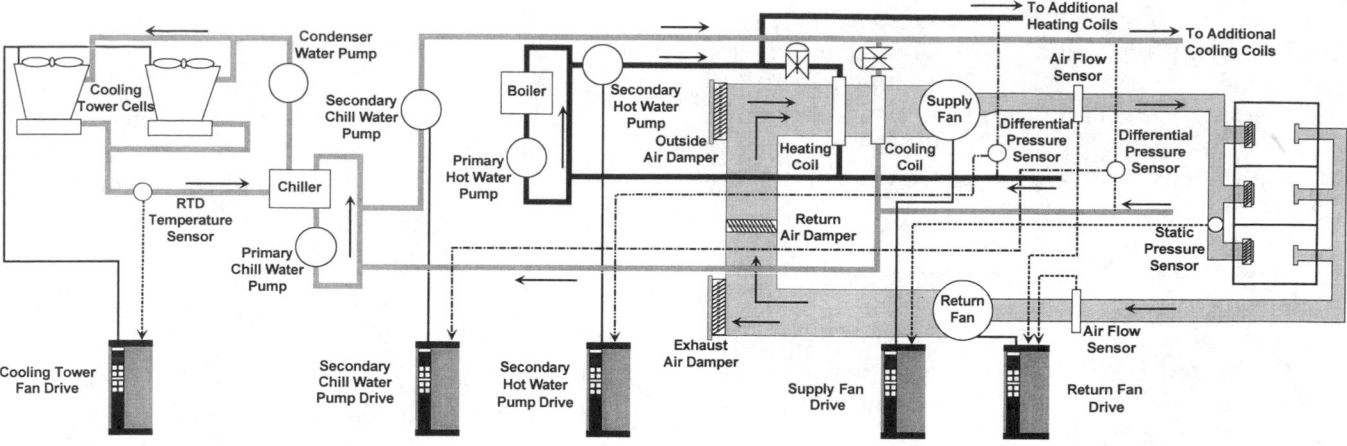

Graham Company

Fig. 4 Typical applications for variable-frequency drives in a comfort conditioning system Consider a variable-speed drive for every motor of significant size that has a variable load. Consider all appropriate types, not just variable-frequency drives.

of the power produced by the VFD is not a good replica of a sine wave. This loss varies widely among different types and models of VFD's, and among different models of motors. For a given VFD and motor, the amount of energy waste depends on the speed reduction and the motor load. Major motor manufacturers now offer models that are designed for reduced losses with VFD's.

VFD's increase losses in the transformers that feed them because of distortion of the input waveform. This effect varies widely between drive types and models, but is usually minor.

The overall system efficiency when using VFD's is somewhat lower than the efficiency typically achieved with modern DC drive systems and variable-pulley drives.

■ Large Turndown Ratio

For equipment driven by AC induction motors or synchronous motors, VFD's offer by far the best turndown ratio. Older types of VFD's offer a turndown ratio of at least 10:1. The best of the new models approach the near-zero-speed capability of DC drives.

■ Ease of Installation and Retrofit

You can often retrofit a VFD easily, with no changes to the existing mechanical installation being needed in most applications. However, some VFD's may cause existing motors to run substantially hotter. With such drives, it is advisable to replace the motor as well.

The electrical components of the drive are contained in an enclosure whose size is proportional to the motor size. Drives for motors of several horsepower typically have the volume of a shoe box. Drives for larger motors are typically freestanding units. The largest are about one or two cubic meters in volume. The drive can be mounted anywhere convenient, even at a considerable distance from the motor. VFD's can be installed in almost any environment, provided that they have an appropriate enclosure.

■ Low Maintenance

Variable-frequency drives have no components that transmit mechanical power, and no moving parts other than pushbuttons. If proper enclosures are selected, the units require no routine maintenance or cleaning.

By the 1990's, the electrical reliability of VFD's improved to the point that electrical failures are rare. In most cases, repairs can be accomplished by exchanging easily removable components. Most manufacturers offer models with self-diagnostic capability. The drives can be adequately protected against electrical faults, either by internal features or by external protective devices.

Motor Limitations in Applications with High Low-Speed Torque

At the outset, it is important to understand that a VFD may not be usable with conventional AC motors in applications where the motor must maintain high torque as the speed is reduced. Over half of all larger motors are used in such applications. Torque requires current, and current produces motor heating. However, conventional AC motors lose their ability to get rid of heat as their speed is reduced.

This is a limitation of the motor, not the drive. These limitations disappear if motors are used that can reject heat effectively at low speed.

■ Torque Categories

The relationship of the torque requirement to speed varies widely with the application. It is conventional to group applications into three torque categories:

- *"variable-torque"* applications, in which the torque requirement increases with speed, typically with the square of speed. The most common examples are centrifugal fans and centrifugal pumps. These applications require the lowest motor current at reduced speeds.

• *"constant-torque"* applications, in which the maximum torque may be required at any speed. Examples are conveyor belts, saws, tumblers, and reciprocating compressors. The description "constant-torque" is not literally correct in many applications, since the torque may vary widely, for example, with a conveyor belt.

• *"constant-power"* or *"constant-horsepower"* applications, in which the torque requirement is related inversely to speed. Constant-power applications require very high starting torque. Examples are drill presses, milling machines, lathes, and some grinding and crushing applications.

■ Motor Cooling Limitations

Motor heat is generated in both the stator and rotor. The amount of heat generated in the stator depends on the motor current. The amount of heat generated in the rotor depends on the stator current and on the percentage of slip. (Slip is what induces circulating currents in the rotor.) Heat from both the stator and the rotor must be able to find its way to the outside.

The rotor has impeller blades on its ends to make it act as a centrifugal fan. In open motors, the fan circulates air through the motor, cooling the stator and rotor directly. In enclosed motors, the fan moves heat toward the inside surface of the exterior shell. The stator can lose heat by direct conduction to the motor shell, but the rotor depends almost entirely on the internal air circulation created by the fan. Many motors also have a

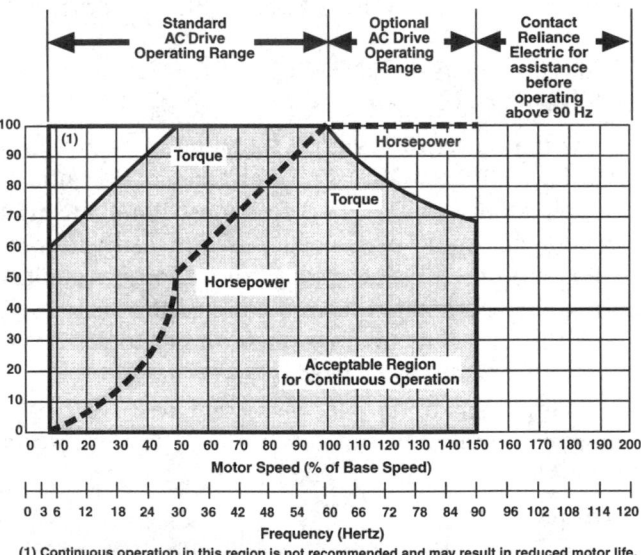

(1) Continuous operation in this region is not recommended and may result in reduced motor life.

Reliance Electric

Fig. 5 Operating envelope for an AC motor designed to be used with a variable-frequency drive The application's torque and speed requirements determine whether this is an appropriate motor. This chart shows that there is danger in operating the motor at low speed with high torque. In this region, the motor cannot get rid of its heat adequately. This motor can be operated up to 50% faster than across-the-line speed. However, be careful to limit the torque at higher speed. Otherwise, the power rating will be exceeded.

fan impeller installed on the external shaft that blows air over the outside of the shell.

As the motor is slowed by the VFD, the internal and external fans circulate less air. In "constant-torque" applications, the amount of heat produced at low speed may not decline much, if any. In "constant-power" applications, heat production may rise radically. The motor cannot get rid of the heat without a large rise in temperature. The rise in temperature may be severe enough to damage insulation and bearings within a short period of time.

■ Motor Ratings for Variable-Speed Operation

AC induction motors were originally designed to operate at a fixed speed, and most are still designed that way. VFD's introduce a mode of operation for which induction motors are not designed. Figure 5 is a graph of the maximum torque and power that is allowable with a typical high-efficiency motor as the speed is reduced. Note that the allowable torque remains constant down to a certain speed, and then drops. (Some motors allow maximum torque down to minimum speed.) The graph also shows that the maximum allowable power drops immediately and inversely with speed.

Ironically, torque and power limitations may be more severe with "high-efficiency" motors. One of the techniques used to reduce the energy consumption of high-efficiency motors is reducing the capacity of the fan. Contrary to popular misconception, "high-efficiency" motors do not necessarily run cooler than conventional motors. In variable-speed applications, they may even run hotter.

This limitation can be removed by the development of motors for high-torque VFD applications that have cooling independent of motor speed. This is not a new idea. DC motors are commonly available with separate fans and filters that maintain a constant flow of outside air through the motors. At the moment, however, a custom AC motor is generally required to provide high torque at low speeds.

■ Adapting to Applications that Require High Torque at Low Speed

You can do several things to avoid motor overheating when using a VFD in an application that requires high torque at low speed:

• *select a motor specifically designated for maximum torque at reduced speed.* At present, such motors allow full-load torque to be maintained down to zero speed, so they can be used in "constant-torque" applications. However, allowable power is still inversely proportional to speed, so they cannot be used in "constant-power" applications.

• *use a motor rated for operation in a high-temperature environment.* Note that this is not the same as buying a motor that has a high rated temperature rise. Such a motor may be a cheap

Table 1. COMPARATIVE CHARACTERISTICS OF VARIABLE-FREQUENCY MOTOR DRIVES

	VVI	CSI	PWM
Efficiency (1)	>90%	>90%	>90%
Turndown Ratio (2)	lower	lower	higher
Power Rating	smallest	largest	growing
Physical Size	middle	largest	smallest
Can Drive Multiple Motors	inherent	option	inherent
Starting Torque Capability	low (3)	high	high
Can Operate Motor Faster Than Across-Line Speed	yes	limited	yes
Regeneration	option	inherent	option
Harmonic Distortion to Line	high	high	low to moderate
Notching	moderate to high (4)	moderate to high (4)	none
Motor Overheating Currents (5)	moderate to high	moderate to high	low to high
Voltage Stress on Motor	minimal	moderate	moderate to high
Motor Cable Resonance	none	none	may occur (6)
High-Frequency EMI	none	none	low to high (7)
Transient EMI	moderate	moderate	none
Audible Motor Noise	minimal	minimal	low to high (8)
Cogging	occurs	occurs	none
Power Factor	low to high (9)	low to high (9)	high
Overload Protection	extra	inherent	extra
Short-Circuit Protection	extra	inherent	extra
Open-Circuit Protection	inherent	extra	inherent

NOTES:
1. Efficiency depends more on the model and on the application than on the type of drive. In recent years, most efficiency development has occurred in PWM drives. Efficiencies of modern drives vary from 93% to 97%.
2. Six-step VVI and CSI drives are generally limited to a turndown ratio of about 10:1 by cogging. PWM drives have nominal ratios up to 100:1. Turndown ratios differ with model and application.
3. VVI drives are limited to equipment, such as fans and pumps, whose torque requirements are proportional to speed.
4. Depends on the type of rectifier and the number of steps.
5. In VVI and CSI drives, motor overheating is caused mainly by low-frequency harmonics. In PWM drives, motor overheating is caused mainly by high-frequency harmonics. Motors differ in their vulnerability to the two types of harmonics.
6. Depends on chopper frequency and cable length.
7. Depends on chopper frequency and other details.
8. Depends on chopper frequency, motor construction, other factors.
9. Depends on the type of devices used in the rectifier.

motor that runs hot. What you want is a motor that normally runs cool, but is rated for a hot environment. Insulation ratings are one factor that determines the ability of a motor to tolerate hot environments. For details, see "Insulation Temperature Ratings" in Measure 10.1.1.

• *use a motor with a high service factor.* "Service factor" is the highest percentage of full load at which a motor can operate continuously under standard test conditions without overheating. Motors used in high-torque applications are usually selected with a high service factor, even without variable-speed operation.

• *use a "high-efficiency" motor.* High-efficiency motors have lower winding resistance and improved magnetic efficiency, so they carry more current and produce higher torque for a given amount of heat generation. Note that the torque ratings of some "high-efficiency" motors are reduced at lower speeds, limiting these motors to variable-torque applications.

• *use an oversized motor.* This may or may not cost a penalty in motor efficiency, depending on the design of the motor and the speed profile of the application.

• *cool the motor externally.* For example, install a fan to blow air over the motor. If the motor is installed in a hot location, duct air to the motor from a cooler location. This method cools the stator by conduction through the motor casing. Stator cooling is most critical in an induction motor because the stator has the windings that are most vulnerable to insulation failure. If you use external cooling, make it dependable. Interlock the motor so it cannot run without external cooling.

• *substitute an open motor for an enclosed motor.* This may be acceptable if the environment is not too dirty. If necessary, clean the motor periodically.

• *use a VFD with a minimum tendency to cause motor heating.* All types of VFD's cause a motor to run hotter, but some add more heat than others. Motor heating caused by VFD's is discussed below.

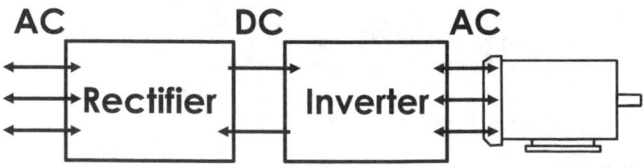

Graham Company

Fig. 6 How all variable-frequency drives work Alternating current from the power system is converted to direct current by the rectifier. The direct current is converted back into alternating current by the inverter, at a frequency that will drive the motor at the desired speed.

• *maintain a minimum speed.* If the application has a minimum speed, set the VFD so that the motor cannot be slowed below this speed. This method requires a fairly high minimum speed. Take steps to keep operators from inadvertently resetting the VFD to a lower speed.

None of these tricks alone may suffice to prevent motor overheating in a high-torque application, but a combination of them may be satisfactory. In the final analysis, there are some applications where a VFD should not be used with the types of motors that are presently available.

Sizes and Prices

There is no longer an upper limit on the motor power that can be controlled with a VFD. Standard models are available in large sizes, and virtually any size is available on a custom basis. The lower size limit of VFD's is determined primarily by their relatively high price in small sizes. Although VFD's are available for motors as small as 0.5 horsepower, prices fall only slowly below about 5 horsepower.

As with all semiconductor devices, the cost of VFD's has declined steeply since their introduction, and cost may continue to drop. In the mid-1990's, the price of large VFD's ranges from $50 to $200 per horsepower (or kilowatt), depending on the type and options. The price of small VFD's ranges from $100 to $400 per horsepower (or kilowatt). Installation cost is typically higher than equipment cost for small- and medium-sized drives, and may be less than equipment cost for large drives.

It may be advisable to replace the motor as well as the drive in some cases. The prices of motors range from $50 to $100 per horsepower (or kilowatt) in sizes larger than 10 horsepower.

Basic Operation of Variable-Frequency Drives

The basic operation of a VFD, as sketched in Figure 6, is simple: (1) a rectifier converts the three-phase alternating current from the power system to single-phase direct, (2) a DC circuit uses inductors and/or capacitors to smooth the direct current, and (3) an inverter switches the direct current on and off using semiconductor devices to simulate alternating current of any desired frequency. The synthetic alternating current (or voltage, depending on the type of drive) is supplied to the motor. The motor turns at a speed that corresponds to the supplied frequency.

The controls of the VFD also regulate the voltage supplied to the motor in proportion to the speed. This is because torque is produced most efficiently if the ratio of voltage to speed remains constant.

(The term "inverter" is also commonly used to mean the entire VFD. In this discussion, we will limit the term to the portion of the VFD that converts DC to artificial AC. The term "converter" is sometimes used for either the rectifier or the inverter.)

Types of VFD's

At present, there are three major VFD designs, along with several hybrids and less common types. The design of the inverter gives each type of VFD its name. The three types are: variable-voltage input (VVI), current source input (CSI), and pulse-width modulation (PWM). Table 1 summarizes their comparative characteristics. This table emphasizes characteristics that distinguish the three types. Refer to the text for more detail, and for characteristics and options that are common to all drive types.

Most manufacturers are now concentrating their development effort on PWM drives, because this is the only type that can accurately simulate a sine wave current in the motor. For this reason, the other two major types are becoming less common. However, they are likely to remain on the market for some time because of their lower cost, and because they avoid some problems that are peculiar to PWM drives. Furthermore, existing units will remain in service for a long time.

The "load-commutated inverter" (LCI) and "cycloconverter" are custom types that have been used on very large applications. We will not discuss them further.

All VFD's use high-power semiconductor switching devices that are capable of controlling current only in an on-off manner. As a result, the alternating current that is provided to the motor is not a smooth sine wave. This makes the motor less efficient. It also causes a variety of problems in the motor and in the rest of the electrical system. The evolution of VFD's is driven largely by the desire to produce motor current that more closely resembles a true sine wave. There is also a major interest in reducing distortion of the input power voltage by the VFD.

As a first step in understanding these problems, let's get acquainted with the switching devices that are causing the design headaches. Then, we will look at the three types of drives in greater detail.

■ Types of Switching Devices

The drive design is related to the types of semiconductor switching devices that are used in the inverter and rectifier. Three types of switching devices are presently used in VFD's.

The silicon controlled rectifier (SCR) is the oldest type, and it still has the highest voltage and current ratings. As the name implies, the SCR is a modification of a high-current rectifier. The main difference is that conduction is inhibited until a starting pulse is delivered to the "gate" terminal of the device. The main disadvantage of SCR's is that they can switch only slowly. Therefore, drives that use them in the inverter can produce only pulses of long duration.

Once current starts to flow in an SCR, it will not stop unless the driving voltage across the device is stopped. Stopping the voltage for this purpose is called "commutation." Commutation of SCR's used in rectifiers is usually simple. Commutation of SCR's used in inverters is more complicated. It is accomplished in a variety of ways, depending on the type of inverter.

The thyristor is similar to the SCR. The main difference is that the current flow can be stopped by applying a turn-off pulse to the "gate" terminal. For this reason, it is sometimes called a "gate turn-off" (GTO) thyristor. The GTO has smaller voltage and current ratings than the SCR, but its response is faster. This allows GTO's to control current in a larger number of steps per cycle. However, GTO's require a fairly large current for the turn-off signal, which produces heat and requires additional circuitry.

(The terminology in the VFD literature is not consistent. "Thyristor" is often used to describe an SCR, and vice versa.)

Power transistors provide faster response and more accurate control of current flow than SCR's and GTO's. These characteristics made it practical to develop the pulse-width modulation (PWM) type of drive. The main disadvantages of transistors has been limited voltage and current capacity, which makes it necessary to use more of them, and to use more complex circuits. The newest type of transistor used in VFD's is the "insulated gate bipolar transistor" (IGBT). This type of transistor has faster response than earlier power transistors, which were called "bipolar junction transistors" "Darlington transistors," or simply "power transistors."

In addition to these three actively controlled devices, diodes are widely used in the rectifier. A diode is a simple, passive device that allows current to flow in only one direction. Diodes start to conduct current as soon as any voltage is applied in the forward direction.

■ Variable Voltage Input (VVI) Drives

The variable-voltage input (VVI) type of drive is the simplest in design and concept. Figure 7 is a diagram of a VVI drive.

The rectifier is controlled to vary the DC voltage appropriately for the motor speed and torque requirements. The inverter switches SCR's or thyristors to produce a square-wave voltage waveform. This crude waveform is called "6-step" or "6-pulse," because it is produced by the successive firing in pairs of the six output devices of the inverter.

The filter in a VVI drive consists of a large capacitor and a relatively small inductor. The filtering in the DC circuit isolates the input and output waveforms from each other.

Applying the abrupt voltage changes of a square wave to the motor produces a ragged current waveform in the motor windings. Motor windings are almost entirely inductive, so the response to the abrupt voltage

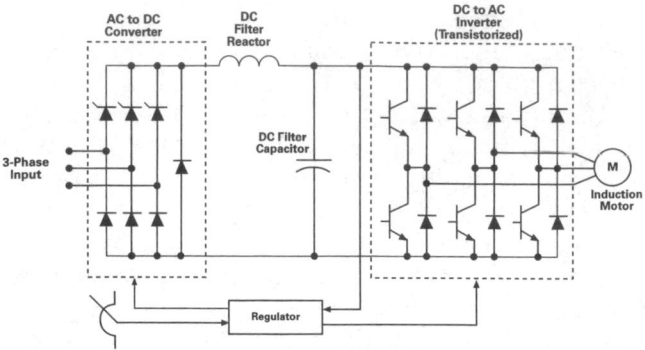

Fig. 7 Variable-voltage-input (VVI) drive This is the simplest type of variable-frequency drive. The output switching devices approximate a sine wave voltage for the motor by a series of square waves at different voltages.

rise is a slow current rise, while the response to the abrupt voltage drop is an "inductive kick" that resists the reduction of current. (In contrast, if a sinusoidal voltage is applied to an inductor, the resulting current is also a sine wave.)

VVI drives yield the lowest overall motor system efficiency and starting torque. They were commonly selected for driving fans and centrifugal pumps, but they are now being displaced by PWM drives.

■ Current Source Input (CSI) Drives

The current-source inverter (CSI) type of drive operates in the same general way as a VVI drive. Figure 8 is a diagram of a CSI drive.

The major difference is that the CSI drive is able to force a square wave of current, rather than voltage, through the motor. This is accomplished by using large inductors in both legs of the DC circuit. Inductors resist change in current, and they are large enough to force a steady flow of current through the motor windings in sequence as the inverter switches the DC circuit from one phase of the motor windings to the next.

■ Pulse Width Modulation (PWM) Drives

In the PWM drive, the inverter uses transistors to switch the direct current at high frequency to deliver a series of voltage pulses to the motor. The width of each pulse is tailored so that the voltage pulses interact with the reactance of the motor windings to produce current flow in the motor that approximates a sine wave. Figure 9 is a diagram of a PWM drive.

Most newer PWM drives also change their output voltage in steps that simulate a sine wave voltage. The filters in the DC circuit isolate the input and output from each other.

The inverter circuit in PWM drives is sometimes called a "chopper." The pulse repetition rate is sometimes called the "carrier frequency." (The latter is a misnomer, but evocative.) In contemporary units, the pulse rate ranges from 1,000 to 12,000 Hertz. The choice of pulse rate is a design compromise between factors discussed below.

The control circuitry of PWM drives is more complex than the control circuitry of VVI and CSI drives. This originally inhibited the development of PWM drives, but no longer. Microprocessors and integrated circuits allow essentially unlimited control complexity without much additional equipment cost. PWM drives were introduced at about the same time as the other two types, and they have been improving steadily.

The basic PWM concept allows more room for improvement than the other two types. Advances in power transistors have improved power handling capacity and the ability to tailor the current waveform in the motor.

PWM drives are not superior in all respects. Their high-frequency voltage pulses create a variety of problems that do not appear in the other two types. Most PWM development has been devoted to minimizing these problems.

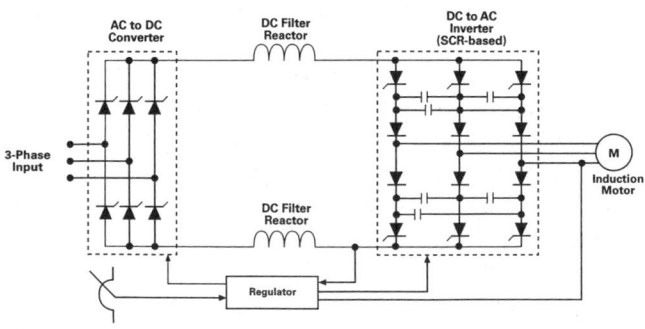

Fig. 8 Current-source-input (CSI) drive The main distinction in this type of drive is that the two inductors in the DC circuit are very large. They are able to force current through the inductance of the motor. Thus, the switching action of the drive simulates a sine wave of current in the motor windings, although the actual waveform is a series of square waves.

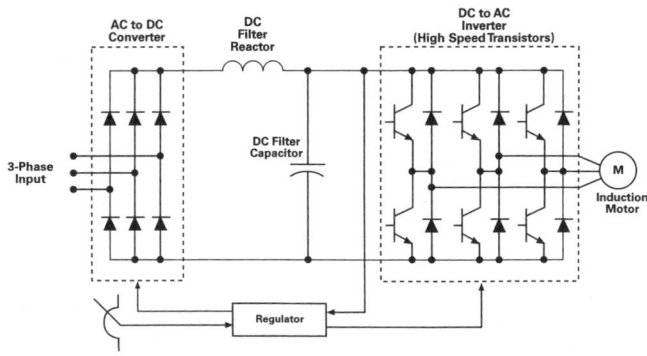

Fig. 9 Pulse width modulation (PWM) drive This drive sends a stream of short voltage pulses to the motor. The voltage pulses react with the inductance of the motor windings to create a sine wave current in the motor that is closer to a true sine wave than is possible with other types of electronic drives.

Waveform Distortion: Introduction

VFD's may create a variety of electrical problems for the motor it drives and for the power supply system that serves it. In other words, a VFD may cause problems in its output and its input.

Motor problems are caused by the fact that the drive does not deliver a true sine wave voltage to the motor. Electric motors are designed to operate with a smooth sine wave voltage input. When such power is provided, the current in the motor windings is also a smooth sine wave. However, if voltage with an irregular waveform is supplied to a motor, the resulting motor current does not follow the voltage waveform. This is because a motor is a huge inductor, and inductors respond differently to different frequency components of the applied voltage. In extreme cases, the motor current waveform may be radically different from that applied voltage waveform.

All VFD's create voltage and current waveforms that are severely different from true sine waves. In turn, this creates very irregular motor current, with a host of resulting problems. Each of the three main types of VFD's has an entirely different set of waveform characteristics. Figure 10 shows the voltage and current waveforms produced by each of the drive types.

Power system problems are caused by the fact that the VFD does not draw current from the system as a sine wave, but in a pattern that is a serious distortion of a sine wave. This distorts the waveform of the power supply system, causing trouble for the system itself and for other equipment that draws power from the system. Figure 11 shows how different models of PWM drives affect the voltage and current waveforms in the power supply system at the input to the drive.

Distortion of the power system waveform is becoming more of a concern because there are more sources of waveform distortion, and because more types of equipment are coming into use that are sensitive to waveform distortion. New sources of waveform distortion include VFD's, electronic ballasts for fluorescent and HID lighting, and computer power supplies. Equipment that is vulnerable to waveform distortion includes electronic equipment that deals with small signals (communications, medical test equipment, etc.), power line carrier communications, transformers, and other induction motors served by the same power system.

■ Input and Output Distortion, Voltage and Current Distortion

When considering the distortion caused by a VFD, be clear about whether the subject is output distortion (to the motor) or input distortion (to the power supply system). Also, be clear about whether the subject is distortion of the voltage waveform or the current waveform. In some cases, the relationship between voltage distortion and current distortion may seem bizarre.

With respect to the motor, the output current waveform is the key issue. The current waveform in the motor windings determines the useful work produced by the motor. It is also a major factor in amount of heat losses.

The waveform of the voltage output supplied to the motor is of interest primarily in its ability to create a desirable current waveform. Contrary to what you might expect, it does not take a sine wave voltage to create a sine wave current in the motor. A reasonably good sine

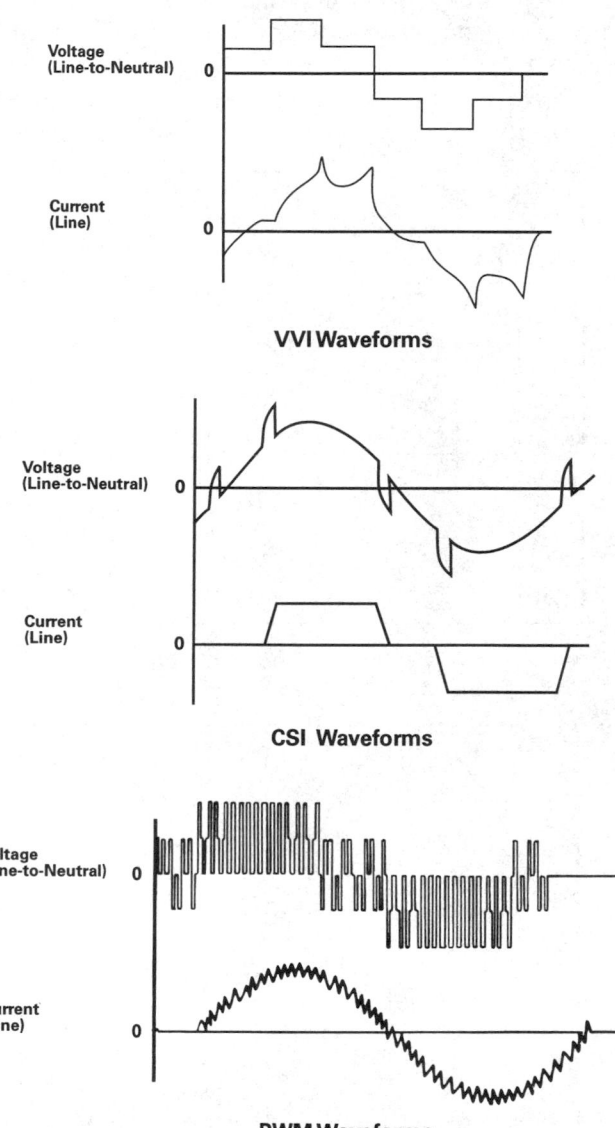

VVI Waveforms

CSI Waveforms

PWM Waveforms

Reliance Electric

Fig. 10　Voltage output waveforms and motor current waveforms　VVI and PWM drives are designed to produce a particular voltage output. The voltage reacts with the inductance of the motor to produce a current waveform. PWM drives produce a current waveform that is much closer to a sine wave. CSI drives use brute force to directly create a square wave of current in the motor.

wave current can be created in the motor windings with a series of voltage pulses, if the timing and amplitude of the pulses are controlled properly. (This works because the motor winding is a big inductor. It accumulates the energy of the incoming pulses in its magnetic field.)

With respect to the power system, distortion of the input voltage waveform is the primary consideration. This distortion causes interference with other equipment served by the same power system, since voltage is what the other equipment sees. Distortion of the input voltage waveform is often called a "power quality" problem. (This term also covers interruptions of service, phase imbalance, and other problems that are not related to VFD's.)

Distortion of the input current waveform can adversely affect the power supply system itself, with effects such as reduced power factor. Also, abrupt distortion of the current waveform causes radiated electrical noise from the power supply system that may interfere with sensitive equipment.

■ Harmonic Analysis of Distortion

A useful fact of mathematics is that any wave shape can be described as a combination of pure sine waves, each of which is an integral multiple of a basic or "fundamental" frequency. Each of these multiples is called a "harmonic." For example, the harmonics of the 60 Hz line frequency are 60 Hz, 120 Hz, 180 Hz, etc. Electrical engineers use this fact to analyze waveform distortion in electrical power systems, which is commonly called "harmonic distortion."

Individual harmonic frequencies or narrow ranges of harmonic frequencies may cause particular problems.

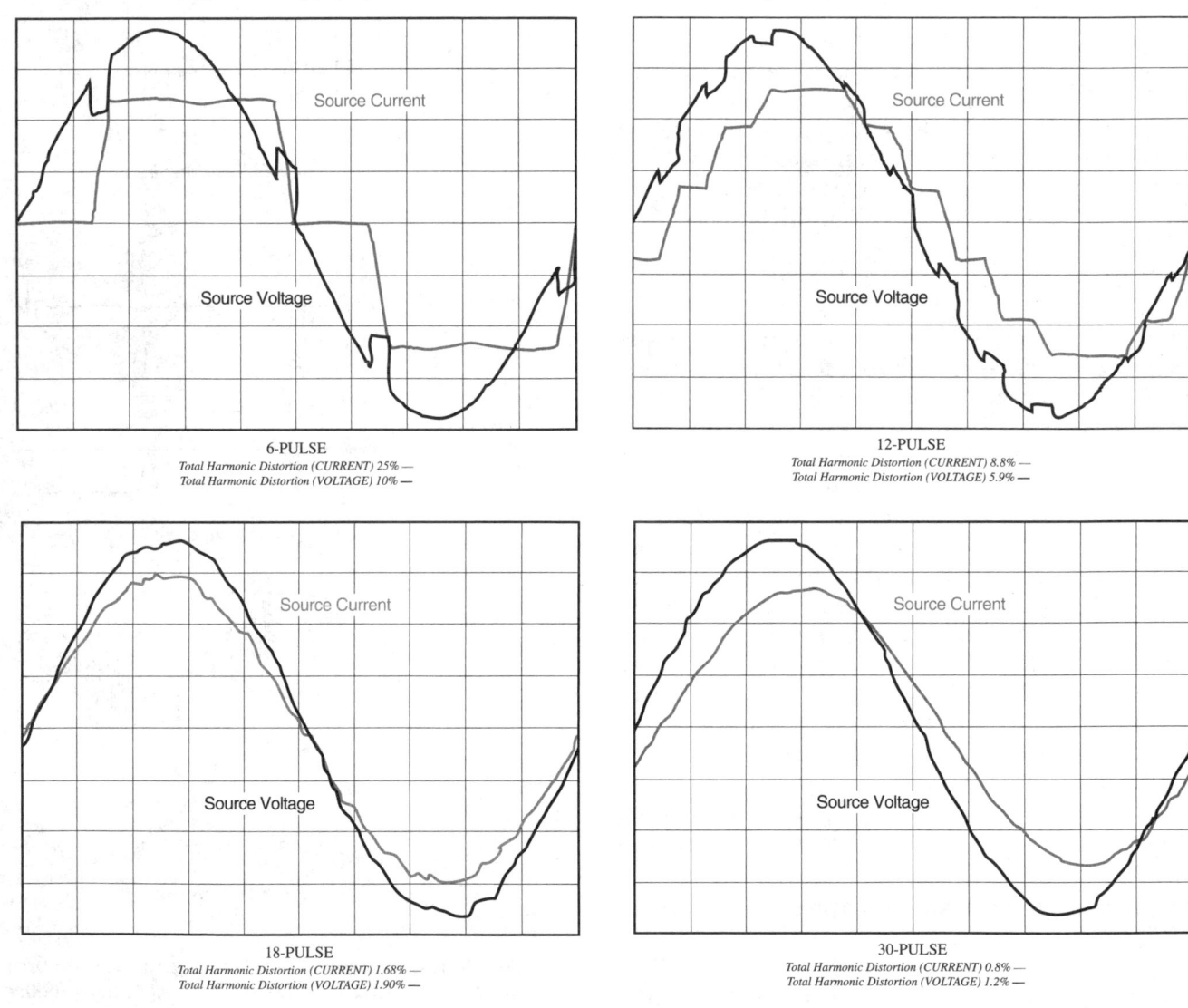

6-PULSE
Total Harmonic Distortion (CURRENT) 25% —
Total Harmonic Distortion (VOLTAGE) 10% —

12-PULSE
Total Harmonic Distortion (CURRENT) 8.8% —
Total Harmonic Distortion (VOLTAGE) 5.9% —

18-PULSE
Total Harmonic Distortion (CURRENT) 1.68% —
Total Harmonic Distortion (VOLTAGE) 1.90% —

30-PULSE
Total Harmonic Distortion (CURRENT) 0.8% —
Total Harmonic Distortion (VOLTAGE) 1.2% —

Robicon

Fig. 11 Input current and voltage distortion   If the drive draws current from the electrical system in a manner that is not proportional to the supply voltage, the voltage of the electrical system will be distorted. In these diagrams, the "source current" is the driving factor, and the "source voltage" is what results. Large, abrupt changes in current draw create the most voltage distortion. These curves are at the input of four different models of PWM drives. The characteristics of the electrical system, especially its "impedance," determine the amount of voltage distortion that appears in other parts of the electrical system.

However, in most cases, the concern is "total harmonic distortion." As the name implies, this is the total waveform distortion caused by all harmonics. This is calculated as the percentage of the total harmonic current compared to the current at the fundamental frequency. "Distortion factor" is another way of measuring total harmonic distortion, expressed as the percentage of the fundamental frequency current compared to the total current.

Any distortion of the basic sine wave could in principle be called "harmonic distortion." However, this term is usually reserved for types of distortion that make a large change in the shape of the waveform. Other types of distortion, such as brief current spikes, may cause serious trouble even though they contribute little to total harmonic distortion.

■ Effect of Power System Characteristics on Waveform Distortion

The characteristics of the power system itself determines how severely a VFD may cause distortion. Relative size is important. Any given VFD produces more voltage distortion if it takes up a larger fraction of system capacity.

Another factor is the "impedance" of the transformer. A VFD produces less voltage distortion if the transformer feeding the system has a low impedance. The impedance of a transformer measures the sensitivity of the transformer's voltage output to its current output. (Impedance is a general electrical term that means a ratio of voltage to current.) Transformers may be rated in terms of "percent impedance," which is defined as the drop in transformer output voltage that occurs in going from no load to the rated transformer load.

For example, if a 460-volt system is served by a transformer that has a 4% impedance, the power system voltage drops about 18 volts at full load, compared to the voltage at no load. If a VFD draws a current pulse equal to half the system capacity, it will lower the system voltage by about 9 volts during the pulse.

Also important is the impedance of the power wiring. In contrast with transformer impedance, this is especially important with abrupt or high-frequency distortion. This is because the inductance and capacitance of the wiring itself can have a filtering or absorbing effect on high-frequency distortion components.

Figure 12 shows large differences in voltage and current input distortion among various drive models. Also, Figure 12 shows that the electrical characteristics of the power system have an enormous effect on the amount of distortion that appears in the power system.

■ How to Calculate Harmonic Distortion in Future VFD Installations

Some VFD manufacturers offer to perform a computer calculation of the harmonic distortion that may be expected from a VFD installation. As input, the program needs the electrical characteristics of the VFD, the motor, and the transformer. It also needs the length and diameter of the wiring between the transformer and the drive, and between the drive and the motor.

This calculation is limited to predicting harmonic distortion. It cannot predict all potential problems, such as electromagnetic interference and motor overheating.

■ How to Measure Distortion in Existing Systems

You can measure and analyze the distortion in an existing VFD installation. A variety of equipment is available for this purpose. Some distortion measuring equipment is expensive, but you can rent it.

Simple handheld analyzers measure total harmonic distortion, along with measurements of basic variables, such as voltage, current, and reactive power (KVAR's). These meters are limited to dealing with problems related to total harmonic distortion.

A harmonic distortion analyzer is a powerful tool that plots a graph of the harmonics that are present. It can indicate the percentage of power represented by each harmonic, making it possible to diagnose the likely cause of problems.

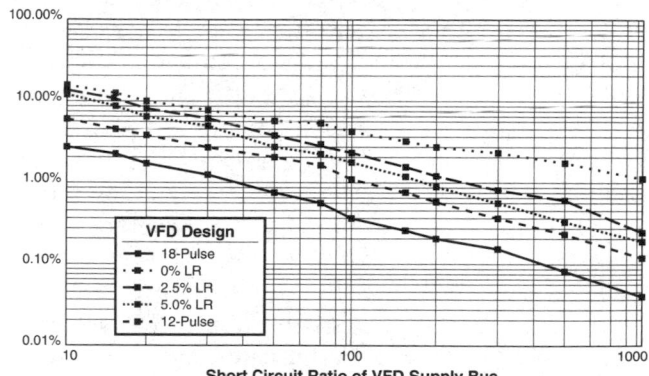

INPUT VOLTAGE TOTAL HARMONIC DISTORTION

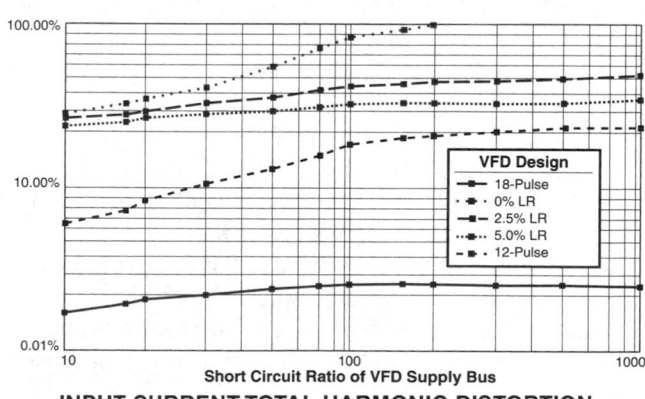

INPUT CURRENT TOTAL HARMONIC DISTORTION

Robicon

Fig. 12 Voltage and current distortion in the electrical power system caused by variable-frequency drives These curves tell us that voltage distortion behaves differently from current distortion. They also tell us that the design of the drive and the characteristics of the electrical system have a large effect on the amount of distortion.

To pinpoint problems related to specific harmonics or transients, especially if they do not make up a large part of total distortion, you may need to use an oscilloscope. This displays the actual voltage and current waveforms.

Take measurements at various points to see how distortion varies throughout the power supply system. The worst distortion appears at the input to the VFD, unless there is a resonance effect in the system (discussed below). Check the distortion at the input to any equipment that may be affected by the operation of the VFD.

Unless you want to become an expert in finding distortion bugs, call in the electric utility or hire a consultant to do the analysis. If you plan to involve the utility, make a commitment to solving any power quality problem that affects their lines. Otherwise, if you attract the utility's attention to a power quality problem that you are causing, they may respond with a rate penalty.

■ Distortion Standards

How much distortion is acceptable in a VFD system? IEEE Standard 519 is intended to answer this question. Standard 519, which is updated occasionally, sets limits for the two worst forms of voltage distortion caused by VFD's. These are harmonic distortion and notching.

IEEE Standard 519 sets a limit on total harmonic distortion based on the type of facility. The present version limits total harmonic distortion to 3% for facilities having sensitive equipment, such as office buildings and hospitals. A limit of 5% is set for general industrial facilities. A limit of 10% is set for specialized industrial systems that can tolerate high distortion.

IEEE Standard 519 also sets a limit on "notching," which is discussed below. As the name implies, notching is a brief, severe cut in voltage that typically occurs several times per cycle. The Standard limits the "area" of the notch, which is the voltage times the duration, expressed in volt-milliseconds.

Unlike most standards, Standard 519 cannot be applied to a particular VFD in isolation. This is because the severity of waveform distortion depends on the characteristics of the power system as well as the characteristics of the VFD. Manufacturers of VFD's may claim that their drives satisfy the requirements of Standard 519, but you have to read the fine print to find out what assumptions are made about the system characteristics. In many cases, VFD's that satisfy Standard 519 will not cause trouble, but the Standard does not guarantee this.

IEEE Standard 519 requires that distortion be measured at the "point of common connection" in the power system. For practical purposes, this usually means the secondary of the transformer. This is not a conservative measurement. A VFD is likely to cause more distortion at points in the power system that are closer to the VFD.

Types of Distortion Caused by VFD's

Distortion is produced by the rectifier and by the inverter. Distortion produced by the rectifier is a problem primarily as distortion fed back to the power supply system. Distortion produced by the inverter is strongest in the direction of the motor and the motor circuit.

In PWM and VVI drives, the filter between the rectifier and inverter tends to prevent rectifier distortion from traveling to the motor, and to prevent inverter distortion from traveling back to the input. In CSI drives, especially those using SCR's in the rectifier, distortion caused by inverter operation can travel back to the power input. These are generalizations. Variations in drive design may greatly reduce the types of distortion produced by the basic types of drives, or they may introduce unexpected modes of distortion.

There is no standard way of classifying the types of distortion produced by VFD's. As mentioned previously, almost any distortion caused by a VFD could be considered as harmonic distortion. However, this is too general to be useful. The following breakdown distinguishes the types of distortion produced by different types of drives, and it provides a basis for the subsequent discussion of problems and solutions.

■ Harmonics of the Line Frequency

Harmonic distortion of the input power waveform occurs when the drive draws current at a rate that is not proportional to the applied voltage. In PWM and VVI drives, input harmonic distortion is caused primarily by the action of the rectifier. In CSI drives, input harmonic distortion may also be caused by the action of the inverter and the inductors of the DC circuit.

The rectifier acts in synchronism with the 3-phase input current, so it produces harmonics of the input frequency. It turns out that rectifier action produces only odd harmonics that are not divisible by three, i.e., the 5th, 7th, 11th, etc. For example, with 60-Hz power, these harmonics are 300 Hz, 420 Hz, 660 Hz, etc.

The harmonics decline in magnitude with higher number. Therefore, this type of harmonic distortion is a low-frequency problem. Fortunately, rectifiers do not produce much third-harmonic distortion. Third harmonics are especially troublesome in 3-phase power distribution because they create large return currents in the neutral leg.

The basic rectifier used by most VFD's consists of a simple full-wave bridge of six identical components. The components may be diodes, SCR's, or thyristors. The rectifier produces substantial harmonic distortion with any of these.

With a diode rectifier, each phase connection of the power system "sees through" the diode to the DC section during each half-cycle of conduction. During the conduction cycle, each phase sees two current pulses as the other two phases apply voltage to the DC section.

These current pulses distort the input voltage waveform to an extent that is determined by the load on the drive and the impedance of the power system. The total harmonic distortion of the current waveform is usually much larger than the distortion of the voltage waveform in percentage terms, but this does not imply that the effects of voltage distortion are less troublesome.

With an SCR or thyristor rectifier, conduction is delayed as load is reduced. This forces current flow even further out of synchronism with the applied voltage, further increasing harmonics. Most PWM drives use diode rectifiers. Some CSI and VVI drives also use diodes to reduce harmonic distortion and notching.

Total harmonic distortion can be reduced greatly by operating the rectifier in a sequence of steps. The so-called "12-step" drive, for example, actually has three input steps per half-cycle of line voltage. Some of the newest PWM drives have reduced harmonic distortion to low levels with a larger number of voltage steps.

■ **Harmonics of the Motor Frequency**

Harmonic distortion of the output waveform of a VFD is caused primarily by the abrupt switching action of the inverter. This occurs in synchronism with the motor speed, so it produces harmonics of the motor frequency. Therefore, this type of harmonic distortion is a low-frequency problem.

Large, low-frequency harmonics occur only in CSI and VVI drives, which deliver current and voltage, respectively, in large pulses. In contrast, PWM drives deliver voltage to the motor in small pulses, and they have high-frequency harmonic problems discussed next.

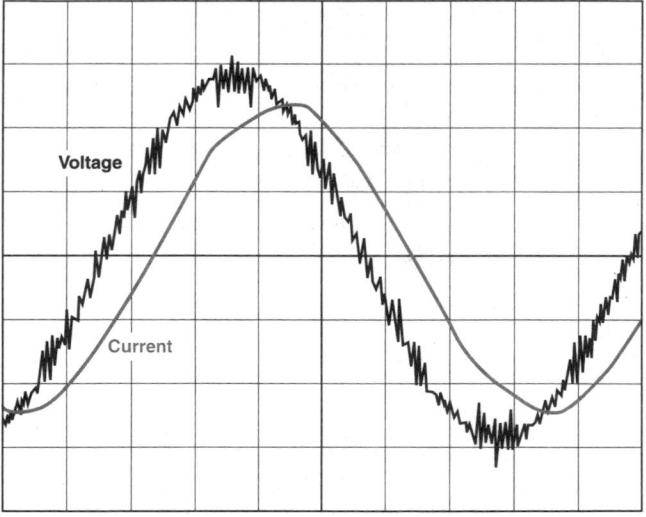

Fig. 13 Phase shift between input current and voltage This occurs with some variable-frequency drives because of delay in the action of the switching devices. The effect is lower power factor in the electrical system.

In CSI drives, depending on design details, the motor-frequency harmonics may be able to feed back to the input, increasing the total harmonic distortion of the input power.

■ **Harmonics of the PWM Inverter Frequency**

The inverters of contemporary PWM drives produce voltage pulses at fixed rates, which range from 1,000 to 12,000 Hertz in different models. The stream of pulses may produce significant harmonics at frequencies up to 100 times the pulse rate.

The pulse rate (or "carrier frequency") is a design compromise. It was soon discovered that the low pulse rates of the original PWM drives create a high level of audible noise in the motors. To reduce this problem, the pulse rate was increased to the high end of the audible range, so that all harmonics would occur above the audible range. Unfortunately, under adverse circumstances, these frequencies have enough higher harmonics to produce a substantial amount of radiated electrical noise, and they may produce standing waves in the motor cables. In addition, PWM harmonics are associated with "voltage stress," a poorly understood phenomenon that causes breakdown of motor insulation.

■ **SCR and Thyristor Rectifier Notching**

"Notching" is a deep, transient reduction of the power system input voltage, so called because it appears as a notch in the voltage waveform. Notching is caused by the action of SCR and thyristor rectifiers. These devices may not be switched on until there is a large voltage across them. Thus, the power system is subjected to a virtual short circuit during the brief period when the voltage in the filter section builds up to match the input voltage.

Notching could be considered a source of harmonic distortion, but it is treated separately in IEEE Standard 519. Notching pulses have very short duration, so they do not contribute much to total harmonic distortion, but their magnitude makes them potentially troublesome.

■ **Voltage-Current Phase Shift**

If the rectifier uses SCR's or thyristors, current draw from the power line is reduced as the motor load falls by delaying the turn-on of the SCR's or thyristors. This causes the average current input to lag the input voltage. Figure 13 shows this effect.

This phase shift is completely different from the lagging power factor of inductive devices, such as motors and fluorescent ballasts. However, it has the same effect of reducing the power factor of the drive system. Power factor is important in some applications, but not in others, as discussed below.

Diode rectifiers do not produce substantial phase shift between the input current and input voltage. This is because diodes draw current as soon as the voltage is applied to them.

Table 2. WAVEFORM-RELATED PROBLEMS AND SOLUTIONS IN VARIABLE-FREQUENCY DRIVES

Potential Problem	Cause	Solutions
MOTOR SYSTEM PROBLEMS		
▪ motor overheating and energy waste	▪ low-frequency harmonics caused by crude current waveform	▪ select PWM drive (1) ▪ select many-step inverter
	▪ high-frequency harmonics produced by PWM inverters	▪ select drive to avoid problem (1) ▪ select "inverter duty" motor ▪ select oversized (derated) motor (2)
▪ electric resonance in drive-to-motor cable • insulation failure • blown fuses	▪ high-frequency harmonics of PWM inverter pulses	▪ limit length of motor cable ▪ change cable length to avoid resonance ▪ select PWM drive with lower chopper frequency (3) ▪ avoid PWM inverter (1)
▪ motor insulation voltage stress	▪ (poorly documented) voltage spikes from PWM inverters and perhaps other types	▪ select "inverter duty" motors ▪ select drive to avoid problem (1)
▪ cogging	▪ large current steps in CSI and VVI drives	▪ select PWM inverter (1) ▪ select many-step inverter ▪ keep motor speed high enough to avoid cogging (typically above 10% speed)
▪ mechanical resonances	▪ steps or distortion in motor current waveform, especially in CSI and VVI drives	▪ select PWM inverter (1) ▪ select many-step inverter ▪ install resonance-absorbing coupling between motor and load ▪ modify driven equipment to avoid mechanical resonances
POWER SYSTEM PROBLEMS		
▪ effects of power system waveform distortion • interference with test equipment, computers, communications gear, etc. • interference with power line carrier communications • overheating and energy waste in transformers, lighting ballasts, other motors connected to the power system • accelerated reading of power meters	▪ line-frequency harmonics caused by rectifier	▪ select drive with many-step diode rectifier
	▪ notching caused by commutation of SCR rectifiers	▪ avoid SCR rectifier ▪ install input filter or isolation transformer ▪ electrically isolate affected equipment
	▪ voltage spikes induced by large current steps in CSI and VVI drives	▪ select drive to avoid problem ▪ install input filter or isolation transformer ▪ electrically isolate affected equipment
	▪ high-frequency harmonics caused by PWM inverter	▪ select drive to avoid problem (1) ▪ install tuned filter in input ▪ install input filter or isolation transformer ▪ electrically isolate affected equipment
▪ radiated electric noise	▪ high-frequency harmonics of PWM inverter pulses, mostly in motor cable	▪ select drive to avoid problem (1) ▪ select PWM drive with lower chopper frequency (3) ▪ limit length of motor cable ▪ install drive and motor cable far from affected equipment ▪ improve grounding of affected equipment ▪ install input filter or isolation transformer
	▪ notching caused by commutation of SCR rectifiers	▪ avoid SCR rectifier ▪ install input filter or isolation transformer ▪ improve grounding of affected equipment
	▪ current spikes in input and output of CSI and VVI drives	▪ select PWM inverter (1) ▪ select many-step inverter ▪ install output filter ▪ install input filter ▪ improve grounding of affected equipment
▪ electric resonance in power supply system • overvoltage damage to connected equipment • wiring insulation failure • blown fuses • blown power factor correction capacitors	▪ triggering of resonance between power factor correction capacitors and system inductance, usually by input harmonics of CSI and VVI drives	▪ select drive with low input harmonics ▪ install input filter ▪ eliminate capacitors served by same transformer (4)
	▪ harmonics of PWM carrier frequency	▪ select drive to avoid problem (1) ▪ install input filter
▪ reduced power factor (5)	▪ voltage-current phase shift caused by SCR and thyristor rectifiers	▪ select VFD having diode rectifier
AUDIBLE NOISE PROBLEMS		
▪ motor noise	▪ high-frequency voltage pulses of PWM inverters	▪ avoid PWM inverter (1) ▪ select PWM inverter with higher chopper frequency ▪ select motor for low noise ▪ acoustically isolate motor
▪ drive noise	▪ vibration of components, especially large coils	▪ select drive for low noise ▪ acoustically isolate drive
▪ filter noise	▪ coil vibration	▪ select drive to avoid need for external filters ▪ acoustically isolate filter

Notes:
1. The best PWM inverters have little input or output harmonic distortion, but they still have potential EMI, audible noise, and "voltage stress" problems.
2. Oversized motors operate with reduced efficiency at higher partial loads. Effect of oversizing motors with VFD's is not well documented.
3. However, higher chopper frequency is desirable to reduce audible motor noise.
4. This leaves the problem of improving power factor by other means, which may be difficult.
5. Power factor drops with motor speed. With centrifugal loads, current and power consumption drop with motor speed, reducing adverse effect.

Waveform-Related Problems and Solutions

The various types of waveform distortion create a variety of problems. Enough experience has been gained to predict where these problems may arise. After an initial period of denial during the 1980's, the VFD industry has become aggressive about reducing waveform distortion and its effects. The development of standards for waveform distortion, especially IEEE Standard 519, has attracted the attention of designers and users to the problem. As a result, some manufacturers have model lines specifically designed for low distortion.

Table 2 summarizes the problems caused by VFD waveform distortion, along with their causes and solutions. Use the Table as a checklist to avoid or minimize the problems that are discussed.

The number of potential problems may appear daunting, but this does not mean that VFD's are a rat's nest of troubles. Not all problems arise in all applications. However, some methods used to correct a particular waveform-related problem may cause a different problem. You can usually avoid problems, or keep them within acceptable limits, by knowing how to select the right equipment for the application, and by being aware of appropriate installation practices. At the same time, avoid the cost of features that you do not need.

■ Motor Overheating and Energy Waste

Motors are designed to operate with pure sine wave voltages at the line frequency. A large part of the output current harmonics created by VFD's cannot produce useful motor power. The unusable energy of the harmonics is converted to heat, which is a waste of energy and raises the temperature of the motor.

The tendency of harmonics to raise motor temperature depends on the application. In variable-torque applications, the reduced motor current at low speed more than compensates for the heat added by harmonics. However, this saves the motor only if it operates at substantially reduced speed most of the time. In "constant-torque" and "constant-power" applications, where motor current remains high at reduced speeds,

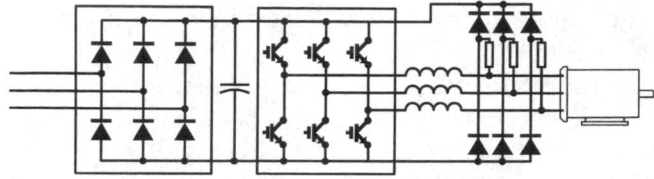

Fig. 15 Output voltage clipper This is a more elaborate method of reducing high-voltage transients in the motor.

heat generated by harmonics is a serious problem at all speeds.

The ideal solution to the problem of harmonics is to select a drive that produces a current waveform in the motor that approaches a true sine wave. Some of the newer PWM drives are capable of this. In retrofit applications, you can afford to pay a higher price for a better VFD if this avoids the need to replace the motor. (However, replacing the existing motor may be desirable as a separate energy conservation measure. See Measure 10.1.1 for details.)

A less satisfactory solution is to install a motor that is designed to use more of the energy of harmonics. Such motors are named "inverter duty" motors. This classification is not the same as "high efficiency" motors, although inverter-duty motors generally have higher efficiency than average. The advantages of inverter-duty motors are not well documented. There is reason to suspect that they perform better with the low-frequency harmonics of CSI and VVI drives, but not much better with the high-frequency harmonics of PWM drives.

"High-efficiency" motors can use more of the low-frequency harmonics because they are designed to reduce the eddy current losses that convert these harmonics into heat. On the other hand, there have been reports of failures of "high-efficiency" motors when used with VFD's. Undoubtedly, some of these failures are due to the motors being used in higher-torque applications for which they were not rated.

■ Motor Insulation Failure with PWM Drives

Motors driven by PWM drives have been subject to a type of insulation failure that has been called "voltage stress." It is widely assumed that this occurs because the high chopper frequencies induce voltage spikes in the motor coils high enough to cause insulation failure.

One solution to this problem is to install an inductive filter between the drive and the motor to reduce the voltage spikes. Figure 14 shows this approach. However, this is a delicate balance, because PWM drives must send pulses to the motor to function properly, and too much filtering interferes with the pulses.

Another solution is to install a voltage limiter between the drive and the motor, as show in Figure 15. This eliminates any voltage at the motor input that exceeds a particular level. However, the large inductance

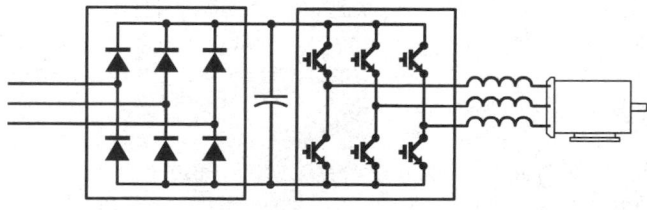

Fig. 14 Output chokes Inductors between the drive and the motor may be useful for reducing several problems, especially high-voltage transients that destroy motor insulation. However, they may also interfere with the efficiency of PWM drives, the type that has the worst problem with transients.

within the motor may create voltage spikes farther inside the motor.

You could minimize this problem by selecting motors that have insulation of high quality. However, you usually cannot judge insulation quality from the motor catalog information. Motors with higher temperature ratings may have insulation that is able to withstand higher voltages, but this is an indirect relationship.

Another possibility is that insulation fails from continuous hammering of the motor's coils by audible vibrations, for which PWM drives are notorious. However, this possibility has not been well investigated.

■ Mechanical Resonances

Especially with CSI and VVI drives, the abrupt steps in motor current produced by the drive cause torque pulses in the motor. If the motor vibrations occur at a resonant frequency of the equipment, they may drive mechanical resonances large enough to make noise, create noticeable vibration, or break equipment.

The most satisfactory solution is to select a drive that has a relatively smooth current waveform. Another solution is to install a vibration-absorbing coupling between the motor and the equipment. Still another is to modify the equipment so that it has no resonances that are triggered by the drive vibration.

■ Cogging in CSI and VVI Drives

"Cogging" is a form of torsional vibration that occurs at very low speed with CSI and VVI drives, typically below about 10% of full speed. This is generally not a resonance phenomenon, but direct hammering that is caused by the large current steps that occur with these drives. Cogging can be avoided completely by selecting PWM drives. With CSI drives, select a model that has stepped output current. Another solution is to set the low-speed limit of the motor above the speed at which cogging becomes noticeable, if the application allows this.

■ Audible Motor Noise with PWM Drives

Some PWM drives may make the motor noisy. The noise occurs because the frequency of the voltage pulses is high enough to induce vibration of the steel laminations that form the magnetic circuit inside the motor. Also, the coils may vibrate. These components are designed to avoid vibration at normal line frequency, but they may resonate easily at the pulse frequencies.

Any possible resonance is likely to be excited because the pulse frequency spectrum is broad. It includes a distribution of harmonics and sub-harmonics of the chopper frequency. Furthermore, the distribution of harmonics changes continuously as the pulse widths are modulated to create the current wave.

The noise problem in PWM drives cannot be solved by using filters between the drive and the motor, because a filter would interfere with transmission of the pulses needed to produce the current waveform. The IGBT type of transistor, which is now common in PWM drives, is claimed to reduce audible noise.

Some manufacturers increase the inverter pulse rate to reduce the noise problem. This produces noise in a less sensitive portion of the range of human hearing, or at frequencies too high to be heard. Also, the motor components may be less excitable at the higher frequencies. This solution is reported to greatly reduce audible noise, but it may create a problem of electric noise, as discussed below.

If motor noise increases substantially at a particular speed, it may be possible to lock out this range of speed. However, this is a clumsy solution. If you plan to replace the motor, shop for a motor that is designed to remain quiet when used with your type of VFD.

Motor noise may be spread through the equipment in which the motor is installed. For example, if a fan motor is located inside an air handling unit, motor noise is transmitted by the ducts.

■ Audible Drive and Filter Noise

The VFD itself may produce audible noise at a level that is objectionable in quiet environments. The large coils used in the DC circuit of CSI are driven to hum by the large currents they carry. Also, PWM and VVI drives may have filter coils that may hum. Large inductors, or coils, may be installed externally to block electrical noise, and these may hum audibly.

The solution is careful selection of equipment. Also, you can isolate noisy equipment by enclosing it or by installing it away from sensitive locations.

■ Conducted Electrical Noise

Another nasty surprise of early VFD's was interference with other electrical equipment served by the same power system. VFD's may backfeed a large amount of electrical noise into the power supply system. This noise may interfere with computers, test equipment, medical instruments, communications equipment, and other electronic equipment involving small voltages. The electrical noise of VFD's is especially crippling to power line carrier communication systems, which use the power wiring of the facility to carry digital or analog signals. This problem was eventually forced into the open, leading to major improvements, but there is still a wide variation in the electrical noise generated by different models.

Most electrical noise is caused by transients and by high-frequency harmonics, rather than by the low-frequency harmonics that are responsible for most of a drive's rated "total harmonic distortion." "Notching" is generally the worst source of electrical noise from VFD's that use SCR's and thyristors, especially in the rectifier. With some PWM drives, the carrier frequency and its harmonics may backfeed to the power system.

VFD's may also produce starting surges, like any other large piece of equipment. However, the effect of VFD's is usually to mute starting surges, compared to operating the motor across the line. Further, some VFD's have a "soft start" that further reduces the starting surge.

You can avoid interference problems by selecting a VFD that does not produce enough noise to cause trouble. As an alternative, install power supply filters at the equipment that is vulnerable to interference.

Or, if the entire power circuit is vulnerable to electrical noise, it is probably cheaper to install a filter at the input to the VFD. Figure 16 shows this approach. Filtering can be expensive because the filter on each drive must handle the entire motor current.

Select the appropriate type of filter for the kind of noise being blocked. In many cases, a simple inductor ("reactor") in the power line suffices. Another option is an isolation transformer, but this is more expensive and may not be significantly better. An isolation transformer is bulky, typically larger than the VFD itself, and it absorbs a few percent of the electrical power. In some cases, a filter that is tuned to an isolated objectionable frequency may be appropriate.

Don't bother trying to use surge suppressors of the kind that are used to protect computers and other equipment against lightning strikes. They will provide no benefit. These devices are intended to shunt high-voltage spikes to ground, whereas the noise signals from VFD's are a fraction of the line voltage.

■ Radiated Electrical Noise

Electrical noise from VFD's may be radiated through space, causing interference with equipment that absorbs the radiation. This type of noise transmission is called "electromagnetic interference" (EMI). EMI can occur at any frequency, from individual pulses to continuous high-frequency interference.

If the frequency of the EMI is high enough, it may be called "radio-frequency interference" (RFI). PWM drives may radiate noise at the low end of the radio frequency band, up to several hundred kilohertz or even up to several megahertz. Although the pulse rate of the drive is only several kilohertz, the abrupt nature of the pulses creates high-order harmonics that may affect sensitive equipment at much higher frequencies.

EMI generally has the same causes as conducted noise, and the same types of equipment are affected. The differences are in the manner of transmission and the preventive measures. EMI spreads by radiation from cables that carry drive input or output current. The cables act as a transmitting antenna. In order for interference to occur, the affected equipment must have unshielded components, such as test leads, that act as a receiving antenna.

The seriousness of EMI depends on the drive installation and the nature of the facility, as well as the drive itself. A problem can arise only if there is equipment in the facility that is sensitive to the EMI produced by the drive. Interference is unlikely to occur if the equipment is far from the drive or affected portions of the power system.

There are several solutions to EMI. One is to select a drive that does not cause the problem. Another is to install filters between the drive and the cables. If necessary, use strict grounding and shielding procedures with the drive, the sensitive equipment, or both.

In the United States, devices that produce RFI are subject to regulation and classification by the Federal Communications Commission. There are two main classes of RFI sources, Class A and Class B. Class A devices are considered satisfactory for general industrial environments, while Class B is a more stringent standard intended for commercial-type environments. Neither classification may be satisfactory for specialized environments that are sensitive to RFI, such as laboratories or communications centers. The threshold frequency for regulation is 10 KHz. PWM drives generally are subject to this regulation, whereas CSI and VVI drives generally are not.

■ Resonance in the Drive-to-Motor Connection with PWM Drives

Under some conditions, PWM drives may create standing waves of voltage and current in the cable that connects the drive to the motor. Pulses produced by the VFD are reflected back into the cable at the motor, and again at the drive. If the cable is a resonant length, standing waves will occur in the cable. This means that each new pulse overlaps and reinforces the reflections of earlier pulses.

At the points in the cable where voltage peaks occur, the voltage may be high enough to break through the insulation. At the points in the cable where the current peaks occur, the current may be high enough to cause localized heating, including blown fuses.

Resonance cannot occur unless the cable is long enough to allow overlap between pulses. The minimum length for resonance depends on the pulse rate of the drive. Higher chopper frequencies allow resonance to occur in shorter cables. Manufacturers of PWM drives recommend a maximum cable length of about 500 feet (160 meters) for pulse rates of several kilohertz and a

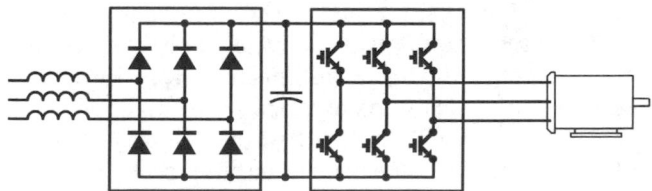

Graham Company

Fig. 16 Input line filter These are large inductors that reduce the spread of electrical noise from the drive into the electrical power system. Inductors can also be used to reduce harmonic distortion.

maximum of cable length of about 200 feet (60 meters) for pulse rates of about 10 kilohertz.

If the motor cable may be long enough to allow a resonance problem to occur, use a drive with a low pulse rate. If a long cable is necessary, it may be possible to avoid resonance by carefully selecting the cable length to avoid any of the specific lengths that allow resonance to occur. (Like an antenna, the cable length must be "tuned" to a multiple of the wavelength to create standing waves. Any length that lies between these multiples will not resonate.) In the extreme case, avoid using a PWM drive.

Don't try to avoid resonance by filtering the output pulses of a PWM drive. This would prevent the motor current waveform from developing properly in the motor.

■ Resonance and Blown Fuses in the Facility Power System

Resonance in the power system may occur at a relatively low harmonic frequency if power factor correction capacitors are installed in the system. This problem is most likely to occur with CSI and VVI drives. The capacitors and the inductance in the system form a tuned circuit that may resonate at one of the low harmonic frequencies produced by such drives. Resonance causes large standing waves of voltage and current to develop. It is not likely that this problem will arise, but the consequences may be severe if it does.

If power factor correction capacitors are a cause of the problem, it may be possible to move them to a different part of the power supply system. Power factor is important primarily to the power company, so it does not matter where the capacitors are installed within the facility. There should be no capacitors on the same side of the transformer that serves the drive, unless the drive has very low harmonic distortion.

If it appears that power factor capacitors are blowing fuses, resonance may not be the cause. The reactance of capacitors is inversely proportional to frequency, so the current at harmonic frequencies passes through the capacitors more easily than the line frequency current. If there is a large amount of harmonic current distortion, the additional harmonic current flow through the capacitors may blow the fuses.

Some PWM drives may create standing waves of voltage and current in the power system wiring, as they may do in the motor cables. The high peak voltages may harm other equipment connected to the same power system, and the high peak currents may overheat wiring and blow fuses. Fortunately, this problem is easier to avoid than resonance in the motor cable. Any chopper pulses are much smaller at the input end of the drive, since they must leak through the rectifier and filter section of the drive to get to the input wiring. The pulses have a much higher frequency than the line frequency, so it is relatively easy to suppress them with a filter

installed at the input end of the drive. Some drive models have such a filter built into the unit.

■ Overheating and Energy Waste in Other Inductive Equipment

Distortion of the power supply system waveform by VFD's causes overheating and energy waste in other inductive devices connected to the system. Affected equipment includes transformers, other motors, and magnetic lighting ballasts. This problem occurs because the harmonics cannot be used by the equipment, and are converted to heat. The general solution to this problem is to select a drive that produces a low value of total harmonic distortion.

Input filters can be used to reduce high-frequency harmonics that cause "skin effect" in wiring and in the coils of affected equipment. Skin effect is an interaction between the current in the wire and its own alternating magnetic field that forces the current toward the outer surface of the wire. Skin effect reduces the effective area of the wire, which increases resistance losses.

■ Accelerated Readings of Induction Power Meters

It is reported that harmonics injected into the power supply system by VFD's can cause electricity meters to run faster. This causes the customer to donate money to the utility.

Application-Related Selection Considerations

Now that we know how to avoid problems caused by waveform distortion, we have a firm basis for understanding drive choices that relate to specific applications. As VFD's have evolved, more features have become standard, or at least common. Nowadays, you rarely need to order a custom VFD. The key to a satisfactory and economical installation is knowing what is available and selecting the appropriate characteristics. You can get information about the following features from the catalogs. If the information is too scanty, call the manufacturer for details.

■ Control Input Provisions

At the outset, limit your search to those manufacturers who offer VFD's having the control capabilities needed for your application. For example, some manufacturers offer only models with manual speed control. For most energy conservation applications, the drive should be equipped to accept standard pneumatic, electric, or electronic control signals, depending on the type of control system that you need or prefer. If your system uses power line carrier communications, select a model that can understand the signal protocol used by your carrier system.

■ Torque and Current Characteristics

Torque is the turning force produced by the motor. Match the torque characteristics of the drive to the torque characteristics of the load, as discussed previously.

Torque is directly related to the motor input current. Therefore, the torque characteristics of a VFD are essentially equivalent to its current-producing characteristics. The maximum torque that a VFD can produce is limited by its current rating.

All three types of VSD's are capable of producing high torque, but torque ratings differ among models intended for different applications. Recall that the inrush current of a motor connected across the line is several times higher than its full-load current. In an application where the motor needs a high starting torque, the drive must be able to deliver much more starting current than the full-load current. For this reason, drives are rated in terms of percentages of motor power. The ratings may be much higher than 100% to provide for starting. Operation at currents higher than "100%" is usually limited to a short period of time.

In variable-torque applications, starting torque is needed primarily to overcome the inertia of the load. With loads having high inertia, such as centrifugal fans, the VFD can operate with a modest torque capability because it can accelerate the load over a period of time, as discussed below.

Many constant-torque applications are jerky, i.e., subject to brief overloads. For such applications, the drive should be rated to provide adequate overcurrent for brief periods of time. A cheaper alternative is a drive that is designed to slow down under brief overloads without tripping out.

■ Current Limits

The maximum output current is usually adjustable. This feature is intended primarily to protect the motor from excessive current. The catalog rating of the drive is usually defined as "100%" current. Drives for variable-torque applications typically have a maximum current limit of about 110%, while drives for constant-torque applications typically have a current limit around 150%.

■ Voltage-Speed Ratio Adjustment

The drive reduces the voltage to the motor as the speed is reduced. This keeps the drive from over-exciting the magnetic circuit of the motor, which would waste energy. Most modern drives provide for a variety of voltage-vs.-speed characteristics. This adapts the drive efficiently to applications with different torque requirements. The voltage-speed characteristic is usually set at the time of installation, and is not changed thereafter.

A fixed ratio of voltage to speed may not provide enough voltage at low speed to overcome resistance losses in the motor windings, resulting in inadequate torque. To compensate for this, the drive increases the voltage-speed ratio at slower speeds. This feature may be called "IR compensation," "current boost," or "voltage boost."

■ Soft Start

Any VFD provides soft starting, compared to starting the motor across the line. This is simply because the drive cannot produce a starting current as strong as the across-line inrush current. In addition to this effect, many models offer a "soft start" option, which provides a gradual or staged build-up of the motor current when starting. This feature further reduces bearing shock, drive belt wear, and the motor heating that occurs from large inrush currents.

■ Acceleration Control

Acceleration control allows the operator to set the time for the drive to reach full speed. Acceleration control is useful for reducing mechanical stress when starting variable-torque loads that have a high moment of inertia, such as large fans. Acceleration control is also useful for providing smooth speed increase with hoisting machinery, conveyors, and other materials handling equipment.

When starting a load with a high static torque, such as a loaded conveyor belt, the load will not start to move at all until the motor torque exceeds the load torque. The acceleration should proceed smoothly from this point. Therefore, acceleration control differs between loads having different starting torque characteristics.

■ Ability to Drive Multiple Motors

PWM and VVI drives are able to drive multiple motors without problems. These two types produce a controlled voltage output that is independent of the motor characteristics. This voltage can be used by any type and number of motors for which the voltage waveform is appropriate, and the motors can be turned on and off individually.

CSI drives may have difficulty driving multiple motors. This is because the reactance of the motor is an integral part of the output circuit of the drive. In particular, the drive must be matched with the motor reactance to provide proper commutation of the SCR's in the inverter. If a CSI drive is connected to more than one motor, and one of the motors is turned off, commutation may not occur. This may burn out the SCR's or cause the drive to be halted by its safety devices. Some CSI drives are advertised for use with multiple motors. These may have additional circuitry to avoid problems.

■ Full- and Part-Load Efficiency

Except near full load, the VFD itself contributes only a relatively small part of the total losses in a VFD drive system. Even at full load, losses in the VFD itself amount to only a few percent of energy input. More important is the effect of the drive on the efficiency of the motor and the electrical power system, as described previously. To review, the drive reduces losses in the motor by making the current flow in the motor as close to a sine wave as possible, and it reduces losses in the power system by minimizing total harmonic distortion.

The efficiency of all drives drops with speed. There are large differences in the low-speed losses of different drive models. Drives that use SCR's waste energy in commutation. This loss is proportional to the square of the speed, and becomes small at low speeds. PWM drives have a fairly constant loss across the speed range. As with most equipment, larger drives are more efficient, other factors being equal. In recent years, most development has been focussed on PWM drives, so the best PWM drives are becoming more efficient than the other types.

Low-speed efficiency tends to be higher in applications where the torque remains high at reduced speed. This is because the losses within the drive are smaller in relation to the amount of power delivered to the motor.

■ Turndown Ratio

With VVI and CSI drives, the minimum speed is determined by "cogging" (explained previously) and by the crudeness of the output waveform. Single-pulse (often called "six-step") outputs limit the turndown ratio to about 10:1. A larger number of steps allows larger turndown ratios.

PWM drives do not suffer from cogging. In principle, they can produce a fairly smooth current waveform of any frequency. Some PWM drives advertise capability to operate at essentially zero speed, but this capability increases cost.

Some less expensive drives simulate operation at very low speed by "jogging." This consists of cycling the drive on and off at the drive's lowest speed or at some other speed that can be selected by the operator.

For a given drive, motors can operate at lower speed with variable-torque applications than with constant-torque applications. This is because the currents are lower.

■ Speed Increase

Most VFD's can operate the motor faster than its rated across-the-line speed. Typical maximum drive speeds range from 110% to 200% of across-line speed. PWM and VVI drives can easily be designed to provide high motor speeds. CSI drives can provide only limited overspeed, because speeds much higher than the input frequency interfere with commutation in the inverter.

Regardless of the drive capability, it is generally not prudent to operate a motor much faster than its design speed. Bearing wear is accelerated, any vibration due to rotor imbalance is increased, and serious overspeeding may cause failure of the rotor from centrifugal force.

VFD's typically cannot increase output power once the motor has reached the across-line speed. This means that torque must decline as the speed increases above the normal motor speed. Most applications call for more power at higher speed, so you may have to increase the power rating of the drive if you want to operate at higher speed. By the same token, motor heating is increased at higher speed.

If you want to increase your motor speed above the line frequency speed, consider replacing the motor with one that is designed for this purpose. Some manufacturers offer motors designed to operate with VFD's at speeds much higher than the line frequency speed. Some of these motors can maintain their torque output at the higher speeds.

■ Speed Precision

For most applications, any type of VFD will control speed with sufficient accuracy. However, certain applications, such as winders in paper mills, require exceptional accuracy in speed control that can be satisfied only by certain combinations of drive and motor types.

Induction motors inherently operate with "slip," which is a loss of speed compared to the speed that corresponds to the frequency of the input power. Standard-torque induction motors (e.g., NEMA Design B) slip no more than a few percent, so variation in speed is less than this amount. High-torque motors have higher slip, up to about 5%, so they have greater speed variation.

Synchronous motors provide the greatest speed precision because they have no slip. The magnetic field of the rotor does not depend on induction from the stator, but is created by its own source of direct current. As a result, the rotor speed is locked to the stator frequency.

Most VFD's control the frequency of their motor current within a fraction of one percent. "Slip compensation" is used in conventional VFD's to improve speed regulation. Even better speed precision is possible with a controller that uses feedback of the motor speed or shaft position from an encoder that is mounted on the motor. "Vector control" is one term for this capability.

■ Fixed Speed Settings

An option with all three types of drives is operation at fixed speeds that may be set by the operator. The drive may be switched to operation at one or more fixed speeds at various times, in addition to normal variable-speed operation. A fixed speed may be used for night temperature setback and air purge cooling in HVAC systems, for smoke purging in fire control, etc.

■ Speed Range Lockout

An option with all three types of drives is the ability to lock out one or more ranges of operating speeds. The speed ranges are set by the operator. This feature is used most commonly to avoid mechanical or electrical resonances related to motor speed. For example, operation of a cooling tower fan at a certain speed may cause serious vibration of the cooling tower structure. Speed lockout is usually a poor alternative to fixing the problem at its source, but it may be the only practical option.

■ Reversing

The ability to start motors in either direction is a simple, inexpensive feature that is standard in many models of VFD's. Simple reversing must occur with the motor at a dead stop, which is sometimes called "static reversing."

Even if reversing is not needed routinely, consider getting the capability for rare occasions when it may be useful. For example, reverse operation of cooling tower fans may be useful for clearing ice from the tower structure during freezing weather. Reverse operation may also be used for clearing jammed equipment, such as conveyors and drill presses.

■ Ability to Start Drive with Motor Turning

Basic VFD's are intended to start motors from a fully stopped condition. Some drives allow the VFD to be started with the motor turning. This is useful with equipment that may be turning while idle, such as fans that are kept moving by air flow. Also, this feature deals with brief interruptions of power. Without this feature, the drive will trip out, the load must be stopped, and the process restarted from zero speed. To provide this function, the drive must be able to sense the motor speed and to synchronize the drive frequency with the motor before starting.

■ Automatic Restart

If the motor system is stopped by a power outage, it may be necessary to restart the VFD manually. To avoid this nuisance, select a model that offers the option of automatic restarting. However, do not get this feature where unexpected restarting of equipment may create a safety hazard.

■ Braking and Regeneration

Some models of VFD's can act as a brake to slow or stop loads such as hoists and conveyors. This requires absorbing power from the load. "Regenerative" braking feeds the power absorbed from the load back to the power supply system, conserving energy. CSI drives with SCR rectifiers have inherent regenerative braking capability. The effectiveness varies with the design of the drive. PWM and VVI drives must have special components and control circuitry to provide regenerative braking. Braking is accomplished in some models by dissipating the absorbed power in resistors.

"Four-quadrant control" is a term sometimes used for the ability to provide braking in both the forward and reverse directions. This is a quaint way of expressing the four possible conditions that may occur, namely: forward direction and forward torque, forward direction and reverse (braking) torque, reverse direction and reverse torque, and reverse direction and forward (braking) torque. Some drives provide "four-quadrant control" with regeneration, and some without it.

■ Emergency Bypass

The drive may offer an internal bypass that allows the motor to be operated directly across the line in the event of drive failure. This feature is useful with applications that allow for emergency motor operation at full speed, such as ventilation fans, cooling tower fans, and chilled water pumps.

If the application makes a bypass useful, try to select a drive that has a "test" position in the bypass. This allows the drive to be connected to the power supply for test purposes independently of the bypassed motor. Any bypass circuit, whether part of the drive or external, must include an appropriate motor starter and fuses for operation of the motor directly across the line.

■ Power Factor

Power factor is important primarily as a possible factor in electricity pricing. It is also a minor factor in energy losses within the power distribution system of the facility. See Reference Note 21, Electricity Pricing, for an explanation of power factor.

When a motor is driven with a VFD, the power factor that appears on the power system depends primarily on the type of rectifier used by the drive. The characteristics of the motor itself have little effect on power factor when the motor is driven by a VFD.

Drives that use diode rectifiers have high power factor at all motor speeds. The power factor of SCR and thyristor rectifiers drops almost linearly with decreasing motor speed, for reasons discussed previously. Diode rectifiers are used on virtually all contemporary PWM drives, so PWM drives have high power factor as a class. Some PWM drives even improve the power factor of the power system to which they are connected, to the extent that their capacity exceeds the momentary requirements of the motor. Diode rectifiers are also used on some models of VVI and CSI drives.

The power factor of a VFD matters most if the utility charges a power factor penalty. If the utility measures power factor at the time of the facility's peak load, the power factor of the VFD may still not matter much. All types of VFD's have high power factor under heavy load, and the absolute amount of "reactive power" (KVAR's) produced by the drive does not increase much with declining load, if at all.

In any event, the drive will not reduce the power factor of the facility unless the VFD is large in relation to the facility's peak load. Utilities do not charge a power factor penalty unless the power factor falls below a stated threshold, which is typically 85%.

For similar reasons, the power factor of a VFD generally does not cause overloading of the power system within the facility.

It may be dangerous to try to improve the power factor of the facility with capacitors if a VFD is installed.

This problem is discussed above, under "Resonance and Blown Fuses in the Facility Power System."

■ Protection from Electrical Faults

Different VFD models can be damaged by different types of electrical faults. All modern drives have the ability to protect themselves against common electrical faults. Some modes of protection may be extra-cost options. Select protective features that are appropriate for your type and model of drive, and for the faults that are likely to occur in your power system.

Drives that use SCR's to control the output to the motor (VVI and CSI drives, generally) are vulnerable to any fault that can interfere with the reverse voltage needed to turn off the SCR's at each power cycle. These include shorts to ground and across the line. If the current is not interrupted, the SCR will burn itself out. CSI drives are fairly easy to protect from faults and grounds because the large inductors in the DC section prevent rapid increase in current. This gives fuses or other protective devices time to work.

Drives using transistors (PWM drives, generally) are less vulnerable because transistors can be shut off simply by turning off their control current. Also, the most modern type of transistor (IGBT) has a faster response time than earlier types of power transistors.

■ Resistance to Tripping of Drive by Transients

A feature commonly called "ride-through" keeps the drive from tripping off when there are brief interruptions of power. Ride-through relies on energy stored in the DC section of the drive, and a separate power supply may be used to keep the DC section charged. The ride-through capability of different models ranges from a small fraction of a second to several seconds, depending on the drive and the load.

■ Diagnostics and Displays

Any electronic device is subject to unpredictable component failure. This makes it desirable to have diagnostic capability that allows the unit to be returned to service quickly. Smaller, cheaper units have indicator lights that warn of a few specific conditions. More sophisticated units have readouts that can indicate most failure conditions. Try to get readouts that display in plain language, rather than in codes.

A useful feature available with larger units is a recording of drive conditions during the time immediately prior to a fault. For example, this would record if an overvoltage occurred prior to a fault. Another useful diagnostic feature is a history of previous faults.

■ Altitude, Cooling, Heating, and Humidity

VFD catalogs commonly list a maximum altitude of 1,000 meters (3,300 feet) without saying why. This is not because the drive needs oxygen to breathe, but because the thinner air at higher altitude is less effective in cooling the drive. You can use a VFD at any altitude, but you have to provide extra cooling at higher altitudes, or limit the drive to a lower output.

A VFD must be kept above a minimum temperature. This keeps electrolytic capacitors from freezing and keeps the electrons from getting stuck in the silicon. A small thermostatically-controlled heater installed inside the housing can take care of this requirement.

Humidity must be limited to prevent condensation on the circuitry. This starts with an appropriate enclosure. The internal heat of the drive should keep the relative humidity low enough during operation. If the drive may become damp when it is shut down, an electric heater may be installed inside the housing to lower the relative humidity prior to starting. If the environment makes this step advisable, install a humidity sensor inside the housing to prevent the drive from starting until the heater reduces the relative humidity.

DIRECT-CURRENT MOTORS AND DRIVES

Before variable-frequency AC motor drives (VFD's) were developed, DC drive systems were the only means of achieving a high turndown ratio while maintaining efficiency and torque. VFD's are now making heavy inroads into the applications that were once dominated by DC drives. In the foreseeable future, DC drives will continue to be the best choice for some specialized applications. There are few retrofit applications, if any, where a DC motor drive is the best choice to replace a constant-speed AC motor. We will treat DC drives lightly, covering only the main features that contrast them with VFD's.

How DC Motors Work

The DC motor is slightly more complex in construction than an AC induction motor. Figure 17 shows a DC motor with all the main parts visible.

A DC motor has two sets of windings. A set of field (stator) windings inside the motor case create a stationary magnetic field. (In small motors, the field may be created by a permanent magnet.) The armature (rotor) has a large number of separate windings distributed around the shaft like a squirrel cage. External current is passed through both sets of windings, creating magnetic fields. Interaction between the two fields creates the force that turns the rotor.

The current flow in the rotating armature windings is switched by the "commutator," which consists of the electrical contacts for all the windings, arranged around one end of the rotor shaft. Stationary carbon brushes mounted on the motor case keep switching current to successive armature windings as the commutator turns under the brushes. Therefore, the windings are always energized at the same position in space, located where the windings are subjected to greatest torque from the magnetic field of the field windings. This allows DC motors to produce high torque at all speeds, and provides rapid response to changes in the applied voltage.

Different methods of connecting the field and armature windings are used to establish the torque and speed control characteristics of DC motors. In a "shunt-wound" motor, voltage is applied to the field and to the armature in parallel. Shunt-wound motors have torque that is relatively low, but the torque remains constant with changing speed. Shunt-wound motors have good speed stability. In "series-wound" motors, the current passes through the field on the way to the armature. This type has high starting torque, but the torque declines sharply with increasing speed. Series-wound motors have poor speed stability. A common hybrid type, "compound wound," has a small series field winding in addition to a larger shunt field coil.

How DC Drives Work

A DC drive unit controls motor speed by varying the voltage supplied to the armature windings, and hence the current. The drive may also vary the voltage and current to the field windings, if this is necessary to achieve the desired torque characteristics.

In the past, a DC drive was a cumbersome affair, as you can tell by visiting the elevator machinery room of an older building. A conventional AC motor was used to drive a DC generator, which produced direct current to drive a DC motor. The DC motor and DC generator were controlled by a bulky electro-mechanical panel. In all, an old DC drive required four big, expensive components.

Modern DC drives are much smaller and simpler, and correspondingly less expensive. The DC drive rectifies and filters the incoming AC power to make DC current. Speed control is entirely electronic. In applications that need precise speed control, the drive uses a tachometer mounted on the motor to provide speed feedback.

A DC drive is similar in size and appearance to a variable-frequency AC drive. The rectifier and filter in a DC drive are similar in function to the first two sections of an AC drive.

Advantages and Disadvantages

DC drives and AC variable-frequency drives are close competitors for some applications, so let's compare the two. The main advantages of DC drives over VFD's are:

- *higher turndown ratios.* DC motors can slow down almost to zero speed with some drives. However, the same capability is appearing in some VFD's.
- *greater speed precision.* DC drive systems are capable of better speed precision than VFD's. DC drives will probably retain this advantage because the motor torque can be controlled more positively

by the drive unit. Figure 18 shows an application where speed control is critical.

- *higher efficiency.* DC drive units have lower losses than VFD's because they have no inverter sections. The overall efficiency of DC drive systems is higher than for VFD systems, especially at lower speeds and percentages of load. This is largely because DC motors use their magnetic fields more efficiently under these conditions.
- *absence of torque-related overheating at low speeds.* DC motors have long been employed to provide high torque at low speeds. They have

WESINC

Fig. 17 Direct-current motor This elevator motor clearly shows the major components of a DC motor. The six stator poles and the rotor windings are energized by electric current. Current reaches the rotor windings through carbon brushes that slide over the commutator on the near end of the rotor. The commutator keeps the magnetic field of the rotor stationary in space as the rotor turns. Note that there is a rotor brush corresponding to each stator pole. The rotor and stator fields remain aligned for maximum torque. Speed and torque are varied by changing the current in the windings.

evolved the ability to survive at low speed with the high currents needed to produce high torque. DC motors may have separate cooling fans, which are independent of motor speed. A similar approach could be used with AC drives, eliminating this as an advantage.

• *relative simplicity.* The rectifier section of a DC drive has about the same design as the rectifier section of a VFD, but it needs no inverter section or complex inverter controls.

The main disadvantages of DC drives, in comparison with AC drives, are:

• *need for motor maintenance.* The commutator rubs against the carbon brushes, creating carbon dust that must be cleaned out of a DC motor often to prevent arcing. At longer intervals, the brushes must be replaced. At still longer intervals, the surface of the commutator must be refinished, which takes the motor out of service for a period of time.

• *difficulty of protecting the motor in severe environments.* The need for access to the commutator and brushes for cleaning makes it awkward to seal a DC motor for use in harsh environments. However, DC motors for harsh environments are available.

• *larger motor size.* The geometry of the field and armature windings requires a DC motor to be somewhat larger than an AC motor of the same power. External cooling systems are also bulky, but where these are used, they allow the DC motor to serve applications that could not be served by an AC motor.

• *power factor decreases with speed.* Unlike AC drives that use diode rectifiers, the power factor of DC drives declines with motor speed.

• *cannot bypass failed drive.* A DC motor cannot be operated across the line in the event of drive failure, because it would have no speed control. (The speed of an AC motor is determined by the frequency of the input power.)

As with AC drives, the cost of a DC drive system is a combination of equipment and labor costs. Comparing the two types, AC drive installations are typically less expensive in smaller sizes, while DC drive installations may be less expensive in larger sizes.

Waveform-Related Problems and Solutions

Electronic DC drives use switching devices similar to those used in VFD's. Therefore, electronic DC drives may create similar problems in the power supply system. The solutions to these problems are also similar. DC drives have not yet reached the same degree of sophistication in avoiding electrical problems as VFD's,

Reliance Electric

Fig. 18 DC motor application This large DC motor in a paper mill operates a winder, which requires precise control of speed and torque. A tachometer is installed on the near end of the shaft as part of the speed control. Note the external cooling ducts, which provide effective cooling at all motor speeds and loads.

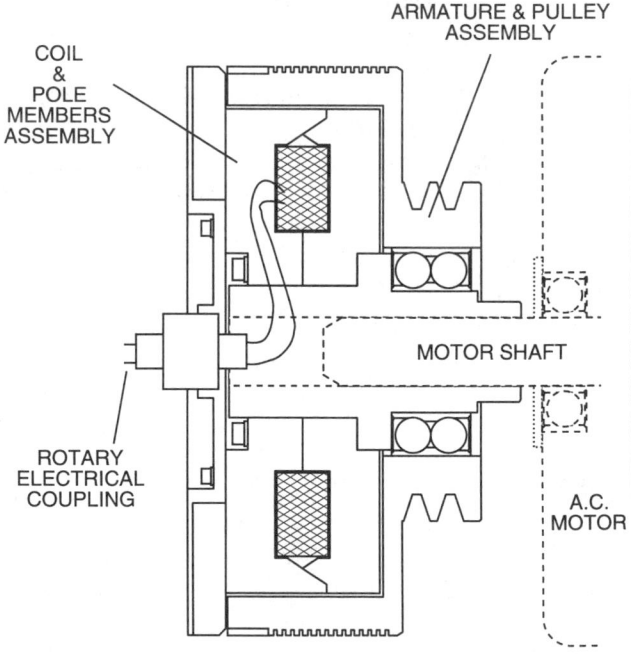

Coyote Electronics, Inc.

Fig. 19 Eddy current clutch combined with a pulley The inner portion of the clutch is fixed to the motor shaft. The outer portion, including a belt pulley, rotates freely on a ball bearing. The outer portion is driven by an induced magnetic field. The coils that induce the field are energized through a rotating electrical contact.

but they presumably will catch up in time. Refer to the previous discussion of VFD's for details.

The smoothness of the DC voltage produced by DC drives depends on the design of the rectifier and filter. It is possible for a DC drive to produce very smooth motor voltage, but not all models do so. As with AC motors, voltage distortion causes inefficiency and overheating in the DC motor.

(Old-style DC drives had none of these problems. The motor-generator set of the DC drive isolated the DC motor from the power supply, and it provided smooth DC power to the motor.)

Application-Related Selection Considerations

Different applications require different DC drive characteristics. These are the most common selection considerations for DC drives:

- control input provisions
- torque characteristics
- acceleration control
- turndown ratio
- speed precision
- braking and regeneration
- diagnostics and displays
- altitude, temperature, and humidity limits.

You may notice that these selection characteristics are similar to those for VFD's, but that there are fewer of them. Refer to the previous discussion of VFD's for the details of these issues.

VARIABLE-PULLEY DRIVES

Variable-pulley drives are belt drives in which the effective diameter of one of the sheaves can be changed as the equipment is operating. The drives use V-belts. The change in effective diameter is achieved by changing the spacing of the cheeks of the sheave. This squeezes the V-belt farther outward, or allows it to fall farther into the groove. Commercially available models are equipped to accept standard control signals.

Variable-pulley drives are among the most efficient methods of varying motor speed. Typically 3% to 6% of input power is lost to the belt drive at all speeds. This is about the same as the belt loss that occurs with constant-speed belt drives. The motor always operates at its design speed, so it suffers no efficiency loss from speed reduction. The motor does suffer the usual loss in efficiency as load is reduced, if the power requirement falls with the speed of the driven equipment.

The turndown ratio of variable-pulley drives is limited to about 3:1. This is sufficient to achieve most of the energy saving potential in variable-torque applications, such as fans and centrifugal pumps. However, the speed range may not be large enough for many industrial applications.

The maximum horsepower of variable-pulley drives is limited by the fact that power is transmitted through a single belt. At present, the largest commercial size is about 50 horsepower.

The variable-pulley mechanisms of different manufacturers differ substantially in bulk. Some are

Coyote Electronics, Inc.

Fig. 20 Eddy current clutch This direct-drive unit is rigidly mounted to the motor case for driving a pump.

quite small, and fit directly on the shaft of the existing motor. Others requires an intermediate belt drive installed on a separate base with the motor. The need to insert the drive unit between the motor and the driven equipment makes retrofit impractical in many cases.

Variable-pulley drives probably deserve greater attention. They have been eclipsed by variable-frequency drives because of the latter's ease of installation, more aggressive marketing, and the popularity of anything that is electronic. Variable-pulley drives are simple and they have few failure modes. They are free of electrical noise problems. Costs are relatively low. Periodic belt replacement is their only major maintenance requirement. Turndown ratio is their only serious performance limitation.

EDDY CURRENT CLUTCH DRIVES

An eddy current clutch is an electromagnetic slip clutch. Figure 19 shows a cross section of an eddy current clutch, for a unit that is an integral part of a belt drive pulley. The motor drives a rotating electromagnet that induces eddy currents in a metal disk or drum connected to the output shaft. The magnetic field of the driven element interacts with the induced eddy currents

to create torque in the driven element. This is similar to the action of an AC induction motor.

The speed of the driven element is controlled by varying the current that creates the magnetic field. This requires an external electrical connection.

Eddy current drives are less efficient than the types covered previously. They were promoted for energy conservation during the 1970's, primarily as fan drives. Their lower efficiency, compared to the emerging AC variable-frequency drives, caused them to fade away. They may be making a limited comeback, primarily because they are free of the electrical problems that have become notorious with variable-frequency drives.

Eddy current clutches have a fairly high turndown ratio. However, efficiency falls rapidly as speed is reduced, so there is little energy saving below a turndown of approximately 3:1.

The eddy current clutch may be mounted directly to the motor frame, as in Figure 20. This provides a strong, compact mounting. Units are also available that mount directly on the shaft of an existing motor. The clutch assembly has about half the volume of the motor, and it requires a small external controller. Maintenance requirements are low because there are few moving parts, and no parts are subject to frictional contact.

WESINC

Fig. 21 Large wound-rotor motor This air handling unit supply fan serves 150,000 square feet of air conditioned space. The wound-rotor motor was installed to allow an adjustment of the fan speed. The speed is reduced by using the resistor bank in the background to dissipate power that is induced in the rotor windings. In addition to wasting motor power, the resistors heat the cooled air. The building owners have been paying for this energy waste for decades. An ordinary motor with drive belts and pulleys would have been cheaper and much more efficient.

WOUND-ROTOR MOTORS

A wound-rotor motor is a variation of an AC induction motor. As with other induction motors, the magnetic field of the rotor is induced by the magnetic field of the stator. The main physical difference is that the wound-rotor motor has more complex windings embedded in the rotor. The rotor windings are divided into phases to match the number of phases of the stator. The rotor windings have connections that are brought to the outside of the motor through slip rings.

The external rotor connections allow the current that is induced in the rotor windings to be reduced by placing resistors in series with the windings. Reducing the rotor field slows the motor. This method allows motor speed to reduced to 50-70% of full speed, below which motor operation becomes unstable.

This method of speed control dissipates a large amount of energy in the resistors. Indeed, the presence of a large external resistor bank is the most obvious clue that a motor is a wound-rotor type. Figure 21 shows an example.

Wound-rotor motors are generally large. They tend to be found in direct-drive applications that lack other means of adjusting speed. However, they could be found in combination with belts or other types of transmissions.

Wound-rotor motors have been rendered obsolete by their poor efficiency and by the availability of better methods of controlling speed. They may get a reprieve from an unusual type of controller that feeds the energy extracted from the rotor back to the power supply system, conserving energy. This is called "slip energy recovery." This type of controller can be retrofitted to existing wound-rotor motors and is claimed to have high efficiency. It creates harmonic distortion of the input power in a manner similar to that of variable-frequency drives.

Setpoints
Matching the Controls to the Equipment
Two-Position Controls
■ **Anticipators**
Stepped Controls

Proportional Controls
■ **Proportional-Integral (PI) Controls**
■ **Proportional-Integral-Derivative (PID) Controls**
■ **Limitations of Integral and Derivative Control**
Analog and Digital Controls
Open-Loop and Closed-Loop Controls

This Note covers the basics of automatic controls. To keep the discussion from being too general, and hence confusing, we will focus on temperature controls to provide examples. However, the same principles apply to any condition that you want to control automatically, for example, pressure in boilers, air flow in variable-volume systems, water flow in variable-flow systems, etc. For system conditions other than temperature, replace "temperature" with "pressure," "flow," or other relevant variable in the following discussion.

Setpoints

The purpose of thermostatic controls and many other types of automatic controls is to keep a selected variable (e.g., temperature, pressure, etc.) close to a particular value. This selected value is called the "setpoint." To do this, the control uses a sensor to measure the controlled variable. The sensor produces an output signal that is sent to the equipment that produces the desired condition. The signal tells the equipment whether to increase or decrease its output.

For example, if the controlled variable is space temperature, the control is a thermostat and the desired room temperature is the setpoint. The thermostat responds to the space temperature by sending a signal to the heating or cooling equipment. The equipment operates in response to the thermostat signal in a way that keeps the space temperature close to the setpoint.

The control usually allows a person to change the relationship between the measured variable and the control signal. This allows the person to change the setpoint, or the sensitivity of the control, or other characteristics. For example, all room thermostats allow a person to change the room temperature setting. Some room thermostats, but not all, allow a person to change the sensitivity of the thermostat to temperature changes.

Matching the Controls to the Equipment

The output of the control must be matched to the characteristics of the equipment being controlled. That is, they must "speak the same language." These characteristics need to be matched:

• *method of transmission.* At present, the most common methods of transmitting control signals are electrical and pneumatic (air pressure). As technology advances, other methods are becoming popular. For example, light beams in fiber optic cables are used for transmitting signals over long distances, and infrared beams are used for transmitting signals over short distances.

• *method of coding information.* The signal may be a simple electrical connection that is turned on or off. For example, this type of control signal is commonly used with small heaters, air conditioners, and water heaters.

The signal may be proportional in strength to the measured condition. This method of coding is called "proportional" or "analog." For example, pneumatic controls are usually analog.

The signal may be coded as a series of digital pulses. All electronic control systems are digital. Most sensors and actuators are inherently analog devices, so an "analog-to-digital converter" is needed at each end of a digital control system.

• *signal range or protocol.* The range of output from the sensor must match the range of input to the controlled equipment. For example, simple electrical on-off thermostats typically operate at 12 or 24 volts, and the controlled equipment is designed to accept these voltages. Pneumatic controls typically operate in a pressure range of 5 to 15 pounds per square inch. Electrical analog control systems typically operate by varying a current between 4 to 20 milliamperes.

Digital signals can be coded in many ways to carry many different kinds of information. Unfortunately, industry has not yet evolved a common set of standards for digital signals. For more about this problem, see Reference Note 13, Energy Management Control Systems.

• *direct- or reverse-acting.* In a proportional control system, if the sensor output increases as the measured variable increases, the control response is called "direct-acting." If the sensor output decreases as the measured variable increases, the response is called "reverse-acting."

Similarly, if an actuator increases its output as the input signal increases, it is called "direct-acting." In the opposite case, it is "reverse-acting."

Combining direct-acting and reverse-acting control components can simplify a control system. For example, a single direct-acting thermostat can simultaneously control cooling with a direct-acting actuator (cooling increases as temperature rises) and heating with a reverse-acting actuator (heating decreases as temperature rises).

• *response curve.* Most analog controls are approximately "linear," which means that the control signal varies proportionally to the measured variable. The output of equipment in response to the control signal is not always linear. For example, the flow of air through a typical damper is not linearly proportional to the position of the damper actuator.

• *sensitivity.* The sensor is limited in its ability to sense changes in the controlled variable. At the other end of the system, the actuator requires a minimum change in signal to make it respond. For example, a sensitive room thermostat may control temperature within 0.2°C, while a less sensitive thermostat may control temperature within 1.0°C. The sensitivity of some control elements is variable. Sensitivity is a factor in control stability.

• *rate of response.* In many cases, it is not desirable for the equipment to respond instantly to the control signal, as this may cause control instability. For example, water valves and air valves can respond so quickly that they can oscillate under adverse conditions. Some equipment, such as baseboard convectors, respond so slowly that instability is not a problem. (Response rate is only one factor among several that affect system response and stability.)

Response rate can be slowed in a variety of ways. One is to select actuators that respond slowly. In pneumatic systems, the signal itself can be delayed by putting a restriction in the air line. Digital systems provide almost unlimited flexibility in tailoring control response. Rate of response is the basis of "derivative" control, which is discussed below.

Two-Position Controls

The simplest type of control turns equipment on and off. Such controls are called "two-position," "make-break," and other names. Most two-position thermostats have a bimetallic temperature sensing element that opens and closes an electrical contact as it changes shape in response to temperature. Two-position thermostats are used to control equipment that is on-off in nature, such as electric heating coils, burners in small furnaces and

boilers, and compressors in small air conditioners and refrigerators.

With two-position control, the temperature cycles between two temperature limits, as shown in Figure 1. The difference between the turn-on and turn-off temperatures is called the "differential." Typical two-position space thermostats have differentials from 1°F to 3°F (0.5°C to 2°C). The need for a substantial temperature differential is a weakness of two-position control. If the differential were too small, the equipment would cycle on and off at an excessive rate.

Note that the temperature shown in Figure 1 is the temperature at the thermostatic element itself. The temperature differential in the space or in the controlled equipment usually is larger than the differential at the thermostat. This is because the thermostat is somewhat isolated, so there is a lag in sensing the space or system temperature. Also, the temperature continues to drift while the heating or cooling equipment starts.

■ Stepped Controls

A stepped, or staged, thermostat consists of several two-position thermostatic elements installed in a single housing, with progressively higher (or lower) temperature settings. Stepped thermostats are used to control multi-stage equipment, such as electric heaters with several elements, boilers with multi-stage burners, and reciprocating cooling compressors with unloading cylinders.

As the load increases and successive thermostatic elements turn on their equipment, the temperature gets further away from the setpoint (the desired temperature). This limits the number of steps that are practical.

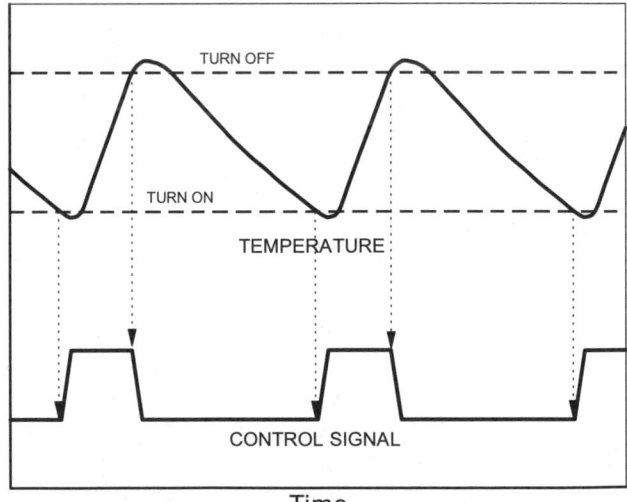

Fig. 1 Two-position control This is the response of a heating system. The difference between the "on" and "off" temperatures is the "differential." In this example, the heating capacity is large, so the temperature rises quickly during the "on" cycle. Also, the heating load is large, so the temperature falls quickly during the "off" cycle.

■ Anticipators

Anticipators are an accessory of two-position thermostats that control space heating equipment that heat space air quickly. The thermostat heats up more slowly than the air in the space. This is because the thermostat is partially enclosed, and because it is mounted on a solid base, which also heats up more slowly than the space air. If no correction were made for this lag in thermostat response, the air in the space would overheat with each heating cycle.

This problem is overcome by installing an "anticipator" in the thermostat. An anticipator is a tiny electric heating element inside the thermostat. It is usually powered by the thermostat control current. Current starts to heat the anticipator when the thermostat circuit is closed by the temperature sensing element, at the same time that the space heating unit starts. The anticipator can be adjusted so that the rate of warming inside the thermostat matches the rate of warming in the space.

Figure 1 in Measure 5.7.4 shows a two-stage heating thermostat that has a separate anticipator for each stage. Each anticipator can be adjusted separately.

Anticipators are used most commonly with forced-air heaters, which heat the air in spaces quickly. Anticipators are used less often in thermostats that control convective and radiant heaters, because these have less tendency to overheat a space. Anticipators are not used in thermostats that are located in the return air stream of fan-forced equipment, because the lag in such thermostats is negligible.

Proportional Controls

A proportional control, as the name implies, provides an increasingly stronger control signal as the temperature gets farther away from the setpoint. Figure 2 shows the response of an ideal proportional control.

Proportional controls are used with equipment that modulates its output, such as hydronic coils, centrifugal and screw chillers, burners in large boilers, pumps in variable-flow systems, etc.

For example, pneumatic thermostats are proportional control devices. Typically, a bimetallic temperature sensing element actuates a tiny valve that bleeds compressed air into a plastic or copper signal line. The air pressure in the line is the output signal of the thermostat.

Instability occurs when the response of a system gets out of phase with actual conditions. Equipment output may increase when it should be decreasing, or vice versa, causing large swings in output. Instability may occur because the system does not respond quickly enough, so the controls continue to require more response when it is not needed. The delay in response may occur because of the travel time of air in ducts, the

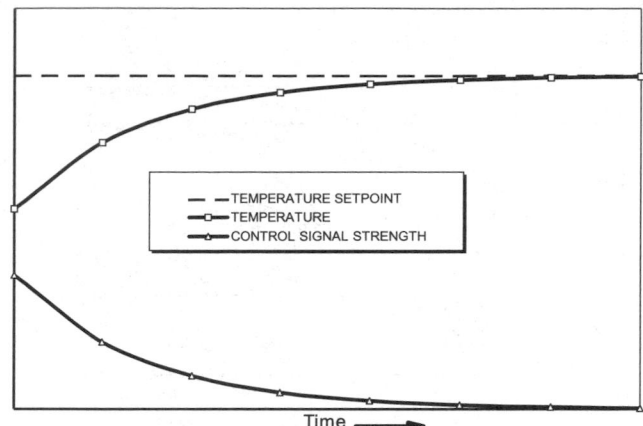

Fig. 2 Proportional control   The strength of the control signal is proportional to the difference between the system output and the setpoint. The response shown here is ideal. Real proportional control systems rarely behave this well.

travel time of liquid in pipes, time delay in signal travel, heat absorption by the mass of space or equipment, etc. By the same token, the tendency toward instability increases if the control responds to conditions too quickly and too strongly, not allowing the controlled equipment to catch up.

A practical problem of simple proportional controls is friction and play in both the control devices and in the controlled equipment. These cause the equipment not to respond fully to the temperature signal, increasing the offset. For example, a loose linkage or a sticky control valve can seriously degrade control accuracy. The same factors also cause "hysteresis," which means that the system stabilizes at different temperatures for a given setpoint, depending on whether the temperature was previously rising or falling.

■ Proportional-Integral (PI) Controls

In a proportional control, the difference between the temperature setting and the actual temperature is called the "offset." (It is sometimes called "drift," "droop," or other names.) Figure 3 illustrates offset.

Offset increases as the load increases. It is the main weakness of simple proportional control. Offset can be reduced by making the system more sensitive, i.e., by producing a stronger control signal in response to the temperature offset. However, excessive sensitivity causes control instability, which is the tendency of the control to oscillate or "hunt."

"Integral" control is a refinement of proportional controls that eliminates the offset error. An older name for this feature is "automatic reset." A proportional-integral (PI) control has an additional feature that senses the difference between the actual temperature and the setpoint. The control changes its response to eliminate this internal difference. Integral control requires additional hardware in the thermostat or the control system.

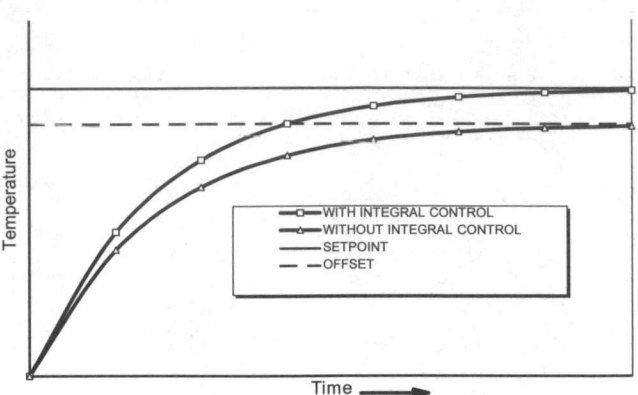

Fig. 3 Proportional-integral (PI) control Integral control senses the difference between the setpoint and the actual system output, and responds to close the gap between the two. The response is relatively slow, which is appropriate for many applications.

"Integral" is a term taken from the mathematical method of calculus. An integral is a summation of a variable over some period. In a PI control, the vigor with which the control system acts to return the system to the zero-offset position is proportional to the accumulated offset, or more precisely, the summation of the offset over time.

■ **Proportional-Integral-Derivative (PID) Controls**

"Derivative" control is a further refinement of proportional controls that is almost always combined with integral control, yielding proportional-integral-derivative (PID) control. Derivative control improves the speed of response, as shown in Figure 4. It is used because integral control alone is rather slow, requiring a summation of the offset over time.

"Derivative" is another calculus term, which means the rate of change of a variable. In derivative control, the rate of change of the control setting is proportional to the offset. If there is a large offset, the control provides a large signal to restore the system to its setpoint.

Derivative control is especially valuable in systems where output is able to change so rapidly that the system response may wander far from the setpoint. Typical applications for PID control are controlling boiler pressure and controlling the flow of air from variable-flow fans.

An unfortunate effect of derivative control is a tendency to overshoot the setpoint, as shown in Figure 4. Instead of proceeding smoothly to the setpoint, the system oscillates around the setpoint. The system is stable if the oscillations damp out quickly. The combination of proportional and integral control provides the desired stability if the controls are adjusted properly.

Derivative control has a further disadvantage in cases where the signal may be noisy, for example, an

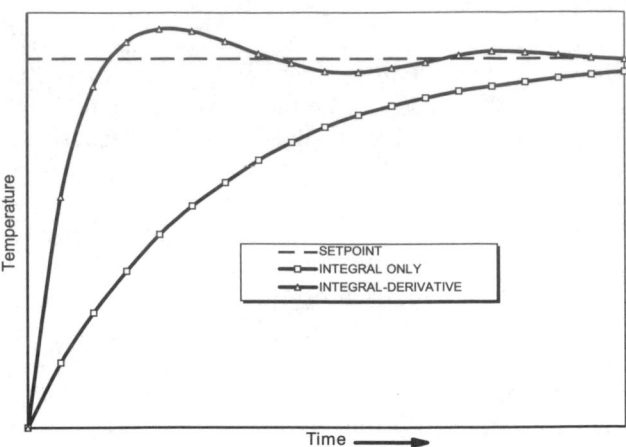

Fig. 4 Proportional-integral-derivative (PID) control A "derivative" characteristic increases the speed of response in a proportional control. Overshoot is an inherent aspect of derivative control. Keeping the system response stable requires a proper combination of integral and derivative characteristics.

analog electrical signal in a wire that picks up electrical noise from the environment. A noise spike tells the derivative control that there is a large offset error, which causes the system to respond strongly, although only momentarily. The same problem arises if the signal is subject to small but rapid fluctuations, for example, the return temperature of a chilled water system. Thus, derivative control may be twitchy.

Integral and derivative control are general methods of signal correction, which are not characteristic of any particular type of controls (i.e., pneumatic, electric, or electronic). They are accomplished differently with different types of controls. Integral and derivative control are standard features of electronic digital controls, but they require additional complications in pneumatic and electric analog controls.

■ **Limitations of Integral and Derivative Control**

Integral and derivative control is not a panacea. It complicates the control system, making it more difficult to diagnose problems. You have to set the rate of response for both the integral and the derivative features of a PID control, based on the speed of response and the lag of the system. For example, the control of a water chiller should take into account the time required for water to circulate through the system, the control characteristics of the equipment that uses chilled water, and the throttling characteristics of the chiller. Improper settings can cause control instability or other problems.

Integral and derivative control can compensate for the effects of play and friction in control linkages, within limits. They cannot compensate for improper thermostat installation, such as placing a space thermostat on an exterior or sun-exposed wall, or installing a chilled water thermostat in the supply line of a reciprocating chiller.

Analog and Digital Controls

All the principles discussed previously apply equally to analog and digital control systems.

Digital controls can provide perfect accuracy in transmitting signals. They also provide perfect accuracy in the internal calculations of the control system. However, the variable being measured (e.g., temperature, pressure, flow, etc.) is usually an analog signal. The sensor technology determines the sensitivity and accuracy of the input signal. Likewise, the output (e.g., a valve setting) is usually an analog variable. The accuracy of the output is determined largely by the controlled device. For example, a sticky valve will not accurately control the flow of water in a coil.

Digital signals are processed with microprocessors, which can be programmed to perform a variety of functions. The response characteristics of a digital control system can be tailored in great detail. These are mixed blessings. See Reference Note 13, Energy Management Control Systems, for more about digital controls.

Open-Loop and Closed-Loop Controls

A closed-loop control system compares the actual system output to the setpoint. The difference between the two creates another signal, called "feedback," that is used by the control system to reduce the difference. Virtually all control systems used in buildings and plants are closed-loop. This includes all types of thermostatic controls.

An open-loop control simply sends a signal to the controlled equipment. The response of the equipment to the control is determined by the magnitude of the control signal, and by the sensitivity and response speed of the controlled equipment. There is no feedback, except perhaps the response of a human operator. A valve on a garden hose is an open-loop control system. A manual damper on a wood stove is another.

What Causes Air Leakage
- Unbalanced Fan Systems
- Wind
- Chimney (Stack) Effect

How to Find Air Leaks
How to Estimate Air Leakage
Vapor Barriers and Air Leakage
Maintenance to Reduce Air Leakage

Air leakage through the skin of a building is a major cause of energy waste in many buildings. This brief Note explains the causes of air leakage, gives methods for finding and estimating leakage, and lays the groundwork for the Measures recommended in Section 6.

We use the term "building air leakage" to mean the passage of air through the outer structural components of the building, including the walls, windows, doors, roof, and floors. The term "infiltration" is often used for air leakage. Strictly speaking, infiltration is air leakage into the building. The less common term "exfiltration" is air leakage out of the building.

The total amount of air entering the building must equal the total amount of air leaving the building, but this does not mean that infiltration equals exfiltration. The most important reason for the difference is the operation of ventilation fans. For example, exhaust fans create a negative pressure in the building that causes infiltration, but no exfiltration. The air exhausted by the fans is not "leakage," as we define the term, although it may carry away a large amount of energy.

What Causes Air Leakage

Air moves only in response to a pressure differential. The following are the three major causes of pressure differential across the building envelope. The amount of air leakage is proportional to the aggregate effect of these three causes. Air leakage is inversely proportional to the resistance of the envelope to leakage.

■ Unbalanced Fan Systems

A pressure differential is created across the building envelope if fans in one part of the building exhaust more or less air than fans in another part of the system bring into the building. Many air conditioning systems are inherently unbalanced, usually because the system has exhaust fans but no arrangements to force outside air through the conditioning units. In some cases, pressure imbalance in conditioning systems can be eliminated with simple adjustments. In other cases, this would require major modifications. See Measure 4.2.1 about balancing intake and exhaust in air handling systems.

A negative pressure is commonly created inside laboratories, hospital rooms, and other isolated spaces to prevent contamination of one part of the building by another. This causes a certain amount of infiltration through the outer envelope.

Conversely, a positive pressure is sometimes created in buildings for the sake of comfort. The positive pressure prevents entry of drafts. It also forces warm air through the walls in cold weather, increasing the wall temperature.

■ Wind

The pressure created by wind against a surface is proportional to the square of the wind speed. For example, a wind of 32 miles per hour has a "velocity pressure" of about 0.5 inches water gauge. This means that if all the energy of wind striking a wall is converted to pressure, the pressure on the wall would be 0.5 inches. This is similar in magnitude to the pressure differentials created by the fans of air conditioning systems.

Wind pressure is positive on the upwind side of a building, and is negative on the downwind side, so air leakage caused by wind tends to flow through the building. The magnitude of negative pressure on the downwind side of the building is much less than the magnitude of the positive pressure on the upwind side. This is because the kinetic energy of the wind is dissipated in turbulence on the downwind side.

Not much can be done about wind pressure on the broad face of a building. However, individual outside air intakes can be shielded from velocity pressure in a fairly effective manner. Refer to Measure 4.2.9 for methods.

■ Chimney (Stack) Effect

If it is cold outside a building, the warm air inside the building is less dense than the air outside, and hence lighter. Therefore, the total weight of the atmosphere acting at ground level creates more pressure outside the building than inside the building. The difference in pressure at the base of the building causes air to flow into the building at its lower levels.

The increase in pressure at the bottom of the building is transmitted to the upper levels of the building, which become higher in pressure than the outside. Thus, chimney effect causes air to leak into the bottom of a building, and to leak out the top of the building. The situation is reversed in a building that is being cooled during warm weather.

The force of chimney effect is proportional to the difference between the inside and outside temperatures. Unfortunately, this makes chimney effect worst when the heating and cooling loads are highest.

Chimney effect is also proportional to the height of the building. In tall buildings, chimney effect can be a strong force, so special methods may be required to limit air leakage. The most visible of these is the revolving door, a hallmark of tall buildings. The revolving door acts as an air lock at ground level.

There is an important fact to apply in minimizing air leakage in tall buildings. If the upper portion of a building is sealed so that air cannot leak out the top, then the force of chimney effect will not induce infiltration at the lower floors. Measures to prevent air leakage from upper floors are recommended in Subsection 6.4. Personnel doors that lead to roofs and penthouses are common paths for leakage. Measures to reduce leakage through these penetrations are recommended in Subsection 6.1.

In principle, chimney effect could be eliminated by isolating each level of the building from the other levels. However, the need for elevators and stairs makes this practically impossible. Also, it might be possible to counteract the negative pressure on lower flows by pressurizing the elevator shafts and other vertical penetrations. However, this approach requires energy to move and condition the outside air that is used for pressurization.

How to Find Air Leaks

Here are some methods of finding air leaks in the building envelope:

- *feel* works well for localized inward leaks, especially during cold weather. Feel is not sensitive enough for outward leaks.
- *small flames* are sensitive to leaks in both directions. Just don't burn the place down. This is how the Three Mile Island nuclear plant meltdown got started. Candles are old fashioned. Use a barbecue lighter.
- *test smoke* is the most effective visual method of finding outward air leakage, but it is poor at detecting inward leakage. Smoke testing has become more practical with the introduction of canisters that can release small puffs. (These are sometimes called "smoke pencils.") Get smoke sources from air conditioning supply houses. Old-fashioned smoke bombs produce far too much smoke, frighten occupants, and set off some smoke alarms.
- *a lightweight cloth* or a *piece of tissue paper* dangled near a suspected leak is sensitive to both

inward and outward flow. This method's only disadvantage is lack of glamor.
- *a lightweight ribbon attached to the end of a stick* works well for finding leaks deep in wall cavities and other places where you cannot reach. Tape a miniature flashlight to the same stick so you can see into the dark hole.
- *infrared thermography* has the unique ability to trace air leakage paths through envelope structures. Scan inside surfaces for incoming leaks, and the outside surfaces for outgoing leaks. Thermography must be done during cold weather, the colder the better. Scans should be done at night, several hours after sunset, to avoid false images caused by solar heating. Thermography is expensive, it cannot get into tight places, and it requires special expertise. For more, see Reference Note 15, Infrared Thermal Scanning.

How to Estimate Air Leakage

It is not possible to estimate the amount of air leakage with precision, even when the locations of leaks are known. The amount of leakage depends on imbalances in the conditioning systems, the average chimney effect, and the average effect of wind, and other factors that cannot be defined accurately. Go ahead and minimize leaks that you find, and forget about making detailed estimates of savings.

ASHRAE has developed formulas and tables for estimating leakage around windows and doors. These methods may be useful if it appears that most leakage occurs at such components. You start by estimating the total opening area of the leakage paths. For example, air leakage around the sash of particular window types can be estimated from ASHRAE tables by measuring the total length of the joints where leakage can occur. These methods are not accurate, but they give you some basis for deciding whether to improve or replace leaky windows or doors.

A significant amount of energy loss can occur from leakage of outside air into the structure of walls and ceilings, rather than into the occupied space. Such leakage bypasses insulation, reducing the effective R-value of the envelope. There is no practical method of measuring the effect of such leakage in the field, but it may be estimated from a thermographic survey. The solution to such leakage is to seal air leakage at the outside surface of the envelope wherever it is practical to do so.

A "blower door" is a device used to test the overall leakage rate in small buildings, such as residential houses. It consists of a panel that fits tightly in a door, window, or other exterior opening. A fan mounted in the panel creates a pressure in the building. A manometer mounted in the panel measures the pressure. If the envelope is tight, the manometer registers a higher pressure than if the building is leaky. The total amount of leakage is derived from the characteristics of the fan.

Blower doors are difficult to use. In order to find dispersed, small envelope leaks, it is necessary to seal all the intentional openings in the building, including furnace flues, hoods, fireplaces, etc. By the same token, the blower door helps to highlight and quantify such major leaks if the blower door test is repeated as each leakage path is sealed. For example, the blower door test may indicate that it is worthwhile to install a low-leakage damper in a kitchen hood system.

Vapor Barriers and Air Leakage

See Reference Note 43, Vapor Barriers, for an explanation of vapor barriers. In brief, a vapor barrier is an impermeable membrane that blocks the flow of atmospheric water vapor through the building envelope. Vapor barriers can also be an effective means of preventing air leakage through walls, ceilings, and other large areas where they are installed.

Vapor barriers provide little benefit in the parts of the building envelope that commonly have the worst air leakage, such as windows and door frames. Vapor barriers are broken where wiring must pass through the surface. For example, this accounts for the drafts that you may feel at electrical switches and receptacles that are installed in outside walls.

Maintenance to Reduce Air Leakage

Leakage seals for doors, windows, and other movable envelope components require occasional replacement. Caulking materials used to seal fixed components lose elasticity and shrink with time. Therefore, maintenance is an essential part of deterring air leakage. The effectiveness of sealing depends on good workmanship and selecting appropriate sealing materials and methods.

A Quick Review: What is "Heat"?
How Insulation Blocks Heat Flow
- Preventing Heat Conduction
- Blocking Heat Radiation
- Preventing Heat Flow by Mass Transfer
- Preventing Convection

External Factors that Affect Insulation
   Performance
Conduction Lag and Heat Storage Effects
Heat-Reflecting "Insulation"

This Reference Note explains how insulation keeps heat from moving. This is the "thermal" part of insulation behavior. With this background as a foundation, Reference Notes 42 through 45 continue with the practical aspects of selecting and installing insulation.

A Quick Review: What is "Heat"?

Since the purpose of insulation is to control heat, let's review what "heat" is. Heat is a form of energy that occurs in two entirely different forms. The first form occurs in any material substance — solid, liquid, or gas. This energy consists of the random vibration of the atoms and molecules that make up the material. The vibration energy increases with temperature. (Conversely, temperature is a measure of the intensity of heat energy. The exact definition of temperature is very theoretical. You don't need to know it for practical work.)

The second form of heat is electromagnetic energy, which travels in empty space. Heat radiation is the same as light, X-rays, and radio waves. It originates from the vibration of the electric charges in matter. Conversely, any electromagnetic radiation that is absorbed by matter is converted to heat in the material.

The energy of all electromagnetic energy can be considered as heat. However, we usually consider heat radiation as the energy emitted by a hot object, such as the sun or a radiator. Heat is also radiated by objects near room temperature, such as the walls of a building. Most heat radiation falls within a broad range of wavelengths that extends beyond the red end of the visible light spectrum. This broad spectrum is called "infrared."

How Insulation Blocks Heat Flow

The purpose of insulation is to keep heat from moving. To understand how insulation works, you need to understand how heat moves. There are three fundamental processes of heat movement: "conduction," "radiation," and "mass transfer." A fourth mode of heat movement, "convection," is driven by the first three. We will summarize how insulation works in relation to each

of these four modes of heat movement. This will give you enough background to interpret the thermal performance specifications of any type of insulation.

■ Preventing Heat Conduction

Most heat loss through building insulation occurs by conduction. Conductive heat transfer is an exchange of kinetic energy between atoms and molecules as they collide with each other. Therefore, conduction occurs only in matter. You can keep heat from traveling by conduction if you create an empty zone that has no atoms and molecules. Stated differently, heat cannot move by conduction across a vacuum. This is the insulating principle used in thermos bottles. The reason why vacuum is not used for building insulation is that no one has yet devised a reliable, inexpensive vacuum container for large surfaces.

Vacuum is not yet available as a practical form of insulation on a large scale, so insulation uses the next best approach. This is to approximate a vacuum by minimizing the amount of matter through which the heat can move. Gases have much lower density than solids, so insulation works by eliminating as much solid material as possible, and replacing it with a gas. In all porous insulation, the surrounding air is the gas that is used.

An empty air space does not work well as insulation, because air is easily moved by convection and pressure differences. Porous insulation holds the air in place by using small quantities of sold material as a matrix to hold the air still. Air adheres weakly to the solid material, so the air is held stationary. (Even so, the insulation must be enclosed effectively to keep air from moving.) The solid material is distributed so it occupies as much of the space as possible, so that each air molecule is close to an attachment point.

The solid material used in porous insulation may be in the form of fibers or granules. Insulation fiber is usually made of glass or glassy slag. Granules can be made of chopped waste paper ("cellulose"), various minerals in expanded form, foam beads, or other materials. Reference Note 43 covers the common types of insulation materials.

Most heat conduction in porous insulation occurs through the trapped air, much less is through the solid material. However, heat conduction through the solid material is significant. For this reason, manufacturers try to make the solid component as thin as possible.

In the case of fiber insulation, conduction through the fibers can be reduced by orienting them perpendicular to the heat flow, so that heat cannot travel along the fibers. Fiber batt and board insulation is made this way. Loose fiber pouring insulation does not have this advantage.

Many gases have better insulating properties than air. Some of these are used in plastic foam insulation, in less common types of insulation, and in windows. Chlorofluorocarbons (CFC's) have been the primary gases used as alternatives to air. However, concern about the effect of CFC's on the atmosphere is motivating a search for other gases to replace CFC's.

All insulation that uses alternative gases requires closed cells to keep the gas from escaping and being replaced by air. This has proven to be a serious weakness of this kind of insulation. Foam insulation, which dominates the market because it is easy to manufacture, holds the gases in tiny plastic bubbles. Unfortunately, the insulating gas is able to slowly diffuse through the plastic.

Manufacturers of foam insulation use various methods of resisting this leakage, such as increasing the thickness of the bubble walls and bonding metal foil to the faces of insulation boards. The rate of leakage may be low in good insulation, but all contemporary plastic foam insulation will eventually lose its special gases. This has spurred the use of other materials to hold the gases, such as glass foam, but no alternative to plastic foam is presently acceptable for wide application.

■ Blocking Heat Radiation

Radiation heat loss and heat gain is not, by itself, a problem in the opaque parts of the building envelope. All opaque construction materials stop heat radiation completely. However, the energy of heat radiation must go somewhere. Radiation that strikes an opaque material is either absorbed or reflected. Absorbed radiation increases the temperature of the material, and the added heat is transferred by conduction. Since insulation blocks heat conduction, it indirectly blocks heat radiation also.

Radiation heat transfer occurs inside insulation materials at a microscopic level, from one molecule to another. However, this effect is small at normal temperatures. For insulation of typical thickness, this process accounts for less than 10% of heat transfer.

Heat radiation is important in transparent parts of the envelope, namely, windows and skylights. In fact, it is so important that Section 8 is set up to deal with it separately. Refer to there for the Measures that deal with glazing.

■ Preventing Heat Flow by Mass Transfer

If you heat some material, then move it somewhere else and let it cool, you have moved heat by mass transfer. The mass is a carrier for the heat energy. Mass transfer is a powerful method of moving heat. In air conditioning systems, air is moved by fans and water is pumped to transfer heat. Unfortunately, air is also effective in moving heat through the building envelope. If the envelope is leaky, loss of heat carried by air may account for much more energy waste than conduction or convection.

Air leakage in buildings is driven by wind, chimney effect, and unbalanced operation of fans. Reference Note 40, Building Air Leakage, gives the details of these processes.

You can eliminate air leakage by making the envelope airtight. In most cases, the insulation itself plays only a secondary role in minimizing leakage. Air movement is stopped primarily by impermeable structural components and by installing "vapor barriers," which are explained in Reference Note 42.

■ Preventing Convection

Convection is movement of air that occurs when the air is heated or cooled. The air adjacent to a heated surface is warmed by conduction, becomes less dense, and rises. If the air encounters a colder surface, it becomes more dense, and falls. In an enclosed space, such as a wall cavity, this cycle continues as long as one side of the cavity is warmer than the other. If the air inside building envelope cavities is free to move, convection is a powerful mode of heat loss.

Insulation prevents convection by holding the air in a cavity still. To do this, the insulation must entirely fill the space. Even a narrow void can seriously undermine the effectiveness of insulation, especially if it runs vertically. For this reason, the workmanship of the installer has a significant effect on insulation performance. Reference Note 44 covers this aspect.

The density of insulation installed in a cavity is an important factor. The optimum density is a compromise between conduction and convection. In practice, the density of commercial insulation is determined by manufacturing limitations and economics. For example, the mineral structure of vermiculite makes it too dense to have a very good insulation value.

On the other hand, glass fiber blanket insulation typically is less dense than optimum because the manufacturer is minimizing material cost. You can improve the performance of fibrous insulation by packing somewhat more of it into the cavity than is specified by the manufacturer. For example, use an R-19 batt where an R-11 batt is specified. This will also help to fill the cavity. However, you should still be careful to fill the cavity as uniformly as possible.

External Factors that Affect Insulation Performance

The insulation characteristics quoted by the manufacturer are measured under a particular set of conditions. If the insulation is used under a different set of conditions, its performance may differ significantly. Perhaps the most important external factor is the temperature of the insulation. The thermal resistance of all thermal insulation improves as the temperature falls. This is a bonus for building insulation in cold climates. Conversely, if insulation is used for applications at much higher temperatures, such as insulating steam pipe, the insulation value may be lower than expected.

Another factor is the orientation of the insulation. This may affect fibrous insulation, reducing the insulation value somewhat if the fibers are oriented vertically.

In most applications, these variations in characteristics are minor. Usually, they will not create any surprises if you select the type of insulation for the particular application. However, if you use a generic type of insulation, such as vermiculite, which may be used in a wide variety of applications, be sure to obtain its characteristics for the conditions under which you plan to use it.

The thermal resistance of most types of insulation is seriously degraded if moisture can condense inside the insulation, or if air is allowed to move through it. Design the installation and select the insulation to avoid both problems. See Reference Note 44, Insulation Integrity, for the details.

Conduction Lag and Heat Storage Effects

Insulation is inherently light in weight because it must contain little mass. Although insulation obstructs heat flow, it does not delay the heat flow that does occur. However, if insulation is combined with the mass of a heavy structural component, such as a masonry wall or a concrete roof slab, the combination structure may have a significant delaying effect on heat flow. The key factor in the time delay is the heat storage capacity of the massive material. All masonry materials have similar specific heats, so their heat storage effect is proportional to the mass of the insulated component.

The mass effect is important primarily with masonry construction. Masonry buildings may have wall heat conduction delays of several hours or longer. A lag in conduction does not occur with metal structures, because heat travels quickly in metal. Thermal lag is not important in frame structures, because the heat storage capacity of the structure is too low to develop much lag.

The lag effect is significant only if the mass of the structure is exposed to outside conditions, i.e., if the insulation is installed inside the structure. If the weather alternates on a daily basis between too hot and too cold, the lag effect is beneficial. This is because the lag effect reduces the average temperature differential across the insulation. However, if the insulation is installed outside the mass, the mass is not subjected to large temperature changes, so the lag effect is greatly reduced.

If the mass of the envelope is exposed to the inside of the building, i.e., if the insulation is installed on the outside, there is an adverse interaction with temperature setback (which is described in Measure 4.3.2). What happens is that a large amount of energy is required to refill the masonry with heat at the end of the setback period. On the other hand, using mass inside the envelope is important with passive solar installations (covered in Subsection 8.4).

Heat-Reflecting "Insulation"

The warm surfaces inside a building radiate heat energy toward the envelope. Knowing this gave someone the idea of installing reflective surfaces in the building envelope to reflect heat back into the space. It is convenient to piggyback this reflecting surface on batt and board insulation, although their functions are independent. (In the case of foam insulation, foil faces are also intended to retard gas leakage, as we discussed previously.)

Be careful when you read the specifications for insulation with reflective backings. The stated R-value is likely to include the effect of the reflective backing when the insulation is installed in an ideal manner. The benefit of the reflective backing increases strongly at greater temperature differentials. Therefore, manufacturers are likely to specify their insulation performance at a large temperature differential, exaggerating the R-value.

The reflective surface does not work unless there is an air space in front of it. This space needs to be a large fraction of an inch (about two centimeters), at least. For example, if board insulation is attached to an inside wall with battens (furring strips), an adequate gap is created when the interior finish is applied over the battens. Another opportunity to use reflective backing exists where the insulation is exposed to the interior space. This occurs in areas where appearance is not critical and the insulation is not exposed to damage, as in ceiling plenums and the exposed ceilings of industrial buildings.

If the design of the building does not include an air gap that would allow you to use reflective insulation, should you change the installation to create an air gap? Probably not. You will probably get better overall performance by filling the space with insulation. Furthermore, creating a gap in an insulated assembly invites serious convective heat loss.

Reflecting internal heat radiation is an important technique for improving the poor thermal characteristics of glazing. So-called "low-emissivity" coatings are used on glazing to reflect heat back into the building. See Section 8 for details.

Heat-reflecting insulation can be valuable for reducing heat loss from very hot surfaces, such as high-pressure steam lines. See Measure 1.11.1 for such applications.

What is a Vapor Barrier?
Purposes of Vapor Barriers
Where Should the Vapor Barrier be Installed?

Vapor Barrier Materials
How to Vent the Water Vapor
House Wraps are Not Vapor Barriers

What is a Vapor Barrier?

A vapor barrier is an impermeable membrane that blocks the flow of air through the building envelope. A vapor barrier is an essential part of the building envelope. Because the purpose of a vapor barrier is not obvious, this important component is often omitted or installed incorrectly.

The main purpose of a vapor barrier is preventing the passage of the water vapor that is contained in air. Vapor barriers and the insulation affect each other. They must both be installed so that they interact beneficially rather than harmfully.

Purposes of Vapor Barriers

The specific functions of vapor barriers are:

• *protecting the envelope structure and insulation from condensation damage.* Many wall materials are permeable to the flow of water vapor from inside to outside, or vice versa. As water vapor from the inside of the building moves outward through a wall on a cold day, it encounters progressively lower temperatures. At the point in the wall where the temperature of the air equals the dew point, the vapor starts to condense, and it keeps condensing from that point outward. Figure 1 illustrates this. Condensation damages all types of envelope structures. It rots wood structures, it rusts steel structural members and steel masonry reinforcements, and it causes freeze cracking of masonry. Installing a vapor barrier on the warm side of the envelope prevents water vapor from traveling through the wall, and thereby prevents condensation.

The protective function of vapor barriers is not inherently related to insulation. Condensation can occur inside the envelope structure whether it is insulated or not. If water vapor condenses inside insulation, the dampness reduces thermal resistance, and may damage the insulation.

• *preventing air leakage through the envelope.* A well-installed vapor barrier prevents or greatly reduces air leakage through the envelope surfaces, although it does not reduce air currents inside the envelope structure itself. At the same time, the vapor barrier reduces air flow through the

insulation, preserving the R-value. For more about this function, see Reference Note 40, Building Air Leakage. (As a matter of perspective, vapor barriers do nothing to reduce air leakage through the major envelope penetrations, such as doors, windows, roof hatches, and fan openings. These penetrations

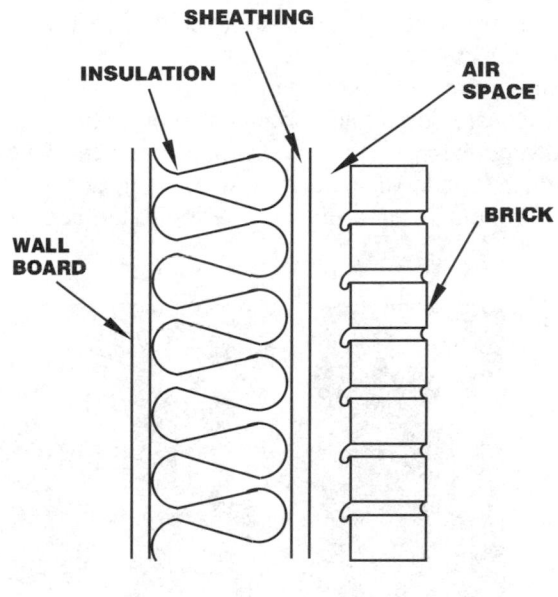

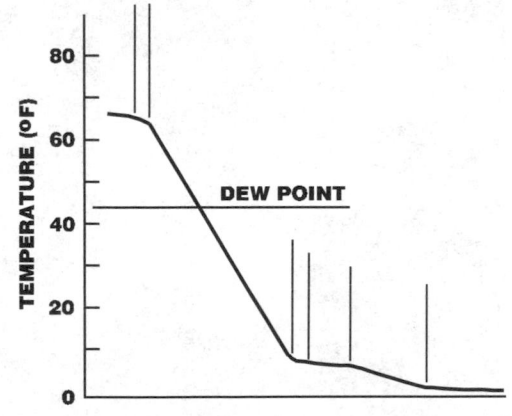

Fig. 1 Why building structures need vapor barriers This is a cross section of an insulated stud wall with an outer brick veneer. Below it is a graph of the temperature inside the wall. It is cold outside and warm inside. It is also humid inside. If water vapor can flow through the wall, it will reach a point at which it condenses. From that point outward, the wall is damp. A vapor barrier on the inner surface keeps the water vapor from flowing through the wall.

account for a majority of air leakage in most buildings.)

- *maintaining interior humidification.* If humidification is used, a vapor barrier reduces the amount of energy and water required to maintain the desired level of humidity.

Where Should the Vapor Barrier be Installed?

Vapor barriers must be installed on the warm side of the insulation. This is because condensation occurs as water vapor moves from the warm side of the wall to the cold side. If a vapor barrier is installed on the cold side, it traps moisture inside the envelope, making moisture problems worse.

This poses a dilemma in climates where the weather can be hot and humid in summer, but cold in winter. The deciding factor is how cold it gets. Where the winters are seriously cold, as in Minnesota, the best compromise is to install the vapor barrier on the inside. In humid climates where winter temperatures are mild, as in Houston, the best compromise probably is not to use a vapor barrier. If this decision is made, the envelope should be made of materials, such as masonry, glass, and aluminum, that withstand periodic dampening.

It might be tempting to solve this problem by installing a vapor barrier on both sides of the envelope. However, this is the worst approach. Vapor barriers on both sides of the envelope would almost certainly trap harmful amounts of moisture.

Vapor Barrier Materials

In principle, a vapor barrier can be any unbroken surface that is impermeable to water vapor. For example, a common vapor barrier material is polyethylene plastic film, typically installed in thicknesses from .002" to .008" (0.05 mm to 0.2 mm). This material is inexpensive, transparent, easy to handle, and is available in wide widths. It can be attached by stapling, mastic, and other means. Figure 2 shows a properly installed vapor barrier using this material.

Vapor barriers can be attached to permeable insulation, such as glass fiber batts or blankets. The vapor barrier is commonly in the form of impregnated kraft paper, sometimes with a thin foil layer. This type of vapor barrier is unreliable because there is no effective way to close the gap between adjacent lengths of insulation. Fold-over strips intended for overlapping the vapor barrier of adjacent batts are generally ignored by installers.

WESINC

Fig. 2 Perfectly installed insulation and vapor barrier   This large room has wood stud walls and a wood rafter cathedral ceiling. Glass fiber batt insulation has been inserted snugly into the stud and rafter spaces, leaving no gaps. The tabs on the paper backing of the insulation are overlapped and closely stapled to the edges of the studs and rafters. A vapor barrier of 0.008" thick polyethylene sheet is stapled over the insulation. The vapor barrier is overlapped several feet at all joints. Plenty of excess plastic material is left in all corners. This slack keeps the plastic from being torn when the wallboard is installed. The vapor barrier is stapled to the window frames, preventing air leakage around the windows.

If the insulation material itself is very impermeable, such as extruded foam board insulation, it may act as its own vapor barrier. This characteristic is useful only if adjacent sheets of insulation are joined to create an unbroken surface. This requires special installation techniques that are difficult to enforce on the job.

Some envelope construction materials, such as asphalt roofing and sheetmetal walls, are impermeable. Therefore, they act as a vapor barrier, whether this is desirable or not. The critical question is whether these impermeable materials are located on the warm side of the insulation. If they are, they can serve as the vapor barrier. If not, they create a moisture venting problem that must be handled properly to prevent damage.

How to Vent the Water Vapor

When installing insulation, create a path for venting water vapor from the insulated cavity. A vapor barrier on the warm side of the envelope must be combined with a venting path on the cold side of the insulation. This is because no vapor barrier is perfect, and because water may get into the structure, typically from rain. In general, the better the vapor barrier and the drier the conditions, the less venting is required.

Effective venting is a challenge with roofs, because they are susceptible to leaks and have an impermeable outer surface. In buildings with attics, a common solution is to vent the attic to the outside. In cathedral ceilings, leave an air space above the insulation to allow water vapor to travel out to the vents, which should be installed along the full lengths of the ridges and eaves.

Venting walls and soffits is just as important, and the same principles apply. If there is an impermeable surface on the cold side of the insulation, such as a sheetmetal outer wall, leave a gap between the cold side of the insulation and this surface. The gap acts as a path for water vapor. In turn, vent this gap to the outside of the cold surface. Walls that can be wetted by precipitation require thorough venting.

At the other extreme, some wall materials are so porous that moisture may vent directly through the wall. Such material is especially vulnerable to rain soaking. Keep insulation away from direct contact with the wall. Generous roof overhangs are an excellent means of keeping walls dry, if the walls are not too tall.

Portions of the building that are located over soil have the problem of moisture migration into the building, rather than outward. The usual solution is to vent the crawl spaces to the outside, as with attics.

All insulated cavities where water may accumulate should have drains, as discussed in Reference Note 44, Insulation Integrity.

House Wraps are Not Vapor Barriers

"House wraps" are an item that goes through episodes of popularity. House wraps tend to be confused with vapor barriers, although their function is entirely

WESINC

Fig. 3 This is not a vapor barrier This is a house wrap. It must be made of appropriate permeable material. Using vapor barrier material as a house wrap would cause moisture damage in the walls. The outer sheathing of this house is plywood. If the plywood is well installed, the house wrap is superfluous.

different. Using vapor barrier material as a house wrap can cause serious moisture damage to the structure. Conversely, house wrap material will not work as a vapor barrier.

The purpose of a house wrap is to prevent air infiltration through the building structure. It is always installed on the outer surface of the building. It plays the same role as a wind breaker in human clothing. As we said previously, one of the benefits of a vapor barrier is preventing air leakage. However, in most climates, vapor barriers are installed on the inner surface of the structure. Therefore, they do not protect the structure, or the insulation inside the structure, from heat loss that is induced by wind.

House wrap material must be permeable. Otherwise, it will act as a vapor barrier that is installed on the wrong side of the surface, and cause moisture damage. A common material used for house wrap is a fiber-reinforced paper. The material is stapled to the outer surface of the structure before installing the outer

weather surface (siding, brick, etc.). Figure 3 shows a typical house wrap installation.

The fact that the material is permeable does not significantly interfere with its wind protection. It is like human clothing that protects from wind while allowing moisture to vent from the body.

We should ask whether house wraps are really needed. If a building is constructed properly, house wrap is superfluous. If the exterior sheathing is installed with sufficient care, it will shield the wall structure from wind. Furthermore, a building should not depend on a structural component that has a reliable life that is less than the life of the building. House wraps are fragile, compared to other structural materials. It is unlikely that they will survive for the life of the building, especially if the exterior surface that protects the house wrap will be replaced during the life of the building.

House wraps are not snake oil, but they have a limited range of useful application. They are most valuable when renewing the exterior of a house with leaky walls.

U-Value and R-Value
Other Insulation Selection Characteristics
- Fire Characteristics
- Material Cost
- Ease of Installation
- Care Required in Installation
- Ability to Fill Voids
- Tendency to Settle
- Resistance to Moisture Damage
- Aging Behavior
- Resistance to Physical Damage
- Temperature Range
- Vermin Resistance
- Growth of Microorganisms
- Moisture Permeability
- Emission of Noxious Chemicals

Contemporary Types of Envelope Insulation
- Glass Fiber
- Mineral Fiber
- Plastic Foam Boards
- Plastic Foam Beads
- Sprayed Plastic Foam
- Loose Dry Cellulose
- Wet Sprayed Cellulose
- Vermiculite
- Perlite
- Lightweight Concrete
- Soil

This Note provides specific information about the types of building insulation that are presently available, along with the factors that you should consider in selecting them.

U-Value and R-Value

The thermal effectiveness of insulation is measured by "U-value" and "R-value," which are two ways of expressing the same information. The formal name for U-value is "thermal conductivity." In English units, U-value is the number of BTU's per hour that pass through one square foot of the material for each degree Fahrenheit of temperature differential across the material.

U-value is used in the well known formula for heat transfer by conduction. If Q is the rate of heat flow, expressed in BTU's per hour, then:

$$Q = U \times (\text{surface area}) \times (\text{temperature differential})$$

Thermal conductivity is also expressed in some tables as "k," "c," or "C." These symbols are not used consistently. "C" usually is the thermal conductivity for a particular thickness that may be specified arbitrarily. "k" is commonly used in two ways, for a thickness of either one inch or one foot. Metric tables use these same symbols with definitions that are expressed in metric units.

The "thermal resistance" of a material is the reciprocal of its thermal conductivity. In English units, thermal resistance is expressed as "R-value." Mathematically, $R = 1/U$.

R-value is handy because it allows you to calculate the total R-value of an assembly, such as a wall or roof, simply by adding the R-values of all its components.

For example, to calculate the total R-value of a stud wall that is filled with insulation, add the R-values of the outer air layer attached to the exterior siding, the exterior siding itself, the air layer between the siding and the outer wall sheathing, the wall sheathing itself, the insulation inside the stud space, the interior sheathing, and the air layer inside the space.

U-values cannot be added. To calculate the heat transfer through an assembly, first calculate the total thermal resistance by adding the R-values. Then, take the reciprocal of the total R-value and use it in the formula above.

The R-values of most insulation materials are fairly independent of external factors. The R-value of insulation increases at lower temperatures, but this effect is not large enough in envelope insulation to favor one type of insulation over another. (In contrast, the R-values of glazing units are strongly influenced by orientation, temperature, internal drafts, and wind. These insulating characteristics of glazing are covered in Section 8.)

Table 1 lists the typical range of R-values for the common types of building insulation. The R-values of some common construction materials are included for comparison.

Be aware that R-value does not imply resistance to air leakage. It is a common mistake to use insulation as caulking material to plug envelope leaks. All types of fibrous and porous insulation are poor at blocking air flow.

Other Insulation Selection Characteristics

The R-value of insulation is by no means the only important selection consideration. There are many types of insulation, each having advantages and disadvantages as a construction material. You need to consider a

Table 1. R-VALUES OF COMMON INSULATION AND CONSTRUCTION MATERIALS

Material	R-Value per inch
Mineral wool*	2.9 to 3.8
Glass fiber batt & blanket*	2.9 to 3.8
Glass fiber board	3.8 to 4.2
Cellulose, shredded paper, new	approx. 3.3
Extruded polystyrene board**	4.0 to 5.2
Polystyrene bead board**	4.0 to 5.0
Polyurethane foam board**	5.0 to 6.0
Polyisocyanurate foam board	5.5 to 8.0
Perlite	approx. 2.7
Vermiculite	approx. 2.3
Lightweight concrete	0.2 to 2.0
Concrete, sand or gravel mix	0.1
Concrete & cinder blocks	0.15 to 0.3
Brick	0.2
Plywood	1.2
Pine	1.2
Acoustical tile, typical	2.5
Fiberboard sheathing, typical	2.5

* Higher density provides better R-value because commercial batt insulation has less than the optimum density of fibers.

** Assumes CFC fill gas. R-value diminishes with age because of gas loss. The figures assume moderate aging.

surprisingly large number of characteristics for each application. If insulation is mismatched to the application, you may have serious problems. Make your selection using reliable product data.

The following characteristics are important for typical building insulation applications. In specialized applications, there may be other insulation characteristics that are important, such as adhesion or the ability to mold the insulation into shapes. The main point is to recognize all the characteristics that are important for each application.

■ Fire Characteristics

The fire characteristics of construction materials are expressed in a variety of ways. Three of the most common are:

• *fuel value.* If the insulation has no fuel value, it cannot burn. In fact, its presence inhibits the spread of fire. However, be aware that noncombustible insulation materials may be combined with materials that have fuel value, such as binders and backing paper. An example is glass fiber batt insulation, which uses an organic binder, and which may have a flammable backing sheet.

• *flame spread,* roughly speaking, is the rate at which fire moves along the material. This factor is significant if the insulation itself is the primary fuel for the fire. It is irrelevant if fire spreads outside the insulation.

• *smoke production* indicates the amount of smoke that the insulation material produces when it burns. Unfortunately, this rating does not address the critical issue of the smoke's toxicity.

Fire safety classifications still fail to adequately define another critical factor, which is ease of ignition. This factor may never be defined satisfactorily, because the ignition of material in a fire depends on many factors, such as the manner of installation, the way the fire surrounds the material, changes in the properties of the insulation before it ignites, etc.

The fire resistance of insulation may be improved considerably by adding retardants, which keep the insulation from catching fire. The performance of retardants varies considerably depending on test conditions. The fire retardant rating assigned to a given insulation material may differ radically from one country to another. The basic problem is that fire retardants cease to function effectively if the fire exceeds a certain temperature and duration. *Any material that has fuel value will eventually burn.*

The importance of fire characteristics depends on where the insulation is installed. For example, fire behavior is less important for insulation installed on a concrete roof deck than for insulation installed under interior paneling.

■ Material Cost

There are large cost differences among different types of insulation. The main reason that more expensive types of insulation are on the market is that they are easier to install.

■ Ease of Installation

Labor is a large component of insulating cost, and labor requirements vary widely with different types of insulation. For example, insulation that can be installed by pouring or blowing requires less labor than insulation that is fitted into place by hand. Using foam board insulation is a very quick way of insulating vertical surfaces.

■ Care Required in Installation

This is not the same as ease of installation. The most common example is fiber batt insulation. Even though batts are easy to install, they are commonly installed in a manner that leaves voids and reduces their effectiveness. Similarly, foam board insulation is exceptionally easy to install, but failing to seal the boards properly at the edges can waste most of their insulating value.

■ Ability to Fill Voids

Voids transfer heat because they allow convection. Aside from open areas that are needed for venting, the entire cavity in the envelope structure should be filled.

Most types of insulation have some tendency to leave voids. Each type of insulation requires its own techniques to eliminate or minimize voids.

■ Tendency to Settle

Loose fill insulation tends to settle, which reduces its R-value. If a flat surface is being insulated, the solution is simply to add more material to compensate. However, if the insulation is used to fill a vertical cavity, such as a stud wall, settling causes the upper portion of the cavity to become completely uninsulated. Cellulose insulation has the worst settling among the common insulation types.

Fiber batt and blanket insulation does not settle if laid flat. If installed vertically, it must be installed snugly in cavity, and it must be supported by a backing sheet. If batt insulation is suspended by a backing sheet alone, it will eventually fall away.

■ Resistance to Moisture Damage

Insulation may be wetted occasionally in ways that are not obvious. Insulation installed in a masonry wall is wetted by rain soaking through the wall. Attic insulation is wetted by occasional roof leaks. Lack of an effective vapor barrier causes permeable insulation to be wetted by condensation during cold weather. Most types of insulation tolerate occasional light moisture, but some do not. Some applications, such as exterior foundation insulation, subject the insulation to continuous soaking. This destroys the R-value of most insulation sooner or later, unless the insulation material is exceptionally resistant to moisture.

■ Aging Characteristics

Inorganic insulation generally does not deteriorate with time. At the other extreme, plastic foam insulation suffers a rapid initial lowering of R-value, followed by a slower long-term degradation of R-value. Cellulose insulation is reported to deteriorate somewhat with time.

■ Resistance to Physical Damage

Insulation may encounter various types of hazards. It can be protected by selecting a type that resists the hazard, or by external protection, or by a combination of both. For example, roof insulation must resist people walking on it. The protection can be provided by selecting a plastic foam insulation that is dense enough to resist the pressure, or by distributing the pressure with a covering of ballast, or by installing a walkway over the insulation. Plastic foam insulation is destroyed rapidly by sunlight, so it must be protected with an opaque surface. Pipe insulation in a machinery room is subject to local impact damage, which can be prevented with suitable guards. And so forth.

■ Temperature Range

Temperature tolerance is not often a factor in envelope insulation, but recognize when it may be. For example, do not use polystyrene foam where it may be exposed to temperatures of 170°F or higher. Such temperatures may occur near flues, ceiling heaters, and incandescent lighting fixtures.

■ Vermin Resistance

Anyone who reads H. P. Lovecraft knows about rats in the walls. Vermin eat cellulose, and they may use other types of loose fill insulation as nesting material. The problem is not primarily loss of insulation value, but proliferation of the vermin.

■ Growth of Microorganisms

Some types of insulation may act as a growth medium for microorganisms, such as mold or fungus. Microorganisms may grow in insulation under some conditions even if the material itself is not organic. These microorganisms may contribute to "sick building syndrome." This is primarily a problem in duct insulation, not in envelope insulation, but envelope insulation may become a problem if it is wet.

■ Moisture Permeability

Preventing moisture migration through insulated surfaces is important in most applications. If the insulation is impermeable, it may serve as a vapor barrier, but only if it is installed in a manner that leaves no gaps. For more about vapor barriers, see Reference Note 42.

Permeability is not an all-or-nothing characteristic. For example, some people assume that plastic insulation is an absolute vapor barrier. This is not true. All plastic insulation is permeable to some degree, and some types of plastic insulation are much more permeable than others.

■ Emission of Noxious Chemicals

Experience with urea formaldehyde wall insulation during the 1970's and 1980's demonstrated that some types of insulation may emit dangerous chemicals. No similar problem is known with other common types of insulation, but the lesson should be remembered. In particular, be cautious about using insulation that requires mixing noxious components on site, or that emits noxious chemical products.

Contemporary Types of Envelope Insulation

The diversity of insulation applications keeps a variety of insulation types on the market. Each application usually has no more than one or two types of insulation that are best suited to it. The following listing includes the major types worth considering for conventional applications. It omits materials that are now obsolete or unacceptable for building insulation, such as cork, asbestos, and urea formaldehyde.

■ Glass Fiber

Glass fiber insulation has good R-value because the fibers are oriented perpendicular to the direction of heat flow, and have little contact with each other. It is available in the form of batts, blankets, semi-rigid boards, and loose fill.

Batts are used for wall cavities. They have a stiff paper backing sheet for attachment and handling, and to act as a vapor barrier. Board insulation is attached to the inside surfaces of walls and roofs. Batts, blankets, and loose fill are used for attic insulation. Batts, blankets, and boards use a binder to hold the fibers together. Loose glass fiber insulation can be installed by blowing.

Contrary to what you might expect, the R-value of fiber insulation is higher if the material is compressed more densely than it is supplied, without going to extremes. The cost of material leads manufacturers to produce batts that have less than optimal density. For this reason, when filling a cavity with blankets or batts, consider using thicker insulation to achieve greater density. This also tends to reduce voids. (But, do not do this if it intrudes on an air space needed for moisture venting.) Semi-rigid fiber boards have better R-value than the fluffy blankets and batts, even though they contain more solid matter.

The kraft paper backing sheets of batt insulation is a mediocre vapor barrier at best. To achieve reliable blocking of water vapor, install a separate vapor barrier, such as large sheets of plastic film, over the insulation.

Glass fiber is inert, which makes it generally safe for occupants of the structure. However, the fibers are a safety hazard during installation. They can penetrate eyes and skin, and there is suspicion that they may cause lung disease. Glass fiber may be a hazard to occupants if it can break loose, as from duct linings,

■ Mineral Fiber

Mineral fiber has essentially the same physical properties as glass fiber and it is used the same way. It is made from dirty raw materials, such as boiler slag, so it is not as pretty as glass fiber. It has shorter fibers, so it is more likely to be used in loose form, as "mineral wool," than in blankets and batts.

■ Plastic Foam Boards

Plastic foam insulation created a revolution in the construction industry because of its high R-value, low moisture permeability, and ease of installation. Most plastic foam is installed in the form of boards, which are used for wall, roof, and subsurface insulation.

Plastic foam has high R-value because the gas is perfectly encapsulated, which prevents gas movement by convection or external pressure differences. The walls of the foam bubbles are thin, which limits conduction through the solid material. The plastic is expanded into a foam by gas injection. CFC's were originally used as the blowing gas, but these are being replaced by other gases because of concern about ozone depletion in the atmosphere.

Unfortunately, the R-value diminishes with age because the insulating gas leaks out and is replaced by air. Even aged foam insulation has a better R-value than other types of insulation, but the advantage may be substantially reduced.

Foam boards are easy and safe to handle. They have almost no weight, they can be cut easily with a box knife, and they do not irritate eyes or skin. The foam can be bonded to structural and finish materials, such as plywood, chipboard, and gypsum board.

The major disadvantage of all plastic foam insulation is its behavior in a fire. Plastic is made from petroleum, and it burns if it is ignited adequately. A great deal of effort has gone into making plastic insulation fire retardant, and some types can resist small fires. However, once ignited by a surrounding fire, plastics produce large amounts of toxic smoke that kills occupants quickly.

For this reason, it is prudent to avoid using plastic insulation in interior applications. Sheathing plastic insulation in non-flammable material, such as gypsum board, may buy time to evacuate a burning space, but this practice cannot keep the plastic from igniting. In a big fire, plastic insulation melts, runs out into the fire, and then ignites.

There are several types of plastic foam used in board insulation. At present, the main types are:

- *extruded polystyrene foam.* Polystyrene is the cheapest of the foam insulation materials, but it is still about twice as expensive as glass fiber for a given surface area. It is limited to temperatures below about 170°F (77°C).

- *polystyrene beadboard.* Polystyrene foam beads can be fused together to form insulating boards. In fact, the beads can be molded into any shape, such as coffee cups, but boards are the most common shape for insulation purposes. If kept dry, beadboard has about the same thermal properties as extruded polystyrene. It is less resistant to long-term moisture penetration. This is important, for example, in insulation that is installed below grade.

- *extruded polyurethane foam.* Polyurethane foam has better R-value than polystyrene. It is more expensive. It tolerates somewhat higher temperatures, around 250°F. It is less resistant than expanded polystyrene to long-term moisture penetration in wet applications, such as foundation insulation below grade.

- *extruded polyisocyanurate foam.* Polyurethane is combined with isocyanurate plastics to create insulation materials that are usable over a wider range of temperatures, from about -290°F to about +300°F.

■ Plastic Foam Beads

Loose plastic foam beads are used for pouring applications, such as the cavities of masonry walls. Polystyrene is the most common material for making foam beads. The fire characteristics discussed for foam boards also apply to foam beads. Foam beads are replacing perlite and vermiculite in applications where fire resistance is not important. They have become popular as an aggregate for lightweight concrete.

■ Sprayed Plastic Foam

Sprayed plastic foam typically is applied to flat roof decks. It requires specialized equipment to create the foam on site. It produces a lumpy surface that can puddle water if it is installed on a flat surface. Foam is destroyed rapidly by sunlight, so exposed surfaces need a durable opaque coating. Since the surface is irregular, it is not practical to apply flat sheathing to protect the foam, and the foam is not quite strong enough to support people walking on it. Therefore, you need separate walkways to protect the foam.

■ Loose Dry Cellulose

"Cellulose" insulation consists of shredded old newspapers treated with chemicals. For example, borax may be added to retard fire, and other chemicals may be added to improve vermin resistance. Cellulose insulation was advocated heavily during the 1970's as a useful way of disposing of waste paper. Its thermal properties are good. Its major weaknesses are settling, fire hazard, degradation if wetted, and the fact that it makes food and nesting material for vermin. Its properties are largely dependent on additives, so quality varies widely among different manufacturers.

Its main advantages are low cost and ease of installation. It typically is blown into place using specialized pneumatic equipment. It is widely used in attics. It can also be blown into walls, but this application is riskier because of its tendency to settle, difficulty in filling voids, vulnerability to moisture, and vulnerability to the skill and honesty of the installer.

Paper burns, so it is important to isolate cellulose insulation from any heat sources, such as flues and incandescent light fixtures.

■ Wet Sprayed Cellulose

Cellulose insulation can also be applied as a slurry of cellulose material mixed with a glue. It can be applied to any orientation, including overhead, although the thickness of each application is limited. When the slush dries, it becomes like papier mache, and it is dimensionally stable.

■ Vermiculite

Vermiculite is a class of minerals consisting primarily of silicates of magnesium, aluminum, and iron that are bound with water. Vermiculite ore has a flat, layered structure that looks like mica, and it has a significant water content. It is converted to insulation by a process called "exfoliation," which is similar to making popcorn. When exposed quickly to high temperature, the water in the ore bursts into steam, expanding the layers and creating hollow granules that look like tiny accordions.

The R-value of vermiculite is lower than that of other insulation materials, and it varies depending on the source of the raw material, the particle size, and the manufacturing process. The main advantages of vermiculite are lack of flammability, ability to withstand high temperatures, benign handling characteristics, and easy pouring. Vermiculite is a good aggregate for lightweight concrete. It is also used as a component of sprayed fire retardant insulation.

■ Perlite

Perlite is made from a certain type of volcanic rock that is high in silica. The material contains a small percentage of water. The rock is first crushed. Then, it is heated quickly to a temperature above 1,600°F (870°C). As the rock melts, the water creates a froth of tiny bubbles, expanding the material by a factor of four to twenty. The cooled material retains the sealed bubbles, making it relatively light and providing significant thermal resistance. The individual particles are physically strong, making the material an effective aggregate for lightweight concrete. It can also be used in loose form, typically to fill the holes in concrete block walls.

■ Lightweight Concrete

Concrete can be made with virtually any material as an aggregate. If a lightweight material is used instead of sand or gravel, both the weight and the thermal conductivity of the concrete may be greatly reduced. Lightweight concrete does not have the high R-value of true insulation, but it insulates much better than conventional concrete.

Plastic foam beads make a concrete that is incredibly light. Perlite and vermiculite produce a concrete that is somewhat heavier and stronger. There is a compromise between R-value and strength. The compromise can be adjusted over a wide range by selecting the type and density of the aggregate material.

■ Soil

In the frenzied days of the 1970's energy crises, "earth sheltered" construction captured the fancy of many, perhaps because the idea of dirt was appealing in its own right. Soil can be considered a very poor grade of expanded mineral insulation. Its only advantage is that it is dirt cheap. However, it costs so much money and energy to modify a building's structure to accommodate soil that using it for insulation is self-defeating. Using soil adds serious problems, including high pressure on the retaining structure, roof weight, rain and ground water leakage, radon entry, visibility limitations, and the esthetics of a prehistoric burial mound.

Minimize Thermal Conduction Paths
Eliminate Convective Paths
Provide Water Drainage, if Necessary
Install the Vapor Barrier Properly

The effectiveness of insulation depends on the way it is installed. This Note covers the main points of installing insulation in a way that maintains its effectiveness.

Minimize Thermal Conduction Paths

Any solid material that bridges the space between the inner and outer surfaces of the envelope acts as a short circuit for heat. The amount of heat loss is proportional to the R-value and thickness of the conducting material. Wall studs, sill plates, sole plates, window and door framing, floor slabs that extend to the exterior structure, and the edges and supporting structures of curtain wall panels are all conductive short circuits. Metal structural elements have high thermal conductivity, and their effect is especially severe if they have wide flanges at the inner and outer envelope surfaces.

In the worst cases, conductive short circuits can severely limit the benefit of adding insulation. An example is insulating a concrete block wall by pouring insulation into the holes in the blocks. This provides some benefit, but a large amount of heat will continue to be conducted through the webs in the blocks.

To avoid conduction short circuits, try to use an insulation method that provides an unbroken layer of insulation. For example, covering the interior or exterior surface of a concrete block wall with insulation board is much more effective than filling the holes in the concrete blocks with insulation.

WESINC

Fig. 1 What is wrong in this picture? The batt thickness is appropriate for the stud widths. However, the insulation is stuffed far inside the cavity, reducing its thickness and creating a large convection path in front. The kraft paper vapor barrier is not effective, because the paper is not joined at the edges. The method of installing the electrical wiring force the insulation installer to stuff the insulation behind the wiring. Fiber insulation is used instead of proper caulking around the window, and under the sole plate. Anything else?

Eliminate Convection Paths

Convection carries heat through voids in the envelope structure. Convection is a powerful mode of heat transfer, so voids should be minimized when installing insulation. This is largely a matter of workmanship and selecting the appropriate type of insulation for the job.

Everyone has seen batt insulation that is stuffed between stud spaces in a manner that leaves large gaps. See Figure 1. A good way to insert tight fitting batt insulation between studs is to use a trowel to tease the edges of the batts past the studs. If electrical wiring is in the way, a large void may be left behind the wiring by sloppy installation. Cut notches or gashes in batts to fit around wiring.

When inserting loose fill insulation into existing walls, there is a risk that the material will snag on obstructions, leaving voids. Preventing voids requires installation methods that are vigorous enough to move the insulation past obstructions. Air blowing, rodding, and vibration are the methods used. Good pouring characteristics may be a major factor in selecting the insulation material.

A major problem with pouring insulation into existing walls is the presence of obstructions that create voids in the insulation. Frame walls may have fire stops, which are horizontal pieces that block the space between studs to retard the spread of fire by convection in the stud space. The gap between the brick and block courses of masonry walls is filled with dried mortar and wall ties. With any wall, it is difficult to get the insulation to fill the space all the way to the top.

If you want to use rigid insulation in a wall cavity, you will have to take special steps to prevent convection. If rigid insulation is placed loosely in a cavity, convection provides a way for heat to bypass the insulation, making the insulation almost useless. There is little hope of blocking convection unless the insulation boards are firmly attached to one of the wall surfaces and are sealed at the edges.

There are two conditions that make it important to leave a space between the insulation and the outer skin of the building to vent water vapor. One is that the outer skin is impermeable. The other is that the outer skin is permeable and can get wet. (This is discussed further in Reference Note 42, Vapor Barriers.) The vented space allows strong convection. From the standpoint of insulation value, the vent space is effectively outside the insulated portion of the envelope.

Provide Water Drainage, if Needed

Anticipate the possibility that water will occasionally get into the envelope structure. The moisture may come from rain, condensation, piping leaks, etc. If the structure is well vented and is not vulnerable to wetting, it may suffice to allow the water to escape by evaporation through the vents.

If water may accumulate in a cavity, install drain holes. Design the drain holes so they will not be plugged by the insulation. This is a problem especially with insulation materials that can flow, such as foam beads, cellulose, and vermiculite.

Install the Vapor Barrier Properly

Installing the vapor barrier is an integral part of the insulation job. Failing to install the vapor barrier properly can reduce the effectiveness of the insulation and can cause structural damage. For example, in Figure 1, the vapor barrier material that is attached to the insulation has been rendered almost useless by bad installation practice. See Reference Note 42, Vapor Barriers, about installing vapor barriers.

This brief Note is intended to clarify the economic aspects of installing insulation. The points are illustrated with a typical example.

Unlike most energy conservation measures, adding insulation is a matter of degree. Furthermore, it is not possible to clearly define a "best" amount of insulation. From the standpoint of energy savings alone, most applications would benefit from installing as much insulation as possible. The total lifetime saving continues to increase as the amount of insulation is increased. However, more is not necessarily better. Beyond a certain amount of insulation, the payback period gets longer.

The economics of adding insulation to an existing building are heavily influenced by the original thermal resistance of the structure. For example, if an attic already has a certain amount of insulation, adding more insulation has a relatively long payback period.

The economically optimum amount of insulation is also affected by the installer's overhead and labor costs. Both of these cost factors tend to be independent of the R-value of the insulation installed. Therefore, high overhead and labor costs argue in favor of adding more insulation.

These points are illustrated in the following example of adding insulation to an uninsulated masonry wall. These are the conditions:

- the bare, uninsulated wall has an R-value of 3.0
- the annual average temperature differential is 30°F
- energy loss costs $10 per million BTU
- overhead cost is fixed at $0.12 per square foot
- base labor cost is $0.20 per square foot
- additional labor cost is $0.01 per R-unit per square foot
- material cost is $0.09 per R-unit per square foot.

From these numbers, the table gives the savings per square foot and the economic return that result from installing different amounts of insulation. The economic return is expressed in two ways, as payback period and as savings-investment ratio. The latter is the inverse of payback period, if no discount factor is included.

In this example, the shortest payback period occurs with an R-value of 6. A wide range of R-values provide payback periods in the same range. However, total savings always increase with additional R-value. An

Total R-Value	Cost, $	Net 20-Year Saving, $	Payback Period, yr.	Savings/ Investment Ratio
3	---	---	---	---
4	0.42	3.96	1.92	10.42
5	0.52	6.48	1.49	13.46
6	0.62	8.14	1.42	14.13
7	0.72	9.30	1.44	13.92
8	0.82	10.14	1.50	13.37
9	0.92	10.76	1.57	12.70
10	1.02	11.24	1.66	12.02
12	1.22	11.92	1.86	10.77
15	1.52	12.50	2.17	9.22
20	2.02	12.98	2.71	7.37
30	3.02	12.74	3.83	5.22

accountant might select the R-value of 6 for best rate of return, whereas a conservation enthusiast might select an R-value of about 20 to maximize long-term savings. The prudent building owner might select a value somewhere between these two. The range in this example is typical of many insulation jobs.

In many cases, such as insulating wall cavities, there is no need to perform such a detailed analysis because space limitations make it impossible to add enough insulation to achieve the shortest payback. In any case, it is difficult to go seriously wrong by adding too much insulation.

This example does not take the time value of money into consideration, nor does it consider changing energy costs. The time value of money is an important consideration from an economic standpoint, but strictly speaking, it is not relevant to energy conservation. In essence, the time value of money is largely the crux of the argument between energy conservation enthusiasts and those who make choices on a strictly economic basis.

It is tempting to ignore the time value of money and to ignore the possibility of cost changes because these factors are unpredictable over the long term, whereas energy consumption is fairly predictable. However, the fact that the economic factors are not predictable does not mean that they are unimportant.

Daylighting Design

**How to Visualize the Performance
of a Daylighting Installation
The Key Aspects of Daylighting Design**
- **Visual Quality**
- **Daylight Penetration**
- **Avoid Visual Desensitization**

- **Control Electric Lighting Efficiently**
- **Understand Solar Cooling Load**
- **Exploit Passive Solar Heating**
A Daylighting Wish List

The amount of sunlight falling on the surface of almost any building, even on a cloudy day, theoretically is adequate to provide all lighting requirements throughout the building. However, methods of distributing sunlight deep within buildings have not progressed much beyond the conceptual stage.

Most of the successful daylighting installations that you can find today use skylights or translucent roof panels. Figures 1, 2, and 3 are examples. Skylights and translucent roofs can provide daylighting throughout a large space, but they are effective only for the floor level immediately below them.

Just making glazed holes in the roof does not provide satisfactory illumination. Failure of skylight installations is common, as illustrated by Figure 4.

Daylighting through windows has been far less successful. With current techniques, daylighting by windows typically cannot extend into the space much farther than the height of the window above the floor. However, even this limited benefit is rarely achieved. Ordinary windows do not convert sunlight into useful illumination. Instead, bare windows primarily cause discomfort. Occupants respond to the discomfort they cause by excluding sunlight from the building. Aside from wasting the energy saving of daylighting, this also forfeits the view and ambiance that windows are supposed to provide. Figure 5 illustrates this failure of contemporary architecture.

There is probably no area of energy conservation in which so many fundamental problems have been ignored as in daylighting. Unsuccessful attempts at daylighting not only fail to provide satisfactory illumination, but they typically increase energy consumption for lighting, heating, and cooling. In other words, most daylighting installations waste more energy than they save. In addition, they often create heating and cooling comfort problems.

Kalwall Corporation

Fig. 1 Daylighting for a school cafeteria The entire roof over this space is translucent, so the roof material must have very low light transmission to avoid glare.

Kalwall Corporation

Fig. 2 Daylighting for a swimming pool This is a classic application for daylighting because the illumination levels and lighting geometry are not critical. It also benefits from a large amount of passive solar heating. The surfaces of the translucent panels are warmed by absorption, which keeps them dry in this humid environment.

Despite this dismal record, there is hope for tapping the potential of daylighting. The key insight is that daylighting is complex. To achieve success, you must know where the pitfalls are hidden. Therefore, this brief introduction to daylighting will show you various examples of daylighting failures, and explain why they occurred.

To create effective daylighting, you must master three subject areas. The first is the human factors of illumination requirements and visual comfort. The second is integrating daylighting with artificial lighting. The third is integrating daylighting with the heating and cooling of the building. None of the individual specialties involved in the design of buildings deals with all these factors. Therefore, you will have to go outside your own professional specialty to learn all you need.

How to Visualize the Performance of a Daylighting Installation

Daylighting design faces an especially difficult problem, which is visualizing the conditions that will exist in the finished project, including visual quality, heating and cooling comfort, and ambiance.

Architects' models are a trap for the unwary. Viewing a model from the outside utterly fails to convey how you would feel on the inside of the real space.

Vistawall Architectural Products

Fig. 3 Daylighting for an atrium   This kind of grandiose application is often used as an element of ambiance. It can save a significant amount of lighting energy, but only if the architect cooperates with the lighting engineer.

The best school for learning daylighting consists of finding good examples of daylighting, and studying them under all operating conditions. At present, you will have to do a lot of searching to find good examples, unless you are lucky. But, there is no other way. Designers need to spend more time outside their offices, and to develop the skills of critical observation in the field.

Once you have developed this base of experience, you may be able to benefit from various computer models and other computational tools. However, using these tools without extensive observation of real, successful installations will only mislead you.

The Key Aspects of Daylighting Design

Daylighting design is demanding because it requires dealing with a large number of factors, some of which are in conflict. These factors can be grouped in the following subject areas.

■ Visual Quality

To make daylighting satisfactory from a visual standpoint, it must have these characteristics:

- *the right intensity,* neither too dim nor too bright. Nightfall, clouds, and other things interfere with sunlight, as illustrated by Figures 6 and 7. Electric lighting must be available, and it must be controlled to provide just the right amount of supplemental lighting when sunlight is inadequate. On the other hand, daylighting can never be allowed to make any area of the visual field too bright, or occupants will block out the daylighting permanently.

- *absence of excessively bright areas within the visual field,* which causes visual desensitization and discomfort. This problem is called "glare." (The term has other meanings as well.) Bright areas may be acceptable if they are behind the viewer or are well above the horizontal. Figure 8 shows an example where daylighting through windows was abandoned because it failed this criterion.

- *the right direction,* which basically means eliminating veiling reflections. This is a seriously neglected aspect of lighting design. In general, veiling reflections are avoided by not having lighting originate in the direction that the viewer is facing. This involves space layout as well as lighting.

See Reference Note 51, Factors in Lighting Quality, for a more complete discussion about each of these factors.

■ Daylight Penetration

If daylighting is to be an important energy conservation measure, it must be able to serve as the primary source of lighting for a large area. Not only must the sunlight reach the activity area, but it must arrive from a direction that is visually effective, as discussed above.

WESINC

Fig. 4 Skylights are not the same as daylighting   Just making glazed holes in the roof does not create satisfactory daylighting, as this abandoned and covered skylight on a classroom building testifies. The problem was excessive glare.

WESINC

Fig. 5 Windows are not the same as daylighting   If direct sunlight can enter a space at any time, the occupants will shut out sunlight entirely, as we see here. The situation is not helped by the vertical blinds, which are an unsatisfactory method of modulating sunlight. There is a nice view, but the occupants can't enjoy it, except on the top floor, which is shaded by a deep overhang. This vast amount of glazing serves little useful purpose, and it increases heating and cooling costs considerably.

WESINC

Fig. 6 Accommodate clouds   A thick cloud passing in front of the sun may reduce the available sunlight by a factor of five. Your daylighting application must be able to tolerate this large, abrupt change. Or, you must install devices to regulate the amount of sunlight that enters the space. And, the electric lighting controls must be able to provide just the right amount of supplemental lighting.

WESINC

Fig. 7 Accommodate shading by external features   Portions of this large hotel are shaded at different times of day by taller adjacent buildings. Plans for daylighting should expect this, including the possibility of later construction nearby.

WESINC

Fig. 8 Daylighting abandoned because of glare Large windows were installed high in the walls of this gymnasium. They have been blocked completely. Put yourself in the position of a basketball player attempting to follow the ball with direct sunlight coming through the windows. It is difficult to imagine any usage of this space for which the windows would provide acceptable daylighting. On the other hand, well designed skylights might have worked well. This space will remain dependent on electric lighting.

WESINC

Fig. 9 Daylighting abandoned because of visual desensitization This classroom has continuous rows of tall windows at the sides and rear. The room is used to train second lieutenants of the U.S. Marine Corps. Even these perfect physical specimens cannot accommodate the large differences in brightness between the windows and the inside of the space. The windows have been covered with draperies, which are too cumbersome to adapt to changes in daylighting and to the changing lighting needs inside the space. The room will remain dependent on electric lighting.

There are many conceptual methods that have been devised for increasing the penetration of sunlight into a space. They all use some combination of reflection, refraction, and diffusion. The Measures in Subsection 8.3 recommend the techniques that are most practical at the present time.

■ Avoid Visual Desensitization

The human eye is able to adapt to an enormous range of illumination levels, over a ratio of approximately a million to one. However, at any given moment, the eye adapts to the brightest illumination levels within the visual field, and cannot see well within the darker areas of the visual field.

The eye adapts to different light levels by three processes: (1) changes in the size of the pupils, (2) changes in the response of the nerves in the visual system, and (3) changes in the amounts of certain chemicals (photopigments) in the light sensing cells of the retina. At normal indoor brightness levels, only the first two of these processes occur, and both of them respond in a time frame of about one second. At normal indoor lighting levels, the delay in adapting is barely noticeable when looking from a well lighted part of a room to a darker part, especially if the range of brightness falls within a ratio of less than 10:1. (This ratio declines with age.)

Electric lighting designed to contemporary standards keeps illumination levels within a narrow range, which allows people to see with little effort of adaptation. In contrast, even the most cleverly designed daylighting applications produce large variations of illumination levels within the space. If daylighting is not well planned, the range of illumination levels within the visual field can far exceed the ability of viewers to adapt to the differences in brightness. Figure 9 shows an example where valuable daylighting was abandoned because windows did not have effective methods of modulating the brightness of daylighting.

People are not forgiving about large variations of brightness within a space. If daylighting causes strong contrasts in illumination, occupants invariably turn on all the electric lights in an attempt to match the illumination levels provided by the daylighting. (The term "fill-in lighting" has been coined by lighting designers to certify this practice, in defiance of the logic of daylighting.) Figure 10 shows an example.

Even worse, unless daylighting is executed very cleverly, the brightness of windows or skylights activates the third adaptation process, depleting the photopigments in the retinal cells, which greatly reduces visual sensitivity. Recovery from this chemical change is slow. Therefore, the eye remains desensitized as long as daylighting is present. This increases the illumination levels needed inside the space, typically to a level several times higher than needed by people whose eyes are adapted to lower lighting levels. This may be acceptable if daylighting is the only source of lighting in the space. However, if a small daylighted perimeter area reduces the visual sensitivity of people farther inside the space, they are likely to turn on all the available lighting equipment.

WESINC

Fig. 10 An expensive daylighting installation that saves no energy A lot of useful space in this office building was given up to create an atrium illuminated by a large skylight. However, note the electric lights that are turned on everywhere around the perimeter of the atrium. These attempt to equalize the illumination levels of the shaded areas with the sunlighted areas. Someone probably won a design award for this.

■ Control Electric Lighting Efficiently

Daylighting saves energy only if it provides a reduction in artificial lighting. Daylighting is useless unless the controls of the lighting fixtures are sophisticated enough and reliable enough to respond to the amount of daylighting that is available from moment to moment. Lack of appropriate lighting controls is a common failing of daylighting installations, as illustrated in Figure 11.

In addition, the electric lighting controls must respond to other variables, such as the occupancy of the spaces, the types of activities being conducted, and individual needs and preferences. For example, the

WESINC

Fig. 11 What's wrong in this picture? This common area of a shopping mall is illuminated effectively by the available daylighting from the skylights. The electric lights, which are all turned on, add virtually nothing on this bright day. There may be a legitimate concern that the HID lighting cannot respond quickly enough to large changes in daylighting caused by clouds. Installing fluorescent lighting instead of HID lighting would have avoided this problem.

WESINC

Fig. 12 The inevitable fate of greenhouse additions to restaurants The restaurant wants a cheap addition that offers daylighting and ambiance. But people are not African violets. The owner must eventually address the discomfort during both hot and cold weather, and covers the greenhouse. This one uses a canvas tarpaulin, which needs replacement every few years. Still, it is better than reflecting films, which are ugly and do not reduce the heat gain and glare sufficiently. Why do architects continue to make this common mistake?

electric lighting in a well designed space may have photoelectric controls for some of the fixtures, time controls for all the fixtures, separate controls for different numbers of lamps within each fixture, override switches for late cleaning crews, and so forth.

People tend to turn lights on, but not to turn them off. This behavior is generally beyond the designer's control. Hoping that occupants will keep unnecessary electric lights turned off is not an acceptable design approach. For specific methods of controlling electric lights to exploit daylighting, see Subsections 9.5, 9.6, and 9.7.

■ Understand Solar Cooling Load

It appears that people who design daylighting installation commonly have no understanding of the amount of solar heat gain that comes along with daylighting. Sooner or later, the cooling load becomes unbearable, in terms of both comfort and energy cost. The result is that all or most of the daylighting installation is abandoned. Figure 12 shows an example.

Some daylighting advocates claim that daylighting reduces cooling load because sunlight is a cooler light source than electric lamps. The basis of this claim is that sunlight has a light-to-heat ratio of about 110 lumens per watt, if the entire solar spectrum enters the space in equal proportions. If all but the visible portion of the sunlight were filtered out, the efficacy exceeds 300 lumens per watt. Currently available glazing can provide daylight that is between these two values. By comparison, the light-to-heat ratio of conventional fluorescent lighting is about 70 lumens per watt.

This theoretical advantage is undermined by the fact that daylighting cannot be distributed as effectively as artificial lighting in real applications. Therefore, an excess of sunlight must be admitted to provide adequate illumination. In a sloppy design, so much extra sunlight may be admitted that the added cooling cost may cancel the saving in lighting cost.

Reflective and absorptive window coatings are widely used as a means of reducing solar cooling load. (These are recommended by Measures 8.1.3 and 8.1.4 for retrofit applications.) These methods have also been used in the vain hope of taming sunlight for illumination purposes. This cannot work, because a direct beam of sunlight is unsatisfactory for illumination, no matter how much it is attenuated. If tinted glass were dark enough to reduce the intensity of a direct beam of sunlight to acceptable levels, it would have to be almost opaque. The illumination provided by direct sunlight is about 6,000 footcandles. Dark tinted glass provides an attenuation of about 80 percent, which still leaves an illumination level of about 1,200 footcandles. Even if the windows are darkly tinted or highly reflective, the response of building occupants is still to close the curtains over all windows that allow direct sunlight.

For maximum efficiency, daylighting should use as little glazing area as possible, which means that the glazing should be clear or translucent. Daylighting forces a compromise between the positive value of free illumination and the penalties of conductive heat loss (while heating) and solar heat gain (while cooling). Any form of window treatment that attenuates the passage of light through the glazing increases the glazed area, along with its undesirable consequences.

Design is complicated by the fact that most climates require both heating and cooling. In such climates, one approach is to base the amount of glazing on the daylighting requirement, and not provide any extra for heating. Another approach is to use shading that can be adjusted continually to provide the best compromise between lighting costs and cooling costs during warm weather. External shading (Measure 8.1.1) is most effective, while interior shading (Measure 8.1.2) may be marginally useful for daylighting.

■ Exploit Passive Solar Heating

Once you have a good understanding of the amount of heating energy in sunlight, try to combine daylighting and passive heating wherever possible. Both involve using solar energy through glazing in a controlled manner. However, there are major differences between daylighting and passive heating. Daylighting can be accomplished with relatively small amounts of glazing, whereas passive heating requires large areas of glazing. Daylighting is desirable any time that sunlight is available, but passive heating is desirable only when there is a heating load in the space.

During the dark hours of cold weather, the glazing used for daylighting is a path for considerable heat loss. Furthermore, cold weather brings many more hours of darkness than of sunlight. It has been argued that continuously exposed glazing collects more energy during the day than it loses at night. In northern locations, this is true only if the glazing is oriented toward the sun and has exceptionally high thermal resistance.

See Reference Note 47, Passive Solar Heating Design, for details.

A Daylighting Wish List

We can make a "wish list" of developments that would make daylighting a powerful and widely applicable energy saving technique. These developments would create a breakthrough, if they were economical and free of serious practical problems:

- light pipes and other devices to move light into the interior of the building in a visually effective manner
- glazing materials with high thermal resistance, for applications where heating is a major cost

- glazing materials with opacity that is variable over a wide range, for applications where daylighting is combined with passive solar
- glazing that filters out the infrared component of sunlight in a non-absorptive manner, for applications in warm climates.

Development continues in all these areas. Check their current status when you have an opportunity to use daylighting.

Passive Solar Heating Design

System Elements
Comparison between Active and Passive Solar
Where to Consider Passive Solar Heating
Energy Saving Potential
- Heating
- Lighting
Primary Design Issues
- Glazing Area
- Glazing Location
- Absorption, Distribution, and Control of Sunlight Within the Building
- Regulating Heat Input
- Limiting Heat Loss

- Amount of Heat Storage Mass
- Mechanical Properties of the Storage Medium
- Location of Heat Storage Mass
- Control of Heat Storage Input and Output
- Coordinate with Daylighting
- Coordinate with Electric Lighting
- Coordinate with Heating and Cooling Equipment
- Longevity of Materials and Installation Methods
- Water Leakage
- Wind
- Snow
- Maintenance
- Esthetics
- Cost Efficiency

Passive solar heating is the direct use of sunlight for space heating. The concept is simple, but creating a successful installation may be complex. Passive solar heating is not a concept for casual experimentation, because failure is almost certain to leave a big mess. In general, the larger the fraction of the building's heating that is provided by passive solar, the more complex the design must be to avoid adverse effects.

Passive heating should include daylighting wherever possible, since both involve the controlled intake of solar energy through glazing. However, combining the two is not easy.

Passive solar heating is a broad concept. The following discussion presents the general principles. Use them as a basis for developing specific applications.

System Components

Figure 1 shows the basic conceptual scheme of a passive solar installation, which includes these components:

- large glazing units, to collect the thinly concentrated energy of sunlight
- variable shading devices, to control solar input
- removable glazing insulation, to limit heat loss during darkness
- devices to absorb sunlight and emit heat at desired locations within the space
- thermal storage mass, to provide heating during periods of darkness
- shading devices to control the flow of sunlight into the storage mass
- adjustable insulation, to control the flow of heat from the storage mass
- light diffusion and distribution devices, to make the incoming sunlight suitable for illumination as well as for heating.

This description is conceptual. Probably no real system would have all these elements. For example, it is simpler to control the release of heat from the thermal storage mass by exploiting the inherent time lag of the material, rather than by using adjustable insulation.

The term "passive" implies the absence of moving parts. However, Figure 1 shows that a passive system may require moving parts. These are potentially the most troublesome part of the system. These components are not presently available as standard equipment, and it may be difficult to fabricate them for many installations. The future evolution of passive solar depends largely on eliminating custom components. This will reduce cost, simplify design, and improve reliability.

Figure 2 shows a rationally designed passive solar installation for a house. It makes an interesting contrast with the comprehensive system depicted in Figure 1.

Comparison between Active and Passive Solar

Active and passive solar systems have almost nothing in common, except for the advantage of collecting free solar energy. Active solar systems are primarily mechanical systems, which have architectural ramifications. Passive solar is primarily an architectural feature, which must be tightly integrated with the building's mechanical systems. Table 1 summarizes the main areas of difference between the two.

Where to Consider Passive Solar Heating

In terms of geography, there is sufficient sunlight for passive solar heating throughout the middle latitudes. The coldest weather in these latitudes is associated with the passage of cold fronts, which are followed by clear skies. So, passive heating is often available when it is needed the most. The value of passive solar is greatly reduced in locations that tend to be cloudy or foggy in winter.

Of course, solar energy is available only during the daytime. Winter days become shorter at higher latitudes. More northerly latitudes tend to be ineligible because they have few hours of sunlight in winter, and because their lower average temperatures cause greater heat loss through the glazing.

In terms of building configuration, passive heating requires exposure to the sun. Therefore, passive solar requires an orientation that is generally toward the south, or facing upward. The acceptable range of orientation is fairly narrow. In winter, the sun rises well to the south of due east, and sets well to the south of due west. Also, the sun remains low in the sky all day. (Reference Note, 24, Characteristics of Sunlight, covers the geometry of solar motion in greater detail.)

Reference Note 46, Daylighting Design, points out that it is difficult to get sunlight to penetrate far into the building interior. This is generally not a limitation for passive heating, because building heat loss occurs through the envelope. Passive heating can be quite effective as a perimeter heating system. However, passive heating is limited to sides of the building (including the roof) that face the sun.

In a building that consists of tall, open space, such as a warehouse, the geometry of the space may allow passive solar to heat the entire building. If sunlight can be delivered from overhead, it is possible to provide widespread daylighting along with heating.

Advocates of passive solar heating tend to promote it for residential and small commercial applications. However, passive solar heating is especially well adapted to many industrial activities, for these reasons:

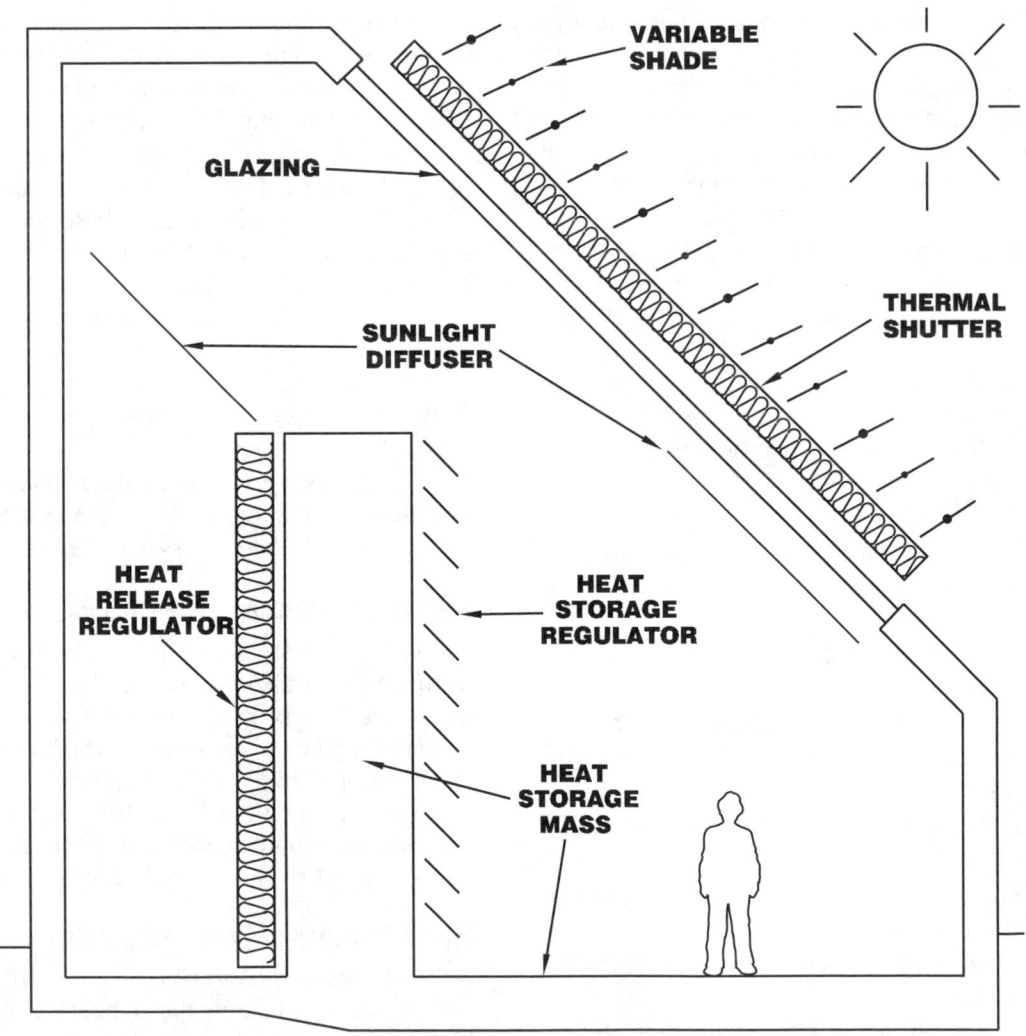

Fig. 1 Complete passive solar heating system This conceptual drawing symbolizes all the functions that any passive heating system should have. Bear in mind that any passive heating system is also a daylighting system, and it must perform well in both roles. In real installations, the clever designer will combine these functions wherever possible, and will exploit the inherent features of the building, such as heat storage mass.

• they are less sensitive to passive solar's lack of precise control of heat gain and illumination. Industrial work involves physical exertion, which makes people less sensitive to small temperature changes that would annoy sedentary workers.

• most industrial tasks are tolerant of a greater range of illumination levels than is office work

• in industrial-type structures, it is often practical to substitute glazing elements for conventional roof and wall materials

• unconventional sunlight control devices are less likely to create an appearance problem in the industrial environment

• industrial facilities tend to have high inherent thermal storage because of the mass of equipment and exposed floor slabs

• availability of skilled maintenance personnel is an important advantage for passive solar installations that have unusual mechanisms.

The need to integrate passive solar heating with both the external and internal design of the structure tends to limit passive solar heating to new buildings, where passive heating is an integral part of the design. This being said, do not overlook the possibility of exploiting passive solar in existing buildings, especially if conditions, such as climate, glazing exposure, and internal layout, are favorable.

Energy Saving Potential

■ Heating

The heating capability of sunlight is weak in comparison with the heat losses that can occur from a poorly insulated building in cold weather. (See Reference Note, 24, Characteristics of Sunlight, for the heat content of sunlight.) Therefore, the effectiveness of solar heating is dominated by the quality of the building envelope. If a building has good envelope insulation and little air leakage, passive solar may provide over half of the heating requirement in most eligible climates. This assumes that the building does not have a large ventilation requirement.

Passive heating does not necessarily have to provide a large part of total heat input to be worthwhile. In fact, passive solar is most economical as a supplemental heat

WESINC

Fig. 2 Rationally designed passive solar installation Two large sunlight collectors are installed in the cathedral ceiling of the living room of this house. Each has an insulated cover, one shown fully open, and the other fully closed. Actuators to control the position of the covers are not yet installed. Before the covers were installed, the surface area proved adequate to maintain the temperature of the space during most winter weather, without other heat. Satisfactory temperature was maintained throughout the night, probably because of heat absorption in the gypsum wallboard. However, the uninsulated glazing sweated profusely at night, causing damage to the frame and floor. The lighting level is bright, but not oppressive, except when reading. Placing the reading chairs in the shaded portion of the room solved this problem. During warm weather, holding the covers in a slightly open position provides very pleasant daylighting with minimal cooling load. The slope of the roof faces southeast, which is not optimum for collecting sunlight in winter. Therefore, the covers are hinged on the right side. They are intended to track the sun so that sunlight will reflect into the space from the white underside of the covers. The actuators, not yet installed here, are the only serious challenge. They must hold the covers rigidly against strong wind in both directions, and they must not be too ugly for the neighborhood. Sliding covers would have been a much easier solution if the roof had faced toward the south.

Table 1. COMPARATIVE CHARACTERISTICS OF ACTIVE AND PASSIVE SOLAR SYSTEMS

	Active Solar	Passive Solar
Heating Capacity	Can be increased indefinitely by adding collector area and storage capacity. In practice, limited by economics and weather patterns.	Limited to faces of the building that have a generally southern orientation, and to adjacent spaces.
Daylighting	Does not provide daylighting.	Inherently provides daylighting in heated spaces. Requires specialized techniques, separate from the control of heating, to make it satisfactory.
Heat Distribution	Heat is distributed with conventional equipment. Space temperature can be controlled precisely.	Heat distribution is dependent on interior space layout and exterior orientation of building. Heat release is subject to relatively wide temperature fluctuations, unless highly unusual features are included.
Dependence of Heat Collection on Temperatures	Cannot absorb solar heat until the temperature of the fluid in the collector rises above the distribution and/or storage temperature of the heating application. Collector performance declines with ambient temperature.	Outside and inside temperatures have virtually no effect on the amount of energy collected. Heat gain through glazing is almost always higher than heat loss during hours of sunlight. Low ambient temperature during dark hours creates major heat loss, unless system has insulating components.
Heat Loss from Spaces	Does not increase heat loss.	Large glazing areas create a serious heat loss path during hours of darkness. Requires troublesome insulating components, especially in colder climates.
Controlling Excess Solar Gain	Excess solar input is easily isolated from the spaces. Protecting collectors from excess temperatures may be troublesome.	Tends to create extreme space cooling load. May require troublesome components to prevent this.
Integration with Building Structure	Collectors are distinct from the building structure. Attaching the collectors to the structure may require strengthening. Cost cannot be reduced by substituting for other building components.	Integral with building structure. Likely to require radical change to structural design. Collectors may substitute for conventional windows and skylights. Other components are extra. Designing external movable components to resist wind and precipitation is a challenge.
Space Requirements	Heat distribution systems are the same as for conventional heat sources. In addition, a large storage tank is needed. Collectors require a large exterior area.	No internal space is needed for equipment. The layout for distributing sunlight and heat within the building may require additional space, and may require unconventional features, such as multi-story spaces adjacent to the walls.
Long-Term Survival	Collector components are subjected to ultraviolet degradation, oxidation, and daily cycles of thermal expansion over a large temperature range. Metal and glass components are durable, but soft components require periodic replacement.	Metal and glass components are durable, but soft components are a potential source of leakage from precipitation. Plastic glazing has limited life. Systems are likely to have unusual moving components requiring maintenance. Movable components can be designed to shield vulnerable components at night and during warm weather.
Appearance	Solar collectors are ugly, but they can be separated visually from the structure. Supporting structures can be hidden.	Components are large and unusual, a major visual element of the structure. Can make the entire building irreparably ugly if the designer is not clever.
Component Availability	Collectors and controls are the only specialized components, and are widely available.	Large glazing elements are the only specialized components available as common items, and these are not entirely satisfactory. Other specialized components must be fabricated on an experimental basis, which makes them expensive and unpredictable.
Maturity of Design Doctrine	Extensive experience has been accumulated with many designs. Extensive documentation exists. Best designs are not yet clearly distinguished. Methods of rejecting excess collector heat are not entirely resolved. Design of active system can evolve largely in isolation from other aspects of building design.	Design doctrine has not yet started to evolve in a serious way. Most design issues are not yet generally recognized. There have been few successful installations, if any, to serve as models for various building types. Evolution of passive design requires difficult interactions with conventional practices of architecture, interior design, lighting design, and HVAC design. Design of custom components requires creativity, along with engineering skills outside the conventional realm of building design.
Economics	Costs are well known. Prices have leveled out. Payback period is long, and will remain so. Variations in design result in relatively small cost differences.	Costs can vary widely. Component costs are high because they are still custom items. Major opportunity for cost reduction remains. Payback period remains to be demonstrated.

source. Probably the best approach is to add as much passive solar capacity as possible without requiring elaborate and expensive features.

■ Lighting

See Reference Note 46 about the energy saving potential of daylighting. If cleverly designed, a passive heating system can provide as much illumination as a system designed primarily for daylighting.

Primary Design Issues

Success with passive solar requires attention to an array of considerations that initially appears bewildering. As with any complex matter, the best approach is to identify all the elements, attack them individually, and then figure out a clever way to combine them. The following are the major issues of passive solar design.

■ Glazing Area

Because of the low energy density of sunlight, passive heating requires large glazing areas, unless the climate is mild. There is a trend of diminishing returns as glazing area is increased. That is, adding more glazing results in a greater number of hours per year when heating capacity exceeds need.

Daylighting requires much less glazing area than passive heating. To avoid excessive brightness, convert most of the sunlight to heat before it is seen by the occupants.

■ Glazing Location

Heat is released into the space at points where the sunlight is absorbed, rather than where it enters. Therefore, consider the location of glazing in relation to the locations of the heat absorbers and the heat storage masses. You may have a great deal of flexibility in making these arrangements. Furthermore, as with conventional heaters, convection can be exploited to distribute heat through the space. You need to tailor the location of glazing more carefully for daylighting than for heating.

Sunlight absorbers, which may consist of nothing more than dark pieces of cloth, can easily be moved to accommodate the glazing. However, the geometrical relationship between the glazing and the heat storage masses is fixed. For example, if you use a masonry wall as a heat storage element, locate the glazing so that sunlight falls primarily on the wall.

■ Absorption, Distribution, and Control of Sunlight Within the Building

Many attempts at passive solar failed because they simply dumped sunlight into the space without regard to consequences. Each location in the space where sunlight is absorbed acts as a heating unit, so these locations must be planned. For example, painting a sunlit wall in a dark color causes heat and light to be absorbed there, whereas painting the wall in a light color causes heat and light to reflect throughout the space.

People cannot work at most indoor tasks in full sunlight. Also, strong sunlight eventually destroys organic materials, such as upholstery, wallpaper, etc. Therefore, provide shading for sensitive areas. For example, install diffusing screens over individual work areas.

Activities within the building may move repeatedly, so design the passive heating system to adapt easily to changing activities.

■ Regulating Heat Input

One of the crippling flaws of passive solar has been failure to provide effective methods of blocking excess sunlight. The most efficient approach is to stop the sunlight outside the building with exterior shading. This may be difficult. External shading devices are large, they must be monumentally strong to resist wind forces and snow loads, and they require mechanisms that are cumbersome and unattractive. In addition to being an engineering challenge, they radically change the appearance of the building.

A less efficient method of controlling excess heat gain is to vent warm air from the space. This method generally is limited to spaces with tall or sloped ceilings. Relying on vents poses the risk of serious heat loss through convective leakage. Vent dampers are the kind of equipment that tends to be forgotten, so they are not operated or maintained properly.

Glazing with controllable opacity may come to market at an acceptable price. This would be a major advance for passive solar, because it would allow control of heat input without the external apparatus.

■ Limiting Heat Loss

Another major problem with passive heating is the high conductive heat loss of glazing, especially during hours of darkness, when there is no heat input and outside temperatures are lowest.

With skylights and other glazing that is slanted, the problem is especially severe. ASHRAE data indicate that multiple-glazed units have two to three times more conductive heat loss when they are installed in a heavily slanted orientation than when they are installed in a vertical orientation. In addition, slanted glazing requires more frame structure and stiffeners to resist sagging, and these components conduct heat.

The low thermal resistance of glazing, especially of skylights, allows the inside surfaces to get cold at night, causing condensation problems. In facilities that have humidification or other moisture sources, skylights may sweat prodigiously. The sweating is unsightly, it grows mildew, and it destroys wooden or steel framing. The condensation may be copious enough to drip on the space below, causing moisture damage inside the space. Condensation usually is not serious during daylight hours, because enough solar heat is absorbed by the glazing to keep it above the dew point of the inside air.

The large, cold surfaces of bare glazing may create discomfort at night, especially if people are close to the glazing.

The heat loss problem was recognized early in the history of passive solar, although it probably was underestimated. Many concepts were devised to add insulation to the glazing during hours of darkness. This conceptual class of insulation has been given a variety of names, including "movable insulation" and "thermal shutters." Unfortunately, all the methods that have been popularized so far have serious practical problems.

At first glance, the easy approach seems to be installing movable insulation inside the glazing. In fact, many types of internal insulation have been tried. These included various types of quilted shades, movable panels of various designs, and insulating shutters. Unfortunately, all these methods fall afoul of condensation problems. Interior insulation keeps the glazing at outside temperature. Moisture infiltrates past the insulation and condenses on the glazing. The insulation traps the moisture against the glazing and its surrounding structure, promoting mildew, rot, and rust.

Another approach is installing movable insulation on the outside of the building. This approach avoids condensation problems. It also protects the glazing from hail, snow, etc. However, the insulating panels must be as large as the glazing. Large external movable panels are an engineering headache. They must be able to withstand wind, they must be designed to prevent air leakage between the panels and the glazing, and they must not look too bizarre for the neighborhood.

A third approach is installing movable insulation inside the glazing, between the panes. For example, one briefly popular concept was blowing foam beads into the space between the panes overnight. This method fell out of favor because the beads stick to the glazing from electrostatic attraction. It is a pity that this did not work. Installing movable insulation between the panes avoids condensation problems, provided that the space is vented to the outside. Adjustable insulation inside glazing units merits more development effort.

Since you need adjustable shading to control heat gains, try to design the exterior movable insulation to act as an adjustable shading device. Needless to say, this involves additional complications.

The oppressive need for movable insulation would disappear entirely if glazing were available for skylights that has a high thermal resistance. Depending on climate, a minimum R-value between 6 and 15 would be sufficient to eliminate the need for movable insulation. For passive heating and daylighting, the glazing would not have to be transparent, only translucent, with a reasonably high light transmission. Such glazing may become available within the foreseeable future.

■ Amount of Heat Storage Mass

Heat storage is a necessary part of almost any passive heating system, because the sun does not shine continuously. With passive systems, storage occurs by the absorption of sunlight in mass. Storage capacity is determined primarily by the amount of mass that is exposed to direct sunlight. Ideally, enough solar energy should be absorbed in the storage mass during the daytime to carry through the hours of darkness.

Buildings are heavy, so a large heat storage potential exists in the building structure. For example, a typical office space may contain several tons of gypsum board that can serve as an effective thermal storage medium if it receives sufficient exposure to sunlight. Concrete floors and masonry walls often have enough mass to provide all the thermal storage mass that may be desired, provided that it is exposed to sunlight. Heavy machinery has a significant amount of heat storage capacity. The clever designer will exploit the mass of the building and its contents as much as possible.

The storage effectiveness of mass is reduced if it is covered by insulating materials, such as carpets, and wall finishes installed over furring strips.

In new construction, the heat absorbing capacity of massive components, such as floor slabs, often can be increased inexpensively by adding more material.

The usual candidate material for thermal storage mass is some type of masonry, such as concrete, brick, stone, tile, etc. Water also has been used. The weight of material required for storage is depends on the material's specific heat. (Specific heat is the heat capacity per unit of weight, in comparison with the heat capacity of water.) The specific heat of water is 1.0, of concrete and most masonry products is between 0.20 and 0.27, of steel is about 0.12.

The volume of the thermal storage mass depends on both the specific heat and the specific gravity of the material. (Specific gravity is the density of a material in comparison with the density of water.) The specific gravity of water is 1.0, of concrete and most construction stone is between 2.0 and 3.0, of steel is about 7.7.

If you multiply the specific gravity of each of these materials by its specific heat, you get the heat storage capacity per unit of volume. By coincidence, water and most bulk construction materials have about the same heat storage capacity on a volumetric basis.

■ Mechanical Properties of the Storage Medium

The heat storage mass is subject to thermal expansion and contraction. Centuries of experience have taught designers how to deal with this in common structural materials.

Masonry tolerates expansion well, but it must be kept loaded in compression, like the bricks in a wall. Tiles cemented to a surface are likely to break loose because of differences in thermal expansion between the tile and the masonry behind it.

Water is cheap. It has exceptionally high specific heat, which reduces structural loading, but it cannot serve as a structural element itself. It has no thermal lag. Convective currents in water prevent thermal lag, and also cause vertical temperature stratification. If water is stored in a transparent container, a dye should be added to it to absorb sunlight.

■ Location of Heat Storage Mass

When masses are releasing heat, they act as huge, low-temperature radiators, which provide comfortable heating throughout a large area. Their location tends not to be critical.

As with any heating system, it is desirable to release the heat near the envelope, to offset the envelope heat losses. From this standpoint, for example, it is better to absorb sunlight in a floor slab close to the exterior wall than to absorb sunlight in an interior wall.

■ Control of Heat Storage Input and Output

Using heat storage efficiently involves three factors: the rate of heat absorption, the rate of heat release, and the timing of heat storage and release. Fortunately, in keeping with the concept of a "passive" system, it is often possible to design these factors into the system without resorting to devices that need to be controlled.

The rate of heat absorption is determined by the surface area exposed to sunlight and by the absorptance of the surface. Mass is effective for heat storage only if it is directly illuminated by sunlight. Therefore, the relative placement of the glazing and storage mass is critical. The motion of the sun causes different parts of the interior to be illuminated throughout the day, which may be advantageous. Warming the space in the morning can be accelerated, at the expense of delaying heat storage, by using internal heat absorbing screens to shade the mass.

The absorptance of the storage mass is determined entirely by its surface. In general, dark colors absorb the most sunlight. "Color" indicates absorptance only in the visible portion of the solar spectrum, which accounts for only about 35% of total solar energy. Absorptance in the infrared portion of sunlight is more difficult to determine. Refer to Measure 8.2.2 for methods of determining the absorptance of particular materials.

The rate of heat output from the storage mass is determined by the surface area and by the emittance of the surface. The emittance of most solid materials used for heat storage is about 0.8, which is satisfactory for the purpose. There is generally no need to tinker with emittance. (Refer to Measure 8.2.2 if you want to learn more about it.)

The timing of heat release in most passive solar systems depends on the thermal lag of the storage mass. Thermal lag is a delay in the release of heat from a mass after the heat has been absorbed. Using thermal lag to control the timing of heat release is usually the preferred method. It is not precise, but it minimizes the need for maintenance or active control.

Thermal lag in a material results from the interplay of its thermal conductivity, heat capacity, and geometry. As heat is absorbed by the sunlit surface of a material, the temperature of the surface is raised, and the rise in temperature forces the heat deeper into the material. When the surface is no longer illuminated, the process reverses. The surface emits heat and becomes cooler, which creates a flow of heat toward the surface. Of course, this is not an on-and-off process. Heat is continuously emitted from the surface, whether it is sunlit or not. As the space cools at night, the increased temperature differential across the surface draws more heat out of the mass.

The thermal lag is much longer and more distinct if the storage mass is heated by the sun on one side and the space being heated is located on the other side. In such cases, there is a distinct peak in heat emission into the space that may occur many hours after sunset.

Calculating thermal lag is somewhat complex. It can be done using some of the more sophisticated energy analysis computer programs. The U.S. National Institute of Standards and Technology has done much of the work in calculating thermal lag in buildings. They offer guidance in this subject.

Water has virtually no thermal lag, because convection keeps transferring heat to the outer surface of the container. The thermal lag of metals is minimal because of their high thermal conductivity.

In some cases, it may be desirable to use adjustable insulation to bottle up heat within the storage mass until needed.

■ Coordinate with Daylighting

Try to exploit daylighting if you use passive heating, since the sunlight is coming into the space anyway. By the same token, you need to consider lighting conditions in a passive solar installation to avoid intolerable brightness and glare.

The methods you use to control sunlight for illumination are quite different from the methods you use for passive heating. Daylighting is desirable any time that sunlight is available, but passive heating is desirable only when there is a heating load in the building. Illumination requires much less glazing area than passive heating, but more careful distribution of sunlight. Daylighting requires still more auxiliary devices, such as light diffusers.

■ Coordinate with Electric Lighting

Refer to Subsection 9.5 for methods of controlling electric lighting to exploit daylighting.

■ Coordinate with Heating and Cooling Equipment

To avoid wasting heating and cooling energy, be sure to design the thermostatic controls of the conditioning equipment so that they do not fight the passive heating system. Passive heating results in swings of temperature. If the passive system is designed properly, the temperature swings remain small enough to avoid discomfort. Design the thermostatic controls to keep heating and cooling equipment turned off as long as temperatures remain within acceptable limits. This is called "deadband." See Measure 4.3.4.2 for details.

■ Longevity of Materials and Installation Methods

Your choice of the materials and installation methods has a major effect on longevity and maintenance requirements. There is a strong temptation, when buying large expanses of glazing, to use short-lived materials to reduce cost. Your successors will curse you for this if you succumb. If you cannot afford the right materials, forget about passive solar. See Measure 8.3.2 about selecting materials for longevity.

■ Water Leakage

Large expanses of glazing tend to be vulnerable to water leakage, especially if the glazing is non-vertical. See Measure 8.3.2.

■ Wind

Shading devices and movable insulation for passive solar systems have a large amount of surface area, which makes it important to design them strongly. In many applications, wind is the greatest impediment to using external devices.

■ Snow

Fortunately, skylights tend to shed snow, provided that they have even a modest slope. Heat loss through the glazing, combined with the insulating property of snow itself, causes the bottom layer of snow to melt and slide off the smooth surface of the glazing. Also, sunlight penetrates snow and warms the surface underneath.

Nonetheless, snow can be very heavy. Skylights should be designed to resist the weight of an overnight wet snowfall. If insulating covers are used, they can be designed to carry the snow load. This requires reliable controls that respond automatically to snowfall. Snow melts and turns to ice, so external mechanisms must be designed to avoid jamming by ice.

■ Maintenance

In theory, passive solar systems should require minimal maintenance. Good design places priority on achieving this ideal. If mechanisms are necessary, such as movable shading devices and thermal shutters, design these for ruggedness and easy maintenance.

■ Esthetics

Daylighting and passive solar heating have a major effect on both the exterior appearance and internal layout. Blending these elements into the design of a building requires imagination and a fine esthetic sense. These qualities are not prevalent in contemporary architecture, and many buildings heated by passive solar are lovely only to their designers. Sadly, many attempts at passive solar design have been so ugly that they degrade the value of the building. The only solution is for owners to be aware of this potential problem, and to cast a critical eye on the esthetic aspects of the design.

■ Cost Efficiency

The clever designer will attempt to satisfy all the functions symbolized in Figure 1, while minimizing the hardware and complexity required. For example, it may be possible to dispense with specific thermal storage devices if the mass of the building is exploited effectively for thermal storage. Considerable thermal storage may be added cheaply by increasing the quantities of inexpensive masonry materials in floor slabs and walls.

In new construction, it may be possible to minimize materials costs by substituting glazing for other roof and wall surfaces.

Lumens
Footcandles and Lux
Candlepower

Brightness
In Summary ...

In doing lighting efficiency work, you need to measure light intensity. You also need to know how to express light intensity for selecting lamps and for laying out the overall lighting configuration. Unfortunately, lighting terminology tends to be confusing and somewhat inconsistent. This brief Note introduces you to the terms that the lighting trade uses to communicate about light intensity, and it points out which of these terms are important to know.

Lumens

"Lumen" is the unit of total light output from a light source. If a lamp or fixture were surrounded by a transparent bubble, the total rate of light flow through the bubble is measured in lumens. Lumens indicate a rate of energy flow. Thus, it is a power unit, like the watt or horsepower.

Typical indoor lamps have light outputs ranging from 50 to 10,000 lumens. You use lumens to order most types of lamps, to compare lamp outputs, and to calculate lamp energy efficiencies (which are expressed as lumens per watt).

Note that lumen output is not related to the light distribution pattern of the lamp. A large fraction of a lamp's lumen output may be useless if it goes in the wrong directions.

Footcandles and Lux

"Footcandles" and "lux" are units that indicate the density of light that falls on a surface. This is what light meters measure. For example, average indoor lighting ranges from 100 to 1,000 lux, and average outdoor sunlight is about 50,000 lux.

The footcandle is an older unit based on English measurements. It is equal to one lumen per square foot. It is being replaced by lux, a metric unit equal to one lumen per square meter. One footcandle is 10.76 lux. Although footcandles are now officially obsolete, they probably will continue to be used because many existing light meters are calibrated in footcandles.

The general term for lux or footcandles is "illuminance." The general term is sometimes used by lighting engineers, but the units of lux or footcandles are more commonly used.

You use footcandles or lux to measure the adequacy of lighting on the task. Footcandles and lux relate only to the task area, not to the lighting equipment or to the geometry of the space. For example, you could create an illumination level of 100 lux on a surface by using a single spotlight located far away, or by using many cove lights nearby.

For energy conservation work in existing facilities, you need a light meter that measures illuminance in footcandles or lux. You will use it continually as you lay out lighting, select fixtures to be delamped, etc. Light meters have become inexpensive, so you can afford to spend the money to get a rugged electronic unit of good quality, rather than the older type that uses a fragile meter movement. Figure 1 shows a footcandle meter.

Candlepower

"Candlepower" is a measure of lighting concentration in a light beam. It is used primarily with lamps that focus, such as spotlights and PAR lamps. In lamps where candlepower is specified, the candlepower rating usually applies only to a small spot in the center of the beam.

WESINC

Fig. 1 Footcandle meter The meter is used to measure "illuminance." It is the only measuring instrument that you need for most applied lighting efficiency work. Being an older model, this meter indicates in units of footcandles. Newer models indicate in units of lux.

The official unit of candlepower is the "candela," which is equal to one lumen per steradian. (A steradian is a fraction of the surface area of a sphere that is equal to the square of the radius divided by the total surface area. This is approximately 8% of the total surface area.) This term is rarely used in practical work. Lamp catalogs usually list "candlepower" rather than candelas. This is like using "horsepower" as both a general term and a specific unit. To confuse matters further, candelas were earlier called "candles."

Brightness

In general, "brightness" is an expression of the amount of light emitted from a surface per unit of area. "Brightness" is not an official term of the lighting trade, and lighting designers may become huffy when you use it. However, the concept is essential for understanding visual quality, especially in relation to contrast and glare.

Brightness does not inherently relate to lamps, or even to light sources. The light could be reflected or transmitted. For example, the bright surface could be the surface of a fluorescent tube, a page of a book, a window with a view of the sky, or a store window with reflections.

The closest official term is "luminance," which is expressed as candelas per square meter of light emitting surface. (Luminance used to be measured in "footlamberts," which is now an obsolete term.) For example, the luminance of a heavily overcast sky is about 1,000 candelas per square meter, and the luminance of a typical frosted light bulb is about 100,000 candelas per square meter.

Luminance is defined in terms of the direction of light emission. The details get technical, and you probably will not need to deal with them. In brief, the brightness of an object usually depends on the direction from which you look at it.

Note that luminance has nothing to do with size of the light emitting surface. The light source could be as small as a lamp filament, or it could be as large as the whole sky, or it could be a task area, such as a desk top.

Measuring brightness ("luminance") is tricky and requires specialized equipment. For practical work, learn how to avoid excessive brightness, so you won't need to measure it. If you do a good job of laying out lighting, people within the space will not be subjected to brightness that is severe enough to cause glare.

Luminance is the converse of illuminance. The former describes the intensity of light that is leaving a surface, whereas the latter describes the intensity of light that is falling on a surface. For light reflected from a surface, luminance equals illuminance multiplied by the percentage of reflectance.

"Brightness" also is used to describe the subjective sensation of light intensity. This sensation largely depends on the overall layout of the scene surrounding the viewer. An uncomfortable level of brightness is described as "glare." (The term "glare" is used in several ways. It is an important concept, but is not precisely defined by the lighting trade. Various types of glare are explained in Reference Note 51, Factors in Lighting Quality.)

In Summary ...

So, here is the overall picture. A lamp produces a certain amount of light, measured in lumens. This light falls on surfaces with a density that is measured in footcandles or lux. A person looking at the scene sees different areas of his visual field in terms of levels of brightness, or luminance, measured in candelas per square meter.

Many characteristics other than light intensity are important in selecting light sources. These include color, operating temperature, starting time, etc. To learn about all of them, see Reference Note 52, Comparative Light Source Characteristics.

Factors in Lighting Quality

Visual Quality = Visual Efficiency + Visual Comfort
Illumination Intensity
- **The Effect of Illumination Level on Task Efficiency**
- **The Effect of Illumination Level on Comfort**

Uniformity of Task Illumination
Self-Shadowing Within the Task
Shadowing by the Viewer and Adjacent Objects
Background Lighting
- **Effect of Background Lighting on Visual Efficiency**
- **Effect of Background Lighting on Visual Comfort and Esthetics**
- **Minimizing Energy Consumption for Background Lighting**

Light Source Glare

Veiling Reflections
- **Where Veiling Reflections Cause Trouble**
- **The Strength of Veiling Reflections**
- **How to Avoid Veiling Reflections**
- **Relationship to Reflected Glare**
- **Polarization**

Color Rendering
- **How Human Beings See Color**
- **Sunlight and Vision**
- **Lamp Spectra and Vision**
- **Color Rendering Affects the Required Illumination Level**
- **How to Accommodate Color Vision Deficiencies**
- **Loss of Color Vision in Low Light**

The most important aspect of lighting is visual quality. As you seek to make lighting more efficient, do it with an understanding of the factors that contribute to visual quality. When you change lighting for the sake of energy conservation, you assume all the responsibilities of lighting design, an activity that even specialists find difficult. You are likely to create bad lighting if you do not understand the factors that affect visual quality. On the other hand, if you deal with each of these factors properly, you can be sure that the lighting will be good.

Visual Quality = Visual Efficiency + Visual Comfort

Good visual quality consists of high visual efficiency combined with visual comfort. Visual efficiency is a measure of a person's ability to perform tasks involving vision. For example, persons can be tested to measure their rate of making proofreading errors under various lighting conditions. Visual efficiency can be measured with objective tests, although it is difficult to make test conditions relevant to typical environments.

Visual efficiency depends both on the nature of objects being viewed and on the nature of the illumination. With respect to the objects themselves, the most important factor is the size of details. For example, large type is easier to read than small type.

With respect to illumination alone, the most important factor in visual efficiency is the illumination level. In brief, higher illumination levels produce greater visual efficiency, up to a point. Illumination levels are determined by lamp output and light distribution, fixture light distribution, and lighting layout.

A third major factor that increases visual efficiency is the contrast within the subject being viewed. Contrast depends on both the nature of the task and the nature of the lighting. You can achieve the best possible contrast by carefully selecting the fixtures and their locations. In the case of colored subjects, you can also improve contrast by selecting lamps for good color rendering.

Less important to visual efficiency is the relative brightness of the task and surroundings, and other factors.

Visual comfort is a matter of feelings and perceptions. These may range from subtle moods to overt discomfort. Visual comfort is related to the ease of seeing the task, which involves the same factors as visual efficiency. In addition, visual comfort is affected by the brightness of the task, the presence of extreme brightness in the visual field, the contrast in brightness between the task and the background, and other factors.

Visual comfort is very subjective. For example, the amount of background lighting may not matter to a person who is concentrating on a task, but it is significant to a person who gazes at the surroundings. Even objective irritations, such as glare, are difficult or impossible to quantify. Attempts to quantify visual comfort have been devised, for example, by measuring "visual comfort probability," but such measurements are imprecise.

Visual efficiency and visual comfort sometimes conflict. For example, intense illumination improves visual efficiency, but it is uncomfortable over an extended period of time.

Illumination Intensity

Illumination intensity, or illumination level, is expressed in footcandles or lux. It is usually the first criterion used in lighting design. Illumination levels usually are selected according to the anticipated types of tasks. Other factors, such as the age of viewers, may also be considered in selecting illumination levels.

The *IESNA Lighting Handbook* is a primary source for recommended illumination levels. Some typical ranges of illumination levels from the *IESNA Lighting Handbook* (1993 edition) are, in footcandles (multiply by 10 for lux):

corridors and stairs	10 - 20
office reception areas	10 - 20
locker rooms	10 - 20
loading docks	20
exterior walkways	5
reading, books	20 - 50
reading, photographs	50 - 100
reading, poor quality copies	50 - 200
proofreading	100 - 200
drafting, high contrast	50 - 100
cartography	100 - 200
work at video monitors	5 - 10
(if shielded from glare)	
dairy barns	10 - 50
meat packing	20 - 50
bakeries	20 - 50
kitchens, commercial	50 - 100
kitchens, residential	20 - 100
automotive repair	50 - 100
autopsy	200 - 500
cloth inspection	1,000 - 2,000

To select within these ranges, the *IESNA Lighting Handbook* provides factors for the age of viewers, the reflectances of the room surfaces, and the speed and accuracy required in the tasks. Typical of contemporary American lighting practice, these lighting levels range between generous and extravagant.

Selecting illumination levels is highly subjective. Recommended levels have changed radically over time. In the years that followed Edison's invention of the light bulb, the emphasis was primarily on increasing lighting levels. The early development of lighting was driven explicitly by the electric utility industry's desire to sell more electricity. In the United States, recommended illumination levels soared between the 1950's and the 1970's. For example, the Illuminating Engineering Society of North America (IESNA) recommended office lighting of 30 to 50 footcandles in the year 1952. Twenty years later, on the eve of the first modern "energy crisis," the same organization recommended lighting levels three times higher.

Research on the effects of lighting on human performance first concentrated on the effect of illumination level. Illumination level interested researchers because it can be measured objectively, it is easy to define, and it can be controlled easily as an experimental variable. Also, illumination level had been the limiting factor in lighting throughout the previous history of humanity, and electric lighting now made it possible to increase illumination level at will. As late as the 1970's, the main thrust of lighting research was to demonstrate that ever higher illumination levels would produce ever better performance on detailed tasks. This research tended to elevate "more is better" to the level of dogma.

Most of the early research was conducted in laboratory settings, for short durations, and typically with young subjects. Long-term visual comfort, esthetics, and energy efficiency were largely ignored. The latter considerations began to receive more attention during the 1970's. Since then, research has suggested that excessive lighting levels and poor placement of fixtures might also be a problem. The combined concerns of energy efficiency and visual comfort halted the rapid escalation of lighting levels, and in some cases, reversed it.

After the energy crises began in 1973, lighting level recommendations in the United States came tumbling back down. Even so, U.S. lighting recommendations today may be several times higher than in other countries. This range of experience demonstrates that human vision does not have narrowly definable requirements for illumination intensity.

■ The Effect of Illumination Level on Task Efficiency

To summarize a rather complex and incomplete field of research, the illumination intensity required to provide good visual efficiency (without regard to comfort) depends strongly on three interrelated factors:

- *the size of details in the task.* For example, less light is needed to read large print efficiently than to read small print.

- *task contrast.* For tasks of high contrast (e.g., black ink on white paper), performance quickly reaches a plateau, above which increased illumination adds little to performance. However, for tasks of low contrast (e.g., machine work, fabric inspection, surgery), visual efficiency continues to improve significantly up to illumination levels that are extremely high. For such tasks, the illumination levels needed to achieve best perception are much too high for long-term comfort.

- *the age of the viewer.* Starting at about the age of 40, visual sensitivity starts to decline significantly. By age 60, visual sensitivity is declining rapidly. In some tests of visual performance, older subjects had substantially lower performance scores than young subjects at all illumination levels. Also, older viewers suffer more serious degradation of performance at lower illumination levels. The extent of this degradation with age varies greatly with the visual quality of tasks. One experiment with tasks of high visual quality showed that viewers of all ages do not suffer much degradation of performance within the range of 10 to 1,000 footcandles.

Higher illumination levels reduce the effect of vision problems that arise commonly in old age, including farsightedness, astigmatism, and cataracts. For example, a person who is becoming farsighted can see closer in brighter light than in dim light. Most of this benefit probably arises from contraction of the pupil, which minimizes the effect of defects in the lens.

It seems clear that these are the three main factors that determine illumination level, but different tests seem to suggest wildly different lighting levels. The reason seems to be that there are factors other than lighting that have a major effect on the performance of visual tasks.

For example, during the 1970's, the Illuminating Engineering Research Institute (IERI) conducted tests that consisted of proofreading mimeographed sheets of good and poor quality. The tests were scored in terms of accuracy and speed. The tests showed little difference in performance in the range between 10 footcandles and 1,000 footcandles, except for older persons reading poor quality material. On the other hand, other tests reported that increased lighting levels significantly increased speed and accuracy, even beyond 1,000 footcandles (10,000 lux).

Office work is an important exception to the findings that higher illumination levels improve work efficiency. There may be several reasons. One is that office work has very high contrast, namely, black type on white paper. Another is that office tasks offer redundant cues to meaning. (For the same reason, the speed of reading ordinary text is not significantly reduced if a few words are misspelled.) Another theory is that perception in office work requires a certain amount of thought, and the eye is quicker than the mind, even in poor light.

Thus, the present practice of specifying illumination levels by generic type of activity is painting with a brush that is too broad. Within a given type of activity, there may be great variability in the visual quality of the subject matter. For example, graphics work involving large, colorful posters requires good color rendering but not high illumination intensity, whereas line engravings require high illumination levels and careful fixture placement, but only modest color rendering.

With tasks that have variable lighting requirements, the key to energy efficiency is supplementary lighting. For example, a dentist uses an intense, narrowly focussed lamp to work inside the patient's mouth, while the rest of the office is illuminated at typical indoor levels.

■ The Effect of Illumination Level on Comfort

Within a normal range, the exact illumination level does not appear to be a major factor in visual comfort. Complaints about visual comfort are usually associated with specific lighting defects, such as glare and excessive task contrast. Experiments and field observations over the years yield these relationships between illumination level and visual comfort:

- There is a broad range of illumination levels that people consider comfortable, for a given activity.
- People prefer higher illumination levels for tasks that are more visually demanding, i.e., that have finer detail or poorer contrast.
- Persons prefer higher illumination when concentrating narrowly on a task.
- Older people prefer higher illumination levels.

It seems that most formal experiments do not make an adequate distinction between the average illumination of the task area and illumination of a narrow area of concentration. For example, a person may prefer an illumination level of 30 footcandles (300 lux) over his desk, but he may want twice this illumination on a piece of paper with poor quality text.

A newer finding, not yet adequately covered by research, is that higher illumination levels become uncomfortable over the long term. Extensive field observations by Wulfinghoff Energy Services, Inc. indicate that 100 to 200 footcandles (1,000 to 2,000 lux) is an upper limit for long-term comfort in office work for most people.

This same experience revealed that people do not appear to recognize when high illumination level is the source of their discomfort. Where excessive artificial lighting is available on an elective basis (usually, with multiple switching), workers tend to keep artificial lighting turned up to a level that causes long-term discomfort, usually in the form of fatigue and headache. When an outside party turns down the lighting, the response of the occupants typically is gratified relief. But also typically, occupants eventually restore the excessive illumination level. (Part of this may be a social response. I.e., people may feel that all available lights are meant to be turned on, regardless of illumination level.)

Low lighting levels do not appear to cause discomfort, provided that the viewer is able to control the lighting level. For example, if a person can read well at 10 footcandles, he may choose to do so for hours on end. Early lighting dogma highlighted low lighting levels as a source of visual discomfort, typically described as "eye strain." However, this problem does not arise unless the illumination level is a fraction of contemporary standards, and only in cases where higher lighting levels are not available at the option of the viewer.

Uniformity of Task Illumination

Illumination within the task area should be as uniform as possible, and differences from one part of the task area to another should be gradual. In other words, lighting should not be spotty.

For purposes of this discussion, the "task area" is defined as the working area within the field of vision. Differences in illumination between the task area and the surroundings may be much greater, as discussed below.

This is not to say that the task area should be uniformly bright overall. The task itself should be brighter than the immediate background in the task area. For example, a desk top should be significantly darker than the paperwork. (Illumination level is determined by the light fixtures, but task brightness is determined by the reflectance of the task surfaces.)

In one test of task lighting conducted by Wulfinghoff Energy Services, Inc., efficient and pleasing illumination was achieved using 30 footcandles on one side of a brown desk where papers were read, with 20 footcandles illuminating a keyboard in the middle, and with 15 footcandles illuminating the side of the desk where a coffee cup and pencil holder resided. (Care was taken to avoid glare and veiling reflections, which are discussed below.)

It may not be possible to achieve reasonably uniform illumination over the task area by using a single fixture. A single fixture would need to be several times farther from the task area than the width of the task area. This condition cannot be met with typical modern ceiling heights. With low fixture heights, it is usually necessary to install two fixtures, on opposite sides of the viewer and with partially overlapping illumination patterns.

Self-Shadowing Within the Task

Tasks that are perfectly flat, such as a drafting table, are rare. Most tasks are three-dimensional to varying degrees. The higher areas of the task tend to cast shadows on the lower areas. This shadowing may be important for creating contrast. An extreme example is metal engraving, where the differences in height in the metal surface are tiny, but the shadows in the surface account for most of the viewer's ability to see the engraving.

On the other hand, too much self-shadowing is unpleasant. For example, one of the most important aspects of lighting in glamor photography is minimizing self-shadowing at surfaces. Almost any type of lighting produces enough shadowing for satisfactory contrast, provided that the fixtures are arranged to avoid glare and veiling reflections. The problem usually is too much shadowing, rather than too little.

Professional photographers describe lighting that produces surface shadowing as "harsh." Generally, the cause of harsh lighting is a light source that has a small surface area in relation to its distance from the task. In other words, as the light source approaches a point source, it has a greater tendency to create shadows within the task surfaces, and the shadows are sharper. If the

task is not flat, shadows cast by one part of the task may mask adjacent areas of the task surface, which reduces visual efficiency.

Even in office work, where most material is flat, concentrated light sources create annoying shadows. For example, turning the pages of a book results in large changes in illumination as each page turns parallel to the light beam. With work that is very three-dimensional, such as most industrial work, the shadows caused by small light sources may seriously hamper performance.

Point-source lighting may be employed to provide a dramatic highlight for some activities, such as displays. However, people should not have to work within the area illuminated in this manner. A prime example is theater stage lighting, which is effective for the audience but is notoriously uncomfortable for the actors. (Even in these cases, a number of distributed point sources are used, rather than a single point source.)

Not only point sources create sharp shadows. Light sources that are small in one dimension create them also. For example, a row of single-tube fluorescent strip fixtures suspended over a work area creates shadows that are strong and sharp-edged in a direction parallel to the lamps.

Photographers long ago discovered the solution to the problems caused by this kind of shadowing, which is to radically increase the surface area of their light sources. If you visit a commercial photographic supply store, you will see many types of large focussing reflectors, and even larger reflecting screens, whose purpose is to increase the effective area of the light source.

Experiments with office task lighting conducted by Wulfinghoff Energy Services, Inc. suggest that the average width of the fixture's radiating surface should be at least one fifth of the distance of the fixture from the task. For example, a round reflector fixture installed five feet above a desk should have a diameter of at least one foot. The surface area required to avoid excessive shadowing varies with the type of task. Generally, it is desirable to make the light source area as large as practical without introducing glare or veiling reflections.

(The extreme of large light source surface area is "equivalent sphere illumination," a theoretical concept that captured the fancy of the lighting trade for a while. In this concept, light is produced uniformly by a hemisphere surrounding the viewer. This geometry effectively eliminates shadowing. However, it is far from ideal, because the part of the hemisphere within the viewer's visual field produces source glare and veiling reflections.)

Contemporary area lighting, which uses closely spaced fluorescent fixtures of large surface area, almost eliminates troublesome shadowing. In moving away from this inefficient type of lighting toward more efficient task lighting, be careful to avoid excessive shadowing. Here again, the main points are effective fixture placement and making the light source area as large as possible.

Light reflected from large surfaces, such as an adjacent wall, also reduces the intensity of shadows. Exploit this fact by making walls and ceilings as reflective as possible, using a diffusing surface finish. In other words, use flat white paint liberally.

Shadowing by the Viewer and Adjacent Objects

Another type of shadowing that diminishes visual quality is shadowing of the task by the viewer's own body or by surrounding objects. The problem is caused by light sources that are located too far behind the viewer, or in the wrong position with respect to other shadowing objects.

To avoid excessive shadowing by the viewer's hands, as when writing, it is necessary to install more than one fixture. The fixtures are located on opposite sides of the task. This positioning reduces the shadows by the crossfire of light, and it avoids source glare and veiling reflections. However, the remaining shadows of the viewer's hands slant inward toward the center of the visual field. Light sources need large surface areas to diffuse these shadows.

Using a number of point sources, such as reflector lamps, is not a satisfactory substitute for a large light source area. Each point source creates a shadow that is sharply defined, although less intense than with a single lamp.

Background Lighting

Background lighting is lighting outside the immediate area of the task, i.e., what you see out of the corner of your eye or what you see when you look up from the task. The issue of concern is how much difference is acceptable between the brightness ("luminance," to purists) of the task and the background. The concept is most relevant to environments where there is a clear separation between an immediate task area, such as a desk or machine tool, and the rest of the space. The distinction blurs in environments where looking around is part of the primary visual activity, as in retail stores and meeting rooms.

The need for brightly lighted surroundings is largely an assumption of current lighting dogma, rather than a firm conclusion based on research. Formal laboratory studies show little agreement about people's preferences for background brightness. However, the inconsistency of the research does not mean that the need for background lighting is fanciful. Many real human needs have not been described or explained scientifically.

■ Effect of Background Lighting on Visual Efficiency

The human eye adapts to different lighting levels by three processes: (1) change in the size of the pupils, (2) change in the response of the nerves in the visual system, and (3) change in the amounts of certain chemicals in the light sensing cells of the retina ("rods" and "cones").

Only the first two of these processes occur at normal indoor brightness levels. Both types of adaptation occur in a time frame of about one second. When looking from a well lighted part of a room to a dark part of a room, the delay in adapting is barely noticeable. This is especially true if the viewer is looking around casually. Therefore, it appears that the intensity of background illumination has little effect on task efficiency as long as it is limited to indoor lighting levels.

Excessive brightness activates the third adaptation process, depleting the light sensitive chemicals in the retinal cells. It may take many minutes after exposure for the chemicals to be replaced, restoring full sensitivity.

The two major sources of excessive brightness in the visual field are daylight and poorly shielded light fixtures. If windows or unshielded fixtures are part of the background, they increase the overall lighting level required to maintain a given level of visual efficiency. At the same time, these bright light sources make the rest of the surroundings appear darker. This may create the false impression that the (apparently) dark background is reducing visual efficiency, leading to the wasteful response of increasing background lighting.

■ Effect of Background Lighting on Visual Comfort and Esthetics

Dark surroundings generally do not cause headaches, sore eyes, or other physical symptoms commonly associated with bad lighting. Physical symptoms of visual discomfort arise only in unusual cases where the surroundings are brighter than the task.

This leaves esthetic preference as the main factor of concern with background illumination. Approach lighting efficiency in a manner that is consistent with the importance of esthetics to the activity. For example, lighting esthetics is important in the lobby of luxury hotel, but it is fairly unimportant in a boiler factory.

For environments where lighting esthetics is important, the *IESNA Lighting Handbook* offers background lighting recommendations related to considerations including "visual clarity," "spaciousness," "relaxation," "privacy," and "pleasantness." These recommendations appear to be bald assertions, which does not mean that they are invalid. In a nutshell, the *IESNA Handbook* recommends high background illumination to convey spaciousness, and low background illumination to convey privacy. No surprises there!

The esthetic importance of background lighting depends on the extent of the viewer's concentration on the task area. For example, if a person is concentrating on a crossword puzzle, he is oblivious to lighting more than a few feet away. However, as soon as the puzzle solver looks up, he becomes conscious of the lighting in all the surroundings that he can see. This suggests, for example, that background lighting is relatively unimportant for detailed assembly work and may be more important for typical office work.

In environments where lighting does not play an explicitly decorative role, the main problem to be avoided is leaving people with the sensation of being stranded in an island of light. The seriousness of this issue is a matter of opinion, as is the amount of background lighting that is considered necessary to avoid it.

Depending on the environment, the minimum acceptable background brightness may range from 5% to 30% of the task brightness. Current lighting dogma in the United States says that background brightness should be between 30% to 100% of the brightness of the task. However, observation of the real world reveals that such extravagance is not necessary for most activities.

■ Minimizing Energy Consumption for Background Lighting

Efficient background lighting is primarily a matter of increasing the brightness of background surfaces, rather than flooding the entire environment with light. The way to maximize the brightness of distant surfaces with minimum energy consumption is to use surface colors that are highly reflective and diffuse. (See Measure 9.7.1 for details.)

If you need background lighting, put it on ceilings and walls, because these fill most of the visual field. In turn, these surfaces should sufficiently illuminate adjacent furnishings. For example, if a person is working alone in a large area, it is most important to put some light on the walls, rather than turning on all the lights between the person and the walls.

The size of the surrounding visible space largely determines whether an "island of light" sensation occurs. In a small room, or in an office cubicle of limited dimensions, the sensation may not arise. On the other hand, a mechanic working at a lighted bench in the middle of a darkened aircraft hangar will certainly experience the sensation. An efficient way to deal with this problem is to visually isolate the activity area. Partitions that extend above head level are very effective. In addition, architects and interior decorators have a variety of tricks to create visual boundaries without solid partitions, such as foliage, changes in decor, etc.

Light Source Glare

"Glare" is an important issue for visual efficiency and comfort. The term is somewhat vague, because it has been used to describe a variety of lighting problems. Glare is perhaps best described as discomfort caused by areas of excessive brightness within the field of view. It is usually not a matter of brightness alone, but also to substantial differences in brightness within the visual field. The eye can adapt to an enormous range of brightness, but when the eye is confronted simultaneously with differing brightness levels, it cannot adapt to them all.

The discomfort may be caused by the excessive brightness itself, or by strain resulting from attempting to see detail in the darker areas, or perhaps by other causes. Glare usually involves sources bright enough to deplete the light sensitive chemicals in the retinal cells, which prevents rapid adaptation from bright to dark.

All conventional light sources have the potential to be major sources of glare if they are not shielded from view. To illustrate this point, here is a comparison of the typical brightness of light sources compared to the brightness of typical office work (in candelas per square meter):

the sun, at high elevations	1,000,000,000
incandescent lamp filament	2,000,000
frosted light bulb	100,000
fluorescent tube	10,000
fluorescent fixture, prismatic diffuser	4,000
overcast sky	2,000
window facing north, viewing clear sky	500
office desk, covered with paperwork, etc.	150

Thus, any light source is much brighter than a typical task area. It is well known that visual efficiency and comfort suffer if there are areas in the background that are brighter than the task. The extent of discomfort and desensitization depends of these characteristics of the background brightness:

- *the size of the light source within the visual field.* For example, the table above shows that incandescent lamp filaments are 10,000 times brighter than typical indoor scenes. Yet clear bulbs that expose their filaments to vision are widely used in chandeliers and other lamps. This is tolerable because the size of the filaments in the visual field, at typical distances, is minuscule. A frosted bulb with the same filament may be just as annoying because it appears larger, even though its surface brightness is less by a factor of 20.

- *the position of the light source in the visual field.* Light sources produce less discomfort the farther they are from the line of sight. Higher positions are less annoying, probably because humans evolved to see by the overhead sun.

- *the duration of exposure.* For example, brief exposure to clear chandelier lamps in a hotel lobby is not annoying, but longer exposure would be.
- *the brightness surrounding the light source.* For example, a light fixture mounted on a white ceiling is perceived as having less glare than the same fixture mounted on a dark ceiling.
- *the age of the viewer.* Sensitivity to glare increases radically with age, mostly beyond mid-life. A variety of changes occur to the eye as it ages, all of which conspire to decrease tolerance to glare.

In general, avoid light source glare by keeping the source out of the visual field of the viewer. This includes any localized bright spots that may be caused by reflection. In spaces that have only a single occupant, such as one-person offices, this may be fairly simple to accomplish.

Difficulty arises in spaces that have more than one viewer, because the fixtures serving one task area may be visible from other task areas. With contemporary lighting, a great deal of effort has been devoted to making fixtures appear less bright to viewers outside the task area. This is generally done by designing the fixture to limit light emission outside the desired area of coverage. Many fixtures designed this way, including downlights with internal baffles, are extremely wasteful of energy. On the other hand, some modern fluorescent fixtures have achieved a much better compromise between light output efficiency and control of light spread.

Another approach, as yet rarely used in contemporary lighting, is shielding the entire fixture from the view of adjacent areas by installing blinders in the line of sight with the fixture. This approach is discussed in Measure 9.7.6, which deals with task lighting. Glare shielding appears to be almost mandatory with task lighting. This method deserves more attention than it presently receives.

The glare of light sources may be reflected from surfaces. If reflection brings the image of a light source more within the line of sight, the reflected glare may be worse than the direct glare from the light source itself. Generally, reflected glare from light sources is not a problem if the lighting is laid out to avoid veiling reflections, as discussed next.

Veiling Reflections

To understand veiling reflections, recall that you are able to see details in an object because they absorb light differently than the areas that surround them. For example, you can read the words on this page because the ink absorbs more light than the paper. If you were looking at a colored picture, you would distinguish the colors because the different dyes absorb different wavelengths from the incoming light.

WESINC

Fig. 1 A work station with serious veiling reflections A person working at this desk is subject to severe veiling reflections from the large light fixture, and from the illuminated area of wall under the light fixture. When the person turns around to work at the computer, he sees strong reflections from the surface of the monitor.

By the same token, if all the light striking a surface were reflected unchanged, you could see no details in the surface. Most surfaces reflect a fraction of the incoming light in this manner. In other words, for a certain fraction of the incoming light, the surface acts like a mirror. The light reflected in this manner is called a "veiling reflection," because it acts as a veil that masks the light that carries information about the task.

For example, veiling reflections in store windows keep you from seeing the merchandise, veiling reflections from the surface of water keep you from seeing the fish, and veiling reflections from glossy paper keep you from seeing the text or pictures.

Veiling reflections reduce visual efficiency by reducing contrast. For example, dull black ink absorbs almost all light, while white paper reflects over 80%. In this example, the reflectance ratio is very high. On the other hand, if ink and paper are glossy, 50% of the incoming light may be reflected as a veiling reflection. This reduces the contrast between the ink and the white background to a ratio of about 2:1. If the inks are light colored, the contrast is degraded by an even greater amount.

Veiling reflections also cause discomfort by making it more difficult to see. This discomfort typically occurs over a long term, and the viewer typically is unaware of the cause.

Veiling reflections waste energy in two ways. First, the light contained in the veiling reflection is entirely wasted. Second, by reducing contrast, veiling reflections increase the overall illumination level that is needed for a viewer to function at a given level of visual efficiency.

Veiling reflections are perhaps the most underestimated problem in lighting design. They tend to be ignored by lighting designers because the locations and orientations of tasks usually are not known at the time the light fixtures are installed. Even if the task locations and orientations are known, it takes a lot of additional effort to lay out the fixtures to avoid veiling reflections.

■ Where Veiling Reflections Cause Trouble

Veiling reflections are limited by the law of reflection, which states the angle of reflection must equal the angle of incidence (arrival). Veiling reflections can cause trouble only if the surfaces in the task area are oriented so that the reflections from a light source are in the viewer's line of sight.   Figure 1 shows a task geometry that has serious veiling reflections.

If veiling reflections can occur from a surface within the task area, they are a potentially serious problem if people must look at the surface for extended periods. Examples are office work, drafting, and sheetmetal work. Veiling reflections are less likely to be a problem if the eyes are continually scanning in different directions, as in work behind a sales counter.

Veiling reflections are usually not a major consideration where activities involve irregularly shaped or moving objects, because the objects reflect light in all directions. However, veiling reflections may be a problem if there are flat surfaces behind the object, as in assembly operations on top of a flat table.

Veiling reflections are almost certain to occur where light fixtures are installed in a blanket manner, without regard to the task location or orientation. In a space with many fixtures, there is a large probability that a viewer will encounter veiling reflections from some fixture in the vicinity. The general solution is to customize the lighting to each task. Task lighting (Measure 9.7.6), which extends this concept to its limits, holds considerable promise for eliminating veiling reflections in the future.

Metals reflect strongly and in a specular (mirror-like) manner, so it is especially important to avoid veiling reflections in industrial operations involving metal surfaces.

Strong veiling reflections can occur with daylighting, whether you are trying to exploit daylighting deliberately or whether it is merely incidental to the presence of windows and skylights. (See Subsection 8.3 about daylighting.) If there is

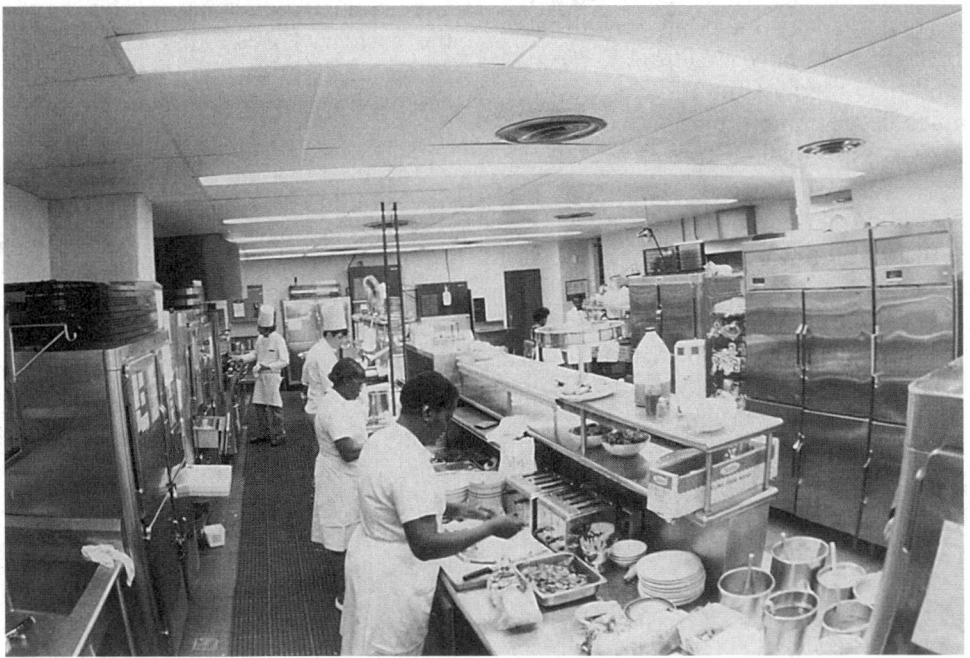

WESINC

Fig. 2 Some lessons about glare and veiling reflections   This hotel kitchen requires a high illumination level throughout the space for food preparation and sanitation. At first glance, the metal surfaces everywhere appear to cause serious glare and veiling reflections. However, the shelves at the rear of the counters shield the work area from most veiling reflections. The work involves a large amount of movement and shifting gaze, which limits the duration of exposure to veiling reflections. The uniformity of the lighting reduces glare by not requiring the cooks' eyes to adapt to different brightness levels as they work. This lighting environment is at the opposite end of the spectrum from those where task lighting is appropriate. With task lighting, each work area must be arranged individually to avoid glare and veiling reflections.

daylighting in a space, you have to arrange activities with respect to the windows and skylights first, because their locations are fixed. Then, arrange the light fixtures.

■ The Strength of Veiling Reflections

Veiling reflections are a surface phenomenon. Their intensity depends primarily on the smoothness of the surface. All fairly smooth surfaces can produce strong veiling reflections. For example, polished wood produces veiling reflections from the very top surface of the lacquer. The grain of the wood is rendered visible by light that penetrates the lacquer deeply enough for the light to be selectively absorbed and reflected by the wood's structural details.

If the surface is very smooth, such as a glass desk top, the veiling reflection is actually an image of the light source. Such images are bright and limited in size. For a surface to produce mirror-like reflection, its surface irregularities need to be smaller than about one wavelength of light (about 0.0005 millimeter).

Even surfaces that are not considered to be optically smooth, such as uncoated paper, can produce significant veiling reflections. The brightness of the veiling reflections is reduced because the incoming light is scattered somewhat. For the same reason, the veiling reflections cover more of the surface than a reflected image of the light source would cover.

Veiling reflections are not formed if the surface is rough enough to break up the incoming light beam completely. This requires the surface to be composed entirely of irregularities much larger than the wavelengths of light. For example, clay brick, blotter paper, and fabrics made from natural fibers produce little veiling reflections.

Surfaces that conduct electricity, such as metals, tend to reflect strongly at all angles of incidence. (This is why glass mirrors depend on a thin metal coating to provide most of their reflective property.) As a result veiling reflections can be very strong in environments where bare metal is within the field of view, as in sheetmetal shops and commercial kitchens (see Figure 2).

With smooth surfaces that do not conduct electricity, such as paper, plastics, and wood, the strength of veiling reflection depends on two factors: the index of refraction of the material, and the angle of incidence. The index of refraction is a physical property of the material related to its density. Veiling reflections are very strong if the light strikes the material within about 30° of the surface. Reflections are weakest if light strikes a surface at an angle near the perpendicular. However, veiling reflections are usually objectionable even in the latter case, because the fixture is then close to the viewer. For example, a light fixture located directly over a desk produces severe veiling reflections.

The strength of veiling reflections does not depend on the color of the material. For example, the veiling reflections from polished black marble and polished white marble are approximately equal, for a given angle of incidence. Therefore, a larger fraction of the light reflected from black marble is in the form of veiling reflections.

■ How to Avoid Veiling Reflections

One foolproof way to avoid veiling reflections is to arrange the layout of the task area so that no smooth surface within the field of view can reflect light from a light source into the eyes of the viewer. In general, install light fixtures off to the side of the viewer's line of sight. At the same time, select the fixture locations to avoid glare and self shadowing, and to provide a uniform illumination level. (These considerations were discussed previously.) Installing the fixtures to the side and slightly behind the viewer is usually best.

If you have a task area where veiling reflections occur, either move the light source, or the task, or the viewer. Or, change the angle of the reflecting surface.

Do not assume that all working surfaces are horizontal. For example, computer monitor screens are approximately vertical, allowing veiling reflections from light fixtures far behind the user. Paper in a typewriter or printer lies at some angle between vertical and horizontal, allowing veiling reflections from a fixture that is overhead or to the rear of the viewer. Locating the fixtures to the side usually avoids veiling reflections from non-horizontal surfaces.

Another way of eliminating veiling reflections is to shield from view the fixtures that create veiling reflections, as discussed previously under "Light Source Glare." This method may be much less cumbersome than rearranging fixtures, especially in a space where task areas are spread out in all directions.

■ Relationship to Reflected Glare

The same layout that makes veiling reflections a problem also invites glare that is caused by reflection of the light source from surfaces in the task area. Glare is a matter of excessive brightness. Whether reflections can cause unacceptable glare depends mainly on the degree of smoothness of the surface. If the task surface is smooth enough to preserve an image of the light source, the reflection will be bright enough to cause desensitization and discomfort. With surfaces that are moderately rough, such as uncoated paper, the reflection of the light source may be scattered enough to reduce its brightness to a tolerable level, even though the veiling reflection is strong enough to reduce contrast seriously.

In general, arranging the lighting to eliminate veiling reflections also eliminates reflected glare, and vice versa.

Figure 2 is an environment that illustrates several points about glare and veiling reflections.

■ Polarization

To be complete, we should mention the polarization of light in veiling reflections. Light waves consist of electric and magnetic fields vibrating perpendicular to the direction of travel. Most light is randomly polarized, which means that the electric fields in the photons of a light beam are oriented randomly around the direction of travel. When light is reflected from a non-conducting surface, the vertical component of the reflected electric field partially cancels the vertical component of the incoming electric field. Such interference cannot occur in the horizontal component. As a result, veiling reflections are polarized parallel to the reflecting surface.

Polarized sunglasses are designed to pass only light that is vertically polarized, and thus they block horizontally polarized veiling reflections from asphalt roads, the surfaces of swimming pools, etc. Therefore, a viewer could avoid most veiling reflections from his desk by wearing polarized sunglasses. This is not practical, so some people have tried the opposite tack, installing vertical polarizers on light sources. This approach has a large energy cost because available polarizing materials absorb a large fraction of the light that passes through them.

Even in principle, trying to use polarization to eliminate veiling reflections is a dead end. If a fixture is located where it can produce veiling reflections, it will also create reflected glare, which is not eliminated by polarization. The only correct way to prevent veiling reflections is by correct fixture placement, perhaps combined with effective shielding.

Color Rendering

A large part of the information and pleasure that people derive from vision is the result of their ability to distinguish colors. Colors are wavelengths of light. The color of an object is the result of selective absorption and reflection of different wavelengths of light by that object. When we design lighting, we want to bring out all the colors in the objects that we see. The ability of lighting to do this is called "color rendering."

Color rendering is determined primarily by the type of lamp or other light source. Selecting lamps is a compromise between color rendering and other factors, especially energy efficiency and cost. To achieve appropriate color rendering, match the color characteristics of the lamp to the color rendering needs of the task and to the physiology of human vision. The penalties for inadequate color rendering are reduced contrast, missing detail, higher illumination requirements, and discomfort.

The quality of color rendering that is required in lighting varies enormously among different activities. In some applications, such as clothing merchandising and medical treatment, color rendering is critical. Other activities, such as paperwork, may involve little or no color vision because there is no color in the task. However, even such activities may require good color rendering in the surroundings for visual comfort. Activities such as warehousing require only mediocre color rendering, even though the articles being handled may be colorful.

■ How Human Beings See Color

There are two types of light receptor cells in the retina of the human eye, called rod cells and cone cells. Only the cone cells can distinguish color. We see color because there are actually three different types of cone cells, each of which has a peak sensitivity at a different wavelength of light. The sensitivity peaks are around 600 nanometers ("red"), 530 nanometers ("green"), and 450 nanometers ("blue"). The red-peaking and green-peaking cone cells are more sensitive than the blue-peaking cells, and they respond to broader ranges of wavelengths.

When light of a particular color enters the eye, the visual portion of the nervous system senses the differences in the responses of the three types of cone cells, and this unique combination of responses is sensed as the color being viewed.

Because vision works this way, the eye may "see" a particular color in two entirely different ways. One way is to receive light of the actual wavelength corresponding to that color. For example, grass looks green because chlorophyll reflects light in the green part of the spectrum. However, the eye may also see "green" as a combination of blue light and yellow light, even though no actual green light enters the eye.

Color printing and color television mimic the eye's method of seeing color. They reproduce all colors by mixing three standard colors. (One system uses red, green, and blue as the primary colors. Another uses cyan, magenta, and yellow.) Color reproduction would be a severe technical challenge if it were necessary to replicate all the actual colors in nature.

Similarly, a lamp can produce a sensation of "white" light in two ways. One is to reproduce the continuous color spectrum of sunlight. The other is to mix a limited number of colors. Only the first method can provide good color rendering of all colors and combinations of colors. In the second case, the quality of color rendering also depends on how colors are mixed in the objects being viewed. If a color in an object being viewed is absent in the light source, that color is not revealed.

Color perception is very subjective, and in most cases, it is very forgiving. The eye compensates considerably for differences in the color distribution of the light source. To a large extent, people see what they expect to see. Thus, for example, incandescent and fluorescent lamps appear to give good color rendering, even though their light output has a very non-uniform distribution of wavelengths. Only when the lamp lacks light output in broad bands of wavelengths, as for example with clear mercury vapor lamps, does the rendering of colors appear distinctly "muddy."

■ Sunlight and Vision

About 35% of solar energy that penetrates to sea level consists of visible wavelengths. This energy is radiated in a continuous spectrum that is strong across the entire wavelength range of human color sensitivity. The eye is most sensitive at the wavelength at which sunlight is strongest, probably because human vision evolved to make the best use of sunlight.

Sunlight is the *de facto* standard against which the color rendering of all lamps is compared. Sunlight is almost ideal as a light source because human color vision is closely matched to the characteristics of sunlight, and because the continuous spectrum of sunlight is able to illuminate all the colors in visible objects.

■ Lamp Spectra and Vision

When we see an object under artificial light, we are seeing the light of the lamp as it is reflected by the object. When we see colors, we are seeing the wavelengths corresponding to those colors reflected strongly, while the other wavelengths of light are absorbed. Therefore, if we are to see the colors in objects, the light source must emit wavelengths corresponding to those colors. Good color rendering requires that the light source should contain colors in roughly the same proportions that they occur in sunlight.

At the present time, there is no type of lamp that closely mimics the visible wavelength characteristics of sunlight. There are several types of lamps that have a high "color rendering index" (described in Reference Note 52), but all these have light output spectra that are very different from the spectrum of the sun.

Incandescent lamps produce a continuous spectrum of wavelengths, so they reveal all colors. This is why incandescent lighting is preferred for surgery, fabric display, graphics work, and other activities where it is critical to see subtle distinctions in color. However, the light output of incandescent lamps is skewed heavily toward the red end of the spectrum, so colors in the blue end of the visible spectrum are revealed relatively weakly.

In contrast, high intensity discharge (HID) lighting produces narrow wavelengths of light, which poses a serious problem for coloring rendering. One type of HID lamp, metal halide, attempts to emulate the solar spectrum by emitting a large number of discrete wavelengths across the visible spectrum. This method works fairly well. However, if an object being viewed has a discrete color component that is absent in the lamp spectrum, that color will not be visible. Thus, a given HID lamp may produce good color rendering for some objects and not for others.

For most general lighting, the best compromise between efficiency and color rendering is achieved with fluorescent lighting. Fluorescent lighting is a two-step process. Light is emitted inside the tube at invisible ultraviolet wavelengths. The ultraviolet light stimulates phosphors to emit light in a continuous band of visible wavelengths. Different phosphors are selected to emit light in different bands. Many different color characteristics are available with fluorescent lamps.

For more detail about the spectra of individual lamp types, see Reference Notes 54 (incandescent lighting), 55 (fluorescent lighting), and 56 (HID and LPS lighting).

■ Color Rendering Affects the Required Illumination Level

A basic limitation of the lumens-per-watt method of rating lamp energy efficiency is that it fails to recognize the role of color in visual efficiency. The extreme example of this weakness is low-pressure sodium lamps, which initially proliferated for many applications because of their high lumen-per-watt ratings. These lamps emit light at only one wavelength, so they fail to provide visual cues that are based on color differences. This is a poor way of revealing robbers in parking lots, for example. A lamp of the same wattage with good color rendering might be much more visually effective, although it produces fewer lumens.

The same principle extends to lamps with mediocre color rendering. For example, one study reported that if lamps having a color rendering index (CRI) of 60 are replaced with lamps having a CRI of 85, the same overall visual quality can be maintained while reducing the lumen level by 25%. However, this would be true only for a scene that depends on color discrimination.

The relationship between lumen levels and color rendering applies even to viewers who are colorblind. This is because objects reflect light only at the wavelengths emitted by the lamp. If the lamp spectrum has serious gaps, the non-illuminated colors in the object cannot be distinguished from the dark background.

■ How to Accommodate Color Vision Deficiencies

Approximately 8% of men and 0.5% of women have reduced ability to distinguish colors. Total color blindness is rare. It is believed that most color blindness is caused by defects in one or more of the three types of cone cells.

For people with partial color vision deficiencies, color vision can be improved by higher illumination levels and better lamp color rendering. In cases of extreme color blindness, nothing can be done with the lighting to improve color vision.

■ Loss of Color Vision in Low Light

The cone cells of the eye are much less sensitive than the rod cells. Therefore, color vision disappears in low light. You can observe this at night. With bright moonlight, you can see outdoor objects fairly well as shades of gray, but you can see colors only under street lights or in a flashlight beam.

This loss of color vision in dim light is not a factor at normal indoor lighting levels.

Comparative Light Source Characteristics

Lumen Output	**Effect of Temperature on Light Output**
Lumen Degradation	**Starting Interval**
Service Life	**Restarting Time**
Efficiency/Efficacy	**Control of Light Distribution**
Ballast Energy Consumption	**Decorative Lamp Options**
Potential for Lamp Substitution and Mismatch	**Mounting Position Limitations**
Dimming Ability	**Acoustical Noise**
Color Rendering Index (CRI)	**Power Factor**
Lamp Color	**Harmonic Distortion**
Starting Temperature	**Electromagnetic Interference**

The energy needed for lighting goes into lamps. Lamps almost exclusively determine the color characteristics of lighting. Lamps are also the limiting factor in lighting efficiency, and they determine all the electrical characteristics of the lighting system. So, selecting the lamps should be your starting point when deciding how to illuminate an activity efficiently. When a lamp is coupled with its auxiliary equipment (e.g., a ballast or ignitor) and installed in a fixture, it becomes the complete light source that is the basic element of the lighting design. Much of energy conservation consists of replacing one light source with another that is more efficient.

Table 1 provides an overview of the major characteristics that distinguish one light source from another. The rest of this Note explains the light source characteristics that are listed in Table 1.

For greater detail about individual lamp types, see Reference Note 54 for incandescent lamps, Note 55 for fluorescent lamps and their accessories, and Note 56 for high-intensity discharge and low-pressure sodium lamps and their accessories. Reference Note 50, Measuring Light Intensity, explains the units that are used to measure the output of light sources.

Lumen Output

Incandescent lamps cover the entire range of capacities that are needed for general lighting. There are no limits on size, except for market demand.

With fluorescent lamps, the power density of the lamp must be kept low to maintain efficiency. Therefore, the size of the lamp must increase with output, and size itself imposes the upper limit on output. The lower size limit is imposed by efficiency and cost, since smaller fluorescent lamps are less efficient but not less expensive. Similar constraints exist with low-pressure sodium lamps.

With HID lamps, there is no hard upper limit except market demand. The lower size limit is efficiency and cost, since smaller lamps are less efficient but not much less expensive.

Lumen Degradation

With all lamp types, output declines with age. Most types of lamps lose output most rapidly at the beginning of their lives. High-pressure sodium lamps are an exception, maintaining output well in the first half of life, but losing output rapidly thereafter. The factors that cause light output to decline are discussed in Reference Notes 54 through 56, respectively, for the different types of lamps.

One way to maintain a higher average light output is to replace lamps before they fail. This is probably not economical with lamps whose light output drops early in life. It may be practical with high-pressure sodium lamps. Consider this as a possibility if you do group relamping.

Service Life

The lighting industry defines service life as the time period during which 50% of lamps fail. For most lamp types, there is a fairly broad distribution, so some lamps fail well before the average. If lamps are replaced on a group basis, this may substantially shorten the replacement interval.

With all lamp types, service life is reduced by increasing the number of starts. Good information does not appear to be available on the effect of frequent cycling. For example, available data on the effect of cycling on fluorescent lamps generally starts with a minimum of 3-hour cycles.

With incandescent lamps, there is a severe compromise between service life and efficiency, since both depend primarily on filament temperature.

Table 1. COMPARATIVE LIGHT SOURCE CHARACTERISTICS

	Conventional Incandescent	Halogen Incandescent	Conventional Fluorescent
Lumen Output (lumens)	10 to 50,000	300 to 40,000	900 to 12,000
Lumen Degradation (percent of initial lumens)	15 to 40	8 to 15	8 to 25
Service Life (hours)	750 to 4,000	2,000 to 6,000	7,000 to 20,000
Efficacy (lumens per watt)	7 to 22	14 to 22	30 to 90
Ballast Energy Consumption (percent of lamp wattage)	none	none	5 (high quality electronic ballasts) to 20 (cheap magnetic ballasts)
Potential for Lamp Substitution and Mismatch	Unlimited substitution wherever the lamp fits the fixture, provided that fixture heat capacity is adequate.	Unlimited substitution wherever the lamp fits the fixture, provided that fixture heat capacity is adequate.	Limited within narrow ranges of wattage by lamp size, socket style, and ballast compatibility.
Dimming Ability	Unlimited.	Unlimited, except that units with diodes cannot be dimmed.	Requires special dimming ballasts.
Color Rendering Index (CRI)	100	100	50 to 95
Lamp Color	Perceived as white, but actually reddish yellow. Continuous spectrum, but concentrated very heavily toward the red end, with little energy at the blue end.	Perceived as white, but actually reddish yellow. Continuous spectrum, somewhat more balanced in color than conventional incandescent.	Almost unlimited color choice by selection of phosphors. Spectra range from very spiky to fair approximation of continuous solar spectrum. All emit strong mercury spectra.
Starting Temperature	no limit	no limit	60°F (15°C) for high efficiency lamps, 50°F (10°C) for conventional lamps. Special ballasts can be installed to provide much lower starting temperatures.
Effect of Temperature on Light Output	Minimal.	Minimal.	Serious loss of light output above and below optimum lamp temperature (about 100°F).
Starting Interval	Instantaneous.	Instantaneous.	Instantaneous for lamps with instant-start ballasts. About one second for rapid-start ballasts. One to several seconds for preheat ballasts.
Restarting Time	Instantaneous.	Instantaneous.	Shorter than cold start.
Control of Light Distribution	Some styles allow very tight focussing.	Some styles allow very tight focussing.	Allows only loose focussing. Most control perpendicular to lamp axis.
Decorative Lamp Options	Wide variety of decorative styles.	Mini-reflectors are often used as decor.	None.
Mounting Position Limitations	None. Some lamps are designed to exploit base-up fixtures.	None.	Long lamps should be mounted nearly horizontal.
Acoustical Noise	Minimal.	Minimal.	All magnetic ballasts produce some noise, and defective noisy units are fairly common. Some electronic ballasts have noticeable noise.
Power Factor	No problem.	No problem.	Ballasts with high power factor are available. Some ballasts have low power factor.
Harmonic Distortion	None.	None.	High distortion occurs primarily in cheaper electronic ballasts.
Electromagnetic Interference	None.	None.	Some problems may occur with any electronic ballast. Problems are more severe with cheaper units.

Compact Fluorescent	Mercury Vapor	Metal Halide	High-Pressure Sodium	Low-Pressure Sodium
250 to 1,800	1,200 to 60,000	4,000 to 160,000	2,000 to 50,000	1,800 to 35,000
15 to 20	35 to 45	30 to 45	25 to 35	
10,000	24,000	5,000 to 20,000	10,000 to 24,000	18,000
25 to 70, including ballast losses	35 to 65	70 to 130	50 to 150	100 to 190
10 (electronic ballasts) to 20 (magnetic ballasts)	8 (large lamps) to 50 (small lamps)	7 (large lamps) to 30 (small lamps)	10 (large lamps) to 35 (small lamps)	ca. 20
Screw-in lamps substitute for each other and for most incandescent lamps, except where they are too large to fit. Cannot be used in dimming fixtures. Other compact lamps have specialized bases that limit substitution.	Substitutions within type highly limited by ballast compatibility. Some mercury vapor lamps substitute for incandescent lamps without external ballasts, but these offer minimal efficacy advantage.	Substitutions within type highly limited by ballast compatibility.	Substitutions within type highly limited by ballast compatibility. Some HPS lamps are designed as direct substitutes for mercury vapor lamps, offering major efficacy improvement but worse color rendering than other HPS lamps.	Substitutions within type highly limited by ballast compatibility and specialized sockets.
Units with integral ballasts cannot be dimmed.	Single-stage dimming to 50% output, only with transformer ballasts.	Single-stage dimming to 50% output, only with transformer ballasts. Major lamp color change.	Single-stage dimming to 30% output, only with transformer ballasts. Major lamp color change.	None presently available.
60 to 85	40 to 50	60 to 70	20 to 85	0 to 20
Similar to conventional fluorescent. Wide color choice, 2,700K to 5,000K.	Distinctly blue-green. Clear lamps concentrate most energy in a few narrow mercury emission lines. Phosphor lamps add additional broader color lines and some continuous spectrum.	Bluish white. Clear lamps emit a large number of narrow lines across the visible spectrum. Phosphor lamps add more continuous spectrum.	Yellowish white. Narrow continuous spectrum centered around 600 nanometers. Units with enhanced color increase output at red end of spectrum. All are deficient in blue half of spectrum.	Monochromatic yellow.
-20°F (-29°C) to 32°F (0°C)	typically -20°F (-29°C)	typically -20°F (-29°C)	typically -20°F (-29°C)	typically -20°F (-29°C)
Serious loss of light output above and below optimum lamp temperature (about 100°F). Lamps that use mercury amalgam maintain light output much better at low temperatures.	Minimal loss of output above -20°F (-29°C).	Minimal loss of output above -20°F (-29°C).	Minimal loss of output above -20°F (-29°C).	Minimal loss of output above -20°F (-29°C).
One to several seconds. Units with mercury amalgam require about one minute to reach full brightness.	4 to 8 minutes	3 to 10 minutes	5 to 10 minutes	7 to 15 minutes
Shorter than cold start.	Similar to cold start.	5 to 10 minutes longer than cold start.	Substantially shorter than cold start.	Immediate.
Allows moderately tight focussing, especially with unconventionally large fixtures.	Allows moderately tight focussing.	Allows moderately tight focussing.	Allows moderately tight focussing.	Allows only loose focussing. Most control perpendicular to lamp axis.
Globes, some specialized diffusers.	None.	None.	None.	None.
None.	Some models have designated mounting positions.	Most models have designated mounting positions.	Not sensitive to mounting position.	
Good units are quiet, cheap units may be noisy.	Ballasts are magnetic, and produce some noise.	Ballasts are magnetic, and produce some noise.	Ballasts are magnetic, and produce some noise.	Ballasts are magnetic, and produce some noise.
Units with high power factor are available. Some have low power factor.	Ballasts with high power factor are available. Some ballasts have low power factor.	Ballasts with high power factor are available. Some ballasts have low power factor.	Ballasts with high power factor are available. Some ballasts have low power factor.	Ballasts with high power factor are available. Some ballasts have low power factor.
All units with electronic ballasts have significant harmonic distortion. Cheaper units have much more than others.	Minor, assuming that the ballasts are magnetic.	Minor, assuming that the ballasts are magnetic.	Minor, assuming that the ballasts are magnetic.	Minor, assuming that the ballasts are magnetic.
Some problems may occur with any models having electronic ballasts, but are likely to be less severe than with conventional fluorescent fixtures. Problems are more severe with cheaper units.	None, assuming that the ballasts are magnetic. No experience with electronic ballasts.	None, assuming that the ballasts are magnetic. No experience with electronic ballasts.	None, assuming that the ballasts are magnetic. No experience with electronic ballasts.	None, assuming that the ballasts are magnetic. No experience with electronic ballasts.

Efficiency/Efficacy

The ability of a light source to turn electricity into light is almost always expressed in terms of "luminous efficacy," or simply, "efficacy." The units of efficacy are lumens per watt. Except for the units, the concept of efficacy is the same as the concept of efficiency. (With efficiency, the input and output units are the same, so that efficiency is expressed as a simple percentage.)

Efficiency, or efficacy, varies widely within each lamp type. The theoretical maximum efficacy of a white light source is about 220 lumens per watt. Table 1 shows that the best modern light sources are approaching this target figure.

For all lamp types, there is a strong correlation between efficacy and size. In Table 1, the higher numbers are for the largest lamps, and the lower numbers for the smallest lamps.

With incandescent lamps, there is a serious compromise between efficacy and lamp life.

With fluorescent and HID lamps, efficacy relates strongly to color rendering. With fluorescent lamps, the most efficient lamps have the best color rendering. Conversely, with HID lamps, good color rendering imposes a severe efficacy penalty.

The rated efficacy of lamps generally is measured under optimum conditions. Efficacy declines with age for all lamp types. Most of the loss consists of a decline in light output without a corresponding drop in energy input. The relationship between the output and efficacy varies with lamp type.

With all lamp types except incandescent, efficacy drops with non-standard ambient temperature and with low supply voltage.

These figures do not include ballast losses, which vary considerably with lamp size and type. One exception is the compact fluorescent fixtures, which include the losses of their integral ballasts.

Ballast Energy Consumption

All conventional fluorescent and HID lamps require ballasts, and some HID lamps require an additional starting device called an "igniter." The purposes of these devices are described in Reference Notes 55 and 56, respectively. Ballasts dissipate a significant amount of energy, which detracts from the overall efficiency of the lamp and fixture.

With fluorescent lighting, ballast energy consumption depends mainly on whether the ballast is magnetic or electronic, the latter being more efficient. There are large efficiency variations within each type. Other features, such as the ability of a ballast to turn off lamp filaments once the lamp has started, also affect energy consumption.

Compact fluorescent fixtures include an integral ballast. With some compact fluorescent fixtures, the tube is detachable from the ballast and can be replaced separately.

With HID lamps, ballast technology is still old and magnetic, and progress toward higher ballast efficiency is slow. Magnetic ballast losses do not increase rapidly with increasing lamp capacity, so the relative losses are much greater with smaller lamps.

Some types of HID lamps have been developed that do not need ballasts, but these lamps are very inefficient.

Potential for Lamp Substitution and Mismatch

Substituting one light source for another is a common step in improving lighting efficiency. You may wish to substitute lamps of a particular type to reduce output, or to increase efficiency, or both. Or, you may wish to substitute a more efficient type of lamp.

Lamp substitution is much easier and more economical if you can switch lamps without the need to replace the fixture or the auxiliary equipment. This is possible with some types of lamps, but not with others.

On the negative side, it is often possible to substitute lamps in a way that wastes energy. Improper substitutions may also shorten lamp life and do other damage to equipment. Similarity of lampholders, or sockets, allows the wrong size, type, or model of lamp to be installed in a fixture. Measure 9.8.2 recommends fixture marking to minimize this problem.

With incandescent lamps, lamp substitution is limited only by the socket configuration and the maximum safe fixture wattage. This is a problem for energy conservation because it allows lamps of higher wattage than needed to be installed in the fixture.

With fluorescent lamps, substitution is confined to a narrow range of wattage by the physical dimensions of the lamp. Also, substitutions are limited by ballast compatibility. Unfortunately, improper substitutions are easy to make because of common socket types.

Substituting compact fluorescent fixtures for incandescent lamps is an important energy conservation measure. Unfortunately, the relative bulkiness of the compact fluorescent units often makes it impossible to install them in existing fixtures.

Most HID lamps for general lighting use medium and mogul bases, but lamp substitutions are limited by the need for ballast compatibility. Some high-pressure sodium lamps are designed as direct substitutes for mercury vapor lamps, providing a significant increase in efficiency, but worse color rendering, than other types of high-pressure sodium lamps.

Some mercury vapor lamps are designed as direct substitutes for incandescent lamps, without external ballasts, but these offer little improvement in efficiency.

Dimming Ability

Dimming is a well established technique for incandescent lighting, but it has only recently become practical for fluorescent lighting. Dimming of HID types is still in its infancy.

Dimming radically reduces the efficiency of incandescent lamps, but modern fluorescent dimming equipment incurs only a modest efficiency penalty.

Dimming incandescent and HID lamps changes their color radically. Dimming fluorescent fixtures does not cause much color change.

Color Rendering Index (CRI)

The relationship between human color vision and the color output of lamps is discussed in Reference Note 51, Factors in Lighting Quality. At present, the "color rendering index" (CRI) is the only selection criterion generally available to indicate how well particular lamps render the colors of illuminated objects. CRI figures are becoming increasingly common in lamp catalogs.

The CRI scale consists of numbers from 0 to 100. The CRI of a particular lamp indicates the amount that it changes color perception of standard colored patterns in comparison with the color perception of these patterns when they are illuminated by incandescent test lamps.

CRI does not tell the whole story about a lamp's color rendering. Roughly speaking, the CRI indicates the ability of a lamp to allow individual colors to be seen and to distinguish between colors. However, two lamps having the same CRI may bring out colors differently. If the CRI is high, the differences may be too subtle to notice in direct vision. However, color photography is sensitive to color differences, so CRI alone is not adequate for selecting lamps for photography.

A high CRI does not imply a smooth distribution of colors, as occurs in incandescent light. In fact, a lamp with a high CRI may have a spiky light distribution curve. (For example, this is true of metal halide lamps that have high CRI's.) This may be satisfactory for general color rendering, but such lamps may not accurately represent pure colors, such as those used in graphics work.

The types of lamps presently being used for general lighting differ radically in their ability to render colors well. At one extreme, incandescent lamps have a continuous color spectrum that brings out all colors well, although their color distribution is skewed toward the red end of the visible spectrum.

At the other extreme are lamps that emit only a few narrow bands of wavelengths, such as mercury vapor lamps. The most extreme case is low-pressure sodium lamps, which emit only one pure color. Low-pressure sodium lighting stands apart from the other types because it has virtually no ability to bring out colors.

In the middle are fluorescent lamps and some types of HID lamps that have been modified for improved color rendering. These lamps have a continuous (but uneven) distribution of wavelengths, upon which is superimposed tall emission spikes at wavelengths that are characteristic of the gases in the lamps.

With HID lamps, there is a severe compromise between color rendering and efficiency. For this reason, HID lamps are available with a large range of CRI's. With fluorescent lamps, there is a compromise between color rendering and cost, also resulting in a wide range of CRI's.

A lamp's own color appearance does not provide reliable guidance to its color rendering ability. The eye cannot judge this characteristic of a lamp. For example, a lamp could appear "white" either by having a continuous output spectrum or by mixing a few colors. The Color Rendering Index is necessary to tell how well the lamp brings out colors in other objects.

Lamp Color

Odd as it may seem, the ability of a lamp to display colors well is not strongly related to the lamp's own average color. A lamp produces good color rendering by emitting light throughout the visible spectrum. The eye tends not to notice variations in the appearance of illuminated objects due to an uneven or skewed wavelength distribution from the lamp. For example, large differences in average lamp color, such as the difference between "cool white" and "warm white" fluorescent lamps, do not strongly affect perceived color rendering.

Lamp color is discussed in the lighting trade mostly in the context of theories about how light affects mood. Specifically, reddish colors are supposed to be "warm" and uplifting, while bluish colors are "cold" and severe. The scientific support for such notions is flimsy, but they provide plenty of advertising copy for lamp manufacturers.

On the other hand, lamp color is critical for photography. Color film is sensitive to lamp color, so good color photographs require careful matching of lamp color to film characteristics. For example, you might not notice the difference between lighting a room with a metal halide or a high-pressure sodium lamp, especially if both lamps have a high CRI. However, color photographs of the room would reveal a stark difference.

A common color specification for lamps is the "correlated color temperature," which is becoming increasing common in lamp catalogs, especially for incandescent and fluorescent lamps. Roughly speaking, this is the temperature of an ideal incandescent source that is closest in color to the lamp.

With incandescent lamps, the color temperature is fairly close to the actual temperature of the filament. With fluorescent and HID lamps, color temperature bears no relationship to the temperature of the lamp because these lamp types do not produce light primarily by a thermal process. In fluorescent lamps, the color temperature represents the average color of all the phosphors in the lamp. In HID lamps, color temperature is a weighted average of the temperatures corresponding to the different wavelengths emitted by the gases in the lamp.

Low color temperatures tend to be reddish, and high color temperatures tend to be bluish. However, this is not a reliable guideline. A fluorescent or HID lamp may have a prominent tint that is quite different from the color that corresponds to the color temperature of the lamp.

Color temperature is expressed in degrees Kelvin, or kelvins, rather than degrees Celsius, so it is sometimes called "Kelvin temperature." (The Kelvin scale uses degrees of the same size as the Celsius scale, but the zero point is absolute zero instead of the freezing temperature of water.)

A more precise description of color is the CIE Chromaticity scale. (CIE are the initials in French of the International Commission on Illumination). This is a two-dimensional color chart that includes all visible colors. A color is specified by two position coordinates on the chart, "x" and "y." The Chromaticity Diagram is much more accurate in describing a particular color than the color temperature can be. Unless you have a lot of experience using the chart, you need to have a copy on hand when reading lamp chromaticity specifications in a catalog.

Neither color temperature nor the Chromaticity Chart can distinguish between a pure color and a color that is the result of mixing other colors. Therefore, neither provides an indication of the color rendering ability of a lamp. On the other hand, the Color Rendering Index (CRI) does not provide information about average lamp color. For example, a high-pressure sodium lamp having a CRI of 60 may produce a overall reddish cast, while a metal halide lamp with the same CRI may produce an overall bluish cast.

Starting Temperature

All lamp types, except incandescent, have minimum starting temperatures that limit the environments in which they can be used. Conventional fluorescent fixtures are not rated to start much below room temperature, a fact that is commonly overlooked. Starting temperatures can be lowered by using special low temperature ballasts, but this does not solve the problem of reduced efficiency at low temperature, discussed next. Compact fluorescent lamps have lower design starting temperatures. HID lamps have even lower starting temperatures, which makes them suitable for outdoor use.

Effect of Temperature on Light Output

The fact that a lamp will start at a low temperature does not imply that the lamp will operate efficiently at that temperature. The light output of fluorescent lamps drops severely with reduced ambient temperature, for reasons covered in Reference Note 55. Compact fluorescent lamps that use a mercury amalgam maintain light output at much lower temperatures than lamps without this feature.

The light output of HID lamps is relatively insensitive to ambient temperature because most have double glass envelopes.

Lamps can operate at lower ambient temperatures if they are installed in enclosed fixtures that retain heat. However, this is a two-edged sword for fluorescent fixtures. When the environment is warm, the lamp will become warmer than its optimum temperature, and its efficiency and life will decline.

The efficiency of fluorescent and HID lamps declines roughly in parallel with the fall in light output.

Starting Interval

Incandescent lamps reach full brightness virtually instantly.

Fluorescent lamps may start almost instantly, or they may need a second or two, depending on the type of starter. They require a few seconds to reach full brightness, but this may not be noticeable. An exception is compact fluorescent lamps that use a mercury amalgam. The mercury evaporates from the amalgam slowly, so brightness grows perceptibly, requiring perhaps a minute to reach maximum.

HID lamps are notorious for long starting times, which limits them to applications that can tolerate a delay of several minutes in reaching full output. Some new types of HID lamps are being introduced that allow rapid starting, but so far these suffer a serious efficiency penalty.

Restarting Time

HID lamps cannot restart until they have cooled somewhat. Mercury vapor and metal halide lamps may take longer to restart than to start from a cold state. As a result, HID lighting is limited to long-duration applications where the need for light is predictable.

Some specialized HID lamps are designed for quicker restarting, but these carry a cost penalty, and possibly an efficiency penalty as well. For example, one type contains two separate arc tubes. If the warm arc tube is turned off, power is transferred to the cold arc tube.

Control of Light Distribution

Control of the light distribution pattern is an essential aspect of lighting efficiency. The lamp and the fixture together form a system that determines the distribution pattern. The geometry of the lamp may limit the ability of the fixture to exercise precise control. In general, the smaller the light source, the better the control. Light from a point source can be focussed precisely, but light from a large source, such as a conventional fluorescent lamp, cannot be focussed except in a broad pattern.

Some incandescent lamps have filaments that are designed for small size to maximize focussing ability. These are not needed for general lighting. For most purposes, ordinary incandescent bulbs can be treated as a point source, with the fixture controlling the distribution pattern. However, details of the lamp matter even for general lighting. For example, the shape of the filament determines the light distribution pattern from the lamp, especially if the lamp envelope is clear.

There are many types of incandescent reflector lamps that provide accurate control of the distribution pattern. Compact fluorescent lamps are also available in reflector fixtures.

Conventional fluorescent tubes have a broad distribution pattern. A bare tube has a pattern that is shaped like a doughnut surrounding the tube. In fluorescent lighting, the fixture plays an important role in controlling the distribution pattern.

HID lamps have relatively small arcs that behave approximately as point sources when the lamps are installed in large fixtures. Low-pressure sodium lamps have a relatively long arc that is similar in dimensions to a short fluorescent tube.

Decorative Lamp Options

Some incandescent lamps are meant to be seen for decorative purposes, and these are available in a large variety of shapes, in clear or coated glass, and with various filament configurations. Fluorescent and HID lamps are not meant to be viewed directly, so they do not offer such options. Compact fluorescent fixtures offer a limited range of decorative configurations, such as globes.

Even with decorative lamps, careful selection of wattage, filament configuration, and other features can reduce energy waste.

Mounting Position Limitations

In most cases, incandescent lamps are not sensitive to mounting position. There are exceptions. For example, some bulbs are designed with straight filaments aligned with the bulb axis to minimize darkening when the lamps are mounted in base-up fixtures.

Conventional long fluorescent tubes should be installed horizontally, or nearly so. If the tubes are tilted and are operated only for short periods, the mercury in the tubes may accumulate at one end, making them hard to start.

Some HID lamps can be operated only in a limited range of mounting positions, for reasons covered in Reference Note 56.

Acoustical Noise

Most acoustical noise in lighting emanates from magnetic ballasts, although electronic ballasts may produce some noise. Selecting your ballasts carefully is the way to avoid this problem.

Power Factor

Low power factor is a problem related primarily to cheaper magnetic ballasts used with fluorescent and HID lighting. Refer to Subsection 9.2 for details.

Harmonic Distortion

Harmonic distortion is a distortion of the waveform in the facility's electrical power system that may cause trouble with other electrical equipment. The problem is related to ballasts, primarily cheaper electronic ballasts. Refer to Subsection 9.2 for details.

Electromagnetic Interference

Electronic ballasts operate at high frequency. The high frequency current in the lighting system may radiate interference into space, and the interference may also propagate in the electrical power system. Refer to Subsection 9.2 for details.

Construction Efficiency Codes
Equipment Efficiency Standards for Manufacturers

In response to the energy shortages of the 1970's, many versions of engineering and legal guidelines for lighting efficiency were put forward. A hodgepodge drawn from these were enacted into law. At this time, lighting efficiency standards are diverse and still evolving. Individual standards are incomplete, unbalanced in emphasis, and sometimes quirky in logical structure. The reason is apparent if you consider how difficult it would be to mandate all the aspects of lighting efficiency that are covered in Section 9.

Construction Efficiency Codes

In the United States, Federal law requires each state to include efficiency standards for lighting and other aspects of construction in its building codes. The underlying source for most of the efficiency standards in the United States, and in some other countries, is ASHRAE Standard 90. The ASHRAE standard may be referenced directly by the state codes, or it may be referenced though model energy codes developed by interstate code organizations, including Building Officials and Code Administrators International (BOCA) and the Council of American Building Officials (CABO).

Alternatively, some states that are particularly aggressive about energy conservation have developed energy conservation codes independently of ASHRAE. U.S. federal law mandates that any State or local codes must meet the minimum requirements of the ASHRAE-derived codes.

Because of practical difficulties, there are no lighting efficiency codes based on a comprehensive approach to lighting efficiency. The leading lighting conservation standards, including ASHRAE Standard 90, are a grab bag of different pragmatic approaches, including the following:

- *illumination level standards.* The first widespread approach to mandating lighting efficiency was defining tables of illumination levels for different types of activities. This method has the serious weakness of failing to account for the efficiency of lamps and fixtures, or for the efficiency of light distribution. Furthermore, specific lighting levels are subject to continuing dispute. Illumination level standards have been almost completely abandoned.

- *lighting power budget.* A newer method of mandating lighting efficiency that goes directly to the question of energy input is the "lighting power budget." This approach specifies the lighting wattage that is allowed for different types of activities. This approach is the major element of ASHRAE Standard 90 and its derivatives. Lighting power budgets have several major weaknesses. They apply only to permanently installed lighting, which ignores the possibility that the efficient lighting fixtures of the future will be easily movable. (For more about movable fixtures, refer to Measure 9.6.2, which deals with task lighting.) Lighting power budgets are virtually impossible to police as lighting is changed in response to changing activities. Also, they do not account for light fixtures that are plugged into receptacles.

- *component efficiency standards.* Some construction codes include efficiency standards for certain components. For example, ASHRAE Standard 90 requires that all ballasts should have a power factor of at least 90%. Some state codes specify lamp and ballast efficiencies, usually following Federal law.

- *control requirements.* The newest development in efficiency codes is requirements for efficient controls. For example, ASHRAE Standard 90 has a rather weak and general section related to manual lighting controls. It also has an oddly illogical provision that allows the installed lighting power to be increased if automatic controls are installed.

Equipment Efficiency Standards for Manufacturers

In the United States, Federal law mandates minimum efficiencies in the manufacture and importation of a limited number of common types of lighting equipment. At present, there are two such Federal laws:

- The National Appliance Energy Conservation Act of 1988 specifies minimum ballast efficacy factors (described in Subsection 9.2) for certain common classes of fluorescent ballasts.

- The National Energy Policy Act of 1992 specifies efficiency standards for the more common types of fluorescent lamps, for compact fluorescent lamps, and for the more common types of incandescent lamps. The same law pushes for vaguely stated improvements in HID lamps and in light fixtures, although specific action in these areas will require enactments at later dates.

Incandescent Lighting

Mother Nature's Basic Light Source
How Incandescent Lighting Works
The Spectrum of Incandescent Lamps
The Color Rendering of Incandescent Lamps
Why Incandescent Lighting is Not Efficient
Practical Effects of Filament Temperature
- Limitations in Filament Materials
- Effect of Filament Temperature on Color
- Effect of Filament Temperature on Service Life
- Ultraviolet Emission

Methods of Improving Incandescent Lamp
  Efficiency
- Heavy Fill Gases
- Tungsten Halogen Lamps
- Infrared Reflecting Coatings
- Low-Voltage Filaments
Filament Configuration Options
Service Life and Lumen Degradation
A Comparison of Incandescent Lamps
How Incandescent Lamps are Named

The simple looking incandescent light bulb is one of the marvels of civilization. Perfecting it was the achievement that made Thomas Edison famous, after many others had failed. This Note will give you a thorough understanding of this wonderful device, along with its inherent efficiency limitations. With this background, you will know where to use incandescent lighting, and where to avoid it. In addition, this Note provides practical information that you need to select the right lamps for applications where incandescent lighting is appropriate.

Mother Nature's Basic Light Source

All light that human beings see is created when electrons lose energy by falling from one energy level to a lower energy level. To get rid of the difference in energy, the electron emits a photon, which is a small package or particle of light. This is the basic process of any kind of light source, whether it is a light bulb, a fluorescent tube, an LED, an electroluminescent night light, a firefly, or a camp fire. In brief, to make a useful light source, we have to get a lot of electrons in one place, excite them to a higher energy level, and let them drop to a lower energy level.

Electrons are a part of every atom, so we have no trouble finding enough electrons to create a light source. Also, there are many ways to excite the electrons. In the three main types of light sources — incandescent, fluorescent, and HID — the electrons are excited by physical collisions. Most of these collisions are with other electrons.

When an electron drops from one energy level to a lower energy level, it has to get rid of the difference in energy. It can do this in several ways, but the most common is to emit a photon. The wavelength of the photon is determined only by its energy. As we said before, the energy of the photon equals the change in energy of the electron as it falls from one level to another.

If the photon has a wavelength that falls within a particular range, our eyes can see the photon. A good light source emits many photons within this visible range. Some light sources (such as low-pressure sodium lamps) emit most of their photons in the visible range. Others (such as candles and incandescent lamps) emit only a small fraction in the visible range. For a light source to be efficient, a large fraction of the energy that we put into the light source should produce photons in the visible range.

We have not yet explained what an electron's "energy level" means, so let's explain it now. An electron is a material particle. Like any moving particle, it has kinetic energy. If the electron is traveling in empty space, all its energy is kinetic. However, electrons in light sources spend a lot of time traveling under the influence of nearby atoms. The electron has a negative charge, and atoms have a positive nucleus, so an electron is attracted to the atom by electrostatic force. The energy stored up in this attraction is called potential energy. An electron's total energy is the sum of its kinetic and potential energies. This total energy is what we mean by the "energy level."

(The energy level actually applies to the system consisting of the electron and the atom or atoms with which it interacts. However, the electron is the main player, and it undergoes most of the change in energy. Therefore, it is common to describe the energy level as belonging to the electron.)

If an electron is attached to a single isolated atom, it moves in a precise standing-wave pattern. Each vibration pattern has a different combination of kinetic and potential energy, with a different total energy. If the atom is isolated, as it is in a low-pressure gas, the electron energy levels are stable and sharply defined. Therefore, when an electron changes its energy level, the photons are emitted at a limited number of precise wavelengths. The light from a low-pressure sodium lamp is an example of this.

If atoms are packed together, as in a solid lamp filament, the electrons of one atom feel the influence of the surrounding atoms, and they collide with other electrons. This interference distorts the vibration patterns of the electrons, changing their energy levels in a random way. Thermal agitation also randomly changes the distance between atoms. As a result, the wavelengths of light emitted by electrons in a hot solid are distributed in a broad, continuous statistical pattern. When light is emitted in this manner, it is called "incandescent."

Now, we have all the ingredients of any light source. In modern lamps, we put energy into the lamp in the form of electricity, which is a stream of electrons. The electrons collide with the electrons in atoms of specially selected materials in the lamp, exciting the latter electrons to higher energy states. The excited electrons fall back to lower energy states, and emit light in the process.

How Incandescent Lighting Works

Incandescent light is the light emitted by a hot, glowing body. Heat is the source of energy that drives incandescent light emission. Heat is the kinetic energy of random motion of the atoms in a substance. Temperature indicates the intensity of this motion. If the temperature is high enough, the collisions between atoms excite many electrons to higher energy states. The electrons fall back to lower energy levels, emitting light, as we just discussed.

It does not matter where the heat energy comes from, in principle. You can create incandescent light by heating a solid object with a blow torch. The hot gas mantle of the sun is heated by nuclear energy. The incandescent gas of a candle flame is heated by chemical energy. The role of electricity in "electric" lighting is simply to heat the filament in a lamp.

An incandescent light source does not have to be a solid. The surface of the sun, which is our most

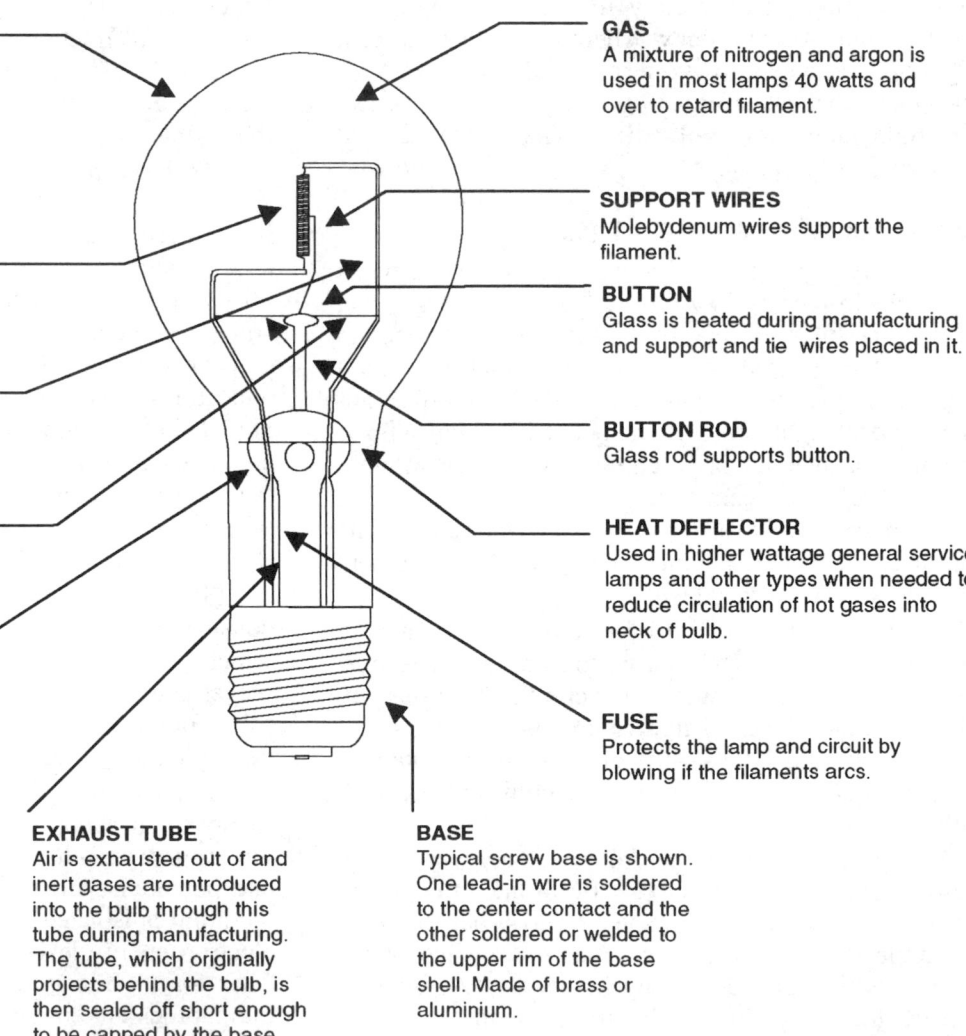

BULB
Soft glass is generally used. Hard glass is used for some lamps to withstand higher bulb temperatures and for added protection against bulb breakage due to moisture. Bulbs are made in various shapes and finishes.

FILAMENT
The filament material generally used is tungsten. The filament may be a straight wire, a coil or a coiled-coil.

LEAD-IN WIRES
Made of copper from base to stem press and nickel-plated copper or nickel from stem press to filament, they carry the current to and from the filament.

TIE WIRES
Molybdenum wires support lead-in wires.

STEM PRESS
The lead-in wires in the glass have an air-tight seal here and are made of a nickel-iron alloy core and a copper sleeve (Dumet wire) to assure about the same coefficient of expansion as the glass.

GAS
A mixture of nitrogen and argon is used in most lamps 40 watts and over to retard filament.

SUPPORT WIRES
Molebydenum wires support the filament.

BUTTON
Glass is heated during manufacturing and support and tie wires placed in it.

BUTTON ROD
Glass rod supports button.

HEAT DEFLECTOR
Used in higher wattage general service lamps and other types when needed to reduce circulation of hot gases into neck of bulb.

FUSE
Protects the lamp and circuit by blowing if the filaments arcs.

EXHAUST TUBE
Air is exhausted out of and inert gases are introduced into the bulb through this tube during manufacturing. The tube, which originally projects behind the bulb, is then sealed off short enough to be capped by the base.

BASE
Typical screw base is shown. One lead-in wire is soldered to the center contact and the other soldered or welded to the upper rim of the base shell. Made of brass or aluminium.

Osram Sylvania Inc.

Fig. 1 How a conventional incandescent lamp is made

important incandescent light source, is a gas in which the atoms are tightly packed by the sun's gravity. The intense thermal agitation prevents the atomic energy levels involved in light emission from settling down in stable states. This causes light to be emitted in the random, "incandescent" manner. The same is true of electric arc lighting.

Virtually all incandescent lighting today is done with electric lamps. Figure 1 shows how the basic type of incandescent lamp is constructed. The light emitting element is a tungsten wire, or filament, which acts as a resistor. Passing electricity through the filament heats it enough to make it a useful source of light.

The Spectrum of Incandescent Lamps

The light emitted by any light source has a characteristic pattern of wavelengths, which is called its "spectrum." The spectrum of any incandescent lamp is a lopsided bell curve. At the long-wavelength (low energy) end of the spectrum, light emission rises slowly, and then levels off at a peak. On the short-wavelength (high energy) side of the peak, emission falls off steeply.

The amount of heat in a material is expressed by its temperature. If the temperature of the source increases, the additional kinetic energy increases both the number and the intensity of the electron excitations. This has two dramatic effects on the energy output pattern:

• the amount of light energy that is emitted increases very rapidly. In fact, the rate of emission is proportional to the fourth power of the temperature. (This fourth-power relationship is called the Stefan-Boltzmann Law.)

• the average wavelength of the light becomes shorter. More specifically, the peak wavelength is inversely proportional to the absolute temperature. (This inverse relationship is called Wien's Displacement Law.)

If an object becomes hot enough, it emits a significant amount of energy at wavelengths that can be seen by human beings. At a temperature of 1,000°F (540°C), a solid emits a dull red glow. Continuing to raise the temperature eventually yields a useful amount of light. The filament temperature of ordinary incandescent bulbs ranges from about 3,800°F (2,100°C) to about 5,000°F (2,800°C). Figure 2 shows the spectrum of incandescent lamps.

The Color Rendering of Incandescent Lamps

An important characteristic of any light source is its ability to reveal the colors in the material that is being illuminated. When you look at a colored object, the colors that you see come from the light source, not from the object itself. Therefore, a light source must emit all colors if it is to have good color rendering. The broad, continuous spectrum of incandescent light emission does provide good color rendering. In fact, the Color

Rendering Index (CRI) of incandescent lamps is defined as 100, or near 100, because it includes all the colors of the visible spectrum. (For more about CRI, see Reference Note 52, Comparative Light Source Characteristics.)

This fact seems to mislead many into believing that incandescent lamps are "ideal" in terms of color rendering. This is far from true. In fact, sunlight is closest to being the ideal light source. This is because sunlight contains a large amount of energy at all wavelengths within the visible spectrum. The same is not true of incandescent lamps. Figure 3 compares the spectrum of sunlight at visible wavelengths to the spectrum of incandescent lamps.

The difference results from the fact that the surface temperature of the sun is about 10,000°F (6,000°C), while the temperature of an incandescent lamp filament is half of this, or less. As Figure 2 shows, visible light emission by an incandescent lamp lies at one edge of its wavelength spectrum. The result is that the wavelength distribution across the visible spectrum is very unbalanced, dropping nearly to zero at the blue end of the spectrum.

Thus, incandescent lamps have very good color rendering from the standpoint of having a continuous spectrum of visible wavelengths, but the color rendering is very poor in terms of the relative strengths of the different wavelengths. Human vision compensates for the latter defect, so we barely notice it when we are looking at colors in person. However, photographic processes reveal the unbalance of wavelengths very strongly.

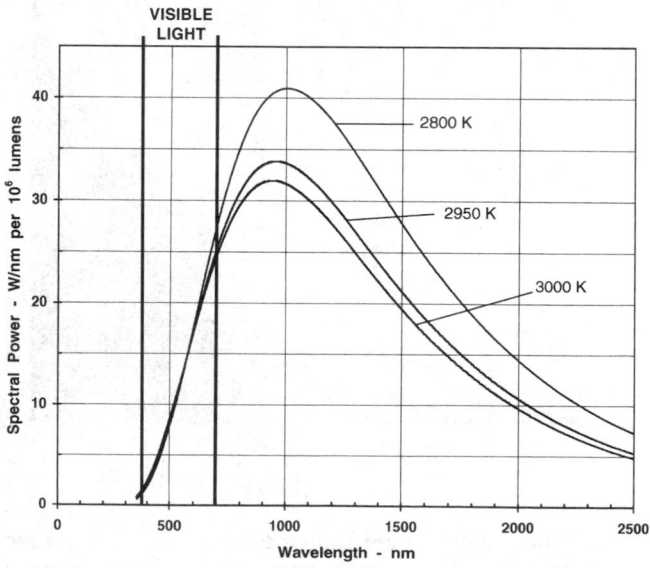

Fig. 2 The spectrum of incandescent lamps Most of the energy output is invisible, lying in the infrared region. Only the shortest wavelengths are visible to the human eye. Higher filament temperatures push more of the radiation into the visible region.

Why Incandescent Lighting is Not Efficient

If human beings could see all wavelengths, incandescent lighting would be highly efficient, because all the energy input to an incandescent lamp is converted to radiation. However, incandescent sources emit over a much broader range of wavelengths than the human eye can see, as shown by Figure 2.

The sun is about the most efficient incandescent light source from the standpoint of human vision, because the peak of the sun's light distribution pattern occurs near the center of the human vision spectrum. Even so, only about 35% of the sun's energy is visible.

Noontime Sunlight

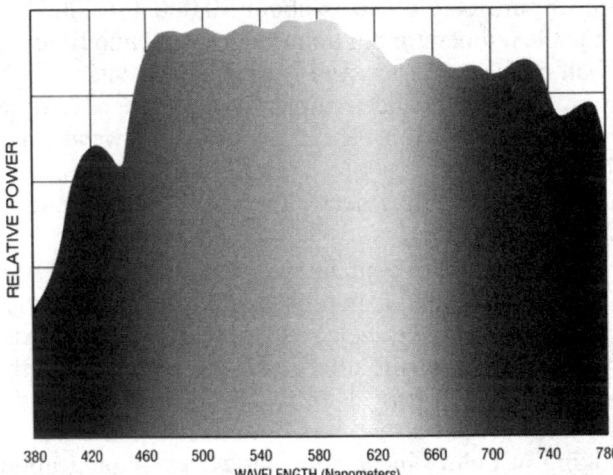

Incandescent

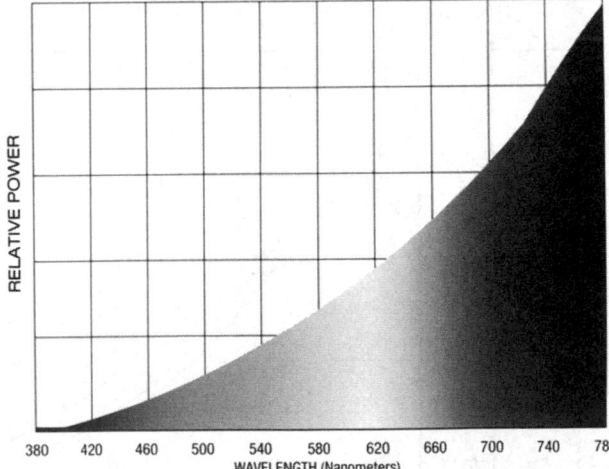

Osram Sylvania Inc.

Fig. 3 The visible spectrum of the sun and the visible spectrum of incandescent lamps The high temperature of the sun causes sunlight to have a fairly uniform distribution of wavelengths across the visible spectrum. In contrast, incandescent lamps barely have a temperature high enough to get into the blue end of the visible spectrum. The distribution of colors is very skewed. The renowned color rendering of incandescent lamps results from their continuous spectrum, not from a good balance of colors.

Incandescent lamps, which have filament temperatures much lower than the surface temperature of the sun, waste most of their energy at infrared wavelengths. These wavelengths are too long for humans to see. Carbon arc lamps operate at temperatures approaching the sun's, so they have correspondingly higher efficiency.

In principle, it is possible to make incandescent lighting efficient by capturing the infrared radiation before it leaves the lamp and using this energy to heat the filament. In this way, the visible light output remains and the input energy is reduced. Some contemporary incandescent lamps employ this technique, but they recover only a small fraction of the heat radiation.

Practical Effects of Filament Temperature

As we have seen, the light emission characteristics of incandescent lamps depend on the filament temperature. The filament temperature also has a strong effect on other aspects of incandescent lighting, which we will cover now.

■ Limitations in Filament Materials

Almost all modern incandescent lighting uses a solid, electrically heated filament as the light source. The development of incandescent light has largely been a search for filament materials that can produce light more efficiently. Primarily, this means finding materials that can survive the high temperatures needed to produce strong light emission. Thomas Edison's main contribution to the invention of the light bulb was finding filament materials that could survive long enough to be useful. His first successful lamps used filaments of carbon, in the form of charred bamboo fiber, charred cotton threads, and other materials. He and his collaborators also experimented with wire filaments made of platinum, osmium, and tantalum.

The breakthrough to modern incandescent lighting occurred in 1908, when Edison's General Electric Company developed a process for making filaments from tungsten, an extremely brittle metal. This improved lamp efficiency by about 250%. Most of this increase resulted from the high melting point of tungsten, along with its low rate of evaporation at high temperature. Also, the light emission pattern of tungsten is skewed toward shorter wavelengths compared to an ideal incandescent source, so more of its emission occurs at visible wavelengths. Tungsten is now used as the filament material for virtually all incandescent lamps.

Unfortunately, the temperature of a tungsten filament is still much too low to provide a wavelength spectrum that matches the sensitivity pattern of the human eye. No materials exist that can survive in a solid state near the high temperature needed to produce maximum incandescent light emitting efficiency.

(It is worth mentioning that arc lighting, the oldest practical form of electric light, operates at a much higher temperature than incandescent lamps that use solid filaments. Arc lighting failed to become popular because the carbon electrodes have short life, and because no one figured out how to make arc lights in small sizes.)

■ Effect of Filament Temperature on Color

Because the filament temperature determines the wavelength distribution of the light, the color characteristics of an incandescent lamp are described by its "color temperature." This is defined as the temperature of an ideal incandescent source that emits at the same peak wavelength. The color temperature is fairly close to the actual filament temperature. Color temperature is explained further in Reference Note 52, Comparative Light Source Characteristics.

■ Effect of Filament Temperature on Service Life

Tungsten melts at about 6,100°F (3,370°C). A tungsten filament heated just below its melting point emits light at an efficacy of about 50 lumens/watt, but a filament survives only briefly at this temperature. ("Efficacy" is lamp efficiency when expressed in terms of lumens per watt.) Metal evaporates from the filament surface at an increasing rate as the temperature approaches the melting point. The loss of metal weakens the filament, and it also blackens the surface of the bulb. Thus, filament temperature is a compromise between efficiency and service life.

The desire for longer service life has created a market for incandescent lamps that operate at reduced voltage in relation to their ratings. For example, "130-volt" lamps are commonly used on 120 volt circuits to achieve long life. Unfortunately, there is no magic in doing this. If a lamp is operated at reduced voltage, light output declines sharply. Also, the color temperature

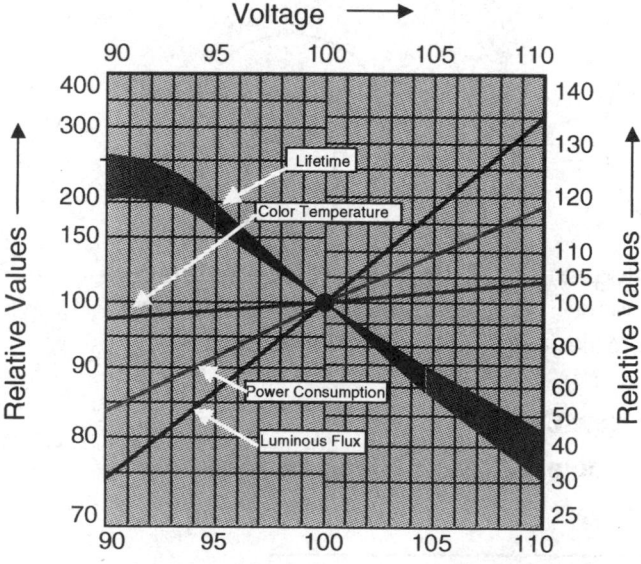

Osram Sylvania Inc.

Fig. 4 Effects of reducing lamp voltage Lamp voltage has a strong effect on filament temperature, which affects light output, service life, color temperature, and efficiency.

drops, further crippling color rendering at the blue end of the spectrum. Figure 4 shows these effects.

■ Ultraviolet Emission

Because filament temperature is limited, the light distribution of incandescent lamps is heavily skewed toward the infrared. An accidental bonus of this characteristic is that most incandescent lamps do not emit enough ultraviolet light to cause trouble. However, incandescent lamps that operate at the high end of the temperature range may emit enough ultraviolet to be damaging in some applications. For example, display lighting may cause fading of fabrics. For such applications, lamps are available with bulbs that block ultraviolet light.

Methods of Improving Incandescent Lamp Efficiency

Filament temperature affects the efficiency of incandescent lamps to an extent that dwarfs all other factors. Virtually all significant improvements to incandescent lamp efficiency, with one exception, are intended to sustain higher filament temperatures. The "high efficiency" incandescent lamps presently on the market use one or more of the following techniques.

■ Heavy Fill Gases

Evaporation of tungsten from the filament can be retarded by filling the lamp with inert gas. Tungsten atoms attempting to leave the metal surface are reflected back by collisions with the gas atoms. The amount of fill gas must be limited, because the gas conducts heat away from the filament. In ordinary lamps, the optimum pressure is near atmospheric. This has the fortunate side effect of reducing the hazard of bulb breakage.

If the fill gas has higher atomic weight, it is more effective in reducing tungsten evaporation, and it conducts less heat away from the filament. Ordinary lamps are filled with nitrogen, which has an atomic weight of 14. Argon, which has an atomic weight of 40, may also be used. Krypton is best of all, with an atomic weight of 84, but it is expensive.

Using krypton increases the cost of an ordinary bulb by about 50%, while providing an efficiency improvement of 5% to 15%. Krypton is more economical to use in halogen lamps. Halogen lamps contain the filament in a small capsule that requires little gas, and their higher filament temperatures allow the krypton to make a bigger difference.

■ Tungsten Halogen Lamps

If the filament is surrounded by a hot gas that is a member of the halogen family (i.e., fluorine, chlorine, bromine, or iodine), tungsten that evaporates from the filament forms a compound with the halogen gas. When the halide compound contacts the filament, the high filament temperature breaks down the compound and the tungsten is returned to the filament. Figure 5 diagrams the halogen lamp cycle.

The regeneration of the filament allows it to run hotter and last longer. Tungsten filaments tend to burn out locally. However, when a part of the filament becomes thinned, the added resistance of that part causes it to become hotter than the rest of the filament. In turn, the higher temperature enhances the breakdown of the tungsten halide gas, tending to repair that part of the filament preferentially.

The halogen gas also combines with tungsten that deposits on the lamp envelope, keeping the lamp clean. The envelope must be very hot to make this process work. This is one reason why halogen lamps should not be operated with dimmers.

In order to keep the gas hot enough, the filament is enclosed by a small transparent capsule made of heat resistant material. The material may be quartz (but is not always), so halogen lamps are often called "quartz" lamps. The material must be durable because the gases inside the capsule, which include the halogen gas, krypton, and perhaps other gases, operate at a pressure of several atmospheres.

The high temperature and pressure inside the quartz capsule are a safety hazard. For this reason, halogen lamps require either an outer glass envelope or special enclosed fixtures. Halogen reflector lamps are made in the PAR configuration because of its ruggedness. Halogen lamps intended as direct replacements for ordinary incandescent bulbs may be exposed, so they have an outer glass shell that is heavier than ordinary bulbs. These features make halogen lamps more expensive to manufacture.

As with other incandescent lamps, halogen lamps must compromise efficiency with service life. If all the advantage goes to efficiency, an efficiency improvement averaging 30% can be achieved. If all the advantage goes to service life, service life is increased by a factor of three to five.

■ Infrared Reflecting Coatings

The inside of the lamp envelope can be coated with a film that selectively reflects light of infrared wavelengths back to the filament, reducing the amount of electrical energy needed to heat the filament. This is the only major improvement in incandescent lamp efficiency that works by a method other than increasing filament temperature.

Figure 6 shows a halogen lamp that uses this efficiency improvement. To date, infrared reflecting coatings are used only with halogen lamps. The small quartz capsule of halogen lamps is ideal as a carrier for the reflective coating because it can accurately focus the heat radiation back on the filament, and because its small size reduces the cost of the coating.

Do not confuse these coatings with similar coatings that are applied to the rear shell of some reflector lamps. Such lamps are used to reduce the amount of heat in the light beam. Infrared radiation from the filament is selectively released through the back of the lamp. To work, these lamps must be installed in a vented fixture. They do not save energy, and they are an invitation to fires because of the additional heat released in the vicinity of the lamp.

■ Low-Voltage Filaments

For a given wattage, low-voltage lamps have thicker filaments, which allows them to run hotter and reduces the need for heat-robbing support wires. As an extreme example, a 12-volt, 25-watt lamp with a rated life of

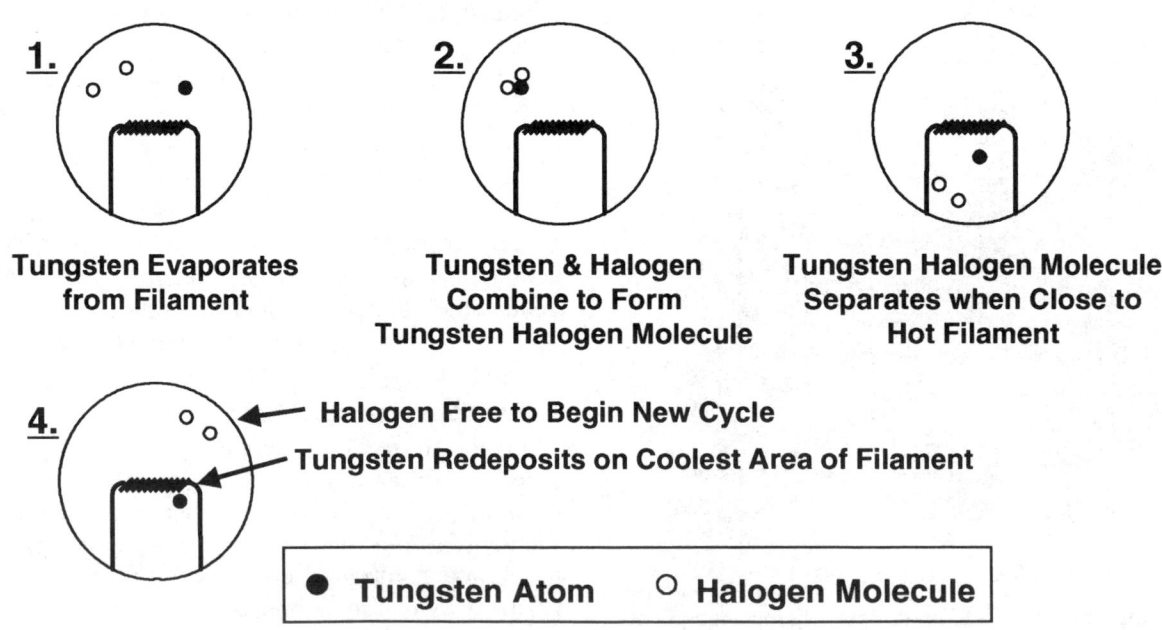

1. Tungsten Evaporates from Filament

2. Tungsten & Halogen Combine to Form Tungsten Halogen Molecule

3. Tungsten Halogen Molecule Separates when Close to Hot Filament

4. ← Halogen Free to Begin New Cycle
Tungsten Redeposits on Coolest Area of Filament

● Tungsten Atom   ○ Halogen Molecule

Osram Sylvania Inc.

Fig. 5 The tungsten halogen lamp cycle

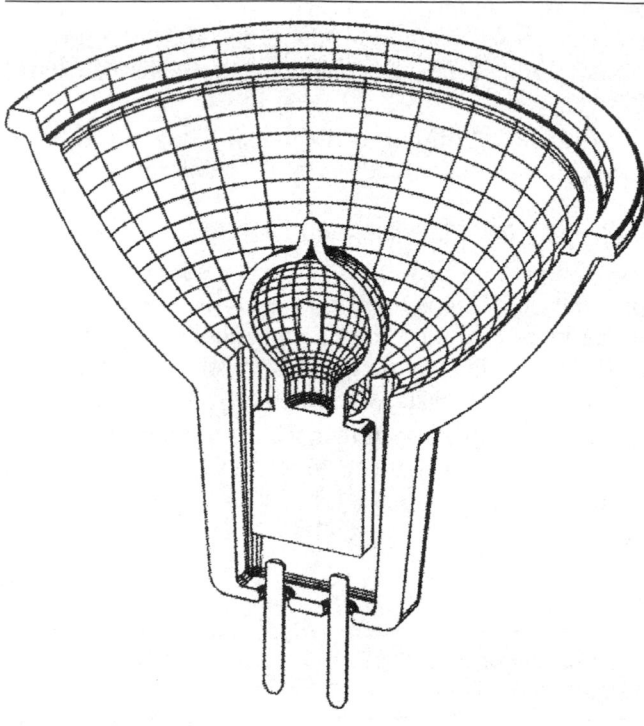

Osram Sylvania Inc.

Fig. 6 Tungsten halogen lamp The inside surface of the spherical inner capsule can be coated to reflect infrared radiation back to the filament. The reflected heat reduces the amount of electricity needed to heat the filament, improving lamp efficiency.

1,000 hours is about 70% more efficient than a comparable 240-volt lamp. The efficiency advantage declines at higher wattages. For example, a 12-volt, 100-watt lamp is about 25% more efficient than a comparable 240-volt lamp. A major disadvantage of this method is the need for a low-voltage power system, which adds cost and its own energy losses.

Filament Configuration Options

The lamp buyer has a choice of filament configurations. This choice affects the following lamp characteristics:

- the *light distribution pattern*, especially in clear bulbs. This can have a major effect on light trapping inside the fixture, as well as the efficiency of light distribution to the activities. See Measure 9.1.3 about how to select lamps for efficient light distribution.

- the *pattern of tungsten deposition on the glass bulb*, as discussed below

- *resistance to shock or vibration* (different designs for each). Filaments designed to be more durable require more support structure for the filament. Unfortunately, the supports conduct heat away from the filament and reduce efficiency.

- *efficiency*. Low-voltage filaments are inherently more efficient, as previously discussed. This is true of thicker filaments in general.

- *tightness of focus*. Smaller filaments are closer to being point sources, so they allow the light to be focussed more tightly.

Service Life and Lumen Degradation

The life of an incandescent lamp is limited by evaporation of material from the filament. The same factor causes a loss of light output as the lamp ages, which is called "lumen degradation" or "lumen depreciation" in the lighting trade.

Tungsten does not evaporate from the filament uniformly. If it did, the life of an individual lamp would be predictable within narrow limits. However, in a batch of similar incandescent lamps, a significant number may fail at half the average life. The reason is that small imperfections in the filament increase resistance and cause the filament to be hotter at those points. Evaporation increases rapidly with temperature, so the metal evaporates faster at the imperfections, further increasing the resistance and the rate of loss. The filament may break at one these points long before there is much tungsten loss overall. These processes are unstable and unpredictable, so the life of any individual lamp is unpredictable.

The resistance of the filament increases from the tungsten loss, which reduces energy input and light output. In addition, the tungsten that evaporates from the filament comes to rest on the inside of the lamp shell, where it absorbs light. Metals are very opaque, so even very thin deposits absorb a significant amount of light.

The absorption loss varies widely. In lamps that have no fill gas ("Class B"), tungsten atoms proceed from the filament to the glass shell with little interference. Therefore, the distribution of metal on the glass is fairly uniform.

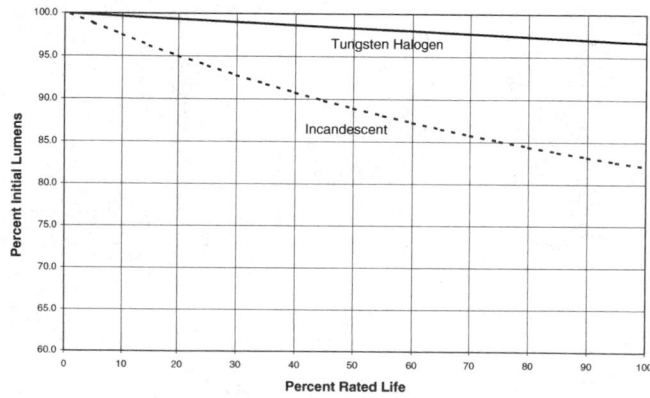

Osram Sylvania

Fig. 7 Lumen depreciation of incandescent lamps Tungsten halogen lamps have longer life, and they also lose much less light output during their life. This is because the filaments do not become as thinned by evaporation, and less tungsten darkens the bulb.

In gas-filled lamps ("Class C"), evaporated tungsten is caught by the fill gas. The heat of the filament causes strong convection currents inside the lamp, which carries the tungsten to the top of the lamp. As a result, the darkening is localized and the effect on light distribution depends on the lamp's mounting position. For example, in a torchere fixture, which projects light out the top, absorption becomes severe toward the end of lamp life. Conversely, the absorption loss is minor in a downlight. One type of lamp filament, the straight axial filament, is specifically for bulbs mounted base-up because most of the lost tungsten is deposited harmlessly in the base of the lamp.

Halogen lamps have longer service lives and lower lumen degradation because they return the evaporated tungsten to the filament. Unfortunately, the halogen process does not return all the tungsten to the thinned spots. Otherwise, these lamps would last virtually forever.

Figure 7 compares the lumen degradation of conventional and halogen incandescent lamps.

A Comparison of Incandescent Lamps

Table 1 lists efficacies and service lives for typical incandescent lamps used in general lighting. Note the trends in the table that illustrate the relationships between efficacy (the lamp efficiency expressed as lumens per watt), wattage, service life, and lamp type.

For a comparison of incandescent lamps with other lamp types, see Reference Note 52, Comparative Light Source Characteristics.

Table 1. TYPICAL INCANDESCENT LAMP CHARACTERISTICS
Efficacy (Lumens per Watt) and Life (Hours)

Watts	Standard Bulb		"R" Reflector		"ER" Reflector		Standard PAR		Halogen PAR		IR Halogen PAR		Halogen Bulb	
15	8	2,500												
25	9	2,500												
30			7	2,000										
40	12	1,500												
42													14	3,500
45			11	2,000										
45									12	2,000				
50					10	2,000								
50											18	2,000		
52													15	3,500
60	15	1,000												
65							10	2,000						
72													16	3,500
75							10	2,000						
75			12	2,000										
75	16	750												
90									14	2,000				
100	14	2,500												
100	17	750												
100											20	3,000		
120							11	2,000						
120					12	2,000								
120			13	2,000										
150	19	750												
150							12	2,000						
150									12	4,000				
150			13	2,000										
250									13	6,000				
300	21	750												
500	17	2,500												
500	21	1,000												
500			16	2,000										
1,000	23	1,000												
1,500	22	1,000												

INCANDESCENTS:

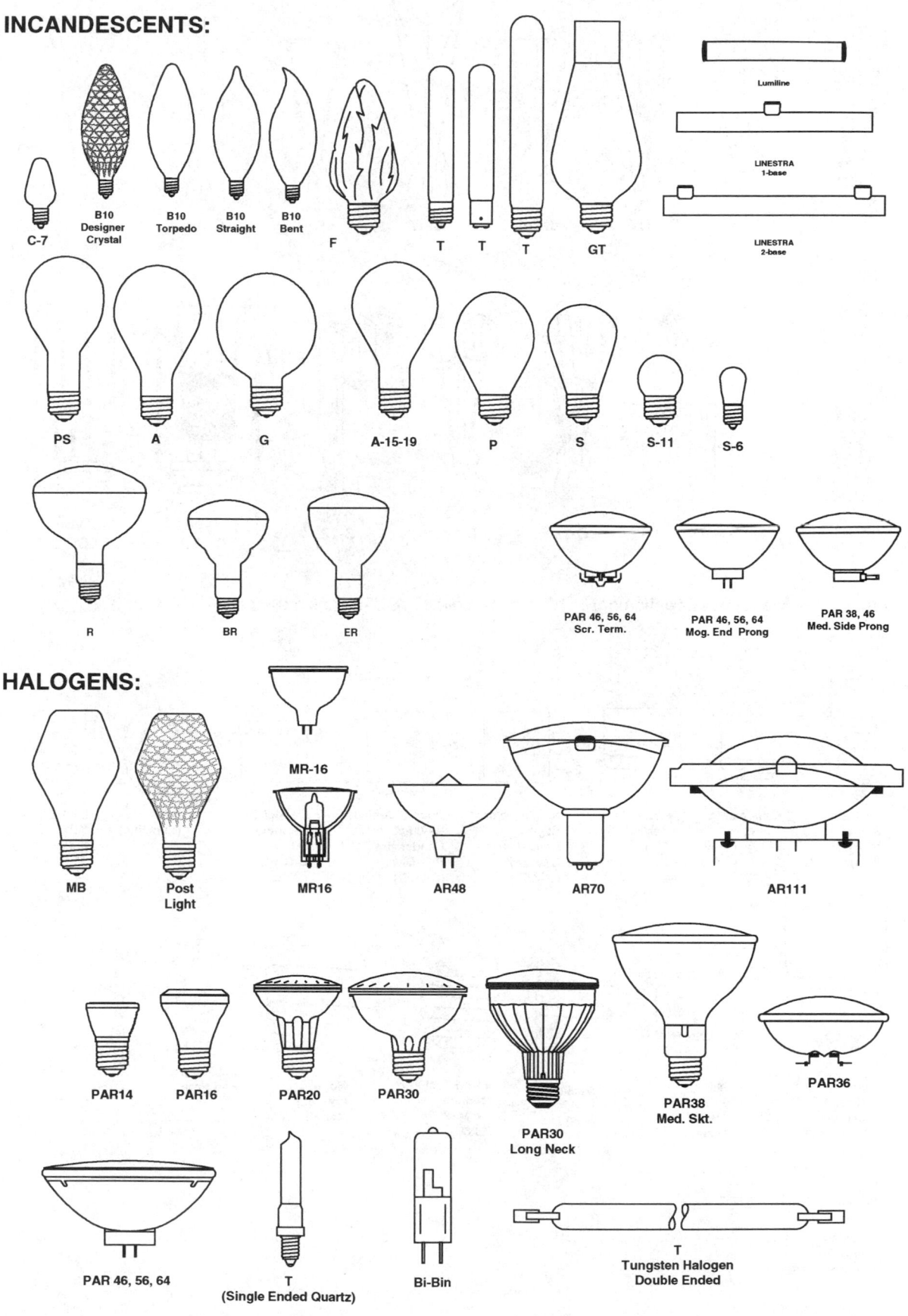

HALOGENS:

Fig. 8 Common incandescent lamp shapes, conventional and halogen

Osram Sylvania Inc.

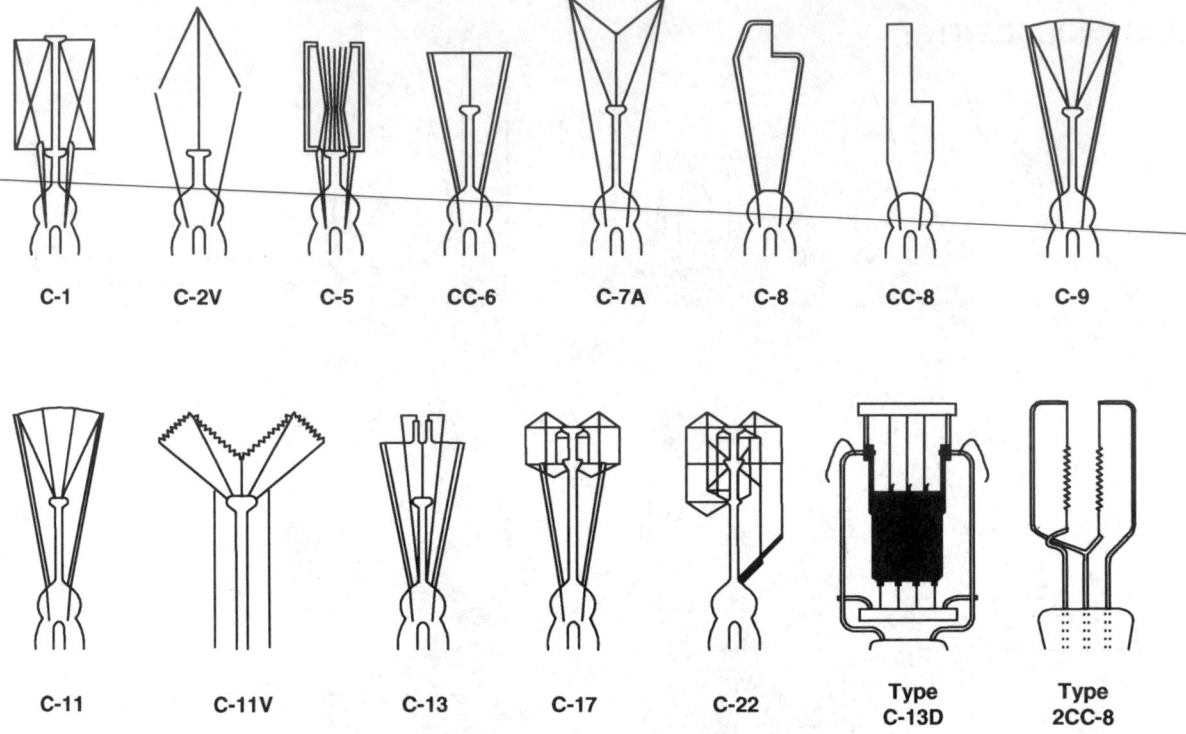

C-1   C-2V   C-5   CC-6   C-7A   C-8   CC-8   C-9

C-11   C-11V   C-13   C-17   C-22   Type C-13D   Type 2CC-8

Osram Sylvania Inc.

Fig. 9 Incandescent lamp filaments   "C" means "coiled." "CC" means "coiled coil."

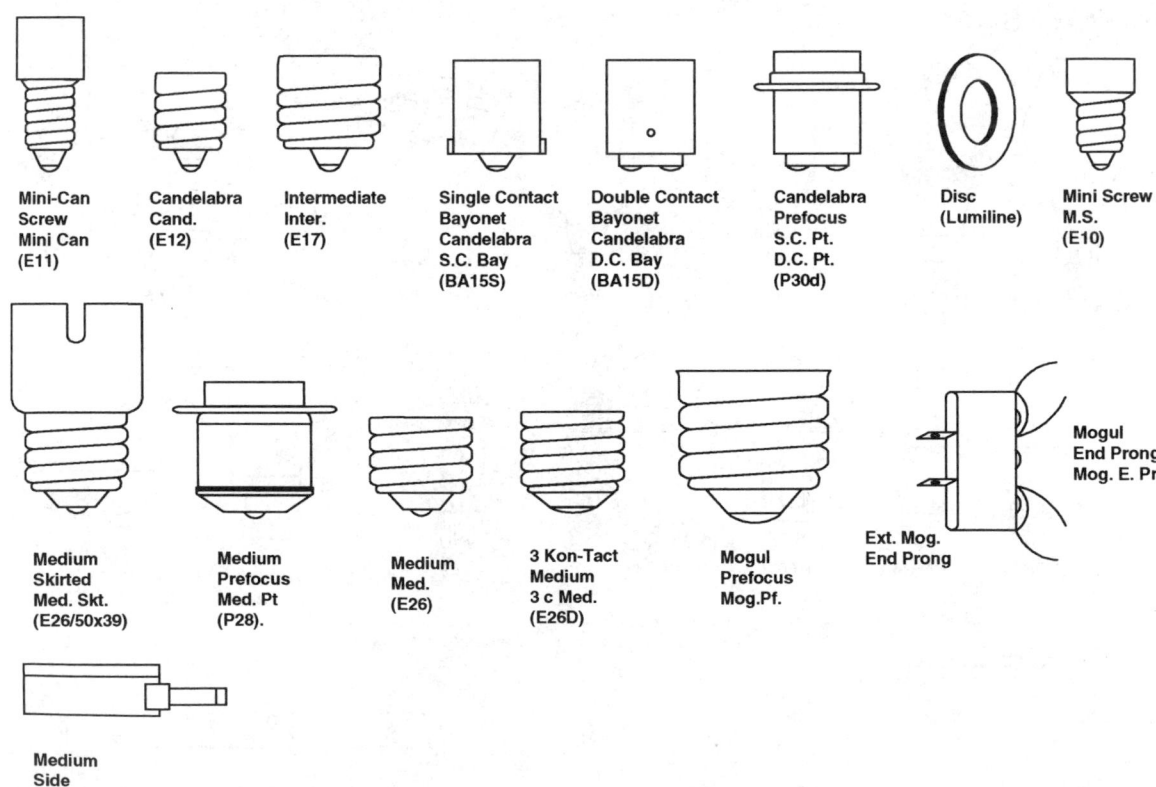

Mini-Can Screw Mini Can (E11)   Candelabra Cand. (E12)   Intermediate Inter. (E17)   Single Contact Bayonet Candelabra S.C. Bay (BA15S)   Double Contact Bayonet Candelabra D.C. Bay (BA15D)   Candelabra Prefocus S.C. Pt. D.C. Pt. (P30d)   Disc (Lumiline)   Mini Screw M.S. (E10)

Medium Skirted Med. Skt. (E26/50x39)   Medium Prefocus Med. Pt (P28).   Medium Med. (E26)   3 Kon-Tact Medium 3 c Med. (E26D)   Mogul Prefocus Mog.Pf.   Ext. Mog. End Prong   Mogul End Prong Mog. E. Pr.

Medium Side Prong

Osram Sylvania Inc.

Fig. 10 Incandescent lamp base types   Most space lighting is done with lamps having "medium" or "mogul" bases. The smaller types are used for decorative or display lighting. Prong bases are used for high-wattage lamps.

How Incandescent Lamps are Named

There is an extraordinary variety of incandescent lamps available, even for general lighting. Special lamps for locomotives, ovens, refrigerators, traffic signals, swimming pools, etc., swell the catalogs. Lamps are available from 3 watts to 1,500 watts in small increments, with extensions in wattage at each end of the range for special applications. Aside from wattage, lamps are identified in these ways:

- *bulb shape and size.* Lamps have dimension codes that consists of letters and a number, such as A-19, PAR-30, G-25, etc. The letter or letters indicate the shape. The number is the approximate maximum diameter in eighths of an inch. For example, an A-19 lamp is a round bulb of approximately 2-3/8" diameter.

  Figure 8 is a chart of the most common lamp shapes. Ordinary incandescent bulbs are "A" shape, except that larger wattages are commonly "PS" (pear-shaped with straight neck) shapes. These typically have higher efficiency than other shapes. Reflector types include "R" (typically, indoor flood), "PAR" (parabolic reflector), "ER" (ellipsoidal reflector), and others. A trip to the hardware store shows that there are many other shapes, including tubular, globes, flame, etc.

- *filament type.* Filament codes consist of one or two letters and a number. The letters usually are "C" (coiled), "CC" (coiled coil), or "S" (straight). The numbers describe different suspension arrangements. Various filament configurations are shown in Figure 9. The shape of the filament and the way it is suspended in the bulb affects the light distribution pattern, resistance to vibration and shock, efficiency, and the pattern of bulb blackening from evaporated filament material.

- *base or socket type.* Most lamps for general lighting have "medium" screw bases, except that larger wattages use "Mogul" screw bases. There are many types of specialty bases, including an entire family just for halogen lamps. Incandescent base types are pictured in Figure 10.

- *class* is either "B" (vacuum) or "C" (gas filled). Smaller lamps usually are class B, and lamps larger than 25 watts usually are class C. This rating does not tell you anything about the type of gas filling.

- *special characteristics,* which could be anything relevant, such as "halogen," "green," "vibration service," "traffic signal," etc.

Fluorescent Lighting

This Note covers the theory and design of fluorescent lighting equipment. It will help you select fluorescent lighting equipment with assurance, and it explains the changes that are occurring in fluorescent lighting technology.

How Fluorescent Lighting Works

A fluorescent lamp consists of a glass tube with electrodes at each end. The inside surface of the tube is coated with one or more types of phosphors. Mercury vapor fills the tube at very low pressure, along with one or more "buffer" gases. Figure 1 shows how a conventional fluorescent tube is made.

Fluorescent lamps produce light through a sequence of physical processes, each of which is related to one or more of the tube's components. The design of the components has been refined over the years to improve efficiency, color rendering, service life, and other properties. You need to understand the purpose of each component to sort through the numerous choices that are available in fluorescent lighting today. Let's examine each one.

■ The Mercury Vapor

Go to Reference Note 54 and read "Mother Nature's Basic Light Source." Then, come back to here.

Light is generated in a fluorescent lamp in two stages. The first stage is an "arc" discharge. The purpose of the mercury vapor is to form the arc.

An arc is a gas or vapor that is carrying an electric current. Free electrons traveling through space are accelerated by an electric field to high speeds. The electrons slam into the mercury vapor atoms, knocking loose more electrons, which then contribute to the electric current. The heavy, ionized mercury atoms are also moved by the electric field, but much more slowly. The electrons are negative and the ions are positive, so they move in opposite directions.

At the extremely low pressure of the mercury vapor, each mercury atom is isolated from the effect of other atoms. Therefore, the energy levels of its electrons are sharply defined. The electrons are excited to higher levels by collisions with electrons that travel freely in the space between the atoms.

To make a fluorescent lamp efficient, it is designed to excite the mercury atoms to their first excited level, and to minimize exciting the atoms to other levels. This concentrates the emitted light energy at a particular wavelength. Having one dominant wavelength makes it possible to design the phosphors for efficient light conversion and accurate control of color.

The light that is emitted when mercury atoms relax from their first excited energy level has a wavelength of 254 nanometers. This wavelength is outside the visible spectrum, far into the ultraviolet. The job of the phosphors is to convert this ultraviolet light to a broad spectrum of visible light, as discussed below.

It is not possible to concentrate all the light emission at 254 nanometers. In typical fluorescent lamps, about 50% of the input energy becomes radiation at this wavelength, and about 10% becomes radiation from the next higher energy level, which produces light at 185 nanometers, even further into the ultraviolet.

Some collisions in the arc stimulate even higher energy levels in the mercury vapor. Ironically, these mostly produce light at longer wavelengths. The reason is that many of the higher states decay in steps, each step emitting a photon of relatively long wavelength. (About 2% of the light emitted by the mercury vapor is in the visible spectrum. This is the light that you see in school laboratory demonstration lamps.) The higher energy states are stimulated by electrons that accumulate too much energy before hitting a mercury atom. Also, a relatively weak electron can raise an already excited atom to a higher energy level.

Electrons are first emitted by the electrodes at both ends of the tube. Fluorescent lamps usually operate from alternating current, so the electrons flow half the time in one direction, and half the time in the other direction. Most of these original electrons are lost to the tube wall before they reach the other end. If current is to keep flowing, the original electrons must be replaced by knocking electrons loose from the mercury atoms.

The process of knocking an electron loose is called "ionization." The energy needed to create the free electrons is mostly wasted because only part of the ionization energy is ever recovered as light.

The mercury atom that is missing one electron is called an "ion." It has a single positive charge, because it is missing the negative charge of the electron. Even though ions are charged particles, they carry little current because they are heavy and slow. However, the heavy ions are important because they heat the electrodes by impact. The impact and heat help to release the electrons from the electrodes. The same impacts also eventually destroy the electrodes.

Ions may re-combine with electrons while they are wandering in the arc, but this is unlikely. If an ion does not reach the negative electrode, it usually hits the tube wall and combines with an electron there.

Fluorescent lighting is limited to low intensity, typically about 10 watts per foot (30 watts per meter) of tube length. This is because the current in the lamp must be limited. If the flow of electrons is too high, the probability increases that an electron will further excite mercury atoms that are already excited. This excites a cascade of light emission at many wavelengths, most of which are too long to be useful.

Mercury is the only light emitting vapor used in fluorescent lighting. It is the only substance yet discovered that has the combination of characteristics needed for efficient generation of light by the fluorescent process. The favorable characteristics of mercury are:

- The density of the vapor is optimum for efficient lamp operation near room temperature. The optimum density for mercury vapor occurs at about 104°F (40°C). Indoor fluorescent lamps operate near this temperature.

- Mercury emits light strongly from its first excited energy level. Concentration of light from the first energy level avoids wasting energy produced at unusable wavelengths. It also allows the lamp design to be optimized to deal with the single primary wavelength.

- The wavelength of mercury's first excited level can be converted to visible light fairly efficiently by phosphors that have other desirable characteristics. These other characteristics include good color rendering, long life, reasonable cost, etc.

- Mercury has a critical interaction with argon, discussed below, that greatly improves the efficiency of ionizing mercury atoms to carry the lamp current.

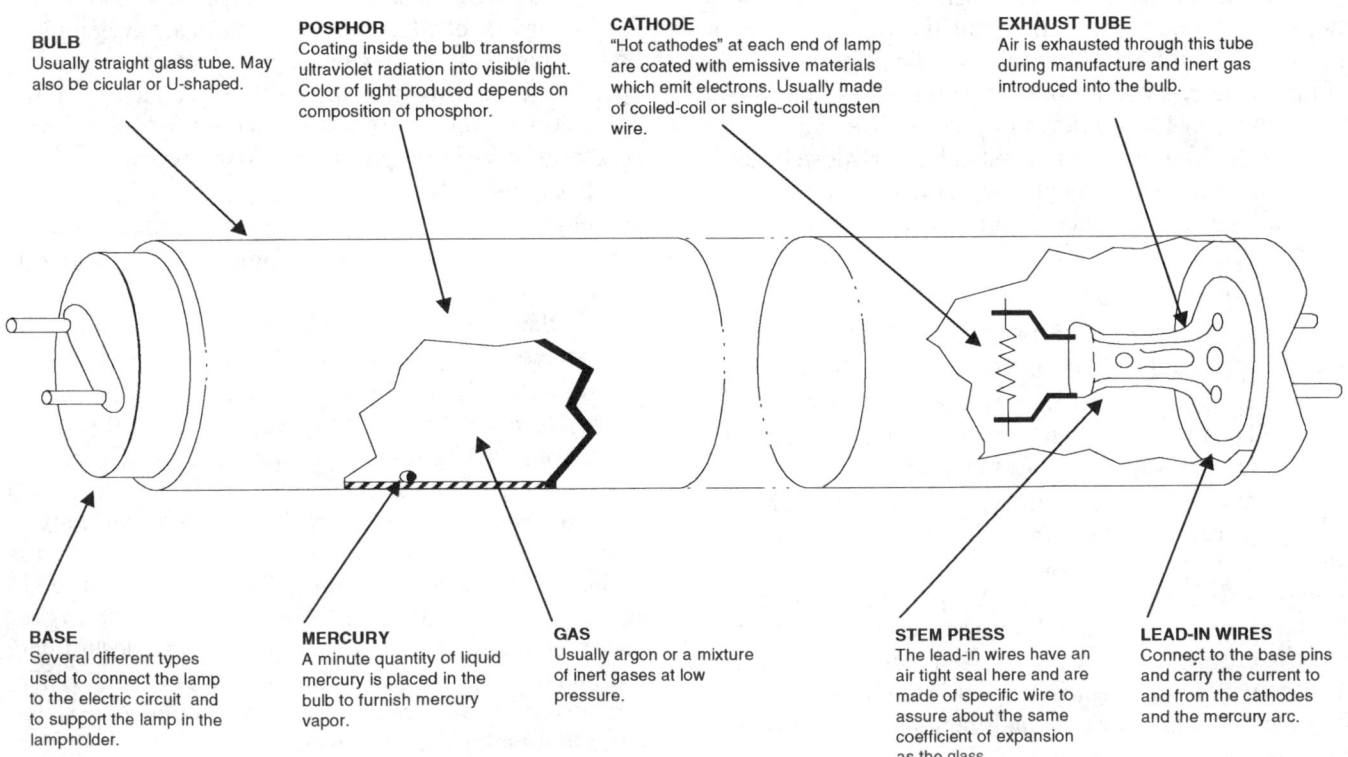

BULB
Usually straight glass tube. May also be cicular or U-shaped.

POSPHOR
Coating inside the bulb transforms ultraviolet radiation into visible light. Color of light produced depends on composition of phosphor.

CATHODE
"Hot cathodes" at each end of lamp are coated with emissive materials which emit electrons. Usually made of coiled-coil or single-coil tungsten wire.

EXHAUST TUBE
Air is exhausted through this tube during manufacture and inert gas introduced into the bulb.

BASE
Several different types used to connect the lamp to the electric circuit and to support the lamp in the lampholder.

MERCURY
A minute quantity of liquid mercury is placed in the bulb to furnish mercury vapor.

GAS
Usually argon or a mixture of inert gases at low pressure.

STEM PRESS
The lead-in wires have an air tight seal here and are made of specific wire to assure about the same coefficient of expansion as the glass.

LEAD-IN WIRES
Connect to the base pins and carry the current to and from the cathodes and the mercury arc.

Osram Sylvania Inc.

Fig. 1 How a conventional fluorescent lamp is made

Phosphors

Phosphors convert the ultraviolet light emitted by the mercury vapor into visible light. Phosphors are crystals, which are regular arrays of atoms. Crystals are literally huge molecules. Like all molecules, they have characteristic energy levels. Instead of having discrete energy levels like individual atoms, crystals have broad energy bands that are created from the energy levels of the closely packed atoms. When the phosphor crystals are excited by the ultraviolet radiation from the mercury vapor, they emit light over a broad spectrum of wavelengths. Energy is lost in this process, so the wavelengths emitted by the phosphors are longer than the wavelength of mercury emission. Phosphors are selected to emit most of their radiation in the visible spectrum.

Some of the mercury radiation is dissipated by generating vibrations in the crystal structure, which is heat. The efficiency of the phosphors in converting the ultraviolet to visible light ranges from 35% to 50%.

Phosphor crystals have additives called "activators" that modify their absorption and emission characteristics. The phosphors and activators are selected to maximize absorption at the 254-nanometer wavelength of mercury emission, and to emit light within the visible spectrum.

The makeup of the phosphors determines the color of a fluorescent lamp. To a large extent, the phosphors also determine the efficacy and cost of the lamp. Selecting a particular model of fluorescent lamp is largely a matter of selecting particular phosphors.

Older fluorescent lamps use a single type of phosphor. The newest phosphor systems combine three different types of phosphors, each of which produces a color distribution with a different peak. The three-phosphor system improves maximum lamp efficacy from its previous maximum of about 75 lumens/watt to about 90 lumens/watt. The color rendering index (CRI) is increased from around 60 to as high as 90. Loss of light output during the life of the lamp is reduced from about 20% to about 7%. The disadvantage of the triple phosphor systems is that are considerably more expensive than the older types.

Figure 2 shows the output spectra of fluorescent lamps with several different types of phosphors. It shows that some lamps emit most of their light at discrete wavelengths, while others emit much of their light as a continuous spectrum. Among both types, light output can be tailored somewhat to emphasize the red or blue end of the spectrum.

Buffer Gases Increase Lamp Efficiency and Output

Fluorescent lamps contain one or more "buffer gases" in addition to the mercury vapor. In fact, fluorescent lamps contain much more buffer gas than mercury vapor, typically at several hundred times higher pressure than the pressure of the mercury vapor (but still only a small fraction of atmospheric pressure). Argon is always the primary buffer gas that is used in fluorescent lighting. Argon is needed to work with the mercury vapor in the lamp, as we shall see.

The buffer gas is sometimes called the "inert gas" or "rare gas," because the gases that are used in fluorescent lamps have these chemical classifications. Their essential characteristic is that they resist excitation. This is because their first energy level is exceptionally high. (This is the main reason why these gases are chemically inert.)

Buffer gases perform several functions. One of these is reducing the loss of the electrons that transmit energy to the mercury atoms. Stated differently, the buffer gas increases the probability that any one electron will stimulate mercury atoms to emit light.

Buffer gases are needed for this purpose because the density of the mercury vapor in the lamp is very low. If the tube contained only mercury vapor, an average electron would travel about two inches (50 mm) before hitting a mercury atom. This distance is so long that the electron would be more likely to hit the tube wall or the positive electrode than to hit a mercury atom. By the same token, few mercury atoms would be stimulated, and light output would be feeble.

Furthermore, the lamp would not work at all with normal voltage, because the loss of electrons would exceed the number of new electrons created by ionization. The only way to make the lamp work would be to use a much higher voltage to create more ionization. This would make fluorescent lighting expensive and dangerous. The lamp would be inefficient because of the high electron loss. Also, it would be inefficient because the high voltage would accelerate the electrons to excessive energies. In turn, the electrons would stimulate the mercury to energy levels that are too high for efficient light emission.

The solution to the problem of low mercury vapor density is to add another gas, called a buffer gas, in sufficient quantity to shorten the distance that an electron travels between collisions. In a fluorescent lamp, the presence of the buffer gas reduces the average distance between collisions to about 0.004 inches (0.1 mm).

(Mercury itself cannot be added at higher pressure, because it exists as a vapor in equilibrium with liquid at room temperature. Attempting to add more mercury vapor would simply cause more of it to condense into the liquid state. Also, in a dense mercury vapor, the light emitted by excited atoms would be absorbed by the surrounding unexcited atoms. This is because an atom absorbs strongly at the same wavelengths it emits.)

The free electrons that carry current through the tube gain energy progressively as they bounce between atoms, accelerated by the electric field of the electrodes. A free electron cannot stimulate an atom to a higher energy level until it gains an amount of energy equal to the

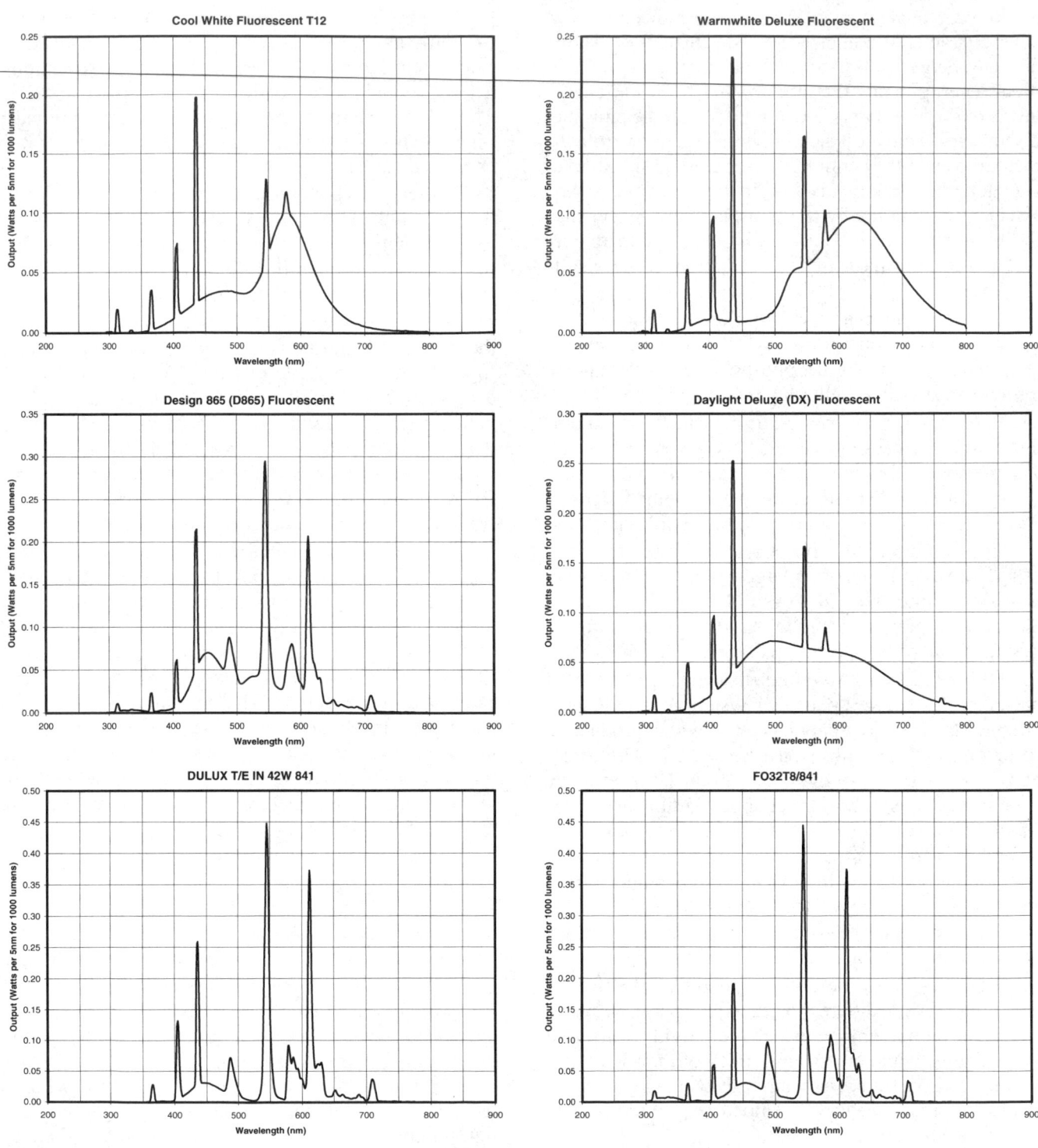

Osram Sylvania Inc.

Fig. 2 Visible light spectra of fluorescent lamps The upper four are from T12 lamps. These emit a large fraction of their light as a continuous spectrum, providing good color rendering. The bottom left spectrum is from a compact fluorescent lamp. The bottom right is from a T8 lamp. These two emit much more of their light at discrete wavelengths.

lowest energy level of any atom it encounters. The buffer gases have high energy levels, whereas mercury does not. This makes it likely that a mercury atom will be the first type of atom stimulated. This is also why the first energy level of mercury is stimulated more than its other energy levels.

To improve efficiency, krypton is added to the argon in some lamps. This reduces the energy loss that occurs from the heating of the buffer gas by electron collisions. Even though most collisions between free electrons and the buffer atoms are elastic (which means that no energy is lost in the collisions), there is a transfer of kinetic energy from the electrons to the atoms. This kinetic energy is not useful, and the atoms eventually dissipate their kinetic energy as heat in collisions with other atoms, the wall of the tube, or an electrode. Krypton is heavier than argon, so it absorbs less energy in collisions.

Krypton is more expensive than argon, so it is used only in premium lamps. Krypton cannot substitute entirely for argon, because it does not have the fortunate matching of energy levels that aids starting. This is discussed next.

■ The Role of Argon in Ionizing Mercury

Argon is always selected as the primary buffer gas because it must perform an entirely different function from the one discussed previously. This function is helping to create enough free electrons to maintain the current within the lamp.

The action of buffer gas discussed previously greatly reduces the loss of the free electrons emitted by the electrodes. Still, most of these electrons do not survive the entire trip through the tube. The lost electrons are replaced by knocking additional electrons loose from the mercury vapor. The difference here is that the electrons are not merely excited, but are knocked completely free of the atom. In other words, the mercury is "ionized."

Electrons are unlikely to ionize mercury directly, because they tend to give up their energy exciting lower mercury energy levels. Instead, a free electron excites an argon atom, which then ionizes a mercury atom. Bear in mind that a free electron is about 100 times more likely to hit an argon atom than a mercury atom, so the electron may accumulate enough energy to ionize a mercury atom before it actually hits one.

Argon is special for this purpose because it has an energy level that is slightly higher (11.6 electron volts) than the ionization energy of mercury (10.4 electron volts). Equally important, the argon atom does not relax quickly from this excited state, so it is likely to collide with a mercury atom first. When the collision occurs, the excitation energy of the argon is immediately transferred to the mercury atom, ionizing the mercury atom. This fortunate interaction between mercury and argon is one of the factors that has made fluorescent lighting practical.

This process is especially important to help the fluorescent tube get started without needing higher voltage. For this reason, the argon in a fluorescent lamp is often called the "starting gas."

■ Buffer Gases Protect the Electrodes and Phosphors

The buffer gas also protects the electrodes and phosphor crystals. The heavy ions are kept at low speed by repeated collisions with the buffer atoms. If this did not happen, the electric field would accelerate the ions to energies that would quickly destroy the electrodes. By the same token, the cloud of buffer gas around the electrodes rebounds evaporated electrode material back to the electrodes.

In much the same way, the buffer gas shields the phosphor crystals from damage by mercury ion bombardment.

■ Electrodes

The electrodes in a fluorescent tube are the gateways through which the current from the ballast enters the arc. The electrons in the metal electrodes don't want to jump out into the vacuum of the tube. They have to be forced to do so. Two methods are used in fluorescent lighting to get the electrons to leave the electrodes. One is to wrench the electrons loose with a strong electric field. This method is used in "instant-start" lamps.

The other method is to heat the electrodes so that the electrons are knocked out of the electrode surface by thermal agitation. (This process is called "thermionic emission.") "Rapid-start" and "preheat" fluorescent lamps use this method. Once the lamp is operating, bombardment of the electrodes by ions and electrons may keep the electrodes hot enough to continue electron emission, so the electric current that heats the filaments can be turned off.

Both methods require energy that produces no light, so they reduce lamp efficiency. To minimize the amount of energy needed to extract electrons from the electrodes, the electrodes are coated with a material that reduces the amount of energy needed to evict the electrons. This material is gradually depleted from the electrode surface during operation, especially when the lamp starts from a cold state. The original amount of the material, and its rate of loss, determine lamp life. The amount of emissive material is limited by the fact that the lost material contaminates the phosphors and darkens the interior of the lamp. This forces a compromise between lamp life and preservation of lumen output.

The loss of energy at the electrodes can be reduced by increasing the frequency of the alternating current from the ballast. Part of the loss in a fluorescent lamp occurs because the arc dissipates with each half cycle of the applied voltage, so it must be continually restarted. Also, during the relatively long time between cycles in 60 Hz current, some of positive mercury ions are attracted to the negative electrode, where they give up their energy without producing light. If the frequency of the input current is greatly increased, the mercury ions do not have time to wander between cycles, and the column of ions does not have time to dissipate completely, so the arc is easier to restart.

The development of electronic ballasts made it possible to supply current to lamps at high frequency. Lamp efficacy rises to a plateau at about 20 KHz, with an improvement of about 10%. By a fortunate coincidence, 20 KHz is the top of the frequency range for human hearing, so electronic ballasts are designed to operate at this frequency or higher. Fortunately, the lamps do not have to be modified to exploit high-frequency power.

The Effect of Temperature on Efficiency and Light Output

Unlike incandescent and HID lighting, the efficiency and the output of fluorescent lighting is sensitive to the temperature surrounding the lamp. The primary reason is that the surface temperature of the lamp determines the vapor pressure of the mercury, which is the primary source of light.

As mentioned previously, the optimum density for mercury vapor occurs at about 104°F (40°C). Accounting for some insulation effect by the glass bulb and phosphors, this means that the exterior surface of the lamp should be at a temperature of about 85°F (30°C). Figure 3 shows that efficiency declines seriously at higher and lower temperatures.

Since the vapor pressure of mercury declines with temperature, the light output declines also.

It is possible to use fluorescent lamps in cooler environments by installing them in sealed fixtures that

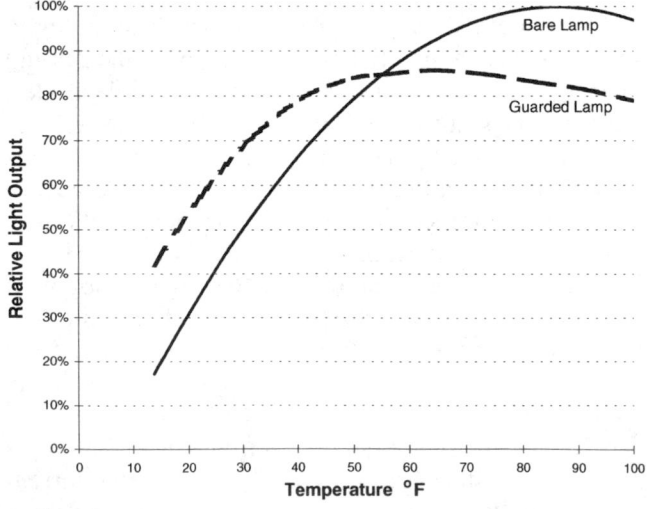

Osram Sylvania Inc.

Fig. 3 The effect of ambient temperature on lamp efficiency
The surface temperature of the lamp controls the vapor pressure of the mercury, which determines light output and efficiency.

retain heat better. However, this simply shifts the optimum temperature. Light loss in the enclosure also reduces the effective maximum efficiency of the lamp.

Compact fluorescent lamps operate at higher temperature because they have higher current densities. As a result, these lamps can be used in cooler environments. However, deviating from the optimum temperature for these lamps also reduces efficiency.

The Effect of Lamp Shape on Efficiency

The length of the tube is a factor in efficiency. The electrodes consume energy without producing light, and light emission in the vicinity of the electrodes is inefficient. These losses are essentially constant, regardless of the length of the tube. Also, the ends of the tube become darkened by evaporated filament material. Therefore, longer tubes tend to be more efficient than shorter tubes. For example, a typical 8-foot tube is about 10% more efficient than a comparable 4-foot tube, which in turn is about 20% more efficient than a comparable 2-foot tube.

Curvature of the tube contributes to inefficiency. Light emission in a given lamp type is less efficient close to the wall of the tube. Bends in the tube force the arc closer to the tube wall, so U-tube and circular configurations tend to be less efficient than straight tubes. For example, a straight 4-foot tube is about 10% more efficient than a comparable U-tube.

The diameter of the tube is a compromise between competing efficiency factors. If the tube is fat, too much radiation from the center of the tube is re-absorbed by the mercury vapor. If the tube is slender, it has more surface area in relation to its gas volume, which increases quenching of the mercury arc at the tube surface. Also, smaller diameter increases the current density for a given wattage, which may reduce efficiency. Since the diameter of a fluorescent lamp affects efficiency in several ways, you cannot judge the efficiency of a fluorescent lamp from the diameter alone.

The high cost of high-efficiency phosphors motivates smaller tube diameters. This factor is largely responsible for the introduction of T8 (1" diameter) tubes for applications that previous used T12 (1.5" diameter) tubes.

Service Life and Number of Starting Cycles

The nominal service life of conventional fluorescent lamps averages about 20,000 hours. The actual service life is affected strongly by the number starting cycles during the life of the lamp. The reason is that starting the lamp causes the electrodes to lose a part of their emissive material.

The nominal service life is usually based on three hours of operation per start. Lamp manufacturers are coy about stating the service life for shorter periods of operation. Going the other way, operating a lamp

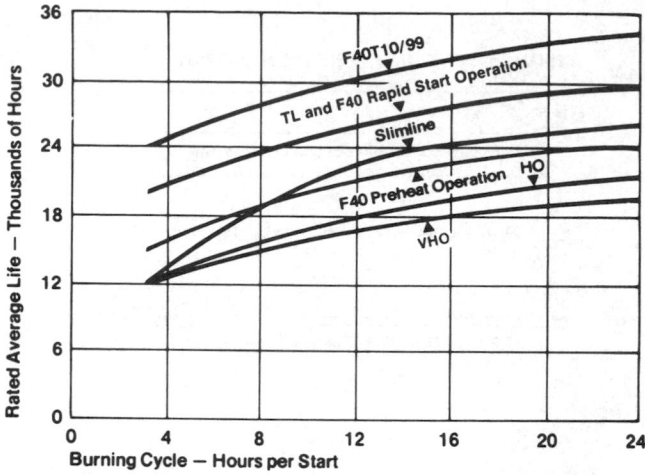

Philips Lighting

Fig. 4 Service life of fluorescent lamps Lamps that operate at higher current, or output, have shorter lives. The number of starting cycles has a major effect of service life. Unfortunately, manufacturers don't like to publish data for starting cycles shorter than three hours.

continuously may increase the service life by about 50%. Figure 4 charts the effect of starting cycles on the service life of several conventional types of fluorescent lamps.

Lumen Degradation

Figure 5 charts the lumen degradation of several types of conventional fluorescent lamps. The figure shows that lumen degradation ranges from 10% to 40%, depending on the type of lamp.

The main causes of lumen degradation in fluorescent lamps are decay of the phosphors and darkening of the lamp by material lost from the electrodes. Phosphor decay is much greater in lamps having higher current.

How Fluorescent Lamps are Named

The features of fluorescent lamps have been changing continually since fluorescent lighting was developed, and the rate of change is now much greater than it was in the past. These features include lamp shape, starting method, wattage, lamp color, color rendering, efficiency, base type, etc. The naming of fluorescent lamps has become a muddle as manufacturers attempt to describe each new type of fluorescent lamp by its lamp code.

At present, the only way to completely decode a fluorescent lamp designation is to use a detailed catalog. Here are some examples:

- "F48T12/CW/WM" is a straight fluorescent tube, 48 inches long, 1.5 inches diameter, "cool white" color, Watt Miser (a General Electric trade mark). The wattage and other characteristics are unspecified.

- "F96T8/SPX41/HO" is a straight fluorescent tube, 96 inches long, 1.0 inch diameter, RE841 phosphor, "high output." The wattage and other features are unspecified.

- "F13DBX23T4/SPX35" is a compact fluorescent tube, 13 watts input, Double Biax (a General Electric trade name) configuration with Type 23 base, 0.5 inches diameter, RE835 phosphor.

You can see that there are some common features here. All fluorescent lamp designations start with the letter "F." The other consistent designation is the lamp diameter. In United States practice, the diameters of all types of fluorescent tubes are designated with the letter "T" in the lamp type code, followed by the diameter in eighths of an inch. For example, a "T4" compact fluorescent lamp has a tube diameter of one half inch.

The designation of lamp length is confused, and it may or may not appear in the lamp code. For example, "F96" means that the lamp is 96 inches long, but "F32" means that the lamp power is 32 watts.

With newer fluorescent lamps, the lamp's output color is usually designated by a phosphor type. However, older types of fluorescent lamps use names for colors, such as "daylight" and "cool white."

Conventional types of fluorescent lamps (that is, not compact types) are classified by starting method, such as "instant start." Starting methods are discussed below. The starting method also implies the type of lamp base, in the case of the older, thick lamps. Most compact fluorescent lamps presently use preheat starting, so the starting method is usually not specified.

Compact fluorescent lamps are evolving with different naming patterns than the older, larger types.

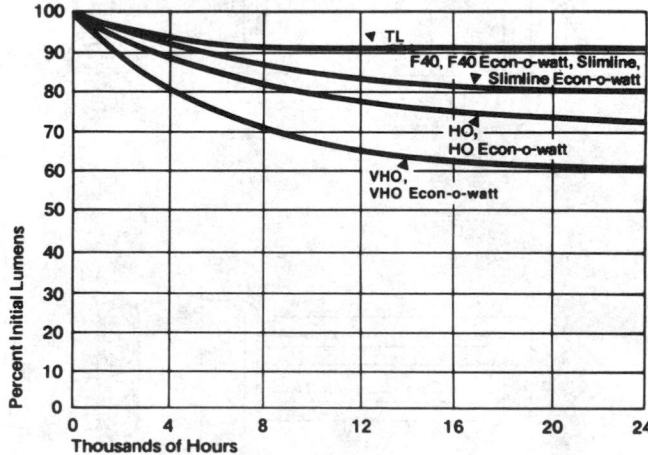

Philips Lighting

Fig. 5 Lumen degradation The loss of light output occurs early in the life of the lamp. Lamps that operate with high current suffer much more loss than the best models. Therefore, "high-output" lamps are not a good bargain, especially in terms of efficiency.

FLUORESCENTS:

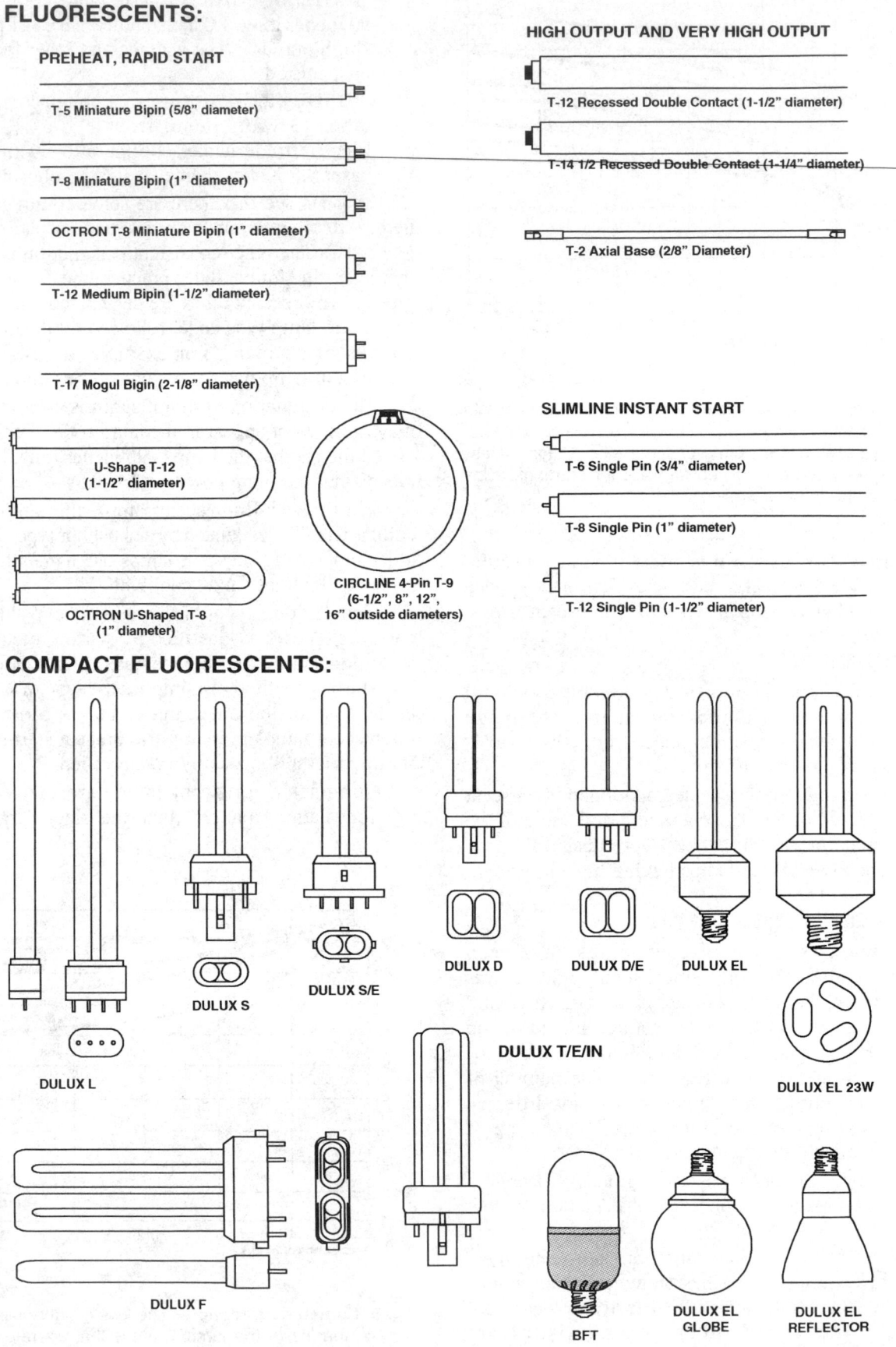

PREHEAT, RAPID START

T-5 Miniature Bipin (5/8" diameter)

T-8 Miniature Bipin (1" diameter)

OCTRON T-8 Miniature Bipin (1" diameter)

T-12 Medium Bipin (1-1/2" diameter)

T-17 Mogul Bigin (2-1/8" diameter)

U-Shape T-12
(1-1/2" diameter)

OCTRON U-Shaped T-8
(1" diameter)

CIRCLINE 4-Pin T-9
(6-1/2", 8", 12",
16" outside diameters)

HIGH OUTPUT AND VERY HIGH OUTPUT

T-12 Recessed Double Contact (1-1/2" diameter)

T-14 1/2 Recessed Double Contact (1-1/4" diameter)

T-2 Axial Base (2/8" Diameter)

SLIMLINE INSTANT START

T-6 Single Pin (3/4" diameter)

T-8 Single Pin (1" diameter)

T-12 Single Pin (1-1/2" diameter)

COMPACT FLUORESCENTS:

DULUX L

DULUX S

DULUX S/E

DULUX D

DULUX D/E

DULUX EL

DULUX T/E/IN

DULUX EL 23W

DULUX F

BFT

DULUX EL
GLOBE

DULUX EL
REFLECTOR

Osram Sylvania Inc.

Fig. 6 Common types of fluorescent lamps

Because compact lamps are spawning many base types, the base type code is likely to be included in the lamp designation. However, unless you always deal with the same type of base, you need a catalog to show you a picture of the base type.

Figure 6 shows most of the older types of fluorescent lamps, and a representative sample of compact fluorescent types.

Why Fluorescent Lamps Need Ballasts

All fluorescent lamps need a ballast to operate. The arc in an operating lamp has almost no electrical resistance. Without some means of regulating the current, the lamp current would be much too high for efficient light emission and the lamp would quickly destroy itself. The ballast regulates the lamp current independently of the lamp's changing resistance characteristics.

In addition, many fluorescent lamps need a voltage higher than the line voltage to get started. (The voltage requirement relates mostly to lamp length and the type of electrode.) The ballast provides the increased starting voltage.

The lamp's characteristics dictate the design of the ballast. The ballast and lamps must be closely matched to achieve best efficiency. Lamps may operate with the wrong ballasts, but not efficiently.

There are two main types of ballasts, magnetic and electronic. Magnetic ballasts have a more efficient variation called a hybrid ballast.

■ How Magnetic Ballasts Work

The original method of regulating lamp current is to use a coil in series with the lamps. The coil restrains alternating current by inductance. This type of ballast is still common. It is called a "magnetic" ballast because the coil produces a magnetic field, although this is irrelevant to the ballast's operation. It is also called an "electromagnetic" or "core and coil" ballast. A magnetic ballast is similar in construction to a transformer. The inductance of the secondary winding determines the lamp current, and the turns ratio between the primary and secondary windings determines the starting voltage.

A simple magnetic ballast is a large inductor, so it causes the lighting system to have a low power factor. This problem can be corrected easily by including a large capacitor in series with the coil.

Two-tube magnetic ballasts also use capacitors for another purpose, to aid starting. A capacitor is connected in parallel with one of the lamps so that full ballast voltage first appears across the other lamp. After that lamp starts, its resistance disappears, the capacitor charges, and starting voltage appears across the first lamp.

Figure 7 shows the inside of a magnetic ballast that has all these features.

Substantial improvements in the efficiency of magnetic ballasts have been made in recent years. These include reducing the resistance of the windings, improving the cores to reduce eddy currents, shaping the wire so that there is less insulation dead space between the wires, and reducing losses in the capacitors.

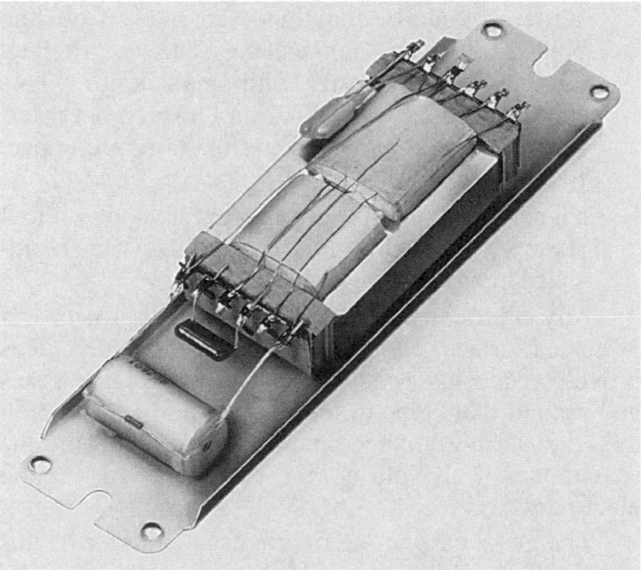

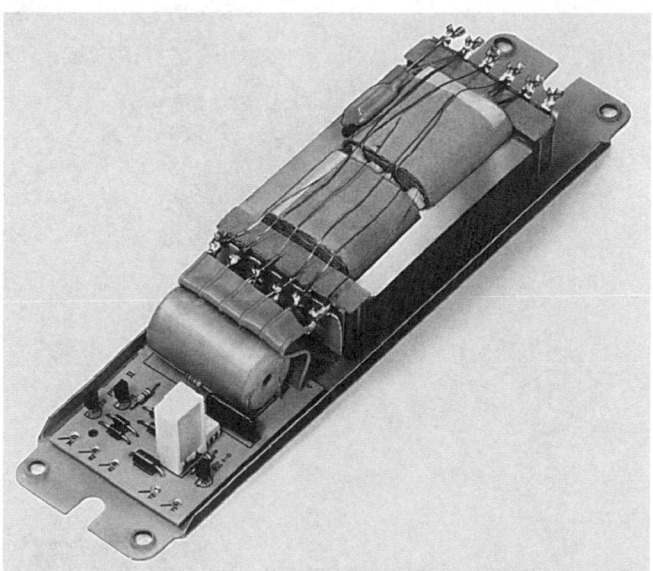

Advance Transformer Co.

Fig. 7 Inside a magnetic ballast Most of this is a transformer, made of two large coils wound on an iron core. The large capacitor to the left of the coils improves power factor. The smaller capacitor next to it aids the starting of two lamps in series. On top is a small thermal switch for fire safety, which makes the ballast "P" rated.

Advance Transformer Co.

Fig. 8 Inside a hybrid ballast This is the same as the magnetic ballast in Figure 7, except that it includes a small electronic switch to turn off the lamp filament after the lamp starts.

"Hybrid" ballasts are magnetic ballasts that include an automatic switch to turn off the filaments of rapid-start lamps after the lamp starts. Figure 8 shows the inside of a hybrid ballast.

■ How Electronic Ballasts Work

Electronic ballasts are rapidly entering the field, although they are still substantially more expensive than magnetic ballasts. They eliminate the large coils and associated heat loss of conventional ballasts because they operate on an entirely different set of principles. Figure 9 shows the inside of a typical unit.

Electronic ballasts regulate lamp current by delivering it in high-frequency pulses. The ballast circuits can tailor the characteristics of the lamp current to optimize a variety of factors that affect efficiency.

Electronic ballasts not only operate more efficiently themselves, they also improve the efficacy of the lamps by about 10 percent, because of the high-frequency operation discussed previously.

The ability of electronic ballasts to tailor the lamp current makes it possible to provide efficient dimming of fluorescent lamps. Special models are needed to provide dimming.

Methods of Starting Fluorescent Lamps

The un-ionized mercury vapor in a cold fluorescent lamp has virtually infinite resistance. To get the lamp started, it is necessary to create ions and free electrons to form an arc and carry the current. The arc is started by ejecting electrons from the electrodes. The impact

Advance Transformer Co.

Fig. 9 Inside an electronic ballast The big coils are gone, replaced by a few small electronic components that drive the lamps at ultrasonic electrical frequency. The ballast itself has smaller losses, and it makes the lamps operate more efficiently. However, electronic devices are potentially less reliable than coils of wire, so expect to replace these ballasts more often.

of these electrons on gas atoms creates ions and more free electrons, as we discussed previously. The arc spreads down the length of the tube as the ions and electrons are accelerated by the electric field of the electrodes. Several methods are used in fluorescent lighting to get the arc started:

- *instant-start* systems use the brute force of high voltage to tear electrons out of the electrode surface. The starting voltage is about three times the operating voltage. Once lamp current starts to flow, bombardment of the electrodes by ions heats the electrodes, which reduces the amount of voltage needed to operate the lamp and improves efficiency. Instant-start tubes with single-pin terminals are often called slimline tubes.

- *rapid-start* systems heat the electrodes to eject electrons by thermal agitation. Rapid-start ballasts have separate windings to supply filament heating voltage.

  Energy can be saved by turning off the filaments after the lamp is started. This can be done by a switch in the ballast or by a switch sealed into the tube itself.

- *preheat* systems also use electrodes in the form of heated filaments. In these lamps, the filaments are heated before voltage is applied across the lamp. Closing a switch in the secondary circuit of the ballast heats the filaments. Releasing the switch simultaneously turns off the filaments and applies an inductive jolt that starts the lamp.

  In fixtures that are very old or very cheap, the switch is manual. Other fixtures use a simple automatic switch called a "starter." With older tube types, the starter is usually contained separately from the ballast in a small replaceable canister. Preheat starting is used mostly with small lamps. This starting method was fading into history, but it has returned as the main method of starting compact fluorescent lamps.

- *trigger-start* ballasts operate preheat lamps without starters by operating the lamps as if they were rapid-start lamps.

Any of these starting systems needs to have a grounded surface near the entire length of the tube. This provides capacitive coupling between the lamp gases and ground that helps to ionize the cold gas. Without this, the starting voltage would have to overcome the resistance of the entire gas column between the electrodes.

The metal case of the fixture usually performs this function. The case needs a good ground connection to start the lamp properly. An alternative method is to install a conducting strip alongside the tube, or to give the tube a conductive transparent coating. Among lighting equipment designers, the grounded surface is called a "starting aid."

Environmental Problems

Mercury is so well attuned to both fluorescent and HID lighting that it almost seems ordained to provide efficient lighting. Unfortunately, mercury is a serious environmental pollutant, and discarded lamps account for a large fraction of the mercury that is released into the environment. Although mercury is a necessary component of virtually all efficient lighting today, its environmental penalty is a strong impetus for design compromises that reduce the quantity of mercury in lamps, and for development of efficient light sources that do not require mercury.

Expect that the mercury content of fluorescent lamps will make disposal increasingly difficult, and that the cost of disposal will grow to become a significant part of the cost of fluorescent lighting.

How HID Lighting Works
HID Lamp Construction
HID Ballasts
HID Lamps that Do Not Need Ballasts
Ignitors

Characteristics of Mercury Vapor Lamps
Characteristics of Metal Halide Lamps
Characteristics of High-Pressure Sodium Lamps
Characteristics of Low-Pressure Sodium Lighting
Status of HID Lamp Efficiency
Status of HID Ballast Efficiency

This Reference Note explains the basics of high-intensity discharge (HID) lighting. This is a category of lighting that presently includes three distinct types of lamps:

- mercury vapor (which we will abbreviate "MV")
- metal halide ("MH")
- high-pressure sodium ("HPS").

There are large differences in efficiency, color rendering, and other important characteristics within this category of lighting. Understanding the factors that cause these differences is essential for selecting HID lamps and the accessories that they need to operate.

An HID lamp of any type must be operated with a ballast, as with fluorescent lighting. All three types are installed in similar fixtures. Figure 1 shows a typical example.

Another type of lighting, low-pressure sodium ("LPS"), is similar in physical characteristics to HID lighting. It is made by manufacturers of HID lighting, and is usually cataloged with HID equipment. For these reasons, we will cover it here also.

How HID Lighting Works

The basic physical processes in HID lighting are similar to the first stage of fluorescent lighting, which is the emission of light from a mercury vapor. See Reference Note 55, Fluorescent Lighting, for details of that process.

Like fluorescent lamps, HID lamps use an excited metal vapor as the main source of light. Beyond that similarity, the two types are different in the way that they produce light. The difference starts with the pressure of the vapor. The pressure of the metal vapors in HID lamps is thousands of times higher than the pressure of the mercury vapor in fluorescent lamps. Mercury vapor and metal halide lamps operate at a pressure of several atmospheres. High-pressure sodium lamps operate around one tenth of atmospheric pressure.

The higher pressure of the metal vapors is achieved by operating the lamp at high temperature. The metal vapors and inert gases are heated by using a relatively high current inside an arc tube that is kept small to concentrate the heat. By the same token, the high density of atoms in HID lamps allows the higher current.

The light output of HID lamps is very intense in relation to the size of the lamp. In fact, the name "high-intensity discharge" originally was intended to distinguish these lamps from fluorescent lamps, which are discharge lamps that produce light at low intensity in relation to their surface area. The light-emitting region of an HID lamp is so small that the lamp can be used in focussing fixtures, like incandescent lamps.

The intense excitation of the gases results in a spectrum of light output that is mostly at visible wavelengths. This is a major difference from fluorescent lamps. In fluorescent lamps, most light emitted by the mercury vapor is in the ultraviolet region, so a phosphor is needed to convert the ultraviolet light to visible light.

Even so, some HID lamps use phosphors. These are used primarily to broaden the color spectrum, and to shift some of the light emission from the vapor toward the red end of the visible spectrum. They may also convert unwanted ultraviolet radiation to visible light.

HID lamps contain inert gases in addition to the metal vapors. The inert gases aid starting, in the same way as in fluorescent lamps. The inert gases are also used to control the electrical characteristics of the arc, which affect wattage and efficiency.

HID lamps do not need inert gases to serve as "buffer" gases, which are used in fluorescent lamps to shorten the average path of electrons between collisions. The high density of the metal vapors makes this function unnecessary.

HID Lamp Construction

HID lamps have a small inner arc tube that contains one or more metal vapors, plus inert gases. The arc tube is generally cylindrical, and it has electrodes at each end. The arc tube is made of materials, such as fused quartz or alumina, that withstand the high operating temperature and the chemical activity of the metal vapors.

Figures 2, 3, and 4 show the parts of each type of lamp. Figure 5 shows the parts of a low-pressure sodium lamp.

Ruud Lighting, Inc.

Fig. 1 Typical HID light fixture The ballast is in the upper left compartment. The capacitor and ignitor (if any) are in the other compartment. The two compartments are separated by an air space to isolate the heat of the ballast. The slot and set screw allow the height of the lamp socket to be changed, as a means of adjusting the light distribution pattern. The reflector should be made of glass if any light is intended to pass through it. Plastic diffusers are darkened by the ultraviolet light and heat of HID lamps.

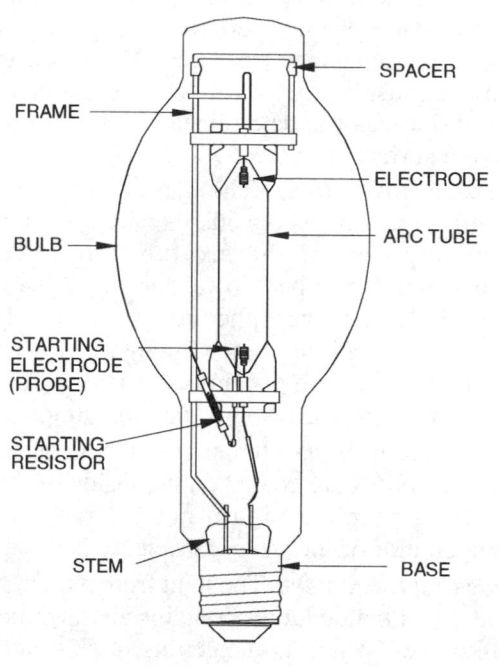

Osram Sylvania Inc.

Fig. 2 Mercury vapor lamp

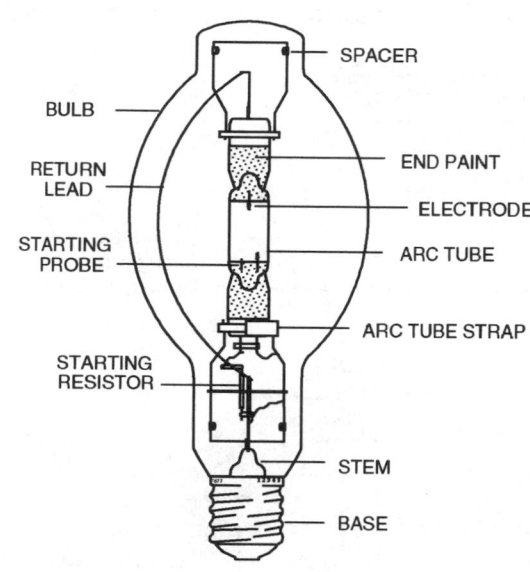

Osram Sylvania Inc.

Fig. 3 Metal halide lamp

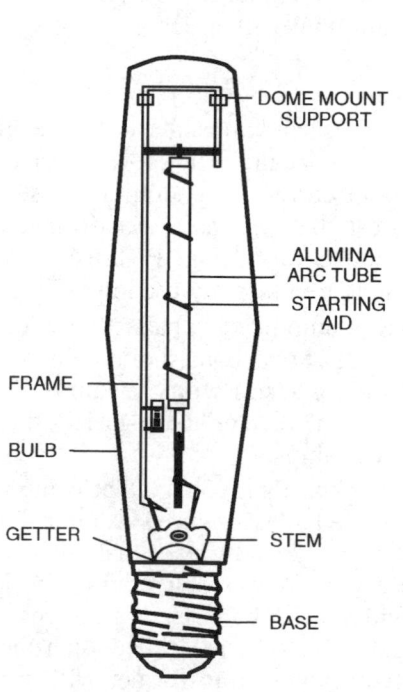

Osram Sylvania Inc.

Fig. 4 High-pressure sodium lamp

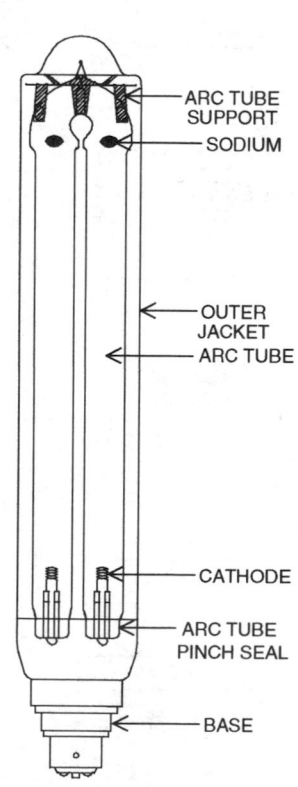

Osram Sylvania Inc.

Fig. 5 Low-pressure sodium lamp

HID lamps do not use filaments to heat the electrodes. Electrode heating is accomplished entirely by the impact of ions and electrons on the electrodes.

Most HID lamps for general lighting have an outer glass shell, which has several important functions:

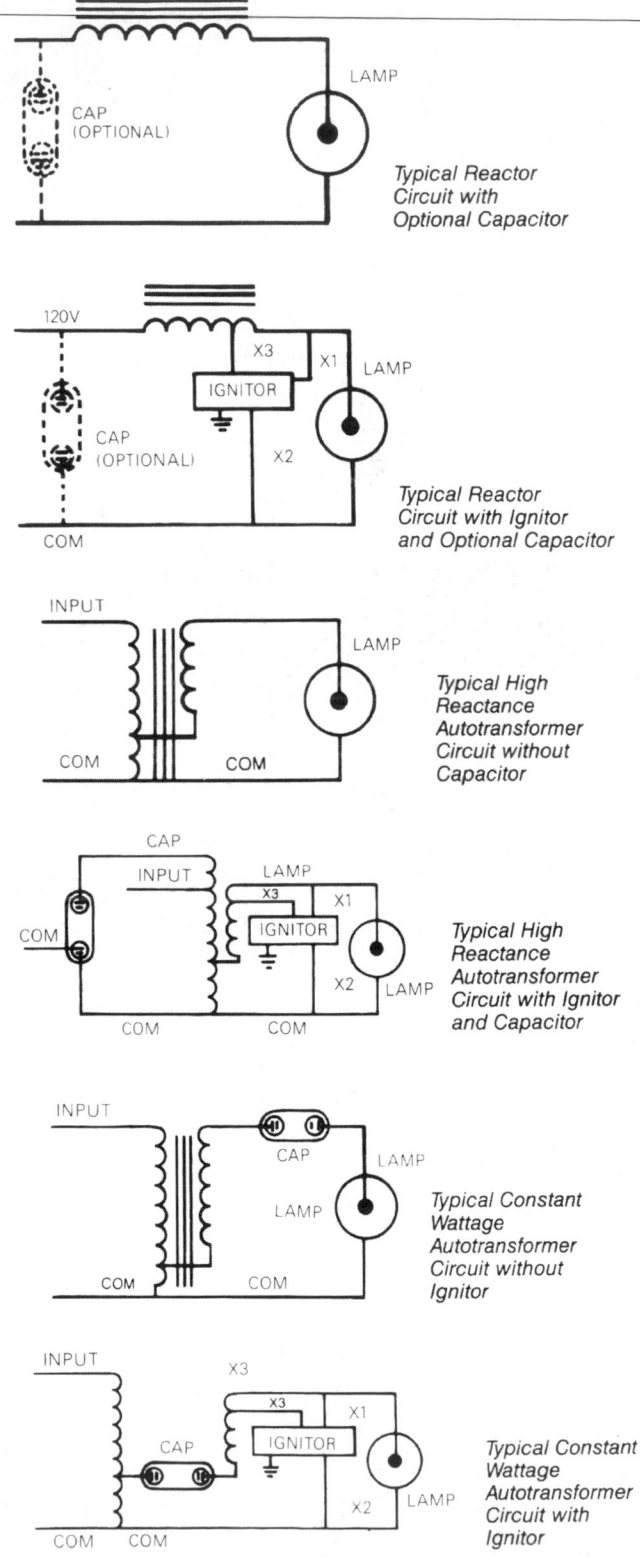

Typical Reactor Circuit with Optional Capacitor

Typical Reactor Circuit with Ignitor and Optional Capacitor

Typical High Reactance Autotransformer Circuit without Capacitor

Typical High Reactance Autotransformer Circuit with Ignitor and Capacitor

Typical Constant Wattage Autotransformer Circuit without Ignitor

Typical Constant Wattage Autotransformer Circuit with Ignitor

Philips Lighting

Fig. 6 HID ballasts

• *maintaining high temperature.* The outer shell reduces heat loss by convection and radiation, and allows the lamp to operate properly over a wide range of ambient temperatures. The entire arc tube surface must be kept warm, because the density of the metal vapor is determined by the coolest point on the surface.

• *physical protection.* The glass bulb protects the fragile components outside the arc tube, including the supports for the arc tube, the electrode connections, and perhaps a starting resistor. The outer bulb also keeps the arc tube clean. Dirt on the arc tube might absorb enough heat to shatter the tube. The bulb generally is filled with an inert gas to prevent corrosion of the hot components.

• *holding phosphors.* In lamps that have phosphors, the phosphors are coated on the inside of the outer shell. Phosphors could not tolerate the high temperature of the inner capsule.

• *blocking ultraviolet.* The light from mercury vapor and metal halide lamps contains a large amount of ultraviolet, which is dangerous to skin and eyes. Ultraviolet also bleaches fabrics and other organic materials, and it may cause other damage. The heat resistant material of the arc tube is transparent to ultraviolet, so the outer bulb is needed to block it.

HID lamps can operate even if the outer shell breaks away, which eliminates the protection against ultraviolet radiation. As a safety feature, some lamps have a fuse link that burns out when envelope breakage brings the link into contact with air. Some specialty lamps lack an outer bulb, so they must be installed in fixtures that include an ultraviolet filter.

HID Ballasts

As in fluorescent lamps, the arc inside an HID is an electrical short circuit. Therefore, the lamp needs a ballast to limit current. In addition, most HID lamps need a ballast that acts as a transformer to provide increased voltage for starting. Figure 6 shows the types of ballasts that are presently used in HID lighting.

If an HID lamp needs a ballast only for regulating current, the ballast may consist of nothing more than an inductor (coil) in series with the lamp. This type of ballast is called a "reactor" ballast. It is the usual type for small HID lamps.

If the lamp needs a voltage that is higher than the line voltage, the ballast is wired as a transformer. This type of ballast has the advantage that fluctuations in line voltage affect the power delivered to the lamp much less than with reactor ballasts. For this reason, ballasts of the transformer type are called "constant wattage" (CW), if they have a transformer with a completely isolated secondary winding, or "constant wattage autotransformer" (CWA), if they use an autotransformer.

CWA ballasts are presently the most common type for medium and large lamps.

Magnetic ballasts have low power factor because they are inductive devices. The power factor can be improved simply by adding a capacitor to the ballast circuit. The power factor can be improved as much as desired, up to 100%.

If a power factor correction capacitor is installed in a reactor ballast, it draws high current if the secondary is disconnected, as when the lamp fails. As a result, this type of ballast allows fewer lamps of a given wattage to be installed on a circuit. A model recently introduced solves this problem by disconnecting the capacitor when the lamp fails.

HID Lamps that Do Not Need Ballasts

An unusual type of HID lamp does not require an external ballast. Lamps of this type use an internal tungsten filament as a resistor to limit current. These lamps can be used as direct replacements for incandescent lamps, but they are no more efficient than incandescent lamps of comparable wattage. Their only advantage is longer life compared to incandescent lamps. Much more efficient alternatives to incandescent lighting are available, so we will say nothing more about them.

Ignitors

With most high-pressure sodium lamps and some low-wattage metal halide lamps, the starting voltage is so much higher than the operating voltage that it does not make sense to design the ballast to start the lamp. With these lamps, a small electronic device called an "ignitor" or "starter" generates high voltage pulses for starting. The ignitor may be a separately replaceable component, or it may be part of the ballast package.

Characteristics of Mercury Vapor Lamps

In a warmed-up mercury vapor lamp, almost all the light is emitted by excited mercury atoms. Unlike fluorescent lamps, most of this radiation is in the visible spectrum. This is because the mercury atoms are hit by electrons so frequently that they are likely to be excited to progressively higher states before they can relax to their lowest state. Mercury tends to relax from its higher energy levels in stages, so the energy of the emitted light is lower and the wavelengths are longer.

Mercury vapor lamps are the least efficient type of HID lamp. They are substantially less efficient than conventional fluorescent lamps.

The color rendering of uncoated mercury vapor lamps is poor, most of the light being concentrated in a few narrow wavelength bands. Clear mercury vapor lamps emit a blue-green light, and they make colors in illuminated objects look muddy. Figure 7 shows the typical spectrum of a mercury vapor lamp.

Mercury vapor lamps may use phosphors to improve color. Usually, the phosphors are selected to create one or more bands of wavelengths in the normally unoccupied red end of the mercury spectrum.

Argon is the usual starting gas for mercury vapor lamps, as in fluorescent lamps. Most mercury vapor lamps have a single starting electrode that is close to one of the main electrodes. The starting electrode is connected to the opposite main electrode through a resistor. Once the lamp starts to operate normally, the low resistance of the arc short-circuits the starting resistor, so little current continues to flow through the starting electrode.

Mercury vapor lamps suffer serious lumen degradation, as shown in Figure 10. Some lamps eventually lose half their light output. Lumen loss occurs most rapidly in the early part of the life cycle.

If a mercury vapor lamp is turned off, it cannot be restarted for several minutes. The reason is that the un-ionized hot mercury vapor is so dense that it prevents the available voltage from ionizing the vapor.

The mounting position of the lamp affects its operation to some extent. Best efficiency is achieved if the arc tube is oriented vertically, which keeps the arc in the center of the arc tube. Some mercury vapor lamps are designated for specific mounting positions.

Characteristics of Metal Halide Lamps

Metal halide lamps evolved from mercury vapor lamps. Their major improvement is using a variety of metal vapors, rather than mercury vapor alone. Some of the selected metals produce light more efficiently than mercury, which improves the overall efficiency of the lamp. Each of the metal vapors produces a different variety of colors, so that the sparse color spectrum of a mercury vapor lamp expands to a thicket of many color bands. Figure 8 shows the light spectrum of a typical clear metal halide lamp.

Metal halide lamps may use phosphors to provide a more uniform color spectrum. Phosphors incur a small efficiency penalty.

The term "metal halide" is used because the additional metals are introduced as compounds with some of the halogen elements (fluorine, chlorine, bromine, or iodine). Iodine is the usual halogen that is used. A combination of sodium, scandium, and lithium halides is common, along with the necessary mercury. Another common combination is holmium, thulium, and dysprosium halides, along with the mercury.

Compounding the metals with a halogen is necessary to vaporize them at the lamp operating temperatures. Also, some of the metals would attack the arc tube materials in their elemental form. In the hot arc, the halide compound breaks down and the excited metal atoms emit light. If the pure metal wanders out of the arc, it recombines with the halogen before it can cause trouble.

Unfortunately, metal halide lamps are not interchangeable with mercury vapor lamps, except in some small sizes. Argon is used for starting, as with mercury vapor lamps, but the starting voltage is higher in relation to the current. This requires a different ballast. Smaller metal halide lamps have arc tubes that are too small to accommodate starting electrodes, so they require ignitors for starting.

Exceptionally high temperature is needed to break down the metal halide compounds in the arc. This creates a number of disadvantages for metal halide lamps:

- the emissive material evaporates more quickly from the electrodes, so lamp life is only about half as long as for mercury vapor and high-pressure sodium lamps
- restarting a hot lamp requires an exceptionally long time
- the high operating temperature (about 1750°F, or 950°C) and pressure of the arc tube pose an explosion hazard. This makes it prudent to install metal halide lamps in special fixtures that can contain the hot fragments of an exploding arc tube.

Metal halide lamps typically suffer somewhat less lumen degradation than mercury vapor lamps. However, they may still lose over a third of their output before failing, as shown in Figure 11. As

with mercury vapor lamps, the lumen loss increases most rapidly in the early part of the life cycle.

Metal halide lamps are especially sensitive to mounting position. Efficiency suffers and color changes if the arc comes in contact with the wall of the arc tube by convection. Some lamps have bimetal switches for their starting electrodes that require the right mounting position to operate reliably. The allowable mounting positions for metal halide lamps are usually coded into the lamp designation. Figure 13 charts the loss of light output from a typical metal halide lamp at various mounting angles.

Characteristics of High-Pressure Sodium Lamps

As the name implies, high-pressure sodium lamps use sodium as the main light emitting element. Unlike mercury, the first excited state of sodium lies almost in the middle of the visible spectrum. The high density and temperature of the lamp do not strongly excite the sodium atoms to higher energy levels, but randomly alter the first excited level. This causes the light to be emitted over a broad, continuous spectrum of visible wavelengths. This spectrum is skewed toward the red. The lamp itself looks very orange, but the illumination may not seem unbalanced to the naked eye. Figure 9 shows the spectrum of a typical lamp.

High-pressure operation reduces lamp efficiency because the atoms do not dissipate all their collision energy in the form of light, and because a certain amount of light is shifted out of the visible spectrum. Ironically,

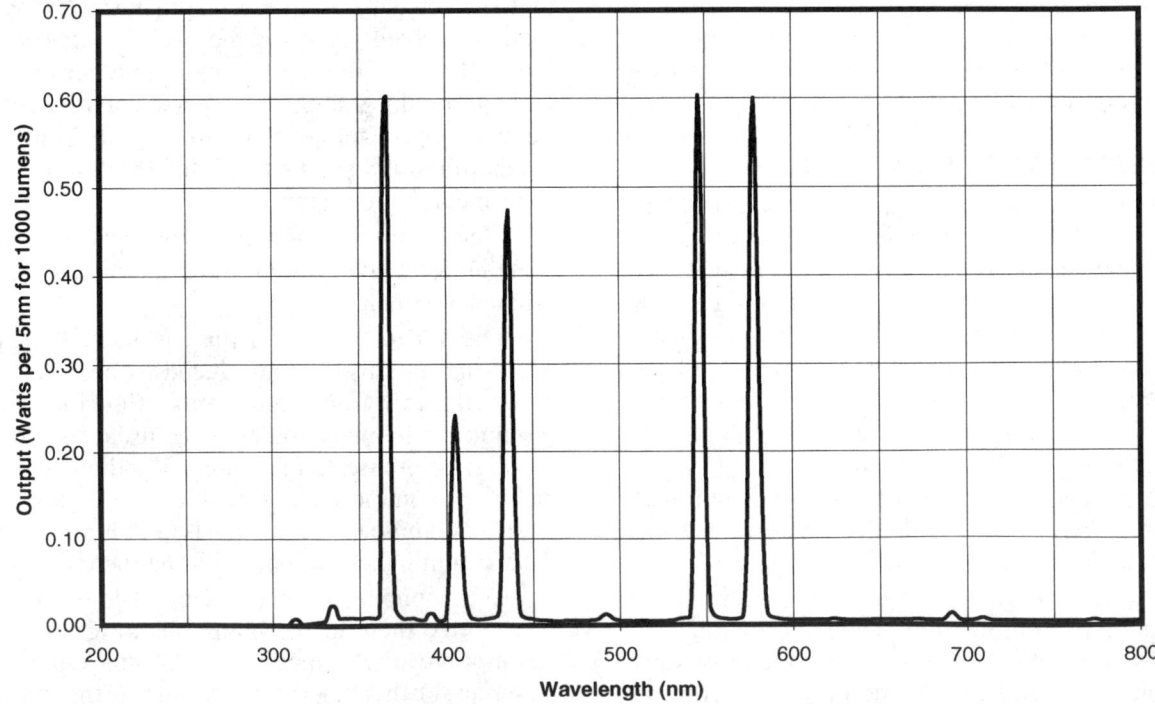

Fig. 7 Spectrum of a clear mercury vapor lamp Virtually all the light output is restricted to these five narrow bands of wavelengths. Any colors in a scene that are not contained in these bands will disappear.

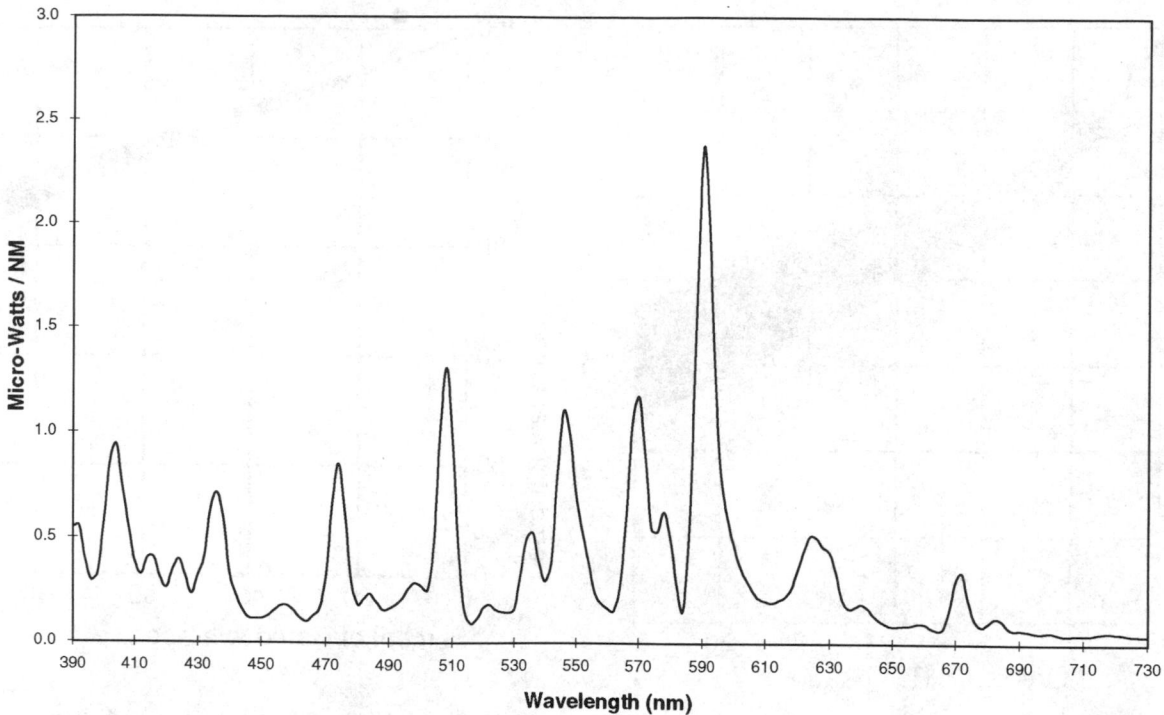

Fig. 8 Spectrum of a clear metal halide lamp This large number of broad bands come from a combination of metal vapors. Together, they provide much more coverage of the visible spectrum than mercury vapor. At least some light is provided at all visible wavelengths. The red end of the spectrum has somewhat less illumination than the rest.

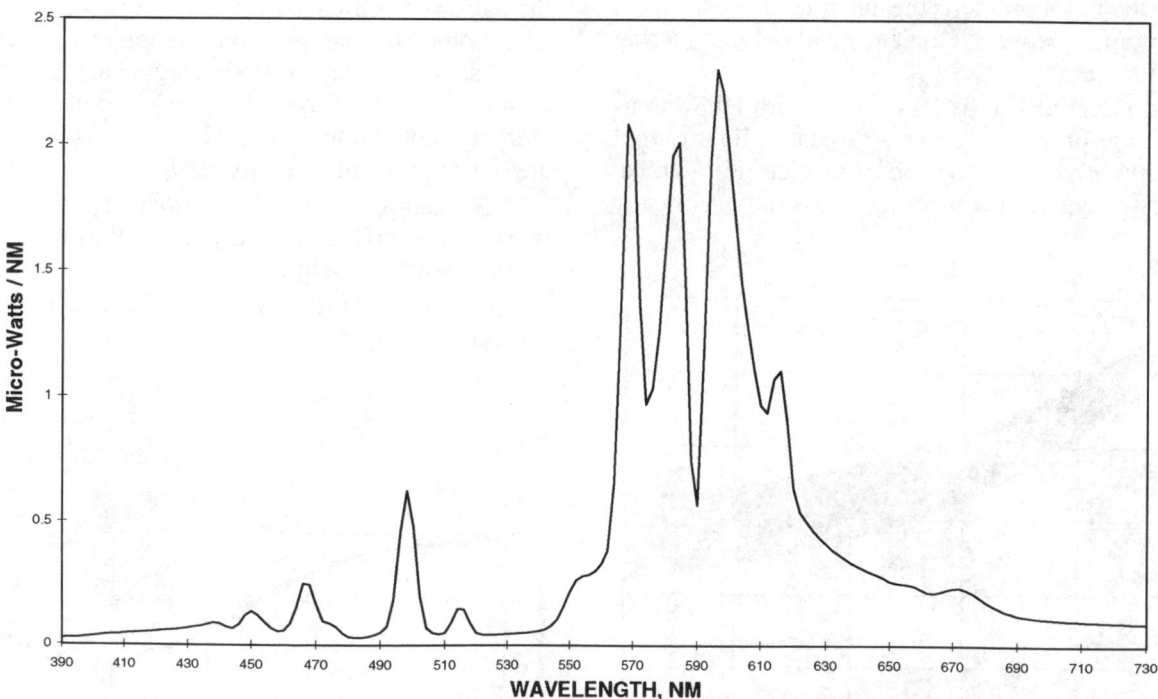

Fig. 9 Spectrum of a high-pressure sodium lamp Most of this spectrum is expanded from a narrow band of wavelengths that occurs in low-pressure sodium at 589 nanometers. Making the sodium vapor hot and dense causes collisions among the sodium atoms, which randomly distort the emission wavelengths. This wastes some energy as heat and as light outside the visible spectrum. The lamp can be designed for different pressures, changing the spectrum and the compromise between color rendering and energy efficiency.

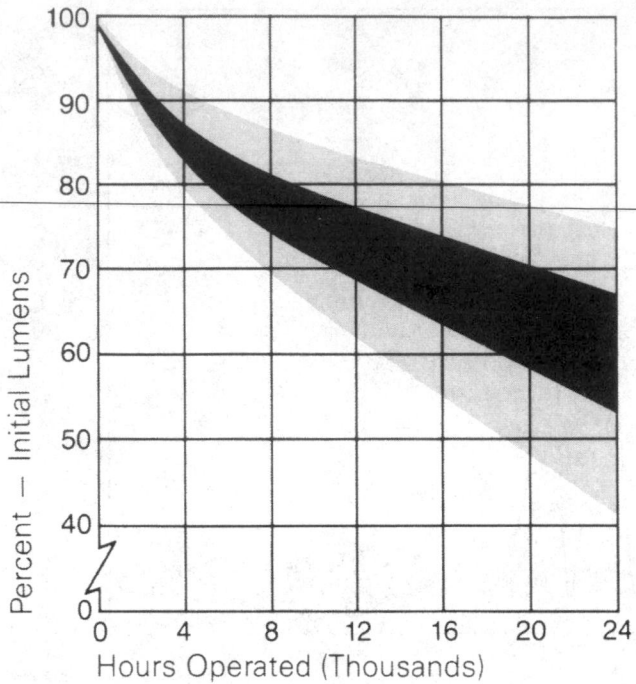

Philips Lighting

Fig. 10 Lumen degradation of mercury vapor lamps

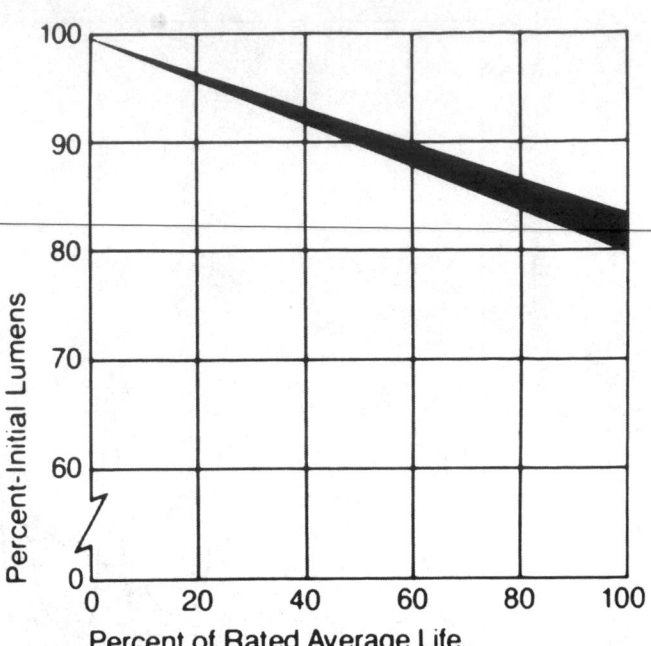

Philips Lighting

Fig. 12   Lumen degradation of high-pressure sodium lamps

light output near the original sodium wavelength is blocked because the dense un-ionized gas surrounding the arc absorbs light strongly at this wavelength. By changing the vapor pressure, the manufacturer can adjust the compromise between efficiency and color rendering over a wide range.

Lumen degradation is less severe with HPS lamps than with the other two types. Typically, light output falls by 20% or less by the end of service life. Unlike the other two types, the rate of light loss remains steady throughout life, which results in a higher average light output. See Figure 12.

Most HPS lamps require an ignitor in addition to the ballast. Such lamps usually use xenon as a starting gas. Some HPS lamps avoid the ignitor and achieve a lower starting voltage by using argon and neon together as starting gases. These lamps can be used to replace mercury vapor lamps without changing ballasts, but they are not as efficient as lamps designed to use ignitors.

HPS lamps use much less mercury than the other two types of HID lamps. As a result, they have virtually no ultraviolet emission.

Most HPS lamps are not sensitive to mounting position.

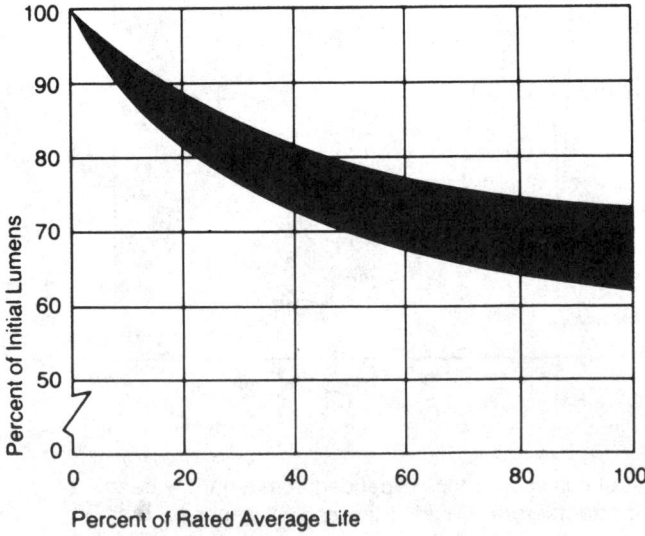

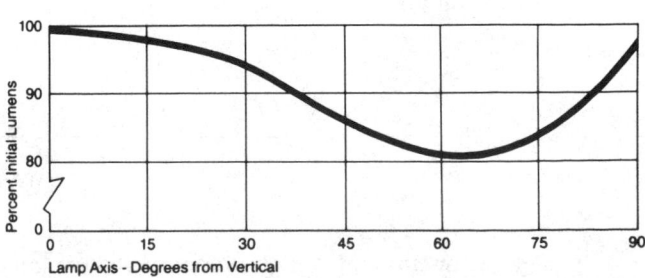

Philips Lighting

Fig. 13 Loss of light output caused by mounting position
This is for a typical metal halide lamp.

Philips Lighting

Fig. 11 Lumen degradation of metal halide lampss

Characteristics of Low-Pressure Sodium Lighting

As the name implies, the main physical difference between low-pressure sodium lighting and high-pressure sodium lighting is the pressure of the sodium vapor in the lamp. LPS is optimized for lumens-per-watt efficacy, but sacrifices output and color rendering to achieve it.

At very low pressure and density, excited sodium vapor emits a large fraction of its energy at one wavelength in the middle of the visible spectrum near 589 nanometers. (Actually, there are two very closely spaced wavelengths.) LPS lamps have very high lumen-per-watt efficiency for three reasons: (1) the high efficiency of light emission, (2) the absence of wasteful conversion processes, and (3) the fact that the wavelength of sodium emission falls in the most sensitive range of human vision.

LPS lamps have achieved an efficacy of 300 lumens/watt in the laboratory, and some commercial lamps claim 180 lumens/watt. (The theoretical maximum efficacy for a white light source is about 220 lumens/watt, which demonstrates that the high efficacy of LPS lamps is achieved partly at the cost of color rendering.)

The operating temperature of the lamp is critical to efficiency because it determines the pressure of the sodium vapor. The optimum temperature is about 500°F (260°C). To maintain this temperature, the lamp has an outer bulb and internal heat reflecting films to concentrate the lamp's waste heat. These features are needed because the current density in LPS lamps is low, so they need fairly large surface areas to produce much light. The large surface area invites heat loss, and it makes the lamp vulnerable to changes in ambient temperature. As a result, LPS lamps operate much of the time at efficacies that are lower than advertised.

Smaller LPS lamps are started by high voltage from a transformer ballast. Some larger lamps have preheat electrodes. In most LPS lamps, a mixture of 99% neon and 1% argon is used as a buffer and starting gas. This combination is used to exploit the same type of energy exchange between the excited states of argon and mercury vapor that is exploited in HID lamps. Helium, xenon, and other gases may also be included to optimize efficiency.

In LPS lamps of lower wattage, the inner arc tube is usually bent in the shape of a hairpin, which allows both electrode connections to be made at one end of the tube. Larger lamps have electrodes at opposite ends.

Because of their specific ballast requirements and unusual lamp sockets, low-pressure sodium lamps are not interchangeable with other lamp types.

Low-pressure sodium lamps require an exceptionally long time to reach full output, up to 15 minutes. This is because the cold sodium metal is being vaporized by a heat source of relatively low intensity. A hot lamp usually restarts immediately, but don't depend on it.

Table 1.　TYPICAL H.I.D.　LAMP EFFICACIES

Watts	CRI	C/P	MV	MH	HPS
100	15	C	38		
	20+	C			95
	25	C	41		
	40+	P	45		
	50+	P	46		
	60+	C		78	
	80+	C			47
	80+	P			25
250	15	C	45		
	20+	C			110
	20+	P			104
	25	C	48		
	40+	P	52		
	50+	P	50		
	60+	C		92	90
	60+	P			80
	80+	P	92		
400	15	C	52		
	20+	C			125
	20+	P			119
	40+	P	57		
	50+	P	55		
	60+	C		100	94
	60+	P			89
	70	C		90	
	70+	P		100	
1,000	15	C	57		
	20+	C			140
	50+	P	62		
	60+	C		125	
	70+	P		125	

NOTES:
- The figures do not include ballast losses.
- "C" indicates clear lamps.
- "P" indicates phosphor coated lamps.

The crippling disadvantage of LPS lighting is its poor color rendering. In fact, if a Color Rendering Index is listed at all for an LPS lamp, it may be zero! People dislike LPS road and parking lot lighting because it makes the world look black-and-white. It is difficult to visualize applications for which LPS is the best choice. In the past, it was used too much for general lighting, and it is now being used less.

Status of HID Lamp Efficiency

There is a wide range of efficiencies among HID lamps. As with other types of lamps, efficiency is called "efficacy" when expressed in terms of lumens per watt. HID and LPS lamps have the following ranges of efficacy, which are compared with the range for fluorescent lamps:

mercury vapor	35 - 65
metal halide	70 - 130
high-pressure sodium	50 - 150
low-pressure sodium	100 - 190
fluorescent	30 - 95

Note that these figures do not include ballast losses. Ballast losses are large in contemporary HID lighting, and they vary considerably with lamp size and type, as discussed next.

This simple listing may be misleading because the higher HID efficacies are available only in very large lamps with poor color rendering. Table 1 displays the relationship between efficacy, color rendering (expressed by the "color rendering index", CRI), and wattage for some typical HID lamps used in general lighting.

Status of HID Ballast Efficiency

As of the mid-1990's, virtually all HID ballasts produced by major manufacturers are the old magnetic type. Some versions of magnetic ballasts with improved efficiency are entering the market. In general, ballast efficiency rises dramatically with increased lamp wattage. Stated differently, the efficiency of ballasts for smaller lamps is relatively poor, so they dissipate a substantial fraction of lighting power. Table 2 displays typical ballast losses as a percentage of lamp wattage, calculated from the listings of a major ballast manufacturer.

Table 2 shows that there is considerable room for improvement, even with the old magnetic technology. There is no doubt that adoption of electronic ballast

Table 2.
TYPICAL H.I.D. MAGNETIC BALLAST LOSSES
As a Percentage of Lamp Wattage

lamp wattage	MV	MH	HPS	LPS
35				72*
50	48*		32*	
70			30*	
90				38*
100	18	29*	30*	
135				32*
150			29*	
175	13	23		
180				22
200		20		
250	14	18	18	
400	13	14	14	
1,000	8	8	10	
1,500		7		

* The figures with asterisks are for high-power-factor reactor ballasts. The other ballasts are CWA units.

technology similar to that now used in fluorescent lighting could offer even greater improvements.

This beckoning market for better HID ballasts will undoubtedly attract manufacturers, so shop around. But, beware. Remember the years of trouble that beset the first electronic fluorescent ballasts. High-efficiency magnetic ballasts are a safer bet for the time being.

The U.S. National Energy Policy Act of 1992 calls for vaguely defined improvements in HID efficiency, but any improvements will not become mandatory until 1999, if at all. This indicates that the industry has not yet decided how to go about making major improvements in HID efficiency.

Most energy losses occur in the winding and the steel core, not in the capacitor or ignitor, so efficiency is not a factor in selecting the latter components (when they are separate from the ballast).

Light Distribution Patterns of Fixtures

Distribution Patterns of Common Fixture Types
Effect of Lamps on Distribution Pattern
General Geometrical Principles

Effect of Diffusers
How to Measure Distribution Patterns

In order to position fixtures rationally, you need to know how the fixtures distribute their light. Light fixtures differ greatly in their light distribution patterns. This Note points out the factors to consider.

Distribution Patterns of Common Fixture Types

The light distribution patterns of many types of fixtures are illustrated in a table in the *IESNA Lighting Handbook*. The table displays most of the common types of fluorescent and HID fixtures.

The table in the *IESNA Handbook* covers only a sampling of incandescent fixture types. The range of incandescent fixture designs is too vast to be covered by any listing, and each design distributes light differently with different types of lamps.

For non-symmetrical fixtures, the *IESNA Handbook* shows the light distribution in two directions, parallel and perpendicular to the long axis of the fixture. The differences in these two patterns should caution you that the orientation of a fixture is important.

The light distribution pattern depends on the distance from the fixture. At distances less than several times the longest dimension of the fixture, the actual pattern may differ substantially from the patterns shown in the *IESNA Handbook*. Light distribution becomes more uniform closer to the fixture, especially if the fixture has a large surface area.

Effect of Lamps on Distribution Pattern

Fluorescent and HID fixtures integrate the lamp's light distribution pattern into the design of the fixture. Therefore, the published light distribution pattern for the fixture tells you all you need to know. Usually, all the lamps that fit a particular fixture will produce the same light distribution pattern.

The opposite is true with incandescent lamps. Their light distribution patterns may vary radically, depending on the filament design, the use of frosting on the bulb, and the use of reflective surfaces built into the lamp. Since different types and wattages of incandescent lamps are interchangeable in fixtures that use common types of sockets, it is your job to select the most efficient type of lamp for the fixture. This is a continuing problem, because incandescent lamps must be replaced continually. See Measure 9.1.3 for details.

General Geometrical Principles

For light sources that are elongated, the laws of geometry require that light is emitted most intensely in a direction perpendicular to the longer axis. A bare fluorescent tube is an extreme example. At a distance several times the length of the tube, its light distribution pattern is shaped like a doughnut surrounding the tube, i.e., the light is emitted most intensely perpendicular to the axis of the tube, and little light is emitted along the axis of the tube. Size is not a factor here. The same pattern applies to a straight lamp filament.

Any significantly unsymmetrical light source, such as a 2-tube fluorescent fixture with a wraparound diffuser, has a significantly unsymmetrical light distribution pattern. Exploit this when moving fixtures around. For example, if fluorescent fixtures are used to illuminate corridors, traffic lanes, tunnels, and other narrow spaces, light is distributed much more uniformly if the fixtures are turned perpendicular to the length of the space. The improved light distribution may allow the number of tubes to be reduced by half or more.

At the same time, distributing the light along the length of the space may produce intense source glare if you are not careful. This is because a viewer looking down the narrow space sees all the fixtures at maximum intensity. You can minimize this problem by using fixtures that block light emission along the line of sight, or by installing baffles near the fixtures.

Similarly, if a single oblong fixture illuminates a desk, turning the fixture so that its long side faces the desk increases the illumination level. It also reduces annoying shadows on the desk.

The rules of geometry apply to light sources of any size. The filaments in incandescent light bulbs have strongly unsymmetrical light distribution patterns, which makes it important to select bulbs for the type and orientation of the fixture. See Measure 9.1.3 for details of selecting incandescent lamps.

Effect of Diffusers

Diffusers radically alter the light distribution pattern of a fixture. In fact, the diffuser may have a greater effect on the light distribution pattern than any other aspect of the fixture. Expect unusual light distribution patterns from fixtures with fresnel lenses, shielding vanes, embossed plastic diffusers, egg crate grilles, etc. When purchasing new fixtures, ask the manufacturer for the light distribution patterns before you do the lighting layout.

How to Measure Distribution Patterns

You can measure the light distribution patterns of installed fixtures by using an ordinary light meter to make a plot. Select a representative fixture that is far away from a wall, disconnect all nearby fixtures, and eliminate daylight. After the lamps are warmed up, measure illumination intensities at the working height, keeping the light meter aimed straight up. (If the fixture is intended to illuminate surfaces that are not horizontal, orient the light meter perpendicular to the illuminated surface.) Make one traverse along the long axis of the fixture, and another perpendicular to the long axis. The results may not meet scientific standards of accuracy, but they will be good enough to guide you in making improvements.

The next step is to measure the effect of overlapping patterns of adjacent fixtures. To do this, simply repeat the measurements with all the fixtures turned on.

CONTRIBUTORS OF ILLUSTRATIONS

Advance Transformer Co.
O'Hare International Center
10275 West Higgins Road
Rosemont IL 60018

Airolite Company
114 Westview Avenue
Marietta OH 45750-0666

American Mill Sales
10764 Noel Street
Los Alamitos CA 90720

ASHRAE
American Society of Heating,
Refrigerating and Air Conditioning
Engineers, Inc.
1791 Tullie Circle, N.E.
Atlanta GA 30329

American Standard, Inc.
240 Princeton Avenue
Trenton NJ 08619

Armstrong International, Inc.
816 Maple Street
P.O. Box 408
Three Rivers MI 49093

Bacharach, Inc.
625 Alpha Drive
Pittsburgh PA 15238

Baltimore Aircoil Company
7595 Montevideo Road
Jessup MD 20794

Blender Products, Inc.
5010 Cook Street
Denver CO 80216

Brayden Automation
1807 East Mulberry
Fort Collins CO 80524

Calmac Manufacturing Corp.
101 W. Sheffield Avenue
Englewood NJ 07631

Carrier Corporation
P.O. Box 4808
Carrier Parkway
Syracuse NY 13221

Celotex Corporation
4010 Boy Scout Boulevard
Tampa FL 33607

Cleaver-Brooks
P.O. Box 421
Milwaukee WI 53201

Construction Specialties, Inc.
P.O. Box 380
Route 405
Muncy PA 17756

Coyote Electronics, Inc.
4701 Old Denton Road
Fort Worth TX 76117

Danfoss Automatic Controls
7941 Corporate Drive
Baltimore MD 21236

Donlee Technologies
693 North Hills Road
York PA 17402

Doucette Industries, Inc.
701 Grantley Road
P.O. Box 2337
York PA 17405

Dow Chemical Company
200 Larkin Center
1605 Joseph Drive
Midland MI 48674

Duo-Gard Industries, Inc.
40442 Koppernick Road
Canton MI 48187

Fuel Efficiency, Inc.
Davis Industrial Park
P.O. Box 271
Clyde NY 14433

GE Lighting
Nela Park
1975 Noble Road
Cleveland OH 44112

Goodway Tools Corporation
420 West Avenue
Stamford CT 06902-6384

Graham Company
8800 W. Bradley Road
P.O. Box 23880
Milwaukee WI 53223

Hi-Fold Door Corporation
N6170 1070th Street
River Falls WI 54022

Huvco, LLC
3416 Prices Distillery Road
Ijamsville MD 21754

IMR Environmental Equipment, Inc.
5401 Central Avenue
St. Petersburg FL 33710

Industrial Combustion
351 21st Street
Monroe WI 53566

Kalwall Corporation
P.O. Box 237
1111 Candia Road
Manchester NH 03105

Kentube Engineered Products
555 West 41st Street
Tulsa OK 74107

Leeson Electric Corporation
2100 Washington Street
Grafton WI 53024

Lennox Industries Inc.
1600 Metrocrest Drive
Carrollton TX 75006

LightScience Corporation
P.O. Box 034084
Indialantic FL 32903

Ludell Manufacturing Company
5200 West State Street
Milwaukee WI 53208

M&I Door Systems
230 Bayview Drive
Units 1-7
Barrie, Ontario L4N 5E9
CANADA

M&I Heat Transfer Products Ltd.
601 The Queensway East
Mississauga, Ontario L5A 3X6
CANADA

Major Industries, Inc.
P.O. Box 306
7120 Stewart Avenue
Wausau WI 54402-0306

Osram Sylvania Inc.
100 Endicott Street
Danvers MA 01923

Owens-Corning Fiberglas
Mail Stop 2B
1 Owens-Corning Parkway
Toledo OH 43659

Paragon Electric Company, Inc.
606 Parkway Boulevard
Two Rivers WI 54241-0028

Parker Boiler Co.
5930 Bandini Boulevard
Los Angeles CA 90040-2999

Paul Mueller Company
1600 W. Phelps Street
Springfield Missouri 65802

Pennsylvania Separator Co.
P.O. Box 340
21 S. Pickering Street
Brookville PA 15825

Perlite Institute
88 New Dorp Plaza
Staten Island NY 10306

Philips Lighting
200 Franklin Square Drive
Somerset NJ 08875

Preferred Utilities Manufacturing
Corporation
P.O. Box 1280
11 South Street
Danbury CT 06813

Pure Water/Clean Air Group, NA
Suite 5C
642 Locust Street
Mt. Vernon NY 10552

Reliance Electric
6065 Parkland Boulevard
Cleveland OH 44124-6106

Resources Conservation Inc.
P.O. Box 71
Greenwich CT 06836

Rite-Hite Corporation
8900 N. Arbon Drive
Milwaukee WI 53223

Robicon
500 Hunt Valley Drive
New Kensington PA 15068

Ruud Lighting, Inc.
9201 Washington Avenue
Racine WI 53406

Schweiss Distributing, Inc.
Box 220
Fairfax MN 55332

Solarex
630 Solarex Court
Frederick MD 21703

Spirax Sarco, Inc.
P.O. Box 119
Allentown PA 18105

Sun Tunnel Skylights
786 McGlincey Lane
Campbell CA 95008

Super Sky Products, Inc.
10301 N. Enterprise Drive
Mequon WI 53092

Todd Combustion, Inc.
15 Progress Drive
P.O. Box 884
Shelton CT 06484-0884

Trane Company
3600 Pammel Creek Road
La Crosse WI 54601-7599

Unenco Electronics, Inc.
Suite 104
1350 South Loop Road
Alameda CA 94502

Vaughn Manufacturing Company
26 Old Elm Street
PO Box 5431
Salisbury MA 01952-5431

Vistawall Architectural Products
803 Airport Road
Terrell TX 75160

Water Technology of Pensacola
3000 W. Nine Mile Road
Pensacola FL 32534

WaterFurnace International, Inc.
9000 Conservation Way
Ft. Wayne IN 46809

Waterless Co.
1233 Camino Del Mar
Del Mar CA 92014

WESINC
Wulfinghoff Energy Services, Inc.
3936 Lantern Drive
Wheaton MD 20902

1488

DISCLAIMER & WARNING

In these times of declining personal responsibility, it has become common for people who make mistakes or suffer injuries to seek someone else to blame. A legal system that feeds this impulse has made it advisable to warn against even the most absurd misuse of products, such as warning users of windows not to fall out of them. In this climate, it has even become advisable for publishers to warn readers that the information contained in books is not perfect and complete.

The *Energy Efficiency Manual* includes suggestions for your consideration. Energy efficiency is still evolving, and many of the concepts have not been tested extensively. The suggestions are based on the knowledge of the author, who does not claim to have complete or perfect knowledge of each topic. This book is not intended to be your sole source of information about any activity. Neither the author nor the publisher can guarantee the accuracy of any information provided.

Many energy conservation activities involve the risks of pioneering. For this reason, we start the book with "The Right Way to Do Energy Conservation." It is a major objective of the *Energy Efficiency Manual* to indicate the potential reliability of each activity, but risks of failure vary with the circumstances of the application and with the skills of the people who do the work. Furthermore, it is impossible to foresee all the conditions that may occur.

Energy efficiency often involves situations that have safety implications or the potential for consequential damage. It is not a purpose of the *Energy Efficiency Manual* to deal with such issues. Safety may be mentioned, but only as an issue that is peripheral to the primary issue of energy efficiency. The ways in which a party may suffer injury are legion, and unpredictable in general. The author and publisher of this book specifically disclaim any liability for damages or injuries that may occur. Again, it remains the responsibility of the reader to select the necessary personnel and equipment to accomplish any activities effectively and safely.

We caution the reader that the *Energy Efficiency Manual* is not an engineering, architectural, or contract document. Where expert or specialized services are required, it is essential for the work to be planned and performed by appropriately qualified practitioners.

The author and later readers will be grateful for any suggestions and for identification of any errors. Please address them to the publisher.

Notice Regarding Patented Devices, Systems, or Concepts

Many of the ideas presented in this book are unusual, or even novel. It is conceivable that some of the concepts may be covered by patents. It is the responsibility of the user to avoid infringing any patent rights that may exist.

INDEX

THREE EASY WAYS TO ORDER

To order by TELEPHONE: call **toll-free 888-280-2665 (888-280-BOOK)**, from the United States and Canada only. All major credit cards are accepted.

To order on the INTERNET: visit our Web site, **www.ENERGYBOOKS.com**, and follow the easy instructions. Order from anywhere on earth for quick delivery.

To order by MAIL: call our toll-free number or visit our Web site for current mail order information.

YOUR SATISFACTION IS GUARANTEED. For a full refund, no questions asked, return the book with proof of purchase within 90 days. You can't lose!

REGISTRATION

To receive free updates, announcements of future editions, and other valuable information, please make a copy of this form, fill it out, and mail it to:

Energy Institute Press
3936 Lantern Drive
Wheaton, Maryland 20902
USA

Name _____

Title _____

Organization _____

Address _____

_____

City _____ State _____

ZIP/Postal Code _____ Country _____

Where Purchased _____

Date _____

This information will not be released to any other organization.

THREE EASY WAYS TO ORDER

To order by TELEPHONE: call **toll-free 888-280-2665 (888-280-BOOK)**, from the United States and Canada only. All major credit cards are accepted.

To order on the INTERNET: visit our Web site, **WWW.ENERGYBOOKS.COM**, and follow the easy instructions. Order from anywhere on earth for quick delivery.

To order by MAIL: call our toll-free number or visit our Web site for current mail order information.

YOUR SATISFACTION IS GUARANTEED. For a full refund, no questions asked, return the book with proof of purchase within 90 days. You can't lose!